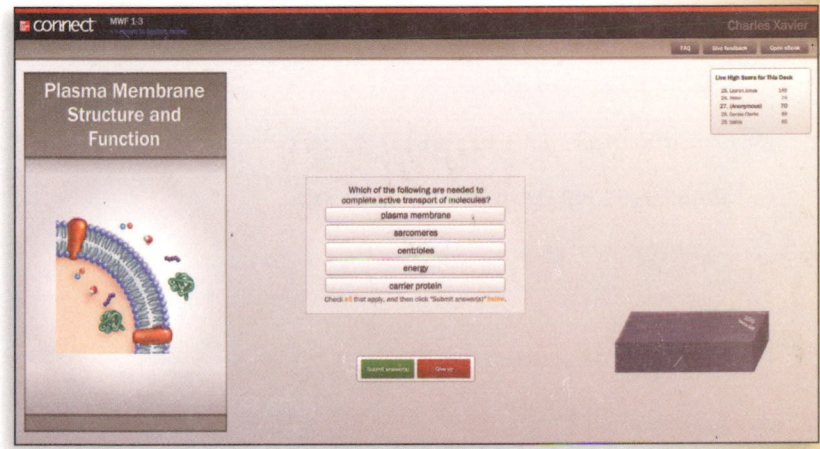

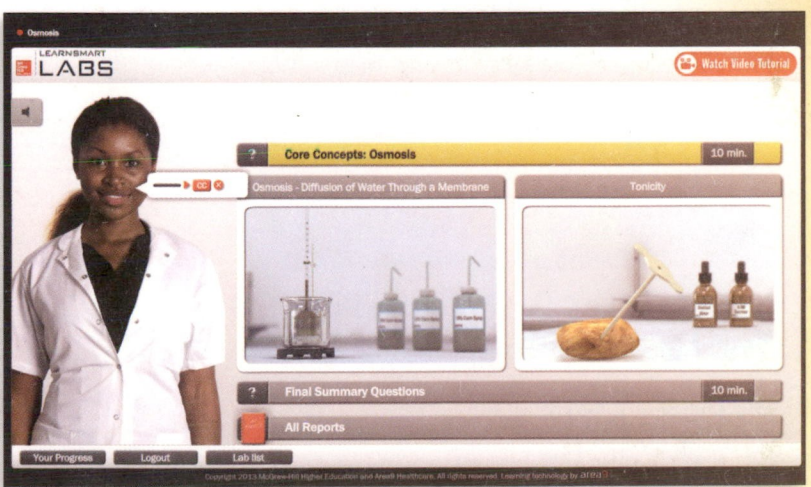

Brief Contents

CONCEPTS
of BIOLOGY

Third Edition

Sylvia S. Mader

Mc
Graw
Hill
Education

CONCEPTS OF BIOLOGY, THIRD EDITION

Published by McGraw-Hill Education, 2 Penn Plaza, New York, NY 10121. Copyright © 2014 by McGraw-Hill Education. All rights reserved.
Printed in the United States of America. Previous editions © 2011 and 2009. No part of this publication may be reproduced or distributed in any
form or by any means, or stored in a database or retrieval system, without the prior written consent of McGraw-Hill Education, including, but
not limited to, in any network or other electronic storage or transmission, or broadcast for distance learning.

Some ancillaries, including electronic and print components, may not be available to customers outside the United States.

This book is printed on acid-free paper.

1 2 3 4 5 6 7 8 9 0 DOW/DOW 10 9 8 7 6 5 4 3

ISBN 978-0-07-352553-2
MHID 0-07-352553-7

Senior Vice President, Products & Markets: *Kurt L. Strand*
Vice President, General Manager, Products & Markets: *Marty Lange*
Vice President, Content Production & Technology Services: *Kimberly Meriwether David*
Managing Director: *Michael S. Hackett*
Director, Biology: *Lynn Breithaupt*
Brand Manager: *Eric Weber*
Director of Development: *Rose Koos*
Senior Development Editor: *Anne Winch*
Marketing Manager: *Chris Loewenberg*
Director, Content Production: *Terri Schiesl*
Content Project Manager: *Jayne Klein*
Senior Buyer: *Sandy Ludovissy*
Senior Designer: *Laurie B. Janssen*
Cover Image: *FLPA/Alamy*
Senior Content Licensing Specialist: *Lori Hancock*
Compositor: *Electronic Publishing Services Inc., NYC*
Typeface: *10/12 Utopia Std*
Printer: *R. R. Donnelley*

All credits appearing on page or at the end of the book are considered to be an extension of the copyright page.

Library of Congress Cataloging-in-Publication Data

Mader, Sylvia S.
  Concepts of biology / Sylvia Mader. — Third edition.
    pages cm
  Includes index.
  ISBN 978-0-07-352553-2 — ISBN 0-07-352553-7 (hard copy : acid-free paper) 1. Biology-Textbooks. I. Title.
  QH308.2.M234 2014
  570-dc23

                                                                                    2013025268

The Internet addresses listed in the text were accurate at the time of publication. The inclusion of a website does not indicate an endorsement by
the authors or McGraw-Hill Education, and McGraw-Hill Education does not guarantee the accuracy of the information presented at these sites.

www.mhhe.com

About the Author

Dr. Sylvia S. Mader has authored several nationally recognized biology texts published by McGraw-Hill. Educated at Bryn Mawr College, Harvard University, Tufts University, and Nova Southeastern University, she holds degrees in both Biology and Education. Over the years she has taught at University of Massachusetts, Lowell, Massachusetts Bay Community College, Suffolk University, and Nathan Mathew Seminars. Her ability to reach out to science-shy students led to the writing of her first text, *Inquiry into Life,* that is now in its fourteenth edition. Highly acclaimed for her crisp and entertaining writing style, her books have become models for others who write in the field of biology.

Although her writing schedule is always quite demanding, Dr. Mader enjoys taking time to visit and explore the various ecosystems of the biosphere. Her several trips to the Florida Everglades and Caribbean coral reefs resulted in talks she has given to various groups around the country. She has visited the tundra in Alaska, the taiga in the Canadian Rockies, the Sonoran Desert in Arizona, and tropical rain forests in South America and Australia. She was thrilled to think of walking in Darwin's steps when she journeyed to the Galápagos Islands with a group of biology educators. Dr. Mader was also a member of a group of biology educators who traveled to China to meet with their Chinese counterparts and exchange ideas about the teaching of modern-day biology.

For My Children   —Sylvia Mader

Preface

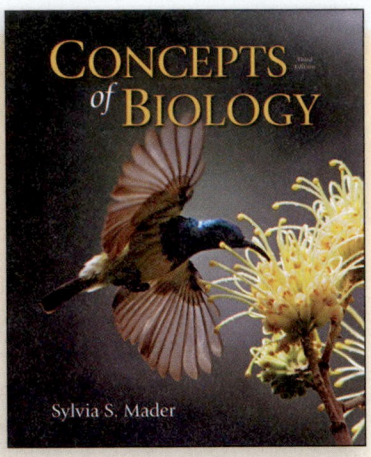

Concepts of Biology, Third Edition, recognizes the value of the traditional approach while still engaging students in the excitement of relevancy to themselves and the world around them. The text abounds with analogies and engaging illustrations as it proceeds from an examination of chemistry to the biosphere.

A significant new feature of this edition is the integration of media assets into the chapter content. Virtually every section of the textbook is now linked to MP3 files, 3D and 2D animations of biological processes, and National Geographic and ScienCentral videos. In addition, McGraw-Hill offers a full suite of adaptive learning tools including LearnSmart, LearnSmart Labs, LearnSmart Prep, LearnSmart Achieve, and SmartBook, all designed to assess a student's existing knowledge base and then adapt to address any deficiencies (see pages xii-xiii).

The conceptual approach of this text is apparent in its organization. *Concepts of Biology* is organized around the five major theories of biology: The Cell Theory, The Gene Theory, The Theory of Homeostasis, The Theory of Ecosystems, and The Theory of Evolution. The evolutionary theme was strengthened this edition to show the relevancy of the evolutionary approach. Natural selection, for example explains how resistance occurs among bacteria as well as pests, and common descent explains why the same genes, such as *Hox* genes, are found in organisms as different as bacteria, plants, and humans. Today, an understanding of evolution is assisting researchers in numerous fields from molecular biology to ecological restoration. To be consistent with this trend, the explanatory power of evolution has been increased in the running text, the chapter introductions, and the applications.

The revised introductions are now entitled "Looking at Life." Their varied topics illustrate how biology pertains to the life of organisms, including humans. The many applications in this text reflect its major themes: evolution, relevancy, and the scientific process. Like all parts of this text, the introductions and applications encourage a conceptual understanding of life.

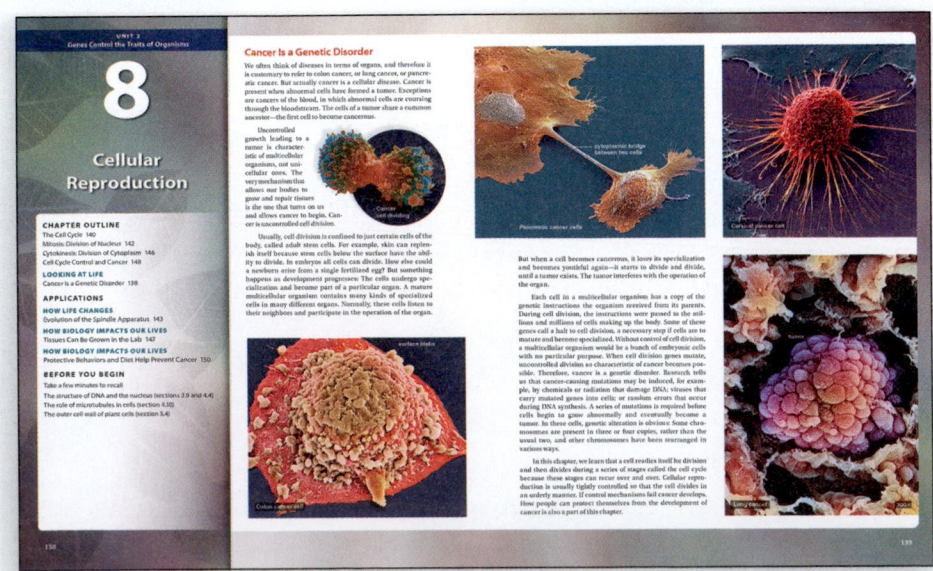

New to this Edition

In this edition, many former one-page sections have been combined into attractive two-page sections with a single title; half page sections have become one-page sections with new titles. This modular approach has the benefit that each illustration is on the same or facing page to its reference and it will rarely be necessary for students to turn a page in order to reach the end of a section. A short introduction ties numbered subsections into a comprehensive whole.

- A new feature in the Chapter Outline "Before You Begin" alerts students to any previous references in the book that pertain to the Learning Outcomes of the chapter.
- Many introductions are new and most contain new photographs and arrangements to be more visually appealing to today's students.
- The Learning Outcomes are new and placed at the start of each numbered section. Connect homework and test bank content are directly tied to these outcomes, allowing instructors to test student comprehension of specific concepts and focus classroom time where it is needed.
- Check Your Progress questions at the end of each section are new and are now better tied to the learning outcomes. All questions are answered in the Appendix.
- All Summaries are new: The number of bullets has been reduced to only one or two for each section.
- Several media tools (MP3 files, Animations, 3D Animations, Videos) are integrated into the running text as icons, alerting students to the availability of these learning tools. These icons become active links in the online eBook version of the textbook.

 MP3 Files
These three- to five-minute audio files serve as a review of the material in the chapter, and they also assist the student in the pronunciation of scientific terms.

 Animations
Drawing on McGraw-Hill's vast library of animations, the author has selected animations that will enhance the student's understanding of complex biological processes.

 3D Animations
For topics such as photosynthesis and cellular respiration, McGraw-Hill has produced a series of dynamic 3D animations that may be used both as presentation tools in the classroom, and as mini-tutorials that can be assigned within Connect or your course management system.

 Videos
Two different types of movies are integrated into this edition of the text. The ScienCentral videos are short news clips on advances in the sciences. The National Geographic videos provide students with a glimpse of the complexity of life that normally would not be possible in the classroom.

 Virtual Labs
These simulated experiments serve as excellent tutorials, allowing students to explore the topics covered in select chapters of the text.

Guided Tutorials
In addition to the assets listed above, a series of 2-minute guided tutorials of some of the more difficult topics in the text are available. A complete list of these tutorials is provided on page vi of the Preface.

Guided Tutorials

The narrated videos below were prepared by the authors of the textbook to assist you in understanding some of the more difficult topics in biology. Each video explores a specific figure in the text and is narrated by Mader series co-author Michael Windelspecht. During the video, important terms and processes are called out, allowing you to focus on the key aspects of the figure. All of these tutorials are embedded within the Connect Plus eBook and are available with assessment in the Connect question bank.

Hydrogen Bonding
Levels of Protein Organization
Endomembrane System
Endosymbiotic Theory
Osmosis and Tonicity
Sodium-Potassium Pump
Activation Energy
Cellular Respiration Overview
Electron Transport Chain
Noncyclic Photosynthesis
Calvin Reactions
Mitosis
Tumor Suppressor Genes
Proto-Oncogenes
Meiosis
Dihybrid Cross
Linkage
DNA Replication
Overview of Gene Expression
Lac Operon
Polymerase Chain Reaction
Hardy-Weinberg Equilibrium
Viral Life Cycle
Zygospore Life Cycle
Angiosperm Life Cycle
Protostomes and Deuterostomes
Primate Classification
Cohesion-Tension Model
Pressure-Flow Model
Alternation of Generations
Negative Feedback
Cardiac Cycle
Capillary Exchange
Blood Clotting
Inflammatory Response
B-Cell Clonal Selection
T-Cell Clonal Selection
HIV Infection Cycle

Hormonal Control of Digestion
External and Internal Respiration
Urine Formation
Neuron Action Potentials
Synaptic Cleft
Skeletal Muscle Contraction
Action of a Peptide Hormone
Action of a Steroid Hormone
Ovarian Cycle
Embryonic Stages of Development
Fetal Circulation
Patterns of Population Growth
Carbon Cycle
Cycling of Energy and Nutrients in an Ecosystem
Factors Influencing the Distribution of Biomes
Global Climate Change

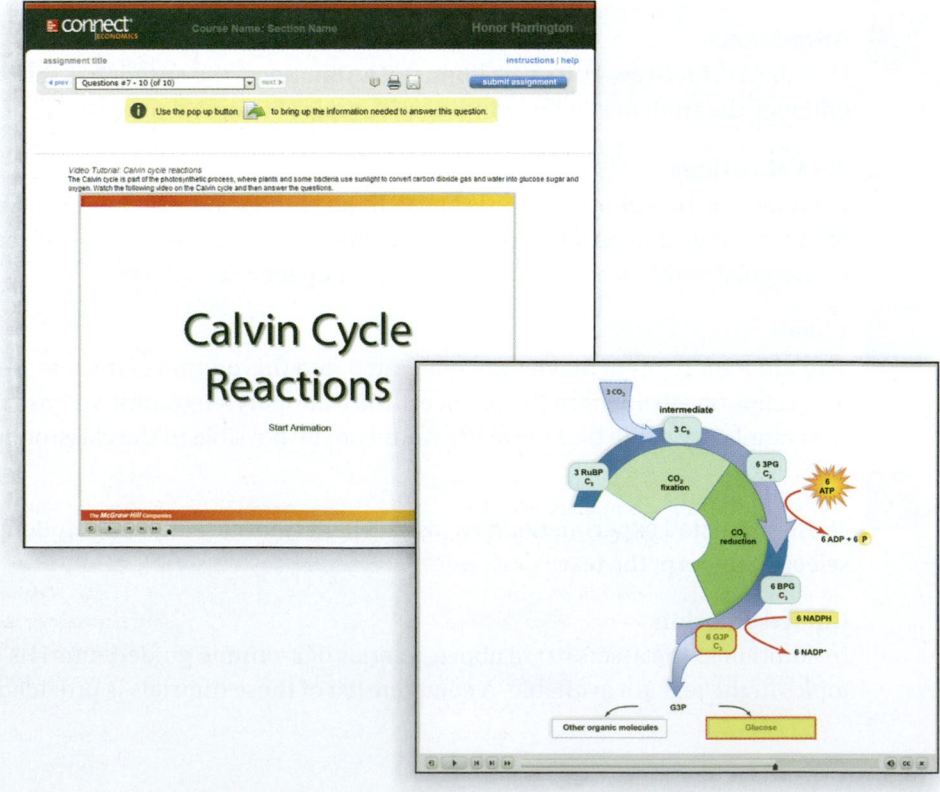

Applications

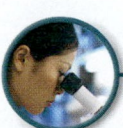

A Student's Guide to
Using This Textbook

Chapter Outline
Lists the concepts and applications that will be discussed in the chapter.

Before You Begin
Links the content of the chapter with material from earlier in the text.

31
Lymph Transport and Immunity

CHAPTER OUTLINE
The Lymphatic System 0
Innate Immunity 0
Adaptive Immunity 0
Immune System Failures 0

LOOKING AT LIFE
HIV/AIDS: A Global Disaster 0

APPLICATIONS
HOW LIFE CHANGES
Evolution of the Immune System 0
HOW SCIENCE PROGRESSES
Monoclonal Antibodies 0
HOW BIOLOGY IMPACTS OUR LIVES
Allergic Reactions 0

BEFORE YOU BEGIN
Take a few minutes to recall
The functions of plasma membrane proteins (section 5.8)
Functions of the lymphatic systems (sections 26.3, 26.7, and 30.6)
Structure and function of white blood cells (section 30.5)

HIV/AIDS: A Global Disaster

AIDS (acquired immunodeficiency syndrome) is caused by a virus known as the human immunodeficiency virus (HIV). HIV infects and reproduces in the cells of the immune system, the very cells that normally keep us free of disease. As the virus bursts forth, its host cell is destroyed. No wonder a person infected with HIV is unable to fight off the onslaught of viruses, fungi, and bacteria that attack the body every day.

At first a person infected with HIV suffers from weight loss, chronic fever, cough, diarrhea, swollen glands, and shortness of breath. Finally, without treatment, the person will have AIDS and develop one or more opportunistic infections. These are illnesses that can occur only because the immune system is so weak. The list is long but a common opportunistic infection is Kaposi sarcoma, a cancer of blood vessels caused by a herpesvirus. Reddish-purple, coin-sized spots and skin lesions occur with this type of cancer. *Pneumocystis* pneumonia, a fungal infection that causes the lungs to become useless as they fill with fluid and debris is another common opportunistic illness.

An HIV infection is acquired by having sex with an infected person or by sharing a needle with an infected person. Therefore,

HIV particles bud from an infected cell

612

Looking at Life
The opening essay illustrates how biology pertains to the life of organisms including humans. Like all parts of this text, the introductions encourage a conceptual understanding of life.

Learning Outcomes
Placed at the start of each numbered section, the learning outcomes provide you with an overview of what you are to know.

31.1 Lymphatic vessels transport lymph

LEARNING OUTCOMES
When you complete this section, you should be able to
1. List four functions of the lymphatic system.
2. Describe the one-way transport of the lymphatic vessels.
3. State the chief functions of four lymphatic organs and three patches of lymphatic tissue.

The **lymphatic system**, which is closely associated with the cardiovascular system, has four main functions that contribute to homeostasis:

Patches of lymphatic tissue in the body include: the **tonsils**, located in the pharynx; **Peyer patches**, located in the intestinal wall; and the vermiform **appendix**, attached to the cecum. These structures encounter pathogens and antigens that enter the body by way of the mouth.

This completes our discussion of the lymphatic system. The next part of the chapter begins our discussion of defenses against disease.

Animation Lymphatic System

▶ **31.1 CHECK YOUR PROGRESS**
1. Summarize how the lymphatic system contributes to homeostasis.
2. Associate lymph nodes with a function of the lymphatic system. Explain this association.

Media Integration
Enhances your study of biology with media. Go to **www.mhhe.com/maderconcepts3** to access animations, videos, and MP3 files referenced throughout this book. Also, ask your instructor about the eBook, LearnSmart™, and related quizzes available through Connect® and ConnectPlus® Biology.

Check Your Progress
Questions at the end of each section help you assess or apply your understanding of the concept.

Connecting the Concepts

Shows how the concepts of the chapter are related, and how they relate to concepts in other chapters. *Analyze and Evaluate* questions allow you to test your reasoning ability. All questions are answered in the Appendix.

Chapter Summary

This illustrated and bulleted summary is organized according to the chapter concepts. Boldface terms are included as an additional aid to help you review the chapter.

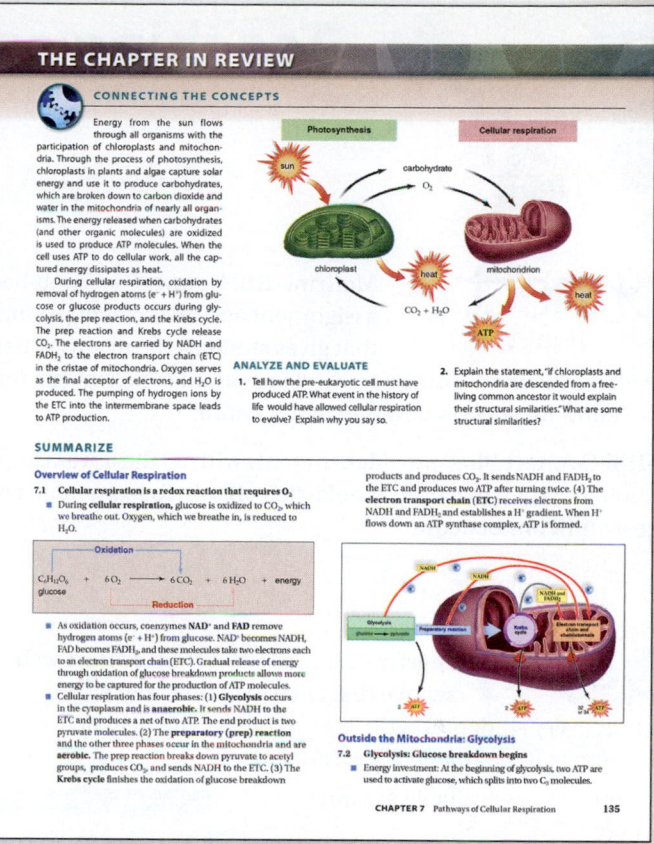

Chapter Review Questions

The end-of-chapter questions offer another way to review the chapter concepts. Included are *Test Yourself* multiple-choice questions and *Thinking Conceptually* questions that ask you to apply your understanding of a concept. *Get Involved* questions give you an opportunity to reason as a scientist.

Virtual Labs

For selected chapters, these online labs can help you better understand the content and provide you with the opportunity to investigate associated topics from a scientific perspective.

Media Study Tools

Provides a link to the *Concepts of Biology* companion website where you can find additional review materials.

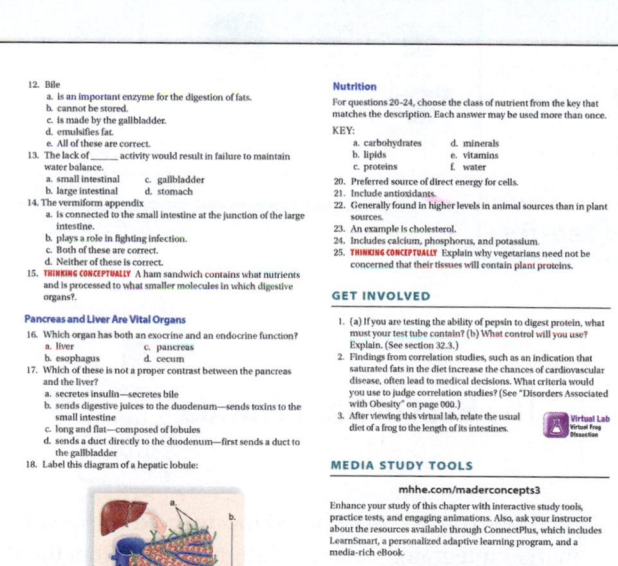

Teaching and Learning Tools

McGraw-Hill Connect® is a web-based assignment and assessment platform that gives students the means to better connect with coursework, instructors, and important concepts that they will need to know for success now and in the future.

McGraw-Hill Connect Plus® provides students with all the advantage of Connect Biology, plus a dynamic, media-rich eBook. To learn more visit **www.mcgrawhillconnect.com**

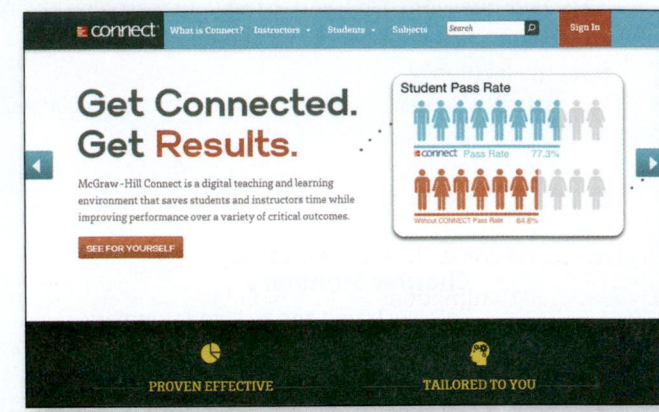

Tegrity Campus® is a lecture capture web service that allows you to easily record every class for every student. Tegrity is available as an integrated feature of Connect Biology and as a standalone.

- Make your classes available anytime, anywhere
- Simple one-click recording. No IT assistance required.
- Students can search a word or phrase and be taken to the exact place in your lecture they need to view.

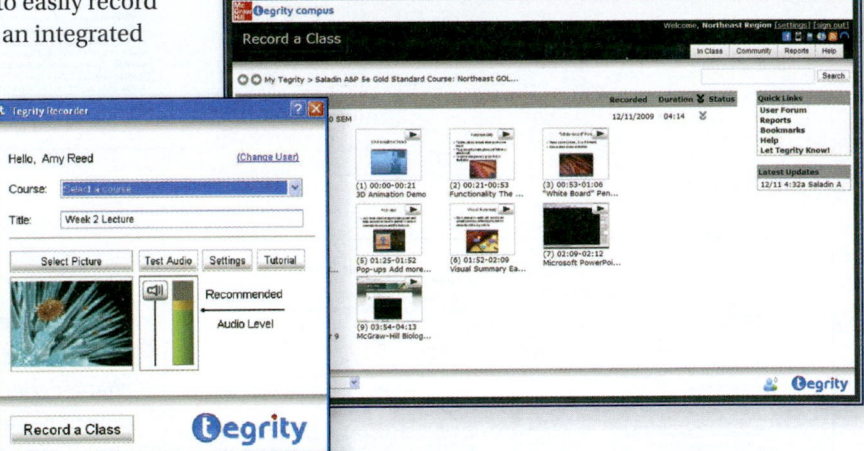

LEARNSMART®

Integrated within Connect and available as a standalone, **McGraw-Hill LearnSmart™** is the premier learning system designed to effectively assess a student's knowledge of course content. Through a series of adaptive questions, LearnSmart intelligently pinpoints concepts the student does not understand and maps out a personalized study plan for success. LearnSmart prepares students with a base of knowledge, allowing instructors to focus valuable class time on higher-level concepts.

New **SmartBook™** facilitates the reading process by identifying what content a student knows and doesn't know through adaptive assessments. As the student reads, the reading material constantly adapts to ensure the student is focused on the content he or she needs the most to close any knowledge gaps.

See pages xii-xiii of the preface for more information about the **LearnSmart Advantage™** suite of adaptive tools or go to **www.LearnSmartAdvantage.com.**

The **Best** of **Both Worlds**

McGraw-Hill and Blackboard Inc. teamed up to deliver the first to market integrated course solution which offers the deepest integration of publisher content within an LMS for Blackboard versions 8.0, 9.0, and 9.1.

Connect Single Sign-on. A single login and single environment provide seamless access to all course resources—all McGraw-Hill's resources are available within the Blackboard Learn platform.

Deep Integration. One-click access to McGraw-Hill Connect assignments and tools—all from within Blackboard Learn™.

One Gradebook. Automatic grade synchronization with Blackboard gradebook. All grades for McGraw-Hill assignments are recorded in the Blackboard gradebook automatically.

Instructor Resources

Connect Biology provides easy access to the following resources:

Presentation Tools

- Enhanced image PowerPoints® with editable art
- Lecture PowerPoints with animations
- Animation PowerPoints
- Labeled and unlabeled JPEG files of art, photos, and tables from the textbook.
- Instructor's Manual containing chapter outlines, lecture enrichment ideas, and discussion questions.
- Laboratory Resource Guide to accompany the *Concepts of Biology Laboratory Manual*

Animations for a New Generation

Dynamic, 3D animations of key biological processes bring an unprecedented level of control to the classroom. Innovative features keep the emphasis on teaching rather than entertaining.

New Guided Tutorials

Prepared by the authors of the textbook to assist students in understanding some of the more difficult topics in biology. Each video explores a specific figure in the text and is narrated by Mader series co-author Michael Windelspecht. During the video, important terms and processes are called out, allowing the student to focus on the key aspects of the figure. A list of tutorials can be found on page vi of the preface.

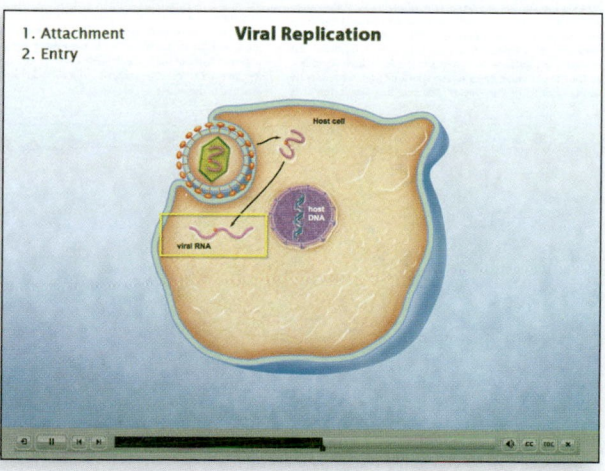

Computerized Test Bank

A comprehensive bank of test questions is provided within a computerized test bank powered by McGraw-Hill's flexible electronic testing program, **EZ Test Online.** A new tagging scheme allows you to sort questions by Bloom's difficulty level, learning outcome, topic, and section. With EZ Test Online, instructors can select questions from multiple McGraw-Hill test banks or author their own, and then either print the test for paper distribution or give it online.

Laboratory Manual

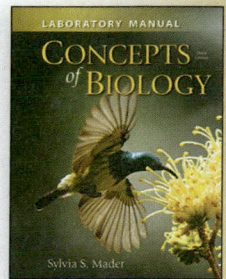

The *Concepts of Biology Laboratory Manual* is written by Dr. Sylvia Mader. Every laboratory has been written to help students learn the fundamental concepts of biology and the specific content of the chapter to which the lab relates, as well as gain a better understanding of the scientific method.

Companion Website

www.mhhe.com/maderconcepts3

The *Concepts of Biology* companion website allows students to access a variety of free digital learning tools that include

- Animations, videos, and MP3 files
- Chapter-level quizzing
- Vocabulary flashcards
- Virtual Labs
- Biology Prep

Create

Design your teaching resources to match the way you teach! With McGraw-Hill Create™ you can easily rearrange chapters, combine material from other content sources, and quickly upload content you have written. Access thousands of leading McGraw-Hill textbooks for content that fits your objectives and arrange your book to fit your teaching style. Create even allows you to personalize your book's appearance by selecting the cover and adding your name, school, and course information. Order a Create book and you'll receive a complimentary print review copy in 3–5 business days or a complimentary electronic review copy (eComp) via email in minutes. Go to **www.mcgrawhillcreate.com** today and register to experience how McGraw-Hill Create™ empowers you to teach *your* students *your* way.

The LearnSmart Advantage

LEARNSMART®

LearnSmart is the only truly adaptive learning system that intelligently identifies course-content students have not yet mastered and maps out personalized study plans for their success.

When LearnSmart has identified a specific subject area where the student is struggling, he or she is given a "time out" and directed to the textbook section and learning objective for remediation.

Dynamically generated reports document student progress and areas for additional reinforcement, offering at-a-glance views of their strengths and weaknesses.

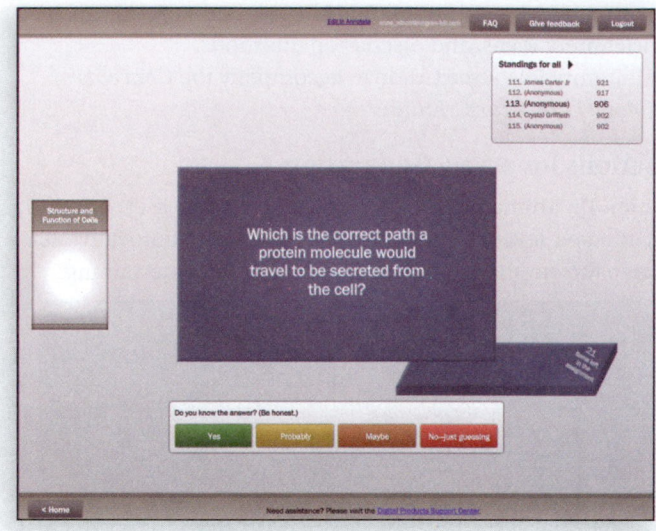

SMARTBOOK™

Powered by an intelligent diagnostic and adaptive engine, **SmartBook** facilitates the reading process by identifying what content a student knows and doesn't know through adaptive assessments.

As the student reads, the reading material constantly adapts to ensure the student is focused on the content he or she needs the most to close any knowledge gaps.

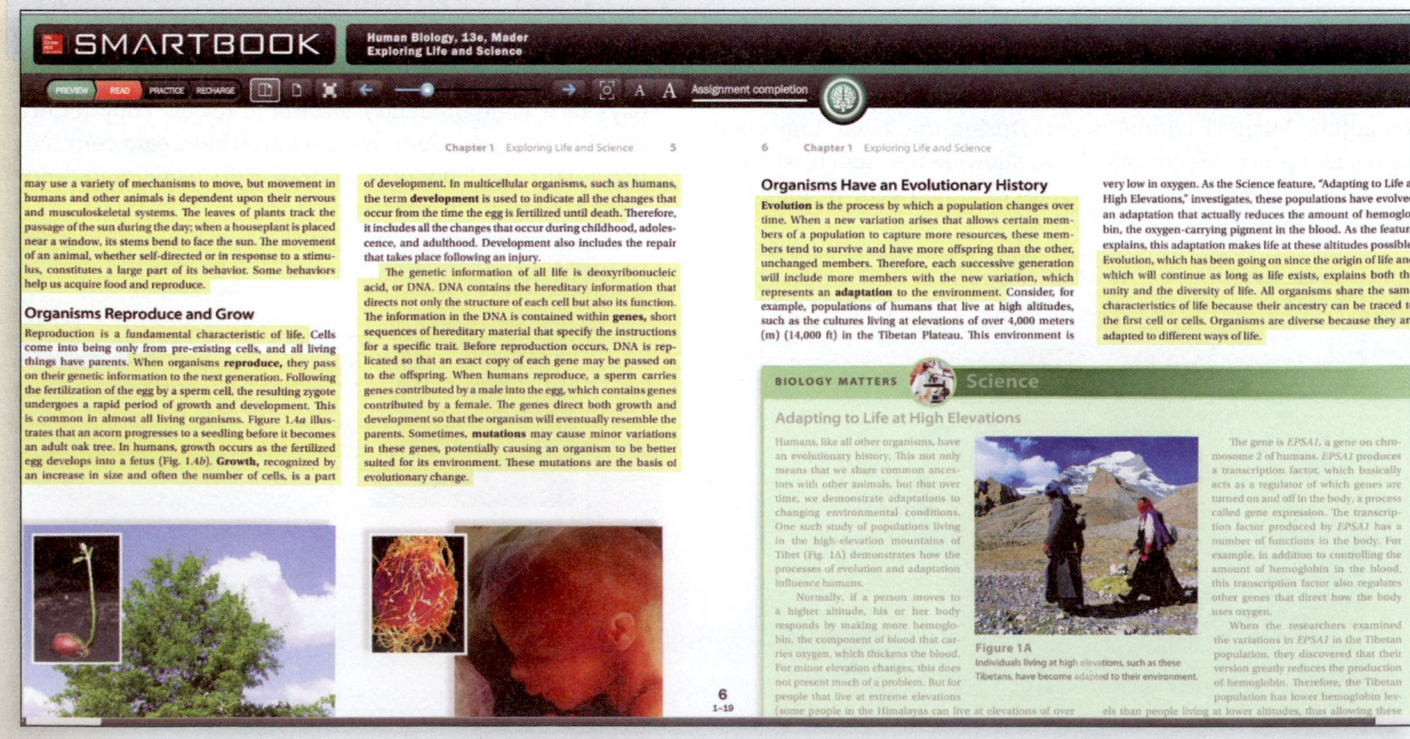

The Evolution of Learning

• MCGRAW-HILL LEARNSMART • SMARTBOOK • LEARNSMART PREP • LEARNSMART LABS • LEARNSMART ACHIEVE • LEARNSMART MASTER

LearnSmart Prep quickly and efficiently prepares students for a college level course. Prep uses a set of diagnostic questions to help identify what a student knows and doesn't know. It then provides a unique learning plan focused on helping the student master the basic skills and concepts he or she needs the most before entering the classroom.

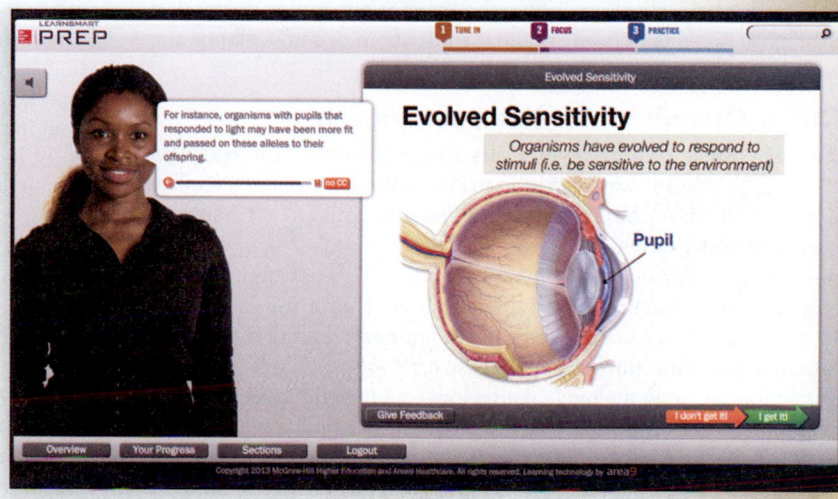

LearnSmart Labs is a super-adaptive simulated lab experience that brings meaningful scientific exploration to students. Through a series of adaptive questions, LearnSmart Labs identifies a student's knowledge gaps and provides resources to quickly and efficiently close those gaps. Once the student has mastered the necessary basic skills and concepts, they engage in a highly realistic simulated lab experience that allows for mistakes and the execution of the scientific method.

Whether your need is to overcome the logistical challenges of a traditional lab, provide better lab prep, improve student performance, or make your online experience one that rivals the real world, LearnSmart Labs accomplishes it all.

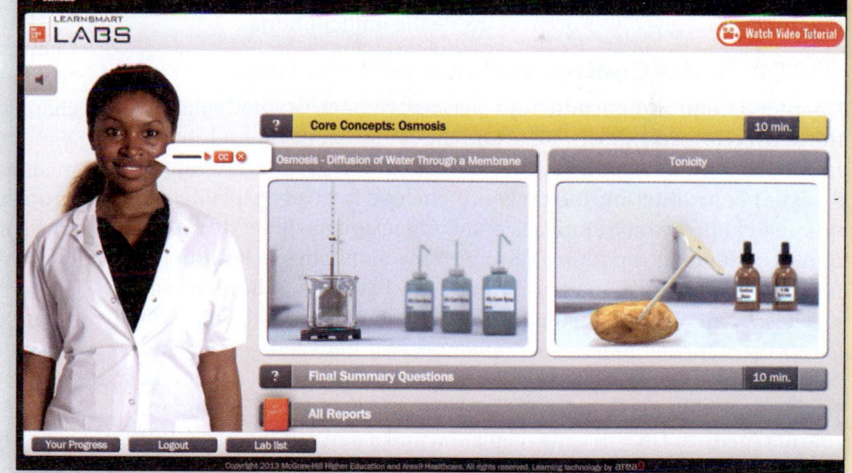

AN INNOVATIVE SUITE OF **ADAPTIVE LEARNING PRODUCTS** FUELED BY INTELLIGENT AND PROVEN LEARNING TECHNOLOGY

www.LearnSmartAdvantage.com

Detailed List of Content Changes

Chapter 1 Biology, The Study of Life Section 1.2 (pages 6-7) was rewritten to help students follow the experiment under discussion and Figure 1.2 was revised to allow students to see at a glance the results of the scientific study.

UNIT 1 Organisms Are Composed of Cell

Chapter 2 Basic Chemistry of Cells The application "The Harmful Effects of Acid Rain" was rewritten to stress the detrimental effects of an abnormal pH on organisms and inanimate structures. **Chapter 3 Organic Molecules of Cells** New introductory photos strikingly reveal relevancy of chemistry to our lives. In section 3.7, a new figure better compares the structure of various amino acids and the discussion directly progresses from amino acids to protein structure. **Chapter 4 Structure and Function of Cells** Figure 4.4B was redone to show the connection between the nuclear envelope and the endoplasmic reticulum. Table 4.11, a summary of the organelles, is now a better study tool for students. **Chapter 5 Dynamic Activities of Cells** This chapter was reorganized to position the plasma membrane discussion after the cell chapter and the study of enzymes prior to the photosynthesis chapter. **Chapter 6 Pathways of Photosynthesis** In section 6.10, modes of photosynthesis was rewritten to clarify and improve accuracy and a new table summarizes this section. **Chapter 7 Pathways of Cellular Respiration** was reorganized into two major sections to emphasize that glycolysis occurs outside the mitochondria.

UNIT 2 Genes Control the Traits of Organisms

Chapter 8 Cellular Reproduction This new chapter, devoted solely to the cell cycle and mitosis, aids student comprehension and better highlights the connection between the cell cycle and cancer. **Chapter 9 Sexual Reproduction** The process of meiosis is clearly explained in this new chapter which emphasizes how meiosis contributes to evolution by producing variations. Chromosome anomalies in section 9.7 illustrate how nondisjunction during meiosis results in human syndromes. **Chapter 11 Molecular Biology of Inheritance** A new section (11.9) dramatically illustrates the connection between genetic and chromosomal mutations and other human syndromes. **Chapter 12 Regulation of Gene Activity** This chapter emphasizes the important role of repetitive DNA in gene regulation and a new application (12B) summarizes the influence of epigenetic inheritance on the phenotype. **Chapter 13 Biotechnology and Genomics** Genomics is an expanding field in biology and the text recognizes this by better explaining the importance of functional and comparative genomics, proteomics, and bioinformatics in sections 13.10–13.12.

UNIT 3 Organisms Are Related and Adapted to Their Environment

Chapter 14 Evidence of Evolution The new title for this chapter emphasizes that Darwin presented evidence for evolution and that more evidence continues to accumulate. A dramatic new illustration (Fig. 14.6) shows how the altered activity of *Hox* genes can contribute to evolution of morphology. **Chapter 16 The Evolutionary History of Life on Earth** The contribution of molecular genetics to recognizing species is exemplified in a new application, "DNA Barcoding of Life." **Chapter 17 Evolution of Microbial Life** Figure 17.6 which outlines the origin of

the first cell is better integrated into the text and later the discussion of autotrophic bacteria is much improved. **Chapter 19 Evolution of Plants and Fungi** In section 19.8 the text suggests that a symbiotic relationship between fungi and plants contributed to the ability of plants to invade land. **Chapter 20 Evolution of Animals** In section 20.11, student interest in echinoderms is promoted by new photos and section 20.14 discusses and illustrates how amphibians evolved from lobe-finned fishes. **Chapter 21 Evolution of Humans** Section 21.4 includes recent finds in South Africa that question whether Lucy gave rise to humans. An expanded Neandertal section (21.5) addresses the question of whether humans interbred with Neandertals.

UNIT 4 Plants are Homeostatic

Chapter 22 Plant Organization and Homeostasis The section "Homeostatic Mechanisms of Plants" was rewritten to sharpen the presentation. **Chapter 23 Transport and Nutrition in Plants** Figures 23.1B and C are colorful additions to the chapter that better describe the structure of xylem and phloem. A relevant application "Plants Can Clean Up Toxic Messes" was rewritten to feature poplar trees and mustard plants. **Chapter 24 Control of Growth and Responses in Plants** In section 24.9, phytochrome structure and function was updated to emphasize its importance to plant physiology.

UNIT 5 Animals Are Homeostatic

Chapter 26 Animal Organization and Homeostasis The application "UV Rays: Too Much or Too Little?" is a new relevant addition to the chapter. **Chapter 29 Locomotion and Support Systems** New section 29.2 highlights the similar functions of a crayfish exoskeleton and a mammalian endoskeleton. A new application (29D) compares the structure and function of fast- and slow-twitch muscle fibers. **Chapter 30 Circulation and Cardiovascular Systems** This chapter has two relevant applications: "How to Prevent Cardiovascular Disease" was revised and the application "What to Know When Giving Blood" is new. **Chapter 31 Lymph Transport and Immunity** The application "Evolution of the Immune System" highlights that macrophages may have evolved from amoebas. **Chapter 34 Osmoregulation and Excretion** Several revised illustrations will make it easier for students to trace the path of urine and learn the functions of kidney tubules. **Chapter 35 Coordination by Hormone Signaling** The section "Hormones affect cellular metabolism" was rewritten to bring the principles of hormone action into sharper focus. The application "Fish Gills and Parathyroid Glands" gives evidence that the parathyroid glands evolved from a pharyngeal pouch.

UNIT 6 Organisms Live in Ecosystems

Chapter 37 Population Ecology The application "Sustainability of the U.S. Population" was revised to reflect recent demographic statistics. **Chapter 38 Behavioral Ecology** A comparative approach based on evolution is now used to discuss sexual selection in an expanded section. **Chapter 39 Community and Ecosystem Ecology** The discussion of ecological succession is much improved by showing appropriate stages in Glacier Bay, Alaska. **Chapter 41 Conservation Biology** A new application explores the efforts to restore the floodplain along the Illinois River in keeping with sustainable principles.

Acknowledgments

Many able and committed individuals assisted in the development of *Concepts of Biology*. I wish to acknowledge the efforts of my closest advisors at McGraw-Hill who helped me bring this book to fruition. In particular, let me thank my developmental editors, Rose Koos and Anne Winch, for advice and guidance during the preparation of the manuscript. The managing director, Michael Hackett and my editor, Eric Weber steadfastly encouraged and supported all those involved. The project managers, Jayne Klein and Vicki Krug, faithfully and carefully steered the book through the publication process. Chris Loewenberg, the marketing manager, is charged with educating the sales representatives on its message. The design of this book is the result of the creative talents of Laurie Janssen and many others who assisted in deciding the appearance of each element in the text. Electronic Publishing Services followed my guidelines as they created and reworked many illustrations, emphasizing pedagogy and beauty to arrive at the best presentation on the page. Evelyn Jo Johnson and Lori Hancock did a superb job of finding just the right photographs and micrographs. I was also very fortunate to have Bea Sussman as my copy editor, and Dawnelle Krouse and Kay J. Brimeyer as my proofreaders. As always, my family was extremely patient with me as I remained determined to meet every deadline on the road to publication. My husband, Arthur Cohen, is also a teacher of biology. The many discussions we have about the minutest detail to the gravest concept are invaluable to me. I am very much indebted to the contributors and reviewers whose suggestions and expertise were so valuable as I developed *Concepts of Biology*.

360° Development

McGraw-Hill's 360° Development Process is an ongoing, never-ending, market-oriented approach to building accurate and innovative print and digital products. It is dedicated to continual large-scale and incremental improvement driven by multiple customer feedback loops and checkpoints. This is initiated during the early planning stages of our new products, and intensifies during the development and production stages, then begins again upon publication in anticipation of the next edition.

This process is designed to provide a broad, comprehensive spectrum of feedback for refinement and innovation of our learning tools, for both student and instructor. The 360° Development Process includes market research, content reviews, course- and product-specific symposia, accuracy checks, and art reviews. We appreciate the expertise of the many individuals involved in this process.

Ancillary Authors

Connect Question Bank: Michael Duus *Appalachian State University;* **Question Bank Reviewer:** Dave Cox, *Lincoln Land Community College;* **Test Bank:** Douglas Darnowski, *Indiana University Southeast,* and Dave Cox, *Lincoln Land Community College;* **Test Bank Reviewer:** Chris Sorenson, *St. Cloud Technical and Community College;* **Tutorial Development:** Krissy Johnson and Sandy Windelspecht, *Ricochet Creative Productions;* **Tutorial Reviewers:** Michael Troyan, *Penn State University,* and Ted Zerucha, *Appalachian State University;* **Companion Website Quizzes:** Alex James, *Ricochet Creative Productions;* **Quiz Reviewer:** Roberta Batorsky, *Middlesex Community College;* **Lecture Outlines:** Stephanie Songer, *North Georgia College and State University;* **Instructor's Manual:** Mark Matusiak, *Rockwood Summit High School;* **LearnSmart Authors:** Patrick Galliart, *North Iowa Area Community College,* Alex James, *Ricochet Creative Productions,* Stacy Zell, *Carroll Community College,* Joy Brookshire, *Kennesaw State University,* Daniel Matusiak, *St. Charles Community College,* Tammy Atchison, *Pitt Community College,* and Sylvester Allred, *Northern Arizona University;* **Reviewers:** Jill Nugent, *University of North Texas,* and MaryJo Witz, *Monroe Community College*

Third Edition Reviewers

Cynthia Anderson, *Georgia Military College*
Jack R. Brook, *Mt. Hood Community College*
Matthew Rex Burnham, *Jones County Junior College*
Erin Christensen, *Middlesex County College*
Robert W. Colburn, *Middlesex County College*
James F. Duke, *Calhoun Community College*
Mark Fairbrass, *Georgia Military College*
Patrick Galliart, *North Iowa Area Community College*
Melissa Greene, *Northwest Mississippi Community College*
Mark Henry, *Illinois Central College*
Tanganika Johnson, *Southern University and A&M College*
John Jones, *Calhoun Community College*
Mary Jane Keleher, *Salt Lake Community College*
Holly Langille, *Northwest Florida State College*
Martin Lowery, *Lonestar College*
Marla Gomez, *Nicholls State University*
Rhana Paris, *College of the Albemarle, Dare Campus*
Subbarayan R. Pochi, *University of Miami*
Adele Register, *Rogers State University*
Vicki Schmidt, *Fox Valley Technical College*
Jennifer Scoby, *Illinois Central College*
Lisa Strong, *Northwest Mississippi Community College*
Sherri Townsend, *North Arkansas College*

Contents

UNIT 6

Organisms Live in Ecosystems 728

CONCEPTS *of* BIOLOGY

1

Biology, the Study of Life

Fire Ants Protect Their Own

Fire ants have a red to reddish-brown color, but even so, they most likely take their name from the ability to STING. Their stinger protrudes from the rear, but in a split second, they can grab a person's skin with their mandibles and position the stinger between their legs to sting from the front. The stinger injects a toxin into the tiny wound, and the result is a burning sensation. The next day, the person has a white pustule at the site of the sting. The success of this defense mechanism is clear because most animals, including humans, try to stay away from bees, wasps, and ants—and any other animal that can sting.

Fire ants usually live in an open, grassy area, and they sting in order to defend their home, which is a mound of soil that they have removed from subterranean tunnels. They use the tunnels to safely travel far afield when searching for food, which they bring back to their nest mates. The queen and many worker ants live in chambers within the mound or slightly below it. The queen is much larger than the other members of

Fire ant mound

the colony, and she has only one purpose: to produce many thousands of small, white eggs. The eggs develop into cream-colored, grublike larvae, which are gently tended by worker ants to keep them clean and well fed. When the larvae become encased by a hard covering, they are pupae. Inside a pupa, an amazing transformation takes place, and eventually an adult ant breaks out. Most of these adults are worker ants, but in the spring, a few are winged "sexuals," which are male and female ants with the ability to reproduce. The sexuals remain inside the colony with nothing to do until the weather is cooperative enough for them to fly skyward to mate. A few of the fertilized females manage to survive the perils of an outside existence long enough to start another colony.

All of the ants in a colony have the same mother, namely the queen ant who produces the eggs. The workers are sterile, closely related sister ants. Because of their close genetic relationship, we can view the members of a colony as a superorganism. The queen serves as the reproductive system, while the workers serve as the digestive and urinary systems, as well as all the other systems that keep the superorganism functioning. What fosters cooperation between the members of the superorganism? The answer is chemicals, pheromones secreted externally that influence the behavior and even the development of the ants. Fire ants, like other ants, produce several different pheromones that send messages when released into the air. The message could be "food is available" or "be alert for possible danger." The queen even releases pheromones that cause workers to attend her.

Why does it work, in a biological sense, for these sisters to spend their lives working away, raising more sterile sisters and defending the colony with little regard for their own safety? It works because the few sexual females that survive their temporary existence on the outside pass the colony's joint genes on to future generations in new and different places. Any social system that allows an organism to pass on its genes is a successful one from an evolutionary point of view.

In this chapter, we will first learn how the scientific understanding of life progresses by making observations and doing experiments. Then we will examine the five scientific theories around which this book is organized. The theory of evolution is examined in particular detail because it is the unifying theory of biology.

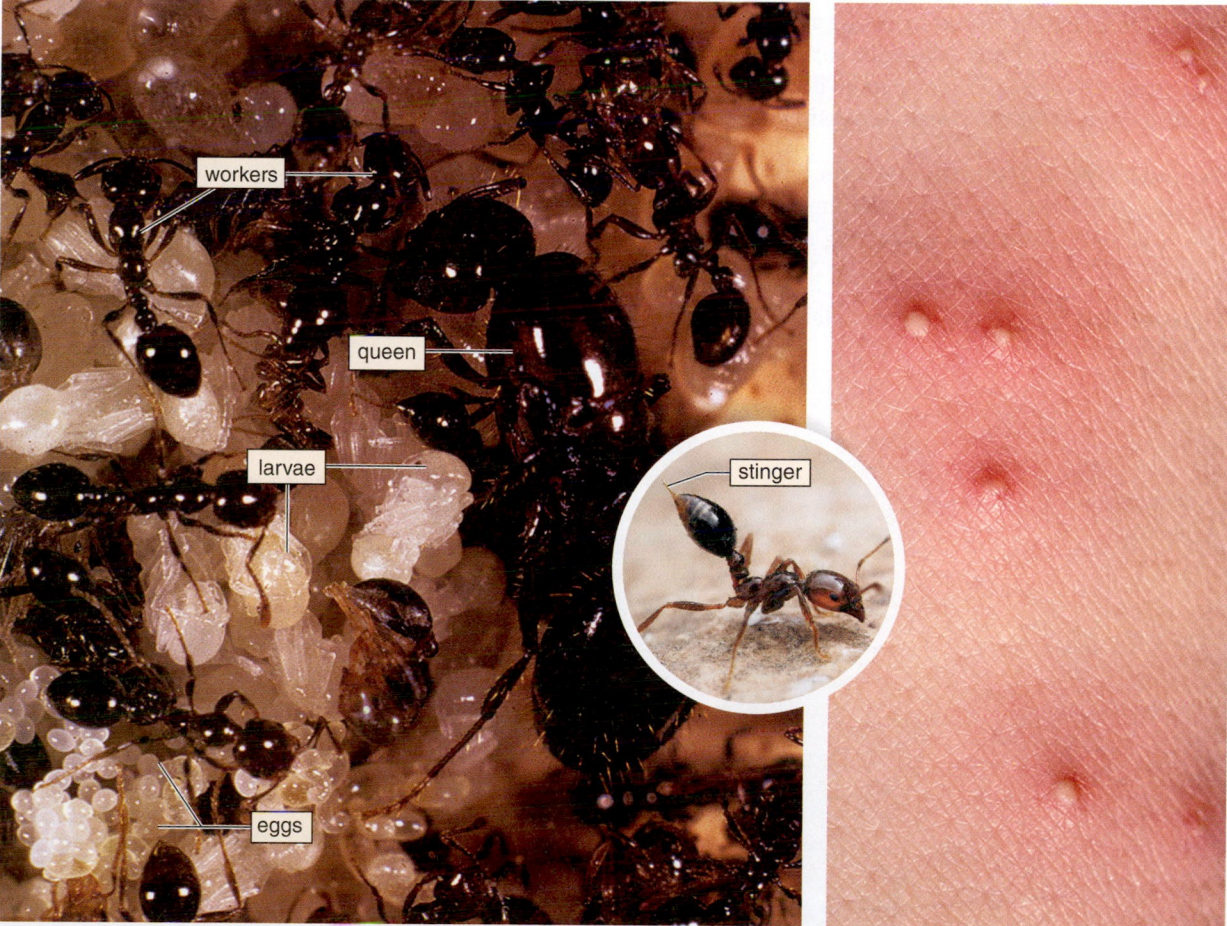

Fire ant colony (*Solenopsis invicta*)

Pustules caused by fire ants

The Process of Science

Biology is the scientific study of life, and therefore it is appropriate for us to first consider what we mean by science. Science is a way of making sense of the natural world around us. Religion, aesthetics, and ethics are all ways that humans can find order in the natural world. Science, unlike these other ways of knowing, is testable. It also leads to improved technology and is responsible for the modern ways in which we travel, communicate, farm, build our houses, and even how we conduct science.

1.1 Scientists use a preferred method

LEARNING OUTCOMES

When you complete this section, you should be able to

1. Describe the four steps of the scientific method.
2. Analyze the five basic theories of biology.

Despite the wide diversity of scientists and what they study (Fig. 1.1A), the usual four steps of the scientific method are (1) making observations, (2) formulating a hypothesis, (3) performing experiments and making observations, and (4) coming to a conclusion (Fig. 1.1B).

Making Observations The scientific method begins with **observation.** We can observe with our noses that dinner is almost ready, observe with our fingertips that a surface is smooth and cold, and observe with our ears that a piano needs tuning. Scientists also extend the ability of their senses by using instruments; for example, the microscope enables them to see objects they could never see with the naked eye. Finally, scientists may expand their understanding even further by taking advantage of the knowledge and experiences of other scientists. For instance, they may look up past studies on the Internet or at the library, or they may write or speak to others who are researching similar topics.

Formulating a Hypothesis After making observations and gathering knowledge about a phenomenon, a scientist uses inductive reasoning. **Inductive reasoning** occurs whenever a person uses creative thinking to combine isolated facts into a cohesive whole. Chance alone can help a scientist arrive at an idea. The most famous case pertains to the antibiotic penicillin, which was discovered in 1928. While examining a petri dish of bacteria that had accidentally become contaminated with the mold *Penicillium,* Alexander Fleming observed an area around the mold that was free of bacteria. Fleming had long been interested in finding cures for human diseases caused by bacteria, and was very knowledgeable about antibacterial substances. So when he saw the dramatic effect of *Penicillium* mold on bacteria, he reasoned that the mold might be producing an antibacterial substance. We call such a possible explanation for a natural event a **hypothesis.** A hypothesis is based on existing knowledge, so it is much more informed than a mere guess. Fleming's hypothesis was supported by further observations. Sometimes a hypothesis is not supported, and must be either modified and subjected to additional study, or rejected.

All of a scientist's past experiences, no matter what they might be, may influence the formation of a hypothesis. But a scientist considers only hypotheses that can be tested by experiments or further observations. Moral and religious beliefs, while very important to our lives, differ between cultures and through time, and are not always testable.

Performing Experiments and Making Observations Scientists often perform an **experiment,** a series of procedures to test a hypothesis. The manner in which a scientist intends to

FIGURE 1.1A Biologists work in a variety of settings.

Biologist in an agricultural field

Biochemist in a laboratory

Ecologist examining an artificial reef

conduct an experiment is called its design. A good experimental design ensures that scientists are testing what they want to test and that their results will be meaningful. When an experiment is done in a laboratory, all conditions can be kept constant except for an **experimental variable,** which is deliberately changed. One or more **test groups** are exposed to the experimental variable, but one other group, called the **control group,** is not. If, by chance, the control group shows the same results as the test group, the experimenter knows the results are invalid.

Scientists often use a **model,** a representation of an actual object. For example, modeling occurs when scientists use software to decide how human activities will affect climate, or when they use mice instead of humans for, say, testing a new drug. Ideally, a medicine that is effective in mice should still be tested in humans. And whenever it is impossible to study the actual phenomenon, a model remains a hypothesis in need of testing. Someday, a scientist might devise a way to test it.

The results of an experiment or further observations are referred to as the **data.** Mathematical data are often displayed in the form of a graph or table. Sometimes studies rely on statistical data. Let's say an investigator wants to know if eating onions can prevent women from getting osteoporosis (weak bones). The scientist conducts a survey asking women about their onion-eating habits and then correlates these data with the condition of their bones. Other scientists critiquing this study would want to know: How many women were surveyed? How old were the women? What were their exercise habits? What criteria were used to determine the condition of their bones? And what is the probability that the data are in error? Even if the data do suggest a correlation, scientists would want to know if there is a specific ingredient in onions that has a direct biochemical or physiological effect on bones. After all, correlation does not necessarily mean causation. It could be that women who eat onions eat lots of vegetables, and have healthier diets overall than women who do not eat onions. In this way scientists are skeptics who always pressure one another to keep investigating.

Coming to a Conclusion Scientists must analyze the data in order to reach a **conclusion** about whether a hypothesis is supported or not. The data can support a hypothesis, but they do not prove it "true" because a conclusion is always subject to revision. On the other hand, it is possible to prove a hypothesis false. Because science progresses, the conclusion of one experiment can lead to the hypothesis for another experiment as represented by the return arrow in Figure 1.1B. In other words, results that do not support one hypothesis can often help a scientist formulate another hypothesis to be tested. Scientists report their findings in scientific journals so that their methodology and data are available to other scientists. Experiments and observations must be *repeatable*—that is, the reporting scientist and any scientist who repeats the experiment must get the same results, or else the data are suspect.

Scientific Theory The ultimate goal of science is to understand the natural world in terms of **scientific theories,** which are accepted explanations (concepts) for how the world works. The results of innumerable observations and experiments support a scientific theory. This text is organized around the following five basic theories of biology:

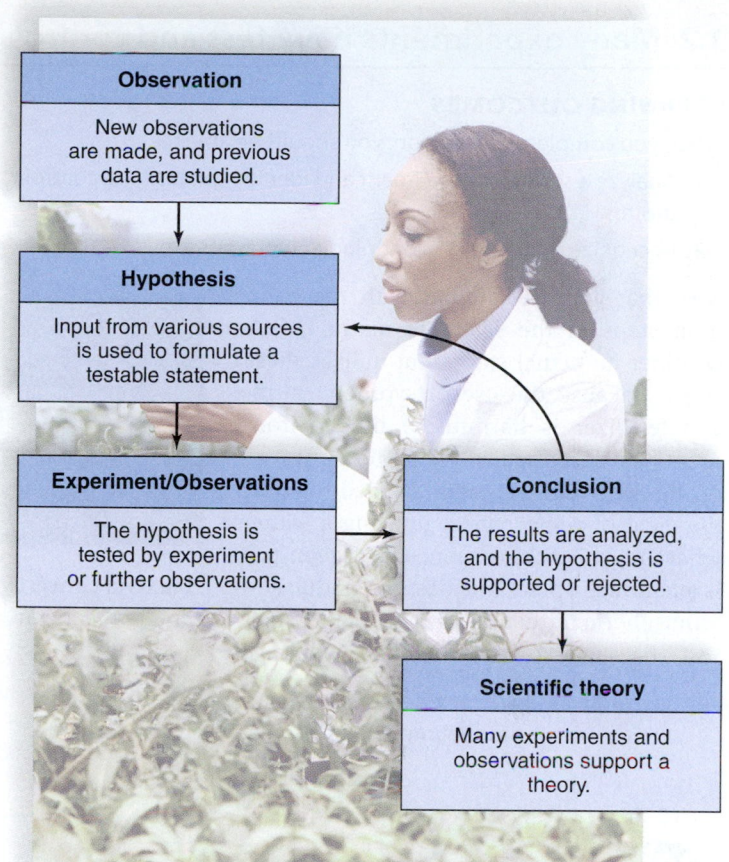

FIGURE 1.1B Flow diagram for the scientific method. On the basis of observations, a scientist formulates and tests a hypothesis. The data either supports or does not support the hypothesis. The back arrow means that a scientist tests many related hypotheses. A scientific theory is supported by many varied studies.

Theory	Concept
Cell	All organisms are composed of cells, and new cells come only from preexisting cells.
Gene	All organisms contain coded information that dictates their form, function, and behavior.
Homeostasis	All organisms have an internal environment that must stay relatively constant within a range protective of life.
Ecosystem	All organisms are members of populations that interact with each other and with the physical environment within a particular locale.
Evolution	All organisms have a common ancestor, but each is adapted to a particular way of life.

We will discuss these theories in detail later in the chapter, but right now let's turn our attention to an example of a scientific experiment.

▶ **1.1 CHECK YOUR PROGRESS**
1. Identify the purpose of each step in the scientific method.
2. Compare how each of the five basic theories of biology pertains to an organism.

1.2 Many experiments have test and control groups

LEARNING OUTCOMES

When you complete this section, you should be able to

1. Analyze a scientific experiment and identify the test and control groups.
2. Recognize that science is ongoing and progressive.

Now that you are familiar with the common steps in the scientific method, let's consider an actual study that utilizes these steps. Because the use of synthetic nitrogen fertilizer is harmful to the environment (as described in "Organic Farming" on this page), researchers decided to study the yield of winter wheat utilizing a winter wheat/pigeon pea rotation. The pigeon pea

nodule

is a **legume,** a plant that has root nodules where bacteria convert atmospheric nitrogen to a form plants such as winter wheat can use. The scientists formulated this hypothesis:

Hypothesis A winter wheat/pigeon pea rotation will cause winter wheat production to increase as well as or better

than an *artificial fertilizer treatment,* which meant that synthetic nitrogen fertilizer would be applied.

This study had a good design because it included test groups and a control group (Fig. 1.2*a*). All environmental conditions for all groups are kept constant, but the test groups are exposed to an experimental variable, the factor being tested. The **control group** is not exposed to the experimental variable. The use of a control group ensures that the data from the test groups are due to the experimental variable and not to some unknown outside influence. Test groups should be as large as possible to eliminate the influence of undetected differences in the test subjects.

The investigators decided to grow the winter wheat in pots and to have three sets of pots:

Control Pots Winter wheat was planted in clay pots of soil that received no fertilization treatment—that is, no artificial fertilizer and no preplanting of pigeon peas.

Test Pots I Winter wheat was grown in clay pots in soil that received the artificial fertilizer treatment.

HOW BIOLOGY IMPACTS OUR LIVES   *Application*

1A Organic Farming

Besides being health conscious, people who buy organic may also be socially conscious. Organic farming is part of a movement to make agriculture sustainable by using farming methods that protect the health of people and ecosystems and preserve the land so that it can be productive for our generation and all future generations.

Modern agricultural methods have been dramatically successful at increasing yield, but at what price? We now know that modern farming practices lead to topsoil depletion and groundwater contamination. Without topsoil, the nutrient-rich layer that nourishes plants, agriculture is impossible, and yet modern farming practices such as tilling the land and allowing it to lie fallow (bare) allow topsoil to erode and disappear. One solution is to use a legume as a ground cover because it both protects and nourishes the soil. The researchers who did the study described in section 1.2 used pigeon peas as a way to enrich the soil between winter wheat plantings.

Instead of growing legumes, farmers are accustomed to making plants bountiful by applying more and more synthetic nitrogen fertilizer. Unfortunately, nitrogen fertilizers pollute wells used for drinking water and also huge bodies of water, such as the Chesapeake Bay, the Gulf of Mexico, and the Great Lakes. Nitrates in the drinking water of infants leads to the "blue-baby" syndrome and possible death due to lack of oxygen in the blood. In adults, nitrates are implicated in causing digestive tract cancers. Certainly they can cause an algal bloom, recognized as a green scum on the water's surface.

In response to these problems, organic farmers severely limit the use of nitrogen fertilizers and instead rely on crop rotation, alternately planting a nitrogen-providing legume and a nitrogen-requiring crop such as wheat. Organic farmers also cut way back on the use of herbicides and pesticides, and this may be the primary reason you and others buy organic. The long-term consumption of these chemicals has been associated with such health problems as birth defects, nerve damage, and cancer. Children may be especially sensitive to health risks posed by pesticides; this is the chief reason lawns sprayed with pesticides carry warning signs. We should all be aware that we too can contribute to an organic lifestyle by limiting the use of synthetic chemicals on our lawns and gardens. In doing so, we improve our health and help preserve the environment for ourselves and future generations.

CONSIDER THESE QUESTIONS

1. The United States exports its current farming technology, with all its long-range problems, to other countries. Should this be continued?
2. Should farmers use organic farming methods even though it may take time for it to be profitable? Why or why not?

 **connect** Explore the concepts through a variety of multimedia assets and question types.

|BIOLOGY

www.mcgrawhillconnect.com

Control pots
no treatment

Test pots I
artificial fertilizer treatment

a. Experiment

Test pots II
pigeon pea treatment

Control Pots
■ = no treatment
Test Pots
■ = artificial fertilizer treatment
■ = pigeon pea treatment

best yield

more yield

least yield

Wheat Yield (grams/pot)

year 1   year 2   year 3

b. Results

FIGURE 1.2 Scientific study. **a.** Experiment contained three types of pots. **b.** Pigeon pea treatment (brown) had poor yield in the first year but by far the best yield by the third year.

Test Pots II Winter wheat was grown in clay pots following pigeon pea plants grown in the summertime. The pigeon pea plants were then turned over in the soil. In other words these pots received a *pigeon pea treatment*.

Results Figure 1.2*b* is a color-coded bar graph that allows you to see at a glance the comparative amount of wheat obtained from each group of pots. After the first year, winter wheat yield was higher in test pots treated with artificial fertilizer than in the control pots. To the surprise of investigators, test pots preplanted with pigeon peas did not produce as high a yield as the control pots.

> **Conclusion** The hypothesis was not supported. Wheat yield following the growth of pigeon peas was not as great as that obtained with artificial fertilizer treatments.

Follow-Up Experiment and Results The researchers decided to continue the experiment, using the same design and the same pots as before, to see whether the buildup of residual soil nitrogen from pigeon peas would eventually increase wheat yield to a greater extent than the use of artificial fertilizer. This was their new hypothesis:

> **Hypothesis** A sustained pigeon pea treatment will eventually cause an increase in winter wheat production.

They predicted that wheat yield following 3 years of pigeon pea treatment would surpass wheat yield following artificial fertilizer treatment.

Analysis of Results After 2 years, the yield from pots treated with artificial fertilizer was less than it had been the first year. Indeed, wheat yield in pots following a summer planting of pigeon peas was the highest of all the treatments. After 3 years, wheat yield in pots treated with artificial fertilizer was greater than in the control pots but not nearly as great as the yield in pots following summer planting of pigeon peas. Compared to the first year, wheat yield increased almost fourfold in pots having a pigeon pea/winter wheat rotation.

> **Conclusion** The hypothesis was supported. At the end of 3 years, the yield of winter wheat following a pigeon pea/winter wheat rotation was much better than for the other types of test pots.

To explain their results, the researchers suggested that the soil was improved by the buildup of the organic matter in the pots as well as by the addition of nitrogen from the pigeon peas. They published their results in a scientific journal,[1] where their experimental method and results would be available to the scientific community.

▶ **1.2 CHECK YOUR PROGRESS**
1. Identify the test and control pots in the winter wheat experiment.
2. Explain why the researchers did a follow-up experiment.

[1]Bidlack, J. E., Rao, S. C., and Demezas, D. H. "Nodulation, nitrogenase activity, and dry weight of chickpea and pigeon pea cultivars using different *Bradyrhizobium* strains," *Journal of Plant Nutrition* 24:549–560 (2001).

FIGURE 1.3A Levels of biological organization.

Biosphere
Regions of the Earth's crust, waters, and atmosphere inhabited by living organisms

↑

Ecosystem
A community plus the physical environment

↑

Community
Interacting populations in a particular area

↑

Population
Organisms of the same species in a particular area

↑

Organism
An individual; complex individuals contain organ systems

↑

Organ System
Composed of several organs working together

↑

Organ
Composed of tissues functioning together for a specific task

↑

Tissue
A group of cells with a common structure and function

↑

Cell
The structural and functional unit of all living things

↑

Molecule
Union of two or more atoms of the same or different elements

↑

Atom
Smallest unit of an element composed of subatomic particles

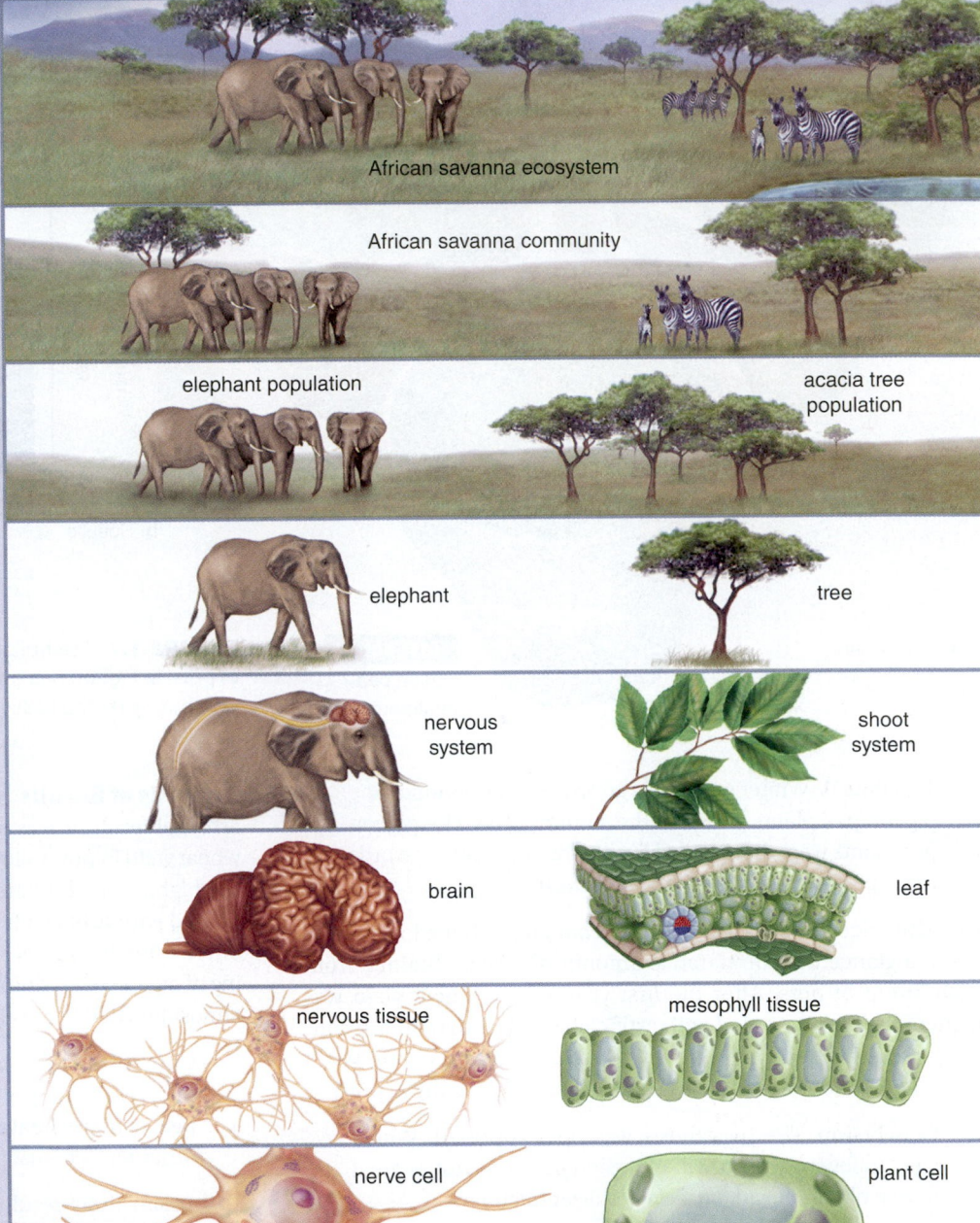

African savanna ecosystem

African savanna community

elephant population

acacia tree population

elephant

tree

nervous system

shoot system

brain

leaf

nervous tissue

mesophyll tissue

nerve cell

plant cell

methane

oxygen

The Cell Theory

From huge menacing sharks to minuscule exotic orchids, life is very diverse. Despite this diversity, biologists have concluded that life can be understood in terms of the five theories that are emphasized in this text. The first theory we will discuss is the cell theory.

1.3 Cells are the fundamental unit of living things

LEARNING OUTCOMES

When you complete this section, you should be able to

1. Identify two major principles of the cell theory.
2. Recognize that cells reproduce and use materials and energy to stay alive.

Figure 1.3A illustrates very well why we will first discuss the **cell theory,** which says that *cells are the fundamental unit of living things.* In a cell, **atoms,** the smallest portions of an element, combine with themselves or other atoms to form **molecules.** Although cells are composed of molecules, cells, and not molecules, are alive. Some cells, such as unicellular paramecia, live independently. Other cells, such as those of the alga *Volvox,* cluster together in microscopic colonies. An elephant is a multicellular organism in which similar cells combine to form a **tissue;** one common tissue in animals is nerve tissue. Tissues make up **organs,** as when various tissues combine to form the brain. Organs work together in **organ systems;** for example, the brain works with the spinal cord and a network of nerves to form the nervous system. Organ systems are joined together to form a complete living thing, or **organism.** Only a microscope can reveal that organisms are composed of cells (Fig. 1.3B).

Later in this chapter, we will consider the higher levels of biological organization shown in Figure 1.3A.

Cells Come from Other Cells Cells come only from a previous cell, and organisms come only from other organisms. In other words, cells and organisms **reproduce.** Every type of living thing can reproduce, or make another organism like itself. Bacteria, protists, and other unicellular organisms simply split in two. In most multicellular organisms, the reproductive process is more complex. It begins with the pairing of two cells—a sperm from one partner and an egg from the other partner. The union of sperm and egg, followed by many cell divisions, results in an immature stage that grows and develops through various stages to become an adult.

Cells Use Materials and Energy Cells and organisms cannot maintain their organization or carry on life's activities without an outside source of nutrients and energy. Nutrients function as building blocks or for energy. **Energy** is the capacity to do work, and it takes work to maintain the organization of the cell and the organism. When cells use nutrients to make their parts and products, they carry out a sequence of chemical reactions. Nerve cells and muscle cells also use energy as organisms move about. The term **metabolism** encompasses all the chemical reactions that occur in a cell.

The ultimate source of energy for nearly all life on Earth is the sun. Plants and certain other organisms are able to capture solar energy and carry on **photosynthesis,** a process that transforms solar energy into the chemical energy of organic nutrients. All life on Earth acquires energy by breaking down nutrients made by photosynthesizers. This applies even to plants.

▶ **1.3 CHECK YOUR PROGRESS**
1. Explain how life has order by referring to Figure 1.3A.
2. Identify how cells reproduce, maintain themselves, and acquire energy.

a.   b.   c.

FIGURE 1.3B Only micrographs (pictures taken microscopically), such as the one in (**c**), can reveal that organisms are composed of cells.

The Gene Theory

The cell theory studied in section 1.3 and the gene theory are intimately connected. Genes are housed in cells, and when cells divide, they pass on genes to the next cell or organism. Genes code for proteins, and it is proteins that directly bring about the traits of organisms.

1.4 Organisms have a genetic inheritance

LEARNING OUTCOMES

When you complete this section, you should be able to

1. Restate the gene theory in your own words.
2. Identify several applications of DNA technology.

A nineteenth-century scientist named Gregor Mendel is often called the father of genetics because he was the first to conclude, following experimentation with pea plants, that units of heredity now called **genes** are passed from parents to offspring. Later investigators, notably James Watson and Francis Crick, discovered that genes are composed of the molecule known as **DNA** (**deoxyribonucleic acid**). The work of these and many other investigators allows us to state the first premise of the **gene theory:** *Genes are hereditary units composed of DNA.* Our increasing knowledge of DNA tells us that genes contain coded information that controls the structure and function of cells and organisms. The spiral staircase structure of DNA contains four different types of molecules called nucleotides, each represented by a different color in Figure 1.4. DNA can **mutate** (undergo permanent changes), and each type of organism has its own particular sequence of these four nucleotides. This is called coded information because a particular nucleotide sequence codes for a particular protein. **Proteins** are cellular molecules that determine what the cell and the organism are like. The second premise of the gene theory is: *Genes control the structure and function of cells and organisms* by coding for proteins.

The gene theory has been extremely fruitful, meaning that it has led to much experimentation and many applications. Every field of biology and most aspects of our lives have changed because of the ability to analyze and manipulate DNA. Here are a few examples:

Basic Genetic Research

We can extract DNA and study metabolism at the molecular level. Therefore, we will soon know how one cell type differs from another.

We can also sequence the nucleotides in DNA and study how the DNA activity is regulated. One day we will know how this makes humans different from chimpanzees, for example.

Medicine

Genetic testing can tell us what diseases we are prone to, and doctors can use this information to prescribe drug therapy or tell us how best to protect ourselves.

Drugs for diabetes, blood disorders, vaccines, and many other diseases are now made by utilizing DNA technology.

Relationship of Species

DNA technology aids in discovering the history of life on Earth—that is, who is related to whom. For example, a recent comparative study concluded that early humans did interbreed with the archaic humans known as Neandertals.

Wildlife biologists use DNA sequence data to determine how best to conserve various species.

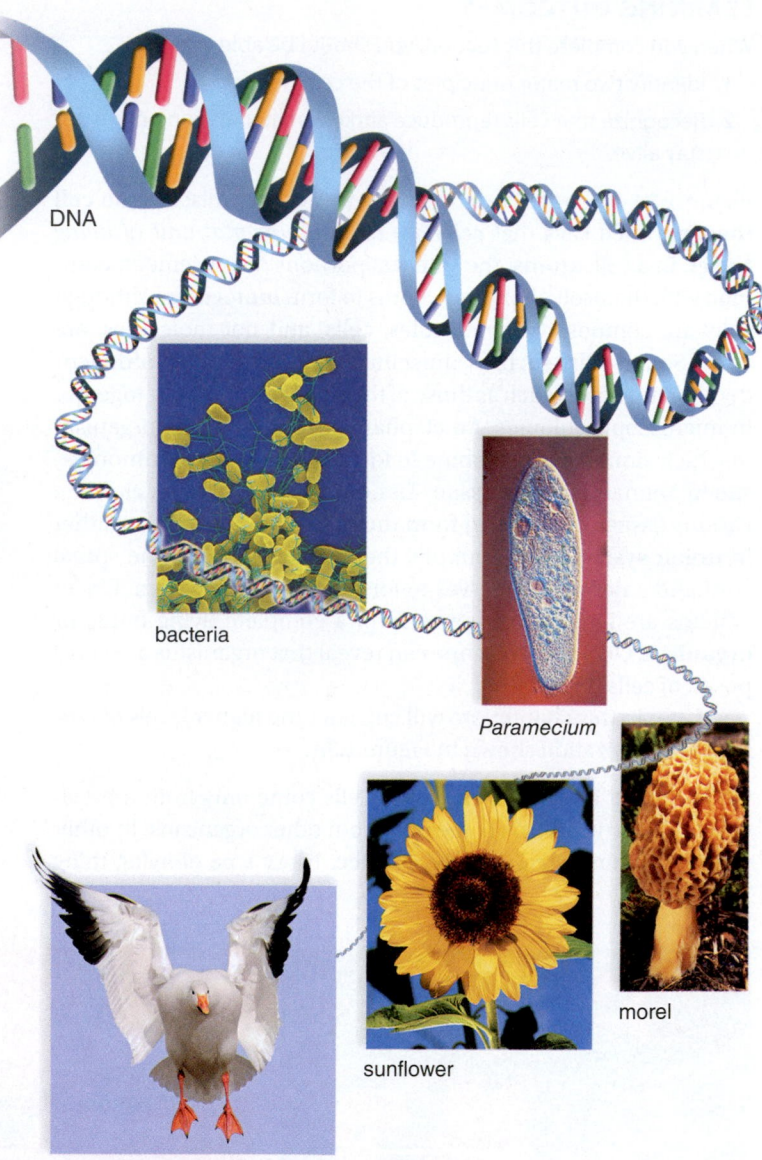

DNA

bacteria

Paramecium

morel

sunflower

snow goose

FIGURE 1.4 DNA differences account for the variety of life on Earth as exemplified by these examples.

Video
Melanoma Marker

▶ 1.4 CHECK YOUR PROGRESS

1. Explain why the organisms in Figure 1.4 produce only their own kind when they reproduce.
2. List several ways in which DNA technology has enhanced biological understanding.

The Theory of Homeostasis

To survive, cells and organisms must maintain a state of biological balance, or **homeostasis.** For example, temperature, moisture level, acidity, and other physiological factors must remain within the tolerance range of cells. Without the ability to respond to internal and external stimuli organisms wouldn't be able to maintain homeostasis. When temperatures fall, complex organisms generate their own warmth and/or seek an environment that will raise their temperature, for example.

1.5 Organisms regulate their internal environment

LEARNING OUTCOMES

When you complete this section, you should be able to

1. Explain the theory of homeostasis.
2. Analyze the relationship between homeostasis and response to a stimulus.

The **theory of homeostasis** tells us that *cells and organisms have an internal environment* and that *living systems regulate this environment so that it stays fairly constant.* While individual cells are homeostatic, most examples of homeostasis involve multicellular organisms. Animals have intricate feedback and control mechanisms that do not require any conscious activity. For example, when a student is so engrossed in her textbook that she forgets to eat lunch, her liver releases stored sugar to keep her blood sugar level within normal limits. In this case, hormones regulate sugar storage and release, but in other instances, the nervous system is involved in maintaining homeostasis.

Many animals depend on behavior to regulate their internal environment. The same student may realize that she is hungry and decide to visit the local diner. A lizard may raise its internal temperature by basking in the sun (Fig. 1.5A) or cool down by moving into the shade. Similarly, fire ants move upward into the mound when the warmth of the sun is needed and move back down into their cooler subterranean passageways when the sun is too hot.

We will see that plants are, to a degree, homeostatic. For example, they bend toward sunlight and have mechanisms that

FIGURE 1.5B Plants respond to light by bending toward it.

contain the damage done by hungry insects to their leaves or infections caused by bacteria and viruses.

Response to Stimuli The ability to respond to stimuli assists the homeostatic ability of organisms. For example, only because they can respond to the presence of predaceous insects can plants protect their integrity. Even unicellular organisms can respond to their environment. For some, the beating of microscopic hairs, and for others, the snapping of whiplike tails move them toward or away from light or chemicals. Multicellular organisms can manage more complex responses. The American turkey vulture uses smell to detect a carcass over a mile away and then soars toward dinner. A monarch butterfly can sense the approach of fall and begin its flight south where resources are still abundant. Bats forage at night. They emit high-pitched sounds and the speed with which the sound bounces back to their large, sometimes over-size, ears allows them to calculate where their insect prey is located.

Animation
Bat Echolocation

When a plant bends toward a source of light (Fig. 1.5B), it acquires the energy it needs for photosynthesis, and when an animal darts safely away from danger, it lives another day. All together, daily activities are termed the behavior of the organism. Organisms display a variety of behaviors as they search and compete for energy, nutrients, shelter, and mates. Many organisms display complex communication, hunting, and defensive behaviors as well. The behavior of an organism often assists homeostasis.

▶ **1.5 CHECK YOUR PROGRESS**
1. Compare various mechanisms for keeping the internal environment fairly constant.
2. Explain how these mechanisms require a response to a stimulus.

FIGURE 1.5A Lizards bask in the sun to raise their body temperature.

The Theory of Ecosystems

The organization of life extends beyond the individual to the **biosphere,** the zone of air, land, and water at the Earth's surface where living organisms are found. Individual organisms belong to a **population,** all the members of a species within a particular area. The populations within a **community** interact among themselves and with the physical environment (soil, atmosphere, etc.), thereby forming an **ecosystem.**

1.6 The biosphere is divided into ecosystems

LEARNING OUTCOMES

When you complete this section, you should be able to

1. Discuss the theory of ecosystems and how ecosystems function.
2. Identify various ways humans threaten the existence of ecosystems such as tropical rain forests and coral reefs.

The **theory of ecosystems** says that *organisms form units in which they interact with the biotic (living) and abiotic (nonliving) components of the environment.* One example of an ecosystem is a North American grassland, which is inhabited by populations of rabbits, hawks, and many other animals, as well as various types of grasses. These populations interact by forming food chains in which one population feeds on another. For example, rabbits feed on grasses, while hawks feed on rabbits and other organisms.

As Figure 1.6 shows, ecosystems are characterized by *chemical cycling* and *energy flow,* both of which begin when plants, such as grasses, take in solar energy and inorganic nutrients to produce food (organic nutrients) by photosynthesis. Chemical cycling (gray arrows) occurs as chemicals move from one population to another in a food chain, until death and decomposition allow inorganic nutrients to be returned to the photosynthesizers once again. Energy (orange arrows), on the other hand, flows from the sun through plants and the other members of the food chain as they feed on one another. The energy gradually dissipates and returns to the atmosphere as heat. Because energy does not cycle, ecosystems could not stay in existence without solar energy and the ability of photosynthesizers to absorb it.

The Biosphere Climate largely determines where different ecosystems are found in the biosphere. For example, deserts exist in areas of minimal rain, while forests require much rain. The two most biologically diverse ecosystems—tropical rain forests and coral reefs—occur where solar energy is most abundant. The human population tends to modify these and all ecosystems for its own purposes. Humans clear forests or grasslands in order to grow crops; later, they build houses on what was once farmland; and finally, they convert small towns into cities. As coasts are developed, humans send sediments, sewage, and other pollutants into the sea.

Tropical rain forests and coral reefs are home to many organisms. The canopy of the tropical rain forest alone supports a variety of organisms, including orchids, insects, and monkeys. Coral reefs, which are found just offshore in the Southern Hemisphere, provide a habitat for many animals, including jellyfish, sponges, snails, crabs, lobsters, sea turtles, moray eels, and some of the world's most colorful fishes. Like tropical rain forests, coral reefs are severely threatened as the human population increases in size. Aside from pollutants, overfishing and collection of coral for sale to tourists destroy the reefs.

Video Coral Reef Ecosystems

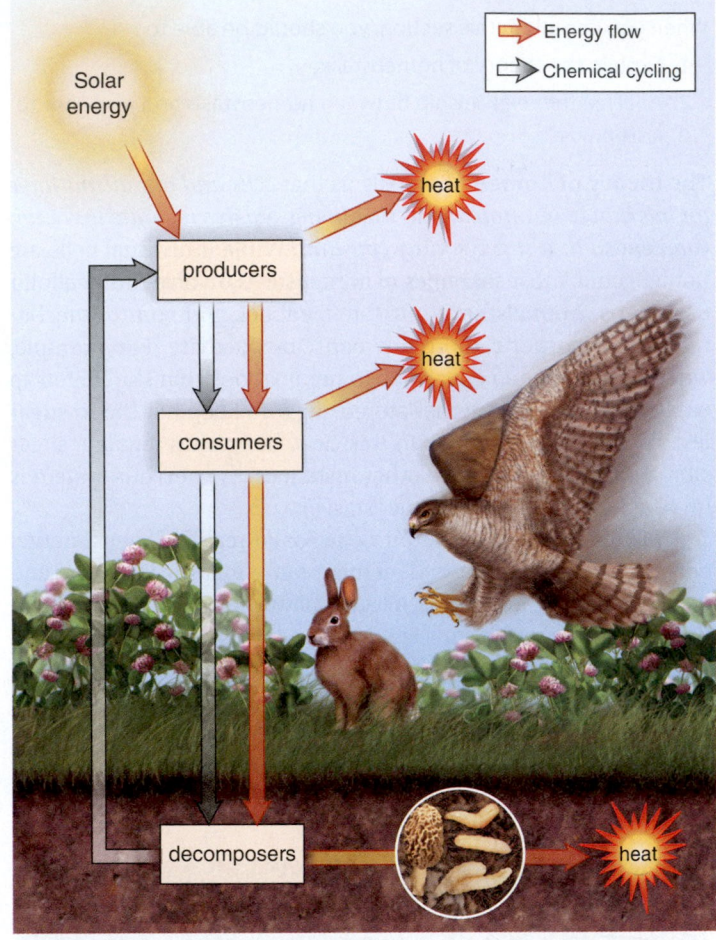

FIGURE 1.6 A grassland is a major ecosystem. Chemicals cycle because decomposers return inorganic nutrients to producers, which provide organic nutrients to consumers including decomposers. With each transfer of nutrients, energy is lost as heat.

Video Tallgrass Prairie Ecology

It has long been clear that humans depend on healthy ecosystems for food, medicines, and various raw materials. We now realize that we depend on them even more for the services they provide. The workings of ecosystems ensure that environmental conditions are suitable for the continued existence of humans.

▶ **1.6 CHECK YOUR PROGRESS**

1. Use Figure 1.6 to explain why chemicals cycle but energy flows through an ecosystem.
2. List several ways humans endanger the continued existence of ecosystems.

The Theory of Evolution

Evolution explains the unity and diversity of life. All organisms share the same characteristics because they are descended from a common source. During descent, however, life changes as different forms become adapted to their environment. Evolution is the unifying concept of biology because it can explain so many aspects of life, including why organisms have shared characteristics despite their great diversity.

1.7 The ancestry of species can be determined

LEARNING OUTCOMES

When you complete this section, you should be able to

1. Relate the theory of evolution to common descent and categorize the data that support this concept.
2. Determine how the branching pattern of an evolutionary tree indicates evolutionary relationships.

The **theory of evolution** says that *organisms have shared characteristics because of common descent.* Just as you and your close relatives can trace your ancestry to a particular pair of great grandparents, so species can trace their ancestry to a common source. An **evolutionary tree** is like a family tree. Just as a family tree shows how a group of people have descended from one couple, an evolutionary tree traces the ancestry of a group to a **common ancestor.** In the same way that one couple can have diverse children, a population can be a common ancestor to several other groups. Over time, diverse life-forms have arisen.

Biologists have discovered that it is possible to trace the evolution of any group—and even life itself—by using molecular data, the fossil record, the anatomy and physiology of organisms, and the embryonic development of organisms. The common ancestors for birds are known from the fossil record, and *Archaeopteryx,* an early bird, clearly has reptile characteristics (Fig. 1.7A). Because the evidence is so clear, birds are now classified as reptiles. Some biologists call them flying dinosaurs. The reptiles that exist today include crocodiles, lizards, snakes, and turtles and birds! The evolutionary tree in Figure 1.7B traces the ancestry of *Archaeopteryx* to an early reptilian ancestor.

In section 1.8 we will examine an evolutionary tree of life and consider how organisms are classified. Then in section 1.9 we will show that natural selection is the mechanism that results in adaptation to the environment, such as the ability of birds to fly. One important thing to remember is that only species (types of organisms) evolve and not individual organisms. Genetics can help you understand why. The genetic makeup you inherited from your parents can mutate during your lifetime and cause cellular changes, but cannot alter the characteristics you pass to your children. On the other hand, mutations that show up in populations can have increased representation in the next generation.

▶ **1.7 CHECK YOUR PROGRESS**
1. Explain the concept of common descent and identify the types of data that support this concept.
2. Identify the information gained from evolutionary trees such as the one in Figure 1.7B.

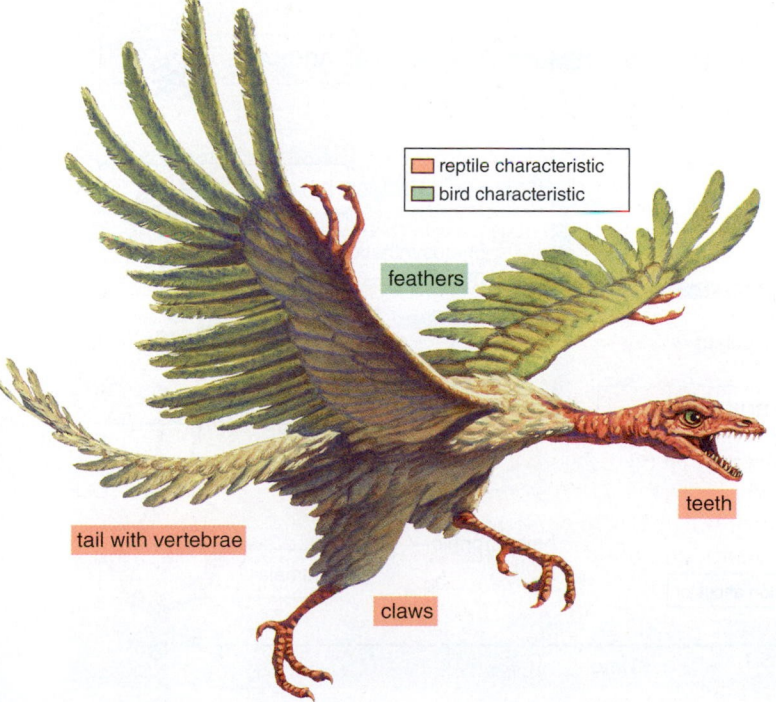

FIGURE 1.7A This depiction of *Archaeopteryx* shows its bird and reptile characteristics.

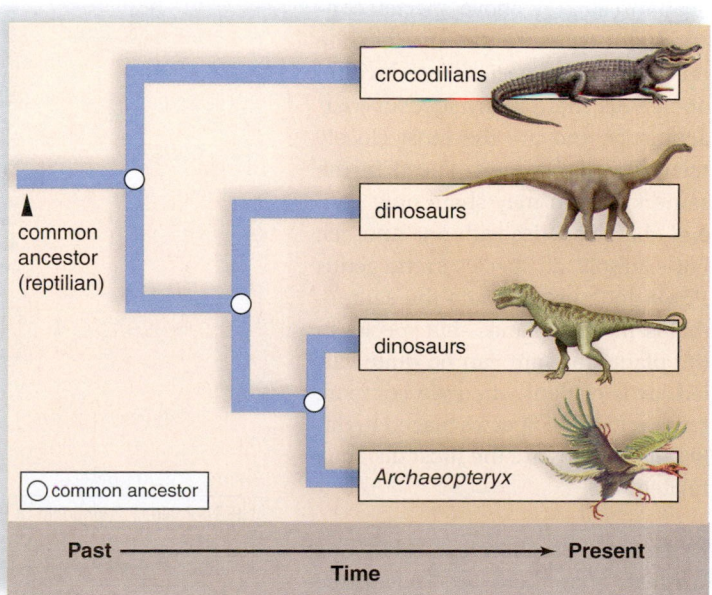

FIGURE 1.7B An evolutionary tree shows how the ancestry of *Archaeopteryx* can be traced to a common ancestor with crocodiles and dinosaurs. Each circle is an intervening common ancestor.

1.8 Evolutionary relationships help biologists group organisms

LEARNING OUTCOMES

When you complete this section, you should be able to

1. List the major categories of classification starting with the least inclusive category.
2. Identify the evolutionary relationship of the domains Bacteria, Archaea, and Eukarya.
3. Recognize the scientific name of organisms and classify organisms to the same extent as in this section.

Despite their diversity, organisms share certain characteristics, and this can be explained by evolution from a common source. For example, all forms of life are composed of cells and use DNA as their genetic material. Figure 1.8A is an evolutionary tree that shows how major groups of organisms are related through evolution.

Organizing Diversity Because life is so diverse, it is helpful to group organisms into categories. **Taxonomy** is the discipline of identifying and grouping organisms according to certain rules. Taxonomy makes sense out of the bewildering variety of life on Earth and is meant to provide valuable insight into evolution. As more is learned about living things, including the evolutionary relationships between species, taxonomy changes. DNA technology is now being used to revise current information and to discover previously unknown relationships between organisms.

The basic classification categories, or taxa, going from least inclusive to most inclusive, are **species, genus, family, order, class, phylum, kingdom,** and **domain** (Table 1.8). The least inclusive category, species, is defined as a group of interbreeding individuals. Each successive classification category above species contains more types of organisms than the preceding one. Species placed within one genus share many specific characteristics and are the most closely related, while species placed in the same kingdom may share only general characteristics with one another. For example, all species in the genus *Pisum* look pretty much the same—that is, like pea plants—but species in the plant kingdom can be quite varied, as is evident when we compare grasses to trees. Species placed in different domains are the most distantly related.

Domains Biochemical evidence suggests that there are only three domains: **Bacteria, Archaea,** and **Eukarya.** Figure 1.8A shows how the domains are related. Both domain Bacteria and domain Archaea

TABLE 1.8	Levels of Classification	
	Human	**Corn**
Domain	Eukarya	Eukarya
Kingdom	Animalia	Plantae
Phylum	Chordata	Anthophyta
Class	Mammalia	Monocotyledones
Order	Primates	Commelinales
Family	Hominidae	Poaceae
Genus	*Homo*	*Zea*
Species*	*H. sapiens*	*Z. mays*

*To specify an organism, you must use the full binomial name, such as *Homo sapiens*.

evolved from the first common ancestor soon after life began. These two domains contain the unicellular **prokaryotes,** which lack the membrane-enclosed nucleus found in the **eukaryotes** of domain Eukarya. However, the DNA of archaea differs from that of bacteria, and their cell surface is chemically more similar to eukaryotes than to bacteria. So, biologists have concluded that eukarya split off from the archaeal line of descent. Prokaryotes are structurally simple but metabolically complex. Archaea (Fig. 1.8B) can live in aquatic environments that lack oxygen or are

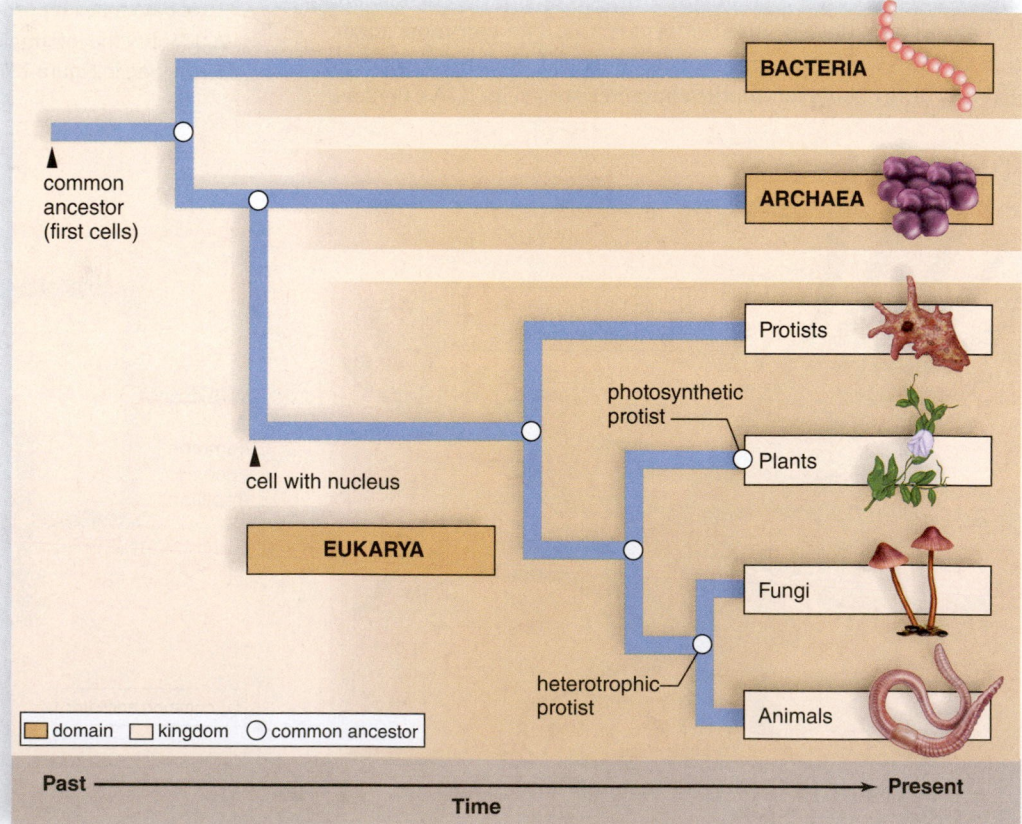

FIGURE 1.8A All species have a common ancestor that existed about 4 billion years ago. Domains Bacteria and Archaea were the first to appear. Domain Eukarya, which includes protists, plants, fungi, and animals, shares an ancestor with domain Archaea.

too salty, too hot, or too acidic for most other organisms. Perhaps these environments are similar to those of the primitive Earth, and archaea are the least evolved forms of life, as their name implies. Bacteria (Fig. 1.8C) are variously adapted to living almost anywhere—in water, soil, and the atmosphere, as well as on our skin and in our mouths and large intestines.

Animation
Three Domains

Taxonomists are in the process of deciding how to categorize the organisms within domains Archaea and Bacteria into several kingdoms. Domain Eukarya, on the other hand, contains four major groups of organisms (Fig. 1.8D). **Protists,** which now comprise a number of kingdoms, range from unicellular forms to a few multicellular ones. Some are photosynthesizers, while others must acquire their food. Common protists include algae, the protozoans, and the water molds. Figure 1.8A shows that plants, fungi, and animals evolved from protists. **Plants** (kingdom Plantae) are multicellular photosynthetic organisms. Examples of plants include azaleas, zinnias, and pines. Among the **fungi** (kingdom Fungi) are the familiar molds and mushrooms that, along with bacteria, help decompose dead organisms. **Animals** (kingdom Animalia) are multicellular organisms that must ingest and process their food. Aardvarks, jellyfish, and zebras are representative animals.

Scientific Names Biologists use binomial nomenclature to assign each living thing a two-part name called its scientific name. For example, the scientific name for mistletoe is *Phoradendron*

tomentosum. The first word is the genus, and the second word is the **specific epithet** of a species within that genus. The genus may be abbreviated (e.g., *P. tomentosum*), and if the species is unknown it may be indicated by sp. (e.g., *Phoradendron* sp.). Scientific names are universally used by biologists to avoid confusion. Common names tend to overlap and are often in the language of a particular country. Scientific names are based on Latin, a universal language that not too long ago was well known by most scholars.

▶ **1.8 CHECK YOUR PROGRESS**
1. Describe the evolutionary relationship between domains.
2. Explain what the fire ant's name *Solenopsis xyloni* tells you.
3. Identify the domain and kingdom for fire ants.

Protists: Several kingdoms

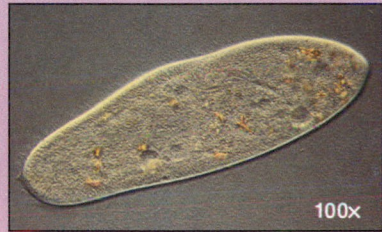

- Protozoans, certain algae, slime molds, and water molds
- Complex single cell (sometimes filaments, colonies, or even multicellular)
- Absorb, photosynthesize, or ingest food

100×

Paramecium, a unicellular protozoan

KINGDOM: Plants

- Certain algae, mosses, ferns, conifers, and flowering plants
- Multicellular, usually with specialized tissues, containing complex cells
- Photosynthesize food

Passiflora, passion flower, a flowering plant

KINGDOM: Fungi

- Molds, mushrooms, yeasts, and ringworms
- Mostly multicellular filaments with specialized, complex cells
- Absorb food

Amanita, a red-cap mushroom

KINGDOM: Animals

- Sponges, worms, insects, fishes, frogs, turtles, birds, and mammals
- Multicellular with specialized tissues containing complex cells
- Ingest food

Vulpes, a red fox

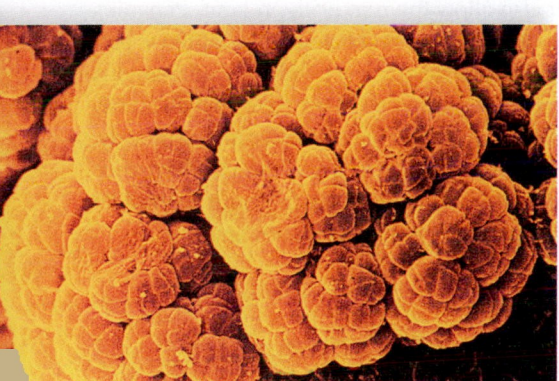

- Prokaryotic cells of various shapes
- Adaptations to extreme environments
- Absorb or chemosynthesize food
- Unique chemical characteristics

Methanosarcina mazei, an archaeon   20,000×

FIGURE 1.8B Domain Archaea.

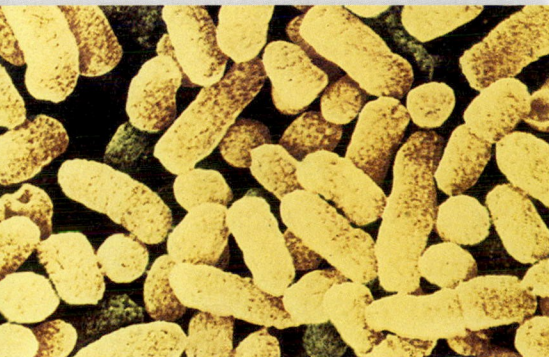

- Prokaryotic cells of various shapes
- Adaptations to all environments
- Absorb, photosynthesize, or chemosynthesize food
- Unique chemical characteristics

Escherichia coli, a bacterium   6,600×

FIGURE 1.8C Domain Bacteria.

FIGURE 1.8D Domain Eukarya.

1.9 Evolution through natural selection results in adaptation to the environment

LEARNING OUTCOME

When you complete this section, you should be able to

1. Explain how the process of natural selection allows a species to become adapted to its environment.

The phrase "common descent with modification" sums up the process of evolution because it means that, as descent occurs from common ancestors, modifications occur that cause these organisms to be adapted (suited) to the environment. Through many observations and experiments, Charles Darwin, the father of evolution, came to the conclusion that natural selection is the process that makes modification—that is, **adaptation**—possible. In other words, the theory of evolution also states that, as evolution occurs, *natural selection brings about adaptation to the environment*. Adaptation to various environments accounts for the diversity of life.

Natural Selection During the process of natural selection, some aspect of the environment selects which traits are more apt to be passed on to the next generation. The selective agent can be an *abiotic* agent (part of the physical environment, such as altitude) or a *biotic* agent (part of the living environment, such as a deer). Figure 1.9A shows how deer could act as a selective agent for a particular mutant. Mutations fuel natural selection because mutations introduce variations among the members of a population. In Figure 1.9A, a plant species generally produces smooth leaves, but a mutation

occurs that causes one plant's leaves to be covered with small extensions or "hairs." The plant with hairy leaves has an advantage because the deer (the selective agent) prefer to eat smooth leaves rather than hairy leaves. Therefore, the plant with hairy leaves survives best and produces more seeds than most of its neighbors. As a result, generations later most plants of this species produce hairy leaves.

As with this example, Darwin realized that although all individuals within a population have the ability to reproduce, not all do so with the same success. Prevention of reproduction can run the gamut from an inability to capture resources, as when long-necked, but not short-necked, giraffes can reach their food source, to an inability to escape being eaten because long legs, but not short legs, can carry an animal to safety. Whatever the example, it can be seen that living things having advantageous traits can produce more offspring than those lacking them. In this way, living things change over time, and these changes are passed on from one generation to the next. Over long periods of time, the introduction of newer, more advantageous traits into a population causes a species to become adapted to an environment.

Video Finches' Natural Selection

Penguins (Fig. 1.9B), for example, are adapted to an aquatic existence in the Antarctic. An extra layer of downy feathers is covered by short, thick feathers that form a waterproof coat. Layers of blubber also keep the birds warm in cold water. Most birds have forelimbs proportioned for flying, but penguins have stubby, flattened wings suitable for swimming. Their feet and tails serve as rudders in the water, but their flat feet also allow them to walk on land. Penguins can also hop from one rock to another and have a bill adapted to eating small shellfish. Penguins also have many behavioral adaptations for living in the Antarctic. They often slide on their bellies across the ice and snow in order to conserve energy when moving quickly. Their eggs—one, or at most two—are carried on the feet, where they are protected by a pouch of skin. This allows the birds to huddle together for warmth while standing erect and incubating eggs.

▶ **1.9 CHECK YOUR PROGRESS**
  1. Create a scenario to show that natural selection can result in adaptation to the environment.

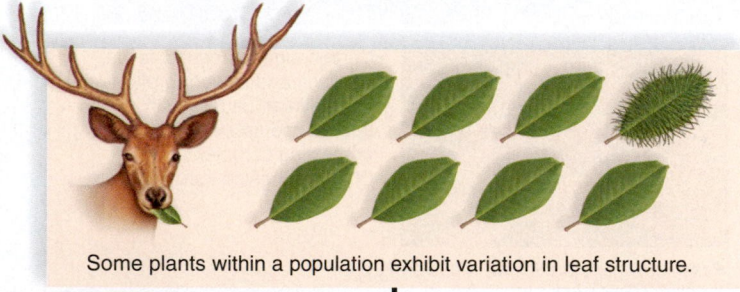

Some plants within a population exhibit variation in leaf structure.

Deer prefer a diet of smooth leaves over hairy leaves. Plants with hairy leaves reproduce more than other plants in the population.

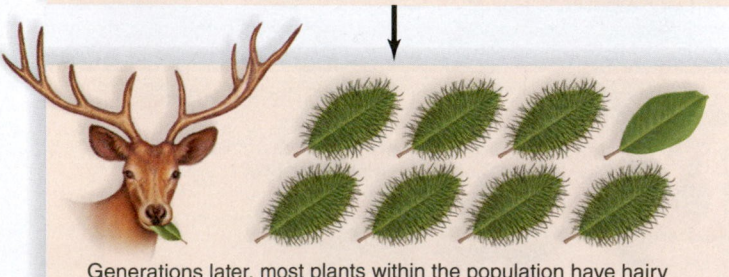

Generations later, most plants within the population have hairy leaves, as smooth leaves are selected against.

FIGURE 1.9A Predatory deer act as a selective agent to bring about change in a plant population.

FIGURE 1.9B Penguins are adapted to sliding across the ice of Antarctica.

1B Evolution's Many Applications

The principles of evolution not only increase our understanding of how the world works but also help us solve practical problems that impact our lives. Many good examples can be cited in the fields of agriculture, medicine, and conservation.

Agriculture

The fruit of the wild banana plant is small and tough with large hard seeds. In contrast, the bananas we eat today are large, soft, sweet,

and for practical purposes seedless. Humans produced this type of banana by using **artificial selection;** in this case, humans were the selective agent and not the environment. Most of the vegetables we eat today, and our domesticated animals including horses, dogs, and cows, were produced in the same way.

Understanding the evolution of our agricultural plants helps us keep them healthy. For example, maize chlorotic dwarf virus (MCDV) causes an infection of young corn plants that makes them sick and reduces yield. However, it's known that our domesticated corn is derived from wild plants called teosinte, and scientists have found teosinte species in the wild that are resistant to several viral diseases, including the one caused by MCDV. This gene has been transferred to corn plants so that they too are resistant.

Farmers use pesticides to protect their crops from insects, or they grow plants that have been engineered to produce the pesticide. However, the pesticide is a selective agent for those members of the insect population that carry genes for the resistant trait. Because these insects reproduce more than nonresistant insects, a large percentage of the insect population becomes resistant. Understanding this process has caused scientists to suggest that farmers make a part of their fields pesticide free. This will allow nonresistant insects to also reproduce, and in this way the percentage of resistant insects in the next generation will be reduced.

Medicine

In the presence of an antibiotic, resistant bacteria are selected to reproduce over and over again, until the entire population of bacteria becomes resistant to the antibiotic. In 1959, a new antibi-

otic called methicillin became available to treat bacterial infections that were already resistant to penicillin. By 1997, 40% of hospital staph infections were caused by MRSA (methicillin-resistant *Staphylococcus aureus*). By now, the same bacteria can spread freely through the general population when people are in close contact. The infection is called CA-MRSA (community-acquired MRSA).

A knowledge of evolution has not only allowed scientists to understand how pathogens (e.g., bacteria and viruses) become resistant to antibiotics, but has also helped them create a process to develop new drugs to kill them. Millions of possible drugs are selected based on their ability to kill a particular pathogen. Then the best of these are tweaked chemically before this new batch of chemicals are tested for their ability to kill the pathogen. This selection process is repeated time and time again until a new drug has been developed. This drug is then tested in another mammal (e.g., mouse) or another primate (e.g., chimpanzee) that is closely related to humans through evolution. If the drug has few harmful side effects, it is prescribed to humans to cure the disease.

Conservation

A knowledge of evolution helps scientists decide which technologies can help save the environment. For example, most of us still fill the tanks of our cars with gasoline derived from oil. Yet oil is a nonrenewable resource that will eventually be depleted. What we need is a renewable resource that can be replaced over and over again. Corn is a renewable resource that can be used to produce ethanol, a fuel that substitutes for gasoline and is somewhat better for the environment. Furthermore, some scientists believe that, instead of using corn, which is food for animals and humans, billions of tons of currently unused waste materials in the United States could be converted to ethanol. By mocking the natural selection process as described above for perfecting a drug, the best bacteria for changing waste to ethanol could be arrived at. Using the natural selection process to achieve the best drug or the best bacterium or to select anything for a particular task is now described as using *directed evolution.*

A knowledge of evolution can also help us save endangered species in the wild. For example, some populations of chinook salmon are listed under the U.S. Endangered Species Act as either threatened or endangered. To save them, it is possible to build hatcheries, breed more fish, and introduce these fish into rivers where small populations of wild chinook salmon now live. However, hatcheries should mimic as much as possible the selection pressures that wild populations are exposed to.

CONSIDER THESE QUESTIONS

1. Is it possible to convince people that evolution has practical value? Why or why not?
2. Should the government require us all to be screened and quarantine carriers of resistant bacteria to protect others? Why or why not?
3. We are told to control bacteria by washing our hands. Should we also use sanitizers? Why or why not?

 **connect** Explore the concepts through a variety of multimedia
| BIOLOGY   assets and question types.

www.mcgrawhillconnect.com

1.10 Evolution from a common ancestor accounts for the characteristics of life

LEARNING OUTCOMES

When you complete this section, you should be able to

1. Use the concept of common descent to explain why organisms share the same characteristics.
2. Identify the basic characteristics of life.

The diversity of life has been mentioned several times by now. With so much diversity, how can we possibly define life? The best way we know to distinguish the living from the nonliving is to list the characteristics shared by all organisms. These characteristics of life must have been present in the original common ancestor or else they would not be present in all organisms.

MP3
Life Characteristics

1. **Life is organized.** The levels of biological organization extend from cells to the biosphere (see page 9). The first living organisms were unicellular, and only later did multicellular forms arise. Once several different types of organisms arose, they interacted among themselves and became the biotic components of ecosystems.

2. **Life uses materials and energy.** The metabolic pathways that allow an organism to maintain its organization and to grow are the same in all organisms (see page 9). We will study these metabolic pathways in future chapters because they are so critical to the lives of organisms.

3. **Life reproduces.** Unicellular organisms simply divide when they reproduce, but in multicellular forms, new life often begins with a fertilized egg that grows and develops into a new organism. When organisms reproduce, genetic differences arise that allow evolution to occur (see pages 9–10).

4. **Life is homeostatic.** Regulatory mechanisms allow cells and organisms to keep their internal environment relatively constant (see page 11). Homeostasis evolved because those members of a population that were homeostatic had more offspring than those that were less homeostatic.

5. **Life responds to stimuli.** Organisms respond to internal stimuli and external stimuli, and this allows them to maintain homeostasis. Response to stimuli also accounts for the behavior of organisms (see page 11). Behavior evolves through natural selection in the same manner as do anatomical features.

6. **Life forms ecosystems.** Interactions are a hallmark of living things. Cells interact within organisms, and populations interact in ecosystems. We could not exist without food produced by plants and without bacteria and fungi that decompose dead remains (see page 12).

7. **Life evolves.** The history of life began with a common source, but as life reproduces it passes on genes that can mutate. Through mutations, advantages arise that are suited to the environment, and through natural selection they become more prevalent in a population. Adaptation to different environments accounts for the variety of life on Earth (see pages 13–16).

▶ **1.10 CHECK YOUR PROGRESS**

1. Describe how the hawk and her offspring illustrate the characteristics of life.

THE CHAPTER IN REVIEW

CONNECTING THE CONCEPTS

The scientific method consists of making observations, formulating a hypothesis, testing the hypothesis, and coming to a conclusion on the basis of the results (data). The conclusions of many studies have allowed scientists to develop the five theories (cell theory, gene theory, theory of homeostasis, theory of ecosystems, and theory of evolution) on which this book is based. Theories are conceptual schemes that tell us how the world works. All theories of biology are related. For example, the gene theory is connected to the theory of evolution because mutations create differences between the members of a population. Better-adapted members have the opportunity through natural selection to reproduce more, and in that way a species becomes adapted to its environment.

Any two theories are related. For example, evolution is also connected to the theory of ecosystems because, as natural selection occurs, species become adapted to living in a particular ecosystem. We can connect this observation to the cell theory because, if a gazelle's nerve cells can conduct nerve impulses faster to its muscle cells than can a lion's nerve cells, the gazelle is more likely to escape capture.

In exploring the theories, we have also discussed the characteristics of life. The cell theory taught us that all organisms are composed of cells and that cells are the fundamental units of life. The theory of homeostasis tells us that all organisms have mechanisms that allow them to keep their internal environment relatively constant. The gene theory tells us all organisms have genes, hereditary units that

undergo mutations leading to the variety of life. Even so, all life-forms share similar characteristics because they can trace their ancestry to a common source as stated by the theory of evolution. All life-forms live in ecosystems where interactions allow them to acquire the materials and energy they need to continue their existence. Humans are also dependent on ecosystems, and when they preserve the biosphere, they are preserving their own existence as well.

ANALYZE AND EVALUATE

1. Give your own example (not taken from this reading) to show that two theories are related.
2. Explain in your own words how bacteria become resistant to an antibiotic.

SUMMARIZE

The Process of Science

1.1 Scientists use a preferred method

- The scientific method consists of four steps: (1) making **observations** using both our senses and special instruments; (2) formulating a possible explanation, called a **hypothesis,** by using inductive reasoning; (3) doing **experiments** that involve an **experimental variable, test groups,** and a **control group** that is not exposed to the experimental variable. (Alternatively, scientists can simply make further observations. When doing experiments, scientists sometimes work with a **model**); and (4) coming to a **conclusion** based on **data** as to whether the hypothesis is supported or not.
- A **scientific theory** is a major concept supported by many observations, experiments, and data.

1.2 Many experiments have test and control groups

- Unlike the test groups that receive experimental treatment, the control group receives no experimental treatment and is used to ensure that results are due to no other factor. In an experiment that tested types of fertilizer, the control group was given no fertilization treatment.

The Cell Theory

1.3 Cells are the fundamental unit of living things

- Biological organization extends from the molecules in cells to the organism and beyond.

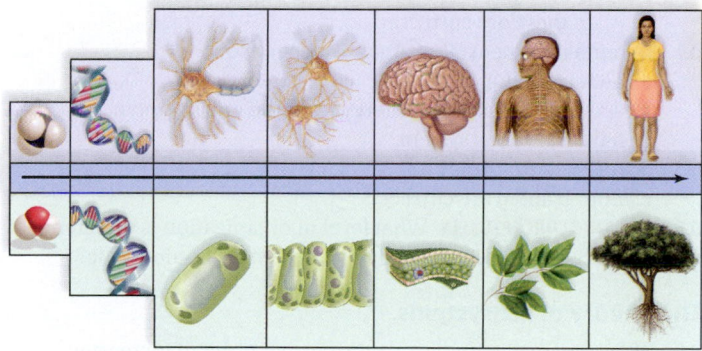

- The **cell theory** tells us that **cells,** the fundamental units of life, come from other cells as **reproduction** of cells and the organism occurs. In cells, **atoms** combine to form **molecules;** similar cells make up a **tissue;** and tissues compose **organs** that work together in **organ systems.** Organ systems work together in an **organism.**
- Cells and organisms acquire materials and **energy** from the environment to maintain their organization. Thereafter, **metabolism** carries out chemical reactions. Plants are able to capture solar energy during **photosynthesis** and produce nutrients that sustain all organisms.

The Gene Theory

1.4 Organisms have a genetic inheritance

- The **gene theory** tells us that **genes** are hereditary units composed of **DNA.** Genes control the structure and function of cells and organisms by coding for cellular molecules (**proteins**). Because genes can **mutate,** each organism has it own particular sequence of DNA nucleotides.
- The gene theory has been very fruitful, yielding many practical applications, such as those listed on page 10.

The Theory of Homeostasis

1.5 Organisms regulate their internal environment

- The **theory of homeostasis** states that organisms and cells have mechanisms that keep the internal environment relatively constant. Only then can life continue.
- Homeostasis involves the use of sense receptors to monitor the external and internal environment. Organisms can respond to changes in the environment. For example, when plants turn toward sunlight, they acquire the energy they need to photosynthesize.

The Theory of Ecosystems

1.6 The biosphere is divided into ecosystems

- The **theory of ecosystems** says that within a local environment: The members of each species are a **population.** Populations form a **community** in which they interact with each other. In a community, chemicals cycle and energy flows but does not cycle. Chemical cycling requires interaction with the physical environment.
- Diverse ecosystems, including tropical rain forests and coral reefs, are being destroyed by human activities.

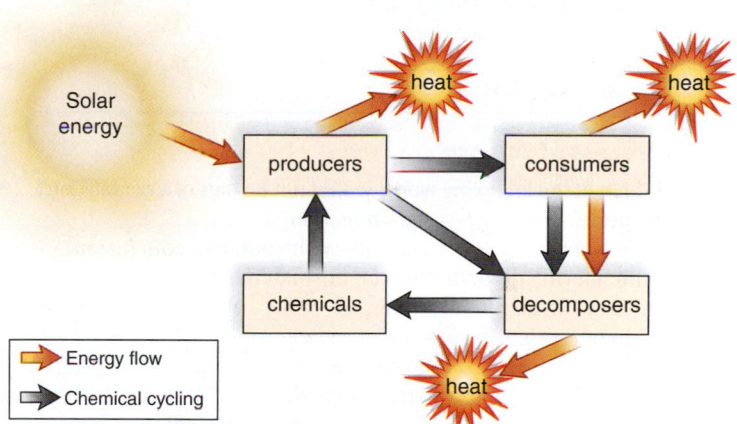

The Theory of Evolution

1.7 The ancestry of species can be determined

- The **theory of evolution** says that all species (living or extinct) can trace their ancestry to a common source. An **evolutionary tree** depicts the pattern of descent by way of **common ancestors.**

1.8 Evolutionary relationships help biologists group organisms

- **Taxonomy** is the classification of organisms according to the evolutionary relationships. The classification categories are **species** (least inclusive), **genus, family, order, class, phylum, kingdom,** and **domain** (most inclusive). There are three domains: **Bacteria, Archaea,** and **Eukarya.**

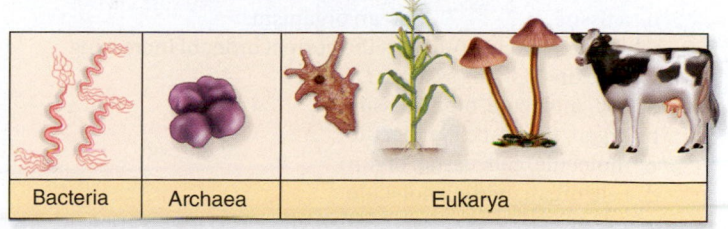

- Domain Archaea and domain Bacteria contain **prokaryotes** (organisms without a membrane-enclosed nucleus). Domain Eukarya contains **eukaryotes** (organisms with a membrane-enclosed nucleus). There are four major groups in domain Eukarya:

  Protists—unicellular to multicellular organisms with various modes of nutrition

  Fungi—molds and mushrooms

  Plants—multicellular photosynthesizers

  Animals—multicellular organisms that ingest food
- To classify an organism, two-part scientific names—binomial nomenclature—are used, consisting of the genus name and the **specific epithet.**

1.9 Evolution through natural selection results in adaptation to the environment

- The theory of evolution also says that modifications are introduced as evolution occurs, and if these modifications assist **adaptations,** they become more common through **natural selection.** The result is a wide variety of life-forms on Earth, each adapted to a different environment.

1.10 Evolution from a common ancestor accounts for the characteristics of life

- Organisms have shared characteristics because of common descent. Life is organized, uses materials and energy, reproduces, is homeostatic, responds to stimuli, forms ecosystems, and evolves.

TEST YOURSELF

The Process of Science

1. Which of the following words would not be part of a conclusion?
   a. proof
   b. support
   c. rejection
   d. All can be part of a conclusion.
2. Which term and definition are mismatched?
   a. data—factual information
   b. hypothesis—the idea to be tested
   c. conclusion—what the data tell us
   d. All of these are properly matched.
3. Which of these describes the control group in the winter wheat experiment? The control group was
   a. planted with pigeon peas.
   b. treated with nitrogen fertilizer.
   c. not treated.
   d. not watered.
   e. Both c and d are correct.
4. **THINKING CONCEPTUALLY** What's the relationship between the scientific method and the five theories on which this book is based?

The Cell Theory

5. The level of organization that includes cells of similar structure and function is
   a. an organ.
   b. a tissue.
   c. an organ system.
   d. an organism.
6. Which sequence represents the correct order of increasing complexity in living systems?
   a. cell, molecule, organ, tissue
   b. organ, tissue, cell, molecule
   c. molecule, cell, tissue, organ
   d. cell, organ, tissue, molecule

7. All of the chemical reactions that occur in a cell are called
   a. homeostasis.
   b. metabolism.
   c. heterostasis.
   d. cytoplasm.
8. The process of turning solar energy into organic chemical energy is called
   a. work.
   b. metabolism.
   c. photosynthesis.
   d. respiration.
9. **THINKING CONCEPTUALLY** Explain why the first step in formulating the cell theory was a microscopic study of plant and animal tissues.

The Gene Theory

10. Genes are
    a. present in eukaryotes but not in prokaryotes.
    b. composed of RNA and DNA.
    c. passed on from cell to cell and from organism to organism.
    d. All of these are correct.
11. Genes
    a. code for proteins.
    b. can mutate.
    c. are always found in cells.
    d. All of these are correct.
12. **THINKING CONCEPTUALLY** What's the relationship between genes and the diversity of life?

The Theory of Homeostasis

13. Which of the following are a part of homeostasis?
    a. Animals keep their internal temperature relatively constant.
    b. Your blood cell count is always about the same.
    c. Certain organs, such as the kidneys, excrete wastes.
    d. Plants are able to turn toward the sun.
    e. All of these are correct.
14. To remain homeostatic, organisms need to
    a. be multicellular.
    b. acquire material and energy from the environment.
    c. have a nervous system.
    d. respond to stimuli.
    e. Both b and d are correct.
15. **THINKING CONCEPTUALLY** What level of organization is threatened first if conditions in the body do not remain homeostatic?

The Theory of Ecosystems

16. Which sequence represents the correct order of increasing complexity?
    a. biosphere, community, ecosystem, population
    b. population, ecosystem, biosphere, community
    c. community, biosphere, population, ecosystem
    d. population, community, ecosystem, biosphere
17. In an ecosystem, energy
    a. flows and nutrients cycle.
    b. cycles and nutrients flow.
    c. and nutrients flow.
    d. and nutrients cycle.
18. An example of chemical cycling occurs when
    a. plants absorb solar energy and make their own food.
    b. energy flows through an ecosystem and becomes heat.
    c. hawks soar and nest in trees.
    d. death and decay make inorganic nutrients available to plants.
    e. we eat food and use the nutrients to grow or repair tissues.
19. Energy is brought into ecosystems by which of the following?
    a. fungi and other decomposers
    b. cows and other organisms that graze on grass
    c. meat-eating animals
    d. organisms that photosynthesize, such as plants
    e. All of these are correct.

20. **THINKING CONCEPTUALLY** How is a college campus, which is composed of buildings, students, faculty, and administrators, like an ecosystem?

The Theory of Evolution

21. Organisms are related because they
    a. all have the same structure and function.
    b. share the same characteristics.
    c. can all trace their ancestry to a common source.
    d. all contain genes.

22. An evolutionary tree
    a. shows common ancestors.
    b. depicts the history of a group of organisms.
    c. is based on appropriate data.
    d. shows how certain organisms are related.
    e. All of these are correct.

23. Classification of organisms reflects
    a. similarities.
    b. evolutionary history.
    c. Neither a nor b is correct.
    d. Both a and b are correct.

24. Which of these exhibits an increasingly more inclusive scheme of classification?
    a. kingdom, phylum, class, order
    b. phylum, class, order, family
    c. class, order, family, genus
    d. genus, family, order, class

25. Humans belong to the domain
    a. Archaea.   c. Eukarya.
    b. Bacteria.   d. None of these are correct.

26. In which group are you most likely to find unicellular organisms?
    a. Protists   c. Plantae
    b. Fungi   d. Animalia

27. The second word of a scientific name, such as *Homo sapiens,* is the
    a. genus.   d. species.
    b. phylum.   e. family.
    c. specific epithet.

28. Modifications that make an organism suited to its way of life are called
    a. ecosystems.   c. adaptations.
    b. populations.   d. None of these are correct.

    For questions 29–32, match each item to a characteristic of life in the key.

KEY:
    a. is organized   d. is homeostatic
    b. uses materials   e. responds to stimuli
       and energy   f. forms ecosystems
    c. reproduces   g. evolves

29. organisms exhibit behavior
30. populations interact
31. giraffes produce only giraffes
32. exhibit common descent with modification
33. **THINKING CONCEPTUALLY** Give evidence to support the phrase "Evolution is the unifying theory of biology."

GET INVOLVED

1. You want to grow large tomatoes and notice that a name-brand fertilizer claims to yield larger produce than a generic brand. How would you test this claim?

2. After viewing this virtual lab, decide whether the experimental variable (see page 5) is the dependent or independent variable.

    Virtual Lab
    Dependent and
    Independent Variables

3. An investigator spills dye on a culture plate and notices that the bacteria live despite exposure to sunlight. He decides to test whether the dye is protective against ultraviolet (UV) light. He exposes to UV light one group of culture plates containing bacteria and dye and another group containing only bacteria. The bacteria on all plates die. Complete the following diagram to identify the steps of his investigation.

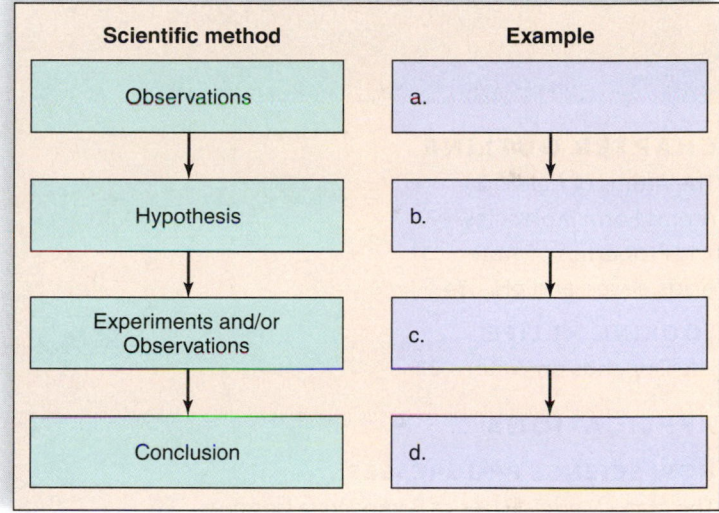

MEDIA STUDY TOOLS

mhhe.com/maderconcepts3

Enhance your study of this chapter with interactive study tools, practice tests, and engaging animations. Also, ask your instructor about the resources available through ConnectPlus, which includes LearnSmart, a personalized adaptive learning program, and a media-rich eBook.

2

Basic Chemistry of Cells

BEFORE YOU BEGIN

Take a few minutes to recall

How the scientific method results in scientific theories (section 1.1)

The first several levels of biological organization (Figure 1.3A)

The tenets of the cell theory (section 1.3)

Life Depends on Water

Scientists and laymen alike have often pondered the question, "Why did life arise on Earth?" It's long been thought that the answer, in part, must involve the presence of water. Three-fourths of the surface of our planet is covered by water. Water is so abundant that if the Earth's surface were absolutely smooth, it would be covered by water. Only land that projects above the seas provides a terrestrial environment.

Cells most likely arose in the oceans. Any living system is 70–90% water, a medium in which chemical reactions can easily occur. A watery environment supports and protects cells while providing an external transport system for chemicals. Homeostasis is also assisted by the ability of water to absorb and give off heat in a way that prevents rapid temperature changes. The abundance of terrestrial life correlates with the abundance of water; therefore, a limited variety of living things is found in the deserts, but much variety exists in the tropics, which receive, by far, the most rain.

Planet Earth has an abundance of water.

The tropics are also warm, and water helps maintain a constant year-round temperature day and night.

Do any of the other planets have life? To answer this question, scientists first look for signs of water on a planet because life as we know it does not exist without water. NASA has long seen signs of water on Mars, and in 2008 the *Phoenix*, a robotic laboratory, landed on Mars and found ice. Also, the soil chemistry of the surrounding area suggested that the planet could have been warmer and wetter sometime within the past few million years. Now, scientists hope that evidence of life on Mars will also be found one day.

The strength of the association between water and organisms is observable in that all animals, whether aquatic or terrestrial, make use of water to reproduce. Animals that live in the sea or in fresh water can simply deposit their eggs and sperm in the water, where they join to form an embryo that develops in the water. The sperm of humans, like those of many other terrestrial animals, are deposited inside the female, where they are protected from drying out. Then, as with most other mammals, the offspring develops within a fluid, called the amniotic fluid, while contained within the uterus. Amniotic fluid cushions the embryo and protects it against possible traumas, while maintaining a constant temperature. Later, it prevents the limbs from sticking to the body and allows the fetus to move about.

This chapter discusses the properties of water that assist living things in maintaining homeostasis. It also covers the basic chemistry necessary to understanding how the cell, and therefore the organism, functions. Some chemicals alter the properties of water and, in that way, threaten the ability of organisms to maintain homeostasis.

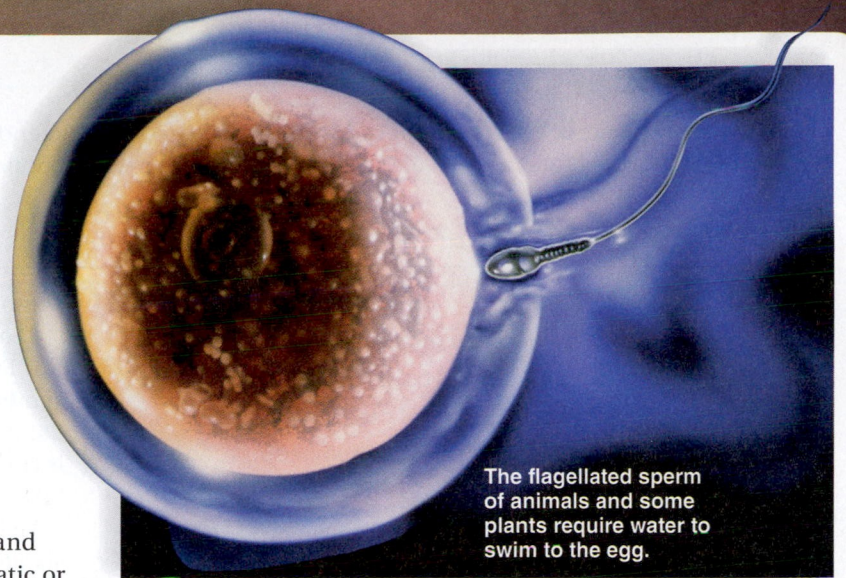

The flagellated sperm of animals and some plants require water to swim to the egg.

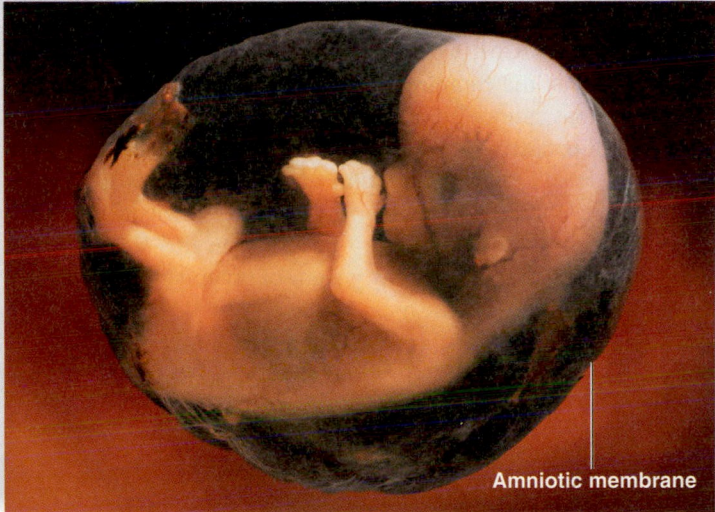

Amniotic membrane

Humans, like other animals, develop in a water environment.

Mercury Venus Earth Mars Ceres Jupiter Saturn Uranus Neptune Pluto and Charon

Of all the planets, only Earth has abundant water and abundant life.

The Atoms of Cells

Cells, the smallest units of life, are composed of molecules (see Fig. 1.3A), and so it is appropriate that we begin our study of life by considering what constitutes a molecule. Our story begins with atoms because atoms join together to make a molecule.

2.1 Six types of atoms are basic to life

LEARNING OUTCOMES

When you complete this section, you should be able to

1. List the six types of atoms basic to cells.
2. Describe the locations and charges of the subatomic particles.
3. Distinguish between the atomic symbol, number, mass, and isotopes.

Both the Earth's crust and all organisms are matter. Turn a page, throw a ball, pat your dog, rake leaves; **matter** refers to anything that takes up space and has mass. Everything from the water we drink to the air we breathe is composed of matter.

Elements and Atoms It is quite remarkable that only 92 elements serve as the building blocks of matter. An **element** is a substance that cannot be broken down by chemical means into a simpler substance. However, an **atom** is the smallest unit of an element that still retains the chemical and physical properties of the element. Only six types of atoms—carbon, hydrogen, nitrogen, oxygen, phosphorus, and sulfur—are basic to life. These atoms make up about 95% of the body weight of organisms, such as the macaws in Figure 2.1A. The macaws have gathered on a salt lick in South America. Salt contains the atoms sodium and chlorine and is a common substance sought by many forms of life.

Every atom has a name and also a symbol. The **atomic symbol** for sodium is Na because *natrium* means sodium in Latin. Chlorine, on the other hand, has the symbol Cl, which is consistent with its English name. Other atoms, however, also take their symbol from Latin. For example, the symbol for iron is Fe because *ferrum* means iron in Latin. The symbols for the six atoms basic to life are C, H, N, O, P, and S. Therefore, we can use the acronym CHNOPS to help us remember these six atoms. As we shall discover in Chapter 3, the properties of the atoms CHNOPS are essential to the uniqueness of cells and organisms. But other atoms are also important to organisms, including sodium, potassium, calcium, iron, and magnesium.

Subatomic Particles Physicists have identified a number of subatomic particles—particles that are less complex than an atom but are components of an atom. The three best-known subatomic particles are positively charged **protons,** uncharged **neutrons,** and negatively charged **electrons.** Protons and neutrons are located within the nucleus of an atom, and electrons move about the

FIGURE 2.1A The table, superimposed on a photograph of macaws on a salt lick, contrasts the proportion of elements in living organisms with those in the Earth's crust.

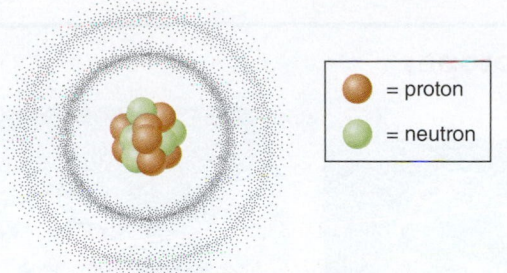

FIGURE 2.1B The stippled area shows the probable location of electrons.

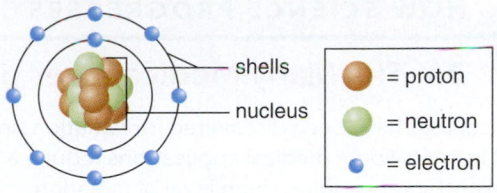

FIGURE 2.1C The shells in this atomic model represent the average location of electrons.

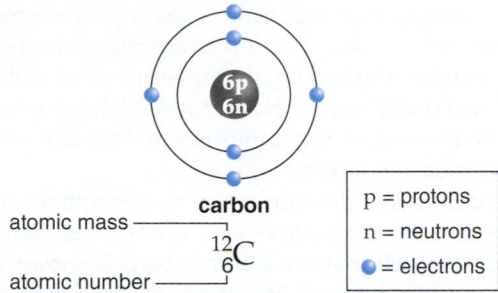

FIGURE 2.1D Atomic model of a carbon atom.

TABLE 2.1	Subatomic Particles		
Particle	**Electric Charge**	**Atomic Mass**	**Location**
Proton	+1	1	Nucleus
Neutron	0	1	Nucleus
Electron	−1	0	Electron shell

nucleus. The stippling in Figure 2.1B shows the probable location of the electrons in an atom that has ten electrons. In Figure 2.1C, the stippling has been converted to circles that represent **electron shells,** the approximate orbital paths of electrons. The inner shell has the lowest energy level and can hold two electrons. The outer shell has a higher energy level and can hold eight electrons. An atom is most stable when the outer shell has eight electrons.

In science, a model is a useful simulation of a structure or process rather than the actual structure or process. Biologists find that the model of an atom shown in Figure 2.1C is sufficient for their purposes, and you will be asked to create such atomic models. Actually, today we know that most of an atom is empty space. If an atom could be drawn the size of a football field, the nucleus would be like a gumball in the center of the field, and the electrons would be tiny specks whirling about in the upper stands. Electrons don't have to always stay within certain shells. In our analogy to a football field, the electrons might very well stray outside the stadium at times.

Animation Atomic Structure

Atomic Number and Mass Number All atoms of an element have the same number of protons. This is called the **atomic number.** For example, the atomic number for a carbon atom is 6. The atomic number not only tells you the number of protons, but it also tells you the number of electrons when the atom is electrically neutral. The model of a carbon atom in Figure 2.1D shows that an electrically neutral carbon atom has six protons and six electrons. Each atom has its own specific mass. The **mass number** of an atom depends on the presence of protons and neutrons, both of which are assigned one atomic mass unit (Table 2.1). Electrons are so small that their mass is considered zero in most calculations.

The mass number for carbon is 12. By convention, when an atom stands alone, the atomic number is written as a subscript to the lower left of the atomic symbol. The mass number is written as a superscript to the upper left of the atomic symbol. Regardless of position, the smaller number is always the atomic number as shown in Figure 2.1D.

Isotopes **Isotopes** are atoms of a single element that differ in their number of neutrons. Isotopes have the same number of protons, but different mass numbers. For example, the element carbon has three common isotopes:

$$^{12}_{6}C \qquad ^{13}_{6}C \qquad ^{14}_{6}C^*$$

*radioactive

Carbon 12 has six neutrons, carbon 13 has seven neutrons, and carbon 14 has eight neutrons. Unlike the other two isotopes, carbon 14 is unstable; it breaks down into atoms with lower atomic numbers. When it decays, it emits radiation in the form of radioactive particles, or radiant energy. Therefore, carbon 14 is called a radioactive isotope. Biologists and other scientists have found many beneficial uses for radiation. For example, Melvin Calvin and his co-workers used carbon 14 to discover the sequence of reactions that occur during the process of photosynthesis.

Animation Half-life

Atomic Mass Elements also have an **atomic mass,** the average mass of all its isotopes. The term *atomic mass* is used rather than *atomic weight,* because mass is constant, whereas weight changes according to the gravitational force of a body. The gravitational force of the Earth is greater than that of the moon; therefore, substances weigh less on the moon, even though their mass has not changed.

▶ **2.1 CHECK YOUR PROGRESS**

1. Use Figure 2.1A to identify the four atoms that are most common in cells.
2. Identify the placement and number of protons and electrons in oxygen (atomic number 8).
3. Determine the difference between the isotopes calcium 40 and calcium 48 (calcium has an atomic number of 20).

2A The Many Medical Uses of Radioactive Isotopes

Many medical uses have been discovered for radiation since it was discovered in 1860. Some medical applications require a low level of radiation and some require a high level of radiation.

Low Levels of Radiation

Radioactive isotopes can be used to image the body's organs and tissues. For example, after a patient drinks a solution containing a minute amount of iodine 131, it is concentrated in the thyroid gland. (The thyroid is the only organ to take up iodine, which it uses to make thyroid gland hormones.) In Figure 2A.1, the missing area (*upper right*) in the X-ray indicates the presence of a tumor that does not take up radioactive iodine.

A procedure called positron-emission tomography (PET) is a way to determine the comparative activity of tissues. Radioactively labeled glucose, which emits a subatomic particle known as a positron, is injected into the body. The radiation given off is detected by sensors and analyzed by a computer. The result is a color image that shows which tissues took up glucose and are metabolically active. In Figure 2A.2, the red areas surrounded by green indicate which areas of the brain are most active. Physicians use PET scans of the brain to evaluate patients who have memory disorders of an undetermined cause and suspected brain tumors or seizure disorders that could possibly benefit from surgery. PET scans of the heart can detect signs of coronary artery disease and low blood flow to the heart muscle. For this procedure, the patient is injected with a radioisotope of the metallic element thallium (thallium 201). The more thallium taken up by the heart muscle, the better the blood supply to the heart.

Video
Nuclear Medicine

High Levels of Radiation

Radioactive substances in the environment can harm cells, damage DNA, and cause cancer. Marie Curie, who helped discover radiation, and many of her co-workers developed cancer. The release of radioactive particles following a nuclear power plant accident can have far-reaching and long-lasting effects on human health. However, high levels of radiation can also be put to good use. Radiation from radioactive isotopes has been used for many years to sterilize medical and dental products, and in the future it may be used to sterilize the U.S. mail in order to free it of possible pathogens, such as anthrax spores.

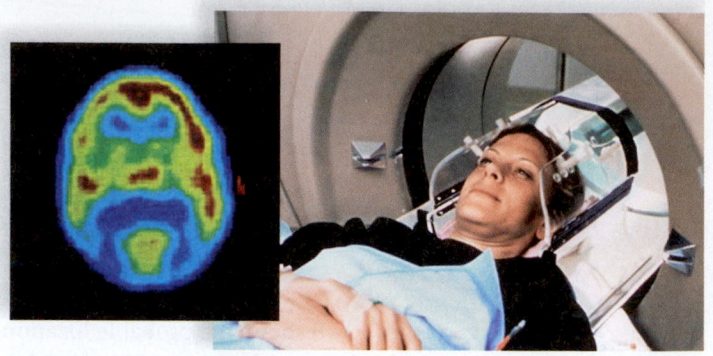

FIGURE 2A.2 Detection of brain activity by doing a PET scan.

Rapidly dividing cells are particularly sensitive to damage by radiation. For this reason, some cancerous growths can be controlled or eliminated by irradiating the area containing the growth. Radiotherapy can be administered externally, as depicted in Figure 2A.3, or it can be given internally. Today, internal radiotherapy allows radiation to destroy only cancer cells, with little risk to the rest of the body. For example, iodine 131 is commonly implanted to treat thyroid cancer, probably the most successful cancer treatment.

CONSIDER THESE QUESTIONS

1. Because it's impossible to tell ahead of time which investigations will prove fruitful, should the government place no restrictions on its support of scientific endeavors?
2. Should anyone, regardless of age and preexisting condition, be permitted to participate in a medical study if they want to?
3. Are you willing to be a guinea pig in experiments that might be harmful to you? Why or why not?

connect
BIOLOGY Explore the concepts through a variety of multimedia assets and question types.

www.mcgrawhillconnect.com

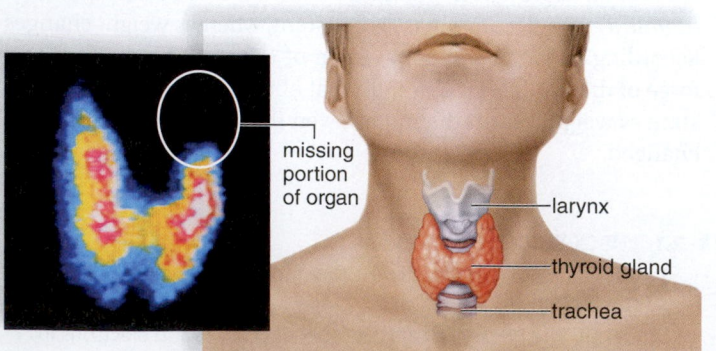

missing portion of organ

larynx

thyroid gland

trachea

FIGURE 2A.1 Detection of thyroid cancer by using radioactive iodine.

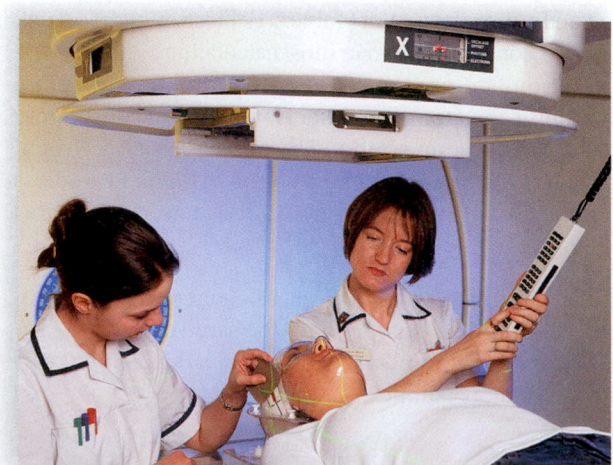

FIGURE 2A.3 Radiotherapy helps cure cancer.

Atoms Form Molecules

The number of electrons in an atom's outer shell largely determines its reactivity. Although some atoms do not react, the biologically important ones do tend to react, and the result is an association called a chemical bond. The two common types of chemical bonds are ionic bonds and covalent bonds. When covalent bonds exhibit polarity, hydrogen bonding may occur.

2.2 After atoms react, they have a completed outer shell

LEARNING OUTCOMES

When you complete this section, you should be able to

1. Use the periodic table of the elements to construct electrically neutral atoms.
2. Determine how many electrons are in the outer shell of a neutral atom when provided only with the atomic number.
3. Discuss the importance of the octet rule.

Once chemists discovered a number of the elements, they arranged them in a periodic table according to their characteristics. Notice in Figure 2.2A that the atoms in the periodic table are arranged according to increasing atomic number. The table also tells you how many shells surround the nucleus and how many electrons are in the outer shell of an atom. The row number tells the number of shells, and the column headed by a Roman numeral tells you the number of electrons in the outer shell. For example, carbon in the second row has two shells, and being in column IV, it has four electrons in the outer shell.

A model can be drawn for each of the atoms in the periodic table. Figure 2.2B illustrates models for the six atoms common to organisms, namely CHNOPS. For these atoms and all the others up through number 20 (calcium), each lower level is filled with electrons before the next higher level contains any electrons. The first shell (closest to the nucleus) can contain two electrons; thereafter, each additional shell can contain eight electrons.

Among CHNOPS, hydrogen is most stable when the outer shell has two electrons, and the others are most stable when the

FIGURE 2.2A A portion of the periodic table of the elements. A column (headed by a Roman numeral) tells the number of electrons in the outer shell; the row tells the number of shells in an atom. Four rows are shown. For a complete table, see inside back cover.

outer shell has eight electrons. Most atoms, therefore, obey the so-called **octet rule:** They will give up, accept, or share electrons in order to have eight electrons in the outer shell. Therefore, the number of electrons in an atom's outer shell, called the **valence shell,** determines its chemical reactivity. The size of an atom can also affect reactivity. Both carbon and silicon have four outer electrons, but only the smaller carbon atom often bonds to other carbon atoms and forms long-chained molecules.

Except for the atoms in column VIII, which already have eight electrons in the outer shell, atoms routinely bond with one another. For example, oxygen does not exist in nature as a single atom, O; instead, two oxygen atoms are joined to form a molecule (O_2). Other naturally occurring molecules include hydrogen (H_2) and nitrogen (N_2). When atoms of two or more elements bond together in fixed proportions, the product is called a **compound.** Water (H_2O) is a compound that contains atoms of hydrogen and oxygen. A **molecule** is the smallest part of a compound that still has the properties of that compound.

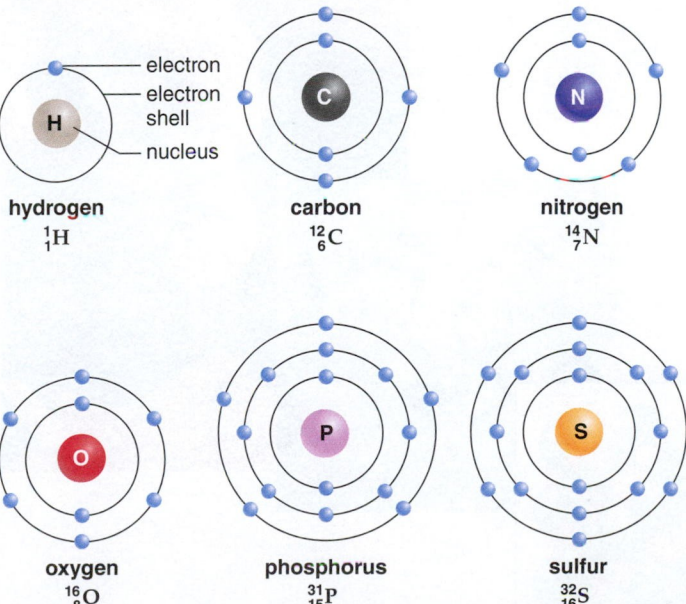

FIGURE 2.2B Models of six atoms before reacting. The first electron shell can contain up to two electrons and, in the atoms depicted, every shell after that can contain eight electrons. Fill the first shell before filling the next.

▶ **2.2 CHECK YOUR PROGRESS**

1. Choose any atom in the periodic table and determine how many shells the atom has and how many electrons are in the outer shell.
2. Explain why phosphorus in Figure 2.2B has only five electrons in its outer shell.
3. Explain why you would expect a water molecule to contain two hydrogen atoms and one oxygen atom.

2.3 Ionic bonds and covalent bonds are common

LEARNING OUTCOME

When you complete this section, you should be able to

1. Recognize and construct electron models of molecules that contain an ionic bond or covalent bonds.

When two toddlers are at play, one might take a toy from the other instead of sharing. Atoms act similarly when one takes (an) electron(s) from the other. Chlorine (Cl) is an atom that has seven electrons in its outer shell, so if it acquires one more, it has a completed outer shell. Sodium (Na), on the other hand, which has only one electron in its third shell, lets chlorine have the electron. It is an electron donor. Once the two atoms react in this way, each has eight electrons in the outer shell (Fig. 2.3A).

MP3
Chemical Bonding

The electron transfer, however, causes a charge imbalance in each atom. The sodium atom has one more proton than it has electrons; therefore, it has a net charge of +1 (symbolized by Na^+). The chlorine atom has one more electron than it has protons; therefore, it has a net charge of –1 (symbolized by Cl^-). Such charged particles are called **ions.**

Ionic Bonds Ionic compounds are held together by an **ionic bond,** which is an attraction between negatively and positively charged ions. When sodium reacts with chlorine, an ionic compound called sodium chloride (NaCl) results. Sodium chloride is an example of a salt; it is commonly known as table salt because it is used to season food (Fig. 2.3A). **Salts** can exist as dry solids, but when salts are placed in water, they release ions as they dissolve. NaCl separates into Na^+ and Cl^-. Ionic compounds are most commonly found in this separated (dissociated) form in

TABLE 2.3		Significant Ions in the Human Body
Name	**Symbol**	**Special Significance**
Sodium	Na^+	Found in body fluids; important in muscle contraction and nerve conduction
Chloride	Cl^-	Found in body fluids
Potassium	K^+	Found primarily inside cells; important in muscle contraction and nerve conduction
Phosphate	PO_4^{3-}	Found in bones, teeth, and the high-energy molecule ATP
Calcium	Ca^{2+}	Found in bones and teeth; important in muscle contraction and nerve conduction
Bicarbonate	HCO_3^-	Important in acid-base balance

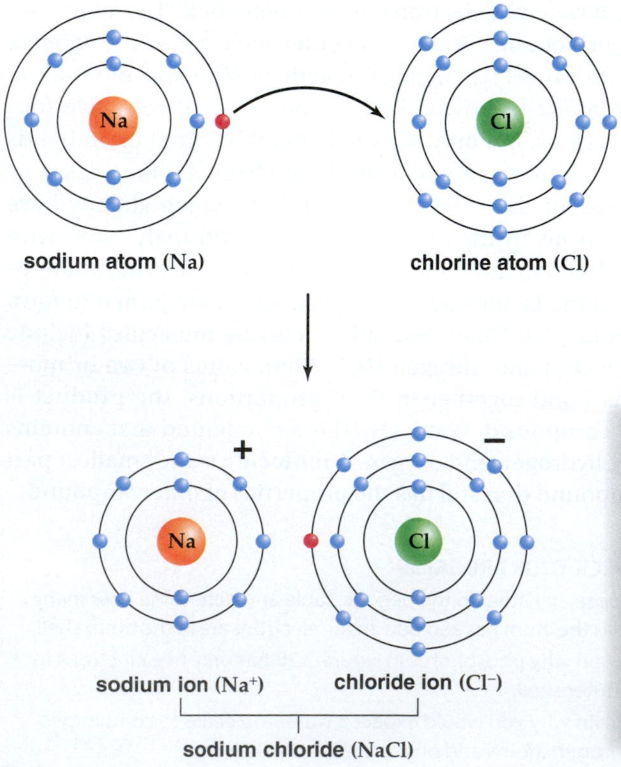

living things because biological systems are 70–90% water.

Animation
Ionic Bonds

Sodium (Na^+) and chloride (Cl^-) are not the only biologically important ions. Some, such as potassium (K^+), are formed by the transfer of a single electron to another atom; others, such as calcium (Ca^{2+}) and magnesium (Mg^{2+}), are formed by the transfer of two electrons. Biologically important ions in the human body are listed in Table 2.3. The balance of these ions in the body is important to our health. Too much sodium in the blood can cause high blood pressure; too much or too little potassium results in heartbeat irregularities; and not enough calcium leads to rickets (bowed legs) in children. Bicarbonate ions are involved in maintaining the acid-base balance of the body.

sodium atom (Na)

chlorine atom (Cl)

sodium ion (Na^+)

chloride ion (Cl^-)

sodium chloride (NaCl)

a.

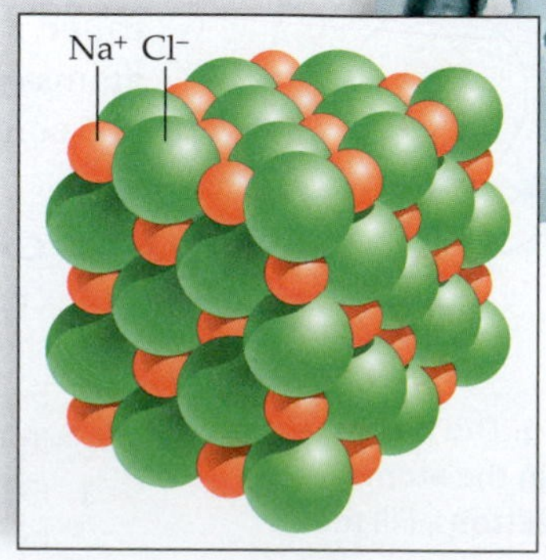

Na^+ Cl^-

b.

FIGURE 2.3A Sodium chloride (table salt). **a.** Formation of sodium chloride. **b.** Sodium chloride in a three-dimensional lattice makes up the crystals found in table salt.

Covalent Bonds Sometimes toddlers do share toys; first one has the toy and then the other. Similarly, some atoms nearly always share electrons, and both atoms are satisfied because the outer shell of each then has eight electrons (or two electrons in the case of hydrogen). If hydrogen is in the presence of a strong electron acceptor, such as oxygen, it gives up its one electron to become a hydrogen ion (H^+) (also called a proton because this ion has no electrons). But hydrogen can also share its electron with another atom. For example, one hydrogen atom can share with another hydrogen atom. When their two orbitals overlap, a pair of electrons is shared within a so-called **covalent bond.** Sharing is illustrated by drawing molecular models called electron models (Fig. 2.3B).

Animation
Ionic Versus
Covalent Bonding

A common way to symbolize that atoms are sharing electrons is to draw a line between the two atoms, as in the structural formula H—H. In a molecular formula, the line is omitted, and the molecule is simply written as H_2 (Fig. 2.3B). Sometimes, atoms share more than one pair of electrons to complete their octets. A double covalent bond occurs when two atoms share two pairs of electrons. To show that oxygen gas (O_2) contains a double bond,

the molecule can be written as O=O. It is also possible for atoms to form triple covalent bonds, as in nitrogen gas (N_2), which can be written as N≡N. Single covalent bonds between atoms are quite strong, but double and triple bonds are even stronger.

The gas methane results when carbon binds to four hydrogen atoms (CH_4). In methane, each bond actually points to one corner of a tetrahedron. A ball-and-stick model is the best way to show this arrangement, while a space-filling model comes closest to showing the actual shape of the molecule (Fig. 2.3C). The shapes of molecules help dictate the roles they play in organisms.

Chemical Reactions Chemical reactions, such as those in photosynthesis, are very important to organisms. An overall equation for the photosynthetic reaction indicates that some bonds are broken and others are formed:

$$6\,CO_2 \;+\; 6\,H_2O \;\longrightarrow\; C_6H_{12}O_6 \;+\; 6\,O_2$$

carbon dioxide — water — glucose — oxygen

This equation says that six molecules of carbon dioxide react with six molecules of water to form one glucose molecule and six molecules of oxygen. The reactants (molecules that participate in the reaction) are shown to the left of the arrow, and the products (molecules formed by the reaction) are shown to the right. Notice that the equation is "balanced"—that is, the same number of each type of atom occurs on both sides of the arrow.

Note the glucose molecule in the equation. It has six atoms of carbon, 12 atoms of hydrogen, and six atoms of oxygen bonded together to form one molecule. The structural formula for glucose is shown in Figure 3.3A.

▶ **2.3 CHECK YOUR PROGRESS**
1. Contrast why you would expect sodium (Na) to give and chlorine (Cl) to accept an electron but you would expect the atoms in a water molecule to share electrons.
2. Knowing that oxygen (O) is able to attract an electron to a greater degree than hydrogen (H), determine the correct charges for the ions that result when water breaks down like this: $H_2O \rightarrow H + OH$.

Electron model	Structural formula	Molecular formula
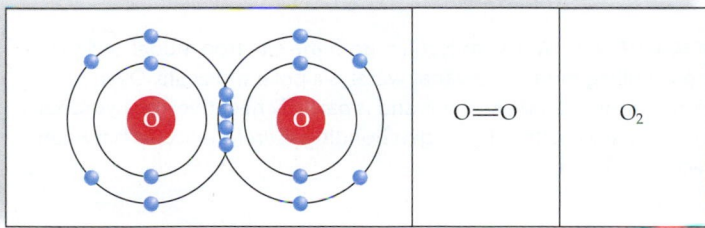	H—H	H_2

hydrogen gas

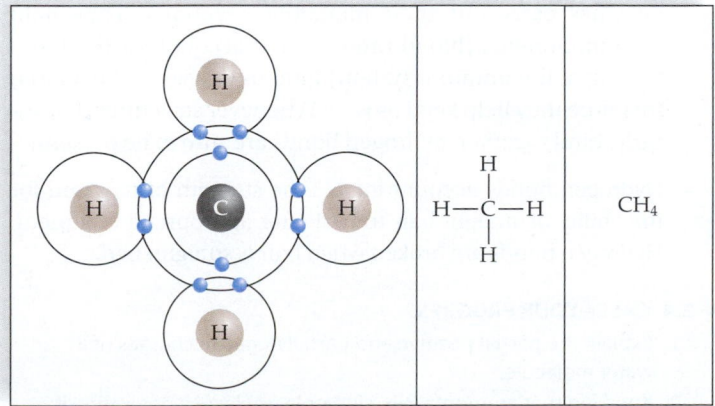	O=O	O_2

oxygen gas

	H | H—C—H | H	CH_4

methane

FIGURE 2.3B Electron models and formulas representing covalently bonded molecules.

Ball-and-stick model	Space-filling model
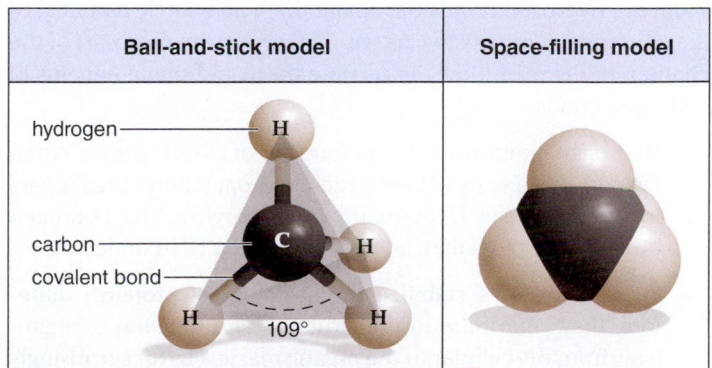	

hydrogen — H
carbon — C
covalent bond — 109°

FIGURE 2.3C Other types of molecular models—in this case, for methane (CH_4).

2.4 A covalent bond can be nonpolar or polar

LEARNING OUTCOMES

When you complete this section, you should be able to

1. Differentiate between a nonpolar and a polar covalent bond.
2. Predict when hydrogen bonding will occur and the effects of these bonds on the molecule.

When the sharing of electrons between two atoms is fairly equal, the covalent bond is said to be a **nonpolar covalent bond.** All the molecules in Figure 2.3B, including methane (CH_4), are nonpolar. However, in a water molecule (H_2O), the sharing of electrons between oxygen and each hydrogen is not completely equal. The attraction of an atom for the electrons in a covalent bond is called its **electronegativity.** The larger oxygen atom, with its greater number of protons, is more electronegative than the hydrogen atom. The oxygen atom can attract the electron pair to a greater extent than each hydrogen atom can. The shape of a water molecule allows the oxygen atom to maintain a slightly negative charge (δ^-, or "delta minus"), and allows the hydrogen atoms to maintain slightly positive charges (δ^+, or "delta plus"). The unequal sharing of electrons in a covalent bond creates a **polar covalent bond,** and in the case of water, the molecule itself is a polar molecule. Figure 2.4 shows the electron model and the space-filling model of a water molecule.

Animation
Electronegativity

Hydrogen Bonds The polarity of water molecules causes the hydrogen atoms in one molecule to be attracted to the oxygen atoms in other water molecules. This attraction, which is weaker than an ionic or covalent bond, is called a **hydrogen bond.** Because a hydrogen bond is easily broken, it is often represented by a dotted line (Fig. 2.4).

Hydrogen bonding is not unique to water. Many biological molecules have polar covalent bonds involving an electropositive hydrogen and usually an electronegative oxygen or nitrogen. In these instances, hydrogen bonds can occur because the partially positive hydrogen atoms are attracted to partially negative atoms. This attraction causes hydrogen bonds to form within the same molecule or between different molecules.

Hydrogen bonds are a bit like Velcro: Each tiny hook and loop is weak, but when hundreds of hooks and loops come together, they are collectively strong. Continuing the analogy, a Velcro fastener is easy to pull apart when needed, and in the same way, hydrogen bonds are disrupted when energy is applied. The versatility of hydrogen bonds has made them absolutely essential to the existence of organisms, as we shall see in the next part of the chapter. In the meantime, let us note these additional benefits of hydrogen bonds:

- Hydrogen bonds hold the two strands of DNA together. When DNA makes a copy of itself, each hydrogen bond breaks easily, allowing the DNA to unzip. Otherwise, the hydrogen bonds, acting together, add stability to the DNA molecule.

- Hydrogen bonds stabilize the structure of proteins; therefore, they contribute to the structure and function of organisms from the cellular to the organismal level. Not surprisingly, the strength of bones and the activity of muscles in animals is dependent in part on hydrogen bonding.

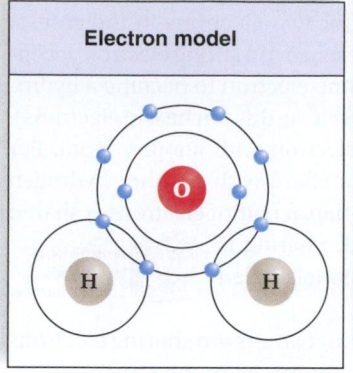

Electron model

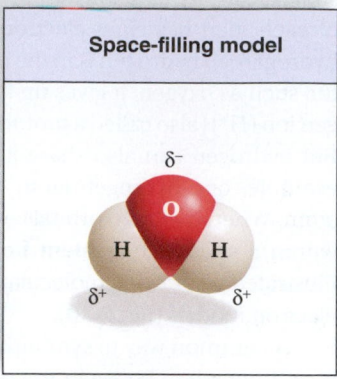

Space-filling model

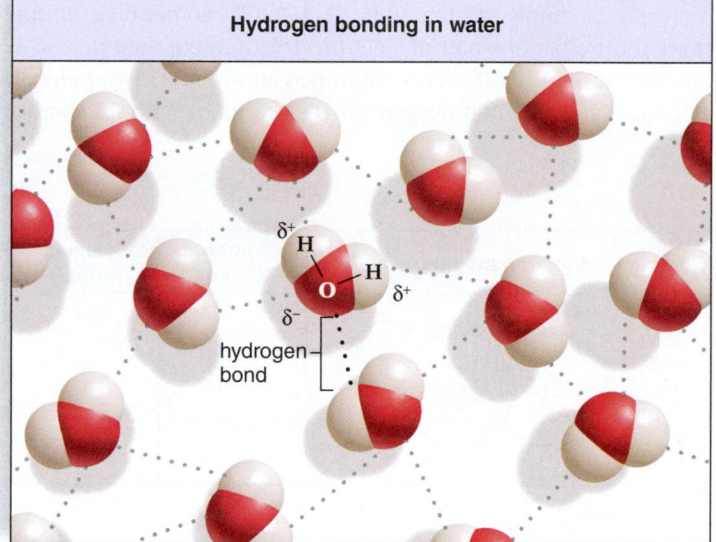

Hydrogen bonding in water

hydrogen bond

FIGURE 2.4 Water molecules. Both the electron model and the space-filling model show that water is a polar molecule. Oxygen attracts the shared electrons and is partially negative; the hydrogens are partially positive. Hydrogen bonding therefore occurs between water molecules.

- Hydrogen bonds not only help enzymes keep their normal shape (without which they cannot function), but also help enzymes carry out their metabolic reactions. They help immunoproteins (blood proteins that account for the functioning of the immune system) bind to viruses and bacteria; therefore they help keep us well. Whenever structures temporarily bind together, hydrogen bonds are sure to be present.

- Hydrogen bonds account for 1/3 the strength of hair and for the ability of straight hair to curl after appropriate treatment. Hydrogen bonds are broken when hair is straightened.

▶ **2.4 CHECK YOUR PROGRESS**

1. Explain the partial positive and partial negative charges of a water molecule.
2. Knowing that ammonia (NH_3) is a polar molecule, determine if hydrogen bonding occurs between ammonia molecules and between ammonia and water molecules.

The Properties of Water

The introduction "Life Depends on Water" on pages 22–23 stresses the close association between water and organisms. In this part of the chapter, we study four properties of water and show how these properties benefit organisms. We will see that water (1) is cohesive and adhesive, (2) expands as it freezes, (3) takes up and gives off heat but changes temperature slowly, and (4) dissolves other polar substances.

2.5 Water molecules stick together and to other materials

LEARNING OUTCOMES

When you complete this section, you should be able to

1. Recognize that hydrogen bonding between water molecules ensures its liquid nature at temperatures suitable to life.
2. Explain why water's hydrogen bonding is essential to the ability of water to serve as a transport medium.

Hydrogen bonding accounts for most of the properties of water that make life possible. For example, without hydrogen bonding, frozen water would melt at –100° Celsius (°C), and liquid water would boil at –91°C, making most of the water on Earth steam, and life unlikely. But because of hydrogen bonding, water is a liquid at temperatures typically found on the Earth's surface. It melts at 0°C and boils at 100°C.

Also because of hydrogen bonding, water molecules exhibit **cohesion** (they stick together) and **adhesion** (they stick to other polar materials). Diving into a pool breaks the hydrogen bonds, but they re-form behind you. Cohesion allows water to flow freely without separating. It also gives water a high surface tension so that a water strider can skip across the top of a pond, for example.

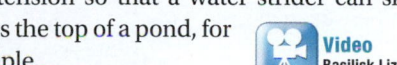

water strider

Video Basilisk Lizard

As a result of cohesion and adhesion, water is an excellent transport medium within vessels in both plants and animals. How is it possible for water to rise to the top of even very tall trees? Water transport in plants is somewhat like sucking water through a straw—or rather, a bundle of straws. Water evaporating from leaves pulls up more water molecules from plant vessels. Cohesion keeps the water column from breaking apart, and adhesion of water molecules to vessel walls prevents the water column from falling backward (Fig. 2.5).

The liquid portion of our blood, which transports dissolved and suspended substances throughout the body, is 90% water. When we drink water, it eventually enters our blood vessels. The cohesive and adhesive properties of water permit blood to flow and help prevent it from falling backward, particularly in the smallest blood vessels, the capillaries (Fig. 2.5).

Animation Properties of Water

> ▶ **2.5 CHECK YOUR PROGRESS**
> 1. Relate the two properties of water that make it an excellent transport medium to hydrogen bonding.
> 2. Cite examples to show that all organisms depend on water as a transport medium.

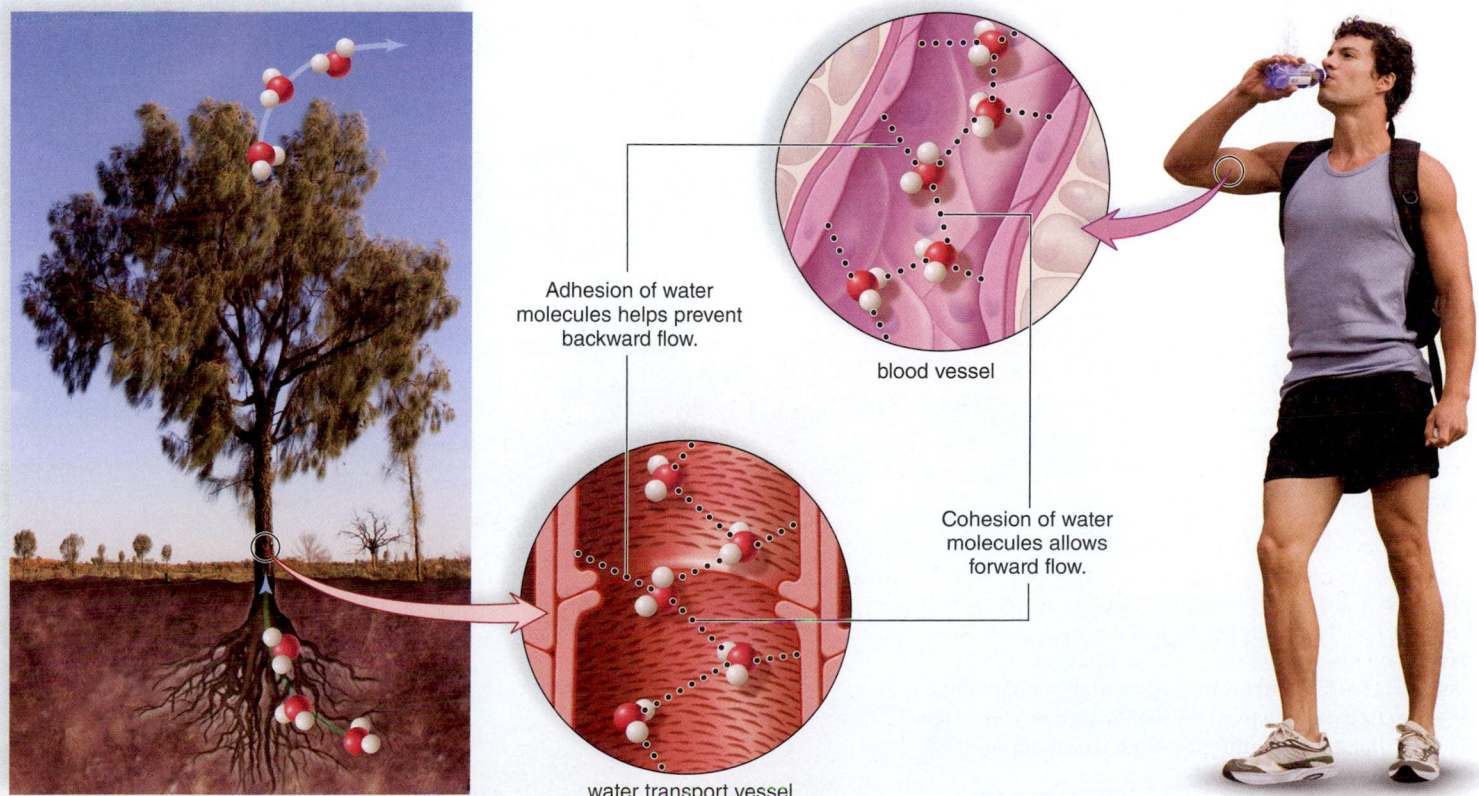

Adhesion of water molecules helps prevent backward flow.

blood vessel

Cohesion of water molecules allows forward flow.

water transport vessel

FIGURE 2.5 Water molecules are cohesive (stick together) and adhesive (stick to vessels) and water makes an excellent transport medium in the vessels of trees and humans.

2.6 Both frozen and liquid water help organisms maintain a normal temperature

LEARNING OUTCOME

When you complete this section, you should be able to

1. Describe why water expands when it freezes and why it warms up and cools down slowly.

Remarkably, water is more dense at 4°C than it is at 0°C. Most substances contract when they solidify, but *water expands when it freezes.* In ice, water molecules form a lattice, in which the hydrogen bonds are farther apart than they are in liquid water. This is why cans of soda burst when placed in a freezer, and why frost heaves make northern roads bumpy in the winter. It also means that ice is less dense than liquid water, and therefore ice floats (Fig. 2.6A).

If ice did not float on water, it would sink, and ponds, lakes, and perhaps even the ocean, would freeze solid, making life impossible in the water and also on land. Instead, bodies of water always freeze from the top down. The ice acts as an insulator to prevent the water below it from freezing and also to prevent the loss of heat to the external environment.

In a pond, the ice protects the protists, plants, and animals so that they can survive the winter (Fig. 2.6B). These animals, except for the otter, are ectothermic, which means they take on the temperature of the outside environment. This might seem disadvantageous; however, water remains relatively warm because of its high heat capacity.

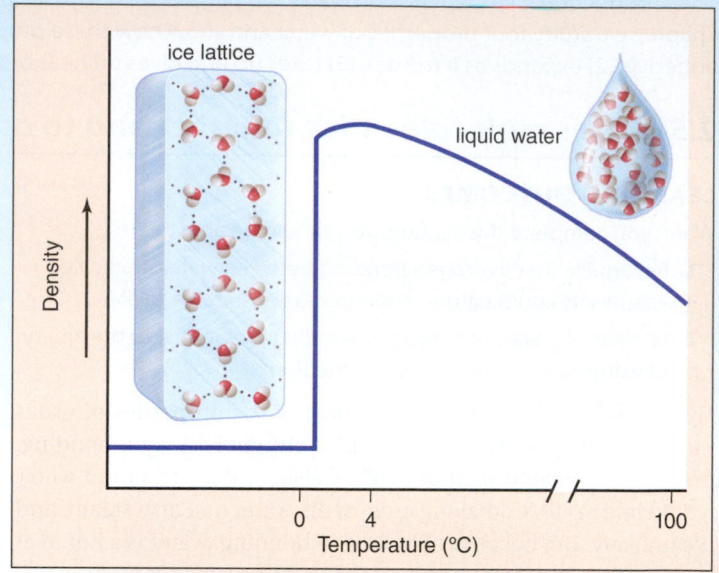

FIGURE 2.6A Ice is less dense than liquid water.

Protists provide food for fish.

River otters visit ice-covered ponds.

Aquatic insects survive in air pockets.

Freshwater fish take oxygen from water.

Common frogs and pond turtles hibernate.

FIGURE 2.6B A pond in winter. During the winter, frogs and turtles hibernate and, in this way, lower their oxygen needs. Insects survive in air pockets. Fish, as you will learn later in this text, have an efficient means of extracting oxygen from the water, and they need less oxygen than the endothermic otter, which depends on muscle activity to warm its liquid blood.

Heat Capacity of Water The many hydrogen bonds that link water molecules help water take up heat, without a great change in temperature. Because *the temperature of water rises slowly,* organisms are better able to maintain their normal internal temperatures and are protected from rapid temperature changes. One calorie is the amount of heat energy needed to raise the temperature of 1 gram (g) of water 1°C. In comparison, other covalently bonded liquids require an input of about one-half calorie to raise the temperature 1°C. Then, too, water holds onto its heat, and *the temperature of water falls slowly* compared to other liquids. This property of water is important, not only for aquatic organisms, but for all living things. Converting 1 g of the coldest liquid water to ice requires the loss of 80 calories of heat energy.

Water also has a high heat of vaporization and *it takes much heat to make water boil* because hydrogen bonds must be broken before water boils and water molecules vaporize (evaporate into the environment). Converting 1 g of the hottest water to a gas requires an input of 540 calories of heat energy. Water's high heat of vaporization gives animals in a hot environment an efficient way to release excess body heat. When an animal sweats or gets splashed, body heat is used to vaporize the water, thus cooling the animal (Fig. 2.6C).

Because of water's high heat capacity and high heat of vaporization, the temperatures along coasts are moderate. During the summer, the ocean absorbs and stores solar heat, and during the winter, the ocean slowly releases it. In contrast, the interior regions of continents experience abrupt changes in temperature.

FIGURE 2.6C The bodies of organisms cool when their heat is used to evaporate water.

▶ **2.6 CHECK YOUR PROGRESS**
   1. Explain why igloos built from blocks of ice keep Eskimos warm.
   2. Explain how evaporative coolers that use a fan to draw hot air through a water-soaked fiber pad work.

2.7 Water dissolves other polar substances

LEARNING OUTCOME

When you complete this section, you should be able to
   1. Identify water as a hydrophilic solvent for polar molecules.

Because of its polarity, *water dissolves a great number of substances.* A **solution** contains both a **solute,** usually a solid, and a **solvent,** usually a liquid. When ionic salts—for example, sodium chloride (NaCl)—are put into water, the negative ends of the water molecules are attracted to the sodium ions, and the positive ends of the water molecules are attracted to the chloride ions. This causes the sodium ions and the chloride ions to dissociate (separate) as it dissolves in water (Fig. 2.7). Water is also a solvent for larger molecules that contain ionized atoms or are polar. When water moves from the roots to the leaves in trees, it serves as a transport vehicle for dissolved minerals. And the transport function of blood is only possible because salts and molecules are dissolved, or suspended, in blood plasma.

Those molecules that can attract water are said to be **hydrophilic.** When ions and molecules disperse in water, they move about and collide, allowing reactions to occur. Therefore, water facilitates chemical reactions. Nonionized and nonpolar molecules that cannot attract water are said to be **hydrophobic.** Gasoline contains nonpolar molecules, and therefore it does not mix with water and is hydrophobic.

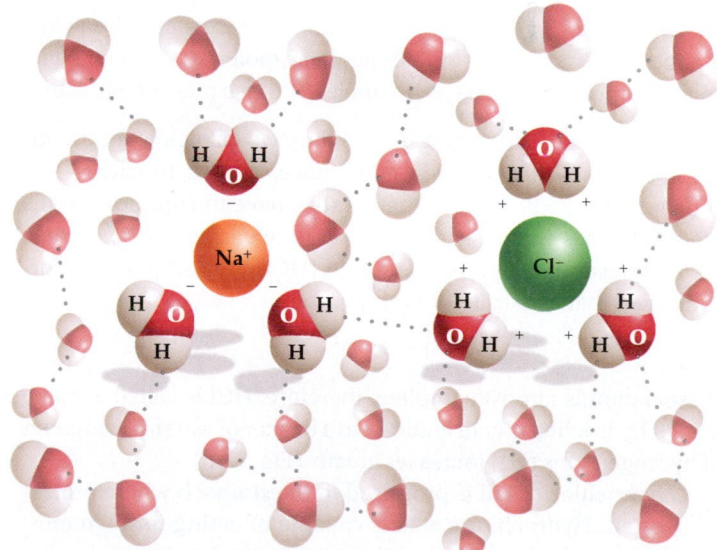

FIGURE 2.7 An ionic salt dissolves in water.

▶ **2.7 CHECK YOUR PROGRESS**
   1. Emulsifiers combine with and cause nonpolar fats to disperse in water. What property must emulsifiers have that fats lack?

Acids, Bases, and pH

Organisms are particularly sensitive to the hydrogen ion concentration [H⁺] of liquids. A pH scale is used to judge any changes caused by the addition of acids or bases to a fluid. Buffers assist in keeping the [H⁺] of body fluids relatively constant.

2.8 Acids and bases affect cells and organisms

LEARNING OUTCOME

When you complete this section, you should be able to

1. Distinguish between acids and bases.

When water dissociates, it releases an equal number of **hydrogen ions (H⁺)** and **hydroxide ions (OH⁻)**:

FIGURE 2.8A Dissociation of water molecules releases equal amount of H⁺ and OH⁻.

$$H—O—H \rightleftharpoons H^+ + OH^-$$
$$\text{water} \quad\quad \text{hydrogen} \quad \text{hydroxide}$$
$$\text{ion} \quad\quad \text{ion}$$

Only a few water molecules at a time dissociate. The actual number of H⁺ is $(1 \times 10^{-7}$ moles per liter [moles/l])[1], and an equal concentration of OH⁻ at $(1 \times 10^{-7}$ moles/l) is also present (Fig. 2.8A).

Acids: Excess Hydrogen Ions When we eat acidic foods, our blood becomes more acidic. Lemon juice, vinegar, tomatoes, and coffee are all acidic foods. What do they have in common? **Acids** are substances that dissociate in water, releasing hydrogen ions (H⁺). For example, hydrochloric acid (HCl) is an important inorganic acid that dissociates in this manner:

$$HCl \longrightarrow H^+ + Cl^-$$

Dissociation is almost complete; therefore, HCl is called a strong acid. If hydrochloric acid is added to a beaker of water, the number of hydrogen ions (H⁺) increases greatly (Fig. 2.8B).

Hydrochloric acid is produced in the stomach where protein is digested. Hydrochloric acid is capable of eating through most metals, and is highly toxic, burning on contact. However, a layer of mucus protects the stomach wall.

Bases: Excess Hydroxide Ions When we take in basic substances, the blood becomes more basic. Milk of magnesia and bicarbonate of soda are basic solutions familiar to most people.

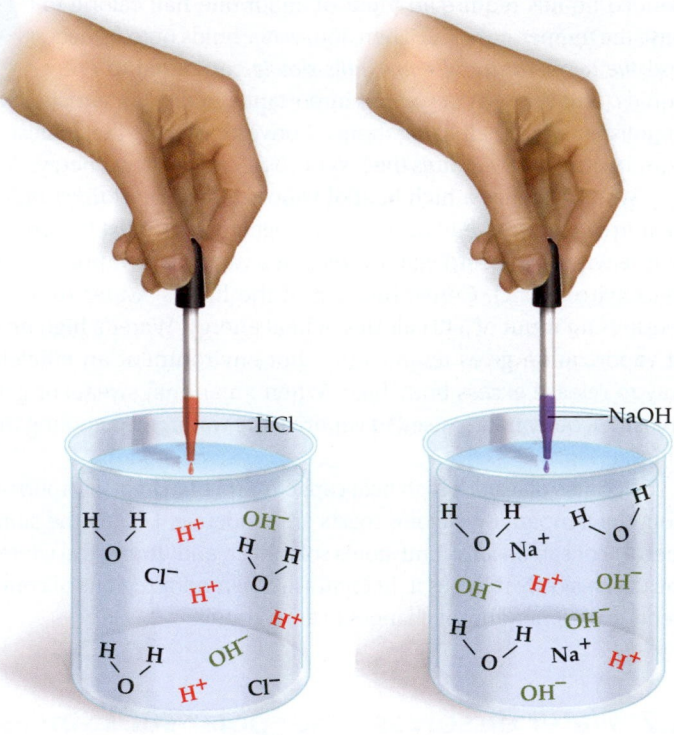

FIGURE 2.8B Acids add H⁺ so that H⁺ is now greater than OH⁻.

FIGURE 2.8C Bases add OH⁻ so that H⁺ is now less than OH⁻.

Bases are substances that either take up hydrogen ions (H⁺) or release hydroxide ions (OH⁻). For example, sodium hydroxide (NaOH) is an important inorganic base that dissociates in this manner:

$$NaOH \longrightarrow Na^+ + OH^-$$

Dissociation is almost complete; therefore, sodium hydroxide is called a strong base. If sodium hydroxide is added to a beaker of water, the number of hydroxide ions increases (Fig. 2.8C).

Sodium hydroxide is also known as lye or caustic soda. It is just as dangerous as a strong acid and can be used to etch aluminum. Contact with strong acids and bases should be avoided. For this reason, containers of these chemicals are marked with warning symbols.

▶ **2.8 CHECK YOUR PROGRESS**
  1. Identify which one, H⁺ or OH⁻, increases when acids are added to water. When bases are added to water.

---

[1]In chemistry, a mole is defined as the amount of matter that contains as many objects (atoms, molecules, ions) as the number of atoms in exactly 12 g of ¹²C.

2.9 The pH scale measures acidity and basicity

LEARNING OUTCOMES

When you complete this section, you should be able to

1. Explain and use the pH scale.

2. Describe a buffer and identify how buffers assist organisms.

The **pH scale** is used to indicate the acidity or basicity (also called alkalinity) of a solution. The pH scale ranges from 0 to 14 (Fig. 2.9). A pH of 7 represents a neutral state, in which the hydrogen ion concentration [H⁺] equals the hydroxide ion concentration [OH⁻]. A pH below 7 is an acidic solution because [H⁺] is greater than [OH⁻]. A pH above 7 is basic because [OH⁻] is greater than [H⁺].

Moving down the pH scale from pH 14 to pH 0, each unit has ten times the [H⁺] of the previous unit. Moving up the scale from 0 to 14, each unit has ten times the [OH⁻] of the previous unit. The pH scale eliminates the use of cumbersome numbers. For example, in the following list, hydrogen ion concentrations are on the left, and the pH is on the right:

[H⁺] (moles per liter)	pH
$0.000001 = 1 \times 10^{-6}$	6
$0.0000001 = 1 \times 10^{-7}$	7
$0.00000001 = 1 \times 10^{-8}$	8

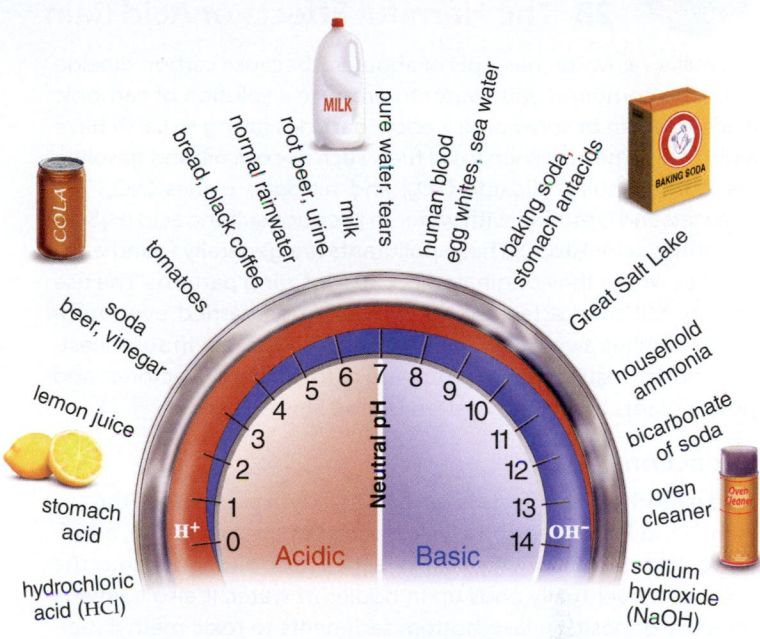

FIGURE 2.9 The pH scale. The dial of this pH meter indicates that pH ranges from 0 to 14. pH 7 is neutral pH (has equal amount of H⁺ and OH⁻). pH 0–7 is acidic with more H⁺ than OH⁻; and pH 7–14 is basic with more OH⁻ than H⁺.

To further illustrate the relationship between [H⁺] and pH, consider the following question: Which of the pH values listed above indicates a higher hydrogen ion concentration [H⁺] than pH 7, and therefore would be an acidic solution? A number with a smaller negative exponent indicates a greater quantity of hydrogen ions; therefore, pH 6 is an acidic solution.

In most organisms, pH needs to be maintained within a narrow range. The pH of human blood is between 7.35 and 7.45; this is the pH at which our proteins, such as cellular enzymes, function properly. To maintain normal pH, blood is buffered.

Buffers A **buffer** resists changes in pH. Many commercial products, such as aspirin, shampoos, or deodorants, are buffered as an added incentive for us to buy them. Blood and many other body fluids are buffered so the pH stays within a certain range. If blood pH rises much above pH 7.45, alkalosis is present, and if the pH lowers much below 7.35, acidosis is present. Weakness, cramping, and irritability are symptoms of alkalosis. Seizures, coma, and even death can result from acidosis. Normally, buffers take up excess hydrogen ions (H⁺) or hydroxide ions (OH⁻), thus preventing these conditions.

Usually a buffer consists of a combination of chemicals. For example, carbonic acid (H_2CO_3) and bicarbonate (HCO_3) are two chemicals present in blood that keep pH within normal limits. Carbonic acid is an acid that releases bicarbonate and H⁺ during this dissociation reaction:

$$H_2CO_3 \rightleftharpoons H^+ + HCO_3^-$$
$$\text{carbonic acid} \qquad \text{bicarbonate ion}$$

Being a weak acid H_2CO_3 tends not to dissociate but will do so upon the addition of a base. When bases add hydroxide ions (OH⁻) to blood, the dissociation reaction occurs because the OH⁻ combines with the H⁺ and water forms. On the other hand, when acids add H⁺ to blood, any dissociated carbonic acid simply re-forms:

$$H^+ + HCO_3^- \longrightarrow H_2CO_3$$

In addition to buffers, breathing helps maintain pH by ridding the body of CO_2, because the more CO_2 there is in the body, the more carbonic acid there is in the blood. As powerful as the buffer and the respiration mechanisms are in maintaining pH, only the kidneys rid the body of a wide range of acidic and basic substances and otherwise adjust the pH. The kidneys are slower acting than other mechanisms, but they have a more powerful effect on pH. If the kidneys malfunction, alkalosis or acidosis may result.

Ecosystems are buffered, but acid deposition can overcome their buffering ability, leading to sterile lakes and dead forests as discussed in "The Harmful Effects of Acid Rain" on page 36.

▶ **2.9 CHECK YOUR PROGRESS**

1. Determine if rainwater with a pH of about 5.6 is acidic or basic. Does it have more or less H⁺ than pH 7?

2. Explain why the bicarbonate ion HCO₃⁻ can help buffer blood.

2B The Harmful Effects of Acid Rain

Normally, rainwater has a pH of about 5.6 because carbon dioxide in the air combines with water to produce a solution of carbonic acid. Acid rain or snow or dry acidic particles falling to Earth have a pH of less than 5. When fossil fuels such as coal, oil, and gasoline are burned, sulfur dioxide (SO_2) and nitrogen oxides (NO_x) are released and combine with water to produce sulfuric acid (H_2SO_4) and nitric acid (HNO_3). These pollutants are generally found eastward of where they originated because of wind patterns. The use of very tall smokestacks causes them to be carried even hundreds of miles away (Fig. 2B). For example, acid rain in southeastern Canada results from the burning of fossil fuels in factories and power plants in the midwestern United States.

Impact on Lakes

Acid rain adversely affects lakes, particularly in areas where the soil is thin and lacks limestone (calcium carbonate, or $CaCO_3$), a buffer to acid deposition. Acid rain leaches toxic aluminum into the soil and it eventually ends up in bodies of water. It also converts mercury deposits in lake bottom sediments to toxic methyl mercury, which accumulates in fish. People are now advised against eating fish from the Great Lakes because of high mercury levels. Hundreds of lakes are devoid of fish in Canada and New England, and thousands have suffered the same fate in the Scandinavian countries. Some of these lakes have no signs of life at all.

Impact on Forests

The leaves of plants are damaged by acid rain so that they can no longer carry on photosynthesis as before. When plants are under stress, they become susceptible to diseases and pests of all types.

Forests on mountaintops receive more rain than those at lower levels; therefore, they are more affected by acid rain (Fig. 2B). Forests are also damaged when toxic chemicals such as aluminum are leached into the soil. These kill soil fungi that assist roots in acquiring the nutrients trees need. In New England, 1.3 million acres of high-elevation forests have been devastated.

Impact on Humans and Structures

Humans may be affected by acid rain. Inhaling dry sulfate and nitrate particles appears to increase the occurrence of respiratory illnesses, such as asthma. Buildings and monuments made of limestone and marble break down when exposed to acid rain. The paint on homes and automobiles is likewise degraded.

CONSIDER THESE QUESTIONS

1. Does it concern you that acid rain might affect the lake where you usually fish or the lumber industry that employs many workers?
2. Would you be willing to drive less to help prevent acid rain? Why or why not?
3. At a time when people are stressed by the economy, should they still be concerned about environmental degradation? Why or why not?

connect |BIOLOGY Explore the concepts through a variety of multimedia assets and question types.

www.mcgrawhillconnect.com

Emissions from power plants and industrial facilities add acids to the atmosphere.

Statues corrode due to acid deposition.

Trees die due to acid deposition.

Lakes become sterile due to acid deposition.

FIGURE 2B Environmental effects of acid rain.

THE CHAPTER IN REVIEW

CONNECTING THE CONCEPTS

It is possible to connect the concepts in this chapter to the structure and properties of the water molecule (H₂O). A water molecule contains the atoms hydrogen and oxygen. For any atom, the atomic number is equal to the number of protons, while the mass number is the number of protons plus the number of neutrons. Because atoms of the same element have isotopes, the atomic mass of an atom is not a whole number.

An atom of hydrogen has one electron in its outer shell, while oxygen has six electrons in its outer shell. When two hydrogens and one oxygen react, covalent bonds form the H₂O molecule. In water, oxygen shares a pair of electrons with each hydrogen, and in that way each hydrogen has two electrons in its outer shell while oxygen has eight electrons in its outer shell.

Because oxygen is more electronegative than a hydrogen atom, the covalent bonds in water are polar bonds. The oxygen carries a slightly negative charge, while each hydrogen carries a positive charge. Therefore, hydrogen bonds form between water molecules. Water, being a polar molecule, can dissolve other polar substances. Hydrogen bonding leads to the other properties of water, such as cohesiveness, the tendency to change temperature slowly, and expansion when it freezes.

Water has a neutral pH, the pH preferred by organisms. If an acid is added to water, the pH decreases; if a base is added to water, the pH increases. Buffers that resist a change in pH can also be added to water.

Water is considered an inorganic substance because it doesn't contain a carbon atom. As we will learn in Chapter 3, organic molecules do contain carbon, and certain organic molecules—carbohydrates, lipids, proteins, and nucleic acids—are unique to living things.

ANALYZE AND EVALUATE

1. Analyze the structure of methane (CH₄) by comparing it to the structure of water. Be sure to explain the formula for methane.

2. Support this statement: The chemical structure of water accounts for its properties that are critical to life.

SUMMARIZE

The Atoms of Cells

2.1 Six types of atoms are basic to life

- **Matter** takes up space; it can be a solid, a liquid, or a gas. All matter is composed of **elements,** substances that cannot be broken down by chemical means. An **atom** is the smallest unit of an element. Each type atom has an **atomic symbol.** Six atoms—carbon, hydrogen, nitrogen, oxygen, phosphorus, and sulfur—play significant roles in all organisms.
- Atoms contain subatomic particles. The best-known subatomic particles are **protons** (positive charge), **neutrons** (uncharged), and **electrons** (negative charge). Electrons are located in **electron shells** that circle the nucleus.
- The **atomic number** is the number of protons in the nucleus of an atom. The **mass number** of an atom is the number of protons plus the number of neutrons in the nucleus.
- The **isotopes** of an element have the same number of protons, but they differ in atomic mass due to different numbers of neutrons. **Atomic mass** is the average mass of an element's isotopes.

Atoms Form Molecules

2.2 After atoms react, they have a completed outer shell

- The **octet rule** states that atoms tend to react to achieve eight electrons in the outer shell (the most stable number) (or two electrons in the case of hydrogen). The number of electrons in an atom's outer shell, called the **valence shell,** determines how it will react.

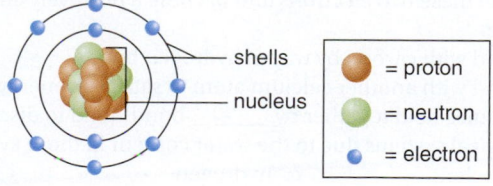

- A **compound** results when atoms of two or more elements are bonded together in fixed proportions. A **molecule** is the smallest part of a compound that has the properties of that compound.

2.3 Ionic bonds and covalent bonds are common

- An **ionic bond** occurs when particles called **ions** are held together by the attraction between negative and positive ions. A **covalent bond** occurs when two atoms share electrons in such a way that each atom has eight electrons in the outer shell (or two in the case of hydrogen).

2.4 A covalent bond can be nonpolar or polar

- A **nonpolar covalent bond** occurs when the sharing of electrons between atoms is fairly equal as in methane. The ability of an atom to attract electrons in a covalent bond is called its **electronegativity.** A **polar covalent bond** occurs when the sharing of electrons is not equal. Unequal sharing can result in a polar molecule (e.g., water).
- Hydrogen bonds can occur between polar molecules. A **hydrogen bond** is a weak attraction between a slightly positive hydrogen atom and a slightly negative atom of another molecule, or between atoms of the same molecule. Hydrogen bonds are individually weak and easily broken, but collectively strong.

The Properties of Water

2.5 Water molecules stick together and to other materials

- Hydrogen bonding is responsible for **cohesion** of water molecules and their **adhesion** to other polar materials. These two properties allow water to serve as a transport medium in living organisms (e.g., trees and humans).

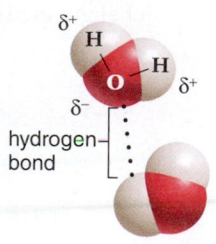

2.6 Both frozen water and liquid water help organisms maintain a normal temperature

- Frozen water expands and floats because hydrogen bonding in ice is more rigid and open. Ice protects water and organisms below it from freezing and the water remains relatively warm because it has a high heat capacity
- Hydrogen bonding causes water to have a high heat capacity and a high heat of vaporization. The temperature of water rises and falls slowly and this protects living things from rapid changes in temperature. Water's high heat of vaporization helps organisms resist overheating.

2.7 Water dissolves other polar substances

- **Solutions** contain a **solvent** (often water) and a **solute** (dissolved substance). The polarity of water makes it a solvent that facilitates chemical reactions. **Hydrophilic** molecules (ionized and/or polar, such as salts) dissolve in water. **Hydrophobic** molecules (nonionized and nonpolar, such as gasoline) do not dissolve in water.

Acids, Bases, and pH

2.8 Acids and bases affect cells and organisms

- When water ionizes, it releases an equal number of **hydrogen ions (H^+)** and **hydroxide ions (OH^-)**. **Acids** add hydrogen ions (H^+) to water. **Bases** add hydroxide ions (OH^-) to water.

2.9 The pH scale measures acidity and basicity

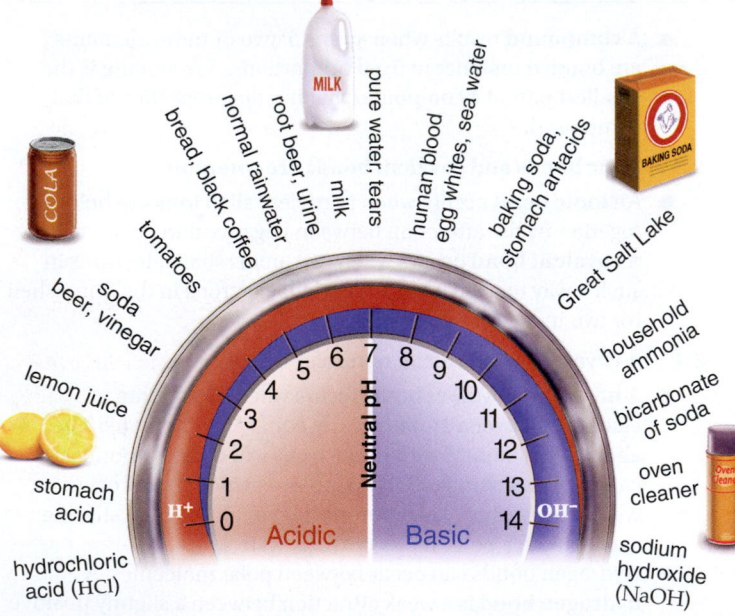

- Pure water has a neutral pH of 7. An acidic solution has a pH below 7: $[H^+]$ is greater than $[OH^-]$. A basic solution has a pH above 7: $[OH^-]$ is greater than $[H^+]$. Most organisms need to maintain pH within a narrow range (e.g., the pH of human blood is about 7.35–7.45).
- A **buffer** is a chemical or combination of chemicals that resists changes in pH and helps keep pH within normal limits.

TEST YOURSELF

The Atoms of Cells

1. CHNOPS are
   a. the only atoms found in nonliving and living things.
   b. the only atoms found in living things.
   c. atoms basic to life.
   d. atoms found in rocks.
   e. Both b and c are correct.
2. Which of the following is not a component of an atom?
   a. proton   c. neutron
   b. lectons   d. electron
3. The atomic number tells you the
   a. number of neutrons in the nucleus.
   b. number of protons in the atom.
   c. atomic mass of the atom.
   d. number of its electrons if the atom has a neutral charge.
   e. Both b and d are correct.
4. Which of the subatomic particles contributes almost no weight to an atom?
   a. protons in the electron shells
   b. electrons in the nucleus
   c. neutrons in the nucleus
   d. electrons at various energy levels
5. Isotopes of the same element differ from each other only by the number of neutrons.
   a. True   b. False
6. The periodic table does not
   a. give information about the various elements.
   b. indicate the number of protons and the number of valence electrons.
   c. indicate whether an element forms ionic or covalent bonds.
   d. play a useful role in chemistry today.
7. **THINKING CONCEPTUALLY** Explain why the title Periodic Table of the Elements is appropriate.

Atoms Form Molecules

8. The rule stating that the outer electron shell is most stable when it contains eight electrons is the
   a. stability rule.   c. octet rule.
   b. atomic rule.   d. shell rule.
9. How many electrons does nitrogen require to fill its outer shell?
   a. zero   c. two
   b. one   d. three
10. Explain why the correct formula for ammonia is NH_3, not NH_4.
11. When an atom gains electrons, it
   a. forms a negatively charged ion.
   b. forms a positively charged ion.
   c. forms covalent bonds.
   d. gains atomic mass.
12. An atom that has two electrons in the outer shell, such as calcium, would most likely
   a. share to acquire a completed outer shell.
   b. lose these two electrons and become a negatively charged ion.
   c. lose these two electrons and become a positively charged ion.
   d. bind with carbon by way of hydrogen bonds.
   e. bind with another calcium atom to satisfy its energy needs.
13. Molecules held together by _____ bonds tend to dissociate in biological systems due to the water content in those systems.
   a. covalent   c. hydrogen
   b. ionic   d. nitrogen

nitrogen
$^{14}_{7}N$

14. Which type of bond results from the sharing of electrons between atoms?
    a. covalent
    c. hydrogen
    b. ionic
    d. neutral

15. In the molecule CH_4,
    a. all atoms have eight electrons in the outer shell.
    b. all atoms are sharing electrons.
    c. carbon could accept more hydrogen atoms.
    d. All of these are correct.

16. In which of these are the electrons always shared unequally?
    a. double covalent bond
    d. polar covalent bond
    b. triple covalent bond
    e. ionic and covalent bonds
    c. hydrogen bond

17. An example of a hydrogen bond would be the
    a. bond between a carbon atom and a hydrogen atom.
    b. bond between two carbon atoms.
    c. bond between sodium and chlorine.
    d. bond between two water molecules.

18. **THINKING CONCEPTUALLY** Explain why an atom with two electrons in its outer shell becomes an ion when it reacts with another atom.

The Properties of Water

19. Water flows freely, but does not separate into individual molecules because water is
    a. cohesive.
    c. hydrophobic.
    b. hydrophilic.
    d. adhesive.

20. Water can absorb a large amount of heat without much change in temperature, and therefore it has
    a. a high surface tension.
    b. a high heat capacity.
    c. ten times as many hydrogen ions.
    d. ten times as many hydroxide ions.

21. Which of these properties of water cannot be attributed to hydrogen bonding between water molecules?
    a. Water stabilizes temperature inside and outside the cell.
    b. Water molecules are cohesive.
    c. Water is a solvent for many molecules.
    d. Ice floats on liquid water.
    e. Both b and c are correct.

Question 22 is based on this graph:

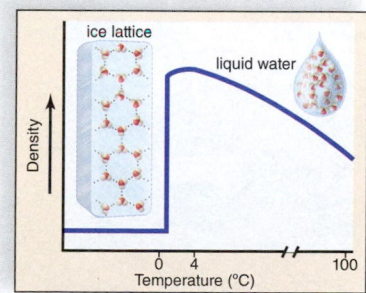

22. This graph illustrates
    a. that ice is less dense than water.
    b. that ice is more dense than water.
    c. why ice floats on water.
    d. why water molecules form hydrogen bonds with other water molecules.
    e. Both a and c are correct.
    f. All of these are correct.

23. **THINKING CONCEPTUALLY** Explain why you would expect the blood of animals to be mostly water.

Acids, Bases, and pH

24. Acids
    a. release hydrogen ions in solution.
    b. cause the pH of a solution to rise above 7.
    c. take up sodium ions and become neutral.
    d. increase the number of water molecules.
    e. Both a and b are correct.

25. Which of these best describes the changes that occur when a solution goes from pH 5 to pH 8?
    a. The hydrogen ion concentration decreases as the solution goes from acidic to basic.
    b. The hydrogen ion concentration increases as the solution goes from basic to acidic.
    c. The hydrogen ion concentration decreases as the solution goes from basic to acidic.

26. When water dissociates, it releases
    a. equal amounts of H^+ and OH^-.
    b. more H^+ than OH^-.
    c. more OH^- than H^+.
    d. only H^+.

27. Rainwater has a pH of about 5.6; therefore, rainwater is
    a. a neutral solution.
    b. an acidic solution.
    c. a basic solution.
    d. It depends on whether the rainwater is buffered.

28. If a chemical accepted H^+ from the surrounding solution, the chemical could be
    a. a base.
    d. None of these are correct.
    b. an acid.
    e. Both a and c are correct.
    c. a buffer.

29. **THINKING CONCEPTUALLY** Compare a chemical buffer such as the bicarbonate ion in the blood, the process of breathing, and kidney function with regard to how they regulate pH in the body.

GET INVOLVED

1. Natural phenomena often require an explanation. Based on Figure 2.7, explain why the oceans don't freeze.
2. Melvin Calvin used radioactive carbon (as a tracer) to discover a series of molecules that form during photosynthesis. Explain why carbon behaves chemically the same, even when radioactive.

MEDIA STUDY TOOLS

mhhe.com/maderconcepts3

Enhance your study of this chapter with interactive study tools, practice tests, and engaging animations. Also, ask your instructor about the resources available through ConnectPlus, which includes LearnSmart, a personalized adaptive learning program, and a media-rich eBook.

3

Organic Molecules of Cells

BEFORE YOU BEGIN

Take a few minutes to recall

The first several levels of biological organization (Figure 1.3A)

The atoms basic to life (section 2.1)

How covalent bonds form (section 2.3)

The significance of hydrogen bonding (section 2.4)

Plants and Animals Are the Same but Different

Certainly plants are similar to animals in many ways. Their basic organization starts with cells, and water moves through the vessels of plants as it does in animals. Without water, they die. Both grow, reproduce, adapt, and exhibit the same characteristics of life.

Still, we might be hard pressed to think of ways that plants and animals are exactly the same until we look inside their cells. Then it becomes obvious that plants are indeed like animals. Vegetarians have no trouble sustaining themselves as long as they include a variety of plants in their diet. That's because plants and animals generally have the same molecules in their cells—namely, carbohydrates, lipids, proteins, and nucleic acids. When we feed on plants, we digest their large biomolecules to smaller molecules, and then we use these smaller molecules to build our own types of carbohydrates, lipids, proteins, and nucleic acids.

"Same but different" will be a common theme in this chapter about the molecules of cells. For example, the genetic material for both plants and animals is the nucleic acid DNA. But each type of plant and animal has its own

particular genes, even though the way genes function in cells is the same in all types of organisms.

Sameness is especially evident when animals acquire vitamins from plants and use them exactly as plants do in their own metabolism. You could go so far as to suggest that an animal's inability to make vitamins is not disadvantageous, as long as it can get the vitamins it needs from plants. That way, an animal is not using up its own energy to make a molecule it can get otherwise. Now it has more energy to use for growth, defense, and reproduction.

Vitamins assist enzymes, the molecules in cells that speed chemical reactions. Plants and animals have to build their own enzymes, but these enzymes function similarly. The enzymes needed to extract energy from nutrient molecules and form ATP, the energy currency of all cells, are the same in plants and animals. Plant cells have many more types of enzymes than do animals because they carry on photosynthesis to form their own food. Plant cells also produce molecules that allow them to protect themselves from predators, maintain an erect posture, and in general, be more colorful than most animals. The pictures on these pages show how animals interact with plants. Plants provide food and they also provide a home for many animals. Humans, when young, love to return to the treetops, the first home for our early ancestors.

In this chapter, we continue our look at basic chemistry by considering the types of molecules unique to organisms. These are the molecules that account for the structure and function of all cells in any type of organism.

Fundamentals of Organic Chemistry

We begin our study of organic molecules by examining the chemistry of carbon, the atom that makes the variety of organic molecules in cells possible. The organic molecules made by cells are called *biomolecules*. Functional groups added to biomolecules increase their variety and allow them to play particular roles in cells. Large biomolecules are modular, and their final size is dependent on how many subunits are joined end to end.

3.1 The chemistry of carbon makes diverse molecules possible

LEARNING OUTCOMES

When you complete this section, you should be able to

1. Discuss how the chemistry of carbon results in a great variety of organic molecules.
2. List the four classes of biomolecules in cells.
3. Identify ways that functional groups can affect the activity of biomolecules.

Carbon is so versatile that an entire branch of chemistry, called **organic chemistry,** is devoted to it. What is there about carbon that makes organic molecules the same and also different? Carbon is quite small, with a total of only six electrons: two electrons in the first shell and four electrons in the outer shell. To acquire four electrons to complete its outer shell, a carbon atom almost always shares electrons with—you guessed it—CHNOPS, the elements basic to organisms.

Because carbon needs four electrons to complete its outer shell, it can share with as many as four other elements, and this spells variety. But even more significant to the shape, and therefore the function, of organic molecules is the fact that carbon often shares electrons with another carbon atom. The C—C bond is quite stable and can result in carbon chains that are quite long. Hydrocarbons are chains of carbon atoms bonded exclusively to hydrogen atoms:

octane, a molecule in gasoline

Branching is possible at any carbon atom, and a hydrocarbon can also turn back on itself to form a ring compound when placed in water. One example is cyclohexane, used as an industrial solvent and in the manufacture of nylon.

Carbon can form double bonds with itself and other atoms. Double bonds restrict the movement of attached atoms, and in that way contribute to the shape of the molecule. Carbon is also capable of forming a triple bond with itself, as in acetylene, H—C≡C—H.

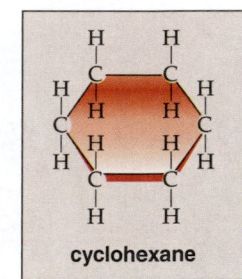

cyclohexane

Biomolecules There are only four classes of biomolecules in any organism: **carbohydrates, lipids, proteins,** and **nucleic acids.** But despite the limited number of classes, the biomolecules in cells are quite varied. A bacterial cell contains some 5,000 different organic molecules, and a plant or animal cell has twice that number. This variety of biomolecules makes the diversity of life possible. For example, each of the organisms in Figure 3.1A uses a carbohydrate as a structural molecule: A cactus uses cellulose to strengthen its cell walls, while a bacterium uses peptidoglycan for that purpose; a crab uses chitin to strengthen its shell.

Carbon is the essential ingredient in all biomolecules. Much as a salad chef first puts lettuce in a bowl, so organic molecules begin with carbon. A chef may then add other ingredients, such as cucumbers or radishes, to the lettuce to make different types of salads. So, the variety of biomolecules comes about when different groups of atoms are added to carbon.

Crab shell contains chitin.

Plant cell walls contain cellulose.

Bacteria cell walls contain peptidoglycan.

FIGURE 3.1A Each of these organisms uses a different type of structural carbohydrate.

Functional groups		
Group	**Structure**	**Compound**
Hydroxyl	$R\!-\!OH$	Alcohol; present in sugars and some amino acids
Carbonyl	$R\!-\!\overset{\displaystyle O}{\underset{\displaystyle H}{C}}$	Aldehyde; present in sugars
	$R\!-\!\overset{\displaystyle O}{\underset{}{C}}\!-\!R$	Ketone; present in sugars
Carboxyl (acidic)	$R\!-\!\overset{\displaystyle O}{\underset{\displaystyle OH}{C}}$	Carboxylic acid; present in fatty acids, amino acids
Amino	$R\!-\!\overset{\displaystyle H}{\underset{\displaystyle H}{N}}$	Amine; present in amino acids
Sulfhydryl	$R\!-\!SH$	Thiol; forms disulfide bonds when present in adjacent amino acids
Phosphate	$R\!-\!O\!-\!\overset{\displaystyle O}{\underset{\displaystyle OH}{P}}\!-\!OH$	Organic phosphate; present in nucleotides and phospholipids

R = rest of molecule

FIGURE 3.1B Functional groups of biomolecules.

Functional Groups The carbon chain of a biomolecule is called its skeleton or backbone. The terminology is appropriate because just as a skeleton accounts for your shape, so does the carbon skeleton of a biomolecule account for its underlying shape. The reactivity of a biomolecule is largely dependent on the attached functional groups. A functional group is a specific combination of bonded atoms that always reacts in the same way, regardless of the particular carbon skeleton. As shown in Figure 3.1B, an *R* can be used to stand for the "rest" of the molecule because only the functional group is involved in reactions.

Notice that functional groups with a particular name and structure are found in certain types of compounds. For example, the addition of an —OH (hydroxyl group) to a carbon skeleton turns that molecule into an alcohol. When an —OH replaces one of the hydrogens in ethane, a 2-carbon hydrocarbon, it becomes ethanol, a type of alcohol that is consumable by humans. Whereas ethane, like other hydrocarbons, is **hydrophobic** (not soluble in water), ethanol is **hydrophilic** (soluble in water) because the —OH functional group is polar. Because cells are 70–90% water, the ability to interact with and be soluble in water profoundly affects the function of organic molecules in cells. Biomolecules containing carboxyl (acid) groups (—COOH) are polar, and when they ionize, they release hydrogen ions, making a solution more acidic:

$$-COOH \longrightarrow -COO^- + H^+$$

testosterone   estrogen

FIGURE 3.1C The functional groups in male and female sex hormones are highlighted.

Functional groups determine the activity of a biomolecule in cells. You will see that alcohols react with carboxyl groups when a fat forms, and that carboxyl groups react with amino groups during protein formation. Notice in Figure 3.1C that the male sex hormone testosterone differs from the female sex hormone estrogen only by its attached groups. Yet, these molecules help bring about the characteristics that determine whether an individual is male or female.

Isomers **Isomers** are organic molecules that have identical molecular formulas but a different arrangement of atoms. In essence, isomers are variations in the architecture of a molecule. Isomers are another example of how the chemistry of carbon leads to variations in organic molecules.

The following two molecules are isomers of one another; they have the same molecular formula but different functional groups. Therefore, we would expect them to react differently in chemical reactions:

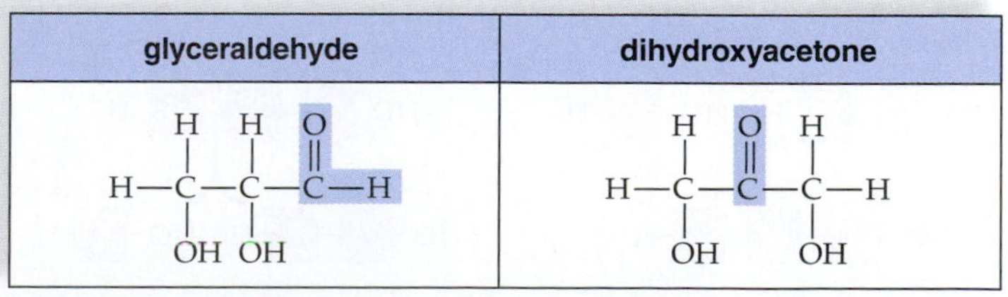

glyceraldehyde	dihydroxyacetone
$\overset{\displaystyle H}{\underset{}{H\!-\!\overset{\displaystyle H}{\underset{\displaystyle OH}{C}}\!-\!\overset{\displaystyle H}{\underset{\displaystyle OH}{C}}\!-\!\overset{\displaystyle O}{\underset{}{C}}\!-\!H}}$	$\overset{}{H\!-\!\overset{\displaystyle H}{\underset{\displaystyle OH}{C}}\!-\!\overset{\displaystyle O}{\underset{}{C}}\!-\!\overset{\displaystyle H}{\underset{\displaystyle OH}{C}}\!-\!H}$

▶ **3.1 CHECK YOUR PROGRESS**
1. Identify ways that carbon alone can result in a variety of molecules.
2. List the four classes of biomolecules discussed in this chapter.
3. Use testosterone and estrogen to exemplify that functional groups can have dramatic effects on the function of biomolecules.

3.2 Molecular subunits can be linked to form varied large biomolecules

LEARNING OUTCOMES

When you complete this section, you should be able to

1. Describe the reactions that allow biomolecules to assemble and disassemble.
2. Name the subunits that are released when you digest carbohydrates, lipids, and proteins.

The largest of the biomolecules are called **polymers** because they are constructed by linking together a large number of the same type of subunit, called **monomers.** A protein can contain hundreds of amino acids, and a nucleic acid can contain thousands of nucleotides. How do polymers get so large? Cells use the modular approach when constructing polymers. Just as a train increases in length when boxcars are hitched together one by one, so a polymer gets longer as subunits bond to one another.

In Figure 3.2A*a*, notice how synthesis (the construction) of a polymer occurs. A cell uses a **dehydration reaction** to synthesize any biomolecule that has subunits. In this reaction, the equivalent of a water molecule consisting of an —OH (hydroxyl group) and an —H (hydrogen atom) is removed as the reaction occurs. After water is removed, a bond exists between the two subunits.

In Figure 3.2A*b*, notice how degradation (breaking down) of a biomolecule occurs. To degrade a biomolecule, our digestive tract, or any cell, uses an opposite type of reaction. During a **hydrolysis reaction,** an —OH group from water attaches to one subunit, and an —H from water attaches to the other subunit. (*Hydro* means "water," and *lysis* means "breaking apart.") In other words, water is used to break the bond holding subunits together. Biologists frequently refer to hydrolysis, sometimes called a hydrolytic reaction, so it is a term that you will want to be familiar with.

In order for these reactions, or almost any other type of reaction, to occur in a cell, an enzyme must be present. An **enzyme** is a molecule that speeds a reaction by bringing reactants together. The enzyme may even participate in the reaction, but it is unchanged by it. Frequently, subunits must be energized before they will bind together because synthesis of a polymer requires energy. On the other hand, hydrolysis of a polymer can release energy.

Breakdown and synthesis of biomolecules is always taking place in your body. For example, food largely contains the classes of biomolecules mentioned earlier: carbohydrates, lipids, and proteins (Fig. 3.2B). Even a late-night pizza, a quick hamburger, or an afternoon snack contains many of these large biomolecules. When you digest your food, these molecules get broken down into their subunits (Table 3.2). For example, digestion of bread releases monosaccharides (e.g., glucose), while digestion of meat re-leases amino acids. Your body then takes these subunits and builds from them the particular carbohydrates and proteins that make up your cells.

FIGURE 3.2B All foods contain carbohydrates, lipids, proteins, and nucleic acids whose subunits are used as building blocks or a source of energy by cells.

Carbohydrates

Lipids   Proteins

TABLE 3.2	Biomolecules	
Category	**Example**	**Subunit(s)**
Carbohydrates*	Polysaccharide	Monosaccharide
Lipids	Fat	Glycerol and fatty acids
Proteins*	Polypeptide	Amino acid
Nucleic acids*	DNA, RNA	Nucleotide

*Polymers

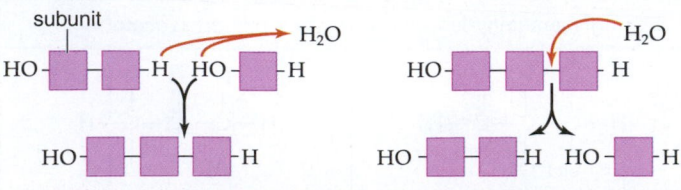

a. Dehydration reaction   **b.** Hydrolysis reaction

FIGURE 3.2A Synthesis and degradation of polymers. **a.** In cells, synthesis often occurs when subunits bond during a dehydration reaction (removal of H_2O). **b.** Degradation occurs when the subunits separate during a hydrolysis reaction (the addition of H_2O).

▶ **3.2 CHECK YOUR PROGRESS**
1. Contrast the result of a dehydration reaction with the result of a hydrolysis reaction.
2. Identify the subunits released when you digest bread (carbohydrate), meat (protein), and butter (fat).

Carbohydrates

The majority of carbohydrates have a carbon to hydrogen to oxygen ratio of 1:2:1. The general formula (CH_2O) explains their name because it indicates that they are hydrates (water) of carbon. Some carbohydrates serve as a source of quick energy and others as energy-storage molecules. Still others are structural components of organisms (see Fig. 3.1A).

3.3 Simple carbohydrates provide quick energy

LEARNING OUTCOME

When you complete this section, you should be able to

1. Distinguish between the structure and function of monosaccharides and disaccharides.

Simple carbohydrates are the sweet-tasting sugars. **Monosaccharides** consist of only a single sugar molecule. A monosaccharide can have a carbon backbone of three to seven carbons. It may also have many hydroxyl groups, and this polar functional group makes it soluble in water. **Glucose,** a monosaccharide with six carbon atoms, has a molecular formula of $C_6H_{12}O_6$. Several ways to represent glucose are shown in Figure 3.3A. When the carbon atoms are included, the formula looks crowded, so a common practice is to omit the carbon atoms, or even to simply show the hexagon shape. You are supposed to imagine the molecule as flat, with the darkened region facing you. Certain atoms bonded to carbon are above the ring, and others are below it, as indicated.

Despite the fact that glucose has several isomers, such as fructose and galactose, we usually think of $C_6H_{12}O_6$ as glucose. This sugar is the major source of cellular fuel for all organisms. Glucose is transported in the blood of animals, and it is the molecule that is broken down in nearly all types of cells to release energy. The liver, however, takes up the other isomers of glucose from the bloodstream and metabolizes them. Nutritionists are now studying whether the extensive use of high-fructose corn syrup to sweeten processed foods causes damage to the liver and comes at a cost to our health.

Ribose and **deoxyribose** are monosaccharides with five carbon atoms. They are significant because they are present respectively in the nucleic acids RNA and DNA, discussed later in this chapter.

A **disaccharide,** which is also a simple sugar, contains two monosaccharides that have joined during a dehydration reaction. Figure 3.3B shows how the disaccharide maltose (an ingredient used in brewing beer) arises when two glucose molecules bond together. When our hydrolytic digestive juices break this bond, the result is two glucose molecules. Sucrose, or table sugar, is another disaccharide of special interest because it is the transport sugar in plants. Plants transport sucrose from cells carrying on photosynthesis to other parts of their

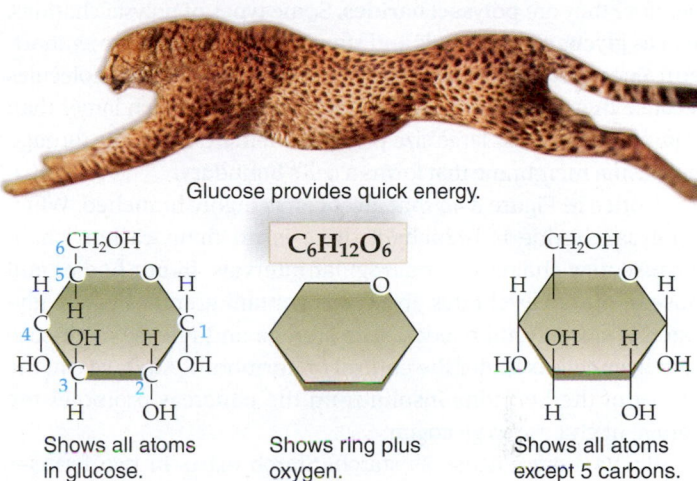

Glucose provides quick energy.

$C_6H_{12}O_6$

Shows all atoms in glucose. Shows ring plus oxygen. Shows all atoms except 5 carbons.

FIGURE 3.3A Three ways to represent glucose, a source of quick energy for this cheetah and for all organisms.

bodies that are not photosynthesizing. Sucrose is also the sugar we use to sweeten our food. We acquire the sugar from plants such as sugarcane and sugar beets. You may also have heard of lactose, a disaccharide found in milk. Lactose is glucose combined with galactose. Lactose is called milk sugar because it is the major carbohydrate component of milk. Individuals who are lactose intolerant cannot break this disaccharide down and therefore experience unpleasant digestive tract symptoms when they consume cow's milk or other products made from cow's milk.

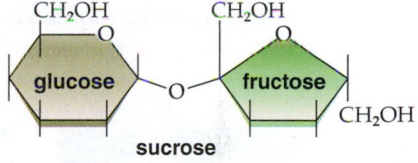

sucrose

MP3
Carbohydrates

▶ **3.3 CHECK YOUR PROGRESS**

1. Contrast the structure and importance of glucose and sucrose in the physiology of plants and animals.

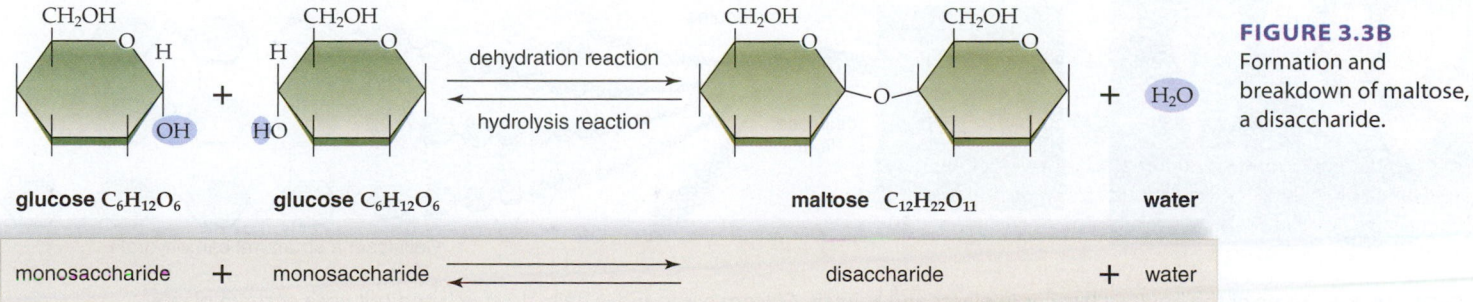

glucose $C_6H_{12}O_6$ glucose $C_6H_{12}O_6$ dehydration reaction / hydrolysis reaction   maltose $C_{12}H_{22}O_{11}$ + H_2O water

monosaccharide + monosaccharide → disaccharide + water

FIGURE 3.3B
Formation and breakdown of maltose, a disaccharide.

3.4 Complex carbohydrates store energy and provide structural support

LEARNING OUTCOME

When you complete this section, you should be able to

1. Compare the structure and varied roles of complex carbohydrates in organisms.

Complex carbohydrates are polymers of monosaccharides, and therefore they are **polysaccharides.** Some types of polysaccharides, such as **glycogen** in animals and **starch** in plants, function as short-term energy-storage molecules. They serve as storage molecules because they are not as soluble in water and are much larger than a simple sugar. Their large size prevents them from passing through the plasma membrane that forms a cell's boundary.

Notice in Figure 3.4*a* that glycogen is highly branched. When a polysaccharide is branched, there is no main carbon chain because new chains occur at regular intervals. In our bodies and those of other vertebrates, liver cells contain granules where glycogen is stored until needed. The storage and release of glucose from liver cells is under the control of hormones. After we eat, the release of the hormone insulin from the pancreas promotes the storage of glucose as glycogen.

Plants store glucose as starch. Starch exists in two forms—nonbranched and branched. The branched form is featured in Figure 3.4*b*. Both the nonbranched and branched forms of starch serve as glucose reservoirs in plants. The cells of a potato contain granules in which starch resides during winter until energy is needed for growth in the spring. Plant cells can hydrolyze starch and therefore can tap into these reservoirs for energy. After animals ingest starch, digestive (hydrolytic) enzymes break it down, thereby releasing glucose that enters their body and cells.

Some types of polysaccharides are structural polysaccharides, such as **cellulose** in plants, **chitin** in animals and fungi, and **peptidoglycan** in bacteria (see Fig. 3.1A). The cellulose subunit is simply glucose, but in chitin, the subunit has an attached amino group. The structure of peptidoglycan is more complex because each subunit also has a peptide chain.

Cellulose is the most abundant carbohydrate and, indeed, the most abundant organic molecule on Earth (Fig. 3.4*c*). Plants produce over 100 billion tons of cellulose each year. Wood and cotton are cellulose plant products. Wood is used for construction, and cotton is used for cloth.

The majority of animals lack the necessary enzymes for digesting cellulose because the glucose subunits are joined as shown in Figure 3.4*c*. But cellulose does serve as dietary fiber, which maintains regularity of elimination. Microorganisms are able to digest the bond between glucose subunits in cellulose. The protozoans in the gut of termites allow them to digest wood. In cows and other ruminants, microorganisms break down cellulose in a special stomach pouch before the "cud" is returned to the mouth for more chewing and reswallowing.

> ▶ **3.4 CHECK YOUR PROGRESS**
>
> 1. List and compare the structure and function of complex carbohydrates in plants and animals. What is the same and what is different about these polysaccharides?

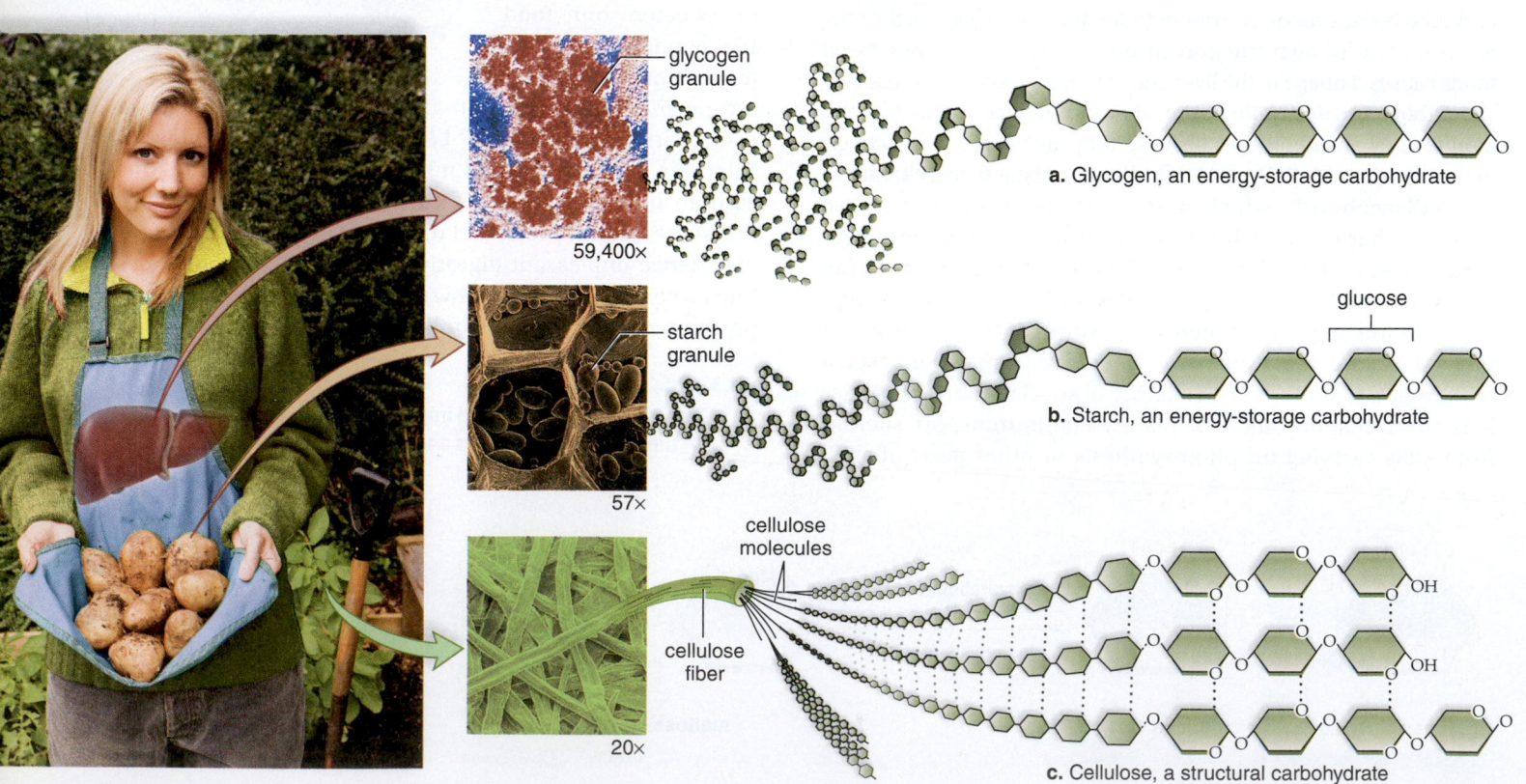

FIGURE 3.4 Some of the polysaccharides in plants and animals. Glycogen is found in animals; starch and cellulose occur in plants.

Lipids

Most lipids are insoluble in water due to a lack of polar groups. Fats and oils, which function as long-term energy-storage molecules, are the best-known lipids. Phospholipids are constructed like fats but have a polar group that makes them soluble in water. Phospholipids are a major part of the plasma membrane that separates a cell from its environment. Steroids are a large class of lipids having a structure different from that of fats.

3.5 Fats and oils are rich energy-storage molecules

LEARNING OUTCOME

When you complete this section, you should be able to

1. Distinguish the makeup of a fat and the difference between saturated and unsaturated fatty acids.

Fats and **oils** contain two types of subunit molecules: glycerol and fatty acids. **Glycerol** is a compound with three —OH groups. A **fatty acid** consists of a long hydrocarbon chain with a —COOH (acid) group at one end. When a fat or oil forms, the acid portions of three fatty acids react with the —OH groups of glycerol during a dehydration reaction (Fig. 3.5). Because there are three fatty acids attached to each glycerol molecule, fats and oils are sometimes called **triglycerides.** Notice that the exposed polar groups in glycerol and fatty acids no longer appear in a triglyceride, and this makes triglycerides insoluble in water. On the other hand, the many C—H bonds of fats and oils make them a richer source of chemical energy than the carbohydrates glycogen and starch.

Fatty acids are primary components of fats and oils. Most of the fatty acids in cells contain 16 or 18 carbon atoms per molecule, although smaller ones are also found. Fatty acids are either saturated or unsaturated. **Saturated fatty acids** have no double bonds between the carbon atoms. The carbon chain is saturated, so to speak, with all the hydrogens it can hold. **Unsaturated fatty acids** have double bonds (see yellow highlight in Figure 3.5) wherever the number of hydrogens in the carbon chain is less than two per carbon atom.

The double bond creates a bend in the fatty acid chain. These kinks (not shown) prevent close packing between the hydrocarbon chains and account for the fluidity of oils. On the other hand, the saturated fatty acid chains in butter can pack together tightly because they have no kinks; therefore, butter is fairly solid. **Trans fats** contain fatty acids that have been partially hydrogenated to make them more saturated, and thus more solid. Complete hydrogenation of oils causes all double bonds to become saturated. Partial hydrogenation does not saturate all bonds. It reconfigures some double bonds, and the hydrogen atoms end up on different sides of the chain. (*Trans* in Latin means across):

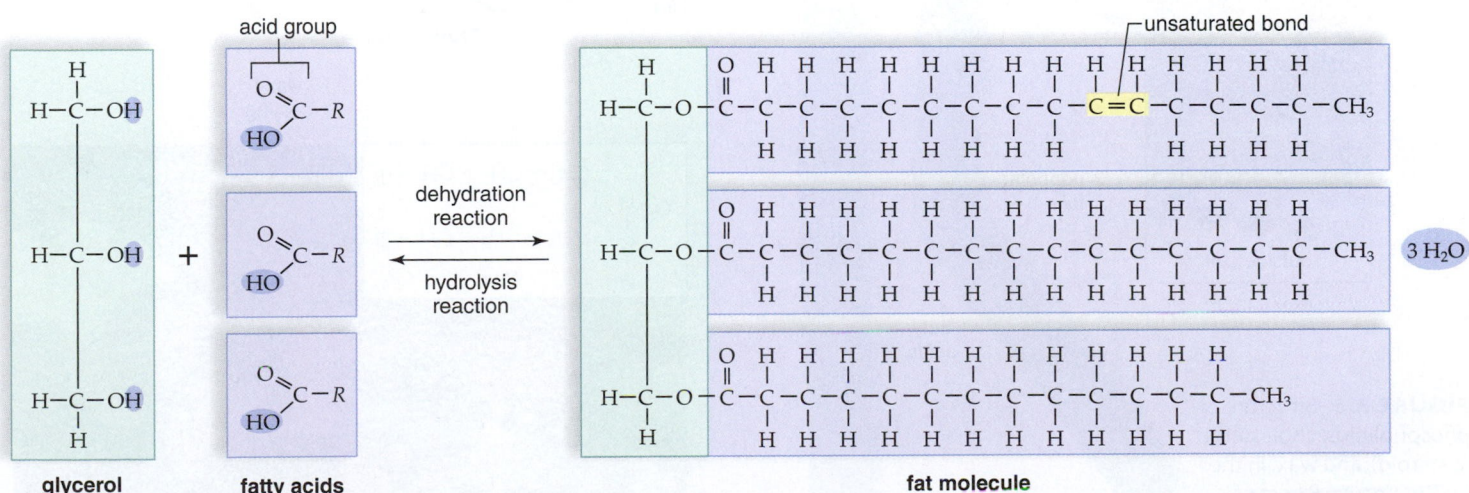

| unsaturated | saturated | trans fats |
| (oils) | (butter) | (hydrogenated oils) |

Trans fats are often found in processed foods—particularly margarine, baked goods, and fried foods. Saturated fats and trans fats contribute to the buildup of abnormal lipid material called plaque inside blood vessels. Plaque leads to high blood pressure. Unsaturated oils, particularly monounsaturated (one double bond) but also polyunsaturated (many double bonds), have been found to protect against abnormal plaque buildup.

MP3
Lipids

▶ **3.5 CHECK YOUR PROGRESS**
   1. Contrast unsaturated fatty acids with saturated fatty acids in structure and health consequences.

FIGURE 3.5 Formation and breakdown of a fat. The *R* (the rest of the molecule) in each fatty acid stands for the long hydrocarbon chains that are depicted in the fat molecule on the right.

3.6 Other lipids have structural, hormonal, or protective functions

LEARNING OUTCOME

When you complete this section, you should be able to

1. Compare the structure and function of phospholipids, steroids, and waxes.

The phospholipids, steroids, and waxes are also important lipids in organisms (Fig. 3.6). Like fats, **phospholipids** contain glycerol and three groups bonded to glycerol. In phospholipids, only two of these groups are fatty acids. After bonding to glycerol, the fatty acid chains form the hydrophobic tails of the molecule. The third group contains a polar phosphate group that becomes the polar head of a phospholipid. In a watery environment, phospholipids naturally form a bilayer in which the hydrophilic heads project outward and the hydrophobic tails project inward. A cell's plasma membrane consists of a phospholipid bilayer, as will be discussed in more detail in Chapter 5. A plasma membrane is absolutely essential to the structure and function of a cell.

Steroids are lipids that have an entirely different structure from that of fat. A steroid molecule has a skeleton of four fused carbon rings. Cholesterol is the steroid that stabilizes an animal's plasma membrane. It is also the precursor of several other steroids, such as the sex hormones testosterone and estrogen (see Fig. 3.1C). Among its many effects, testosterone is responsible for the generally greater muscle development of human males. For this reason, athletes of both sexes sometimes take anabolic steroids—testosterone or steroids that resemble testosterone—in an attempt to improve their athletic performance. This use of steroids is now banned by most athletic organizations. Anabolic steroid abuse can create serious health problems involving the kidneys and the cardiovascular system, as well as changes in sexual characteristics. Females take on male characteristics, and males become feminized.

Like saturated fats, cholesterol also participates in the formation of plaque along cardiovascular walls. Plaque can restrict blood flow and result in heart attacks or strokes.

In **waxes,** long-chain fatty acids bond with long-chain alcohols. Waxes are solid at normal temperatures. Being hydrophobic, they are also waterproof and resistant to degradation. In many plants, waxes, along with other molecules, form a protective coating that retards the loss of water from all exposed parts. In many animals, waxes are involved in skin and fur maintenance. In humans, wax is produced by glands in the outer ear canal. Earwax contains cerumin, an organic compound that at the very least repels insects, and in some cases even kills them. It also traps dust and dirt, preventing them from reaching the eardrum.

Honeybees produce beeswax in glands on the underside of their abdomen. They then use this beeswax to make the six-sided cells of the comb where their honey is stored. Honey contains the sugars fructose and glucose, breakdown products of sucrose.

▶ **3.6 CHECK YOUR PROGRESS**
1. Compare the structure of cholesterol to that of phospholipids and waxes and give a function for each.

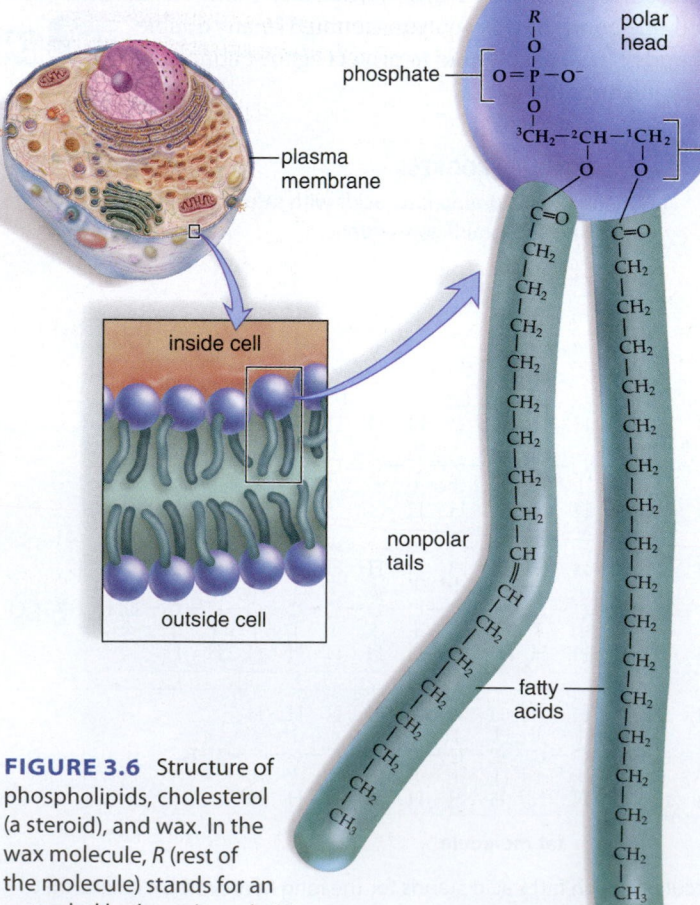

FIGURE 3.6 Structure of phospholipids, cholesterol (a steroid), and wax. In the wax molecule, *R* (rest of the molecule) stands for an extended hydrocarbon chain.

Phospholipid

Cholesterol

Wax

wax-covered fruit

3A Controlling Obesity

Obesity, an excess accumulation of body fat, is a serious medical condition that afflicts over 30% of adults and over 16% of children and adolescents in the United States today. Obesity is judged by a person's body mass index (BMI)—weight in relation to height—as described in Figure 3A.

Disorders Associated with Obesity

Two serious illnesses are associated with obesity: type 2 diabetes and cardiovascular disease. When a person has type 1 diabetes, the pancreas no longer produces insulin, a hormone that stimulates cells to take in glucose. When a person has type 2 diabetes, the pancreas produces insulin, but the cells do not respond to it. In both instances, glucose builds up in the blood and spills over into the urine. Therefore, urinalysis detects when a person has diabetes. How might diet contribute to the occurrence of type 2 diabetes? Simple sugars in foods, such as candy and ice cream, immediately enter the bloodstream, as do sugars from the digestion of starch within white bread and potatoes. When the blood glucose level rises rapidly, the pancreas produces an overload of insulin to bring the level under control. Chronically high insulin levels due to diet apparently lead to insulin resistance, a high blood fatty acid level, and type 2 diabetes.

Cardiovascular disease due to arteries blocked by plaque is another condition seen in obese individuals. Plaque contains saturated fats and also cholesterol. Cholesterol is carried in the blood by two types of lipoproteins: low-density lipoprotein (LDL) and high-density lipoprotein (HDL). LDL is thought of as "bad" because it carries cholesterol from the liver to the cells, whereas HDL is thought of as "good" because it carries cholesterol from the cells to the liver, which takes it up and converts it to bile salts. Limiting cholesterol (present in cheese, egg yolks, shrimp, and lobster) in the diet may be helpful. Beef, dairy foods, and coconut oil are rich sources of saturated fat, which tends to raise LDL levels. Further, processed foods made with or fried in partially hydrogenated oils (e.g., vegetable shortening and stick margarine) are sources of trans fats that contribute to plaque formation.

Eating Sensibly

Before turning to more drastic measures, an overweight or obese person should first attempt to lose weight by lowering their caloric intake and increasing their caloric output through exercise. Only then will the body metabolize its stored fat for energy needs, allowing the person to lose weight. There are no quick and easy solutions for losing weight. The typical fad diet is nutritionally unbalanced and difficult to follow over the long term. Weight loss and weight maintenance require permanent lifestyle changes, such as increasing the level of physical activity and reducing portion sizes. Once body weight is under control, it needs to be maintained by continuing to eat sensibly. Complex carbohydrates, such as those in whole-grain breads and cereals, are preferable to simple carbohydrates, such as candy and ice cream, because they do not cause a spike in blood insulin level

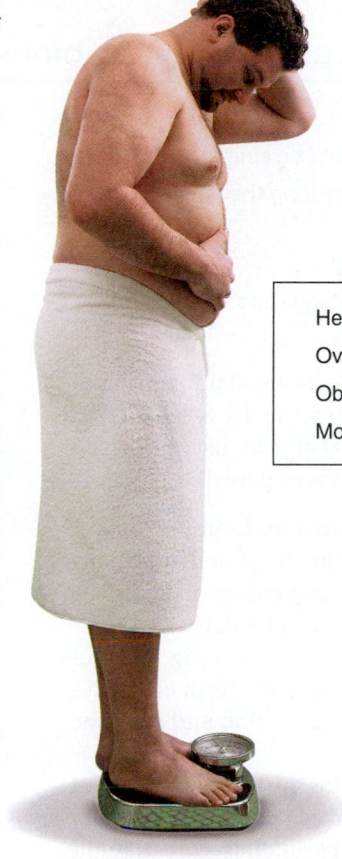

$$BMI = \frac{body\ weight\ (in\ kilograms)}{height^2\ (in\ meters)}$$

or

$$\frac{weight\ (pounds) \times 703.1}{height^2\ (inches)}$$

Healthy BMIs	=	18.5 to 24.9
Overweight BMIs	=	25.0 to 29.9
Obese BMIs	=	30.0 to 39.9
Morbidly obese BMIs	=	40.0 or more

FIGURE 3A You can determine your own body mass index (BMI) using either of the formulas shown here. If you are obese, you should try to reduce your BMI with diet and exercise.

and they contain dietary fiber plus vitamins and minerals. We also need to watch the type of fat in our diet. Unsaturated fatty acids in olive and canola oils, most nuts, and cold-water fish tend to lower LDL cholesterol levels. Furthermore, cold-water fish (e.g., herring, sardines, tuna, and salmon) contain polyunsaturated fatty acids and especially monounsaturated fatty acids that can reduce the risk of cardiovascular disease.

CONSIDER THESE QUESTIONS

1. Today restaurants usually serve large portions of food and drinks. Should they be required to serve only smaller portions to help prevent obesity?
2. Should food manufacturers be required to stop using high-fructose corn syrup, which has been implicated in childhood obesity?

 Explore the concepts through a variety of multimedia assets and question types.

www.mcgrawhillconnect.com

Proteins

Proteins are polymers of amino acids. Amino acids have the same general structure, but they differ by the side chain attached to a central carbon. The particular sequence of amino acids determines the final shape of a protein. The shape of a protein determines its structure and function in cells. The functions of proteins are quite varied but it is possible in general to categorize the various functions of proteins.

3.7 Proteins are the most versatile of life's biomolecules

LEARNING OUTCOMES

When you complete this section, you should be able to

1. Describe the versatility of proteins by listing their many functions.
2. Explain the structure of an amino acid.
3. Identify the peptide bond within a peptide and explain its polarity.

Proteins, which are polymers of amino acids, are of primary importance to each type of cell. As much as 50% of the dry weight of a cell consists of proteins. Presently, over 100,000 proteins have been identified. Here are some of their many functions in animals:

Support Some proteins are structural proteins. Examples include the silk protein in spider webs; keratin, the protein that makes up hair (Fig. 3.7A*a*) and fingernails; and collagen, the protein that lends support to skin, ligaments, and tendons.

Metabolism Some proteins are enzymes. They bring reactants together and thereby speed chemical reactions in cells. They are specific for one particular type of reaction and can function at body temperature.

Transport Channel and carrier proteins in the plasma membrane allow substances to enter and exit cells. Other proteins transport molecules in the blood of animals—for example, **hemoglobin** is a complex protein that transports oxygen (Fig. 3.7A*b*).

Defense Proteins called antibodies combine with disease-causing agents to prevent them from destroying cells and upsetting homeostasis, the relative constancy of the internal environment.

Regulation Hormones are regulatory proteins. They serve as intercellular messengers that influence cell metabolism. For example, the hormone insulin regulates the content of glucose in the blood and in cells, while growth hormone determines the height of an individual.

Motion The contractile proteins actin and myosin allow parts of cells to move and cause muscles to contract. Muscle contraction enables animals to move from place to place (Fig. 3.7A*c*).

Proteins are such a major part of organisms that tissues and cells of the body can sometimes be characterized by the proteins they contain or produce. For example, muscle cells contain large amounts of actin and myosin for contraction; red blood cells are filled with hemoglobin for oxygen transport; and support tissues, such as ligaments and tendons, are composed of tough fibers made from collagen.

Amino Acids The subunit of a protein is an amino acid. Amino acids have a unique carbon skeleton in which a central carbon atom bonds to a hydrogen atom, two functional groups, and a side chain, or *R* group. The name amino acid is appropriate because one of two functional groups is an —NH$_2$ (amino group), and another is a —COOH (acid group). The *R* group is the rest of the molecule:

amino group		acid group
	H	
H$_2$N—	C	—COOH
	R	

R = rest of molecule

Amino acids differ from one another according to their particular *R* group, a side chain that lacks color in Figure 3.7B. The *R* groups range in complexity from a single hydrogen atom to a complicated ring compound. Some *R* groups are polar and some are not. Also, the amino acid cysteine has an *R* group that ends

FIGURE 3.7A Proteins have many functions in humans.

a. Hair is a protein.

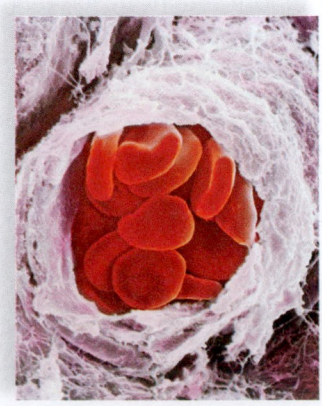

b. Hemoglobin is a protein.

c. Muscle contains protein.

Nonpolar R groups

valine (Val) methionine (Met) phenylalanine (Phe)

Nonpolar R groups

Polar R groups

cysteine (Cys) asparagine (Asn) serine (Ser)

Polar R groups

Ionized (polar) R groups

glutamic acid (Glu) lysine (Lys) aspartic acid (Asp)

Ionized (polar) R groups

FIGURE 3.7B A sampling of amino acids.

with a —SH group; two such groups can form a disulfide bond, —S—S—. Several other amino acids commonly found in cells are shown in Figure 3.7B.

Peptides Proteins are large biomolecules composed of **amino acid** subunits. Figure 3.7C shows how two amino acids join by a dehydration reaction between the carboxyl group of one and the amino group of another. A **peptide** is two or more amino acids bonded together, and a **polypeptide** is a chain of many amino acids joined by peptide bonds. A protein may contain more than one polypeptide chain; therefore, a single protein may have a very large number of amino acids. In 1953, Frederick Sanger developed a method to determine the amino acid sequence of a polypeptide. We now know the sequences of thousands of polypeptides, and it is clear that each type of polypeptide has its own particular sequence.

The covalent bond between two amino acids is called a **peptide bond.** The atoms associated with the peptide bond share the electrons unevenly because oxygen attracts electrons more than nitrogen. Therefore, the hydrogen attached to the nitrogen has a slightly positive charge (δ^+), while the oxygen has a slightly negative charge (δ^-).

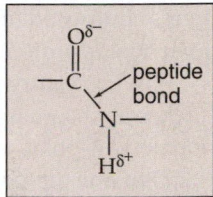

The polarity of the peptide bond means that hydrogen bonding is possible between the —CO of one amino acid and the —NH of another amino acid in the same or a different polypeptide.

▶ **3.7 CHECK YOUR PROGRESS**
1. Describe the major functions of proteins in organisms.
2. Contrast the ways in which amino acids are the same but different.
3. Determine why you would expect hydrogen bonding to occur in a protein.

amino group ⎡ ⎤ acid group ⎡ ⎤ peptide bond

amino acid + amino acid → (dehydration reaction / hydrolysis reaction) → dipeptide + H_2O water

FIGURE 3.7C Formation and breakdown of a peptide.

3.8 The shape of a protein is necessary to its function

LEARNING OUTCOME

When you complete this section, you should be able to

1. Describe how the four levels of a protein's structure result in a particular shape.

The shape of a protein is suited to its function. For example, collagen has a super-coiled helical shape that allows it to lend mechanical strength to bones, tendons, and ligaments and also to support the body's tissues, including the skin. Hemoglobin has a globular shape that allows it to travel inside blood vessels as it carries oxygen throughout the body. However, environmental

conditions, such as extremes of temperature and pH, can *denature* a protein—that is, alter its shape—so that it can no longer perform its usual function. For example, cooking an egg causes the protein albumin in egg white to coagulate. Stomach acids denature any protein we ingest, including hormones, enzymes, and the muscle proteins in meat. Denaturation unravels a protein's usual shape and makes it more susceptible to hydrolysis by digestive enzymes.

Animation
Protein Denaturation

Alcohol, including both wood alcohol (methanol) and drinking alcohol (ethanol), and the salts of heavy metals, such as lead (Pb), mercury (Hg), and silver (Ag), also denature proteins. A

70% alcohol solution is used as a disinfectant because it destroys bacteria by unraveling their proteins, and silver nitrate ($AgNO_3$) acts similarly when it is routinely used to disinfect the eyes of newborns.

The final shape of a protein is dependent on its sequence of amino acids, as discussed next.

Levels of Protein Organization The structure of a protein has at least three levels of organization and can have four levels (Fig. 3.8). The first level, called the *primary structure,* can be likened to a string of beads because it is the linear sequence of the amino acids joined by peptide bonds. Each particular polypeptide has its own sequence of amino acids. Just as an alphabet of 26 letters can form the sequence of many different words, so can 20 amino acids form the sequence of many different proteins. In the next part of this chapter (page 55), we learn that genes composed of DNA specify the sequence of amino acids in a particular protein.

The *secondary structure* of a polypeptide comes about when it takes on a certain orientation in space. As mentioned, the peptide bond is polar, and hydrogen bonding is possible between the —CO of one amino acid and the —NH of another amino acid in a polypeptide. Due to hydrogen bonding, two possible shapes can occur even within the same protein: a right-handed spiral, called an alpha helix, and a folding of the chain, called a pleated sheet. Fibrous proteins, such as those in hair and nails, exist as helices or as pleated sheets.

Globular proteins have a *tertiary structure* as their final three-dimensional shape. In muscles, myosin molecules have a rod shape ending in globular (globe-shaped) heads. In enzymes, the polypeptide bends and twists in different ways. Invariably, the hydrophobic portions are packed mostly on the inside, and the hydrophilic portions are on the outside, where they can make contact with fluids. The tertiary shape of a polypeptide is maintained by various types of bonding between the *R* groups; covalent disulfide bonds, ionic, and hydrogen bonding all occur.

Some proteins have only one polypeptide, and others have more than one polypeptide, each with its own primary, secondary, and tertiary structures. These separate polypeptides are arranged to give some proteins a fourth level of organization, termed the *quaternary structure*. Hemoglobin is a complex protein having a quaternary structure; most enzymes also have a quaternary structure.

MP3 Proteins

▶ **3.8 CHECK YOUR PROGRESS**
1. Explain why you would expect to find globular proteins but not fibrous proteins in plant cells.

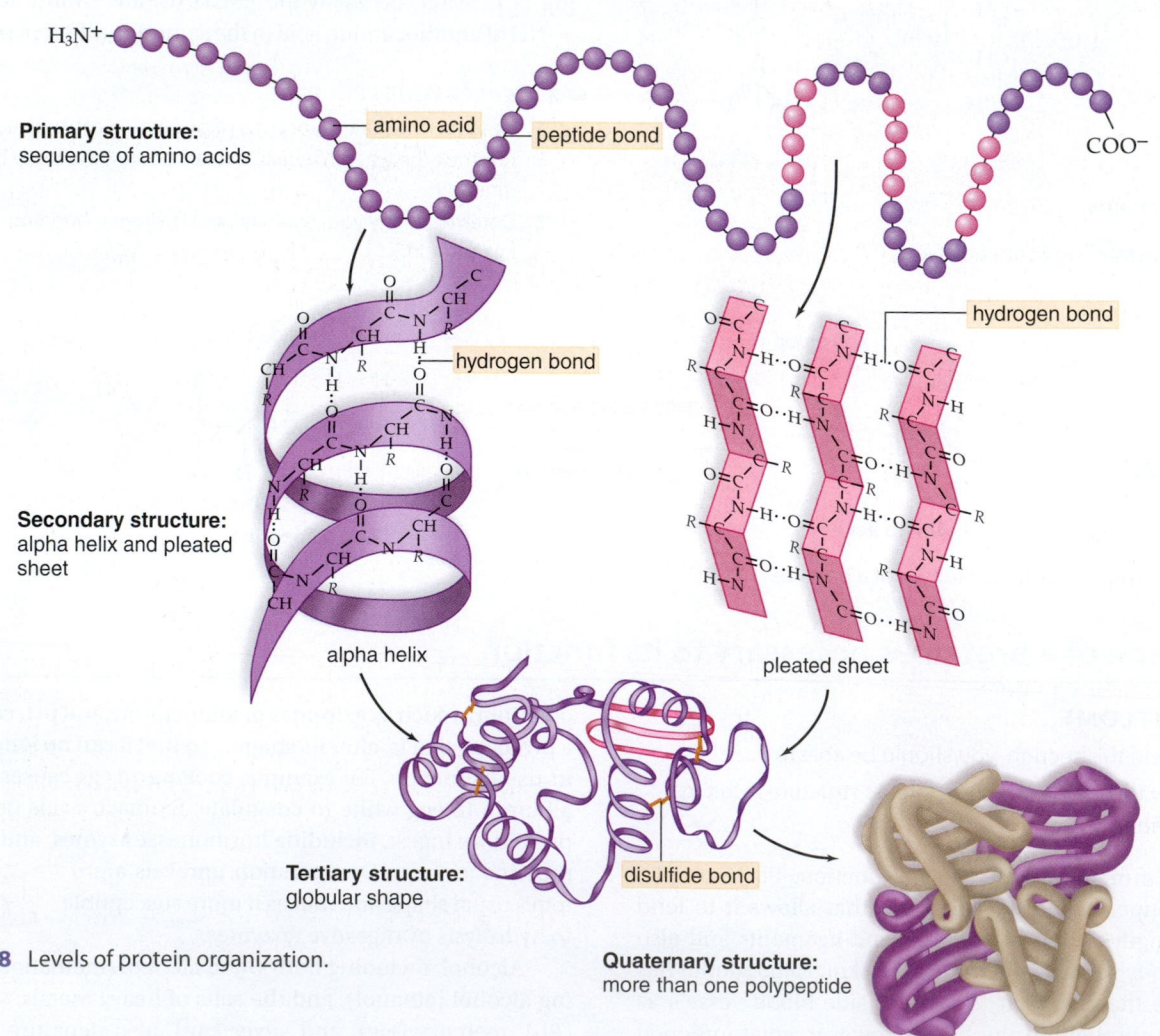

Primary structure: sequence of amino acids

amino acid · peptide bond

Secondary structure: alpha helix and pleated sheet

hydrogen bond

hydrogen bond

alpha helix

pleated sheet

Tertiary structure: globular shape

disulfide bond

Quaternary structure: more than one polypeptide

FIGURE 3.8 Levels of protein organization.

3B Molecular Evolution—A New Endeavor

Molecular genetics has been a critical part of biology ever since Watson and Crick discovered the structure of DNA in 1959. By now, biologists have determined the nucleotide sequence in the genes of many different organisms. Comparing this information has helped us to determine how organisms are related to each other and also to decipher the history of life. The application of molecular genetics to the study of evolution has been invaluable. For example, the tree of life depicted in Figure 1.8A is based on sequence differences in both DNA and RNA. It is not a stretch to say that our knowledge of taxonomy and evolution at every level of classification has been transformed by molecular studies.

Ultimately, molecular evolutionists want to associate new gene mutations with specific advantages that are selected by the environment.

Bacterial Laboratory Studies

As a first step, some biologists are simply studying the occurrence of mutations in populations of bacteria. Mutations are "the raw material of evolution" because, without them, a variety of organisms would not be possible. In one study conducted by Richard Lenski at Michigan State University, bacteria were the organisms of choice because they mutate frequently due to a reproductive rate of only a few hours. In 1988, genetically identical *Escherichia coli* bacteria were placed in 12 flasks; the environment has been held constant and no new bacteria have ever been added. By now, the original bacteria have produced 45,000 generations each. Periodically, a sample of bacteria is removed from each flask and cultured on agar in petri dishes (note the petri dishes piled up in Figure 3B) to detect any genetic mutations. The greatest change detected so far occurred after about 31,500 generations when some bacteria from a particular flask tested positive for the ability to digest citrate, something no other *E. coli* do. Since then, the citrate-eaters, which are considered a new species, have increased in number presumably due to a selective advantage. The student featured in Figure 3B participated in testing trillions of bacteria for their ability to eat citrate.

Studies in the Wild

Another approach was undertaken by Douglas W. Schemski of Michigan State University and H. D. Bradshaw, Jr., of the University of Washington. They wanted to test a molecular evolutionary change in the wild and decided to use two species of monkey flowers as their experimental material. A species with pink flowers is mainly pollinated by bumblebees, and a species with red flowers is mainly pollinated by hummingbirds. A difference in

FIGURE 3B The student in the foreground is meditating after helping to plate trillions of bacteria in the petri dishes piled up behind him.

the gene called *YUP* was found to be primarily responsible for the color difference and therefore the pollinator difference—bees prefer the pink color, and hummingbirds prefer the red color. Among other experiments, these investigators substituted "pink" *YUP* genes for "red" *YUP* genes in certain plant embryos. When these altered embryos became plants in the wild, the gene substitutions were found to have a dramatic effect on pollinator visits. The plants had orange flowers instead of red flowers and this caused the number of bumblebee visits to increase dramatically. This experiment provides insight into how evolution can be linked to a particular gene. It is hypothesized that pink flowers originated before red flowers, so the "newly evolved" flowers produced by the investigators are a throwback.

Scientists expect that the number of molecular evolution studies seeking to link specific DNA mutations with particular adaptive changes will expand and ever increase our understanding of the evolutionary process in years to come. Stay tuned for exciting developments!

CONSIDER THESE QUESTIONS

1. Sometimes major scientific advances have humble beginnings. Therefore, should all scientific studies be funded whether they seem worthwhile or not?
2. Biologists know of a gene that allows tomato plants to thrive when watered with salty water. If this gene is inserted into tomato plants, have they evolved? Why or why not?

Pink monkey flower

Red monkey flower

 **connect**
|BIOLOGY

Explore the concepts through a variety of multimedia assets and question types.

www.mcgrawhillconnect.com

Nucleic Acids

DNA is the heredity material for all species on planet Earth, and yet life is very diverse. The structure of DNA and its relationship to RNA and proteins explain how it can be the basis for such diversity. DNA provides the information and ATP, an energy carrier in cells, provides the energy that allow cells to build proteins, the molecules that make all humans the same and yet different. ATP is a nucleotide as are the subunits of DNA.

3.9 DNA stores coded information

LEARNING OUTCOMES

When you complete this section, you should be able to

1. Compare the structure and function of DNA and RNA in cells.
2. Account for the role of mutations in evolution and the history of life.

In Chapter 1 we learned that genes are hereditary units composed of **DNA** (**deoxyribonucleic acid**) and that genes control the structure and function of cells and organisms by coding for proteins. However, only by understanding DNA's structure can

we understand how it functions (Fig. 3.9A). Like a computer program whose coded information formats the appearance of text or moves cartoon figures across a screen, DNA dictates a cell's activities, even though it does not participate in those activities.

DNA is the hereditary material in both the girl and the rabbit but the two are different because the sequence of the four types of nucleotides in their DNA is different.

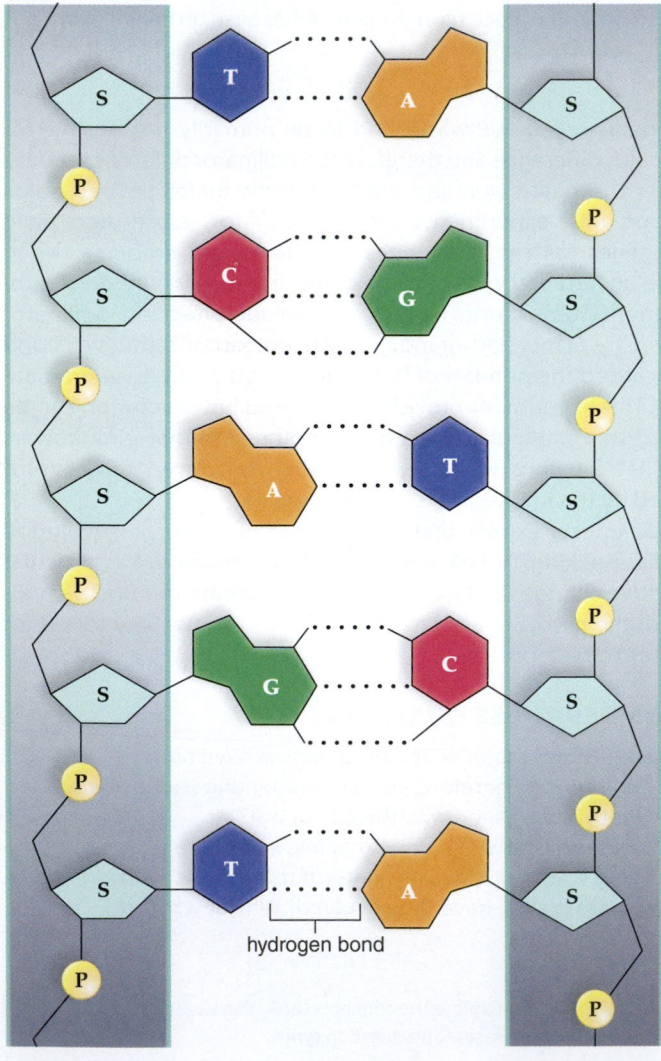

DNA: ladder configuration

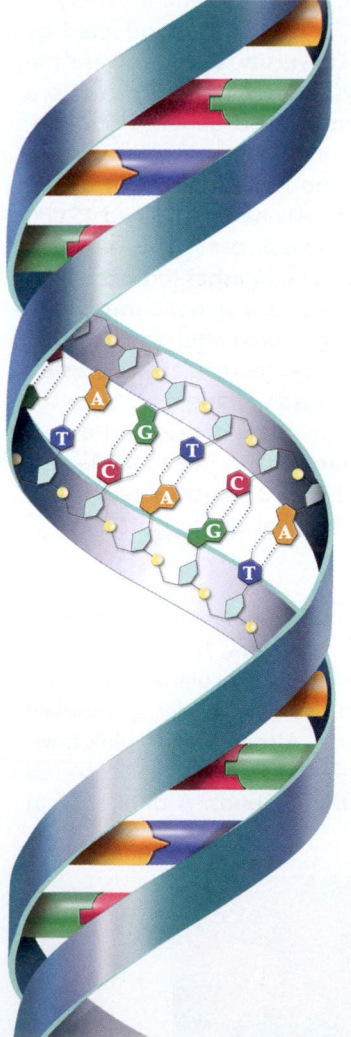

hydrogen bond

DNA: double helix

DNA: space-filling model

FIGURE 3.9A DNA structure at three levels of complexity.

A **nucleotide** is a molecular complex of three types of molecules: a phosphate (phosphoric acid), a pentose (5-carbon) sugar, and a nitrogen-containing base. These molecules are called bases because their presence raises the pH of a solution. A **nucleic acid** is a polymer of nucleotides.

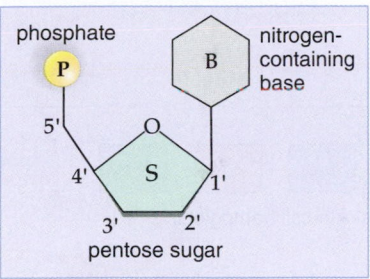

One nucleotide

Both DNA and its helper, **RNA** (**ribonucleic acid**), are nucleic acids. Early investigators called them nucleic acids because they were first detected in the nuclei of cells.

MP3
Nucleic
Acids

Structure of DNA The nucleotides in DNA contain the pentose sugar deoxyribose, accounting for its name—deoxyribonucleic acid. DNA is double-stranded, as shown in Figure 3.9A. The ladder structure of DNA is so called because the sugar and phosphate molecules make up the sides of a ladder, and hydrogen-bonded bases make up the rungs of the ladder. The hydrogen bonds are represented by dotted lines. The bases can be in any order, but between strands, thymine (T) is always paired with adenine (A), and guanine (G) is always paired with cytosine (C). These are, therefore, called **complementary bases.** Every organism has a particular sequence of paired bases, and after many years of research, biologists became quite adept at determining the sequence manually. Now, equipment is available that can sequence DNA automatically and rapidly. A DNA molecule can have thousands—even millions—of bases, and a particular **gene** is only a segment of these. By now, we know the DNA base sequence of all the genes (called the **genome**) in innumerable organisms, including rabbits and humans.

We will see in Chapter 11 that base pairing is absolutely essential when DNA replicates and a copy of the genome is passed from generation to generation of cells and organisms. It is also essential when RNA is synthesized.

Flow of Genetic Information Figure 3.9B shows how the nucleotides are arranged in RNA, a single-stranded molecule. Notice how the sugar and phosphate molecules form the backbone of the molecule, while the bases project to the side. The 5-carbon sugar molecule is ribose, and this accounts for its name—ribonucleic acid. The bases in RNA are guanine (G), adenine (A), cytosine (C), and uracil (U), which replaces thymine in DNA (Table 3.9).

When an RNA is synthesized, RNA nucleotides pair complementarily with those of a DNA strand. At that time, the base uracil in RNA pairs with adenine in DNA. Several types of RNA are available in cells, but **messenger RNA** (**mRNA**) is of interest now because it is a copy of a gene that codes for proteins. After mRNA is synthesized, it moves to where proteins are made in the cell. The sequence of bases in mRNA determines the sequence of amino acids in a protein. Therefore, genetic information flows from DNA to mRNA to protein:

$$DNA \longrightarrow mRNA \longrightarrow protein$$

Just as a monitor is needed to see the results of a computer program, so determining the proteins in a cell allows us to see the results of a cell's genetic information.

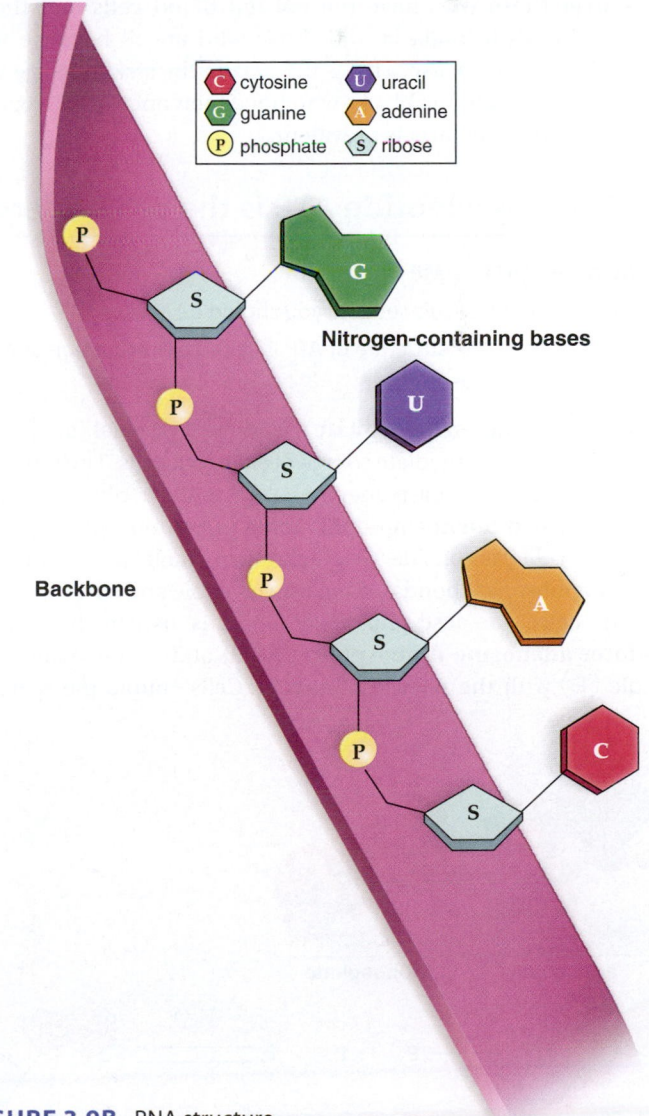

Nitrogen-containing bases

Backbone

C	cytosine	U	uracil
G	guanine	A	adenine
P	phosphate	S	ribose

FIGURE 3.9B RNA structure.

TABLE 3.9	DNA Structure Compared to RNA Structure	
	DNA	**RNA**
Sugar	Deoxyribose	Ribose
Bases	Adenine, guanine, thymine, cytosine	Adenine, guanine, uracil, cytosine
Strands	Double-stranded with base pairing	Single-stranded
Helix	Yes	No

Genetic Mutations We now know that a **genetic mutation** is a change in the sequence of bases in a gene and a mutation can result in an altered amino acid sequence in a protein. Without mutations, evolution would be impossible because genetic changes can result in adaptive changes. Individuals with an adaptive change are able to secure more resources and therefore to reproduce to a greater extent than those without the adaptive change.

The genetic disorder sickle-cell disease exemplifies how even a seemingly harmful mutation can be adaptive. In sickle-cell disease, an individual's red blood cells are sickle-shaped because at one particular spot the amino acid valine (Val) appears in hemoglobin instead of the amino acid glutamate (Glu) (Fig. 3.9C). When a person inherits a double mutation, red blood cells lose their normally round flexible shape and become hard and jagged. When these abnormal red blood cells go through small blood vessels, they may clog the flow of blood, causing pain, organ damage, and a low red blood cell count.

However, when a person inherits a single mutation, the red blood cells become sickle-shaped only on occasion, such as when they are invaded by a malarial parasite. The sickle shape causes the parasite to die and gives these people an advantage over those who have normal red blood cells and therefore succumb to malaria and those who are ill because they have sickle-cell disease. This is the reason the sickle-cell gene is more common among blacks who trace their ancestry to regions of Africa where malaria is prevalent.

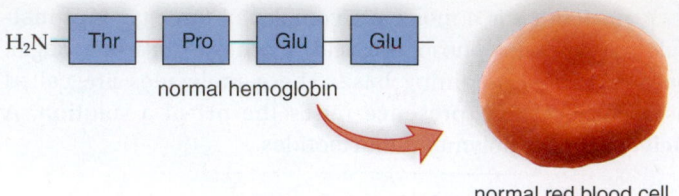

normal hemoglobin

normal red blood cell

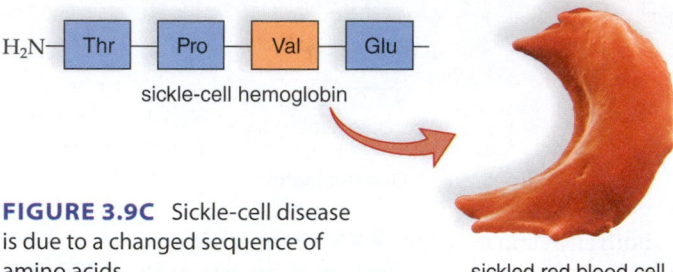

sickle-cell hemoglobin

FIGURE 3.9C Sickle-cell disease is due to a changed sequence of amino acids.

sickled red blood cell

▶ **3.9 CHECK YOUR PROGRESS**
1. Determine the complementary base sequence to GATCCA in DNA.
2. Use Table 3.9 to identify the differences in structure between DNA and RNA.
3. Explain why a mutation would result in an altered messenger RNA molecule.

3.10 The nucleotide ATP is the cell's energy carrier

LEARNING OUTCOME

When you complete this section, you should be able to

1. Explain how the structure of ATP is suited to its function as a carrier of usable energy.

ATP (**adenosine triphosphate**) is a very special nucleotide because it is the immediate source of energy in cells. The molecule is composed of the base adenine and the sugar ribose—a compound termed **adenosine**—plus three linked phosphate groups (Fig. 3.10). ATP is considered a high-energy molecule because the last two phosphate bonds (in red) are unstable and easily broken.

In cells, the last phosphate bond is usually hydrolyzed to form **adenosine diphosphate** (**ADP**) and a phosphate molecule Ⓟ with the release of energy. Cells couple the released

energy to energy-requiring processes such as the synthesis of DNA. In muscle cells, the energy is used for muscle contraction, and in nerve cells, it is used to conduct nerve impulses. Just as you spend money when you pay for a product or a service, cells "spend" ATP when they need something done. ATP is a usable energy molecule, not an energy-storage molecule as are carbohydrates and fat. Therefore, ATP is called the energy currency of cells. We will discuss more about ATP in Chapter 7.

MP3
ATP

▶ **3.10 CHECK YOUR PROGRESS**
1. Knowing that glucose breakdown leads to ATP buildup, determine how it benefits your body for the liver to store glycogen.

FIGURE 3.10 ATP (adenosine triphosphate) hydrolysis releases energy.

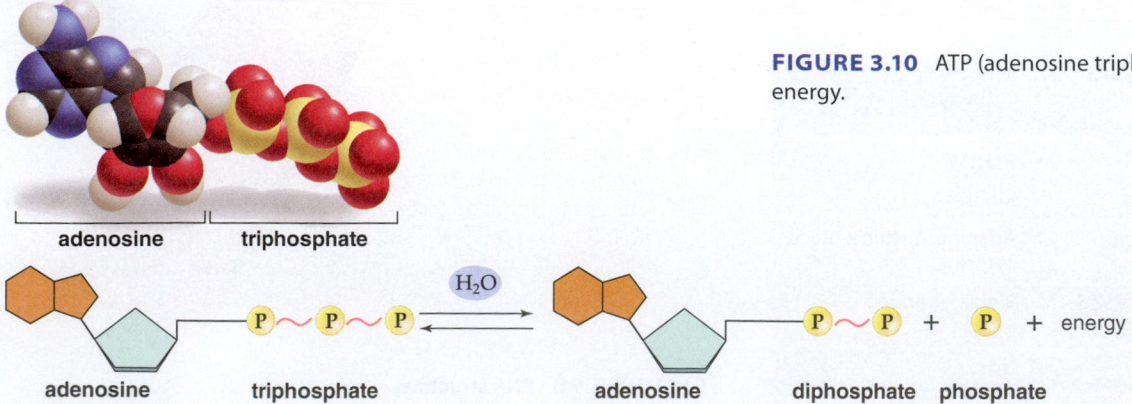

adenosine   triphosphate

adenosine   triphosphate   adenosine   diphosphate   phosphate

THE CHAPTER IN REVIEW

CONNECTING THE CONCEPTS

The large biomolecules—carbohydrates, fats, proteins, and nucleic acids—are present in all cells and organisms, but the particular type varies with the organism. For example, while animals store glucose as glycogen, plants use starch. Plants store unsaturated oils in their seeds, and animals store saturated fats in their bodies. DNA, the hereditary (genetic) material, has a sequence of nucleotide bases that varies according to the species and indeed even among members of the same species. The sequence of bases is coded information that determines what types of proteins will be in each organism. All cells and organisms have their particular mix of proteins, and while the function of proteins varies from species to species, proteins serve as enzymes in all organisms.

While DNA specifies the sequence of amino acids in proteins, it does not participate in the making of proteins. However, through complementary base pairing, messenger RNA receives the coded information from DNA and participates in protein synthesis so that each protein has its own sequence of amino acids. Proteins then take on a particular shape that allows them to carry out their specific function. When DNA mutates, its sequence of bases, and thus the shape of the resulting protein, changes. This is the source of the genetic variety that allows evolution to occur.

ANALYZE AND EVALUATE

1. Divide the particular biomolecules studied in this chapter according to these categories: (a) energy storage and use, (b) genetic information storage, and (c) ongoing activities of the cell. Explain your reasoning.
2. It's currently estimated that humans have about 20,000 genes. Conceivably, then, they have at least this number of proteins. Explain how a single molecule (DNA) can code for so much variety of proteins.
3. Humans have about 210 different cell types. Concentrating just on proteins, in general how would you expect these cell types to differ biochemically?

SUMMARIZE

Fundamentals of Organic Chemistry

3.1 The chemistry of carbon makes diverse molecules possible

- Carbon needs four electrons to complete its outer shell; it can share with as many as four other elements. The carbon—carbon bond is very stable, and carbon chains can be very long. Organic molecules (having carbon and hydrogen) are a major component of organisms and thus are called **biomolecules.**
- **Functional groups** are a specific combination of atoms that always react in the same way. They can make a biomolecule **hydrophobic** or **hydrophilic. Isomers** have identical molecular formulas but a different arrangement of atoms (or functional groups).

3.2 Molecular subunits can be linked to form varied large biomolecules

- **Dehydration reactions** form **polymers** which have many subunits, called **monomers. Hydrolysis reactions** break them apart. The large biomolecules in cells are carbohydrates, lipids, proteins, and nucleic acids. The carbohydrate subunits are monosaccharides; lipid subunits are glycerol and fatty acids; the protein subunits are amino acids; and the nucleic acid subunits are nucleotides.

Carbohydrates

3.3 Simple carbohydrates provide quick energy

- **Monosaccharides** (each composed of a single sugar molecule) are simple sugars. **Glucose** is a simple sugar and a major source of quick energy. **Ribose** and **deoxyribose** are 5-carbon sugars in RNA and DNA, respectively. **Disaccharides** are formed from two monosaccharides joined during dehydration. Sucrose (table sugar) is a disaccharide.

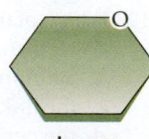

glucose

3.4 Complex carbohydrates store energy and provide structural support

- **Polysaccharides** (polymers of monosaccharides that can be broken down to sugar molecules for energy) include **glycogen** (stored glucose in animals) and **starch** (stored glucose in plants). Polysaccharides used for structural support include **cellulose** (in plants), **chitin** (in animals and fungi), and **peptidoglycan** (in bacteria).

Lipids

3.5 Fats and oils are rich energy-storage molecules

- **Fats** and **oils (triglycerides)** contain three **fatty acids** attached to a **glycerol** molecule. **Saturated fatty acids** (no double bonds) are characteristic of solid fats found in animals. **Unsaturated fatty acids** (double bonds) are characteristic of liquid oils found primarily in plant seeds. **Trans fats** found in processed foods contain partially hydrogenated fatty acids and are particularly harmful to our health.

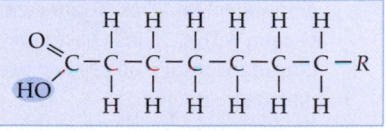

fatty acid

3.6 Other lipids have structural, hormonal, or protective functions

- In **phospholipids,** the major component of the plasma membrane, glycerol is bound to two fatty acids and a polar group containing phosphate. **Steroids,** which serve as a plasma membrane component (cholesterol) or have a hormonal function (estrogen and testosterone), are composed of four fused rings. In **waxes,** which prevent water loss in plants and assist in skin and fur maintenance in animals, long chain fatty acids bond with long chain alcohols.

Proteins

3.7 Proteins are the most versatile of life's biomolecules

- **Proteins** (polymers of amino acids) have the following functions:
  Support (structural proteins)
  Metabolism (speed chemical reactions)
  Transport (**hemoglobin** transports oxygen)
  Defense (antibodies combine with antigens to remove them)
  Regulation (hormones)
  Motion (muscle contraction)
- **Amino acids** have a central carbon attached to an amino group (—NH$_2$), an acid group (—COOH), and an *R* group. Amino acids differ according to their *R* group.
- A **peptide** consists of two or more amino acids bonded together. A **peptide bond** is the covalent bond between two amino acids. A **polypeptide** is a chain of amino acids joined by peptide bonds.

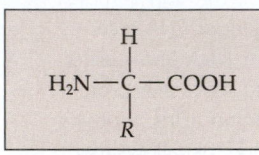

amino acid

3.8 The shape of a protein is necessary to its function

- The primary structure is the linear sequence of amino acids in a protein. The secondary structure is the particular way a polypeptide folds—alpha helix or pleated sheet. The tertiary structure is a globular protein's final three-dimensional shape. Proteins that consist of more than one polypeptide may have a quaternary structure.

Nucleic Acids

3.9 DNA stores coded information

- A **gene** is a segment of DNA that codes for a protein. The **genome** is all of an organism's genes. The **nucleic acids, DNA** (**deoxyribonucleic acid**) and **RNA** (**ribonucleic acid**), are polymers of nucleotides. A **nucleotide** is composed of a phosphate, a pentose sugar, and a nitrogen-containing base. The nucleotides in DNA contain the sugar deoxyribose and the bases thymine (T), adenine (A), guanine (G), and cytosine (C). DNA is double-stranded and helical. In DNA, T pairs with A; G pairs with C. These bases are called **complementary bases** because their pairing is specific.

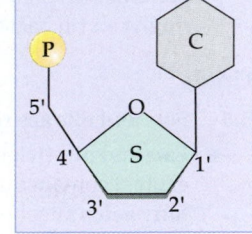

nucleotide

- RNA compares to DNA in these ways:

	DNA	RNA
Sugar	Deoxyribose	Ribose
Bases	Adenine, guanine, thymine, cytosine	Adenine, guanine, uracil, cytosine
Strands	Double-stranded with base pairing	Single-stranded
Helix	Yes	No

- In RNA, the sugar is ribose and the bases are U (for uracil), A, G, and C. RNA is single-stranded and not helical. When **messenger RNA** (**mRNA**) forms, DNA's coded information is passed to messenger RNA through complementary base pairing. (U pairs with A.) The sequence of bases in messenger RNA determines the sequence of amino acids in a protein.
- A **genetic mutation** is an altered sequence of bases, which in turn can result in an altered protein. Mutations are necessary to the evolutionary process and help account for the diversity of life.

3.10 The nucleotide ATP is the cell's energy carrier

- **ATP** (**adenosine triphosphate**) is composed of adenine and ribose (**adenosine**) plus three phosphate groups (triphosphate). The last phosphate bond is hydrolyzed to form **adenosine diphosphate** (**ADP**)+P, with the release of energy. Energy from ATP breakdown is used for biomolecule synthesis, muscle contraction, and nerve conduction.

TEST YOURSELF

Fundamentals of Organic Chemistry

1. Which of the following is an organic molecule?
   - a. CO
   - b. H$_2$O
   - c. C$_6$H$_{12}$O$_6$
   - d. O$_2$
   - e. All of these are correct.
2. Which of these is not a characteristic of carbon?
   - a. forms four covalent bonds
   - b. bonds with other carbon atoms
   - c. is sometimes ionic
   - d. can form long chains
   - e. sometimes shares two pairs of electrons with another atom
3. Organic molecules containing carboxyl groups are
   - a. nonpolar.
   - b. acidic.
   - c. basic.
   - d. More than one of these is correct.
4. Subunits are attached together to create polymers when a hydroxyl group and a hydrogen atom are _____ in a _____ reaction.
   - a. added, dehydration
   - b. removed, dehydration
   - c. added, hydrolysis
5. **THINKING CONCEPTUALLY** Based on the position of silicon in the periodic table of the elements, how is silicon like and how is it different from carbon? Explain why silicon does not form the same variety of molecules as carbon.

For questions 6–9, match the subunit with one of the biomolecules in the key.

KEY:
- a. carbohydrate
- b. lipid
- c. protein
- d. nucleic acid

6. Glycerol
7. Glycogen
8. Nucleotide
9. Amino acid

Carbohydrates

10. Which of the following is a disaccharide?
    - a. glucose
    - b. ribose
    - c. fructose
    - d. maltose
11. Plants store glucose as
    - a. maltose.
    - b. glycogen.
    - c. starch.
    - d. None of these are correct.
12. Which of these makes cellulose nondigestible in humans?
    - a. a polymer of glucose subunits
    - b. a fibrous protein
    - c. the linkage between the glucose molecules
    - d. the peptide linkage between the amino acid molecules
    - e. ionization of the carboxyl groups
13. **THINKING CONCEPTUALLY** After examining Figure 3.4, give reasons why cellulose rather than starch is a structural component of plant cell walls.

Lipids

14. A triglyceride contains
    a. glycerol and three fatty acids.
    b. glycerol and three sugars.
    c. protein and three fatty acids.
    d. protein and three sugars.
15. A fatty acid is unsaturated if it
    a. contains hydrogen.
    b. contains carbon—carbon double bonds.
    c. contains a carboxyl (acidic) group.
    d. bonds to glycogen.
    e. bonds to a nucleotide.
16. Saturated fatty acids and unsaturated fatty acids differ in
    a. the number of double bonds present.
    b. their consistency at room temperature.
    c. the number of hydrogen atoms present.
    d. All of these are correct.
17. _____ is the precursor of _____.
    a. Estrogen, cholesterol
    b. Cholesterol, glucose
    c. Testosterone, cholesterol
    d. Cholesterol, testosterone and estrogen
18. Which of these is not a lipid?
    a. steroid   d. wax
    b. fat   e. phospholipids
    c. polysaccharide
19. **THINKING CONCEPTUALLY** Explain why phospholipids lend themselves to forming a bilayer membrane.

Proteins

20. Nearly all _____ are _____.
    a. proteins, enzymes   c. enzymes, proteins
    b. sugars, monosaccharides
21. The difference between one amino acid and another is found in the
    a. amino group.   d. peptide bond.
    b. carboxyl group.   e. carbon atoms.
    c. R group.
22. The joining of two adjacent amino acids is called a
    a. peptide bond.   c. covalent bond.
    b. dehydration reaction. d. All of these are correct.
23. Covalent bonding between R groups in proteins is associated with the _____ structure.
    a. primary   c. tertiary
    b. secondary   d. None of these are correct.
24. The three-dimensional structure of a protein that contains two or more polypeptides is the
    a. primary structure.   c. tertiary structure.
    b. secondary structure.   d. quaternary structure.
25. **THINKING CONCEPTUALLY** Why would it be appropriate to say that proteins (and not their DNA) make species different from one another?

Nucleic Acids

26. Nucleotides
    a. are composed of a sugar, a nitrogen-containing base, and a phosphate group.
    b. are the monomers of fats and polysaccharides.
    c. join together by covalent bonding between the bases.
    d. are present in both DNA and RNA.
    e. Both a and d are correct.
27. Which of the following pertains to an RNA nucleotide, not a DNA nucleotide?
    a. contains the sugar ribose
    b. contains a nitrogen-containing base
    c. contains a phosphate molecule
    d. becomes bonded to other nucleotides by condensation
28. ATP
    a. is an amino acid.
    b. has a helical structure.
    c. is a high-energy molecule that can break down to ADP and phosphate.
    d. provides enzymes for metabolism.
29. **THINKING CONCEPTUALLY** Early chemists noted the parallel construction of a single-stranded nucleic acid and a protein. What did they mean by "parallel construction" in this context?

GET INVOLVED

1. You hypothesize that the unsaturated oil content of temperate plant seeds will help them survive freezing temperatures better than the saturated oil content of tropical plant seeds. (a) How would you test your hypothesis? (b) If your hypothesis is supported, what might you also conclude?
2. Chemical analysis reveals that an abnormal form of an enzyme contains a polar amino acid while the normal form contains a nonpolar amino acid. Formulate a hypothesis you would want to test regarding the abnormally constructed enzyme.

MEDIA STUDY TOOLS

mhhe.com/maderconcepts3

Enhance your study of this chapter with interactive study tools, practice tests, and engaging animations. Also, ask your instructor about the resources available through ConnectPlus, which includes LearnSmart, a personalized adaptive learning program, and a media-rich eBook.

4

Structure and Function of Cells

BEFORE YOU BEGIN

Take a few minutes to recall

The role of glucose and ATP as energy sources
    (sections 3.3 and 3.10)
How phospholipids form membrane (section 3.6)
The metabolic function of proteins in cells (section 3.7)
The function of DNA and RNA in cells (section 3.9)

Cells: What Are They?

Imagine that you have never taken a biology course, and you are alone in a laboratory with a bunch of slides of plant and animal tissues and a microscope. The microscope is easy to use, and soon you are able to focus it and begin looking at the slides.

Your assignment is to define a cell. In order not to panic, you idly look at one slide after another, letting your mind wander. Was this the way Robert Hooke felt back in the seventeenth century, when he coined the word "cell"? What did he see? Actually, Hooke was using a light microscope, as you are, when he happened to look at a piece of cork. He drew what he saw like this:

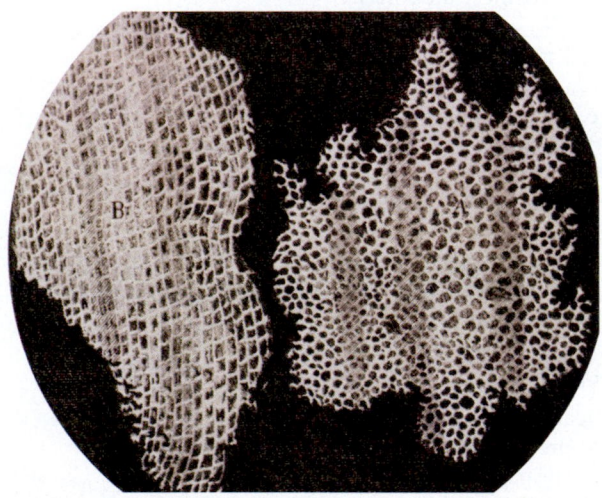

Hooke saw almost nothing except for outlines, which we know today are the cell walls of plant cells. Similarly, you can make out the demarcations between onion root cells in the micrograph on page 61. After comparing these to the nerve cells below, you might conclude that a cell is an entity, a unit of a larger whole.

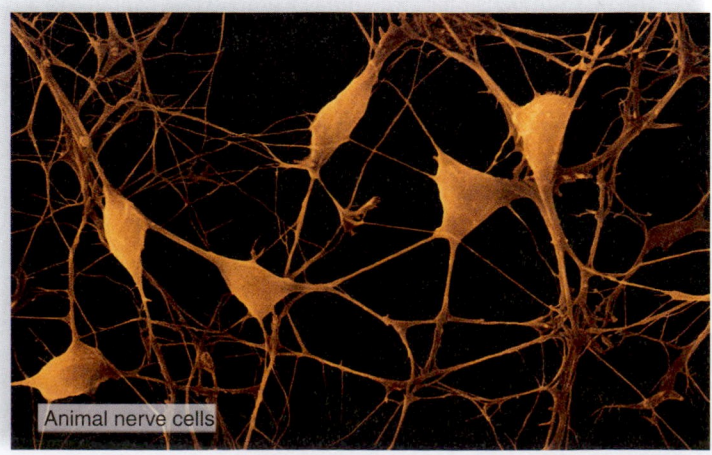

Animal nerve cells

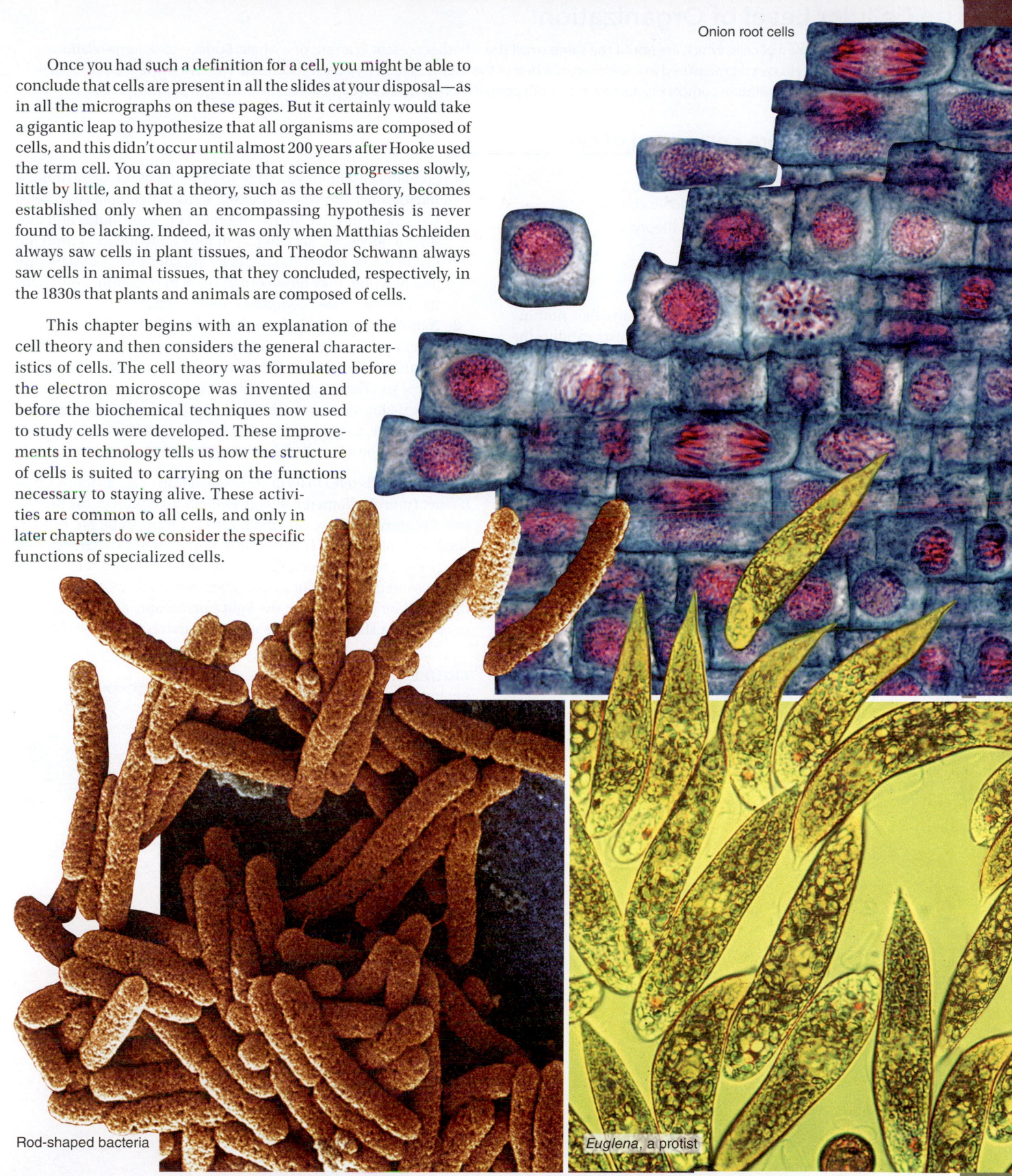

Onion root cells

Once you had such a definition for a cell, you might be able to conclude that cells are present in all the slides at your disposal—as in all the micrographs on these pages. But it certainly would take a gigantic leap to hypothesize that all organisms are composed of cells, and this didn't occur until almost 200 years after Hooke used the term cell. You can appreciate that science progresses slowly, little by little, and that a theory, such as the cell theory, becomes established only when an encompassing hypothesis is never found to be lacking. Indeed, it was only when Matthias Schleiden always saw cells in plant tissues, and Theodor Schwann always saw cells in animal tissues, that they concluded, respectively, in the 1830s that plants and animals are composed of cells.

This chapter begins with an explanation of the cell theory and then considers the general characteristics of cells. The cell theory was formulated before the electron microscope was invented and before the biochemical techniques now used to study cells were developed. These improvements in technology tells us how the structure of cells is suited to carrying on the functions necessary to staying alive. These activities are common to all cells, and only in later chapters do we consider the specific functions of specialized cells.

Rod-shaped bacteria

Euglena, a protist

The Cellular Level of Organization

All organisms are composed of cells, which are about the same small size whether present in an ant or a whale. Surface-to-volume relationships explain why most cells can be measured in micrometers, a unit of the metric system (see back endsheet). The two major types of cells—prokaryotic and eukaryotic—differ in complexity, but even so both contain DNA and have a cytoplasm enclosed by a plasma membrane.

4.1 All organisms are composed of cells

LEARNING OUTCOMES

When you complete this section, you should be able to

1. List and explain three tenets of the cell theory.
2. Evaluate why cells are so small.

The **cell theory** states the following:

1. ***A Cell Is the Basic Unit of Life*** This means that nothing smaller than a cell is alive. A unicellular organism exhibits the characteristics of life we discussed in Chapter 1. No smaller unit exists that is able to reproduce, respond to stimuli, remain homeostatic, grow and develop, take in and use materials from the environment, and adapt to the environment. In short, life has a cellular nature. On this basis, we can make two other deductions.

2. ***Organisms Are Made Up of Cells*** While it may be apparent that a unicellular organism is a cell, what about more complex organisms? Lilacs and rabbits as well as other visible organisms are multicellular. Figure 4.1A illustrates that a lilac leaf is composed of cells, and Figure 4.1B illustrates that the intestinal lining of a rabbit is composed of cells. Is there any tissue in these organisms that is not composed of cells? For example, you might be inclined to say that bone does not contain cells. But if you were to examine bone tissue under a microscope, you would see that it, too, is composed of cells. Cells have distinct forms—a bone cell looks quite different from a nerve cell, and they both look quite different from the cell of a lilac leaf. Although cells are specialized in structure and function, they have certain parts in common. This chapter discusses those common components.

3. ***New Cells Arise Only from Preexisting Cells*** This statement wasn't readily apparent to early investigators, who believed that organisms could arise from dirty rags, for example. Today, we know you cannot get a new lilac bush or a new rabbit without preexisting lilacs and rabbits. When lilacs, rabbits, or humans reproduce, a sperm cell joins with an egg cell to form a zygote, which is the first cell of a new multicellular organism.

Cell Size Cells tend to be quite small. A frog's egg, at about 1 millimeter (mm) in diameter, is large enough to be seen by the human eye. But most cells are far smaller than 1 mm; some are even as small as 1 micrometer (μm)—one thousandth of a millimeter. Cell structures and biomolecules that are smaller than a micrometer are measured in terms of nanometers (nm). Figure 4.1C outlines the visual range of the eye, the light microscope, and the electron microscope using units of the metric system (see back endsheet).

FIGURE 4.1B Rabbit, with a photomicrograph of its intestinal lining below.

Rabbit, an animal

Micrograph of intestine reveals cells.   140×

Lilac, a plant

Micrograph of leaf reveals cells.   80×

FIGURE 4.1A Lilac leaf, with a photomicrograph below.

"Microscopes Allow Us to See Cells" on page 64 explains why the electron microscope allows us to see so much more detail than the light microscope does.

Why are cells so small? To answer this question, consider that a cell needs a surface area large enough to allow sufficient nutrients to enter and to rid itself of wastes. Small cells, not large cells, are more likely to have this adequate surface area per volume. Consider a balloon: The air in the balloon is the volume, and the balloon's skin is its surface area. A larger balloon has more volume, as you can appreciate by trying to blow up a large balloon compared to a small balloon. How might you appreciate the amount of surface area per volume? Figure 4.1D shows one way because it calculates the surface area per volume for different-sized cubes. Cutting a large cube into smaller cubes provides a lot more surface area per volume. The calculations show that a large cube has limited surface area per volume compared to a large cube composed of many individual cubes.

We would expect, then, that actively metabolizing cells would have to remain small. A chicken's egg is several centimeters in diameter, but the egg is not actively metabolizing. Once the egg is incubated and metabolic activity begins, the egg divides repeatedly without growth. Cell division restores the amount of surface area needed for adequate exchange of materials.

Further, cells that specialize in absorption have modifications that greatly increase the surface-area-to-volume ratio of the cell. The cells along the surface of the intestinal wall have surface foldings called microvilli (sing., microvillus) that increase their surface area. Nerve cells and some large plant cells are long and thin, and this increases the ratio of plasma membrane to cytoplasm. Nerve cells are shown on page 60.

▶ **4.1 CHECK YOUR PROGRESS**

1. Apply the cell theory to the human body.
2. Relate a high surface-area-to-volume ratio to the efficiency of cells.

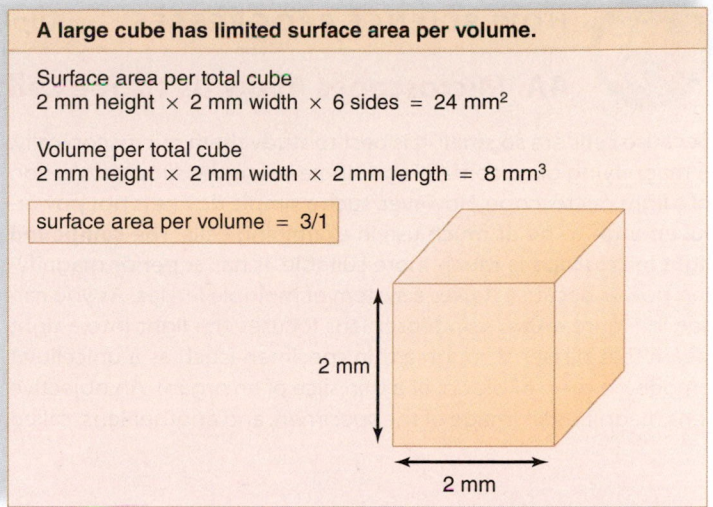

A large cube has limited surface area per volume.

Surface area per total cube
2 mm height × 2 mm width × 6 sides = 24 mm^2

Volume per total cube
2 mm height × 2 mm width × 2 mm length = 8 mm^3

surface area per volume = 3/1

2 mm

2 mm

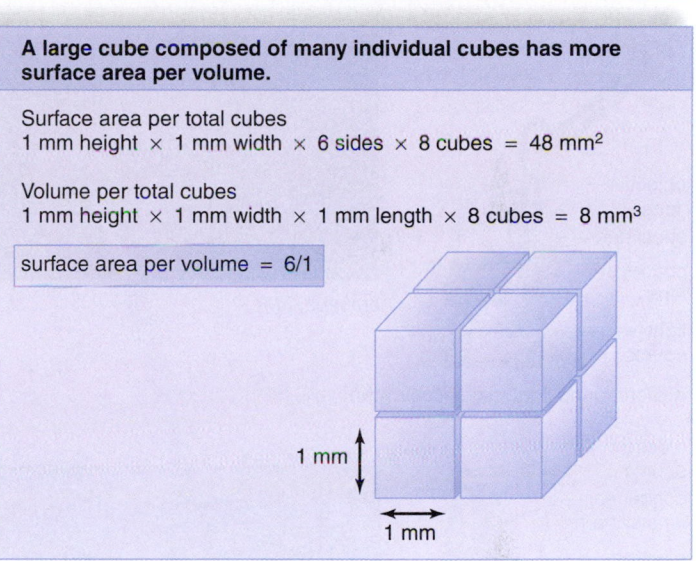

A large cube composed of many individual cubes has more surface area per volume.

Surface area per total cubes
1 mm height × 1 mm width × 6 sides × 8 cubes = 48 mm^2

Volume per total cubes
1 mm height × 1 mm width × 1 mm length × 8 cubes = 8 mm^3

surface area per volume = 6/1

1 mm

1 mm

FIGURE 4.1D Surface-area-to-volume relationships.

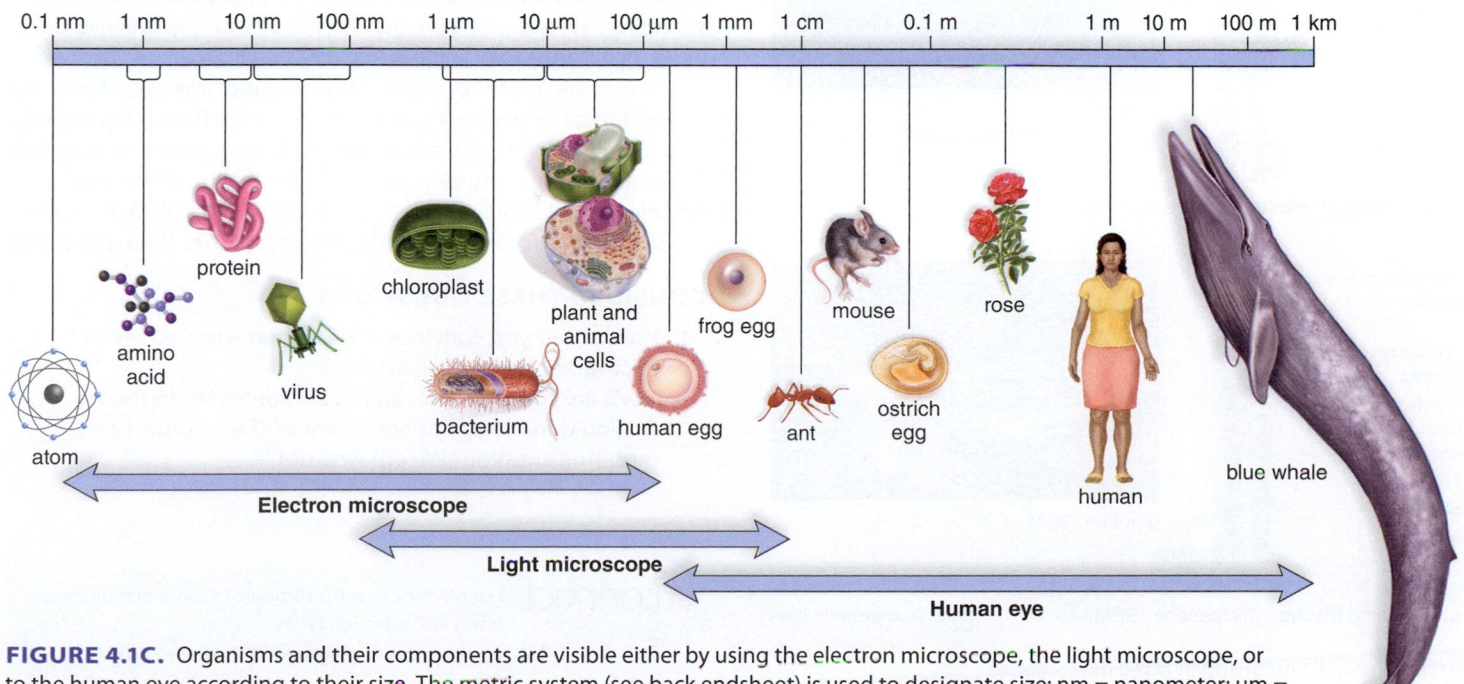

0.1 nm   1 nm   10 nm   100 nm   1 μm   10 μm   100 μm   1 mm   1 cm   0.1 m   1 m   10 m   100 m   1 km

protein

chloroplast

amino acid

virus

plant and animal cells

bacterium

frog egg

human egg

mouse

ant

ostrich egg

rose

human

blue whale

atom

Electron microscope

Light microscope

Human eye

FIGURE 4.1C. Organisms and their components are visible either by using the electron microscope, the light microscope, or to the human eye according to their size. The metric system (see back endsheet) is used to designate size: nm = nanometer; μm = micrometer; mm = millimeter; cm = centimeter; m = meter; km = kilometer.

4A Microscopes Allow Us to See Cells

Because cells are so small, it is best to study them microscopically. A magnifying glass containing a single lens is the simplest version of a light microscope. However, such a simple device is not powerful enough to be of much use in examining cells. The **compound light microscope** is much more suitable. It has superior magnifying power because it uses a system of multiple lenses. As you can see in Figure 4A*a*, a condenser lens focuses the light into a tight beam that passes through a thin specimen (such as a unicellular amoeba, a drop of blood, or a thin slice of an organ). An objective lens magnifies an image of the specimen, and another lens, called the ocular lens, magnifies it yet again. It is the image from the ocular lens that is viewed with the eye. The most commonly used compound light microscope is called a bright-field microscope, because the specimen, which is typically stained, appears dark against a light background.

The compound light microscope is widely used in research, clinical, and teaching laboratories. However, the use of light to produce an image reduces the ability to view two objects as separate—the resolution—is not as good as with an electron microscope. The resolution limit of a compound light microscope is 0.2 μm, which means that objects less than 0.2 μm apart appear as a single object. Although there is no limit to the magnification that could be achieved with a compound light microscope, there is a definite limit to the resolution.

An electron microscope can produce finer resolution than a light microscope because, instead of using light, it fires a beam of electrons at the specimen. Electrons have a shorter wavelength than does light. The essential design of an electron microscope is similar to that of a compound light microscope, but its lenses are made of electromagnets, instead of glass. Because the human eye cannot see the images produced by electron microscopes, they are projected onto a screen or viewed on a television monitor.

There are two types of electron microscopes: the transmission electron microscope and the scanning electron microscope. These microscopes didn't become widely used until about 1970. A **transmission electron microscope** passes a beam of electrons through a specimen (Fig. 4A*b*). Because electrons do not have much penetrating ability, the section must be very thin—usually between 50 and 150 nm. The transmission electron microscope can discern fine details, with a limit of resolution around 1.0 nm and a magnifying power up to 200,000 times larger than the actual size. A **scanning electron microscope** does not pass a beam through a specimen; rather, it collects and focuses electrons that are scattered from the specimen's surface and generates an image with a distinctive three-dimensional appearance (Fig. 4A*c*).

Scientists often preserve microscopic images; these are referred to as micrographs. A captured image from a light microscope is termed a light micrograph (LM), or a photomicrograph. There are also transmission electron micrographs (TEMs) and scanning electron micrographs (SEMs). The latter two are black-and-white in their original form, but computers can colorize them for clarity.

CONSIDER THESE QUESTIONS

1. How would you convince a friend that what we see in micrographs actually exists?
2. TEMs are colorless but can have color added to them. Do you think color enhancement of TEMs borders on misrepresentation? Why or why not?

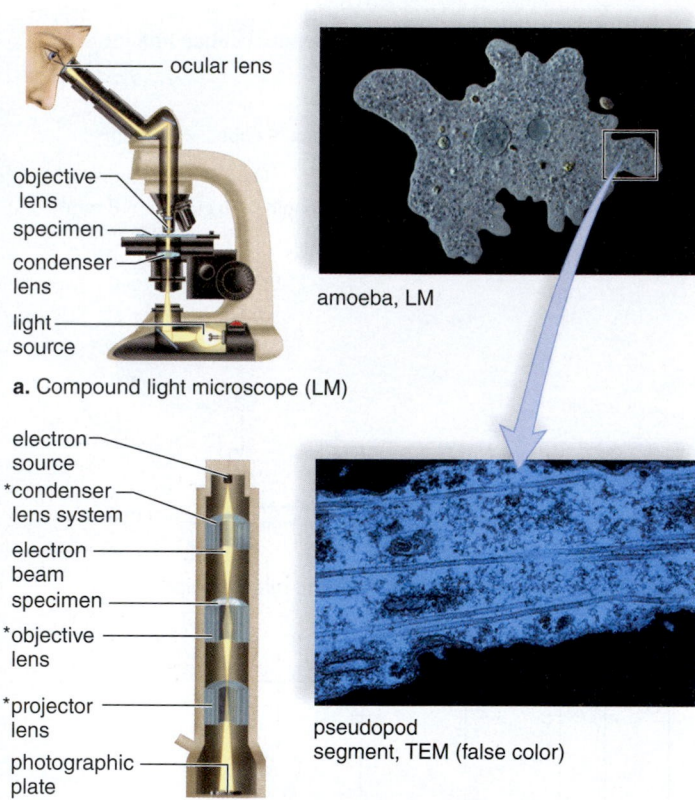

amoeba, LM

a. Compound light microscope (LM)

pseudopod segment, TEM (false color)

b. Transmission electron microscope (TEM)

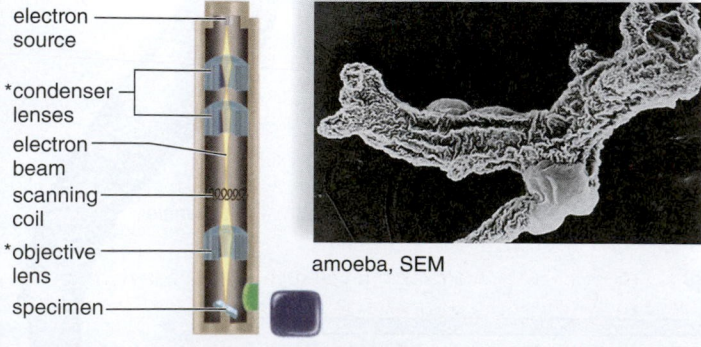

amoeba, SEM

c. Scanning electron microscope (SEM)   *electromagnetic lens

FIGURE 4A Comparison of three microscopes.

 **connect**
BIOLOGY   Explore the concepts through a variety of multimedia assets and question types.

www.mcgrawhillconnect.com

4.2 Prokaryotic cells evolved first

LEARNING OUTCOMES

When you complete this section, you should be able to

1. Understand the classification of prokaryotes into two domains.
2. Identify and give a function for each component of a bacterium.

Fundamentally, two different types of cells exist. **Prokaryotic cells** (*pro,* before, and *karyon,* nucleus) are so named because they lack a membrane-enclosed nucleus. The other type of cell, called a eukaryotic cell, has a nucleus. Prokaryotic cells are minuscule in size compared to eukaryotic cells (Fig. 4.2A). Prokaryotes are present in great numbers in the air, in bodies of water, in the soil, and also in and on other organisms.

As discussed on page 14, prokaryotic cells are divided into two groups, largely based on biochemical, including DNA, evidence. These two groups are so biochemically different that they have been placed in separate domains, called domain **Bacteria** and domain **Archaea.**

Figure 4.2B shows the generalized structure of a bacterium. Like a eukaryotic cell, a bacterium is full of a semifluid substance called **cytoplasm** that is enclosed by a **plasma membrane.** The plasma membrane is a phospholipid bilayer (see Fig. 3.6) with embedded proteins:

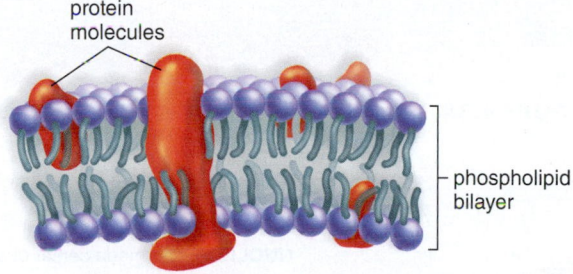

The plasma membrane has the important function of regulating the entrance and exit of substances into and out of the cytoplasm. After all, the cytoplasm has a normal composition that needs to be maintained. It contains thousands of **ribosomes** where proteins are produced. A long, looped, threadlike strand of DNA, the **chromosome** of a prokaryotic cell, is located within a region of the cytoplasm known as a **nucleoid.** When bacteria reproduce by splitting in two, each new cell gets a copy of the chromosome. Cyanobacteria (*cyan,* blue-green) are able to photosynthesize in the same manner as plants because they have light-absorbing chlorophyll on internal membranes.

In addition to the plasma membrane, bacteria have a **cell wall,** which helps maintain the shape of the cell. The cell wall may in turn be surrounded by a **capsule.** Many short, hollow protein rods called **pili** project through the cell wall. Pili attach the cell to solid substances and produce a slime that coats your teeth, rocks at the bottom of lakes, and the hulls of ships, for example. Motile bacteria usually have long, very thin flagella (sing., flagellum), which rotate like propellers, rapidly moving the bacterium in a fluid medium.

Bacteria are well known for causing serious diseases, such as tuberculosis, anthrax, tetanus, throat infections, and gonorrhea. However, they are important to the environment because they decompose the remains of dead organisms and contribute to the cycling of chemicals in ecosystems. Also, their great ability to synthesize molecules can be put to use for the manufacture of all sorts of products, from industrial chemicals to foodstuffs and drugs.

Video E. Coli Wars

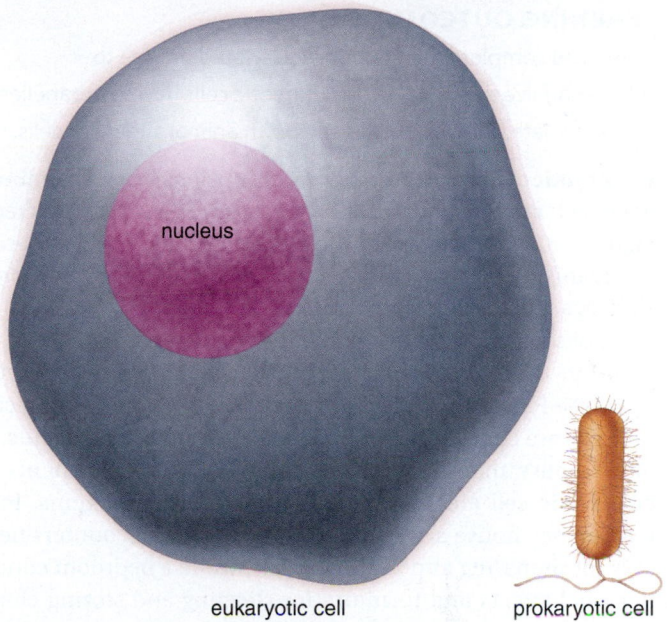

FIGURE 4.2A Eukaryotic cells are much larger than prokaryotic cells, as shown in this proportional drawing.

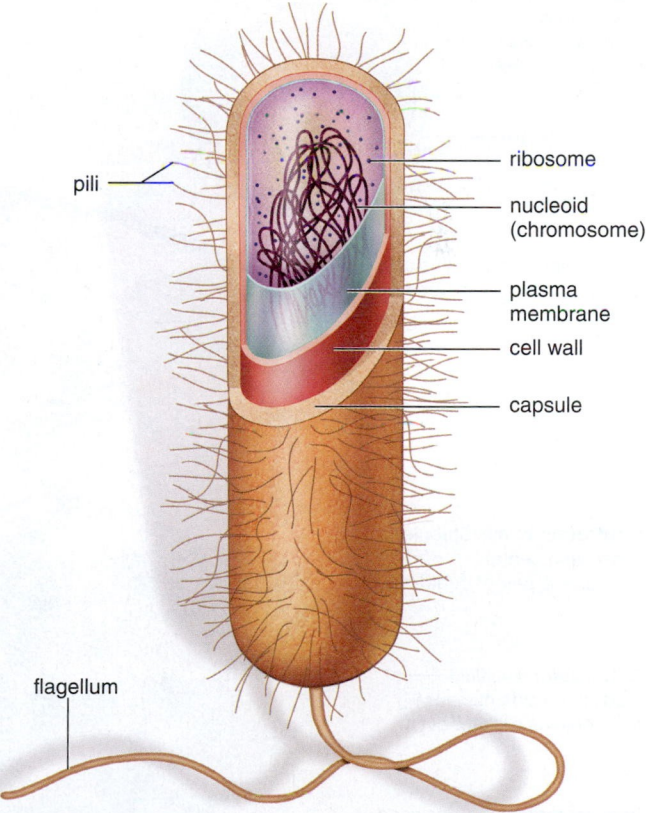

FIGURE 4.2B A prokaryotic cell such as this bacterium is structurally simple but metabolically complex.

▶ **4.2 CHECK YOUR PROGRESS**

1. Explain the classification of prokaryotes into two domains.
2. Identify a function for each labeled structure in Figure 4.2B.

4.3 Eukaryotic cells contain specialized organelles: An overview

Eukaryotic cells (*eu,* true, and *karyon,* nucleus) have a membrane-enclosed **nucleus,** which houses their DNA. As depicted in Figure 1.8D, protists, fungi, plants, and animals are the groups of organisms that have eukaryotic cells and are in the domain **Eukarya,** the third domain of life.

Eukaryotic cells are much larger than prokaryotic cells, and therefore they have less surface area per volume than prokaryotic cells (see Fig. 4.1D). This disadvantage has been solved because the cells are compartmentalized—they have compartments. Just like a house that has separate rooms, the compartments of a eukaryotic cell are specialized for particular functions. In the kitchen of a house are utensils, appliances, and counters necessary for preparing and serving meals, while a bedroom contains personal effects and furniture for sleeping and storing clothes.

Similarly, a cell contains **organelles** (meaning "little organs") that are specialized and perform only specific functions. The organelles are located within the cytoplasm, a semifluid interior enclosed by a plasma membrane. As in prokaryotic cells, the plasma membrane is a phospholipid bilayer that contains proteins, shown in the circular blowup of Figure 4.3A.

Eukaryotic cells are rich in membrane, and most organelles are membranous. Originally, the term organelle referred only to membranous structures, but we will use it to include any well-defined subcellular structure. By that definition, the little particles called ribosomes are also organelles. At first it might seem difficult to learn the names and functions of all the structures in plant and animal cells. One technique that will help is to have a mental image of the structure and then discover its function. So in Figures 4.3A and 4.3B, first look at the structure and then follow the leader back to its name and function. A well-known truism in biology states, "Structure suits function." Why might that be? In the course of evolution, those organisms whose cells possessed organelles

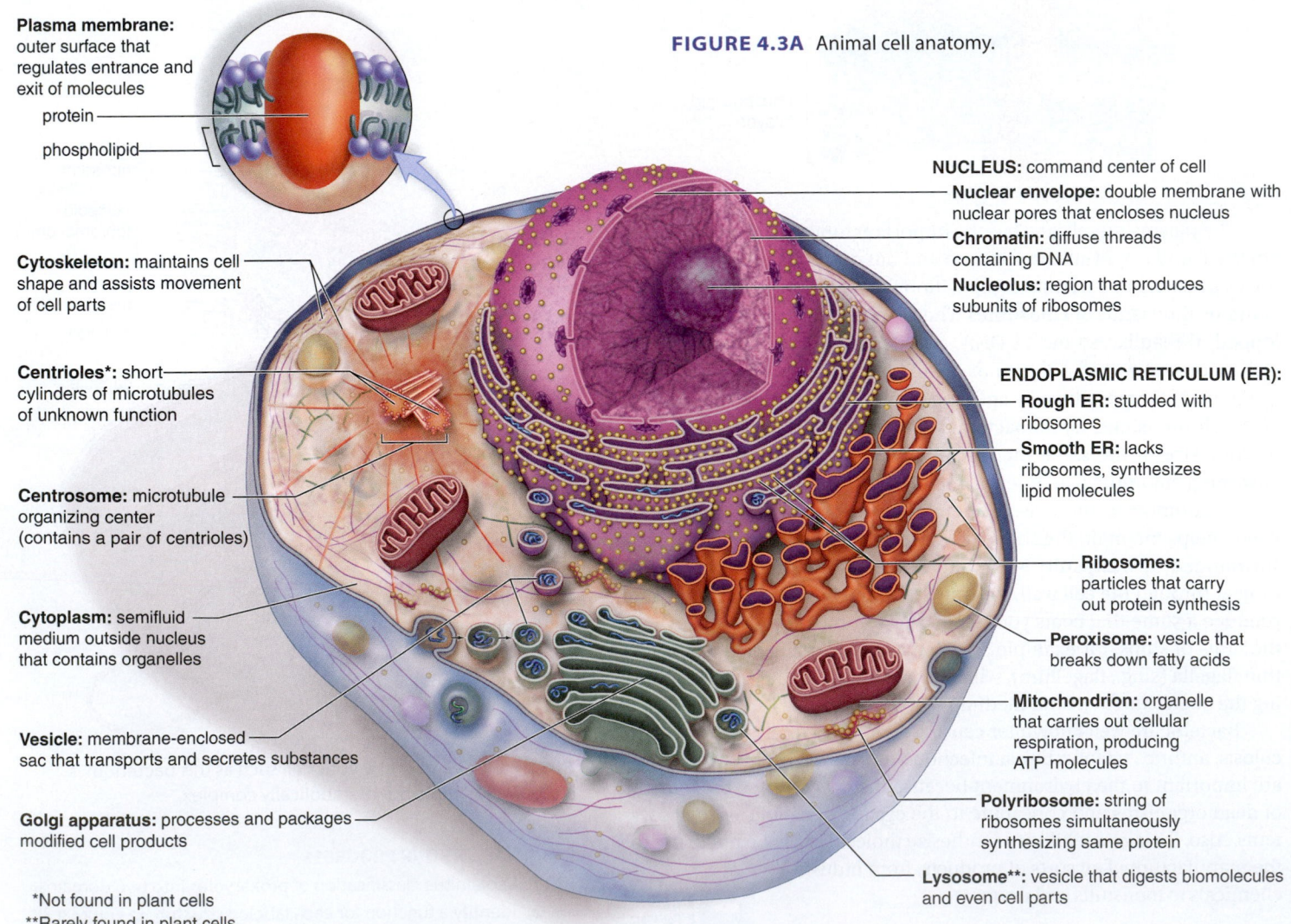

FIGURE 4.3A Animal cell anatomy.

Plasma membrane: outer surface that regulates entrance and exit of molecules

protein

phospholipid

Cytoskeleton: maintains cell shape and assists movement of cell parts

Centrioles*: short cylinders of microtubules of unknown function

Centrosome: microtubule organizing center (contains a pair of centrioles)

Cytoplasm: semifluid medium outside nucleus that contains organelles

Vesicle: membrane-enclosed sac that transports and secretes substances

Golgi apparatus: processes and packages modified cell products

NUCLEUS: command center of cell

Nuclear envelope: double membrane with nuclear pores that encloses nucleus

Chromatin: diffuse threads containing DNA

Nucleolus: region that produces subunits of ribosomes

ENDOPLASMIC RETICULUM (ER):

Rough ER: studded with ribosomes

Smooth ER: lacks ribosomes, synthesizes lipid molecules

Ribosomes: particles that carry out protein synthesis

Peroxisome: vesicle that breaks down fatty acids

Mitochondrion: organelle that carries out cellular respiration, producing ATP molecules

Polyribosome: string of ribosomes simultaneously synthesizing same protein

Lysosome:** vesicle that digests biomolecules and even cell parts

*Not found in plant cells
**Rarely found in plant cells

suited to their function were more likely to have surviving off-spring, and slowly over time all organisms of that group had such cells and organelles.

In this chapter we are going to concentrate on aspects of structure and function common to both animal and plant cells. Both types of cells have a nucleus that houses chromatin (DNA) and ribosomes that produce proteins in the same manner, for example. The fundamental aspects of cellular organization and function do not vary between the two types of cells. Still, we have an opportunity in this chapter to point out how the two types of cells differ as listed in Table 4.3. The cell wall of plants is covered in Chapter 5, which concerns the structure and function of cell surfaces. However, Table 4.3 will assist you in learning the major differences between animal and plant cells. The other cell structures (plasma membrane, nucleus, centrosome, endoplasmic reticulum, ribosomes, Golgi, peroxisomes, cytoskeleton) are present in both plant and animal cells.

The various cells in your body have the structures depicted in Figure 4.3A but many of your cells have additional structures and modifications to carry on particular functions. Similarly, the plants in your garden and the trees in your yard have cells with the structures shown in Figure 4.3B but they also have cells that are specialized in different ways. Multicellular organisms in particular have specialized cells and this leads to their diversity in form and capabilities.

TABLE 4.3	Animal and Plant Cell Differences	
Structure	**Animal Cell**	**Plant Cell**
Cell wall	No	Yes
Chloroplast	No	Yes
Lysosomes	Yes	Rarely present
Centrioles	Yes	No
Large central vacuole	No	Yes
Shape	Round	Rectangular

It's good to keep in mind as you study cell structures and their functions that despite their small size cells display all the characteristics of life we studied in Chapter 1. This chapter tells you how they can accomplish this feat.

▶ **4.3 CHECK YOUR PROGRESS**
1. Identify how a large eukaryotic cell benefits from having organelles.
2. Explain why you would expect plant cells to have chloroplasts and both plant and animal cells to have mitochondria.

FIGURE 4.3B Plant cell anatomy.

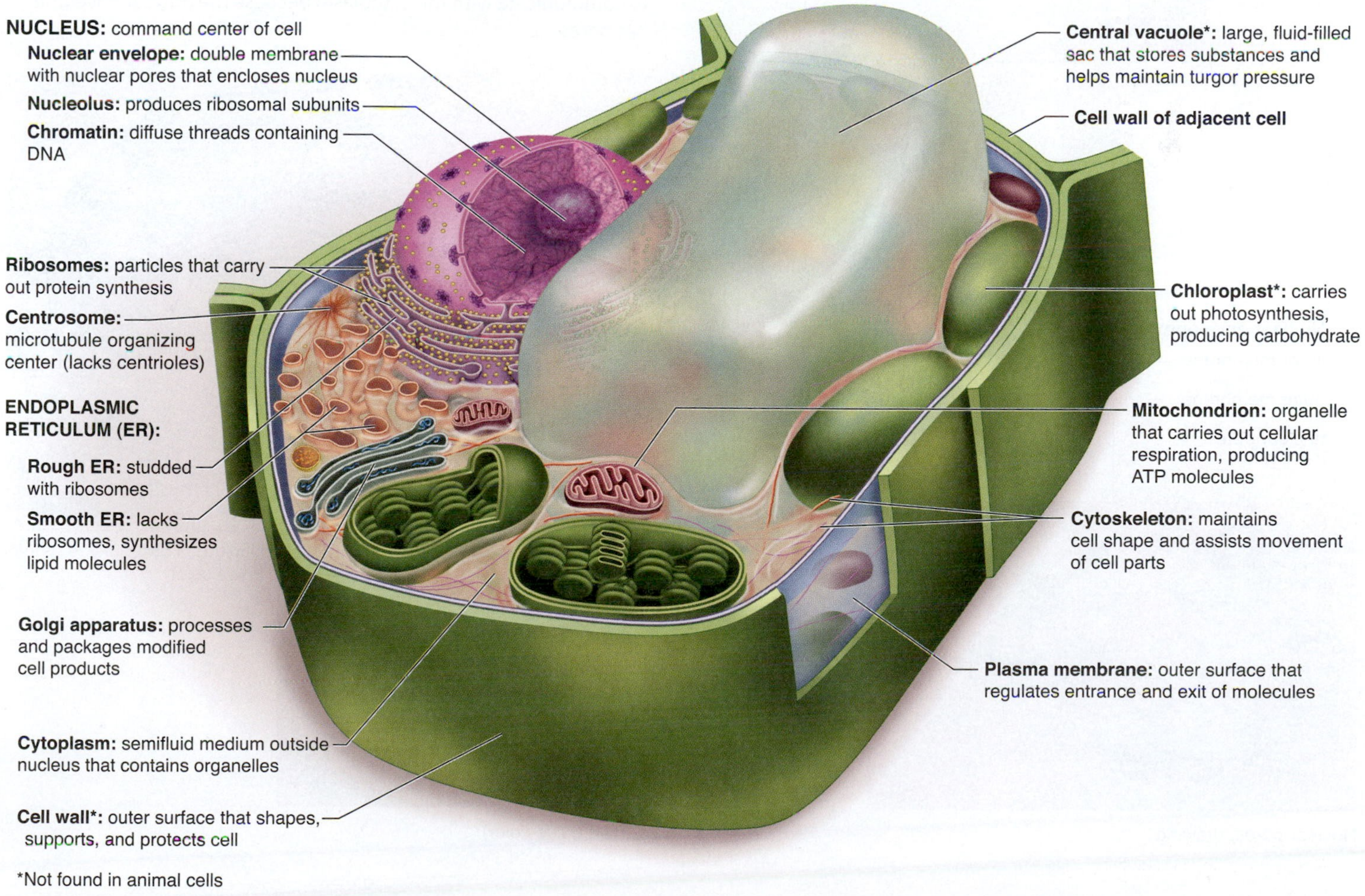

NUCLEUS: command center of cell
Nuclear envelope: double membrane with nuclear pores that encloses nucleus
Nucleolus: produces ribosomal subunits
Chromatin: diffuse threads containing DNA

Ribosomes: particles that carry out protein synthesis
Centrosome: microtubule organizing center (lacks centrioles)

ENDOPLASMIC RETICULUM (ER):
Rough ER: studded with ribosomes
Smooth ER: lacks ribosomes, synthesizes lipid molecules

Golgi apparatus: processes and packages modified cell products

Cytoplasm: semifluid medium outside nucleus that contains organelles

Cell wall*: outer surface that shapes, supports, and protects cell

Central vacuole*: large, fluid-filled sac that stores substances and helps maintain turgor pressure

Cell wall of adjacent cell

Chloroplast*: carries out photosynthesis, producing carbohydrate

Mitochondrion: organelle that carries out cellular respiration, producing ATP molecules

Cytoskeleton: maintains cell shape and assists movement of cell parts

Plasma membrane: outer surface that regulates entrance and exit of molecules

*Not found in animal cells

The Nucleus and the Endomembrane System

This part of the chapter discusses certain organelles of eukaryotic cells—namely, the nucleus, the ribosomes, the endoplasmic reticulum, and the Golgi apparatus, which are all involved in producing proteins that may serve necessary functions in the cell or may be secreted out of the cell.

4.4 The nucleus is a control center

LEARNING OUTCOMES

When you complete this section, you should be able to

1. Identify and give a function for each component of the nucleus.
2. Determine how the nucleus controls the functioning of the ribosomes.

The nucleus is a prominent structure in a eukaryotic cell (Fig. 4.4A). It generally has an oval shape and is located near the center of a cell. The nucleus contains DNA, the genetic material that is passed from cell to cell and from generation to generation. DNA dictates which proteins a cell is to synthesize and these proteins determine the cell's structure and functions; therefore, the nucleus is the command center of a cell.

At the time of cell division, DNA and proteins are organized into the several **chromosomes** of a eukaryotic cell. Following cell division, the chromosomes become extended into **chromatin**, which looks grainy, but actually is a network of fine strands. A **nucleolus** is a dark region of chromatin where the subunits of ribosomes are produced.

The nucleus is separated from the cytoplasm by a double membrane known as the **nuclear envelope.** Even so, the nucleus communicates with the cytoplasm. The nuclear envelope has **nuclear pores** of sufficient size to permit the passage of ribosomal subunits out of the nucleus into the cytoplasm, and the passage of proteins from the cytoplasm into the nucleus. High-power electron micrographs show nonmembranous components associated with the pores that form a nuclear pore complex.

The nuclear envelope is a part of an **endomembrane system,** which is composed of membranous structures that are either directly connected or communicate by way of transport vesicles.

FIGURE 4.4A The nucleus is the control center of the cell. It is able to communicate with the cytoplasm because the nuclear envelope has pores.

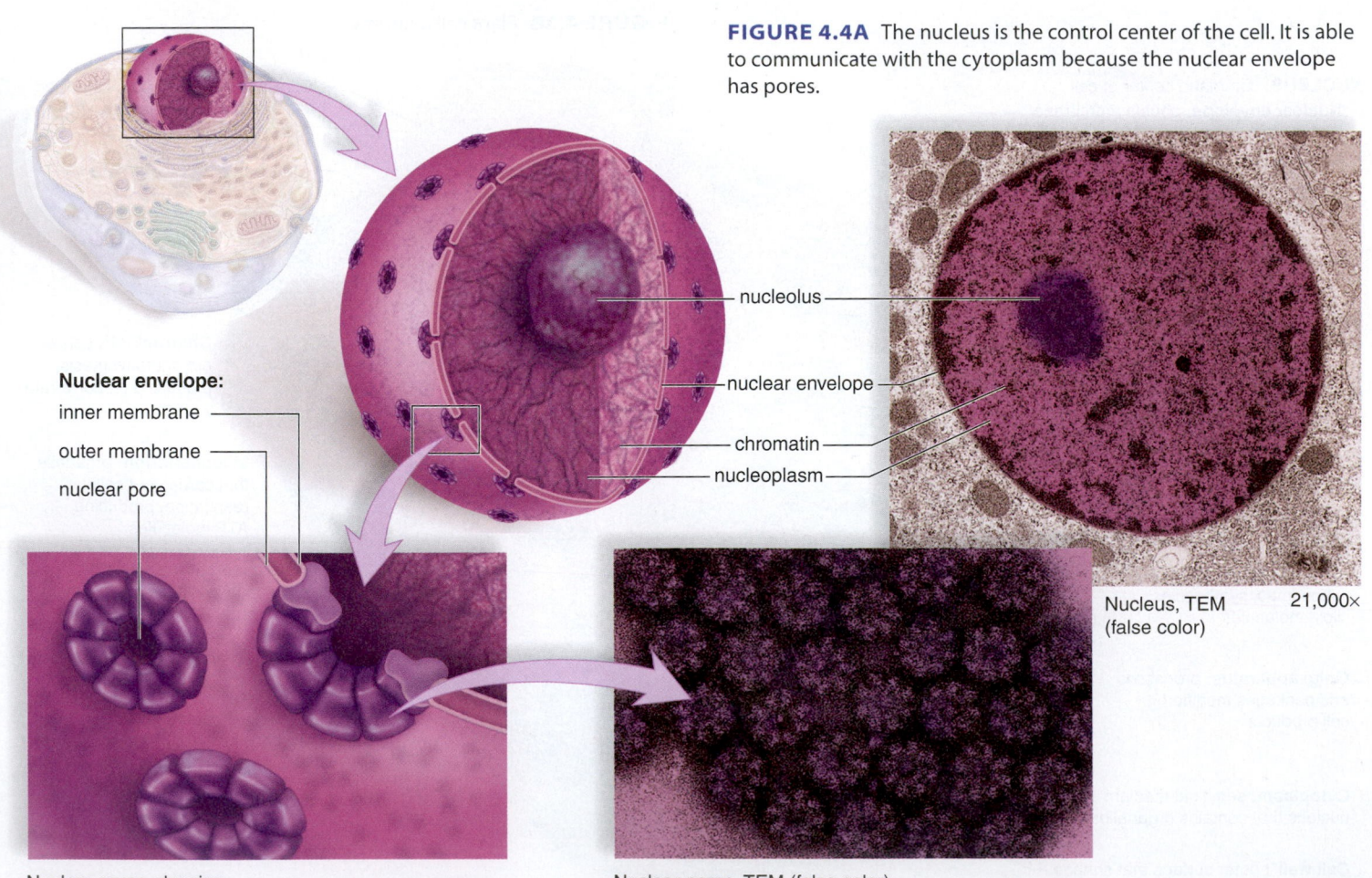

nucleolus

nuclear envelope

chromatin

nucleoplasm

Nuclear envelope:
inner membrane
outer membrane
nuclear pore

Nucleus, TEM   21,000×
(false color)

Nuclear pores, drawing

Nuclear pores, TEM (false color)

Ribosomes Ribosomes are particles that produce plentiful proteins. When you are going to make something, you usually need a surface on which to do your work. In the same manner, a cell uses ribosomes as a workbench for producing proteins.

Ribosomes are measured in nanometers, which means they are quite small; eukaryotic ribosomes are slightly larger than those in prokaryotes. In both types of cells, ribosomes are composed of two subunits, one large and one small.

In eukaryotic cells, some ribosomes occur freely within the cytoplasm, either singly or in groups called polyribosomes. Other ribosomes attach to the **endoplasmic reticulum** (**ER**), a membranous system of flattened saccules (small sacs) and tubules that is contiguous with the nuclear envelope (Fig. 4.4B, *left*).

The nucleus is the control center of the cell because the genes specify the sequence of amino acids in proteins. When a protein is needed, an RNA copy of a gene called messenger RNA (mRNA) leaves the nucleus by way of a nuclear pore and becomes attached to a ribosome. A ribosome uses the sequence of nucleotides in the mRNA as a code to produce a protein with the correct order of its amino acids (see page 200). Just as a dressmaker uses a pattern and directions to make a garment, so a ribosome uses the coded information provided by mRNA to produce a particular protein.

Attachment of Ribosomes to the ER Proteins produced by cytoplasmic ribosomes often enter a membranous organelle. Those produced by ribosomes attached to the ER end up in the interior of the ER. As shown in Figure 4.4B, *right*, ❶ after the mRNA leaves the nucleus and enters the cytoplasm, ❷ it becomes attached to a ribosome, and ❸ the ribosome becomes attached to the ER. The newly produced protein enters the interior of the ER. The protein folds into the correct shape inside the ER. Recall from Figure 3.8 that a protein can have up to four levels of organization. The shape of a protein is very important to its functioning appropriately.

▶ **4.4 CHECK YOUR PROGRESS**
1. How does the nuclear envelope permit RNA to exit the nucleus?
2. Identify the relationship between ribosomes and the nucleus and also the ER.

FIGURE 4.4B mRNA is formed in the nucleus of a gene and moves to cytoplasm where protein synthesis occurs at a ribosome. Attachment of ribosome to endoplasmic reticulum (ER) makes it rough ER.

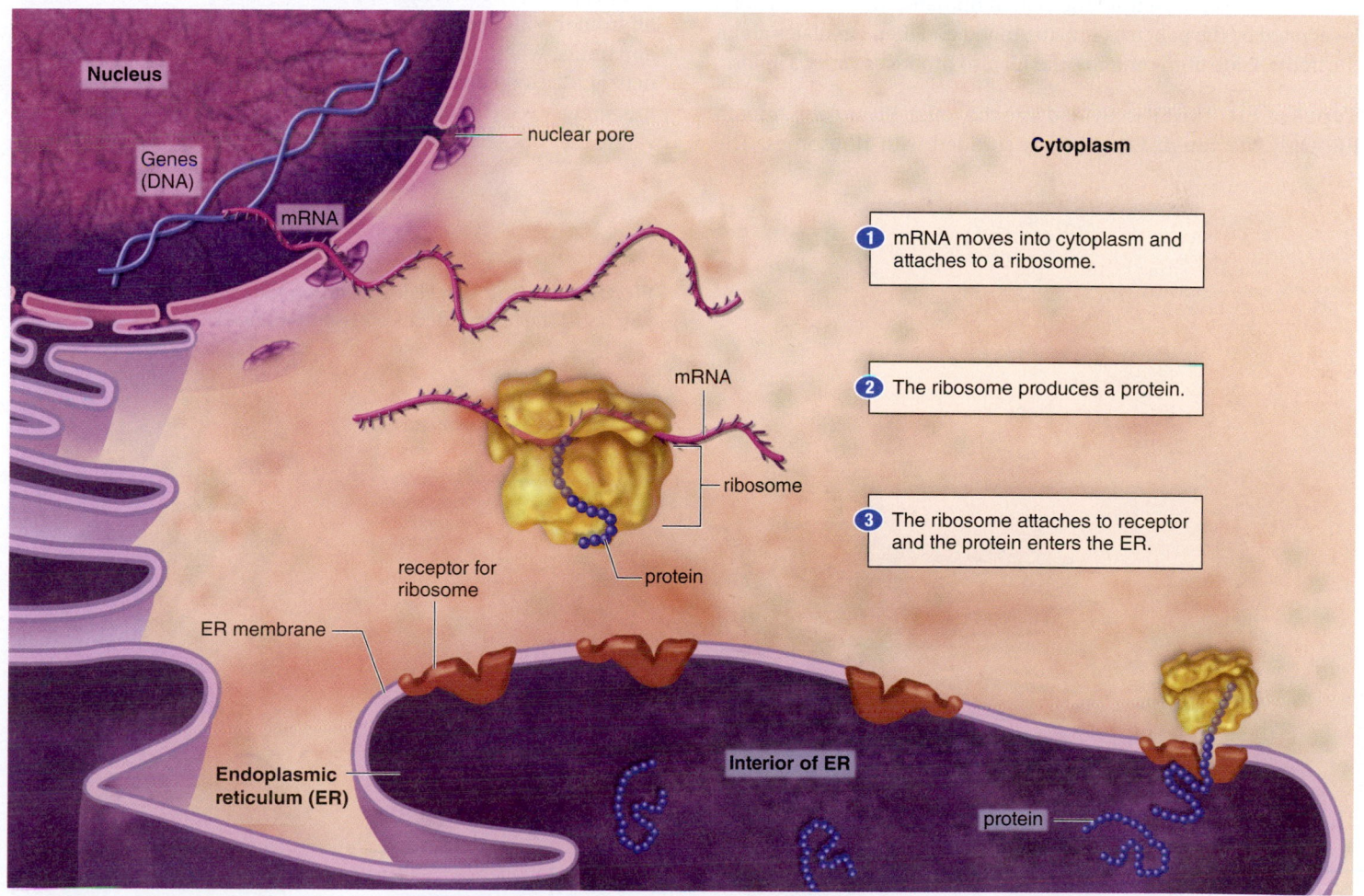

4.5 The ER produces and transports proteins and lipids to the Golgi apparatus

The term endoplasmic reticulum is a difficult one but becomes simpler if we break it down. *Endoplasmic* means "within the plasm" of the cell and *reticulum* is an elegant way of saying "network." Just as a long street can be lined by different neighborhoods and have different names according to the neighborhood, the outer membrane of nuclear envelope becomes the membrane of the ER. The membranous tubules and flattened sacs of the ER typically account for more than half of the total membrane within an average animal cell. Its twists and turns as it courses through the cytoplasm like a long snake enclose a single internal space. This space will be termed the interior of the ER (Fig. 4.4B). If you compare Figure 4.4B to Figure 4.5A, you can see that Figure 4.4B shows only a small portion of the ER found in a cell.

Because many ribosomes attach themselves to the ER, it becomes the location where all the proteins are produced for the many membranes inside a eukaryotic cell as well as most of the proteins that are secreted from the cell. In humans, the protein insulin is secreted by the pancreas into the blood and then circulates about the body. Aside from proteins, the ER also produces various lipids.

Types of ER The ER is divided into the rough ER and the smooth ER. Only the **rough ER** (**RER**) is studded with ribosomes. The ribosomes are attached to the side of the membrane that faces the cytoplasm. Figure 4.4B shows how a protein enters the interior of the ER from an attached ribosome. Once inside, a protein undergoes the process of folding into its final shape.

Smooth ER (**SER**), which is continuous with rough ER, does not have attached ribosomes. Therefore, it has a smooth appearance in electron micrographs and more important it does not participate in protein production. Smooth ER is abundant in gland cells, where it synthesizes lipids of various types. For example, cells that synthesize steroid hormones from cholesterol have much SER. In the liver, SER, among other functions, adds lipid to proteins, forming the lipoproteins that carry cholesterol in the blood. Also, the SER of the liver increases in quantity when a person consumes alcohol or takes barbiturates on a regular basis, because SER contains the enzymes that detoxify these molecules.

The RER and SER, working together, produce membrane, which is composed of phospholipids and various types of proteins, including those that have carbohydrate chains (see Fig. 5.1A). Proteins to be secreted from the cell remain in the interior of the ER, but the ones destined to become membrane constituents become embedded in its membrane. Because the ER produces membrane, it can form the transport vesicles by which it communicates with the Golgi apparatus. **Transport vesicles** pinch off from the ER and carry protein and lipids, notably to the Golgi apparatus, where they undergo further modification. The products of the Golgi apparatus are utilized by the cell or repackaged in secretory vesicles that make their way to the plasma membrane where they are secreted (see Fig. 4.5B).

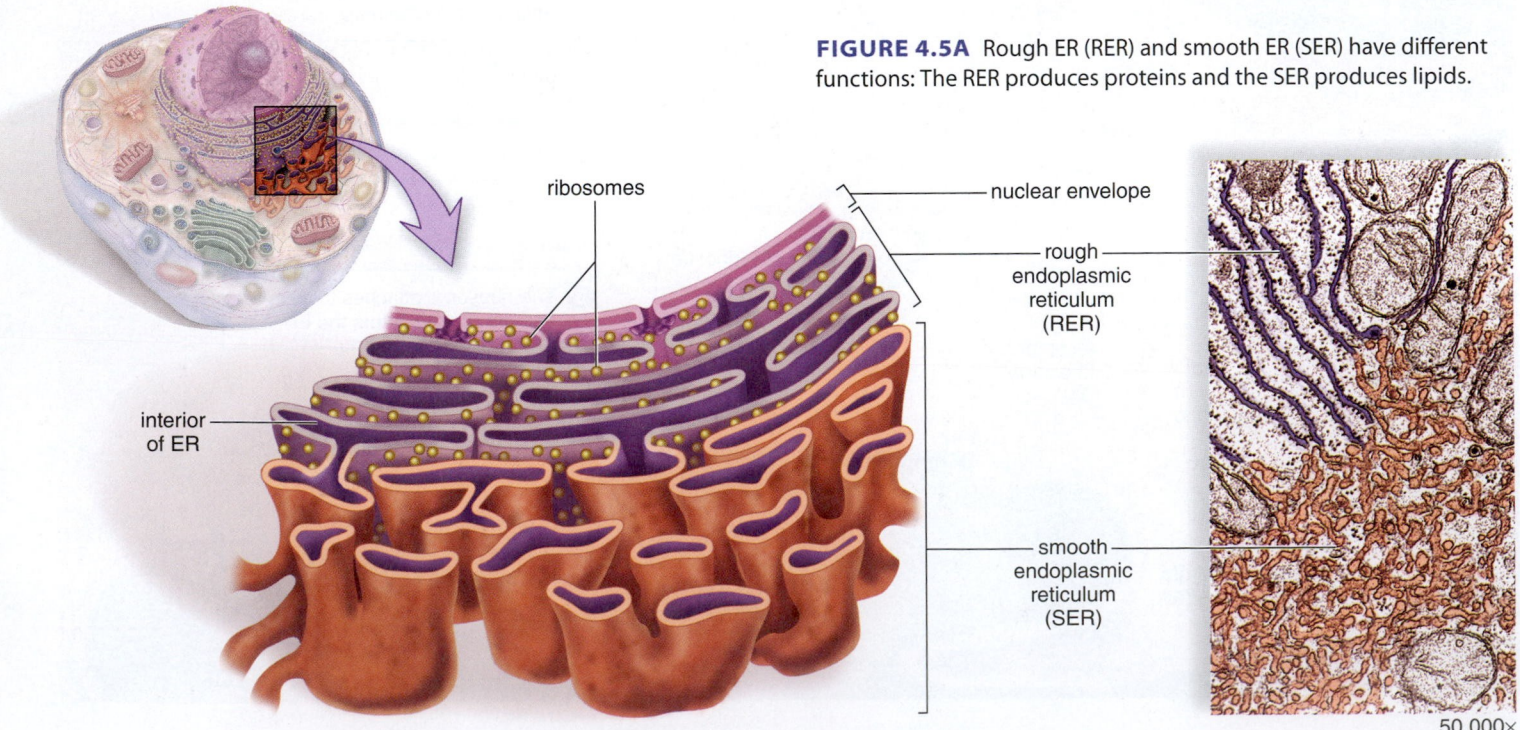

FIGURE 4.5A Rough ER (RER) and smooth ER (SER) have different functions: The RER produces proteins and the SER produces lipids.

ribosomes

interior of ER

nuclear envelope

rough endoplasmic reticulum (RER)

smooth endoplasmic reticulum (SER)

50,000×

Golgi Apparatus The **Golgi apparatus** is named for Camillo Golgi, who discovered its presence in cells in 1898. The Golgi apparatus, or simply the Golgi, typically consists of a stack of three to twenty slightly curved, flattened saccules whose appearance can be compared to a stack of pancakes (Fig. 4.5B). One side of the stack is directed toward the ER, and the other side is directed toward the plasma membrane. Vesicles can frequently be seen at the edges of the saccules.

The Golgi receives, processes, and packages proteins and lipids, so that they may be sent to their final destination in the cell. In particular, it readies proteins for secretion. Protein-filled vesicles that bud from the rough ER and lipid-filled vesicles that bud from the smooth ER are received by the Golgi. Thereafter, the Golgi alters these substances as they move through its saccules. For example, proteins may have attached carbohydrate chains and the Golgi contains enzymes that modify these chains. In some cases, the modified carbohydrate chain serves as a signal molecule that determines the protein's final destination in the cell.

The Golgi sorts and packages proteins and lipids in vesicles that carry them to their final destination. Secretory vesicles proceed to the plasma membrane, where they stay until a signal molecule triggers the cell to release them. Then their membrane becomes part of the plasma membrane as they discharge their contents during **secretion.** Digestive enzymes, for example, are secreted into a tube that carries them to the digestive tract, where they break down food to nutrient molecules.

Lysosomes are another type of vesicle produced by the Golgi apparatus. They have a very low internal pH and contain powerful hydrolytic digestive enzymes. Lysosomes have significant digestive functions inside the cell. For example, they are important in recycling cellular material and digesting worn-out organelles, such as old mitochondria (Fig. 4.5C).

Sometimes biomolecules are engulfed (brought into a cell by vesicle formation) at the plasma membrane. When a lysosome fuses with such a vesicle, its contents are digested by lysosomal enzymes into simpler subunits that then enter the cytoplasm. Some white blood cells defend the body by engulfing bacteria, which are then enclosed within vesicles. When lysosomes fuse with these vesicles, the bacteria are digested.

A cell can have dozens of lysosomes and this perhaps suggests their importance in helping to maintain homeostasis inside a cell. Unfortunately there are many different types of lysosomal storage diseases, so called because a molecule builds up in a lysosome because the enzyme needed to break it down is absent or nonfunctional. Tay-Sachs disease is one such condition, in which a newborn appears healthy but then gradually becomes nonresponsive, deaf, and blind before dying within a few months. The brain cells are filled with particles containing a type of lipid that cannot be digested by lysosomes.

Animation
Lysosomes

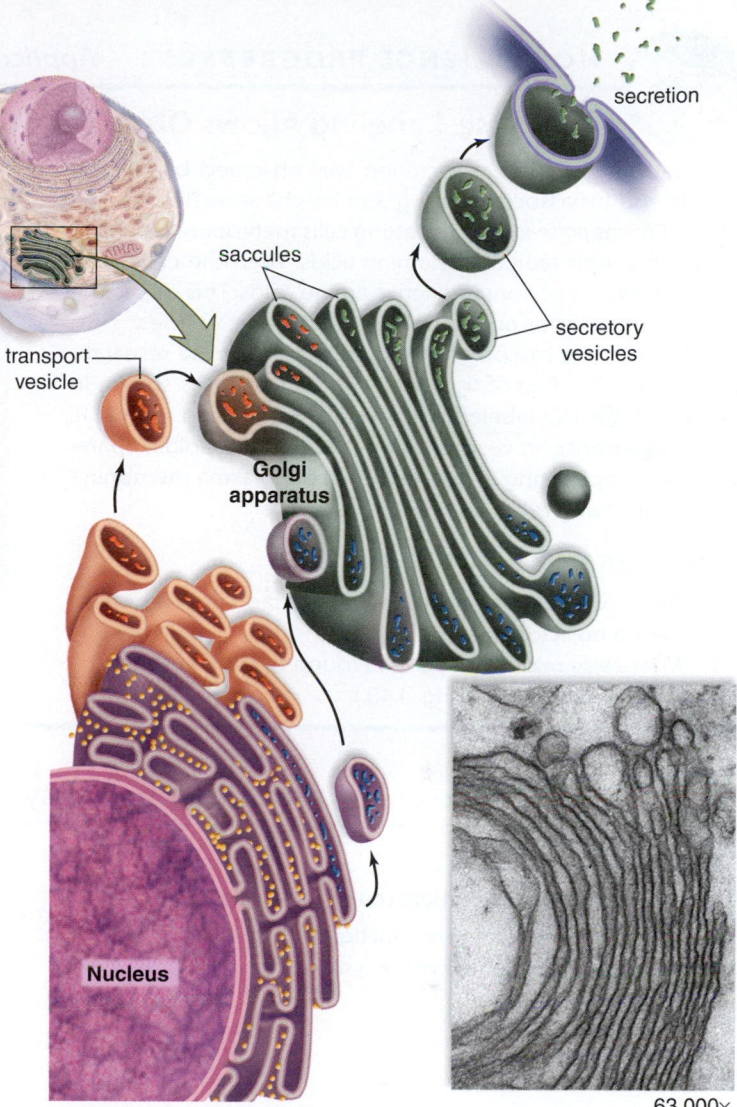

transport vesicle

saccules

secretion

secretory vesicles

Golgi apparatus

Nucleus

63,000×

FIGURE 4.5B The Golgi apparatus receives molecules for modification in vesicles from both rough and smooth ER. After processing these molecules, it packages them for secretion at the plasma membrane.

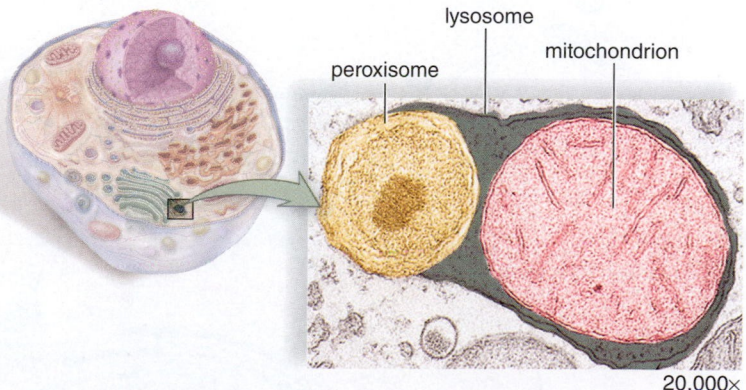

peroxisome

lysosome

mitochondrion

20,000×

FIGURE 4.5C Lysosomes fuse with incoming vesicles and digest their contents. They also take in and destroy spent organelles, as shown here.

▶ **4.5 CHECK YOUR PROGRESS**

1. Identify the path of a protein from production to secretion from a cell.
2. Contrast the roles of lysosomes and secretory vesicles in cells and suggest a possible way the Golgi is able to produce both types of vesicles.

4B Pulse-Labeling Allows Observation of the Secretory Pathway

The pathway of protein secretion was observed by George Palade and his associates using a pulse-chase technique. The rough ER was *pulse-labeled* by letting cells metabolize for a very short time with radioactive amino acids. Then the cells were given an excess of nonradioactive amino acids. This *chased* the labeled amino acids out of the ER into transport vesicles.

Electron microscopy techniques allowed these researchers to trace the fate of the labeled amino acids, as shown in Figure 4B: ❶ The labeled amino acids were found in the ER, then in ❷ transport vesicles, and then in ❸ the Golgi apparatus, before appearing in ❹ vesicles at the plasma membrane and finally being released.

CONSIDER THESE QUESTIONS

1. Why would Palade have labeled sulfur and not carbon in the amino acids? (See Fig. 3.7A.)
2. Where else might Palade have found the labeled amino acids in the cell? (See Fig. 4.4B.)

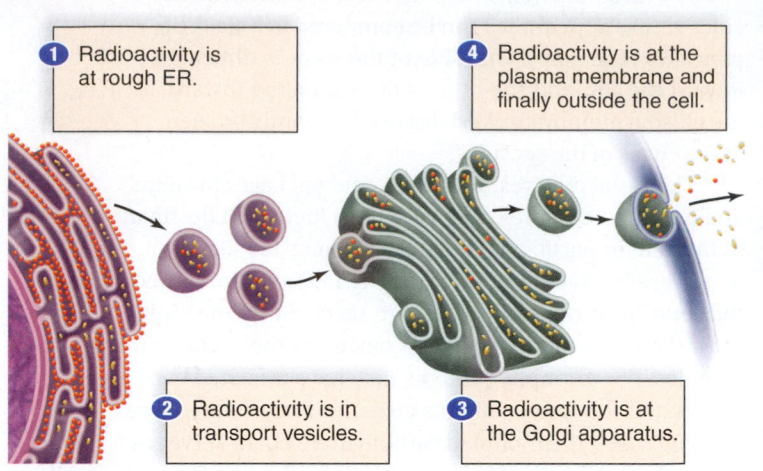

❶ Radioactivity is at rough ER.

❹ Radioactivity is at the plasma membrane and finally outside the cell.

❷ Radioactivity is in transport vesicles.

❸ Radioactivity is at the Golgi apparatus.

FIGURE 4B The secretory pathway.

4.6 The organelles of the endomembrane system work together

LEARNING OUTCOME

When you complete this section, you should be able to

1. Describe the structure and function of the organelles that belong to the endomembrane system.

The **endomembrane system** includes the nuclear envelope, the endoplasmic reticulum (ER), the Golgi apparatus, lysosomes, and the transport vesicles.

Figure 4.6 shows how the components of the endomembrane system work together: ❶ In the secretory pathway, proteins produced in the rough ER are carried in ❷ transport vesicles to ❸ the Golgi apparatus, which sorts the proteins and packages them into vesicles that transport them to various cellular destinations.

❹ Secretory vesicles take the proteins to the plasma membrane, where they exit the cell when the vesicles fuse with the membrane. For example, secretion into ducts occurs when the salivary glands produce saliva or when the pancreas produces digestive enzymes. Similarly, lipids move from the smooth ER to the Golgi apparatus and can eventually be secreted.

❺ In animal cells, a lysosomal pathway occurs. Lysosomes produced by the Golgi apparatus ❻ fuse with incoming vesicles from the plasma membrane and digest biomolecules and debris. White blood cells are well known for engulfing pathogens (e.g., disease-causing viruses and bacteria) that are then broken down in lysosomes.

FIGURE 4.6 The organelles of the endomembrane system.

secretion

plasma membrane

incoming vesicle

incoming vesicle

❻

lysosome

❺

transport vesicle

lipid

smooth endoplasmic reticulum

ribosome

Nucleus

❹ **secretory vesicle**

❸ **Golgi apparatus**

protein

❷ **transport vesicle**

❶ **rough endoplasmic reticulum**

▶ **4.6 CHECK YOUR PROGRESS**

1. Identify which organelles in the endomembrane system produce, modify, or break down a biomolecule.

Vacuoles and Vesicles

Cells have various membranous sacs that look the same in electron micrographs but have different functions. Lysosomes, as discussed previously, contain powerful hydrolytic enzymes that digest biomolecules and even cell parts. Peroxisomes are more specialized and assist mitochondria by breaking down lipids, among other functions. Some of the vacuoles in protists and plants are unique to them and not found in other eukaryotes.

4.7 Vacuoles are common in plant cells

LEARNING OUTCOME

When you complete this section, you should be able to

1. Contrast the function of vacuoles in protists with their function in plants.

Like vesicles, **vacuoles** are membranous sacs, but vacuoles are larger than vesicles. The vacuoles of some protists are quite specialized, including contractile vacuoles for ridding the cell of excess water and digestive vacuoles for breaking down nutrients. Vacuoles usually store substances. Few animal cells contain vacuoles, but fat cells contain a very large lipid-engorged vacuole that takes up nearly two-thirds of the volume of the cell!

Plant vacuoles contain not only water, sugars, and salts but also water-soluble pigments and toxic molecules. The pigments are responsible for many of the red, blue, or purple colors of flowers and some leaves. The toxic substances help protect a land plant from feeding insects.

Typically, plant cells have a large **central vacuole** that may take up to 90% of the volume of the cell. The vacuole is filled with a watery fluid called cell sap that gives added support to the cell (Fig. 4.7). **Turgor pressure,** which is the pressure of cell contents against the plant cell wall, is maintained by the central vacuole. A plant cell can rapidly increase in size by enlarging its vacuole. Eventually, a plant cell also produces more cytoplasm.

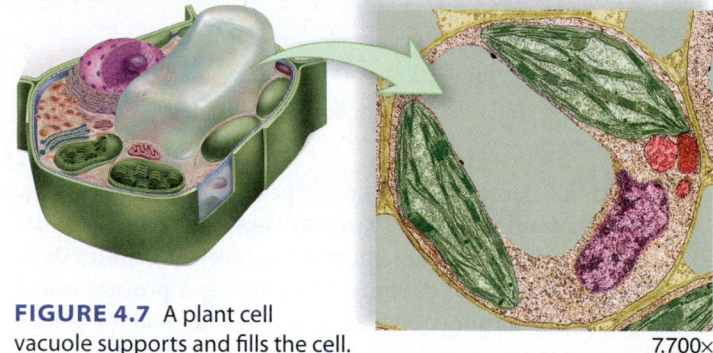

FIGURE 4.7 A plant cell vacuole supports and fills the cell.

7,700×

The central vacuole functions in storage of both nutrients and wastes. Metabolic waste products are pumped across the vacuole membrane and stored permanently in the central vacuole. As organelles age and become nonfunctional, they fuse with the vacuole, where digestive enzymes break them down. This is a function carried out by lysosomes in animal cells.

▶ **4.7 CHECK YOUR PROGRESS**
1. Explain why the central vacuole of plants is more than a bag of water.

4.8 Peroxisomes have many different functions

LEARNING OUTCOME

When you complete this section, you should be able to

1. Discuss in general the functions of peroxisomes.

Peroxisomes, similar to lysosomes, are membrane-enclosed vesicles that contain enzymes. A concentration of enzymes resulting in a protein crystal is characteristic of peroxisomes (Fig. 4.8). Peroxisomes originate at the ER and thereafter they import enzymes from the cytoplasm. Some of these enzymes synthesize needed substances and several of them break down toxic substances. All peroxisomes contain enzymes whose actions result in hydrogen peroxide (H_2O_2):

$$RH_2 + O_2 \longrightarrow R + H_2O_2$$

Hydrogen peroxide, a toxic molecule, is immediately broken down to water and oxygen by another peroxisomal enzyme called catalase. When hydrogen peroxide is applied to a wound, bubbling occurs as catalase breaks it down.

Peroxisomes have varied functions but are especially prevalent in cells that are synthesizing and breaking down lipids. In the liver, some peroxisomes produce bile salts from cholesterol, and others break down fats. In a 1992 movie, *Lorenzo's Oil,* the peroxisomes in a boy's cells lack a membrane protein needed to import a specific enzyme and/or long-chain fatty acids from the cytoplasm. As a result, long-chain fatty acids accumulate in his

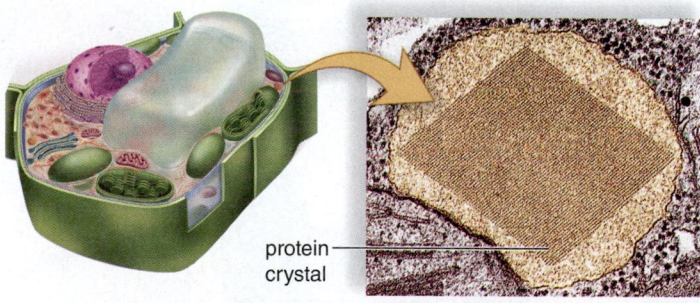

protein crystal

155,000×

FIGURE 4.8 Peroxisomes contain so many enzymes, they sometimes form a protein crystal.

brain, and he suffers neurological damage. This disorder is known as adrenoleukodystrophy.

Plant cells also have peroxisomes. In germinating seeds, they oxidize fatty acids into molecules that can be converted to sugars needed by the growing plant. In leaves, peroxisomes can assist photosynthesis.

▶ **4.8 CHECK YOUR PROGRESS**
1. Identify a function that would immediately distinguish peroxisomes from lysosomes.

Chloroplasts and Mitochondria

Chloroplasts transform solar energy into the energy of carbohydrates, which serve as organic food for themselves and all organisms in the biosphere. Mitochondria transform the energy of carbohydrates to that of ATP molecules. All cells use ATP molecules as a source of energy for metabolic reactions and processes.

4.9 Chloroplasts and mitochondria have opposite functions

LEARNING OUTCOME

When you complete this section, you should be able to

1. Contrast the structure and function of chloroplasts with that of mitochondria.

We learned in Chapter 1 that all organisms must acquire energy and nutrients from their environment. Plants, however, can use solar energy and the inorganic nutrients water and carbon dioxide to make energy-rich carbohydrates during a process called **photosynthesis.** Carbohydrates serve as organic food for plants; therefore, we say that plants make their own food. Photosynthesis not only takes in carbon dioxide but also releases oxygen.

Plants can photosynthesize because their cells contain organelles called **chloroplasts.** Each plant cell may contain as many as 100 chloroplasts, and a square millimeter of leaf can contain up to 500,000 chloroplasts (Fig. 4.9). The green pigment chlorophyll, and other pigments as well, are responsible for the ability of chloroplasts to absorb solar energy. Within a chloroplast, chlorophyll is located in the membrane of flattened sacs called thylakoids.

Chloroplasts are of great significance to the biosphere, including humans, because they are the ultimate source of all food for organisms. Consider that you either feed directly on plants or on animals that have fed on plants. Another source of food for the biosphere is carbohydrates made by cyanobacteria and algae, because they also use pigments to absorb solar energy and photosynthesize in the same manner as plants.

How would you know that chloroplasts produce carbohydrates when the sun is shining? One way is to look for starch grains to accumulate in plant cells when the sun is out. Set a plant in the dark and the starch grains disappear.

In contrast to chloroplasts, nearly all organisms and types of cells, including both plant and animal cells, contain mitochondria (Fig. 4.9). **Mitochondria** are indispensable to cells because they carry on **cellular respiration,** the process that transforms the energy of carbohydrates to that of ATP molecules. It's called cellular respiration because mitochondria take in oxygen and give off carbon dioxide. Because mitochondria produce ATP, they are called the powerhouse of a cell.

Cellular respiration and photosynthesis are opposite reactions:

carbohydrate + oxygen $\leftrightarrows$ carbon dioxide + water + energy

For cellular respiration, read left to right and replace energy with ATP. For photosynthesis, read right to left and replace energy with solar energy.

Cells use ATP, not glucose, as a direct source of metabolic energy—using a molecule of glucose would be energy-inefficient and wasteful. You use change, not a dollar bill, to buy something that costs five cents. In the same manner, an organism converts

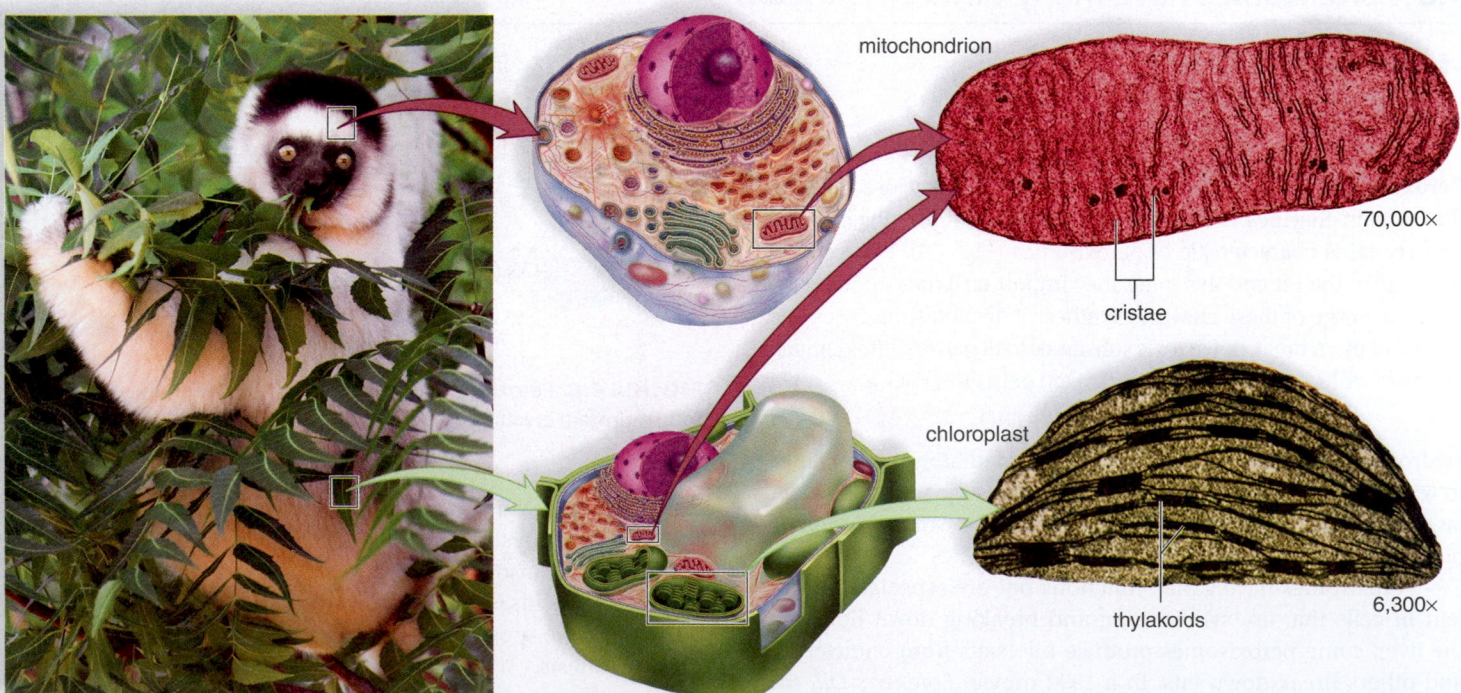

FIGURE 4.9 Plant cells carry on photosynthesis in green leaves where chloroplasts absorb solar energy because they contain the green pigment chlorophyll in thylakoid membranes. Mitochondria in plant and animal cells carry on cellular respiration, a process that produces ATP on the membranous invaginations called cristae.

carbohydrates to many molecules of ATP and uses them as a source of energy for individual reactions, such as linking amino acids during protein synthesis. Mitochondria are most abundant in human cells that carry out energy-intensive activities. For example, ATP provides the energy for muscle contraction and nerve conduction. Mitochondria are not as complex as chloroplasts but their inner membrane does fold back and forth, forming the cristae that act as shelves where ATP is formed. ATP exits mitochondria and enters the cytoplasm where it is utilized as an immediate energy source.

It is of great interest to scientists that both chloroplasts and mitochondria provide evidence that they were once free-living prokaryotes. For example, they have their own DNA in a nucleoid region, and they make some of their own proteins. For a more thorough discussion of this topic, see "How the Eukaryotic Cell Evolved" on this page.

▶ **4.9 CHECK YOUR PROGRESS**
   1. Describe how it is possible to trace the energy of the sun to ATP molecules.

HOW LIFE CHANGES *Application*

4C How the Eukaryotic Cell Evolved

Life's history is written in the fossil record, which includes the remains of past life, often encased by stone (see Section 16.1). The fossil record tells us that the prokaryotic cell was present about 3.5 BYA (billion years ago); the eukaryotic cell evolved in stages (Fig. 4C). The nuclear envelope and nucleus may have arisen around 2 BYA from an infolding of the plasma membrane, but what about the organelles such as mitochondria and chloroplasts? Much evidence supports the proposal that these organelles were once free-living prokaryotes that were either prey to or parasites of a eukaryotic cell. Their outer double membrane tells us that the eukaryotic cell engulfed them—the outer membrane is derived from the host plasma membrane, and the inner membrane was their own outer surface. By now, they are endosymbionts—organisms that live inside a host cell and are indispensable to their host, the cell that engulfed them.

Mitochondria and chloroplasts have their own DNA—circular like that of a prokaryote—and they carry on protein synthesis in the same manner as bacteria. Then, too, they reproduce by splitting as do bacteria, and their reproduction occurs independently of host cell reproduction.

The theory of endosymbiosis explains that because all cells have mitochondria, aerobic bacteria entered the host cell first, perhaps just when oxygen began to rise in the atmosphere due to the advent of photosynthesis by free-living cyanobacteria (see Fig. 17.11C). A host cell with an endosymbiont that used oxygen and produced ATP molecules would have been a distinct evolutionary advantage. Later, a cyanobacterium entered certain cells, and these cells became capable of photosynthesis. Being able to make your own food does away with the need to find it elsewhere. Eventually the relationship between host cells and endosymbionts became so beneficial that by now they cannot live separately from one another!

Animation
Endosymbiosis

CONSIDER THESE QUESTIONS

   1. Explain the phraseology "the host had an evolutionary advantage." Be sure to mention comparative number of offspring in your explanation.
   2. If you compared the structure of a cyanobacterium with that of a chloroplast, what similarities would you expect to find?

 Explore the concepts through a variety of multimedia
|BIOLOGY assets and question types.
www.mcgrawhillconnect.com

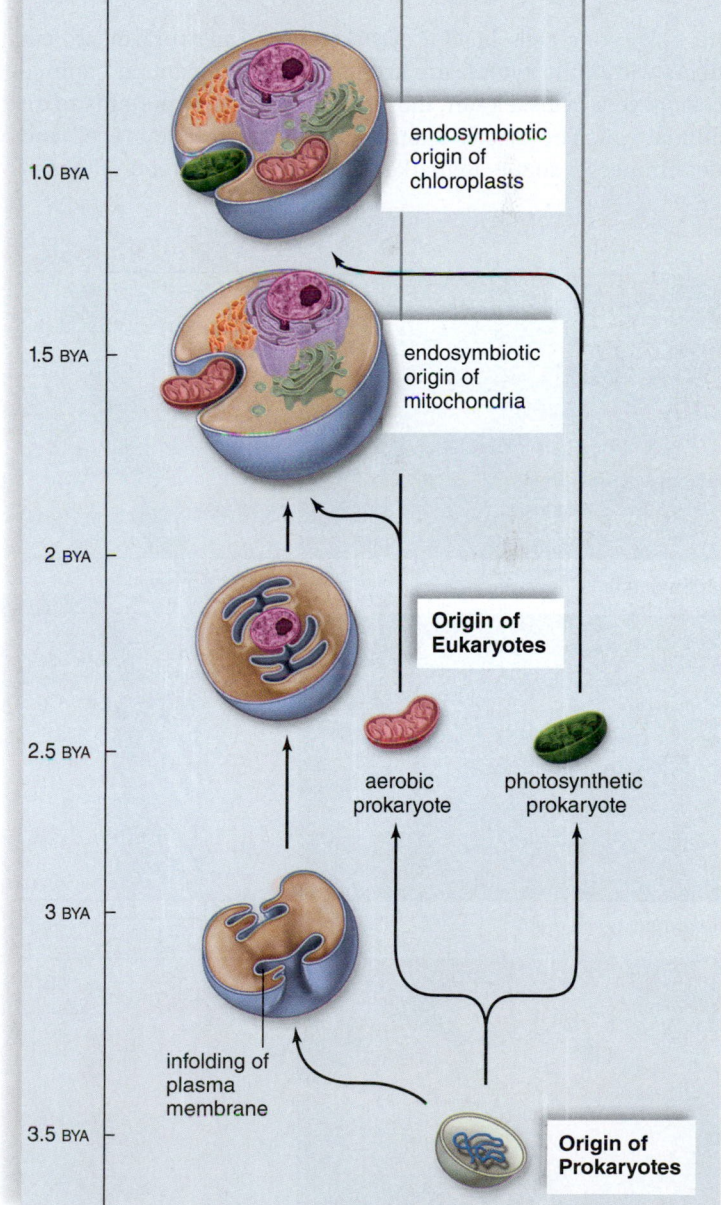

FIGURE 4C The eukaryotic cell was fully formed when a nucleated cell engulfed prokaryotes that became endosymbionts.

The Cytoskeleton and Cell Movement

As you know, bones and muscles give an animal structure and produce movement. Similarly, the fibers of the cytoskeleton maintain cell shape and cause the cell and its organelles to move. Cilia and flagella are also instrumental in producing movement, so they are included in this part of the chapter as well.

4.10 The cytoskeleton maintains cell shape and assists movement

LEARNING OUTCOMES

When you complete this section, you should be able to

1. Identify and contrast the structure and function of cytoskeletal fibers and flagella.
2. Describe the possible relationship between centrioles and flagella.
3. Understand that motor molecules enable cell parts and also cilia and flagella to move.

All eukaryotic cells have a **cytoskeleton,** a network of protein fibers within the cytoplasm. Even though both plant and animal cells have a cytoskeleton, most of our discussion pertains to an animal cell. The cytoskeleton (1) supports the animal cell and determines its shape. Remarkably, however, the protein fibers of the cytoskeleton can assemble and disassemble their subunits rapidly, and this accounts for why the shape of some animal cells can change from moment to moment. Similarly, the cytoskeleton (2) anchors the organelles in place but can also allow them to move, as when a vesicle moves from the Golgi to the plasma membrane. Because the cytoskeleton has the dual function of support and movement, it is appropriately described as the "skeleton and muscles" of an animal cell.

If you could look closely at the cytoskeleton in an animal cell, you would note the three different types of fibers shown in Figure 4.10A. Because you can't see these fibers with a light microscope, scientists prepare fluorescent antibodies, each of which attaches to only one type of fiber and then they photograph the cells under fluorescent light. Actin filaments and microtubules have binding sites for **motor molecules,** which are proteins able to break down ATP and use the resulting energy to change shape in order to move from one binding site to the next. Motor molecules are either kinesins, dyneins, or myosins. As they move along a cytoskeletal fiber, they pull their cargo (lysosomes, chromosomes, vesicles, etc.) with them.

Actin filaments are so named because they contain two twisted strands of actin, a fibrous protein. Bundles of actin filaments support the plasma membrane and other structures, such as the microvilli (short projections) of intestinal cells. However, you can primarily associate actin filaments with movement. For example, actin interacts with myosin when muscle contraction occurs. Just as you do when participating in a tug of war, a myosin head attaches, detaches, and reattaches further along an actin filament, and this pulls the actin filament in the opposite direction:

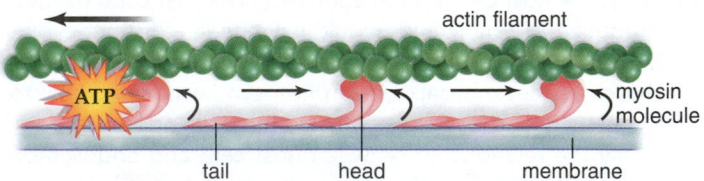

Also in conjunction with myosin, actin filaments act like purse strings to pinch off and separate cells during cell division. Then, too, they are responsible for the ability of white blood cells to crawl and to engulf disease-causing agents such as viruses and bacteria.

Intermediate filaments have a diameter that is intermediate between actin filaments and microtubules. Although the specific protein composition varies with the type of cell, intermediate filaments always have a ropelike structure that provides mechanical strength. Some intermediate filaments support the nuclear envelope, whereas others support the plasma membrane and take part in the formation of cell-to-cell junctions. Intermediate filaments made of the protein keratin strengthen skin cells.

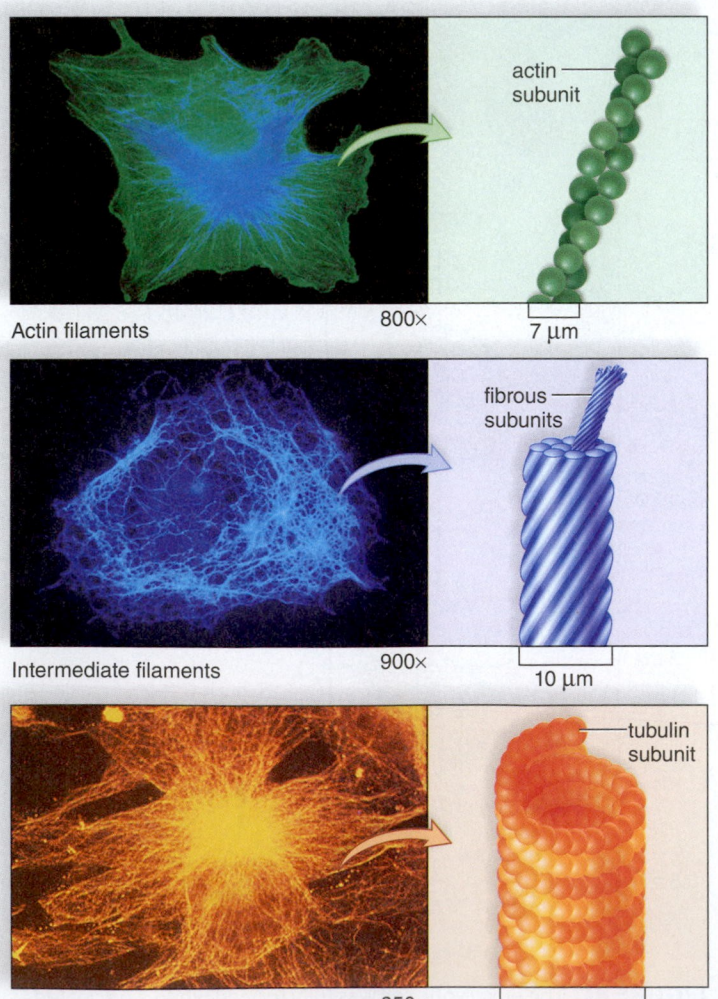

FIGURE 4.10A The cytoskeleton contains these three types of fibers that support the cell.

Microtubules are cylindrical structures composed of 13 rows of a protein called tubulin. Assembly is under the control of a microtubule organizing center (MTOC) located in the **centrosome** (see Fig. 4.3A). Microtubules radiate from the centrosome, helping to maintain the shape of the cell and acting as tracks along which organelles can move (Fig. 4.10B). Before a cell divides, microtubules disassemble and then reassemble into a structure called a spindle, which distributes chromosomes in an orderly manner.

Cilia and Flagella Cilia and flagella (sing., cilium, flagellum) are whiplike projections of cells. **Cilia** move stiffly, like an oar, and **flagella** move in an undulating, snakelike fashion. Cilia are short (2–10 μm), and flagella are longer (usually no more than 200 μm). Unicellular protists utilize cilia or flagella to move about. The ciliated cells that line our respiratory tract sweep debris trapped within mucus back up into the throat, which helps keep the lungs clean. Similarly, ciliated cells move an egg along the oviduct, where it can be fertilized by a flagellated sperm cell (Fig. 4.10C, *left*).

A cilium and a flagellum have the same organization of microtubules within a plasma membrane covering (Fig. 4.10C, *right*). Dynein side arms, powered by ATP, allow the microtubules in cilia and flagella to interact and bend, and thereby to move.

A particular genetic disorder illustrates the importance of normal cilia and flagella. Some individuals have an inherited defect that leads to malformed microtubules in cilia and flagella. Not surprisingly, they suffer from recurrent and severe respiratory infections, because the ciliated cells lining their respiratory passages fail to keep their lungs clean. They are also infertile due to the lack of ciliary action to move the egg in a female, or the lack of flagellar action by sperm in a male.

Centrioles Located in the centrosome, **centrioles** are short, barrel-shaped organelles composed of microtubules. It's possible that centrioles give rise to **basal bodies,** which lie at the base of

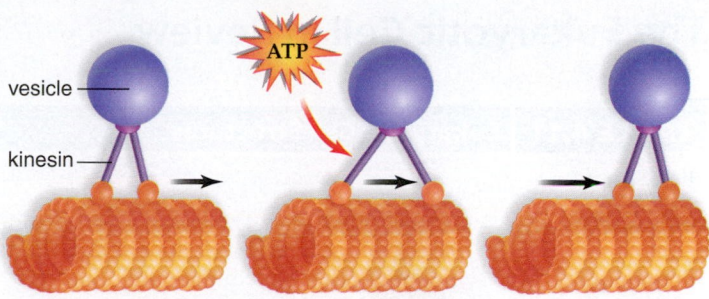

vesicle moves, not microtubule

FIGURE 4.10B The motor molecule kinesin is moving a vesicle along a microtubule track.

and organize the microtubules in cilia and flagella (Fig. 4.10C, *far right*). It's also possible that centrioles help organize the spindle, mentioned earlier, which is so necessary to cell division. However, plant cells lack centrioles in their centrosomes yet they do form a spindle during cell division.

▶ **4.10 CHECK YOUR PROGRESS**

1. Identify with examples the two general functions of cytoskeletal fibers.
2. Compare the structure and function of actin filaments and microtubules.
3. Explain how motor molecules function to allow cell parts and flagella to move.

FIGURE 4.10C The motor molecule dynein and the microtubule organization in cilia and flagella allows these cellular projections that arise from basal bodies to move.

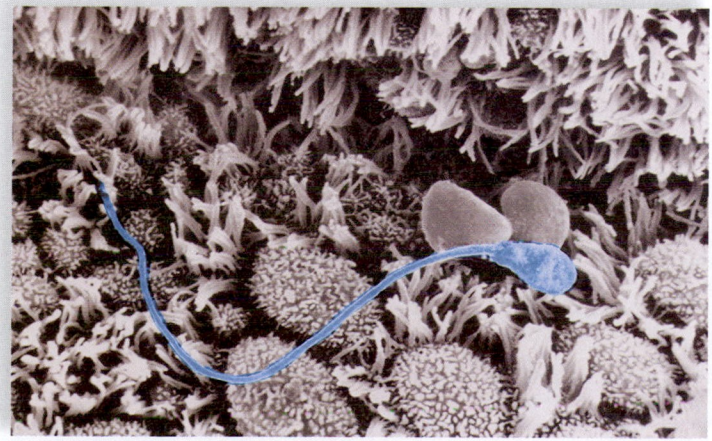

Flagellated sperm in oviduct lined by ciliated cells   7,200×

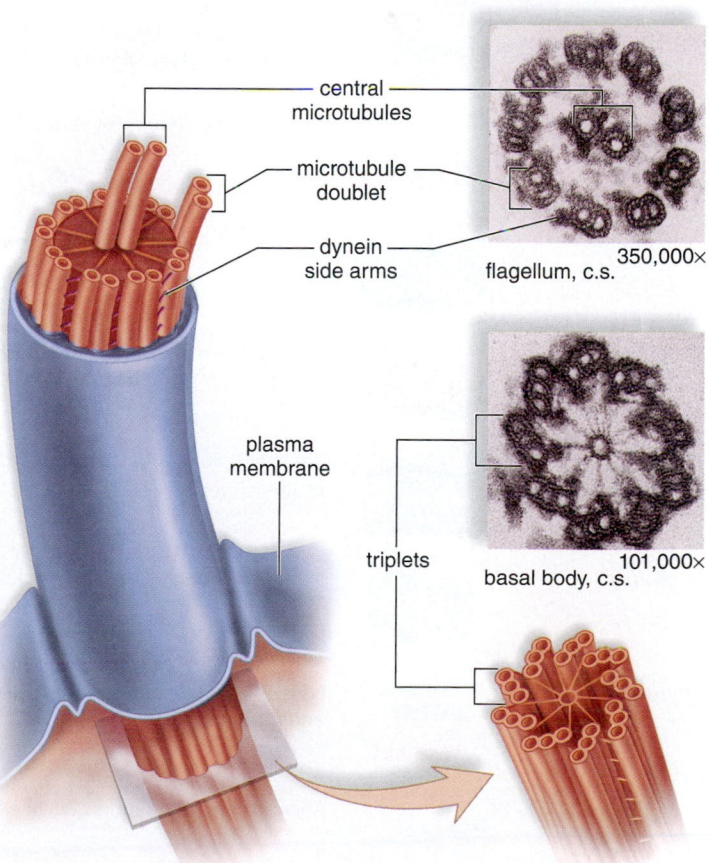

Flagellum   **Basal body**

The Eukaryotic Cell in Review

TABLE 4.11 — Eukaryotic Cell Structures

Cell Structure	Description	Function
Nucleus	Enclosed by nuclear envelope with pores; contains nucleolus and chromatin where DNA is located	Controls cell because it houses genetic information; nucleolus produces ribosomal subunits
Ribosome	Small particle having two subunits composed of RNA and protein	Produces proteins according to the coded information supplied by mRNA nucleotides
Endoplasmic reticulum (ER)	Network of membranous tubules and flattened sacs Rough ER: studded with ribosomes Smooth ER: lacks ribosomes	 Produces proteins; forms transport vesicles Produces lipids; forms transport vesicles
Golgi Apparatus	Stack of several flattened saccules	Carries out processing and packaging of proteins and lipids; forms secretory vesicles and lysosomes
Vesicle	Tiny membranous sac containing cell products	Isolates and transports biomolecules
Vacuole	Small to large membranous sac—plant cells have large central vacuole featured here	Stores varied biomolecules and wastes; in plant cells produces turgor pressure
Lysosome	Vesicle produced by Golgi apparatus that contains hydrolytic digestive enzymes	Breaks down debris, large biomolecules, and even organelles
Peroxisome	Vesicle with enzymes for varied functions	Breaks down fatty acids and poisons such as hydrogen peroxide
Chloroplast	Enclosed by a double membrane; interior has membranous thylakoids that contain chlorophyll	Absorbs solar energy; carries on photosynthesis and produces carbohydrates
Mitochondrion	Enclosed by a double membrane; inner membrane forms cristae	Carries on cellular respiration with the breakdown of carbohydrates to produce ATP
Plasma Membrane	Phospholipid bilayer with embedded proteins	Regulates the passage of molecules into and out of cell
Cell Wall	In plant cells, outer layer of cellulose	Helps maintain shape of a plant cell, protects and supports the cell
Cytoskeleton	Network of actin and intermediate filaments and microtubules	Maintains cell shape, assists transport of organelles within cell
Flagella (and Cilia)	Microtubule-containing cellular extensions	Move the cell; cilia move substances along its surface

THE CHAPTER IN REVIEW

Our knowledge of cell anatomy has in part been gathered by studying micrographs of cells. This has allowed cytologists (biologists who study cells) to arrive at a generalized picture of cells, such as those depicted for an animal and a plant cell in section 4.3. Eukaryotic cells, taken as a whole, contain several types of organelles, and the learning outcomes throughout the chapter suggest that you should know the structure and function of each one. A concept to keep in mind is that "structure suits function." For example, ribosomal subunits move from the nucleus to the cytoplasm; therefore, it seems reasonable that the nuclear envelope has pores. Finding relationships between structure and function will give you a deeper understanding of the cell and boost your memory capabilities.

Also, realizing that the organelles work together is helpful. If you wanted to describe the involvement of cell parts to make a protein, you would start with the nucleus because

chromosomes contain DNA, which specifies the order of amino acids in a particular protein. From there, you would mention the ribosomes at the rough endoplasmic reticulum (RER), transport vesicles, the Golgi apparatus, and a possible final destination for the protein. Analogies can help you remember the structure and function of organelles. For example, the endomembrane system can be compared to a post office: Proteins (the letters) are deposited into the RER (the local post office), which sends them to a Golgi (the regional sorting center) from which they are sent to their correct destinations. The pulse-labeling technique, described in section 4B, provides evidence to support this analogy.

Table 4.11 suggests ways for you to group the organelle for study and understanding. Lysosomes and peroxisomes are vesicles with digestive functions: Lysosomes digest various biomolecules while peroxisomes break down lipids. The origin of the eukaryotic cell links

together what you know about the structure of prokaryotic and eukaryotic cells because the endosymbiotic theory says that mitochondria and chloroplasts were once free-living prokaryotes.

In Chapter 5, we continue our general study of the cell by considering some of the functions common to all cells. For example, all cells exchange substances across the plasma membrane, and they also carry out enzymatic metabolic reactions, which either release or require energy.

ANALYZE AND EVALUATE

1. Use the structure of the prokaryotic cell to support the endosymbiotic theory.
2. Explain how the structure of the endoplasmic reticulum suits its function.
3. Microtubules are a part of the cytoskeleton and are found in cilia and flagella. What function of the cytoskeleton is consistent with the presence of microtubules in these structures?

SUMMARIZE

The Cellular Level of Organization

4.1 All organisms are composed of cells

- The **cell theory** states the following:
  A cell is the basic unit of life.
  All organisms are made up of cells.
  New cells arise only from preexisting cells.
- Most cells are quite small (measured in micrometers). A small size gives them a **surface-area-to-volume ratio** that is large enough to allow adequate exchanges with the environment.

4.2 Prokaryotic cells evolved first

- **Prokaryotic cells** are classified in the domains **Archaea** and **Bacteria.** They are simpler in structure and much smaller than eukaryotic cells and lack a membrane-enclosed nucleus. The single chromosome is in a region called the **nucleoid.**
- Like all cells, prokaryotic cells have a **plasma membrane** surrounding a **cytoplasm** (semifluid interior). The prokaryotic cytoplasm contains plentiful **ribosomes,** protein-producing particles. In addition, prokaryotic cells are protected by a **cell wall** and a **capsule. Pili** allow bacteria to attach to solid substances.

4.3 Eukaryotic cells contain specialized organelles: An overview

- The cells of **eukaryotic organisms** in the domain **Eukarya** have a membrane-enclosed **nucleus** and **organelles,** which are structures specialized to perform specific

functions. (See Figs. 4.3A and 4.3B.) The organization of eukaryotic cells allows them to be larger than prokaryotic cells.

The Nucleus and the Endomembrane System

4.4 The nucleus is a control center

- The nucleus contains genes, composed of DNA, located within **chromatin** that becomes organized as **chromosomes** when nuclear division occurs. The **nucleolus** produces ribosomal subunits. The nucleus is enclosed by a double membrane called the **nuclear envelope** with **nuclear pores** that permit communication between the nucleus and the cytoplasm.
- The nuclear membrane is a part of the endomembrane system, which is composed of membranous structures that are either directly connected or communicate by way of transport vesicles.
- An RNA copy of a gene called messenger RNA (mRNA) exits the nucleus and attaches to a ribosome that then produces a protein. When the ribosome attaches to the **endoplasmic reticulum (ER),** the protein enters the ER where the protein takes on its final shape.

4.5 The ER produces and transports proteins and lipids to the Golgi apparatus

- The outer membrane of the nuclear envelope is directly connected to the endoplasmic reticulum, a system of tubules and flattened saccules (small sacs) that join and form one whole. The **rough ER (RER)** has attached ribosomes and therefore it produces proteins (rough ER); the **smooth ER (SER)** produces lipids.

- **Transport vesicles** from the ER carry proteins and lipids to the **Golgi apparatus,** a stack of flattened saccules which modify and package proteins and lipids for distribution. In the secretory pathway, transport vesicles leave the Golgi apparatus and travel to the plasma membrane, where **secretion** occurs.
- The Golgi apparatus also produces **lysosomes,** membranous vesicles that digest biomolecules and cell parts because they contain hydrolytic digestive enzymes. In the lysosomal pathway, lysosomes fuse with vesicles and digest any biomolecules or particles entering the cell by way of the plasma membrane.

4.6 The organelles of the endomembrane system work together
- The nuclear envelope, the ER, Golgi apparatus, lysosomes, and transport vesicles make up the **endomembrane system.** The nucleus and these organelles work together to produce proteins and lipids that function within the cell or are secreted for specific purposes in the organism. For example, they produce the proteins and lipids found in the plasma membrane and the regulatory proteins that leave the cell.

Vacuoles and Vesicles
4.7 Vacuoles are common in plant cells
- **Vacuoles,** like vesicles, are membranous sacs but they are larger in size and specialize in storage. Plant cells have a large **central vacuole** that stores watery cell sap and maintains **turgor pressure.**

4.8 Peroxisomes have many different functions
- **Peroxisomes** are vesicles that contain enzymes and when the concentration rises, the proteins crystallize. Unlike lysosomes, peroxisomes arise at the ER and absorb enzymes from the cytoplasm. They synthesize some substances and break down toxins such as hydrogen peroxide.

Chloroplasts and Mitochondria
4.9 Chloroplasts and mitochondria have opposite functions
- **Chloroplasts** have internal membranes called thylakoids, where chlorophyll absorbs solar energy and produces carbohydrates during **photosynthesis.**
- **Mitochondria** have membranous shelves called cristae, where **cellular respiration** converts carbohydrate energy into the energy of ATP molecules.

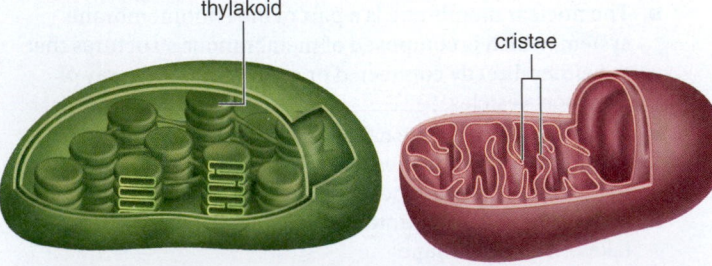
thylakoid

cristae

The Cytoskeleton and Cell Movement
4.10 The cytoskeleton maintains cell shape and assists movement
- The cytoskeleton is a network of protein fibers within the cytoplasm. **Actin filaments** are organized in bundles or networks. **Intermediate filaments** are ropelike assemblies of polypeptides. **Microtubules** are made of the globular protein tubulin. **Motor molecules** such as dynein move vesicles along microtubules. They act as tracks for organelle movement. The

microtubule organizing center (MTOC) regulates microtubule assembly and is located in the **centrosome.**
- **Cilia** (short) and **flagella** (long) are whiplike projections from cells that allow the cell to move. Cilia and flagella grow from **basal bodies,** perhaps derived from **centrioles.**

The Eukaryotic Cell in Review
- See Table 4.11.

TEST YOURSELF

The Cellular Level of Organization
1. The cell theory states
   a. cells form as organelles and molecules become grouped together in an organized manner.
   b. the normal functioning of an organism depends on its individual cells.
   c. the cell is the basic unit of life.
   d. only eukaryotic organisms are made of cells.
2. When you examine a cell using a light microscope, which of these can you usually see?
   a. the nucleus only
   b. the nucleus and the nucleolus
   c. the nucleus, the nucleolus, and the threads of chromatin
   d. all of these plus the DNA double helix
3. The small size of cells best correlates with
   a. their ability to reproduce.
   b. their prokaryotic versus eukaryotic nature.
   c. an adequate surface area for exchange of materials.
   d. their vast versatility.
   e. All of these are correct.
4. Which of the following structures are found in both plant and animal cells?
   a. centrioles   d. mitochondria
   b. chloroplasts   e. All of these are found in both types
   c. cell wall   of cells.
5. Eukaryotic cells compensate for a low surface-to-volume ratio by
   a. taking up materials from the environment more efficiently.
   b. lowering their rate of metabolism.
   c. compartmentalizing their activities into organelles.
   d. reducing the number of activities in each cell.
6. The cell wall and capsule of bacteria
   a. are located inside the plasma membrane.
   b. compensate for the lack of a plasma membrane.
   c. provide easy access to the cytoplasm.
   d. have projections called pili.
   e. Both b and c are correct.

The Nucleus and the Endomembrane System
7. What is synthesized by the nucleolus?
   a. mitochondria   c. transfer RNA
   b. ribosomal subunits d. DNA
8. The organelle that can modify a protein and determine its destination in the cell is the
   a. ribosome.   c. Golgi apparatus.
   b. vacuole.   d. lysosome.
9. Which of these is not involved in protein production and secretion?
   a. smooth ER   c. plasma membrane
   b. nucleus   d. All of these are correct.

10. **THINKING CONCEPTUALLY** Communication is critical in cells. How does the nucleus communicate with the cytoplasm, and how does the rough ER communicate with the Golgi?

11. _____ are produced by the Golgi apparatus and contain _____.
    a. Lysosomes, DNA
    b. Mitochondria, DNA
    c. Lysosomes, enzymes
    d. Nuclei, DNA

12. Vesicles from the ER most likely are on their way to
    a. the rough ER.
    b. the lysosomes.
    c. the Golgi apparatus.
    d. the plant cell vacuole only.
    e. the location suitable to their size.

13. Which organelle in the endomembrane system is incorrectly matched with its function?
    a. Nucleus—contains genetic information regarding the sequence of amino acids in proteins
    b. Transport vesicles—the way the nucleus communicates with the ER
    c. Golgi apparatus—involved in modification and packaging of proteins
    d. Lysosomes—digest biomolecules and cell parts
    e. All of these associations are correct.

14. **THINKING CONCEPTUALLY** A concept is an encompassing idea tested by the scientific method. The concept of the endomembrane system is based on what data?

Vacuoles and Vesicles

15. The central vacuole of plant cells may contain
    a. flower color pigments.
    b. toxins that protect plants against herbivorous animals.
    c. sugars.
    d. All of these are correct.

16. Peroxisomes
    a. are the same as lysosomes.
    b. are made by the Golgi apparatus.
    c. accumulate as we age.
    d. contain enzymes involved in lipid metabolism.

Chloroplasts and Mitochondria

17. Mitochondria
    a. are involved in cellular respiration.
    b. break down ATP to release energy for cells.
    c. contain stacks of thylakoid membranes.
    d. are present in animal cells but not in plant cells.
    e. All of these are correct.

18. The products of photosynthesis are
    a. glucose and oxygen.
    b. oxygen and water.
    c. carbon dioxide and water.
    d. glucose and water.

19. Why are mitochondria but not chloroplasts called the powerhouses of the cell?
    a. Mitochondria form glucose, but chloroplasts break it down.
    b. Mitochondria but not chloroplasts have their own genetic material.
    c. Mitochondria but not chloroplasts capture solar energy.
    d. Mitochondria but not chloroplasts directly provide ATP to the cell.
    e. Both a and b are correct.

20. **THINKING CONCEPTUALLY** Both chloroplasts and mitochondria are critical to your existence. How so?

The Cytoskeleton and Cell Movement

21. Which of these are involved in movement of the cell or the cell contents?
    a. actin filaments
    b. microtubules
    c. basal bodies
    d. All of these are correct.

22. Which of these statements is not true?
    a. Actin filaments are found in muscle cells.
    b. Microtubules radiate from the ER.
    c. Intermediate filaments sometimes contain keratin.
    d. Motor molecules that are moving organelles use microtubules as tracks.

23. Plant cells lack centrioles, and this correlates with their lack of
    a. mitochondria.
    b. flagella.
    c. a large central vacuole.
    d. All of these are correct

The Eukaryotic Cell in Review

For questions 24–28, match the functions to the organelles in the key.

KEY:
    a. endoplasmic reticulum and Golgi apparatus
    b. peroxisomes and lysosomes
    c. chloroplast and mitochondria
    d. centrosome and microtubules
    e. nucleus and ribosomes

24. Carbohydrate metabolism resulting in ATP formation
25. Contain enzymes for breaking down substances
26. Protein formation and secretion
27. Protein production as DNA dictates
28. Movement of the cell and its parts

GET INVOLVED

1. Utilizing Palade's procedure, described on page 72, you decide to label and trace the base uracil. What type of molecule are you labeling, and where do you expect to find it in Figure 4B?

2. After publishing your study from question 1, you are criticized for failing to trace uracil from mitochondria. Why might you have looked for uracil in mitochondria, and what comparative difference between the nuclear envelope and the mitochondrial double membrane might justify your study as is?

MEDIA STUDY TOOLS

mhhe.com/maderconcepts3

Enhance your study of this chapter with interactive study tools, practice tests, and engaging animations. Also, ask your instructor about the resources available through ConnectPlus, which includes LearnSmart, a personalized adaptive learning program, and a media-rich eBook.

5

Dynamic Activities of Cells

BEFORE YOU BEGIN

Take a few minutes to recall

How energy functions in organisms and ecosystems (sections 1.3 and 1.6)

How phospholipids form membrane (section 3.6)

The basic structure of the plasma membrane (section 4.3)

Life Is Organized

A cheetah captures its food.

On the African plain, a cheetah swiftly chases and captures an impala. Then the cheetah gets down to the business of lunch. As the cheetah dashed toward its prey, nerve cells caused muscle cells to contract. All the cells in the cheetah contributed toward the final goal of acquiring materials and energy from the environment—the impala.

The cells that exist in the cheetah, you, and me have a long evolutionary history that traces them back to even the first cells to evolve. The cheetah could not exist today without evolution of photosynthesizing cells that capture solar energy. Their ability is centered in the organization of membrane—membrane marks the boundary and forms the internal structure of a cell. Chlorophyll located in membrane captures solar energy.

Cattle stand and graze to get their food.

A cheetah feeds on the impala, a grazing animal.

Less than 1% of the solar energy that strikes the Earth is taken up by photosynthesizing plants, algae, and bacteria but this is enough to power the biological world. You, like the cheetah, are dependent on photosynsthesis because you either consume plants or animals that have fed on plants. As you acquire biomolecules from your food, some are used to build your own tissues, and some are used as a source of energy. This energy is eventually converted to heat, which escapes into the environment.

The illustrations on this page give examples of energy flow. The one involving humans goes like this: from the sun, to corn plants, to cattle, to humans out for a run with their dog. Your gnawing stomach makes you aware of the need to eat food every day, but you may not realize why, like all organisms, humans are dependent on a constant flow of energy from the sun. The answer is: "Energy dissipates." When muscle contraction is over, energy escapes into the body of an animal and then into the environment. The heat given off when your muscles contract is put to good use. It keeps you warm, but it is no longer usable by photosynthesizers for chemical reactions. It is too diffuse. However, solar energy is concentrated enough to allow plants to keep on photosynthesizing and, in that way, provide the biosphere with organic food.

This introduction gives you an overview of why energy flows through the biological world and cannot accumulate. Energy is an important part of metabolism, and so are enzymes, the proteins that speed chemical reactions. Without enzymes, the cheetah and you would not be able to make use of biomolecules to maintain your bodies and carry on such activities as muscle contraction. Enzymes work in conjunction with membrane, and membrane is the first topic of this chapter. The plasma membrane permits cells to exist in the first place.

Hot gasses surrounding the surface of the sun

Corn plants make their own food.

Humans eat plants or animals to get their food.

The Plasma Membrane Structure and Function

In this section, we will study the structure of the plasma membrane and the functions of the many types of proteins found within it. The plasma membrane is not a passive boundary for the cell. Rather, it has many varied functions that are often dependent on the many proteins present in the membrane.

5.1 The plasma membrane is a phospholipid bilayer with embedded proteins

LEARNING OUTCOMES

When you complete this section, you should be able to

1. Describe the structure of a plasma membrane using the fluid-mosaic model.
2. Differentiate six types of protein molecules in the plasma membrane.

The plasma membrane marks the boundary between the outside and the inside of a cell. Its integrity and function are necessary to the life of the cell. Not only does it serve as the boundary of the cell, but it also regulates the passage of molecules into and out of the cell, and it allows a cell to communicate with its neighbors.

In both bacteria and eukaryotes, the plasma membrane is a **phospholipid bilayer** that has the consistency of olive oil. Recall that the polar head of a phospholipid is hydrophilic, while the nonpolar tails are hydrophobic. The polar heads of the phospholipids face toward the outside of the cell and toward the inside of the cell, where there is a watery medium. The nonpolar tails face inward toward each other, where there is no water.

3D Animation
Membrane Transport:
Lipid Bilayer

The membrane contains embedded proteins that lie within it and peripheral proteins that lie along the inside. The embedded proteins have a hydrophobic region within the membrane and hydrophilic regions that extend beyond the surface of the membrane. The presence of these regions prevents embedded proteins from flipping, but they can move laterally.

The **fluid-mosaic model** states that the protein molecules embedded in the membrane have a pattern (form a mosaic) within the fluid phospholipid bilayer (Fig. 5.1A). The pattern varies according to the particular membrane and also within the same membrane at different times. **Cholesterol** molecules are steroids present in animal cell membrane. Cholesterol interacts with the adjacent fatty acid chain of a phospholipid and in that way reduces membrane fluidity and permeability. Other steroids perform this function in the plasma membranes of plants.

Both phospholipids and proteins can have attached carbohydrate (sugar) chains. Molecules carrying such chains are called **glycolipids** and **glycoproteins,** respectively. Because the carbohydrate chains occur only on the outside surface, and because peripheral proteins occur only on the inner surface of the membrane, the two sides of the membrane are not identical.

MP3
Membrane Structure

In animal cells, the carbohydrate chains project into the extracellular matrix (ECM) shown and discussed in section 5.4. The ECM protects the cell and has various other functions. For

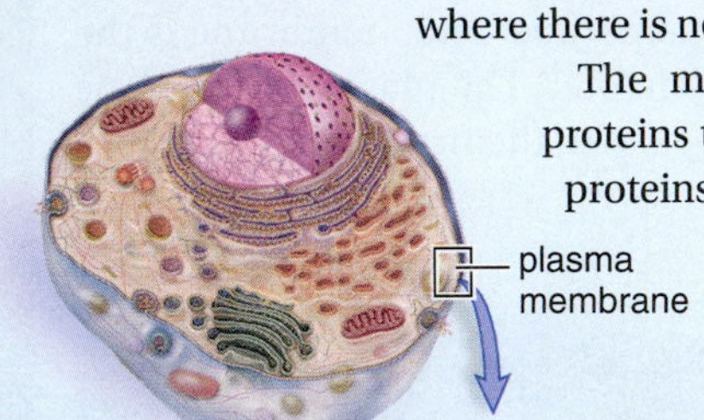

plasma membrane

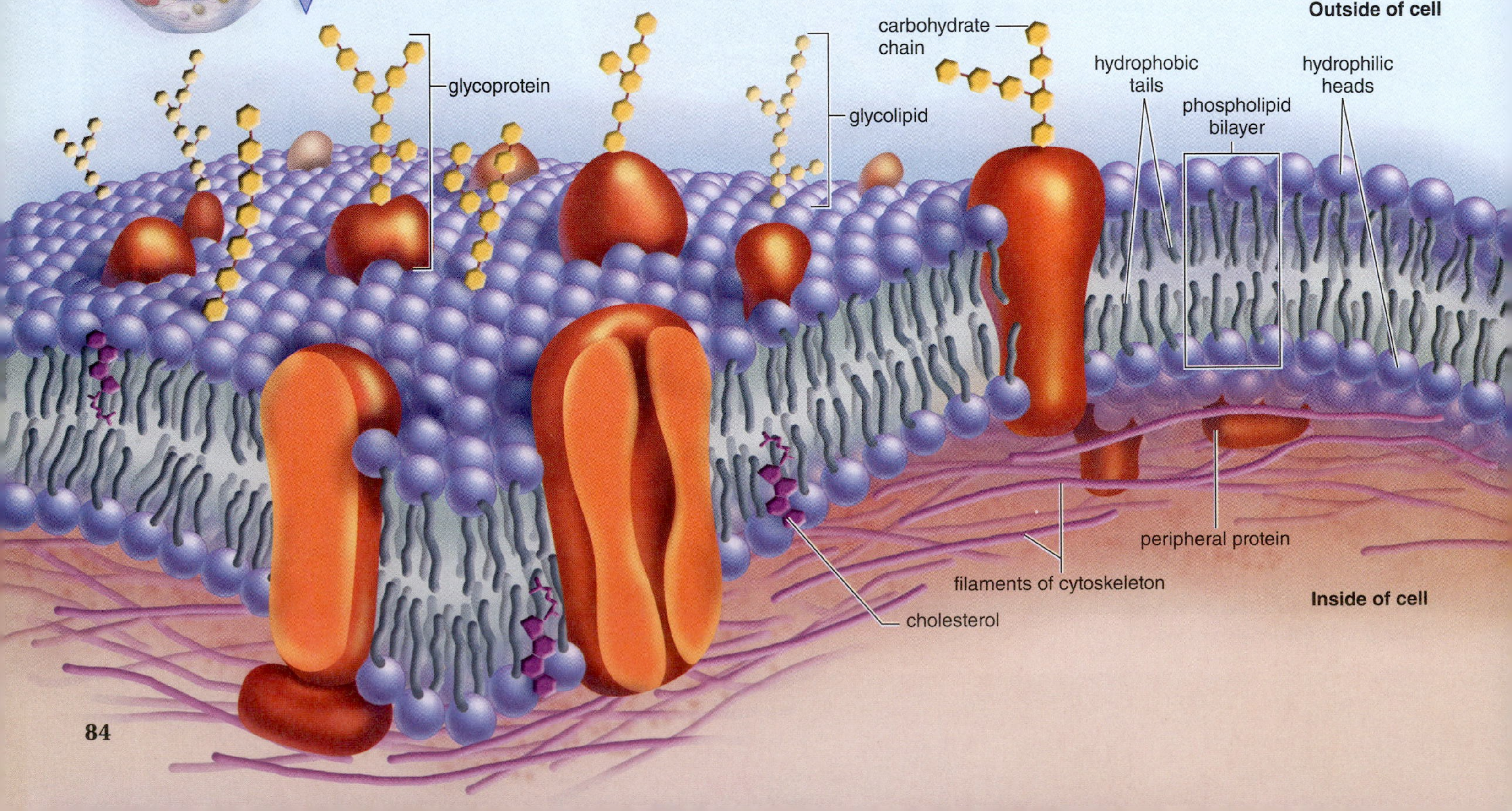

FIGURE 5.1A The plasma membrane is a phospholipid bilayer with embedded proteins (red). The proteins have various functions.

Outside of cell

glycoprotein

carbohydrate chain

glycolipid

hydrophobic tails

phospholipid bilayer

hydrophilic heads

peripheral protein

filaments of cytoskeleton

cholesterol

Inside of cell

example, it facilitates adhesion between cells and the reception of signaling molecules that influence the behavior of the cell.

The Proteins The plasma membranes of different cells and the membranes of various organelles each have their own particular collections of these specific types of embedded proteins. The six types of embedded proteins described here and depicted in Figure 5.1B help a membrane fulfill its functions. Decide whether each of these proteins helps the membrane regulate the passage of molecules into and out of the cell or helps the cell communicate with its environment and/or its neighbors:

Channel proteins Channel proteins have a channel that, when open, allows molecules to simply move across the membrane. For example, a channel protein allows hydrogen ions to flow across the inner mitochondrial membrane. Without this movement of hydrogen ions, ATP would never be produced.

Carrier proteins Carrier proteins are different from channel proteins because they combine with a substance and help it move across the membrane. For example, a carrier protein transports sodium and potassium ions across a nerve cell membrane. Without this carrier protein, nerve conduction would be impossible.

Cell recognition proteins Cell recognition proteins are glycoproteins. Foreign cells bear their own glycoproteins that enable the immune system to recognize them and mount a defense.

Without this recognition, harmful organisms (pathogens) would be able to freely invade the body.

Receptor proteins Receptor proteins have a binding site for a specific molecule. The binding of this molecule causes the protein to change its shape and, thereby, bring about a cellular response. The coordination of the body's organs is totally dependent on signaling molecules that bind to receptors. For example, the liver stores glucose after it is signaled to do so by insulin.

Enzymatic proteins Some plasma membrane proteins are enzymatic proteins that carry out metabolic reactions directly. Without the presence of enzymes, some of which are attached to the various membranes of the cell, a cell would never be able to perform the metabolic reactions necessary for its proper function.

Junction proteins As discussed in section 5.4, proteins are also involved in forming various types of junctions between cells. The junctions assist cell-to-cell communication.

▶ **5.1 CHECK YOUR PROGRESS**
1. Explain why the structure of the plasma membrane is referred to as a mosaic.
2. Identify four types of proteins in the plasma membrane that do not have a transport function.

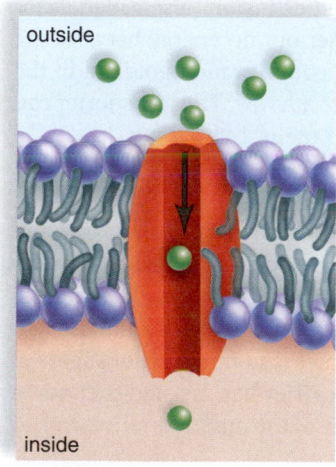

channel protein

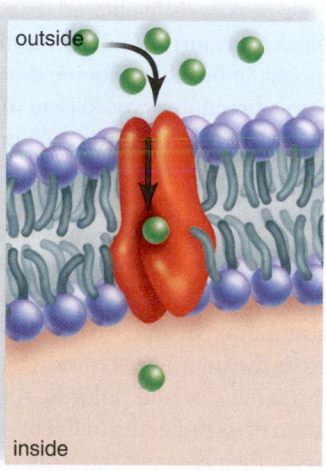

carrier protein

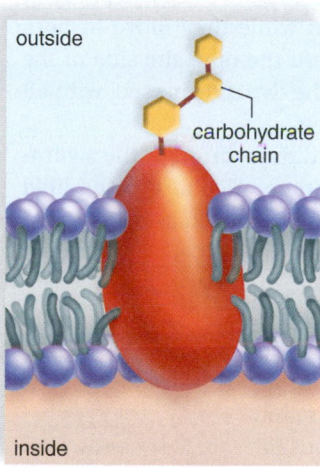

carbohydrate chain

cell recognition protein

FIGURE 5.1B The proteins (red) in a membrane have many varied functions.

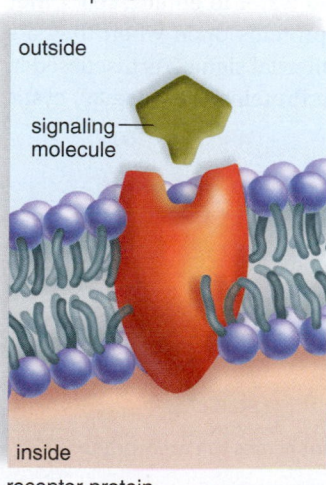

signaling molecule

receptor protein

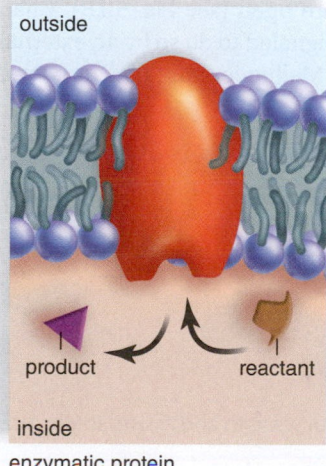

product   reactant

enzymatic protein

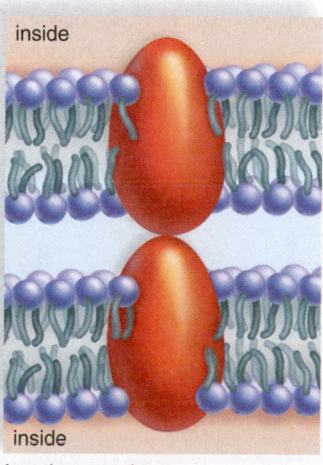

junction proteins

The Passage of Molecules Into and Out of Cells

The plasma membrane is **differentially permeable,** meaning that only certain substances can freely diffuse across the membrane and others cannot. For the latter, a transporter and/or energy are involved (Table 5.2). Facilitated diffusion moves substances toward a lower concentration, a transporter is required but no energy. Active transport moves substances toward a higher concentration and both a transporter and energy are required. Bulk transport, which moves substances by vesicle formation, is not affected by concentration gradients.

5.2 Diffusion across a membrane requires no energy input

LEARNING OUTCOMES

When you complete this section, you should be able to

1. Compare and contrast simple diffusion with facilitated diffusion.
2. Predict the effect of osmotic conditions on animal versus plant cells.

During **diffusion,** a molecule moves from a high concentration to a low concentration until it is distributed equally. In other words, the molecule follows its **concentration gradient** until the gradient disappears. Diffusion is a physical process that can be observed with any type of molecule, any place. For example, if you release a perfume in one corner of a room, the scent spreads out into all corners by diffusion.

Diffusion occurs because molecules are in motion, and it is a *passive* form of transport because no energy need be added for it to happen. Diffusion can take place across a membrane if the substance is able to cross the membrane. In Figure 5.2A, a red dye is added to water on one side of the membrane. The dye particles are able to cross the membrane, and while they move in both directions, the net movement is toward the opposite side of the membrane (long arrow). Eventually, the dye is dispersed, with no net movement of dye in either direction.

Very few molecules can simply diffuse through the hydrophobic portion of the plasma membrane. Alcohols, being lipid soluble, can diffuse across, and so can gases such as oxygen (O_2) and carbon dioxide (CO_2). Diffusion allows O_2 to enter the blood from the air sacs of the lungs, and CO_2 to move in the opposite direction. No energy is required because the gases are simply following their concentration gradient. Similarly, cellular respiration in cells sets up favorable concentrations for gas exchange to occur by diffusion.

 **MP3** Diffusion

 **3D Animation** Membrane Transport: Diffusion

Animation How Diffusion Works

Facilitated Diffusion Certain molecules (e.g., water, glucose, and amino acids) and ions (Na⁺, Cl⁻, Ca²⁺) cross plasma membranes at a rate faster than expected based on their size and polarity because

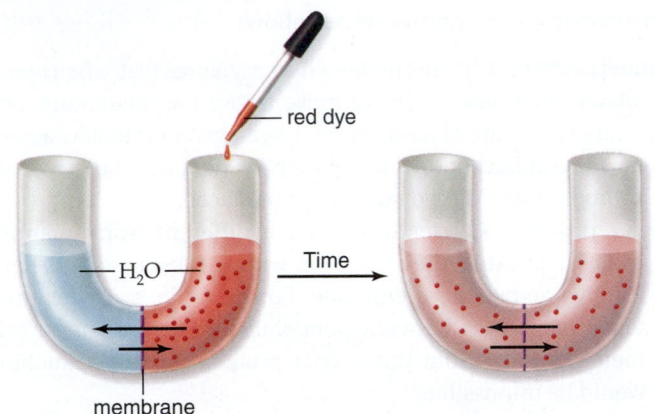

FIGURE 5.2A Some molecules can move freely by simply diffusing across a membrane.

their passage is facilitated. **Facilitated diffusion** (also called facilitated transport) requires a transporter but no energy because the molecule or ion is moving down its concentration gradient in the same direction it would tend to move anyway. The transporter can be a carrier protein or a channel protein, but each is *specific* because it assists the passage of only its own particular substance.

Animation How Facilitated Diffusion Works

A carrier protein is slower acting than a channel protein because it combines with a molecule and then deposits it on the other side (Fig. 5.2B). A carrier protein is highly specific, and the abundance of glucose transporters accounts for why glucose can cross the membrane hundreds of times faster than other sugars of the same size and polarity. After a carrier has assisted the movement of its molecule to the other side of the membrane, it is free to assist the passage of another one.

Channel proteins allow ions and water to enter a cell if they are open (see Fig. 5.1B). Channel proteins open when they are signaled to do so by an external or internal signal. As discussed in "Malfunctioning Plasma Membrane Proteins" on page 89, cystic

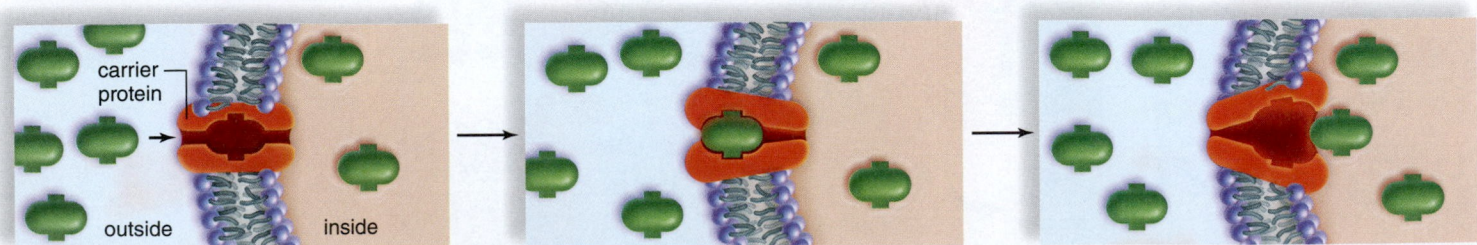

FIGURE 5.2B During this type of facilitated diffusion, a carrier protein combines with and assists solute movement across the membrane.

fibrosis is due to the inability of a Cl⁻ channel to open. A channel protein is faster acting than a carrier protein because the substance simply flows through it. Thousands of millions of water molecules per second pass through a single channel protein, named an **aquaporin** because of its specificity.

Osmosis The diffusion of water across the plasma membrane due to concentration differences is called **osmosis.** When osmosis occurs, the solute is unable to cross the plasma membrane whereas water, the solvent, is able to freely cross the membrane. Which way the water moves is dependent on the solute versus water concentration on both sides of the membrane. In the laboratory, cells are normally placed in **isotonic solutions** (*iso,* same as) in which the cell neither gains nor loses water because the concentration of solute versus water is the same on both sides of the membrane (Fig. 5.2C). In medical settings, a 0.9% solution of sodium chloride (NaCl) is known to be isotonic to red blood cells; therefore, intravenous solutions usually have this concentration.

Animation How Osmosis Works

MP3 Osmosis

Cells placed in a **hypotonic** (*hypo,* less than) **solution** cause the cell to gain water. Outside the cell, the concentration of solute is less, and the concentration of water is greater, than inside the cell. Animal cells placed in a hypotonic solution expand and sometimes burst. The term *lysis* refers to disrupted cells; *hemolysis,* then, is disruption of red blood cells (*hemo,* blood). Organisms that live in fresh water have to prevent their internal environment from gaining too much water. Many protozoans, such as paramecia, have contractile vacuoles that rid the body of excess water. Freshwater fishes excrete a large volume of dilute urine and take in salts at their gills, ensuring that their internal fluids don't become hypotonic to their cells.

Animation Hemolysis and Crenation

Video Contractile Vacuoles

When a plant cell is placed in a hypotonic solution, the large central vacuole gains water, and the plasma membrane pushes against the rigid cell wall as the plant cell becomes *turgid.* The plant cell does not burst because the cell wall does not give way. **Turgor pressure** in plant cells is extremely important in maintaining the plant's erect position. If you forget to water your plants, they wilt due to decreased turgor pressure.

Cells placed in a **hypertonic** (*hyper,* more than) **solution** lose water. Outside the cell, the concentration of solute is more, and the concentration of water is less, than inside the cell. Animal cells placed in a hypertonic solution shrivel (*crenate*). Marine fishes

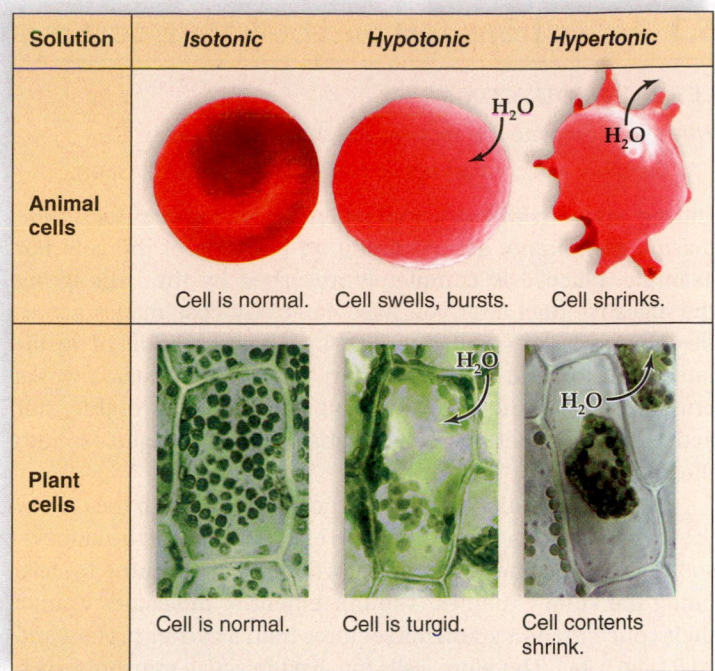

Solution	*Isotonic*	*Hypotonic*	*Hypertonic*
Animal cells	Cell is normal.	Cell swells, bursts.	Cell shrinks.
Plant cells	Cell is normal.	Cell is turgid.	Cell contents shrink.

FIGURE 5.2C When cells are in an isotonic solution, they remain normal. When cells are in a hypotonic solution, they gain water, and when they are in a hypertonic solution, they lose water to the environment.

prevent their internal environment from becoming hypertonic to their cells by excreting salts across their gills. Hypertonicity can be put to good use. For example, meats are sometimes preserved by being salted. Bacteria are killed, not by the salt, but by the lack of water in the meat.

Animation How Osmosis Works

3D Animation Membrane Transport: Osmosis

When a plant cell is placed in a hypertonic solution, the plasma membrane pulls away from the cell wall as the large central vacuole loses water. This is an example of *plasmolysis,* shrinking of the cytoplasm due to osmosis.

Animation Plasmolysis

► **5.2 CHECK YOUR PROGRESS**
   1. Identify the reason facilitated diffusion does not require energy.
   2. Explain why a red blood cell is subject to destruction when it is in a hypotonic or hypertonic solution.

TABLE 5.2		Passage of Molecules Into and Out of the Cell		
	Name	**Direction**	**Requirements**	**Examples**
Energy Not Required	Diffusion	Toward lower concentration	No transporter, no energy	Lipid-soluble molecules, and gases
	Facilitated diffusion	Toward lower concentration	Transporter, no energy	Some sugars, amino acids
Energy Required	Active transport	Toward higher concentration	Transporter, energy	Sugars, amino acids, ions
	Bulk transport		Vesicle formation	Polypeptides, polysaccharides, and particles

5.3 Active transport and bulk transport require energy input

LEARNING OUTCOME

When you complete this section, you should be able to

1. Compare and contrast active transport with bulk transport.

During **active transport,** molecules or ions move across the plasma membrane, accumulating on one side of the cell. For example, glucose is completely absorbed by the cells lining the digestive tract after you have eaten. Glucose moves across the lining of the small intestine by a combination of facilitated diffusion and active transport. Facilitated diffusion works only as long as the concentration gradient is favorable, but active transport permits cells to absorb all of the glucose into the body.

Most of the iodine that enters the body collects in the cells of the thyroid gland for the production of a hormone. In the kidneys, sodium can be almost completely withdrawn from urine by cells lining the kidney tubules. The movement of molecules against their concentration gradients requires both a carrier protein and ATP (Fig. 5.3A). Therefore, cells involved in active transport, such as kidney cells, have a large number of mitochondria near their plasma membranes to generate ATP.

Proteins engaged in active transport are often called *pumps.* The **sodium-potassium pump,** vitally important to nerve function, undergoes a change in shape when it hydrolyzes ATP, and this allows it to combine alternately with sodium ions and potassium ions to move them across the membrane.

3D Animation
Membrane Transport: Active Transport

Animation
How the Sodium-Potassium Pump Works

Bulk transport occurs when fluid or particles are brought into a cell by vesicle formation, called **endocytosis** (Fig. 5.3B), or out of a cell by evagination, called **exocytosis.** To imagine exocytosis, reverse the arrows in Figure 5.3B.

Animation
Endocytosis and Exocytosis

Polypeptides, polysaccharides, and particles are too large to be moved by carrier proteins. Instead, endocytosis takes them into a cell, and exocytosis takes them out. If a particle taken in is large, such as a food particle or another cell, the process is called **phagocytosis.** Phagocytosis is common in unicellular organisms, such as amoebas. It also occurs in humans. Certain types of human white blood cells are amoeboid—that is, they are mobile like an amoeba, and are able to engulf debris such as worn-out red blood cells or bacteria. When an endocytic vesicle fuses with a lysosome in the cell, digestion occurs (see Fig. 4.5C).

Pinocytosis occurs when vesicles form around a liquid or around very small particles. Cells that use pinocytosis to ingest

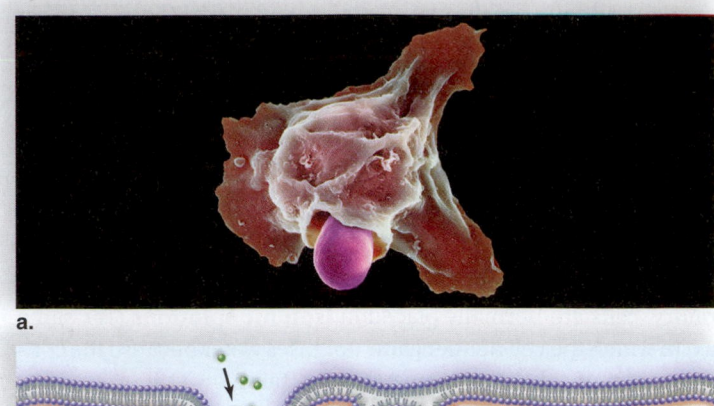

a.

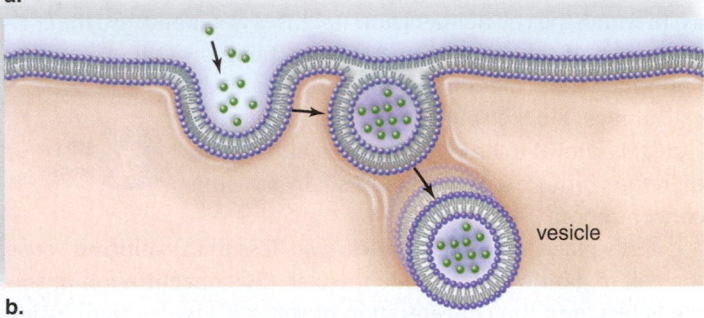

b.

FIGURE 5.3B Bulk transport into the cell is by endocytosis. **a.** A white blood cell is engulfing and destroying a fungal cell. **b.** Drawing shows that bulk transport involves vesicle (or vacuole) formation.

substances include white blood cells, cells that line the kidney tubules and the intestinal wall, and plant root cells.

During **receptor-mediated endocytosis,** receptors for particular substances are found at one location in the plasma membrane. This location is called a coated pit because there is a layer of protein on its intracellular side. Receptor-mediated endocytosis is selective and much more efficient than ordinary pinocytosis. It is involved when substances move from maternal blood into fetal blood at the placenta, for example.

In contrast to endocytosis, digestive enzymes and hormones are transported out of the cell by exocytosis. In cells that synthesize these products, secretory vesicles accumulate near the plasma membrane. The vesicles release their contents only when the cell is stimulated by a signal received at the plasma membrane, a process called regulated secretion.

▶ **5.3 CHECK YOUR PROGRESS**

1. Compare active transport to a turnstile that allows people to pass through after they have paid.
2. Describe how you would redirect the arrows in Figure 5.3B to illustrate exocytosis.

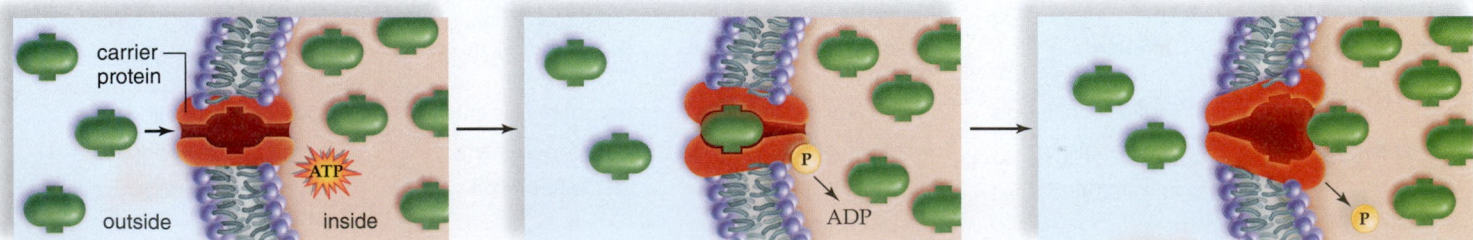

FIGURE 5.3A During active transport, a protein carrier assists solute movement toward a higher concentration, and energy is required.

5A Malfunctioning Plasma Membrane Proteins

Suppose you went to the doctor for a particular medical condition. Would you expect to hear that your condition was due to your plasma membrane? That is what might happen with certain illnesses. Take type 2 diabetes, for example.

Type 2 Diabetes

The typical type 2 diabetes patient is somewhat, or even grossly, overweight. The symptoms include unusual hunger and/or thirst, excessive fatigue, blurred vision, sores that do not heal, and frequent urination, especially at night. The doctor does a urinalysis and finds sugar in the urine, and yet the blood test shows insulin in the blood. Usually when we eat sugar, the pancreas, a gland that lies near the stomach, releases the hormone insulin into the bloodstream, and it travels to the cells, where it binds to its receptor protein. The binding of insulin signals a cell to send carrier proteins (often called transporters) to the plasma membrane that will carry glucose into the cells. In the case of type 2 diabetes, the insulin binds to its receptor protein, but the number of carrier proteins sent to the plasma membrane for glucose is not enough. The result is too much glucose in the blood, which spills over into the urine.

Patients can prevent, or at least control, type 2 diabetes by switching to a healthy diet and engaging in daily exercise.

Color Blindness

If you have ever found yourself accidentally wearing socks of two different colors, you may have endured a little teasing about being color blind. Color vision is dependent on the action of cone cells present in the retina, the part of the eye that allows us to see. People with normal color vision have three types of cones: blue, green, and red, each activated by different wavelengths of visible light. The perception of color often requires activation of a combination of these three types of cone cells.

When a cone cell receives the wavelength of light to which its particular photopigment is sensitive, a signal is sent to close sodium ion channels in its plasma membrane. However, some people, mostly males, have inherited a mutation that results in a lack of functional red or green photopigment proteins and therefore an inability to close sodium ion channels in the plasma membrane. Such individuals have what is termed "red-green color blindness" and have difficulty distinguishing these two colors. In a much less common situation, both red and green photopigments are missing; such people may lack all color vision and see a monochromatic world.

Cystic Fibrosis (CF)

The typical CF patient is a child, usually younger than 3 years of age, who has experienced repeated lung infections or poor growth. The doctor orders a test that measures the amount of salt (NaCl) in the child's sweat, because children with CF have more salt in their sweat than normal children. Usually, chloride ions (Cl⁻) pass easily through a plasma membrane channel protein, but when their passage is not properly regulated, a thick mucus appears in the lungs and pancreas (Fig. 5Aa). The mucus clogs the lungs, causing breathing problems. It also provides fertile ground for bacterial growth. The result is frequent lung infections, which

FIGURE 5A a. Cystic fibrosis is due to a defective *CF* gene and defective Cl⁻ channel proteins. **b.** Treatment includes use of a percussion vest to loosen the excess mucus in the lungs.

chromosome 7

defective *CF* gene

DNA

mRNA

H_2O Cl⁻   — cytoplasm

Cl⁻ H_2O Cl⁻   — Chloride ions and water are trapped inside cell.
            H_2O

— Defective Cl⁻ channel protein does not allow chloride ions to pass through.

— Thick, sticky mucus results.

a. Cause of cystic fibrosis

b. Child undergoing percussion treatment

eventually damage the lungs and contribute to an early death. Also, thick digestive fluids may clog ducts leading from the pancreas to the small intestine. This prevents enzymes from reaching the small intestine, where they are needed to digest food. Digestive problems and slow growth result. Various treatments have been devised to extend the lives of CF patients (Fig. 5Ab).

Video Good Poison

CONSIDER THESE QUESTIONS

1. How can you convince people that they can prevent type 2 diabetes with good nutrition and exercise?
2. Should textbooks use strong contrasts in color—even ones jarring to other people—because color-blind individuals cannot otherwise see a contrast?

 Explore the concepts through a variety of multimedia
|BIOLOGY assets and question types.

www.mcgrawhillconnect.com

The Plasma Membrane and Cell Communication

Both plant and animal cells have extracellular material previously produced by the cell and transported across its plasma membrane. Plant cells have a fairly rigid cell wall that gives them shape. Channels called plasmodesmata extend between adjacent cells and allow them to communicate. Some animal cells have junctions that allow them to communicate and others have a complex extracellular matrix that both supports the cell and permits communication.

5.4 Extracellular material allows cells to join together and communicate

LEARNING OUTCOMES

When you complete this section, you should be able to

1. Compare the cell wall of plant cells to the extracellular matrix (ECM) of animal cells.
2. List and describe three types of junctions that join animal cells.

Plant Cells All plant cells have a primary cell wall in which cellulose microfibrils are held together by noncellulose carbohydrates such as glycan and pectin. The middle lamella, a packing material between cells, is made of pectin, a water-attracting material.

Living plant cells are connected by **plasmodesmata** (sing., plasmodesma), numerous narrow, membrane-lined channels that pass through the cell wall (Fig. 5.4A). Cytoplasmic strands within these channels allow direct exchange of some materials between adjacent plant cells and, ultimately, all the cells of a plant.

Animal Cells In animals, certain tissues have junctions between their cells that allow them to behave in a coordinated manner (Fig. 5.4B). In the heart, stomach, and bladder where tissues get stretched, the cells are joined together by **anchoring junctions.** In these junctions, intercellular filaments attached to internal cytoplasmic deposits are held in place by intermediate filaments of the cytoskeleton.

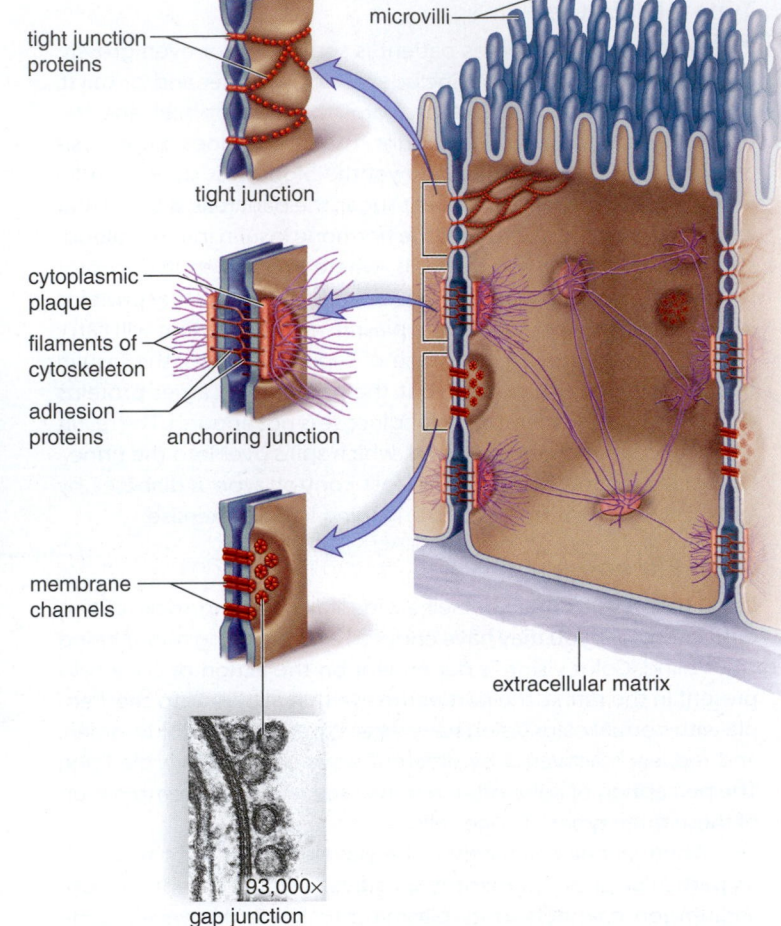

FIGURE 5.4B In certain tissues, animal cells are joined by three possible types of junctions. Microvilli are small cellular extensions that increase the surface area for absorption.

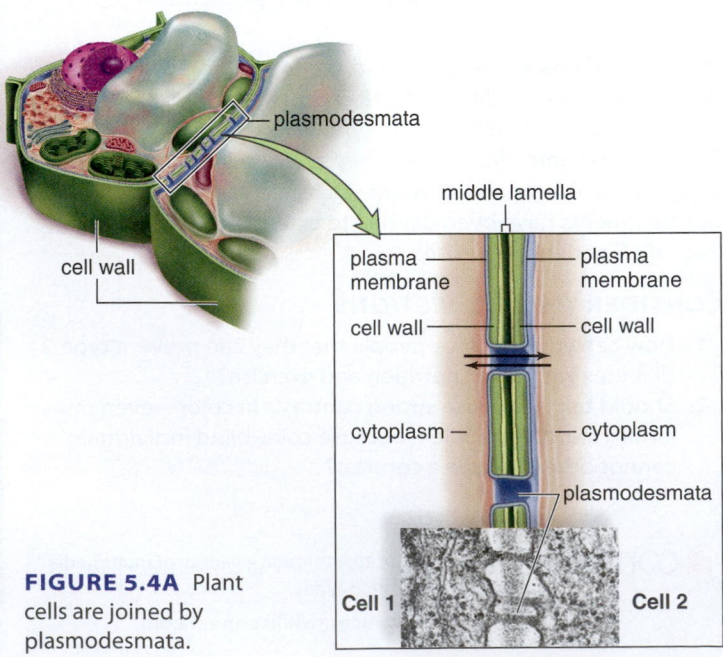

FIGURE 5.4A Plant cells are joined by plasmodesmata.

In other tissues, cells are even more closely joined by **tight junctions,** in which plasma membrane proteins actually attach to each other, producing a zipperlike fastening. Tight junctions in the cells of the intestine prevent digestive juices from leaking into the abdominal cavity, and in the kidneys, urine stays within the kidney tubules, because the cells are joined by tight junctions.

Gap junctions, on the other hand, form when two identical plasma membrane channels join. The channel of each cell is lined by six plasma membrane proteins. A gap junction lends strength to the cells, but it also allows small molecules and ions to pass between them. In heart muscle and smooth muscle, cells contract in a coordinated manner because they permit ions to flow between the cells.

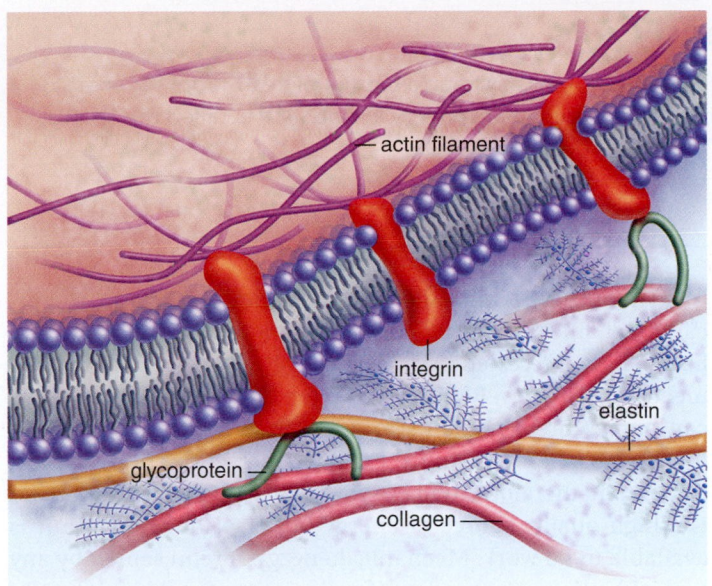

FIGURE 5.4C Extracellular matrix of animal cell.

Extracellular Matrix Where junctions are not present, animal cells have a protein-rich **extracellular matrix** (**ECM**) between cells, which varies from quite flexible, as in cartilage, to rock solid, as in bone. Collagen and elastin fibers are two well-known structural proteins in the ECM (Fig. 5.4C). Collagen gives the matrix strength, and elastin gives it resilience.

An elaborate mixture of glycoproteins (proteins with short chains of sugars attached to them) is also present in the matrix. One glycoprotein attaches to a plasma membrane protein called integrin. Integrin spans the membrane and internally attaches to actin filaments of the cytoskeleton. Because of these connections, the matrix can influence cell migration patterns during growth and development and in many ways help coordinate the behavior of cells.

▶ **5.4 CHECK YOUR PROGRESS**
1. What features do the plasmodesmata of plant cells and the gap junctions of animal cells have in common?
2. Contrast the types of biomolecules in plant cell walls with the types found in the ECM of animal cells.

HOW LIFE CHANGES *Application*

5B Evolution of the Plasma Membrane

The plasma membrane must have arisen early in the evolution of life because both bacteria and eukaryotic cells have the same type of plasma membrane. A plasma membrane serves as a barrier to keep the contents of the cell different from the extracellular environment while still allowing molecules to enter and exit the cell. A prokaryote with a plasma membrane that functioned selectively would have reproduced more than one lacking that advantage. Therefore, through the natural selection process, all cells have the same type of selectively permeable plasma membrane.

The phospholipid component of the plasma membrane naturally forms a boundary, as discussed in section 3.6, and much of the specialty work of the plasma membrane is done by its protein components (see section 5.1). The form—and therefore the function—of proteins is ultimately controlled by genes. This means that during the history of life many variations in plasma membrane structure and function could have been tried out and most likely natural selection is still shaping how the plasma membrane functions.

The evolution of multicellular organisms (fungi, plants, and animals) would have been impossible without a cell's ability to produce extracellular material that glues them together and allows them to signal one another. It often happens in evolution that an early change permits a later change. Scientists hypothesize that prokaryotes evolved signaling mechanisms, which were inherited by the first eukaryotic cells. Multicellular organisms use them to keep their cells functioning for the benefit of the entire organism. Typically, cell signaling occurs when a signaling molecule binds to a receptor protein in a target cell's plasma membrane. The signal causes the receptor protein to initiate a series of reactions within a signal transduction pathway (Fig. 5B). The end product of the pathway (not the signal) directly affects the

metabolism of the cell. We will revisit cell signaling in subsequent chapters, particularly when we study control of cell division, the development of cancer, and how hormones function.

CONSIDER THIS QUESTION

1. Does it seem reasonable that the plasma membrane of your cells is the same as the membrane that surrounds a bacterium. Why or why not?

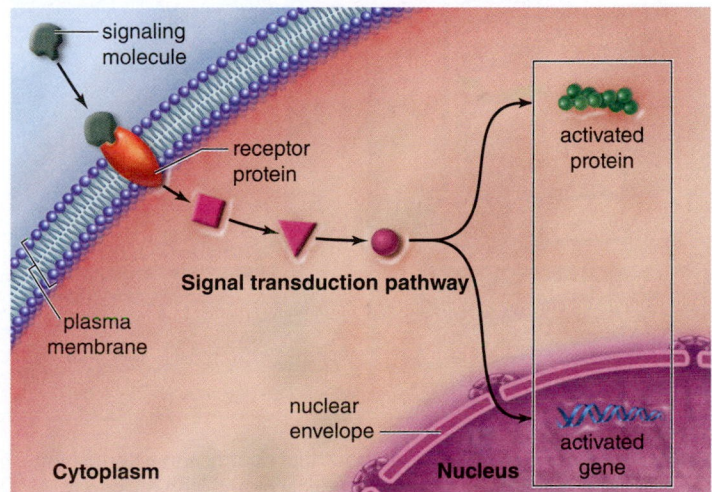

FIGURE 5B A cell signal received at the plasma membrane initiates a signal transduction pathway that activates a gene or a protein.

Energy Transformations in Organisms and Cells

Cells readily convert—that is, transform—one form of energy into another, but even so they need a continual fresh supply of energy. Two energy laws will be used to explain this paradox. ATP is the preferred form of immediate energy in cells. ATP is called the "energy currency" of cells because when cells do any kind of work, they "spend" ATP.

5.5 Energy makes things happen

LEARNING OUTCOMES

When you complete this section, you should be able to

1. Explain the difference between potential energy and kinetic energy.
2. State two energy laws and apply them to organisms.

Organisms are highly ordered, and energy is needed to maintain this order. Organisms acquire energy, store energy, and release energy, and only by transforming one form of energy into another form can organisms continue to stay alive. Despite its importance to organisms and society, energy is a commodity that cannot be seen. Most authorities define **energy** as the capacity to do work—to make things happen.

Potential energy is stored energy of position or configuration.

Kinetic energy is energy of motion.

Potential energy is being converted to kinetic energy.

diving board

FIGURE 5.5A Food has stored potential energy and a nourishing lunch supplied this diver with the chemical energy she is using to mount the steps to a diving board. Climbing steps is energy of motion or kinetic energy. At the diving board this diver possesses the stored energy of position, a form of potential energy. Potential energy is converted to kinetic energy once she dives.

There are five specific forms of energy: radiant, chemical, mechanical, electrical, and nuclear. In this book, we are particularly interested in radiant energy, chemical energy, and mechanical energy. Radiant energy, in the form of solar energy, can be captured by plants to make food for themselves and for the biosphere. Chemical energy is present in organic molecules, and therefore, food is a source of chemical energy. Food is a high-quality source of energy because the energy in it is available to do work. Mechanical energy is represented by any type of motion—the motion of a diver, as well as the motion of atoms, ions, or molecules. The latter is better known as **heat.** Heat is low-quality energy because it is too dispersed to do useful work. Usually heat is used to warm something, such as the human body, but we learned in Chapter 2 that excess heat can evaporate sweat, and in that way, lower the temperature of the body.

All the specific types of energy we have mentioned are either potential energy or kinetic energy. **Potential energy** is stored energy, and **kinetic energy** is energy in action. Potential energy is constantly being converted to kinetic energy, and vice versa.

Animation
Energy Conversion

Let's look at the example in Figure 5.5A. The chemical energy in the food a diver had for lunch contains potential energy of configuration. Energy is in the bonds and the relationships between atoms. When the diver climbs the ladder to the diving board, the potential energy of food is converted to the kinetic energy of motion. Once she reaches the diving board, kinetic energy has been converted to the potential energy of location (greater altitude). As she makes her dive, this potential energy is converted to kinetic energy again. The diver could not continue this cycle for long without a new supply of chemical energy. Chemical energy supplies the potential energy that organisms need to stay in existence and to perform work.

Both potential and kinetic energy are important to organisms because cells constantly store energy and then gradually release it to do work. To take an example, liver cells store energy as glycogen, and then they break down glycogen in order to make ATP molecules, which carry on the work of the cell.

It is important to have a way to measure energy. A **calorie** is the amount of heat required to raise the temperature of 1 g of water by 1°C. This isn't much energy, so the caloric value of food is listed in nutrition labels and in diet charts in terms of **kilocalories** (1,000 calories). In this text, we use Calorie (C) to mean 1,000 calories.

Energy Laws Two laws, called the laws of thermodynamics, govern the use of energy. These laws were formulated by early researchers who studied energy relationships and exchanges. Neither nonliving objects nor organisms can circumvent these laws.

The first law of thermodynamics—the law of conservation of energy—states that energy cannot be created or destroyed, but it can be changed from one form to another.

Figure 5.5B shows how this law applies to organisms. Grass is able to convert solar energy to chemical energy, and a horse, like all animals including humans, is able to convert chemical energy into the energy of motion. However, notice that with every energy transformation, some energy is lost as heat. The word "lost" recognizes that when energy has become heat, it is no longer usable to perform work.

The second law of thermodynamics states that energy cannot be changed from one form to another without a loss of usable energy.

Let's look at Figure 5.5B in a bit more detail. When grass photosynthesizes, it uses solar energy to form carbohydrate molecules from carbon dioxide and water. Carbohydrates are energy-rich molecules, while carbon dioxide and water are energy-poor molecules. Not all of the captured solar energy becomes carbohydrates; some becomes heat:

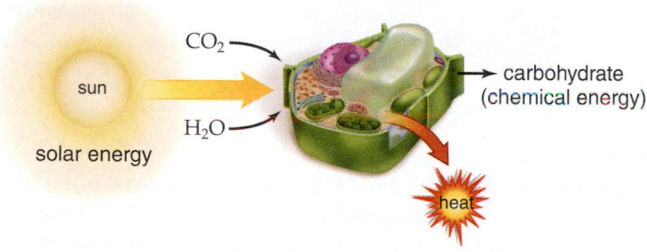

Plant cells do not create or destroy energy in this process—the sun is the energy source, and the unusable heat is still a form of energy. Similarly, as a horse uses the energy derived from carbohydrates to power its muscles, none is destroyed, but some becomes heat, which dissipates into the environment:

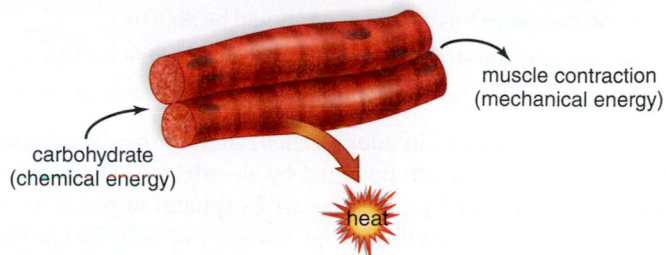

With transformation upon transformation, eventually all of the captured solar energy becomes heat that is lost to the environment. Therefore, we say that energy flows through organisms and does not cycle. Therefore, all organisms are dependent on a constant supply of solar energy because photosynthesizers use solar energy to synthesize organic molecules.

The second law of thermodynamics tells us that as energy conversions occur, disorder increases because it is difficult to use heat to perform more work. The word **entropy** is often used to describe this disorder. Energy transformations can occur, but they always increase entropy because no conversion of energy is ever 100% efficient. For example, the gasoline engine in an automobile is between 20% and 30% efficient in converting chemical energy into mechanical energy. The majority of energy is lost as heat. Cells are capable of about 40% efficiency, with the remaining energy given off to the surroundings as heat.

MP3 Laws of Thermodynamics

▶ **5.5 CHECK YOUR PROGRESS**
1. Describe how both potential and kinetic energy are important to organisms.
2. Apply the two laws of thermodynamics to a horse that is running away.

FIGURE 5.5B The grass converts solar energy to the chemical energy of nutrient molecules. The horse converts a portion of this chemical energy to the mechanical energy of motion. Eventually, all solar energy absorbed by the plant dissipates as heat. (These are Icelandic horses, native to Iceland.)

5.6 Cellular work is powered by ATP

LEARNING OUTCOMES

When you complete this section, you should be able to

1. Discuss why the structure of ATP is suited to its role in cells.
2. Explain how ATP breakdown can drive an endergonic reaction.

Many of the appliances in your kitchen, such as the dishwasher, stove, and refrigerator, are powered by electricity. Cells, as mentioned earlier, use **ATP** (adenosine triphosphate) to power reactions. ATP is often called the energy currency of cells. Just as you use cash to purchase all sorts of products, a cell uses ATP to carry out nearly all of its activities, including synthesizing biomolecules, transporting ions across plasma membranes, and causing organelles and cilia to move.

ATP is a nucleotide, the type of molecule that serves as a monomer for the construction of DNA and RNA. Its name, adenosine triphosphate, means that it contains the sugar ribose, the nitrogen-containing base adenine, and three phosphate groups (Fig. 5.6A). The three phosphate groups are negatively charged and repel one another. It takes energy to overcome their repulsion, and thus these phosphate groups make the molecule unstable.

ATP easily loses the last phosphate group because the breakdown products, ADP (adenosine diphosphate) and a separate phosphate group symbolized as $\circledP$, are more stable than ATP. This reaction is written as: ATP $\longrightarrow$ ADP + $\circledP$. ADP can also lose a phosphate group to become AMP (adenosine monophosphate).

The continual breakdown and regeneration of ATP is known as the ATP cycle (Fig. 5.6A). As soon as ATP forms, it is used in a reaction that requires energy. Then ATP is rebuilt from ADP + $\circledP$. Each ATP molecule undergoes about 10,000 cycles of synthesis and breakdown every day. Our bodies use some 40 kilograms (kg) of ATP daily, and the amount on hand at any one moment is sufficiently high to meet only current metabolic needs.

ATP's instability, the very feature that makes it an effective energy donor, keeps it from being an energy-storage molecule. Instead, the many H—C bonds of carbohydrates and fats make them the energy-storage molecules of choice. Their energy is extracted during cellular respiration and used to rebuild ATP, mostly within mitochondria. Cellular respiration, during which glucose is broken down, is called an **exergonic reaction** because this process gives up energy. In other words, energy *exits* from cellular respiration and is used to build up ATP. However, only 40% of the potential energy of glucose is converted to the potential energy of ATP; the rest is lost as heat. The production of ATP is still worthwhile for the following reasons:

1. ATP is suitable for use in many different types of cellular reactions that only occur if energy is supplied. Such reactions are called **endergonic reactions.** In other words, energy must *enter* in order for these reactions to occur.

2. When ATP becomes ADP + $\circledP$, the amount of energy released is more than the amount needed for a biological purpose, but not overly wasteful. Section 5.7 explains how cells capture the energy of ATP breakdown to perform work.

Figure 5.6B illustrates how exergonic reactions can drive endergonic reactions so that energy can be captured for a particular

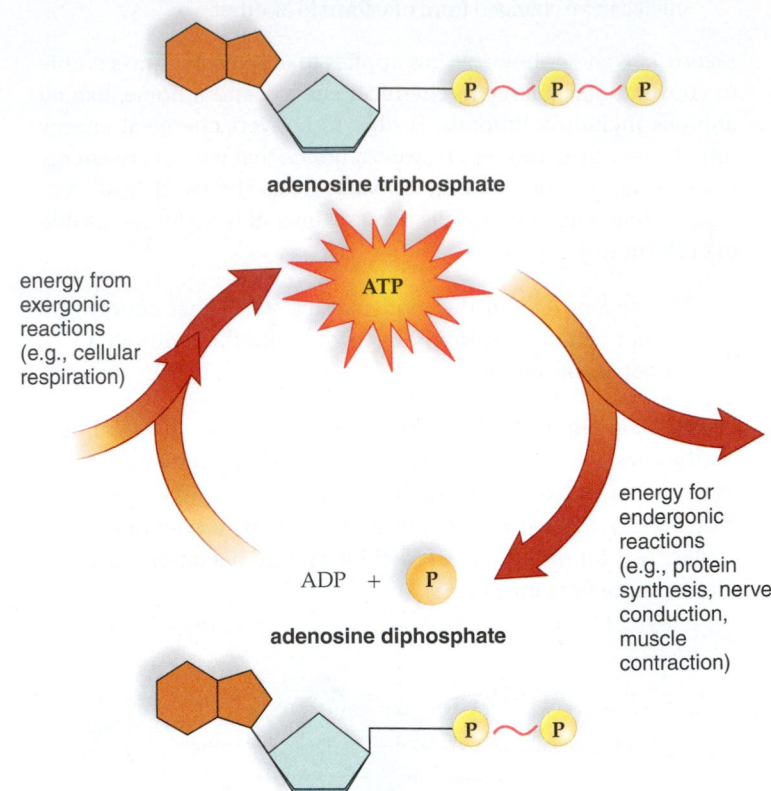

FIGURE 5.6A ATP cycle. First, ATP is produced in mitochondria, and then it is used for energy-requiring reactions in the cell. Therefore, ATP carries energy between exergonic reactions (in mitochondria) and endergonic reactions in the rest of the cell. When a phosphate group is removed by hydrolysis, ATP releases the appropriate amount of energy for most metabolic reactions.

FIGURE 5.6B This person is working and releasing energy (exergonic) and constructing a building which is energy-requiring (endergonic). Similarly there are also exergonic and endergonic reactions.

purpose. To bring about a structure such as a building or an organism, energy must be captured.

How can the energy released by ATP hydrolysis be transferred to a reaction that requires energy so that the reaction will occur? In other words, how does ATP act as a carrier of chemical energy? The answer is that ATP breakdown is coupled to the energy-requiring reaction. **Coupled reactions** are reactions that occur in the same place, at the same time, and in such a way that an energy-releasing (exergonic) reaction drives an energy-requiring (endergonic) reaction. Usually the energy-releasing reaction is the hydrolysis of ATP. Because the cleavage of ATP's phosphate group releases more energy than the amount consumed by the energy-requiring reaction, entropy increases, and both reactions proceed. The simplest way to represent a coupled reaction is like this:

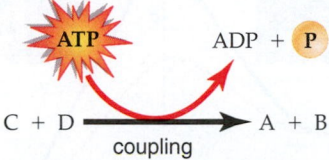

$$ATP \longrightarrow ADP + \textcircled{P}$$
$$C + D \xrightarrow{\text{coupling}} A + B$$

This reaction tells you that coupling occurs, but it does not show how coupling is achieved. A cell has two main ways to couple ATP hydrolysis to an energy-requiring reaction: ATP is used to energize a reactant, or ATP is used to change the shape of a reactant. Both can be achieved by transferring a phosphate group to the reactant.

For example, when iodine collects inside a thyroid cell, ATP is hydrolyzed, and instead of the last phosphate group floating away, an enzyme attaches it to a protein. This causes the protein to undergo a change in shape that allows it to move more iodine into the cell. As a contrasting example, when a polypeptide is synthesized at a ribosome, an enzyme transfers a phosphate group from ATP to each amino acid in turn, and this transfer supplies the energy that allows an amino acid to bond with another amino acid.

Figure 5.6C shows how ATP hydrolysis provides the necessary energy for muscle contraction. During muscle contraction, myosin filaments pull actin filaments to the center of the cell, and the muscle shortens. ❶ A myosin head at the end of a filament combines with ATP (three connected green triangles) and takes on its resting shape. ❷ When the ATP breaks down to ADP (two green triangles) plus ⓟ, the myosin head attaches to actin. ❸ The release of ADP plus ⓟ from the myosin head causes it to change shape and pull on the actin filament. The cycle begins again at ❶, when a myosin head combines with ATP and takes on its resting shape once more. During this cycle, chemical energy has been transformed to mechanical energy, and entropy has increased.

Animation
Breakdown of ATP and Cross-Bridge Movement During Muscle Contraction

Through coupled reactions, ATP drives forward energetically unfavorable processes that must occur if the high degree of order essential for life is maintained. Biomolecules must be made and organized to form cells and tissues; the internal composition of the cell and the organism must be sustained; and movement of cellular organelles and the organism must occur if life is to continue.

This completes our discussion of energy transformations in cells. In the next section of the chapter, we will study metabolism in general.

▶ **5.6 CHECK YOUR PROGRESS**
1. Identify reasons why ATP is called the energy currency of cells.
2. Use Figures 5.3A and 5.6C to show how ATP hydrolysis can be coupled to energy-requiring reactions.

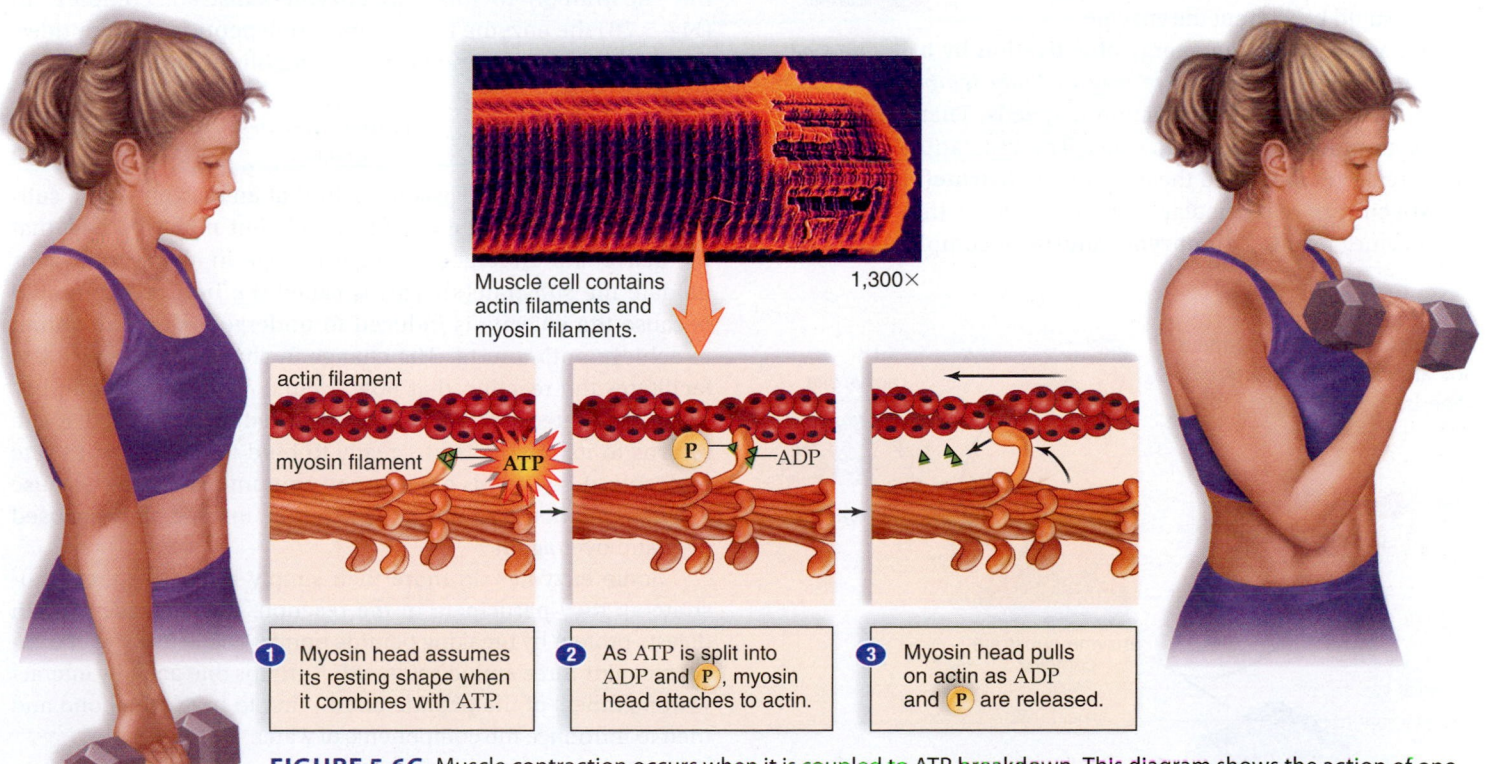

Muscle cell contains actin filaments and myosin filaments.

1,300×

actin filament

myosin filament — ATP

ⓟ — ADP

❶ Myosin head assumes its resting shape when it combines with ATP.

❷ As ATP is split into ADP and ⓟ, myosin head attaches to actin.

❸ Myosin head pulls on actin as ADP and ⓟ are released.

FIGURE 5.6C Muscle contraction occurs when it is coupled to ATP breakdown. This diagram shows the action of one myosin head, but actually many myosin heads work in unison. The electron micrograph is of muscle tissue.

Enzyme Function and Metabolic Pathways

The metabolism of a cell is governed by the enzymes present, because few reactions occur in a cell unless an enzyme is present. Enzymes lower the energy of activation by bringing specific reactants together in a way that causes them to react. Therefore, the study of metabolism involves a study of enzymes and how their activity is regulated by local conditions and by the influence of inhibitors.

5.7 Enzymes speed reactions

LEARNING OUTCOMES

When you complete this section, you should be able to

1. Explain how enzymes speed chemical reactions.
2. List conditions that affect enzyme speed and amount of product.
3. Explain two ways an inhibitor of enzymes can function.

The food on your dinner plate doesn't break down into nutrient molecules until it enters your digestive tract, where it encounters enzymes. An **enzyme** is typically a protein molecule that functions as an organic catalyst to speed a chemical reaction without itself being affected by the reaction. Just like your digestive tract, cells contain many types of enzymes. Regardless of where they are, enzymes cause reactions to occur. However, enzymes can only speed reactions that would occur anyway, not energetically unfavorable reactions.

MP3 Enzymes

Energy of Activation Imagine the graph in Figure 5.7A as a roller coaster ride. To get the ride started, you have to push the car (the reactants) to the top of an incline. Then, just as the car will naturally fall, the reaction will occur. In the lab, heat is often used to increase the effective collisions between molecules so that the reaction can occur. When an enzyme is present, the **energy of activation** (E_a) is lower than it would be without the enzyme.

Animation Energy of Activation

Enzymes lower the energy of activation by bringing reactants together in an effective way *at body temperature*. Each enzyme is specific to the reaction it speeds. That is why a cell needs so many different enzymes. The reactants in an enzymatic reaction are called the enzyme's **substrate**(s). Substrates are specific to a particular enzyme because they bind with an enzyme, forming an enzyme-substrate complex. Only one

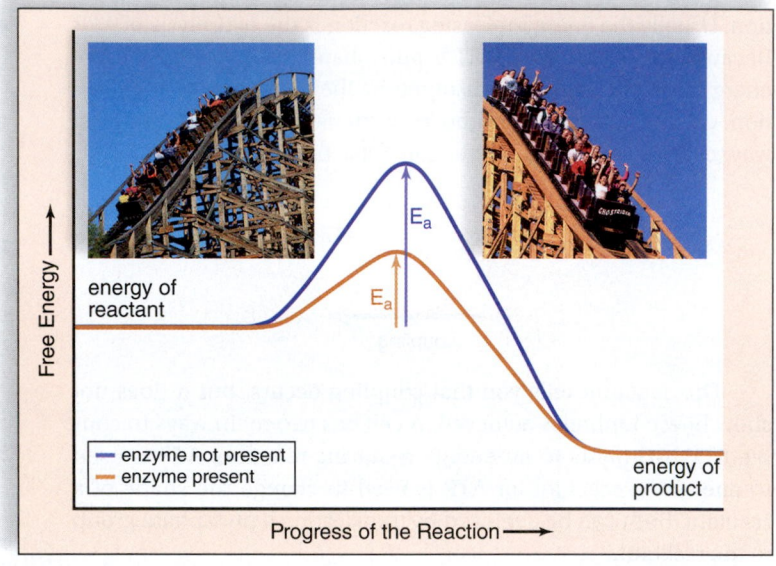

FIGURE 5.7A The energy of activation (E_a) (as when a roller coaster rises) allows the reaction to take place (roller coaster descends).

small part of the enzyme, called the **active site,** binds with the substrate(s) to form an enzyme-substrate complex. In (Fig. 5.7B) the enzyme is digesting a polypeptide to dipeptides, the product. This reaction can be symbolized as:

$$\underset{\text{enzyme}}{E} + \underset{\text{substrate}}{S} \longrightarrow \underset{\substack{\text{enzyme-substrate} \\ \text{complex}}}{ES} + \underset{\text{enzyme}}{E} + \underset{\text{product}}{P}$$

At one time, biologists thought that an enzyme and a substrate fit together like a key fits a lock, but now we know that the active site undergoes a slight change in shape to accommodate the substrate(s). This is called the **induced fit model** because the enzyme is induced to undergo a slight alteration to achieve optimum fit. The change in shape of the active site facilitates the reaction that now occurs. After the reaction has been completed, the product is released, and the active site returns to its original state, ready to bind to another substrate molecule. A cell needs only a small amount of enzyme because enzymes are not used up by the reaction; instead, they are used over and over again.

Some enzymes do more than simply bind with their substrate(s); they participate in the reaction. For example, trypsin digests protein by breaking peptide bonds. The active site of trypsin contains three amino acids with *R* groups that actually interact with members of the peptide bond—first to break the bond and then to introduce the components of water.

Factors Affecting Enzyme Speed The speed of a reaction is the amount of product produced per unit time. Generally, enzymes

FIGURE 5.7B An enzyme is not altered by the reaction it speeds.

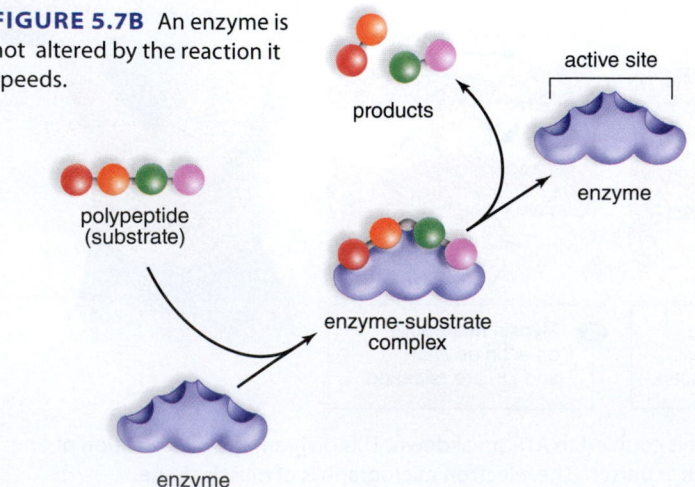

work quickly, and in some instances they can increase the reaction rate more than 10 million times. To achieve the maximum rate, enough substrate should be available to fill the active sites of all enzyme molecules most of the time. Increasing the amount of substrate, providing an adequate temperature, an optimal pH, and cofactors increase the rate of an enzymatic reaction.

Animation
How Enzymes Work

1. **Substrate Concentration** Molecules must come together in order to react. Generally, enzyme activity increases as substrate concentration increases because there are more chance encounters between substrate molecules and the enzyme. As more substrate molecules fill active sites, more product results per unit time. But when the enzyme's active sites are filled almost continuously with substrate, the enzyme's rate of activity cannot increase any more. Maximum rate has been reached.

   Just as the amount of substrate can increase or limit the rate of an enzymatic reaction, so the amount of active enzyme can also increase or limit the rate of an enzymatic reaction.

2. **Temperature** Typically, as temperature rises, enzyme activity increases (Fig. 5.7C). This occurs because warmer temperatures cause more effective encounters between enzyme and substrate.

   The body temperature of an animal seems to affect whether it is normally active or inactive. It has been suggested that the often cold temperature of a reptile's body (Fig. 5.7D) hinders metabolic reactions and may account for why mammals are more prevalent today. The generally warm temperature of a mammal's body (Fig. 5.7E) allows its enzymes to work at a rapid rate despite a cold outside temperature.

   In the laboratory, if the temperature rises beyond a certain point, enzyme activity eventually levels out and then declines rapidly because the enzyme has been **denatured.** An enzyme's shape changes during denaturation, and then it can no longer bind its substrate(s) efficiently. Nevertheless, some prokaryotes can live in hot springs because their enzymes do not denature.

3. **pH** Each enzyme also has an optimal pH at which the rate of the reaction is highest. At this pH value, these enzymes

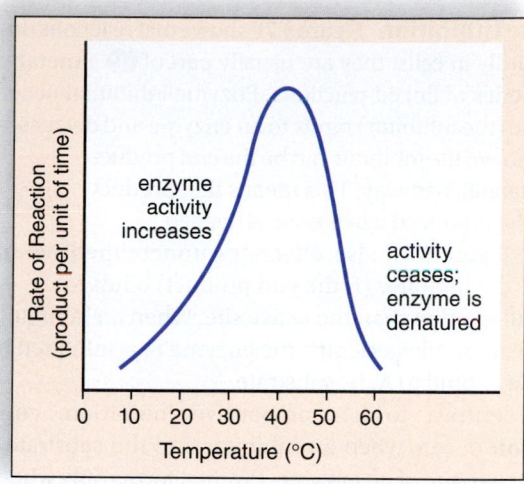

FIGURE 5.7C Temperature affects the rate of an enzymatic reaction.

have their normal configurations. The globular structure of an enzyme is dependent on interactions, such as hydrogen bonding, between *R* groups. A change in pH can alter the ionization of these side chains and disrupt normal interactions; under extreme conditions of pH, denaturation eventually occurs. Again, the enzyme's shape has been altered so that it is unable to combine efficiently with its substrate.

4. **Cofactors** Many enzymes require the presence of an inorganic ion, or a nonprotein organic molecule, in order to be active; these necessary ions or molecules are called **cofactors.** The inorganic ions are metals such as copper, zinc, or iron. The nonprotein organic molecules are called **coenzymes.** These cofactors assist the enzyme and may even accept or contribute atoms to the reactions.

   Vitamins are relatively small organic molecules that are required in trace amounts in the diets of humans and other animals for synthesis of coenzymes. The vitamin becomes part of a coenzyme's molecular structure. If a vitamin is not available, enzymatic activity decreases, and the result is a vitamin-deficiency disorder. For example, niacin deficiency results in a skin disease called pellagra, and riboflavin deficiency results in cracks at the corners of the mouth.

FIGURE 5.7D If body temperature tends to be cold, as in reptiles, reaction rates are slow.

FIGURE 5.7E If body temperature tends to be warm, as in mammals, reaction rates increase.

Enzyme Inhibition Figure 5.7F shows that reactions do not occur haphazardly in cells; they are usually part of **❶ a metabolic pathway,** a series of linked reactions. Enzyme inhibition occurs when a molecule (the inhibitor) binds to an enzyme and decreases its activity. As shown, the inhibitor can be the end product of a metabolic pathway. This means that product will not be produced when none is needed.

Animation
Biochemical Pathways

❷ Figure 5.7F also illustrates **noncompetitive inhibition** because the inhibitor (F, the end product) binds to the enzyme E_1 at a location other than the active site. When an inhibitor is at this site, called the allosteric site, the enzyme E_1 is inhibited because it is unable to bind to A, its substrate.

In contrast to noncompetitive inhibition, **competitive inhibition** occurs when an inhibitor and the substrate compete for the active site of an enzyme. Product forms only when the substrate, not the inhibitor, is at the active site. In this way, the amount of product is regulated.

Normally, enzyme inhibition is reversible, and the enzyme is not damaged by being inhibited. When enzyme inhibition is irreversible, the inhibitor permanently inactivates or destroys an enzyme. Many metabolic poisons are irreversible enzyme inhibitors.

Animation
Feedback Inhibition of Biochemical Pathways

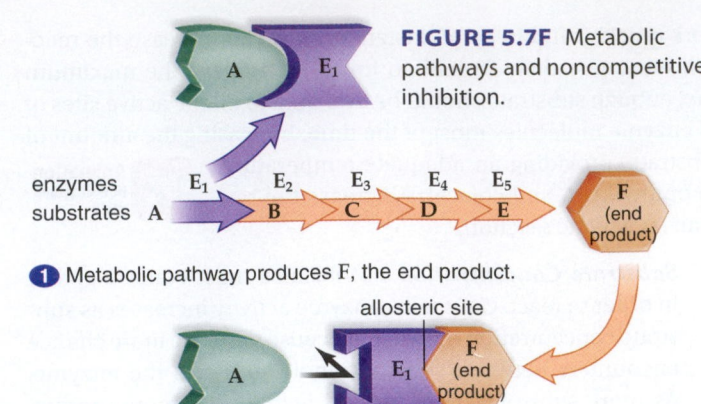

FIGURE 5.7F Metabolic pathways and noncompetitive inhibition.

enzymes
substrates A

❶ Metabolic pathway produces F, the end product.

❷ A cannot bind to E_1; the enzyme has been inhibited by F.

▶ **5.7 CHECK YOUR PROGRESS**
1. Identify the significance of an enzyme's active site to its function.
2. Explain why it is helpful to know the conditions under which an enzyme functions best.
3. Evaluate the benefit and the drawback to enzyme inhibition.

HOW BIOLOGY IMPACTS OUR LIVES   *Application*

5C Enzyme Inhibitors Can Spell Death

Cyanide gas was formerly used to execute people. How did it work? Cyanide can be fatal because it binds to a mitochondrial enzyme necessary for the production of ATP. MPTP (1-methyl-4-phenyl-1,2,3,6-tetrahydropyridine) is another enzyme inhibitor that stops mitochondria from producing ATP. The toxic nature of MPTP was discovered in the early 1980s, when a group of intravenous drug users in California suddenly developed symptoms of Parkinson disease, including uncontrollable tremors and rigidity. All of the drug users had injected a synthetic form of heroin that was contaminated with MPTP.

Sarin is a chemical that inhibits an enzyme at neuromuscular junctions, where nerves stimulate muscles. When the enzyme is inhibited, the signal for muscle contraction cannot be turned off, so the muscles are unable to relax and become paralyzed. Sarin can be fatal if the muscles needed for breathing become paralyzed. In 1995, terrorists released sarin gas on a subway in Japan. Although many people developed symptoms, only 17 died (Fig. 5C).

A fungus that contaminates and causes spoilage of sweet clover produces a chemical called warfarin. Cattle that eat the spoiled feed die from internal bleeding because warfarin inhibits a crucial enzyme for blood clotting. Today, warfarin is widely used as a rat poison. Unfortunately, it is not uncommon for warfarin to be mistakenly eaten by pets and even very small children, with tragic results. Still, following surgical procedures, patients are sometimes prescribed a medicine called Coumadin, the brand name for warfarin, to prevent inappropriate blood clotting.

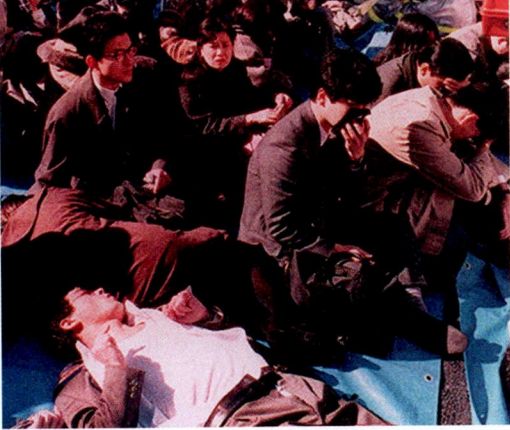

FIGURE 5C The aftermath when sarin, a nerve gas that results in the inability to breathe, was released by terrorists in a Japanese subway in 1995.

CONSIDER THESE QUESTIONS
1. Should all poisons be banned, or are they a necessary evil?
2. Should countries have poison detectors to warn people of a biological attack before they get sick? Why or why not?

connect
BIOLOGY   Explore the concepts through a variety of multimedia assets and question types.

www.mcgrawhillconnect.com

CONNECTING THE CONCEPTS

The dynamic plasma membrane is quite appropriately called the gatekeeper of the cell because its numerous proteins allow only certain substances to enter or exit. In a person who has diabetes, cystic fibrosis, or high cholesterol, knowing that a protein within the plasma membrane of a cell is malfunctioning is the first step toward curing the condition.

Multicellular organisms require mechanisms to join their cells and allow them to communicate. The plasma membrane excretes materials that support cells and also function in cell-to-cell communication. All the cells contain the same glycoproteins and glycolipids that mark the cell as belonging to that organism.

A cell is also dynamic because it carries out enzymatic reactions, many of which release or require energy. Energy is the ability to do work, to bring about change, and to make things

happen, whether it's a leaf growing or a human running. Exchanges across the plasma membrane allow the cell to continue to perform its usual reactions. Few reactions occur in a cell without the presence of an enzyme because enzymes lower the energy of activation by bringing substrates together. Enzymes are proteins, and as such they are sensitive to environmental conditions, including pH, temperature, and any inhibitors present. In a cell, noncompetitive inhibition regulates the activity of an enzyme

ATP, the universal energy "currency" of life, makes energy-requiring (endergonic) reactions go. Most often in cells, the exergonic breakdown of carbohydrates drives the buildup of ATP molecules. The metabolic pathways inside cells use the chemical energy of ATP to synthesize molecules, to cause muscle contraction, and even to allow you to read these words.

In Chapter 6, we will see how photosynthesis inside chloroplasts transforms solar energy into the chemical energy of carbohydrates. Then, in Chapter 7, we will discuss how carbohydrate products are broken down in mitochondria as ATP is built up. Chloroplasts and mitochondria are the cellular organelles that permit energy to flow from the sun through all organisms.

ANALYZE AND EVALUATE

1. How does enzyme structure result in a lowering of activation barriers?
2. In a cell, what environmental conditions must be met for an endergonic reaction such as polypeptide synthesis to occur?
3. Which types of proteins in the plasma membrane are involved in cell-to-cell recognition and communication?

SUMMARIZE

The Plasma Membrane Structure and Function

5.1 The plasma membrane is a phospholipid bilayer with embedded proteins

- The plasma membrane and all cell membranes are **phospholipid bilayers.** The current model for the plasma membrane is called the **fluid-mosaic model** because proteins form a mosaic pattern in the fluid phospholipid bilayer. **Cholesterol** is a steroid that lends stability to animal plasma membranes. Phospholipids and proteins with attached carbohydrate chains are called **glycolipids** and **glycoproteins,** respectively.
- Proteins in the plasma membrane have numerous functions. Channel proteins allow passage of molecules, and carrier proteins

assist the passage of molecules through the membrane. Cell recognition proteins are glycoproteins that help the body recognize its own cells. Receptor proteins bind specific signaling molecules. Enzymatic proteins carry out metabolic reactions. Certain junction proteins assist cell-to-cell communication.

The Passage of Molecules Into and Out of Cells

5.2 Diffusion across a membrane requires no energy input

- A plasma membrane is **differentially permeable:** Few molecules can diffuse through it. During **diffusion,** molecules in solution move down a **concentration gradient** until equally distributed. In **facilitated diffusion,** a transporter, either a carrier or channel protein, passively assists the passage of a solute. Channel proteins such as **aquaporins,** provide an open conduit; carrier proteins are not as fast because they combine with the substance.
- **Osmosis** is diffusion of water across a semipermeable membrane (e.g., plasma membrane). Cells placed in an **isotonic solution** neither gain nor lose water. Cells placed in a **hypotonic solution** gain water. Cells placed in a **hypertonic solution** lose water. Expansion of a cell due to gain of water in a hypotonic solution is called **turgor pressure.** Shrinking of a cell due to loss of water in a hypertonic solution is called **plasmolysis.**

5.3 Active transport and bulk transport require energy input

- During **active transport,** a substance moves against its concentration gradient. Protein carriers involved in active transport are called pumps (e.g., **sodium-potassium pump**).

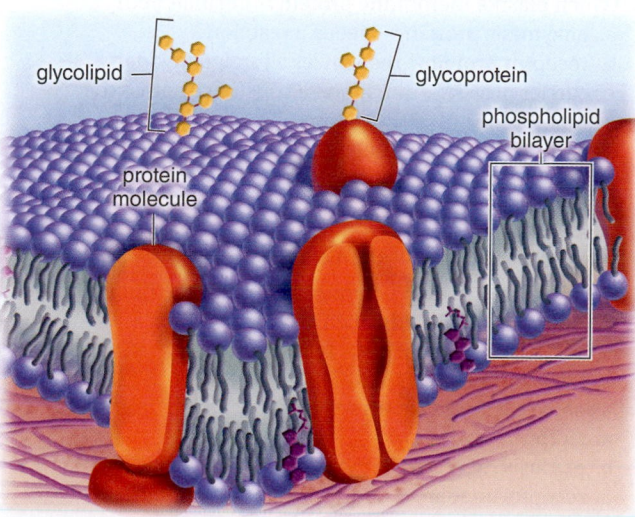

glycolipid

glycoprotein

phospholipid bilayer

protein molecule

- Bulk transport involves vesicle formation. In **endocytosis,** vesicles transport substances into the cell. **Phagocytosis** occurs when amoeboid cells engulf debris or bacteria. **Pinocytosis** occurs when vesicles form around a liquid or very small particles. **Receptor-mediated endocytosis** is selective and more efficient than ordinary pinocytosis. In **exocytosis,** vesicles transport substances (e.g., digestive enzymes, hormones) out of the cell.

The Plasma Membrane and Cell Communication

5.4 Extracellular material allows cells to join together and communicate

- Plant cells have cell walls; plant cells are joined by **plasmodesmata.** Animal cells are joined by junctions, such as **anchoring junctions** (cells are joined by filaments); **tight junctions** (function like a zipper); and **gap junctions** (channels are present). Animal cells have an **extracellular matrix (ECM)** that plays a role in cell signaling.

Energy Transformations in Organisms and Cells

5.5 Energy makes things happen

- **Energy** is the capacity to do work. The five forms of energy are radiant, chemical, mechanical, electrical, and nuclear. **Heat** is diffuse energy of motion that has limited usefulness but can be used to warm substances. **Potential energy** is stored energy, while **kinetic energy** is energy in action.
- A **calorie** is used to measure energy. A calorie is the amount of heat needed to raise the temperature of 1 g of water by 1°C. The caloric value of food is measured in **kilocalories.**
- Two laws apply to energy and its use. (1) First law of thermodynamics: Energy cannot be created or destroyed, but it can be changed from one form to another. (2) Second law of thermodynamics: Energy cannot be changed from one form to another without a loss of usable energy. When energy is transformed, some is always lost as heat, and **entropy** increases.

5.6 Cellular work is powered by ATP

- When **ATP** breaks down to ADP + P, energy is released. Continual hydrolysis and regeneration of ATP is called the ATP cycle. An **exergonic reaction** releases energy. An **endergonic reaction** requires energy to occur.
- ATP breakdown is coupled to energy-requiring reactions. A **coupled reaction** allows ATP breakdown (an exergonic reaction) to drive an endergonic reaction. ATP is not stored and is continually made during cellular respiration. As soon as it forms, ATP is used to drive an endergonic reaction in the cell.

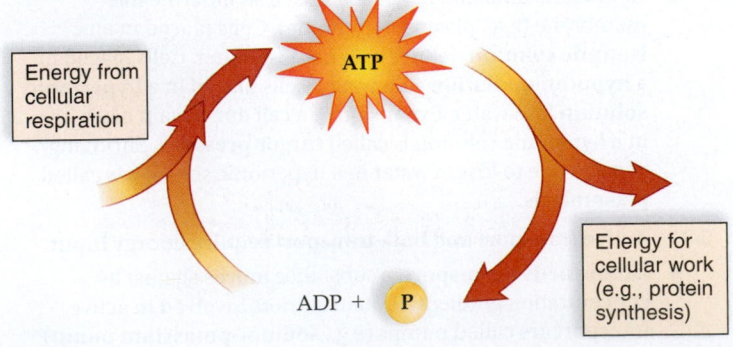

Energy from cellular respiration

ATP

ADP + P

Energy for cellular work (e.g., protein synthesis)

Enzyme Function and Metabolic Pathways

5.7 Enzymes speed reactions

- **Enzymes** are typically protein molecules that speed chemical reactions without being affected by the reactions. An enzyme lowers the **energy of activation** by bringing reactants together on its surface.
- Enzymes are specific, and the reactants in an enzymatic reaction are called its **substrate**(s). The **active site** is a small part of the enzyme that binds with the substrate(s), forming an enzyme-substrate complex. According to the **induced fit model,** the shape of the active site changes slightly to accommodate the **substrate**(s).
- Enzyme speed is affected by local conditions. Enzyme activity increases as substrate concentration increases until all active sites have been filled. As temperature rises, enzymatic activity increases until the temperature gets too high. Then the enzyme is **denatured,** and activity levels off and declines. Each enzyme has an optimal pH and may require **cofactors** or **coenzymes** for optimal reaction. **Vitamins** are required for synthesis of coenzymes.
- Enzymes can be inhibited noncompetitively and competitively. A **metabolic pathway** contains a number of sequential enzymes, and if one is inhibited, the pathway can shut down. Enzyme inhibition occurs when a substance binds to an enzyme and decreases its activity. In **noncompetitive inhibition,** the inhibitor, often the end product of a metabolic pathway, binds to an enzyme at the allosteric site. In **competitive inhibition,** an inhibitor and a substrate compete for the enzyme's active site.

TEST YOURSELF

The Plasma Membrane Structure and Function

1. In the fluid-mosaic model, which component of the membrane is fluid?
   a. the glycoprotein   d. All of these are fluid.
   b. only cholesterol   e. Both b and c are correct.
   c. only the lipid bilayer
2. Cholesterol is found in the _____ of _____ cells.
   a. cytoplasm, plant
   b. plasma membrane, animal
   c. plasma membrane, plant
   d. None of these are correct.
3. Which plasma membrane protein is mismatched?
   a. enzymatic protein—speeds a reaction
   b. receptor protein—recognizes a foreign invader
   c. carrier protein—transports a substance across the membrane
   d. channel protein—transports a substance across the membrane
   e. All of these are correct.

The Passage of Molecules Into and Out of Cells

4. Cells involved in active transport have a large number of _____ near their plasma membrane.
   a. vacuoles   c. actin filaments
   b. mitochondria   d. lysosomes
5. A coated pit is associated with
   a. simple diffusion.
   b. osmosis.
   c. receptor-mediated endocytosis.
   d. pinocytosis.

6. Phagocytosis is common in
   a. amoebas.
   c. red blood cells.
   b. white blood cells.
   d. Both a and b are correct.
7. When a cell is placed in a hypotonic solution,
   a. solute exits the cell to equalize the concentration on both sides of the membrane.
   b. water exits the cell toward the area of lower solute concentration.
   c. water enters the cell toward the area of higher solute concentration.
   d. solute exits and water enters the cell.
   e. Both c and d are correct.
8. **THINKING CONCEPTUALLY** Human blood always has a greater concentration of solutes than does tissue fluid. Why is that important to blood's transport function?

For questions 9–12, match the items to the transport methods in the key. Each question may have more than one answer.

KEY:
   a. simple diffusion
   b. facilitated diffusion
   c. osmosis
   d. active transport

9. Movement of molecules, from high concentration to low concentration
10. Requires a membrane
11. Requires energy input
12. Requires a transporter

The Plasma Membrane and Cell Communication

13. Which type of junction holds neighboring cells together so tightly that fluids cannot pass between them?
    a. anchoring
    c. plasmodesmata
    b. gap
    d. tight
14. Communication between cells often involves
    a. cell signaling.
    b. a signal transduction pathway.
    c. only enzymes and energy.
    d. the passage of molecules by channel proteins.
    e. All but c are correct.

Energy Transformations in Organisms and Cells

15. A diver at a diving board has _____ energy, and during the dive, the diver represents _____ energy.
    a. potential, potential
    c. potential, kinetic
    b. kinetic, kinetic
    d. kinetic, potential
16. The _____ of energy would break a law of thermodynamics.
    a. creation
    c. Both a and b are correct.
    b. transformation
    d. Neither a nor b is correct.
17. As a result of energy transformations,
    a. entropy increases.
    b. entropy decreases.
    c. heat energy is gained.
    d. energy is lost in the form of heat.
    e. Both a and d are correct.
18. When ATP breaks down, what type of reaction has occurred?
    a. hydrolysis
    d. Both a and c are correct.
    b. endergonic
    e. All of these are correct.
    c. exergonic

19. Coupling two reactions
    a. allows ATP to drive an endergonic reaction.
    b. allows energy to be created so that a transformation will occur.
    c. allows potential energy to become mechanical energy.
    d. opposes the energy laws governing energy use.
    e. Both a and c are correct.
20. **THINKING CONCEPTUALLY** How is the bulk of food you bring home from the store like glycogen, and how is ATP like the meal you prepare?

Enzyme Function and Metabolic Pathways

21. The current model for enzyme action is called the
    a. induced fit model.
    c. lock-and-key model.
    b. activation model.
    d. active substrate model.
22. The active site of an enzyme
    a. is identical to that of any other enzyme.
    b. is the part of the enzyme where its substrate can fit.
    c. can be used over and over again.
    d. is not affected by environmental factors such as pH and temperature.
    e. Both b and c are correct.
23. Vitamins can be components of
    a. coreactants.
    c. coenzymes.
    b. cosugars.
    d. None of these are correct.
24. Temperature
    a. usually inhibits an enzyme because a boiling temperature can denature an enzyme.
    b. can affect the speed of an enzymatic reaction.
    c. is the type of energy that combines with an allosteric site.
    d. All of these are correct.

GET INVOLVED

1. Use the first and second laws of thermodynamics to explain why each ecosystem needs a continuous supply of solar energy.
2. You have been asked to design an experiment to illustrate enzyme function. What factors, including environment, should you consider?
3. For a detailed examination of the processes involved in the movement of molecules across the plasma membrane, watch McGraw-Hill's 3D Animation "Membrane Transport."

3D Animation
Membrane Transport

MEDIA STUDY TOOLS

mhhe.com/maderconcepts3

Enhance your study of this chapter with interactive study tools, practice tests, and engaging animations. Also, ask your instructor about the resources available through ConnectPlus, which includes LearnSmart, a personalized adaptive learning program, and a media-rich eBook.

6

Pathways of Photosynthesis

BEFORE YOU BEGIN

Take a few minutes to recall

The role of producers in an ecosystem (section 1.6)

The function of chloroplasts (section 4.9)

How energy conversions occur (section 5.5) and the ATP cycle (section 5.6)

The Sun Drives Photosynthesis

The sun's rays power photosynthesis, the marvelous process that directly produces food for plants and indirectly for nearly all organisms on Earth. Plants are colorful; they contain pigments that can absorb a portion of the light energy that floods our world every day. Some of this energy is stored in the covalent bonds of newly manufactured sugar molecules.

White or visible light contains different colors of light, from violet to blue, green, yellow, orange, and finally red. Land plants use all the colors except green—and that's why we see them as green! It's easy to show that plants do not absorb green light for photosynthesis. Simply put a sprig of the plant elodea in a glass jar, fill it with water (elodea lives in water), and shine a green light on it. NOTHING HAPPENS. But switch to a bright white light, and watch the bubbling. Bubbling is caused by oxygen gas escaping from the water. Plants give off oxygen when they photosynthesize and this is the very gas that we and other animals breathe in to stay alive.

Algae are also photosynthesizers. Red algae live deep in the ocean, and despite their name, some forms of red algae are dark-colored, almost black, which means that they are able to use

all the different colors in white light for photosynthesis. Does this mean that if plants use green light in addition to all the other colors when they photosynthesize, they would appear black to us? Yes, it does. Look out the window and imagine that all the plants you see are black instead of green.

Land plants evolved from aquatic green algae, which typically live on the surface of ponds and oceans where light is readily available. They contain the green pigment chlorophyll, which absorbs the blue and red ranges of light. Plants are green and our world is beautiful because their primary photosynthetic pigment doesn't absorb green light! The neglect of green light by chlorophyll in a habitat where light is abundant is not a significant drawback to them.

Green leaves and variously colored eukaryotic algae carry on photosynthesis in chloroplasts, as discussed in this chapter. We will see that photosynthesis consists of two connected types of metabolic pathways: the light reactions and the Calvin cycle reactions. The absorption of solar energy during the light reactions splits water and releases oxygen and also drives the Calvin cycle reactions, which produce a carbohydrate. The conclusion is that plants provide the oxygen we inhale and the glucose we use as a source of energy when they photosynthesize to provide food for themselves.

If land plants used all colors of light to photosynthesize, they would be black in color.

green alga, *Chlorella*

100×

Overview of Photosynthesis

Photosynthesizers not only produce food for the biosphere, but they are also the source of the fossil fuel our society burns to maintain its standard of living. The overall equation for photosynthesis shows the starting reactants and the end products. But we will see that, in actuality, photosynthesis requires two metabolic pathways: the light reactions that occur in thylakoid membranes and the Calvin cycle reactions that occur in the stroma of a chloroplast.

6.1 Photosynthesizers are autotrophs that produce their own food

LEARNING OUTCOME

When you complete this section, you should be able to

1. Identify and describe the anatomy of the leaf tissues and cellular structure where photosynthesis occurs.

Photosynthesis converts solar energy into the chemical energy of a carbohydrate. Photosynthetic organisms, including plants, algae, and cyanobacteria, are called **autotrophs** because they produce their own organic food (Fig. 6.1A). Photosynthesis produces an enormous amount of carbohydrate. So much that, if it could be instantly converted to coal and the coal loaded into standard railroad cars (each car holding about 50 tons), the photosynthesizers of the biosphere would fill more than 100 cars with coal *per second*.

No wonder photosynthetic organisms are able to sustain themselves and all other organisms on Earth. With few exceptions, it is possible to trace any food chain back to plants and algae. In other words, producers, which have the ability to synthesize carbohydrates, feed not only themselves but also consumers, which must take in preformed organic molecules. Collectively, consumers are called **heterotrophs.** Both autotrophs and heterotrophs use organic molecules produced by photosynthesis as a source of building blocks for growth and repair and as a source of chemical energy for cellular work.

Our analogy about photosynthetic products becoming coal is apt because the bodies of many ancient plants became the coal we burn today. This process began several hundred million years ago, and that is why coal is called a fossil fuel. Today we use coal in large part to produce electricity. In the future, it's possible not only agricultural crops (e.g., corn), but also grasses, algae, and wastes, such as wood chips, will be used to produce biofuels (alcohols, methane, or diesel gas) and these will run power plants or your car.

Video Plants

FIGURE 6.1A Photosynthetic organisms include cyanobacteria, algae, and plants.

Euglena, a green flagellate

Kelp, an alga

Trees, deciduous and evergreen plants

Gloeocapsa, a cyanobacterium

Diatoms, an alga

Sunflower, a garden plant

Moss, a plant

Chloroplasts Chloroplasts occur in algae and plants. Photosynthesis takes place in the green portions of plants, particularly the leaves (Fig. 6.1B). **1** The leaves of a plant contain mesophyll tissue in which cells are specialized for photosynthesis. The raw materials for photosynthesis are water and carbon dioxide. The roots of a plant absorb water, which then moves in vascular tissue up the stem to a leaf where it is distributed by way of **2** the leaf veins. **3** Carbon dioxide in the air enters a leaf through small openings called **stomata** (sing., stoma).

After entering a leaf cell, carbon dioxide and water diffuse into **chloroplasts,** the organelles that carry on photosynthesis. **4** A double membrane surrounds a chloroplast and its fluid-filled interior, called the **5 stroma.** A different membrane system within the stroma forms flattened sacs called **thylakoids,** which in some places are stacked to form **6 grana** (sing., granum), so called because they looked like piles of seeds to early microscopists.

7 The space of each thylakoid is thought to be connected to the space of every other thylakoid within a chloroplast, thereby forming an inner compartment within chloroplasts called the thylakoid space. The thylakoid membrane contains **chlorophyll** and other pigments that are capable of absorbing solar energy. This is the energy that drives photosynthesis. The stroma contains a metabolic pathway where carbon dioxide is first attached to an organic compound and then converted to a carbohydrate. Therefore, it is proper to associate the absorption of solar energy with the thylakoid membranes making up the grana and to associate the conversion of carbon dioxide to a carbohydrate with the stroma of a chloroplast.

3D Animation
Photosynthesis: Structure of a Chloroplast

Humans, and indeed nearly all organisms, release carbon dioxide into the air. This is some of the same carbon dioxide that enters a leaf through the stomata and is converted to a carbohydrate. Carbohydrate, in the form of glucose, is the chief energy source for most organisms.

▶ **6.1 CHECK YOUR PROGRESS**
   1. Identify two major parts of a chloroplast and tell how each contributes to photosynthesis.

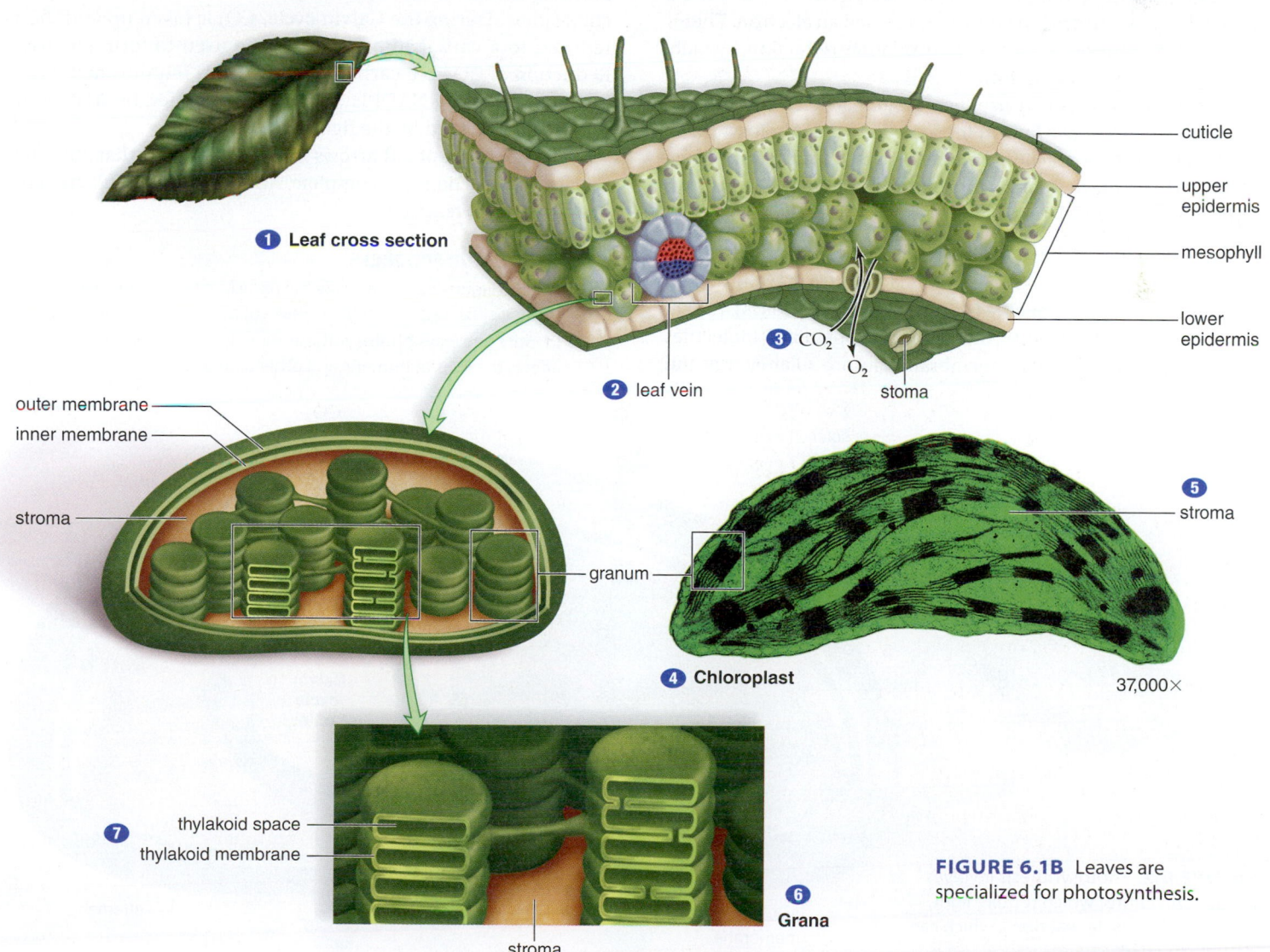

1 Leaf cross section

cuticle

upper epidermis

mesophyll

lower epidermis

3 CO_2

O_2

2 leaf vein

stoma

outer membrane
inner membrane

stroma

granum

5
stroma

4 Chloroplast

37,000×

7

thylakoid space
thylakoid membrane

6
Grana

stroma

FIGURE 6.1B Leaves are specialized for photosynthesis.

6.2 Photosynthesis involves two sets of reactions: The light reactions and the Calvin cycle reactions

LEARNING OUTCOMES

When you complete this section, you should be able to

1. Describe how the overall equation for photosynthesis is a redox equation.
2. Identify how the light reactions drive the Calvin cycle reactions.

During photosynthesis, hydrogen atoms are transferred from water to carbon dioxide with the release of O_2 and the formation of glucose:

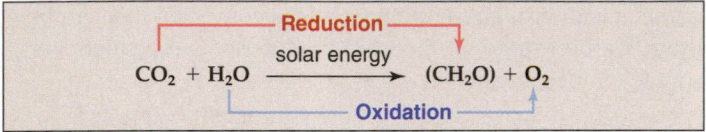

$$CO_2 + H_2O \xrightarrow{\text{solar energy}} (CH_2O) + O_2$$

Reduction →
← Oxidation

In chemistry, **oxidation** is the loss of electrons, and **reduction** is the gain of electrons. In cells, oxidation is instead the loss of hydrogen atoms, and reduction is the gain of hydrogen atoms. This difference is easily explained because a hydrogen atom contains one electron and one hydrogen ion ($e^- + H^+$); when a molecule loses a hydrogen atom, it has lost an electron, and when a molecule gains a hydrogen atom, it has gained an electron. Therefore, photosynthesis is indeed an oxidation-reduction, usually shortened to a **redox reaction.**

Even though we often use the overall reaction for photosynthesis, researchers have known for some time that photosynthesis involves two sets of reactions. These two sets of reactions can be associated with the two parts of a chloroplast—namely, the stacks of thylakoids (grana) and the stroma, as shown in Figure 6.2.

Light Reactions The **light reactions** (sometimes called the light-dependent reactions) are so named because they only occur when solar energy is available (during the daylight hours). At that time, the chlorophyll molecules and other pigment molecules located within the thylakoid membranes absorb solar energy and use it to energize electrons taken from water, which splits, releasing oxygen:

$$H_2O \longrightarrow e^- + 2H^+ + \tfrac{1}{2} O_2$$

The energy of these electrons empowered by the sun is captured and later used for ATP production.

Energized electrons are also taken up by a coenzyme called NADP$^+$ (nicotinamide adenine dinucleotide phosphate). After NADP$^+$ accepts electrons, it combines with an H$^+$ derived from water. The reaction that reduces NADP$^+$ is as follows:

$$NADP^+ + 2\,e^- + H^+ \longrightarrow NADPH$$

The lower set of red arrows in Figure 6.2 show that the NADPH and ATP produced by the light reactions are sent to the Calvin cycle reactions.

Calvin Cycle Reactions The Calvin cycle reactions (sometimes called the light-independent reactions because they can occur both day and night) are named for Marvin Calvin who discovered them and their cyclical arrangement. The enzymatic reactions of the Calvin cycle occur in the stroma of a chloroplast. During the Calvin cycle, CO_2 is taken up and then reduced to a carbohydrate that can be used to form glucose. Reduction of CO_2 to a carbohydrate (CH_2O) requires hydrogen atoms supplied by NADPH and energy supplied by ATP, both molecules provided by the light reactions.

The upper set of red arrows in Figure 6.2 show that after the Calvin cycle reactions are complete, ADP + ⓅP and NADP$^+$ are sent back to the light reactions.

▶ **6.2 CHECK YOUR PROGRESS**

1. Refer to Figure 6.2 and identify the participating role for each molecule that appears in the overall equation for photosynthesis.
2. Identify two ways photosynthesis directly and/or immediately affects the lives of humans and other animals?

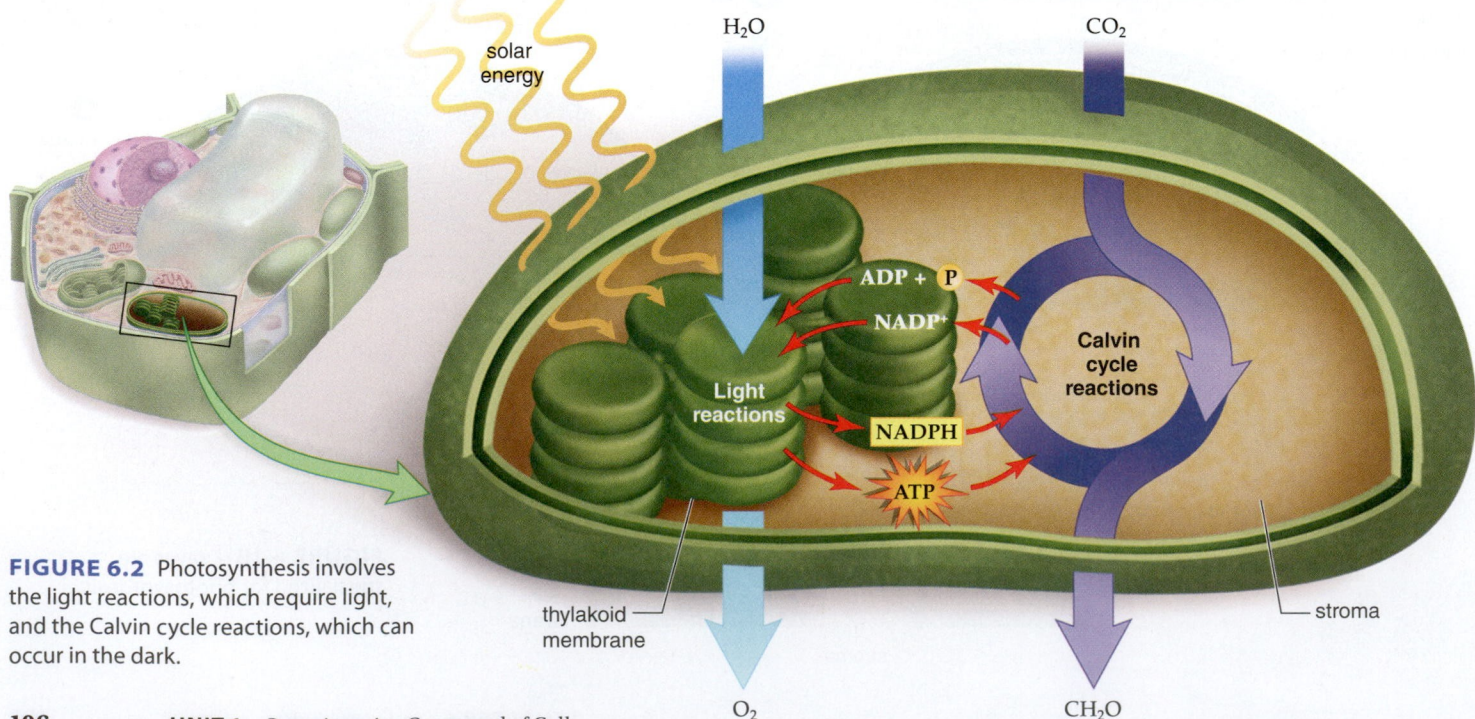

FIGURE 6.2 Photosynthesis involves the light reactions, which require light, and the Calvin cycle reactions, which can occur in the dark.

The Light Reactions—Harvesting Energy

The light reactions begin when pigments in the thylakoid membrane absorb certain wavelengths of light. The thylakoid membrane is highly organized for absorbing solar energy and producing ATP and NADPH, needed by the Calvin cycle reactions to reduce carbon dioxide to a carbohydrate in the stroma.

6.3 Solar energy is absorbed by pigments

LEARNING OUTCOME

When you complete this section, you should be able to

1. List the types of photosynthetic pigments and briefly explain their functions.

Solar energy (radiant energy from the sun) can be described in terms of its wavelength and its energy content. Figure 6.3A lists the different types of radiant energy, from the shortest wavelength, gamma rays, to the longest, radio waves. We are most interested in white light, or *visible light,* because it is the type of radiation used for photosynthesis and for vision.

When visible light is passed through a prism, we can observe that it is made up of various colors. (Actually, of course, our brain interprets these wavelengths as colors.) The colors in visible light range from violet (the shortest wavelength) to blue, green, yellow, orange, and red (the longest wavelength). The energy content is highest for violet light and lowest for red light.

The pigments within most types of photosynthesizing cells are **chlorophylls *a*** and ***b*** and **carotenoids.** These pigments are capable of absorbing various portions of visible light. The absorption spectrum for these pigments is shown in Figure 6.3B. Both chlorophyll *a* and chlorophyll *b* absorb violet, blue, and red light better than the light of other colors. Because green light is reflected and only minimally absorbed, leaves appear green to us. The yellow or orange carotenoids are able to absorb light in the violet-blue-green range. Only in the fall is it obvious that pigments other than chlorophyll assist in absorbing solar energy. In the spring and summer, plant cells mask the instability of chlorophyll by using ATP molecules to rebuild it. As the hours of sunlight lessen in the fall, sufficient energy to rebuild chlorophyll is not available. Further, enzymes are working at a reduced speed because of the lower temperatures. Therefore, the amount of chlorophyll in leaves slowly disintegrates. When that happens, we begin to see yellow and orange pigments in the leaves.

In some trees, such as maples, certain pigments accumulate in acidic vacuoles, leading to a brilliant red color. The brown color of certain oak leaves is due to wastes left in the leaves.

3D Animation
Photosynthesis: Properties of Light

▶ **6.3 CHECK YOUR PROGRESS**
  1. Explain why white light drives photosynthesis better than any of the individual colors of light.

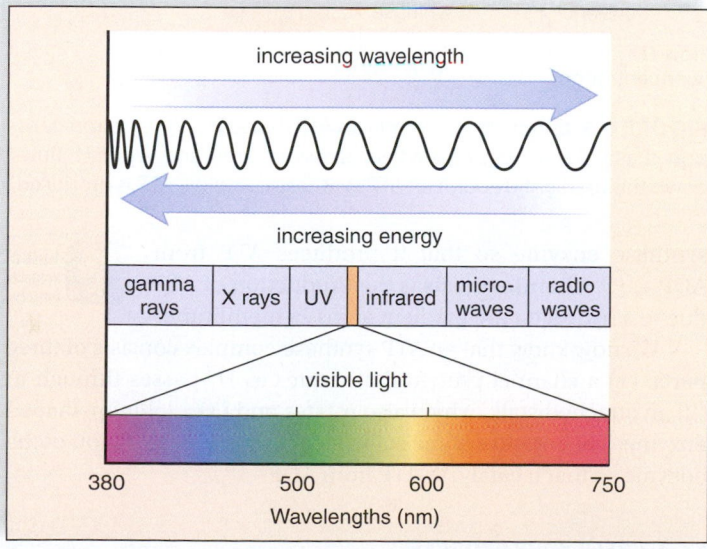

FIGURE 6.3A The electromagnetic spectrum includes visible light, which drives photosynthesis. Here, visible light is expanded to show its component colors.

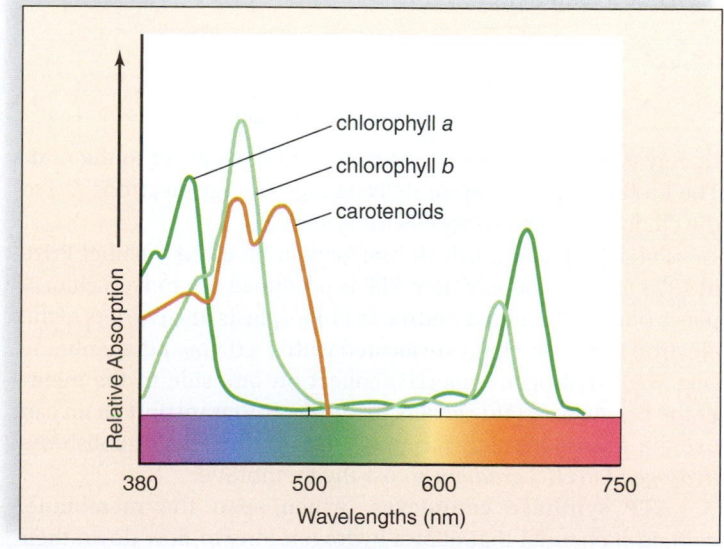

FIGURE 6.3B Absorption spectrum of photosynthetic pigments. Notice that these pigments do not absorb green light—that's why the leaves of plants appear green to us.

6.4 Solar energy boosts electrons to a higher energy level

LEARNING OUTCOME

When you complete this section, you should be able to

1. Explain the function of each portion of a photosystem.

In the thylakoid membrane, photosystems contain (1) a pigment complex that absorbs solar energy and (2) an electron-acceptor molecule (Fig. 6.4). Each pigment complex consists of antenna molecules and a reaction center. Antenna molecules are light-absorbing pigments, such as chlorophyll *a* and *b* and carotenoid pigments. They are called antenna molecules because they absorb light energy just as a radio antenna absorbs radio waves. The antenna molecules pass all their energy on to the reaction center, which contains a particular chlorophyll *a* molecule. The reaction center excites electrons and sends them on to the electron acceptor. Remember the old-fashioned "use a mallet to ring the bell, win a prize" carnival game? Similarly, solar energy has been used to launch electrons from the reaction center all the way up to the energy level of the electron acceptor.

Two types of photosystems, called **photosystem I** (**PS I**) and **photosystem II** (**PS II**), participate in the light reactions. The reaction center in PS I absorbs light with a wavelength of 700 nm, and the one in PS II absorbs light with a wavelength of 680 nm.

Animation
Absorption of Light

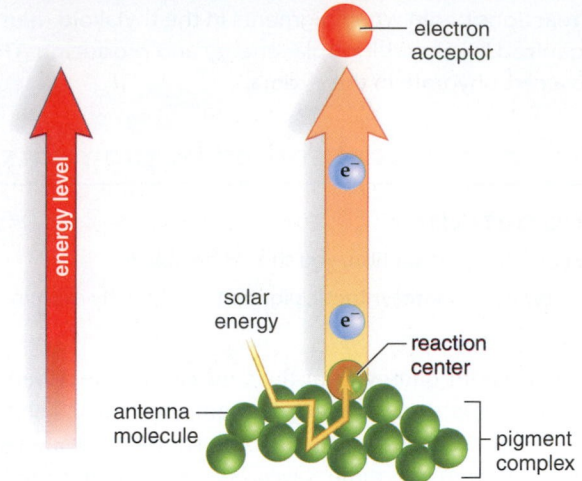

FIGURE 6.4 A general model of a photosystem.

▶ **6.4 CHECK YOUR PROGRESS**
  1. Describe the function of a pigment complex in a photosystem.

6.5 Solar energy is converted to the chemical energy of ATP

LEARNING OUTCOME

When you complete this section, you should be able to

1. Recognize how solar energy is converted to the chemical energy of ATP during the light reactions.

Solar energy energizes electrons in PS II and this energy is converted to that of ATP. The process is complicated and involves the **electron transport chain** (**ETC**), a series of carriers (colored boxes) located in the thylakoid membrane that pass electrons from one to another.

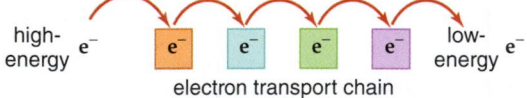

electron transport chain

Energy is released as electrons pass from one carrier to the next: The PS II electron acceptor delivers high-energy electrons (e⁻) to the chain, and low-energy electrons leave it.

Peter Mitchell, a British biochemist, received a Nobel Prize in 1978 for his model of how ATP is produced not only in chloroplasts but also in mitochondria. In chloroplasts, the carriers of the electron transport chain are located within a thylakoid membrane (Fig. 6.5). Hydrogen ions (H^+) collect on one side of the membrane because certain carriers of the electron transport chain can use the released energy to pump them there. This establishes a *hydrogen ion* (H^+) *gradient* across the membrane.

ATP synthase complexes, which span the membrane, contain a channel that allows hydrogen ions to flow down their concentration gradient. The flow of hydrogen ions through the channel is a form of kinetic energy that alters the active site of the ATP

Animation
Electron Transport and ATP Synthesis

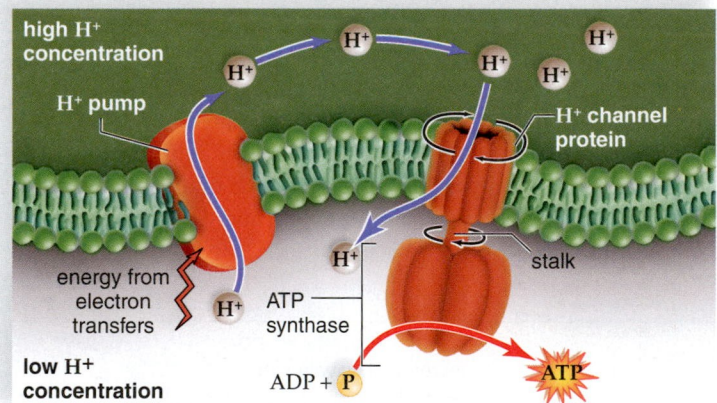

FIGURE 6.5 Chemiosmosis produces ATP molecules. An electron transport chain establishes a H^+ gradient across a membrane. When H^+ flows down this gradient through an ATP synthase complex, ATP is produced.

synthase enzyme so that it produces ATP from ADP + Ⓟ. **Chemiosmosis** is the production of ATP due to a hydrogen ion gradient across a membrane.

Video
Spinach Battery

We now know that an ATP synthase complex consists of three parts: (1) a channel protein that rotates as H^+ passes through it; (2) an attached stalk, which also rotates and (3) a lollipop-shaped enzyme. The rotation of the stalk changes the configuration of the enzyme so that it catalyzes ATP from ADP + Ⓟ.

▶ **6.5 CHECK YOUR PROGRESS**
  1. Describe, as if to a friend, how the energy captured by PS II results in ATP production.

6.6 The noncyclic flow of electrons produces ATP and NADPH

When you complete this section, you should be able to

1. Describe the noncyclic electron pathway during the light reactions and give a function for each participant.

During the light reactions in the thylakoid membrane, electrons follow a *noncyclic pathway* that begins with PS II (Fig. 6.6). ❶ The absorbed solar energy in PS II is passed from one pigment to the other in the pigment complex, until it is concentrated in the reaction center, which is a special chlorophyll *a* molecule. ❷ Electrons (e⁻) become so energized that they escape from the reaction center and ❸ move to a nearby electron acceptor.

PS II would disintegrate without replacement electrons; thus, ❹ electrons are removed from water, which splits, releasing oxygen to the atmosphere. Notice that with the loss of electrons, water has been oxidized, and that indeed, ❺ the oxygen released during photosynthesis comes from water. The oxygen escapes and exits into the atmosphere by way of a leaf's stomata (see Fig. 6.1B). ❻ However, the hydrogen ions (H⁺) are trapped inside a thylakoid.

❼ The electron acceptor sends the energized electrons down an electron transport chain and ATP is produced. (We now know that as the electrons pass from one carrier to the next, certain carriers pump hydrogen ions across the thylakoid membrane. In this way, energy is captured and stored in the form of a hydrogen ion (H⁺) concentration gradient. Later, as H⁺ flows down this gradient through an ATP synthase complex, ATP is produced (see Fig. 6.5).

When the PS I pigment complex absorbs solar energy, energized electrons leave its reaction center and are captured by a different ❽ electron acceptor. (Low-energy electrons from the electron transport chain adjacent to PS II replace those lost by PS I.) This electron acceptor passes its electrons on to ❾ NADP⁺ molecules. ❿ Each NADP⁺ molecule accepts two electrons and an H⁺ to become a reduced form of the molecule—that is, NADPH. ATP and NADPH from the light reactions are used by the Calvin cycle reactions to reduce carbon dioxide to a carbohydrate.

▶ **6.6 CHECK YOUR PROGRESS**
1. Explain how the capture of solar energy leads to the oxidation of water and release of oxygen during photosynthesis.

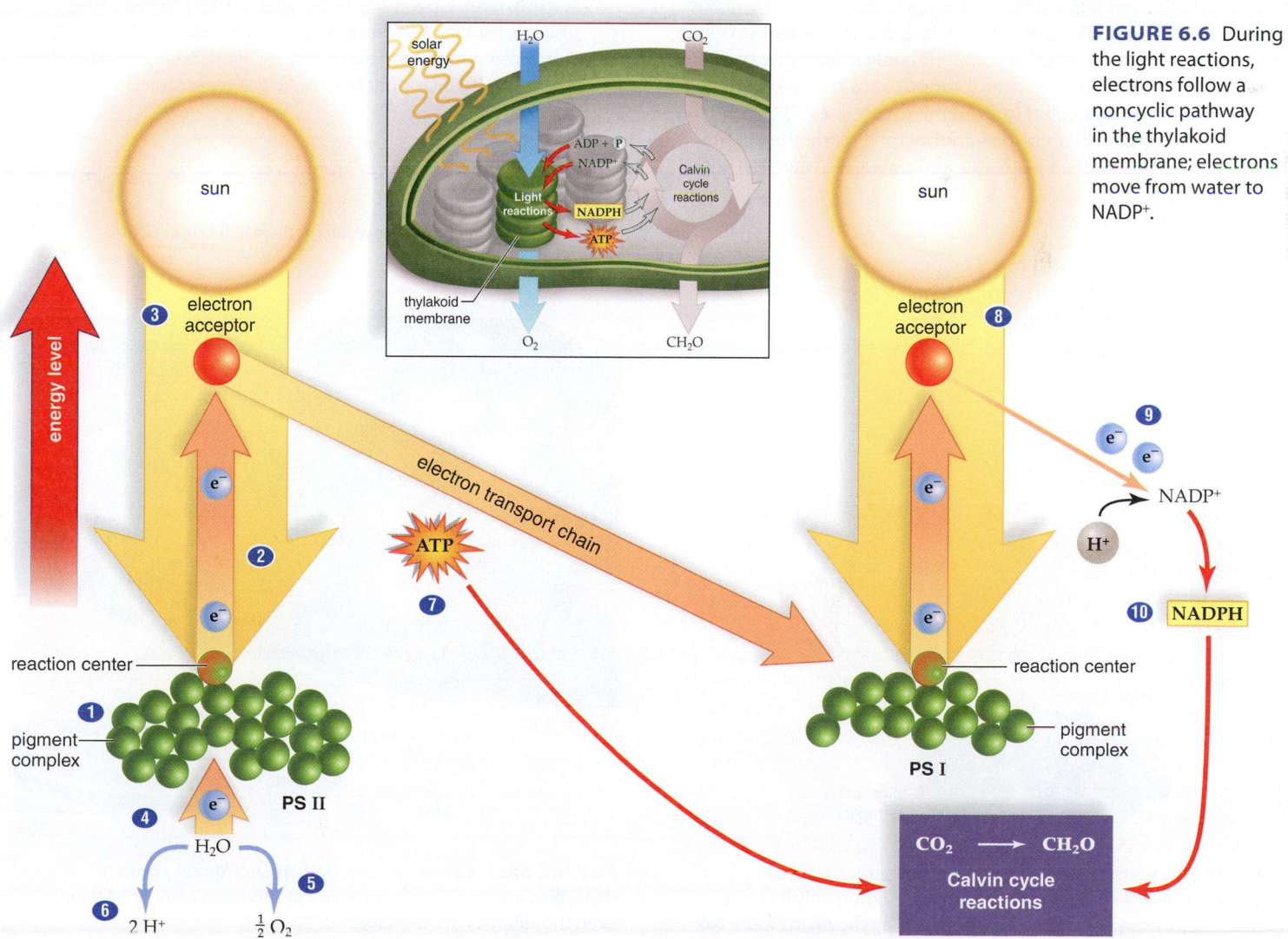

FIGURE 6.6 During the light reactions, electrons follow a noncyclic pathway in the thylakoid membrane; electrons move from water to NADP⁺.

6A Photosystem I Evolved Before Photosystem II

Most bacteria that can produce their own food do not utilize the noncyclic flow of electrons we have just described and instead make use of a cyclic flow of electrons, as illustrated in Figure 6A.1. The cyclic electron pathway begins when the antenna complex of PS I absorbs solar energy and energized electrons are taken up by an electron acceptor. The acceptor sends the electrons down a transport chain, and ATP is produced. The pathway is cyclic because the electrons return to PS I, where they are energized once more. Notice that the cyclic electron pathway produces ATP but no NADPH. To reduce carbon dioxide to a carbohydrate, both ATP and hydrogen atoms are required. PS I cannot split water to acquire hydrogen atoms because it absorbs solar energy at a lower level than PS II (680 nm compared to 700 nm). Instead, bacteria that only use the cyclic electron pathway use H_2S (also present in the early atmosphere) as a hydrogen source to reduce carbon dioxide to a carbohydrate. They release elemental sulfur (S) and not oxygen. In fact, they live in O_2-free environments.

Although cyanobacteria, unique among bacteria, are able to utilize the noncyclic electron pathway and release oxygen, they can revert to a cyclic pathway when necessary. When they revert, they use hydrogen sulfide (H_2S) as a hydrogen source to reduce carbon dioxide just like other bacteria groups that can produce their own food. However, cyanobacteria , unlike the other bacteria, contain thylakoids flattened membranous sacs arranged adjacent to the surface of the cell (Fig. 6A.2). We can speculate that the evolution of thylakoids among cyanobacteria made the evolution of the noncyclic electron pathway and oxygenic photosynthesis possible. Just as you need a place to set up your electronic hardware, so photosynthesizers need a location for the noncyclic electron pathway. Thylakoid membrane provides this location.

Cyanobacteria were the first organisms to have thylakoids and to release oxygen into the atmosphere. We tend to think of oxygen as a gift to organisms because most make use of oxygen in mitochondria to produce ATP. However, this mindset is not completely accurate. Oxygen is actually dangerous to organisms because it damages organic molecules. Organisms get rid of oxygen by using it during cellular respiration. Unique among organisms, cyanobacteria are able to perform both oxygenic photosynthesis and cellular respiration within the same compartment—their prokaryotic cell. The evidence that cyanobacteria became the chloroplasts of algae and plants once they were taken up by pre-eukaryotic cells is already extensive (see page 75). However, as further evidence, biologists are busy finding remnants of the cellular respiration capabilities of cyanobacteria in chloroplasts.

CONSIDER THESE QUESTIONS

1. Evolution often proceeds by "adding on" rather than starting over. How is the evolution of oxygen-releasing photosynthesis an example of this concept?
2. During evolution, attributes can be "lost." Today, what ability do free-living cyanobacteria have that chloroplasts do not have?

connect | BIOLOGY   Explore the concepts through a variety of multimedia assets and question types.

www.mcgrawhillconnect.com

FIGURE 6A.1 Photosynthesis evolved in stages. Scientists hypothesize that a cyclic electron pathway utilizing only PS I evolved before the noncyclic electron pathway shown in Figure 6.6.

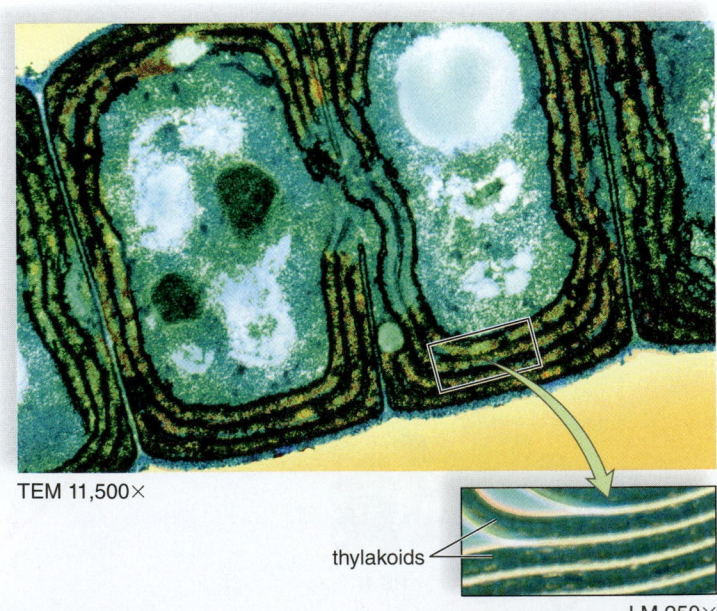

FIGURE 6A.2 Cyanobacteria, such as *Oscillatoria,* contain thylakoids and can photosynthesize in the same manner as plants even though they are prokaryotes.

6.7 A thylakoid is highly organized for its task

LEARNING OUTCOME

When you complete this section, you should be able to

1. Explain the division of the molecular complexes in the thylakoid membrane into those that "get ready" and those that represent the "payoff."

Let's divide the molecular complexes in the thylakoid membrane (Fig. 6.7) into those that "get ready" and those that represent the "payoff."

Get Ready

1 PS II consists of a pigment complex that absorbs solar energy and passes electrons on to an electron-acceptor molecule. **2** PS II receives replacement electrons from water, which splits, releasing H$^+$ and oxygen (O$_2$).

3 The electron transport chain, consisting of a series of electron carriers such as cytochrome complexes, passes electrons from PS II to PS I. (Notice, therefore, that PS I receives replacement electrons from the electron transport chain.) **4** Members of the electron transport chain also pump H$^+$ from the stroma into the thylakoid space. Eventually, an electron gradient is present: The thylakoid space contains much more H$^+$ than the stroma.

5 PS I consists of a pigment complex that absorbs solar energy and sends excited electrons on to an electron-acceptor molecule, which passes them to an enzyme called NADP$^+$ reductase.

Payoff

6 NADP$^+$ reductase, an enzyme, receives electrons and reduces NADP$^+$. NADP$^+$ combines with H$^+$ and becomes **7** NADPH.

8 H$^+$ flows down its concentration gradient through a channel in an ATP synthase complex. This complex contains an enzyme that then enzymatically binds ADP to (P), producing **9** ATP.

This method of producing ATP is called **chemiosmosis** because ATP production is tied to an H$^+$ gradient across a membrane.

We have now concluded our study of the light reactions. In the next part of the chapter, we will study the Calvin cycle reactions.

▶ **6.7 CHECK YOUR PROGRESS**
1. Identify how the production of NADPH differs from the production of ATP during the light reactions.

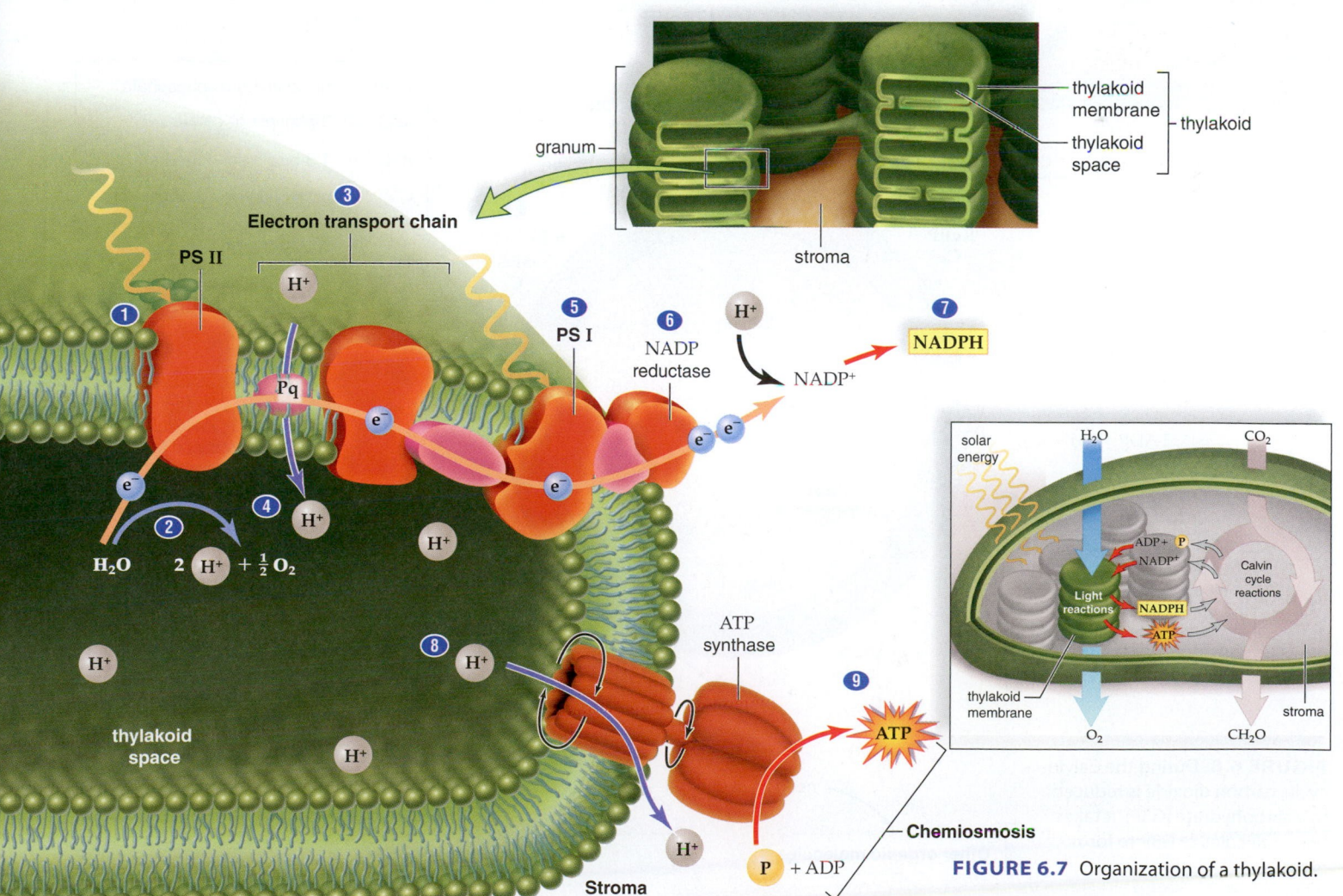

FIGURE 6.7 Organization of a thylakoid.

The Calvin Cycle Reactions—Making Sugars

From the overall equation for photosynthesis, associate carbon dioxide and carbohydrate (CH_2O) with the Calvin cycle reactions. The Calvin cycle reactions are light-independent because they can occur in the light or the dark, but even so they could not occur without the ATP and NADPH from the light reactions because they are used to reduce CO_2 to a carbohydrate.

6.8 ATP and NADPH from the light-dependent reactions are needed to produce a carbohydrate

LEARNING OUTCOME

When you complete this section, you should be able to

1. List the three stages of the Calvin cycle and describe the major event that occurs during each stage.

The Calvin cycle is a series of reactions that produces carbohydrate before returning to the starting point once more (Fig. 6.8). Melvin Calvin and his colleagues used the radioactive isotope ^{14}C as a tracer to discover its many reactions. The reactions can be divided into these phases: CO_2 fixation, CO_2 reduction, and finally regeneration of RuBP, the molecule that combines with and fixes CO_2 from the atmosphere. The steps in the Calvin cycle are multiplied by three for reasons that will be explained.

CO_2 Fixation ❶ During the first phase of the Calvin cycle, CO_2 from the atmosphere combines with RuBP, a 5-carbon molecule, and a C_6 (6-carbon) molecule results. The enzyme that speeds this reaction, called **RuBP carboxylase,** is a protein that makes up 20–50% of the protein content in chloroplasts. The reason for its abundance may be that it is unusually slow (it processes only a few molecules of substrate per second compared to thousands per second for a typical enzyme), and so there has to be a lot of it to keep the Calvin cycle going.

❷ The C_6 molecule immediately splits into two C_3 molecules. (Remember that all molecules are multiplied by three in Figure 6.9.) This C_3 molecule is called 3PG.

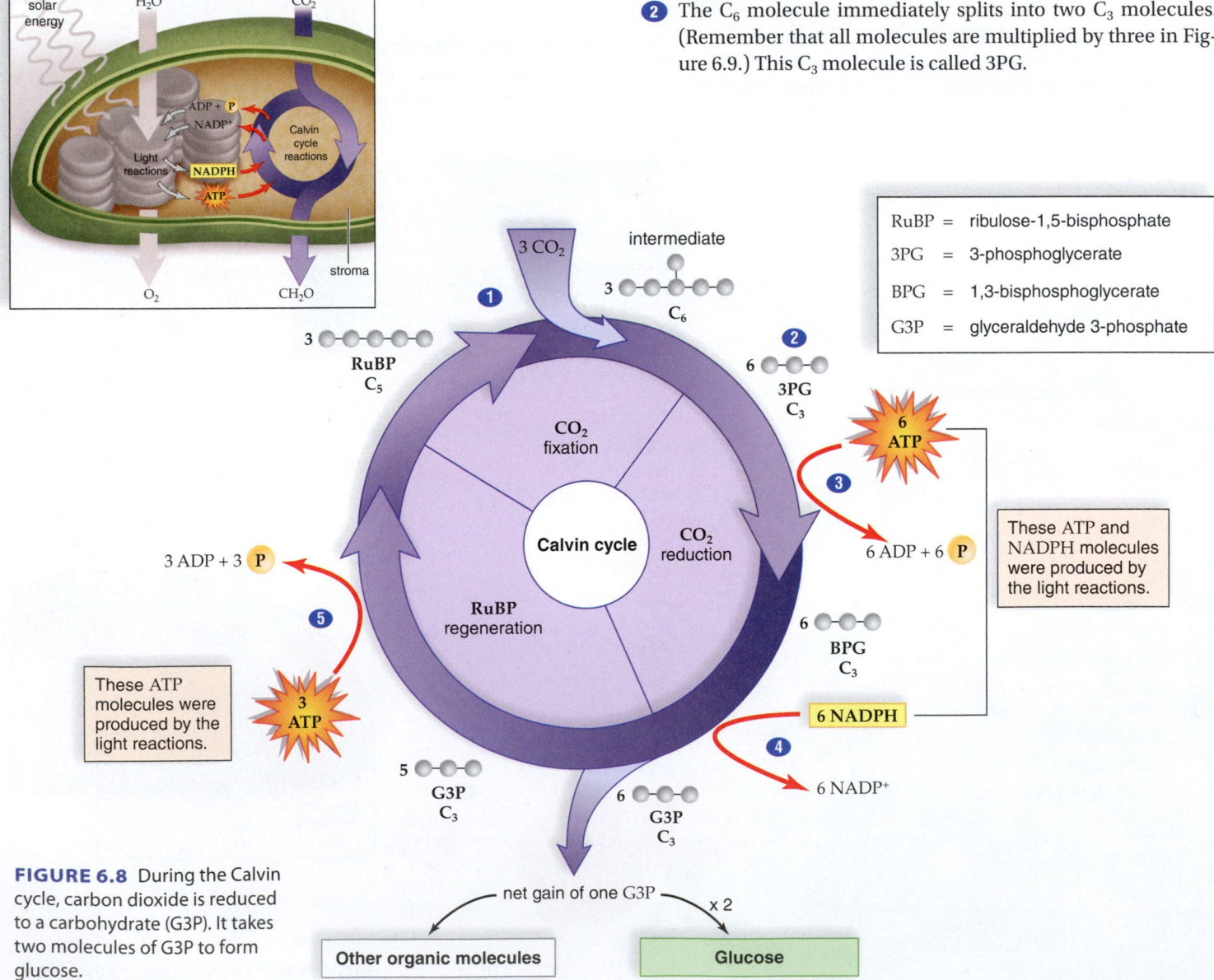

FIGURE 6.8 During the Calvin cycle, carbon dioxide is reduced to a carbohydrate (G3P). It takes two molecules of G3P to form glucose.

CO₂ Reduction ③ and ④ Each of the 3PG molecules undergoes reduction to G3P in two steps:

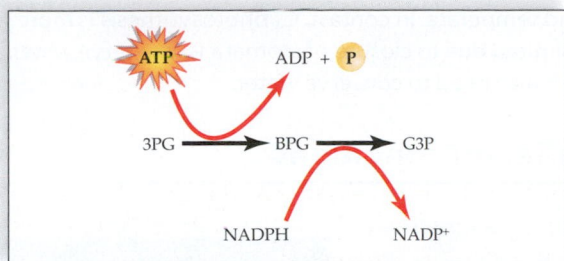

During this phase, CO_2 is reduced to a carbohydrate because $R-CO_2$ (3PG) has become $R-CH_2O$ (G3P). Energy and electrons are needed for this reduction reaction, and these are supplied by ATP and NADPH, which become ADP + Ⓟ, and $NADP^+$, respectively.

RuBP Regeneration The reactions in Figure 6.8 are multiplied by three because it takes three turns of the Calvin cycle to allow one G3P to exit. Why? Because, for every three turns of the Calvin cycle, five molecules of G3P are used to re-form three molecules of

RuBP, and the cycle continues. Notice that 5 × 3 (carbons in G3P) = 3 × 5 (carbons in RuBP):

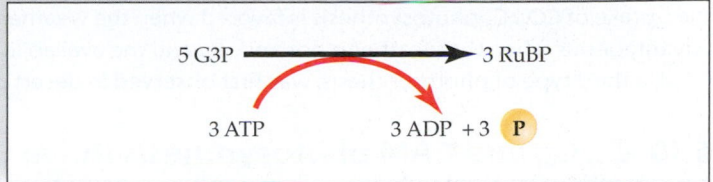

⑤ As this ATP produced by the light reactions breaks down, ADP + Ⓟ results.

The ADP + Ⓟ and $NADP^+$ that result from the Calvin cycle return to the light reactions so that they can continue to produce ATP and NADPH for the light-independent reactions.

▶ **6.8 CHECK YOUR PROGRESS**
1. Distinguish the use of ATP and NADPH during the Calvin cycle reactions.
2. Explain the expression that "plants use solar energy to produce sugar."

6.9 In photosynthesizers, a carbohydrate is the starting point for other molecules

LEARNING OUTCOME

When you complete this section, you should be able to

1. Summarize how the output of the Calvin cycle is used to make other molecules a plant cell needs.

G3P (glyceraldehyde-3-phosphate) is the product of the Calvin cycle that can be converted to all sorts of organic molecules. Compared to animal cells, algae and plants have greater biochemical capabilities. They use G3P for the purposes described in Figure 6.9.

① Notice that glucose phosphate is among the organic molecules that result from G3P metabolism. This is of interest to us because glucose is the molecule that plants and other organisms most often metabolize to produce the ATP molecules they require. Glucose is blood sugar in humans.

Glucose (with the removal of the phosphate) can be combined with ② fructose (with the removal of phosphates) to form ③ sucrose, the transport form of sugar in plants.

④ Glucose phosphate is also the starting point for the synthesis of ⑤ starch and ⑥ cellulose. Starch is the storage form of glucose. Some starch is stored in chloroplasts, but most starch is stored in roots. Cellulose is a structural component of plant cell walls and becomes fiber in our diet because we are unable to digest it.

⑦ A plant can use the hydrocarbon skeleton of G3P to form fatty acids and glycerol, which are combined in plant oils such as the corn oil, sunflower oil, and olive oil used in cooking. ⑧ Also, when nitrogen is added to the hydrocarbon skeleton derived from G3P, plants can make any type of amino acid from scratch.

▶ **6.9 CHECK YOUR PROGRESS**
1. Identify the benefit of including Figure 6.9 in this chapter.

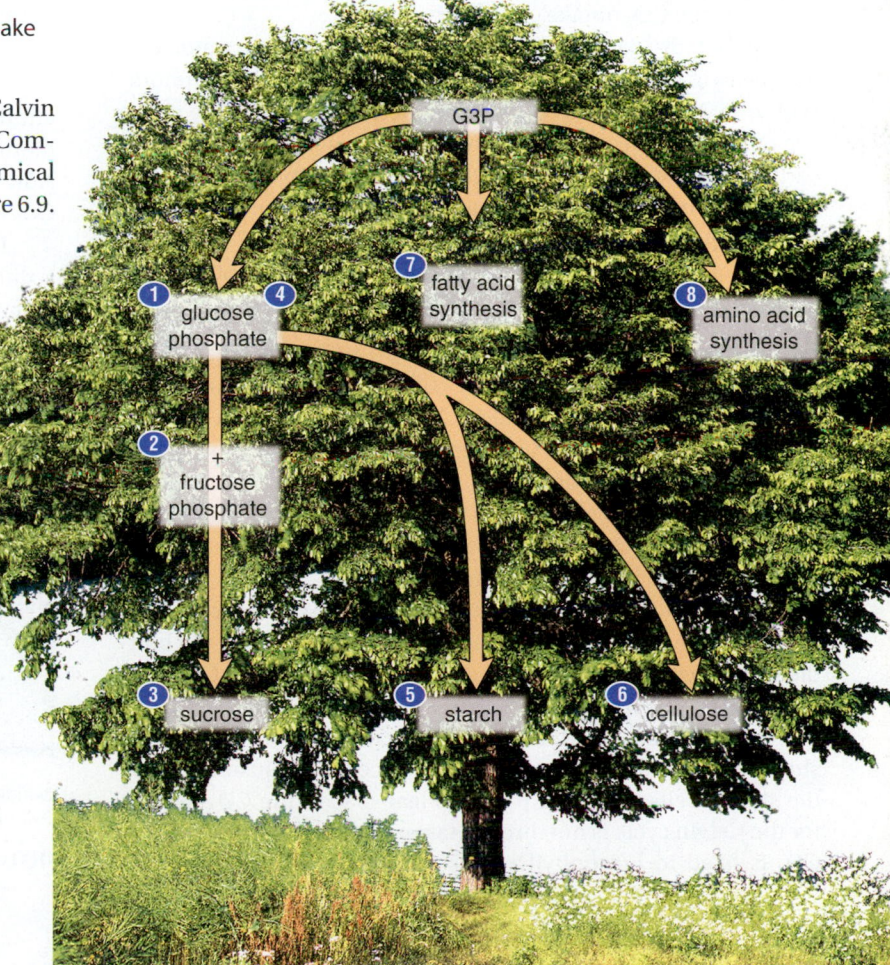

FIGURE 6.9 Trees and all photosynthesizers have the metabolic pathways that are able to convert G3P to all the molecules noted.

Types of Photosynthesis

Thus far, we have been observing C_3 photosynthesis, named for the number of carbons in the first observable molecule following the uptake of CO_2. C_3 photosynthesis is favored when the weather is moist and temperate. In contrast, C_4 photosynthesis is more advantageous when the weather is hot and dry and the availability of CO_2 is limited due to closure of stomata to conserve water. CAM, a third type of photosynthesis, was first observed in desert plants, which also need to conserve water.

6.10 C_3, C_4, and CAM photosynthesis thrive under different conditions

LEARNING OUTCOMES

When you complete this section, you should be able to

1. Compare the leaf structure of a C_3 plant with that of a C_4 plant.
2. Describe the advantages and disadvantages of C_3 and C_4 photosynthesis.
3. Explain CAM photosynthesis and the environmental conditions to which it is suited.

C_3 Plants In regions where temperature and rainfall tend to be moderate, plants efficiently carry on **C_3 photosynthesis,** and are therefore called C_3 plants. In a C_3 plant, the first stable molecule after CO_2 fixation is a C_3 molecule, namely 3PG (Fig. 6.10A). Look again at the Calvin cycle (see Fig. 6.8), and notice that after CO_2 combines with RuBP, the resulting C_6 molecule immediately breaks down to PGA, a C_3 molecule. The use of a radioactive tracer allows you to determine that this molecule is the first detectable one following CO_2 uptake.

On hot, dry days, when stomata close to conserve water, CO_2 decreases in leaf spaces and O_2 increases due to continued photosynthesis. Now, high levels of O_2 compete with CO_2 for the active site of RuBP carboxylase and less C_3 is produced. Further, if RuBP carboxylase combines with oxygen, it becomes an oxygenase that breaks down components of the Calvin cycle in a process called **photorespiration** because it uses O_2 and produces CO_2. Photorespiration leads to a severe decrease in yield that is of concern because many food crops are C_3 plants.

C_4 Plants Some plants have evolved an adaptation that allows them to be successful in hot, dry conditions. These plants carry out **C_4 photosynthesis** instead of C_3 photosynthesis. In a C_4 plant, the first detectable molecule following CO_2 fixation is a C_4 molecule having four carbon atoms. The anatomy of a C_4 plant is different from that of a C_3 plant. In a C_3 leaf, mesophyll cells, arranged in parallel rows, contain well-formed chloroplasts. In a C_4 leaf, chloroplasts are located in the mesophyll cells, but they are also located in bundle sheath cells, which surround the leaf vein. Further, mesophyll cells are arranged concentrically around the bundle sheath cells, shielding the bundle sheath cells from O_2 in leaf spaces. (Fig. 6.10B).

In C_4 plants, the Calvin cycle occurs only in the bundle sheath cells and not in the mesophyll cells. Because the bundle sheath cells are not exposed to leaf spaces, the CO_2 needed for the Calvin cycle is not directly taken from leaf spaces. Instead, CO_2 is fixed in mesophyll cells by an enzyme that has a high

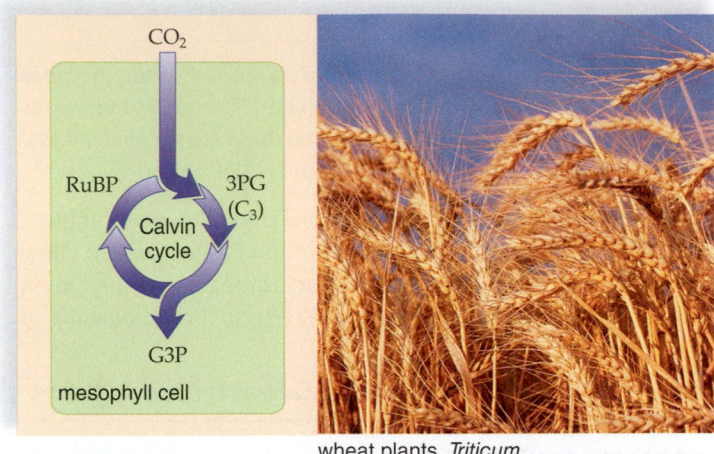

wheat plants, *Triticum*

FIGURE 6.10A Carbon dioxide fixation in C_3 plants as exemplified by these wheat plants.

affinity for CO_2 and the result is a C_4 molecule. The C_4 molecule is modified and then pumped into bundle sheath cells (Fig. 6.10C). Now CO_2 enters the Calvin cycle. This represents partitioning of CO_2 fixation and the Calvin cycle in space.

It takes energy to make C_4 molecules and to pump them to a different part of the cell; therefore, you would think that the C_4 pathway would be disadvantageous. Yet in hot, dry climates,

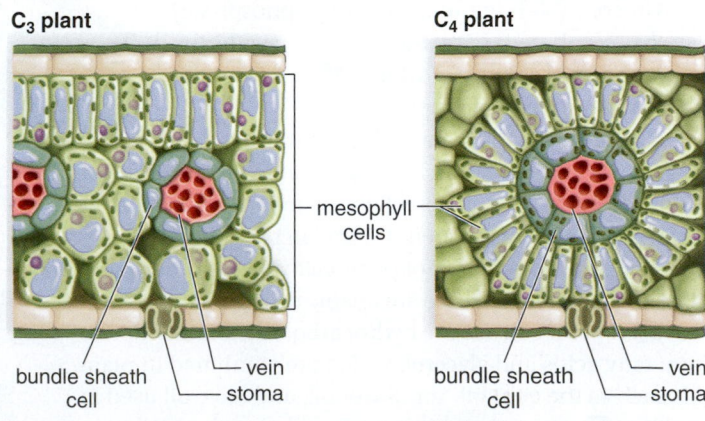

FIGURE 6.10B Anatomy of a C_3 plant compared to a C_4 plant.

the net photosynthetic rate of C_4 plants such as sugarcane, corn, and Bermuda grass is two to three times that of C_3 plants such as wheat, rice, and oats. Why do C_4 plants enjoy such an advantage? The answer is that when the weather is hot and dry and the stomata close, RuBP carboxylase is not exposed to O_2, and yield is maintained.

When the weather is moderate, C_3 plants ordinarily flourish but when the weather becomes hot and dry, C_4 plants have their chance to take over, and they begin to predominate. In the early summer, lawns exhibit mainly C_3 plants such as Kentucky bluegrass and creeping bent grass in the cooler parts of the United States, but by midsummer, crabgrass, a C_4 plant, begins to take over.

CAM Plants Another type of photosynthesis is called **CAM photosynthesis.** CAM stands for crassulacean-acid metabolism. It gets its name from the Crassulaceae, a family of flowering succulent (water-containing) plants that live in warm, arid regions of the world. CAM was first discovered in these plants, but now it is known to be prevalent among most succulent plants that grow in desert environments, including cacti.

Whereas a C_4 plant represents partitioning in space—that is, CO_2 fixation occurs in mesophyll cells, and the Calvin cycle occurs in bundle sheath cells—CAM is partitioning by the use of time. During the night, CAM plants use C_3 molecules to fix CO_2, forming C_4 molecules. These molecules are stored in large vacuoles in mesophyll cells. During the day, the C_4 molecules release CO_2 to the Calvin cycle when NADPH and ATP are available from the light reactions (Fig. 6.10D).

The primary advantage of this partitioning again relates to the conservation of water. CAM plants open their stomata only at night, and this reduces water loss significantly. During the day, the stomata close, limiting the amount of carbon dioxide that enters the plant. Although CAM photosynthesis does not permit rapid growth, it is a very successful adaptation to stressful desert conditions.

Comparison C_4 plants most likely evolved in, and are adapted to, areas of high light intensities, high temperatures, and limited rainfall. However, C_4 plants are more sensitive to cold, and C_3 plants probably do better than C_4 plants below 25°C. CAM plants, on the other hand, compete well with either type of plant when the environment is extremely arid. Surprisingly, CAM is quite widespread and has evolved in 30 families of flowering plants, including cacti, stonecrops, orchids, and bromeliads. It is also found among nonflowering plants, such as some ferns and cone-bearing trees.

corn plants, *Zea*

FIGURE 6.10C Carbon dioxide fixation in C_4 plants as exemplified by corn plants.

pineapple, *Ananas*

FIGURE 6.10D Carbon dioxide fixation in CAM plants as exemplified by pineapple plants.

▶ **6.10 CHECK YOUR PROGRESS**
1. Relate the leaf structure of a C_4 plant to C_4 photosynthesis.
2. Identify the advantage of a C_3 plant when the weather is cool and moist.
3. Explain why CAM photosynthesis is "partitioning in time."

TABLE 6.1	Types of Photosynthesis		
	C_3 photosynthesis	**C_4 photosynthesis**	**CAM photosynthesis**
Weather	Cool, moist	Hot, dry	Hot, dry
First detectable molecule	C_3	C_4	C_4
CO_2 fixation	Mesophyll	Mesophyll	Mesophyll
Calvin cycle	Mesophyll	Bundle sheath	Mesophyll
Limitation	Photorespiration	ATP cost	ATP cost, CO_2 availability
Plants	Cereals, soybeans, most trees, lawn grasses	Corn, sugarcane, crabgrass, sorghum	Orchids, stonecrops, cactuses, bromeliads

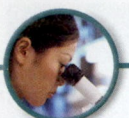

6B Tropical Rain Forests and Global Climate Change

Tropical rain forests occur near the equator, wherever temperatures are above 26°C and rainfall is regular and heavy (100–200 cm per year). Huge trees with buttressed trunks and broad, undivided, dark-green leaves predominate. Nearly all of the plants in a tropical rain forest are woody, and woody vines are also abundant. There is no undergrowth except at clearings. Instead, orchids, ferns, and bromeliads live in the branches of the trees.

Despite the fact that tropical rain forests have dwindled from an original 14% to 6% of land surface today, they still make a substantial contribution to global CO_2 fixation. Taking into account all ecosystems, marine and terrestrial, photosynthesis produces organic matter that is 300–600 times the mass of the people currently living on Earth. Tropical rain forests contribute greatly to the uptake of CO_2 and the productivity of photosynthesis because they are the most efficient of all terrestrial ecosystems.

We have learned that organic matter produced by photosynthesizers feeds all organisms and that photosynthesis releases oxygen (O_2), a gas needed to complete the process of cellular respiration. Does photosynthesis by tropical rain forests provide any other service that has significant worldwide importance? A rise in the average global temperature during the twenty-first century due to the introduction of certain gases, chiefly carbon dioxide, into the atmosphere is expected (Fig. 6B). The process of photosynthesis and also the oceans act as a sink for carbon dioxide. For at least a thousand years prior to 1850, atmospheric CO_2 levels remained fairly constant at 0.028%. Since the 1850s, when industrialization began, the amount of CO_2 in the atmosphere has increased to 0.038%.

Much like the panes of a greenhouse, CO_2 in our atmosphere traps radiant heat from the sun and warms the world. Therefore, CO_2 and other gases that act similarly are called *greenhouse gases*. Without any greenhouse gases, Earth's temperature would be about 33°C cooler than it is now. Therefore, it is hypothesized that increasing the concentration of these gases will cause an increase

in global temperatures that will disrupt climate patterns recognizable by such events as heatwaves, droughts, and storms including an increased number of hurricanes and tornadoes.

Burning fossil fuels adds CO_2 to the atmosphere. Have any other factors contributed to an increase in atmospheric CO_2? Between 10 and 30 million hectares of rain forests are lost every year to ranching, logging, mining, and other means of developing forests for human needs. The clearing of forests often involves burning them, which is double trouble for global climate change. Each year, deforestation of tropical rain forests accounts for 20–30% of all the carbon dioxide in the atmosphere. At the same time, burning removes trees that would ordinarily absorb CO_2 (Fig. 6B).

Some investigators hypothesized that an increased amount of CO_2 in the atmosphere will cause photosynthesis to increase in the remaining portion of the forest. To study this possibility, they measured atmospheric CO_2 levels, daily temperature levels, and tree girth in La Selva, Costa Rica, for 16 years. The data collected demonstrated relatively *lower* forest productivity at higher temperatures. These findings suggest that, if temperatures rise, tropical rain forests may add to ongoing atmospheric CO_2 accumulation rather than the reverse. Therefore it behooves us to do what we can to limit carbon dioxide emissions into the atmosphere and prevent a rise in global temperatures.

CONSIDER THESE QUESTIONS

1. What are some other advantages to preserving tropical rain forests aside from helping to prevent any further rise in CO_2?
2. What can the countries of the world that have no tropical rain forests do to help preserve them?

 Explore the concepts through a variety of multimedia assets and question types.

www.mcgrawhillconnect.com

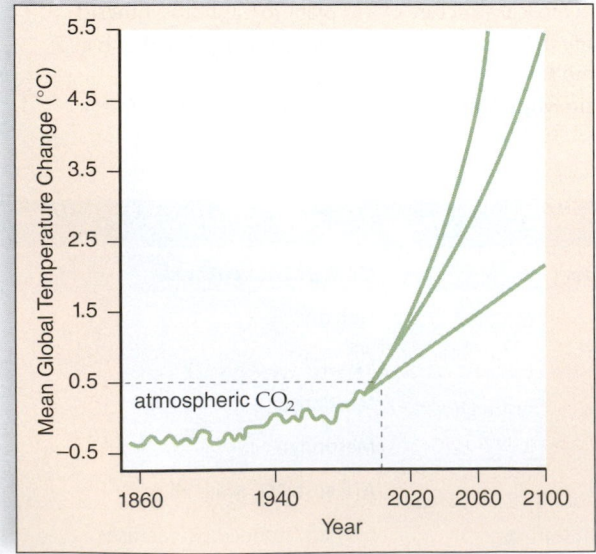

FIGURE 6B This graph shows how rising atmospheric CO_2 affects global temperature. As shown the global temperature is now 0.05°C higher worldwide than in the twentieth century. Burning forests adds to atmospheric CO_2.

THE CHAPTER IN REVIEW

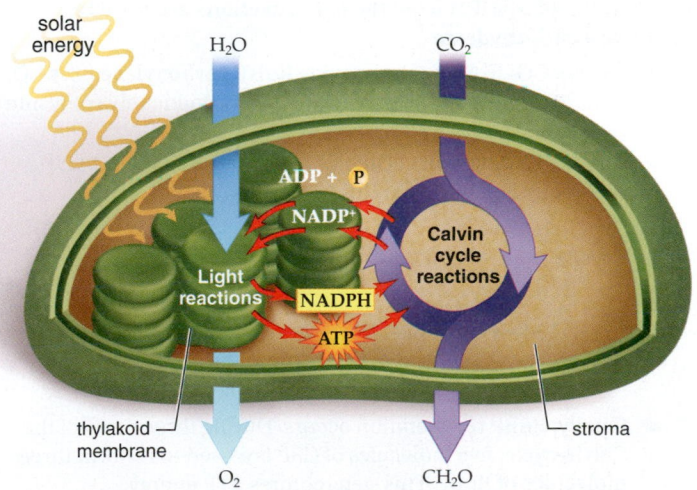

CONNECTING THE CONCEPTS

The overall reaction for photosynthesis:

$$CO_2 + H_2O \longrightarrow (CH_2O) + O_2$$

takes place in chloroplasts. This equation does not reflect that photosynthesis requires two separate sets of reactions: the light reactions (take place in thylakoid membrane) and the Calvin cycle reactions (take place in stroma).

The light reactions absorb solar energy and convert it into chemical forms of energy that drive the Calvin cycle reactions. As a result of the noncyclic flow of electrons, NADPH carries electrons, and ATP provides energy to reduce carbon dioxide to a carbohydrate during the Calvin cycle.

Photosystem I, which is capable of a cyclic flow of electrons, evolved before photosystem II. Some photosynthesizing bacteria today utilize only photosystem I. They must live under anaerobic conditions because they die in the presence of oxygen released by photosynthesizers who utilize both photosystem I

and photosystem II. Cyanobacteria do utilize both photosystems, and they carry on cellular respiration, which soaks up oxygen.

C_3 photosynthesis was the first form of photosynthesis to evolve. We can tell it evolved when oxygen was in limited supply because RuBP carboxylase is inefficient in the presence of oxygen. By now, two other forms of photosynthesis have evolved—C_4 photosynthesis (partitioning in space) and CAM photosynthesis (partitioning in time). Both of the alternative forms of photosynthesis are means of supplying the active site of RuBP carboxylase with CO_2, while limiting its exposure to oxygen produced by the plant.

The details of photosynthesis should not cause us to lose sight of its great contribution to the biosphere. It keeps the biosphere functioning because it supplies energy, in the form of carbohydrates, to all organisms. Most organisms have a way to tap into the energy provided

by carbohydrates. It's called cellular respiration and is the subject of our next chapter. Cellular respiration is completed within mitochondria. Mitochondria are called the powerhouses of the cell because they convert the energy of carbohydrates (and other organic molecules) to that of ATP molecules, the energy currency of cells.

ANALYZE AND EVALUATE

1. What is the significance of cyanobacteria in the history of life?
2. Prepare an overview of photosynthesis that includes a simplified version of the light reactions (Fig. 6.6) and the Calvin cycle reactions (Fig. 6.8). Compare your diagram to that produced by another group and critique both diagrams.
3. Why would you predict that C_4 photosynthesis evolved after C_3 photosynthesis?

SUMMARIZE

Overview of Photosynthesis

6.1 Photosynthesizers are autotrophs that produce their own food

- **Photosynthesis** converts solar energy to the chemical energy of a carbohydrate. Producers (**autotrophs**) produce food for themselves and for consumers (**heterotrophs**).
- In eukaryotes, **chloroplasts** carry out photosynthesis. **Chlorophyll** and other pigments within the membranes of thylakoids of a **granum** absorb solar energy. Conversion of CO_2 to a carbohydrate occurs in the **stroma**, the enzyme-containing interior of chloroplasts. CO_2 enters a leaf through small openings called **stomata.**

6.2 Photosynthesis involves two sets of reactions: The light reactions and the Calvin cycle reactions

- The overall reaction for photosynthesis shows that CO_2 is reduced (**reduction** is gain of electrons) and water is oxidized (**oxidation** is loss of hydrogen atoms), resulting in a carbohydrate and oxygen. This is a **redox reaction:**

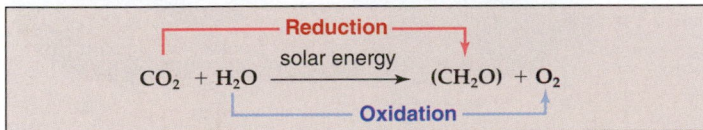

- The **light reactions** only occur in thylakoids during the day when solar energy is available. The light reactions split water, releasing O_2, and send NADPH and ATP to the Calvin cycle reactions. The **Calvin cycle reactions** are enzymatic

reactions that occur in the stroma anytime, day or night. The Calvin cycle reactions use the NADPH and ATP from the light reactions to reduce CO_2 to a carbohydrate (CH_2O).

The Light Reactions—Harvesting Energy

6.3 Solar energy is absorbed by pigments

- **Chlorophylls *a* and *b*** and **carotenoids** absorb violet, blue, and red light better than other portions of visible light. Because these pigments don't absorb green light, leaves appear green to us.

6.4   Solar energy boosts electrons to a higher energy level

- Within thylakoid membranes, pigment complexes in **photosystem I (PS I)** and **photosystem II (PS II)** absorb solar energy. Energized electrons are passed by a reaction center chlorophyll *a* molecule to an electron acceptor.

6.5   Solar energy is converted to the chemical energy of ATP

- Solar energy energizes electrons in a PS II and this energy is converted to that of ATP. (1) The electrons release energy as they pass down an **electron tranport chain (ETC)** and certain carriers use this energy to establish an H^+ gradient across the thylakoid membrane. (2) When the hydrogen ions flow through the channel of an **ATP synthase complex,** this enzyme produces ATP from ADP + $\widehat{P}$. The process is called **chemiosmosis.**

6.6   The noncyclic flow of electrons produces ATP and NADPH

- During a noncyclic electron pathway, electrons move from PS II down an electron transport chain to PS I and on to $NADP^+$. Replacement electrons are removed from water, which splits, releasing O_2 and H^+. An electron transport chain between PS II and PS I stores energy in the form of an H^+ gradient and then this gradient generates ATP. $NADP^+$ receives e^- and H^+ and becomes NADPH.
- The light reactions send NADPH and ATP to the Calvin cycle reactions.

6.7   A thylakoid is highly organized for its task

- In a get-ready phase, PS II absorbs solar energy and water splits, releasing H^+ and O_2. Members of the electron transport chain pump H^+ from the stroma to the thylakoid space and an H^+ gradient results. PS I absorbs solar energy, and electrons are picked up by $NADP^+$ reductase.
- In a payoff phase, $NADP^+$ reductase passes electrons to $NADP^+$, and NADPH results. **Chemiosmosis** occurs and the ATP synthase complex produces ATP.

The Calvin Cycle Reactions—Making Sugars

6.8   ATP and NADPH from the light reactions are used to produce a carbohydrate

- During CO_2 fixation, the enzyme **RuBP carboxylase** fixes CO_2 to RuBP, producing a C_6 molecule that immediately splits into two C_3 molecules (3PG).
- Then CO_2 reduction occurs. Each 3PG is reduced to a G3P molecule. This step requires ATP and NADPH:

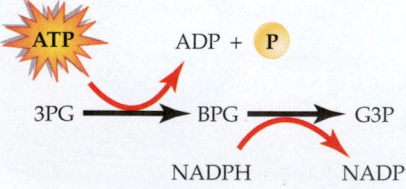

- Finally, RuBP regeneration occurs. During three turns of the Calvin cycle, five molecules of G3P are used to re-form three molecules of RuBP. This step requires ATP energy.

6.9   In photosynthesizers, a carbohydrate is the starting point for other molecules

- **G3P** (phosphoglyceraldehyde-3-phosphate) can be converted to all organic molecules needed by a plant. It takes two G3P molecules to make one glucose molecule.

6.10 C_3, C_4, and CAM photosynthesis thrive under different conditions

- **C_3 photosynthesis** is adapted to conditions of moderate temperature and rainfall. When weather is hot and dry, RuBP carboxylase uses O_2 as a substrate instead of CO_2 because the CO_2 supply is low, due to stomata closure. Now **photorespiration** occurs, reducing yield.
- **C4 photosynthesis,** which is adapted to hot, dry weather, prevents RuBP from combining with O_2. CO_2 fixation occurs in mesophyll and results in a C_4 molecule which is pumped to bundle sheath cells, where the Calvin cycle occurs and RuBP carboxylase is not exposed to O_2. This is partitioning photosynthesis in space.
- During **CAM photosynthesis,** which is adapted to desert conditions, stomata are closed during the day, and this conserves water. CO_2 fixation at night results in a C_4 molecule. CO_2 does not enter the Calvin cycle until the next day. This is partitioning photosynthesis in time.

TEST YOURSELF

Overview of Photosynthesis

1. The raw materials for photosynthesis are
   a. oxygen and water.
   b. oxygen and carbon dioxide.
   c. carbon dioxide and water.
   d. carbohydrates and water.
   e. carbohydrates and carbon dioxide.
2. During photosynthesis, _____ is reduced to _____.
   a. CO_2, oxygen
   b. oxygen, CO_2
   c. water, oxygen
   d. CO_2, a carbohydrate
3. The light reactions
   a. take place in the stroma.
   b. consist of the Calvin cycle.
   c. Both a and b are correct.
   d. Neither a nor b is correct.
4. The function of the light reactions is to
   a. obtain CO_2.
   b. make carbohydrate.
   c. convert light energy into usable forms of chemical energy.
   d. regenerate RuBP.
5. Label the following diagram of a chloroplast.

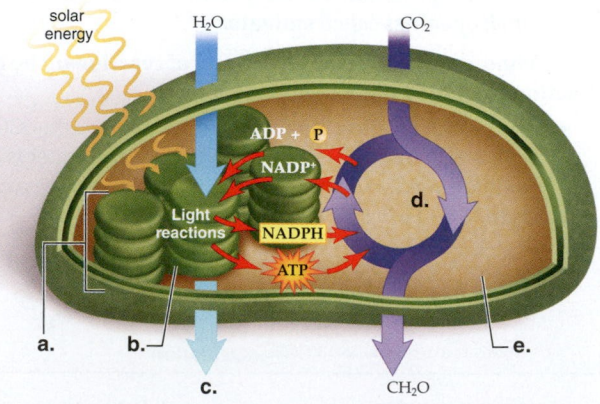

6. **THINKING CONCEPTUALLY** In order for the biosphere to have animal life, some animals must eat plants. Explain.

The Light Reactions—Harvesting Energy

7. When leaves change color in the fall, _____ light is absorbed for photosynthesis.
   - a. orange range
   - b. red range
   - c. violet-blue-green range
   - d. None of these are correct.

8. A photosystem contains
   - a. pigments, a reaction center, and an electron receiver.
   - b. ADP, ⓅP, and hydrogen ions (H^+).
   - c. protons, photons, and pigments.
   - d. cytochromes only.
   - e. Both b and c are correct.

9. PS I, PS II, and the electron transport chain are located in the
   - a. thylakoid membrane.
   - b. stroma.
   - c. outer chloroplast membrane.
   - d. cell's nucleus.

10. The final acceptor of electrons during the noncyclic electron pathway is
    - a. PS I.
    - b. PS II.
    - c. water.
    - d. ATP.
    - e. $NADP^+$.

11. When electrons in the reaction center of PS II are passed to an electron acceptor, they are replaced by electrons that come from
    - a. oxygen.
    - b. glucose.
    - c. carbon dioxide.
    - d. water.

12. During the light reactions of photosynthesis, ATP is produced when hydrogen ions (H^+) move
    - a. down a concentration gradient from the thylakoid space to the stroma.
    - b. against a concentration gradient from the thylakoid space to the stroma.
    - c. down a concentration gradient from the stroma to the thylakoid space.

The Calvin Cycle Reactions—Making Sugars

13. The Calvin cycle reactions
    - a. produce carbohydrates.
    - b. convert one form of chemical energy into a different form of chemical energy.
    - c. regenerate more RuBP.
    - d. use the products of the light reactions.
    - e. All of these are correct.

14. The Calvin cycle requires _____ from the light reactions.
    - a. carbon dioxide and water
    - b. ATP and NADPH
    - c. carbon dioxide and ATP
    - d. ATP and water
    - e. NADH and water

15. **THINKING CONCEPTUALLY** The overall equation for photosynthesis doesn't include ATP. Why is ATP needed?

Types of Photosynthesis

16. C_4 photosynthesis
    - a. occurs in plants whose bundle sheath cells contain chloroplasts.
    - b. takes place in plants such as wheat, rice, and oats.
    - c. is an advantage when the weather is warm.
    - d. Both a and c are correct.

17. CAM photosynthesis
    - a. is the same as C_4 photosynthesis.
    - b. is an adaptation to cold environments in the Southern Hemisphere.
    - c. is prevalent in desert plants that close their stomata during the day.
    - d. stands for chloroplasts and mitochondria.

18. The different types of photosynthesis are dependent upon the timing and location of
    - a. CO_2 fixation.
    - b. nitrogen fixation.
    - c. H_2O fixation.
    - d. All of these are correct.

GET INVOLVED

1. The graph shows the absorbance of colored light rays by a filament of green alga. Aerobic bacteria were added to the experimental setup, and they congregated in the regions of greatest absorbance. What hypotheses are supported by this experiment?

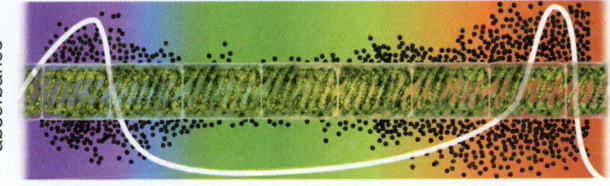

oxygen-seeking bacteria

absorbance

filament of green alga

2. Photosynthesis makes use of light in the visual range, the same range that is suitable to our vision. Use the theory of evolution (specifically adaptation) to explain this observation.

3. For an interactive exploration of photosynthesis events take a moment to view McGraw-Hill's 3D animation "Photosynthesis."

MEDIA STUDY TOOLS

mhhe.com/maderconcepts3

Enhance your study of this chapter with interactive study tools, practice tests, and engaging animations. Also, ask your instructor about the resources available through ConnectPlus, which includes LearnSmart, a personalized adaptive learning program, and a media-rich eBook.

7

Pathways of Cellular Respiration

BEFORE YOU BEGIN

Take a few minutes to recall

The chemical structure of carbohydrates (sections 3.3
    and 3.4)

The structure of a mitochondrion (section 4.9)

The ATP cycle (section 5.6)

ATP Is Universal

ATP (adenosine triphosphate) is ancient—a molecular fossil, really—and it is universal. ATP was present 3.5 billion years ago when life began. Like other nucleotides, it probably formed in the early atmosphere and then rained down into the oceans to become incorporated into the first cells that evolved. Because all living cells are related to the first cell(s), ATP became universal. Whether you are a snake hanging from a tropical tree, an octopus out for a swim in the ocean, bacteria with undulating flagella, or a tourist taking a swamp walk in the Everglades—your cells are making and using ATP, and so are the cells of the cypress trees and all other plants lining your swamp walk. When cells require energy to do work, they split ATP.

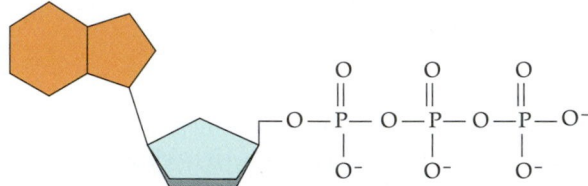

adenosine triphosphate

What is the secret of ATP? Why is ATP suited to be the energy currency of cells? ATP is a nucleotide with three phosphate groups, and therein lies the secret of its ability to supply energy. The phosphates repel one another because each has a negative charge. The breakdown of ATP to ADP relieves the repulsion and releases a significant amount of energy. Cells are able to couple the energy of a triphosphate nucleotide to reactions that require energy. Without coupling, all the energy of breakdown would be lost as heat, and life would not exist.

We can imagine that ATP became incorporated into the first cells and that its breakdown releases energy. But then what happens? When ATP becomes ADP + ⓟ, it has to become ATP again or else cells die. Enter the mitochondria. Mitochondria are aptly called the powerhouses of the cell because most ATP is made there from ADP + ⓟ. Without mitochondria, a eukaryotic

Emerald tree boas, *Corallus caninus*

cell has a limited capacity to rebuild its ATP. The power plants of modern human society use all sorts of fuels—fossil fuels, nuclear energy, and renewable energy sources such as wind and sun—to produce electricity. Similarly, mitochondria use glucose, fatty acids, and even proteins as fuels to produce ATP.

ATP is unique among the cell's storehouse of chemicals; no other molecule performs its functions in such a solitary manner. Amino acids must join to make a protein, and nucleotides must join to make DNA or RNA, but ATP has the same structure and function in all cells. It provides the energy that makes all forms of life possible. Whether you go skiing, take an aerobics class, or just hang out, ATP molecules provide the energy needed for your muscles to contract. ATP is also needed for nerve conduction, protein synthesis, and any other reaction in a cell that requires energy. ATP molecules are produced during cellular respiration, the process by which cells harvest the energy stored in organic compounds. Cellular respiration is the topic of this chapter.

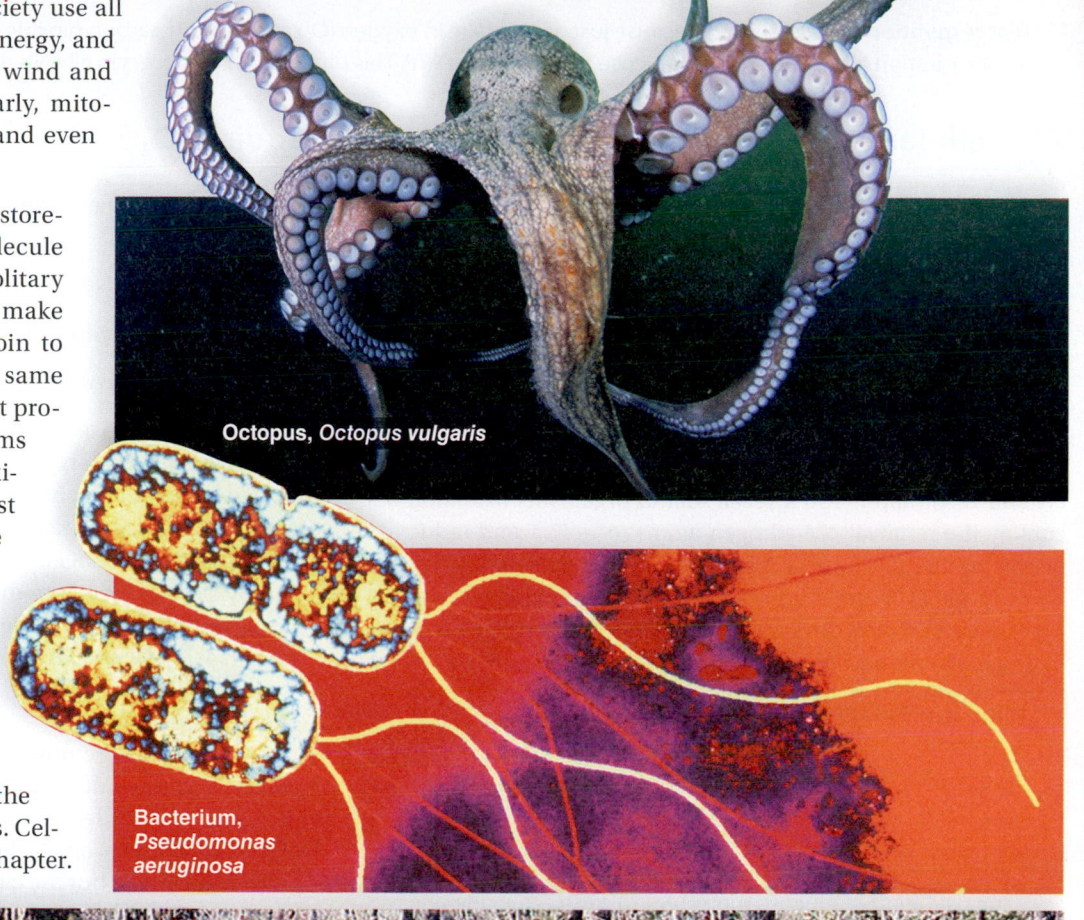

Octopus, *Octopus vulgaris*

Bacterium, *Pseudomonas aeruginosa*

Common bald cypress, *Taxodium distichum*

Overview of Cellular Respiration

Cellular respiration is aptly named because just as you take in oxygen (O_2) and give off carbon dioxide (CO_2) during breathing, so does cellular respiration as it breaks down glucose and builds up ATP. This process, which occurs in all cells of the body, is the reason you breathe.

7.1 Cellular respiration is a redox reaction that requires O_2

LEARNING OUTCOMES

When you complete this section, you should be able to

1. Recognize that the overall breakdown of glucose is a redox equation.
2. Explain the function of NAD^+ and FAD during cellular respiration.
3. Identify the four phases of cellular respiration and the location of each within the cell.

Photosynthesis is a redox reaction (see section 6.2), and so is cellular respiration. You'll recall that in organisms oxidation occurs by the removal of hydrogen atoms ($e^- + H^+$), and reduction occurs by the addition of hydrogen atoms. As cellular respiration occurs, hydrogen atoms are removed from glucose, and the result is carbon dioxide (CO_2). On the other hand, oxygen receives hydrogen atoms and becomes water:

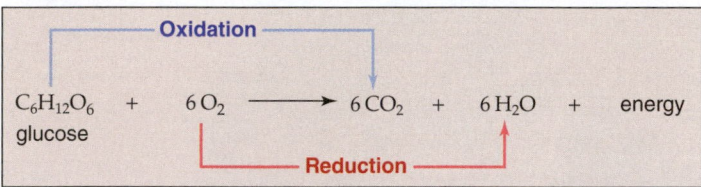

The oxidation of glucose molecules to carbon dioxide releases energy, which is then used for ATP production (Fig. 7.1A).

NAD^+ and FAD The coenzymes **NAD^+** (nicotinamide adenine dinucleotide) and **FAD** (flavin adenine dinucleotide) assist the enzymes involved in the oxidation of glucose breakdown products.

NAD^+ accepts two electrons ($2\ e^-$) and one hydrogen ion (H^+) so that NADH results:

$$NAD^+ + 2\ e^- + H^+ \longleftrightarrow NADH$$

FAD accepts two hydrogen atoms ($2\ e^- + 2\ H^+$) to become $FADH_2$:

$$FAD + 2\ e^- + 2\ H^+ \longleftrightarrow FADH_2$$

The electrons received by NADH and $FADH_2$ are *high-energy electrons*, which they carry to an **electron transport chain** (**ETC**), located within a mitochondrion (Fig. 7.1B). As the electrons move down the ETC, their energy is captured for ATP production. Oxygen is the final acceptor of electrons ($2\ e^-$) at the end of the ETC, and it quickly takes on H^+ to become water. NAD^+ and FAD can be likened to shuttle buses that travel between two designated places because, as soon as NAD^+ and FAD have discharged their "passengers" at the ETC, they return and pick up more electrons.

Animation
How the NAD^+ Works

The gradual removal of electrons from glucose requires many steps. How does the cell benefit by the gradual oxidation of glucose? Consider that if the engine of your car were to burn all the gasoline at once, there would be a big explosion and a great deal of heat, but the car wouldn't go far. Instead, the engine burns gasoline in spurts and captures the energy little by little so that your car can go several miles to the gallon. In the same way, the cell oxidizes glucose in a stepwise manner allowing several ATP molecules to be produced per respiration of one glucose molecule.

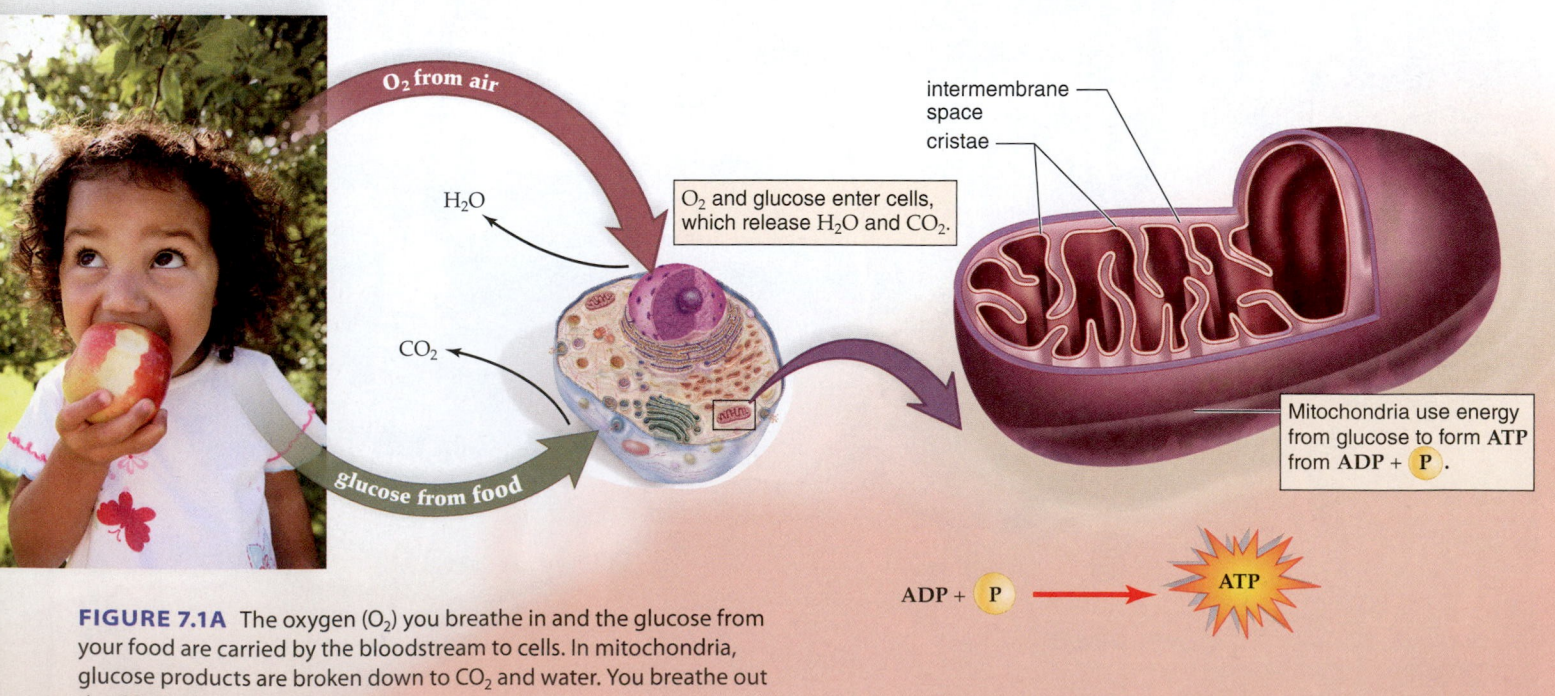

FIGURE 7.1A The oxygen (O_2) you breathe in and the glucose from your food are carried by the bloodstream to cells. In mitochondria, glucose products are broken down to CO_2 and water. You breathe out the CO_2.

Phases of Cellular Respiration Figure 7.1B shows the four phases of cellular respiration that take place in succession. Glycolysis, the first phase, occurs in the cytoplasm and does not require oxygen, so glycolysis is called an **anaerobic** process. The other phases of cellular respiration occur inside the mitochondria, where oxygen is a critical player. Because they require oxygen, these phases are called **aerobic.**

Glycolysis and two more phases accomplish the breakdown of glucose to carbon dioxide and water. Aside from producing NADH and $FADH_2$, two of these phases produce ATP as shown at the bottom of Figure 7.1B. These are the four phases of cellular respiration:

- Glycolysis breaks down glucose to two molecules of pyruvate. Glycolysis produces NADH but also two molecules of ATP outside the ETC.
- The preparatory (prep) reaction, which takes place in the matrix of the mitochondrion, breaks down pyruvate with the release of CO_2 and formation of NADH.
- The Krebs cycle (sometimes called the citric acid cycle) also takes place in the matrix of a mitochondrion. As further oxidation of substrates occurs, more CO_2 is released and both NADH and $FADH_2$ result. The Krebs cycle is also able to produce two ATP per glucose molecule outside the ETC.

- The fourth phase of cellular respiration (the electron transport chain [ETC] and chemiosmosis) produces most of the ATP per glucose breakdown. The details are as we have already outlined for photosynthesis (section 6.5). Passage of electrons down the ETC releases energy that is used to create a H^+ gradient. As H^+ flows down its gradient through an ATP synthase complex, ATP is produced. The number of ATP produced varies with the tissue, but as much as 32–34 ATP per glucose molecule is possible (Fig. 7.1B).

Thus far, we have not accounted for the role of oxygen and the production of water during cellular respiration. In Figure 7.1B the diffusion of oxygen stops at the ETC because oxygen is the final acceptor of electrons (e^-) from the ETC. After oxygen receives electrons, it combines with H^+ and becomes water. Free oxygen doesn't stay very long in cells (if it did even DNA stands the chance of getting oxidized). Instead oxygen is immediately converted to water.

MP3
Cellular Respiration

▶ **7.1 CHECK YOUR PROGRESS**
1. Explain the role of NAD^+ and FAD during the oxidation of glucose.
2. Identify which two of the four phases of glucose breakdown release carbon dioxide.

FIGURE 7.1B The four phases of cellular respiration consist of glycolysis, which occurs in the cytoplasm, as well as the preparatory reaction, the Krebs cycle, and the electron transport chain (ETC), which occur in a mitochondrion. NADH and $FADH_2$ shuttle electrons from the first three phases to the ETC. The end result of cellular respiration is a total of 36–38 molecules of ATP.

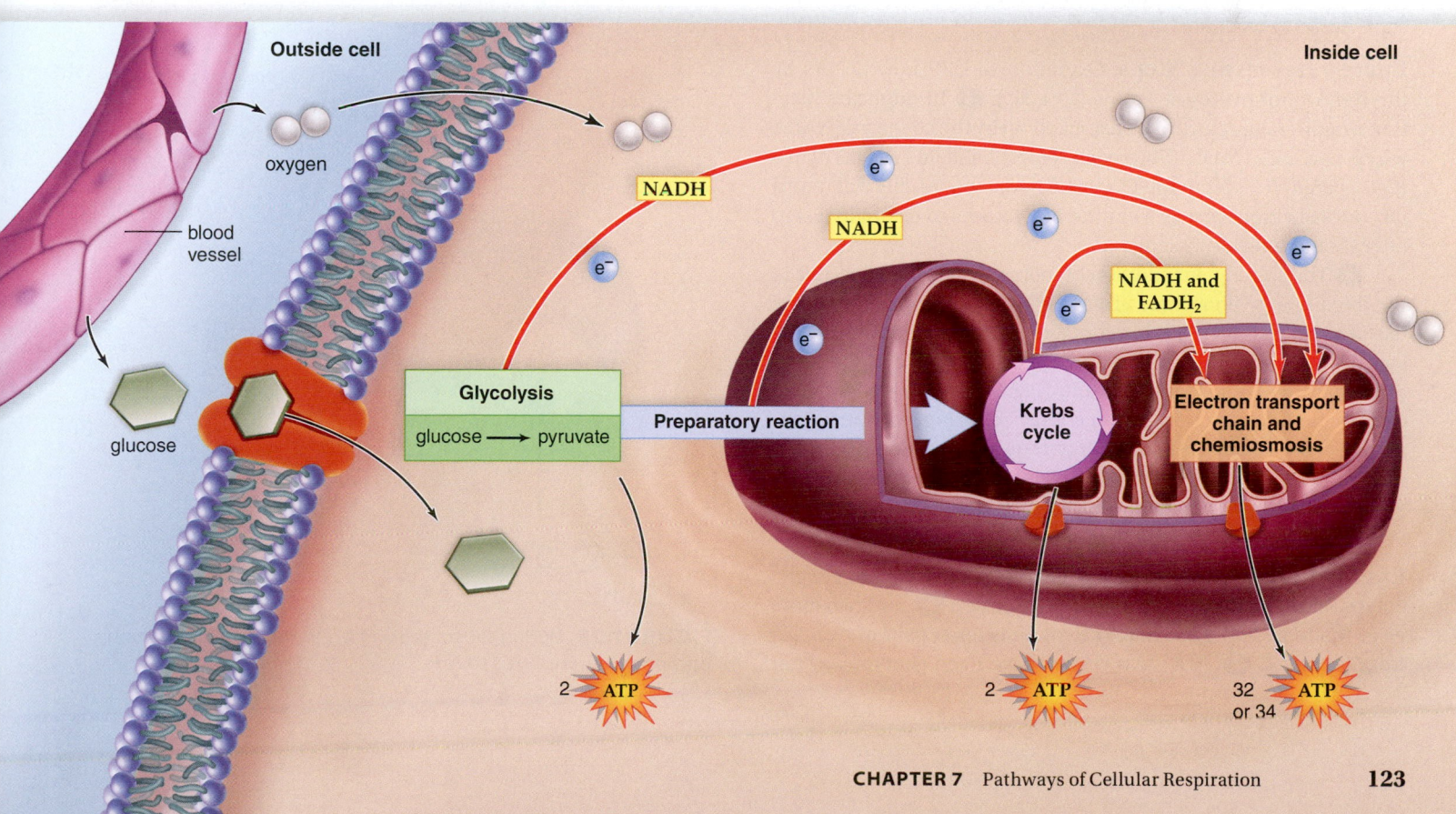

Outside the Mitochondria: Glycolysis

Glycolysis is a good example of a metabolic pathway, in which the first reaction leads to the next reaction, which leads to the next reaction until there is an end product. Keep in mind that each reaction in a metabolic pathway requires its own enzyme that speeds the reaction by combining with its substrate.

7.2 Glycolysis: Glucose breakdown begins

LEARNING OUTCOMES

When you complete this section, you should be able to

1. List the inputs and outputs of glycolysis.
2. Explain substrate-level ATP synthesis.
3. Distinguish between the energy-investment steps and the energy-harvesting steps of glycolysis.

Glycolysis, which takes place within the cytoplasm outside the mitochondria, is the breakdown of glucose to two pyruvate molecules (Fig. 7.2A, on facing page). Since glycolysis occurs universally in organisms, it most likely evolved before the Krebs cycle and the electron transport chain. This may be why glycolysis occurs in the cytoplasm and does not require oxygen. There was no free oxygen in the early atmosphere of the Earth.

Glycolysis is a long series of reactions (Fig. 7.2A), and just as you would expect for a metabolic pathway, each step has its own enzyme. The pathway can be conveniently divided into the energy-investment steps and the energy-harvesting steps. During the energy-investment steps, ATP is used to "jump-start" glycolysis. This deficit is made up for as glycolysis proceeds.

Energy-Investment Steps ❶ As glycolysis begins, two ATP are used to activate glucose, a 6-carbon (C_6) molecule that ❷ splits into two C_3 molecules, known as G3P. Each G3P has a phosphate group. From this point on, each C_3 molecule undergoes the same series of reactions.

Energy-Harvesting Steps Oxidation of G3P now occurs by the removal of hydrogen atoms (e^- + H^+). ❸ In duplicate reactions, high-energy electrons are picked up by the coenzyme NAD^+, which becomes NADH. Now, each NADH molecule will carry two *high-energy electrons* to the ETC. The NAD^+ that results will return for more electrons just as a shuttle bus returns to its original stop after discharging its passengers.

❹ The subsequent addition of inorganic phosphate results in two high-energy phosphate groups. ❺ Each phosphate group is used to synthesize an ATP, by a process called **substrate-level ATP synthesis.** During this process, an enzyme passes a high-energy phosphate to ADP so that ATP results (Fig. 7.2B). This is an example of a coupled reaction because an energy-releasing reaction is happening at the same time as an energy-requiring reaction. ❻ Oxidation occurs again, but by removal of water (H_2O). ❼ Substrate-level ATP synthesis occurs again, and ❽ two molecules of pyruvate result. Glycolysis is now complete.

Energy Benefit of Glycolysis Two ATP were used to get glycolysis started, but two ATP were made in step ❺, and two more were made in step ❼. Therefore, there is a net gain of two ATP from glycolysis ❽. Also, two NADH have been generated. After

these NADH have delivered their high-energy electrons to the ETC, more ATP will be produced as a result of glycolysis.

Inputs and Outputs of Glycolysis Altogether, the inputs and outputs of glycolysis are as follows:

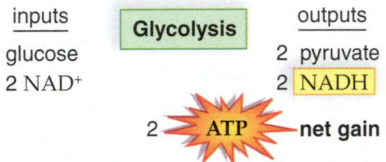

Notice that so far, we have accounted for only two of the 36 or 38 ATP formed per glucose molecule during cellular respiration. If oxygen is available, pyruvate enters mitochondria, where it is oxidized to CO_2 during the prep reaction and the Krebs cycle. Only if pyruvate enters mitochondria can 36–38 ATP be produced as a result of glucose breakdown.

3D Animation
Cellular Respiration: Glycolysis

▶ 7.2 CHECK YOUR PROGRESS

1. Explain what happens in general terms so that glycolysis results in two NADH and a net gain of two ATP.
2. Identify the type of bond a substrate must have in order to react with ADP to form ATP.

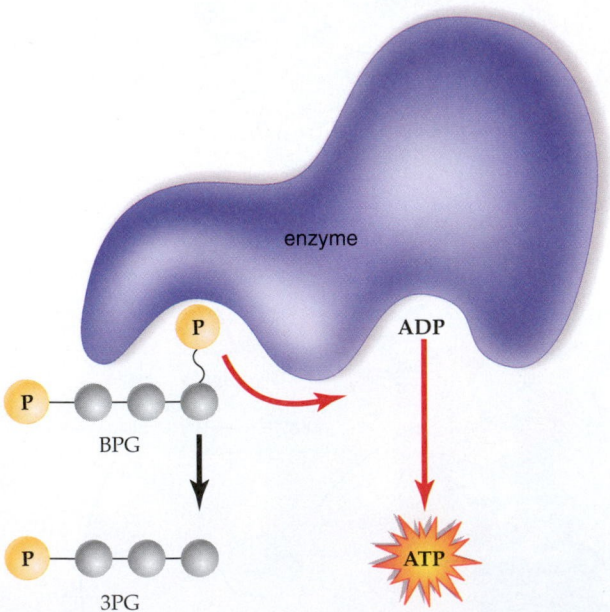

FIGURE 7.2B During substrate-level ATP synthesis, a high-energy phosphate is transferred to ATP.

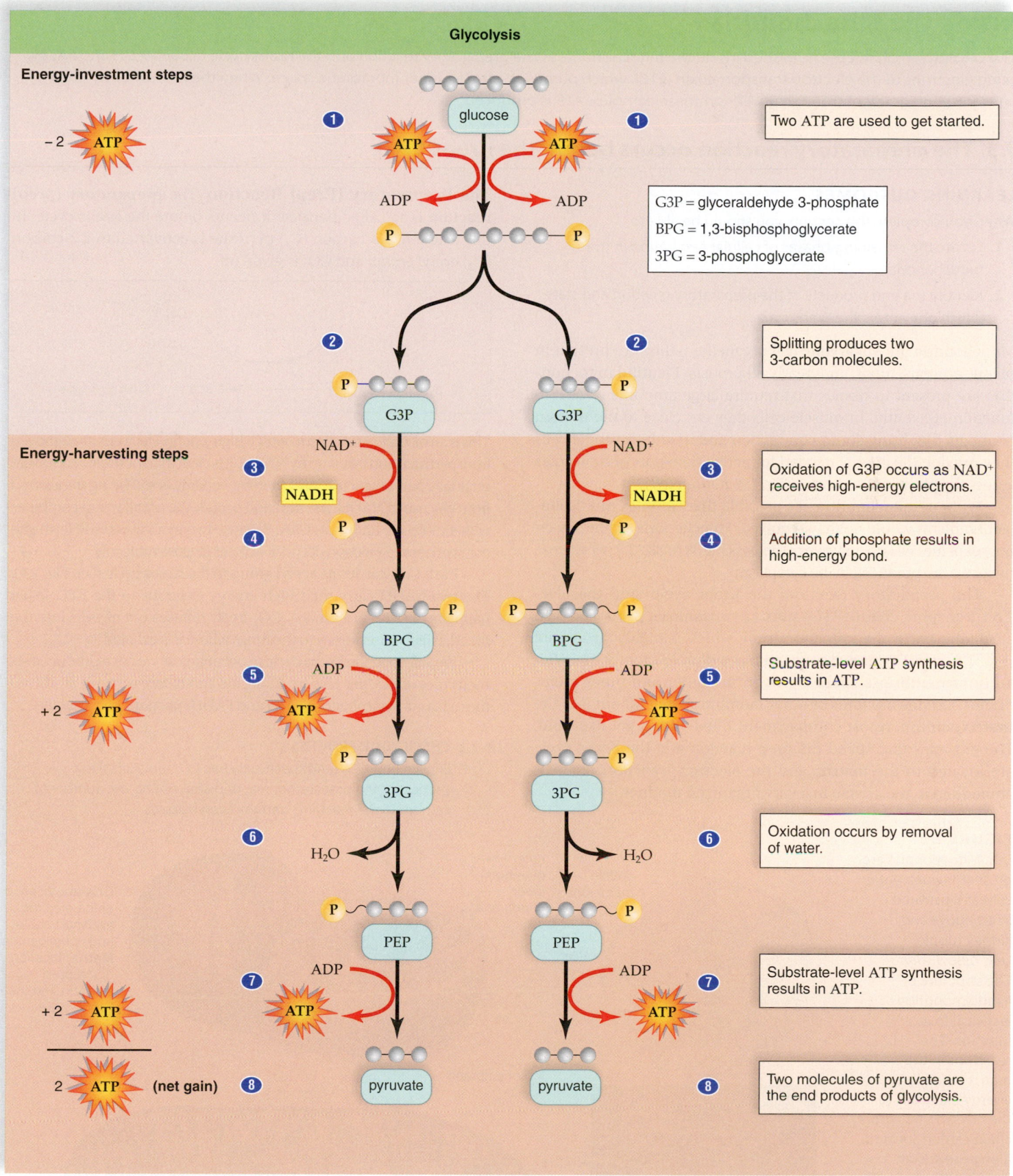

Glycolysis

Energy-investment steps

– 2 ATP

❶ glucose
❶ Two ATP are used to get started.

G3P = glyceraldehyde 3-phosphate
BPG = 1,3-bisphosphoglycerate
3PG = 3-phosphoglycerate

❷ ❷ Splitting produces two 3-carbon molecules.

Energy-harvesting steps

G3P G3P

NAD⁺ NAD⁺
❸ NADH NADH ❸ Oxidation of G3P occurs as NAD⁺ receives high-energy electrons.

P P
❹ ❹ Addition of phosphate results in high-energy bond.

BPG BPG

ADP ADP
❺ ATP ATP ❺ Substrate-level ATP synthesis results in ATP.

+ 2 ATP

3PG 3PG

❻ H₂O H₂O ❻ Oxidation occurs by removal of water.

PEP PEP

ADP ADP
❼ ATP ATP ❼ Substrate-level ATP synthesis results in ATP.

+ 2 ATP

2 ATP (net gain) ❽ pyruvate pyruvate ❽ Two molecules of pyruvate are the end products of glycolysis.

FIGURE 7.2A Glycolysis. Each gray ball represents one carbon atom. Therefore glucose, a C_6 molecule, has six gray balls.

Inside the Mitochondria

The breakdown of glucose is finished inside the mitochondria, as the preparatory reaction and the Krebs cycle occur. NADH and $FADH_2$ bring electrons to the electron transport chain (ETC), which pumps H^+ into the inner membrane space. When the H^+ flow down their concentration gradient through an ATP synthase complex, ATP is produced.

7.3 The preparatory reaction occurs before the Krebs cycle

LEARNING OUTCOMES

When you complete this section, you should be able to

1. Locate the remaining phases of cellular respiration in the mitochondria.
2. Identify the end products of the preparatory reaction and state which one enters the Krebs cycle.

Mitochondria are eukaryotic cell organelles—they are present in plants, animals, fungi, and nearly all protists. Plentiful mitochondria are present in tissues that require large amounts of ATP to function efficiently. In muscle cells they are close to the myosin filaments that bring about contraction and in kidney tubule cells they provide the energy needed for active transport of molecules across the plasma membrane (Fig. 7.3, *bottom left*).

Without mitochondria, it's possible that complex multicellular organisms would not have evolved. Mitochondria are the powerhouses of the cell and only two of the possible 36 or 38 ATP per glucose molecule are produced in the cytoplasm.

The preparatory reaction, the Krebs cycle, and the electron transport chain (ETC) plus chemiosmosis are located in particular parts of a mitochondrion, which is highly structured (Fig. 7.3, *right*). A mitochondrion has a double membrane, with an **intermembrane space** between the outer and inner membranes. **Cristae** are folds of inner membrane that jut out into the **matrix,** an innermost compartment filled with a gel-like fluid. The enzymes that speed the prep reaction and the Krebs cycle are located in the matrix, and the electron transport chain is embedded in the cristae in a very organized manner.

The Preparatory (Prep) Reaction The **preparatory (prep) reaction** is so called because it occurs before the Krebs cycle. In this reaction, a 3-carbon (C_3) pyruvate is converted to a 2-carbon (C_2) *acetyl group,* and CO_2 is given off:

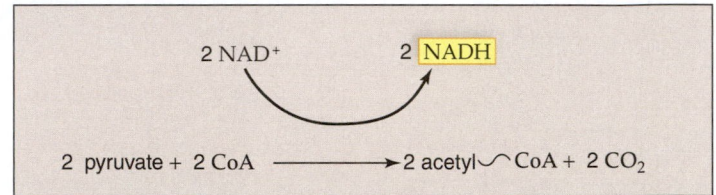

This is an oxidation reaction in which high-energy electrons are removed from pyruvate and taken up by NADH. It is clear that these electrons were originally part of glucose, and therefore energy is still being transferred from glucose to NADH. As oxidation occurs, the acetyl group combines with a molecule known as CoA. Because two pyruvates result from glycolysis, the prep reaction occurs twice per glucose molecule.

Each CoA carries its acetyl group to the Krebs cycle (see Fig. 7.4). The two NADH carry their high-energy electrons to the ETC. What happens to CO_2? In humans, CO_2 freely diffuses out of cells into the blood, which transports it to the lungs where it is exhaled.

We still need the Krebs cycle to occur before we have produced six molecules of CO_2 per glucose molecule and before we have produced the maximum number of NADH and $FADH_2$ per glucose molecule.

▶ **7.3 CHECK YOUR PROGRESS**
1. Explain how CO_2 (and specifically two CO_2 molecules) is a product of the preparatory reaction after noting the number of carbons in the reactants and the acetyl groups.

FIGURE 7.3

Left: Mitochondria are present in eukaryotic cells as illustrated in art above and kidney tubule cells below. *Right:* The structure of a mitochondrion is critical to its function. The reactions in the matrix feed high-energy electrons to the electron transport chain carriers located along the cristae. Only in this way is ATP eventually produced.

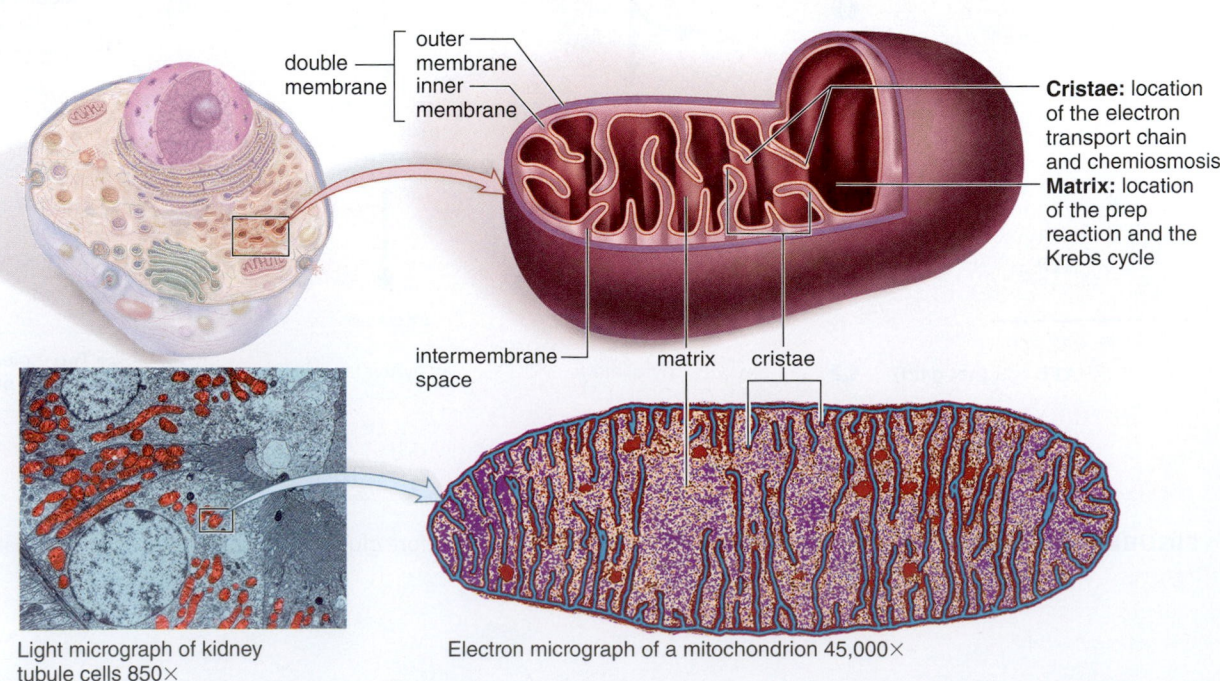

Light micrograph of kidney tubule cells 850×

Electron micrograph of a mitochondrion 45,000×

7.4 The Krebs cycle generates much NADH

LEARNING OUTCOME

When you complete this section, you should be able to

1. List and explain the outputs of the Krebs cycle.

The **Krebs cycle,** named for the person who discovered it, is a metabolic pathway located in the matrix of mitochondria. The cycle consists of a series of reactions that return to their starting point, namely a molecule of citrate (citric acid) (Fig. 7.4). This molecule is the reason the pathway is also called the citric acid cycle.

Steps of the Cycle ❶ At the start of the Krebs cycle, an enzyme speeds the removal of an acetyl group (C_2) from CoA. ❷ This acetyl group joins with a 4-carbon (C_4) molecule, forming a 6-carbon (C_6) citrate molecule. ❸, ❹ Following formation of citrate, an oxidation reaction occurs twice over; both times NADH forms and CO_2 is released. ❺ Per turn, enough energy remains to form a molecule of ATP from ADP + Ⓟ. This is substrate-level ATP synthesis because the energy that drives the reaction comes from the breakdown of substrates. ❻ Additional oxidation reactions produce

a FADH$_2$ and another NADH. ❼ The cycle has now returned to its starting point and is ready to receive another acetyl group.

 **Animation** How the Krebs Cycle Works

Because the Krebs cycle turns twice for each original glucose molecule, the inputs and outputs of the cycle per glucose molecule are as follows:

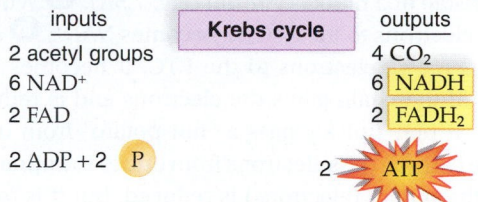

inputs	Krebs cycle	outputs
2 acetyl groups		4 CO_2
6 NAD$^+$		6 NADH
2 FAD		2 FADH$_2$
2 ADP + 2 Ⓟ		2 ATP

Note that the six carbon atoms originally located in a glucose molecule have now become six molecules of CO_2. The prep reaction produces two CO_2, and the Krebs cycle produces four CO_2 per glucose molecule. We have already mentioned that this is the CO_2 we breathe out. In other words, the CO_2 we exhale comes from the mitochondria within our cells.

Energy Benefit of the Krebs Cycle The outputs show that, per glucose molecule, the Krebs cycle produces two ATP, six NADH, and two FADH$_2$. The NADH and FADH$_2$ carry their high-energy electrons to the ETC where they collectively result in 22 ATP per glucose molecule. Clearly the Krebs cycle is critically important in capturing the energy of glucose.

3D Animation Cellular Respiration: Citric Acid Cycle

▶ **7.4 CHECK YOUR PROGRESS**

1. Describe how the Krebs cycle is the workhorse of cellular respiration by comparing its outputs to those of glycolysis and the prep reaction.

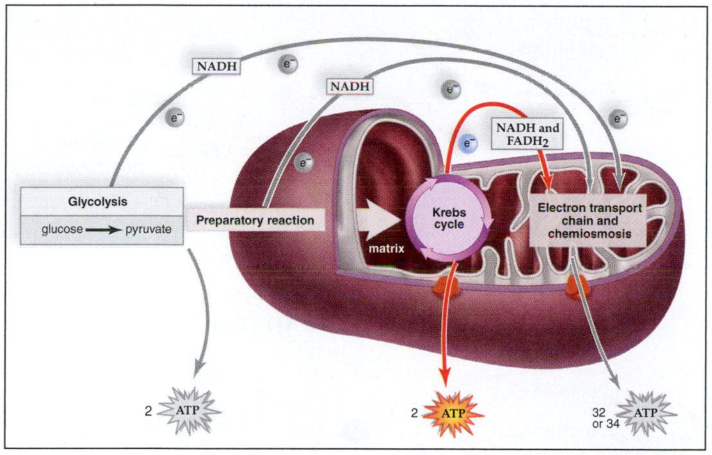

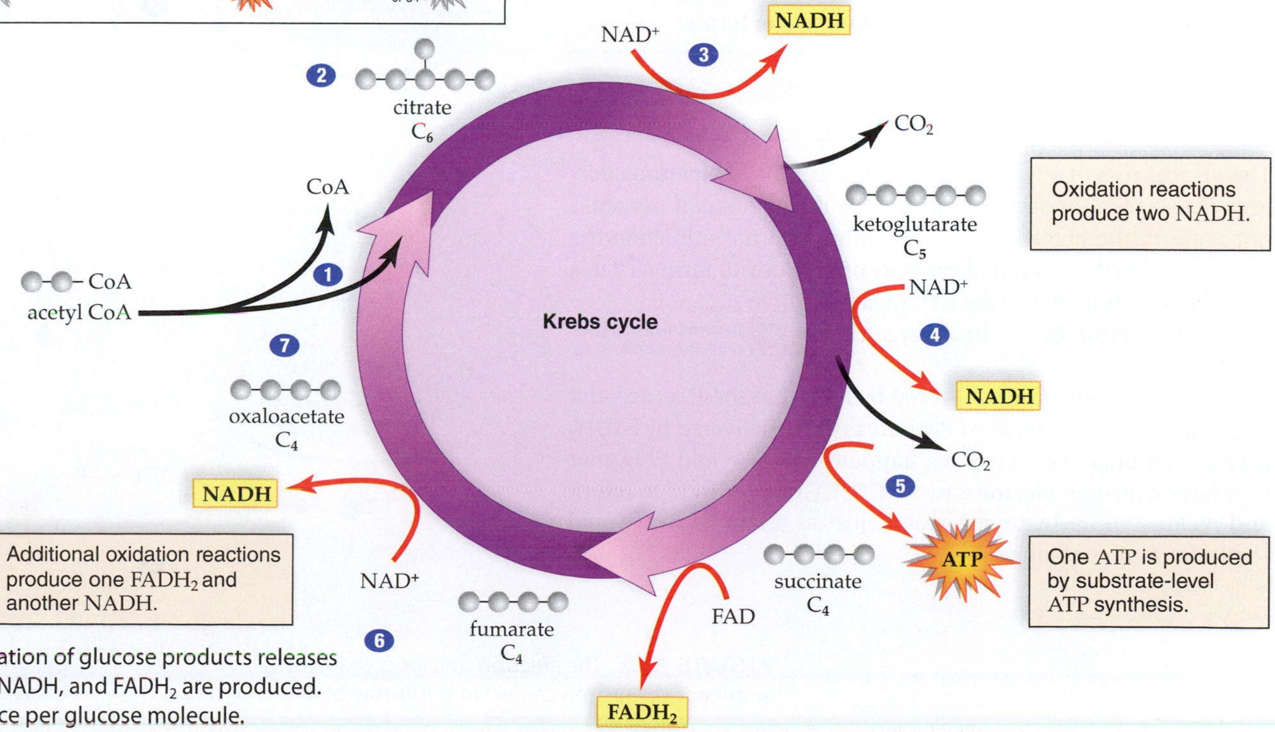

FIGURE 7.4 Oxidation of glucose products releases energy so that ATP, NADH, and FADH$_2$ are produced. This cycle turns twice per glucose molecule.

Oxidation reactions produce two NADH.

One ATP is produced by substrate-level ATP synthesis.

Additional oxidation reactions produce one FADH$_2$ and another NADH.

Krebs cycle

citrate C_6

ketoglutarate C_5

oxaloacetate C_4

succinate C_4

fumarate C_4

acetyl CoA

7.5 The electron transport chain captures energy and ATP synthase produces ATP

LEARNING OUTCOME

When you complete this section, you should be able to

1. Describe the organization of mitochondrial cristae and how mitochondria produce ATP with water as a side product.

An **electron transport chain (ETC)** is a series of electron carriers in the cristae of a mitochondrion (Fig. 7.5A). **1a** When NADH gives up its electrons to the ETC, it becomes NAD$^+$; **1b** and when FADH$_2$ gives up its electrons to the ETC, it becomes FAD. The first carrier of the chain gains the electrons and is reduced. Just as children at play quickly pass a "hot potato" from one to the other, so the carriers pass electrons from one to the other. A carrier that holds the potato (electrons) is reduced, but it is oxidized as soon as it passes the electrons to the next carrier. This oxidation-reduction process continues, and each of the carriers, in turn, becomes reduced and then oxidized as the electrons move down the chain. The electrons that enter the ETC are high-energy electrons (a hot potato) carried by NADH and FADH$_2$. A hot potato loses heat as it is passed along, and in the same way electrons have lost their energy by the time they leave the ETC.

2 Several of the carriers in the ETC are cytochrome molecules. A **cytochrome** is a protein that has a tightly bound heme group with a central atom of iron, just as hemoglobin does. When the iron accepts electrons, it becomes reduced, and when it gives them up, it becomes oxidized. A number of poisons, such as cyanide, cause death by binding to and blocking the function of cytochromes. **3** As the pair of electrons is passed from carrier to carrier, energy is captured and eventually used to form ATP molecules.

What is the role of oxygen in cellular respiration and the reason we take in oxygen by breathing? Oxygen is the final acceptor of electrons from the ETC. Oxygen receives the energy-spent electrons from the last of the carriers. **4** After receiving electrons, oxygen combines with hydrogen ions, and water forms:

$$\tfrac{1}{2}O_2 + 2\,e^- + 2\,H^+ \longrightarrow H_2O$$

The critical role of oxygen as the final acceptor of electrons during cellular respiration is exemplified by the fact that, if oxygen is not present, the chain does not function, and the mitochondria produce no ATP. The limited capacity of the body to form ATP in a way that does not involve the ETC means that death eventually results if oxygen is not available.

Animation
Electron Transport System and ATP Synthesis

Notice that electrons delivered by NADH to the ETC account for the production of three ATP, whereas those delivered by FADH$_2$ account for only two ATP. What happens to NAD$^+$ and FAD after they have delivered electrons to the ETC? They are "free" to return and pick up more hydrogen atoms, just as a shuttle bus would return for more passengers.

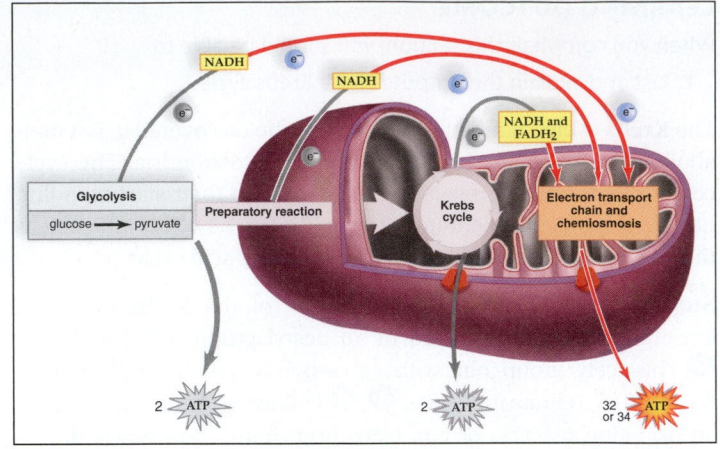

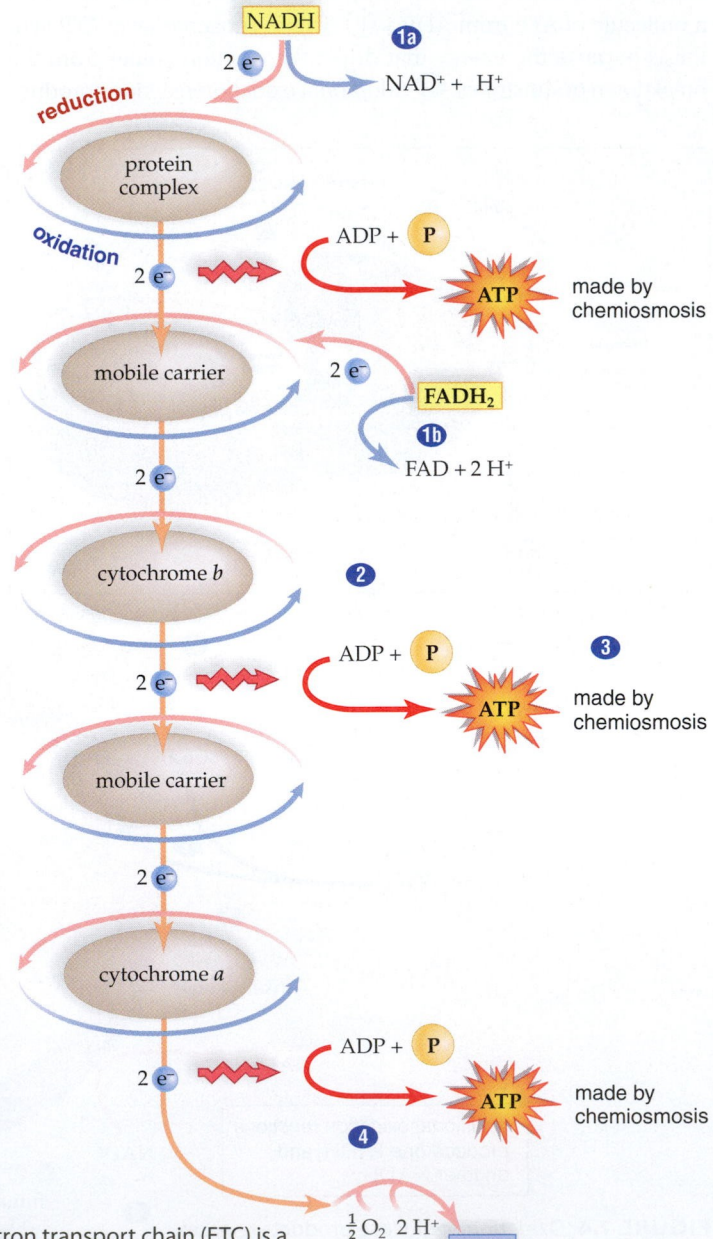

FIGURE 7.5A The electron transport chain (ETC) is a sequence of electron carriers in the cristae of mitochondria.

Organization of Cristae At this juncture it is important to recall the structure of a mitochondrion. As described in section 7.3, the cristae of a mitochondrion separate the matrix from the intermembrane space.

The cristae of mitochondria contain the ETC and the ATP synthase complex. Most of the carriers of the ETC occur in three stationary protein complexes, but two are mobile carriers that ferry electrons between the complexes (Fig. 7.5B). Following the example of section 6.7, we will divide our discussion into the "get ready" phase and the "payoff" phase.

Get Ready As we have discussed, NADH and FADH$_2$ carry electrons to the ETC and thereafter the carriers of the ETC pass electrons from one to the other. What happens to the hydrogen ions (H$^+$) carried by NADH and FADH$_2$? Certain complexes of the ETC use the released energy to pump hydrogen ions from the matrix into the intermembrane space of a mitochondrion. Vertical blue arrows in Figure 7.5B (*top*) show that the protein complexes of the ETC pump H$^+$ into the intermembrane space. This establishes a strong gradient; there are about ten times as many hydrogen ions in the intermembrane space as in the matrix.

Payoff The hydrogen ions then flow down their concentration gradient from the intermembrane space into the matrix. As hydrogen ions flow from high to low concentration through an ATP synthase complex (Fig. 7.5B, *bottom*), the enzyme **ATP synthase** synthesizes ATP from ADP + Ⓟ. This process is called **chemiosmosis** because ATP production is tied to the establishment of an H$^+$ gradient. The action of certain poisons supports the chemiosmotic model. For example, a poison that inhibits ATP synthesis causes the H$^+$ gradient to become larger than usual.

ATP production is comparable to a hydroelectric power plant. Water is trapped behind a dam, and when released, its motion is used to generate electricity. Similarly, hydrogen ions are trapped inside the intermembrane space, and when they pass through an ATP synthase complex, this energy is used to generate ATP.

Once formed, ATP moves out of mitochondria and is used to perform cellular work, during which it breaks down to ADP and Ⓟ. Then these molecules are returned to the mitochondria for recycling. At any given time, the amount of ATP in muscles would sustain a strenuous activity for only about 15 seconds (Fig. 7.5C); therefore, ATP synthase must continually produce ATP. It is estimated that the mitochondria produce our body weight in ATP every day.

How many ATP are produced per glucose molecule by means of chemiosmosis? See section 7.6, which calculates the number of ATP per each phase of cellular respiration and then totals the number of ATP per glucose molecule.

3D Animation
Cellular Respiration:
Electron Transport Chain

▶ **7.5 CHECK YOUR PROGRESS**
1. Explain by summarizing cellular respiration where and how the children playing tag in Figure 7.5C get their ATP.

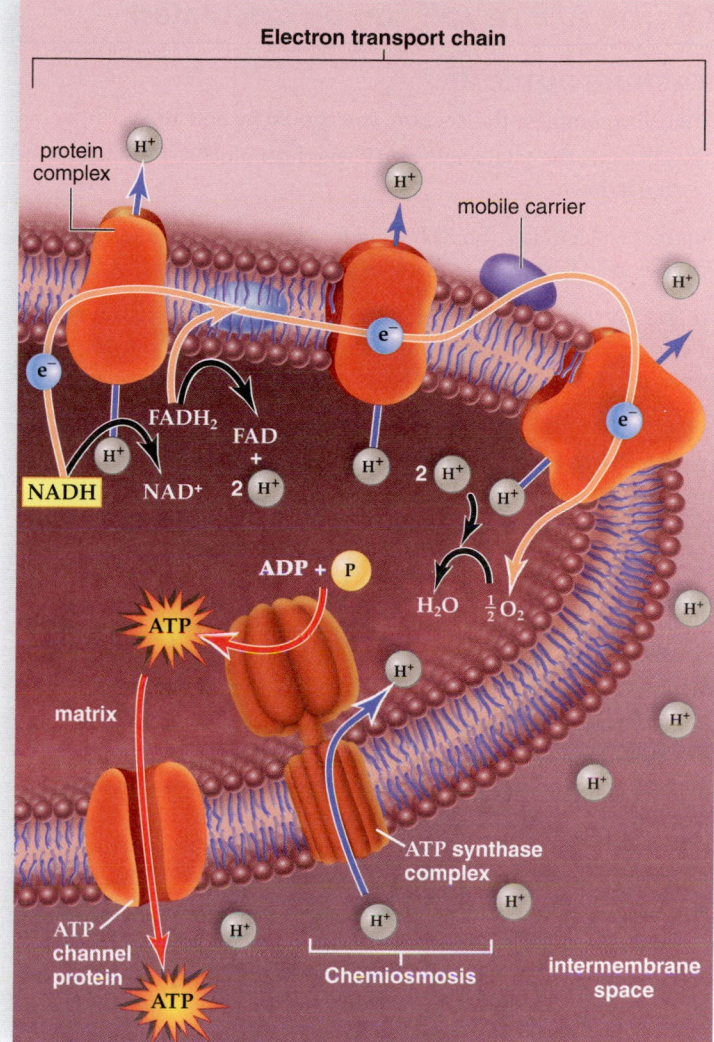

FIGURE 7.5B Protein complexes of the ETC use the released energy of electrons to pump hydrogen ions into the intermembrane space (*top*). When the hydrogen ions flow down their concentration gradient through an ATP synthase complex, ATP is produced (*bottom*).

FIGURE 7.5C Continual ATP production is a necessity. Muscles typically have enough ATP on hand for a child to play tag for about 15 seconds.

7.6 The ATP payoff can be calculated

LEARNING OUTCOME

When you complete this section, you should be able to

1. Calculate the total energy (ATP) yield per glucose molecule breakdown.

Figure 7.6 calculates the ATP yield for the complete breakdown of glucose to CO_2 and H_2O during cellular respiration.

In the Cytoplasm ❶ Per glucose molecule, there is a net gain of two ATP from glycolysis, which takes place in the cytoplasm. These two ATP are produced by substrate-level ATP synthesis (see Fig. 7.2B).

In the Mitochondrion ❷ The prep reaction does not generate any ATP molecules directly. ❸ The Krebs cycle, which occurs in the matrix of mitochondria, directly accounts for two ATP per glucose molecule. These two ATP form by substrate-level ATP synthesis. ❹ This adds up to a subtotal of four ATP produced by substrate-level ATP synthesis.

❺ Most ATP is produced by the electron transport chain (ETC) and chemiosmosis. Per glucose molecule, ten NADH and two $FADH_2$ take electrons to the ETC.

❻ In many animal cells, NADH formed outside a mitochondrion by glycolysis cannot cross the inner mitochondrial membrane, but a "shuttle" mechanism allows its electrons to be delivered to the ETC inside the mitochondrion. The cost to the cell is one ATP for each NADH that is shuttled to the ETC. This reduces the overall count of ATP produced as a result of glycolysis in some animal cells to four, instead of six, ATP.

For each NADH formed *inside* the mitochondrion by ❼ the prep reaction and ❽ the Krebs cycle, three ATP result, but ❾ for

each $FADH_2$ from the Krebs cycle only two ATP are produced. As Figure 7.5A shows, $FADH_2$ delivers its electrons to the ETC after NADH, and therefore its two electrons account for the production of only two ATP.

❿ Altogether, 32 or 34 ATP are made as a result of electrons passing down the ETC. ⓫ Therefore, the total number of ATP produced as a result of complete glucose breakdown is 36 or 38.

Efficiency of Cellular Respiration It is interesting to calculate how much of the energy in a glucose molecule eventually becomes available to the cell. The difference in energy content between the reactants (glucose and O_2) and the products (CO_2 and H_2O) is 686 kcal. An ATP phosphate bond has an energy content of 7.3 kcal, and if 36 ATP are produced during glucose breakdown, the yield is 263 kilocalories (kcal). Therefore, 263/686, or 39%, of the available energy is usually transferred from glucose to ATP. The rest of the energy is lost in the form of heat.

It is good to remind ourselves why cells go to the bother of transforming the energy within glucose to that in ATP. In the end, it's more efficient to spend a penny (ATP) for something you want instead of a dollar (glucose) when no change will be forthcoming. The "something" a cell wants is any cellular process that requires energy.

3D Animation
Cellular Respiration: Summary

▶ **7.6 CHECK YOUR PROGRESS**
1. Compare the number of ATP produced outside to that produced inside the mitochondria per glucose molecule.
2. Identify the part of Figure 7.6 that evolved in eukaryotic cells before the other part. Explain your answer.

FIGURE 7.6
Energy yield per glucose molecule.

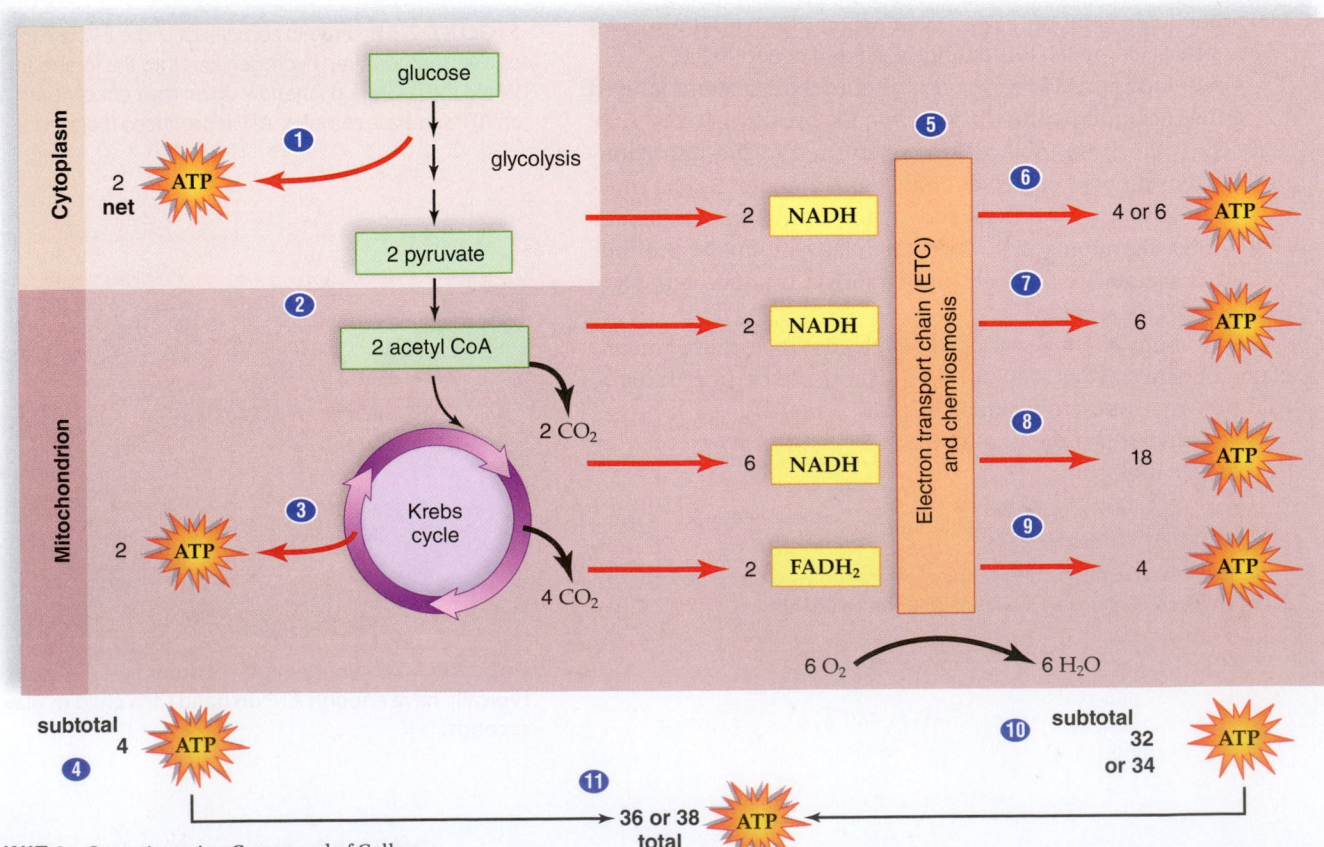

Fermentation

Complete glucose breakdown demands an input of oxygen to accept electrons from the electron transport chain (ETC). If oxygen is not available, the ETC stops working, but cells then turn to fermentation, an anaerobic process. Fermentation produces only two ATP per glucose molecule, instead of 36 or 38, but even so, it is useful to a cell when oxygen is not available.

7.7 When oxygen is in short supply, the cell switches to fermentation

LEARNING OUTCOMES

When you complete this section, you should be able to

1. Compare and contrast fermentation to glycolysis.
2. Explain why fermentation is a benefit to the body.

Fermentation produces two ATP per glucose molecule when oxygen is not available. During fermentation, pyruvate (from glycolysis) is the final acceptor for electrons instead of oxygen (Fig. 7.7). In animal cells, including those of humans, the pyruvate is reduced to lactate. Other types of organisms instead produce alcohol with the release of CO_2. Bacteria vary as to whether they produce an organic acid, such as lactate, or an alcohol and CO_2. Yeasts are good examples of organisms that generate ethyl alcohol and CO_2 as a result of fermentation.

In Figure 7.7, notice that the NADH formed during glycolysis becomes NAD^+ once more after it has given its two electrons to pyruvate. Only because NAD^+ is replenished can glycolysis and substrate-level ATP synthesis continue in the absence of oxygen.

Benefits Versus Drawbacks of Fermentation Despite its low yield of only two ATP, fermentation is essential to organisms because it is anaerobic. When our muscles are working vigorously over a short period of time, as when we run, fermentation is a way to produce ATP even though oxygen is temporarily in limited supply.

Fermentation products are toxic to cells. Yeasts ferment the sugar of grapes in the absence of oxygen, but they die from the alcohol they produce. In humans, blood carries away the lactate formed in muscles. Eventually, however, lactate begins to build up, changing the pH and causing the muscles to "burn," and eventually to fatigue so that they no longer contract. When we stop running, our bodies are in **oxygen debt,** as signified by the fact that we continue to breathe very heavily for a time.

Grapes coated with yeast.

Comparison of Yields Fermentation only produces two ATP per glucose molecule, instead of the 36 or 38 possible with cellular respiration. Therefore, following fermentation, most of the potential energy in a glucose molecule is still waiting to be released. Fortunately, the liver converts lactate back to pyruvate so that cellular respiration can pick up where it left off:

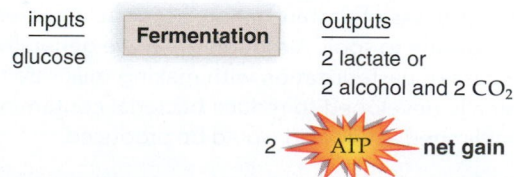

inputs	Fermentation	outputs
glucose		2 lactate or
		2 alcohol and 2 CO_2

2 ATP net gain

▶ **7.7 CHECK YOUR PROGRESS**
1. Compare the role of NAD^+ in glycolysis to that of NAD^+ in fermentation.

glucose

−2 ATP   2 ATP

2 ADP

2 P—P—P

G3P

2 P   2 NAD^+

2 NADH

2 P—•—•—• P

BPG

4 ADP

+4 ATP   4 ATP

2 •—•—•

pyruvate

or

2 ATP (net gain)

2 CO_2

2 lactate   or   2 alcohol

FIGURE 7.7 Fermentation consists of glycolysis followed by a reduction of pyruvate by NADH. "Free" NAD^+ returns to pick up more electrons.

7A Fermentation Helps Produce Numerous Food Products

Common grocery items such as bread, yogurt, soy sauce, pickles, and maybe even wine are among the many foods produced when microorganisms ferment (break down sugar in the absence of oxygen). Foods produced by fermentation last longer because the fermenting organisms have removed many of the nutrients that would attract other organisms. As mentioned, the products of fermentation can even be dangerous to the very organisms that produced them, as when yeasts are killed by the alcohol they produce.

Fermenting Yeasts Leaven Bread and Produce Alcohol

Baker's yeast, *Saccharomyces cerevisiae,* is added to bread for the purpose of leavening—the dough rises when the yeasts give off CO_2. The ethyl alcohol produced by the fermenting yeast evaporates during baking. The many different varieties of sourdough breads obtain their leavening from a starter composed of fermenting yeasts along with bacteria from the environment. Depending on the community of microorganisms in the starter, the flavor of the bread ranges from sour and tangy, as in San Francisco–style sourdough, to a milder taste, such as that produced by most Amish friendship bread recipes.

Ethyl alcohol is desired when yeasts are used to produce wine and beer. When yeasts ferment the carbohydrates of fruits, the end result is wine. If they ferment grain, beer results. A few specialized varieties of beer, such as traditional wheat beers, have a distinctive sour taste because they are produced with the assistance of lactic acid–producing bacteria, such as those of the genus *Lactobacillus.* Stronger alcoholic drinks (e.g., whiskey and vodka) require distillation to concentrate the alcohol content.

The acetic acid bacteria, including *Acetobacter aceti,* spoil wine because these bacteria convert alcohol to acetic acid (vinegar). Until the renowned nineteenth-century scientist Louis Pasteur invented the process of pasteurization, acetic acid bacteria commonly caused wine to spoil. Although today we generally associate the process of pasteurization with making milk safe to drink, it was originally developed to reduce bacterial contamination in wine so that limited acetic acid would be produced.

Fermenting Bacteria That Produce Acid

Yogurt, sour cream, and cheese are produced through the action of various lactic acid bacteria that cause milk to sour. Milk contains lactose, which these bacteria use as a substrate for fermentation. Yogurt, for example, is made by adding lactic acid bacteria, such as *Streptococcus thermophilus* or *Lactobacillus bulgaricus,* to milk and then incubating it to encourage the bacteria to act on lactose.

During the production of cheese, an enzyme called rennin must also be added to the milk to cause it to coagulate and become solid.

Old-fashioned brine cucumber pickles, sauerkraut, and kimchi are pickled vegetables produced by the action of acid-producing, fermenting bacteria from the genera *Lactobacillus* and *Leuconostoc,* which can survive in high-salt environments. Salt is added to draw liquid out of the vegetables and aid in their preservation. The bacteria need not be added because they are already present on the surfaces of the vegetables.

Soy Sauce Is a Combination Product

Soy sauce is traditionally made by adding a mold, *Aspergillus,* and a combination of yeasts and fermenting bacteria to soybeans and wheat. The mold breaks down starch, supplying the fermenting microorganisms with sugar they can use to produce alcohol and organic acids.

CONSIDER THESE QUESTIONS

1. Use the products on this page to demonstrate that seemingly irrelevant scientific findings can be used to benefit humankind.
2. Do the products on this page negate the statement that our food is ultimately derived from plants? Why or why not?

connect | BIOLOGY   Explore the concepts through a variety of multimedia assets and question types.

www.mcgrawhillconnect.com

Fermentation helps make the products shown on this page.

Intersections of Metabolic Pathways

Carbohydrates, proteins, and fats can be metabolized by entering degradative (catabolic) pathways at different locations. Catabolic pathways also provide metabolites needed for the synthesis (anabolism) of various important substances. Therefore, catabolism and anabolism both use the same pools of metabolites.

7.8 Organic molecules can be broken down and synthesized as needed

LEARNING OUTCOME

When you complete this section, you should be able to

1. Recognize the relationship between protein, carbohydrate, and fat metabolism.

Certain substrates recur in various key metabolic pathways, and therefore they form a **metabolic pool.** In the metabolic pool, these substrates serve as entry points for the degradation or synthesis of larger molecules (Fig. 7.8). Degradative reactions break down molecules and collectively participate in **catabolism.** Synthetic reactions build up molecules and collectively participate in **anabolism.**

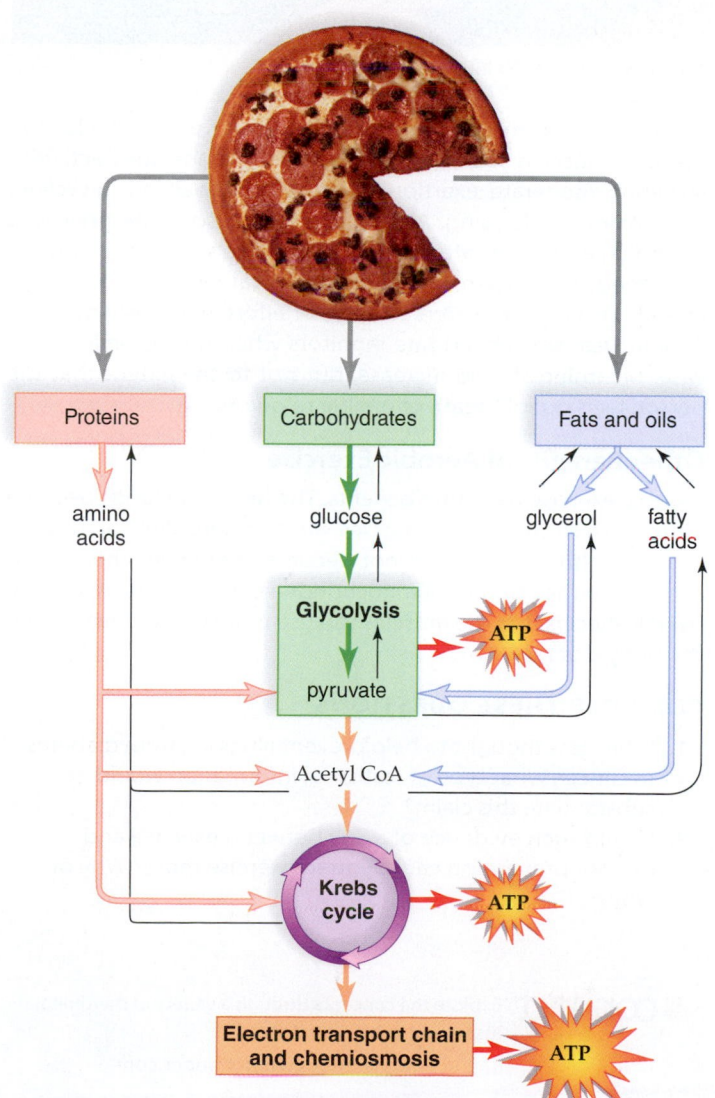

FIGURE 7.8 The metabolic pool concept.

Catabolism We already know that glucose is broken down during cellular respiration. When a fat is used as an energy source, it is hydrolyzed to glycerol and three fatty acids. Then, as Figure 7.8 indicates, glycerol can enter glycolysis. The fatty acids are converted to acetyl CoA, which can enter the Krebs cycle. An 18-carbon fatty acid results in nine acetyl CoA molecules. Calculation shows that respiration of these can produce a total of 108 ATP molecules. For this reason, fats are an efficient form of stored energy—there are three long-chain fatty acids per fat molecule.

The carbon skeleton of amino acids can enter glycolysis, be converted to acetyl CoA, or enter the Krebs cycle directly. The carbon skeleton is produced in the liver when an amino acid undergoes **deamination,** removal of the amino group. The amino group becomes ammonia (NH_3), which enters the urea cycle and becomes part of urea, the primary excretory product of humans. Just where the carbon skeleton begins degradation depends on the length of the R group, as this determines the number of carbons left after deamination.

Anabolism The degradative reactions of catabolism drive the synthetic reactions of anabolism because they release energy that is required for anabolism to occur. For example, we have already mentioned that ATP breakdown drives synthetic reactions. But catabolism is also related to anabolism in another way. The substrates making up the pathways in Figure 7.8 can be used as starting materials for synthetic reactions. In other words, polymers can be oxidized to monomers that can be used to synthesize other biomolecules. In this way, carbohydrate intake can result in the formation of fat: G3P from glycolysis can be converted to glycerol, and acetyl groups from the prep reaction can be joined to form fatty acids. Fat synthesis follows. This explains why you gain weight from eating too much candy, ice cream, or cake.

Some substrates of the Krebs cycle can be converted to amino acids through transamination. During transamination, one amino acid donates its amino group to an organic acid that lacks one. Regardless of transamination, plants are able to synthesize all the amino acids they need. Animals, however, lack some of the enzymes necessary for synthesis of all 20 amino acids. Adult humans can synthesize eleven of the common amino acids, but they cannot synthesize the other eight. These are the *essential amino acids* because it is quite possible for humans to suffer from protein deficiency if their diets do not contain adequate quantities of these amino acids.

▶ **7.8 CHECK YOUR PROGRESS**

1. Consult Figure 7.8 and explain the central position of acetyl CoA in metabolism.

7B Exercise Burns Fat

Combining exercise with a sensible diet that satisfies the body's nutritional requirements appears to be the best long-term approach to weight management. Any form of exercise is preferable to inactivity because all types increase your energy use—in other words, they all "burn calories" at a higher rate than the body does at rest. However, some forms of exercise may be more effective for weight loss than others.

Exercise Offers the Opportunity to Burn Fat

Recall that a fat molecule is also called a triglyceride because it contains three fatty acids. The graph in Figure 7B suggests that muscle cells store both ❶ fat and ❹ glycogen, but the longer we exercise, the more depleted these stores become. Now muscle cells begin to burn either ❷ fatty acids or ❸ glucose from the blood. The fatty acids were deposited in the blood by adipose tissue, which makes us look fat because it does indeed store fat in the body. Figure 7B also suggests that the longer we exercise at about 70% effort, the more fatty acids are burned instead of glucose.

As mentioned in section 7.8, fats are very energy-rich molecules. In fact, the amount of energy released by breaking down fat is more than double that derived from breaking down carbohydrate or protein. Simply taking a quick walk for a minimum of 30 minutes on most days can help burn fat. But since fatty acids store so much energy, prolonged exercise is the best way to use up some of this stored energy.

Prolonged Aerobic Exercise Burns Fat

The burning of fats requires oxygen; it cannot be done anaerobically. When a fat is broken down in an enzyme-catalyzed reaction, the glycerol can be used in glycolysis. But the fatty acids break down to form acetyl CoA, which feeds into the Krebs cycle. The Krebs cycle continues only as long as oxygen is present to receive electrons from the electron transport chain. If oxygen runs out, as occurs when exercise is extremely vigorous, cells switch to fermentation, which is anaerobic, and fat burning decreases.

Aerobic exercise burns fat.

For this reason, people who want to lose weight usually have the most success with aerobic exercise. This includes activities requiring moderate exertion, such as brisk walking, bicycling, swimming, and jogging. Aerobic exercise is so named because breathing and heart rate increase in order to supply the muscles with adequate oxygen. In fact, the heart rate is sometimes used as an indicator of the correct range of effort, which is why some people wear small heart rate monitors when they exercise. Likewise, breathing should increase, but not to the extent that the person feels out of breath or unable to go on.

Other Benefits of Aerobic Exercise

Aerobic exercise has other benefits. The heart and lungs become better able to supply the muscles with oxygen. Skeletal muscles change—although they do not become very bulky, their blood supply and the number of mitochondria in the cells increase. The overall effect is that the muscles can use more energy, even when the body is at rest.

CONSIDER THESE QUESTIONS

1. Exercise is thought to help prevent illnesses, from diabetes to cancer. What type of statistical information would substantiate this claim?
2. Would such evidence of a link between exercise and disease prevention cause you to exercise more? Why or why not?

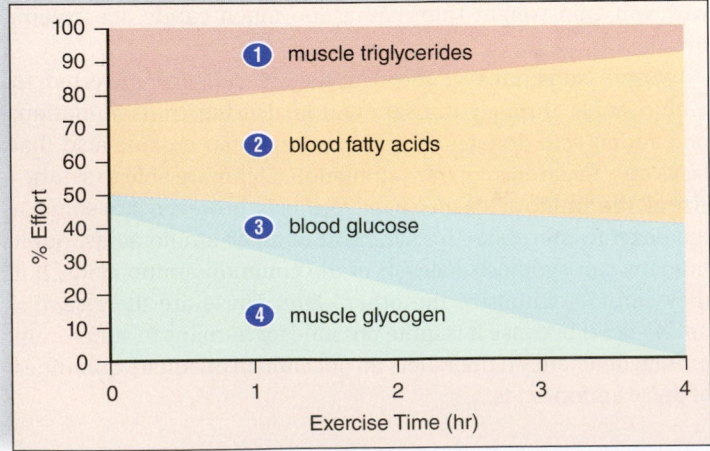

FIGURE 7B The longer you exercise, the more likely you are to burn fatty acids.

 **connect** |BIOLOGY Explore the concepts through a variety of multimedia assets and question types.

www.mcgrawhillconnect.com

THE CHAPTER IN REVIEW

CONNECTING THE CONCEPTS

Energy from the sun flows through all organisms with the participation of chloroplasts and mitochondria. Through the process of photosynthesis, chloroplasts in plants and algae capture solar energy and use it to produce carbohydrates, which are broken down to carbon dioxide and water in the mitochondria of nearly all organisms. The energy released when carbohydrates (and other organic molecules) are oxidized is used to produce ATP molecules. When the cell uses ATP to do cellular work, all the captured energy dissipates as heat.

During cellular respiration, oxidation by removal of hydrogen atoms (e^- + H^+) from glucose or glucose products occurs during glycolysis, the prep reaction, and the Krebs cycle. The prep reaction and Krebs cycle release CO_2. The electrons are carried by NADH and $FADH_2$ to the electron transport chain (ETC) in the cristae of mitochondria. Oxygen serves as the final acceptor of electrons, and H_2O is produced. The pumping of hydrogen ions by the ETC into the intermembrane space leads to ATP production.

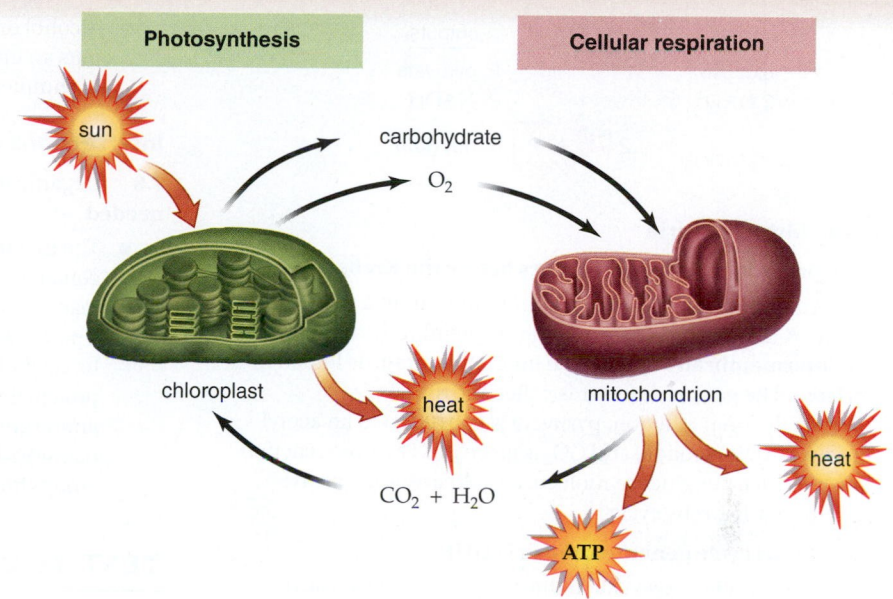

ANALYZE AND EVALUATE

1. Tell how the pre-eukaryotic cell must have produced ATP. What event in the history of life would have allowed cellular respiration to evolve? Explain why you say so.

2. Explain the statement, "if chloroplasts and mitochondria are descended from a free-living common ancestor it would explain their structural similarities." What are some structural similarities?

SUMMARIZE

Overview of Cellular Respiration

7.1 Cellular respiration is a redox reaction that requires O_2

- During **cellular respiration,** glucose is oxidized to CO_2, which we breathe out. Oxygen, which we breathe in, is reduced to H_2O.

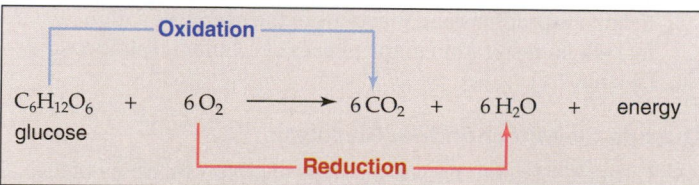

- As oxidation occurs, coenzymes **NAD+** and **FAD** remove hydrogen atoms (e^- + H^+) from glucose. NAD+ becomes NADH, FAD becomes $FADH_2$, and these molecules take two electrons each to an electron transport chain (ETC). Gradual release of energy through oxidation of glucose breakdown products allows more energy to be captured for the production of ATP molecules.
- Cellular respiration has four phases: (1) **Glycolysis** occurs in the cytoplasm and is **anaerobic.** It sends NADH to the ETC and produces a net of two ATP. The end product is two pyruvate molecules. (2) The **preparatory (prep) reaction** and the other three phases occur in the mitochondria and are **aerobic.** The prep reaction breaks down pyruvate to acetyl groups, produces CO_2, and sends NADH to the ETC. (3) The **Krebs cycle** finishes the oxidation of glucose breakdown

products and produces CO_2. It sends NADH and $FADH_2$ to the ETC and produces two ATP after turning twice. (4) The **electron transport chain (ETC)** receives electrons from NADH and $FADH_2$ and establishes a H^+ gradient. When H^+ flows down an ATP synthase complex, ATP is formed.

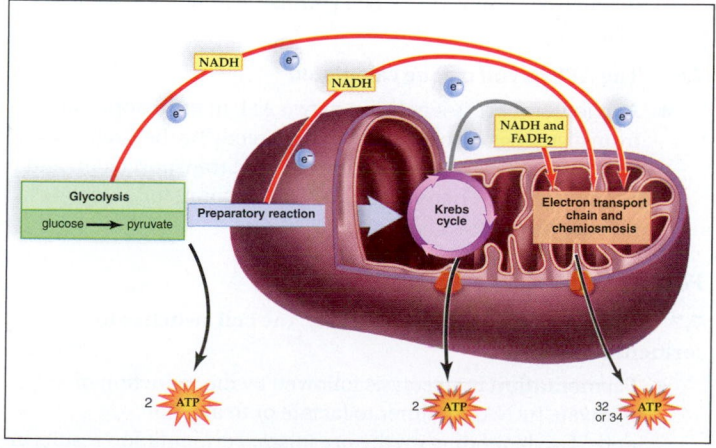

Outside the Mitochondria: Glycolysis

7.2 Glycolysis: Glucose breakdown begins

- Energy investment: At the beginning of glycolysis, two ATP are used to activate glucose, which splits into two C_3 molecules.

Energy harvesting: Removal of hydrogen atoms results in two NADH and four high-energy phosphate bonds. Four ATP result from **substrate-level ATP synthesis.** Therefore, the net gain is two ATP—two used, four made—per glucose molecule. Glycolysis begins with C_6 glucose and ends with two C_3 pyruvate molecules:

inputs   **Glycolysis**   outputs

glucose   2 pyruvate

2 NAD⁺   2 NADH

2   **ATP**   net gain

Inside the Mitochondria

7.3   The preparatory reaction occurs before the Krebs cycle

- The other phases of cellular respiration occur in a mitochondrion. A mitochondrion has a double membrane separated by an **intermembrane space.** The inner membrane is folded into **cristae.** The gel-like interior is called the **matrix.**
- During the prep reaction, pyruvate is converted to an acetyl group, NADH is made, and CO_2 is given off. The prep reaction occurs twice per glucose molecule. CoA carries the acetyl groups to the Krebs cycle.

7.4   The Krebs cycle generates much NADH

- The Krebs cycle is also called the citric acid cycle because citrate begins and ends the cycle. Six NADH and two $FADH_2$ are made per glucose as CO_2 is released. Two ATP per glucose results from substrate-level ATP synthesis.

7.5   The electron transport chain captures energy and ATP synthase produces ATP

- NADH and $FADH_2$ bring electrons to the electron transport chain which contains several **cytochrome** molecules. Three ATP are made per NADH, and two ATP are made per $FADH_2$. As electrons move from one complex to another, H⁺ are pumped into the intermembrane space, creating an H⁺ gradient. Oxygen (we breathe in) is the final acceptor of electrons from the electron transport chain. Oxygen is reduced to water.
- **Chemiosmosis:** As H⁺ flow down the H⁺ gradient from the intermembrane space through the ATP synthase complex, to the matrix, the enzyme **ATP synthase** synthesizes ATP from ADP + Ⓟ.

7.6   The ATP payoff can be calculated

- Altogether there is a net gain of two ATP in the cytoplasm from glycolysis, two ATP are made directly by the Krebs cycle, and 32–34 ATP are made by the electron transport chain and chemiosmosis. Therefore, the total ATP count for complete glucose breakdown is 36 or 38 ATP.

Fermentation

7.7   When oxygen is in short supply, the cell switches to fermentation

- **Fermentation** is glycolysis followed by the reduction of pyruvate by NADH either to lactate or to alcohol and CO_2, depending on the organism. Fermentation results in only two ATP molecules, but provides ATP when oxygen is not available.

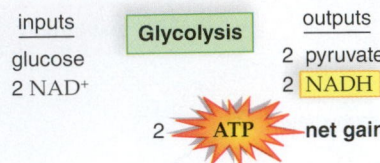

inputs   **Fermentation**   outputs

glucose   2 lactate or

                    2 alcohol and 2 CO_2

2   **ATP**   net gain

- Alcohol or lactate buildup is toxic to cells. In humans, lactate puts an individual into **oxygen debt** because oxygen is needed to completely metabolize it.

Intersections of Metabolic Pathways

7.8   Organic molecules can be broken down and synthesized as needed

- The **metabolic pool** concept recognizes that metabolism contains enzymatic pathways that can intersect. Degradative reactions are a part of **catabolism,** and synthetic reactions are a part of **anabolism.** Carbohydrates, fats, and proteins can all be catabolized to produce ATP molecules. Amino acids from protein first have to undergo **deamination,** the removal of the amino group. During anabolism, molecules from the breakdown pathways can be used to build molecules. Example: Acetyl groups from carbohydrate breakdown can be used to build fat.

TEST YOURSELF

Overview of Cellular Respiration

1. During cellular respiration, _____ is oxidized and _____ is reduced.
   - a. glucose, oxygen
   - b. glucose, water
   - c. oxygen, water
   - d. water, oxygen
   - e. oxygen, carbon dioxide
2. The products of cellular respiration are energy and
   - a. water.
   - b. oxygen.
   - c. water and carbon dioxide.
   - d. oxygen and carbon dioxide.
   - e. oxygen and water.
3. The correct order of the phases of cellular respiration is
   - a. Krebs cycle, glycolysis, prep reaction, ETC.
   - b. glycolysis, prep reaction, Krebs cycle, ETC.
   - c. prep reaction, Krebs cycle, glycolysis, ETC.
   - d. None of these are correct.
4. Relate the products and reactants of the overall equation for cellular respiration to the phases of cellular respiration (section 7.1).

Outside the Mitochondria: Glycolysis

5. During the energy-harvesting steps of glycolysis, which of the following chemicals are produced?
   - a. ATP and NADH
   - b. ADP and NADH
   - c. ATP and NAD
   - d. ADP and NAD
6. Which of the following is needed for glycolysis to occur?
   - a. pyruvate
   - b. glucose
   - c. NAD⁺
   - d. ATP
   - e. All of these are needed except a.
7. Which of these are carrying high-energy electrons?
   - a. NAD⁺ and NADH
   - b. FAD and $FADH_2$
   - c. water and carbon dioxide
   - d. NADH and $FADH_2$

Inside the Mitochondria

8. Acetyl CoA is the end product of
   a. glycolysis.
   b. the preparatory reaction.
   c. the Krebs cycle.
   d. the electron transport chain.

9. Which of these is not true of the prep reaction? The prep reaction
   a. begins with pyruvate and ends with acetyl CoA.
   b. produces more NADH than does the Krebs cycle.
   c. occurs in the mitochondria.
   d. occurs after glycolysis and before the Krebs cycle.

10. The final electron acceptor in the electron transport chain is
    a. NADH.   c. oxygen.
    b. $FADH_2$.   d. water.

11. The H^+ released by NADH and $FADH_2$
    a. passes down the ETC along with electrons.
    b. is pumped into the intermembrane space.
    c. passes through the ATP synthase complex.
    d. Both b and c are correct.

12. What is the name of the process that results in ATP production using the flow of hydrogen ions?
    a. substrate-level ATP synthesis
    b. fermentation
    c. reduction
    d. chemiosmosis

13. How many ATP molecules are produced when one NADH donates electrons to the ETC?
    a. 1   c. 36
    b. 3   d. 10

14. Label this diagram of a mitochondrion, and state a function for each portion indicated.

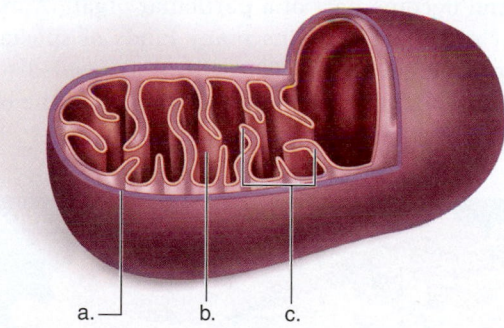

a. ___   b. ___   c. ___

15. **THINKING CONCEPTUALLY** If carbohydrate breakdown supplies the energy for ATP buildup, why does entropy increase as required by the second law of thermodynamics (see section 5.5)?

Fermentation

16. Fermentation occurs in the absence of
    a. CO_2.   c. oxygen.
    b. H_2O.   d. sodium.

17. Which are possible products of fermentation?
    a. lactate   c. CO_2
    b. alcohol   d. All of these are correct.

18. Which of the following is not true of fermentation? Fermentation
    a. has a net gain of only two ATP.
    b. occurs in the cytoplasm.
    c. donates electrons to the electron transport chain.
    d. begins with glucose.
    e. is carried on by yeast.

19. Fermentation does not yield as much ATP as cellular respiration does because fermentation
    a. generates mostly heat.
    b. makes use of only a small amount of the potential energy in glucose.
    c. creates by-products that require large amounts of ATP to break down.
    d. creates ATP molecules that leak into the cytoplasm and are broken down.

Intersections of Metabolic Pathways

20. Fatty acids are broken down to
    a. pyruvate molecules, which take electrons to the electron transport chain.
    b. acetyl groups, which enter the Krebs cycle.
    c. glycerol, which is found in fats.
    d. amino acids, which excrete ammonia.
    e. All of these are correct.

21. Deamination of an amino acid results in
    a. CO_2.   c. NH_2.
    b. H_2O.   d. O_2.

22. **THINKING CONCEPTUALLY** Why would you expect both photosynthesis and cellular respiration to use an electron transport chain to produce ATP? (See section 4.9.)

GET INVOLVED

1. You discover that mitochondria can still produce ATP when placed in certain solutions without their outer membranes. What is the pH of the solution—acidic or alkaline—and why?

2. You are working with acetyl CoA molecules that contain only radioactive carbon. They are incubated with all the components of the Krebs cycle long enough for one turn of the cycle. Examine Figure 7.4 and explain why the carbon dioxide given off is radioactive.

3. For an interactive exploration of the phases of cellular respiration, watch McGraw-Hill's 3D animation "Cellular Respiration."

MEDIA STUDY TOOLS

mhhe.com/maderconcepts3

Enhance your study of this chapter with interactive study tools, practice tests, and engaging animations. Also, ask your instructor about the resources available through ConnectPlus, which includes LearnSmart, a personalized adaptive learning program, and a media-rich eBook.

8

Cellular Reproduction

CHAPTER OUTLINE

LOOKING AT LIFE

APPLICATIONS

HOW LIFE CHANGES

HOW BIOLOGY IMPACTS OUR LIVES

HOW BIOLOGY IMPACTS OUR LIVES

BEFORE YOU BEGIN

Take a few minutes to recall

The structure of DNA and the nucleus (sections 3.9 and 4.4)
The role of microtubules in cells (section 4.10)
The outer cell wall of plant cells (section 5.4)

Cancer Is a Genetic Disorder

We often think of diseases in terms of organs, and therefore it is customary to refer to colon cancer, or lung cancer, or pancreatic cancer. But actually cancer is a cellular disease. Cancer is present when abnormal cells have formed a tumor. Exceptions are cancers of the blood, in which abnormal cells are coursing through the bloodstream. The cells of a tumor share a common ancestor—the first cell to become cancerous.

Uncontrolled growth leading to a tumor is characteristic of multicellular organisms, not unicellular ones. The very mechanism that allows our bodies to grow and repair tissues is the one that turns on us and allows cancer to begin. Cancer is uncontrolled cell division.

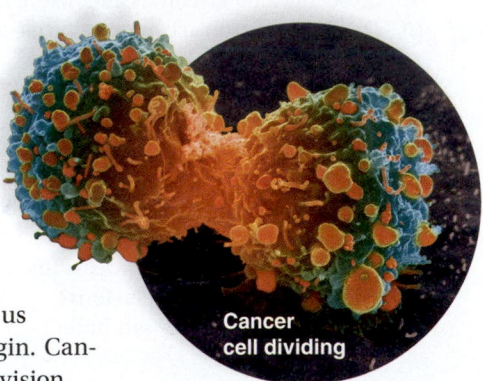
Cancer cell dividing

Usually, cell division is confined to just certain cells of the body, called adult stem cells. For example, skin can replenish itself because stem cells below the surface have the ability to divide. In embryos all cells can divide. How else could a newborn arise from a single fertilized egg? But something happens as development progresses: The cells undergo specialization and become part of a particular organ. A mature multicellular organism contains many kinds of specialized cells in many different organs. Normally, these cells listen to their neighbors and participate in the operation of the organ.

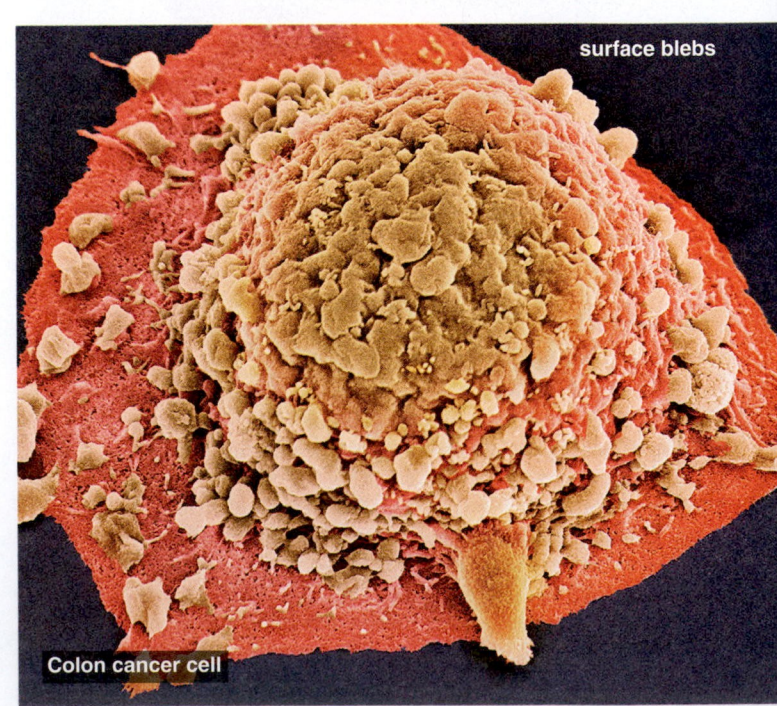
surface blebs

Colon cancer cell

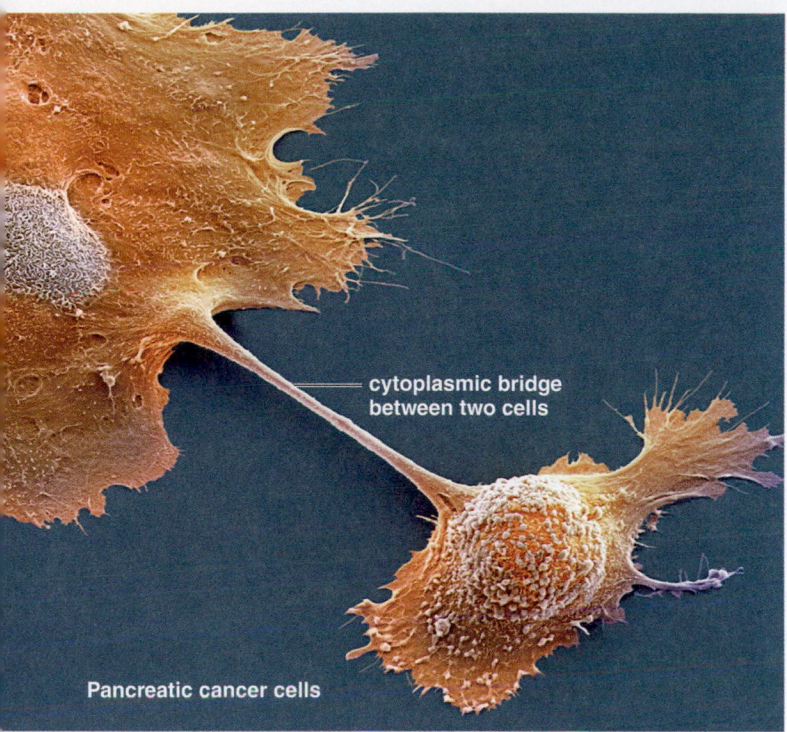

cytoplasmic bridge
between two cells

Pancreatic cancer cells

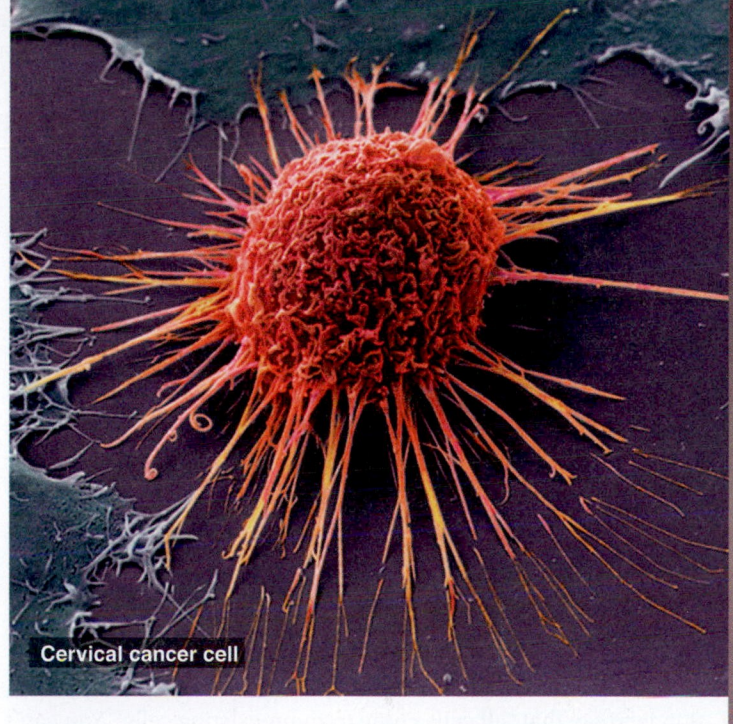

Cervical cancer cell

But when a cell becomes cancerous, it loses its specialization and becomes youthful again—it starts to divide and divide, until a tumor exists. The tumor interferes with the operation of the organ.

Each cell in a multicellular organism has a copy of the genetic instructions the organism received from its parents. During cell division, the instructions were passed to the millions and millions of cells making up the body. Some of these genes call a halt to cell division, a necessary step if cells are to mature and become specialized. Without control of cell division, a multicellular organism would be a bunch of embryonic cells with no particular purpose. When cell division genes mutate, uncontrolled division so characteristic of cancer becomes possible. Therefore, cancer is a genetic disorder. Research tells us that cancer-causing mutations may be induced, for example, by chemicals or radiation that damage DNA; viruses that carry mutated genes into cells; or random errors that occur during DNA synthesis. A series of mutations is required before cells begin to grow abnormally and eventually become a tumor. In these cells, genetic alteration is obvious: Some chromosomes are present in three or four copies, rather than the usual two, and other chromosomes have been rearranged in various ways.

In this chapter, we learn that a cell readies itself for division and then divides during a series of stages called the cell cycle because these stages can recur over and over. Cellular reproduction is usually tightly controlled so that the cell divides in an orderly manner. If control mechanisms fail cancer develops. How people can protect themselves from the development of cancer is also a part of this chapter.

tumor

Lung cancer

300×

The Cell Cycle

The cell cycle occurs in complex organisms as they grow and repair their tissues. The process requires a number of stages during which the cell gets ready to divide by doubling its DNA and organelles and then it undergoes division.

8.1 Cellular reproduction depends on the cell cycle

LEARNING OUTCOMES

When you complete this section, you should be able to

1. Recognize the importance of cellular reproduction and the cell cycle.
2. List and describe the stages of the cell cycle.

We humans, like other multicellular organisms, begin life as a single cell called a zygote. In 9 short months, however, we become trillions of cells because cellular reproduction has occurred over and over again (Fig. 8.1A*a*). Even after we are born, cellular reproduction doesn't stop—it continues as we grow (Fig. 8.1A*b*), and when we are adults, it replaces worn-out or damaged tissues (Fig. 8.1A*c*). Right now, your body is producing thousands of new red blood cells, skin cells, and cells that line your respiratory and digestive tracts. If you suffer a cut, cellular reproduction helps repair the injury.

One way to emphasize the importance of cellular reproduction is to say that "all cells come from preexisting cells." You can't have a new cell without a preexisting cell, and you can't have a new organism without a preexisting cell. Cellular reproduction is necessary for the production of both new cells and new organisms.

Cellular reproduction involves the cell cycle (Fig. 8.1B). Cell division is only a small part of the **cell cycle,** an orderly set of stages that takes place between the time a eukaryotic cell divides and the time the resulting daughter cells also divide. The amount of time it takes for a cell to complete the cycle varies widely, but adult mammalian cells can usually finish the cell cycle in about 24 hours.

Interphase For most of the cell cycle, the cell is in **interphase,** defined as the period of time between cell divisions. During interphase, a cell is performing its normal work of communicating with other cells, secreting substances, and carrying out cellular respiration. In addition, the cell may be preparing for cell division. Interphase has three stages: G_1, S, and G_2. Years ago, when cell biologists picked the terms G_1, S, and G_2, they said that G stood for "gap," but this now seems inappropriate because the cell is actually busy during both G phases. So today, it is better to think of G as standing for "growth." Protein synthesis is very much a part of these growth stages.

During the **G_1 stage,** the cell first recovers from the previous division. Then it may make a commitment to divide again. If not, a cell can enter **G_0,** which is a substage of the G_1 stage. During G_0, a cell continues to perform normal everyday processes, but no preparations are being made for cell division. However, cells can exit the G_0 stage upon receiving proper signals from other cells and other parts of the body. Some types of cells, such as nerve and muscle cells, are more apt to be in G_0 than other types of cells.

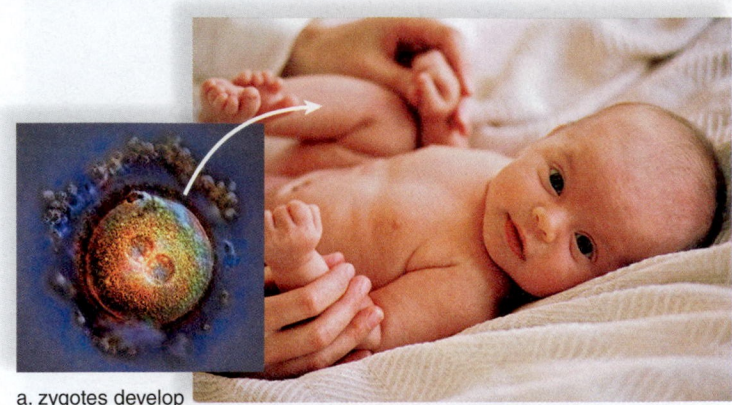
a. zygotes develop

b. children grow

c. tissues repair

FIGURE 8.1A Cellular reproduction occurs when **(a)** embryos develop, **(b)** organisms grow larger, and **(c)** injuries heal.

Embryonic cells and also adult stem cells do not enter G_0. **Adult stem cells** are relatively nonspecialized cells whose job is to divide and produce cells that will become mature. In red bone marrow, the division of stem cells gives rise to all the types of blood cells in the body.

As soon as a cell makes a commitment to divide, during G_1 it increases in size, doubles its organelles (such as mitochondria and ribosomes), and accumulates molecules that will be used for DNA synthesis.

The **S stage** of interphase follows the G_1 stage. The "S" stands for synthesis, and this is the stage during which DNA replicates—makes a copy of itself. When a eukaryotic cell is not undergoing division, the DNA (and associated proteins) within the nucleus is a tangled mass of thin threads called chromatin. A human cell contains an astonishing 2 m of chromatin, and moving this large quantity of DNA during cell division would be no easy task. However, just before cell division, DNA and associated proteins are packaged into a set of **chromosomes**. Once the chromosomes are visible, it is possible to photograph and count them. Each species has a characteristic chromosome number; for instance, human cells contain 46 chromosomes, corn has 20 chromosomes, and a goldfish has 94.

The **G_2 stage** follows the S stage. It extends from the completion of DNA replication to the onset of cell division. Organelle replication continues during this stage, and the cell synthesizes proteins that will assist cell division. For example, it makes the proteins that form microtubules. Microtubules are used during the mitotic stage to form a spindle apparatus that helps nuclear division occur.

M Stage A nuclear division, called **mitosis,** occurs during the **M stage.** Mitosis maintains the chromosome number and each daughter nucleus has the same number of chromosomes as the parent cell had. This is possible because DNA replication was organized and each chromosome prior to mitosis has two identical halves called **sister chromatids;** in other words, it is a duplicated chromosome. Each sister chromatid contains an identical double DNA helix. The two chromatids are held together at a constricted area called a **centromere** and during mitosis the centromeres split and a daughter nucleus receives a copy of every chromosome the parent cell had.

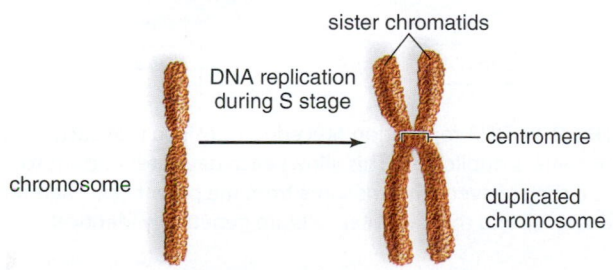

Following mitosis the parent cell and daughter cells are genetically identical. **Cytokinesis,** or division of the cytoplasm, which starts even before mitosis is finished, is also part of the M stage of the cell cycle (Fig. 8.1C).

Animation
Overview of Cell Division

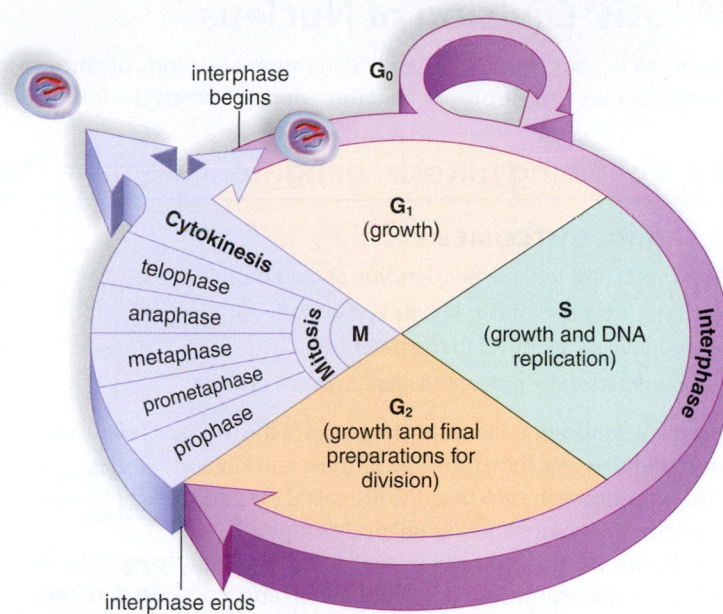

FIGURE 8.1B During the cell cycle a cell gets ready to divide (interphase) and then it divides (mitosis and cytokinesis). A cell can break out of the cell cycle and become specialized or it can enter a resting stage known as G_0.

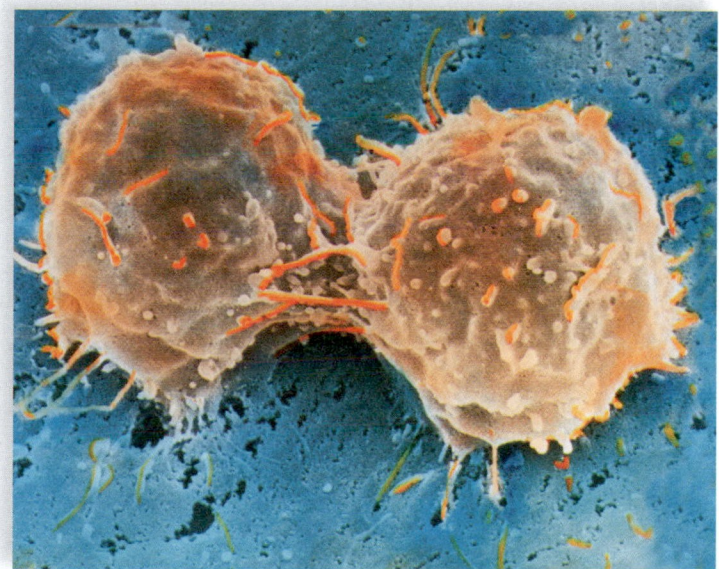

FIGURE 8.1C Cytokinesis is a noticeable part of the cell cycle because division of the cytoplasm occurs. These are animal cells, which divide by a process called furrowing.

▶ **8.1 CHECK YOUR PROGRESS**
1. Explain why cellular reproduction and the cell cycle are essential to humans and all organisms.
2. Identify the three stages of the cell cycle that occur as a cell gets ready to divide and the two events that occur as they do divide.

Mitosis: Division of Nucleus

Every cell in your body has the same number and kinds of chromosomes. The cell cycle ensures that this will happen because the chromosomes first duplicate, and then during mitosis the identical parts separate and go into daughter nuclei.

8.2 Following mitosis, daughter cells have the same chromosome count as the parent cell

LEARNING OUTCOMES

When you complete this section, you should be able to

1. Describe how mitosis results in daughter cells that are genetically identical to the parent cell.
2. Summarize the role of the spindle in cell division.

Mitosis is duplication division. The nuclei of the two new cells, called the **daughter cells,** have the same number and kinds of chromosomes as the cell that divides, called the **parent cell.** How this comes about is relatively simple.

MP3
Mitosis

At the start of mitosis, as you know from studying the cell cycle, each chromosome is duplicated and composed of two identical parts, called sister chromatids. This diagram shows you how one duplicated chromosome can give rise to two identical chromosomes:

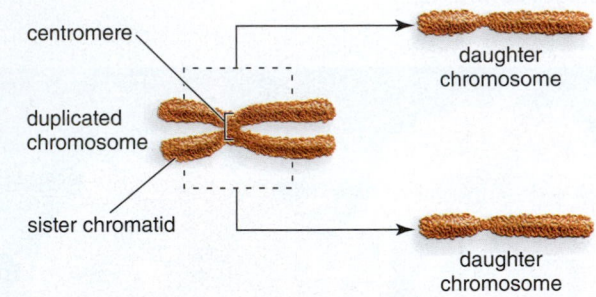

It is important when studying cell division to determine the number of chromosomes by counting the number of independent centromeres. For example, you can tell that all the cells in Figure 8.2 have the same number of chromosomes because they have the same number of centromeres. Since the chromosomes occur in pairs, the cells are **diploid,** symbolized as **2n.** Pairs are determined by shape and size; therefore the cells in Figure 8.2 have two pairs—one pair is long and the other is short. The difference in color signifies that one of each pair (e.g., blue) was inherited from the father and the other (e.g., red) was inherited from the mother. If the chromosomes are not in pairs, the cell is **haploid,** symbolized as **n.** In animal cells, the cell is diploid, but in many protists and fungi, the cell is haploid before and after mitosis.

The Spindle Apparatus A **centrosome** is the microtubule organizing center of the cell, which divides at the start of a nuclear division. The daughter centrosomes produce the spindle fibers of a **spindle apparatus,** which assists the separation of the chromatids as they move toward the opposite poles of the spindle. As soon as the chromatids separate, they are called daughter chromosomes.

In animal cells, the centrioles are short cylinders of microtubules located in centrosomes. It seems doubtful that centrioles help form the spindle, because plant cells don't have centrioles but

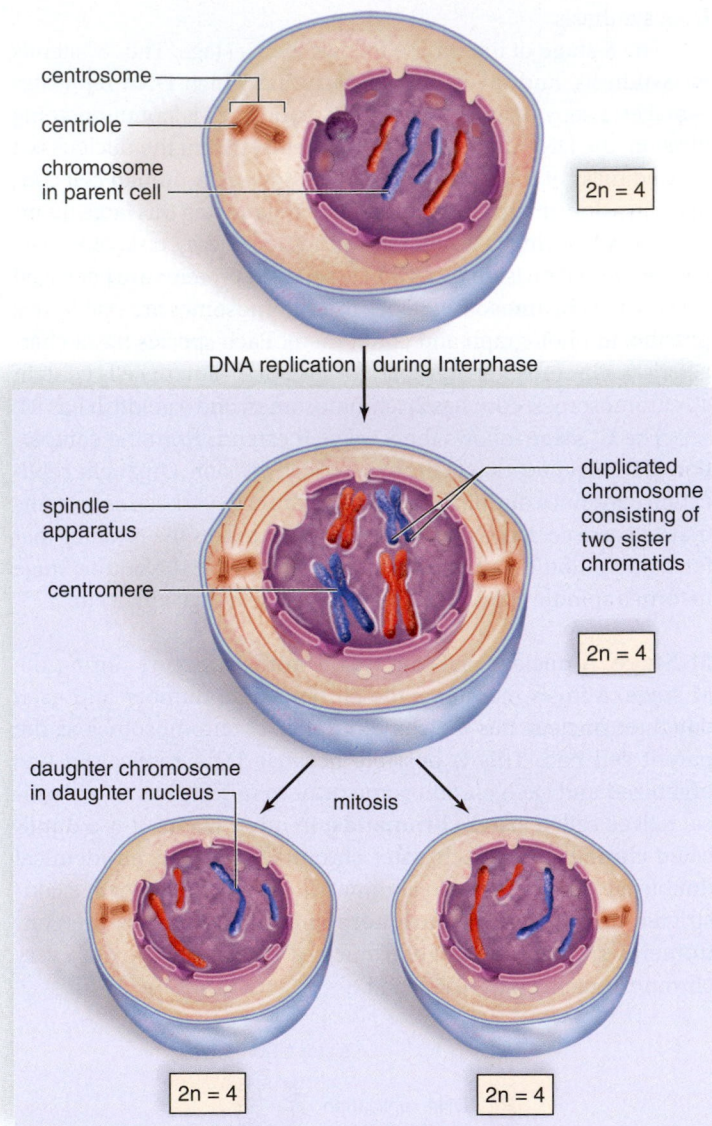

FIGURE 8.2 DNA replication precedes mitosis so that each chromosome is duplicated. This allows each daughter nucleus to receive a copy of every chromosome from the parent cell. Therefore, the parent cell and the daughter cells are genetically identical.

they do have a spindle apparatus. "Evolution of the Spindle Apparatus" on the next page discusses how the spindle fibers function and how the spindle may have evolved.

▶ **8.2 CHECK YOUR PROGRESS**
1. Explain why DNA replication must precede mitosis.
2. Contrast the chromosomal makeup of a parent cell that has one long red and one long blue chromosome, with the chromosomes in the two daughter nuclei, following mitosis.

HOW LIFE CHANGES *Application*

8A Evolution of the Spindle Apparatus

In plant and animal cells and most fungi, mitosis involves a spindle apparatus composed of microtubules organized into spindle fibers of two types: kinetochore fibers and polar fibers. The nuclear envelope has fragmented and the spindle takes up the entire cell (Fig. 8A.1, *center*).

During mitosis, each chromatid has a kinetochore (a special assembly of proteins) in the region of the centromere. You can think of a kinetochore as an engine that runs along a train track, where the train track is a kinetochore spindle fiber. As the kinetochore moves up the fiber, the fiber disassembles. In this way, the daughter chromosomes are seemingly pulled apart (Fig. 8A.1, *left*). The spindle apparatus also contains polar fibers that overlap at the equator of the spindle. When the polar spindle fibers lengthen, the poles are pushed apart (Fig. 8A.1, *right*). It's said that the daughter chromosomes separate by a push-pull system.

The origin of the spindle apparatus is an evolutionary puzzle that researchers have been investigating for quite some time. When bacteria, which lacks a nucleus, reproduce, the daughter chromosomes are attached to a plasma membrane site, and they separate as the cell elongates. No microtubules are involved in this process, which is called binary fission (Fig. 8A.2, *left*). In some way, mitosis must have evolved from binary fission, but it is unlikely that mitosis developed in a straight-line manner. Instead, the evolution of the spindle apparatus must have involved numerous dead ends and variations that finally resulted in the spindle used by animals, plants, and most fungi. The fossil record is unlikely

to help discover the evolutionary pathway because soft, pliable cells don't make good fossils. Therefore, researchers studying the origin of the spindle apparatus have turned to living unicellular protists to see how the spindle may have evolved. Two significant groups of unicellular organisms offer clues:

1. In dinoflagellates (Fig. 8A.2, *center*), the nuclear envelope does not fragment, and microtubules merely stabilize the nuclear envelope when mitosis occurs. The daughter chromosomes are attached to the nuclear envelope as the nucleus elongates and divides.
2. In diatoms and yeast cells (Fig. 8A.2, *right*), the nuclear envelope does not fragment; the spindle forms inside the nucleus and functions as in plant and animal cells.

These studies result in a hypothesis that once the eukaryotic cell arose, spindle fibers became more and more involved in the process of chromosome separation so that today the nuclear envelope fragments and the spindle apparatus fills the cell during mitosis (Fig. 8A.1, *center*).

CONSIDER THESE QUESTIONS

1. Bacteria make a protein related to tubulin which is found in eukaryotic microtubules. In what way does this help trace the evolution of the spindle?
2. An opportunistic person makes use of any benefits that come their way. Why could it be said that evolution is opportunistic?

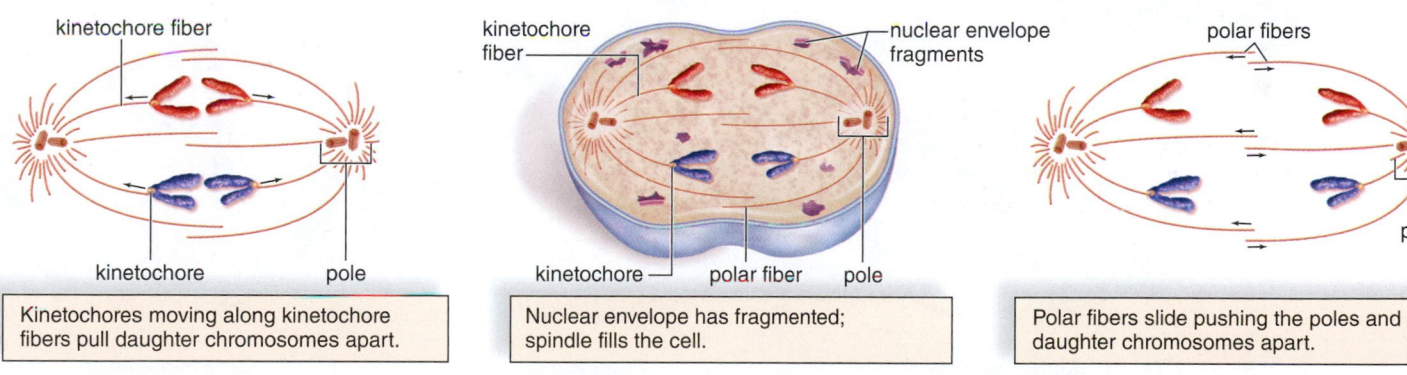

Kinetochores moving along kinetochore fibers pull daughter chromosomes apart.

Nuclear envelope has fragmented; spindle fills the cell.

Polar fibers slide pushing the poles and daughter chromosomes apart.

FIGURE 8A.1 Structure and function of spindle in animal cells.

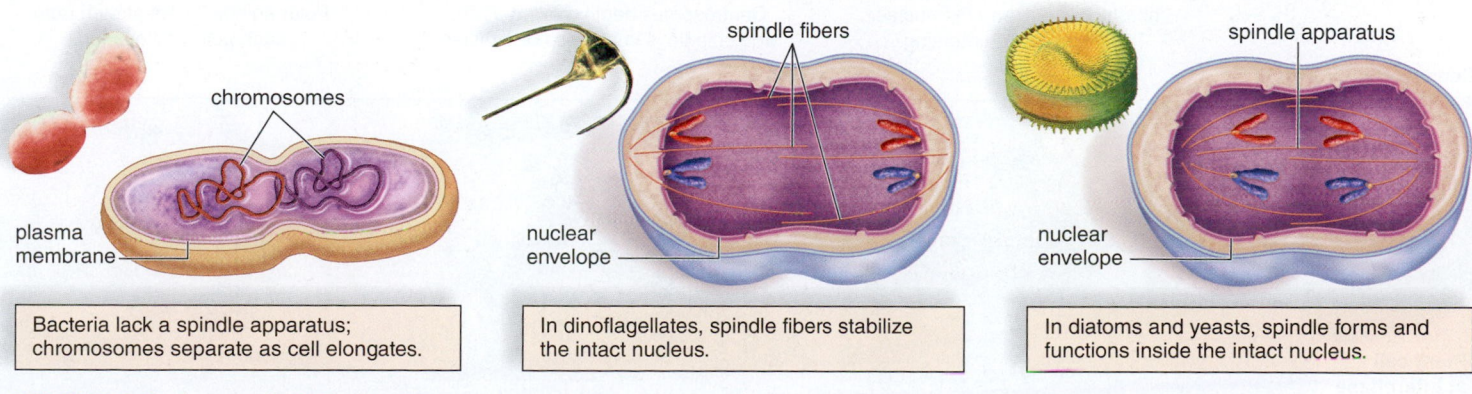

Bacteria lack a spindle apparatus; chromosomes separate as cell elongates.

In dinoflagellates, spindle fibers stabilize the intact nucleus.

In diatoms and yeasts, spindle forms and functions inside the intact nucleus.

FIGURE 8A.2 Survey of other cells with regard to spindle.

LEARNING OUTCOME

When you complete this section, you should be able to

1. Describe and contrast the events in each phase during mitosis.

Figure 8.3A describes mitosis in animal cells (*top*) and in plant cells (*bottom*). The poles of animal cells are prominent because short spindle fibers, called an **aster**, radiate from the pair of centrioles located within each centrosome. Plant cells don't have centrioles or an aster as noted in Table 8.3.

Mitosis is one stage of the cell cycle; however, it is composed of a number of phases simplified as **prophase**, **metaphase**, **anaphase**, and **telophase**. Each phase can be further divided into "early" and/or "late." Because you count the number of centromeres to determine the number of chromosomes, all the cells in Figure 8.3A have four chromosomes. During mitosis, kinetochores develop in the region of centromeres and these facilitate attachment of chromosomes to the spindle apparatus. Keep in mind that, although mitosis is divided into phases, it is a continuous process.

The process of mitosis ensures that all the cells of an individual have the same chromosomes and the same genes. All living cells require a copy of the same genes that allow them to carry out their normal activities, such as protein synthesis, cellular respiration, and yes, to divide.

3D Animation
Cell Cycle and Mitosis: Mitosis

Video
Mitosis

FIGURE 8.3A Phases of mitosis in animal cells and plant cells.

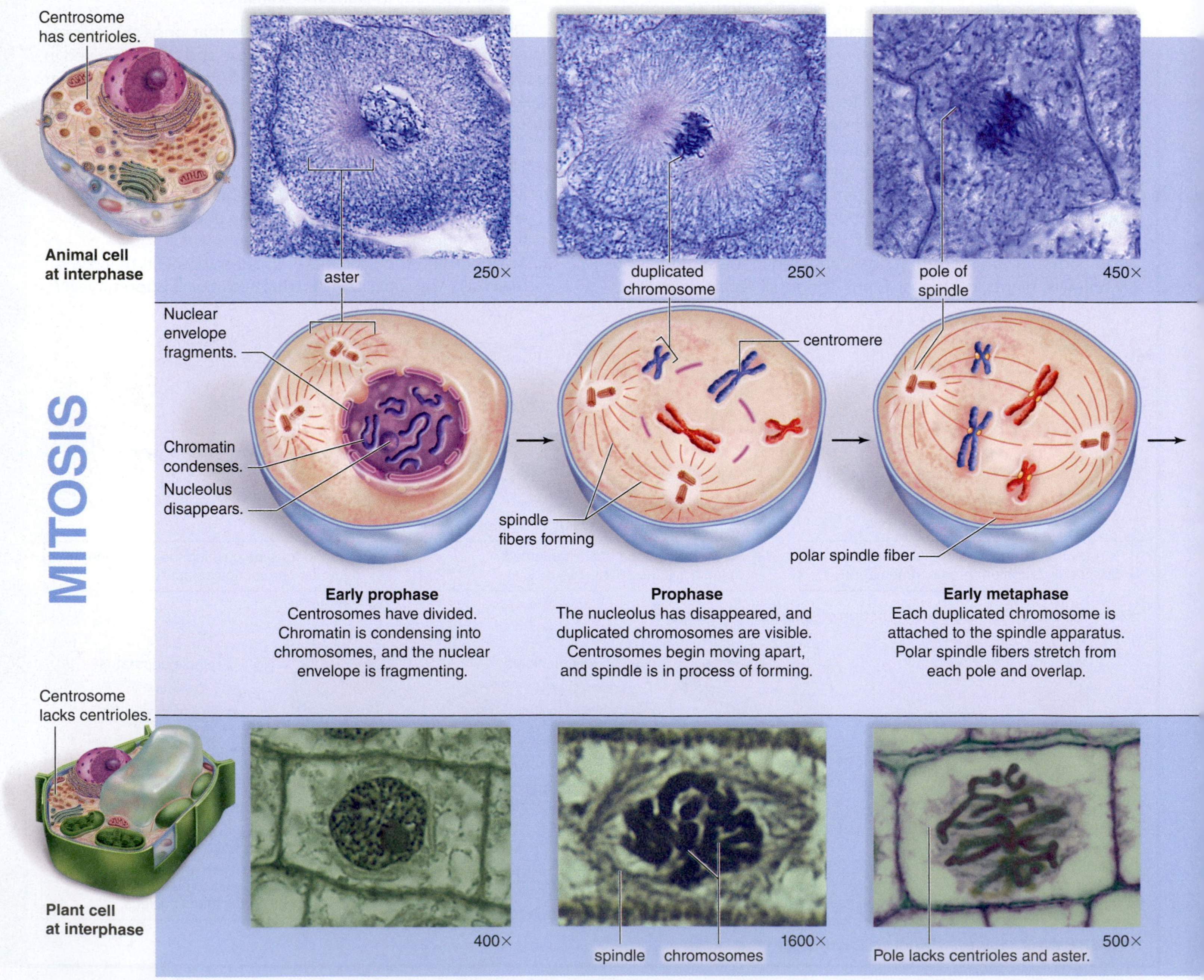

MITOSIS

Centrosome has centrioles.

Animal cell at interphase

aster   250×

duplicated chromosome   250×

pole of spindle   450×

Nuclear envelope fragments.

Chromatin condenses.

Nucleolus disappears.

centromere

spindle fibers forming

polar spindle fiber

Early prophase
Centrosomes have divided. Chromatin is condensing into chromosomes, and the nuclear envelope is fragmenting.

Prophase
The nucleolus has disappeared, and duplicated chromosomes are visible. Centrosomes begin moving apart, and spindle is in process of forming.

Early metaphase
Each duplicated chromosome is attached to the spindle apparatus. Polar spindle fibers stretch from each pole and overlap.

Centrosome lacks centrioles.

Plant cell at interphase

400×

spindle   chromosomes   1600×

Pole lacks centrioles and aster.   500×

TABLE 8.3 Comparison of Mitosis in Animal and Plant Cells

Structure/Function	Animal Cells	Plant Cells
Nuclear division	Yes	Yes
Centrosome present	Yes	Yes
Centrioles in centrosome	Yes	No
Spindle pole has aster	Yes	No
Spindle apparatus separates sister chromatids	Yes	Yes
Events during mitotic phases	Same	Same
Maintains chromosome number	Yes	Yes
Occurrence	Embryonic and adult stem cells	Root tip and shoot tip cells

▶ 8.3 CHECK YOUR PROGRESS

1. Describe metaphase and cite a benefit to the position of the chromosomes during this phase.

2. Explain how it's possible for each of your body cells to have the same number of chromosomes.

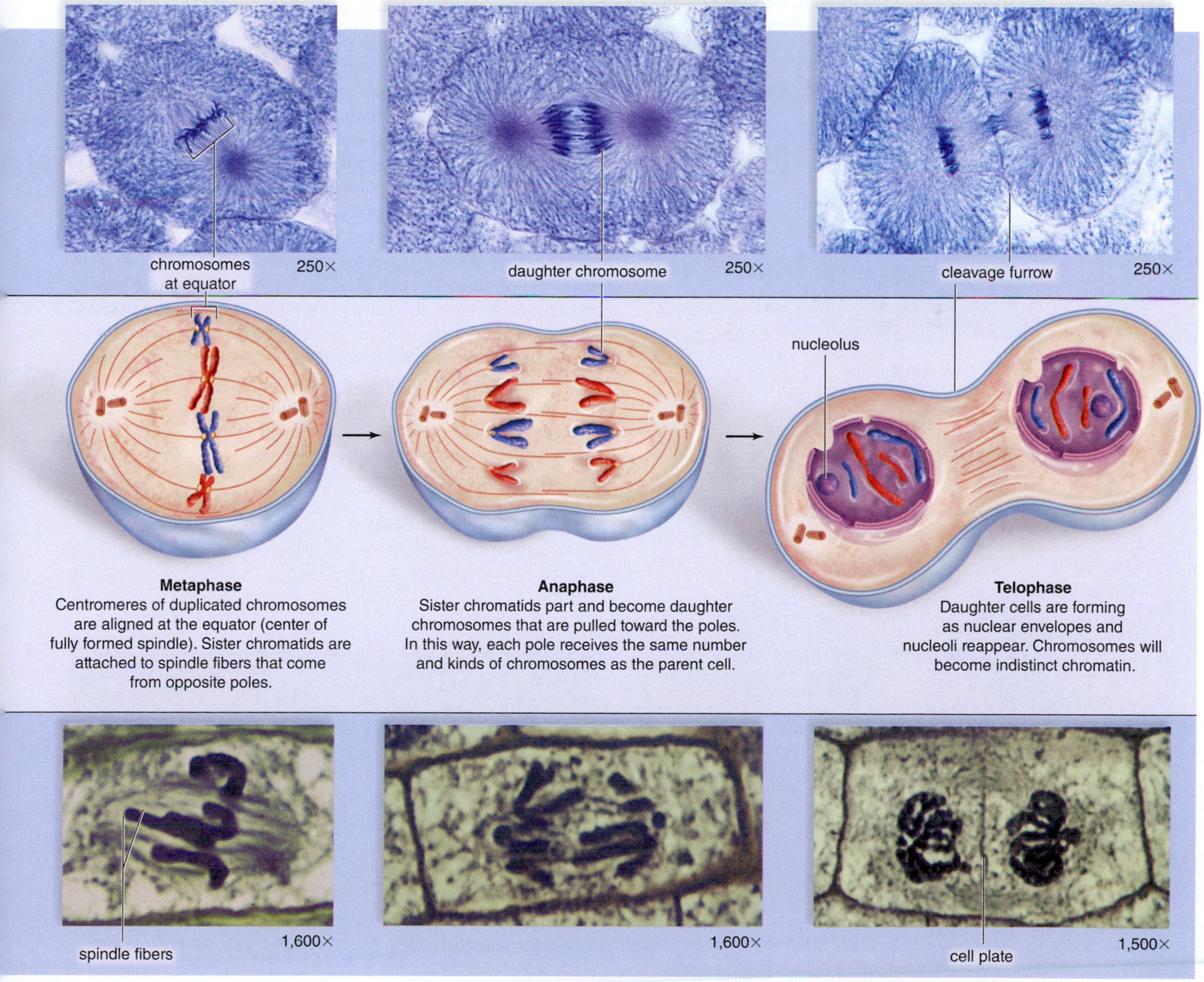

chromosomes at equator   250×

daughter chromosome   250×

cleavage furrow   250×

nucleolus

Metaphase
Centromeres of duplicated chromosomes are aligned at the equator (center of fully formed spindle). Sister chromatids are attached to spindle fibers that come from opposite poles.

Anaphase
Sister chromatids part and become daughter chromosomes that are pulled toward the poles. In this way, each pole receives the same number and kinds of chromosomes as the parent cell.

Telophase
Daughter cells are forming as nuclear envelopes and nucleoli reappear. Chromosomes will become indistinct chromatin.

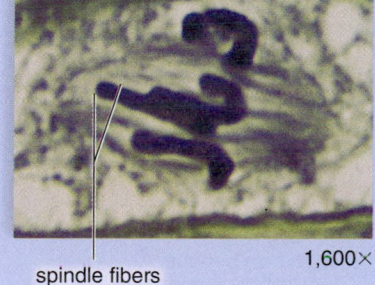

spindle fibers   1,600×

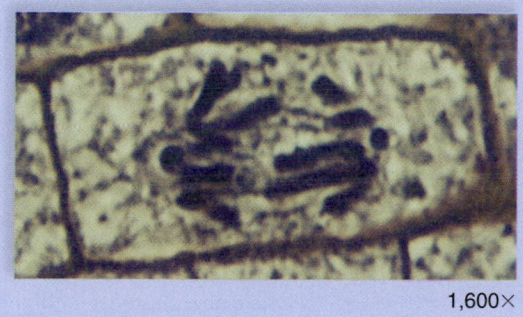

1,600×

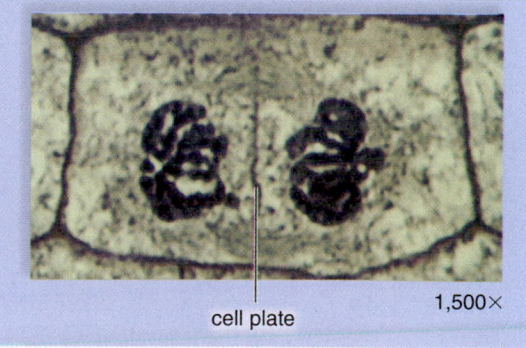

cell plate   1,500×

Cytokinesis: Division of Cytoplasm

Mitosis is division of the nucleus and cytokinesis is division of the cytoplasm. Without cytokinesis, cells would have multiple nuclei and become quite large.

8.4 Cytokinesis follows mitosis

LEARNING OUTCOME

When you complete this section, you should be able to

1. Contrast cytokinesis in an animal cell and a plant cell.

We should always remember that mitosis means nuclear division and that cell division involves not only division of the nucleus but also division of the cytoplasm. Cytokinesis, meaning division of the cytoplasm, follows mitosis in most cells, but not all of them. When mitosis occurs but cytokinesis does not occur, the result is a multinucleated cell. For example, you will see later in this book that skeletal muscle cells in vertebrate animals and, at one point, the embryo sac in a flowering plant are multinucleated.

Ordinarily, cytokinesis begins during telophase and continues after the nuclei have formed until there are two daughter cells.

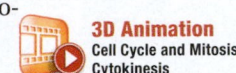 **3D Animation**
Cell Cycle and Mitosis: Cytokinesis

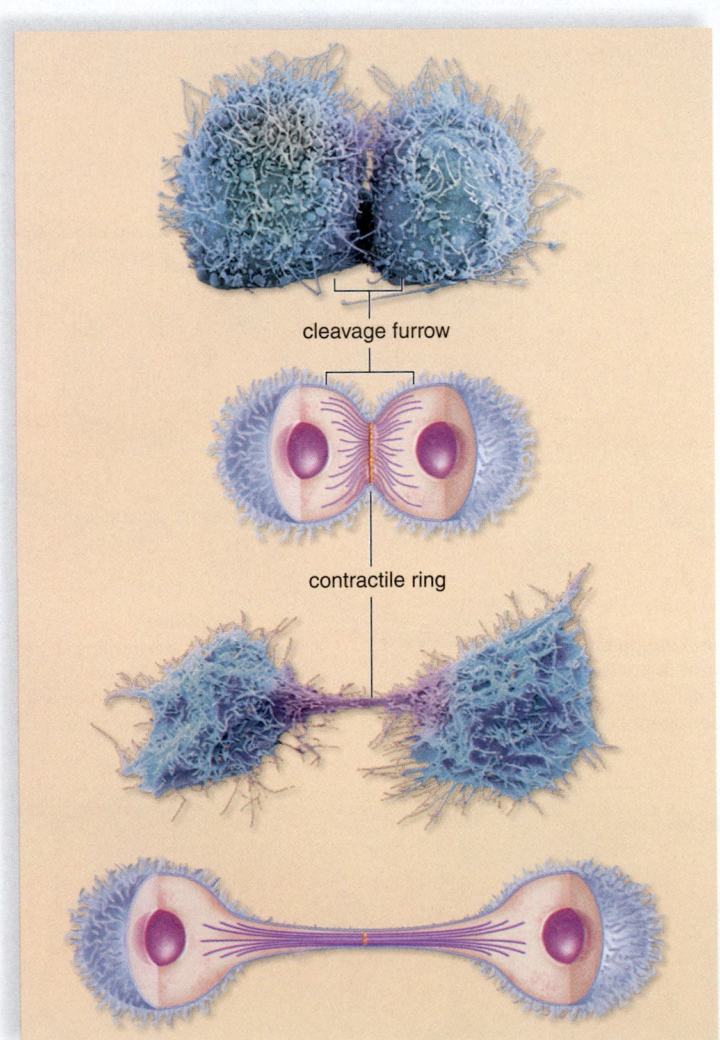

cleavage furrow

contractile ring

4,000×

FIGURE 8.4A Cytokinesis in an animal cell involves a cleavage furrow and a contractile ring that pinches off the two cells.

Animal Cell Cytokinesis In animal cells, a **cleavage furrow,** which is an indentation of the membrane between the two daughter nuclei, begins at the start of telophase. The cleavage furrow deepens when a band of actin filaments, called the contractile ring, slowly forms a circular constriction between the two daughter cells. The action of the contractile ring can be likened to pulling a drawstring ever tighter about the middle of a balloon. A narrow bridge between the two cells is visible during telophase, and then the contractile ring continues to separate the cytoplasm until there are two independent daughter cells. First, the cleavage furrow appears, and then a contractile ring tightens the constriction (Fig. 8.4A).

Plant Cell Cytokinesis In plant cells, cytokinesis occurs by a process different from that seen in animal cells. The rigid cell wall that surrounds plant cells does not permit cytokinesis by furrowing. Instead, cytokinesis in plant cells involves the building of new plasma membranes and cell walls between the daughter cells.

Cytokinesis is apparent when a small, flattened disk appears between the two daughter plant cells. Electron micrographs reveal that the disk is composed of vesicles (Fig. 8.4B). The Golgi apparatus produces these vesicles, which move along microtubules to the region of the disk. As more vesicles arrive and fuse, a cell plate can be seen. The **cell plate** is simply newly formed plasma membrane that expands outward until it reaches the old plasma membrane and fuses with it. The new membrane releases molecules that form the new plant cell walls. These cell walls are later strengthened by the addition of cellulose fibrils.

We have completed our study of mitosis and cytokinesis. In the next part of the chapter, we study cell cycle control, because when it falters, cancer develops.

 **Animation**
Cytokinesis

▶ **8.4 CHECK YOUR PROGRESS**
1. Explain why plant cells and not animal cells have a cell plate during cytokinesis.

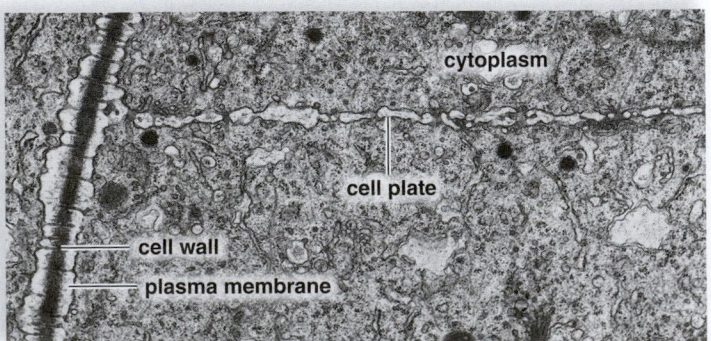

cytoplasm

cell plate

cell wall

plasma membrane

24,000×

FIGURE 8.4B Cytokinesis in plant cells involves a cell plate where new plasma membrane and cell wall form between the two daughter cells.

8B Tissues Can Be Grown in the Lab

Most people are now aware that stem cells can undergo the cell cycle and generate tissues for the cure of devastating human diseases, such as diabetes, cancer, brain disorders, and heart ailments (Fig. 8B). For many years, scientists have known about two types of stem cells: embryonic stem (ES) cells and adult stem cells.

Embryonic stem cells are simply the cells of an early embryonic stage. These cells can stay alive longer and are better at producing different tissues than adult stem cells, but to acquire them a human embryo must be destroyed. Embryos are sometimes "left over" at fertility clinics, but even so many people reject the use of ES cells because it means the destruction of a potential human life. Adult stem cells are difficult to glean from the human body, and they do not multiply readily in the laboratory. Also, their

potential to become all different types of tissues is not as great as that of ES cells. One drawback to both ES cells and adult stem cells is the danger of rejection by the recipient. Remember the many different types of proteins that occur in the plasma membrane? Some of those mark the cell as belonging to us, and if a transplanted tissue or organ carries different markers, our body works against them until they die. This is called rejection of the transplant.

Breakthrough

By now, scientists are experienced at coaxing stem cells to become specialized cells, but research would really benefit from an unlimited source of stem cells in order to achieve the goal of replacing diseased or damaged tissues in the human body. The scientific community is now hopeful that such a source has been found, thanks to a little-known Japanese scientist who worked alone for 10 years in a tiny laboratory. Through patient research, Shinya Yamanaka was able to discover why ES cells are *pluripotent*—able to become any type of tissue in the body. He hypothesized that pluripotent cells produce certain proteins that specialized cells do not produce. Yamanaka worked with mouse skin cells until he knew that only four particular genes do the trick of making cells pluripotent. In 2006 he published his results in the journal *Cell*. Just 5 months later, United States scientists induced human skin cells to become pluripotent by supplying them with active forms of the four genes. These skin cells are termed iPS (induced pluripotent stem) cells. For every cell that became pluripotent, thousands of skin cells are treated. But the inefficiency doesn't matter because scientists have access to millions of skin cells. Such cells can even be obtained by simply swabbing the inside of a person's mouth! Researchers are still improving their technique and resolving various safety issues, but they feel confident they will be able to make tissues for human transplant. If replacement tissues are produced using the patient's own skin cells, rejection should not be a problem. However, scientists hope that eventually labs can stockpile so many different types of tissues, a good match will be available for most every person. Because spinal cord injuries should be treated within a few hours, there isn't time to use the patient's own skin cells to produce replacement nerve cells.

CONSIDER THESE QUESTIONS

1. Currently, the main safety issue with iPS cells is that they might cause cancer. If you were 75 and had Alzheimer disease, would you be willing to take the chance of cancer in order to correct this condition?
2. Imagine that you are a scientist who worked all alone for 10 years to reach a breakthrough. Should you be allowed to patent your "invention," or should it be available to everyone?

 Explore the concepts through a variety of multimedia assets and question types.

www.mcgrawhillconnect.com

human embryo   human skin

Stem cells

embryonic
stem cells (ES)   induced pluripotent
stem cells (iPS)

**Different
treatments**

**Types of
tissues**

pancreatic tissue   nervous tissue   cardiac tissue

FIGURE 8B ES (embryonic stem) cells and iPS (induced pleuripotent stem) cells both produce many different types of specialized cells and tissues in the lab. Safety issues need to be resolved, but eventually scientists believe that iPS tissues will be available to cure human ills.

Cell Cycle Control and Cancer

The cell cycle's control system ensures that the cell cycle occurs in an orderly manner. Cancer develops when the cell cycle control system is not functioning as it should. A cell cycle out of control accounts for the abnormal characteristics of cancer cells.

8.5 Cell cycle control depends on checkpoints

LEARNING OUTCOMES

When you complete this section, you should be able to

1. Explain how various checkpoints control the cell cycle.
2. Describe the role of apoptosis in cells.

In order for the body to remain healthy, the cell cycle must be controlled. The method of cell cycle control can be understood by comparing it to the events that occur in an automatic washing machine. The washer's control system starts to wash only when the tub is full of water, does not spin until the water has been emptied, delays the most vigorous spin until rinsing has occurred, and so forth. Similarly, the cell cycle's control system ensures that the G_1, S, G_2, and M stages occur in order and only when the previous stage has been successfully completed. The cell cycle has checkpoints that can delay the cycle until all is well.

The cell cycle has many checkpoints, but we will consider only three: G_1, G_2, and M (mitotic) (Fig. 8.5).

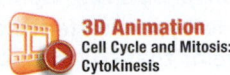 **3D Animation**
Cell Cycle and Mitosis: Cytokinesis

① The G_1 checkpoint is especially significant, because if the cell cycle passes this checkpoint, the cell is committed to divide. If the cell does not pass this checkpoint, it can enter G_0, during which it performs specialized functions but does not divide. If the DNA is damaged beyond repair, the internal signaling protein **p53** can stop the cycle at this checkpoint. First, p53 attempts to initiate DNA repair, but if that is not possible, it brings about the death of the cell by **apoptosis**, defined as programmed cell death.

② The cell cycle hesitates at the G_2 checkpoint, ensuring that DNA has replicated. This prevents the initiation of the M stage unless the chromosomes are duplicated. Also, if DNA is damaged, as from exposure to solar radiation or X-rays, arresting the cell cycle at this checkpoint allows time for the damage to be repaired, so that it is not passed on to daughter cells. If repair is not possible, apoptosis occurs.

③ The M checkpoint occurs during the mitotic stage. The cycle hesitates at the M checkpoint to make sure the chromosomes are going to be distributed accurately to the daughter cells. The cell cycle does not continue until every duplicated chromosome is ready for the chromatids to separate.

Apoptosis During apoptosis, the cell progresses through a typical series of events that bring about its destruction. The cell rounds up and loses contact with its neighbors. The nucleus fragments, and the plasma membrane develops blisters. Finally, the cell breaks into fragments, and its bits and pieces are engulfed by white blood cells and/or neighboring cells. We now know that cells routinely harbor the enzymes, now called *caspases*, that bring about apoptosis. These enzymes are ordinarily held in check by inhibitors, but are unleashed by either internal or external signals.

Cell division and apoptosis are two opposing processes that keep the number of cells in the body at an appropriate level. They are normal parts of growth and development. An organism begins as a single cell that repeatedly undergoes the cell cycle to produce many cells, but eventually some cells must die in order for the organism to take shape. For example, when a tadpole becomes a frog, the tail disappears as apoptosis occurs. In humans, the fingers and toes of an embryo are at first webbed, but later the webbing disappears as a result of apoptosis, and the fingers are freed from one another. Apoptosis is also helpful when it destroys precancerous cells. Otherwise, a tumor might develop.

▶ **8.5 CHECK YOUR PROGRESS**
1. Distinguish between the G_1 and G_2 checkpoint.
2. Identify the functions of apoptosis in cells.

FIGURE 8.5
Checkpoints control and keep the cell cycle occurring normally.

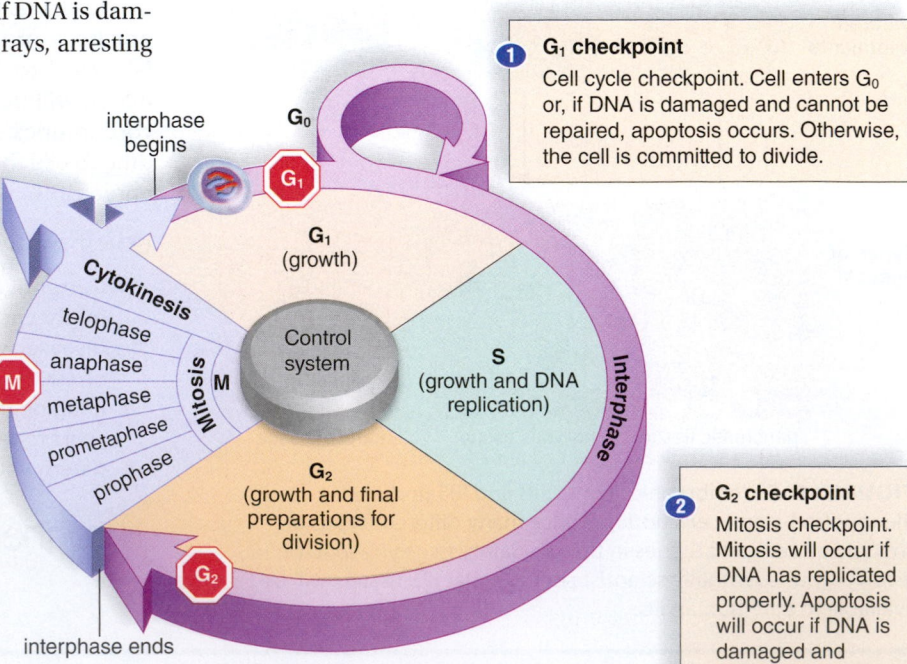

① G_1 checkpoint
Cell cycle checkpoint. Cell enters G_0 or, if DNA is damaged and cannot be repaired, apoptosis occurs. Otherwise, the cell is committed to divide.

③ M checkpoint
Spindle assembly checkpoint. Mitosis will not continue if chromosomes are not properly aligned.

② G_2 checkpoint
Mitosis checkpoint. Mitosis will occur if DNA has replicated properly. Apoptosis will occur if DNA is damaged and cannot be repaired.

interphase begins

G_0

G_1

G_1 (growth)

Control system

S (growth and DNA replication)

Interphase

G_2 (growth and final preparations for division)

G_2

interphase ends

Cytokinesis

telophase

anaphase

metaphase

prometaphase

prophase

Mitosis

M

8.6 Cancer cells are distinctly abnormal

LEARNING OUTCOMES

When you complete this section, you should be able to

1. Describe the characteristics of cancer cells.
2. Relate the characteristics of cancer cells to a cell cycle out of control.

Development of Cancer As explained in the introduction to this chapter, **mutations** (permanent DNA changes) due to different environmental assaults can result in abnormal growth of cells and eventually cancer. Any tissue that already has a high rate of cell division is inherently more susceptible to **carcinogenesis,** the development of cancer, because division gives cells the opportunity to undergo a series of genetic mutations, each one making the next generation of cells more abnormal. Once the cell cycle is out of control, its checkpoints are not working and apoptosis is not occurring. Yet, the cells live on and keep dividing even though they are abnormal.

Cancers are classified according to tissue of origin. *Carcinomas* are cancers of the tissue type that lines organs; *sarcomas* are cancers arising in muscle or bone and cartilage; and *leukemias* are cancers of the blood. In general, cancer cells have the following characteristics:

Cancer cells form tumors Normal cells anchor themselves to a substrate and/or adhere to their neighbors. Then they exhibit contact inhibition and stop dividing. But when cancer is present, cells have lost all restraint; they pile on top of one another and grow in multiple layers, forming a **tumor.** Normal cells respond to signals from their neighbors telling them when to grow and when to stop growing. Cancer cells have no need for stimulatory signals, and they do not respond to inhibitory signals. As cancer develops, the most aggressive cell becomes the dominant cell of the tumor.

Cancer cells undergo angiogenesis and metastasis To grow larger than a million cells (about the size of a pea), a tumor must have a well-developed capillary network to bring it nutrients and oxygen. **Angiogenesis** is the formation of new blood vessels. The low oxygen content in the middle of a tumor may turn on genes for secretions that diffuse into the nearby tissues and cause new vessels to form. Due to mutations, cancer cells tend to be motile because they have a disorganized internal cytoskeleton and lack intact actin filament bundles. To metastasize, cancer cells must make their way across the extracellular matrix and invade a blood vessel or lymphatic vessel. Invasive cancer cells are odd-shaped (Fig. 8.6) and don't look at all like normal cells. Cancer cells produce proteinase enzymes that degrade the membrane and allow them to invade underlying tissues. When these cells begin new tumors far from the primary tumor, **metastasis** has occurred. Not many cancer cells achieve this feat (maybe 1 in 10,000), but those that successfully metastasize spread the cancer throughout the body.

Cancer cells have abnormal nuclei The nuclei of cancer cells are enlarged and may contain an abnormal number of chromosomes. For example, the nuclei of the cervical cancer cells shown in Figure 8.6 have increased to the point that they take up most of the cell. The chromosomes are also abnormal; some parts may be

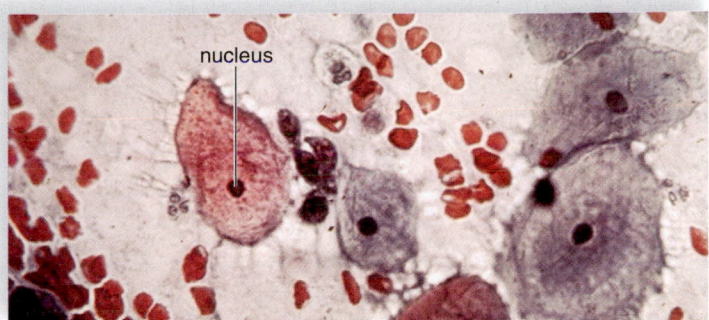
Normal cervical cells

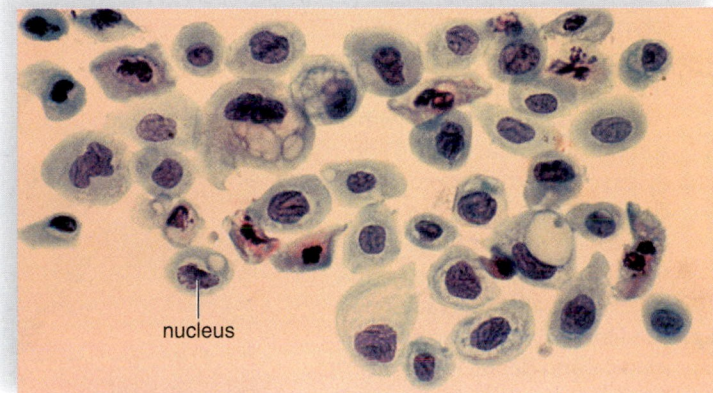
Precancerous cervical cells

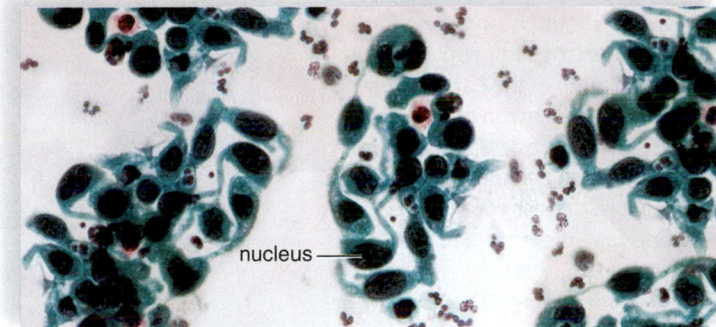
Cancerous cervical cells

FIGURE 8.6 These micrographs show a progression toward cancer of the cervix in a female. Cervical cancer is often caused by the human papillomavirus (HPV), a sexually transmitted virus. Therefore, all young women should be vaccinated to protect themselves against any new HPV infections.

duplicated, or some may be deleted. In addition, gene amplification (extra copies of specific genes) is seen much more frequently in cancer cells than in normal cells.

Cancer cells lack differentiation Cancer cells are nonspecialized and do not contribute to the functioning of a body part. A cancer cell does not look like a specialized skin, muscle, nerve, or liver cell; instead, it looks distinctly abnormal. Normal cells enter the cell cycle about 70 times, and then they die. Cancer cells can enter the cell cycle repeatedly, and in this way they are immortal.

▶ **8.6 CHECK YOUR PROGRESS**

1. Identify characteristics of cancer cells that indicate the cell cycle is out of control.

8C Protective Behaviors and Diet Help Prevent Cancer

Evidence suggests that the risk of certain types of cancer can be reduced by adopting protective behaviors and the right diet.

Protective Behaviors

The following behaviors help prevent cancer:

Don't smoke People who smoke two or more packs of cigarettes per day have lung cancer mortality rates 15–25 times greater than those of nonsmokers. Smokeless tobacco (chewing tobacco or snuff) increases the risk of cancers of the mouth, larynx, throat, and esophagus.

Use sunscreen Almost all cases of skin cancer are considered sun-related. Use a sunscreen with a sun protection factor (SPF) of at least 15, and wear protective clothing if you are going to be out during the brightest part of the day. Don't sunbathe on the beach or in a tanning salon.

Avoid radiation Even though most medical and dental X-rays are adjusted to deliver the lowest dose possible, unnecessary X-rays should be avoided. Radon gas from the radioactive decay of uranium in the Earth's crust can accumulate in houses and increase the risk of lung cancer, especially in cigarette smokers. It is best to test your home and take the proper remedial actions.

Be tested for cancer Do the shower check for breast cancer or testicular cancer. Have other exams done regularly by a physician.

Be aware of occupational hazards Exposure to several different industrial agents (nickel, chromate, asbestos, vinyl chloride, etc.) and/or radiation increases the risk of various cancers. The risk from asbestos is greatly increased when combined with cigarette smoking.

Carefully consider hormone therapy A large study conducted by the Women's Health Initiative found that combined estrogen-progestin therapy prescribed to ease the symptoms of menopause increased the incidence of breast cancer. Also, the risk outweighed the possible decrease in the number of colorectal cancer cases sometimes attributed to hormone therapy.

The Right Diet

Statistical studies have suggested that people who follow the following dietary guidelines are less likely to have cancer:

Sunscreen with SPF 15 minimizes skin cancer.

Increase consumption of foods rich in vitamins A and C Beta-carotene, a precursor of vitamin A, is found in carrots, fruits, and dark-green, leafy vegetables. Vitamin C is present in citrus fruits. Vitamin C also prevents the conversion of nitrates and nitrites into carcinogenic nitrosamines in the digestive tract.

Limit consumption of salt-cured, smoked, or nitrite-cured foods Consuming salt-cured or pickled foods may increase the risk of stomach and esophageal cancers. Smoked foods, such as ham and sausage, contain chemical carcinogens similar to those in tobacco smoke. Nitrites, sometimes added to processed meats (e.g., hot dogs and cold cuts) and other foods to protect them from spoilage, are associated with the development of cancer.

Include vegetables from the cabbage family in the diet The cabbage family includes cabbage, broccoli, brussels sprouts, kohlrabi, and cauliflower. These vegetables may reduce the risk of gastrointestinal and respiratory tract cancers.

Be moderate in the consumption of alcohol The risks of cancer development rise as the level of alcohol intake increases. The strongest associations are with oral, pharyngeal, esophageal, and laryngeal cancer, but cancer of the breast and liver are also implicated. People who both drink and smoke greatly increase their risk for developing cancer.

Maintain a healthy weight The risk of cancer (especially colon, breast, and uterine cancers) is 55% greater among obese women, and the risk of colon cancer is 33% greater among obese men, compared to people of normal weight.

CONSIDER THESE QUESTIONS

1. What mental processes might cause people to sunbathe even when they know skin cancer could result from this behavior? What could you say to change their mind? Are these the same mental processes that cause people to smoke and drink?

2. It can take years to acquire cancer by neglecting this list of do's and don'ts. How much does that affect people's behavior today?

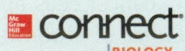 Explore the concepts through a variety of multimedia assets and question types.

www.mcgrawhillconnect.com

CONNECTING THE CONCEPTS

Each type of eukaryotic organism has a characteristic number of chromosomes in the nucleus; humans have 46 chromosomes. During cellular reproduction each new cell receives a copy of each chromosome and the cell cycle brings this about. During interphase (the time between nuclear divisions), DNA replication occurs so that by the time chromatin has compacted the visible chromosomes are duplicated and composed of two sister chromatids. It is separation of the chromatids—each a double DNA helix—during mitosis that assures the daughter nuclei have the same number of chromosomes and the same genes as the parent cell had. Without the orderly events of

mitosis dependent as it is on the spindle apparatus, eukaryotic organisms could not continue to exist. Again, we come to the realization that the cytoskeleton is of primary importance to the life of the cell and the organism.

The cell cycle has a number of pauses, checkpoints that determine if the DNA is normal and capable of functioning properly. If by chance the DNA appears to have been damaged in some way, it is repaired if possible. If this is not possible, then the cell dies by a process called apoptosis. The lack of apoptosis is associated with the development of cancer in which distinctly abnormal cells are capable of dividing over and over again until a tumor develops.

The human life cycle contains two types of nuclear divisions: mitosis and meiosis. Mitosis, the topic of this chapter, is essential to growth and repair of the organism. The type of nuclear division discussed in the next chapter, meiosis, is essential to reproduction of the organism.

ANALYZE AND EVALUATE

1. How is DNA "packaged" in order to ensure that every new cell receives a complete copy of the organism's genetic material?
2. How does the spindle apparatus function so that each daughter nucleus receives a copy of each chromosome?

SUMMARIZE

The Cell Cycle

8.1   Cellular reproduction depends on the cell cycle

- Cellular reproduction (and the cell cycle) occur repeatedly as complex organisms develop, grow, and repair their tissues. The cell theory tells us that new cells come only from preexisting cells.
- In the **cell cycle, interphase** (G_1, S, G_2 stages) precedes the **M stage,** which includes **mitosis** and **cytokinesis.** In G_1 **stage** cells grow and can make a commitment to divide; in **S stage** DNA replication results in duplicated chromosomes; and in G_2 **stage** growth occurs and proteins are made to form microtubules.

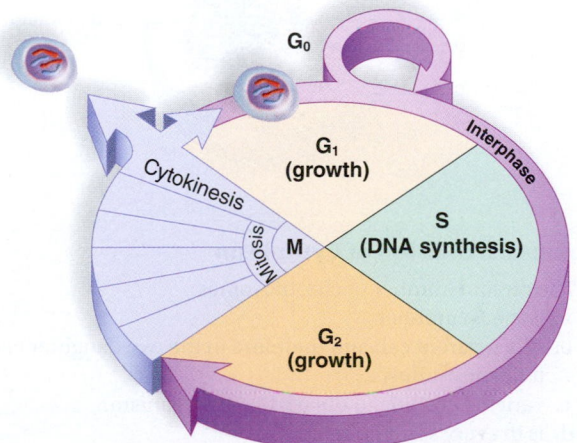

- Embryonic cells and **adult stem cells** divide all the time; cells in the G_0 **stage** have dropped out of the cell cycle and do not divide until stimulated to do so.

Mitosis: Division of Nucleus

8.2   Following mitosis, daughter cells have the same chromosome count as the parent cell

- Mitosis is duplication division; the **parent cell** and the **daughter cells** all have the same number and kinds of chromosomes because the identical chromatids of each duplicated chromosome separate and become daughter chromosomes.
- The parent cell can be **diploid (2n)** or **haploid (n),** depending on the species. The number of centromeres equals the number of chromosomes a cell has.
- Centrosomes form the **spindle apparatus,** which helps ensure orderly separation of chromatids.

8.3   Mitosis has a set series of phases

- During mitosis, the spindle poles of animal cells have centrioles and an **aster.** Plant cells have poles but no centrioles or asters. For a comparison of mitosis in animal and plant cells, see Table 8.3.
- The stages of mitosis are: **prophase**—the nuclear envelope fragments, chromosomes are visible, and a spindle apparatus appears; **metaphase**—the chromosomes attach to spindle fibers and align at the equator; **anaphase**—sister chromatids separate and daughter chromosomes move to the poles of the spindle; **telophase**—daughter nuclei form and cytokinesis begins.

Cytokinesis: Division of Cytoplasm

8.4   Cytokinesis follows mitosis

- In animal cells, cytokinesis involves a **cleavage furrow.** In plant cells, cytokinesis involves the formation of a new plasma membrane and cell wall at a **cell plate.**

Cell Cycle Control and Cancer

8.5 Cell cycle control depends on checkpoints

■ Checkpoint **G₁** ensures the cell is normal before the cell cycle continues; checkpoint G₂ ensures that DNA replicated properly; checkpoint M ensures that chromosomes will distribute accurately to daughter cells. In general, if the cell cycle is unable to continue, apoptosis occurs. **Apoptosis** initiated by **p53** is programmed cell death orchestrated by unleashed enzymes.

8.6 Cancer cells are distinctly abnormal

■ Due to **mutations, carcinogenesis** occurs and cancer is present when cells divide uncontrollably and a **tumor** develops. Cell cycle control and apoptosis are lacking.

■ Cancer cells have abnormal characteristics: They lack differentiation, have abnormal nuclei, form tumors, promote **angiogenesis** (formation of new blood vessels), and undergo **metastasis** (formation of tumors distant from primary tumor).

TEST YOURSELF

The Cell Cycle

1. Which of these statements is incorrect? Just before mitosis in a eukaryotic cell,
   a. homologous pairs of chromosomes can be seen.
   b. each chromosome has two sister chromatids.
   c. one chromatid came from the father and one came from the mother.
   d. each sister chromatid carries the same genes.

2. In the cell cycle,
   a. mitosis cannot occur without interphase.
   b. the single event during interphase is chromosome duplication.
   c. cells are metabolically inactive during interphase.
   d. a DNA double helix divides in two.

For questions 3–6, match the descriptions to a stage in the key.

KEY:
   a. G₁ stage
   b. S stage
   c. G₂ stage
   d. M (mitotic) stage

3. At the end of this stage, each chromosome consists of two attached chromatids.
4. During this stage, daughter chromosomes are distributed to two daughter nuclei.
5. The cell doubles its organelles and accumulates the materials needed for DNA synthesis.
6. The cell synthesizes the proteins needed for cell division.

7. Interphase
   a. is the same as prophase, metaphase, anaphase, and telophase.
   b. is composed of G₁, S, and G₂ stages.
   c. requires the use of polar spindle fibers and kinetochore spindle fibers.
   d. is the majority of the cell cycle.
   e. Both b and d are correct.

For questions 8–10, match the descriptions to the terms in the key.

KEY:
   a. centrosome
   b. chromosome
   c. centromere
   d. cyclin

8. Point of attachment for sister chromatids
9. Found at a spindle pole in the center of an aster
10. Coiled and condensed chromatin

Mitosis: Division of Nucleus

11. The two identical halves of a duplicated chromosome
    a. always stay together.
    b. are different sizes.
    c. become daughter chromosomes.
    d. are called homologues.

For questions 12–15, match each description to a phase of mitosis in the key.

KEY:
   a. prophase
   b. metaphase
   c. anaphase
   d. telophase

12. The nucleolus disappears, and the nuclear envelope breaks down.
13. The spindle disappears, and the nuclear envelopes form.
14. Sister chromatids separate.
15. Chromosomes are aligned on the spindle equator.

16. Mitosis in animal cells but not plant cells
    a. maintains the chromosome number.
    b. uses a spindle apparatus.
    c. has centrioles at the poles.
    d. produces two unequal daughter cells.

17. Label this diagram of a cell in early prophase of mitosis:

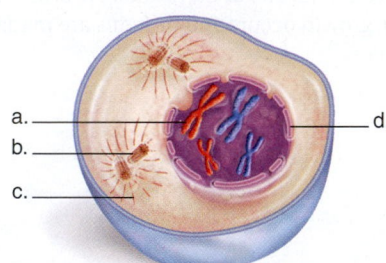

Cytokinesis: Division of Cytoplasm

18. The diploid number of chromosomes
    a. is the 2n number.
    b. is in a parent cell and therefore in the two daughter cells following mitosis.
    c. varies according to the particular organism.
    d. is in every somatic cell.
    e. All of these are correct.

19. During which mitotic phases are duplicated chromosomes present?
   a. all but telophase
   b. prophase and anaphase
   c. all but anaphase and telophase
   d. only during metaphase at the metaphase plate
   e. Both a and b are correct.
20. At the metaphase plate during metaphase of mitosis, there are
   a. single chromosomes.
   b. duplicated chromosomes.
   c. G_1 stage chromosomes.
   d. always 23 chromosomes.
21. If a parent cell has 14 chromosomes prior to mitosis, how many chromosomes will each daughter cell have?
   a. 28 because each chromatid is a chromosome
   b. 14 because the chromatids separate
   c. only 7 after mitosis is finished
   d. any number between 7 and 28
   e. 7 in the nucleus and 7 in the cytoplasm, for a total of 14

Cell Cycle Control and Cancer

22. Which of these is an incorrect statement?
   a. Checkpoints allow the cell to continue if all is normal.
   b. A DNA abnormality can cause apoptosis to occur.
   c. The cell cycle stages can occur out of order and all will be well.
   d. Mutations can cause the cell cycle to occur repeatedly.
23. Which of the following is typical of normal cells, but not typical of cancer cells?
   a. Cell cycle control is always present.
   b. The cells have enlarged nuclei.
   c. The cells stimulate the formation of new blood vessels.
   d. The cells are capable of traveling through blood and lymph.
24. Which of the following is not characteristic of cancer cells?
   a. Cancer cells often undergo angiogenesis.
   b. Cancer cells tend to be nonspecialized.
   c. Cancer cells undergo apoptosis.
   d. Cancer cells often have abnormal nuclei.
   e. Cancer cells can metastasize.
25. When cancer occurs,
   a. cells cannot pass the G_1 checkpoint.
   b. control of the cell cycle is impaired.
   c. apoptosis has occurred.
   d. the cells can no longer enter the cell cycle.
   e. All of these are correct.

GET INVOLVED

1. The 3D animation "Cell Cycle and Mitosis" provides a visual exploration of the events and stages of cell division.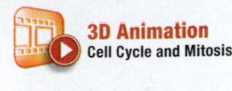
2. A student is microscopically examining cells undergoing mitosis in a plant root tip. He decides to determine the relative length of time taken for each phase by counting the number of cells among 100 that are in each phase. Hypothesize his results and give a reason for your hypothesis.
3. Which of the cancer cell characteristics on page 149 is substantiated by this virtual lab?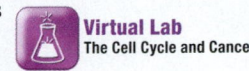

MEDIA STUDY TOOLS

mhhe.com/maderconcepts3

Enhance your study of this chapter with interactive study tools, practice tests, and engaging animations. Also, ask your instructor about the resources available through ConnectPlus, which includes LearnSmart, a personalized adaptive learning program, and a media-rich eBook.

9

Sexual Reproduction

CHAPTER OUTLINE

BEFORE YOU BEGIN

Take a few minutes to recall

The sequence of bases in DNA can differ (section 3.9)

DNA replicates once per cell cycle (section 8.1)

How the spindle apparatus interacts with chromosomes (section 8.3)

SEX—What is it?

Male elephant seals fight until they're bloody, birds flap their wings and prance about, human males buy a convertible and otherwise, too, expend a lot of effort in order to have sex. When we consider sex, we tend to think of behaviors that bring the sexes together and also the union they so desire, but the actual purpose of sex, beyond having an offspring, is rarely considered. A couple of biological examples that strip away any unnecessary concerns might reveal to us the true meaning of sex.

Let's take frogs, for example. When it's spring and the time of year for sex, male frogs puff themselves up and croak incessantly to attract a female. But the sex act itself is quite simple. Males clasp a female and they both shed their gametes at approximately the same time in the water. Hundreds of flagellated sperm fertilize the same number of visible eggs and development of new frogs begins. This example suggests that the essence of sex is not their behavior but the union of sperm and egg, right? Well that's a logical conclusion, but to arrive at a more in-depth meaning of sex let's look at another example.

If we allow ourselves to search beyond the animal kingdom, we will find that sex is present even among protists that don't produce an egg and sperm. Consider the life cycle of *Chlamydomonas,* unicellular green protists that live in freshwater ponds. When the weather is good, *Chlamydomonas* repeatedly divide in two to reproduce asexually

Human sexual behavior is complex.

(without sex). But when the weather turns bad—winter is coming or the pond is drying up—*Chlamydomonas* turn to sex and reproduce sexually. Then, *Chlamydomonas* become gametes which fuse two at time to form a diploid zygote. The zygote has a thick wall and can overwinter in order to survive an unfavorable environment. In the spring, the zygote releases cells whose genetic makeup will vary one from the other.

Aha! this could very well be the significance of sex. Asexual reproduction, which utilizes mitosis, produces individuals that have the same genetic makeup as the parent. Asexual reproduction is a quick and easy way to populate an environment with offspring that are already adapted to the environment. However, a particular off-spring with a genetic makeup different from that of a parent might be more suited to a new or changing environment. Sexual reproduction increases genetic variability because gametes receive various com-binations of genes and then, any two of these gametes, one from each parent, donate chromosomes to the offspring. Some new combinations of genes may make an offspring particularly suited to an environment that is different from that of the parent. But before we go on discussing evolution, let's turn our attention to the topic of this chapter, namely meiosis, the type of cell division that occurs during the production of gametes in animals. We want to explore exactly how meiosis is able to reduce the chromosome number and introduce genetic variability among the gametes at the same time. Because gam-etes fuse during fertilization, a reduced num-ber keeps the species chromosome number constant generation after generation and because meiosis introduces genetic variability among gametes, the spe-cies is more likely to become adapted to a new environment.

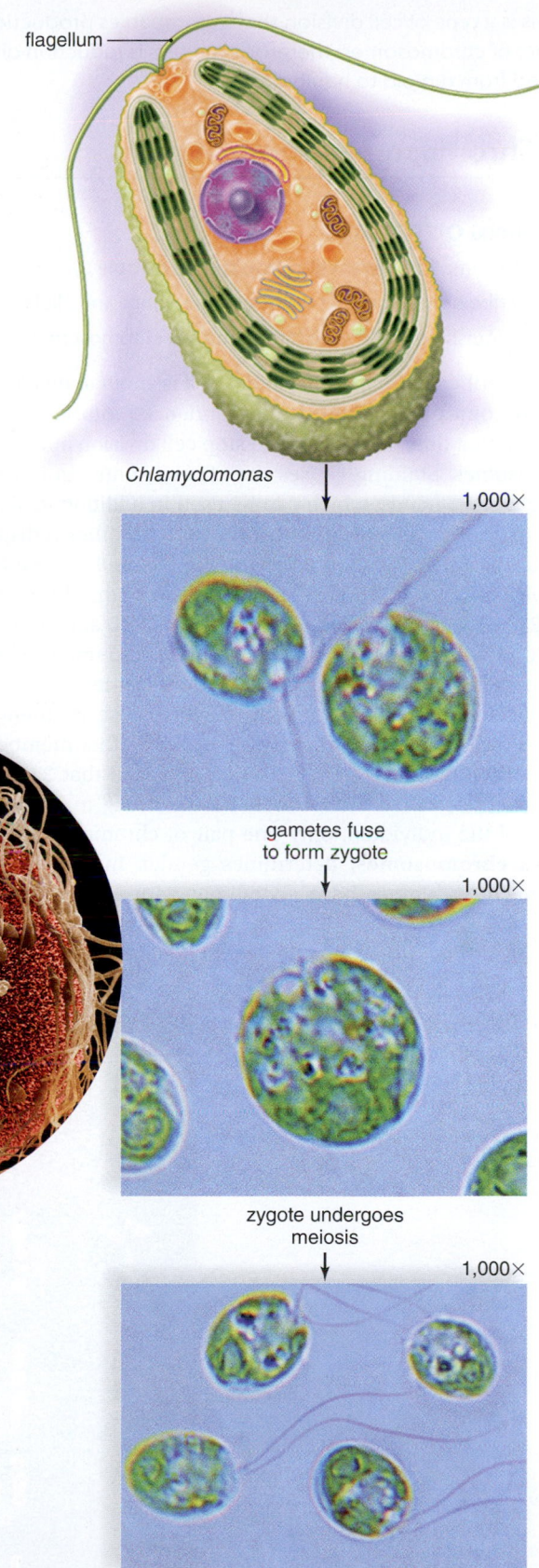

flagellum

Chlamydomonas

1,000×

gametes fuse
to form zygote

1,000×

zygote undergoes
meiosis

1,000×

These *Chlamydomonas* cells have a varied genetic makeup.

gametes

When frogs mate, their gametes are visible.

The Basics of Meiosis

Meiosis is a type of cell division that accompanies production of gametes, such as the sperm and egg. The gametes have the haploid number of chromosomes; therefore, meiosis is reduction division. In this section, we concentrate on how the chromosome number is reduced from diploid to haploid.

9.1 Chromosomes come in pairs

LEARNING OUTCOMES

When you complete this section, you should be able to

1. Describe homologues in terms of appearance and alleles.
2. Distinguish between autosomes and sex chromosomes.

As we described in Chapter 8, when a cell is about to divide, chromatin becomes highly coiled and condensed into the chromosomes. At that time, it's possible to stop cell division and view the chromosomes. Staining causes the chromosomes to have dark and light cross-bands, which can be used in addition to size and shape to distinguish one chromosome from the other. A display of the chromosomes is called a **karyotype;** Figure 9.1 is the karyotype of a human male. This karyotype shows that human cells have 23 pairs of chromosomes, for a total of 46 altogether. This is the diploid (2n) number of chromosomes in humans. Half this number is the haploid (n) number of chromosomes.

Why do chromosomes come in pairs? Because one member of each pair was donated by the mother and the other member was donated by the father. Also, it should be noted that 22 pairs of chromosomes, called **autosomes,** have nothing to do with the gender of the individual; only one pair of chromosomes, called the **sex chromosomes,** determines gender. In human males, the sex chromosomes have a different appearance; the larger one is called an **X chromosome** and the smaller is called a **Y chromosome.** Females have two X chromosomes. We will be discussing the sex chromosomes in more detail later. Just now, let's continue our discussion of the autosomes.

The members of an autosome chromosome pair are called **homologues,** or homologous chromosomes, because not only do they look alike, but they also carry genes for the same traits, such as type of hairline, length of fingers, or type of earlobe. Just as your mother may have long fingers and your father may have short fingers, each homologue may have different versions of a gene. Alternate versions of a gene for a particular trait (e.g., finger length) are called **alleles.** Alleles occur at the same location on each homologue. Your mother's allele for long fingers is on one homologue, and your father's allele for short fingers is on the other homologue.

Recall that following DNA replication during the cell cycle, a chromosome has two parts, called **sister chromatids.** Each sister chromatid is a double helix. Because the sister chromatids are duplicates of each other the alleles on sister chromatids are exactly alike.

▶ 9.1 CHECK YOUR PROGRESS

1. Compare and contrast homologues with sister chromatids.

FIGURE 9.1 This karyotype of a normal male shows 23 pairs of homologous chromosomes. These chromosomes are duplicated, and each one is composed of two sister chromatids

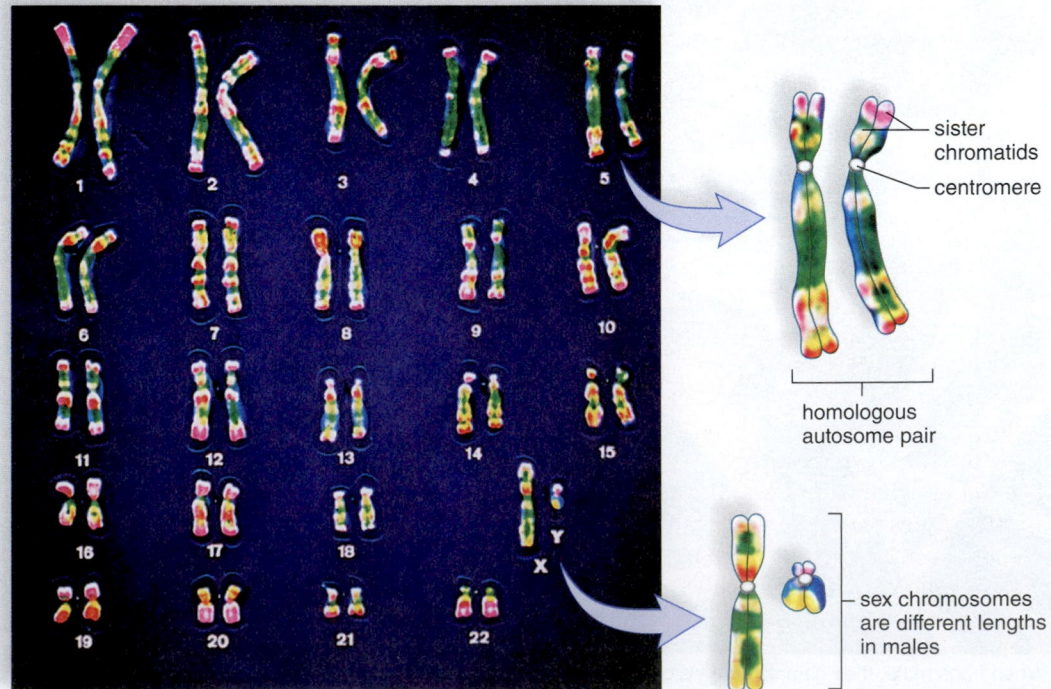

9.2 Homologues separate during meiosis

LEARNING OUTCOME

When you complete this section, you should be able to

1. Understand the role of synapsis in reducing the chromosome number during meiosis I.

Meiosis is reduction division. Because meiosis involves two divisions, four daughter cells result. Each of these daughter cells has one of each kind of chromosome and, therefore, half as many chromosomes as the parent cell. Usually, the daughter cells mature into **gametes** (sex cells—sperm and egg) that fuse during fertilization. **Fertilization** restores the diploid number of chromosomes in the zygote, the first cell of the new individual. If the gametes carried the diploid instead of the haploid number of chromosomes, the chromosome number would double with each fertilization. After several generations, the zygote would be nothing but chromosomes. For example, in humans with a diploid number of 46 chromosomes, in five generations the chromosome number would increase to 1,472 chromosomes (46×2^5). In 10 generations this number would increase to a staggering 47,104 chromosomes (46×2^{10}).

MP3
Meiosis

Animation
How Meiosis
Works

In Figure 9.2, the diploid (2n) number of chromosomes is four, and there are two pairs of chromosomes. The short chromosomes are one pair, and the long chromosomes are another. The haploid (n) number of chromosomes for this cell is two. A haploid cell has only one of each kind of chromosome and therefore lacks pairs of chromosomes.

Prior to the first division, called **meiosis I,** DNA replication has occurred, and the chromosomes are duplicated. During meiosis I, the homologues come together and line up side by side. This so-called **synapsis** results in an association of four chromatids that stay in close proximity during the first two phases of meiosis I. Also, because of synapsis, there are pairs of homologues at the equator during meiosis I. (Keep in mind that only during meiosis I is it possible to observe paired chromosomes at the equator.) Synapsis leads to a reduction in the chromosome number because it permits orderly separation of homologues. The daughter nuclei are haploid when they receive only one member of each pair. The haploid (n) nature of each daughter cell can be verified by counting its centromeres. Each chromosome, however, is still duplicated, and no replication of DNA occurs between meiosis I and meiosis II. The period of time between meiosis I and meiosis II is called **interkinesis.**

During **meiosis II,** the sister chromatids of each chromosome separate, becoming daughter chromosomes that are distributed to daughter nuclei. In the end, each of four daughter cells has the n, or haploid, number of chromosomes, and each chromosome consists of one chromatid. In "Life Cycles Are Varied" on page 162 we learn that only in the life cycle of animals do these haploid daughter cells become gametes. In plants they become spores. Spores are haploid cells capable of developing into a haploid organism. Plants have a life cycle that alternates between a diploid generation and a haploid generation. In protists such as *Chlamydomonas,* the zygote is the only diploid phase of the life cycle and it undergoes meiosis to produce haploid cells, each of which becomes a mature organism.

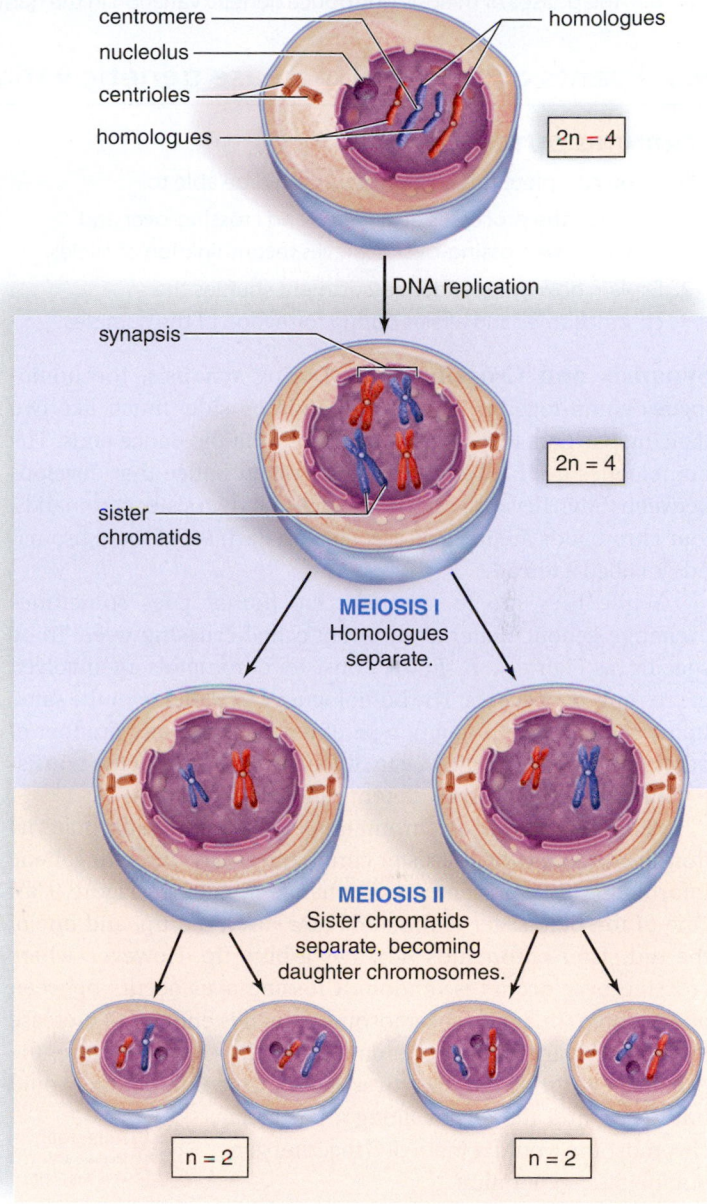

FIGURE 9.2 Meiosis produces daughter cells that are genetically different from the parent cell. Four daughter cells result because meiosis includes two divisions: During meiosis I, the homologues separate, and during meiosis II the chromatids separate, becoming daughter chromosomes.

▶ 9.2 CHECK YOUR PROGRESS

1. Discuss how DNA replication results in duplicated chromosomes and explain the term duplicated.
2. Justify, as if to a fellow student, why the daughter cells following meiosis I are termed haploid.
3. Explain why you would not expect to find both a long red chromosome and a long blue chromosome in a daughter cell following meiosis.

The Phases of Meiosis

The same sequence of phases we saw during mitosis occur during both meiosis I and meiosis II. In addition to reducing the chromosome number, the phases of meiosis I introduce genetic variation in the gametes and, therefore, increase the genetic variability of the next generation.

9.3 Events of meiosis I increase genetic variation among the gametes

LEARNING OUTCOMES

When you complete this section, you should be able to

1. Describe the processes of synapsis and crossing-over and explain how crossing-over achieves recombination of alleles.
2. Explain how independent assortment shuffles the chromosomes and alleles during formation of the gametes.

Synapsis and Crossing-Over During **synapsis,** the homologues come together and line up side by side, much like two dancing partners who will stay together until the dance ends. The homologues are held in place by a protein lattice that develops between them. Because each homologue has two sister chromatids, four chromatids are in close association. Each set of four chromatids is called a **tetrad.**

While they are in synapsis, the homologues sometimes exchange genetic material, an event called **crossing-over.** To be specific, as Figure 9.3A shows, nonsister chromatids are involved in crossing-over events. The homologues carry alleles for the same traits, such as finger length, type of hairline, and any number of other traits, but their alleles can differ. This means that the nonsister chromatids can carry different genetic information.

After the nonsister chromatids exchange genetic material during crossing-over, the sister chromatids carry different genetic information as represented by a change in color in Figure 9.3A: One of the blue sister chromatids now has a red tip, and one of the red sister chromatids now has a blue tip. However, where crossing-over occurs is random. Crossing-over occurs between one to three times per chromosome, which is enough to increase the genetic variability of the daughter cells—and therefore, the gametes. Without crossing-over genetic recombination would not occur, and the alleles along a particular chromosome would remain tied together generation after generation.

▶ **Animation**
Meiosis and Crossing-Over

Independent Assortment Meiosis I not only shuffles the alleles during crossing-over, it also shuffles the chromosomes. Maternal and paternal chromosomes do not necessarily stay together during meiosis because the homologues align randomly at the equator. In Figure 9.3B, ❶ the parent cell has two pairs of homologues, which undergo synapsis soon after meiosis I begins. ❷ Notice that two orientations are possible at the equator because either homologue can face either pole of the spindle. In the simplest of terms, with reference to Figure 9.3B, the red chromosomes don't have to be on the left, and the blue chromosomes don't have to be on the right. They randomly *align* at the equator.

❸ The homologues and their attached alleles separate, so that one chromosome from each pair goes to each daughter nucleus, and the daughter cells are haploid. ❹ All possible combinations of chromosomes and alleles can occur among the gametes. Therefore, **independent assortment** of homologues occurs during meiosis. In the simplest of terms, any short chromosome

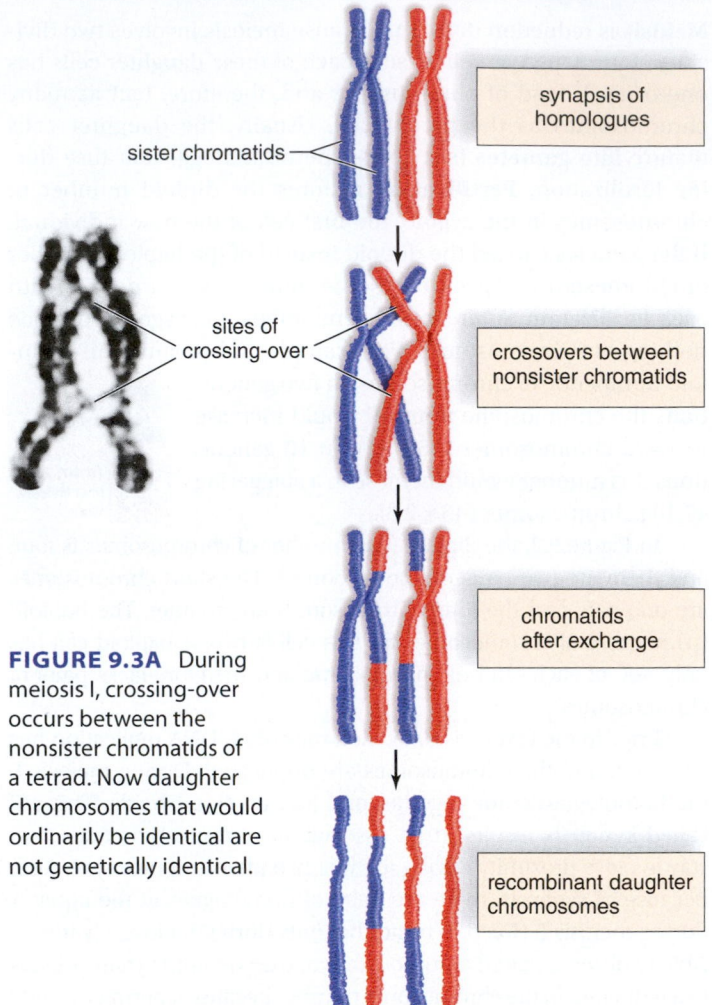

FIGURE 9.3A During meiosis I, crossing-over occurs between the nonsister chromatids of a tetrad. Now daughter chromosomes that would ordinarily be identical are not genetically identical.

(blue or red) can be with any long chromosome (blue or red). The genetic variation brought about by independent assortment of chromosomes is increased by crossing-over.

▶ **Animation**
Random Orientation of Chromosomes During Meiosis

Fertilization The union of male and female gametes during fertilization produces a **zygote,** the first cell of the new individual. As we have seen, the gametes produced by individuals, such as humans, have the same number of chromosomes, but the chromosomes may carry different genetic information due to independent assortment and crossing-over. In humans, each gamete has 23 chromosomes. Considering the fusion of unlike gametes due to independent assortment, it means that $(2^{23})^2$, or 70,368,744,000,000, chromosomally different zygotes are possible, even assuming no crossing-over. If crossing-over occurs once, then $(4^{23})^2$, or 4,951,760,200,000,000,000,000,000,000, genetically different zygotes are possible for every couple. Keep in mind that crossing-over can occur several times between homologues.

Genetic Variation The process of sexual reproduction brings about genetic variation among members of a population. Therefore, if the environment changes, genetic variability among offspring, introduced by sexual reproduction, may be advantageous (Fig. 9.3C). In other words, some offspring may have a better chance of survival and reproductive success than others in a population. For example, suppose the ambient temperature were to rise due to global climate change. A dog with genes for the least amount of fur may have an advantage over other dogs of its generation.

3D Animation
Meiosis: Genetic Diversity

Animation
Genetic Diversity

In a changing environment, sexual reproduction, with its shuffling of genetic information due to meiosis and fertilization, is expected to give at least a few offspring a better chance of survival when environmental conditions change.

FIGURE 9.3C The puppies in this litter differ in appearance because crossing-over and independent assortment occurred during meiosis, and fertilization brought different gametes together.

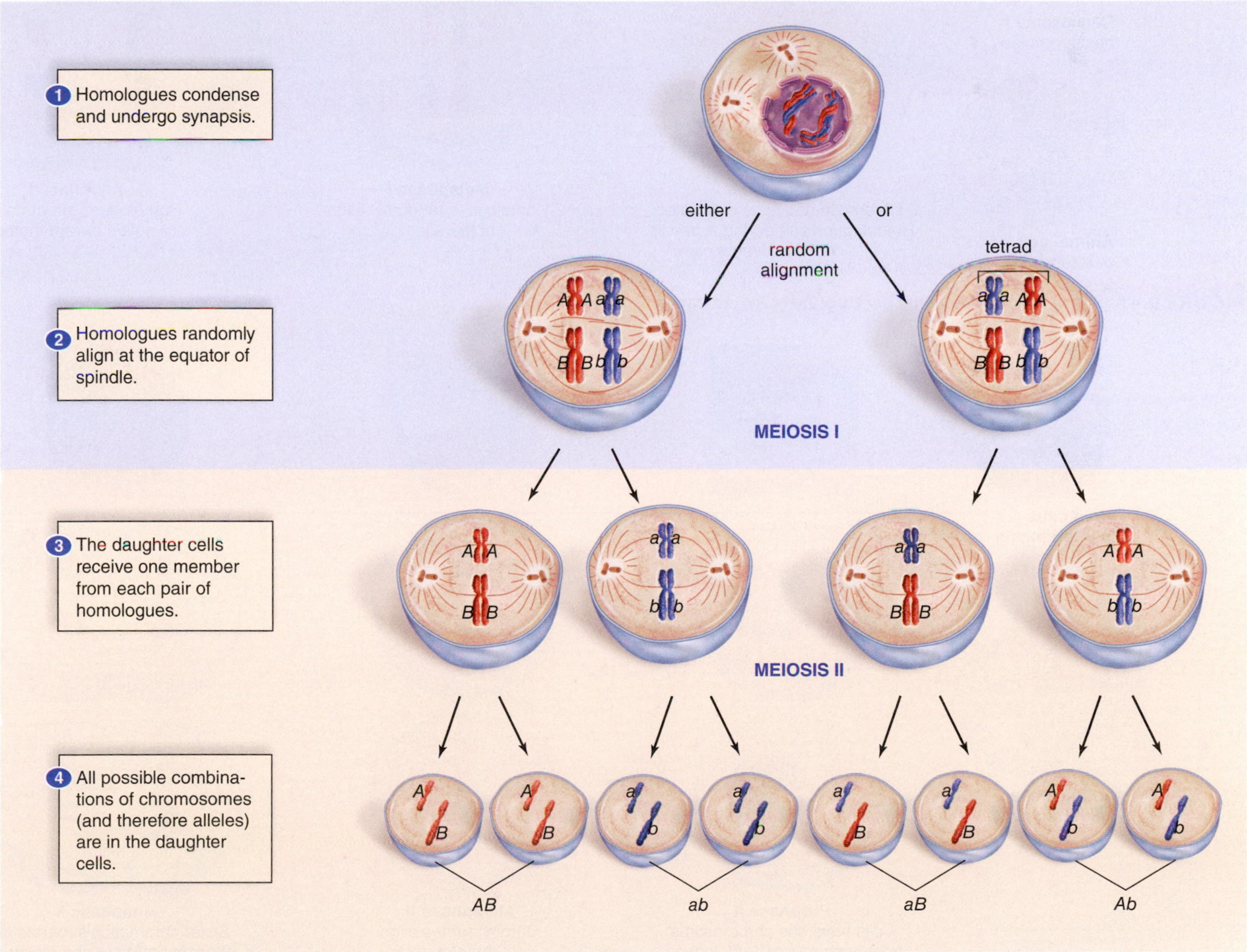

▶ **9.3 CHECK YOUR PROGRESS**

1. Identify why crossing-over occurs only between nonsister chromatids.
2. Determine the chromosomes of the daughter cells if mother (i.e., red) and father (i.e., blue) chromosomes did not randomly separate during meiosis I.

❶ Homologues condense and undergo synapsis.

either   or

random alignment   tetrad

A A a a   *a a A A*

B B b b   *B B b b*

MEIOSIS I

❷ Homologues randomly align at the equator of spindle.

❸ The daughter cells receive one member from each pair of homologues.

A A   *a a*   *a a*   *A A*

B B   *b b*   *B B*   *b b*

MEIOSIS II

❹ All possible combinations of chromosomes (and therefore alleles) are in the daughter cells.

A *A*   *a* *a*   *a* *a*   *A* *A*

B *B*   *b* *b*   *B* *B*   *b* *b*

AB   *ab*   *aB*   *Ab*

FIGURE 9.3B Independent assortment of chromosomes and alleles increases genetic variation. Each chromosome carries many alleles, but for the sake of simplicity, the homologues have been assigned only one allele. Blue background = 2n; tan background = n.

9.4 Both meiosis I and meiosis II have four phases

LEARNING OUTCOME

When you complete this section, you should be able to

1. List the phases of meiosis and briefly explain what events occur during each phase.

The same four phases of mitosis—prophase, metaphase, anaphase, and telophase—occur during both meiosis I (Fig. 9.4A) and meiosis II (Fig. 9.4B). During **prophase I,** the nuclear envelope fragments, the nucleolus disappears as the spindle appears, and the condensing homologues undergo synapsis. The formation of tetrads helps prepare the homologues for separation; it also allows crossing-over to occur between nonsister chromatids. During **metaphase I,** tetrads are present and homologues align randomly at the spindle equator. Following separation of the homologues during **anaphase I** and re-formation of the nuclear envelopes during **telophase I,** the daughter nuclei are haploid: Each daughter cell contains only one chromosome from each pair of homologues. The chromosomes are duplicated, and each still has two sister chromatids.

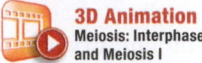

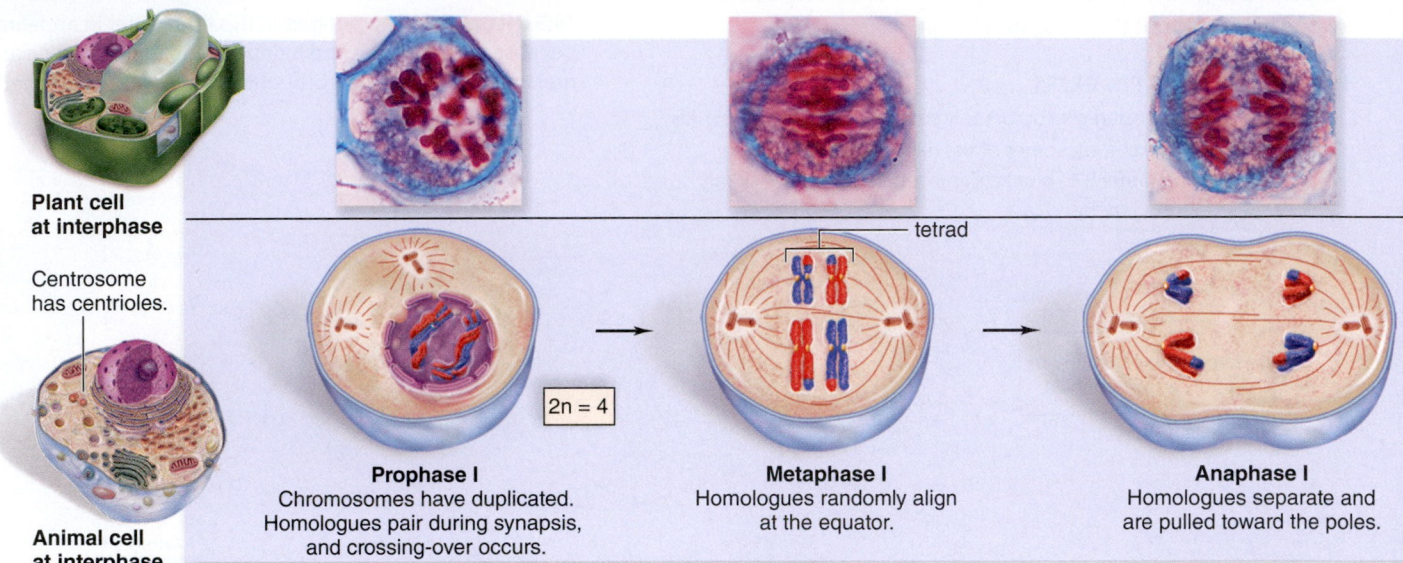

FIGURE 9.4A Phases of meiosis I. Blue background = 2n; tan background = n.

Prophase I
Chromosomes have duplicated. Homologues pair during synapsis, and crossing-over occurs.

Metaphase I
Homologues randomly align at the equator.

Anaphase I
Homologues separate and are pulled toward the poles.

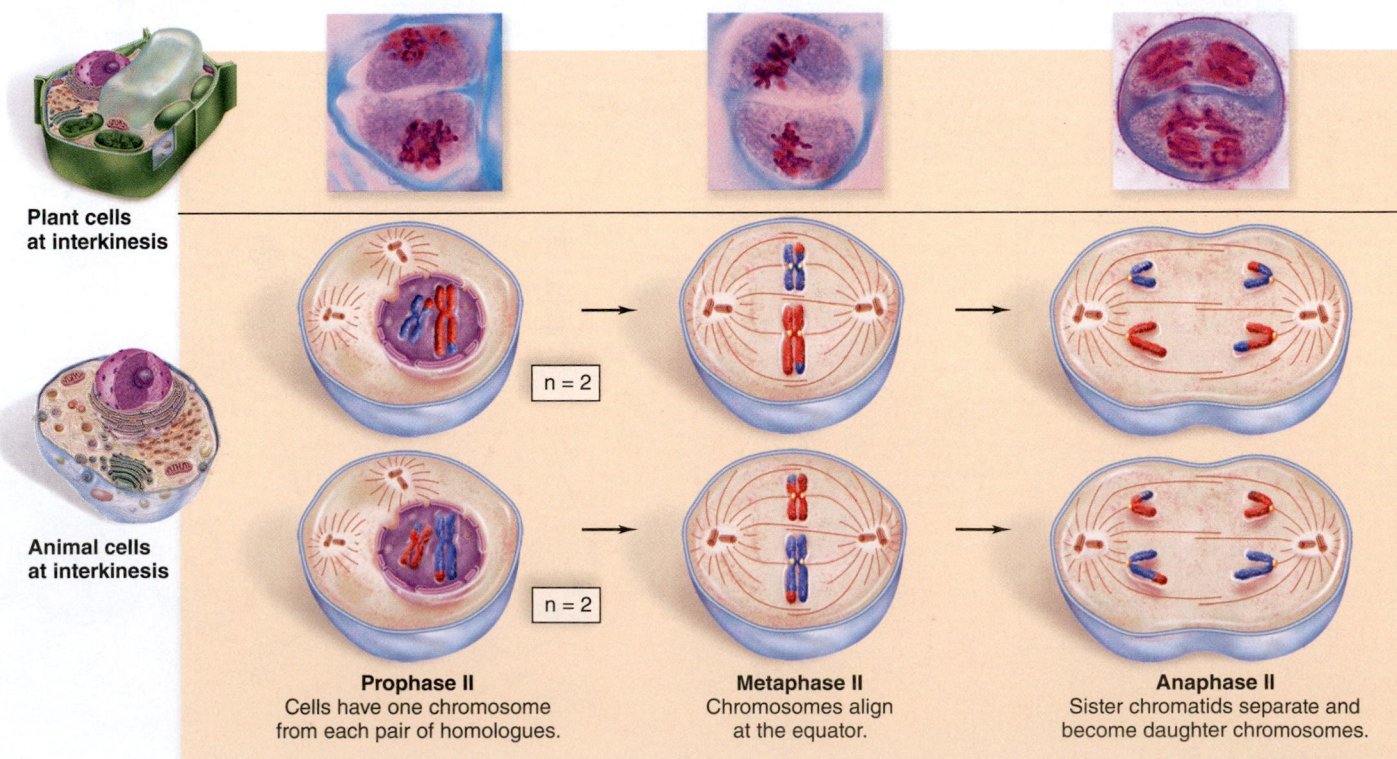

Prophase II
Cells have one chromosome from each pair of homologues.

Metaphase II
Chromosomes align at the equator.

Anaphase II
Sister chromatids separate and become daughter chromosomes.

FIGURE 9.4B Phases of meiosis II. Tan background = n.

No replication of DNA occurs during interkinesis, the period of time between meiosis I and meiosis II. When you think about it, the events of meiosis II are the same as those for mitosis, except the cells are haploid. At the beginning of prophase II, a spindle appears, while the nuclear envelope fragments and the nucleolus disappears. Duplicated chromosomes (one from each homologue) are present, and each attaches to the spindle. During metaphase II, the duplicated chromosomes line up at the spindle equator. During anaphase II, sister chromatids separate and move toward the poles. Each pole receives the same number and kinds of chromosomes.

3D Animation
Meiosis: Meiosis II

In telophase II, the spindle disappears as nuclear envelopes re-form.

"Life Cycles Are Varied" on page 162 discusses when meiosis occurs in the different life cycles of eukaryotes.

▶ **9.4 CHECK YOUR PROGRESS**
1. Identify two reasons why tetrad formation is so critical during prophase I.
2. Identify whether anaphase I or anaphase II has duplicated chromosomes at the poles and explain.

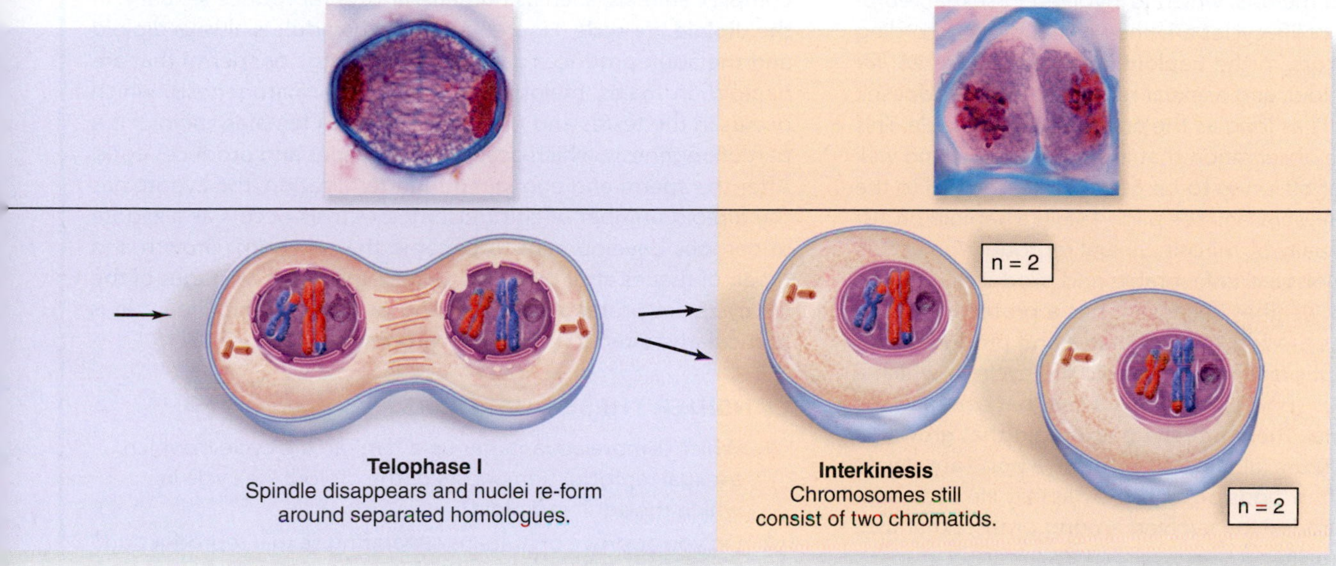

Telophase I
Spindle disappears and nuclei re-form around separated homologues.

Interkinesis
Chromosomes still consist of two chromatids.

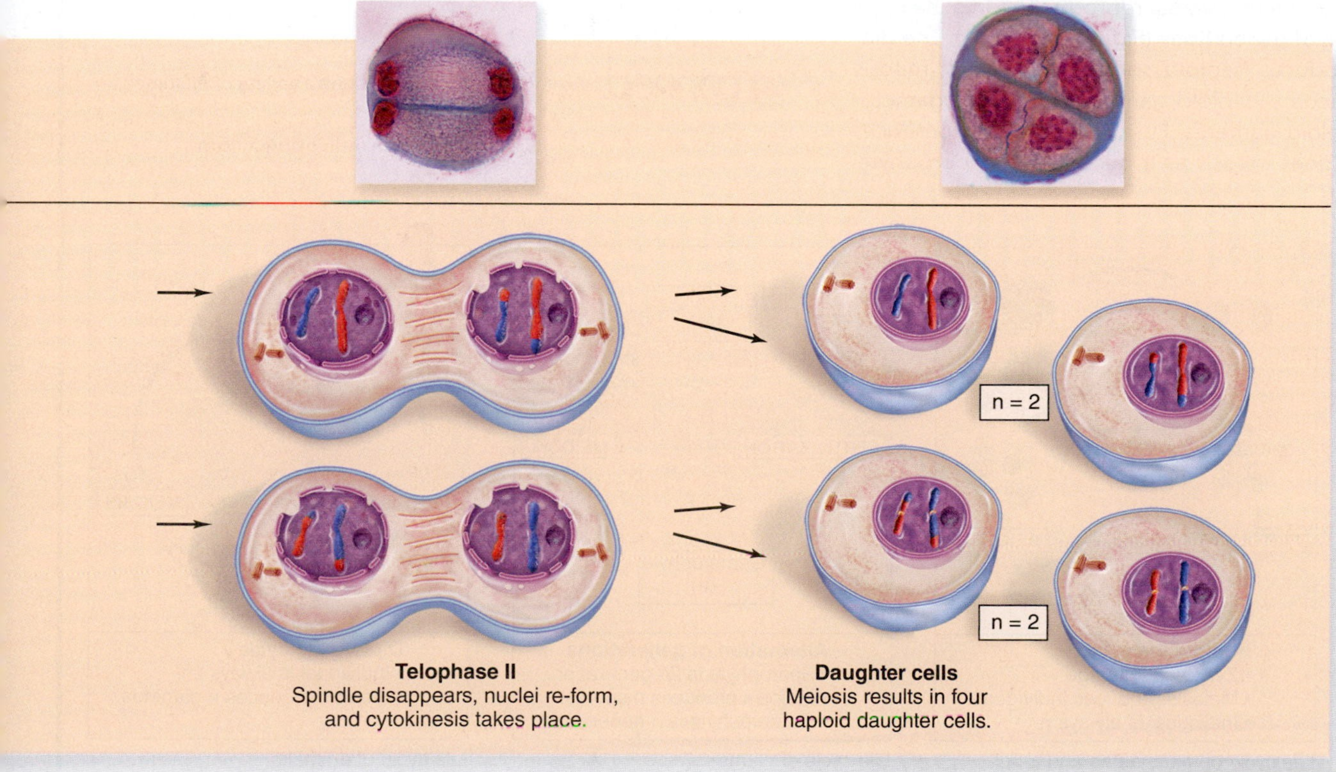

n = 2

n = 2

Telophase II
Spindle disappears, nuclei re-form, and cytokinesis takes place.

Daughter cells
Meiosis results in four haploid daughter cells.

n = 2

n = 2

9A Life Cycles Are Varied

A **life cycle** includes the reproductive events that occur from one generation to the next. Prokaryotes and a few protists reproduce asexually by means of binary fission. During asexual reproduction, there is only one parent, and the offspring are genetically identical to that parent. Asexual reproduction is advantageous when an organism is already adapted to the environment because it allows the organism to quickly colonize a suitable environment.

The life cycle of eukaryotes usually involves both mitosis, which is asexual, and meiosis, which is involved in sexual reproduction. Three types of life cycles are known that differ according to when meiosis occurs. In the **haploid life cycle** (Figure 9A, *far left*), the adult is haploid, and asexual reproduction that doesn't involve meiosis occurs as long as the environment is stable. This is consistent with the observation that if the parent is doing well it is advantageous for offspring to be genetically identical to the parent. The alga *Chlamydomonas* (see page 155) is a haploid protist that reproduces asexually by mitosis. Sexual reproduction occurs when growth conditions are unfavorable and produces a zygote that can survive bad weather because it has a protective covering. Meiosis occurs as the zygote germinates and produces haploid individuals. Fungi practice the haploid life cycle in which a sporangium produces spores. A spore is a cell that can give rise to a haploid individual. Therefore, the black mold that grows on bread and the green scum (algae) that floats on a pond are always haploid. Evolution often "adds on" and the haploid life cycle suggests that sexual reproduction evolved among protists that usually reproduced asexually.

Plants have various means of reproducing asexually. As an example, the "eye" of a sweet potato will produce an entire plant. But, in general, plants practice sexual reproduction in a life cycle known as **alternation of generations** (Figure 9A, *center*). The diploid sporophyte produces haploid spores by meiosis; mitosis occurs as spores become haploid gametophytes. The gametophyte produces haploid gametes. Fusion of gametes produces a zygote that undergoes mitosis as it becomes the sporophyte.

The majority of plants, including pines, corn, and pea plants, are diploid most of the time, and the haploid generation is short-lived. It's been suggested that this life cycle originally enabled the gametophyte to exploit one environment and the sporophyte to exploit another. We will see that in plants fully adapted to the land environment, the gametophyte generation is dependent on the sporophyte generation for its very existence.

Asexual reproduction does occur in the animal kingdom, but complex animals, such as humans, always reproduce sexually. In the **diploid life cycle** (Figure 9A, *right*). the adult is always diploid and the adult produces gametes (either eggs or sperm) that are haploid. In males, meiosis is a part of spermatogenesis, which occurs in the testes and produces sperm. In females, meiosis is a part of oogenesis, which occurs in the ovaries and produces eggs. After the sperm and egg join during fertilization, the zygote has the diploid number of chromosomes. Mitosis occurs as a zygote undergoes development to become the newborn. Growth and repair of tissues after birth also require mitosis. If all stages of the life cycle are protected from drying out, the diploid life cycle is particularly suited to the land environment.

CONSIDER THESE QUESTIONS

1. Which is more advantageous, a haploid life cycle in which asexual reproduction occurs or the diploid life cycle in which the adult is diploid?

2. Do you approve of humans separating sexual reproduction from the sex act as when they reproduce by in vitro fertilization? Why or why not?

 Explore the concepts through a variety of multimedia assets and question types.

www.mcgrawhillconnect.com

FIGURE 9A
Common life cycles. Blue background = 2n; tan background = n.

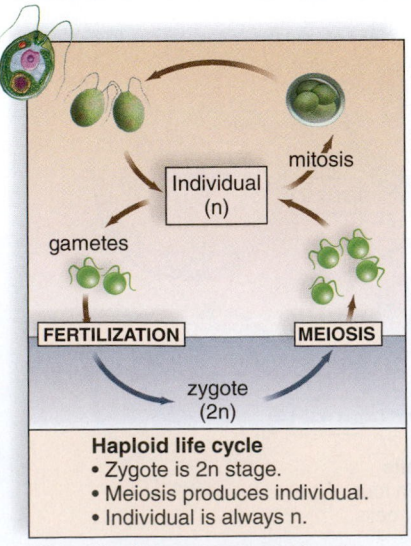

Haploid life cycle
• Zygote is 2n stage.
• Meiosis produces individual.
• Individual is always n.

Life cycle of many algae and fungi

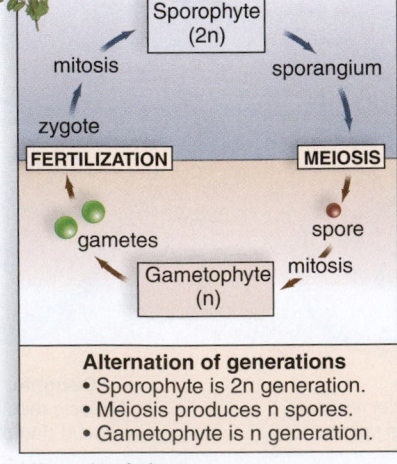

Alternation of generations
• Sporophyte is 2n generation.
• Meiosis produces n spores.
• Gametophyte is n generation.

Life cycle of plants

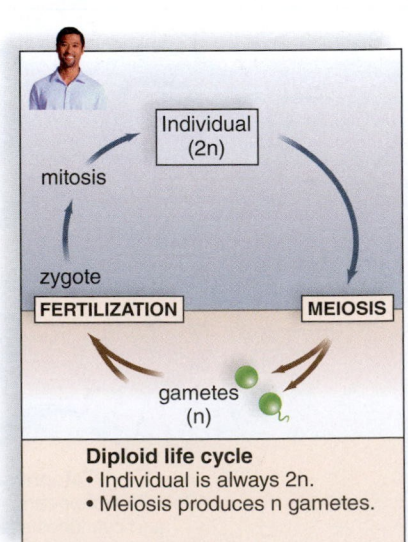

Diploid life cycle
• Individual is always 2n.
• Meiosis produces n gametes.

Life cycle of animals

9.5 Meiosis can be compared to mitosis

LEARNING OUTCOME

When you complete this section, you should be able to

1. Contrast the outcome and events of meiosis I and meiosis II with the events of mitosis.

Figure 9.5 compares meiosis to mitosis. Notice the following:

- Meiosis requires two nuclear divisions, but mitosis requires only one nuclear division.
- Meiosis results in four daughter cells. Mitosis results in two daughter cells.
- Following meiosis, the four daughter cells are haploid, meaning that they have half the chromosome number of the parent cell. Following mitosis, the daughter cells are diploid, having the same chromosome number as the parent cell.
- Following meiosis, the daughter cells are genetically dissimilar to each other and to the parent cell. Following mitosis, the daughter cells are genetically identical to each other and to the parent cell.

These differences between meiosis and mitosis are due to certain events:

- During meiosis I, tetrads form, and crossing-over occurs during prophase I. These events do not occur during mitosis.
- During metaphase I of meiosis, tetrads are at the equator. The homologues align at the spindle equator independently. During metaphase in mitosis, duplicated chromosomes align at the spindle equator.
- During anaphase I of meiosis, homologues separate, and duplicated chromosomes (with centromeres intact) move to opposite poles. During anaphase of mitosis, sister chromatids separate, becoming daughter chromosomes that move to opposite poles.

The events of meiosis II are just like those of mitosis except that in meiosis II, the daughter cells have the haploid number of chromosomes. Could abnormal meiosis cause the inheritance of a chromosome anomaly? Section 9.6 shows how this is possible.

Animation
Comparison of Meiosis and Mitosis

▶ **9.5 CHECK YOUR PROGRESS**

1. Identify the event that allows mitosis to maintain the chromosome number and the event that allows meiosis to reduce the chromosome number.

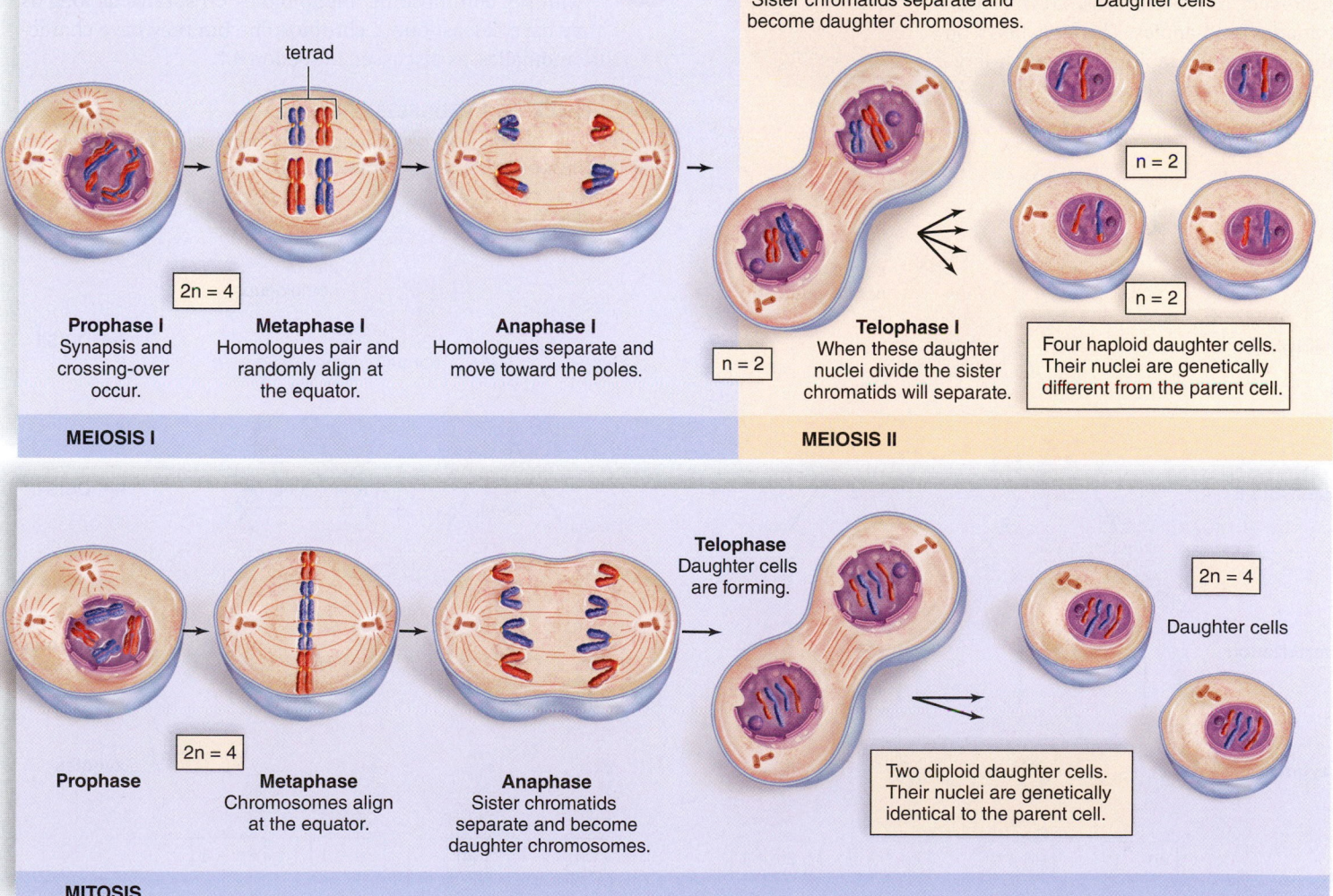

Sister chromatids separate and become daughter chromosomes.

Daughter cells

tetrad

2n = 4

n = 2

n = 2

n = 2

Prophase I
Synapsis and crossing-over occur.

Metaphase I
Homologues pair and randomly align at the equator.

Anaphase I
Homologues separate and move toward the poles.

Telophase I
When these daughter nuclei divide the sister chromatids will separate.

Four haploid daughter cells. Their nuclei are genetically different from the parent cell.

MEIOSIS I

MEIOSIS II

Telophase
Daughter cells are forming.

2n = 4

Daughter cells

2n = 4

Prophase

Metaphase
Chromosomes align at the equator.

Anaphase
Sister chromatids separate and become daughter chromosomes.

Two diploid daughter cells. Their nuclei are genetically identical to the parent cell.

MITOSIS

FIGURE 9.5 Meiosis (top) compared to mitosis (bottom). Blue background = 2n; tan background = n.

Chromosome Number Anomalies

When meiosis does not occur normally, a chromosome number anomaly can be inherited. The anomaly is most likely due to nondisjunction, which can occur during either meiosis I or meiosis II.

9.6 Nondisjunction causes chromosome number anomalies

LEARNING OUTCOMES

When you complete this section, you should be able to

1. Distinguish between a polyploid and an aneuploid.
2. Explain how nondisjunction may bring about an abnormal chromosome number in a gamete.

Changes in chromosome number include polyploidy and aneuploidy. When a eukaryote has three or more complete sets of chromosomes, it is called a **polyploid.** More specifically, triploids (3n) have three of each kind of chromosome, tetraploids (4n) have four sets, pentaploids (5n) have five sets, and so on. Although polyploidy is not often seen in animals, it is a major evolutionary mechanism in plants, including many of our most important crops—wheat, corn, cotton, and sugarcane, as well as fruits such as watermelons, strawberries, bananas, and apples. The strawberry on the left is an octaploid and much larger than the diploid one on the right. Also, many attractive flowers, including chrysanthemums and daylilies, are polyploids.

The usual cause of chromosome number anomalies in animals is nondisjunction. **Nondisjunction** occurs during meiosis I when homologues fail to separate and both homologues go into the same gamete (Fig. 9.6A) or during meiosis II when the sister chromatids fail to separate and both daughter chromosomes go into the same gamete (Fig. 9.6B). If one of these gametes happens to fuse with a normal gamete and development continues, an individual with a chromosome anomaly known as **aneuploidy** results. An extra chromosome creates a condition known as a **trisomy** because one of the chromosomes is present in triplet. The absence of a chromosome is called a **monosomy** because one chromosome is present only once. The absence of an autosomal chromosome is lethal in human embryos, and only one autosomal trisomy (trisomy number 21, called Down syndrome) survives for any length of time. Those with sex chromosome aneuploidies do survive as long as they have at least one X chromosome but they have characteristic anomalies, as discussed in section 9.7.

> **9.6 CHECK YOUR PROGRESS**
> 1. Describe two ways nondisjunction can result in an egg with one too many chromosomes.

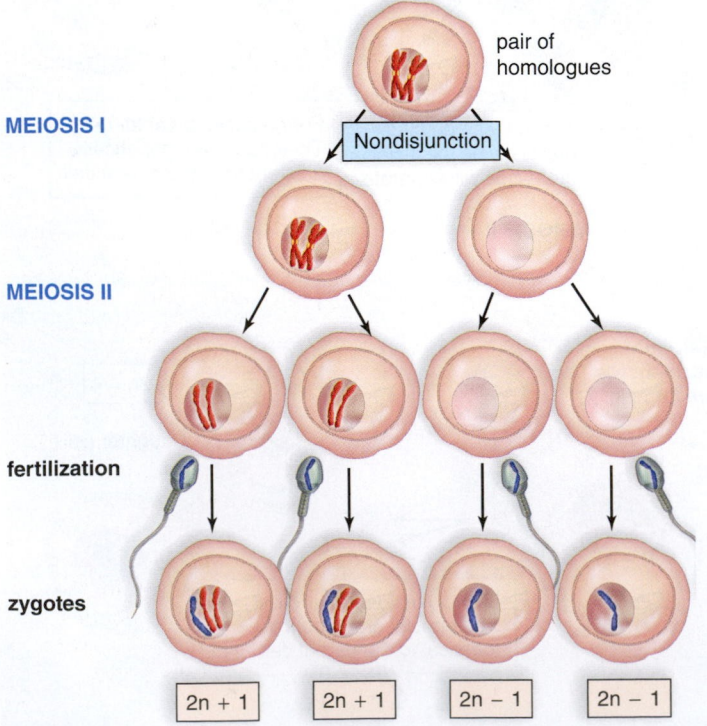

FIGURE 9.6A Nondisjunction of chromosomes during meiosis I of oogenesis, followed by fertilization with normal sperm.

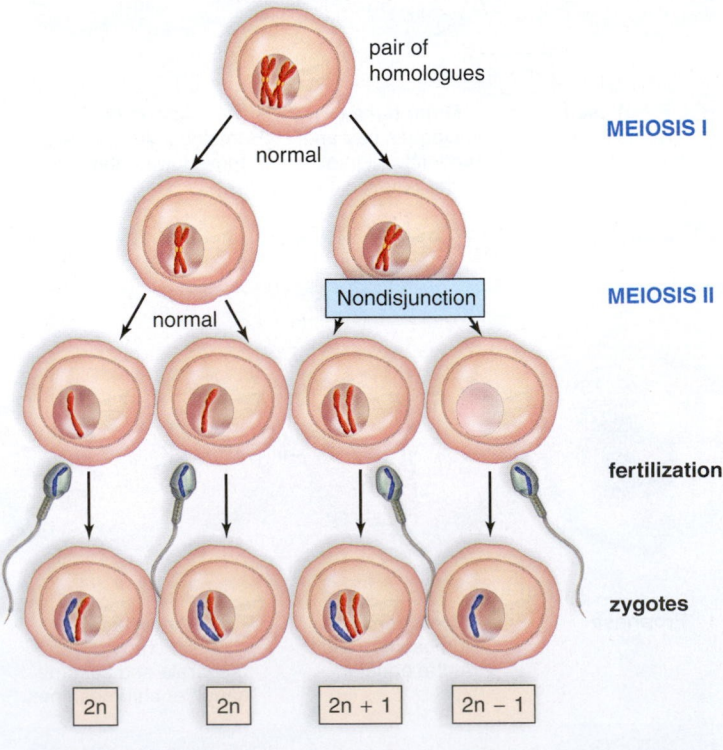

FIGURE 9.6B Nondisjunction of chromosomes during meiosis II of oogenesis, followed by fertilization with normal sperm.

9.7 Chromosome number anomalies cause syndromes

LEARNING OUTCOME

When you complete this section, you should be able to

1. Describe Down syndrome, Turner syndrome, and Klinefelter syndrome.

When an individual inherits an aneuploidy, certain characteristics commonly appear that together are called a **syndrome.**

Trisomy 21 (Down Syndrome) The most common autosomal trisomy among humans is trisomy 21, also called **Down syndrome.** This syndrome is easily recognized by these characteristics: short stature, eyelid fold, flat face, stubby fingers, wide gap between the first and second toes, large, fissured tongue, round head, distinctive palm crease, heart problems, and intellectual disability, which can sometimes be severe. In addition, these individuals have an increased chance of developing Alzheimer disease later in life.

Over 90% of individuals with Down syndrome have three copies of chromosome 21. Usually, two copies are contributed by the egg; however, in 8% of cases, the sperm contributes the extra chromosome. The chances of a woman having a child with Down syndrome increase rapidly with age. In women aged 20–30, 1 in 1,400 births have Down syndrome, while in women 30–35, about 1 in 750 births have Down syndrome. It is thought the longer oocytes are dormant in the ovaries, the greater the chances of a nondisjunction event.

HOW BIOLOGY IMPACTS OUR LIVES

Application

9B Hope for Down Syndrome

Chris Burke (Fig. 9Ba) was born with Down syndrome, and his parents were advised to put him in an institution. But Chris's parents didn't do that. They gave him the same loving care and attention they gave their other children, and it paid off. Chris is remarkably talented. He is a playwright, actor, and musician. he starred in *Life Goes On* (1989–1993), a TV series written just for him, and he is sometimes asked to be a guest star in other TV shows. His love of music and collaboration with other musicians have led to the release of several albums—like Chris, the songs are uplifting and inspirational. You can read more about this remarkable individual in his autobiography, *A Special Kind of Hero.*

The genes that cause Down syndrome are located on the bottom third of chromosome 21 (Fig. 9Bb). Extensive investigative work has been directed toward discovering the specific genes responsible for the characteristics of the syndrome. Thus far, investigators have discovered several genes that may account for various conditions seen in persons with Down syndrome. For example, they have located genes most likely responsible for the increased tendency toward leukemia, cataracts, accelerated rate of aging, and intellectual disability. The gene for intellectual disability, dubbed the *Gart* gene, causes an increased level of purines in the blood, a finding associated with mental retardation. One day, it may be possible to control the expression of the *Gart* gene even before birth so that at least this symptom of Down syndrome does not appear.

CONSIDER THESE QUESTIONS

1. Do you believe that individuals with a trisomy should reproduce? Why or why not?
2. If you had a trisomy, would you want to receive genetic therapy in an attempt to cure it? Why or why not?

Explore the concepts through a variety of multimedia assets and question types.

www.mcgrawhillconnect.com

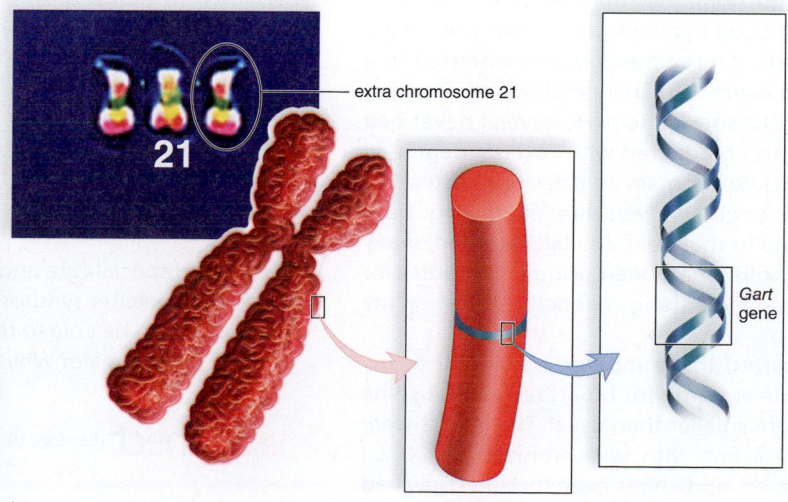

extra chromosome 21

21

Gart gene

FIGURE 9B Down syndrome. a. Chris Burke was born with Down syndrome. **b.** The Down syndrome karyotype shows an extra chromosome 21. An extra copy of the *Gart* gene, which leads to a high level of purines in the blood, may account for the intellectual disability seen in persons with Down syndrome.

Although an older woman is more likely to have a Down syndrome child, most babies with Down syndrome are born to women younger than age 40 because this is the age group having the most babies. A karyotype of the individual's chromosomes can detect a Down syndrome child. However, young women are not routinely encouraged to undergo the procedures necessary to get a sample of fetal cells because the risk of complications is greater than the risk of having a Down syndrome child. Fortunately, a test based on substances in maternal blood can help identify fetuses who may need to be karyotyped.

Sex Chromosome Number Anomalies Two sex chromosome anomalies are of special interest. In **Turner syndrome,** females have only a single X chromosome. They tend to be short, with a broad chest and widely spaced nipples. These individuals also have a low posterior hairline and neck webbing. Their ovaries, oviducts, and uterus are very small and underdeveloped. Turner females do not undergo puberty or menstruate, and their breasts do not develop. However, some have given birth following in vitro fertilization using donor eggs.

A male with **Klinefelter syndrome** has two or more X chromosomes in addition to a Y chromosome. The extra X chromosomes become inactivated. In Klinefelter males, the testes and prostate gland are underdeveloped, and facial hair is lacking. There may be some breast development. Affected individuals have large hands and feet and very long arms and legs. They are usually slow to learn but not severely disabled intellectually, unless they inherit more than two X chromosomes. No matter how many X chromosomes are present, an individual with a Y chromosome is a male.

▶ **9.7 CHECK YOUR PROGRESS**

1. Explain why Turner syndrome and Klinefelter syndrome have sexual characteristic anomalies but Down syndrome does not.

HOW BIOLOGY IMPACTS OUR LIVES   *Application*

9C Living with Klinefelter Syndrome

In 1996, at the age of 25, I was diagnosed with Klinefelter Syndrome (KS). Being diagnosed has changed my life for the better.

I was a happy baby, but when I was still very young, my parents began to believe that there was something wrong with me. I knew something was different about me, too, as early on as 5 years old. I was very shy and had trouble making friends. One minute I'd be well behaved, and the next I'd be picking fights and flying into a rage. Many psychologists, therapists, and doctors tested me because of school and social problems and severe mood changes. Their only diagnosis was "learning disabilities" in such areas as reading comprehension, abstract thinking, word retrieval, and auditory processing. No one could figure out what the real problem was, and I hated the tutoring sessions I had. In the seventh grade, a psychologist told me that I was stupid and lazy, I would probably live at home for the rest of my life, and I would never amount to anything. For the next 5 years, he was basically right, and I barely graduated from high school.

I believe, though, that I have succeeded because I was told that I would fail. I quit the tutoring sessions when I enrolled at a community college; I decided I could figure things out on my own. I received an associate degree there, then transferred to a small liberal arts college. I never told anyone about my learning disabilities and never sought special help. However, I never had a semester below a 3.0, and I graduated with two B.S. degrees. I was accepted into a graduate program but decided instead to accept a job as a software engineer even though I did not have an educational background in that field. As I later learned, many KS'ers excel in computer skills. I had been using a computer for many years and had learned everything I needed to know on my own, through trial and error.

Around the time I started the computer job, I went to my physician for a physical. He sent me for blood tests because he noticed that my testes were smaller than usual. The results were conclusive: Klinefelter syndrome with sex chromosomes XXY. I initially felt denial, depression, and anger, even though I now had an explanation for many of the problems I had experienced all my life. But then I decided to learn as much as I could about the condition and treatments available. I now give myself a testosterone injection once every 2 weeks, and it has made me a different person, with improved learning abilities and stronger thought processes in addition to a more outgoing personality.

I found, though, that the best possible path I could take was to help others live with the condition. I attended my first support group meeting 4 months after I was diagnosed. By spring 1997, I had developed an interest in KS that was more than just a part-time hobby. I wanted to be able to work with this condition and help people forever. I have been very involved in KS conferences and have helped to start support groups in the United States, Spain, and Australia.

Since my diagnosis, it has been my dream to have a son with KS, although when I was diagnosed, I found out it was unlikely that I could have biological children. Through my work with KS, I had the opportunity to meet my fiancee Chris. She has two wonderful children: a daughter, and a son who has the same condition that I do. There are a lot of similarities between my stepson and me, and I am happy I will be able to help him get the head start in coping with KS that I never had. I also look forward to many more years of helping other people seek diagnosis and live a good life with Klinefelter syndrome.

Stefan Schwartz
stefan13@mail.ptd.net

CONSIDER THESE QUESTIONS

1. Should special care and consideration be given to a child with Klinefelter syndrome? Why or why not?
2. Would you be able to treat a male with Klinefelter syndrome as a normal male? Why or why not?

 Explore the concepts through a variety of multimedia assets and question types.

www.mcgrawhillconnect.com

THE CHAPTER IN REVIEW

CONNECTING THE CONCEPTS

Meiosis is the type of nuclear division that reduces the chromosome number in plants and animals. It occurs in the life cycle of nearly all eukaryotes. In animals it is a part of gamete production that precedes fertilization. Fertilization in both plants and animals restores the full or diploid number of chromosomes.

While reducing the chromosome number, meiosis also introduces variability by the twin means of crossing-over and independent assortment of chromosomes. The genetic makeup of offspring in animals is different from that in both parents due to meiosis.

The number of chromosomes in succeeding generations would constantly double without the process of meiosis to reduce the number. Without the introduction of variability, adaptation to the present or a future environment would not occur. For natural selection to work, diversity must be present. It is a remarkable revelation to know that the diversity of life that accompanies the process of evolution would not be as rich without meiosis. The contribution of meiosis to genetic diversity in each new generation will also be apparent when we study Mendelian genetics in the next chapter.

ANALYZE AND EVALUATE

1. Synapsis during meiosis is necessary to crossing-over and independent assortment of chromosomes. Explain.
2. What other process, aside from meiosis, results in increased variation among offspring?
3. Create a scenario by which meiosis could have evolved in cells accustomed to undergoing mitosis.

SUMMARIZE

The Basics of Meiosis

9.1 Chromosomes come in pairs

- A **karyotype** shows that eukaryotes have pairs of chromosomes called **homologues**. Humans have 22 pairs of **autosomes** and one pair of **sex chromosomes**. A **Y chromosome** is shorter than an **X chromosome**. Males are XY and females are XX.
- Following DNA replication, each chromosome has two **sister chromatids** held together at a **centromere**. Homologues have **alleles** for the same trait—e.g., type of hairline. Sister chromatids have exact alleles—e.g., widow's peak.

9.2 Homologues separate during meiosis

- **Meiosis** is reduction division. Each of four daughter cells has only one of each kind of chromosome. Meiosis requires one DNA replication and two cell divisions, called **meiosis I** and **meiosis II.** The period of time between meiosis I and meiosis II is called **interkinesis.**
- Homologues come together during **synapsis** and then separate during meiosis I; sister chromatids separate during meiosis II. The daughter cells are haploid. In animals, the daughter cells become **gametes** (egg and sperm) with the haploid number of chromosomes. The diploid number is restored with **fertilization.**

The Phases of Meiosis

9.3 Events of meiosis I increase genetic variation among the gametes

- During meiosis I, synapsis (pairing of homologues to form a **tetrad**) and **crossing-over** (exchange of genetic material) between nonsister chromatids occurs. Crossing-over recombines genetic information and increases the variability of genetic inheritance on the chromosomes. The daughter cells contain all possible combinations of chromosomes because of **independent assortment.** Independent assortment occurs because the homologues align during meiosis I

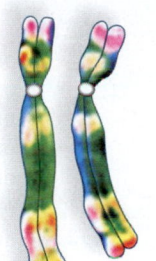

with either homologue facing either pole. Fertilization brings together genetically different gametes that fuse to form a **zygote.**

9.4 Both meiosis I and meiosis II have four phases

- Meiosis I: **prophase I**—homologues pair and crossing-over occurs; **metaphase I**—homologue pairs align at equator independently; **anaphase I**—homologues separate; **telophase I**—daughter cells are haploid.
- Interkinesis is the time period between meiosis I and meiosis II. No DNA replication occurs. Meiosis II: During stages designated by the Roman numeral II, the chromatids of duplicated chromosomes from meiosis I separate, producing a total of four daughter cells for meiosis.

9.5 Meiosis can be compared to mitosis

- See Figure 9.5 and note that homologues only pair during metaphase I of meiosis and that four haploid daughter cells result from meiosis but not mitosis.

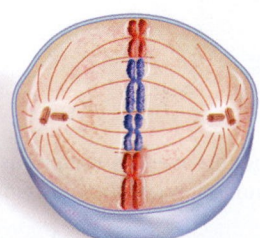 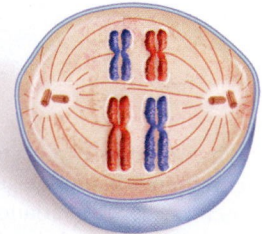

Metaphase during mitosis   Metaphase I during meiosis

Chromosome Number Anomalies

9.6 Nondisjunction causes chromosome number anomalies

- A **polyploid** has a multiple of the haploid number of chromosomes; an **aneuploid** is a **monosomy** (2n–1) or a **trisomy** (2n+1).

- Aneuploidy is due to **nondisjunction** when homologues do not separate during meiosis I or when chromatids do not separate during meiosis II.

9.7   Chromosome number anomalies cause syndromes

- A **syndrome** is due to the inheritance of a set of physical characteristics that can be overcome with proper medical care and support. **Down syndrome** is an autosomal trisomy. **Turner syndrome** and **Klinefelter syndrome** result from sex chromosome anomalies.

TEST YOURSELF

The Basics of Meiosis

1. Which of these statements is incorrect? Following DNA replication in a eukaryotic cell,
   a. homologues of chromosomes can be seen.
   b. each chromosome has two sister chromatids.
   c. one chromatid came from the father and one came from the mother.
   d. each sister chromatid carries the same genes.

2. Which is a correct contrast between autosomes and sex chromosomes in humans?
   a. 22 pairs—one pair
   b. control gender—control enzymes
   c. are always duplicated—are always single
   d. are always visible—are never visible

3. Which is a best description of chromosomes in the four daughter cells following meiosis? Each daughter cell is
   a. diploid and genetically the same as the original mother cell.
   b. haploid but each chromosome is duplicated.
   c. diploid but each chromosome has one chromatid.
   d. haploid and each chromosome consists of one chromatid.

4. **THINKING CONCEPTUALLY** Consider this mother cell, and explain why each daughter cell will have only one short chromosome of either color with one long chromosome of either color.

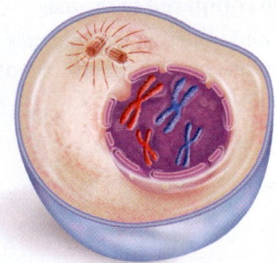

5. Why would you expect to see crossing-over between nonsister chromatids of homologues and not sister chromatids in order to increase genetic variation?
   a. Both sister and nonsister chromatids carry the same alleles.
   b. Only sister chromatids would carry alternate alleles.
   c. Only nonsister chromatids would carry alternate alleles.
   d. Homologues carry the same alleles.

6. **THINKING CONCEPTUALLY** If cousins reproduce, meiosis may not be able to increase the genetic variability of the next generation. Why not?

The Phases of Meiosis

7. At the equator during metaphase I of meiosis, there are
   a. single chromosomes.
   b. unpaired duplicate chromosomes.
   c. homologues.
   d. always 23 chromosomes.

8. At the equator during metaphase II of meiosis, there are
   a. single chromosomes.
   b. unpaired duplicated chromosomes.
   c. homologue pairs.
   d. always 23 chromosomes.

9. During which phase of meiosis do homologues separate?
   a. prophase II   d. anaphase I
   b. telophase I   e. anaphase II
   c. metaphase I

10. Which of these helps to ensure that genetic diversity will be maintained?
    a. independent alignment during metaphase I
    b. crossing-over during prophase I
    c. fusion of sperm and egg nuclei during fertilization
    d. All of these are correct.

For questions 11–15, match the statements to the items in the key. Answers may be used more than once.

KEY:
   a. prophase I of meiosis
   b. metaphase I of meiosis
   c. metaphase II of meiosis
   d. prophase of mitosis
   e. metaphase of mitosis

11. Tetrads form as synapsis occurs.
12. Diploid number of chromosomes align individually at the equator.
13. Haploid number of chromosomes are at the equator.
14. Crossing-over occurs.
15. Homologue pairs are at the equator.

16. Which is an incorrect comparison between meiosis and mitosis?
    a. four daughter cells—two daughter cells
    b. crossing-over occurs—crossing-over does not occur
    c. homologues separate—chromatids separate
    d. daughter cells are diploid—daughter cells are haploid

For questions 17–20, match the statements to the items in the key. Answers may be used more than once, and more than one answer may be used.

KEY:
   a. mitosis
   b. meiosis I
   c. meiosis II
   d. Both meiosis I and meiosis II are correct.
   e. All of these are correct.

17. A parent cell with ten duplicated chromosomes will produce daughter cells with five duplicated chromosomes each.
18. A parent cell with five duplicated chromosomes will produce daughter cells with five chromosomes consisting of one chromatid each.
19. A parent cell with ten duplicated chromosomes will produce daughter cells with ten chromosomes consisting of one chromatid each.

20. Involved in growth and repair of tissues.
21. Knowing that all sexual life cycles contain mitosis, meiosis, and fertilization, fill in all the horizontal boxes. Fill in the square boxes with the notations n or 2n.

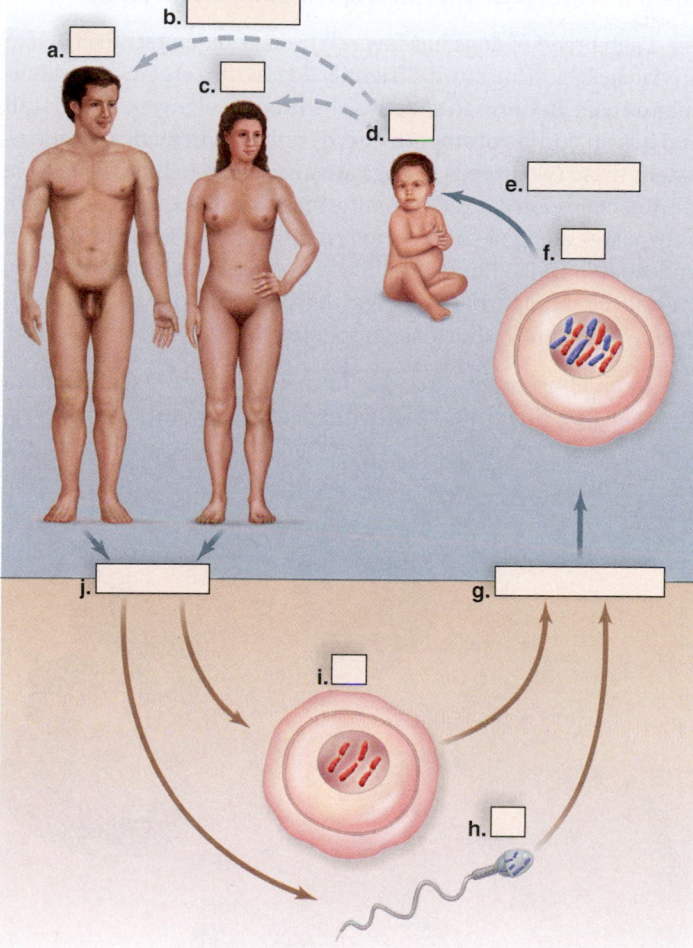

22. A male with underdeveloped testes and some breast development most likely has
    a. Down syndrome.   c. Turner syndrome.
    b. Jacobs syndrome.   d. Klinefelter syndrome.
23. Turner syndrome (X0) can only result if nondisjunction occurred during
    a. mitosis.   c. meiosis II.
    b. meiosis I.   d. Both b and c are correct.
24. Which way(s) can an egg receive the same two chromosomes during meiosis?
    a. Both homologues go into the same daughter cell following meiosis I
    b. Both chromatids go into the same daughter cell following meiosis II
    c. Both chromatids go into the same daughter cell following meiosis I
    d. Both a and b are correct.

GET INVOLVED

1. Genetic testing shows that Mary has only 46 chromosomes, but both members of one homologue pair came from her father. In which parent(s) did nondisjunction occur? Explain.
2. Criticize the hypothesis that it would be possible to clone an individual by using an egg and a sperm with the exact genetic makeup as those that produced the individual.
3. The 3D animation "Meiosis" provides a visual exploration of the events during meiosis.

MEDIA STUDY TOOLS

mhhe.com/maderconcepts3

Enhance your study of this chapter with interactive study tools, practice tests, and engaging animations. Also, ask your instructor about the resources available through ConnectPlus, which includes LearnSmart, a personalized adaptive learning program, and a media-rich eBook.

10

Patterns of Genetic Inheritance

BEFORE YOU BEGIN

Take a few minutes to recall

The role of DNA and proteins in cells (section 3.9)

How homologues carry different alleles (section 9.1)

The independent separation of homologues during meiosis (section 9.3)

Troubles with Dog Breeding

When dogs—or people—reproduce, they pass on their genes, units of heredity that determine what the offspring will be like. Dog breeders rely on this common knowledge when they choose close relatives with like characteristics to reproduce with one another. Only in this way has it been possible to produce over 150 different dog breeds from an original common ancestor.

Each breed of dogs has favored traits. Golden retrievers were bred to be beautiful, sturdy, friendly dogs with a cream- to golden-colored coat. In German shepherds, the body is longer than it is tall and has an outline of smooth curves rather than angles. Unfortunately, these two breeds of dogs are also prone to hip dysplasia, a painful condition caused by malformed hip joints. Inbreeding not only passes along desirable traits, it also is more likely to pass on undesirable traits. English bulldogs and pugs look cute, but they may develop breathing and digestive problems due to a protruding lower jaw and a shortened upper jaw.

In an effort to improve dog breeding today, reformers point out that dog breeding became especially organized and systematic in the nineteenth century and by now some pedigree

English bulldog

dog breeds are so inbred they might cease to exist. One cause of genetic diversity loss is the use of "super-sires," dogs that have won prizes at dog shows and thereafter are used through artificial insemination to produce many litters. The Kennel Club of Great Britain, which sets the standards for awarding winners of dog shows, has put 12 breeds on what it calls its "worry list." They now recognize that inbreeding limits the gene pool so that a breed loses vigor and is unable to maintain itself. Now it is up to responsible breeders to sometimes mate their dogs to unrelated dogs in order to introduce new genes that will help maintain the overall health of the breed.

As in human genetics, tracing the causes of genetic disorders in dogs is complicated, but modern genetics can help. Breeders can keep careful records of matings and the results of those matings. From these records, they should be able to determine the pattern of a disorder's inheritance and which dogs should not be used for breeding. It is also possible today to screen for many genetic diseases, so that dogs with the same but hidden (recessive) genetic faults are not mated to each other. In other words proper dog breeding can avoid producing dogs with physical deformities. Modernization can also include the successful use of gene therapy in collies and Briard sheepdogs to cure blindness, another condition resulting from continual inbreeding. Researchers injected a harmless virus carrying the corrective gene beneath the retina and waited several months. When tested, the dogs could see through the eye that was treated, but not through the eye that was not treated. The researchers hope their work will be a step toward curing blindness in humans one day.

Dog owners can also help. Researchers who have studied hip dysplasia in dogs tell us this condition is not a birth defect.

Collie

The dogs are born with what appear to be normal hips and then develop the disease later. In addition, some dogs with the genetic tendency do not develop the condition, while the degree of hip dysplasia in others can vary. It appears that multiple genetic factors, plus environmental factors, are involved in determining the degree of hip dysplasia. This is also true of many human disorders—both genes and the environment seem to play a role. Researchers have identified the possible contributing factors for hip dysplasia. A test group of Labrador retrievers, who were fed 25% less than normal, showed less hip dysplasia than a control group allowed to eat as much as they wanted. The researchers concluded that rapid weight gain can contribute to the development of hip dysplasia.

This chapter begins our study of inheritance by taking a look at the work of Gregor Mendel, who is often called the father of genetics. After discussing how genes function in peas, we will turn our attention to humans. The same laws of heredity function in plants, dogs, humans, and every other organism. By understanding these laws we can prevent and cure genetic disorders in dogs and humans.

German shepherds are subject to hip dysplasia.

Dog shows influence breeding.

Mendel's Laws

The experiments performed by Gregor Mendel with garden peas refuted the blending model of inheritance prevalent at the time. In contrast to the blending model, Mendel's work showed that inheritance is particulate and, therefore, traits such as tall or short height always recur in future generations.

10.1 Mendel developed a particulate model of inheritance

LEARNING OUTCOMES

When you complete this section, you should be able to

1. Contrast the blending model and the particulate model of inheritance.
2. Identify pea plant characteristics that make them good experimental subjects.

Like begets like—zebras always produce zebras, never camels; pumpkins always produce seeds for pumpkins, never watermelons. It is apparent to anyone who observes such phenomena that parents pass hereditary information to their offspring. However, an offspring can be markedly different from either parent. For example, black-coated mice occasionally produce white-coated mice. The science of genetics founded by Gregor Mendel, an Austrian monk (Fig. 10.1A), provides explanations about not only the stability of inheritance, but also the variations observed between generations and among organisms.

Various hypotheses about heredity had been proposed before Mendel began his experiments in the 1860s. In particular, investigators had been trying to support a blending concept of inheritance. Most plant and animal breeders acknowledged that both sexes contribute equally to a new individual, and they felt that parents of contrasting appearance always produce offspring of intermediate appearance. According to this blending concept, a cross between plants with red flowers and plants with white flowers would yield only plants with pink flowers. When red and white flowers reappeared in future generations, the breeders mistakenly attributed this to an instability in the genetic material.

The blending model of inheritance had offered little help to Charles Darwin, the father of evolution. If populations contained only intermediate individuals and normally lacked variations, how could diverse forms evolve? However, Mendel's model of inheritance does account for the presence of variations among the members of a population, generation after generation.

FIGURE 10.1A Gregor Mendel examining a pea plant.

Although Darwin was a contemporary of Mendel, Darwin never learned of Mendel's work and it went unrecognized until 1900. Therefore, Darwin was never able to make use of Mendel's research to support his theory of evolution, and his treatise on natural selection lacked a strong genetic basis.

Design of Mendel's Experiments Mendel had a background suitable to his task. Previously, he had studied science and mathematics at the University of Vienna, and at the time of his genetic research, he was a substitute natural science teacher at a local high school. Aside from theoretical knowledge, Mendel knew how to cultivate plants. Most likely, his knowledge of mathematics prompted him to use a statistical basis for his breeding experiments. He prepared for his experiments carefully and conducted preliminary studies with various animals and plants. He then chose to work with the garden pea, *Pisum sativum.*

The garden pea was a good choice. The plants are easy to cultivate, have a short generation time, and produce many offspring. A pea plant normally self-pollinates because the reproductive organs in the flower are completely enclosed by petals (Fig. 10.1B), **1** As in all flowering plants, the reproductive organs in peas are the stamen and the carpel. A stamen produces sperm-bearing pollen in the anther, and the carpel produces egg-bearing ovules in the ovary. When Mendel wanted the plants to self-fertilize, he covered the flowers with a bag to ensure that only the pollen of that flower would reach the carpel of that flower. Even though pea plants normally self-fertilize, they can be cross-pollinated by an experimenter who manually transfers pollen from an anther to the carpel. **2** Mendel prevented self-fertilization by cutting away the anthers before they produced any pollen. **3** Then he dusted that flower's carpel with pollen from another plant. **4** Afterwards, the carpel developed into a pod containing peas. In the cross illustrated here, pollen from a plant that normally produces yellow peas was used to fertilize the eggs of a plant that normally produces green peas. These plants produced only yellow peas. In the next, or F_2, generation, a few peas were green again.

Many varieties of pea plants were available, and Mendel chose 22 of them for his experiments. When these varieties self-fertilized, they were *true-breeding,* meaning that the offspring were like the parent plants and like each other. In contrast to his predecessors, Mendel studied the inheritance of relatively simple, clear-cut, and easily detected traits, such as seed shape, seed color, and flower color, and he observed no intermediate characteristics among the offspring (Fig. 10.1C).

As Mendel followed the inheritance of individual traits, he kept careful records. Then he used his understanding of the mathematical laws of probability to interpret his results and to arrive at a theory that has been supported by innumerable experiments since. It is called a *particulate model of inheritance* because it is based on the existence of minute

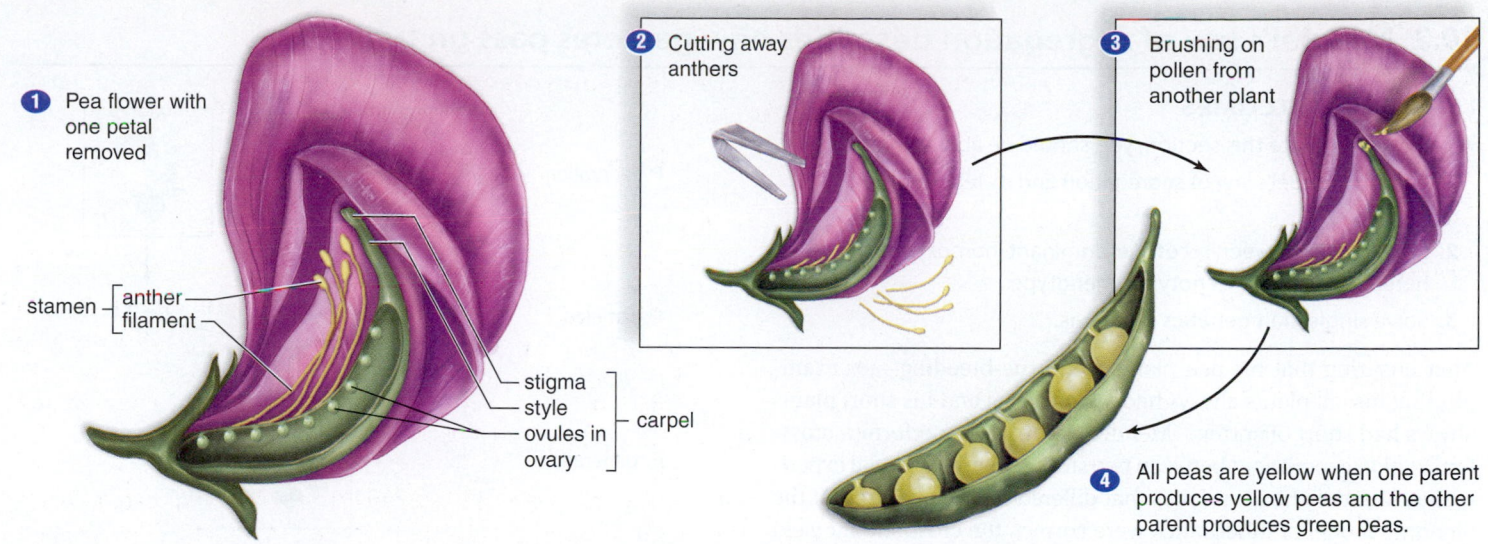

① **Pea flower with one petal removed**

stamen { anther, filament }

stigma
style
ovules in ovary } carpel

② **Cutting away anthers**

③ **Brushing on pollen from another plant**

④ All peas are yellow when one parent produces yellow peas and the other parent produces green peas.

FIGURE 10.1B Garden pea anatomy and the cross-pollination procedure Mendel used.

Trait	Characteristics				F₂ Results		
	Dominant		Recessive		Dominant	Recessive	Ratio
Stem length	Tall		Short		787	277	2.84:1
Pod shape	Inflated		Constricted		882	299	2.95:1
Seed shape	Round		Wrinkled		5,474	1,850	2.96:1
Seed color	Yellow		Green		6,022	2,001	3.01:1
Flower position	Axial		Terminal		651	207	3.14:1
Flower color	Purple		White		705	224	3.15:1
Pod color	Green		Yellow		428	152	2.82:1
				Totals:	14,949	5,010	2.98:1

FIGURE 10.1C Garden pea traits and crosses studied by Mendel. His F₂ results allowed him to deduce the first of his laws of heredity.

particles, or hereditary units, that we now call genes. Inheritance involves the reshuffling of the same genes from generation to generation.

Mendel clearly stated conclusions that are now considered laws of heredity which are stated in sections 10.2 and 10.3.

▶ **10.1 CHECK YOUR PROGRESS**

1. Contrast the expected results using the blending model and the particulate model of inheritance when a tall pea plant is crossed with a short pea plant.

CHAPTER 10 Patterns of Genetic Inheritance   **173**

LEARNING OUTCOMES

When you complete this section, you should be able to

1. Explain Mendel's law of segregation and its relationship to meiosis.
2. Distinguish between recessive/dominant, homozygous/heterozygous, and genotype/phenotype.
3. Solve single-trait genetics problems.

After ensuring that his pea plants were true-breeding—for example, that his tall plants always had tall offspring and his short plants always had short offspring—Mendel was ready to perform a cross-fertilization experiment between two strains. For these initial experiments, Mendel chose varieties that differed in only one trait. If the blending model of inheritance were correct, the cross should yield offspring with an intermediate appearance compared to the parents. For example, the offspring of a cross between a tall plant and a short plant should be intermediate in height.

Mendel called the original parents the **P generation** and the first batch of offspring the **F$_1$ (for filial) generation.** The final batch of offspring became the **F$_2$ generation.** He performed reciprocal crosses: First, he dusted the pollen of tall plants onto the stigmas of short plants, and then he dusted the pollen of short plants onto the stigmas of tall plants. In both cases, all F$_1$ offspring resembled the tall parent.

Certainly, these results were contrary to those predicted by the blending model of inheritance. Rather than being intermediate, the F$_1$ plants resembled only one parent. Did these results mean that the other characteristic (i.e., shortness) had disappeared permanently? Apparently not, because when Mendel allowed the F$_1$ plants to self-pollinate, ¾ of the F$_2$ generation were tall, and ¼ were short, a 3:1 ratio (Fig. 10.2A). Therefore, the F$_1$ plants were able to pass on a factor for shortness—it didn't just disappear.

Mendel counted many offspring. For this particular cross, he counted a total of 1,064 offspring, of which 787 were tall and 277 were short. In all the crosses he performed, he found just about a 3:1 ratio in the F$_2$ generation. The characteristic that had disappeared in the F$_1$ generation reappeared in almost ¼ of the F$_2$ offspring (see Fig. 10.1C).

Mendel's mathematical approach led him to interpret his results differently than previous breeders. He knew that the same rounded off ratio was obtained among the F$_2$ generation time and time again for the crosses he was studying. Eventually, Mendel arrived at this explanation: A 3:1 ratio among the F$_2$ offspring was possible if the F$_1$ parents contained two separate copies of each hereditary factor, one of these being dominant and the other recessive. The factors separated when the gametes were formed, and each gamete carried only one copy of each factor; random fusion of all possible gametes occurred upon fertilization. Only in this way would shortness recur in the F$_2$ generation.

After doing many F$_1$ crosses, called **monohybrid crosses** because they examine only one trait, Mendel arrived at the first of his laws of inheritance—the law of segregation, which is a cornerstone of his particulate theory of inheritance.

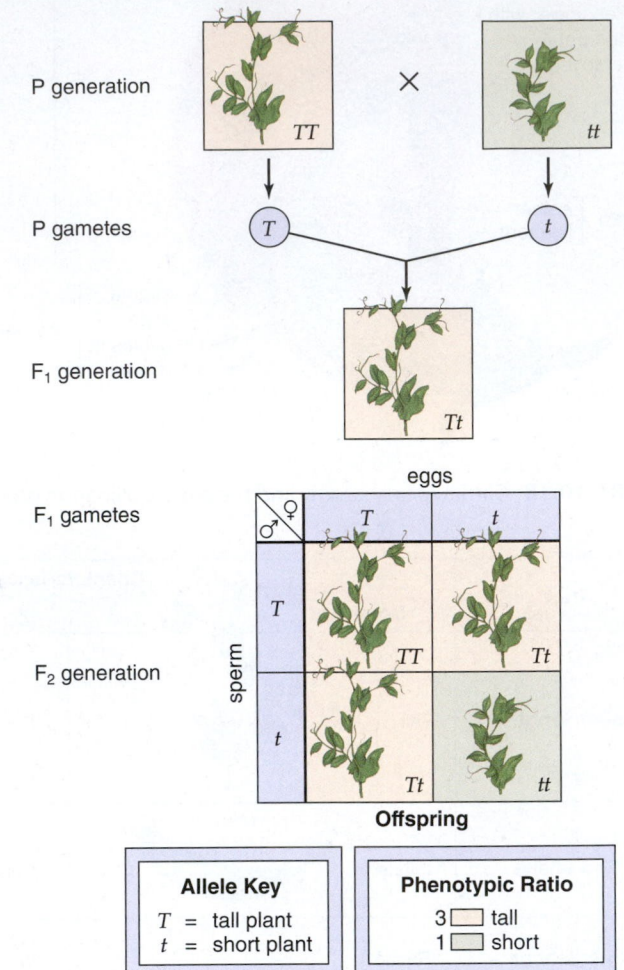

FIGURE 10.2A Monohybrid cross performed by Mendel.

The **law of segregation** states the following:
- Each individual has two factors for each trait.
- The factors segregate (separate) during the formation of the gametes.
- Each gamete contains only one factor from each pair of factors.
- Fertilization gives each new individual two factors for each trait.

It is important to understand Figure 10.2A because all the crosses in this chapter are presented in the same way. Notice that the gametes (blue) have only one letter. The individuals (tan and green boxes) have two letters. The individuals in tan boxes all have at least one capital, namely, a T. Only the individuals in green boxes have two small letters, namely, tt. The drawings show you the appearance of the individuals. The results of the cross, the ratio 3:1, is based on the appearance of the individuals. You will learn more about this cross, including the correct terminology to describe it, on the next page.

Genotype Versus Phenotype Mendel said that organisms received "factors" from their parents, but today we use the term "genes." Traits are controlled by **alleles,** alternative forms of a gene. The alleles occur on homologues at a particular **gene locus** (Fig. 10.2B). The **dominant allele** is so named because of its ability to mask the expression of the other allele, called the **recessive allele.** (Therefore, dominant does not mean the normal or most frequent condition.) The dominant allele is identified by a capital letter, and the recessive allele by the same letter, but lowercase. Usually, the letter chosen has some connection to the trait itself. For example, when considering stem length in peas, the allele for tallness is *T,* and the allele for shortness is *t.*

As you learned in Figure 9.2, meiosis is the type of cell division that reduces the chromosome number. During meiosis I, homologues each having sister chromatids separate. During meiosis II, the chromatids separate. Therefore, the process of meiosis explains Mendel's law of segregation and why there is only one allele for each trait in a gamete.

In Mendel's cross (see Fig. 10.2A), the original parents (P generation) were true-breeding; therefore, the tall plants had two copies of the same allele for tallness (*TT*), and the short plants had two copies of the same allele for shortness (*tt*). When an organism has two identical alleles, as these had, we say it is **homozygous.** Because the parents were homozygous, all gametes produced by the tall plant contained the allele for tallness (*T*), and all gametes produced by the short plant contained the allele for shortness (*t*).

After cross-fertilization, all the individuals in the resulting F$_1$ generation had one allele for tallness and one for shortness (*Tt*). When an organism has two different alleles at a gene locus, we say that it is **heterozygous.** Although the plants of the F$_1$ generation had one of each type of allele, they were all tall. The allele that is expressed in a heterozygous individual is the dominant allele. The allele that is not expressed in a heterozygote is the recessive allele.

TABLE 10.2	Genotype Versus Phenotype
Genotype	**Phenotype**
TT, homozygous dominant	Tall plant
Tt, heterozygous	Tall plant
tt, homozygous recessive	Short plant

You can see that two organisms with different allelic combinations for a trait can have the same outward appearance. For example, *TT* and *Tt* pea plants are both tall. For this reason, it is necessary to distinguish between the alleles present in an organism and the appearance of that organism.

The word **genotype** refers to the alleles an individual receives at fertilization. Genotype may be indicated by letters or by short, descriptive phrases. Genotype *TT* is called homozygous dominant, genotype *tt* is called homozygous recessive, and genotype *Tt* is called heterozygous.

The word **phenotype** refers to the physical appearance of the individual. The homozygous dominant (*TT*) individual and the heterozygous (*Tt*) individual both show the dominant phenotype and are tall, while the homozygous recessive (*tt*) individual shows the recessive phenotype and is short. Table 10.2 compares genotype with phenotype.

Continuing with the discussion of Mendel's cross (see Fig. 10.2A), the F$_1$ plants produce gametes in which 50% have the dominant allele *T* and 50% have the recessive allele *t.* During the process of fertilization, we assume that all types of sperm (*T* or *t*) have an equal chance to fertilize all types of eggs (*T* or *t*). When this occurs, a cross between heterozygotes always produces just about a 3:1 (dominant to recessive) ratio among the offspring. Figure 10.1C gives Mendel's results for several crosses between heterozygotes, and you can see that the phenotypic ratio was always close to 3:1.

Linkage All of the alleles on any chromosome form a **linkage group** and will be inherited together unless crossing-over occurs. Only when we are dealing with more than one trait does linkage become a consideration, however.

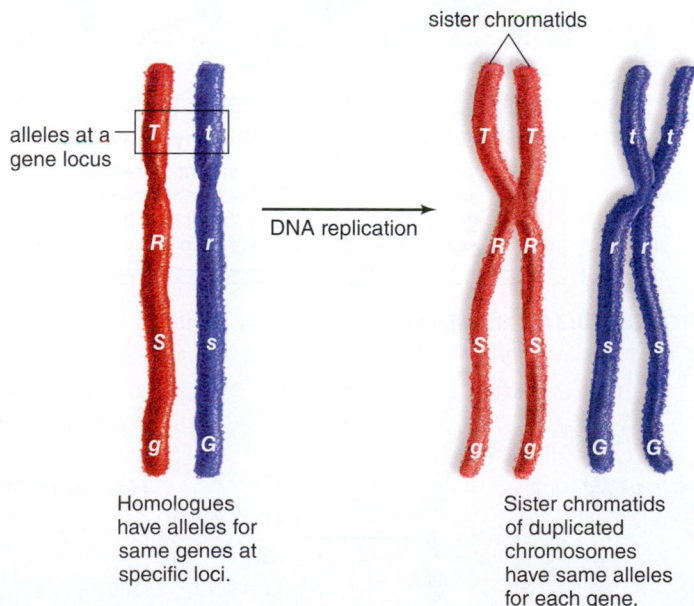

FIGURE 10.2B Occurrence of alleles on homologues.

▶ **10.2 CHECK YOUR PROGRESS**

1. Identify the genotype of a plant (after choosing an appropriate letter) that is (**a**) homozygous for round seeds, a dominant characteristic; and (**b**) heterozygous for round seeds.
2. List all possible gametes for each of the following genotypes and note the proportion of each gamete: (**a**) *WW*, (**b**) *Rr*, (**c**) *Tt*, and (**d**) *TT*.
3. Identify which of these genotypes (*Bb, BB, bb*) a white rabbit could have if *B* = dominant black allele and *b* = recessive white allele.
4. Predict the phenotypic ratio among the offspring if a heterozygous rabbit reproduces with one of its own kind. If there are 120 rabbits, how many would you expect to be white?

10.3 Mendel's law of independent assortment describes inheritance of multiple traits

LEARNING OUTCOMES

When you complete this section, you should be able to

1. Explain Mendel's law of independent assortment and its relationship to meiosis.
2. Predict the results of two-trait genetics problems when the parents are homozygous or heterozygous for two traits.

Mendel performed a second series of crosses in which true-breeding plants differed in two traits. For example, he crossed tall plants having green pods with short plants having yellow pods (Fig. 10.3). The F_1 plants showed both dominant characteristics. As before, Mendel then allowed the F_1 plants to self-pollinate. These F_1 crosses are called **dihybrid crosses** because they are examining two traits. Mendel reasoned that two possible results could occur in the F_2 generation:

1. If the dominant factors (TG) always segregate into the F_1 gametes together, and the recessive factors (tg) always stay together, two phenotypes would occur among the F_2 plants—tall plants with green pods and short plants with yellow pods.
2. If the four factors segregate into the F_1 gametes independently, four phenotypes would occur among the F_2 plants—tall plants with green pods, tall plants with yellow pods, short plants with green pods, and short plants with yellow pods.

Figure 10.3 shows that Mendel observed four phenotypes among the F_2 plants, supporting the second hypothesis. Therefore, Mendel knew that the gametes for a dihybrid cross always consist of the two dominants (such as TG), the two recessives (such as tg), and ones that have a dominant and a recessive (such as Tg and tG). When doing genetics problems, it may be helpful to use the diagram in Figure 10.3B to form the gametes for a dihybrid cross.

Because Mendel always observed a phenotypic ratio of 9:3:3:1 for the particular type of dihybrid cross shown in Fig. 10.3A, he knew that the gametes contained all possible combinations of the involved factors (called alleles, today).

The **law of independent assortment** states the following:
- Each pair of factors separates (assorts) independently (without regard to how the others separate).
- All possible combinations of factors can occur in the gametes.

As long as the alleles are not linked, the process of meiosis explains why Mendel's F_1 plants produced every possible type of gamete, and therefore why four phenotypes appear among the F_2 generation of plants. As was explained in Figure 9.3B, homologues are randomly aligned at the spindle's equator—either homologue can face either spindle pole. Because of this, the daughter cells from meiosis I (and also meiosis II) have all possible combinations of unlinked alleles.

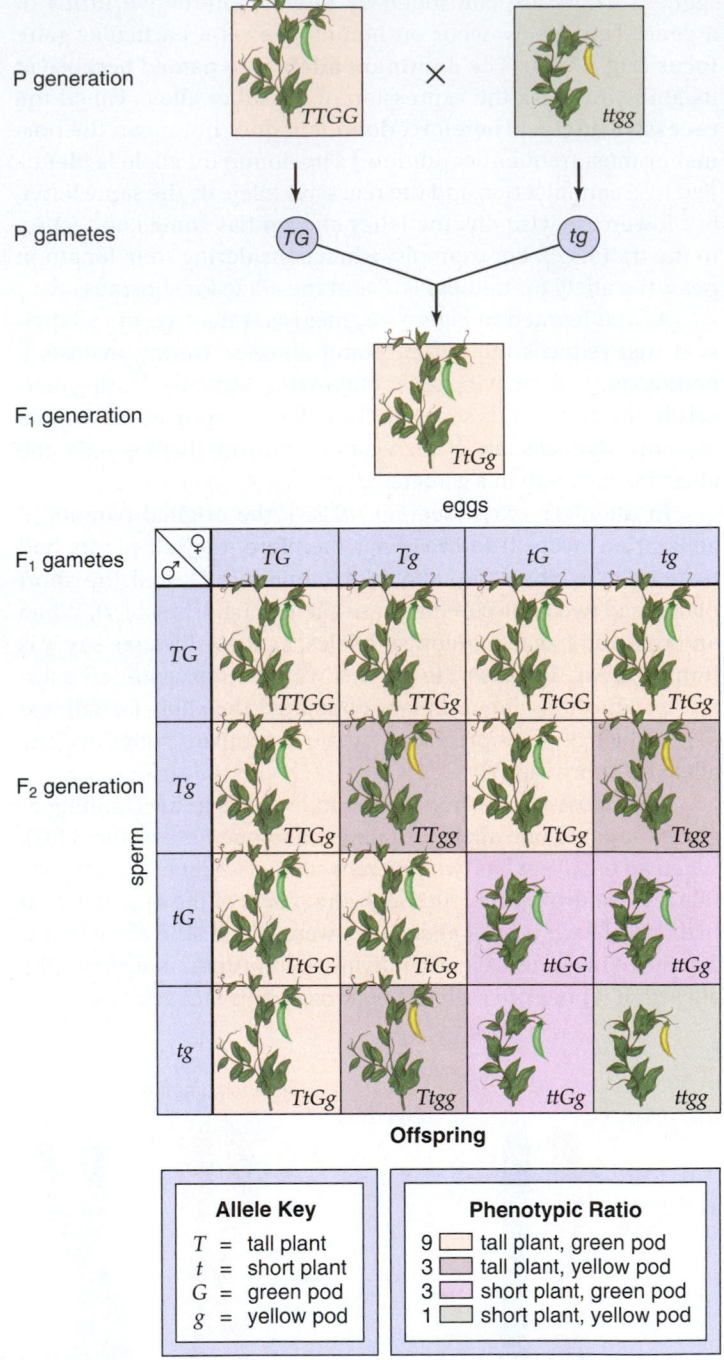

FIGURE 10.3A Dihybrid cross performed by Mendel.

Allele Key	Phenotypic Ratio
T = tall plant	9 ☐ tall plant, green pod
t = short plant	3 ☐ tall plant, yellow pod
G = green pod	3 ☐ short plant, green pod
g = yellow pod	1 ☐ short plant, yellow pod

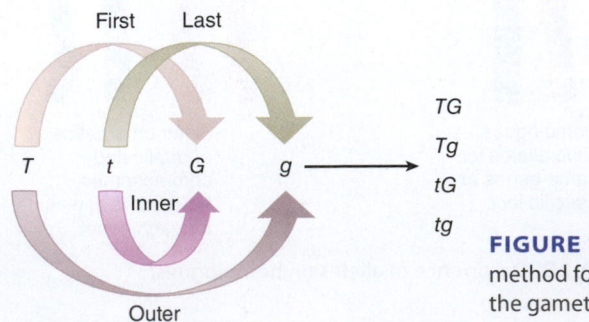

FIGURE 10.3B FOIL method for forming the gametes.

► **10.3 CHECK YOUR PROGRESS**
1. A plant is homozygous for round seeds and homozygous for purple flowers. If such flowers are crossed, what phenotype will all the offspring have?
2. Identify the gametes of an animal whose genotype is WwSs.

10A The Theory of Natural Selection

It's amazing to learn that Darwin didn't know any genetics and was unaware of Mendel's studies with garden peas. Both men lived at the same time, although in different countries—Mendel in Austria and Darwin outside London, England. Mendel actually visited a London expedition in 1862 when he was 40 and Darwin was 53, but they never met. Darwin's scientific theory of natural selection was based on observations of phenotypes, such as the many different types of horses, and not on their genotype. He was very much aware, however, that traits are inherited. He even performed **artificial selection** in which the breeder selects individuals with the desired traits (Fig. 10A.1). Because the traits are passed from parents to offspring, the next generation has a higher proportion of the desired traits than the previous generation. Artificial selection is the way humans have produced many of the crops that sustain us today, including wheat, corn, bananas, and tomatoes.

Despite having a basic knowledge of genetics, it wasn't until the mid-twentieth century that biologists introduced the concept of alleles into Darwin's theory of natural selection in this way:

- The alleles of genes are responsible for the traits of an individual.
- As we have learned with peas, individuals pass their alleles to their offspring, and the alleles on separate chromosomes are shuffled with each generation due to the process of meiosis. (During meiosis, you'll recall, recombination of alleles occurs as a result of crossing-over, and independent assortment of chromosomes occurs because homologue pairs align at the equator randomly. See Figure 9.3B.)
- Mutations are the raw material of evolution because they introduce new traits. Because beneficial mutations are bound to be selected, their effect is greater than neutral or harmful mutations. Indeed, without beneficial mutations, new species could not arise.

FIGURE 10A.2 Due to natural selection, the bill of a verditer flycatcher can catch insects; that of an oystercatcher can pry open the shell of a mussel. The talons of an osprey are adapted to catching fish.

- The new combination of alleles, plus any mutations, will likely make some individuals more suited to acquiring food in their environment and therefore better able to survive and reproduce than other members of a population (Fig. 10A.2). To take another example, a plant better able to survive drought would have an advantage over other members of a plant population in a dry environment.
- In this way, each generation becomes better adapted to the environment than the previous generation. Adaptation to the environment makes natural selection very different from artificial selection, which often does not consider adaptation. Race horses, for example, are bred for speed without considering whether the trait is useful for survival.

Darwin's theory of natural selection has stood the test of time, as is witnessed by its consistency with the principles of Mendelian genetics described in this chapter as well as with those of molecular genetics (Chapter 11).

CONSIDER THESE QUESTIONS

1. The compatibility of the gene theory with the theory of evolution lends support to the theory of evolution. How so?
2. Why is it significant that natural selection but not artificial selection is expected to result in adaptation to the environment?
3. Why did it work for Mendel and Darwin to rely only on the phenotypes of organisms to deduce their hypotheses?

FIGURE 10A.1 Due to artificial selection, draft horses are large enough to do heavy lifting; quarter horses are agile enough to do maneuvers; thoroughbred horses are fast enough to race.

connect | BIOLOGY   Explore the concepts through a variety of multimedia assets and question types.

www.mcgrawhillconnect.com

10.4 Support for Mendel's laws is various

LEARNING OUTCOMES

When you complete this section, you should be able to

1. Explain why Mendel performed testcrosses and why they still might be done today.
2. Understand how to use the results of a cross to deduce the genotype of a parent.
3. Understand that Mendel's laws are consistent with those of probability.
4. Predict the results of two-trait genetics problems when you are given or can deduce the genotype of the parents.

One-trait Testcrosses To confirm that the F₁ of his one-trait crosses were heterozygous, Mendel crossed his F₁ generation plants with true-breeding, short (homozygous recessive) plants. Mendel performed these so-called **testcrosses** because they allowed him to support the law of segregation. For the cross in Figure 10.4A, he reasoned that half the offspring should be tall and half should be short, producing a 1:1 phenotypic ratio. His results supported the hypothesis that alleles segregate when gametes are formed. In Figure 10.4A, the homozygous recessive parent can produce only one type of gamete (*t*), and so the Punnett square has only one column. When a heterozygous individual is crossed with one that is homozygous recessive, the probable results are always a 1:1 phenotypic ratio.

Today, a one-trait testcross is used to determine if an individual with the dominant phenotype is homozygous dominant (e.g., *TT*) or heterozygous (e.g., *Tt*). Since both of these genotypes produce the dominant phenotype, it is not possible to determine the genotype by observation. Figure 10.4B shows that if the individual is homozygous dominant, all the offspring will be tall. Each parent has only one type of gamete, and therefore a Punnett square is not required to determine the results.

Two-trait Testcrosses When doing a two-trait testcross, an individual with the dominant phenotype for both traits is crossed with one having the recessive phenotype for both traits. Suppose when doing a testcross you are working with fruit flies in which:

L = long wings	*G* = gray body
l = vestigial (short) wings	*g* = black body

 You wouldn't know by examination whether the fly on the left was homozygous or heterozygous for wing and body color. The fly could have these genotypes:

LLGG	*LLGg*
LlGG	*LlGg*

and still have long wings and gray body.

In order to find out the genotype of the test fly, you cross it with the one on the right. You know by examination that this vestigial-winged and black-bodied fly is homozygous recessive for both traits. In other words this fly has the genotype *llgg*.

If the test fly is homozygous dominant for both traits with the genotype *LLGG*, it will form only one gamete: *LG*. Therefore, all the offspring from the proposed cross will have long wings and

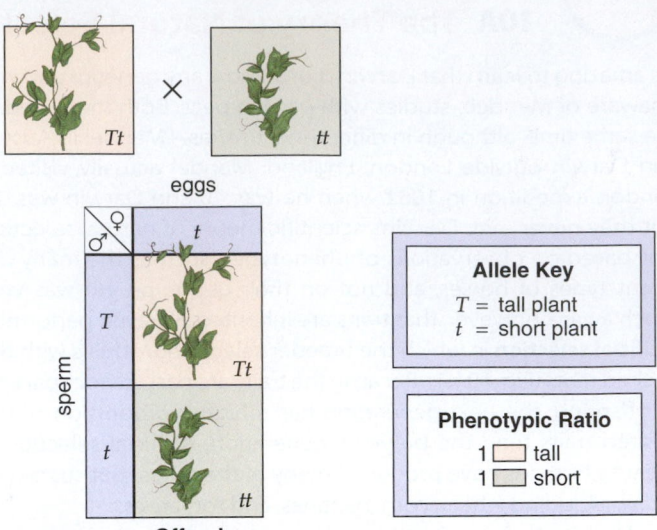

FIGURE 10.4A One-trait testcross, when the individual with the dominant phenotype is heterozygous.

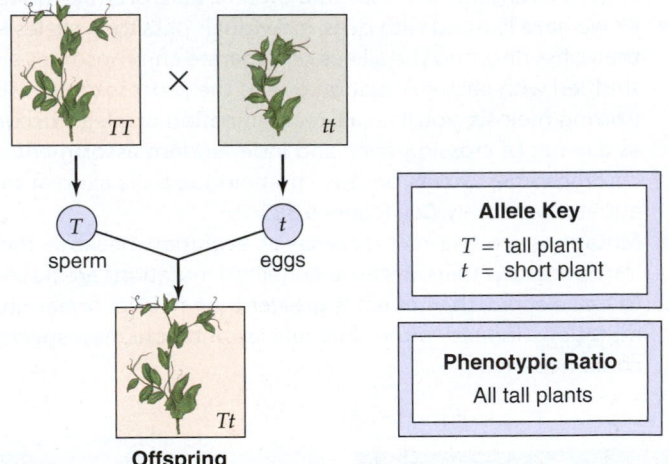

FIGURE 10.4B One-trait testcross, when the individual with the dominant phenotype is homozygous.

a gray body despite the genotype of its partner. What will be the results for each of the genotypes listed?

If the test fly is heterozygous for both traits, having the genotype *LlGg*, it will form four different types of gametes:

Gametes: *LG*   *Lg*   *lG*   *lg*

and have four different offspring:

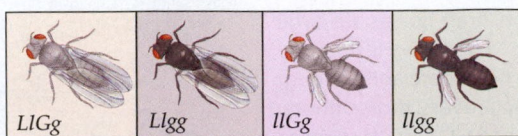

The presence of the offspring with vestigial wings and a black body shows that the test fly is heterozygous for both traits and has the genotype *LlGg*. Otherwise, it could not have this offspring. In general, you will want to remember that when an individual heterozygous for two traits is crossed with one that is recessive for the traits, the offspring have a 1:1:1:1 phenotypic ratio.

Laws of Probability The diagram we have been using to calculate the results of a cross is called a **Punnett square.** The Punnett square allows us to easily calculate the chances, or the probability, of genotypes and phenotypes among the offspring.

Consider a one-trait cross. Like flipping a coin each gamete (sperm or egg) in Figure 10.4C can bear an *E* or *e*. Therefore, as illustrated in the Punnett square, an offspring of the cross has a 50% (or ½) chance of receiving an *E* for unattached earlobe or an *e* for attached earlobe from each parent:

The chance of *E* = ½
The chance of *e* = ½

How likely is it that an offspring will inherit a specific set of two alleles, one from each parent? To answer this question we have to turn to the rules of probability. The product rule of probability tells us that we have to multiply the chances of independent events to get the answer:

1. The chance of *EE* = ½ × ½ = ¼
2. The chance of *Ee* = ½ × ½ = ¼
3. The chance of *eE* = ½ × ½ = ¼
4. The chance of *ee* = ½ × ½ = ¼

The Punnett square does this for us because we can easily see that each genotype occurs in ¼ of the total number of squares. How do we get the phenotypic results? The sum rule of probability tells us that when the same event can occur in more than one way, we add the results. Because 1, 2, and 3 all result in unattached earlobes, we add them up to know that the chance of unattached earlobes is ¾, or 75%. The chance of attached earlobes is ¼, or 25%. The Punnett square doesn't do this for us—we have to add the results ourselves.

Know that in genetics "chance has no memory." So, if a couple has four children, each child has a 25% chance of having attached earlobes. This may not be significant if we are considering earlobes. But it does become significant if we are considering a recessive genetic disorder, such as cystic fibrosis, a debilitating respiratory illness. If a heterozygous couple has four children, each child has a 25% chance of inheriting two recessive alleles, and all four children could have cystic fibrosis.

Consider a two-trait cross. We can use the product rule and the sum rule of probability to predict the results of a dihybrid cross, such as the one shown in Figure 10.3. The Punnett square carries out the multiplication for us, and we add the results to find that the phenotypic ratio is 9:3:3:1. We expect these same results for each and every dihybrid cross. Therefore, it is not necessary to do a Punnett square over and over again for either a monohybrid or a dihybrid cross. *Instead, we can simply remember the probable results of 3:1 and 9:3:3:1.* But we have to remember that the 9 represents the two dominant phenotypes together, the 3's are a dominant phenotype with a recessive, and the 1 stands for the two recessive phenotypes together. This tells you the probable phenotypic ratio among the offspring, but not the chances for each possible phenotype. Because the Punnett square has 16 squares, the chances are $^9/_{16}$ for the two dominants together, $^3/_{16}$ for the dominants with each recessive, and $^1/_{16}$ for the two recessives together. To take an example, what is the chance of a tall plant with a yellow pod in Figure 10.3? How many possible genotypes are there for this type plant? You are correct if you

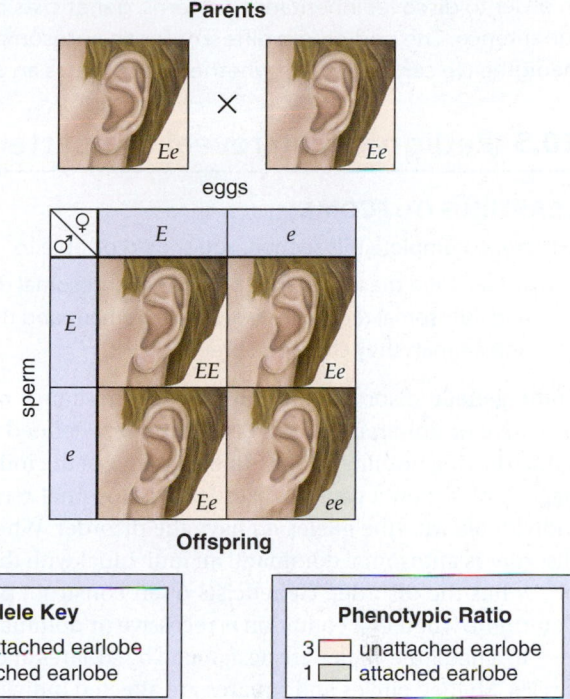

Allele Key
E = unattached earlobe
e = attached earlobe

Phenotypic Ratio
3 ⬜ unattached earlobe
1 ⬜ attached earlobe

FIGURE 10.4C Use of a Punnett square to calculate probable results; in this case a 3:1 phenotypic ratio.

answered the chances of this phenotype is $^3/_{16}$ but the number of possible genotypes is only two. What are the probable results for a one-trait testcross when the test subject is heterozygous and when the subject is heterozygous in two traits?

Why did Mendel do so many crosses? Mendel counted the results of many similar crosses to get the probable results, and in the laboratory, we too have to count the results of many individual crosses to get the probable results for a cross. Why? Consider that each time you toss a coin, you have a 50% chance of getting heads or tails. If you tossed the coin only a couple of times, you might very well have the same result both times. However, if you toss the coin many times, you are more likely to finally achieve 50% heads and 50% tails.

▶ **10.4 CHECK YOUR PROGRESS**

1. Predict the results when a heterozygous fruit fly (*LlGg*) is crossed with a homozygous recessive (*llgg*). What are the chances of offspring with long wings and a black body? In fruit flies, long wings (*L*) is dominant over vestigial (short) wings (*l*), and gray body (*G*) is dominant over black body (*g*).

2. Identify the genotype of both parents and their offspring when a trotter is mated to a pacer and the offspring is a pacer. In horses, trotter (*T*) is dominant over pacer (*t*).

3. Predict the chance of each child having freckles when a man with freckles reproduces with a woman with freckles and as yet, none of their children have freckles. In humans, freckles is dominant over no freckles.

4. Identify the probable phenotypic ratio among the offspring of the dihybrid cross *LlGg* × *LlGg* without doing a Punnett square. What are the chances of an offspring with short wings and a black body?

Mendel's Laws Apply to Humans

In order to discover inheritance patterns, geneticists construct a family tree called a pedigree to determine a trait's pattern of inheritance. The pedigree is different for an autosomal dominant versus an autosomal recessive trait. Therefore, by constructing a pedigree we can determine whether a disorder is an autosomal dominant disorder or an autosomal recessive disorder.

10.5 Pedigrees can reveal the patterns of inheritance

LEARNING OUTCOME

When you complete this section, you should be able to

1. Understand the inheritance patterns of autosomal dominant and autosomal recessive disorders in humans and the plants and animals they choose to breed.

Some genetic disorders are due to the inheritance of abnormal recessive or dominant alleles on autosomes, defined as chromosomes having nothing to do with the gender of the individual (see page 156). When a genetic disorder is autosomal recessive, only individuals with the alleles *aa* have the disorder. When a genetic disorder is autosomal dominant, an individual with the alleles *AA* or *Aa* has the disorder. Geneticists often construct **pedigrees** to determine whether a condition is recessive or dominant.

In a pedigree, males are designated by squares and females by circles. Shaded circles and squares are affected individuals. A line between a square and a circle represents a union. A vertical line going downward leads to offspring. (If there is more than one offspring, they are placed off a horizontal line.) A pedigree shows the pattern of inheritance for a particular condition. Consider these two possible patterns of inheritance:

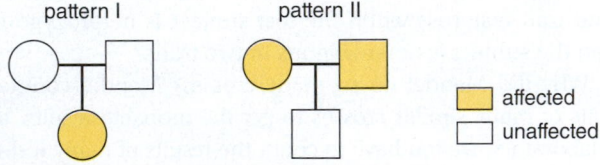

pattern I   pattern II

affected
unaffected

Which pattern of inheritance (I or II) do you think pertains to an autosomal dominant characteristic, and which pertains to an autosomal recessive characteristic?

In pattern I, the child is affected, but neither parent is; this can happen if the condition is recessive and the parents are *Aa*. Notice that the parents are **carriers** because they appear normal but are capable of having a child with the genetic disorder. In pattern II, the child is unaffected, but the parents are affected. This can happen if the condition is dominant and the parents are *Aa*.

Figure 10.5A shows other ways to recognize an autosomal recessive pattern of inheritance, and Figure 10.5B shows other ways to recognize an autosomal dominant pattern of inheritance. In these pedigrees, generations are indicated by Roman numerals placed on the left side. Notice in the third generation of Figure 10.5A that two closely related individuals have produced three children, two of whom have the affected phenotype. This illustrates that reproduction between closely related persons increases the chances of children inheriting two copies of a potentially harmful recessive allele.

The inheritance pattern of alleles on the X chromosome follows different rules than those on the autosomal chromosomes, as discussed in section 10.9.

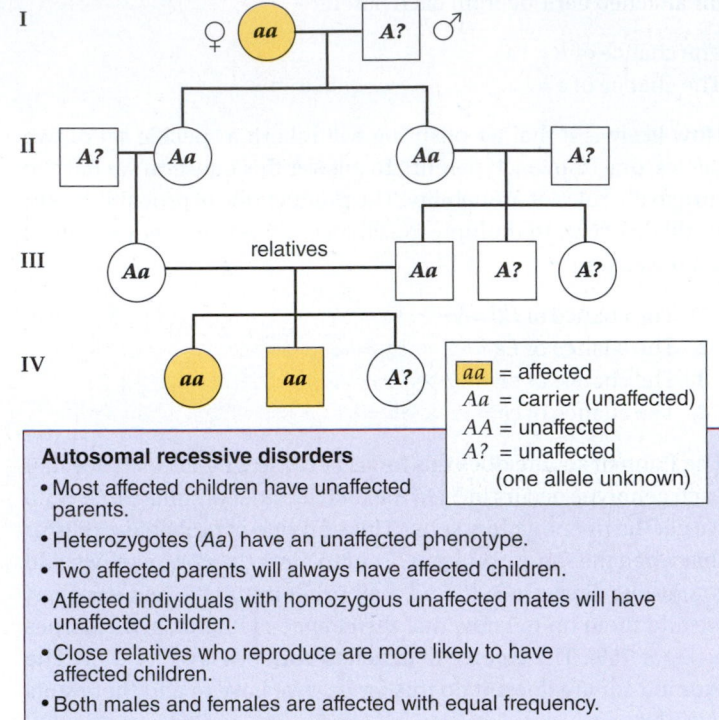

Autosomal recessive disorders
- Most affected children have unaffected parents.
- Heterozygotes (*Aa*) have an unaffected phenotype.
- Two affected parents will always have affected children.
- Affected individuals with homozygous unaffected mates will have unaffected children.
- Close relatives who reproduce are more likely to have affected children.
- Both males and females are affected with equal frequency.

aa = affected
Aa = carrier (unaffected)
AA = unaffected
A? = unaffected (one allele unknown)

FIGURE 10.5A Autosomal recessive pedigree.

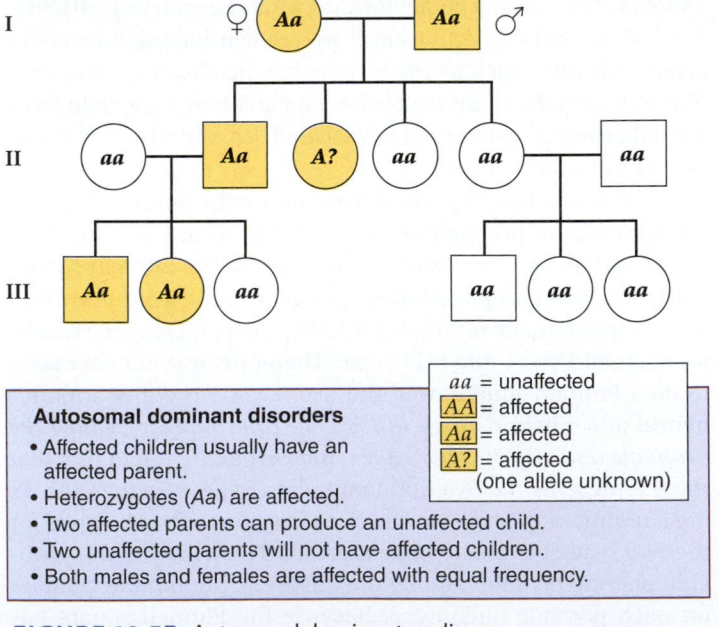

Autosomal dominant disorders
- Affected children usually have an affected parent.
- Heterozygotes (*Aa*) are affected.
- Two affected parents can produce an unaffected child.
- Two unaffected parents will not have affected children.
- Both males and females are affected with equal frequency.

aa = unaffected
AA = affected
Aa = affected
A? = affected (one allele unknown)

FIGURE 10.5B Autosomal dominant pedigree.

▶ **10.5 CHECK YOUR PROGRESS**

1. Identify how Figure 10.5A demonstrates that dog breeders should keep careful pedigree records.

10B Genetic Disorders May Now Be Detected Early On

A variety of procedures are available to test for genetic disorders. Some of these involve testing after pregnancy, and others involve testing before pregnancy. It's important to realize that a woman is not pregnant until the embryo implants itself in the womb.

Testing After Implantation

During **amniocentesis,** a long needle is passed through the abdominal and uterine walls to withdraw a small amount of the fluid that surrounds the fetus and contains a few fetal cells. Thereafter, genetic tests can be done on this fluid and on fetal chromosomes from the cells. During chorionic villus sampling (CVS), a long, thin tube is inserted through the vagina into the uterus. Then a sampling of fetal cells is obtained by suction. The cells do not have to be cultured, as they must be following amniocentesis, and testing can be done immediately. These procedures carry a slight risk of miscarriage. If the fetus is found to have a genetic disorder some may find it unacceptable for ethical, psychosocial, or religious reasons to terminate the pregnancy.

Preimplantation Genetic Diagnosis (PGD)

PGD is expected to play an ever-more-significant role in the control and prevention of genetic disease. Because it does not involve the termination of a pregnancy there is a reduced risk of miscarriage, and parents can avoid the heartbreak of having a child with a genetic disorder. The prospective parents have two choices: They can elect to test the embryo, or they can elect to test the egg.

If the embryo is to be tested (Fig. 10B.1), development begins in laboratory glassware through the process of in vitro fertilization (IVF). A physician obtains eggs from the prospective mother and sperm from the prospective father, and places them in the same receptacle, where fertilization occurs. Then the zygote (fertilized egg) begins dividing. A single cell is removed from the 8-celled embryo and subjected to PGD. Removing a single cell does not affect the developing embryo. Only healthy embryos that test negative for the genetic disorders of interest are placed in the mother's uterus, where they hopefully implant and continue developing.

For testing the egg (Fig. 10B.2) you need to know that meiosis in females results in a single egg and at least two nonfunctional cells called polar bodies. Polar bodies, which later disintegrate, receive very little cytoplasm, but they do receive a haploid number of chromosomes. When a woman is heterozygous for a recessive genetic disorder, about half the polar bodies have received the mutated allele, and in these instances the egg received the normal allele. Therefore, if a polar body tests positive for a mutated allele, the egg received the normal allele. Only normal eggs are then used for IVF. Even if the sperm should happen to carry the mutation, the zygote will, at worst, be heterozygous. But the phenotype will appear normal. Testing the egg may be favored over testing the embryo because it does not involve discarding rejected embryos.

CONSIDER THESE QUESTIONS

1. Do you agree that PGD is an acceptable procedure to avoid having a child with a genetic disorder, whereas terminating a pregnancy is not acceptable?
2. Is IVF acceptable to you, and if so do you prefer testing the egg over testing the embryo? Explain your reasoning.

 Explore the concepts through a variety of multimedia assets and question types.

www.mcgrawhillconnect.com

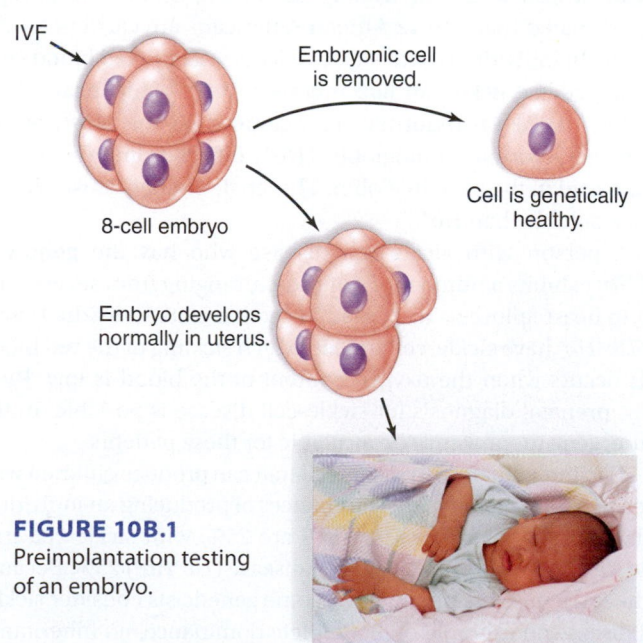

FIGURE 10B.1
Preimplantation testing of an embryo.

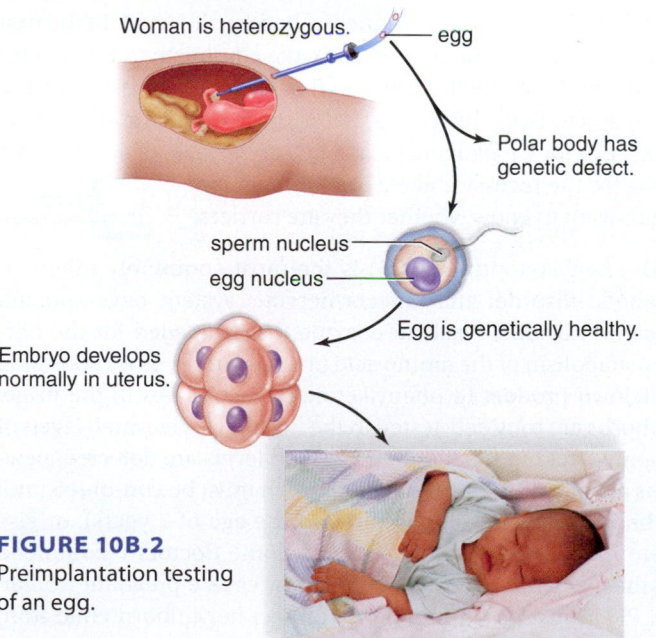

FIGURE 10B.2
Preimplantation testing of an egg.

LEARNING OUTCOME

When you complete this section, you should be able to

1. List some common autosomal genetic disorders, state the symptoms, and describe the inheritance pattern.

Autosomal Recessive Disorders In humans, a number of genetic disorders are controlled by a single pair of alleles. Four of the best-known autosomal recessive disorders are Tay-Sachs disease, cystic fibrosis, phenylketonuria, and sickle-cell disease. Individuals can be carriers for these diseases.

Tay-Sachs Disease In a baby with Tay-Sachs disease, development begins to slow down between 4 and 8 months of age, and neurological impairment and psychomotor difficulties then become apparent. The child gradually becomes blind and helpless, develops uncontrollable seizures, and eventually becomes paralyzed prior to dying. Tay-Sachs disease results from a lack of the lysosomal enzyme Hex A and the subsequent storage of its substrate, a lipid, in lysosomes. As the lipid builds up in the lysosomes, it crowds the organelles and impairs their function, especially in the brain.

Carriers of Tay-Sachs disease have about half the level of Hex A activity found in homozygous dominant individuals but they appear to be normal. Prenatal diagnosis of the disease is possible following either amniocentesis or chorionic villus sampling. The gene for Tay-Sachs disease is located on chromosome 15.

Cystic Fibrosis Cystic fibrosis (CF) is the most common lethal genetic disease among Caucasians in the United States. Abnormal secretions related to the chloride ion channel characterize this disorder. One of the most obvious symptoms in CF patients is extremely salty sweat. In children with CF, the mucus in the bronchial tubes and pancreatic ducts is particularly thick and viscous, interfering with the function of the lungs and pancreas. To ease breathing, the thick mucus in the lungs has to be loosened periodically, but still the lungs frequently become infected. In the past few years, new treatments, including the administration of antibiotics by means of a nebulizer and the use of a percussion vest to loosen mucus in the lungs (Fig. 10.6A), have raised the average life expectancy for CF patients to as much as 35 years of age. Genetic testing for the recessive allele is possible if individuals want to know whether they are carriers.

Video
Good Poison

PKU Phenylketonuria (PKU) is the most commonly inherited metabolic disorder that affects nervous system development. Affected individuals lack an enzyme that is needed for the normal metabolism of the amino acid phenylalanine, so an abnormal breakdown product (a phenylketone) accumulates in the urine. Newborns are routinely tested in the hospital for elevated levels of the amino acid in the blood. If elevated levels are detected, newborns are placed on a special diet, which must be continued until the brain is fully developed (around the age of 7 years), or else severe intellectual disability occurs. Some doctors recommend that the diet continue for life, but in any case, a pregnant woman with PKU must be on the diet to protect her unborn child from harm. Many diet products, such as soft drinks, have warnings that the product contains the amino acid phenylalanine.

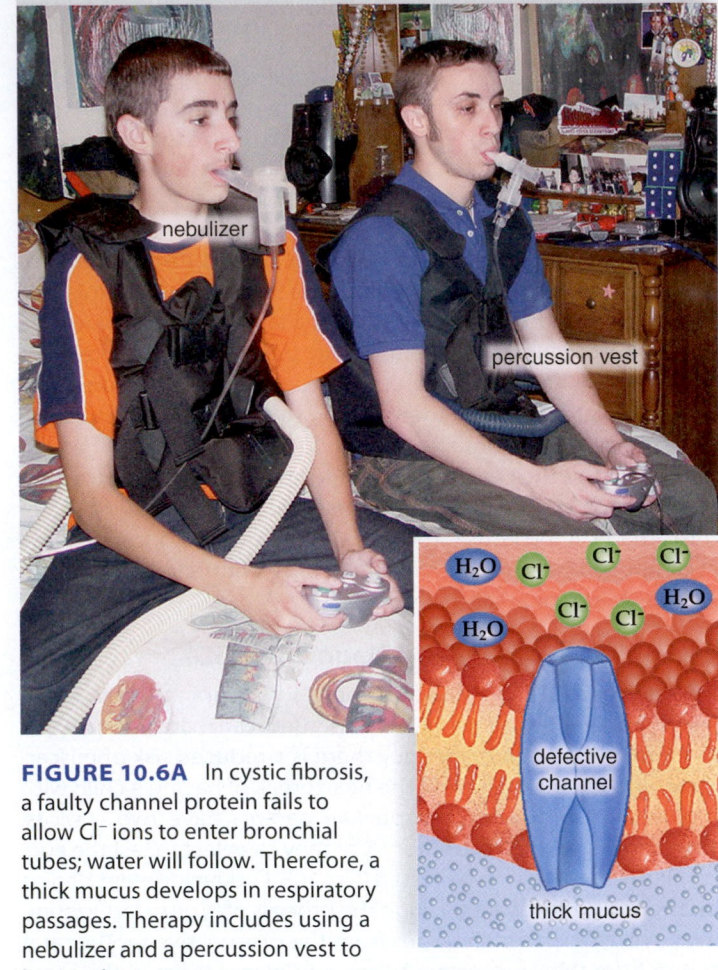

FIGURE 10.6A In cystic fibrosis, a faulty channel protein fails to allow Cl^- ions to enter bronchial tubes; water will follow. Therefore, a thick mucus develops in respiratory passages. Therapy includes using a nebulizer and a percussion vest to loosen the mucus.

Sickle-Cell Disease Sickle-cell disease occurs among people of African descent and is not usually seen among other racial groups. It is estimated that 1 in 12 African Americans are carriers for the disease. In individuals with sickle-cell disease, the red blood cells are shaped like sickles, or half-moons, instead of biconcave discs (see Fig. 10.8B). An abnormal hemoglobin molecule (Hb^S) causes the defect. Normal hemoglobin (Hb^A) differs from Hb^S by one amino acid in the protein globin. The single change causes Hb^S to be less soluble than Hb^A.

A person with sickle-cell disease who has the genotype Hb^SHb^S exhibits a number of symptoms, ranging from severe anemia to heart failure as discussed in section 10.8. Individuals who are Hb^AHb^S have sickle-cell trait, in which sickling of the red blood cells occurs when the oxygen content of the blood is low. Presently, prenatal diagnosis for sickle-cell disease is possible. In the future, gene therapy may be available for these patients.

Two individuals with sickle-cell trait can produce children with three possible phenotypes. The chances of producing an individual with a normal genotype (Hb^AHb^A) are 25%, with sickle-cell trait (Hb^AHb^S) 50%, and with sickle-cell disease (Hb^SHb^S) 25%. Because of the three possible phenotypes, some geneticists consider sickle-cell disease an example of incomplete dominance, an inheritance pattern discussed in section 10.7.

Autosomal Dominant Disorders A number of autosomal dominant disorders have been identified in humans. Three relatively common ones are neurofibromatosis, Huntington disease, and achondroplasia.

Neurofibromatosis Neurofibromatosis, sometimes called von Recklinghausen disease, is one of the most common genetic disorders and is seen equally in every racial and ethnic group throughout the world, many times in families with no history of the disorder. Once it appears in a family, it becomes a recurring trait.

At birth, or later, the affected individual may have six or more large, tan spots on the skin. Such spots may increase in size and number and get darker. Small, benign tumors (lumps) called neurofibromas, which arise from the fibrous coverings of nerves, may develop. In most cases, symptoms are mild, and patients live a normal life. In some cases, however, the effects are severe and include skeletal deformities, such as a large head, and eye and ear tumors that can lead to blindness and hearing loss. Many children with neurofibromatosis have learning disabilities and are hyperactive.

In 1990, researchers isolated the gene for neurofibromatosis and learned that it controls the production of a protein called neurofibromin, which normally blocks growth signals leading to cell division. Any number of mutations can lead to a neurofibromin that fails to block cell growth, and the result is the formation of tumors. Some mutations are caused by inserted DNA bases that do not belong in their present location.

Huntington Disease Huntington disease is a neurological disorder that leads to progressive degeneration of brain cells. Figure 10.6B shows that a portion of the brain involved in motor control atrophies. This, in turn, causes severe muscle spasms that worsen with time (Fig. 10.6C). The disease is caused by a single mutated copy of the gene for a protein called huntingtin. Most patients appear normal until they are of middle age and have already had children, who may eventually also be stricken. Occasionally, the first sign of the disease in the next generation is seen in teenagers or even younger children. There is no effective treatment, and death comes 10–15 years after the onset of symptoms.

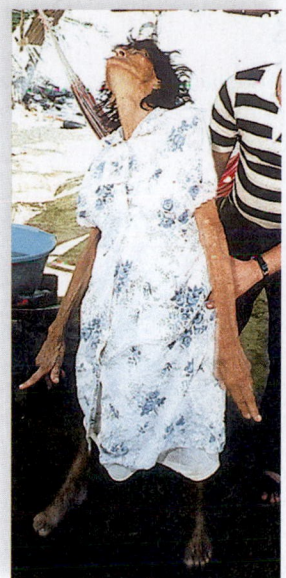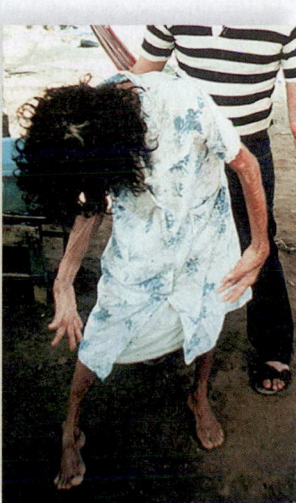

FIGURE 10.6C Patients with Huntington disease have neuromuscular spasms.

Several years ago, researchers found that the gene for Huntington disease is located on chromosome 4. A test was developed to detect the presence of the gene, but few people want to know if they have inherited the gene because there is no cure. At least now we know that the disease stems from a mutation that causes huntingtin to have too many copies of the amino acid glutamine. The normal version of the huntingtin protein has stretches of between 10 and 25 glutamines. If the huntingtin protein has more than 36 glutamines, it changes shape and forms large clumps inside neurons. Even worse, it attracts and causes other proteins to clump with it. One of these proteins, called CBP, ordinarily helps nerve cells survive. Researchers hope they may be able to combat the disease by boosting normal CBP levels.

Another possible treatment is to transplant stem cells into the brain. Scientists hypothesize that these will replace some of the damaged neurons. Animal studies show promising results.

Achondroplasia Achondroplasia is a common form of dwarfism associated with a defect in the growth of long bones. Individuals with achondroplasia have short arms and legs and a swayback, but a normal torso and head. About 1 in 25,000 people have achondroplasia. The condition arises when a gene on chromosome 4 undergoes a spontaneous mutation. Individuals who have achondroplasia are heterozygotes (*Aa*). The homozygous recessive (*aa*) genotype yields normal-length limbs. The homozygous dominant condition (*AA*) is lethal, and death generally occurs shortly after birth.

Preimplantation Genetic Diagnosis, discussed on page 181, offers a way to have a child free of a genetic disease that runs in the family.

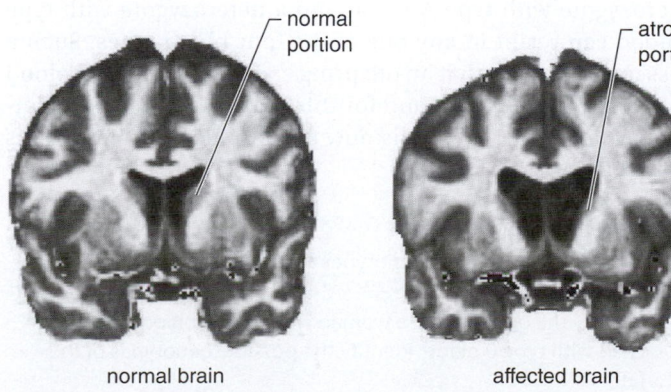

normal portion   atrophied portion

normal brain   affected brain

FIGURE 10.6B A normal brain compared to the brain of a patient affected by Huntington disease.

▶ **10.6 CHECK YOUR PROGRESS**

1. Identify the genotype of normal parents who have a child with cystic fibrosis.
2. Predict the chances that two blind collies could have an offspring that was not blind if the allele for blindness is dominant and the parents are heterozygous.

Beyond Mendel's Laws

In this part of the chapter, we see that Mendelian genetics also applies to complex patterns of inheritance, such as incomplete dominance, multiple alleles, polygenic inheritance, and pleiotropy. You will want to recognize each of these patterns of inheritance and be able to solve genetics problems concerning them.

10.7 Variations in the recessive/dominant allele relationship occur

LEARNING OUTCOMES

When you complete this section, you should be able to

1. Contrast the inheritance pattern for a condition that is incompletely dominant with one that is codominant.
2. Explain the inheritance pattern for ABO blood type.

When the heterozygote has an intermediate phenotype between that of either homozygote, **incomplete dominance** is exhibited. In a cross between a true-breeding, red-flowered four-o'clock strain and a true-breeding, white-flowered strain (Fig. 10.7), ① the offspring have pink flowers. But this is not an example of the blending model of inheritance. When the plants with pink flowers self-pollinate, ② the offspring have a phenotypic ratio of 1 red-flower: 2 pink-flower: 1 white-flower. The reappearance of all three phenotypes in this generation makes it clear that flower color in this instance is controlled by a single pair of alleles.

It would appear that in R_1R_1 individuals, a double dose of pigment results in red flowers; in R_1R_2 individuals, a single dose of pigment results in pink flowers; and because the R_2R_2 individuals produce no pigment, the flowers are white.

In humans, familial hypercholesterolemia (FH) is an example of incomplete dominance. An individual with two alleles for this disorder develops fatty deposits in the skin and tendons and may have a heart attack as a child. An individual with one normal allele and one FH allele may suffer a heart attack as a young adult, and an individual with two normal alleles does not have the disorder.

Perhaps the inheritance pattern of other human disorders should be considered as incomplete dominance. For example, to detect the carriers of cystic fibrosis and Tay-Sachs disease, it is customary to determine the amount of enzyme activity of the gene in question. When the activity is one-half that of the dominant homozygote, the individual is a carrier. In other words, at the level of gene expression, the homozygotes and heterozygotes differ in the same manner as four-o'clock plants.

Multiple Alleles When a trait is controlled by **multiple alleles,** the gene exists in several allelic forms. But each person usually has only two of the possible alleles. For example, a person's ABO blood type is determined by multiple alleles. The following alleles determine the presence or absence of antigens on red blood cells:

I^A = A antigen on red blood cells

I^B = B antigen on red blood cells

i = Neither A nor B antigen on red blood cells

The possible phenotypes and genotypes for blood type are as follows:

Phenotype	Genotype
A	I^AI^A, I^Ai
B	I^BI^B, I^Bi
AB	I^AI^B
O	ii

The inheritance of the ABO blood group in humans is also an example of **codominance** because both I^A and I^B are fully expressed in the presence of the other. Therefore, a person inheriting one of each of these alleles will have type AB blood.

This inheritance pattern differs greatly from Mendel's findings, because more than one allele is fully expressed. Both I^A and I^B are dominant over i. There are two possible genotypes for type A blood and two possible genotypes for type B blood. Use a Punnett square to confirm that reproduction between a heterozygote with type A blood and a heterozygote with type B blood can result in any one of the four blood types. Such a cross makes it clear that an offspring can have a different blood type from either parent, and for this reason, DNA fingerprinting, instead of blood type, is now used to identify the parents of an individual.

▶ **10.7 CHECK YOUR PROGRESS**

1. Identify the potential phenotypes for the children of two FH carriers.
2. Identify the genotype of a woman with type A blood who has a child with type 0 blood. Identify the possible genotypes of the father.

FIGURE 10.7
Incomplete dominance.

eggs

Offspring

10.8 Variations in the one gene–one trait relationship occur

LEARNING OUTCOME

When you complete this section, you should be able to

1. Contrast the inheritance pattern for a condition that is pleiotropic with one that is governed by polygenes.

Pleiotropy When an allele pair has more than one effect on the body, **pleiotropy** exists. For example, persons with Marfan syndrome have disproportionately long arms, legs, hands, and feet; a weakened aorta; poor eyesight; and other characteristics (Fig. 10.8Aa). All of these characteristics are due to the production of abnormal connective tissue (Fig. 10.8Ab). Marfan syndrome has been linked to a mutated gene (FBN_1) on chromosome 15 that ordinarily specifies a functional protein called fibrillin. Fibrillin is essential for the formation of elastic fibers in connective tissue (Fig. 10.8Ab). Without the structural support of normal connective tissue, the aorta can burst, particularly if the person is engaged in a strenuous sport, such as volleyball or basketball. Flo Hyman may have been the best American woman volleyball player ever, but she fell to the floor and died at the age of only 31 because her aorta gave way during a game. Now that coaches are aware of Marfan syndrome, they are on the lookout for it among very tall basketball players. Chris Weisheit, whose career was cut short after he was diagnosed with Marfan syndrome, said, "I don't want to die playing basketball."

Many other disorders, including porphyria and sickle-cell disease, are examples of pleiotropic traits. Porphyria is caused by a chemical insufficiency in the production of hemoglobin, the pigment that makes red blood cells red. The symptoms of porphyria are photosensitivity, strong abdominal pain, port-wine-colored urine, and paralysis in the arms and legs. In the

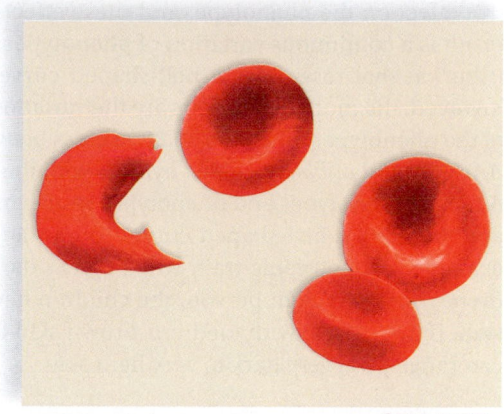

FIGURE 10.8B
In persons with sickle-cell disease, the red blood cells are sickle-shaped.

SEM, 7,000×

late 1700s and early 1800s, many members of the British royal family suffered from this disorder, which can lead to epileptic convulsions, bizarre behavior, and coma.

In a person suffering from sickle-cell disease with the genotype Hb^SHb^S, as described in section 10.6, the cells are sickle-shaped (Fig. 10.8B). The abnormally shaped cells slow down blood flow and clog small blood vessels. In addition, sickled red blood cells have a shorter life span than normal red blood cells. Affected individuals may exhibit a number of symptoms, including severe anemia, physical weakness, poor circulation, impaired mental function, pain, high fever, rheumatism, paralysis, spleen damage, low resistance to disease, and kidney and heart failure.

Although sickle-cell disease is a devastating disorder, it provides heterozygous individuals with a survival advantage. People who have sickle-cell trait are resistant to the protozoan parasite that causes malaria. The parasite spends part of its life cycle in red blood cells feeding on hemoglobin, but it cannot complete its life cycle when sickle-shaped cells form and break down earlier than usual.

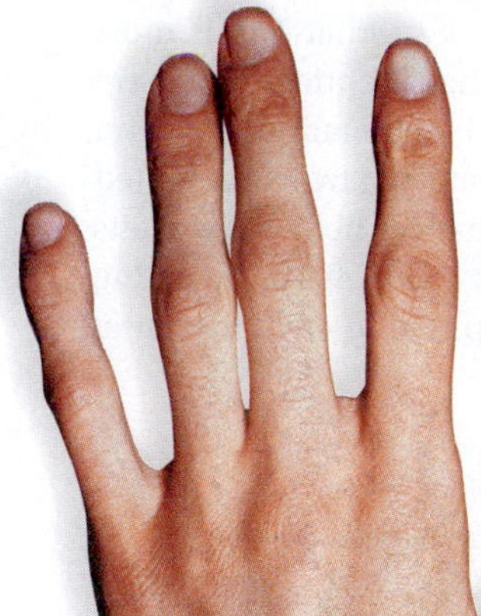

FIGURE 10.8A a. Very long fingers and toes are characteristic of Marfan syndrome. **b.** Defective connective tissue leads to the various characteristics of Marfan syndrome.

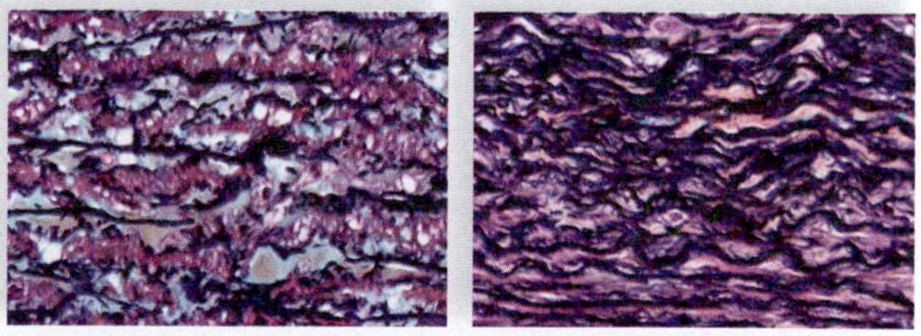

Marfan syndrome   Normal

b. Connective tissue contrast

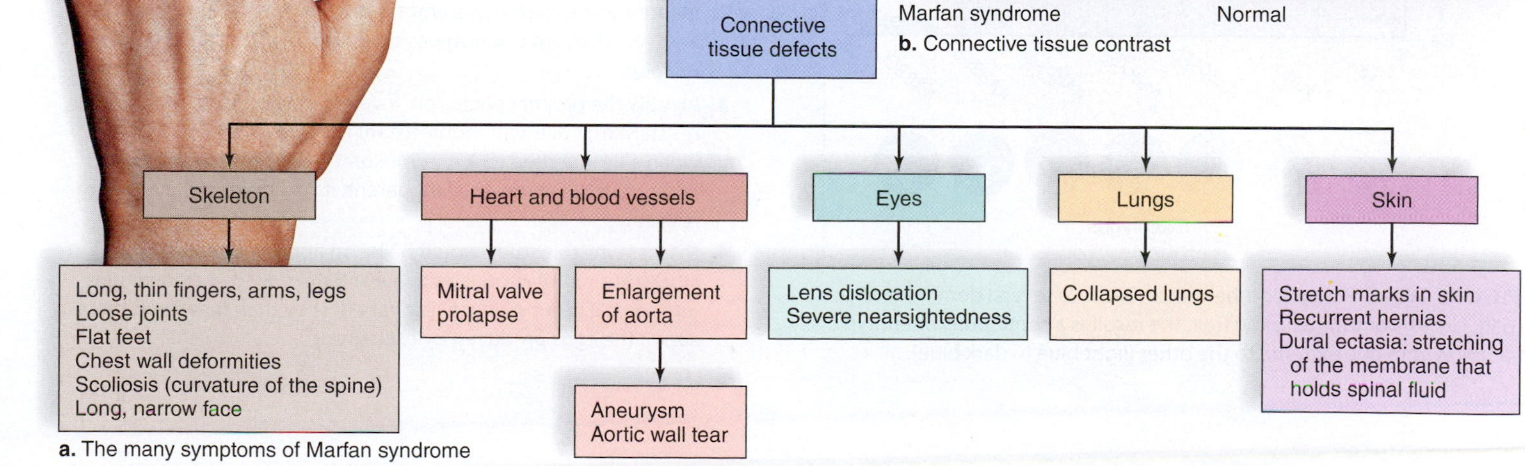

Connective tissue defects

| Skeleton | Heart and blood vessels | Eyes | Lungs | Skin |

| Long, thin fingers, arms, legs
Loose joints
Flat feet
Chest wall deformities
Scoliosis (curvature of the spine)
Long, narrow face | Mitral valve prolapse | Enlargement of aorta | Lens dislocation
Severe nearsightedness | Collapsed lungs | Stretch marks in skin
Recurrent hernias
Dural ectasia: stretching of the membrane that holds spinal fluid |

Aneurysm
Aortic wall tear

a. The many symptoms of Marfan syndrome

Polygenic Inheritance Several allelic pairs govern a trait controlled by polygenes. Each dominant allele has a quantitative effect on the phenotype, and these effects are additive. The result is a continuous variation of phenotypes, resulting in a distribution that resembles a bell-shaped curve. The more genes involved, the more continuous are the variations and distribution of the phenotypes. In Figure 10.8C, a cross between the genotypes *AABBCC* and *aabbcc* yields F₁ hybrids with the genotype *AaBbCc*. A range of genotypes and phenotypes results in the F₂ generation, and therefore a bell-shaped curve. To give an easily understood example of a polygenic trait, when a very dark person has children with a very light person, the children have medium-brown skin. If two people with medium-brown skin have children, they can range from very dark to very light skin.

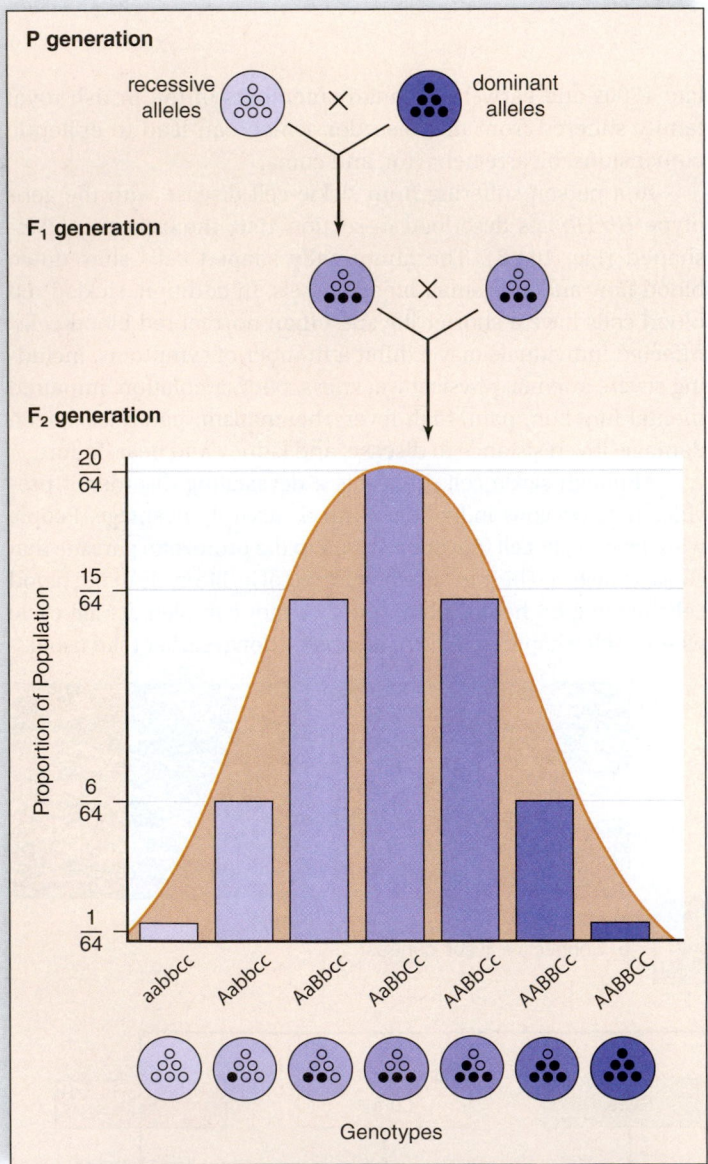

FIGURE 10.8C Polygenic inheritance: Because several dominant alleles (dark circles) contribute to the trait, the result is a continuous phenotypic variation from one extreme to the other (light blue to dark blue).

Multifactorial traits are controlled by polygenes but are also subject to environmental influences. Recall that rapid weight gain possibly contributes to the occurrence of hip dysplasia in dogs, as discussed in the introduction to this chapter. In humans, skin color and disorders such as cleft lip and/or palate, clubfoot, congenital dislocations of the hip, hypertension, diabetes, schizophrenia, and even allergies and cancers are likely due to the combined action of many genes plus environmental influences. In Figure 10.8C brown shading was added to show that, for example, skin color is influenced by exposure to the sun.

Reports have surfaced in recent years that all sorts of behaviors, including alcoholism, phobias, and even suicide, can be associated with particular genes. No doubt, behavioral traits are somewhat controlled by genes, but it is impossible at this time to determine to what degree. And very few scientists would support the idea that these behavioral traits are predetermined by our genes.

The relative importance of genetic and environmental influences on the phenotype can vary, but in some instances the environment seems to have an extreme effect. One interesting study showed that cardiovascular disease is more prevalent among offspring whose biological *or adoptive* parents had cardiovascular disease. Can you suggest environmental ways that adoptive parents can bring on cardiovascular disease in their children, based on your study of fats in Chapter 3?

These examples lend support to the belief that human traits controlled by polygenes are likely also subject to environmental influences. Therefore, many investigators are trying to determine what percentage of various traits is due to nature (inheritance) and what percentage is due to nurture (the environment). Some studies use twins separated since birth, because if identical twins in different environments share the same trait, that trait is most likely inherited. These studies suggest that identical twins are more similar in (1) intellectual talents, (2) personality traits, and (3) levels of lifelong happiness than are fraternal twins separated at birth. However, biologists conclude that all behavioral traits are partly heritable, and that genes exert their effects by acting together in complex combinations susceptible to environmental influences.

▶ **10.8 CHECK YOUR PROGRESS**

1. Discuss why cystic fibrosis (CF) should also be considered a pleiotropic disorder.

2. Identify seven genotypes among the offspring that will result in seven different phenotypes when *AaBbCc* is crossed with *AaBbCc*.

3. Identify the proper conclusion: Investigators crossed insecticide resistant fruit flies with nonresistant fruit flies. They found that when a fly inherited its X chromosome and chromosomes 2 and 3 from its insecticide resistant parent, it had maximal insecticide resistance.

4. Discuss why this study would be be an approved method for studying a multifactorial trait: Investigators divided cloned hens into several groups whose diet varied. They then measured the size of the eggs produced by each group.

Sex-Linked Inheritance

The trait white eye in *Drosophila,* the fruit fly, was the first allele to be definitively assigned to a chromosome—in this case, the X chromosome. X-linked alleles have an unusual inheritance pattern because the Y chromosome does not have a corresponding allele. Some human disorders have the X-linked pattern of inheritance.

10.9 Traits transmitted via the X chromosome have a unique pattern of inheritance

LEARNING OUTCOME

When you complete this section, you should be able to

1. Understand X-linked recessive inheritance and be able to describe the genotype of males and females for an X-linked trait.

By the early 1900s, investigators had noted the parallel behavior of chromosomes and genes during meiosis (see Fig. 9.3B), but they were looking for further data to support their belief that the genes were located on the chromosomes. A Columbia University group, headed by Thomas Hunt Morgan, performed the first experiments definitely linking a gene to a chromosome. This group worked with fruit flies (*Drosophila*). Fruit flies are even better subjects for genetic studies than garden peas: They can be easily and inexpensively raised in simple laboratory glassware; females mate and then lay hundreds of eggs during their lifetimes; and the generation time is short, taking only about 10 days when conditions are favorable.

Drosophila flies have the same sex chromosome pattern as humans, and this facilitates our understanding of a cross performed by Morgan. Morgan took a newly discovered mutant male with white eyes and crossed it with a red-eyed female. All of the offspring had red eyes; therefore, he knew that red eyes are the dominant characteristic and white eyes are the recessive characteristic. He then crossed the F_1 flies. The F_2 generation showed the expected 3 red-eyed: 1 white-eyed ratio, but it struck Morgan as odd that all of the white-eyed flies were males:

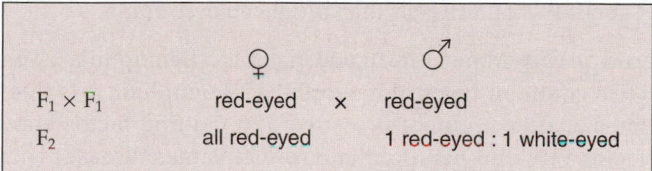

	♀		♂
$F_1 \times F_1$	red-eyed	×	red-eyed
F_2	all red-eyed		1 red-eyed : 1 white-eyed

Obviously, a major difference between the male flies and the female flies was their sex chromosomes. Could it be possible that an allele for eye color was on the Y chromosome but not on the X? This idea could be quickly discarded because usually females have red eyes, and they have no Y chromosome. But perhaps an allele for eye color was on the X chromosome, and not on the Y chromosome. Figure 10.9 indicates that this explanation matches the results obtained in the experiment. Therefore, the white-eye allele must be on the X chromosome. Once investigators had discovered a number of genes on one chromosome, they used crossing-over frequencies to determine the sequence of alleles on a chromosome, even human chromosomes. Today the accepted procedure for "mapping the chromosomes" is purely molecular (see Chapter 13).

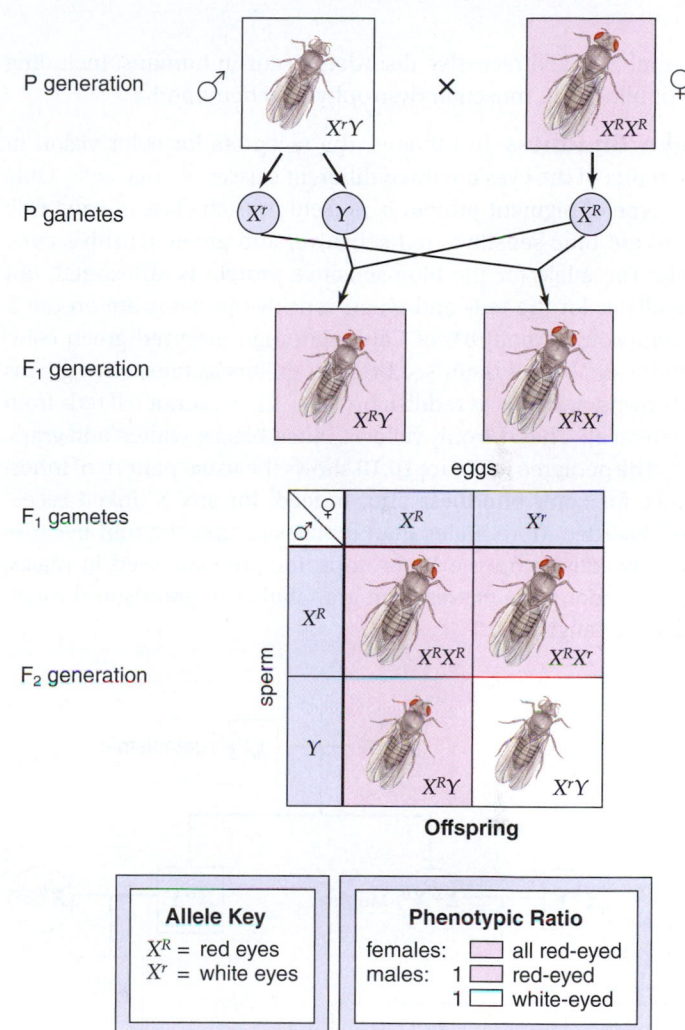

FIGURE 10.9 X-linked recessive inheritance.

Pattern of Inheritance Notice that X-linked alleles have a different pattern of inheritance than alleles on the autosomes because the Y chromosome lacks these alleles, and the inheritance of a Y chromosome cannot offset the inheritance of an X-linked recessive allele. For the same reason, affected males always receive an X-linked recessive mutant allele from the female parent—they receive the Y chromosome from the male parent.

▶ **10.9 CHECK YOUR PROGRESS**

1. Identify the genotype of a white-eyed male? A white-eyed female?

LEARNING OUTCOMES

When you complete this section, you should be able to

1. Distinguish between a pedigree of autosomal and X-linked disorders.
2. List some common X-linked disorders and state the symptoms of each.

Several X-linked recessive disorders occur in humans, including color blindness, muscular dystrophy, and hemophilia.

Color Blindness In humans, the receptors for color vision in the retina of the eyes are three different classes of cone cells. Only one type of pigment protein is present in each class of cone cell; there are blue-sensitive, red-sensitive, and green-sensitive cone cells. The allele for the blue-sensitive protein is autosomal, but the alleles for the red- and green-sensitive proteins are on the X chromosome. About 8% of Caucasian men have red-green color blindness. Most of them see brighter greens as tans, olive greens as browns, and reds as reddish browns. A few cannot tell reds from greens at all. They see only yellows, blues, blacks, whites, and grays.

The pedigree in Figure 10.10 shows the usual pattern of inheritance for color blindness and, indeed, for any X-linked recessive disorder. More males than females exhibit the trait because recessive alleles on the X chromosome are expressed in males. The disorder often passes from grandfather to grandson through a carrier daughter.

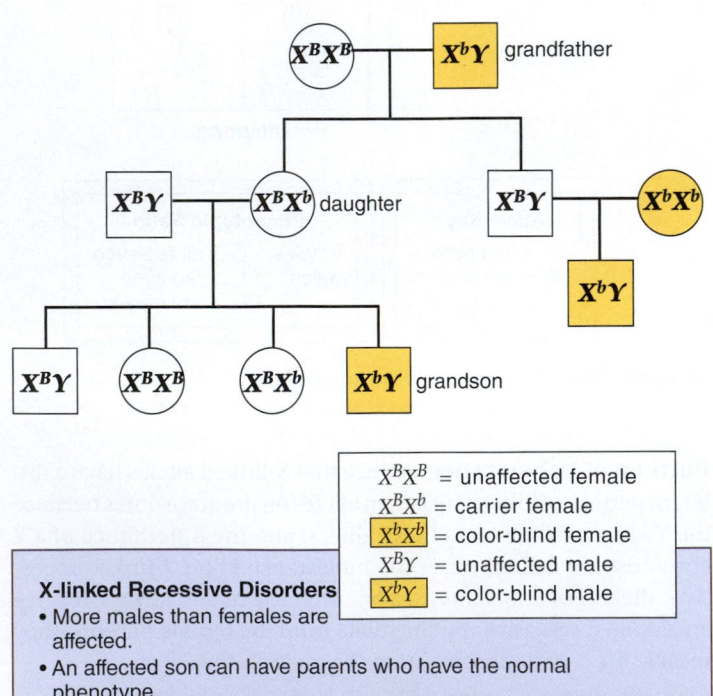

$X^B X^B$ = unaffected female
$X^B X^b$ = carrier female
$X^b X^b$ = color-blind female
$X^B Y$ = unaffected male
$X^b Y$ = color-blind male

X-linked Recessive Disorders
- More males than females are affected.
- An affected son can have parents who have the normal phenotype.
- For a female to have the characteristic, her father must also have it. Her mother must have it or be a carrier.
- The characteristic often skips a generation from the grandfather to the grandson.
- If a woman has the characteristic, all of her sons will have it.

FIGURE 10.10 X-linked recessive pedigree.

Muscular Dystrophy Muscular dystrophy, as the name implies, is characterized by wasting away of the muscles. The most common form, Duchenne muscular dystrophy, is X-linked and occurs in about 1 out of every 3,600 male births. Symptoms, such as waddling gait, toe walking, frequent falls, and difficulty rising, may appear as soon as the child starts to walk. Muscle weakness intensifies until the individual is confined to a wheelchair. Death usually occurs by age 20; therefore, affected males are rarely fathers. The recessive allele remains in the population through passage from carrier mother to carrier daughter.

The allele for Duchenne muscular dystrophy has been isolated, and researchers have discovered that the absence of a protein called dystrophin causes the disorder. Much investigative work determined that dystrophin is involved in the release of calcium from the endoplasmic reticulum in muscle fibers. The lack of dystrophin causes calcium to leak into the cell, which promotes the action of an enzyme that dissolves muscle fibers. When the body attempts to repair the tissue, fibrous tissue forms, and this cuts off the blood supply so that more and more cells die. In the meantime, the calves of the legs enlarge.

A test is now available to detect carriers of Duchenne muscular dystrophy. Also, various treatments have been tried. Immature muscle cells have been injected into muscles, and for every 100,000 cells injected, dystrophin production occurs in 30–40% of muscle fibers. The allele for dystrophin has been inserted into thigh muscle cells, and about 1% of these cells then produced dystrophin.

Calves enlarge due to fibrous tissue.

Hemophilia About 1 in 10,000 males is a hemophiliac. There are two common types of hemophilia: Hemophilia A is due to the absence or minimal presence of a clotting factor known as factor VIII, and hemophilia B (or Christmas disease) is due to the absence of clotting factor IX. Hemophilia is called the bleeder's disease because the affected person's blood either does not clot or clots very slowly. Although hemophiliacs bleed externally after an injury, they also bleed internally, particularly around joints. Hemorrhages can be stopped with transfusions of fresh blood (or plasma) or concentrates of the clotting protein. Also, clotting factors are now available as biotechnology products.

▶ **10.10 CHECK YOUR PROGRESS**

1. Identify the genotypes of all the individuals involved when a female with normal vision and a color-blind male have a daughter who is color blind.
2. Identify the possible genotypes of offspring when a homozygous red-eyed *Drosophila* female reproduces with a white-eyed male.
3. Identify the *Drosophila* cross that would produce white-eyed males: (a) $X^R X^R \times X^r Y$ or (b) $X^R X^r \times X^R Y$. In what ratio?

CONNECTING THE CONCEPTS

Working with the garden pea and using the laws of probability, Mendel gave us two laws of genetics that apply to all organisms, including humans. The first of Mendel's laws tells us that an individual has two alleles, but the gametes have only one allele for every trait. The second law tells us that the gametes have all possible combinations of alleles. This increases variability among the offspring. Today we know that Mendel's laws are consistent with our knowledge of chromosome behavior during meiosis as long as the alleles are on separate chromosomes (see Fig. 9.3B). Mendel was fortunate to be working with nonlinked genes because we now know that the alleles on one chromosome tend to stay together during the process of meiosis, except when crossing-over occurs.

Certain patterns of inheritance, such as polygenic inheritance and X-linked inheritance, do not negate, but rather extend, the range of Mendelian analysis. Males are more apt than females to display an X-linked disorder because they receive only one set of X-linked alleles from their mother. Their father gives them a Y chromosome, which is blank for these alleles.

Just as Mendelian genetics proposes, genes have loci on the chromosomes, but today we know that genes are composed of DNA and that the genes are on the chromosomes. In Chapter 11, we will learn that the sequence of the bases in each gene determines the sequence of amino acids in a protein. It is our proteins that make us who we are.

ANALYZE AND EVALUATE

1. Considering that Mendel applied the laws of probability to his crosses, why did he count so many F_2 results?
2. Modern genetics also uses meiosis to understand Mendel's results. Why does understanding meiosis help?
3. Of what value is Mendelian genetics when it does not tell us how genes actually function?

SUMMARIZE

Mendel's Laws

10.1 Mendel developed a particulate model of inheritance

- According to the blending concept, parents of contrasting appearance always produce offspring of intermediate appearance. In contrast, Mendel developed a particulate model of inheritance. He used the garden pea, an excellent experimental organism because it was easy to grow, had a short generation time, and produced many offspring. Mendel said inheritance involves factors now called genes and a reshuffling of the same genes to offspring.

10.2 Mendel's law of segregation describes how gametes pass on traits

- Mendel first did **monohybrid crosses** that involved only one trait. He began with a **P generation,** and that was followed by an **F_1 generation** and then an **F_2 generation.**
- Mendel's mathematical approach allowed him to formulate the **law of segregation:** An individual has two factors for each trait, which separate during gamete formation so that each gamete contains only one factor from each pair.
- Today we say that each trait has two **alleles** (alternate forms of a gene), and the **dominant allele** masks expression of the **recessive allele.** A **homozygous** dominant individual has two copies of the dominant allele; a homozygous recessive individual has two copies of the recessive allele. A **heterozygous** organism has one of each type of allele at a **gene locus.** Genotype refers to the genes of an individual. **Phenotype** refers to an individual's physical appearance. Based on the law of segregation, $Tt \times Tt$ gives these results:

Gametes	Phenotypic Ratio
T = tall plant	3 ☐ tall
t = short plant	1 ☐ short

- All the alleles on one chromosome form a **linkage group,** which can be broken by crossing-over.

10.3 Mendel's law of independent assortment describes inheritance of multiple traits

- Mendel's two-trait crosses, called **dihybrid crosses,** allowed him to formulate the **law of independent assortment.** Each pair of factors on separate chromosomes assorts (separates) independently (see Fig. 9.3B). Therefore, all possible combinations of factors can occur in the gametes daughter cells, so that if $TtGg \times TtGg$, then:

Gametes	Phenotypic Ratio
TG	9 ☐ tall plant, green pod
Tg	3 ☐ tall plant, yellow pod
tG	3 ☐ short plant, green pod
tg	1 ☐ short plant, yellow pod

10.4 Support for Mendel's laws is various

- Consistent with Mendel's laws, a heterozygous individual crossed with a homozygous recessive individual results in a 1:1 phenotypic ratio among the offspring. This **testcross** determines whether a dominant phenotype is homozygous dominant or heterozygous.
- An individual heterozygous for two traits crossed with an individual recessive for those traits results in a 1:1:1:1 phenotypic ratio.
- Mendel's results are consistent with the rules of probability. The probability of genotypes and phenotypes in the next generation can be calculated using a **Punnett square.**

Mendel's Laws Apply to Humans

10.5 Pedigrees can reveal the patterns of inheritance

- **Pedigrees** show the patterns of inheritance for particular conditions. Most alleles are on the **autosomal chromosomes** (nonsex chromosomes). When a trait is autosomal recessive, the child may be affected while the parents are not. When a trait is autosomal dominant, the child may be unaffected even though a parent is affected. **Carriers** appear normal but are capable of parenting a child with a genetic disorder.

10.6 Some human genetic disorders are autosomal recessive and some are autosomal dominant

- Tay-Sachs disease, cystic fibrosis, PKU, and sickle-cell disease are examples of autosomal recessive genetic disorders controlled by a single pair of alleles on a pair of autosomal chromosomes. Neurofibromatosis, Huntington disease, and achondroplasia are examples of autosomal dominant genetic disorders controlled by a single pair of alleles on a pair of autosomal chromosomes.

Beyond Mendel's Laws

10.7 Variations in the recessive/dominant allele relationship occur

- In **incomplete dominance,** a heterozygote has the intermediate phenotype between its homozygous parents (e.g., pink color in four o'clocks). In the F_2 generation, all three genotypes reappear.
- **Multiple alleles** control a trait when the gene exists in several allelic forms. The inheritance of the ABO blood group in humans is an example of multiple alleles and **codominance.**

10.8 Variations in the one gene–one trait relationship occur

- **Pleiotropy** occurs when a single gene has more than one effect, and often leads to a syndrome. Examples are Marfan syndrome and sickle-cell disease.

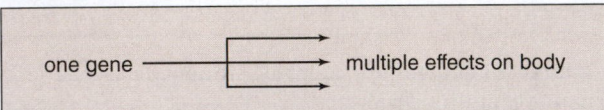

- In **polygenic inheritance,** a trait is governed by two or more sets of alleles, and continuous variation of phenotypes results in a bell-shaped curve.

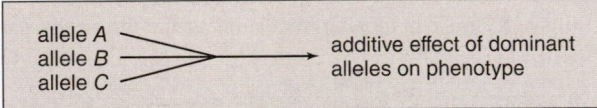

- **Multifactorial traits** are controlled by polygenes subject to environmental influences.

Sex-Linked Inheritance

10.9 Traits transmitted via the X chromosome have a unique pattern of inheritance

- In X-linkage, females have two X chromosomes and therefore can be homozygous or heterozygous. Males have only one X chromosome (the Y lacks X-linked alleles), and therefore a recessive allele on the X chromosome always shows itself in males. Therefore, more males than females are affected in the pedigree.

10.10 Humans have X-linked recessive disorders

- X-linked recessive disorders in humans include color blindness, muscular dystrophy, and hemophilia. More males than females have X-linked disorders because the recessive allele cannot be masked by a dominant allele. An X-linked recessive pedigree indicates that the trait often passes from grandfather through a carrier daughter to a grandson.

TEST YOURSELF

Mendel's Laws

1. Peas are good for genetics studies because they
   a. cannot self-pollinate.
   b. have a long generation time.
   c. are easy to grow.
   d. have fewer traits than most plants.
2. **THINKING CONCEPTUALLY** Explain how a blending model of inheritance would not provide the variation needed for evolution to occur, but particulate inheritance, as revealed by Mendel, does support evolution.
3. Which of the following is *not* consistent with the law of segregation?
   a. Each gamete contains one factor from each pair of factors in the parent.
   b. Factors segregate during gamete formation.
   c. Following fertilization, the new individual carries two factors for each trait.
   d. Each individual has one factor for each trait.
4. A fruit fly heterozygous for long wings is crossed with another of the same genotype. What is the genotype of both flies?
   a. *LL*   c. *Ll*
   b. *ll*   d. *LlLl*
5. What phenotypic ratio is expected among the offspring of the flies described in question 4.
   a. 2:1   c. 4:1
   b. 1:1   d. 3:1
6. A testcross involving a heterozygous fly from question 4 gave a phenotypic ratio of 1:1. Which of these describes the ratio?
   a. one long-winged fly per every short-winged fly
   b. one homozygous dominant fly per every homozygous recessive fly
   c. one recessive fly per every parental fly
   d. All of these are correct.
7. If two parents with short fingers (dominant) have a child with long fingers, what is the chance their next child will have long fingers?
   a. 25%   c. 75%
   b. 50%   d. 100%
8. In humans, pointed eyebrows (*B*) are dominant over smooth eyebrows (*b*). Mary's father has pointed eyebrows, but she and her mother have smooth. What is the genotype of the father?
   a. *BB*   d. *BbBb*
   b. *Bb*   e. Any one of these may be correct.
   c. *bb*
9. **THINKING CONCEPTUALLY** How does Figure 10.2B demonstrate that genes belong to linkage groups?
10. According to the law of independent assortment,
    a. all possible combinations of factors can occur in the gametes.
    b. only the parental combinations of gametes can occur in the gametes.
    c. only the nonparental combinations of gametes can occur in the gametes.
11. Which of these is the genotype of a dihybrid?
    a. *LlGG*   c. *LLgg*
    b. *LlGg*   d. None of these are correct.

12. A fly heterozygous for two traits is crossed with another of the same genotype. What results are expected?
    a. 3:1
    b. 1:1
    c. 9:3:3:1
    d. 1:1:1:1
13. A testcross could be a cross between
    a. *aaBB* × *AABB*
    b. *AABb* × *A?Bb*
    c. *A?B?* × *aabb*
14. Determine the probability that an *aabb* individual will be produced from an *AaBb* × *aabb* cross.
    a. 50%
    b. 25%
    c. 75%
    d. 100%
    e. 0%
15. **THINKING CONCEPTUALLY** A couple is concerned about preserving human life once fertilization has occurred. Which procedure would you recommend (fetal, embryonic, or egg testing) to detect a genetic disorder? Explain. (See Application 10B on page 181.)

Mendel's Laws Apply to Humans

For questions 16–20, match each description to a condition in the key.

KEY:
    a. Tay-Sachs disease
    b. cystic fibrosis
    c. phenylketonuria (PKU)
    d. sickle-cell disease
    e. Huntington disease

16. Abnormal hemoglobin
17. The most common lethal genetic disorder among U.S. Caucasians
18. Results from lack of the enzyme hex A, resulting in the storage of its substrate in lysosomes
19. Results from the inability to metabolize the amino acid phenylalanine
20. Late-onset neuromuscular genetic disorder
21. Two affected parents have an unaffected child. The trait involved is
    a. autosomal recessive.
    b. incompletely dominant.
    c. controlled by multiple alleles.
    d. autosomal dominant.

Beyond Mendel's Laws

22. If a man of blood group AB marries a woman of blood group A whose father was type O, what phenotypes could their children be?
    a. A only
    b. A, AB, B, and O
    c. AB only
    d. A, AB, and B
    e. O only
23. An anemic person has a number of problems, including lack of energy, fatigue, rapid pulse, pounding heart, and swollen ankles. This could be an example of
    a. pleiotropy.
    b. sex-linked inheritance.
    c. polygenic inheritance.
    d. codominance.

Sex-Linked Inheritance

24. Using *A* and *a* for the alleles, what is the genotype of the starred individual.

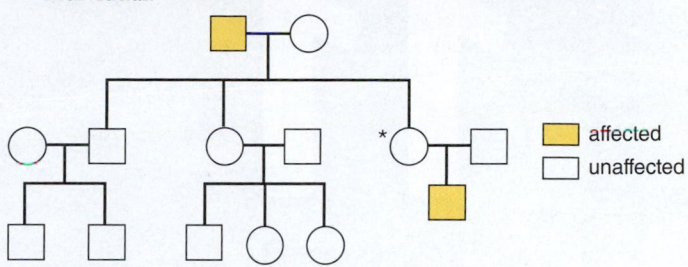

GET INVOLVED

1. You want to determine whether a newly found *Drosophila* characteristic is dominant or recessive. Would you wait to cross this male fly with another with this characteristic or cross it now with a fly that lacks the characteristic?
2. You want to test whether the leaf pattern of a plant is influenced by the amount of fertilizer in the environment. What would you do?
3. The virtual lab "Punnett Squares" provides an interactive environment in which to test your understanding of the principles of Mendelian inheritance and your ability to do monohybrid problems.

MEDIA STUDY TOOLS

mhhe.com/maderconcepts3

Enhance your study of this chapter with interactive study tools, practice tests, and engaging animations. Also, ask your instructor about the resources available through ConnectPlus, which includes LearnSmart, a personalized adaptive learning program, and a media-rich eBook.

11

Molecular Biology of Inheritance

BEFORE YOU BEGIN

Take a few minutes to recall

The levels of protein organization (section 3.8)

The structure of DNA and RNA and the flow of genetic information in a cell (section 3.9)

How the nucleus and the ribosomes work together (section 4.4)

Some genetic disorders of humans (sections 10.6 and 10.10)

Arabidopsis Is a Model Organism

Arabidopsis thaliana is a small flowering plant related to cabbage and mustard plants. *Arabidopsis* has no commercial value—in fact, it is a weed! However, it has become a model organism for the study of plant molecular genetics. Other model organisms in genetics are Mendel's peas and Morgan's fruit flies. Work with these models produces results that apply to many different organisms. For example, Mendel's two laws are generally applicable to all organisms, and Morgan's discovery of X-linkage is applicable to nearly all sexually reproducing animals. *Arabidopsis* is a very useful model organism for these reasons:

- It is small, so many hundreds of plants can be grown in a small amount of space. *Arabidopsis* consists of a flat rosette of leaves from which grows a flower stalk.
- Generation time is short. It takes only 5–6 weeks for plants to mature, and each one produces about 10,000 seeds!
- *Arabidopsis* normally self-pollinates, but it can easily be cross-pollinated. This feature facilitates gene mapping and the production of strains with multiple mutations.
- The number of base pairs in its DNA is larger than you might suppose: 115,409,949 base pairs are distributed in 5 chromosomes (2n = 10) and 23,000 genes.

When Mendel worked with peas toward the end of the nineteenth century, he merely hypothesized that parents must pass genetic factors (now

Arabidopsis thaliana

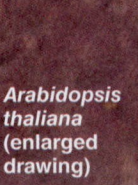

Arabidopsis thaliana (enlarged drawing)

called alleles) to their offspring. A progression of genetic studies in the twentieth century first established that DNA is the genetic material. Then scientists discovered the structure of DNA and how it functions in a cell. Today, the emphasis is on studying the specific function of individual genes, which are segments of DNA.

Every organism has its own sequence of DNA bases, and we know the order of the bases for many organisms, including humans and *Arabidopsis*. Irradiating the seeds of *Arabidopsis* causes a mutation, a change in the normal sequence of bases. The creation of *Arabidopsis* mutants plays a significant role in revealing what each of its genes do. For example, if a mutant plant lacks stomata (openings in leaves), we know that the affected gene influences the formation of stomata.

Think of Mendel working in an abbey garden, and then think of today's laboratory, where researchers study model organisms with the aid of advanced, high-speed equipment. Amazing, too, is the recognition that, just like Mendel's peas, the work with *Arabidopsis* can assist our understanding of how human genes function. In this chapter, we begin our study of molecular genetics by examining the structure of DNA with the surety that structure suits function. Our ultimate goal is to show that DNA meets the criteria for the genetic material: (1) DNA is variable between species and able to store the information that makes species different. (2) DNA is constant within a species and able to be replicated with high fidelity during cell division. (3) DNA is able to undergo rare changes, called mutations, that provide the raw material needed for evolution to occur.

Arabidopsis flower

Mutated flower

Mutated flower

A flat of *Arabidopsis*

Lab

DNA and RNA Structure and Function

Watson and Crick deduced the structure of DNA from the data gathered by other scientists. Chemists told them that nucleic acids are polymers of nucleotides and that the bases of DNA can be in any order but the base adenine symbolized as A always paired with T (thymine) while C (cytosine) is always paired with G (guanine). X-ray diffraction told them that DNA is a double helix. The structure of DNA explains how it can have a distinctive sequence of paired bases in each species and yet replicate faithfully.

11.1 DNA and RNA are polymers of nucleotides

LEARNING OUTCOMES

When you complete this section, you should be able to

1. Differentiate between the structure of DNA and RNA.
2. Understand how the structure of DNA provides both variability and constancy.

It wasn't until the early 1950s that James Watson, a postdoctoral student, and Francis Crick, a biophysicist, deduced the double helical structure of DNA after studying the data that had been gathered by previous scientists. They can be likened to a person who has the parts of a bicycle laid out on the grass and is trying to figure out how to put the parts together to make the bicycle run. Their achievement was to create a complete structure of DNA from knowing only its parts. Once the structure of DNA was known, biologists could determine how it functioned. What *did* Watson and Crick know when they started to build their model of DNA?

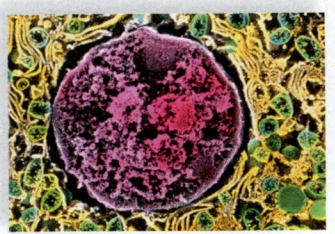
False colored EM of nucleus

DNA and RNA We want to keep in mind that the structure of DNA suits its function, and so it is fitting to point out that an early investigator found DNA in the nucleus of cells! In 1869, the Swiss physician Johann Miescher removed nuclei from pus cells and found they contained a chemical he called nuclein. Nuclein, he said, was rich in phosphorus and had no sulfur, and these properties distinguished it from protein. Later, other chemists working with nuclein found that it had acidic properties. Therefore, they decided to call the molecule **nucleic acid.**

Nucleotides Early in the 1900s, researchers discovered that nucleic acids contain only nucleotides. Each generalized **nucleotide** has three parts: a pentose sugar (5-carbon sugar), a phosphate, and a nitrogen-containing base. In this representation of a nucleotide, the carbons in the sugar are numbered:

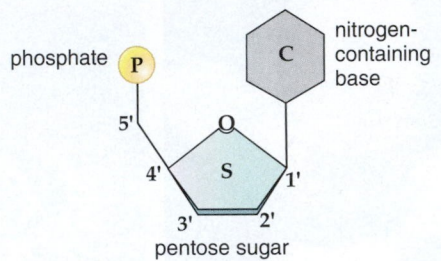

The particular sugar and base of a nucleotide can differ. For example, from Figure 11.1Aa, you can choose one of two sugars. If you substitute the sugar deoxyribose, you have constructed a DNA nucleotide, but if you substitute the sugar ribose, you have constructed an RNA nucleotide. To complete your DNA nucleotide, you have a choice of four bases, symbolized as **C, T, A,** and **G.**

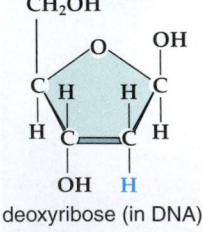

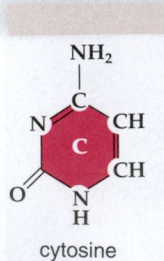
deoxyribose (in DNA) ribose (in RNA)

a. Sugars

FIGURE 11.1A a. A DNA nucleotide always has the sugar deoxyribose while a RNA nucleotide always has the sugar ribose. Notice that deoxyribose lacks an oxygen found in ribose. **b.** In DNA nucleotides, the four bases are symbolized as C, T, A, and G. In RNA nucleotides the four bases are C, U, A, and G. Notice that pyrimidines have only one organic ring while purines have two organic rings.

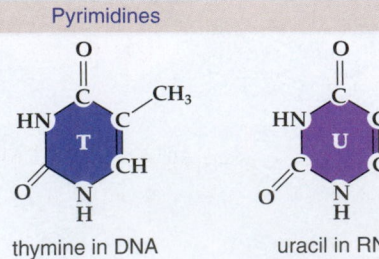

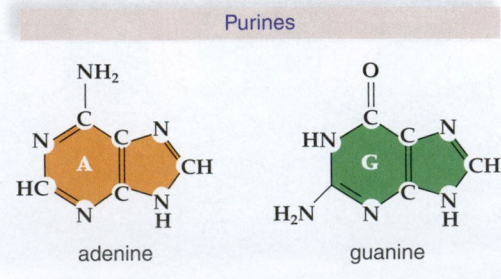

b. Bases

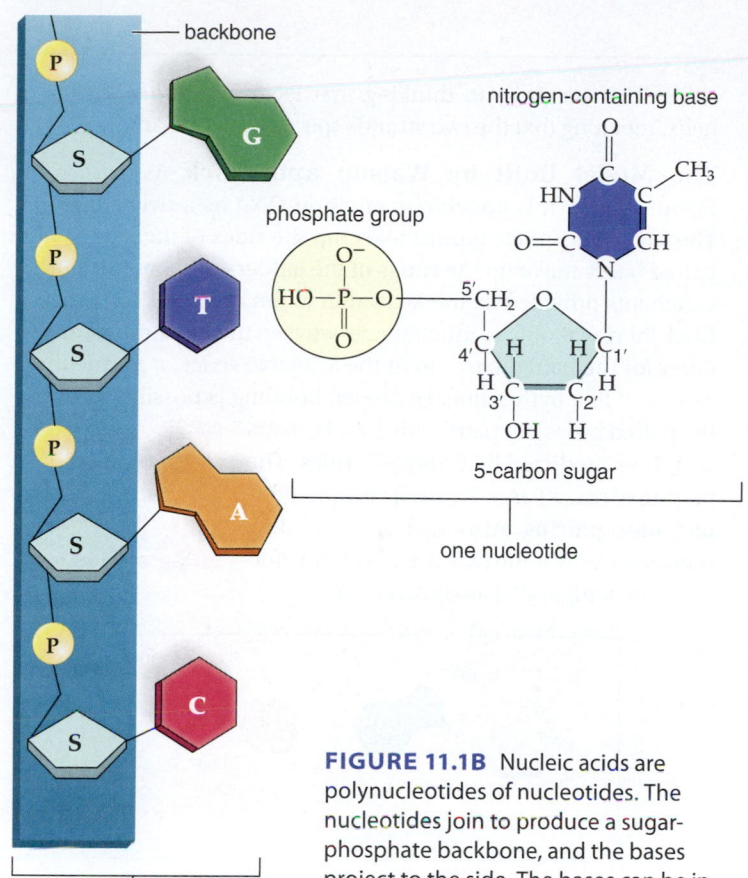

FIGURE 11.1B Nucleic acids are polynucleotides of nucleotides. The nucleotides join to produce a sugar-phosphate backbone, and the bases project to the side. The bases can be in any order.

FIGURE 11.1C The sequence of DNA bases differs in plants, the researcher Dian Fossey, and the gorillas. This is how DNA stores the genetic information that accounts for the phenotypic differences between species.

An RNA nucleotide is the same except for a **U** in place of the T. If you see a T in a nucleic acid, as in Figure 11.1B, you know you are dealing with DNA, but if you see a U, the molecule has to be an RNA. Some bases mentioned are pyrimidines (one ring), and some are purines (two rings) (Fig. 11.1A*b*).

Polymers In a nucleic acid, the nucleotides join in a particular way. Notice in Figure 11.1B how the sugars are joined by phosphate molecules. In the backbone of both DNA and RNA polymers, a sugar group continuously alternates with a phosphate group, and the bases project to the side. The dark bar in Figure 11.1B represents the backbone of the molecule. An important feature discovered by the chemist Erwin Chargaff is that the bases can be in any order. Without this aspect of its structure, DNA wouldn't have the variability needed to be a genetic material (Fig. 11.1C).

Chargaff's Rules With the development of new chemical techniques in the 1940s, it became possible for Erwin Chargaff to analyze the base content of the DNA of different species. At the time, researchers thought that perhaps the four bases repeated the same in all species, like this: ATGC, ATGC, ATGC.... If that were the case, all species would have the same sequence of bases. But Chargaff discovered that the base composition of DNA varies between species. A sample of his data is given in Figure 11.1D. You can see that,

DNA Composition in Various Species (%)				
Species	A	T	G	C
Homo sapiens (human)	31.0	31.5	19.1	18.4
Drosophila melanogaster (fruit fly)	27.3	27.6	22.5	22.5
Zea mays (corn)	25.6	25.3	24.5	24.6
Neurospora crassa (fungus)	23.0	23.3	27.1	26.6
Escherichia coli (bacterium)	24.6	24.3	25.5	25.6

FIGURE 11.1D Chargaff's data showed that the DNA base composition of various species differs. For example, in humans the A and T percentages are about 31%, but in fruit flies these percentages are about 27%. However, in all organisms, the amount of A = T and the amount of G = C.

while some species have approximately 25% of each type of nucleotide, most do not. Instead, the percentage of each type of nucleotide differs from species to species, as you would expect if each species has its own distinctive sequence of bases. Most important for Watson and Crick, Chargaff also discovered that, despite DNA's variability, the amount of A = T and the amount of G = C.

▶ **11.1 CHECK YOUR PROGRESS**

1. Contrast the chemical components of DNA and RNA.
2. Explain in what way DNA differs from one species to another species.

11.2 DNA is a double helix

When you complete this section, you should be able to

1. Analyze the significance of DNA's X-ray pattern to Watson and Crick.

2. Understand that complementary base pairing is dependent on hydrogen bonding.

In the mid-1950s, researchers were racing against each other to discover the structure of DNA. Much had already been learned. They knew that DNA is a polymer of nucleotides and, based on Chargaff's data, that the amount of base A = T and the amount of base C = G. But exactly how was the molecule put together? Watson and Crick, working at Cambridge, were much aided by the findings of Rosalind Franklin and Maurice H. F. Wilkins, who were at King's College of London.

X-ray Diffraction of DNA Franklin and Wilkins studied the structure of DNA using X-ray crystallography. Franklin found that if a concentrated, viscous solution of DNA is made, it can be separated into fibers. Under the right conditions, the fibers are enough like a crystal (a solid substance whose atoms are arranged in a definite manner) to produce an X-ray diffraction pattern (Fig. 11.2A, *left*). The X-ray diffraction pattern of DNA suggested to Watson and Crick that DNA is a **double helix.** The helical shape is indicated by the crossed (X) pattern in the center of the photograph in Figure 11.2A, *right*. The dark portions at the top and bottom of the photograph indicate that some portion of the double helix is repeated.

> **Video**
> DNA Dark Lady

Watson and Crick also knew that the structure of some proteins is helical and maintained by hydrogen bonding between amino acids. This set them to thinking that DNA is a *double-stranded* helix, meaning that the two strands spiral about one another.

The Model Built by Watson and Crick As shown in Figure 11.2B*a*, it is possible to envision DNA as a twisted ladder. The sugar-phosphate groups make up the sides of the ladder, and paired bases make up the rungs of the ladder. Mathematical measurements provided by the X-ray diffraction data told Watson and Crick there was only so much space between the two strands, and in order for the paired bases to fit the available space, a purine must be linked to a pyrimidine. Hydrogen bonding is possible between the paired bases if A pairs with T and C pairs with G as predicted by Chargaff's rules. The pairing of these bases is called **complementary base pairing.** Also, hydrogen bonding is possible only if the two strands of the molecule are antiparallel as shown here:

Animation
DNA Structure

3D Animation
DNA Replication:
DNA Structure

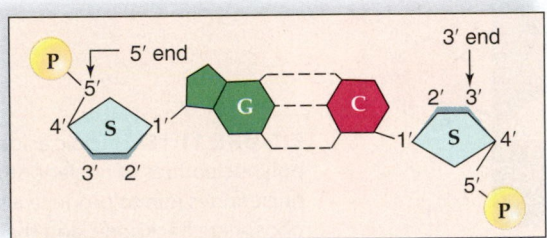

Notice in Figure 11.2B*b* that the sugars in the right strand are upside down with respect to the sugars in the left strand.

▶ **11.2 CHECK YOUR PROGRESS**

1. Discuss why DNA from *Arabidopsis* and humans have the same X-ray diffraction pattern.

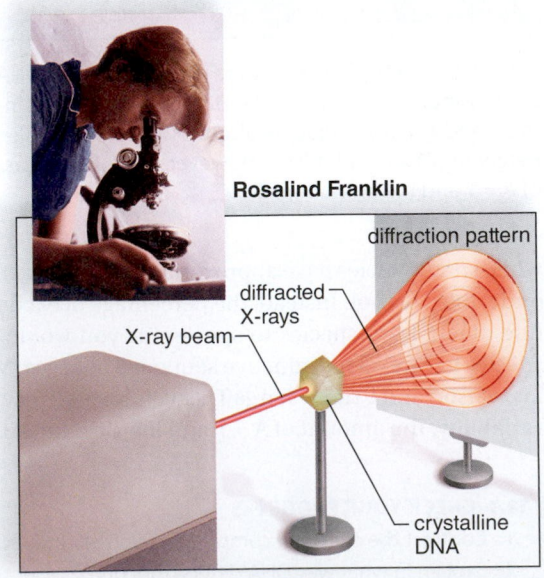

Rosalind Franklin

diffraction pattern

diffracted X-rays

X-ray beam

crystalline DNA

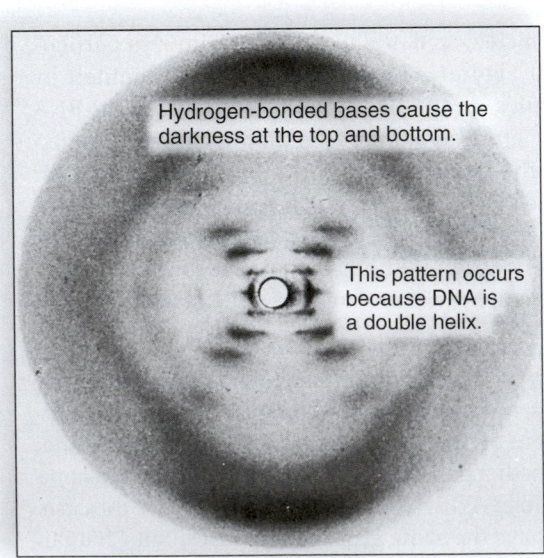

Hydrogen-bonded bases cause the darkness at the top and bottom.

This pattern occurs because DNA is a double helix.

FIGURE 11.2A *Left:* When a crystal is X-rayed, the way the beam is diffracted reflects the pattern of the molecules in the crystal. *Right:* DNA's diffraction pattern had an X in the center, telling Watson and Crick that DNA is a helix. The dark portions at the top and bottom told them that some feature is repeated over and over. Watson and Crick determined that the feature was the hydrogen-bonded bases.

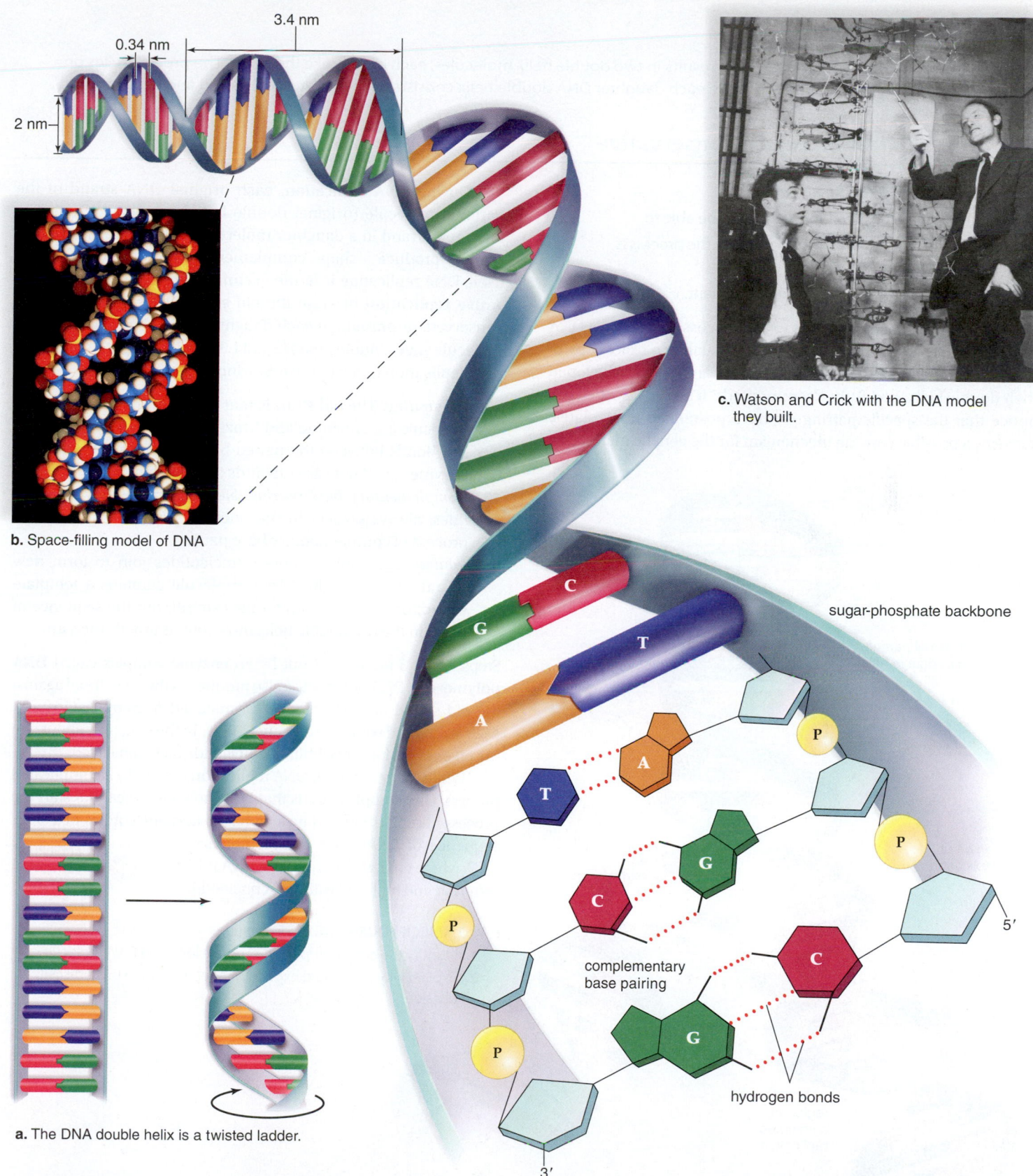

b. Space-filling model of DNA

c. Watson and Crick with the DNA model they built.

sugar-phosphate backbone

C

G

T

A

complementary base pairing

hydrogen bonds

5′

3′

a. The DNA double helix is a twisted ladder.

FIGURE 11.2B a. The ladder configuration for DNA. Twisting the ladder produces the double helix. **b.** Space-filling model and enlargement of the double helix. Each strand of the molecule has a sugar-phosphate backbone and among the bases, A is always hydrogen bonded to T, and C is always bonded to G. **c.** Watson and Crick used the X-ray diffraction pattern of DNA to build their double helix model.

DNA Replication

DNA replication is a duplication because it results in two double helix molecules, each exactly like the other. The process relies on complementary base pairing, and therefore each daughter DNA double helix consists of a template (old) strand and a new strand.

11.3 DNA replication is semiconservative

LEARNING OUTCOMES

When you complete this section, you should be able to

1. Summarize how DNA replicates and why the process is semiconservative.
2. Explain why replication errors are infrequent.

The term **DNA replication** refers to the process of copying a DNA molecule. Following replication, there is usually an exact copy of the DNA double helix. As soon as Watson and Crick developed their double helix model, they commented, "It has not escaped our notice that the specific pairing we have postulated immediately suggests a possible copying mechanism for the genetic material."

During DNA replication, each original DNA strand of the parental molecule (original double helix) serves as a template for a new strand in a daughter molecule. A **template** is a pattern used to produce a shape complementary to itself. DNA replication is termed **semiconservative replication** because the old strand is conserved, or present, in each daughter DNA molecule (new double helix) (Fig. 11.3).

Animation Meselson and Stahl Experiment

Animation DNA Replication

Replication requires the following steps:

1. *Unwinding.* The old strands that make up the parental DNA molecule are unwound and "unzipped" (i.e., the weak hydrogen bonds between the paired bases are broken). A special enzyme called helicase unwinds the molecule.
2. *Complementary base pairing.* New complementary nucleotides, always present in the nucleus, are positioned by the process of complementary base pairing.
3. *Joining.* The complementary nucleotides join to form new strands. Each daughter DNA molecule contains a template strand, or old strand, and a new strand, but the sequence of bases in the two double helix molecules is exactly the same.

Steps 2 and 3 are carried out by an enzyme complex called **DNA polymerase.** DNA polymerase also proofreads the new strand against the old strand and detects any mismatched nucleotides; usually, each is replaced with a correct nucleotide. In the end, only about one mistake occurs for every 1 billion nucleotide pairs replicated.

While easily outlined, DNA replication is actually a complicated process. One complication is that DNA polymerase cannot start the process of replication and instead has to add nucleotides to a short sequence of RNA bases complementary to DNA where replication starts. The RNA is later removed and replaced with DNA nucleotides.

3D Animation DNA Replication

▶ **11.3 CHECK YOUR PROGRESS**
1. Identify the critical steps that produce identical DNA molecules.
2. Explain why you would expect DNA replication to precede cell division.

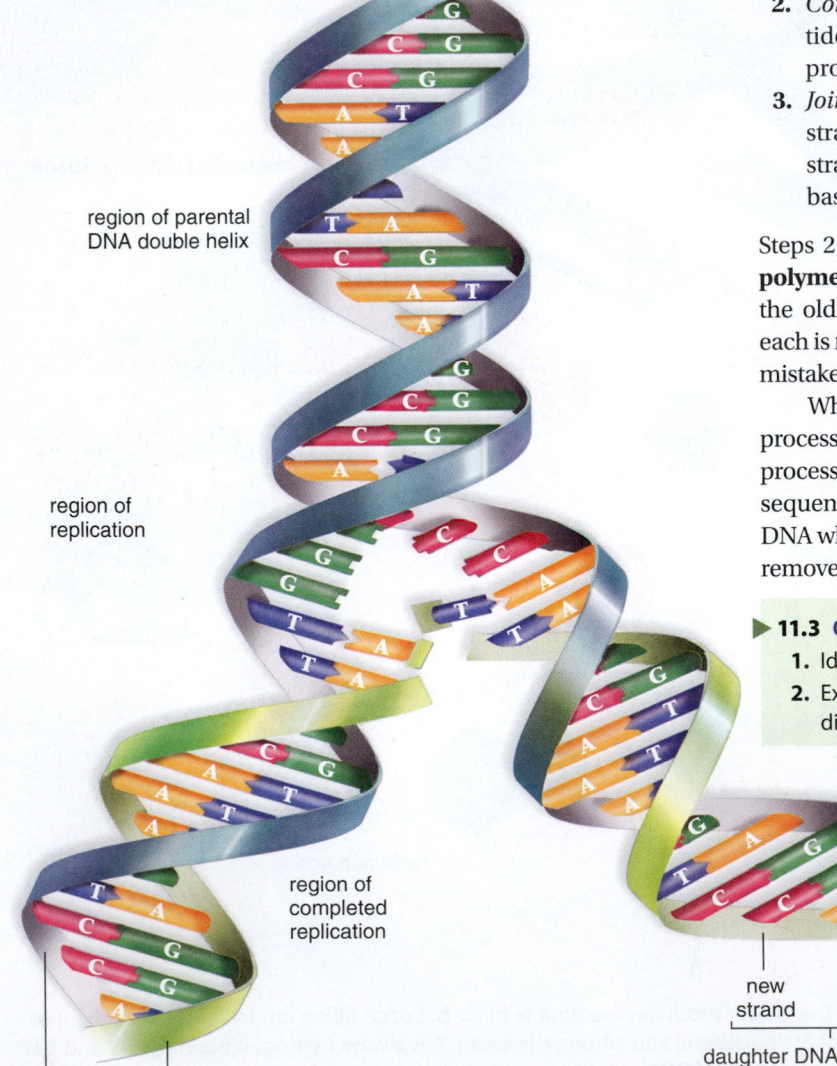

region of parental DNA double helix

region of replication

region of completed replication

template strand

new strand

daughter DNA double helix

new strand

template strand

daughter DNA double helix

FIGURE 11.3 Semiconservative replication. Notice that each new double helix contains a template (old) strand and a new strand.

11A DNA Replication in a Test Tube

The polymerase chain reaction (PCR) is a powerful molecular way to select and copy a particular segment of DNA in a test tube (Fig. 11A). It's called the polymerase chain reaction because the enzyme DNA polymerase is bringing about replication of a chosen segment over and over again. Once it starts, a chain of replications occurs until there are thousands of copies of the same DNA segment. The beauty of the process is its precision in selecting only a tiny portion of DNA to be copied. The primers used can seek out and bind to a segment of viral DNA, human DNA, plant DNA—or any particular type of DNA, as long as it is in the test tube. Analysis of the copied segment produces visual results.

Animation
Polymerase Chain
Reaction

The expression **"DNA fingerprinting"** refers to the use of PCR to identify a specific individual. Just like a traditional fingerprint, the analytical results are unique to the individual. There is only a slight chance a genetic fingerprint will be the same for any two people. To guard against even this possibility, however, PCR is done twice, using two different target segments for replication.

The applications for DNA fingerprinting are extremely varied, and some of them are quite dramatic. For example, Arthur Whitfield had already served 22 years for rape when DNA fingerprinting of a saved semen sample taken from the vagina of the rape victim showed he was not the rapist. On the basis of that evidence, the state of Virginia set him free. Any type of body sample, such as a drop of blood or even a single hair root from a crime scene, can be used to identify a suspect—or to clear a suspect. Investigators identified the victims of the September 11, 2001, terrorist attacks in the United States by comparing the DNA of body remains with that of a few cells taken from a personal object, such as a toothbrush. Parental testing can match an individual to close relatives. For example, the biological father of Anna Nicole Smith's daughter, Danielynn, was found to be photographer Larry Birkhead through the use of DNA fingerprinting.

Applications of PCR are limited only by our imaginations. Both vegetables and meats can be tested to detect specific bacterial or viral contamination. Medical applications are numerous. When the segment of DNA chosen for replication matches that of a virus or mutated gene, we know that a viral infection, a genetic disorder, or cancer is present. PCR-based diagnostic tests are available for various sexually transmitted diseases and also tuberculosis. Because it detects the presence of an infective agent itself, PCR testing produces fast, reliable results that are helping to make the nation's blood supply safe. It has also been noted that a negative PCR-based test for various infections of spinal fluid, such as spinal meningitis, can be relied on with confidence. PCR analysis is essential to preimplantation genetic diagnosis, which requires testing a polar body or embryonic cell for a particular genetic mutation. Similarly, prospective parents can be tested for being genetic carriers, or their children can be tested for actually having a particular genetic disorder.

Biologists have also found many uses for PCR. PCR analysis can be carried out on mummies that are thousands of years old. Hair, muscle, skin, or bone samples have been used to determine the gender of mummies, for example. An interesting phylogenetic question is the evolutionary relationship of Neandertals to modern humans. PCR studies have been done and we have some genetic evidence based on mitochondrial DNA that humans and Neandertals did interbreed at least in some places where their populations both existed at the same time. Ecologists have found that PCR analysis of droppings can help them estimate the size of rare animal populations in a particular area and to infer the animals' eating habits.

CONSIDER THESE QUESTIONS

1. The chief limitation of PCR is inadvertent contamination with stray DNA in the test tube. How could this concern be addressed, or should PCR not be used?
2. Should the DNA fingerprints of convicted felons be kept on file for easy identification in the case of future crimes?
3. Should PCR procedures be available for any and all purposes, or should they be regulated. If so, by whom?

connect
|BIOLOGY

Explore the concepts through a variety of multimedia assets and question types.

www.mcgrawhillconnect.com

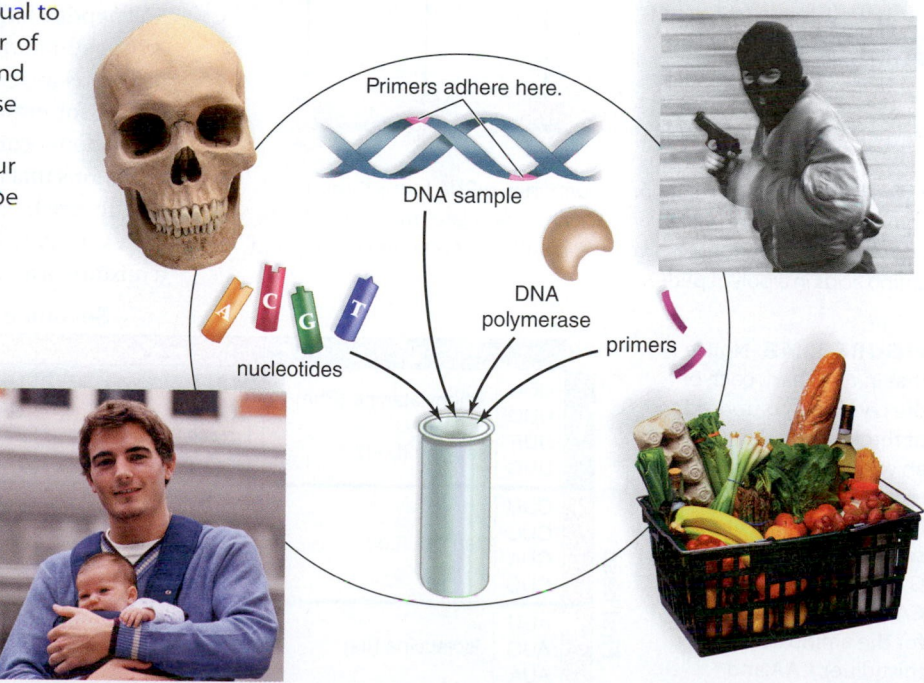

FIGURE 11A PCR requirements are minimal, but the applications are many and varied. Evolutionary relationships can be clarified, criminals can be identified, foods can be tested for contamination, and parents can be identified.

Gene Expression

Producing a cellular protein requires two steps: During transcription, DNA is a template (mold) for RNA formation, and during translation, the sequence of bases in a messenger RNA (mRNA) codes for the sequence of amino acids in a polypeptide. Transfer RNA (tRNA) and ribosomal RNA (rRNA) are also active during translation. Once a protein product is present in a cell, a gene has been expressed.

11.4 Transcription is the first step in gene expression

LEARNING OUTCOMES

When you complete this section, you should be able to

1. Explain how the mRNA molecule bears a genetic code.
2. Describe how mRNA is formed and how it is processed.

Figure 11.4A gives an overview of gene expression, which is complete once a new protein has been made. First, during **transcription,** a strand of RNA forms that is complementary to the DNA template strand. Transcription means to "make a faithful copy," and in this case, a sequence of nucleotides in DNA is copied into a sequence of nucleotides in mRNA. The mRNA molecule that forms is a transcript of a gene. Second, gene expression also requires the process of **translation.** Translation means "to put information into a different language." In this case, a sequence of nucleotides is translated into a sequence of amino acids. This is possible only if the bases in DNA and mRNA code for amino acids. This code is called the **genetic code.**

MP3 Transcription

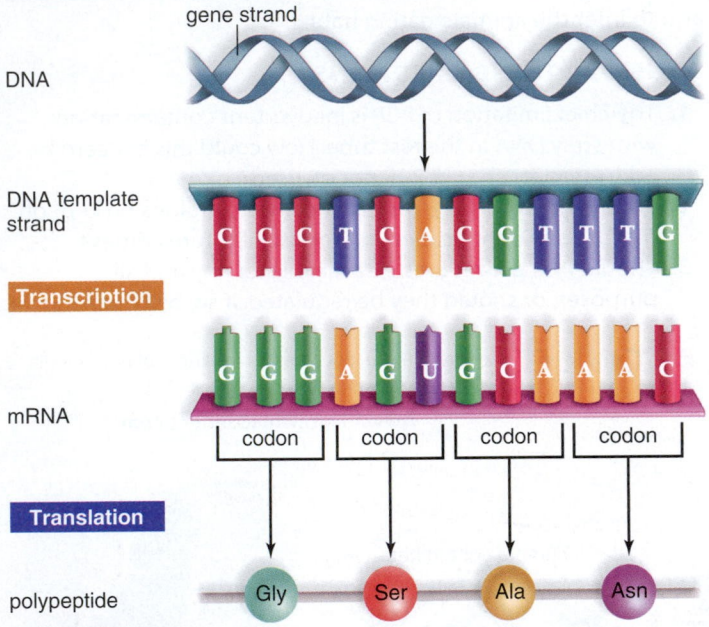

FIGURE 11.4A Transcription occurs when mRNA forms through complementary base pairing with the DNA template strand. Translation occurs when the sequence of codons in mRNA specifies the sequence of amino acids in a polypeptide.

The Genetic Code Recognizing that there must be a genetic code, investigators wanted to know how four bases (A, C, G, U) could provide enough combinations to code for 20 amino acids. If the code were a singlet code (one base standing for one amino acid), only four amino acids could be encoded. If the code were a doublet (any two bases standing for one amino acid), it would still not be possible to code for 20 amino acids. But if the code were a triplet, then the four bases could supply 64 different triplets, far more than needed to code for 20 different amino acids. It should come as no surprise then to learn that the genetic code is a **triplet code.**

Each three-letter (base) unit of an mRNA molecule is called a **codon.** The translation of all 64 mRNA codons has been determined (Fig. 11.4B). Sixty-one triplets correspond to a particular amino acid; the remaining three are **stop codons,** which signal polypeptide termination. The one codon that stands for the amino acid methionine is also a **start codon,** signaling polypeptide initiation. Notice, too, that most amino acids have more than one codon; for example, leucine, serine, and arginine have six different codons. This offers some protection against possibly harmful mutations that change the sequence of the bases.

To crack the code, a cell-free experiment was done: Artificial RNA was added to a medium containing bacterial ribosomes and a mixture of amino acids. Comparison of the bases in the RNA with

FIGURE 11.4B Notice that in this chart, each of the codons is composed of three letters. As an example, find the rectangle where C is the first base and A is the second base. U, C, A, or G can be the third base. CAU and CAC are codons for the amino acid His (histidine); CAA and CAG are codons for Gln (glutamine).

First base	Second base: U	Second base: C	Second base: A	Second base: G	Third base
U	UUU phenylalanine (Phe) UUC phenylalanine (Phe) UUA leucine (Leu) UUG leucine (Leu)	UCU serine (Ser) UCC serine (Ser) UCA serine (Ser) UCG serine (Ser)	UAU tyrosine (Tyr) UAC tyrosine (Tyr) UAA *stop* UAG *stop*	UGU cysteine (Cys) UGC cysteine (Cys) UGA *stop* UGG tryptophan (Trp)	U C A G
C	CUU leucine (Leu) CUC leucine (Leu) CUA leucine (Leu) CUG leucine (Leu)	CCU proline (Pro) CCC proline (Pro) CCA proline (Pro) CCG proline (Pro)	CAU histidine (His) CAC histidine (His) CAA glutamine (Gln) CAG glutamine (Gln)	CGU arginine (Arg) CGC arginine (Arg) CGA arginine (Arg) CGG arginine (Arg)	U C A G
A	AUU isoleucine (Ile) AUC isoleucine (Ile) AUA isoleucine (Ile) AUG methionine (Met) *(start)*	ACU threonine (Thr) ACC threonine (Thr) ACA threonine (Thr) ACG threonine (Thr)	AAU asparagine (Asn) AAC asparagine (Asn) AAA lysine (Lys) AAG lysine (Lys)	AGU serine (Ser) AGC serine (Ser) AGA arginine (Arg) AGG arginine (Arg)	U C A G
G	GUU valine (Val) GUC valine (Val) GUA valine (Val) GUG valine (Val)	GCU alanine (Ala) GCC alanine (Ala) GCA alanine (Ala) GCG alanine (Ala)	GAU aspartic acid (Asp) GAC aspartic acid (Asp) GAA glutamic acid (Glu) GAG glutamic acid (Glu)	GGU glycine (Gly) GGC glycine (Gly) GGA glycine (Gly) GGG glycine (Gly)	U C A G

the resulting polypeptide allowed investigators to decipher the code. For example, an mRNA with a sequence of repeating guanines (GGGGGG...) would encode a string of glycine amino acids.

The genetic code is just about universal in organisms. This suggests that the code dates back to the very first organisms on Earth and that all organisms are related.

The Process of Transcription Suppose you have a woodworking encyclopedia in your bookcase and you want to make a step stool. Rather than taking the entire set of encyclopedias out to the workshop, you might copy the instructions to make a step stool onto a sheet of paper and take just that to your workshop. We can liken the DNA in a nucleus to the encyclopedia containing the instructions for all sorts of wood products (polypeptides). The sheet of paper becomes an mRNA molecule that has instructions for the step stool, which represents the particular polypeptide to be made.

During transcription, an RNA molecule is transcribed off the DNA template strand (Fig. 11.4C). Although all three classes of RNA are formed by transcription, we will focus on transcription to form **messenger RNA** (**mRNA**). Transcription begins when the enzyme RNA polymerase opens up the DNA helix just in front of it so that complementary base pairing can occur. Then RNA polymerase joins the RNA nucleotides, and an mRNA molecule results. When mRNA forms, it has a sequence of bases complementary to the DNA template strand; wherever A, T, G, or C is present in the DNA template strand, U, A, C, or G is incorporated into the mRNA molecule. Now mRNA is a faithful copy of a **gene strand.** A gene is a portion of DNA that codes for a product, often a protein product.

> ▶ **Animation**
> Stages of Transcription

> ▶ **3D Animation**
> Molecular Biology of the Gene: Transcription

Processing mRNA In eukaryotes, the newly synthesized primary-mRNA must be processed before it enters the cytoplasm. During processing, which occurs in the nucleus, one end of the primary-mRNA is modified by the addition of a cap that is composed of an altered guanine nucleotide (Fig. 11.4D). At the other end, there is a poly-A tail, a series of adenosine nucleotides. These modifications provide stability to the mRNA; only those that have a cap and tail remain active in the cell.

Most genes in humans are interrupted by segments of DNA that do not code for protein. These portions are called **introns** because they are intervening segments. The other portions of the gene, called **exons,** contain the protein-coding portion of the gene. In primary-mRNA splicing, the introns are removed by enzymes and the exons are joined together. Only the mature mRNA molecule consisting of continuous exons is ready to be translated.

Processing can utilize all the exons of a gene. In some instances, however, only certain exons form the mature RNA transcript and are utilized. The result will be a different protein product, depending on the exons used. In other words, so-called **alternative mRNA splicing** can potentially increase the possible number of protein products that can be made from a single gene.

> ▶ **3D Animation**
> Molecular Biology of the Gene: mRNA Modifications

▶ **11.4 CHECK YOUR PROGRESS**
1. Relate the terms *genetic code* and *codon* to the structure of mRNA.
2. Analyze the significance of transcription to gene expression.
3. Explain how alternative mRNA splicing affects gene expression.

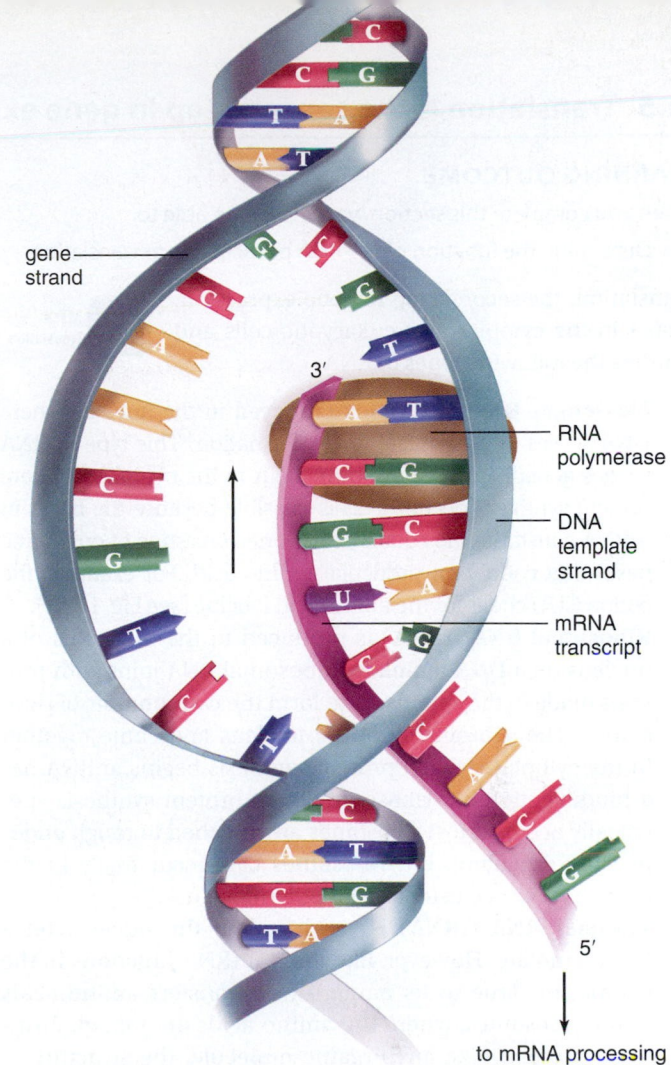

FIGURE 11.4C During transcription, complementary RNA is made off the DNA template strand. A portion of DNA unwinds and unzips at the point of attachment of RNA polymerase. A strand of mRNA is produced when complementary bases join in the order dictated by the sequence of bases in the DNA template strand.

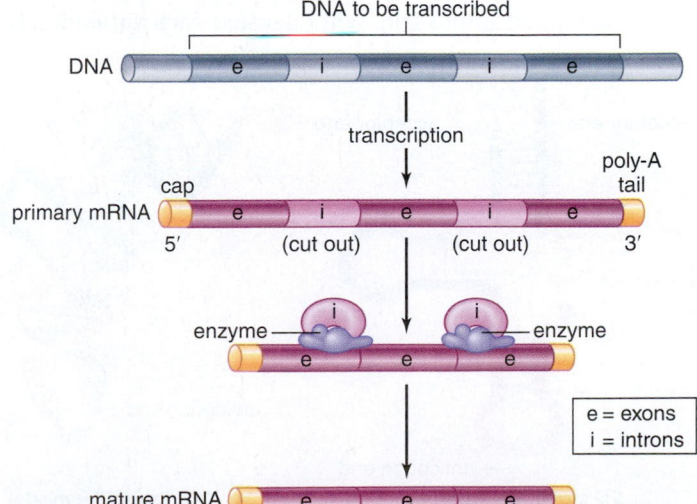

FIGURE 11.4D A cap and a poly-A tail are attached to the ends of the primary RNA transcript, and the introns are removed so that only the exons remain. This mature mRNA molecule moves into the cytoplasm of the cell, where translation occurs.

LEARNING OUTCOME

When you complete this section, you should be able to

1. Determine the function of the RNA participants in translation.

Translation, the second step in gene expression, occurs in the cytoplasm of eukaryotic cells and requires the following types of RNA:

 **MP3** Translation

- **Messenger RNA (mRNA)** is produced in the nucleus where DNA serves as a template for its formation. This type of RNA carries genetic information from DNA to the cytoplasm where protein synthesis occurs. This is possible because the bases in mRNA constitute the codons, each one consisting of only three bases that code for a particular amino acid. For example, the codon CUU codes for the amino acid leucine (see Fig. 11.4B).

- **Ribosomal RNA (rRNA)** is produced in the nucleolus of a nucleus off a DNA template. Ribosomal RNA joins with proteins made in the cytoplasm to form the two subunits of ribosomes. The subunits leave the nucleus and come together in the cytoplasm when protein synthesis begins. mRNA has a binding site on a ribosome where protein synthesis specifically occurs. Most ribosomes are attached to rough endoplasmic reticulum, but ribosomes can occur freely in the cytoplasm or in clusters called polyribosomes.

- **Transfer RNA (tRNA)** is produced in the nucleus off a DNA template. However, like mRNA, tRNA functions in the cytoplasm. True to its name, tRNA transfers amino acids to the ribosomes, where the amino acids are joined, forming a protein. Like any organic molecule, the structure of tRNA can be represented by various models, each depicting a specific aspect of the actual molecule (Fig. 11.5A). The space-filling model shows the molecule's three-dimensional shape, but the cloverleaf model shows clearly that tRNA contains regions where intramolecular base pairing occurs. It's this pairing that gives tRNA two ends: The **acceptor end** binds to an amino acid, and the **anticodon end** binds to a codon for that amino acid. Why? Because each **anticodon** is complementary to the codon for its amino acid. For example, an anticodon for leucine would be GAA because GAA binds to CUU, the codon for leucine. (Recall that U occurs in RNA instead of T.)

The Process of Translation In our previous analogy, DNA was a woodworking encyclopedia, and mRNA was a copy of a page for making a step stool. Continuing that comparison, a ribosome is a table in the workshop where the step stool is made. The term translation is appropriate because DNA and RNA are made of nucleotides, and polypeptides are made of amino acids. In other words, one language (nucleic acids) gets translated into another language (protein).

During translation, the sequence of codons in the mRNA at a ribosome directs the sequence of amino acids in a polypeptide. This works because the tRNA-amino acid complexes bind to the mRNAs in the order dictated by the sequence of the codons. There can be several tRNAs for each amino acid, depending on the number of codons for that amino acid (see Fig. 11.4B). As mentioned, this helps ensure that, despite changes in DNA base sequences, the sequence of amino acids will remain the same.

Let's revisit Figure 11.4B to show how translation works. If the codon sequence is ACC, GUA, and AAA, what will be the anticodons and sequence of amino acids in a portion of the polypeptide? Check your answers against the following chart:

Codon	Anticodon	Amino Acid
ACC	UGG	Threonine
GUA	CAU	Valine
AAA	UUU	Lysine

Function of a Ribosome Both prokaryotic and eukaryotic cells contain thousands of ribosomes per cell because they are needed for protein synthesis. Ribosomes have a binding site for mRNA and three binding sites for transfer RNA (tRNA) molecules (Fig. 11.5B*a*). These binding sites are called the **A site,** the **P site,** and the **E site.** The tRNA at the A site bears only an *a*mino acid, while

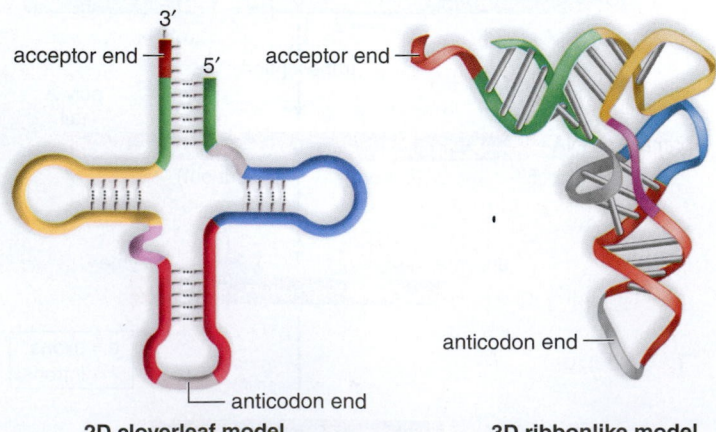

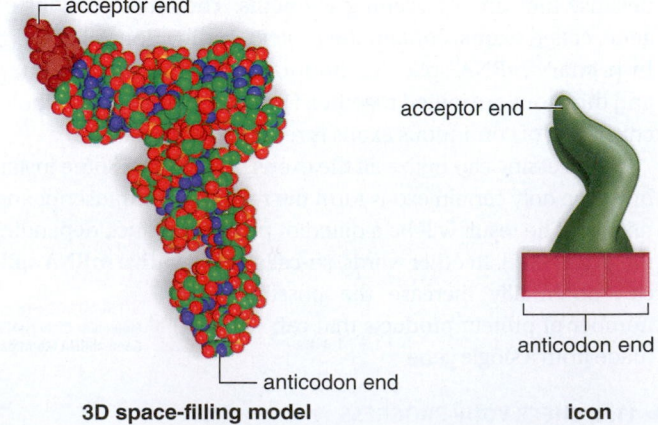

2D cloverleaf model   **3D ribbonlike model**   **3D space-filling model**   **icon**

FIGURE 11.5A In the cloverleaf model of tRNA, the base pairing within the molecule that creates the anticodon loop and the acceptor end where the amino acid attaches are obvious. The ribbonlike model gives a more realistic view of the polynucleotide chain making up the molecule. The 3D space-filling model shows the actual shape of the molecule. The icon for tRNA used in this book is based on the space-filling model.

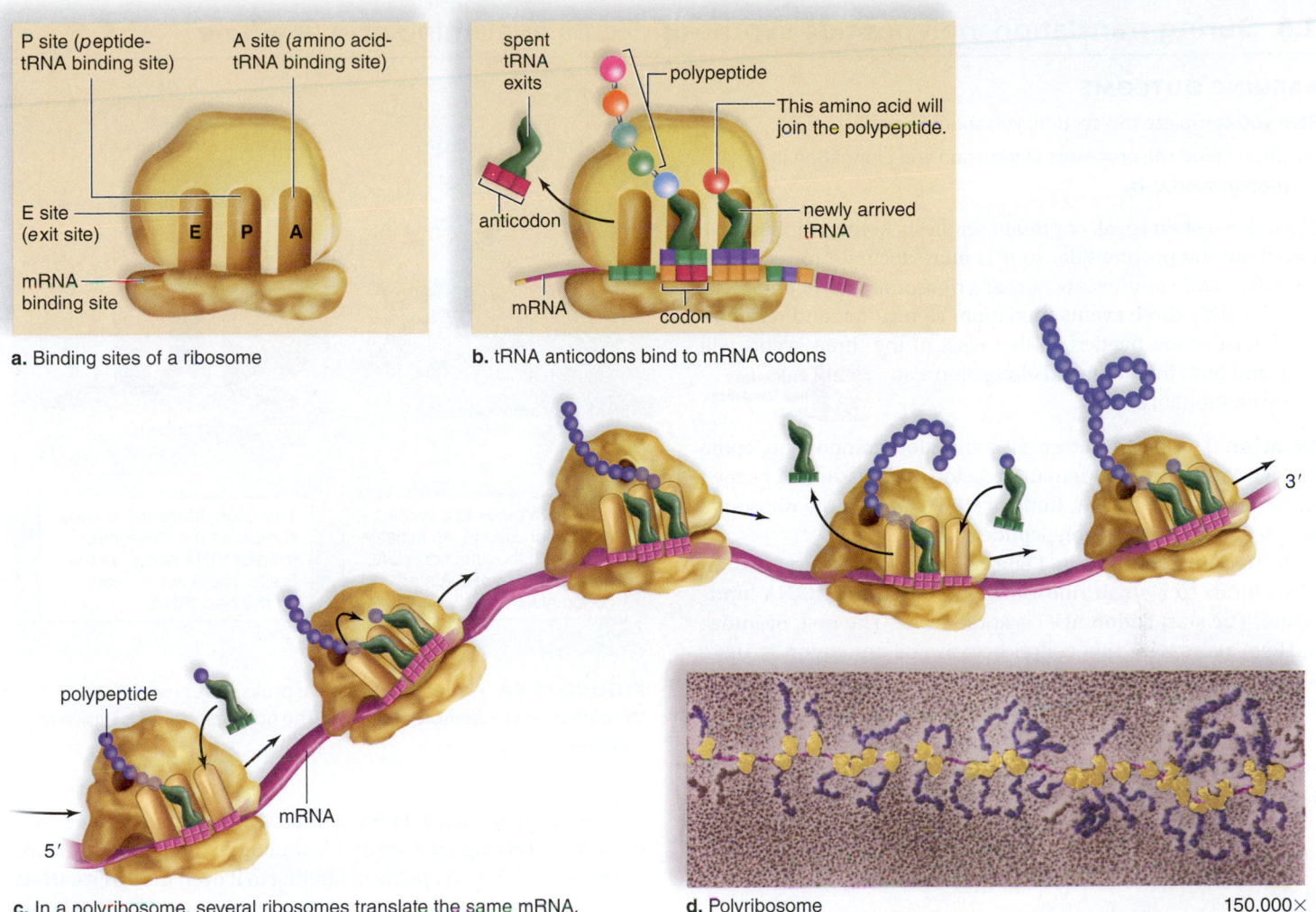

a. P site (*p*eptide-tRNA binding site) A site (*a*mino acid-tRNA binding site)

E site (*e*xit site)

mRNA binding site

a. Binding sites of a ribosome

b. spent tRNA exits

polypeptide

This amino acid will join the polypeptide.

anticodon

newly arrived tRNA

mRNA

codon

b. tRNA anticodons bind to mRNA codons

polypeptide

mRNA

5′

3′

c. In a polyribosome, several ribosomes translate the same mRNA.

d. Polyribosome

150,000×

FIGURE 11.5B **a.** A ribosome has three binding sites where (**b**) tRNA anticodons bind to codons. **c., d.** A polyribosome is a number of ribosomes all translating the same mRNA transcript. This makes translation more efficient.

the tRNA at the P site bears a *p*eptide. The tRNA that leaves the *e*xit site bears neither, and this is why it leaves the ribosome. The tRNA binding sites facilitate complementary base pairing between tRNA anticodons and mRNA codons.

Conceivably three tRNAs could momentarily be at the ribosome at the same time, but more likely an amino acid-tRNA is just coming and/or an empty tRNA is just leaving (Fig. 11.5Bb). Just as a parent might drop off a student and wait to make sure the student joins a line-up to enter the school, so a tRNA takes an amino acid to a ribosome and leaves after the amino acid has taken its place in a growing polypeptide. The methodology to get the amino acids lined up properly in a polypeptide is discussed in section 11.6. Also, translation begins with the start codon and terminates at a stop codon.

What is the function of ribosomes? They help ensure that the amino acids in a polypeptide are sequenced according to the order originally specified by DNA. This comes about because the anticodons of the tRNAs bind to the codons in a particular sequence.

A **polyribosome** is several ribosomes attached to and translating the same mRNA (Fig. 11.5Bc, d). As soon as the initial portion of mRNA has been translated by one ribosome, and the ribosome

has begun to move down the mRNA, another ribosome attaches to the mRNA. A polyribosome greatly increases the efficiency of translation. The average speed of protein synthesis is 20 peptide bonds per second. But when you consider that many ribosomes may be synthesizing the same protein, the speed per protein may be much higher, even up to 2,000 identical proteins per second. Not surprisingly, it has been calculated that *E. coli* spends 90% of its energy on supplying the substrates and the means to carry out protein synthesis.

Efficient antibiotics attack a bacterium where it is most vulnerable. The antibiotic tetracycline blocks the A site of bacterial ribosomes so tRNAs cannot bind to this site, and the antibiotic streptomycin causes a misreading of the genetic code so that a faulty protein is produced. Either of these events leads to death of the bacterium.

▶ **11.5 CHECK YOUR PROGRESS**

1. Determine the last three amino acids in a polypeptide attached to the tRNA at the P site if their RNA codons are CCA, UAC, and AGA.

11.6 During translation, polypeptide synthesis occurs one amino acid at a time

LEARNING OUTCOME

When you complete this section, you should be able to

1. Understand the processes of initiation and elongation in protein synthesis.

Although we often speak of protein synthesis, some proteins have more than one polypeptide, so it is more accurate to recognize that polypeptide synthesis occurs at a ribosome. Polypeptide synthesis involves three events: initiation, elongation, and termination. Enzymes are needed so that each of the three events will occur, and both initiation and elongation also require an input of energy.

Animation
How Translation Works

Initiation During **initiation** all translation components come together. Proteins called initiation factors help assemble a small ribosomal subunit, mRNA, initiator tRNA, and a large ribosomal subunit for the start of a polypeptide synthesis.

Initiation is shown in Figure 11.6A. In prokaryotes, an mRNA binds to a small ribosomal subunit at the mRNA binding site. The start codon AUG is at the P site. The first, or initiator, tRNA pairs with this codon because its anticodon is UAC. As you can see by examining Figure 11.4B, AUG is the codon for methionine. Methionine is always the first amino acid of a polypeptide. After the small ribosomal unit has attached, a large ribosomal subunit joins to the small subunit. Although similar in many ways, initiation in eukaryotes is much more complex.

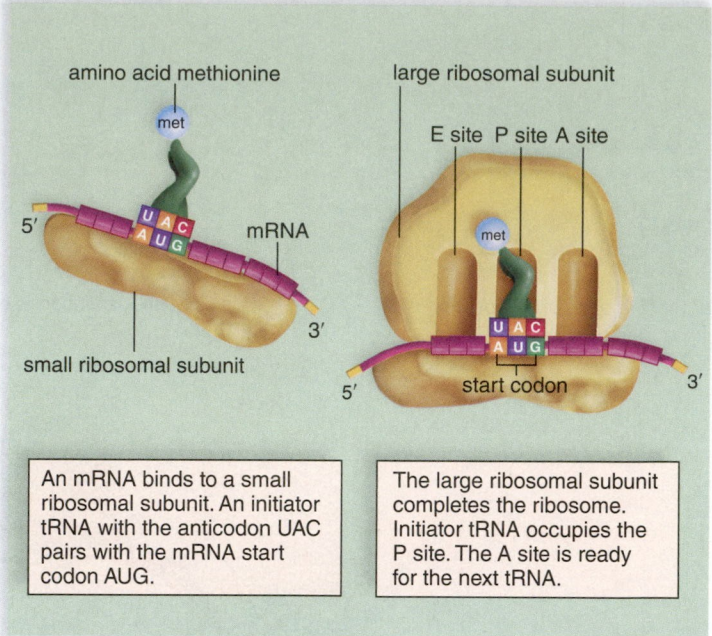

An mRNA binds to a small ribosomal subunit. An initiator tRNA with the anticodon UAC pairs with the mRNA start codon AUG.

The large ribosomal subunit completes the ribosome. Initiator tRNA occupies the P site. The A site is ready for the next tRNA.

FIGURE 11.6A An initiation event in prokaryotes requires that the participants assemble as shown. The first amino acid is typically methionine.

As shown in Figure 11.6A, a ribosome has three binding sites for tRNAs. Although the second is ordinarily for a peptide-tRNA, the initiator tRNA is capable of binding to it even though it carries only the amino acid methionine. The next amino acid-tRNA binds to the A site.

Elongation and Termination During **elongation,** a polypeptide increases in length, one amino acid at a time. In addition to the participation of tRNAs, elongation requires elongation factors, which facilitate the binding of tRNA anticodons to mRNA codons at a ribosome.

Elongation is shown in Figure 11.6B, where ❶ a tRNA with an attached peptide is already at the P site, and a tRNA carrying its appropriate amino acid is just arriving at the A site. ❷ Once the next tRNA is in place at the A site, the peptide will be transferred to this tRNA. This transfer requires energy and a ribozyme. **Ribozymes,** located in the larger ribosomal subunit, are enzymes composed of RNA instead of protein. Ribozymes join peptides (from the P sites) to amino acids at the A sites. The bond that joins them together is a peptide bond (see Fig. 3.7C). ❸ After peptide bond formation occurs, the peptide is one amino acid longer than it was before. ❹ Finally, **translocation** occurs: The ribosome *moves forward,* and the peptide-tRNA is now at the P site of the ribosome. The used tRNA exits from the E site. A new codon is at the A site, ready to receive another tRNA. Eventually, the ribosome reaches a stop codon, and **termination** occurs, during which the polypeptide is released.

3D Animation
Molecular Biology of the Gene: Translation

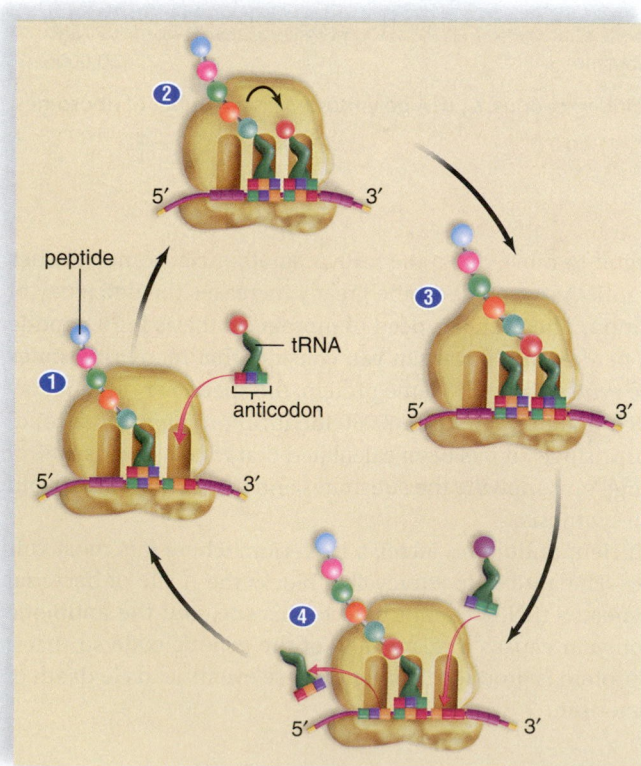

FIGURE 11.6B During elongation, a polypeptide increases by one amino acid at a time. As a result of #1 and #2, the peptide at the P site will pass to the amino acid that just arrived at the A site. Then between #3 and #4 translocation occurs: The ribosome moves forward so that the peptide-tRNA is now at the P site. The empty tRNA exits.

▶ **11.6 CHECK YOUR PROGRESS**

1. Describe how elongation sequences the amino acids in the correct order.

11.7 Let's review gene expression

LEARNING OUTCOME

When you complete this section, you should be able to

1. Describe gene expression step by step.

Gene expression requires two steps, called transcription and translation. Figure 11.7 shows that in a eukaryotic cell, transcription occurs in the nucleus and translation occurs in the cytoplasm. ❶ and ❷ mRNA is produced and processed before leaving the nucleus. ❸–❻ After mRNA becomes associated with ribosomes, polypeptide synthesis occurs one amino acid at a time. Table 11.7 reviews the participants in gene expression.

Many ribosomes can be translating the same section of DNA at a time, and collectively these ribosomes are called a polyribosome. As discussed earlier, some ribosomes remain free in the cytoplasm, and others become attached to rough ER. ❼ After the polypeptide enters the lumen of the ER by way of a channel, it is folded and further processed by the addition of sugars, phosphates, or lipids. ❽ When the ribosome reaches a stop codon and termination occurs, ribosomal units and mRNA are separated from one another, and the polypeptide is released.

We have finished our examination of gene expression (the making of a protein). In the next part of the chapter, we will study the biochemistry of mutations.

TABLE 11.7 | Participants in Gene Expression

Name of Molecule	Significance
DNA	Sequence of DNA bases that constitutes the genetic information of an organism
mRNA	Sequence of RNA bases, complementary to DNA, that constitutes the codons
tRNA	Having a series of three RNA bases (called an anticodon) that are complementary to a codon
rRNA	Structural RNA present in a ribosome where a polypeptide is synthesized
Amino acid	Subunit of a polypeptide that is transported to a ribosome by a tRNA
Polypeptide	Series of amino acids specified by DNA and performing an enzymatic or structural function in a cell

▶ **11.7 CHECK YOUR PROGRESS**

1. Describe in detail what a cell does with the genetic information stored by DNA.
2. Suggest a fate for a protein that is finalized inside the ER (Hint: See 4.6, p. 72).

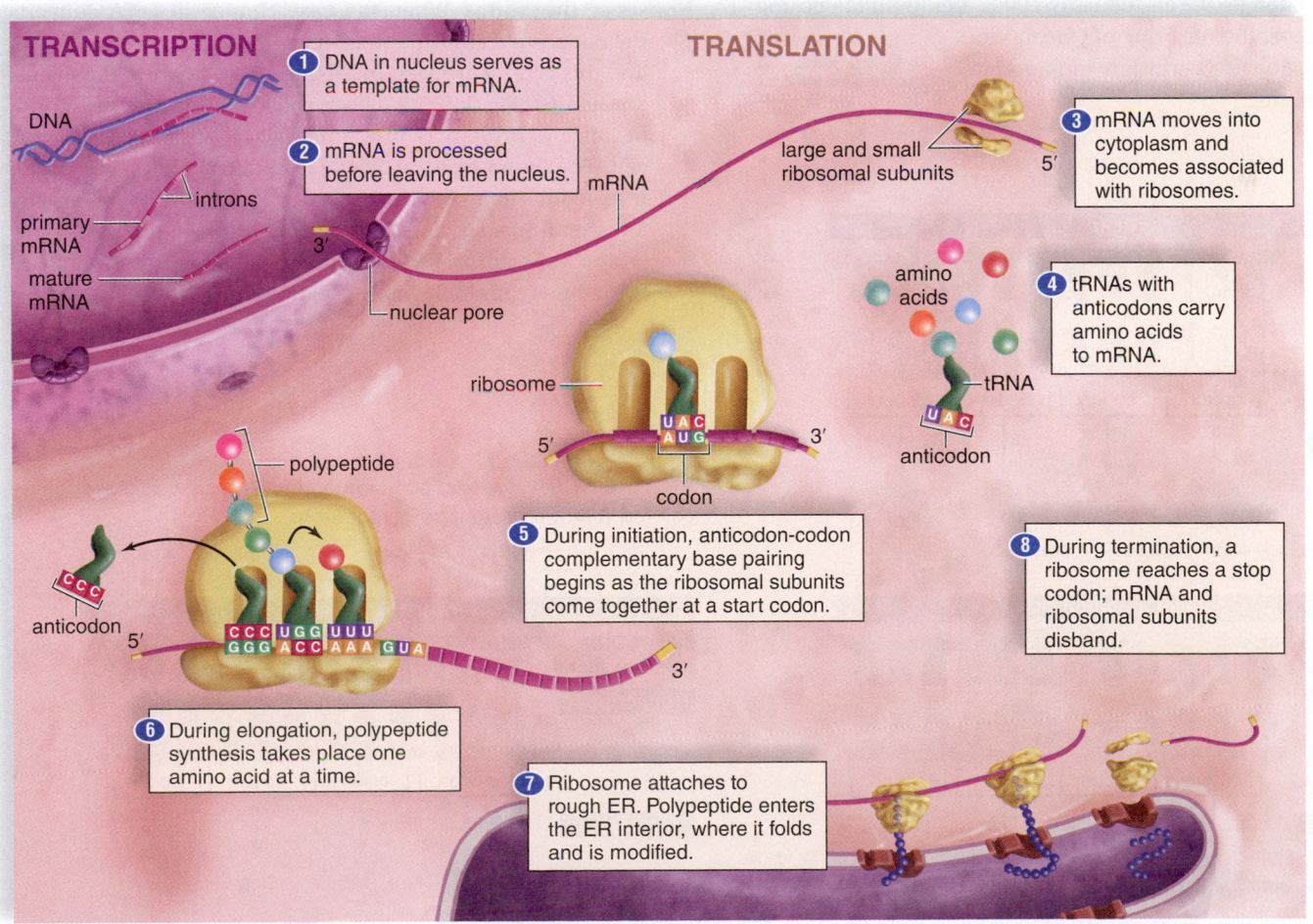

FIGURE 11.7 Summary of gene expression in eukaryotes.

Mutations: Altered Gene Expression

Mutations can cause an altered gene expression and disorders. For example, the human disorders we studied in Chapter 10, such as Huntington disease and Marfan syndrome, are due to mutations. Even so, mutations are the raw material for evolution because without them diversity would not occur among the members of populations. Not all mutations are harmful; natural selection is possible only when members of a population exhibit beneficial variations. We will be examining both genetic mutations and chromosomal mutations in this section.

11.8 Genetic mutations affect a specific allele

LEARNING OUTCOMES

When you complete this section, you should be able to

1. Describe the effects of the various types of point mutations.
2. Identify inheritance characteristics of a triplet repeat expansion.

A **genetic mutation** is a permanent change in the base sequence of a gene. We are particularly interested in mutations that lead to an altered gene expression and disorders. We will be mentioning these disorders as we continue our discussion. Some mutations are spontaneous—they happen for no apparent reason—while others are induced by mutagens. Various forms of radiation, such as radioactive elements, X-rays, and ultraviolet (UV) light, and certain organic chemicals, such as pesticides and compounds in cigarette smoke, are mutagens that cause changes in genes and chromosomes. A base change in the DNA of body cells can result in cancer. A base change during the production of gametes, the offspring of the individual may be affected. Rarely is it possible to trace such a change, but in the early 1900s, hemophilia became prevalent among the offspring of Queen Victoria possibly due to a germ-line mutation in her or her parents.

Animation
Addition and
Deletion Mutations

Point Mutations It's remarkable that a change in a single DNA base pair, called a **point mutation,** can lead to a change in the sequence of amino acids and therefore a change in the shape of a protein. Recall that proteins function properly only when they have the correct shape.

Biologists who have studied point mutations have categorized them as shown in Figure 11.8A like this:

Silent mutations In Figure 11.8A*b*, a single base change has occurred in the DNA compared to the normal sequence found in Figure 11.8A*a*. However, because of the redundancy of codons—multiple codons code for the same amino acid—no change occurs in the amino acid sequence; therefore, the mutation is a **silent mutation.** Investigators have discovered that the human genome contains many such changes called single nucleotide polymorphisms (SNPs), and the particular pattern of SNPs can be used to trace a person's ancestry. It's hoped that knowing a person's ethnicity will assist physicians in prescribing the correct medications for that person.

Nonsense mutations When the base change in DNA converts the codon to a stop codon instead of one for an amino acid, and this is not corrected by DNA repair enzymes, the results are serious (Fig. 11.8A*c*). This is called a **nonsense mutation,** because a protein is now missing or nonfunctional in a cell. To continue our analogy (see page 201), imagine if the instructions for making a stool (DNA for a gene) were missing a paragraph, you would not be able to make the stool (protein). Cystic fibrosis and Duchenne muscular dystrophy are sometimes caused by nonsense mutations.

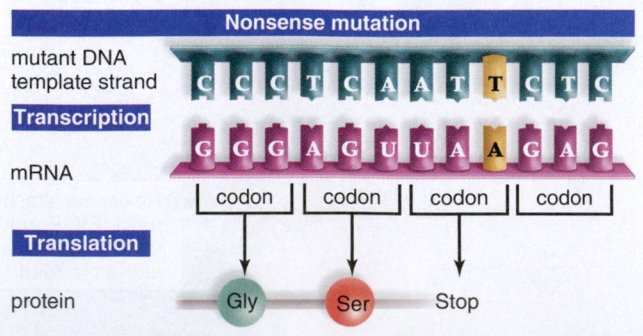

FIGURE 11.8A Types of point mutations. Note that the base pairs colored yellow represent point mutations.

Normal

normal DNA template strand: C C C T C A A T G C T C

Transcription

mRNA: G G G A G U U A C G A G

codon | codon | codon | codon

Translation

protein: Gly — Ser — Tyr — Glu

a. Normal sequence of amino acids in a protein.

Silent mutation

mutant DNA template strand: C C C T C G A T G C T C

Transcription

mRNA: G G G A G C U A C G A G

codon | codon | codon | codon

Translation

protein: Gly — Ser — Tyr — Glu

b. Silent mutations cause no change in sequence of amino acids.

Nonsense mutation

mutant DNA template strand: C C C T C A A T T C T C

Transcription

mRNA: G G G A G U U A A G A G

codon | codon | codon | codon

Translation

protein: Gly — Ser — Stop

c. Nonsense mutations cut short the sequence of amino acids.

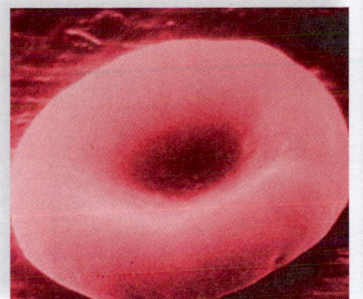
Normal cell

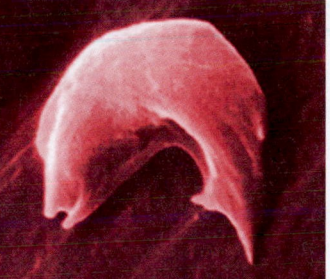
Sickled cell

FIGURE 11.8B Normal red blood cells compared to sickled red blood cells.

FIGURE 11.8C The mother and daughter with 100 repeats have mild myotonic dystrophy, whereas the infant with 1,000 repeats has severe myotonic dystrophy.

Missense mutations In a **missense mutation** (Fig. 11.8A*d*), a protein results but due to a single base change in DNA, the protein can have the wrong shape. In red blood cells a missense mutation occurs when valine (Val) appears at one spot in hemoglobin instead of glutamic acid (Glu). This change from hydrophilic glutamic acid to the hydrophobic valine causes hemoglobin to form semirigid rods and cells become sickle-shaped. The person is subject to various ills because the malformed cells clog blood vessels and die off more quickly than normal-shaped cell (Fig. 11.8B).

Frameshift mutations We expect an altered protein when one or more nucleotides are either inserted or deleted from DNA because the result is a completely new sequence of amino acids. This mutation is called a **frameshift mutation** (Fig. 11.8A*e*) because the reading frame has changed. The sequence of codons is read from a specific starting point, as in this sentence: THE CAT ATE THE RAT. If the letter C is deleted from this sentence and the reading frame shifts, we read THE ATA TET HER AT—something doesn't make sense. Cystic fibrosis is known to sometimes occur due to a frameshift mutation because just like a nonsense mutation, a particular protein is not performing its usual function.

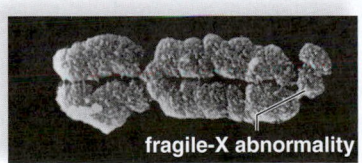
fragile-X abnormality

Triplet Repeat Expansion
Fragile X syndrome is one of the most common genetic causes of intellectual impairment. The term fragile X syndrome originated because diagnosis used to be dependent upon observation of an X chromosome whose tip is attached to the rest of the chromosome by only a thin thread. This condition results from repeats of CGG. Some genes normally have a repeat of three nucleotides (**triplet repeat**), such as CTGCTGCTG. Perhaps due to a replication error prior to meiosis, the number of repeats can expand (increase in number) from one generation to another. Eventually, an offspring can inherit such a large number of repeats that illness results. Huntington disease is known to be due to CAG repeat, and myotonic dystrophy characterized by muscle weakness and myotonia (slow relaxation of muscles after contraction) is due to a CTG repeat (Fig. 11.8C).

▶ **11.8 CHECK YOUR PROGRESS**

1. Analyze the cause for a genetic disorder when there is a change in the DNA from ATA to ATT (Hint: See Fig. 11.4B).
2. Compare and contrast a silent mutation with a frameshift mutation.

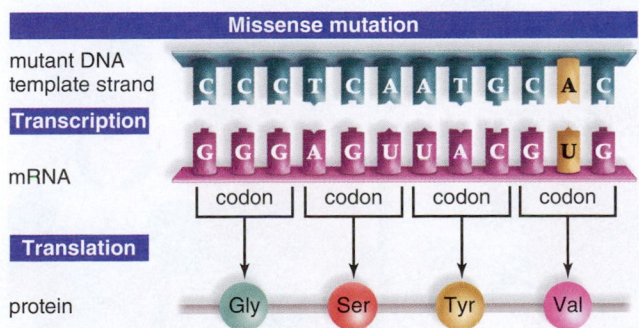
d. Missense mutations cause a change in sequence of amino acids.

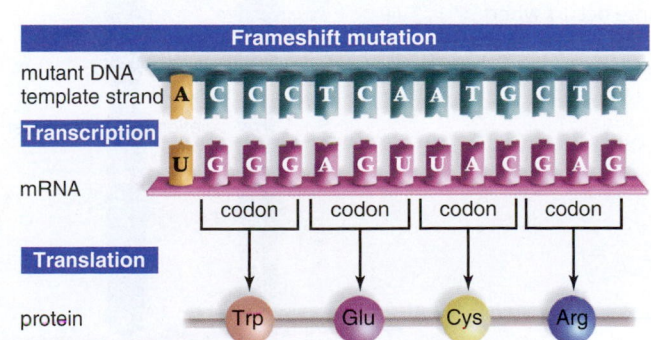
e. Frameshift mutations cause a completely new sequence of amino acids.

LEARNING OUTCOMES

When you complete this section, you should be able to

1. Describe a chromosomal deletion, duplication, translocation, and inversion.
2. List disorders caused by changes in chromosome structure.

Chromosomal mutations include changes in chromosome number and changes in chromosome structure. Section 9.7 discussed a few of the possible human syndromes due to a change in chromosome number. This section discusses human syndromes associated with changes in chromosome structure. A syndrome is a collection of symptoms that occur together and characterize a particular condition. The syndromes discussed here are often due to the same environmental agents—radiation, certain organic chemicals—we mentioned previously. Even viruses can cause chromosomes to break apart. Ordinarily, when breaks occur, the segments reunite to give the same sequence of alleles. But their failure to reunite correctly can result in one of several types of chromosomal mutations: deletion, duplication, inversion, or translocation.

Chromosomal mutations are quite characteristic of cells as they become cancerous. Chromosomal mutations can also occur during meiosis, and if the offspring inherits the abnormal chromosome, a syndrome may develop. The examples that follow show that chromosomal mutations have medical significance, but as then discussed at the end of the section, evidence is accumulating that chromosomal mutations have also been important in the evolution of life.

Deletions, Duplications, and Inversions A **deletion** occurs when a single break causes a chromosome to lose an end piece or when two simultaneous breaks lead to the loss of an internal chromosomal segment. The chromosome is shorter than it was before because certain alleles have been deleted from the chromosome. An individual who inherits a normal chromosome from one parent and a chromosome with a deletion from the other parent no longer has a pair of alleles for each trait, and a syndrome can result. Williams syndrome occurs when chromosome 7 loses a tiny end piece (Fig. 11.9A). Children who have this syndrome look like pixies because they have turned-up noses, wide mouths, small chins, and large ears. Although their academic skills are poor, they exhibit excellent verbal and musical abilities. One

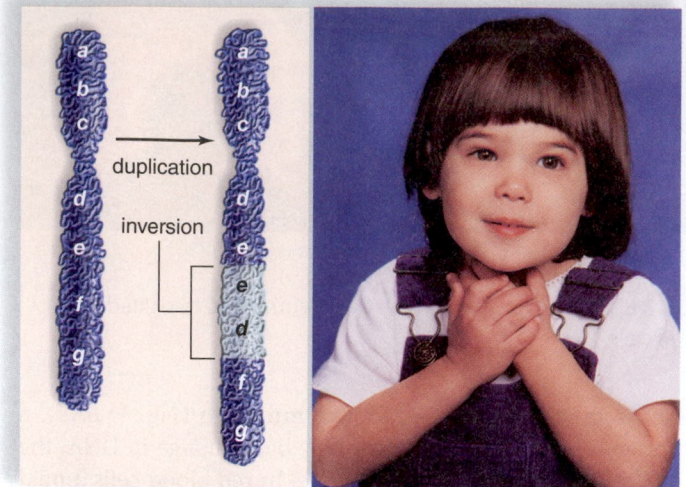

FIGURE 11.9B Autistic characteristics appear when a duplication and inversion is present on chromosome 15.

allele that governs the production of the protein elastin is missing, which affects the health of the cardiovascular system and causes their skin to age prematurely. Such individuals are very friendly but need an ordered life, perhaps because of the loss of the gene for a protein that is normally active in the brain. Cri du chat (cat's cry) syndrome is seen when chromosome 5 is missing an end piece. The affected individual has a small head, is intellectually impaired, and has facial abnormalities. Abnormal development of the glottis and larynx results in the most characteristic symptom— the infant's cry resembles that of a cat.

In a **duplication,** a chromosomal segment is repeated in the same chromosome or in a nonhomologous chromosome. The affected chromosome is longer than it was before because it now has multiple copies of an allele, and the individual will have more than two alleles for a certain trait. An inverted duplication is known to occur in chromosome 15. **Inversion** means that a segment joins in the direction opposite from normal. Children with this syndrome, called inv dup 15 syndrome, have poor muscle tone, intellectual impairment, seizures, a curved spine, and autistic characteristics, including poor speech, hand flapping, and lack of eye contact (Fig. 11.9B).

FIGURE 11.9A Williams syndrome occurs when chromosome 7 has lost an end piece.

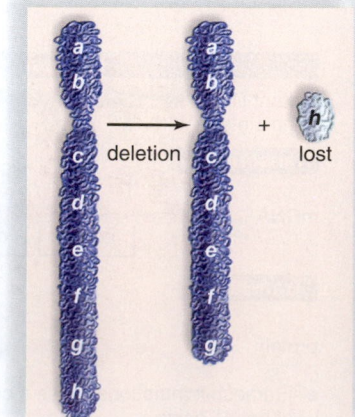

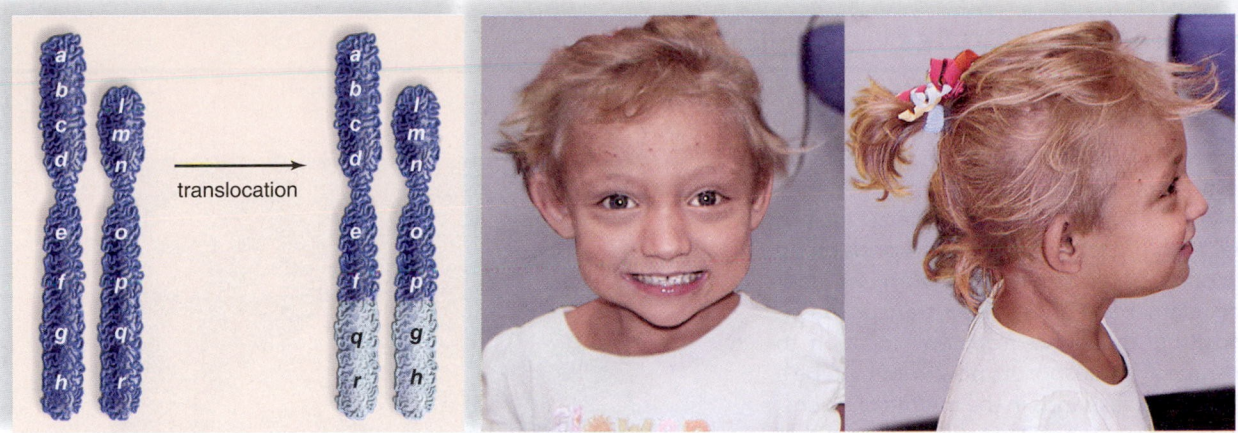

FIGURE 11.9C If a translocation between chromosome number 2 and 20 disrupts an allele on chromosome 20, Alagille syndrome can result.

Translocation A **translocation** is the exchange of chromosomal segments between two nonhomologous chromosomes. A person who has both of the involved chromosomes has the normal amount of genetic material and is healthy, unless the chromosome exchange breaks an allele into two pieces. The person who inherits only one of the translocated chromosomes will have only one copy of certain alleles and three copies of certain other alleles.

In 5% of cases, a translocation that occurred in a previous generation between chromosomes 21 and 14 is the cause of one type of Down syndrome. The affected person inherits two normal chromosomes 21 and an abnormal chromosome 14 that contains a segment of chromosome 21. In these cases, Down syndrome is not related to the age of the mother, but instead tends to run in the family of either the father or the mother. People with Down syndrome tend to have similar characteristics, including an upward slant to the eyes, flat face, large fissured tongue, low muscle tone, deep crease across the center of the palm, and an increased risk of cardiac defects and Alzheimer disease. There is apt to be some degree of intellectual disability, but most individuals with syndrome have IQs in the mild to moderate disability range (see "Hope for Down Syndrome" on page 165).

Figure 11.9C shows a girl with the distinctive face of someone with Alagille syndrome. People with Alagille syndrome also have abnormalities of the eyes, internal organs, and severe itching. Alagille syndrome may be caused by a translocation between chromosomes 2 and 20. Because it is sometimes caused by a deletion on chromosome 20, it can be deduced that the translocation disrupted an allele on chromosome 20. The symptoms of Alagille syndrome range from mild to severe, so some people may not be aware they have the syndrome.

Inversion An inversion occurs when a segment of a chromosome is turned 180 degrees. You might think this is not a problem because the same genes are present, but the reverse sequence of the alleles can lead to altered gene activity. Crossing-over between an inverted chromosome and the noninverted homologue can lead to recombinant chromosomes that have both duplicated and

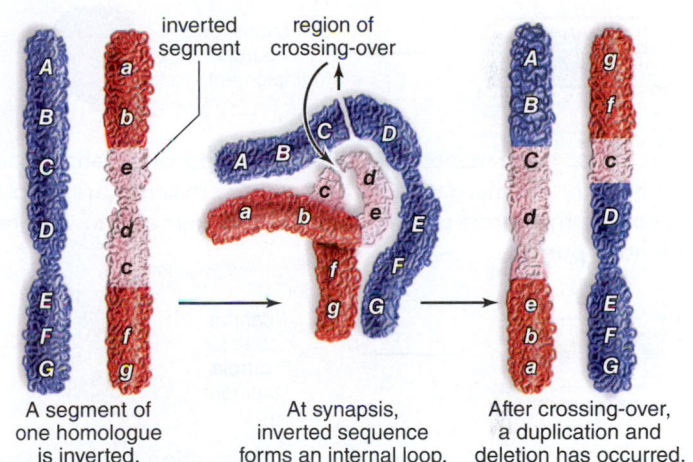

| A segment of one homologue is inverted. | At synapsis, inverted sequence forms an internal loop. | After crossing-over, a duplication and deletion has occurred. |

FIGURE 11.9D Inversion.

deleted segments. This happens because alignment between the two homologues is possible only when the inverted chromosome forms a loop (Fig. 11.9D).

Chromosomal Mutations and Evolution Humans have a variety of hemoglobin genes with slightly different functions. Perhaps a duplication of an original hemoglobin gene occurred, and this was followed by different small point mutations in each copy so that each began to function differently. Another example of a chromosomal mutation and evolution is known. Our chromosomes are different from those of a chimpanzee, perhaps due to a translocation that occurred in a common ancestor. Chimpanzees have 24 pairs of chromosomes, and we have 23 due to a fusion of two chromosomes.

► **11.9 CHECK YOUR PROGRESS**
  1. Explain why a person who inherits a chromosome carrying a duplication will have three copies of certain genes.
  2. Identify the deletion that would occur if an individual inherited only the right chromosome of the translocated ones shown in Figure 11.9C.

11B Transposons Cause Mutations

In 1983 Barbara McClintock, shown in Figure 11B*a*, received a Nobel Prize in Physiology or Medicine for her work in genetics. When she began studying inheritance in corn (maize) plants, geneticists believed that each nucleotide had a fixed position on a chromosome. In the course of her studies with corn, she concluded that "controlling elements"—later called transposons—could undergo transposition and move from one location to another on the chromosome. If a transposon, now known to be a short sequence of DNA nucleotides, lands in the middle of a gene, it prevents the expression of that gene. Dr. McClintock said that because transposons are capable of suppressing gene expression, they could account for the pigment pattern of the corn strain popularly known as Indian corn (Fig. 11B*b*).

Suppose, for example, that the expression of a normal gene results in a corn kernel that is purple:

What happens if transposition causes a transposon to land in the middle of this normal gene? The cells of the corn kernel are unable to produce the purple pigment, and the corn kernel is now white, instead of purple:

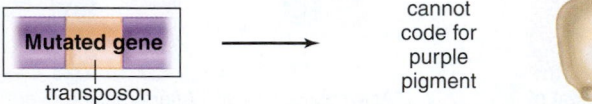

While mutations are usually stable, a transposition is very unstable. When the transposon jumps to another chromosomal location, some cells regain the ability to produce the purple pigment, and the result is a corn kernel with a speckled pattern, as shown in Figure 11B*b*.

When McClintock first published her results in the 1950s, the scientific community paid little attention. Years later, when molecular genetics was well established, transposons were also discovered in bacteria, yeasts, plants, fruit flies, and humans. We now know that transposons make up 45% of all the DNA in a human cell! So the study of transposons is now of tremendous importance. Unfortunately, transposons can have harmful effects:

1. Transposons cause genetic mutations when they block transcription. They are known to cause human diseases, including hemophilia and muscular dystrophy.
2. Transposons cause chromosomal mutations, including translocations, deletions, and inversions, because they often carry a copy of certain host genes with them when they jump. Transposons also cause duplications if they leave copies of themselves and certain host genes before jumping.
3. Transposons in bacteria encourage the spread of human infectious diseases because they contain one or more genes that make a bacterium resistant to antibiotics.

a.

b.

FIGURE 11B a. Barbara McClintock discovered transposons. Her experimental material was the maize (corn) plant. **b.** Transposons are responsible for the speckled or striped patterns in Indian corn.

Because of their ability to strongly affect the genotype, transposons are thought likely to have played a significant role in evolution and the development of organisms. In her acceptance speech for the Nobel Prize, the 81-year-old scientist proclaimed that "it might seem unfair to reward a person for having so much pleasure over the years, asking the maize plant to solve specific problems, and then watching its responses."

Animation
Transposons: Shifting Segments of the Genome

CONSIDER THESE QUESTIONS

1. Transposons are generally considered "selfish DNA parasites." Why is this expression warranted? In what sense is all DNA selfish?
2. One family of transposons in *Drosophila melanogaster,* called P elements, probably appeared in the twelfth century and since has spread to all populations of the *Drosophila* species. Create a scenario by which this would be possible.

 **connect** |BIOLOGY   Explore the concepts through a variety of multimedia assets and question types.

www.mcgrawhillconnect.com

CONNECTING THE CONCEPTS

A genetic material should meet three requirements: It must be variable, accounting for the differences between species; it must be able to replicate; and it must be able to undergo mutations. The ability of DNA to fulfill these requirements lies in the sequence of its bases. Furthermore, DNA contains coded information stored from one generation to the next that permits the synthesis of particular proteins, and it is the difference in proteins between species that makes one species different from another. By studying the activity of genes in cells, geneticists have confirmed that proteins are the link between genotype and phenotype. In other words, you have blue, or brown, or hazel eye pigments because of the types of enzymes (proteins) contained within the cells of your eyes.

Polypeptide synthesis is a two-step process. Three types of RNA (mRNA, rRNA, and tRNA) participate in polypeptide synthesis, but only mRNA carries the coded information to the ribosomes containing rRNA. mRNA contains a faithful copy of a gene—a portion of the DNA—and this copy is used during transcription to build a polypeptide. Just as an office saves the master copy of a form so it can always make more copies, the eukaryotic cell keeps its master copy of the genes safely stored in the nucleus. The tRNAs capture amino acids and bring them to the ribosome where binding of tRNA to the correct codons ensures that the amino acids will be sequenced in the correct order.

Mutations arise when the DNA (not the protein machinery) undergoes a permanent change. After all, DNA and not RNA is passed to the next generation. Faithful copy of the genetic material is top priority for a cell because even a single change in the base sequence of a gene can have a profound effect on the protein product. Genetic mutations and even chromosomal mutations do occur, however, and indeed are a necessity for the process of evolution.

We now know the sequence of bases in human DNA, but it turns out that humans have far fewer genes than expected. A complicated organism such as a human can make do with fewer genes if each gene has more than one function, according to how it is regulated. Regulation of gene activity, to be discussed in Chapter 12, has become the focal point of modern-day research.

ANALYZE AND EVALUATE

1. Show that DNA fulfills the criteria for a genetic material.
2. Mature cells are specialized for particular functions and contain their own specific set of proteins. What does this say about the activity of protein-coding genes in specialized cells?
3. Drugs are usually molecules that affect the activity of proteins in cells. Futuristic drugs might affect which part(s) of the protein synthesis procedure?

SUMMARIZE

DNA and RNA Structure and Function

11.1 DNA and RNA are polymers of nucleotides

- **Nucleotides** in the nucleic acid DNA contain the sugar deoxyribose and the bases adenine (**A**), thymine (**T**), cytosine (**C**), and guanine (**G**). The nucleotides in **RNA** contain the sugar ribose and the bases A, C, G, and uracil (**U**) instead of T.

- Chargaff discovered that the base composition of DNA varies between species, showing that DNA does have the variability to be a genetic material. In all species, the percentage of A always equals the percentage of T, and the percentage of G equals the percentage of C.

11.2 DNA is a double helix

- Watson and Crick constructed the first **double helix** model of DNA, using all the data mentioned in section 11.1 as well as Franklin and Wilkins's X-ray diffraction data. In the model, A pairs with T and G pairs with C. This is called **complementary base pairing.** Each species has its own sequence of bases. The double helix model suggests how the replication of DNA occurs.

DNA Replication

11.3 DNA replication is semiconservative

- **Semiconservative replication** means that each new double helix contains an old strand that acted as a template and a new strand made off the template. The steps in replication are unwinding, complementary base pairing, and joining. **DNA polymerase** is used in pairing and joining.

Gene Expression

11.4 Transcription is the first step in gene expression

- Gene expression (the making of a protein) has two steps: **transcription** and **translation.** During transcription, a sequence of DNA nucleotides is transcribed into a sequence of RNA nucleotides. The nucleotides in a **messenger RNA (mRNA)** are complementary to those in a protein-encoding **gene.** A gene is a portion of the DNA molecule. The **genetic code** is a **triplet code:** Every three bases in mRNA, called a **codon,** stands for an amino acid. The genetic code has **start** and **stop codons,** and most amino acids have more than one codon.

- In eukaryotes, an mRNA is processed before leaving the nucleus. mRNA receives a cap and a tail. During splicing of primary mRNA, **introns** are removed, and all or some **exons** are used to form mature mRNA. **Alternative mRNA splicing** means that several types of proteins can be encoded by the same segment of DNA.

11.5 Translation is the second step in gene expression: An overview

- All three types of mRNA are involved in translation: (1) mRNA contains a sequence of codons complementary to the bases in DNA. (2) Ribosomal RNA (**rRNA**) and proteins to form the ribosomes. (3) Each **transfer RNA (tRNA)** brings a specific amino acid to the ribosomes. A tRNA has an **acceptor end** where its amino acid binds and also an **anticodon end.** The **anticodon** is a sequence of three bases that are complementary to a codon for that amino acid.

- A ribosome has binding sites for mRNA and three tRNAs at the **A site,** the **P site,** and the **E site.** An amino acid-tRNA is at the A site, and a peptide-tRNA is at the P site. A tRNA that lacks an attached amino acid exits from the E site. A **polyribosome** is composed of several ribosomes attached to and translating the same mRNA. This increases the efficiency of protein synthesis.

11.6 During translation, polypeptide synthesis occurs one amino acid at a time

- During **initation** in prokaryotes, ribosomal subunits, mRNA, and initiator tRNA come together. Initiation is more complicated in eukaryotes. The first amino acid–tRNA is one that carries the amino acid methionine (Met) to the ribosome.
- During **elongation,** as described in Figure 11.6B, the peptide increases one amino acid at a time in this way: The peptide at the P site binds to the amino acid at the A site. A **ribozyme** speeds the formation of a peptide bond between the two.
- Now **translocation** occurs: The ribosome moves forward, and the peptide-tRNA is now at the P site. The spent tRNA exits from the E site. This process occurs over and over again. At **termination,** the ribosome reaches a stop codon, and the polypeptide is released.

11.7 Let's review gene expression

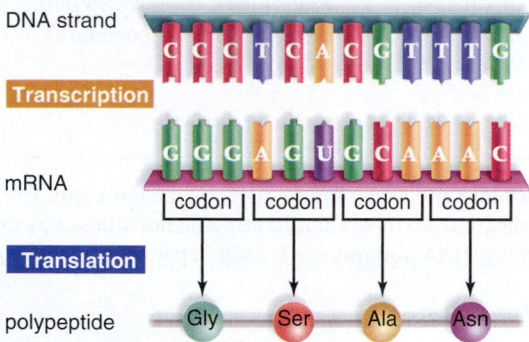

Mutations: Altered Gene Expression

11.8 Genetic mutations affect a specific allele

- A **genetic mutation** is a permanent change in the base sequence of a gene. A **point mutation** is only one base pair change; in a **silent mutation** the base pair change does not affect the sequence of amino acids; in a **nonsense mutation** a stop codon is introduced that stops protein synthesis; in a **missense mutation,** an amino acid substitution occurs and in a **frameshift mutation** all amino acids change.
- In a **triplet repeat expansion** (three bases are repeated over and over again) each generation receives more of the repeat with ever greater chance of disorder developing.

11.9 Chromosomal mutations affect several alleles

- **Chromosomal mutations** are either changes in chromosome number (see section 9.6 and 9.7) or changes in chromosome structure that affect the function of alleles. In **deletions** a portion of a chromosome is missing; in **duplications** a segment of a chromosome is duplicated; in an **inversion** a segment is joined in an opposite direction; in a **translocation** there is an exchange of segments between two nonhomologous chromosomes.

TEST YOURSELF

DNA and RNA Structure and Function

1. If 30% of an organism's DNA is thymine, then
   a. 70% is purine.   d. 70% is pyrimidine.
   b. 10% is guanine.   e. Both c and d are correct.
   c. 30% is adenine.

2. In a DNA molecule, the
   a. backbone is sugar and phosphate molecules.
   b. bases are covalently bonded to the sugars.
   c. sugars are covalently bonded to the phosphates.
   d. bases are hydrogen-bonded to one another.
   e. All of these are correct.

3. Which of these characteristics of DNA is not paired with a proper explanation for that characteristic?
   a. variable between species—sequence of bases can vary
   b. store information—sugar-phosphate backbones never vary
   c. constant within a species—can be replicated by complementary base pairing
   d. able to undergo mutations—sequence of bases can change

4. Which of these statements is true concerning DNA structure?
   a. A sugar bonds to phosphate and to a base.
   b. A sugar bonds only to two phosphate groups.
   c. U is present in DNA but absent in RNA.
   d. Sugars, being ring structures, hydrogen bond together.

DNA Replication

5. Because each daughter molecule contains one old strand of DNA, DNA replication is said to be
   a. conservative.   c. semidiscontinuous.
   b. preservative.   d. semiconservative.

6. During DNA replication, the parental strand ATTGGC would code for the daughter strand
   a. ATTGGC.   c. TAACCG.
   b. CGGTTA.   d. GCCAAT.

7. DNA polymerase carries out replication, except it cannot
   a. carry out complementary base pairing.
   b. unwind the double helix.
   c. join the nucleotides together.
   d. proofread the polymer for accuracy of base pairing.

8. **THINKING CONCEPTUALLY** Azidothymicline (AZT), the well-known medicine for HIV infection, is a DNA base analogue that hinders DNA replication. Explain why it works.

Gene Expression

For questions 9–13, match each molecule to its special significance in gene expression as listed in the key.

KEY:
   a. stores genetic information from generation to generation
   b. sequence of three RNA bases complementary to those in DNA
   c. has an anticodon
   d. located in ribosomes
   e. the gene product

9. rRNA
10. mRNA codon
11. Protein
12. DNA
13. tRNA
14. Transcription produces _____, while translation produces _____.
   a. DNA, RNA   c. polypeptides, RNA
   b. RNA, polypeptides   d. RNA, DNA

15. Which of the following statements does not characterize the process of transcription? Choose more than one answer if correct.
    a. During transcription, RNA nucleotides base pair to the DNA template strand.
    b. To make RNA, the base uracil pairs with adenine.
    c. The enzyme RNA polymerase synthesizes RNA.
    d. RNA is made in the cytoplasm of eukaryotic cells.
16. Because there are more codons than amino acids,
    a. some amino acids are specified by more than one codon.
    b. some codons do not specify any amino acid.
    c. some amino acids do not have codons.
    d. Both a and b are correct
17. If the sequence of bases in the DNA template strand is TAGC, then the sequence of bases in RNA will be
    a. ATCG.   d. GCTA.
    b. TAGC.   e. Both a and b are correct.
    c. AUCG.
18. mRNA processing
    a. is the same as transcription.
    b. is an event that occurs after RNA is transcribed.
    c. is the rejection of old, worn-out RNA.
    d. pertains to the function of transfer RNA during protein synthesis.
    e. Both b and d are correct.
19. Label this diagram showing the participants in translation:

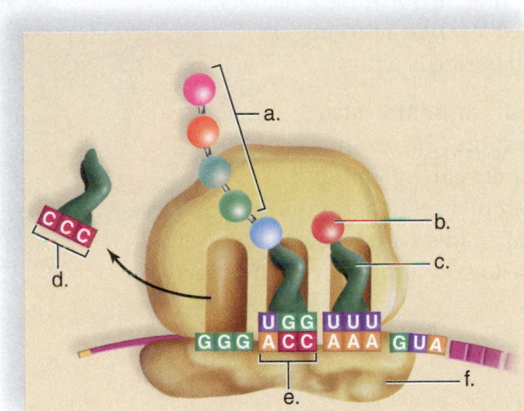

20. During protein synthesis, the anticodon of transfer RNA (tRNA) pairs with
    a. DNA nucleotide bases.
    b. ribosomal RNA (rRNA) nucleotide bases.
    c. messenger RNA (mRNA) nucleotide bases.
    d. other tRNA nucleotide bases.
    e. Any one of these pairings can occur.
21. Following is a segment of a DNA molecule. (Remember that only the template strand is transcribed.) What are (a) the RNA codons, (b) the tRNA anticodons, and (c) the sequence of amino acids in a protein?

22. **THINKING CONCEPTUALLY** What type information does DNA store, and how does it store the information?

Mutations: Altered Gene Expression

23. How would you know that a genetic mutation has occurred?
    a. A person has an infectious disease.
    b. A person inherits a genetic disorder.
    c. Transcription and translation occur.
    d. All of these are correct.

For questions 24–26, match the resulting event with the genetic mutation in the key.

KEY:
    a. silent mutation
    b. missense mutation
    c. nonsense mutation
    d. frameshift mutation

24. One amino acid in the sequence is substituted for another.
25. A stop codon appears before protein synthesis is complete.
26. The sequence of amino acids has completely changed.

For questions 27–29, match the resulting event with the chromosomal mutation in the key.

KEY:
    a. duplication
    b. deletion
    c. inversion
    d. translocation

27. Exchange of segments occurs between nonhomologous chromosomes.
28. A segment of a chromosome is present over again.
29. A segment of the chromosome runs in the opposite direction.
30. **THINKING CONCEPTUALLY** Mutations can cause cancer, but on the other hand, it is important for DNA to mutate. Explain this contradiction.

GET INVOLVED

1. How would you test your hypothesis that the genetic condition neurofibromatosis is due to a transposon?
2. Knowing that you can clone plants from a few cells in tissue culture, how would you determine if an isolated *Arabidopsis* allele causes a particular mutation?
3. The virtual lab "DNA and Genes" provides an interactive tutorial about the influence of mutations on the sequence of amino acids in a protein as illustrated in Figure 11.8A.

Virtual Lab
DNA and Genes

MEDIA STUDY TOOLS

mhhe.com/maderconcepts3

Enhance your study of this chapter with interactive study tools, practice tests, and engaging animations. Also, ask your instructor about the resources available through ConnectPlus, which includes LearnSmart, a personalized adaptive learning program, and a media-rich eBook.

12

Regulation of Gene Activity

BEFORE YOU BEGIN

Take a few minutes to recall

A simplified overview of gene expression (Figure 4.4B)

The function of transcription and translation in gene expression (section 11.7)

How cell cycle control is missing in cancer cells (section 8.6)

Moth and Butterfly Wings Tell a Story

After you set an ornate moth free, all that is left on your hand is a smudge of dustlike residue. The residue is actually composed of many scales, the units of moth and butterfly wings. The multitude of scale colors and patterns in moths and butterflies is awe-inspiring. Each individual scale is a particular color and may vary completely from a neighboring scale. The color of the scales is due to the presence of particular pigments that transmit, absorb, and reflect certain colors of light.

Most specialists who study insects agree that scales evolved from the bristles of an ancestor to moths and butterflies. Over time, the bristles became wide and flat and lost any sensory function. You might think that scales have an accessory and unnecessary function, but evidence suggests otherwise. For example, the easy detachment of scales may have made it easier for ancient moths and butterflies to escape from spiderwebs and other predators. The possible protective function of scales is strengthened by their role in forming eyespots, a rounded eyelike marking on moth and butterfly wings.

Other animals also have eyespots. For example, eyespots can be found on the tail of a redfish, on the bodies of spiders and caterpillars, and occasionally on the back of a lynx's

Leafwing butterfly

eyespot

Bull's-eye moth

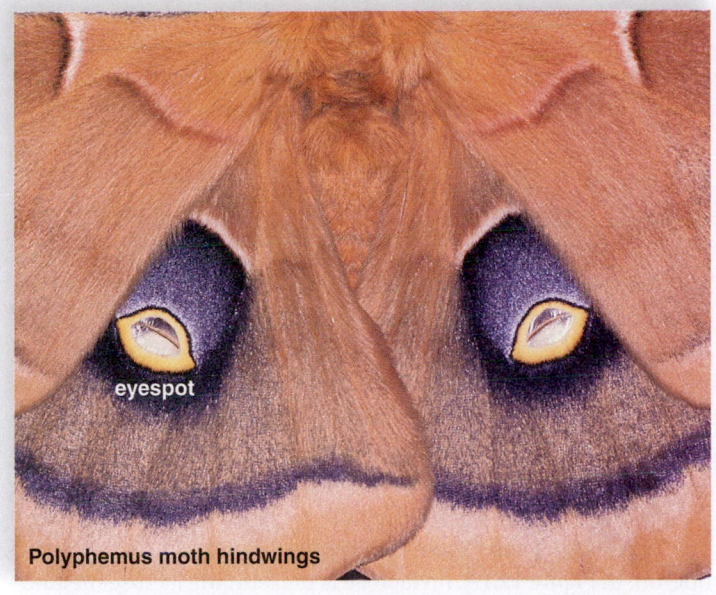

eyespot

Polyphemus moth hindwings

retractable antennae

eyespot

Caterpillar of citrus swallowtail butterfly

eyespots

Indian spectacled cobra

ear or the back of a cobra's hood. Eyespots confuse a potential predator and may divert attacks to body margins, thereby saving most of the animal from damage. Certainly an animal as delicate as a moth or butterfly needs all the help it can get to keep its body from being attacked.

Still more evidence suggests the importance of scales to the life of a moth or butterfly. Developmental biologists specializing in evolution have discovered that the same regulatory genes involved in building insect limbs are also involved in determining eyespot patterns. Eyespots are the result of ***Hox* genes,** the master regulatory genes that lie at the center of development itself. Like that of any animal, the body organization of a moth or a butterfly is determined by these very ancient genes. They are developmental regulatory genes that act as important switches for organizing embryonic cells into specific structures. This means that the amazing scales of moths and butterflies are a product of the genes that make them what they are. Without developmental regulatory genes, there would be no scales and indeed no moths and butterflies. The two are intertwined. The next time a moth or butterfly flits by—think about it.

Gene regulation plays an important role in the development of an organism before birth and in its health and welfare after birth. This chapter surveys the field of gene regulation in prokaryotes and eukaryotes before taking another look at cancer, which is caused by the lack of proper gene regulation.

Transcriptional Control

Gene regulation was first discovered in prokaryotes, and the study of operons in prokaryotes offers an opportunity to understand what is meant by gene regulation. This part of the chapter also introduces the concept that DNA-binding proteins are involved in regulating gene expression whether the topic involves chromosome structure or development of the embryo. The use of proteins to control gene expression in eukaryotes requires you to visually imagine the production of a protein in the cytoplasm that doubles back to bind to DNA in the nucleus.

12.1 DNA-binding proteins usually turn genes off in prokaryotes

LEARNING OUTCOME

When you complete this section, you should be able to

1. List and state the function of the various components of an operon and also the specific role of the repressor.

As you know, the process of mitosis ensures that every cell in your body contains the same number of chromosomes and genes. Yet, the body is composed of different tissues, such as muscle tissue, nervous tissue, and bone tissue. How is this diversity possible if all cells have a copy of every gene? Gene regulation is the answer to this conundrum: Only certain genes are active in each type of cell, and the protein products of these cells cause the cells to become specialized in structure and function. But how does gene regulation work? The first clue came from studies with the bacteria *Escherichia coli,* which live in the human intestine. The nutrients available to *E. coli* are very much dependent on the diet of their host, and it would be wasteful for bacteria to exert energy producing enzymes they do not need. In 1961, the French microbiologists François Jacob and Jacques Monod showed that *Escherichia coli* are capable of producing only the enzymes they need because they can turn off the expression of genes that are not needed. In other words, bacteria control which genes are expressed.

The Operon Model Jacob and Monod proposed the **operon** model to explain gene regulation in prokaryotes. Later, they received a Nobel Prize for their investigations. Note in Figure 12.1A that the operon model has these components:

- A **regulatory gene** is located outside the operon. The regulatory gene codes for a DNA-binding protein that acts as a **repressor.** When active, the repressor binds to an operator, and this turns genes off.
- A **promoter** is a short sequence of DNA where RNA polymerase first attaches when genes are to be transcribed. Basically, the promoter signals the start of an operon.

- An **operator** is a short portion of DNA where an active repressor binds. When an active repressor is bound to the operator, RNA polymerase cannot attach to the promoter, and transcription does not occur. The operator therefore acts as an off switch for transcription of the **structural genes** which are transcribed as a unit. (In Figure 12.1A, the structural genes are the lactose metabolizing genes.)

Figure 12.1A and Figure 12.1B apply this information to the *lac* operon. In Figure 12.1A, no lactose is present; the lactose metabolizing genes are turned off because the repressor is bound to the operator. In Figure 12.1B, RNA polymerase has room to attach to the promoter, and the lactose metabolizing genes are turned on. The production of mRNA transcripts and lactose enzymes follows. What happened? Lactose is present and has combined with

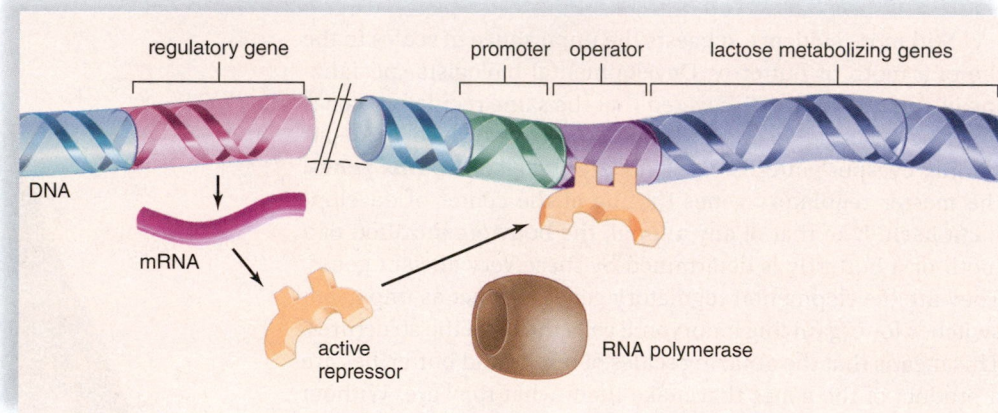

FIGURE 12.1A Inactive *lac* operon: The genes are turned off, and lactose enzymes are not produced.

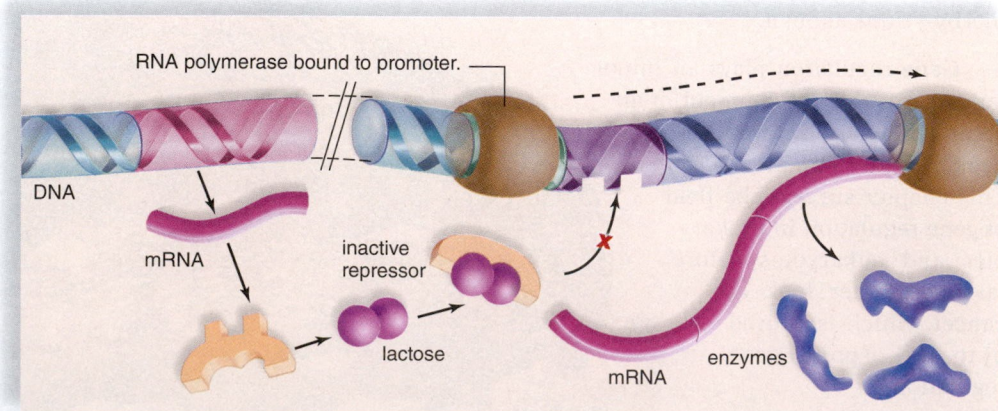

FIGURE 12.1B Active *lac* operon: The genes are being transcribed, and lactose enzymes are produced.

the repressor, inactivating it. In other words, a change in the shape of the repressor allows the operon to be active.

Animation
Combination of Switches: The *Lac* Operon

The *lac* operon is usually turned off because lactose is usually absent and glucose, the preferred sugar to metabolize, is usually present. Some other operons are usually active in *E. coli*. For example, the enzymes needed to make the amino acid tryptophan are always present, unless of course, tryptophan is present in the environment. Then, *E. coli* absorbs tryptophan from the environment and turns the operon off. Turning the operon off again requires that the repressor bind to the operator, but in this instance the repressor becomes active after it combines with tryptophan (Fig. 12.1C).

Animation
The Tryptophan Repressor

In summary, we have seen that regulatory genes code for proteins and these proteins affect the expression of genes after

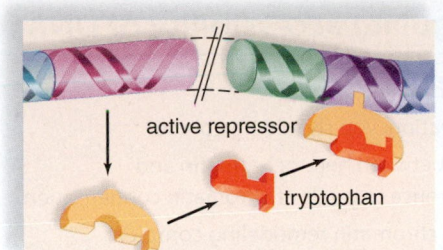

FIGURE 12.1C
When tryptophan combines with the repressor, it is active.

binding to DNA. In prokaryotes, regulatory DNA binding proteins must be inactive for transcription to occur.

▶ **12.1 CHECK YOUR PROGRESS**
   1. Explain when and how a *lac* repressor is inactivated in a bacterium.

12.2 DNA-binding proteins usually turn genes on in eukaryotes

LEARNING OUTCOME

When you complete this section, you should be able to

1. Identify the DNA-binding proteins that turn genes on in eukaryotes.

Some genes in a eukaryotic cell, such as a human cell, are called housekeeping genes because they govern functions that are common in many types of cells, such as glucose metabolism. Housekeeping genes tend to be active all the time. However, genes that account for the specialization of cells, such as hemoglobin in red blood cells, insulin in pancreatic cells, and myosin in muscle cells, are active only in certain cells.

Animation
Control of Gene Expression in Eukaryotes

Although some operons like those of prokaryotic cells have been found in eukaryotic cells, transcription is usually controlled by DNA-binding proteins that turn genes on (Fig. 12.2). Chief among them are

Animation
Transcription Factors

❶ **transcription factors,** proteins that assist (instead of preventing) the binding of RNA polymerase to a promotor. A cell has many different types of transcription factors, and the absence of one can prevent transcription from occurring. Even if all transcription factors are present, transcription may not begin without the assistance of a transcription activator. ❷ **Transcription activators** bind to regions of DNA called **enhancers.** A different transcription activator is believed to regulate the activity of any particular gene. ❸ Enhancers can be quite a distance from the promoter, but bending of DNA can bring the transcription activator attached to the enhancer into

the vicinity of the transcription factors. Once contact is made between the transcription activator and the regulatory protein complex at the promoter, transcription begins and the result is a pre-mRNA transcript.

The need for so many regulatory proteins to activate transcription means that regulation of gene expression can differ between cells and that fine-tuning is possible. The switching on of a gene requires that certain regulatory proteins be present. As we shall see in section 12.3, the initiation (start) of transcription must also take into account the accessibility of the gene to be transcribed.

▶ **12.2 CHECK YOUR PROGRESS**
   1. Distinguish between a transcription factor and transcription enhancer.

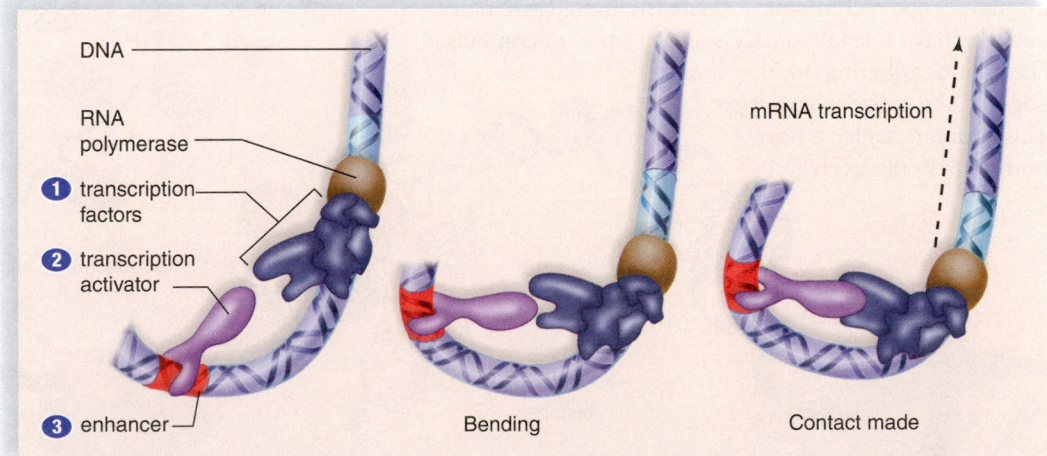

DNA
RNA polymerase
❶ transcription factors
❷ transcription activator
❸ enhancer

mRNA transcription

Bending   Contact made

FIGURE 12.2 Transcription in eukaryotic cells requires that transcription factors bind to the promoter and transcription activators bind to an enhancer. The enhancer is far from the promoter, but the DNA loops so that the activator can bind to the factors. Only then does transcription begin so that a pre-mRNA transcript results.

12.3 Histones regulate accessibility of DNA for transcription

LEARNING OUTCOMES

When you complete this section, you should be able to

1. Explain the difference between heterochromatin and euchromatin with reference to levels of chromatin condensation.
2. Identify the role of the chromatin remodeling complex.

The DNA in eukaryotes is always associated with plentiful proteins, and together they make **chromatin,** a stringy material that can be observed in the interphase nucleus. **Histones** are proteins that form spools around which DNA winds, and they participate in gene regulation. Without histones, DNA would not be able to fit inside a nucleus. For example, a human cell contains about 2 m of DNA, and a nucleus is only 5–8 μm in diameter. At the time of cell division, chromatin is compacted to form the chromosomes (Fig. 12.3A).

Levels of Chromatin Structure Previously, we mentioned that chromatin is condensed to form chromosomes that are visible during cell division. To achieve this compacted form of chromatin, DNA is first periodically wound around a core of eight histone molecules so that it looks like beads on a string. Each bead is called a **nucleosome.** The string is further compacted when it folds into a zigzag fiber, which loops back and forth. At this level of compaction, chromatin is called **euchromatin.** Euchromatin may be accessed by RNA polymerase and transcription factors that are needed to promote transcription. Euchromatin is not darkly stained.

Under a microscope, we often observe darkly stained chromatin (Fig. 12.3B*a*). These areas of highly coiled chromatin are called **heterochromatin.** Heterochromatin is considered inactive chromatin because the genes it contains are infrequently transcribed, if at all.

Prior to cell division, a protein scaffold helps condense chromatin into a form characteristic of metaphase chromosomes. No doubt, compact chromosomes are easier to move about than more extended chromatin.

Heterochromatin Is Not Transcribed A dramatic example of heterochromatin is seen in mammalian females. Females have a small, darkly staining mass of condensed chromatin adhering to the inner edge of the nuclear envelope. This structure, called a **Barr body** after its discoverer,

Murray Barr, is an inactive X chromosome. On a random basis, one of the X chromosomes undergoes inactivation in the cells of female embryos. The inactive X chromosome does not produce gene products, and therefore female cells have a reduced amount of product from genes on the X chromosome.

How do we know that Barr bodies are inactive X chromosomes that are not producing gene product? Suppose 50% of the cells have one X chromosome active and 50% have the other X chromosome active. Wouldn't the body of a heterozygous female be a mosaic, with "patches" of genetically different cells? This is exactly what investigators have discovered. Human females who

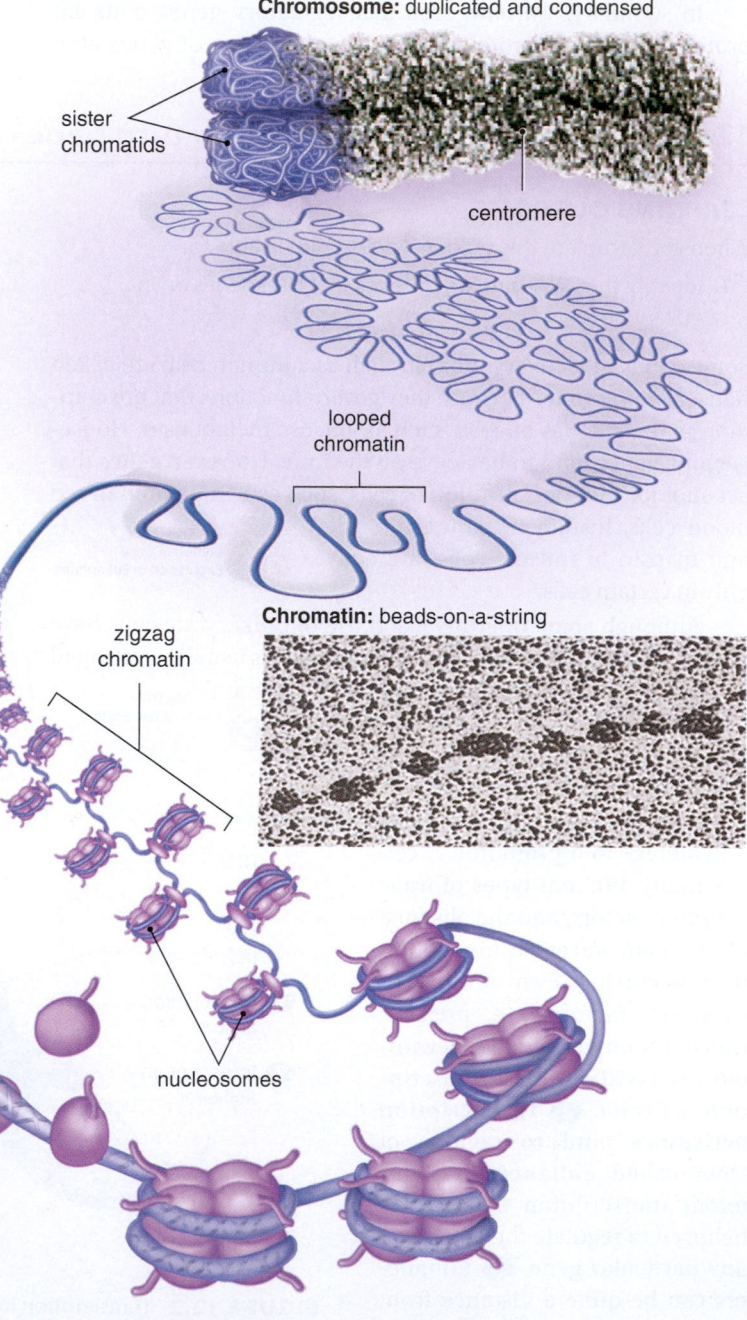

Chromosome: duplicated and condensed

sister chromatids

centromere

looped chromatin

zigzag chromatin

Chromatin: beads-on-a-string

nucleosomes

histones

FIGURE 12.3A Levels of chromatin condensation from a double DNA helix to a metaphase chromosome.

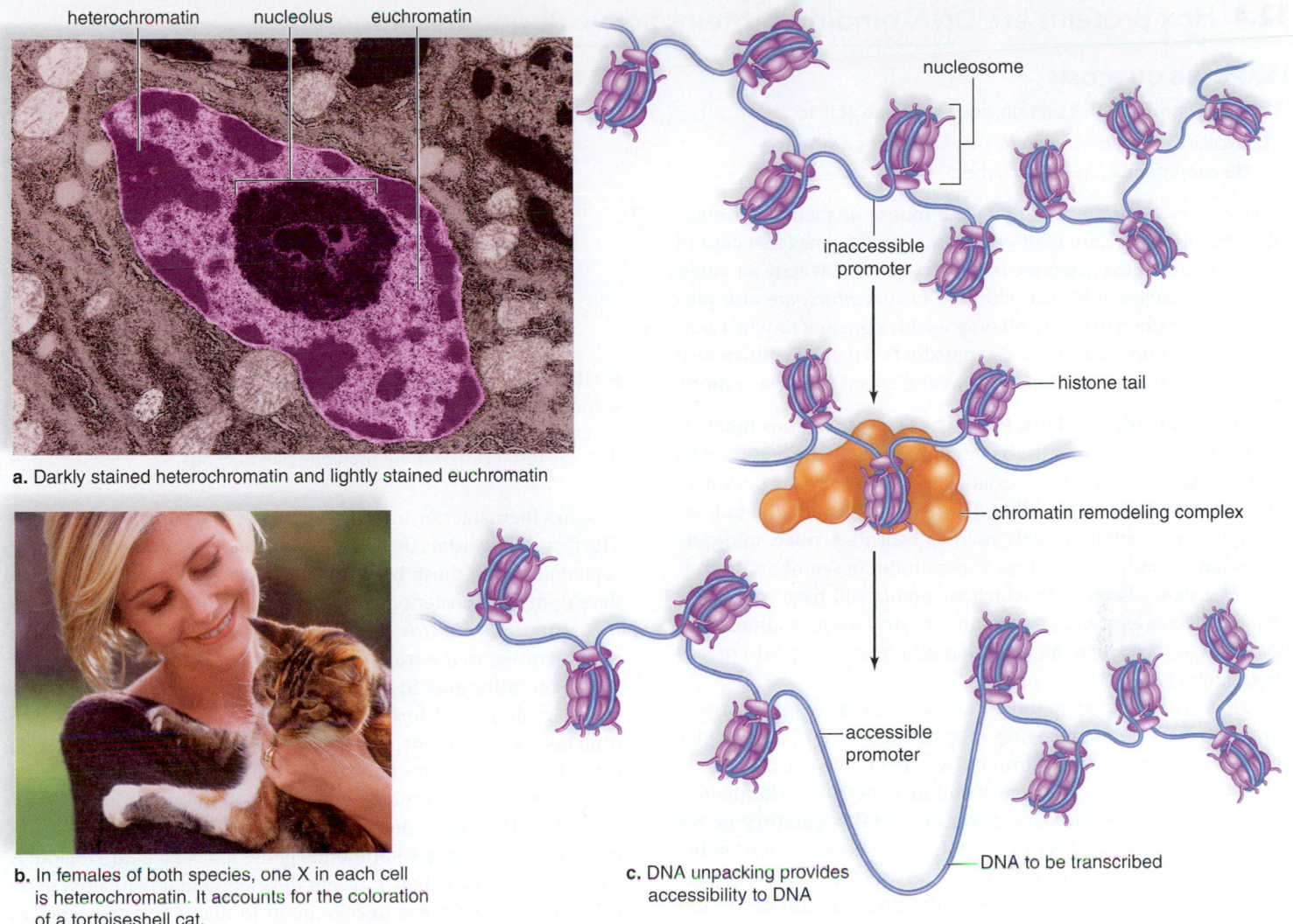

a. Darkly stained heterochromatin and lightly stained euchromatin

b. In females of both species, one X in each cell is heterochromatin. It accounts for the coloration of a tortoiseshell cat.

c. DNA unpacking provides accessibility to DNA

FIGURE 12.3B **a.** A nucleus contains heterochromatin which is inactive (not being transcribed) and euchromatin which is active (being transcribed). **b.** Tortoiseshell cats are females which have patches of both black and orange. Black appears when X chromosomes carrying an allele for orange are inactivated and orange appears when X chromosomes carrying an allele for black are inactivated. **c.** When euchromatin is transcribed, a chromatin remodeling complex pushes aside the histone portions of nucleosomes so that RNA polymerase and transcription factors have access to the gene to be transcribed.

are heterozygous for an X-linked recessive form of ocular albinism have patches of pigmented and nonpigmented cells at the back of the eye. Women heterozygous for Duchenne muscular dystrophy have patches of normal muscle tissue and degenerative muscle tissue (the normal tissue increases in size and strength to make up for the defective tissue). And women who are heterozygous for X-linked hereditary absence of sweat glands have patches of skin lacking sweat glands. The female tortoiseshell cat also provides dramatic support for a difference in X-inactivation in its cells. In these cats, an allele for black coat color is on one X chromosome, and a corresponding allele for orange coat color is on the other X chromosome. The patches of black and orange in the coat can be related to which X chromosome is in the Barr bodies of the cells found in the patches (Fig. 12.3B*b*).

Euchromatin Is Transcribed Active genes in eukaryotic cells are associated with more loosely compacted euchromatin. His-tones regulate accessibility to DNA, and euchromatin becomes genetically active when histones no longer bar access to DNA.

When DNA in euchromatin is transcribed, a so-called **chromatin remodeling complex** pushes aside the histone portion of a nucleosome so that access to DNA is not barred and transcription can begin (Fig. 12.3B*c*). After **unpacking** occurs, many decondensed loops radiate from the central axis of the chromosome. These chromosomes have been named lampbrush chromosomes because their feathery appearance resembles the brushes that were once used to clean kerosene lamps.

What regulates whether chromatin exists as euchromatin or heterochromatin? Histone molecules have tails, strings of amino acids that extend beyond the main portion of the nucleosome (Fig. 12.3B*c*). In euchromatin, the histone tails tend to have attached acetyl groups ($-COCH_3$); in heterochromatin, the histone tails tend to bear methyl groups ($-CH_4$).

▶ **12.3 CHECK YOUR PROGRESS**
1. Characterize the type of chromatin that is available for transcription in the nucleus.

LEARNING OUTCOME

When you complete this section, you should be able to

1. Explain what it means to say that *Hox* genes are "master developmental regulatory genes."

Investigators using the fruit fly or the mouse as their experimental material have begun to discover the types of genes that control development. These genes are regulatory genes that code for either signaling proteins or for transcription factors which are also proteins. As described in Figure 5B on page 91, signaling proteins activate transduction pathways and a transduction pathway ends with a transcription factor that binds to DNA and affects gene expression.

Hox Genes Code for Hox Proteins The *Hox* genes function during development after the basic coordinates of the body and its various segments have been established. (Coordinates designate which end of the animal will be the head and which the tail, for example.) Segmental animals such as fruit flies, mice, and ourselves have a body divided into a set number of segments. In fruit flies, *Hox* genes determine which segments will have wings and which will have legs. *Hox* gene mutations may result in abnormalities, such as legs where antennae should be or wings where legs should be (Fig. 12.4A).

Hox genes can be found in many different types of organisms, and they can be recognized by the presence of a particular sequence of bases called a **homeobox** (Fig. 12.4A). The homeobox codes for a particular sequence of amino acids called a homeodomain. *Hox* genes are **master developmental regulatory genes** because they code for Hox proteins which are transcription factors that bind to and activate other regulatory genes. The homeodomain is the DNA-binding portion of the transcription factor and a variable sequence determines which target genes are affected. The target genes are directly involved in pattern formation, or what the animal will look like.

The importance of *Hox* genes is underscored by the finding that they are highly conserved, being present in the genomes of many organisms, including mammals such as mice and even humans. In both flies and mammals, the position of the *Hox* on the chromosome

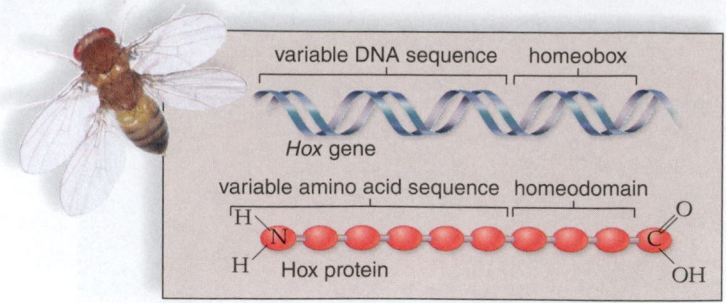

FIGURE 12.4A *Hox* genes are master developmental genes identified in many different species by the presence of a homeobox that codes for a homeodomain. If a *Hox* gene malfunctions in fruit flies, the result can be more than one pair of wings.

matches their anterior-to-posterior expression pattern in the body. The first gene clusters determine the final development of anterior segments, while those later in the sequence determine the final development of posterior segments of the animal's body.

Mutations in *Hox* genes in the mammalian body have effects similar to the transformations observed in fruit flies. For instance, mutations in two adjacent *Hox* genes in the mouse result in shortened forelimbs that are missing the radius and ulna bones. In humans, mutations in a different *Hox* gene cause synpolydactyly, a rare condition characterized by extra digits (fingers and toes), some of which are fused to their neighbors.

While *Hox* genes determine the final development of a segment, other master developmental genes affect an entire region. A surprising finding is that a master developmental regulatory gene called *Pax6* triggers eye development in many different types of organisms, including a fly, a human, and a squid, even though the structure of their eyes is entirely different (Fig. 12.4B).

▶ **12.4 CHECK YOUR PROGRESS**

1. Explain why a master developmental regulatory gene can greatly affect the body.

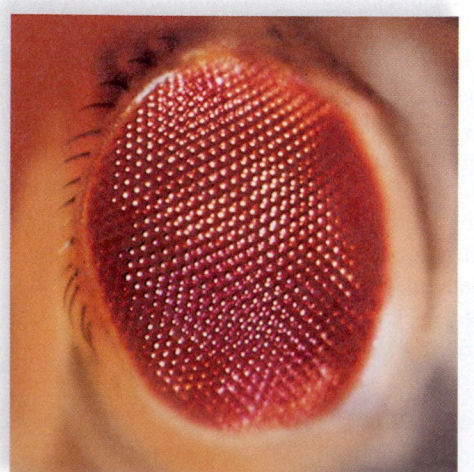
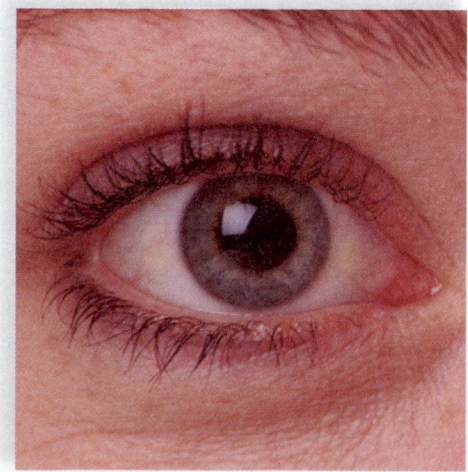
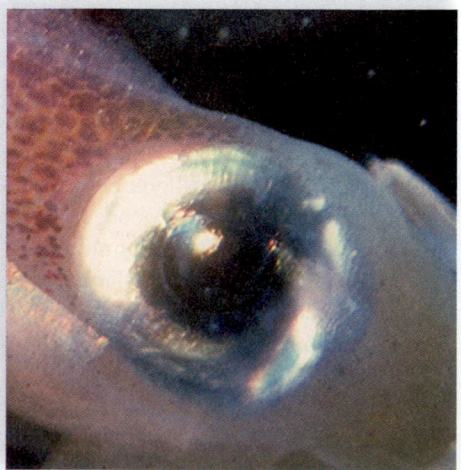

FIGURE 12.4B *Pax6* is a master developmental regulatory gene that turns on eye development in a fly, a human, and a squid.

12A   Regulatory Genes and the Origin of the Genus *Homo*

Fossils of immediate ancestors to the genus *Homo* indicate that they spent part of their time climbing trees and that they retained many apelike traits. In some, the arms, like those of an ape, were long compared to the length of the legs. Then, too, our ancestors had strong wrists and long, curved fingers and toes. These traits would have served well for climbing, and these predecessors to humans probably climbed trees for the same reason that chimpanzees do today: to gather fruits and nuts in trees and to sleep aboveground at night in order to avoid predatory animals, such as lions and hyenas.

Whereas our brain is about the size of a grapefruit, that of our predecessors was about the size of an orange—and only slightly larger than that of a chimpanzee. There is no evidence that they manufactured stone tools; presumably, they were not smart enough to do so.

Several years ago, Stephen Stanley of Johns Hopkins University concluded that the genus *Homo* could not have evolved if our immediate ancestors lived in trees. The obstacle relates to the way we, members of *Homo,* develop our large brain. Unlike apes, we retain a high rate of fetal brain growth through the first year after birth. (That is why a 1-year-old child has a very large head in proportion to the rest of its body.) The brain of apes grows rapidly before birth, but immediately after birth the brain grows more slowly. As a result, an adult human brain is more than three times as large as that of an adult chimpanzee.

A continuation of the high rate of fetal brain growth in our ancestors eventually allowed the genus *Homo* to evolve. But the continued brain growth is linked to underdevelopment of the body as a whole. Although the human brain eventually becomes more complex, human babies are remarkably weak and uncoordinated. Such helpless infants must be carried about and tended. Human babies are unable to cling to their mothers the way chimpanzee babies can.

The origin of the genus *Homo* entailed a great evolutionary compromise. Humans gained a large brain, but they were saddled with a long period of infantile helplessness. The positive value of a large brain must have outweighed the negative aspects of infantile helplessness, such as the inability of adults to climb trees while holding an infant, or else genus *Homo* wouldn't have evolved. Having a larger brain meant that humans were able to outsmart or ward off predators with weapons they were clever enough to manufacture.

If regulatory genes were involved, as presumably they were, very few genetic changes were required to delay maturation and produce the large brain of *Homo.* The mutation of a master developmental regulatory gene, such as a *Pax* gene, that controls one or more other genes most likely could have been all that was needed. As we learn more about the human genome, we will eventually uncover the particular gene or gene combinations that caused early *Homo* to have a large brain, and this will be a very exciting discovery. As of now, we know that changes in gene expression occur more often in other organs (such as the liver) than they do in the brain, but any such changes could have produced a dramatic effect. This correlates with the unique pattern of human brain development. The human brain expands as nerve cells arise and make contact with one another. A gene called protocadherin has recently been singled out as a possible candidate for a master regulatory gene in the brain. Perhaps this will be the gene that functions differently in the human brain compared to the ape brain.

A young chimpanzee can cling to its mother, leaving her hands free to climb a tree.

A human baby cannot cling and has to be carried.

CONSIDER THESE QUESTIONS

1. Should researchers spend much time and resources discovering what makes our brain different from that of the apes? Why or why not?

2. *Mosaic evolution* is a term used for evolution of one part of the body at the expense of another. Why does this term apply to the aspects of human evolution discussed in this reading?

3. During human development, life begins as a single cell that divides many times, and these cells slowly take on the shape and function of specialized tissues and organs. Does it make sense to you that regulatory genes play an important role in development? Why or why not?

 **connect**
|BIOLOGY   Explore the concepts through a variety of multimedia assets and question types.

www.mcgrawhillconnect.com

Posttranscriptional Control

The role of RNAs in gene expression is only now being recognized and given greater significance. Alternative mRNA splicing has been known for some time, but the presence and activity of small RNA sequences in regulating gene expression is a new finding. Biologists are asking: Is RNA simply DNA's helper, or is RNA the regulator that controls what the cell and the organism will be like?

12.5 Alternative mRNA splicing results in varied gene products

LEARNING OUTCOME

When you complete this section, you should be able to

1. Describe how alternative mRNA splicing controls the specific protein products of a gene.

You'll recall that during pre-mRNA processing, introns (noncoding regions) are excised, and exons (expressed regions) are spliced together. When introns are removed, **pre-mRNA alternative splicing** of exons can occur, and this affects gene expression. For example, an exon that is normally included in an mRNA transcript may be skipped, and it is excised along with the flanking introns (Fig. 12.5). The resulting mature mRNA has an altered sequence, and the protein product differs. Sometimes introns remain in an mRNA transcript; when this occurs, the protein coding sequence also changes.

Examples of alternative pre-mRNA splicing abound. Both the hypothalamus and the thyroid gland produce a protein hormone called calcitonin, but the mRNA that leaves the nucleus is not the same in both types of cells. This causes the thyroid to release a slightly different version of calcitonin than the hypothalamus. Evidence of alternative mRNA splicing is also found in other cells, such as those that produce neurotransmitters, muscle regulatory proteins, and antibodies. This process allows the cells of humans and other complex organisms to recombine genes in many new and novel ways to create the great variety of proteins found in these organisms.

Researchers are busy determining how introns and other types of small RNAs can affect mRNA processing and translation as discussed in section 12.6.

Aside from alternative splicing, posttranscriptional control of gene expression can also be achieved by modifying the speed of transport of mRNA from the nucleus into the cytoplasm. Evidence indicates there is a difference in the length of time it takes various mRNA molecules to pass through a nuclear pore, affecting the amount of gene product realized per unit time following transcription.

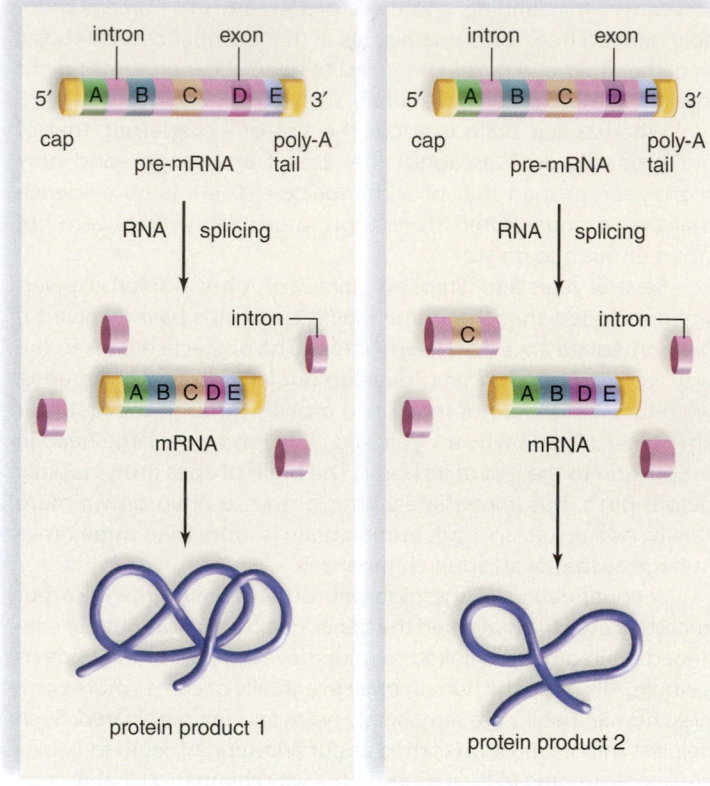

FIGURE 12.5 Because the pre-mRNAs are processed differently in these two cells, distinct proteins result. This is a form of posttranscriptional control of gene expression.

▶ **12.5 CHECK YOUR PROGRESS**
1. Identify the components of the mRNA molecule in Figure 12.5 if splicing did not occur.

12.6 Small RNA (sRNA) molecules fine-tune gene expression

LEARNING OUTCOME

When you complete this section, you should be able to

1. Explain the role of small RNAs in regulating the amount of a gene's product.

Once scientists understood gene expression, they were faced with a dilemma: The nucleus contains vastly more DNA than is needed to account for the number of proteins in the cell. The DNA that did not code for protein was initially termed "junk" DNA because its function was unknown, but scientists have now begun to understand the role of this DNA in the cell. Their work is on the cutting edge of modern genetic research!

We now know that noncoding DNA is used to form small RNA (sRNA) molecules that function in a number of ways to regulate gene expression. How does a cell form sRNA molecules and how do they regulate gene expression? Notice in

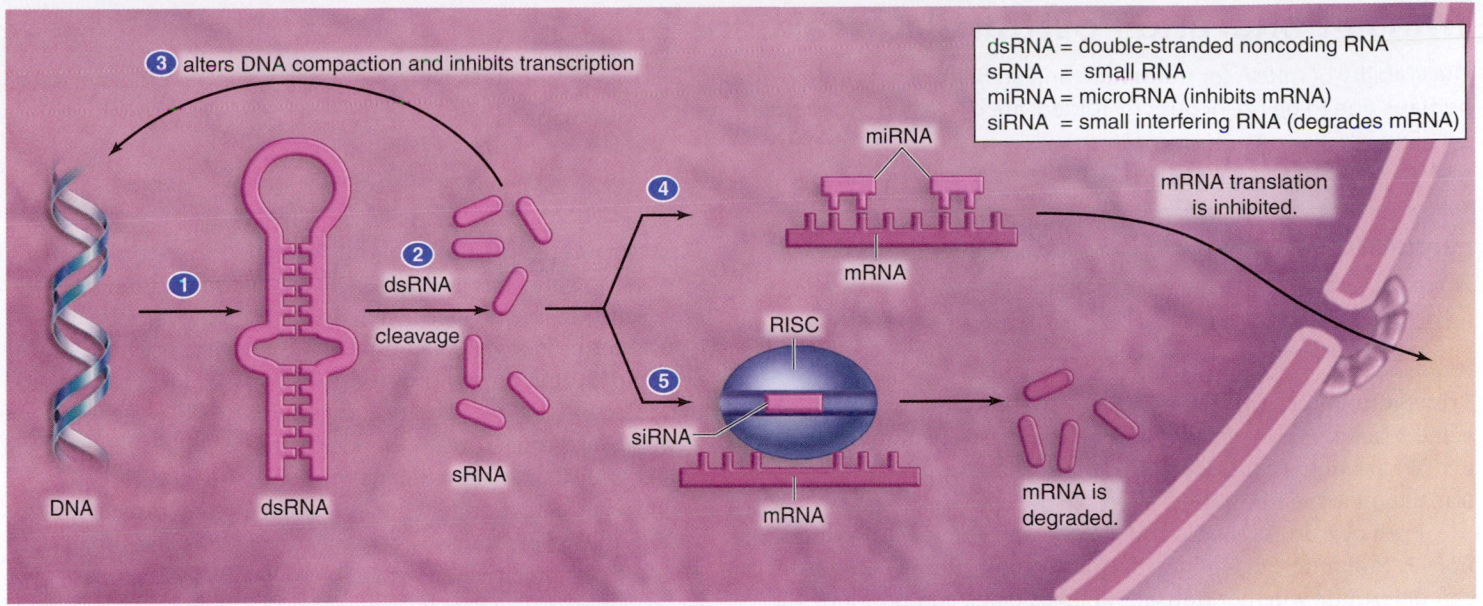

FIGURE 12.6 Transcription of the DNA ① may lead to looped and double-stranded RNA (dsRNA). ② The cleavage of the dsRNA produces many small RNA (sRNA) molecules. ③ An sRNA can double-back to increase DNA compaction, or may become an miRNA or siRNA. ④ miRNA reduces translation by binding to complementary mRNA transcripts. ⑤ siRNA help form a complex called RISC, which then degrades any mRNA transcript with a sequence of bases that are complementary to the siRNA. Translation of this mRNA transcript does not occur.

Figure 12.6 that transcribed RNA can sometimes form loops as hydrogen bonding occurs between its bases. Enzymes in the cell dice up this double-stranded (dsRNA) to form **small RNAs** (**sRNAs**). Some of these sRNA molecules regulate transcription, while others are involved in the regulation of translation. Specifically, three ways have been found by which sRNA may regulate gene expression:

- Small RNAs can alter the compaction of DNA so that some genes become inaccessible for transcription.

- Small RNAs are the source of **microRNAs** (**miRNAs**), small snippets of RNA that can bind to and lessen the translation of an mRNA in the cytoplasm. In that case, the amount of gene product made is regulated.

- Small RNAs are also the source of **small interfering RNAs** (**siRNAs**) that join with an enzyme to form an *RNA Induced Silencing Complex* (RISC). The complex targets certain mRNAs for breakdown.

By using a combination of miRNA and siRNA molecules, a cell can fine-tune the amount of protein product in a cell, much like a dimmer switch allows a light to be on maximally, to varying degrees of brightness, or not at all. Previously, scientists did not know that RNA could be this active in gene regulation and have now begun to hypothesize on the significance of this finding. For example, small RNAs may play a role in the regulation of development and help account for why one species is different from another. Because of their abundance in vertebrates compared to invertebrates, some researchers hypothesize that small RNAs

might be responsible for the evolution of various vertebrates, including humans. Other scientists have turned their attention to applying these new findings in various ways.

Uses for siRNAs and miRNAs Some viruses have RNA as their genetic material, and it's possible that siRNA evolved as a way to prevent these viruses from infecting the cell. Today, siRNA is an important way for cells to defend themselves against viral infections, and it's possible that they could be employed to help defend us against emerging viral diseases.

Physicians are looking into the possibility of using siRNAs as therapeutic agents to silence the expression of genes that cause Alzheimer disease, cardiovascular disease, or cancer. The 2006 Nobel Prize in Physiology or Medicine was awarded jointly to Andrew Fire and Craig Mello for constructing the first artificial siRNA to suppress the expression of a specific gene.

Recent research indicates that miRNAs may be useful to help genetically modify plants. When *Arabidopsis* plants are in great need of phosphate, an increased number of phosphate carrier proteins appear in root cells because an miRNA causes limited production of an enzyme that degrades the carriers! This raises the possibility that miRNA can help scientists develop plant strains for various stressful plant environments.

► **12.6 CHECK YOUR PROGRESS**
1. Distinguish between the activity of siRNAs and miRNAs in cells.
2. Identify reasons why scientists are very interested in the recent discovery of small RNAs and their functions.

Translational and Posttranslational Control

The stability of mRNA for translation and stability of the protein product following translation can vary. Proteins that are old, or unused perhaps due to being incorrectly folded, are degraded by a proteasome complex.

12.7 Both the activity of mRNA and the protein product are regulated

LEARNING OUTCOME

When you complete this section, you should be able to

1. List ways in which regulation of gene expression can occur in the cytoplasm.

Translational control begins when the processed mRNA molecule reaches the cytoplasm and before there is a protein product. Translational control involves the availability of an mRNA for translation at the ribosome.

Presence or absence of the 5′ cap and the length of the 3′ poly-A (adenine nucleotide) tail of a mature mRNA transcript can determine whether translation takes place and how long the mRNA is active. If histone proteins are not needed because DNA replication is not occurring, histone mRNA is degraded by nucleases within minutes due to the loss of the poly-A tail. On the other hand, the long life of mRNAs that code for hemoglobin in mammalian red blood cells is attributed to the persistence of their caps and their long poly-A tails.

Translation can also be inhibited either by the presence of miRNAs attached to the mRNA or by the presence of translation repressor proteins. For example, ferritin, an iron-storing protein, is not made unless iron is present. If iron is not present, a **translation repressor protein** attaches to the beginning of the ferritin mRNA, making it impossible for mRNA to bind to a ribosome. When iron enters the cell, the iron combines with the repressor protein and it dissociates from the ferritin mRNA. Now translation begins. Unlike the use of repressor proteins in prokaryotes, this is an example of translational control by the use of a repressor protein.

Posttranslational control begins once a protein has been synthesized and has become active. Posttranslational control represents the last chance a cell has for influencing gene expression. Some proteins are not immediately active after synthesis. For example, at first bovine proinsulin is a single, long polypeptide that folds into a three-dimensional structure. Cleavage results in two smaller chains that are bonded together by disulfide (S—S) bonds. Only then is active insulin present.

Just how long a protein remains active in a cell is usually regulated by the use of **proteases,** enzymes that break down proteins. To protect a cell, proteases are typically confined to lysosomes or special structures called **proteasomes** (Fig. 12.7). For a protein to enter a proteasome, it has to be tagged with a signaling protein that is recognized by the proteasome cap. When the cap recognizes the tag, it opens and allows the protein to enter the core of the structure where the protein is digested to peptide fragments. Proteasomes contribute to the development of cancer if they digest regulatory proteins, including p53 (see page 226), which help control the cell cycle and prevent cancer. Drugs that inhibit proteasome activity and thereby raise the level of p53 are now being investigated as a way to treat cancer. The inability of proteasomes to degrade other proteins is implicated in Alzheimer, Parkinson, and mad cow disease. Notice that proteasomes function in regulating gene expression because they help control the amount of protein product in the cytoplasm.

▶ **12.7 CHECK YOUR PROGRESS**
1. Distinguish between translational control and posttranslational control by giving an example of each type of control.

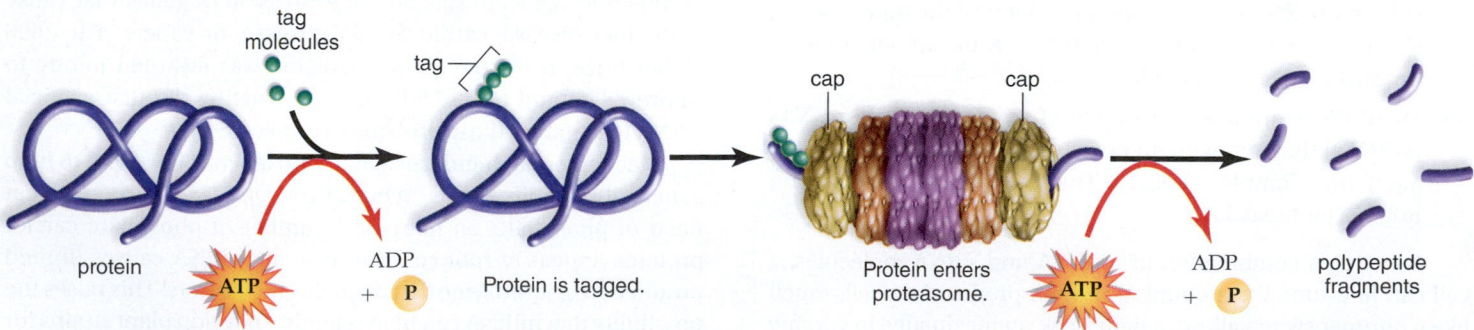

FIGURE 12.7 Proteins to be degraded are first tagged with a signaling molecule. Then they enter a proteasome, a large, cylindrical complex that contains a protease. The protease enzymes digest the protein to polypeptide fragments. The process requires energy in the form of ATP.

12B Epigenetic Inheritance and Who You Are

The many regulatory mechanisms we have discussed in this chapter are a part of epigenetic (beyond the genotype) inheritance. This statement recognizes that you may inherit a particular gene from your parents but it may not be expressed. Let's take some examples as we review the levels of genetic control featured in Figure 12B.1.

Histones and small RNAs play a role in chromatin packing ❶ and heterochromatin formation, which make genes inaccessible for transcription. There is even evidence that heterochromatin formation is an important way to turn genes off as specialization occurs during development.

70×   75×

12B.2 The fly on the left is normal, but the one on the right has legs where antennae should be because a Hox protein malfunctioned during development.

Transcriptional control ❷ depends on DNA-binding proteins, including transcription factors and transcription activators, which initate transcription. We learned that Hox proteins can particularly influence the phenotype because they are master regulatory proteins that turn on and activate many regulatory genes, sometimes inappropriately (Fig. 12B.2).

Posttranscriptional control ❸ also occurs in the nucleus. Pre-mRNA alernative splicing can affect gene expression by determining what protein products a cell makes. Also, RISC (RNA-induced silencing complex) can degrade an mRNA, and attached miRNA snippets can dampen the degree to which it is translated once it reaches the cytoplasm.

Translational control ❹ occurs in the cytoplasm. Various molecules in the cytoplasm determine whether translation begins. For example, specific translation repressor proteins can make it impossible for an mRNA to bind to a ribosome.

Posttranslational control ❺ occurs in the cytoplasm after protein synthesis is complete. As mentioned, the cell has giant protein complexes, called proteasomes, which carry out the task of destroying proteins! Only functioning proteins determine the phenotype.

The conclusion is that there are a host of regulatory molecules (both RNA molecules and proteins) that play a role in determining the phenotype. These molecules are a part of our epigenetic inheritance and they, too, aside from particular genes, determine our inheritance and also may play an important role in growth, aging, and cancer, as future studies will determine. The possible contribution of epigenetic inheritance to the process of evolution cannot be overestimated.

CONSIDER THESE QUESTIONS

1. Do you think epigenetic inheritance could be as important as genetic inheritance? Why or why not?
2. Why might it be advantageous to have many levels of genetic control as in Figure 12B.1? Why might it be disadvantageous?

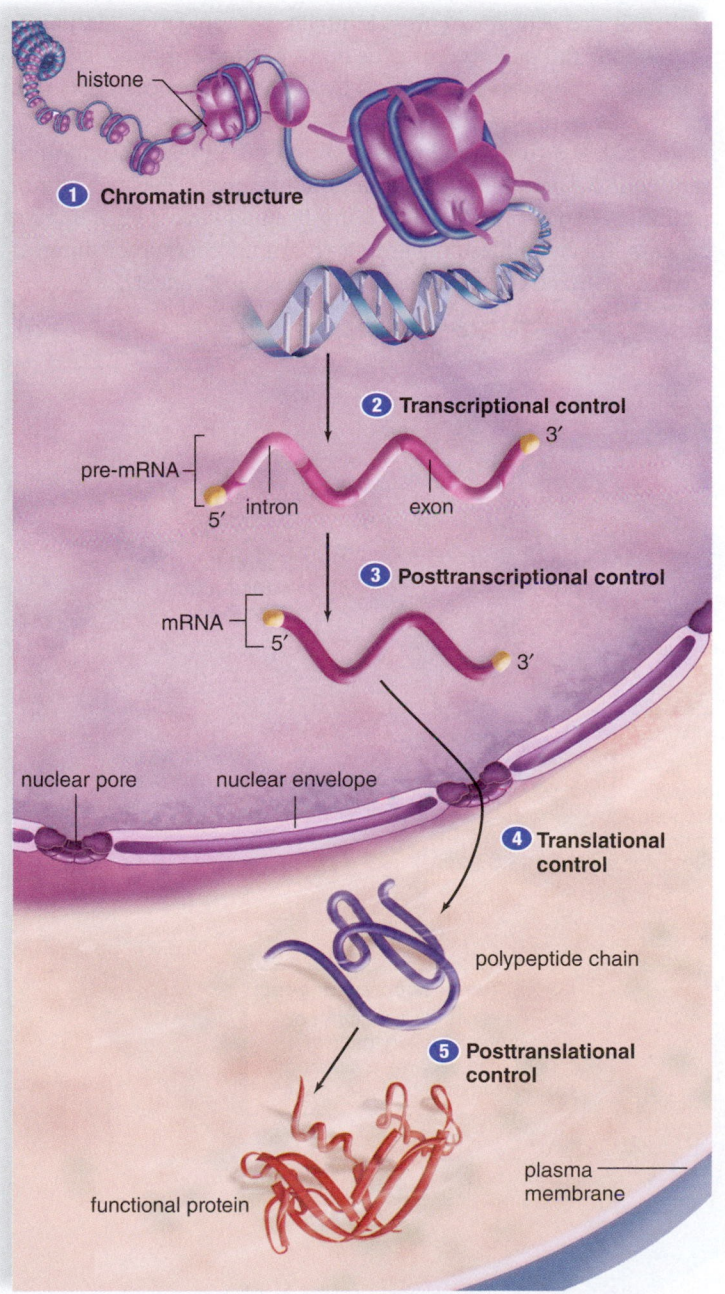

 Explore the concepts through a variety of multimedia assets and question types.

www.mcgrawhillconnect.com

12B.1 Levels of epigenetic inheritance in eukaryotic cells.

Cancer: Lack of Cell Cycle Control

In this part of the chapter, we study genes, called proto-oncogenes and tumor suppressor genes, that control the cell cycle. These genes are part of regulatory pathways that stretch from the plasma membrane to the nucleus where they are located. Several proteins in these pathways must malfunction before cancer develops; therefore, it takes several years for cancer to develop.

12.8 Two types of genes ordinarily control the cell cycle

LEARNING OUTCOMES

When you complete this section, you should be able to

1. Contrast the signal transduction pathways of proto-oncogenes and tumor suppressor genes.
2. Describe the development of cancer as a multistep process.

Recall that the cell cycle consists of interphase followed by mitosis. Two types of genes that ordinarily keep the cell cycle functioning as it should are proto-oncogenes and tumor suppressor genes:

1. **Proto-oncogenes** are regulatory genes that promote the cell cycle and inhibit apoptosis. They are often likened to the gas pedal of a car because they keep the cell cycle going. When proto-oncogenes mutate, they become cancer-causing genes, called **oncogenes,** that overstimulate the cell cycle. Also apoptosis is inhibited to a greater extent.
2. **Tumor suppressor genes** are regulatory genes that inhibit the cell cycle and promote apoptosis. They are often likened to the brakes of a car because they inhibit the cell cycle and stop cells from dividing. When tumor suppressor genes mutate, tumors are more likely to develop and apoptosis is inhibited.

Animation
How Tumor Suppressor Genes Block Cell Division

These genes are a part of signal transduction pathways that extend from the plasma membrane to the nucleus (Fig. 12.8A). A normal stimulatory pathway ends with a proto-oncogene. The pathway consists of a stimulatory growth factor, the receptor, signaling proteins, and a transcription factor. When a growth factor binds to the receptor, the stimulatory pathway is activated. The transcription factor is necessary for turning on a proto-oncogene that codes for a protein that is part of the stimulatory pathway. For example, Ras proteins are a part of the stimulatory pathway. When a proto-oncogene mutates, the result could be a Ras protein that overstimulates the cell cycle. A Ras protein has been implicated in about one-third of all cancers in humans. A mutation in any one of the genes that code for a signaling protein in a stimulatory pathway can lead to overstimulation of the cell cycle and excessive inhibition of apoptosis.

The normal inhibitory pathway also contains signaling proteins, a receptor, and a transcription factor. In this instance, however, the external signal is an inhibiting growth factor and the transcription factor turns on a tumor suppressor gene. The protein p53 is a transcription factor instrumental in stopping the cell cycle and activating chromosomal repair enzymes. If repair is impossible, the p53 protein goes on to promote apoptosis. You'll recall that apoptosis is a process by which cell death occurs due to the release of particular enzymes inside a cell (see section 8.5, page 148). If a tumor suppressor gene undergoes a mutation, it may code for a protein that cannot inhibit the cell cycle or cannot promote apoptosis. Lack of p53 is implicated in over half of human cancers. The retinoblastoma protein (RB) is another tumor suppressor protein that is dysfunctional in many types of cancer.

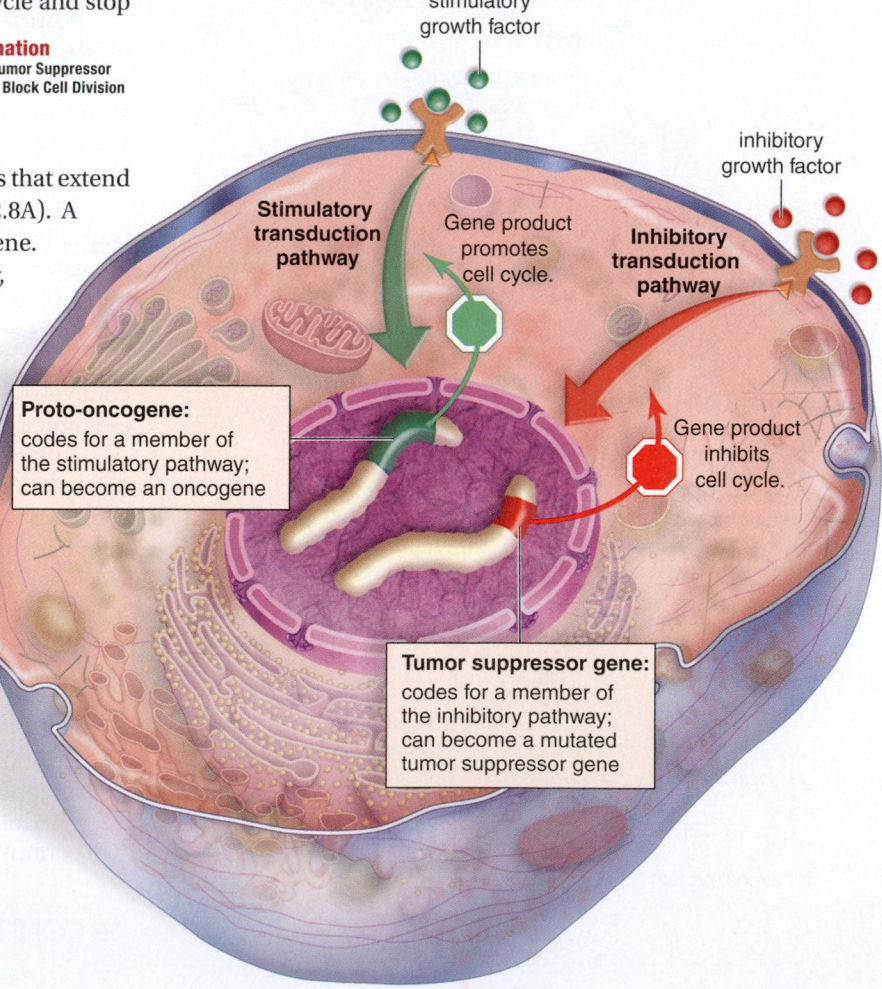

stimulatory growth factor

inhibitory growth factor

Stimulatory transduction pathway

Gene product promotes cell cycle.

Inhibitory transduction pathway

Gene product inhibits cell cycle.

Proto-oncogene: codes for a member of the stimulatory pathway; can become an oncogene

Tumor suppressor gene: codes for a member of the inhibitory pathway; can become a mutated tumor suppressor gene

FIGURE 12.8A The cell cycle is regulated through control of gene expression. A stimulatory signal transduction pathway ends in stimulation of a proto-onocogene (green arrow) An inhibiting signal transduction pathway ends in stimulation of a tumor suppressor gene (red arrow).

Development of Cancer As we have seen, two types of signal transduction pathways are of fundamental importance to normal operation of the cell cycle and control of apoptosis. Therefore, it is not surprising that inherited or acquired defects in these pathways contribute to **carcinogenesis,** the development of cancer (Fig. 12.8B). Some of the inherited cancer-causing genes are known:

BRCA1 and *BRCA2* In 1990, DNA linkage studies of large families in which females tended to develop breast cancer identified the first gene allele associated with breast cancer. Scientists named this gene *breast cancer 1*, or *BRCA1* (pronounced brakuh). Later, they found that breast cancer in other families was due to a faulty allele of another gene they called *BRCA2*. Both alleles are mutant tumor suppressor genes that are inherited in an autosomal recessive manner. If one mutated allele is inherited from either parent, a mutation in the other allele is required before the predisposition to cancer is increased. Because the first mutated gene is inherited, it is present in all cells of the body, and then cancer is more likely wherever the second mutation occurs. If the second mutation occurs in the breast, breast cancer may develop. If the second mutation is in the ovary, ovarian cancer may develop if additional cancer-causing mutations occur.

RB **gene** The *RB* gene is also a tumor suppressor gene. It takes its name from its association with an eye tumor called a retinoblastoma, which first appears as a white mass in the retina. A tumor in one eye is most common because it takes mutations in both alleles before cancer can develop. Children who inherit a mutated allele are more likely to have tumors in both eyes.

RET **gene** An abnormal allele of the *RET* gene, which predisposes a person to thyroid cancer, can be passed from parent to child. *RET* is a proto-oncogene known to be inherited in an autosomal dominant manner—only one mutated allele is needed for an increased predisposition to cancer. The remaining mutations necessary for thyroid cancer to develop are acquired (not inherited).

Figure 12.8B shows that carcinogenesis requires several mutations. First, a single cell undergoes a mutation that causes it to begin to divide repeatedly. Among the progeny of this cell, one cell mutates further and can start a tumor whose cells have further selective advantages. A tumor is present, but it is called cancer in situ because it is contained within its place of origin. To grow larger than a pea, a tumor must have a well-developed capillary network to bring it nutrients and oxygen. The tumor cells release growth factors that lead to **angiogenesis,** the formation of new blood vessels. One treatment for cancer uses drugs that break up the network of new capillaries in the vicinity of a tumor.

New mutations cause the tumor cells to have a disorganized internal cytoskeleton and to lack intact actin filament bundles. By now they are motile cells and can invade underlying tissues because they produce proteinase enzymes that degrade their extracellular matrix. Other mutations give cancer cells the ability to invade lymphatic vessels and blood vessels, which take them to other parts

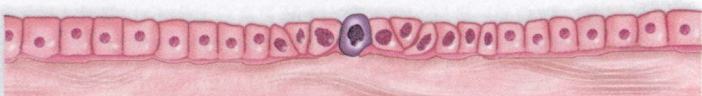
Mutation arises in a tumor suppressor gene. Mutation also inactivates chromosome repair enzymes.

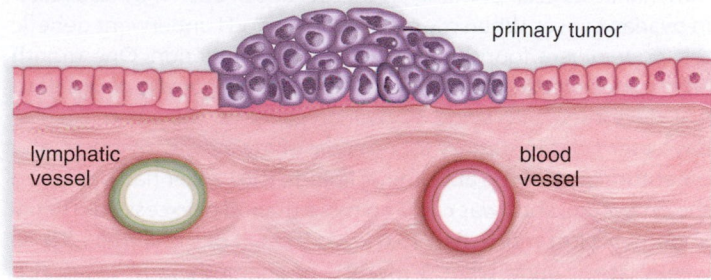
primary tumor

lymphatic vessel

blood vessel

Tumor has started to form. More mutations arise in tumor suppressor genes and mutations also occur in pronto-oncogenes. Cancer in situ is present.

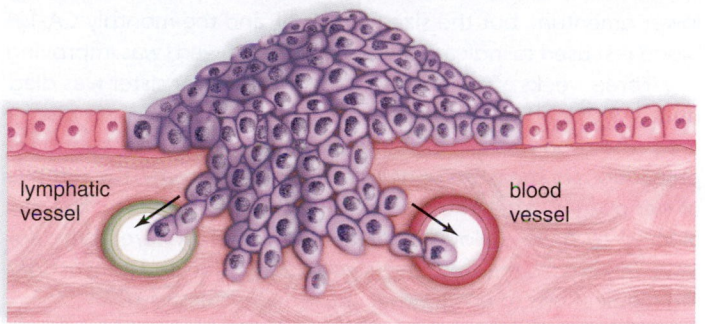
lymphatic vessel

blood vessel

Apoptosis does not occur and tumor becomes aggressive—invades lymphatic and blood vessels.

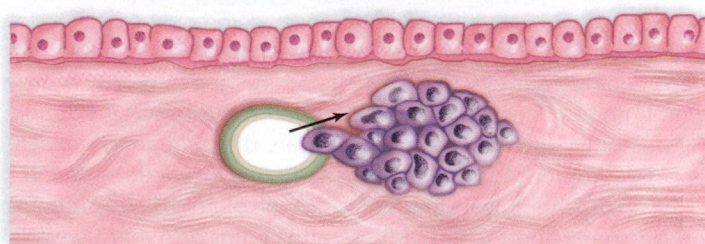
New metastatic tumors are found some distance from the primary tumor.

FIGURE 12.8B The development of cancer requires a series of mutations leading first to a localized tumor and then to metastatic tumors. With each successive step toward cancer, the most genetically altered and aggressive cell becomes the dominant type of tumor. The cells take on characteristics of embryonic cells in that they are not differentiated, and they can divide uncontrollably.

of the body. **Malignancy** is present when cancer cells are found in nearby lymph nodes. When cancer cells initiate new tumors far from the primary tumor, **metastasis** has occurred. Not many cancer cells achieve this feat (maybe 1 in 10,000), but those that successfully metastasize make the probability of complete recovery doubtful.

▶ **12.8 CHECK YOUR PROGRESS**
1. Contrast a mutation in a proto-oncogene with a mutation in a tumor suppressor gene.
2. Discuss three examples of mutated genes that can lead to cancer and tell whether they are tumor suppressor genes or oncogenes.

12C   A *BRCA* Female Tells Her Story

I am a *BRCA1* female [as are the sisters in Figure 12C]. The *BRCA1* mutation manifests itself primarily as breast cancer, but it may also result in ovarian cancer. When my mother, sister, and I underwent genetic testing we were found to carry the *BRCA1* mutation. One month after genetic testing and after speaking with a genetic counselor, I was diagnosed with stage 3C ovarian cancer at the age of 35. Five days later, I was in surgery for removal of my ovaries, oviducts, and omentum (part of the abdominal lining). The cancer had devoured the omentum, and I was carrying more than 7 l of excess fluid in my abdomen. I lost more than 13.6 kg as a result of the surgery. Three weeks later, I underwent 3 months of chemotherapy. I was also diagnosed with deep vein thrombosis in my left thigh and had to take warfarin for 6 months. CT scans indicated some cancer in the lower omentum, but the size was trivial, and the monthly CA-125 blood test used to indicate ovarian cancer showed I was improving.

Three weeks after my diagnosis, my younger sister was diagnosed with stage 1 breast cancer. The breast cancer was discovered when she was undergoing a routine exam before taking the precautionary measure of having an oophorectomy, the removal of her ovaries. After the discovery of the breast cancer, she also elected to have a double mastectomy and hysterectomy, followed by chemotherapy. She is now cancer-free and is investigating medications to ensure that she remains that way.

We knew we had a history of reproductive cancer in our family: My maternal grandmother and aunt both died of ovarian cancer in their forties, and my mother was diagnosed with early stage cancer of the oviducts (a type of *BRCA1* cancer commonly mistaken for ovarian cancer) at age 46. She underwent chemotherapy, and then had additional surgery to remove her uterus and to biopsy nearby lymph nodes. It was also recommended that she have a double mastectomy. However, we were unaware that our family's experience with reproductive cancer was a result of a genetic mutation.

The psychological effects of cancer on our families resulted in helplessness, depression, sadness, and confusion. Our husbands, while very supportive, dealt with the news and treatments differently. Our children were initially shocked, but recovered quickly from the news of the cancer. They have been positive role models for the entire family and their resiliency has been inspiring. We received information from the Mayo Clinic inviting us to participate in a "Familial Breast and Ovarian Cancer Study." This is exciting news for us because participating in case studies and research groups helps us contribute to finding cures for these cancers. This particular study will help pinpoint any other contributing genetic and environmental factors regarding ovarian and breast cancer, as well as other minor cancers that may be caused by *BRCA1* or *BRCA2* gene mutations.

SEM of breast cancer cell

My sister and I now have concerns that our children may also carry the mutation. Consultation with several doctors and genetic counselors has revealed that children should be aware of the issues associated with the mutation, but that the information should be age-appropriate. It is recommended that children have genetic counseling when they are in their late teens or early twenties, and that testing should be at their own discretion. Usually, testing can wait until age 25. By this age, woman have or know if they are going to have children and are more likely to have a relationship with a doctor and make regular gynecological appointments. Mammograms and breast MRIs should be started annually at age 30 if a person has a *BRCA1* or *BRCA2* mutation.

This revelation also necessitates many changes in lifestyle. A healthy diet and a daily exercise routine are highly recommended. I exercise 20–30 minutes each day and constantly try new activities to keep my interest piqued. My sister is on a stringent diet and fitness routine. In fact, at her best she runs a 5K road race in under 30 minutes. My mother did not establish a regular exercise routine, but she rides her bicycle several times a week with my father. The challenges my family endured as a result of the cancer and *BRCA1* genetic mutation may seem daunting, but genetic testing, support from family, friends, and physicians, and healthy lifestyle changes made all the difference in our attitudes toward dealing with this chronic disease.

CONSIDER THESE QUESTIONS

1. Should young people be made aware that they could carry a cancer gene? Why or why not?
2. Just as all women are advised to have mammograms, should everyone be advised to be tested for cancer-causing alleles? Why or why not?

connect | BIOLOGY   Explore the concepts through a variety of multimedia assets and question types.

www.mcgrawhillconnect.com

FIGURE 12C   These three sisters have all had breast cancer. Genetic tests can identify women at risk for breast cancer so that they can choose to have frequent examinations to allow for early detection.

THE CHAPTER IN REVIEW

Gene regulation involves control at different levels of gene expression. Anyone who has cooked a meal knows how important control can be if all is to go well. The meal should be balanced, and the quantity of each food should be appropriate to the number of diners. Similarly, a cell needs to make only the proteins that are immediately needed and in the amount needed. In both prokaryotes and eukaryotes, the products of regulatory genes are DNA-binding proteins that determine which genes will be transcribed. In eukaryotes, if the gene is located in euchromatin, transcription factors and transcription activators initiate transcription, and the result is a pre-mRNA molecule. This is the first step toward specialization of a cell, but the importance of posttranscriptional control is only now being recognized. Alternative pre-mRNA splicing can affect the particular protein product of cells, and small RNAs (miRNAs and siRNAs)

determine which mRNAs will be translated and in what quantity. Does this mean that small RNAs have the upper hand when it comes to the protein constituency of the cell? Some scientists think so.

In the cytoplasm, gene regulation is still needed to make sure that translation is appropriately timed. After all, for protein synthesis to occur as needed, all necessary materials must be available and mRNA must be active only as long as needed. A cell couldn't continue to exist if all the proteins ever made couldn't be disposed of. Returning to our meal analogy, there are always some leftovers that go into the disposal. Similarly, proteasomes keep the content of the cell up-to-date by degrading previously made proteins. You can see, then, that regulation of gene activity is an absolute necessity in cells. Also, knowledge of how genes are regulated can help explain not only the specialization of cells but also how basic genetic

differences have arisen among species, such as between humans and chimpanzees.

In Chapter 13, you will see how our molecular knowledge has contributed to a biotechnology revolution. We now know how to isolate and move genes between organisms of the same species and even different species. We have sequenced the DNA of humans and have much to report about how the present state of our understanding can be used to benefit our well-being.

ANALYZE AND EVALUATE

1. How does the evolution of the genus *Homo* warrant an understanding of regulatory genes?
2. How can a knowledge of regulation be used to cure diseases?
3. Make up an analogy to illustrate the importance of genetic regulation in cells.

SUMMARIZE

Transcriptional Control

12.1 DNA-binding proteins usually turn genes off in prokaryotes

- The **operon** model explains gene regulation in prokaryotes. In prokaryotes, a **regulatory gene** codes for a **repressor,** a DNA-binding protein. An operon also includes a **promoter,** an **operator** (*off switch*) and **structural genes.** All of these are portions of DNA. The important concept is that a DNA-binding protein, the active repressor, ordinarily turns the structural genes off by binding to the operator. Then, RNA polymerase cannot bind to the promoter.

- In the *lac* operon, the repressor is usually active and the structural genes are usually turned off—lactose is not a preferred food. If lactose is present and needs to be digested, the repressor becomes inactive after binding to lactose. In the *trp* operon, the repressor must be activated by tryptophan. When tryptophan is present, the enzymes to make tryptophan (coded for by structural genes) are not needed.

12.2 DNA-binding proteins usually turn genes on in eukaryotes

- In eukaryotes, regulatory genes code for **transcription factors** and **transcription activators,** proteins that act together to turn on genes one by one. Transcription factors bind to the promoter and assist the binding of RNA polymerase. Transcription may still not begin until a transcription activator binds to an **enhancer** some distance away. The chromosome bends to bring the enhancer near the transcription factors.

12.3 Histones regulate accessibility of DNA for transcription

- In **nucleosomes,** DNA is wound around **histones.** Nucleosomes zigzag and then coil as chromatin condenses.

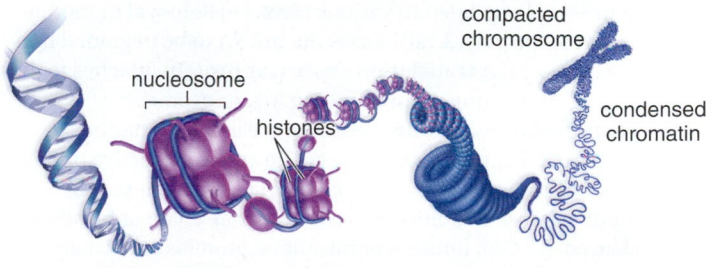

- Highly condensed **chromatin** is called **heterochromatin. Barr bodies** are inactive X chromosomes. The genes in Barr bodies are not expressed. Because one X chromosome is inactive in each cell, females are mosaics.

- **Euchromatin** consists of loosely condensed chromatin. Genes in euchromatin are accessible to RNA polymerase and transcription factors after a **chromatin remodeling complex** pushes aside the histones to expose the DNA. This is called **unpacking** the DNA.

12.4 Hox proteins are DNA-binding proteins active during development

- *Hox* genes are **master developmental regulatory genes** that code for **Hox proteins.**

- *Hox* genes contain a homeobox that codes for a homeodomain. The **homeobox** can be used to identify *Hox* genes and they are found in many different segmental animals. Hox proteins are transcription factors that turn on genes directing the development of segments.
- Other master developmental genes affect the development of an entire region. Investigators were surprised to discover that the ***Pax6*** gene directs the formation of an eye in diverse animals.

Posttranscriptional Control

12.5 Alternative mRNA splicing results in varied gene products

- Posttranscriptional control of gene expression occurs in the nucleus and involves mRNA processing and the speed at which mRNA leaves the nucleus. **Pre-mRNA alternate processing** by which different exons make up an mRNA can cause the particular protein product in cells to be different.

12.6 Small RNA (sRNA) molecules fine-tune gene expression

- **Small RNAs (sRNAs)** are transcribed from a DNA template located in what use to be called junk DNA because it did not code for protein. Small RNAs can double-back to affect compaction of DNA so that some genes become inaccessible for transcription. Small RNAs are the source of **microRNAs (miRNAs)**, small snippets of RNA that can bind to and dampen the translation of an mRNA in the cytoplasm. Small RNAs are also the source of **small interfering RNAs (siRNAs)** that join with an enzyme to form a RISC (RNA-induced silencing complex). RISC degrades any mRNA that has a base sequence complementary to that in siRNA.

Translational and Posttranslational Control

12.7 Both the activity of mRNA and the protein product are regulated

- **Translational control** begins when processed mRNA reaches the cytoplasm before there is a protein product. Translation of mRNA is regulated in various ways: (1) Removal or modification of the poly-A tail causes the mRNA to be degraded by nucleases. (2) A **translation repressor protein** attaches to the mRNA, preventing it from binding to a ribosome.
- **Posttranslational control** begins once a protein has been synthesized and becomes active. The amount of active protein product is regulated in various ways: (1) A protein may be modified by the addition of a carbohydrate chain or by being cleaved. (2) Old, unused, or misfolded proteins are usually degraded by a **proteasome,** a structure that contains proteases.

Cancer: Lack of Cell Cycle Control

12.8 Two types of genes ordinarily control the cell cycle

- **Proto-oncogenes** promote the cell cycle and inhibit apoptosis. A stimulatory signal transduction pathway turns on a proto-oncogene whose product stimulates the cell cycle. When cancer occurs, the product of an **oncogene** leads to overstimulation of the pathway and the cell cycle and excessive inhibition of apoptosis.
- **Tumor suppressor genes** inhibit the cell cycle and promote apoptosis. An inhibitory signal transduction pathway turns on a tumor suppressor gene whose product inhibits the cell cycle and promotes apoptosis.
  - When cancer occurs, the product of a mutated tumor suppressor gene fails to stop the cell cycle and fails to promote apoptosis.

- **Carcinogenesis** refers to tumor formation due to repeated mutations. **Angiogenesis** (formation of new blood vessels) provides nutrients to a growing tumor. Malignancy is present when motile cells invade lymphatic and blood vessels. **Metastasis** has occurred when a new tumor forms far from the first tumor.

TEST YOURSELF

Transcriptional Control

1. Label this diagram of an operon:

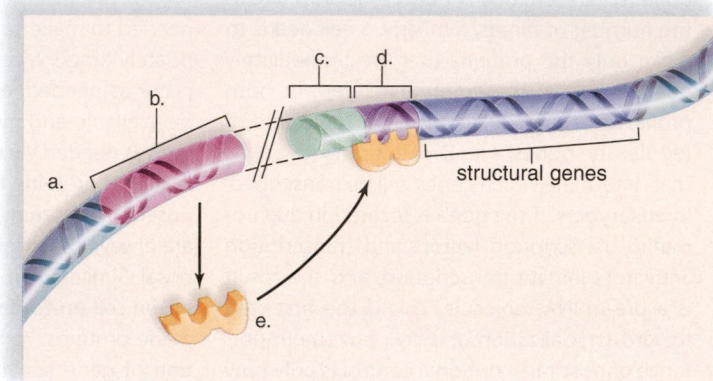

2. In operon models, the function of the promoter is to
   a. code for the repressor protein.
   b. bind with RNA polymerase.
   c. bind to the repressor.
   d. code for the regulatory gene.
3. Which of these correctly describes the function of a regulatory gene for the *lac* operon?
   a. prevents transcription from occurring
   b. a sequence of DNA that codes for the repressor
   c. prevents the repressor from binding to the operator
   d. keeps the operon off until lactose is present
   e. Both b and d are correct.
4. Which of these associations is mismatched?
   a. loosely packed chromatin—gene can be active
   b. transcription factors—gene is inactivated
   c. mRNA—translation can begin
   d. proteasomes—protein is inactive
5. The genes that determine which body parts form on each body segment of a fruit fly are called _____ genes.
   a. promoter   c. intron
   b. exon   d. *Hox*
6. Which of the following is part of a transcription factor that binds to DNA?
   a. homeobox   c. homeotic gene
   b. homeodomain   d. Bicoid protein
7. **THINKING CONCEPTUALLY** Compare transcriptional control in prokaryotes and eukaryotes pointing out similarities and differences.

Posttranscriptional Control

8. Investigators were surprised to find that
   a. transcriptional regulation occurs.
   b. small RNAs affect the activity of mRNA.
   c. proteasomes digest proteins.
   d. chromatin compacting occurs.
   e. All of these are correct.

9. Only
   a. miRNA binds to complementary sequences.
   b. siRNA helps degrade mRNA.
   c. miRNA is made from a DNA template.
   d. siRNA acts in the nucleus.
10. Alternative pre-mRNA splicing can
    a. produce miRNA and siRNA.
    b. result in different protein products.
    c. produce transcription factors.
    d. occur in the cytoplasm and the nucleus.
11. **THINKING CONCEPTUALLY** How does regulation by small RNAs affect what the organism will be like?

Translational and Posttranslational Control

12. Which of these is a true statement?
    a. Once an mRNA is made, a product always follows.
    b. Translation of mRNA can alter the protein product.
    c. Translational control can alter the amount of a protein product.
    d. Proteasomes degrade mRNA molecules.
13. Proteasomes
    a. accept only proteins that have been properly tagged.
    b. are cell structures.
    c. contain catalytic enzymes.
    d. are protein complexes.
    e. All of these are correct.
14. **THINKING CONCEPTUALLY** Give examples to show that all stages of regulatory gene control ultimately affect whether translation occurs or not.

For questions 15–19, match the examples to the gene expression control mechanisms in the key. Each answer can be used more than once.

KEY:
   a. chromatin structure
   b. transcriptional control
   c. posttranscriptional control
   d. translational control
   e. posttranslational control
15. Insulin does not become active until 30 amino acids are cleaved from the middle of the molecule.
16. The mRNA for vitellin is longer-lived if it has a poly-A tail.
17. Genes in Barr bodies are inactivated.
18. Calcitonin is produced in both the hypothalamus and the thyroid gland, but in different forms due to exon splicing.
19. DNA-binding proteins are active.
20. **THINKING CONCEPTUALLY** A variety of mechanisms regulate gene expression in eukaryotic cells. What are the benefits and the drawbacks of this arrangement?

Cancer: Lack of Cell Cycle Control

For questions 21–24, choose two answers for each type of gene.
KEY:
   a. Cell cycle is promoted and apoptosis is inhibited.
   b. Cell cycle is inhibited and apoptosis is promoted.
   c. Signal transduction pathway contains normal proteins.
   d. Signal transduction pathway contains abnormal proteins.

21. Tumor suppressor gene
22. Proto-oncogene
23. Mutated tumor suppressor gene
24. Oncogene
25. Sequence these events that lead to the development of cancer:
    a. Cells gain the ability to invade underlying tissues.
    b. Metastatic tumors occur.
    c. Cell division leads to a tumor.
    d. Blood vessels arise and service the tumor.

GET INVOLVED

1. You receive much criticism for your conclusion that development in the mouse and fruit fly is similar because you have found several homeoboxes in the genes of both organisms. Why? (See section 12.4.)
2. You are a skilled cytologist and want to show that a particular environmental pollutant causes a cell to divide uncontrollably. How will you do this? (See section 12.8.)

MEDIA STUDY TOOLS

mhhe.com/maderconcepts3

Enhance your study of this chapter with interactive study tools, practice tests, and engaging animations. Also, ask your instructor about the resources available through ConnectPlus, which includes LearnSmart, a personalized adaptive learning program, and a media-rich eBook.

13

Biotechnology and Genomics

BEFORE YOU BEGIN

Take a few minutes to recall

Prokaryotic cell structure (section 4.2)

Cell cycle events (section 8.1)

DNA structure and function (sections 11.1–11.7)

Witnessing Genetic Engineering

Genetic engineering has been around since 1973, so by now many genetically modified organisms (GMOs) have been produced. Fish and cows are now expressing foreign genes that make them grow larger. Pigs have been engineered to make their organs acceptable for transplant into humans. Strawberry and potato plants don't freeze, and soybeans are resistant to viral, bacterial, and fungal pathogens—all because they have been genetically engineered. Bacteria produce human insulin as well as other important medicines. And gene therapy in humans, which is the insertion of normal human genes to make up for ones that do not function properly, is a reality.

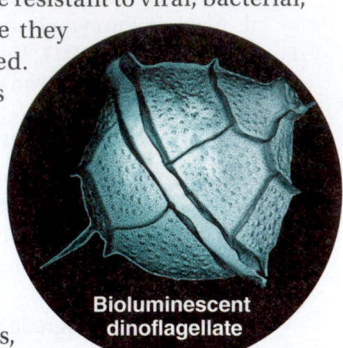

Bioluminescent dinoflagellate

With so many examples of GMOs, you might think it would be easy to "prove" to a friend that it is possible to transfer a gene from one organism to another—but how would you go about it? Well, first you need a gene that makes its appearance known visibly. How about a gene for bioluminescence? Some organisms, including fireflies, jellyfish, glowworms, beetles, and various fishes, can create their own light because they are bioluminescent. The advantages of bioluminescence are varied. Glowworms use their light to attract their prey, and fireflies use the ability to glow to attract mates. The gene for bioluminescence in jellyfish codes for a protein called green fluorescent protein (GFP), and when this gene is transferred to another organism, it glows!

Bioluminescent fish

The basic technique you would use to genetically engineer an organism is relatively simple. For example, to transfer the jellyfish gene for bioluminescence to a pig, first locate the gene among all the others in a jellyfish genome. Then fragment the DNA, and introduce the fragment that contains the bioluminescence gene into the embryo of a pig, mouse, or rabbit, for example. The result is a "glow-in-the-dark" organism.

Genes have no difficulty crossing the species barrier. Mammalian genes work just as well in bacteria, and an invertebrate gene, such as the bioluminescence gene, has no trouble functioning in mammals. The genes of any organism are composed of DNA, and the manufacture of a protein (and indeed, the function of that protein) is similar, regardless of the DNA source. Glowing pigs, mice, and rabbits are certainly living proof that genes can be transferred and also that all cells use basically the same machinery to produce proteins.

Is it ethical to give a mouse a gene that makes it glow? Advocacy groups have even graver concerns about creating genetically modified organisms. Some worry that modified bacteria and plants might harm the environment. Others fear that products produced by GMOs might not be healthy for humans. Perhaps terrorists could use biotechnology to produce weapons of mass destruction. Finally, to what extent is it proper to improve the human genome? All citizens should be knowledgeable about genetics and biotechnology so that they can participate in deciding these issues.

This chapter looks at genetic engineering by first considering the cloning of organisms and genes and then the production of products and modified organisms for various purposes. Studying the human genome has progressed from knowing its base sequence to figuring out how the genome functions. Applying the findings for the benefit of the natural world, including humans, is a completely modern endeavor that cannot help but expand in the years to come. Comparative genomics is expected to shed light on our relationship to other animals. Proteomics and bioinformatics are new fields very much dependent on computer technologies. The news media said Craig Venter had created a cell when he used the computer to produce an entire bacterial genome in the lab. The genome worked perfectly when he inserted it into a bacterial cell cleared of its own DNA!

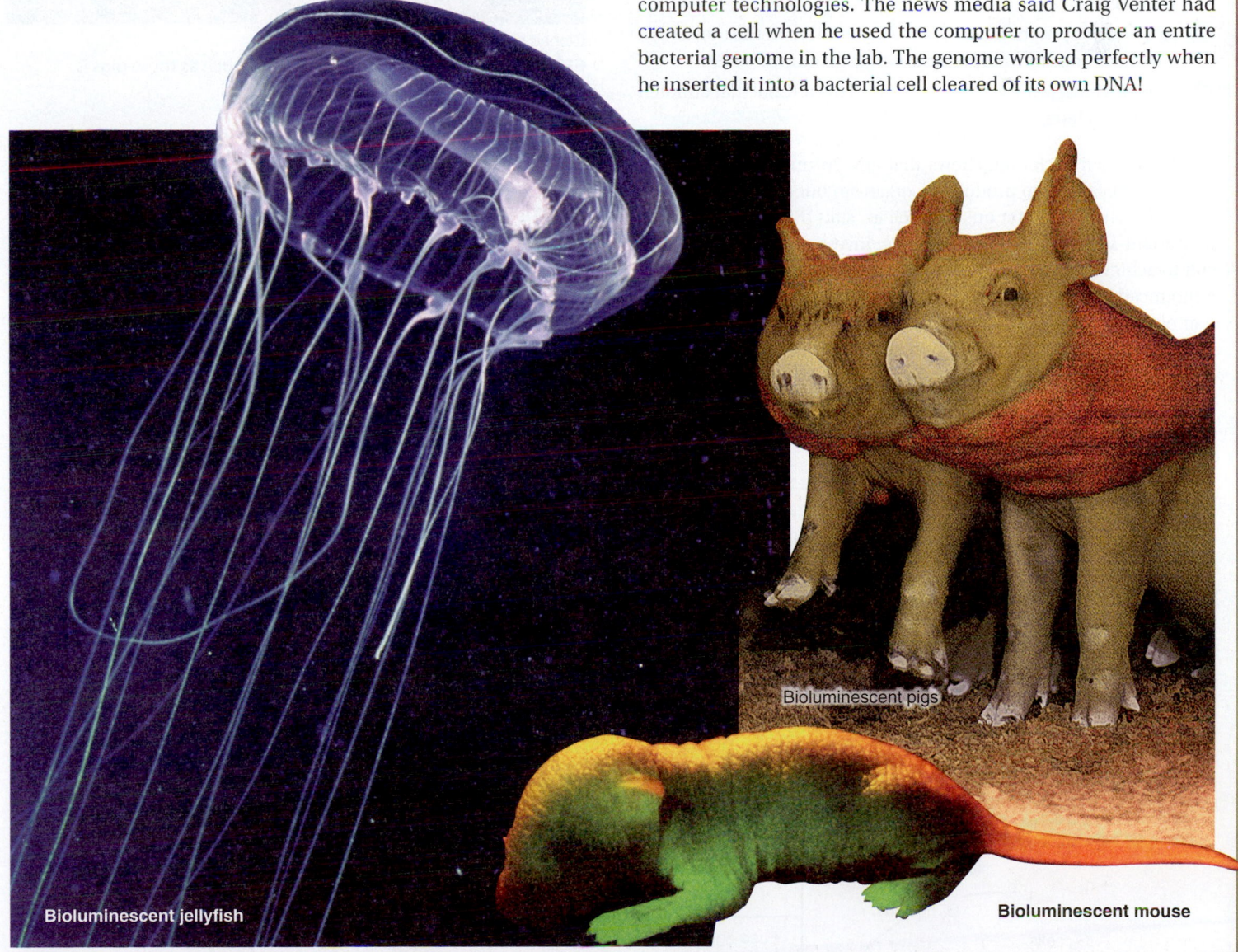

Bioluminescent pigs

Bioluminescent jellyfish

Bioluminescent mouse

Cloning of Organisms, Genes, and DNA Segments

Biotechnology is the use of a natural biological system to make a product or achieve an end desired by humans. Today, we are able to clone organisms, genes, and segments of DNA. When an organism is cloned, the product has the appearance of the organism that donated the nucleus. When a gene is cloned, a copy of it can be inserted into an organism to change the genotype. And when a DNA segment is cloned, the segments can be used to identify the individual from which they came.

13.1 Animals can be cloned using surrogate mothers

LEARNING OUTCOME

When you complete this section, you should be able to

1. Explain the procedure for cloning an organism.

With **reproductive cloning,** the desired end is an individual that is exactly like the original individual. Figure 13.1A shows the steps necessary to clone an animal:

1 After choosing an animal to be cloned, remove a cell to serve as the source for a 2n nucleus.

2 Place this nucleus in an egg after its own nucleus has been removed.

3 After development begins, place the embryo in the uterus of a surrogate mother so development will continue to term.

4 The newborn animal will be a clone of the animal that donated the 2n nucleus.

Reproductive cloning shows that any 2n nucleus contains all the genes necessary to produce an organism but all the genes must be turned on in order for an adult cell to "start over" and overcome its present specialization. At one time, investigators found it difficult to achieve this hurdle. But in March 1997, Scottish investigators announced they had successfully cloned a Dorset sheep, which they named Dolly. How was their procedure different from all the others that had been attempted? Again, an adult nucleus was placed in an enucleated egg cell. However, the donor cell had been starved, which caused it to stop dividing and go into a resting stage (the G_0 stage of the cell cycle) (Fig. 13.1A). The G_0 nucleus was amenable to cytoplasmic signals for initiation of development.

Today it is common practice to clone farm animals that have desirable traits, and even to clone rare animals that might otherwise become extinct. However, the cloning of farm animals is not yet

Cloned pigs

FIGURE 13.1B Cloning of farm animals such as these pigs is commonplace today.

efficient. In the case of Dolly, out of 29 clones, only one was successful. Also, cloned animals may not be healthy. Dolly was put down by lethal injection in 2003 because she was suffering from lung cancer and crippling arthritis. She had lived only half the normal life span for a Dorset sheep. In the United States, no federal funds can be used for experiments to reproductively clone humans.

Since Dolly, a number of mammals have been successfully cloned, including mice, rabbits, cats, dogs, deer, horses, cattle, mules, and rhesus monkeys. But the debate regarding animal cloning is intense. Some people who are unopposed to eating cloned plants object to eating cloned pigs like the ones shown in Figure 13.1B. Still, cloning may be the only way at present to save endangered species from extinction.

▶ **13.1 CHECK YOUR PROGRESS**

1. Contrast the maturity of the nucleus in Figure 13.1A with the maturity of the cytoplasm used.

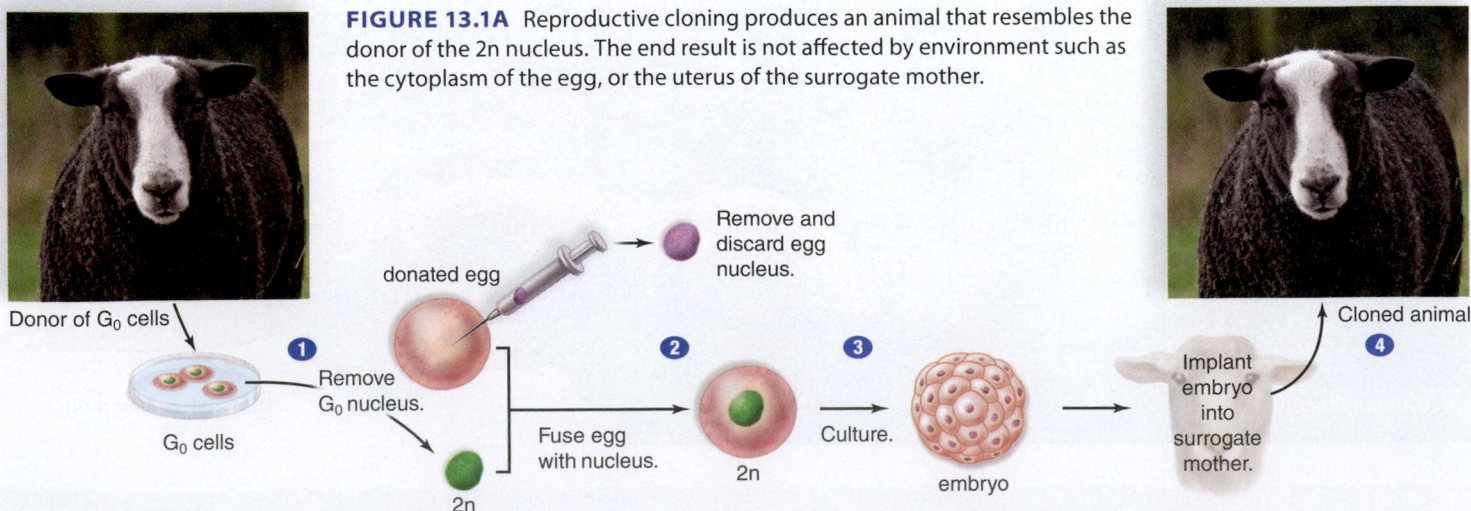

FIGURE 13.1A Reproductive cloning produces an animal that resembles the donor of the 2n nucleus. The end result is not affected by environment such as the cytoplasm of the egg, or the uterus of the surrogate mother.

Donor of G_0 cells

G_0 cells

donated egg

Remove and discard egg nucleus.

1 Remove G_0 nucleus.

2n

Fuse egg with nucleus.

2 2n

3 Culture.

embryo

Implant embryo into surrogate mother.

Cloned animal

4

13.2 Genes can be cloned in bacteria

LEARNING OUTCOME

When you complete this section, you should be able to

1. Describe the use of recombinant DNA technology to clone a gene.

Gene cloning is done to produce many identical copies of the same gene. Gene cloning requires **recombinant DNA (rDNA)**, which contains DNA from two or more different sources. To create rDNA, a technician needs a **vector,** by which the gene of interest will be introduced into a host cell, which is often a bacterium. One common vector is a **plasmid,** a small accessory ring of DNA found in bacteria. The ring is not part of the bacterial chromosome and replicates on its own.

Figure 13.2 traces the steps in cloning a gene:

1 A **restriction enzyme** is used to cleave the plasmid. Hundreds of restriction enzymes occur naturally in bacteria, where they cut up any viral DNA that enters the cell. They are called restriction enzymes because they *restrict* the growth of viruses, but they also act as molecular scissors to cleave any piece of DNA at a specific site. For example, the restriction enzyme called *Eco*RI always cuts double-stranded DNA at this sequence of bases and in this manner:

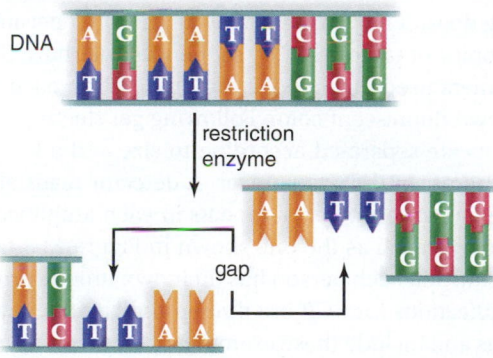

Notice that there is now a gap into which a piece of foreign DNA can be placed if it begins and ends in bases complementary to those exposed by the restriction enzyme. To ensure this, it is only necessary to cleave the foreign DNA; for example, a human chromosome that contains the gene for insulin (or a jellyfish chromosome that contains the gene for green fluorescent protein [GFP]) is cleaved with the same type of restriction enzyme.

Animation
Restriction Endonucleases

2 The enzyme **DNA ligase** is used to seal foreign DNA into the opening created in the plasmid. The single-stranded, but complementary, ends of a cleaved DNA molecule are called "sticky ends" because they can bind a piece of DNA by complementary base pairing. Sticky ends facilitate the pasting of the plasmid DNA with the DNA of the inserted gene. The use of both restriction enzymes and ligase allows researchers to cut and paste DNA strands at will. Now the vector is complete, and an rDNA molecule has been prepared.

3 Some of the bacteria take up a recombinant plasmid, especially if the bacteria have been treated to make them more permeable.

4a Gene cloning occurs as the plasmid replicates on its own. Scientists clone genes for a number of reasons. They might want to determine the base sequence of the gene and compare it to other cloned genes. Or, they might use the genes to genetically modify other organisms.

4b The bacterium has been **genetically engineered** and is a **genetically modified organism (GMO)** that can make a product (e.g., insulin or GFP) it could not make before.

▶ **13.2 CHECK YOUR PROGRESS**
1. Explain why the human DNA (containing the gene) fits into the "gap" in the plasmid created by the restriction enzyme (Fig. 13.2).

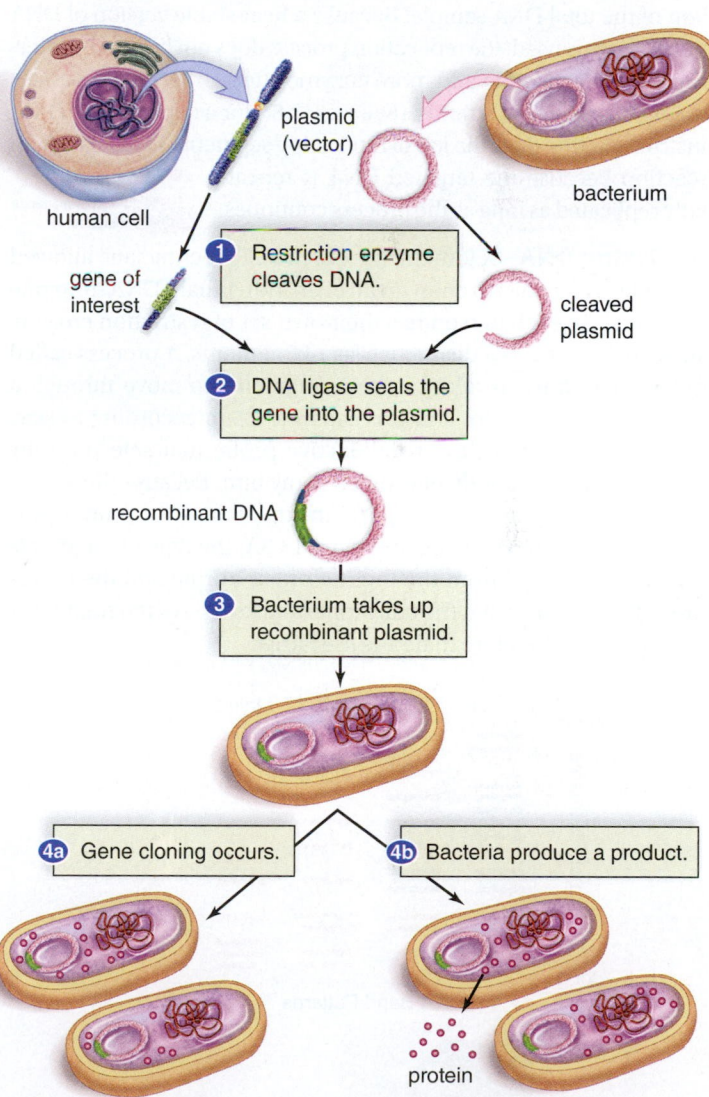

FIGURE 13.2 Cloning a gene that produces a product. Human DNA and plasmid DNA are cleaved by a specific type of restriction enzyme. Then the human DNA, containing the gene of interest, is spliced into a plasmid by the enzyme DNA ligase. Gene cloning is achieved after a bacterium takes up the recombinant plasmid. If the gene functions normally as expected, the protein product may also be retrieved.

13.3 DNA segments can be cloned in a test tube

LEARNING OUTCOMES

When you complete this section, you should be able to

1. Describe the polymerase chain reaction and how DNA segments are analyzed.
2. List several applications of PCR that benefit human society.

The **polymerase chain reaction** (**PCR**) is widely used in biotechnology laboratories to create copies of DNA segments quickly in a test tube. We previously discussed the reaction in the context of DNA replication (see "DNA Replication in a Test Tube" on page 199) but now we want to consider the process in the context of biotechnology. Through the use of primers to mark the location of a DNA sequence, the enzyme DNA polymerase amplifies (makes copies of) only the targeted segment. The segment can be less than one part in a million of the total DNA sample! Because a heat-stable version of DNA polymerase is used, the replication process does not have to be interrupted by the need to add more enzyme after a high heat separates the newly replicated strands. (See Fig. 13.8A for a depiction of a PCR instrument that does the job of replicating segments.) PCR is a chain reaction because the targeted DNA is repeatedly replicated as long as the process continues.

Animation
Polymerase Chain Reaction

Analyzing DNA Before the advent of PCR, technicians allowed restriction enzymes to chop up all of an individual's DNA from one cell. Because each person has their own set of restriction enzyme sites, the result was a distinctive set of fragments. A process called **gel electrophoresis** allowed DNA fragments to move through a gel in an electric field, and this separate them according to size. It was customary to use a radioactive probe that selected only certain fragments for display on an X-ray film. Because the visible pattern that resulted was unique to the individual, the process was called **DNA fingerprinting.** In Figure 13.3A, the type of fragments the child inherited from the mother are in purple and the others are in red. It's clear that the baby inherited its type of red fragments from male 1. Therefore male 1 is the father.

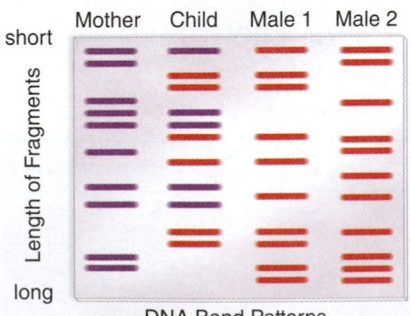

FIGURE 13.3A DNA fingerprinting using targeted STRs and gel electrophoresis can establish paternity. The profile of the child matches that of the mother and male 1.

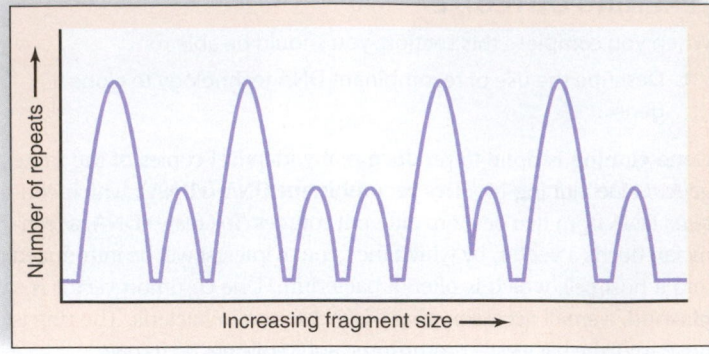

FIGURE 13.3B In this automated system of doing a DNA fingerprint, the STRs are fluorescently labeled, and a detector reads the number of repeats per fluorescent color.

Now with the advent of PCR, the method of choice to achieve a DNA fingerprint is called short tandem repeat (STR) profiling. **Short tandem repeats** are DNA sequences that contain a number of base repeats as in TCG′TCG′TCG′TCG. This example has four repeats, but many sequences have more repeats than this. The number of repeats is inherited and unique to the individual. STR profiling doesn't require analyzing all the DNA because PCR can make copies of targeted DNA sequences that have STR repeats. The primers used during PCR are fluorescent; each type repeat has its own fluorescent color. Following gel electrophoresis, DNA fragments are separated according to size and a laser is used to excite fluorescent STR sequences. A detector reads all the colors and displays the number of repeats in each amplified sequence. The printout, such as the one shown in Figure 13.3B, is the DNA fingerprint, and each person has their own unique printout.

Applications for PCR are limited only by the imagination of scientists and include these examples:

- A viral infection, a genetic disorder, or cancer can be confirmed when the amplified DNA sequence matches that of a known virus or mutated gene.
- DNA fingerprinted from blood or tissues at a crime scene has been successfully used in screening suspects, convicted criminals, and in exonerating those wrongly convicted.
- DNA fingerprinting through STR profiling has been extensively used to identify the victims of natural disasters, such as the tsunamis in Indonesia and Japan.
- Relatives can be found, paternity suits can be settled (Fig. 13.3A), genetic disorders can be detected, and illegally poached ivory and illegally hunted wildlife meat can be recognized using this technology. **Video** World Trade Center DNA
- PCR has also shed new light on evolutionary studies by comparing DNA extracted from ancient specimens with that of living organisms.

▶ **13.3 CHECK YOUR PROGRESS**

1. Contrast the purpose and process of recombinant DNA technology with that of PCR.
2. List several useful applications for the PCR process.

Biotechnology Products and Genetically Modified Organisms

Bacteria, genetically modified using recombinant DNA technology, can be used as a source of the cloned gene; to produce a biotechnology product; or to perform various services. Plants and animals and also humans can be genetically modified. The process in humans is called gene therapy.

13.4 Genetically modified bacteria produce useful products

LEARNING OUTCOME

When you complete this section, you should be able to

1. Discuss the many services of genetically modified bacteria.

Genetically modified bacteria can be used in various ways:

1. **GM bacteria clone a gene that can be isolated and used in basic research.** For example, a gene that helps plants keep track of the daily cycle of light and dark was cloned and is now being studied by scientists at Scribbs Research Institute to see how it works. Similarly, scientists at the University of Toronto cloned a gene for early-onset Alzheimer disease as a first step toward studying how the gene is implicated in this condition.

2. **GM bacteria can be used to produce commercial products.** Of more direct interest to you, perhaps, are the **biotechnology products** now being produced by GM bacteria. For example, if the gene in Figure 13.2 is the one for insulin from human cells, and if the gene is expressed in the bacteria, then the protein being made by the GM bacterium is insulin. GM bacteria are grown in huge vats called bioreactors, and the gene product is collected from the medium, packaged, and sold as a commercial product.

Animation
Early Genetic Engineering Experiments

The products featured in Figure 13.4 may be of special interest to you. GM bacteria assist in making and washing our clothes, in keeping us safe from terrorists, and in making medical biotechnology products. Aside from the insulin being injected by the young girl, GM bacteria produce other medical products such as clotting factor VIII, human growth hormone, t-PA (tissue plasminogen activator), and hepatitis B vaccine.

Not shown in Figure 13.4 are several other products. For example, an eel-like fish, the ocean pout, produces a natural antifreeze protein that is now made commercially by GM bacteria. The product is readily available to all, even ice cream manufacturers who want their product to be free of ice crystals. In the past, the cheese-making industry was dependent upon a substance called rennet, which was collected from the stomach lining of calves. With the decline in the veal industry (veal is a meat derived from calves), a rennet shortage resulted, and no satisfactory substitute could be found. The essential ingredient in rennet for making cheese is chymosin, and the cheese industry was dramatically rescued when the chymosin gene was isolated from calf cells and cloned in bacteria.

3. **GM bacteria can be used to modify plants.** Many other uses have been found for genetically modified bacteria, aside from the production of products. For example, bacteria that normally live on plants and encourage the formation of ice crystals have been changed from frost-plus to frost-minus bacteria. As a result, new crops such as frost-resistant strawberries are being developed. Also, a bacterium that normally colonizes the roots of corn plants has now been endowed

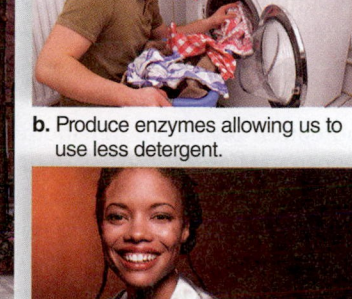

a. Help produce biodegradable dog waste bags and dye for blue jeans.

b. Produce enzymes allowing us to use less detergent.

c. Produce spider web silk.

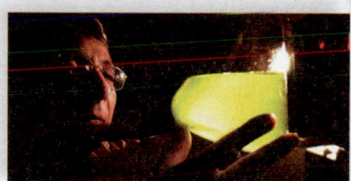

d. Glow in dark if chemical warfare agent is present.

e. Produce vaccines.

f. Produce insulin.

FIGURE 13.4 Genetically modified bacteria are useful in these various ways.

with genes (from another bacterium) that code for an insect toxin. The toxin protects the roots from insects.

4. **GM bacteria can be used to prevent environmental pollution.** GM bacteria can also perform various services. Bacteria can be selected for their ability to degrade a particular substance, and this ability can then be enhanced by genetic engineering. For instance, naturally occurring bacteria that eat oil can be genetically engineered to do an even better job of cleaning up beaches after oil spills. Bacteria can also remove sulfur from coal before it is burned and help clean up toxic waste dumps. One such strain was given genes that allowed it to clean up levels of toxins that would have killed other strains. Further, these bacteria were given "suicide" genes that caused them to self-destruct when the job was done.

▶ 13.4 CHECK YOUR PROGRESS

1. List and describe four ways that our society can make use of genetically modified bacteria.

13.5 Plants can be genetically modified

LEARNING OUTCOME

When you complete this section, you should be able to

1. List ways in which crops have been genetically modified to increase yield and improve their nutritional value.

Corn, potato, soybean, and cotton plants have been engineered to be resistant to either insect predation or widely used herbicides. In Figure 13.5A, the *B.t.t.+* potato plant produces an insecticide protein and is resistant to the Colorado potato beetle. The expected benefits of Bt potatoes are increased yield, reduced postharvest losses, and reduced human exposure to pesticides because the plants will no longer need externally applied pesticides. Some corn and cotton plants are now both insect- and herbicide-resistant. In 2006, GM crops were planted on more than 252 million acres worldwide, an increase of over 13% from the previous year. If crops are resistant to a broad-spectrum herbicide and weeds are not, the herbicide can be used to kill the weeds. When herbicide-resistant plants were planted, weeds were easily controlled, less tillage was needed, and soil erosion was minimized.

Other forms of resistance have also been accomplished. Potato blight due to a fungus infection is the most serious potato disease in the world. About 150 years ago, it was responsible for the Irish potato famine, which caused the deaths of millions of people. By placing a gene from a naturally blight-resistant wild potato into a farmed variety, researchers have now made potato plants that are invulnerable to a range of blight strains.

Genetic modification of crops can also improve their agricultural value. For example, crop production is currently limited by the effects of salinization on about 50% of irrigated lands. Salt-tolerant crops would increase yield in these agricultural fields. To produce a salt-tolerant tomato, scientists first identified a gene coding for a channel protein that transports Na^+ across the vacuole membrane. Sequestering the Na^+ in a vacuole prevents it from interfering with plant metabolism. Then the scientists cloned the gene and used it to genetically engineer plants that overproduce the channel protein. The modified plants thrived when watered with a salty solution.

Some progress has also been made to increase the nutritional value of crops. Soybeans have been developed that mainly produce the monounsaturated fatty acid oleic acid, a change that may improve human health. Golden Rice, so-named because it is literally yellow in color, is an example of how a common food can be genetically modified to

FIGURE 13.5A The potato plant on the left is nonresistant to the Colorado potato beetle, while the plant on the right is resistant.

fulfill a definite need. Currently, millions of children in the Southern Hemisphere whose diet contains mainly rice suffer from vitamin A deficiency, which can lead to blindness. It took about 10 years of research that began at the Institute of Plant Sciences, Zurich, to genetically engineer a rice that is golden because it contains significant amounts of beta-carotene, the precursor to vitamin A. Expected to be commercially available in 2013, eating this improved rice will raise the level of vitamin A in the body and remove the threat of blindness. Initially, three genes taken from corn (maize), bacteria, and daffodils were inserted into the genome of common white rice in order to make it golden and contain beta-carotene (Fig. 13.5B). Still, people have two basic concerns about genetically modified foods: food safety and environmental impact.

Genetic engineering of plants has also developed many products for medical use, such as human hormones, clotting factors, and antibodies. One type of antibody made by corn can deliver radioisotopes to tumor cells, and another made by soybeans may be developed to treat genital herpes.

▶ 13.5 CHECK YOUR PROGRESS

1. Identify ways GM crops can have an improved yield or improved nutritional value.

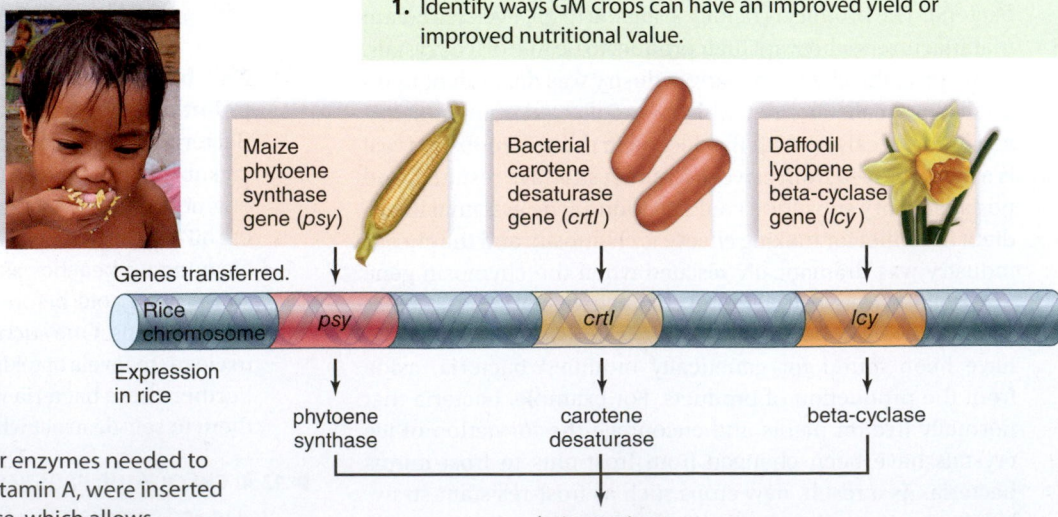

FIGURE 13.5B Three genes coding for enzymes needed to produce beta-carotene, a precursor to vitamin A, were inserted into white rice. The result was Golden Rice, which allows children to produce vitamin A, thereby preventing blindness.

13A Are Genetically Engineered Foods Safe?

People who are much concerned about the safety of transgenic foods grant that GM crops allow traditional farmers to use less chemical pesticides and herbicides—but, even so, they ask these questions:

Can Transgenic Foods Make People Sick?

Before a GM food makes it to the market, the government's Food and Drug Administration (FDA) must approve it for human consumption. The FDA tests the food to make sure it is not poisonous and it does not contain an allergen—a substance that some people would be allergic to. It could be that more and better tests are needed for allergens, but right now it appears that transgenic foods, such as Bt corn, do not contain allergens that would make people sick.

Bt corn is so-called because it contains a foreign gene taken from the bacterium *Bacillus thuringiensis*. The gene allows corn plants to resist destruction by corn borers. About a dozen Bt varieties, including corn, potato, and even a tomato, are approved for human consumption. The planting of Bt corn has steadily increased over the years (Fig. 13A). To prevent allergens from entering foods, scientists have purposefully been removing genes from soybeans and other crops that code for allergens. And it appears that Bt corn is actually safer for humans because it is not attacked by a fungus that puts toxins in damaged cereal crops. The fungal toxins can cause cancer and birth defects.

Activists would like GM foods clearly labeled so the consumer could choose not to buy them. Unfortunately, this has proven difficult to achieve because, for example, most cornmeal is derived from both conventional and genetically engineered corn. So far, one type of food source is not separated from another on a large scale. However, some health-food stores do label foods that do not contain GMOs.

Video
GM Food Safety

Are GM Crops Environmentally Safe?

Many studies have been done to answer this question, and these studies conclude that GM plants are not harmful to the environment simply because they are transgenic. What about the possibility that GM plants might reproduce with wild plants to produce monster weeds that will be resistant to insects and herbicides? So far this has not happened. In 2003, ecologists at the University of Tennessee crossed Bt oilseed rape plants with their wild relative; the hybrid plants were poor competitors that did not take over the area. Is it possible that, like traditional pesticides, GM crops might kill off nontarget insects, even pollinators of the crop? Several years ago, a study found that monarch butterfly larvae were harmed if they were raised in the laboratory and fed only pollen from Bt corn plants. More recent studies conclude that monarch larvae do not ordinarily consume enough pollen to cause them harm. When mature, the butterflies themselves feed on flower parts that do not contain the Bt insecticide. Still, like traditional pesticides, resistance to the Bt insecticide can occur among insects. To prevent this, researchers are reproducing transgenic strains that carry more than one type of toxin gene, and this will make it more difficult for resistance to develop.

Are Transgenic Crops Really Needed?

In 2009, the Union of Concerned Scientists released a study that concluded that GM crops do not result in increased yield when compared with traditional crops. Those that support growing GM crops reply that GM crops prevent a loss of yield to pests during storage, and they allow farmers to practice no-till farming, which saves fossil fuels and prevents soil erosion. The use of less insecticides and herbicides makes farming easier, more profitable, and more sustainable.

CONSIDER THESE QUESTIONS

1. Are you in favor of banning GM crops—even if it means food prices would rise? Do you approve of ecoterrorists who burn GM crops and destroy biotechnology labs?
2. What data would allow to decide that the the rice described in Figure 13.5B should be sold because it prevents childhood blindness in less-developed countries?

Bt corn resistant to corn borer

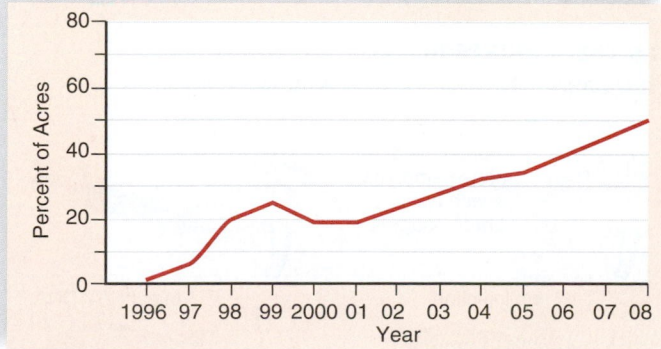

Bt corn usage since 1996

FIGURE 13A Bt corn (*top*) has been genetically modified to express a bacterial gene that allows the plants to resist insects such as corn borers (*see insert*), which decrease yield. The graph shows how the planting of Bt corn has increased over the years.

Explore the concepts through a variety of multimedia assets and question types.

www.mcgrawhillconnect.com

13.6 Animals can be genetically modified

LEARNING OUTCOME

When you complete this section, you should be able to

1. List and describe three uses for GM animals.

Techniques have been developed to insert genes into the eggs of animals. It is possible to microinject foreign genes into eggs by hand, but another method uses vortex mixing. DNA and eggs are placed in an agitator with silicon-carbide needles. The needles make tiny holes, through which the DNA can enter. When these eggs are fertilized, the resulting offspring are GM animals.

Increased yield is a desired end for genetic modification of animals. When animal eggs acquire the gene for bovine growth hormone, the fishes, cows, pigs, rabbits, and sheep that result are larger than before.

Gene pharming, the use of GM animals to produce pharmaceuticals, is being pursued by a number of firms. Genes that code for therapeutic and diagnostic proteins are incorporated into an animal's DNA, and the proteins appear in the animal's milk (Fig. 13.6A). Plans are under way to produce medicines for the treatment of cystic fibrosis, cancer, blood diseases, and other disorders by this method. Figure 13.6A outlines the procedure for producing GM mammals. **1** The gene of interest (in this case, for human growth hormone) is microinjected into donor eggs. **2** Following in vitro fertilization, the zygotes are placed in host females, where they develop. **3** After the GM female offspring mature, the product is secreted in their milk. Then, cloning can be used to produce many animals that produce the same product: **4** Donor enucleated eggs are fused with 2n GM nuclei. The eggs are coaxed to begin development in vitro. **5** Development continues in host females until the clones are born. **6** The female offspring are clones that have the same product in their milk.

Video
Firefly Rx

Video
Cloned Milk

GM animals are also used for basic research. Figure 13.6B shows how researchers demonstrated that a section of DNA called *SRY* (sex-determining region of the Y chromosome) produces a male animal. The *SRY* DNA was cloned, and then one copy was injected into one-celled mouse embryos with two X chromosomes. Injected embryos developed into males, but any that were not injected developed into females. GM mice have also been created to study human diseases. An allele such as the one that causes cystic fibrosis can be cloned and inserted into mice embryonic stem cells, and occasionally a mouse embryo homozygous for cystic fibrosis will result. This embryo develops into a mutant mouse that has a phenotype similar to that of a human with cystic fibrosis. New drugs for the treatment of cystic fibrosis can then be tested in these mice.

Xenotransplantation is the use of animal organs, instead of human organs, in transplant patients. Scientists have chosen to work with pigs because they are prolific and have long been raised as a meat source. Pigs are being genetically modified to make their organs less likely to be rejected by the human body. The hope is that one day a pig organ will be as easily accepted by the human body as a blood transfusion from a person with the same blood type.

▶ **13.6 CHECK YOUR PROGRESS**

1. Distinguish between the various uses for GM animals.

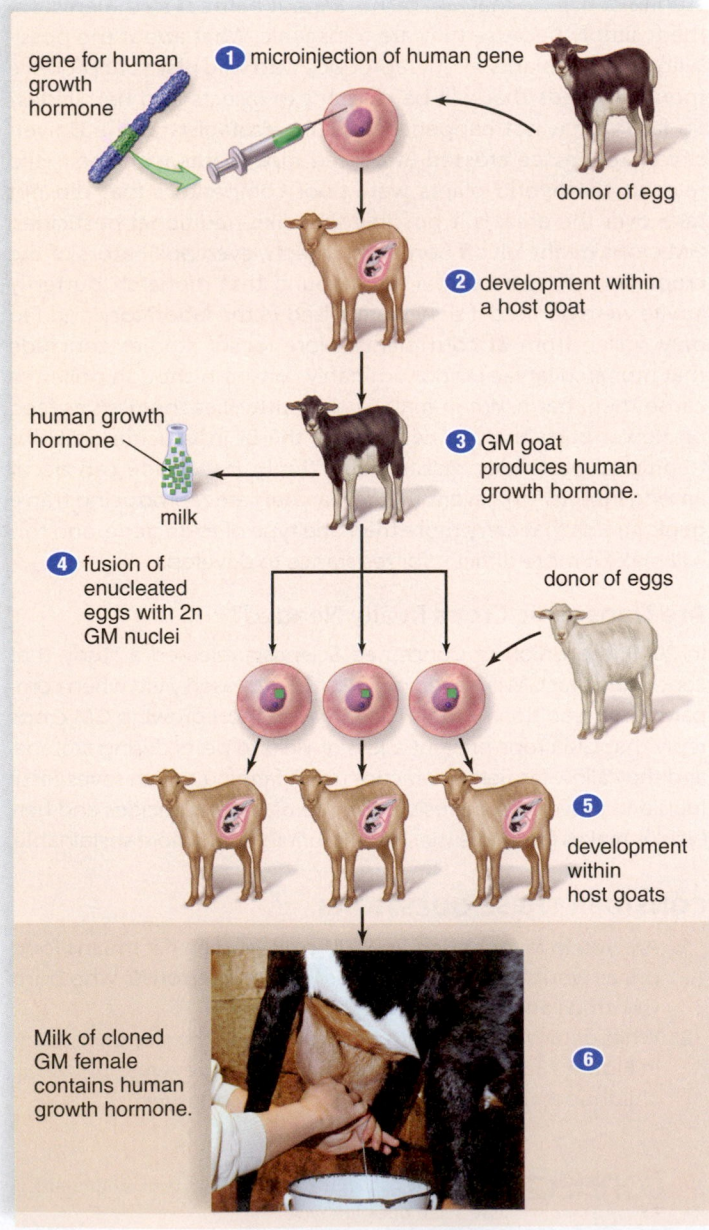

FIGURE 13.6A Procedure for producing many female clones that yield the same product.

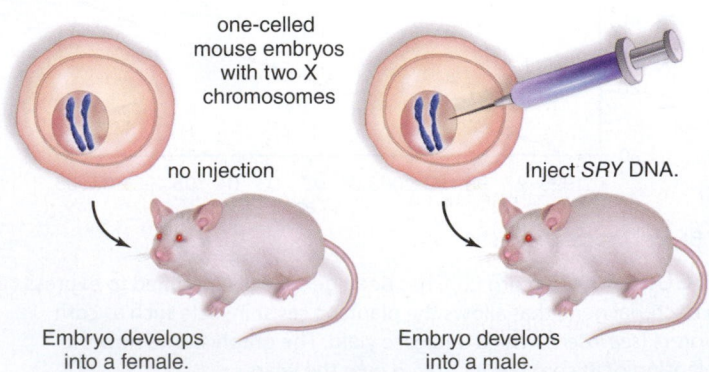

FIGURE 13.6B GM mice showed that maleness is due to *SRY* DNA.

13.7 A person's genome can be modified

LEARNING OUTCOME

When you complete this section, you should be able to

1. Explain the difference and give examples of ex vivo gene therapy and in vivo gene therapy.

The manipulation of an organism's genes can be extended to humans in a process called gene therapy. **Gene therapy** is the insertion of a foreign gene into human cells for the treatment of a disorder. Gene therapy has been used to cure inborn errors of metabolism as well as more generalized disorders, such as cardiovascular disease and cancer. *Ex vivo gene therapy* means the gene is inserted into cells that have been removed and then returned to the body while *in vivo gene therapy* means the gene is delivered directly into the body. Figure 13.7 shows regions of the body that have received copies of normal genes by various methods of gene transfer. Viruses genetically modified to be safe can be used to ferry a normal gene into the body, and so can liposomes, which are microscopic globules of lipids specially prepared to enclose the normal gene. On the other hand, sometimes the gene is injected directly into a particular region of the body.

Ex Vivo Gene Therapy Children who have SCID (severe combined immunodeficiency) lack the enzyme ADA (adenosine deaminase), which is involved in the maturation of white blood cells. Therefore, these children are prone to constant infections and may die without treatment. To carry out gene therapy, bone marrow stem cells are removed from the bone marrow of the patient and infected with a virus that carries a normal gene for the enzyme. Then the cells are returned to the patient, where it is hoped they will divide to produce more blood cells with the same genes.

Familial hypercholesterolemia is a condition in which high levels of blood cholesterol make patients subject to fatal heart attacks at a young age. Through ex vivo gene therapy, a small portion of the liver is surgically excised and then infected with a virus containing a normal gene for the receptor before being returned to the patient. Patients are expected to experience lowered serum cholesterol levels following this procedure.

Investigators are also working on a cure for phenylketonuria (PKU), an inherited condition that can cause intellectual impairment. If detected early enough, the child can be placed on a special diet for the first few years of life, but this is very inconvenient. These investigators believe they will be able to inject the gene directly into the DNA of excised liver cells, which will then be returned to the patient.

In Vivo Gene Therapy Cystic fibrosis patients lack a gene that codes for the transmembrane carrier of the chloride ion. They often suffer from numerous and potentially deadly infections of the respiratory tract. In gene therapy trials, the gene needed to cure cystic fibrosis is sprayed into the nose or delivered to the lower respiratory tract by a virus or by liposomes, artificial vesicles made from phospholipids. So far, this treatment has resulted in limited success.

Genes are also being used to treat medical conditions such as poor coronary circulation. Scientists have known for some time that VEGF (vascular endothelial growth factor) can cause the growth of new blood vessels. The gene that codes for this growth factor can be

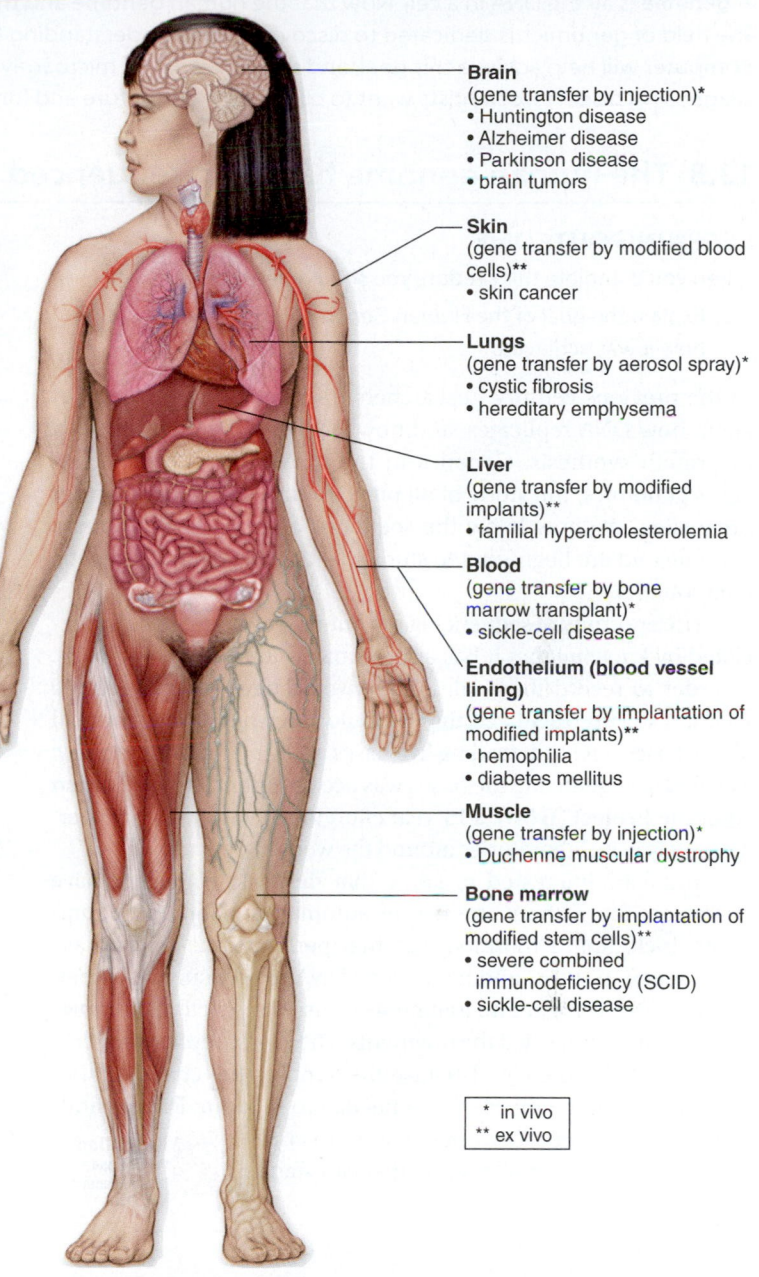

Brain
(gene transfer by injection)*
• Huntington disease
• Alzheimer disease
• Parkinson disease
• brain tumors

Skin
(gene transfer by modified blood cells)**
• skin cancer

Lungs
(gene transfer by aerosol spray)*
• cystic fibrosis
• hereditary emphysema

Liver
(gene transfer by modified implants)**
• familial hypercholesterolemia

Blood
(gene transfer by bone marrow transplant)*
• sickle-cell disease

Endothelium (blood vessel lining)
(gene transfer by implantation of modified implants)**
• hemophilia
• diabetes mellitus

Muscle
(gene transfer by injection)*
• Duchenne muscular dystrophy

Bone marrow
(gene transfer by implantation of modified stem cells)**
• severe combined immunodeficiency (SCID)
• sickle-cell disease

* in vivo
** ex vivo

FIGURE 13.7 Sites of ex vivo and in vivo gene therapy to cure the conditions noted.

injected alone or within a virus into the heart to stimulate branching of coronary blood vessels. Patients report that they have less chest pain and can run longer on a treadmill.

Gene therapy is increasingly being applied as a part of cancer therapy. Genes are used to make healthy cells more tolerant of chemotherapy and to make tumors more vulnerable to chemotherapy. The gene *p53* brings about apoptosis, and there is much interest in introducing it into cancer cells and, in that way, killing them off.

▶ **13.7 CHECK YOUR PROGRESS**

1. Explain one of the examples of gene therapy in Figure 13.7.

The Human Genome and How It Functions

A **genome** is all the DNA in a cell. Now that the human genome and the genomes of many other organisms have been base-sequenced, the field of genomics is dedicated to discovering and understanding the function of the genome and the proteins in a cell. The computer will help achieve this goal, and so will the use of microarrays. Microarrays are also of interest to the medical community in diagnosing diseases. Scientists want to compare the structure and function of the human genome to that of other organisms.

13.8 The human genome has been sequenced

LEARNING OUTCOME

When you complete this section, you should be able to

1. Explain the goal of the Human Genome Project and describe how it was achieved.

In the previous century, researchers discovered the structure of DNA, how DNA replicates, and how DNA and RNA are involved in protein synthesis. Genetics in the twenty-first century concerns genomics, the study of all our DNA as well as that of other organisms. We now know the sequence of bases in the human genome and are beginning to study the function of all its genetic components.

The enormity of sequencing the human genome can be appreciated by knowing that it has approximately 3.2 billion base pairs. In order to record them all, a book would have to have 500,000 pages. It's been calculated that it would take you 60 years to read all the bases out loud, reading five bases a second for 8 hours a day. The feat of sequencing the bases was accomplished by the **Human Genome Project** (**HGP**), a 13-year effort that involved both university and private laboratories around the world.

You'll be interested to know that the task would not have been possible without the use of automated laboratory equipment, including a thermocycler that performs the polymerase chain reaction (PCR) and high-speed DNA sequencing machines. Human DNA was cut into fragments by using restriction enzymes, and then PCR amplified the fragments. The thermocycler shown in Figure 13.8A is so-called because the temperature changes automatically. A high temperature is needed to separate DNA strands prior to the use of a DNA polymerase to copy target DNA. A DNA polymerase that can stand

Animation
Polymerase
Chain Reaction

FIGURE 13.8B An automated DNA sequencer produces a pattern that can be read by a computer to tell the sequence of the bases.

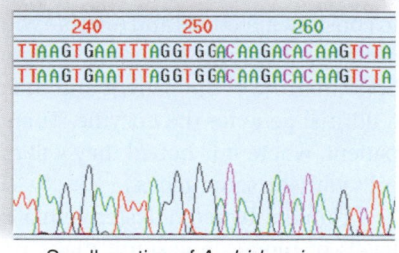

240   250   260
TTAAGTGAATTTAGGTGGACAAGACACAAGTCTA
TTAAGTGAATTTAGGTGGACAAGACACAAGTCTA

Small section of *Arabidopsis* genome

this high temperature was extracted from the bacterium *Thermus aquaticus,* which lives in hot springs. Therefore it is called *Taq DNA polymerase.*

Years before, investigators had perfected a manual method (requiring many copies of the same DNA) to decipher a base sequence, but naturally they began to use automated sequencers when they became available (Fig. 13.8B). Over the 13-year span of the HGP, DNA sequencers were constantly improved, and now modern instruments can automatically analyze up to 3 billion base pairs of DNA in a 24-hour period. Where did the DNA come from? Sperm DNA was the material of choice because it has a much higher ratio of DNA to protein than other types of cells. However, white cells from the blood of female donors were also used. The DNA donors were of European, African, American (both North and South), and Asian ancestry.

▶ **13.8 CHECK YOUR PROGRESS**

1. Identify two pieces of laboratory equipment essential to sequencing the human genome.

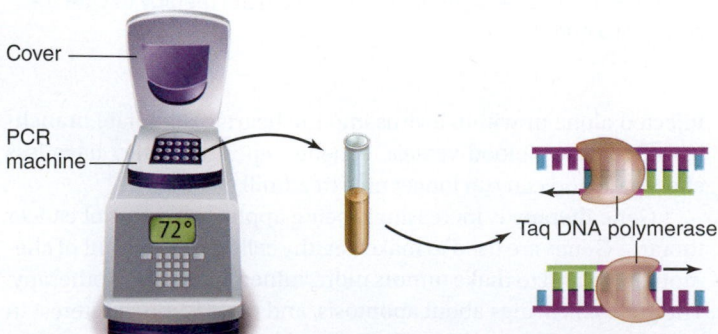

Cover

PCR machine

72°

Taq DNA polymerase

FIGURE 13.8A PCR is done in a thermocycler that changes the temperature automatically. Target DNA (purple) between the primers (green) is copied at a temperature of 72°C. Each time the target DNA is copied, the amount of target DNA doubles.

LEARNING OUTCOME

When you complete this section, you should be able to

1. Distinguish between protein-coding DNA and the various types of noncoding DNA.

Coding DNA While we often speak of protein-coding genes, Figure 13.9 reminds us that only exons are incorporated into mRNA and go on to participate in protein synthesis at a ribosome. An amazing discovery was that only 2% of our genome consists of protein-coding exons. The number of our protein-coding genes is 23,000. A much larger number was expected, considering that the genome has 3 billion base pairs. The genome can contain more than one copy of each protein-coding gene. Multiple copies of the same gene can allow an organism to produce a great deal of product within a short time. Also, a duplicate gene doubles the possibility of a gene undergoing an advantageous mutation that could contribute to evolution. The term **pseudogene** is used for copies of genes that are nonfunctional because of a mutation.

Noncoding DNA The genome has a tremendous amount (98%) of noncoding DNA. You might think that noncoding DNA could vary between parent and child, but it does not. Each parent passes on a haploid set of chromosomes that contains both coding and noncoding DNA in a particular order. The fact that noncoding DNA is often transcribed into RNA raises the possibility that a heretofore-overlooked regulatory network may be what allows eukaryotes, including humans, to achieve a structural complexity far beyond anything seen in the prokaryotic world.

Among the noncoding DNA, 24% consists of **introns.** Once regarded as merely intervening sequences, introns may very well be regulators of gene expression. The presence of introns allows exons to be put together in various sequences so that different mRNAs and proteins can result from a single gene. In general, more complex organisms have more and larger introns. Introns can also be a source of small RNAs.

The rest of the noncoding DNA is colored red in Figure 13.9. Of this, **repetitive DNA** is the much larger piece (59%). This is the portion of our genome that was once specifically called "junk DNA." Although many scientists still dismiss repetitive DNA as having no function, others point out that the centromeres and the ends of chromosomes are composed of repetitive elements and, therefore, repetitive DNA may not be as useless as once thought. For example, repetitive elements of the centromere could possibly help with segregating the chromosomes during cell division.

Repetitive DNA can occur as tandem repeats and interspersed repeats. **Tandem repeat** means that the repeated sequences are next to each other on the chromosome. For example, ACTGACTGACTG would be a tandem repeat. The number and types of tandem repeats may vary significantly from one individual to another, making them invaluable as indicators of heritage. One type of tandem repeat sequence, referred to as short tandem repeats, or STRs, has become a standard method in forensic science

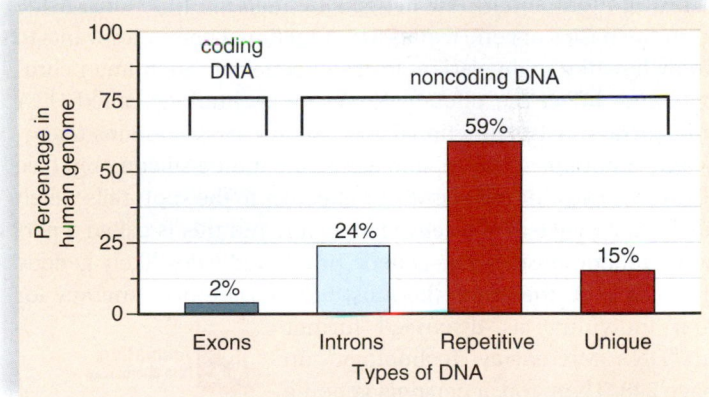

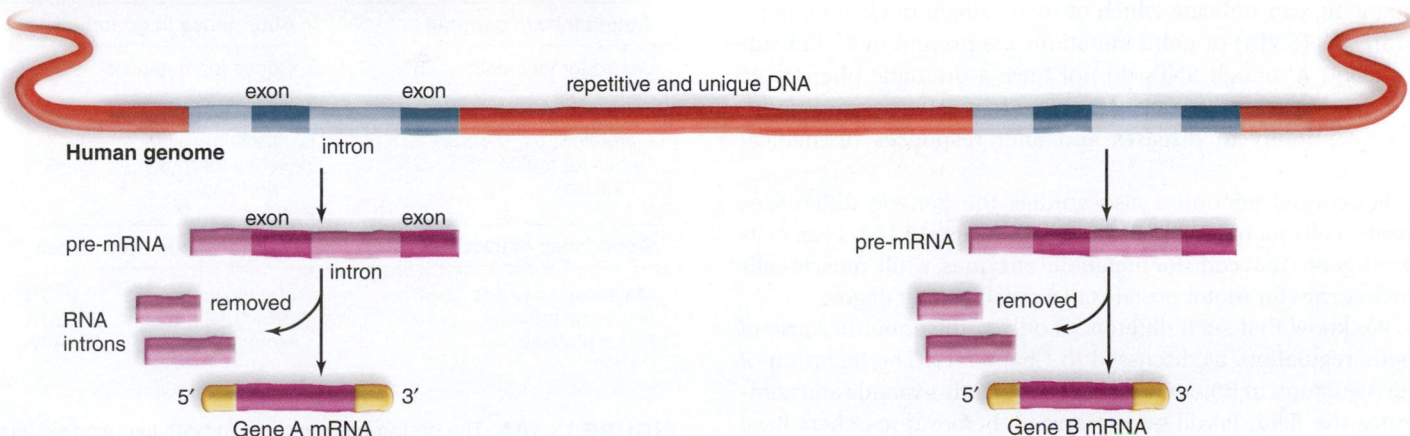

FIGURE 13.9 A genome contains protein-coding DNA called exons (dark blue) and noncoding DNA, including introns (light blue) and other intergenic sequences (red). Only the exons are present in mRNA and specify protein synthesis. Children inherit both coding and noncoding DNA from their parents.

for distinguishing one individual from another through DNA fingerprinting. The second type of repetitive DNA is called an **interspersed repeat,** meaning that the repetitions may be placed intermittently along a single chromosome, or across multiple chromosomes. Because of their common occurrence, interspersed repeats are thought to play a role in the evolution of new genes.

Transposons are specific repetitive DNA sequences, called elements, that have the remarkable ability to move within and between chromosomes. Because it can move, a repetitive DNA sequence known as the *Alu element* is interspersed on multiple chromosomes once per every 5,000 base pairs in human DNA. Transposons have by now been discovered in most organisms. Barbara McClintock received a Nobel Prize in 1983 for her discovery of transposons in corn (see "Transposons Cause Mutations" on page 210).

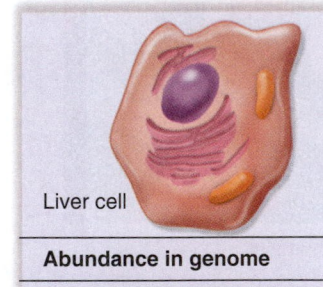
Animation
Transposons: Shifting Segments of the Genome

The movement of transposons to a new location sometimes alters neighboring genes, particularly by decreasing their expression. In other words, a transposon sometimes acts like a regulatory gene. The movement of transposons throughout the genome is thought to be a driving force in the evolution of organisms. In fact, many scientists now think that many repetitive DNA elements are or were originally derived from transposons.

Little is known about **unique noncoding DNA,** and it's possible that this region along with repetitive DNA produces the regulatory small RNAs discussed in Chapter 12. What was once thought to be a vast "junk DNA" wasteland may instead play an active role in regulating protein-coding gene expression. This previously overlooked RNA signaling network may even be what allows humans to achieve their structural complexity. Altogether, we now envision a more complex and dynamic genome than was apparent merely a few decades ago.

▶ **13.9 CHECK YOUR PROGRESS**
1. Describe three types of noncoding DNA.
2. Identify which type of noncoding DNA occupies the largest percentage of the genome and functions it might serve.

13.10 Functional and comparative genomics analyze the human genome

LEARNING OUTCOMES

When you complete this section, you should be able to

1. Differentiate between functional and comparative genomics.
2. Discuss the type of studies being done by researchers in these fields.

Functional Genomics After researchers determined the sequence of bases in the human genome, they were interested in knowing more about how the genetic material functions. Aside from the basics of gene expression, functional genomics wants to know how the many variant forms of genes on the human chromosomes affect the phenotype. A new technology called DNA microarrays assists functional genetics. An array contains microscopic amounts of base sequences spotted on a silicon chip (see Fig. 13B). The binding of mRNA molecules to the spots tells which alleles are active in the cells of a person and this is called a person's genetic profile. The genetic profile indicates likely genetic illnesses and, therefore, the most appropriate drug therapy for that individual as discussed further in "DNA Microarray Technology" on page 248. Then, too, a personal genome

Animation
DNA Microarray

sequencing can indicate which of many single nucleotide polymorphisms (SNPs) or point mutations are present in an individual's DNA. Although SNPs do not have a dramatic phenotypic effect, it is hypothesized that particular SNPs patterns can account for susceptibility to diseases and alter responses to medical treatments.

Functional genomics also studies the genetic differences between cells such as those illustrated in Figure 13.10A. Liver cells express genes that code for metabolic enzymes, while muscle cells express genes for motor proteins to a much greater degree.

We know that such differences only come about because of genetic regulation, as discussed in Chapter 12. The inclusion of gene regulation in functional genomics greatly expands and complicates the field. It will be some years before researchers have completely revealed the previously hidden ways in which genetic expression is regulated.

Comparative Genomics The aim of comparative genomics is to compare the human genome to the genome of other organisms such as those listed in Table 13.10 and indeed to compare the genomes of all species one to the other. Because researchers are busy sequencing the genomes of so many different organisms comparative genomics will expand greatly in the years to come. The organisms listed in Table 13.10 are model organisms, so-called because the work done with any one of them can be applied to all the other forms of life.

Note that first line in Table 13.10 compares genome size of these organisms by giving the number of nucleotide base pairs (A=T; G=C) in their particular genomes. It would be reasonable to

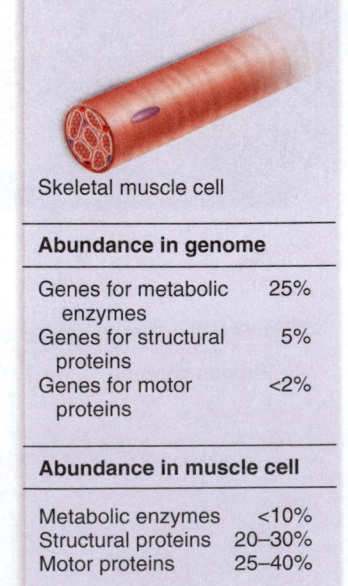

Liver cell		
Abundance in genome		
Genes for metabolic enzymes		25%
Genes for structural proteins		5%
Genes for motor proteins		<2%
Abundance in liver cell		
Metabolic enzymes		>50%
Structural proteins		<10%
Motor proteins		<5%

Skeletal muscle cell		
Abundance in genome		
Genes for metabolic enzymes		25%
Genes for structural proteins		5%
Genes for motor proteins		<2%
Abundance in muscle cell		
Metabolic enzymes		<10%
Structural proteins		20–30%
Motor proteins		25–40%

FIGURE 13.10A The genome is the same in both liver and skeletal muscle cells, but the expressed genes are different as shown by the abundance of proteins in the cells.

TABLE 13.10 Genome Sizes of Humans and Some Model Organisms

Organism	Homo sapiens (human)	Mus musculus (mouse)	Drosophila melanogaster (fruit fly)	Arabidopsis thaliana (flowering plant)	Caenorhabditis elegans (roundworm)	Saccharomyces cerevisiae (yeast)
Estimated Size	3,200 million bases	2,500 million bases	180 million bases	125 million bases	97 million bases	12 million bases
Estimated Number of Genes	~23,000	~30,000	13,600	25,500	19,100	6,300
Average Gene Density	1 gene per 100,000 bases	1 gene per 100,000 bases	1 gene per 9,000 bases	1 gene per 4,000 bases	1 gene per 5,000 bases	1 gene per 2,000 bases

assume that the complexity of an organism could be judged by the size of its genome. For example, the size of the human genome is much larger than that of the bacterium. However, if we take a look at the bar graph in Figure 13.10B we will notice some unexpected inconsistencies. The bar graph in Figure 13.10 tells us that plants have more nucleotide base pairs in their DNA than do mammals! What could bring this difference about? Even the two related plants have difference genome sizes: *Echinops bannaticus* has double the amount of base pairs than does *E. nanus*. Perhaps this is a clue that polyploidy brings about these genomic differences. Polyploidy, which is common among plants, is known to account for why many plants have more genetic material than animals and why some plants have more genetic material than other plants.

The next line in Table 13.10 compares the number of genes these organisms have. Again some surprises appear that require an explanation. Evidence suggests that the large number of genes in the plant *Arabidopsis* could be due to duplicate genes. If the same gene is present in duplicate, it is still counted as two genes. (If one gene is present in duplicate, not all of them have to be.) Humans have a few genes in duplicate, as exemplified by the various hemoglobin genes: fetal hemoglobin in the fetus, myoglobin in muscles, and regular hemoglobin in red blood cells. Table 13.10 tells us that a mouse has slightly more genes than a human has. Whether these genes code for the same proteins and are regulated in the same way as in humans is a matter that will eventually be scrutinized in detail. Researchers are very interested in comparing the human genome to that of the great apes because it is known that their genomes are so similar. The application "We Are Closely Related to Chimpanzees" on page 250 compares the human chromosomes to those of chimpanzees.

We naturally assume that all the genes referred to in Table 13.10 are protein-coding genes. If they are, biologists are embarrassed to say that in some cases they do not know what proteins they code for. So far, it seems that some protein coding genes apparently code for unknown amino acid sequences in unknown proteins! Another critical question is why the number of genes in mammals is so few considering the large number of bases in their genomes. As we know, much of the human genome consists of repetitive and noncoding sequences (see Fig. 13.9) that nevertheless have a critical function. They are involved in regulating the expression of protein-coding genes and only the final result of gene expression—the various proteins in the cells—determine the phenotype. On the other hand, researchers have discovered that the same type of genes can be present in a wide variety of organisms from bacteria to humans. These genes are said to be conserved genes because as the process of evolution occurred, these genes stayed as they were and were passed on to nearly all organisms with only minor changes. A well-known example is the developmental *Hox* genes that occur in organisms across the spectrum of life despite different developmental outcomes.

Much remains to be discovered in functional and comparative genomics, which is a dynamic and active field of biology today.

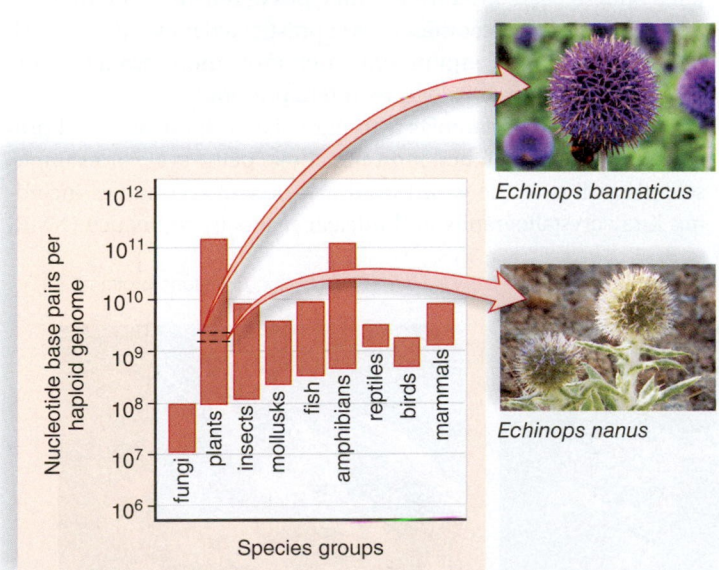

Echinops bannaticus

Echinops nanus

FIGURE 13.10B The size of plant genomes tends to be larger than that of mammals, most likely because of polyploidy (duplication of genetic material). This could also explain why *Echinops bannaticus* has double the amount of base pairs than does *E. nanus*.

▶ **13.10 CHECK YOUR PROGRESS**

1. Identify the type of research that would be consistent with the goals of functional genomics and comparative genomics.

LEARNING OUTCOMES

When you complete this section, you should be able to

1. Explain why the study of the proteome requires much work and new technologies.
2. Summarize the many functions of proteins in a proteome.

The information stored in a genome allows a cell to produce proteins in the cytoplasm. A cell could not continue to exist without a constant supply of proteins. **Proteomics** is a new field of endeavor that studies the structure and function, and interaction of cellular proteins within a cell or organism. The term **proteome** refers to the entire collection of a species' proteins. The number of proteins that occur in a cell can number in the hundreds of thousands and therefore usually far outnumbers the coding genes in the genome.

Two processes account for why the proteome is larger than the genome: alternative pre-mRNA splicing in the nucleus and posttranslational modification in the cytoplasm. When mRNA is first made off of a DNA template strand, it contains a mix of exons (protein-coding regions) and introns (nonprotein-coding regions). Only the exons will be in mature RNA, and the introns, which occur in between the exons, will be spliced out. An opportunity arises for alternative splicing because a different selection of exons can appear in mature RNA. Different selections of exons means different proteins occur in the cell. The other mechanism that accounts for the many proteins in a cell, namely posttranslational modification, occurs in the rough endoplasmic reticulum (RER) and the Golgi apparatus. Polypeptides enter the interior space of the RER as they are being produced by ribosomes and then vesicles transport the proteins to the Golgi apparatus (Fig. 13.11A). During their stay in the RER and Golgi apparatus, polypeptides gain side chains of various sorts and these can be rearranged in different ways. Because any particular polypeptide is individually modified according to the needs of the cell, the number of different proteins that exit the Golgi apparatus is larger than the number that entered it.

Proteins have a particular structure and that structure has to be suitable to the particular function they will have in the cell (Fig. 13.11B). The large number of potential proteins in the cytoplasm presents a formidable task for biologists who wish to study both their structure and function. The enormity of the task can be appreciated by imagining that you have been given the task of studying and reporting to others in a logical way the many different

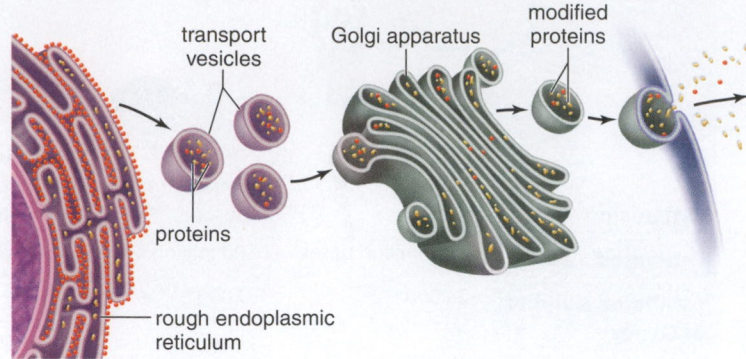

FIGURE 13.11A Proteins made at the RER are modified and therefore diversified in the Golgi apparatus.

types of objects for sale in department stores. Your task would be complicated, too, by knowing that not all department stores sell the same items. We can liken this to knowing that any particular cell has their own mix of proteins (see Fig. 13.10A). You might think that it would be possible to use mRNA and DNA microarray chips to determine what proteins are present in a cell (see page 248). But this is unreliable because the presence of mRNA does not always mean the protein is present nor does it tell you the quantity of the protein present. Protein microarray chips can be prepared which theoretically work on the same principle as DNA chips. Hundreds or even thousands of different proteins from an organism are spotted on the chip at known locations. Fluorescent tags are applied to proteins from a cell and when the chip is exposed to cellular proteins some will bind to particular spots, right? Well no, because proteins don't bind by complementary base pair as do nucleotides. They vary greatly in their chemistry, some being hydrophobic and some hydrophilic, for example. Proteins bind to each other by several types of noncovalent intereactions. Therefore, proteomics uses sophisticated methods and high powered laboratory equipment to confirm the presence of the protein and the quantity of the protein present.

One goal of proteomics is to understand the structure of proteins and how they function in metabolic pathways. For example, scientists all over the world used a variety of techniques including X-ray crystallography and nuclear magnetic resonance (NMR)

FIGURE 13.11B Building 3D protein models represented here by an enzyme (*left*) and histone proteins in a nucleosome (*right*) helps scientists understand how the structure of a protein can affect its interactions and functions.

Gluconate kinase, an enzymatic protein

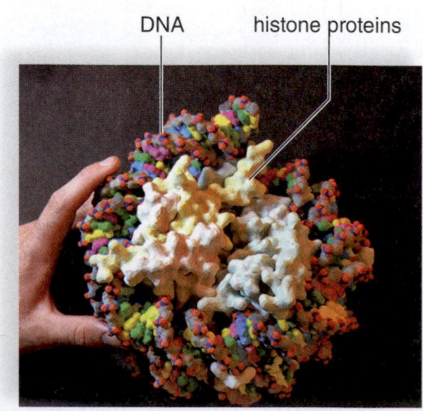

Nucleosome: the core is histone proteins

TABLE 13.11 Types of Proteins in the Proteome

Function		Examples	Function		Examples
Metabolism		In cells, enzymes bring reactants together and thereby speed chemical reactions. Enzymes are specific to a reaction and function at body temperatures.	Transport		Channel and carrier proteins in the plasma membrane allow substances to enter and exit cells. Hemoglobin is a complex protein that transports oxygen.
Signaling		Hormones serve as intercellular messengers. For example, insulin influences target cell metabolism. Growth hormone determines the height of an individual.	Gene Expression Regulation		Transcription in cells is largely controlled by DNA-binding proteins including transcription factors, activators, and repressors. Ribosomal proteins assist translation.
Movement		The contractile proteins actin and myosin allow parts of cells to move and cause muscles to contract. Muscle contraction allows animals to move from place to place.	Defense		Proteins called antibodies combine with disease-causing agents and prevent them from destroying cells and upsetting homeostasis.
Support		Keratin is a tough fibrous protein that makes up fingernails. Other support proteins include collagen in the extracellular matrix and silk protein in spiderwebs.	Osmolarity		Plasma proteins are involved in osmolarity, even though some specialize in signaling, defense, coagulation, and transport. Researchers have found 325 distinct proteins in blood plasma.

spectroscopy to collect data stored in the Protein Data Bank which is now a worldwide consortium. A special printer (called the Z corporation printer) can use these data to build a 3D model of over 37,136 proteins (see Fig. 13.11B). Such models make it easier to study the shapes of proteins and how they possibly interact with other molecules. After many protein models are made, it's possible that scientists will be able to predict final three-dimensional shape of a protein from knowing its sequence of amino acids and even predict the effects of DNA mutations on the protein's shape and function. Then drugs can be designed to interfere with the action of the protein if advisable. Also, it may be possible one day to tailor drug treatment to the particular proteome of the individual.

Biologists have a pretty good handle on the general functions of proteins in a cell (Table 13.11). Every reaction in a cell requires an enzyme to bring the reactants together so that the reaction takes place with minimum energy input. This means that enzymes are specific to their substrates. The plasma membrane contains various proteins and many are involved in cell signaling because they are receptors for molecules that affect the metabolism of the cell. A cell could not continue to exist without exchanges with its environment. Some plasma membrane proteins are channels or carriers that allow molecules to enter and/or exit the cell. While plants support themselves by producing cellulose and other nonprotein molecules, animals rely largely on the protein collagen in the extracellular matrix and in muscle and bone for support. Actin and myosin interact within muscle to produce movement. In a cell, motor molecules composed of protein moving along microtubules allow vesicles to travel between organelles! DNA binding proteins regulate transcription in prokaryotes and eukaryotes alike. In complex animals, antibody production provides a specific defense against disease-causing antigens. The other plasma proteins also play important functions such as maintaining the volume and therefore the pressure of blood.

► **13.11 CHECK YOUR PROGRESS**
1. Analyze why a cell contains many more proteins than there are protein-coding genes.
2. Identify two possible ways to study the structure and function of a protein.

13B DNA Microarray Technology

With advances in robotic technology, it is now possible to place the entire human genome onto a single microarray (Fig. 13B). The mRNA from the organism or the cell to be tested is labeled with a fluorescent dye and added to the chip. When the mRNAs bind to spots on the microarray, a fluorescent pattern results that is recorded by a computer. Now the investigator knows what DNA is active in that cell or organism. A researcher can use this method to determine the difference in gene expression between two different cell types, such as between liver cells and muscle cell, for example.

Animation
DNA Microarray

Genetic Profiles

A mutation microarray, the most common type, can be used to generate a person's genetic profile. A mutation microarray contains hundreds of thousands of known disease-associated mutant gene alleles. Genomic DNA from the individual to be tested is labeled with a fluorescent dye, and then added to the microarray. The spots on the microarray fluoresce if the individual's DNA binds to the mutant alleles on the chip, indicating that the individual may have a particular disorder or is at risk for developing it later in life. This technique can generate a genetic profile much more quickly and inexpensively than older methods involving DNA sequencing.

Diseased Tissues

DNA microarrays also promise to hasten the identification of genes associated with diseased tissues such as a tumor or a diseased connective tissue. First, mRNA derived from the diseased tissue and the corresponding normal tissue are labeled with different fluorescent dyes, say red and green. The normal tissue serves as a control. The investigator applies the mRNA from the both normal and abnormal tissue to the microarray. The relative intensities of fluorescence from a spot on the microarray indicate the amount of mRNA originating from that gene in the diseased tissue relative to the normal tissue. If a gene is inactive in the diseased tissue, more copies of mRNA will bind to the microarray from the control tissue, and the spot will appear more red than green.

Genomic microarrays are also used to identify links between disease and chromosome mutations. Each spot on the microarray is a known chromosome mutation. Labeled genomic DNA from diseased and control tissues bind to the DNA on the chip, and the relative floruescence from both dyes is determined. A difference in fluorescence of the two dyes indicates which chromosomal mutations may be associated with the disease being studied.

Microarray Applications

The microarray applications rival those of PCR (polymerase chain reaction). A microarray can

- Detect genetic differences between cells. In how many ways is a liver cell genetically different from a muscle cell or how genetically different are two types of epithelial tissues?

DNA microarray

Fluorescent mRNA did bind to DNA on chip.

DNA on chip

fluorescent mRNA

Fluorescent mRNA did not bind to DNA on chip.

testing subject's mRNA

FIGURE 13B DNA microarray technology.   A DNA microarray contains many microscopic samples of DNA bound to known locations on a silicon chip. A fluorescently labeled mRNA from a tissue or organism binds to the DNA on the chip by complementary base pairing. The fluorescent spots indicate that binding has occurred and that the gene functions in that cell.

- Create a genetic profile for an individual. A genetic profile lists all the disease-related mutations that occur in a person's genome.
- Identify genes that are active or even chromosome mutations that occur in diseased tissues, such as a tumor. Genetic differences can indicate how a tumor should be treated.
- Detect a specific genetic disease when a mutant allele does not bind to a spot on a microarray as well as a normal allele.
- Distinguish species on the basis of genetic differences. Such differences can be critical for classifying bacteria, for example.
- Detect cell responses to a particular environment. For example, does an environment that offers only fructose as a sugar cause detectable genetic differences in liver cells?

CONSIDER THESE QUESTIONS

1. Why might a researcher want to know what genes are being expressed in different cell types?
2. How might the information from a DNA microarray be used to develop new drugs to treat disease?

 Explore the concepts through a variety of multimedia assets and question types.

BIOLOGY

www.mcgrawhillconnect.com

13.12 Bioinformatics assists the study of genomics and proteomics

An enormous amount of data was collected in order to sequence the bases in the human genome. Similarly, much data has now been collected regarding the many proteins in the proteome. Fortunately, today high-powered computers are available to store all these data. A **database** has data stored in a single location and organized for easy accessibility. We have already had the opportunity to mention the Protein Data Bank, which is a worldwide database for protein information. Similar datbases (e.g., GenBank) make gene sequence information available per chromosome (Fig. 13.12).

The application of computer technologies, specifically developed software, and statistical techniques to study biological information in a database, is called **bioinformatics.** The first software system to find genes and make an initial decision about their functions was developed to study the genome of the bacterium *Haemophilus influenzae* in 1995. Since then software has been developed to find both normal and mutated genes in the human genome and to sequence the amino acids in the polypeptide they encode. A database can also be searched to determine how closely a human gene or protein matches that in other organisms. For example, the genome data now available for the Neandertals can be compared to that of the human genome. This information is extremely helpful to those studying the evolutionary relationship between the two groups. The latest studies suggest that modern humans did mate with Neandertals to some extent in places where their populations overlapped.

The human genome has 3 billion known base pairs, and without the computer it would be almost impossible to make sense of these data. BLAST, which stands for Basic Local Alignment Search Tool, is a computer program that can even identify homologous genes among the genomic sequences of various organisms. **Homologous genes** are genes that code for the same proteins, although the base sequence may be slightly different. Such differences can help trace the history of evolution among a group of organisms. BLAST and other computer software are expected to help with constructing a very complex tree of life, for example. In proteomics, BLAST enables researchers to find homologous proteins. Modeling homologous proteins can help researchers predict the structure of other proteins with similar sequences of amino acids.

Bioinformatics can be helpful in any study involving diverse populations around the globe. For example, humans inherit patterns of base sequence differences now called haplotypes (from the terms haploid and genotype). If a chromatid has a G rather than a C at a particular location, this change might be accompanied by other base differences near the G. Researchers want to discover the most common haplotypes among African, Asian, and European populations and to do this they will have to collect, store, and compare data from thousands of people. If different ethnicities have different haplotypes it could mean that they will respond differently to particular medical treatments. Such information would be very helpful to the medical community. Bioinformatics also has various other applications in the field of

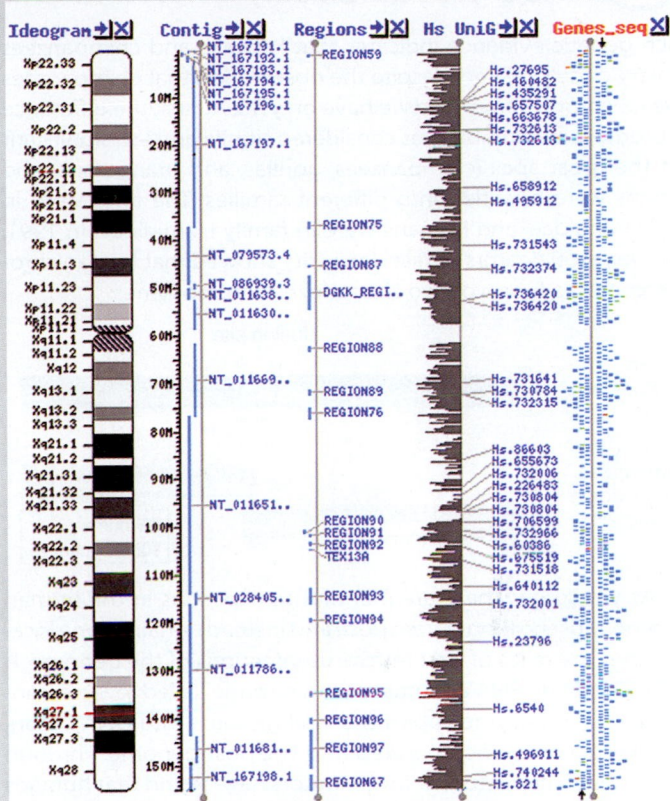

FIGURE 13.12 These X chromosome data are available at the National Center for Biotechnology Information (NCBI) website.

medicine. For example, researchers found the function of the protein that causes cystic fibrosis by using the computer to search for genes in model organisms that have the same sequence. Model organisms, you'll recall, are organisms that are easily studied and these studies apply to all other organisms, see "*Arabidopsis* is a Model Organism" on pages 192–193. Because they knew the function the cystic fibrosis gene in model organisms, researchers could deduce the function in humans. This was a necessary step toward developing specific treatments for cystic fibrosis. And the computer can help correlate genomic differences among large numbers of people with a specific disease. For example, we now know that an individual's genome often contains multiple copies of a gene. But individuals may differ in the number of copies, called copy number variations. Now it seems that the number of copies in a genome can be associated with specific diseases.

It is safe to say that without bioinformatics, our progress would be extremely slow in determining the function of DNA sequences, comparing our genome to model organisms, and knowing how genes and proteins interact in cells. Instead, with the help of bioinformatics, progress should proceed rapidly in these and other areas.

13C We Are Closely Related to Chimpanzees

Much genetic evidence indicates that humans and chimpanzees are very closely related, despite the observation that chimpanzees have 48 chromosomes and we have only 46. At first, the difference in chromosome number was considered significant—so significant that the great apes (chimpanzees, gorillas, and orangutans) and humans were classified into different families. The apes were in family Pongidae, and humans were in family Hominidae. In 1991, however, investigators at Yale University showed that human chromosome 2 is a fusion of two chimpanzee chromosomes:

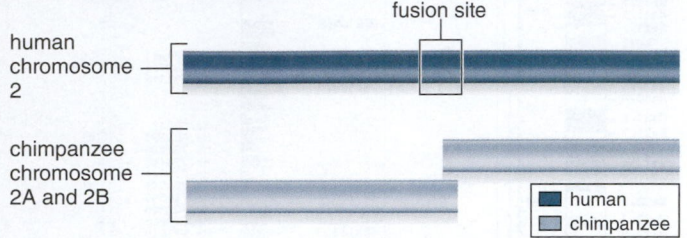

As you know, there are plentiful transposons in the human genome. Most no longer transpose and instead remain at one location. They are relics of past retrovirus infections of the genome. A retrovirus has an RNA genome and an enzyme called reverse transcriptase. This enzyme copies the viral genome into DNA before the DNA becomes incorporated into the host genome. Through the course of many studies, investigators have found that humans and chimpanzees have similar patterns of transposons in their genomes. Here is an example showing how the Alu element (a transposon) inserted itself into the human genome and the chimpanzee genome in the vicinity of hemoglobin genes:

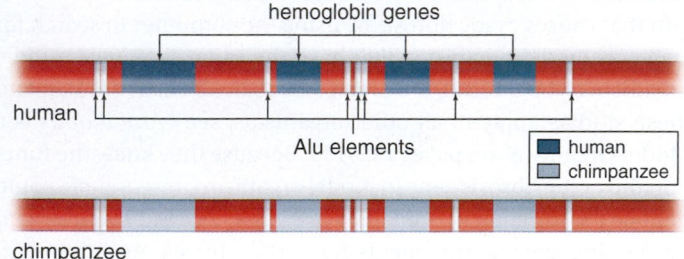

Pseudogenes are nonfunctional copies of genes that were active in the past. Pseudogenes are presently not active due to a mutation that prevents them from coding for a functional protein. The pseudogene pattern in human DNA is most similar to that in chimpanzee DNA and to a lesser extent to that of the other great apes. These and other studies have caused a reclassification of the primates most closely related to us. All of the great apes are now in the same family as humans; chimpanzees are in the same subfamily (Homininae).

Modern genomic data show that the base sequence of chimpanzees and human DNA differs only by 1.5% of their bases. By now, we take as a given that humans and chimpanzees are genetically similar, and it has become the task of geneticists to determine what genetic factors make us different from chimpanzees. Even though our base sequences are quite similar, significant differences do exist. For example, when we compare the chimpanzee and the human genome, many DNA stretches (about 5 million) are absent in one or

Now that we know chimpanzees are closely related to us, studies have shifted to finding the genetic differences that make us so anatomically different.

the other genome. We know that in the evolution of mammals there was an explosion in the amount of noncoding sequences relative to the number of coding genes. So a common mammalian ancestor may have had a larger quantity of noncoding DNA, and each mammalian species lost different DNA stretches since their lines of descent separated. Many examples tell us that noncoding DNA is not the mere junk as was largely thought only a few years ago, and so it could very well be that primate anatomical differences could be due to which stretches of noncoding DNA were retained by the particular primate. Microarray chip data support the conclusion that certain genes are not expressed in chimpanzees to the same degree as in humans. Which genes might these be? Evidence is gaining that the genes controlling the development of the brain may have been affected the most by the gaps present in the chimpanzee genome compared to the human genome. This may account for why our brains are larger than those of chimpanzees today.

CONSIDER THESE QUESTIONS

1. This reading concludes that chimpanzees and humans differ most in gene regulation. Would you have expected gene regulation to account for major differences in evolution? Why or why not?

2. Would you be willing to participate in a comparative human study to determine what types of genomes are prone to what illnesses? Why or why not?

3. Would you expect your genome to differ from that of your siblings? Why or why not?

 Explore the concepts through a variety of multimedia assets and question types.

www.mcgrawhillconnect.com

CONNECTING THE CONCEPTS

This chapter is divided into first a study of biotechnology and then a study of genomics. Biotechnology is the use of organisms for the benefit of humans, and includes many modern techniques to manipulate the genome. Cloning can be done to make copies of organisms, genes, or DNA segments. Recombinant DNA (rDNA) technology allows researchers to use bacteria to clone a human gene and produce protein products such as vaccines, hormones, and growth factors. Being able to transfer genes from one organism to another has led to the production of GM plants and animals and gene therapy in humans. Today, plants and animals can be engineered to make a product or to possess desired characteristics. Gene therapy offers the promise of curing human genetic disorders such as muscular dystrophy, cystic fibrosis, hemophilia, and many others.

We now know the sequence of the bases in the human genome, thanks to the use of PCR automation and DNA base sequencers. Our genome contains only 23,000 protein-coding genes, even though it is 3 billion base pairs long. Research indicates that 98% of our genome is noncoding (for proteins), and consists largely of introns and repetitive DNA, which includes the transposons. Researchers are now scrambling to discover the usefulness of noncoding DNA, much of which is apparently transcribed into RNA. Could it be that a vast RNA regulatory network accounts for the complexity of humans?

Sequencing the human genome has led to new fields of endeavor. Functional and comparative genomics analyze the human genome; proteomics studies the many proteins in cells, and bioinformatics assists the study of genomics and proteomics. Each of these fields makes medical contributions.

Evolution, which is sometimes defined as an alteration of species as a result of genetic changes, is the topic of the next unit of this text. Charles Darwin, who knew nothing about genes, was the first to present significant evidence that evolution occurs.

ANALYZE AND EVALUATE

1. Biotechnology is correctly named because it is a form of applied biology. Give examples to support this statement.
2. Biotechnology products benefit humans in what ways?
3. Sequencing the human genome has led to much research. Give three examples.

SUMMARIZE

Cloning of Organisms, Genes, and DNA Segments

13.1 Animals can be cloned using surrogate mothers

■ The desired result of **reproductive cloning** is a new individual exactly like the original individual. Many farm animals are cloned today, and it may be the only way to save some endangered species from extinction.

13.2 Genes can be cloned in bacteria

■ **Gene cloning** using **recombinant DNA** (rDNA) produces identical copies of a particular gene. To make **recombinant DNA,** a **plasmid** is a **vector** prepared to accept a foreign gene when both the plasmid and foreign DNA are cleaved by the same **restriction enzyme. DNA ligase** seals the foreign DNA into the plasmid, and the plasmid carries the foreign gene into the host cell—in this case, a bacterium. Now the bacterium has been **genetically engineered** and is a **genetically modified organism (GMO).**

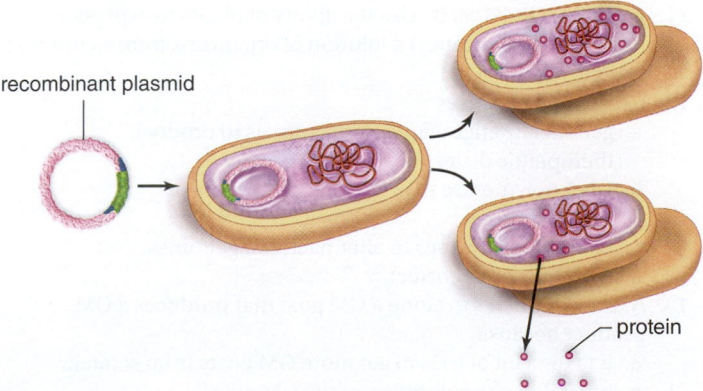

recombinant plasmid

protein

■ A gene is cloned as the plasmid replicates, and the bacterium will produce a new and different protein product.

13.3 DNA segments can be cloned in a test tube

■ The **polymerase chain reaction** (**PCR**) uses the enzyme DNA polymerase to quickly make copies over and over again (hence, chain reaction) of a specific fragment of DNA. Previously, to accomplish DNA fingerprinting, restriction enzymes were used to fragment the entire genome, now PCR makes copies of **short tandem repeats** (e.g., ATGATGATG). **Gel electrophoresis** (fragments migrate through a gel in an electrical field) can compare fragment sizes but today the amount of fluorescence given off can indicate the number of repeats. PCR has many applications to assisting genomic research.

Biotechnology Products and Genetically Modified Organisms

13.4 Genetically modified bacteria produce useful products

■ **GM bacteria** produce **biotechnology products,** either medical (e.g., vaccines, growth hormone) or commercial (e.g., products that clean up oil spills). GM bacteria have many other uses including cleaning up oil spills and protecting crops from frost and pests.

13.5 Plants can be genetically modified

■ Certain crops have been modified to resist disease, insects, or herbicides. Genetic engineering is being used to improve the agricultural and food qualities of crops. Some plants have been engineered to manufacture medical products.

13.6 Animals can be genetically modified

■ Genes can be inserted into the eggs of animals. Through **gene pharming,** GM animals produce pharmaceuticals. GM mice are bred for research. **Xenotransplantation** is the use of modified animal organs in human transplant patients.

13.7 A person's genome can be modified

- **Gene therapy** can be done in two ways: Using ex vivo therapy, cells or tissues are removed from the body, given a normal gene, and then reinserted into the body. Using in vivo therapy, a gene is delivered directly into the body.

The Human Genome and How It Functions

13.8 The human genome has been sequenced

- The **Human Genome Project** (**HGP**) determined the order of bases in the human genome. The **polymerase chain reaction** and automated equipment (PCR thermocycler, DNA sequencers) facilitated the project.

13.9 The genome contains coding and noncoding DNA

- Coding DNA (exons) is transcribed into mRNA that binds to a ribosome where protein synthesis occurs. Some protein-coding genes occur in multiple copies; a **pseudogene** is a nonfunctional copy. Noncoding DNA contains (1) **introns,** (2) **repetitive DNA,** and (3) **unique noncoding DNA.** Surprisingly, this DNA may be transcribed into RNA, and this RNA is thought to have a regulatory function. Repetitive DNA contains **tandem repeats, interspersed repeats,** and **transposons.** The function of unique noncoding DNA is still to be discovered.

13.10 Functional and comparative genomics analyze the human genome

- **Functional genomics** studies how genomes differ; for example by the presence of **single nucleotide polymorphisms** (**SNPs**). **DNA microarrays,** which contain DNA fragments on a silicon chip, can be used to determine the genetic activity in a cell or an individual. This is called the **genetic profile** of an individual. **Comparative genomics** focuses on determining how species are related and how their genes compare to those of humans. **Conserved genes** were present in the earliest of ancestors and did not change as they were passed from ancestor to ancestor.

13.11 Proteomics analyzes the proteins in human cells

- **Proteomics** is the study of the structure, function, and interaction of cellular proteins. The human **proteome** is the complete collection of proteins that humans produce.

13.12 Bioinformatics assists the study of genomics and proteomics

- **Bioinformatics** is the application of computer technologies to the study of the genome, including genomics and proteomics. **Databases** help scientists detect **homologous genes** in various species.

TEST YOURSELF

Cloning of Organisms, Genes, and DNA Segments

1. During reproductive cloning, a(n) _____ is placed into a(n) _____.
   a. enucleated egg cell, adult cell nucleus
   b. adult cell nucleus, enucleated egg cell
   c. egg cell nucleus, enucleated adult cell
   d. enucleated adult cell, egg cell nucleus
2. The major challenge to cloning animals is
   a. finding surrogate mothers.
   b. obtaining enough enucleated eggs.
   c. the health of cloned animals.
   d. keeping embryos alive in culture.

3. Which of the following is not a clone?
   a. a colony of identical bacterial cells
   b. identical quintuplets
   c. a forest of identical trees
   d. eggs produced by oogenesis
   e. copies of a gene produced through PCR
4. Put the lettered phrases in the correct order to form a plasmid-carrying recombinant DNA.
   a. Use restriction enzymes.
   b. Use DNA ligase.
   c. Remove plasmid from parent bacterium and acquire donor eukaryotic DNA.
   d. Introduce plasmid into new host bacterium.
5. Restriction enzymes found in bacterial cells are ordinarily used
   a. during DNA replication.
   b. to degrade the bacterial cell's DNA.
   c. to degrade viral DNA that enters the cell.
   d. to attach pieces of DNA together.
6. Recombinant DNA technology is used
   a. for gene therapy.
   b. to clone a gene.
   c. to acquire GM bacteria.
   d. Both b and c are correct.
7. The restriction enzyme *Eco*RI has cut double-stranded DNA in the following manner. The piece of foreign DNA to be inserted begins and ends with what base pairs?

8. **THINKING CONCEPTUALLY** What is the value of cloning organisms, genes, and DNA segments?

Biotechnology Products and Genetically Modified Organisms

9. Which of these would you not expect to be a biotechnology product?
   a. phospholipid   c. modified enzyme
   b. protein hormone   d. clotting enzyme
10. Which of these are desired ends of GM plants?
    a. salt-tolerant crops
    b. drought-tolerant crops
    c. protein-enriched foods
    d. crops that can resist herbicides
    e. All of these are correct.
11. **THINKING CONCEPTUALLY** Use the ability of plants to express a human gene to support evolution of organisms from a common source.
12. Gene pharming uses
    a. genetically engineered farm animals to produce therapeutic drugs.
    b. DNA polymerase to produce many copies of targeted genes.
    c. restriction enzymes to alter bacterial genomes.
    d. All of these are correct.
13. A farmer decides to clone a GM goat that produces a GM product because
    a. it takes a lot of tries to get more GM goats from scratch.
    b. he prefers the coloration of the GM goat.
    c. it's more likely to produce GM goats than by breeding them.
    d. Both a and c are correct.

14. The use of both ex vivo and in vivo gene therapy shows that
    a. only one type of therapy will be successful.
    b. there is more than one way to introduce beneficial genes into the body.
    c. human cells will take up beneficial genes directly or by using a viral vector.
    d. researchers can't make up their minds.
    e. Both b and c are correct.
15. **THINKING CONCEPTUALLY** Explain why gene therapy researchers prefer to genetically modify the stem cells of white blood cells, as opposed to the white blood cells themselves.

The Human Genome and How It Functions

16. Which of the following is not required for the polymerase chain reaction?
    a. DNA polymerase   c. a DNA sample
    b. RNA polymerase   d. nucleotides
17. Today, the polymerase chain reaction (PCR)
    a. uses RNA polymerase.
    b. takes place in huge bioreactors.
    c. uses a heat-tolerant enzyme.
    d. makes lots of nonidentical copies of DNA.
18. **THINKING CONCEPTUALLY** You have 30 dinosaur genes. Explain why it would be impossible to create a dinosaur, even by using PCR to increase the quantity of each gene.
19. Because of the Human Genome Project, we know
    a. the sequence of the base pairs of our DNA.
    b. the sequence of all genes along the human chromosomes.
    c. all the mutations that lead to genetic disorders.
    d. All of these are correct.
    e. Both a and c are correct.
20. Which of these pairs is mismatched?
    a. repetitive DNA—tandem repeats and interspersed repeats
    b. tandem repeats—ACGACGACG
    c. transposons—jumping genes
    d. introns—same as exons
    e. Both a and c are mismatched.
21. Repetitive DNA could
    a. be the source of small RNAs that play a regulatory role in cells.
    b. account for an RNA network necessary to our complexity.
    c. be more important than formerly thought.
    d. All of these are correct.
22. Which of these pairs is mismatched?
    a. genome—all the genes of an individual
    b. proteome—all the proteins in an individual
    c. bioinformatics—all the genetic information present in the organism
    d. genetic profile—includes the mutations of an individual

23. The field of comparative genomics is not concerned with
    a. the function of gene products.
    b. the number of genes in various organisms.
    c. who is related to whom.
    d. how to cure human genetic diseases.
24. Which of these is a true statement?
    a. The complexity of an organism does not necessarily correlate with the number of coding genes.
    b. Genomes do not contain both coding and noncoding DNA.
    c. Bioinformatics is not used as a tool for studying genomes.
    d. Alternative splicing of existing genes is not possible.
25. If we knew the sequence of genes on the chromosomes,
    a. it would reveal which genes are active in which cells.
    b. it would reveal how genes are regulated in a cell.
    c. more genes could be isolated and used for gene therapy.
    d. All of these are correct.

GET INVOLVED

1. Using this virtual lab to help you, design an experiment based on Figure 13.6B that would allow you to determine where a dominant gene for "tail" is located on mouse chromosome 10.
2. When doing a gene therapy study, what is the advantage of utilizing an ex vivo instead of an in vivo procedure? (See section 13.7.)
3. What experimental steps reviewed in this virtual lab are needed to place a foreign gene into the genome of another organism?

MEDIA STUDY TOOLS

mhhe.com/maderconcepts3

Enhance your study of this chapter with interactive study tools, practice tests, and engaging animations. Also, ask your instructor about the resources available through ConnectPlus, which includes LearnSmart, a personalized adaptive learning program, and a media-rich eBook.

14

Evidence of Evolution

BEFORE YOU BEGIN

Take a few minutes to recall

The "Vice Versa" of Animals and Plants

Adaptations provide powerful evidence for evolution. Bacteria that are able to survive and reproduce in the presence of an antibiotic have become adapted to their environment. Penguins are birds adapted to swimming in the ocean, and bats are mammals that can fly due to wings made of skin stretched over long fingers.

Insects are adapted to taking nectar from particular plants. It might seem as if bees go to all flowers, but they don't. They prefer sweet-smelling flowers with ultraviolet shadings that lead them to where nectar, a surgery liquid that serves as food, can be found. The bee feeding apparatus is a long, specialized tongue, called a proboscis, that is just the right size to reach down into a narrow floral tube where the nectar is located. As the bee goes about the business of feeding, pollen clings to its hairy body, and then as the bee moves from flower to flower of the species to which it is adapted, the pollen is distributed. Why does a flower provide the bee with nectar? By providing bees with nectar, flowers are helping to ensure their reproduction.

The orchid *Ophrys elegans* has a unique appearance that causes a bumblebee to visit it. The center of the flower looks like a female bumblebee is resting there. Actually, this is due to a petal that resembles a bumblebee. Occasionally, a male bee tries to mate with the petal, and when it does, it gets dusted with pollen, which the male bee takes to the next flower of this species.

Butterflies tend to feed from colorful composite flowers that provide them with a flat landing platform. Each individual flower of the composite has a floral tube that allows the long, thin butterfly proboscis to reach the nectar. Hummingbirds flap their wings rapidly—called hovering—in order to remain in one spot while they feed during the day from odorless, red

proboscis of honeybee

landing platform

floral tube

Honeybee-pollinated flower

packet of pollen

petal resembles
a female bumblebee

Bumblebee-pollinated flower

flat landing
platform

Butterfly-pollinated
flower

hummingbirds
hover

long thin beak

floral tube with
curved back margins

Hummingbird-pollinated flower

proboscis

hawk moth hovers

Hawk moth-pollinated flower

flowers that curve backward. A hummingbird's long, thin beak can access the nectar through a slender floral tube.

Moth-pollinated flowers are white, pale yellow, or pink—colors that are visible at night, when moths are active. The flowers give off a strong, sweet perfume that attracts moths, which hover as they extend a long, thin proboscis to gather the nectar at the base of a floral tube. The Madagascar star orchid (*Angraecum sesquipedale*) has a very long floral tube that holds its nectar much like a long, thin goblet would hold a drink. When Darwin first saw a picture of this orchid, he exclaimed, "What insect could suck it?" Later, he said in his book on orchids, "In Madagascar there must be moths with proboscises capable of an extension to ten and eleven inches [25.4 cm–27.7 cm]!" Many were skeptical, but Darwin was vindicated when in 1903, the zoologists Lionel Walter Rothschild and Karl Jordan discovered a large hawk moth living in Madagascar that has a proboscis 25–30 cm in length. As the hawk moth approaches the flower, it unrolls its proboscis and inserts it into the floral tube in order to feed.

We begin our study of evolution in this chapter by examining the work of Charles Darwin, who provided evidence that evolution consists of descent from a common ancestor and adaptation to the environment. Further, Darwin offered a mechanism for evolution he called natural selection. He called it natural selection because the environment, in a sense, chooses which members of a population reproduce, and in that way, adaptation to the environment is eventually achieved.

Darwin Collected Evidence of Evolution

Other scientists before Darwin hypothesized that evolution occurs, but developed no mechanism. Darwin concluded that evolution occurs when taking a trip around the world as a naturalist aboard the HMS *Beagle*. After studying artificial selection and the work of Thomas Malthus, Darwin—and later Alfred Wallace—suggested natural selection as a mechanism for evolution.

14.1 Observations support and give evidence for evolution

LEARNING OUTCOMES

When you complete this section, you should be able to

1. Identify two early biologists who suggested evolution occurs and criticize the mechanism proposed by Lamarck.
2. Describe Darwin's trip and some of the observations he made.

Biologists before Darwin had suggested that evolution occurs. Georges Cuvier, who founded the science of **paleontology,** the study of fossils, knew that fossils showed a succession of different life-forms through time (Fig. 14.1A). He hypothesized that a series of past catastrophes (local extinctions) had occurred and that after each one, a region was repopulated by species from surrounding areas. The result of all these catastrophes was change appearing over time.

In contrast to Cuvier, Jean-Baptiste de Lamarck, an invertebrate biologist, concluded on the basis of fossil evidence that more complex organisms are descended from less complex organisms. To explain the process of adaptation to the environment, Lamarck offered the idea of *inheritance of acquired characteristics,* which proposes that use and disuse of a structure can bring about inherited change. One example Lamarck gave—and the one for which he is most famous—is that the long neck of a giraffe developed over time because animals stretched their necks to reach food high in trees and then passed on a longer neck to their offspring (Fig. 14.1B).

Neither Cuvier nor Lamarck arrived at a satisfactory explanation for the evolutionary process. The inheritance of acquired characteristics has never been substantiated by experimentation. For example, if acquired characteristics were inherited, people who use tanning machines would have tan children, and people who have LASIK surgery to correct their vision would have children with perfect vision.

However, in December 1831, a new chapter in the history of biology began. A 22-year-old naturalist, Charles Darwin (1809–1882), set sail on the journey of a lifetime aboard the British naval vessel HMS *Beagle* (Fig. 14.1C). Darwin's primary mission on this journey around the world was to expand the navy's knowledge of natural resources in foreign lands. The captain of the *Beagle,* Robert Fitzroy, also hoped that Darwin would find evidence to support the biblical account of creation. Contrary to Fitzroy's wishes, Darwin amassed observations that would eventually support another way of thinking and change the history of science and biology forever.

During the trip, Darwin made numerous observations. For example, he noted that the rhea of South America was suited to living on a plain and looked like the ostrich that lived in Africa. However, the rhea was not an ostrich. Why not? Because the rhea evolved in South America, while the ostrich evolved in Africa. Darwin also found that species varied according to whether they lived in the Patagonian desert or in a lush tropical rain forest. Then, too, unique animals lived only on the Galápagos Islands, located off the coast of South America, not on the mainland. A marine iguana had large claws that allowed it to cling to rocks and a snout that enabled it to eat algae off rocks. One type of finch, lacking the long bill of a woodpecker, used a cactus spine to probe for insects. Why were these animals found only in the Galápagos Islands? Had they evolved there?

When Darwin explored the region that is now Argentina, he saw raised beaches for great distances along the

FIGURE 14.1B Lamarck thought the long neck of a giraffe was due to continued stretching in each generation.

FIGURE 14.1A One of the animals that Cuvier reconstructed from fossils was the mastodon.

Rhea

Patagonian desert

Earth's strata contain fossils

Charles Darwin, age 31

Tropical rain forest

Woodpecker finch

Marine iguana

FIGURE 14.1C *Middle:* Charles Darwin took a trip around the world aboard the HMS *Beagle*. The map shows where Darwin observed unique animals and ecosystems.

coast. He thought it would have taken a long time for such massive movements of the Earth's crust to occur. While Darwin was making geologic observations, he also collected fossils that showed today's plants and animals resemble, but are not exactly like, their forebears. Darwin had brought Charles Lyell's *Principles of Geology* on the *Beagle* voyage. This book said that weathering causes erosion and that, thereafter, dirt and rock debris are washed into the rivers and transported to oceans. When these loose sediments are deposited, layers of soil called **strata** (sing., stratum) result. The strata, which often contain fossils, are uplifted from below sea level to form land. Lyell's book went on to support a **uniformitarianism**

hypothesis, which states that geologic changes occur at a uniform rate. This idea of slow geologic change is still accepted today, although modern geologists have concluded that rates of change have not always been uniform. Darwin was convinced that the Earth's massive geologic changes are the result of slow processes and that, therefore, in contrast to thought at that time, the Earth was old enough to have allowed *evolution* to occur.

► **14.1 CHECK YOUR PROGRESS**
   1. Explain the significance of Darwin relying on evidence to arrive at a conclusion.

LEARNING OUTCOMES

When you complete this section, you should be able to

1. Explain how it benefited Darwin to know about artificial selection and also Wallace's conclusion.

2. List four essential components of Darwin's theory of natural selection.

Darwin made a study of **artificial selection,** a process by which humans choose, on the basis of certain traits, the animals and plants that will reproduce. For example, foxes are very shy and normally shun the company of people, but in 40 years time, Russian scientists have produced silver foxes that now allow themselves to be petted and even seek attention (Fig. 14.2A). They did this by selecting the most docile animals to reproduce. The scientists noted that some physical characteristics changed as well. The legs and tails became shorter, the ears became floppier, and the coat color patterns changed. Artificial selection is only possible because the original population exhibits a range of characteristics, allowing humans to select which traits they prefer to perpetuate.

To take another example, several varieties of vegetables can be traced to a single ancestor that exhibits various characteristics. Chinese cabbage, brussels sprouts, and kohlrabi are all derived from one species of wild mustard (Fig. 14.2B). Cabbage was produced by selecting for reproduction only plants that had overlapping leaves; brussels sprouts came from crossing only plants with certain types of buds; and kohlrabi was produced by crossing only the plants that had enlarged stems.

Darwin thought that a process of selection might occur in nature without human intervention. Using the process of artificial selection helped him arrive at the mechanism of natural selection, which allows evolution to occur.

Natural Selection Darwin was very much impressed by an essay written by Thomas Malthus about the reproductive potential of humans. Malthus had proposed that death and famine

Chinese cabbage   Brussels sprouts   Kohlrabi

Wild mustard

FIGURE 14.2B These three vegetables came from the wild mustard plant through artificial selection.

are inevitable because the human population tends to increase faster than the supply of food. Darwin applied this concept to all organisms and saw that available resources were insufficient for all members of a population to survive. For example, he calculated the reproductive potential of elephants. Assuming a life span of about 100 years and a breeding span of 30–90 years, a single female probably bears no fewer than six young. If all these young survive and continue to reproduce at the same rate, after only 750 years, the descendants of a single pair of elephants would number about 19 million! Each generation has the same reproductive potential as the previous generation. Therefore, Darwin hypothesized, there is a constant struggle for existence, and only certain members of a population survive and reproduce in each generation. What members might those be? The members that have some advantage and are best able to compete successfully for limited resources.

Applying Darwin's thinking to giraffes, we can see that long-necked giraffes would be better able to feed off leaves in trees than short-necked giraffes. The longer neck gives giraffes an advantage that, in the end, would allow them to produce more offspring than short-necked giraffes. So, eventually, all the members of a giraffe population (individuals of a species in one locale) would have long necks. Or, what about bacteria living in an environment of antibiotics? The few bacteria that can survive in this environment have a tremendous advantage, and therefore their offspring will make up the next generation of bacteria, and this strain of bacteria will be resistant to the antibiotic.

FIGURE 14.2A Artificial selection has produced domesticated foxes.

Darwin called the process by which organisms with an advantage reproduce more than others of their kind **natural selection** because some aspect of the environment acts as a **selective agent** and chooses the members of the population with the advantageous phenotype to reproduce more than the other members.

Natural selection has these essential components:

- *The members of a population have inheritable variations.* For example, a wide range of differences exists among the members of a population. Many of these variations are inheritable. Inheritance of variations is absolutely essential to Darwin's hypothesis, even though he did not know the means by which inheritance occurs.
- *A population is able to produce more offspring than the environment can support.* The environment contains only so much food and water, places to live, potential mates, and so forth. The environment can't support all the offspring that a population can produce, and each generation is apt to be too large for the environment to support.
- *Only certain members of the population survive and reproduce.* Certain members have an advantage suited to the environment that allows them to capture more resources than other members, as when long-necked giraffes are better able to browse on tree leaves. This advantage allows these members of the population to survive and produce more offspring. Today, this is called differential reproduction or **fitness** because the better adapted an individual is, the more likely they are to have offspring.
- *Natural selection results in a population adapted to the local environment.* In each succeeding generation, an increasing proportion of individuals will have the adaptive characteristics—the characteristics suited to surviving and reproducing in that environment (Fig. 14.2C).

Now it is possible to form a definition of evolution. **Evolution** consists of changes in a population over time due to the accumulation of inherited characteristics. Evolution explains the unity and diversity of organisms. "Unity" means organisms share the same characteristics of life because they share a common ancestry, traceable even to the first cell or cells. "Diversity" comes about because each type of organism (each species) is adapted to one of the many different environments in the biosphere (e.g., oceans, deserts, mountains).

Independently, Alfred Wallace also arrived at natural selection as a mechanism for evolution, as explained next.

Alfred Russel Wallace Like Darwin, Alfred Russel Wallace (1823–1913) was a naturalist. While he was a schoolteacher at Leicester in 1844–1845, he met Henry Walter Bates, a biologist who interested him in insects. Together, they went on a collecting trip to the Amazon that lasted several years. Wallace's knowledge of the world's flora and fauna was further expanded by a tour he made of the Malay Archipelago from 1854–1862. Later, he divided the islands into a western group and an eastern group on the basis of their different plants and animals. The dividing line between these islands is a narrow but deep strait now known as the Wallace line.

FIGURE 14.2C The brightly colored tree frog can hide among tropical plants where the large red eyes confuse predators. The frog climbs trees and other plants assisted by toes with suction cups.

Just as Darwin had done, Wallace wrote articles and books that clearly showed his belief that species change over time and it was possible for new species to evolve. Later, he said he had pondered for many years about a mechanism to explain the origin of a species. He, too, had read Malthus's essay on human population increases, and in 1858, while suffering an attack of malaria, the idea of "survival of the fittest" came upon him. He quickly completed an essay outlining a natural selection process, which he chose to send to Darwin for comment. Darwin was stunned upon its receipt. Here before him was the hypothesis he had formulated as early as 1844, but never published. Darwin told his friend and colleague Charles Lyell that Wallace's ideas were so similar to his own that even Wallace's "terms now stand as heads of my chapters" in the book he had begun in 1856.

Darwin suggested that Wallace's paper be published immediately, even though he himself had nothing in print yet. Lyell and others who knew of Darwin's detailed work substantiating the process of natural selection suggested that a joint paper be read to the Linnean Society. The title of Wallace's section was "On the Tendency of Varieties to Depart Indefinitely from the Original Type." Darwin allowed the abstract of a paper he had written in 1844 and an abstract of his book *On the Origin of Species* to be read. This book was published in 1859.

By now, evolution by natural selection has been supported by so many observations and experiments that it is considered a theory rather than a hypothesis. Modern investigators have shown that it is possible to observe the process of natural selection, as described in the application "Natural Selection Can Be Witnessed" on page 263.

▶ **14.2 CHECK YOUR PROGRESS**
  1. Explain in four steps how a population becomes adapted to the environment.

More Evidence of Evolution

The evidence for evolution is categorized according to its source. Evidence for common descent is based on fossils, comparative anatomy, biogeography, and molecular observations.

14.3 Fossils provide powerful evidence for common descent

LEARNING OUTCOME

When you complete this section, you should be able to

1. Explain why transitional fossils provide powerful evidence for common descent.

The best evidence for evolution comes from **fossils,** the physical remains of organisms that lived on Earth between 10,000 and billions of years ago. Usually, when an organism dies, the soft parts are either consumed by scavengers or decomposed by bacteria. This means that most fossils consist of hard parts, such as shells, bones, or teeth, because these are usually not consumed or destroyed. Fossils are also the traces of past life, such as trails, footprints, burrows, worm casts, or even preserved droppings. Or fossils can also be such items as pieces of bone, impressions of plants pressed into shale, organisms preserved in ice, and even insects trapped in tree resin (which we know as amber).

More and more fossils have been found because paleontologists have been out in the field looking for them (Fig. 14.3A*a*). Weathering and erosion of rocks produces an accumulation of particles that vary in size and nature and are called sediment. This process, called sedimentation, has been going on since the Earth was formed, and can take place on land or in bodies of water. Sediment becomes a stratum, a recognizable layer in a sequence of layers. Any given stratum is older than the one above it and younger than the one immediately below it (Fig. 14.3A*b*). For a fossil to be encased by rock, the remains are first buried in sediment; then the hard parts are preserved by a process called mineralization; and finally, the surrounding sediment hardens to form rock.

FIGURE 14.3B Fossils are carefully cleaned, and organisms are reconstructed.

Usually, **paleontologists** remove fossils from layers of sediment called strata to study them in the laboratory (Fig. 14.3B), and then they may decide to exhibit them. The **fossil record** is the history of life recorded by fossils and the most direct evidence we have that evolution has occurred. The species found in ancient sedimentary rock are not the species we see about us today.

Darwin relied on fossils to formulate his theory of evolution, and today we have a far more complete record than was available to Darwin. The record tells us that, in general, life has progressed from the simple to the complex. Unicellular prokaryotes are

FIGURE 14.3A a. Paleontologists carefully remove fossils for further study. **b.** The deeper the strata (sing., stratum), the older the fossils found there.

a.

b.

strata

the first signs of life in the fossil record, followed by unicellular eukaryotes and then multicellular eukaryotes. Among the latter, fishes evolved before terrestrial plants and animals. On land, non-flowering plants preceded the flowering plants, and amphibians preceded the reptiles, including the dinosaurs. Dinosaurs are directly linked to the evolution of birds, but only indirectly linked to the evolution of mammals, including humans.

Transitional Fossils Darwin used the phrase "descent with modification" to explain evolution. Because of descent, all organisms can trace their ancestry to an original source. For example, you and your cousins have a **common ancestor** in your grandparents and also in your great grandparents, and so forth. In the end, one couple can give rise to a great number of descendants that look quite different.

A **transitional fossil** is either a common ancestor for two different groups of organisms or an individual closely related to the common ancestor for these groups. Transitional fossils allow us to trace the descent of organisms. Even in Darwin's day, scientists knew of the *Archaeopteryx lithographica* fossil, which is an

FIGURE 14.3C Fossil of *Archaeopteryx* (top) and an artist's representation (bottom).

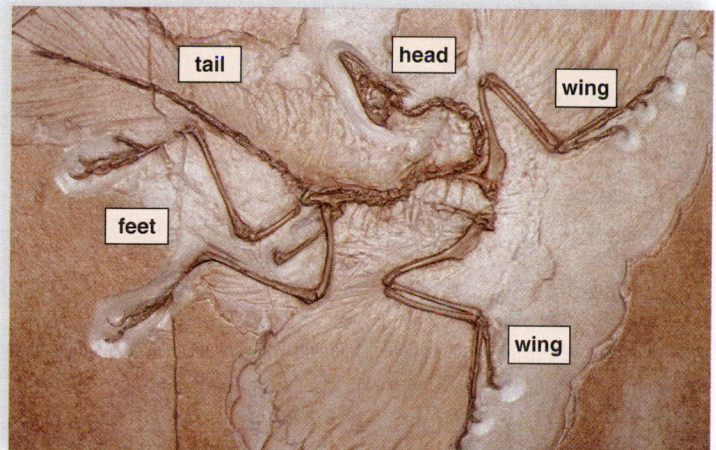

Archaeopteryx fossil

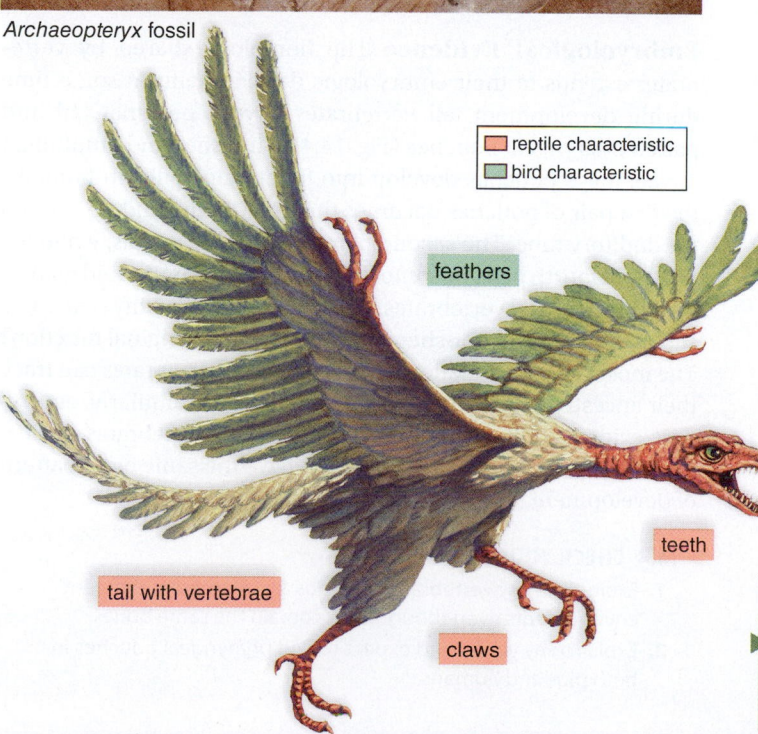

reptile characteristic
bird characteristic

feathers

tail with vertebrae

claws

teeth

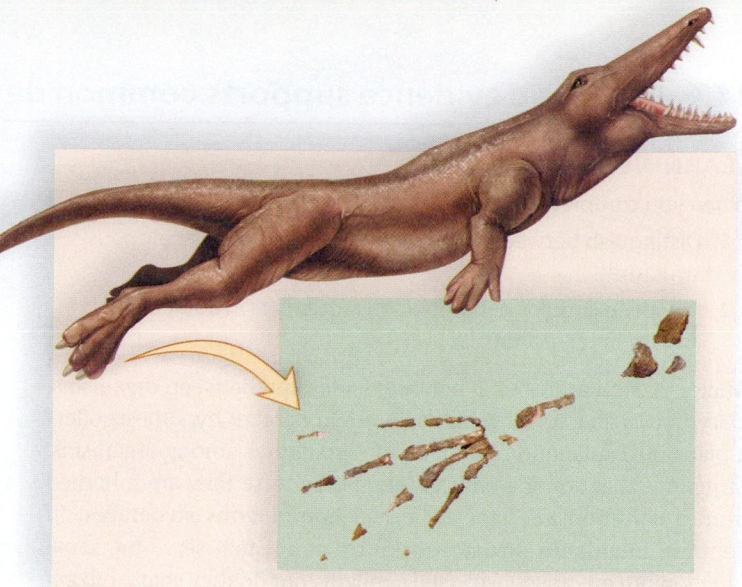

FIGURE 14.3D *Ambulocetus natans,* an ancestor of the modern toothed whale, and its fossil remains.

intermediate between reptiles and birds. The dinosaur-like skeleton of these fossils has reptilian features, including jaws with teeth, and a long, jointed tail, but *Archaeopteryx* also had feathers and wings. Figure 14.3C shows a fossil of *Archaeopteryx* along with an artist's representation of the animal based on the fossil remains. Many more prebird fossils have been discovered in China. These fossils are progressively younger than *Archaeopteryx:* The skeletal remains of *Sinornis* suggest it had wings that could fold against its body like those of modern birds, and its grasping feet had an opposable toe, but it still had a tail. Another fossil, *Confuciusornis,* had the first toothless beak. A third fossil, called *Iberomesornis,* had a breastbone to which powerful flight muscles could attach. Such fossils show how the bird of today evolved.

Scientists had always thought whales had terrestrial ancestors. Now, fossils have been discovered that support this hypothesis (see Fig. 15.1A). *Ambulocetus natans* (meaning "the walking whale that swims") was the size of a large sea lion, with broad, webbed feet on its forelimbs and hindlimbs that enabled it to both walk and swim. It also had tiny hoofs on its toes and the primitive skull and teeth of early whales. Figure 14.3D is an artist's re-creation, based on fossil remains of *Ambulocetus,* which lived in freshwater streams. An older genus, *Pakicetus,* was primarily terrestrial, and yet also had the dentition of an early toothed whale. A younger genus, *Rodhocetus,* had reduced hindlimbs that would have been no help for either walking or swimming, but may have been used for stabilization during mating.

The origin of mammals is also documented. The synapsids are proto-mammals whose descendants diversified into different types of premammals. Slowly, mammal-like fossils acquired skeletal features that adapted them to live more efficiently on land. For example, the legs projected downward rather than to the side as in reptiles. The earliest true mammals were shrew-sized creatures that have been unearthed in fossil beds about 200 million years old.

▶ **14.3 CHECK YOUR PROGRESS**
1. Explain what "common descent" means.
2. Identify two transitional fossils that support common descent.

14.4 Anatomic evidence supports common descent

LEARNING OUTCOMES

When you complete this section, you should be able to

1. Distinguish between homologous structures and analogous structures.
2. Identify examples of homologous structures in adult animals and embryonic animals.

Anatomic similarities exist between fossils and between organisms. Darwin was able to show that a common descent hypothesis offers a plausible explanation for anatomic similarities among organisms. Structures that are anatomically similar because they are inherited from a common ancestor are called **homologous structures.** In contrast, **analogous structures** are structures that serve the same function but are not constructed similarly, nor do they share a *recent* common ancestor. The wings of birds and insects and the eyes of octopuses and humans are analogous structures. The presence of homology, not analogy, is evidence that organisms are closely related. Studies of comparative anatomy and embryologic development and increasingly genetic data reveal homologous structures. For example, we now know that the gene *Pax6* (see section 12.4, page 220.) initiates the development of the eye in diverse organisms. Does this mean that all eyes are homologous structures instead of analogous ones?

Comparative Anatomy Vertebrate forelimbs are used for flight (birds and bats), orientation during swimming (whales and seals), running (horses), climbing (arboreal lizards), or swinging from tree branches (monkeys). Yet, despite their dissimilar functions all vertebrate forelimbs contain the same sets of bones organized in similar ways (Fig. 14.4A). The most plausible explanation for this unity is that the basic forelimb plan belonged to a common ancestor for all vertebrates, and then the plan was modified as each type of vertebrate continued along its own evolutionary pathway.

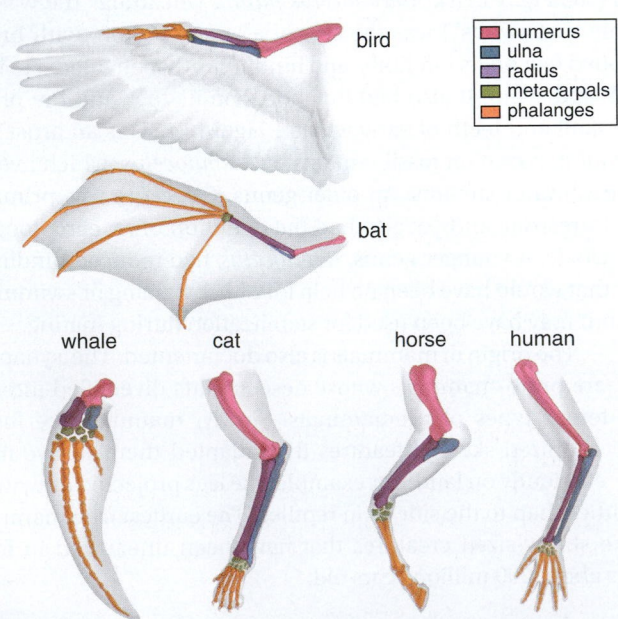

FIGURE 14.4A Despite differences in structure and function, vertebrate forelimbs have the same bones.

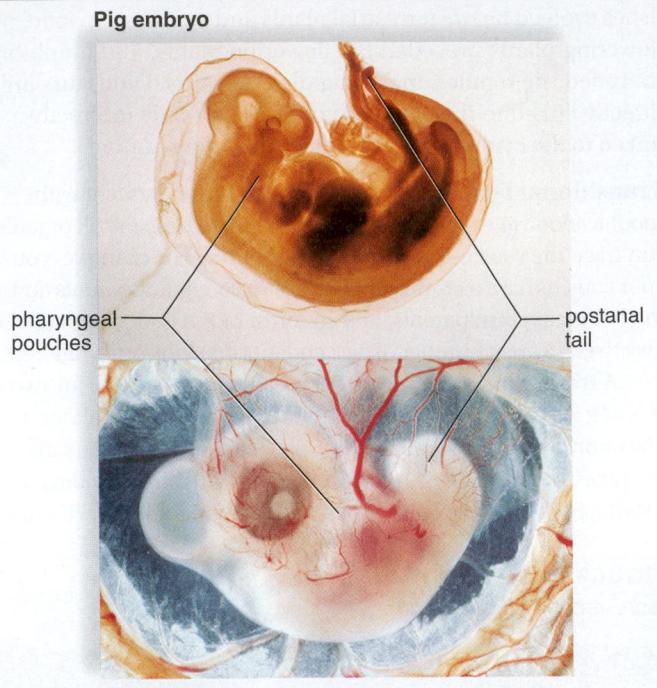

Pig embryo

pharyngeal pouches — postanal tail

Chick embryo

FIGURE 14.4B Vertebrate embryos have features in common, even though they have different appearances as adults.

Vestigial structures are fully developed in one group of organisms but reduced and possibly nonfunctional in similar groups. For example, modern whales have a vestigial pelvic girdle and legs because their ancestors walked on land. Similarly, snakes have no use for hindlimbs, and yet some have remnants of a pelvic girdle and legs. Humans have a tailbone but no tail. The presence of vestigial structures can be explained by the common descent hypothesis: Vestigial structures occur because organisms inherit their anatomy from their ancestors; they are traces of an organism's evolutionary history.

Embryological Evidence The homology shared by vertebrates extends to their embryologic development. At some time during development, all vertebrates have a postanal tail and paired pharyngeal pouches (Fig. 14.4B). In fishes and amphibian larvae, these pouches develop into functioning gills. In humans, the first pair of pouches becomes the cavity of the middle ear and the auditory tube. The second pair becomes the tonsils, while the third and fourth pairs become the thymus and parathyroid glands. Why do terrestrial vertebrates develop and then modify structures such as pharyngeal pouches that have lost their original function? The most likely explanation is that terrestrial vertebrates can trace their ancestry to amphibians and then to fishes. Similarly, embryonic evidence tells us that echinoderms and vertebrates share a common ancestor because they both have the same early pattern of development (see Fig. 20.2A).

▶ **14.4 CHECK YOUR PROGRESS**

1. Exemplify how vertebrate forelimbs are adapted to different environments even though they contain the same bones.
2. Explain why you would expect to find pharyngeal pouches in both pigs and humans.

14A Natural Selection Can Be Witnessed

Darwin formed his natural selection hypothesis, in part, by observing the adaptations of tortoises and finches on the Galápagos Islands. Tortoises with domed shells and short necks live on well-watered islands, where grass is available. Those with shells that flare up in front have long necks and are able to feed on tall cacti. They live on arid islands, where treelike prickly-pear cactus is the main food source. Similarly, the islands are home to many different types of finches (Fig. 14A.1). The heavy beak of the large, ground-dwelling finch is suited to a diet of seeds. The beak of the warbler-finch is suited to feeding on insects found among ground vegetation or caught in the air. The longer, somewhat more pointed beak and split tongue of the cactus-finch are suited for probing cactus flowers for nectar.

Video Galápagos Finches

Beak Size and Natural Selection

Today, investigators, such as Peter and Rosemary Grant of Princeton University, are actually watching natural selection as it occurs. In 1973, the Grants began a study of the various finches on Daphne Major, near the center of the Galápagos Islands. The weather swung widely back and forth from wet years to dry years, and they found that the beak size of the medium ground finch, *Geospiza fortis*, adapted to each weather swing, generation after generation (Fig. 14A.2). These finches like to eat small, tender seeds that require a smaller beak, but when the weather turns dry, they have to eat larger, drier seeds, which are harder to crush. Then, the birds that have a larger beak depth have an advantage and produce more offspring. Dry weather acts as a selective agent for a *G. fortis* beak size that has more depth than in the previous generation.

Video Finches Natural Selection

Silent Crickets and Natural Selection

A research team led by Marlene Zuk, a professor of biology, reported that prior to 2001 the Hawaiian field cricket population (*Teleogryllus oceanicus*) on the island of Kauai contained very few silent males. Chirping males have a wing structure that produces the chirping sound that attracts females. By 2006, over 90% of male crickets were silent because their wings were flat and unable to produce the chirping sound. In just 20 generations, the population had undergone this dramatic evolutionary change due to a particular selective agent that caused the silent phenotype to be advantageous. A deadly parasitic fly (*Ormia ochracea*) uses the male crickets' chirping as a way to locate them. The fly deposits her eggs on a male cricket's back, and they develop into maggots (Fig. 14A.3). Over a week's time, the maggots eat the cricket's internal organs and then emerge from its dead body to undergo metamorphosis into adult flies. The silent males were not parasitized and increased in number because many of them mated with females and passed on their genes. How did they do it? The silent males wait near any remaining chirping males and intercept incoming females. Normally, female crickets will not accept a male until he completes a final mating song, but even that is beginning to change and females will now accept silent males, thereby allowing them to increase in number.

CONSIDER THESE QUESTIONS

1. Cite some examples of natural selection that involve resistance to drugs or pesticides. Now that we know how powerful

A ground-dwelling finch feeds on seeds.

A cactus-finch probes flowers for nectar.

FIGURE 14A.1 Finches on the Galápagos Islands.

A warbler-finch feeds on insects.

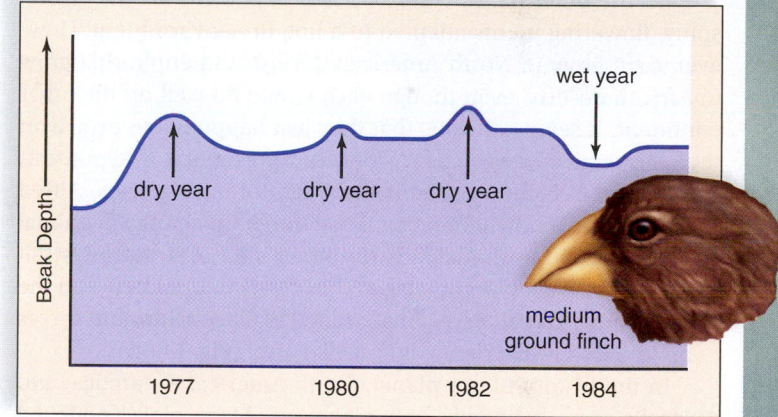

FIGURE 14A.2 The beak size of a ground finch varies from generation to generation, according to the weather.

FIGURE 14A.3 Maggots feeding on a cricket.

natural selection can be, how can we keep resistance from happening?

2. Today's drug designers biochemically create drugs for a particular purpose, and then they keep testing and selecting which ones to improve. Is this natural selection at work?

connect **BIOLOGY**   Explore the concepts through a variety of multimedia assets and question types.

www.mcgrawhillconnect.com

14.5 Biogeographic evidence supports common descent

LEARNING OUTCOME

When you complete this section, you should be able to

1. Explain how biogeography supports common descent.

Biogeography is the study of the distribution of plants and animals in different places throughout the world. Because organisms evolve in particular locales, you expect a different mix of plants and animals whenever geography separates continents, islands, or seas. As mentioned, Darwin noted that South America lacked rabbits. He concluded that rabbits evolved somewhere else and had no means of reaching South America. Instead, the Patagonian hare is adapted to the environment of a rabbit and resembles a rabbit in anatomy and behavior but has the face of a guinea pig, from which it probably evolved in South America. To take another example, on the Galápagos Islands in Darwin's day, short-necked tortoises inhabited islands with lush vegetation, and long-necked tortoises inhabited dry islands where cacti are abundant (Fig. 14.5A). Their long necks helped them feed on the cacti. When Darwin observed these tortoises, he wondered if these two types of tortoises adapted to different isolated environments had evolved from the same mainland tortoise.

On the other hand, both cacti and euphorbia are succulent, spiny, flowering plants adapted to a hot, dry environment. However, cacti grow in North American deserts and euphorbia grow in African deserts, even though each would do well on the other continent. It seems obvious that they just happened to evolve on their respective continents. Aside from presenting a hypothesis that natural selection explains the origin of new species, Alfred Wallace is well known for recognizing the sharp change in animal species inhabiting the islands on either side of a narrow strait bisecting the Malay Archipelago. This deep channel between the Oriental and Australian regions is called Wallace's line, and serves as an impassable barrier to animal dispersal (Fig. 14.5B).

In the history of our planet, South America, Antarctica, and Australia were originally one continent. Marsupials (pouched mammals) arose at around the time Australia separated and drifted away on its own. Isolation allowed marsupials to diversify

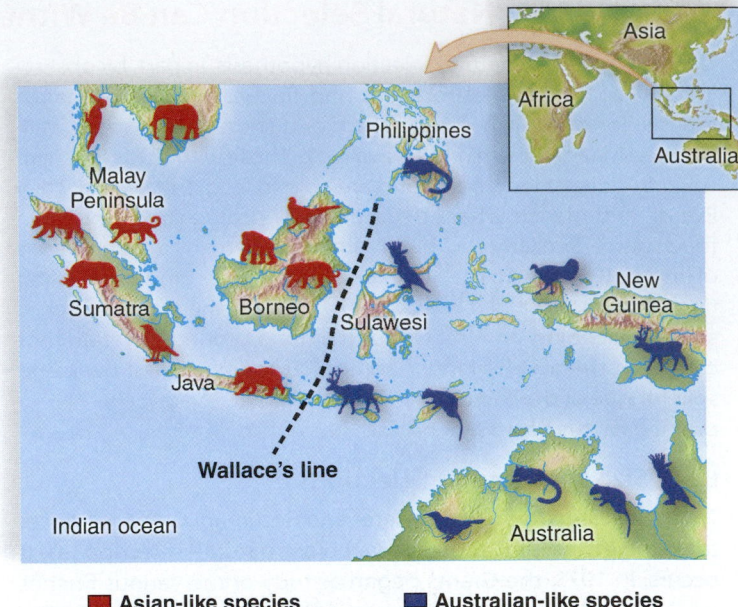

■ **Asian-like species**   ■ **Australian-like species**

FIGURE 14.5B Wallace's line (a deep channel) divides the Malay Archipelago into two regions, whose animal populations evolved separately when exposed to different environmental pressures.

into many different forms suited to various environments of Australia. They were free to do so because there were few, if any, placental (modern) mammals in Australia. In South America, where there are placental mammals, marsupials are present but they are not as diverse. These examples support the hypothesis that evolution is influenced by the physical environment and the mix of plants and animals that happen to be there—that is, by biogeography.

▶ **14.5 CHECK YOUR PROGRESS**
1. Explain the occurrence of two unrelated species with the same characteristics on separate continents.

a.

b.

FIGURE 14.5A In the Galápagos Islands, (**a**) tortoises with short necks evolved on islands with lush vegetation, and (**b**) tortoises with long necks evolved on dry islands where cacti are the chief source of food.

LEARNING OUTCOMES

When you complete this section, you should be able to

1. Identify three types of molecular homologies.
2. Explain why sequencing proteins can show relatedness.
3. Understand how *Hox* genes can function effectively in so many different types of organisms.

With the expansion of molecular studies in the past and present century, the molecular evidence for common descent has grown exponentially. Molecules that are similar because they are inherited from a common ancestor are now called **molecular homologies.** Here we will divide our examples of molecular homologies into three categories:

Similar molecules occur in all cells All cells contain the same types of biomolecules and they interact similarly. This makes it possible for researchers to study biochemical pathways in various model organisms and be assured their findings will apply to most other organisms. For example, almost all organisms use DNA to store genetic information and the same triplet code which specifies the sequence of amino acids in a protein is passed onto mRNA.

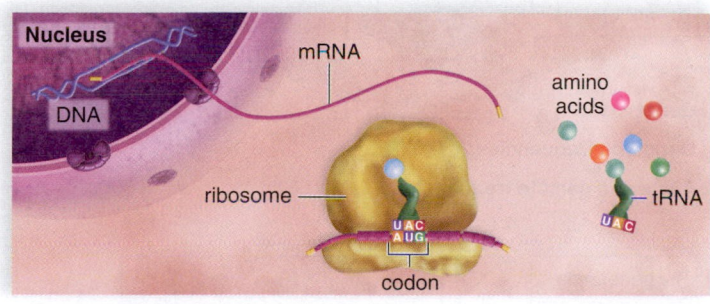

Cellular respiration depends on nearly the same enzymes in all organisms to break down glucose, and the released energy is captured by the same electron transport chain before ATP molecules are produced by ATP synthase. This provides evidence that these processes evolved early in the history of life and were thereafter passed from common ancestor to common ancestor to all life-forms.

Similar genes and metabolic proteins occur in diverse species Now that we can sequence genes and proteins, we are aware that many different types of organisms contain the same genes called homologous genes and proteins. Cytochrome *c* is a small protein in the electron transport chain of organisms. Researchers find that the sequence of amino acids in cytochrome *c* is similar from yeast to humans but not exactly because mutations occurred in each line of descent. The number of amino acid differences correlates how long ago organisms shared a common ancestor and therefore signifies their degree of relatedness. For example, the number of amino acid differences between humans and yeast is much larger than between humans and monkeys. To take another example, the p53 protein you will recall helps prevent cancer by initiating repair of damaged DNA during the cell cycle. Two species of monkeys differ from human p53 protein by the same number of amino acid differences, showing that their common ancestor was recent. The number of amino acid differences in the p53 protein correlates with anatomical data that indicate the degree of relatedness between animals including humans.

Similar developmental genes occur in diverse species *Hox* genes are master developmental regulatory genes that were first discovered in fruit flies. They are master regulatory genes because they turn on other developmental genes, and in this way they have a significant effect on the outcome of development.

Much to the surprise of researchers, the same *Hox* genes have been found in diverse invertebrates and also in diverse mammals, including humans. *Hox* genes, which can be recognized by the presence of a short sequence of exact DNA bases, even occur in plants! The presence of *Hox* genes in so many different types of organisms clearly supports the concept of descent from a common source. *Hox* genes must have evolved in a very early common ancestor, but their activity has been modified, accounting for the many different body plans among organisms. Figure 14.6 shows that the neck of a giraffe did not get longer by adding more vertebrae; instead each bone grew longer due to the continued expression of genes that control bone formation.

▶ **14.6 CHECK YOUR PROGRESS**

1. Identify the way that sequencing genes or proteins can show relatedness.
2. Explain how master development genes can bring about different effects during the development of different types of organisms.

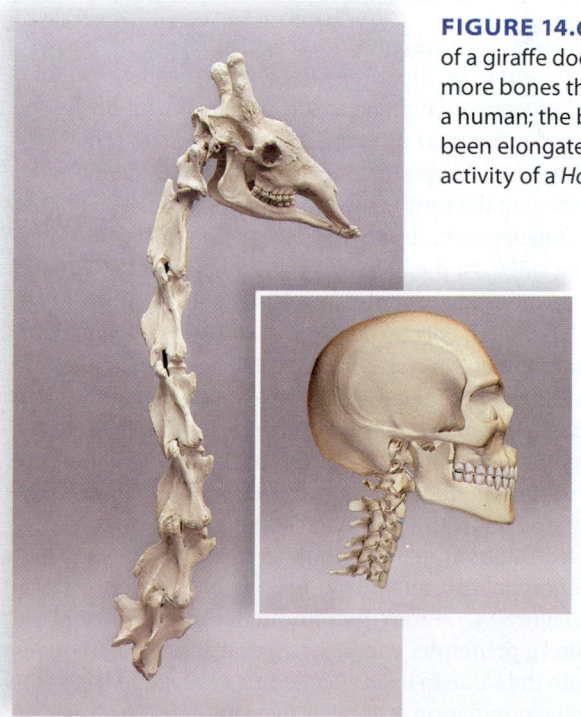

FIGURE 14.6 The neck of a giraffe doesn't have more bones than that of a human; the bones have been elongated due to the activity of a *Hox* gene.

Evidence Through Study of Populations

Microevolution is evolution beneath the species level. The Hardy-Weinberg principle states that gene pool frequencies in a population, calculated by using the equation $p^2 + 2pq + q^2$, will stay constant generation after generation, unless evolution occurs.

14.7 Gene pool frequency changes determine evolution

LEARNING OUTCOMES

When you complete this section, you should be able to

1. Explain how the Hardy-Weinberg principle tells us when microevolution has occurred.
2. Identify five conditions that must be met or else evolution occurs.

Not until the 1930s were population geneticists able to apply the principles of genetics to populations and thereafter develop a way to recognize when microevolution has occurred. The **gene pool** of a population is composed of all the alleles in all the individuals making up the population. *When the allele frequencies for a population change, microevolution has occurred.* Microevolution does not necessarily result in a visible change but let's take an example that does.

A peppered moth can be light-colored or dark-colored. Suppose you decide that D = dark color and d = light color. Furthermore, you find that in one Great Britain population, only 4% (0.04) of the moths are homozygous dominant (DD), 32% (0.32) are heterozygous (Dd), and 64% (0.64) are homozygous recessive (dd). From these genotype frequencies, you can calculate the allele frequencies in the population:

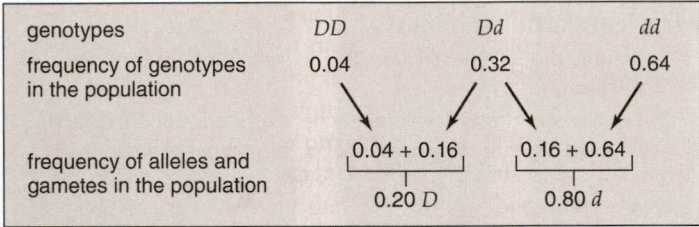

The frequency of the gametes (sperm and egg) produced by this population will necessarily be the same as the allele frequencies. And, we can use these gamete frequencies to calculate the ratio of genotypes in the next generation by using a Punnett square (Fig. 14.7). The results show that the genotype frequencies (and therefore the allele frequencies) in the next generation are the same as they were in the previous generation. In other words, the homozygous dominant moths are still 0.04; the heterozygous moths are still 0.32; and the homozygous recessive moths are still 0.64 of the population. This remarkable finding tells us that *sexual reproduction alone cannot bring about a change in genotype and allele frequencies.* Also, the dominant allele need not increase from one generation to the next. Dominance does not cause an allele to become a common allele.

The potential constancy, or equilibrium state, of gene pool frequencies was independently recognized in 1908 by G. H. Hardy, an English mathematician, and W. Weinberg, a German physician. They used the binomial equation ($p^2 + 2pq + q^2 = 1$) to calculate the genotype and allele frequencies of a population, as illustrated in Figure 14.7. From their findings, they formulated the **Hardy-Weinberg principle,** which states that gene pool frequencies will remain the same in each succeeding generation of a sexually reproducing population as long as five conditions are met:

1. *No mutations.* Allele changes do not occur, or changes in one direction are balanced by changes in the opposite direction.
2. *No gene flow.* Migration of alleles into or out of the population does not occur.
3. *Random mating.* Individuals pair by chance, not according to their genotypes or phenotypes.
4. *No genetic drift.* The population is very large, and changes in allele frequencies due to chance alone are insignificant.
5. No natural selection. No selective agent favors one genotype over another.

These conditions are rarely met, and gene pool frequencies do change from one generation to the next.

▶ **14.7 CHECK YOUR PROGRESS**
1. Identify when microevolution has occurred and the possible causes.

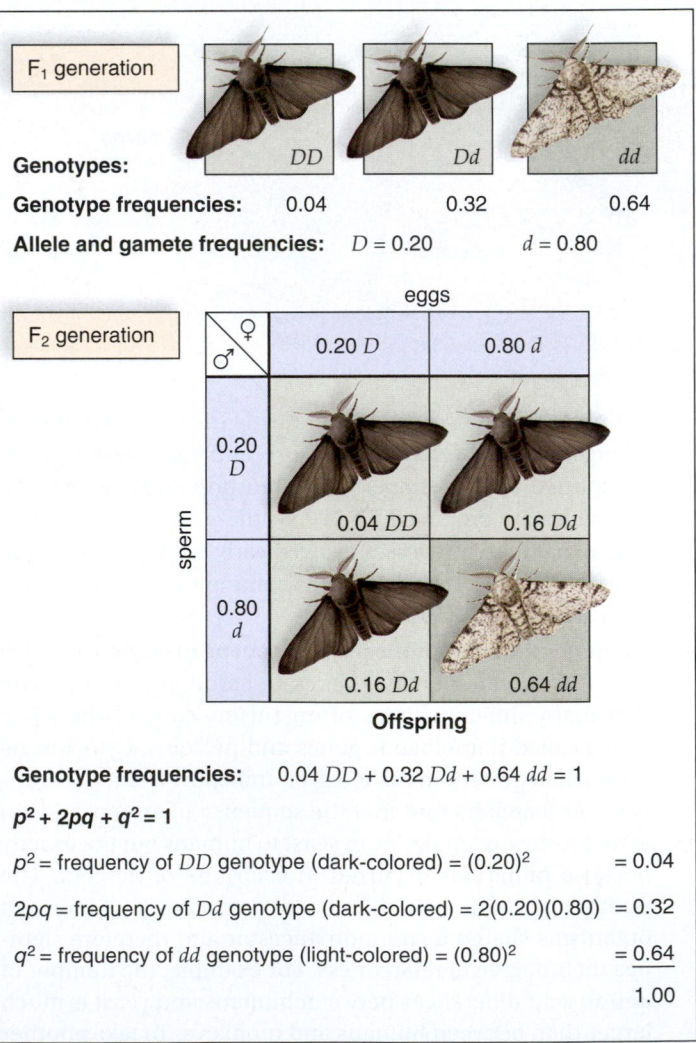

Genotype frequencies: 0.04 DD + 0.32 Dd + 0.64 dd = 1

$p^2 + 2pq + q^2 = 1$

p^2 = frequency of DD genotype (dark-colored) = $(0.20)^2$ = 0.04

$2pq$ = frequency of Dd genotype (dark-colored) = 2(0.20)(0.80) = 0.32

q^2 = frequency of dd genotype (light-colored) = $(0.80)^2$ = 0.64

1.00

FIGURE 14.7 Calculating gene pool frequencies for F_1 and F_2.

14.8 Both mutations and sexual recombination produce variations

LEARNING OUTCOME

When you complete this section, you should be able to

1. Recognize that mutations and sexual recombination are necessary to the process of evolution.

Mutations, which are permanent genetic changes, are the raw material for evolutionary change because, without mutations, there could be no inheritable phenotypic variations among members of a population. The rate of mutations is generally very low—on the order of 1 per 100,000 cell divisions. Also, it is important to realize that evolution is not goal-oriented, meaning that no mutation arises because the organism "needs" one. For example, the mutation that causes bacteria to be resistant was already present before antibiotics appeared in the environment.

Mutations are the primary source of genetic differences among prokaryotes that reproduce asexually. Generation time is so short that many mutations can occur quickly, even though the rate is low, and because these organisms are haploid, any mutation that results in a phenotypic change is immediately tested by the environment.

In diploid organisms, a recessive mutation can remain hidden and become significant only when a homozygous recessive genotype arises. The importance of recessive alleles increases if the environment is changing; it's possible that the homozygous recessive genotype could be helpful in a new environment, if not the present one. It's even possible that natural selection will maintain a recessive allele if the heterozygote has advantages (see section 14.13).

In sexually reproducing organisms, sexual recombination is just as important as mutation in generating phenotypic differences, because sexual recombination can bring together a new and different combination of alleles. This new combination might produce a more successful phenotype. Success, of course, is judged by the environment and counted by the relative number of healthy offspring an organism produces.

▶ **14.8 CHECK YOUR PROGRESS**

1. Identify which of the natural selection components (page 259) depend on mutations and sexual recombination.

14.9 Nonrandom mating and gene flow change gene pool frequencies

LEARNING OUTCOME

When you complete this section, you should be able to

1. Identify why nonrandom mating and gene flow can change gene frequencies.

Random mating occurs when individuals pair by chance. You make sure random mating occurs when you do a genetic cross on paper or in the lab, and cross all possible types of sperm with all possible types of eggs. **Nonrandom mating** occurs when only certain genotypes or phenotypes mate with one another. **Assortative mating** is a type of nonrandom mating that occurs when individuals mate with those having the *same* phenotype with respect to a certain characteristic. For example, flowers such as the garden pea usually self-pollinate—therefore, the same phenotype has mated with the same phenotype (Fig. 14.9A). Assortative mating can also be observed in human society. Men and women tend to marry individuals with characteristics such as intelligence and height that are similar to their own. Assortative mating causes homozygotes for certain gene loci to increase in frequency and heterozygotes for these loci to decrease in frequency.

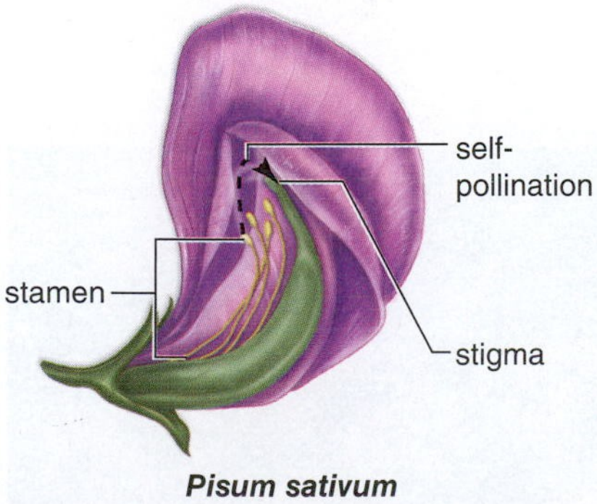

Gene flow, also called gene migration, is the movement of alleles between populations. When animals move between populations or when pollen is distributed between species (Fig. 14.9B), gene flow has occurred. When gene flow brings a new or rare allele into the population, the allele frequency in the next generation changes. When gene flow between adjacent populations is constant, allele frequencies continue to change until an equilibrium is reached. Therefore, continued gene flow tends to make the gene pools similar and reduce the possibility of allele frequency differences between populations.

▶ **14.9 CHECK YOUR PROGRESS**

1. Predict the effect of nonrandom mating and gene flow on the gene pool of populations.

FIGURE 14.9A The anatomy of the garden pea (*Pisum sativum*) ensures self-pollination and nonrandom mating.

self-pollination

stamen

stigma

Pisum sativum

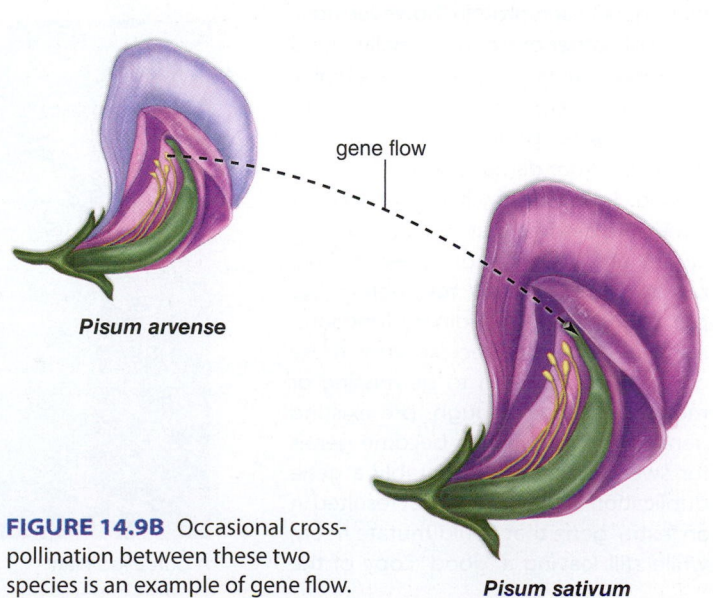

gene flow

Pisum arvense

Pisum sativum

FIGURE 14.9B Occasional cross-pollination between these two species is an example of gene flow.

14B Sometimes Mutations Are Beneficial

Imagine trying to redesign a vital mechanical part of an airplane, while still keeping that plane in flight. Sounds nearly impossible, doesn't it? This was one of the puzzles facing early evolutionary biologists. After all, mutations are the main way in which new traits and features arise during evolution, and yet most mutations cause damage. If a feature is important, how can it be altered while still allowing an organism and its offspring to survive?

Geneticists have shown one possible way mutations can accumulate without impairing present function: gene duplication (Fig. 14B.1). An extra (and possibly unused) copy of a gene may result from errors during cell division, efforts to repair breakage to DNA, or other mechanisms. The surprising idea here is that these seeming accidents actually can provide raw material for natural selection. Particularly in plants, many examples of gene duplication have been found—for example, the wild mustard plant has undergone at least two duplications of *all* its chromosomes in the past, as well as duplication of several individual genes at various times in history.

An intriguing example of gene duplication involves the sweet-tasting proteins. Of the thousands of proteins studied so far, most have no noticeable flavor—but about half a dozen have an intensely sweet taste. These rare, sweet-tasting proteins are found in plants and plant products from several different continents: The protein "curculin" is found in the fruit of a Malaysian herb (Fig. 14B.2); "mabinlin" can be extracted from a traditional Chinese herb; "thaumatin" is found in the fruit of a West African rain forest shrub; and "brazzein" comes from a fruit that grows wild in Gabon, Cameroon, and Zaire. Each of these proteins tastes sweet only to humans and certain monkeys. From the plant's point of view, the proteins likely provided an advantage: Sweeter fruits would be eaten more often and their seeds distributed more widely, ensuring the growth of more plants with genes for making sweet proteins. A question still remains: How did these unusual proteins come about?

No one yet knows exactly how these proteins originated, but gene duplication is a likely answer. The proteins look nothing alike, are found in unrelated plants, and clearly did not come from some ancient shared plant gene. Each protein, however, does resemble other proteins normally found in healthy plants. Brazzein and mabinlin, for example, closely resemble "proteinase inhibitors," proteins that can help prevent further damage when a plant is injured. Interestingly, however, neither sweet protein has that function. Similar stories are true of most sweet-tasting proteins: They closely resemble other plant proteins with ordinary functions, but the sequences necessary for those other functions seem to be missing or mutated. It's as though pre-existing genes were recycled to become genes for sweet proteins. Presumably a gene duplication in the distant past resulted in an "extra" gene that could mutate freely, while still leaving a "good" copy of the

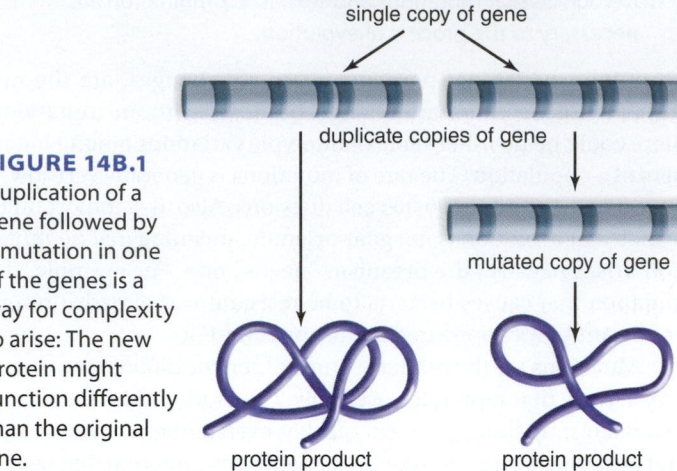

FIGURE 14B.1
Duplication of a gene followed by a mutation in one of the genes is a way for complexity to arise: The new protein might function differently than the original one.

single copy of gene

duplicate copies of gene

mutated copy of gene

protein product   protein product

gene to support the plant's functions. In time, the extra copy of the gene acquired mutations that happened to provide a sweet taste, and plants with that mutation gained a special appeal for local diners.

CONSIDER THESE QUESTIONS

1. Humans and perhaps apes and monkeys like sweet foods. How does this benefit plants containing sweet proteins?
2. Are humans influencing the evolution of plants when they propagate them? When they genetically modify them and then propagate them?
3. In what way is artificial selection harmful to the plants and animals selected to reproduce?

Explore the concepts through a variety of multimedia assets and question types.

www.mcgrawhillconnect.com

A *Curculigo* plant

The fruits develop at base of leaves.

FIGURE 14B.2 The sweet protein curculin is present in the fruit of a *Curculigo* plant.

LEARNING OUTCOMES

When you complete this section, you should be able to

1. Explain why small populations are more likely to be affected by genetic drift.
2. Contrast the bottleneck and founder effect.

Genetic drift refers to changes in the allele frequencies of a gene pool due to chance rather than selection by the environment. Therefore, genetic drift does not necessarily result in adaptation to the environment, as does natural selection. For example, in California, there are a number of cypress groves, each a separate population. The phenotypes within each grove are more similar to one another than they are to the phenotypes in the other groves. Some groves have longitudinally shaped trees, and others have pyramidally shaped trees. The bark is rough in some colonies and smooth in others. The leaves are gray to bright green or bluish, and the cones are small or large. The environmental conditions are similar for all the groves, and no correlation has been found between phenotype and the environment across groves. Therefore, scientists hypothesize that these variations among the groves are due to genetic drift.

Animation
Simulation of
Genetic Drift

Small Versus Large Populations Although genetic drift occurs in populations of all sizes, a smaller population is more likely to show the effects of drift. Suppose the allele *B* (for brown) occurs in 10% of the members in a population of frogs. In a population of 50,000 frogs, 5,000 will have the allele *B*. If a hurricane kills off half the frogs, the frequency of allele *B* may very well remain the same among the survivors. On the other hand, 10% of a population with ten frogs means that only one frog has the allele *B*. Under these circumstances, a natural disaster could very well do away with that one frog, should half the population perish. Or, let's suppose that five green frogs out of a ten-member population die. Now the frequency of allele *B* will increase from 10–20% (Fig. 14.10A).

Bottleneck and Founder Effects When a species is subjected to near extinction because of a natural disaster (e.g., hurricane,

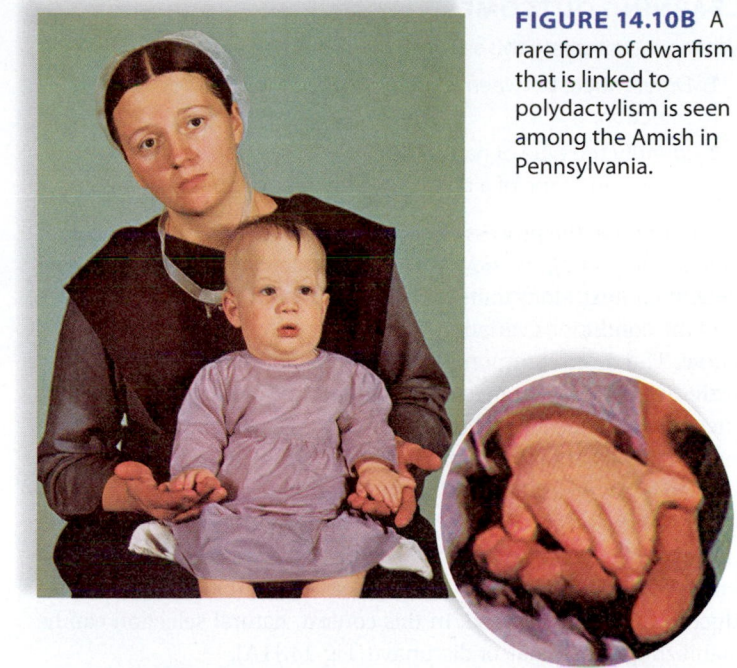

earthquake, or fire) or because of overhunting, overharvesting, and habitat loss, it is as if most of the population has stayed behind and only a few survivors have passed through the neck of a bottle. This so-called **bottleneck effect** prevents the majority of genotypes from participating in the production of the next generation. The extreme genetic similarity found in cheetahs is believed to be due to a bottleneck effect. In a study of 47 different enzymes, each of which can have several different forms, the sequence of amino acids in the enzymes was exactly the same in all the cheetahs. What caused the cheetah bottleneck is not known, but today cheetahs suffer from relative infertility because of the intense inbreeding that occurred after the bottleneck. Even if humans were to intervene and the population were to increase in size, the cheetah could still become extinct without genetic variation. Other organisms pushed to the brink of extinction suffer a plight similar to that of the cheetah.

The **founder effect** is an example of genetic drift in which rare alleles, or combinations of alleles, occur at a higher frequency in a population isolated from the general population. Founding individuals could contain only a fraction of the total genetic diversity of the original gene pool. Which alleles the founders carry is dictated by chance alone. The Amish of Lancaster County, Pennsylvania, are an isolated group that was begun by German founders. Today, as many as 1 in 14 individuals carries a recessive allele that causes an unusual form of dwarfism (affecting only the lower arms and legs) and polydactylism (extra fingers) (Fig. 14.10B). In the general population, only 1 in 1,000 individuals has this allele.

▶ **14.10 CHECK YOUR PROGRESS**

1. Explain why natural disasters are apt to cause gene pool frequency changes in small rather than large populations.
2. Differentiate between a bottleneck and a founder effect.

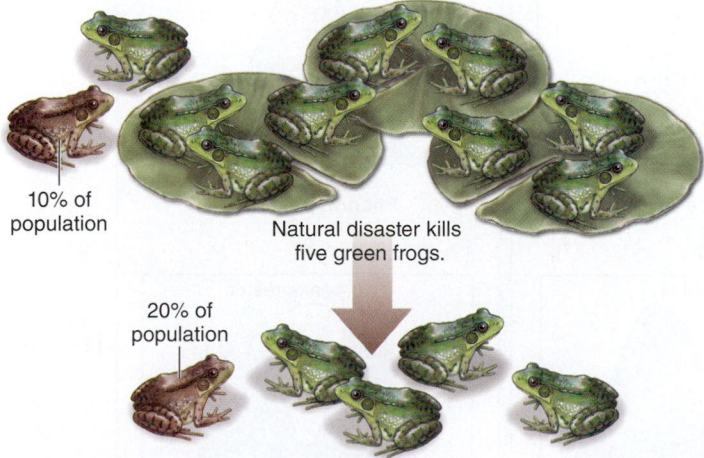

10% of population

Natural disaster kills five green frogs.

20% of population

FIGURE 14.10A Chance events can cause allele frequency changes and genetic drift.

LEARNING OUTCOMES

When you complete this section, you should be able to

1. Differentiate between stabilizing, directional, and disruptive selection.

2. Identify the type of natural selection affecting a trait by a change in shape of a phenotype range over time.

After outlining the process of natural selection earlier in this chapter (see section 14.2), we now wish to consider natural selection in a genetic context. Many traits are polygenic (controlled by many genes), and the continuous variation in phenotypes results in a bell-shaped curve. The most common phenotype is intermediate between two extremes. When this range of phenotypes is exposed to the environment, natural selection favors the one that is most adaptive under the present environmental circumstances. Natural selection acts much the same way as a governing board that decides which applying students will be admitted to a college. Some students will be favored and allowed to enter, while others will be rejected and not allowed to enter. Of course, in the case of natural selection, the chance to reproduce is the prize awarded. In this context, natural selection can be stabilizing, directional, or disruptive (Fig. 14.11A).

Stabilizing selection occurs when an intermediate phenotype is favored (Fig. 14.11A). It can improve adaptation of the population to those aspects of the environment that remain constant. With stabilizing selection, extreme phenotypes are selected against. As an example, consider that when Swiss starlings lay four to five eggs, more young survive than when the female lays more or less than this number. Genes determining physiological characteristics, such as the production of yolk, and behavioral characteristics, such as how long the female will mate, are involved in determining how many eggs are laid.

Human birth weight is another example of stabilizing selection. Through the years, hospital data have shown that human infants born with an intermediate birth weight (3–4 kg) have a better chance of survival than those at either extreme (either much less or much greater than usual). When a baby is small, its systems may not be fully functional, and when a baby is large, it may have experienced a difficult delivery. Stabilizing selection reduces the variability in birth weight in human populations (Fig. 14.11B).

Directional selection occurs when an extreme phenotype is favored, and the distribution curve shifts in that direction. Such a shift can occur when a population is adapting to a changing environment (see Fig. 14.11A).

Two investigators, John Endler and David Reznick, both at the University of California, conducted a study of guppies, which are known for their bright colors and reproductive potential. These investigators

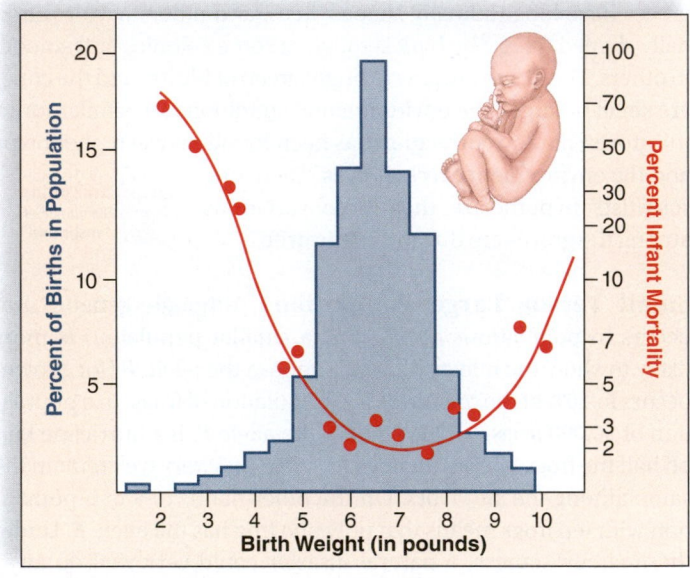

FIGURE 14.11B Stabilizing selection as exemplified by human birth weight.

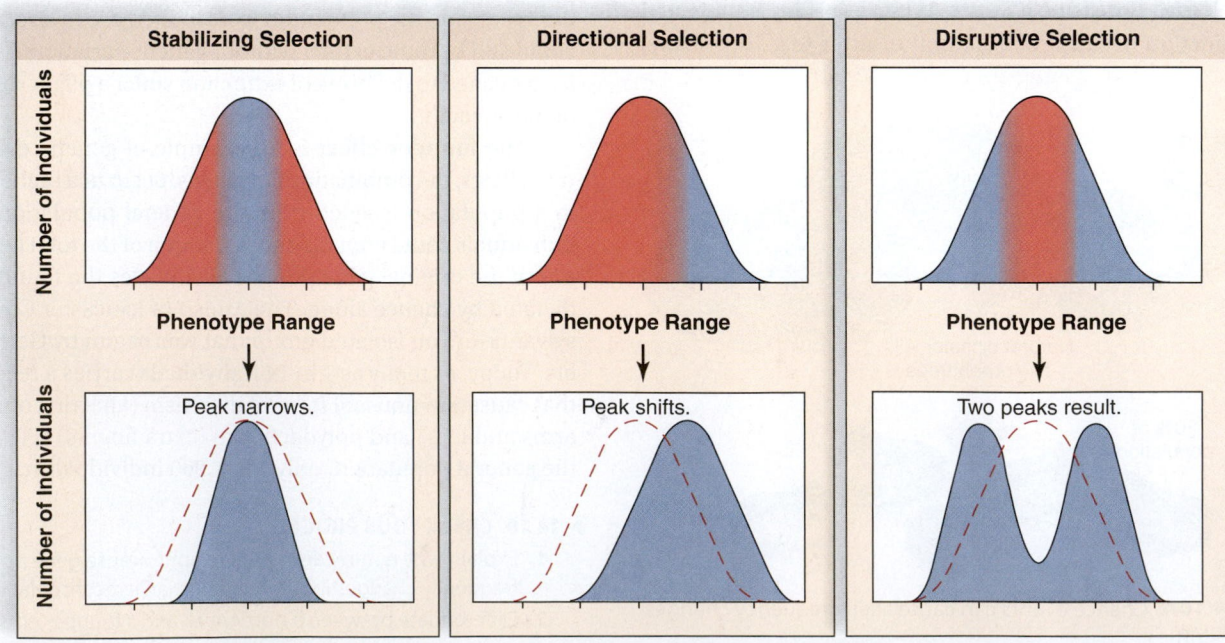

FIGURE 14.11A Phenotype ranges before and after three types of selection. Blue represents selected phenotype(s).

a. Experimental site

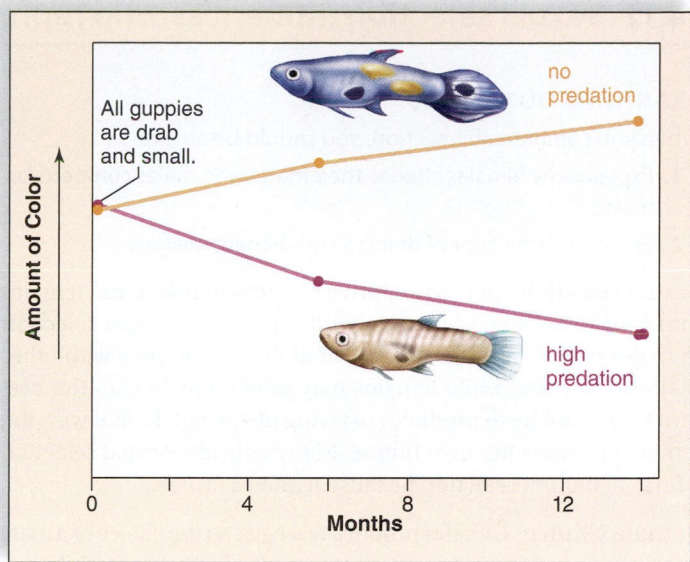

b. No predation results in colorful guppies.

FIGURE 14.11C Directional selection in guppies.

noted that on the island of Trinidad, when male guppies are subjected to high predation by other fish, they tend to be drab in color and to mature early and at a smaller size. The drab color and small size are most likely protective against being found and eaten. On the other hand, when male guppies are exposed to minimal or no predation, they tend to be colorful, to mature later, and to attain a larger size.

Endler and Reznick performed many experiments of particular interest. They took a supply of guppies from a high-predation area (below a waterfall) and placed them in a low-predation area (above a waterfall) (Fig. 14.11Ca). The waterfall prevented the predator fish (pike) from entering the low-predation area. They monitored the guppy population for 12 months, and during that year, the guppy population above the waterfall underwent directional selection (Fig. 14.11Cb). The male members of the population became colorful and large in size. The members of the guppy population below the waterfall (the control population) remained drab and small.

In **disruptive selection,** two or more extreme phenotypes are favored over any intermediate phenotype (see Fig. 14.11A, *right*). For example, British land snails (*Cepaea nemoralis*) have a wide habitat range that includes low-vegetation areas (grass fields and hedgerows) and forests. In forested areas, thrushes feed mainly on light-banded snails, and the snails with dark shells become more prevalent. In low-vegetation areas, thrushes feed mainly on snails with dark shells, and light-banded snails become more prevalent.

Therefore, these two distinctly different phenotypes are found in the population (Fig. 14.11D).

Stabilizing selection, discussed in section 14.13, maintains the heterozygote, especially if it has an advantage over the homozygote, as seen in sickle-cell disease.

▶ **14.11 CHECK YOUR PROGRESS**
1. Identify the type of selection if the peak of the phenotype range (**a**) shifts in one direction, (**b**) narrows, or (**c**) two peaks result.
2. Give an example of each type of selection mentioned in your answer to the first question.

FIGURE 14.11D Disruptive selection in snails.

Forested areas

Low-lying vegetation

When you complete this section, you should be able to

1. Explain why females choose their mates and males compete for mates.
2. Explain the concept of doing a cost-benefit analysis.

Sexual selection refers to adaptive changes in males and females that lead to an increased chance to reproduce. Sexual selection in males may result in an increased ability to compete with other males for a mate, while females may select a male with the best fitness (the ability to produce surviving offspring). In that way, the female increases her own fitness. Many consider sexual selection a form of natural selection because it affects fitness.

Female Choice Females produce few eggs, so the choice of a mate becomes a serious consideration. In a study of satin bowerbirds, two opposing hypotheses regarding female choice were tested: 1. Good genes hypothesis: Females choose mates on the basis of traits that improve the chance of survival. 2. Runaway hypothesis: Females choose mates on the basis of traits that improve male appearance.

The term runaway pertains to the possibility that the trait will be exaggerated in the male until its mating benefit is checked by the trait's unfavorable survival cost.

As investigators observed the behavior of satin bowerbirds, they discovered that aggressive males were usually chosen as

FIGURE 14.12A Sexual dimorphism. In the Raggiana Bird of Paradise, males have brilliantly colored plumage. The drab females tend to choose flamboyant males as mates.

mates by females. It could be that inherited aggressiveness does improve the chance of survival, or it could be that aggressive males are good at stealing blue feathers from other males. Females prefer blue feathers as bower decorations. Therefore, the data did not clearly support either hypothesis.

The Raggiana Bird of Paradise is remarkably dimorphic, meaning that males and females differ in size and other traits. The males are larger than the females and have beautiful orange flank plumes. In contrast to males, the females are drab (Fig. 14.12A). Female choice can explain why male birds are more ornate than females. Consistent with the two hypotheses, it is possible that the remarkable plumes of the male signify health and vigor. Or, it's possible that females choose the flamboyant males on the basis that their sons will have an increased chance of being selected by females. Some investigators have hypothesized that extravagant male features could indicate that they are relatively parasite-free. In barn swallows, females also choose those with the longest tails, and investigators have shown that males that are relatively free of parasites have longer tails than otherwise.

Male Competition Males can father many offspring because they continuously produce sperm in great quantity. We expect males to compete in order to inseminate as many females as possible. Cost-benefit analyses have been done to determine if the benefit of access to mating is worth the cost of competition among males.

Baboons, a type of Old World monkey, live together in a troop. Baboons travel three to six miles a day hunting for food and sleeping in trees at night. As the troop moves along, the dominant males along with new mothers and infants are in the center surrounded by other females. Less dominant males are on the fringes. If the troop is threatened, the dominant males leave the center and move toward the danger. They cover the troop as it retreats and attack when necessary. Both males and females have separate **dominance hierarchies** in which a higher-ranking animal has greater access to resources than a lower-ranking animal. Dominance is established by psychological contests between individuals in which it is determined who will give way to whom. Dominance is obvious in two aspects of troop life: grooming and presenting. Grooming occurs when one member of the troop picks through the hair of another and removes parasites and bits of debris with fingers or teeth. The dominant males are groomed by other members of the troop and, in general, a less dominant animal grooms another. Presenting occurs when one baboon takes a position with head down on the ground and the buttocks raised toward another. Presenting demonstrates submission because it puts the presenting animal in a very vulnerable position. The establishment of dominance is useful to the troop because it often prevents actual fighting. When a disturbance occurs within the troop, it is usually quieted by a dominant male.

Baboons are dimorphic; the males are larger than the females, and they can threaten other members of the troop with their long, sharp canines. One or more males become dominant by frightening the other males and their willingness to fight if psychology doesn't work (Fig. 14.12B). As might be expected, a male baboon pays a cost for his dominant position. Being larger means that he needs more food, and being willing and able to fight other males

FIGURE 14.12B A male olive baboon displaying full threat. In olive baboons, *Papio anubis,* males are larger than females and have enlarged canines. Competition between males establishes a dominance hierarchy for the distribution of resources.

and predators means that he may get hurt. Is there a reproductive benefit to his behavior? Yes, in that dominant males do indeed monopolize females when they are most fertile. Nevertheless, there may be other ways to father offspring. A male may act as a helper to a female and her offspring; then, the next time she is in estrus, she may mate preferentially with him instead of a dominant male. Or subordinate males may form a friendship group that opposes a dominant male, making him give up a receptive female.

A **territory** is an area that is defended against competitors. Scientists are able to track an animal in the wild in order to determine its home range or territory. Territoriality includes the type of aggressive behavior needed to defend a territory. Baboons travel within a home range, and dominant males decide where and when

the troop will move. If the troop is threatened, dominant males not only protect the troop as it retreats, they attack intruders when necessary. Vocalization and displays, rather than outright fighting, however, may be sufficient to defend a territory. In songbirds, for example, males use singing to announce their willingness to defend a territory. Other males of the species become reluctant to make use of the same area.

Red deer stags (males) on the Scottish island of Rhum compete to be the harem master of a group of hinds (females) that mate only with them. The reproductive group occupies a territory that the harem master defends against other stags. Harem masters first attempt to repel challengers by roaring. If the challenger remains, the two lock antlers and push against one another (Fig. 14.12C). If the challenger then withdraws, the master pursues him for a short distance, roaring the whole time. If the challenger wins, he becomes the harem master.

A harem master can father two dozen offspring at most, because he is at the peak of his fighting ability for only a short time. And there is a cost to being a harem master. Stags must be large and powerful in order to fight; therefore, they grow faster and have less body fat. During bad times, they are more likely to die of starvation, and in general, they have shorter lives. Harem master behavior will persist in the population only if its cost (reduction in the potential number of offspring because of a shorter life) is less than its benefit (increased number of offspring due to harem access).

► **14.12 CHECK YOUR PROGRESS**
1. Discuss why female choice can increase male fitness but male competition can lead to early death.
2. Identify the cost and the benefit of bright male plumage and keeping a harem of females.

a. Males roar to frighten other males.

b. Males compete by fighting one another.

FIGURE 14.12C Competition between male red deer. **a.** Roaring alone may frighten off a challenger, but (**b**) outright fighting may be necessary.

14.13 How variations are maintained in a population

LEARNING OUTCOMES

When you complete this section, you should be able to

1. Identify ways in which diversity is maintained in a population.
2. Explain why sickle-cell disease is maintained in a population subject to malaria.

Variations are maintained in a population for any number of reasons. Mutation still creates new alleles, and recombination still recombines these alleles during gametogenesis and fertilization. Gene flow might still occur. If the receiving population is small and mostly homozygous, gene flow can be a significant source of new alleles. Genetic drift also occurs, particularly in small populations, and the end result may be contrary to adaptation to the environment. Natural selection never starts from scratch and therefore the result is often a compromise. An erect posture freed the hands of humans but subjected the spine to injury because it is imperfectly adapted. But the benefit of freeing the hands must have been worth the risk of spinal injuries or it would not have evolved. An inefficient selective agent can play a role in maintaining diversity; predatory birds never catch all the white moths when pollutants darken the vegetation. A changing environment retains the ability of the medium ground finch on the Galápagos Islands to change its beak size as appropriate to the food supply. Clearly, the maintenance of variation among a population has survival value for the species. Here, we consider that heterozygote superiority in a particular environment can assist the maintenance of genetic, and therefore phenotypic, variations in future generations.

Sickle-Cell Disease Sickle-cell disease can be a devastating condition. Patients may have severe anemia, physical weakness, poor circulation, impaired mental function, pain and high fever, rheumatism, paralysis, spleen damage, low resistance to disease, and kidney and heart failure. In these individuals, the red blood cells are sickle-shaped and tend to pile up and block flow through tiny capillaries. The condition is due to an abnormal form of hemoglobin

(Hb), the molecule that carries oxygen in red blood cells. People with sickle-cell disease (Hb^SHb^S) tend to die early and leave few offspring, due to hemorrhaging and organ destruction. Interestingly, however, geneticists studying the distribution of sickle-cell disease in Africa have found that the recessive allele (Hb^S) has a higher frequency in regions (purple color) where the disease malaria is also prevalent (Fig. 14.13). Malaria is caused by a protozoan parasite that lives in and destroys the red blood cells of the normal homozygote (Hb^AHb^A). Individuals with this genotype have fewer offspring, due to an early death or to debilitation caused by malaria.

People who are heterozygous (Hb^AHb^S) have an advantage over both homozygous genotypes because they don't die from sickle-cell disease and they don't die from malaria. The parasite causes any red blood cell it infects in these individuals to become sickle-shaped. Sickle-shaped red blood cells lose potassium, and this causes the parasite to die. **Heterozygote advantage** causes all three alleles to be maintained in the population. It's as if natural selection were a store owner balancing the advantages and disadvantages of maintaining the recessive allele Hb^S in the warehouse. As long as the protozoan that causes malaria is present in the environment, it is advantageous to maintain the recessive allele, as shown in the following table:

Genotype	Phenotype	Result
Hb^AHb^A	Normal	Dies due to malarial infection
Hb^AHb^S	Sickle-cell trait	Lives due to protection from both
Hb^SHb^S	Sickle-cell disease	Dies due to sickle-cell disease

Heterozygote advantage is also an example of stabilizing selection because the genotype Hb^AHb^S is favored over the two extreme genotypes, Hb^AHb^A and Hb^SHb^S. In the parts of Africa where malaria is common, one in five individuals is heterozygous (has sickle-cell trait) and survives malaria, while only 1 in 100 is homozygous, Hb^SHb^S, and dies of sickle-cell disease.

What happens in the United States where malaria is not prevalent? As you would expect, the frequency of the Hb^S allele is declining among African Americans because the heterozygote has no particular advantage in this country.

Cystic Fibrosis Stabilizing selection is also thought to have influenced the frequency of other alleles. Cystic fibrosis is a debilitating condition that leads to lung infections and digestive difficulties. In this instance, the recessive allele, common among individuals of northwestern European descent, causes the person to have a defective plasma membrane protein. The agent that causes typhoid fever can use the normal version of this protein, but not the defective one, to enter cells. Here again, heterozygote superiority caused the recessive allele to be maintained in the population.

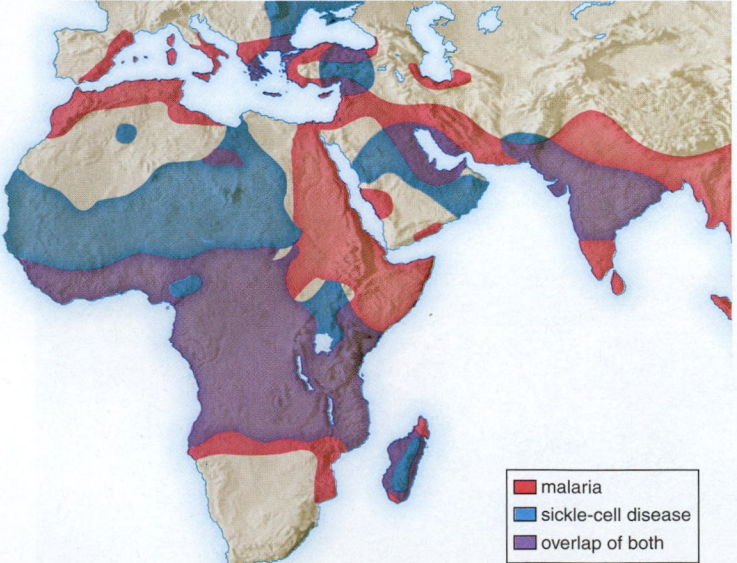

- ■ malaria
- ■ sickle-cell disease
- ■ overlap of both

FIGURE 14.13 Sickle-cell disease is more prevalent in areas of Africa where malaria is more common.

▶ **14.13 CHECK YOUR PROGRESS**

1. Identify two ways variations can be maintained in a population.
2. Demonstrate how sickle-cell disease is an example of stabilizing selection.

THE CHAPTER IN REVIEW

CONNECTING THE CONCEPTS

Darwin took a trip around the world as the naturalist aboard the HMS *Beagle*. During his trip, he collected fossils and made several observations that made him think evolution occurs. Darwin was aware of artificial selection, and he had read an essay by Malthus suggesting that the members of a population compete with one another for resources. Darwin began to see that a competitive edge would allow certain members of a population to survive and reproduce more than other members of the population. Assuming that advantageous traits are inheritable, future generations would eventually acquire adaptations to the local environment. Darwin called this process, by which a population adapts to its environment, natural selection because nature selects which members of a population will reproduce to a greater extent. Natural selection is like artificial selection except the environment instead of a breeder selects which plants or animals will reproduce.

Evolution explains the unity and diversity of life. Life is unified because of common descent, and it is diverse because of adaptations to particular environments. Darwin used the expression "descent with modification" to explain evolution. Support for common descent includes transitional fossils, anatomic features (homologous structures, vestigial structures, and embryologic similarities), biogeographic data, and molecular evidence.

In the 1930s, biologists developed a way to apply the principles of genetics to evolution. Populations would be in a Hardy-Weinberg equilibrium (allele frequencies stay the same) if mutation, gene flow, nonrandom mating, genetic drift, and natural selection did not occur. However, these events do occur, and they are the agents of evolutionary change that lead to microevolution, recognizable by allele frequency changes. Mutations provide the raw material for evolution. Genetic drift results in allele frequency changes due to a chance event, as when only a few members of a population are able to reproduce because of a natural disaster or because they have founded a colony. Natural selection is the only agent of evolution that results in adaptation to the environment.

Chapter 15 concerns macroevolution, the manner in which new species arise. The origin of new species is essential to the history of life on Earth, which we consider in Chapter 16.

ANALYZE AND EVALUATE

1. We now know that evolution by natural selection can be observed over a short period of time (years, months). Give examples.
2. Why would you expect evolution to have a genetic basis? Use industrial melanism to support the genetic basis of evolution.
3. Why would it be *incorrect* to say that bacteria became resistant in order to escape being killed by antibiotics?

SUMMARIZE

Darwin Collected Evidence of Evolution

14.1 Observations support and give evidence for evolution

- Before Darwin, Cuvier founded the science of **paleontology** and said that catastrophes cause evolution to occur. Also, Lamarck proposed the inheritance of acquired characteristics as a mechanism of evolution.
- On his trip around the world, Darwin observed that species change from place to place and through time. In his book Lyell explained how fossils come to be in **strata** and suggested the idea of **uniformitarianism** (geologic changes occur at a uniform rate). The book and Lyell's observations convinced Darwin that the Earth had existed long enough for evolution to have occurred.

14.2 Natural selection is a mechanism for evolution

- During **artificial selection,** humans (not the environment) select certain characteristics to perpetuate. During **natural selection,** an environmental **selective agent** selects which organisms will reproduce.
- Natural selection has several components: (1) The members of a population have inheritable variations. (2) A population is able to produce more offspring than the environment can support. (3) Members of a population differ in **fitness.** Certain members are able to survive and reproduce more than other members because they have an advantage suited to the environment. (4) Natural selection results in a population adapted to its environment.

- **Evolution** can be defined as changes in a population over time due to an accumulation of inherited differences. Wallace was a naturalist who had also read Malthus and arrived at conclusions similar to those of Darwin.

More Evidence of Evolution

14.3 Fossils provide powerful evidence for common descent

- **Fossils** are hard parts of organisms or other traces of life found in sedimentary rock. **Paleontologists** study the **fossil record,** which indicates that life has progressed from simple to complex. **Transitional fossils** have the characteristics of two different groups and thus provide clues to the evolutionary relationships between organisms.
- During evolution, organisms share **common ancestors** just as you and your cousins share grandparents.

14.4 Anatomic evidence supports common descent

- **Homologous structures** are anatomical similarities due to common ancestry. **Analogous structures** have the same functions in different organisms but are not anatomically similar. Only homologous structures (not analogous structures) indicate that organisms have a *recent* common ancestor.

bat   bird

- Other types of anatomic evidence exist. Organisms have **vestigial structures** despite their being reduced and nonfunctional because they were once functional in an ancestor. All vertebrates share the same embryonic features, such as pharyngeal pouches and a post-anal tail, but they are later modified for different purposes.

14.5 Biogeographic evidence supports common descent
- **Biogeography** is the study of the distribution of organisms around the globe. Plants and animals evolved in particular locations, and therefore widely separated similar environments contain different but similarly adapted organisms.

14.6 Molecular evidence supports common descent
- **Molecular homologies** occur in organisms related by common descent. To illustrate molecular homologies consider (1) All cells use DNA to store genetic information and the same enzymes during cellular respiration to build ATP. (2) The same type enzymes occur in metabolic pathways. The number of amino acid differences in metabolic molecules such as p53 tell how closely related two groups are. (3) Similar developmental genes such as *Hox* genes occur in many different types of organisms, suggesting that they are all related through common descent.

Evidence Through Study of Populations

14.7 Gene pool frequency changes determine evolution
- **Microevolution** is evidenced by changes in **gene pool** frequencies. Hardy and Weinberg showed that it was possible to calculate the genotype and allele frequencies of a population by using the following equation:

$$p^2 + 2pq + q^2 = 1$$

  This equation predicts a Hardy-Weinberg equilibrium. The **Hardy-Weinberg principle** states that microevolution does not occur as long as mutations, gene flow, nonrandom matings, genetic drift, and natural selection do not occur.
- Generally, allele frequencies do change between generations, and microevolution does occur. For example, dark moths become prevalent in moth populations when trees become dark due to pollution. This is called **industrial melanism.**

14.8 Both mutations and sexual recombination produce variations
- **Mutations** are the primary source of genetic differences in prokaryotes. Sexual recombination and mutations are equally important in eukaryotes.

14.9 Nonrandom mating and gene flow change gene pool frequencies
- **Nonrandom mating** occurs when only certain genotypes or phenotypes mate with one another. **Assortative mating** is a type of nonrandom mating in which individuals mate with those that have the same phenotype for a particular characteristic.
- **Gene flow** results when alleles move between populations due to migration.

14.10 Genetic drift is more likely to alter gene pool frequencies in small populations
- **Genetic drift** refers to changes in allele frequency in a gene pool due to chance. The **bottleneck effect** prevents the majority of genotypes from participating in production of the next generation. The **founder effect** occurs when rare alleles contributed by the founders of a population occur at a higher frequency in isolated populations.

14.11 Natural selection can be stabilizing, directional, or disruptive
- In **stabilizing selection,** extreme phenotypes are selected against while intermediate phenotypes are favored. In **directional selection,** an extreme phenotype is favored. In **disruptive selection,** two or more extreme phenotypes are favored over the intermediate phenotype.

14.12 Sexual selection influences who reproduces
- **Sexual selection** is increased reproduction resulting from the ability to obtain mates. The process can result in **sexual dimorphism,** a difference in the appearance of males and females. Females who produce few eggs over a lifetime choose certain males over others, and males, who produce many sperm, compete with one another for mates. A **cost-benefit analysis** considers whether the increased number of offspring is worth the cost of being dominant or defending a **territory.**

14.13 How variations are maintained in a population
- **Heterozygote advantage** causes the sickle-cell allele to be maintained in Africa, even though the homozygous recessive is lethal because the heterozygote is protective against malaria. The recessive allele for cystic fibrosis is believed to have been maintained because a faulty membrane protein doesn't allow the typhoid bacterium to enter cells.

TEST YOURSELF

Darwin Collected Evidence of Evolution
1. Why was it helpful to Darwin to learn that Lyell had concluded the Earth was very old?
   a. An old Earth has more fossils than a new Earth.
   b. It meant there was enough time for evolution to have occurred slowly.
   c. It meant there was enough time for the same species to spread into all continents.
   d. Darwin said artificial selection occurs slowly.
   e. All of these are correct.
2. Which of these pairs is mismatched?
   a. Charles Darwin—natural selection
   b. Cuvier—series of catastrophes explains the fossil record
   c. Lamarck—uniformitarianism
   d. All of these are correct.
3. Which is most likely to be favored during natural selection, but not artificial selection?
   a. fast seed germination rate
   b. short generation time
   c. efficient seed dispersal
   d. lean pork meat production
4. Which of these is/are necessary to natural selection?
   a. variations
   b. differential reproduction
   c. inheritance of differences
   d. All of these are correct.

5. **THINKING CONCEPTUALLY** The adaptive results of natural selection cannot be determined ahead of time. Explain.

More Evidence of Evolution

6. The fossil record offers direct evidence for common descent because you can
   a. see that the types of fossils change over time.
   b. sometimes find common ancestors.
   c. trace the ancestry of a particular group.
   d. trace the biological history of organisms.
   e. All of these are correct.
7. Which of the following is not an example of a vestigial structure?
   a. human tailbone   c. pelvic girdle in snakes
   b. ostrich wings   d. dog kidney
8. If evolution occurs, we would expect different biogeographic regions with similar environments to
   a. all contain the same mix of plants and animals.
   b. each have its own specific mix of plants and animals.
   c. have plants and animals with similar adaptations.
   d. have plants and animals with different adaptations.
   e. Both b and c are correct.
9. DNA nucleotide differences between organisms
   a. indicate how closely related organisms are.
   b. indicate that evolution occurs.
   c. explain why there are phenotypic differences.
   d. are to be expected.
   e. All of these are correct.

For questions 10–13, match the evolutionary evidence in the key to the description. Choose more than one answer if correct.

KEY:
   a. biogeographic evidence   c. molecular evidence
   b. fossil evidence   d. anatomic evidence

10. Islands have many unique species not found elsewhere.
11. All vertebrate embryos have pharyngeal pouches.
12. Distantly related species have more amino acid differences in cytochrome c.
13. Transitional links have been found between major groups of animals.
14. **THINKING CONCEPTUALLY** Why can researchers make decisions about who is related to whom using only DNA base sequence data? (See section 14.6.)

Evidence Through Study of Populations

For questions 15 and 16, consider that about 75% of white North Americans can taste the chemical phenylthiocarbamide. The ability to taste is due to the dominant allele T. Nontasters are tt. Assume this population is in Hardy-Weinberg equilibrium.

15. What is the frequency of t?
    a. 0.25   d. 0.09
    b. 0.70   e. 0.60
    c. 0.55
16. What is the frequency of heterozygous tasters?
    a. 0.50   c. 0.2475
    b. 0.21   d. 0.45
17. When a population is small, there is a greater chance of
    a. gene flow.   d. mutations occurring.
    b. genetic drift.   e. sexual selection.
    c. natural selection.

18. The offspring of better-adapted individuals are expected to make up a larger proportion of the next generation. The most likely explanation is
    a. mutations and nonrandom mating.
    b. gene flow and genetic drift.
    c. mutations and natural selection.
    d. mutations and genetic drift.
19. The northern elephant seal went through a severe population decline as a result of hunting in the late 1800s. The population has rebounded but is now homozygous for nearly every gene studied. This is an example of
    a. negative assortative mating.   d. a bottleneck.
    b. migration.   e. disruptive selection.
    c. mutation.
20. Natural selection is the only process that results in
    a. genetic variation.
    b. adaptation to the environment.
    c. phenotypic change.
    d. competition among individuals in a population.
21. Which of these is an example of stabilizing selection?
    a. Over time, *Equus* developed strength, intelligence, speed, and durable grinding teeth.
    b. British land snails mainly have two different phenotypes.
    c. Swiss starlings usually lay four or five eggs, thereby increasing their chances of more offspring.
    d. Drug resistance increases with each generation; the resistant bacteria survive, and the nonresistant bacteria get killed off.
    e. All of these are correct.
22. **THINKING CONCEPTUALLY** Explain why evolutionists are more interested in mutations and natural selection than in the other causes for microevolution.

GET INVOLVED

1. You decide to repeat the guppy experiment described in section 14.11 because you want to determine what genotype changes account for the results. What might you do to detect such changes?
2. A cotton farmer applied a new insecticide against the boll weevil for several years. At first, the treatment was successful, but then the insecticide became ineffective and the boll weevil rebounded. Did evolution occur? Explain.

MEDIA STUDY TOOLS

mhhe.com/maderconcepts3

Enhance your study of this chapter with interactive study tools, practice tests, and engaging animations. Also, ask your instructor about the resources available through ConnectPlus, which includes LearnSmart, a personalized adaptive learning program, and a media-rich eBook.

15

Speciation and Evolution

BEFORE YOU BEGIN

Take a few minutes to recall

"Epigenetic Inheritance and Who You Are" 225
Fossil, anatomical, and molecular evidence for evolution (sections 14.3–14.6)
Conditions that cause microevolution (section 14.7)

Hybrid Animals Do Exist

The immense liger, an offspring of a lion father and a tiger mother, really impressed Brian. Upon returning from the show, he immediately began researching more information. To his surprise, he found that ligers are one of many hybridized species that have been recorded. His search led him to common hybrid websites that discussed mules, zorses, zonkeys, and beefalos. He also discovered several strange hybrids, such as the wolphin, a cross between a false killer whale and a dolphin; a grolar, a cross between a grizzly bear and a polar bear; and a cama, a cross between a camel and a llama. Usually, in naming hybrids, the name of the male parent is used first. Thus, a zorse has a zebra father and a horse mother.

A hybrid results from breeding two closely related, but distinct, species. Lions and tigers meet this criterion, but a hybrid between a cat and a rabbit would not exist because these animals are not closely related. Hybrids are usually the result of human activities, either by direct intervention or by placing related species in the same setting. For example, humans have mated female donkeys and male horses to develop mules for centuries. The vast majority of hybrids have been born in zoos as a result of bringing together related species from different continents. After all, lions

Liger

live in Africa and tigers live in Asia. Therefore, most ligers are born in captivity. Reports of ligers in zoos can be traced back to the early 1800s.

Brian found that ligers are much larger than their parental stock. In fact, they are the largest felines in the world, measuring up to 4 m tall when standing on their hind legs and weighing as much as 454 kg. Their coat color is usually tan with tiger stripes on the back and hindquarters and lion cub spots on the abdomen. A liger can produce both the "chuff" sound of a tiger and the roar of a lion. Male ligers may have a modest lion mane or no mane at all. Most ligers have an affinity for water and love to swim. Generally, ligers have a gentle disposition; however, considering their size and heritage, handlers should be extremely careful. Brian also discovered tigons, rare animals that have a tiger father and a lion mother. Generally, tigons are smaller than their parental stock.

Most hybrids are sterile. Mules inherit an uneven number of chromosomes that cannot pair up during meiosis, making it rarely possible for them to form gametes. Ligers have an even number of chromosomes but males are infertile presumedly because the pairing of the Y from a lion and the X from the tiger is impossible during meiosis. Female ligers are fertile and Li-ligers (both parents are ligers), Li-tigons (father is a liger, mother is a tigon), Ti-ligers (father is a tiger, mother is a liger), and Ti-tigons (father is a tiger, mother is a tigon) have been produced. These unusual hybrids display a variety of lion and tiger traits.

This chapter is about speciation, the origin of species. Without the origin of new species there would have been no history of life on Earth. In this chapter we will see how speciation occurs and how it may be observed during present times and in the fossil record.

Tigon

Mule

Zorse

Diversity Requires Speciation

Macroevolution is the evolution of species. A species can be defined using the evolutionary species or the biological species concept. In order for a species to be biologically distinct, reproductive barriers are needed to maintain its genetic differences from other species. These barriers consist of prezygotic and postzygotic isolating mechanisms.

15.1 Species have been defined in more than one way

LEARNING OUTCOME

When you complete this section, you should be able to

1. Compare and contrast the evolutionary species concept with the biological species concept.

In Chapter 14, we concluded that **microevolution** was any allele frequency change within the gene pool of a population. Macroevolution, which is observed best within the fossil record, requires the origin of species, also called speciation. **Speciation** is the splitting of one species into two or more species, or the transformation of one species into a new species over time. Speciation is the final result of changes in gene pool allele and genotype frequencies. The diversity of life we see about us is absolutely dependent on speciation, so it is important to be able to define a species and to know when speciation has occurred. In Chapter 1 we defined a species as a type of organism, but now we want to characterize a species in more depth.

The **evolutionary species concept** recognizes that every species has its own evolutionary history, at least part of which is in the fossil record. As an example, consider that the species depicted in Figure 15.1A are a part of the evolutionary history of toothed whales. Binomial nomenclature, discussed in section 1.5, was used to name these ancestors of killer whales as well as the other species of toothed whales living today. The two-part scientific name when translated from the Latin often tells you something about the organism. For example, the scientific name of the dinosaur *Tyrannosaurus rex* means "tyrant-lizard king."

The evolutionary species concept relies on traits, called diagnostic traits, to distinguish one species from another. As long as these traits are the same, fossils are considered members of the same species. Abrupt changes in these traits, such as hindlimbs, indicate the evolution of a new species in the fossil record. In summary, the evolutionary species concept states that members of a species share the same distinct evolutionary pathway and that species can be recognized by diagnostic trait differences.

One advantage of the evolutionary species concept is that it applies to both sexually and asexually reproducing organisms. However, a major disadvantage can occur when anatomic traits are used to distinguish species. The presence of variations, such as size differences in male and female animals, might make you think you are dealing with two species instead of one, and the lack of distinct differences could cause you to conclude that two fossils are the same species when they are not.

The evolutionary species concept necessarily assumes that the members of a species are reproductively isolated. If members of different species were to reproduce with one another, their evolutionary history would be mingled, not separate. By contrast, the **biological species concept** relies primarily on reproductive isolation rather than trait differences to define a

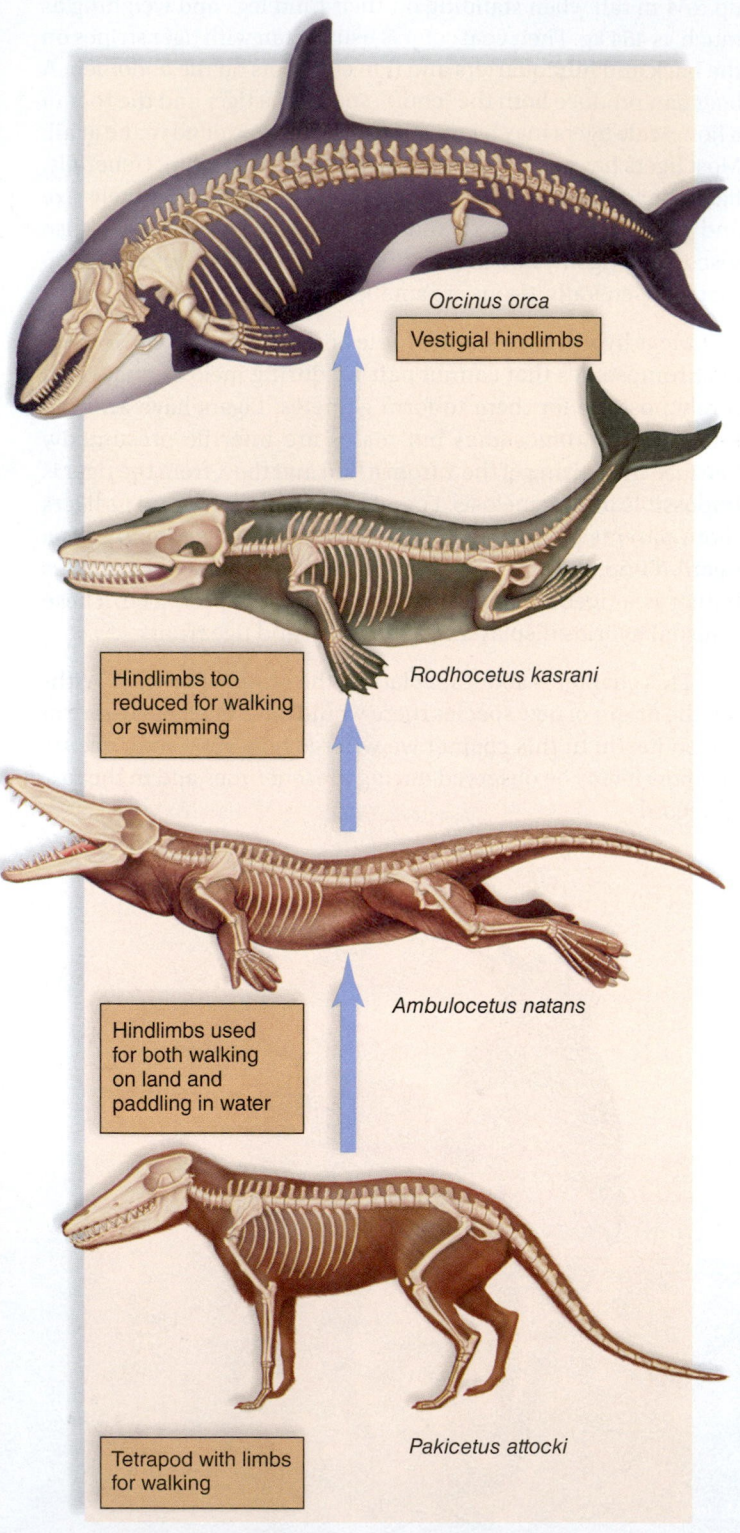

Orcinus orca

Vestigial hindlimbs

Rodhocetus kasrani

Hindlimbs too reduced for walking or swimming

Ambulocetus natans

Hindlimbs used for both walking on land and paddling in water

Pakicetus attocki

Tetrapod with limbs for walking

FIGURE 15.1A Evolution of the modern toothed whale.

pit-see
Acadian flycatcher, *Empidonax virescens*

fitz-bew
Willow flycatcher, *Empidonax trailli*

che-bek or che-bek
Least flycatcher, *Empidonax minimus*

FIGURE 15.1B Three species of flycatchers. The call of each bird is given on the photograph.

species. In other words, although traits can help us distinguish species, the most important criterion, according to the biological species concept, is reproductive isolation—the members of a species have a single gene pool. While useful, the biological species concept cannot be applied to asexually reproducing organisms, to organisms known only by the fossil record, or to species that interbred when they lived near one another. The benefit of the concept is that it can designate species even when trait differences may be difficult to find. The flycatchers in Figure 15.1B are very similar, but they do not reproduce with one another; therefore, they are separate species. They live in different habitats. The Acadian flycatcher inhabits deciduous forests and wooded swamps, especially beeches; the willow flycatcher inhabits thickets, bushy pastures, old orchards, and willows; and the least flycatcher inhabits open woods, orchards, and farms. They also have different calls. Conversely, when anatomic differences are apparent, but reproduction is not deterred, only one species is present. Despite noticeable variations, humans from all over the world can reproduce with one another and belong to one species. The Massai of East Africa and the Eskimos of Alaska are kept apart by geography, but we know that, should they meet, reproduction between male and female would be possible (Fig. 15.1C).

The biological species concept gives us a way to know when speciation has occurred, without regard to anatomic differences. As soon as descendants of a group of organisms are able to reproduce only among themselves, speciation has occurred.

The biological species concept is supplemented by our knowledge of molecular genetics. DNA base sequence data and differences in proteins can indicate the relatedness of groups of organisms but it can't indicate when speciation has occurred. Macroevolution starts with speciation but anatomic data such as the differences in the whales in Figure 15.1A play a prominent role in deciphering the history of life.

FIGURE 15.1C The Massai of East Africa (*left*) and the Eskimos of Alaska (*right*) belong to the same species.

▶ **15.1 CHECK YOUR PROGRESS**
  1. Contrast an advantage of the evolutionary species concept with an advantage of the biological species concept.

15.2 Reproductive barriers maintain genetic differences between species

LEARNING OUTCOME

When you complete this section, you should be able to

1. List and give examples of five prezygotic isolating mechanisms and three postzygotic isolating mechanisms.

As mentioned in the previous section, for two species to be separate, they must be reproductively isolated—that is, gene flow must not occur between them. Isolating mechanisms that prevent successful reproduction from occurring are reproductive barriers (Fig. 15.2A). In evolution, reproduction is successful only when it produces fertile offspring.

Prezygotic (before the formation of a zygote) **isolating mechanisms** are those that prevent reproductive attempts and make it unlikely that fertilization will be successful if mating is attempted. Scientists have identified several types of isolation that make it highly unlikely for particular genotypes to contribute to a population's gene pool:

Habitat isolation When two species occupy different habitats, even within the same geographic range, they are less likely to meet and attempt to reproduce. This is one of the reasons that the flycatchers in Figure 15.1B do not mate, and that red maple and sugar maple trees do not exchange pollen. In tropical rain forests, many animal species are restricted to a particular level of the forest canopy, and in this way they are isolated from similar species.

Temporal isolation Several related species can live in the same locale, but if each reproduces at a different time of year, they do not attempt to mate. Five species of frogs of the genus *Rana* are all found at Ithaca, New York. The species remain separate because the period of most active mating is different for each (Fig. 15.2B), and because whenever there is an overlap, different breeding sites are used. For example, wood frogs breed in woodland ponds or shallow water, leopard frogs in lowland swamps, and pickerel frogs in streams and ponds on high ground.

Behavioral isolation Many animal species have courtship patterns that allow males and females to recognize one another. The male blue-footed boobie in Figure 15.2C does a dance unique to the species. Male fireflies are recognized by females of their species by the pattern of their flashings; similarly, female crickets recognize male crickets by their chirping. Many males recognize females of their species by sensing chemical signals called pheromones. For example, female gypsy moths have special abdominal glands from which they secrete pheromones (see section 28.2, p. 556) that are detected downwind by receptors on the antennae of males.

Video Flirting Flies

Mechanical isolation Inaccessibility of pollen to certain pollinators can prevent cross-fertilization in plants, and the sexes of many insect species have genitalia that do not match. When animal genitalia or plant floral structures are incompatible,

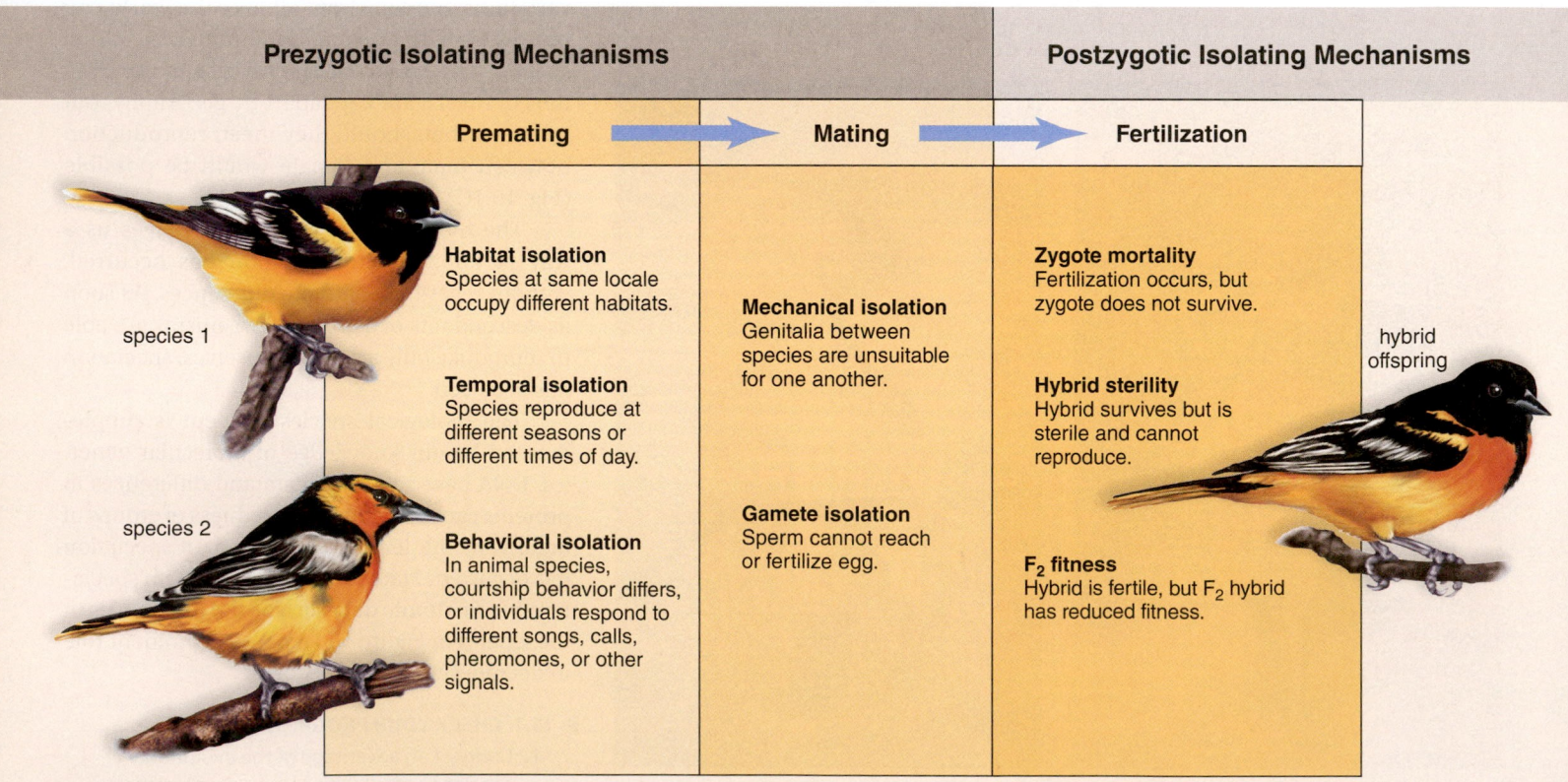

Prezygotic Isolating Mechanisms			Postzygotic Isolating Mechanisms
Premating	**Mating**	**Fertilization**	

Habitat isolation
Species at same locale occupy different habitats.

Temporal isolation
Species reproduce at different seasons or different times of day.

Behavioral isolation
In animal species, courtship behavior differs, or individuals respond to different songs, calls, pheromones, or other signals.

Mechanical isolation
Genitalia between species are unsuitable for one another.

Gamete isolation
Sperm cannot reach or fertilize egg.

Zygote mortality
Fertilization occurs, but zygote does not survive.

Hybrid sterility
Hybrid survives but is sterile and cannot reproduce.

F$_2$ fitness
Hybrid is fertile, but F$_2$ hybrid has reduced fitness.

species 1

species 2

hybrid offspring

FIGURE 15.2A Reproductive barriers.

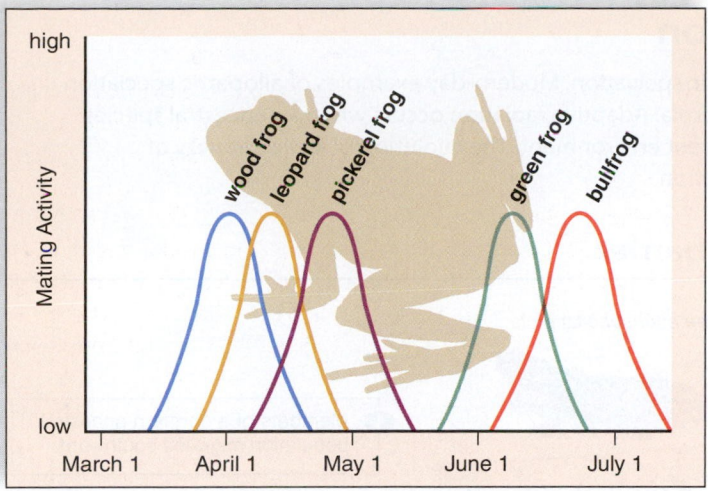

FIGURE 15.2B Mating activity peaks at different times of the year for these species of frogs.

FIGURE 15.2C Male blue-footed boobie doing a courtship dance for a female.

reproduction cannot occur. Other characteristics can also make mating impossible. For example, male dragonflies have claspers that are suitable for holding only the females of their own species.

Gamete isolation Even if the gametes of two different species meet, they may not fuse to become a zygote. In animals, the sperm of one species may not be able to survive in the reproductive tract of another species, or the egg may have receptors only for sperm of its species. In plants, pollen grains are species-specific and will not form a pollen tube for another species. Without a pollen tube, the sperm cannot successfully reach the egg.

Postzygotic (after the formation of a zygote) **isolating mechanisms** prevent hybrid offspring from developing or breeding, even if reproduction attempts have been successful.

Zygote mortality A hybrid zygote may not be viable, and so it dies. A zygote with two different chromosome sets may fail to go through mitosis properly, or the developing embryo may receive incompatible instructions from the maternal and paternal genes so that it cannot continue to exist.

Hybrid sterility The hybrid zygote may develop into a sterile adult. As is well known, a cross between a male horse and a female donkey produces a mule, which is usually sterile—it cannot reproduce (Fig. 15.2D). Sterility of hybrids generally results from complications in meiosis that lead to an inability to produce viable gametes. A cross between a cabbage and a radish produces offspring that cannot form gametes, most likely because the cabbage chromosomes and the radish chromosomes cannot align during meiosis (see section 15.5).

F₂ fitness Even if hybrids can reproduce, their offspring may be unable to reproduce. In some cases, mules are fertile, but their offspring (the F_2 generation) are not fertile.

▶ **15.2 CHECK YOUR PROGRESS**

1. Give an example of a prezygotic and postzygotic isolating mechanism and tell how each keeps species separate.

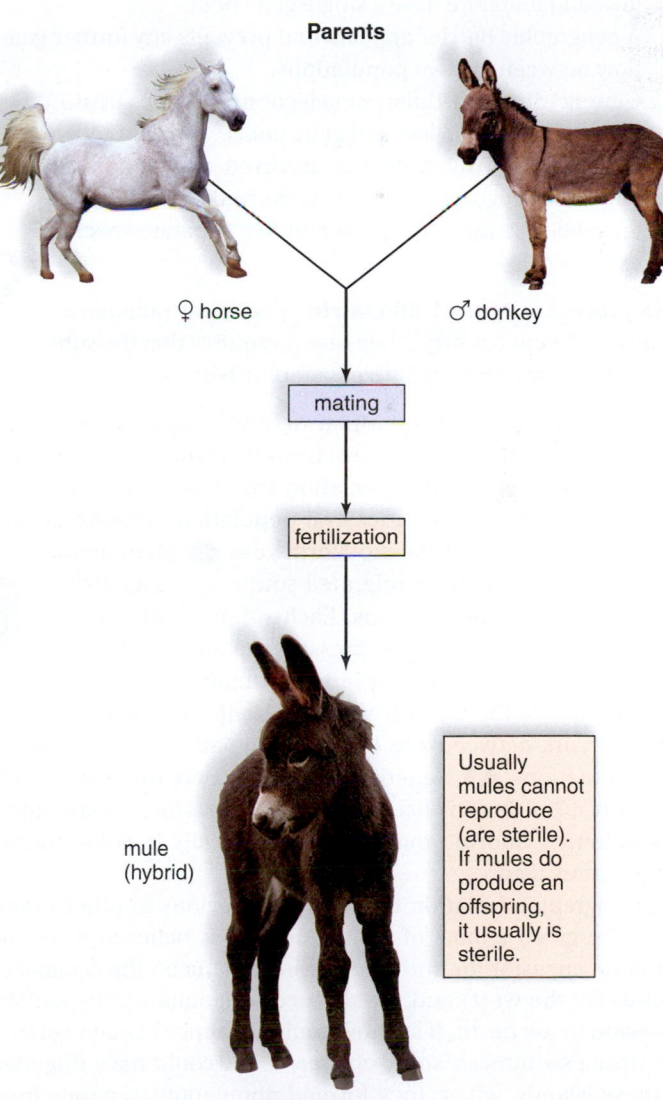

Parents

♀ horse   ♂ donkey

mating

fertilization

mule (hybrid)

Usually mules cannot reproduce (are sterile). If mules do produce an offspring, it usually is sterile.

Offspring

FIGURE 15.2D Mules cannot reproduce due to chromosome incompatibility.

Speciation Due to Geographic Separation

Geographic isolation fosters the genetic changes that result in allopatric speciation. Modern-day examples of allopatric speciation include the evolution of distinct forms of *Ensatina* salamanders in California. Adaptive radiation occurs when an ancestral species evolves into several new and different species, each adapted to a different environment. The evolution of a wide variety of honeycreepers on the Hawaiian Islands is an example of adaptive radiation.

15.3 Allopatric speciation utilizes a geographic barrier

LEARNING OUTCOMES

When you complete this section, you should be able to

1. List five steps that result in allopatric speciation.
2. Describe real-life examples of allopatric speciation.

In 1942, Ernst Mayr, an evolutionary biologist, published the book *Systematics and the Origin of Species,* in which he proposed the biological species concept and this process by which speciation could occur:

- Two subpopulations of a species are experiencing gene flow and therefore have a single gene pool.
- A geographic barrier appears and prevents any further gene flow between the two populations.
- Genetic drift and different selection pressures cause divergence between the isolated gene pools.
- Reproductive isolation has occurred and continues even when the geographic barrier is removed.
- Speciation is now complete and two separate species exist.

This process is termed **allopatric speciation** (allopatric means "different country") because it requires that the subpopulations be separated by a geographic barrier.

***Ensatina* Salamanders** Much data in support of allopatric speciation have since been discovered. Figure 15.3A features an example of allopatric speciation that has been extensively studied in California. An ancestral population of *Ensatina* salamanders lives in the Pacific Northwest. ❶ Members of this ancestral population migrated southward, establishing a series of subpopulations. Each subpopulation was exposed to its own selective pressures along the coastal mountains and the Sierra Nevada mountains. ❷ Due to the presence of the Central Valley of California, gene flow rarely occurs between the eastern populations and the western populations. ❸ Genetic differences increased from north to south, resulting in distinct forms of *Ensatina* salamanders in southern California that differ dramatically in color and no longer interbreed.

Geographic isolation is even more obvious in other examples. The green iguana of South America is believed to be the common ancestor for both the marine iguana on the Galápagos Islands (to the west) and the rhinoceros iguana on Hispaniola, an island to the north. If so, how could it happen? Green iguanas are strong swimmers, so by chance, a few could have migrated to these islands, where they formed populations separate from each other and from the parent population back in South America. Each population continued on its own evolutionary path as new mutations, genetic drift, and different selection pressures

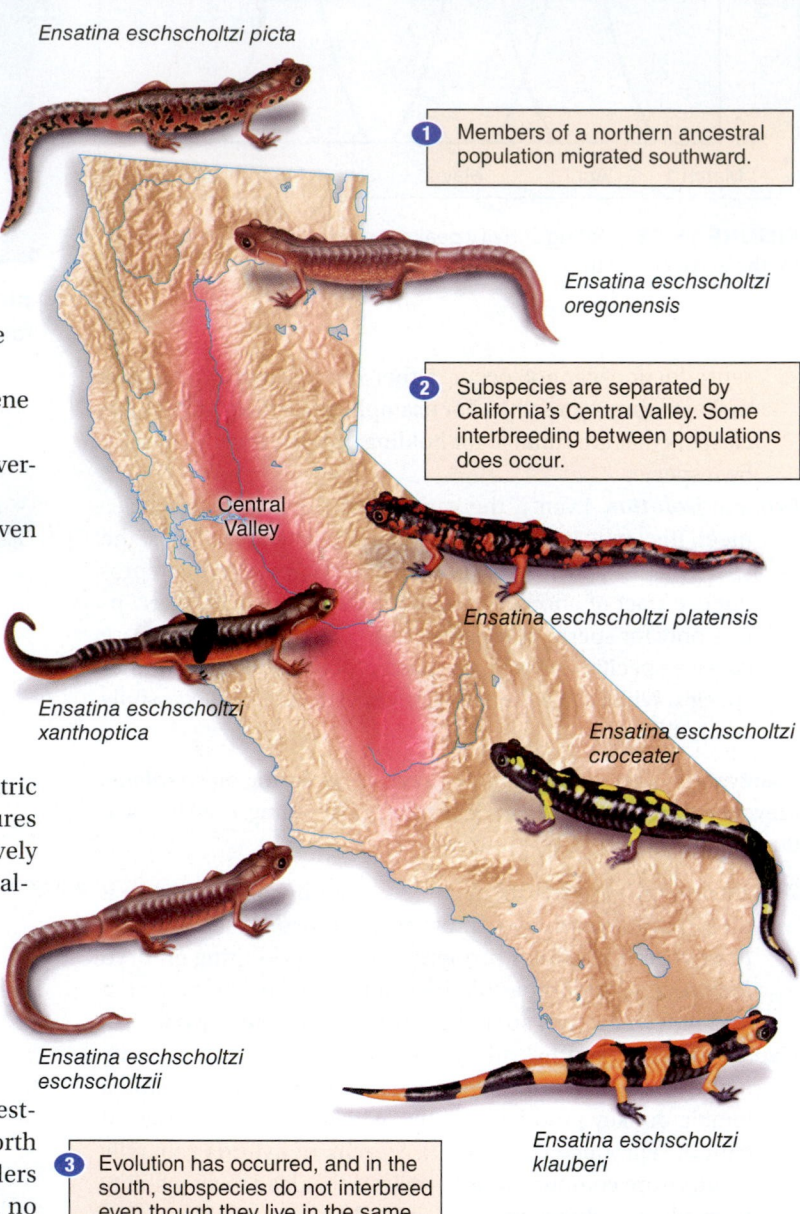

Ensatina eschscholtzi picta

1 Members of a northern ancestral population migrated southward.

Ensatina eschscholtzi oregonensis

2 Subspecies are separated by California's Central Valley. Some interbreeding between populations does occur.

Central Valley

Ensatina eschscholtzi platensis

Ensatina eschscholtzi xanthoptica

Ensatina eschscholtzi croceater

Ensatina eschscholtzi eschscholtzii

3 Evolution has occurred, and in the south, subspecies do not interbreed even though they live in the same environment.

Ensatina eschscholtzi klauberi

FIGURE 15.3A Allopatric speciation among *Ensatina* salamanders.

occurred. Eventually, reproductive isolation developed, and the result was three species of iguanas that are reproductively isolated from each other.

Sockeye Salmon and *Anolis* Lizards A more detailed example of allopatric speciation involves sockeye salmon in Washington state. In the 1930s and 1940s, hundreds of thousands

FIGURE 15.3B Sockeye salmon at Pleasure Point Beach, Lake Washington.

FIGURE 15.3C Sockeye salmon in the Cedar River. The river connects with Lake Washington.

of sockeye salmon were introduced into Lake Washington. Some colonized an area of the lake near Pleasure Point Beach (Fig. 15.3B). Others migrated into the Cedar River (Fig. 15.3C). Andrew Hendry, a biologist at McGill University, is able to tell Pleasure Point Beach salmon from Cedar River salmon because they differ in shape and size due to the demands of reproducing in the river. In the river, males tend to be more slender than those along the beach. A slender body is better able to turn sideways in a strong current, and the courtship ritual of a sockeye salmon requires this maneuver. On the other hand, the females tend to be larger than those along the beach. This larger body helps them dig slightly deeper nests in the gravel beds on the river bottom. Their deeper nests are not disturbed by river currents and will remain warm enough for egg viability.

Hendry has another way to tell beach salmon from river salmon. Ear stones called otoliths reflect variations in water temperature while a fish embryo is developing. Water temperatures at Pleasure Point Beach are relatively constant compared to Cedar River temperatures. By checking otoliths in adults, Hendry found that a third of the sockeye males at Pleasure Point Beach had grown up in the river. Yet the distinction between male and female shape and size according to the two locations remains. Therefore, these males are not successful breeders along the beach. In other words, reproductive isolation has occurred.

dewlap

This example shows that a side effect to adaptive changes can be reproductive isolation. Another example is seen among *Anolis* lizards, which court females by extending a colorful flap of skin, called a "dewlap." The dewlap must be seen in order to attract mates. Therefore, populations of *Anolis* in a dim forest tend to evolve light-colored dewlaps that reflect light, while populations in open habitats evolve dark-colored dewlaps. This change in dewlap color causes the populations to be reproductively isolated, because females distinguish males of their species by the color of the dewlap.

***Ficedula* Flycatchers** As populations become reproductively isolated, postzygotic isolating mechanisms may arise before prezygotic isolating mechanisms. Postzygotic isolating mechanisms can keep species separate but they represent a large investment of energy to no avail. For example, the production of a hybrid requires an investment of energy that does not result in the passage of genes to future generations. Therefore, natural selection would favor the evolution of prezygotic isolating mechanisms over postzygotic isolating mechanisms. The term *reinforcement* is given to the process of natural selection favoring variations that lead to prezygotic reproductive isolation. An example of reinforcement has been seen in *Ficedula* flycatchers of the Czech Republic and Slovakia. When the pied and collared flycatchers occur in close proximity, the pied flycatchers have evolved a different coat color from that of the collared flycatchers. The difference in color helps the two species recognize and mate with their own species.

▶ **15.3 CHECK YOUR PROGRESS**

1. Describe how the green iguana may have given rise to both the marine iguana and the rhinoceros iguana by expanding on the text description.

15.4 Adaptive radiation produces many related species

When you complete this section, you should be able to

1. Describe and give examples of adaptive radiation.

Adaptive radiation occurs when a single ancestral species gives rise to a variety of species, each adapted to a specific environment. An *ecological niche* is where a species lives and how it interacts with other species. When an ancestral finch arrived on the Galápagos Islands, its descendants spread out to occupy various niches. Geographic isolation of the various finch populations caused their gene pools to become isolated. Because of natural selection, each population adapted to a particular habitat on its island. In time, the many populations became so genotypically different that now, when by chance they reside on the same island, they do not interbreed, and are therefore separate species. The finches use beak shape to recognize members of the same species during courtship. Rejection of suitors with the wrong type of beak is a behavioral type of prezygotic isolating mechanism.

Video
Finches Adaptive
Radiation

Similarly, on the Hawaiian Islands, a wide variety of honeycreepers are descended from a common goldfinch-like ancestor that arrived from Asia or North America about 5 million years ago. Today, honeycreepers have a range of beak sizes and shapes for feeding on various food sources, including seeds, fruits, flowers, and insects, as shown in Figure 15.4, which uses common names for the species.

FIGURE 15.4 Adaptive radiation in Hawaiian honeycreepers.

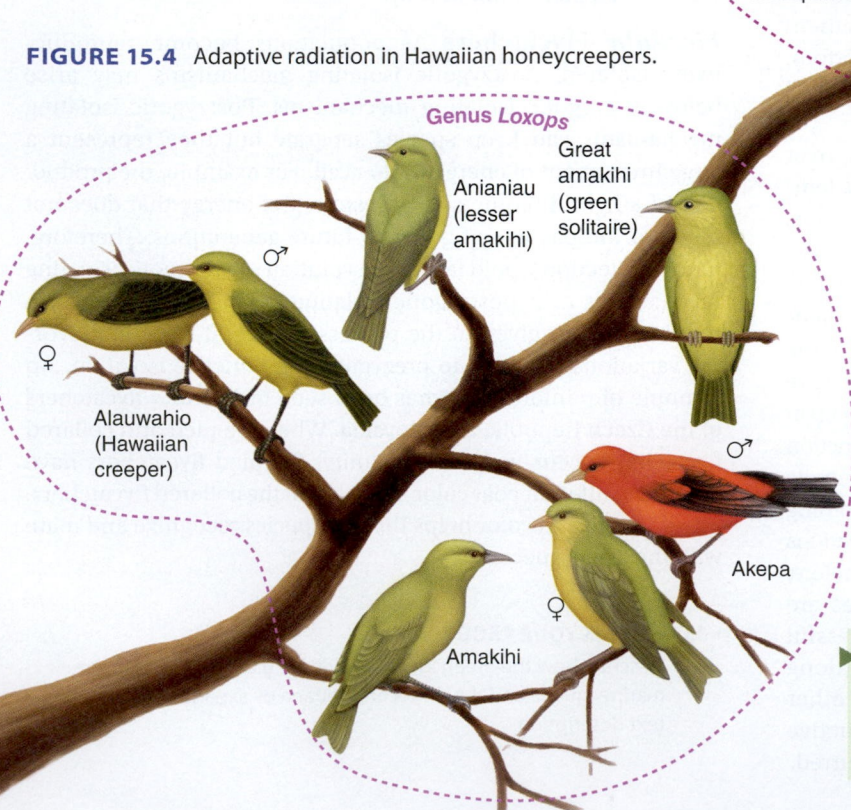

* Extinct species or subspecies

Adaptive radiation has occurred in both plants and animals throughout the history of life on Earth when a group of organisms exploits a new environment. For example, with the demise of the dinosaurs about 66 million years ago, mammals underwent adaptive radiation as they exploited niches previously occupied by the dinosaurs.

This completes our discussion of allopatric speciation. The next part of the chapter discusses speciation when there is no geographic barrier.

▶ **15.4 CHECK YOUR PROGRESS**

1. Explain why the many finch species on the Galápagos Islands are an example of adaptive radiation and how you would know they are separate species.

Speciation Without Geographic Separation

Speciation without the presence of a geographic barrier does occur, and the best examples are due to chromosome number changes in plants. Hybridization, followed by doubling of the chromosome number, can occur naturally or as a result of artificial selection. Such events must have occurred during the artificial selection of corn over the years.

15.5 Speciation occasionally occurs without a geographic barrier

LEARNING OUTCOMES

When you complete this section, you should be able to

1. Relate sympatric speciation in plants to polyploidy.
2. Distinguish between autoploidy and alloploidy.

Speciation without the presence of a geographic barrier is termed **sympatric speciation.** Sympatric speciation has been difficult to substantiate in animals. For example, two populations of the Meadow Brown butterfly, *Maniola jurtina,* have different distributions of wing spots. The two populations are both in Cornwall, England, and they maintain the difference in wing spots, even though there is no geographic boundary between them. But, as yet, no reproductive isolating mechanism has been found. In contrast, we know of instances in plants by which a postzygotic isolating mechanism has given rise to a new species within the range and habitat of the parent species. In other words, no geographic barrier was required. All instances in plants involve **polyploidy,** additional sets of chromosomes beyond the diploid (2n) number. Sympatric speciation is more common in flowering plants than in animals due to self-pollination. A polyploid plant can reproduce only with itself, and cannot reproduce with the parent (2n) population because not all the chromosomes would be able to pair during meiosis. Two types of polyploidy are known: autoploidy and alloploidy.

Speciation through **autoploidy** is seen in diploid plants when nondisjunction occurs during meiosis and a diploid species produces diploid gametes. If this diploid gamete fuses with a haploid gamete, a triploid plant results. A triploid (3n) plant is sterile and cannot produce offspring because the chromosomes cannot pair during meiosis. Humans have found a use for sterile plants because they produce fruits without seeds. Figure 15.5A contrasts a diploid banana with seeds to today's polyploid banana that produces no seeds. If two of the diploid gametes fuse, the plant is a tetraploid (4n) and the plant is fertile, as long as it reproduces with another of its own kind. The fruits of polyploid plants are much larger than those of diploid plants. The huge strawberries of today are produced by octaploid (8n) plants.

FIGURE 15.5A Autoploidy: The small, diploid-seeded banana is contrasted with the large, polyploid banana that produces no seeds.

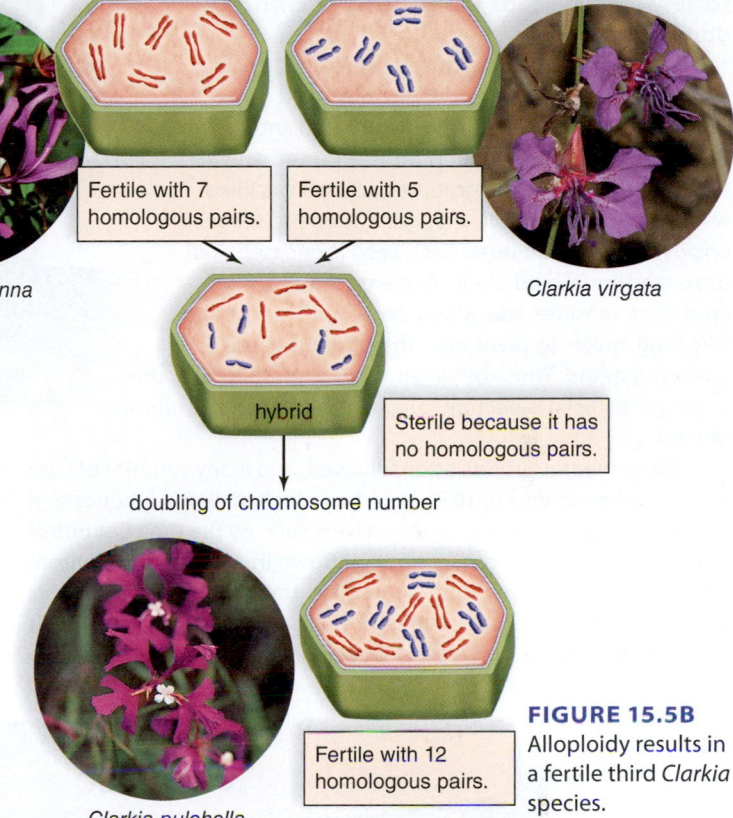

Clarkia concinna

Fertile with 7 homologous pairs.

Fertile with 5 homologous pairs.

Clarkia virgata

hybrid

Sterile because it has no homologous pairs.

doubling of chromosome number

Clarkia pulchella

Fertile with 12 homologous pairs.

FIGURE 15.5B Alloploidy results in a fertile third *Clarkia* species.

Speciation through **alloploidy** requires two steps. The prefix *allo-,* which means "different," is appropriate because the process begins when two different but related species of plants hybridize. First, when two different but related species of plants hybridize, the hybrid is sterile because it has no homologous pairs of chromosomes. Second, if and when a doubling of the chromosome number occurs a new fertile plant results. Figure 15.5B gives an example of alloploidy. The Western wildflower, *Clarkia concinna,* has seven (7) pairs of homologous chromosomes; a related species, *C. virgata,* has five (5) pairs of chromosomes. The hybrid has twelve chromosomes but is sterile because it has no homologous pairs and therefore meiosis of gametogenesis cannot occur. However, researchers have located a plant now called *C. pulchella* that is fertile due to doubling of the chromosome number, which allows the twelve (12) chromosomes to pair during meiosis. Alloploidy also occurred during the evolution of the wheat plant, which is commonly used today to produce bread.

▶ **15.5 CHECK YOUR PROGRESS**

1. Explain which isolating mechanism keeps the three species of *Clarkia* separate.

15A The Many Uses of Corn, an Allotetraploid

When the world record for eating corn on the cob was set at 33½ ears in 12 minutes, the last thing on anyone's mind was the evolution of corn. Corn, also known as maize (*Zea mays*), represents one of the most remarkable plant-breeding achievements in the history of agriculture. Today, modern society literally reaps the benefits of corn as a domestic product.

Modern corn bears little resemblance to its ancient ancestor, an inconspicuous wild grass called teosinte from southern Mexico. Teosinte is a drought-tolerant grass that produces reproductive spikes fairly close to the ground. Each spike is filled with two rows of small, triangular kernels (seeds) enclosed in a tough husk. Each seed is encased and protected by a hard shell. Ancient peoples discovered that teosinte was a source of food and began selecting spikes to plant near their homes, close to irrigation systems. Thus, between 4000 and 3000 B.C., the hand of artificial selection began to shape the evolution of corn.

Experimental hybridization followed, and many varieties of corn were developed. By A.D. 1070, corn had reached North America and was being grown by the Iroquois in New York. By the time Columbus visited the Americas, corn was being grown in a number of environments. Columbus even commented on the fields of corn and its great taste. We now know that corn is an allotetraploid, meaning it is 4n. Hybridization between two related species must have been

Teosinte
(*Zea mexicana*)

followed by doubling of the chromosomes, accounting for why ears of corn are now so large (Fig. 15A).

Today corn is America's number-one field crop, yielding approximately 9.5 billion bushels yearly. It is an important food source for both humans and livestock. Corn is a component of over 3,000 grocery products, including cereals, corn syrup, cornstarch, ice cream, soft drinks, chips, snack foods, and even peanut butter. It is also used in making glue, shoe polish, ink, soaps, and synthetic rubber. Now, corn is also a source for the production of ethanol to fuel our vehicles. The uses of corn seem to be limited only by our imaginations (Fig. 15A).

CONSIDER THESE QUESTIONS

1. Using corn to produce ethanol raises the cost of corn because it makes less available for food and feed in the United States and abroad. Should we continue to convert corn to ethanol? Why or why not?

2. Much corn is grown as feed for animals. The sewage produced by animals can be a threat to our health and eating beef can lead to circulatory problems. Should we stop eating beef?

connect |BIOLOGY Explore the concepts through a variety of multimedia assets and question types.

www.mcgrawhillconnect.com

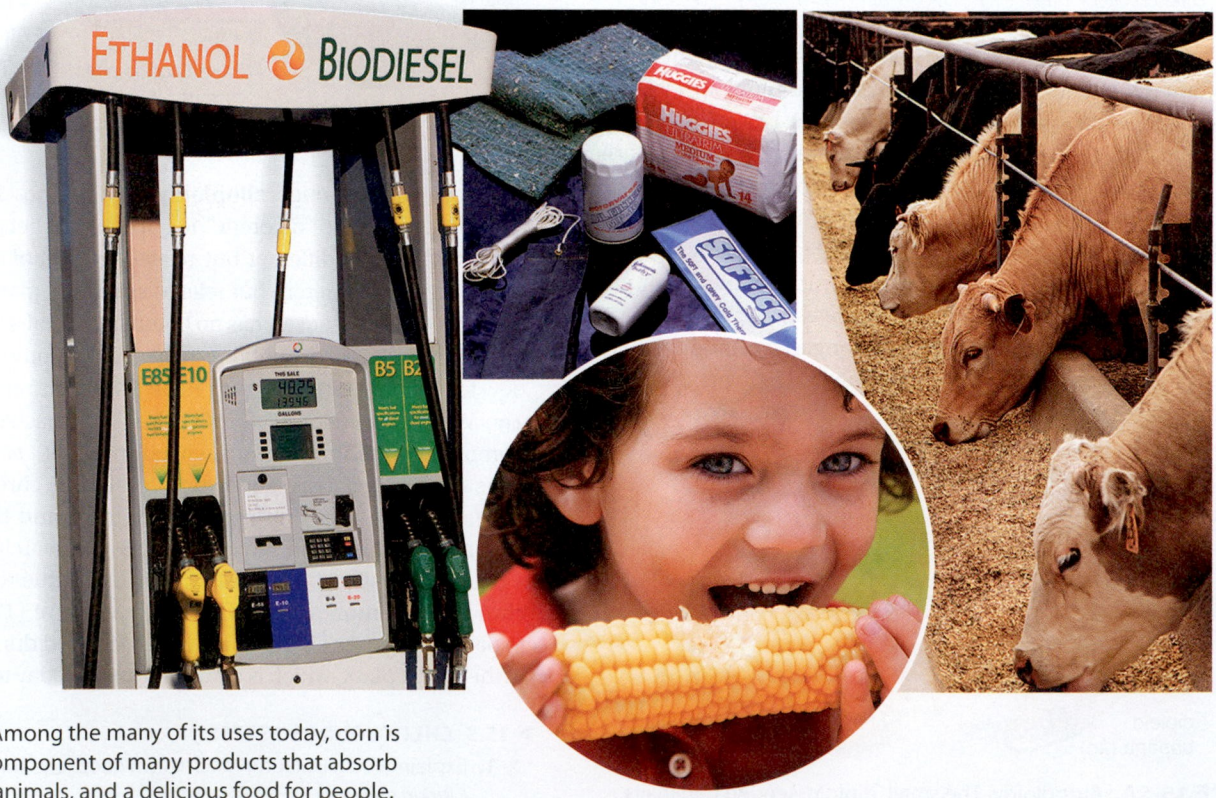

FIGURE 15A Among the many of its uses today, corn is a fuel source, a component of many products that absorb water, a feed for animals, and a delicious food for people.

Principles of Macroevolution

The gradualistic model of speciation can be contrasted with the punctuated equilibrium model. The gradualistic model predicts transitional links that may fossilize while the punctuated equilibrium model predicts few, if any, transitional links in the fossil record. Studies of developmental regulatory *Hox* genes show us that developmental changes could contribute to rapid speciation. Still, any phenotypic change, even if it occurs rapidly, is subject to natural selection, and when we view, say, the evolution of the horse, we see no constant pattern toward a goal.

15.6 Speciation occurs at different tempos

LEARNING OUTCOME

When you complete this section, you should be able to

1. Compare and contrast the gradualistic model of speciation with the punctuated equilibrium model.

Many evolutionists conclude, as Darwin did, that evolutionary changes occur gradually. Therefore, these evolutionists support a *gradualistic model,* which proposes that speciation occurs after populations become isolated, with each group continuing slowly on its own evolutionary pathway. These evolutionists often show the history of groups of organisms by drawing the type of diagram shown in Figure 15.6A. Note that in this diagram, an ancestral species has given rise to two separate species, represented by a slow change in plumage color. The gradualistic model suggests that it is difficult to indicate when speciation occurred because there would be so many transitional links. However, in some cases, it has been possible to trace the evolution of a group of organisms by finding transitional links.

After studying the fossil record, some paleontologists tell us that species can appear quite suddenly, and then they remain essentially unchanged phenotypically until they undergo extinction. Based on these findings, they developed a *punctuated equilibrium model* to explain the pace of evolution. This model says that periods of equilibrium (no change) are punctuated (interrupted) by speciation. Figure 15.6B shows this way of representing the history of evolution over time. This model suggests that transitional links are less likely to become fossils and less likely to be found. Moreover, speciation is apt to involve an isolated population at one locale, because a favorable genotype could spread more rapidly within such a population. Only when this population expands and replaces other species is it apt to show up in the fossil record.

A strong argument can be made that it is not necessary to choose between these two models of evolution and that both could very well assist us in interpreting the fossil record. In a stable environment, a species may be kept in equilibrium by stabilizing selection for a long period. If environmental change is rapid, a new species may arise suddenly before the parent species goes on to extinction. Because geologic time is measured in millions of years, the "sudden" appearance of a new species in the fossil record could actually represent many thousands of years. Using only a small rate of change (0.0008/year), two investigators calculated that the brain size in the human lineage could have increased from 900–1,400 cm^3 in only 135,000 years. This would be a very rapid change in the fossil record. Actually, the record indicates that brain enlargement took about 500,000 years, indicating that the real pace was slower than it might have been.

▶ **15.6 CHECK YOUR PROGRESS**

1. Explain what needs to be determined before the many new and varied excavated human skulls in East Africa can be used to support either model of speciation.

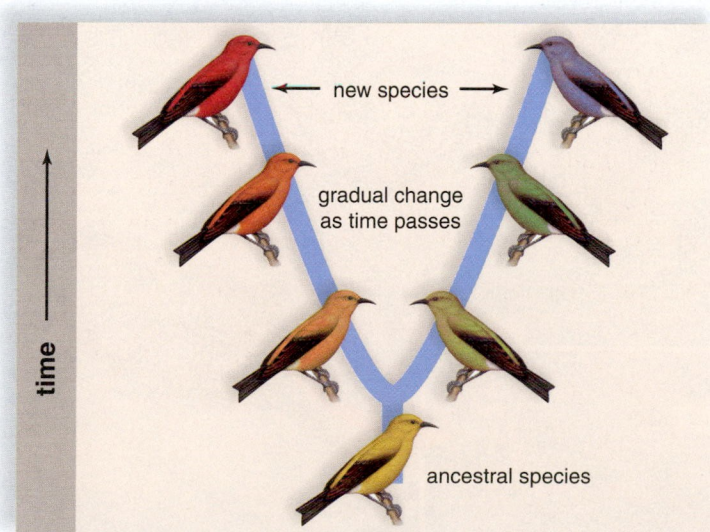

FIGURE 15.6A Gradualistic model of speciation.

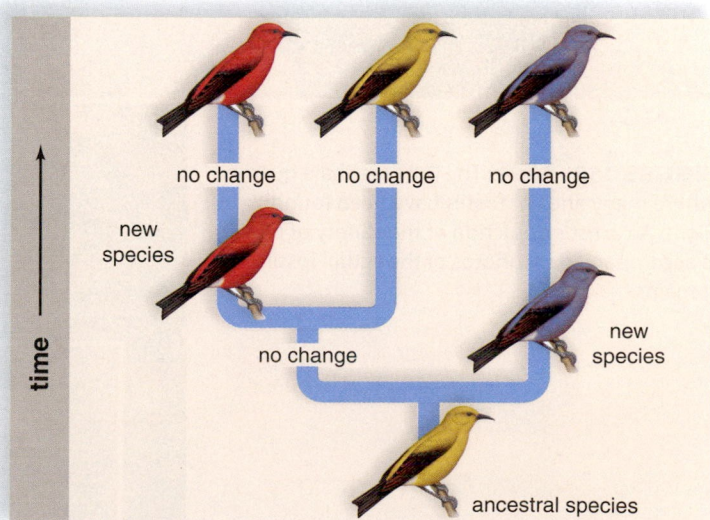

FIGURE 15.6B Punctuated equilibrium model of speciation.

15B The Burgess Shale Hosts a Diversity of Life

Finding the Burgess Shale, a rock outcropping in Yoho National Park, British Columbia, was a chance happening. In 1909, Charles Doolittle Walcott of the Smithsonian Institute was out riding when his horse stopped in front of a rock made of shale. He cracked the rock open and saw the now-famous fossils of the animals depicted in Figure 15B. Walcott and his team began working the site and continued on their own for quite a few years. Around 1960, other paleontologists became interested in studying the Burgess Shale fossils.

As a result of uplifting and erosion, the intriguing fossils of the Burgess Shale are relatively common in that particular area. However, the highly delicate impressions and films found in the rocks are very difficult to remove from their matrix. Early attempts to remove the fossils involved splitting the rocks along their sedimentary plane and using rock saws. Unfortunately, these methods were literally "shots in the dark," and many valuable fossils were destroyed in the process. Using ultraviolet light to see the fossils and diluted acetic acid solutions to remove the matrix was more successful in freeing the fossils.

The fossils tell a remarkable story of marine life some 540 MYA (millions of years ago). In addition to fossils of organisms that had external skeletons, many of the fossils are remains of soft-bodied invertebrates; these are a great find because soft-bodied animals rarely fossilize. During this time, all organisms lived in the sea, and it is believed the barren land was subject to mudslides, which entered the ocean and buried the animals, killing them. Later, the mud turned into shale, and later still, an upheaval raised the shale. Before the shale formed, fine mud particles filled the spaces in and around the organisms so that the soft tissues were preserved and the fossils became somewhat three-dimensional.

The fossils tell us that the ancient seas were teeming with weird-looking, invertebrate animals. All of today's groups of animals can trace their ancestry to one of these strange-looking forms, which include sponges, arthropods, worms, and tribolites, as well as spiked creatures and oversized predators. The animals featured in Figure 15B have been assigned to these genera because they are believed to be the type of animal mentioned:

FIGURE 15B *Above:* The Burgess Shale quarry, where many ancient fossils have been found. *Right:* An artist's depiction of the variety of fossils is accompanied by photos of the actual fossilized remains.

Opabinia

Genus	Type of
Opabinia	crustacean
Thaumaptilon	sea pen
Vauxia	sponge
Wiwaxia	segmented worm

The vertebrates, including humans, are descended from *Pikaia,* the only one of the fossils that has a supporting rod called a notochord. (In vertebrates, the notochord is replaced by the vertebral column during development.)

Unicellular organisms have also been preserved at the Burgess Shale site. They appear to be bacteria, cyanobacteria, dinoflagellates, and other protists. Fragments of algae are preserved in thin, shiny carbon films. A technique has been perfected that allows the films to be peeled off the rocks.

Anyone can travel to Yoho National Park, look at the fossils, and get an idea of the types of animals that dominated the world's oceans for nearly 300 million years. Some of the animals had external skeletons, but many were soft-bodied. Interpretations of the fossils vary. Some authorities hypothesize that the great variety of animals in the Burgess Shale evolved within 20–50 million years, and therefore the site supports the hypothesis of punctuated equilibrium. Others believe that the animals started evolving much earlier and that we are looking at the end result of an adaptive radiation requiring many more millions of years to accomplish. Some investigators present evidence that all the animals are related to today's animals and should be classified as such. Others believe that several of them are unique creatures unrelated to the animals of today. Regardless of the controversies, the fossils tell us that speciation, diversification, and eventual extinction are part of the history of life.

CONSIDER THESE QUESTIONS

1. How should the scientific community react when confronted with a phenomenon like the Burgess Shale?
2. Should students be exposed to phenomena that cannot as yet be fully explained? Why or why not?

connect | BIOLOGY   Explore the concepts through a variety of multimedia assets and question types.

www.mcgrawhillconnect.com

Thaumaptilon

Vauxia

Wiwaxia

15.7 Development plays a role in speciation

LEARNING OUTCOME

When you complete this section, you should be able to

1. Explain how regulatory genes could have contributed to speciation of higher classification categories.

Whether slow or fast, how could evolution have produced the myriad of animals in the Burgess Shale and, indeed, in the history of life? Or, to ask the question in a genetic context, how can genetic changes bring about such major differences in form? It has been suggested since the time of Darwin that the answer must involve developmental processes. In 1917, D'Arcy Thompson asked us to imagine an ancestor in which all parts are developing at a particular rate. A change in regulatory gene expression could stop a developmental process or continue it beyond its normal time. For instance, if the growth of limb bones were stopped early, the result would be shorter limbs, and if it were extended, the result would be longer limbs compared to those of an ancestor. Or, if the whole period of growth were extended, a larger animal would result, accounting for why some species of horses are so large today.

Using new kinds of microscopes and the modern techniques of cloning and manipulating genes, investigators have indeed discovered genes whose *differential expression* can bring about changes in body shapes (Fig. 15.7A). More surprisingly, these same regulatory genes occur in all organisms. This finding suggests that these genes must date back to a common ancestor that lived more than 600 MYA (before the Burgess Shale animals), and that despite millions of years of divergent evolution, all animals share the same regulatory switches for development. Previously in section 12.4, page 220, we pointed out how the same regulatory gene, *Pax6*, turns on eye development even though the animal kingdom contains many different types of eyes, and it was long thought that each type of eye would require its own set of genes. Flies, crabs, and other arthropods have compound eyes composed of hundreds of individual

> Despite millions of years of divergent evolution, all animals share the same regulatory genes for development.

visual units. Humans and all other vertebrates have a camera-type eye with a single lens. So do squids and octopuses. Humans are not closely related to either flies or squids, so wouldn't it seem likely that all three types of animals evolved "eye" genes separately? Not so. In 1994, Walter Gehring and his colleagues at the University of Basel, Switzerland, discovered that *Pax6* is required for eye formation in all animals tested. Mutations in the *Pax6* gene lead to failure of eye development in both people and mice, and remarkably, the mouse *Pax6* gene can cause an eye to develop on the leg of a fruit fly.

Increase in Complexity The regulatory genes called *Hox* genes have been much studied, and investigators tell us that the number of these genes increased twice during the evolution of animals. Both expansions are associated with an increase in complexity, defined by the appearance of different cell types. One expansion occurred during the evolution of vertebrates. Invertebrates have 13 *Hox* genes, while vertebrates, including humans, have four copies of the 13 *Hox* gene set. It appears that the set underwent a series of duplications, and some of the duplicate genes may have taken on new functions, a process mentioned in "Sometimes Mutations Are Beneficial" on page 268. Similarly, some other sets of regulatory genes that operate in development were duplicated when vertebrates evolved. This increase in the number of regulatory gene numbers may have contributed to the evolution of vertebrates and to their complexity.

eye on fruit fly leg

The limbs of these terrestrial mammals are shaped for running (or walking).

The limbs of birds are shaped for flight.

FIGURE 15.7A Differential expression of the same regulatory genes during development can account for differences in vertebrate limbs.

Development of Limbs Wings and arms are very different, but both humans and birds express the *Tbx5* regulatory gene in developing limb buds. *Tbx5* codes for a transcription factor that turns on the genes needed to make a limb. What seems to have changed as birds and mammals evolved are the genes that *Tbx5* turns on. Perhaps in an ancestral tetrapod, the Tbx5 protein triggered the transcription of only one gene. In mammals and birds, a few genes are expressed in response to Tbx5 protein, but the particular genes are different.

Hindlimb reduction has occurred during the evolution of other mammals. For example, as whales and manatees evolved from land-dwelling ancestors into fully aquatic forms, the hindlimbs became greatly reduced in size. Similarly, legless lizards have evolved many times. A stickleback fish study has shown how natural selection can lead to major skeletal changes in a relatively short time. The three-spined stickleback fish occurs in two forms in North American lakes. In the open waters of a lake, long pelvic spines help protect the stickleback from being eaten by large predators. But on the lake bottom, long pelvic spines are a disadvantage because dragonfly larvae seize and feed on young sticklebacks by grabbing them by their spines. The presence of short spines in bottom-dwelling stickleback fish can be traced to a reduction in the development of the pelvic-fin bud in the embryo, and this reduction is due to the altered expression of a regulatory gene.

Development of Overall Shape Vertebrates have repeating segments, as exemplified by the vertebral column. In general, *Hox* genes control the development of repeated structures along the main body axes of vertebrates. Shifts in how long *Hox* genes are expressed per segment in embryos are responsible for why the snake has hundreds of rib-bearing vertebrae and essentially no neck in contrast to other vertebrates, such as a chick (Fig. 15.7B).

> Animal diversity is due in large part to variations in the expression of ancient regulatory genes during development.

Changes in the timing of *Hox* gene expression can also account for the evolution of four legs rather than a fin. In vertebrates with legs, the *Hox* genes are turned on again in a later phase of development. This phase is associated with the further growth outward of the limb bones to form the limb and digits where a fin formerly existed. On the other hand, the inability of a *Hox* gene to turn on other regulatory genes in certain segments can explain why insects have just six legs, and other arthropods, such as crayfish, have ten legs. In general, the study of *Hox* genes has shown how animal diversity is due to variations in the expression of ancient genes rather than to wholly new and different genes.

Many investigators consider *Hox* genes to be **master regulatory genes** because they turn on other genes that directly control the shape and structure of an organism. There seems to be a relatively small number of genes that can be called master regulatory genes of development.

Human Evolution The sequencing of genomes has shown us that our DNA base sequence is very similar to that of chimpanzees, mice, and, indeed, all vertebrates. Based on this knowledge and the work just described, investigators no longer expect to find new genes to account for the evolution of humans. Instead, they predict that differential gene expression and/or new functions for "old" genes will explain how humans evolved.

We have to keep in mind that developmental changes result in a phenotype that is subject to natural selection. We would expect that during the history of life many changes in phenotype were not advantageous and therefore did not become prevalent in future generations.

▶ **15.7 CHECK YOUR PROGRESS**

1. Discuss how the study of regulatory genes supports the possibility of rapid speciation in the fossil record.

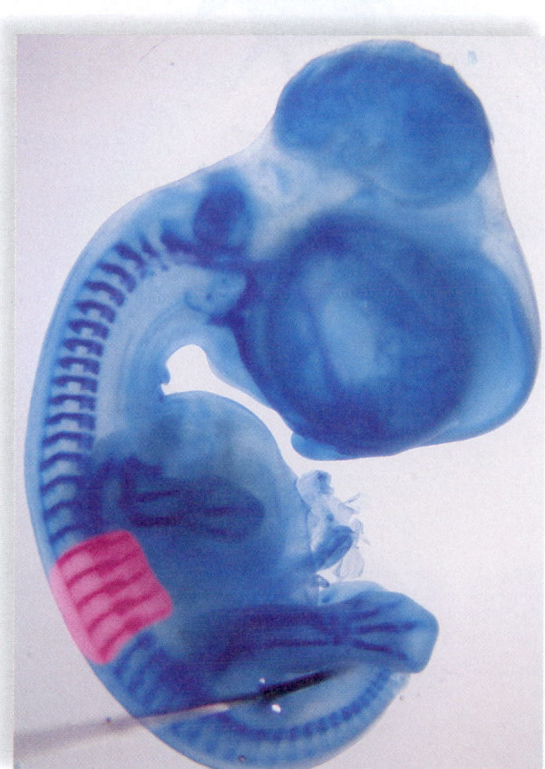

a. Chick vertebrae

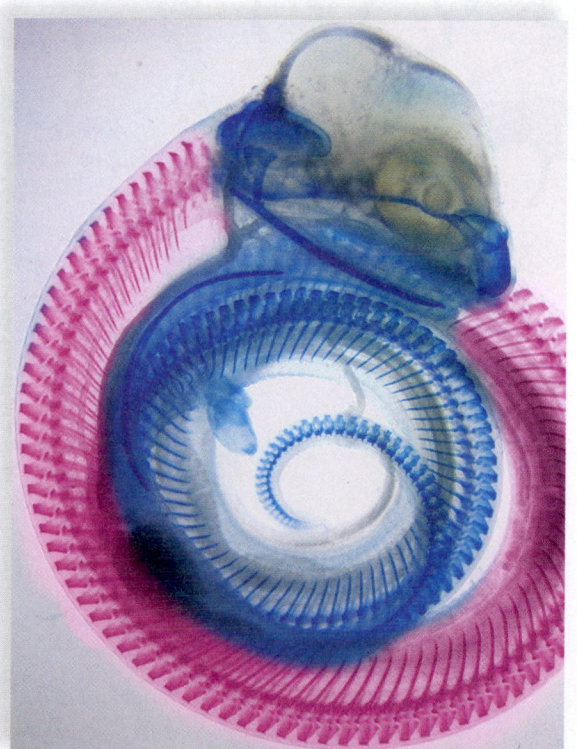

b. Snake vertebrae

FIGURE 15.7B
Differential expression of a *Hox* gene causes (**a**) a chick to have fewer vertebrae than (**b**) a snake in a particular region (colored pink) of the spine.

Burke, A. C. 2000. *Hox* genes and the global patterning of the somitic mesoderm. In Somitogenesis. C. Ordahl (ed.) *Current Topics in Developmental Biology,* Vol. 47. Academic Press.

15.8 Speciation is not goal-oriented

LEARNING OUTCOME

When you complete this section, you should be able to

1. Discuss the evolution of the horse to show that evolution is not goal-oriented.

The evolution of the horse, *Equus*, has been studied since the 1870s, and at first the ancestry of this genus seemed to represent a model for gradual, straight-line evolution until its goal, the modern horse, had been achieved. Three trends were particularly evident during the evolution of the horse: increase in overall size, toe reduction, and change in tooth size and shape.

By now, however, many more fossils have been found, making it easier to tell that the lineage of a horse is complicated by the presence of many ancestors with varied traits. The family tree in Figure 15.8 is an oversimplification because each of the names is a genus that contains several species, and not all past genera in the horse family are included. It is apparent, then, that the ancestors of *Equus* form a thick bush of many equine species and that straight-line evolution did not occur. Because *Equus* alone remains and the other genera have died out, it might seem as if evolution was directed toward producing *Equus,* but this is not the case. Instead, each of these ancestral species was adapted to its environment. Adaptation occurs only because the members of a population with an advantage are able to have more offspring than other members. Natural selection is opportunistic, not goal-oriented.

Fossils named *Hyracotherium* have been designated as the first probable members of the horse family, living about 57 MYA. These animals had a wooded habitat, ate leaves and fruit, and were about the size of a dog. Their short legs and broad feet with several toes would have allowed them to scamper from thicket to thicket to avoid predators. *Hyracotherium* was obviously well adapted to its environment because this genus survived for 20 million years.

The family tree of *Equus* tells us once more that speciation, diversification, and extinction are common occurrences in the fossil record. The first adaptive radiation of horses occurred about 35 MYA. The weather was becoming drier, and grasses were evolving. Eating grass requires tougher teeth, and an increase in size and longer legs would have permitted greater speed to escape enemies. The second adaptive radiation of horses occurred about 15 MYA and included *Merychippus* as a representative of these groups of speedy grazers that lived on the open plain. By 10 MYA, the horse family had become quite diversified. Some species were large forest browsers, some were small forest browsers, and others were large plains grazers. Many species had three toes, but some had one strong toe. (The hoof of the modern horse includes only one toe.)

Modern horses evolved about 4 MYA from ancestors who had features adaptive for living on an open plain, such as large size, long legs, hoofed feet, and strong teeth. The other groups of horses prevalent at the time became extinct, no doubt for complex reasons.

▶ **15.8 CHECK YOUR PROGRESS**

1. Describe what the evolutionary tree of the modern horse would look like if evolution were goal-oriented.

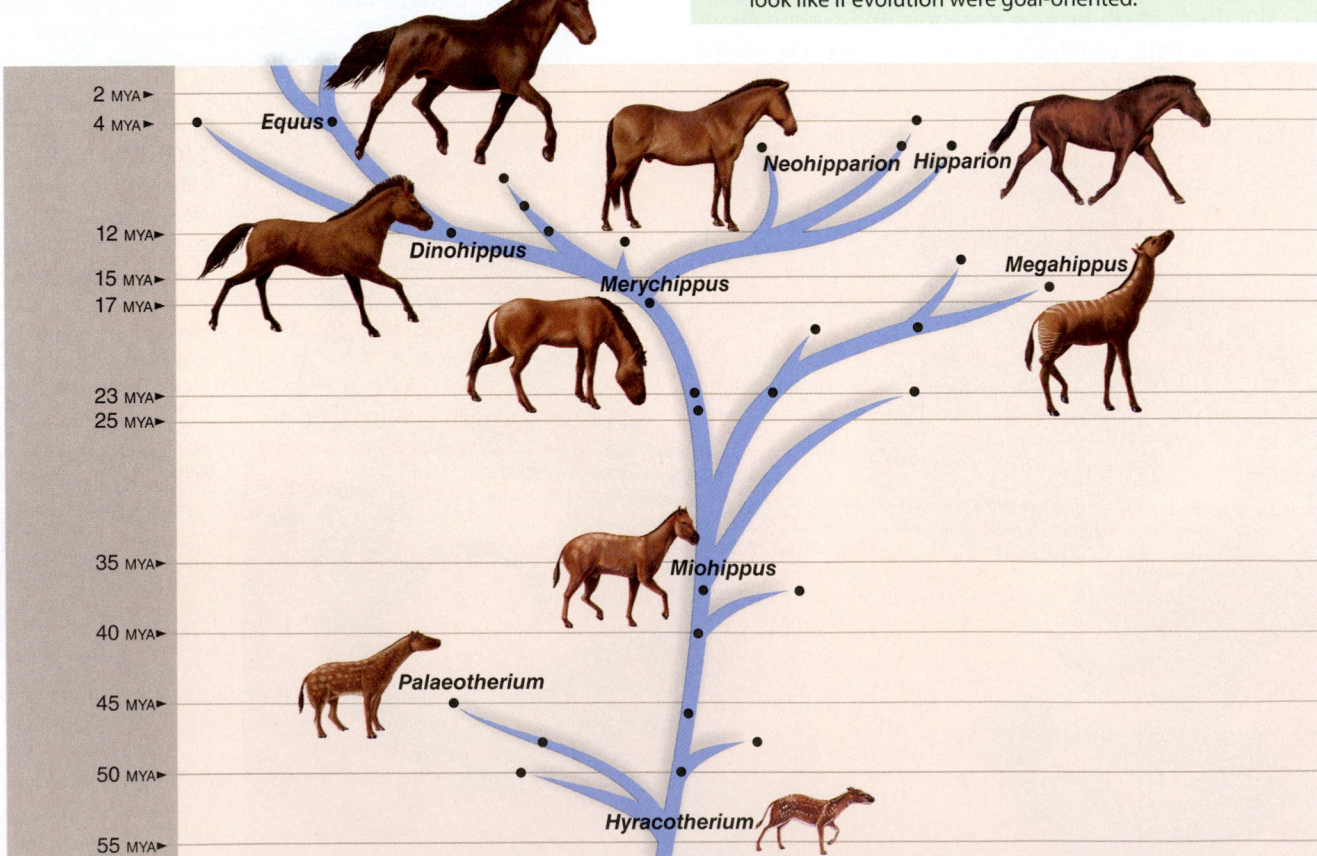

FIGURE 15.8 Simplified family tree based on the fossil record of *Equus*. Every dot represents a genus.

CONNECTING THE CONCEPTS

Macroevolution, the study of the origin and history of the species on Earth, is the subject of this chapter and the next. The biological species concept states that the members of a species have an isolated gene pool and can only reproduce with one another.

The origin of species is called speciation. Speciation usually occurs after two populations derived from a larger one are separated geographically. If the members of a salamander population are suddenly divided by a barrier, each new population becomes adapted to its particular environment over time. Eventually, the two populations may become so genetically different that even if members of each population come into contact, they will not be able to produce fertile offspring. Because gene flow between the two populations is no longer possible, the

salamanders are considered separate species. Aided by geographic separation, multiple species can repeatedly arise from an ancestral species, as when a common ancestor from the mainland led to many species of Hawaiian honeycreepers, each adapted to its own particular environment.

Does speciation occur gradually, as Darwin supposed, or rapidly (in geologic time), as described by the punctuated equilibrium model? The fossils of the Burgess Shale support the punctuated equilibrium model. How can genetic changes bring about such major changes in form, whether fast or slow? Investigators have now discovered ancient regulatory genes (e.g., *Hox* genes), whose differential expression can bring about changes in body shapes and organs.

Evolution is not directed toward any particular end, and the traits of the species alive

today arose through common descent with adaptations to a local environment. The subject of Chapter 16 is the evolutionary history and classification of organisms today.

ANALYZE AND EVALUATE

1. Scientists make observations and then formulate testable hypotheses to explain the observations. What testable hypotheses have biologists made about speciation? (*Hint:* How might species arise? What might cause them to arise?)
2. Paleontologists suggested the punctuated equilibrium model. What data did they use?
3. The study of evolution is a scientific endeavor. Review the scientific method (Figure 1.1B) and explain this statement.

SUMMARIZE

Diversity Requires Speciation

15.1 Species have been defined in more than one way

- **Macroevolution** depends on speciation. **Speciation** occurs when one species splits into two or more species or when one species becomes a new species over time. According to the **evolutionary species concept,** every species has its own evolutionary history, and a species can be recognized by diagnostic traits. According to the **biological species concept,** members of a species are reproductively isolated from members of other species. They can only reproduce with members of their own species.

15.2 Reproductive barriers maintain genetic differences between species

- **Prezygotic isolating mechanisms** prevent reproductive attempts. **Postzygotic isolating mechanisms** prevent zygote development or F_1 and F_2 hybrid offspring from breeding.

Prezygotic Isolating Mechanisms		Postzygotic Isolating Mechanisms
Premating →	Mating →	Fertilization
Habitat isolation	Mechanical isolation	Zygote mortality
Temporal isolation		Hybrid sterility
	Gamete isolation	
Behavioral isolation		F_2 fitness

Speciation Due to Geographic Separation

15.3 Allopatric speciation utilizes a geographic barrier

- **Allopatric speciation** begins when populations derived from a larger one are separated by a barrier and they start to differ genetically and phenotypically. Following separation, postzygotic mechanisms followed by prezygotic mechanisms can develop over time.

15.4 Adaptive radiation produces many related species

- When **adaptative radiation** occurs, several new species evolve from an ancestral species, and they adapt to fill different niches separated by geographic barriers. The many types of Hawaiian honeycreepers are a result of adaptive radiation.

Speciation Without Geographic Separation

15.5 Speciation occasionally occurs without a geographic barrier

- **Sympatric speciation** occurs without a geographic barrier. Speciation through **polyploidy** occurs when a plant acquires an additional set of chromosomes beyond the diploid (2n) number. A polyploidy cannot reproduce with a 2n plant. Speciation through **autoploidy** occurs when a diploid gamete fuses with a haploid gamete, resulting in a triploid plant, which is sterile. Speciation through **alloploidy** occurs when two different but related species of plants hybridize, and then the chromosome number doubles making the hybrid fertile.

Principles of Macroevolution

15.6 Speciation occurs at different tempos

- According to the gradualistic model, speciation occurs gradually, perhaps due to a gradually changing environment. According to the punctuated equilibrium model, periods of equilibrium are interrupted by rapid speciation. Perhaps, if the environment changes rapidly, new species may suddenly arise.
- On occasion, fossil record data may fit one model of speciation, and on another occasion, it may fit the other model. Speciation, diversification, and eventual extinction are part of the history of life.

15.7 Development plays a role in speciation

- Despite millions of years of divergent evolution, all animals share the same regulatory genes for development. *Hox* gene duplications could have brought about increased complexity in animals. Many investigators consider *Hox* genes to be **master regulatory genes** because they turn on other genes that control the shape and structure of the organism.
- Eye development, limb development, and shape determination are controlled by the same regulatory genes in different animals. But differential expression can account for differences in outcome. Investigators hypothesize that differential gene expression during development coupled with natural selection can account for the process of evolution, including human evolution.

15.8 Speciation is not goal-oriented

- In horses, each ancestral species adapted to its environment, but due to a changing environment, only *Equus* survived. The fossil record of *Equus* shows speciation, diversification, and extinction. The family tree shows that at least two major adaptive radiations occurred in the past.
- Natural selection is opportunistic, not goal-oriented; adaptation occurs because members with an advantage can have more offspring.

TEST YOURSELF

Diversity Requires Speciation

1. A biological species
   a. always looks different from other species.
   b. always has a different chromosome number from that of other species.
   c. is reproductively isolated from other species.
   d. never occupies the same niche in different environments.

For questions 2–7, indicate the type of isolating mechanism described in each scenario in the key.

KEY:

a. habitat isolation	e. gamete isolation
b. temporal isolation	f. zygote mortality
c. behavioral isolation	g. hybrid sterility
d. mechanical isolation	h. low F_2 fitness

2. Males of one species do not recognize the courtship behaviors of females of another species.
3. One species reproduces at a different time than another species.
4. A cross between two species produces a zygote that always dies.
5. Two species do not interbreed because they occupy different areas.
6. The sperm of one species cannot survive in the reproductive tract of another species.
7. The offspring of two hybrid individuals exhibit poor vigor.

8. Which of these is a prezygotic isolating mechanism?
   a. habitat isolation
   b. temporal isolation
   c. hybrid sterility
   d. zygote mortality
   e. Both a and b are correct.
9. Male moths recognize females of their species by sensing chemical signals called pheromones. This is an example of
   a. gamete isolation.
   b. habitat isolation.
   c. behavioral isolation.
   d. mechanical isolation.
   e. temporal isolation.
10. Which of these is mechanical isolation?
    a. Sperm cannot reach or fertilize an egg.
    b. Courtship patterns differ.
    c. The organisms live in different locales.
    d. The organisms reproduce at different times of the year.
    e. Genitalia are unsuited to each other.
11. **THINKING CONCEPTUALLY** Regardless of how speciation occurs or how species are defined, what is required for separate species to be present?

Speciation Due to Geographic Separation

12. Complete the following diagram illustrating allopatric speciation by using these phrases: genetic changes (used twice), geographic barrier, species 1, species 2, species 3.

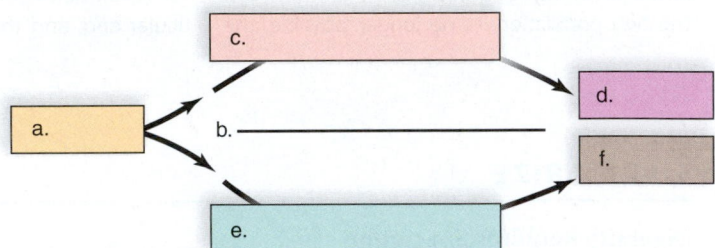

13. The creation of new species without the need of a geographic barrier is called
    a. isolation speciation.
    b. allopatric speciation.
    c. allelomorphic speciation.
    d. sympatric speciation.
    e. symbiotic speciation.
14. The many species of Galápagos finches are each adapted to eating different foods. This is the result of
    a. gene flow.
    b. adaptive radiation.
    c. sympatric speciation.
    d. genetic drift.
    e. All of these are correct.
15. **THINKING CONCEPTUALLY** The Hawaiian Islands are some distance from any mainland, and the plants and animals on each island are unique. Only short distances separate the Florida Keys from each other and the mainland. The mainland and the Keys all contain the same species. Explain.

Speciation Without Geographic Separation

16. Allopatric, but not sympatric, speciation requires
    a. reproductive isolation.
    b. geographic isolation.
    c. spontaneous differences in males and females.
    d. prior hybridization.
    e. rapid rate of mutation.
17. Which of the following is not a characteristic of plant alloploidy?
    a. hybridization
    b. chromosome doubling
    c. self-fertilization
    d. All of these are characteristics of plant alloploidy.

18. Corn is an allotetraploid, which means that its
    a. chromosome number is 4n.
    b. development resulted from hybridization.
    c. development required a geographic barrier.
    d. Both a and b are correct.

Principles of Macroevolution

19. Transitional links are least likely to be found if evolution proceeds according to the
    a. gradualistic model.
    b. punctuated equilibrium model.
    c. Both a and b are correct.
    d. None of these are correct.
20. Adaptive radiation is only possible if evolution is punctuated.
    a. true
    b. false
21. Why are there no fish fossils in the Burgess Shale?
    a. The habitat was not aquatic.
    b. Fish do not fossilize easily because they do not have shells.
    c. The fossils of the Burgess Shale predate vertebrate animals.
    d. There are fish fossils in the Burgess Shale.
22. Which of the following can influence the rapid development of new species of animals?
    a. the influence of molecular clocks
    b. a change in the expression of regulatory genes
    c. the sequential expression of genes
    d. All of these are correct.
23. Which of the following does not seem to influence speciation?
    a. the evolution of different types of *Hox* genes
    b. the evolution of new types of genes that control development
    c. Only the environment and not genes influence speciation.
    d. Both a and b are correct.
24. Which gene is incorrectly matched to its function?
    a. *Hox*—body shape
    b. *Pax6*—body segmentation
    c. *Tbx5*—limb development
    d. All of these choices are correctly matched.

25. **THINKING CONCEPTUALLY** Explain the statement that "*Hox* genes are ancient genes."
26. In the evolution of the modern horse, which was the goal of the evolutionary process?
    a. large size
    b. single toe
    c. Both a and b are correct.
    d. Neither a nor b is correct.
27. Which of the following does not pertain to *Hyracotherium,* an ancestral horse genus?
    a. small size
    b. single toe
    c. wooded habitat
    d. All of these are characteristics of *Hyracotherium.*

GET INVOLVED

1. You want to decide what definition of a species to use in your study. What are the advantages and disadvantages of the evolutionary and biological species concepts?
2. You decide to create a hybrid by crossing two species of plants. If the hybrid is a fertile plant that produces normal-sized fruit, what conclusion is possible?

MEDIA STUDY TOOLS

mhhe.com/maderconcepts3

Enhance your study of this chapter with interactive study tools, practice tests, and engaging animations. Also, ask your instructor about the resources available through ConnectPlus, which includes LearnSmart, a personalized adaptive learning program, and a media-rich eBook.

16

The Evolutionary History of Life on Earth

BEFORE YOU BEGIN

Take a few minutes to recall

What an evolutionary tree is (section 1.7)

Life's characteristics (section 1.10)

Fossil and biogeographic evidence for evolution (sections 14.3 and 14.5)

Motherhood Among Dinosaurs

Because dinosaurs are classified as reptiles, paleontologists at first assumed that dinosaurs behaved in the same manner as today's reptiles. For example, the female American alligator lays her eggs in a bowl-shaped nest made from vegetation and mud. She also covers the eggs with vegetation, which protects and keeps the eggs warm as it decays. The mother stays nearby, and when the young call out from inside the eggs, she opens the nest, allowing them to escape. The Nile crocodile is known for further helping her young by carrying the hatchlings down to the water in her mouth.

Birds also construct bowl-shaped nests. The structural evidence that birds are dinosaurs is strong. Gerald Mayr of the Senckenberg Research Institute in Frankfurt, Germany, has stated that unique traits shared by *Archaeopteryx* and other early birdlike fossils are also present in dinosaurs, such as *Microraptor,* a gliding dinosaur. Beyond the structural evidence, we now have behavioral evidence that at least some dinosaurs nested in the same manner as birds, giving us additional indications that birds are the last surviving dinosaurs.

Within the past 30 years, similar bowl-shaped nests containing dinosaur eggs have been found in Mongolia, Argentina, and Montana. Paleontologists Jack Horner and Bob Makela of Montana State University discovered an entire colony of bowl-shaped nests in Montana. The nests contained fossilized eggs and bones along with eggshell fragments. The space between the nests was large enough for an adult parent to stand lengthwise. From this evidence, these researchers concluded that the baby dinosaurs stayed in the nest after hatching until they had grown large enough to walk around and fragment the eggshells. The spacing between the nests suggested that perhaps the mother fed the young. Such behavior would be more like that of a bird than an alligator or a crocodile.

The nests belonged to a dinosaur the paleontologists called *Maiasaura,* meaning "good mother lizard." *Maiasaura* was a large dinosaur with a head that looked like that of a horse. The head had

Microraptor, a winged gliding dinosaur

Fossil bones of *Maiasaura* hatchlings have been found.

Fossilized remains of *Maiasaura* eggs (boulder-shaped) have been found.

Today's birds also build bowl-shaped nests.

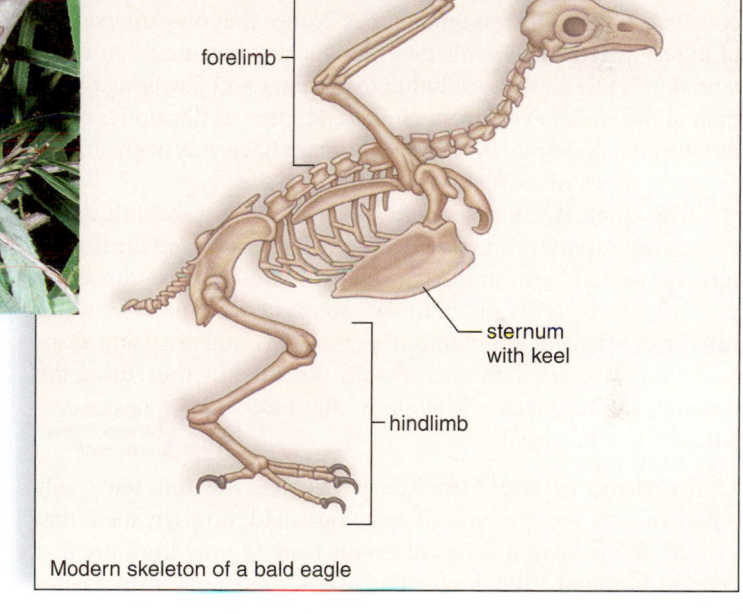

forelimb

sternum with keel

hindlimb

Modern skeleton of a bald eagle

a skull crest, which could have served as a resonating chamber when the dinosaurs communicated with one another. *Maiasaura* could stand and walk on either two or four legs and had a heavy, muscular tail. Perhaps the tail was used for defense, or perhaps these dinosaurs were protected by their herd behavior. The remains of an enormous herd of *Maiasaura,* found by Horner and colleagues, are estimated to consist of nearly 30 million bones, representing 10,000 animals, in an area measuring about 2.6 km². Herd behavior and a means of communication allowed Horner to conclude that these particular dinosaurs may have had some sort of social structure.

Since 1978, when Horner and Makela found the nests of *Maiasaura,* others have made similar discoveries. For example, in 1993, Mark Norrell of the American Museum of Natural History found nests in Mongolia that belonged to a dinosaur called *Oviraptor,* which was much smaller and more birdlike than *Maiasaura.* *Oviraptor* was less than 1.8 m long, lightly built, and fast-moving on long legs. Furthermore, the fossilized female parent had been sitting on her eggs like a chicken! Researchers believe this parent and her nest were buried by a fast-moving sandstorm.

Further evidence that birds are surviving dinosaurs would be to show that extinct dinosaurs were endothermic (warm blooded)

as birds are today. A new field called **paleophysiology** has found ways to use the genes of extinct animals to reconstruct their physiological molecules, such as hemoglobin, in order to answer such questions. Paleophysiology is an exciting new field that should eventually give us physiological evidence in addition to anatomic evidence about the evolutionary relationships of organisms. If a living cell of an extinct animal can be found—say in one of these fossilized eggs—it might be possible to use its nucleus to clone the animal using the method described in Figure 13.1A.

In this chapter, we will examine the history of life, as revealed by the fossil record, before considering how evolutionary relationships are determined by traditionalists and cladists. Scientists use data regarding evolutionary relationships to construct diagrams depicting these relationships.

Fossils Tell a Story

The geologic timescale gives a brief outline of the history of life on Earth, based on a complex fossil record. Meteorite bombardment and severe climate changes (due to continental drift) most likely played a role in the mass extinctions that have occurred during the history of life.

16.1 The geologic timescale is based on the fossil record

LEARNING OUTCOME

When you complete this section, you should be able to

1. Use the geologic timescale to trace macroevolution in broad outline.

Because all life-forms evolved from the first cell or cells, life has a history, and this history is revealed by the fossil record. The **geologic timescale,** which was developed by both geologists and paleontologists, depicts the history of life based on the fossil record. We will be referring to the geologic timescale in future chapters as we study the evolution of various groups of organisms; therefore, it would be beneficial for you to become familiar with it now (Table 16.1).

The timescale divides the history of Earth into eras, then periods, and then epochs. The three eras (the Paleozoic, the Mesozoic, and the Cenozoic eras) span the greatest amounts of time, and the epochs have the shortest time frames. Notice that only the periods of the Cenozoic era are divided into epochs, meaning that more attention is given to the evolution of primates and flowering plants than to the earlier evolving organisms. Modern civilization is given its own epoch, despite the fact that humans have only been around for about 0.04% of the history of life.

The timescale provides both relative dates and absolute dates. For example, when you say, "Flowering plants evolved during the Jurassic period," you are using relative time, because flowering plants evolved earlier or later than groups in other periods. If you use the dates given in millions of years (MYA), you are using absolute time. Absolute dates are usually obtained by measuring the amount of a radioactive isotope in the rocks surrounding the fossils.

> **Animation**
> Geologic History of the Earth

Limitations of the Timescale Because the timescale tells when various groups evolved and flourished, it might seem that evolution has been a series of events leading only from the first cells to humans. This is not the case; for example, prokaryotes never declined and are still the most abundant and successful organisms on Earth. Even today, they constitute up to 90% of the total weight of organisms in the oceans.

Then, too, the timescale lists mass extinctions, but it doesn't tell when specific groups became extinct. **Extinction** is the total disappearance of a species or a higher group; a **mass extinction** occurs when a large number of species disappear in a few million years or less. For lack of space, the geologic timescale can't depict in detail what happened to the members of every group of organisms mentioned.

If we could trace the descent of all the millions of groups ever to have evolved, the entirety would resemble a dense bush. Some lines of descent would be cut off close to the base; some would continue in a straight line up to the present; and others would split,

producing two or even several groups. The geologic timescale can't show the many facets, twists, and turns of the history of life.

How to Read the Timescale Using Table 16.1, you can trace the history of life by beginning with ❶ Precambrian time at the bottom of the timescale. The timescale indicates that the first cells (the prokaryotes) arose some 3,500 MYA (millions of years ago). The prokaryotes evolved before any other group.

The Precambrian time was very long, lasting from the time the Earth first formed until 542 MYA. The fossil record during the Precambrian time is meager, but the fossil record from the Cambrian period onward is rich, as we know from our study of the Burgess Shale (see "The Burgess Shale Hosts a Diversity of Life" on page 290). This helps explain why the timescale usually is not broken down into periods until the Cambrian period of the Paleozoic era. (British geologists decided which strata formed an era or period and then named them. The names most often refer to places and ancient tribes in England. For example, the Ordovician period is named after an ancient tribe called the Ordovices.)

We can also use the timescale to check when certain groups evolved and/or flourished. For example, during ❷ the Ordovician period, the first simple plants appeared on land, and the first jawless and jawed fishes appeared in the sea. A mass extinction occurred at the end of the Ordovician period, as is often the case at the end of a period. The reasons for mass extinctions are diverse and will be examined in section 16.3.

On the timescale, note ❸ the Carboniferous period. Rich coal deposits formed world-wide during this period because the climate conditions were perfect for coal formation. Coal formed in England, North America, and elsewhere. This is the very coal we burn today to fuel our modern way of life. The timescale tells us that reptiles appeared during the Carboniferous period, but of course it does not mention that reptiles are especially well adapted to living on land because they do not have to return to water to reproduce. This will be discussed in Chapter 20.

❹ The evolution of dinosaurs was a significant event during the Mesozoic era. The dinosaurs (but not the mammals, which also appeared during the Mesozoic era) perished during the mass extinction at the end of the Cretaceous period. Why didn't mammals become extinct? Perhaps their small size and lack of specialization helped them survive. Once the dinosaurs departed, ❺ mammals underwent adaptive radiation in the Tertiary period to fill the niches left empty by the dinosaurs.

> ▶ **16.1 CHECK YOUR PROGRESS**
> 1. Identify the correct era and period of the geologic timescale for the first appearance of fishes, amphibians, reptiles, and mammals.

TABLE 16.1	The Geologic Timescale: Major Divisions of Geologic Time and Some of the Major Evolutionary Events of Each Time Period					

Era	Period	Epoch	Millions of Years Ago (MYA)	Plant Life	Animal Life	
Cenozoic ⑤		Holocene	(0.01–0)	Human influence on plant life	Age of *Homo sapiens*	
			Significant Mammalian Extinction			
	Quaternary	Pleistocene	(1.80–0.01)	Herbaceous plants spread and diversify.	Presence of Ice Age mammals. Modern humans appear.	
		Pliocene	(5.33–1.80)	Herbaceous angiosperms flourish.	First hominids appear.	
		Miocene	(23.03–5.33)	Grasslands spread as forests contract.	Apelike mammals and grazing mammals flourish; insects flourish.	
		Oligocene	(33.9–23.03)	Many modern families of flowering plants evolve.	Browsing mammals and monkeylike primates appear.	
	Tertiary	Eocene	(55.8–33.9)	Subtropical forests with heavy rainfall thrive.	All modern orders of mammals are represented.	
		Paleocene	(65.5–55.8)	Flowering plants continue to diversify.	Primitive primates, herbivores, carnivores, and insectivores appear.	
Mesozoic ④			**Mass Extinction: Dinosaurs and Most Reptiles**			
	Cretaceous		(145.5–65.5)	Flowering plants spread; conifers persist.	Placental mammals appear; modern insect groups appear.	
	Jurassic		(199.6–145.5)	Flowering plants appear.	Dinosaurs flourish; birds appear.	
			Mass Extinction			
	Triassic		(251–199.6)	Forests of conifers and cycads dominate.	First mammals appear; first dinosaurs appear; corals and molluscs dominate seas.	
Paleozoic			**Mass Extinction**			
③	Permian		(299–251)	Gymnosperms diversify.	Reptiles diversify; amphibians decline.	
	Carboniferous		(359.2–299)	Age of great coal-forming forests; ferns, club mosses, and horsetails flourish.	Amphibians diversify; first reptiles appear; first great radiation of insects.	
			Mass Extinction			
	Devonian		(416–359.2)	First seed plants appear. Seedless vascular plants diversify.	First insects and first amphibians appear on land.	
	Silurian		(443.7–416)	Seedless vascular plants appear.	Jawed fishes diversify and dominate the seas.	
			Mass Extinction			
②	Ordovician		(488.3–443.7)	Nonvascular plants appear on land.	First jawless and then jawed fishes appear.	
	Cambrian		(542–488.3)	Marine algae flourish.	All invertebrate phyla present; first chordates appear.	
Precambrian Time ①			630	Soft-bodied invertebrates		
			1,000	Protists diversify.		
			2,100	First eukaryotic cells		
			2,700	O_2 accumulates in atmosphere.		
			3,500	First prokaryotic cells		
			4,570	Earth forms.		

16.2 Continental drift has affected the history of life

LEARNING OUTCOME

When you complete this section, you should be able to

1. Explain how the drifting of the continents influenced the distribution of fossil remains.

In the past, scientists thought that the Earth's crust was immobile, that the continents had always been in their present positions, and that the ocean floors were only a catch basin for the debris that washed off the land. But in 1920, Alfred Wegener, a German meteorologist, presented data from a number of disciplines to support his hypothesis of **continental drift.** He proposed that the continents are not fixed; instead, their positions and the positions of the oceans have changed over time. Wegener's hypothesis was finally confirmed in the 1960s. During the Paleozoic era, the continents joined to form one supercontinent that Wegener called Pangaea (Fig. 16.2). First, Pangaea divided into two large subcontinents, called Gondwana and Laurasia, and then these also split to form the continents of today. Presently, the continents are still drifting in relation to one another.

Continental drift explains why the coastlines of several continents are mirror images of each other—for example, the outline of the west coast of Africa matches that of the east coast of South America. The same geologic structures are also found in many of the areas where the continents touched. A single mountain range runs through South America, Antarctica, and Australia. Continental drift also explains the unique distribution patterns of several fossils. For example, fossils of *Lystrosaurus,* a reptile abundant during the Triassic period, have now been discovered in Antarctica, far from Africa and Southeast Asia, where the reptile also occurs. With mammalian fossils, the situation is different: Australia, South America, and Africa all have their own distinctive mammals because they evolved after the continents separated. Why are marsupials primarily in Australia? They started evolving in the Americas and were able to reach Australia while the southern continents were still joined. Once Australia separated off, marsupials were able to diversify because placental mammals on that continent offered little competition. In contrast, only a few marsupials can be found in the Americas.

Plate Tectonics Why do the continents drift? A branch of geology known as **plate tectonics** has concluded that the Earth's crust is fragmented into slablike plates that float on a lower hot mantle layer. The continents and the ocean basins are a part of these rigid plates, which move like conveyor belts.

At ocean ridges, seafloor spreading occurs as molten mantle rock rises and material is added to the ocean floor. Seafloor spreading causes the continents to move a few centimeters per

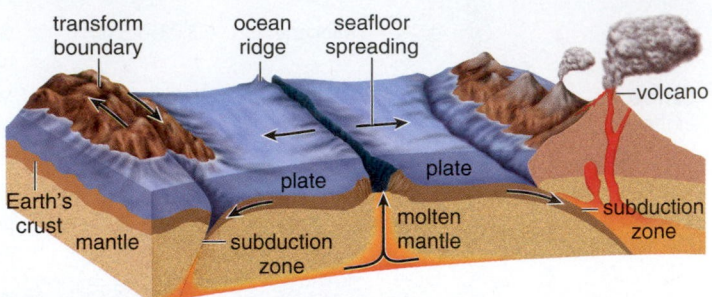

year on the average. At *subduction zones,* the forward edge of a moving plate sinks into the mantle and is destroyed, forming deep ocean trenches bordered by volcanoes or volcanic island chains. The Earth isn't getting bigger or smaller, so the amount of oceanic crust being formed is equal to the amount being destroyed. When two continents collide, the result is often a mountain range; for example, the Himalayas resulted when India collided with Eurasia. The place where two plates meet and scrape past one another is called a *transform boundary*. The San Andreas fault in Southern California is at a transform boundary, and the movement of the two plates is responsible for the many earthquakes in that region. Earthquakes are visible evidence that the plates are moving.

▶ **16.2 CHECK YOUR PROGRESS**

1. Explain why you would expect to find dinosaur bones on all continents since they evolved during the early Mesozoic era.

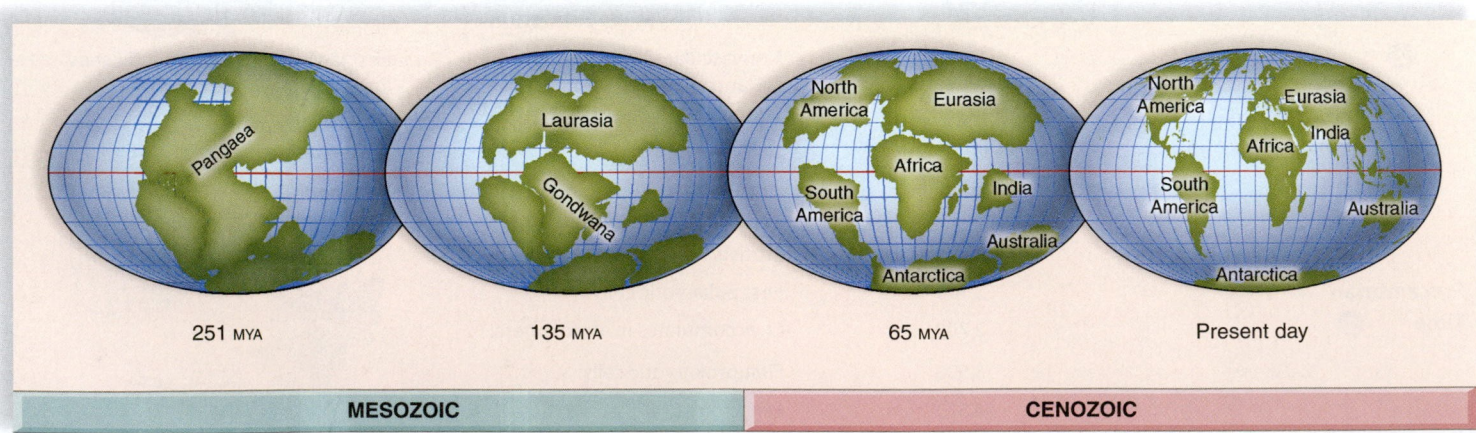

FIGURE 16.2 According to the **continental drift** hypothesis, about 251 MYA, all the continents were joined into a supercontinent called Pangaea. During the Mesozoic era, the joined continents of Pangaea began moving apart, forming two large continents called Laurasia and Gondwana. Then all the continents began to separate. Presently, North America and Europe are drifting apart at a rate of about 2 cm per year.

16.3 Mass extinctions have affected the history of life

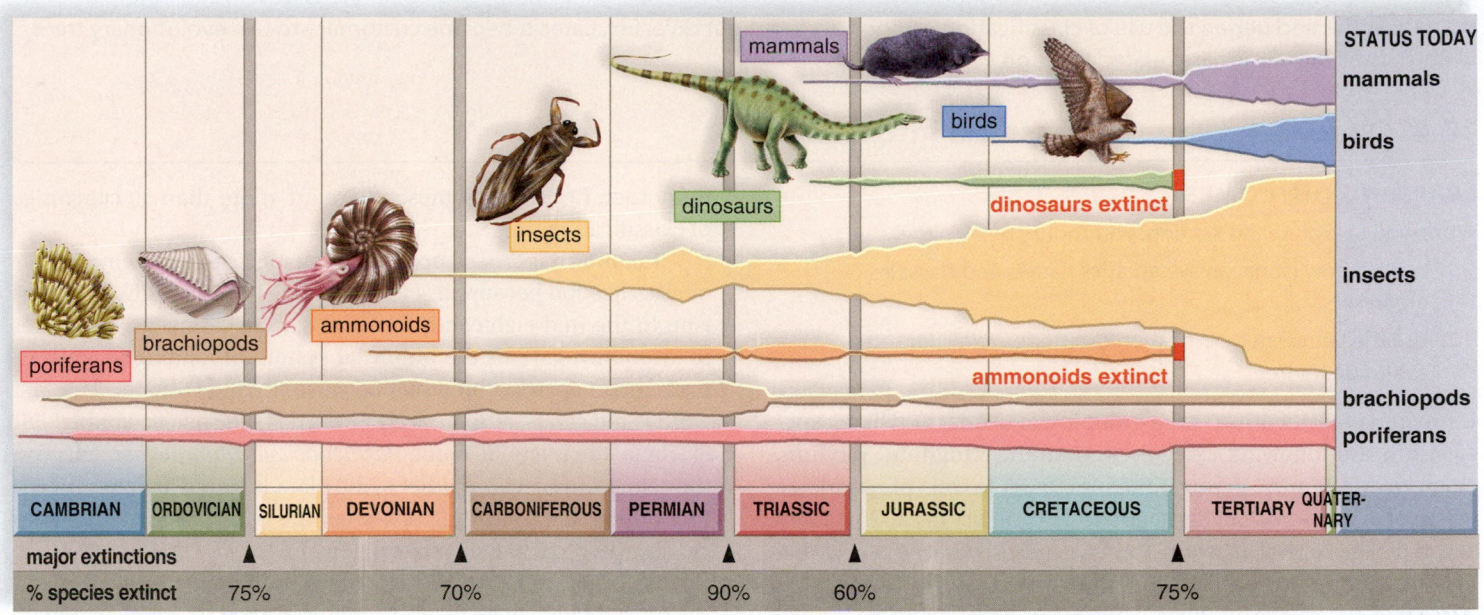

FIGURE 16.3 Five significant mass extinctions and their effects on the abundance of certain forms of marine and terrestrial life. The width of the horizontal bars indicates the varying abundance of each life-form considered. MYA = million years ago.

LEARNING OUTCOME

When you complete this section, you should be able to

1. Identify two possible explanations for the mass extinctions noted in the geologic timescale.

At least five mass extinctions have occurred throughout history of the Earth: at the ends of the Ordovician, Devonian, Permian, Triassic, and Cretaceous periods (Fig. 16.3; see Table 16.1). Is a mass extinction due to some cataclysmic event, or is it a more gradual process brought on by environmental changes, including tectonic, oceanic, and climatic fluctuations?

Certainly, continental drift contributed to the Ordovician extinction. This extinction occurred after Gondwana arrived at the South Pole. Immense glaciers, which drew water from the oceans, chilled even the once-tropical land. Marine invertebrates and coral reefs, which were especially hard hit, didn't recover until Gondwana drifted away from the pole and warmth returned.

An astronomical event may have caused the mass extinction at the end of the Devonian period which saw an end to 70% of marine invertebrates. A bolide is an asteroid (minor planet) that explodes, producing meteorites that fall to Earth. Iridium is a rare element in the Earth's crust but more common in asteroids and meteorites. Helmont Geldsetzer of Canada's Geological Survey notes that iridium has been found in Devonian rocks in Australia, suggesting that a bolide was involved. Some scientists believe that this mass extinction could have been due to movement of Gondwana back to the South Pole, however.

The extinction at the end of the Permian period was quite severe; 90% of species disappeared. The latest hypothesis attributes the Permian extinction to excess carbon dioxide. When Pangaea formed, there were no polar ice caps to initiate ocean currents. The lack of ocean currents caused organic matter to stagnate at the bottom of the ocean. Then, as the continents drifted

into a new configuration, ocean circulation switched back on. Now, the extra carbon on the seafloor was swept up to the surface where it became carbon dioxide, a deadly gas for sea life. The trilobites became extinct, and the crinoids (sea lilies) barely survived. Excess carbon dioxide on land led to global warming that altered the pattern of vegetation. Areas that had been wet and rainy became dry and warm, and vice versa. Burrowing animals that could escape land surface changes seemed to have the best chance of survival.

The extinction at the end of the Triassic period has been attributed to the environmental effects of a meteorite collision with Earth. Central Quebec has a crater half the size of Connecticut that some believe is the impact site. The dinosaurs may have benefited from the mass extinction at the end of the Triassic because this is when the first of the gigantic dinosaurs took charge of the land.

The Cretaceous ended, however, with the extinction of the dinosaurs. In 1977, Walter and Luis Alvarez said that the Cretaceous extinction was due to a bolide because Cretaceous clay contains an abnormally high level of iridium. The result of a large meteorite striking Earth could have been similar to that of a worldwide atomic bomb explosion: A cloud of dust would have mushroomed into the atmosphere, blocking out the sun and causing plants to freeze and die. A layer of soot has been identified in the strata alongside the iridium, and a huge crater that could have been caused by a meteorite was found in the Caribbean–Gulf of Mexico region on the Yucatán peninsula.

▶ **16.3 CHECK YOUR PROGRESS**
1. Compare two main reasons for mass extinctions in the geologic timescale.
2. Explain how mass extinctions might contribute to the ongoing process of evolution.

Linnaean Systematics

Systematics is the study of the diversity of organisms at all levels of biological organization. Carolus Linnaeus developed binomial nomenclature and began the use of classification categories to sort out diversity. Later, it became customary to use evolutionary trees to show how organisms are related through evolution.

16.4 Organisms can be classified into categories

LEARNING OUTCOMES

When you complete this section, you should be able to

1. Explain how Linnaean systematics names and classifies organisms.
2. Use the Linnaean classification categories to explain an evolutionary tree.

Linnaean classification is the grouping of extinct and living species into the following categories: **domain, kingdom, phylum, class, order, family, genus,** and **species.** A **taxon** (pl., taxa) is a group of organisms that fills a particular category of classification. There can be several species within a genus, several genera within a family, and so forth—the higher the category, the more inclusive it is. Thus, the categories form a hierarchy. The organisms that fill a particular classification category are distinguishable from other organisms by a set of characteristics, or simply characters, that they share. A **character** is any trait, whether structural, molecular, reproductive, or behavioral, that distinguishes one group from another. Organisms in the same domain have general characters in common; those in the same species have quite specific characters in common.

In most cases, categories of classification can be divided into additional categories, such as superorder, order, suborder, and infraorder. Considering these, there are more than 30 categories of classification.

Taxonomy is the science of naming species. Taxonomy is a part of classification because a scientific name helps classify an organism. In the mid-eighteenth century, Carolus Linnaeus, the father of taxonomy, gave us the **binomial system** of naming organisms. Each name is called binomial because it has two parts. The first word is the genus, and the second word is the specific epithet. For example, the scientific name *Parthenocissus quinquefolia* tells you the genus and specific epithet of the plant featured in Figure 16.4A. From there, the ascending boxes tell the other categories used to classify this plant in the Linnaean system.

Why is it preferable to use the scientific name, *Parthenocissus quinquefolia,* instead of the common name, Virginia creeper? Common names often differ between countries and even within the same country, but scientific names are always based on Latin, a universal language for scientists. Classification categories necessarily hypothesize the evolutionary relationships between species.

Evolutionary Trees One goal of systematics is to determine **phylogeny,** or the evolutionary history of a group of organisms. Traditionally, these relationships are represented by an **evolutionary (phylogenetic) tree,** a diagram that indicates common

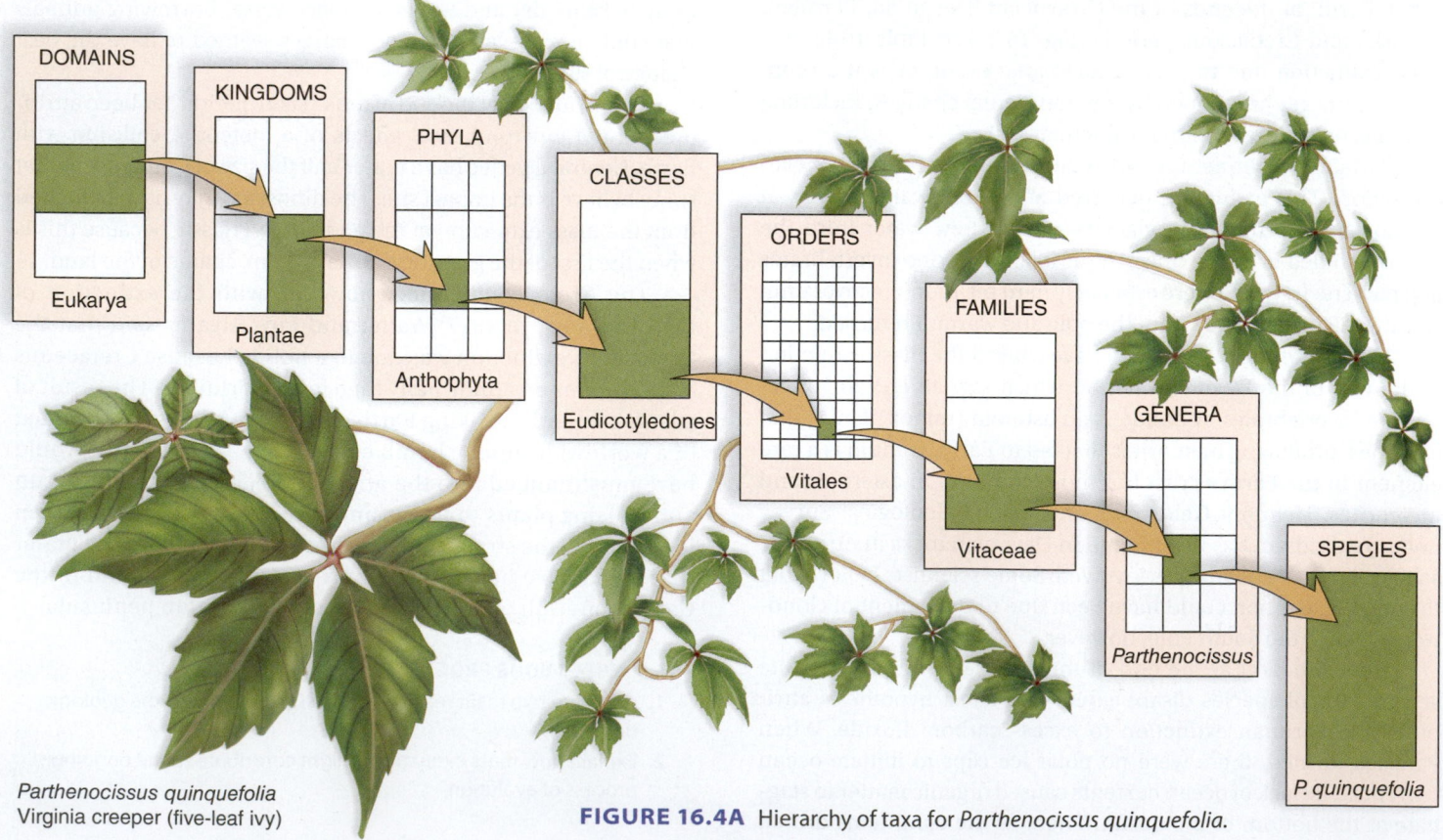

Parthenocissus quinquefolia
Virginia creeper (five-leaf ivy)

FIGURE 16.4A Hierarchy of taxa for *Parthenocissus quinquefolia.*

ancestors and lines of descent (lineages). Each branch point in a phylogenetic tree is a divergence from a **common ancestor**, a species that gives rise to two new groups. For example, this portion of an evolutionary tree shows that monkeys and apes share a common primate ancestor:

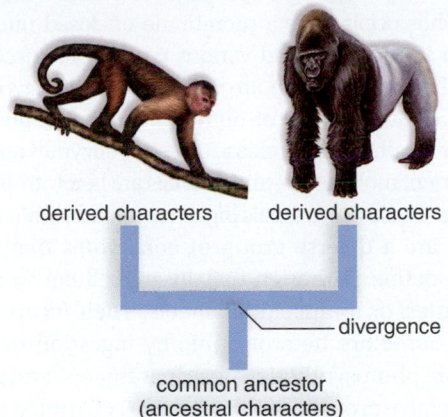

Divergence is presumed because monkeys and apes have their own individual characteristics (often called **derived characters**). For example, skeletal differences allow an ape to swing from limb to limb of a tree, while monkeys run along the tops of tree branches. The common primate ancestor to both monkeys and apes has **ancestral characters** that are shared by the ancestor as well as by monkeys and apes. For example, the common primate ancestor must have been able to climb trees, as both monkeys and apes can do.

A phylogenetic tree has many branch points, and a tree can show that it is possible to trace the ancestry of a group of organisms farther and farther back in the past. For example, reindeer, monkeys, and apes give birth to live young, and they all have a common ancestor that was a placental mammal:

Classification is a part of systematics because classification categories list the unique characters of each taxon, which ideally reflect phylogeny. A species is most closely related to other species in the same genus, then to genera in the same family, and so forth, from order to class to phylum to kingdom. When we say that two species (or genera, families, etc.) are closely related, we mean that they share a more recent common ancestor with each other than they do with members of other taxa. For example, all the animals in Figure 16.4B are related because we can trace their ancestry back

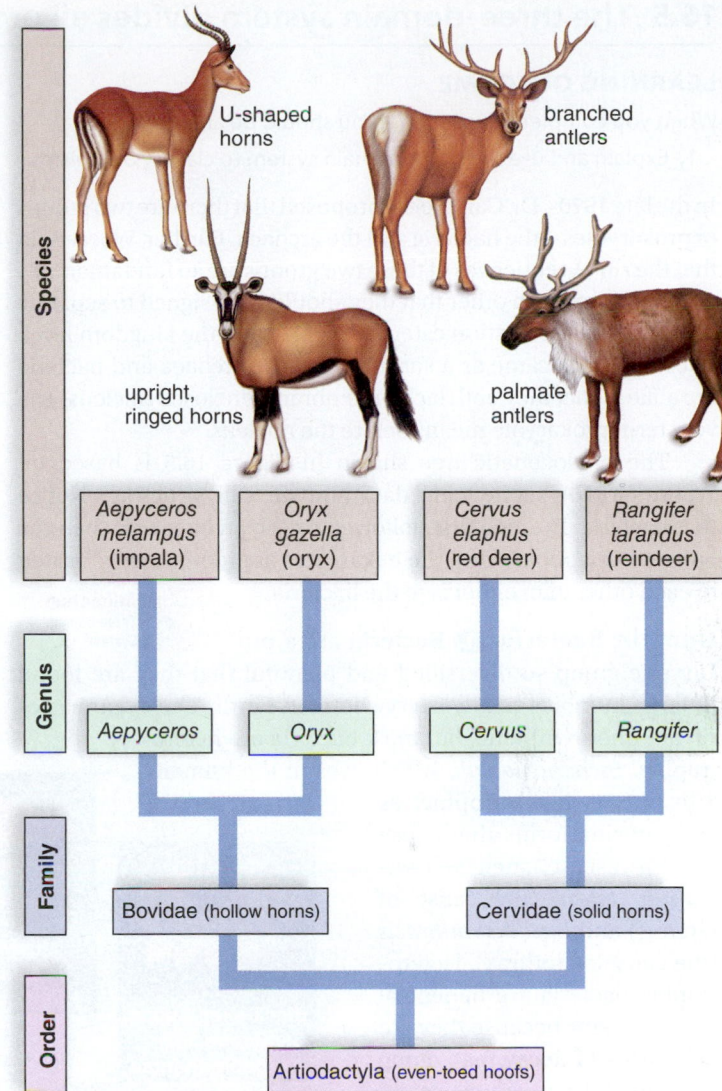

FIGURE 16.4B Linnaean classification and phylogeny. This shows that the reindeer is more closely related to the red deer than to the impala and the oryx.

to the same order. The animals in the order Artiodactyla all have even-toed hoofs. Animals in the family Cervidae have solid horns, called antlers, but the horns in red deer (genus *Cervus*) are highly branched, while those in reindeer (genus *Rangifer*) are palmate (shaped like a hand). In contrast, animals in the family Bovidae have hollow horns, and unlike the Cervidae, both males and females have horns, although they are smaller in females.

▶ **16.4 CHECK YOUR PROGRESS**

1. List the Linnaean classification categories from the most inclusive (domain) to the least inclusive (species) and tell which of these categories is used in the scientific name of an organism.

2. Explain why more inclusive categories are at the bottom of an evolutionary tree and least inclusive categories at the top of the tree.

LEARNING OUTCOME

When you complete this section, you should be able to

1. Explain and use the three-domain system to classify organisms.

In the late 1970s, Dr. Carl Woese proposed that there are two groups of prokaryotes—the bacteria and the archaea. Further, Woese said that the rRNA sequences of these two groups are so fundamentally different from each other that they should be assigned to separate domains, a classification category higher than the kingdom level. These findings came as a surprise because archaea and bacteria are alike in that they both lack a membrane-enclosed nucleus. The very term prokaryote means before the nucleus.

The phylogenetic tree shown in Figure 16.5 is based on Woese's rRNA sequencing data and on cell structure. Notice that bacteria diverged first, followed by the archaea and then the eukaryotes. Archaea and the eukaryotes are more closely related to each other than either is to the bacteria.

Animation
Three Domain System

Domain Bacteria ❶ **Bacteria** are a prokaryotic group so diversified and plentiful that they are found in large numbers nearly everywhere on Earth. The cyanobacteria are photosynthetic, but most bacteria are heterotrophic. *Escherichia coli,* which lives in the human intestine, is heterotrophic, as are parasitic forms that cause human disease, such as *Clostridium tetani* (the cause of tetanus) and *Bacillus anthracis* (the cause of anthrax). Heterotrophic bacteria are beneficial in ecosystems because they are organisms of decay that, along with fungi, keep chemicals cycling so that plants always have a source of inorganic nutrients.

Domain Archaea ❷ **Archaea** are prokaryotes that do not look different from bacteria under the microscope. They are known for living under extreme conditions, and they are difficult to culture. For example, the methanogens live in anaerobic environments, such as swamps and marshes and the guts of animals. The halophiles are salt-lovers, living in bodies of water such as the Great Salt Lake in Utah. The thermoacidophiles thrive in extremely hot, acidic environments, such as hot springs and geysers. The branched nature of diverse

lipids in the archaeal plasma membrane could possibly help them live in extreme conditions.

Domain Eukarya ❸ **Eukaryotes** are unicellular to multicellular organisms whose cells have a membrane-enclosed nucleus. Sexual reproduction is common, and various types of life cycles are seen. Aside from the protists which are very diverse, each type of eukaryote has a characteristic type of nutrition. Plants are photosynthetic, fungi are saprotrophic—they release digestive enzymes and absorb the resulting nutrient molecules—and animals are heterotrophic by ingestion. Even the smallest of animals digest their food within a cavity.

Protists are a diverse group of eukaryotes that are hard to classify and define. Although usually unicellular, some are filaments, colonies, or multicellular sheets. Their forms of nutrition are diverse; some are heterotrophic by ingestion or absorption and others are photosynthetic. Algae, paramecia, and slime molds are representative protists. Biologists have concluded that the protists do not share one common ancestor and therefore the group should be divided into many different kingdoms. The number of kingdoms is still being determined, illustrating that classification changes as new data become available.

Fungi are eukaryotes that form spores, lack flagella, and have cell walls containing chitin. Most are multicellular. Fungi are heterotrophic by absorption—they secrete digestive enzymes and then absorb nutrients from decaying organic matter. Mushrooms, molds, and yeasts are representative fungi. Despite appearances, molecular data suggest that fungi and animals are more closely related to each other than either group is to plants.

Plants are nonmotile eukaryotic multicellular organisms. They possess true tissues and the organ system level of organization. Plants are autotrophic and carry on photosynthesis. Examples include cacti, ferns, and cypress trees.

Animals are motile, eukaryotic, multicellular organisms. Most also have true tissues and the organ system level of organization. Animals are heterotrophic by ingestion. Worms, whales, and insects are examples of animals.

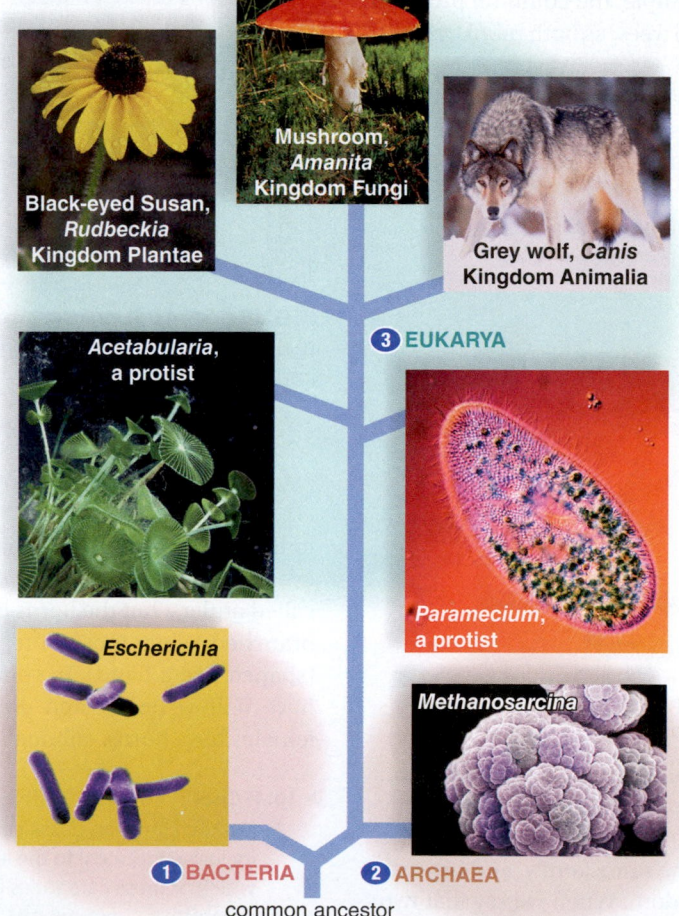

Black-eyed Susan, *Rudbeckia* Kingdom Plantae

Mushroom, *Amanita* Kingdom Fungi

Grey wolf, *Canis* Kingdom Animalia

Acetabularia, a protist

❸ EUKARYA

Paramecium, a protist

Escherichia

Methanosarcina

❶ BACTERIA　　❷ ARCHAEA

common ancestor

FIGURE 16.5 The three domain phylogenetic tree shows that domain Eukarya is more closely related to the Archaea than either is to the Bacteria.

▶ **16.5 CHECK YOUR PROGRESS**

1. Identify the basis for proposing the three-domain system and tell what type organisms are in each domain.

16A   DNA Barcoding of Life

Traditionally, taxonomists have relied on anatomic data to tell species apart, but laypeople don't have a lot of anatomic data to refer to when they want to identify, say, a bug in their house or a plant when taking a walk in the woods. Enter the Consortium for the Barcode of Life (CBOL), which proposes that any person with the proper handheld scanner could identify any organism on Earth.

So far, scientists have identified only about 1.5 million species out of a potential 30 million. And there is no central database that keeps track of the known species. The consortium is hoping to change that. The idea of using barcodes to identify species is not new, but Paul Hebert and his colleagues at the University of Guelph in Canada are the first to suggest it would be possible to use the base sequence in DNA to develop a barcode for each living thing. The order of DNA's nucleotides—A, T, C, and G—within a particular gene common to all organisms in each kingdom would fill the role taken by numbers in the scanning devices used in warehouses and stores (Fig. 16A.1).

Speedy DNA barcoding would not only be a boon to taxonomists, but it would also benefit farmers who need to identify a pest attacking their crops, doctors who need to know the correct antivenin for snakebite victims, and college students who are expected to identify the plants, animals, and protists on an ecological field trip. Already, the CBOL has accumulated hundreds of thousands of DNA barcodes representing species across the diversity of life.

FIGURE 16A.2 Kate Stoeckle and Louisa Strauss used DNA barcodes to discover mislabeled fish sold in Manhattan.

The CBOL initiative has the potential to be a powerful tool for conservation biologists and wildlife officials worldwide. A DNA barcode can be used to identify illegal trade in endangered species and for the early detection of invasive species that are brought into a country as a consequence of global transportation. In 2008, a pair of New York City high school students found a commercial application for the CBOL database. Kate Stoeckle and Louisa Strauss (Fig. 16A.2) collected over 60 fish samples from various Manhattan restaurants and grocery stores, and they sent them off to have their DNA barcodes compared to a global library of fish barcodes representing nearly 5,500 fish species. They discovered that several of the samples were mislabeled, and that less expensive fish were being sold as more expensive fish. In one case, they even found that an endangered Acadian redfish was being sold as red snapper.

CONSIDER THESE QUESTIONS

1. Would the ability to identify any species on Earth benefit society? Why or why not?
2. What would it take to convert this method of identifying organisms to a process for identifying criminals? How would it work?
3. If your proposed procedure in question 2 worked, would you try to make money from it?

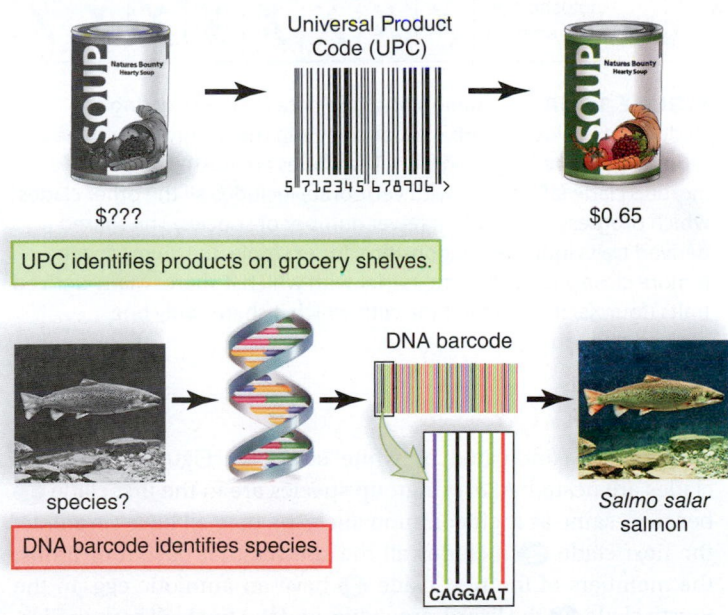

FIGURE 16A.1 The Universal Product Code (UPC), which uses numbers to identify products sold in stores, versus a DNA barcode, which uses DNA base sequencing to identify organisms.

connect BIOLOGY Explore the concepts through a variety of multimedia assets and question types.
www.mcgrawhillconnect.com

Modern Systematics: Cladistics

For many years, biologists relied on Linnaean systematics to classify organisms and construct evolutionary trees. A new method called cladistics utilizes only objective data based on shared traits to determine relationships among organisms. The objective data come from the fossil record, from comparative anatomy, and increasingly from molecular findings. Cladists display their data in phylogenetic trees called cladograms.

16.6 Cladograms reflect evolutionary history

LEARNING OUTCOME

When you complete this section, you should be able to

1. Understand the process that cladists use to construct a cladogram.

Biologists are always seeking new and improved ways to discover the evolutionary history of life on Earth. Tracing evolutionary history would be easy if similarities alone could be used, but this is not the case because evolution is quite variable and sometimes even reverses to a former state. For example, some vertebrates have teeth and some do not, and therefore we need a methodology that will tell us which is the ancestral state—teeth or no teeth. In this instance, the fossil record tells us that possession of teeth is an early characteristic of vertebrates. But if a fossil is unavailable, we need some other method. Today, the most commonly used method to determine evolutionary relationships when a complete fossil record is not available is called phylogenetic cladistics, or simply **cladistics.**

How to Construct a Cladogram Cladistics, which is based on the work of Willi Hennig, is a way to trace the evolutionary history of a group by using shared traits, derived from a common ancestor, to determine which species are most closely related. These traits are then used to construct phylogenetic trees called cladograms. A **cladogram** depicts the evolutionary history (phylogeny) of a group based on the available data. Let's see how it works.

The first step when constructing a cladogram is to draw up a table that summarizes the ancestral and derived traits of the species being compared (Fig. 16.6A). An **outgroup** (in this case, lancelets) has at least one trait (i.e., notochord in embryo) that is shared by all the other species. This is a **shared ancestral trait.** The other traits, called **shared derived traits** (e.g., vertebrae), occur only in members of the **ingroup.**

Only the shared derived traits are used to construct the cladogram, so why do we need an outgroup? The outgroup helps us know which traits are the shared derived traits. Any trait not found in the outgroup must be a shared derived trait. We could go to the fossil record to discover which traits are shared derived traits, but the fossil record is rarely complete enough to use exclusively.

The shared derived traits are indicated in a cladogram (Fig. 16.6B). A cladogram contains several **clades;** each clade includes the common ancestor and all its descendants that share one or more traits. The clades get smaller because the number of species with the same shared derived traits gets smaller. Because the outgroup does not have vertebrae and all the species in the ingroup do have vertebrae, we know that the first common ancestor for the ingroup was a vertebrate—it had vertebrae. Common

Traits		Species							
		ingroup							lancelet (outgroup)
		chimpanzee	terrier	finch	crocodile	lizard	frog	tuna	
6	hair, mammary glands	X	X						
5	gizzard			X	X				
4	epidermal scales			X	X	X			
3	amniotic egg	X	X	X	X	X			
2	four limbs	X	X	X	X	X	X		
1	vertebrae	X	X	X	X	X	X	X	
	notochord in embryo	X	X	X	X	X	X	X	X

FIGURE 16.6A This table shows the data for constructing the cladogram in Figure 16.6B. The lancelet is in the outgroup and the other species are in the ingroup. The clades are nested; the first ingroup clade (all species with vertebrae) includes all the other clades which progressively contain fewer number of species. The shared derived traits indicate relationships: for example, the chimpanzee is more closely related to the terrier with which it shares many derived traits (four Xs) than to the tuna with which it shares only one derived trait (one X).

ancestors are indicated by white circles in Figure 16.6B. The clades are nested: All the ingroup species are in the first clade ❶ because, same as their common ancestor, they all have vertebrae; the next clade ❷ includes all the species that have four limbs; the members of the next clade ❸ have an amniotic egg. In the fourth clade ❹ the lizard, crocodile, and the finch all have epidermal scales, but only the crocodile and finch are in the next clade ❺ because they have a gizzard, which the lizard lacks. Feathers do not help decide a clade because only the finch has feathers.

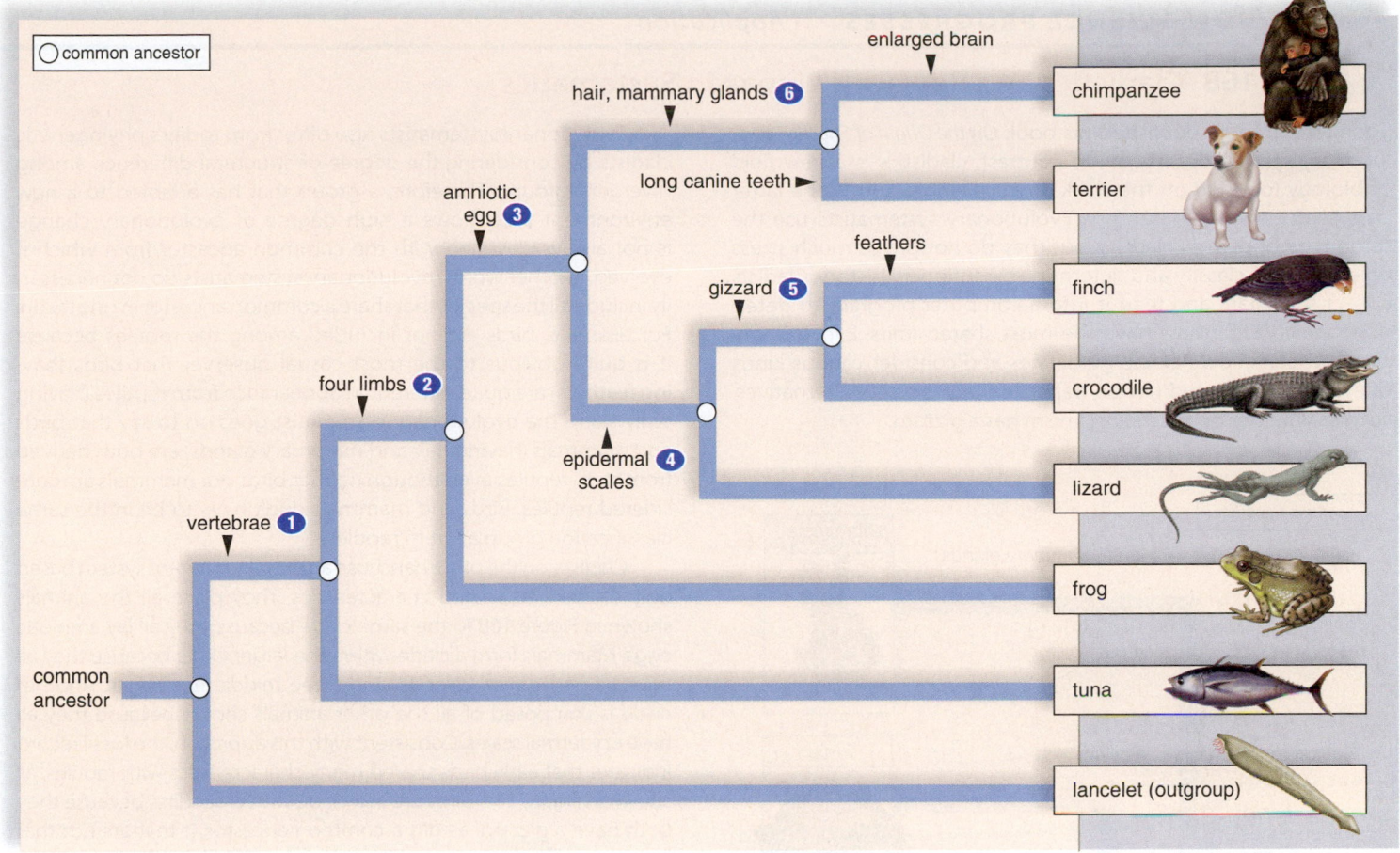

FIGURE 16.6B Based on the data in Figure 16.6A, the ingroup in this cladogram has six clades which are nested: The first clade **1** includes all the others; the last clade **6** contains only the chimpanzee and the terrier. Each clade contains a common ancestor with a derived trait that is shared by all members of the clade. For example only the chimpanzee and terrier are in clade **6** because only they have hair and mammary glands.

The last clade **6** in our cladogram includes the terrier and the chimpanzee because, same as their common ancestor, they both have hair and mammary glands. Similiar to feathers, the canine teeth of the dog and the enlarged brain of the chimp are not helpful in deciding clades because they are not shared by any other species in our ingroup.

The terms you need to learn to understand cladistics are given in Table 16.6.

TABLE 16.6	Terms Used in Cladistics
Outgroup	Species that define(s) which study group trait is oldest
Ingroup	Species that will be placed into clades in a cladogram
Ancestral trait	Traits present in both the outgroup and the ingroup
Clade	Evolutionary branch of a cladogram that contains a common ancestor and all its descendant species
Shared derived traits	Traits that distinguish a particular clade
Parsimony	Results in the simplest cladogram possible

How to Judge a Cladogram In order to tell if a cladogram has produced the best hypothesis, cladists are often guided by the principle of **parsimony,** which states that the minimum number of assumptions is the most logical. That is, they construct the cladogram that leaves the fewest number of shared derived characters unexplained or that minimizes the number of evolutionary changes. The rule of parsimony works best for traits that evolve at a slower rate than the frequency of speciation events. A problem with parsimony can arise when DNA sequencing is used to help construct cladograms. Mutations, especially in noncoding DNA, can be quite frequent, and if so, base changes are not reliable data for distinguishing clades. For this reason, some systematists have begun using statistical tools rather than parsimony to construct phylogenetic trees. This branch of systematics is called statistical phylogenetics. In any case, the reliability of a cladogram is dependent on the knowledge and skill of the particular investigator gathering the data and doing the character analysis.

► **16.6 CHECK YOUR PROGRESS**
   1. Describe the table that cladists use as the first step toward constructing a cladogram.

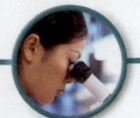

16B Cladistics Has Replaced Linnaean Systematics

Soon after Darwin published his book *On the Origin of Species,* evolutionary systematics began. In contrast, cladistics is a new field of biology founded on the work of Willi Hennig during the latter part of the twentieth century. Evolutionary systematists use the same type of data as cladists, but they do not put as much stress on ancestry to classify and determine evolutionary history. Cladists collect their data and feed it into a computer program to determine which organisms have the most shared traits. Evolutionary systematists follow flexible guidelines and consider various kinds of evidence that need not be translated into discrete alternatives, such as whether or not the organism has a gizzard.

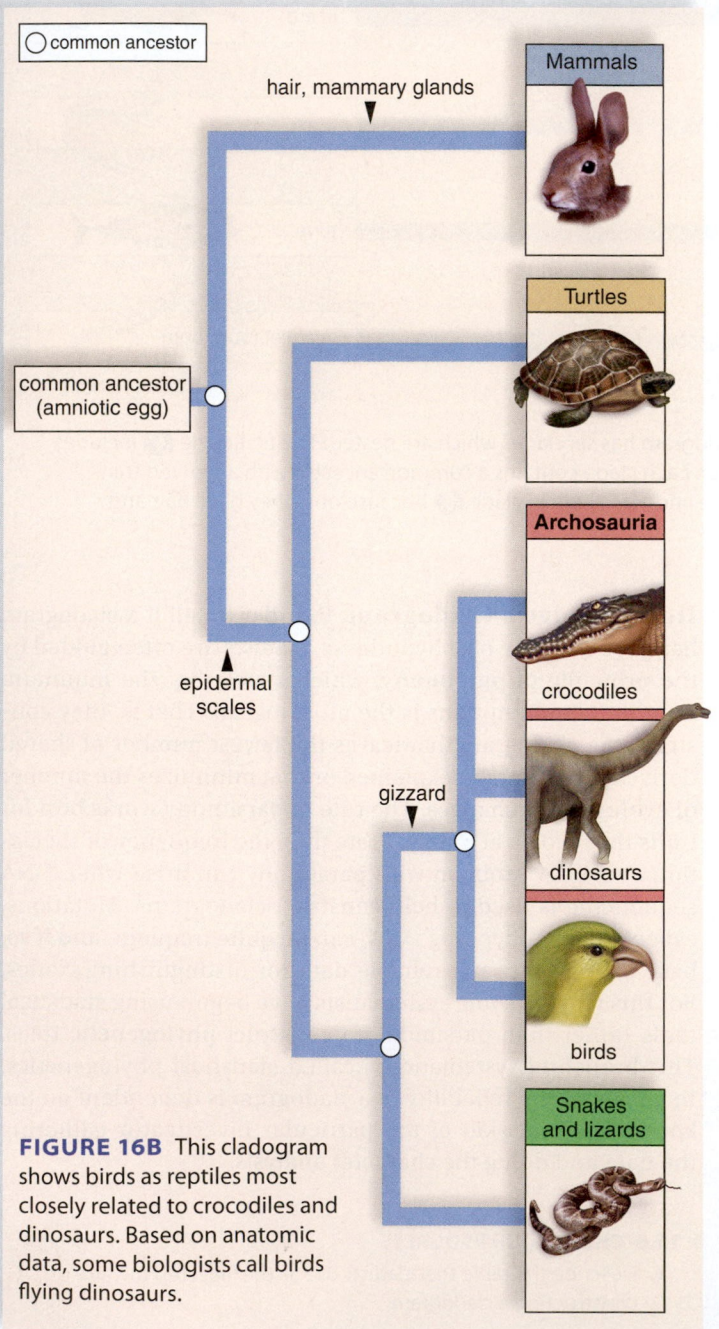

FIGURE 16B This cladogram shows birds as reptiles most closely related to crocodiles and dinosaurs. Based on anatomic data, some biologists call birds flying dinosaurs.

Evolutionary systematists also differ from today's phylogenetic cladists by considering the degree of structural difference among divergent groups. Therefore, a group that has adapted to a new environment and shows a high degree of evolutionary change is not always classified with the common ancestor from which it evolved. In other words, evolutionary systematists do not necessarily include all the species that share a common ancestor in one taxon. For example, birds are not included among the reptiles because it is quite obvious to the most casual observer that birds (having feathers) are quite different in appearance from reptiles (having scaly skin). The evolutionary systematist goes on to say that birds and mammals (having hair and mammary glands) are both derived from stem reptiles even though neither birds nor mammals are considered reptiles. Birds and mammals don't have to be in the same classification group as stem reptiles.

Cladists, on the other hand, use an entirely different system based only on derived shared characteristics. They place all the animals shown in Figure 16B in the same clade because they all lay amniotic eggs. Mammals form a clade within the larger clade because they all have hair, mammary glands, and three middle ear bones. Another clade is composed of all the other animals shown because they all have epidermal scales. Consistent with this approach, the fossil record indicates that early birds shared many characteristics with reptiles, as shown in Figure 16B. Birds are in a clade with crocodiles because they both have a gizzard, as did a common ancestor. It matters not that birds are now adapted to a different environmental niche. To indicate that crocodiles and birds, along with the dinosaurs, are closely related, a cladist would place them in a clade of their own called Archosaurs.

Because Linnaean classification is not consistent with the approach of phylogenetic cladistics, some cladists have proposed a different system of classification, called the International Code of Phylogenetic Nomenclature, or PhyloCode. PhyloCode sets forth rules to follow in naming clades. Other biologists are hoping to modify Linnaean classification to be consistent with the principles of cladistics.

Two major problems may be unsolvable: (1) Clades are hierarchical, as are Linnaean categories. However, there may be more clades than Linnaean taxonomic categories, and therefore it is difficult to equate clades with taxons. (2) The taxons are not necessarily equivalent in the Linnaean system. To take an example, the family taxon within kingdom Plantae may not be equivalent to the family taxon within kingdom Animalia. Because of such problems, some cladists recommend abandoning Linnaean classification altogether.

CONSIDER THESE QUESTIONS

1. Why is cladistics considered more objective than evolutionary systematics? Discuss this as a possible reason why cladistics has been well received by the biology community.
2. Should science be willing to always change, or are there reasons to stay with the traditional way? What reasons might those be?

 Explore the concepts through a variety of multimedia assets and question types.

www.mcgrawhillconnect.com

LEARNING OUTCOME

When you complete this section, you should be able to

1. Summarize the types of data used by systematists to trace phylogeny.

Systematists use fossil, morphological, and molecular data to determine the correct sequence of common ancestors in any particular group of organisms.

Fossil Record Data One of the advantages of fossils is that they can be dated, but unfortunately it is not always possible to tell to which group, living or extinct, a fossil is related. For example, at present, paleontologists are discussing whether fossil turtles indicate that turtles are distantly or closely related to crocodiles. On the basis of his interpretation of fossil turtles, Oliver C. Rieppel of the Field Museum of Natural History in Chicago is challenging the conventional interpretation that turtles are ancestral (have traits seen in a common ancestor to all reptiles) and are not closely related to crocodiles, which evolved later. His interpretation is being supported by molecular data that show turtles and crocodiles are closely related.

If the fossil record were more complete, there might be fewer controversies about the interpretation of fossils. One reason the fossil record is incomplete is that most fossils represent only the harder body parts, such as bones and teeth. Soft parts are usually eaten or decayed before they have a chance to be buried. This may be one reason it has been difficult to discover when angiosperms (flowering plants) first evolved. By the time angiosperm fossils are found in the Cretaceous, they have already diversified so angiosperms must have evolved earlier but no fossil has been found to support this hypothesis.

Morphological Data Comparative anatomy, including developmental evidence such as that shown in Figure 16.7A, provides information regarding homology. **Homologous structures** are similar to each other because of common descent. For example, the forelimbs of vertebrates contain the same bones, organized just as they were in a common ancestor, despite adaptations to different environments. As Figure 14.4A shows, even though a horse has but a single digit and toe (the hoof) while a bat has four lengthened digits that support its wing, a horse's forelimb and a bat's forelimb contain the same bones.

In contrast to homologous structures, **analogous structures** have the same function in different groups but do not have a common ancestry. For example, both cacti and spurges are adapted similarly to a hot, dry environment, and both are succulent (thick, fleshy) with spiny leaves. However, the details of their flower structure indicate that these plants are not closely related. The construction of phylogenetic trees is dependent on discovering homologous structures and avoiding the use of analogous structures to decide ancestry.

Molecular Data Speciation occurs when mutations bring about changes in the base-pair sequences of DNA. Systematists, therefore, assume that the more closely species are related, the fewer changes there will be in DNA base-pair sequences. Because DNA codes for amino acids, it also follows that the more closely species are related, the fewer differences will exist in the amino acid sequences within their proteins. Software breakthroughs have made it possible to analyze nucleotide sequences or amino acid sequences quickly and accurately using a computer. Also, these analyses are available to anyone doing comparative studies through the Internet, so each investigator doesn't have to start from scratch. The combination of accuracy and availability of past data has made molecular systematics a standard way to study the relatedness of groups of organisms today. To take an example, cytochrome c is a protein present in all aerobic organisms, so its

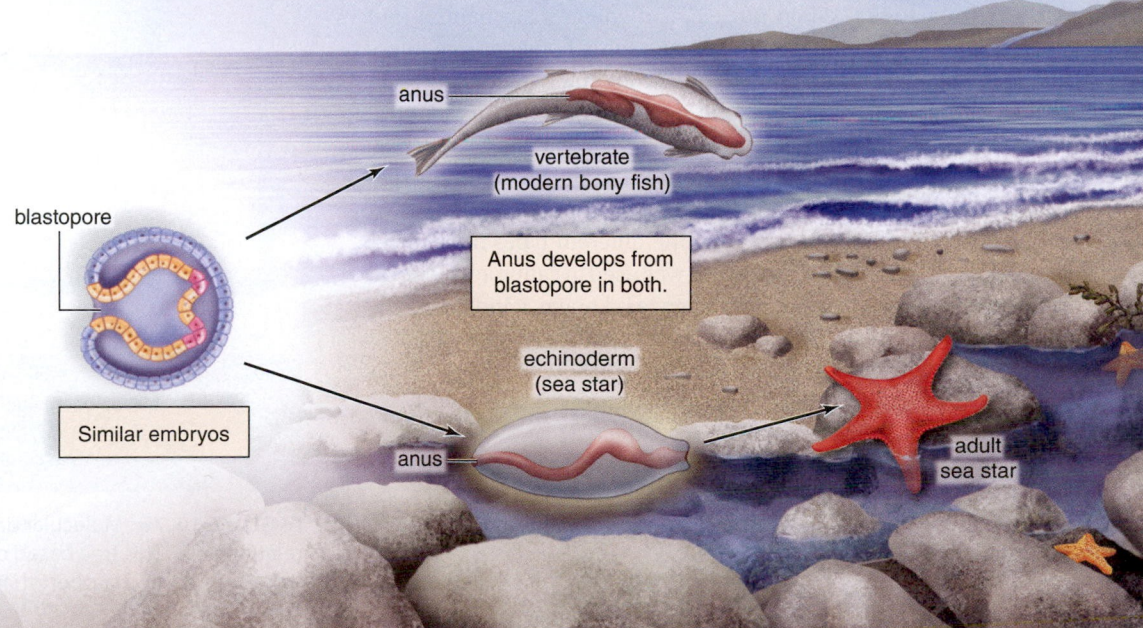

FIGURE 16.7A Development reveals homologies. Echinoderms (e.g., sea star) and vertebrates develop similarly—in both, the embryonic blastopore becomes the anus. Later, the echinoderm, but not the vertebrate, becomes radially symmetrical.

blastopore

Similar embryos

anus

vertebrate
(modern bony fish)

Anus develops from blastopore in both.

echinoderm
(sea star)

anus

adult sea star

sequence has been determined for a number of different organisms. The amino acid difference in cytochrome *c* between chickens and ducks is only 3, but between chickens and humans there are 13 amino acid differences. From this data you can conclude that, as expected, chickens and ducks are more closely related than are chickens and humans. Since the number of proteins available for study in all organisms at all times is limited, most new studies today focus on differences in RNA and DNA.

All cells have ribosomes, which are essential for protein synthesis. Further, the genes that code for ribosomal RNA (rRNA) have changed very slowly during evolution in comparison to other genes. Therefore, it is believed that comparative rRNA sequencing provides a reliable indicator of the similarity between organisms. Ribosomal RNA sequencing helped investigators conclude that all organisms can be divided into the three domains, as discussed in section 16.5. The same RNA and DNA comparisons can be used in other ways, aside from deciphering evolutionary relationships. For example, conservationists can use them to determine that a species is rare and endangered.

DNA differences can substantiate data, help trace the course of macroevolution, and fill in the gaps of the fossil record. The phylogenetic tree of primates shown in Figure 16.7B is based on DNA differences. These data suggest that humans share a recent common ancestor with chimpanzees, and therefore according to the rules of cladistics, should be classified with them, as we now are to the level of subfamily (see Fig. 21.2A).

Mitochondrial DNA (mtDNA) mutates ten times faster than nuclear DNA. Therefore, when determining the phylogeny of closely related species, investigators often choose to sequence mtDNA instead of nuclear DNA. One such study concerned North American songbirds. It had long been suggested that these birds diverged into eastern and western subspecies due to retreating glaciers some 250,000–100,000 years ago. Sequencing of mtDNA allowed investigators to conclude that groups of North American songbirds diverged from one another an average of 2.5 million years ago (MYA). Since the old hypothesis based on glaciation is apparently flawed, a new hypothesis is required to explain why eastern and western subspecies arose among these songbirds.

Molecular Clocks When nucleic acid changes are neutral (not tied to adaptation) and accumulate at a fairly constant rate, these changes can be used as a kind of **molecular clock** to indicate relatedness and evolutionary time. The researchers doing comparative mtDNA sequencing used their data as a molecular clock when they equated a 5.1% nucleic acid difference among songbird subspecies to 2.5 MYA. In Figure 16.7B, the researchers used their DNA sequence data to suggest how long the different types of primates have been separate. The fossil record was used to calibrate the clock: When the fossil record for one divergence is known, it indicates how long it probably takes for each nucleotide pair difference to occur. When the fossil record and molecular clock data agree, researchers have more confidence that the proposed phylogenetic tree is correct.

Animation
Molecular Clock

▶ **16.7 CHECK YOUR PROGRESS**

1. Explain the increasing use of DNA analysis to construct evolutionary trees.

FIGURE 16.7B Molecular data helps determine evolutionary relationships. This tree based on molecular data correlates with fossil evidence, lending support that this tree is accurate.

human

common chimpanzee

white-handed gibbon

rhesus monkey

green monkey

capuchin monkey

lesser bushbaby

| 60 | 50 | 40 | 30 | 20 | 10 | | PRESENT |

Million years ago (MYA)

← Increased difference in DNA

THE CHAPTER IN REVIEW

CONNECTING THE CONCEPTS

Understanding the evolutionary history of life is of concern to all biologists. The geologic timescale only describes, in general, the history of life on Earth. If we could trace the descent of all the millions of groups ever to have evolved, the entirety would resemble a dense bush.

The species alive today are the end product of all the changes that occurred on Earth as life evolved. There have been five mass extinctions, brought on chiefly by climate change and meteorite collisions. The continents drift and this has resulted in climate changes. Would the history of life have been different if these events had not occurred? For example, if the continents had not separated 65 MYA, what types of mammals, if any, would be alive today? Given a different sequence of environments, a different mix of plants and animals might very well have resulted.

Linnaean taxonomy gives every known species a two-part name consisting of a genus and a specific epithet. Most biologists today

have adopted the three-domain system of classifying species. The archaea are structurally similar to bacteria, but their rRNA differs from that of bacteria and is instead similar to that of eukaryotes. The eukaryotes consist of protists, fungi, plants, and animals. The classification categories assist traditionalists in constructing evolutionary trees. Traditionalists, often called evolutionary systematists, also use the degree of divergence to decide how to group organisms and display them in the tree. A famous example is the reluctance of traditionalists to group birds with reptiles because birds have feathers, a unique characteristic.

Phylogenetic cladistics is by now a widely accepted way to determine evolutionary relationships. Organisms that share derived traits are placed, along with their common ancestor, in the same clade and these relationships are depicted in a cladogram. To a cladist it is clear that birds are reptiles because they share derived traits with reptiles. Cladists study the diversity of life and use the fossil record, anatomic data,

and molecular data to decide how organisms are related.

Chapter 17 begins our detailed review of the history of life with a look at the evolution of the first cell(s) and bacteria and archaea. The rest of the chapters in this part review protists, plants, fungi, and animals. The ultimate aim of systematics is a tree of life that will show how all organisms are related through evolution.

ANALYZE AND EVALUATE

1. Do you think a tree of life that shows how all organisms are related will increase your desire to preserve all species and the biosphere? Why or why not?
2. Do you look forward to tracing the ancestry of humans to the very first living source? Why or why not?
3. Does it surprise you that humans are related to all other organisms on the planet? Why or why not?

SUMMARIZE

Fossils Tell a Story

16.1 The geologic timescale is based on the fossil record

- Eras, periods, and epochs divide up the **geologic timescale.** These divisions can be used to indicate the relative timing of events, but the MYA dates provide the absolute timing. **Extinctions,** the total disappearance of a species, have been common during the history of life. **Mass extinction** is the disappearance of many groups within a few million years or less.

16.2 Continental drift has affected the history of life

- **Continental drift**: The positions of continents and oceans have changed over time and are still changing. Continental drift can help explain the distribution of organisms on Earth, such as why marsupials are more prevalent in Australia.
- **Plate tectonics** explains the movements of Earth's crust. Oceanic ridges form when molten mantle lava rises and material is added to plates. Subduction zones occur where a plate sinks into the mantle and is destroyed. The place where two plates meet and scrape past one another is called a transform boundary.

16.3 Mass extinctions have affected the history of life

- Mass extinctions occurred at the ends of the Ordovician, Devonian, Permian, Triassic, and Cretaceous periods. Climate change due to continental drift and meteorite collisions can help explain mass extinctions.

Linnaean Systematics

16.4 Organisms can be classified into categories

- **Systematics** is the study of organism diversity at all levels of organization. Classification permits an organized study of diversity. The main **Linnaean classification** categories (**taxons**) are **domain, kingdom, phylum, class, order, family, genus,** and **species.**
- **Taxonomy** begins when species are named because in the **binomial system** a scientific name tells the genus and the specific epithet. Organisms in the same species have specific **characters** in common; those in the same domain have general characters in common.
- **Phylogeny** is the evolutionary history of a group of organisms. A **phylogenetic (evolutionary) tree** indicates **common ancestors** and lines of descent. **Derived characters** are the particular characteristics of a group, while **ancestral characters** are shared with a common ancestor.

16.5 The three-domain system divides all organisms into three large groups

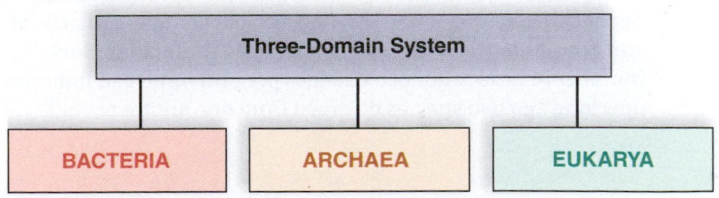

In domain Bacteria, the **bacteria** are a diverse and plentiful group of prokaryotes. In domain Archaea, the **archaea** encompass prokaryotes that are chemically different from bacteria and thrive in extreme environments. In domain Eukarya, the **eukaryotes** comprise a wide variety of unicellular to multicellular organisms that all have a membrane-enclosed nucleus but differ in nutrition and life cycles. Eukaryotes include protists, fungi, plants, and animals.

Modern Systematics: Cladistics

16.6 Cladograms reflect evolutionary history

■ **Cladistics** uses shared derived characters and common ancestry to construct **cladograms** and to group organisms in clades:

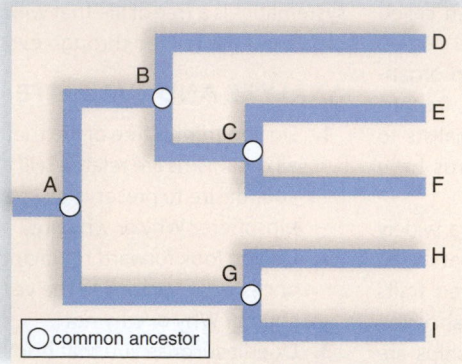

○ common ancestor

■ A **clade** is an evolutionary branch in a cladogram that includes the common ancestor and all descendant species. In the preceding cladogram, A, B, C, and G are common ancestors. Clades are nested, and therefore they vary in size. In the preceding cladogram, all the species (A–I) are in the largest clade. Species B–F are in a smaller clade, and species C, E, F and G–I form clades also. The members in each clade have **shared derived traits** that do not appear in members of an **outgroup** (not shown in the preceding cladogram). Any traits that are in both the outgroup and the **ingroup** (species A–I) are **shared ancestral traits** and are not used to tell clades apart. Each clade other than the largest one (A–I) has certain shared derived traits that define the clade. The rule of **parsimony** says that the simplest cladogram that can be constructed for an ingroup is the preferred one.

16.7 Certain types of data are used to trace phylogeny

■ The fossil record provides data that can trace the history of life and lineages. Fossil data are much prized because they are direct data.

■ Comparative anatomic data are used to trace phylogeny. **Homologous structures** are similar in structure (but not necessarily in function) because they were inherited from a common ancestor. **Analogous structures** are similar in function because they are adaptations to a similar environment, and they were not inherited from a common ancestor. Therefore, investigators must avoid using analogous structures.

■ Molecular data are used to trace phylogeny. The fewer the changes in DNA and amino acid sequences, the more closely related are two species. When nucleic acid changes are neutral and accumulate at a fairly constant rate, a **molecular clock** (number of base sequence changes per unit time) can indicate how long ago two species diverged from one another.

TEST YOURSELF

Fossils Tell a Story

For questions 1–4, match the phrases with a division of geologic time in the key.

KEY:
- a. Cenozoic era
- b. Mesozoic era
- c. Paleozoic era
- d. Precambrian time

1. Dinosaur diversity; evolution of birds and mammals
2. Prokaryotes abound; eukaryotes evolve and become multicellular
3. Mammalian diversification
4. Invasion of land by plants
5. Continental drift helps explain
   - a. mass extinctions.
   - b. the distribution of fossils on the Earth.
   - c. geologic upheavals such as earthquakes.
   - d. climate changes.
   - e. All of these are correct.
6. Which of the following would be a correct statement? Humans appeared
   - a. early in the history of the Earth.
   - b. at the same time as the dinosaurs.
   - c. after the apes appeared.
   - d. quite recently in the history of the Earth.
   - e. Both c and d are correct.
7. **THINKING CONCEPTUALLY** Australia, South America, and Africa, which are separate continents, all have mammals but the specific ones are different. Explain this observation.

Linnaean Systematics

8. Which is the scientific name of an organism?
   - a. *Rosa rugosa*
   - b. *Rosa*
   - c. *rugosa*
   - d. *rugosa rugosa*
   - e. Both a and d are correct.
9. Classification of organisms reflects
   - a. anatomical and molecular similarities
   - b. evolutionary history.
   - c. Neither a nor b is correct.
   - d. Both a and b are correct.
10. Which of these is sequenced from less inclusive to more inclusive?
    - a. kingdom, phylum, class, order
    - b. phylum, class, order, family
    - c. class, order, family, genus
    - d. genus, family, order, class
11. **THINKING CONCEPTUALLY** The adoption of the three-domain system emphasizes the increased use of genetic similarities and differences to classify organisms. Explain.

Modern Systematics: Cladistics

12. Use the data from the following table to fill in the phylogenetic tree for vascular plants (plants having vascular transport tissue):

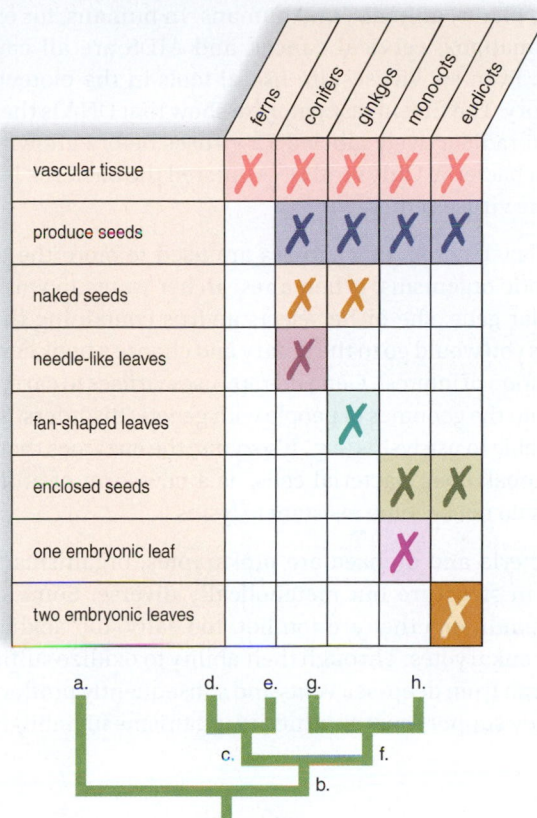

	ferns	conifers	ginkgos	monocots	eudicots
vascular tissue	X	X	X	X	X
produce seeds		X	X	X	X
naked seeds		X	X		
needle-like leaves		X			
fan-shaped leaves			X		
enclosed seeds				X	X
one embryonic leaf				X	
two embryonic leaves					X

a.   d. e. g.   h.

c.   f.

b.

vascular tissue

13. One benefit of the fossil record is
    a. that hard parts are more likely to fossilize.
    b. fossils can be dated.
    c. its completeness.
    d. that fossils congregate in one place.
    e. All of these are correct.
14. Which pair is mismatched?
    a. homology—character similarity due to a common ancestor
    b. molecular data—DNA strands match
    c. fossil record—bones and teeth
    d. homology—functions always differ
    e. molecular data—molecular clock
15. The discovery of common ancestors in the fossil record, the presence of homologies, and nucleic acid similarities help scientists decide
    a. how to classify organisms.
    b. the proper cladogram.
    c. how to construct phylogenetic trees.
    d. how evolution occurred.
    e. All of these are correct.
16. Molecular clock data are based on
    a. common adaptations among animals.
    b. DNA dissimilarities in living species.
    c. DNA fingerprinting of fossils.
17. In cladistics,
    a. a clade must contain the common ancestor plus all its descendants.
    b. derived characters help construct cladograms.
    c. data for the cladogram are presented.

d. the species in a clade share homologous structures.
    e. All of these are correct.
18. In evolutionary systematics, birds are assigned to a different group from reptiles because
    a. they evolved from reptiles and couldn't be a clade.
    b. they are adapted to a different ecological niche compared to reptiles.
    c. feathers came from scales, and feet came before wings.
    d. all classes of vertebrates are only related by way of a common ancestor.
    e. All of these are correct.
19. **THINKING CONCEPTUALLY** DNA differences are expected to be consistent with evolutionary trees based on structure. Explain.
20. The three-domain classification system is based on
    a. mitochondrial biochemistry and plasma membrane structure.
    b. cellular structure and rRNA sequence data.
    c. plasma membrane and cell wall structure.
    d. nuclear and mitochondrial biochemistry.
21. Which of these are domains? Choose more than one answer if correct.
    a. Bacteria   d. animals
    b. Archaea   e. plants
    c. Eukarya
22. Which of these are eukaryotes? Choose more than one answer if correct.
    a. Bacteria   d. animals
    b. Archaea   e. plants
    c. Eukarya
23. Which of these is a true statement?
    a. Eukaryotes are more closely related to bacteria than they are to archaea.
    b. Fungi, animals, and plants have different means of acquiring nutrients.
    c. The close relationship between bacteria and archaea places them in their own domain.
    d. Archaea evolved during the Precambrian time, but bacteria evolved during the Cambrian period.
    e. All of these are correct.

GET INVOLVED

1. Suppose that you were asked to supply an evolutionary tree of life and decided to use Figure 16.5. How is this tree consistent with evolutionary principles?
2. Explain the occurrence of living fossils, such as horseshoe crabs, that closely resemble their ancestors in the fossil record.

MEDIA STUDY TOOLS

mhhe.com/maderconcepts3

Enhance your study of this chapter with interactive study tools, practice tests, and engaging animations. Also, ask your instructor about the resources available through ConnectPlus, which includes LearnSmart, a personalized adaptive learning program, and a media-rich eBook.

17

Evolution of Microbial Life

BEFORE YOU BEGIN

Take a few minutes to recall

Life's characteristics (section 1.10)
Prokaryotic cell structure (section 4.2)
The three domains of life (section 16.5)

At Your Service: Viruses and Bacteria

Viruses are noncellular entities responsible for a number of diseases in plants, animals, and humans. In humans, for example, polio, smallpox, cervical cancer, and AIDS are all caused by viruses. Even so, viruses are useful tools in the biotechnology laboratory. The first investigators to show that DNA is the genetic material radioactively labeled T2 viruses before allowing them to infect bacteria. Only viral DNA entered the bacteria, but many complete viruses emerged.

Today, recombinant viruses are used to store the genes of eukaryotic organisms. When a researcher wants to work with a particular gene, she or he selects a virus containing that gene, much as you would go to the library and choose a book containing information of interest. Gene therapy uses viruses to carry normal genes into the genomes of people with genetic disorders. Soon, we may be able to use lysins (e.g., lysozyme) the enzymes that viruses use to break open bacterial cells, as a new form of antibiotic—bacteria do not become resistant to lysins.

Bacteria and archaea are prokaryotes, organisms that are simple in structure but metabolically diverse. Some can live under conditions that are too hot, too salty, too acidic, or too cold for eukaryotes. Through their ability to oxidize sulfides that spew forth from deep-sea vents and subsequently produce nutrients, they support communities of organisms in habitats where

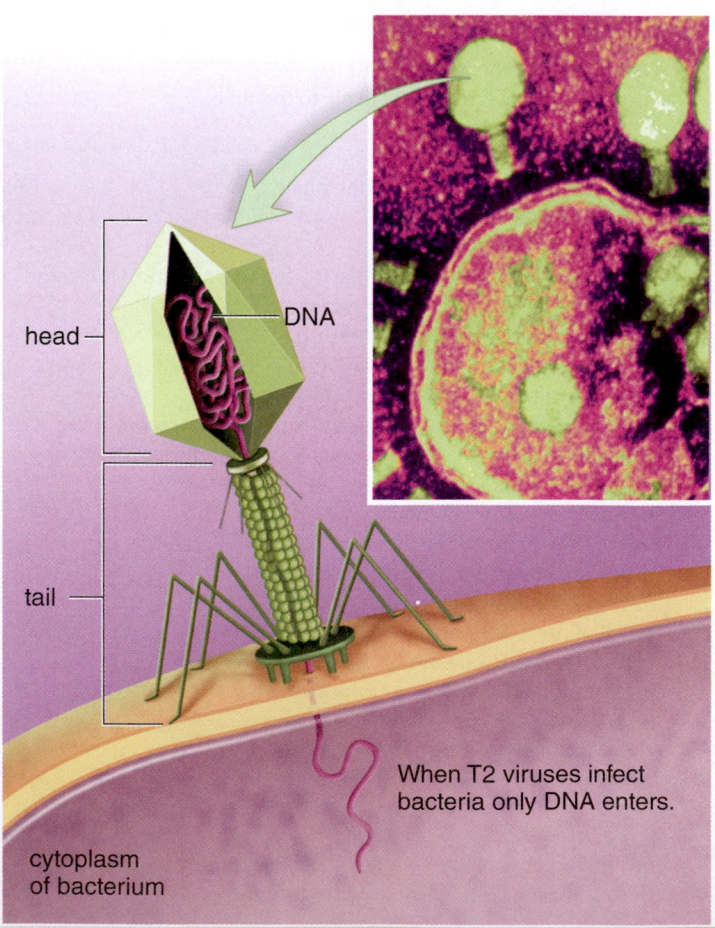

head

DNA

tail

When T2 viruses infect bacteria only DNA enters.

cytoplasm
of bacterium

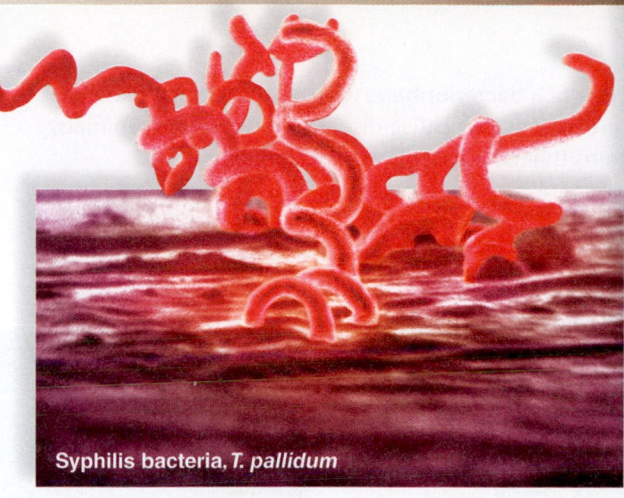
Syphilis bacteria, *T. pallidum*

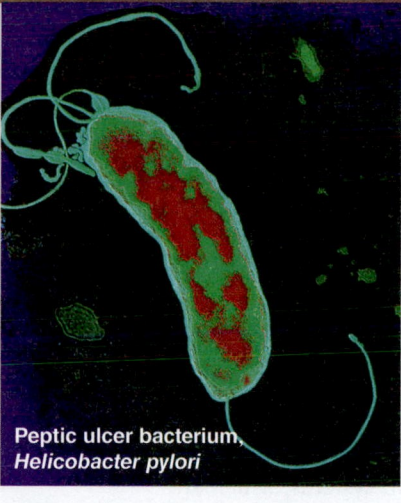
Peptic ulcer bacterium, *Helicobacter pylori*

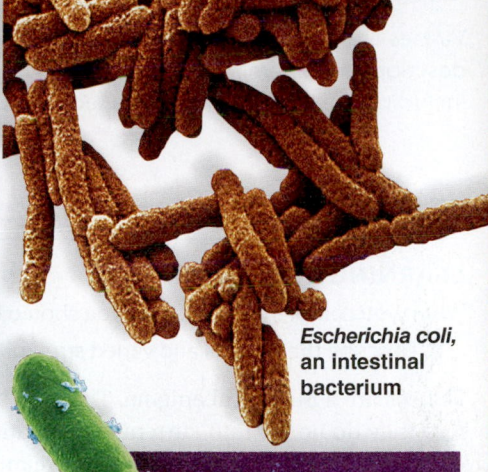
Escherichia coli, an intestinal bacterium

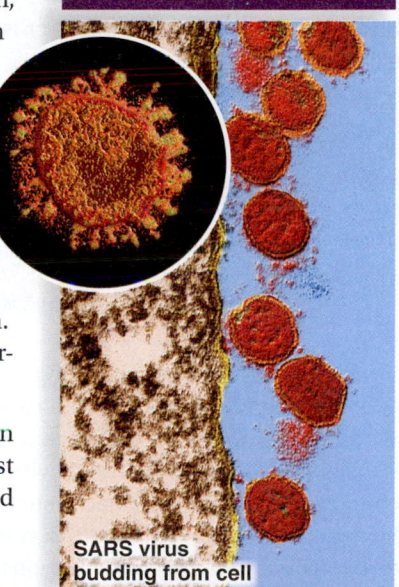
Anthrax bacteria, *Bacillus anthracis*

the sun never shines. Humans use bacteria in the environment to mine minerals, degrade sewage, and improve the soil because they fix atmospheric nitrogen in plant nodules. In addition, bacteria's vast ability to break down almost any substance has been applied to bioremediation, the biological cleanup of harmful chemicals called pollutants. Bacteria have been used to clean up oil spills, and some strains have helped remove Agent Orange, a potent herbicide, from soil samples. Dual cultures of two types of bacteria have been shown to degrade polychlorinated biphenyls (PCBs), chemicals formerly used as coolants and lubricants.

Although some bacteria cause deadly diseases, others produce antibiotics, such as streptomycin, that can help cure such illnesses. Due to the ease with which they can be grown and manipulated in the laboratory, bacteria are used to study basic life processes. The details of how DNA specifies the order of amino acids in a protein were discovered by studying the process in *Escherichia coli*. Through genetic engineering, bacteria produce many commercial products, including human insulin for diabetics. And now new research tells us that the many bacteria living on and in our digestive system perform services for us that have too long gone unrecognized. A new term—*microbiome*—has been coined to refer to these microbes that interact with human cells and keep us healthy. As an example, although the bacterium *Helicobacter pylori* sometimes causes ulcers, researchers now tell us that it more often plays a role in keeping our weight under control. Other types of bacteria produce vitamins, allow us to digest whole grains and prevent inflammation. A new mindset and much more research is needed for us to become fully aware of the preventive services of the microbes that inhabit our bodies.

This chapter will introduce you to the *microbes*, so-called because they are too small to be seen without the aid of a microscope. Even though viruses are noncellular, it's possible they were the first to contain a genetic material. Prokaryotes including both archaea and bacteria lack a nucleus and are closely related to the first cell(s) to have evolved.

SARS virus budding from cell

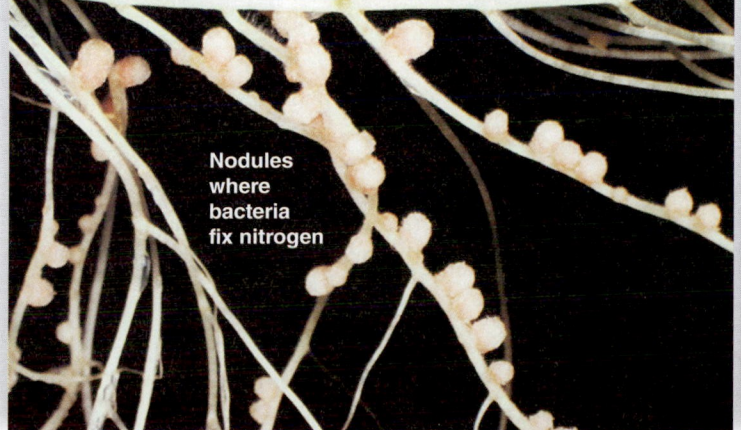
Nodules where bacteria fix nitrogen

before   after

Bioremediation of an oil spill

Viruses, Viroids, and Prions

Viruses are noncellular, and today they reproduce only inside living cells. The life cycle of a bacteriophage (viruses that attack bacteria) describes, in general, the life cycle of a virus in any type of cell. Of special concern are viruses that cause diseases in plants and animals, including humans. Viroids are naked strands of RNA and prions are misfolded proteins that cause diseases when present in cells.

17.1 Viruses have a simple structure

LEARNING OUTCOME

When you complete this section, you should be able to

1. Describe the general and varied structure of a virus.

Viruses are a biological enigma. They are noncellular, and therefore they do not fit into current classification systems, which are devoted to categorizing the cellular organisms on Earth. How viruses might fit into the origin of the first cell is still being considered. Many viruses can be purified and crystallized, and the crystals may be stored just as chemicals are stored. Still, viral crystals become infectious when the viral particles they contain are given the opportunity to invade a specific host cell. Viruses have a DNA or RNA genome, but they can reproduce only by using the metabolic machinery of a host cell.

The size of a virus is comparable to that of a large protein macromolecule, ranging from 0.2–2 μm. Therefore, viruses are best studied through electron microscopy. The following diagram summarizes viral structure:

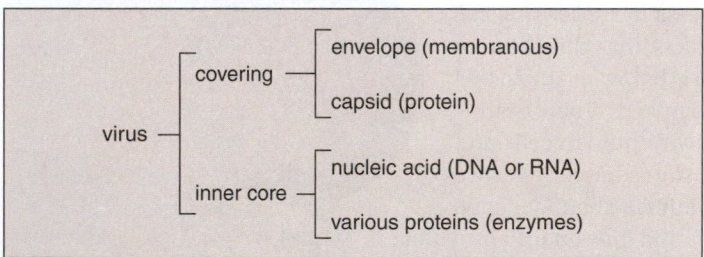

Viruses possess an outer covering. They all have a **capsid** which is composed of protein and the capsid can be surrounded by an outer membranous envelope. If there is no envelope, the virus is said to be naked. Naked viruses can be transmitted by contact with inanimate objects, such as desktops. Viruses also have an inner core which at the very least contains a nucleic acid (DNA or RNA) genome. The observation that some viruses have an RNA genome is considered significant for reasons we will be discussing. A viral genome has as few as 3 and as many as 100 genes; a human cell contains about 23,000 genes. The inner core of some viruses also includes various proteins (e.g., enzymes).

Figure 17.1A shows an example of a naked virus, while Figure 17.1B is an example of an enveloped virus. The envelope is actually a piece of the host's plasma membrane that also contains viral glycoprotein spikes. The spikes assist the virus in entering the host cell. Enveloped viruses are usually transmitted by direct contact with an infected individual. Aside from its genome, a viral particle may also contain various proteins, especially enzymes such as the polymerases, which are needed to produce viral DNA and/or RNA.

Viruses are categorized by (1) their type of nucleic acid, which can be DNA or RNA, and whether it is single-stranded or

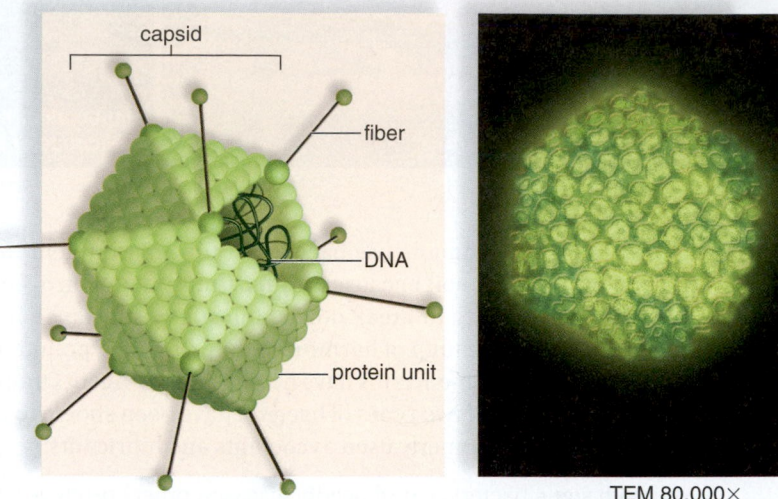

FIGURE 17.1A Adenovirus, a naked virus, with a polyhedral capsid and a fiber at each corner.

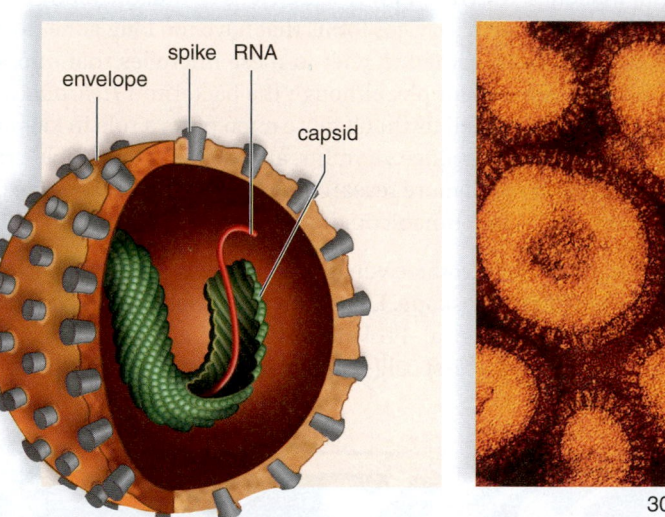

FIGURE 17.1B Influenza virus, surrounded by an envelope with spikes.

double-stranded; (2) their size and shape (the capsid can have projecting fibers); and (3) the presence or absence of an outer envelope.

▶ **17.1 CHECK YOUR PROGRESS**
1. Identify the two features shared by all viruses.

17.2 Some viruses reproduce inside bacteria

LEARNING OUTCOME

When you complete this section, you should be able to

1. Differentiate the ways a virus can reproduce inside a bacterium.

All sorts of cells, whether prokaryotic or eukaryotic, are susceptible to a viral infection. Viruses are specific. Specificity extends even to the type of cell infected by the virus. For example, tobacco mosaic virus especially infects tobacco leaves, and adenoviruses attach to cells in our respiratory tract, causing colds. Specificity of a virus arises because viruses enter a cell when either an envelope spike or a capsid protein combines with a particular receptor protein in the plasma membrane of the host cell.

Video
How Viruses Attack

Bacteriophages, or simply phages, are viruses that parasitize bacteria. There are two types of bacteriophage life cycles, termed the lytic cycle and the lysogenic cycle (Fig. 17.2). Most types of bacteriophages have the lytic cycle, and only a few types are lysogenic. In the **lytic cycle,** viral reproduction occurs, and the host cell undergoes lysis, by which the cell breaks open to release the viral particles. In the **lysogenic cycle,** viral reproduction does not immediately occur, but reproduction may take place sometime in the future.

Lytic Cycle Figure 17.2 shows that the lytic cycle may be divided into five stages: attachment, penetration, biosynthesis, maturation, and release. ❶ During *attachment,* portions of the capsid combine with a receptor on the rigid bacterial cell wall in a lock-and-key manner. ❷ₐ During *penetration,* a viral enzyme digests away part of the cell wall, and viral DNA is injected into the bacterial cell. ❸ *Biosynthesis* of viral components begins after the virus brings about inactivation of host genes not necessary to viral replication. The virus takes over the machinery of the cell in order to carry out viral DNA replication and produce multiple copies of the capsid protein. ❹ During *maturation,* viral DNA and capsids assemble to produce several hundred viruses. Lysozyme, an enzyme coded for by a viral gene, is produced; this disrupts the cell wall, and ❺ the *release* of new viruses occurs. The bacterial cell dies as a result.

Video
Virus Lytic Cycle

Lysogenic Cycle With the lysogenic cycle, the infected bacterium does not immediately produce phages but may do so sometime in the future. In the meantime, the phage is *latent*—not actively replicating. ❷ᵦ Following attachment and penetration, *integration* occurs. Viral DNA becomes incorporated into bacterial DNA with no destruction of host DNA. While latent, the viral DNA is called a *prophage.* The prophage is replicated along with the host DNA, and all subsequent cells, called lysogenic cells, carry a copy of the prophage. Certain environmental factors, such as ultraviolet radiation, can induce the prophage to enter the lytic stage of biosynthesis, followed by maturation and release.

▶ **17.2 CHECK YOUR PROGRESS**

1. Compare the lytic and lysogenic cycles that occur when a virus infects a cell.

FIGURE 17.2 The lytic and lysogenic cycles occur when a virus infects a bacterium.

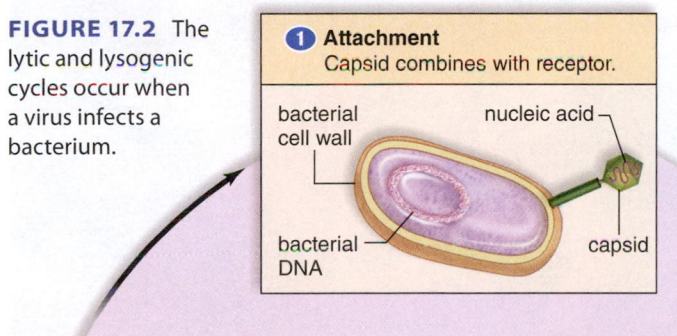

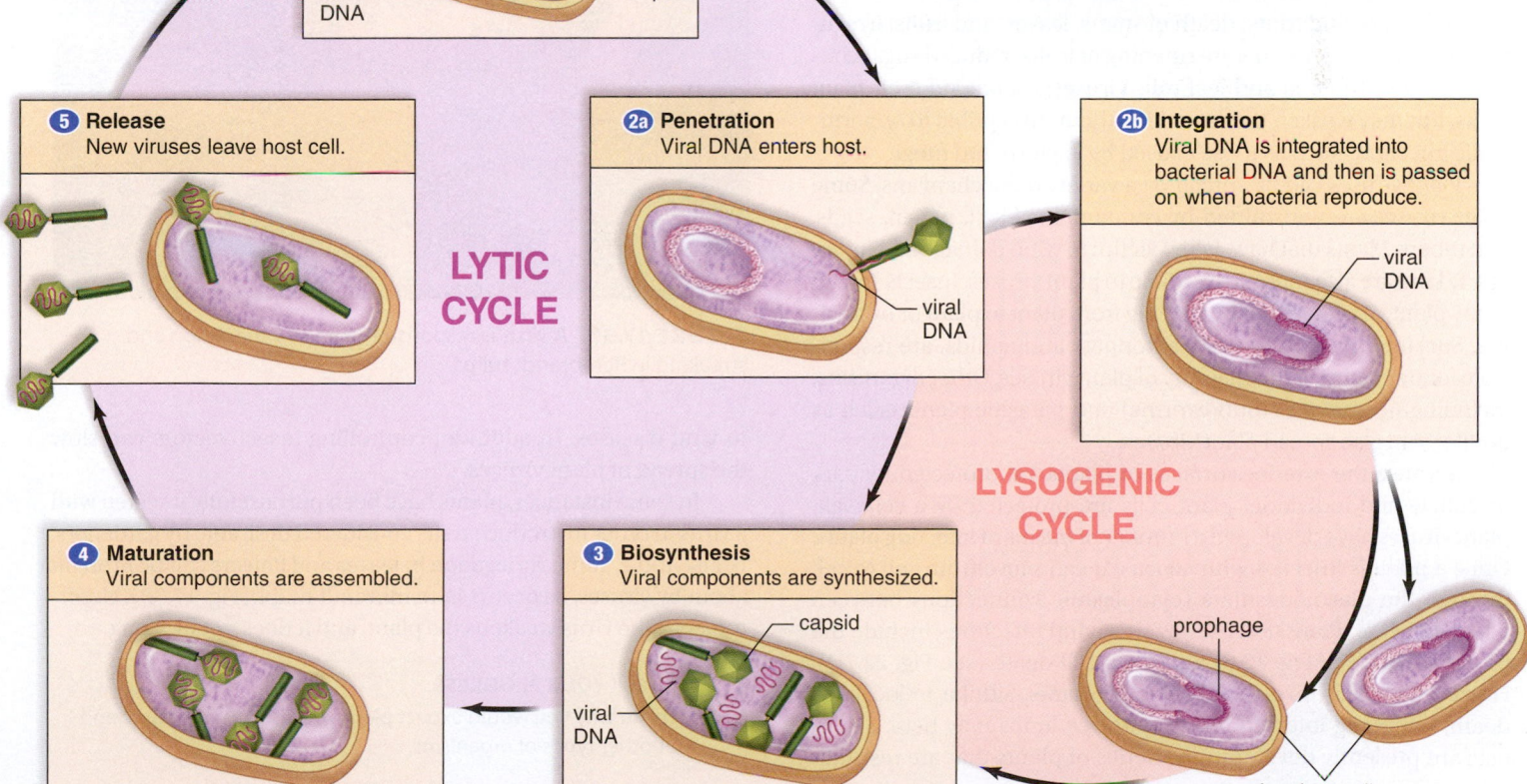

17.3 Viruses are responsible for a number of plant diseases

LEARNING OUTCOME

When you complete this section, you should be able to

1. Discuss the significance of viral infections in plants.

Approximately 2,000 kinds of plant diseases have been attributed to viruses. Plant viruses are responsible for the loss of over $60 billion annually by reducing the world-wide yield of important agricultural and horticultural crops.

Tobacco Mosaic Virus (TMV) In 1898, Martinus Beijerinck, working with a disease prevalent in tobacco plants, determined that the disease was not caused by a bacterium but by a virus. We now call this virus tobacco mosaic virus (TMV). In the 1930s, Wendell Stanley was studying TMV when he discovered the structure of viruses.

TMV can infect a number of plants, including orchid, potato, tobacco, and tomato plants. The virus can remain viable for years on dried plant debris and is extremely tolerant of very high temperatures. TMV enters plants through wounds sustained during transplanting or pruning. It spreads rapidly once it is in the host. The virus interferes with chlorophyll production, and the infected plant develops unsightly light-green, yellow, or white spots on its leaves and fruits (Fig. 17.3A). Tobacco products are the most common source of infection, and smokers can pass on the virus to plants by handling them. In fact, at one time, smokers were not allowed to work in ketchup factories because their touch could ruin tomatoes.

Transmission and Infection Plant viruses do not differ significantly in size and shape from bacteriophages and animal viruses. With the exception of three groups of DNA viruses, all of the other plant viruses are RNA viruses. The generalized symptoms of plants infected with a virus include: stunted growth; discoloration of leaves, flowers, and fruits; death of stems, leaves, and fruits; irregularities in fruit size; premature ripening of fruits; reduced sugar content of fruits; tumors; and leaf roll. Viruses seldom kill their plant hosts, but they weaken them, making them susceptible to opportunistic infections such as those caused by bacteria and fungi.

Plant viruses can be spread by a variety of mechanisms. Some plant viruses are transmitted by contaminated soil, pollen, seeds, and tubers. Plants that have fallen victim to wind damage and injury of any kind are also more susceptible to plant viruses. Insects spread many plant viruses, either by moving from plant to plant or by feeding. Sucking insects, such as leafhoppers and aphids, are responsible for transmitting the majority of plant viruses. Other organisms, including nematodes (roundworms) and parasitic plants, such as dodder, can also spread plant viruses.

Because the exterior surfaces of plants are protected by bark or cuticle and individual plant cells are protected by a cell wall, plant viruses have developed a number of means of infecting plants. Once a plant is infected with a virus, it can move from cell to cell through the plasmodesmata (cytoplasmic connections between adjacent cells). Plant defenses against viral infections include the use of siRNA (see Fig. 12.6) and local cell death (see Fig. 22.7D). However, there is no cure for infected plants. Similar to local cell death, removing infected leaves and tree limbs may help. Scientists are presently developing varieties of plants that are resistant

FIGURE 17.3A The tobacco mosaic virus (TMV) is responsible for discoloration in the leaves of tobacco plants.

FIGURE 17.3B A virus is responsible for the variegation and streaking in Rembrandt tulips.

to viral diseases. In addition, controlling insect vectors can slow the spread of plant viruses.

In some instances, plants have been purposefully infected with a virus in order to produce traits considered desirable by gardeners. For example, some variegation in leaves and flowers can be brought about by viruses, as occurs in Rembrandt tulips (Fig. 17.3B). Unfortunately, the virus weakens the plant, and it does not live long.

▶ **17.3 CHECK YOUR PROGRESS**

1. Explain why you would expect plant viruses to infect plants and not other types of organisms.

17A Humans Suffer from Emergent Viral Diseases

Emergent diseases are ones that we newly recognize as causing infections in people. Sometimes the human disease existed before but has begun to spread widely. The modern age of travel makes it convenient for a person to wake up in Bangkok in the morning and sleep in Los Angeles that night. It also provides pathogens with unprecedented mobility. No longer are outbreaks necessarily limited to a small geographic region.

Very often today, an emergent disease was originally confined to a specific group of animals, such birds, mice, and apes, but then it became capable of infecting humans. Viruses are constantly in a state of evolutionary flux and can acquire new spikes (surface proteins) that allow them to attach to and enter human cells. At first, the immune system is unable to recognize the new threat and does not mount a defense in time to defeat the pathogen. Thus, the virus successfully completes its life cycle, causes a disease, and spreads to other victims. A virus that cannot pass from human to human after jumping from an animal host will not be capable of causing an epidemic, a large-scale infection of many persons. So far, this has been the case with bird flu, a disease that will soon be discussed.

Some emergent diseases are transmitted by vectors, usually insects that carry pathogens from an infected individual or animal reservoir to a healthy individual. Mosquitoes serve as a common vector for several viral diseases, among them **West Nile disease,** which began in 1937 in Uganda, Africa. Many times, extensive efforts are required to control and eradicate the vector or the reservoir of the virus. For example, a **hantavirus** that causes a highly fatal hemorrhagic infection is transmitted to humans through fecal droppings of deer mice. Control of the deer mice population helps control the disease.

Vaccines are presently or soon will be available for many emerging diseases. It has been difficult to produce a vaccine against **HIV** (see Fig. 17.4B and "HIV/AIDS: A Global Disaster" on pages 612–613) because HIV often mutates and frequently changes its spikes. Such changes help a virus remain infectious. A vaccine is now available for the **H1N1 virus,** which causes fever of about 38°C, cough, sore throat, headache, chills, muscle aches, diarrhea, and/or vomiting. A few of the people infected develop pneumonia and respiratory failure. Originally called swine flu, it is now named after its spikes, dubbed H1 and N1. It is these spikes that allow the virus to infect human cells. H1N1 emerged in Mexico and has now spread around the globe. To avoid catching or passing on the virus, cover your nose and mouth when you cough,

wash your hands frequently, avoid touching any part of your face, avoid close contact with sick people, and stay home if you are sick.

Researchers are working on vaccines for these emerging diseases:

The **severe acute respiratory syndrome** (**SARS**) virus causes high fever, body aches, and pneumonia. In 2003, its path of death was easily traced from Southeast Asia to Toronto, Canada. Its apparent mode of transmission is by droplet infection and direct contact, so some people protected themselves from SARS by wearing surgical masks (Fig. 17A.1).

Avian influenza (or **bird flu**) was of special concern a few years ago. This disease arose in Southeast Asia, where markets are crowded with humans and animals, particularly domesticated chickens. Human symptoms include cough, sore throat, muscle aches, eye infections, respiratory distress, and life-threatening complications. So far, the disease does not often spread from chickens to humans, nor is it efficiently transmitted among humans. Nevertheless, in some areas, chickens have been exterminated as a precaution (Fig. 17A.2).

Video
Killer Flu Recreated

Ebola is one of a number of viruses that can cause hemorrhagic fever. These extremely diabolical pathogens are highly contagious and can quickly cause intolerable fever and extensive tissue damage, leading to profuse internal bleeding, multiple organ failure, and certain death. The vector and animal reservoir for Ebola virus are unknown. The Zaire and Sudan strains were found in Africa, but the Reston strain was discovered in a 1989 shipment of 100 crab-eating monkeys, imported from the Philippines to Reston, Virginia. The movie thriller *Outbreak* was based on a book, *The Hot Zone,* whose story line concerned these monkeys.

Video
Virus Crisis

CONSIDER THESE QUESTIONS

1. What special concerns might people have about catching a newly discovered emergent disease?
2. Is it a service or a disservice for the media to publicize emergent diseases as much as they do?
3. Are we too critical or not critical enough of developing countries where emergent diseases are apt to originate?

connect
BIOLOGY

Explore the concepts through a variety of multimedia assets and question types.

www.mcgrawhillconnect.com

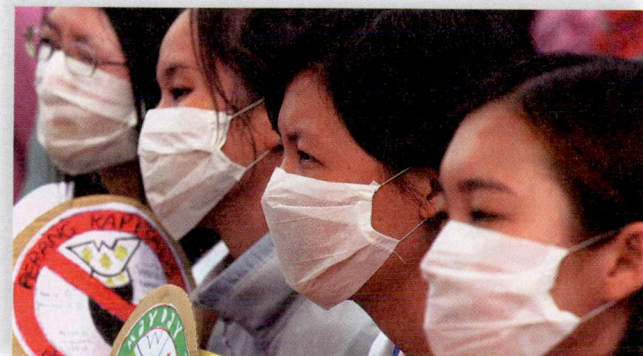

FIGURE 17A.1 Surgical masks provide protection against the transmission of SARS from person to person.

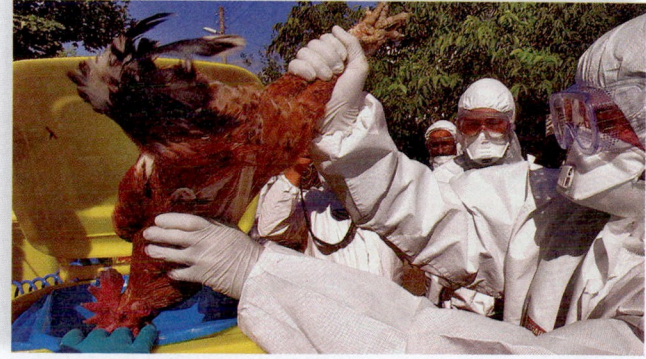

FIGURE 17A.2 Exterminating infected chickens provides protection against the transmission of bird flu from chickens to humans.

17.4 Viruses also reproduce inside animal cells and cause animal diseases

LEARNING OUTCOME

When you complete this section, you should be able to

1. Describe how viruses, including HIV, invade and reproduce inside animal cells.

Viruses are responsible for a number of diseases in humans (Table 17.4). Rabies affects many species of mammals, including humans. Viruses are responsible for several childhood maladies, including measles, mumps, chickenpox, warts, and viral pinkeye. More serious human viral diseases include polio, yellow fever, dengue fever, type 1 herpes (fever blisters), type 2 herpes (genital herpes), shingles, mononucleosis, hepatitis, HIV infection, smallpox, rubella, and various forms of flu and colds. In the past, several emergent viruses have captured the public's attention, including H1N1 and bird flu (see "Humans Suffer from Emergent Viral Diseases" on page 321).

Life Cycle of a DNA Virus Although different in shape, host range, and genetic composition, DNA viruses that invade human and animal cells use reproductive strategies similar to those of bacteriophages. As illustrated in Figure 17.4A, replication of an animal virus with a DNA genome involves these steps:

1. *Attachment.* Glycoprotein spikes projecting through the envelope allow the virus to bind only to host cells having specific receptor proteins on their surface.

2. *Penetration.* After the virus is brought into the host cell, uncoating—removal of the viral capsid—follows, and viral DNA is released into the host.

   Biosynthesis. **3a** The capsid and other proteins are synthesized by host cell ribosomes according to viral mRNA instructions. The viral spikes are modified in the ER before becoming part of the plasma membrane. **3b** The host cell's enzymes make many copies of the viral DNA.

4. *Maturation.* Viral proteins and DNA replicates are assembled to form new viruses.

5. *Release.* In an enveloped virus, budding occurs, and the virus gains its envelope, which usually consists of the host's plasma membrane components and glycoprotein spikes that were coded for by the viral DNA.

Animation
Entry of an Animal Virus into Host Cell

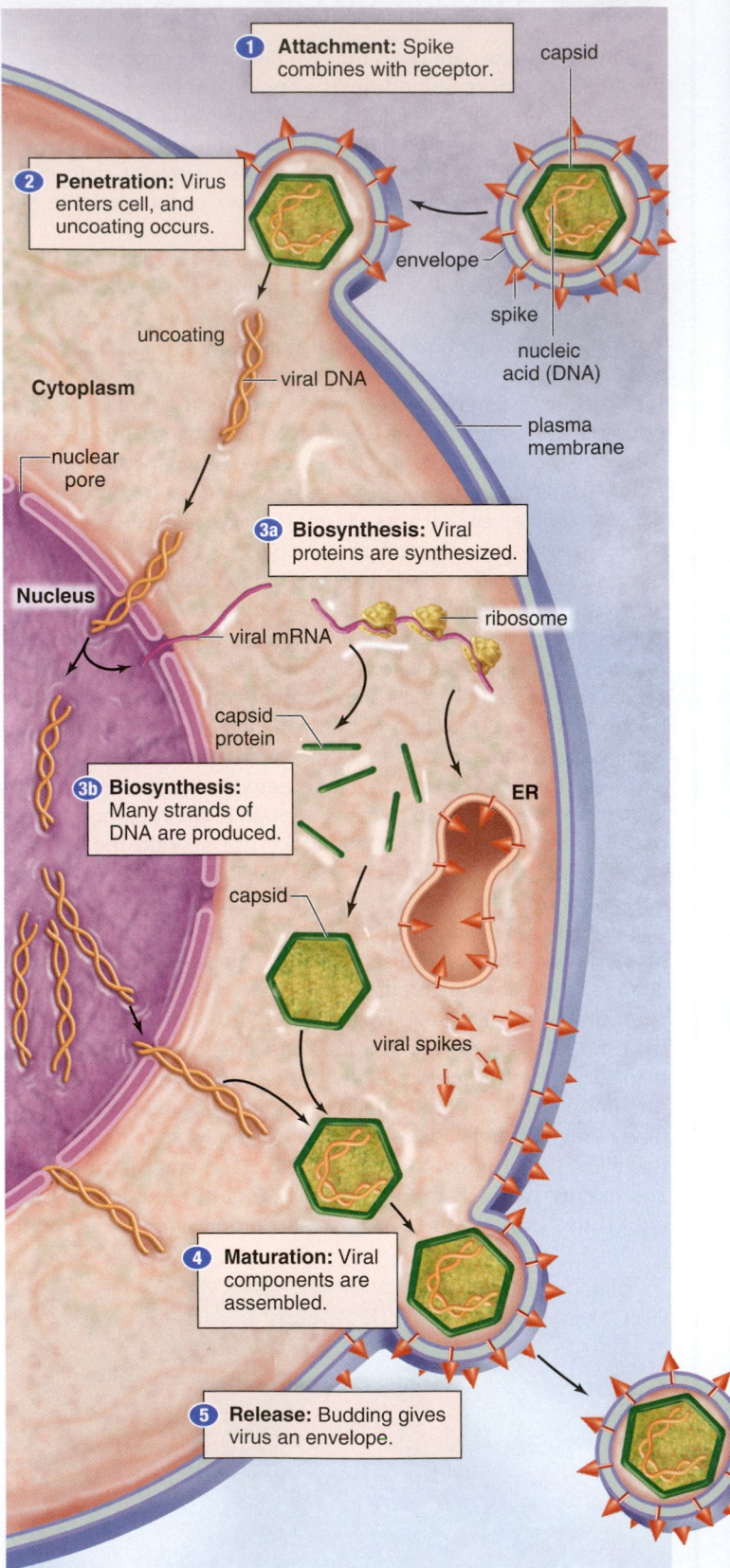

FIGURE 17.4A Reproduction of an animal DNA virus.

TABLE 17.4	Viral Diseases in Humans
Category	**Disease**
Sexually transmitted diseases	AIDS (HIV infection), genital warts, genital herpes
Childhood diseases	Mumps, measles, chickenpox, German measles
Respiratory diseases	Common cold, influenza, severe acute respiratory syndrome (SARS)
Skin diseases	Warts, fever blisters, shingles
Digestive tract diseases	Gastroenteritis, diarrhea
Nervous system diseases	Poliomyelitis, rabies, encephalitis
Other diseases	Smallpox, hemorrhagic fevers, cancer, hepatitis, mononucleosis, yellow fever, dengue fever, conjunctivitis, hepatitis C

HIV The genome for **HIV** (the human immunodeficiency virus that causes AIDS) consists of RNA, instead of DNA. In addition, HIV is a **retrovirus,** meaning that it uses reverse transcription from RNA into DNA in order to insert a copy of its genome into the host's genome. Figure 17.4B shows the steps for retrovirus reproduction:

Animation
Replication Cycle
of a Retrovirus

① *Attachment.* During attachment, the HIV virus binds to the plasma membrane. HIV has a spike that allows the virus to bind to a receptor in the host-cell plasma membrane.

② *Penetration.* After attachment, the HIV virus fuses with the plasma membrane, and the virus enters the cell. A process called uncoating removes the capsid, and RNA is released.

③ *Reverse transcription.* This event in the reproductive cycle is unique to retroviruses. The enzyme called **reverse transcriptase** makes a DNA copy (cDNA) of the retrovirus's RNA genetic material. Usually, in cells DNA is transcribed into RNA. Retroviruses can do the opposite only because they have this unique enzyme, from which they take their name. (*Retro* in Latin means reverse.)

④ *Integration.* Following DNA replication, double-stranded DNA is integrated into host DNA. The term **provirus** refers to viral DNA integrated into host DNA. HIV is usually transmitted to another person by means of cells that contain proviruses. Also, proviruses serve as a latent reservoir for HIV during drug treatment. Even if drug therapy results in an undetectable viral load, investigators know that there are still proviruses inside infected cells.

⑤ *Biosynthesis.* When the provirus is activated, perhaps by a new and different infection, the normal cell machinery directs the production of more viral RNA, some of which is used as mRNA. Viral mRNA brings about the synthesis of capsid proteins and viral spikes. The spikes are modified in the ER and then enter the plasma membrane.

⑥ *Maturation.* Capsid proteins, viral enzymes, and RNA can now be assembled to form new viral particles.

⑦ *Release.* During budding, the virus gets its envelope and spikes coded for by the viral genetic material.

Animation
HIV Replication

The characteristics for an HIV infection are discussed in "HIV/AIDS: A Global Disaster" on pages 612–613.

Viroids and Prions

Viroids and prions are subviral particles that can also cause diseases. **Viroids** are naked strands of RNA that cause diseases in plants. They may be introns (coding for ribosomal components) that have gone amuck and interfere with gene expression in general. **Prions** (*pro*tein *in*fectious particles) are misfolded proteins whose presence causes other proteins to also become misfolded. Rare but serious brain diseases, such as mad cow disease, are caused by prions.

Animation
How Prions Arise

▶ **17.4 CHECK YOUR PROGRESS**

1. (**a**) Identify which retrovirus reproductive steps (Fig. 17.4B) differ from those of a DNA animal virus (Fig. 17.4A). (**b**) What does the enzyme reverse transcriptase do?

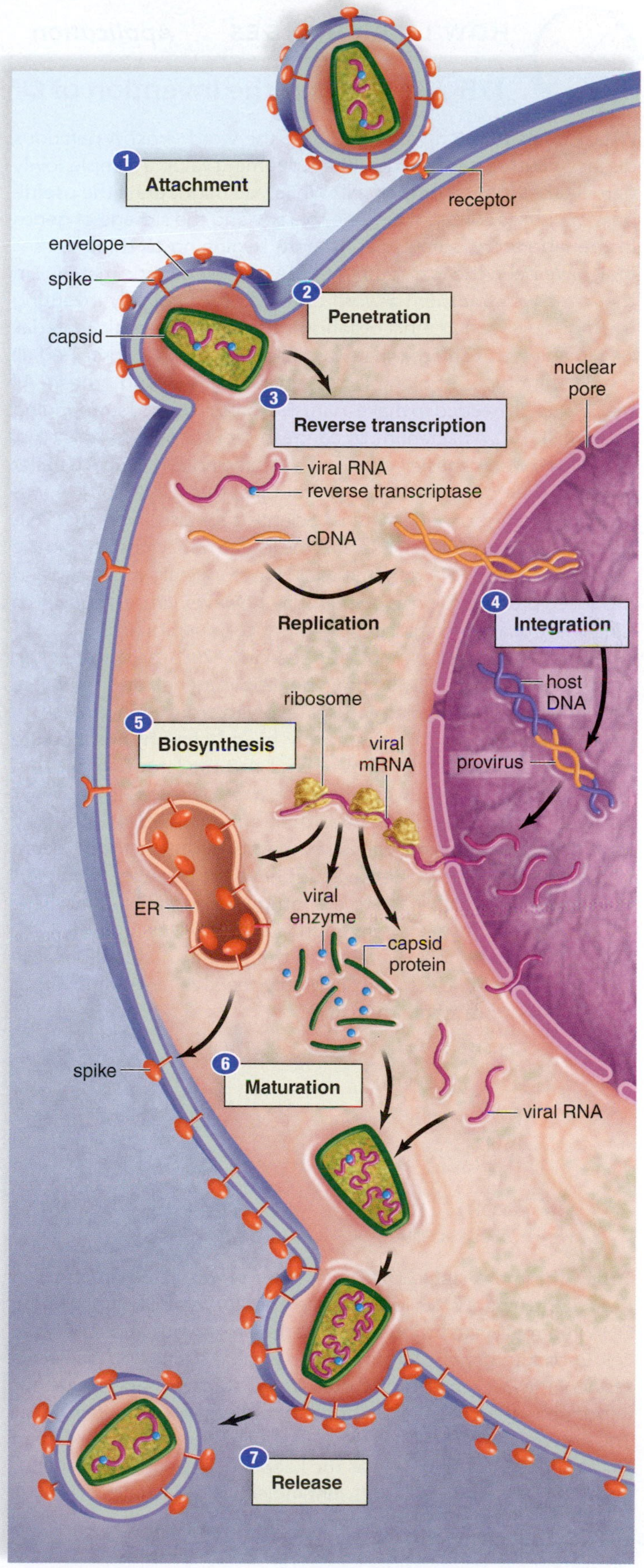

FIGURE 17.4B Reproduction of HIV.

17B Viruses and the Invention of DNA

What comes to mind when you hear the word virus? A nefarious disease-causing particle with little redeeming value? A simple, lowly particle that is not a proper cell? While you might think the usefulness of viruses is limited to gene therapy and the storage of genes in the laboratory, some scientists are beginning to think viruses may have been the first to utilize DNA as their genetic material. In that case, we owe them a word of thanks for originally inventing the molecule we and most organisms use as their genetic material.

Imagine an early sea that was lifeless although it contained all the different types of biomolecules, including RNA and protein. All that would be needed to make a virus is a protein capsid surrounding RNA (Fig. 17B.1). A bacterium, by contrast, is much more complex (Fig. 17B.2). Why would the first virus have RNA and not DNA? RNA is easier to make abiotically (in the lab, without life) than DNA, so RNA was probably the first genetic material. Scientists are also aware that RNA can be both a substrate and an enzyme (called ribozymes), raising the possibility that an RNA genome could have originally replicated and produced proteins all on its own. (Today, however, viruses require the machinery of a cell to replicate and produce their proteins.)

This is the best part of our scenario: Viruses of today can use either RNA or DNA as their genetic material, and evidence suggests that DNA-based viruses evolved from RNA viruses. In addition, it appears that there may have been an intermediate form between the earlier RNA viruses and the later DNA viruses. This intermediate form was, of course, a retrovirus, as is HIV. Why? Because retroviruses use RNA as their genetic material, but they have that handy enzyme called reverse transcriptase that makes a DNA copy of their RNA. Retroviruses have the capacity to convert RNA into DNA, which leads to the intriguing possibility that these viruses are responsible for inventing DNA.

DNA has a benefit over RNA as the genetic material because DNA is more stable. Oxygen is highly reactive, and removing it makes DNA more stable than RNA:

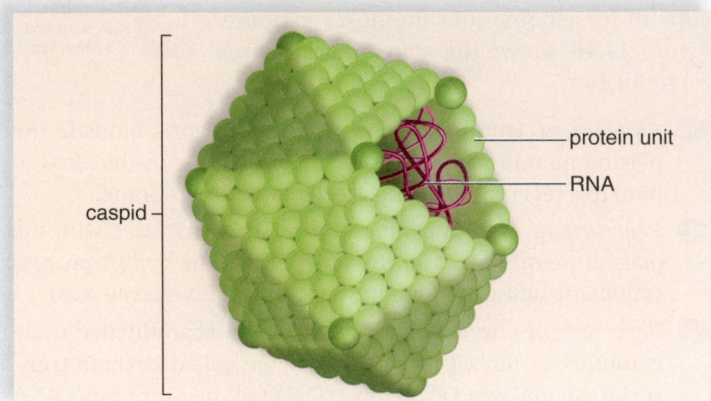

FIGURE 17B.1 A virus is easier to make than a prokaryote because all it takes is a capsid surrounding a genetic material. Like this one, some viruses use RNA as their genetic material.

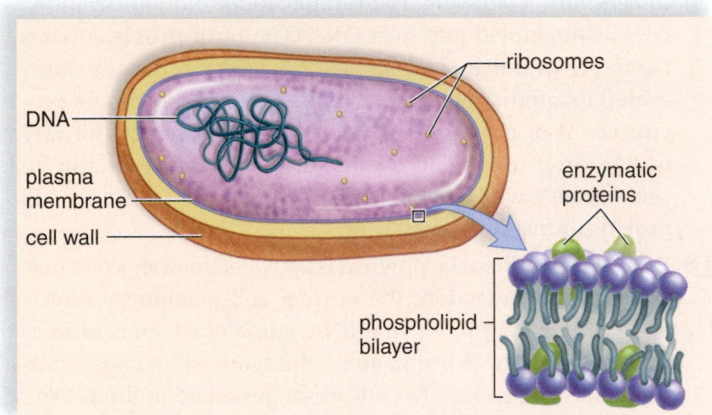

FIGURE 17B.2 A prokaryote is more complex, both metabolically and structurally, than a virus. Like this one, prokaryotes always use DNA as their genetic material.

ribose in RNA   deoxyribose in DNA

Today, the more stable DNA effectively serves as a hard copy of genetic instructions, while RNA is a temporary vehicle for transferring DNA information to various places in the prokaryotic and eukaryotic cell. Encoding genetic information in a more stable DNA molecule results in fewer replicating mistakes and surer passage of genetic information than if RNA continued to serve as the genetic material. The present variety of life on Earth, including humans, is a testament to the success of using DNA as the keeper of genetic information—and it may all be due to the ancient molecular capabilities of viruses.

CONSIDER THESE QUESTIONS

1. Today, viruses are obligate parasites, and they depend on the machinery of their host to replicate their genetic material and produce their proteins. Also, viruses do not carry on cellular respiration as prokaryotes do in their cytoplasm. Using this information and Figures 17B.1 and 17B.2, create a series of proposed and hypothetical evolutionary stages between the first viruses and bacteria. (*Hint:* The stages should include both gain of metabolic capabilites and the increased structural complexity needed.)

2. Speculate that the viruses of today are degenerate forms of viruses that gave rise to cells. What functions did they lose? Why would the parasitic way of life have caused them to lose these functions?

3. You will learn when we study animals that a parasitic way of life emphasizes reproduction over all other capabilities. Why would that be?

 Explore the concepts through a variety of multimedia assets and question types.

www.mcgrawhillconnect.com

Origin of Life

The first organic molecules could have originated in the atmosphere, or perhaps at hydrothermal vents. The discovery that some viruses have an RNA genome and that RNA enzymes (ribozymes) function in cells suggests that the first genes were composed of RNA. Today, however, cells use DNA to store genetic information and specify the sequence of amino acids in proteins.

17.5 Experiments show how small organic molecules may have first formed

LEARNING OUTCOME

When you complete this section, you should be able to

1. Summarize two hypotheses for the origin of small organic molecules.

Two different hypotheses have been developed to explain how the organic molecules of cells could have formed from inorganic molecules.

Prebiotic Soup Hypothesis Early Earth had an atmosphere, but it was not the same as today's atmosphere. When the Earth formed, intense heat, produced by gravitational energy and radioactivity, resulted in several stratified layers. Heavier atoms of iron and nickel became the molten liquid core, and dense silicate minerals became the semiliquid mantle. Massive volcanic eruptions produced the first crust and the first atmosphere.

Gases from volcanoes consist mainly of these inorganic chemicals: water vapor (H_2O), nitrogen (N_2), and carbon dioxide (CO_2), with only small amounts of hydrogen (H_2), methane (CH_4), hydrogen sulfide (H_2S), and carbon monoxide (CO). Notice that this atmosphere contains no oxygen and, therefore, is a reducing atmosphere. This would have been fortuitous on early Earth because oxygen (O_2) attaches to organic molecules, preventing them from joining to form larger molecules. In support of the hypothesis that small inorganic molecules such as these could have produced the first organic molecules, Stanley Miller and Harold Urey performed the ingenious experiment diagrammed in Figure 17.5. ❶ They placed a mixture resembling a strongly reducing atmosphere—methane (CH_4), ammonia (NH_3), hydrogen (H_2), and water (H_2O)—in a closed system, and ❷ circulated it past an electric spark (simulating lightning). ❸ After condensing the gases to a liquid, ❹ they heated the liquid. ❺ After

a week's run, Miller and Urey withdrew the liquid and discovered that a variety of amino acids and organic acids had been produced.

Without such a highly reducing mixture of gases the Miller-Urey experiment does not work as well. Even so, these experiments lend support to the hypothesis that the Earth's first atmospheric gases could have reacted with one another to produce small organic molecules. Energy would have been required, but early Earth had abundant sources of energy in the form of lightning, volcanic activity, and intense radiation from a sun that had just formed. We know that there would have been plenty of time for synthesis to occur because the Earth is some 4.6 billion years old and the first cells are about 3.6 billion years old. Neither oxidation (there was no free oxygen) nor decay (there were no bacteria) would have destroyed the first molecules, and rainfall would have washed them into the ocean, where they accumulated for hundreds of millions of years. Therefore, the oceans would have been a thick, warm, prebiotic soup.

Iron-Sulfur World Hypothesis Other investigators are concerned that Miller and Urey used ammonia as one of the atmospheric gases. They point out that, whereas inert nitrogen gas (N_2) would have been abundant in the primitive atmosphere, ammonia (NH_3) would have been scarce. Where might NH_3 have been abundant? A team of researchers at the Carnegie Institution in Washington, D.C., believe they have found the answer: in hydrothermal vents on the ocean floor. These vents line the huge ocean ridges, where molten magma wells up and adds material to the ocean floor. Cool water seeping through the vents is heated to a temperature as high as 350°C, and when it spews back out, it contains various mixed iron and nickel sulfides that can act as catalysts to change N_2 to NH_3. Even today, these conditions produce nutrient molecules for microorganisms that support a diverse community of other organisms, including huge clams and tube worms living in the vicinity of the ocean ridges (see Fig. 17.10A).

A laboratory test of the iron-sulfur hypothesis worked perfectly. Under ventlike conditions, 70% of various N_2 sources were converted to NH_3 within 15 minutes. Two German organic chemists, Gunter Wachtershaüser and Claudia Huber, have gone one more step. They have shown that organic molecules will react and amino acids will form peptides in the presence of iron and nickel sulfides under ventlike conditions.

plume of hot water rich in iron sulfides

hydrothermal vent

Animation
Miller-Urey

FIGURE 17.5 Laboratory re-creation of a chemical evolution in the atmosphere. Atmospheric gases in the ocean may have reacted to form the first small organic molecules.

❷ electrode

❶ stopcock for adding gases

electric spark

CH_4
NH_3
H_2
H_2O — gases

❺ stopcock for withdrawing liquid

hot water out

❸ condenser

cool water in

liquid droplets

boiler

❹ heat

small organic molecules

▶ **17.5 CHECK YOUR PROGRESS**

1. Identify the main differences between the two hypotheses for the origin of small organic molecules.

17.6 RNA may have been the first macromolecule

LEARNING OUTCOMES

When you complete this section, you should be able to

1. Discuss why RNA, instead of DNA or protein, may have been the first macromolecule.

2. Describe the steps by which a protobiont may have evolved into the first cell.

3. Explain the division of Figure 17.6A into a chemical evolution followed by a biological evolution.

Figure 17.6A gives an overview for the origin of life. So far we have discussed how during a chemical evolution, the first small organic molecules formed abiotically (without rather than within an organism). When these small organic molecules joined together the first macromolecules would have formed. The RNA-first hypothesis states that RNA was the first macromolecule to form. If so, a protobiont would have formed when plasma membrane surrounded RNA and then the protobiont became a true cell when DNA took over as the genetic material. "Viruses and the Invention of DNA" on page 324 makes this same supposition, except

it suggests the transition occurred in viruses. In Figure 17.6A it is hypothesized that protobionts were living precursors to living cells and therefore the transformation from protobionts to living cells is a part of biological evolution.

RNA-First Hypothesis For several decades, scientists have been studying which of the three macromolecules—RNA, DNA, or protein—led to the origin of the first protobiont. Certainly, amino acids were available to form proteins, and the necessary molecules to form nucleotides were also present in the prebiotic soup. Nucleotides could have then polymerized to form nucleic acids.

The possibility that proteins alone led to the first protobiont is rejected by some because, as you know, proteins do not store genetic information. The genetic material must be able to (1) store genetic information and (2) replicate in order to transmit the genetic information to daughter cells. Scientists have discovered that RNA, not DNA, could have performed both functions by itself. Today, RNA stores genetic information within the sequence of its bases when it participates in the formation of proteins.

During the past three decades, scientists have been able to show that RNA also acts as an enzyme (called a ribozyme), and therefore an RNA genome could have possibly replicated on its own. Ribozymes also join amino acids during protein synthesis. Thomas Cech and Sidney Altman shared a Nobel Prize in 1989 because they discovered that RNA can be both a substrate and an enzyme.

It would be absolutely critical that the first genome be able to both replicate and carry out protein synthesis. The researchers at the Whitehead Institute for Biomedical Research discovered that ribozymes can perform almost any metabolic function, including replication of RNA. They determined this by placing about 1,000 ribozymes in a test tube and selecting the ones that perform an enzymatic function the best. After selecting only the best for a particular function multiple times, the final group of ribozymes is much more efficient than the original group.

These findings have led these researchers to say that it was an "RNA world" some 4 billion years ago (BYA), and that RNA molecules were the first form of life! In this world, RNA molecules competed with each other for free nucleotides and were subject to natural selection. Only the most efficient RNA molecules survived and became the genetic material for the protobiont.

Other Hypotheses In contrast to the RNA first hypothesis, Sidney Fox of the University of Miami suggested that once amino acids were present in the oceans, they collected in shallow puddles along the rocky shore. There, the heat of the sun could have caused them to congregate and become proteinoids. When Fox simulated the formation of proteinoids in the lab and returned them to water, they formed structures with cell-like properties. They resembled bacteria, could divide, and perhaps were subject to a selection process during which they acquired other properties.

First and foremost, the protobiont would have had to acquire an outer membrane. The plasma membrane of today's cells

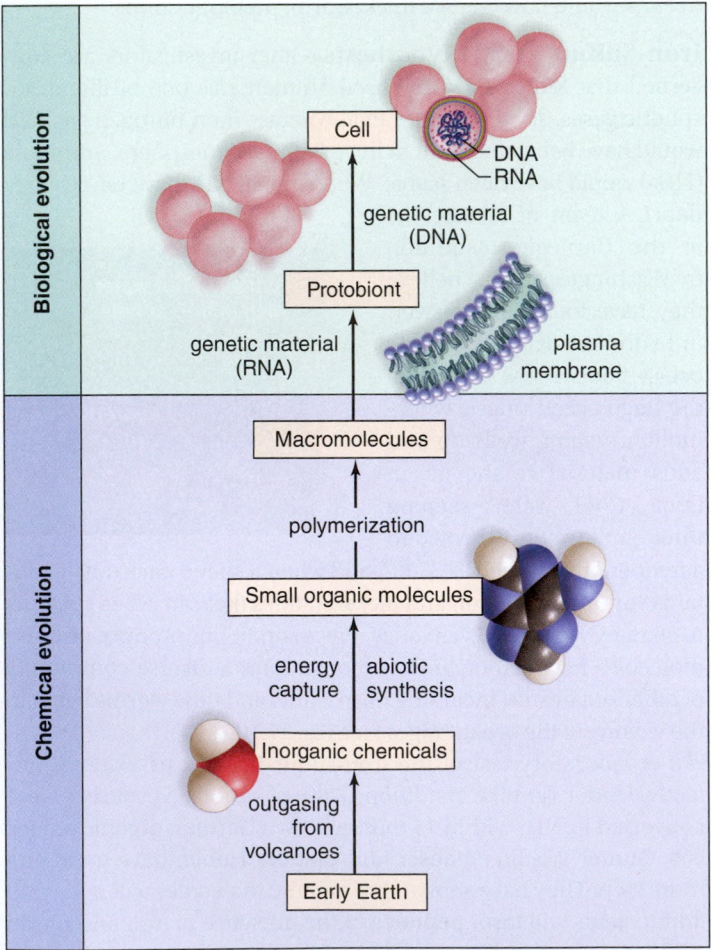

FIGURE 17.6A The origin of the first cell(s) can be broken down into these steps. How viruses might fit into this scenario is currently being considered.

separates the living interior from the nonliving exterior. There are two hypotheses about the origin of the first plasma membrane. Sidney Fox was able to show that if lipids are made available to proteinoids, they acquire a lipid-protein outer membrane (Fig. 17.6B). In contrast, Alec Bangham of the Animal Physiology Institute in Cambridge, England discovered that when he extracted lipids from egg yolks and placed them in water, the lipids naturally organized themselves into double-layered bubbles, roughly the size of a cell. Bangham's bubbles soon became known as **liposomes** (Fig. 17.6C). Later, Bangham, along with biophysicist David Deamer of the University of California, realized that liposomes might have provided life's first boundary. Perhaps liposomes with a phospholipid membrane engulfed early RNA molecules that had enzymatic abilities. The liposomes would have protected the molecules from their surroundings and concentrated them so they could react (and evolve) quickly and efficiently.

The protobiont would have needed enzymes to carry on nutrition so that it could grow. If organic molecules formed in the atmosphere and were carried by rain into the ocean, nutrition would have been no problem because simple organic molecules could have served as food. This hypothesis suggests that the protobiont was a heterotrophic fermenter that took in preformed food. On the other hand, if the protobiont evolved at hydrothermal vents, it may have carried out chemosynthesis as discussed in section 17.5. Chemosynthetic bacteria obtain energy for synthesizing organic molecules by oxidizing inorganic compounds, such as hydrogen sulfide (H_2S), a molecule that is abundant at the vents. Only with the advent of photosynthesis was oxygen added to the atmosphere, making aerobic cellular respiration possible.

A protobiont became a true cell when it contained a DNA information system: DNA $\longrightarrow$ RNA $\longrightarrow$ protein. How did it arise? Those who support an RNA don't hypothesize that viral reverse transcriptase made the first DNA molecule (see section 17.4). Instead, they suggest that (1) A ribozyme could have acted in the same manner as the enzyme reverse transcriptase to produce DNA. Then DNA took over the function of storing genetic information, and RNA became its helper for specifying the sequence of amino acids in proteins. (2) Ribozymes synthesized these proteins that took over most of the enzymatic functions in cells.

Sidney Fox suggested instead that the first proteins making up proteinoids may have had enzymatic properties. When these first enzymatic proteins were exposed to the selective pressures mentioned earlier only the most efficient remained to synthesize DNA and RNA.

Recap for the Origin of Life Let's review our hypotheses of how life began (see Fig. 17.6A). (1) An abiotic synthesis process created small organic molecules such as amino acids and nucleotides, perhaps in the atmosphere or at hydrothermal vents as discussed in section 17.5. (2) These subunits joined together to form polymers along the shoreline (warm seaside rocks or clay) or at the vents. The first polymers could have been RNA or proteins, or they could have evolved together. (3) A plasma membrane produced a protobiont, which had enzymatic properties such that

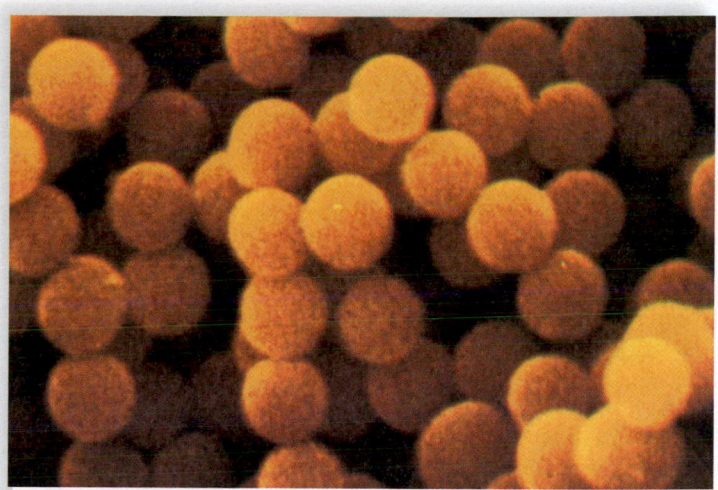

FIGURE 17.6B Proteinoids are made of protein and could have acquired an outer lipid-protein membrane during the origin of protobionts.

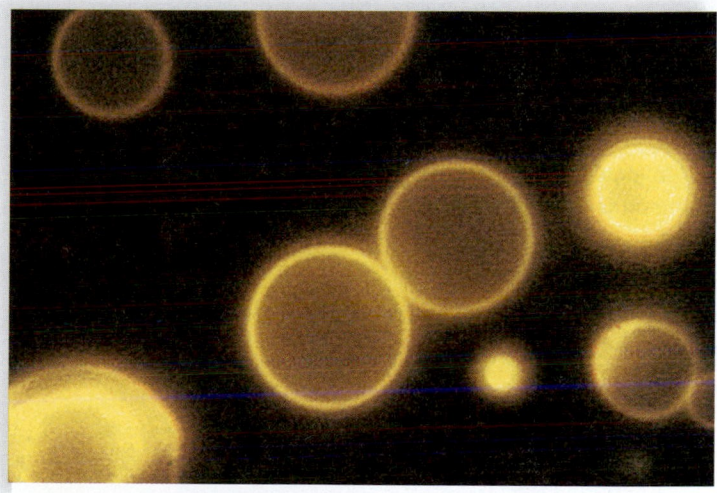

FIGURE 17.6C Liposomes, which are composed of lipids, have a double-layered outer membrane.

it could grow. If the protobiont developed in the ocean, it was a heterotrophic fermenter; if it developed at hydrothermal vents, it was a chemoautotroph. (4) Once the protobiont contained DNA genes, a true cell had evolved. The first genes may have been RNA molecules, but later DNA became the information storage molecule of heredity. Biological evolution—and the history of life—had begun!

▶ **17.6 CHECK YOUR PROGRESS**

1. Identify why RNA could have replicated on its own and discuss the significance of this ability.
2. Identify the cellular components that would have allowed a protobiont to (**a**) metabolize and (**b**) reproduce.
3. Explain why photosynthesis would have preceded aerobic cellular respiration during biological evolution.

The Prokaryotes

Prokaryotes lack a true nucleus and other membranous organelles. They all reproduce by binary fission but have varied means of nutrition. The photosynthetic cyanobacteria release oxygen into the atmosphere, and the archaea are well known for living in extreme environments.

17.7 Prokaryotes have unique structural features

LEARNING OUTCOME

When you complete this section, you should be able to

1. Describe the structure of prokaryotes, including possible shapes and types of envelopes and appendages.

Bacteria (domain Bacteria) and archaea (domain Archaea) are prokaryotes, but each is placed in its own domain because of molecular and certain cellular differences. Both are unicellular organisms that generally range in size from 1–10 μm in length and from 0.7–1.5 μm in width. The term prokaryote means "before a nucleus," and these organisms lack a nucleus. They also do not have membranous organelles (Fig. 7.7A and 4.2B). Although bacteria lack mitochondria, they still carry on cellular respiration; their enzymes are attached to the inside of the plasma membrane. As a group, bacteria have far greater metabolic capabilities than animal cells; for example, they can use almost any organic compound as a food source. Thousands of ribosomes, which are smaller than eukaryotic ribosomes, carry out protein synthesis.

A dense area called a nucleoid contains a single chromosome consisting of a circular strand of DNA. Many bacteria also have accessory rings of DNA called *plasmids*. Plasmids can be extracted and used as vectors to carry foreign DNA into host bacteria during genetic engineering processes. The outer envelope of a bacterium consists of a plasma membrane and a cell wall that is strengthened by *peptidoglycan,* a complex molecule containing a unique amino disaccharide. The cell wall may be surrounded by a layer of polysaccharides. A well-organized layer is called a capsule, while a loosely organized one is called a slime layer that protects the bacterium from host defenses.

The appendages of bacteria include pili and flagella. Pili are short, bristlelike fibers that allow bacteria to adhere to surfaces. The pili of *Neisseria gonorrhoeae* enable it to attach to host cells and cause the sexually transmitted disease gonorrhea. If present, flagella (rotating filaments of the protein flagellin) allow movement.

Animation
Bacterial
Locomotion

Common Shapes of Bacteria Bacteria have three basic shapes (Fig. 17.7B): Cocci (sing., coccus) are round or spherical; bacilli (sing., bacillus) are rod-shaped; and spirilla (sing., spirillum) are spiral- or helical-shaped.

▶ **17.7 CHECK YOUR PROGRESS**
   1. Identify how the outer envelope of a bacterium can vary.

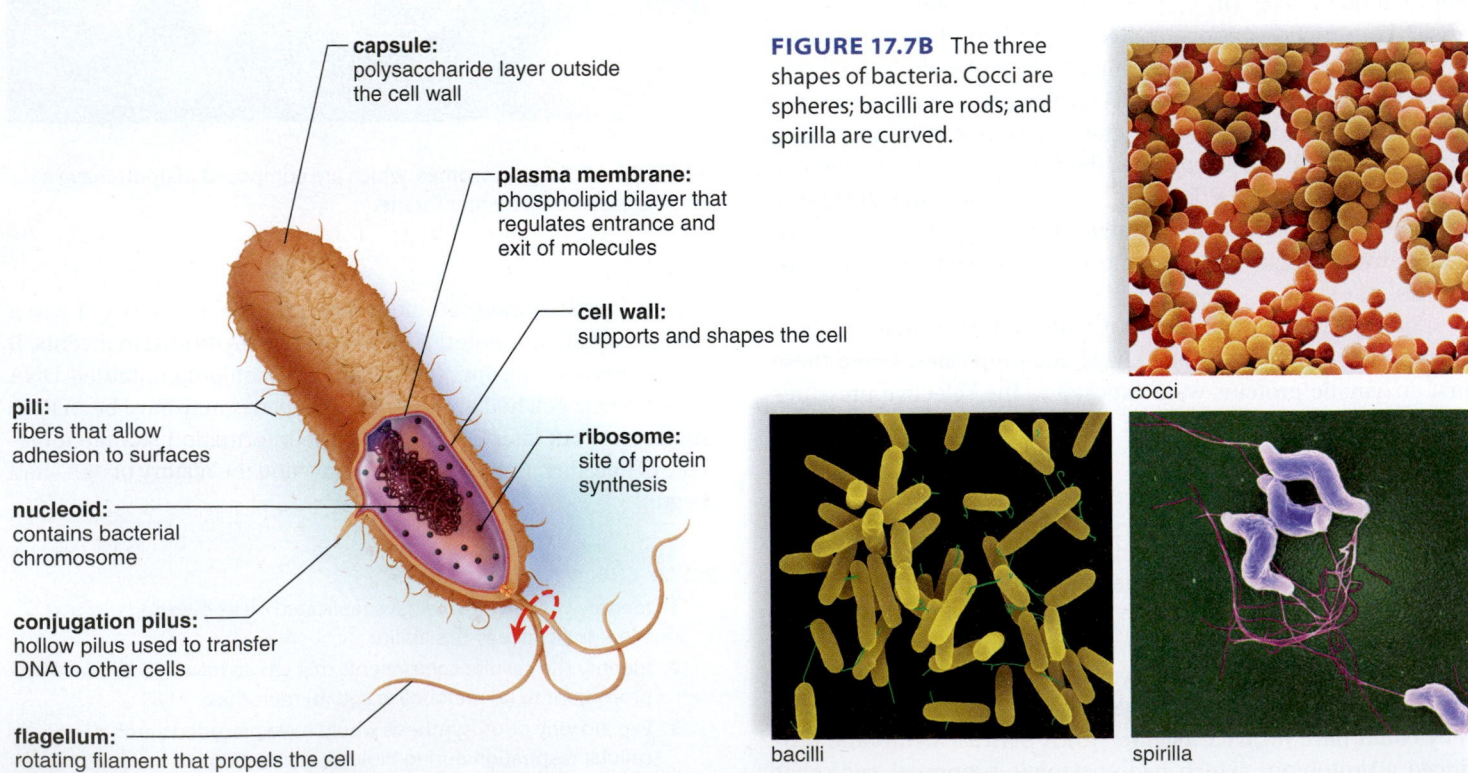

capsule:
polysaccharide layer outside
the cell wall

plasma membrane:
phospholipid bilayer that
regulates entrance and
exit of molecules

cell wall:
supports and shapes the cell

pili:
fibers that allow
adhesion to surfaces

ribosome:
site of protein
synthesis

nucleoid:
contains bacterial
chromosome

conjugation pilus:
hollow pilus used to transfer
DNA to other cells

flagellum:
rotating filament that propels the cell

FIGURE 17.7A Anatomy of a bacterium.

FIGURE 17.7B The three shapes of bacteria. Cocci are spheres; bacilli are rods; and spirilla are curved.

cocci

bacilli

spirilla

LEARNING OUTCOME

When you complete this section, you should be able to

1. Describe the reproductive strategy of prokaryotes.

Mitosis, which requires formation of a spindle apparatus, does not occur in prokaryotes. Prokaryotes, as exemplified by bacteria, reproduce asexually by means of **binary fission,** a process that results in two prokaryotes of nearly equal size. The single circular chromosome replicates and then the two copies separate as the cell enlarges. Newly formed plasma membrane and cell wall separate

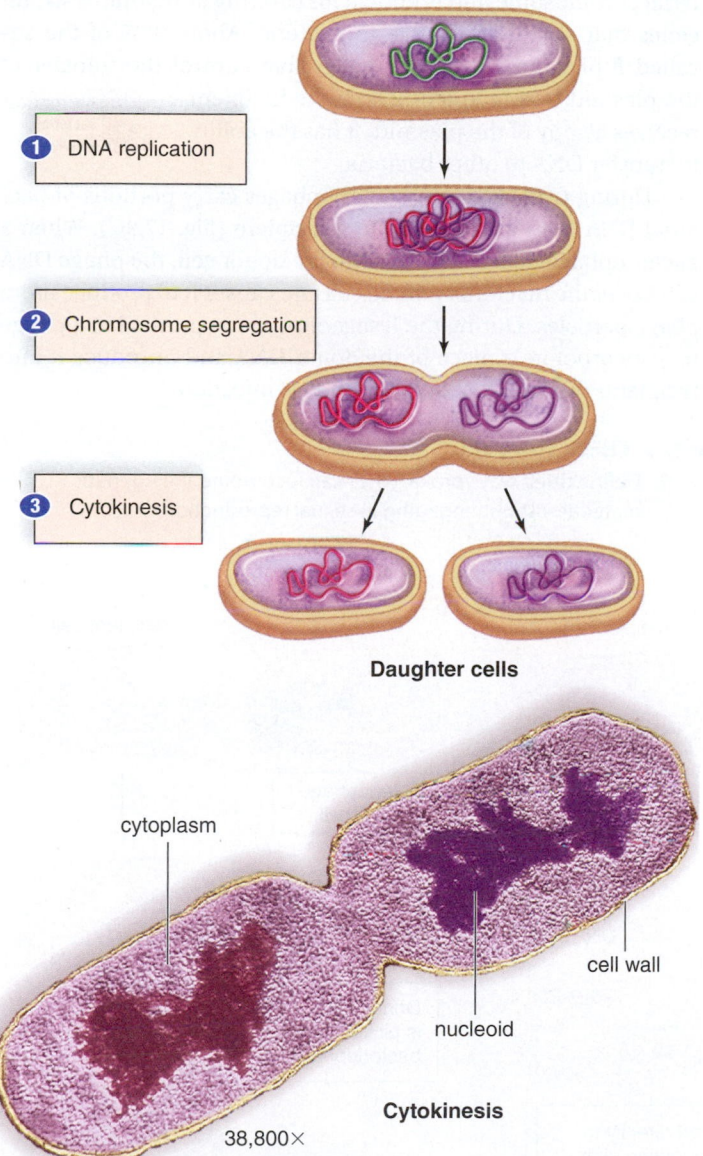

Daughter cells

cytoplasm

cell wall

nucleoid

Cytokinesis

38,800×

FIGURE 17.8 Binary fission requires these three steps and results in two identical bacteria. The photos illustrate that even in bacteria the genetic material is lengthy and must be condensed to fit inside the cell.

the cell into two cells. In other words, binary fission requires three steps: DNA replication, chromosome segregation, and cytokinesis. Figure 17.8 shows the process in bacteria. ❶ DNA replication is the first step. DNA replication begins at one location as the DNA double strand unzips. Each strand acts as a template for synthesis of a daughter strand by semiconservative DNA replication, the same as in eukaryotes. Each circular DNA strand is attached to the plasma membrane. ❷ Chromosome segregation is the second step. The cell elongates, and the chromosomes segregate (separate). ❸ Cytokinesis, which is illustrated in both art and by micrograph, is the third step. Cytokinesis requires that the plasma membrane invaginate to divide the cytoplasm. New cell wall formation also occurs.

Binary fission is asexual, and the offspring are at first genetically identical to the parent cell. But bacterial DNA has a relatively high mutation rate, and bacteria have a generation time as short as 12 minutes under favorable conditions. Therefore, mutations are generated and passed on to offspring more quickly than in eukaryotes. Also, prokaryotes are haploid—each new cell gets only one copy of each gene—and so mutations are immediately subjected to natural selection, which determines any possible adaptive benefit.

Animation
Binary Fission

Formation of Endospores in Bacteria When faced with unfavorable environmental conditions, some bacteria form **endospores** sometimes simply called spores. A portion of the cytoplasm and a copy of the chromosome dehydrate and are then encased by a heavy, protective spore coat. In some bacteria, the rest of the cell deteriorates, and the endospore is released. Spores survive in the harshest of environments—desert heat and dehydration, boiling temperatures, polar ice, and extreme ultraviolet radiation. They also survive for very long periods. When anthrax endospores 1,300 years old germinate, they can still cause a severe infection (usually seen in cattle and sheep). Anthrax spores can be used as a bioterrorism weapon (see "Disease-Causing Microbes Can Be Biological Weapons" on page 331). In 2001, 22 cases of anthrax, including five deaths, occurred after spores were purposely sent through the mail. Humans also fear a deadly, but uncommon, type of food poisoning called botulism that is caused when endospores germinate inside cans of food (see section 17.12). To germinate, the endospore absorbs water and grows out of the spore coat. Within a few hours, it becomes a typical bacterial cell, capable of reproducing once again by binary fission. Endospore formation is not a means of reproduction, but it does allow bacteria to survive and to disperse to new places.

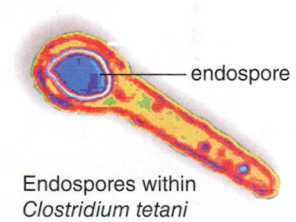

endospore

Endospores within
Clostridium tetani

▶ **17.8 CHECK YOUR PROGRESS**

1. Identify the essential steps of binary fission in bacteria.

17.9 How genes are transferred between bacteria

LEARNING OUTCOME

When you complete this section, you should be able to

1. Describe the three mechanisms for genetic recombination among bacteria.

Much of what we know about molecular genetics has come from studying the processes of DNA replication and gene expression in bacteria such as *E. coli,* which lives in the human intestine. Section 17.7 discussed the structure of bacteria. We learned that bacteria, unlike viruses, are cellular and that they have a single chromosome composed of double-stranded DNA. The chromosome is actually quite large and condensed into a region called the nucleoid. Section 17.8 reviewed how bacteria reproduce by binary fission. At that time, DNA replicates and is pulled apart as the cell elongates. Formation of a new plasma membrane and cell wall divide the cell. Now we will consider how a bacterium can acquire new genes from others of its own kind.

In eukaryotes, genetic recombination occurs as a result of sexual reproduction. But even though sexual reproduction does not occur among bacteria, three means of gene transfer take place: transformation, conjugation, and transduction. In all three mechanisms, the donor is the cell that provides the genetic material for transfer, and the recipient is the cell that receives the material. The genes that allow bacteria to be resistant to antibiotics can be transferred by any one of these methods.

Transformation occurs when a recipient bacterium picks up (from its surroundings) free pieces of DNA secreted by live bacteria or released by dead bacteria (Fig. 17.9A). Transformation was discovered by Frederick Griffith in 1928 while working with two strains of *Streptococcus pneumoniae:* The R strain was not infectious while the S strain was infectious. When living R bacteria picked up a substance from dead S bacteria, they too became infectious. Griffith determined that the R strain had received DNA from the S strain. Transformation usually incorporates only a small amount of donor DNA into the chromosome of the recipient. Cells that do not naturally undergo transformation can be induced to do so by treatments that disrupt the cell wall.

Animation
Bacterial
Transformation

Conjugation occurs between bacteria when the donor cell passes DNA to the recipient by way of a conjugation pilus, which temporarily joins the two bacteria. Figure 17.9B illustrates that a donor cell with a plasmid can initiate conjugation. A **plasmid** is a small circle of DNA that can replicate independently of the bacterial chromosome and is known for carrying antibiotic-resistant genes that confer resistance to bacteria. About 35% of the so-called F plasmid consists of genes that control the transfer of the plasmid to a recipient. Once the recipient receives a copy of the plasmid, it has the ability to transfer DNA to other bacteria.

Animation
Bacterial
Conjugation

During **transduction,** bacteriophages carry portions of bacterial DNA from a donor cell to a recipient (Fig. 17.9C). When a bacteriophage injects its DNA into the donor cell, the phage DNA takes over the machinery of the cell and causes it to produce more phage particles. During the lysogenic cycle in particular, a phage may incorporate a piece of the donor DNA and introduce it into recipients during subsequent rounds of infection.

▶ **17.9 CHECK YOUR PROGRESS**
1. Define three ways prokaryotes can recombine their genetic material without engaging in sexual reproduction.

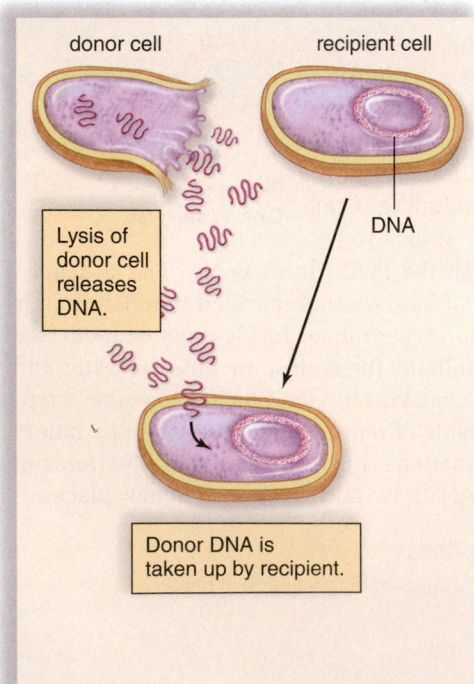

FIGURE 17.9A Gene transfer by transformation.

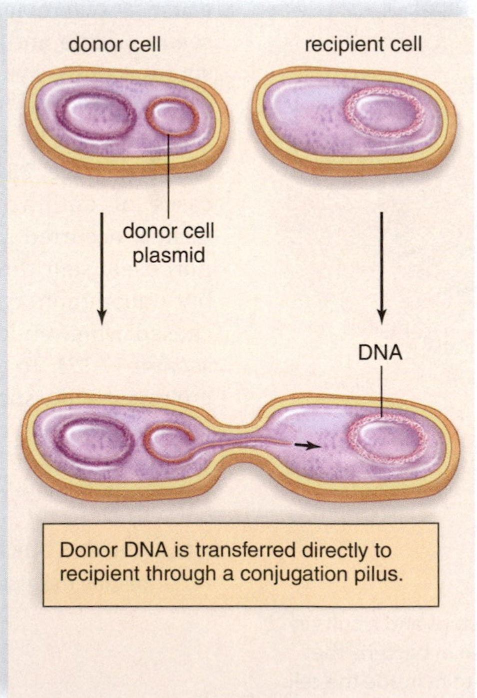

FIGURE 17.9B Gene transfer by conjugation.

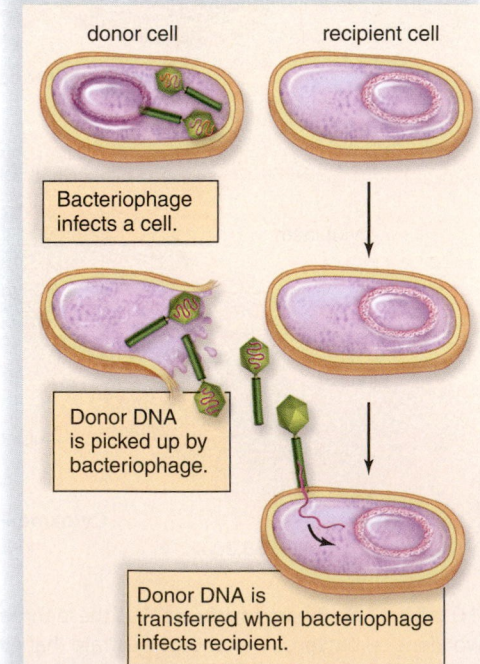

FIGURE 17.9C Gene transfer by transduction.

17C Disease-Causing Microbes Can Be Biological Weapons

Biological warfare is the use of viruses and bacteria, or their toxins, as weapons of war. In recent years, several nations have used genetic engineering to produce biological warfare agents that have an enhanced resistance to antimicrobial drugs and an altered pathogenic effect and incubation period. Bioterrorists prefer pathogens that are highly contagious, consistently produce a desired detrimental effect on a population, have a short incubation period, and are easy to disseminate and deliver to a population. When dealing with these dangerous agents, protective clothing must be worn (Fig. 17C).

In addition to humans, valuable animals and crops can be the targets of biological attacks. An attack upon a nation's cattle, pigs, or other domesticated animals could have serious consequences on the nation's food, animal products (wool, hides), and medicinal supplies produced by GM animals. Agents that could be used against animals include anthrax, glanders, swine fever, hog cholera, foot-and-mouth disease, and fowl pest disease. Valuable crops, as well as commercially and medically important plants, can also be the targets of biological warfare. Several species of fungi and insects could be devastating to targeted plants. In addition, several herbicides are considered biological weapons because they serve as bioregulators in plants.

The likely microbial agents for use by bioterrorists are these:

Anthrax, caused by the bacterium *Bacillus anthracis,* is a biological agent that is easy to acquire, grow, and disseminate. Anthrax occurs in three forms: inhalation anthrax, cutaneous anthrax, and gastrointestinal anthrax. Inhalation anthrax is the deadliest form to humans. It begins with flulike symptoms and, if not treated promptly, can lead to death in 24–36 hours after the onset of respiratory distress.

Smallpox is caused by the variola virus, a very dangerous and highly contagious airborne virus. Through diligent vaccination, smallpox has not been recorded since 1980. However, many young people have not received the vaccination, and in others the vaccine may have lost its effectiveness. After a 7–17-day incubation period, the disease begins with a fever, headache, and malaise. Patients are the most contagious 3–6 days after onset of the fever. Because many physicians have not seen smallpox, it can easily be misdiagnosed until the telltale lesions develop. Eventually, pus-filled lesions and blisters form on the patient's body. The pustules crust and form deep scars in survivors. In an unvaccinated population, smallpox has a 30% mortality rate.

Botulism, caused by the toxin of the anaerobic bacterium *Clostridium botulinum,* can be a lethal foodborne agent. Recently aerosol forms of the toxin have been developed. Botulinum toxins are some of the most lethal toxins known. They are easy to manufacture and weaponize, and represent a major threat to human populations. Initial symptoms are blurred vision, difficulty swallowing, and muscle weakness. The mortality rate from botulism is high, and death usually results from respiratory failure.

Plague, caused by the bacterium *Yersinia pestis,* has been called the Black Death and bubonic plague in the past, and has been responsible for millions of deaths. In a biological warfare scenario, the plague bacterium can be delivered by infected fleas, causing traditional bubonic plague, or by airborne droplets, causing the more deadly pneumonic plague.

Tularemia is caused by the bacterium *Francisella tularensis.* The disease has several forms, with the inhaled form being the most likely biological warfare candidate. Just 10–50 organisms inhaled by a human can cause an infection. A variety of symptoms may accompany tularemia, including fever, chills, headache, weakness, abdominal pain, vomiting, chest pain, and cutaneous ulcers. Pneumonia may develop in many victims, and death can result.

Hemorrhagic fevers, caused by several types of viruses, are characterized by high fever and severe, uncontrollable bleeding from several organs. Four deadly hemorrhagic fevers caused by virulent viruses are Crimean-Congo fever, Rift Valley fever, Marburg fever, and Ebola fever. Ebola is the best-known of these viruses.

The bacterial diseases anthrax, plague, botulism, and tularemia usually respond to specific antibiotics. Hemorrhagic fevers, if diagnosed soon enough, may respond to specific antiviral drugs, but these may be in short supply. Therefore, vaccines and preventives may be the best way to counter biological agents.

CONSIDER THESE QUESTIONS

1. Should flyovers at sports events be discontinued to prevent bioterrorism?
2. Should the United States retaliate in kind in case of a bioterrorism attack?

FIGURE 17C Bioterrorism represents a threat to our health.

 Explore the concepts through a variety of multimedia assets and question types.

www.mcgrawhillconnect.com

LEARNING OUTCOME

When you complete this section, you should be able to

1. Compare the metabolic diversity of prokaryotes in terms of their need for oxygen and their means of acquiring food.

As exemplifed by bacteria, prokaryotes require nutrients just as eukaryotic cells do. One difference, however, concerns their need for oxygen. Some bacteria are **obligate anaerobes,** meaning that they are unable to grow in the presence of free oxygen. A few serious illnesses—such as botulism, gas gangrene, and tetanus—are caused by anaerobic bacteria. Other bacteria, called **facultative anaerobes,** are able to grow in either the presence or the absence of gaseous oxygen. Most bacteria, however, are **aerobic** and, like animals, require a constant supply of oxygen to carry out cellular respiration.

Heterotrophic Bacteria Most bacteria are heterotrophs that take in organic nutrients. They are aerobic **saprotrophs,** which means that they secrete digestive enzymes into the environment for the breakdown of large organic molecules to smaller ones that can be absorbed. There is probably no natural organic molecule that cannot be digested by at least one prokaryotic species. In ecosystems, saprotrophic bacteria are called *decomposers*. They play a critical role in recycling matter and making inorganic molecules available to photosynthesizers.

The vast metabolic capabilities of heterotrophic prokaryotes have long been exploited by humans. Bacteria are used commercially to produce chemicals, such as ethyl alcohol, acetic acid, butyl alcohol, and acetones. Prokaryotic action is also involved in the production of butter, cheese, sauerkraut, rubber, cotton, silk, coffee, and cocoa. Even antibiotics are produced by some bacteria.

Heterotrophs may be either free-living or symbiotic, meaning that they form mutualistic, commensalistic, or parasitic relationships. *Mutualism* exists when both partners benefit. For example, mutualistic bacteria live in the root nodules of soybean, clover, and alfalfa plants, where they receive organic nutrients and assist their hosts by reducing atmospheric nitrogen (N_2) for incorporation into organic compounds. Other mutualistic bacteria that live

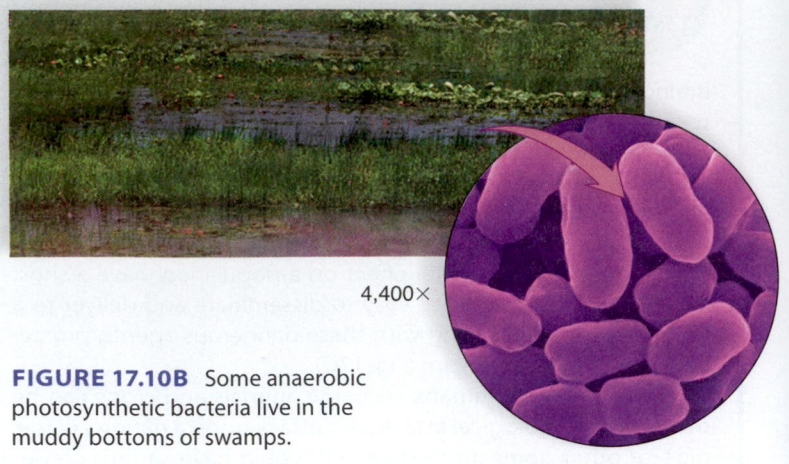

4,400×

FIGURE 17.10B Some anaerobic photosynthetic bacteria live in the muddy bottoms of swamps.

in human intestines release vitamins K and B_{12}, which we can use to produce blood components. In the stomachs of cows and goats, special mutualistic prokaryotes digest cellulose, enabling these animals to feed on grass.

Commensalism often occurs when one population modifies the environment in such a way that a second population benefits. For example, obligate anaerobes can live in our intestines only because the bacterium *Escherichia coli* uses up the available oxygen. The *parasitic* bacteria cause disease, including human diseases.

Autotrophic Bacteria Some bacteria and archaea are **chemosynthetic.** They oxidize an inorganic chemical to acquire electrons and energy to reduce carbon dioxide to an organic compound. The nitrifying bacteria oxidize ammonia (NH_3) and nitrites (NO_2^-) and their metabolic abilities keep nitrogen cycling through ecosystems. Other prokaryotes oxidize hydrogen sulfide (H_2S) spewing from hydrothermal vents 2.5 km below sea level. The archaea called methanogens oxidize hydrogen gas (H_2) and give off methane. These chemosynthetic bacteria support the growth of a community of organisms found at vents (Fig. 17.10A). This discovery lends support to the suggestion that the first cells originated at hydrothermal vents (see section 17.5). Some bacteria are photosynthetic and use solar energy and hydrogen atoms to reduce carbon dioxide to organic compounds. There are two types of photosynthetic bacteria: those that evolved first and do not give off oxygen (O_2), and those that evolved later and do give off O_2. The green and purple sulfur bacteria are photosynthetic bacteria that contain a different type of chlorophyll than plants. They do not give off O_2, because they do not use water as an electron donor; instead, they can, for example, use hydrogen sulfide (H_2S):

$$\text{solar energy} + CO_2 + 2\,H_2S \longrightarrow (CH_2O) + 2\,S + H_2O$$

These anaerobic bacteria live in the muddy bottoms of bogs and marshes, where there is no O_2 (Fig. 17.10B). The cyanobacteria (Fig. 17.10C) contain chlorophyll *a*, as do plants, and they carry on photosynthesis in the same manner as plants:

$$\text{solar energy} + CO_2 + H_2O \longrightarrow (CH_2O) + O_2$$

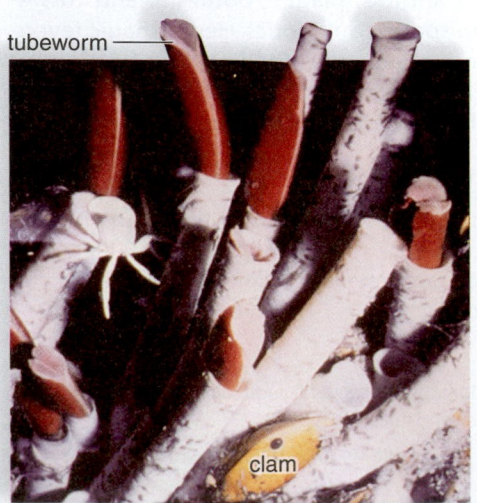

tubeworm

clam

FIGURE 17.10A At hydrothermal vents, a community that contains huge tubeworms and clams depends on chemosynthetic bacteria and archaea.

Cyanobacteria Formerly, the **cyanobacteria** were called blue-green algae and were classified with eukaryotic algae, but now they are known to be bacteria. Cyanobacteria are named for the blue-green pigment called phycocyanin they contain. But they can also have other pigments that make them appear red, yellow, brown, or black, rather than blue-green. These pigments occur in the membrane of flattened disks called **thylakoids.** Cyanobacteria photosynthesize in the same manner as plants and are believed to be responsible for first introducing oxygen into the early atmosphere. Without this event, animal evolution, as we know it, would not have occurred. Ancient seas contained stromatolites, which look like strange boulders but contain cyanobacteria. Some stromatolites are living today, and some contain fossils dated 2 BYA.

Cyanobacterial cells are rather large, ranging from 1–50 mm in width. They can be unicellular, filamentous, or colonial. (Fig. 17.10C). Cyanobacteria lack any visible means of locomotion, although some glide when in contact with a solid surface, and others oscillate (sway back and forth). Some cyanobacteria have a special advantage because they possess **heterocysts,** which are thick-walled cells without nuclei, where nitrogen fixation occurs. The ability to photosynthesize and also to fix atmospheric nitrogen (N_2) means that their nutritional requirements are minimal. They can serve as food for heterotrophs in ecosystems that are otherwise nutrient poor. Being bacteria, cyanobacteria reproduce by binary fission and can produce endospores that resist freezing and drying out. Endospore formation means that cyanobacteria can come back when a dry lake receives water once again.

Cyanobacteria are common in fresh and marine waters, in soil, and on moist surfaces. But they are also found in harsh habitats, such as deserts, the frozen lakes of Antarctica, extremely acidic, basic, or salty water, and even hot springs, where water temperatures approach 75°C. Cyanobacteria are the first photosynthetic organisms to appear on cooled lava after a volcanic eruption. Scientists hypothesize that they were the first colonizers of land during the course of evolution.

Cyanobacteria are symbiotic with a number of organisms, including some protists, plants, and animals. When living in these organisms, cyanobacteria often lose their cell walls and essentially function as chloroplasts inside the cells of their host. In association with fungi, they form **lichens** that can grow on rocks, buildings, and trees. A lichen is a symbiotic relationship in which the cyanobacterium provides organic nutrients to the fungus, while the fungus possibly protects and furnishes inorganic nutrients to the cyanobacterium. It is also possible that the fungus is parasitic on the cyanobacterium. Lichens help transform rocks into soil; other forms of life may then follow. Some lichens serve as bioindicators of air pollution.

Cyanobacteria are ecologically important in still another way. If care is not taken in disposing of industrial, agricultural, and human wastes, phosphates drain into lakes and ponds, resulting in a "bloom" of these organisms. The surface of the water becomes turbid, and light cannot penetrate to lower levels. When a portion of the cyanobacteria dies off, the decomposing prokaryotes use up the available oxygen, causing fish to die from lack of oxygen.

▶ **17.10 CHECK YOUR PROGRESS**

1. Identify the major difference between heterotrophs and autotrophs and the major difference between chemosynthetic and photosynthetic bacteria.

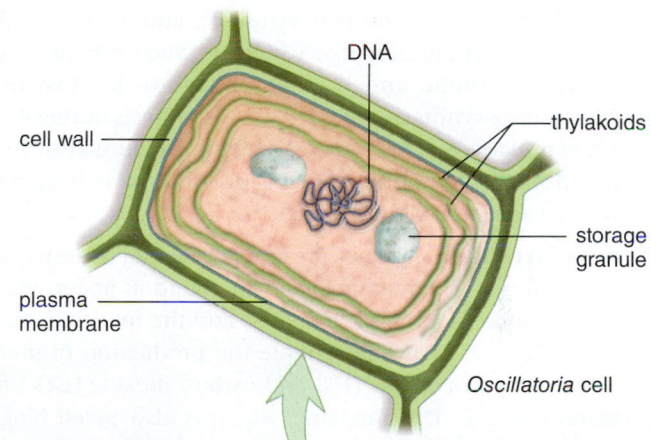

Oscillatoria cell

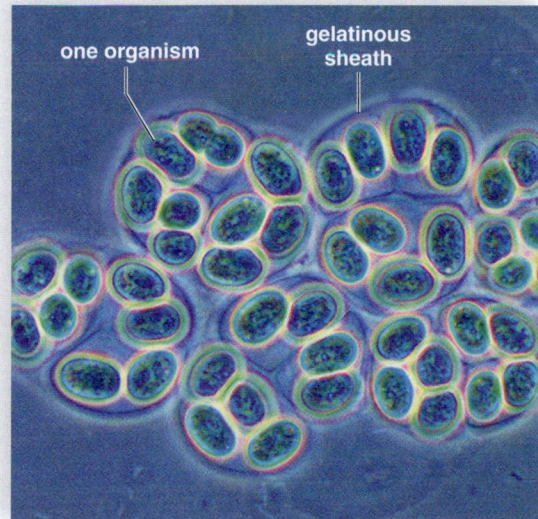

Gloeocapsa, a unicellular form   250×

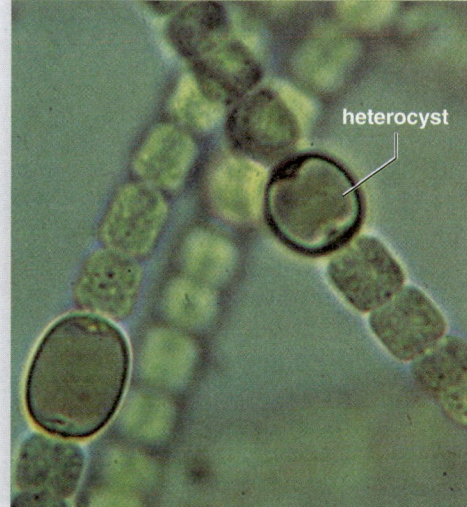

Anabaena, a colonial form   800×

Oscillatoria, a filamentous form   100×

FIGURE 17.10C Diversity among the cyanobacteria.

17.11 Some archaea live in extreme environments

LEARNING OUTCOME

When you complete this section, you should be able to

1. Describe the structure and function of archaea, and name three types.

As previously mentioned, scientists currently propose that the tree of life contains three domains: Archaea, Bacteria, and Eukarya. Because **archaea** and some bacteria are found in extreme environments (hot springs, thermal vents, and salt basins), they may have diverged from a common ancestor relatively soon after life began. Later, the eukarya are believed to have split off from the archaeal line of descent. Archaea and eukaryotes share some of the same ribosomal proteins (not found in bacteria), initiate transcription in the same manner, and have similar types of tRNA.

Structure and Function The plasma membranes of archaea contain unusual lipids that allow them to function at high temperatures. The archaea have also evolved diverse cell wall types, which facilitate their survival under extreme conditions. The cell walls of archaea do not contain peptidoglycan, as do the cell walls of bacteria. In some archaea, the cell wall is largely composed of polysaccharides, and in others, the wall is pure protein. A few have no cell wall.

Archaea have retained primitive and unique forms of metabolism. Most archaea are chemosynthetic, and a few are photosynthetic. The **methanogens,** which produce methane (CH_4), are chemosynthetic and the halophiles are photosynthetic. Perhaps chemosynthesis predated photosynthesis during the evolution of prokaryotes. Archaea are sometimes mutualistic or even commensalistic, but none are parasitic—that is, archaea are not known to cause infectious diseases.

Types of Archaea Archaea are often discussed in terms of their unique habitats. The methanogens are found in anaerobic environments, such as swamps, marshes, and the intestinal tracts of animals (Fig. 17.11A). They couple the production of methane (CH_4) from hydrogen gas (H_2) and carbon dioxide (CO_2) to the formation of ATP. This methane, which is also called biogas, is released into the atmosphere, where it contributes to the greenhouse effect and global warming. About 65% of the methane in our atmosphere is produced by methanogenic archaea.

The **halophiles** are adapted to living in high salt concentrations (usually 12–15%; by contrast, the ocean is about 3.5% salt). Halophiles have been isolated from highly saline environments, such as the Great Salt Lake in Utah, the Dead Sea in the Mideast, solar salt ponds, and hypersaline soils (Fig. 17.11B). These archaea have evolved a number of mechanisms to thrive in high-salt environments. They depend on a pigment related to the rhodopsin in our eyes to absorb light energy to pump out chloride, and they use another similar type of pigment to synthesize ATP.

A third major type of archaea are the **thermoacidophiles** (Fig. 17.11C). These archaea are isolated from extremely hot and acidic environments, such as hot springs, geysers, hydrothermal vents, and around volcanoes. They reduce sulfur to sulfides, producing acidic sulfates, and these archaea grow best at a pH of 1 to 2. Thermoacidophiles survive best at temperatures above 80°C; some can even grow at 105°C (remember that water boils at 100°C)!

Methanosarcina mazei

FIGURE 17.11A Methanogen habitat and structure.

Great Salt Lake, Utah

Halobacterium salinarium

FIGURE 17.11B Halophile habitat and structure.

Boiling springs and geysers in Yellowstone National Park

Sulfolobus acidocaldarius

FIGURE 17.11C Thermoacidophile habitat and structure.

▶ **17.11 CHECK YOUR PROGRESS**

1. List and describe the three types of archaea distinguished by their unique habitats.

17.12 Prokaryotes have medical and environmental importance

LEARNING OUTCOME

When you complete this section, you should be able to

1. Discuss the medical and environmental importance of prokaryotes.

The vast majority of bacterial species are not pathogenic to humans. However, several have had a tremendous impact on human health since antiquity and continue to plague us today. Several of the diseases caused by bacteria are listed in Table 17.12. Many species of pathogenic bacteria invade and destroy the tissues of their host; others produce powerful **toxins** (poisons).

More different types of human disease are caused by bacteria from the genus *Streptococcus* than any other type of bacteria. Most of the infections are respiratory in nature, but this genus also contributes to tooth decay and skin diseases. Scarlet fever results from a strep infection that leads to rheumatic fever and heart damage. By releasing enzymes that destroy connective and muscle tissues, *Streptococcus* infections can lead to amputation and death.

Tuberculosis (TB) kills more people worldwide than any other infectious disease. Tuberculosis is a chronic disease caused by *Mycobacterium tuberculosis,* closely related to the causative agent of leprosy. A TB infection generally affects the lungs but can occur elsewhere.

Food poisoning is caused by a number of bacterial species. *E. coli* is capable of producing exotoxins (secreted outside the cell) that cause food poisoning and some forms of traveler's diarrhea. *Clostridium botulinum* produces a neurotoxin that is perhaps the most toxic substance on Earth. When canning, people may fail to heat food above the boiling point of water. Under these conditions, *Clostridium* can produce endospores that survive the canning process. These spores then germinate, and the cells that grow in the airless environment of the can produce the exotoxin that causes botulism. This is the same substance that is used in Botox (botulinum toxin A), a cosmetic injection that reduces the severity of wrinkles. The endotoxins (a part of the cell) of *Salmonella* can cause food poisoning, and another species is responsible for typhoid fever.

Video
E. Coli Wars

TABLE 17.12	Bacterial Diseases in Humans
Category	**Disease**
Sexually transmitted diseases	Syphilis, gonorrhea, chlamydia
Respiratory diseases	Strep throat, scarlet fever, tuberculosis, pneumonia, Legionnaires disease, whooping cough, inhalation anthrax
Skin diseases	Erysipelas, boils, carbuncles, impetigo, acne, infections of surgical or accidental wounds and burns, leprosy (Hansen disease)
Digestive tract diseases	Gastroenteritis, food poisoning, dysentery, cholera, peptic ulcers, dental caries
Nervous system diseases	Botulism, tetanus, leprosy, spinal meningitis
Systemic diseases	Plague, typhoid fever, diphtheria
Other diseases	Tularemia, Lyme disease

Chlamydia trachomatis is a small, intracellular parasite that causes a variety of human diseases, including trachoma, the leading cause of preventable blindness in the world. *Chlamydia* is also the leading sexually transmitted disease in the United States. In rare cases, the infection can develop into life-threatening pelvic inflammatory disease.

Treatment Sometimes **vaccines** are available that can make people immune to bacterial diseases so that a future infection does not occur. For example, successful vaccines have been prepared for the bacterial diseases whooping cough and diphtheria. For the most part, however, the medical profession must rely on antibiotics to cure a bacterial disease rather than on vaccines to prevent them from occurring in the first place. Unfortunately, pathogenic bacteria seem to be winning the evolutionary arms race against antibiotics. Whereas penicillin was once 100% effective against hospital strains of *Staphylococcus aureus,* today it is far less effective. New strains, such as MRSA (methicillin-resistant *Staphylococcus aureus*), are becoming serious health threats because some strains are now transmitted from person to person. Penicillin and tetracycline, long used to cure gonorrhea, now have a failure rate of more than 40% against certain strains of *Neisseria.* Pulmonary tuberculosis is on the rise, particularly among AIDS patients, the homeless, and the rural poor, and the strains are resistant to the usual combined antibiotic therapy. Presently scientists are seeking new and innovative ways to kill bacteria such as lysins, the enzymes that viruses use to break open bacterial cells. In the meantime, following these guidelines should help you avoid acquiring any resistant diseases:

- Get vaccinated when vaccines are available. Take antibiotics exactly and only as a doctor prescribes. Never take an antibiotic for a viral infection, such as a cold or the flu.
- Always handle food safely by keeping your hands, utensils, and countertops clean and by keeping raw meat, poultry, and fish from contacting other foods. Cook foods thoroughly and refrigerate them promptly.
- Wash hands frequently and thoroughly. Exercise, eat right, drink lots of water, and get plenty of sleep.

Prokaryotes Were and Are Environmentally Important Ancient photosynthetic cyanobacteria altered Earth's primitive atmosphere by releasing copious amounts of their waste gas, oxygen. The presence of oxygen in the atmosphere, in turn, led to the evolution of cellular respiration and the rise of the diversity of life on Earth. Today, the descendants of these cyanobacteria continue to add valuable oxygen to the atmosphere. On the surface of the Earth, many species of bacteria serve as the principal creators of soil fertility and help recycle the nutrients tied up in leaf clutter and animal corpses. Bacteria that live on the roots of certain plants, such as legumes (peanuts, clover), acacias, and the tiny aquatic fern *Azolla,* fix atmospheric nitrogen and make it available to plants. Prokaryotes play an essential role in the carbon, nitrogen, sulfur, and phosphorus environmental cycles.

▶ **17.12 CHECK YOUR PROGRESS**

1. Discuss how you can protect yourself against bacterial infections.

17D Why Can You Catch Gonorrhea Over and Over Again?

Each year, as many as 1 million persons in the United States catch gonorrhea from a sexual partner. Various parts of the body can be affected. Diagnosis in men is not difficult. The patient complains of pain during urination and has a milky urethral discharge 3–5 days after contact with the pathogen (Fig. 17D.1*a*). Inflammation of the testicles is also possible.

Women are less apt to have symptoms, but may have pain upon urination and a discharge that can be bloody. If gonorrhea is suspected, patients should be tested for the presence of the infection because if untreated in men or women, gonorrhea can lead to pelvic inflammatory disease (PID), a painful condition that causes reduced fertility. In women, PID can also lead to tubal pregnancies. Because of scarring, the fertilized egg remains in and starts developing in the tube that's suppose to conduct it to the uterus. The result can be a burst tube which is an emergency medical condition. A newborn who contacts the bacteria in the birth canal can become blind. Gonorrhea can even spread to other internal organs and result in crippling arthritis and fatal heart disease.

Gonorrhea proctitis, an infection of the anus characterized by anal pain and blood or pus in the feces, also occurs. Oral/genital contact can cause infection of the mouth, throat, and tonsils. If, by chance, the person touches infected genitals and then touches his or her eyes, a severe eye infection can result (Fig. 17D.1*b*). A person with gonorrhea has a greater risk of also becoming infected with HIV, the virus that causes AIDS.

The chances of getting a gonorrheal infection from an infected partner are good. Women have a 50–60% risk, while men have a 20% risk of contracting the disease after even a single exposure to an infected partner. Gonorrhea is usually curable with antibiotic therapy, but resistance to antibiotics is becoming more common, and 40% of all strains are now known to be resistant to therapy. Rates of

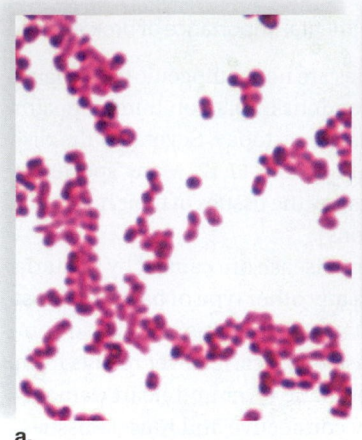
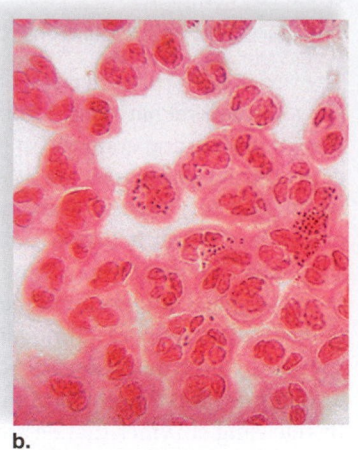

a.   **b.**

FIGURE 17D.2 a. *Neisseria gonorrhoeae,* the organism that causes gonorrhea. The organism is a coccus that occurs in pairs. **b.** Microscopic examination of a discharge containing the organism.

infection remain high among adolescents, young adults, and African Americans. Also, women using the birth control pill have a greater risk of contracting gonorrhea because hormonal contraceptives cause the genital tract to be more receptive to pathogens. There is no effective vaccine to prevent a gonorrheal infection. Various types have been formulated but to no avail. The same person can catch the disease again within a few days of being cured—and often does.

Why can we catch gonorrhea over and over again, and why isn't there a vaccine for it? Gonorrhea is caused by the bacterium *Neisseria gonorrhoeae* (Fig. 17D.2), but unlike some bacterial diseases, a number of different strains exist. When there is only one variety of the organism, the immune system builds up antibodies that attack the pathogen should it enter the body again. With so many strains to choose (or suffer!) from, a typical adult could catch a different strain of gonorrhea each time, build up immunity to it, and continue to become infected with different strains for the rest of his or her life.

The best way to prevent catching a sexually transmitted disease (STD) is by (1) practicing abstinence, (2) having a monogamous (always the same partner) sexual relationship with someone who does not have an STD and is not an intravenous drug user, and (3) always using a female or male latex condom in the proper manner. You should also avoid oral/genital contact—just touching the genitals can transfer an STD in some cases.

CONSIDER THESE QUESTIONS

1. What would you tell a friend about gonorrhea in order to persuade her or him to practice safe sex?
2. At what age should we instruct children about gonorrhea? Defend your answer.

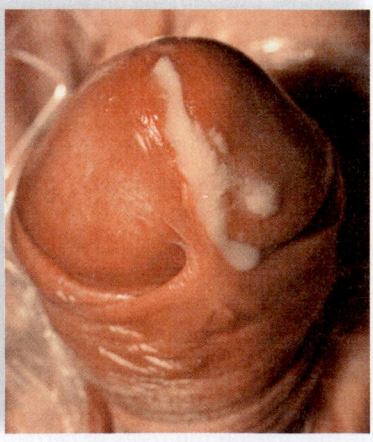
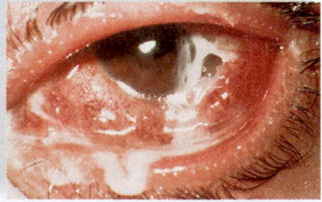

a.   **b.**

FIGURE 17D.1 a. Gonorrhea, also called the clap, is a sexually transmitted disease that is easier to diagnose in males because they have a milky discharge and pain when urinating. **b.** Touching the infected genitals and then the eyes can cause the eyes to become infected. Newborns who come in contact with the bacteria in the birth canal can become blind.

 Explore the concepts through a variety of multimedia assets and question types.

www.mcgrawhillconnect.com

THE CHAPTER IN REVIEW

CONNECTING THE CONCEPTS

Viruses are noncellular, disease-causing agents. As such, the medical significance of viruses cannot be underestimated. Nevertheless, humans use viruses for gene research and even for gene therapy after disarming their capability to cause disease. Some scientists speculate that retroviruses may have contributed to the evolution of DNA.

How viruses could have contributed to the evolution of the first cell is very much up in the air. But some viruses use RNA as their genetic material, and many scientists support the hypothesis that RNA was the first genetic material, especially because RNA can be both a substrate and a ribozyme that carries out many of the same functions as enzymatic proteins do. Once a protobiont evolved, biological evolution began. The protobiont would have carried on the same metabolic activities as cells do today and would have presumably evolved into a prokaryotic cell by acquiring DNA as its genetic material.

Prokaryotes are cellular, but their structure is simpler than that of eukaryotes—they lack a nucleus and membranous organelles. Although there are significant structural differences between prokaryotes and eukaryotes, many biochemical similarities do exist between the two. Therefore, the details of protein synthesis, first worked out in bacteria, are applicable to all cells, including those of humans. Today, GM bacteria routinely make products and otherwise serve the needs of humans.

Many prokaryotes can live in environments that may resemble the habitats available when the Earth first formed. We find prokaryotes in such hostile habitats as swamps, the Dead Sea, and hot sulfur springs. The fossil record suggests that the prokaryotes evolved before the eukaryotes. Many investigators have performed experiments that suggest how a protobiont may have preceded the evolution of the true cell which would have been a prokaryotic cell.

Cyanobacteria are believed to have introduced oxygen into the Earth's ancestral atmosphere, and they may have been the first colonizers of the terrestrial environment. Most bacteria are decomposers that recycle nutrients in both aquatic and terrestrial environments. Clearly, humans are dependent on the past and present activities of prokaryotes.

All living things trace their ancestry to the prokaryotes, which contributed to the evolution of the eukaryotic cell. The mitochondria and chloroplasts of the eukaryotic cell are derived from bacteria that took up residence inside a nucleated cell, as we will discuss in Chapter 18. The rest of Unit 3 pertains to the evolution of protists, plants, fungi, and animals, which are all eukaryotic organisms.

ANALYZE AND EVALUATE

1. Despite the generalized term "pathogen" to refer to disease-causing viruses and bacteria, bacteria in particular have redeeming features. Explain.

2. Explain the scenario described in Figure 17.6A by using either the iron-sulfur world hypothesis or the prebiotic soup hypothesis from section 17.5.

3. One of the most important events in the history of the Earth is the evolution of photosynthesis. Which organism first carried on photosynthesis, and why was it significant?

SUMMARIZE

Viruses, Viroids, and Prions

17.1 Viruses have a simple structure

- A virus is composed of an outer **capsid** made of protein and an inner nucleic acid core. Viruses reproduce by using the metabolic machinery of the host cell.
- Viruses are categorized by whether their nucleic acid is DNA or RNA and single- or double-stranded; by their size and shape; and by the presence or absence of an outer envelope.

17.11 Some viruses reproduce inside bacteria

- Viruses are specific: Each infects a preferred organism or tissue. **Bacteriophages** are viruses that parasitize bacteria.
- The **lytic cycle** has five stages: attachment, penetration, biosynthesis, maturation, and release. In the **lysogenic cycle**, integration also occurs, as shown here:

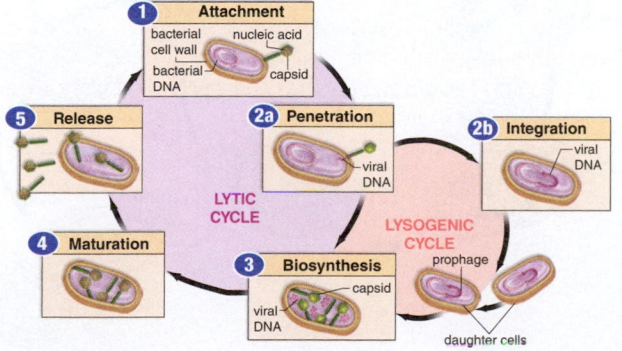

17.3 Viruses are responsible for a number of plant diseases

- The first virus discovered was the tobacco mosaic virus, which especially infects tobacco leaves but also can infect a number of different plants. Plant viruses are spread, particularly among injured plants, via contaminated tools, soil, pollen, seeds, or tubers, or by insects, nematodes, or parasitic plants.

17.4 Viruses also reproduce inside animal cells and cause animal diseases

- The reproductive stages of an animal virus are similar to those of a bacteriophage: attachment, penetration, biosynthesis, maturation, and release. Diseases caused by viruses include parvovirus in dogs, rabies in mammals, measles in children, and some cancers.
- A **retrovirus,** such as **HIV** (the AIDS virus), has an RNA genome and uses **reverse transcriptase** (from RNA to DNA) to make a copy of its genome that is inserted into the host genome. A latent **provirus** now exists. A provirus can become an active virus.
- Disease-causing agents also include **prions** (protein infectious particles) that cause rare brain diseases and **viroids** (bits of naked RNA) that cause plant diseases.

Origins of Life

17.5 Experiments show how small organic molecules may have first formed

- The Miller and Urey experiment supports the prebiotic soup hypothesis: Gases from the early Earth's reducing atmosphere could have reacted to produce small organic molecules.
- Other experiments support the iron-sulfur world hypothesis: Iron (and nickel) sulfides at hydrothermal vents could have catalyzed reactions necessary to the formation of small organic molecules.

17.6 RNA may have been the first polymer

- The finding that RNA can be both a substrate and enzyme, supports the hypothesis that RNA was the first macromolecule and original genome.
- **Protobionts** arose when macromolecules were surrounded by a plasma membrane. Either proteinoids or **liposomes** could possibly have acquired a plasma membrane. A protobiont may have had an RNA genome and became a true cell when it contained a DNA information system.

The Prokaryotes

17.7 Prokaryotes have unique structural features

- Bacteria have a nucleoid that contains a single chromosome consisting of a circular strand of DNA. Ribosomes carry out protein synthesis.
- The outer envelope consists of a plasma membrane and a cell wall strengthened by peptidoglycan. Pili and/or flagella may be present.
- Bacteria occur in three shapes: They may be round, rod-shaped, or spiral.

17.8 Prokaryotes reproduce by binary fission

- Prokaryotes reproduce asexually by **binary fission. Endospore** formation in bacteria occurs during unfavorable conditions.

17.9 How genes are transferred between bacteria

- Genes are transferred by **transformation** (pick up DNA from a medium), **conjugation** (receive DNA, sometimes a **plasmid,** via a conjugation pilus), and **transduction** (receive DNA via a virus).

17.10 Prokaryotes have various means of nutrition

- **Obligate anaerobes** cannot tolerate the presence of oxygen, while **facultative anaerobes** can. **Aerobic** bacteria require oxygen.
- Most bacteria are heterotrophs, which act as decomposers because they are **saprotrophs. Chemosynthetic** prokaryotes oxidize inorganic compounds to acquire electrons needed to reduce CO_2 to a carbohydrate. Photosynthetic prokaryotes use solar energy to produce carbohydrates; some do not release oxygen.
- **Cyanobacteria** are photosynthetic and do release oxygen; their pigments are located in **thylakoids.** Cyanobacteria sometimes fix atmospheric nitrogen in **heterocysts,** and form **lichens.** They are common in many aquatic and terrestrial habitats, including harsh environments.

17.11 Some archaea live in extreme environments

- **Archaea** are prokaryotes more closely related to eukaryotes than bacteria are. (1) **Methanogens** produce methane in anaerobic environments.
  (2) **Halophiles** live in high-salt environments.
  (3) **Thermoacidophiles** inhabit extremely hot, acidic environments.

17.12 Prokarotes have medical and environmental importance

- Prokaryotes cause many diseases (gonorrhea, strep throat, tuberculosis, food poisoning, botulism, diphtheria) sometimes by producing a **toxin. Vaccines** are available to prevent some bacterial diseases, but for the most part the medical profession relies on antibiotics to cure bacterial disease. Many bacteria have become resistant to antibiotics. Certain prokaryotes produce oxygen and play roles in the carbon, nitrogen, sulfur, and phosphorus cycles, which are essential to life.

TEST YOURSELF

Viruses, Viroids, and Prions

1. A virus contains a
   a. cell wall.
   b. plasma membrane.
   c. nucleic acid.
   d. cytoplasm.
   e. More than one of these are correct.
2. Some scientists consider viruses nonliving because
   a. they do not locomote.
   b. they cannot reproduce independently.
   c. their nucleic acid does not code for protein.
   d. they are acellular.
   e. Both b and d are correct.
3. Which of these are found in all viruses?
   a. envelope, nucleic acid, capsid
   b. DNA, RNA, and proteins
   c. proteins and a nucleic acid
   d. proteins, nucleic acids, carbohydrates, and lipids
   e. tail fibers, spikes, and a rod shape
4. The five stages of the lytic cycle occur in this order:
   a. penetration, attachment, release, maturation, biosynthesis
   b. attachment, penetration, release, biosynthesis, maturation
   c. biosynthesis, attachment, penetration, maturation, release
   d. attachment, penetration, biosynthesis, maturation, release
   e. penetration, biosynthesis, attachment, maturation, release
5. Capsid proteins are synthesized during which phase of viral replication?
   a. replication   d. proteination
   b. biosynthesis   e. All of these are correct.
   c. assembly
6. **THINKING CONCEPTUALLY** Label the life cycles and give a reason why each type of life cycle is advantageous to the virus.

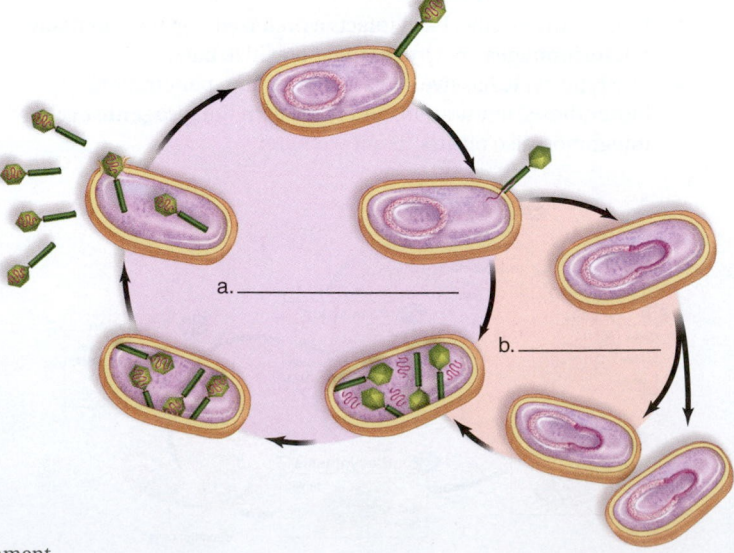

a. _____

b. _____

7. RNA retroviruses have a special enzyme that
   a. disintegrates host DNA.
   b. polymerizes host DNA.
   c. transcribes viral RNA to DNA.
   d. translates host DNA.
   e. repairs viral DNA.
8. Retroviruses
   a. parasitize only plant cells.
   b. have a reverse life cycle.
   c. include HIV.
   d. carry on anaerobic cellular metabolism.
   e. All of these are correct

Origin of Life

9. The atmosphere in which life arose lacked
   a. carbon.   c. oxygen.
   b. nitrogen.   d. hydrogen.
10. The RNA-first hypothesis for the origin of cells is supported by the discovery of
    a. ribozymes.   c. polypeptides.
    b. proteinoids.   d. nucleic acid polymerization.
11. A true cell came into being when ribozymes
    a. produced the first DNA genome.
    b. carried out protein synthesis.
    c. digested DNA polymerase.
    d. substituted for DNA ligase.
    e. Both a and b.
12. Liposomes (phospholipid droplets) are significant because they show that
    a. the first plasma membrane contained protein.
    b. a plasma membrane could have easily evolved.
    c. a biological evolution produced the first cell.
    d. there was water on the early Earth.
    e. the protobiont had organelles.
13. Protobionts probably obtained energy as
    a. photosynthetic autotrophs.   c. heterotrophs.
    b. chemosynthetics.   d. None of these are correct.
14. Which of these is an incorrect statement?
    a. The chemicals that Miller and Urey used to show that chemical evolution occurred in the atmosphere included ammonia (NH_3).
    b. Other experiments showed that nickel sulfides can act as a catalyst to change N_2 in the atmosphere to NH_3.
    c. Iron and nickel sulfides were abundant in the early atmosphere.
    d. Both a and b are incorrect.

The Prokaryotes

15. **THINKING CONCEPTUALLY** What data did researchers apply to hypothesize that RNA and not DNA was the original genetic material?
16. Bacterial cells contain
    a. ribosomes.   d. vacuoles.
    b. nuclei.   e. More than one of these are correct.
    c. mitochondria.
17. Which is not true of prokaryotes? They
    a. are living cells.
    b. lack a nucleus.

c. all are parasitic.
d. include both archaea and bacteria.
e. evolved early in the history of life.
18. Bacterial endospores function in
    a. reproduction.
    b. survival.
    c. protein synthesis.
    d. storage.
19. Archaea differ from bacteria in that they
    a. have a nucleus.
    b. have membrane-enclosed organelles.
    c. have peptidoglycan in their cell walls.
    d. are often photosynthetic.
    e. None of these are correct.

For questions 20–24, determine which type of organism listed in the key is being described. Each answer in the key may be used more than once.

KEY:
   a. bacteria
   b. archaea
   c. both bacteria and archaea
   d. neither bacteria nor archaea

20. Peptidoglycan in cell wall
21. Methanogens
22. Sometimes parasitic
23. Contain a nucleus
24. Plasma membrane containing lipids
25. **THINKING CONCEPTUALLY** On what basis would you decide (aside from their name) that cyanobacteria are bacteria?

GET INVOLVED

1. While a few drugs are effective against some viruses, they often produce a number of side effects by impairing the function of body cells. Most antibiotics (antibacterial drugs) do not cause side effects. Why would antiviral medications be more likely to produce side effects?
2. The bacterium *E. coli* is a model organism. What characteristics make *E. coli* particularly useful in genetic experiments?

MEDIA STUDY TOOLS

mhhe.com/maderconcepts3

Enhance your study of this chapter with interactive study tools, practice tests, and engaging animations. Also, ask your instructor about the resources available through ConnectPlus, which includes LearnSmart, a personalized adaptive learning program, and a media-rich eBook.

18

Evolution of Protists

Protists Cause Disease Too

Many people relate disease to viruses, bacteria, and an occasional fungus. Little do they realize that protists cause disease too.

Malaria is caused by a protist that may have infected humans ever since they evolved. The infection causes recurring cycles of chills, fever, and sweating every few days. These symptoms are due to bursting of the red blood cells where the parasite's spores exist for a part of its life cycle. The protozoan that causes malaria was identified and named *Plasmodium* in 1880, and researchers learned in 1898 that the *Anopheles* mosquito transmits the protozoan from person to person. Even so, the administrators charged with building the Panama Canal in the early 1900s had other explanations. The name malaria means "bad air," and they still believed that breezes coming off swamps caused malaria. The best protection against malaria, they said, was a morally correct lifestyle. Malaria was finally brought under control in Panama when a young physician, Dr. William C. Gorgas, was given the resources to prevent *Anopheles* from breeding. Unfortunately, malaria still affects millions around the world, particularly in South America and Africa.

Amoebic dysentery, which is characterized by bloody diarrhea, is caused by *Entamoeba histolytica*, an amoeboid protozoan. This infection is more likely to occur in the tropics, where the parasite is prevalent. It may be associated with poor hygiene because it is spread by food or water contaminated with feces.

The bite of the *Anopheles* mosquito transmits malaria.

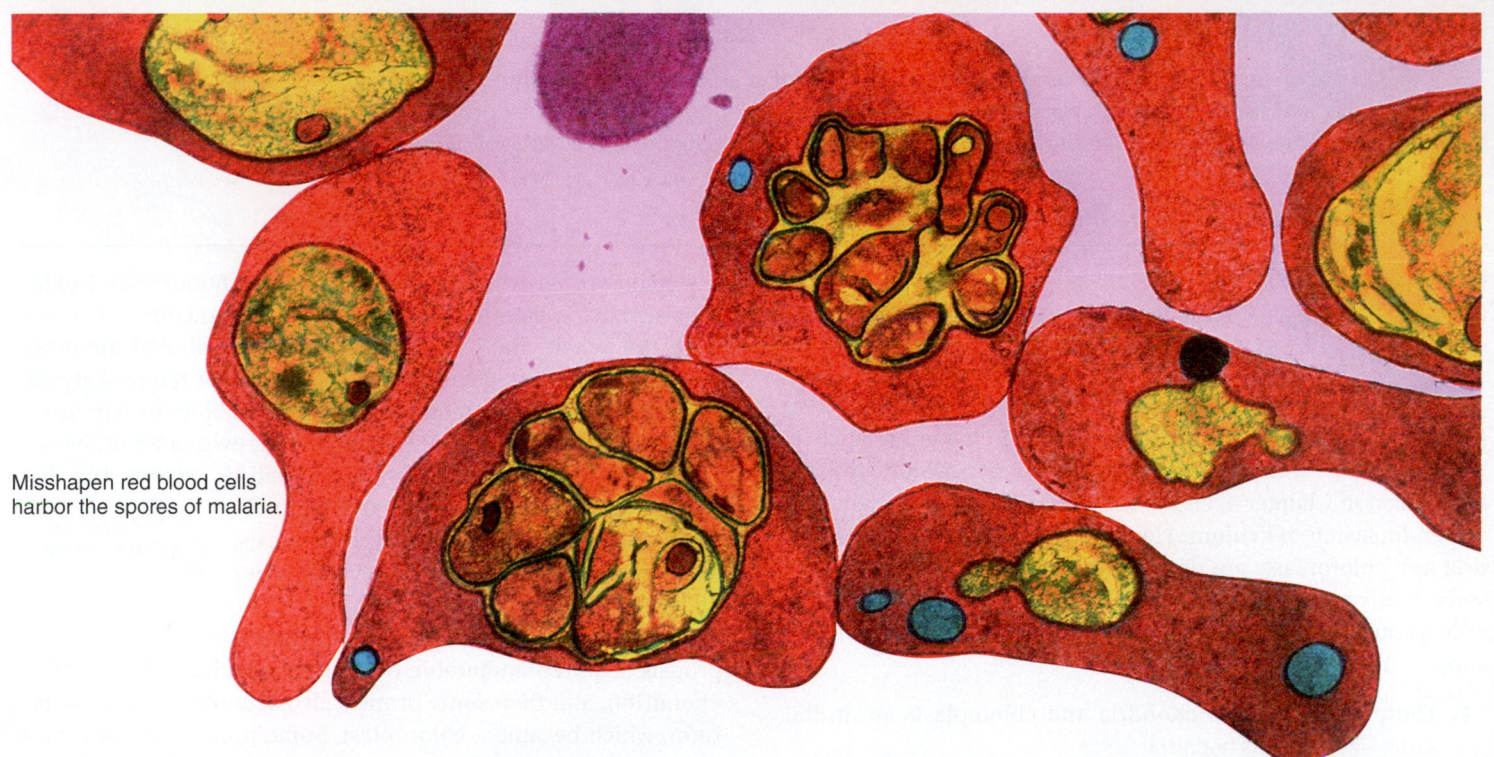

Misshapen red blood cells harbor the spores of malaria.

Giardiasis is caused by a multiflagellated protozoan that adheres to the human intestinal lining by means of a sucking disk. The primary symptom of infection is extreme diarrhea. *Giardia* does not have a vector; instead, the protozoan is taken into the body by drinking contaminated water. Persons who drink shallow well water or water from a stream while camping or hiking, or those who accidentally ingest pool water while swimming, are subject to possible infection. Since 1971, *Giardia* has been the most commonly identified waterborne pathogen in the United States. You can protect yourself by only drinking water that has been properly filtered.

Flagellated protozoans in the genus *Trypanosoma* are well known for causing tropical diseases. Each disease is transmitted by a specific insect vector. Chagas disease occurs in Central and South America after an insect commonly called the "kissing bug" deposits feces containing the parasite *Trypanosoma cruzi* in its bite. Symptoms of Chagas disease include localized swelling, loss of strength, bone pain, anemia, and possible heart failure. African sleeping sickness, caused by *Trypanosoma brucei,* has reemerged as a serious health and economic problem in sub-Saharan Africa despite eradication efforts. The vector is the large, brown, and stealthy tsetse fly, named for the sound it makes while flying. Fever, lymph node swelling, and general malaise occur before the parasite makes its way to the brain. Neurological complications result in a stupor that accounts for the name sleeping sickness. Few recover from this disease, which occurs only in Africa.

Biting female sand fly transmits leishmaniasis

Still other trypanosomes cause leishmaniasis, a disease transmitted by the bite of an infected female sand fly. This vector, about one-third the size of a mosquito, is a noiseless flyer that usually bites at night. The disease is common in tropical and subtropical countries, and therefore rare in the United States. Usually, leishmaniasis manifests itself as skin sores that heal within a few months, leaving noticeable scars. A more serious form spreads to the internal organs and is potentially fatal if untreated. Cutaneous strains from North and South America may destroy nasal and cheek mucosa and cause extreme facial disfigurement. Twenty cases of the cutaneous form and 12 cases of visceral infection were reported in soldiers during Operation Desert Storm from 1990–1991.

In this chapter, we will examine the many types of protists, including their diverse forms and lifestyles. We begin by describing how protists evolved from prokaryotes.

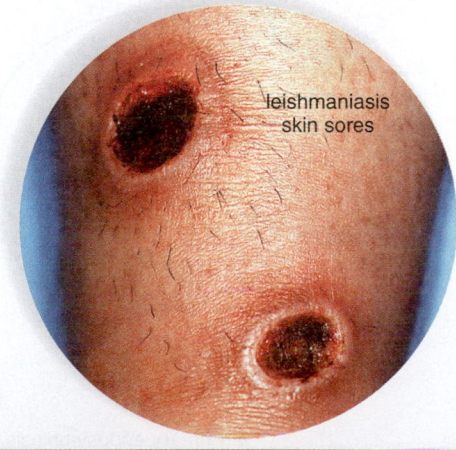

leishmaniasis skin sores

Diversity of Protists

The name protist means "first," and indeed the protists were the first eukaryotes. The multicellular groups of eukaryotes—namely, the plants, animals, and fungi—can each trace their ancestry to a particular group of unicellular protists. The diversity of protists reflects the many different ways a unicellular organism can be structured and fulfill the characteristics of life.

18.1 Eukaryotic organelles arose by endosymbiosis

LEARNING OUTCOME

When you complete this section, you should be able to

1. Describe the origin of mitochondria and chloroplasts in eukaryotes.

Protists are eukaryotes. The eukaryotic cell contains a nucleus and various membranous organelles. The **endosymbiotic theory,** introduced in Chapter 4, shows the evolution of the nucleus and the endoplasmic reticulum. Here we review that the mitochondria and chloroplasts are derived from independent prokaryotic cells. A eukaryotic cell having a flagellum (containing microtubules) could have engulfed these prokaryotes. Observational data support this theory:

1. The present-day mitochondria and chloroplasts are in the same size range as bacteria.
2. Mitochondria and chloroplasts have their own DNA and make some of their own proteins.
3. Mitochondria and chloroplasts divide independently by binary fission, just as bacteria do.
4. Mitochondria and chloroplasts are enclosed by a double membrane. The outer membrane resembles that of a eukaryotic cell, and the inner membrane resembles that of a bacterial cell.

Endosymbiosis accounts for why mitochondria and chloroplasts have a double membrane; the outer membrane represents the vesicle that brought them into the cell, and the inner membrane is the original plasma membrane of the prokaryote. Although the endosymbionts retained their ability to reproduce by binary fission and to make some of their own proteins, evolutionary change brought about the mutualistic relationship that exists today. The host cell provides a protective home for mitochondria and chloroplasts, and they in turn serve as the energy organelles of the cell.

 **Animation** Endosymbiosis

Figure 18.1*a* shows that during *primary endosymbiosis,* all protists acquired an aerobic bacterium, which became a mitochondrion, and then some protists also acquired a cyanobacterium, which became a chloroplast. Some protists actually have plastids with a total of four membranes, suggesting that during *secondary endosymbiosis,* these chloroplasts were originally part of independent protists (Fig. 18.1*b*)! Certainly, endosymbiosis was a common occurrence during the evolution of eukaryotes.

▶ **18.1 CHECK YOUR PROGRESS**
1. Explain the role of mitochondria in a photosynthetic protist.

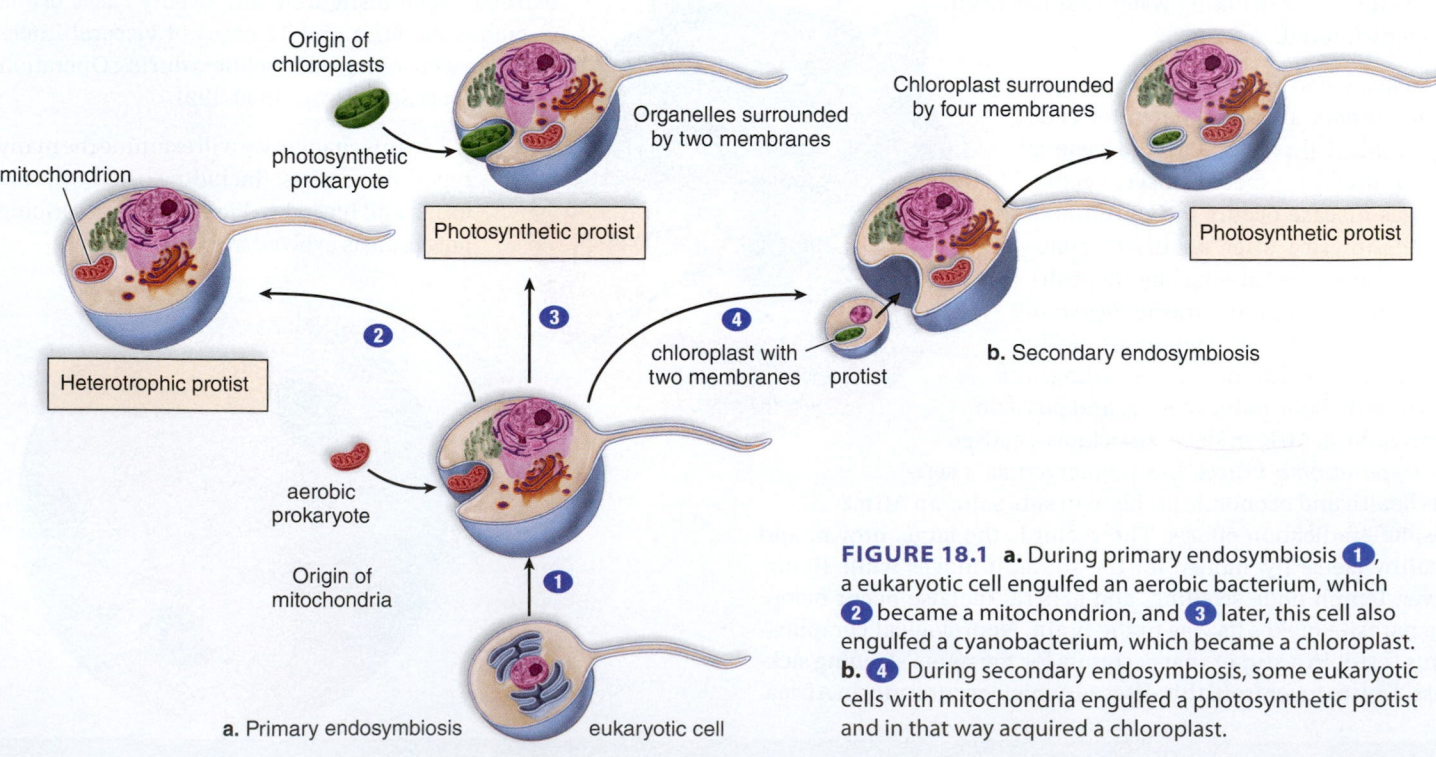

FIGURE 18.1 a. During primary endosymbiosis ❶, a eukaryotic cell engulfed an aerobic bacterium, which ❷ became a mitochondrion, and ❸ later, this cell also engulfed a cyanobacterium, which became a chloroplast. **b.** ❹ During secondary endosymbiosis, some eukaryotic cells with mitochondria engulfed a photosynthetic protist and in that way acquired a chloroplast.

18.2 Protists differ in many ways

LEARNING OUTCOME

When you complete this section, you should be able to

1. Summarize how protists are diverse by comparing size, mode of nutrition, reproduction, and symbiotic relationships.

Protists are very diverse, and sequencing their genomes is now the more standard way to determine their relationships than is comparing their structure and physiology. Because of their great diversity, the term *protist* is useful only if we realize that in general it refers to mostly unicellular eukaryotes that are neither animals, fungi, nor plants. Actually, some protists are more closely related to animals and plants than they are to each other. Only by considering the protists can we trace evolution from the prokaryotes to higher forms of life.

Protists are usually aquatic but can live in moist locations on land. Their size ranges from microscopic (as small as 1 μm) to 200 m (a football field is about 101 m). While the smaller protists are indeed unicellular, the larger ones tend to be multicellular. Even the unicellular ones are complex because each is a complete organism and its one cell has to carry on all the functions of life. The amoeboids and ciliates possess unique organelles, such as a contractile vacuole that assists in water regulation and digestive vacuoles that break down their food (Fig. 18.2).

Asexual reproduction by mitosis is common among protists, but there are exceptions. Sexual reproduction involving meiosis and perhaps aquatic spore formation is seen, especially when the environment turns hostile. **Spores** are haploid reproductive cells that are often resting cells resistant to adverse conditions, and able to survive until favorable conditions return once more. Some protozoans form cysts, another type of resting stage. In parasites, a cyst often serves as a means of transfer to a new host.

Protists carry on different modes of nutrition. You already know from Figure 18.1 that some have chloroplasts and are therefore photosynthetic. The term **algae** (sing., alga) refers to photosynthetic protists. Despite having the same mode of nutrition, the photosynthesizers need not be closely related. After all, they became photosynthetic by simply engulfing a photosynthetic prokaryote or a photosynthetic protist. Even though we recognize that they need not be closely related, we will still discuss the algae in section 18.10 of this chapter. The protists that lack chloroplasts are heterotrophic and either ingest their food or absorb organic molecules from the environment. The heterotrophic protists are generally known as **protozoans,** a term that means "before the animals." Some protists can go back and forth between making their own food and absorbing it, depending on the availability of light and organic food.

As you know from the introduction to this chapter, protists of medical importance are parasitic and cause disease. Photosynthetic protists can be endosymbionts that do not cause disease. They assist their hosts by supplying them with nutrients. For example, the growth of coral reefs in warm tropical waters is dependent on the protists that live within their bodies. A helpful endosymbiont need not be photosynthetic. A termite lives by ingesting wood only because a protist living in its gut contains a prokaryote that can digest wood!

Protists are of great ecological importance. Being aquatic, the photosynthesizers give off oxygen and function as a source of food in both freshwater and saltwater ecosystems. They are a major component of **plankton,** organisms that are suspended in the water and serve as food for heterotrophic protists and animals. The heterotrophic protists are themselves food for fishes and other animals that live in the sea.

One proposed evolutionary tree of protists, based on molecular sequence data, is shown in Figure 18A.1 (p. 344). Each group of organisms (**1**–**5**) is in its own kingdom. Some authorities have suggested many more kingdoms for the protists, and the designation kingdom Protista is no longer taxonomically acceptable.

▶ 18.2 CHECK YOUR PROGRESS

1. Discuss the diversity of protists in terms of structure and nutrition.

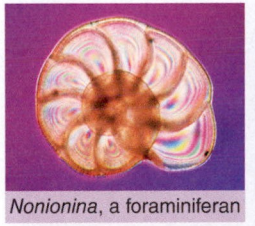

Nonionina, a foraminiferan

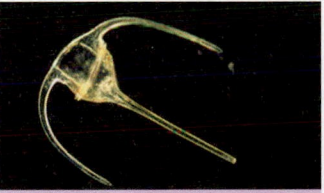

contractile vacuole / digestive vacuoles

Blepharisma, a ciliate with visible vacuoles

Ceratium, an armored dinoflagellate

Acetabularia, a single-celled green alga

Bossiella, a coralline red alga

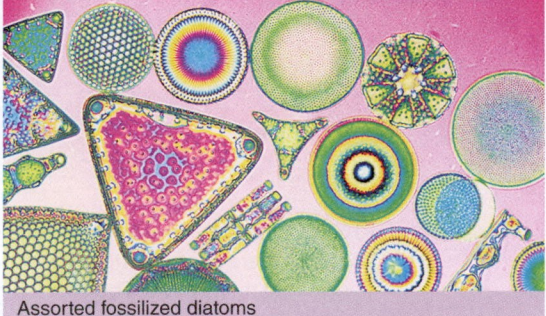

Assorted fossilized diatoms

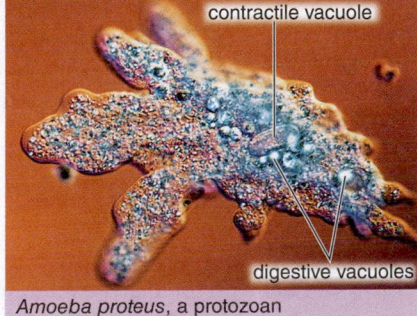

contractile vacuole / digestive vacuoles

Amoeba proteus, a protozoan

FIGURE 18.2 Some of the many living protists. Diatoms, dinoflagellates, green algae, and red algae are all photosynthesizers. Foraminiferans, ciliates, and amoebas are heterotrophs.

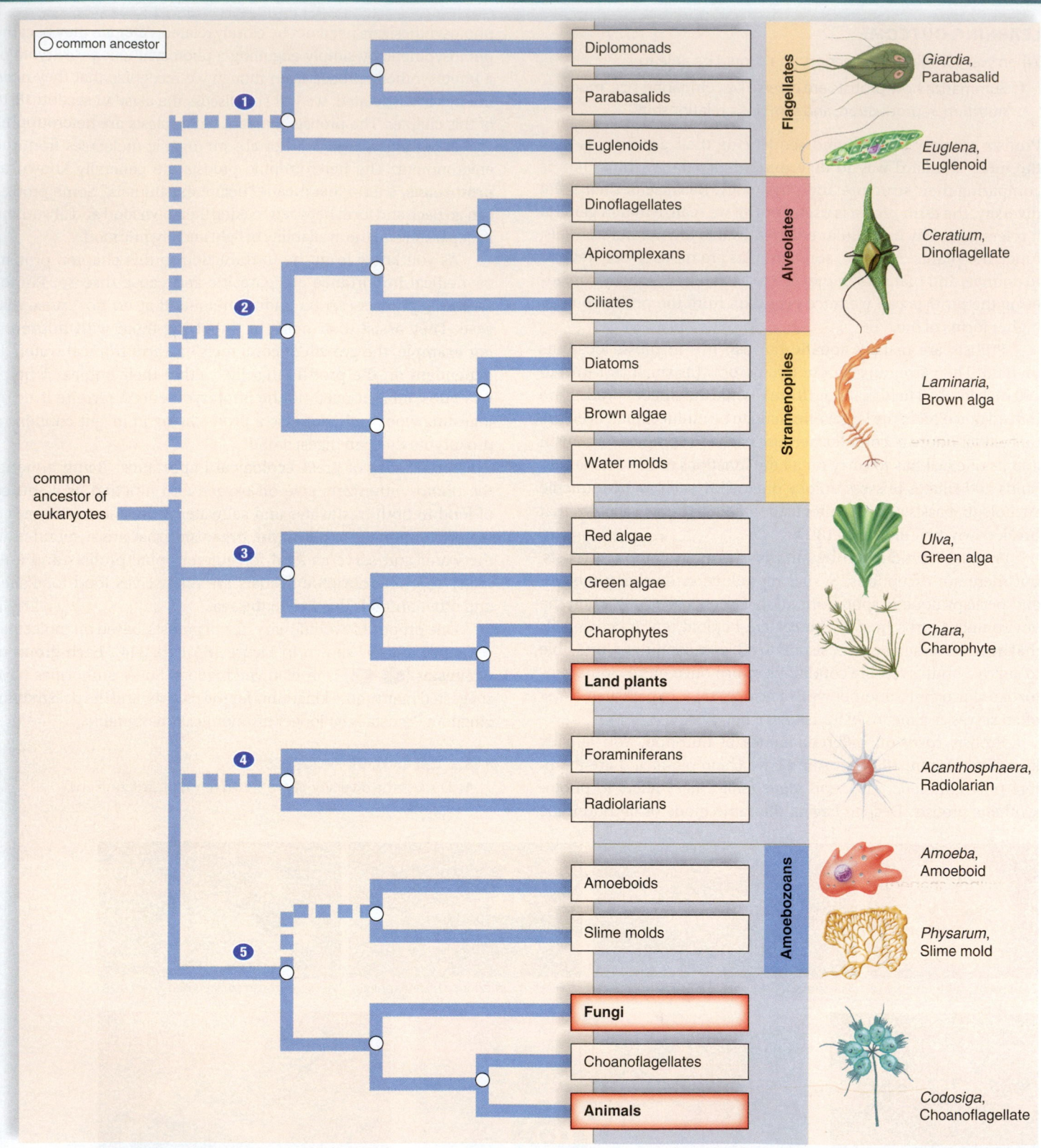

○ common ancestor

Diplomonads — *Giardia*, Parabasalid

Parabasalids

Euglenoids — *Euglena*, Euglenoid

Flagellates

Dinoflagellates — *Ceratium*, Dinoflagellate

Apicomplexans

Ciliates

Alveolates

Diatoms — *Laminaria*, Brown alga

Brown algae

Water molds

Stramenopiles

common ancestor of eukaryotes

Red algae — *Ulva*, Green alga

Green algae

Charophytes — *Chara*, Charophyte

Land plants

Foraminiferans — *Acanthosphaera*, Radiolarian

Radiolarians

Amoeboids — *Amoeba*, Amoeboid

Slime molds — *Physarum*, Slime mold

Amoebozoans

Fungi

Choanoflagellates

Animals — *Codosiga*, Choanoflagellate

Significant Groups in the Evolutionary Tree:
Flagellates have whiplike flagella.
Alveolates have sacs beneath the plasma membrane.
Stramenopiles have ancestor with hairy flagella.
Amoebozoans move by pseudopods (or cytoplasmic
streaming) and engulf their food.

FIGURE 18A.1 This proposed evolutionary tree of protists suggests there are five kingdoms of protists; some authorities have designated many more. The dotted lines on the left are tentative hypotheses. On the right are representative members of each group.

18A The Eukaryotic Big Bang

Take a look at the evolutionary tree in Figure 18A.1 (p. 344). Notice that all five of the numbered groups in the tree apparently evolved at the same time. Ordinarily researchers would be able to find a *root* of a tree—in this case, a living protist group that is closest to the common ancestor for all the protists. At first biologists proposed that the flagellates be placed at the root of the tree. This suggestion was supported by molecular data and by the lack of mitochondria in the parasitic members of this group. After all, if they lack mitochondria, they never participated in the first endosymbiotic event, right? Well, wrong because by now mitochondrial genes have been found in these organisms. So, apparently flagellates did have mitochondria at one time, but have since lost them perhaps due to their parasitic way of life.

As of this writing, the eukaryotic tree is left without a root, giving rise to a *eukaryotic big-bang hypothesis*. This hypothesis states that all the different groups of protists evolved so quickly from the original eukaryote that there is no known root to their tree. Another way to represent the eukaryotic big-bang hypothesis is shown in Figure 18A.2, *right,* which indeed does look like a tree (Fig. 18A.2, *left*). Like some trees, multiple trunks are coming from an original trunk as also shown in the photograph.

Unfortunately, the Precambrian fossil record is very poor. Fossils that are clearly related to modern protistan groups are dated only 1.2–1.7 billion years ago, even though evidence found in Australian shales indicates that eukaryotes were present 2.7 billion years ago. If the fossil record was more complete, we might be able to definitively determine whether any of the protists are at the root of the tree. As you know, the fossil record is much better if organisms have hard parts, and of course, a unicellular organism such as an amoeba has no hard parts. If we did have good fossil record data, researchers would be closer to saying which of the protists are at the bottom of the tree. Without it, molecular biologists are hard at work determining the relationships between the different types of protists purely on the basis of DNA sequence data, and the results can be confusing to them and to us. Some of the proposed groupings seem very strange based on the structure and physiology of the protists in that group. Water molds (saprotrophic and threadlike) are related to diatoms (photosynthetic and pillbox-shaped). Ciliates (heterotrophic with cilia) and dinoflagellates (photosynthetic with flagella) are related by way of a common ancestor. But of course these protists may have changed quite a lot from the common ancestor for the group.

This much we know for sure on the basis of molecular data: Lumping all the single-celled eukaryotes (protists) into a single kingdom is artificial and does not represent how evolution actually occurred. But should they be placed in several kingdoms, thereby increasing the number of kingdoms in the domain Eukarya? Should some of them be placed in the other kingdoms—plants, animals, or fungi—or should new kingdoms be created that would include, say, the plants and some of the protists, or include the animals and some of the protists? For example, on the basis of molecular data, some authorities group the green algae called stoneworts with the plants. Notice also that in Figure 18A.1 (p. 344) ⑤ the fungi and animals are related by way of a common ancestor. Should they be in one kingdom because they share a common ancestor along with the choanoflagellates, which are the living protists most closely related to animals? Much remains to be decided about eukaryotic classification.

CONSIDER THESE QUESTIONS

1. Does uncertainty about how evolution exactly occurred call into question the theory that evolution does occur? Why or why not?
2. Which are more reliable data to decide evolutionary lineages: genetic similarities and differences between protist groups or structural similarities and differences between protist groups? Explain.
3. How can scientists avoid mistaken conclusions based on their observational data?

connect BIOLOGY Explore the concepts through a variety of multimedia assets and question types.
www.mcgrawhillconnect.com

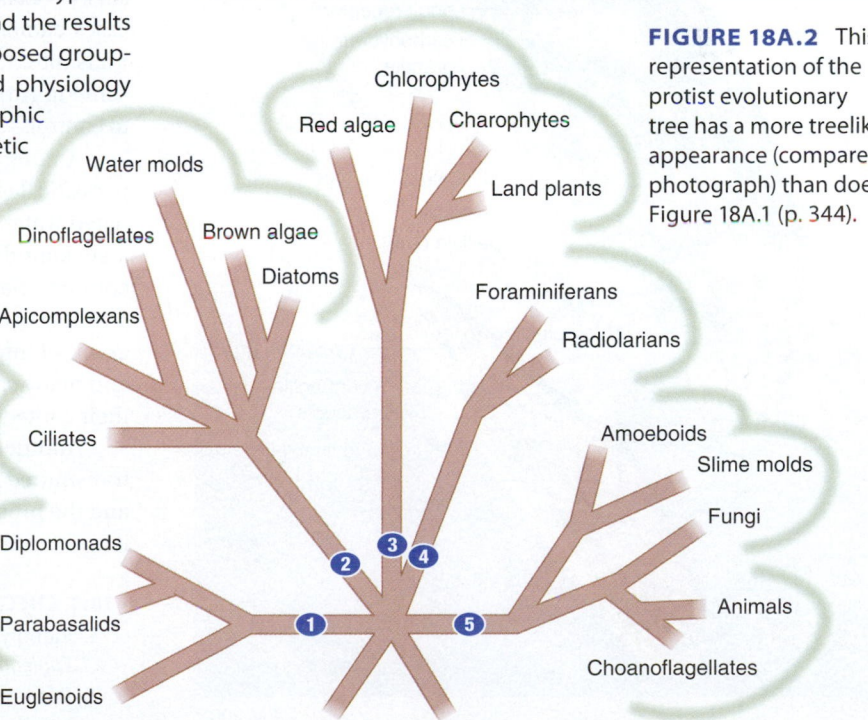

FIGURE 18A.2 This representation of the protist evolutionary tree has a more treelike appearance (compare to photograph) than does Figure 18A.1 (p. 344).

Protozoan Protists

As discussed, the designation protozoans, while useful, is not used to classify protists. Similarly, we will discuss protozoans by their means of locomotion. Flagellates move by flagella, amoeboids move by pseudopods, ciliates move by cilia, and apicomplexans are not motile.

18.3 Protozoans called flagellates move by flagella

LEARNING OUTCOMES

When you complete this section, you should be able to

1. Use euglenoids to describe how nutrition can be variable in a protist.
2. List the diseases caused by flagellates and tell whether each is common to a temperate or a tropical zone.

All the protists discussed in this section are in a group called **flagellates** (see Fig. 18A.1 p. 344) because they have a whiplike flagellum.

Euglenoids The **euglenoids** include about 1,000 species of small, flagellated (10–500 µm), freshwater, unicellular organisms that typify the problem of classifying protists by structure. One-third of all genera have chloroplasts; the rest do not. This may not be surprising when we consider that their chloroplasts are like those of green algae and are probably derived from them through endosymbiosis. A pyrenoid is a special region of the chloroplast where polysaccharides form. Euglenoids produce an unusual type of polysaccharide called paramylon. Euglenoids that lack chloroplasts ingest or absorb their food.

Euglenoids have two flagella, one of which is typically much longer than the other and projects from an anterior, vase-shaped invagination. It is called a tinsel flagellum because there are hairs on it. Near the base of this flagellum is an eyespot, which shades a photoreceptor for detecting light. Because euglenoids are enclosed by a flexible pellicle composed of protein bands lying side by side, they can assume different shapes as the underlying cytoplasm undulates and contracts. A common euglenoid is *Euglena deces,* an inhabitant of freshwater ditches and ponds. A contractile vacuole allows this protist to rid its body of excess water (Fig. 18.3).

Parasitic Flagellates Diplomonads, Fig. 18A.1, are parasitic flagellates. We will consider three examples. (1) *Trypanosoma brucei,* transmitted by the bite of the tsetse fly, is the cause of African sleeping sickness in humans. The area of the bite becomes an open sore, from which the trypanosomes move toward the lymphatic glands or remain in the bloodstream, where they divide every 5–7 hours. Weight loss and recurrent attacks of fever occur during this phase of the disease. The trypanosomes invade the central nervous system, leading to the typical symptoms of sleeping sickness—disturbed sleep cycle, change in personality, and coma. Many thousands of cases of human sleeping sickness are diagnosed each year. Fatalities or permanent brain damage are common. (2) Another trypanosome, *Trypanosoma cruzi,* causes Chagas disease in humans in Central and South America. Approximately 45,000 people die yearly from the severe cardiac and digestive problems caused by this parasite. (3) Leishmaniasis, characterized by skin sores and in some cases damage to the internal organs, is caused by a trypanosome transmitted by sand flies. These diseases are particularly troublesome in Africa and South America, and so far have been difficult to control.

Among the parabasalids (Fig. 18A.1), *Giardia lamblia* is a parasitic flagellate whose cysts are transmitted by way of contaminated water. It attaches to the human intestinal wall by means of a sucking disk and causes severe diarrhea. *Giardia* is the most common flagellate in the human digestive tract and also lives in a variety of other mammals. Beavers seem to be an important reservoir of infection in the mountains of the western United States, and many cases of infection have been acquired by hikers who fill their canteens at beaver ponds.

Another parabasalid, *Trichomonas vaginalis* is a sexually transmitted flagellate that infects the vagina and urethra of women and the prostate, seminal vesicles, and urethra of men. Therefore, it is a common cause of vaginitis in the United States.

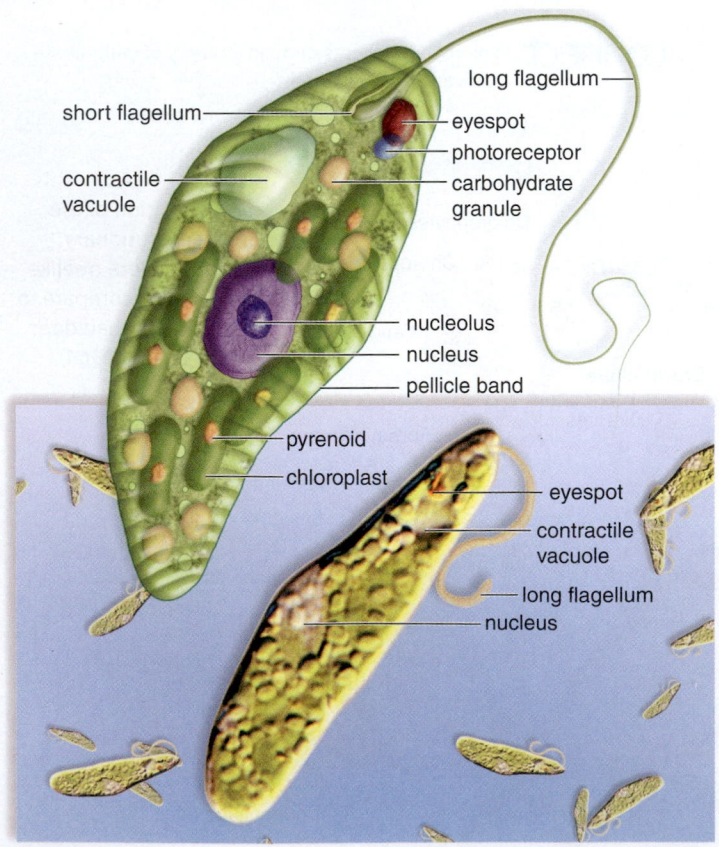

short flagellum
long flagellum
eyespot
photoreceptor
carbohydrate granule
contractile vacuole
nucleolus
nucleus
pellicle band
pyrenoid
chloroplast
eyespot
contractile vacuole
long flagellum
nucleus

LM 200×

FIGURE 18.3 *Euglena,* a flagellate.

> **18.3 CHECK YOUR PROGRESS**
> 1. Identify the benefit of a photoreceptor to photosynthetic *Euglena*.
> 2. Determine why you might not want to fill a canteen at a beaver pond.

18.4 Protozoans called amoeboids move by pseudopods

When you complete this section, you should be able to

1. Identify how amoeboids move, feed, and reproduce.
2. Distinguish amoebas from foraminiferans and from radiolarians.

As you know, we are using the term protozoan to denote any unicellular heterotrophic protist, although the term is not in the evolutionary tree of protists. In this section, all the protists discussed produce **pseudopods,** extensions that form when cytoplasm streams in a particular direction.

Video
Amoeba
Locomotion

In Figure 18A.1 (p. 344), **amoeboids** are in a larger group called **Amoebozoans.** In oceans and freshwater lakes and ponds, the amoeboids are a part of the **zooplankton,** microscopic suspended organisms that feed on other organisms. They use pseudopods to move and to engulf their food. Hundreds of species of amoeboids have been classified. *Amoeba proteus* is a commonly studied freshwater member of this group (Fig. 18.4A). When amoeboids feed, **phagocytosis** occurs as pseudopods surround and engulf their prey, which may be algae, bacteria, or other protists. Digestion then occurs within a *food vacuole.* Some white blood cells in humans also move by pseudopods and phagocytize debris and parasites which they digest in lysosomes. Freshwater amoeboids have *contractile vacuoles,* where excess water from the cytoplasm collects before the vacuole appears to "contract," releasing the water through a temporary opening in the plasma membrane. Amoeboids reproduce asexually. The process of mitosis and cytokinesis results in two amoebas whereas before there was only one amoeba.

Entamoeba histolytica is a parasitic amoeboid that lives in the human large intestine and causes amoebic dysentery. The ability of the organism to survive as cysts allows amoebic dysentery to be transmitted to a new host. Complications arise when this parasite invades the intestinal lining and reproduces there. If the parasites enter the body proper, liver and brain involvement can be fatal.

On the basis of molecular sequence data, **foraminiferans** and **radiolarians** are not in a clade with amoeboids (see Fig. 18A.1). These protists have shells called tests, which are intriguing and beautiful. In the foraminiferans, the calcium carbonate test is often multichambered. Threadlike pseudopods extend through openings in the test, which covers the plasma membrane. Deposits of foraminiferans for millions of years, followed by geologic upheaval, formed the White Cliffs of Dover along the southern coast of England (Fig. 18.4B). In radiolarians, the glassy silicon test is internal and usually has a radial arrangement of spines (Fig. 18.4C). The pseudopods are external to the test.

The tests of dead foraminiferans and radiolarians form a deep layer (700–4,000 m) of sediment on the ocean floor. The radiolarians lie deeper than the foraminiferans because their glassy test is insoluble at greater pressures. The presence of either or both is used as an indicator of oil deposits on land and sea. Their fossils date as far back as Precambrian times and are evidence of the antiquity of the protists. Because each geologic period has a distinctive form of foraminiferan, they can be used as index fossils to date sedimentary rock. The great Egyptian pyramids are built of foraminiferan limestone. One foraminiferan test found in the pyramids is about the size of a silver dollar. This species, known as *Nummulites,* has been found in deposits worldwide, including in Mississippi.

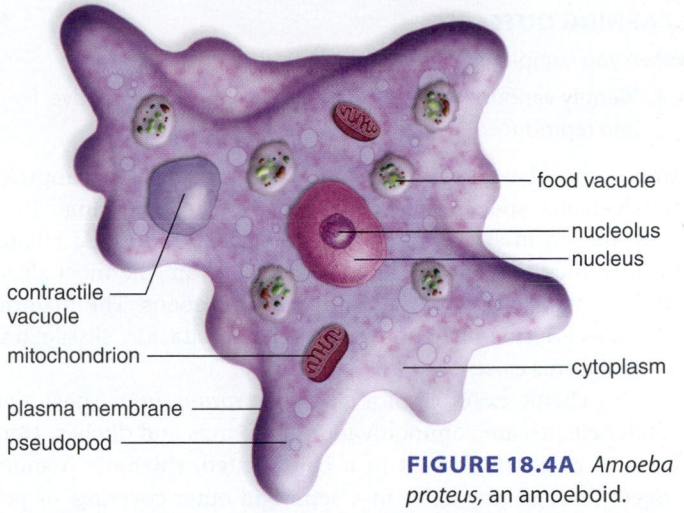

FIGURE 18.4A *Amoeba proteus,* an amoeboid.

food vacuole
nucleolus
nucleus
cytoplasm
contractile vacuole
mitochondrion
plasma membrane
pseudopod

160×

FIGURE 18.4B Foraminiferans, such as *Globigerina,* built the White Cliffs of Dover, England.

SEM 200×

FIGURE 18.4C Radiolarian tests.

▶ **18.4 CHECK YOUR PROGRESS**

1. Discuss why you might consider an amoeba to be a scavenger.
2. List ways foraminiferans and radiolarians are alike.

LEARNING OUTCOME

When you complete this section, you should be able to

1. Identify various types of ciliates and tell how ciliates move, feed, and reproduce.

Among the Alveolates (Fig. 18.1A), **ciliates** consist of approximately 8,000 species of unicellular protists that range from 10–3,000 μm in size. Members of this group are called ciliates because they move by means of cilia. They are the most structurally complex and specialized of all protozoans. The majority of ciliates are free-living; however, several parasitic, sessile, and colonial forms exist.

The classic example of a ciliate is *Paramecium*. These unicellular ciliates are commonly found in ponds and ditches. Hundreds of cilia, which beat in a coordinated, rhythmic manner, project through tiny holes in a semirigid outer covering, or pellicle (Fig. 18.5A). Numerous oval capsules lying in the cytoplasm just beneath the pellicle contain **trichocysts.** Upon mechanical or chemical stimulation, trichocysts discharge long, barbed threads that are useful for defense and for capturing prey. Toxicysts are similar, but they release a poison that paralyzes prey.

When a paramecium feeds, food particles are swept down a gullet, below which food vacuoles form. Following digestion, the soluble nutrients are absorbed by the cytoplasm, and the nondigestible residue is eliminated at the anal pore.

Ciliates reproduce both asexually and sexually. Asexual reproduction involves both mitosis and cytokinesis by transverse fission. Ciliates have two types of nuclei: a large macronucleus and one or more small micronuclei. The macronucleus controls the cell's normal metabolism, while the micronuclei are concerned with reproduction. Sexual reproduction involves conjugation (Fig. 18.5B). In each ciliate, the macronucleus disintegrates and the micronuclei undergo meiosis, producing eight cells but only one remains to undergo

mitosis. *The two ciliates exchange a micronucleus.* Each ciliate now has two micronuclei that fuse and undergo sufficient mitosis to form a new macronucleus and two new micronuclei.

Ciliates are a diverse group of protozoans. Barrel-shaped didiniums expand to consume paramecia much larger than themselves. *Suctoria* have an even more dramatic way of getting food. They rest quietly on a stalk until a hapless victim comes along. Then they promptly paralyze it and use their tentacles like straws to suck it dry. *Stentor* may be the most elaborate ciliate, resembling a giant blue vase decorated with stripes (Fig. 18.5C). *Ichthyophthirius,* a ciliate, is responsible for a common disease in fishes called "ich." If left untreated, it can be fatal.

▶ **18.5 CHECK YOUR PROGRESS**

1. Discuss why you might consider ciliates to be predators.

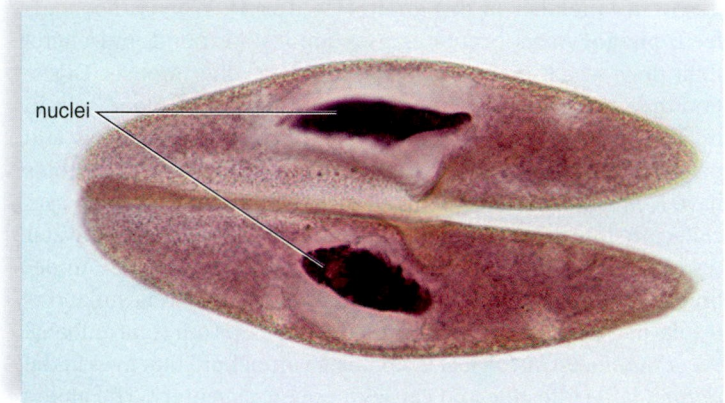

100×

FIGURE 18.5B During conjugation, two paramecia first unite at their oral areas and then exchange micronuclei.

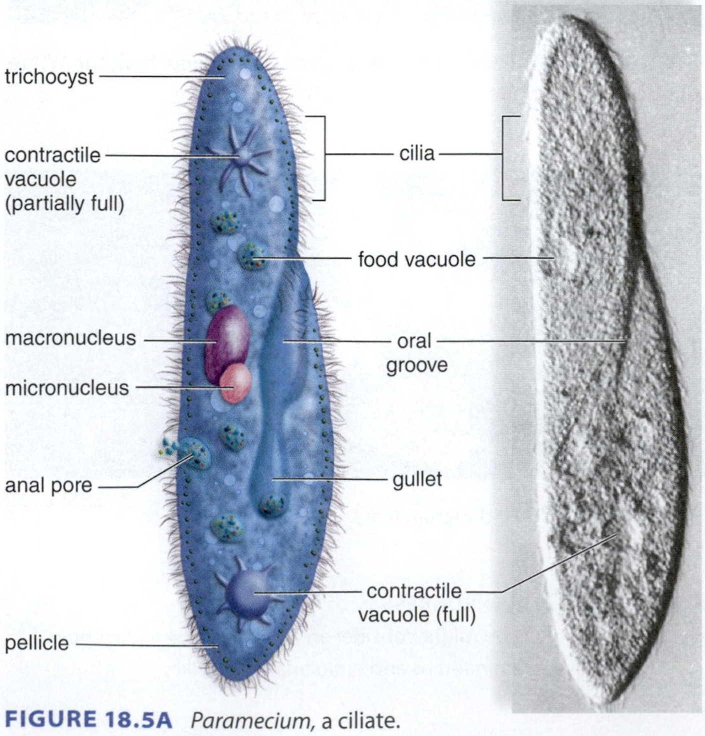

FIGURE 18.5A *Paramecium,* a ciliate.

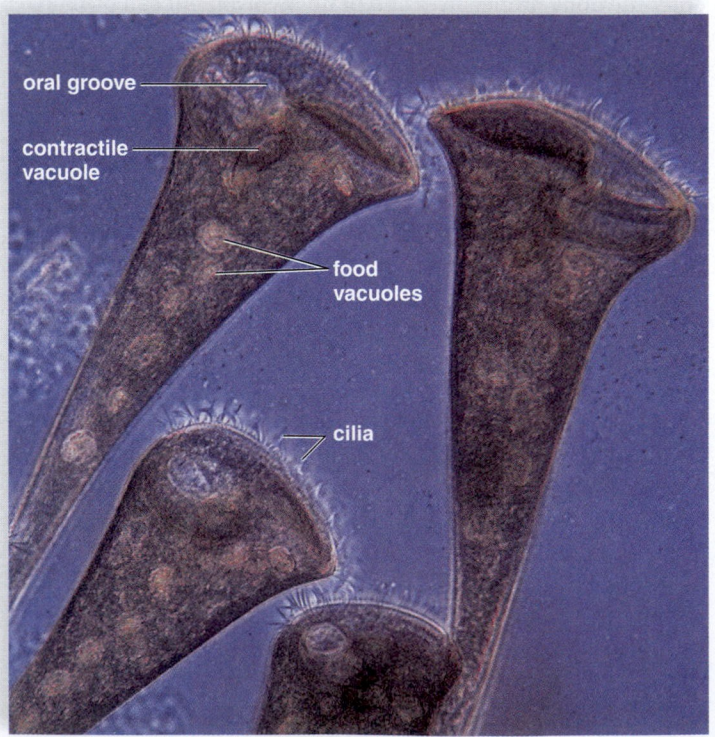

125×

FIGURE 18.5C *Stentor,* a ciliate.

LEARNING OUTCOME

When you complete this section, you should be able to

1. Describe the life cycle of *Plasmodium vivax*.

The parasites discussed in this section are called **apicomplexans** (see Fig. 18A.1) because they have an apical complex of microtubules within the cell. Apicomplexans number nearly 3,900 species of nonmotile, parasitic, spore-forming protozoans. Many apicomplexans have multiple hosts.

Pneumocystis jiroveci causes the type of pneumonia seen primarily in AIDS patients. During its sexual reproduction, thick-walled cysts form in the lining of pulmonary air sacs. The cysts contain spores that successively divide until the cyst bursts and the spores are released. Each spore becomes a new mature organism that can reproduce asexually but may also enter the sexual stage and form cysts.

Today, approximately 1 million people die each year from **malaria,** a chronic disease with potentially lifelong effects, that is caused by four parasites in the genus *Plasmodium*. The disease is spread by a mosquito vector that passes the parasite to humans. International travel coupled with new resistant forms of the vector and parasite are causing malaria to increase in frequency.

The life cycle of the sporozoan *Plasmodium vivax,* a common cause of malaria, is shown in Figure 18.6. The female *Anopheles* mosquito bites humans and other animals to acquire the protein she needs to produce eggs. ❶ If the mosquito picks up the parasite, the sexual phase of the parasite's life cycle occurs in her body. ❷ The next human the mosquito bites will acquire the parasite, which ❸ begins its asexual phase in the liver. ❹ The chills and fever of malaria appear after red blood cells are infected and burst, ❺ releasing parasites and toxic substances into the blood. ❻ Some of these parasites become gametocytes that will be taken up by a mosquito.

Toxoplasma gondii, another apicomplexan, causes toxoplasmosis, particularly in cats, but also in people. In pregnant women, the parasite can infect the fetus and cause birth defects and intellectual disability; in AIDS patients, it can infect the brain and cause neurological symptoms.

▶ **18.6 CHECK YOUR PROGRESS**

1. Identify why an insecticide is used to keep malaria under control.

female gamete

male gamete

food canal

Sexual phase in mosquito

zygote

sporozoite

salivary glands

FIGURE 18.6 Life cycle of *Plasmodium vivax,* the cause of one type of malaria.

❶ In the gut of a female *Anopheles* mosquito, gametes fuse, and the zygote undergoes many divisions to produce sporozoites, which migrate to her salivary gland.

❷ When the mosquito bites a human, the sporozoites pass from the mosquito salivary glands into the bloodstream and then the liver of the host.

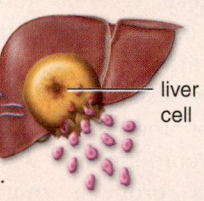

❸ Asexual spores (merozoites) produced in liver cells enter the bloodstream and then the red blood cells, where they feed as trophozoites.

liver cell

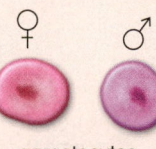

❻ Some merozoites become gametocytes, which enter the bloodstream. If taken up by a mosquito, they become gametes.

♀ ♂

gametocytes

Asexual phase in humans

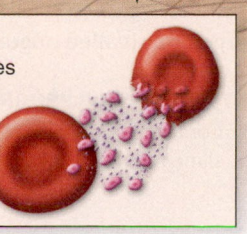

❹ When the red blood cells rupture, merozoites invade and reproduce asexually inside new red blood cells.

❺ Merozoites and toxins pour into the bloodstream when the red blood cells rupture.

Protists Called "Molds"

In forests and woodlands, slime molds phagocytize, and thereby help dispose of bacteria and dead plant material. In contrast water molds are parasitic both in the water and on land. Both slime molds and water molds are heterotrophic but beyond that they have little in common. Their designation as "molds" is due only to their general appearance and they differ significantly in structure from the molds classified as fungi.

18.7 The diversity of protists includes slime molds and water molds

LEARNING OUTCOME

When you complete this section, you should be able to

1. Compare and contrast slime molds with water molds.

For the sake of convenience only, we discuss slime molds and the water molds in this same section.

Slime Molds **Slime molds** are of two types, known as plasmodial slime molds and cellular slime molds. Usually, **plasmodial slime molds** exist as a plasmodium, a diploid, multinucleated, cytoplasmic mass enveloped by a slimy sheath that creeps along, phagocytizing decaying plant material in a forest or agricultural field. Approximately 500 species of plasmodial slime molds have been described. Many species are brightly colored.

Video Decomposers

Plasmodium, *Physarum*   1×

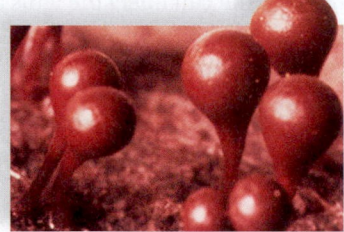
Sporangia, *Hemitrichia*   10×

At times that are unfavorable to growth, such as during a drought, the plasmodium develops many **sporangia,** reproductive structures that produce spores. The spores can survive until moisture is sufficient for them to germinate. In plasmodial slime molds, spores release a haploid flagellated cell or an amoeboid cell. Eventually, two of them fuse to form a zygote that feeds and grows, producing a multinucleated plasmodium once again (Fig. 18.7).

In keeping with their name, **cellular slime molds** exist as individual amoeboid cells. They are common in soil, where they feed on bacteria and yeasts. Their small size prevents them from being seen. Nearly 70 species of cellular slime molds have been described.

As the food supply runs out or unfavorable environmental conditions develop, the cells release a chemical that causes them to aggregate into a pseudoplasmodium. The pseudoplasmodium stage is temporary and eventually gives rise to a fruiting body, in which sporangia produce spores. When favorable conditions return, the spores germinate, releasing haploid amoeboid cells, and the asexual cycle begins again.

Water Molds The **water molds** usually live in the water, where they form furry growths when they parasitize fishes or insects and decompose remains. Despite their common name, some water molds live on land and parasitize insects and plants. Nearly 500 species of water molds have been described. The water mold *Phytophthora infestans* was responsible for the 1840s potato famine in Ireland. However, most water molds are saprotrophic and live off dead organic matter. Another well-known water mold is *Saprolegnia*, which is often seen as a white, cottonlike mass on dead organisms.

Water molds have a filamentous body like that of fungi, but their cell walls are largely composed of cellulose, whereas fungi have cell walls of chitin. The life cycle of water molds also differs from that of fungi. During asexual reproduction, water molds produce motile spores (2n zoospores), which have flagella covered by fine hairs. The organism is diploid (not haploid as in the fungi), and meiosis produces gametes. The phylum name Oomycota refers to the enlarged tips (called oogonia) where eggs are produced.

dead insect

filaments of water mold   10×

FIGURE 18.7 Life cycle of a plasmodial slime mold.

zygote

mature plasmodium

sporangia formation begins

diploid (2n)

haploid (n)

FERTILIZATION

MEIOSIS

amoeboid cells

germinating spore

or

flagellated cells

▶ **18.7 CHECK YOUR PROGRESS**

1. Contrast where you might find and how you would recognize a slime mold and a water mold.

Algal Protists

Algae are the photosynthetic protists. Our survey of algae includes the golden brown diatoms and the variously colored dinoflagellates, which are major producers in the oceans; the red algae and the brown algae, which are multicellular; and the green algae, which are ancestral to land plants. Some authorities classify the green algae with the plants.

18.8 The diatoms and dinoflagellates are significant algae in the oceans

LEARNING OUTCOME

When you complete this section, you should be able to

1. Contrast the anatomy of diatoms and dinoflagellates and tell why they are significant algae in the oceans.

Diatoms Most **diatoms** (approximately 11,000 species) are free-living, photosynthetic cells that inhabit aquatic and marine environments. Diatoms are the most numerous unicellular algae in the oceans and freshwater environments. Diatoms are a significant part of the **phytoplankton,** photosynthetic organisms suspended in the water in both freshwater and marine ecosystems, where they serve as an important source of food and oxygen for heterotrophs.

The structure of a diatom is often compared to a hat box because the cell wall has two halves, or valves, with the larger valve acting as a "lid" that fits over the smaller valve (Fig. 18.8A). When diatoms reproduce asexually, each receives one old valve. The new valve fits inside the old one; therefore, new diatoms are smaller than the original ones. When they reproduce sexually, the size returns to normal.

The cell wall of a diatom has an outer layer of silica, a common ingredient in glass. The valves are covered with a great variety of striations and markings that form beautiful patterns when observed under the microscope. These are actually depressions or pores through which the organism makes contact with the outside environment. The remains of diatoms, called diatomaceous earth, accumulate on the ocean floor and are mined for use as filtering agents, soundproofing materials, components of reflective paints, and gentle polishing abrasives such as those in silver polish and toothpaste.

Dinoflagellates The **dinoflagellates** (about 4,000 species) are usually enclosed by protective cellulose plates impregnated with silicates (Fig. 18.8B). Typically, the organism has two flagella; one lies in a longitudinal groove with its distal end free, and the other lies in a transverse groove that encircles the organism. The longitudinal flagellum acts as a rudder, and the beating of the transverse flagellum causes the cell to spin as it moves forward.

The chloroplasts of a dinoflagellate vary in color from yellow-green to brown, and some species, such as *Noctiluca,* are capable of bioluminescence (producing light). Being a part of the phytoplankton, the dinoflagellates are an important source of food for small animals in the ocean. They also live within the bodies of some invertebrates as symbionts. Symbiotic dinoflagellates lack cellulose plates and flagella and are called zooxanthellae. Corals, members of the animal kingdom, usually contain large numbers of zooxanthellae, which provide their hosts with organic nutrients while the corals in turn provide wastes that fertilize the algae. Some dinoflagellates lack chloroplasts and are heterotrophic; some of these are parasitic.

1,785×

FIGURE 18.8A *Cyclotella,* a diatom. Diatoms live in "glass houses" because the outer visible valve, which fits over the smaller inner valve, contains silica.

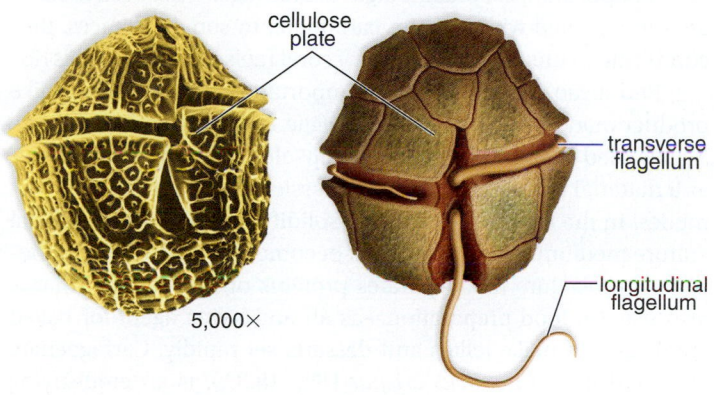

cellulose plate

transverse flagellum

longitudinal flagellum

5,000×

FIGURE 18.8B *Gonyaulax,* a dinoflagellate. This dinoflagellate is responsible for the poisonous "red tide" that sometimes occurs along the coasts.

Like the diatoms, dinoflagellates are one of the most important groups of producers in marine environments. Occasionally, however, particularly in polluted waters in late summer, they undergo a population explosion and become more numerous than usual. At these times, their density can equal 30,000 in a single milliliter. When dinoflagellates, such as *Gonyaulax,* increase in number, they may cause a phenomenon called **red tide.** Massive fish kills can occur as the result of a powerful neurotoxin produced by these dinoflagellates. Humans who consume shellfish that have fed during a *Gonyaulax* outbreak may experience shellfish poisoning, which paralyzes the respiratory organs.

Video Good Poison

Dinoflagellates usually reproduce asexually by mitosis and a splitting of the cell lengthwise. During sexual reproduction, the zygote is a resting stage that eventually undergoes meiosis. Only one of the four haploid cells becomes another dinoflagellate.

▶ **18.8 CHECK YOUR PROGRESS**

1. Contrast the external protection of a diatom with that of a dinoflagellate.

18.9 Red algae and brown algae are multicellular

Red Algae The **red algae** include more than 5,000 species of multicellular organisms closely related to the green algae discussed in section 18.10. These algae live primarily in warm seawater, both shallow and deep. Some grow attached to rocks in the intertidal zone, where they are exposed at low tide. Others can grow at depths exceeding 200 m, where light barely penetrates. Red algae are usually fairly small and delicate, although some species can exceed a meter in length.

Some forms of red algae are simple filaments, but most have complex branches with a feathery, flat, or expanded, ribbonlike appearance. Coralline algae are red algae whose cell walls are impregnated with calcium carbonate. In some instances, they contribute as much to the growth of coral reefs as coral animals do.

Red algae are economically important. Agar is a gelatin-like product made primarily from the algae *Gelidium* and *Gracilaria*. Agar is used commercially to make capsules for vitamins and drugs, as a material for making dental impressions, and as a base for cosmetics. In the laboratory, agar is a solidifying agent for a bacterial culture medium. When purified, it becomes the gel for electrophoresis, a procedure that separates proteins or nucleotides. Agar is also used in food preparation—as an antidrying agent for baked goods and to make jellies and desserts set rapidly. Carrageenan, extracted from *Chondrus crispus* (Fig. 18.9A), is an emulsifying agent (causes fat to disperse in water) for the production of chocolate and cosmetics. The reddish-black wrappings around sushi rolls consist of processed blades from *Porphyra*, another red alga.

Brown Algae The **brown algae** (over 1,500 species) range from small forms with simple filaments to large, multicellular forms that may reach more than 100 m in length. The vast majority of brown algae, such as rockweed (*Fucus*) live in cold ocean waters

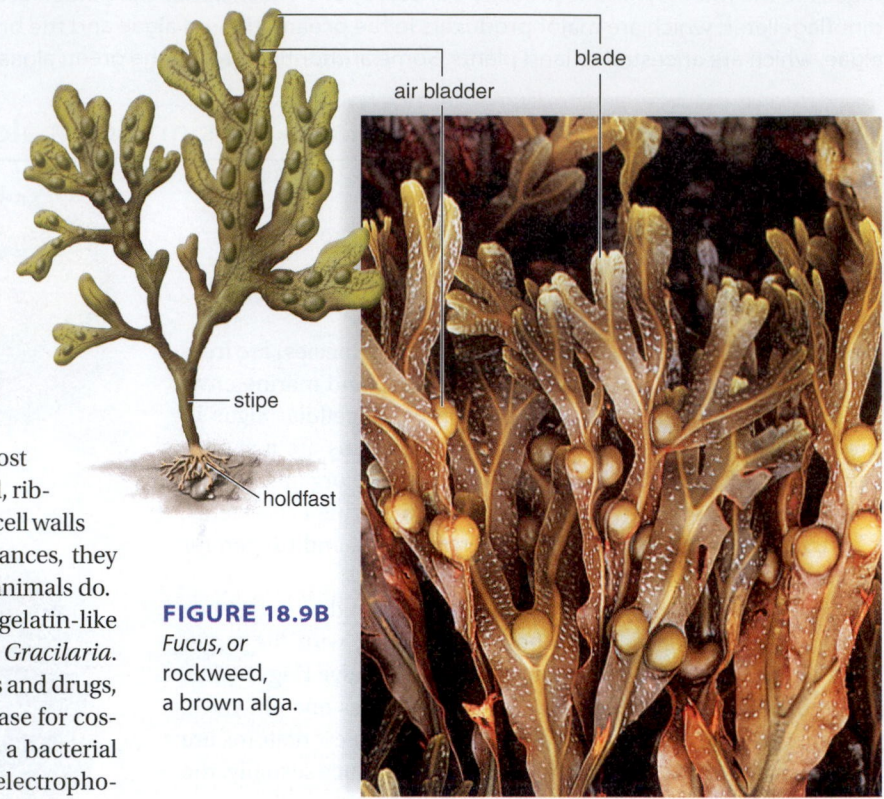

FIGURE 18.9B *Fucus,* or rockweed, a brown alga.

(Fig. 18.9B). The brown algae have chlorophylls *a* and *c* in their chloroplasts and a type of carotenoid pigment (fucoxanthin) that gives them their characteristic color. Reserve food is stored as a carbohydrate called *laminarin*.

Laminaria, commonly called kelp, are brown algae that, like *Fucus,* grow along the shoreline. In deeper waters, the giant kelps (*Macrocystis* and *Nereocystis*) often grow extensively in vast beds. Individuals of the genus *Sargassum* sometimes break off from their holdfasts and form floating masses. Brown algae not only provide food and habitat for marine organisms, but also are harvested for human food and for fertilizer in several parts of the world. *Macrocystis* is the source of alginate (algin), a pectinlike material that is added to ice cream, sherbet, cream cheese, and other products to give them a smooth, stable consistency.

Laminaria is unique among the protists because members of this genus show tissue differentiation—for example, they transport organic nutrients by way of a tissue that resembles the phloem in land plants. Most brown algae have the same life cycle as plants, but some species of *Fucus* are unique in that meiosis produces gametes, and the adult is always diploid, as in animals (see Fig. 9A on page 162).

The multicellular forms of green, red, and brown algae are called **seaweeds,** a common term for any large, complex alga. Brown algae are often observed along the rocky coasts in the north temperate zone, where they are pounded by waves as the tide comes in and are exposed to dry air as the tide goes out. They dry out slowly, however, because their cell walls contain a mucilaginous, water-retaining material.

FIGURE 18.9A *Chondrus crispus,* a red alga.

▶ **18.9 CHECK YOUR PROGRESS**

1. Explain how rockweed (*Fucus*) got its name. What are its air bladders for?

18.10 Green algae are ancestral to plants

When you complete this section, you should be able to

1. Contrast the anatomy of five types of green algae, and describe how they reproduce.

Some biologists classify green algae as plants because they have chlorophyll *a* and *b*, store excess carbohydrates as starch, and have cellulose in their cell walls. However, the green algae do not develop from an embryo protected by the organism, as do plants adapted to living on land.

The **green algae** include approximately 7,500 species. Although green algae contain chlorophyll, they are not always green; some possess pigments that give them an orange, red, or rust color. They inhabit a variety of environments, including oceans, freshwater environments, snowbanks, the bark of trees, and the backs of turtles. The green algae also form symbiotic relationships with fungi and animals. As discussed in section 19.8, they associate with fungi in lichens. The majority of green algae are unicellular; however, filamentous and colonial forms exist. Some multicellular green algae are seaweeds that resemble lettuce leaves.

Chlamydomonas An actively moving unicellular green alga called *Chlamydomonas* inhabits still, freshwater pools. Its fossil ancestors date back over a billion years. It has a definite cell wall and a single, large, cup-shaped chloroplast that contains a *pyrenoid,* a dense body where starch is synthesized. In many species, a bright red eyespot, or stigma, exists on the chloroplast, which is sensitive to light and helps bring the organism into the light, where photosynthesis can occur. Two long, whiplike flagella project from the anterior end of this alga and operate with a breaststroke motion.

Chlamydomonas has a life cycle typical of protists, which is called the **haploid life cycle** because all the stages are haploid except for the zygote and the zygospore (Fig. 18.10A). During asexual reproduction, mitosis produces as many as 16 daughter cells still within the haploid parent cell wall. Each daughter cell then secretes a cell wall and acquires flagella. The daughter cells escape by secreting an enzyme that digests the parent cell wall. *Chlamydomonas* only occasionally reproduces sexually when growth conditions are unfavorable. Gametes of two different mating types come into contact and join to form the zygote. A heavy wall forms around the zygote, and it becomes a resistant zygospore that undergoes a period of dormancy. When a zygospore germinates, it produces four haploid zoospores by meiosis. Each zoospore becomes an adult *Chlamydomonas*.

Spirogyra Filamentous green algae have end-to-end chains of cells that form after cell division occurs in only one plane. In some algae, the filaments are branched, and in others they are unbranched. *Spirogyra* is an unbranched, filamentous green alga. Filamentous green algae often grow epiphytically on (not taking nutrients from) aquatic flowering plants; they also attach to rocks or other objects under water. Some filaments are suspended in the water.

Spirogyra is found in green masses on the surfaces of ponds and streams. It has ribbonlike, spiralled chloroplasts (Fig. 18.10B). During sexual reproduction, *Spirogyra* undergoes

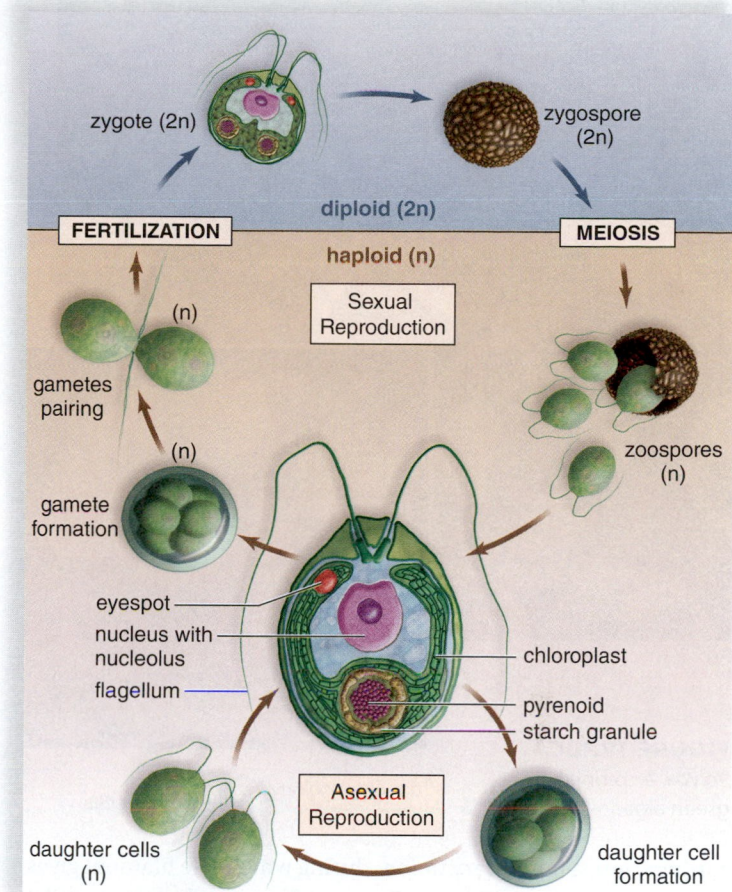

FIGURE 18.10A Reproduction in *Chlamydomonas*, a motile green alga.

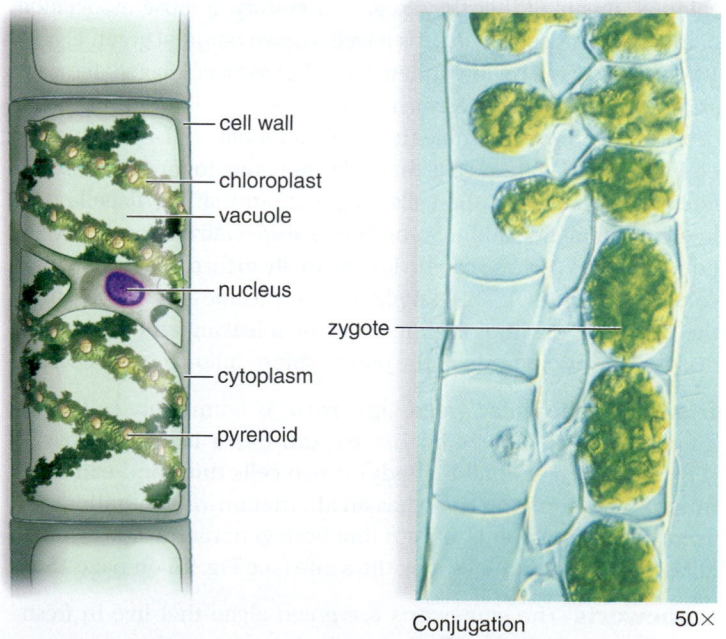

Conjugation 50×

FIGURE 18.10B Cell anatomy and conjugation in *Spirogyra*, a filamentous green alga.

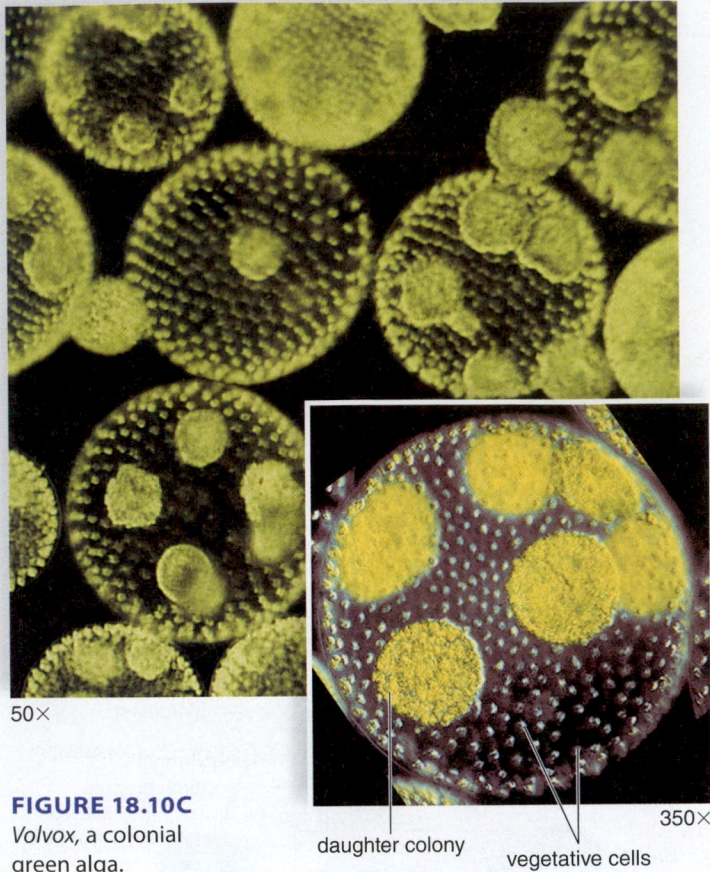

50×

FIGURE 18.10C
Volvox, a colonial
green alga.

daughter colony   vegetative cells   350×

Ulva, several individuals

One individual

FIGURE 18.10D *Ulva,* a multicellular green alga.

conjugation, a temporary union, during which one filament gives genetic material to the other. The two filaments line up parallel to each other, and the cell contents of one filament move into the cells of the other filament, forming diploid zygotes. Resistant zygospores survive the winter, and in the spring, they undergo meiosis to produce new haploid filaments.

Volvox Among the flagellated green algae, a number of forms are colonial, meaning that they exist in a **colony,** a loose association of independent cells. *Volvox* is a well-known colonial green alga. A *Volvox* colony is a hollow sphere with thousands of flagellated cells arranged in a single layer surrounding a watery interior. Each cell of a *Volvox* colony resembles a *Chlamydomonas* cell—perhaps it is derived from daughter cells that fail to separate following zoospore formation. In *Volvox,* the cells cooperate in that the flagella beat in a coordinated fashion. Some cells are specialized for reproduction, and each of these can divide asexually to form a new daughter colony (Fig. 18.10C). This daughter colony resides for a time within the parent colony, but then it leaves by releasing an enzyme that dissolves away a portion of the parent colony, allowing it to escape.

Ulva A multicellular green alga, *Ulva,* is commonly called sea lettuce because it lives in the sea and has a leafy appearance (Fig. 18.10D). The thallus (body) is two cells thick and can be as much as a meter long. *Ulva* has an alternation-of-generations life cycle like that of plants, except that both generations look exactly alike and all the gametes look the same (see Fig. 9A on page 162).

Stoneworts The stoneworts are green algae that live in freshwater lakes and ponds. They are called stoneworts because some species, such as *Chara,* are encrusted with calcium carbonate

Chara, several individuals

One individual

branch

main axis

node

FIGURE 18.10E *Chara,* a stonewort.

deposits (Fig. 18.10E). The main axis of the alga, which can be over a meter long, is a single file of very long cells. Whorls of branches occur at multicellular nodes, regions between the giant cells of the main axis. Each of the branches is also a single file of cells.

Stoneworts basically have the same life cycle as *Chlamydomonas* (see Fig. 18.10A). However, during sexual reproduction, they do produce male and female multicellular reproductive structures at the nodes. The male structure produces flagellated sperm, and the female structure produces a single egg. The diploid zygote is retained until it is enclosed by tough walls. DNA sequencing data suggest that among green algae, the stoneworts are most closely related to plants (see Fig. 18A.1).

▶ **18.10 CHECK YOUR PROGRESS**
   1. Determine how a molecular biologist would decide which terrestrial plants are most closely related to stoneworts.

THE CHAPTER IN REVIEW

CONNECTING THE CONCEPTS

While certain structures present in eukaryotic cells may have been unique to them alone, others appear to be endosymbionts, present only because they were engulfed by a much larger cell. Mutualism is a powerful force that shaped the eukaryotic cell and also shapes all sorts of relationships in the living world. For example, we have already mentioned that mutualism between flowers and their pollinators has contributed to the success of flowering plants.

The protists are a bewildering collection that may represent the first eukaryotes to have evolved, and they can be studied using different approaches. Molecular data tell us how they might be related, and medical and ecological data tell us of their importance in the world today. Both approaches are meaningful. We expect plants, animals, and fungi to have a common ancestor with some particular living protist group, and we are closer today to discovering which living protist is most closely related to which multicellular group. A green charophyte alga is most closely related to land plants; a flagellate is most closely related to animals. The ultimate goal is to construct a tree of life that shows how all organisms are related.

Recall that "Life Cycles Are Varied" on page 162 reviews the three major types of life cycles. Although asexual reproduction is common among protists, some reproduce sexually using the haploid life cycle; others use the alternation-of-generations life cycle (i.e., *Ulva*) as do plants; and still others use the diploid life cycle (i.e., *Fucus*) the same as animals. The photosynthetic algae (e.g., diatoms, dinoflagellates, and green algae) are important generators of oxygen and producers of food for the biosphere. The heterotrophic protozoans (e.g., amoebas and ciliates) participate in food chains in both fresh and marine bodies of water.

All possible forms of reproduction and nutrition are present among the protists, but each of the other eukaryotic groups specializes in a particular type of reproduction and a particular method of acquiring needed nutrients.

We will see that fungi reproduce by means of windblown spores during both asexual and sexual life cycles, and they are saprotrophic. Plants have the alternation-of-generations life cycle and are photosynthetic. Animals have the diploid life cycle and are heterotrophic. Chapter 19 pertains to the evolution of plants and fungi, while Chapter 20 discusses the evolution of animals.

ANALYZE AND EVALUATE

1. Many scientists appreciate protists for their many different ways of life and the services they provide for the biosphere. Support this approach to studying the protists.

2. Other scientists are determined to discover how the protists are related through evolution. Support this approach to studying the protists.

3. Explain why the term kingdom Protista is no longer used but the term protists is still used.

SUMMARIZE

Diversity of Protists

18.1 Eukaryotic organelles arose by endosymbiosis

- **Protists** were the first eukaryotes. Protists are mostly unicellular organisms and therefore are not considered animals, fungi, or plants. Actually, however, some protists are more closely related to animals and plants than they are to each other. The current classification of protists is subject to change in the future as more data becomes available.

- The **endosymbiotic theory** tells us that during primary endosymbiosis, aerobic bacteria were engulfed by a eukaryotic cell and became the mitochondria of protists. Some cells also engulfed cyanobacteria, and these became the chloroplasts of photosynthetic protists. During secondary endosymbiosis, some protists engulfed photosynthetic protists, and in this way they became photosynthetic.

18.2 Protists differ in many ways

- Molecular data are now used to group protists and to tell which ones are more closely related by way of common ancestors. The current classification of protists is subject to change in the future as more data becomes available.

- Protists are diverse in cellular organization, nutrition, reproduction, and locomotion. During reproduction, they may form **spores,** which are haploid, often resting cells.

- **Algae** are photosynthetic; **protozoans** are heterotrophic by ingestion or absorption, and some are parasitic. Slime molds are phagocytic, and water molds are heterotrophic by absorption.

- Protists are a part of **plankton,** suspended organisms in aquatic environments that serve as food for larger organisms.

Protozoan Protists

18.3 Protozoans called flagellates move by flagella

- **Euglenoids** are flagellates with a flexible body wall; many specimens contain chloroplasts. The group flagellates also includes the trypanosomes (causing such diseases as African sleeping sickness and leishmaniasis), *Giardia,* and *Trichomonas.*

18.4 Protozoans called amoeboids move by pseudopods

- **Pseudopods** are extensions that form when cytoplasm streams forward. When pseudopods engulf material **phagocytosis** has occurred. Examples of **amoeboids** in the **zooplankton** include *Amoeba* and *Entamoeba.*

- **Foraminiferans** and **radiolarians** have pseudopods and tests (shells) that build up on the ocean floor. A geologic upheaval produced the White Cliffs of Dover, which are composed of foraminifera tests.

18.5 Protozoans called ciliates move by cilia

- **Ciliates** are unicellular but have a complex internal structure. *Paramecium,* a well-known ciliate, is found in ponds. Paramecia have capsules called **trichocysts** that discharge barbed threads for defense or to capture prey. Paramecia reproduce asexually by mitosis with transverse fission or sexually by conjugation.

18.6 Some protozoans are not motile

■ The **apicomplexans** having an apical complex of microtubules include spore-forming and parasitic protozoans. *Plasmodium* causes **malaria,** a widespread disease in tropical countries. *Toxoplasma gondii* causes toxoplasmosis in AIDS patients and pregnant women.

Protists Called "Molds"

18.7 The diversity of protists includes slime molds and water molds

■ The slime molds and water molds are both heterotrophs but they differ in structure from each other and from fungi. **Plasmodial slime molds** exist as a plasmodium that phagocytizes decaying plant material and produces a spore-forming **sporangium** when the weather turns dry. **Cellular slime molds** exist as individual amoeboid cells. **Water molds** form furry growths on insects or fishes. Their bodies are filamentous, and they asexually produce zoospores that have hairy flagella.

Algal Protists

18.8 The diatoms and dinoflagellates are significant algae in the oceans

■ Both **diatoms** and **dinoflagellates** are significant producers of oxygen in the biosphere. Diatoms have a cell wall, and dinoflagellates have protective cellulose plates impregnated with silica.

■ Diatoms and dinoflagellates are marine producers. As part of the **phytoplankton,** they are an important source of food for heterotrophs. Dinoflagellates are responsible for a toxic bloom called the **red tide.**

18.9 Red algae and brown algae are multicellular

■ **Red algae** and **brown algae** are not closely related, but both are multicellular **seaweeds.** Red algae live in warm seawater, and coralline red algae are a significant part of coral reefs. Uses of red algae include serving as a source of agar and as a wrapping for sushi rolls. Brown algae live along northern rocky coasts; they provide food and habitat for marine organisms and are used for food and fertilizer by humans.

18.10 Green algae are ancestral to plants

■ **Green algae** photosynthesize in the same manner as green plants, have cellulose cell walls, and store carbohydrates as starch. However, they are not adapted to reproducing on land.

■ Green algae diversity is exemplified by (1) *Chlamydomonas,* flagellated and unicellular, with a **haploid life cycle;** (2) *Spirogyra,* filamentous, with a spiral chloroplast; **conjugation** produces a zygote; (3) *Volvox,* a **colony** of flagellated cells; daughter colonies develop inside the adult; (4) *Ulva,* multicellular; called sea lettuce; reproduce in the same manner as plants; and (5) stoneworts, a main axis has whorls of filamentlike branches encrusted with calcium carbonate; DNA sequencing places these green algae closest to plants.

Ulva

TEST YOURSELF

Diversity of Protists

1. Which of these sequences depicts an endosymbiotic evolutionary scenario?
   a. cyanobacteria—mitochondria
   b. Golgi—mitochondria
   c. mitochondria—cyanobacteria
   d. cyanobacteria—chloroplasts

2. Which of the following pairs is matched correctly if the first is ancestral to the second?
   a. slime mold—fungus   c. green algae—plants
   b. water mold—animals   d. protozoan—prokaryotes

3. Which of the following are photosynthetic?
   a. algae   d. protozoans
   b. slime molds   e. More than one answer is correct.
   c. water molds

4. Determining how protists evolved will allow us to better understand the origin of
   a. plants.   d. bacteria.
   b. animals.   e. a, b, and c are all correct choices.
   c. fungi.

5. **THINKING CONCEPTUALLY** Why would you predict that mitochondria contain DNA that codes for mitochondrial proteins?

Protozoan Protists

6. What structure exemplifies the difficulty of using structure to classify euglenoids?
   a. flagella   d. mitochondrion
   b. nucleus   e. c and d are both correct.
   c. chloroplast

7. Which of the following moves by means of flagella?
   a. *Paramecium*   d. Both a and b are correct.
   b. *Euglena*   e. None of the choices are correct.
   c. amoeba

8. Contractile vacuoles are found in _____ and function in _____.
   a. amoeboids, feeding
   b. amoeboids, water regulation
   c. ciliates, feeding
   d. ciliates, reproduction
   e. apicomplexans, attachment to host cells

9. Ciliates
   a. can move by pseudopods.
   b. are not as varied as other protists.
   c. have a gullet for food gathering.
   d. are closely related to the radiolarians.

10. List the four means of protozoan locomotion. Then, for each type, name an organism that uses this means of locomotion and give a unique characteristic of its group.

11. **THINKING CONCEPTUALLY** Why might you predict that among the protists, animals would be most closely related to a protozoan?

Protists Called "Molds"

12. Which characteristic of slime molds makes them like amoebas?
    a. have nonmotile spores
    b. phagocytize their food
    c. form a zygote
    d. photosynthesize
    e. All of these are correct.

13. Which type of mold is a saprotrophic protist, as are fungi?
    a. cellular slime mold
    b. plasmodial slime mold
    c. water mold
14. **THINKING CONCEPTUALLY** Give an evolutionary explanation for why water molds, slime molds, and fungi have some characteristics in common even though they are unrelated. (Hint: See page 16.)

Algal Protists

15. Which of these statements is correct about algae?
    a. They are all multicellular.
    b. They all photosynthesize.
    c. They are all about the same size.
    d. They all cause diseases.
16. Which pair is properly matched?
    a. water mold—amoeboid   c. *Plasmodium vivax*—mold
    b. trypanosome—flagellate d. amoeboid—algae
17. Which of the following statements is incorrect?
    a. Unicellular protists can be quite complex.
    b. Euglenoids are motile but have chloroplasts.
    c. Plasmodial slime molds are amoeboid but have sporangia.
    d. *Volvox* is colonial but box-shaped.
    e. Both b and d are incorrect.
18. Dinoflagellates
    a. usually reproduce sexually.
    b. have protective cellulose plates.
    c. are insignificant producers of food and oxygen.
    d. have cilia instead of flagella.
    e. tend to be larger than brown algae.

For questions 19–23, match each organism to a characteristic in the key. Answers can be used more than once, and each organism can have more than one answer.

KEY:
    a. photosynthetic   d. closely related to plants
    b. protozoan   e. closely related to animals
    c. cause disease

19. Red algae
20. Ciliates
21. Brown algae
22. Amoeboids
23. Green algae

24. Which of these is not a green alga?
    a. *Volvox*   d. *Chlamydomonas*
    b. *Fucus*   e. *Ulva*
    c. *Spirogyra*
25. Which is not a characteristic of brown algae?
    a. multicellular
    b. chlorophylls *a* and *b*
    c. live along rocky coasts
    d. harvested for commercial reasons
    e. contain a brown pigment
26. Which of these protists do not have any type of flagella?
    a. *Volvox*   d. *Chlamydomonas*
    b. *Spirogyra*   e. trypanosomes
    c. dinoflagellates

27. Which pair is not properly matched?
    a. paramecia—trichocysts
    b. euglenoids—parasitic
    c. diatoms—live in glass houses
    d. *Ulva*—multicellular
28. All of the following descriptions are true of brown algae except that they
    a. range in size from small to large.
    b. are a type of seaweed.
    c. live on land.
    d. are photosynthetic.
    e. are usually multicellular.
29. Which of the following statements is false?
    a. Slime molds and water molds are protists.
    b. Some algae have flagella.
    c. Amoeboids have pseudopods.
    d. Among protists, only green algae ever have a sexual life cycle.
    e. Conjugation occurs among some green algae.
30. In the haploid life cycle (e.g., *Chlamydomonas*),
    a. meiosis occurs following zygote formation.
    b. the adult is diploid.
    c. fertilization is delayed beyond the diploid stage.
    d. the zygote produces sperm and eggs.
31. Which alga is most closely related to plants?
    a. *Euglena*   d. diatoms
    b. *Ulva*   e. stoneworts
    c. brown algae
32. Give a reason why diatoms, dinoflagellates, red algae, and brown algae are useful, or otherwise significant, to humans.
33. **THINKING CONCEPTUALLY** Considering your study of this chapter, what do you find surprising about the organisms grouped together as stramenopiles (see Fig. 18A.1)?

GET INVOLVED

1. While studying a unicellular alga, you discover a mutant in which the daughter cells do not separate after mitosis. This gives you an idea about how filamentous algae may have evolved. Explain your idea.
2. You are an investigator trying to discover a cure for malaria. Why might you decide to target human red blood cells? What might you want to learn about the merozoite stage of infection that is not known now (see #4 in Fig. 18.6)?

MEDIA STUDY TOOLS

mhhe.com/maderconcepts3

Enhance your study of this chapter with interactive study tools, practice tests, and engaging animations. Also, ask your instructor about the resources available through ConnectPlus, which includes LearnSmart, a personalized adaptive learning program, and a media-rich eBook.

19

Evolution of Plants and Fungi

BEFORE YOU BEGIN

Take a few minutes to recall

The roles of plants and fungi in an ecosystem (Fig. 1.6)

How to read an evolutionary tree (section 1.7B)

Plant evolutionary history as revealed by fossil evidence (Table 16.1)

The general characteristics of plants and fungi (section 16.5)

Some Plants Are Carnivorous

We think of plants as largely minding their own business as they quietly photosynthesize their food. So it may come as a surprise that some plants are carnivorous—they feed on insects, or even on amphibians, birds, and mammals. Carnivorous plants are adapted to living in bogs, swamps, and marshes, where water collects, oxygen is limited, and decomposers are inhibited from recycling nutrients. These plants can survive where other plants cannot because they feed on animals, usually insects, as a source of nitrogen. We can think of carnivorous plants as a part of the great adaptive radiation of flowering plants into all sorts of environments on planet Earth. Let's look at three plant species among the 600 or so that are carnivorous.

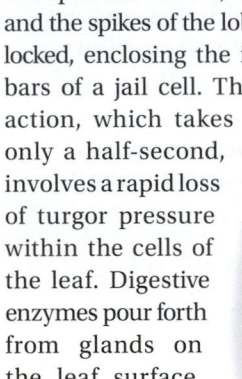

Venus flytrap with fly

The narrow green leaves of a Venus flytrap (*Dionaea muscipula*) end with two reddish lobes on either side of a midrib. The lobes are fringed by spikes and have a few isolated trigger hairs on their upper surface. An insect, most likely a fly, is lured to the leaves because the spikes are lined by a band of sweet-smelling nectar glands. When the fly touches one trigger hair twice or two hairs in rapid succession, the trap is sprung, and the spikes of the lobes become interlocked, enclosing the insect like the bars of a jail cell. This action, which takes only a half-second, involves a rapid loss of turgor pressure within the cells of the leaf. Digestive enzymes pour forth from glands on the leaf surface,

spiked fringe

insect

leaf

trigger hairs

Venus flytrap, *Dionaea muscipula*

Venus flytrap flower

breaking down the helpless victim. As a part of this remarkable adaptation, a small insect—not worth the energy to digest—walks free by just exiting between the spikes.

Sundew plants (e.g., *Drosera capensis*) are rather low-growing, so they are able to capture crawling insects as well as flying ones. The leaves are visually attractive, covered with hairs tipped with knobs that sparkle like dew in the sun. The insect gets stuck on the sticky hairs, and the knobs secrete mucuslike juices, which break down the insect. Rolling from the tip, the leaves enclose the prey, preventing it from escaping and hastening the digestive process.

Among the pitcher plants, the yellow trumpet pitcher (*Sarracenia flava*) stands over 91 cm tall, and its leaves form a pitcher. Just like a pitcher in your kitchen, this one is also filled with water—containing digestive juices, of course. The pitcher has a hood covered with glands that secrete nectar to attract insects, such as ants. Any inquisitive insect that leaves the hood to investigate the pitcher is greeted by downward-pointing hairs. And because the sides of the pitcher are slippery, the insect loses its grip, tumbling into the lethal waters.

The carnivorous plants, like most plants, are adapted to living on land. Of all things, their flowers are pollinated by insects! The flowers produce seeds within fruits. In the three species just discussed, the fruit is a dry capsule that contains rather small seeds distributed by wind. Unfortunately, today they would find few habitats that would suit their needs for growth. While our description of carnivorous plants may make you think that they are plentiful in the wild, such is not the case. They are sentinels of the health of the environment and their present day restricted distribution is a call to action that many have responded to. Carnivorous plant societies in many states tell us that the wetland habitats of carnivorous plants have been filled in for agricultural purposes and housing developments. Only a watchful society can prevent the disappearance of carnivorous plants from the wild. In the past they have even been excessively collected for sale to the general public. Many of them, however, can be propagated in greenhouses and therefore any one who wishes to grow them should be sure the plants have been commercially cultivated. A love of carnivorous plants can lead to a life dedicated to the preservation of wildlife in general.

This chapter emphasizes the evolution of plants leading to their ability to reproduce and be prevalent in so many different habitats on land. Fungi, also discussed in this chapter, have ecological, economic, and medical importance.

pitcher interior filled with bugs

hood with nectar-producing glands

pitcher with pitfall trap

bulbs release digestive enzymes

Sundew leaf enfolds prey

sticky hairs

narrow leaf form

Cape sundew plant, *Drosera capensis*

Yellow trumpet pitcher flower

Yellow trumpet pitcher plant, *Sarracenia flava*

Evolution of Plants

Land plants have to deal with the constant threat of desiccation (drying out). In order to reproduce, they must protect sperm, egg, and embryo from drying out, and they require a suitable means of dispersing their offspring on land. We will trace the evolution of land plants and explore how they manage to live on land, which presents them with many challenging problems, such as transporting water through an erect body.

19.1 Plants have a green algal ancestor

LEARNING OUTCOMES

When you complete this section, you should be able to

1. List key structural innovations that occurred during the evolution of various plant groups.
2. Compare the traits of charophytes and land plants.

Plants are multicellular, photosynthetic eukaryotes that range in size from the diminutive duckweed to the giant coastal redwoods of California. The land plants evolved from a freshwater green algal species some 500 million years ago (MYA)(Fig. 19.1A). As evidence for a green algal ancestry, scientists have known for some time that all green algae and plants (1) contain chlorophylls *a* and *b* and various accessory pigments; (2) store excess carbohydrates as starch; and (3) have cellulose in their cell walls.

Video Plants

It is now customary for molecular systematists to compare DNA and RNA base sequences between organisms. The results suggest that among the green algae, land plants are most closely related to freshwater green algae, known as **charophytes.** Fresh water, of course, exists in bodies of water on land, and natural selection would have favored those specimens best able to make the transition to the land itself. The land environment at the time was barren and represented a vast opportunity for any photosynthetic plants that were able to leave the water and take advantage of the new environment.

There are several types of charophytes—*Spirogyra,* for example, is a charophyte. But botanists tell us that among living charophytes, Charales (an order with 300 macroscopic species) and *Coleochaete* (a genus with 30 microscopic species) are most like land plants. First, let's take a look at these filamentous green algae (Fig. 19.1B). The Charales (e.g., *Chara*) are commonly known as stoneworts because some species are encrusted with calcium carbonate deposits. The body consists of a single file of very long cells anchored in mud by thin filaments. Whorls of branches occur at multicellular nodes, regions between the giant cells of the main axis. Male and female reproductive structures

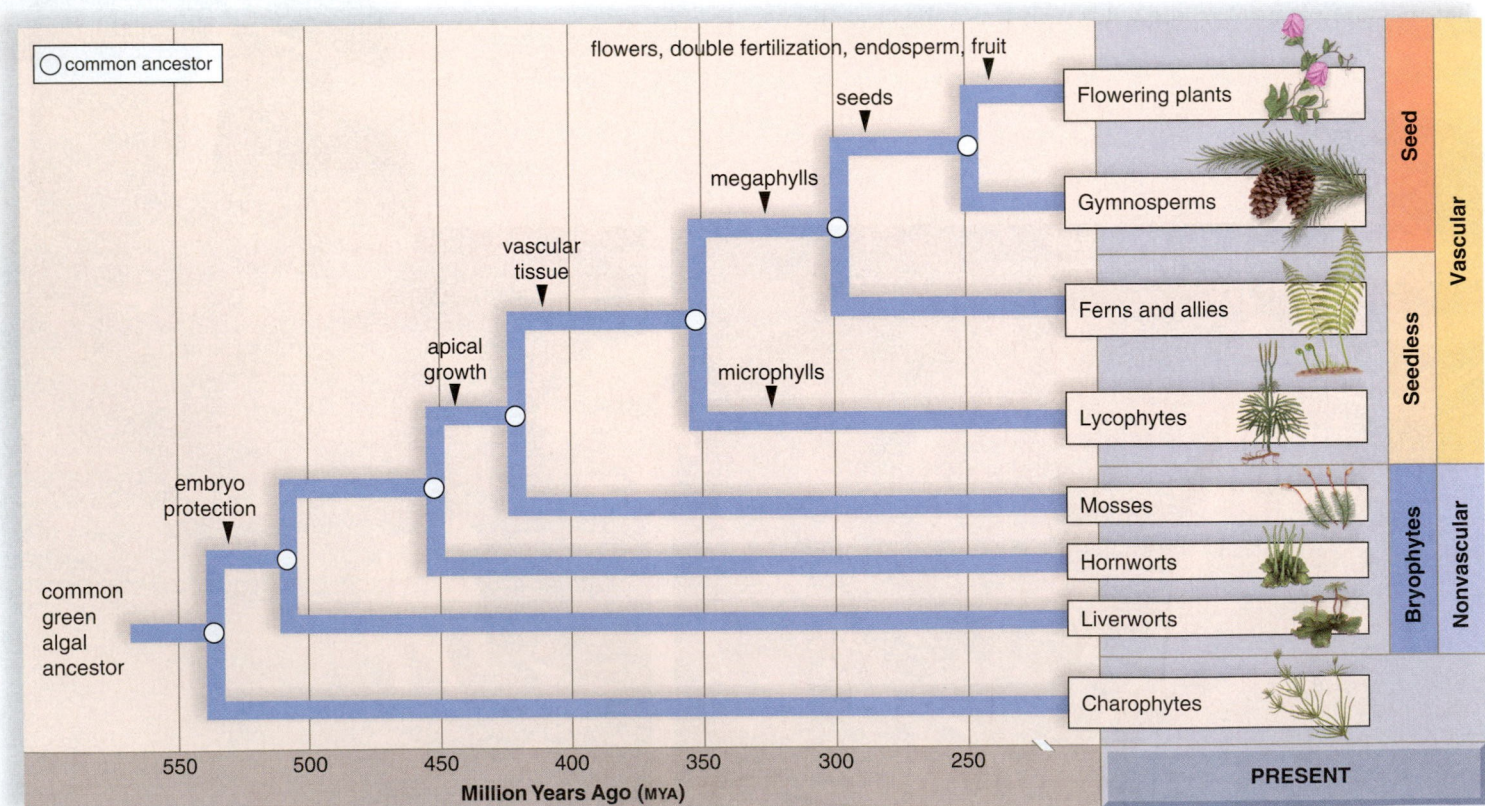

FIGURE 19.1A The evolution of plants involves these significant innovations. In particular, protection of a multicellular embryo was seen in the first plants to live on land. Vascular tissue permits the transport of water and nutrients. The evolution of the seed increased the chance of survival for the next generation.

node

Chara

Coleochaete

FIGURE 19.1B Close algal relatives of plants.

grow at the nodes. A *Coleochaete* looks like a flat pancake, but the body is actually composed of elongated branched filaments of cells that spread flat across the substrate or form a three-dimensional cushion.

These two groups of charophytes have several features that would have promoted the evolution of the complex land plants listed in the table to the right. The following features are present in charophytes and/or improved upon in land plants today:

1. The cellulose cell walls of charophytes and land plants are laid down by the same unique type of cellulose-synthesizing complexes. The mechanism of cell wall formation in charophytes is nearly identical to that of land plants.

    In land plants, a strong cell wall helps them stay erect. The leaves of land plants also have a waxy cuticle that prevents water loss.

2. The apex (top) of charophytes produce cells that allow their filaments to increase in length. At the nodes, other cells can divide asymmetrically to produce reproductive structures.

    Land plants are noted for their use of apical tissue to produce new organs, such as additional leaves, as they increase in height. Apical growth in mosses also led to the branching that is seen in plants that have vascular tissue—a means to deliver water and nutrients to cells. Large leaves (megaphylls) as in ferns provide a greater surface for the absorption of solar energy.

3. The plasmodesmata of charophytes pass through cell walls and provide a means of communication between neighboring cells.

    Land plants also have plasmodesmata, which may have played a role in the evolution of specialized tissues, but this is not known for certain.

4. Charophytes have a haploid life cycle. However, the haploid adult protects and passes nutrients to the zygote. This feature is preadaptive for living on land.

    Land plants not only protect the zygote, but they also protect the embryo from drying out. Therefore, an alternate name for the land plant clade is **embryophyta.**

DOMAIN: Eukarya
KINGDOM: Plants

CHARACTERISTICS

Multicellular, usually with specialized tissues; photosynthesizers that became adapted to living on land; most have alternation-of-generations life cycle.

Charophytes
Live in water; haploid life cycle; share certain traits with the land plants.

LAND PLANTS (embryophytes)
Alternation-of-generations life cycle; protect a multicellular sporophyte embryo; organs produce gametes; apical tissue produces complex tissues; waxy cuticle prevents water loss.

Bryophytes (hornworts, liverworts, mosses)
Low-lying, nonvascular plants that prefer moist locations: Dominant gametophyte produces flagellated sperm; unbranched, dependent sporophyte produces windblown spores.

VASCULAR PLANTS (lycophytes, ferns and their allies, seed plants)
Dominant, branched sporophyte has vascular tissue: Lignified xylem transports water, and phloem transports organic nutrients; typically has roots, stems, and leaves; and gametophyte is eventually dependent on sporophyte.

Lycophytes (club mosses)
Leaves are microphylls with a single, unbranched vein; sporangia borne on sides of leaves produce windblown spores; independent and separate gametophyte produces flagellated sperm.

Ferns and Allies (pteridophytes)
Leaves are megaphylls with branched veins; dominant sporophyte produces windblown spores in sporangia borne on leaves; and independent and separate gametophyte produces flagellated sperm.

SEED PLANTS (gymnosperms and angiosperms)
Leaves are megaphylls; dominant sporophyte produces heterospores that become dependent male and female gametophytes. Male gametophyte is pollen grain, and female gametophyte occurs within ovule, which becomes a seed.

Gymnosperms (cycads, ginkgoes, conifers, gnetophytes)
Usually large; cone-bearing; existing as trees in forests. Sporophyte bears pollen cones, which produce windblown pollen (male gametophyte), and seed cones, which produce seeds.

Angiosperms (flowering plants)
Diverse; live in all habitats. Sporophyte bears flowers, which produce pollen grains, and ovules within ovary. Following double fertilization, ovules become seeds that enclose a sporophyte embryo and endosperm (nutrient tissue). Fruit develops from ovary.

Referring to Figure 19.1A and the table above, you can see that the bryophytes were the first land plants and they protect the embryo. A dramatic change in the alternation-of-generations life cycle led to the protection of all phases of the life cycle from drying out and the formation of seeds in the seed plants.

▶ **19.1 CHECK YOUR PROGRESS**

1. List the innovations that allowed plants to adapt to a land environment.
2. Identify features of charophytes that helped give rise to land plants.

LEARNING OUTCOME

When you complete this section, you should be able to

1. Evaluate differences in the alternation-of-generations life cycle among land plants.

All plants have a life cycle that includes **alternation of generations.** In this life cycle, two multicellular individuals alternate, each producing the other (Fig. 19.2A). The two individuals are (1) a sporophyte, which represents the diploid generation, and (2) a gametophyte, which represents the haploid generation.

The **sporophyte** (2n) is named for its production of spores by meiosis. A **spore** is a haploid reproductive cell that develops into a new organism without needing to fuse with another reproductive cell. In the plant life cycle, a spore undergoes mitosis and becomes a gametophyte.

The **gametophyte** (n) is named for its production of gametes. In plants, eggs and sperm are produced by mitotic cell division. A sperm and egg fuse, forming a diploid zygote that undergoes mitosis and becomes the sporophyte.

Two observations are in order. First, meiosis produces haploid spores. This is consistent with the realization that the sporophyte is the diploid generation and spores are haploid reproductive cells. Second, mitosis occurs both as a spore becomes a gametophyte and again as a zygote becomes a sporophyte. Indeed, it is the occurrence of mitosis at these times that results in two generations.

Dominant Generation Plants differ as to which generation is dominant—that is, more conspicuous. The appearances of the gametophyte and the sporophyte in each group of plants are shown in Figure 19.2B. Only the sporophyte has vascular tissue for transporting water and nutrients, and only plants with a dominant sporophyte attain significant height.

Notice that as the sporophyte gains in dominance, the gametophyte becomes microscopic. Microscopic size allows the gametophyte to be dependent on and protected by the generation that has vascular tissue. As the gametophyte becomes smaller among vascular plants, its dependence on the sporophyte increases.

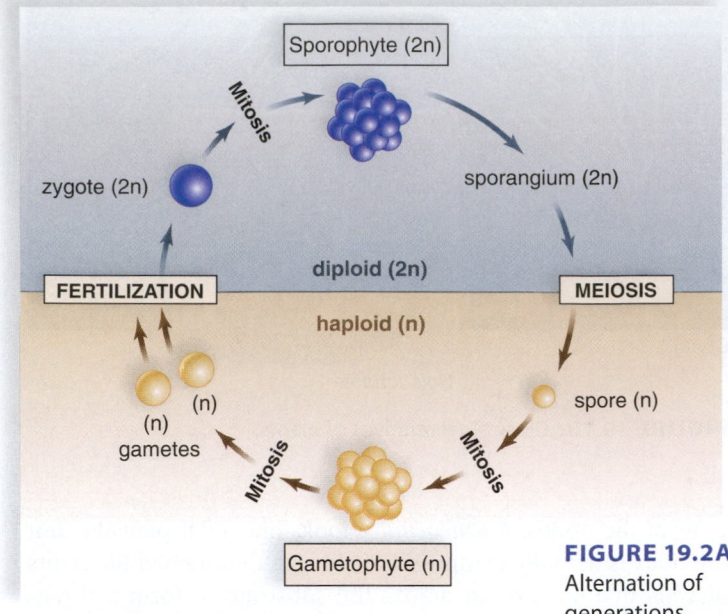

FIGURE 19.2A Alternation of generations.

Reproductive Adaptation to the Land Environment

Sporophyte dominance can be associated with increasing adaptation for reproduction in a dry, terrestrial environment. To emphasize this concept, we will contrast features of fern adaptation to those of flowering plant adaptation. Ferns are seedless vascular plants with a dominant sporophyte. In ferns:

1. The sporophyte produces spores that disperse (scatter) separate gametophytes.
2. The gametophyte is a small, heart-shaped structure that has no vascular tissue and can dry out if the environment is not moist.
3. Each **archegonium** (pl., archegonia) on the surface of a gametophyte produces an egg that is fertilized by a *flagellated sperm,* which must swim to the archegonium in a film of external water (Fig. 19.2C*a*).

FIGURE 19.2B The size of the gametophyte is progressively reduced as the sporophyte becomes more dominant.

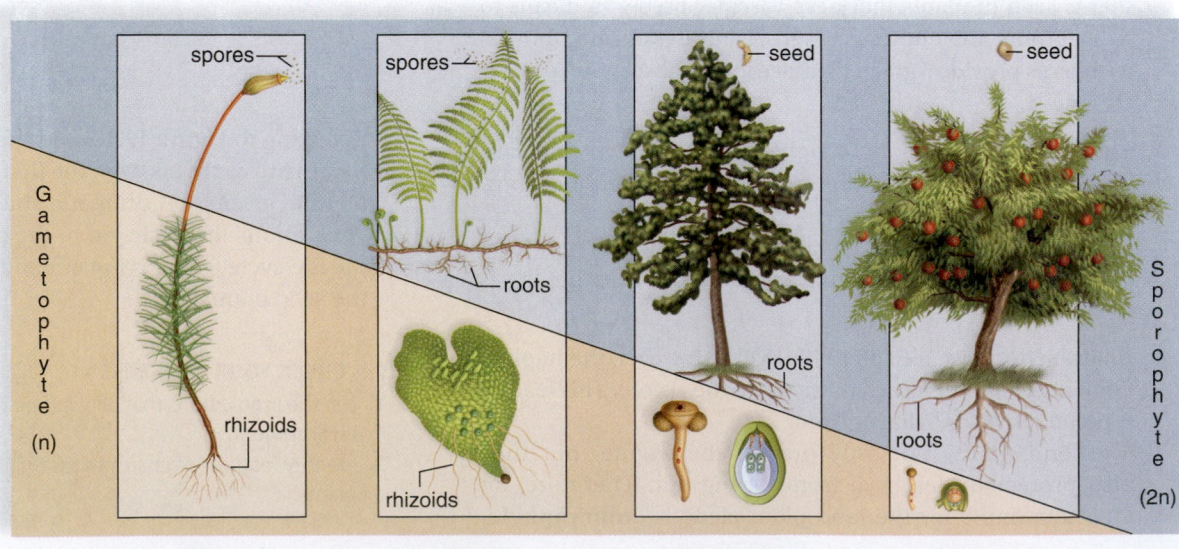

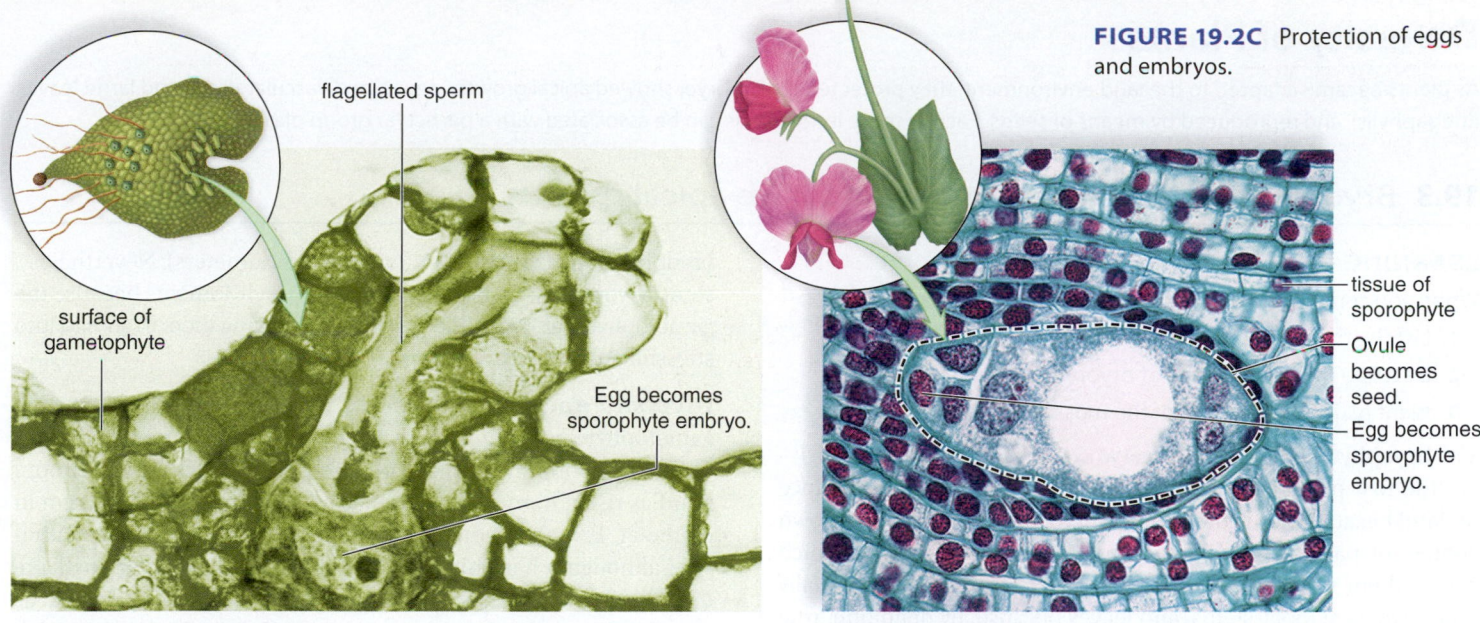

flagellated sperm

surface of gametophyte

Egg becomes sporophyte embryo.

a. Archegonium in seedless plants

FIGURE 19.2C Protection of eggs and embryos.

tissue of sporophyte

Ovule becomes seed.

Egg becomes sporophyte embryo.

b. Ovule in seed plants

The water-dependent gametophyte makes it more difficult for ferns and related plants to spread to and live in dry environments.

Flowering plants are seed plants with a dominant sporophyte. In flowering plants:

1. The sporophyte produces seeds that disperse the new sporophytes generation protected by seed coats.
2. The female gametophyte is microscopic and retained and protected within an **ovule,** a sporophyte structure located within the sporophyte tissue of a flower (Fig. 19.2C*b*).
3. The male gametophytes are **pollen grains.** Pollen grains are transported by wind, insects, or birds; therefore, they do not need external water to reach the egg. Following fertilization, the ovule becomes a seed.

In seed plants, all reproductive structures are protected from drying out in the terrestrial environment.

Other Adaptations to the Land Environment Sporophyte dominance is accompanied by adaptations for water and nutrient transport and also for preventing water loss. The sporophyte is protected against drying out in ways other than the ability to transport water. The leaves and other exposed parts of the sporophyte plant are covered by a waxy cuticle (Fig. 19.2D, *left*). The **cuticle** is relatively impermeable and provides an effective barrier to water loss, but it also limits gas exchange. Leaves and other photosynthesizing organs have little openings called **stomata** (sing., stoma) that let carbon dioxide enter while allowing oxygen and water to exit (Fig. 19.2D, *right*). A

stoma is bordered by guard cells that regulate whether it is open or closed. A stoma closes when the weather is hot and dry, and this keeps water loss to a minimum.

Having given a broad overview of this chapter, we will now discuss each group of plants in turn. Section 19.3 discusses the bryophytes.

► **19.2 CHECK YOUR PROGRESS**
1. Identify the role of each generation in the alternation-of-generations life cycle.
2. Compare the haploid generation of a fern to the haploid generation of a seed plant.

FIGURE 19.2D Leaf features of vascular plants.

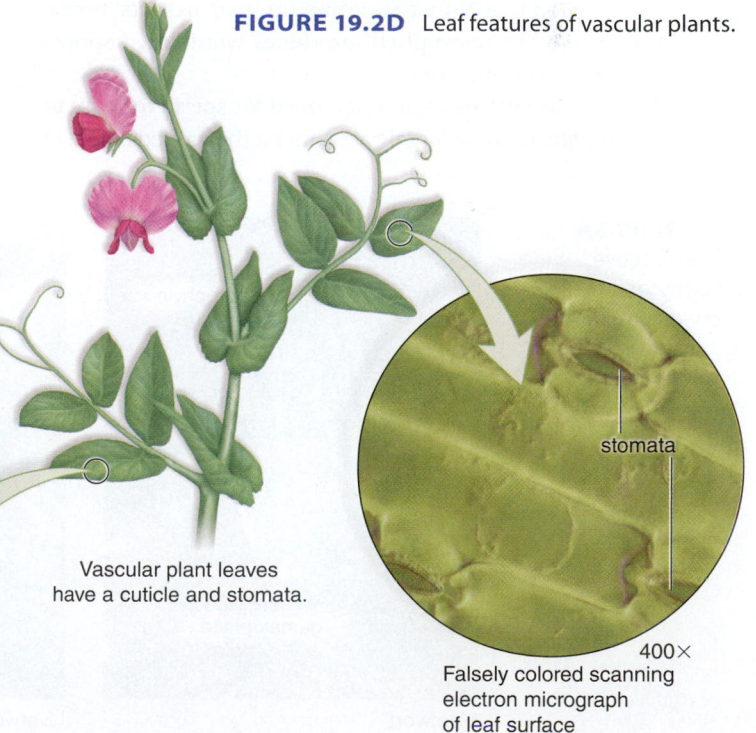

cuticle

Vascular plant leaves have a cuticle and stomata.

Stained photomicrograph of a leaf cross section

stomata

400×

Falsely colored scanning electron micrograph of leaf surface

Diversity of Plants

As plants became adapted to the land environment, they protected the embryo, showed apical growth, developed vascular tissue and large leaves (megaphylls), and reproduced by means of seeds. Each of these innovations can be associated with a particular group of plants.

19.3 Bryophytes protect the embryo and have apical growth

LEARNING OUTCOMES

When you complete this section, you should be able to

1. List the traits that help classify a plant as a bryophyte.
2. Describe the three groups of bryophytes.
3. Identify the components of the moss life cycle.

The **bryophytes**—hornworts, liverworts, and mosses (Fig. 19.3A)—are the first plants to colonize land. They successfully reproduce on land because they protect the embryo and produce wind-blown spores. In particular, mosses demonstrate apical growth, which allows them to produce complex tissues. They only superficially appear to have roots, stems, and leaves because, by definition, true roots, stems, and leaves must contain vascular tissue, which the bryophytes lack. Therefore, bryophytes are often called the **non-vascular plants.** Lacking vascular tissue, which is specialized for transporting water and organic nutrients throughout the body of a plant, most bryophytes remain low-lying, and even mosses, which do stand erect, only reach a maximum height of about 20 cm.

Bryophyte Reproduction In bryophytes, the gametophyte is the dominant generation, meaning the generation we recognize as the plant. The female gametophyte produces eggs in archegonia, and the male gametophyte produces flagellated sperm in antheridia. The sperm swim to the vicinity of the egg in a continuous film of water. Following fertilization, the zygote becomes a sporophyte embryo that is protected from drying out within the archegonium. The embryo develops into a sporophyte that is attached to, and derives its nourishment from, the photosynthetic gametophyte. This can be likened to a child that grows up in its parents' house and never leaves. The sporophyte produces windblown spores that are resistant to drying out.

The lack of vascular tissue and the need for sperm to swim to archegonia in a film of water largely account for the limited height of bryophytes (usually no taller than a few centimeters). Nevertheless, some bryophytes compete well in harsh environments because the gametophyte can reproduce asexually, allowing them to spread into stressful and even dry habitats.

Diversity and Importance of Bryophytes The **hornwort** gametophyte usually grows as a thin rosette or ribbonlike thallus. The small sporophytes of a hornwort resemble tiny, green broom handles rising from a thin gametophyte, usually less than 2 cm in diameter. Like the gametophyte, a sporophyte can photosynthesize, although it has only one chloroplast per cell. **Liverworts** are divided into two groups. One group has a flattened body, known as a thallus. The name liverwort refers to the lobes of a thallus, which to some resemble those of the liver. The majority of liverwort species are actually leafy and resemble a moss.

Mosses are the largest phyla of nonvascular plants, with over 12,000 species. There are three distinct groups of mosses: peat mosses, granite mosses, and true mosses. Although most prefer damp, shaded locations in the temperate zone, some survive in deserts, and others inhabit bogs and streams. In forests, they frequently form a mat that covers the ground and rotting logs. In dry environments, they may become shriveled, turn brown, and look completely dead. As soon as it rains, however, the plant becomes green and resumes metabolic activity.

Bryophytes are of great ecological significance. They contribute to the lush beauty of rain forests, the stability of dunes, and the conversion of mountain rocks to soil because of their ability to trap and hold moisture, retain metals in the soil, and tolerate desiccation. The cell walls of peat moss have a tremendous ability to absorb water, and so peat moss is often used in gardening. One percent of the Earth's surface is peatlands, where dead *Sphagnum,* in particular, accumulates and does not decay. This material, called peat, can be extracted and used for various purposes, such as fuel and building materials.

FIGURE 19.3A
Representative bryophytes. In liverworts, gemmae (groups of cells) can detach and start a new plant.

Hornwort

Liverwort female gametophyte

Moss gametophyte

Once scientists know which genes give *Sphagnum* the ability to resist chemical reagents and decay, as well as animal attacks, these qualities could be transferred to other plants through genetic engineering.

Life Cycle of Mosses Figure 19.3B describes the life cycle of a typical temperate-zone moss. ❶ The mature gametophyte consists of shoots that bear antheridia and archegonia. ❷ An antheridium has an outer layer of sterile cells and an inner mass of cells that become flagellated sperm. An archegonium, which looks like a vase with a long neck, has an outer layer of sterile cells with a single egg located at the base. ❸ After fertilization, the sporophyte embryo is protected from drying out because it is located within the archegonium.

❹ The mature sporophyte lacks vascular tissue and is dependent on the gametophyte. It consists of a foot (not shown) enclosed in female gametophyte tissue, a stalk, and an upper capsule (the **sporangium**), where windblown spores are produced by meiosis. In some species, the sporangium can produce as many as 50 million spores. ❺ The spores disperse the gametophyte generation. ❻ A spore germinates into an alga-like, branching filament of cells that precedes and produces the upright leafy shoots.

Having discussed the first plants on land, namely the bryophytes, we will now take a look at the lycophytes, the first plants to have vascular tissue.

▶ **19.3 CHECK YOUR PROGRESS**
1. List several key features common to all bryophytes.
2. Identify an advantage and disadvantage of the manner in which bryophytes reproduce on land.

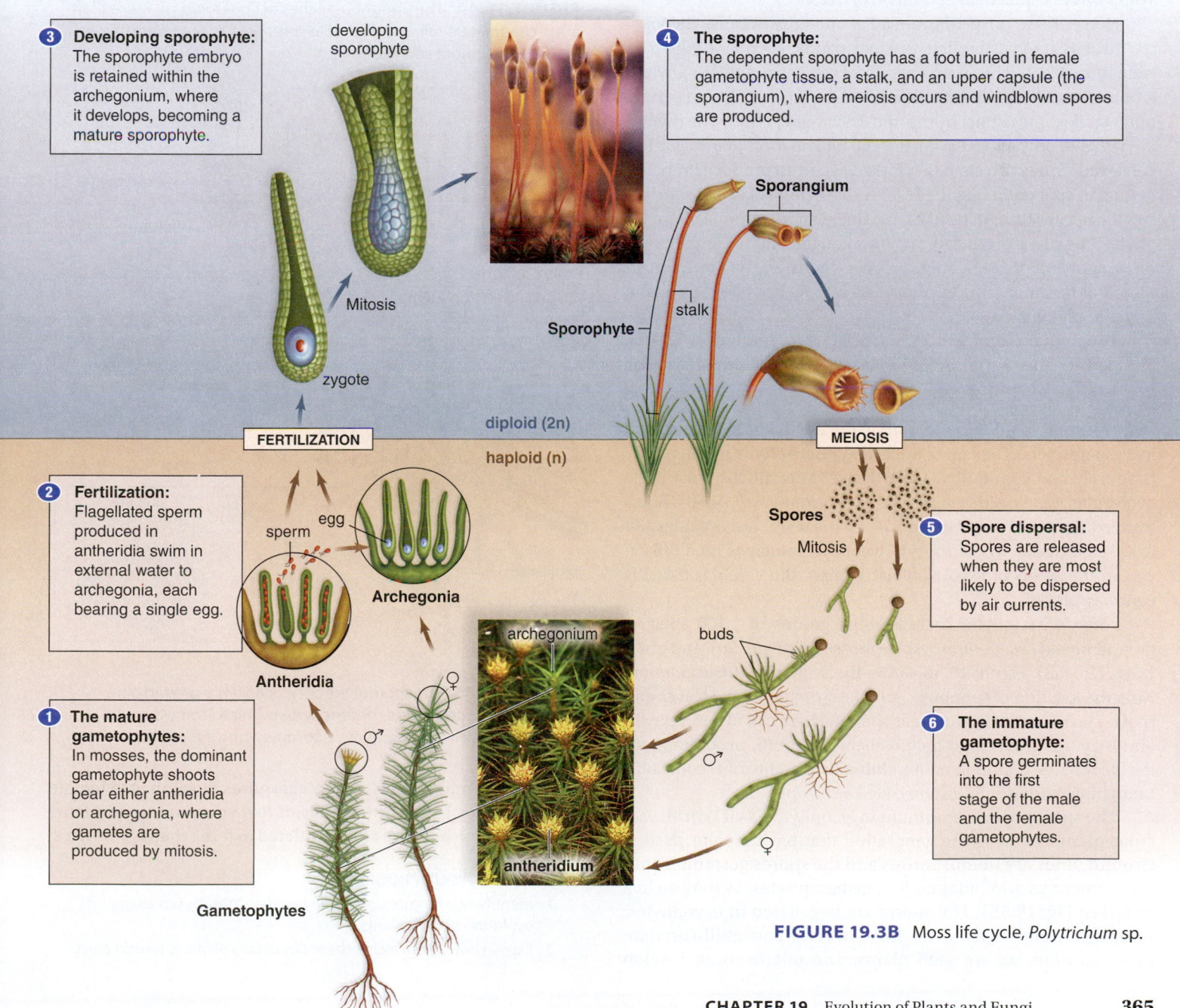

3 Developing sporophyte: The sporophyte embryo is retained within the archegonium, where it develops, becoming a mature sporophyte.

developing sporophyte

4 The sporophyte: The dependent sporophyte has a foot buried in female gametophyte tissue, a stalk, and an upper capsule (the sporangium), where meiosis occurs and windblown spores are produced.

Mitosis

zygote

Sporangium

Sporophyte — stalk

FERTILIZATION

diploid (2n)

haploid (n)

MEIOSIS

2 Fertilization: Flagellated sperm produced in antheridia swim in external water to archegonia, each bearing a single egg.

sperm egg

Archegonia

Antheridia

Spores

Mitosis

5 Spore dispersal: Spores are released when they are most likely to be dispersed by air currents.

buds

archegonium

antheridium

1 The mature gametophytes: In mosses, the dominant gametophyte shoots bear either antheridia or archegonia, where gametes are produced by mitosis.

6 The immature gametophyte: A spore germinates into the first stage of the male and the female gametophytes.

Gametophytes

FIGURE 19.3B Moss life cycle, *Polytrichum* sp.

19.4 Lycophytes have vascular tissue for transport

Today, **vascular plants** dominate the natural landscape in nearly all terrestrial habitats. Trees are vascular plants that achieve great height because they have roots that absorb water from the soil and a vascular tissue called **xylem** that transports water through the stem to the leaves. (Another conducting tissue, called **phloem,** transports nutrients in a plant.) Further, the cell walls of the conducting cells in xylem contain **lignin,** a material that strengthens plant cell walls; therefore, the evolution of xylem was essential to the evolution of trees.

Animation
Vascular Tissue

However, we are getting ahead of our story because the fossil record tells us that the first vascular plants, such as *Cooksonia*, were more like bushes than trees. *Cooksonia* was a rhyniophyte, a group of vascular plants that flourished during the Silurian period, but then became extinct by the mid-Devonian period. The rhyniophytes were only about 6.5 cm tall and had no roots or leaves. They consisted simply of a stem that forked evenly to produce branches ending in sporangia (Fig. 19.4A). The branching of *Cooksonia* was significant because, instead of the single sporangium as in bryophytes, the plant produced many sporangia, and therefore many more spores. For branching to occur, meristem has to be positioned at the apex (tip) of stems and also its branches, as it is in vascular plants today.

The sporangia of *Cooksonia* produced windblown spores, and because it was not a seed plant, it was a **seedless vascular plant,** as are lycophytes and ferns.

Lycophytes In addition to a stem, the first **lycophytes** also had leaves and roots. The leaves are called **microphylls** because they have only one strand of vascular tissue. Microphylls most likely evolved as simple side extensions of the stem (Fig. 19.4B). Roots evolved simply as lower extensions of the stem; the organization of vascular tissue in the roots of lycophytes today is much like it was in the stems of fossil vascular plants—the vascular tissue is centrally placed.

Today's lycophytes include three groups of 1,150 species: ground pines (*Lycopodium*), spike mosses (*Selaginella*), and quillworts (*Isoetes*). Figure 19.4B shows the structure of *Lycopodium*. Note the location of sporangia, and the structure of the leaves and roots. The microphylls that bear sporangia are called sporophylls, and they are grouped into club-shaped **strobili,** accounting for the lycophytes' common name, **club mosses.** The roots come off a branching, underground stem called a rhizome.

The sporophyte is dominant in lycophytes, as it is in all vascular plants. (This is the generation that has vascular tissue!) Ground pines are homosporous and the spores germinate into inconspicuous and independent gametophytes, as they do in a fern (see Fig. 19.5E). The sperm are flagellated in bryophytes, lycophytes, and ferns. (The spike mosses and quillworts are heterosporous, as are seed plants, and microspores develop

FIGURE 19.4A The upright branches of *Cooksonia*, no more than a few centimeters tall, terminated in sporangia as seen here in the drawing and photo of a fossil.

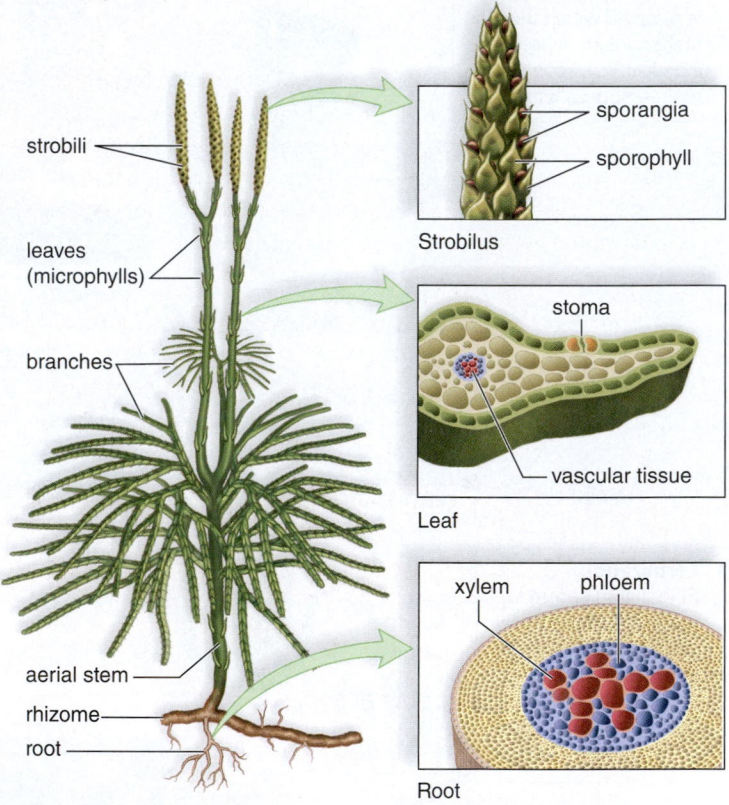

FIGURE 19.4B The sporophyte of *Lycopodium*, a ground pine, develops an underground rhizome system. This rhizome is an undergound stem that produces true roots along its length.

into male gametophytes, and megaspores develop into female gametophytes.) Yet on the basis of leaf structure, seed plants are believed to be more closely related to ferns than lycophytes.

19.5 Ferns have large leaves called megaphylls

Ferns and their allies—horsetails, and whisk ferns—are seedless vascular plants. However, like the seed plants, which are studied in section 19.6, ferns have megaphylls. **Megaphylls** are broad leaves with several strands of vascular tissue. Figure 19.5A shows the difference between microphylls and megaphylls, and how megaphylls could have evolved.

Megaphylls, which evolved about 370 MYA, allow plants to efficiently collect solar energy, leading to the production of more food and possibly more offspring than plants without megaphylls. Therefore, the evolution of megaphylls made plants more fit. Recall that fitness, in an evolutionary sense, is judged by the number of living offspring an organism produces in relationship to others of its own kind.

Ferns and their allies were dominant from the late Devonian period through the Carboniferous period as were the lycophytes. Today, the lycophytes are quite small, but some of the extinct relatives of today's club mosses were 35 m tall and dominated the Carboniferous swamps. The horsetails, at 18 m, and ancient tree ferns, at 8 m, also contributed significantly to the great swamp forests of the time (see "Carboniferous Forests Became the Coal We Use Today" on page 372).

Horsetails Today, **horsetails** consist of one genus, *Equisetum,* and approximately 15 species of distinct seedless vascular plants. Most horsetails inhabit wet, marshy environments around the globe. About 300 MYA, horsetails were dominant plants and grew as large as modern trees. Today, horsetails have a rhizome that produces hollow, ribbed aerial stems and reaches a height of 1.3 m

FIGURE 19.5B Horsetail (*Equisetum*) has whorls of branches at the nodes of the stem. Spore-producing sporangia are borne in strobili.

(Fig. 19.5B). The whorls of slender, green side branches at regular intervals along the stem make the plant bear a resemblance to a horse's tail. The leaves may have been megaphylls at one time, but now they are reduced and encircle the nodes. Many horsetails have strobili at the tips of all stems; others send up special buff-colored stems that bear the strobili. The spores germinate into inconspicuous and independent gametophytes.

FIGURE 19.5A a. Microphylls have a single strand of vascular tissue, which explains why they are quite narrow. In contrast, megaphylls have several branches of vascular tissue and are more broad. **b.** These steps show the manner in which megaphylls may have evolved. All vascular plants, except lycophytes, bear megaphylls, which can gather more sunlight and produce more organic food than microphylls.

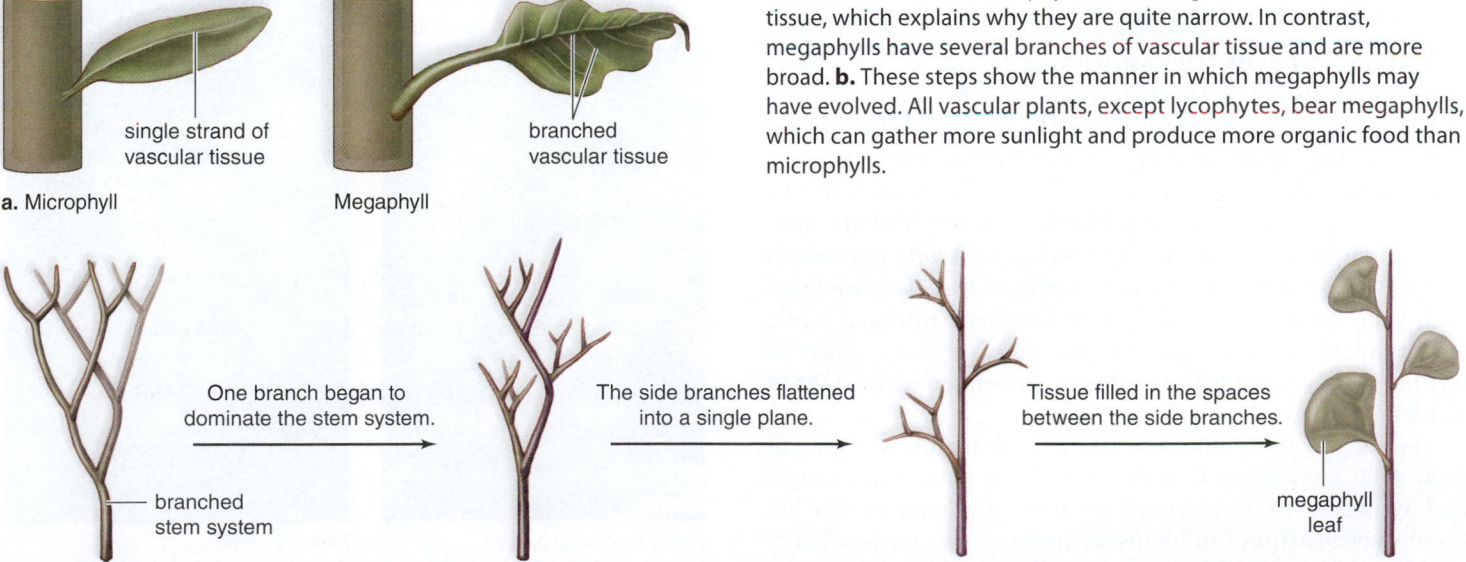

The stem and branches are tough and rigid because of silica deposited in the cell walls. Early Americans, in particular, used horsetails for scouring pots and called them "scouring rushes." Today, they are still used as ingredients in a few abrasive powders.

Whisk Ferns The **whisk ferns** are represented by the genera *Psilotum* and *Tmesipteris.* Both genera live in southern climates as epiphytes (plants that live in/on trees), or they can also be found on the ground. The two *Psilotum* species resemble a whisk broom (Fig. 19.5C) because they have no leaves. A horizontal rhizome gives rise to an aerial stem that repeatedly forks. The sporangia are borne on short side branches. The two to five species of *Tmesipteris* have appendages that some botanists maintain are reduced megaphylls.

Ferns The **ferns** include approximately 11,000 species. Ferns are most abundant in warm, moist, tropical regions, but they can also be found in temperate regions and as far north as the Arctic Circle. Several species live in dry, rocky places, and others have adapted to an aquatic life. Ferns range in size from minute aquatic species less than 1 cm in diameter to giant tropical tree ferns that exceed 20 m in height.

The megaphylls of ferns, called **fronds,** can be divided into leaflets (Fig. 19.5D). The sporangia of a fern can be located on the underside of the leaflets or on fertile fronds, as in the cinnamon fern. The windblown spores disperse the gametophyte, the generation that lacks vascular tissue. The royal fern has fronds that stand about 1.8 m tall; those of the hart's tongue fern are straplike and leathery; and those of the maidenhair fern are broad, with subdivided leaflets (Fig. 19.5D). In nearly all ferns, the leaves first appear in a curled-up form called a fiddlehead, which unrolls as it grows.

The two generations of a fern are considered separate and independent of one another. The separate heart-shaped gametophyte produces flagellated sperm that swim in a film of water from the antheridium to the egg within the archegonium, where fertilization occurs. Eventually, the gametophyte disappears and the sporophyte is independent. This can be likened to a child who grows up in its parents' house, and then goes out on its own. Many ferns can reproduce asexually by fragmentation of the fern rhizome.

The Uses of Ferns The ostrich fern (*Matteuccia truthiopteris*) is the only edible fern to be traded as a food, and it comes to the table in North America as "fiddleheads." In tropical regions of Asia, Africa, and the western Pacific, dozens of types of ferns are taken from the wild and used as food, but they are not traded commercially. The fern *Azolla* harbors *Anabaena,* a nitrogen-fixing cyanobacteria, and *Azolla* is grown in rice paddies, where it fertilizes rice plants. It is estimated that each year, *Azolla* converts more atmospheric nitrogen into a form available for plant growth than all the legumes. And like legumes, its use avoids the problems associated with applying artificial fertilizer.

Ferns and their allies are used as medicines in China to treat such conditions as boils and ulcers, whooping cough, and dysentery. Extracts from ferns have also been used to kill insects because they inhibit insect molting. Also, ferns beautify gardens, and horticulturists may use them in floral arrangements. Vases, small boxes and baskets, and also jewelry are

FIGURE 19.5C The whisk fern *Psilotum* has no leaves—the branches carry on photosynthesis. The sporangia are yellow.

Cinnamon fern, *Osmunda cinnamomea*

Hart's tongue fern, *Campyloneurum scolopendrium*

Maidenhair fern, *Adiantum pedatum*

FIGURE 19.5D Diversity of fern fronds.

made from the trunk of tree ferns. Their black and very hard vascular tissue provides many interesting patterns.

Life Cycle of Ferns The life cycle of a typical temperate-zone fern is shown in Figure 19.5E. **1** The dominant sporophyte produces windblown spores by meiosis within **2** sporangia. A cluster of sporangia forms a sorus (pl., sori) on the underside of the leaflets. **3** The windblown spores disperse **4** the gametophyte, the generation that lacks vascular tissue.

5 The separate heart-shaped gametophyte produces flagellated sperm that swim in a film of water from the antheridium to the egg within the archegonium, where fertilization occurs. **6** The sporophyte embryo is protected within the archegonium, where it gradually develops into a mature sporophyte. As stated, the gametophyte disappears, and the sporophyte is independent.

▶ **19.5 CHECK YOUR PROGRESS**
1. Identify two ways in which the fern life cycle is dependent on external water.
2. Compare key differences between the life cycle of a fern and a moss.

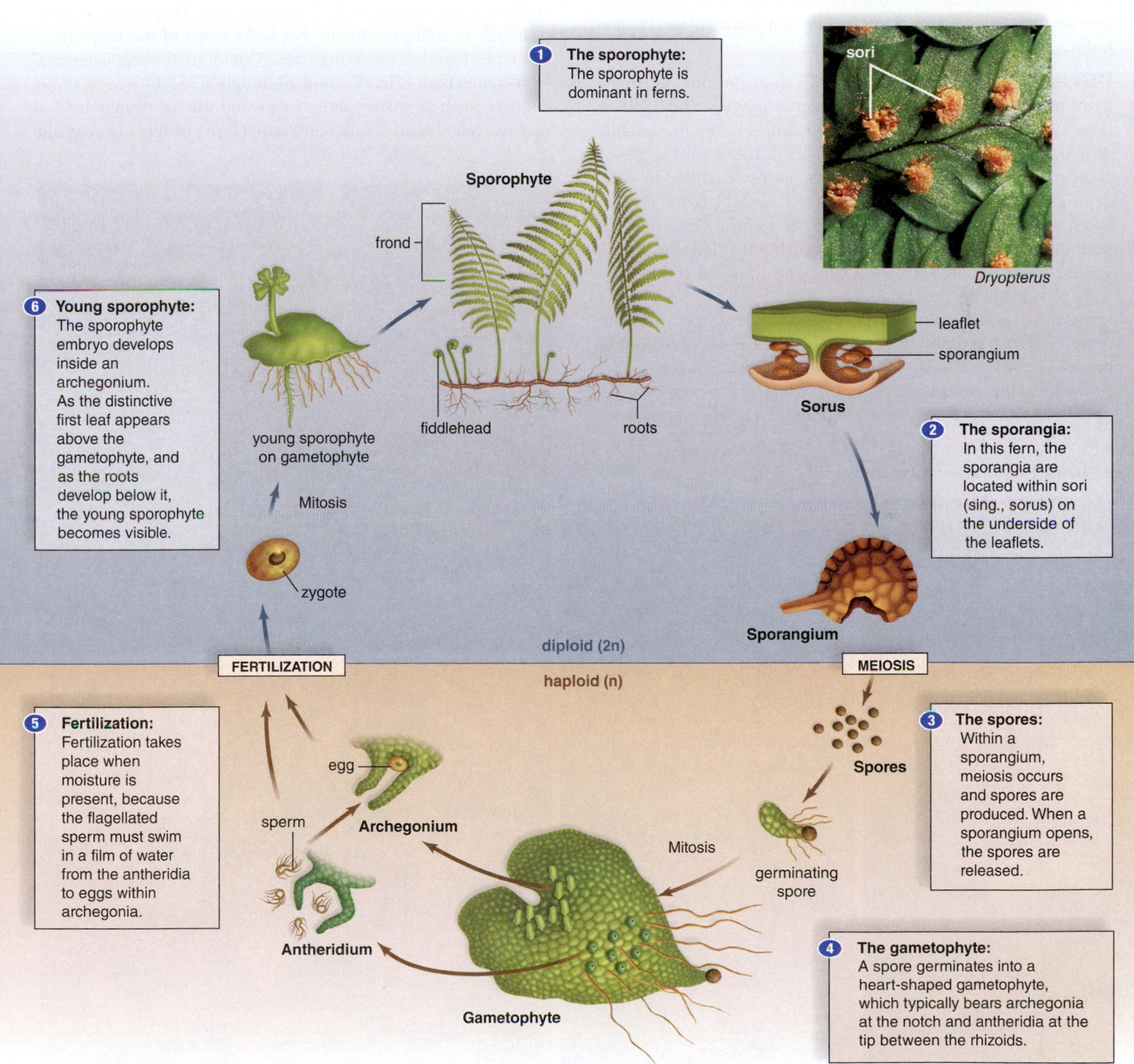

FIGURE 19.5E Fern life cycle.

19.6 Most gymnosperms bear cones on which the seeds are "naked"

LEARNING OUTCOMES

When you complete this section, you should be able to

1. Describe the four groups of living conifers.
2. Identify the components of the pine life cycle.

The evolution of the seed was the next significant innovation in the evolution of plants. Gymnosperms and angiosperms are **seed plants.** The **seed** contains a sporophyte generation, along with stored food, within a protective seed coat. The ability of seeds to survive harsh conditions until the environment is again favorable for growth largely accounts for the dominance of seed plants today.

Diversity of Gymnosperms The four groups of living **gymnosperms** are **cycads, ginkgoes, gnetophytes,** and **conifers** (Fig. 19.6A). All of these plants have ovules and subsequently develop seeds that are exposed on the surface of cone scales or analogous structures. (Since the seeds are not enclosed by fruit, gymnosperms are said to have "naked seeds.") Early gymnosperms were present in the swamp forests of the Carboniferous period, and they became dominant during the Triassic period. Today, living gymnosperms are classified into 780 species, the most plentiful being the conifers.

Conifers consist of about 125 species of trees, many of them evergreens such as pines, spruces, firs, cedars, hemlocks, redwoods, cypresses, yews, and junipers. The name *conifer* signifies plants that bear cones, but other gymnosperm phyla are also cone-bearing. Vast areas of northern temperate regions are covered in evergreen coniferous forests. The tough, needlelike leaves of pines conserve water because they have a thick cuticle and recessed stomata.

The coastal redwood (*Sequoia sempervirens*), a conifer native to northwestern California and southwestern Oregon, is the tallest living vascular plant; it may attain 115 m in height. Another conifer, the bristlecone pine (*Pinus longaeva*) of the White Mountains of California, is the oldest living individual tree (as opposed to clonal tree). One specimen is 4,900 years of age.

Economic Value of Conifers The wood of pines and other conifers is used extensively in construction. This wood consists primarily of transport tissue that lacks some of the more rigid cell types found in flowering trees. Therefore, it is considered a "soft" rather than a "hard" wood. Although called softwoods, some conifers, such as yellow pine, have wood that is actually harder than so-called hardwoods. The foundations of the 130-year-old

FIGURE 19.6A Gymnosperm diversity.

Cycad, *Encephalartos humilis*
Female plant with large seed cones

Ginkgo, *Ginkgo biloba*
Female maidenhair tree with seeds

Gnetophyte, *Ephedra*
Branched shrub with scalelike leaves

Conifer, *Picea*
Spruce tree with pollen cones and seed cones

Brooklyn Bridge are made of southern yellow pine. Resin, produced naturally by pines to prevent insect and fungal invasion, is harvested commercially for a derived product called turpentine.

Life Cycle of Pines Figure 19.6B shows the life cycle of a typical conifer. ❶ Pine trees have two types of cones and produce two types of spores, an innovation by seed plants. This innovation leads to the production of pollen grains and seeds. ❷ A megaspore mother cell within an ovule produces four megaspores by meiosis. Only one of these becomes a microscopic and dependent **female gametophyte.** Microspore mother cells produce microspores by meiosis, and they become the **male gametophytes,** which are windblown pollen grains. ❸ During **pollination,** pollen grains are transported by wind to female gametophytes, and ❹ the sperm they contain fertilize the eggs of the female gametophytes. *Note that no external water is needed to accomplish fertilization in a seed plant.*

❺ The sporophyte embryo is enclosed within the ovule, which becomes a "naked" seed on the scale of the seed cone. Blown by the wind, the winged seeds disperse the sporophyte, the generation that has vascular tissue.

Observe that in the life cycle of seed plants, the ovule holds in turn the megaspore mother cell, surviving megaspore, haploid female gametophyte, diploid zygote, sporophyte embryo, and finally develops into a seed. The early gymnosperms were dominant and enjoyed great height during the Carboniferous period, as discussed in "Carboniferous Forests Became the Coal We Use Today" on page 372.

▶ **19.6 CHECK YOUR PROGRESS**

  1. Explain how the pine life cycle is independent of external water.
  2. Compare key differences between the life cycle of a pine and a fern.

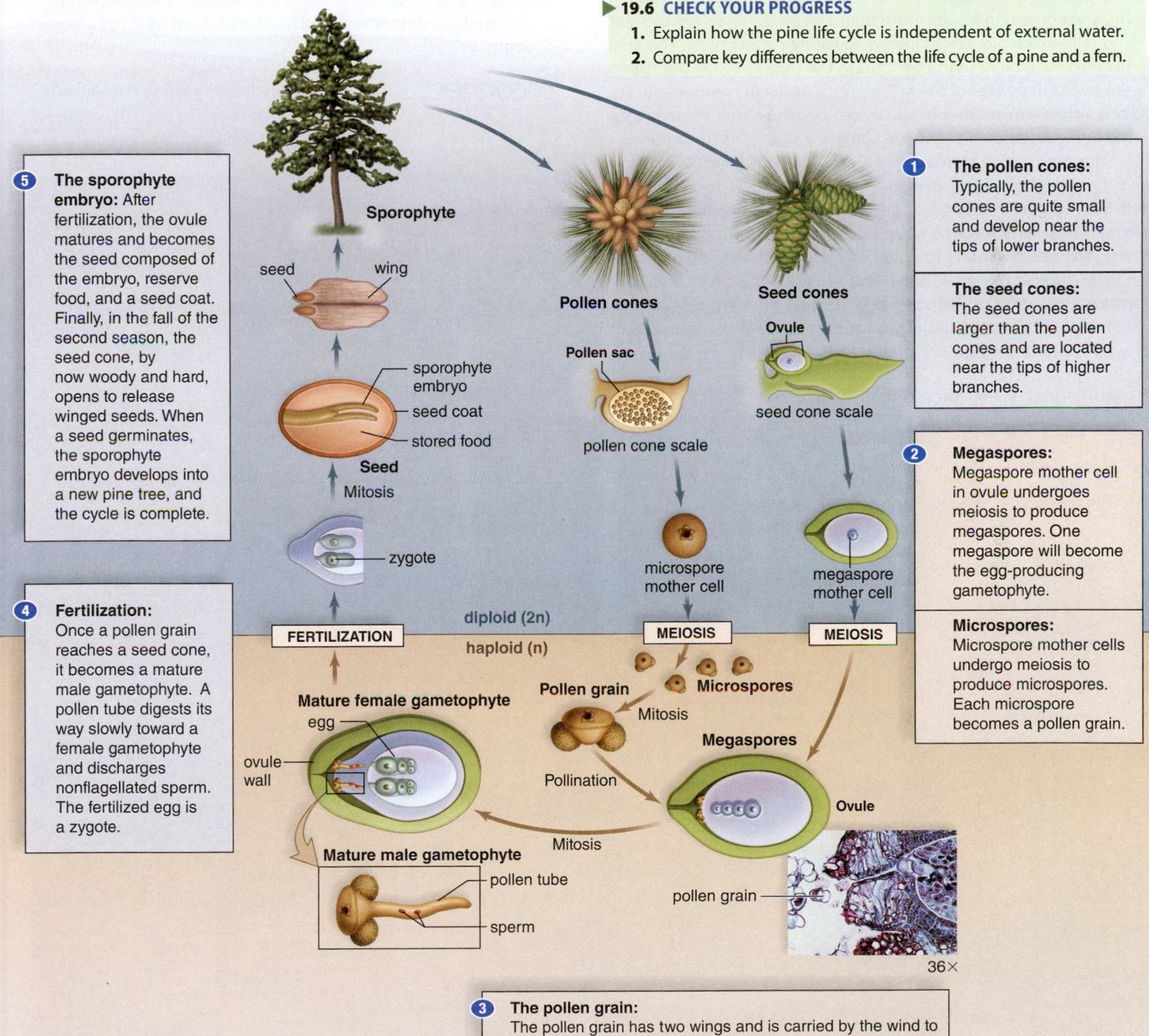

5 **The sporophyte embryo:** After fertilization, the ovule matures and becomes the seed composed of the embryo, reserve food, and a seed coat. Finally, in the fall of the second season, the seed cone, by now woody and hard, opens to release winged seeds. When a seed germinates, the sporophyte embryo develops into a new pine tree, and the cycle is complete.

4 **Fertilization:** Once a pollen grain reaches a seed cone, it becomes a mature male gametophyte. A pollen tube digests its way slowly toward a female gametophyte and discharges nonflagellated sperm. The fertilized egg is a zygote.

1 **The pollen cones:** Typically, the pollen cones are quite small and develop near the tips of lower branches.

The seed cones: The seed cones are larger than the pollen cones and are located near the tips of higher branches.

2 **Megaspores:** Megaspore mother cell in ovule undergoes meiosis to produce megaspores. One megaspore will become the egg-producing gametophyte.

Microspores: Microspore mother cells undergo meiosis to produce microspores. Each microspore becomes a pollen grain.

3 **The pollen grain:** The pollen grain has two wings and is carried by the wind to the seed cone during pollination.

Sporophyte
seed — wing
sporophyte embryo
seed coat
stored food
Seed
Mitosis
zygote
FERTILIZATION
diploid (2n)
haploid (n)
Mature female gametophyte
egg
ovule wall
Mature male gametophyte
pollen tube
sperm

Pollen cones
Pollen sac
pollen cone scale
microspore mother cell
MEIOSIS
Microspores
Pollen grain
Mitosis
Pollination

Seed cones
Ovule
seed cone scale
megaspore mother cell
MEIOSIS
Megaspores
Mitosis
Ovule
pollen grain
36×

FIGURE 19.6B Pine life cycle.

19A Carboniferous Forests Became the Coal We Use Today

Our industrial society runs on fossil fuels, such as **coal.** The term fossil fuel might seem odd at first, until you realize that it refers to the remains of organic material from ancient times. During the Carboniferous period, more than 300 MYA, a great swamp forest (Fig. 19A) encompassed what is now northern Europe, the Ukraine, and the Appalachian Mountains in the United States. The weather was warm and humid, and the trees grew very tall. These were not the trees we know today; instead, they were related to today's seedless vascular plants: the club mosses, horsetails, and ferns! Club mosses today may stand as high as 30 cm, but their ancient relatives were 35 m tall and 1 m wide. The spore-bearing cones were up to 30 cm long, and some had leaves more than 1 m long. Horsetails too—at 18 m tall—were giants compared to today's specimens. The tree ferns were also taller than tree ferns in the tropics today, and there were two other types of trees: seed ferns and early gymnosperms. "Seed fern" is a misnomer because we now know that these plants, which only resemble ferns, were actually a type of gymnosperm.

The amount of biomass in a Carboniferous forest was enormous, and occasionally the swampy water rose and the trees fell. Submerged trees do not decompose well, and their partially decayed remains became covered by sediment that sometimes changed to sedimentary rock. Sedimentary rock applied pressure, and the organic material then became coal, a fossil fuel. This process continued for millions of years, resulting in immense deposits of coal. Subsequent geologic upheavals raised the deposits to the level where they can be mined today.

FIGURE 19A Swamp forest of the Carboniferous period.

With a change of climate, the trees of the Carboniferous period became extinct, and only their much smaller relatives survived to our time. Without these ancient forests, our life today would be far different because coal helped bring about our industrialized society.

CONSIDER THESE QUESTIONS

1. Just as children are dependent on the hard work of their parents, modern-day humans are dependent on the organisms that preceded them. For example, how are we dependent on the plants present in an ancient swamp forest?
2. Would you predict that the great swamp forests occurred when the continents drifted and coalesced near the equator? Explain.

 Explore the concepts through a variety of multimedia assets and question types.

www.mcgrawhillconnect.com

Fossil seed fern

club mosses

horsetail

seed fern

early gymnosperm

fern

19.7 Angiosperms have flowers in which the seeds are "covered"

LEARNING OUTCOMES

When you complete this section, you should be able to

1. Relate the evolution of the flower to other unique features of angiosperms.
2. List and describe the parts of a flower.
3. Identify the components of the angiosperm life cycle.

Angiosperms are the flowering plants. Evidence suggests that angiosperms evolved in the mesozoic some 200 MYA. The gymnosperms were abundant at that time but declined during the mass extinction at the end of the Cretaceous period. Angiosperms survived and underwent an early adaptive radiation during the Tertiary period. They have gone on to become the dominant plants of modern times. The exact ancestral past of angiosperms has remained a mystery. To find the plant in existence today that might be most closely related to the first angiosperms, botanists have turned to DNA comparisons. Gene-sequencing data singled out *Amborella trichopoda* as having ancestral traits. Its flowers are only about 4–8 mm in diameter, and the petals and sepals look the same; therefore, they are called tepals. Plants bear either male or female flowers, with a variable number of stamens or carpels.

Amborella trichopoda

The flower is an innovation of angiosperms and so, too, is the fruit. Whereas the flower attracts a specific pollinator prior to seed formation, the fruit serves as a means by which animals disperse seeds, as we shall discuss in Chapter 25. Angiosperms are an exceptionally large and successful group of plants, with 250,000 known species—six times the number of all other plant groups combined. Angiosperms live in all sorts of habitats, from fresh water to deserts, and from the frigid north to the torrid tropics. It would be impossible to exaggerate the importance of angiosperms in our everyday lives. Angiosperms include all the hardwood trees of temperate deciduous forests and all the broadleaved evergreen trees of tropical forests. Also, all herbaceous (nonwoody) plants, such as grasses and most garden plants, are flowering plants. This means that all the fruits, vegetables, nuts, herbs, and grains that are the staples of the human diet are angiosperms. As discussed in "Flowering Plants Provide Many Services" on pages 376–377 angiosperms provide us with clothing, food, medicines, and many other commercially valuable products.

The flowering plants are called angiosperms because their ovules, unlike those of gymnosperms, are always enclosed within sporophyte tissues. In the Greek derivation of their name, *angio* (vessel) refers to the ovary, which develops into a fruit, a unique angiosperm product that contains the seeds.

Angiosperm Diversity Most flowering plants belong to one of two groups: monocotyledons, often shortened to simply the **monocots** (about 65,000 species), and eudicotyledons, shortened to **eudicots** (about 175,000 species). Monocots and eudicots are named for their number of **cotyledons.** Cotyledons are seed leaves

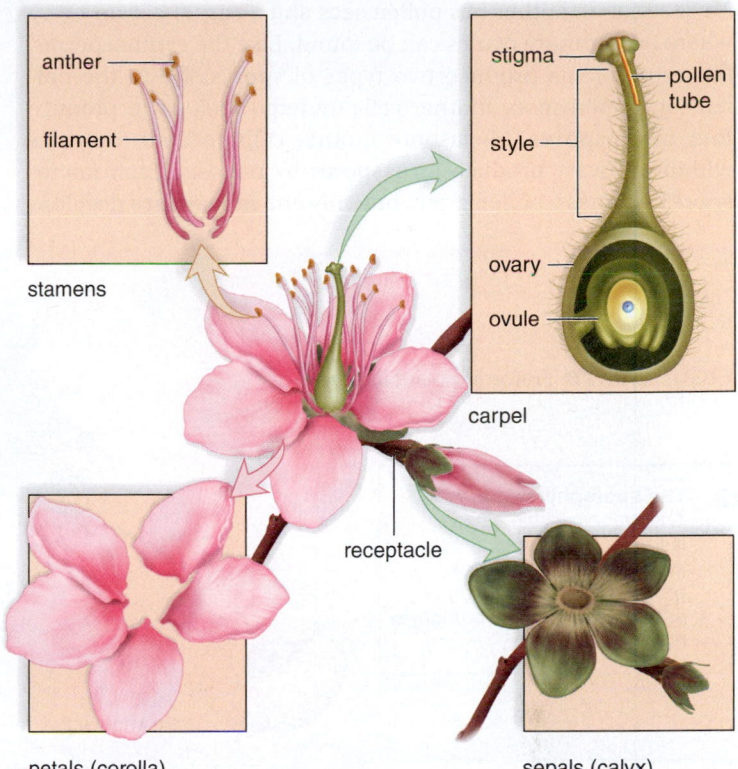

stamens

petals (corolla)

carpel

receptacle

sepals (calyx)

FIGURE 19.7A A generalized flower.

that contain nutrients and nourish the plant embryo. Monocots have only one cotyledon in their seeds. Common monocots include corn, tulips, pineapple, bamboo, and sugarcane. Monocot flower parts, such as petals, occur in threes or multiples of three. Eudicots possess two cotyledons in their seeds. Common eudicots include cacti, strawberries, dandelions, poplars, and beans. Eudicot flower parts occur in fours or fives, or multiples thereof. The flower shown in Figure 19.7A is a eudicot.

The Flower Although flowers vary widely in appearance, most have certain structures in common. The flower stalk expands slightly at the tip to form a **receptacle,** which bears the other flower parts in whorls (circles) (Fig. 19.7A).

1. The **sepals,** collectively called the calyx, protect the flower bud before it opens. The sepals may drop off or may be colored like the petals. Usually, however, sepals are green and remain attached to the receptacle.
2. The **petals,** collectively called the corolla, are quite diverse in size, shape, and color. The petals often attract a particular pollinator.
3. Next are the **stamens.** Each stamen consists of two parts: the anther, a saclike container, and the filament, a slender stalk. The anther contains **pollen sacs.**
4. At the very center of a flower is the **carpel,** a vaselike structure with three major regions: the **stigma,** an enlarged sticky knob; the **style,** a slender stalk; and the **ovary,** an enlarged base that encloses one or more ovules. The ovule becomes the seed, and the ovary becomes the fruit.

Figure 19.7B depicts the life cycle of a typical flowering plant. ❶ The life cycle involves the stamens and the carpel of a flower. Notice that an anther has pollen sacs and a carpel has an ovary where one or more ovules can be found. Like the gymnosperms, flowering plants produce two types of spores. ❷ In the pollen sacs, microspore mother cells undergo meiosis to produce four **microspores.** Megaspore mother cells located in ovules within an ovary produce megaspores by meiosis. Each microspore becomes a pollen grain, but only one **megaspore** develops into an egg-bearing female gametophyte called the **embryo sac.** ❸ In most angiosperms, the embryo sac has seven cells; one of these is an egg, and another contains two polar nuclei, so called because they came from opposite ends of the embryo sac.

During pollination, a pollen grain is transported by various means from the anther to the stigma of a carpel, where it germinates. The **pollen tube** carries two sperm to the female gametophyte in the ovule. ❹ During **double fertilization,** one sperm unites with an egg, forming a diploid zygote, and the other unites

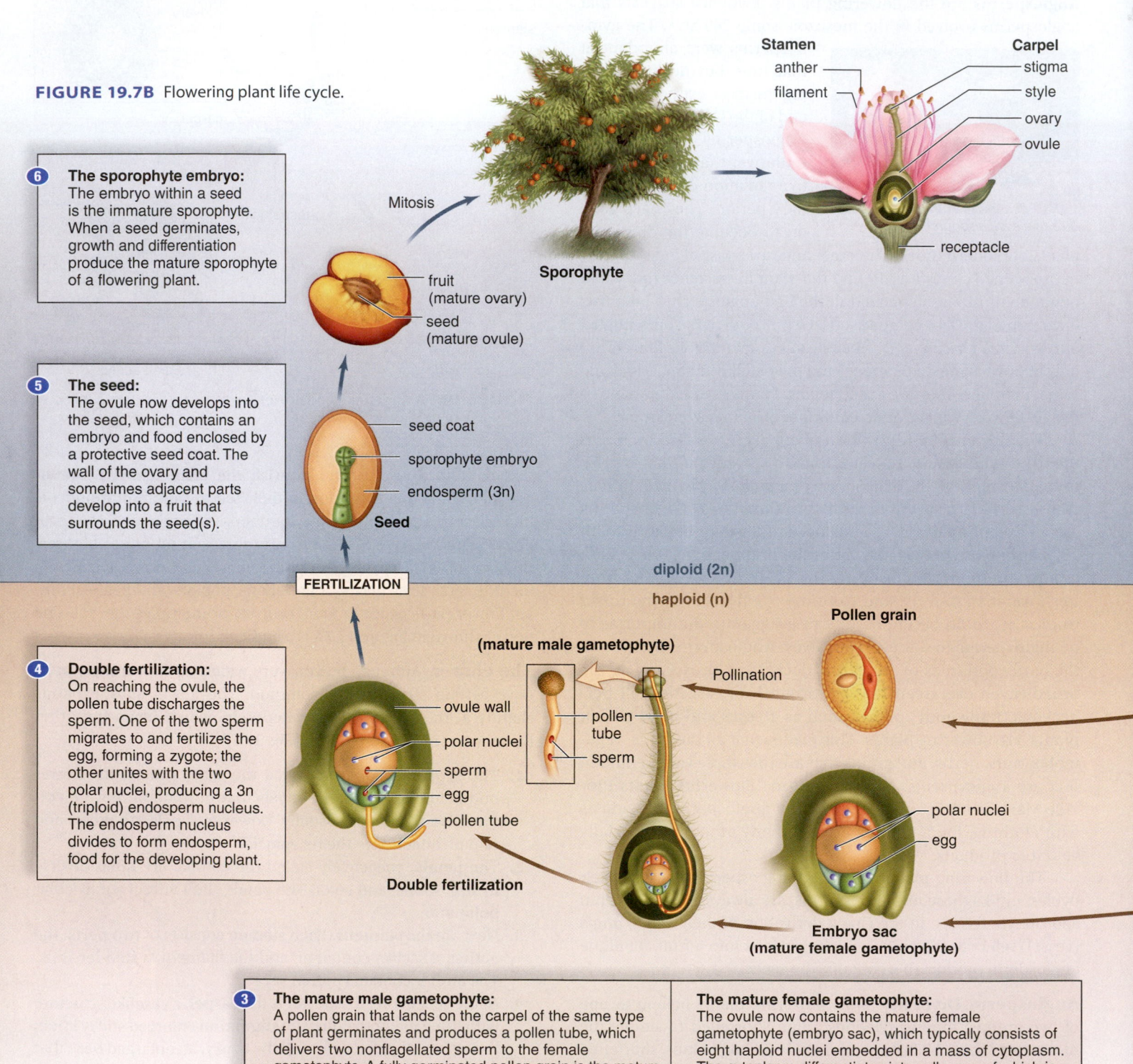

FIGURE 19.7B Flowering plant life cycle.

❻ **The sporophyte embryo:** The embryo within a seed is the immature sporophyte. When a seed germinates, growth and differentiation produce the mature sporophyte of a flowering plant.

❺ **The seed:** The ovule now develops into the seed, which contains an embryo and food enclosed by a protective seed coat. The wall of the ovary and sometimes adjacent parts develop into a fruit that surrounds the seed(s).

❹ **Double fertilization:** On reaching the ovule, the pollen tube discharges the sperm. One of the two sperm migrates to and fertilizes the egg, forming a zygote; the other unites with the two polar nuclei, producing a 3n (triploid) endosperm nucleus. The endosperm nucleus divides to form endosperm, food for the developing plant.

❸ **The mature male gametophyte:** A pollen grain that lands on the carpel of the same type of plant germinates and produces a pollen tube, which delivers two nonflagellated sperm to the female gametophyte. A fully germinated pollen grain is the mature male gametophyte.

The mature female gametophyte: The ovule now contains the mature female gametophyte (embryo sac), which typically consists of eight haploid nuclei embedded in a mass of cytoplasm. The cytoplasm differentiates into cells, one of which is an egg and another of which contains two polar nuclei.

with the polar nuclei, forming a **triploid endosperm** that will be food for the embryo. Ultimately, the ovule becomes a seed that contains the sporophyte embryo and stored food enclosed within a seed coat.

⑤ In angiosperms, seeds are covered by a **fruit,** which is derived from an ovary and possibly accessory parts of the flower. Some fruits, such as apples and tomatoes, provide a fleshy covering for seeds, and other fruits, such as pea pods and acorns, provide a dry covering. ⑥ When a seed germinates, it becomes the mature sporophyte, a flowering plant.

Advantages of Fruits The fruits of flowers protect and aid in the dispersal of seeds. Dispersal occurs when seeds are transported by wind, gravity, water, or animals to another location. Fleshy fruits may be eaten by animals, which transport the seeds possibly some distance away before depositing them when they defecate. Because animals live in particular habitats or have particular migration patterns, they are apt to deliver the fruit-enclosed seeds to a suitable location for seed germination and development of the plant.

Advantages of Seeds Dispersal by seeds offers advantages in a dry environment because the plant that germinates has the rudiments of vascular tissue. Seeds are protected by a strong seed coat and in one experiment geranium seeds even 129 years old germinated and the plant survived. The endosperm/cotyledons of seeds provide nourishment that helps the new plant survive until it starts photosynthesizing.

Video Fruit Bat Seed Dispersal

Video Dung Seed Dispersal

Flowers and Diversification As discussed in "The 'Vice-Versa' of Animals and Plants" on page 254–255, plants and their specific pollinators, such as bees, wasps, flies, butterflies, moths, and even bats, are mutualistic. Glands located in the region of the ovary produce nectar, a nutrient that pollinators gather as they go from flower to flower. The pollinator has mouthparts that are able to obtain the nectar from the base of a particular type of flower. The pollinator disperses pollen only to this flower type.

Today, there exist some 250,000 species of flowering plants and over 700,000 species of insects, suggesting that the success of angiosperms has contributed to the success of insects, and vice versa. In recent years, the populations of bees and other pollinators have been declining worldwide. Consequently, some plants are endangered because they have lost their normal pollinator. The decline in pollinator populations has been caused by a variety of factors, including pollution, habitat loss, and emerging diseases. Although insecticides should not be applied to crops that are blooming, they frequently are. While beekeepers can quickly move their beehives, wild bees have no protection whatsoever. Then, too, widespread aerial applications to control mosquitoes, medflies, grasshoppers, gypsy moths, and other insects leave no region where wild insect pollinators can reproduce and repopulate. Too often, a "chem-lawn" philosophy advocates treating lawns with pesticides and regards dandelions and clover, favored by bees, as weeds. Others encourage farms, suburbs, and cities to provide a habitat where bees and butterflies can live. Migratory pollinators, such as monarch butterflies and some hummingbirds, also need nectar corridors in order to survive.

Animation Pollinators

▶ **19.7 CHECK YOUR PROGRESS**
1. Identify what unique angiosperm features should be associated with the flower.
2. Explain how the angiosperm life cycle involves the stamen and the carpel.

---

① **The stamen:**
An anther at the top of each stamen has four pollen sacs.

The carpel:
The ovary at the base of a carpel contains one or more ovules. The contents of an ovule change during the flowering plant life cycle.

stigma

style

Anther

Carpel — **ovule**

ovary

pollen sac

microspore mother cell

megaspore mother cell

MEIOSIS   MEIOSIS

Mitosis

Microspores   **Megaspores**

Mitosis

degenerating megaspores

Ovule

② **Microspores:**
Microspore mother cells undergo meiosis to produce microspores. Each microspore becomes a pollen grain.

Megaspores:
Megaspore mother cell inside ovule undergoes meiosis to produce megaspores. One megaspore will become the egg-producing female gametophyte.

19B Flowering Plants Provide Many Services

Plants define the features of and are the producers in most ecosystems. Humans derive most of their sustenance from three flowering plants: wheat, corn, and rice (Fig. 19B.1). All three of these plants are in the grass family and are collectively, along with other species, called grains. Most of the Earth's over 7 billion people have a simple way of life, growing their food on family plots. A virus or other disease could hit any one of these three plants and cause massive loss of life from starvation.

Wheat, corn, and rice originated and were first cultivated in different parts of the globe. Wheat is commonly used in the United States to produce flour and bread. It was first cultivated in the Middle East (Iran, Iraq, and neighboring countries) about 8000 B.C.; hence, it is thought to be one of the earliest cultivated plants. Wheat was brought to North America in 1520 by early settlers; now the United States is one of the world's largest producers of wheat.

Corn, or what is properly called maize, was first cultivated in Central America about 7,000 years ago. Maize developed from a plant called teosinte, which grows in the highlands of central Mexico. By the time Europeans were exploring Central America, over 300 varieties were already in existence—growing from Canada to Chile. We now commonly grow six major varieties of corn: sweet, pop, flour, dent, pod, and flint.

Rice originated several thousand years ago in southeastern Asia, where it grew in swamps. Today we are familiar with brown rice and white rice. Brown rice results when the seeds are threshed to remove the hulls, but the seed coat and complete embryo remain. If the seed coat and embryo are removed, leaving only the starchy endosperm, white rice results. Because the seed coat and embryo are a good source of vitamin B and fat-soluble vitamins, brown rice is the healthier choice. Today, rice is grown throughout the tropics and subtropics where water is abundant. It is also grown in some parts of the western United States by flooding diked fields with irrigation water.

Do you have an "addiction" to sugar? This simple carbohydrate comes almost exclusively from two plants—sugarcane (grown in South America, Africa, Asia, the southeastern United States, and the Caribbean) and sugar beets (grown mostly in Europe and North America). Today most processed foods are sweetened with high-fructose corn syrup made by changing corn starch into a mixture of slightly more fructose than glucose.

Many foods are bland or tasteless without spices. In the Middle Ages, wealthy Europeans spared no cost to obtain spices from the Middle and Far East. In the fifteenth and sixteenth centuries, major expeditions were launched in an attempt to find better and cheaper routes for spice importation. The explorer Christopher Columbus convinced the queen of Spain that he would find a shorter route to the Far East by traveling west by ocean rather than east by land. Columbus's idea was sound, but he encountered a little barrier, the New World. Nevertheless, this discovery later provided Europe with a wealth of new crops, including corn, potatoes, peppers, and tobacco.

Our most popular drinks—coffee, tea, and cola—also come from flowering plants. Coffee originated in Ethiopia, where it was first used (along with animal fat) during long trips for sustenance and to relieve fatigue. Coffee as a drink was not developed until the thirteenth century in Arabia and Turkey, and it did not catch on in Europe until the seventeenth century. Tea is thought to have been developed somewhere in central Asia. Its earlier uses were almost exclusively medicinal, especially among the Chinese, who still drink tea for medical reasons. The drink as we now know it was not developed until the fourth century. By the mid-seventeenth century, it had become popular in Europe. Cola is a common ingredient in tropical drinks and was used around the turn of the century, along with the drug coca (used to make cocaine), in the "original" Coca-Cola.

Rubber is another plant that has many uses today. The product was first made in Brazil from the thick, white sap (latex) of

FIGURE 19B.1 Species of grains important to humans.

Wheat plants, *Triticum*

Corn plants, *Zea*

Rice plants, *Oryza*

b. Cotton, *Gossypium*, for cloth

c. Palm leaves, *Arecaceae*, for thatched roofs

a. Rubber, *Hevea*, for auto tires

d. Tulips, *Tulipa*, for beauty

FIGURE 19B.2 Uses of plants.

the rubber tree (Fig. 19B.2*a*). Once collected, the sap is placed in a large vat, where acid is added to coagulate the latex. When the water is pressed out, the product is formed into sheets or crumbled and placed into bales. Much stronger rubber, such as that in tires, was originally made by adding sulfur and heating, a process called vulcanization; this produces a flexible material less sensitive to temperature changes. Today, though, much rubber is synthetically produced.

Before the invention of synthetic fabrics, cotton and other natural fibers were our usual source of clothing (Fig. 19B.2*b*). China is now the largest producer of cotton. The cotton fiber itself comes from filaments that grow on the seed. In sixteenth-century Europe, cotton was a little-understood fiber known only from stories brought back from Asia. Columbus and other explorers were amazed to see the elaborately woven cotton fabrics in the New World. But by 1800, Liverpool, England, was the world's center of cotton trade. Over 30 species of native cotton now grow around the world, including the United States.

Plants have been used for centuries to construct a number of important household items, including the house itself. Lumber, the major structural portion in buildings, comes mostly from a variety of conifers: pine, fir, and spruce, among others. In the tropics, trees and even nonwoody plants provide important components for houses. In rural parts of Central and South America, palm leaves are preferable to tin for roofs, because they last as long as 10 years and are quieter during a rainstorm (Fig. 19B.2*c*). In the Middle East, numerous houses along rivers are made entirely of reeds.

An actively researched area today involves medicinal plants, and about 50% of all pharmaceutical drugs have their origins from plants. The treatment of some cancers appears to rest in the discovery of new plants. Indeed, the National Cancer Institute (NCI) and most pharmaceutical companies have spent millions

of dollars to send botanists out to collect and test plant samples around the world.

Over the centuries, malaria has caused far more human deaths than any other disease. After European scientists became aware that malaria can be treated with quinine, which comes from the bark of the cinchona tree, a synthetic form of the drug, chloroquine, was developed. Widespread use of the drug in certain areas caused the malaria parasite that lives in red blood cells to become resistant to chloroquine. Fortunately, the drug mefloquine is available to prevent malaria in these areas.

Numerous plant extracts continue to be misused for their hallucinogenic or other effects on the human body: Coca is used to produce cocaine and crack, opium poppy for morphine, and wild yam for steroids.

In addition to all these uses of plants, we should not forget or neglect their aesthetic value (Fig. 19B.2*d*). Flowers brighten any yard, ornamental plants accent landscaping, and trees provide cooling shade during the summer and protect us from the winter wind. Plants also produce oxygen, which is so necessary for all animals and also for the plants themselves.

CONSIDER THESE QUESTIONS

1. Why is it unfortunate that humans are so dependent on only wheat, corn, and rice for their sustenance? How would it be possible to get people to eat more varied grains?

2. In what ways aside from food are people dependent on plants? Would it be possible for us to live without them? If so, how; if not, why not?

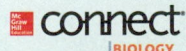 Explore the concepts through a variety of multimedia assets and question types.

www.mcgrawhillconnect.com

Evolution and Diversity of Fungi

Fungi feed on organic materials, and their structure is adaptive to their saprotrophic mode of nutrition. Their lifestyle allows land fungi to form mutualistic relationships with algae in lichens and with plant roots in mycorrhizae. Land fungi reproduce by means of windblown spores, but the appearance of the spore-bearing structure for each major group of land fungi differs.

19.8 Fungi differ from plants and animals

LEARNING OUTCOMES

When you complete this section, you should be able to

1. Describe the body of a fungus and how fungi are adapted to living on land.
2. Evaluate the benefits derived from the ecological relationships of fungi.
3. Explain how land fungi may be related to the aquatic group called the chytrids.

The **fungi,** listed in the table to the right, are diverse group of eukaryotes whose body is usually a mass of filaments called a **mycelium** (Fig. 19.8A*a*). Each of the filaments is a **hypha.** Some fungi have cross walls (called septa) that divide a hypha into a chain of cells. These hyphae are called septate (Fig. 19.8A*b*). Septa have pores that allow cytoplasm and even organelles to pass from one cell to the other along the length of the hypha. Nonseptate fungi have no cross walls, and their hyphae are multinucleated. Hyphae give the mycelium quite a large surface area per volume of cytoplasm, and this facilitates absorption of nutrients into the body of a fungus. Hyphae grow at their tips, and the mycelium absorbs and then passes nutrients on to the growing tips.

Fungal cells are quite different from plant cells, not only because they lack chloroplasts, but also because their cell wall contains chitin rather than cellulose. Chitin, like cellulose, is a polymer of glucose, but in chitin, a nitrogen-containing amino group is attached to each glucose molecule. Chitin is the major structural component of the exoskeleton of arthropods, such as insects, lobsters, and crabs. The energy reserve of fungi is not starch, but glycogen, as in animals.

Fungi are strict heterotrophs, but they release digestive enzymes into the external environment and digest their food outside the body, while animals in contrast ingest their food

and digest it internally. Most fungi are saprotrophs; they decompose the corpses of plants and animals; therefore, along with the bacteria of decay, fungi play an important role in ecosystems by returning inorganic nutrients to the food producers—that is, photosynthesizers. Fungi can degrade even cellulose and lignin in the woody parts of trees. It is common to see fungi (brown rot or white rot) on the trunks of fallen trees. The body of a fungus can become large enough to cover acres of land. In fact, an 8,500-year-old fungus covers nearly 10 km² of forest floor in northeast Oregon (and has been called "the humongous fungus among us").

Fungi form mutualistic relationships, called **mycorrhizae,** with the roots of most plants, helping them grow more successfully in dry or poor soils. The fungal hyphae greatly increase the surface area of roots for the absorption of water and nutrients, and in return the plant provides organic carbon to the fungus. Early plant fossils indicate that the relationship between fungi and plant roots is an ancient one, and therefore it may have helped plants adapt to life on dry land.

A **lichen** is a mutualistic association between a particular land fungus and a cyanobacterium or alga (see Fig. 19.8D). In a lichen, a thin, tough upper layer and a loosely packed lower layer, both formed by fungal hyphae, shield a middle layer of photosynthetic cells.

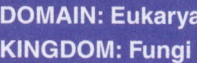

DOMAIN: Eukarya
KINGDOM: Fungi

CHARACTERISTICS
Usually multicellular without flagella; absorb food; haploid life cycle with windblown spores during sexual and asexual reproduction.

Chytrids (chytridiomycota)
Aquatic with flagellated spores and gametes

Zygospore fungi (zygomycota)
For example, terrestrial black bread molds with thick-walled zygospore and nonmotile spores produced in sporangia

Sac fungi (ascomycota)
For example, terrestrial cup fungi that produce nonmotile spores in asci; many sac fungi reproduce asexually by producing spores called conidia.

Club fungi (basidiomycota)
For example, terrestrial mushrooms that produce nonmotile spores in basidia

water mold
black bread mold
cup fungus
mushroom

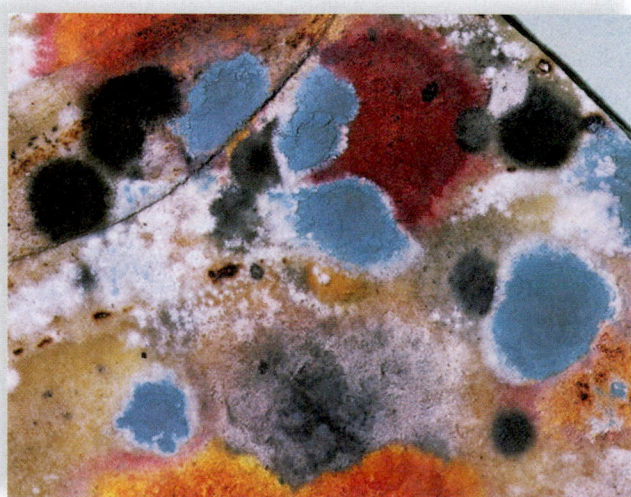

a. Fungal mycelia on a corn tortilla

nuclei
septum
cell wall
nonseptate hypha   septate hypha

b. Cell structure of hyphae

FIGURE 19.8A Fungal mycelia and hyphae.

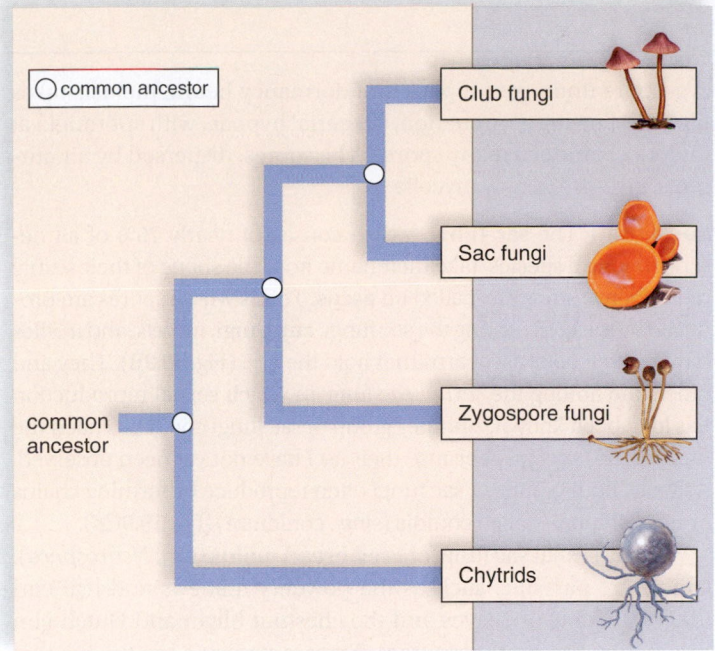

FIGURE 19.8B Evolution of fungi. The land fungi (zygospore, sac, and club fungi) are related as shown. All three types of land fungi reproduce windblown spores during sexual reproduction but they can be differentiated by the type of spore-forming structure.

The algae provide the fungus with organic food, while the fungus transfers minerals and water to the photosynthesizer and also offers protection from predation and desiccation. The organic acids given off by fungi can release minerals from rocks, and therefore lichens are ecologically important because they create new soil, allowing plants to invade the area. Lichens can reproduce asexually by releasing reproductive units that contain hyphae and an algal cell.

Groups of Fungi Means of reproduction distinguish the four different groups of fungi, which evolved as shown in Figure 19.8B. Notice that the aquatic group the **chytrids,** sometimes called water molds, are closest to the root of the evolutionary tree. Unlike the other groups of fungi, they have flagellated spores and gametes; perhaps the common ancestor for all the fungi had these characteristics before adaptation to land occurred.

Fungi adapted to living on land, called the **land fungi,** are nonmotile and do not have flagella at any stage in their life cycle. They move toward a food source by growing toward it. Hyphae can grow as much as a kilometer a day! Land fungi produce windblown spores during both asexual and sexual reproduction (Fig. 19.8C). In fungi, spores germinate into new mycelia. Among the three types of land fungi, sac fungi form the largest group and they are fungi most often found in mycorrhizae and lichens (Fig. 19.8D). Sexual reproduction in fungi involves conjugation of hyphae from two different mating types (usually designated + and –). Often, the

FIGURE 19.8C The fungus earthstar, releasing hordes of spores. Fungi release spores during both asexual reproduction and sexual production. Earthstar is a sac fungus.

haploid nuclei from the two hyphae do not immediately fuse to form a zygote. Eventually, the nuclei do fuse and the zygote undergoes meiosis in a structure unique to the particular group of fungi, as discussed in section 19.9.

Video Spore Dispersal

▶ **19.8 CHECK YOUR PROGRESS**

1. Describe the body of a fungus and how it acquires nutrients and grows.
2. Explain the ecological role of fungi and how they form mutualistic relationships with plant roots and algae.
3. Identify the possible evolutionary relationships among the various groups of fungi.

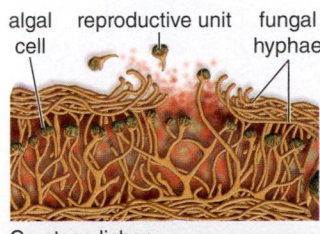

Crustose lichen

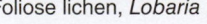

Foliose lichen, *Lobaria*

Fruticose lichen, *Cladonia*

FIGURE 19.8D Lichen structure and examples.

LEARNING OUTCOMES

When you complete this section, you should be able to

1. Describe sexual reproduction in each group of land fungi, naming the structure in which meiosis occurs.

2. Explain how zygospore fungi and sac fungi reproduce asexually.

The fungi adapted to living on land are the zygospore fungi, the sac fungi, and the club fungi.

Zygospore Fungi The zygospore fungi are mainly saprotrophs living off plant and animal remains in the soil or on bakery goods in the pantry. Some are parasites of small soil protists or worms, and even of insects, such as the housefly. The **black bread mold,** *Rhizopus stolonifer* (Fig. 19.9A), is well known to many of us. In *Rhizopus,* the hyphae are specialized: Some are horizontal and exist on the surface of the bread; others grow into the bread to anchor the mycelium and carry out digestion; and ❶ still others are stalks that bear sporangia. As in plants, a sporangium in fungi is a capsule that produces spores.

The name of this group of fungi refers to the zygospore, which is seen during sexual reproduction. When *Rhizopus* reproduces sexually, ❷ the ends of + strain and – strain hyphae join, ❸ haploid nuclei fuse, and ❹ a thick-walled **zygospore** results. The

zygospore undergoes a period of dormancy before meiosis takes place. Following germination, ❺ aerial hyphae, with sporangia at their tips, produce many spores. The spores, dispersed by air currents, give rise to new mycelia.

Sac Fungi The **sac fungi,** which consist of nearly 75% of all described fungal species, take their name from the shape of their sexual reproductive structure, called an **ascus.** This is where spores are produced by meiosis. Among the sac fungi, **cup fungi,** morels, and truffles have conspicuous ascocarps that hold the asci (Fig. 19.9B). They and others are among the *sexual sac fungi* in which sexual reproduction has long been known. Another group of sac fungi could be called the *asexual sac fungi* just because their asci have not yet been observed. Actually, both groups of sac fungi often reproduce by forming chains of asexual spores called conidia (sing., conidium) (Fig. 19.9Ca).

Many sexual sac fungi are **red bread molds** (e.g., *Neurospora*). Others are parasitic, such as the powdery mildews and leaf curl fungi that grow on leaves and the chestnut blight and Dutch elm disease that destroy these trees. Ergot, a parasitic sac fungus that infects rye, produces hallucinogenic compounds.

The term **yeast** is generally applied to the unicellular sac fungi. *Saccharomyces cerevisiae* is a budding yeast (Fig. 19.9Cb). Unequal binary fission occurs, and a small cell gets pinched off and then grows

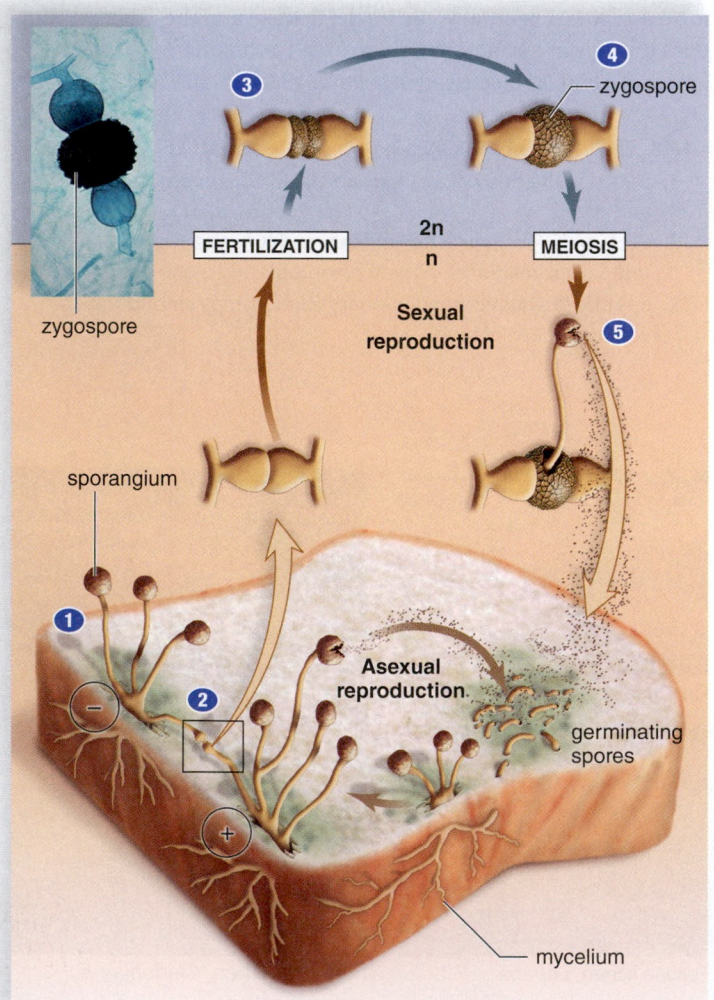

FIGURE 19.9A Black bread mold, *Rhizopus stolonifer.*

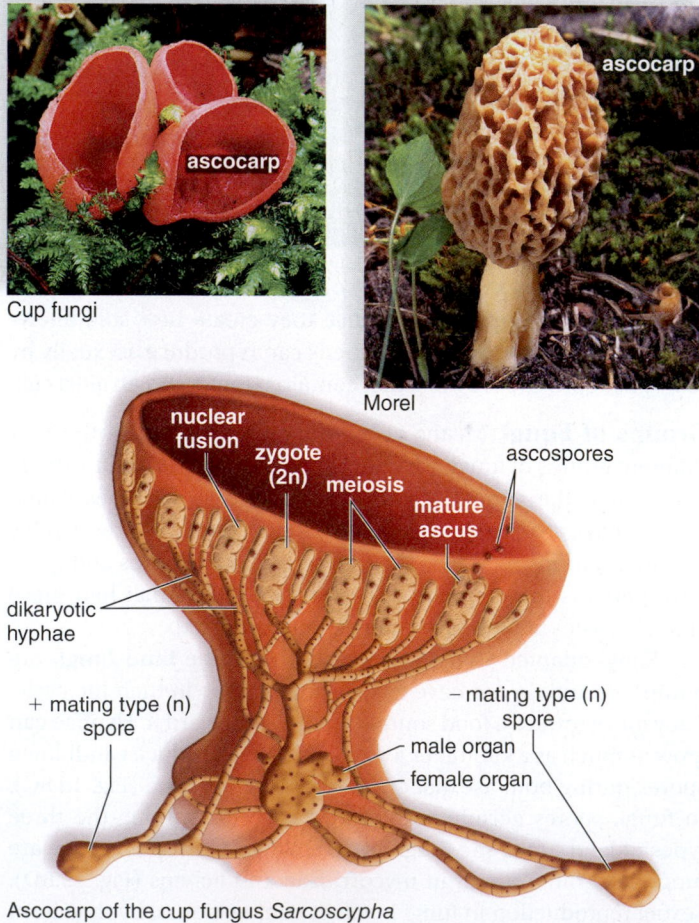

Cup fungi

Morel

Ascocarp of the cup fungus *Sarcoscypha*

FIGURE 19.9B Sexual reproduction in sac fungi.

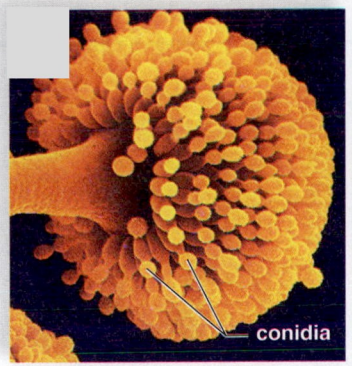

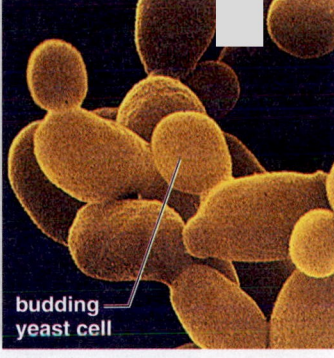

a.

b.

FIGURE 19.9C Asexual reproductive structures in sac fungi.

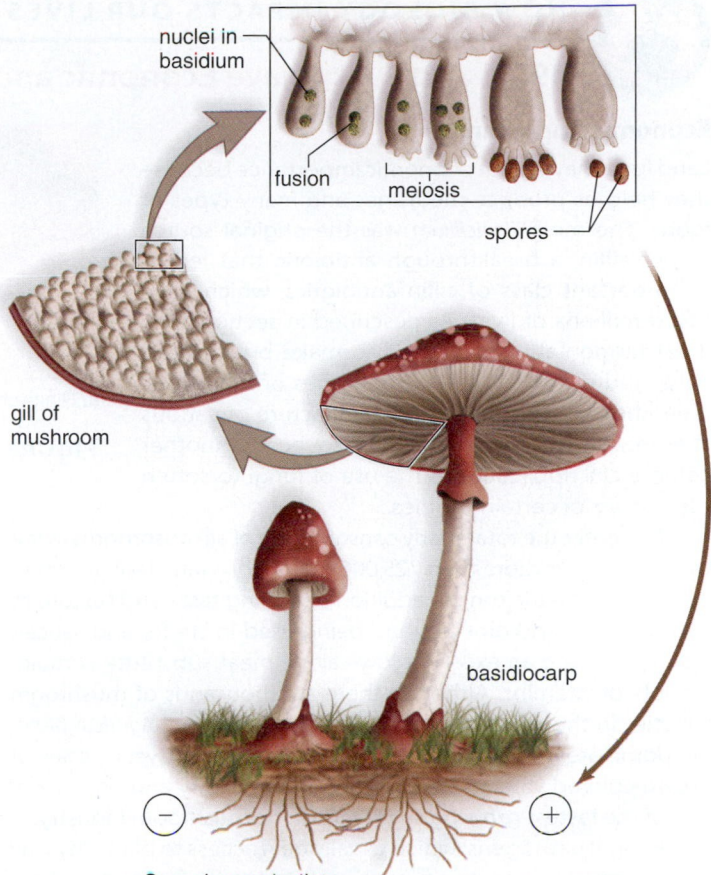

Sexual reproduction

to full size. Sexual reproduction, which occurs when the food supply runs out, results in the formation of asci and ascospores.

When some yeasts ferment, they produce ethanol and carbon dioxide. In the wild, yeasts grow on fruits, and historically, the yeasts already present on grapes were used to produce wine. Today, selected yeasts are added to relatively sterile grape juice in order to make wine. Also, yeasts are added to prepare grains to make beer. Both the ethanol and carbon dioxide are retained in beers and sparkling wines; carbon dioxide is released from other wines. In breadmaking, the carbon dioxide produced by yeasts causes the dough to rise; the gas pockets are preserved as the bread bakes.

Formerly some sac fungi were called the deuteromycota, or the imperfect fungi, because their means of sexual reproduction was unknown. However, on the basis of molecular data and structural characteristics, these fungi have now been identified as sac fungi. The asexual sac fungi include the yeast *Candida* and the molds *Aspergillus. Penicillium,* the source of penicillin was considered asexual until the asci of *Penicillium* were discovered, and therefore, it has officially been renamed *Talaromyces.*

Club Fungi The club fungi take their name from the shape of the sexual reproductive structure, called a **basidium** (pl., basidia), where spores are produced by meiosis. The basidia are located within a basidiocarp (Fig. 19.9D), which forms aboveground after – and + mycelia located in the soil fuse. **Mushrooms** sometimes develop in a ring, called a fairy ring, on the outer living fringes of a 2n mycelium located underground.

When you eat a mushroom, you are consuming a basidiocarp. Shelf or bracket fungi found on dead trees are also basidiocarps. Less well-known basidiocarps are puffballs, bird's nest fungi, and stinkhorns. In puffballs, spores are produced inside parchmentlike membranes and then released through a pore or when the membrane breaks down. Stinkhorns resemble a mushroom with a spongy stalk and a compact, slimy cap. Stinkhorns emit an incredibly disagreeable odor. Flies are attracted by the odor, and when they linger to feed on the sweet jelly, they pick up spores that they later distribute.

Although club fungi occasionally produce conidia asexually, they usually reproduce sexually by forming basidia. However, the rusts and smuts that parasitize cereal crops, such as corn, wheat, oats, and rye, don't form basidiocarps. They remain hidden within the plant until the plant matures, and then they produce numerous small spores within basidia. "Land Fungi Have Economic and Medical Importance" on page 382 emphasizes the economic and medical importance of fungi.

Mushroom

Shelf fungi

Giant puffball

FIGURE 19.9D Sexual reproduction in club fungi involves a basidiocarp of which three types are shown.

▶ **19.9 CHECK YOUR PROGRESS**

1. Contrast sexual reproduction in black bread mold with sexual reproduction in a mushroom.
2. Contrast asexual reproduction in black bread mold with asexual reproduction in a sac fungus.

19C Land Fungi Have Economic and Medical Importance

Economic Importance

Land fungi have great economic importance because they help us produce medicines and many types of foods. The mold *Penicillium* was the original source of penicillin, a breakthrough antibiotic that led to an important class of cillin antibiotics, which have saved millions of lives. As described in section 19.9, yeast fermentation is utilized to make bread, beer, wine, and distilled spirits. Other types of fungal fermentation contribute to the manufacture of various cheeses, as well as soy sauce from soybeans. Another commercial application is the use of fungi to soften the centers of certain candies.

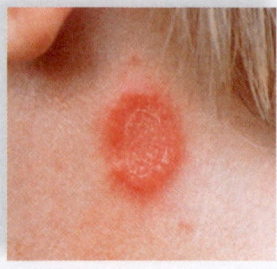
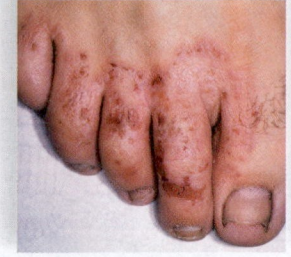
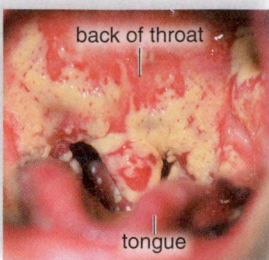

a. Ringworm **b.** Athlete's foot **c.** Thrush

FIGURE 19C.2 Human fungal diseases.

Presently, the total yearly consumption of all mushrooms in the United States is more than 725,000 metric tons (an elephant often weighs one metric ton). In addition to adding taste and texture to soups, salads, and omelets, and being used in stir-fry and sauces, mushrooms are an excellent low-calorie meat substitute containing lots of vitamins. Although there are thousands of mushroom varieties in the world, the white button mushroom, *Agaricus bisporus,* dominates the U.S. market. However, in recent years, sales of brown-colored variants have surged in popularity and have been one of the fastest-growing segments of the mushroom industry.

Fungal pathogens, which usually gain access to plants by way of the stomata or a wound, are a major concern for farmers. Serious crop losses occur each year due to fungal disease (Fig. 19C.1). As much as one-third of the world's rice crop is destroyed each year by rice blast disease. Corn smut is a major problem in the midwestern United States. Various rusts attack grains, and leaf curl is a fungal disease of fruit trees.

Medical Importance

Certain mushrooms are poisonous, and so wild mushrooms for eating should be carefully chosen. *Amanita* spp. are a deadly wild mushroom known as the death angel. Ergot, a fungus that grows on grain, can cause ergotism in a person who eats contaminated bread. Ergotism is characterized by hysteria, convulsions, and sometimes death.

Mycoses are diseases caused by fungi. Mycoses have three customary levels of invasion: Cutaneous mycoses only affect the skin; subcutaneous mycoses affect a deeper layer; and systemic mycoses spread throughout the body by traveling in the bloodstream. Fungal diseases that can be contracted from the environment include rose gardener's disease from thorns, Chicago disease from old buildings, and basketweaver's disease from grass cuttings. A new concern is *Cryptococcus gattii,* a rare form of airborne yeast that causes a lung infection after inhalation. This infection has a 25% chance of causing death! Several opportunistic fungal infections now seen in AIDS patients stem from fungi that are always present in the body but take the opportunity to cause disease when the immune system becomes weakened.

Tineas are infections of the skin caused for the most part by fungi. Ringworm is a cutaneous infection contracted from soil. The fungal colony does not penetrate the skin but grows outward, forming a ring of inflammation. The center of the lesion begins to heal, producing a characteristic red ring surrounding an area of healed skin (Fig. 19C.2*a*). Athlete's foot is a form of tinea that affects the foot, causing itching and peeling of the skin between the toes (Fig. 19C.2*b*).

Candida albicans causes the widest variety of fungal infections. Disease occurs when antibacterial treatments kill off the microflora community, allowing *Candida* to proliferate. Vaginal *Candida* infections in women are commonly called "yeast infections." Oral thrush is a *Candida* infection of the mouth common in newborns and AIDS patients (Fig. 19C.2*c*). In individuals with inadequate immune systems, *Candida* can move throughout the body, causing a systemic infection that can damage the heart, brain, and other organs.

CONSIDER THESE QUESTIONS

1. Make a list of the many indirect ways that fungi benefit humans. How many can you think of?
2. Should we try to label organisms such as fungi good or bad? Why or why not? Give examples for your reasoning.
3. Plants, for the most part, are not parasites. Why is that? What does a parasite such as the fungus that causes athlete's foot have in common with a predator such as a lion?

 Explore the concepts through a variety of multimedia assets and question types.

www.mcgrawhillconnect.com

FIGURE 19C.1 Plant fungal disease.

THE CHAPTER IN REVIEW

CONNECTING THE CONCEPTS

Land plants and fungi most likely evolved from aquatic ancestors and went on to adapt to the land environment. The land environment is not all that friendly to life because of the constant threat of drying out. Adaptations are required, and in the case of land plants a comparative study shows what innovations led to the extreme success of angiosperms in the land environment. In plants, gametophyte dependence on a sporophyte with large leaves and vascular tissue works well. Further, in seed plants, no external water is needed for male gametophytes (pollen grains) to be transported to the female gametophyte. The evolution of the flower and the recruitment of animals to help with seed and fruit dispersal was also totally suitable to the land environment. Angiosperms are the most widely dispersed of the land plants and can live in a wide variety of land habitats.

Humans are also adapted to living on land, and it is interesting to compare their reproductive adaptations with those of angiosperms. Humans require no external water to pass sperm to the female because males have a penis to fulfill this function. This works for animals that have the power to move about as they wish. Whereas the female portion of the flower does not release the offspring until it is suitably packaged in a seed, the human female retains the embryo until it can survive on its own in the land environment. This comparison shows the type of adaptations that are necessary to reproduce in the dry land environment.

Land fungi are adapted to the land environment because they produce windblown spores within both asexual and sexual life cycles. Whereas plants are photosynthetic, fungi are saprotrophic. They release enzymes into the environment to digest organic remains and absorb the resultant nutrient molecules. Without photosynthesis carried out by algae and plants and without decomposition carried out by bacteria and fungi, animals could not exist. Animals are not essential to the biosphere, but plants and fungi are!

Animal evolution is the topic of Chapter 20. Even today, more groups of animals live in water than on land. Certain molluscs (e.g., snails), certain arthropods (e.g., insects), and many vertebrates (examples can be found among the amphibians, reptiles, birds, and mammals) live on land. Birds and mammals are the most successful of these groups today. Most people agree that birds are a continuation of dinosaur evolution, while mammals filled the ecological niches of the dinosaur groups that died out at the end of the Cretaceous period, as discussed in Chapter 16.

ANALYZE AND EVALUATE

1. Think of the way ferns reproduce. Why is it a limitation for the gametophyte to be independent, instead of dependent on the sporophyte as seed plants are?

2. Support the suggestion that "DNA is selfish" and that all the energy an organism such as a plant expends on reproduction is merely a way to perpetuate the organism's DNA.

3. Do you approve of comparing reproduction in humans to reproduction in plants? Why or why not?

SUMMARIZE

Evolution of Plants

19.1 Plants have a green algal ancestor

- Plants and green algae contain chlorophylls *a* and *b*, store carbohydrates as starch, and have cellulose in their cell walls.
- Molecular data indicate that **charophytes** and land plants have the following characteristics in common: (1) Both produce their cell walls in the same way. (2) Both show apical growth, which allows plants to produce new organs. (3) Both have plasmodesmata. (4) Both have an alternation-of-generations life cycle. Charophytes protect the zygote. Land plants also protect the embryo, and thus an alternate name for the land plant clade is **embryophyta.**

19.2 Plants have an alternation-of-generations life cycle

- In the **alternation-of-generations** life cycle, the **sporophyte** is the diploid generation; it produces haploid spores by meiosis. A **spore** is a haploid reproductive cell that develops into a new generation. The **gametophyte** is the haploid generation; it produces gametes (eggs and sperm) by mitosis.
- During the evolution of plants, the sporophyte became more dominant as plants became increasingly adapted to life on land. An independent gametophyte and flagellated sperm, which fertilize the egg produced by an **archegonium,** prevent ferns from reproducing sexually in dry environments. A female gametophyte protected within an **ovule** and **pollen grains** carried by insects allow flowering plants to reproduce sexually in dry environments. The sporophyte of flowering plants is protected from drying out by a waxy **cuticle** interrupted by **stomata.**

Diversity of Plants

19.3 Bryophytes protect the embryo and have apical growth

- **Hornworts, liverworts,** and **mosses** are **bryophytes.** The bryophytes are the **nonvascular plants.**
- In the moss life cycle, the dependent sporophyte produces windblown spores in a **sporangium.** The dominant gametophyte produces flagellated sperm (see Fig. 19.3B).

19.4 Lycophytes have vascular tissue for transport

- The **vascular plants** have vascular tissue consisting of **lignin**-strengthened **xylem** for water transport and **phloem** for nutrient transport.
- The lycophytes are one of the first types of plants to have vascular tissue. Lycophytes and ferns are **seedless vascular plants.** The leaves of lycophytes are **microphylls. Lycophytes** are called **club mosses;** the sporangia are produced in club-shaped **strobili.**

19.5 Ferns have large leaves called megaphylls

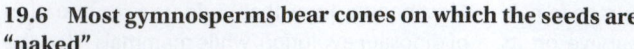

- **Ferns** (and their allies) were among the first to have **megaphylls,** large leaves. Fern leaves are called **fronds. Horsetails** (whorls of branches at nodes) and **whisk ferns** (no leaves) may have had megaphylls in the past.
- In the fern life cycle, the dominant sporophyte produces windblown spores; the separate gametophyte produces flagellated sperm (see Fig. 19.5E).

19.6 Most gymnosperms bear cones on which the seeds are "naked"

- Gymnosperms and angiosperms are the **seed plants. A seed** contains a sporophyte generation stored within a protective seed coat. **Gymnosperms** include **cycads, ginkgoes, gnetophytes,** and **conifers** (evergreen trees).
- In the pine life cycle, the sporophyte is dominant (see Fig. 19.6B). On the surface of pollen cones, within pollen sacs, microspores become **male gametophytes** (pollen grains). During **pollination,** pollen grains are transported by wind to the **female gametophyte.** On the surface of seed cones, following fertilization, ovules, which contain female gametophytes, become seeds *not* enclosed by fruit.

19.7 Angiosperms have flowers in which the seeds are "covered"

- **Monocots** have one **cotyledon** (seed leaf) and flower parts in threes or multiples of three. **Eudicots** have two cotyledons and flower parts in fours, fives, or multiples thereof.
- The main parts of a flower are the **receptacle** (bears other flower parts) and the **sepals, petals, stamens, carpel, stigma, style,** and **ovary** (see Fig. 19.7A).
- In the **angiosperm** life cycle, within an ovule, the **megaspore** develops into the **embryo sac** (the female gametophyte). In the **pollen sacs** of the anther, **microspores** become pollen grains (the male gametophyte) (see Fig. 19.7B). Pollen is transported by wind or animals to the stigma of a carpel. A **pollen tube** carries two sperm to the female gametophyte in the ovule. **Double fertilization** produces the **triploid endosperm** and a zygote. Following fertilization, the ovule becomes a seed, which is covered by a **fruit** derived from the ovary wall and perhaps from accessory parts of the flower.

Evolution and Diversity of Fungi

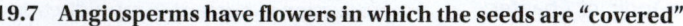

19.8 Fungi differ from plants and animals

- A **mycelium,** which is a mass of filaments called **hyphae,** makes up the fungal body. **Land fungi** produce windblown spores during both sexual and asexual reproduction. **Chytrids** are aquatic fungi that produce flagellated spores and gametes. Most **fungi** are saprotrophs that decompose dead remains.
- Mycorrhizal fungi form a mutualistic relationship called **mycorrhizae** with plant roots, thereby increasing the root surface area as well as the water and nutrient absorption of the plant. A **lichen** is a mutualistic association between a fungus and cyanobacteria or green algae. Types of lichens are crustose, fruticose, and foliose.

19.9 Land fungi occur in three main groups

- During sexual reproduction, zygospore fungi (**black bread mold**) have a **zygospore;** sac fungi (**cup fungi, red bread mold, yeast**) have an **ascus;** and club fungi (**mushrooms**) have a **basidium.**

TEST YOURSELF

Evolution of Plants

1. Land plants (but not charophytes)
   a. protect the zygote and embryo.
   b. have the alternation-of-generations life cycle.
   c. are multicellular.
   d. All of these are correct.
2. Which of the following is not a plant adaptation to land?
   a. recirculation of water
   b. protection of embryo in maternal tissue
   c. development of flowers
   d. creation of vascular tissue
   e. seed production
3. Plant spores are
   a. haploid and genetically different from each other.
   b. haploid and genetically identical to each other.
   c. diploid and genetically different from each other.
   d. diploid and genetically identical to each other.
4. Sporophyte dominance should be associated with
   a. windblown spores.
   b. independent gametophytes.
   c. pollen grains.
   d. a female gametophyte protected by a sporophyte.
   e. Both c and d are correct.
5. The gametophyte is the dominant generation in
   a. ferns.
   b. mosses.
   c. gymnosperms.
   d. angiosperms.
   e. More than one of these are correct.
6. **THINKING CONCEPTUALLY** The evolution of the vascular system allowed land plants to grow tall. What are the advantages of height in land plants?

Diversity of Plants

7. In bryophytes, sperm usually move from the antheridium to the archegonium by
   a. swimming.
   b. flying.
   c. insect pollination.
   d. worm pollination.
   e. bird pollination.
8. A fern sporophyte will develop on which region of the gametophyte?
   a. near the notch, where the archegonia are located
   b. near the tip, where the antheridia are located
   c. anywhere on the gametophyte
   d. All of these are correct.
9. Which of the following is a seedless vascular plant?
   a. gymnosperm
   b. angiosperm
   c. fern
   d. monocot
   e. eudicot

10. How are ferns different from mosses?
    a. Only ferns produce spores for reproduction.
    b. Ferns have vascular tissue.
    c. In the fern life cycle, the gametophyte and the sporophyte are both independent.
    d. Ferns do not have flagellated sperm.
    e. Both b and c are correct.
11. In the life cycle of the pine tree, the ovules are found on
    a. needlelike leaves.   d. root hairs.
    b. seed cones.   e. All of these are correct.
    c. pollen cones.
12. Label the parts of the flower in the following illustration.

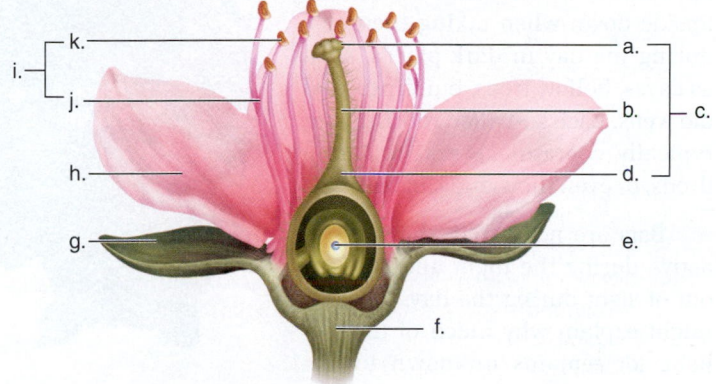

13. A seed contains a mature
    a. embryo.   c. ovary.
    b. ovule.   d. pollen grain.
14. Endosperm is produced by the union of the
    a. egg and polar nuclei.
    b. egg and sperm nuclei.
    c. polar and sperm nuclei.
    d. polar and egg nuclei.
15. Which of these plants contributed the most to our present-day supply of coal?
    a. bryophytes
    b. seedless vascular plants
    c. gymnosperms
    d. angiosperms
    e. Both b and c are correct.
16. **THINKING CONCEPTUALLY** Pollen cones are located on a tree's lower branches. They do not occur next to seed cones on the upper branches of the same tree. What are the possible advantages to this arrangement?

Evolution and Diversity of Fungi

17. A mushroom is like a plant because it
    a. is a multicellular eukaryote.
    b. produces spores.
    c. is adapted to a land environment.
    d. is photosynthetic.
    e. All but d are correct.
18. Which feature is best associated with hyphae?
    a. strong, impermeable walls
    b. rapid growth
    c. large surface area
    d. pigmented cells
    e. Both b and c are correct.

19. Symbiotic relationships of fungi include
    a. athlete's foot.
    b. lichens.
    c. mycorrhizae.
    d. Only b and c are correct.
    e. All three examples are correct.
20. A fungal spore
    a. contains an embryonic organism.
    b. germinates into an organism.
    c. is always windblown.
    d. is most often diploid.
    e. Both b and c are correct.
21. Conidia are formed
    a. asexually at the tips of special hyphae.
    b. during sexual reproduction.
    c. by all types of fungi except water molds.
    d. only when it is windy and dry.
    e. as a way to survive a harsh environment.
22. Which of the following diseases is/are caused by *Candida?*
    a. oral thrush
    b. athlete's foot
    c. vaginal yeast infection
    d. ringworm
    e. Both a and c are correct.
23. **THINKING CONCEPTUALLY** Pine seedlings grow more vigorously if they are transplanted with some of their native soil. Explain.

GET INVOLVED

1. Which of the following experimental group(s) would you expect to complete alternation of generations and why? (a) Mosses that are provided with water only to their rhizoids; or (b) mosses that are provided with water to their rhizoids, as well as water sprayed into the air.
2. An orchid produces flowers that attract particular male moths because the flowers resemble females of the same moth species. Why would you expect this to be an effective pollen dispersal strategy?

MEDIA STUDY TOOLS

mhhe.com/maderconcepts3

Enhance your study of this chapter with interactive study tools, practice tests, and engaging animations. Also, ask your instructor about the resources available through ConnectPlus, which includes LearnSmart, a personalized adaptive learning program, and a media-rich eBook.

20

Evolution of Animals

BEFORE YOU BEGIN

Take a few minutes to recall

The Secret Life of Bats

Believe it or not, bats are closely related to humans. They are mammals with hair and mammary glands. Usually, a single bat is born at a time; the mother carries a newborn around for a while, and then leaves it behind when it is not nursing. The young begin to fly in a few days. There are other mammals that can glide, but bats are the only mammal that can truly fly. Their wings extend from elongated fingers, all the way down to the feet. The feet have claws, which enable bats to hang upside down when taking their ease during the day in dark places, such as caves, hollow trees, buildings, and old wells. Each so-called roosting site typically contains several to hundreds, or even thousands, of bats.

Bats are nocturnal; they are active during the night and stay out of sight during the day. This might explain why much of their behavior remains unknown to us. Or, it could be that we are simply overwhelmed by their variety. There are more than 1,000 species of bats, and bats are second only to rodents in the number of species and individuals. One out

Black flying fox bat

Vampire bat, *Desmodus rotundus*

of every four mammalian species on Earth is a type of bat! Bats are ecologically important, and that is one reason an effort to conserve bats is under way. Some bats feed on fruit, nectar, and pollen, and in so doing, they disperse pollen and also seeds, which pass through their digestive tract. Other bats feed on insects, greatly reducing the numbers of those that flit about during the night. These bats offer a way to biologically control the insect population at no trouble to ourselves and without using pesticides.

Vampire bat feeding off a sow

The face of a bat varies greatly; many species have odd-looking appendages on the snout and very large, elaborately convoluted ears. These modifications help them emit and receive sounds at a higher frequency than is audible to the human ear. After sending out "ultrasounds," the returning echoes tell them the location of any nearby object. In other words, many bats use echolocation to find their prey.

Some bats feed on blood, and perhaps because of our interest in vampires, we know more about them. The common vampire bat, *Desmodus rotundus,* is about the size of an adult human's thumb, and weighs less than 43 grams, but its wingspan is nearly 20.5 cm. The bat finds its prey using echolocation, smell, and heat, and then uses its limbs as crutches to catapult toward a sleeping animal. Once the vampire bat finds a victim, it uses special sensors in its nose to locate a superficial vein. Contrary to popular myth, vampire bats do not suck the blood from their victims. Rather, using razor-sharp incisors, they painlessly open a small wound in their prey. A numbing chemical in the bat's saliva keeps the victim from waking up. The bat then uses its tongue to lap up the blood as it oozes from the wound or, if need be, repeatedly darts its tongue in and out of the wound. Typically, a vampire bat needs 30 milliliters (30 ml) of blood per day to survive, but can consume up to 60% of its body weight during a 20-minute feeding. After feeding, the bat returns to its roost. The highly specialized stomach of the vampire bat shunts the blood plasma to the kidneys for elimination, and only the red blood cells are used for nourishment. The saliva of vampire bats contains the most powerful anticoagulant known. In recent years, desmoteplase, a genetically engineered drug derived from the saliva of a vampire bat, has been successfully used in heart attack and stroke victims. Humans are rarely the victims of these infamous blood-eating winged parasites. Vampire bats usually feed upon the blood of cattle, pigs, horses, and large birds.

Bats are just one tiny part of the animal kingdom, the focus of this chapter. We begin by examining the characteristics that distinguish animals from other types of eukaryotes.

wing

Bat eating cactus fruit

Gnome fruit bat

Animals and How They Evolved

Animals have their own particular way of life that differs from that of plants and fungi. They evolved from a protistan ancestor and thereafter diversified into various groups. The present phylogenetic (evolutionary) tree is based on molecular and anatomic data. The anatomic data of interest to us concerns number of tissue layers, type of symmetry, and developmental homologies. Developmental homologies strongly reveal relationships between animals. For example, a similar type larva (immature form) tells us that molluscs, annelids, and flatworms are related and a similar pattern of development tells us that vertebrates are most closely related to the echinoderms among the invertebrates.

20.1 Animals have distinctive characteristics

LEARNING OUTCOMES

When you complete this section, you should be able to

1. Discuss the characteristics of animals.
2. Describe the colonial flagellate hypothesis.
3. Differentiate between protostomes and deuterostomes on the basis of embryonic development.

Animals have distinctive characteristics that make them different from plants and fungi. Like both of these groups, animals are *multicellular* eukaryotes, but unlike plants, which make their own food through photosynthesis, animals are *heterotrophs* and must acquire nutrients from an external source. Fungi are also heterotrophs, but fungi digest their food externally and absorb the breakdown products. Free-living animals *ingest* (*eat*) *their food* and digest it internally. Some parasitic animals absorb nutrient molecules from their host.

Animals usually carry on *sexual reproduction* and begin life only as a fertilized diploid egg. From this starting point, they undergo a series of

Adult frog

developmental stages to produce an organism that has specialized tissues within organs that have specific functions. Two types of tissues in particular—*muscles and nerves*—characterize animals. The presence of these tissues allows an animal to perform a variety of flexible movements that help it search actively for food and prey on other organisms. Coordinated movements also allow animals to seek mates, shelter, and a suitable climate—behaviors that have resulted in the vast diversity of animals. The more than 30 animal phyla we recognize today are believed to have evolved from a single ancestor.

Figure 20.1A illustrates that a frog, like other animals, goes through a number of embryonic stages to become a larval form (the tadpole) with specialized organs, including muscular and nervous systems that enable it to swim. A larva is an immature stage that typically lives in a different habitat and feeds on different foods than the adult. By means of a change in body form called metamorphosis, the larva, which only swims, turns into a sexually mature adult frog that swims and hops. The aquatic tadpole lives on plankton, and the terrestrial adult typically feeds on insects and worms. A large African bullfrog will try to eat just about anything, including other frogs, as well as small fish, reptiles, and mammals.

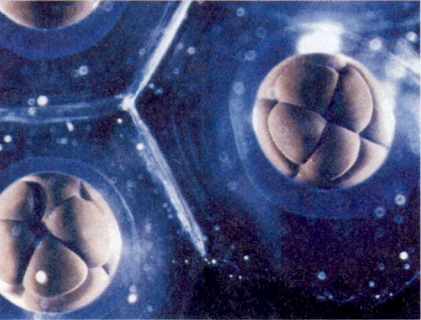  

Embryonic stages produce a tadpole (*bottom left*).

A tadpole undergoes metamorphosis to become a frog (*top left*).

FIGURE 20.1A Developmental stages of a frog.

Protistan Ancestor In Chapter 19, we discussed evidence that plants most likely share a green algal ancestor with the charophytes. What about animals? Did they also evolve from a protist, most likely a particular motile protozoan? The **colonial flagellate hypothesis** states that animals are descended from an ancestor that resembled a hollow spherical colony of flagellated cells. Figure 20.1B shows how the process would have begun with ❶ an aggregate of a few flagellated cells. From there ❷ a colony of cells could have formed a hollow sphere. ❸ Individual cells within the colony would have become specialized for particular functions, such as reproduction. (A *Volvox* is a spherical colony of cells and certain cells have only a reproductive function.) ❹ Two tissue layers could have arisen by the infolding of certain cells into a hollow sphere. Tissue layers arise in this manner during the development of animals today. The colonial flagellate hypothesis is also attractive because it implies that radial symmetry preceded bilateral symmetry in the history of animals, as is probably the case. We will see that among animals, the cnidarians, as represented by *Hydra* (Fig. 20.4B), are radially symmetrical animals that have only two tissue layers. Most animal groups have three tissue layers and the presence of only two tissue layers is evidence that the cnidarians evolved early in the history of animal evolution. In a radially symmetrical animal, any longitudinal cut produces two identical halves; in a bilaterally symmetrical animal, only one longitudinal cut yields two identical halves:

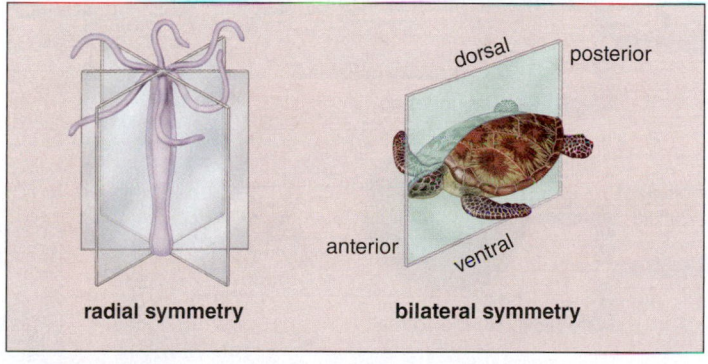

radial symmetry   bilateral symmetry

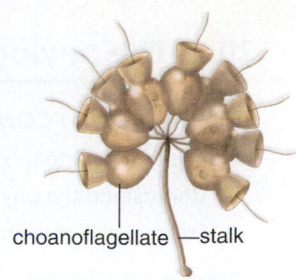

choanoflagellate —stalk

Among the protists, the choanoflagellates (collared flagellates) most likely resemble the unicellular ancestor of animals, and molecular data tell us that they are the closest living relatives of animals! A choanoflagellate is a single cell, 3–10 μm in diameter, with a flagellum surrounded by a collar of 30–40 microvilli. Movement of the flagellum creates water currents that pull the protist along. As the water moves through the microvilli, they engulf bacteria and debris from the water. Interesting to our story, choanoflagellates also exist as a colony of cells. Several can be found together at the end of a stalk or simply clumped together like a bunch of grapes.

Evolution of Animal Body Plans We are unable to trace the evolution of animals from a protozoan ancestor because, as discussed in Chapter 15, representatives of all animal phyla appeared at once in geologic terms around 540 MYA, the start of the Cambrian period. To review the many types of fossils found in the so-called **Cambrian explosion,** take a look at Figure 15B on page 290.

Biologists have long speculated on what could have caused so many different animal body plans to suddenly occur. Was it due to the emergence of a predatory lifestyle or a sudden buildup of oxygen at this time? Most biologists are satisfied that genetic studies have now answered this question. The many varieties caused by the number, position, size, and patterns of body parts is mainly due to the activity of master regulatory genes, in particular the *Hox* genes. Section 15.7, pages 290–293 of this text, discusses in detail how, despite millions of years of divergent evolution, all animals share the same *Hox* genes for development. Data have allowed biologists to conclude that animal diversity is due in large part to differential expression of these ancient regulatory genes during development.

▶ **20.1 CHECK YOUR PROGRESS**
1. List three characteristics of all animals.
2. Explain how tissues could arise from a colony of cells.

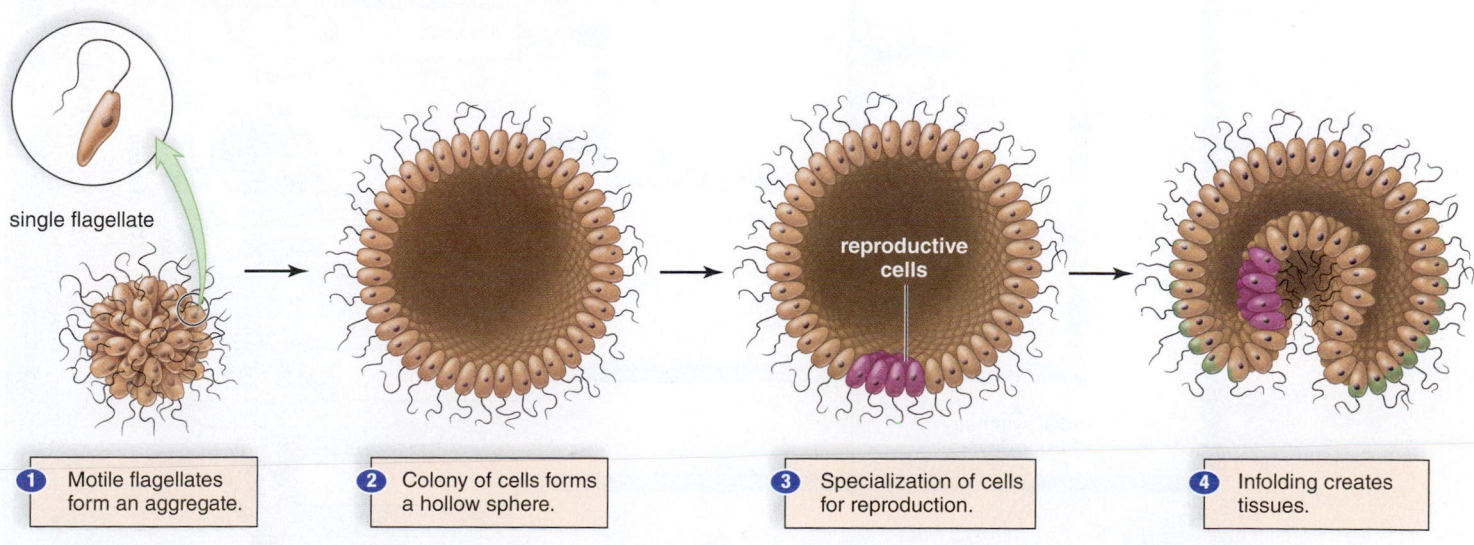

| ❶ Motile flagellates form an aggregate. | ❷ Colony of cells forms a hollow sphere. | ❸ Specialization of cells for reproduction. | ❹ Infolding creates tissues. |

single flagellate

reproductive cells

FIGURE 20.1B The colonial flagellate hypothesis.

20.2 The phylogenetic tree of animals is based on molecular and anatomic data

LEARNING OUTCOMES

When you complete this section, you should be able to

1. Understand the phylogenetic tree of animals.
2. Distinguish between protostome and deuterostome development.

There is no adequate fossil record by which to trace the early evolution of animals. Therefore, the phylogenetic tree of animals shown in Figure 20.2A is based on molecular and anatomic data, including homologies that become apparent during the development of animals. When utilizing molecular data, it is assumed that the more closely related two organisms are, the more DNA base sequences they will have in common.

Refer to the tree in Figure 20.2A and note the animal phyla on the far right that we will be discussing in this chapter. Note further that sponges are multicellular but they alone have no tissues; next to evolve, the cnidarians have only two tissue layers and radial symmetry. All the other animal phyla have bilateral symmetry, three tissue layers, and a body cavity. The animals in five phyla have protostome development and animals in only two phyla have deuterostome development. Suppose you wanted to trace the evolution of echinoderms back to an ancestral protist.

Working backward, you would mention the ancestor with deuterostome development that they share with chordates. Then, they share ancestors with an ever increasing number of phyla until finally you reach a protist ancestor for all animal phyla. Now let's look at the anatomic characteristics of animals in more detail.

Type of Symmetry Three types of symmetry exist in the animal world: (1) *Asymmetry* is seen in sponges that have no particular body shape. (2) Cnidarians are *radially symmetrical*—they are organized circularly, similar to a wheel, and two identical halves are obtained, no matter how the animal is sliced longitudinally. (3) The rest of the animals are *bilaterally symmetrical* as adults. They have a definite left and right half, and only a longitudinal cut down the center of the animal produces two equal halves.

Radially symmetrical animals are sometimes attached to a substrate—that is, they are **sessile.** This type of symmetry is useful because it allows these animals to reach out in all directions from one center. This advantage also applies to floating animals with radial symmetry, such as jellyfish. Bilaterally symmetrical animals tend to be active and to move forward from an anterior end. During the evolution of animals, bilateral symmetry is accompanied

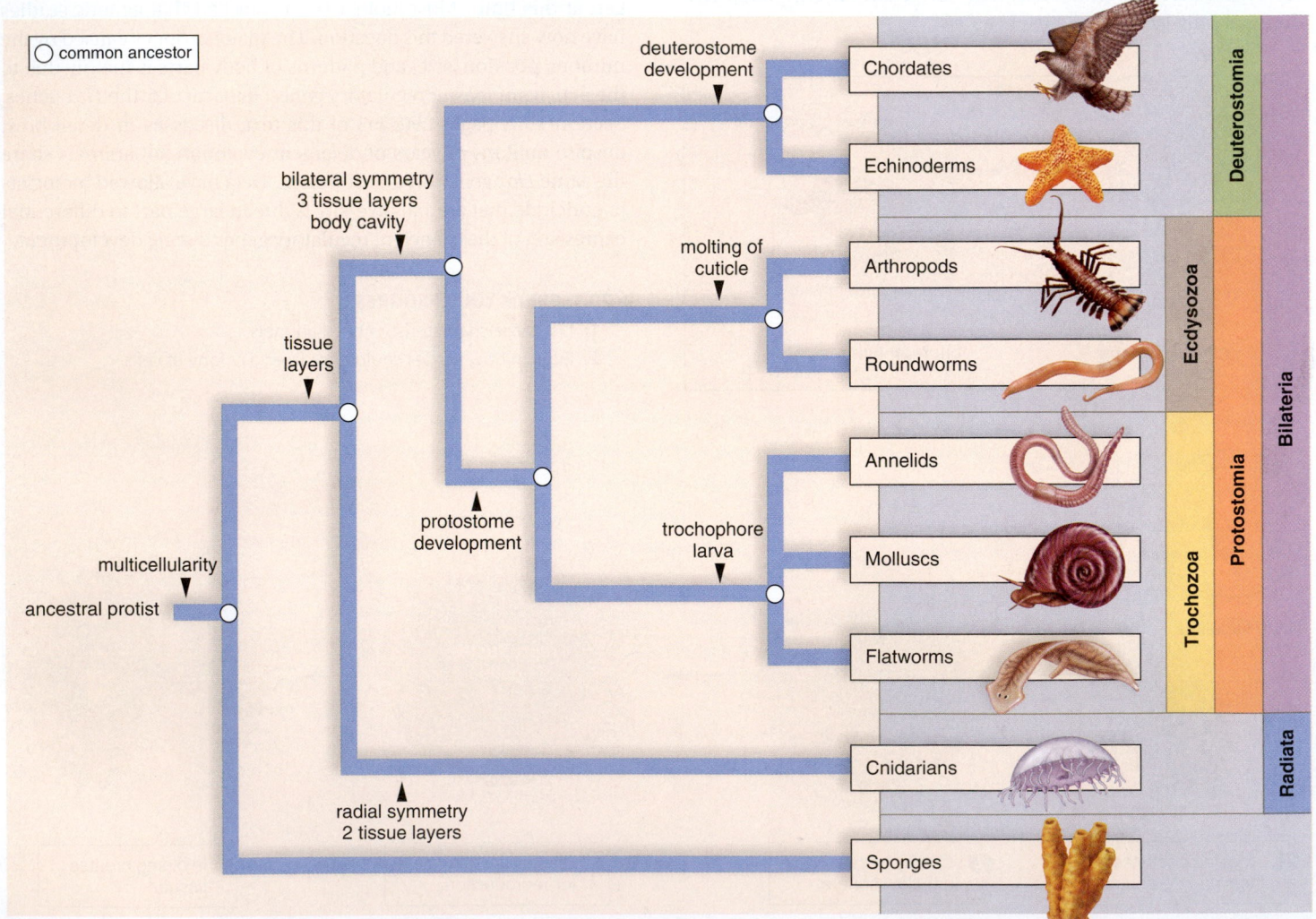

FIGURE 20.2A Phylogenetic tree of animals.

by **cephalization,** localization of a brain and specialized sensory organs at the anterior end of an animal.

Embryonic Development Like all animals, sponges are multicellular, but they do not have true tissues as the other animals do. Therefore, sponges have the cellular level of organization. True tissues appear in the other animals as they undergo embryonic development. The first three tissue layers are often called **germ layers** because they give rise to the organs and organ systems of complex animals. Animals such as cnidarians, which have only two tissue layers (ectoderm and endoderm) as embryos, have the tissue level of organization. Animals that develop further and have all three tissue layers (ectoderm, mesoderm, and endoderm) as embryos have the organ level of organization. Notice in the phylogenetic tree in Figure 20.2A that the animals with three tissue layers are either **protostomes** or **deuterostomes.**

Figure 20.2B shows that protostome and deuterostome development are differentiated by three major events:

1. Cleavage, the first event of development, is cell division without cell growth. In protostomes, spiral cleavage occurs, and daughter cells sit in grooves formed by the previous cleavages. The fate of these cells is fixed and determinate in protostomes; each can contribute to development in only one particular way. In deuterostomes, radial cleavage occurs, and the daughter cells sit right on top of the previous cells. The fate of these cells is indeterminate—that is, if they are separated from one another, each cell can go on to become a complete organism.

2. As development proceeds, a hollow sphere of cells, called a blastula, forms, and the indentation that follows produces an opening called the blastopore. In protostomes, the mouth appears at or near the blastopore; in deuterostomes, the anus appears at or near the blastopore, and only later does a second opening form the mouth.

   Video Blastocyst Formation

3. Certain protostomes and all deuterostomes have a body cavity completely lined by mesoderm, and it is called a **true coelom,** but more often simply shortened to a coelom. The coelom develops differently in the two groups. In most protostomes, the mesoderm arises from cells located near the embryonic blastopore, and a splitting produces the coelom. In deuterostomes, the coelom arises as a pair of mesodermal pouches from the wall of the primitive gut. The pouches enlarge until they meet and fuse.

The deuterostomes include the echinoderms and the chordates, two groups of animals that we will examine in detail later. The protostomes are divided into the ecdysozoans and the trochozoans. The **ecdysozoans** include the roundworms and the arthropods. Both of these types of animals molt, meaning that they shed their outer covering as they grow. Ecdysozoan means molting animals. The **trochozoans** either presently have—or their ancestors had—a trochophore larva (see section 20.5, page 395). The trochophore larva is translucent and pear-shaped with one or more circlets of cilia that allow it to swim freely and acquire food in its marine planktonic habitat. Posterior and anterior tufts of cilia are often present. Although not often stressed, this larva does have a fully formed digestive tract and other organs.

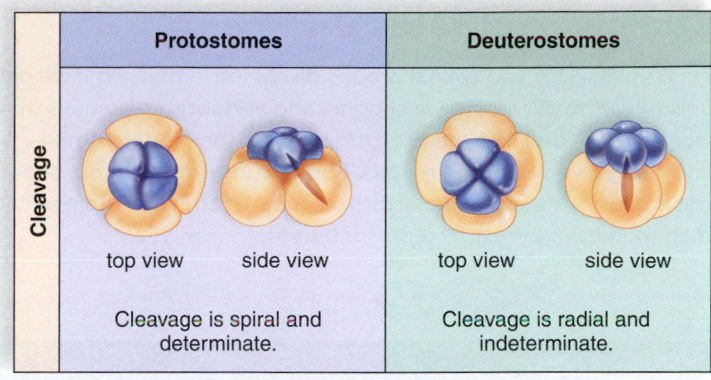

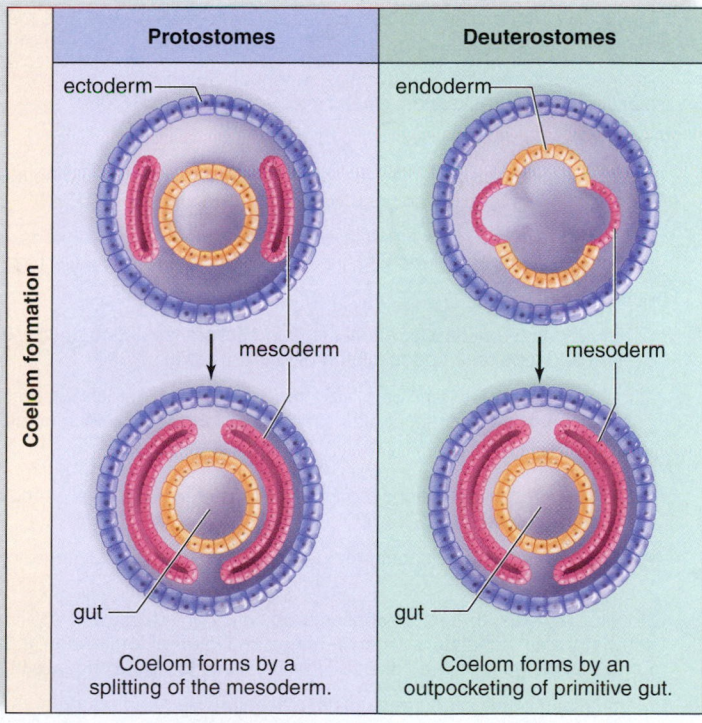

FIGURE 20.2B Protostomes and deuterostomes may be contrasted in these ways: how cleavage occurs, the fate of the embryonic blastopore, and how the coelom, if present, develops.

▶ **20.2 CHECK YOUR PROGRESS**

1. List three basic anatomic features that distinguish the different groups of animals.
2. Differentiate between protostome and deuterostome development in three ways.

The Invertebrates

For convenience, the animal groups discussed in this text have been divided into **invertebrates** (animals that do not have an endoskeleton of cartilage and bone) and **vertebrates** (animals that do have an endoskeleton). The invertebrates with three tissue layers are divided into the protostomes and the deuterostomes. A few chordates are invertebrates, but most are vertebrates. As you know, animals evolved in the sea, and as surprising as it may seem, most animals still live in the water. Among the invertebrates, only the molluscs, annelids, and arthropods have terrestrial representatives. Among the vertebrates, the amphibians, reptiles, birds, and mammals have terrestrial representatives.

DOMAIN: Eukarya
KINGDOM: Animals

CHARACTERISTICS
Multicellular, usually with specialized tissues; ingest or absorb food; diploid life cycle.

INVERTEBRATES

Sponges (bony, glass, spongin): *Asymmetrical, saclike body perforated by pores; internal cavity lined by choanocytes; spicules serve as internal skeleton. 5,150[†]

Radiata

Cnidarians (hydra, jellyfish, corals, sea anemones): Radially symmetrical with two tissue layers; sac body plan; tentacles with nematocysts. 10,000[†]

Protostomia (trochozoans)

Flatworms (planarians, tapeworms, flukes): *Bilateral symmetry with cephalization; *three tissue layers and organ systems; acoelomate with incomplete digestive tract that can be lost in parasites; hermaphroditic. 20,000[†]

Molluscs (chitons, clams, snails, squids): *Coelom; all have a foot, mantle, and visceral mass; foot is variously modified; in many, the mantle secretes a calcium carbonate shell as an exoskeleton; all organ systems. 110,000[†]

Annelids (polychaetes, earthworms, leeches): Segmented, with body rings and setae; cephalization in some polychaetes; hydroskeleton; closed circulatory system. 16,000[†]

Protostomia (ecdysozoans)

Roundworms (*Ascaris*, pinworms, hookworms, filarial worms): Pseudocoelom and hydroskeleton; complete digestive tract; free-living forms in soil and water; parasites common. 25,000[†]

Arthropods (crustaceans, spiders, scorpions, centipedes, millipedes, insects): Chitinous exoskeleton with jointed appendages undergoes molting; insects—most have wings—are most numerous of all animals. 1,000,000[†]

Deuterostomia

Echinoderms (sea stars, sea urchins, sand dollars, sea cucumbers): Radial symmetry as adults; unique water vascular system and tube feet; endoskeleton of calcium plates. 7,000[†]

Chordates (tunicates, lancelets, vertebrates): All have notochord, dorsal tubular nerve cord, pharyngeal pouches, and postanal tail at some time; contains mostly vertebrates in which notochord is replaced by vertebral column. 56,000[†]

VERTEBRATES

Fishes (jawless, cartilaginous, bony): *Endoskeleton, jaws, and paired appendages in most; internal gills; single-loop circulation; usually scales. 28,000[†]

Amphibians (frogs, toads, salamanders): Jointed limbs; lungs; three-chambered heart with double-loop circulation; moist, thin skin. 6,900[†]

Reptiles (snakes, turtles, crocodiles): Amniotic egg; rib cage in addition to lungs; three- or four-chambered heart typical; scaly, dry skin; copulatory organ in males and internal fertilization. 8,000[†] Birds (songbirds, waterfowl, parrots, ostriches): Endothermy, feathers, and skeletal modifications for flying; lungs with air sacs; four-chambered heart. 10,000[†]

Mammals (monotremes, marsupials, placental): Hair and mammary glands. 4,800[†]

*After these characters are listed, they are present in the rest, unless stated otherwise.
[†]Number of species.

20.3 Sponges are multicellular animals

When you complete this section, you should be able to

1. Discuss the body, skeleton, and reproduction of a sponge, and how sponges differ from other animals.

While all animals are multicellular, **sponges** (phylum Porifera, 7,000 species) are the only animals to lack true tissues and to have a *cellular level of organization*. Sponges, unlike other animals, are asymmetrical, meaning that they have no particular symmetry. Actually, they have few cell types and no nerve or muscle cells to speak of. Sponges may be more closely related to protists than to the multicellular animals.

Body of a Sponge Sponges are in phylum Porifera because their saclike bodies are perforated by many pores (Fig. 20.3). Sponges are aquatic, largely marine animals that vary greatly in size, shape, and color. But they all have a canal system of varying complexity that allows water to move through their bodies.

The interior of the canals is lined with flagellated cells called collar cells (choanocytes). The beating of the flagella produces water currents that flow through the pores into the central cavity and out through the osculum, the upper opening of the body. Even a simple sponge only 10 cm tall is estimated to filter as much as 100 l of water each day. It takes this much water to supply the needs of the sponge. A sponge is a stationary **filter feeder,** also called a suspension feeder, because it filters suspended particles from the water by means of a straining device—in this case, the pores of the walls and the microvilli making up the collar of collar cells. Microscopic food particles that pass between the microvilli are engulfed by the collar cells and digested by them in food vacuoles.

Skeleton The skeleton of a sponge prevents the body from collapsing. All sponges have fibers of spongin, a modified form of collagen; a bath sponge is the dried spongin skeleton from which all living tissue has been removed. Today, however, commercial "sponges" are usually synthetic. Typically, the skeleton of sponges also contains spicules—small, needle-shaped structures with one to six rays. Traditionally, the type of spicule has been used to classify sponges as three types: bony, glass, or spongin. The success of sponges, which have existed longer than many other animal groups, can be attributed to their spicules. Sponges have few predators because a mouth full of spicules is an unpleasant experience. Also, sponges produce a number of foul-smelling and toxic substances that discourage predators.

Reproduction Sponges can reproduce both asexually and sexually. Asexual reproduction occurs by *budding* or by fragmentation. During budding, a small protuberance appears and gradually increases in size until a complete organism forms. Budding produces colonies of sponges that can become quite large. During sexual reproduction, eggs and sperm are released into the central cavity, and the zygote develops into a flagellated larva that may swim to a new location. If a sponge is fragmented into small parts, each part will regenerate into a complete sponge. If the cells of a sponge are mechanically separated, they will reassemble into a complete and functioning organism!

▶ **20.3 CHECK YOUR PROGRESS**

1. List general and specific ways in which sponges differ from other animals.

Yellow tube sponge

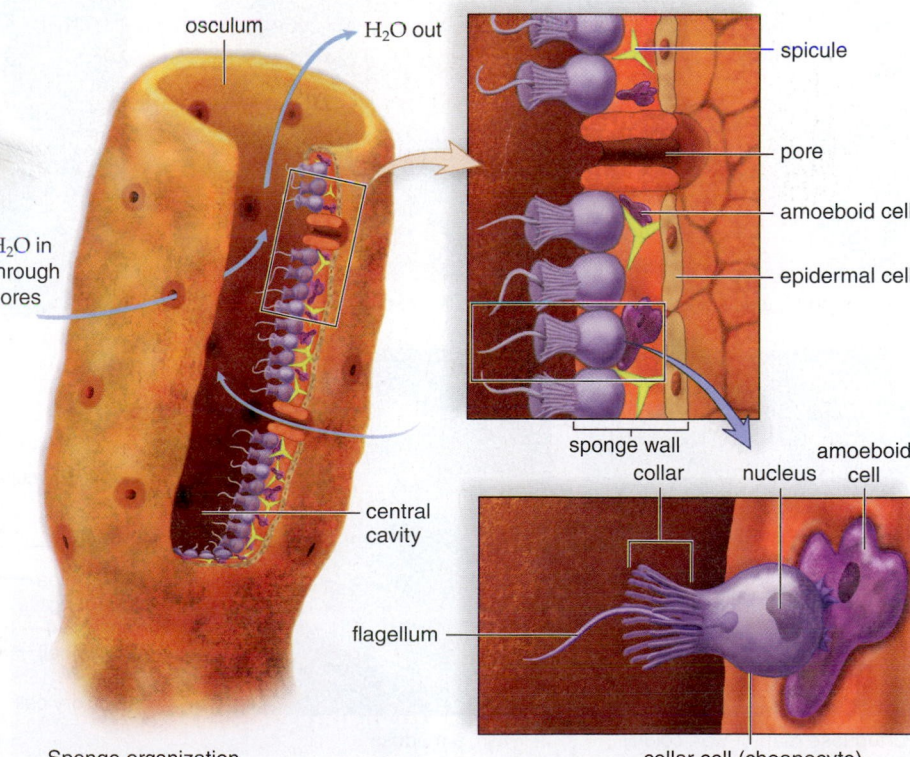

Sponge organization

FIGURE 20.3 Sponge anatomy.

20.4 Cnidarians are radially symmetrical with two tissue layers

LEARNING OUTCOME

When you complete this section, you should be able to

1. Discuss the unique features of cnidarians.

Cnidarians (phylum Cnidaria, 10,000 species) are an ancient group of invertebrates with a rich fossil record (Fig. 20.4A). Most cnidarians live in the sea, but a few freshwater species exist. Although stationary, or at best slow-moving, cnidarians have an effective means of capturing prey. They are *radially symmetrical* and capture their prey with a ring of tentacles that bear specialized stinging cells, called cnidocytes (Fig. 20.4B). Each cnidocyte has a capsule called a **nematocyst** containing a long, spirally coiled, hollow thread. When the trigger of the cnidocyte is touched, the nematocyst is discharged. Some nematocysts merely trap a prey or predator; others have spines that penetrate the prey's body and inject paralyzing toxins before the prey is captured. Once caught, the prey is drawn into a **gastrovascular cavity** that has only one opening, a mouth. Cnidarians can digest a prey of fairly large size because extracellular digestion occurs in this cavity. A complete digestive tract has both a mouth and anus; therefore, cnidarians have an *incomplete digestive tract*.

During development, cnidarians acquire only two germ layers (ectoderm and endoderm), and as adults, they have the *tissue level of organization*. Cnidarians are capable of coordinated movements because the ectodermal cells have contractile fibers that are stimulated by nerve cells, which form a nerve net. Sensory cells, which receive external stimuli, also communicate with the nerve net.

Two basic body forms are seen among cnidarians—the polyp and the medusa. The mouth of the polyp is directed upward from the substrate, while the mouth of the medusa is directed downward. In any case, cnidarians have a *sac body plan* with only one opening. A medusa has much jellylike packing material, called **mesoglea,** and is commonly called a "jellyfish." Polyps are tubular and generally attach to a rock with some, but not as much, mesoglea (Fig. 20.4B).

Cnidarians, as well as other marine animals, have been the source of medicines, particularly drugs that counter inflammation. Like the cnidarians, the other groups of animals to be studied also have true tissues.

▶ **20.4 CHECK YOUR PROGRESS**

1. List ways that the two body plans of cnidarians are the same and how they differ.

Sea anemone, a solitary polyp

Coral, a colonial polyp

Portuguese man-of-war, colony of modified polyps and medusae

Jellyfish, a medusa

FIGURE 20.4A Cnidarian diversity.

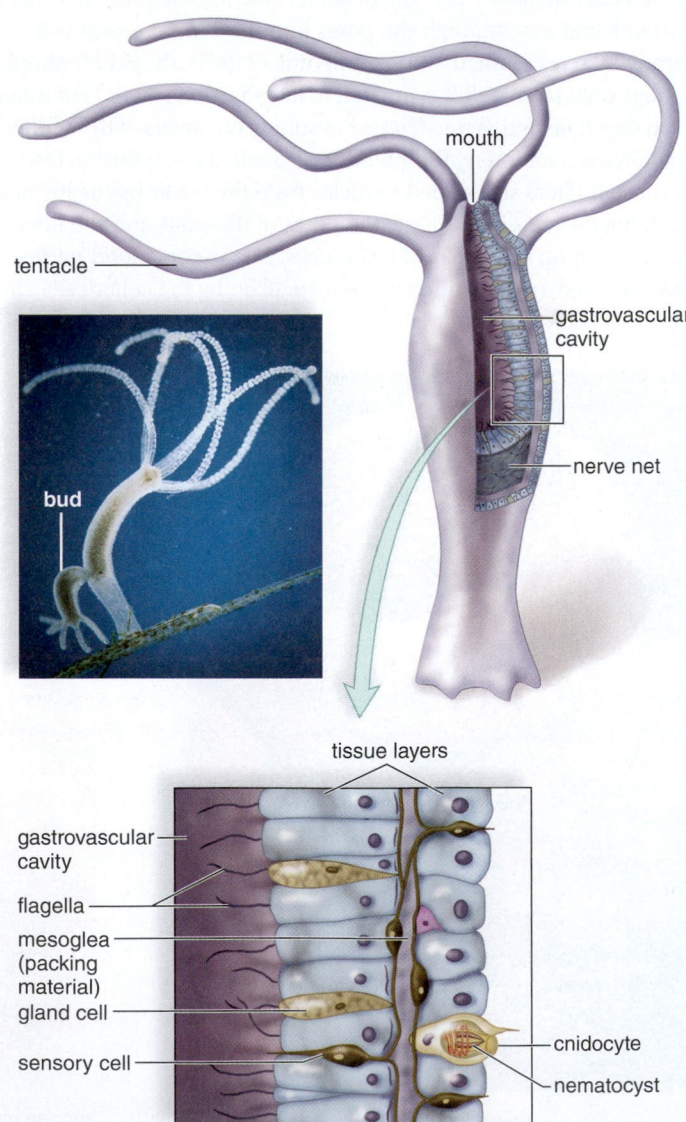
FIGURE 20.4B Anatomy of *Hydra,* a polyp.

LEARNING OUTCOMES

When you complete this section, you should be able to

1. Distinguish ways in which free-living flatworms are the same as cnidarians and ways in which they differ.
2. Understand the anatomy and life cycle of tapeworms and flukes.

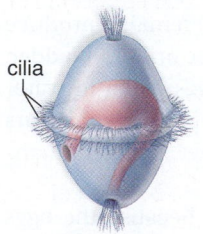

cilia

trochophore larva

Among the protostomes, the trochozoans either have a **trochophore larva** (molluscs and annelids) or have an ancestor that had one sometime in the past (flatworms). A **larva** is an immature stage that can live independently.

The **flatworms** (phylum Platyhelminthes, 20,000 species) have *bilateral symmetry.* Flatworms also have three germ layers: ectoderm, endoderm, and mesoderm. These germ layers are responsible for the development of various organ systems. The presence of mesoderm, in addition to ectoderm and endoderm, gives bulk to the animal and leads to more organ formation: Flatworms have the organ system level of organization. Nevertheless, flatworms have no body cavity. In the other animals to be studied, the organs lie in a body cavity called a coelom that is lined by mesoderm. Because the flatworms have no body cavity, they are called **acoelomates.**

Free-living flatworms, called **planarians,** have several body systems, including a digestive system (Fig. 20.5A). The animal captures food by wrapping itself around the prey, entangling it in slime, and pinning it down. Then the planarian extends a muscular pharynx and, by a sucking motion, tears up and swallows its food. The pharynx leads into a three-branched *gastrovascular cavity* where digestion begins. Digestion is finished inside the cells that line the gastrovascular cavity. The digestive tract is *incomplete* because it has only one opening, and undigested food passes out through the pharynx. This is the sac body plan.

Living in fresh water, planarians have a well-developed excretory system composed of a series of interconnecting canals that run the length of the body on each side. Flame cells contain cilia that move back and forth, bringing water into the canals that empty at pores. The beating of the cilia reminded an early investigator of the flickering of a flame; therefore, he called them **flame cells.**

Planarians are **hermaphrodites,** meaning that they have both male and female sex organs. The worms practice cross-fertilization: The penis of one is inserted into the genital pore of the other, and a reciprocal transfer of sperm takes place. The fertilized eggs hatch in 2–3 weeks as tiny worms. Development of the larva introduces bilateral symmetry; in other words, the larva was bilaterally symmetrical, and this type of symmetry is retained by the adult.

Planarians have a **ladderlike nervous system.** A small anterior brain and two lateral nerve cords are joined by cross-branches called transverse nerves. Planarians exhibit cephalization; aside from a brain, the "head" end has light-sensitive organs (the eyespots) and chemosensitive organs located on the auricles. The presence of mesoderm permits the development of three muscle layers—an outer circular layer, an inner longitudinal layer, and a diagonal layer—that allow for varied movement. A ciliated epidermis allows planarians to glide along a film of mucus.

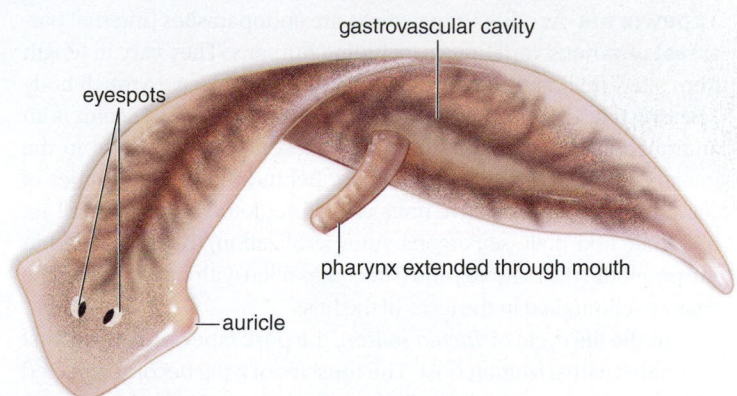

Digestive system

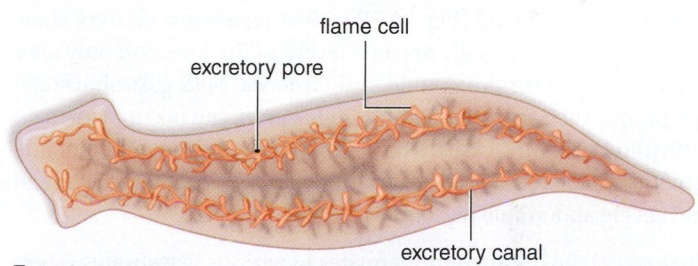

Excretory system

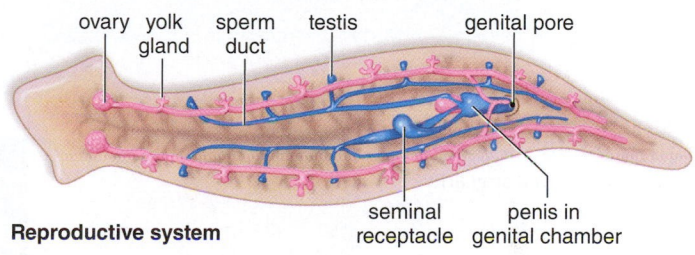

Reproductive system

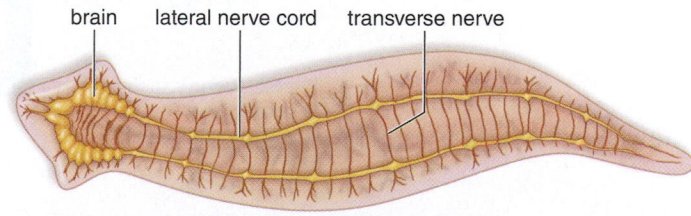

Nervous system

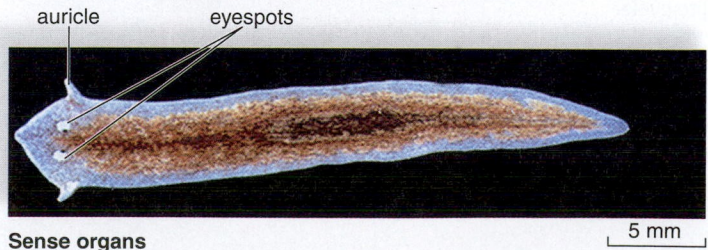

Sense organs

FIGURE 20.5A Planarians have a three-branched gastrovascular cavity, an excretory system with flame cells, the reproductive system of a hermaphrodite, and a ladderlike nervous system.

Parasitic Flatworms Tapeworms and flukes are two types of parasitic flatworms.

Tapeworms As adults, tapeworms are endoparasites (internal parasites) of various vertebrates, including humans. They vary in length from a few millimeters to nearly 20 m. Tapeworms have a tough body covering that is resistant to the host's digestive juices. The scolex is an anterior region that bears hooks and suckers for attachment to the intestinal wall of the host (Fig. 20.5B). Behind the scolex, a series of self-contained reproductive units called proglottids contain a full set of female and male sex organs. After fertilization, the organs within the proglottid disintegrate, and it becomes filled with mature eggs. The eggs are eliminated in the feces of the host.

In the life cycle of *Taenia solium,* the pork tapeworm, a pig host alternates with a human host. The muscles of a pig become infected with bladder worms when pigs eat food contaminated with egg-containing feces. When humans eat infected pork that has not been thoroughly cooked, a bladder worm becomes a tapeworm attached to their intestinal wall (Fig. 20.5B). Most tapeworm carriers show no symptoms and usually become aware of the infection only after noticing tapeworm proglottids in their feces. Mild gastrointestinal symptoms, such as nausea or abdominal pain, can occur in infected individuals. In rare cases where tapeworm proglottids enter the appendix, pancreas, or bile duct, a person may experience sudden and severe abdominal discomfort.

Flukes All flukes are endoparasites of various vertebrates. Their flattened and oval-to-elongated body is covered by a protective body wall. The anterior end of the animal has an oral sucker and at least one other sucker used for attachment to the host. Flukes are named for where they live in the body. Blood flukes (*Schistosoma* spp.) occur predominantly in the Middle East, Asia, and Africa. Adults are small (approximately 2.5 cm long) and may live for years in their human hosts. Nearly 800,000 persons die each year from an infection called schistosomiasis.

Adult humans become infected when they expose their skin to water that contains *Schistosoma* larvae released from a snail (Fig. 20.5C). Male and female flukes live in the veins of the human abdominal cavity. Here, they mate, and the females produce eggs. When the eggs penetrate the intestine or urinary bladder, they leave the body in feces or urine. The eggs hatch in water and become larvae that infect snails. Asexual reproduction occurs within the snails, and then the larval form that infects humans escapes the snails and enters the water.

Schistosomiasis is a debilitating disease because the eggs cause much tissue damage when they penetrate the walls of the veins of the small intestine or urinary bladder. The tissues hemorrhage, so that blood often appears in urine or feces. Even worse, many of the eggs produced by the female worms do not leave the veins, but are swept up in the circulatory system and deposited in the host's liver, where they are encapsulated.

▶ **20.5 CHECK YOUR PROGRESS**
1. List ways in which flatworms are more complex than cnidarians.
2. Summarize which organs are affected when a person has a tapeworm infection. A blood fluke infection?

FIGURE 20.5B Tapeworm (*Taenia solium*) anatomy and life cycle.

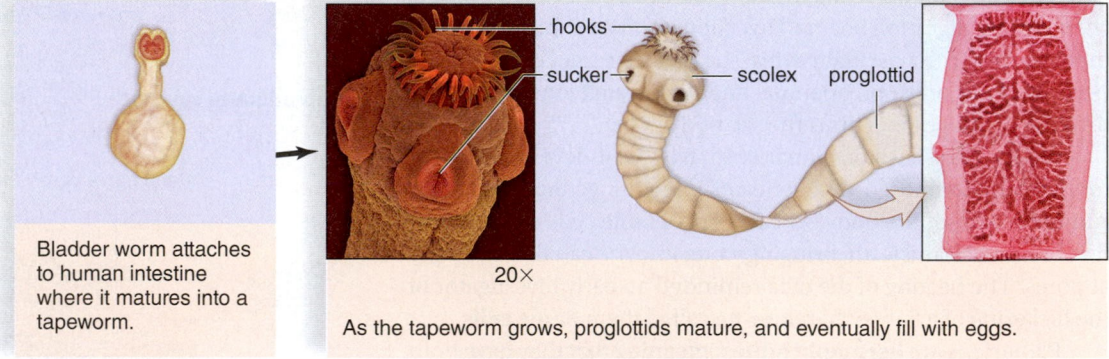

hooks · sucker · scolex · proglottid

20×

Bladder worm attaches to human intestine where it matures into a tapeworm.

As the tapeworm grows, proglottids mature, and eventually fill with eggs.

FIGURE 20.5C Sexual portion of blood fluke (*Schistosoma* spp.) life cycle.

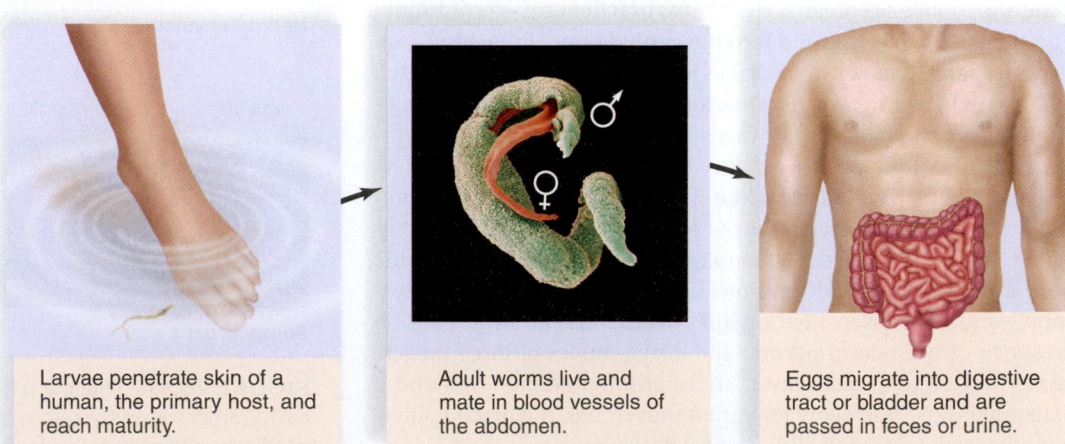

Larvae penetrate skin of a human, the primary host, and reach maturity.

Adult worms live and mate in blood vessels of the abdomen.

Eggs migrate into digestive tract or bladder and are passed in feces or urine.

20A Nemertine Worms Are Closely Related to Whom?

Suppose your biologist friend drags you along for a walk along a muddy tidal flat (Fig. 20A.1). He wants to show you all the long, flat worms that suddenly appear as the sun goes down and the tide goes out. You stand dumbfounded when long tubes shoot out with explosive force from the anterior end of the worms and wrap around small prey he calls ampipods but look like shrimp. Astonished, you ask, "What are these worms?" "They're called ribbon worms, or more correctly nemertine worms," he answers, "and I have to capture some and take them back to the lab. Easy does it, or they might fall apart," he cautions.

Back at the lab, a cursory examination makes the two of you think the worms are a type of flatworm (Fig. 20A.2). After all, they are flat, although much longer than a flatworm, and they use cilia to glide in the slime they secrete. Except for the long, blind (no opening) tube called a proboscis, the animal feeds like a flatworm. When the proboscis retracts, it brings prey to a mouth and then an esophagus thrusts out to suck the prey into a digestive tract.

After a careful microscopic examination your friend is not so sure they are flatworms. The worms do have an excretory system consisting of a flame cell system like that of flatworms. But they also have a complete digestive system with both a mouth and an anus. And even more striking, the proboscis is in a body cavity, called the rhynchocoel, in a handy textbook (Fig. 20A.3).

Well, it's time to look and see if any molecular studies have been done to classify ribbon worms, your friend suggests. In an issue of *Molecular Biology and Evolution,* you find an article[1] in which the authors compared sequences of DNA coding for an rRNA taken from a nemertine worm and that taken from seven known protostomes and two deuterostomes. The authors assumed that

the greater the similarity in base sequence, the closer the evolutionary relationship between the groups being compared. To construct a phylogenetic tree, they used three different computer techniques that provide a statistical indication of how likely it is that any given relationship is in fact the correct conclusion for their data. The results were clear and very similar for all three analyses. Nemertine worms are more closely related to molluscs and annelids (see sections 20.7 and 20.8) than they are to the flatworms. These two groups of protostomes are coelomates, and the correct interpretation is that the rhynchocoel is indeed the coelom for the ribbon worms.

CONSIDER THESE QUESTIONS

1. In a world that had no DNA data, how did biologists decide who was related to whom?
2. Why is it important for anatomic data to be backed up by DNA data and DNA data to be backed up by anatomic data?
3. Should biologists go out in the field to study organisms, or doesn't it matter? Explain.

 Explore the concepts through a variety of multimedia assets and question types.

www.mcgrawhillconnect.com

[1]Turbeville, J. M., et al. "Phylogenetic Position of Phylum Nemertini, Inferred from 18S rRNA Sequences: Molecular Data as a Test of Morphological Character Homology," *Molec. Biol. Evol.* 9:235–249 (1992) .

FIGURE 20A.1 Walking a mudflat.

Tropical nemertine worms are colorful; those in the temperate zone tend to be drab.

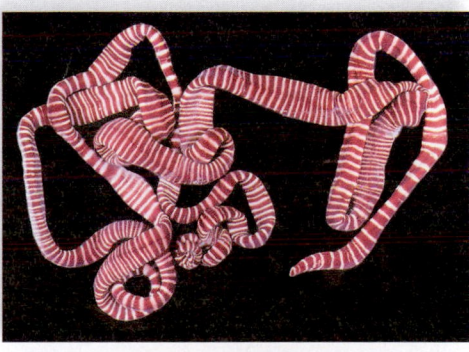

FIGURE 20A.2 Studying nemertine anatomy in the lab.

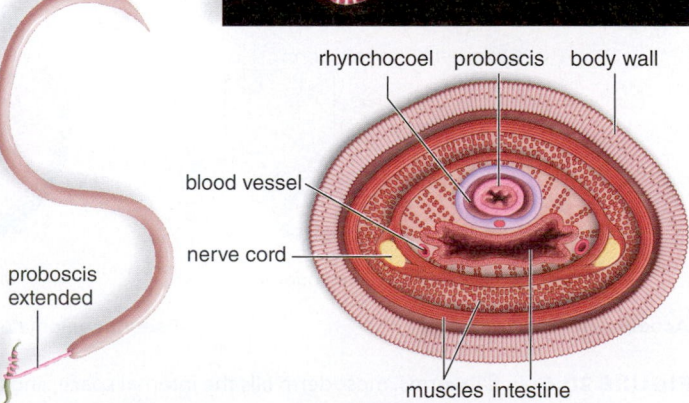

proboscis extended

rhynchocoel   proboscis   body wall

blood vessel

nerve cord

muscles intestine

FIGURE 20A.3 Nemertine anatomy

LEARNING OUTCOMES

When you complete this section, you should be able to

1. Distinguish between an acoelomate, pseudocoelomate, and a coelomate and name a type of animal of each.
2. Discuss the possible advantages of a coelom.

Early on, all animals pass through a developmental stage in which they are a hollow sphere of cells called a blastula. An indentation that follows produces an opening called the blastopore. In protostomes (*proto,* before; *stome,* mouth), the mouth appears at or near the blastopore; in deuterostomes (*deutero,* second), the anus appears at or near the blastopore, and only later does a new opening form the mouth.

Another defining characteristic in animals is the presence or absence of a body cavity, called a coelom. Among protostomes, flatworms do not have any type of body cavity for their internal organs, and roundworms have a pseudocoelom rather than a true coelom. A **pseudocoelom** is a body cavity incompletely lined by mesoderm while a true coelom is completely lined by mesoderm (Fig. 20.6). All other protostomes (molluscs, annelids, arthropods) have a true coelom. Among deuterostomes, both echinoderms and chordates have a true coelom. The term true coelom can be shortened to simply coelom.

In a pseudocoelom, the mesoderm lines only the body wall, and in a coelom, mesoderm lines both the body wall and the digestive tract. The coelom develops differently in protostomes and deuterostomes, as described in Figure 20.2B. In most protostomes, solid mesoderm masses arise from cells located on either side of the embryonic blastopore, and a splitting occurs that produces the coelom. Because of the splitting, it is called a schizocoelom. In deuterostomes, the coelom arises as a pair of mesodermal pouches from the wall of the primitive gut. The pouches enlarge until they meet and fuse, forming a coelom in which mesoderm lines both the body wall and the digestive tract. This is called an enterocoelom.

Advantages of a Coelom A coelom offers many advantages. Body movements are freer because the outer wall can move independently of the enclosed organs. In animals such as the annelids (see Fig. 20.8A), which lack a skeleton, the fluid-filled coelom acts as a hydrostatic skeleton by offering support and resistance to the contraction of muscles so that the animal can move. As analogies, consider that a garden hose stiffens when filled with water and that a water-filled balloon changes shape when squeezed at one end. Similarly, an animal with a hydrostatic skeleton can change shape and perform a variety of movements.

In coelomates, the ample space of a coelom allows complex organs and organ systems to develop. For example, the digestive tract can coil and provide a greater surface area for absorption of nutrients. Coelomic fluid protects internal organs against damage and prevents marked temperature changes. It can even help store and transport substances.

As complex animals, coelomates have the *organ system of organization.* Like other animals, they evolved in the sea, but many now live successfully on land. Terrestrial existence requires breathing air, preventing desiccation, and having means of locomotion and reproduction that are not dependent on external water. The excretory system may be modified for excretion of a solid nitrogenous waste to help conserve water.

► **20.6 CHECK YOUR PROGRESS**
1. Explain the benefit of a coelom in more complex animals.

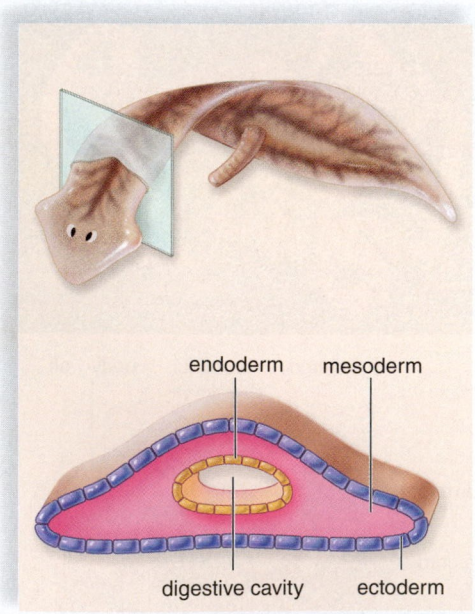

Acoelomate (flatworms)

Pseudocoelomate (roundworms)

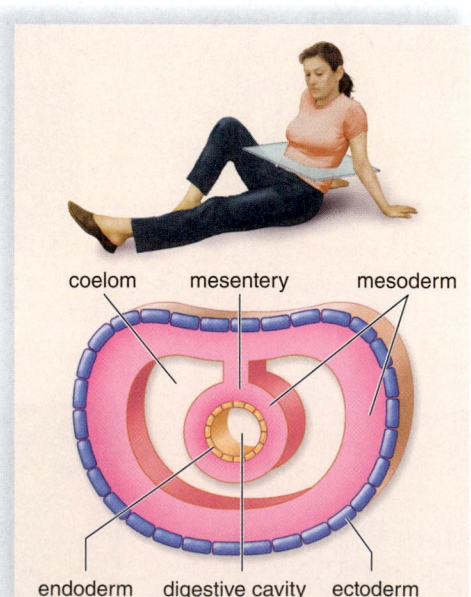

Coelomate (molluscs, annelids, arthropods, echinoderms, chordates)

FIGURE 20.6 In flatworms, mesoderm fills the internal space, and there is no coelom; thus they are acoelomates. In roundworms, mesoderm lines only the body wall; they are pseudocoelomates. In the other animal groups, mesoderm lines both the body wall and the digestive tract; they are coelomates.

20.7 Molluscs have a three-part body plan

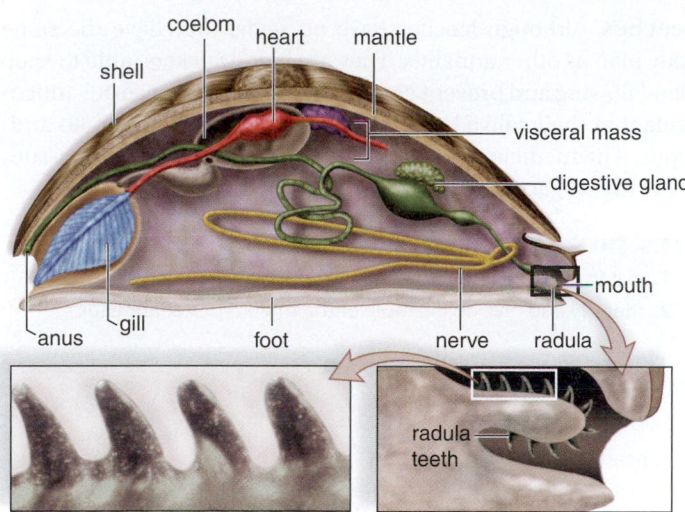

FIGURE 20.7A Hypothetical ancestral mollusc.

FIGURE 20.7B Three groups of molluscs.

Land snail

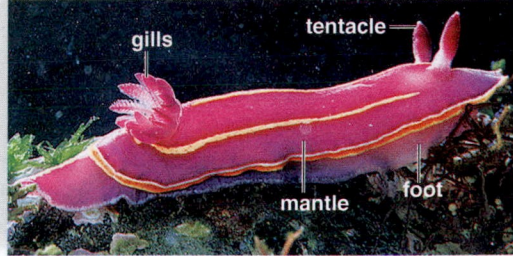

Three-stripe doris nudibranch

① **Gastropods**

Chambered nautilus   Two-spotted octopus

② **Cephalopods**

Blue mussel   Bay scallop

③ **Bivalves**

LEARNING OUTCOMES

When you complete this section, you should be able to

1. Describe the general characteristics shared by all molluscs.
2. Identify and give an example of the three major types of molluscs.

All **molluscs** (phylum Mollusca, 110,000 species) have a body composed of at least three distinct parts: (1) The *foot* is the strong, muscular portion used for locomotion. (2) The *visceral mass* is the soft-bodied portion that contains internal organs. (3) The *mantle* is a membranous, or sometimes muscular, covering that envelops the visceral mass (Fig. 20.7A). The mantle may secrete an exoskeleton called a *shell*. If a foreign body is placed between the mantle and the shell of an oyster (a mollusc), concentric layers of shell are deposited about the particle to form a pearl. Another feature often present in molluscs is a rasping, tonguelike *radula,* an organ that bears many rows of teeth and is used to obtain food.

As shown in Figure 20.7B, three common groups of molluscs are gastropods, cephalopods, and bivalves. **①** In **gastropods** (meaning stomach-footed), including snails and nudibranchs, the animal moves by muscle contractions that pass along its ventrally flattened foot. In terrestrial snails, the mantle is richly supplied with blood vessels and functions as a lung. **②** In **cephalopods** (meaning head-footed), including octopuses, squids, and nautiluses, the foot has evolved into tentacles about the head. The tentacles seize prey, and then a powerful beak and a radula tear it apart. Cephalopods possess well-developed nervous systems and complex sensory organs. Rapid movement and the secretion of a brown or black pigment from an ink gland help cephalopods escape their enemies. Octopuses have no shell, and squids have only a remnant of one concealed beneath the skin. **③** Clams, oysters, scallops, and mussels are called **bivalves** because their shells have two parts. A muscular foot projects ventrally from the shell. In a clam, such as the freshwater clam, the calcium carbonate shell has an inner layer of mother-of-pearl. The clam is a filter feeder. Food particles and water enter the mantle cavity by way of a siphon; mucous secretions cause smaller particles to adhere to the gills; and ciliary action sweeps them toward the mouth. Spent fluid exits the mantle cavity by way of another siphon.

Video Snail's Pace

Video Clam Locomotion

▶ **20.7 CHECK YOUR PROGRESS**

1. Describe the modified foot of a snail, an octopus, and a scallop.

20.8 Annelids are the segmented worms

When you complete this section, you should be able to

1. Recognize the several ways in which annelids exhibit segmentation.
2. Describe with an example the three types of annelids.

Annelids (phylum Annelida, 12,000 species) are segmented, as can be seen externally by the rings that encircle the body of an earthworm (Fig. 20.8A). **Segmentation** assists the animal as a whole. For example, because septa divide the well-developed, fluid-filled coelom into segments, it can be used as a hydrostatic skeleton to facilitate the worm's movement. (In the same way, our segmented vertebral column allows us flexibility of movement.) Movement is also assisted in that the nervous system consists of a brain connected to a ventral nerve cord, with ganglia in each segment. The excretory system consists of **nephridia,** which are tubules in most segments that collect waste material and excrete it through an opening in the body wall. The complete digestive tract is not segmented and there are many specialized organs, from the mouth to the anus. The crop is a food storage organ.

Phylum Annelida contains oligochaetes, polychaetes, and leeches.

Oligochaetes The earthworm is an oligochaete because it has few setae per segment. **Setae** are bristles that anchor the worm or help it move. Earthworms do not have a well-developed head, and they reside in soil, where there is adequate moisture to keep the body wall moist for gas exchange. They are scavengers that feed on leaves or any other organic matter, living or dead, that can conveniently be taken into the mouth along with dirt.

Polychaetes Most annelids are polychaetes (having many setae per segment) that live in marine environments. Figure 20.8B*a* shows a stationary polychaete, with tentacles that form a funnel-shaped fan. This animal is also known as a tube worm because it secretes and lives in a tube, from which it emerges to filter-feed. Water currents created by the action of cilia trap food particles that are directed toward the mouth. The clam worm *Nereis* is a polychaete with a pair of strong, chitinous jaws that extend with a part of the pharynx. In support of its predatory way of life, *Nereis* has a well-defined head region, with eyes and other sense organs (Fig. 20.8B*b*).

Leeches Although leeches have no setae, they have the same body plan as other annelids. They are blood suckers, able to keep blood flowing and prevent clotting by means of a powerful anticoagulant in their saliva known as *hirudin*. Hirudin is also an antiseptic. The medicinal leech is used to remove blood from tissues following surgery (Fig. 20.8B*c*).

▶ **20.8 CHECK YOUR PROGRESS**
1. List three ways annelids exhibit segmentation.
2. Identify and give an example of the three types of annelids.

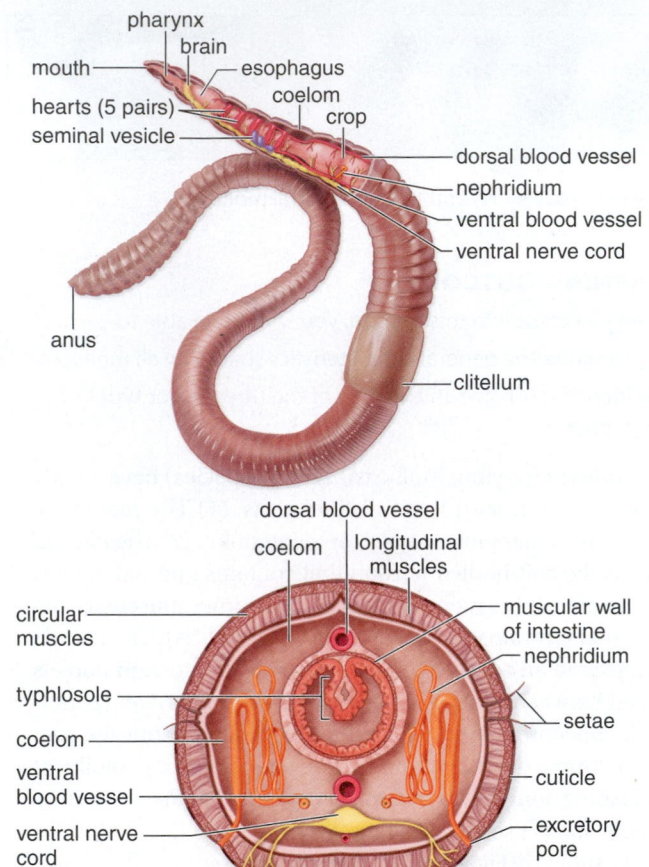

FIGURE 20.8A Earthworm anatomy.

a. Christmas tree worm

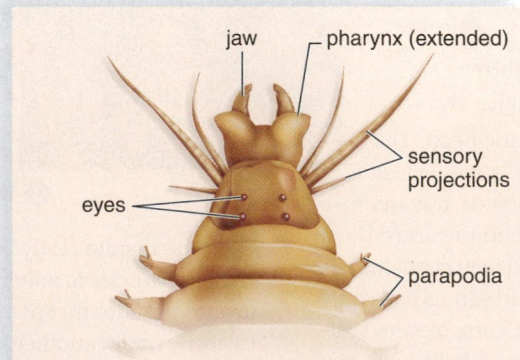

b. Clam worm

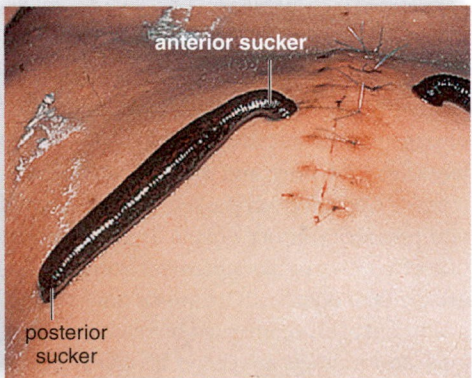

c. Medicinal leech

FIGURE 20.8B Other annelids.

20.9 Roundworms are nonsegmented and plentiful

LEARNING OUTCOME

When you complete this section, you should be able to

1. Understand that several types of roundworms cause parasitic infections.

Among the protostomes, roundworms and arthropods are ecdysozoans, meaning that they are the **molting** animals. When they molt, roundworms and arthropods shed their outer covering.

Roundworms (phylum Nematoda, 20,000 species) are nonsegmented, meaning that they have a smooth outside body wall. Roundworms are generally colorless and less than 5 cm in length, and they occur almost everywhere—in the sea, in fresh water, and in the soil—in such numbers that thousands of them can be found in a small area. Many are free-living and feed on algae, fungi, microscopic animals, dead organisms, and plant juices, causing great agricultural damage. Parasitic roundworms live anaerobically in every type of animal and many plants. Several parasitic roundworms infect humans.

Ascaris Humans become infected with a roundworm called *Ascaris* (Fig. 20.9) when eggs enter the body via uncooked vegetables, soiled fingers, or ingested fecal material and hatch in the intestines. The juveniles make their way into the cardiovascular system and are carried to the heart and lungs. From the lungs, the larvae travel up the trachea, where they are swallowed and eventually reach the intestines. There, the larvae mature and begin feeding on intestinal contents. A female *Ascaris* is very prolific, producing over 200,000 eggs daily. The eggs are eliminated with host feces.

Other Roundworm Parasites Trichinosis is a fairly serious human infection rarely seen in the United States. Humans acquire the disease when they eat meat that contains encysted larvae.

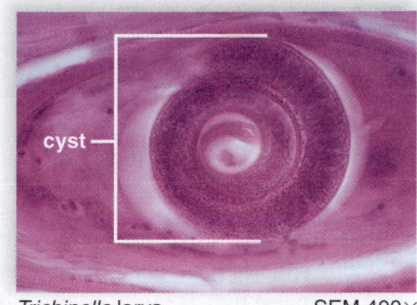

Trichinella larva SEM 400×

Once in the digestive tract, the cysts release the larvae, which develop into adult worms. The female then burrows into the wall of the host's small intestine, where she deposits live larvae that are carried by the bloodstream to the skeletal muscles, where they encyst. The presence of adults in the small intestine causes digestive disorders, fatigue, and fever. After the larvae encyst in muscles, the symptoms include aching joints, muscle pain, and itchy skin.

Elephantiasis is caused by a roundworm called a filarial worm, which utilizes mosquitoes as a secondary host. The adult worms reside in human lymphatic vessels, which normally take up excess tissue fluid but are prevented from doing so by the presence of the worms. The limbs of an infected human can swell to an enormous size, even resembling those of an elephant, hence the name of the disease. More common still is a disabling swelling of the scrotum in men. When a mosquito bites an infected person, it can transport larvae to a new host.

Other roundworm infections are more common in the United States. Children frequently acquire pinworm infections, and hookworm is seen in the southern states, as well as worldwide. A hookworm infection can be very debilitating because the worms attach to the intestinal wall and feed on blood. Good hygiene, proper disposal of sewage, thorough cooking of meat, and regular deworming of pets usually protect people from parasitic roundworms. A common fatal roundworm infection in dogs is due to the heartworm. Mosquitoes serve as the vector.

▶ **20.9 CHECK YOUR PROGRESS**

1. Identify several types of illnesses caused by parasitic roundworms.

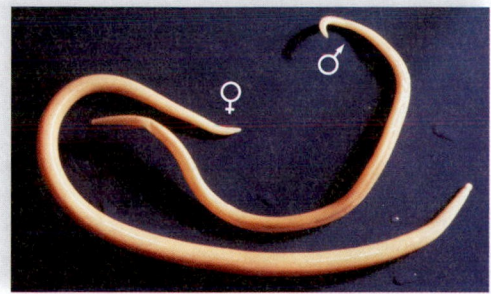

a. *Ascaris*

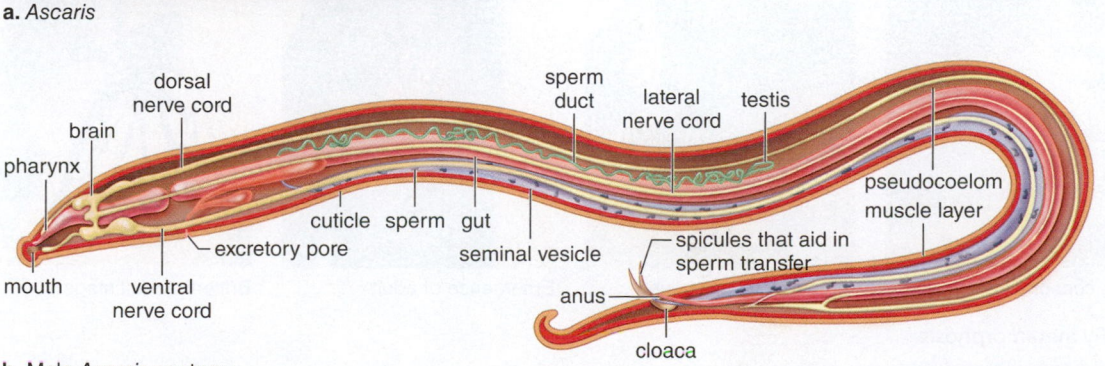

b. Male *Ascaris* anatomy

dorsal nerve cord · brain · pharynx · mouth · ventral nerve cord · cuticle · sperm · gut · excretory pore · seminal vesicle · anus · cloaca · spicules that aid in sperm transfer · sperm duct · lateral nerve cord · testis · pseudocoelom · muscle layer

FIGURE 20.9 a. The roundworm *Ascaris*. **b.** Roundworms such as *Ascaris* have a pseudocoelom and a complete digestive tract with a mouth and an anus.

20.10 Arthropods have jointed appendages

LEARNING OUTCOMES

When you complete this section, you should be able to

1. Identify six characteristics that account for the success of arthropods.
2. Identify the characteristics of crustaceans, arachnids, and insects with examples.

Arthropods (phylum Arthropoda) are extremely diverse. Over 1,150,000 species have been discovered and described, but some experts suggest that as many as 30 million arthropods may exist—most of them insects. The success of arthropods can be attributed to the following six characteristics:

1. *Jointed appendages.* Jointed appendages, which are basically hollow tubes moved by muscles, have become adapted to different means of locomotion, food gathering, and reproduction. Examples are the walking legs of a crayfish shown

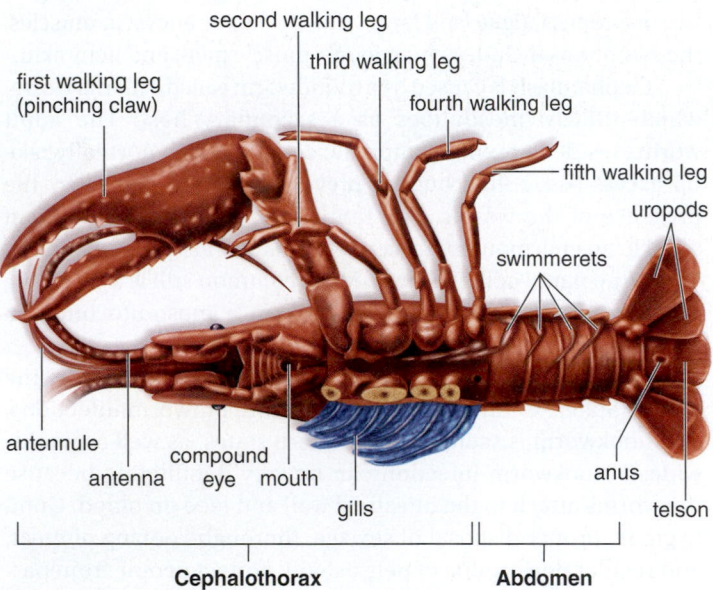

FIGURE 20.10A Exoskeleton and jointed appendages of a crayfish, a crustacean and a well-known arthropod.

in Figure 20.10A. Modifications of appendages account for much of the diversity of arthropods.

2. *Exoskeleton.* A rigid but jointed exoskeleton is composed primarily of **chitin,** a strong, flexible, nitrogenous polysaccharide. The exoskeleton serves many functions, including protecting the body, preventing desiccation, serving as an attachment site for muscles, and aiding locomotion.

   Because an exoskeleton is hard and nonexpandable, arthropods must undergo molting, or shedding of the exoskeleton, as they grow larger. During molting, arthropods are vulnerable and are attacked by many predators.

3. *Segmentation.* Arthropods are segmented animals and have repeating units of the body called segments. In some arthropods, each segment has a pair of jointed appendages; in others, the segments are fused into a head, thorax, and abdomen.

4. *Well-developed nervous system.* Arthropods have a brain and a ventral nerve cord. The head bears various types of sense organs, including compound and simple eyes. Many arthropods also have well-developed touch, smell, taste, balance, and hearing capabilities. Arthropods display many complex behaviors and communication skills.

5. *Adaptation of respiratory organs.* Marine forms utilize gills; terrestrial forms have book lungs (e.g., spiders) or air tubes called **tracheae.** Tracheae serve as a rapid way to transport oxygen directly to the cells. The circulatory system is open, with the dorsal heart pumping blood into various sinuses throughout the body.

6. *Reduced competition through metamorphosis.* Many arthropods undergo a change in form and physiology as a larva becomes an adult. **Metamorphosis** allows the larva to have a different lifestyle than the adult. For example, among insects such as butterflies, the caterpillar feeds on leafy vegetation, while the adult feeds on nectar (Fig. 20.10B). Larval crabs live among and feed on plankton, while adult crabs are bottom dwellers that catch live prey or scavenge dead organic matter.

| Caterpillar, eating stage | Pupa, cocoon stage | Metamorphosis occurs | Emergence of adult | Butterfly, adult stage |

FIGURE 20.10B Monarch butterfly metamorphosis.

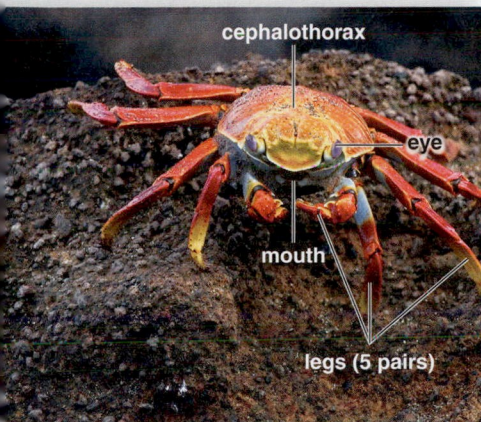

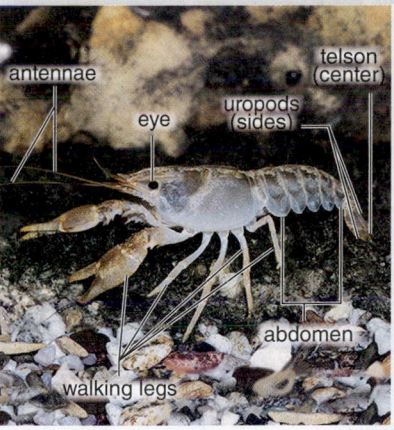
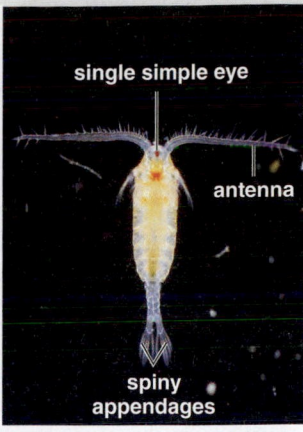

Sally lightfoot crab   Gooseneck barnacles   Crayfish   Copepod

FIGURE 20.10C Crustacean diversity.

Crustaceans The name **crustaceans** is derived from their hard, crusty exoskeleton. Crustaceans (47,000 species) are a group of largely marine arthropods (Fig. 20.10C). Although crustacean anatomy is extremely diverse, both lobsters and crayfish have a head, a part of the cephalothorax that bears a pair of compound eyes, and five pairs of appendages. The first two pairs of appendages, called antennae and antennules, respectively, lie in front of the mouth and have sensory functions. The other three pairs are mouthparts used in feeding. In a crayfish (see Fig. 20.10A), the thorax bears five pairs of walking legs. The first walking leg is a pinching claw. The *gills* are situated above the walking legs. The abdominal segments are equipped with swimmerets, small, paddlelike structures. The last two segments bear the uropods and the telson, which make up a fan-shaped tail.

Crustaceans play a vital role in the food chain. The small shrimp-like krill and especially the tiny copepods are a major source of food for baleen whales, seabirds, and seals. Many species of lobsters, crabs, and shrimp are important in the seafood industry.

Video Voice of the Lobster

Other groups of arthropods live on land: centipedes, with a pair of appendages on every segment, are carnivorous, while millipedes, with two pairs of legs on most segments, are herbivorous (Fig. 20.10D). The head appendages of these animals are similar to those of insects.

The **arachnids** include spiders, scorpions, ticks, mites, and horseshoe crabs (Fig. 20.10E). Spiders have a narrow waist that separates the cephalothorax, which has four pairs of legs, from the abdomen. Spiders use silk threads for all sorts of purposes, from lining their nests to catching prey. The internal organs of spiders also show how they are adapted to a terrestrial way of life. Invaginations of the inner body wall form lamellae ("pages") of spiders' so-called book lungs. Scorpions are the oldest terrestrial arthropods (Fig. 20.10E). Ticks and mites are parasites. Ticks suck the blood of vertebrates and sometimes transmit diseases, such as Rocky Mountain spotted fever or Lyme disease. Like other arachnids, the first pair of appendages in a horseshoe crab is a pinching structure used for feeding and defense (Fig. 20.10E).

Video Lyme Disease

Centipede   Millipede

FIGURE 20.10D Centipede and millipede.

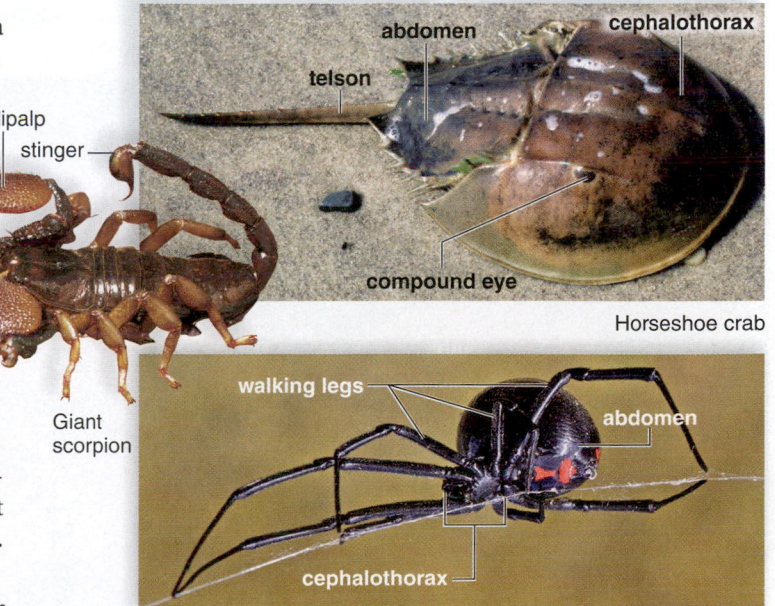

Horseshoe crab

Giant scorpion

Black widow spider

FIGURE 20.10E A spider and two of its relatives.

Insects The **insects** (1,000,000+ described species) are so numerous and so diverse that the study of this one group is a major specialty in biology called entomology (Fig. 20.10F). Some insects show remarkable behavior adaptations, as exemplified by the social systems of bees, ants, termites, and other colonial insects.

Insects are adapted to an active life on land, although some have secondarily invaded aquatic habitats. The body is divided into a head, a thorax, and an abdomen. The head usually bears a pair of sensory antennae, a pair of compound eyes, and several simple eyes. The mouthparts are adapted to each species' particular way of life: A grasshopper has mouthparts that chew, and a butterfly has a long tubular proboscis for siphoning the nectar from flowers. The abdomen contains most of the internal organs; an inefficient circulatory system contains a heart that pumps colorless blood into open spaces called sinuses; an efficient respiratory system consisting of small tubules called tracheae delivers air directly to wing muscles. Wings enhance an insect's ability to survive by providing a way of escaping enemies, finding food, facilitating mating, and dispersing the offspring. The exoskeleton of an insect is lighter and contains less chitin than that of many other arthropods. Insects always have three pairs of legs attached to the thorax.

The male has a penis, which passes sperm to the female. The female, as in the grasshopper, may have an ovipositor for laying the fertilized eggs. Some insects, such as butterflies, undergo complete metamorphosis, involving a drastic change in form (see Fig. 20.10B).

▶ **20.10 CHECK YOUR PROGRESS**

1. List several characteristics of arthropods after examining Figure 20.10A.
2. Explain how you know that the organism in Figure 20.10A is a crustacean.
3. Identify three features of insects that show they are adapted to life on land.

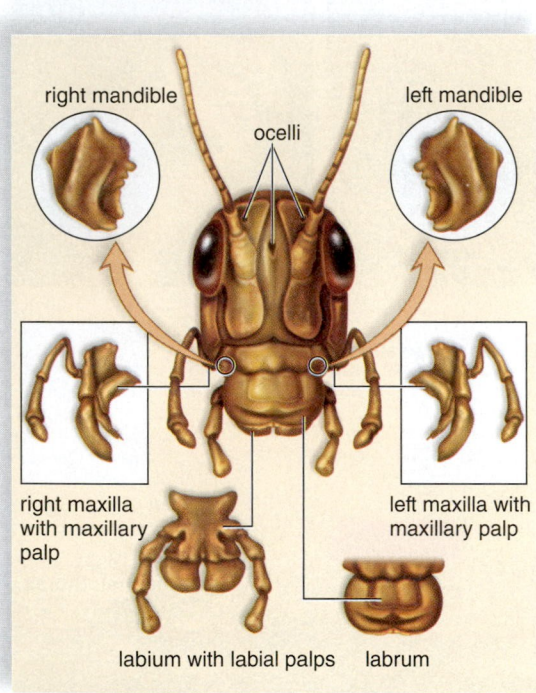

Mouthparts of a grasshopper

Mealybug

Grasshopper

Dragonfly

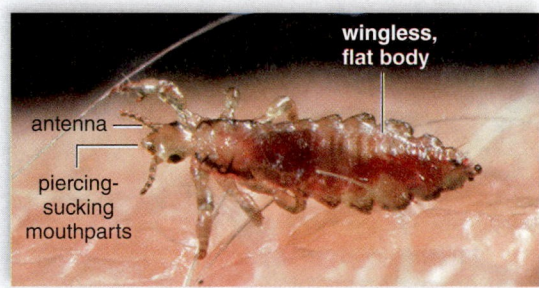

Head louse

Wasp

Butterfly

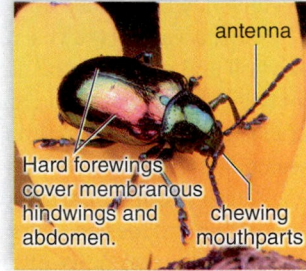

Beetle

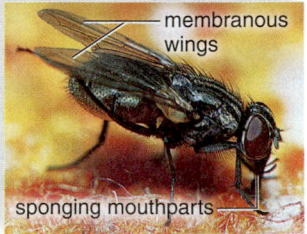

Housefly

Leafhopper

FIGURE 20.10F Insect diversity.

20.11 Echinoderms are radially symmetrical as adults

LEARNING OUTCOMES

When you complete this section, you should be able to

1. Describe how an echinoderm locomotes.
2. Explain how echinoderms are less complex than arthropods.

Echinoderms (phylum Echinodermata, 6,000 species) lack features associated with vertebrates, and yet we know they are related to chordates because both are deuterostomes. The echinoderms are radially, not bilaterally, symmetrical as adults (Fig. 20.11). However, their larva is a free-swimming filter feeder with bilateral symmetry—it metamorphoses into the radially symmetrical adult. Also, adult echinoderms do not have a head, a brain, or segmentation. Their nervous system consists of nerves in a ring around the mouth extending outward radially.

Echinoderm locomotion depends on a **water vascular system.** In the sea star, water enters this system through a sieve plate (Fig. 20.11). Eventually, it is pumped into many tube feet, expanding them. When the foot touches a surface, the center withdraws, producing suction that causes the foot to adhere to the surface. By alternating the expansion and contraction of its many tube feet, a sea star moves slowly along.

Echinoderms don't have a complex respiratory, excretory, or circulatory system. Fluids within the coelomic cavity and the water vascular system carry out many of these functions. For example, gas exchange occurs across the skin gills and the tube feet. Nitrogenous wastes diffuse through the coelomic fluid and the body wall.

Brittle stars are infrequently seen but actually are a common type of echinoderm characterized by a small central disk with five snakelike arms. The arms are flexible and brittle; they break off but can regrow. In contrast, the remains of sand dollars are commonly seen on the beach. In a living sand dollar, skin gills emerge from pores and account for the petal-like pattern on the aboral side. Sea urchins have a spiny shell, and their tube feet are longer than the spines! Both spines and tube feet help with locomotion and capturing food. Sea cucumbers have a shape that really does resemble a cucumber; five rows of tube feet run the length of the animal and at one end are modified as a feeding apparatus. Feather stars have a cup-shaped body; cirri (appendages) attached to the underside of the cup cling to sponges, corals, or rocks. Their feathery arms, in multiples of five, are coated with a sticky substance that helps capture food. Most echinoderms feed variously on organic matter in the sea or substrate, but sea stars prey upon crustaceans, molluscs, and other invertebrates.

Video Sea Urchin Reproduction

▶ **20.11 CHECK YOUR PROGRESS**
1. Identify several types of echinoderms and what they have in common.

FIGURE 20.11 Echinoderm structure and diversity.

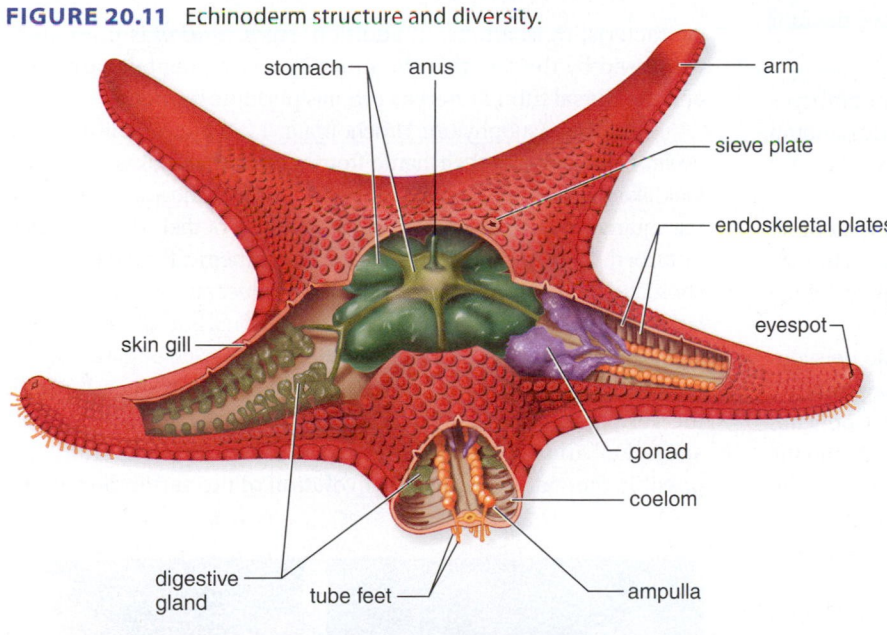

Sea star (starfish) anatomy

Sea lily

Brittle star

Sea cucumber

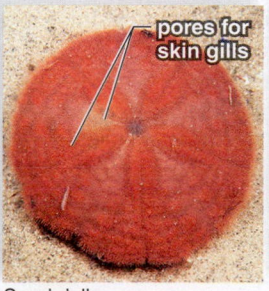

Sea urchin

Sand dollar

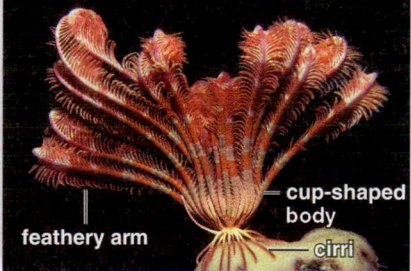

Feather star

The Vertebrates

Four characteristics distinguish the chordates from the other animal phyla. One of these is the presence of the notochord. The invertebrate chordates have a notochord as adults, but in the **vertebrates,** the notochord is replaced by vertebrae. This and the other three characteristics of all chordates pertain to developmental features that may be modified or even disappear as development proceeds. For example, a postanal tail appears during the development of humans but then largely disappears. Aside from these four features, we can mention that chordates have an internal skeleton. An internal skeleton increases in size as the animal grows larger. After examining the key features and evolution of vertebrates in section 20.12, we describe each group of vertebrates in the order they evolved. See Figure 20.12C.

20.12 Four features characterize chordates

LEARNING OUTCOMES

When you complete this section, you should be able to

1. Describe four features all chordates have in common.
2. Describe two features for each of the two groups of nonvertebrate chordates.
3. Identify six derived traits that evolved during the course of vertebrate evolution.

Chordates (phylum Chordata, 63,500 species), like echinoderms, have the deuterostome pattern of development. Most chordates are vertebrates, whose skeleton is internal with muscles attached to its outer suface. (In invertebrates, the skeleton is external, and the muscles are attached to an inner surface.) The placement in vertebrates allows them to enjoy freedom of movement and to attain a larger size than invertebrates. The chordates have the four characteristics depicted in Figure 20.12A and listed here:

1. *The **notochord,** a dorsal support rod,* is present in embryos and sometimes for life. Vertebrates have an endoskeleton of cartilage or bone, including a vertebral column, that has replaced the notochord during development.
2. *A dorsal tubular nerve cord* contains a canal filled with fluid. In vertebrates, the nerve cord is protected by the vertebrae. Therefore, it is called the spinal cord because the vertebrae form the spine.
3. *Pharyngeal pouches* are seen only during embryonic development in most vertebrates. In the invertebrate chordates, the fishes, and some amphibian larvae, the pharyngeal pouches become functioning *gills.* Water passing into the mouth and the pharynx goes through the gill slits, which are supported by *cartilaginous gill arches.* In terrestrial vertebrates that breathe with lungs, the pouches are modified for various purposes. In humans, the first pair of pharyngeal pouches become the auditory tubes. The second pair become the tonsils, while the third and fourth pairs become the thymus gland and the parathyroid glands.
4. *A postanal tail* extends beyond the anus.

Invertebrate Chordates In the **invertebrate chordates,** the notochord is never replaced by the vertebral column. **Lancelets** (subphylum Cephalochordata, 23 species) are marine chordates only a few centimeters long. They look like a lancet, a small, two-edged surgical knife. Lancelets are found in the shallow water along most coasts, where they usually lie partly buried in sandy or muddy substrates with only their anterior mouth and gill apparatus exposed (Fig. 20.12B*a*). They feed on microscopic particles filtered out of the constant stream of water that enters the mouth and exits through the gill slits. Lancelets retain the four chordate

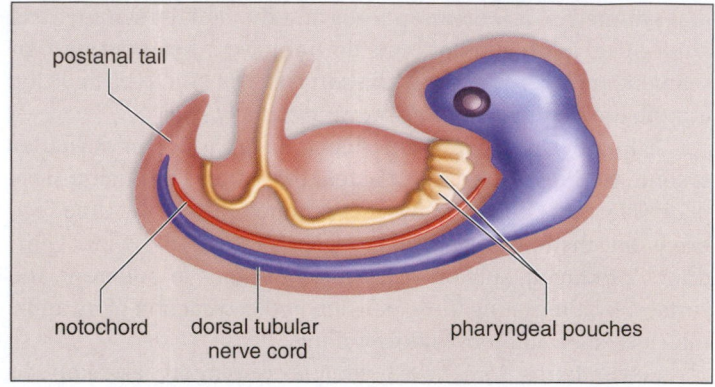

FIGURE 20.12A The four chordate characteristics.

characteristics as adults. In addition, *segmentation* is present, as witnessed by the fact that the muscles are segmentally arranged and the dorsal tubular nerve cord has periodic branches.

Tunicates (subphylum Urochordata, 1,250 species) live on the ocean floor and take their name from a tunic that makes the adults look like thick-walled, squat sacs (Fig. 20.12B*b*). They are also called sea squirts because they squirt water from one of their siphons when disturbed. The tunicate larva is bilaterally symmetrical and has the four chordate characteristics. Metamorphosis produces the sessile adult that lacks a notochord, a nerve cord, and a postanal tail but cilia move water into the pharynx and out numerous gill slits supported by gill arches.

The invertebrate chordates share a common ancestor with the vertebrates (Fig. 20.12C). In contrast to the sudden appearance of all animal phyla at the start of the Cambrian period, it is possible thereafter to trace the evolution of the vertebrates in the

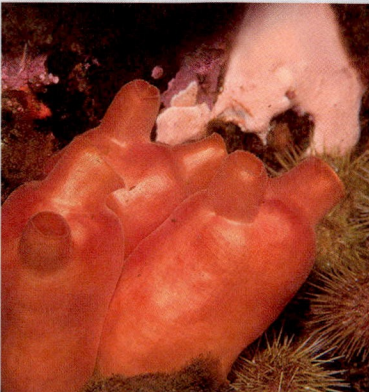

a. A lancelet **b.** Three tunicates

FIGURE 20.12B The invertebrate chordates.

fossil record. Therefore, the phylogenetic tree of the chordates in Figure 20.12C is supported by the fossil record.

Vertebrates The vertebrates include the four types of fishes shown in Figure 20.12C, the amphibians, the reptiles, and the mammals. The designation of birds as reptiles is supported by much evidence such as that presented in section 1.7 and "Motherhood Among Dinosaurs" on pages 298–299. The several types of mammals, our closest relatives, are discussed in section 20.16.

In Figure 20.12C, the fishes were the first type of vertebrate to evolve. Fishes, which live in both marine and fresh water, have more species than any of the other types of vertebrates. A few of today's fishes lack jaws and thus have to suck and otherwise engulf their prey. But most fishes have *jaws,* which are a more efficient means of grasping and eating prey. The jawed fishes and all the other vertebrates are gnathostomes—animals with jaws.

Certain of the early fishes not only had jaws, but also had a *bony skeleton, lungs,* and *fleshy fins.* These characteristics were preadaptive for a land existence, and the amphibians, the first vertebrates to live on land, evolved from these fishes. The amphibians were the first vertebrates to have *limbs.* The terrestrial vertebrates are tetrapods because they have four limbs. Some, such as snakes, no longer have four limbs, but their evolutionary ancestors had four limbs.

Many amphibians, such as the frog, reproduce in an aquatic environment. This means that, in general, amphibians are not fully adapted to living on land. However, reptiles are fully adapted to life on land because, among other features, they produce an *amniotic egg.* The amniotic egg is so named because the embryo is surrounded by an amniotic membrane that encloses the amniotic fluid. Therefore, it is obvious that amniotes develop within an aquatic environment of their own making. In placental mammals, such as humans, the fertilized egg develops inside the female, where the embryo and later the fetus is surrounded by an amniotic membrane. Reproduction in mammals also includes the ability to nurse their young for a period of time after birth because females have mammary glands that produce milk following the birth of a newborn.

▶ **20.12 CHECK YOUR PROGRESS**

1. Identify the four features all chordates have in common.
2. Explain why adult tunicates are classified as chordates even though they look like thick-walled, squat sacs.
3. Identify each of the derived traits in Figure 20.12C with a particular group of vertebrates.

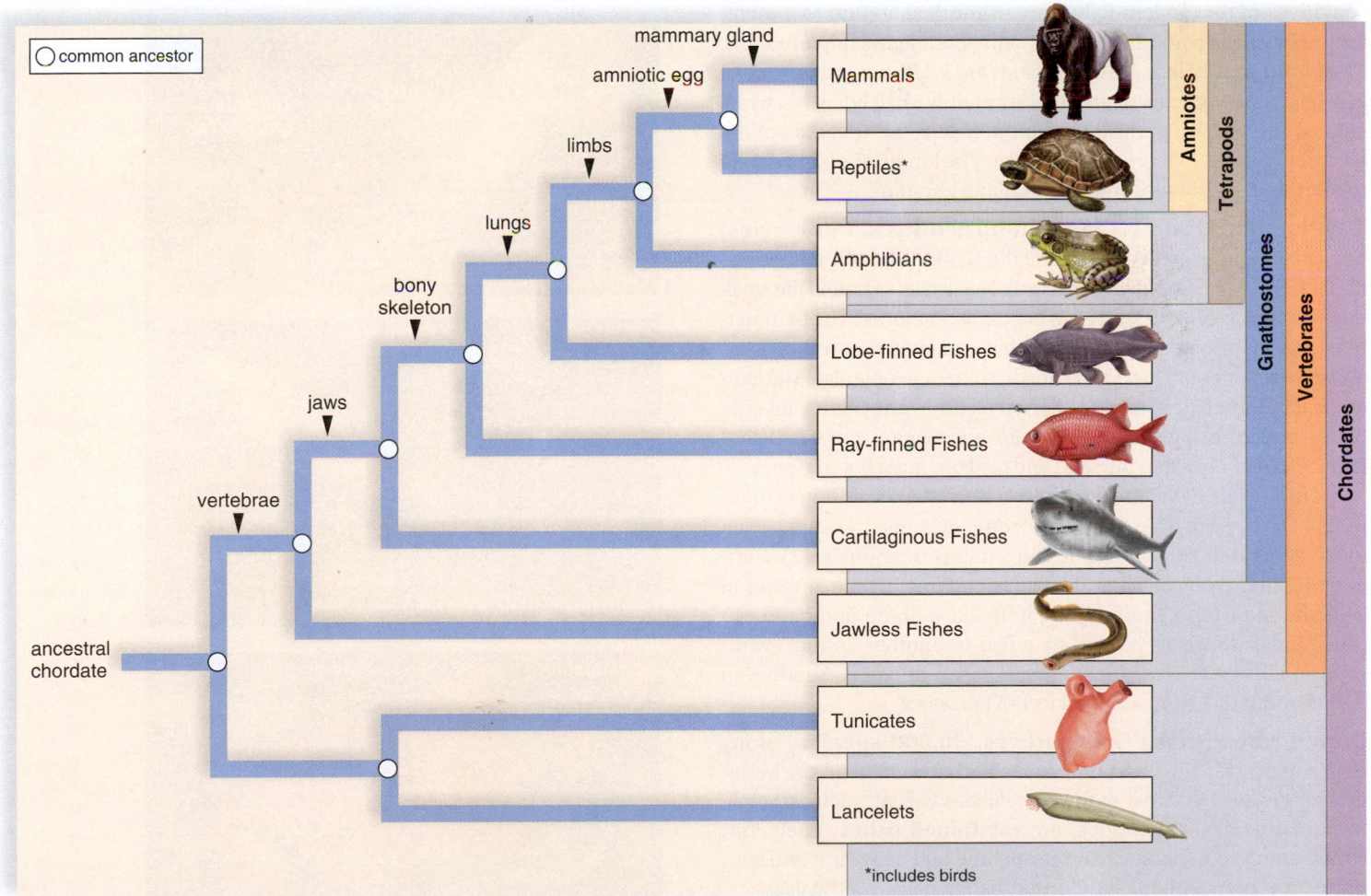

FIGURE 20.12C Phylogenetic tree of the chordates. All vertebrates are chordates (orange and purple bars to right). The vertebrates that have jaws are called gnathostomes (blue bar); the vertebrates that have four legs are tetrapods (brown bar); and the vertebrates whose embryo is surrounded by an amnion (see page 410) are amniotes (tan bar). The amniotic egg made reproduction on land possible.

20.13 Jaws, a bony skeleton, and lungs evolved among the fishes

LEARNING OUTCOME

When you complete this section, you should be able to

1. Identify and discuss the three types of fishes.

The first vertebrates were jawless fishes, which wiggled through the water and sucked up food from the ocean floor. Today, there are three living groups of fishes: jawless fishes, cartilaginous fishes, and bony fishes. The two latter groups have *jaws,* tooth-bearing bones of the head. Jaws are believed to have evolved from the first pair of gill arches, structures that ordinarily support gills.

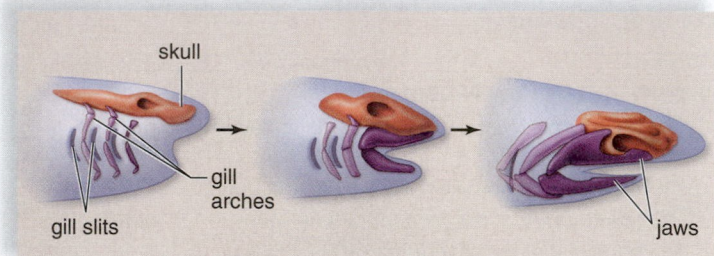

The presence of jaws permits a predatory way of life.

Jawless Fishes (Class Agnatha, 65 species) Living representatives of the **jawless fishes** are cylindrical and up to a meter long. They have smooth, scaleless skin and no jaws or paired fins. The two groups of living jawless fishes are *hagfishes* and *lampreys*. The hagfishes are scavengers, feeding mainly on dead fishes, while some lampreys are parasitic. When parasitic, the oral disk of the lamprey (Fig. 20.13) serves as a sucker. The lamprey attaches itself to another fish and taps into its circulatory system.

Cartilaginous Fishes (Class Chondrichthyes, 750 species) The **cartilaginous fishes,** including the sharks (Fig. 20.13), the rays, and the skates, have skeletons of cartilage, instead of bone. The small dogfish shark is often dissected in biology laboratories. The hammerhead shark is an aggressive predator that usually feeds on other fishes and invertebrates but has been known to attack people also. The largest sharks, the whale sharks, feed on small fishes and marine invertebrates and do not attack humans. Skates and rays are rather flat fishes that live partly buried in the sand and feed on mussels and clams.

Three well-developed senses enable sharks to detect their prey: (1) They are able to sense electric currents in water—even those generated by the muscle movements of animals. (2) They, and all other types of fishes, have a lateral line system, a series of cells that lie within canals along both sides of the body and can sense pressure waves caused by a fish or another animal swimming nearby. (3) They have a keen sense of smell. Sharks can detect about one drop of blood in 115 l of water.

Bony Fishes (Class Osteichthyes, 30,000 species) Bony **fishes** are by far the most numerous and diverse of all the vertebrates (Fig. 20.13). Most of the bony fishes we eat, such as perch, trout, salmon, and haddock, are **ray-finned fishes.** Their fins, which are used to balance and propel the body, are thin and supported by bony spikes. Ray-finned fishes have various ways of life. Some, such as herring, are filter feeders; others, such as trout, are opportunists; and still others, such as piranhas and barracudas, are predaceous carnivores.

Video Cichlid Specialization

Animation Bony Fishes

Ray-finned fishes have a swim bladder, which usually serves as a buoyancy organ. The streamlined shape, fins, and muscle action of ray-finned fishes are all suited to locomotion in the water. Their skin is covered by bony scales that protect the body but do not prevent water loss. To respire, the gills are kept continuously moist by the passage of water through the mouth and out the gill slits. As the water passes over the gills, oxygen is absorbed by the blood, and carbon dioxide is given off. Ray-finned fishes have a single-circuit circulatory system. The heart is a simple pump, and blood flows through its chambers, including a nondivided atrium and ventricle, to the gills. O_2-rich blood leaves the gills and goes to the body proper, eventually returning to the heart for recirculation.

Another type of bony fish is called the **lobe-finned fishes.** Ancestral lobe-finned fishes not only had fleshy appendages that could be adapted to land locomotion, but most also had a lung, which was used for respiration.

▶ **20.13 CHECK YOUR PROGRESS**

1. Identify three ways ray-finned fishes are adapted to live in the water.

FIGURE 20.13 Diversity of fishes.

20.14 Amphibians are tetrapods that can move on land

LEARNING OUTCOME

When you complete this section, you should be able to

1. Understand why the term amphibian appropriately describes many members of this group.

Amphibians (class Amphibia, 6,400 species), whose class name means living on both land and in the water, are represented today by frogs (including toads) and salamanders (including newts). Aside from jointed limbs, amphibians have other features not seen in bony fishes: eyelids for keeping their eyes moist, ears adapted to picking up sound waves, and a voice-producing larynx. The brain is larger than that of a fish. Adult amphibians usually have small lungs. Air enters the mouth by way of nostrils, and when the floor of the mouth is raised, air is forced into the relatively small lungs. Respiration is supplemented by gas exchange through the smooth, moist, and glandular skin. The amphibian heart has only three chambers, compared to the four chambers in a mammalian heart. Mixed blood flows to all parts of the body; some is sent to the skin, where it is further oxygenated.

Most members of this group lead an amphibious life—that is, the larval stage lives in the water, and the adult stage is adapted to living on land. Figure 20.1A illustrates how the frog tadpole undergoes metamorphosis into an adult before taking up life on land. However, the adult usually returns to the water to reproduce. Figure 20.14A compares the appearance of a frog to that of another amphibian, a salamander. In a frog, the head and trunk are fused, and the long hindlimbs are specialized for jumping. The tree frogs have adhesive toepads that allow them to climb trees, while others, the spadefoots, have hardened spades that act as shovels enabling them to dig into the soil. Frogs have smooth skin, and they live in or near fresh water; toads have stout bodies and warty skin, and they live in dark, damp places away from the water. Most salamanders have limbs that are set at right angles to the body and resemble the earliest fossil amphibians. They move like a fish, with a side-to-side, S-shaped motion.

 **Video** Tadpole Development

Video Frog Reproduction

Paleontologists have recently found a well-preserved transitional fossil from the Late Devonian period in arctic Canada that represents an intermediate between lobe-finned fishes and amphibians with limbs. The name of the fossil is *Tiktaalik roseae* (Fig. 20.14B, *left*). This fossil provides unique insights into how the legs of tetrapods arose (Fig. 20.14B, *right*).

 **Animation** Early Vertebrates

▶ **20.14 CHECK YOUR PROGRESS**

1. How does metamorphosis benefit amphibians?

FIGURE 20.14A Frogs and salamanders are well-known amphibians.

a. European edible frog

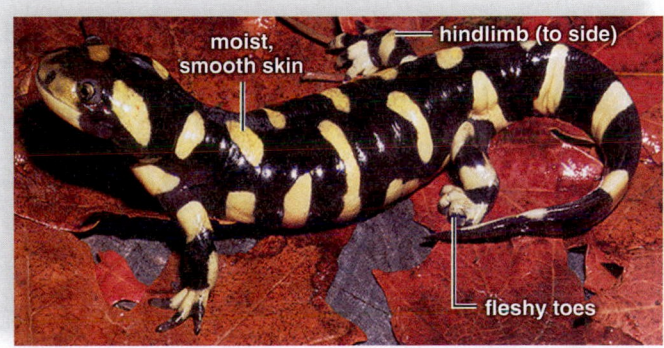

b. Barred tiger salamander

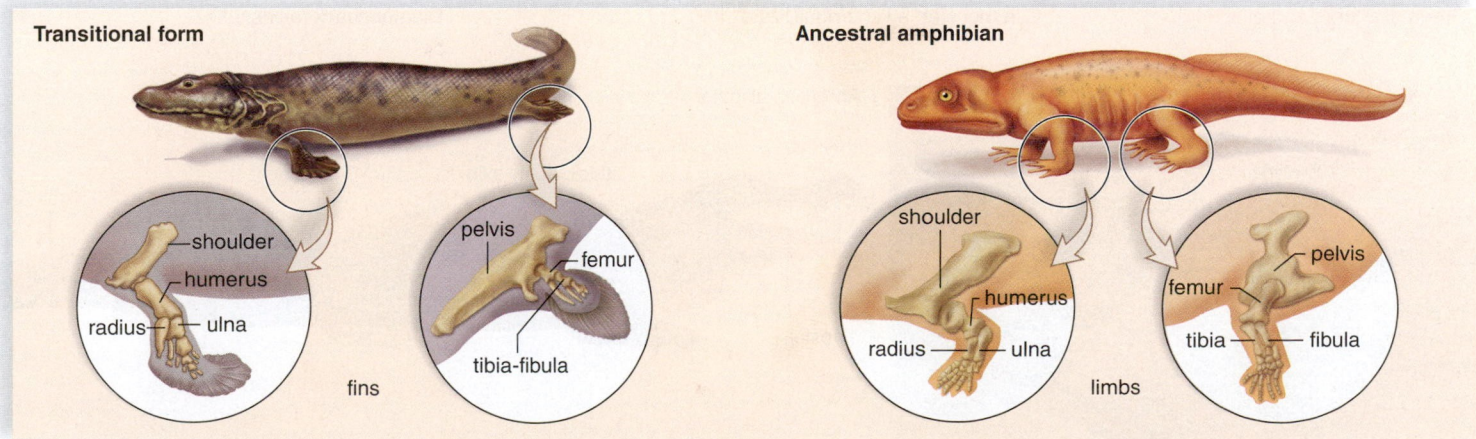

FIGURE 20.14B This transitional form links the fins of ancestral lobe-finned fishes to the limbs of ancestral amphibians.

20.15 Reptiles have an amniotic egg and can reproduce on land

LEARNING OUTCOMES

When you complete this section, you should be able to

1. Identify the adaptations of reptiles, including birds, to a dry land environment.
2. Explain how birds are adapted to the ability to fly.

Reptiles (class Reptilia, 8,000 species) diversified and were abundant during the Permian period and the whole of the Mesozoic era. These animals included the dinosaurs, which became extinct, except for those that evolved into birds. Some dinosaurs are remembered for their great size. *Brachiosaurus*, a herbivore, was about 23 m long and about 17 m tall. *Tyrannosaurus rex*, a carnivore, was 5 m tall when standing on its hind legs. The bipedal stance of some dinosaurs was preadaptive for the evolution of wings in birds.

The reptiles living today are mainly alligators, crocodiles, turtles, snakes, lizards, and tuataras (Fig. 20.15A). The body of a reptile is covered with hard, keratinized scales, which protect the animal from desiccation and from predators. Reptiles have well-developed lungs enclosed by a protective and functional rib cage. The heart has four chambers, but the septum that divides the two halves is incomplete in certain species; therefore, some mixing of O_2-rich and O_2-poor blood occurs.

 Video Leaf-Tailed Geckos

Video Basilisk Lizards

Perhaps the most outstanding adaptation of the reptiles is their means of reproduction, which is suitable to a land

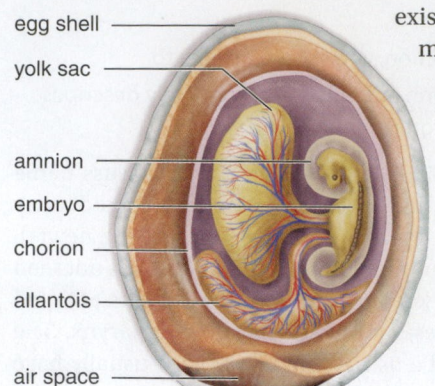

egg shell
yolk sac
amnion
embryo
chorion
allantois
air space

Amniotic egg

existence. The penis of the male passes sperm directly to the female. Fertilization is internal, and the female lays leathery, flexible, shelled eggs. The **amniotic egg** made development on land possible and eliminated the need for a swimming larval stage during development. The amniotic egg has extraembryonic membranes that provide the developing embryo with atmospheric oxygen (chorion) and food (yolk sac). It also removes nitrogenous wastes (allantois) and protects the embryo from drying out and from mechanical injury (amnion).

Fishes, amphibians, and reptiles are **ectotherms,** meaning that their body temperature matches the temperature of the external environment. If it is cold externally, they are cold internally; if it is hot externally, they are hot internally. Most reptiles regulate their body temperatures by exposing themselves to the sun if they need warmth or by hiding in the shadows if they need cooling off.

shell
beak
flipper

Green sea turtle and offspring

clawed foot

Gila monster, a venomous lizard

venom gland
fang
rattle

Diamondback rattlesnake

third eye (not visible)
scaly skin
tail

Tuatara, a living fossil

American alligator

thick, scaly skin
tail
tongue
nostril

FIGURE 20.15A Reptilian diversity.

Birds (class Aves, 10,000 species) are characterized by the presence of *feathers,* which are modified reptilian scales. (Perhaps you have noticed the scales on the legs of a chicken.) Ample data today indicate that birds are closely related to bipedal dinosaurs and that they are reptiles. However, birds lay a hard-shelled amniotic egg, rather than a leathery egg as other reptiles do.

Nearly every anatomic feature of a bird can be related to its ability to fly. The forelimbs are modified as wings. Bird flight requires an airstream and a powerful wing downstroke for lift, a force at right angles to the airstream (Fig. 20.15B). The hollow, very light bones are laced with air cavities. A horny beak has replaced jaws equipped with teeth, and a slender neck connects the head to a rounded, compact torso. Respiration is efficient, because the lobular lungs form anterior and posterior air sacs. The presence of these sacs means that the air moves one way through the lungs, and gases are continuously exchanged across respiratory tissues (see Fig. 32.5c). Another benefit of air sacs is that they lighten the body and aid flying.

Birds have a four-chambered heart that completely separates O_2-rich blood from O_2-poor blood. Unlike other reptiles, birds are **endotherms,** meaning that they generate internal heat. Many endotherms can use metabolic heat to maintain a constant internal temperature. This may be associated with their efficient nervous, respiratory, and circulatory systems. Also, their feathers provide insulation. Birds have no bladder and excrete uric acid in a semidry state.

downstroke

upstroke

FIGURE 20.15B Bird flight.

Birds have particularly acute vision and well-developed brains. Their muscle reflexes are excellent. These adaptations are suited to flight. An enlarged portion of the brain seems to be the area responsible for instinctive behavior. A ritualized courtship often precedes mating. Many newly hatched birds require parental care before they are able to fly away and seek food for themselves. A remarkable aspect of bird behavior is the seasonal migration of many species over very long distances. Birds navigate by day and night, whether it's sunny or cloudy, by using the sun and stars and even the Earth's magnetic field to guide them.

Video
Harris Hawk

Video
Bird Radar

The majority of birds, including eagles, geese, and mockingbirds, can fly. However, some birds, such as emus, penguins, kiwis, and ostriches, are flightless. Traditionally, the classification of birds was based on type of beak (Fig. 20.15C) and foot, and to a lesser extent on habitat and behavior. A bald eagle's beak tears prey apart; a woodpecker's beak can drill in wood; a flamingo's beak strains food from water; a vulture's beak can grasp flesh; and a cardinal's beak can crack tough seeds.

Video
Finches Adaptive
Radiation

▶ **20.15 CHECK YOUR PROGRESS**
  1. Explain why sea turtles reproduce on land and breathe air.
  2. List two characteristics of birds that show they are adapted for flight.

Bald eagle

Pileated woodpecker

FIGURE 20.15C Types of bird beaks.

Flamingo

Turkey vulture

Cardinal

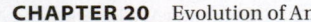

LEARNING OUTCOMES

When you complete this section, you should be able to

1. Discuss the features that mammals have in common and identify three major types of mammals.
2. Identify ten types of placental mammals.

Mammals (class Mammalia, 5,500 species) appeared during the Triassic period, (251–199 MYA), about the same time as the first dinosaurs. The first mammals were small, about the size of mice. During all the time the dinosaurs flourished (199–165 MYA), mammals were a minor group that changed little. Some of the earliest mammalian groups are still represented today by the monotremes and marsupials, but they are not abundant. The placental mammals that evolved later went on to live in many habitats.

The two chief characteristics of mammals (class Mammalia) are *hair* and *milk-producing mammary glands*. Mammals are endotherms, and many of their adaptations are related to temperature control. Hair, for example, provides insulation against heat loss and mammals are active, even in cold weather.

Mammary glands enable females to feed (nurse) their young without leaving them to find food. Nursing also creates a bond between mother and offspring that helps ensure parental care while the young are helpless. In most mammals, the young are born alive after a period of development in the uterus, a part of the female reproductive system. Internal development shelters an offspring and allows the female to move actively about while awaiting its birth.

Monotremes (Fig. 20.16A*a*) are mammals that, like birds, have a cloaca, a terminal region of the digestive tract serving as a common chamber for feces, excretory wastes, and sex cells. They also lay *hard-shelled amniotic eggs.* They are represented by the spiny anteater and the duckbill platypus, both of which live in Australia. The female duckbill platypus lays her eggs in a burrow in the ground. She incubates the eggs, and after hatching, the young lick up milk that seeps from mammary glands on her abdomen. The spiny anteater has a pouch on the belly side formed by swollen mammary glands and longitudinal muscle. Hatching takes place in this pouch, and the young remain there for about 53 days. Then they stay in a burrow, where the mother spiny anteater periodically visits and nurses them.

The offspring of **marsupials** (Fig. 20.16A*b, c*) begin their development inside the female's body, but they are born in a very immature condition. Newborns crawl up into a *pouch on their mother's abdomen.* Inside the pouch, they attach to the nipples of mammary glands and continue to develop. Frequently, more are born than can be accommodated by the number of nipples, and it's "first come, first served."

In Australia, marsupials underwent adaptive radiation for several million years without competition. Thus, marsupial mammals are now found mainly in Australia, with some in Central and South America as well. The Virginia opossum is the only marsupial that occurs north of Mexico. Among the herbivorous marsupials, koalas are tree-climbing browsers, and kangaroos are grazers. The Tasmanian wolf or tiger, thought to be extinct, was a carnivorous marsupial about the size of a collie dog.

The vast majority of living mammals are **placental mammals** (Fig. 20.16B). Developing placental mammals are dependent on the **placenta,** an organ of exchange between maternal blood and fetal blood. Nutrients are supplied to the growing offspring, and wastes are passed to the mother for excretion. While the fetus (developing offspring) is clearly parasitic on the female, in exchange she is able to freely move about while the fetus develops.

Placental mammals are adapted to live on land (some have returned to the water or can fly) and have limbs that allow them to move rapidly. In fact, an evaluation of mammalian features leads us to the obvious conclusion that they lead active lives. The brain is well developed and enlarged due to the expansion of the cerebral hemisphere; the lungs are expanded not only by the action of the rib cage but also by the contraction of the diaphragm, a horizontal muscle that divides the thoracic cavity from the abdominal cavity; and the heart has four chambers. They are endotherms with a warm internal temperature. Hair, when abundant, helps insulate the body. The brain is not fully developed until after birth, and young learn to take care of themselves during a period of dependency on their parents.

These are some of the groups of placental mammals:

The *ungulates* are hoofed mammals, which comprise about a third of all living placentals. The hoofed mammals have a reduced number of toes and are divided into two groups, depending on whether an odd number of toes remains (e.g., horses, zebras, tapirs, rhinoceroses) or whether an even number of toes remains (e.g., pigs, cattle, deer, hippopotamuses, buffaloes,

a. Duckbill platypus, a monotreme of Australian streams

b. Virginia opossum, the only American marsupial

c. Koala, a tree-dwelling Australian marsupial

FIGURE 20.16A Monotremes and marsupials.

White-tailed deer, a forest-dwelling ungulate

African lioness, a grassland-dwelling carnivore

Human, a ground-dwelling primate

Killer whale, a sea-dwelling cetacean

FIGURE 20.16B
Placental mammals.

giraffes). Many of the hoofed animals have elongated limbs and are adapted for running, often across open grasslands. Both groups of ungulates are herbivorous and have large, grinding teeth.

The *carnivores* (e.g., dogs, cats, bears, raccoons, skunks, lions) are predaceous meat eaters with large, conical-shaped canine teeth. Some carnivores are aquatic (e.g., seals, sea lions, and walruses) but must return to land to reproduce.

Video
Mom Grizzly
Teaches Her Cubs

The *primates* are tree-dwelling fruit eaters (e.g., lemurs, monkeys, gibbons, chimpanzees, gorillas, and humans). Humans are ground dwellers, well known for their opposable thumb and well-developed brain.

The *cetaceans* are well-known marine whales and dolphins, which have very little hair or fur. Baleen whales feed by straining large quantities of water containing plankton. Toothed whales feed mainly on fish and squid.

The *chiroptera* are the flying mammals (i.e., bats) that can hang by their feet. Their wings consist of two layers of skin and connective tissue stretched between the elongated bones of all forelimb digits but the first. Many species use echolocation to navigate at night and to locate their usual insect prey. But there are also bird-, fish-, frog-, plant-, and blood-eating bats.

Video
Bat Echolocation

The *rodents* are most often small plant eaters (e.g., mice, rats, squirrels, beavers, and porcupines). The incisors of these gnawing animals suffer heavy wear and tear, and they grow continuously.

The *proboscideans*, the herbivorous elephants, are the largest living land mammals. Their upper lip and nose have become elongated and muscularized to form a trunk.

The *lagomorphans* are the rodentlike jumpers (e.g., rabbits, hares, and pikas). These herbivores have two pairs of continually growing incisors, and their hind legs are longer than their front legs.

The *insectivores* are the small burrowing mammals (e.g., shrews and moles), which have short snouts and live primarily underground.

At one time, researchers thought that insectivores were most like the original placentals. However, more recent analysis suggests that the *edentates* (anteaters) and *pangolins* (scaly anteaters) are the more primitive groups of living placentals.

▶ **20.16 CHECK YOUR PROGRESS**

1. Identify three types of mammals and give examples of each type.

20B Many Vertebrates Provide Medical Treatments for Humans

Hundreds of pharmaceutical products come from other vertebrates, and even those that produce poisons and toxins provide medicines that benefit us.

Natural Products with Medical Applications

The Thailand cobra paralyzes its victim's nerves and muscles with a potent venom that eventually leads to respiratory arrest. However, that venom is also the source of the drug tetrachlorodecaoxide (Immunokine), which has been used for over a decade in multiple sclerosis patients. Immunokine, which is almost without side effects, actually protects the patient's nerve cells from destruction by their immune system.

A compound known as ABT-594, derived from the skin of the poison-dart frog (Fig. 20B*a*), is approximately 50 times more powerful than morphine in relieving chronic and acute pain without addicitve properties.

The southern copperhead snake and the fer-de-lance pit viper are two of the unlikely vertebrates that either serve as the source of pharmaceuticals or provide a chemical model for the synthesis of effective drugs in the laboratory. These drugs include anticoagulants ("clot busters"), painkillers, antibiotics, and anticancer drugs.

A variety of friendlier vertebrates produce proteins that are similar enough to human proteins to be used for medical treatment. Until 1978, when recombinant DNA human insulin was produced, diabetics injected purified insulin from pigs. Currently, the flu vaccine is produced in fertilized chicken eggs. The production of these drugs, however, is often time-consuming, labor intensive, and expensive. Approximately 600 million chicken eggs are used each year in vaccine production.

Animal Pharming

Some of the most powerful applications of genetic engineering can be found in the development of drugs and therapies for human diseases. In fact, this new biotechnology has actually led to a new industry: animal pharming. Animal pharming uses genetically modified (GM) vertebrates, such as mice, sheep, goats, cows, pigs, and chickens, to produce medically useful pharmaceutical products.

The human gene for some useful product is inserted into the embryo of the vertebrate. That embryo is implanted into a foster mother, which gives birth to the GM animal, which contains genes from the two sources. An adult GM vertebrate produces large quantities of the pharmed product in its blood, eggs, or milk, from which the product can be easily harvested and purified.

An example of a pharmed product advanced in development and approved by the FDA for high-risk situations is ATIII, a bioengineered form of human antithrombin. This medication is important in the treatment of individuals who have a hereditary deficiency of this protein and are at high risk for life-threatening blood clots during surgery, childbirth procedures, or the treatment of thromboembolism.

Xenotransplantation

Xenotransplantation, the transplantation of nonhuman vertebrate tissues and organs into humans, is another benefit of genetically altered animals. There is an alarming shortage of human donor organs to fill the need for hearts, kidneys, and livers. In 1963, doctors transplanted chimpanzee kidneys into 13 human patients, 12 of which lived for between 9 and 60 days. One patient survived for 9 months. In the late 1990s, two patients were kept alive using a pig liver outside of their body to filter blood until a human organ was available for transplantation.

Although baboons are phylogenetically closer to humans than pigs, pigs are generally healthier, produce more offspring in a shorter time, and are already farmed for food (Fig. 20B*b*). Despite the fears of some, scientists think that viruses unique to pigs are unlikely to cross the species barrier and infect the human recipient. Currently, pig heart valves and skin are routinely used for treatment of humans (Fig. 20B*c*). Miniature pigs, whose heart size is similar to humans, are being genetically engineered to make their tissues less foreign to the human immune system, in order to avoid rejection.

CONSIDER THESE QUESTIONS

1. Could a viral AIDS-like epidemic be unleashed by cross-species transplantation? What other unseen health consequences might there be?
2. Is it ethical to change the genetic makeup of vertebrates in order to use them as drug or organ factories?

 Explore the concepts through a variety of multimedia assets and question types.

www.mcgrawhillconnect.com

a. Poison-dart frog, source of a medicine

b. Pig, source of organs

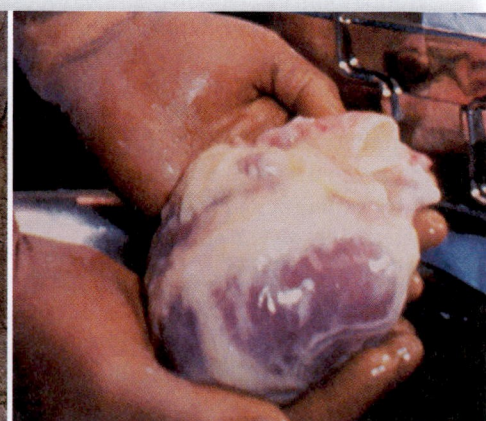

c. Pig heart for transplantation

FIGURE 20B Vertebrates used for medical purposes.

THE CHAPTER IN REVIEW

CONNECTING THE CONCEPTS

Our speedy review of the animal kingdom serves to exemplify the kingdom's diversity. We learned that animals differ in complexity and that these differences can be used to substantiate what DNA base sequencing tells us about their relationships. The wealth of anatomic information gathered by so many past investigators is needed to flesh out what the molecular data tell us. Still, some surprises are evident. For example, it comes as a surprise that roundworms, which are not segmented, are closely related to arthropods, which are segmented. This and other associations will be supported or rejected by future studies.

The great diversity in the animal kingdom extends to animals' different ways of acquiring food. Some are filter feeders that glean their particulate food from the sea around them, while others are predators that can attack and swallow their food in chunks. To define an animal, therefore, we could state only the essential characteristics: Animals are multicellular heterotrophs that acquire their food from the environment. Typically they go through developmental stages from a zygote to a final form that lives freely or as a parasite taking nourishment from others. Animals are very much dependent on the other groups of organisms, including the prokaryotes, protists, plants, and fungi. They could not live in a world alone, and animals did not invade the terrestrial environment until after other organisms had done so. This should be a humbling realization for us.

Evolution builds on what went before, so of course there is a progression among animals from the multicellular level of organization to the tissue level, to the organ level, and finally to the organ system level. Evidence of nerve fibers in a hydra appears before evidence of a brain in a planarian and before the complex brain of a mammal. Still, we have to realize that evolution is a bush and that each type of organism is on its own evolutionary pathway. Sponges, for example, have been around since animals first burst on the scene. Evolution proceeds through common ancestors rather than a straight line from one organism to another organism.

Much can be learned about the process of evolution by studying the diversity of the animal kingdom.

ANALYZE AND EVALUATE

1. Trace the evolution of humans using the evolutionary trees in this chapter (see Figs. 20.2A and 20.12C). How many steps did you write down?
2. Several types of animals are parasites. Recall how evolution judges "fitness," and state the advantages and disadvantages of the parasitic way of life.
3. We have known for some time that humans and echinoderms are both deuterostomes. What can you learn about evolution by considering the relatedness between these two groups of animals?

SUMMARIZE

Animals and How They Evolved

20.1 Animals have distinctive characteristics

- Animals are multicellular heterotrophs that ingest their food; usually reproduce sexually; undergo developmental stages that produce specialized tissues and organs; and have muscles and nerves.
- The **colonial flagellate hypothesis** has been supported by molecular data: Animals are descended from an ancestor that resembled a colony of flagellated cells.
- Animal diversity, which suddenly appeared during the **Cambrian explosion,** can be explained by the differential activity of *Hox* genes during development.

20.2 The phylogenetic tree of animals is based on molecular and anatomic data

- Radially symmetrical animals tend to be **sessile;** bilaterally symmetrical animals tend to have undergone **cephalization.** The latter animals produce three tissue layers (**germ layers**) during development. These tissue layers are called ectoderm, endoderm, and mesoderm.
- Protostomes develop differently from deuterostomes. In **protostomes,** including both **ecdysozoans** and **trochozoans,** the embryonic blastopore becomes the mouth. In **deuterostomes,** the blastopore becomes the anus. If a **true coelom** is present, in protostomes the coelom develops by splitting of the mesoderm. In deuterostomes, the coelom develops from an outpocketing of the gut.

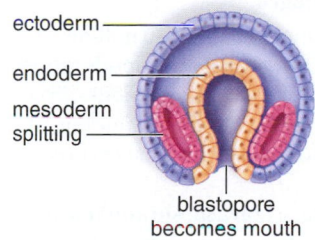
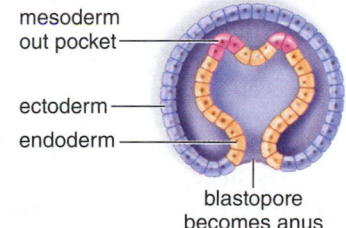

The Invertebrates

20.3 Sponges are multicellular animals

- **Sponges** are multicellular but lack tissue layers and are **filter feeders.**

20.4 Cnidarians are radially symmetrical with two tissue layers

- **Cnidarians** have radial symmetry, an incomplete digestive tract, two tissue layers separated by **mesoglea,** and the tissue level of organization. Two body forms are present: polyp (mouth above) and medusa (mouth below). Unique features are **nematocysts** and a **gastrovascular cavity** (sac body plan).

20.5 Flatworms are trochozoans without a coelom

- Trochozoans have a **trochophore larva** either now (annelids, molluscs) or in a past ancestor (flatworms). **Flatworms,** such as **planarians,** are **acoelomates** with three tissue layers but no body cavity and a sac body plan. Their body systems include an excretory system that contains **flame cells,** a **ladderlike**

nervous system, and an incomplete digestive tract. They are **hermaphrodites.**

- Tapeworms are parasites, they attach to the wall of the digestive tract. Flukes, another type of parasite, are named for their location in the body; there are liver flukes and blood flukes, for example.

20.6 A coelom gives complex animals certain advantages

- Flatworms are acoelomates; roundworms have a **pseudocoelom;** and the other animal groups have a true coelom, a body cavity completely lined with mesoderm. A coelom provides a space for organs to become specialized and have freedom of movement. If a skeleton is lacking, the coelom acts as a hydrostatic skeleton.

20.7 Molluscs have a three-part body plan

- All **molluscs** have a foot, a mantle, and a visceral mass. In **gastropods** (e.g., snails), the foot is flattened; in **cephalopods** (e.g., squids), the foot evolved into tentacles; and in **bivalves** (e.g., clams), the foot projects from the shell.

20.8 Annelids are the segmented worms

- In **annelids, segmentation** is obvious: Rings encircle the body, septa divide the coelom, and paired **nephridia** and blood vessels occupy each segment.
- Oligochaetes (earthworms) have few **setae** per segment, reside in soil, and do not have a well-developed head. They are hermaphroditic. Polychaetes (marine worms) have many setae per segment. They have jaws and sense organs, and are predaceous. Leeches have no setae and are blood suckers.

20.9 Roundworms are nonsegmented and plentiful

- Both roundworms and arthropods are ecdysozoans, the **molting** animals.
- **Roundworms** are nonsegmented and have a pseudocoelom. Roundworms exist in both terrestrial and marine habitats. Some are also parasites, causing trichinosis (muscle infection), elephantiasis (lymph vessel infection), or other infections.

20.10 Arthropods have jointed appendages

- **Arthropods** are very diverse and include about 30 million species, mostly insects. Six characteristics of arthropods are jointed appendages, an exoskeleton of **chitin** that molts, segmentation, a well-developed nervous system including compound eyes, a variety of respiratory organs (insects use **tracheae**), and **metamorphosis.**
- Among **crustaceans,** some (lobsters, crayfish, shrimp) have a cephalothorax in which the head bears sense organs and the thorax has five pairs of walking legs. **Arachnids** (spiders, scorpions) have a cephalothorax in which the thorax has four pairs of walking legs. **Insects** (grasshoppers, flies, butterflies, beetles) usually have wings for flying, but some are aquatic. Their body is divided into a head, thorax, and abdomen, with three pairs of legs attached to the thorax.

20.11 Echinoderms are radially symmetrical as adults

- **Echinoderms** are marine animals with no head, brain, or segmentation. A **water vascular system** provides locomotion and helps carry out respiratory, excretory, and circulatory functions.

The Vertebrates

20.12 Four features characterize chordates

- During their life cycle, **chordates** have a **notochord,** a dorsal tubular nerve cord, pharyngeal pouches, and a postanal tail.

- Among **invertebrate chordates,** all four chordate features are present in an adult **lancelet** and a **tunicate** (sea squirt) larva. Adult sea squirts retain only the gill slits.
- Vertebrate evolution is marked by the evolution of vertebrae, jaws, a bony skeleton, lungs, limbs, and the amniotic egg.

20.13 Jaws, a bony skeleton, and lungs evolved among the fishes

- **Jawless fishes** (hagfishes and lampreys) were the first vertebrates. **Cartilaginous fishes** (sharks and rays) have jaws and a skeleton made of cartilage. **Bony fishes** (e.g., **ray-finned fishes** such as trout, salmon, haddock) have jaws and fins supported by bony spikes. Certain bony fishes are **lobe-finned fishes.** Ancestral lobe-finned fishes also had lungs. Both the lobed fins and the lungs were preadaptive for a land environment.

20.14 Amphibians are tetrapods that can move on land

- **Amphibians** (frogs and salamanders) have jointed limbs, eyelids, ears, a voice-producing larynx, and lungs to help the adult stage live on land. Many amphibians (e.g., frogs) have to return to the water to reproduce. A larval stage undergoes metamorphosis into the adult form that can live on land.
- A transitional fossil shows how lobed fins of lobe-finned fishes were like the limbs of an amphibian.

20.15 Reptiles have an amniotic egg and can reproduce on land

- **Reptiles** (including turtles, snakes, lizards, which are **ectotherms,** and birds, which are **endotherms**) lay an **amniotic egg,** which contains extraembryonic membranes.

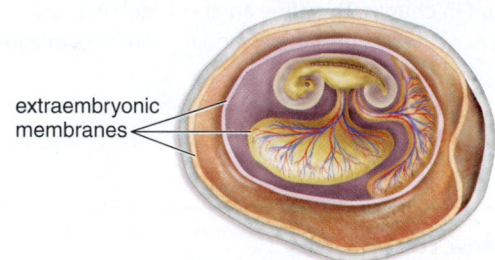

extraembryonic membranes

- **Birds** are adapted for flight. They have well-developed sense organs and lay hard-shelled amniotic eggs.

20.16 Mammals have hair and mammary glands

- **Monotremes** (duckbill platypus, spiny anteater) lay a hard-shelled amniotic egg. **Marsupials** (kangaroo, opossum) have a pouch in which the newborn matures. **Placental mammals** (deer, dogs, cats, whales, humans) have a **placenta** and retain their offspring inside a uterus until birth.

TEST YOURSELF

Animals and How They Evolved

1. Which of these is not a characteristic of animals?
   - a. heterotrophic
   - b. diploid life cycle
   - c. contracting fibers
   - d. single cells or colonial
   - e. lack of chlorophyll
2. The phylogenetic tree of animals shows that
   - a. three germ layers evolved before a coelom.
   - b. both molluscs and annelids are protostomes.
   - c. some animals have radial symmetry.
   - d. sponges were the first to evolve from an ancestral protist.
   - e. All of these are correct.

3. Which of these descriptions does not pertain to both protostomes and deuterostomes?
   a. three germ layers, bilateral symmetry, first opening is mouth
   b. bilateral symmetry, first opening is mouth, all have a true coelom
   c. spiral cleavage, first opening is anus, true coelom develops by splitting of mesoderm
   d. bilateral symmetry, three germ layers, second opening is mouth
   e. None pertain to both protostomes and deuterostomes.

The Invertebrates

4. Animals with a cellular level of organization have
   a. cells only.   c. cells and organs.
   b. cells and tissues.   d. cells, tissues, and organs.
5. Which of these sponge characteristics is not typical of animals?
   a. They practice sexual reproduction.
   b. They have the cellular level of organization.
   c. They have various symmetries.
   d. They have flagellated cells.
   e. Both b and c are not typical.
6. Cnidarians are considered to be organized at the tissue level because they contain
   a. ectoderm and endoderm.   d. endoderm and mesoderm.
   b. ectoderm.   e. mesoderm.
   c. ectoderm and mesoderm.
7. Unlike flatworms, roundworms have
   a. an internal skeleton and an incomplete digestive tract.
   b. an external skeleton and a complete digestive tract.
   c. an internal skeleton and a body cavity.
   d. an external skeleton and a body cavity.
   e. a complete digestive tract and a body cavity.
8. Compared to an animal species that lacks a coelom, one that has a coelom
   a. is more flexible.
   b. has more complex organs.
   c. is more likely to tolerate temperature variations.
   d. Both a and b are correct.
9. A mollusc's shell is secreted by the
   a. foot.   c. visceral mass.
   b. head.   d. mantle.
10. Which of the following is not a feature of an insect?
   a. compound eyes   d. an exoskeleton
   b. eight legs   e. jointed legs
   c. antennae
11. Sea stars move by
   a. expansion and contraction of their tube feet.
   b. producing jets of water.
   c. using their feet as legs and hopping.
   d. taking advantage of water currents to carry them.
12. **THINKING CONCEPTUALLY** What traits of free-living flatworms would be advantageous to an internal parasite?

The Vertebrates

13. Which of the following is not a chordate characteristic?
   a. dorsal supporting rod,   c. pharyngeal pouches
      the notochord   d. postanal tail
   b. dorsal tubular nerve cord e. vertebral column
14. Bony fishes include which two groups?
   a. hagfishes and lampreys
   b. sharks and ray-finned fishes
   c. ray-finned fishes and lobe-finned fishes
   d. jawless fishes and cartilaginous fishes

15. Amphibians arose from
   a. tunicates and lancelets.   d. ray-finned fishes of today.
   b. cartilaginous fishes.   e. bony fishes with lungs.
   c. jawless fishes.
16. What indicates that birds are reptiles?
   a. Birds have scales as well as feathers that are modified scales.
   b. Birds are ectothermic, as are reptiles.
   c. Birds lay leathery, shelled eggs, as do reptiles.
   d. Birds neglect their offspring as do all reptiles.
17. Which of the following is not an adaptation for flight in birds?
   a. air sacs   d. acute vision
   b. modified forelimbs   e. well-developed bladder
   c. bones with air cavities
18. Which of the following is a true statement? Choose more than one answer if correct.
   a. In all mammals, offspring develop completely within the female.
   b. All mammals have hair and mammary glands.
   c. All mammals give birth to one offspring at a time.
   d. All mammals are land-dwelling forms.
   e. All of these are true.
19. Which of the following animals does not produce an amniotic egg? Choose more than one answer if correct.
   a. bat   d. robin
   b. duckbill platypus e. frog
   c. snake
20. **THINKING CONCEPTUALLY** Of what special significance are transitional fossils such as *Tiktaalik,* which preceded the amphibians?

GET INVOLVED

1. For your senior project, you have decided to present evidence that sponges are animals. Describe the procedure you will use.
2. Most investigators today use what type of data to determine relationships among animals? They might go on to substantiate it with what other type of data?
3. After utilizing the virtual lab "Earthworm Dissection" tell what anatomic feature(s) would tell you that an earthworm is not a chordate.

Virtual Lab
Earthworm Dissection

MEDIA STUDY TOOLS

mhhe.com/maderconcepts3

Enhance your study of this chapter with interactive study tools, practice tests, and engaging animations. Also, ask your instructor about the resources available through ConnectPlus, which includes LearnSmart, a personalized adaptive learning program, and a media-rich eBook.

21

Evolution of Humans

BEFORE YOU BEGIN

Take a few minutes to recall

Transitional fossils give evidence of evolution (section 14.3)

Species can sometimes be determined from fossil evidence (section 15.1)

Humans are primates, a type of mammal (section 20.16)

Meet Ardi

Ardi—well, actually *Ardipithecus ramidus,* your long-lost relative—is known only from its fossil remains and reconstructions such as the one shown here. Ardi was found in East Africa not far from where Lucy (*Australopithecus afarensis,* see page 426) was uncovered some 50 years ago. You may recall that Lucy, who was named after a Beatles' song, was previously the oldest fossil for which we had good data. Now, Lucy has to forsake this distinction in favor of Ardi because Ardi (named for a town near where the remains were discovered) is dated at 4.4 million years ago (MYA)— a full million years before Lucy.

It took years of field work to find and then put together not only Ardi's skull and pelvis, but also the arms and hands, the legs and feet, and yes, of course, the teeth. The bones were so fragile and in such tiny soft pieces that paleontologists used computerized tomography (CT) imaging to make molds to show how the pieces fit together. Some of the results are displayed on these pages. Although an upright stance makes Ardi seem large, she was not. Compared to most modern humans, Ardi was small—being about 120 cm tall and 55 kg compare to 177 cm tall and 68 kg for modern humans. Males and females were about the same height and weight. Ardi is usually spoken of as female, but actually she is a composite of over 100 skeletons!

The paleontologists who pieced together Ardi's bones tell us her anatomy is most surprising. Whereas the skeleton of Lucy had some chimplike qualities, Ardi's is more humanlike! Lucy must be on another branch of our family tree, because Ardi stood fully erect—a small bone in the foot and her pelvic structure made sure of that. Still, look at those feet; the big toe splays out, signifying that Ardi could grasp tree limbs with her feet as apes can. That prehensile big toe tells us that Ardi was very much at home in the trees, and her flexible wrists and restricted shoulder movement indicate that she moved carefully along tree limbs on all fours, much as the earliest apes did. On the ground, Ardi didn't knuckle-walk the way chimps do, and in trees she didn't hang and swing from branches as apes do today. Ardi definitely seems more humanlike and more distantly related to chimps than Lucy, and it makes you wonder what changes may be needed in the current primate phylogenetic tree.

The size of Ardi's skull tells us she had a small brain, but she had humanlike teeth and not the long canine teeth found in chimpanzees and gorillas. Without large canine teeth, how did the ardipithecines defend themselves? The males definitely didn't use a threat display to frighten off other males for access to females. Their teeth wouldn't frighten anyone. As a possible motive for walking on the ground, perhaps males gathered food from afar and brought it back to bribe a receptive female for sex. Otherwise, males and females would have been happy to remain in the trees. Ardi lived in a woodland habitat, as evidenced by the existence of trees all around her. This evidence and the finding of other fossils also make it seem unlikely that hominins (humans and their closest relatives) began to walk erect because they lived in a habitat sparsely populated by trees. In other words, they wouldn't have stood up because it was advantageous to see over tall grasses.

Paleontologists and comparative anatomists are studying what the discovery of Ardi does to our early family tree. So far we have only fragmentary data on a few fossils older than Ardi, but hopefully we will learn more about them in the near future. Such information will no doubt help us make conclusions about the root of the hominin tree. This chapter uses the current data to trace the evolution of humans from the earliest primates to the first modern humans. It explains the occurrence of the large human population that now stresses the biosphere.

Reconstructed hand of Ardi

Reconstructed teeth of Ardi

Reconstructed foot of Ardi

Evolution of Primates

Humans are primates and they share certain characteristics with all the other primates. Molecular data tell us that humans are most closely related to chimpanzees and evolutionists have set a date for a common ancestor at about 7 MYA. But so far fossil data have not confirmed the molecular data. Paleontologists are still looking for fossil evidence, and in the meantime Ardi's anatomy (see "Meet Ardi" on pages 418–419) is so surprising.

21.1 Primates are adapted to live in trees

LEARNING OUTCOMES

When you complete this section, you should be able to

1. List and describe all the various types of primates.
2. Discuss four traits common to primates.

The order **Primates** includes prosimians, monkeys, apes, and humans (Fig. 21.1A). In contrast to other types of mammals, primates are adapted for an **arboreal** life—that is, a life spent in trees. The evolution of primates is characterized by trends toward mobile limbs, grasping (prehensile) hands, a flattened face with binocular vision, a large, complex brain, and a reduced reproductive rate. These traits are particularly useful for living in trees.

Mobile Forelimbs and Hindlimbs Primates tend to have prehensile hands and feet, often with opposable thumbs and toes (Fig. 21.1B). In most primates, flat nails have replaced the claws of ancestral primates, and sensitive pads on the undersides of fingers and toes assist the grasping of objects. All primates have a thumb, but it is only truly opposable in Old World monkeys, great apes, and humans. Because an **opposable thumb** can touch each of the other fingers, the grip becomes both powerful and precise. In all but humans, primates with an opposable thumb also have an opposable toe.

The evolution of the primate limb was a very important adaptation for their life in trees. Mobile limbs with clawless, opposable digits allow primates to freely grasp and release tree limbs. They also enable primates to easily reach out and bring food, such as fruit, to the mouth.

FIGURE 21.1A Primate diversity.

Prosimians

Ring-tailed lemur, *Lemus catta*

Tarsier, *Tarsius bancanus*

New World Monkey

White-faced monkey, *Cebus capucinus*

Old World Monkey

Anubis baboon, *Papio anubis*

Asian Apes

Orangutan, *Pongo pygmaeus*

White-handed gibbon, *Hylobates lar*

Stereoscopic Vision A foreshortened snout and a relatively flat face are also evolutionary trends in primates. These may be associated with a general decline in the importance of smell and an increased reliance on vision. In most primates, the eyes are located in the front, where they can focus on the same object from slightly different angles (Fig. 21.1C). The result is **stereoscopic vision** (three-dimensional vision) with good depth perception that permits primates to make accurate judgments about the distance and position of adjoining tree limbs.

Some primates, humans in particular, have color vision and greater visual acuity because the retina contains cone cells in addition to rod cells. Rod cells are activated in dim light, but the blurry image is in shades of gray. Cone cells require bright light, but the image is sharp and in color. The lens of the eye focuses light directly on the fovea, a region of the retina where cone cells are concentrated.

Large, Complex Brain Sense organs are only as beneficial as the brain that processes their input. The evolutionary trend among primates is toward a larger and more complex brain. This is evident when comparing the brains of prosimians, such as lemurs and tarsiers, with those of apes and humans. The portion of the brain devoted to smell is smaller, and the portions devoted to sight have increased in size and complexity. Also, more of the brain is devoted to controlling and processing information received from the hands and the thumb. The result is good hand-eye coordination. A larger portion of the brain is devoted to communication skills, which support primates' tendency to live in social groups.

Reduced Reproductive Rate One other trend in primate evolution is a general reduction in the rate of reproduction, associated with increased age at sexual maturity and extended life spans. Gestation is lengthy, allowing time for forebrain development. One birth at a time is the norm in primates; it is difficult to care for several offspring while moving from limb to limb in trees. The juvenile period of dependency is extended, and learned behavior and complex social interactions are emphasized.

Humans are a type of hominin. The next section gives an overview of the evolution of hominins from primate ancestors.

▶ **21.1 CHECK YOUR PROGRESS**
1. Identify the various types of primates and the traits they have in common.

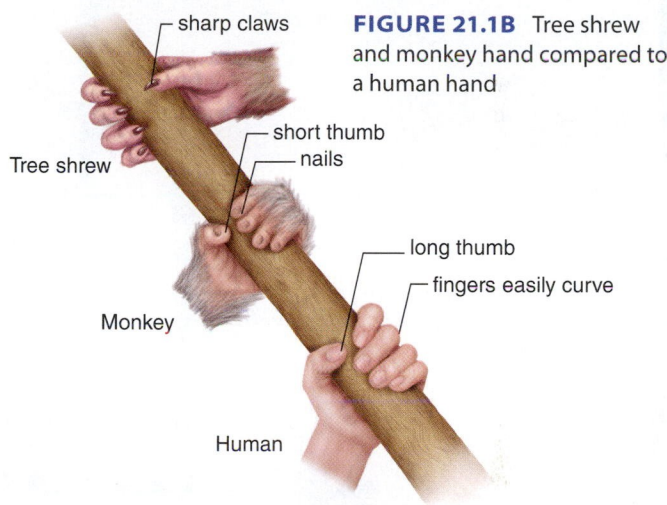

FIGURE 21.1B Tree shrew and monkey hand compared to a human hand

African Apes

Chimpanzee, *Pan troglodytes*   Western lowland gorilla, *Gorilla gorilla*

Humans

Humans, *Homo sapiens*

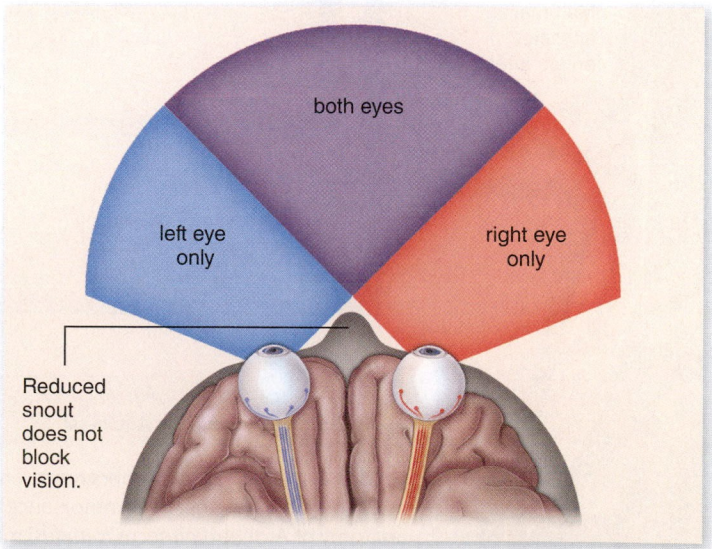

FIGURE 21.1C Stereoscopic vision.

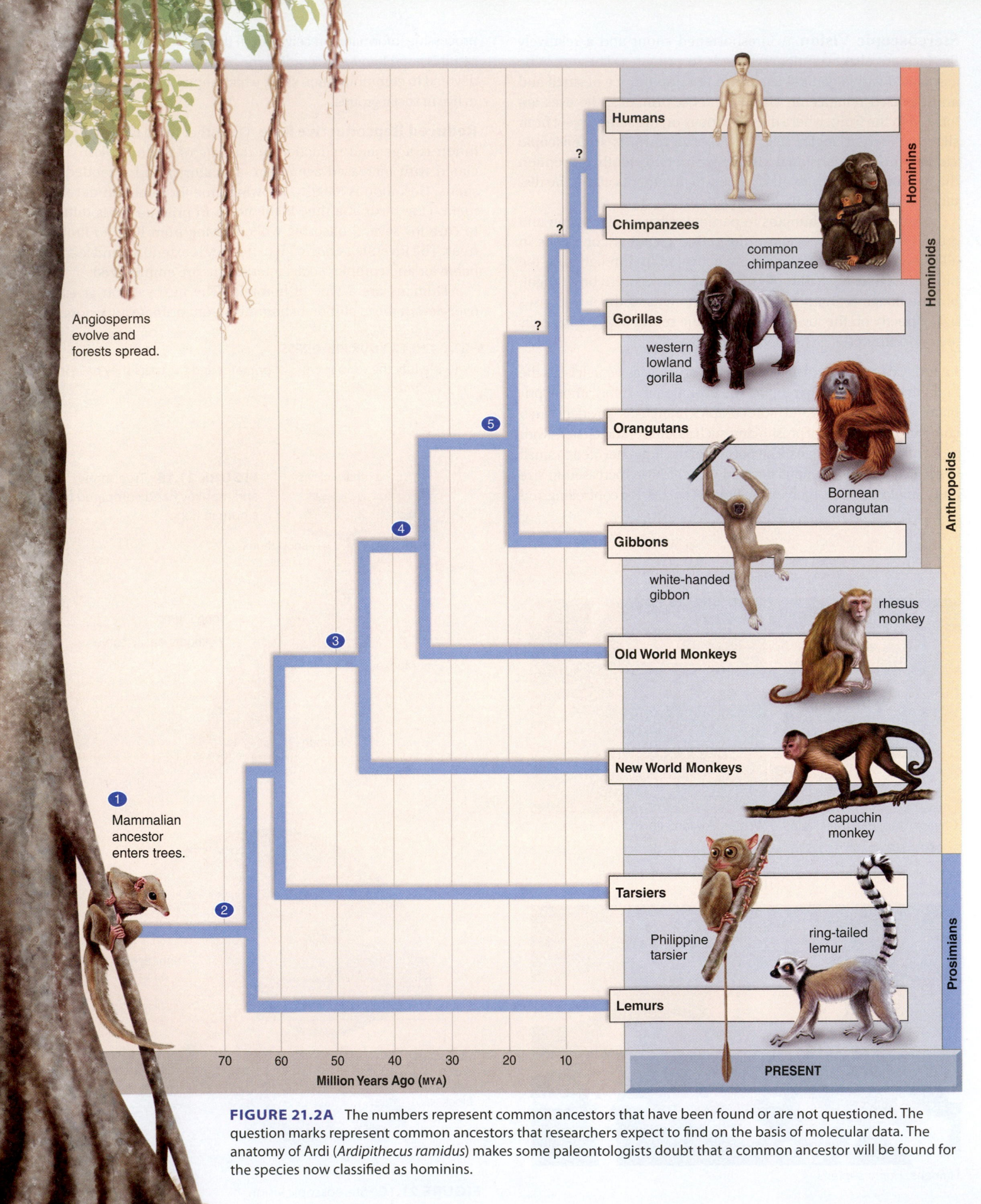

FIGURE 21.2A The numbers represent common ancestors that have been found or are not questioned. The question marks represent common ancestors that researchers expect to find on the basis of molecular data. The anatomy of Ardi (*Ardipithecus ramidus*) makes some paleontologists doubt that a common ancestor will be found for the species now classified as hominins.

21.2 All primates evolved from a common ancestor

LEARNING OUTCOME

When you complete this section, you should be able to

1. Discuss an evolutionary tree that shows the relationships of primates.

Figure 21.2A illustrates the sequence of primate evolution during the Cenozoic era. This evolutionary tree shows that all primates share **1** a common mammalian ancestor and that the other types of primates diverged from the main line of descent (called a lineage) over time. **2** Notice that **prosimians,** represented by lemurs and tarsiers, were the first types of primates to diverge.

3 An **anthropoid** was a common ancestor to all the other primates (monkeys, apes, and humans). New World monkeys often have long, prehensile (grasping) tails and flat noses, and Old World monkeys, which lack such tails, have protruding noses. Two well-known New World monkeys are the spider monkey and the capuchin, the "organ grinder's monkey." Some of the better-known Old World monkeys are the ground-dwelling baboon and the rhesus monkey, which has been used in medical research.

The earliest primate fossils similar to monkeys are found in Africa, dated about 45 MYA. At that time, the Atlantic Ocean would have been too expansive for some of them to easily make their way to South America, where the New World monkeys live today. Scientists hypothesize that a common ancestor to both New World and Old World monkeys arose much earlier, when a narrower Atlantic made crossing much more feasible. The New World monkeys evolved in South America, and the Old World monkeys evolved in Africa.

Dated about 35 MYA, **4** *Proconsul* (means before Consul, a famous performing chimpanzee) fossils probably represent a transitional link between the monkeys and the apes. *Proconsul* was about the size of a baboon, and the size of its brain was also comparable at about 65 cc. *Proconsul* didn't have the tail of a monkey (Fig. 21.2B), but walked as a quadruped on top of tree limbs as monkeys do. Primarily a tree dweller, *Proconsul* may have also spent time exploring nearby environs for food.

Proconsul was probably ancestral to **5** the **dryopithecines,** a common ancestor to hominoids (apes and humans). About 10 MYA, Africarabia (Africa plus the Arabian Peninsula) joined with Asia, and the apes migrated into Europe and Asia. In 1966, Spanish paleontologists announced the discovery of a specimen near Barcelona they named *Dryopithecus,* dated at 9.5 MYA. The anatomy of these bones clearly indicates that *Dryopithecus* was a tree dweller and locomoted by swinging from branch to branch as apes do today.

Common Ancestors Between Apes and Humans Molecular data tell us common ancestors will likely be found where the question marks appear on Figure 21.2A. Today, humans (including their ancestors) and chimpanzees are classified as **hominins**. Molecular clock data suggest a shared common ancestor as late as 7 MYA. Whether humans and chimpanzees shared a common ancestor is doubted by some in particular because Ardi, an early ancestor, does not locomote on the ground or in trees in the same manner as chimpanzees.

Animation
Molecular Clock

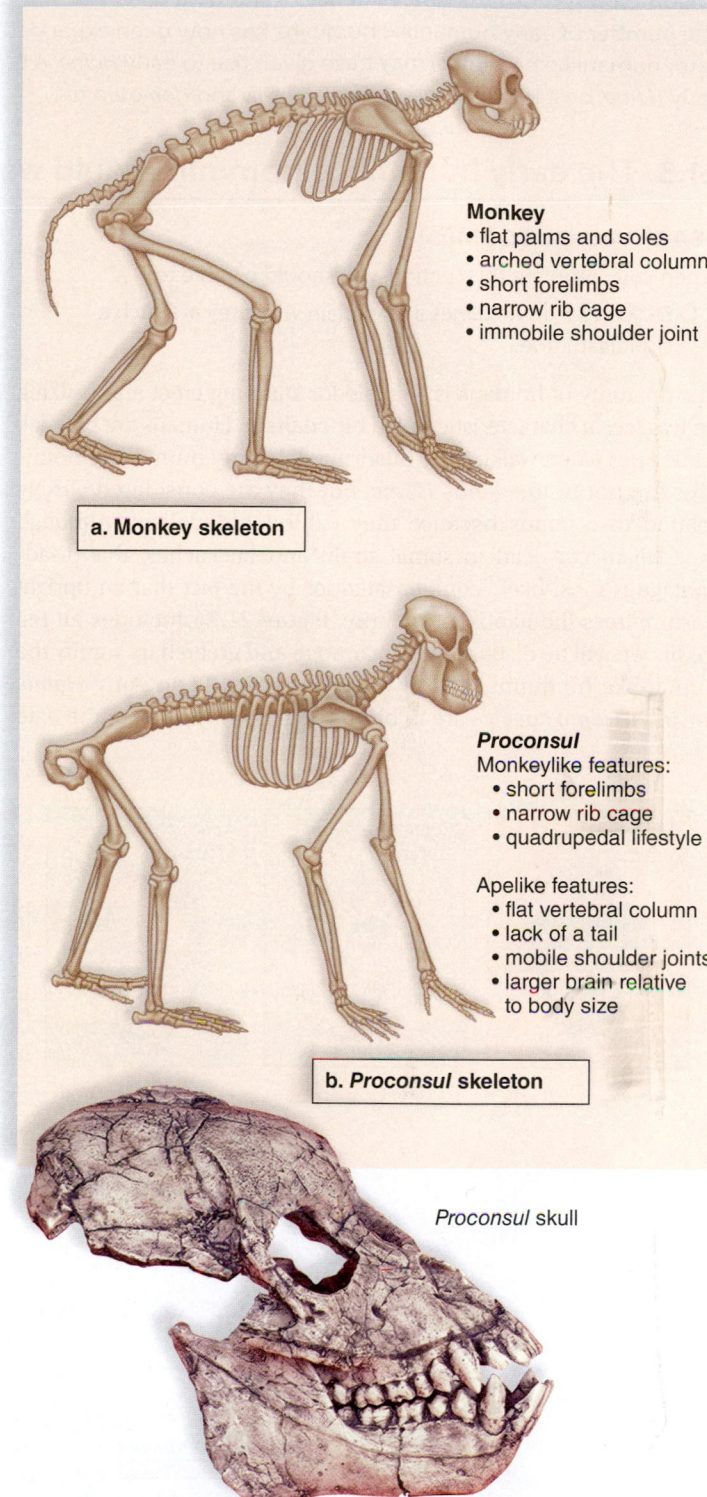

Monkey
- flat palms and soles
- arched vertebral column
- short forelimbs
- narrow rib cage
- immobile shoulder joint

a. Monkey skeleton

Proconsul
Monkeylike features:
- short forelimbs
- narrow rib cage
- quadrupedal lifestyle

Apelike features:
- flat vertebral column
- lack of a tail
- mobile shoulder joints
- larger brain relative to body size

b. *Proconsul* skeleton

Proconsul skull

FIGURE 21.2B Comparison of a monkey skeleton (**a**) with that of *Proconsul* (**b**) shows various dissimilarities, indicating that *Proconsul* is more related to today's apes than to today's monkeys.

▶ **21.2 CHECK YOUR PROGRESS**
1. With the help of Figure 21.2A, trace the path of evolution from a mammalian ancestor to humans.

Evolution of Humans

Paleontologists currently use evidence of standing erect as a way to distinguish early humanlike hominins from apes in the fossil record. The number of early humanlike hominins has now been expanded to include Ardi (*Ardipithecus ramidus*). One of the australopithecines (later humanlike hominins) may have given rise to early *Homo*. A brain size of at least 600 cc is needed for a fossil to be considered an early *Homo*, best represented by *Homo habilis* and *Homo erectus*.

21.3 The early humanlike hominins could walk upright

LEARNING OUTCOME

When you complete this section, you should be able to

1. Describe ardipithecines and explain why they are such a significant find.

The anatomy of humans is suitable for standing erect and walking on two feet, a characteristic called **bipedalism.** Humans are bipedal, while apes when walking are quadrupedal. Early humanlike hominins are not in the genus *Homo*, but they are considered closely related to humans because they exhibit bipedalism. Although bipedalism can lead to spinal strain and backaches, this disadvantage is most likely compensated for by the fact that an upright posture frees the hands for tool use. Figure 21.3A includes all the fossils we will be discussing. The orange and green bars signify the humanlike hominins, and lavender bars signify an early *Homo;* the later *Homo* species are in blue. The bars extend from the date

of a species' appearance in the fossil record to the date it became extinct.

Paleontologists have now found several fossils dated around the time the ape lineage and the human lineage are believed to have split. One of these is *Sahelanthropus tchadensis.* Only the braincase has been found and dated at 7 MYA. But a point at the back of the skull where the neck muscles would have attached suggests bipedalism. The skull of this fossil is very like that of *Ardipithecus ramidus,* discussed next.

Ardipithecines Two species of **ardipithecines** (ardipiths for short) have been uncovered, but only teeth and a few bone bits have been found for one species; therefore, we will concentrate on *Ardipithecus ramidus,* which is affectionately known as a female called Ardi. Actually, the reconstruction of Ardi shown in "Meet Ardi" on pages 418–419 is based on a compilation of more than

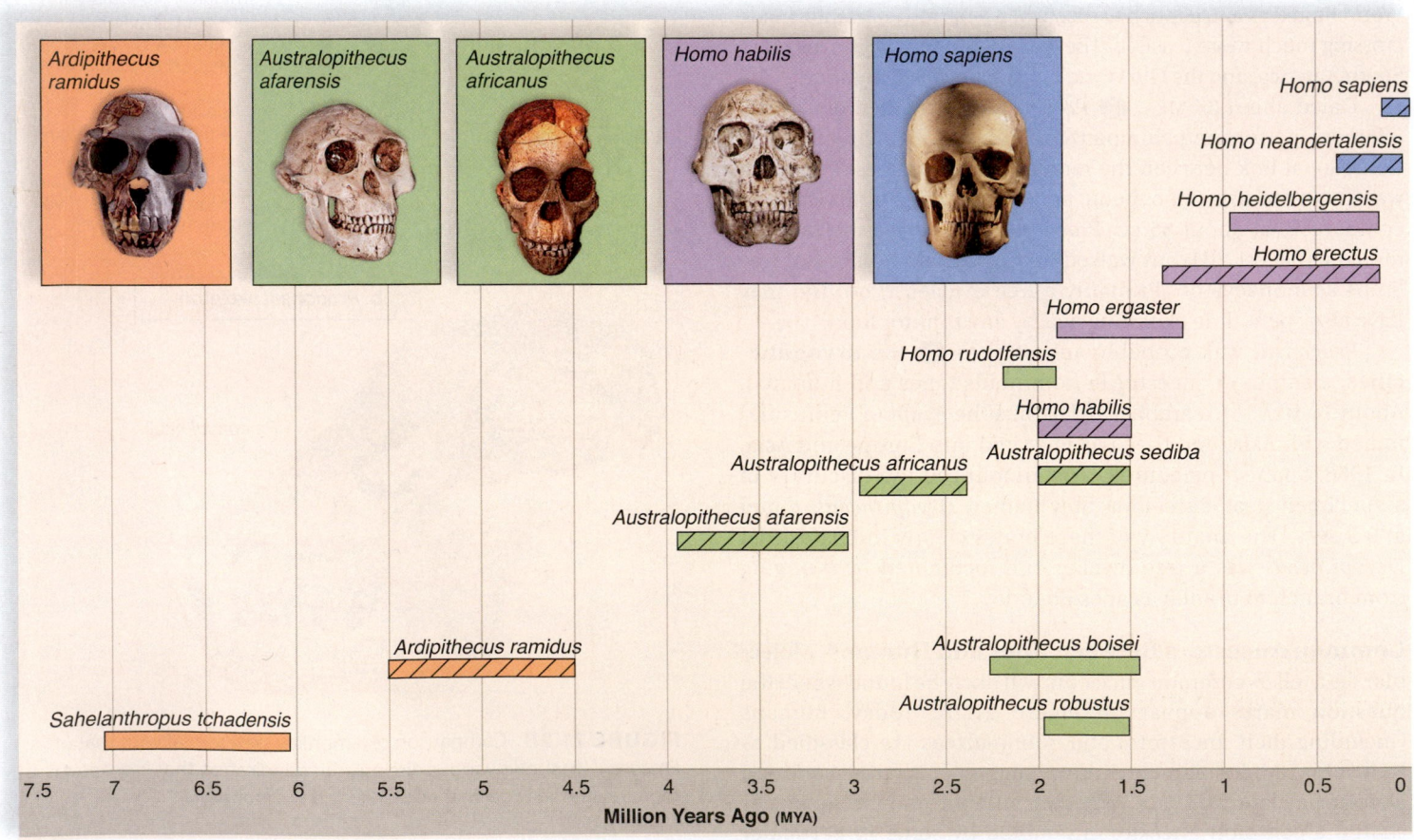

FIGURE 21.3A Several groups of extinct hominins preceded the evolution of modern humans. These groups have been divided into the early humanlike hominins (orange), later humanlike hominins (green), early *Homo* species (lavender), and finally the later *Homo* species (blue). Only modern humans are classified as *Homo sapiens.*

100 different skeletons found at the same site, near the town of Aramis in Ethiopia, East Africa. Some of Ardi's features are primitive—like that of an ape such as *Dryopithecus* (see section 21.2)—but others are like that of a human.

Ardi, which is dated at 4.4 MYA, was about the size of a chimpanzee, standing about 120 cm tall and weighing about 55 kg. Males and females were about the same size.

Ardi had a small head compared to the size of her body; the skull has the same features but is smaller than that of *Sahelanthropus tchadensis*. The brain size was around 300–350 cc, slightly less than the size of a female chimp brain and much smaller than that of a modern human, whose brain size is about 1,360 cc. The muzzle projects, and the forehead is low with heavy eyebrow ridges, a combination that makes the face more primitive than that of the australopithecines. However, the projection of the face is less than that of a chimpanzee because Ardi's teeth were small and like those of an omnivore. She lacked the strong sharp canines of a chimpanzee, and her diet probably consisted mostly of soft rather than tough plant material.

Ardi could walk erect, but she spent a lot of time in trees. Notice in Figure 21.3B that in both human and Ardi skeletons, the spine exits from the center of the skull. Also, the femurs angle (red arrows) toward the knees. These skeletal features assist walking erect by placing the trunk's center of gravity squarely over the feet. Also, Ardi's pelvis and hip joint are broad enough to keep her from swaying from side to side (as chimps do) as she walked. The knee joint in both humans and Ardi is modified to support the body's weight because the bones broaden at this joint. Ardi's feet have a bone, missing in the apes, that would keep her feet squarely on the ground, a sure sign that she was bipedal and did not knuckle-walk as apes do. Nevertheless, she has an opposable big toe the same as apes. An opposable toe allows an animal's feet to grab hold of a tree limb. The wrists of Ardi's hands were flexible, and most likely she moved along tree limbs on all fours, as ancient apes did. Modern apes brachiate—use their arms to swing from limb to limb. Ardi did not do this, but her shoulders were flexible enough and her arms long enough to allow her to reach for limbs to the side or over her head. The conclusion is that Ardi

locomoted *carefully* in trees. While the top of her pelvis is like that of a human and probably served for the attachment of muscles needed for walking, the bottom of the pelvis served for the attachment of strong muscles needed for climbing trees.

Until now, it's been suggested that bipedalism evolved when a dramatic change in climate caused forests to be replaced by grassland. But Ardi lived in the woods, as evidenced by the environment that existed around her. What advantage would walking erect have afforded? Well, males didn't frighten each other in order to have access to females—their canine teeth wouldn't scare anyone. So maybe males who took to the ground and foraged for food had an advantage when they carried it back to bribe females for sex. Or it's possible bipedalism may be associated with the evolution of a helpless infant that has to be hand carried from place to place as discussed in "Regulatory Genes and the Origin of the Genus *Homo*" on page 221.

▶ **21.3 CHECK YOUR PROGRESS**

1. List the differences that you can observe between Ardi and *Homo sapiens* in Figure 21.3B.

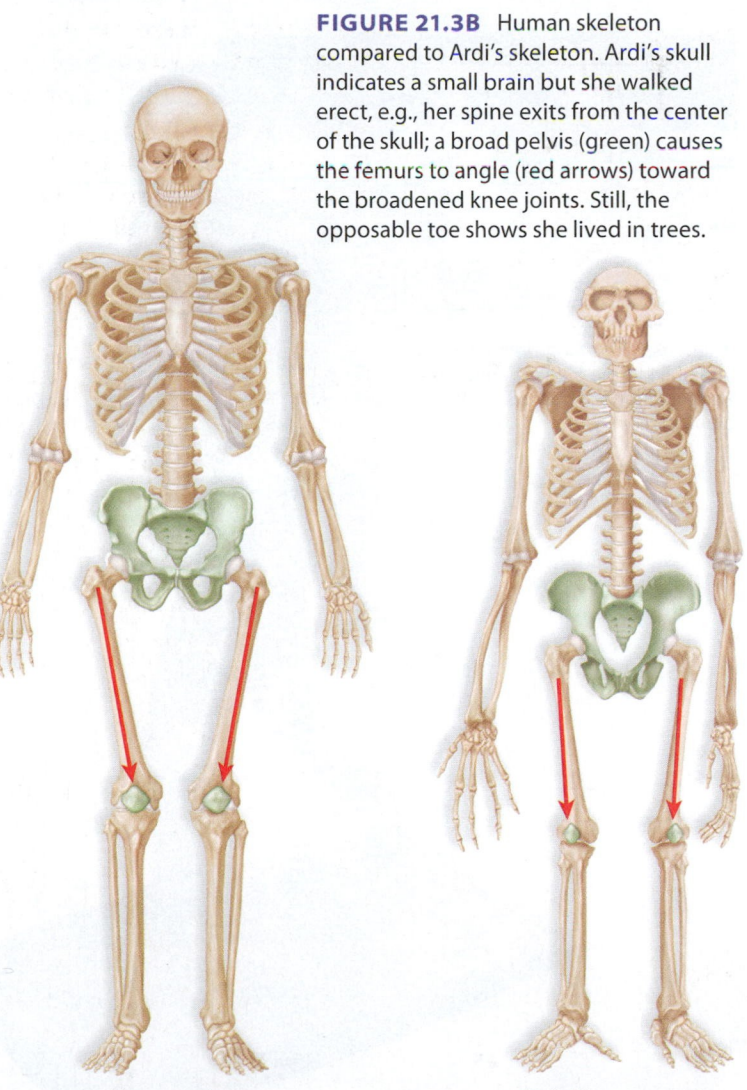

FIGURE 21.3B Human skeleton compared to Ardi's skeleton. Ardi's skull indicates a small brain but she walked erect, e.g., her spine exits from the center of the skull; a broad pelvis (green) causes the femurs to angle (red arrows) toward the broadened knee joints. Still, the opposable toe shows she lived in trees.

Homo sapiens   *Ardipithecus ramidus*

LEARNING OUTCOMES

When you complete this section, you should be able to

1. Exemplify that the australopithecines exhibited mosaic evolution.
2. Discuss the advancements associated with *Homo erectus*.

The **australopithecines** (australopiths for short) are a group of later humanlike hominins that evolved and diversified in Africa from 4 MYA until about 2 MYA. In Figure 21.3A (see p. 424), the australopiths are represented by green bars. The australopiths had a slightly larger brain (370–515 cm³) than the ardipiths and stood a little taller (100–115 cm). Still, like the ardipiths, they exhibited **mosaic evolution,** meaning that as the human form evolved, body parts changed at different rates and, therefore, at different times. Both the ardipiths and the australopiths had a small brain (apelike characteristic) and walked erect (humanlike characteristic).

Among the australopiths, males were distinctly larger than females. Also, some australopithecines were slight of frame and termed gracile (slender). Others were robust (powerful) and tended to have massive jaws because of their large grinding teeth. Their

well-developed chewing muscles were anchored to a prominent bony crest along the top of the skull. The gracile types most likely fed on soft fruits and leaves, while the robust types had a more fibrous diet that may have included hard nuts. Therefore, the australopithecines show adaptations to different ways of life.

Fossil remains of australopithecines have been found in both southern and eastern Africa.

Fossils from East Africa *Australopithecus afarensis* (3.2 MYA), affectionately known as Lucy after the title of a Beatles' song, is the most significant fossil from East Africa. Lucy's forehead was low; the face projected forward with prominent canine teeth and was broader than that of the ardipithecines (Fig. 21.4A). Although the brain was quite small (400 cc), Lucy's skeleton indicates that she stood upright and walked bipedally. She stooped a bit like a chimpanzee, and the arms were somewhat proportionately longer than the legs. This suggests brachiation as a possible mode of locomotion in trees. Otherwise, the skeleton was humanlike, even though the pelvis lacked refinements that would have allowed Lucy to walk with a striding gate as humans do. Even better evidence of bipedal locomotion comes from a trail of fossilized footprints in Laetoli dated about 3.7 MYA. The larger prints are double, as though a smaller-sized being was stepping in the footprints of another—and there are additional small prints off to the side, within hand-holding distance (Fig. 21.4A).

A. afarensis, a gracile type, is believed to be ancestral to the robust types found in eastern Africa such as *Astralopithecus boisei. A. boisei* had a powerful upper body and the largest molars of any hominin. Some 30 years after Lucy was discovered, paleontologists uncovered Selam, the skeleton of a child some called Lucy's baby, even though it is dated 0.1 million years older than Lucy. The face looks more like that of the ardipiths, and the semicircular ear canals suggest that Selam was not as agile a walker as Lucy.

Fossils from South Africa The first australopithecine to be discovered was unearthed in southern Africa by Raymond Dart in the 1920s. This hominin, named ***Australopithecus africanus,*** is a gracile type. A second southern African specimen, *Astralopithecus robustus,* is a robust type also found some time ago.

In 2008, an American anthropologist, Lee Berger, discovered in South Africa the bones of australopiths he named *Australopithecus sediba* (or "wellspring"). Berger hypothesizes that this species gave rise to *Homo*. The brain is small (420 cc) but there is evidence of frontal lobe development including an area devoted to language. The apelike shoulders and tree-climbing arms end with a hand that has a long thumb and short fingers, suggesting the ability to hold and use tools. The pelvis is less flared and more bowl-shaped like that of a human; the legs are long but a modern style ankle rests on a very primitive heel. The skull of a boy suggests a projecting nose and dentition similar to the early *Homo* species.

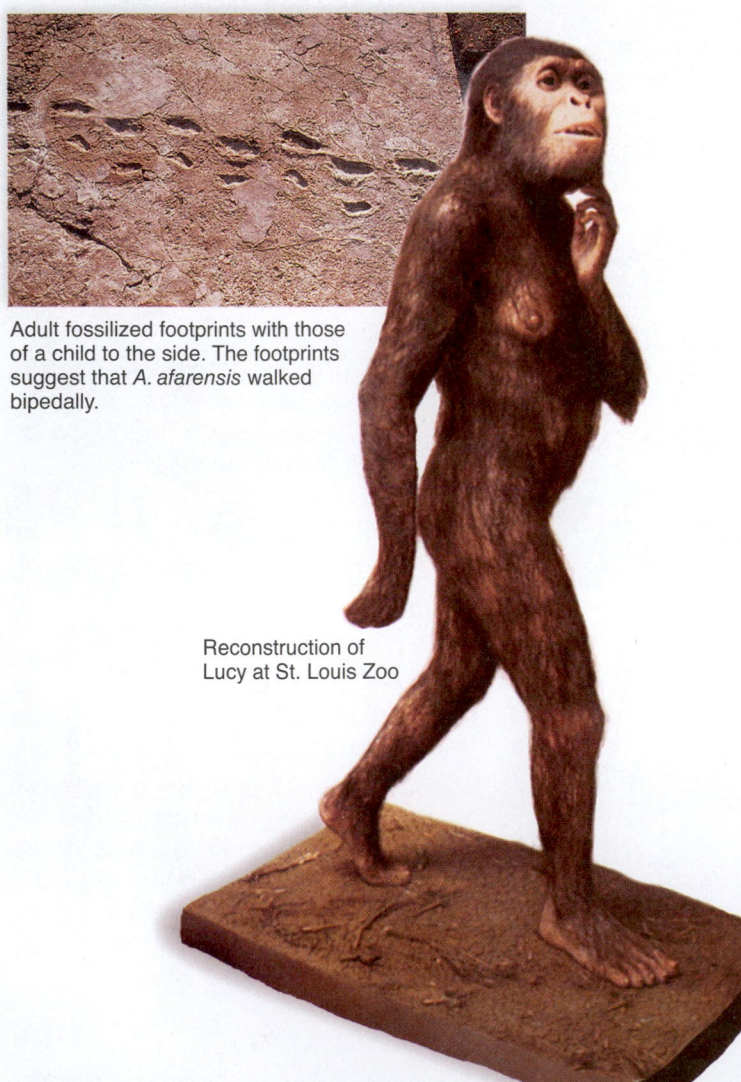

Adult fossilized footprints with those of a child to the side. The footprints suggest that *A. afarensis* walked bipedally.

Reconstruction of Lucy at St. Louis Zoo

FIGURE 21.4A *Australopithecus afarensis.*

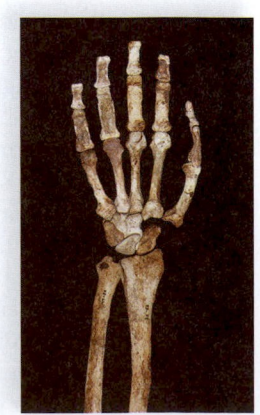

The hand of *A. sediba* (palmar view)

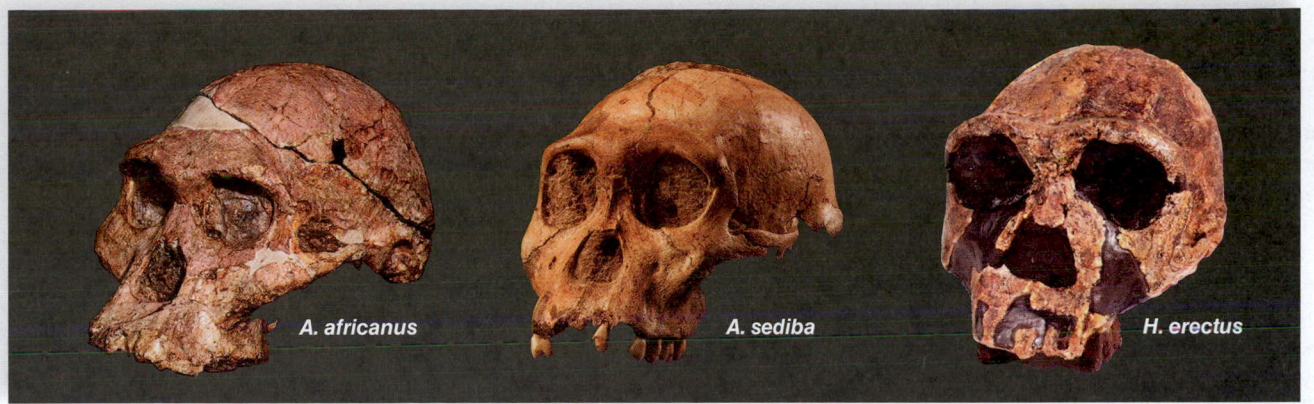

FIGURE 21.4B Skulls from South Africa together with a *Homo erectus* skull from East Africa. Some suggest evolution proceeded in this manner.

Until recently, it was generally agreed that *A. afarensis* (Lucy) was ancestral to the genus *Homo* but now *A. sediba* is vying with Lucy for this spot (Fig. 21.4B). A contradiction remains, however: *A. sediba* is dated at 1.9 MYA, the same as *H. habilis,* discussed next. Meave Leakey suggests: "It is likely that the South African hominins are a separate radiation that took place in the south of the continent."

Early *Homo* Fossils designated as early *Homo* species are represented by lavender bars in Figure 21.3A. These fossils appear in the fossil record somewhat earlier or later than 2 MYA. They all have a brain size of 600 cc or greater, their jaws and teeth resemble those of humans, and tool use is in evidence.

Homo habilis and related fossils (e.g., *Homo rudolfensis*) have a more primitive anatomy than the other *Homo* species. Although the height of *H. habilis* did not exceed that of the australopith, some of this species' fossils have a brain size as large as 800 cc, which is considerably larger than that of an australopith. The cheek teeth of these humans tend to be smaller than even those of the gracile australopithecines. Therefore, it is likely that they were omnivorous and ate meat in addition to plant material. These *Homo* species were found with plentiful tools about them. Their name means handyman.

Homo erectus and related fossils (e.g., *Homo ergaster*) are dated between 1.8 and 0.2 MYA. A Dutch anatomist named Eugene Dubois was the first to unearth *H. erectus* bones in Java in 1891, and since that time many other fossils belonging to this species have been found in Africa and Asia. The recovery of an almost complete skeleton of a 10-year-old boy indicates that *H. erectus* was much taller than the hominins discussed thus far (Fig. 21.4C). Males were 1.8 m tall, and females were 1.55 m. The members of this species had the body proportions of a modern human; the legs were long and the arms did not reach to the knees, providing evidence that they were ground dwellers and did not live in trees. Indeed, they most likely had a striding gait as do humans and a much larger cranium and brain (about 1,000 cc) than *H. habilis.* Although they had prominent brow ridges, the nose projected. This type of nose is adaptive for a hot, dry climate because it permits water to be removed before air leaves the body.

The size of the birth canal indicates that infants were born in an immature state that required an extended period of care

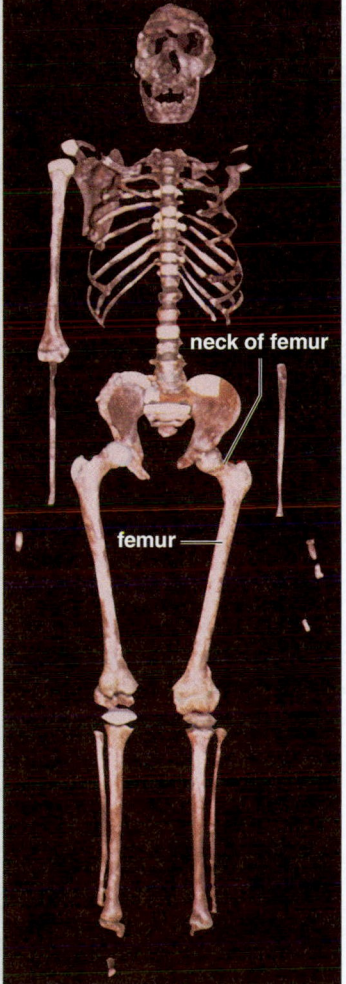

FIGURE 21.4C *Homo erectus* skeleton from East Africa. Notice that the femur is angled because its neck is quite long. (Compare to Fig. 21.3B.)

neck of femur

femur

while the brain developed (see "Regulatory Genes and the Origin of the genus *Homo*" on p. 221). They also had a knowledge of fire and may have been the first to cook meat and have a home base. *H. erectus* first appeared in Africa but then migrated into Europe and Asia sometime between 2 MYA and 1 MYA. Such an extensive population movement is a first in the history of humankind and a tribute to the intellectual and physical skills of the species.

Homo floresiensis In 2004, scientists announced the discovery of the fossil remains of *Homo floresiensis,* another early *Homo* species. The 18,000-year-old fossil of a 1-m tall, 25-kg adult female was discovered on the island of Flores in the South Pacific. The specimen was the size of a 3-year-old and had a braincase only one-third the size of a modern human. Researchers suspect that this diminutive hominin species evolved from a normal-sized, island-hopping *H. erectus* that reached Flores about 840,000 years ago. Apparently, *H. floresiensis* used tools and fire.

▶ **21.4 CHECK YOUR PROGRESS**

1. List ways that the australopithecines exhibited mosaic evolution.
2. Identify advances that account for the significant place of *H. erectus* in human evolution.

21A Biocultural Evolution Began with *Homo*

Culture encompasses human activities and products that are passed on from one generation to another outside of direct biological inheritance. *Homo habilis* could make the simplest of stone tools, called Oldowan tools after a location in Africa where the tools were first found. The main (core) tool could have been used for hammering, chopping, and digging. A flake tool was a type of knife sharp enough to scrape away hide and remove meat from bones. The diet of *H. habilis* most likely consisted of collected plants. But they probably had the opportunity to eat meat scavenged from kills abandoned by lions, leopards, and other large predators in Africa.

Homo erectus, who lived in Eurasia, also made stone tools, but the flakes were sharper and had straighter edges. They are called Acheulian tools for a location in France where they were first found. Their so-called multipurpose handaxes were large flakes with an elongated oval shape, a pointed end, and sharp edges on the sides. Supposedly they were hand-held, but no one knows for sure. In addition, *H. erectus* could have also made many other implements out of wood or bone and even grass, which can be twisted together to make string and rope. Excavations of *H. erectus* campsites dated 400,000 years ago have uncovered literally tens of thousands of tools.

H. erectus, like *H. habilis,* also gathered plants as food. However, *H. erectus* may have harvested large fields of wild plants that were growing naturally. The members of this species were not master hunters, but aside from scavenging meat, they could have hunted a bit. The bones of all sorts of animals litter the areas where they lived. Apparently, they ate pigs, sheep, rhinoceroses, buffalo, deer, and many other smaller animals. *H. erectus* lived during the last Ice Age, but even so, moved northward. No wonder *H. erectus* is believed to have used fire. A campfire would have protected them from wild beasts and kept them warm at night. Also, the ability to cook would have made meat easier to eat. Plants can't provide much food in the dead of winter in northern climates, and so meat must have become a substantial part of the diet when necessary. It's even possible that the campsites of *H. erectus* were "home bases" where people gathered and shared food. If so, these people may have been the first **hunter-gatherers** (Fig. 21A)—that is, they hunted animals and gathered plants. This was a successful way of life that caused the hominin populations to increase from a few

thousand australopithecines in Africa 2 MYA to hundreds of thousands of *H. erectus* by 0.3 MYA. The hunting and gathering way of life doesn't permit a population explosion, however. Children have to be carried long distances, and the men were frequently not around to father children.

Hunting does most likely encourage the development and spread of culture between individuals and generations. Those who could speak a language would have been able to cooperate better as they hunted and even as they sought places to gather food. The ability to speak is now associated with a transcription factor coded for by a *FOXP2* gene. Among animals, only humans have a complex language that allows them to communicate their experiences symbolically. Words stand for objects and events that can be pictured in the mind. The cultural achievements of *H. erectus* essentially began a new phase of human evolution, called **biocultural evolution,** in which natural selection is influenced by cultural achievements rather than by anatomic phenotype. *H. erectus* succeeded in new, colder environments because these individuals occupied caves, used fire, and became more capable of obtaining and eating meat as a substantial part of their diet.

CONSIDER THESE QUESTIONS

1. There is probably nothing you do that doesn't show humans have culture. Name some of these many evidences of culture in your own life.
2. What aspects of your life do not have anything to do with culture? Explain.
3. Some people compare our modern way of life to that of the hunter-gatherers. Show that this is possible. Do such comparisons have any benefit?

 Explore the concepts through a variety of multimedia assets and question types.

www.mcgrawhillconnect.com

FIGURE 21A The *Homo erectus* people may have been hunter-gatherers.

Evolution of Modern Humans

The replacement model of human evolution says that modern humans originated only in Africa and, after migrating into Europe and Asia, replaced the archaic *Homo* species found there. The Neandertals, a group of archaic humans, lived in Europe and Asia. Cro-Magnon is the name often given to modern humans. The human ethnic groups of today differ in ways that can be explained in part by adaptation to the environment.

21.5 Cro-Magnons replaced the other *Homo* species

LEARNING OUTCOME

When you complete this section, you should be able to

1. Understand and use the replacement model to explain the survival of only one species of modern humans (*Homo sapiens*).

Two species are considered the later *Homo* species. They are *H. neandertalensis* in Europe and *H. sapiens* in Africa (blue bars in Fig. 21.3A). Only **H. sapiens** is considered a modern human.

Neandertals The **Neandertals** are an intriguing species of archaic humans that lived between 200,000 and 30,000 years ago. Neandertal fossils have been found in the Middle East and throughout Europe. Neandertals take their name from Germany's Neander Valley, where one of the first Neandertal skeletons, dated some 200,000 years ago, was discovered. The Neandertal Project refers to a group of scientists who are busy sequencing the neandertal genome.

The Neandertal brain was, on the average, slightly larger than that of *Homo sapiens* (1,400 cc, compared with 1,360 cc in most modern humans). The Neandertals had large brow ridges and wide noses. They also had a forward-sloping forehead and a receding lower jaw. Their nose, jaws, and teeth protruded far forward (Fig. 21.5A). Physically, the Neandertals were powerful and heavily muscled, especially in the shoulders and neck. The bones of Neandertals were shorter and thicker than those of modern humans. New fossils show that the pubic bone was long compared to that of modern humans. The Neandertals lived in Europe and Asia during the last Ice Age, and their sturdy build could have helped them conserve heat.

Archaeological evidence suggests that Neandertals were culturally advanced. Some Neandertals lived in caves; however, others probably constructed shelters. They manufactured a variety of stone tools, including spear points, which they could have used for hunting, and scrapers and knives, which would have helped in food preparation. They most likely hunted bears, woolly mammoths, rhinoceroses, reindeer, and other contemporary animals. They used and could control fire, which probably helped them cook frozen meat and keep warm. They even buried their dead with flowers and tools and may have had a religion.

Replacement Model Various early *Homo* species (including the Neandertals) who lived in Europe and Africa around 0.5 MYA are often grouped together as **archaic humans**. The replacement model proposes that the archaic humans in Africa gave rise to *Homo sapiens*. *Homo sapiens* replaced the archaic species in Africa and then migrated out of Africa to Europe and Asia to largely replace the archaic species there also. The replacement model is supported by the fossil record. The earliest remains of modern humans (Cro-Magnon), dated at least 200,000 years BP, have been found only in Africa. Modern humans do not appear in Asia until 100,000 years

BP and not in Europe until 60,000 years BP. The migration of *Homo sapiens* out of Africa is also supported by DNA data in that mitochondrial DNA of Africans is more diverse than the DNA of the people in Europe and elsewhere.

Genome studies have also been done to determine if modern humans interbred with Neandertals once they reached Europe. These studies show that the DNA of Neandertals shares more genetic variations with European populations than with African populations. This supports the hypothesis that some modern humans did breed with Neandertals after migrating out of Africa and that some of your DNA was contributed by the Neandertals. Consistent with the replacement model (Fig. 21.5B), however, modern humans went on to replace the Neandertals.

FIGURE 21.5A A Neandertal female.

▶ **21.5 CHECK YOUR PROGRESS**

1. Identify how the replacement model explains the disappearance of archaic populations in Europe and Asia.

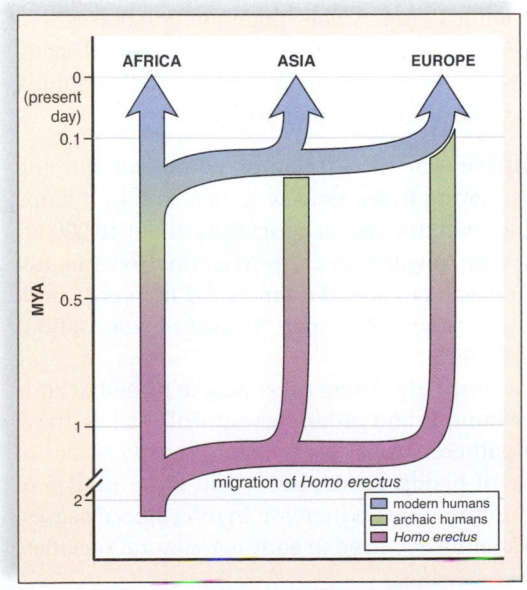

FIGURE 21.5B The replacement model tells us that modern humans (blue) evolved in Africa and then replaced archaic humans in Asia and Europe.

21.6 Cro-Magnons were socially advanced

LEARNING OUTCOME

When you complete this section, you should be able to

1. Relate brain development to the advancement of culture among Cro-Magnons.

Cro-Magnons are the oldest fossils to be designated *Homo sapiens*. In keeping with the replacement model, the Cro-Magnons, who are named after a fossil location in France, were the modern humans who entered Asia from Africa about 100,000 years BP and then spread to Europe. They probably reached western Europe about 40,000 years ago. Cro-Magnons had a thoroughly modern appearance (Fig. 21.6A). They had lighter bones, flat high foreheads, domed skulls housing brains of 1,590 cc, small teeth, and a distinct chin. They were hunter-gatherers in the same manner as *H. erectus*, except that they hunted more efficiently.

Tool Use in Cro-Magnons Cro-Magnons designed and manipulated tools and weapons of increasing sophistication. They made advanced stone tools, including compound tools, as when fitting a stone flake to a wooden handle. Cro-Magnons may have been the first to make knifelike blades and to throw spears, enabling them to kill animals from a distance. They were such accomplished hunters that some researchers believe they may have been responsible for the extinction of many larger mammals, such as the giant sloth, the mammoth, the saber-toothed tiger, and the giant ox, during the late Pleistocene epoch. This event is known as the Pleistocene overkill.

Language and Cro-Magnons A more highly developed brain may have also allowed Cro-Magnons to perfect a language composed of patterned sounds. Language greatly enhanced the possibilities for cooperation and a sense of cohesion within the small bands that were the predominant form of social organization, even for the Cro-Magnons.

The Cro-Magnons were extremely creative. They sculpted small figurines and jewelry out of reindeer bones and antlers. These sculptures could have had religious significance or been seen as a way to increase fertility. The most impressive artistic achievements of the Cro-Magnons were cave paintings, realistic and colorful depictions of a variety of animals, from woolly mammoths to horses, that have been discovered deep in caverns in southern France and Spain. These paintings suggest that Cro-Magnons had the ability to think symbolically, as would be needed in order to speak.

Rise of Agriculture The Cro-Magnons combined hunting and fishing with gathering fruits, berries, grains, and root crops that grew in the wild. With the rise of agriculture about 10,000 BP, modern humans are no longer called Cro-Magnon. However, full dependency on domestic crops and animals did not occur until humans started making tools of bronze, instead of stone, about 4,500 years BP (Fig. 21.6B).

Anthropologists formerly thought that people turned to agriculture because the hunting and gathering way of life had its drawbacks. But hunter-gatherers most likely had a nutritious diet of thousands of types of plants, seeds, fruits, and nuts. Studies of skeletal evidence indicate that an increase in infectious diseases, malnutrition, and anemia occurred in early agricultural societies,

FIGURE 21.6A Cro-Magnon people are the first to be designated *Homo sapiens*. Their tool-making ability and other cultural attributes, such as their artistic talents, are legendary.

FIGURE 21.6B The advent of agriculture affected the way people lived and interacted with one another.

compared to those of hunter-gatherers. Why, then, did people turn to agriculture? Perhaps the Pleistocene overkill had made hunting less productive, and as the weather warmed, the glaciers retreated and left behind fertile soil with rivers and streams full of fish. In suitable locations, such as the fertile crescent in Mesopotamia, fishing villages may have sprung up and caused people to settle down. As people became more sedentary, they may have had more children, especially because the men were home more often. A population increase may have tipped the scales and caused them to adopt agriculture full-time, especially if agriculture could be counted on to provide food for hungry mouths. Once food was readily available, they began to specialize in other ways of life in towns and then cities.

▶ **21.6 CHECK YOUR PROGRESS**

1. Identify modern attributes exhibited by Cro-Magnons not seen before in the evolutionary history of humankind.

21B Migration Patterns Start with Africa

Humans are one of the most widely distributed species on Earth, and human populations have a long history on six of the seven continents. (Humans are not found on Antarctica, which is too dry and cold for human habitation.) Clearly, modern humans are an adaptable species, considering their amazingly broad distribution across the planet.

As discussed on the previous pages, the evolution of humans is very much centered in Africa, and modern humans evolved there around 200,000 BP. From Africa, how did *Homo sapiens* disperse to the other continents? While there is no physical evidence of the original dispersal event, there is one thing that these early human migrants took with them and gave to each of their descendants: their DNA.

Researchers studying the migration history of the human population use mitochondrial DNA (mtDNA) to trace migration patterns. Whereas the zygote receives chromosomes from both parents, the organelles, such as mitochondria, are usually inherited only from the egg because it contains so much more cytoplasm than a sperm. In other words, you received mtDNA only from your mother, who received it from her mother, and so forth to an original ancestor. This process of inheriting mtDNA is called **maternal inheritance** because it assumes that fathers had nothing to do with it. mtDNA, like nuclear DNA, is subject to occasional harmless mutations. After several generations, a particular mutation is present in almost all the men and women where it arose. When people leave a region, they carry the mutation with them. Therefore, by tracing mutations from one local to another, researchers have determined migration patterns.

The results of migration studies are shown in Figure 21B. The frame at the top shows that Cro-Magnon people remained in Africa until about 100,000 BP. The second frame shows a migration through the Middle East into South Asia at about 100,000 BP that may have included Australia by 40,000 BP. The third frame shows a migration from the Middle East to Europe, during a later time frame. Finally, in the last frame, migration reached East Asia by 30,000 BP and continued into the Americas around 15,000 BP. All this colonization occurred during the last Ice Age when glaciation had caused a significant drop in sea level. As a result, land bridges to North America and Australia were available.

To understand why mtDNA studies are appropriate, imagine small groups of colonists migrating out of the main population in Africa and establishing new populations in these other areas. The colonist populations would be expected to have only a small sample of the genetic variation present in their original source population. Geneticists refer to this as a founder effect. As the populations migrated farther from their source population, no exchange of genetic variation (gene flow) took place between the old and new populations. Over time, the new colonist populations can be genetically differentiated from the original source population because of newly acquired mutations in their mtDNA.

CONSIDER THESE QUESTIONS

1. How far back can you trace your genealogy? Does it involve a migration pattern?
2. Do you think migration patterns plus the founder effect plus natural selection can explain the various ethnic groups of humans? Explain how.
3. Do migration patterns show that all humans are related to one another? Explain.

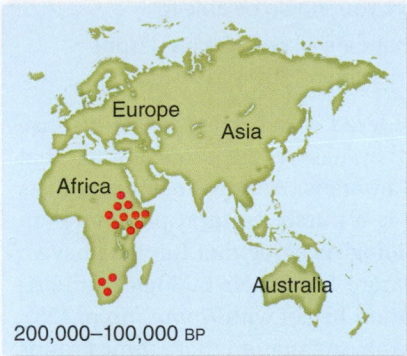

FIGURE 21B *Homo sapiens* (red dots) migrated out of Africa through the Middle East to other continents.

200,000–100,000 BP

Homo sapiens remained in Africa until 100,000 BP.

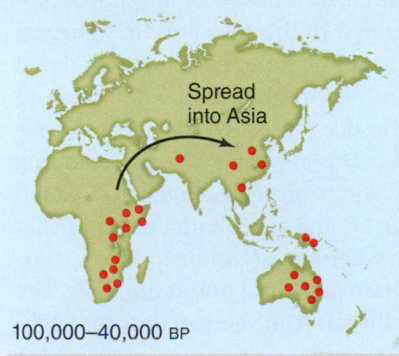

100,000–40,000 BP

Homo sapiens migrated through the Middle East to Asia starting at about 100,000 BP.

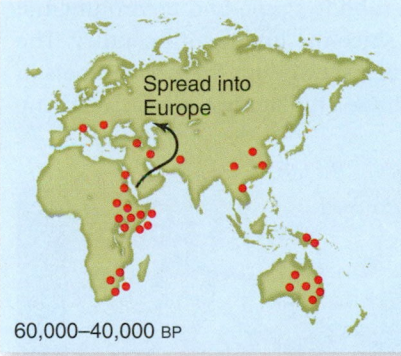

60,000–40,000 BP

Homo sapiens migrated through the Middle East to Europe starting at about 60,000 BP.

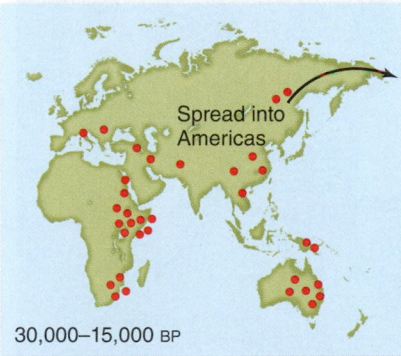

30,000–15,000 BP

Homo sapiens migrated from East Asia to the Americas starting at about 30,000 BP.

 Explore the concepts through a variety of multimedia assets and question types.

www.mcgrawhillconnect.com

LEARNING OUTCOME

When you complete this section, you should be able to

1. Identify evidence that humans are all one species despite adaptation to different environmental conditions.

Many authorities have considered why humans have a sense of ethnicity—meaning that they have a sense of heritage that sets them apart from other humans. Among social scientists some suggest that ethnicity is a purely cultural phenomenon based on shared customs. Biologists agree that humans have a cultural heritage that is separate from their biological inheritance (see "Biocultural Evolution Began with *Homo*" on p. 428). Other social scientists prefer the explanation that ethnic groups have arisen due to a sense of kinship based on a group's biological heritage. Biologists agree with this also and they suggest that because modern human populations were once isolated from one another in different environments, phenotypic and genotypic variations are now noticeable. Therefore people have different ethnicities (Fig. 21.7). One obvious difference among people is skin color. A darker skin is protective against the high UV intensity of bright sunlight. On the other hand, a whiter skin ensures vitamin D production when the UV intensity is low. Harvard University geneticist Richard Lewontin points out, however, that a hypothesis concerning the survival value of dark and light skin has never been tested.

Two correlations between body shape and environmental conditions have been noted since the nineteenth century. The first, known as Bergmann's rule, states endothermic animals in colder regions of their range have a bulkier body build, allowing them to produce more heat. The second, known as Allen's rule, states that animals in colder regions of their range have shorter limbs, digits, and ears, allowing them to conserve heat. The lower surface-area-to-volume ratio in cold climates means that the animal will lose less heat to the environment. On the other hand, longer limbs and appendages increase the surface area of an animal, allowing it to lose more heat in warmer climates. In keeping with these rules, we can note that the Massai of East Africa tend to be slightly built with elongated limbs, while the Eskimos, who live in northern regions of the world, are bulky and have short limbs (see Figure 15.1C).

Other anatomic differences among ethnic groups, such as hair texture, a fold on the upper eyelid (common in Asian peoples), or the shape of lips, cannot be explained as adaptations to the environment. Perhaps these features became fixed in different populations due simply to genetic drift. As far as intelligence is concerned, no significant disparities have been found among different ethnic groups.

Origin of Ethnic Groups The replacement model for the evolution of humans, discussed in section 21.5, pertains to the origin of ethnic groups. This hypothesis proposes that all modern humans have a relatively recent common ancestor—that is, Cro-Magnon, who evolved in Africa and then spread into other regions. Paleontologists tell us that the variation among modern human populations is considerably less than among archaic human populations some 250,000 years ago. If so, all ethnic groups evolved from the same single, ancestral population.

A comparative study of mitochondrial DNA shows that the differences among human populations are consistent with their having a common ancestor no more than a million years ago. Lewontin has also found that the genotypes of different modern populations are extremely similar. He examined variations in 17 genes, including blood groups and various enzymes, among major groups such as African, Asian, and European populations. The results indicated that the great majority of genetic variation—85%—occurs *within* ethnic groups, not among them. In other words, the amount of genetic variation between individuals of the same ethnic group is greater than the variation between ethnic groups.

FIGURE 21.7 Some of the differences between the various ethnic groups in the United States may stem from adaptations to different original environments.

> ▶ **21.7 CHECK YOUR PROGRESS**
> 1. List two adaptations to environmental conditions that might explain the different ethnic groups of humans.
> 2. Discuss evidence that all modern humans are members of the same species.

CONNECTING THE CONCEPTS

One of the most unfortunate misconceptions concerning human evolution is the belief that humans evolved from apes. On the contrary, we can trace our ancestry back to a common ancestor with the apes and then back to the original source of all organisms. Today's apes are our cousins, and we couldn't have evolved from our cousins because we are all contemporaries—living on Earth at the same time. Humans' relationship to apes is analogous to you and your first cousins being descended from your grandparents. When did our common ancestor with the apes live? Was it an ancestor that lived as late as 7 MYA, so that we are closely related to chimpanzees as comparison DNA studies suggest? Do we share a later ancestor with just the chimpanzees as suggested by australopithecine characteristics but not ardipithecine characteristics?

Aside from various anatomic differences related to human bipedalism and intelligence, a cultural evolution separates us from the apes. A hunter-gatherer society evolved when humans became able to make and use tools. That society then gave way to an agricultural economy about 10,000 years ago. The agricultural period extended from that time to about 200 years ago, when the Industrial Revolution began. Now, most people live in urban areas. Perhaps as a result, modern humans are for the most part divorced from nature and often endowed with the philosophy of exploiting and controlling nature.

Our cultural evolution has had far-reaching effects on the biosphere, especially because the human population has expanded to the point that it is crowding out many other species. Our degradation and disruption of the environment threaten the continued existence of many species, including our own. As discussed in Unit VI of this text, however, we now realize that we must work with, rather than against, nature if biodiversity is to be maintained and our own species is to continue to exist.

Before we examine the environment and the role of humans in ecosystems, we will study plant biology and the various organ systems of the human body. Humans need to keep themselves and the environment fit so that they and their species can endure.

ANALYZE AND EVALUATE

1. This part of the text has traced human evolution from the first protists to Cro-Magnon. Draw an evolutionary tree to support this statement.
2. Why would you suggest that biocultural evolution is more influential today than human biological evolution?
3. Suggest ways we could use education to shape our biocultural evolution and perhaps save biodiversity.

SUMMARIZE

Evolution of Primates

21.1 Primates are adapted to live in trees

- Prosimians, monkeys, apes, and humans, are **primates** adapted to an **arboreal** life in trees. Primates are characterized by prehensile hands and feet, including an **opposable thumb; stereoscopic vision;** a large, complex brain; and a reduced reproductive rate.

21.2 All primates evolved from a common ancestor

- The **prosimians** were the first primates to diverge from the main line of descent; the other primates (monkeys, apes, and humans) share an **anthropoid** ancestor (e.g., *Proconsul*); a **dryopithecine** is a common ancestor for the hominoids (gibbons, orang-utans, and hominins).

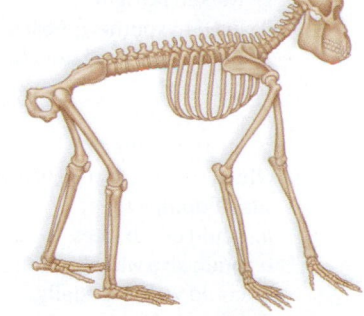
Proconsul

- The designation **hominin** includes chimpanzees, humans, and their closest relatives (humanlike hominins). Ardi has skeletal characteristics—she could not brachiate and did not knuckle-walk—that suggest the hominin ancestor was not chimplike.

Evolution of Humans

21.3 The early humanlike hominins could walk upright

- Humans exhibit **bipedalism,** and therefore this trait has been selected as a necessity to be considered a hominin.

- **Ardipithecines** (7–4.9 MYA) had these traits: (1) small size (122 cm, 50 kg); males and females same size; (2) evidence of bipedalism but with an opposable toe; (3) small brain size (300–350 cc); (4) small teeth including the canines; and (5) lived in trees but did not brachiate.

Ardipithecus ramidus

21.4 Mosaic evolution continues and produces the first humans

- **Australopithecines,** which lived in Africa from 4–1.5 MYA had these traits: (1) not as tall as ardipithecines (100 cm) but brain somewhat larger (400 cc); males larger than females; and (2) robust and gracile types of australopithecines adapted to different ways of life.

- *Australopithecus afarensis* (Lucy) (3.2 MYA) had these traits: (1) somewhat larger teeth with prominent canines; (2) stooped a bit and was an agile walker but lacked a striding gait; (3) lived in trees and may have used brachiation (judged by length of arms); and (4) possible direct ancestor to humans.

- *Australopithecus africanus,* discovered first in South Africa, stood upright and was bipedal and gracile.

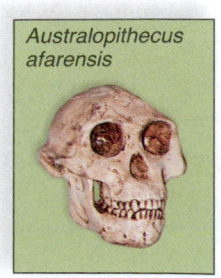
Australopithecus afarensis

- *A. sediba* (1.9 mya) had these traits: (1) brain small but with possible frontal lobe development; (2) tree-climbing arms but tool-grasping hands; (3) long legs with modern-style ankle rests on primitive heel;

(4) projecting nose and dentition like *Homo* species; and (5) possible ancestor to humans.

Homo habilis

- Genus *Homo* had at least a 600-cc brain size and jaws and teeth resembling those of modern humans; tool use is evident. **Homo habilis** was omnivorous with a brain size of 800 cc. Tool use is definite. **H. erectus** (1.8–3 MYA) had these traits: (1) body size increase (1.8 m in males; 1.55 m in females); (2) larger brain (1,100 cc), prominent brow ridges, rounded jaw, and projecting nose; (3) striding gait; (4) born in an immature state; (5) **hunter-gatherers** with tool use; (6) could use fire; and (7) migrated out of Africa.
- **H. floresiensis,** a diminutive version of *Homo sapiens,* discovered in 2004, had a small brain but could use tools and fire.

Evolution of Modern Humans

21.5 Cro-Magnons replaced the other *Homo* species

- The **archaic humans** include all living *Homo* species in Africa, Asia, and Europe around 0.5 MYA. They gave rise to the Neandertals (200,000 BP) in Europe and to *H. sapiens* (100,000 BP) in Africa. Neandertals are also archaic but *Homo sapiens* are modern humans.
- The **Neandertals** (*H. neandertalensis*) lived in Europe and the Near East; they had a brain size of 1,400 cc, were larger-boned than *H. sapiens,* built shelters, used stone tools, hunted successfully, used fire, and had burial ceremonies.
- The replacement model hypothesizes that modern humans evolved in Africa and replaced archaic humans in Asia and Europe.

21.6 Cro-Magnons were socially advanced

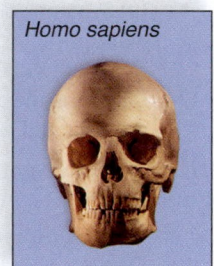
Homo sapiens

- The **Cro-Magnons** (*H. sapiens*) entered Asia and Europe from Africa; they had a brain size of 1,590 cc, were modern in appearance, and subsisted as hunter-gatherers. The Cro-Magnons made advanced stone tools and were good artists. Their highly developed brain facilitated language, social organization, and artistic accomplishments. Agriculture came in about 10,000 BP, and from that point these people are simply called *Homo sapiens.*

21.7 Humans belong to one species

- Humans have a wide geographic distribution and many phenotypic and genotypic variations. The replacement hypothesis proposes that all modern humans have a recent common ancestor. Genetic variations among individuals of the same ethnic group are greater than variations between ethnic groups.

TEST YOURSELF

Evolution of Primates

1. Which of the following lists the correct order of divergence from the main primate line of descent?
   a. prosimians, monkeys, gibbons, orangutans, African apes, humans
   b. gibbons, orangutans, prosimians, monkeys, African apes, humans
   c. monkeys, gibbons, prosimians, African apes, orangutans, humans
   d. African apes, gibbons, monkeys, orangutans, prosimians, humans
   e. *H. habilis, H. ergaster, H. neandertalensis,* Cro-Magnon
2. Stereoscopic vision is possible in primates due to
   a. the presence of cone cells.   c. a shortened snout.
   b. an enlarged brain.   d. None of these are correct.
3. Using the side bars for guidance, name the common ancestors from the earliest to the latest that appear in Figure 21.2A.
4. Which of these is not a feature of *Proconsul*?
   a. prehensile tail
   b. larger brain than a monkey's
   c. mobile shoulder joint
   d. quadrupedal
5. **THINKING CONCEPTUALLY** Explain why some investigators now doubt that humans have a recent common ancestor with chimpanzees.

Evolution of Humans

6. The first humanlike feature to evolve in the hominins was
   a. a large brain.
   b. massive jaws.
   c. a slender body.
   d. bipedal locomotion.
7. The last common ancestor for African apes and hominins
   a. has been found, and it resembles a gibbon.
   b. was probably alive around 7 MYA.
   c. has been found, and it has been dated at 30 MYA.
   d. is not expected to be found because humans evolved before apes.
   e. is now believed to have lived in Asia, not Africa.
8. Which of the following is not a characteristic of robust australopithecines?
   a. massive chewing muscles attached to a bony skull crest
   b. larger brain size than Lucy
   c. walked upright
   d. lived in southern Africa
   e. Both a and c are not characteristics of robust types.
9. Lucy is a(n)
   a. early *Homo.*   c. ardipithecine.
   b. australopithecine.   d. modern human.
10. Which of these is true of ardipithecines but not australopithecines?
    a. could climb trees
    b. brain size was about the same as that of a chimpanzee
    c. could walk bipedally
    d. big toe was opposable
11. Which of these characteristics is not consistent with the genus *Homo?*
    a. large brain size
    b. prolonged infancy
    c. life in the trees
    d. increased intelligence
    e. All of these are characteristics of genus *Homo.*
12. This first species to use fire was probably
    a. *Homo habilis.*
    b. *Homo erectus.*
    c. *Homo sapiens.*

d. *Australopithecus robustus.*
  e. *Australopithecus afarensis.*
13. **THINKING CONCEPTUALLY** How do biocultural evolution and Darwinian evolution by natural selection differ?

Evolution of Modern Humans

14. If the replacement hypothesis is correct,
  a. *Homo erectus* migrated out of Africa.
  b. no hominins and no *Homo* species migrated out of Africa.
  c. humans evolved in and migrated out of Africa.
  d. Both b and c are correct.
  e. Both a and c are correct.
15. Mitochondrial DNA data support the replacement model by showing that
  a. humans evolved in Africa.
  b. only early hominins evolved in Africa.
  c. Neandertals were never replaced and are alive today.
  d. people today are more distantly related than previously thought.
16. Which of these pairs is not correctly matched?
  a. *H. erectus*—made tools
  b. Neandertal—good hunter
  c. *H. habilis*—controlled fire
  d. Cro-Magnon—good artist
  e. *A. robustus*—fibrous diet
17. The first *Homo* species to use art appears to be
  a. *H. neandertalensis.*
  b. *H. erectus.*
  c. *H. habilis.*
  d. *H. sapiens.*
18. The increased reliance on agriculture in some early societies led to increases in which of the following?
  a. infectious disease
  b. malnutrition
  c. anemia
  d. All of these are correct.
19. Complete the key to explain the replacement model depicted below.

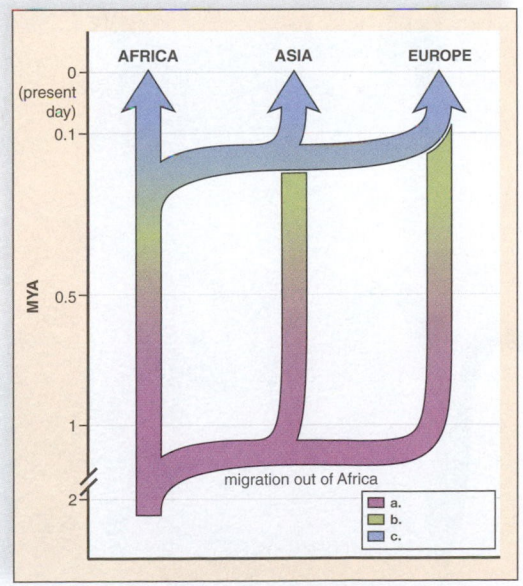

20. Which human characteristic is not thought to be an adaptation to the environment?
  a. bulky bodies of Eskimos
  b. long limbs of Africans
  c. light skin of northern Europeans
  d. hair texture of Asians
  e. Both a and b are correct.
21. **THINKING CONCEPTUALLY** How does genetic variation between ethnic groups compare to genetic variation within ethnic groups? What does this tell us about humans in general?

GET INVOLVED

1. Bipedalism has many selective advantages, including the increased ability to spot predators and prey. However, bipedalism has one particular disadvantage—upright posture leads to a smaller pelvic opening, which makes giving birth to an offspring with a large head very difficult. This situation results in a higher percentage of deaths (of both mother and child) during birth in humans compared to other primates. How can you explain the selection for a trait, such as bipedalism, that has both positive and negative consequences for fitness?
2. You have been provided with DNA from Neandertals and modern humans. What type of study would you do and how might your data support or not support the hypothesis that humans interbred with Neandertals?

MEDIA STUDY TOOLS

mhhe.com/maderconcepts3

Enhance your study of this chapter with interactive study tools, practice tests, and engaging animations. Also, ask your instructor about the resources available through ConnectPlus, which includes LearnSmart, a personalized adaptive learning program, and a media-rich eBook.

22

Plant Organization and Homeostasis

BEFORE YOU BEGIN

Take a few minutes to recall

The cellular structures unique to plants (Fig. 4.3B)
How plants carry on photosynthesis (section 6.2–6.8)
How flowering plants reproduce (Fig. 19.7B)

Plants Fuel You (and Soon Your Car)

Grasses are among the most successful plant groups, and they have long produced food crops for us such as corn, wheat, and rice. Now, a particular grass called switchgrass (*Panicum virgatum*) could become the biofuel that will drive your car. A biofuel is a liquid fuel, such as ethanol, diesel fuel, or even gasoline, made from plant materials. The technology is not here yet, but it is being developed.

In the past, switchgrass was a significant part of the tallgrass prairie that covered much of the Midwest before settlers turned it into farms. Switchgrass doesn't look like the grass in your front lawn. It's strong and tough, as it would have to be because it can be 30.80 cm high. The roots are extensive and spread out into the soil to collect inorganic nutrients and water. Bioengineering is making switchgrass even more hardy and able to grow on marginal land that is too dry and nutrient-poor for

Switchgrass can be 10 ft high.

other plants. Because switchgrass can grow on land no longer suitable for farming, bringing it back in quantity will not crowd out food crops. Once established, switchgrass can be harvested every year for at least 10 years before replanting is needed. After harvesting the stems and leaves, the carbon-retaining root remains in the ground ready to bring forth another shoot system the next year. Best of all, switchgrass takes in more CO_2 than it gives off. Because it requires limited fertilizer and irrigation, it can return as much as five times as much energy as is needed to grow it.

A good source of biofuel (1) is renewable, (2) is not a food source for humans, (3) can grow on land that is not farmable, (4) will produce a new crop every year, (5) returns more energy than is needed to produce it, and (6) takes in more CO_2 while growing than it gives off when burned. All plants are renewable because you can grow a new supply each year, but switchgrass is much better than corn as a biofuel source with regard to the rest of these criteria. Other possibilities, such as cornstalks, (instead of ears of corn) are being considered—but wouldn't it be better to keep the stalks on the ground where they can prevent erosion? Wood chips are sometimes used, but trees do not regrow from their roots within one growing season. Only when wood has been harvested for other reasons is it sensible to use wood chips as a source for biofuel.

The consensus is that switchgrass is an energy crop that can supply as much biofuel in 1 year as the 1970 peak oil production in the United States. Plants already do so much for us, and now they may even allow us to continue our way of life without the threat of climate change. Why? Because certain plants, particularly those that regrow from their roots as switchgrass does, take in more CO_2 than they give off when they are burned! This chapter teaches you about the anatomy of plants. Just as the structure of switchgrass is suited to being the source of biofuel, so plant organs are adapted to the role they play in the life of a plant. We will be discussing the adaptations of plant organs as we explore their anatomy.

Switchgrass is a good potential source of biofuel.

Switchgrass can be baled.

Switchgrass can be used to fuel a power plant.

Organs and Tissues of Flowering Plants

Each vegetative organ of a plant is suited to its usual functions as well as to specialized functions. The two main groups of flowering plants, monocots and eudicots, differ in their anatomy.

22.1 Flowering plants typically have roots, stems, and leaves

LEARNING OUTCOME

When you complete this section, you should be able to

1. Discuss how the structures of roots, stems, and leaves are suited to their respective functions.

From cacti living in a hot desert to water lilies growing in a nearby pond, the flowering plants, or angiosperms, are extremely diverse. (Other plant types, such as mosses, ferns, and gymnosperms, are not considered in this chapter.) Despite their great diversity in size and shape, flowering plants share many common structural features. Most flowering plants have a shoot system (above ground) and a root system (below ground) (Fig. 22.1A). The **shoot system** consists of the stem, the branches, the leaves, and the flowers, which are organs of sexual reproduction. The **root system** consists of the main root and its branches. The stem, the leaf, and the root—called the three vegetative organs—perform functions that allow a plant to live and grow. The flower is discussed in section 25.1.

The Stem A **stem** performs four main functions that are necessary to the life of a plant: support of leaves and flowers, growth of the stem, transport of water and nutrients between the leaves and roots, and sometimes food storage. A stem must be strong enough to lift the leaves so that they can absorb sunlight and take in CO_2 for photosynthesis. Stems grow and produce new tissues throughout the entire lifetime of the plant!

A stem can grow because it has a **terminal bud** in the so-called shoot tip. The activity of a terminal bud produces tissues that become new leaves and new axillary (lateral) buds. In turn, the axillary buds can produce new branches of the stem (or flowers). The part of a stem where new leaves and axillary buds are found is called a **node.** An **internode** is the region between the nodes. The nodes are at first close together, but as the stem increases in length, the nodes get farther apart. Nodes occur only on a stem, and their presence allows you to know you are observing a stem.

A stem not only supports the leaves so that they are better exposed to sunlight, but also contains the vascular tissue that brings water to them and takes away the products of photosynthesis. These products are used as an energy source by other plant parts or are stored. The transport function of a stem is possible only because its vascular tissue is continuous with that of the root system and the leaves. What part of a stem produces the cells that become new vascular tissue as the plant grows taller? The terminal bud, of course.

Stems can store the products of photosynthesis, and some are particularly good at it. For example, the stem of a potato plant grows above and also below ground, where swollen regions we call potatoes store food. How do you know that a white potato is part of an underground stem? The "eyes" of a white potato are actually nodes capable of new growth. If you cut out an eye and place it in water, it will give rise to a complete plant. Other plants have a fleshy horizontal underground stem called a **rhizome,** which survives the winter and from which they can regrow in the spring. Like switchgrass, these plants are called rhizomatous perennials.

Stems can be highly modified. In a cactus, the stem is modified for photosynthesis and also for water storage; the leaves are reduced to spines in order to minimize water loss (Fig. 22.1B).

Leaves Usually, **leaves,** attached to the stem and its branches, are the chief organs of photosynthesis. Photosynthesis, which produces food for plant growth, repair, and reproduction, requires a supply of solar energy, carbon dioxide, and water. Foliage leaves are flattened; this shape gives them a maximum surface area for the collection of solar energy and absorption of carbon dioxide

FIGURE 22.1A Organization of a plant body.

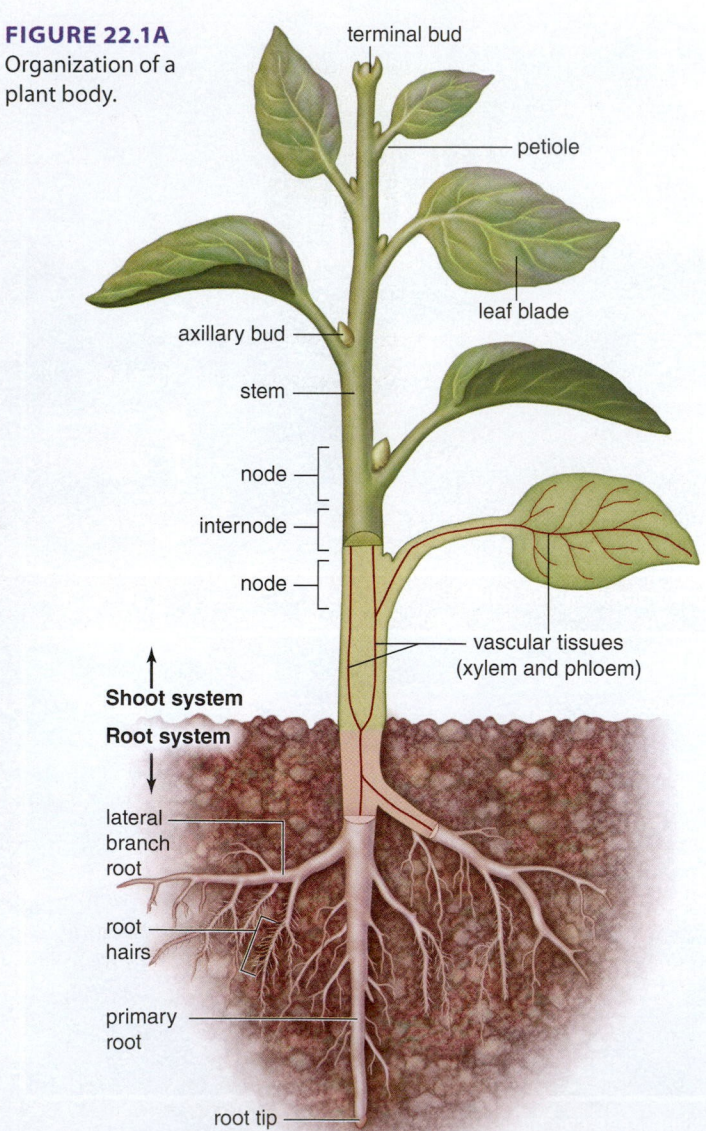

terminal bud

petiole

leaf blade

axillary bud

stem

node

internode

node

vascular tissues (xylem and phloem)

Shoot system

Root system

lateral branch root

root hairs

primary root

root tip

stem

leaves

tendril

Spines are the leaves of a cactus

Tendrils are modified leaves of a cucumber

Leaves of a Venus flytrap capture insects

FIGURE 22.1B Modified leaves adapt to a plant's environment.

through pores. The vascular tissue in a leaf is continuous with that of the stem and functions to bring leaves the water needed for photosynthesis.

The wide portion of a foliage leaf which absorbs solar energy is called the **blade.** The blade is held away from the stem by a stalk called a **petiole,** and therefore a petiole helps expose a leaf to sunlight. Some leaves, such as those of switchgrass, do not have petioles and are instead attached directly to the stem. Leaves without petioles are called sessile leaves.

Leaves are adapted not only to carry on photosynthesis but also to environmental conditions (Fig. 22.1B). Desert plants, such as a cactus, tend to minimize water loss by having reduced leaves. Division of labor occurs in cucumber leaves because some of its leaves are climbing leaves. The climbing leaves are modified into tendrils that attach to nearby objects and help lift the foliage leaves. The leaves of a few plants are specialized for catching insects. The Venus flytrap has hinged leaves that snap shut and interlock when an insect triggers sensitive hairs projecting from inside the leaves. How do many plants cope with winter in the temperate zone? They are **deciduous,** and lose their leaves only to have them grow back in the spring when warmth returns.

Roots Like shoot tips, root tips can grow their entire lives. **Roots** have three main functions: They anchor the plant in the soil, absorb water and minerals from the soil, and produce hormones. In addition, roots often store food. The structure of a root is suited to anchoring a plant and in that way helping to keep it upright. The main part of a root is often a larger and stronger taproot, but a taproot gives off branches that give off still more branches. These many branches spread out and keep a plant solidly fixed in the soil.

The cylindrical shape and slimy surface of a taproot and its branches help them penetrate soil as they grow. This shape also permits water and minerals to be absorbed from all sides. The high surface-to-volume ratio of a root system is increased even more by the presence of **root hairs,** which are delicate extensions of mature root cells. Root hairs are so numerous that they increase the absorptive surface of the root system tremendously. It has been estimated that a single rye plant has about 14 billion root-hair

Taproot

Fibrous root system

FIGURE 22.1C Taproot system (*left*) compared to a fibrous root system (*right*).

cells, and if placed end-to-end, the root hairs would stretch 10,626 km. Root hairs are constantly being replaced, and this same rye plant forms about 100 million new root-hair cells every day.

Roots produce hormones that ensure the root system will be a size appropriate to the needs of the shoot system. A huge shoot system needs a much larger root system than a small one. Hormones are chemical messengers often distributed by the vascular system. Just like the human couriers of a delivery service, hormones deliver messages that keep the parts of a plant well coordinated.

Figure 22.1C shows the two main types of roots. A dandelion has a taproot, which still has many branches. Grasses such as switchgrass have fibrous roots that lack a main taproot, but despite this difference in anatomy, a root system anchors a plant and absorbs water and minerals. Without roots, plants cannot survive for long, as you know from observing the short life span of cut flowers.

▶ **22.1 CHECK YOUR PROGRESS**
  1. Identify a significant function of the stem, the leaf, and the root.

22.2 Flowering plants are either monocots or eudicots

LEARNING OUTCOME

When you complete this section, you should be able to

1. List and describe five differences between monocots and eudicots.

Plants manage to grow from seeds before they have fully developed organs because seeds have cotyledons. The seeds of grasses and many other plants (e.g., tulips and daffodils) have one cotyledon. These plants are known as monocotyledons, or **monocots.** The seeds of kidney beans, peas, lima beans, and many other plants have two cotyledons. These plants are known as eudicotyledons, or **eudicots.** The cotyledons of eudicots supply nutrients when a seedling begins to grow, but the cotyledons of monocots act largely as a transfer tissue for nutrients derived from the endosperm, a storage tissue, before the true leaves begin photosynthesizing.

The term eudicot is a new one for botanists, who formerly compared monocots to a group called dicots. Recent findings about plant evolution have revealed that some of the plants formerly called dicots, such as water lilies, are so ancient that they arose before the angiosperms split into the monocots and eudicots. Therefore, the term dicotyledon is now obsolete.

Adult monocots and eudicots have other structural differences (Fig. 22.2). Some of these differences are observable with the unaided eye, while others require a microscope. For example, in the monocot root, vascular tissue occurs in a ring encircling a core of cells that comprise the pith. In the eudicot root, xylem, which transports

water and minerals, has a star shape while phloem, which transports organic nutrients, is located between the arms of the xylem.

In the monocot stem, the vascular bundles, which contain vascular tissue surrounded by a sheath, are scattered throughout the ground tissue. In a eudicot stem, the vascular bundles occur in a ring. This ring of vascular bundles divides the ground tissue into a cortex and the centrally located pith.

Visible to the naked eye, leaf veins are the vascular bundles within a leaf. Monocots, such as grasses, exhibit parallel venation, and eudicots, such as maples, exhibit netted venation. Plants also differ by the number of flower parts. Monocot flower parts are arranged in multiples of three, and eudicot flower parts occur in multiples of four or five. Not shown in Figure 22.2, pollen grains differ by the number of apertures (thin areas in the wall). Monocot pollen grains usually have one aperture, and eudicot pollen grains usually have three apertures.

Although the division between monocots and eudicots may seem of limited importance, it does in fact affect many aspects of their structure. The eudicots are the larger group and include some of our most familiar flowering plants—from dandelions to oak trees. The monocots include grasses, lilies, orchids, and palm trees, as well as some of our most significant food sources —rice, wheat, and corn.

▶ **22.2 CHECK YOUR PROGRESS**

1. Identify how the vascular bundles are arranged in the stem if the flower parts of a plant occur in threes.

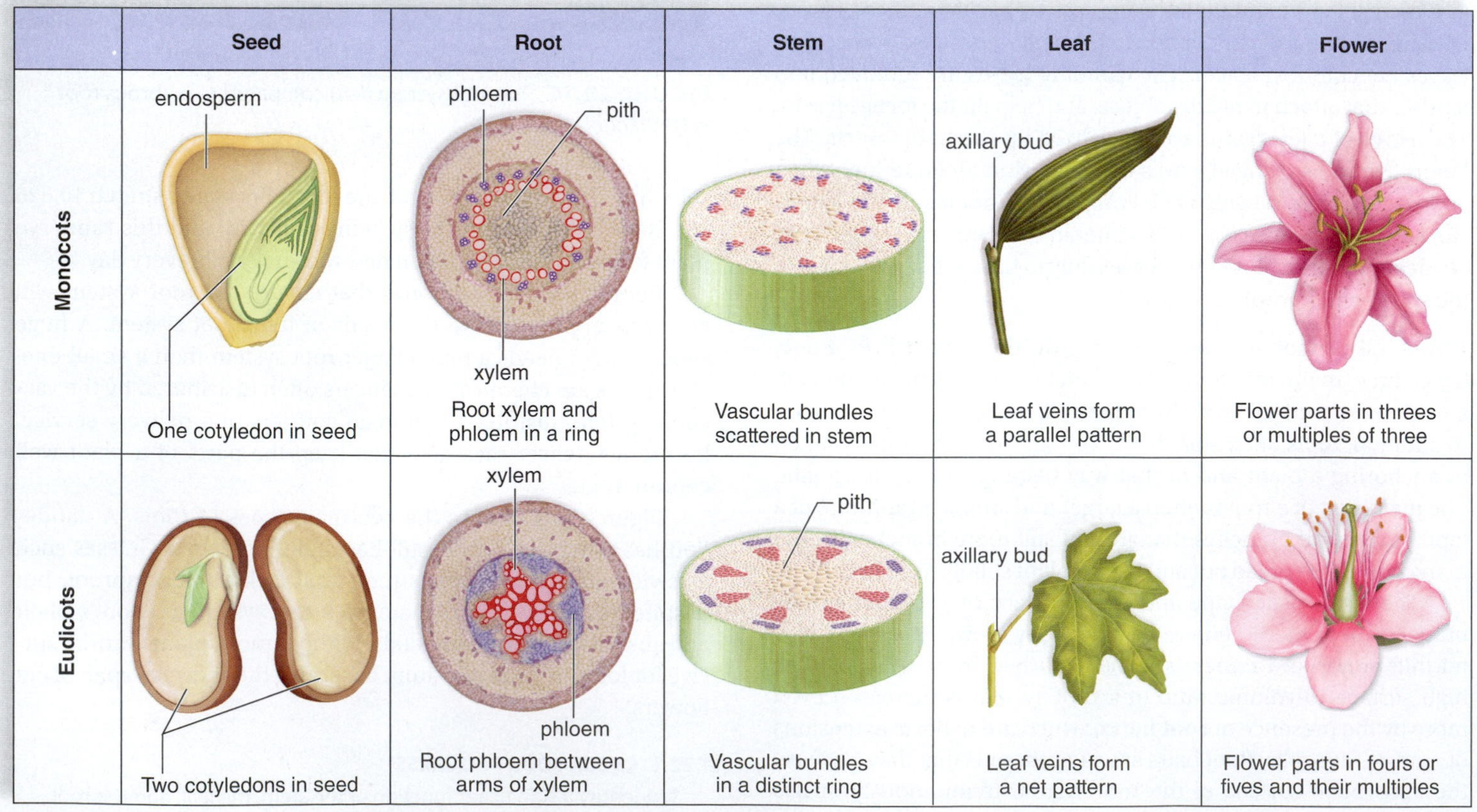

	Seed	Root	Stem	Leaf	Flower
Monocots	One cotyledon in seed	Root xylem and phloem in a ring	Vascular bundles scattered in stem	Leaf veins form a parallel pattern	Flower parts in threes or multiples of three
Eudicots	Two cotyledons in seed	Root phloem between arms of xylem	Vascular bundles in a distinct ring	Leaf veins form a net pattern	Flower parts in fours or fives and their multiples

FIGURE 22.2 Monocots and eudicots differ structurally in several ways.

22A Monocots Serve Humans Well

Although the monocots are a small group compared to the eudicots, they have great importance. From cereal grains to alcoholic beverages, monocots play a significant role in all our lives. The domestication of monocot plants included selective breeding in order to accumulate certain desirable traits in crops. For example, you would probably never recognize "wild" corn because, through selective breeding, we have encouraged large, fleshy, starchy kernels that in no way resemble ancestral corn.

Cereal grasses produce the cereal grains such as rice, wheat, corn, and barley (Fig. 22A) which are a chief source of calories for the majority of the world's people and their livestock. It is remarkable that wheat, corn, and rice are associated with different major cultures or civilizations—wheat with Europe and the Middle East, corn or maize with the Americas, and rice with the Far East. Corn is by far the most important crop plant in the United States, where about 80% of the corn produced goes to feed livestock. However, people in many developing countries rely on corn for as much as 30% of the calories in their diets. Three of the world's four most populous nations are rice-based societies—China, India, and Indonesia. Over 50% of the world's people depend on rice for about 80% of their calorie requirements. A diverse food, rice can be cooked and eaten as is, or can be used to produce breakfast cereals, desserts, rice cakes, and rice flour.

Beer is also produced from cereal grain. The exact origin of beer is unclear, but it is believed to be over 10,000 years old. While most beer uses barley malt (dried young seedlings) to supply enzymes, the specific grain to be fermented varies geographically among all the beer-producing cultures. For example, wheat is used in Mesopotamia, rice in Asia, and sorghum in Africa. Sake, although called "rice wine," is actually beer produced from fermented rice (as opposed to a true wine, which is produced from grapes and other fruits). In the United States, various grains are used to make beer.

Bamboo, the common name for about 1,000 species of grass, ranges in height from 15 cm to 30–35 m. Depending on the species, bamboo can grow up to a foot a day. Eaten not only by pandas, young bamboo is also consumed by humans as a vegetable (yes, those are actual bamboo shoots in Chinese food). Older bamboo is much tougher, and therefore harvested for making musical instruments, furniture, and acupuncture needles, as well as for roofing, flooring, and drainage pipes. In fact, about 73% of the population of Bangladesh lives in bamboo houses.

Finally, many of the flowers bought and sold are monocots, including tulips, daffodils, and lilies. Floriculture, the cultivation and management of ornamental and flowering plants, is a multi-billion-dollar industry in the United States alone.

CONSIDER THESE QUESTIONS

1. When you eat a grain, you are consuming an embryo. Do you find that disturbing? Why or why not?
2. Cattle raised on grain return far less energy to a consumer than the grain would. And cattle are polluting. Would you mind eating only grains so that more people could be fed?

Rice plants, *Oryza*

Wheat plants, *Triticum*

Corn plants, *Zea mays*

Barley

FIGURE 22A Monocot variety.

Explore the concepts through a variety of multimedia assets and question types.

www.mcgrawhillconnect.com

LEARNING OUTCOME

When you complete this section, you should be able to

1. Describe the adaptations of epidermal tissue, ground tissue, and vascular tissue that allow them to perform their respective functions.

Unlike humans, flowering plants grow in size their entire life because they have meristematic (embryonic) tissue composed of cells that divide. **Apical meristem** is located in the terminal bud of the shoot system and in the root tip. When apical meristem cells divide, some of the daughter cells differentiate into one of three primary meristems (see section 22.5). Each primary meristem produces a particular tissue type during the life of a plant:

1. **Epidermal tissue** contains flat epidermal cells, which form the outer protective covering of a plant but are modified for various other functions.
2. **Ground tissue** contains large, thin-walled parenchyma cells that fill the interior of a plant and serve metabolic and other functions.
3. **Vascular tissue** contains xylem and phloem, whose cells form strong, continuous pipelines that transport water and sugar in a plant and provide support.

Epidermal Tissue The entire body of a plant is covered by a layer of closely packed epidermal cells called the **epidermis.** The walls of epidermal cells that are exposed to air are covered with a waxy **cuticle** to minimize water loss. In leaves, the epidermis often contains **stomata** (sing., stoma) (Fig. 22.3A*a*). A stoma is a small opening surrounded by two modified epidermal cells called guard cells. When the stomata are open, CO_2 uptake and water loss occur. As mentioned in section 22.1, roots have long, slender projections of epidermal cells called root hairs (Fig. 22.3A*b*). These hairs increase the surface area of the root for absorption of water and minerals.

In the trunk of a tree, the epidermis is replaced by periderm, which contains cork. New cork cells are made by a meristem called cork cambium. As the new cork cells mature, they increase slightly in volume and become encrusted with *suberin,* a lipid material that both waterproofs them and causes them to die. These nonliving cells protect the plant and help it resist attack by fungi, bacteria, and animals. Lenticels, which are breaks in the periderm (cork plus cork cambium), function in gas exchange in some trees (Fig. 22.3A*c*).

Ground Tissue Ground tissue forms the bulk of stems, leaves, and roots. Ground tissue contains three types of cells (Fig. 22.3B). **Parenchyma cells** are the least specialized of the cell types and are present in all the organs of a plant. When in a leaf, they typically contain chloroplasts that allow them to carry on photosynthesis; in the cortex or pith region, they contain colorless plastids that store the products of photosynthesis. **Collenchyma cells** have thicker primary walls than parenchyma cells. The thickness is uneven, with the thicker areas usually found in the corners of the cell. Collenchyma cells particularly provide structural support in nonwoody plants. The familiar strands in celery stalks are composed mostly of collenchyma cells. **Sclerenchyma cells** have thick secondary cell walls impregnated with *lignin,* a substance that

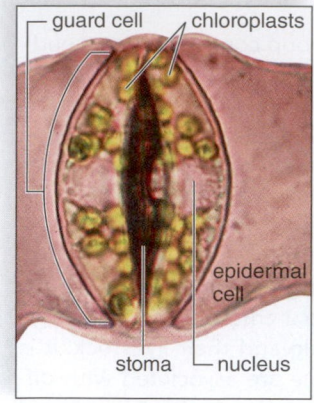

a. Stoma of leaf

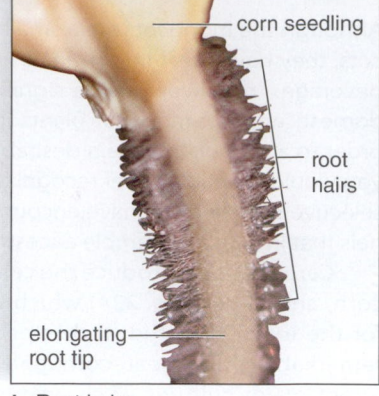

b. Root hairs

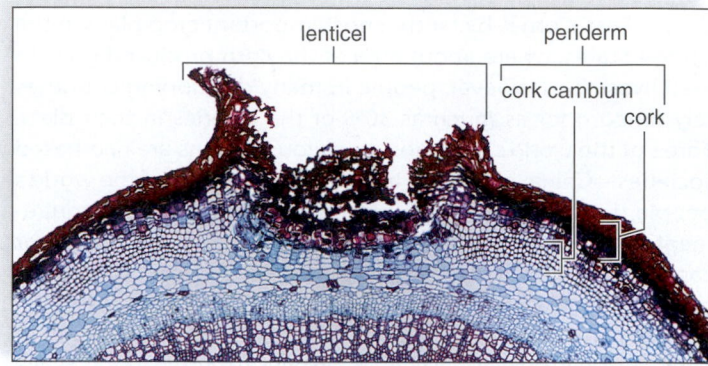

c. Cork of older stem

FIGURE 22.3A Modifications of epidermal tissue.

FIGURE 22.3B Ground tissue cells.

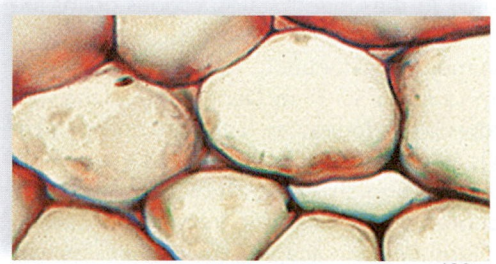

Parenchyma cells with thin walls   100×

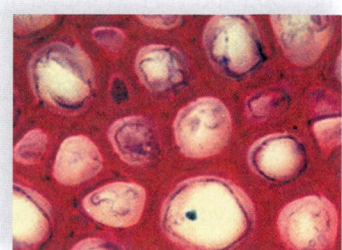

Collenchyma cells   255×
with thicker walls

Sclerenchyma cells   340×
with very thick walls

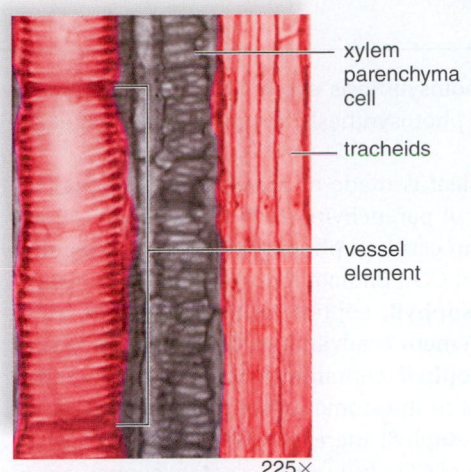

- xylem parenchyma cell
- tracheids
- vessel element

225×

a. Xylem micrograph

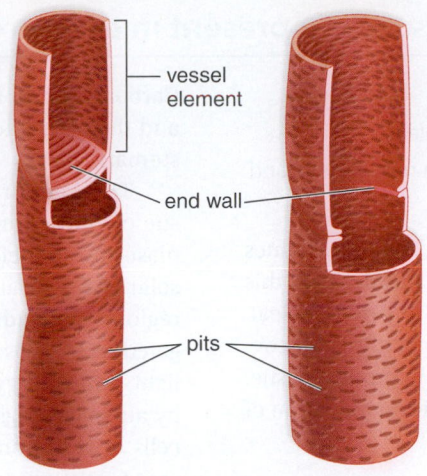

- vessel element
- end wall
- pits

b. Two types of vessels

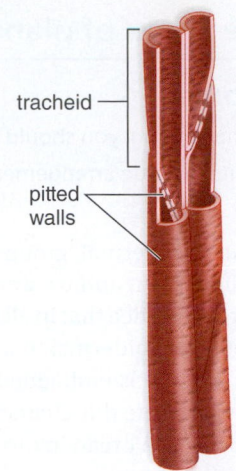

- tracheid
- pitted walls

c. Tracheids

FIGURE 22.3C
Xylem structure **a.** Organization of xylem **b.** Drawing of two types of vessels: On the right, the vessel elements have barred end walls and on the left, the end walls have no bars. **c.** Drawing of tracheids.

makes plant cell walls tough and hard. If we compare a cell wall to reinforced concrete, cellulose fibrils would play the role of steel rods, and lignin would be analogous to the cement. Most sclerenchyma cells are nonliving at maturity; their primary function is to support the mature living regions of a plant. There are two main types of sclerenchyma cells: fibers and sclerids (also known as stone cells). Fibers in plants make them useful for a number of purposes. For example, cotton and flax fibers can be woven into cloth, and hemp fibers can make strong rope.

Vascular Tissue The xylem and phloem of vascular tissue have different functions. **Xylem** transports water and minerals from the roots to the leaves. Xylem contains conducting cells called vessel elements and tracheids (Fig. 22.3C). Both of these are hollow and nonliving, but tracheids are elongated and narrow, while vessel elements are wider and shorter. The end walls of **vessel elements** even if barred are arranged to form a **vessel,** a continuous pipeline for water and mineral transport. The end walls and side walls of **tracheids** have pits, depressions that allow water to move from one tracheid to another.

Phloem transports sugar, in the form of sucrose, and other organic compounds, such as hormones, usually from the leaves to the roots. The conducting cells of phloem are **sieve-tube**

members, arranged to form a continuous sieve tube (Fig. 22.3D). Sieve-tube members contain cytoplasm but no nuclei. The term *sieve* refers to a cluster of pores in the end walls, collectively known as a sieve plate. Sieve-tube members have companion cells which do have a nucleus. Sieve-tube members and their companion cells are joined by numerous plasmodesmata (strands of cytoplasm that extend through pores), and the nuclei of the companion cells maintain the life of the sieve-tube members. The companion cells are also believed to be involved in the transport function of phloem.

Vascular tissue, consisting of both xylem and phloem, extends from the root through the stem to the leaves and vice versa (see Fig. 22.1A). In the root, the vascular tissue is located in a central cylinder; in the stem, it is in vascular bundles supported by fibers; and in the leaves, the vascular bundles are referred to as leaf veins.

Animation
Vascular System of Plants

▶ **22.3 CHECK YOUR PROGRESS**

1. Identify how the epidermis is modified in the root, stem, and leaf.
2. Distinguish between the function of the phloem and the xylem.

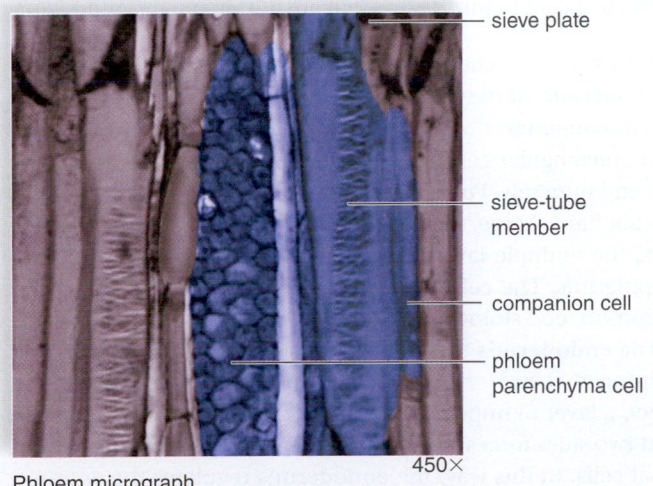

- sieve plate
- sieve-tube member
- companion cell
- phloem parenchyma cell

450×

Phloem micrograph

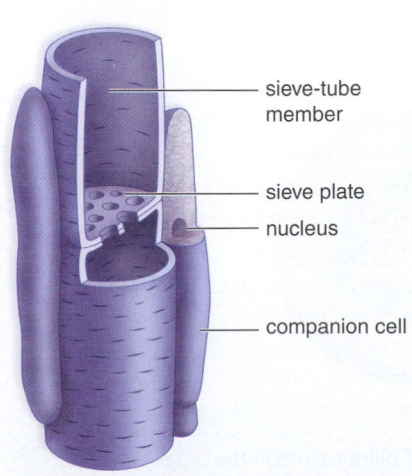

- sieve-tube member
- sieve plate
- nucleus
- companion cell

Sieve-tube member and companion cells

FIGURE 22.3D Phloem structure. Sieve-tube members, which transport sugars and other substances, have companion cells that carry out metabolic functions.

LEARNING OUTCOME

When you complete this section, you should be able to

1. Explain the benefits of tissue arrangement in a root, a stem, and a leaf.

Figure 22.4A shows how epidermal, ground, and vascular tissues are arranged in a eudicot plant and we want to consider how this is adaptive for each organ. Notice that in all three organs—the leaf, the stem, and the root—the epidermal tissue forms the outer covering, and the vascular tissue is embedded within ground tissue. In order to further demonstrate this characteristic organization of plant tissues, let's look at each organ in turn.

Leaf A typical eudicot leaf of a temperate-zone plant is shown in longitudinal section at the top of Figure 22.4B*a*.

The upper epidermis often bears protective hairs and/or glands that secrete irritating substances to provide some protection against predation. These appendages and chemicals discourage insects from eating leaves where insects are apt to land. The upper and lower epidermis has an outer, waxy cuticle, which prevents water loss but also prevents gas exchange because it is not gas permeable. This problem is solved by having stomata located in the lower epidermis. Here the stomata are less likely to serve as an avenue for water loss and an entrance for parasites.

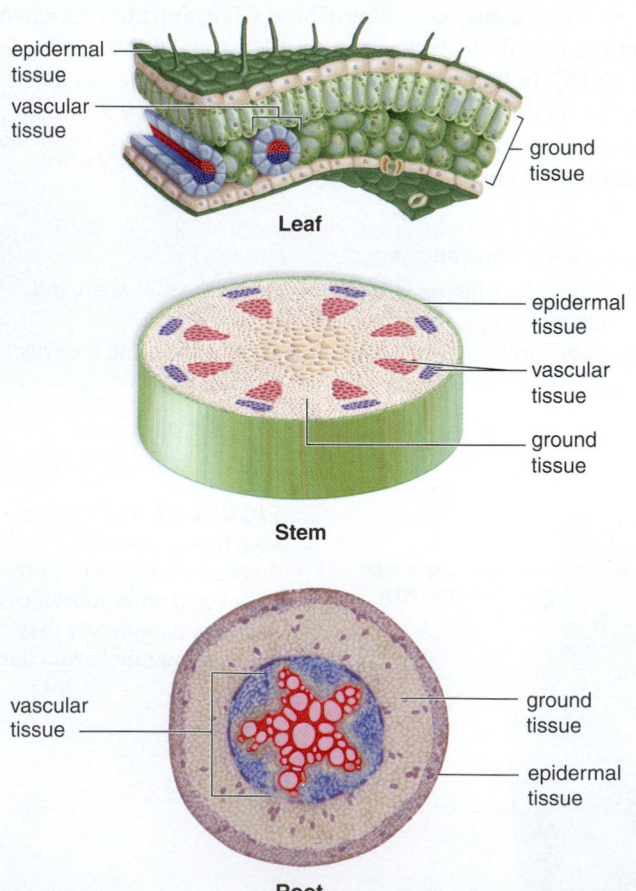

FIGURE 22.4A Arrangement of plant tissues in the organs of eudicots. See also Figure 22.4B.

Carbon dioxide for photosynthesis enters a leaf at the stomata, and the by-product of photosynthesis, oxygen, exits a leaf at the stomata.

The interior of a leaf is made of **mesophyll,** a ground tissue composed mostly of parenchyma cells that contain chloroplasts and therefore can carry on photosynthesis which requires solar energy, water, and CO_2. Eudicot mesophyll has two distinct regions: **palisade mesophyll,** containing tightly packed, elongated cells whose placement is advantageous for absorbing sunlight, and **spongy mesophyll,** containing irregular cells bounded by air spaces right next to the stomata. The loose packing of the cells in the spongy mesophyll increases the amount of surface area for the absorption of CO_2. But water loss from spongy mesophyll can be regulated by closure of the stomata.

Leaf veins branch and terminate in the mesophyll, where xylem can deliver water to the cells that are photosynthesizing. Terminations also collect the product of photosynthesis, a sugar, by way of phloem.

Stem Plants that have nonwoody stems, such as zinnias and daisies, are termed **herbaceous** plants. The ground tissue in the stem consists of the **cortex,** a narrow ring of parenchyma cells beneath the epidermis, and the central **pith,** which stores water and the products of photosynthesis. The cortex is sometimes green and carries on photosynthesis.

Herbaceous stems have distinctive vascular bundles containing xylem and phloem. In the herbaceous eudicot stem, the vascular bundles are arranged in a distinct ring. In the herbaceous monocot stem, the vascular bundles are scattered throughout the ground tissue and pith is not distinguishable. Figure 22.4B*b, c* contrasts herbaceous eudicot and monocot stems.

The vascular tissue of a stem supports the shoot system so that the leaves can collect solar energy for photosynthesis. The strong walls of vessel elements and tracheids are bolstered by the presence of many sclerenchyma cells so that the stem can support the shoot system as it increases in length. Further, the vascular bundles bring water and minerals to the leaf veins, which distribute them to the mesophyll of leaves, the primary organs of photosynthesis in most plants. The vascular bundles also distribute the products of photosynthesis to the root system, where they may be stored as needed, and also to any immature leaves that are not yet photosynthesizing.

Root In a cross section of a mature root (Fig. 22.4B*d*), the following specialized tissues are identifiable: The epidermis, which forms the outer layer of the root, usually consists of only a single layer of rectangular cells that are thin-walled and able to absorb water and minerals. Further, when mature, many epidermal cells have root hairs. Large, thin-walled parenchyma cells make up the cortex, the multiple layer of ground tissue cells located beneath the epidermis. The cells contain starch granules and the cortex functions in food storage.

The **endodermis** is a single layer of rectangular cells that fit snugly together and form the outer tissue of the **vascular cylinder.** Further, a layer of impermeable suberin (the Casparian strip) on all but two sides forces water and minerals to pass through endodermal cells. In this way, the endodermis regulates the entrance

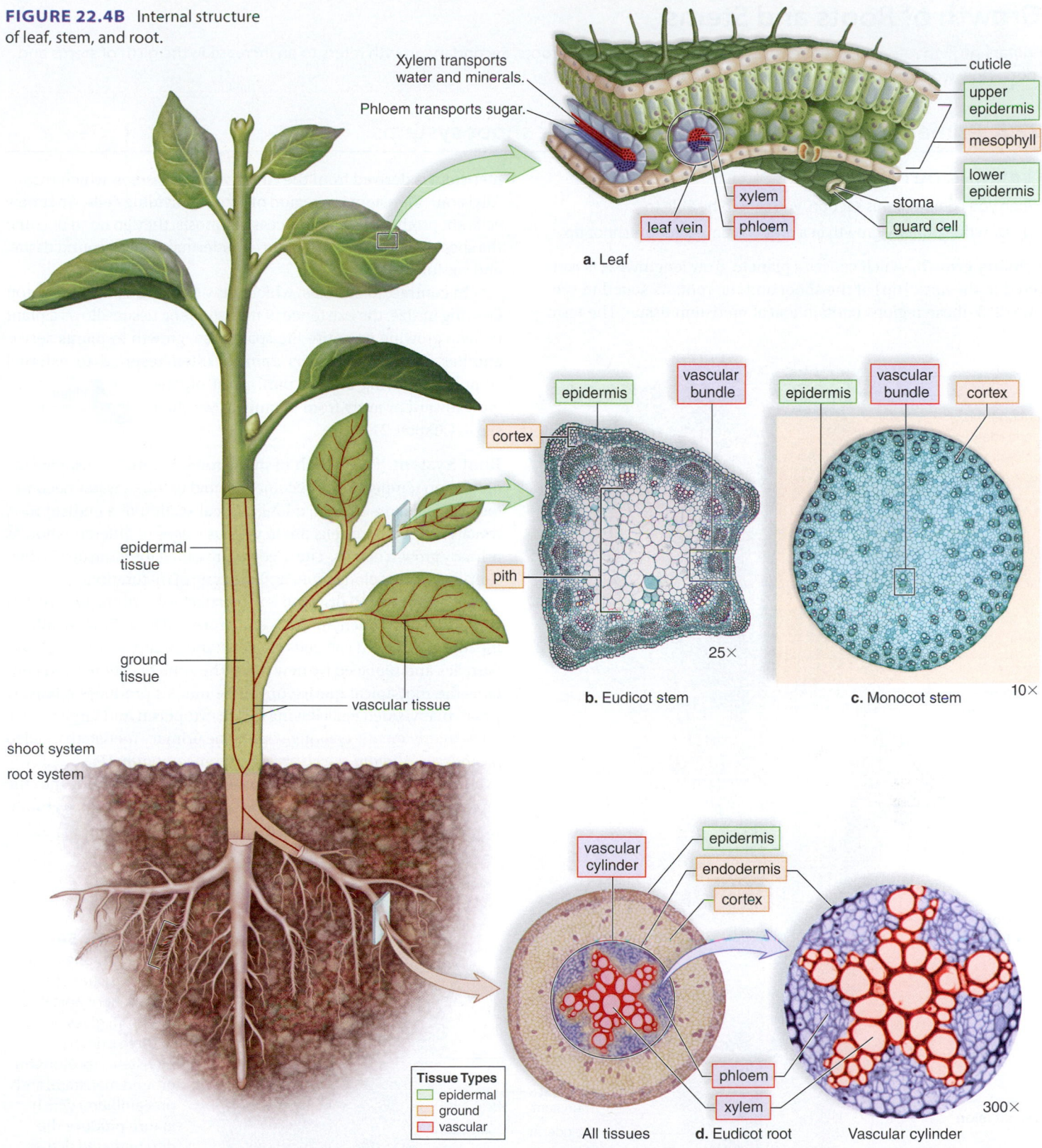

FIGURE 22.4B Internal structure of leaf, stem, and root.

Xylem transports water and minerals.

Phloem transports sugar.

cuticle

upper epidermis

mesophyll

lower epidermis

stoma

guard cell

leaf vein

xylem

phloem

a. Leaf

epidermis

vascular bundle

cortex

pith

25×

b. Eudicot stem

epidermis

vascular bundle

cortex

10×

c. Monocot stem

epidermal tissue

ground tissue

vascular tissue

shoot system

root system

vascular cylinder

epidermis

endodermis

cortex

phloem

xylem

300×

Tissue Types
- epidermal
- ground
- vascular

All tissues

d. Eudicot root

Vascular cylinder

of minerals into the vascular cylinder of the root. Just inside the endodermis layer is the **pericycle,** a layer of actively dividing cells from which lateral branch roots arise (see Fig. 22.5A). The star shape of xylem tissue in a root facilitates the absorption of water across the endodermis from the cortex. Once water and minerals are taken up by xylem, they are transported to the stem. The star shape of xylem means that phloem occurs in separate regions between the arms of the xylem.

▶ **22.4 CHECK YOUR PROGRESS**

1. Identify the location of ground tissue and how it acquires water in the leaf, stem, and root.

Growth of Roots and Stems

Primary growth refers to an increase in the length of stems and roots. Secondary growth refers to an increase in the girth of stems and roots. Only woody plants undergo secondary growth.

22.5 Primary growth lengthens the root and shoot systems

LEARNING OUTCOME

When you complete this section, you should be able to

1. Describe primary growth in a root tip compared to a shoot tip.

Primary growth, which causes a plant to grow lengthwise, is centered in the apex (tip) of the shoot and the root. As stated in section 22.3, these regions contain apical meristem tissue. The term meristem is derived from the Greek word *merismos,* which means "division." Meristem is a region of actively dividing cells. After new cells are produced by the process of mitosis, they go on to become the specialized tissues of a plant—epidermal tissue, ground tissue, and vascular tissue.

In contrast to animals, which grow to maturity and then stop growing in size, the existence of meristematic tissue allows a plant to keep growing its entire life span. Also, growth in plants serves another function. Whereas animals often respond to external stimuli by moving a body part, plant organs grow toward or away from stimuli, as we shall see in Chapter 24.

Video
Seedling Growth

Root System The growth of many roots is continuous, pausing only when temperatures become too cold or when water becomes too scarce. Figure 22.5A*a*, a longitudinal section of a eudicot root, reveals zones where cells are in various stages of differentiation as primary growth occurs. These zones are called the zone of cell division, the zone of elongation, and the zone of maturation.

The **zone of cell division** is protected by the **root cap,** which is composed of parenchyma cells and protected by a slimy sheath. As the root grows, root cap cells are constantly removed by rough soil particles and replaced by new cells. The zone of cell division contains the root apical meristem, where mitosis produces relatively small, many-sided cells having dense cytoplasm and large nuclei. These meristematic cells give rise to the primary meristems, called *protoderm, ground meristem,* and *procambium* (Fig. 22.5A*b*). Eventually, the primary meristems develop, respectively, into the three mature tissue types discussed previously: epidermis, ground tissue (i.e., cortex), and vascular tissue.

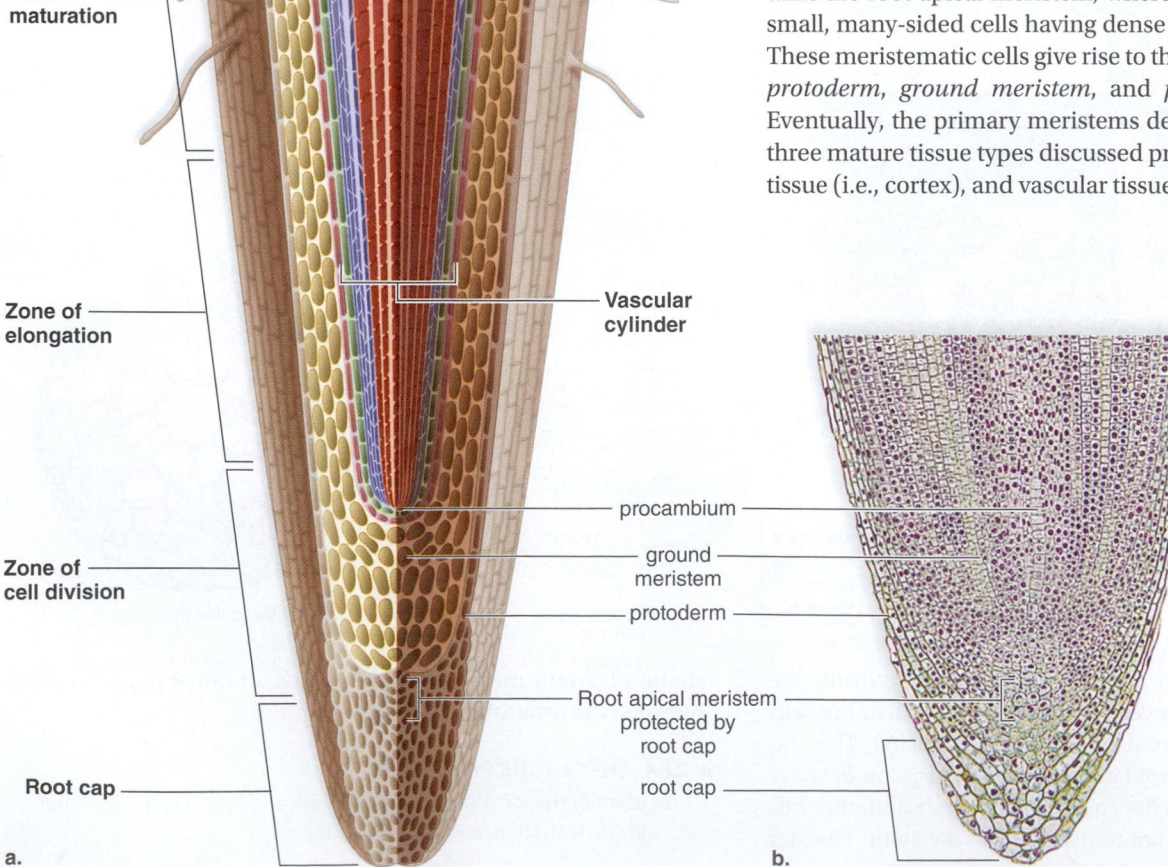

Labels (left to right, top to bottom): epidermis, cortex, endodermis, pericycle, phloem, xylem, root hair, Zone of maturation, Zone of elongation, Vascular cylinder, Zone of cell division, procambium, ground meristem, protoderm, Root apical meristem protected by root cap, Root cap, root cap

FIGURE 22.5A
a. Growth zones and mature tissues of a root. **b.** Root apical meristem gives rise to three primary meristems (protoderm, ground meristem, and procambium) which in turn produce the differentiated tissues of a root. Xylem and phloem develop from the procambium.

a.

b.

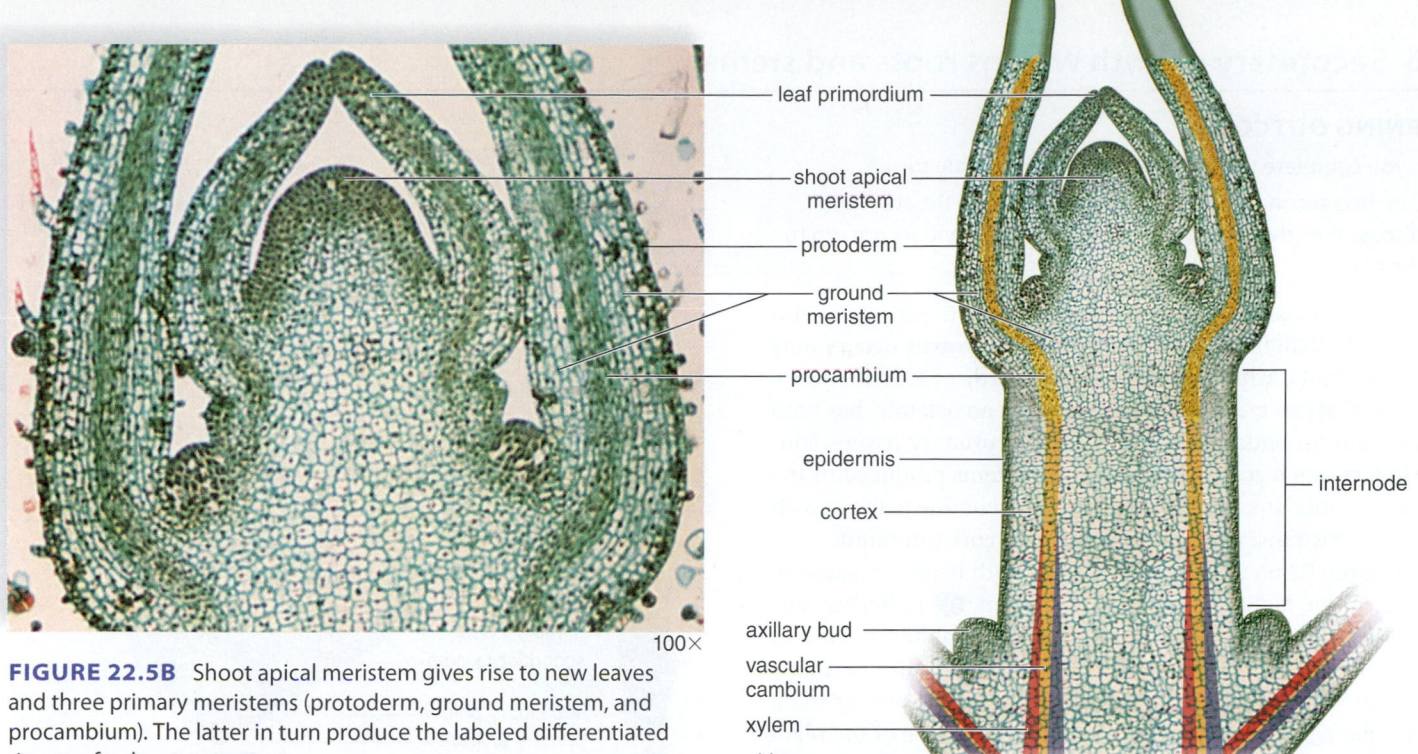

FIGURE 22.5B leaf primordium, shoot apical meristem, protoderm, ground meristem, procambium, epidermis, cortex, internode, axillary bud, vascular cambium, xylem, phloem

FIGURE 22.5B Shoot apical meristem gives rise to new leaves and three primary meristems (protoderm, ground meristem, and procambium). The latter in turn produce the labeled differentiated tissues of a shoot system.

100×

The **zone of elongation** is the region where the root increases in length due to elongation of cells. In the zone of elongation, the cells lengthen, but they are not yet fully specialized. This region also gets longer due to the addition of cells produced by the zone of cell division.

The **zone of maturation** is the region that does contain fully differentiated tissues. This region is recognizable because many of the epidermal cells bear root hairs. Here, also, the mature tissues of a root can be readily distinguished from one another (see Fig. 22.5A*a*).

Shoot System The terminal bud includes the shoot apical meristem and also leaf primordia (young leaves) that differentiate from cells produced by the shoot apical meristem (Fig. 22.5B). The activity of the terminal bud would be hindered by a protective covering comparable to the root cap. Instead, the leaf primordia fold over the apical meristem, providing protection. At the start of the season, the leaf primordia are, in turn, covered by terminal bud scales (scale-like leaves), but these drop off, or abscise, as growth continues.

The shoot apical meristem produces everything in a shoot: leaves, axillary buds, additional stem, and sometimes flowers. In the process, it gives rise to the same primary meristems as in the root. These primary meristems, in turn, develop into the differentiated tissues of a shoot system. The *protoderm* becomes the epidermis of the stem and leaves. *Ground meristem* produces parenchyma cells that become the cortex and pith in the stem and mesophyll in the leaves.

Procambium differentiates into the xylem and phloem of a vascular bundle. Certain cells become tracheids and others become vessel elements. The first sieve-tube members of a vascular bundle do not have companion cells and are short-lived (some live only a day before being replaced). Mature vascular bundles contain fully differentiated xylem, and phloem, and a lateral meristem called vascular cambium, which is responsible for secondary growth. Vascular cambium is discussed more fully in section 22.6.

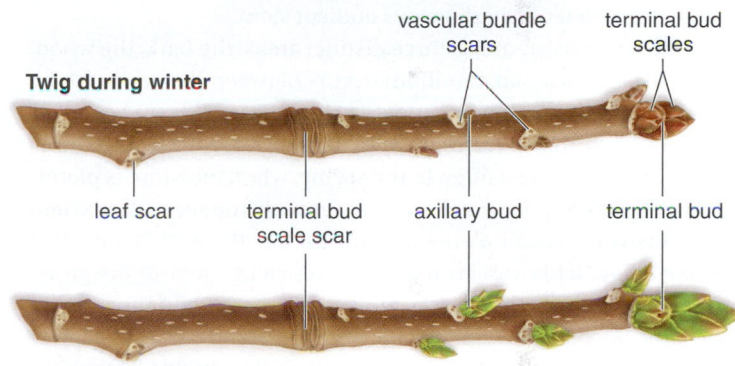

Twig during winter — vascular bundle scars, terminal bud scales, leaf scar, terminal bud scale scar, axillary bud, terminal bud

Twig during spring

FIGURE 22.5C Winter twig showing stem organization.

Winter Twig The anatomy of a winter twig shows the changes that allow a stem to overwinter (Fig. 22.5C). The terminal bud contains the apical meristem and leaf primordia of the shoot tip that will allow the stem to regrow in the spring. The terminal bud is protected by leaves modified to be small, dry, terminal bud scales. The leaves and flowers have abcised; *leaf scars* and *vascular bundle scars* mark the spot of **abscission** (dropoff). Dormant axillary buds in this region can give rise to branches or flowers in the spring. When growth resumes, terminal bud scales fall off and leave a scar that resembles a compact series of concentric circles. You can tell the age of a stem by counting these **terminal bud scale scars** because there is one for each year's growth.

▶ **22.5 CHECK YOUR PROGRESS**
   1. Identify any differences in primary growth of the root tip and shoot tip and associate them with the functions of the root and stem.

22.6 Secondary growth widens roots and stems

LEARNING OUTCOME

When you complete this section, you should be able to

1. Describe secondary growth with emphasis on the stem and discuss the advantages and drawbacks of secondary growth to the plant.

As we just discussed, primary growth occurs in all plants and increases the length of a plant. **Secondary growth** occurs only in woody plants, where it increases the girth of trunks, stems, branches, and roots. A woody plant, such as an oak tree, has both primary and secondary tissues. In a stem, primary tissues continue to form each year from primary meristems produced by the shoot apical meristem. Secondary tissues occur due to the growth of lateral meristems: vascular cambium and cork cambium.

As Figure 22.6A shows, secondary growth begins because of a change in the activity of vascular cambium. ❶ In herbaceous plants, the lateral meristem called vascular cambium is present between the xylem and phloem of each vascular bundle. In woody plants, the vascular cambium develops into a ring of meristem. ❷ Vascular cambium divides parallel to the surface of the plant and produces secondary (new) xylem and phloem each year. Eventually, a woody eudicot stem has an entirely different organization from that of a herbaceous eudicot stem.

❸ A woody stem has three distinct areas: the bark, the wood, and the pith. Vascular cambium occurs between the bark and the wood. **Wood** is actually secondary xylem that builds up year after year. In trees that have a growing season, vascular cambium is dormant during the winter. In the spring, when moisture is plentiful and leaves require much water for growth, the secondary xylem contains wide vessel elements with thin walls. In this so-called *spring wood,* wide vessels transport sufficient water to the growing leaves. Later in the season, moisture is scarce, and the wood at this time becomes *summer wood.* Summer wood contains a lower proportion of vessels and a higher number of tracheids. Tracheids, with their thicker walls and smaller diameters, are less susceptible to water scarcity than the larger, less rigid vessel elements. At the end of the growing season, just before the cambium becomes dormant again, only heavy fibers containing sclerenchyma cells with especially thick secondary walls may develop. When the trunk of a tree has spring wood followed by summer wood, the two together make up one year's growth, or an **annual ring.** You can tell the age of a tree by counting the annual rings.

The **bark** of a tree contains periderm and phloem. The **periderm** is a secondary growth tissue that contains cork and cork cambium. **Cork cambium** lies beneath the epidermis, but later it is part of the periderm, which replaces epidermis. Cork cambium divides and produces the cork cells that give bark its characteristic appearance. Cork cells are impregnated with suberin, a waxy layer that makes them waterproof but also causes them to die. This is protective because it makes the stem less edible. The impermeable cork may be interrupted by **lenticels,** which are pockets of loosely arranged cork cells not impregnated with suberin. Gas exchange for processes such as photosynthesis can be accomplished at the lenticels after stems undergo secondary growth and the epidermis is replaced by periderm. Secondary phloem does not build up as xylem does, and only that year's phloem is found in the bark. The

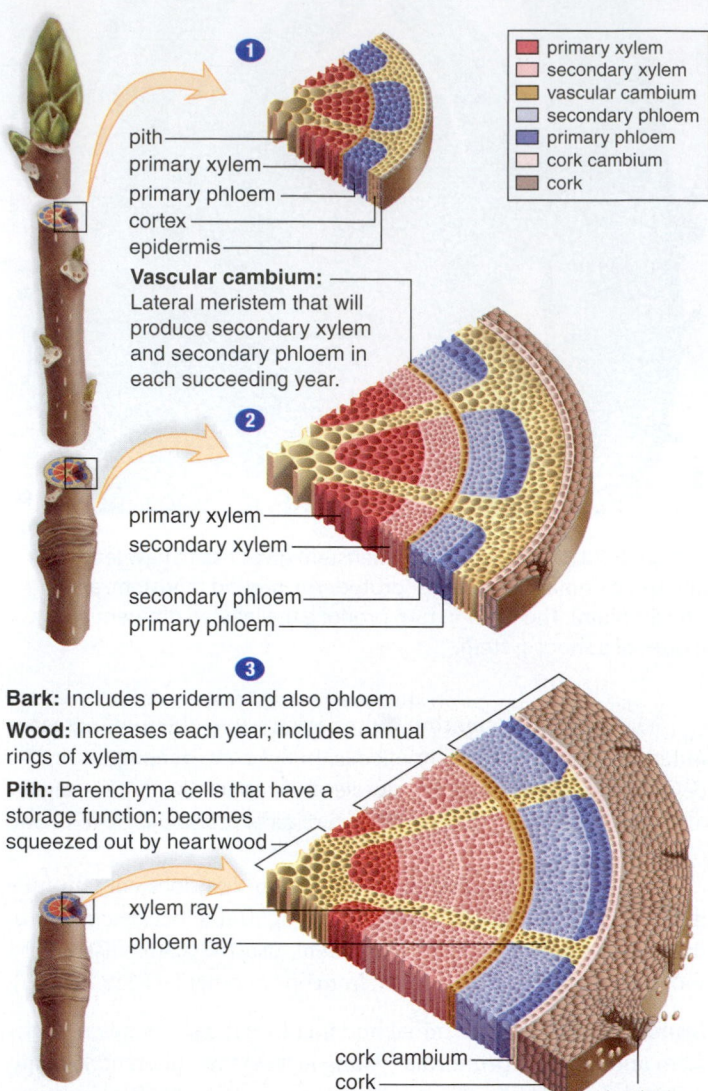

■	primary xylem
■	secondary xylem
■	vascular cambium
■	secondary phloem
■	primary phloem
■	cork cambium
■	cork

❶
pith
primary xylem
primary phloem
cortex
epidermis

Vascular cambium: Lateral meristem that will produce secondary xylem and secondary phloem in each succeeding year.

❷
primary xylem
secondary xylem
secondary phloem
primary phloem

❸
Bark: Includes periderm and also phloem
Wood: Increases each year; includes annual rings of xylem
Pith: Parenchyma cells that have a storage function; becomes squeezed out by heartwood

xylem ray
phloem ray

cork cambium
cork
lenticel

FIGURE 22.6A In a tree, vascular cambium produces secondary xylem and phloem each year. Wood consists of secondary xylem. Also, bark (contains cork, cork cambium, and secondary phloem) protects the trunk.

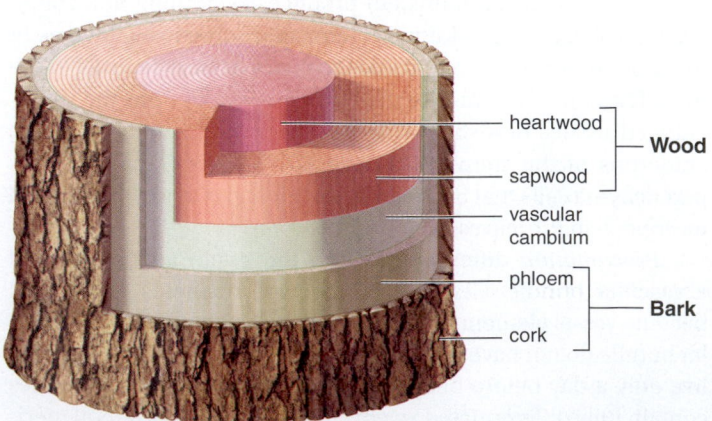

heartwood — **Wood**
sapwood
vascular cambium
phloem — **Bark**
cork

FIGURE 22.6B Heartwood has no transport function and serves as a depository for various substances such as resins and tannin.

bark of a tree should not be removed because, without phloem, sugar cannot be transported. Removal of a ring of bark from a tree can be lethal to the tree. Today, bark is a source of medicines, as seems reasonable because it must possess chemicals that keep it from being eaten by herbivores. But because a tree cannot live without sufficient bark, one branch of research focuses on how much bark can be harvested without harming a tree.

Figure 22.6A also shows the *xylem rays* and *phloem rays* in the cross section of a woody stem. Rays consist of parenchyma cells that permit lateral conduction of substances from the pith to the cortex and some storage of food. After annual rings build up, the center of a tree becomes heartwood that has no transport functions as sapwood does (Fig. 22.6B).

Advantages and Disadvantages of Being Woody Woody plants evolved around 380 MYA and have become very successful, evolving into many species that dominate the landscape. Therefore, there must be some advantages to being woody. To understand

what they might be, consider a herbaceous plant. The longer it lives, the more leaves and roots it has. But it still has only a limited amount of xylem and phloem and may not be able to service a large shoot and root systems adequately. A woody plant is more likely to have enough xylem and phloem to adequately satisfy the needs of its expanding shoot and root systems now and in the future. Woody plants also have reduced competition because, as long as they occupy the same site, no other plant can take it from them. Still, being woody has its disadvantages. It is metabolically expensive to produce wood and the energy could otherwise be used for reproduction. Woody plants also need more defense mechanisms because they are bigger and represent a bigger target for parasites and pathogens.

Animation
Woody Dicot

▶ **22.6 CHECK YOUR PROGRESS**
1. Identify what portion of a woody stem is in fact wood and discuss advantages and drawbacks to producing wood.

HOW LIFE CHANGES   *Application*

22B The First Forests

In 2007, scientists pieced together the fossil remains of a woody tree that lived in the Devonian period (over 385 MYA) before animals had invaded land (Fig. 22B.1). They named the tree *Eospermatopteris* and hypothesized that it stood 900 m tall. They decided that *Eospermatopteris* looked much like a palm tree, indicating that Gilboa, New York, where the fossil remains were found, must have been tropical at that time (Fig. 22B.2).

Wood fossilizes well, and botanists have been able to determine how ancient wood evolved the vessel elements of today's wood. In ancient wood, xylem was composed of long cells with tapered ends, much like the tracheids of today's wood. Later, the cells became shorter, and the tapered ends became flatter and developed perforated end walls. When the diameter of these cells increased, the vessel elements of modern xylem had evolved. The fibers that give wood its structural strength were also seen alongside xylem in ancient wood.

As remarkable as it seems, the trees of the first forests lowered the local temperature and removed vast quantities of carbon dioxide from the atmosphere when they photosynthesized. They also altered local hydrologic cycles by evaporating large amounts of water. The result was the creation of habitats that enabled animals to invade the land environment. In addition, trees provide unique habitats

 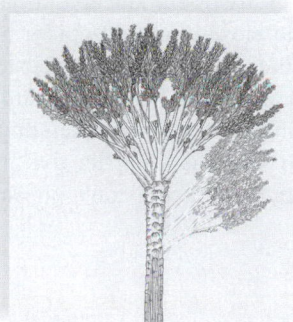 

FIGURE 22B.1 (*Left to right*) *Eospermatopteris* fossil stump, reconstruction, and fossil crown.

FIGURE 22B.2
A living palm tree.

for both plants and animals above ground due to their multiple layers of branches and leaves.

The lesson to take home is that as soon as modern vascular tissue evolved, so did trees that could grow taller and wider because they had the means to transport materials throughout a larger body. Also, wood provides the structural support a larger plant needs to remain erect. Finally, trees help preserve the ecological conditions that allow animals such as humans to live on land.

CONSIDER THIS QUESTION
1. Give evidence that trees play a necessary role in the life of humans and support the hypothesis that they will continue to do so in the future.

 **connect** |BIOLOGY Explore the concepts through a variety of multimedia assets and question types.

www.mcgrawhillconnect.com

Homeostatic Mechanisms of Plants

Plants, like other organisms, have an organization suited to maintain **homeostasis,** the relative constancy of the internal environment. Regulation is usually required to keep internal conditions within tolerable limits, and in this section, we present evidence that plants regulate their internal environment.

22.7 Leaves carry on photosynthesis and help maintain homeostasis

LEARNING OUTCOMES

When you complete this section, you should be able to

1. Explain how the structure of a leaf is suited to carrying on photosynthesis.
2. Analyze the benefits of photosynthesis as a mechanism to maintain homeostasis.
3. List and describe regulating mechanisms in plants that help maintain homeostasis.

Leaves have an entirely different structure from stems and roots. Figure 22.7A shows the cross section of a generalized foliage leaf of a temperate-zone eudicot plant. As mentioned previously, epidermal tissue, called the **epidermis**, is located on the leaf's upper and lower surfaces, where it is interrupted particularly on the lower surface by stomata, openings that allow gas exchange and water evaporation when they are open. The epidermal layers are protective in two ways. The closely packed epidermal cells secrete an outer, waxy cuticle that helps prevent water loss and serves a barrier to parasites and predators. It also bears little hairs, called **trichomes,** which discourage predation by secreting irritating substances. Trichomes also serve as a barrier that prevents parasites from entering the stomata and cuts down on water loss when stomata are open.

As you also know, in a leaf, ground tissue called mesophyll fills the space between epidermal layers. The elongated cells of the palisade mesophyll carry on most of the photosynthesis and are situated to allow its chloroplasts to efficiently absorb solar energy. The irregular cells of spongy mesophyll are bounded by air spaces that receive CO_2 when stomata are open. Their loosely packed arrangement increases the amount of surface area for the absorption of CO_2 and evaporation of water. When water evaporates from a leaf, it causes the water column in xylem to rise and enter a leaf from the leaf vein. Aside from carbon dioxide and water, solar energy is needed for photosynthesis and foliage leaves are generally flat and thin—this shape allows solar energy to penetrate the entire width of the leaf.

Leaf veins not only bring water and minerals to a leaf, they also contain phloem, which distributes the products of photosynthesis to the rest of the plant. **Bundle sheaths** contain parenchyma cells and possibly also sclerenchyma cells that surround vascular tissue. The parenchyma cells in bundle sheaths help regulate the entrance and exit of material into and out of leaf veins.

Our discussion of leaf anatomy allows us to conclude that leaves have a structure that is adaptive to carrying on photosynthesis. The food they provide supplies the energy and building blocks a plant needs to maintain its existence and therefore leaves are organs of homeostasis.

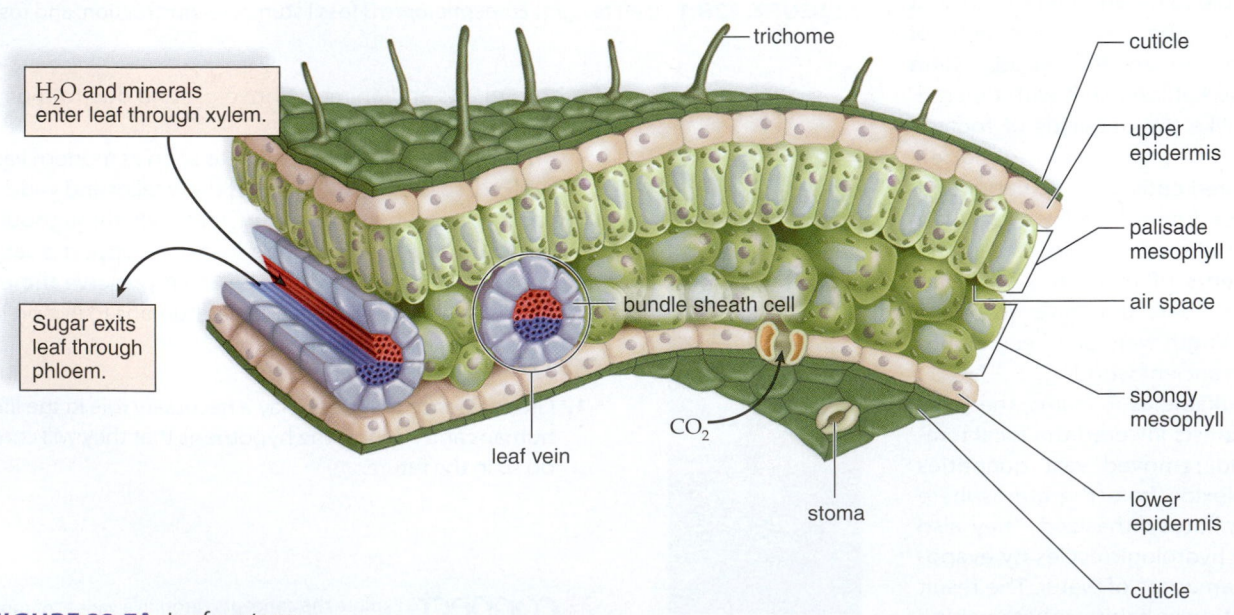

H₂O and minerals enter leaf through xylem.

Sugar exits leaf through phloem.

leaf vein

bundle sheath cell

CO_2

stoma

trichome

cuticle

upper epidermis

palisade mesophyll

air space

spongy mesophyll

lower epidermis

cuticle

FIGURE 22.7A Leaf anatomy.

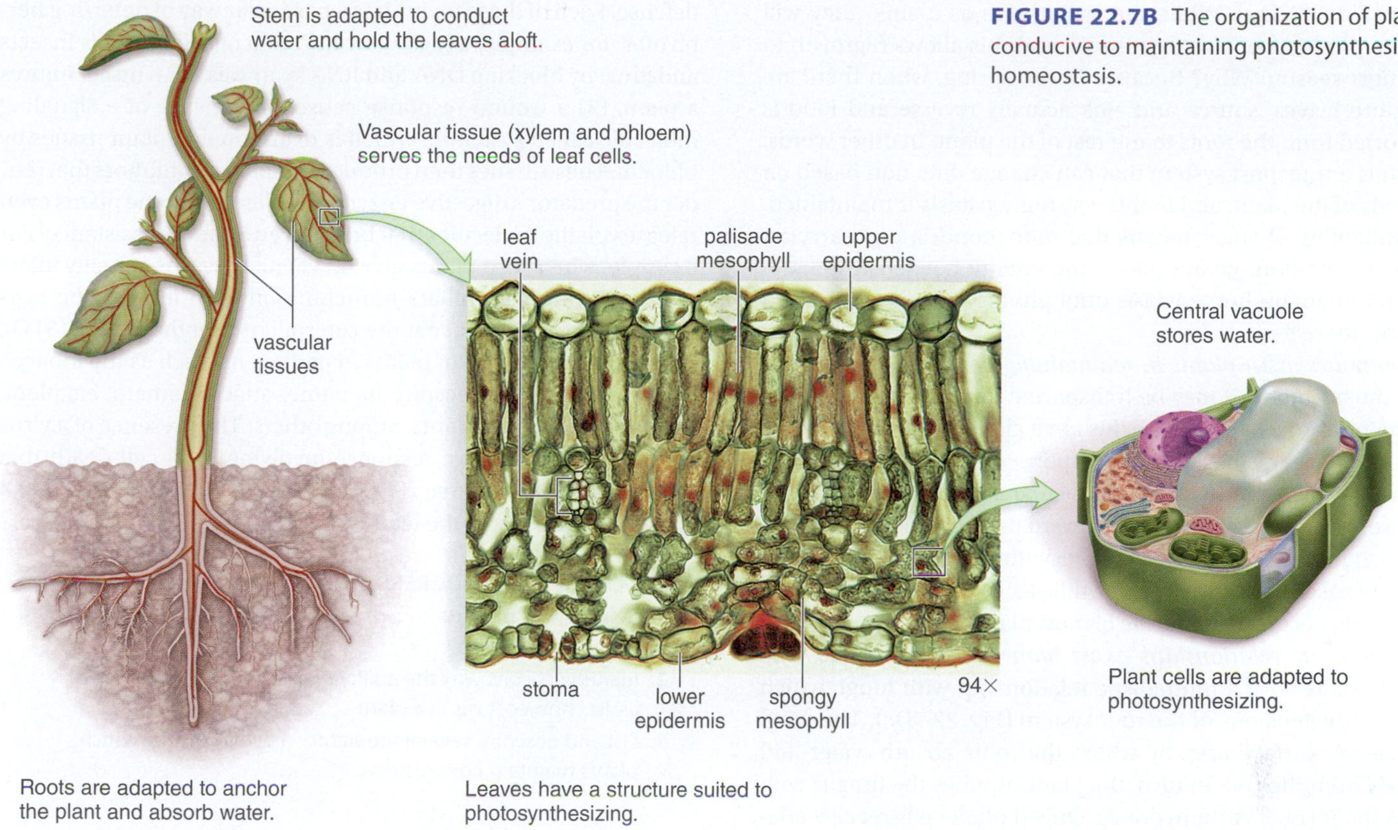

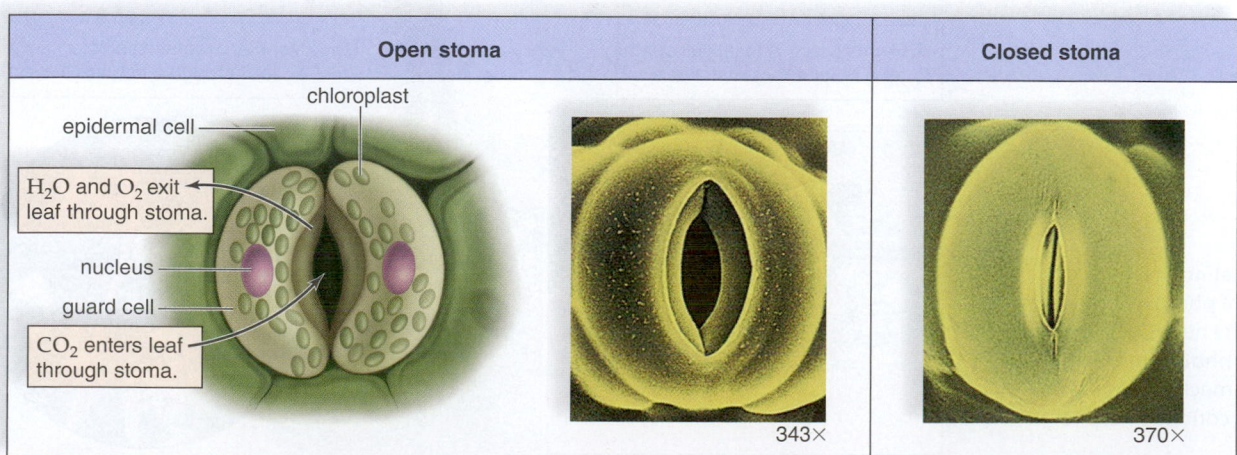

Stem is adapted to conduct water and hold the leaves aloft.

Vascular tissue (xylem and phloem) serves the needs of leaf cells.

vascular tissues

leaf vein   palisade mesophyll   upper epidermis

stoma   lower epidermis   spongy mesophyll   94×

FIGURE 22.7B The organization of plants is conducive to maintaining photosynthesis and homeostasis.

Central vacuole stores water.

Plant cells are adapted to photosynthesizing.

Roots are adapted to anchor the plant and absorb water.

Leaves have a structure suited to photosynthesizing.

Mechanisms of Homeostasis in Plants Figure 22.7B dramatically shows that the functions of the root and stem in making water and minerals available to leaves are absolutely essential for photosynthesis to continue. Plants use various mechanisms to keep the internal conditions of their cells relatively normal so that photosynthesis can continue.

Prevention of dehydration is necessary to homeostasis. Without adequate water, a nonwoody plant first becomes limp and then dies. A plant has two ways of preventing water loss at the leaves: The cuticle, which has a waxy surface, keeps the leaves from drying out; and the stomata have the ability to close. When a plant is water-stressed, a hormone is released that causes the guard cells to change shape and close the stomata (Fig. 22.7C).

This action conserves water, but how does closure of stomata affect photosynthesis? When stomata are closed, CO_2 cannot enter the leaf, and therefore photosynthesis may be put on hold. True, mitochondria give off CO_2, but without a source of oxygen from either chloroplasts or via the stomata, cellular respiration is also on hold. Despite the need for stomata to remain open for both photosynthesis and cellular respiration to occur, stomata close when a plant is water-stressed.

Phloem transport is critical to homeostasis. The distribution is from source to sink. In the summer, the *source* includes the mature leaves that are producing sugar. The *sink* includes the plant parts, such as flowers and roots that are not photosynthesizing. Also, immature leaves may not be photosynthesizing enough to allow

Open stoma		Closed stoma

chloroplast

epidermal cell

H_2O and O_2 exit leaf through stoma.

nucleus

guard cell

CO_2 enters leaf through stoma.

343×   370×

FIGURE 22.7C When open stomata allow CO_2, needed for photosynthesis, to enter a leaf. When a plant is water stressed, stomata close, conserving water. This homeostatic mechanism prevents the plant from drying out.

growth (Fig. 22.7Da). While the roots serve as a sink, they will store the products of photosynthesis, and this allows regrowth to begin next season. Why? Because in the spring, when there are no mature leaves, source and sink actually reverse and food is transported from the roots to the rest of the plant. In other words, phloem is a transport system that can change direction based on the needs of the plant, and in this way homeostasis is maintained. The availability of sugar means that mitochondria can carry on cellular respiration, giving plants the energy needed to remain alive and maintain homeostasis until photosynthesis begins and produces more food.

Hormones assist plants in maintaining homeostasis. Depending on the hormone, it may be transported through the xylem, the phloem, or from cell to cell. Auxins are a class of hormones, some of which are responsible for tropic responses. A **tropism** is a growth response toward or away from a particular stimulus. Among other responses, plants are able to bend toward the light and, in this way, better acquire solar energy for photosynthesis (Fig. 22.7Db). The capture of solar energy and photosynthesis, as we have established, is absolutely necessary to the life of most plants.

Mutualistic relationships assist homeostasis in plants. The roots of plants have a mutualistic relationship with fungi, which function as extensions of the root system (Fig. 22.7Dc). The fungi increase the surface area by which the roots absorb water and minerals from the soil. In turn, the plant supplies the fungus with food in the form of carbohydrates. Only if plant cells receive adequate amounts of minerals is homeostasis possible. For example, a plant needs phosphorus in order to produce adenosine triphosphate (ATP), the energy currency of cells.

Animation
Root Nodules
of Plants

Defense mechanisms help a plant maintain homeostasis. We have already mentioned that (1) leaf cells are covered by a thick, impervious cuticle and that epidermal projections, including thorns and hairs, can discourage hungry insects. If these defense mechanisms fail, many plants rely on (2) chemical toxins—alkaloids such as morphine, quinine, taxol, and caffeine—as the next level of

defense. Each of these toxins has a particular way of deterring herbivores; for example, a high concentration of caffeine kills insects and fungi by blocking DNA and RNA synthesis. If an insect injures a plant, (3) a wound response causes the release of a signaling molecule called systemin, which is distributed to plant tissues by phloem. These tissues then produce proteinase inhibitors that render the predator's digestive enzymes useless. (4) Some plants even release volatile molecules that bring on an animal's assistance! For example, when corn plants give off volatiles, wasps actually inject their eggs into caterpillars munching on corn leaves. The eggs develop into larvae that eat the caterpillars, not the leaves. (5) Or, consider the response of plants to pathogens, such as the tobacco mosaic virus, which despite its name, attacks tomato, eggplant, pepper, and spinach plants, among others. The presence of a virus causes a hypersensitive response involving local cell death that seals off the infested area (Fig. 22.7Dd). Now photosynthesis can continue in the rest of the leaf.

▶ **22.7 CHECK YOUR PROGRESS**
1. Discuss how the structure of a leaf is suited to carrying on photosynthesis.
2. Identify reasons why the ability of a leaf to photosynthesize assists homeostasis in a plant.
3. List and describe several regulatory mechanisms by which plants maintain homeostasis.

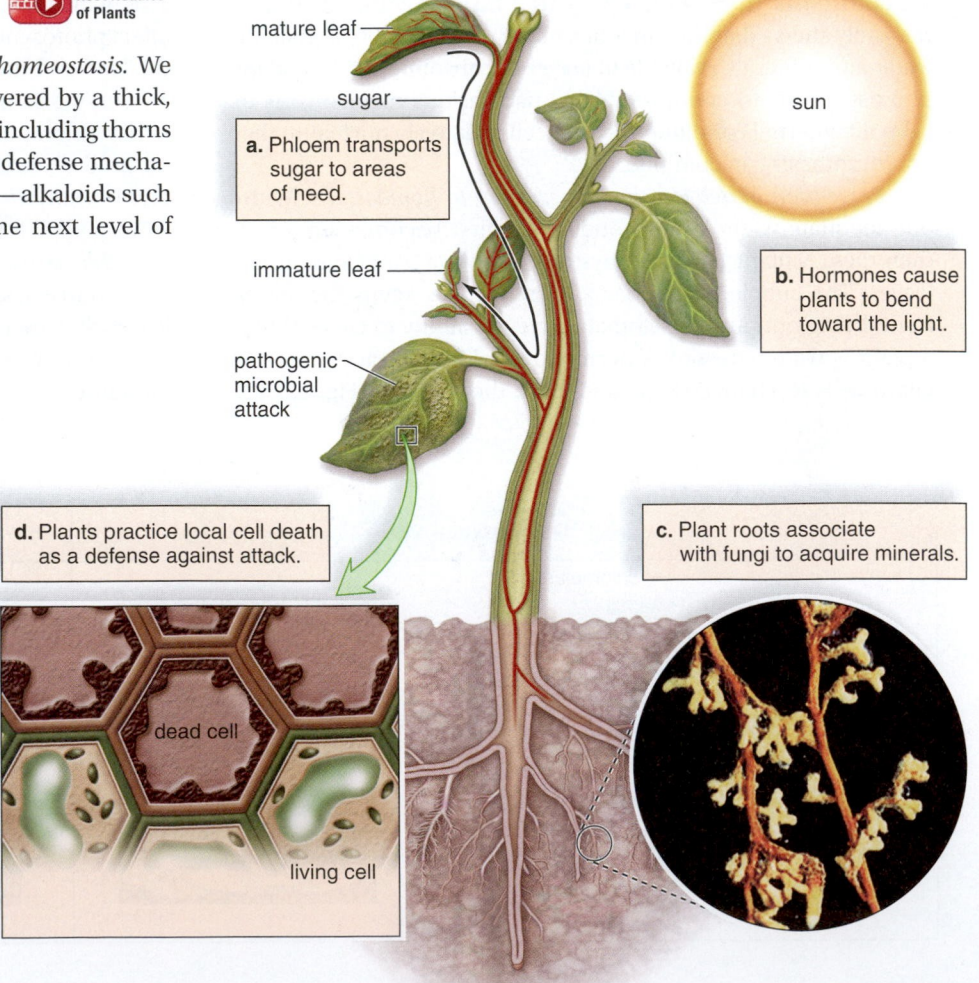

mature leaf

sugar

a. Phloem transports sugar to areas of need.

immature leaf

pathogenic microbial attack

sun

b. Hormones cause plants to bend toward the light.

d. Plants practice local cell death as a defense against attack.

dead cell

living cell

c. Plant roots associate with fungi to acquire minerals.

FIGURE 22.7D Regulating mechanisms include (**a**) distribution of photosynthetic products according to the needs of the plant; (**b, c**) ability to promote photosynthesis in these two ways; and (**d**) mechanisms that allow photosynthesis to continue despite pathogen attack.

THE CHAPTER IN REVIEW

CONNECTING THE CONCEPTS

Flowering plants are adapted to living on land. In Chapter 19, we saw how plants became adapted to reproducing on land. Many other types of adaptations are also required to prosper on land. Because even humid air is drier than a living cell, the prevention of water loss is critical for land plants. The epidermis and the cuticle it produces help prevent water loss and overheating in sunlight. Gas exchange in leaves depends on the presence of stomata, which close when a plant is water-stressed. The cork of woody plants is especially protective against water loss, but where cork is interrupted by lenticels, gas exchange is still possible.

In an aquatic environment, water buoys up organisms and keeps them afloat, but on land, plants had to evolve a way to oppose the force of gravity. The stems of plants contain strong-walled sclerenchyma cells, tracheids, and vessel elements. The accumulation of secondary xylem in particular allows a tree to grow in diameter and offers more support.

In an aquatic environment, water is available to all cells, but on land it is adaptive to have a means of water uptake and transport. In plants, the roots absorb water and have special extensions called root hairs that facilitate water uptake. Xylem transports water to all plant parts, including the leaves. In Chapter 23 we will see how the drying effect of air allows water to move from the roots to the leaves. Roots are buried in soil, where they can absorb water but cannot photosynthesize. In that chapter, we will also see how the properties of water allow phloem to transport sugars from the leaves to the roots and to any other plant part in need of sustenance. We will then discuss the inorganic nutrient needs of plants, which consist of substances they garner from their environment.

ANALYZE AND EVALUATE

1. What evidence convinces you the most that the vegetative organs of plants are adapted to a land existence?
2. A bush is larger than a herbivorous garden plant. How does a bush provide more vascular tissue for the increased number of leaves?
3. Rhizomes are underground stems that spread beneath the surface and give off shoots at nodes. What are the advantages to this type of organization?

SUMMARIZE

Organs and Tissues of Flowering Plants

22.1 Flowering plants typically have roots, stems, and leaves

- A flowering plant has a **shoot system** and a **root system;** both are adapted to the roles they play in the life of a plant.
- In the shoot system, **stems** support leaves; grow longer because they have a **terminal bud** and produce branches because they have **axillary buds;** conduct materials to and from roots and leaves through their vascular tissue and help store water or plant products. The part of a stem where leaves and axillary buds are attached is a **node.** Between nodes are **internodes. Rhizomes** are horizontal stems that give rise to aboveground shoots at nodes.
- **Leaves,** attached to a stem by a **petiole,** are the chief organs of photosynthesis. The leaf **blade** is adapted to absorb solar energy from a flattened surface, absorb CO_2 through pores, and receive water from leaf veins. **Deciduous** trees lose their leaves in the winter.
- **Roots,** being tubular and pointed, anchor a plant, often by means of a taproot and branches. **Root hairs** can absorb water and minerals from all sides. The production of hormones ensures that the size of the root system is appropriate to that of the shoot system. Roots store the products of photosynthesis—for example, in a taproot.

22.2 Flowering plants are either monocots or eudicots

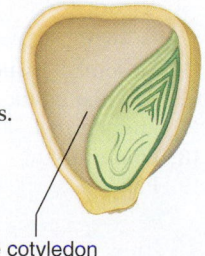

one cotyledon

- **Monocots** (e.g., grasses, lilies) have one cotyledon (seed leaf). **Eudicots** (e.g., dandelions, oak trees) have two cotyledons. Monocots and eudicots have structural differences in their roots, stems, leaves, and number of flowering parts.

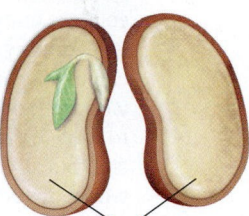

two cotyledons

22.3 Plants have specialized cells and tissues

- **Apical meristem** produces the primary meristems that become **epidermal tissue,** which protects; **ground tissue,** which metabolizes; and **vascular tissue,** which transports. Epidermal tissue contains epidermal cells (protected by a waxy **cuticle**). In leaves, the **epidermis** contains **stomata.**
- Ground tissue contains **parenchyma cells, collenchyma cells** (thick primary walls at corners), and **sclerenchyma cells** (fibers with thick secondary walls).
- Vascular tissue contains **xylem,** which transports water and minerals from roots to leaves, and **phloem,** which often transports sugar from leaves to roots. Xylem contains **vessels** (stacked **vessel elements**) and **tracheids.** Phloem contains **sieve-tube members** and **companion cells.**

22.4 The three types of plant tissues are present in each organ

- A leaf has an upper and lower epidermis (with stomata), **mesophyll** (ground tissue) composed mostly of parenchyma cells, and **leaf veins. Palisade mesophyll** is strategically located to absorb solar energy, and **spongy mesophyll** has open spaces where CO_2 enters the leaf and can be absorbed.
- The stem of a **herbaceous** plant has epidermis, ground tissue in the **cortex** and **pith,** and vascular bundles that are scattered (monocot) or in a ring (eudicot).
- A eudicot root has an outer layer of epidermis, ground tissue in the cortex, and a **vascular cylinder** (contains xylem and phloem). In addition, an impermeable tissue layer called **endodermis** regulates the entrance of minerals into the vascular cylinder; **pericycle** contains actively dividing cells that give rise to branch roots.

Growth of Roots and Stems

22.5 Primary growth lengthens the root and shoot systems

- A root tip has a **zone of cell division**, a **zone of elongation**, and a **zone of maturation** which contains the differentiated tissues of a root. Root apical meristem (protected by **root cap**) produces the primary meristems called protoderm, ground meristem, and procambium which respectively produce the epidermis, cortex, and vascular tissue of a root.
- Shoot apical meristem produces all the structures of a shoot: leaves, axillary buds, additional stem, and sometimes flowers. It gives rise to the primary meristems which in turn produce the differentiated tissues of a stem including epidermis, cortex, and the vascular bundles.
- A winter twig shows that plants are adapted to overwinter. Leaf scars and vascular bundle scars mark the spot of **abscission** (leaf dropoff), and the terminal bud is protected by **terminal bud scales.**

22.6 Secondary growth widens roots and stems

- Lateral meristems, called **vascular cambium** (produces secondary xylem and secondary phloem) and **cork cambium** (produces cork) are responsible for **secondary growth** (increase in girth).
- The trunk of a woody plant contains bark, wood, and pith. Pith is later replaced by nonfunctional heartwood. Only sapwood is functional. **Bark** contains phloem, cork cambium, and cork. **Periderm** (cork cambium and cork) replaces epidermis. Cork cells are impermeable, but **lenticels** permit gas exchange. **Wood** is secondary xylem that builds up year after year; **annual rings** can tell the age of a tree and also its environmental history. A woody plant has some advantages (enough xylem and phloem to service a large shoot and root system, retention of a home site) but also some disadvantages (has to use energy to produce wood, offers a larger target for parasites and predators).

Homeostatic Mechanisms of Plants

22.7 Leaves carry on photosynthesis and help maintain homeostasis

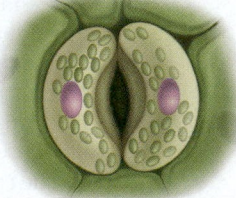

Stoma and guard cells

- The **epidermis** at top and bottom has a protective waxy cuticle. Stomata, most often on the lower surface, allow CO_2 to enter but can be closed by guard cells to prevent water loss. **Trichomes** help deter parasites and predators and also prevent water loss.
- Mesophyll forms the body of a leaf and carries on photosynthesis. Elongated palisade mesophyll cells are arranged for maximum energy absorption, and the open spaces of spongy mesophyll allow its irregularly shaped cells to take up CO_2. Leaf veins surrounded by **bundle sheaths** transport water to and remove sugars from a leaf.
- Some plant organs, such as stems, assist **homeostasis** by supplying leaf cells with the materials they need for photosynthesis. Other mechanisms include the following: (1) Stomata open and close according to water availability; they close to prevent water loss and, therefore, death of the plant. (2) Phloem has the ability to transport sugar between source and sink. (3) Plant **hormones,** such as auxin, cause tropic responses. A **tropism** is a growth response toward or away from a stimulus. (4) Plant defense mechanisms include a waxy cuticle, thorns, the ability to produce toxins, and local cell death. (5) Fungal roots help a plant acquire water and minerals; the plant provides the fungus with food.

TEST YOURSELF

Organs and Tissues of Flowering Plants

1. It is possible to distinguish between a stem and a root by looking for
   a. petioles on stems.   c. nodes on stems.
   b. petioles on roots.   d. nodes on roots.
2. If the epidermal cells of roots had a thick waxy cuticle, they would not
   a. be the primary site of photosynthesis.
   b. give rise to new leaves and flowers.
   c. be able to absorb water and minerals.
   d. be able to anchor the plant.
3. In order for foliage leaves to carry on photosynthesis, it is best if they
   a. have a waxy cuticle at both top and bottom.
   b. are flat and thin.
   c. stack on the stem, one on top of the other.
   d. All of these are correct.
4. Which of these is an incorrect contrast between monocots (stated first) and eudicots (stated second)?
   a. one cotyledon—two cotyledons
   b. leaf veins parallel—net-veined leaves
   c. vascular bundles in a ring—vascular bundles scattered
   d. flower parts in threes—flower parts in fours or fives
   e. All of these are correct.
5. Monocots are the smaller group compared to eudicots but are of extreme importance because all the cereal grains are monocots.
   a. true   b. false
6. Describe the function of apical meristem and axillary buds.
7. A cereal grain is what stage in the life cycle of a plant?
8. Sclerenchyma, parenchyma, and collenchyma cells occur in what type of plant tissue?
   a. epidermal   d. meristem
   b. ground   e. All but d are correct.
   c. vascular
9. Which of these cells in a plant is adapted for strength?
   a. parenchyma cell   d. epidermal cell
   b. collenchyma cell   e. guard cell
   c. sclerenchyma cell

For questions 10–14, match each function to the cell types in the key.
KEY:
   a. meristem   d. parenchyma
   b. sclerenchyma   e. epidermal tissue
   c. vessel element

10. Transport
11. Support
12. Cell division
13. Photosynthesis or storage
14. Protection

For questions 15–19, match each tissue to an organ in the key. Answers can be used more than once.
KEY:
   a. root   c. leaf
   b. stem   d. All of these are correct.

15. Mesophyll tissue
16. Vascular tissue
17. Endodermis
18. Epidermis
19. Pith

20. Which of these is not a contrast between xylem and phloem? A characteristic of xylem is mentioned first.
    a. vessel elements—sieve-tube members
    b. companion cell—sclerenchyma cell
    c. dead cells—cells contain cytoplasm
    d. transports water—transports sugar
21. Cortex is found in
    a. roots, stems, and leaves.   d. stems and leaves.
    b. roots and stems.   e. roots only.
    c. roots and leaves.
22. Explain why mesophyll, cortex, and pith are all designated as ground tissue.
23. **THINKING CONCEPTUALLY** If they lacked epidermal tissue, how would roots, stems, and leaves be affected?

Growth of Roots and Stems

24. New plant cells originate from the
    a. parenchyma.   d. base of the shoot.
    b. collenchyma.   e. apical meristem.
    c. sclerenchyma.
25. Root hairs are found in the zone of
    a. cell division.   d. apical meristem.
    b. elongation.   e. All of these are correct.
    c. maturation.
26. During secondary growth, a tree adds more xylem and phloem through the activity of the
    a. apical meristems.   c. intercalary meristems.
    b. vascular cambium.   d. cork cambium.
27. Between the cork and the phloem in a woody stem is a layer of meristem called
    a. cork cambium.   d. the zone of cell division.
    b. vascular cambium.   e. procambium preceding bark.
    c. apical meristem.
28. Bark contains which of the following?
    a. phloem   d. cork cambium
    b. cork   e. All of these are correct.
    c. periderm
29. Annual rings are the number of
    a. internodes in a stem.
    b. rings of vascular bundles in a monocot stem.
    c. layers of xylem in a stem.
    d. bark layers in a woody stem.
    e. Both b and c are correct.
30. What part of a woody twig allows you to determine the amount of growth last year?
    a. leaf scar   c. vascular bundle scar
    b. terminal bud scale scar d. axillary bud
31. Contrast primary growth with secondary growth by naming the tissue responsible for each and the results of the growth.
32. Annual rings represent an adaptation for the advantage of
    a. telling the age of a tree.
    b. protecting a plant from parasites and predators.
    c. providing enough vascular tissue for a large shoot system.
    d. dividing a trunk into heartwood and sapwood.
    e. Both b and c are correct.
33. **THINKING CONCEPTUALLY** Explain the advantage of having xylem and not phloem add to the girth of trees on the basis of their comparative structure.

Homeostatic Mechanisms of Plants

34. Name an advantage to the plant for
    a. a flat, thin blade.
    b. an epidermis composed of closely packed cells covered by cuticle.

    c. palisade mesophyll containing chloroplasts next to upper epidermis.
    d. spongy mesophyll that is loosely packed.
    e. stomata that can open.
35. **THINKING CONCEPTUALLY** Why is photosynthesis necessary to homeostasis in plants?

For questions 36–40, match the method of regulation to the functions in the key.

KEY:
    a. prevent pathogen invasion
    b. regulate growth
    c. aid metabolism and growth
    d. regulate water loss
    e. distribute sugar according to need

36. Phloem transport
37. Stomata
38. Plant hormones
39. Defense mechanisms
40. Mutualistic relationships
41. You would expect a plant living in a desert to maintain homeostasis by having
    a. deep roots.
    b. thick cuticles with sunken stomata.
    c. narrow leaves.
    d. spongy mesophyll lacking open spaces.
    e. All of these are correct.

GET INVOLVED

1. Utilizing an electron microscope, how might you confirm structurally and biochemically that a companion cell communicates with its sieve-tube member? (See section 22.3 and "Microscopes Allow Us to See Cells" on p. 64.)
2. For your senior project, you decide to make microscopic tissue slides confirming that root tips do have the three zones shown in Figure 22.6A. After growing many seedlings, what would you like your root tip slides to show?

MEDIA STUDY TOOLS

mhhe.com/maderconcepts3

Enhance your study of this chapter with interactive study tools, practice tests, and engaging animations. Also, ask your instructor about the resources available through ConnectPlus, which includes LearnSmart, a personalized adaptive learning program, and a media-rich eBook.

23

Transport and Nutrition in Plants

BEFORE YOU BEGIN

Take a few minutes to recall

The conductive properties of water (section 2.5)

The makeup of plant vascular tissue (section 22.3)

The structure of wood (section 22.6)

The mutualistic relationship of fungi with plants (section 22.7)

Reach for the Stars

Redwood trees, named for the color of their bark but scientifically called *Sequoia sempervirens,* live along the Pacific coast in California. They prefer the coastal ranges of California where the climate is mild and a dense fog drips moisture. Redwoods are gymnosperms, a unique deciduous cypress tree that likes to regrow from a damaged stump. They are giants—the Statue of Liberty provides a mental image of how tall and wide a redwood tree can be. The tallest reach a height of 115 m and are easily wide enough at ground level for a car to pass through. As of 2010, some redwood trees have lived 2,200 years, longer than the length of time since Rome fell. The trees are experts at surviving disasters. Their thick (30 cm) bark protects them from fire, and their root system simply grows anew, above the old one, after a flood. The root system is shallow but wide, and the trees remain erect only because their roots intertwine with those of their neighbors. In the same way, humans are less likely to fall when they hold hands than when they stand alone.

As a growing tree increases in height and width, the amount of its xylem, the vascular tissue that delivers water even up to leaves fluttering in the crown of the tree, also increases. Only tracheids are present in the xylem of gymnosperms, but their thick walls have pits that allow water to move from one to the other all the way to the top of the tree. But how can water travel to such a great height without a pumping device to move it? The cells of your body require water and your muscular heart is a pump that uses up ATP to keep your blood moving. Redwood trees require vast quantities of water

Redwood trees are so named because of their red bark.

to remain viable and carry on photosynthesis. Yet no energy is required and the method of moving water through xylem is passive. It works because water molecules are like little magnets that stick together due to their polarity, and they also stick to the polar walls of tracheids, making it less likely water will fall back. The method of moving water up a tree is simple: As water evaporates from the surface of the leaves, more water moves up from below to replace it.

Scientists go everywhere, and some have even climbed to the top of redwood trees to take measurements and decide how tall a redwood tree could be using this passive system of moving water. They took their instruments with them, and while in the treetops, they measured photosynthetic rates and the amount of pull on the water in tracheids at that great height. They found that the pull on the water column due to evaporation was as great as the pull exerted by the weight of the water to drop back. Beyond this point, the water column breaks apart, and gas bubbles prevent any flow of water in xylem. However, the thin water column in tracheids is less apt to break than the wide one in vessel elements. In addition, the scientists found that little photosynthesis was taking place at the tops of redwoods. They concluded that photosynthesis would stop altogether above 130 m, and therefore no redwood tree, nor any other type of tree, will ever be found taller than this. Trees, including redwood trees, will never reach the stars.

In this chapter, we will see that plants require only inorganic nutrients in order to make the organic molecules that provide the building blocks and energy they need to grow. We will examine how these various substances are transported to their cells.

Scientist using a climbing rope to ascend a redwood.

A car can pass through the trunk of a redwood tree.

Xylem and Phloem Structure

Just as your blood vessels go to all parts of your body, vascular tissue goes to all parts of a plant. Vascular tissues transport water plus minerals and sugar (sucrose) to all plant organs.

23.1 Transport begins in both the roots and leaves of a plant

LEARNING OUTCOME

When you complete this section, you should be able to

1. Describe the continuous pipelines for water transport and sugar transport in plants.

Flowering plants are well adapted to living in a terrestrial environment. Their leaves, which carry on photosynthesis, are positioned to catch the rays of the sun because they are held aloft by the stem (Fig. 23.1A). Carbon dioxide enters leaves at the stomata, but water, the other main requirement for photosynthesis, is absorbed by the roots. Water must be transported from the roots through the stem to the leaves. Vascular plants have a transport tissue, called **xylem,** that moves water and minerals from the roots to the leaves. Xylem, with its strong-walled, nonliving cells, gives trees much-needed internal support. Xylem contains two types of conducting cells: tracheids and vessel elements (see Fig. 23.1B). **Tracheids** are tapered at both ends. The ends overlap with those of adjacent tracheids, and pits allow water to pass from one tracheid to the next. **Vessel elements** are long and tubular with perforation plates at each end. Vessel elements placed end to end form a completely hollow pipeline from the roots to the leaves.

The process of photosynthesis results in sugars, which are used as a source of energy and building blocks for other organic molecules throughout a plant. **Phloem** is the type of vascular tissue that transports organic nutrients to all parts of the plant. Roots buried in the soil cannot possibly carry on photosynthesis, but they still require a source of energy so that they can carry on cellular metabolism. In flowering plants, phloem consists of two types of cells: sieve-tube members and companion cells (Fig. 23.1C). **Sieve-tube members** are the conducting cells of phloem. The end walls are called sieve plates and have numerous pores; strands of cytoplasm extend from one sieve-tube member to another through the pores. Sieve-tube members lack nuclei and other organelles involved in protein synthesis. **Companion cells,** which have nuclei, provide proteins to sieve-tube members.

Sieve-tube members form a continuous *sieve tube* for organic nutrient transport throughout the plant. Knowing that vascular plants are structured in a way that allows materials to move from one part to another does not tell us the mechanisms by which these materials move. Plant physiologists have performed numerous experiments to determine how water and minerals rise to the tops of very tall trees in xylem, and how organic nutrients move in the opposite direction in phloem. We might expect that these

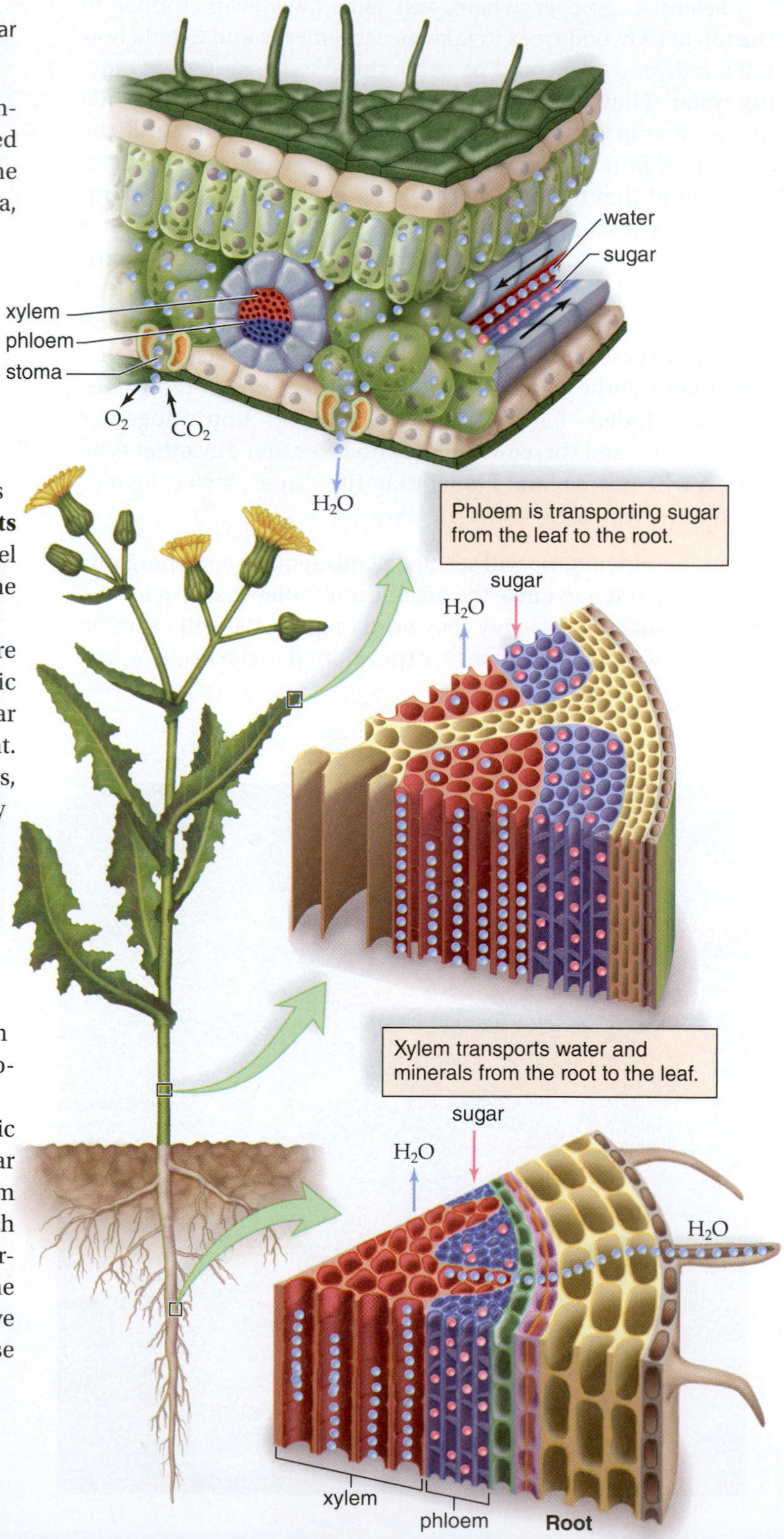

water
sugar

xylem
phloem
stoma

O_2 CO_2

H_2O

Phloem is transporting sugar from the leaf to the root.

sugar
H_2O

Xylem transports water and minerals from the root to the leaf.

sugar
H_2O
H_2O

xylem
phloem   **Root**

FIGURE 23.1A A plant's transport system (blue = phloem; pink = sugar; red = xylem; light blue = water).

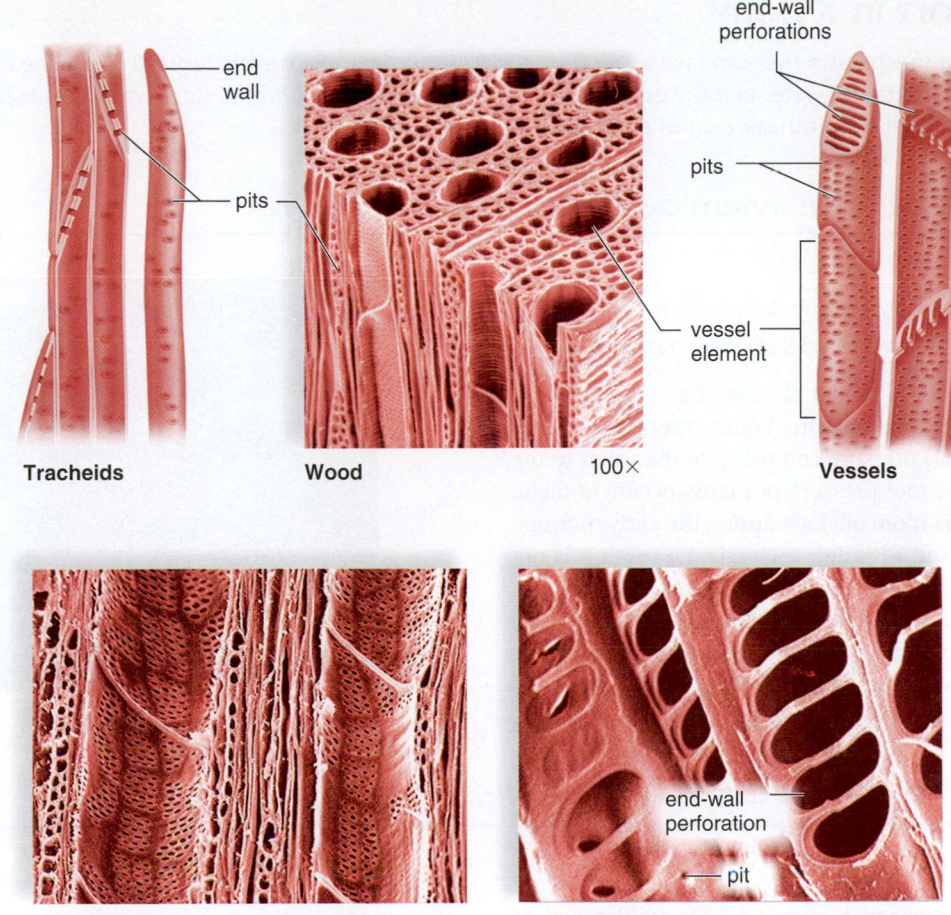

Tracheids

end wall

pits

Wood

100×

Vessels

end-wall perforations

pits

vessel element

Vessel elements in a walnut tree

520×

End-wall perforations

end-wall perforation

pit

3,200×

FIGURE 23.1B Xylem, which makes up wood, contains two types of conducting cells: tracheids and vessel elements. Tracheids (*top left*) are long and slender. Vessel elements (*top right*) have perforation plates for their end walls. A vessel contains many vessel elements.

processes are mechanical in nature and are based on the properties of water, because water is a large part of both **xylem sap** and **phloem sap.**

3D Animation
Plant Transport

In organisms, water molecules diffuse freely across plasma membranes from the area of higher concentration to the area of lower concentration. Other chemical properties of water are also important in movement of xylem sap. The polarity of water molecules and the hydrogen bonding between water molecules allow water to fill xylem cells.

▶ **23.1 CHECK YOUR PROGRESS**

1. Explain why vessel elements would have no need for companion cells.

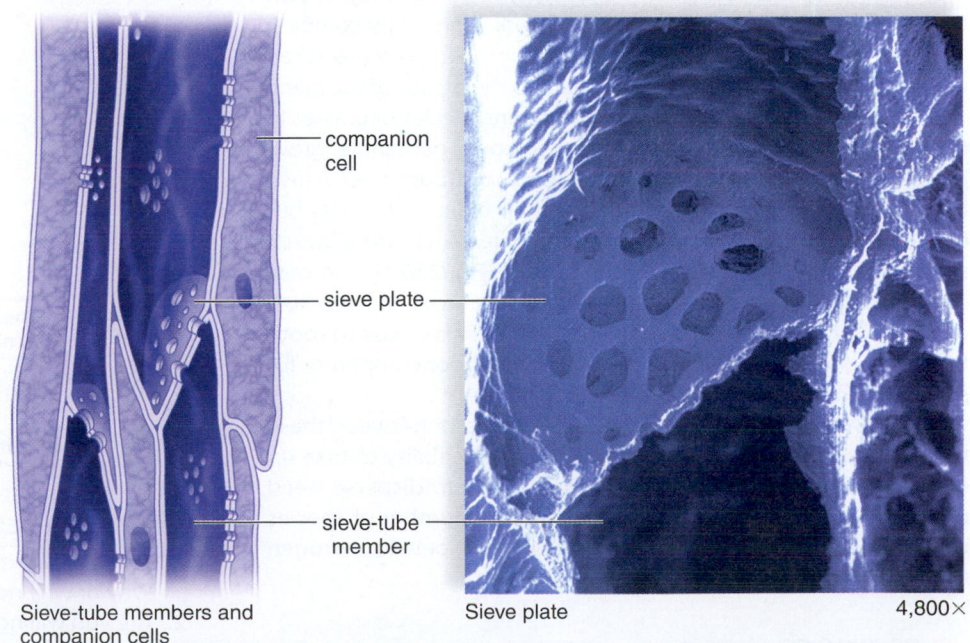

companion cell

sieve plate

sieve-tube member

Sieve-tube members and companion cells

Sieve plate

4,800×

FIGURE 23.1C The conducting cells of phloem are called sieve-tube members because their end walls are perforated in the same manner as a sieve.

Water Transport in Xylem

Plants require a transport mechanism that can take water from the roots to the tops of even very tall trees. The cohesion-tension model relies on the chemical properties of water and the evaporation of water at stomata. Stomata close when plants become water stressed, but then carbon dioxide for photosynthesis cannot enter leaves.

23.2 Water is pulled up in xylem by evaporation from leaves

LEARNING OUTCOME

When you complete this section, you should be able to

1. Describe the cohesion-tension model of xylem transport.

How is it possible for water to rise to the top of a very tall plant? One contributing factor is root pressure. Water entering root cells creates a positive (internal) pressure compared to the water in the surrounding soil. Because root pressure primarily occurs at night, this water accumulation is more obvious during the early morning hours and may be confused with dew. Actually, however, it is not dew but guttation, during which drops of water are forced out of vein endings along the edges of leaves (Fig. 23.2A). Guttation is the result of root pressure. Although root pressure may contribute to the upward movement of water in some instances, it is not believed to be the mechanism by which water can rise to the tops of very tall trees.

FIGURE 23.2A Guttation.

HOW SCIENCE PROGRESSES   *Application*

23A Competition for Resources Is One Aspect of Biodiversity

In college, the biological issues that intrigued G. David Tilman most were concerned with the relationships of species to their environment, including the effects of interactions such as competition or predation. It seemed to him that biodiversity might relate to these interactions: He chose to study the Minnesota grasslands because they often harbor more than 100 plant species within an area of only a few hectares. Long-term experiments had already shown that all these plant species are held in check (limited) by competition for the same resource—namely, soil nitrogen, which is taken up by the roots and transported in xylem to the tissues of the plant.

The researchers planted over 50 different species of grasses in serveral plots and studied how well they could germinate, grow, and reproduce there. The best competitors for nitrogen were native bunchgrasses, which allocated 85% of their biomass to roots but only 0.5% of their biomass to seeds. Little bluestem (*Schizachyrium scoparium*) is an example of a bunchgrass (Fig. 23A.1). The best dispersers (plants that can move to new locations) allocated 30% of their biomass to seeds, but only 40% of their biomass to roots. Bent grass (*Agrostis scabra*) is an example of a poor competitor for soil nitrogen but a good disperser (Fig. 23A.2).

Tilman concluded that coexistence can occur between these two types of plants because they differ in their ability to take up nitrogen (more root biomass) and their ability to disperse (seed production). He found that it is possible for a number of species to coexist when they differ in their ability to compete for nitrogen uptake and their ability to disperse.

FIGURE 23A.1 Little bluestem.   **FIGURE 23A.2** Bent grass.

CONSIDER THESE QUESTIONS

1. Would you expect redwood trees to be good competitors? Why or why not?
2. Are you willing to eat new grains in order for farmers to increase biodiversity? Why or why not?

Explore the concepts through a variety of multimedia assets and question types.

www.mcgrawhillconnect.com

Cohesion-Tension Model Once water enters xylem, it must be transported to all parts of the plant. Transporting water can be a daunting task, especially for some plants, such as redwood trees, which can exceed 115 m in height.

The **cohesion-tension model** of xylem transport, outlined in Figure 23.2B, describes a mechanism for xylem transport that requires no expenditure of energy by the plant and is dependent on the properties of water. The term *cohesion* refers to the tendency of water molecules to cling together. Because of hydrogen bonding, water molecules interact with one another and form a continuous **water column** in xylem, from the leaves to the roots, that is not easily broken. In addition to cohesion, another property of water, called *adhesion,* plays a role in xylem transport. Adhesion refers to the ability of water, a polar molecule, to interact with the molecules making up the walls of the vessels in xylem. Adhesion gives the water column extra strength and prevents it from slipping back.

What Happens in the Leaf? Consider the structure of a leaf, as shown in Figure 23.2B. ❶ The stomata (sing., stoma), while also found elsewhere, are most concentrated in the lower epidermis of a eudicot leaf, where they open directly into the spongy layer of the mesophyll. When the stomata are open, the cells of the spongy layer are exposed to the air, which contains less water than the leaf does. Water then evaporates as a gas or vapor from the spongy layer into the intercellular spaces. Evaporation of water through leaf stomata is called **transpiration.** At least 90% of the water taken up by the roots is eventually lost by transpiration. This means that the total amount of water lost by a plant over a period of time is surprisingly large. An average-sized birch tree with over 200,000 leaves will transpire up to 3,700 l of water per day during the growing season.

The water molecules that evaporate from cells into the spaces of spongy mesophyll are replaced by other water molecules from the leaf veins. Because water molecules are cohesive, transpiration exerts a pulling force, or *tension,* that draws the water column through the xylem to replace the water lost by leaf cells.

Note that the loss of water by transpiration is also the mechanism by which minerals are transported throughout the plant body. Furthermore, the evaporation of water helps keep the temperature of leaf cells within normal limits through evaporative cooling.

What Happens in the Stem? Figure 23.2B shows that ❷ the tension in xylem created by evaporation of water at the leaves pulls the water column in the stem upward. Usually the water column in the stem is continuous because of the cohesive property of water molecules. The water molecules also adhere to the sides of the vessels. What happens if the water column within xylem breaks? The water column "snaps back" down the xylem vessel away from the site of breakage, making it more difficult for conduction to occur. Next time you use a straw to drink a soda, notice that pulling the liquid upward is fairly easy as long as there is liquid at the end of the straw. When the soda runs low and you begin to get air, it takes considerably more suction to pull up the remaining liquid. When putting flowers in a vase, you should always cut the stems under water to preserve an unbroken water column and the life of the flowers.

What Happens in the Root? ❸ Due to the active transport of minerals into the root (see Fig. 23.6B), water enters root hairs passively by osmosis, and from there it enters xylem. The water column in xylem extends from the leaves down to the root. Water is pulled upward from the roots due to the tension in xylem created by the evaporation of water at the leaves.

3D Animation
Plant Transport: Water Transport in Xylem

▶ **23.2 CHECK YOUR PROGRESS**
  1. Identify the properties of water that assist water transport in xylem.
  2. Describe why water is under tension from the leaves to the roots of plants.

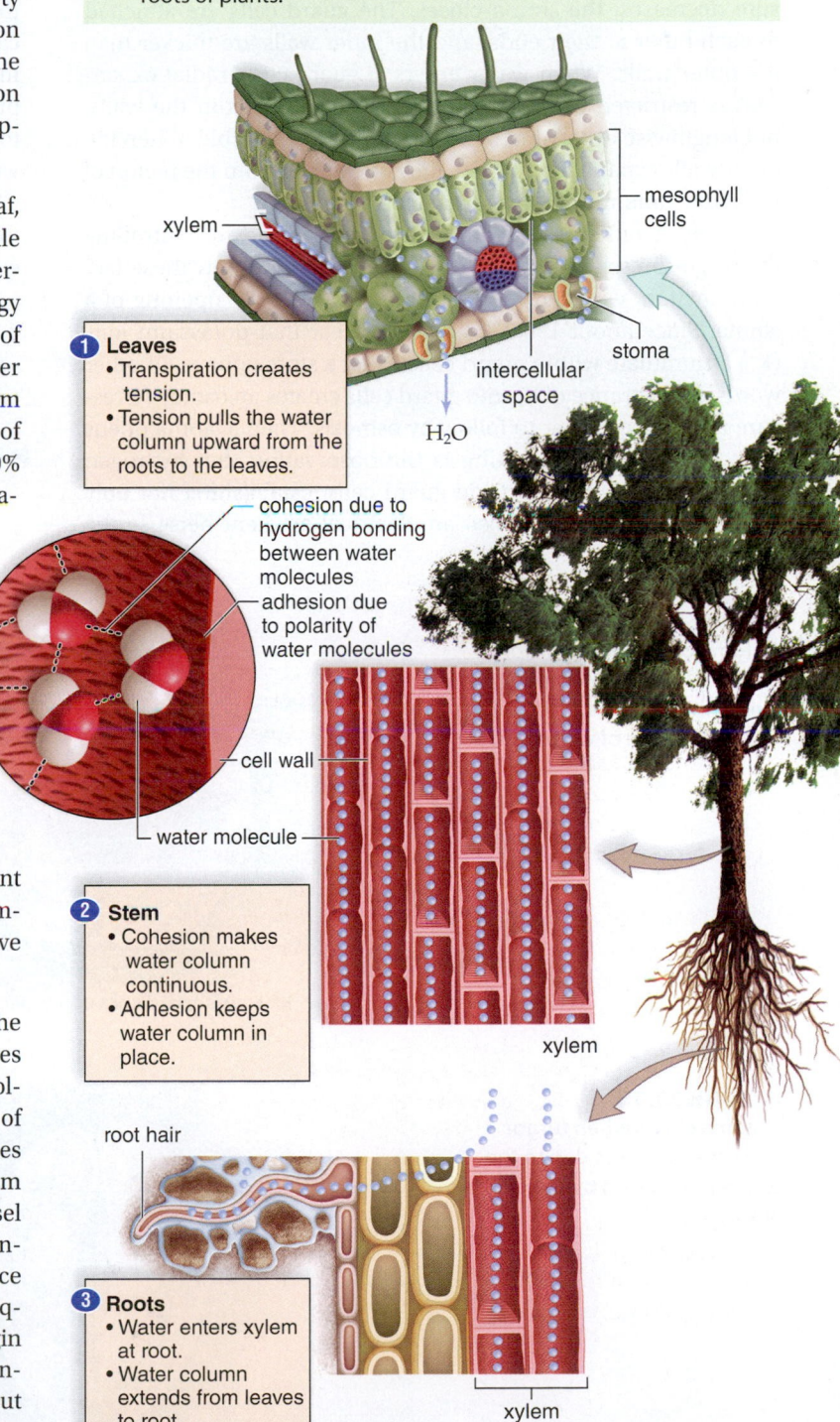

❶ **Leaves**
• Transpiration creates tension.
• Tension pulls the water column upward from the roots to the leaves.

xylem

mesophyll cells

stoma

intercellular space

H_2O

cohesion due to hydrogen bonding between water molecules

adhesion due to polarity of water molecules

cell wall

water molecule

❷ **Stem**
• Cohesion makes water column continuous.
• Adhesion keeps water column in place.

xylem

root hair

❸ **Roots**
• Water enters xylem at root.
• Water column extends from leaves to root.

xylem

FIGURE 23.2B Cohesion-tension model of xylem transport.

23.3 Guard cells regulate water loss at leaves

LEARNING OUTCOMES

When you complete this section, you should be able to

1. Describe how guard cells cause stomata to open.
2. List and describe factors that cause stomata to open or close.

Each **stoma** (pl., stomata) is a small pore in the leaf epidermis bordered by modified epidermal cells called **guard cells.** When water enters the guard cells and turgor pressure increases, the stoma opens; when water exits the guard cells and turgor pressure decreases, the stoma closes. The guard cells are attached to each other at their ends, and the inner walls are thicker than the outer walls. When water enters, a guard cell's radial expansion is restricted because of cellulose microfibrils in the walls, but lengthwise expansion of the outer walls is possible. When the outer walls expand lengthwise, they buckle out from the region of their attachment, and the stoma opens.

Other factors aside from water are involved in controlling the turgor pressure of guard cells. To understand how these factors function, we need to look more closely at the opening of a stoma. Since about 1968, it has been clear that potassium ions (K^+) accumulate within guard cells when a stoma opens. In other words, the entrance of K^+ into guard cells creates an osmotic pressure that causes water to follow by osmosis. Then a stoma opens (Fig. 23.3A). Also interesting is the observation that hydrogen ions (H^+) accumulate outside guard cells, establishing not only a chemical gradient but also an electrical gradient because the inside of the cell is now negative. The electrical gradient allows K^+ to enter by way of a channel protein, and water follows by osmosis.

On the other hand, a stoma closes when turgor pressure decreases due to the exit of K^+ followed by the exit of water (Fig. 23.3B).

Three other factors, aside from water availability, regulate whether stomata open or close. (1) The presence of light causes stomata to open. Evidence suggests that the absorption of blue light by a flavin protein sets in motion the cytoplasmic response that leads to opening of stomata. (2) A high concentration of CO_2 causes stomata to close. Perhaps there is a receptor in the plasma membrane of guard cells that brings about inactivation of a H^+ pump when CO_2 concentration rises, as might happen when photosynthesis ceases. (3) Abscisic acid (ABA), produced by cells in wilting leaves, can cause stomata to close (see section 23.5). Like CO_2, ABA most likely promotes the inhibition of a H^+ pump and, in that way, hinders the creation of conditions that cause water to enter guard cells by osmosis.

Interestingly enough, when plants are kept in the dark, stomata open and close just about every 24 hours. Circadian rhythms (a behavior that occurs nearly every 24 hours) and biological clocks that can keep time are areas of intense investigation.

▶ **23.3 CHECK YOUR PROGRESS**
1. Identify why stomata close and what factors can influence the opening and closing of stomata.

FIGURE 23.3A
A stoma opens when turgor pressure increases in guard cells due to the entrance of K^+ followed by the entrance of water.

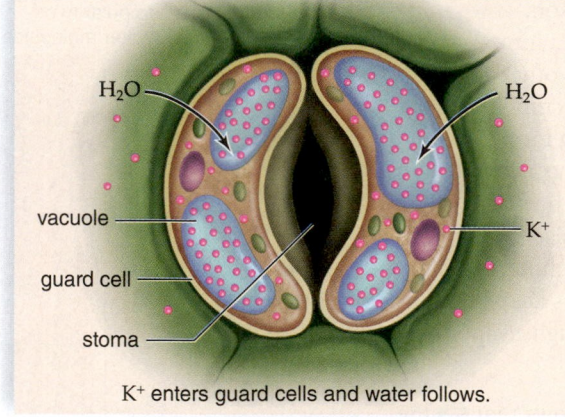
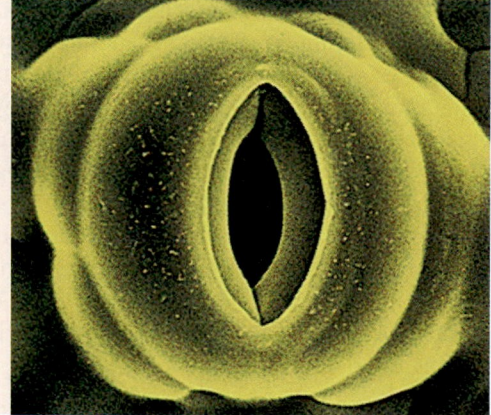

H_2O H_2O vacuole guard cell stoma K^+

K^+ enters guard cells and water follows.

343×

FIGURE 23.3B
A stoma closes when turgor pressure decreases due to the exit of K^+ followed by the exit of water.

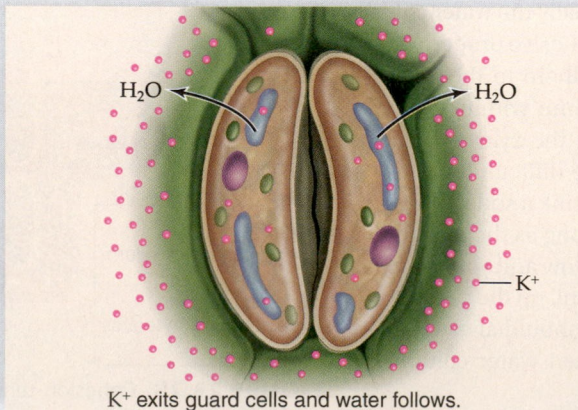
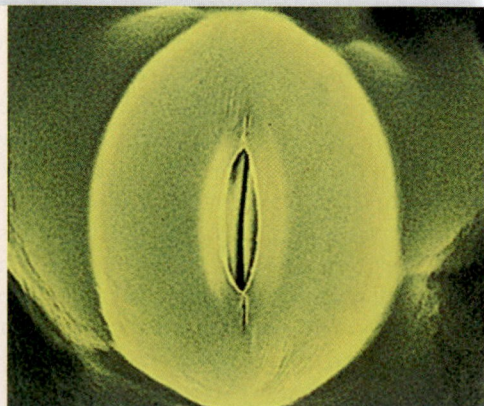

H_2O H_2O K^+

K^+ exits guard cells and water follows.

370×

23B Plants Can Clean Up Toxic Messes

Plants are well known as a source of food, medicines, and firewood. We may also recognize that plants slow the rate of global warming and clean up urban waste water but we may not be aware that plants can clean soil of toxic messes including nitrates which reduce the oxygen capacity of blood, heavy metals that lead to brain damage, and organic chemicals that cause cancer. Plants can even be harvested and the metals reclaimed for industrial use.

Scientists have found that poplar trees and mustard plants are particularly useful to carry out *phytoremediation,* the use of plants to restore the environment to its former natural state.

Poplar Trees A 1.6-km stand of spindly poplars outside Amana, Iowa, were planted there by Louis Licht when he was a University of Iowa graduate student. The trees take up nitrate-laden runoff from fertilized cornfields before the runoff reaches a nearby brook—and perhaps other waters. Nitrate runoff into the Mississippi River from midwestern farms is a major cause of the large "dead zone" of oxygen-depleted water that develops each summer in the Gulf of Mexico. Before the poplar trees were planted, the brook's nitrate level was as much as ten times the amount considered safe.

Gary Pierzynski of Kansas State University and Jerry Schnoor of the University of Iowa used poplars at a site in Dearing, Kansas, to stabilize soils and decrease the movement of metals to ground water. Nothing had grown on waste piles at this site since it was abandoned for over 50 years because of high concentrations of lead and zinc. After they planted 3,100 hybrid poplar trees with soil amendments on 0.80 hectare (0.80 ha), the growth of the trees exceeded expectations and the risk to humans due to windblown dust was greatly reduced.

Then, too, on the basis of results from a pilot site, a University of Washington group in conjunction with the Oregon Department of Environmental Quality began planting poplars about 10 years ago in expectation of remediating an aquifer that was contaminated with trichloroethane, an organic contaminant.

Mustard Plants Canola, a mustard plant, is grown in California's San Joaquin Valley to soak up excess selenium in the soil and help prevent an environmental catastrophe like the one that occurred there in the 1980s. Irrigation caused naturally occurring selenium to rise to the soil surface and excess water caused the selenium to flow off into drainage ditches and eventually end up in Kesterson National Wildlife Refuge. The selenium in ponds at the refuge accumulated in plants and fish and subsequently deformed and killed waterfowl, says Gary Bañuelos, a plant scientist with the U.S. Department of Agriculture who helped remedy the problem (Fig. 23B). He recommended that farmers add selenium–accumulating canola plants to their crop rotations. As a result, selenium levels in runoff are being managed. Although the underlying problem of excessive selenium in soils has not been solved, says Bañuelos, "This is a tool to manage mobile selenium and prevent another unlikely selenium-induced disaster."

Edenspace Systems Corporation of Reston, Virginia, has demonstrated the power of a phytoremediation at a Superfund site on an Army firing range in Aberdeen, Maryland. The company successfully used mustard plants to remove uranium from the firing range, at about 10% of the cost of traditional cleanup

FIGURE 23B Scientist Gary Bañuelos in a field of canola.

methods. Depending on the contaminant involved, traditional cleanup costs can run as much as $1 million per 0.40 ha according to experts. Mustard plants are also capable of accumulating high concentrations of zinc, nickel, cadmium, lead, copper, and cobalt in their bodies. *Phytoextraction* is possible because the metals can later be extracted from the plants. It's estimated that 20 tons of biomass yields approximately $1,000 worth of metals. Scientists are hoping to improve a plant's uptake of metals by genetically modifying them through conventional breeding or by genetic engineering practices.

One possible drawback to phytoremediation is its slow pace. Depending on the contaminant, it can take several growing seasons to clean a site—much longer than by conventional methods. "We normally give phytoremediation a target of one to three years to clean a site," notes Edenspace's Mike Blaylock. "People won't want to wait much longer than that." Even so, we can conclude that phytoremediation makes it possible to treat the most serious types of environmental damage in a way that not only restores the environment but also recovers resources that are valuable to industry.

CONSIDER THESE QUESTIONS

1. Make a list of all the ways plants help us.
2. Should phytoremediation be used if pets get sick feeding on plants full of pollutants?
3. If by chance only an expensive plant could feed on a pollutant, would you be inclined to use it anyway?

 Explore the concepts through a variety of multimedia assets and question types.

www.mcgrawhillconnect.com

Organic Nutrient Transport in Phloem

Plants require a transport mechanism capable of taking sugar to all the areas of a plant that need it, be they leaves, flowers, or roots. The pressure-flow model shows how bulk flow can explain the transport of organic food in a plant.

23.4 Pressure flow explains phloem transport

LEARNING OUTCOME

When you complete this section, you should be able to

1. Describe the pressure-flow model of phloem transport.

Plants transport the organic molecules resulting from photosynthesis to the parts of plants that need them. This includes young leaves that have not yet reached their full photosynthetic potential, flowers that are in the process of making seeds and fruits, and roots, whose location in the soil prohibits them from carrying on photosynthesis.

If a strip of bark is removed all the way around a tree below the level of the majority of its leaves, the bark swells just above the cut, and sugar accumulates in the swollen tissue. We know today that when a tree is girdled like this, the phloem is removed, but the xylem is left intact. Therefore, the results of this girdling experiment tell us that phloem, not xylem, is the tissue that transports sugars.

Radioactive tracer studies with carbon 14 (^{14}C) have confirmed that phloem transports organic molecules. When ^{14}C-labeled CO_2 is supplied to mature leaves, radioactively labeled sucrose is soon found moving down the stem into the roots. It's difficult to get samples from phloem sap without injuring the phloem, but this problem is solved by using aphids, small insects that are phloem feeders. The aphid drives its stylet, a sharp mouthpart that functions like a hypodermic needle, between the epidermal cells, and sap enters its body from a sieve-tube member (Fig. 23.4A). If the aphid is anesthetized using ether, its body can be carefully removed, leaving the stylet. A researcher can then collect and analyze the phloem sap. The use of radioactive tracers and aphids has revealed that sap movement through phloem can be as fast as 60–100 cm per hour and possibly up to 300 cm per hour.

The Pressure-Flow Model The pressure-flow model is a current explanation for the movement of organic materials throughout the plant in phloem. Consider an experiment in which two bulbs are connected by a glass tube. The first bulb contains sucrose at a higher concentration than the second bulb. Each bulb is bounded by a differentially permeable membrane, and the entire apparatus is submerged in distilled water, which lacks ions. As shown in Figure 23.4B, **1** distilled water flows into the first bulb because it has the higher solute concentration. **2** The entrance of water creates a positive *pressure,* and water *flows* toward the second bulb. In this experiment, **3** the pressure was great enough to move out through the membrane of the second bulb—even though the second bulb contained a higher concentration of solute than the distilled water. In the same way, **pressure flow** (movement of water and solutes together due to a pressure gradient) can explain the flow of phloem sap in plants. As an analogy, recall that the entrance of water into a garden hose at a faucet forces water through the hose to a sprinkler. Water exits at the sprinkler where the pressure is always less than the faucet end.

In plants, the sieve tubes of phloem are analogous to the glass tube that connects the two bulbs. Sieve tubes are composed of sieve-tube members, each of which has a companion cell. It is possible that the companion cells assist the sieve-tube members in some way. The sieve-tube members align end to end, and at maturity the wide pores in the sieve-tube plates allow phloem sap to move from one cell to the other. Sieve tubes, therefore, form a continuous pathway for organic nutrient transport throughout a plant.

Flow Is from a Source to a Sink During the growing season, photosynthesizing leaves are producing sugar, which supplies the energy plants need to grow, repair tissues, and to reproduce. Therefore, photosynthesizing leaves are a **source** of sugar (e.g., sucrose). Figure 23.4C shows what happens at the source.

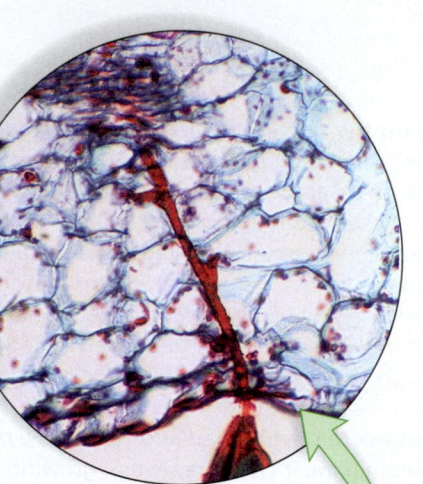

FIGURE 23.4A
Aphids are small insects that remove nutrients from phloem by means of a needlelike mouthpart called a stylet. Excess phloem sap appears as a droplet after passing through the aphid's body.

Under microscope

waste due to feeding on phloem sap

An aphid feeding on a plant stem

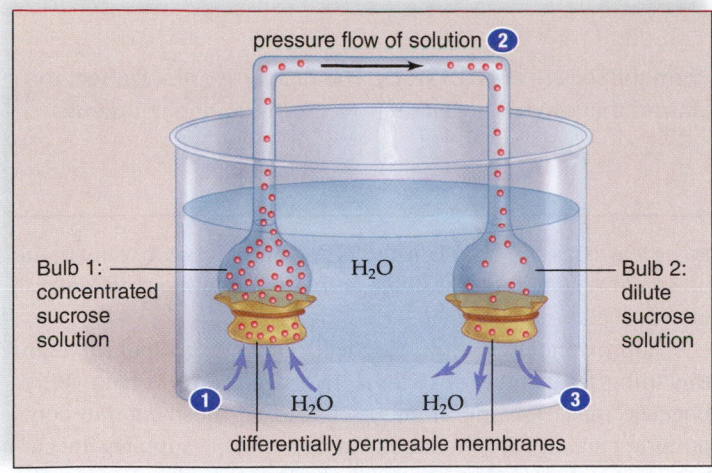

FIGURE 23.4B Bulk flow due to a pressure gradient.

1 Sugar is actively transported into the phloem in this manner: First an energy-requiring proton pump establishes an electrochemical gradient—more H^+ is outside than inside phloem. Sugar then travels across the membrane in conjunction with hydrogen ions (H^+), which are moving down their concentration gradient (see Fig. 23.6B*b*). **2** After sugar enters the sieve tubes, water follows passively by osmosis. **3** The buildup of water within sieve tubes creates the positive pressure that starts the bulk flow of the phloem contents. **4** The roots (and other growth areas) are a **sink** for sugar, meaning that they are removing sugar and using it for cellular respiration. Also, some plants, such as the sugar beet, the carrot, and the white potato, have modified organs for carbohydrate storage. After sugar is actively transported out of sieve tubes, water exits the phloem passively by osmosis and **5** is taken up by xylem. **6** Xylem transports water to the mesophyll of the leaf where it is used for photosynthesis. **7** Although up to 90% of water is transpired, some is used for photosynthesis, and some reenters the phloem by osmosis **2**.

Transport of Sugar The pressure-flow model of phloem transport can account for any direction of flow in sieve tubes if we consider that the direction of flow is always from source to sink. In other words, phloem contents can move either up or down as appropriate for the plant at a particular time in its life cycle. For example, recently formed leaves can be a sink, and they will receive sucrose until they begin to photosynthesize maximally.

3D Animation
Plant Transport:
Translocation in Phloem

▶ **23.4 CHECK YOUR PROGRESS**

1. Explain the process in which sugars move from source to sink in a plant.

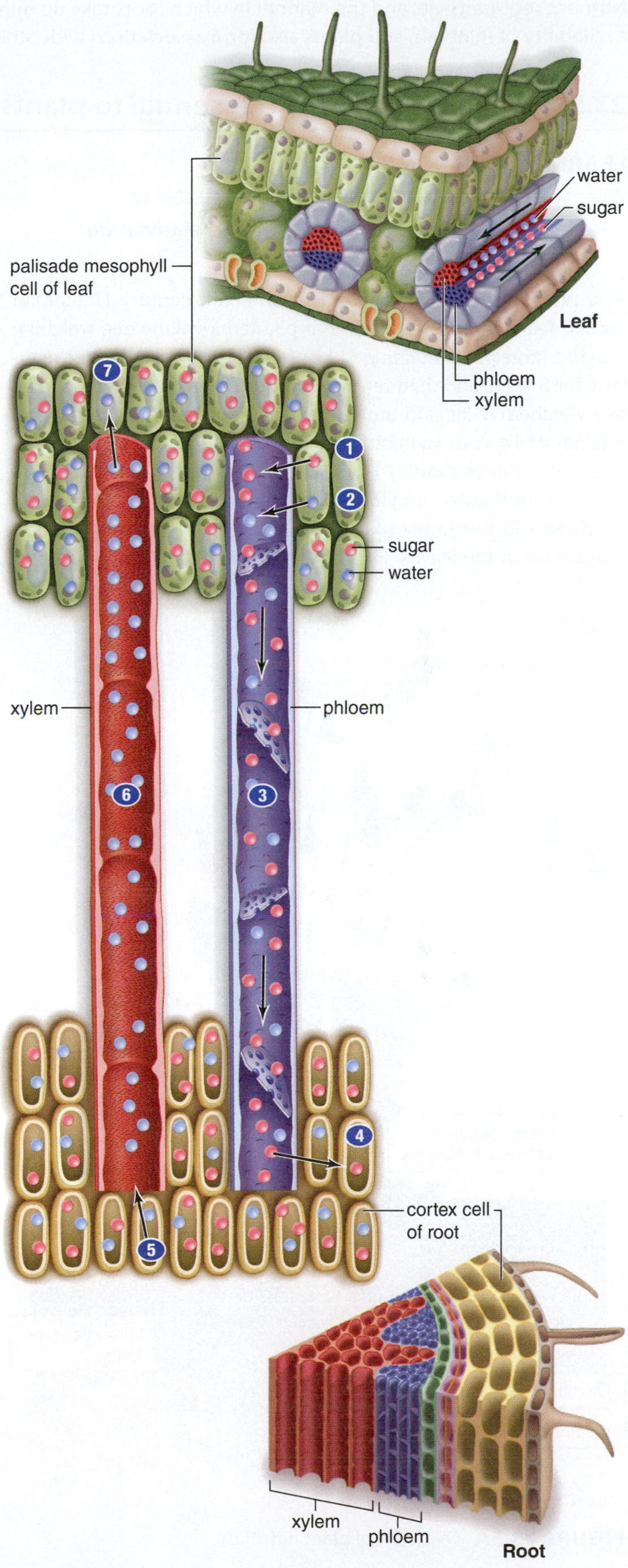

FIGURE 23.4C The pressure-flow model of phloem transport says that sugar is transported in phloem from a source (e.g., photosynthesizing leaves) to a sink (e.g., actively metabolizing roots).

Plant Nutrition and Soil

Nutrient requirements and the manner in which roots take up minerals from the soil are areas of study. The properties of soil affect the availability of minerals, and plants also form associations with other organisms that can enhance their success in acquiring nutrients.

23.5 Certain nutrients are essential to plants

LEARNING OUTCOME

When you complete this section, you should be able to

1. Identify the macronutrients and micronutrients that are required by plants.

In a now-famous experiment, a seventeenth-century Dutchman named Jean-Baptiste van Helmont planted a willow tree weighing 2 kg in a large pot containing 41 kg of soil. He watered the tree regularly for 5 years and then reweighed both the tree and the soil. The tree weighed 77 kg, and the soil weighed only a little less than the original 41 kg. Van Helmont concluded that the tree's increase in weight was due primarily to the addition of water.

Although water is a vitally important nutrient for a plant, van Helmont did not know that most of the water entering a plant evaporates at the leaves. He was also unaware that CO_2 (taken in at the leaves, as shown in Figure 23.5A) combines with water in the presence of sunlight to produce carbohydrates, the chief organic matter of plants.

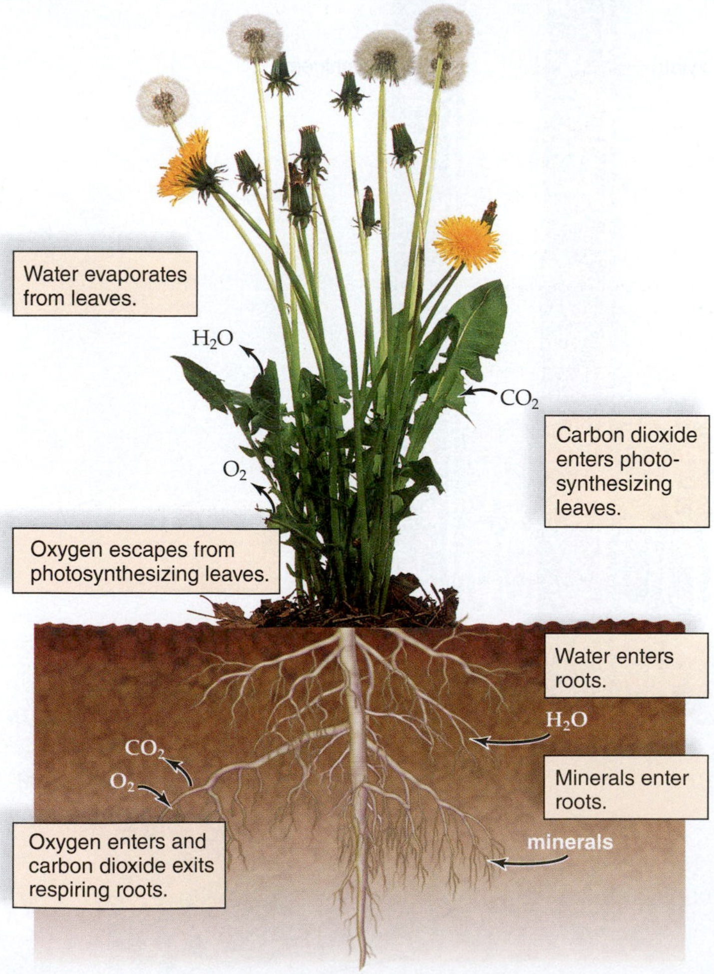

Approximately 95% of a typical plant's dry weight (weight excluding free water) is carbon, hydrogen, and oxygen. Why? Because these are the elements found in most organic compounds, such as carbohydrates. Carbon dioxide supplies the carbon, and water (H_2O) supplies the hydrogen and oxygen for the organic compounds of a plant.

Minerals as Nutrients In addition to carbon, hydrogen, and oxygen, plants require certain other nutrients that are absorbed as minerals by the roots. A **mineral** is an inorganic substance usually containing two or more elements. Minerals are needed to help build molecules. For example, nitrogen is a major component of nucleic acids and proteins; magnesium is a component of chlorophyll, the main photosynthetic pigment; and iron is a building block of cytochrome molecules, which carry electrons during photosynthesis and cellular respiration. The major functions of various **essential nutrients** for plants are listed in Table 23.5. A nutrient is essential if (1) it has an identifiable role, (2) no other nutrient can substitute and fulfill the same role, and (3) a deficiency of this nutrient causes a plant to die or fail to complete its reproductive cycle. Essential nutrients are divided into **macronutrients** (needed in large quantity) and **micronutrients** (needed in trace amounts).

Beneficial nutrients are another category of elements taken up by plants. Beneficial nutrients either are required for or enhance the growth of a particular plant. For example, horsetails require silicon as a mineral nutrient, and sugar beets show enhanced growth in the presence of sodium. Nickel is a beneficial nutrient in soybeans when root nodules are present. Some ferns use aluminum, and locoweeds, which are toxic to livestock, take up selenium.

When a plant is burned, its nitrogen component is given off as ammonia and other gases, but most other essential minerals remain in the ash. Still, the absence of nitrogen shows that the absence of a mineral from the ash cannot be relied on to determine whether a plant needs that mineral. The preferred method for determining the mineral requirements of a plant was developed at the end of the nineteenth century by the German plant physiologists Julius von Sachs and Wilhem Knop. The method, called water culture or **hydroponics,** allows plants to grow well in water, instead of soil, if they are supplied with all the nutrients they need. The investigator omits a particular mineral and observes the effect on plant growth. If growth suffers, the omitted mineral is an essential nutrient (Fig. 23.5B). This method has been more successful for macronutrients than for micronutrients. For studies involving the latter, the water and the mineral salts used must be absolutely pure, but purity is difficult to attain, because even

FIGURE 23.5A Overview of plant nutrition.

Labels on figure:
Water evaporates from leaves.
H_2O
CO_2
O_2
Carbon dioxide enters photosynthesizing leaves.
Oxygen escapes from photosynthesizing leaves.
Water enters roots.
H_2O
CO_2
O_2
Minerals enter roots.
minerals
Oxygen enters and carbon dioxide exits respiring roots.

TABLE 23.5	Some Essential Inorganic Nutrients in Plants			
Element	**Symbol**	**Form**	**Major Functions**	
Macronutrients				
Carbon	C	CO_2		
Hydrogen	H	H_2O	Major component of organic molecules	
Oxygen	O	O_2		
Phosphorus	P	$H_2PO_4^-$ HPO_4^{2-}	Part of nucleic acids, ATP, and phospholipids	
Potassium	K	K^+	Cofactor for enzymes; functions in water balance and openings of stomata	
Nitrogen	N	NO_3^-	Part of nucleic acids, proteins, chlorophyll, and coenzymes	
Sulphur	S	SO_4^{2-}	Part of amino acids, some coenzymes	
Calcium	Ca	Ca^{2+}	Regulates responses to stimuli and movement of substances through plasma membrane; involved in formation and stability of cell walls	
Magnesium	Mg	Mg^{2+}	Part of chlorophyll; activates a number of enzymes	
Micronutrients				
Iron	Fe	Fe^{2+} Fe^{3+}	Part of cytochrome needed for cellular respiration; activates some enzymes	
Boron	B	BO_3^{3-} $B_4O_7^{2-}$	Role in nucleic acid synthesis, hormone responses, and membrane function	
Manganese	Mn	Mn^{2+}	Required for photosynthesis; activates some enzymes such as those of the Krebs cycle	
Copper	Cu	Cu^{2+}	Part of certain enzymes, such as redox enzymes	
Zinc	Zn	Zn^{2+}	Role in chlorophyll formation; activates some enzymes	
Chlorine	Cl	Cl^-	Role in water-splitting step of photosynthesis and water balance	
Molybdenum	Mo	MoO_4^{2-}	Cofactor for enzyme used in nitrogen metabolism	

Solution lacks nitrogen   Complete nutrient solution

Solution lacks phosphorus   Complete nutrient solution

Solution lacks calcium   Complete nutrient solution

FIGURE 23.5B Effects of nutrient deficiencies.

instruments and glassware can introduce micronutrients. It is also possible that the element in question may already be present in the seedling used in the experiment. These factors complicate the process of determining essential plant micronutrients by means of hydroponics.

▶ **23.5 CHECK YOUR PROGRESS**

1. Define a macronutrient and name the macronutrients (aside from C, H, and O) that are required by a plant to make proteins and nucleic acids.

23.6 Roots are specialized for the uptake of water and minerals

LEARNING OUTCOMES

When you complete this section, you should be able to

1. Describe how roots absorb water and minerals from the soil.
2. Explain a simplified soil profile and the benefits of topsoil.

Land plants depend on soil to supply them with water and inorganic nutrients. **Soil** is a mixture of mineral particles (sand, silt, and clay), decaying organic material, organisms, air, and water, which together support the growth of plants. The mineral particles vary in size: Sand particles are the largest; silt particles have an intermediate size; and clay particles are the smallest. Because sandy soils have many large particles, they have large spaces, and the water drains readily through the particles. In contrast to sandy soil, a soil composed mostly of clay particles has small spaces that fill completely with water. The type of soil called loam is composed of roughly one-third each sand, silt, and clay particles. This combination sufficiently retains water and nutrients, while still allowing the drainage necessary to provide air spaces. Loam is one of the most productive soils.

Soil that contains a high percentage of decomposing organic material is called **humus.** Humus mixes with the top layer of soil particles and augments beneficial soil characteristics. Plants do well in soils that contain 10–20% humus.

Humus causes soil to have a loose, crumbly texture that allows water to soak in without doing away with air spaces. After a rain, the presence of humus decreases the chances of runoff. Humus swells when it absorbs water and shrinks as it dries. This action helps aerate soil. Soil that contains humus is nutritious for plants because when bacteria and fungi break down the organic matter in humus, inorganic nutrients are returned to plants.

In a good agricultural soil, the components come together so that there are spaces of air and water. It is best if the soil contains particles of different sizes because only then will there be spaces for air. Roots take up oxygen from air spaces within the soil. Ideally, water clings to soil particles by capillary action and does not fill the spaces. That is why you should not overwater your houseplants!

A **soil profile** is a vertical section of soil, from the ground surface to the unaltered rock below. Usually, a soil profile has parallel layers known as **horizons.** Mature soil generally has three horizons (Fig. 23.6A*a*). The A horizon is the uppermost (or topsoil) layer. It contains leaf litter, humus, and soil organisms, but minerals have drained into the B horizon. The B horizon has two parts: the zone of leaching and the subsoil. Minerals drain from the zone of leaching into the subsoil, which accumulates both inorganic nutrients and organic materials. The C horizon is a layer of weathered and shattered rock.

Root hairs absorb minerals (and also water) from the A and B horizons (Fig. 23.6A*b*). Clay particles have a benefit that sand particles do not have. Some minerals are negatively charged, and others are positively charged. Clay particles are negative, and they can retain positively charged minerals such as calcium (Ca^{2+}) and potassium (K^+), preventing these minerals from being washed away by leaching. Because clay particles are unable to retain negatively charged NO_3^- and $H_2PO_4^-$, the nitrogen and phosphate content of soil is often replenished by applying fertilizers.

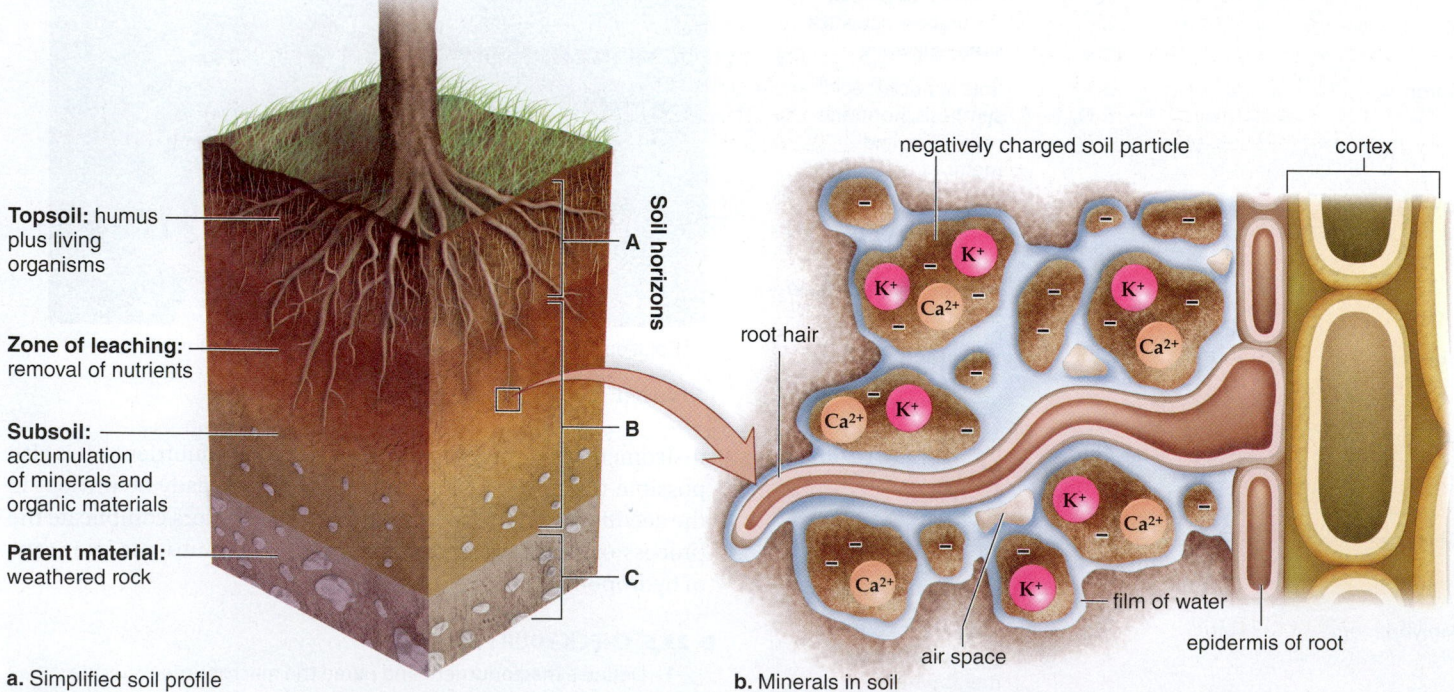

Topsoil: humus plus living organisms

Zone of leaching: removal of nutrients

Subsoil: accumulation of minerals and organic materials

Parent material: weathered rock

Soil horizons

A

B

C

a. Simplified soil profile

negatively charged soil particle

cortex

root hair

K^+ Ca^{2+}

film of water

air space

epidermis of root

b. Minerals in soil

FIGURE 23.6A **a.** Topsoil (A horizon) is a source of mineral (inorganic) nutrients released when microbes digest organic molecules in the soil. Water (delivered by rain or irrigation) leaches minerals into the B horizon. Consequently, plants absorb water and minerals from the A and B horizons. **b.** Negatively charged clay particles bind positively charged minerals such as Ca^{2+} and K^+. Plants release these minerals from soil by exchanging H^+ for them.

Water and Mineral Absorption by Roots Absorption of water and minerals is one of the main functions of a root. In order for water to reach the xylem of a root, it must pass through the cortex in one of two ways. Using the *apoplast route,* water along with minerals, can enter the root and travel through the cortex simply by passing between the porous cell walls (Fig. 23.6B*a*). However, once the water reaches the endodermal layer, it can no longer pass between cells because bands of water-proofing suberin (called the **Casparian strip**) encircle the individual endodermal cells, and this forces water to pass through the endodermal cells before reaching xylem. Alternately, using the *symplast route,* water can enter epidermal cells at root hairs and then progress through cells across the cortex and endodermis of a root by means of cytoplasmic strands within plasmodesmata. Regardless of the pathway, water enters root cells by osmosis or by **aquaporins,** which are channel proteins, as discussed in section 5.2 on page 86. Aquaporins allow water to move in the same direction as it would move by osmosis, but more quickly. Aquaporins are important in moving water into xylem.

Mineral Uptake In contrast to water, which passively crosses plasma membranes, minerals are actively taken up by plant cells. Plants possess an astonishing ability to concentrate minerals—that is, to take up minerals until they are many times more concentrated in the plant than in the surrounding medium. The concentration of certain minerals in roots is as much as 10,000 times greater than in the surrounding soil. Following their uptake by root cells, minerals move into the xylem and are transported into

the leaves by the upward movement of water. Along the way, minerals can exit the xylem and enter those cells that require them. By what mechanism do minerals cross plasma membranes?

Recall that plant cells absorb minerals in the ionic form. For example, although the atmosphere is 79% nitrogen gas, plants require nitrogen in the form of nitrate (NO_3^-) or ammonium (NH_4^+). Phosphorus, another nutrient requirement for plants, is absorbed as phosphate (HPO_4^{2-}); potassium is absorbed as potassium ions (K^+); and so forth. While water can passively (no energy needed) cross a plasma membrane, ions need to be actively transported into plant cells because they are unable to cross the plasma membrane on their own.

It has long been known that plant cells expend energy to actively take up and concentrate mineral ions. Figure 23.6B*b* describes how plants manage to do this. ❶ A proton pump hydrolyzes ATP and uses the energy released to transport hydrogen ions (H^+) out of the cell. This sets up an *electrochemical gradient.* ❷ The electrochemical gradient then drives positively charged ions such as K^+ through a channel protein into the cell. ❸ Negatively charged mineral ions (symbolized as I^-) are transported, along with H^+, by carrier proteins. Since H^+ is moving down its concentration gradient, no energy is required.

▶ **23.6 CHECK YOUR PROGRESS**
1. Explain the significance of the Casparian strip to mineral uptake.
2. Identify several reasons why plants prefer topsoil for growth.

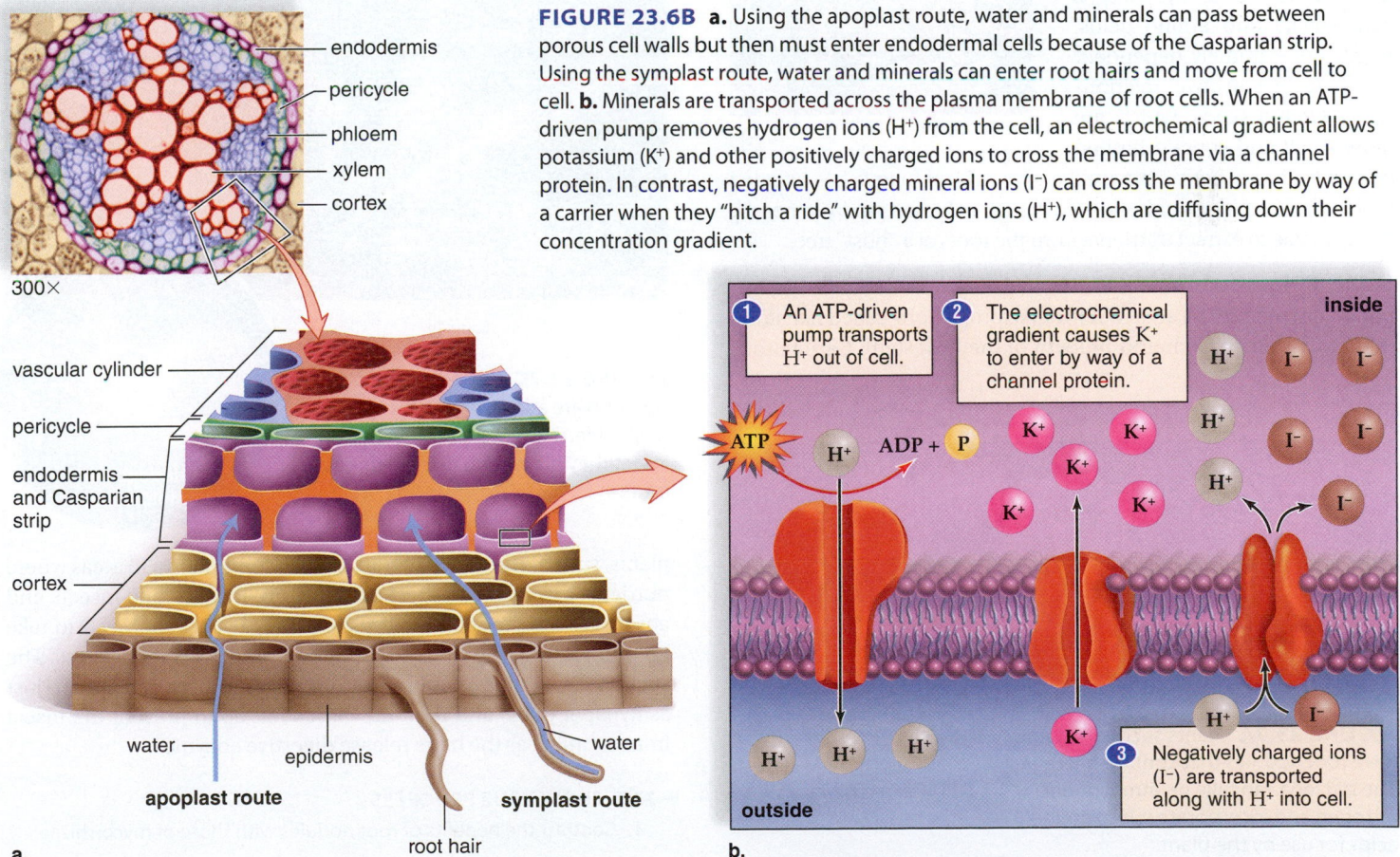

FIGURE 23.6B a. Using the apoplast route, water and minerals can pass between porous cell walls but then must enter endodermal cells because of the Casparian strip. Using the symplast route, water and minerals can enter root hairs and move from cell to cell. **b.** Minerals are transported across the plasma membrane of root cells. When an ATP-driven pump removes hydrogen ions (H^+) from the cell, an electrochemical gradient allows potassium (K^+) and other positively charged ions to cross the membrane via a channel protein. In contrast, negatively charged mineral ions (I^-) can cross the membrane by way of a carrier when they "hitch a ride" with hydrogen ions (H^+), which are diffusing down their concentration gradient.

Labels on micrograph: endodermis, pericycle, phloem, xylem, cortex — 300×

Labels on apoplast/symplast diagram: vascular cylinder, pericycle, endodermis and Casparian strip, cortex, water, epidermis, water, **apoplast route**, **symplast route**, root hair

Labels on membrane diagram:
1 An ATP-driven pump transports H^+ out of cell.
2 The electrochemical gradient causes K^+ to enter by way of a channel protein.
3 Negatively charged ions (I^-) are transported along with H^+ into cell.
inside, outside, ATP, ADP + P

23.7 Adaptations of plants help them acquire nutrients

LEARNING OUTCOME

When you complete this section, you should be able to

1. Describe root nodules, mycorrhizae, and other mutualistic relationships that help plants acquire nutrients.

Root Nodules Some plants, such as legumes, soybeans, and alfalfa, have roots colonized by *Rhizobium* bacteria, which can fix atmospheric nitrogen (N_2). They break the $N \equiv N$ bond and reduce nitrogen to NH_4^+ for incorporation into organic compounds. The bacteria live in **root nodules** and are supplied with carbohydrates by the host plant (Fig. 23.7A). The bacteria, in turn, furnish their host with nitrogen compounds, making this a mutualistic relationship.

Animation
Root Nodule
Formation

Mycorrhizae A second type of mutualistic relationship, called **mycorrhizae,** involves fungi and almost any type of plant root (Fig. 23.7B). Only a small minority of plants do not have mycorrhizae, and these plants are most often limited as to the environment in which they can grow. The fungus increases the surface area available for mineral and water uptake and breaks down organic matter in soil, releasing nutrients that the plant can use. In return, the root furnishes the fungus with sugars and amino acids. Plants are extremely dependent on their mycorrhizal relationships. Orchid seeds, which are quite small and contain limited nutrients, do not germinate until a mycorrhizal fungus has invaded their cells. Nonphotosynthetic plants, such as Indian pipe, use nearby mycorrhizae to extract nutrients from the roots of a "host" tree.

dodder (brown)

Other Relationships Parasitic plants, such as dodders, broomrapes, and pinedrops, send out rootlike projections called haustoria that tap into the xylem and phloem of the host stem. Carnivorous

root   nodule

bacteria

portion of infected cells

FIGURE 23.7A Plants such as legumes have root nodules that contain bacteria. The bacteria can take up atmospheric nitrogen and incorporate it into amino acids for use by the plant.

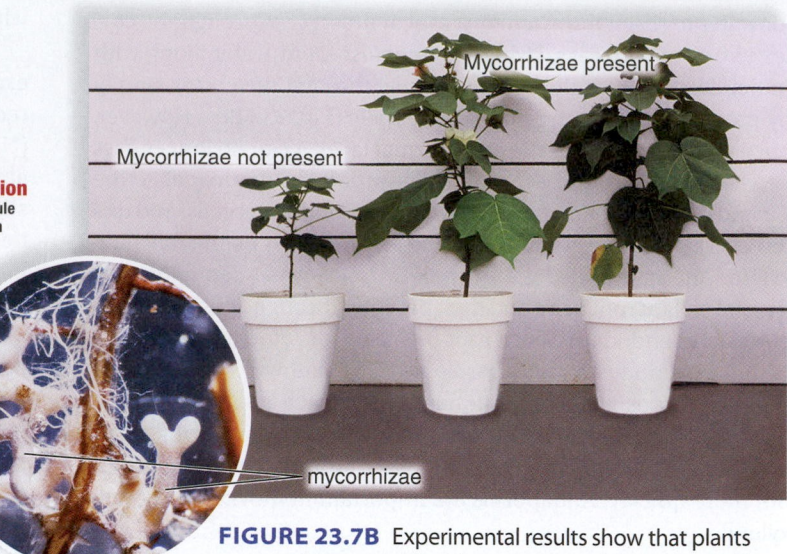

mycorrhizae

FIGURE 23.7B Experimental results show that plants grown with mycorrhizae (*two plants on right*) grow much larger than a plant (*left*) grown without mycorrhizae.

Mycorrhizae present

Mycorrhizae not present

sticky hairs

narrow leaf form

Bulbs release digestive enzymes.

Sundew leaf enfolds prey.

FIGURE 23.7C Plants such as sundews are adapted to a carnivorous way of life. Once they capture an insect they digest it to acquire amino acids.

plants, such as the Venus flytrap and the sundew, live in areas where nutrients are lacking, especially nitrogen. By digesting insects and absorbing the products of digestion, they bypass the need to take up nitrates from the soil and produce amino acids themselves. The sticky hairs on the leaves of a sundew (Fig. 23.7C) trap an insect (just as flypaper does) and then the leaves roll up to prevent the insect from escaping as the hairs release digestive enzymes.

▶ **23.7 CHECK YOUR PROGRESS**

1. Contrast the benefits of root nodules with those of mycorrhizae.

THE CHAPTER IN REVIEW

The land environment offers many advantages to plants, including greater availability of light and carbon dioxide for photosynthesis. (Water, even if clear, filters out light, and carbon dioxide concentration and rate of diffusion are lower in a water environment.) The evolution of a transport system with strong walls was critical, however, in order for plants to make full use of the advantages offered by a land environment. Strong walls help elevate plant leaves so that they are better exposed to solar energy and carbon dioxide in the air. A transport system brings water and minerals from the roots to the leaves and takes the products of photosynthesis down to the roots beneath the soil, where their cells depend on an input of organic molecules to remain alive. An efficient transport system also allows roots to penetrate deeply into the soil to absorb water and minerals.

In addition, a transport system allows materials to be distributed to those parts of the plant body that are growing most rapidly. For example, new leaves and flower buds would grow rather slowly if they had to depend on their own rate of photosynthesis. Height in vascular plants, due to the evolution of their transport system, has other benefits aside from elevation of leaves. First, it is adaptive to have reproductive structures located where the wind can better distribute pollen and seeds. Second, once animal pollination came into existence, it was beneficial for flowers to be located where they would be more easily seen by animals.

Another benefit of a transport system is distribution of hormones that regulate plant responses to the environment. Hormones are involved in growth, reproduction, and development of plants. The next chapter explores the topic of plant hormones.

ANALYZE AND EVALUATE

1. Why do plants require inorganic nutrients from the environment? Why do they require the food they make through photosynthesis?
2. In general terms, explain how plants are able to concentrate minerals and thereby make them available to animals.
3. Give reasons why it is beneficial to have xylem and phloem in the same vascular bundle (stems) and veins (leaves).

SUMMARIZE

Xylem and Phloem Structure

23.1 Transport begins in both the roots and the leaves of plants

- **Xylem** moves water and minerals in **xylem sap** from roots to leaves. Its conducting cells are **tracheids** and **vessel elements.** Vessel elements join end to end and form a vessel.
- **Phloem** transports sugars from in photosynthesizing leaves in **phloem sap** to all plant parts. Its conducting cells are **sieve-tube members,** which have **companion cells.** Sieve-tube members join end to end and form a sieve tube.

Water Transport in Xylem

23.2 Water is pulled up in xylem by evaporation from leaves

- The **cohesion-tension model** of xylem transport explains the movement of water upward. (1) In leaves, **transpiration** generates a tension that moves water upward from roots to leaves. (2) In the stem, cohesion due to hydrogen bonding makes the **water column** continuous, while adhesion keeps the water column from falling back. (3) Water enters the xylem at the root, and tension created due to evaporation of water at the leaves pulls water from the roots.

cohesion
adhesion
cell wall

23.3 Guard cells regulate water loss at leaves

- A **stoma** opens when turgor pressure increases in the **guard cells** due to the entrance of K^+ and then water. A stoma closes when turgor pressure decreases in the guard cells due to the exit of K^+ and water.
- The presence of light causes stomata to open. Abundant CO_2 and the presence of abscisic acid (ABA) causes them to close.

Organic Nutrient Transport in Phloem

23.4 Pressure flow explains phloem transport

- Phloem sap contains sugars, hormones, and some amino acids. Phloem transport begins when (1) sugar is actively transported into phloem at a **source,** and water follows by osmosis. (2) The resulting increase in pressure creates a flow, which moves water and sugar to a **sink.** (3) Sugar is actively transported out of sieve tubes at a sink and water exits also. (4) Later some of this water enters xylem and later still it enters phloem again.

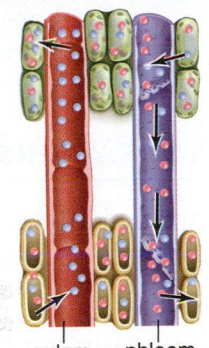

xylem   phloem

Plant Nutrition and Soil

23.5 Certain nutrients are essential to plants

- Plants need only inorganic nutrients such as **minerals** to build all the organic compounds that make up the plant body.
- **Essential nutrients** are either **macronutrients** (needed in large quantity) or **micronutrients** (needed in trace amounts). **Beneficial nutrients** are either required for or enhance plant growth.
- **Hydroponics** can help determine the essential nutrients in growing plants.

23.6 Roots are specialized for the uptake of water and minerals

- **Soil** is composed of mineral particles, organic matter, living organisms, air, and water. **Humus** is soil containing a high amount of decomposing organic matter. A **soil profile** shows **horizons** in a vertical section of soil, from the surface to the rock layer. Plants absorb water and minerals from the A and B horizons. Negatively charged soil retains only positively charged ions. Therefore, soil is apt to be poor in negatively charged nitrate and phosphate.

- The root system absorbs water and minerals. Using the symplast route they enter root hairs and move from cell to cell to xylem. If water and minerals take the apoplast route via cell walls, the **Casparian strip** causes them to enter endodermal cells before reaching xylem. **Aquaporins** (water channel proteins) play a significant role in moving water into xylem.

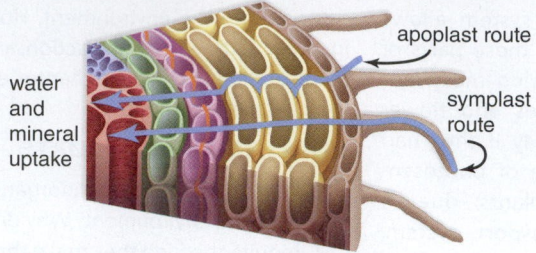

- In mineral uptake by root cells, an ATP-driven proton pump transports H^+ out of the cell; an electrochemical gradient causes K^+ to enter through a channel protein; and negatively charged ions are transported along with H^+ into the cell.

23.7 Adaptations of plants help them acquire nutrients

- Among adaptations are: (1) Nitrogen-fixing bacteria that live in **root nodules** take up atmospheric nitrogen and incorporate it into amino acids that are available to the plant host. The host makes sugar available to the bacteria. (2) **Mycorrhizae** (fungal roots) increase the root surface area and otherwise assist in water and mineral uptake. (3) Some plants obtain nutrients by parasitizing other plants. (4) Other plant leaves are modified to capture and digest insects, providing the plant with nutrients, especially nitrogen-containing amino acids.

TEST YOURSELF

Xylem and Phloem Structure

1. Xylem includes all of the following except
   - a. companion cells.
   - b. vessels.
   - c. tracheids.
   - d. dead cells.

2. **THINKING CONCEPTUALLY** Why would you expect leaves to be involved in the transport of both water and sugar?

Water Transport in Xylem

3. The process responsible for guttation is
   - a. evaporation.
   - b. cohesion.
   - c. root pressure.
   - d. transpiration.

4. What role do cohesion and adhesion play in xylem transport?
   - a. Like transpiration, they create a tension.
   - b. Like root pressure, they create a positive pressure.
   - c. Like sugars, they cause water to enter xylem.
   - d. They create a continuous water column in xylem.
   - e. All of these are correct.

5. What is the significance of the experiment depicted here?

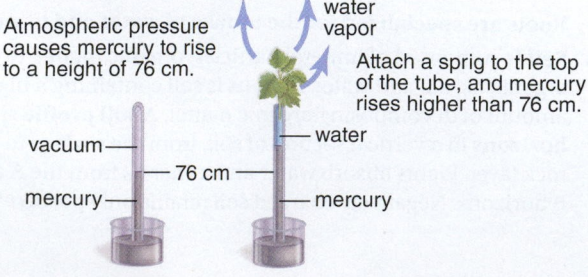

6. An opening in the leaf that allows gas and water exchange is called the
   - a. lenticel.
   - b. hole.
   - c. stoma.
   - d. guard cell.
   - e. accessory cell.

7. What main force drives absorption of water, creates tension, and draws water through the plant?
   - a. adhesion
   - b. cohesion
   - c. tension
   - d. transpiration
   - e. absorption

8. Stomata are usually open
   - a. at night, when the plant requires a supply of oxygen.
   - b. during the day, when the plant requires a supply of carbon dioxide.
   - c. day or night if there is excess water in the soil.
   - d. during the day, when transpiration occurs.
   - e. Both b and d are correct.

9. By which process does phytoremediation work?
   - a. Plants break down harmful chemicals.
   - b. Microbes surrounding plants break down harmful chemicals.
   - c. Plants take up and store harmful chemicals.
   - d. Both a and b are correct.
   - e. a, b, and c are all correct.

10. **THINKING CONCEPTUALLY** Support the hypothesis that the transport of water during the day causes the stomata to open.

Organic Nutrient Transport in Phloem

11. The sugar produced by mature leaves moves into sieve tubes by way of _____, while water follows by _____.
    - a. osmosis, osmosis
    - b. active transport, active transport
    - c. osmosis, active transport
    - d. active transport, osmosis

12. After sucrose enters sieve tubes,
    - a. it is removed by the source.
    - b. water follows passively by osmosis.
    - c. it is driven by active transport to the source, which is usually the roots.
    - d. stomata open so that water flows to the leaves.
    - e. All of these are correct.

13. In contrast to transpiration, the transport of organic nutrients in the phloem
    - a. requires energy input from the plant.
    - b. always flows in one direction.
    - c. results from tension.
    - d. does not require living cells.

14. The pressure-flow model of phloem transport states that
    - a. phloem contents always flow from the leaves to the root.
    - b. phloem contents always flow from the root to the leaves.
    - c. water flow brings sucrose from a source to a sink.
    - d. water pressure creates a flow of water.
    - e. Both c and d are correct.

15. What is the significance of the experiment depicted here?

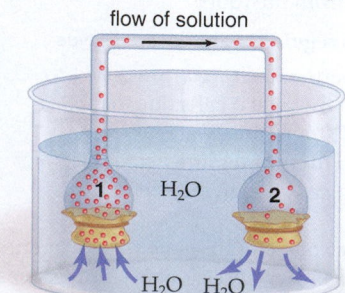

flow of solution

16. **THINKING CONCEPTUALLY** If a plant is water stressed, what do you predict would happen to phloem transport? Support your prediction.

Plant Nutrition and Soil

17. Which of these is not a mineral ion?
    a. NO_3^-
    d. Al^{3+}
    b. Mg^+
    e. All of these are correct.
    c. CO_2

18. Which of these molecules is not a nutrient for plants?
    a. water
    d. nitrogen gas
    b. carbon dioxide gas
    e. None of these are nutrients.
    c. mineral ions

19. A nutrient element is considered essential if
    a. plant growth increases when the concentration of the element is reduced.
    b. plant growth suffers in the absence of the element.
    c. plants can substitute a similar element for the missing element with no ill effects.
    d. the element is a positive ion.

20. Which process is responsible for moving water from the soil into root hairs?
    a. active transport
    c. endocytosis
    b. diffusion only
    d. osmosis and aquaporins

21. The Casparian strip affects
    a. how water and minerals move into the vascular cylinder.
    b. vascular tissue composition.
    c. how soil particles function.
    d. how organic nutrients move into the vascular cylinder.
    e. Both a and d are correct.

22. Plants expend energy in order to take up
    a. carbon dioxide.
    d. Both a and b are correct.
    b. minerals.
    e. a, b, and c are all correct.
    c. water.

23. Which is a component of soil?
    a. mineral particles
    d. air and water
    b. humus
    e. All of these are correct.
    c. organisms

24. Soils rich in which type of soil particle have a high water-holding capacity?
    a. sand
    b. silt
    c. clay
    d. All soil particles hold water equally well.

25. Negatively charged clay particles attract
    a. K^+.
    d. Both a and b are correct.
    b. NO_3^-.
    e. Both a and c are correct.
    c. Ca^+.

26. Plants with mycorrhizae form a mutualistic relationship with
    a. algae.
    d. protozoans.
    b. bacteria.
    e. Both a and b are correct.
    c. fungi.

27. **THINKING CONCEPTUALLY** What evidence aside from Figure 23.7B would support the conclusion that a mycorrhizae association is mutualistic?

GET INVOLVED

1. Based on the data presented in "Competition for Resources Is One Aspect of Biodiversity" on page 460, why would biologists suggest to farmers that polyculture (planting several species at a time) instead of monoculture (planting a single species) would better protect their fields from environmental assaults, such as insect attacks and drought?

2. Using hydroponics, design an experiment to determine if calcium is an essential plant nutrient. State the possible results.

3. After doing the virtual lab "Plant Transpiration," discuss the weather conditions that assist or hinder transpiration.

Virtual Lab
Plant Transpiration

MEDIA STUDY TOOLS

mhhe.com/maderconcepts3

Enhance your study of this chapter with interactive study tools, practice tests, and engaging animations. Also, ask your instructor about the resources available through ConnectPlus, which includes LearnSmart, a personalized adaptive learning program, and a media-rich eBook.

24

Control of Growth Responses in Plants

BEFORE YOU BEGIN

Take a few minutes to recall

All organisms respond to external stimuli (section 1.5)

Diffusion, osmosis, and turgor pressure (section 5.2)

Plasma membrane receptors are involved in cell signaling
("Evolution of the Plasma Membrane" on page 91)

Regulation of gene activity by DNA-binding proteins
(section 12.2)

Plastic Trees

Imagine walking into a field, burying your feet in the ground, and spending the remainder of your life there. Food and water will be provided, but you will have to survive the elements. There will be hot days, frigid nights, severe storms, drought, biting insects, and countless other threats to your survival. This is the environment plants endure every day. Animals can escape from adverse conditions, but plants must acclimate to them. Plants acclimate, in large part, by altering their growth form. This ability of a given plant to change its growth form in response to different environments is called phenotypic plasticity.

In order for a plant to respond to its environment, it must first sense it. Plants do not have the five senses of higher animals, but they are capable of sensing the environment in ways that are meaningful to them. As you will learn in this chapter, a plant can sense touch, day length, light intensity, and the direction of gravity and light. The plant can then adjust its growth form in a way that will enhance its chance of survival. Plants are phenotypically plastic for a number of traits, including anatomical modifications of growth form, changes associated with the seasons or alterations in reproductive timing. For example, the growth form of a pine tree is entirely different when the tree is exposed to a steady and strong wind; a maple tree loses its leaves in preparation for winter; and leaf lettuce bolts and flowers when the weather turns hot and dry. In general, if a plant finds itself growing in a stressful environment, it may allocate energy to reproductive growth at the expense of vegetative growth to increase its probability of producing offspring. A plant growing in a dry, windy area may also reduce evaporative water loss by producing thicker leaves covered by more leaf hairs than it

Normal (right) versus bolting lettuce (above)

Normal (*left*) versus wind-swept Jeffrey Pine (*right*)

would in a sheltered, humid area. In low light conditions, a plant grows tall and spindly to enhance its chances of reaching full sunlight (see Fig. 24.9B). This growth pattern is commonly seen in houseplants that do not receive enough light.

Plants have evolved a number of defense responses to survive attacks by fungi, bacteria, insects, and other animals. Some plants produce biochemical compounds, such as toxins and other antimicrobial chemicals, in response to wounding by fungal or bacterial pathogens. Similarly, in response to herbivore saliva, some plants produce chemical deterrents to prevent the animals from eating any more leaves. (See section 24.11.)

Phenotypic plasticity can even be seen within a plant, especially in trees. At the top of a tree, the "sun leaves" are smaller (to reduce evaporative water loss) and thicker (with more layers of photosynthetic tissue) than the "shade leaves" lower in the tree, which are broad (to intercept low levels of sunlight) and thin (because light cannot penetrate thick leaves). In addition, sun leaves have a thicker cuticle than shade leaves in order to reduce evaporative water loss.

A plant may find itself exposed to a number of different, competing stresses at the same time. It can't exhibit a maximum phenotypic response to all of them at once, so what should it do? The best solution is to be plastic and work out a compromise. Some species are capable of more plasticity than others. With the fast pace of environmental change experienced by plants today, phenotypic plasticity is likely to become an essential survival strategy.

In this chapter, we will see that hormones affect plant growth and movement by changing the rate and direction of cell expansion, differentiation, and division. They also influence seed germination and the growth and development of a plant, including flowers and fruit. We will explore the many roles of plant hormones and some plant responses to environmental stimuli that exhibit their plasticity.

Maple tree in summer

Maple tree in fall

Plant Hormones

Plant hormones, sometimes called phytohormones, are signaling molecules that influence plant growth and development. The action of certain hormones, namely auxins, gibberellins, cytokinins, abscisic acid, and ethylene, are well known and will be discussed. Auxins, gibberellins, and cytokinins are often called stimulatory hormones because they promote growth, while abscisic acid and ethylene are called inhibitory hormones because they promote activities associated with the end of a life cycle. Each hormone affects different aspects of plant responses to stimuli.

24.1 Plant hormones act by utilizing signal transduction pathways

LEARNING OUTCOME

When you complete this section, you should be able to

1. Explain how hormones allow a plant to respond to a stimulus.

Plant hormones are small organic molecules that regulate growth and development at very low concentrations. A hormone is produced and stored in one part of the plant, but it can travel within the vascular system or from cell to cell to another part of the plant. As we shall see, each hormone may cause a variety of responses and may work with other hormones to bring about a specific response suitable to the particular environment.

Being able to respond to environmental stimuli is a beneficial adaptation because it leads to organisms' longevity and ultimately the survival of the species. Flowering plants detect and react to a variety of environmental stimuli. Some examples include light, gravity, carbon dioxide levels, pathogen infection, drought, and touch. Their responses can be short term, as when stomata open and close in response to light levels, or long term, as when plants respond to gravity with the downward growth of the root and the upward growth of the stem.

Although we think of responses in terms of a plant structure change, the mechanism that brings about a response occurs at the cellular level. Research has shown that plant cells respond to stimuli by utilizing signal transduction that begins when a hormone, the molecular "signal" binds to a receptor in the plasma membrane. You first encountered the concept of signal transduction when studying the functions of proteins in the plasma membrane, and "Evolution of the Plasma Membrane" on page 91 explored the concept in further depth.

Figure 24.1 provides a generalized model for a signal transduction pathway, which involves the following:

Reception of Signal In response to an environmental stimulus such as light or gravity, a protein usually located in the plasma membrane is activated by a specific signal. The signal can be one of the hormones you will be studying in this chapter. Plant cells sometimes utilize photoreceptors that have a pigment component sensitive to light. Phototropin located in the plasma membrane is sensitive to blue light, while phytochrome located in the cytoplasm is sensitive to red light. In any case, when a receptor is activated, it begins the transduction process.

Transduction Pathway A change of the signal into a different form. Transduction can involve a series of relay proteins

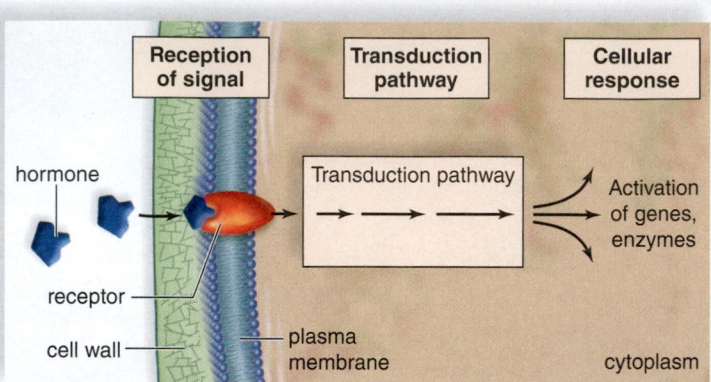

FIGURE 24.1 A typical signal transduction pathway in plants begins when a hormone, the signal, binds to a receptor in the plasma membrane. Only specific cells have receptors for this particular hormone. The binding of the signal activates a transduction pathway consisting of proteins that can bring about a desired cellular response to the presence of a hormone.

that amplify the signal and transform it to one that affects the machinery of the cell. Sometimes the receptor activates a molecule called a second messenger. Ca^{2+} is a well-known second messenger that forms a complex with a protein in a transduction pathway. Or, the end result of transduction can be a protein that enters the nucleus.

Cellular Response Often the Ca^{2+}-protein complex activates a cellular enzyme, and this enzyme proceeds to bring about a structural change in the plant such as a stomata closing or a stem that turns toward the light. A protein that enters the nucleus can have a dramatic effect on plant structure if it binds to a developmental regulatory gene that turns on other genes.

Plant hormones and other types of signals allow a plant to respond to both abiotic and biotic stimuli because they bring about a cellular response leading to an observable macroscopic change.

▶ 24.1 CHECK YOUR PROGRESS

1. List and describe the three components of a signal transduction pathway that lead to a visually observable change in plant structure.

24.2 Auxins promote growth and cell elongation

When you complete this section, you should be able to

1. Understand that auxins are significant plant hormones that affect the growth of the plant.

Auxins are a group of plant hormones that affect many aspects of plant growth and development. Here we will discuss how auxins (1) maintain apical dominance and (2) cause stems to bend toward the light. The most common naturally occurring auxin is indoleacetic acid (IAA), produced in shoot apical meristem and also found in young leaves, in flowers, and in fruits.

Auxins and Apical Dominance Apically produced or applied auxin stimulates terminal bud activity and prevents the growth of axillary buds, a phenomenon called **apical dominance.** When a terminal bud is removed deliberately or accidentally, the nearest axillary buds begin to grow, and the plant branches. Pruning the top (apical meristem) of a plant generally achieves a fuller look. This removes the source of auxin and, therefore, the apical dominance. More branching of the main body of the plant then occurs.

terminal bud removed

branches

Horticulturists apply IAA as a paste to plant cuttings to stimulate vigorous root formation. Auxin production by seeds promotes the growth of fruit. As long as auxin is concentrated in leaves or fruits rather than in the stem, leaves and fruits do not fall off. Therefore, trees can be sprayed with auxin to keep mature fruit from falling to the ground.

Auxins and Bending of Stems The role of auxin in causing stems to bend toward a light source (called **phototropism**) has been studied for quite some time. A **coleoptile** is a protective sheath for the young leaves of the seedling. In 1881, Charles Darwin and his son found that phototropism will not occur if the coleoptile tip of a seedling is removed or covered by a black cap. They concluded that some influence that causes curvature is transmitted from the coleoptile tip to the rest of the shoot.

In a now-famous 1926 experiment, Frits Went cut off the coleoptile tips and placed them on agar (a gelatin-like material). Auxin diffused from the tips into the agar, and then Went divided the agar. He put a small agar block to one side of a tipless coleoptile and found that the shoot curved away from that side. The bending occurred even though the seedlings were not exposed to light. Why? Because (1) the agar block released auxin to one side of the shoot, and (2) only the cells on that side experienced elongation, resulting in curvature of the shoot. The experiment showed that bending occurs because auxin is present on only one side of the stem.

Curvature of shoot occurs beneath the block.

How Auxin Causes Stems to Bend When a stem is exposed to unidirectional light, events occur that cause a stem to bend toward the light (Fig. 24.2). ❶ auxin moves to the shady side, where it activates a proton (H^+) pump. ❷ The resulting acidic conditions loosen the cell wall because hydrogen bonds are broken and cellulose fibrils are weakened by enzymatic action. ❸ The end result of these activities is elongation of the stem on the shady side so that it bends toward the light.

▶ **24.2 CHECK YOUR PROGRESS**

1. Explain how auxin causes stems to bend.

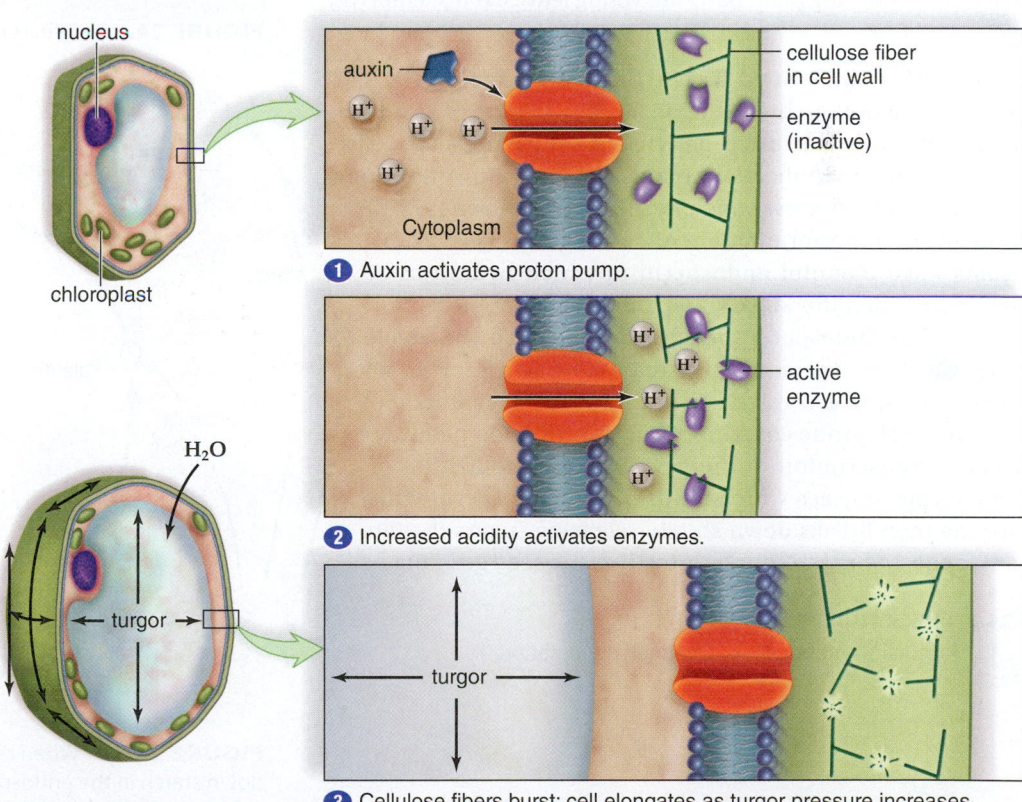

nucleus

auxin

cellulose fiber in cell wall

enzyme (inactive)

Cytoplasm

❶ Auxin activates proton pump.

chloroplast

active enzyme

❷ Increased acidity activates enzymes.

H_2O

turgor

turgor

❸ Cellulose fibers burst; cell elongates as turgor pressure increases.

FIGURE 24.2 A stem elongates on the shady side due to the effects of auxin on its cells.

LEARNING OUTCOME

When you complete this section, you should be able to

1. Understand that gibberellins are significant plant hormones with a known mode of action.

Gibberellins are growth-promoting hormones that (1) cause stems to elongate and (2) seeds to germinate. We know of about 136 gibberellins, and they differ chemically only slightly. The most common of these is gibberellic acid, GA_3 (the subscript distinguishes it from other gibberellins).

Gibberellins and Stem Elongation When gibberellins are applied externally to plants, the most obvious effect is stem elongation. In Figure 24.3A*a,* the plant on the *left* was not treated with gibberellins, while the plant on the *right* was. Gibberellins can cause dwarf plants to grow, cabbage plants to become 2 m tall, and bush beans to become pole beans. In Figure 23.3A*b,* the plant that produced the grapes on the *right* was treated with gibberellins. This treatment causes an increase in the space between the grapes, allowing them to grow larger.

Gibberellins were discovered in 1926 when Ewiti Kurosawa, a Japanese scientist, was investigating a fungal disease of rice plants called "foolish seedling disease." The plants elongated too quickly, causing the stem to weaken and the plant to collapse. Kurosawa found that a fungus infecting the plants was producing an excess of a certain substance. Later, investigators isolated the substance and called it gibberellin, after the name of the fungus, *Gibberella fujikuroi.* It wasn't until 1956 that gibberellic acid was isolated from a flowering plant, rather than from a fungus. Sources of gibberellin in flowering plant parts are young leaves, roots, embryos, seeds, and fruits.

Gibberellins and Seed Germination Commercially, gibberellins are used to break the **dormancy** (a time of low metabolic activity and arrested growth) of seeds and buds. Also, after gibberellin application, plants begin to grow, flowering occurs, or flowers grow larger. Research has shown how GA_3 is involved in the germination of cereal grains (Fig. 24.3B). Grains have plentiful endosperm in which hydrolysis of starch provides the sugars and therefore the energy for growth. The endosperm has a covering called the aleurone. ❶ When the embryo is ready to begin growth, it releases GA_3, which travels to and binds to receptors on cells in the aleurone covering. ❷ A transduction pathway leads to transcription of the gene and ❸ production of the enzyme amylase which diffuses into endosperm. This enzyme then breaks down starch, releasing sugars that are sent to the embryo to spur its growth as the seed germinates.

24.3 CHECK YOUR PROGRESS

1. Explain how gibberellins assist seed germination.

a. Stems elongate, as in plant on *right*.

b. Grape size increases, as in bunch on *right*.

FIGURE 24.3A Effects of gibberellic acid (GA_3).

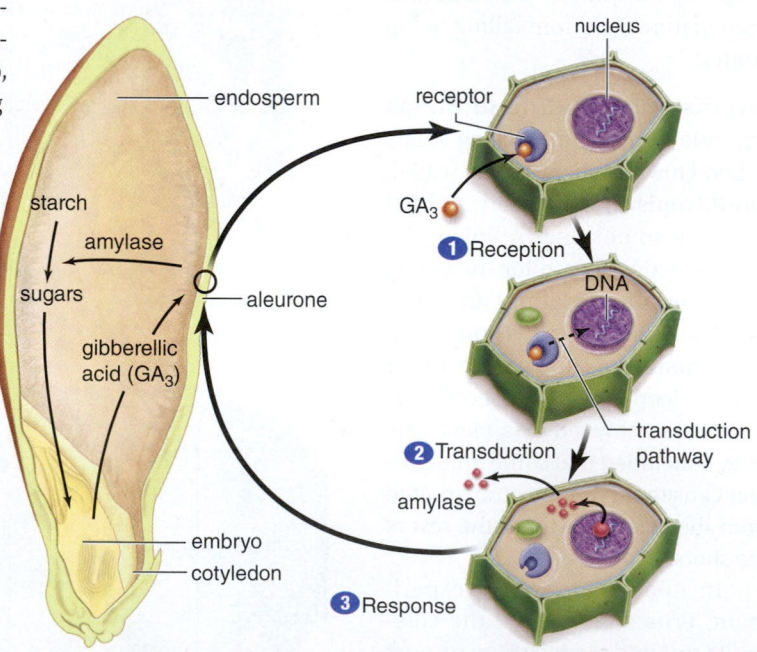

FIGURE 24.3B When cereal grains germinate, amylase breaks down starch in the endosperm. The amylase is produced in aleurone cells in response to the hormone GA_3.

24.4 Cytokinins stimulate cell division and differentiation

LEARNING OUTCOME

When you complete this section, you should be able to

1. Explain that cytokinins are significant plant hormones whose actions can be associated with cell division.

Cytokinins are a class of plant hormones that (1) promote cell division and (2) in combination with auxin bring about differentiation of tissues and (3) prevent senescence. Cytokinins are compounds with a structure resembling adenine, one of the purine bases in DNA and RNA.

Cytokinins and Cell Division The cytokinins were discovered as a result of attempts to grow plant tissue and organs in culture vessels in the 1940s. Researchers found that cell division occurred when coconut milk (a liquid endosperm) and yeast extract were added to the culture medium. Although the effective agent or agents could not be isolated, they were collectively called cytokinins because *cytokinesis* means cell division. A naturally occurring cytokinin was not isolated until 1967. Because it came from the kernels of maize (*Zea mays*), it was called *zeatin*.

Cytokinins have been isolated from various plants, where they occur in the actively dividing tissues of roots and also in seeds and fruits. The cytokinins are produced in root apical meristem and then transported in xylem throughout the plant. A synthetic cytokinin called kinetin also promotes cell division. Cytokinins have been used to prolong the life of flower cuttings, as well as the freshness of vegetables in storage.

Cytokinins and Tissue Differentiation Plant tissue culture is the process of growing a plant from cells or tissues in laboratory glassware, rather than from the germination of seeds. The interactions of hormones are well exemplified by observing how the varying ratios of auxin and cytokinins affect the differentiation of plant tissues in culture. Researchers are well aware that the ratio of auxin to cytokinin and the acidity of the culture medium determine the differentiation of plant tissue (Fig. 24.4). Depending on the proportion of auxin to cytokinin, cultured plant tissue can ❶ remain an undifferentiated mass called a callus or the callus can go on to ❷ produce roots, or ❸ vegetative shoot and leaves, or ❹ floral shoots.

These effects illustrate that each plant hormone rarely acts alone; it is the relative concentrations of hormones and their interactions that produce an effect. Modern researchers studying plant growth responses look for an interplay of hormones. They have reported that chemicals called *oligosaccharins* (fragments of short-chained sugars released from the cell wall) are effective in directing differentiation. They hypothesize that auxin and cytokinins are part of a signal transduction pathway, which leads to the activation of enzymes that release these fragments from the cell wall.

Cytokinins and Senescence When a plant organ, such as a leaf, loses its natural color, it is most likely undergoing an aging process called **senescence.** During senescence, large molecules within the leaf are broken down and transported to other parts of the plant. Senescence need not affect the entire plant at once; for example, as some plants grow taller, they naturally shed their lower leaves, and ripened fruits routinely separate from the parent plant.

Experimenters have shown that if mature bean pods are allowed to remain on a bean plant instead of picking them, the plant undergoes senescence. Why might that be? The parent plant undergoes senescence because bean pods left on a plant for any length of time secrete hormones that change the allocation of nutrients so that the parent plant literally starves to death.

Senescence of leaves, however, can be prevented by applying cytokinins. Not only can cytokinins prevent the death of leaves, but they can also initiate leaf growth. Axillary buds begin to grow despite apical dominance when cytokinins are applied to them.

▶ **24.4 CHECK YOUR PROGRESS**

1. Identify three effects of cytokinins that can be associated with cell division.

❶ Callus only   ❷ Callus produces roots.   ❸ Callus produces vegetative shoots and leaves.   ❹ Callus produces floral shoots.

FIGURE 24.4 The proportion of auxin to cytokinin determines the results when plant tissues are cultured.

LEARNING OUTCOME

When you complete this section, you should be able to

1. Identify reasons why abscisic acid is sometimes called the stress hormone.

Abscisic acid (**ABA**) is produced by any "green tissue" (tissue containing chloroplasts). ABA is also produced in monocot endosperm and roots, where it is derived from carotenoid pigments. Abscisic acid is sometimes called the stress hormone because it (1) initiates and maintains seed and bud dormancy and (2) brings about the closure of stomata.

Researchers once believed that ABA functioned in **abscission,** the dropping of leaves, fruits, and flowers from a plant. But although the external application of ABA promotes abscission, this hormone is no longer believed to function naturally in this process. Instead, the hormone ethylene seems to bring about abscission.

Abscisic Acid and Dormancy Recall that dormancy is a period of low metabolic activity and arrested growth. Dormancy occurs when a plant organ readies itself for adverse conditions by ceasing to grow (even though conditions at the time may be favorable for growth). For example, it is believed that ABA moves from leaves to vegetative buds in the fall, and thereafter these buds are converted to winter buds. A winter bud is covered by thick, hardened scales (Fig. 24.5A). A reduction in the level of ABA and an increase in the level of gibberellins are believed to break seed and bud dormancy. Then seeds germinate, and buds send forth leaves. Corn kernels have begun to germinate on the developing cob in Figure 24.5B, because this maize mutant is deficient in ABA. Abscisic acid is needed to maintain the dormancy of seeds.

FIGURE 24.5B Corn kernels start to germinate on the cob (see arrows) due to low abscisic acid.

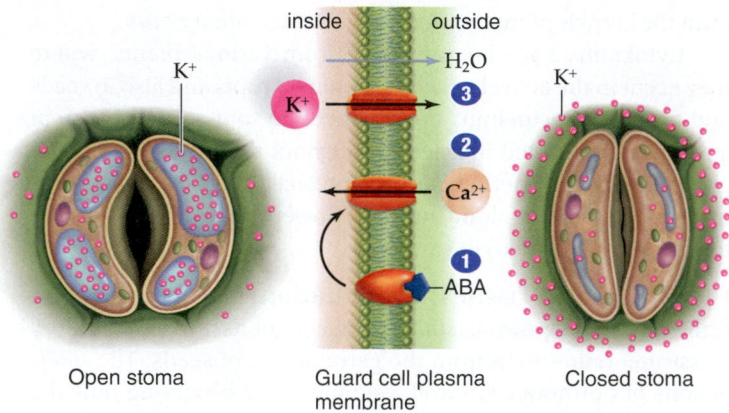

FIGURE 24.5C Abscisic acid promotes closure of stomata.

FIGURE 24.5A
Abscisic acid promotes the formation of winter buds.

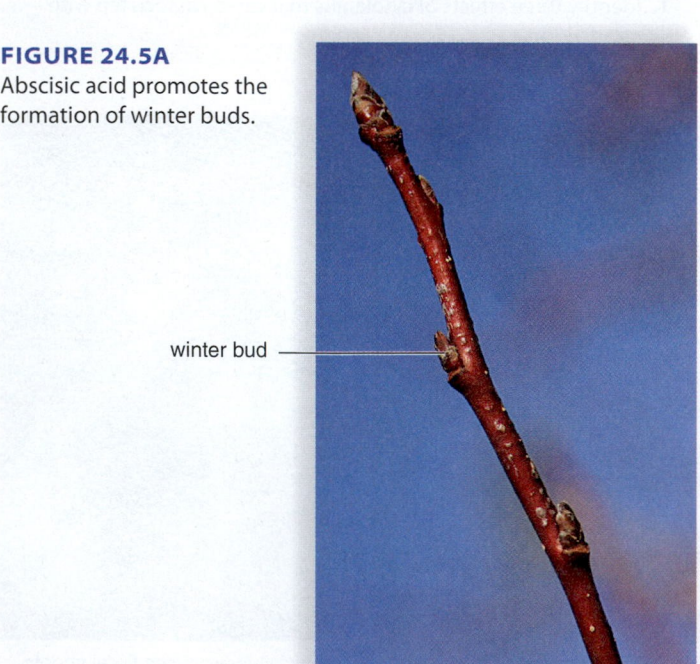

winter bud

Abscisic Acid and Stomata Closure Abscisic acid brings about the closure of stomata when a plant is under water stress, as shown in Figure 24.5C:

1. The stoma is open.
2. When ABA (the first messenger) binds to its receptor in the guard cell plasma membrane, the second messenger (Ca^{2+}) enters. Now, K^+ channels open, and K^+ exit the guard cells. After K^+ exit, so does water.
3. The stoma closes.

Investigators have also found that ABA induces rapid depolymerization of actin filaments and formation of a new type of actin that is randomly oriented throughout the cell. This change in actin organization may be part of the signal transduction pathways involved in stomata closure.

► **24.5 CHECK YOUR PROGRESS**
1. List and explain two reasons abscisic acid is sometimes called the stress hormone.

24.6 Ethylene stimulates the ripening of fruits

Ethylene is a gas formed from the amino acid methionine. This hormone is involved in (1) abscission and (2) the ripening of fruits.

Ethylene and Abscission The presence of auxin, and perhaps gibberellin, probably initiates abscission. But once abscission has begun, ethylene stimulates certain enzymes, such as cellulase, which helps cause leaf, fruit, or flower drop. In Figure 24.6A, a ripe apple, which gives off ethylene, is under the bell jar on the right, but not under the bell jar on the left. As a result, only the holly plant on the right loses it leaves.

Leaf and fruit drop increase due to ethylene production when a plant is stressed by exposure to toxic chemicals, a drastic change in temperature, extreme drought conditions, or increased viral or insect activity. The damage leaves experience when a plant is exposed to ozone has also been attributed to ethylene production.

Ethylene Ripens Fruit In the early 1900s, it was common practice to prepare citrus fruits for market by placing them in a room with a kerosene stove. Only later did researchers realize that an incomplete combustion product of kerosene, namely **ethylene,** ripens fruit. It does so by increasing the activity of enzymes that soften fruits. For example, it stimulates the production of cellulase, which weakens plant cell walls. It also promotes the activity of enzymes that produce the flavor and smell of ripened fruits. And it breaks down chlorophyll, inducing the color changes associated with fruit ripening.

Ethylene moves freely through a plant by diffusion, and because it is a gas, ethylene also moves freely through the air. That is why a barrel of ripening apples can induce ripening of a bunch of bananas some distance away. Ethylene is released at the site of a plant wound due to physical damage or infection (which is why one rotten apple spoils the whole bushel).

The use of ethylene in agriculture is extensive. It is used to hasten the ripening of green fruits such as melons and honeydews, and is also applied to citrus fruits to attain pleasing colors before marketing. Normally, tomatoes ripen on the vine because the plants produce ethylene (Fig. 24.6B). Today, tomato plants can be genetically modified so that they do not produce ethylene. This facilitates shipping because green tomatoes are not subject to as much damage. Once the tomatoes have arrived at their destination, they can be exposed to ethylene so that they ripen (Fig. 24.6C).

Other Effects of Ethylene Ethylene has other varied effects in plants, For example, ethylene, in addition to auxin is involved in axillary bud inhibition. Auxin, transported down from the apical meristem of the stem, stimulates the production of ethylene, and this hormone suppresses axillary bud development. Ethylene also suppresses stem and root elongation, even in the presence of other hormones.

▶ **24.6 CHECK YOUR PROGRESS**

1. Describe the role of ethylene in abscission and fruit ripening.

No abscission   Abscission

FIGURE 24.6A Ethylene promotes abscission.

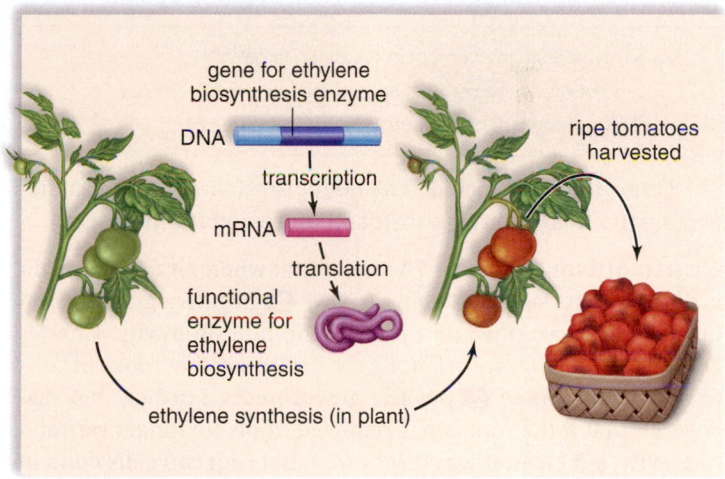

FIGURE 24.6B Wild-type tomatoes ripen on the vine after producing ethylene.

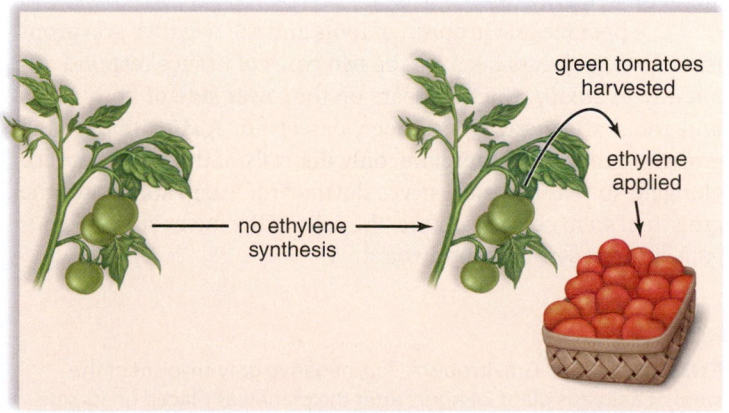

FIGURE 24.6C Tomatoes are genetically modified to produce no ethylene and stay green for shipping.

Plant Responses to the Abiotic Environment

Plant responses often involve growth and special attention to the growth response for particular stimuli. Growth toward or away from a stimulus such as gravity or light is called a tropism. Light is also involved in the circadian rhythms exhibited by plants, but light merely resets a biological clock and its presence is not necessary for the rhythm to be present.

24.7 Tropisms occur when plants respond to stimuli

LEARNING OUTCOME

When you complete this section, you should be able to

1. Identify how tropisms, turgor movements, and sleep movements are responses to particular abiotic stimuli.

Growth toward or away from a stimulus, such as gravity or light, is a **tropism.** Growth toward a stimulus is called a *positive tropism,* and growth away from it is called a *negative tropism.* For example, roots are positively gravitropic because they grow in the direction of gravity. Tropisms are due to differential growth—one side of an organ elongates faster than the other, and the result is a curving toward or away from the stimulus (see section 24.2). A number of tropisms have been observed in plants, the three best-known being **gravitropism, phototropism,** and **thigmotropism:**

> Gravitropism: movement in response to gravity
> Phototropism: movement in response to a light stimulus
> Thigmotropism: movement in response to touch

Other tropisms include chemotropism (chemicals), traumotropism (trauma), skototropism (dark), and aerotropism (oxygen).

Gravitropism Figure 24.7A shows that when an upright plant is placed on its side, the stem displays ❶ negative gravitropism because it grows upward, opposite the pull of gravity. Charles Darwin and his son were among the first to say that roots, in contrast to stems, show ❷ positive gravitropism. Further, they discovered that if the root cap is removed, roots no longer respond to gravity. ❸ Later, it was discovered that root cap cells contain sensors called **statoliths,** which are starch grains located within amyloplasts, a type of plastid. Perhaps gravity causes the amyloplasts to settle to a lower part of the cell, where they come in contact with the endoplasmic reticulum (ER). The ER then releases stored calcium ions (Ca^{2+}), and this leads to the influence of auxin on cell growth.

The positive gravitropism of roots and the negative gravitropism of stems occurs because the two types of tissues respond differently to auxin, which appears on the lower side of both stems and roots after gravity has been perceived. Auxin inhibits the growth of root cells; therefore, only the cells of the upper surface elongate so that the root curves downward. Auxin stimulates the growth of stem cells; therefore, the cells of the lower surface elongate, and the stem curves upward.

FIGURE 24.7A Gravitropism. *Top,* negative gravitropism of the stem of a Coleus plant 24 hours after the plant was placed on its side. *Mid,* positive gravitropism of a root emerging from a corn kernel. *Bottom,* amyloplasts settle in response to gravity in root cap cells.

❶ Negative gravitropism of stem

❷ Positive gravitropism of root

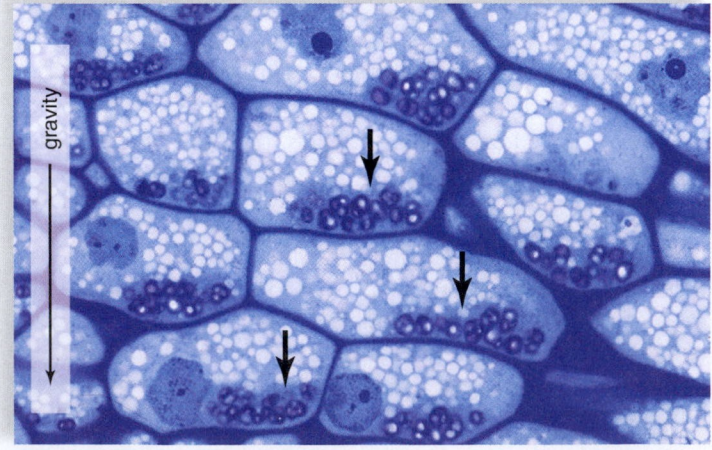

❸ Sedimentation of amyloplasts (arrows)

FIGURE 24.7B Positive phototropism in stems.

FIGURE 24.7D Coiling response of a morning glory plant, *Ipomoea*, illustrates thigmotropism.

Phototropism As discussed in section 24.2, positive phototropism of stems occurs because the cells on the shady side of the stem elongate due to the presence of auxin (Fig. 24.7B). Curving away from light is called negative phototropism. Roots, depending on the species examined, are either insensitive to light or exhibit negative phototropism.

Through the study of mutant plants, we now know that phototropism occurs because plants respond to blue light (Fig. 24.7C). ❶ When blue light is absorbed, the pigment portion of a photoreceptor, called phototropin (phot), undergoes a conformation change. ❷ This change results in the transfer of a phosphate group from ATP (adenosine triphosphate) to a protein portion of the photoreceptor. ❸ The phosphorylated photoreceptor triggers a transduction pathway that, in some unknown way, leads to the migration of auxin to the shady side of a stem. It is clearly adaptive for plants to have a way to increase their photosynthetic efficiency by bending and exposing their leaves to light.

Thigmotropism Unequal growth due to contact with solid objects is called thigmotropism. An example of this response is the coiling of the tendrils or the stems of plants, such as morning glory (Fig. 24.7D). These growth changes occur when a plant part touches a solid object such as a trellis. The cells in contact with the object grow less, while those on the opposite side elongate. Thigmotropism can be quite rapid; a tendril has been observed to encircle an object within 10 minutes. The response also endures; a couple of minutes of touching can bring about a response that lasts for several days. But sometimes the response can be delayed; tendrils touched in the dark may not respond until they are illuminated. ATP, rather than light, can cause the response. Therefore, the need for light may simply be a need for ATP. Also, the hormones auxin and ethylene may be involved because they can induce curvature of tendrils even in the absence of touch.

 **Video** Plant Tactile Responses

▶ **24.7 CHECK YOUR PROGRESS**
  1. Distinguish between positive and negative geotropism, and phototropism and thigmotropism.

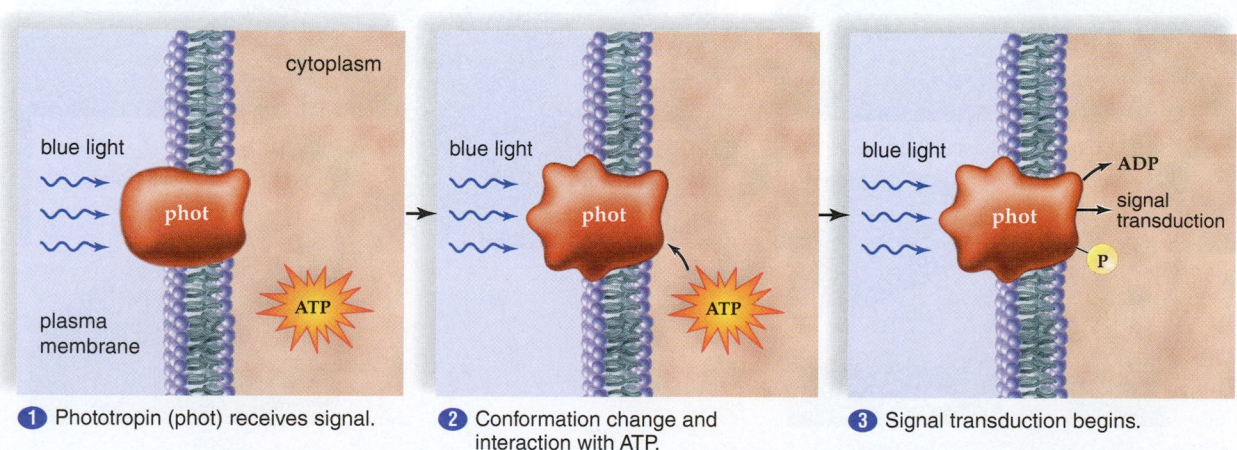

❶ Phototropin (phot) receives signal.

❷ Conformation change and interaction with ATP.

❸ Signal transduction begins.

FIGURE 24.7C In the presence of blue light, a photoreceptor called phototropin (phot) initiates a signal transduction pathway.

LEARNING OUTCOME

When you complete this section, you should be able to

1. Distinguish between turgor movements and sleep movements.

Recall that a plant cell exhibits turgor when it fills with water:

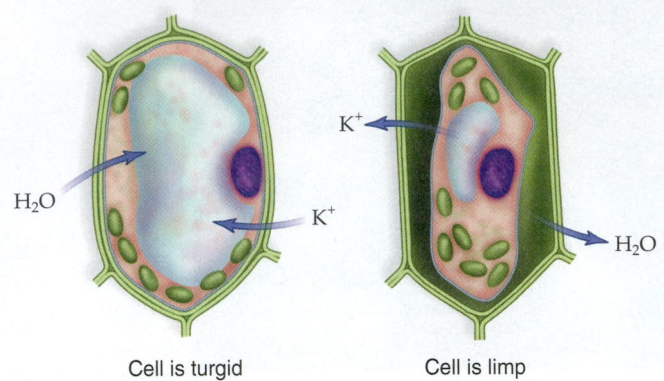

Cell is turgid   Cell is limp

In general, if water exits the many cells of a leaf, the leaf goes limp. Conversely, if water enters a limp leaf and cells exhibit turgor, the leaf moves as it regains its former position. **Turgor movements** are dependent on turgor pressure changes in plant cells. In contrast to tropisms, turgor movements do not involve growth and are not related to the source of the stimulus.

Turgor movements can result from touch, shaking, or thermal stimulation. The sensitive plant, *Mimosa pudica,* has compound leaves, meaning that each leaf contains many leaflets. Touching one leaflet collapses the whole leaf (Fig. 24.8A). *Mimosa*

is remarkable because the progressive response to the stimulus takes only a second or two.

The portion of a plant involved in controlling turgor movement is a thickening called a pulvinus at the base of each leaflet. A leaf folds when the cells in the lower half of the pulvinus, called the motor cells, lose potassium ions (K^+), and then water follows by osmosis. When the pulvinus cells lose turgor, the leaflets of the leaf collapse. An electrical mechanism may cause the response to move from one leaflet to another. The speed of an electrical charge has been measured, and the rate of transmission is about 1 cm per second.

A Venus flytrap closes its trap in less than one second when an insect touches three hairs at the base of the trap, called the trigger hairs. When the trigger hairs are stimulated by the insect, an electrical charge is propagated throughout the lobes of a leaf. Two possible causes of this electrical charge are being studied. (1) Perhaps the cells located near the outer region of the lobes rapidly secrete hydrogen ions into their cell walls, loosening them and allowing the walls to swell rapidly by osmosis; or (2) perhaps the cells in the inner portion of the lobes

Venus flytrap

and the midrib rapidly lose ions, leading to loss of water by osmosis and collapse of these cells. In any case, it appears that turgor movements are involved.

Before

pulvinus   vascular tissue

After

cells retaining turgor   cells losing turgor

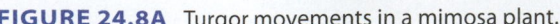

FIGURE 24.8A Turgor movements in a mimosa plant.

Prayer plant (morning)   Prayer plant (night)

Morning glory (morning)   Morning glory (night)

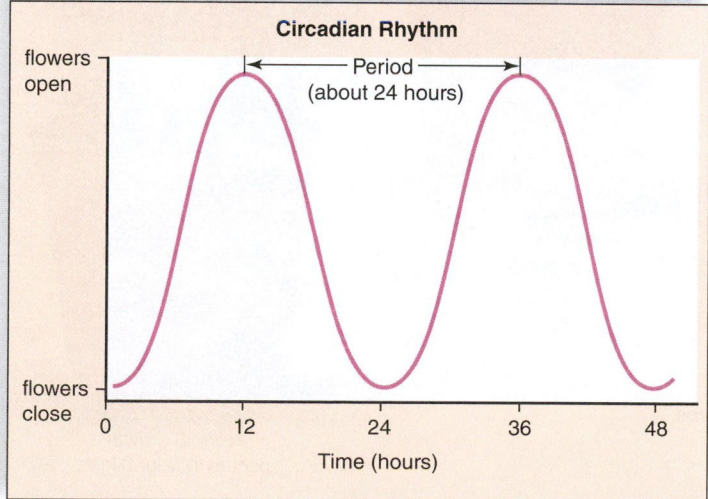

Circadian Rhythm

Period (about 24 hours)

flowers open

flowers close

Time (hours)

FIGURE 24.8B Circadian rhythms occur just about every 24 hours.

Sleep Movements and Circadian Rhythms Leaves that close at night are said to exhibit sleep movements. Activities such as sleep movements that occur regularly in a 24-hour cycle are called **circadian rhythms.** One of the most common examples occurs in a houseplant called the prayer plant (*Maranta leuconeura*) because at night the leaves fold upward into a shape resembling hands at prayer (Fig. 24.8B). This movement is also due to changes in the turgor pressure of motor cells in a pulvinus located at the base of each leaf.

To take a few other examples, morning glory (*Ipomoea leptophylla*) opens its flowers in the early part of the day and closes them at night. In most plants, stomata open in the morning and close at night, and some plants secrete nectar at the same time of the day or night. Figure 24.9B, *bottom*, shows how a circadian rhythm would appear if graphed for a morning glory plant.

To qualify as a circadian rhythm, the activity must (1) occur every 24 hours; (2) take place in the absence of external stimuli, such as in dim light; and (3) be able to be reset if external cues are provided. For example, if you take a transcontinental flight, you will likely suffer jet lag because your body remains attuned to the day-night pattern of its previous environment. But after several days, you will most likely have adjusted and will be able to go to sleep and wake up according to your new time.

Biological Clock The internal mechanism by which a circadian rhythm is maintained in the absence of appropriate environmental stimuli is termed a **biological clock.** If organisms are sheltered from environmental stimuli, their biological clock keeps the circadian rhythms going, but the cycle extends. In prayer plants, for example, the sleep cycle changes to 26 hours when the plant is kept in constant dim light, as opposed to 24 hours when in traditional day/night conditions. Therefore, scientists suggest that biological clocks are synchronized by external stimuli to 24-hour rhythms. The length of daylight compared to the length of darkness, called the photoperiod, sets the clock. Temperature has little or no effect. This is adaptive because the photoperiod indicates seasonal changes better than temperature changes. Spring and fall, in particular, can have both warm and cold days.

Work with *Arabidopsis* and other organisms suggests that the biological clock involves the transcription of a small number of "clock genes." One model proposes that the information-transfer system from DNA to RNA to enzyme to metabolite, with all its feedback controls, is intrinsically cyclical and could be the basis for biological clocks. In *Arabidopsis,* the biological clock involves about 5% of the genome. These genes control sleep movements, the opening and closing of stomata, the discharge of floral fragrances, and the metabolic activities associated with photosynthesis. The biological clock also influences seasonal cycles that depend on day-night lengths, including the regulation of flowering.

While circadian rhythms are outwardly very similar in all species, the clock genes that have been identified are not the same in all species. It would seem, then, that biological clocks have evolved several times to perform similar tasks.

▶ **24.8 CHECK YOUR PROGRESS**

1. Identify evidence that suggests turgor movements close the trap of a Venus flytrap.
2. Identify evidence that suggests sleep movements are due to a biological clock.

LEARNING OUTCOME

When you complete this section, you should be able to

1. Explain the structure and function of phytochrome.

Many physiological changes in flowering plants are related to a seasonal change in day length. Such changes include seed germination, the breaking of bud dormancy, and the onset of senescence. A physiological response prompted by changes in the length of day or night in a 24-hour daily cycle is called **photoperiodism.** In some plants, photoperiodism influences flowering; for example, violets and tulips flower in the spring, and asters and goldenrod flower in the fall. Photoperiodism requires the participation of a biological clock, which can measure time, and it also requires the activity of a plant photoreceptor called phytochrome.

Phytochrome is involved in all sorts of signal transduction pathways that lead to photomorphogenesis (change in structure) and phototropisms as well. A phytochrome system arose in algae and is present in plants and their immediate ancestor. Phytochrome is composed of two identical proteins (Fig. 24.9Aa). Each protein has a larger portion where a light-sensitive region is located. The smaller portion of phytochrome is a kinase that can phosphorylate itself and other proteins to activate them.

Phytochrome can be said to act like a light switch because it can be in the down (inactive) position or in the up (active) position. The active form of phytochrome is designated as P_{fr} because it absorbs far-red light. Far-red light is prevalent in the evening, and it serves to change P_{fr} to P_r, the inactive form of phytochrome. Red light prevalent in daylight activates phytochrome, and it assumes its active conformation known as P_{fr} (Fig. 24.9Ab). P_{fr} can bind to other members of a transduction pathway in the cytoplasm and it can help form a transcription complex in the nucleus when it binds to specific proteins, such as a transcription factor. The complex turns on the expression of master regulatory genes that manage the activity of light-responsive genes that control photomorphogenesis and phototropisms.

Animation
Phytochrome Signaling

Effects of Phytochrome   The phytochrome conversion cycle is now known to control various growth functions in plants. P_{fr} promotes seed germination and inhibits shoot elongation, for example. The presence of P_{fr} indicates to some seeds that sunlight is present and conditions are favorable for germination. This is why some seeds must be only partly covered with soil when planted. Germination of other seeds, such as those of *Arabidopsis,* is inhibited by light, so they must be planted deeper. Following germination, the presence of P_{fr} indicates that sunlight is available, and the seedlings begin to grow normally—the leaves expand and become green and the stem branches. Seedlings that are grown in the dark etiolate—that is, the shoot increases in length, and the leaves remain small (Fig. 24.9B). Only when P_r is converted to P_{fr} does the seedling grow normally.

▶ **24.9 CHECK YOUR PROGRESS**

1. Explain how a plant "knows" that sunlight is present.

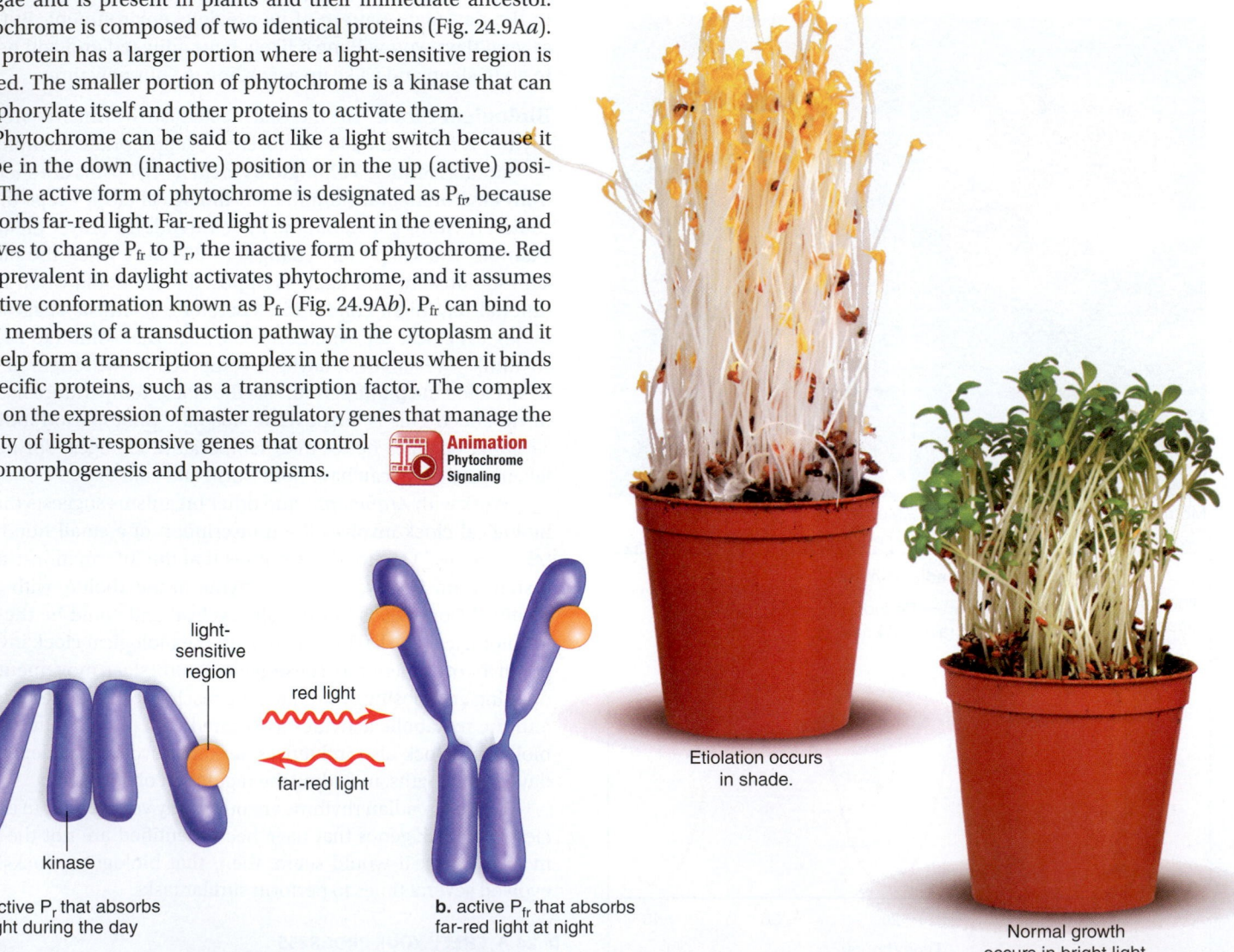

light-sensitive region

red light

far-red light

a. inactive P_r that absorbs red light during the day

b. active P_{fr} that absorbs far-red light at night

kinase

FIGURE 24.9A Phytochrome conversion cycle. Phytochrome responds to light by changing its shape to P_{fr}. P_{fr} interacts with proteins in transduction pathways and can become part of a transcription complex that turns on light-response genes.

Etiolation occurs in shade.

Normal growth occurs in bright light.

FIGURE 24.9B Etiolation allows a plant to seek light while saving energy at the same time. When a plant is in the light it takes on a normal shape as phytochrome becomes P_{fr}.

24.10 Flowering is a response to the photoperiod in some plants

LEARNING OUTCOME

When you complete this section, you should be able to

1. Distinguish between short-day (long-night) plants and long-day (short-night) plants.

The photoperiod, that is the relative length of day and night in a 24-hour period, influences flowering in certain plants. Therefore some plants flower in the fall as days become shorter and others flower in the spring as days become longer.

In the 1920s, when U.S. Department of Agriculture scientists were trying to improve tobacco, they decided to grow plants in a greenhouse, where they could artificially alter the photoperiod. They came to the conclusion that plants can be divided into three groups:

1. **Short-day plants** flower when the day length is shorter than a critical length. (Examples are cocklebur, goldenrod, poinsettia, and chrysanthemum.)

2. **Long-day plants** flower when the day length is longer than a critical length. (Examples are wheat, barley, rose, iris, clover, and spinach.)

3. **Day-neutral plants** are not dependent on day length for flowering. (Examples are tomato and cucumber.)

The criterion for designating plants as short-day or long-day is not an absolute number of hours of light, but a critical number that either must be or cannot be exceeded. Spinach is a long-day plant that has a critical length of 14 hours; ragweed is a short-day plant with the same critical length. Spinach, however, flowers in the summer when the day length increases to 14 hours or more, and ragweed flowers in the fall, when the day length shortens to 14 hours or less. In addition, we now know that some plants require a specific sequence of day lengths in order to flower.

Soon after the three groups of flowering plants were discovered, researchers began to experiment with artificial lengths of light and dark that did not necessarily correspond to a normal 24-hour day. These investigators discovered that the cocklebur, a short-day plant, will *not* flower if a required long dark period is interrupted by a brief flash of white light. (Interrupting the light period with darkness has no effect.) On the other hand, a long-day plant will flower if an overly long dark period is interrupted by a brief flash of white light. They concluded that the length of the dark period, not the length of the light period, controls flowering. Of course, in nature, short days always go with long nights, and vice versa.

To recap, let's consider the illustration on this page:

- Cocklebur is a short-day plant (Fig. 24.10, *left*).
  1 When the night is longer than a critical length, cocklebur flowers. **2** The plant does *not* flower when the night is shorter than the critical length. **3** Cocklebur also does *not* flower if the longer-than-critical-length night is interrupted by a flash of light. This has the same effect as a short night.
- Clover is a long-day plant (Fig. 24.10, *right*). **4** When the night is shorter than a critical length, clover flowers. **5** The plant does *not* flower when the night is longer than a critical length. **6** Clover does flower when a slightly longer-than-critical-length night is interrupted by a flash of light. This has the same effect as a short night.

> ▶ **24.10 CHECK YOUR PROGRESS**
> 1. Explain why the cocklebur plant in Figure 24.10 #3 does not flower.

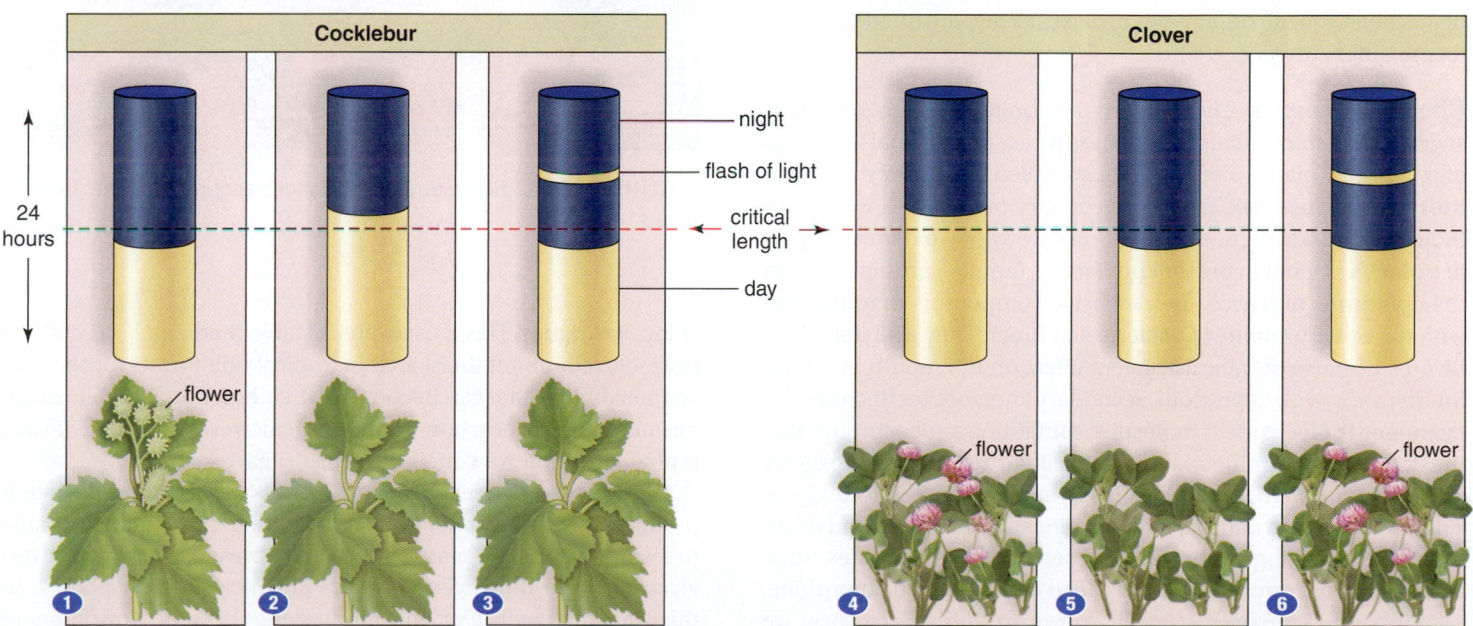

FIGURE 24.10 These experiments support the hypothesis that the length of the dark period and not the length of the day period controls flowering. *Left*: Cocklebur, a long-night plant, will *not* flower if a longer-than-critical-length night is interrupted by a flash of light. *Right*: Clover, a short-night plant, will flower if a longer-than-critical-length night is interrupted by a flash of light.

Plant Responses to the Biotic Environment

Plants have evolved a number of defense responses to survive attacks by fungi, bacteria, insects, and other animals. These responses include physical barriers, chemical toxins, systemic mechanisms, and relationships with animals.

24.11 Plants respond to other organisms in the environment

LEARNING OUTCOME

When you complete this section, you should be able to

1. Describe ways in which plants respond to their biotic environment, including physical barriers, chemical toxins, systemic mechanisms, and relationships with animals.

Plants are always under attack by animals that eat plants, such as insects, and parasites, including fungi (Fig. 24.11A). Fortunately, plants have an arsenal of defense mechanisms.

Physical Barriers A plant's cuticle-covered epidermis and its bark, if present, do a good job of discouraging attackers. The thorns of roses and the spines of cacti are examples of other surface features that deter herbivores. Small hairs called trichomes, which project from the epidermis, may contain poisons. For example, under the slightest pressure, the stiff trichomes of the stinging nettle lose their tips, forming "hypodermic needles" that shoot a stinging chemical into an intruder.

Unfortunately, herbivores have ways around a plant's first line of defense. A fungus can invade a leaf by way of the stomata and set up shop inside a leaf, where it feeds on nutrients meant for the plant. Underground nematodes have sharp mouthparts to break through the epidermis of a root and establish a parasitic relationship, sometimes by way of a single cell, which enlarges and transfers carbohydrates to the animal. Similarly, alfalfa plant bugs have sucking mouth parts and tiny insects called aphids have styletlike mouthparts that allow them to tap into the phloem of a nonwoody stem (see Fig. 23.4A). These examples illustrate why plants need several other types of defenses that are not dependent on the outer surface.

Chemical Toxins The primary metabolites of plants, such as sugars and amino acids, are necessary to the normal workings of a cell, but plants also produce so-called **secondary metabolites** as a defense mechanism. Secondary metabolites were once thought to be waste products, but now we know that they are part of a plant's arsenal to prevent predation. Tannins, present in or on the epidermis of leaves, are defensive compounds that interfere with the outer proteins of bacteria and fungi. They also deter herbivores because of their astringent effect on the mouth and their interference with digestion. Secondary metabolites include the **cyanogenic glycosides** (molecules containing a sugar group) that break down to cyanide and inhibit cellular respiration. Foxglove (*Digitalis purpurea*) produces deadly cardiac and steroid glycosides, which cause nausea, hallucinations, convulsions, and death in animals that ingest them. Some secondary metabolites, such as bitter nitrogenous substances called **alkaloids** (e.g., morphine, nicotine, and caffeine), are well known to humans because we use them for our own purposes. The seedlings of coffee plants contain caffeine at a concentration high enough to kill insects and fungi by blocking DNA and RNA synthesis. The application

Alfalfa plant bug

Fungus infection

Monarch caterpillar and butterfly

FIGURE 24.11A Some organisms directly acquire nutrients from plants.

"Eloy Rodriguez Has Discovered Many Medicinal Plants" on page 490 tells about the search in tropical rain forests for secondary metabolites that can become medicines for humans. Taxol, an unsaturated hydrocarbon from the Pacific yew (*Taxus brevifolia*), is now a well-known cancer-fighting drug.

Predators can overcome the defenses of a plant even when it produces secondary metabolites. Monarch caterpillars are able to feed on milkweed plants, despite the presence of a poisonous glycoside, and they even store the chemical in their bodies. In this way, the caterpillar and the butterfly become poisonous to their own predators (Fig. 24.11A). Birds that become sick after eating a monarch butterfly know to leave them alone thereafter.

Video Cranberries Versus Bacteria

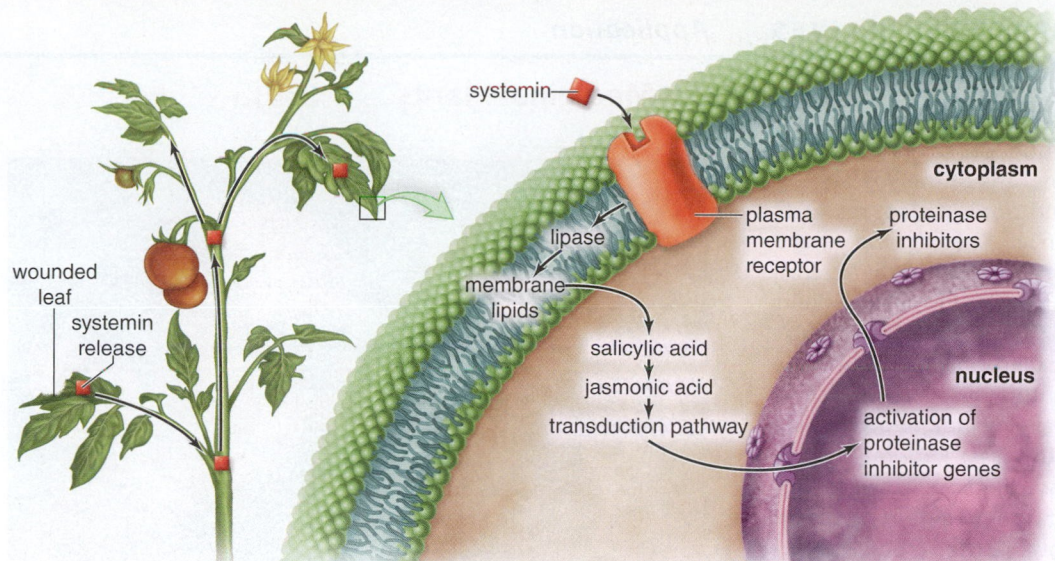

FIGURE 24.11B In response to a wound (*far left*) a leaf releases systemin, which binds to plasma membrane receptors. Activation of a transduction pathway leads to the production of proteinase inhibitors. Proteinase inhibitors destroy the digestive enzymes of insects feeding on the leaves.

Systemic Mechanisms Wound responses illustrate that plants can use signal transduction pathways to produce chemical defenses only when they are needed. After a leaf is chewed or injured, a plant produces proteinase inhibitors, chemicals that inhibit the digestive enzymes of a predator feeding on them. The proteinase inhibitors are produced throughout the plant, not just at the wound site. The growth regulator that brings about this effect is a small peptide called **systemin** (Fig. 24.11B). Systemin is produced in the wound area in response to the predator's saliva, but then it travels between cells to reach phloem, which distributes it about the plant. When systemin binds to its receptors, cells produce proteinase inhibitors. A chemical called jasmonic acid, and also possibly a chemical called salicylic acid, are part of a transduction pathway in these cells. Salicylic acid (the chemical in aspirin) has been known since the 1930s to bring about a phenomenon called systemic acquired resistance (SAR), the activation of genes that lead to the production of antiherbivore chemicals.

On occasion, plants produce a specific gene product that binds (like a key fits a lock) to a viral, bacterial, or fungal gene product made within the cell. This combination offers a way for the plant to "recognize" a particular pathogen. A transduction pathway now ensues, and the final result is a **hypersensitive response** (**HR**) that seals off the infected area and also initiates the wound response just discussed.

Some defenses of plants are called indirect because they do not kill or discourage an herbivore outright. For example, female butterflies are less likely to lay their eggs on plants that already have butterfly eggs. So, because the leaves of some passion flowers (genus *Passiflora*) display physical structures resembling the yellow eggs of *Heliconius* butterflies, these butterflies do not lay eggs on this plant. Other plants produce hormones that prevent caterpillars from metamorphosing into adults and laying more eggs.

Certain plants attract the natural enemies of caterpillars feeding on them. They produce volatile molecules that diffuse into the air and advertise that food is available for a specific carnivore (an animal that eats other animals). For example, lima beans produce volatiles that attract carnivorous mites only when the plants are being damaged by a spider mite. Corn and cotton plants release volatiles that attract wasps, which then inject their eggs into caterpillars munching on their leaves. The eggs develop into larvae that eat the caterpillars, not the leaves. The combined effect of a wide range of volatiles, some that attract predators of plant pests and some that simply prevent egg laying, can result in as much as a 90% reduction in the number of viable eggs on leaves.

Relationships with Animals Mutualism is a relationship between two species in which both species benefit. As evidence that a mutualistic relationship can help protect a plant from predators, consider the bullhorn acacia tree, which provides a home for ants of the species *Pseudomyrmex ferruginea*. Unlike other acacias, this species has swollen thorns with a hollow interior where ant larvae can grow and develop. In addition to housing the ants, acacias provide them with food. The ants feed from nectaries at the base of leaves and eat fat- and protein-containing nodules called Beltian bodies, which are found at the tips of the leaves. In return, the ants constantly protect the plant by attacking and stinging any would-be herbivores because, unlike other ants, they are active 24 hours a day. Indeed, when the ants on experimental trees were removed, the acacia trees died.

Mutualism between ants and a plant

Video
Thorn Tree Ants

Hypersensitive response to fungus invasion

▶ **24.11 CHECK YOUR PROGRESS**
1. Describe several methods that flowering plants use to protect themselves from insect predators.

24A Eloy Rodriguez Has Discovered Many Medicinal Plants

Eloy Rodriguez (Fig. 24A), a Mexican-American biochemist and biology professor, typically spends 7 months of the year at Cornell University teaching biodiversity and tropical plant research, plus doing chemical research in his own lab. Then, for the other 5 months he does field research in the rain forests of South America and the Caribbean and the deserts of Africa. He involves students in all his activities. Rodriguez has spent 25 years traveling through the jungles and deserts to learn about medicinal plants used by native healers. A leader of twentieth-century American ethnobotany (the study of plants used traditionally by indigenous people for food, medicine, shelter, and other purposes), Rodriguez is one of the first modern scientists to extract medicinal compounds from jojoba and candelilla in the laboratory. Yet, he reminds his students that these two plants were probably used medicinally by ancient desert dwellers as well.

Rodriguez points out that, without the participation of native peoples, Americans don't know where to begin to look for medicines in a tropical rain forest that contains 5,000 plant species. Indigenous people can tell us which plants contain potential medicines because their ancestors, for many generations, have been using these plants to heal diseases. He is concerned that only a small percentage of plants recognized as medicines have been studied by western pharmacologists. Of the entire 250,000 flowering plant species, only about 3–5% have been investigated for medicinal purposes, according to Rodriguez.

Rodriguez is working vigorously to save endangered plants, while filling the gap of knowledge between age-old native use of plants and their scientific investigation in modern labs. As he told a *Wildlife Conservation* journalist in 1991, "We just scratched the surface to discover the first drug against malaria (quinine), and the first drug against the cough (codeine), and the first drug against cancer (vincristine)." Quinine was discovered in 1640 when Spanish colonists noticed Peruvians using the bark of the cinchona tree to treat malaria. Subsequently, quinine was isolated in 1820 and synthesized in 1944. Codeine is one of 10 alkaloids produced by the immature seed capsule of the opium poppy; morphine is another. Eli Lilly and Co. introduced the vinca alkaloids, a class of anticancer drugs, in the 1960s. These chemicals have all come from the Madagascar periwinkle.

By combining modern science with the age-old observations of indigenous people around the world, Rodriguez has learned that creosote (*Larrea tridentata*), a plant of the Sonoran desert that natives call "hedionda" or "bad little smeller," contains over 1,000 potential drugs. Native Americans have used creosote for generations to treat colds, chest infections, intestinal problems, menstrual pain, dandruff, toothaches, and other ills. Referring to its chemical properties, Rodriguez calls creosote a "botanical superstar."

Other potential superstars come from Africa. A colleague of Rodriguez, Harvard University anthropologist Richard Wrangham, discovered that sick chimps in Tanzania would chew on the pith of *Vernonia* plants. Subsequently, it was found that *Vernonia* pith contains chemicals with antiparasitic activity against microorganisms that infect both chimps and humans. The chemicals suppress the movement and egg-laying abilities of the parasitic worm *Schistosoma japonicum*. Rodriguez is delighted that one of

FIGURE 24A Eloy Rodriguez tells us that other animals use plants to cure their illnesses and that we can learn to do the same.

the drugs discovered in studying the apes turned out to be an effective drug in humans. Rodriguez and Wrangham coined the word "zoopharmacognosy" to refer to animals' deliberate use of medicinal plants to treat their illnesses. As Rodriguez explained in *National Wildlife*:

> Wild apes five or six million years ago were already using plants. . . . And as the human line evolved, we obviously learned from animals by observing them. It gives us a peek into how we [first started] selecting medicinal plants. (1994. *National Wildlife* 32 [1]:46.)

Rodriguez's major contributions are unique among biologists because he discovers medicinal plants in the field by interacting with native peoples and by observing the practices of apes. He follows this up in his lab by determining which cellular chemicals are responsible for the medicinal effects of the plants and he makes this knowledge available to pharmacology companies.

CONSIDER THESE QUESTIONS

1. Discuss the various ways humans are dependent on plants, and include the way that is discussed in this application.
2. Some chemotherapeutic drugs for cancer are derived from chemicals produced by a plant to deter insect predators. Explain.

 Explore the concepts through a variety of multimedia assets and question types.

www.mcgrawhillconnect.com

THE CHAPTER IN REVIEW

Behavior in plants can be understood in terms of three different levels of organization. On the population level, plant responses that promote survival and reproductive success have evolved through natural selection. At the organism level, hormones coordinate the growth and development of plant parts. And at the cellular level, hormones influence cellular metabolism.

We can illustrate these three perspectives by answering the question, Why do plants bend toward the light? At the population level, plants that bend toward the light will be able to produce more organic food and will have more offspring. At the organism level, light can cause the movement of auxin in certain plant parts, and when auxin moves from the lit side of a stem to the shady side, elongation occurs; thereafter,

the plant bends toward the light. At the cellular level, after auxin is received by a plant cell, cellular activities cause its walls to expand.

The response of both an organism and a cell is dependent on these three steps: (1) reception of signal, (2) transduction of the signal into a different form, and (3) a cellular response. In this chapter, we concentrate on cellular reception and response to a signal. For example, auxin is received by proton pumps that then acidify the cytoplasm. The response is a weakened cell wall that allows the stem to elongate and bend toward the light.

In Chapter 25, we stress the organism level of plants by discussing how they reproduce on land. Certainly we know that the manner in which flowering plants reproduce is an adaptation to the land environment. But we will digress

somewhat to consider the steps that permit plants to reproduce asexually. Sexual reproduction involves seed formation and embryo development. Asexual reproduction of plants in the lab has become especially important because it permits the introduction of improved traits by means of genetic modification.

ANALYZE AND EVALUATE

1. Give a population, organism, and cellular answer to this question: Why do stomata open in the morning and close at night?
2. In general terms, apply the three parts of the signal transduction pathway to the role phytochrome plays in flowering.
3. Give examples to show that each of the hormones studied either affects the growth of a plant positively or negatively.

SUMMARIZE

Plant Hormones

24.1 Plant hormones act by utilizing signal transduction pathways

- **Plant hormones** are small organic molecules that are produced in one part of a plant and travel to other parts, where they affect plant growth and development.
- A signal transduction pathway consists of reception of signal, transduction pathway, and cellular response:

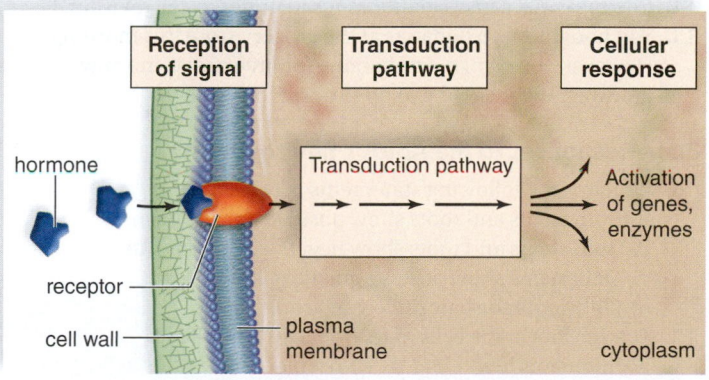

24.2 Auxins promote growth and cell elongation

- IAA (indoleacetic acid) is a natural **auxin.** Auxin encourages the growth of the terminal bud and accounts for **apical dominance** over the growth of axillary buds. The **coleoptile** (a seedling sheath) produces auxin that moves to the shady side of a plant where cells elongate, causing the stem to curve toward the light (called phototropism). Activation of a proton (H⁺) pump leads to elongation of cells on the shady side.

24.3 Gibberellins control stem elongation

- **Gibberellins** are growth-promoting hormones that cause stems to elongate and break the **dormancy** of seeds and buds and bring about seed germination. Experiments with cereal grains show that during germination, reception of GA_3 leads to production of amylase, an enzyme that breaks down starch.

24.4 Cytokinins stimulate cell division and differentiation

- **Cytokinins** promote cell division and tissue differentiation and prevent **senescence. Plant tissue culture** shows that hormone concentrations and their interactions determine tissue differentiation.

24.5 Abscisic acid suppresses growth of buds and closes stomata

- **ABA** (**abscisic acid**) promotes seed dormancy and formation of winter buds, and closes stomata when a plant is water stressed.

24.6 Ethylene stimulates the ripening of fruits

- **Ethylene** causes **abscission** (shedding of leaves, flowers, fruits), ripens fruit, and also suppresses stem and root elongation.

Plant Responses to the Abiotic Environment

24.7 Tropisms occur when plants respond to stimuli

- A **tropism** is plant growth toward or away from a unidirectional stimulus, such as light or gravity:

Gravitropism: movement in response to gravity
Phototropism: movement in response to a light stimulus
Thigmotropism: movement in response to touch

- Auxin is involved in negative **gravitropism** displayed by a stem; and positive gravitropism displayed by roots possibly due to **statoliths**. In **phototropism,** auxin causes stems to curve toward blue light received by a phototropin that initiates a signal transduction pathway. In **thigmotropism,** unequal growth (i.e., coiling of tendrils) results from contact with a solid object. Auxin and ethylene may be involved.

24.8 Turgor and sleep movements are complex responses

- **Turgor movements** (touch, shaking, thermal stimulation) depend on turgor changes. A **circadian rhythm** consists of periodic fluctuations corresponding to a 24-hour cycle. A **biological clock** is an internal mechanism that maintains a circadian rhythm in the absence of stimuli but can be reset by stimuli such as light/dark sequences.

24.9 Response to the photoperiod requires phytochrome

- **Photoperiodism** refers to a physiological response to changes in the length of day or night. **Phytochrome** is a photoreceptor that responds to daylight. P_r absorbs red light; P_{fr} absorbs far-red light. P_{fr} is involved in signal transduction pathways that lead to activation of light-response genes. Light-response genes are responsible for various responses of plants to the photoperiod.

24.10 Flowering is a response to the photoperiod in some plants

- **Short-day plants** flower when day length is shorter than a critical length. **Long-day plants** flower when day length is longer than a critical length. **Day-neutral plants** are not dependent on day length for flowering.

Plant Responses to the Biotic Environment

24.11 Plants respond to other organisms in the environment

- Plants' first line of defense includes a cuticle-covered epidermis, thorns, spines, and trichomes. Chemical defenses are **secondary metabolites** produced by plants that help prevent predation (e.g., tannins, **alkaloids, cyanogenic glycosides**).
- Wound responses involve **systemin** and a signal transduction pathway that leads to activation of genes for proteinase inhibitors. A **hypersensitive response** (**HR**) seals off the area infected by a virus, bacterium, or fungus. Indirect responses include the production of volatile molecules that attract carnivores to kill off herbivores or discourage egg laying on leaves.
- Mutualistic relationships form between certain plants and animals (e.g., the bullhorn acacia tree and ants).

TEST YOURSELF

Plant Hormones

1. During which step of the signal transduction pathway is a second messenger released into the cytoplasm?
   a. reception
   b. response
   c. transduction
2. Which of the following plant hormones causes apical dominance?
   a. auxin
   b. gibberellins
   c. cytokinins
   d. abscisic acid
   e. ethylene

3. Internode elongation is stimulated by
   a. abscisic acid.
   b. ethylene.
   c. cytokinin.
   d. gibberellin.
   e. auxin.
4. **THINKING CONCEPTUALLY** In experiment A, application of GA_3, a gibberellin, causes a dwarf plant to grow as expected. In experiment B, application of GA_3 does not cause a different dwarf plant to grow. Which plant most likely is unable to produce gibberellin on its own, and which one might have a defective receptor for gibberellin? Explain your answer.
5. _____ always promotes cell division.
   a. Auxin
   b. Phytochrome
   c. Cytokinin
   d. None of these are correct.
6. In the absence of abscisic acid, plants may have difficulty
   a. forming winter buds.
   b. closing the stomata.
   c. Both a and b are correct.
   d. Neither a nor b is correct.
7. Ethylene
   a. is a gas.
   b. causes fruit to ripen.
   c. is produced by the incomplete combustion of fuels such as kerosene.
   d. All of these are correct.
8. Which of the following plant hormones is responsible for a plant losing its leaves?
   a. auxin
   b. gibberellins
   c. cytokinins
   d. abscisic acid
   e. ethylene
9. Which is not a plant hormone?
   a. auxin
   b. cytokinin
   c. gibberellin
   d. All of these are plant hormones.

For questions 10–14, match each statement with a hormone in the key.

KEY:
   a. auxin
   b. gibberellin
   c. cytokinin
   d. ethylene
   e. abscisic acid

10. One rotten apple can spoil the barrel.
11. Cabbage plants bolt (grow tall).
12. Stomata close when a plant is water-stressed.
13. Plants point toward the sun.
14. Coconut milk causes plant tissues to undergo cell division.
15. You bought green bananas at the grocery store this morning. However, you want a ripe banana for breakfast tomorrow morning. What could you do to accomplish this?

Plant Responses to the Abiotic Environment

16. Which of the following statements is correct?
   a. Both stems and roots show positive gravitropism.
   b. Both stems and roots show negative gravitropism.
   c. Only stems show positive gravitropism.
   d. Only roots show positive gravitropism.
17. The sensors in the cells of the root cap are called
   a. mitochondria.
   b. central vacuoles.
   c. statoliths.
   d. chloroplasts.
   e. intermediate filaments.

18. A student places 25 pea seeds in a large pot and allows the seeds to germinate in total darkness. Which of the following growth or movement activities might the seedlings exhibit?
    a. gravitropism, as the roots grow down and the shoots grow up
    b. phototropism, as the shoots bend toward the light
    c. thigmotropism, as the tendrils coil around other seedlings
    d. Both a and c are correct.
19. Circadian rhythms
    a. require a biological clock.
    b. do not exist in plants.
    c. are involved in the tropisms.
    d. are involved in sleep movements.
    e. Both a and d are correct.
20. Label this diagram:

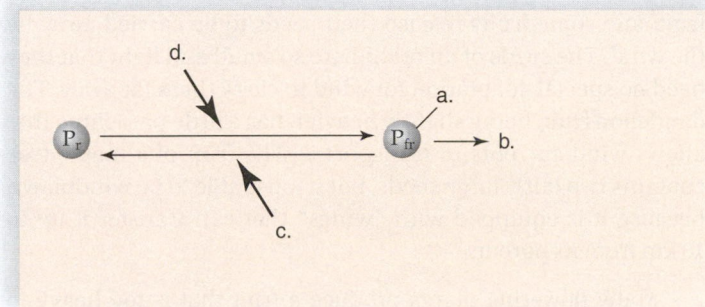

21. Plants that flower in response to long nights are
    a. day-neutral plants.   c. short-day plants.
    b. long-day plants.   d. impossible.
22. Short-day plants
    a. are the same as long-day plants.
    b. are apt to flower in the fall.
    c. do not have a critical photoperiod.
    d. will not flower if a short day is interrupted by bright light.
    e. All of these are correct.
23. A plant requiring a dark period of at least 14 hours will
    a. flower if a 14-hour night is interrupted by a flash of light.
    b. not flower if a 14-hour night is interrupted by a flash of light.
    c. not flower if the days are 14 hours long.
    d. not flower if the nights are longer than 14 hours.
    e. Both b and c are correct.
24. Phytochrome plays a role in
    a. flowering.   c. seed germination.
    b. stem growth.   d. All of these are correct.
25. Phytochrome
    a. is a plant pigment.
    b. is present as P_{fr} during the day.
    c. activates DNA-binding proteins.
    d. is a photoreceptor.
    e. All of these are correct.

26. **THINKING CONCEPTUALLY** Knowing that enzymes are necessary to the production of hormones, create a scenario to show that the activation of phytochrome can possibly lead to the production of hormones.

Plant Responses to the Biotic Environment

27. Primary metabolites are needed for _____, while secondary metabolites are produced for _____.
    a. growth, signal transduction
    b. normal cell functioning, defense
    c. defense, growth
    d. signal transduction, normal cell functioning
28. Which of the following is a plant secondary metabolite used by humans to treat disease?
    a. morphine   d. penicillin
    b. codeine   e. All but d are correct.
    c. quinine
29. See section 28.2 to hypothesize how volatiles released by plants could signal carnivores the availability of food.

GET INVOLVED

1. Based on the data from sections 24.5 and 24.8, you hypothesize that abscisic acid (ABA) is responsible for petiole turgor pressure changes that permit plants to track the sun as shown on right. What observations could you make to support your hypothesis?
2. You formulate the hypothesis that the negative gravitropic response of stems is greater than the positive phototropism of stems. How would you test your hypothesis?

MEDIA STUDY TOOLS

mhhe.com/maderconcepts3

Enhance your study of this chapter with interactive study tools, practice tests, and engaging animations. Also, ask your instructor about the resources available through ConnectPlus, which includes LearnSmart, a personalized adaptive learning program, and a media-rich eBook.

25

Reproduction in Plants

BEFORE YOU BEGIN

Take a few minutes to recall

With a Little Help

There are two times in the life cycle of a flowering plant when it might need a little help. The first is the time of pollination. How can a plant get its pollen from the male part of one flower to the female part of another flower? Some plants, such as the oak, rely on the wind. But others, such as roses, attract pollinators by the color of their petals and their sweet smell, which advertise the availability of nectar (see section 25.2). The second time a flowering plant might need a little help is with the dispersal of its seeds. During dispersal, seeds are carried away from the parent plant to a site where they might have better growing conditions. In addition to achieving more room to grow, taking up residence in a new place may ensure that the species will survive should a disaster, such as a fire, devastate the plants in other locations.

In flowering plants, seeds are enclosed within a protective fruit, and some fruits release their seeds to be carried away by the wind. The seeds of an orchid are so small and light that they need no special adaptation for wind to carry them far away. The dandelion fruit, being slightly heavier, has a little parachute that allows wind currents to transport it. The fruit of a maple tree contains two fairly large seeds, but it too is able to be windblown because it is equipped with "wings" that can transfer it up to 10 km from its parent.

Many flowering plants produce a fruit that is too heavy to be carried by wind. These plants need an alternative dispersal

method, and again animals are willing to oblige. A fleshy fruit such as a berry encloses small seeds, while the fleshy part of, say, a peach encloses a single large seed, sometimes called a stone. Berries, peaches, apples, and cherries tend to be green and hidden from view by green leaves while they are developing. But when they become ripe, they take on an attractive color and scent. These changes entice an animal to eat the fruit, and later the seeds pass through the animal's digestive tract and are deposited some distance from the parent plant. Birds—and even animals as large as bears—enjoy eating berries such as blueberries, huckleberries, and rose hips. Raccoons eat fleshy fruits with large pits, and deer are known to eat crab apples.

Animals also eat nuts, in which a hard covering encloses the seed contents. When the seed is eaten, so is the embryo of the next generation. However, squirrels, as well as birds such as blue jays, store acorns and other nuts during the autumn to tide them over the long winter. Sometimes they forget where to go looking for the nuts, and in the meantime, the seeds enclosed by fruit have been dispersed.

The dry fruits of violets and trillium, among other plants, split open to release seeds that have a cap rich in oil and vitamins. The caps are prized by ants, which set about dragging the seeds back to the nest. Once there, the ants eat only the caps, and the rest of the seed stays intact until it germinates. Up to one-third of all the herbs in a deciduous forest of the United States are dispersed by ants!

Animals also disperse seeds when the hooks and spines of clover, burdock, and cocklebur attach to their fur and are carried some distance away.

This chapter explores in some detail how the plant sexual life cycle is modified to permit the production of seeds enclosed by fruits, an evolutionary event that helps explain the success of flowering plants in a terrestrial environment. We will also discuss the development of the embryo within the seed and the structure of fruits. Finally, we will see that asexual reproduction permits humans to clone plants and their tissues for commercial purposes.

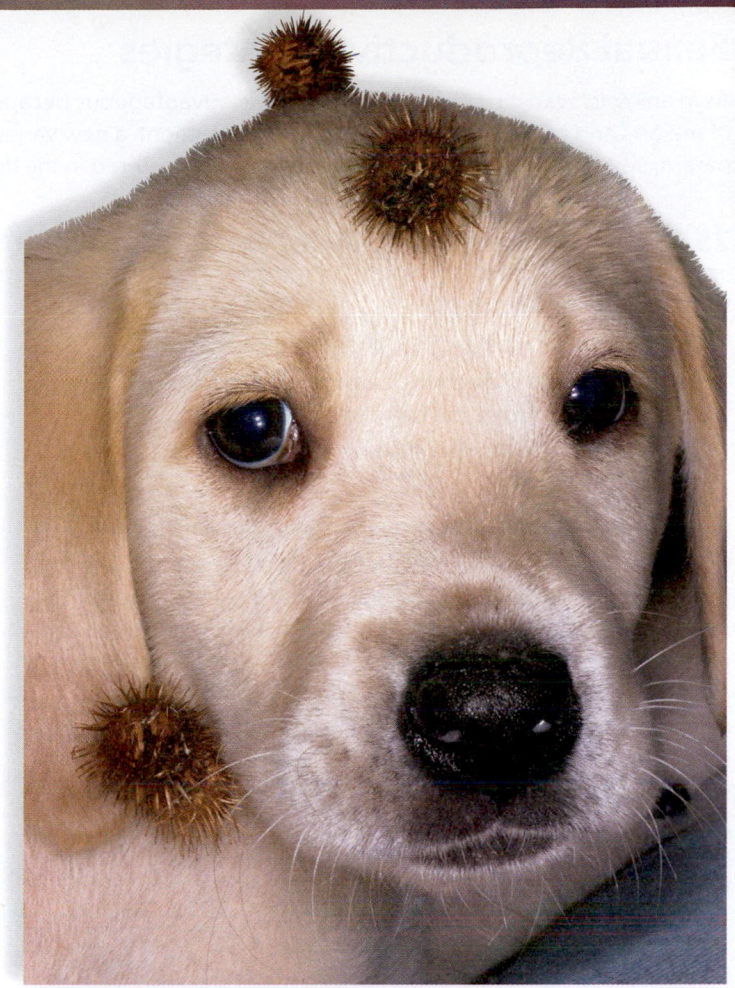

Sexual Reproductive Strategies

As in animals, sexual reproduction in plants is advantageous because it can generate variety among the offspring due to the processes of meiosis and fertilization. In a changing environment, a new variety may be better adapted for survival and reproduction than either parent. Sexual reproduction in flowering plants is centered in the flower, which produces seeds enclosed by a fruit.

25.1 Plants have a sexual life cycle called alternation of generations

LEARNING OUTCOMES

When you complete this section, you should be able to

1. Explain an overall diagram of the flowering plant life cycle, with emphasis on adaptation to the land environment.
2. Label a diagram of a flower and state a function for each part labeled.

Before beginning this chapter, it would serve you well to read "Evolution of Seed Plants" on page 498, which traces the evolution of seed plants, especially the flowering plants. You'll want to realize from an examination of the alternation-of-generations life cycle in early plants that the *sporophyte* produces spores and the *gametophyte* produces gametes during the plant life cycle. Seed plants protect all stages of the life cycle by a modification of the cycle, resulting in the production of two types of spores (microspores and megaspores) and two types of gametes (male gametophyte and female gametophyte) as illustrated in the art accompanying "Evolution of Seed Plants." As a result of the evolutionary process, the life cycle of flowering plants includes these features (Fig. 25.1):

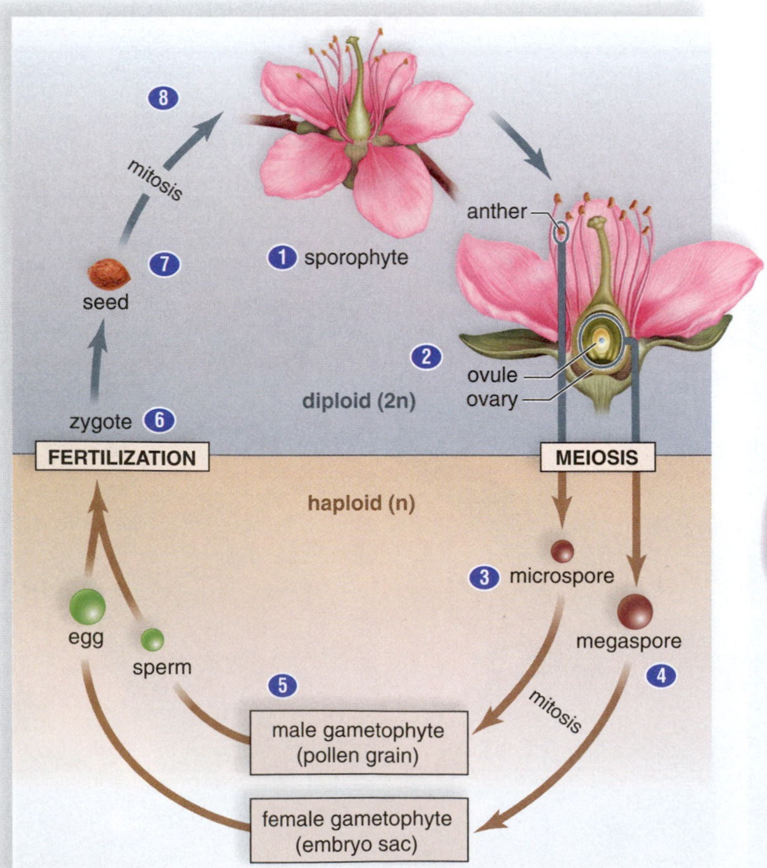

1 In flowering plants, the diploid **sporophyte** is dominant, and it is the generation that bears flowers. **2** A **flower,** which is the reproductive structure of angiosperms, produces two types of spores by meiosis: microspores and megaspores. **3** In the anther, each microspore undergoes mitosis and becomes a **pollen grain,** a **male gametophyte,** which is either windblown or carried by an animal to the vicinity of the female gametophyte. **4** In the meantime, in each ovule of an ovary, a megaspore has undergone mitosis to become the **female gametophyte,** an **embryo sac. 5** At maturity, a pollen grain contains nonflagellated sperm, which travel by way of a pollen tube to the embryo sac. **6** Once a sperm fertilizes an egg, the zygote becomes an embryo, still within an ovule. **7** The **ovule** develops into a **seed,** which contains the embryo and stored food surrounded by a seed coat. The **ovary** becomes a **fruit,** which aids in dispersing the seeds. **8** When a seed germinates, a new sporophyte emerges and, through mitosis and growth, becomes a mature organism.

Notice that the sexual life cycle of flowering plants is adapted to a land existence. The microscopic female gametophytes develop completely within the sporophyte and are thereby protected from desiccation. Pollen grains (male gametophytes) are not released until they develop a thick wall. No external water is needed to bring about fertilization in flowering plants. Instead, the pollen tube provides passage for a sperm to reach an egg. Following fertilization, the embryo and its stored food are enclosed within a protective seed coat until external conditions are favorable for germination.

Flowers Flowers are unique to angiosperms (Fig. 25.1B). Aside from producing the spores (microspores and megaspores) and protecting the gametophytes, flowers often attract pollinators,

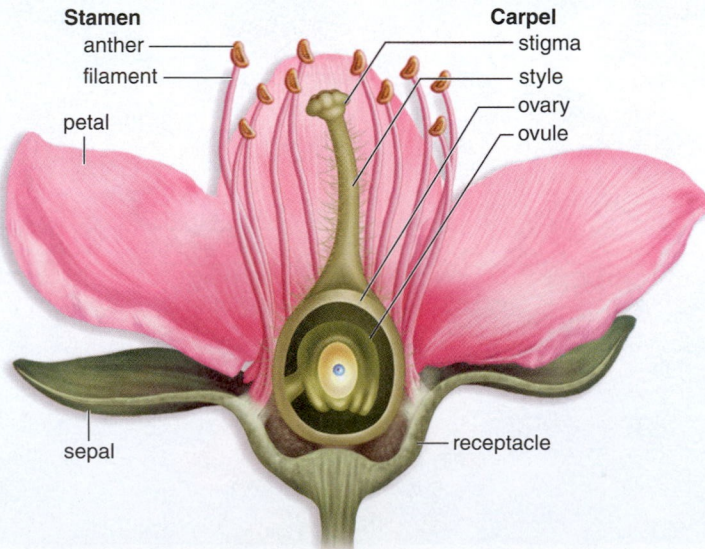

FIGURE 25.1A Alternation of generations in flowering plants.

FIGURE 25.1B Anatomy of a flower.

FIGURE 25.1C **a.** Daylilies are monocots. Monocots have flower parts usually in threes. **b.** Azaleas are eudicots. Eudicots have flower parts in fours or fives (p = petal; s = sepal)

which aid in transporting pollen from plant to plant. Flowers also produce the fruits that enclose the seeds. The evolution of the flower was a major factor leading to the success of angiosperms, with over 240,000 species. Flowering is often a response to environmental signals, such as the length of the day (see section 24.10, p. 487). In many plants, a flower develops when shoot apical meristem that previously formed leaves suddenly stops producing leaves and starts producing a flower enclosed within a bud. In other plants, axillary buds develop directly into flowers. In monocots, flower parts occur in threes and multiples of three; in eudicots, flower parts are in fours or fives and multiples of four or five (Fig. 25.1C).

A typical flower has four whorls of modified leaves attached to a peduncle, which is a receptacle at the end of a flower stalk (see Fig. 25.1B). These are the modified leaves of flowers:

1. The **sepals,** which are the most leaflike of all the flower parts, are usually green, and they protect the bud as the flower develops within. Collectively, the sepals are called the calyx.
2. An open flower next has a whorl of **petals,** whose color accounts for the attractiveness of many flowers. The size, the shape, and the color of petals are attractive to a specific

pollinator. Wind-pollinated flowers may have no petals at all. Collectively, the petals are called the corolla.
3. **Stamens** are the "male" portion of the flower. Each stamen has two parts: the **anther,** a saclike container, and the filament, a slender stalk. Pollen grains develop from the microspores produced in the anther.
4. At the very center of a flower is the **carpel,** a vaselike structure that represents the "female" portion of the flower. A carpel usually has three parts: the stigma, an enlarged sticky knob; the style, a slender stalk; and the ovary, an enlarged base that encloses one or more ovules.

Ovules play a significant role in the production of megaspores and, therefore, female gametophytes. A flower can have a single carpel or multiple carpels. Sometimes several carpels are fused into a single structure, in which case the ovary is compound and has several chambers, each of which contains ovules. For example, an orange develops from a compound ovary, and every section of the orange is a chamber.

Variations in Flower Structure We have space to mention only a few variations in flower structure. Not all flowers have sepals, petals, stamens, or carpels. Those that do are said to be complete and those that do not are said to be incomplete. Flowers that have both stamens and carpels are called perfect (bisexual) flowers; those with only stamens and those with only carpels are imperfect (unisexual) flowers. If staminate flowers and carpellate flowers are on one plant, as in corn, the plant is monoecious (Fig. 25.1D). If staminate and carpellate flowers are on separate plants, the plant is dioecious. Holly trees are dioecious, and if red berries are desired, it is necessary to acquire a plant with staminate flowers and another plant with carpellate flowers.

Staminate flowers   Carpellate flowers

FIGURE 24.1D A corn plant is monoecious. The staminate flowers produce pollen and the carpellate flowers produce ears of corn after receiving pollen.

▶ **25.1 CHECK YOUR PROGRESS**
1. Identify how the gametophytes are protected from drying out in the flowering plant life cycle.
2. Explain where the ovary in a flower is located and what role it plays in the flowering plant life cycle.

25A Evolution of Seed Plants

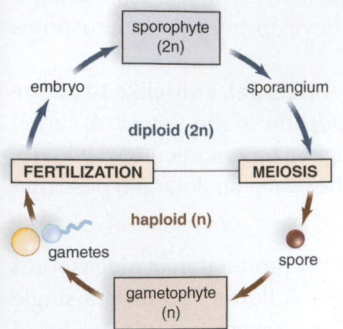

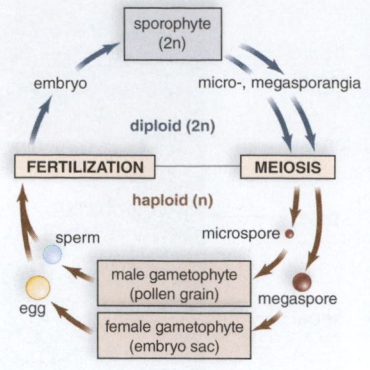

In Chapter 19 we saw that plants are multicellular photosynthetic organisms that have become adapted to living on land. A large part of this adaptation is the ability to protect all stages of the life cycle from drying out. Because plants utilize an alternation-of-generations life cycle, this means that eventually both the gametophyte generation, which produces the gametes (sperm and egg), and the resulting embryo are protected from the dry air of the land environment.

To trace this evolutionary achievement, we begin with the bryophytes, exemplified in Figure 25A by the mosses. Bryophytes differ from all other land plants because the gametophyte is dominant. So when you take a walk in the woods and happen to notice moss plants, you are most likely looking at the gametophyte generation. In bryophytes, the egg is protected, but the sperm are flagellated and must swim to the egg. This is a distinct disadvantage for reproduction on land. Still, the sporophyte generation attached to the gametophyte does produce haploid windblown spores in its capsule, a sporangium where meiosis occurs.

The other plants in Figure 25A have vascular tissue, and the sporophyte is dominant—it's the generation that evolved vascular tissue. In ferns, the gametophyte, a tiny almost microscopic heart-shaped structure, is independent and must fend for itself on the forest floor. It too produces flagellated sperm that require moisture to reach the protected egg. Again, the sporangia, which in our example of a fern are located on the underside of fronds, produce windblown spores.

The last step in our story is the evolution of seed plants. A dramatic change occurs in the life cycle of seed plants, which are either gymnosperms (e.g., a pine tree) or flowering plants (e.g., a fruit tree). Possible evolutionary links between ferns and progymnosperms of the Devonian period have been found but not confirmed. Therefore, we do not know the exact steps by which seed plants evolved, but we do know that today the sporophyte of seed plants produces two types of spores and they develop into two types of gametophytes (compare the diagrams left and right). Microspores become pollen grains that produce sperm. In other words, a pollen grain is a microscopic male gametophyte. Megaspores located in ovules become microscopic female gametophytes that produce eggs. Pollen grains are not released, for example from the pollen cones of gymnosperms, until they have a hard coat. This coat protects the sperm from drying out while they are transported by wind to the female gametophyte (embryo sac), which is always protected within an ovule on the parent plant. The cones of gymnosperms that bear ovules are called seed cones because once the pollen grain has arrived and fertilization has occurred the ovule develops into a seed that contains and protects the sporophyte embryo. The seed protects and disperses the 2n sporophyte embryo. (How different is that from utilizing a spore to disperse a fragile, water-dependent gametophyte, as in ferns?) In seed plants, all stages of the life cycle are protected from drying out, and this wouldn't be possible without the production of two types of spores and two types of gametophytes.

CONSIDER THESE QUESTIONS

1. Hypothesize how the nonseed life cycle could have evolved into the seed life cycle.
2. Show that humans, like seed plants, are adapted to protect all stages of reproduction in a dry environment.
3. How is a seed cone in a pine tree like and different from a flower in an angiosperm?

connect | **BIOLOGY** Explore the concepts through a variety of multimedia assets and question types.

www.mcgrawhillconnect.com

FIGURE 25A As evolution occurred, the gametophyte became reduced in size and dependent on the sporophyte. Seed plants protect all stages of the life cycle from drying out.

25B Evolution of Insect Pollination

Although we generally associate insect pollination only with angiosperms (flowering plants), the practice may have evolved first among gymnosperms (see Fig. 19.6A). Today, beetles are attracted to male cycad cones by the odorous chemicals they secrete. While there, beetles eat the pollen and get covered by it before they leave to possibly visit female cones. In this way beetles are pollinators of cycads. The fossil record tells us that cycads and beetles were both prevalent during the Triassic period (250–205 MYA), long before angiosperms evolved. It's possible, therefore, that cycads and beetles began their relationship back in the Triassic, paving the way for beetles to also visit odorous flowers once they evolved. Even today, beetles are active pollinators of flowers.

During the Cretaceous period, both flowers and insects diversified greatly, perhaps because by then they had forged their great alliance. By being adapted to drink nectar from only certain flowers, pollinators are more likely to deliver pollen to another member of that species. Although most flowering plants depend on several different pollinators, some good examples are used to show that certain flowers and their pollinators are adapted to each other (Fig. 25B). For example, in order of their evolution it seems that

- Beetles are attracted to large, open, bowl-shaped flowers that are often white or greenish. These flowers tend to have a strong sweet smell and offer copious amounts of pollen (Fig. 25B*b*).
- Flies are attracted to "carrion flowers," which have a brownish-red color and a strong stench like that of putrefying flesh. These traits attract flies who are looking for a place to lay their eggs. Instead, they become covered with pollen, which they transport to another flower with the same attraction (Fig. 25B*d*).
- Butterflies are usually attracted to brightly colored, sweet-smelling flowers. These flowers offer nectar in deep tubes that butterflies can access with their long mouthparts (Fig. 25B*a*).
- Bees prefer brightly colored, sweet-smelling, blue or yellow flowers that have a landing platform to receive them. They

have a long tube and tongue for sucking up nectar and pollen baskets on their hind legs. Bees can see ultraviolet shadings that tell them where the nectar is. Even more amazing, researchers now tell us that flowers have different electric fields and bees can detect them!
- Bird-pollinated flowers tend to be bright red or orange, colors only birds notice. The flowers often have a tube shape that accommodates only the shape of a specific species' beak (Fig. 25B*c*).
- Bat-pollinated flowers have a strong sweet, fermented, or musky scent, and the flowers are pale or drab—bats pollinate at night, so bright colors would be wasted on them. Numerous anthers dust pollen onto the bat while it laps at nectar hidden in the flower (Fig. 25B*e*).

Adaptations between flowers and pollinators can be highly specific. For example, some orchids produce flowers that mimic the appearance and scent of females in certain wasp and bee species. The males are drawn to the "fake females," and in the process of attempting to mate with them, they receive a pollen load that they will transport to another flower when they unsuccessfully attempt to mate with it!

CONSIDER THESE QUESTIONS

1. It's quite a controversy today whether pollinators visit one type or different types of flowers. Design an experiment that could help decide this question.
2. Gymnosperms produce much pollen, and it is usually distributed by wind. What does this suggest about the efficiency of wind pollination? Support your hypothesis.

connect |BIOLOGY   Explore the concepts through a variety of multimedia assets and question types.

www.mcgrawhillconnect.com

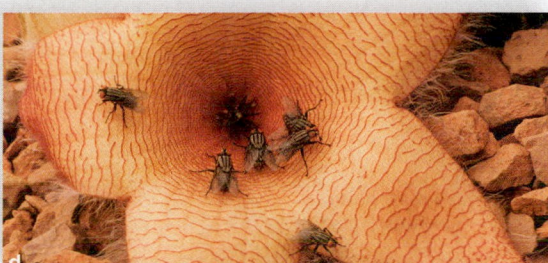

FIGURE 25B Butterflies, birds, and bats are adapted for acquiring nectar from certain flowers. Flowers that attract beetles produce much pollen and those that attract flies have the smell of rotting flesh.

25.2 Pollination and fertilization bring gametes together during sexual reproduction

LEARNING OUTCOME

When you complete this section, you should be able to

1. Describe the events of the flowering plant life cycle with emphasis on gamete maturation and double fertilization.

In section 19.7 (see p. 373), we studied sexual reproduction in flowering plants. Sexual reproduction involves (1) the production of pollen grains (male gametophytes) in the anthers of stamens and (2) the production of an embryo sac (female gametophyte) in an ovule located within the ovary of a carpel (Fig. 25.2A). The production of these two different gametophytes is paramount to the adaptation of seed plants to reproduction on land. It led to their ability to protect all stages of the life cycle from drying out in the land environment.

In this section, we discuss two other important events in the life cycle of flowering plants: pollination and double fertilization.

These events also help explain why flowering plants are able to disperse so well on land.

Pollination During **pollination,** pollen is transferred from the anther to the stigma so that an egg within the female gametophyte is fertilized. Self-pollination occurs if the pollen and stigma are from the same plant. Cross-pollination occurs when the pollen is from a member of the same species but not the same flower. Cross-pollination offers the best chance of the offspring having a different genotype from that of the parent. Plants have various means of achieving cross-pollination.

Some species of flowering plants—for example, the grasses and grains—rely on wind pollination (Fig. 25.2B), as do the gymnosperms, the other type of seed plant. Much of the plant's energy goes into making pollen to ensure that some pollen grains actually reach a stigma. Even the amount successfully transferred

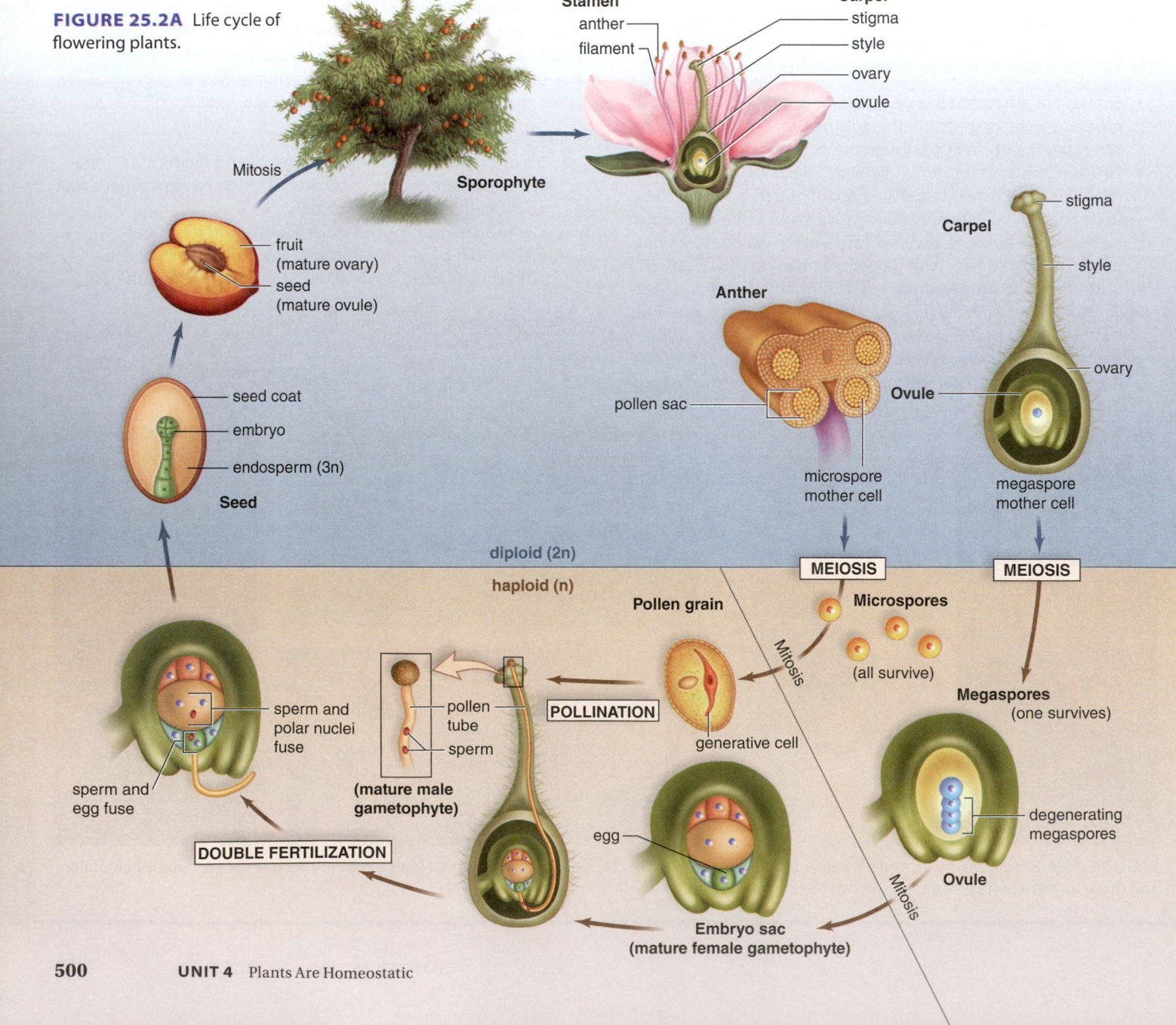

FIGURE 25.2A Life cycle of flowering plants.

FIGURE 25.2B Wind pollination of a grass, with SEM of pollen grains.

Double Fertilization The process of **double fertilization** is also unique in angiosperms. It results in not only a zygote but also a food source for the developing zygote. Note the mature male gametophyte in Figure 25.2A. The generative cell has divided to produce two sperm cells, and a pollen tube is in the process of lengthening. This is in keeping with our expectation of the male gametophyte because, in plants, the gametophyte generation produces gametes. A pollen tube, which is an outgrowth of the inner wall of a pollen grain, digests its way through the tissue of the stigma and style of the carpel. The interval between pollen tube initiation and the time when the tube reaches the embryo sac in an ovule is quite variable, taking from a few hours to a few days.

When the pollen tube reaches the entrance of the embryo sac, double fertilization occurs. Remember that the embryo sac is the female gametophyte, and as such, it produces an egg. As expected, one of the sperm unites with the egg, forming a 2n zygote. Unique to angiosperms, the other sperm unites with two polar nuclei centrally placed in the embryo sac, forming a 3n endosperm cell. This cell eventually develops into the **endosperm,** a nutritive tissue that the developing embryonic sporophyte will use as an energy source. Now the ovule begins to develop into a seed. One important aspect of seed development is formation of the **seed coat** from the ovule wall. A mature seed contains (1) the embryo, (2) stored food, and (3) the seed coat. Figure 25.2D shows a eudicot seed in which the embryo has already formed. The endosperm has been taken up by the cotyledons, or seed leaves.

is staggering: A single corn plant may produce from 20–50 million grains a season. In corn, the flowers tend to be monoecious, and clusters of tiny male flowers move in the wind, freely releasing pollen into the air (see Fig. 25.1D).

Most angiosperms rely on animals—be they insects (e.g., bumblebees, flies, butterflies, and moths), birds (e.g., hummingbirds), or mammals (e.g., bats)—to carry out pollination. The use of animal pollinators is unique to flowering plants, and it helps account for why these plants are so successful on land. By the time flowering plants appear in the fossil record some 135 MYA, insects had long been present. Thus, for millions of years plants and their animal pollinators have coevolved. **Coevolution** means that as one species changes, the other changes too, so that in the end, the two species are suited to one another. Plants with flowers that attracted a pollinator enjoyed an advantage because, in the end, they produced more seeds. Similarly, pollinators that were able to find and remove food from the flower were more successful. Today, we see that the reproductive parts of the flower are positioned so that the pollinator can't help but pick up pollen from one flower and deliver it to another. On the other hand, the mouthparts of the pollinator are suited to gathering the nectar from particular plants.

Many examples of adaptations between flowers and their pollinators are given in "Evolution of Insect Pollination" on p. 499. Here, we can note that, in addition, bees respond to ultraviolet markings called nectar guides that help them locate nectar. Humans do not use ultraviolet light in order to see, but bees are sensitive to ultraviolet light (Fig. 25.2C). A bee has a feeding proboscis that allows it to collect nectar from certain flowers and a pollen basket on its hind legs that allows it to carry pollen back to the hive. Today, the number of bees is declining due to disease and the use of pesticides, causing concern because many fruits and vegetables depend on bee pollination.

▶ **25.2 CHECK YOUR PROGRESS**
1. Describe the mature pollen grain.
2. Describe the products of double fertilization in angiosperms.
3. List the ways flowering plants, in general, are adapted to reproduce on land.

As we see it

nectar guides
As a bee sees it

FIGURE 25.2C Ultraviolet markings guide bees to nectar.

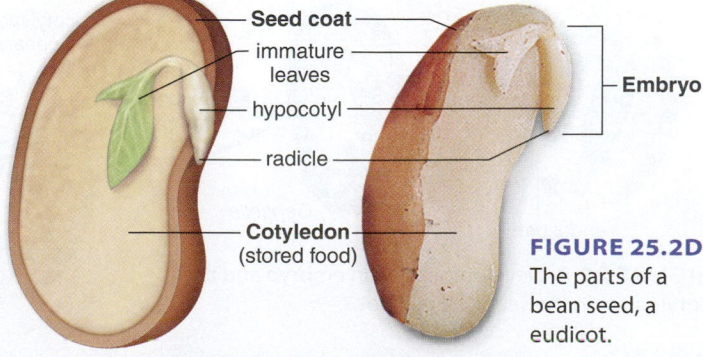

Seed coat
immature leaves
hypocotyl
radicle
Embryo
Cotyledon (stored food)

FIGURE 25.2D The parts of a bean seed, a eudicot.

Seed Development and Growth

This part of the chapter describes the stages of embryo development within a seed, the structure of fruits, and seed germination. With seed germination, the life cycle of flowering plants has come full circle.

25.3 A sporophyte embryo and its cotyledons develop as a seed matures

LEARNING OUTCOME

When you complete this section, you should be able to

1. Divide development of the embryo into six stages and label the three main parts of a seed.

Figure 25.3 shows the stages of development for a eudicot embryo. **1** The zygote stage is the beginning stage. **2** Then, the zygote divides repeatedly in different planes, forming several cells called a **proembryo.** Also formed is an elongated structure called a suspensor that has a basal cell. The suspensor, which anchors the embryo and transfers nutrients to it from the sporophyte plant, will disintegrate later.

3 During the **globular stage,** the proembryo is largely a ball of cells. The root-shoot axis of the embryo is already established at this stage because the embryonic cells near the suspensor will become a root, while those at the other end will ultimately become a shoot.

4 During the **heart stage,** the cotyledons appear as a result of rapid, local cell division, giving the **embryo** a heart shape. Monocots have one **cotyledon,** which in addition to storing certain nutrients, absorbs other nutrient molecules from the endosperm and passes them to the embryo. In eudicots, the cotyledons usually store the nutrient molecules the embryo uses.

5 As the embryo continues to enlarge and elongate, it takes on a torpedo shape, a period known as the **torpedo stage.** Now the root and shoot apical meristems are functional and will allow the seedling to grow and produce all its various specialized tissues once germination occurs (see Fig. 25.5A, B).

6 In the **mature embryo** (or simply the embryo) of the final stage, the epicotyl is the portion between the cotyledons that contributes to shoot development. The hypocotyl is the portion below the cotyledons. It contributes to stem development and terminates in the radicle, or embryonic root. The cotyledons are quite noticeable in a eudicot embryo and may fold over. The embryo stops developing and becomes dormant within its seed coat (derived from the ovule wall).

> ▶ **25.3 CHECK YOUR PROGRESS**
> 1. Identify the origin of each of the three parts of a seed.
> 2. Describe the structure and function of the cotyledon.

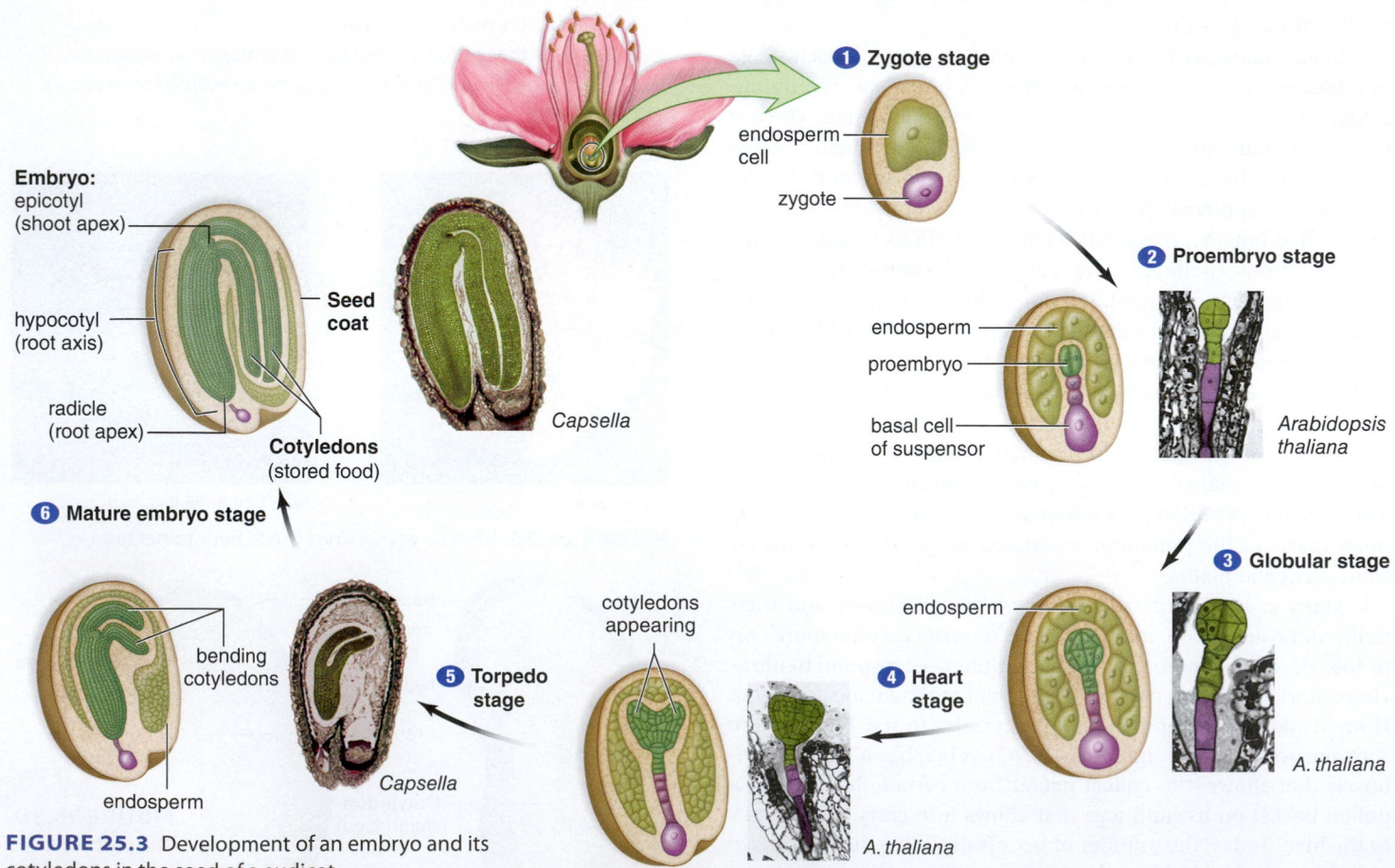

FIGURE 25.3 Development of an embryo and its cotyledons in the seed of a eudicot.

25.4 The ovary becomes a fruit, which assists in sporophyte dispersal

LEARNING OUTCOMES

When you complete this section, you should be able to

1. Give examples of fleshy and dry fruits.
2. Distinguish between simple, accessory, and compound fruits.

A fruit—derived from an ovary and sometimes other flower parts—protects and helps disperse the next 2n sporophyte generation. How does a fruit help in seed dispersal? Often, fruits are an attractive and nutritious package that animals like to eat. Then they deposit the seeds some distance away.

Video Fruit Bat Seed Dispersal

As a fruit develops, the ovary wall thickens to become the **pericarp,** as labeled on the pea pod in Figure 25.4. The pericarp can have as many as three layers that encircle the seed: exocarp, mesocarp, and endocarp.

Fleshy Versus Dry Fruits Figure 25.4 shows diverse types of fruit. ❶ A pea pod is a dry fruit. Peas and beans, being legumes, have a **dehiscent** fruit: it splits open when mature. In other words, pea pods break open to release their seeds. ❷ The fruit of a maple tree is dry and **indehiscent**—it does not split open.

The cereal grains of wheat, rice, and corn are small dry fruits. Sometimes the fruits of cereal grains are mistaken for seeds because a dry pericarp adheres to the seed within. These dry fruits are indehiscent. Humans gather grains before they are released from the plant and then process them to acquire their nutrients.

Some fruits such as berries, apples, and peaches are fleshy. The "flesh" is derived variously. In a tomato, the pericarp is fleshy and in peaches only the mesocarp is fleshy. Peaches have a hard, stony endocarp and are often, therefore, called stone fruits,

although botanically they are known as drupes. This type of endocarp protects the seed so it can pass through the digestive system of an animal and remain unharmed.

Simple Versus Aggregate and Multiple Fruits Fruits can be simple or compound. A **simple fruit** is derived from a single ovary, which can have one or more chambers. A tomato is an example of a simple fruit that has several chambers. When you slice a tomato, you can observe the chambers. **Accessory fruits** are fruits that form from other flower parts, in addition to the ovary. ❸ A strawberry is an accessory fruit because the bulk of the fruit is not from the ovary, but from the receptacle. Similarly, only the core of an apple is derived from the ovary. If you cut an apple crosswise, it is obvious the ovary of an apple flower, like that of a tomato, has several chambers.

A **compound fruit** is derived from a group of ovaries (Fig. 25.4) If a single flower has multiple ovaries, as in a raspberry ❹, then it produces an **aggregate fruit.** In contrast, a pineapple ❺ comes from many ovaries of separate flowers, but they shared a common receptacle. As the fruit developed, the ovaries fused to form a large, **multiple fruit.**

▶ **25.4 CHECK YOUR PROGRESS**

1. Compare the structure and dispersal methods of dry and fleshy fruits.
2. Explain how an apple is both a simple and an accessory fruit.
3. Identify the advantage of fruit production in flowering plants.

FIGURE 25.4 Fruit diversity.

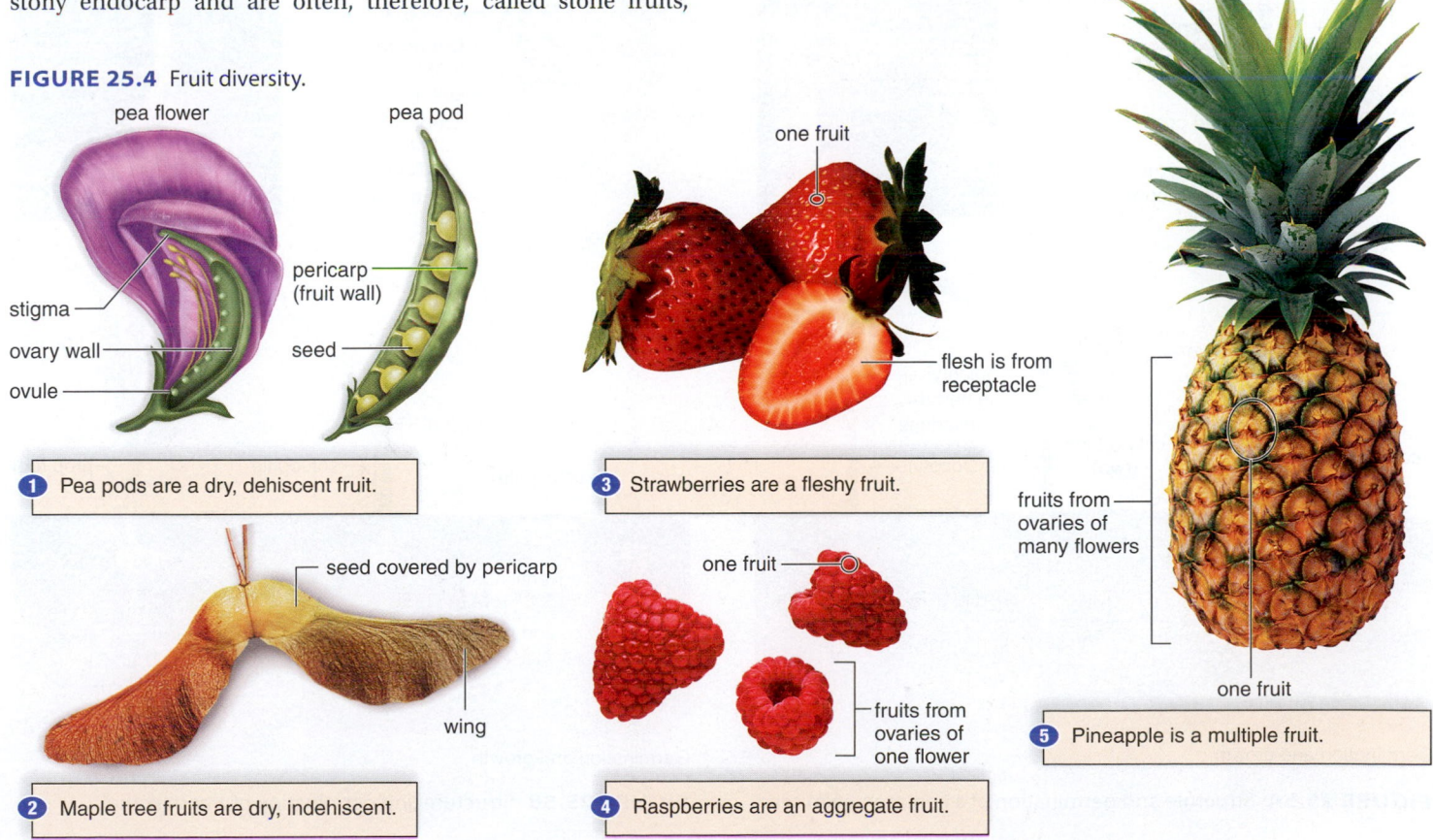

pea flower

pea pod

stigma

ovary wall

ovule

pericarp (fruit wall)

seed

❶ Pea pods are a dry, dehiscent fruit.

seed covered by pericarp

wing

❷ Maple tree fruits are dry, indehiscent.

one fruit

flesh is from receptacle

❸ Strawberries are a fleshy fruit.

one fruit

fruits from ovaries of one flower

❹ Raspberries are an aggregate fruit.

fruits from ovaries of many flowers

one fruit

❺ Pineapple is a multiple fruit.

25.5 With seed germination, the life cycle is complete

LEARNING OUTCOMES

When you complete this section, you should be able to

1. Compare a bean seed to a corn kernel.
2. Compare and contrast germination of a bean plant and a corn plant.

Following dispersal, if conditions are right, seeds may **germinate** to form a seedling. Germination doesn't usually take place until there is sufficient water, warmth, and oxygen to sustain growth. These requirements help ensure that seeds do not germinate until the most favorable growing season has arrived. Some seeds do not germinate until they have been dormant for a period of time. For seeds, *dormancy* is the time during which no growth occurs, even though conditions may be favorable for growth. In the temperate zone, seeds often have to be exposed to a period of cold weather before dormancy is broken. Fleshy fruits (e.g., apples, pears, oranges, and tomatoes) contain inhibitors so that germination does not occur while the fruit is still on the plant. For seeds to take up water, bacterial action and even fire may be needed. Once water enters, the seed coat bursts and the seed germinates. Gibberellins are involved in supplying energy for the embryo to grow (see section 24.3, p. 478).

While the cotyledons of a bean seed are important sources of food for the embryo, it is the embryo that gives rise to the mature plant. The hypocotyl becomes the shoot system which will include the plumule (first leaves). The radicle becomes the root system. If the two cotyledons of a bean seed are parted, the rudimentary plant with immature leaves is exposed (Fig. 25.5A). As the eudicot seedling starts to grow, the shoot is hook-shaped to protect the immature leaves as they emerge from the soil. The cotyledons provide the new seedlings with enough energy to straighten and form true leaves. As the true leaves of the plant begin photosynthesizing, the cotyledons shrivel up.

A corn kernel is actually a fruit, and therefore its outer covering is the pericarp and seed coat combined (Fig. 25.5B). Inside is the single cotyledon. Notice that the monocot embryo also has a plumule (first leaves) and a radicle that will become the root system. However, the immature leaves and the root are covered, respectively, by a coleoptile and a coleorhiza. These sheaths are discarded when the seedling begins to grow.

▶ 25.5 CHECK YOUR PROGRESS

1. Identify ways in which a bean seed and corn kernel are the same and how they are different.
2. Compare the protective methods monocot and eudicot seedlings use to protect their first leaves.

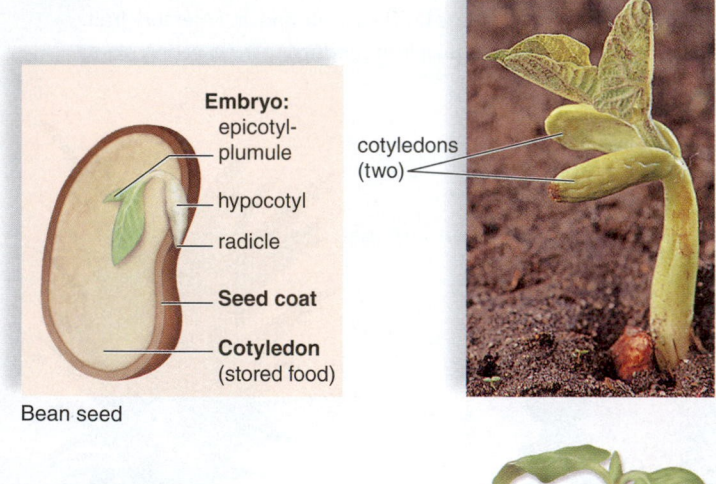

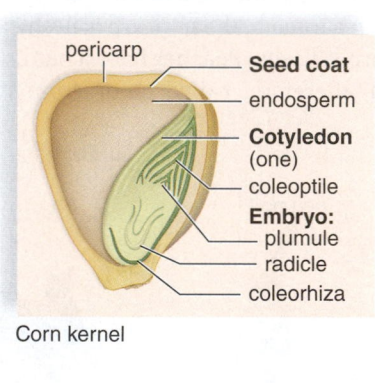

Bean seed

Corn kernel

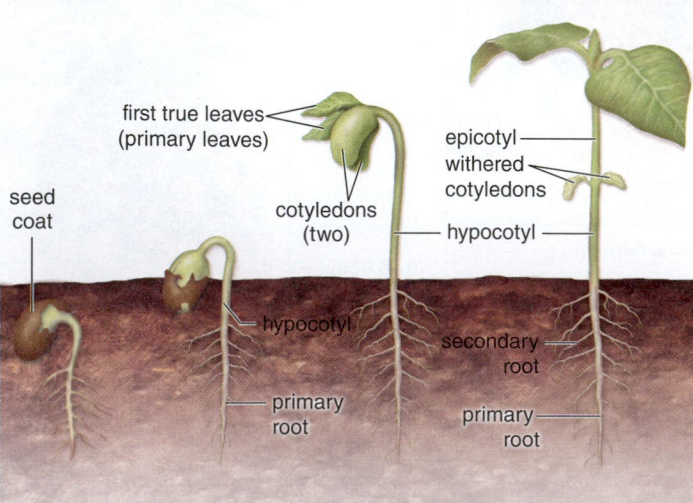

Germination and growth

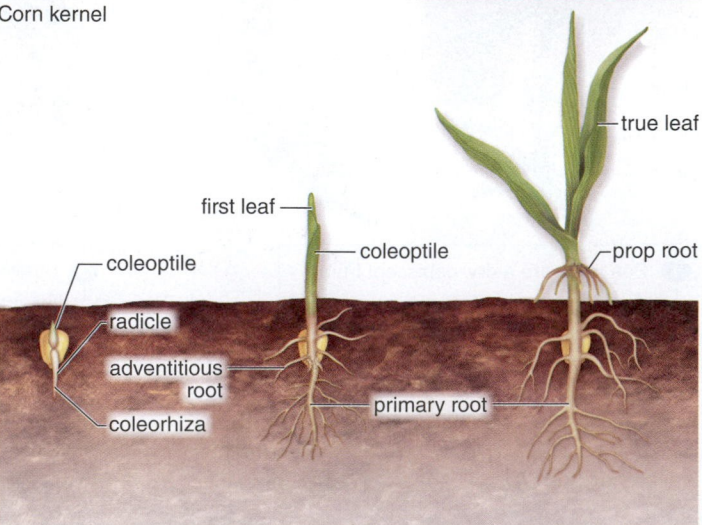

Germination and growth

FIGURE 25.5A Structure and germination of a common bean seed.

FIGURE 25.5B Structure and germination of a corn kernel.

Asexual Reproductive Strategies

We can observe the asexual reproduction of plants in the environment as well as in the laboratory, where cloning of plants in tissue culture is commonplace. Cloning produces identical plants or plant tissues with highly desirable traits for agricultural and commercial purposes.

25.6 Plants have various ways of reproducing asexually

LEARNING OUTCOMES

When you complete this section, you should be able to

1. Understand the benefits of sexual reproduction.
2. Identify ways that plants in the wild reproduce asexually.

Asexual reproduction is the production of an offspring identical to a single parent. Often the offspring is attached in some way to the parent, which gives it support while the offspring is developing (Fig. 25.6). Asexual reproduction is less complicated in flowering plants because pollination and seed production are not required. Therefore, it can be advantageous when a parent is already well adapted to a particular environment that is not subject to change.

You may already be familiar with the examples of asexual reproduction in plants given in Figure 25.6. Plants can grow from the axillary buds of aboveground horizontal stems and various types of underground stems. Aboveground horizontal stems, called **stolons,** run along the ground. Complete strawberry plants can grow from axillary buds that appear at the nodes of a stolon.

Underground horizontal stems, called **rhizomes,** may be long and thin, as in sod-forming grasses, or thick and fleshy, as in irises. Rhizomes survive the winter and contribute to asexual reproduction because each node bears a bud. Irises grow from the buds of rhizomes, as do violets and many grasses.

Some rhizomes have enlarged portions called **tubers** that function in food storage. Potatoes are tubers, in which the eyes are axillary buds that mark the nodes. A bud has the potential to produce a new potato plant if it is planted with a portion of the swollen tuber.

Corms are bulbous underground stems that lie dormant during the winter, just as rhizomes do. They also produce new plants in the next growing season. Gladiolus corms are called bulbs by laypersons, but botanists reserve the term *bulb* for a structure composed of modified leaves attached to a short, vertical stem. Onions grow from bulbs, as do lilies, tulips, and daffodils.

Many different plants can be propagated from stem cuttings since the discovery that an application of the plant hormone auxin can cause stems to produce roots.

▶ **25.6 CHECK YOUR PROGRESS**
1. List three benefits of asexual reproduction in wild plants.
2. Describe methods of asexual reproduction in wild plants.

FIGURE 25.6 Asexual reproduction in plants.

Rhizome — Tuber — Corm

rhizome
branch
adventitious roots
axillary bud
rhizome
tuber
axillary bud
papery leaves
corm
adventitious roots

Parent plant

stolon — Asexually produced offspring

25.7 Cloning of plants in tissue culture assists agriculture

When you complete this section, you should be able to

1. Describe how tissue culture can be used to clone plants with desirable traits.

Tissue culture is the growth of a tissue in artificial liquid or on solid culture medium. Three methods of cloning plants—somatic embryogenesis, meristem tissue culture, and anther tissue culture—are methods of cloning plants due to the ability of plants to grow from single cells. Many plant cells are **totipotent,** which means that each one has the genetic capability of becoming an entire plant.

During *somatic embryogenesis,* hormones cause plant tissues to generate small masses of cells, from which many new genetically identical plants may grow. Thousands of little "plantlets" can be produced by using this method of plant tissue culture (Fig. 25.7A). Many important crop plants, such as tomato, rice, celery, and asparagus, as well as ornamental plants such as lilies, begonias, and African violets, have been produced using somatic embryogenesis. Plants generated from somatic embryos are not always genetically identical clones. They can vary because of mutations that arise spontaneously during the production process. These mutations, called *somaclonal variations,* are another way to produce new plants with desirable traits. Somatic embryos can be encapsulated in hydrated gel, creating artificial "seeds" that can be shipped anywhere.

Meristem tissue culture can also be used to produce plant cells. In this case, the resulting products are clonal plants that always have the same traits. In Figure 25.7B, culture flasks containing meristematic orchid tissue are rotated under lights. If the correct proportions of hormones are added to the liquid medium, many new shoots develop from a single shoot tip. When these are removed, more shoots form. Another advantage to producing identical plants from meristem tissue is that the plants are virus-free. (The presence of plant viruses weakens plants and makes them less productive.)

Anther tissue culture is a technique in which the haploid cells within pollen grains are cultured in order to produce haploid plantlets. Conversely, a diploid (2n) plantlet can be produced if chemical agents are added to the anther culture to encourage chromosomal doubling. Anther tissue culture is a direct way to produce plants that are certain to have the same characteristics.

Cell Suspension Culture A technique called **cell suspension culture** allows scientists to extract chemicals (i.e., secondary metabolites) from plant cells in high concentrations and without having to over-collect wild-type plants growing in their natural environments. These cells produce the same chemicals the entire plant produces. For example, cell suspension cultures of *Cinchona ledgeriana* produce quinine, which is used to treat leg cramping, a major symptom of malaria. And those of several *Digitalis* species produce digitalis, digitoxin, and digoxin, which are useful in the treatment of heart disease.

▶ **25.7 CHECK YOUR PROGRESS**

1. Identify the tissues that can become whole plants in tissue culture.

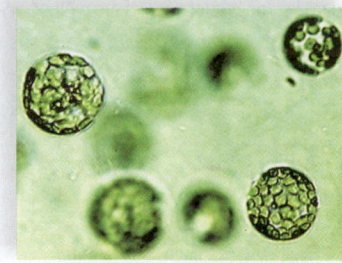
a. Protoplasts, naked cells

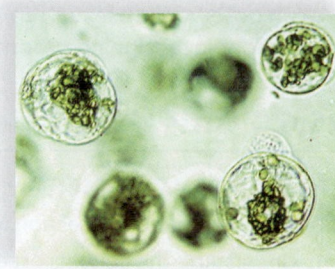
b. Cell wall regeneration

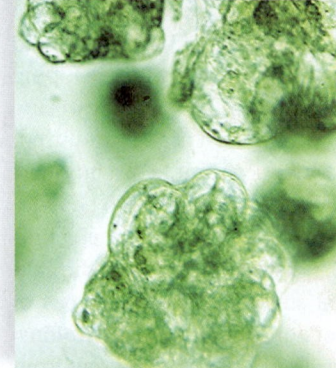
c. Aggregates of cells

d. Callus, undifferentiated mass

e. Somatic embryo

f. Plantlet

FIGURE 25.7A Somatic embryogenesis.

FIGURE 25.7B Producing whole plants from meristem tissue.

THE CHAPTER IN REVIEW

CONNECTING THE CONCEPTS

With this chapter, we bring to a close our study of plants. It is fitting that we end with a look at the reproduction of flowering plants. Life, as we know it, would not be possible without vascular plants—and specifically flowering plants, which now dominate the biosphere. *Homo sapiens* evolved in a world already dominated by flowering plants, and therefore, humans do not know a world without them. The earliest humans were mostly herbivores; they relied on foods they could gather for survival—fruits, nuts, seeds, tubers, roots, and so forth. Plants also provided protection from the environment, offering shelter from heavy rains and noonday sun. Later on, human civilizations could not have begun without the development of agriculture. Most of the world's population still relies primarily on three flowering plants—corn, wheat, and rice—for the majority of its sustenance. Sugar, coffee, spices of all kinds, cotton, rubber, and tea are plants that

have even led to wars due to their importance to some countries' economies.

Although we now live in an industrialized society, we are still dependent on plants and use them for many purposes. In fact, plants may be even more critical to our lives today than they were to our early ancestors on the African plains. For millions of urban dwellers, plants are their major contact with the natural world.

We grow plants not only for food and shelter, but also for their simple beauty. Today, plants even produce the substances needed to lubricate the engines of supersonic jets and to make cellulose acetate for films.

Currently, half of all pharmaceutical drugs are derived from plants. The world's major drug companies are engaged in a frantic rush to collect and test plants from the rain forests for their drug-producing potential. Why the hurry? Because the rain forests may be gone before all the possible cures for cancer, AIDS, and other diseases have been found. Wild

plants cannot only help cure human ills, but may also serve as a source of genes for improving the quality of the plants that support our way of life.

In Unit 5, we study the animal systems, which may seem more familiar to you because humans are animals. However, we should not forget the dependence of animals on plants, a theme that returns in Unit 6 of this text.

ANALYZE AND EVALUATE

1. People don't eat the spores of nonseed plants but do eat the seeds of angiosperms. What is there about seeds that makes them so nutritious?

2. What two mutualistic relationships between animals and flowers have benefited reproduction in flowers? How have they benefited animals?

3. Compared to a fern life cycle, what stages of a flowering plant life cycle are dimorphic?

SUMMARIZE

Sexual Reproductive Strategies

25.1 Plants have a sexual life cycle called alternation of generations

- In flowering plants, the dominant **sporophyte** (2n) produces two types of spores by meiosis: (1) A microspore develops into a **male gametophyte** (**pollen grain**), which produces sperm. (2) A megaspore develops into an egg-producing **female gametophyte** (**embryo sac**) contained within an **ovule**. The **ovule** becomes a **seed**; the **ovary** becomes a **fruit**. (See also section 19.7, p. 373.)

- The sexual life cycle occurs in the **flower**. Within the outer **sepals** and **petals**, the male portion of a flower consists of **stamens**. Each stamen has an **anther** and a filament. The female portion of a flower consists of one or more **carpels**. Each carpel consists of a stigma, a style, and an ovary. Ovules are located in the ovary.

25.2 Pollination and fertilization bring gametes together during sexual reproduction

- **Pollination** transfers pollen from the anther to the stigma of a carpel. Wind pollination sometimes occurs, but pollination by animals is more common. **Coevolution** occurred between flowers and their pollinators.

- In **double fertilization,** two sperm reach the embryo sac. One sperm unites with the egg, and the other unites with two polar nuclei to form **endosperm,** a nutrient substance. The ovule wall becomes the **seed coat** that encloses the multicellular embryo and endosperm (stored food).

embryo sac

Seed Development and Growth

25.3 A sporophyte embryo and its cotyledons develop as a seed matures

- The stages of embryonic development are **proembryo, globular, heart, torpedo,** and **mature embryo. Cotyledons** are embryonic leaves that store food (originally in the endosperm) until the first leaves become functional. Monocots have one cotyledon; eudicots have two.

25.4 The ovary becomes a fruit, which assists in sporophyte dispersal

- The ovary becomes a fruit. The ovary wall becomes the **pericarp.** Fruits may be grouped into various categories: (1) Dry fruits are exemplified by pea pods (**dehiscent**) and cereal grains (**indehiscent**). (2) Fleshy fruits are exemplified by berries, tomatoes, apples, and peaches. (3) **Simple fruits** develop from a

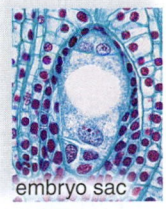

fleshy fruit

flower with a single ovary (grape, bean, wheat, maple). The ovary can have chambers (tomato, apple). (4) In **accessory fruits,** flower parts other than the ovary contribute to the development of the fruit. (5) In **aggregate fruits,** many separate ovaries are from a single flower (strawberry, raspberry). (6) In **multiple fruits,** the ovaries are from separate flowers (pineapple) with one receptacle.

25.5 With seed germination, the life cycle is complete

- **Germination** is regulated by water, warmth, and oxygen availability, among other factors. The embryo breaks out of the seed coat and becomes a seedling with leaves, stems, and roots.

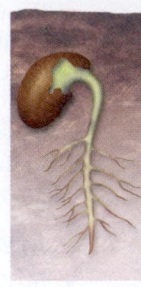 

eudicot   monocot

Asexual Reproductive Strategies

25.6 Plants have various ways of reproducing asexually

- Asexual reproduction has benefits: (1) avoidance of pollination and seed formation, (2) offspring has same adaptations as the parent, and (3) support of the parent. Axillary buds on stems (either aboveground or underground) sometimes give rise to entire plants. Examples of asexual reproductive structures include **stolons, rhizomes, tubers,** and **corms.** Many plants can be propagated from stem cuttings especially because an application of auxin causes roots to sprout.

25.7 Cloning of plants in tissue culture assists agriculture

- **Tissue culture** refers to the growth of tissue in an artificial liquid or solid culture medium. A **totipotent** plant cell has the genetic capability of becoming an entire plant. Cloning methods include somatic embryogenesis, meristem tissue culture, and anther culture. **Cell suspension culture** is a way to obtain secondary metabolites directly from plant cells.

TEST YOURSELF

Sexual Reproductive Strategies

1. Ovule is to carpel as anther is to
   a. sepal.   c. ovary.
   b. stamen.   d. style.
2. Which of the following is not a component of the carpel?
   a. stigma   d. ovule
   b. filament   e. style
   c. ovary
3. The flower part that becomes a fruit is the
   a. carpel.   d. ovary.
   b. stamen.   e. seed.
   c. sepal.
4. Carpels
   a. are the female part of a flower.
   b. contain ovules.
   c. are the innermost part of a flower.
   d. are absent in some flowers.
   e. All of these are correct.

5. Label this diagram of a flower:

6. In plants,
   a. gametes become a gametophyte.
   b. spores become a sporophyte.
   c. both sporophyte and gametophyte produce spores.
   d. only a sporophyte produces spores.
   e. Both a and b are correct.
7. In plants, meiosis directly produces
   a. new xylem.   d. an egg.
   b. phloem.   e. sperm.
   c. spores.
8. In the life cycle of flowering plants, a microspore develops into
   a. a megaspore.   d. an ovule.
   b. a male gametophyte.   e. an embryo.
   c. a female gametophyte.
9. A pollen grain is
   a. a haploid structure.
   b. a diploid structure.
   c. first a diploid and then a haploid structure.
   d. first a haploid and then a diploid structure.
10. The megaspore is similar to the microspore in that both
    a. have the diploid number of chromosomes.
    b. become an embryo sac.
    c. become a gametophyte that produces a gamete.
    d. are necessary to seed production.
    e. Both c and d are correct.
11. The megaspore and the microspore
    a. both produce pollen grains.
    b. both divide meiotically.
    c. both divide mitotically.
    d. produce pollen grains and embryo sacs, respectively.
    e. All of these are correct.
12. The embryo of a flowering plant can be found in the
    a. pollen.   c. microspore.
    b. anther.   d. seed.
13. Which is the correct order of the following events?
    (1) megaspore becomes embryo sac, (2) embryo forms,
    (3) double fertilization, (4) meiosis
    a. 1, 2, 3, 4   c. 4, 3, 2, 1
    b. 4, 1, 3, 2   d. 2, 3, 4, 1

14. Double fertilization refers to the formation of a _____ and a(n) _____.
    a. zygote, zygote
    b. zygote, pollen grain
    c. zygote, megaspore
    d. zygote, endosperm
15. Which of these pairs is incorrectly matched?
    a. polar nuclei—plumule
    b. egg and sperm—zygote
    c. ovule—seed
    d. ovary—fruit
    e. stigma—carpel
16. **THINKING CONCEPTUALLY** Would you expect a wind-pollinated plant or an animal-pollinated plant to produce more pollen? Explain.

Seed Development and Growth

17. A seed is a mature
    a. embryo.
    b. ovule.
    c. ovary.
    d. pollen grain.
18. Globular, heart, and torpedo refer to
    a. embryo development.
    b. sperm development.
    c. female gametophyte development.
    d. seed development.
    e. Both b and d are correct.
19. A seed contains
    a. a seed coat.
    b. an embryo.
    c. stored food.
    d. cotyledon(s).
    e. All of these are correct.
20. The function of the flower is to _____, and the function of fruit is to _____.
    a. produce fruit; provide food for humans
    b. aid in seed dispersal; attract pollinators
    c. attract pollinators; assist in seed dispersal
    d. produce the ovule; produce the ovary
21. Fruits
    a. nourish embryo development.
    b. help with seed dispersal.
    c. signal gametophyte maturity.
    d. attract pollinators.
    e. signal when they are ripe.
22. In an apple, the bulk of the fruit is from the
    a. ovary.
    b. style.
    c. pollen.
    d. receptacle.
23. Which of these is not a fruit?
    a. walnut
    b. pea
    c. green bean
    d. peach
    e. All of these are fruits.
24. **THINKING CONCEPTUALLY** Seed germination sometimes requires exposure to cold temperatures. Explain the benefit to the plant.

Asexual Reproductive Strategies

25. Asexual reproduction in flowering plants
    a. is unknown.
    b. is a rare event.
    c. is common.
    d. produces seeds also.
    e. is no fun.
26. Plant tissue culture takes advantage of
    a. a difference in flower structure.
    b. sexual reproduction.
    c. gravitropism.
    d. phototropism.
    e. totipotency.
27. The term totipotent means
    a. that each plant cell can become an entire plant.
    b. hormones control all plant growth.
    c. all cells develop from the same tissue.
    d. None of these are correct.
28. **THINKING CONCEPTUALLY** Under what environmental conditions would it be advantageous for a plant to carry out asexual reproduction?

GET INVOLVED

1. You notice that a type of wasp has been visiting a particular flower type in your garden. What data about the wasp and flower would allow you to hypothesize that this wasp is a pollinator for this flower type?
2. You are a laboratory scientist who has discovered an unusual lettuce type and wants to propagate it. What could you do if no seeds are available?

MEDIA STUDY TOOLS

mhhe.com/maderconcepts3

Enhance your study of this chapter with interactive study tools, practice tests, and engaging animations. Also, ask your instructor about the resources available through ConnectPlus, which includes LearnSmart, a personalized adaptive learning program, and a media-rich eBook.

26

Animal Organization and Homeostasis

BEFORE YOU BEGIN

Take a few minutes to recall

That organisms are composed of cells (sections 1.3 and 4.1)
The levels of biological organization (Figure 1.3A)
That organisms are homeostatic (section 1.5)

Staying Warm, Staying Cool

Animals have ways of keeping their body temperature within normal limits. Invertebrates, such as insects, and vertebrates, such as reptiles, usually live in warmer climates, where external temperatures can help them maintain a warm body temperature. Basking in the sun while lying on a hot rock allows them to use radiant energy from the sun and heat from the rock to get warm. Some insects fan their wings in the early morning to warm the flight muscles so they can take off. The bodies of many nocturnal moths are covered with scales that help retain the heat of muscle contraction, so that they can be active at night when there is no sun. In contrast to these ectotherms that rely on the sun to warm their bodies, endotherms, exemplified by mammals and birds, have a high metabolic rate—heat from the inside—to stay warm. They also have various structural modifications to help them keep warm, including the hair of mammals and the feathers of birds. So why do chimpanzees and other hairy apes (gorilla, orangutan, and gibbon) live in tropical forests? You guessed it. At some time in their past evolutionary history they must have relied on their hair to help them stay warm. Primates evolved as nocturnal animals in northern broadleaf evergreen forests some 55 MYA. Hair may have been an asset as they foraged

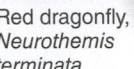

Red dragonfly,
*Neurothemis
terminata*

Chimpanzee, *Pan troglodytes*

These animals are adapted to living in a warm climate.

Red-eyed tree frog,
Agalychnis callidryas

Acklins ground iguana,
Cyclura rileyi nuchalis

Galápagos giant tortoise,
Geochelone nigra

after the sun set. Later, primates followed their habitat when colder northern temperatures caused broadleaf evergreen trees to retreat to the tropics.

Mammals, especially, are noted for being able to live where it is really cold, such as the Arctic Circle. Their stocky bodies, compared to those of close relatives living in warmer climates, give them a reduced surface-area-to-volume ratio that helps keep the heat inside. For example, the snowshoe hare of Canada has a more compact body than the jackrabbit of the American Southwest. Birds and some mammals (e.g., caribou and foxes) have a vascular modification that keeps them from losing heat through exposed limbs. Arteries that carry blood to the limbs are surrounded by veins that bring blood back to the body core. Heat passes from outgoing to incoming vessels, rather than escaping from exposed surfaces. Then, too, lipids in plasma membranes are polyunsaturated and, thus, less apt to freeze. Penguins' high density of feathers and a subdermal fat layer keeps the birds warm despite subzero temperatures.

Endotherms are also known to use behavior modification as a way to stay warm. After all, when birds migrate south in the fall, they are relocating to warmer climates, and mammals such as groundhogs and black bears hibernate during the winter months. During hibernation, animals go into a very deep, sleeplike state in which their heartbeat slows drastically. When groundhogs come out of their burrows, they are checking to see if it is time to stop hibernating. (However, it's a superstition that, if a groundhog does not see its shadow, winter is over and spring is on the way.)

Adaptations to stay warm help determine where various animals reside and help account for why we do not find polar bears in the tropics or iguanas in the Arctic Circle! But animals have adaptations to stay cool, also. Reptiles and other invertebrates move to shady locations when the sun gets too hot. Endotherms, such as humans, become flushed as blood rushes to the skin, where cooling breezes and active sweat glands can dissipate heat. Evaporation of water at the tongue and mouth of dogs serves the same purpose. In this chapter, we focus on how animals maintain a stable internal environment, regardless of the external environment. But first we lay the groundwork for that discussion by describing the overall organization of animal bodies.

These animals are adapted to living in a cold climate.

Caribou, *Rangifer tarandus*

Polar bear, *Ursus maritimus*

Emperor penguins, *Aptenodytes forsteri*

Types of Tissues

A review of the biological levels of organization serves as a backdrop to this chapter, which will consider the structure and function of various tissues, organs, and organ systems. Four different tissue types make up the body of a complex animal and contribute to maintaining homeostasis. These tissues are termed epithelial, connective, muscular, and nervous tissues.

26.1 Levels of biological organization are evident in animals

LEARNING OUTCOME

When you complete this section, you should be able to

1. State the levels of biological organization with reference to a particular organ system.

Figure 26.1 depicts the levels of biological organization in the body of a mammal. **①** The trillions of cells in a mammal's body are its cellular level of organization. Although there are about 100 different types of cells, only like cells make up a tissue. **②** A **tissue** is a group of specialized cells performing a similar function. We will see that the four basic types of tissue in an animal's body are epithelial tissue, connective tissue, muscular tissue, and nervous tissue. **③** An **organ** contains various types of tissues arranged in a particular fashion. In other words, the structure and function of an organ are dependent on the tissues it contains. The many examples given throughout this chapter will help you understand why it is sometimes said that tissues, not organs, are the structural and functional units of the body. **④** Several organs make up an **organ system,** and the organs of the system work together to perform necessary functions for **⑤** the **organism.**

The role of the urinary system is to produce, store, and rid the body of metabolic wastes. These vital functions are dependent on its organs, which in turn are dependent on the tissues making up the organs. The urinary system contains these organs: two kidneys, two ureters, a bladder, and a urethra (Fig. 26.1 **④**). The kidneys remove waste molecules from the blood and produce urine. The tubular shape of the ureters is suitable for passing urine to the bladder, which can store urine because its tissue lining is expandable. Urine passes out of the body by way of the tubular urethra.

Other examples show that an organ's function is dependent on a tissue's specialized cells. In the digestive system, the intestine absorbs nutrients. The cells of the tissue lining the lumen (cavity) of the small intestine have microvilli that increase the available surface area for absorption. Within muscular tissue, muscle cells shorten

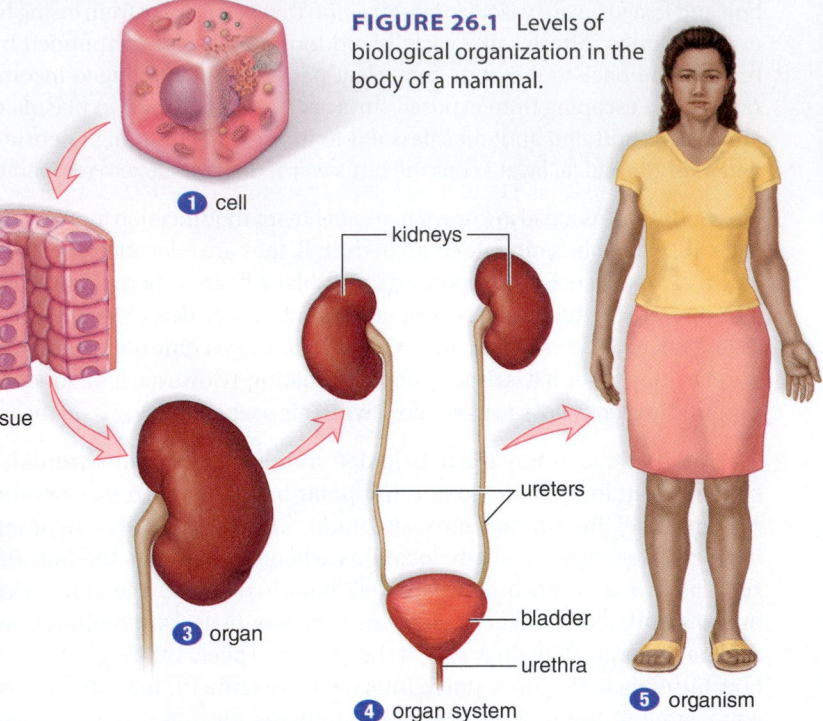

FIGURE 26.1 Levels of biological organization in the body of a mammal.

① cell
② tissue
③ organ
④ organ system
kidneys
ureters
bladder
urethra
⑤ organism

when they contract because they have intracellular components that move past one another. Within nervous tissue, nerve cells have long, slender projections that carry impulses to distant body parts. The biological axiom that "structure suits function," and vice versa, begins with the specialized cells within a tissue. And thereafter, this truism applies also to organs and organ systems.

▶ **26.1 CHECK YOUR PROGRESS**

1. State the levels of biological organization with reference to the urinary system in a complex animal.

26.2 Epithelial tissue covers organs and lines body cavities

LEARNING OUTCOME

When you complete this section, you should be able to

1. Compare the different types of epithelial tissues on the basis of structure and function.

Epithelial tissue, also called *epithelium,* forms the external and internal linings of many organs and covers the surface of the body. Therefore, in order for a substance to enter or exit the body at the digestive tract, the lungs, or the urinary tract, it must cross an epithelial tissue. Epithelial cells adhere to one another, but

are generally only one cell thick. This characteristic enables them to fulfill a protective function and yet allows substances to pass through to a tissue beneath them.

MP3
Epithelial Tissue

Epithelial cells are connected to one another by three types of junctions (see Fig. 5.4B): tight junctions, adhesion junctions, and gap junctions. Because epithelial cells are joined by tight junctions in the intestine, the gastric juices stay out of the body, and the same holds true in the kidneys, where tight junctions cause the urine to stay within the kidney tubules. Adhesion junctions allow epithelial cells in the skin to stretch and bend, while gap junctions permit molecules to pass between adjacent cells. Epithelial cells are exposed on one side, but on the other side they have a basement membrane. The **basement membrane** is simply two thin layers of proteins that anchor the epithelium to underlying connective tissue.

The cells of epithelial tissue differ in shape (Fig. 26.2). **1** Simple **squamous epithelium,** such as that lining the air spaces of the lungs and the central cavity of blood vessels, is composed of a single layer of flattened cells attached to the basement membrane. **2** Simple **cuboidal epithelium,** which lines the lumen of the kidney tubules, contains cube-shaped cells. **3** Simple **columnar epithelium** is a single layer of cells resembling rectangular pillars or columns, with nuclei usually located near the bottom of each cell. Simple columnar epithelium lines the lumen of the digestive tract. The cells have tiny projections called microvilli that increase the absorptive surface area.

Aside from cell type, epithelial tissue is classified according to the number of layers in the tissue. An epithelium is simple when it has one layer of cells and stratified when it has several layers of cells piled on top of one another. As we shall see, the protective outer layer of skin is **stratified** squamous epithelium, but these cells have been reinforced by keratin, a protein that provides

strength. When an epithelium is pseudostratified, it appears to be layered, but true layers do not exist because each cell touches the baseline. An example is **4** **pseudostratified ciliated columnar epithelium,** which lines the trachea (windpipe). Along the trachea, goblet cells produce mucus that traps foreign particles, and the upward motion of cilia carries the mucus to the back of the throat (pharynx), where it may be either swallowed or expelled. Smoking can cause a change in mucus secretion and inhibit ciliary action, resulting in an inflammatory condition called chronic bronchitis.

When an epithelium secretes a product, it is said to be glandular. A **gland** can be a single epithelial cell, such as a mucus-secreting goblet cell, or a gland can contain many cells. Glands that secrete their product into ducts are called exocrine glands, and those that secrete their products into the bloodstream are called endocrine glands.

▶ **26.2 CHECK YOUR PROGRESS**

1. Describe how the structure of squamous epithelium is appropriate to its function.
2. Compare the structure and function of simple columnar epithelial tissue to that of pseudostratified ciliated columnar tissue.

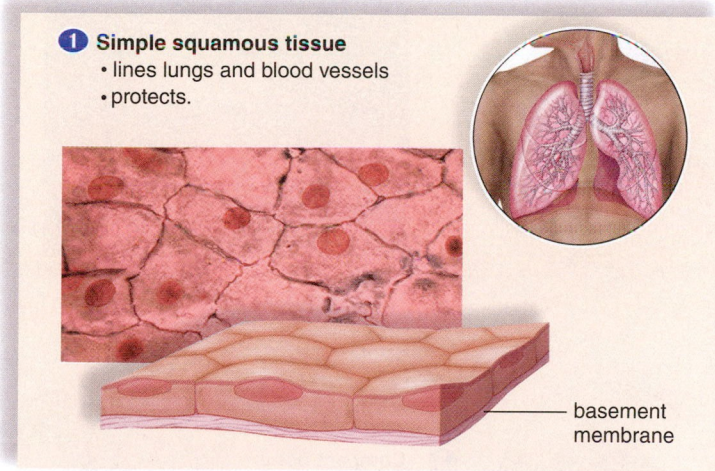

1 Simple squamous tissue
• lines lungs and blood vessels
• protects.

basement membrane

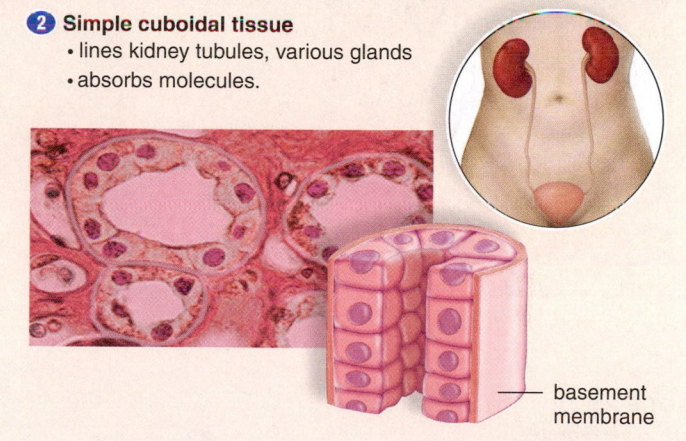

2 Simple cuboidal tissue
• lines kidney tubules, various glands
• absorbs molecules.

basement membrane

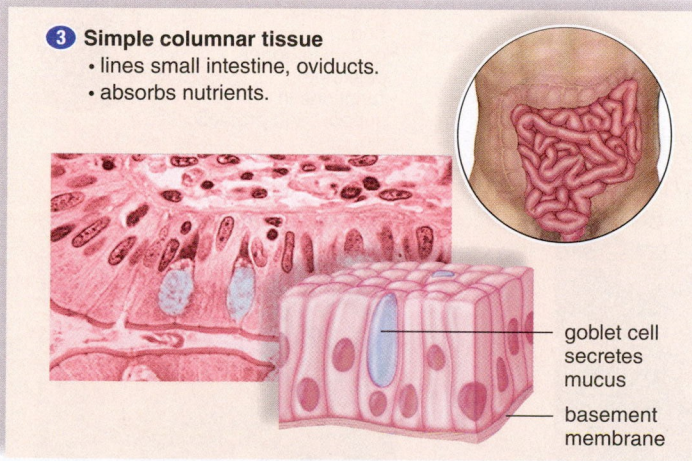

3 Simple columnar tissue
• lines small intestine, oviducts.
• absorbs nutrients.

goblet cell secretes mucus

basement membrane

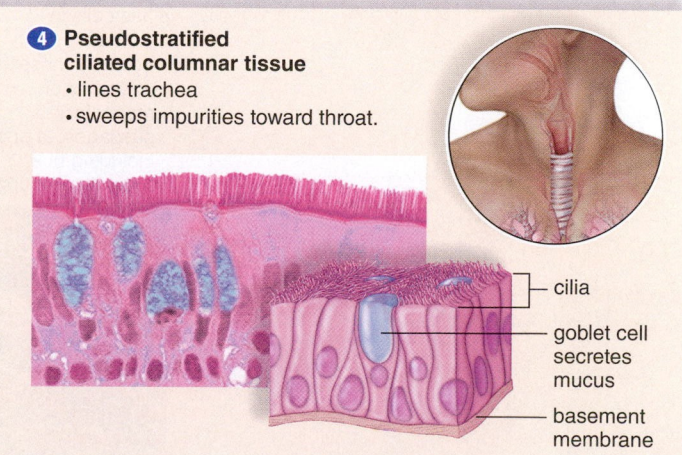

4 Pseudostratified ciliated columnar tissue
• lines trachea
• sweeps impurities toward throat.

cilia

goblet cell secretes mucus

basement membrane

FIGURE 26.2 Types of epithelial tissue in vertebrates.

26.3 Connective tissue connects and supports other tissues

LEARNING OUTCOME

When you complete this section, you should be able to

1. Compare the different types of connective tissues on the basis of structure and function.

Connective tissue is the most abundant and widely distributed tissue in vertebrates. The many different types of **connective tissue** are all involved in binding organs together and providing support and protection. As a rule, connective tissue cells are widely separated by a **matrix,** a noncellular material that varies from solid to semifluid to fluid. The matrix usually has fibers, notably collagen fibers. Collagen is the most common protein in the human body, which gives you

some idea of how prevalent connective tissue is. The fibers lend support and also make connective tissue resilient (able to adapt to changes).

Loose Fibrous and Related Connective Tissues Let's consider **loose fibrous connective tissue** first, and then compare the other types to it (Fig. 26.3A). ❶ This tissue occurs beneath an epithelium and connects it to other tissues within an organ. It also forms a protective covering for many internal organs, such as muscles, blood vessels, and nerves. Its cells are called **fibroblasts** because they produce a matrix that contains fibers, including collagen fibers and elastic fibers. The presence of loose fibrous

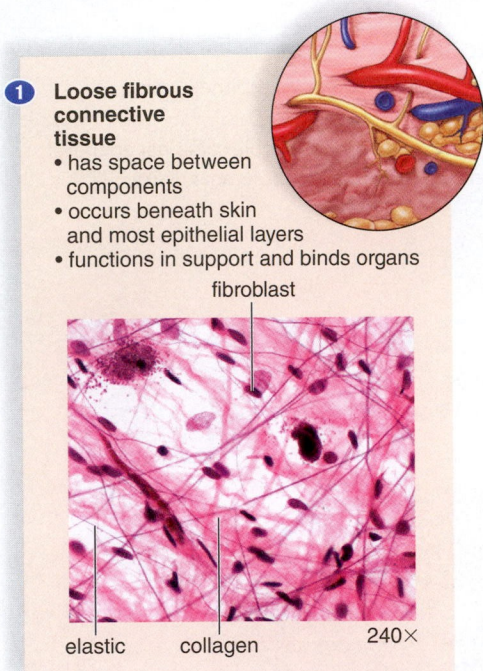

❶ **Loose fibrous connective tissue**
- has space between components
- occurs beneath skin and most epithelial layers
- functions in support and binds organs

fibroblast

elastic   collagen   240×

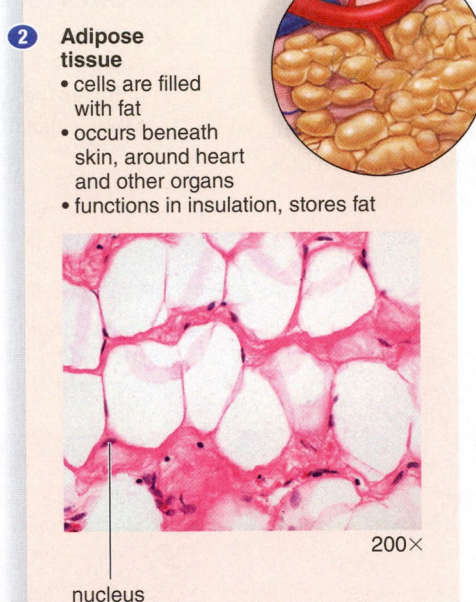

❷ **Adipose tissue**
- cells are filled with fat
- occurs beneath skin, around heart and other organs
- functions in insulation, stores fat

200×

nucleus

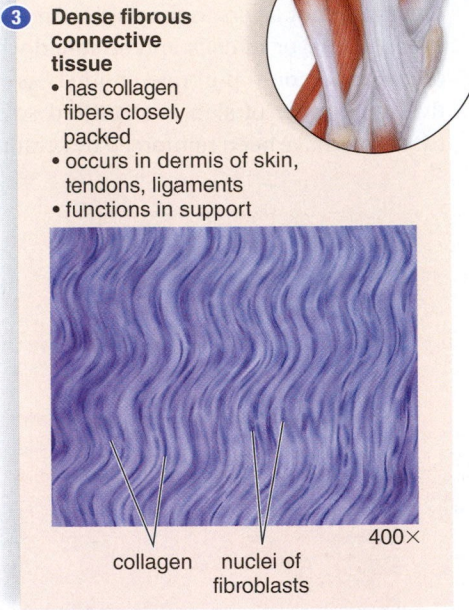

❸ **Dense fibrous connective tissue**
- has collagen fibers closely packed
- occurs in dermis of skin, tendons, ligaments
- functions in support

400×

collagen   nuclei of fibroblasts

FIGURE 26.3A Types of connective tissue in vertebrates.

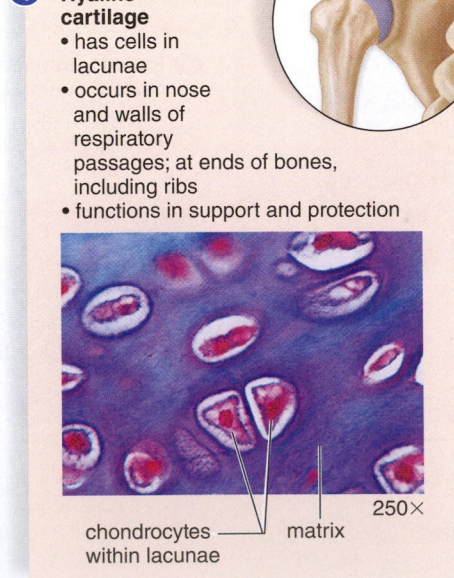

❹ **Hyaline cartilage**
- has cells in lacunae
- occurs in nose and walls of respiratory passages; at ends of bones, including ribs
- functions in support and protection

250×

chondrocytes within lacunae   matrix

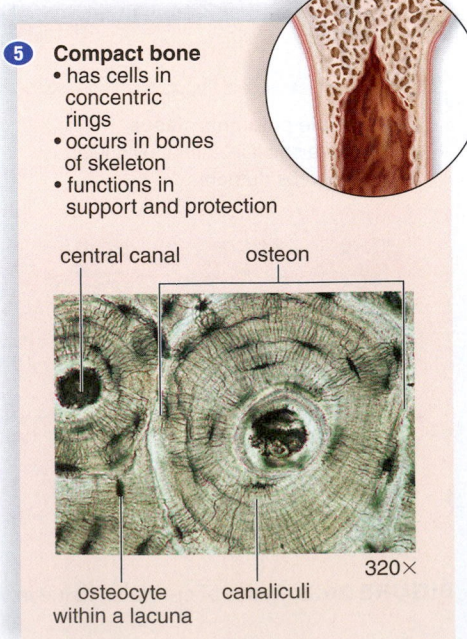

❺ **Compact bone**
- has cells in concentric rings
- occurs in bones of skeleton
- functions in support and protection

central canal   osteon

320×

osteocyte within a lacuna   canaliculi

connective tissue in the walls of the lungs and the arteries allows these organs to expand.

2 Adipose tissue is a type of loose connective tissue in which the fibroblasts enlarge and store fat, and there is limited matrix. Adipose tissue is located beneath the skin and around organs, such as the heart and the kidneys. The body uses this stored fat for energy, insulation, and organ protection.

Compared to loose fibrous connective tissue, **3 dense fibrous connective tissue** contains more collagen fibers, and they are packed closely together. This type of tissue has more specific functions than does loose fibrous connective tissue. For example, dense fibrous connective tissue is found in the dermis of the skin; in **tendons,** which connect muscles to bones; and in **ligaments,** which connect bones to other bones at joints.

In **cartilage,** the cells lie in small chambers called lacunae, separated by a matrix that is solid yet flexible. Unfortunately, because this tissue lacks a direct blood supply, it heals very slowly. **4 Hyaline cartilage,** the most common type of cartilage, contains only very fine collagen fibers. The matrix has a white, translucent appearance. Hyaline cartilage is found in the nose and at the ends of the long bones and the ribs, and it forms rings in the walls of respiratory passages. The human fetal skeleton is also made of this type of cartilage, which is later replaced by bone. Cartilaginous fishes, such as sharks, have a cartilaginous skeleton throughout their lives.

Bone is the most rigid connective tissue. It consists of an extremely hard matrix of inorganic salts, notably calcium salts, deposited around collagen fibers. The inorganic salts give bone rigidity, and the collagen fibers provide elasticity and strength, much as steel rods do in reinforced concrete. **5 Compact bone,** the most common type, consists of cylindrical structural units called osteons. Rings of hard matrix surround the central canal, which contains blood vessels. Bone cells (osteocytes) are located in lacunae between the rings of matrix. Blood vessels in the central canal carry nutrients that allow bone to renew itself. The nutrients can reach all of the cells because minute canals (canaliculi) containing thin extensions of the osteocytes connect osteocytes with one another and eventually with the central canal. These connections allow the osteocytes to have a constant supply of blood and nutrients.

Blood Blood is composed of several types of cells suspended in a liquid matrix called plasma (Fig. 26.3B). Blood is unlike other types of connective tissue in that the matrix (i.e., plasma) is not made by the cells. Some people do not classify blood as connective tissue; instead, they suggest a separate tissue category called vascular tissue.

Blood serves the body well. It transports nutrients and oxygen to cells and removes their wastes. It helps distribute heat and plays a role in fluid, ion, and pH balance. Also, various components of blood help protect us from disease, and blood's ability to clot prevents fluid loss.

Blood contains two types of cells. **Red blood cells** are small, biconcave, disk-shaped cells without nuclei. The presence of the red pigment hemoglobin makes the cells red and, in turn, makes the blood red. Hemoglobin combines with oxygen, and in this way, red blood cells transport oxygen. At a crime scene, the pigment portion of hemoglobin makes it difficult for the perpetrator to remove traces of blood. Forensic specialists can perform tests to confirm that a stain is due to hemoglobin. One test

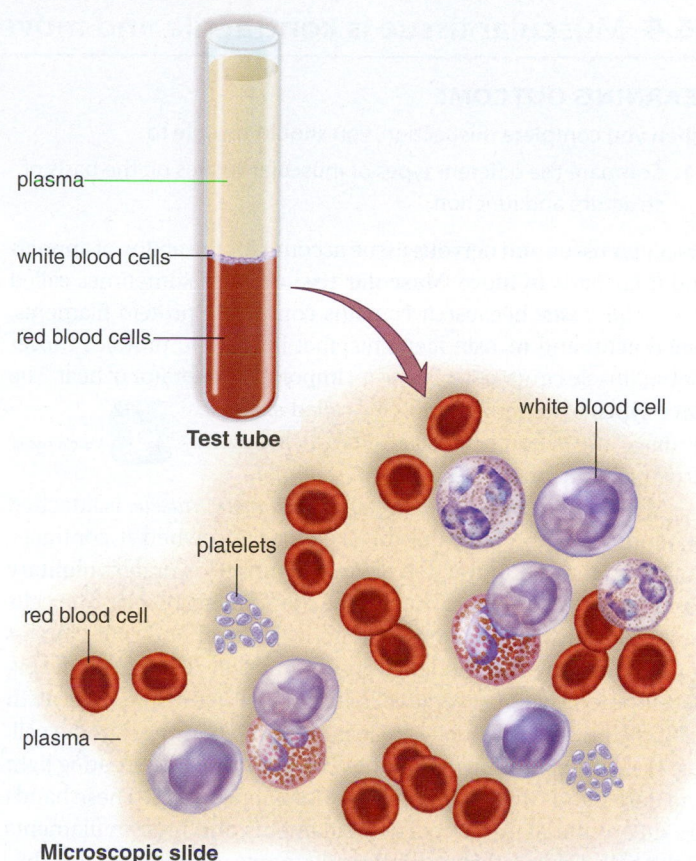

FIGURE 26.3B Composition of blood, a liquid connective tissue.

involves spraying the stain with luminal, a chemical that binds with blood and then glows in the dark. More refined tests can identify the blood as belonging to a specific person.

White blood cells may be distinguished from red blood cells by the fact that they are usually larger, have a nucleus, and without staining, would appear translucent. White blood cells fight infection in two primary ways: Some white blood cells are phagocytic and engulf infectious pathogens, while other white blood cells produce antibodies, molecules that combine with foreign substances and inactivate them.

Platelets are another component of blood, but they are not complete cells; rather, they are fragments of giant cells present only in bone marrow. When a blood vessel is damaged, platelets form a plug that seals the vessel, and injured tissues release molecules that help the clotting process.

Lymph is a liquid connective tissue located in lymphatic vessels. Lymph is derived from **tissue fluid** (water and various dissolved solutes) that bathes the cells. Excess tissue becomes lymph when it is taken up by lymphatic vessels. Lymph nodes, composed of fibrous connective tissue, occur along the length of lymphatic vessels. Lymph is cleansed as it passes through lymph nodes, particularly because white blood cells congregate there.

▶ **26.3 CHECK YOUR PROGRESS**

1. Compare the structure and function of loose fibrous connective tissue to that of compact bone tissue.
2. Explain why blood is classified as a connective tissue.

LEARNING OUTCOME

When you complete this section, you should be able to

1. Compare the different types of muscular tissues on the basis of structure and function.

Muscular tissue and nervous tissue account for the ability of animals and their parts to move. **Muscular tissue** is also sometimes called *contractile tissue* because it contains contractile protein filaments, called actin and myosin filaments, that interact to produce movement. Muscle contraction is also an important generator of heat. The three types of vertebrate muscles called skeletal, cardiac, and smooth muscle differ in location, structure, and function (Fig. 26.4).

MP3 Muscle Tissue

1 **Skeletal muscle,** also called *voluntary muscle,* is attached by tendons to the bones of the skeleton, and when it contracts, the bones move. Contraction of skeletal muscle is under voluntary control and occurs faster than in the other muscle types. The cells of skeletal muscle, called fibers, are cylindrical and quite long—sometimes they run the length of the muscle. They arise during development when several cells fuse, resulting in one fiber with multiple nuclei. The nuclei are located at the periphery of the cell, just inside the plasma membrane. The fibers have alternating light and dark bands that give them a **striated** appearance. These bands are due to the placement of actin filaments and myosin filaments in the cell. The interaction of these filaments accounts for the ability of all three types of muscles to contract (see Fig. 5.6C, p. 95).

2 **Cardiac muscle** is found only in the walls of the heart, and its contraction pumps blood and accounts for the heartbeat. Like skeletal muscle, cardiac muscle has striations, but the contraction of the heart is autorhythmic and involuntary. Cardiac muscle cells also differ from skeletal muscle cells in that they have a single, centrally placed nucleus. The cells are branched and seemingly fused with one another, and the heart appears to be composed of one large, interconnecting mass of muscle cells. Actually, cardiac muscle cells are separate and individual, but they are bound end-to-end at **intercalated disks**, areas where adhesion and gap junctions in folded plasma membranes allow the contraction impulse to spread from one cell to the other.

3 **Smooth muscle** is so named because the cells lack striations. The spindle-shaped cells form layers in which the thick middle portion of one cell is opposite the thin ends of adjacent cells. Consequently, the nuclei form an irregular pattern in the tissue. Smooth muscle is not under voluntary control, and therefore is said to be *involuntary*. Smooth muscle is also sometimes called *visceral muscle* because it is found in the walls of the viscera (intestine, stomach, and other internal organs) and blood vessels. Smooth muscle contracts more slowly than skeletal muscle but can remain contracted for a longer time. When the smooth muscle of the intestine contracts, food moves along its lumen (central cavity). When the smooth muscle of the blood vessels contracts, blood vessels constrict, helping to raise blood pressure. Small amounts of smooth muscle are also found in the iris of the eye, and in the skin, accounting for the ability of hair to stand up.

▶ **26.4 CHECK YOUR PROGRESS**

1. Compare the structure and function of skeletal muscular tissue to that of smooth muscular tissue.
2. Explain why it is beneficial for smooth muscle to be involuntary.

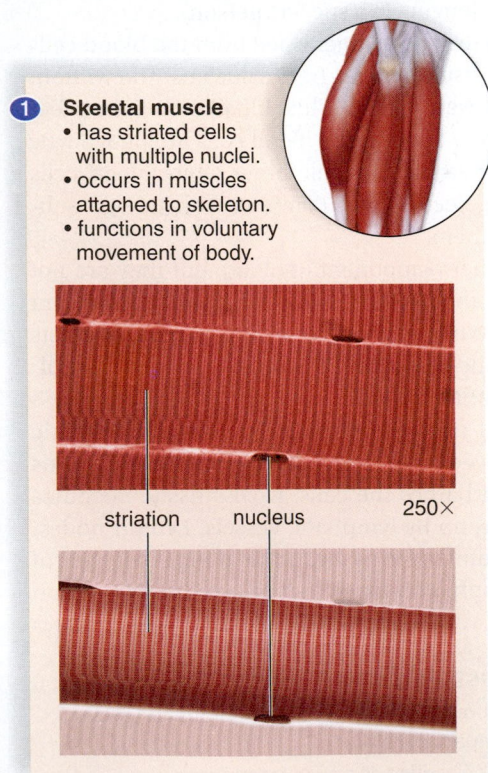

1 **Skeletal muscle**
- has striated cells with multiple nuclei.
- occurs in muscles attached to skeleton.
- functions in voluntary movement of body.

striation   nucleus   250×

2 **Cardiac muscle**
- has branching, striated cells, each with a single nucleus.
- occurs in the wall of the heart.
- functions in the pumping of blood.
- is involuntary.

intercalated disk   nucleus   250×

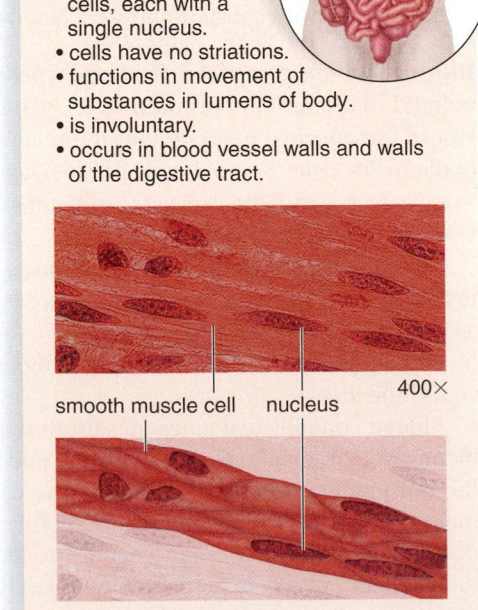

3 **Smooth muscle**
- has spindle-shaped cells, each with a single nucleus.
- cells have no striations.
- functions in movement of substances in lumens of body.
- is involuntary.
- occurs in blood vessel walls and walls of the digestive tract.

smooth muscle cell   nucleus   400×

FIGURE 26.4 Types of muscular tissue.

26.5 Nervous tissue communicates with and regulates the functions of the body's organs

LEARNING OUTCOMES

When you complete this section, you should be able to

1. Identify the parts of a neuron and their specific functions.
2. In general, describe the structure and function of neuroglia.

Nervous tissue coordinates body parts and allows an animal to respond to the environment. Nerves conduct impulses from sensory receptors to the spinal cord and the brain, where integration occurs. The phenomenon called sensation occurs only in the brain. However, nerves also conduct nerve impulses away from the spinal cord and brain to the muscles and glands, causing them to contract and secrete, respectively. In this way, a coordinated response to both internal and external stimuli is achieved.

 **MP3** Nervous Tissue

A nerve cell is called a **neuron.** Every neuron has three parts: dendrites, a cell body, and an axon (Fig. 26.5). A *dendrite* is an extension that conducts signals toward the cell body. The *cell body* contains the major concentration of the cytoplasm and the nucleus of the neuron. An *axon* is an extension that conducts nerve impulses. The brain and spinal cord contain many neurons, whereas **nerves** contain only the axons of neurons. The dendrites and cell bodies of these neurons are located in the spinal cord or brain, depending on whether it is a spinal nerve or a cranial nerve. A nerve is like a land-based telephone trunk cable, because just like telephone wires, each axon is a communication channel independent of the others.

Neuroglia Neuroglia are cells that outnumber neurons as much as 50 to 1, and take up more than half the volume of the brain. Although the primary function of neuroglia is to support and nourish neurons, research is currently being conducted to determine how much they directly contribute to brain function. Neuroglia do not have axons, but even so, researchers are now beginning to gather evidence that they do communicate among themselves and with neurons! Neuroglia also form a plasma-like solution called **cerebrospinal fluid,** which supports and nourishes the brain and spinal cord. Various types of neuroglia are present in the brain. Microglia, astrocytes, and oligodendrocytes are shown in Figure 26.5. Microglia, in addition to supporting neurons, engulf bacterial and cellular debris. Astrocytes provide nutrients to neurons and produce a hormone known as glia-derived growth factor, which someday might be used as a cure for Parkinson disease and other conditions caused by neuron degeneration. (Parkinson is a movement disorder characterized by tremor, rigidity, slowness, and poor balance.) Oligodendrocytes form **myelin,** a whitish, fatty substance that forms a sheath and acts much like the insulation of a telephone cable.

Mature neurons have little capacity for cell division and seldom form tumors. The majority of brain tumors in adults involve actively dividing neuroglia. Most brain tumors have to be treated with surgery or radiation therapy because tight junctions between cells lining the blood vessels do not permit drugs to pass through to the brain.

▶ **26.5 CHECK YOUR PROGRESS**
1. Compare the structure and function of the dendrites to the axon of a motor neuron.
2. Explain the benefit of a long axon in motor neurons.

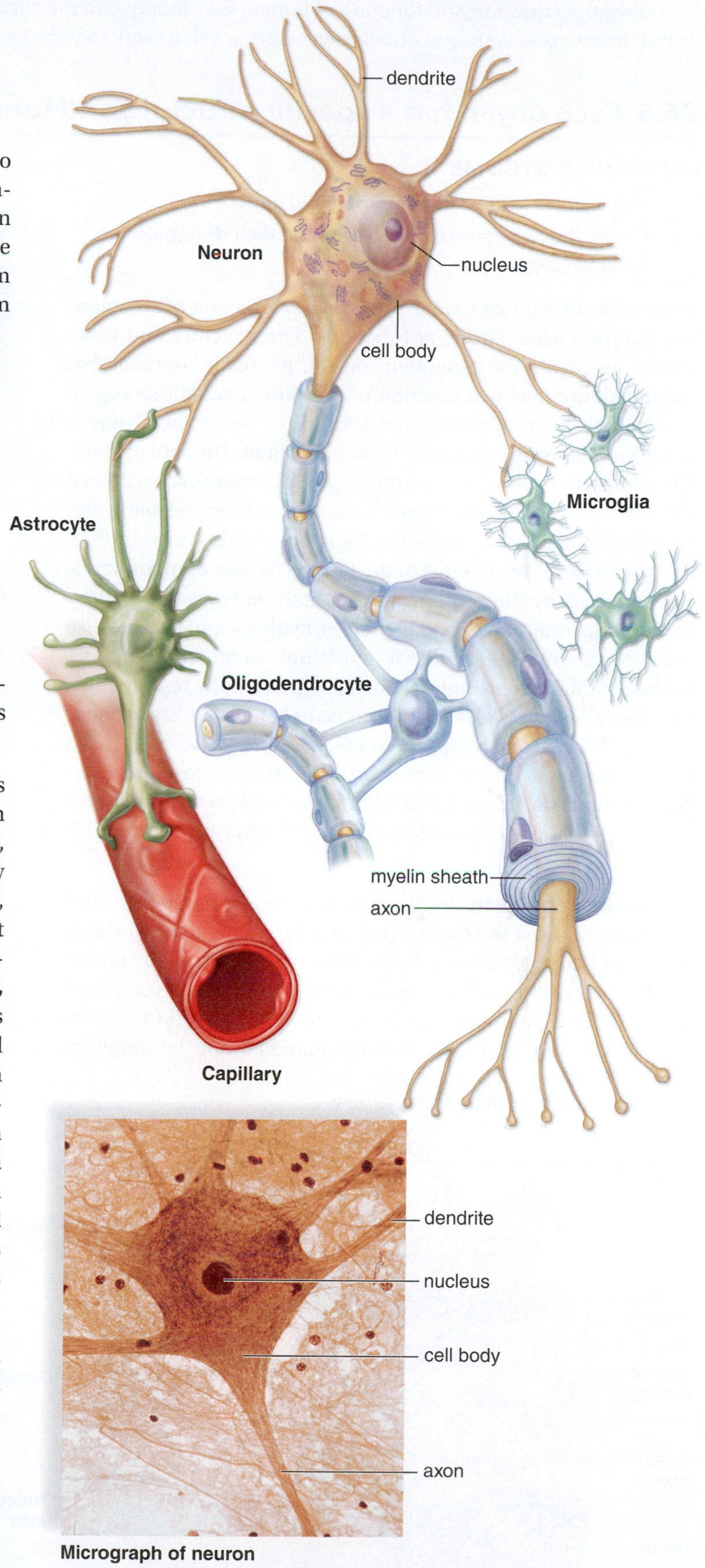

Neuron — dendrite
nucleus
cell body
Astrocyte
Microglia
Oligodendrocyte
myelin sheath
axon
Capillary

dendrite
nucleus
cell body
axon

Micrograph of neuron

FIGURE 26.5 Parts of a neuron and types of neuroglia.

Organs and Organ Systems

A study of the structure and function of human skin, along with the subcutaneous layer, exemplifies how the structure of an organ suits its function. Organ systems contain several organs that work together as they accomplish particular functions in the body.

26.6 Each organ has a specific structure and function

LEARNING OUTCOME

When you complete this section, you should be able to

1. Identify the two main regions of skin and their distinguishing characteristics.

As stated in section 26.1, an organ is a structural unit of an organism that performs particular functions. The structure and function of an organ are dependent on the tissues it contains. For example, the protective function of the **skin,** a vertebrate organ, is enhanced by a thickened and keratinized epidermis, covered by different derivatives according to the animal: The skin of fishes has numerous bony scales; amphibians have smooth skin covered with mucous glands; most reptiles have a skin of epidermal scales that vary in color and shape but birds have scales only on their legs—feathers cover the rest of the body. The skin of mammals is characterized by the presence of hair and derivative structures, such as fingernails and toenails. Scales, feathers, and hair are also involved in controlling body temperature, as are sweat glands. Sweating in mammals not only helps regulate body temperature, but also allows the skin to excrete salts and even nitrogenous wastes. The sensory receptors in the skin provide us with much knowledge about the outside world. In addition to these functions of the skin, mammalian skin cells manufacture precursor molecules that are converted to vitamin D after exposure to UV (ultraviolet) light.

Regions of the Skin In humans, skin has two regions, called the epidermis and the dermis (Fig. 26.6A). A subcutaneous layer, also called the hypodermis, lies between the skin and any underlying structures, such as muscle or bone. **Epidermis** is composed of stratified squamous epithelium. New cells derived from stem cells in the germinal layer become flattened and hardened as

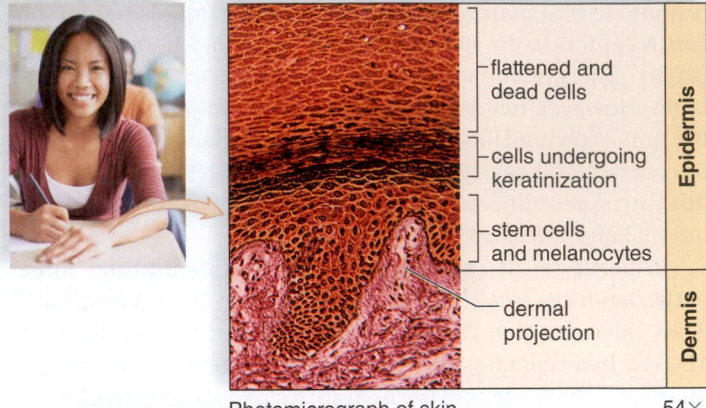

Photomicrograph of skin   54×

FIGURE 26.6B Photomicrograph of skin. Stem cells are a source of new epidermal cells to replace those that get sloughed off.

they are pushed to the surface. Hardening takes place because the cells produce keratin, a waterproof protein. A thick layer of dead keratinized cells arranged in concentric spirals forms fingerprints and footprints. Specialized cells in the epidermis called melanocytes produce melanin, the pigment responsible for skin color (Fig. 26.6B). In a light-skinned person, tanning signifies that melanocytes are trying to protect the skin from the dangerous rays of the sun. Too much ultraviolet radiation can lead to skin cancer.

Video Skin Color Gene

The **dermis** is a region of dense fibrous connective tissue beneath the epidermis. With age and exposure to the sun, the number of collagen and elastic fibers in the dermis decreases, resulting in wrinkles. Botox is a diluted form of a bacterial toxin that blocks innervation of muscles and reduces wrinkling (see Fig. 29C on p. 585). In addition to fibers, the dermis contains blood vessels, hair follicles, oil and sweat glands, and sensory receptors for touch, pressure, pain, heat, and cold (see Fig. 26.6A).

Technically speaking, the **subcutaneous layer** beneath the dermis is not part of the skin. It is composed of loose connective tissue and adipose tissue, which stores fat. A well-developed subcutaneous layer gives the body a rounded appearance, provides protective padding against external assaults, and reduces heat loss. Excessive development of the subcutaneous layer accompanies obesity.

MP3 Human Skin

FIGURE 26.6A Human skin anatomy.

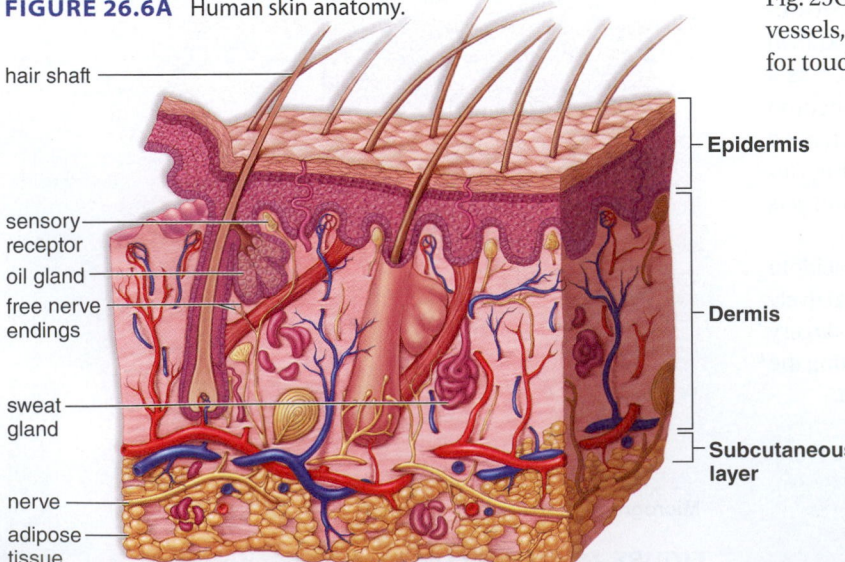

- hair shaft
- sensory receptor
- oil gland
- free nerve endings
- sweat gland
- nerve
- adipose tissue
- Epidermis
- Dermis
- Subcutaneous layer

▶ 26.6 CHECK YOUR PROGRESS

1. Identify the region of skin that contains stratified squamous epithelium and describe its structure and function.
2. Explain how the structure of skin is an adaptation to the land environment.

26A UV Rays: Too Much or Too Little?

The sun is the major source of energy for life on Earth. Without the sun, plants and then most organisms would quickly die out. But the sun's energy can also be damaging. In addition to visible light, the sun emits ultraviolet (UV) radiation, which is more powerful than visible light. UV radiation occurs as three types: long-wavelength UVA, medium-wavelength UVB, and short-wavelength UVC; only UVA and UVB reach the Earth's surface; because UVC is filtered out by the atmosphere, scientists have developed a UV index to measure the strength of UV radiation under various environmental circumstances. In general, the more southern the city, the higher the UV index and the greater the risk of skin cancer. Maps are available that indicate the daily risk levels for various U.S. regions and these can be viewed at http://www.weather.gov/os/uv/.

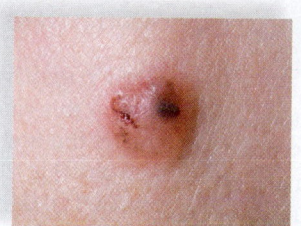 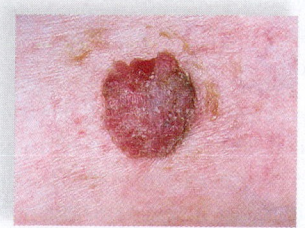 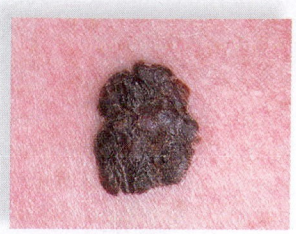

a. Basal cell carcinoma starts in basal cells of epidermis; often occurs on face

b. Squamous cell carcinoma starts in top layer of skin; occurs in areas exposed to sun

c. Melanoma causes pigmented cells to proliferate; spreads to other body parts

FIGURE 26A Skin cancer.

Both UVA and UVB rays can damage skin cells. Too much exposure to UVB rays can directly damage skin cell DNA, leading to mutations that can cause skin cancer. UVB is also responsible for the pain and redness characteristic of a sunburn. Tanning results from UVA rays which penetrate deeply into the skin, damage collagen fibers and induce premature aging of the skin. Tanning occurs when melanin granules increase in keratinized cells at the surface of the skin as a way to prevent further damage by UV rays. UVA rays can also promote skin cancer by causing the production of highly reactive chemicals that indirectly damage DNA.

Skin cancer is the most commonly diagnosed type of cancer in the United States, outnumbering cancers of the lung, breast, prostate, and colon combined. The most common type of skin cancer is basal cell carcinoma (Fig. 26A*a*), which is rarely fatal, but can be disfiguring if allowed to grow. Squamous cell carcinoma (Fig. 26A*b*), the second most common type, is fatal in about 1% of cases. The most deadly form is melanoma, which occurs in adolescents and young adults as well as in older people. If detected early, over 95% of patients survive at least 5 years, but if the cancer cells have already spread throughout the body, only 10–20% can expect to live this long.

Video Melanoma Marker

Melanoma affects pigmented cells and often has the appearance of an unusual mole (Fig. 26A*c*). Any moles that become malignant are removed surgically. If the cancer has spread, chemotherapy and a number of other treatments are also available. In March 2011, the FDA approved a drug called ipilimumab (Yervoy), after it was shown to extend the lives of patients with advanced, late-stage melanoma.

According to the Skin Cancer Foundation, about 90% of nonmelanoma skin cancers, and 65% of melanomas, are associated with exposure to UV radiation from the sun. So how can we protect ourselves? First, try to minimize sun exposure between 10 A.M. and 4 P.M. (when the UV rays are most intense) by wearing protective clothing, hats, and sunglasses. Second, use sunscreen. Sunscreens generally do a better job of blocking UVB than UVA rays. In fact, the SPF, or sun protection factor printed on sunscreen labels, refers only to the degree of protection against UVB. Many sunscreens don't provide as much protection against UVA, and because UVA doesn't cause sunburn, people may have a false sense of security. Some sunscreens do a better job of blocking UVA—look for those that contain zinc oxide, titanium dioxide, avobenzone, or Mexoryl SX. Unfortunately, tanning salons use lamps that emit UVA rays that are two to three times more powerful than the UVA rays emitted by the sun. Because of the potential damage to deeper layers of skin, most medical experts recommend avoiding indoor tanning salons altogether.

Because UV light is potentially damaging, why haven't all humans developed the more protective dark skin that is common to humans living in tropical regions? It turns out that vitamin D is produced in the body only when UVB rays interact with a form of cholesterol found mainly in the skin. This "sunshine vitamin" serves several important functions in the body, including keeping bones strong, boosting the immune system, and reducing blood pressure. Certain foods also contain vitamin D, but it can be difficult to obtain sufficient amounts through diet alone. Therefore, in more temperate areas of the planet, lighter-skinned individuals have the advantage of being able to synthesize sufficient vitamin D. Interestingly, dark-skinned people living in such regions may be at increased risk for vitamin D deficiency.

So how much sun exposure is enough in temperate regions of the world? During the summer months, an average fair-skinned person will synthesize plenty of vitamin D after exposure to 10–15 minutes of midday sun. During winter months, however, anyone living north of Atlanta probably receives too few UV rays to stimulate vitamin D synthesis, and therefore they must fulfill their requirement through their diet.

CONSIDER THESE QUESTIONS

1. Which is more important to you—having a "healthy-looking" tan now, or preserving your skin's health later in life? Why?
2. There is considerable controversy about the level of dietary vitamin D intake that should be recommended. What factors can affect your need for vitamin D?

 **connect** | BIOLOGY

Explore the concepts through a variety of multimedia assets and question types.

www.mcgrawhillconnect.com

LEARNING OUTCOMES

When you complete this section, you should be able to

1. List one or more major organs within each system.
2. Describe the general function(s) for each system.

An organ system is a collection of organs that work together to perform related roles in the organism. The functions of an organ system are dependent on its organs. To take an example, the function of the urinary system is to produce urine, store it, and then transport it out of the body. As stated, the kidneys produce urine, and then tubes called ureters transport it to the urinary bladder for storage until it is released from the body by way of a tube called the urethra.

For the sake of discussion, we will group the organ systems of mammals (e.g., humans) according to these functions: control, protective and motor activity, transport, maintenance, and reproductive. All organ systems are ultimately involved in maintaining homeostasis, the stability of the body's internal environment. An organism has both an external environment and an internal environment. Vertebrates are able to regulate the internal environment so that it remains constant, despite fluctuations in the external environment.

The Control Systems Both the nervous and endocrine systems coordinate and regulate the functions of the body's other systems. The **nervous system** (Fig. 26.7A, *left*) consists of the brain, the spinal cord, and associated nerves. Nerves conduct nerve impulses from sensory receptors (present for example in the eyes and ears) to the brain and spinal cord. They also conduct nerve impulses from the brain and spinal cord to the muscles and glands, allowing us to respond to both external and internal stimuli.

The **endocrine system** (Fig. 26.7A, *right*) consists of the hormonal glands, which secrete chemicals that serve as messengers between body parts. The endocrine system also helps maintain the proper functioning of the male and female reproductive organs.

Both the nervous system and the endocrine system coordinate body parts. However, the nervous system is fast-acting, while the endocrine system is slower but has more lasting effects.

The Protective and Motor Activity Systems The three systems in this category (integumentary, skeletal, and muscular) all protect the body and are involved in motor activity. The **integumentary** (Fig. 26.7B, *left*) **system** consists of the skin and accessory organs. The skin protects the body and contains sensory receptors that initiate nerve impulses that go to the brain. The brain may then order muscles to contract and/or glands to secrete. For example, the skin contains sensory receptors that detect a cold temperature and the response can be shivering. Section 26.9 discusses more completely how the skin is involved in temperature regulation.

The **skeletal system** and the **muscular system** (Fig. 26.7B, *middle, right*) enable the body and its parts to move as a result of nerve stimulation. The skeleton, as a whole, serves as a place of attachment for the skeletal muscles. Contraction of muscles in the muscular system accounts for the actual movement of body parts.

The skeletal and muscular systems, along with the integumentary system, also protect and support the body. The bones of the skeleton protect body parts. For example, the skull forms a protective encasement for the brain, as does the rib cage for the heart and lungs. The skin and muscles assist in this endeavor because they are exterior to the bones.

Bones serve other functions as well. Red bone marrow produces blood cells, and bones serve as storage areas for calcium and phosphate salts.

The Transport Systems Two systems, the cardiovascular system and the lymphatic system, work together to maintain the fluidity of blood as blood, in tiny vessels called capillaries. makes exchanges in the lungs, kidneys, and the digestive tract. These exchanges permit blood to maintain its normal composition so that it can continue to serve the body's cells.

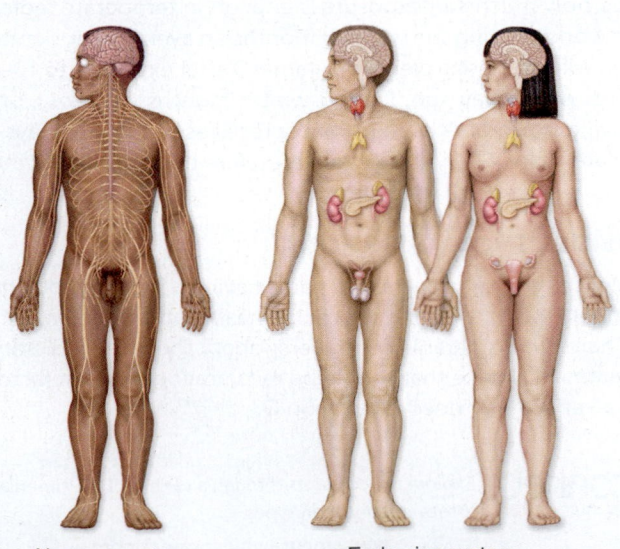

Nervous system   Endocrine system

FIGURE 26.7A The control systems.

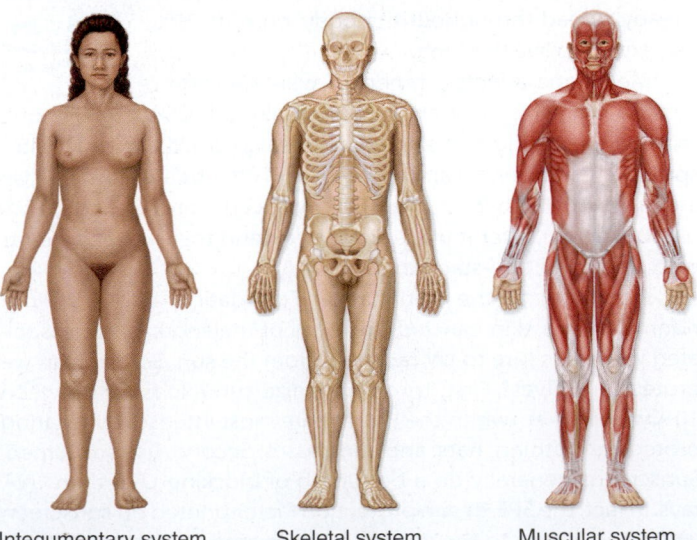

Integumentary system   Skeletal system   Muscular system

FIGURE 26.7B The protective and motor activity systems.

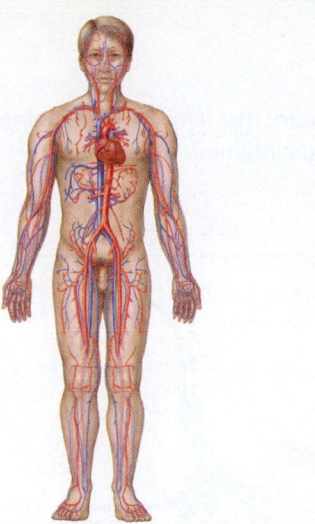 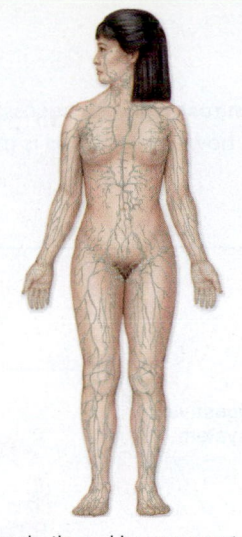

Cardiovascular system   Lymphatic and immune systems

FIGURE 26.7C The body's transport systems.

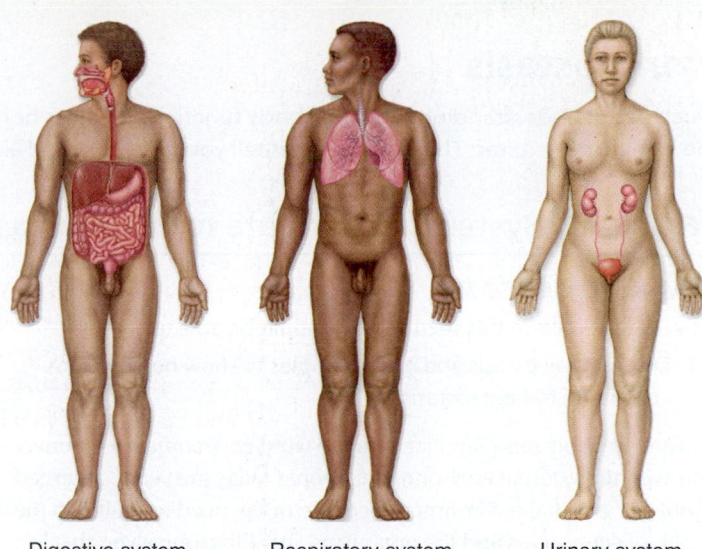

Digestive system   Respiratory system   Urinary system

FIGURE 26.7D The body's maintenance systems.

The **cardiovascular system** (Fig. 26.7C, *left*) consists of the heart and the blood vessels that carry blood throughout the body. The pumping action of the heart propels blood into two circuits; one circuit takes blood to the lungs, and the other circuit takes blood to the body proper. Blood transports nutrients and oxygen to tissue fluid for the cells, and removes waste molecules excreted by cells from the tissue fluid. The internal environment of the body consists of the blood within the blood vessels and the tissue fluid that surrounds the cells.

The **lymphatic system** (Fig. 26.7C, *right*) consists of lymphatic vessels, which carry lymph, and lymphatic organs, including lymph nodes. Lymphatic vessels absorb fat from the digestive system and collect excess tissue fluid, which is returned to the blood in the cardiovascular system. In this way, the lymphatic system assists the cardiovascular system in maintaining blood pressure. The lymphatic organs have various other functions, but all are involved in defending the body against disease. Again, the cardiovascular system assists with this function. Certain blood cells in the lymph and blood are part of an immune system, which specifically protects the body from disease.

The Maintenance Systems Three systems (digestive, respiratory, and urinary) maintain the body by adding substances to and/or removing substances from the blood. If the composition of the blood remains constant, so does that of the tissue fluid.

The **digestive system** (Fig. 26.7D, *left*) consists of the various organs along the digestive tract together with associated organs, such as teeth, salivary glands, the liver, and the pancreas. The accessory organs produce digestive enzymes (salivary glands and pancreas) and bile (liver), which are sent to the digestive tract by way of ducts. The digestive system receives food and digests it into nutrient molecules that enter the blood. Nutrient molecules carried by the bloodstream sustain the body.

The **respiratory system** (Fig. 26.7D, *middle*) consists of the lungs and the tubes that take air to and from the lungs. The body has a way to store energy but no way to store oxygen. Therefore, the respiratory system brings oxygen into the body and takes carbon dioxide out of the body through the lungs. It also exchanges gases with the blood.

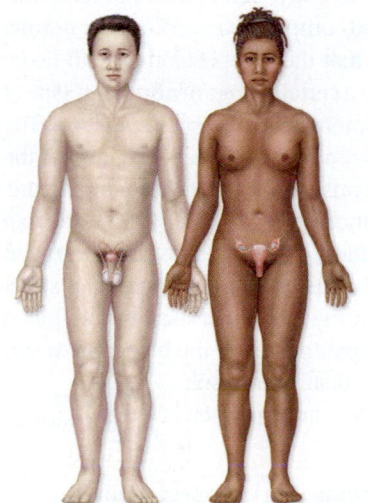

Reproductive system

FIGURE 26.7E
The reproductive systems.

The **urinary system** (Fig. 26.7D, *right*) contains the kidneys and the urinary bladder along with tubes that transport urine. This system rids the blood of wastes and also helps regulate the fluid level and chemical content of the blood.

The Reproductive Systems Different organs are in the **reproductive systems** (Fig. 26.7E) of males and females. The male reproductive system consists of the testes, other glands, and various ducts that conduct semen to and through the penis. The testes produce sex cells called sperm. The female reproductive system consists of the ovaries, oviducts, uterus, vagina, and external genitals. The ovaries produce sex cells called eggs. When a sperm fertilizes an egg, an offspring begins to develop.

▶ **26.7 CHECK YOUR PROGRESS**
   1. Describe the makeup and function of the respiratory system, the digestive system, and the cardiovascular system.

Homeostasis

Much of our understanding of animal body functions involves homeostasis. **Homeostasis** is so necessary that if it is not maintained, the body soon expires. The next sections tell you generally, and also specifically, how homeostasis is maintained.

26.8 Body Systems Contribute to Homeostasis

LEARNING OUTCOME

When you complete this section, you should be able to

1. Define homeostasis and give examples to show how systems contribute to homeostasis.

You are probably most familiar with the word *environment* in connection with the external environment. People today are very concerned about our external environment because of the need to maintain the health of ecosystems and the organisms, including ourselves, that live in them. Your body, however, is composed of many cells, which live in a liquid environment called tissue fluid that is constantly renewed by exchanges with the blood (Fig. 26.8). Therefore, blood and tissue fluid are the internal environment of the body. Tissue fluid remains relatively constant only as long as blood composition remains near normal levels. Relatively constant means that the composition of both tissue fluid and blood usually falls within a certain range of normality. One of the most obvious examples of homeostasis is body temperature. The temperature of the human body is maintained near 37°C, even if the surrounding temperature varies considerably from this temperature. Other examples of homeostasis include regulation of the water–salt content, carbon dioxide concentration, pH, and glucose concentration of blood. If you eat acidic foods, the pH of your blood still stays about 7.4, and even if you eat a candy bar, the amount of sugar in your blood remains at just about 0.1%. If the parameters of the blood fail to stay within a normal range, coma and death can result because cells can only continue to function when their needs are being met.

MP3 Homeostasis

Body Systems and Homeostasis Let us consider how the systems of the body contribute to maintaining homeostasis. The cardiovascular system conducts blood to and away from capillaries, where exchange occurs. The heart pumps the blood, and thereby keeps it moving toward the capillaries. Red blood cells transport oxygen and participate in the transport of carbon dioxide. White blood cells fight infection, and platelets participate in the clotting process. The lymphatic system is accessory to the cardiovascular system. Lymphatic capillaries collect excess tissue fluid and return it via lymphatic vessels to the cardiovascular system. Lymph nodes help purify lymph and keep it free of pathogens.

The digestive system takes in and digests food, providing nutrient molecules that enter the blood to replace those that are constantly being used by the body's cells. The respiratory system removes carbon dioxide from and adds oxygen to the blood. The chief regulators of blood composition are the kidneys and the liver. Urine formation by the kidneys is extremely critical to the body, not only because it rids the body of metabolic wastes, but also because the kidneys carefully regulate blood volume, salt balance, and pH. The liver, among other functions, regulates the glucose concentration of the blood. Immediately after glucose enters the blood, the liver removes the excess for storage as glycogen. Later, glycogen is broken down to replace the glucose that was used by body cells. In this way, the glucose composition of the

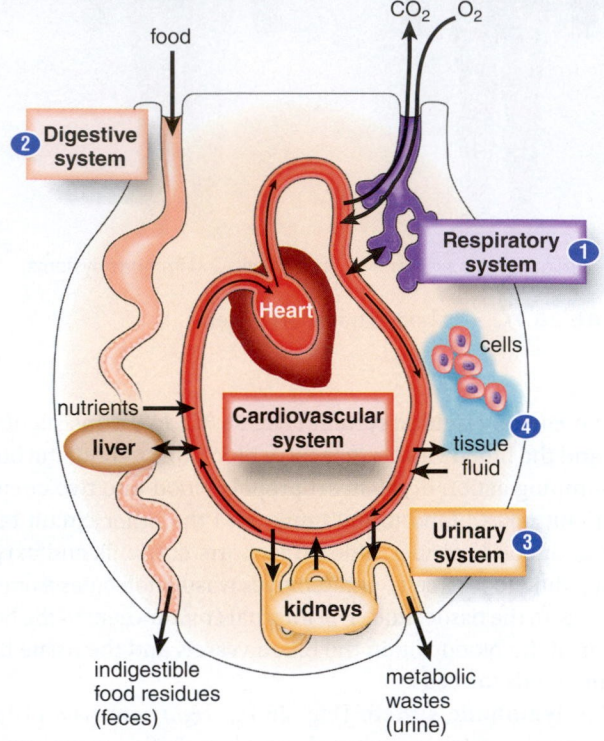

FIGURE 26.8 Process of achieving relative constancy in the internal environment (blood and tissue fluid). **1** The respiratory system exchanges gases with the external environment and with the blood. **2** The digestive system takes in food and adds nutrients to the blood, and **3** the urinary system removes metabolic wastes from the blood and excretes them. **4** When blood exchanges nutrients and oxygen for carbon dioxide and other wastes with tissue fluid, the composition of the tissue fluid stays within normal limits.

blood remains constant. The liver also removes toxic chemicals, such as ingested alcohol and other drugs. The liver makes urea, a nitrogenous end product of protein metabolism.

The nervous system and the endocrine system work together to control body systems so that homeostasis is maintained. For example, they can cause the breathing rate to speed up or slow down. Likewise, they can speed the action of the heart or slow it down. In mechanisms involving the nervous system, sensory receptors send nerve impulses to control centers in the brain, which then direct effectors to become active. Effectors can be muscles or glands. Muscles bring about an immediate change. Endocrine glands secrete hormones that bring about a slower, more lasting change that keeps the internal environment relatively stable.

▶ **26.8 CHECK YOUR PROGRESS**

1. Explain why homeostasis is critical to the health of an animal.
2. Describe how the digestive system contributes to homeostasis and how the cardiovascular system assists the respiratory system.

26.9 Homeostasis is achieved through negative feedback mechanisms

LEARNING OUTCOME

When you complete this section, you should be able to

1. Define negative feedback and describe the essential components of a negative feedback system.

Negative feedback mechanisms reverse a change so that conditions are returned to normalcy. The model for negative feedback shown in Figure 26.9A has two components: a sensor and a control center. The sensor detects a change in the internal environment (a stimulus); the control center initiates an effect that brings conditions back to normal again. Now the sensor is no longer activated. In other words, a negative feedback mechanism is present when the output of the system dampens the original stimulus.

Let's take a simple example. When the pancreas detects that the blood glucose level is too high, it secretes insulin, a hormone that causes cells to take up glucose. Now the blood sugar level returns to normal, and the pancreas is no longer stimulated to secrete insulin.

When conditions exceed their limits and negative feedback mechanisms cannot compensate, illness results. For example, if the pancreas is unable to produce insulin, the blood sugar level becomes dangerously high, and the individual can become seriously ill. The study of homeostatic mechanisms is, therefore, medically important. **Animation** Feedback Mechanism

Complex Examples A home heating system is often used to illustrate how a more complex negative feedback mechanism works. You set the thermostat at, say, 20°C. This is the *set point*. The thermostat contains a thermometer, a sensor that detects when the room temperature is above or below the set point. The thermostat also contains a control center; it turns the furnace off when the room is warm and turns it on when the room is cool. When the furnace is off, the room cools a bit, and when the furnace is on, the room warms a bit. In other words, a negative feedback system results in controlled fluctuation above and below the set point.

In humans, the thermostat for body temperature is located in a part of the brain called the hypothalamus. When the core body temperature becomes higher than normal, the control center directs (via nerve impulses) the blood vessels of the skin to dilate (Fig. 26.9B, *left*). More blood is then able to flow near the

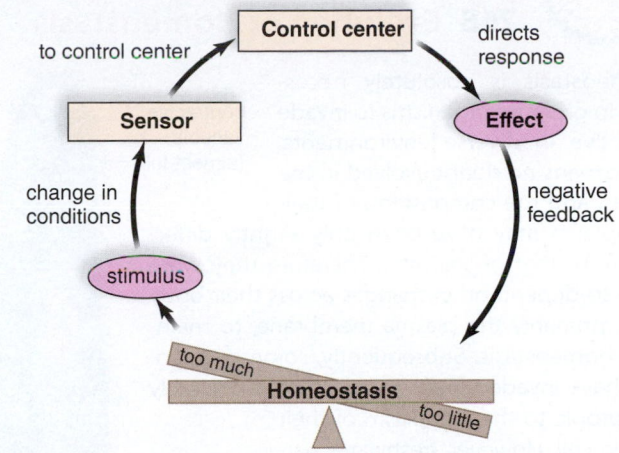

FIGURE 26.9A Model of negative feedback.

surface of the body, where heat can be lost to the environment. In addition, the nervous system activates the sweat glands, and the evaporation of sweat helps lower body temperature. Gradually, body temperature decreases to 37°C.

When the core body temperature falls below normal, the control center directs the blood vessels of the skin to constrict (Fig. 26.9B, *right*). This action conserves heat. If the core body temperature falls even lower, the control center sends nerve impulses to the skeletal muscles, and shivering occurs. Shivering generates heat, and gradually, body temperature rises to 37°C. When the temperature rises to normal, the control center is inactivated.

Notice that a negative feedback mechanism prevents change in the same direction—in other words, body temperature does not get warmer and warmer, because warmth brings changes that decrease body temperature. Also, body temperature does not get colder and colder, because a body temperature below normal causes changes that bring the body temperature up. **MP3** Temperature Regulation

▶ **26.9 CHECK YOUR PROGRESS**

1. Explain the negative feedback mechanism that maintains body temperature.

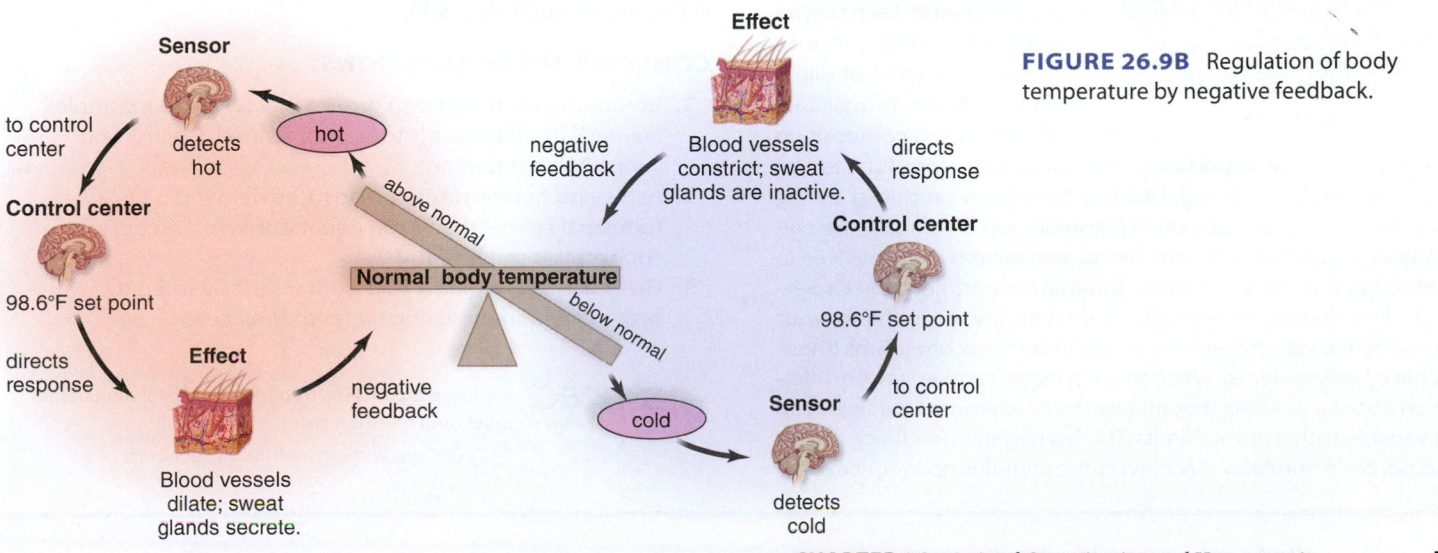

FIGURE 26.9B Regulation of body temperature by negative feedback.

26B Evolution of Homeostasis

Homeostasis is absolutely necessary in order for organisms to invade and live in diverse environments. Protozoans no doubt evolved in the ocean and the composition of their cytoplasm may have been only slightly different from that of seawater. Therefore they were able to depend on exchanges across their body wall, primarily the plasma membrane, to maintain homeostasis. Subsequently, some protozoans have invaded fresh water, which is distinctly hypotonic to the cytoplasm of their single cell. However, freshwater paramecia have two water vacuoles, which are filled by tubules connecting to and receiving water from the endoplasmic reticulum. When the vacuoles are full, they contract, expelling the excess water so that the osmolarity (water-salt composition) of cytoplasm is maintained. As we have seen previously, evolution works with what is available to allow an organism to adapt to a particular environment—in this case, a modification of the endoplasmic reticulum.

The first multicellular organisms to evolve were marine and possibly similar to *Hydra* which exhibits limited coordination between parts, signifying that most likely their cells largely take care of their own homeostatic needs. Not so in molluscs, annelids, and arthropods, which have fully functioning organs, including a circulatory system and a nervous system. The internal environment of these invertebrates is composed of blood and other fluids, as it is in humans.

The water-salt composition of blood is therefore of paramount importance because if it stays constant, so does the water-salt composition of the cells that are serviced by the blood. To take an example, let's consider the nephridia of an earthworm. Each day, an earthworm may produce copious amounts of dilute urine in order to maintain the water-salt composition of blood. Of equal importance, the osmoregulatory function of earthworm nephridia is regulated by hormones produced by the brain! So we see that, early on in the evolution of animals, the nervous system and the endocrine (hormone) system were involved in maintaining the constancy of the internal environment. We also expect to find that negative feedback systems involving the nervous and endocrine systems will be present in complex organisms (those that have body systems). When these systems receive sensory information about a variable, they initiate the mechanisms that will keep that variable within normal limits. This is a negative feedback system because, once normalcy is achieved, the stimulus no longer exists.

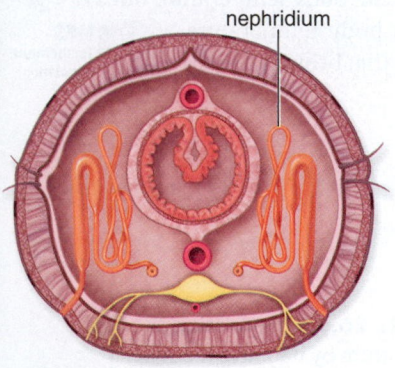

Paramecium

nephridium

Earthworm

Exhaled air is cooled and dried in long convoluted air passages.

Animal fur prevents evaporative loss of water at skin.

Urine is the most hypertonic known among animals.

Fecal pellets are dry.

FIGURE 26B A kangaroo rat is adapted to a dry environment and minimizes water loss in the many ways noted. The only source of water available to the rat is metabolic water derived from cellular respiration.

Invading the land environment requires many adaptations that are not seen in earthworms, which actually live in films of water within the soil. No organism exemplifies the adaptations of a mammal to a dry land environment better than the kangaroo rat, which maintains its blood water-salt balance despite living in a desert. Its fur prevents loss of water to the air, and during the day, it remains in a cool burrow. Exhaled air is usually full of moisture, which is why you can "see your breath" on cold winter mornings—the moisture in exhaled air is condensing. But the kangaroo rat's nasal passage has a highly convoluted mucous membrane surface that captures condensed water from air that is exhaled.

As we shall see in Chapter 34, humans mainly conserve water by producing urine that is hypertonic to blood plasma (the liquid portion of blood). The kangaroo rat forms a very concentrated urine—20 times more concentrated than its blood plasma. Also, its fecal material is almost completely dry. Then, too, the kangaroo rat does not need to drink water; it is so adapted to conserving water that it can survive by using metabolic water derived from cellular respiration (Fig. 26B).

CONSIDER THESE QUESTIONS

1. Communication between organs is essential in a complex animal if homeostasis is to be maintained. Discuss and support this statement.
2. Achieving homeostasis can be more demanding in a terrestrial animal than in an aquatic animal. Discuss and support this statement.
3. Give examples to show that humans are adapted to preserving and providing water to their cells.

 **connect** | **BIOLOGY**   Explore the concepts through a variety of multimedia assets and question types.

www.mcgrawhillconnect.com

THE CHAPTER IN REVIEW

CONNECTING THE CONCEPTS

This chapter serves as an introduction to Unit 5 because it reviews not only the types of tissues in the animal body, but also the organ systems we will be discussing. Two themes are introduced that we will refer to time and time again: (1) structure suits function and (2) homeostasis. Structure suits function means that the function of an organ is reflected in its structure. For example, the function of the heart is to pump blood, and it contains chambers for receiving blood and has thick, muscular walls for pumping blood.

Homeostasis refers to the relative constancy of the internal environment. In complex animals, the internal environment is blood and tissue fluid. Let's take a familiar example of homeostasis in humans. After eating, the

hormone insulin is released, and glucose is removed from the blood and stored in the liver as glycogen. In between eating, the hormone glucagon regulates glycogen breakdown so that the blood glucose level remains at just about 0.1%. In this way, glucose is constantly available to cells for the process of cellular respiration.

In complex animals, the nervous and endocrine systems act together to regulate the actions of organs so that homeostasis is maintained. Sensory receptors gather information from the external and internal environments and convert it to a form that can be processed by the nervous system. Then the nervous system can direct the action of the other systems, including the endocrine system, so that homeostasis is maintained. Movement brought

about by the skeletal and muscular systems may be required to achieve homeostasis. These and other examples of homeostasis will be discussed throughout the chapters of Unit 5. Chapters 27–29 discuss the nervous, sensory, skeletal, and muscular systems.

ANALYZE AND EVALUATE

1. Explain the statement, "Movement brought about by the skeletal and muscular systems may be required to achieve homeostasis."
2. People who take a trip to the South Pole or who walk in space take precautions to help their bodies maintain homeostasis. What precautions do they take?
3. Do you think that blood and tissue fluid (fluid around cells) should be called the internal environment? Why or why not?

SUMMARIZE

Types of Tissues

26.1 Levels of biological organization are evident in animals

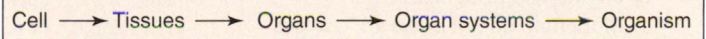

Cell ⟶ Tissues ⟶ Organs ⟶ Organ systems ⟶ Organism

- A **tissue** is composed of specialized cells performing a similar function.
- The four types of animal tissues are shown here:

Epithelial	Connective	Muscular	Nervous

- An **organ** is formed by different types of tissues arranged in a certain way. An **organ system** is made up of several organs that work together to perform necessary functions for the **organism.**

26.2 Epithelial tissue covers organs and lines body cavities

- By being one cell thick, **epithelial tissue** can be protective, and yet also allow substances to pass through. One or more epithelial cells make up the **glands.** Epithelial cells connect to each other by tight junctions, adhesion junctions, and gap junctions. A **basement membrane** anchors epithelium to underlying connective tissue.
- Cells of the epithelium may be **squamous, cuboidal,** or **columnar** in shape; they also may be simple (one layer) or **stratified** (more than one layer). In **pseudostratified ciliated columnar epithelium,** the columnar cells are ciliated and appear to be stratified but are not because each one touches the baseline.

26.3 Connective tissue connects and supports other tissues

- **Connective tissue** cells are separated by a **matrix,** which varies from solid (as in **compact bone**), to semifluid (as in cartilage), to fluid (as in blood).

- The various types of connective tissue are (1) **Loose fibrous connective tissue** (contains **fibroblasts**), including **adipose tissue;** (2) **dense fibrous** connective tissue (**tendons and ligaments**); (3) **cartilage** (**hyaline** is most abundant) and **bone** (matrix for cartilage is more flexible than that for bone); and (4) **blood** (liquid portion is plasma; formed elements are **red blood cells, white blood cells,** and **platelets**) and **lymph** (liquid portion is excess **tissue fluid;** cells are white blood cells).

26.4 Muscular tissue is contractile and moves body parts

- **Skeletal muscle,** attached by tendons to bones, is **striated** and voluntary. **Cardiac muscle,** found in the walls of the heart, is striated and involuntary. **Intercalated disks** bind the cells together. **Smooth muscle,** found in the walls of viscera, is involuntary and not striated.

26.5 Nervous tissue communicates with and regulates the functions of the body's organs

- **Nervous tissue** coordinates body parts and allows animals to respond to their environment. The nervous system receives sensory input, integrates data, and brings about a response.
- A **neuron** has dendrites, a cell body, and an axon. **Nerves** are composed of axons. **Neuroglia** support and nourish neurons and produce **cerebrospinal fluid.** Types of neuroglia include microglia, astrocytes, and oligodendrocytes.

Organs and Organ Systems

26.6 Each organ has a specific structure and function

- An organ performs functions that tissues alone cannot. The **skin** in vertebrates is an organ composed of an **epidermis** (stratified epithelium) and a **dermis** (dense fibrous connective tissue). A **subcutaneous layer** (loose fibrous connective tissue) lies between the skin and underlying structures.

26.7 Several organs work together to carry out the functions of an organ system

- The control systems: The organs of the **nervous system** and **endocrine system** regulate the other systems.
- The protective and motor activity systems: The **integumentary, skeletal,** and **muscular systems** protect the internal organs. The skeletal and muscular systems account for body movements.
- The transport systems: The **cardiovascular system** transports nutrients to cells and wastes from cells. The **lymphatic system** absorbs excess tissue fluid and functions in immunity.
- The maintenance systems: The **digestive system** takes in food and adds nutrients to the blood. The **respiratory system** carries out gas exchange with the external environment and blood. The **urinary system** (i.e., the kidneys) removes metabolic wastes and regulates the pH and salt content of the blood.
- The reproductive systems: Each gender has its own set of organs and therefore its own **reproductive system.**

Homeostasis

26.8 Homeostasis is the constancy of the internal environment

- The internal environment of the body consists of blood and tissue fluid. Tissue fluid stays constant due to exchange of nutrients and wastes with the blood:

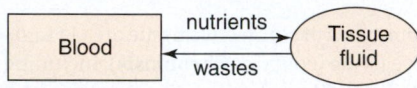

- Examples of homeostasis include a relatively constant body temperature, water content, O_2 and CO_2 concentrations, pH, and glucose concentration. All body systems contribute to homeostasis. Examples include: Cardiovascular system circulates the blood; digestive system provides nutrients; respiratory system breathes in O_2 and breathes out CO_2; urinary system maintains pH and water-salt concentration of blood.

26.9 Homeostasis is achieved through negative feedback mechanisms

- **Negative feedback** mechanisms reverse conditions to within normal limits. Example: Warm body temperature promotes dilation of skin blood vessels; cold body temperature promotes constriction of skin blood vessels.
- Negative feedback mechanisms contain (1) a sensor: detects a change above or below a set point; communicates with control center, often the brain and (2) a control center: brings about an effect that reverses the change and brings conditions back to normal.

TEST YOURSELF

Types of Tissues

1. A grouping of similar cells that perform a specific function is called a
   a. sarcoma.
   c. tissue.
   b. membrane.
   d. None of these are correct.
2. The microvilli on the cells lining the small intestine are adaptations to promote
   a. absorption.
   c. movement.
   b. digestion.
   d. secretion.
3. Which tissue is more apt to line a space?
   a. epithelial tissue
   c. nervous tissue
   b. connective tissue
   d. muscular tissue
4. Tight junctions are most often associated with
   a. connective tissue.
   c. cartilage.
   b. adipose tissue.
   d. epithelium.
5. Which tissue has cells in lacunae?
   a. epithelial tissue
   d. smooth muscle
   b. cartilage
   e. Both b and c are correct.
   c. bone
6. Blood is a(n) _____ tissue because it has a _____.
   a. connective, gap junction
   b. muscular, matrix
   c. epithelial, gap junction
   d. connective, matrix
7. A reduction in red blood cells would cause problems with
   a. fighting infection.
   c. blood clotting.
   b. carrying oxygen.
   d. None of these are correct.
8. White blood cells fight infection by
   a. producing antibodies.
   b. producing toxins.
   c. engulfing pathogens.
   d. More than one of these are correct.
   e. All of these are correct.
9. **THINKING CONCEPTUALLY** Use the concept of "structure suits function" to discuss the structure of the neuron shown in Figure 26.5.

For questions 10–12, match each type of muscle to as many terms in the key as possible.

KEY:
   a. voluntary
   e. spindle-shaped cells
   b. involuntary
   f. branched cells
   c. striated
   g. long, cylindrical cells
   d. nonstriated

10. Skeletal muscle
11. Smooth muscle
12. Cardiac muscle
13. Which characteristic is true of both cardiac and skeletal muscle?
   a. striated
   b. single nucleus per cell
   c. multinucleated cells
   d. involuntary control

14. The main part of a neuron that conducts nerve impulses is the
    a. astrocyte.
    b. axon.
    c. cell body.
    d. dendrite.
15. Which is a true statement?
    a. Cell bodies are primarily located in the brain and spinal cord.
    b. Axons can be very long, even reaching the length of your leg.
    c. Nerve cells have no need of nutrients locked away in the brain as they are.
    d. Both a and b are correct.
    e. All statements are incorrect.
16. **THINKING CONCEPTUALLY** Many cancers develop from epithelial tissue. What two attributes of this tissue type make cancer more likely to develop?

Organs and Organ Systems

17. Which of the following is a function of skin?
    a. temperature regulation
    b. protection against water loss
    c. collection of sensory input
    d. protection from invading pathogens
    e. All of these are correct.
18. Without melanocytes, skin would
    a. be too thin.
    b. lack nerves.
    c. lack color.
    d. None of these are correct.
19. What is the primary feature of human skin that makes it a good protection against most infectious agents, such as bacteria and viruses?
20. The major function of tissue fluid is to
    a. provide nutrients and oxygen to cells and remove wastes.
    b. maintain cell shape.
    c. prevent cells from touching each other.
    d. provide flexibility by allowing cells to slide over each other.
21. Which of these body systems plays the biggest role in fluid balance?
    a. cardiovascular   c. digestive
    b. urinary   d. integumentary
22. Which of the following body systems does not add or remove substances from the blood?
    a. digestive   d. respiratory
    b. cardiovascular   e. nervous
    c. urinary
23. The skeletal system functions in
    a. blood cell production.   c. movement.
    b. mineral storage.   d. All of these are correct.

Homeostasis

24. Which of these body systems contribute to homeostasis?
    a. digestive and urinary systems
    b. respiratory and nervous systems
    c. nervous and endocrine systems
    d. immune and cardiovascular systems
    e. All of these are correct.

25. Which of the following act slowly as effectors in a negative feedback system?
    a. muscles   d. red blood cells
    b. epidermal cells   e. senses
    c. endocrine glands
26. The correct order for a negative feedback mechanism is:
    a. sensory detection, control center, effect brings about a change
    b. control center, sensory detection, effect brings about a change
    c. sensory detection, control center, effect causes no change
    d. None of these are correct.
27. Which of the following is an example of negative feedback?
    a. Air conditioning goes off when room temperature lowers.
    b. Insulin decreases blood sugar levels as the day progresses.
    c. Heart rate lowers when blood pressure drops.
    d. All of these are examples of negative feedback.
28. When a human is cold, the blood vessels
    a. dilate, and the sweat glands are inactive.
    b. dilate, and the sweat glands are active.
    c. constrict, and the sweat glands are inactive.
    d. constrict, and the sweat glands are active.
    e. contract so that shivering occurs.
29. How does shivering increase body temperature?
30. **THINKING CONCEPTUALLY** Both the cardiovascular system and the nervous system pervade the body. Which one would you expect to have a pump, and why?

GET INVOLVED

1. You are a histologist (a person who specializes in tissues) working in the laboratory of a hospital. You receive some tissues taken from a person who died of AIDS. Your task is to identify the types of tissues. What will you do?
2. You hypothesize that more cases of skin cancer occur among people who frequent tanning salons than in others. Describe the study you would do to test your hypothesis.

MEDIA STUDY TOOLS

mhhe.com/maderconcepts3

Enhance your study of this chapter with interactive study tools, practice tests, and engaging animations. Also, ask your instructor about the resources available through ConnectPlus, which includes LearnSmart, a personalized adaptive learning program, and a media-rich eBook.

27

Coordination by Neural Signaling

BEFORE YOU BEGIN

Take a few minutes to recall

How active transport of ions occurs across plasma
    membranes (section 5.3)

The major components of the nervous system (section 26.7)

The contribution of the nervous system to homeostasis
    (sections 26.8 and 26.9)

Getting a Head

How do you define a head? Whether you describe a head as an anterior demarcation of the body or an anterior region containing a brain and sense organs, it is clear that echinoderms, such as sea urchins, don't have one. Sea urchins depend on all those spines to protect them, rather than running away quickly on their many tube feet. What about the clam compared to the octopus? The inactive clam spends its life digging in the sand and does not have a head. But the octopus, in addition to all those creeping arms, does have a head. You might mistake the visceral hump for part of the head, but it is not. Its head is where you see the eyes, because that is where the brain is. An octopus needs its brain, eyes, and arms to go after its prey. A head is a definite advantage to a predator. Good eye-appendage coordination helps a lot too.

Do whales have a head? Marine vertebrates, whether a fish, a dolphin, or a duckbill platypus,

Clam

Sea urchin

Octopus

tend to have a head region that is not distinct from the rest of the body. If you were asked to cut off the head of a fish, where would you cut—the choice is yours! The more streamlined their bodies, the better aquatic animals can move in the water. If we compare the crab to a grasshopper, we can see that the land animal has a more definite head. The scavenger crab uses its claws to bring food that might float or swim away to its mouth, but the grasshopper has no claws. The herbivorous grasshopper has a head that can bob up and down to reach its stationary food on land. It uses all its legs for hopping. So, it appears that lifestyle within a particular environment plays a role in whether an animal has a head or not.

What makes animals have heads is an interesting study, but the real question is, what type of nervous system does an animal have? Some animals have a few nerve cells here or there, as in a hydra, and they cannot do much of anything. But with a good brain, protected by a skull and connected to associated nerves and sense organs, an animal can lead a more complex life. It can receive stimuli from the environment, interpret those stimuli, and respond in an appropriate manner.

The vertebrate brain is divided into the hindbrain, the midbrain, and the forebrain. Fishes and amphibians rely on their midbrain to carry out complex behaviors, but in mammals especially, this function has been taken over by the forebrain. Humans have the best-developed forebrain of all the animals, and when you map it, you can see that an inordinate amount of space is allotted to the hands! When humans came down out of the trees and stood on two legs, they exposed their bellies, but freed their hands. With good eye-hand coordination and the ability to use tools, humans can flourish in almost any terrestrial environment. They can also decide on the best course of action in a wide variety of circumstances.

In this chapter, we discuss the vertebrate nervous system in an evolutionary context. A comparison of animal nervous systems shows how the vertebrate nervous system may have evolved.

Dolphin

Grasshopper

Rainbow trout

Crab

Student

Nervous Systems

In this chapter the application "Evolution of the Nervous System" (p. 536) gives evidence that the vertebrate nervous system is the product of the evolutionary process. In vertebrates, the central nervous system (CNS) specializes in information processing and integration, while the peripheral nervous system (PNS) serves as an avenue of communication between the CNS and the rest of the body. Both systems are composed of neurons, which communicate with each other and with muscles and glands at regions called synapses.

27.1 Vertebrates have well-developed central and peripheral nervous systems

LEARNING OUTCOMES

When you complete this section, you should be able to

1. Relate the three main functions of the nervous system to the CNS and the PNS.
2. Describe the basic structure of a neuron and compare the functions of the three types of neurons.

The **central nervous system** (**CNS**) consists of the brain and spinal cord, as shown in Figure 27.1A*a, b.* The brain is enclosed in the skull, and the spinal cord is housed in the vertebral column. The **peripheral nervous system** (**PNS**) consists of all the nerves and ganglia that lie outside the CNS. All signals enter and leave the CNS through nerves; those that connect to the spinal cord are called *spinal nerves,* whereas those attached to the brain are *cranial nerves.* Nerves contain a number of **nerve fibers.** The CNS and the PNS work together to accomplish these three functions:

1. *Receive sensory input.* Sensory receptors respond to external and internal stimuli by generating nerve impulses that travel to the CNS by way of sensory fibers in the PNS. For example,

suppose you are playing catch with your nephew and it's his turn to throw the ball. Your eyes communicate the trajectory of the ball to your brain by way of cranial nerves. Spinal nerves continually inform your brain about the present position of your limbs. (See Fig. 27.1A*b, left.*)

2. *Perform integration.* The CNS sums up the input it receives from your eyes and limbs and decides what moves will be necessary to catch the ball.

3. *Generate motor output.* Nerve fibers from the CNS go by way of spinal nerves in the PNS to the muscles in your legs, feet, arms, and hands so that you make the moves decided on by your brain for catching the ball. (See Fig. 27.1A*b, right.*)

The CNS also receives information, for example, about blood pressure from visceral sensory receptors and sends motor commands via the autonomic system to increase or decrease pressure as necessary. The autonomic system is further divided into sympathetic and parasympathetic divisions, which will be discussed in section 27.9. The structural components of the human nervous system are complex, and the CNS and PNS must work in harmony to carry out its three functions.

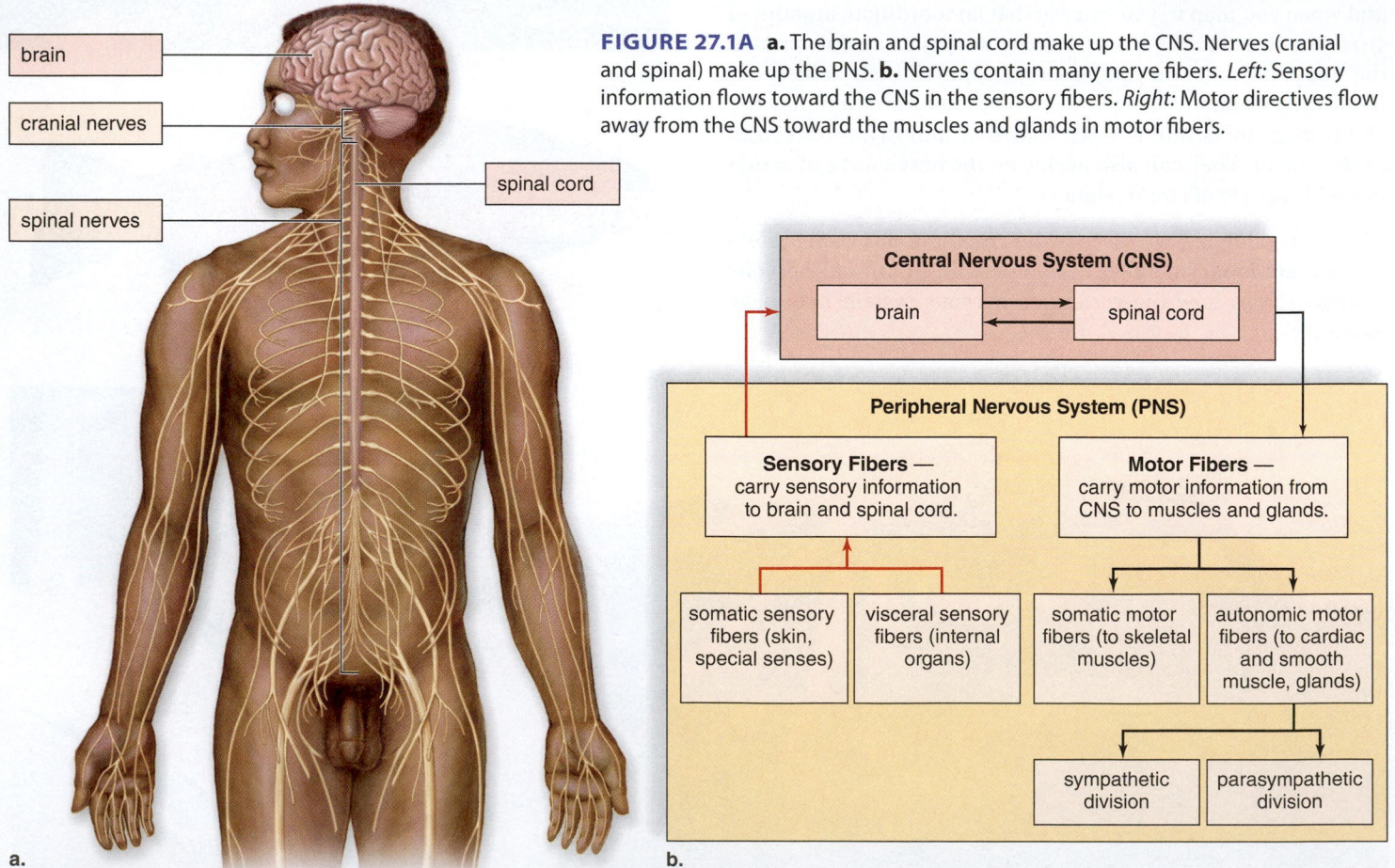

FIGURE 27.1A a. The brain and spinal cord make up the CNS. Nerves (cranial and spinal) make up the PNS. **b.** Nerves contain many nerve fibers. *Left:* Sensory information flows toward the CNS in the sensory fibers. *Right:* Motor directives flow away from the CNS toward the muscles and glands in motor fibers.

brain

cranial nerves

spinal nerves

spinal cord

Central Nervous System (CNS)

brain   spinal cord

Peripheral Nervous System (PNS)

Sensory Fibers — carry sensory information to brain and spinal cord.

Motor Fibers — carry motor information from CNS to muscles and glands.

somatic sensory fibers (skin, special senses)

visceral sensory fibers (internal organs)

somatic motor fibers (to skeletal muscles)

autonomic motor fibers (to cardiac and smooth muscle, glands)

sympathetic division

parasympathetic division

a.

b.

Neurons and Neuroglia **Neurons** are the functional units of the nervous tissue and **neuroglia** are cells that support and care for the neurons. Neurons are specialized to conduct nerve impulses and nerve impulses allow sensory information to reach the CNS and allow the CNS to send out motor directives in nerve fibers. Just as you receive information or give directives by speaking, so neurons use nerve impulses for these purposes.

Neurons vary in appearance, depending on their function and location. They consist of three major parts: a cell body, dendrites, and an axon (Fig. 27.1B). The **cell body** contains a nucleus and a variety of organelles. The **dendrites** are short, highly branched processes that receive signals from the sensory receptors or other neurons and transmit them to the cell body. The **axon** is the portion of the neuron that conducts nerve impulses to another neuron or to other cells. Axons become the nerve fibers of nerves. Many axons are covered by a white insulating layer called the **myelin sheath.**

Video
Making Brain Cells

Neuroglia, or glial cells, greatly outnumber neurons in the brain. There are several different types in the CNS, each with specific functions. Some (microglia) help remove bacteria and debris; others (astrocytes) provide metabolic and structural support directly to the neurons. The myelin sheath is formed from the membranes of tightly spiraled neuroglia. In the PNS, **Schwann cells** perform this function, leaving gaps called **nodes of Ranvier,** or neurofibril nodes. In the CNS, another type of neuroglia, called an oligodendrocyte, performs this function.

Types of Neurons Neurons can be classified according to their function and shape. The axons of **motor neurons** carry nerve impulses from the CNS to muscles or glands. The neuron shown in Figure 27.1B is a motor neuron. Like all motor neurons, it has many dendrites and a single axon. Motor neurons cause muscle fibers to contract or glands to secrete, and therefore they are said to innervate these structures.

The axons of **sensory neurons** take nerve impulses from sensory receptors to the CNS. The sensory receptor may be the end of a sensory neuron itself (a pain or touch receptor), or it may be a specialized cell that forms a synapse with a sensory neuron (e.g., the hair cells of the inner ear). In sensory neurons, the process that extends from the cell body divides into a branch that extends to the periphery and another that extends to the CNS (Fig. 27.1C). Because both of these extensions are long and myelinated and transmit nerve impulses, it is now generally accepted to refer to them as an axon.

Interneurons, also known as association neurons, occur entirely within the CNS. Interneurons parallel the structure of motor neurons (Fig. 27.1D) and their axons conduct nerve impulses between various parts of the CNS. Some lie between sensory neurons and motor neurons, and some take impulses from one side of the spinal cord to the other, or from the brain to the spinal cord and vice versa. They also form complex pathways in the brain where the processes accounting for thinking, memory, and language occur.

MP3
Cells of the Nervous System

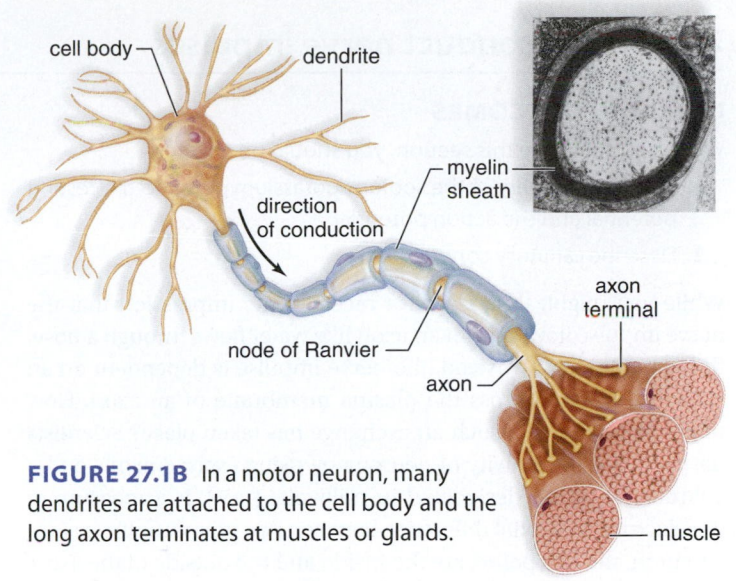

FIGURE 27.1B In a motor neuron, many dendrites are attached to the cell body and the long axon terminates at muscles or glands.

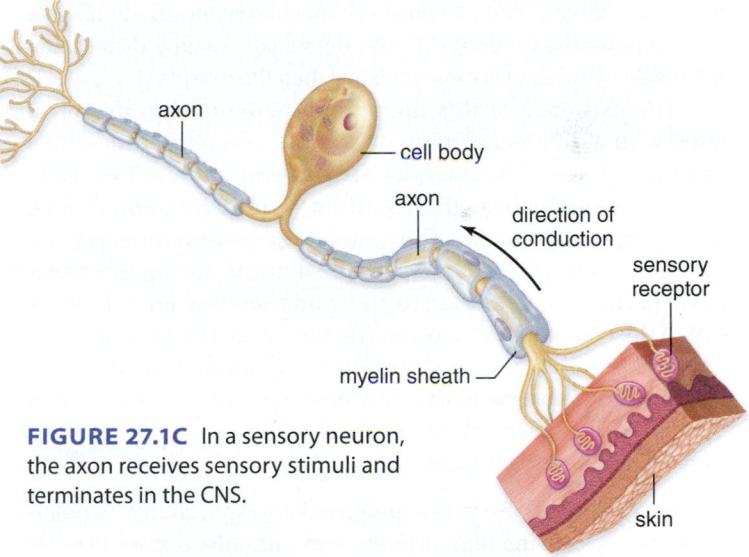

FIGURE 27.1C In a sensory neuron, the axon receives sensory stimuli and terminates in the CNS.

FIGURE 27.1D Interneurons have a short or long axon depending on their location in the CNS.

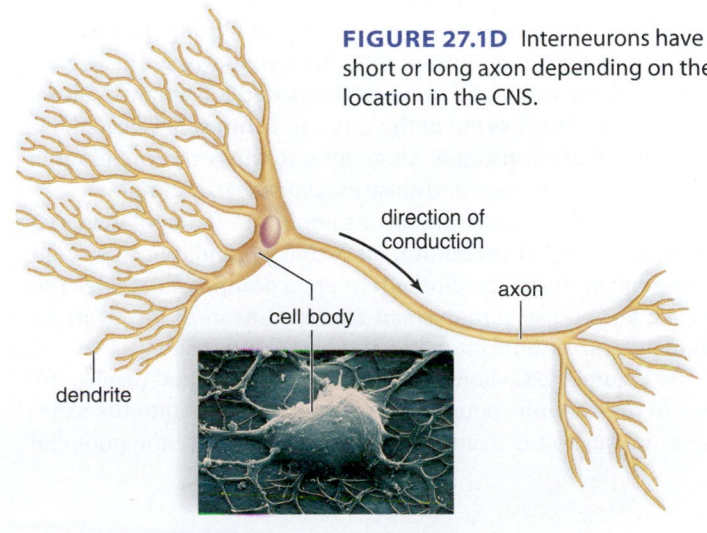

27.2 Axons conduct nerve impulses

LEARNING OUTCOMES

When you complete this section, you should be able to

1. Relate the activity of the sodium-potassium pump to the resting potential and the action potential.
2. Describe saltatory conduction.

While you might, thus far, have received the impression that the nerve impulse flows within an axon like water flows through a hose, this is not the case. Instead, the nerve impulse is dependent on an exchange of ions across the plasma membrane of an axon. How might you judge that such an exchange has taken place? Scientists have studied the activity of neurons by using excised axons and a voltmeter. Voltage, designated in millivolts (mV), is a measure of the electrical potential difference between two points. In the case of a neuron, the two points are the inside and the outside of the axon. When an electrical potential difference exists between the inside and outside of a cell (i.e., across the membrane), it is called a *membrane potential,* and we can say that there is *polarity* in the distribution of electrical charges. When a neuron is not conducting an impulse, its **resting potential** is about –70 mV; the negative sign indicates that the inside of the axon is more negative than the outside (Fig. 27.2A).

The existence of this membrane potential can be correlated with a difference in ion distribution on either side of the axon membrane. The unequal distribution of these ions is, in part, due to the activity of the **sodium-potassium pump,** which moves three sodium ions (Na^+) out of the neuron for every two potassium ions (K^+) it moves into the neuron. The membrane is more permeable to K^+ than to Na^+, and therefore more K^+ leaks out of the cell than Na^+ leaks in. There are also large, negatively charged proteins in the cytoplasm of the axon. All cells maintain a membrane potential, but neurons are unusual in that they alter their membrane potential when they transmit nerve impulses.

Animation
Sodium-Potassium Pump

Action Potential An **action potential** is a rapid change in polarity across an axon membrane as the nerve impulse occurs. In order to visualize the rapid fluctuations in voltage during the action potential, researchers generally find it useful to graph the voltage changes over time (Fig. 27.2B). An action potential uses two types of gated ion channels in the axon membrane. During the first part of an action potential, a gated ion channel allows sodium (Na^+) to pass into the axon, and then another gated ion channel allows potassium (K^+) to pass out of the axon. In contrast to ungated ion channels, which constantly allow ions to cross the membrane, gated ion channels open and close in response to a stimulus.

If a stimulus causes the axon membrane to depolarize to a certain level, called **threshold,** an action potential occurs in an *all-or-none manner.* The strength of an action potential does not change; an intense stimulus can cause an axon to fire (start an action potential) more often in a given time interval.

As Figure 27.2C shows, when an action potential begins, the gates of the sodium channels open, and Na^+ flows into the axon. As Na^+ moves to the inside of the axon, the membrane potential

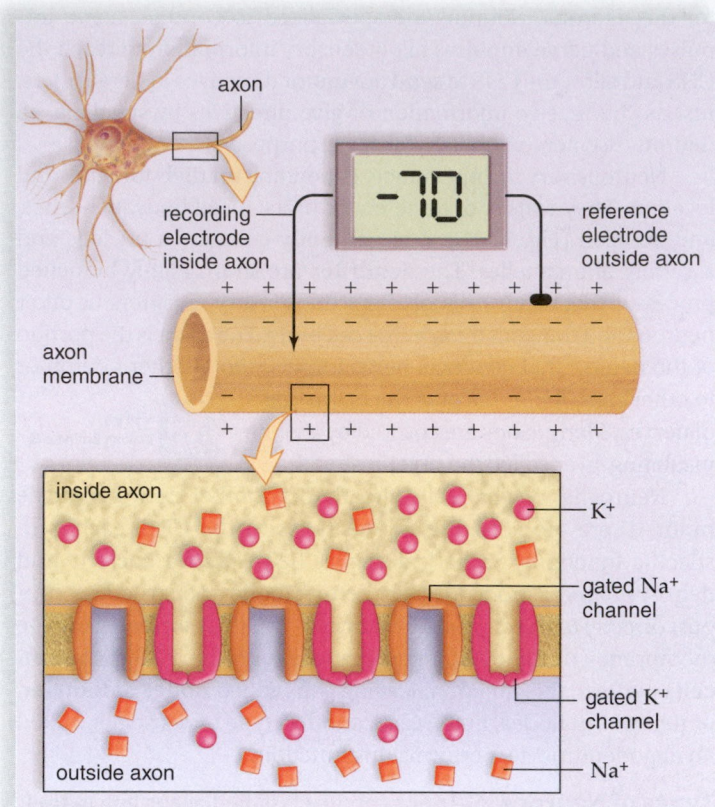

FIGURE 27.2A Resting potential: Note the uneven distribution of Na^+ and K^+ across the membrane. Inside is –70 mV, relative to the outside.

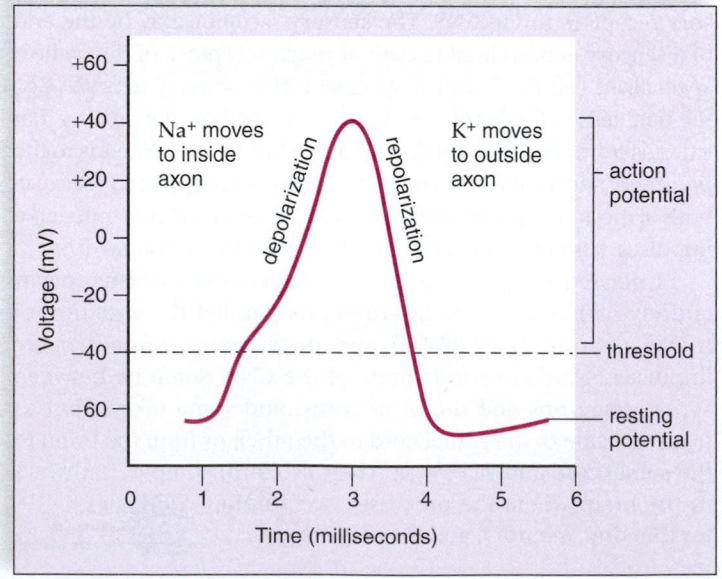

FIGURE 27.2B An action potential can be visualized if voltage changes are graphed over time.

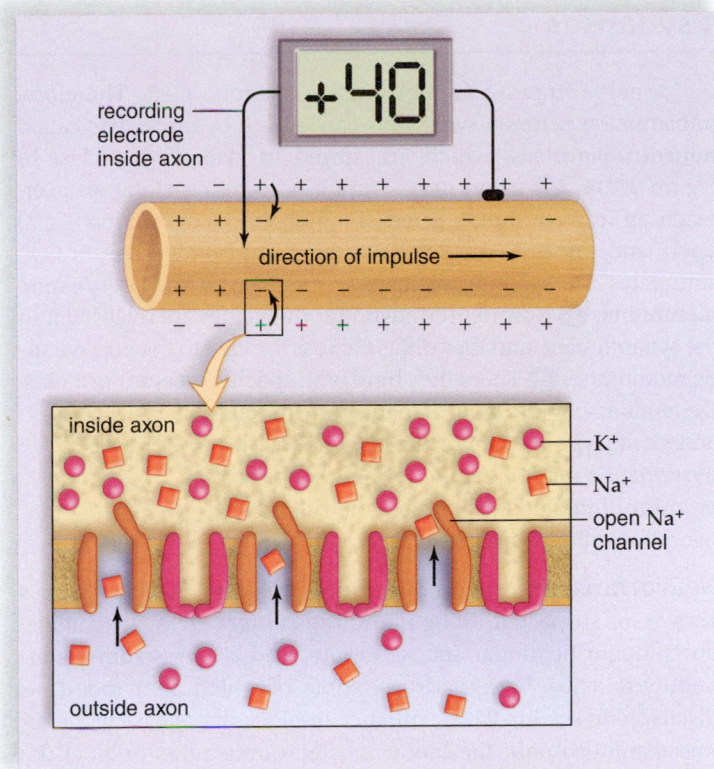

FIGURE 27.2C Action potential begins: Depolarization to +40 mV as Na⁺ gates open and Na⁺ moves to inside the axon.

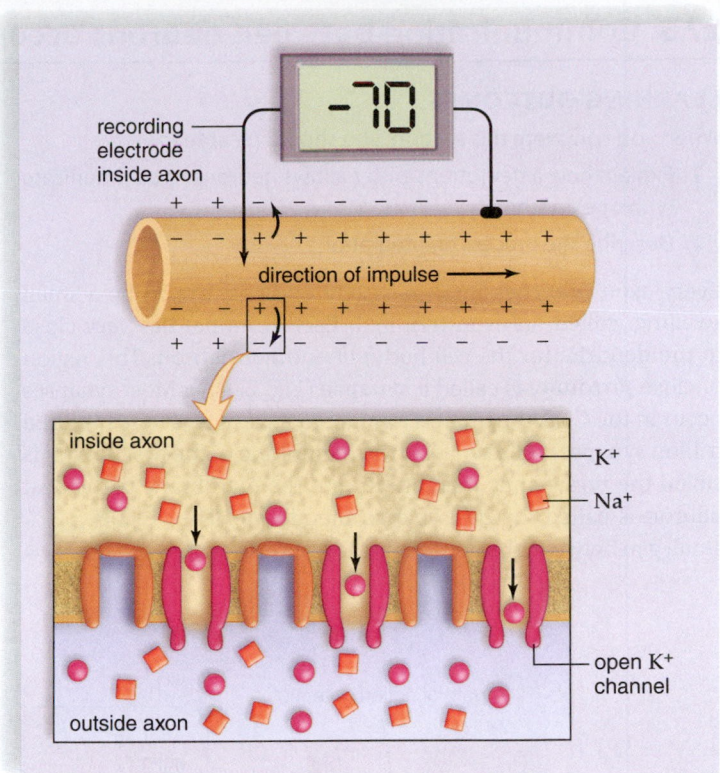

FIGURE 27.2D Action potential ends: Repolarization to −70 mV as K⁺ gates open and K⁺ moves to outside the axon.

changes from −70 mV to +40 mV. This is **depolarization** because the charge inside the axon changes from negative to positive (Fig. 27.2C). This reversal in polarity causes the sodium channels to close and the potassium channels to open.

As Figure 27.2D shows, next, the gates of the potassium channels open, and K⁺ flows out of the axon; the action potential changes from +40 mV back to −70 mV. This is **repolarization** because the inside of the axon becomes negative again as K⁺ exits the axon (Fig. 27.2D).

Aspects of Conduction In nonmyelinated axons, the action potential travels down an axon one small section at a time, at a speed of about 1 m/second. As soon as an action potential has moved on, the previous section undergoes a **refractory period,** during which the Na⁺ gates are unable to open. Notice, therefore, that the action potential cannot move backward and instead always moves down an axon toward its terminals. When the refractory period is over, the sodium-potassium pump has restored the previous ion distribution by pumping Na⁺ to outside the axon and K⁺ to inside the axon.

In myelinated axons, the gated ion channels that produce an action potential are concentrated at the nodes of Ranvier. Just as taking giant steps during a game of "Simon Says" is most efficient, so ion exchange only at the nodes makes the action potential travel faster than in nonmyelinated axons. *Saltar* in Spanish means "to

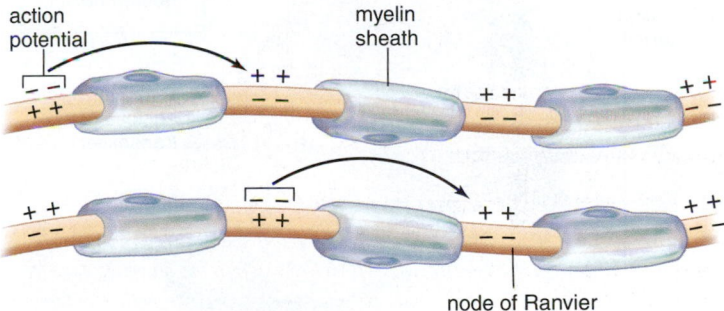

FIGURE 27.2E Saltatory conduction.

jump," and so this mode of conduction is called **saltatory conduction,** meaning that the action potential "jumps" from node to node (Fig. 27.2E). Speeds of 200 m per second have been recorded.

Animation
Nerve Impulse

The symptoms of multiple sclerosis, characterized by impaired motor skills, are due to demyelination.

▶ **27.2 CHECK YOUR PROGRESS**
1. Explain how the activity of the sodium-potassium pump is necessary for both depolarization and repolarization during an action potential.

LEARNING OUTCOMES

When you complete this section, you should be able to

1. Explain how a neurotransmitter allows neurons to communicate with one another.
2. Describe the process of integration.

Every axon branches into many fine endings, tipped by a small swelling, called an axon terminal. Each terminal lies very close to the dendrite (or the cell body) of another neuron. This region of close proximity is called a **synapse** (Fig. 27.3A). Most synapses occur in the CNS; the brain is thought to have at least one hundred trillion synapses. At a synapse, the membrane of the first neuron is called the *pre*synaptic membrane, and the membrane of the next neuron is called the *post*synaptic membrane. The small gap between the neurons is the **synaptic cleft.**

MP3 Synapses

A nerve impulse cannot cross a synaptic cleft. Therefore, transmission across a synapse is carried out by molecules called **neurotransmitters,** which are stored in synaptic vesicles. In Figure 27.3A, ❶ when nerve impulses traveling along an axon reach an axon terminal, gated channels for calcium ions (Ca^{2+}) open, and calcium enters the terminal. The sudden rise in Ca^{2+} stimulates the synaptic vesicles to merge with the presynaptic membrane. ❷ Now neurotransmitter molecules are released into the synaptic cleft, and they diffuse across the cleft to the postsynaptic membrane. ❸ There they bind with specific receptor proteins. Depending on the type of neurotransmitter and/or the type of receptor, the response of the postsynaptic neuron can be toward excitation or toward inhibition. Na^+ enters the cell if the neurotransmitter is excitatory.

Animation Action Potential Propagation

Animation Chemical Synapse

Neurotransmitters Among the more than 100 substances known, or suspected, to be neurotransmitters are acetylcholine, norepinephrine, dopamine, serotonin, and GABA (gamma aminobutyric acid). Various drugs, some of which alter mood, as discussed in section 27.10, enhance or block the release of a neurotransmitter, mimic the action of a neurotransmitter or block the receptor, or interfere with the removal of a neurotransmitter from the synaptic cleft.

Acetylcholine (**ACh**) and **norepinephrine** (**NE**) are frequently mentioned and well-known neurotransmitters in both the CNS and the PNS. Alzheimer disease is associated with a deficiency of ACh in the CNS. In the PNS, ACh is active at neuromuscular junctions (see Fig. 29.7). It excites skeletal muscle but inhibits cardiac muscle. Also, ACh has either an excitatory or an inhibitory effect on smooth muscle or glands, depending on the particular organ. Botulism is a rare kind of food poisoning caused by the bacterium *Clostridium botulinum,* which produces botulin

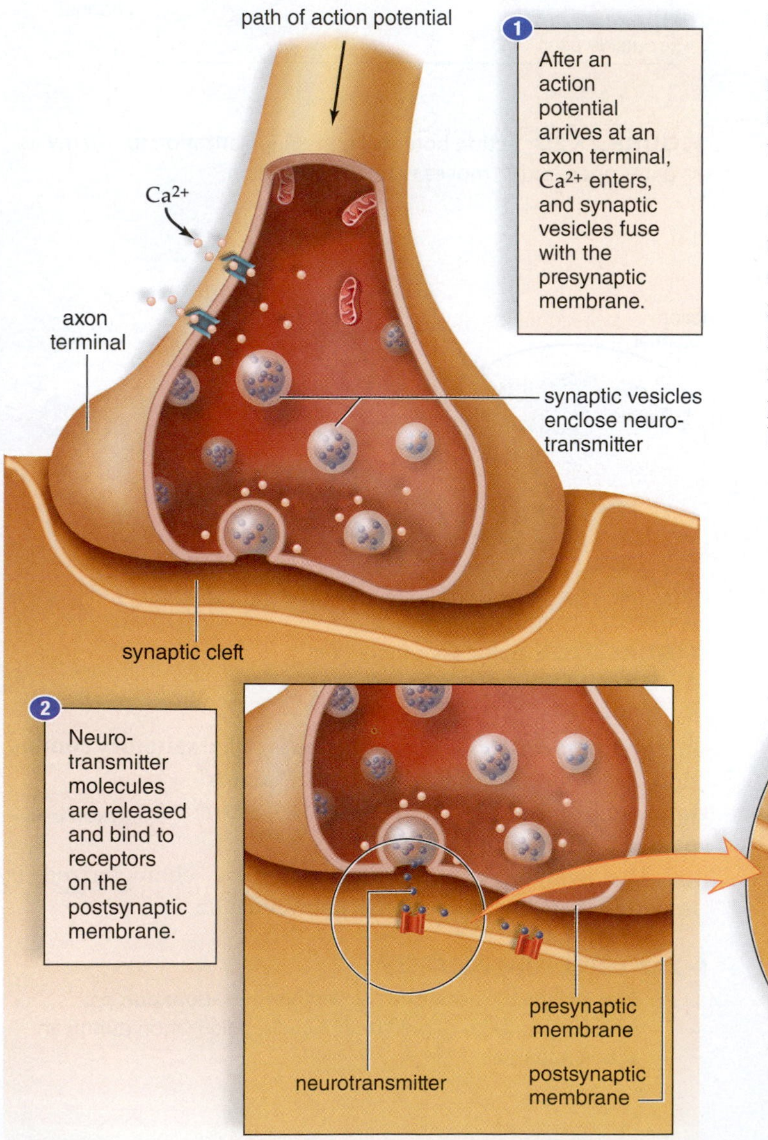

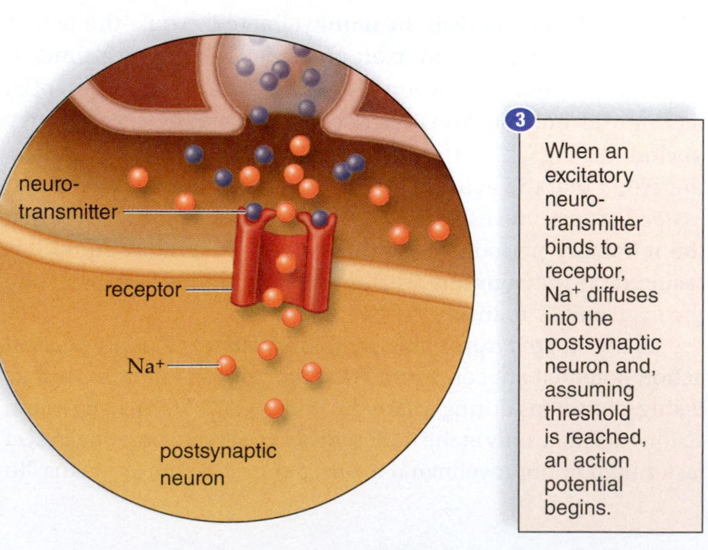

path of action potential

❶ After an action potential arrives at an axon terminal, Ca^{2+} enters, and synaptic vesicles fuse with the presynaptic membrane.

Ca^{2+}

axon terminal

synaptic vesicles enclose neurotransmitter

synaptic cleft

❷ Neurotransmitter molecules are released and bind to receptors on the postsynaptic membrane.

neurotransmitter

presynaptic membrane

postsynaptic membrane

neurotransmitter

receptor

Na^+

postsynaptic neuron

❸ When an excitatory neurotransmitter binds to a receptor, Na^+ diffuses into the postsynaptic neuron and, assuming threshold is reached, an action potential begins.

FIGURE 27.3A Transmission of a nerve impulse across a synapse.

toxin. The toxin blocks the release of ACh at neuromuscular junctions. Six hours to 8 days after eating contaminated food, usually canned improperly, the person may feel the effects of the toxin and may die if the respiratory muscles are affected. Botulinum toxin type A is used as the drug Botox to paralyze facial skeletal muscles and thus reduce the appearance of wrinkles.

In the CNS, NE is important to dreaming, waking, and mood. **Serotonin,** another neurotransmitter, is involved in thermoregulation, sleeping, emotions, and perception. Reduced levels of NE and serotonin seem to be linked to depression, and the antidepressant drug fluoxetine (Prozac) blocks the removal of serotonin from a synapse. In the PNS, NE generally excites smooth muscle.

Dopamine, found primarily in the CNS, is involved in emotions, control of motor function, and attention. Parkinson disease is associated with a lack of dopamine in the brain. Many of the drugs that affect mood act by interfering with or increasing the effect of dopamine. GABA is an abundant inhibitory neurotransmitter in the CNS. The drug diazepam (Valium) binds to the receptors of GABA, thereby increasing its effects.

Neuromodulators are molecules that block the release of a neurotransmitter or modify a neuron's response to a neurotransmitter. The caffeine in coffee, chocolate, and tea keeps us awake by interfering with the effects of inhibitory neurotransmitters in the brain. Two well-known neuromodulators are substance P and endorphins. Substance P is released by sensory neurons when pain is present. **Endorphins** block the release of substance P and, therefore, serve as natural painkillers. They are thought to be associated with the "runner's high" experienced by joggers and to be produced by the brain, not only in the presence of physical stress, but also emotional stress. The opiates—namely, codeine, heroin, and morphine—function similarly to endorphins, and like them, reduce pain and produce a feeling of well-being.

Clearing of Neurotransmitter from a Synapse Once a neurotransmitter has been released into a synaptic cleft and has initiated a response, it is removed from the cleft. The short existence of neurotransmitters at a synapse prevents continuous stimulation (or inhibition) of postsynaptic membranes.

In some synapses, the postsynaptic membrane contains enzymes that rapidly inactivate the neurotransmitter. For example, the enzyme **acetylcholinesterase (AChE)** breaks down acetylcholine. The enzyme GABA transaminase converts GABA into an inactive compound. In other synapses, the presynaptic membrane rapidly reabsorbs the neurotransmitter, possibly for repackaging in synaptic vesicles or for molecular breakdown.

Integration Even though each individual synapse is excitatory or inhibitory, it is important to realize that a single neuron can have many synapses all over its dendrites and the cell body, as seen in the micrograph in Figure 27.3B*a*. Some neurons have as many as 10,000 synapses. Therefore, a neuron is on the receiving end of many excitatory and inhibitory signals. An excitatory neurotransmitter produces a signal that drives the neuron closer to threshold, and an inhibitory neurotransmitter produces a signal that drives the neuron further from threshold (see Fig. 27.2B). Excitatory signals have a depolarizing effect, and inhibitory signals have a hyperpolarizing effect.

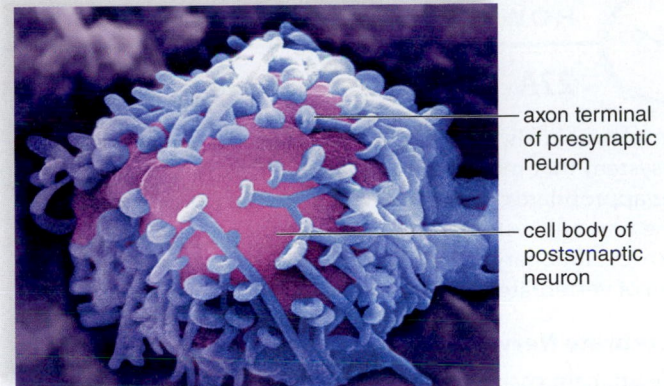

a.

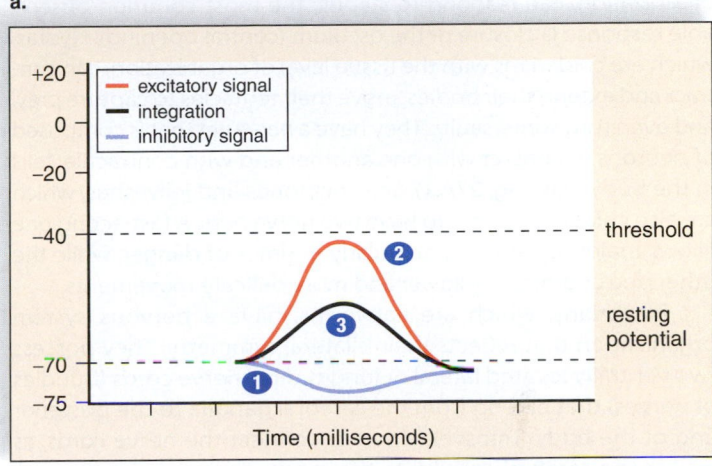

b.

FIGURE 27.3B *Top:* Many neurons synapse with a cell body. *Bottom:* Both inhibitory signals (blue) and excitatory signals (red) are summed up in the dendrite and cell body of the postsynaptic neuron. Only if the combined signals cause the membrane potential to rise above threshold does an action potential occur. In this example, threshold was not reached.

Neurons integrate these incoming signals. **Integration** is the summing up of excitatory and inhibitory signals. If a neuron receives many excitatory signals (either from different synapses or from one synapse at a rapid rate), chances are the axon will transmit a nerve impulse. On the other hand, if a neuron receives both inhibitory and excitatory signals, the summing up of these signals may prohibit the axon from reaching threshold and firing.

In Figure 27.3B*b*, ❶ the inhibitory signals received outweighed ❷ the excitatory signals received by the neuron, and ❸ threshold was never reached following integration. Threshold, as mentioned, must be reached, or else a nerve impulse does not start.

▶ **27.3 CHECK YOUR PROGRESS**
  1. Describe step by step how a nerve impulse is transmitted across a synapse.
  2. Identify ways synapses are essential to integration.

27A Evolution of the Nervous System

In complex animals, the ability to survive is dependent on a nervous system that monitors internal and external conditions and makes appropriate changes to maintain homeostasis. By comparing the nervous system organization of simpler animals, we can discern evolutionary trends that may have led to the nervous system of vertebrates.

Invertebrate Nervous Organization

Simple animals, such as sponges, which have the cellular level of organization, can respond to stimuli; the most common observable response is closure of the osculum (central opening). Hydras, which are cnidarians with the tissue level of organization, can contract and extend their bodies, move their tentacles to capture prey, and even turn somersaults. They have a *nerve net* that is composed of neurons in contact with one another and with contractile cells in the body wall (Fig. 27A.1). Sea anemones and jellyfishes, which are also cnidarians, seem to have two nerve nets. A fast-acting one allows major responses, particularly in times of danger, while the other one coordinates slower and more delicate movements.

Planarians, which are flatworms, have a nervous system organization that reflects their bilateral symmetry. They possess two ventrally located lateral or longitudinal nerve cords (bundles of nerves) that extend from the cerebral ganglia to the posterior end of the body. Transverse nerves connect the nerve cords, as well as the cerebral ganglia, to the eyespots. The entire arrangement is a *ladderlike nervous system*. *Cephalization* has occurred, as evidenced by the concentration of ganglia and sensory receptors in a head region. A cluster of neurons is called a *ganglion* (pl., ganglia), and the anterior cerebral ganglia receive sensory information from photoreceptors in the eyespots and sensory cells

in the auricles. The two lateral nerve cords allow rapid transfer of information from the cerebral ganglia to the posterior end, and the transverse nerves between the nerve cords keep the movement of the two sides coordinated. Bilateral symmetry plus cephalization are two significant trends in the development of a nervous organization that is adaptive for an active way of life. Also, the nervous organization in planarians foreshadows the organization of the nervous system in vertebrates.

In annelids (e.g., earthworm), arthropods (e.g., crab), and molluscs (e.g., squid), the nervous system shows further advances. The annelids and arthropods have the typical invertebrate nervous system. It consists of a brain and a ventral nerve cord that has a ganglion in each segment. The brain, which normally receives sensory information, controls the activity of the ganglia and assorted nerves so that the muscle activity of the entire animal is coordinated. The crab and squid have a well-defined brain and well-developed sense organs, such as eyes. The presence of a brain and other ganglia in the body of all these animals indicates an increase in the number of neurons (nerve cells) among more complex invertebrates.

Vertebrate Nervous Organization

In vertebrates (e.g., cat), cephalization, coupled with bilateral symmetry, results in several types of paired sensory receptors, including the eyes, ears, and olfactory structures, that allow the animal to gather information from the environment. Paired cranial and spinal nerves contain numerous nerve fibers. Vertebrates have many more neurons than do invertebrates. For example, an insect's entire nervous system may contain a total of about 1 million neurons, while a vertebrate's nervous system may contain

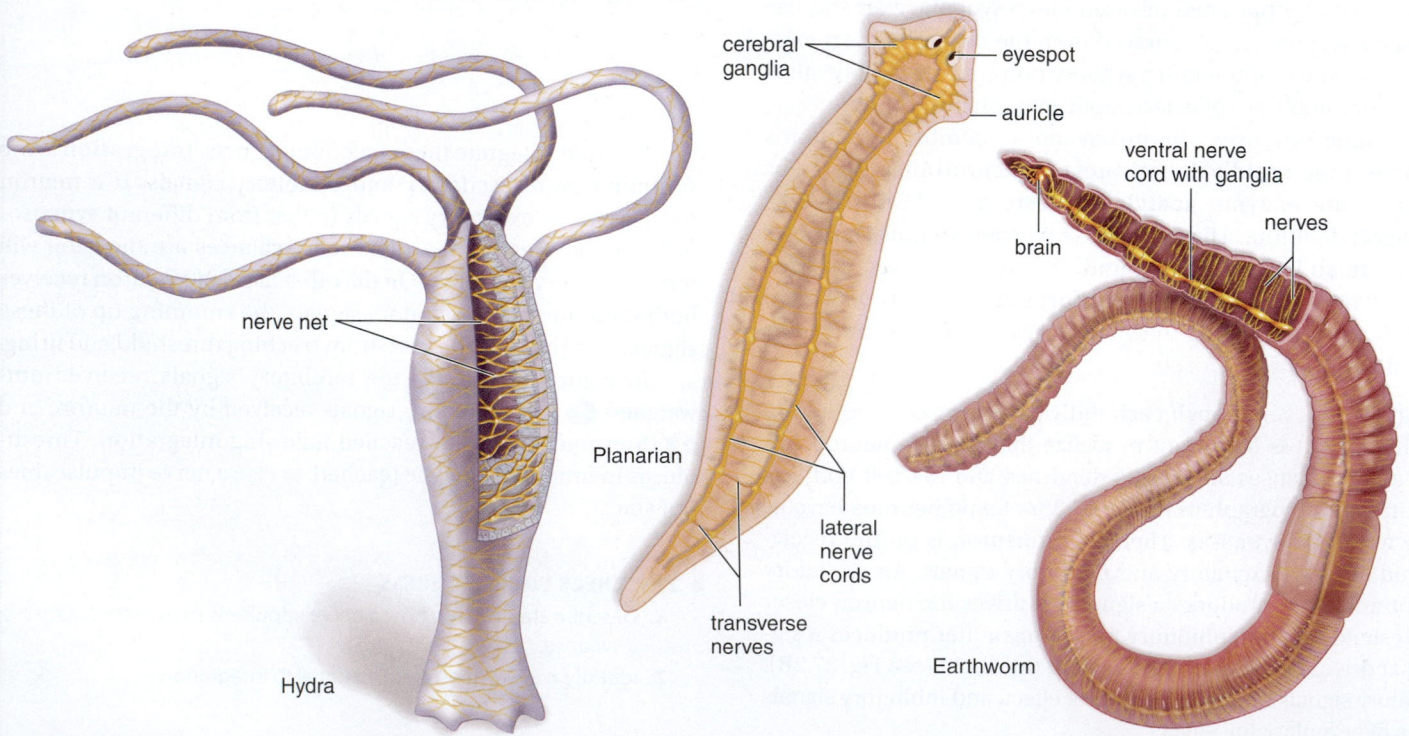

FIGURE 27A.1 Evolution of the nervous system.

many thousand to many billion times that number. A vertebrate's central nervous system (CNS), consisting of a spinal cord and brain, develops from an embryonic dorsal neural tube. The spinal cord is continuous with the brain because the embryonic neural tube becomes the spinal cord posteriorly, while the vertebrate brain is derived from the enlarged anterior end of the neural tube. Ascending tracts carry sensory information to the brain, and descending tracts carry motor commands to the neurons in the spinal cord that control the muscles.

It is customary to divide the vertebrate brain into the hind-brain, midbrain, and forebrain (Fig. 27A.2). The hindbrain is the most ancient part of the brain. Nearly all vertebrates have a well-developed hindbrain that regulates motor activity below the level of consciousness. In humans, for example, the lungs and heart function even when we are sleeping. The medulla oblongata contains control centers for breathing and heart rate. Coordination of motor activity associated with limb movement, posture, and balance eventually became centered in the cerebellum.

The optic lobes are part of the midbrain, which was originally a center for coordinating reflexes involving the eyes and ears. Starting with the amphibians and continuing in the other vertebrates, the forebrain processes sensory information. Originally, the forebrain was concerned mainly with the sense of smell. Later, the thalamus evolved to receive sensory input from the midbrain and the hindbrain and to pass it on to the cerebrum, the anterior part of the forebrain in vertebrates. In the forebrain, the hypothalamus is particularly concerned with homeostasis, and in this capacity, the hypothalamus communicates with the medulla oblongata and the pituitary gland.

The cerebrum, which is highly developed in mammals, integrates sensory and motor input and is particularly associated with higher mental capabilities. In humans, the outer layer of the cerebrum, called the cerebral cortex, is especially large and complex.

MP3
Nervous System
Organization

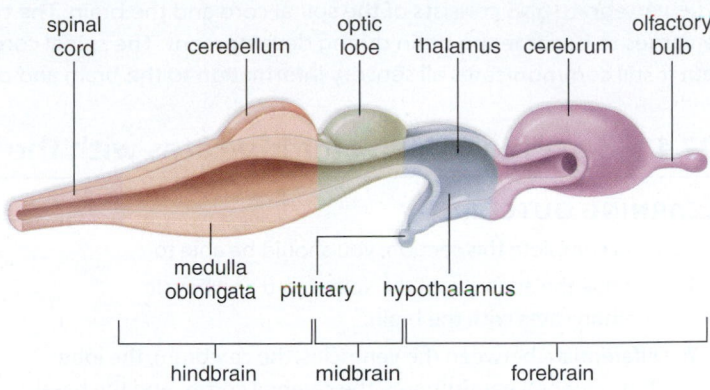

FIGURE 27A.2 Organization of the vertebrate brain.

CONSIDER THESE QUESTIONS

1. Compare the human nervous system (see Fig. 27.1) to that of a planarian.
2. How does the organization of the nervous system in vertebrates differ from that of arthropods?
3. Would you expect the structure and function of a cat's nervous system to be different from that of a human? How?

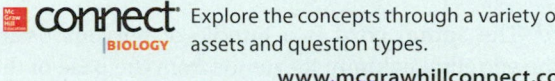

connect
|BIOLOGY   Explore the concepts through a variety of multimedia assets and question types.

www.mcgrawhillconnect.com

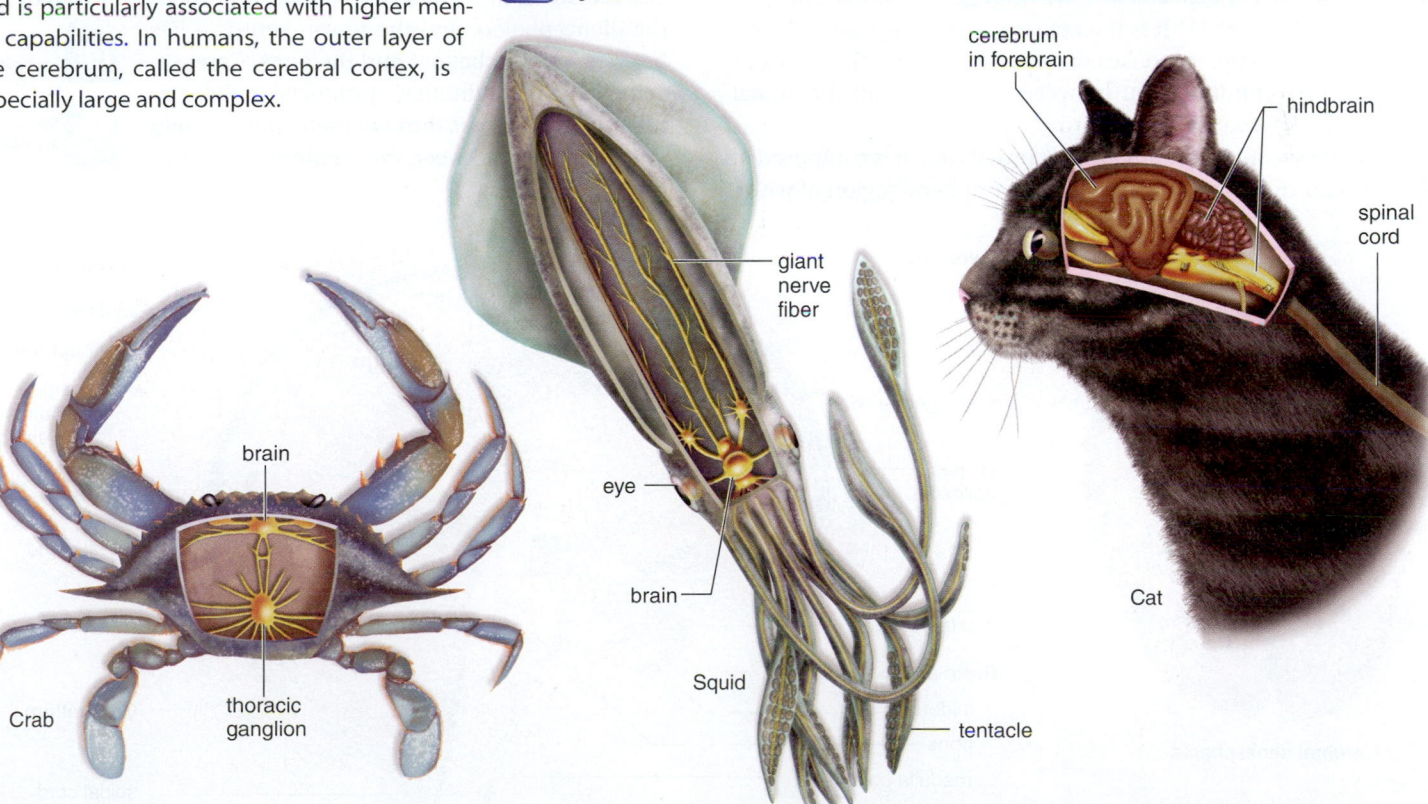

Central Nervous System (CNS)

The vertebrate CNS consists of the spinal cord and the brain. The two are connected because the anterior end of the neural tube enlarges to become the brain during development. The spinal cord can participate in reflex actions without checking with the brain, but it still communicates all sensory information to the brain and only the brain makes higher-level motor and judgmental decisions.

27.4 The spinal cord communicates with the brain

LEARNING OUTCOMES

When you complete this section, you should be able to

1. Describe the anatomy of the spinal cord and how it communicates with the brain.
2. Differentiate between the ventricles, the cerebrum, the lobes of the cerebral hemispheres, the cerebral cortex, and the basal nuclei.

The CNS consists of the spinal cord and the brain, where sensory information is received and motor control is initiated. The spinal cord and the brain are both protected by bone; the spinal cord is surrounded by vertebrae (see Fig. 27.7), and the brain is enclosed by the skull. Both the spinal cord and the brain are wrapped in three protective membranes known as **meninges.** *Meningitis* (inflammation of the meninges) is a serious disorder caused by a number of bacteria or viruses that invade the meninges. The spaces between the meninges are filled with **cerebrospinal fluid,** which cushions and protects the CNS. Cerebrospinal fluid is contained in the central canal of the spinal cord and within the **ventricles** of the brain, which are interconnecting spaces that produce and serve as reservoirs for cerebrospinal fluid.

Spinal Cord The **spinal cord** is a bundle of nervous tissue enclosed in the vertebral column; it extends from the base of the brain to the vertebrae just below the rib cage. The spinal cord has two main functions: (1) It is the center for many **spinal reflexes,** which are automatic responses to external stimuli. (2) It provides a means of communication between the brain and the spinal nerves, which leave the spinal cord.

A cross section of the spinal cord reveals that it is composed of a central portion of **gray matter** and a peripheral region of **white matter.** The gray matter consists of cell bodies and unmyelinated fibers. It is shaped like a butterfly, or the letter H, with two dorsal (posterior) horns and two ventral (anterior) horns surrounding a central canal. The gray matter contains portions of sensory neurons and motor neurons, as well as short interneurons that connect sensory and motor neurons (see Fig. 27.8).

Myelinated long fibers of interneurons that run together in bundles called **tracts** give white matter its color. These tracts connect the spinal cord to the brain. Like a busy superhighway, these tracts continuously pass information between the brain and the rest of the body. Dorsally, the tracts are primarily ascending, taking information *to* the brain; ventrally, the tracts are primarily descending, carrying information *from* the brain. Because the tracts at one point cross over, the left side of the brain controls the right side of the body, and the right side of the brain controls the left side of the body.

If the spinal cord is severed as the result of an injury, paralysis results. If the injury occurs in the cervical (neck) region, all four limbs are usually paralyzed, a condition known as quadriplegia. If the injury occurs in the thoracic region, the lower body may be paralyzed, a condition called paraplegia.

Brain Ventricles The brain contains four interconnected chambers called ventricles (Fig. 27.4A). The two lateral ventricles are inside the cerebrum. The third ventricle is surrounded by the diencephalon, and the fourth ventricle lies between the cerebellum and the pons. Cerebrospinal fluid is continuously produced in the ventricles and circulates through them; it then flows out of the brain between the meninges.

MP3 The Brain

Video Brain Bank

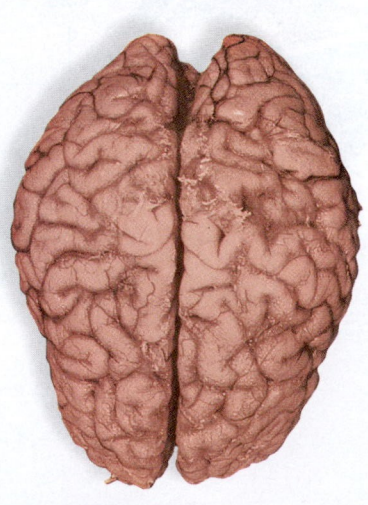

Cerebral hemispheres

FIGURE 27.4A The human brain.

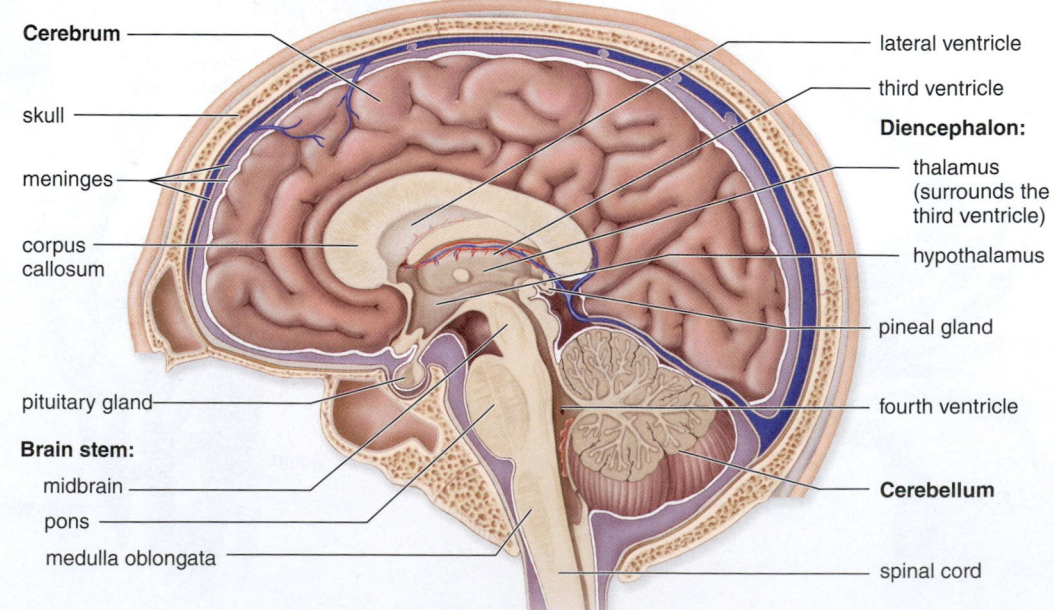

Cerebrum

skull

meninges

corpus callosum

pituitary gland

Brain stem:
  midbrain
  pons
  medulla oblongata

lateral ventricle

third ventricle

Diencephalon:
  thalamus (surrounds the third ventricle)

  hypothalamus

pineal gland

fourth ventricle

Cerebellum

spinal cord

Cerebrum The **cerebrum** is the largest portion of the brain in humans. The cerebrum is the last center to receive sensory input and carry out integration before commanding voluntary motor responses. It communicates with and coordinates the activities of the other parts of the brain.

The cerebrum is divided into two halves, called **cerebral hemispheres** (see Fig. 27.4A). A deep groove called the longitudinal fissure divides the cerebrum into the right and left hemispheres. Each hemisphere receives information from and controls the opposite side of the body. Although the hemispheres appear the same, the right hemisphere is associated with artistic and musical ability, emotion, spatial relationships, and pattern recognition. The left hemisphere is more adept at mathematics, language, and analytical reasoning. The two cerebral hemispheres are connected by a bridge of tracts within the corpus callosum.

Shallow grooves called sulci (sing., sulcus) divide each hemisphere into the **cerebral lobes** (Fig. 27.4B). A *frontal lobe* is the anterior portion of a hemisphere and is associated with motor control, memory, reasoning, and judgment. For example, the frontal lobe enables you to decide how to dress if the temperature plummets to subzero or whether to escape a fire via the stairs or a window. The frontal lobe on the left side contains the motor speech *(Broca) area,* which organizes motor commands to produce speech.

The *parietal lobes* lie posterior to the frontal lobe and are concerned with sensory reception and integration, as well as taste. A *primary taste area* in the parietal lobe accounts for taste sensations.

The *temporal lobe* is located laterally. A primary auditory area in the temporal lobe receives information from our ears. The *occipital lobe* is the most posterior lobe. A *primary visual area* in the occipital lobe receives information from our eyes.

The **cerebral cortex** is a thin (less than 5 mm thick), but highly convoluted, outer layer of gray matter that covers the cerebral hemispheres. The convolutions increase the surface area of the cerebral cortex. The cerebral cortex contains tens of billions of neurons and is the region of the brain that accounts for sensation, voluntary movement, and all the thought processes required for learning and memory as well as for language and speech.

Two regions of the cerebral cortex are of particular interest. The **primary motor area** is in the frontal lobe just ventral to (before) the central sulcus. Voluntary commands to skeletal muscles begin in the primary motor area, and each part of the body is controlled by a certain section. The size of the section indicates the precision of motor control. For example, the face and hand take up a much larger portion of the primary motor area than does the entire trunk. The **primary somatosensory area** is just dorsal to the central sulcus in the parietal lobe. Sensory information from the skin and skeletal muscles arrives here, where each part of the body is sequentially represented in a manner similar to the primary motor area.

While the bulk of the cerebrum beneath the cerebral cortex is composed of white matter, masses of gray matter are located deep within the white matter. These so-called **basal nuclei** (formerly termed basal ganglia) integrate motor commands, ensuring that proper muscle groups are activated or inhibited. **Huntington disease** and **Parkinson disease,** which are both characterized by uncontrollable movements, are believed to be due to malfunctioning basal nuclei.

▶ **27.4 CHECK YOUR PROGRESS**
1. Identify various functions of the cerebrum that illustrate its primary importance to humans.
2. Use Figure 27.4B to point out the cerebral lobes and give a function of each.

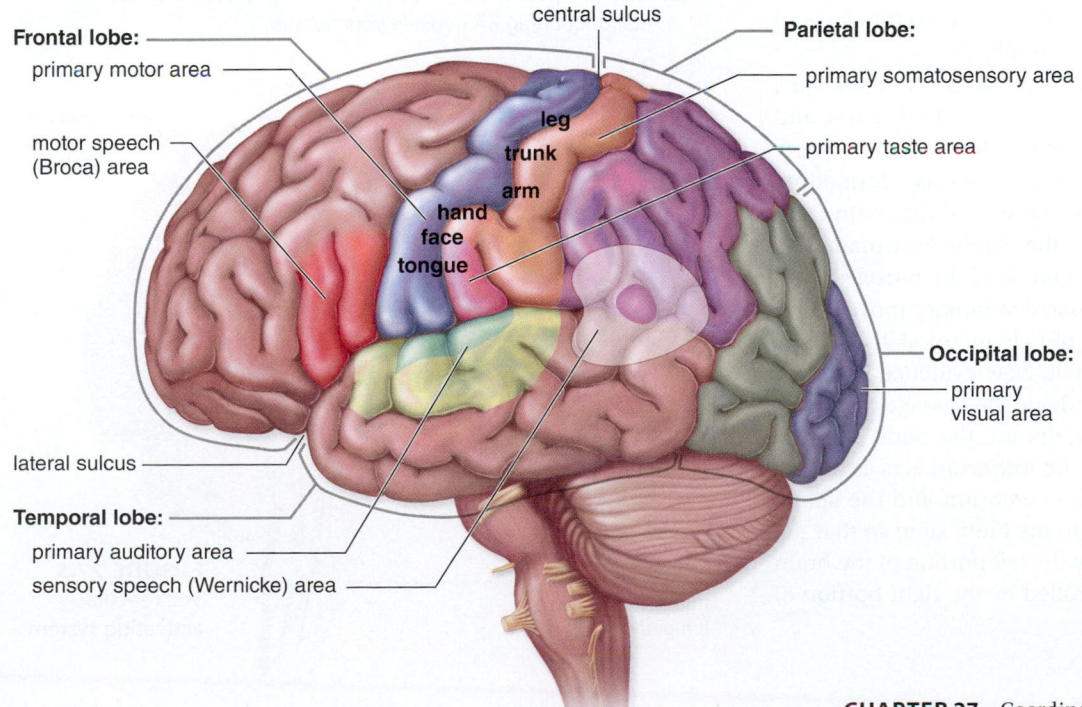

Frontal lobe:
— primary motor area

motor speech (Broca) area

hand
face
tongue

leg
trunk
arm

central sulcus

lateral sulcus

Temporal lobe:
— primary auditory area
— sensory speech (Wernicke) area

Parietal lobe:
— primary somatosensory area
— primary taste area

Occipital lobe:
— primary visual area

FIGURE 27.4B Among the cerebral lobes, the frontal lobe carries out higher level thinking and initiates motor impulses; the other lobes have specialized sensory areas for receiving and integrating sensory information.

27.5 The other parts of the brain have specialized functions

LEARNING OUTCOMES

When you complete this section, you should be able to

1. Identify and state functions for the hypothalamus, cerebellum, midbrain, and the medulla oblongata.
2. Define and exemplify how the reticular activating system (RAS) functions.

The hypothalamus and the thalamus are in the **diencephalon,** a region that encircles the third ventricle (see Fig. 27.4). The **hypothalamus** forms the floor of the third ventricle. It is an integrating center that helps maintain homeostasis by regulating hunger, sleep, thirst, body temperature, and water balance. The hypothalamus controls the pituitary gland and, thereby, serves as a link between the nervous and endocrine systems.

The **thalamus** consists of two masses of gray matter located in the sides and roof of the third ventricle. It is on the receiving end of all sensory input except smell. Visual, auditory, and somatosensory information arrives at the thalamus via the cranial nerves and tracts from the spinal cord. The thalamus integrates this information and sends it on to the appropriate portions of the cerebrum. For this reason, the thalamus is often referred to as the "gatekeeper" for sensory information en route to the cerebral cortex. The thalamus is involved in arousal of the cerebrum, and it also participates in higher mental functions such as memory and emotions.

The **pineal gland,** which secretes the hormone melatonin, is located in the diencephalon. Presently, there is much interest in the role of melatonin in our daily rhythms; some researchers believe it may be involved in jet lag and insomnia. Scientists are also interested in the possibility that this hormone regulates the onset of puberty.

Video
Winter Mood

The **cerebellum** lies under the occipital lobe of the cerebrum and is separated from the brain stem by the fourth ventricle. It is the largest part of the hindbrain. The cerebellum has two portions, which are joined by a narrow central portion. Each portion is primarily composed of white matter, which in longitudinal section has a treelike pattern. Overlying the white matter is a thin layer of gray matter that forms a series of complex folds.

The cerebellum receives sensory input from the eyes, ears, joints, and muscles about the present position of body parts, and it also receives motor output from the cerebral cortex about where these parts should be located. After integrating this information, the cerebellum sends motor impulses by way of the brain stem to the skeletal muscles. In this way, the cerebellum maintains posture and balance. It also ensures that all of the muscles work together to produce smooth, coordinated voluntary movements. The cerebellum assists the learning of new motor skills such as playing the piano or hitting a baseball. New evidence indicates that the cerebellum is important in judging the passage of time.

The **brain stem** contains the midbrain, the pons, and the medulla oblongata (see Fig. 27.4A). The **midbrain** acts as a relay station for tracts passing between the cerebrum and the spinal cord or cerebellum. The tracts cross in the brain stem so that the right side of the body is controlled by the left portion of the brain and the left side of the body is controlled by the right portion of the brain.

The brain stem also has reflex centers for visual, auditory, and tactile responses. The word **pons** means "bridge" in Latin, and true to its name, the pons contains bundles of axons traveling between the cerebellum and the rest of the CNS. In addition, the pons functions with the medulla oblongata to regulate breathing rate, and has reflex centers concerned with head movements in response to visual and auditory stimuli.

The **medulla oblongata** contains a number of reflex centers for regulating heartbeat, breathing, and blood pressure. It also contains the reflex centers for vomiting, coughing, sneezing, hiccuping, and swallowing. The medulla oblongata lies just superior to the spinal cord, and it contains tracts that ascend or descend between the spinal cord and higher brain centers.

The Reticular Activating System The reticular formation is a complex network of nuclei (masses of gray matter) and nerve fibers that extend the length of the brain stem (Fig. 27.5). The reticular formation is a major component of the **reticular activating system** (RAS), which receives sensory signals that it sends up to higher centers, and motor signals that it sends to the spinal cord.

The RAS arouses the cerebrum via the thalamus and causes a person to be alert. Apparently, the RAS can filter out unnecessary sensory stimuli, explaining why you can study with the TV on. If you want to awaken the RAS, surprise it with a sudden stimulus, like splashing your face with cold water; if you want to deactivate it, remove visual and auditory stimuli. General anesthetics function by artificially suppressing the RAS. A severe injury to the RAS can cause a person to be comatose, from which recovery may be impossible.

Several parts of the brain work together in the limbic system, discussed in the next section.

▶ **27.5 CHECK YOUR PROGRESS**

1. Identify which part of the brain helps you play baseball and which two parts are most involved in homeostasis. Justify your choices.
2. Trace the path of the RAS and explain the possible effect of a malfunctioning RAS. Justify your answer.

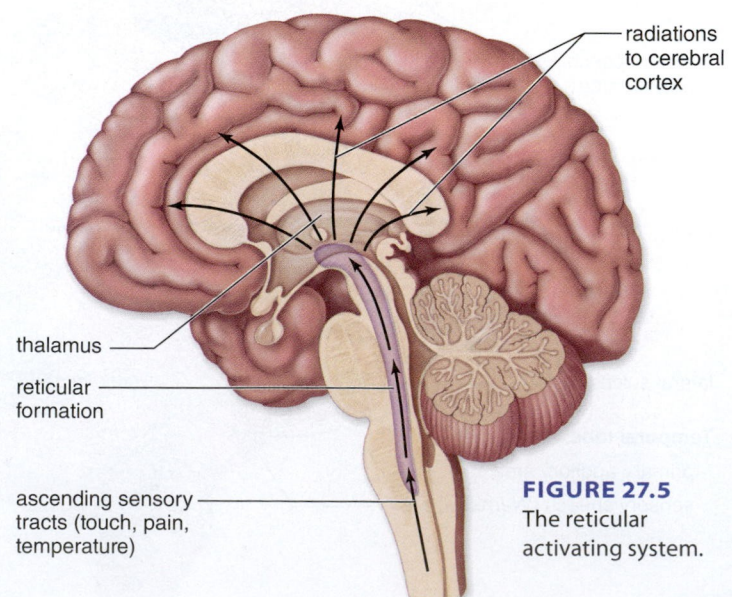

radiations to cerebral cortex

thalamus

reticular formation

ascending sensory tracts (touch, pain, temperature)

FIGURE 27.5
The reticular activating system.

27B If You Don't Snooze, You'll Lose!

Sleep is a bit of an evolutionary conundrum. In humans, with the exception of college students at exam time, sleep typically occupies a third of a person's life. Other mammals have various sleep requirements: Giraffes need only 2 hours, while brown bats need 20. Nonetheless, all mammals and birds sleep, and during the time that they are sleeping, they are vulnerable to predators, thus putting themselves at a huge evolutionary disadvantage. What, then, is the biological reason for engaging in this ostensibly high-risk behavior known as sleep?

An emerging scientific consensus is that sleep is needed to increase the strength of the neural circuits in the brain. In other words, during sleep, your brain is hard at work enhancing and expanding the synaptic connections among neurons. This appears to help you create and strengthen memories as well as process and understand information. It is no surprise that humans require more sleep when the brain is actively developing, such as during embryonic development and early childhood.

As a person falls asleep, drowsiness gives way to light sleep and finally stage 4, which is deep sleep (Fig. 27B). The majority of the sleep cycle in mammals and birds is spent in deep sleep in which responsiveness to stimuli is reduced, as are metabolic rates and skeletal muscle tone. Periodically, the sleeper backtracks; the brain is more active as REM sleep occurs. REM sleep is so named because *r*apid *e*ye *m*ovement occurs. Dreams that occur during REM sleep are more vivid than in the other stages of sleep. Although reptiles other than birds have a resting period during which slower brain activity is correlated with quiescent behavior, they don't have REM sleep. Other vertebrates, such as amphibians and fish, likewise show behavior changes that are associated with resting metabolism, but they too lack REM sleep. Even insects exhibit behavioral and physiological changes that can be equated with rudimentary sleep.

The capacity for sleep has even been observed in the roundworm. *Caenorhabditis elegans,* which is often used in scientific research, has a sleeplike behavioral state called lethargus during which the worm is much less responsive than when it is fully "awake." If the worm is deprived of sleep, it will fall asleep faster and sleep deeper than if it is not sleep-deprived first. Researchers have identified genes in *C. elegans* that regulate sleep in this species. Using *C. elegans* as a model, they are now testing drugs that target these genes as therapies for sleep disorders.

In higher vertebrates, control of sleep is governed by the hypothalamus; damage to the back part of the organ caused drowsiness, while damage to the front part of the organ had the opposite effect, producing insomnia. The hypothalamus is also involved in temperature regulation in endotherms, whose body temperature does drop somewhat during sleep, suggesting that sleep may have evolved from primitive temperature-regulation mechanisms. In cats, the link between temperature regulation and sleep is explicit as both sleep and body temperature are regulated by the same neurons. Could this be evidence that certain genes originally regulated only body temperature and then became involved in controlling sleep also?

One thing we know for sure: Sleep deprivation has negative consequences. Symptoms in humans range from general crankiness to weight loss and immunosuppression. While our lifestyles might be inconvenienced by the requirement for sleep, our brains need down time. You can't perform at your peak if your synapses aren't being revived during sleep. So during your next final exam week, get some sleep!!

CONSIDER THESE QUESTIONS

1. Why does the brain need sleep and can't be awake to enhance and expand synaptic connections? Answer in terms of the possible energy required.
2. If deep sleep were needed to enhance and expand synaptic connections, what might be happening during REM sleep?
3. People use the expression "sleep on it" because sometimes solutions to complex problems come to them while they sleep. Give an explanation consistent with our need for sleep.

 Explore the concepts through a variety of multimedia assets and question types.

www.mcgrawhillconnect.com

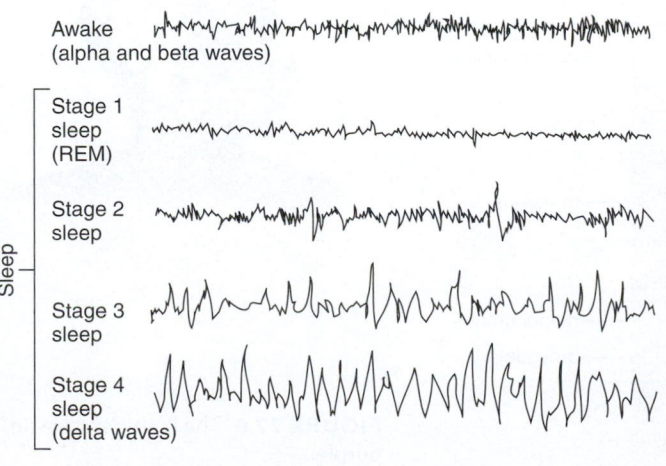

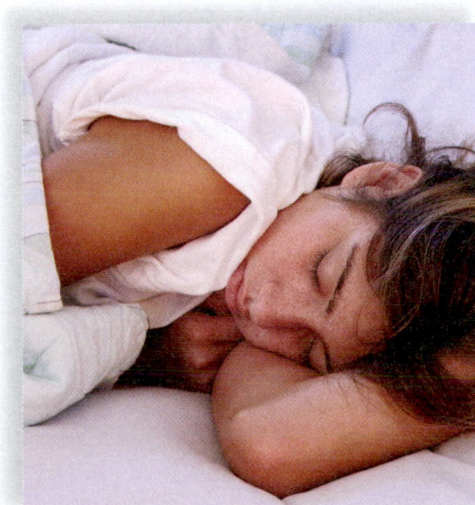

FIGURE 27B REM (rapid eye movement) sleep occurs when our dreams are vivid.

LEARNING OUTCOMES

When you complete this section, you should be able to

1. Understand the structure and function of the limbic system.
2. Discuss the contributions of the hippocampus and the amygdala to the functioning of the limbic system, learning, and memory.

The **limbic system** is a complex network of tracts and nuclei that incorporates portions of the cerebral lobes, the basal nuclei, and the diencephalon (Fig. 27.6). The limbic system blends higher mental functions and primitive emotions into a united whole. It accounts for why activities such as sexual behavior and eating seem pleasurable and also why, say, mental stress can cause high blood pressure.

Two significant structures within the limbic system are the hippocampus and the amygdala, which are essential for learning and memory. The **hippocampus,** a seahorse-shaped structure that lies deep in the temporal lobe, is well situated in the brain to make the frontal lobe aware of past experiences stored in various sensory areas. The **amygdala,** in particular, adds emotional overtones. The smell of smoke not only warns us that the hotel is on fire, but also creates great anxiety. Because the frontal lobe is part of the limbic system, we may be able to calmly analyze the situation and walk to the nearest exit.

Learning and Memory The ability to hold a thought in mind or recall events from the past is termed **memory.** Memories range from a word we learned only yesterday to an early emotional experience that has shaped our lives. Learning takes place when we retain and use past memories.

The frontal lobe is active during short-term memory, as when we temporarily recall a telephone number. Some telephone numbers go into long-term memory. Think of a telephone number you know by heart, and see if you can bring it to mind without also thinking about the place or person associated with that number.

Most likely, you cannot because, typically, long-term memory is a mixture of what is called semantic memory (numbers, words, etc.) and episodic memory (persons, events, etc.). Skill memory is a type of memory that can exist independent of episodic memory. Skill memory enables us to perform motor activities, such as riding a bike or playing ice hockey.

What parts of the brain are functioning when you remember something from long ago? As mentioned, our long-term memories are stored in bits and pieces throughout the sensory areas of the cerebral cortex. The hippocampus gathers this information together for use by the frontal lobe when we remember Uncle Frank or our summer holiday. Why are some memories so emotionally charged? Again, the amygdala is responsible for fear conditioning and associating danger with sensory information received from the thalamus and the cortical sensory areas.

In addition to studying what regions of the brain are involved in memory formation, neurobiologists want to know what exactly neurons are doing when we store memories and bring them back. They have discovered a chemical process caled *long-term potentiation* (*LTP*), which is an enhanced response at synapses within the hippocampus. LTP seems to mainly involve glutamate, an amino acid that can function as a neurotransmitter. To show this, investigators at the Massachusetts Institute of Technology produced mice that lacked the receptor for glutamate only in the hippocampus. Unlike control mice, the defective mice could not learn to run the maze, demonstrating an important role for this chemical in memory.

▶ **27.6 CHECK YOUR PROGRESS**

1. Identify ways in which the limbic system is necessary to our survival.
2. Contrast the memory function of the hippocampus and the amygdala.

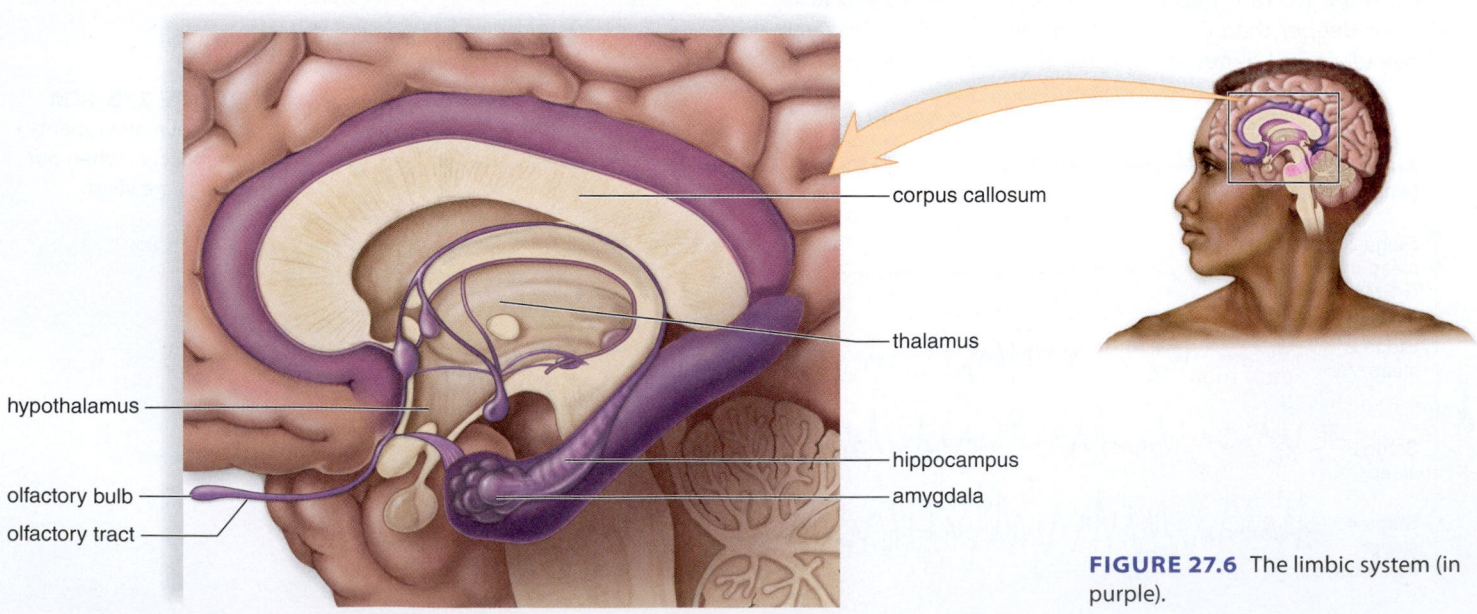

corpus callosum

thalamus

hypothalamus

hippocampus

amygdala

olfactory bulb

olfactory tract

FIGURE 27.6 The limbic system (in purple).

Peripheral Nervous System (PNS)

The peripheral nervous system consists of pairs of nerves that emerge from the brain and spinal cord and serve all parts of the body. Most nerves are mixed nerves, which contain the axons of both sensory and motor neurons. Review the organization of the nervous system in Figure 27.1 and note the two types of motor fibers; some belong to a somatic system and some belong to an autonomic system.

27.7 The peripheral nervous system contains cranial and spinal nerves

LEARNING OUTCOME

When you complete this section, you should be able to

1. Describe the anatomy of the PNS, including the cranial nerves and spinal nerves.

The peripheral nervous system (PNS) lies outside the central nervous system and contains **nerves,** which are bundles of axons. Axons within nerves are also called **nerve fibers:**

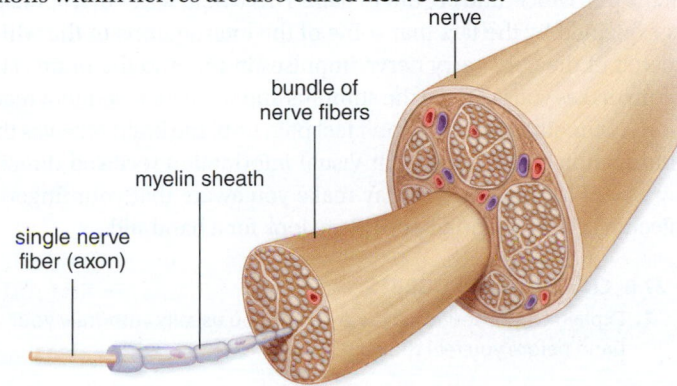

Nerves are designated as cranial nerves when they arise from the brain and as spinal nerves when they arise from the spinal cord. In any case, all nerves take impulses to and from the CNS. So, right now, your eyes are sending messages by way of a cranial nerve to the brain, allowing you to read this page, and your brain, by way of the spinal cord and a spinal nerve, will direct the muscles in your fingers to turn to the next page.

The **cranial nerves** are attached to the brain (Fig. 27.7A). Some of these are sensory nerves—that is, they contain only sensory nerve fibers. Some are motor nerves that contain only motor fibers, and others are mixed nerves that contain both sensory and motor fibers. Cranial nerves largely serve the head, neck, and face. The vagus nerve, which arises from the brain stem, has branches to the pharynx and larynx and to most of the internal organs.

The **spinal nerves** are attached to the spinal cord. Each spinal nerve emerges from the spinal cord by two short branches, or roots (Fig. 27.7B). A spinal nerve separates the axons of sensory neurons from the axons of motor neurons. At the cord, the dorsal root contains the axons of sensory neurons, which conduct impulses to the spinal cord from sensory receptors. The cell body of a sensory neuron is in the **dorsal root ganglion.** A **ganglion** is a cluster of cell bodies outside the CNS. The cell body of a motor neuron is in the cord. The ventral root of a spinal nerve contains the axons of motor neurons, which conduct impulses away from the spinal cord to effectors that are muscle fibers or glands. These two roots join to form a spinal nerve. All spinal nerves are mixed nerves that contain many sensory and motor fibers. Each spinal nerve serves the particular region of the body in which it is located. For example, the intercostal muscles of the rib cage are innervated by the thoracic nerves.

MP3 Organization of the Nervous System

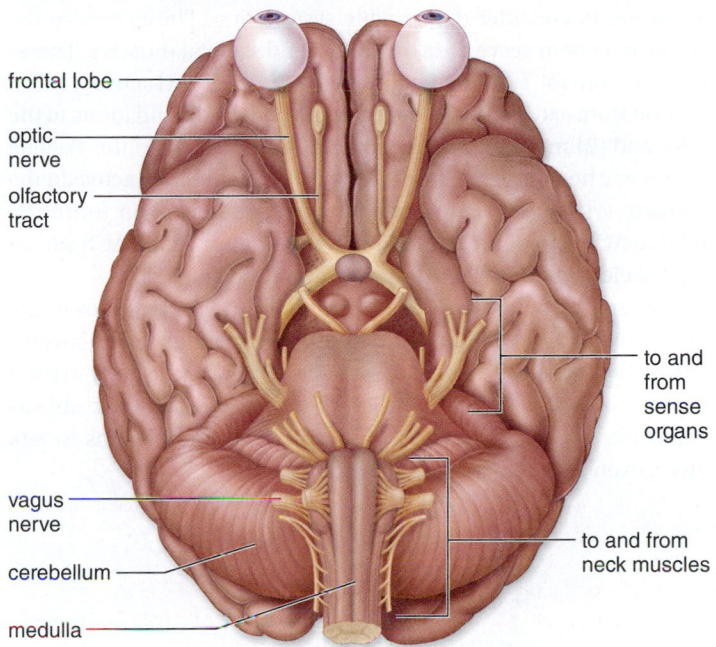

FIGURE 27.7A Ventral surface of brain showing the attachment of the cranial nerves (yellow).

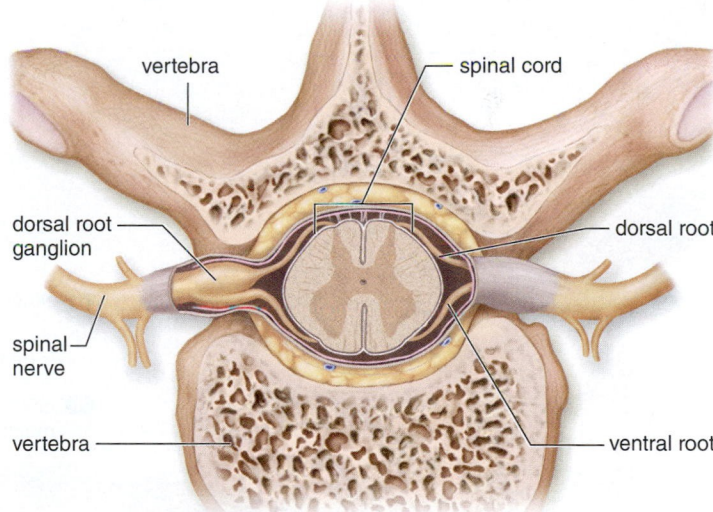

FIGURE 27.7B Cross section of the vertebral column and spinal cord, showing a spinal nerve.

▶ **27.7 CHECK YOUR PROGRESS**

1. Describe the structure of a nerve.
2. Distinguish between location and function of cranial nerves and spinal nerves.

27.8 The somatic system can respond quickly to stimuli

LEARNING OUTCOME

When you complete this section, you should be able to

1. Define the somatic division, and describe the events of a reflex action.

The PNS has two portions—somatic and autonomic—and we are going to consider the somatic system first. The nerves in the **somatic system** serve the skin, joints, and skeletal muscles. Therefore, the somatic system includes nerves that take (1) sensory information from external sensory receptors in the skin and joints to the CNS and (2) motor commands away from the CNS to the skeletal muscles. The neurotransmitter acetylcholine (ACh) is active in the somatic system. In Chapter 29, we will see how axon terminals release ACh into neuromuscular junctions, after which ACh stimulates skeletal muscle fibers to contract (see section 29.7).

Voluntary control of skeletal muscles always originates in the brain. Involuntary responses to stimuli, called **reflexes,** can involve either the brain or the spinal cord. Reflexes enable the body to react swiftly to stimuli that could disrupt homeostasis. Flying objects cause our eyes to blink, and sharp pins cause our hands to jerk away, even without us having to think about it.

The Reflex Arc Figure 27.8 illustrates the path of a spinal reflex that involves only the spinal cord. ❶ If your hand touches a sharp pin, sensory receptors in the skin generate nerve impulses that move along sensory axons through a ❷ dorsal root ganglion toward the spinal cord. ❸ Sensory neurons that enter the cord dorsally pass signals on to many interneurons in the gray matter of the spinal cord. ❹ Some of these interneurons synapse with motor neurons. The short dendrites and the cell bodies of motor neurons are also in the spinal cord, but their axons leave the cord ventrally. Nerve impulses travel along motor axons to an effector, which brings about a response to the stimulus. ❺ In this case, a muscle contracts so that you withdraw your hand from the pin.

Various other reactions are possible—you will most likely look at the pin, wince, and cry out in pain. This whole series of responses is explained by the fact that some of the interneurons in the white matter of the cord carry nerve impulses in tracts to the brain. The brain makes you aware of the stimulus and directs subsequent reactions to the situation. You don't feel pain until the brain receives the information and interprets it! Visual information received directly by way of a cranial nerve may make you aware that your finger is bleeding. Then you might decide to look for a band-aid.

▶ **27.8 CHECK YOUR PROGRESS**

1. Explain why, if you touch a hot stove, you usually withdraw your hand before you feel the pain.

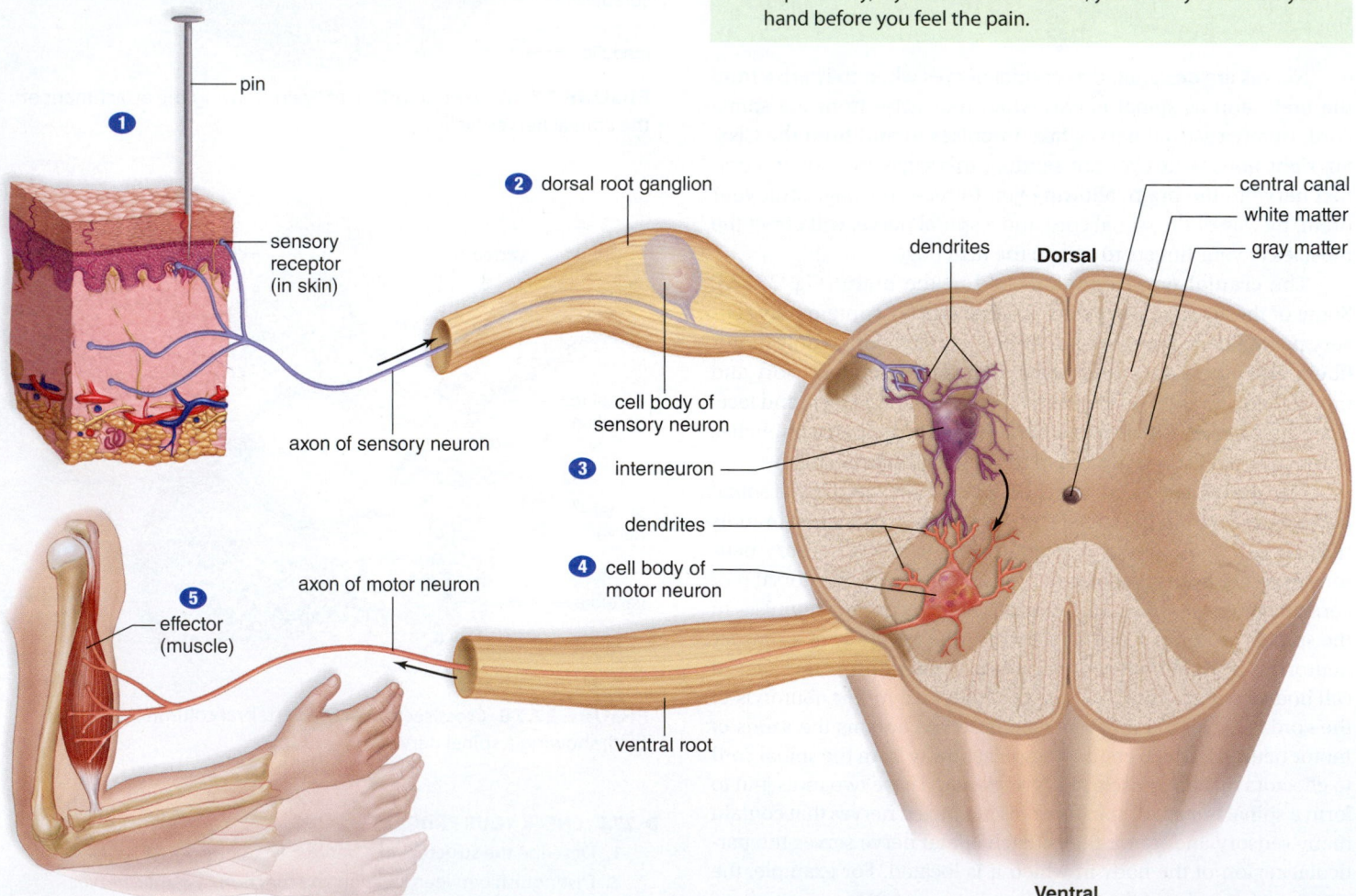

FIGURE 27.8 A reflex arc showing the path of a spinal reflex.

LEARNING OUTCOME

When you complete this section, you should be able to

1. Contrast the structure and function of the sympathetic and parasympathetic divisions of the autonomic nervous system.

The **autonomic system** of the PNS automatically and involuntarily regulates the activity of glands and cardiac and smooth muscle. The system is divided into the parasympathetic and sympathetic divisions (Fig. 27.9). Activation of these systems generally causes opposite responses.

Although their functions are different, the parasympathetic and sympathetic divisions share these same features: (1) They function automatically and usually in an involuntary manner. (2) They innervate all internal organs. (3) They utilize two motor neurons and one ganglion for each outgoing message. The first neuron with a cell body within the CNS has a preganglionic fiber. The second neuron with a cell body within a ganglion has a post-ganglionic fiber.

Reflex actions, such as those that regulate blood pressure and breathing rate, are especially important to the maintenance of homeostasis. These reflexes begin when the sensory neurons in contact with internal organs send information to the CNS. They are completed by motor neurons within the autonomic system.

Parasympathetic Division The **parasympathetic division** includes a few cranial nerves (e.g., the vagus nerve) as well as axons that arise from the last portion of the spinal cord. The parasympathetic division, sometimes called the "housekeeping division," promotes all the internal responses we associate with a relaxed state. For example, it causes the pupil of the eye to constrict, promotes digestion of food, and retards the heartbeat. Some call the parasympathetic division the *rest-and-digest system*. The parasympathetic division utilizes the neurotransmitter ACh.

Sympathetic Division Axons of the **sympathetic division** arise from the other portions of the spinal cord. The sympathetic division is especially important during emergency situations and is associated with *fight or flight*. If you need to fend off a foe or flee from danger, active muscles require a ready supply of glucose and oxygen. The sympathetic division accelerates the heartbeat and dilates the bronchi, while at the same time it inhibits the digestive tract, because digestion is not an immediate necessity if you are under attack. The sympathetic division utilizes the neurotransmitter norepinephrine, which has a structure like that of epinephrine (adrenaline), an adrenal medulla hormone that usually increases heart rate and contractility.

▶ **27.9 CHECK YOUR PROGRESS**

1. Give two examples to show that the sympathetic and parasympathetic divisions have an opposite effect on the body's organs.

FIGURE 27.9 Autonomic system.

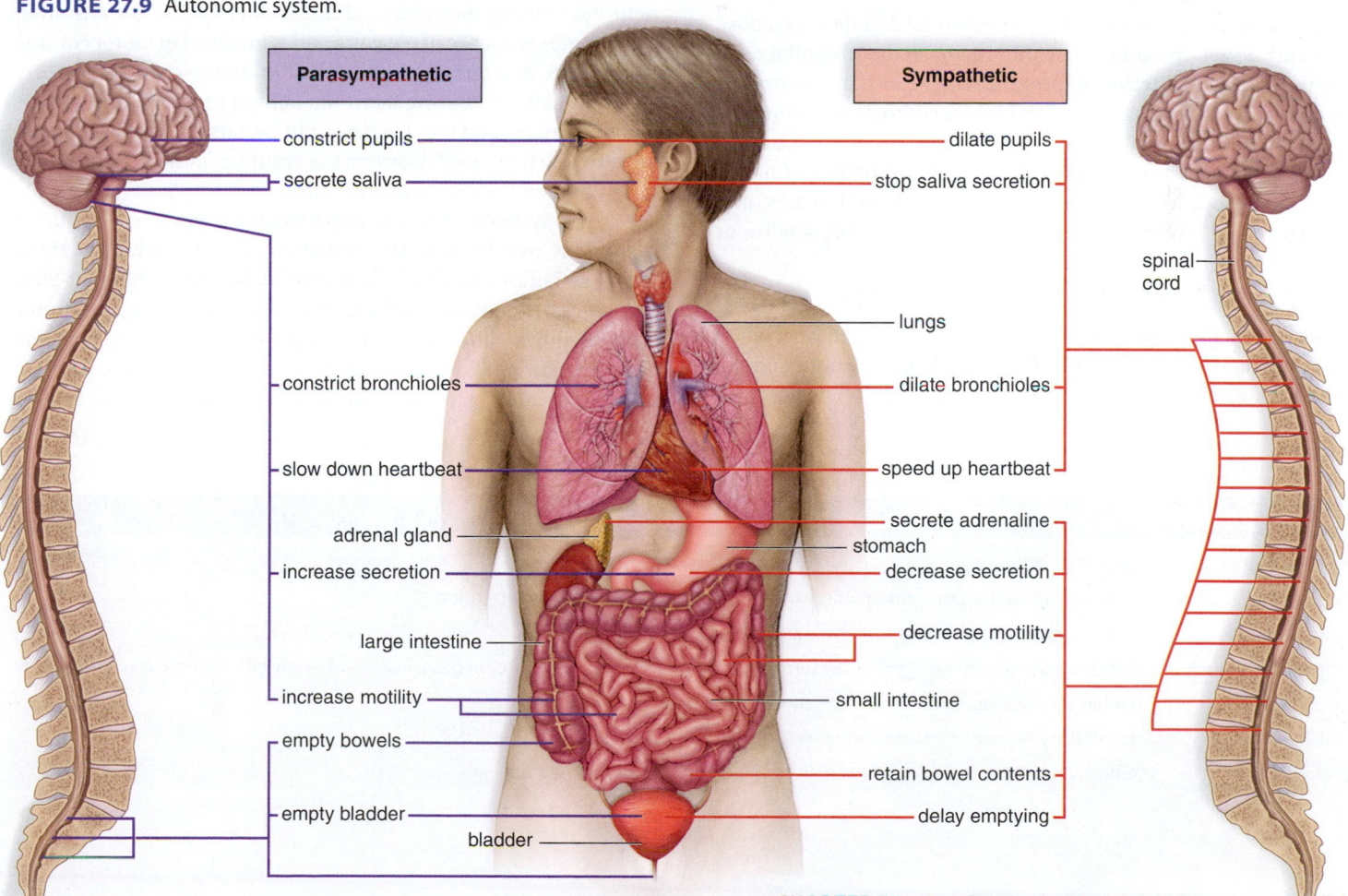

Drug Abuse

Many human neurological problems that are associated with neurotransmitter imbalances are now being treated by administering drugs to correct the imbalance. People often abuse drugs in order to experience a change in mood, leading to euphoria, a feeling of elation.

27.10 Certain neurotransmitters are known to affect behavior and emotional attributes

LEARNING OUTCOME

When you complete this section, you should be able to

1. Classify drugs as to whether they have a depressant, stimulant, or psychoactive effect on the body.

It is amazing to realize that all the thoughts, feelings, and actions of a human are dependent on nerve impulses and neurotransmitters at synapses. By modifying or controlling synaptic transmission, a wide variety of drugs with neurological activity—both legal pharmaceuticals and illegal drugs of abuse—can alter mood, emotional state, behavior, and personality. Some of the more common and better-known neurotransmitters are discussed in section 26.5 and listed in Table 27.10.

Like mental illness, drug abuse is also linked to neurotransmitter levels. Dopamine plays a central role in the working of the brain's limbic system. The limbic system ordinarily promotes healthy, pleasurable activities, such as consuming food. It's possible to abuse behaviors, such as eating, spending, or gambling, because the behavior stimulates the system and makes us feel good. Drug abusers take drugs that artificially affect the limbic system to the point that they neglect their basic physical needs in favor of continued drug use.

Drug abuse is present when a person takes a drug at a dose level and under circumstances that increase the potential for a harmful effect. **Addiction** is present when more of the drug is needed to get the same effect, and when withdrawal symptoms occur when the user stops taking the drug. This is true not only for teenagers and adults, but also for newborn babies of mothers who abuse and are addicted to drugs. Alcohol, drugs, and tobacco can all adversely affect the developing embryo, fetus, or newborn.

The following are some specific drugs of abuse:

Alcohol Alcohol consumption is the most socially accepted form of drug use worldwide. Approximately 65% of adults in the United States consume alcohol on a regular basis. Of those, 5% say they are "heavy drinkers." Notably, 80% of college-age young adults drink (Fig. 27.10A). Unfortunately, so-called binge drinking has resulted in the deaths of many college students.

Alcohol (ethanol) acts as a *depressant* on many parts of the brain, where it affects neurotransmitter release or uptake. For example, alcohol increases the action of GABA, which inhibits motor neurons, and it also increases the release of endorphins, which, as discussed, are natural painkillers. Depending on the amount consumed, the effects of alcohol on the brain can lead to a feeling of relaxation, lowered inhibitions, impaired concentration and coordination, slurred speech, and vomiting. If the blood level of alcohol becomes too high, coma or death can occur.

Chronic alcohol consumption can damage the frontal lobes, decrease overall brain size, and increase the size of the ventricles. Brain damage is manifested by permanent memory loss, amnesia, confusion, apathy, disorientation, or lack of motor coordination. Prolonged alcohol use can also permanently damage the liver, the major detoxification organ of the body, to the point that a liver transplant may be required.

Nicotine When a person smokes tobacco, nicotine is rapidly delivered to the CNS, especially the midbrain. There it binds to neurons, causing the release of dopamine, the neurotransmitter that promotes a sense of pleasure and is involved in motor control. In the PNS, nicotine also acts as a *stimulant* by mimicking acetylcholine and increasing heart rate, blood pressure, and muscle activity. Fingers and toes become cold because blood vessels have constricted. Increased digestive tract motility may account for the weight loss sometimes seen in smokers.

The physiologically and psychologically addictive nature of nicotine is well known. The addiction rate of smokers is about 70%. The failure rate in smokers who try to quit is about 80–90%. Withdrawal symptoms include irritability, headache, insomnia, poor cognitive performance, the urge to smoke, and weight gain. Ways to quit smoking include applying nicotine skin patches, chewing nicotine gum, or taking oral drugs that block the actions

TABLE 27.10	Common CNS Neurotransmitters and Their Actions
Neurotransmitter	**Actions of Interest**
Acetylcholine	Involved in arousal, enhancement of sensory perceptions, and sustaining attention
Norepinephrine	Creates a sense of well-being; low levels are associated with depression
Dopamine	Creates a sense of well-being; affects motor control and attention; deficiency in basal nuclei is associated with Parkinson disease
Serotonin	An inhibitory neurotransmitter that promotes sleep
GABA	An inhibitory neurotransmitter that works contrary to dopamine
Endorphins	Inhibitory neuromodulators; reduce pain by inhibiting substance P release

FIGURE 27.10A Why do people use drugs? Social motivations may be at play if a person smokes and drinks only in the presence of others. Other factors may be involved if the activity occurs to excess when the person is alone.

of acetylcholine. The effectiveness of these therapies is variable. An experimental therapy involves "immunizing" the brain of smokers against nicotine. Injections cause the production of antibodies that bind to nicotine and prevent it from entering the brain. The effectiveness of this new therapy is not yet known.

Club and Date Rape Drugs Methamphetamine and Ecstasy are considered club or party drugs.

Methamphetamine (commonly called meth or speed) is a synthetic drug made by the addition of a methyl group to amphetamine, a stimulant. Because the addition of the methyl group is fairly simple, methamphetamine is often produced from amphetamine in clandestine, makeshift laboratories in homes, motel rooms, or campers. The number of toxic chemicals used to prepare the drug makes a former meth lab site hazardous to humans and to the environment. Over 9 million people in the United States have used methamphetamine at least once in their lifetime. It is available as a powder or as crystals (crystal meth or ice).

The structure of methamphetamine is similar to that of dopamine, and its *stimulatory* effect mimics that of cocaine. It reverses the effects of fatigue, maintains wakefulness, and temporarily elevates the user's mood. The initial rush is typically followed by a state of high agitation that, in some individuals, leads to violent behavior. Chronic use can result in what is called "amphetamine psychosis," characterized by paranoia, auditory and visual hallucinations, self-absorption, irritability, and aggressive, erratic behavior. Excessive intake can lead to hyperthermia, convulsions, and death.

Video
Meth and the Brain

Ecstasy is the street name for MDMA (methylenedioxymethamphetamine), a drug with effects similar to those of methamphetamine. Also referred to as E, X, or the lung drug, it is taken as a pill that looks like an aspirin or candy. Many people using Ecstasy believe that it is totally safe if used with lots of water to counter

its effect on body temperature. A British teen, Lorna Spinks, died after taking two high-strength Ecstasy pills, which caused her body temperature to rise to a fatal level. Also, Ecstasy has an over-stimulatory effect on neurons that produce serotonin, which, like dopamine, elevates our mood. Most of the damage to these neurons can be repaired when the use of Ecstasy is discontinued, but some damage appears to be permanent.

Date rape or predatory drugs include Rohypnol (roofies, flunitrazepam), Gamma-hydroxybutyric acid (GHB), and Special K (ketamine). These drugs can be given to an unsuspecting person, who then becomes vulnerable to sexual assault after the drug takes effect. Relaxation, amnesia, and disorientation occur after taking these drugs, which are popular at clubs because they enhance the effect of heroin and Ecstasy.

Cocaine Cocaine is an alkaloid derived from the shrub *Erythroxylon coca*. Approximately 35 million Americans have used cocaine by sniffing/snorting, injecting, or smoking. Cocaine is a powerful *stimulant* in the CNS that interferes with the re-uptake of dopamine at synapses. The result is a rush of well-being that lasts from 5–30 minutes. People on cocaine sprees (or binges) take the drug repeatedly and at ever-higher doses. The result is sleeplessness, lack of appetite, increased sex drive, tremors, and "cocaine psychosis," a condition that resembles paranoid schizophrenia. During the crash period, fatigue, depression, and irritability are common, along with memory loss and confused thinking.

"Crack" is the street name for cocaine that is processed to a free base for smoking. The term *crack* refers to the crackling sound heard when smoking. Smoking allows extremely high doses of the drug to reach the brain rapidly, providing an intense and immediate high, or "rush." Approximately 8 million Americans use crack. Long-term use is expected to cause brain damage.

Cocaine is highly addictive; related deaths are usually due to cardiac and/or respiratory arrest. The combination of cocaine and alcohol dramatically increases the risk of sudden death.

Heroin Heroin is derived from the resin or sap of the opium poppy plant, which is widely grown—from Turkey to Southeast Asia and in parts of Latin America. Heroin is a highly addictive drug that acts as a *depressant* in the nervous system. Drugs derived from opium are called opiates, a class that also includes morphine and codeine, both of which have painkilling effects.

Heroin is the most abused opiate—it travels rapidly to the brain, where it is converted to morphine, and the result is a rush sensation and a feeling of euphoria. Opiates depress breathing, block pain pathways, cloud mental function, and sometimes cause nausea and vomiting. In addition to addiction, long-term effects of heroin use are hepatitis, HIV/AIDS, and various bacterial infections due to sharing needles with other addicts (Fig. 27.10B). As with other drugs of abuse, addiction is common, and heavy users may experience convulsions and death by respiratory arrest.

Heroin can be injected, snorted, or smoked. Abusers typically inject heroin up to four times a day. It is estimated that 4 million Americans have used heroin at some time in their lives, and over 300,000 people use heroin annually.

FIGURE 27.10B Drug use.

anandamide belong to a class of chemicals called cannabinoids. Receptors that bind cannabinoids are located in the hippocampus, cerebellum, basal ganglia, and cerebral cortex, brain areas that are important for memory, orientation, balance, motor coordination, and perception.

When THC reaches the CNS, the person experiences mild euphoria, along with alterations in vision and judgment. Distortions of space and time can also occur in occasional users. In heavy users, hallucinations, anxiety, depression, rapid flow of ideas, body image distortions, paranoia, and psychotic symptoms can result. The terms "cannabis psychosis" and "cannabis delirium" describe such reactions to marijuana's influence on the brain. Regular usage of marijuana can cause cravings that make it difficult to stop.

Treatment for Addictive Drugs Presently, treatment for drug addiction consists mainly of behavior modification. Heroin addiction can be treated with synthetic opiate compounds, such as methadone or suboxone, that decrease withdrawal symptoms and block heroin's effects. Unfortunately, inappropriate methadone use can be dangerous, as demonstrated by celebrity deaths associated with methadone overdose or taking methadone along with other drugs.

New treatment techniques include the administration of antibodies to block the effects of cocaine and methamphetamine. These antibodies would make relapses by former drug abusers impossible and could be used to treat overdoses. A vaccine for cocaine that would stimulate antibody production is being tested.

Marijuana The dried flowering tops, leaves, and stems of the Indian hemp plant, *Cannabis sativa,* contain and are covered by a resin that is rich in THC (tetrahydrocannabinol). The names *cannabis* and *marijuana* apply to either the plant or THC. Marijuana can be consumed, but usually it is smoked in a cigarette called a "joint." An estimated 14.6 million Americans use marijuana. Although the drug was banned in the United States in 1937, several states have legalized its use for medical purposes, such as lessening the effects of chemotherapy.

It seems that THC may mimic the actions of anandamide, a neurotransmitter not discovered until 1992. Both THC and

▶ **27.10 CHECK YOUR PROGRESS**
  1. Explain why methamphetamine and cocaine have similar physiological effects.

THE CHAPTER IN REVIEW

CONNECTING THE CONCEPTS

The human nervous system has just three functions: sensory input, integration, and motor output. Nerve impulses are the same in all neurons, so how is it that stimulation of the eyes causes us to see, and stimulation of the ears causes us to hear? Essentially, the central nervous system (CNS) carries out the function of integrating incoming data. The brain has particular parts that receive and integrate particular sensory data. The brain allows us to perceive our environment, to reason, and to remember. After sensory data have been processed by the CNS, motor output occurs. Muscles and glands are the effectors that allow us to respond to the original stimuli. Without the muscular and skeletal systems, discussed in Chapter 29, we would never be able to respond to a danger detected by our eyes and ears.

Similar to the wiring of a modern office building, the human peripheral nervous system (PNS) contains nerves that carry sensory input to the CNS and motor output to the muscles and glands. There is a division of labor among the nerves. The cranial nerves for the most part serve the face, teeth, and mouth. All body movements are controlled by spinal nerves, and this is why paralysis may follow a spinal injury. Except for the vagus nerve (a cranial nerve), only spinal nerves make up the autonomic system, which controls the internal organs.

You might argue that sense organs, such as the eyes and ears discussed in Chapter 28, should be considered part of the nervous system, because there would be no sensory nerve impulses without their ability to generate them.

Our view of the world is dependent on the sense organs, which are sensitive to external and internal stimuli.

ANALYZE AND EVALUATE

1. In general, discuss how the nervous system is organized and how it functions.
2. Do you take exception to this statement: "All the thoughts, feelings, and actions of a human are dependent on the nerve impulse and the neurotransmitters at a synapse." Explain why you agree or disagree with this statement.
3. If all behavior is due to muscle contraction controlled by the CNS, then how can the limbic system ever be involved in behavior—it has no motor output.

SUMMARIZE

Nervous Systems

27.1 Vertebrates have well-developed central and peripheral nervous systems

- The nervous system carries out three functions: receive sensory input; perform integration; generate motor output. The **central nervous system** (**CNS**) is composed of the spinal cord and the brain. The **peripheral nervous system** (**PNS**) consists of the nerves and ganglia outside the CNS. This diagram summarizes the structure and function of the two systems:

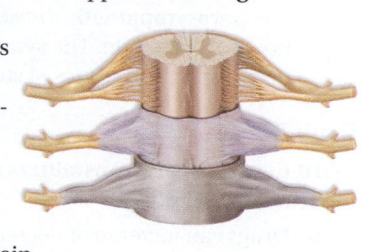

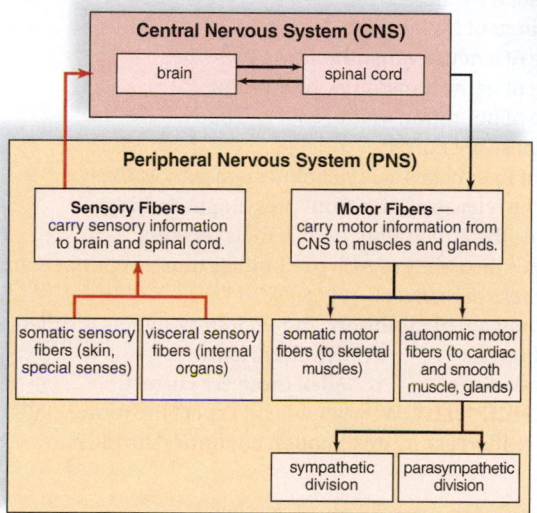

- **Neurons** are the functional units of a nervous system. **Neuroglia** support and nourish neurons. **Schwann cells** form **myelin sheaths** with gaps called **nodes of Ranvier.** A neuron has three parts: a **cell body, dendrites,** and an **axon,** also called a **nerve fiber.** The three types of neurons are **motor, sensory,** and **interneurons.**

27.2 Axons conduct nerve impulses

- In a **resting potential,** the axon is not conducting the impulse; there is more Na^+ outside the axon and more K^+ inside the axon. Above **threshold,** an **action potential** is a rapid change in polarity across the axon membrane as the nerve impulse occurs: Na^+ gates open, and Na^+ moves to inside the axon; K^+ gates open, causing **depolarization,** and K^+ moves to outside the axon, causing **repolarization.**
- Following an action potential, each axon segment undergoes a **refractory period** of recovery. In **saltatory conduction,** ion exchange occurs only at nodes, and the action potential jumps from node to node.

27.3 Communication between neurons occurs at synapses

- A **synapse** is a region of close proximity between an axon terminal and a dendrite. When a **neurotransmitter** is released into a **synaptic cleft,** transmission of a nerve impulse follows.
- Binding of a neurotransmitter to receptors in the postsynaptic membrane causes stimulation or inhibition. The inhibitors **serotonin** and **dopamine** and the neuromodular **endorphin** are primarily found in the CNS. In the PNS, **acetylcholine** (**ACh**) stimulates skeletal muscle; (before it is broken down by **acetylcholinesterase,** or **AChE**) **norepinephrine** (**NE**) generally stimulates smooth muscle.

- A neuron receives signals, which can be excitatory or inhibitory. During **integration,** an excitatory signal drives a neuron closer to threshold because it has a depolarizing effect. An inhibitory signal drives a neuron further from threshold because it has a hyperpolarizing effect.

Central Nervous System (CNS)

27.4 The spinal cord communicates with the brain

- The **spinal cord** and the brain are wrapped in **meninges. Cerebrospinal fluid** fills the spaces between the meninges and the **ventricles** of the brain. The spinal cord participates in **reflex actions** (cell bodies form **gray matter**) and contains **tracts** (form **white matter**) that take messages to and from the brain.

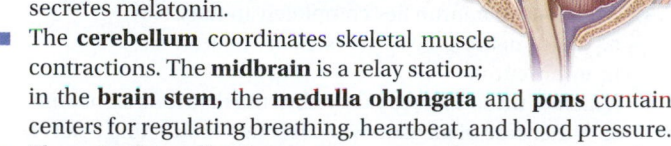

- The brain is well organized. Sensation, reasoning, learning, memory, language, and speech take place in the **cerebrum.** The cerebrum has two **cerebral hemispheres** and each hemisphere is divided into four lobes: frontal, parietal, occipital, and temporal. Each lobe has a particular function.
- The **cerebral cortex,** the convoluted outer covering of the cerebrum, contains a **primary motor area** in the frontal lobe (sends out voluntary motor commands to skeletal muscle) and a primary **somatosensory area** in the parietal lobe (receives sensory information from the skin and skeletal muscles). The temporal lobe receives information from the ears and the occipital lobe receives information from the eyes.
- **Basal nuclei** in the white matter integrate motor commands, and malfunctions result in **Huntington disease** and **Parkinson disease.**

27.5 The other parts of the brain have specialized functions

- In the **diencephalon,** the **hypothalamus** controls homeostasis, while the **thalamus** sends sensory input to the cerebrum. The **pineal gland** secretes melatonin.
- The **cerebellum** coordinates skeletal muscle contractions. The **midbrain** is a relay station; in the **brain stem,** the **medulla oblongata** and **pons** contain centers for regulating breathing, heartbeat, and blood pressure.
- The **reticular activating system** arouses the cerebrum via the thalamus, causing alertness.

27.6 The limbic system is involved in memory and learning as well as in emotions

- The **limbic system** affects mood and emotions. In the limbic system, the **hippocampus** is involved in storing and retrieving a **memory.** The **amygdala** adds emotional overtones.

Peripheral Nervous System (PNS)

27.7 The peripheral nervous system contains cranial and spinal nerves

- Nerves are bundles of axons (**nerve fibers**). Many **cranial nerves** contain either sensory or motor fibers but **spinal nerves** contain both sensory and motor fibers. The ventral roots of spinal nerves contain only motor fibers; motor neurons have their cell body inside the cord. **Dorsal root**

ganglia contain the cell bodies of sensory neurons. A **ganglion** is a cluster of cell bodies outside the CNS.

27.8 The somatic system can respond quickly to stimuli

- The PNS is divided into the **somatic system** and the autonomic system. Nerves in the somatic system serve the skin, joints, and skeletal muscles. Some actions are due to **spinal reflexes** (begin in sensory receptor and end at muscle effector, see Fig. 27.8), which are automatic and involuntary.

27.9 The autonomic system controls the actions of internal organs

- The **parasympathetic division** governs responses during times of relaxation. The **sympathetic division** is in charge of responses during times of stress.

Drug Abuse

27.10 Certain neurotransmitters are known to affect behavior and emotional attributes

- Drugs can increase or decrease the amount of a neurotransmitter at a synapse, and some can mimic the action of a neurotransmitter. **Drug abuse** occurs when the dose level is harmful. **Addiction** is present when more of a drug is needed to get the same effect.
- Among specific drugs of abuse, alcohol is a depressant, while GABA is a stimulant; methamphetamine and Ecstasy are considered club or date rape drugs; cocaine is a powerful stimulant in the CNS; heroin is a depressant that converts to morphine in the brain; and marijuana can alter vision and judgment.

TEST YOURSELF

Nervous Systems

1. Which of the following neuron parts receive(s) signals from other neurons?
   a. cell body   c. dendrites
   b. axon   d. Both a and c are correct.
2. Which of these would be covered by a myelin sheath?
   a. short dendrites   d. interneurons
   b. globular cell bodies e. All of these are correct.
   c. long axons
3. What type of neuron lies completely in the CNS?
   a. motor neuron   c. sensory neuron
   b. interneuron
4. Which of these correctly describes the distribution of ions on either side of an axon when it is not conducting a nerve impulse?
   a. more sodium ions (Na^+) outside and more potassium ions (K^+) inside
   b. more K^+ outside and less Na^+ inside
   c. charged protein outside; Na^+ and K^+ inside
   d. Na^+ and K^+ outside and water only inside
   e. chloride ions (Cl^-) outside and K^+ and Na^+ inside
5. When the action potential begins, sodium gates open, allowing Na^+ to cross the membrane. Now the polarity changes to
   a. negative outside and positive inside.
   b. positive outside and negative inside.
   c. neutral outside and positive inside.
   d. There is no difference in charge between outside and inside.

6. Repolarization of an axon during an action potential is produced by
   a. inward diffusion of NA^+.
   b. outward diffusion of K^+.
   c. inward active transport of Na^+.
   d. active extrusion of K^+.
7. What function is served by the nodes of Ranvier?
   a. The nodes pass the signal on to the next neuron.
   b. Presence of nodes speeds the nerve impulse.
   c. Because they occur in the limbic system they affect mood.
   d. They are vestigial organs having no effect.
8. **THINKING CONCEPTUALLY** From an evolutionary perspective, why is it not surprising that the nerve impulse makes use of a potential difference present in all plasma membranes?
9. Transmission of the nerve impulse across a synapse is accomplished by the
   a. movement of Na^+ and K^+.
   b. release of a neurotransmitter by a dendrite.
   c. release of neurotransmitter by an axon.
   d. release of neurotransmitter by a cell body.
   e. All of these are correct.
10. A drug that inactivates acetylcholinesterase most likely
    a. stops the release of ACh from presynaptic endings.
    b. prevents the attachment of ACh to its receptor.
    c. increases the ability of ACh to stimulate muscle contraction.
    d. All of these are correct.
11. The summing up of inhibitory and excitatory signals is called
    a. excitation.   c. depolarization.
    b. integration.   d. All of these are correct.
12. **THINKING CONCEPTUALLY** Why would you expect the motor skills of a child to increase as myelination continues during early childhood?

Central Nervous System (CNS)

13 Membranes surrounding the CNS are collectively called
   a. meninges.   c. vesicles.
   b. myelin.   d. None of these are correct.
14. Which of the following cerebral lobes is not correctly matched with its function?
   a. occipital lobe—vision
   b. parietal lobe—somatosensory area
   c. temporal lobe—primary motor area
   d. frontal lobe—motor speech (Broca) area
15. The cerebellum
   a. coordinates skeletal muscle movements.
   b. receives sensory input from the joints and muscles.
   c. receives motor input from the cerebral cortex.
   d. All of these are correct.
16. The hypothalamus does not
   a. control skeletal muscles.
   b. regulate thirst.
   c. control the pituitary gland.
   d. regulate body temperature.
17. The limbic system
   a. involves portions of the cerebral lobes and the diencephalon.
   b. is responsible for our deepest emotions, including pleasure, rage, and fear.
   c. is a system necessary to memory storage.
   d. is not directly involved in language and speech.
   e. All of these are correct.

18. Which of these functions is not correctly matched to a part of the brain?
    a. medulla oblongata—regulates heartbeat, breathing, and blood pressure
    b. cerebellum—coordinates voluntary muscle movements
    c. thalamus—regulates daily body rhythms
    d. midbrain—acts as a reflex center for visual, auditory, and tactile responses
19. **THINKING CONCEPTUALLY** Explain why you would expect learning to result in an increase in the number of synapses.

Peripheral Nervous System (PNS)

20. Label this diagram:

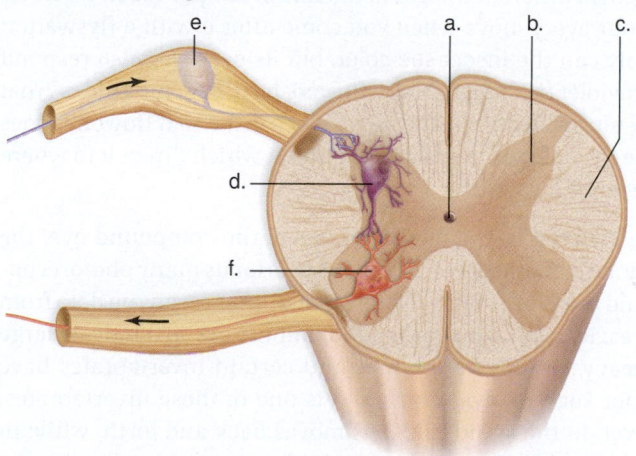

21. The organs of the peripheral nervous system are the
    a. cranial and spinal nerves.
    b. brain and spinal cord.
    c. nerves and spinal cord.
    d. cranial nerves and brain.
22. Which of these are the first and last elements in a spinal reflex?
    a. axon and dendrite
    b. sensory receptor and muscle effector
    c. central horn and dorsal horn of spinal cord
    d. brain and skeletal muscle
    e. Both b and d are correct.
23. Somatic is to skeletal muscle as autonomic is to
    a. cardiac muscle.   c. gland.
    b. smooth muscle.   d. All of these are correct.
24. Which of these statements are incorrect regarding the autonomic nervous system (ANS)?
    a. The ANS includes mainly cranial nerves.
    b. The ANS is divided into a sympathetic and parasympathetic divisions.
    c. Most organs under ANS control are under dual innervation by both sympathetic and parasympathetic divisions.
    d. The ANS affects the activity of cardiac muscle, smooth muscles, and glands.

25. Which of these statements about autonomic neurons is correct?
    a. Autonomic neurons are motor neurons.
    b. Preganglionic neurons have cell bodies in the CNS.
    c. Postganglionic neurons innervate smooth muscle, cardiac muscle, and glands.
    d. All of these are correct.
26. The sympathetic division of the autonomic system will
    a. increase heart rate and digestive activity.
    b. decrease heart rate and digestive activity.
    c. cause pupils to constrict.
    d. None of these are correct.
27. The parasympathetic system does not
    a. stimulate urination.
    b. constrict bronchioles.
    c. promote digestion
    d. speed the heart.
    e. constrict the pupils.

Drug Abuse

28. Which of these drugs creates a sense of well being?
    a. acetylcholine
    b. norepinephrine
    c. dopamine
    d. Both b and c are correct
29. **THINKING CONCEPTUALLY** Physiologically, what do the drugs nicotine, alcohol, cocaine, methamphetamine, heroin, and marijuana have in common?

GET INVOLVED

1. Knowing that the fight-or-flight response is initiated by the release of the neurotransmitter norepinephrine, how might it be possible to control the response in people who are stressed? What complications might ensue?
2. Hypothesize why a man with an amputated leg still feels pain as though it were coming from the missing limb.

MEDIA STUDY TOOLS

mhhe.com/maderconcepts3

Enhance your study of this chapter with interactive study tools, practice tests, and engaging animations. Also, ask your instructor about the resources available through ConnectPlus, which includes LearnSmart, a personalized adaptive learning program, and a media-rich eBook.

28

Sense Organs

BEFORE YOU BEGIN

Take a few minutes to recall

How the PNS and CNS are involved in an animal's response to stimuli (section 27.1)

That integration of sensory data occurs in the cerebral hemispheres (section 27.4)

That nerves conduct sensory data to the cerebrum (section 27.7)

The role of sensory receptors in a reflex arc (section 27.8)

The Eyes Have It

A bumblebee gathering nectar at a flower has an entirely different type of eye from that of a vertebrate. The eye of an insect is called compound because it has many visual units, each sending its own data to the brain, where a mosaic (compound) image is produced. The compound eye produces a crude image, but it is a taskmaster at detecting motion. The unusually rapid recovery of its light receptors makes this possible. The human eye can distinguish only about 24 images per second; after that, the images are fused into one. In contrast, the eye of an insect can distinguish different images at the rate of 330 per second. The fly sees your every move when you come after it with a flyswatter! Not only can the insect see color, but its eyes can also respond to ultraviolet (UV) rays. Many flowers have "nectar guides" that reflect ultraviolet light, and when a bee visits that flower, it does not see the color as much as the guides, which direct it to where the nectar is located.

In contrast to the mosaic image of the compound eye, the vertebrate camera-type eye has one lens for its many photoreceptors, and the brain forms a single image after receiving data from the eyes. Resolving power is good, but the eye is relatively large and heavy, so only vertebrates and certain invertebrates have room for such an eye. The squid is one of these invertebrates. However, in the squid, the lens moves back and forth, while in human eyes, the lenses change shape to accommodate for the distance of an object. Our eyes contain the sensory receptors for light, called the rods and the cones. The rods function well in

Visual Unit

photo-receptor cells

optic nerve fibers

Mosaic image of compound eye.

An insect has a compound eye with many visual units.

dim light but produce an image that is indistinct and lacks color. Turn off the lights tonight, wait a few minutes, and a shadowy-gray world will appear. Flip on the light, and your cones will take over, showing you a distinct, colorful world! Cones have terrific resolving power. The eyes of a hawk contain more than a million cones per cubic millimeter, allowing it to detect a tiny mouse scurrying among the underbrush from a great height. Birds and humans are known for seeing color. Humans have cones of only three different colors (blue, green, and red), but see different shades, depending on which of these is stimulated. Birds, with cones of four to five different colors, have superior color vision to that of humans. The attributes of sensory receptors correlate with an animal's life style. Rods have replaced cones in the eyes of tarsiers, a rat-sized primate that is active only at night.

Previously, scientists thought that the compound eye and the camera-type eye evolved separately, and perhaps many times over, in the animal kingdom. But now evo-devo geneticists (those who study development from an evolutionary perspective) tell us that the same genes are active whether an animal has a compound eye or a camera-type eye. Surprisingly, they have concluded that all image-forming eyes can be traced to an original eye-bearing ancestor.

Predatory mammals tend to have stereoscopic vision. As in humans both eyes face forward and each eye views the object from a different angle. Animals with eyes facing sideways, such as rabbits and zebras, don't have stereoscopic vision, but they do have panoramic vision, meaning that the visual field is very wide. Panoramic vision is useful to prey animals because it helps detect a predator sneaking up on them.

This chapter discusses the major types of animal sense organs from an evolutionary perspective. It stresses the chemoreceptors, the photoreceptors, and the mechanoreceptors such as those in the ears of humans.

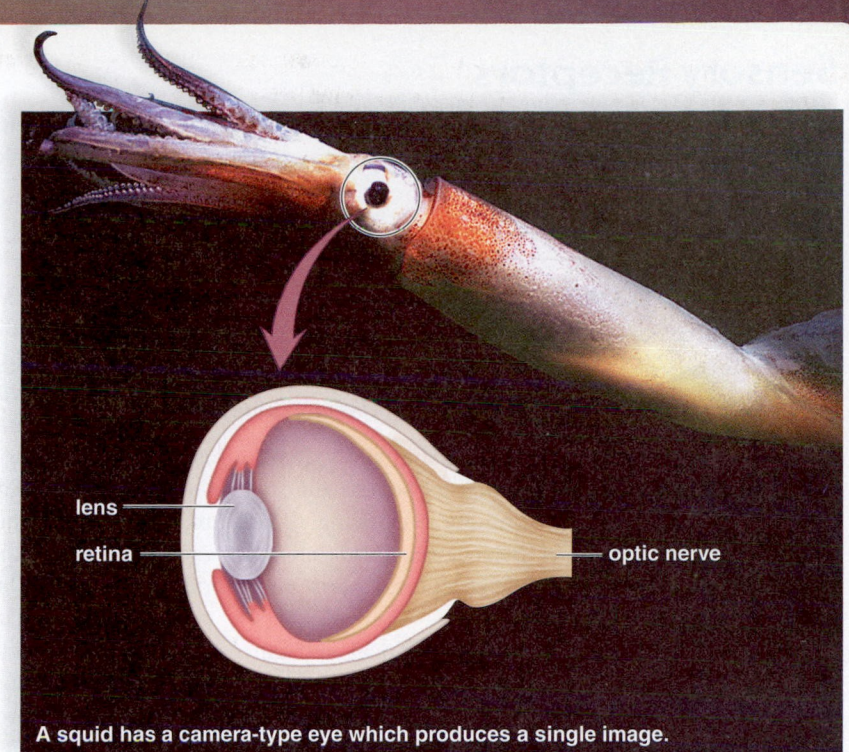

lens
retina
optic nerve

A squid has a camera-type eye which produces a single image.

Eyes look forward.

Eyes look to the side.

Sensory Receptors

Sensory receptors detect certain types of stimuli, including chemical, pain, electromagnetic, temperature, and mechanical. By way of nerves, sensory receptors communicate with the central nervous system, which integrates nerve impulses and directs a response.

28.1 Sensory receptors differ but function similarly

LEARNING OUTCOMES

When you complete this section, you should be able to

1. List and describe five common types of sensory receptors.
2. Explain how the activity of sensory receptors results in sensation.

All animals have **sensory receptors** that allow them to respond to stimuli. Stimuli are environmental signals that tell us about the external or internal environment. Surprisingly, there are only five common categories of sensory receptors: chemoreceptors, pain receptors, electromagnetic receptors, thermoreceptors, and mechanoreceptors.

Chemoreceptors respond to chemical substances in the immediate vicinity. *Taste* and *smell* depend on this type of sensory receptor, but certain chemoreceptors in various other organs are sensitive to internal conditions. For example, chemoreceptors that monitor blood pH are located in the carotid arteries and aorta of humans. If the pH lowers, the breathing rate increases. As more carbon dioxide is expired, the blood pH rises.

Pain receptors (also called nociceptors) are sometimes classified as a type of chemoreceptor. However, pain receptors respond to excessive temperature and mechanical pressure in addition to a range of chemicals, some of which are released by damaged tissues. *Pain* is protective because it alerts us to possible danger. For example, without the pain of appendicitis, we might never seek the medical help needed to avoid a ruptured appendix.

Electromagnetic receptors, responsible for *vision,* are stimulated by changes in electromagnetic waves. Photoreceptors, present in the eyes of most animals, are sensitive to visible light energy. As mentioned in the introduction to this chapter, our eyes contain photoreceptors known as rod cells that result in black-and-white vision, while stimulation of photoreceptors known as cone cells

results in color vision. The eyes of insects can detect ultraviolet radiation, and this helps them notice flowers, particularly the location of nectar. Several types of animals, including gray whales (Fig. 28.1A), migrate long distances from feeding to breeding areas and are believed to use the Earth's magnetic field as a type of compass to orient themselves.

Thermoreceptors are stimulated by changes in *temperature.* Humans have thermoreceptors located in the hypothalamus and the skin; those that respond when temperatures rise are called heat receptors, and those that respond when temperatures lower are called cold receptors. Some snakes, such as pythons, have thermoreceptors located in pits near the mouth that detect the body heat of their prey up to 1–2 m away (Fig. 28.1B). Some researchers classify thermoreceptors as electromagnetic receptors because they equate heat with infrared energy, which is part of the electromagnetic spectrum.

Mechanoreceptors are stimulated by mechanical forces, which most often result in pressure of some sort. When we *hear,* airborne sound waves are converted to fluid-borne pressure waves that can be detected by mechanoreceptors in the inner ear. The external ears of some bats are large for their size (Fig. 28.1C), and their inner ears are able to detect ultrasounds, sounds above the range humans can hear. These bats make ultrasonic clicking noises, and the echo of these sounds tells them where their prey is located in the dark.

The sense of *balance* and the sense of *touch* depend on mechanoreceptors. Also, in humans, pressoreceptors located in certain arteries detect changes in blood pressure, and stretch receptors in the lungs detect the degree of lung inflation. Proprioceptors are mechanoreceptors that respond to the stretching of muscle fibers, tendons, joints, and ligaments, making us aware of the *position* of our limbs.  **MP3** Sensations and Receptors

FIGURE 28.1A Electromagnetic receptors in gray whales help them migrate.

thermoreceptor

FIGURE 28.1B Thermoreceptors that are sensitive to infrared energy help pythons find their prey.

Function of Sensory Receptors In complex animals, sensory receptors transform stimuli into nerve impulses that go to the cerebral cortex of the brain either directly or by way of the spinal cord (Fig. 28.1D). We learned in Chapter 27, sensory receptors are the first element in a reflex arc. But, we are only aware of a reflex action once input has reached the cerebral cortex. At that time, the brain integrates information received from various sensory receptors and **sensation** (conscious detection of stimuli) occurs. If you burn yourself and quickly remove your hand from a hot stove, the brain receives information not only from your skin, but also from your eyes, nose, and all sorts of other sensory receptors, and this input is sent to the frontal lobe of the cerebral cortex before a response can occur.

Some sensory receptors are associated with neurons. They may be free nerve endings or encapsulated nerve endings, or specialized cells attached to neurons. When the receptor proteins in the plasma membrane of a chemoreceptor such as taste receptors bind to certain molecules, ions flow across the plasma membrane. If the stimulus is sufficient, nerve impulses begin and are carried by a sensory nerve fiber to the CNS (Fig. 28.1D). The stronger the stimulus, the greater the frequency of nerve impulses. In other instances, nerve impulses reach the spinal cord first and are conveyed to the brain by ascending tracts. If nerve impulses finally reach the cerebral cortex, sensation occurs.

All sensory receptors initiate nerve impulses; the resulting sensation depends on the part of the brain receiving the impulses. Nerve impulses that begin in the optic nerve eventually reach the occipital lobe of the cerebral cortex, and then we see objects. Nerve impulses that begin in the auditory nerve eventually reach the temporal lobe of the cerebral cortex, and then we hear sounds. If it were possible to switch these nerves, stimulation of the eyes would result in hearing! On the other hand, when a blow to the eye stimulates photoreceptors, we "see stars" because nerve impulses from the eyes can only result in sight.

Before sensory receptors initiate nerve impulses, they carry out some **integration,** the summing up of signals. One type of integration is called **sensory adaptation,** a decrease in

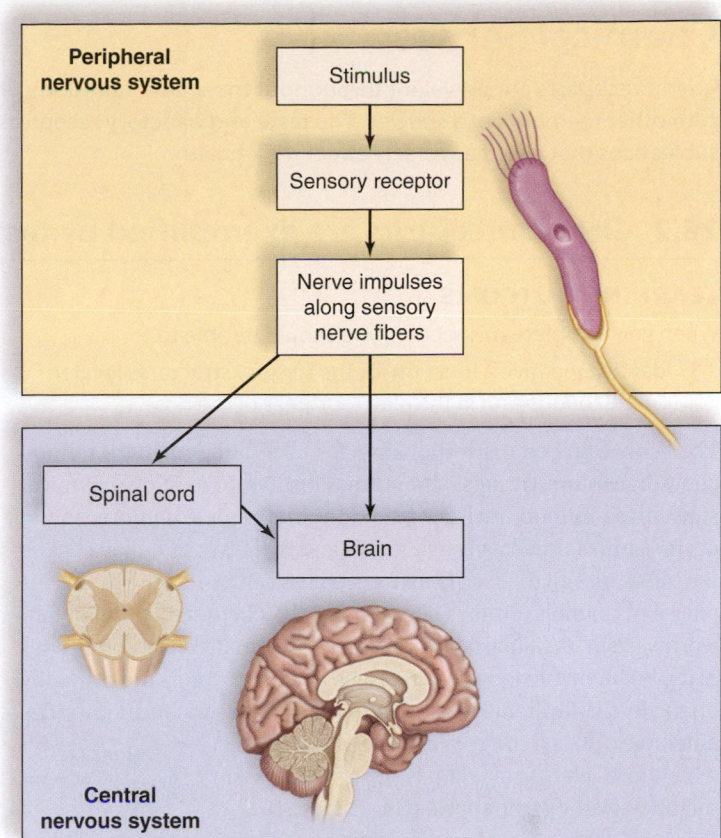

FIGURE 28.1D Nerve impulses from sensory receptors result in sensation in the brain.

the response to a stimulus. We have all had the experience of smelling an odor when we first enter a room and then later not being aware of it at all. Some authorities believe that when sensory adaptation occurs, sensory receptors have stopped sending impulses to the brain. Others believe that the reticular activating system (RAS) has filtered out the ongoing stimuli. You will recall that the RAS conveys sensory information from the brain stem, through the thalamus, to the cerebral cortex. The thalamus acts as a gatekeeper and only passes on information of immediate importance. Just as we can gradually become unaware of particular environmental stimuli, we can suddenly become aware of stimuli that may have been present for some time. This can be attributed to the workings of the RAS, which has synapses with many ascending sensory tracts.

The functioning of sensory receptors makes a significant contribution to homeostasis. Without sensory input, we would not receive information about our internal and external environments. This information, passed through nerves to the spinal cord to the brain, or directly to the brain, leads to appropriate reflex and voluntary actions that keep the internal environment constant.

▶ **28.1 CHECK YOUR PROGRESS**

1. Describe the major function(s) of the five categories of sensory receptors.
2. Use Figure 28.1D to trace the path of sensory input to the brain. If vision results, which lobe of the brain has been stimulated (see Fig. 27.4B)?

FIGURE 28.1C Large ears help some bats use echolocation to find their prey.

Chemical Senses

Chemoreceptors are prevalent throughout the animal kingdom. Pheromones are chemicals that provide a means of communicating with other members of a species. The taste and olfactory receptors of mammals provide chemical information about food sources and substances that could adversely affect their health.

28.2 Chemoreceptors are exemplified by taste buds and olfactory cells

LEARNING OUTCOME

When you complete this section, you should be able to

1. Identify and give a function for the sensory structures located on the tongue and in the olfactory areas of the nose.

The sensory receptors responsible for taste and smell are termed chemoreceptors because they are sensitive to certain chemical substances in food, including liquids, and in air. Chemoreception occurs almost universally in animals and is, therefore, believed to be the most primitive sense. Chemoreceptors can be located in various places in animals. Studies suggest that the chemoreceptors of flatworms, such as planarians, are located in the auricles on the sides of the head. In the housefly, an insect, chemoreceptors are primarily on the feet. A fly literally tastes with its feet instead of its mouth. The antennae of insects detect airborne **pheromones,** which are chemical signals passed between members of the same species (Fig. 28.2A).

> **Video** Mosquitos and Sweat

In vertebrates, such as amphibians, chemoreceptors are located in the nose, mouth, and skin. Snakes and other vertebrates possess vomeronasal organs (VNO), a pair of pitlike organs located in the roof of the mouth. When a snake flicks its forked tongue, pheromones are carried to the VNO, which sends nerve impulses to the brain for interpretation. In mammals, the receptors for taste are located in the mouth; the receptors for smell—and perhaps a VNO for detecting pheromones—are located in the nose.

Taste Buds In mammals, including humans, a receptor for taste is a type of chemoreceptor located in **taste buds** (Fig. 28.2B). Adult humans have approximately 3,000 taste buds, located primarily on the tongue **1**. Many taste buds lie along the walls of the

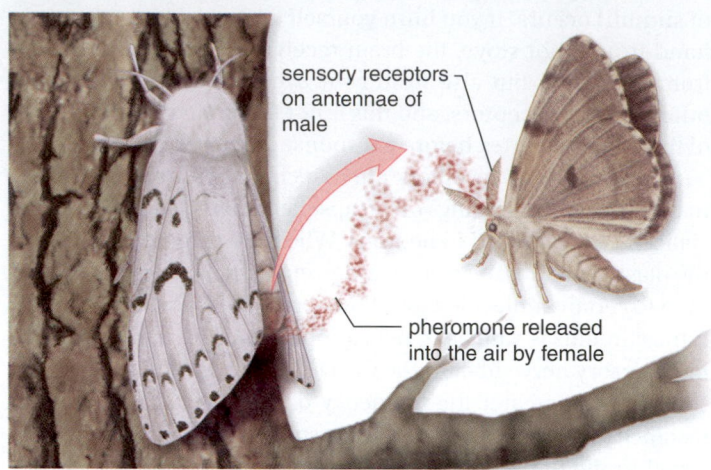

FIGURE 28.2A A male moth responds to a species-specific pheromone acting as a sex attractant.

papillae **2**, the small elevations on the tongue that are visible to the unaided eye. Isolated taste buds are also present on the hard palate, the pharynx, and the epiglottis.

Taste buds **3** have supporting cells and a number of elongated taste receptor cells that end in microvilli at a taste pore **4**. The microvilli, which project from the taste cells into the taste pore, bear receptor proteins for certain molecules. When molecules dissolved in solution bind to receptor proteins, nerve impulses are generated in associated sensory nerve fibers. These nerve impulses go to the brain, including cortical areas that interpret them as tastes.

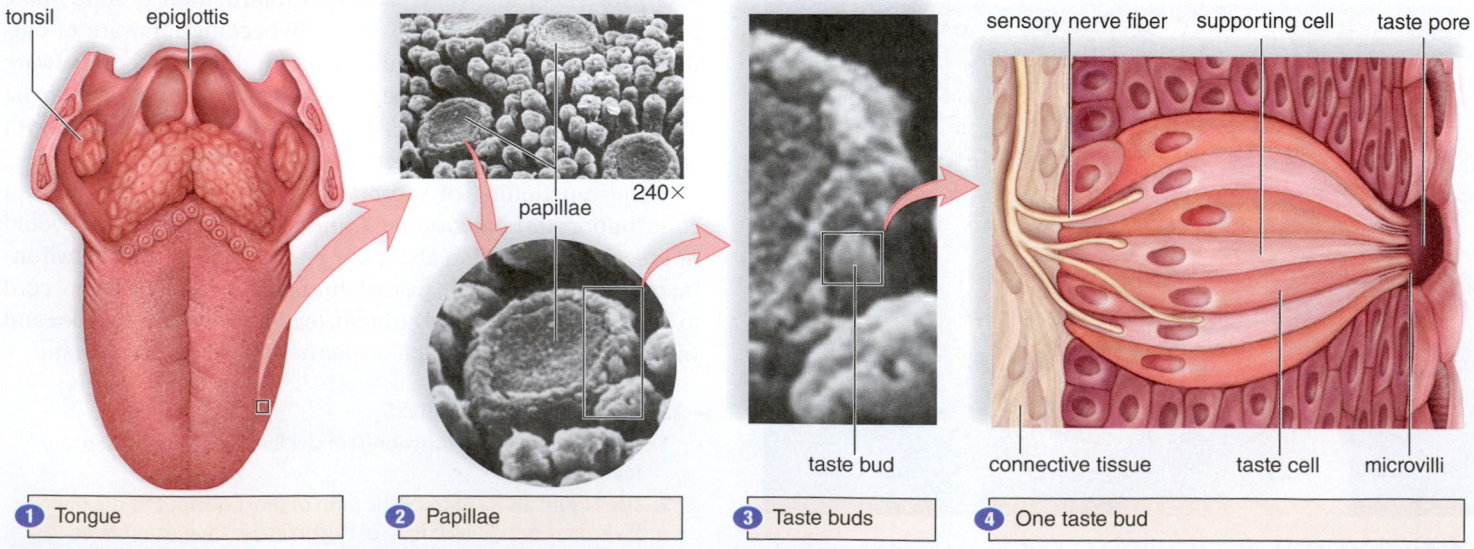

FIGURE 28.2B Taste buds in humans.

Ice cream tastes sweet.

In addition to the familiar tastes—bitter, sweet, salty, and sour—a fifth taste called umami is particularly triggered by glutamate, an amino acid. Umami receptors are responsible for the savory flavor of meats, cheeses, and sauces, such as soy sauce. Receptors for the five tastes are scattered in the taste buds and taste buds do not specialize in any particular taste. A particular food can stimulate receptors for more than one taste. The brain appears to survey the overall pattern of incoming sensory impulses and to take a "weighted average" of their taste messages as the perceived taste.

Olfactory Cells The chemical sense of smell is well developed in mammals, especially in carnivores such dogs, cats, and hyenas, which use it to track down their prey. In humans and other mammals, the sense of smell, or olfaction, is dependent on between 10 and 20 million **olfactory cells.** These structures are located within an olfactory epithelium high in the roof of the nasal cavity (Fig. 28.2C). Olfactory cells are modified neurons. Each cell ends in a tuft of about five olfactory cilia that bear receptor proteins for odor molecules. Each olfactory cell has only 1 out of 1,000 different types of receptor proteins. Nerve fibers from similar olfactory cells lead to the same neuron in the **olfactory bulb,** an extension of the brain. An odor contains many odor molecules that activate a characteristic combination of receptor proteins. A rose might stimulate certain olfactory cells, designated by blue and green in Figure 28.2C, while a gardenia might stimulate a different combination. An odor's signature in the olfactory bulb is determined by which neurons are stimulated. When the neurons communicate this information via the olfactory tract to the olfactory areas of the cerebral cortex, we know we have smelled a rose or a gardenia.

Have you ever noticed that a certain aroma vividly brings to mind a certain person or place? A whiff of perfume may remind you of a former friend, or the smell of boxwood may remind you of your grandfather's farm. The olfactory bulbs have direct connections with the limbic system and its centers for emotions and memory. One investigator showed that when subjects smelled an orange while viewing a painting, they not only remembered the painting when asked about it later, but they also had many deep feelings about the painting.

MP3
Taste and Smell

▶ **28.2 CHECK YOUR PROGRESS**
1. Compare and contrast the function of the chemoreceptors on the tongue and in the nose.
2. Explain (in general) how smelling smoke can result in an investigation of the cause.

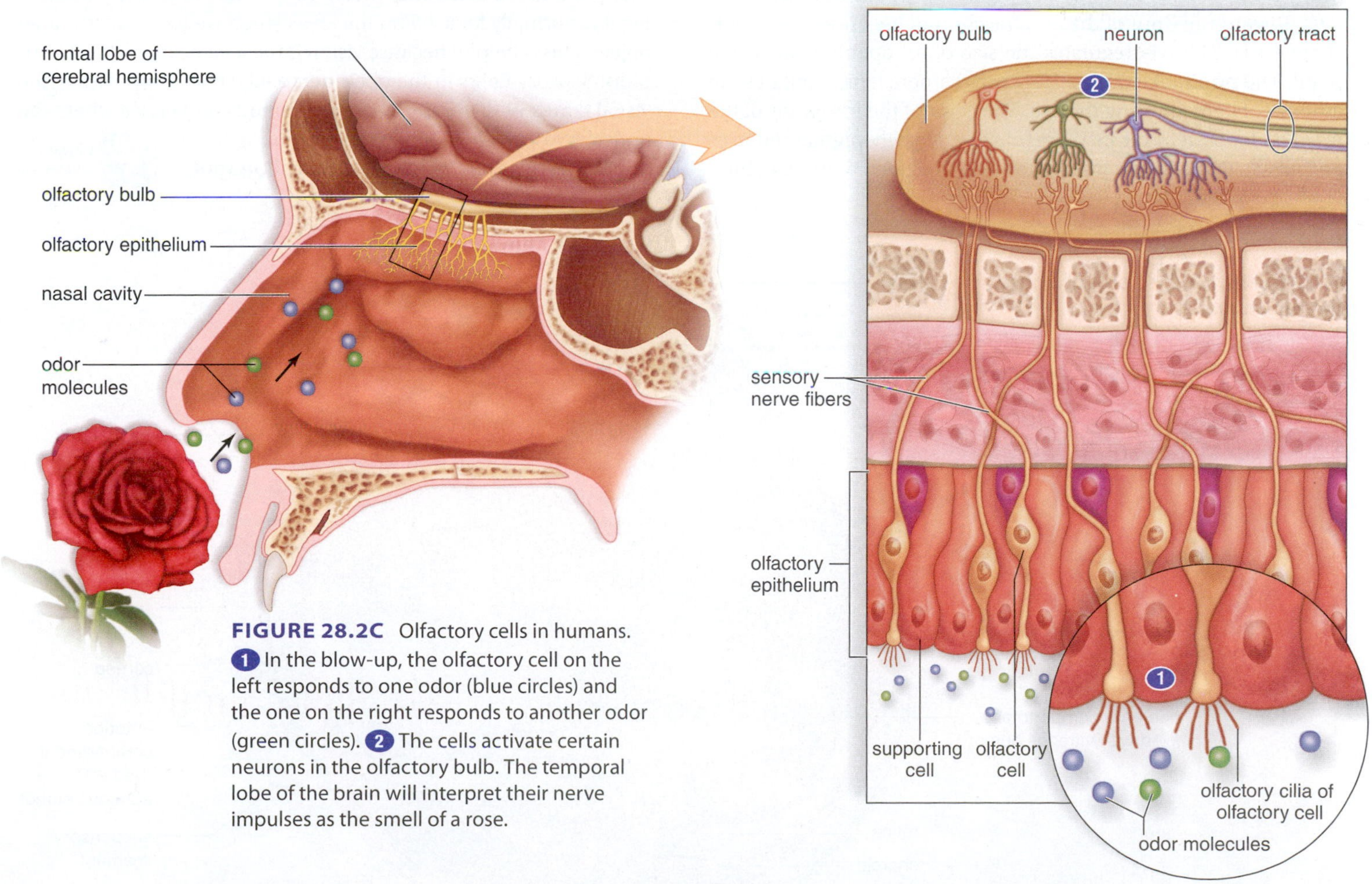

frontal lobe of cerebral hemisphere

olfactory bulb

olfactory epithelium

nasal cavity

odor molecules

olfactory bulb   neuron   olfactory tract

sensory nerve fibers

olfactory epithelium

supporting cell   olfactory cell

olfactory cilia of olfactory cell

odor molecules

FIGURE 28.2C Olfactory cells in humans. ❶ In the blow-up, the olfactory cell on the left responds to one odor (blue circles) and the one on the right responds to another odor (green circles). ❷ The cells activate certain neurons in the olfactory bulb. The temporal lobe of the brain will interpret their nerve impulses as the smell of a rose.

Sense of Vision

The introduction to this chapter compared the compound eye of arthropods to the camera-type eye of a few invertebrates and all vertebrates. This part of the chapter discusses the structure and function of the vertebrate eye and how humans can protect their vision.

28.3 The vertebrate eye is a camera-type eye

LEARNING OUTCOMES

When you complete this section, you should be able to

1. Identify and give a function for the structures of the human eye.
2. Explain how the eye focuses on near and far objects.

The human eye is an elongated sphere about 2.5 cm in diameter that has three layers, or coats: the sclera, the choroid, and the retina (Fig. 28.3A). The outer layer, the **sclera,** is an opaque, white, fibrous layer that covers most of the eye. A mucous membrane called the **conjunctiva** covers the exposed surface of the sclera and lines the inside of the eyelids. In front of the eye, the sclera becomes the **cornea.** The cornea is transparent, being composed of connective tissue with few cells and no blood vessels. Light rays pass through the cornea into the rest of the eye, and therefore the cornea is called the window of the eye. However, the cornea plays an active role in vision by helping to focus light rays. Damage to the cornea is a frequent cause of blindness, but a damaged cornea is replaceable by a corneal transplant.

The middle, thin, dark-brown layer, the **choroid,** contains many blood vessels and a brown pigment that absorbs stray light rays. Toward the front of the eye, the choroid becomes the donut-shaped iris. The **iris** regulates the size of an opening called the pupil. The **pupil,** like the aperture on a camera lens, regulates the amount of light entering the eye. The color of the iris (color of the eyes) is dependent on its pigmentation. Heavily pigmented eyes are brown, while lightly pigmented eyes are green or blue. Behind the iris, the choroid thickens and forms the circular ciliary body. The **ciliary body,** consisting of many radiating folds, contains the ciliary muscles, which control the shape of the lens for near and far vision.

The **lens,** within a membranous capsule, lies directly behind the iris and the pupil. Attached to the ciliary body by suspensory ligaments, the lens divides the cavity of the eye into two compartments: The one in front of the lens is the anterior compartment, and the one behind the lens is the posterior compartment. A basic, watery solution called aqueous humor fills the anterior compartment. The aqueous humor provides a fluid cushion as well as nutrient and waste transport for the eye.

The third layer of the eye, the **retina,** is located in the posterior compartment, which is filled with a clear, gelatinous material called the vitreous humor. The retina contains the photoreceptors; namely, the rod cells and the cone cells. The rod cells, which are very sensitive to light but do not see color, are scattered throughout the retina. The cone cells, which require bright light, give us the ability to distinguish color. The retina has a very special region called the **fovea centralis,** where cone cells are densely packed. Light is normally focused on the fovea when we look directly at an object. This is helpful because vision is most acute in the fovea centralis. Sensory fibers in the **optic nerve** take nerve impulses to the visual cortex of the brain. There are no rods and cones where the optic nerve exits the retina. Therefore, no vision is possible in this area, which is called a **blind spot.**

Video Artificial Eye

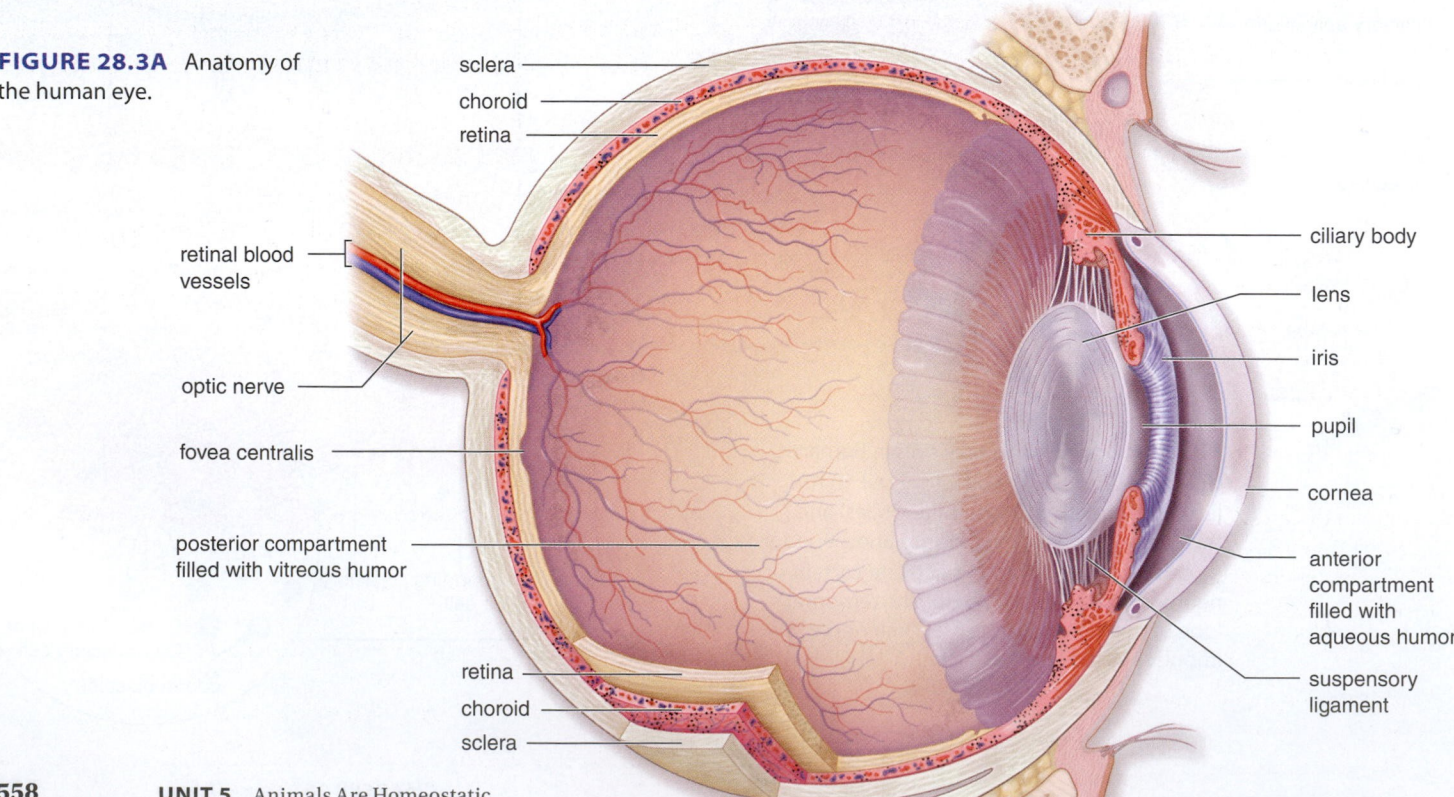

FIGURE 28.3A Anatomy of the human eye.

sclera
choroid
retina
retinal blood vessels
optic nerve
fovea centralis
posterior compartment filled with vitreous humor
retina
choroid
sclera
ciliary body
lens
iris
pupil
cornea
anterior compartment filled with aqueous humor
suspensory ligament

The Lens When we look at an object, light rays pass through the pupil and focus on the retina. The image produced is much smaller than the object because light rays are bent (refracted) when they are brought into focus. Focusing mostly occurs at the cornea as light passes from an air medium to a fluid medium. The lens, however, provides additional focusing power as **visual accommodation** occurs for close vision. The shape of the lens is controlled by the **ciliary muscle** within the ciliary body. When we view a distant object, the ciliary muscle is relaxed, causing the suspensory ligaments attached to the ciliary body to be taut; therefore, the lens remains relatively flat (Fig. 28.3B*a*). When we view a near object, the ciliary muscle contracts, releasing the tension on the suspensory ligaments, and the lens becomes more round due to its natural elasticity (Fig. 28.3B*b*). Because close work requires contraction of the ciliary muscle, it very often causes muscle fatigue known as eyestrain. With normal aging, the lens loses its ability to accommodate for near objects; therefore, people frequently need reading glasses once they reach middle age.

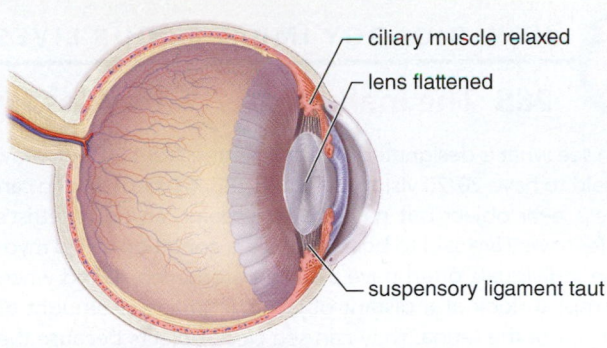

ciliary muscle relaxed
lens flattened
suspensory ligament taut

a.

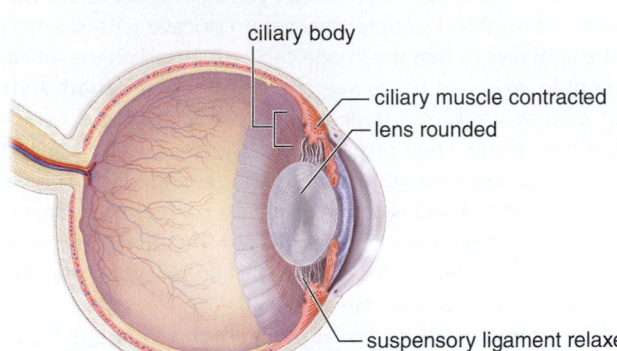

ciliary body
ciliary muscle contracted
lens rounded
suspensory ligament relaxed

b.

FIGURE 28.3B a. Focusing on a distant object. **b.** Focusing on a near object.

▶ **28.3 CHECK YOUR PROGRESS**
1. List and give a specific function for the three coats of the eye.
2. Explain why focusing on a near object is more likely to cause eyestrain.

HOW BIOLOGY IMPACTS OUR LIVES   *Application*

28A Protect Your Eyes from the Sun

The three most frequent causes of blindness are retinal disorders, glaucoma, and cataracts, in that order. Retinal disorders include diabetic retinopathy and macular degeneration. During retinopathy, capillaries to the retina burst, and blood spills into the vitreous humor. Careful regulation of blood glucose levels may protect against this condition. In macular degeneration, the cones are destroyed because thickened choroid vessels no longer function as they should. Glaucoma occurs when fluid

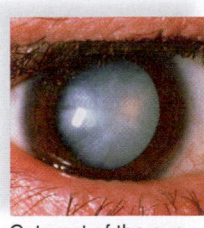

Cataract of the eye

builds up in the compartments and destroys the nerve fibers responsible for peripheral vision. People who have experienced acute glaucoma report that the eyeball feels as heavy as a stone. In cataracts, cloudy spots on the lens of the eye eventually pervade the whole lens. The milky, yellow-white lens scatters incoming light and blocks vision.

Currently, surgery is the only viable treatment for cataracts. First, a surgeon opens the eye near the rim of the cornea and removes the lens from its capsule, using any one of several possible procedures. Then an artificial lens that can correct the patient's near and/or distant vision is inserted into the original lens capsule.

Accumulating evidence suggests that both macular degeneration and cataracts, which tend to occur in the elderly, are caused by long-term exposure to the ultraviolet rays of the sun. It is recommended, therefore, that everyone, especially people who live in sunny climates or work outdoors, wear sunglasses that absorb ultraviolet light. Large lenses worn close to the eyes offer further protection.

The Sunglass Association of America has devised the following system for categorizing sunglasses:

- Cosmetic lenses absorb 20% of UV-A (the type of radiation that reaches the Earth's surface) and 60% of visible light. Such lenses are worn for comfort, rather than protection.
- General-purpose lenses absorb at least 60% of UV-A, and 60–92% of visible light. They are good for outdoor activities in temperate regions.
- Special-purpose lenses block at least 60% of UV-A and 20–97% of visible light. They are good for bright sun combined with sand, snow, or water.

Healthcare providers have noted an increased incidence of cataracts in heavy cigarette smokers. The risk of cataracts doubles in men who smoke 20 cigarettes or more a day and in women who smoke 35 cigarettes or more a day. A possible reason is that smoking reduces the delivery of blood, and therefore nutrients, to the lens.

CONSIDER THESE QUESTIONS

1. Can any part of the eye be damaged and vision not be affected? Why or why not?
2. Are you willing to make any life changes such as always wearing sunglasses and/or quit smoking, in order to protect your eyes?

 Explore the concepts through a variety of multimedia
BIOLOGY   assets and question types.
www.mcgrawhillconnect.com

28B The Inability to Form a Clear Image Can Be Corrected

If you can see what is designated as size 20 letters from 20 feet away, you are said to have 20/20 vision. In Figure 28B, **①** people who can easily see a near object but have trouble seeing an optometrist's chart 20 feet away are said to be *nearsighted,* a condition called myopia. These individuals often have an elongated eyeball, and when they attempt to look at a distant object, the image is brought to focus in front of the retina. They can see close objects because the lens can compensate for the elongated eyeball. In order to see distant objects, nearsighted people may wear concave lenses, which diverge the light rays so that the image can be focused on the retina.

② People who can easily see the optometrist's chart 20 ft away but cannot easily see near objects are *farsighted,* a condition called hyperopia. They often have a shortened eyeball, and when they try to see near objects, the image is focused behind the retina. When the object is distant, the lens can compensate for the short eyeball. To see near objects, these individuals may wear a convex lens that increases the bending of light rays so that the image can be focused on the retina.

③ When the cornea or lens is uneven, the image is fuzzy. This condition, called astigmatism, can be corrected by wearing an unevenly ground lens to compensate for the uneven cornea.

LASIK Surgery

Rather than wearing glasses or contact lenses, many people are now choosing to undergo LASIK eye surgery. LASIK stands for laser-assisted in-situ keratomileusis, which results in reshaping of the cornea. Typically, adults affected by common vision problems (nearsightedness, farsightedness, or astigmatism) respond well to LASIK. An eye exam determines the present thickness and shape of the cornea and how much the cornea needs to be reshaped to achieve 20/20 vision. During the LASIK procedure, a small flap of conjunctiva is first lifted to expose the cornea. Then the laser is used to remove tissue from the cornea. Each pulse of the laser removes a small amount of corneal tissue, allowing the surgeon to flatten or otherwise change the shape of the cornea. After the procedure, the flap of conjunctiva is put back in place and allowed to heal on its own. Most patients achieve vision that is close to 20/20, but the chances for improved vision are based, in part, on the condition of the eyes before surgery.

 Explore the concepts through a variety of multimedia assets and question types.

www.mcgrawhillconnect.com

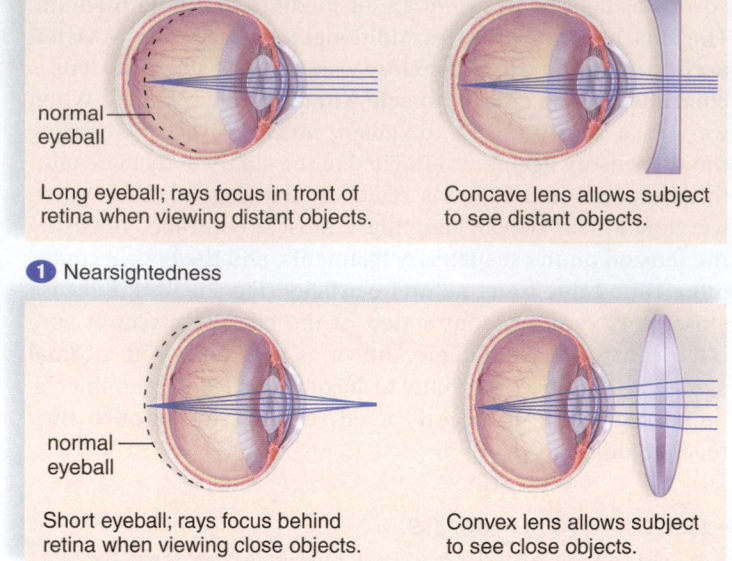

Long eyeball; rays focus in front of retina when viewing distant objects.

Concave lens allows subject to see distant objects.

① Nearsightedness

Short eyeball; rays focus behind retina when viewing close objects.

Convex lens allows subject to see close objects.

② Farsightedness

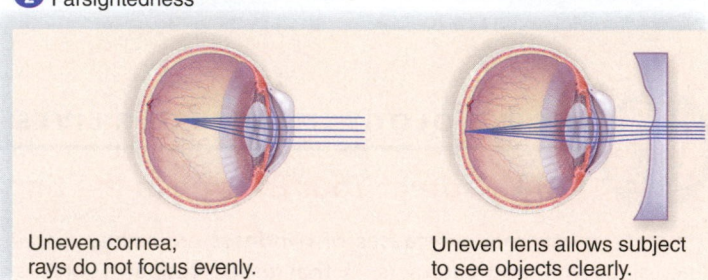

Uneven cornea; rays do not focus evenly.

Uneven lens allows subject to see objects clearly.

③ Astigmatism

FIGURE 28B Common abnormalities of the eye and the possible corrective lens for each.

CONSIDER THESE QUESTIONS

1. People sometimes feel stigmatized because they wear glasses. Is this reasonable? Why or why not?
2. Should people bear the expense of LASIK surgery and a slight chance of worsening their vision for purely cosmetic reasons?

28.4 The retina sends information to the visual cortex

LEARNING OUTCOMES

When you complete this section, you should be able to

1. Describe the role of the rods and the cones in vision.
2. Understand the location and function of bipolar and ganglion cells in the retina.

So far, we have studied how the eye focuses an image on the retina. Now we will consider how vision is achieved. We will see that the visual system of humans does not merely record bits of light and dark like a camera. Instead, it constructs an image that helps us function in the environment.

In a **rod cell** (Fig. 28.4, *right*) the membrane disks of an outer segment contain numerous visual pigment molecules that absorb light. The plasma membrane contains ion channels. Synaptic vesicles are located at the synaptic endings of the inner segment. The visual pigment in rods is a deep-purple pigment called rhodopsin. **Rhodopsin** is a complex molecule made up of the protein opsin and a light-absorbing molecule called *retinal,* which is a derivative

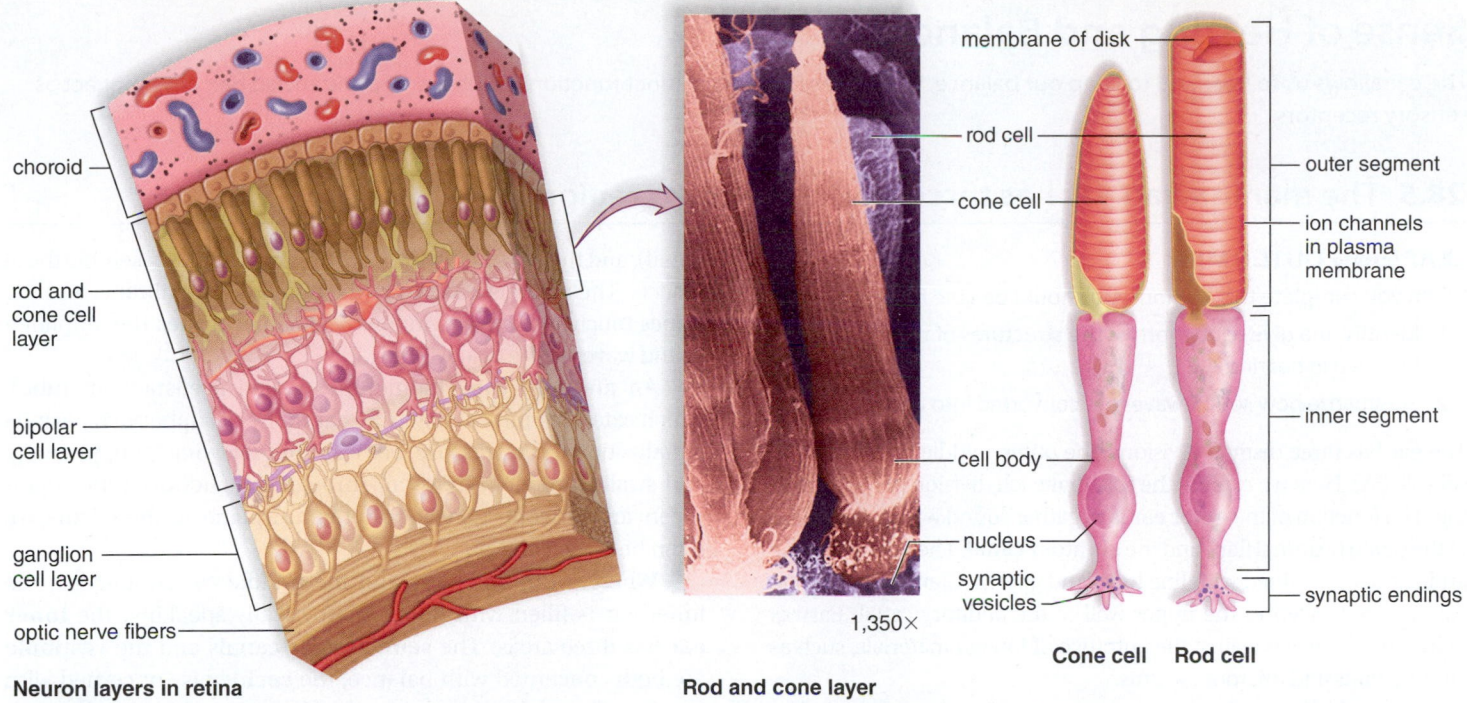

choroid

rod and
cone cell
layer

bipolar
cell layer

ganglion
cell layer

optic nerve fibers

Neuron layers in retina

membrane of disk

rod cell

cone cell

cell body

nucleus

synaptic
vesicles

1,350×

Rod and cone layer

outer segment

ion channels
in plasma
membrane

inner segment

synaptic endings

Cone cell   Rod cell

FIGURE 28.4 Structure and function of the retina.

of vitamin A. When a rod absorbs light, rhodopsin splits into opsin and retinal, leading to a cascade of reactions and the closure of ion channels in the rod cell's plasma membrane. The release of inhibitory transmitter molecules from the rod's synaptic vesicles ceases. Then nerve impulses go to the visual areas of the cerebral cortex.

Rods are very sensitive to light and, therefore, are suited to night vision. (Since carrots are rich in vitamin A, it is true that eating carrots can improve your night vision.) Rod cells are plentiful in the peripheral region of the retina; therefore, they also provide us with peripheral vision and perception of motion. **Cone cells,** on the other hand, are located primarily in the fovea centralis and are activated by bright light. They allow us to detect the fine detail and the color of an object.

Color vision depends on three different kinds of cones, which contain pigments called the B (blue), G (green), and R (red) pigments. Each pigment is made up of opsin and retinal, but there is a slight difference in the opsin structure of each, which accounts for their individual absorption patterns. Various combinations of cones are believed to be stimulated by in-between shades of color. For example, the color yellow is perceived when green cones are highly stimulated, red cones are partially stimulated, and blue cones are not stimulated. In color blindness, usually one type of cone is defective or deficient in number. The most common mutation is the inability to see the colors red and green so that, for example, the individual cannot see the number 74 in the illustration in the center of this page.

Figure 28.4, *left*, shows the retina has three layers of neurons: the **rod cell and cone cell layer,** the **bipolar cell layer,** and the **ganglion cell layer.** The rods and cones are in the layer closest to the choroid. The rods and cones synapse with the bipolar cells,

which in turn synapse with ganglion cells, whose axons are optic nerve fibers. Notice in Figure 28.4 that there are many more rods and cones than ganglion cells. In fact, the retina has as many as 150 million rod cells and 6 million cone cells, but only 1 million ganglion cells. The sensitivity of cones versus rods is mirrored by how directly they connect to ganglion cells. As many as 150 rods may activate the same ganglion cell. No wonder stimulation of rods results in blurred, indistinct vision. In contrast, some cones in the fovea centralis activate only one ganglion cell. This explains why cones, especially in the fovea centralis, provide us with a sharper, more delineated image of an object.

As signals pass to bipolar cells and ganglion cells, integration occurs. Therefore, considerable processing takes place in the retina before ganglion cells generate nerve impulses, which are carried in the optic nerve to the visual cortex. Additional integration occurs in the visual cortex, where a meaningful image is achieved. For example, due to the placement of the eyes in humans, each side of the brain receives data for only half the field of vision. To see the complete object, communication is needed between the two sides of the brain, thereby our vision is stereoscopic (three-dimensional). Also, because the image is inverted and reversed, it must be righted in the brain for us to correctly perceive the visual field.

MP3
Sense
of Vision

▶ **28.4 CHECK YOUR PROGRESS**

1. Contrast the location of the rods and the cones in the retina and the role they play in vision.

2. Use Figure 28.4 to explain why cone cells produce a sharper image of an object.

Sense of Hearing and Balance

The ear allows us to hear and to keep our balance. While these are two distinct functions, both are dependent on hair cells that act as sensory receptors.

28.5 The mammalian ear has three well-developed regions

LEARNING OUTCOMES

When you complete this section, you should be able to

1. Identify and give a function for the structures of the ear that are involved in hearing.
2. Summarize how sound waves are converted into nerve signals.

The ear has three distinct divisions: the outer, middle, and inner ear (Fig. 28.5A). Here we explore the role that each division plays in hearing. The function of the **outer ear** is to gather sound waves. It consists of the pinna (external flap) and the **auditory canal.** The opening of the auditory canal is lined with fine hairs and glands. Glands that secrete earwax are located in the upper wall of the auditory canal. Earwax helps guard the ear against the entrance of foreign materials, such as air pollutants and microorganisms.

The **middle ear** begins at the **tympanic membrane** (often called the eardrum) and ends at a bony wall containing two small openings covered by membranes. These openings are called the *oval window* and the *round window.* The function of the middle ear is to amplify sound waves. Three small bones lie between the tympanic membrane and the oval window. Collectively called the **ossicles,** individually they are the *malleus* (hammer), the *incus*

(anvil), and the *stapes* (stirrup) because their shapes resemble these objects. The malleus adheres to the tympanic membrane, and the stapes touches the oval window. The stapes passes the amplified sound waves to the oval window.

An **auditory tube** (sometimes called a eustachian tube), which extends from each middle ear to the nasopharynx, permits equalization of air pressure. For example, chewing gum, yawning, and swallowing help move air through the auditory tubes upon ascent and descent in an elevator or an airplane. As this occurs, we often hear the ears "pop."

Whereas the outer ear and the middle ear contain air, the inner ear is filled with fluid. Anatomically speaking, the **inner ear** has three areas: The **semicircular canals** and the **vestibule** are both concerned with balance; the **cochlea** is concerned with hearing. The cochlea resembles the shell of a snail because it spirals. Receptors that respond to sound waves are housed in the cochlea.

Hearing To understand how we hear, first trace the path of sound waves through the outer and middle ear in Figure 28.5A. Just as ripples travel across the surface of a pond, sound waves travel by

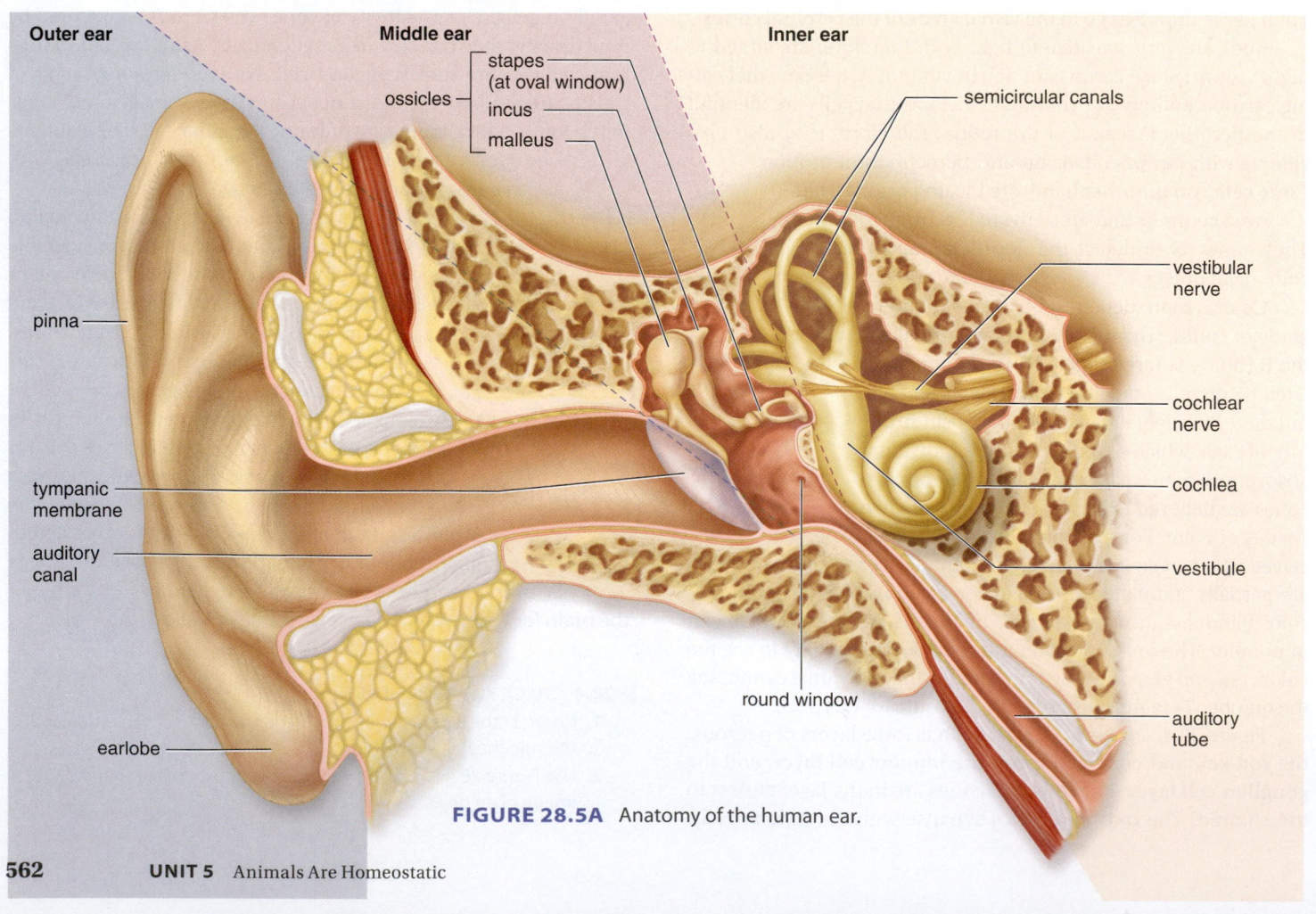

FIGURE 28.5A Anatomy of the human ear.

the successive vibrations of molecules. Ordinarily, sound waves do not carry much energy, but when a large number of waves strike the tympanic membrane, it moves back and forth (vibrates) ever so slightly. The malleus takes the pressure from the tympanic membrane and passes it, by means of the incus, to the stapes, which strikes the oval window. The stapes vibrates the membrane of the **oval window** with a force that has been multiplied about 20 times by the movement of the ossicles. This force allows sound waves to become fluid pressure waves in the inner ear. Eventually, the pressure waves disappear at the **round window.**

Figure 28.5B shows the receptors for hearing at increasing levels of magnification. ❶ These receptors are located in the snail-shaped cochlea, a major part of the inner ear. ❷ A cross section of the cochlea reveals that the receptors are located in the cochlear canal, one of three canals in the cochlea. When the stapes strikes the oval window, pressure waves move through the fluid of these canals. ❸ The **organ of Corti,** by which we hear, consists of little hair cells that occur along the length of the basilar membrane. The hair cells have extensions called stereocilia, which are embedded in the gelatinous tectorial membrane. ❹ When we hear, the fluid pressure waves in the inner ear cause the basilar membrane to vibrate and the stereocilia to bend because they are trapped in the tectorial membrane. The hair cells now generate nerve impulses that travel in the **cochlear nerve** to the brain stem. When these impulses reach the auditory areas of the cerebral cortex, they are interpreted as sound.

Animation
Effects of Sound Waves on Cochlear Structure

Sensory Coding Each part of the organ of Corti is sensitive to different wave frequencies, or *pitch.* Think of the basilar membrane as a rope stretched between two posts. If you pluck the rope at one end, a wave of vibration travels down its length. Similarly, a sound causes a wave in the basilar membrane. If the wave reaches the tip of the organ of Corti, the brain interprets this as a low pitch, such as that of a tuba. If the wave remains near the base, the brain interprets this as a high pitch, such as that of a whistle. Thus, the pitch sensation we experience depends on which region of the basilar membrane vibrates and which area of the brain is stimulated. Researchers believe the brain interprets the tone of a sound based on the distribution of the hair cells stimulated.

Volume is a function of the size (amplitude) of sound waves. Loud noises cause the fluid within the vestibular canal to exert more pressure and the basilar membrane to vibrate to a greater extent. The brain interprets the resulting increased stimulation as volume.

MP3
Sense of Hearing and Equilibrium

▶ **28.5 CHECK YOUR PROGRESS**
1. Identify the location and function of the tympanic membrane and the oval window.
2. Identify the specific sense organ for hearing and tell how it functions.
3. Explain how hearing would be affected if the stereocilia of the cells in the organ of Corti disappeared.

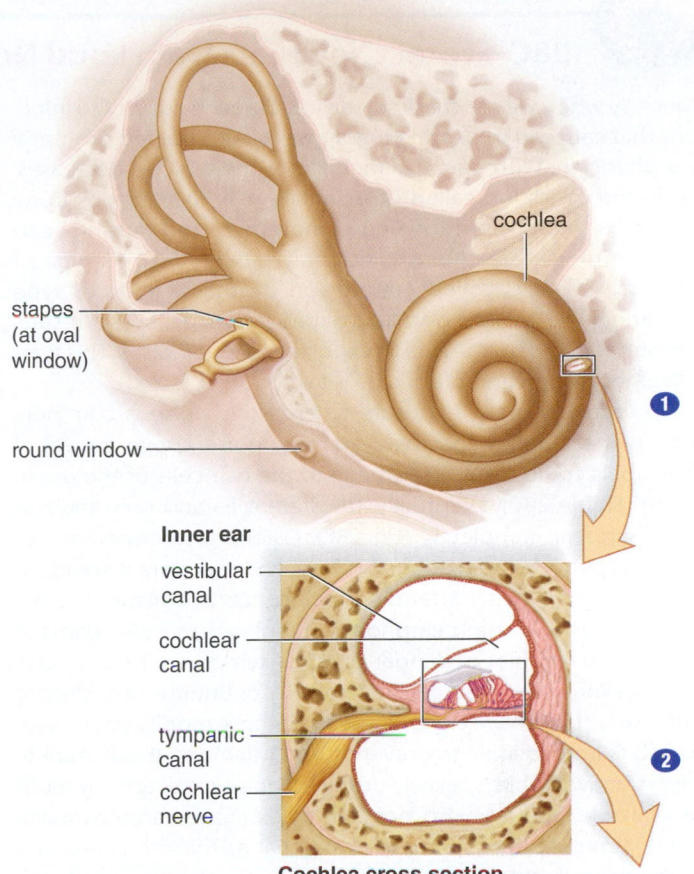

Inner ear

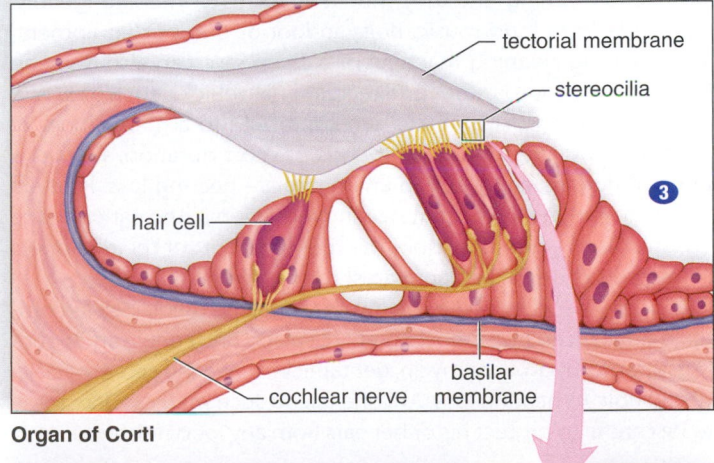

Cochlea cross section

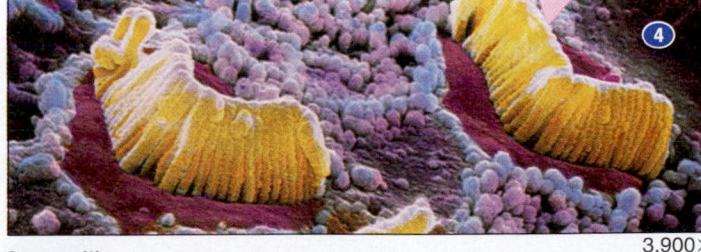

Organ of Corti

Stereocilia

3,900×

FIGURE 28.5B The sense organ for hearing, the organ of Corti, is located in the cochlea and consists of hair cells, whose stereocilia are embedded in the tectorial membrane. Vibration of the membrane causes the stereocilia to bend.

28C Protect Your Ears from Loud Noises

Especially when we are children, the middle ear is subject to infections that can lead to hearing impairment if not treated promptly by a physician. With age, the mobility of the ossicles decreases, and in the condition called otosclerosis, new filamentous bone grows over the stirrup, impeding its movement and causing hearing loss. Surgical treatment is the only remedy for this type of deafness, which is called conduction deafness. Another type of hearing loss, called age-associated nerve deafness, results from stereocilia damage due to exposure to loud noises. This type of deafness is preventable, if care is taken.

In today's society, exposure to excessive noise is common. Noise is measured in decibels, and any noise above a level of 80 decibels could result in damage to the hair cells of the organ of Corti (Table 28C). Eventually, the stereocilia and then the hair cells disappear completely (Fig. 28C). Listening to city traffic for extended periods can damage hearing, and therefore it stands to reason that frequently attending rock concerts, constantly playing loud music, or using earphones at high volume also damage hearing. The first hint of danger could be temporary hearing loss, a "full" feeling in the ears, muffled hearing, or tinnitus (e.g., ringing in the ears). If you have any of these symptoms, modify your listening habits immediately to prevent further damage. If exposure to noise is unavoidable, specially designed noise reduction earmuffs are available, and it is also possible to purchase earplugs made from a compressible, spongelike material at the drugstore or a sporting-goods store. These earplugs are not the same as those worn for swimming, and they should not be used interchangeably.

Aside from loud music, noisy indoor or outdoor equipment, such as a rug-cleaning machine or a chain saw, can also damage hearing. Even motorcycles and recreational vehicles such as snowmobiles and motocross bikes can contribute to a gradual loss of hearing. Exposure to intense sounds of short duration, such as a burst of gunfire, can result in an immediate hearing loss. Hunters may experience a significant hearing reduction in the ear opposite the shoulder where they hold the rifle. (The butt of the rifle offers some protection to the ear nearest the gun when it is shot.)

Finally, people need to be aware that some medicines are ototoxic. Anticancer drugs, most notably cisplatin, and certain antibiotics (e.g., streptomycin, kanamycin, gentamicin) make the ears especially susceptible to hearing loss. Anyone taking such medications needs to be careful to protect his or her ears from any loud noises.

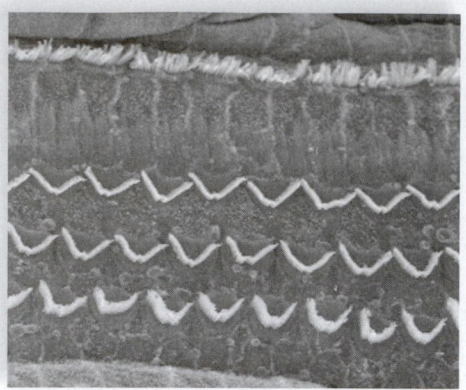

a.

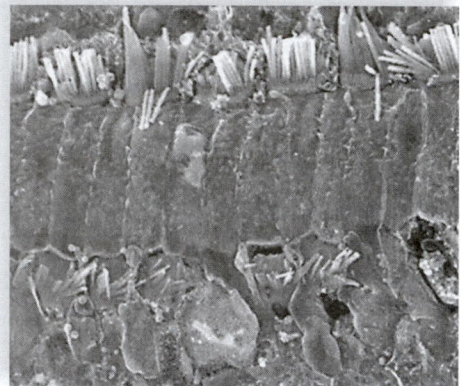

b.

FIGURE 28C
Excessive noise can cause **a.** normal hair cells in the organ of Corti to become damaged, as shown in **b.**

CONSIDER THESE QUESTIONS

1. Are you willing to forgo rock concerts and exercise classes at which the music is too loud in order to save your hearing? Why or why not?
2. Older people who can't hear annoy us. Should they have been more careful to protect their hearing when they were younger? Explain.

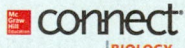 Explore the concepts through a variety of multimedia assets and question types.

www.mcgrawhillconnect.com

TABLE 28C	Noises That Affect Hearing	
Type of Noise	**Sound Level (decibels)**	**Effect**
"Boom car," jet engine, shotgun, rock concert	Over 125	Beyond threshold of pain; potential for hearing loss high
Nightclub, thunderclap	Over 120	Hearing loss likely
Earbuds in external auditory canal	110–120	Hearing loss likely
Chain saw, pneumatic drill, jackhammer, symphony orchestra, snowmobile, garbage truck, cement mixer	100–200	Regular exposure of more than 1 min risks permanent hearing loss
Farm tractor, newspaper press, subway, motorcycle	90–100	Fiteen minutes of unprotected exposure potentially harmful
Lawn mower, food blender	85–90	Continuous daily exposure for more than 8 hrs can cause hearing damage
Diesel truck, average city traffic noise	80–85	Annoying; constant exposure may cause hearing damage

HOW LIFE CHANGES   *Application*

28D Evolution of the Mammalian Ear

Only mammals have an ear like the one depicted in Figure 28.5A, but the invertebrates have body parts that resemble those in the mammalian ear. For example, insects have tympanic membranes in various places, including their front legs! Stereociliated hair cells located beneath the membrane generate nerve impulses when sound waves cause the membrane to vibrate. Also, gravitational balance organs called statocysts are found in cnidarians, molluscs, and arthropods. When movement occurs, a particle called a statolith deflects hair cells, which then generate nerve impulses that go to the brain (Fig. 28D.1).

The outer ear of birds and mammals has a recessed tympanic membrane, and the inner ear contains a utricle and a saccule, which are gravitational balance organs that respond to the movement of particles called otoliths. Fishes have semicircular canals and a utricle and saccule that function as balance organs and are also involved in hearing. In fact, the saccule has an extension called the lagena that increases in size from amphibian to reptile to mammal, in whom it becomes the coiled cochlea. Fishes also have a lateral line system that admits water and contains hair cells enclosed by a gelatinous cap called a cupula (Fig. 28D.2). When the cupula bends in response to pressure waves caused by nearby objects, the stereocilia of the hair cells bend and they generate nerve impulses. The lateral line system of fishes helps them locate other fish, including predators, prey, and mates. This unique system, which can be likened to "seeing in the dark," has no equivalent in terrestrial vertebrates. Still, we know that the hair cells in the organ of Corti do respond to pressure waves, as do the hair cells in the ampullae of the semicircular canals.

Mammals have a unique middle ear due to the presence of three ossicles—the malleus, the incus, and the stapes. The other terrestrial vertebrates have only one ossicle, a stapes (Fig. 28D.3). The question becomes, how did the other ossicles (malleus and incus) arise? Paleontologists offer evidence that the malleus and incus are derived from the jawbone of a reptile! They have found fossils of tiny ancient mammals (about 12 cm long) from the Mesozoic era, in which these ossicles are still connected to the jaw. In later-appearing mammals, these ossicles are present in the middle ear and no longer have any connection to the jaw. The presence of three ossicles instead of one means that sound waves can be better amplified and hearing is enhanced. They also allow better resolution of high frequencies and reduce the transmission of low frequencies.

CONSIDER THESE QUESTIONS

1. What evolutionary principle is exhibited by the presence of structures resembling reptilian jawbones in the middle ear of mammals?
2. Give an evolutionary explanation for why the inner ear of mammals makes use of fluid pressure waves.
3. Mammals don't use a lateral line for detecting the presence of other organisms. Why not—and what do they use? Does your answer depend on the presence of three ossicles in the middle ear? Explain.

 Explore the concepts through a variety of multimedia assets and question types.

www.mcgrawhillconnect.com

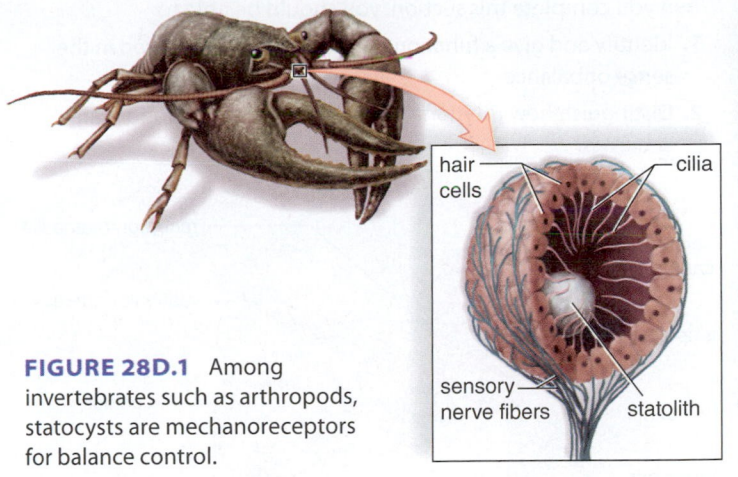

FIGURE 28D.1 Among invertebrates such as arthropods, statocysts are mechanoreceptors for balance control.

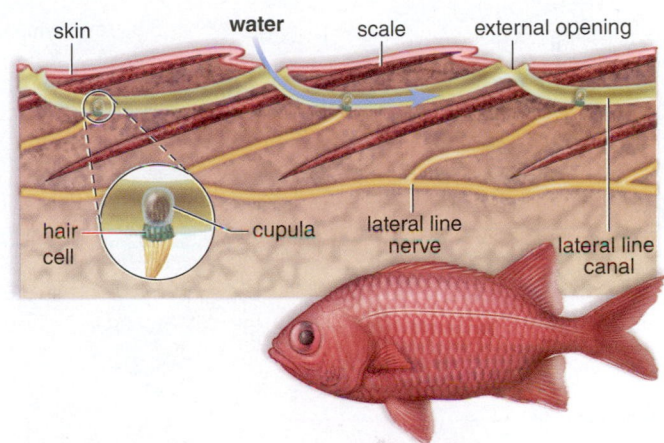

FIGURE 28D.2 A lateral line system, which helps fish detect the presence of other organisms by the pressure waves they generate, is an adaptation unique to fishes.

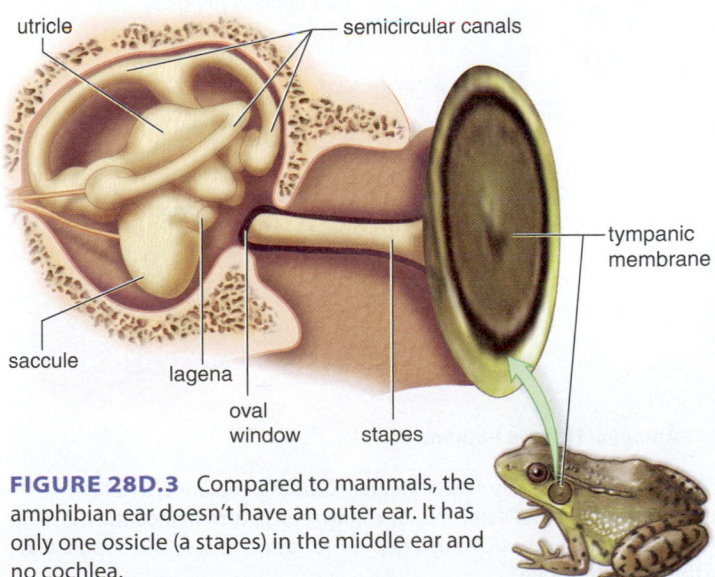

FIGURE 28D.3 Compared to mammals, the amphibian ear doesn't have an outer ear. It has only one ossicle (a stapes) in the middle ear and no cochlea.

When you complete this section, you should be able to

1. Identify and give a function for the structures involved in the sense of balance.

2. Distinguish how rotational and gravitational equilibrium are achieved.

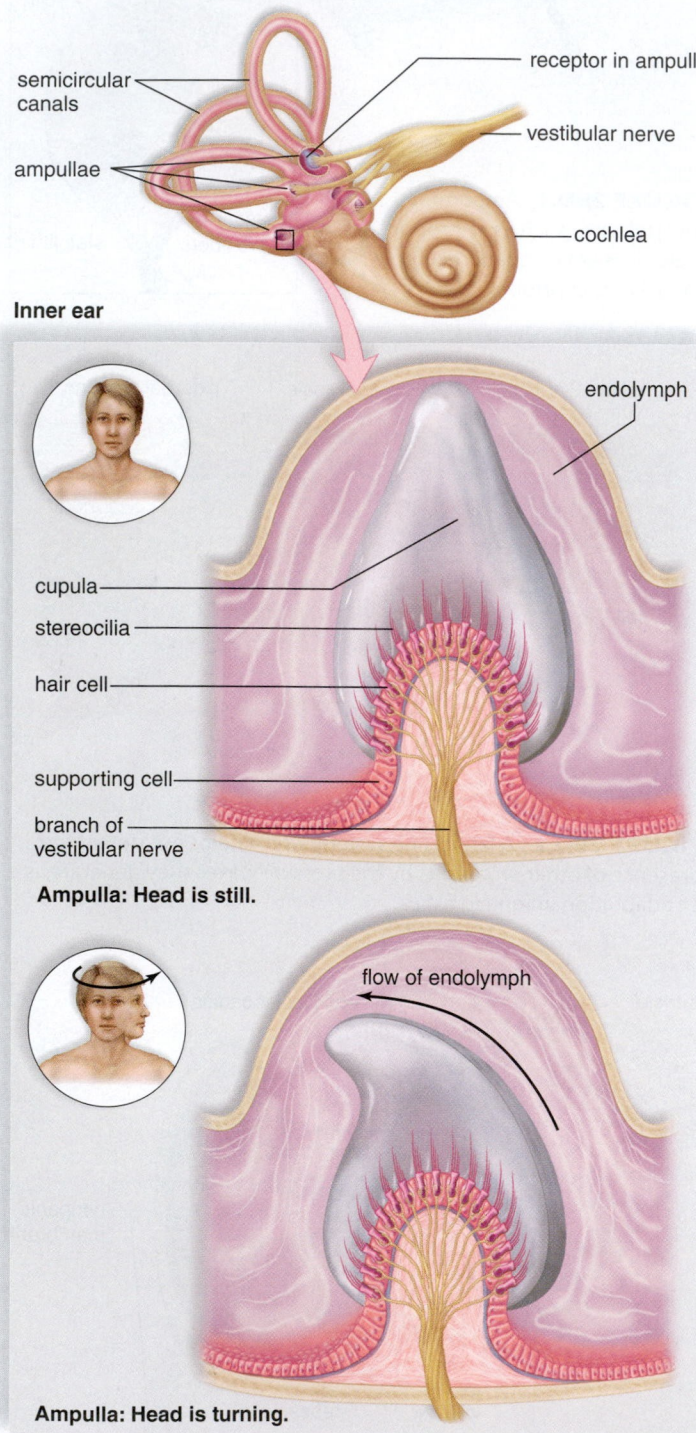

semicircular canals

receptor in ampulla

vestibular nerve

ampullae

cochlea

Inner ear

endolymph

cupula

stereocilia

hair cell

supporting cell

branch of vestibular nerve

Ampulla: Head is still.

flow of endolymph

Ampulla: Head is turning.

FIGURE 28.6A A sense organ for rotational balance, located in the ampullae of the semicircular canals, consists of hair cells, whose stereocilia are embedded in a cupula. Rotation causes the cupula and stereocilia to move.

The receptors for balance detect rotational and/or angular movement of the head (**rotational balance**) and also straight-line movement of the head in any direction (**gravitational balance**).

Rotational Balance In the inner ear, the three semicircular canals are arranged so that there is one in each dimension of space. The base of each of the three canals, called the **ampulla,** is slightly enlarged (Fig. 28.6A). Little hair cells, whose stereocilia are embedded within a gelatinous cap called a cupula, are located within the ampullae. Each ampulla responds to head rotation in a different plane of space because of the way the semicircular canals are arranged. As fluid within a semicircular canal flows over and displaces a cupula, the stereocilia of the hair cells bend. This changes the pattern of signals carried by the vestibular nerve to the brain. The brain uses information from the hair cells within each ampulla of the semicircular canals to maintain balance. The brain generates appropriate motor output to various skeletal muscles to correct our present position in space as needed.

Why does spinning around cause you to become dizzy? When we spin, the cupula slowly begins to move in the same direction we are spinning, and bending of the stereocilia causes hair cells to send messages to the brain. As time goes by, the cupula catches up to the rate we are spinning, and the hair cells no longer send messages to the brain. When we stop spinning, the slow-moving cupula continues to move in the direction of the spin and the stereocilia bend again, indicating that we are moving. Yet, the eyes know we have stopped. The mixed messages sent to the brain cause us to feel dizzy (Fig. 28.6B).

Certain types of movement are likely to cause motion sickness. Seasickness may result from the rocking or swaying of a boat; if the water becomes rough enough, almost anyone can develop this type of motion sickness. Airsickness can occur due to abrupt changes in altitude during a plane trip or the jerky motions of turbulence. Motion sickness is also quite common for passengers in cars, buses, or trains. Motion sickness is the most common cause

FIGURE 28.6B Using playground equipment, children love to test their rotational equilibrium. Adults are usually not too interested.

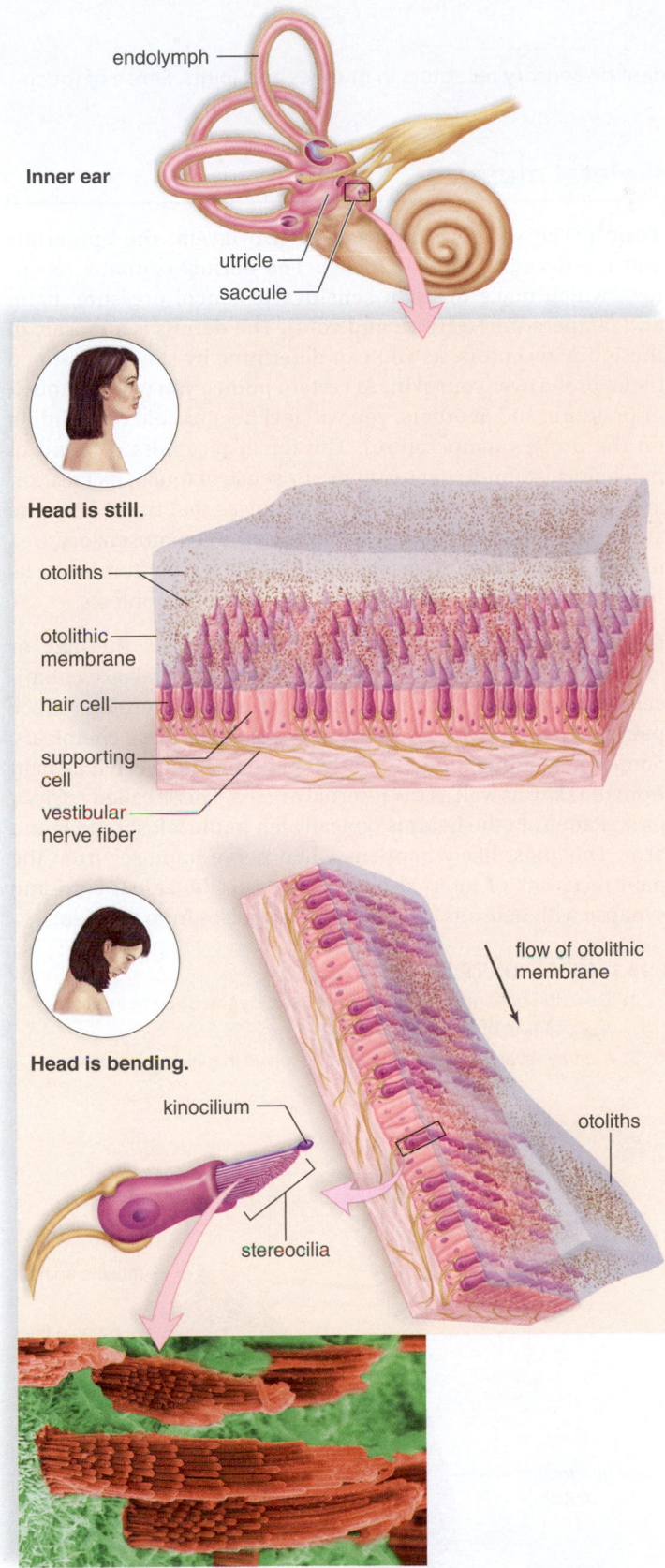

Inner ear

endolymph

utricle

saccule

Head is still.

otoliths

otolithic membrane

hair cell

supporting cell

vestibular nerve fiber

Head is bending.

kinocilium

stereocilia

flow of otolithic membrane

otoliths

FIGURE 28.6C The receptors for gravitational balance are in the utricle and the saccule of the vestibule. When the head bends, otoliths are displaced, causing the membrane to sag and the stereocilia to bend.

of vertigo; other symptoms include nausea, vomiting, and cold sweats. Fortunately, motion sickness subsides soon after the triggering motion ceases. Motion sickness occurs when the brain is bombarded with conflicting sensory input. For example, suppose you are trying to read a book while riding on a bus. As the vehicle goes up and down hills and around curves, starts and stops, and hits the occasional pothole, your inner ear balance receptors send information about all these changes in position to your brain. So do somatic receptors in your joints and muscles and skin (see section 28.7). However, since your eyes are fixed on the pages of your book, the visual information that your brain receives says your position has not changed. The brain becomes overwhelmed, and motion sickness ensues.

One way to avoid motion sickness is to avoid reading while you are in motion; watch the scenery instead. (This is why many people who routinely become carsick as passengers experience no symptoms while driving.) If you are on a train, try to avoid facing backward. If you are on a plane, reserve a seat over the wings, where the plane is most stable. For the same reason, if you travel by ship, pick a cabin near the middle of a deck.

Gravitational Balance Receptors in the utricle and saccule detect movement of the head in the vertical or horizontal planes. The **utricle** and the **saccule** are two membranous sacs located in the inner ear near the semicircular canals. Both of these sacs contain little hair cells, whose stereocilia are embedded within a gelatinous material called an otolithic membrane (Fig. 28.6C). Calcium carbonate ($CaCO_3$) granules, or **otoliths,** rest on this membrane. The utricle is especially sensitive to horizontal (back-forth) movements and bending of the head, while the saccule responds best to vertical (up-down) movements.

When the body is still, the otoliths in the utricle and the saccule rest on the otolithic membrane above the hair cells. When the head bends or the body moves in the horizontal and vertical planes, the otoliths are displaced. The otolithic membrane sags, bending the stereocilia of the hair cells beneath. If the stereocilia move toward the largest stereocilium, called the kinocilium, nerve impulses increase in the vestibular nerve. If the stereocilia move away from the kinocilium, nerve impulses decrease in the vestibular nerve. The frequency of nerve impulses in the vestibular nerve indicates whether your head is moving up or down.

These data reach the cerebellum, which uses them to determine the direction of the movement of the head at that moment. It's important to remember that the cerebellum (p. 538) is vital to maintaining balance and gravitational equilibrium. The cerebellum processes information from the inner ear (the semicircular canals, utricle, and saccule) as well as visual and proprioceptive inputs. In addition, the motor cortex in the frontal lobe of the brain signals where the limbs should be located at any particular moment. After integrating all these nerve inputs, the cerebellum coordinates skeletal muscle contraction to correct our position in space if necessary.

MP3 Sense of Hearing and Equilibrium

▶ **28.6 CHECK YOUR PROGRESS**

1. Contrast how rotational and gravitational balance are achieved.

Somatic Senses

Proprioception, an awareness of the body's position in space, is dependent on sensory receptors in muscles and joints. Sense of touch is dependent on sensory receptors in the skin.

28.7 Mammalian proprioceptors are located in skeletal muscles

LEARNING OUTCOME

When you complete this section, you should be able to

1. Distinguish between proprioceptors and cutaneous receptors with regard to function.

Proprioceptors The receptors that maintain muscle tone and also help maintain the body's balance and posture are called **proprioceptors.** Proprioceptors tell the position of our body parts in space by detecting the degree of muscle contraction (tension), the stretch of tendons, and the movement of ligaments.

Muscle spindles, built into a muscle, act to increase muscle contraction and tension. Golgi tendon organs, found in tendons, decrease muscle tension. The result is a muscle that has the proper length and tension, or muscle tone. Figure 28.7 illustrates the activity of a muscle spindle. In a muscle spindle, sensory nerve endings are wrapped around thin muscle fibers within a connective tissue sheath. The information sent by muscle spindles to the spinal cord and from there to the cerebellum is used to help maintain the body's balance and posture. Proper balance and body position are maintained, despite the force of gravity always acting upon the skeleton and muscles.

When a muscle relaxes and its length increases, the muscle spindle is stretched, and nerve signals to the spinal cord are generated. The more the muscle stretches, the faster the muscle spindle sends its signals. A reflex action then occurs, which results in contraction of muscle fibers adjoining the muscle spindle. The knee-jerk reflex, which involves muscle spindles, offers an opportunity for physicians to test this reflex action.

Touch The skin is composed of two layers: the epidermis and the dermis (see Fig. 26.6A). The dermis contains receptors, which make the skin sensitive to touch, pressure, pain, and temperature (warmth and cold). The dermis is a mosaic of these tiny receptors, as you can determine by slowly passing a metal probe over your skin. At certain points, you will feel touch or pressure, and at others, you will feel heat or cold (depending on the probe's temperature). The touch receptors respond to mechanical stimuli that result in our sense of touch, its location and intensity. They initiate nerve impulses that travel from the skin to the spinal cord and from there to the somatosensory area of the cerebral cortex (see Fig. 27.4B). Touch receptors assist us in knowing where our body is in relation to other objects.

Pain When damage occurs due to mechanical, thermal, or electrical stimulation, or a toxic substance, cells release chemicals that stimulate pain receptors. Aspirin and ibuprofen reduce pain by inhibiting the synthesis of one class of these chemicals. Sometimes, stimulation of internal pain receptors is felt as pain from the skin, as well as the internal organs. This is called *referred pain.* Pain from the heart is typically felt in the left shoulder and arm. This most likely happens when nerve impulses from the pain receptors of internal organs travel to the spinal cord and synapse with neurons also receiving impulses from the skin.

▶ 28.7 CHECK YOUR PROGRESS

1. Describe how proprioceptors are used by the body to indicate the position of the arms and legs.
2. Identify several senses of the body involved in maintaining balance.

FIGURE 28.7 During a knee jerk, a muscle spindle sends ➊ sensory nerve impulses to the spinal cord. ➋ Motor nerve impulses from the spinal cord result in muscle fiber contraction so that muscle tone is maintained.

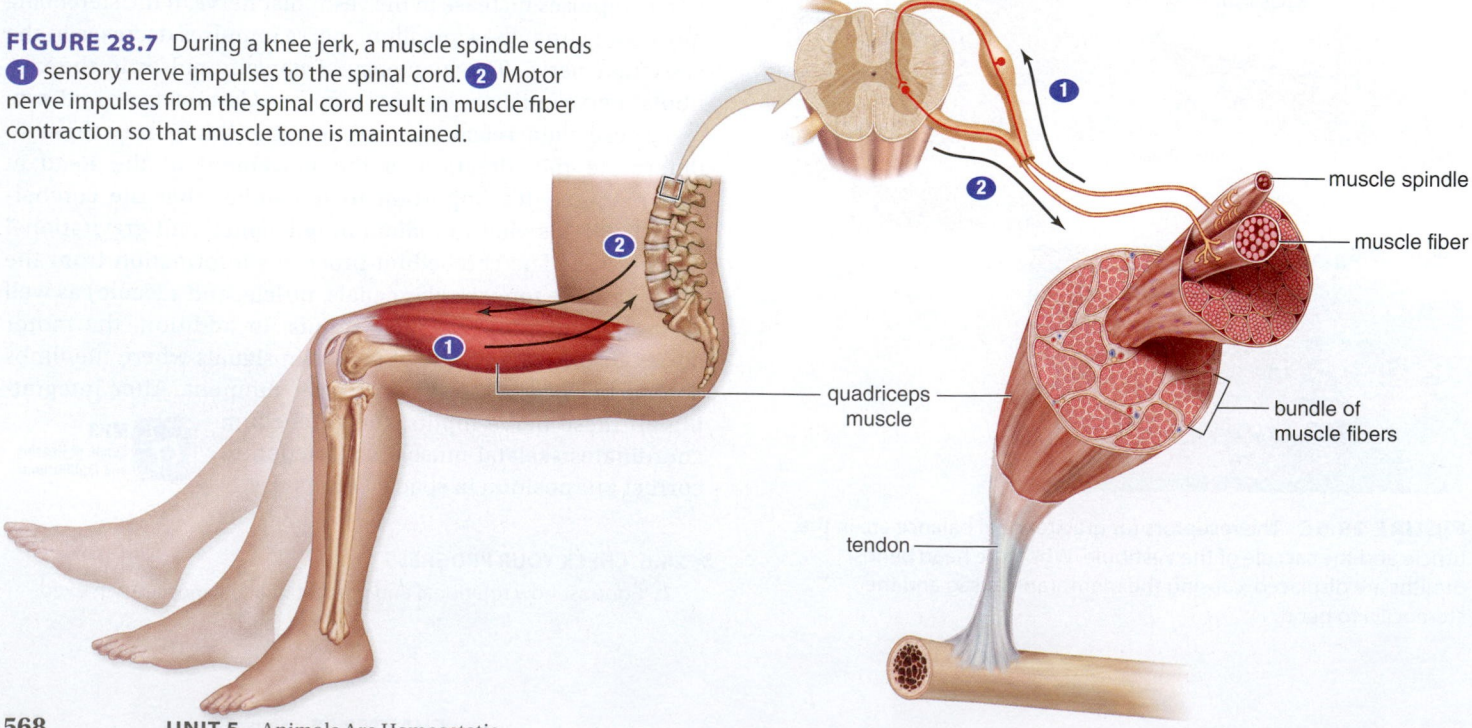

muscle spindle

muscle fiber

quadriceps muscle

bundle of muscle fibers

tendon

THE CHAPTER IN REVIEW

CONNECTING THE CONCEPTS

An animal's information exchange with the internal and external environments is dependent upon just a few types of sensory receptors. In this chapter, we have examined chemoreceptors, such as taste cells and olfactory cells; electromagnetic receptors, such as eyes; and mechanoreceptors, such as the hair cells for hearing and balance.

The senses are not equally developed in all animals. For instance, male moths have chemoreceptors on the filaments of their antennae to detect minute amounts of an airborne pheromone released by a female. This is certainly more efficient than searching for a mate by sight. Birds that live in forested areas signal that a territory is occupied by singing, because it is difficult to see a bird in a tree,

as most birders know. On the other hand, hawks have such a keen sense of sight that they are able to locate a small mouse in a field far below them. Insectivorous bats have an unusual adaptation for finding prey in the dark. They send out a series of sound pulses and listen for the echoes that come back. The time it takes for an echo to return indicates the location of an insect. A unique adaptation is found among the so-called electric fishes of Africa and Australia. They have electroreceptors that can detect disturbances in an electrical current they emit into the water. These disturbances indicate the location of obstacles and prey.

Through the evolutionary process, animals tend to rely on those stimuli and senses that are

adaptive to their particular environment and way of life. In all cases, sensory receptors generate nerve impulses that travel to the brain, where integration results in sensation.

ANALYZE AND EVALUATE

1. What senses are most keen in humans? Why would this be adaptive for a primate?
2. What senses do parents use to bond with their child? What senses are birds apt to use?
3. If you had to give up one of your senses, which one would you choose and why?

SUMMARIZE

Sensory Receptors

28.1 Sensory receptors differ but function similarly

- **Chemoreceptors, thermoreceptors,** and **mechanoreceptors** are familiar types of **sensory receptors. Pain receptors** respond to excessive temperature or pressure and various chemicals. **Electromagnetic receptors** are stimulated by changes in electromagnetic waves (e.g., photoreceptors, UV radiation).
- Sensory receptors initiate nerve impulses that are transmitted to the spinal cord and/or brain. **Sensation** occurs when nerve impulses reach the cerebral cortex. **Sensory adaptation** is a type of **integration** by which response to a stimulus gradually decreases.

Chemical Senses

28.2 Chemoreceptors are exemplified by taste buds and olfactory cells

- Chemoreceptors are responsible for taste and smell. Chemoreceptors are located in various places in various animals. **Pheromones** are chemical signals passed between members of the same species.
- Approximately 3,000 **taste buds** are present on the tongue. The five tastes are sweet, sour, salty, bitter, and umami.
- **Olfactory cells** are modified neurons located high in the nasal cavity that communicate with neurons in the olfactory bulb. Taste and smell depend on the combination of receptors stimulated.

Sense of Vision

28.3 The vertebrate eye is a camera-type eye

- The three layers of the eye are the **sclera,** the **choroid,** and the **retina.** The **conjunctiva** is a mucous membrane that covers the **sclera** including the **cornea** and lines the inside of the eyelids. The choroid forms the **iris,** which has an opening called the **pupil.** In the retina, the **fovea centralis** is

a concentration of cone cells. The **blind spot** is the area containing no rods or cones, where the **optic nerve** exits the retina.

- The **lens,** regulated by the **ciliary body,** lies directly behind the pupil. Light passes through the pupil and the lens focuses it on the retina. In **visual accommodation,** the lens rounds up to allow sight of near objects. At that time, the **ciliary muscle** is contracted. With aging, the lens loses its ability to accommodate.

28.4 The retina sends information to the visual cortex

- The first layer of the retina contains rods and cones: **Rod cells** contain **rhodopsin** and are the sensory receptors for dim light. **Cone cells** are the sensory receptors for bright light and color. In color blindness, one type of cone is defective or deficient in number. The other two layers of the retina are composed of the **bipolar cells** and the **ganglion cells,** which integrate stimuli before sending nerve impulses to the brain.

Sense of Hearing and Balance

28.5 The mammalian ear has three well-developed regions

- The **outer ear** contains the pinna and the **auditory canal.** The **middle ear** houses the **tympanic membrane** and three **ossicles;** the **auditory tube** permits equalization of air pressure; and the **inner ear** is the site of the **semicircular canals** and **vestibule** which are involved in balance control and the **cochlea** where the receptors for hearing are located.

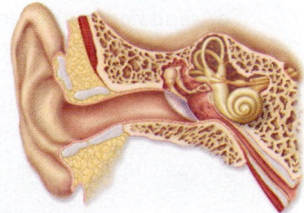

- For hearing to occur, the tympanic membrane and ossicles amplify sound waves that strike the **oval window** membrane and start pressure waves in the fluid within the cochlea. The pressure disappears at the **round window.** In response to the pressure waves, the stereocilia of hair cells on the basilar membrane of the **organ of Corti** bend because they are embedded in the tectorial membrane. Bending causes the hair cells to send impulses by way of the **cochlear nerve** to the cerebral cortex, where they are interpreted as sound.

28.6 Mammalian balance receptors are in the inner ear

- For **rotational balance,** receptors in each **ampulla** of the semicircular canals detect rotational and/or angular movement of the head. The receptors are hair cells whose stereocilia are embedded in a cupula, and when the cupula moves, the hair cells send nerve impulses to the brain.
- For **gravitational balance,** receptors in the **utricle** and **saccule** detect head movements in the vertical or horizontal plane. The receptors are hair cells whose stereocilia are bent when **otoliths** are displaced in the otolithic membrane.

Somatic Senses

28.7 Proprioceptors are located in skeletal muscles

- **Proprioceptors** are receptors involved in reflex actions that maintain muscle tone. They also help us maintain balance because the information they send to the spinal cord is relayed to the cerebellum.
- Touch receptors are a type of receptor in the skin. They send messages to the spinal cord about mechanical stimuli that result in our sense of touch, its location, and intensity. The spinal cord sends these messages on to the somatosensory area of the cerebral cortex.

TEST YOURSELF

Sensory Receptors

1. Chemoreceptors are involved in
   a. hearing.   d. vision.
   b. taste.   e. Both b and c are correct.
   c. smell.
2. Being conscious of stimuli from the internal and external environment is called
   a. responsiveness.   c. sensation.
   b. interpretation.   d. accommodation.
3. **THINKING CONCEPTUALLY** Explain the expression "The sensory receptors are the window of the brain."

Chemical Senses

4. The chemical senses
   a. respond to specific molecules in air and water.
   b. include the taste and smell receptors.
   c. depend on plasma membrane receptors.
   d. send nerve impulses to the brain.
   e. All of these are correct.

Sense of Vision

5. Label this diagram:

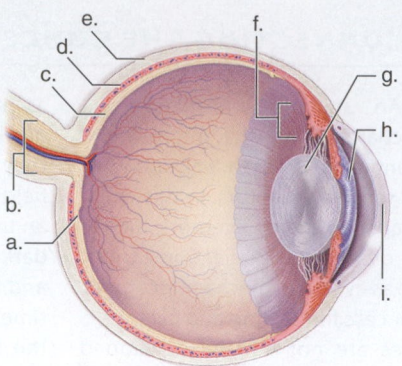

6. Is the eye in question 5 viewing a near or a distant object? How do you know?
7. The thin, darkly pigmented layer that underlies most of the sclera is the
   a. conjunctiva.   c. retina.
   b. cornea.   d. choroid.
8. Which of these sequences describes the correct path for light rays entering the human eye?
   a. sclera, retina, choroid, lens, cornea
   b. fovea centralis, pupil, aqueous humor, lens
   c. cornea, pupil, lens, vitreous humor, retina
   d. cornea, fovea centralis, lens, choroid, rods
   e. optic nerve, sclera, choroid, retina, humors
9. A blind spot occurs where the
   a. iris meets the pupil.
   b. retina meets the lens.
   c. optic nerve meets the retina.
   d. cornea meets the retina.
10. Which part of the eye is incorrectly matched with its function?
    a. pupil—admits light
    b. choroid—absorbs stray light rays
    c. fovea centralis—makes night vision possible
    d. optic nerve—transmits impulses to brain
    e. iris—regulates light entrance
11. During accommodation,
    a. the suspensory ligaments must be pulled tight.
    b. the lens needs to become more rounded.
    c. the ciliary muscle will be relaxed.
    d. All of these are correct.
12. Retinal is
    a. a derivative of vitamin A.
    b. sensitive to light energy.
    c. a part of rhodopsin.
    d. found in both rods and cones.
    e. All of these are correct.
13. A color-blind person has an abnormal type of
    a. rod.   d. cone.
    b. cochlea.   e. None of these are correct.
    c. cornea.

14. Which abnormality of the eye is incorrectly matched with its cause?
    a. astigmatism—either the lens or the cornea is not even
    b. farsightedness—eyeball is shorter than usual
    c. nearsightedness—image focuses behind the retina
    d. color blindness—genetic disorder in which certain types of cones may be missing
15. **THINKING CONCEPTUALLY** What specific physiological defect can be corrected by both cataract surgery and LASIK surgery? Explain.

Sense of Hearing and Balance

16. Label this diagram:

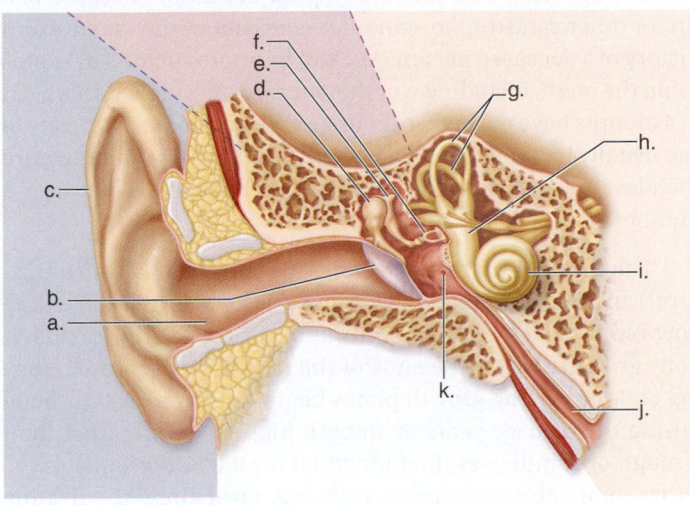

17. The middle ear communicates with the inner ear at the
    a. oval window.
    b. tympanic membrane.
    c. round window.
    d. Both a and c are correct.
18. The ossicle that articulates with the tympanic membrane is the
    a. malleus.   c. incus.
    b. stapes.   d. All of these are correct.
19. Which one of these wouldn't you mention if you were tracing the path of sound vibrations?
    a. auditory canal   d. cochlea
    b. tympanic membrane   e. ossicles
    c. semicircular canals
20. Which one of these correctly describes the location of the organ of Corti?
    a. between the tympanic membrane and the oval window in the inner ear
    b. in the utricle and saccule within the vestibule
    c. between the tectorial membrane and the basilar membrane in the cochlear canal
    d. between the outer and inner ear within the semicircular canals

21. Our perception of pitch is dependent upon the region of the _____ vibrated and the regions of the _____ stimulated.
    a. cochlea, spinal cord
    b. basilar membrane, spinal cord
    c. basilar membrane, auditory cortex
    d. cochlea, cerebellum
22. Loud noises generally lead to hearing loss due to damage to the
    a. outer ear.
    b. middle ear.
    c. inner ear.
23. Which part of the ear is incorrectly matched to its location?
    a. semicircular canals—inner ear
    b. utricle and saccule—outer ear
    c. auditory canal—outer ear
    d. cochlea—inner ear
    e. ossicles—middle ear
24. Which of these structures would assist you in knowing that you were upside down, even if you were in total darkness?
    a. utricle and saccule   c. semicircular canals
    b. cochlea   d. tectorial membrane
25. The lateral line system in fishes allows them to
    a. detect sound.
    b. locate other fish.
    c. maintain gravitational equilibrium.
    d. maintain rotational equilibrium.

Somatic Senses

26. **THINKING CONCEPTUALLY** Use the example of a blind person sitting in a chair to show that proprioception and touch are critical to knowing where the body is located.

GET INVOLVED

1. Suggest a hypothesis that would explain why some people have perfect pitch. How would you test your hypothesis?
2. How does LASIK surgery support the hypothesis that the cornea, not the lens, provides most of the focusing when we see clearly?

MEDIA STUDY TOOLS

mhhe.com/maderconcepts3

Enhance your study of this chapter with interactive study tools, practice tests, and engaging animations. Also, ask your instructor about the resources available through ConnectPlus, which includes LearnSmart, a personalized adaptive learning program, and a media-rich eBook.

29

Locomotion and Support Systems

BEFORE YOU BEGIN

Take a few minutes to recall

The microscopic structure of bone and cartilage, tendons, and ligaments (section 26.3)

The microscopic structure of the three types of muscular tissue (section 26.4)

How acetylcholine functions as a neurotransmitter (section 27.3)

The ability of the frontal lobe and spinal cord to generate motor commands (sections 27.4 and 27.8)

Skeletal Remains Reveal All

Dr. Sandra Bullock, a forensics expert, and her assistant Tom were standing in the tall grass of the empty lot at the corners of Marion and Washington Streets. The bones they were examining had obviously been bleached by the sun and scattered by passing dogs. "Let's get as many bones as we can find into the lab and try to identify them as one of the missing persons reported within the past year," said Sandra.

It is better for forensics if many bones, in good condition, are found. But even bones that are in poor condition, because of a fire or other catastrophe, can offer clues about the identity and history of a deceased person. Age can be approximated by examining the teeth, including whether any are missing. Infants aged 0–4 months have no teeth, of course; children about 6–10 years of age usually have missing "baby teeth"; and young adults acquire their last "wisdom teeth" around age 20. Older adults may have a number of missing or broken teeth.

The condition of the long bones (do they have a healed fracture?) and the joints (are they arthritic?) can also assist in telling how old a person was at the time of death. A thin, cartilaginous growth plate at the ends of the long bones is present during childhood. The growth plates begin to be replaced by bone during the teenage years. A smooth hipbone joint, rather than a rough one, indicates the individual is an adult of some years. Other joints also deteriorate with age. Over time, the hyaline cartilage capping the long bones becomes worn, yellowed, and brittle, and the amount of cartilage lessens. Also, as we age, the pads between the vertebrae are more apt to show damage.

If the skeletal remains include the individual's pelvic bones, these provide the best method for determining an adult's gender. To accommodate a fetus during pregnancy, the pelvis of a female is shallower and wider than that of a male; a wider outlet allows a baby's head to pass when birth occurs. The long bones of the limbs give information about gender as well. In males, the

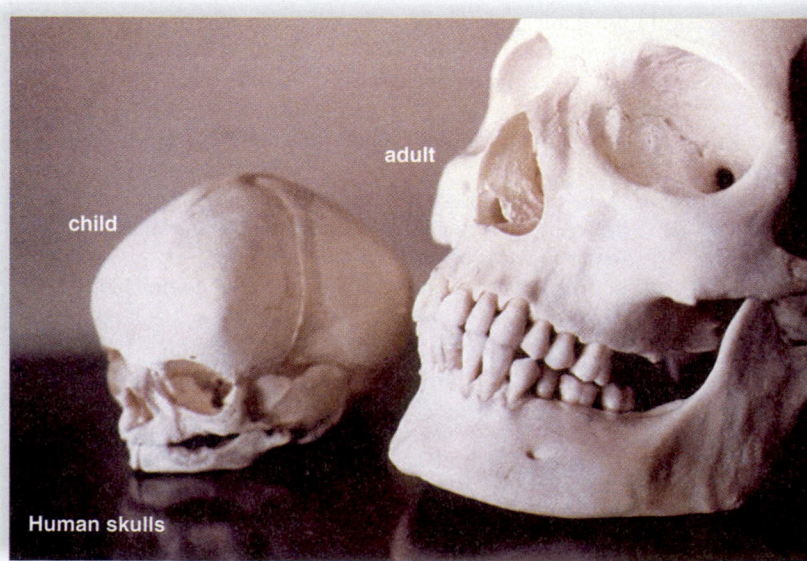

Human skulls

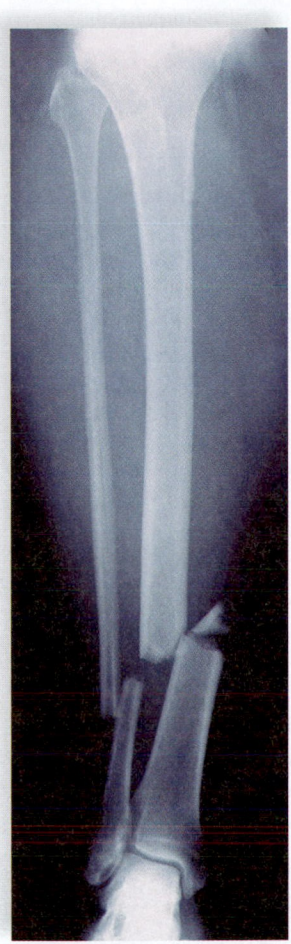

Fracture of both
the tibia and fibula bones

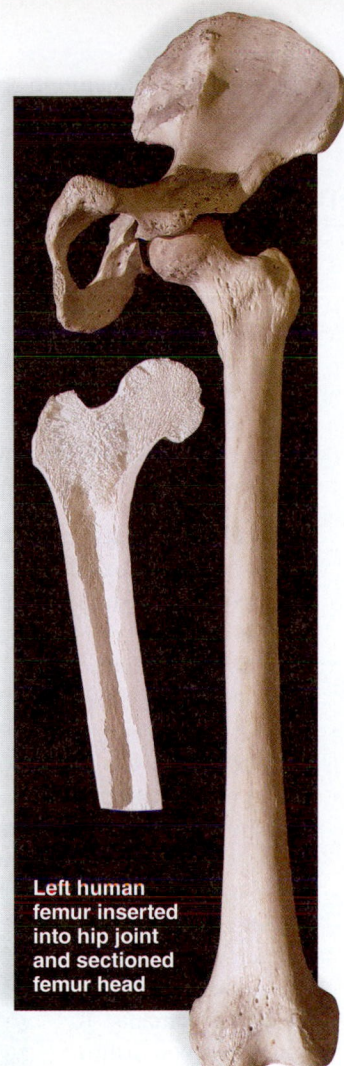

Left human
femur inserted
into hip joint
and sectioned
femur head

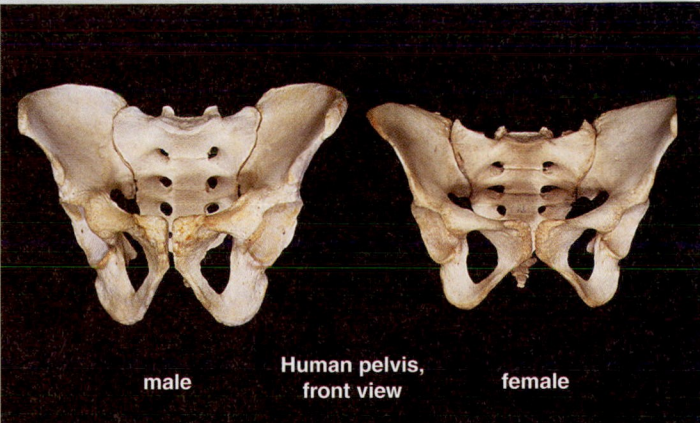

male | Human pelvis, front view | female

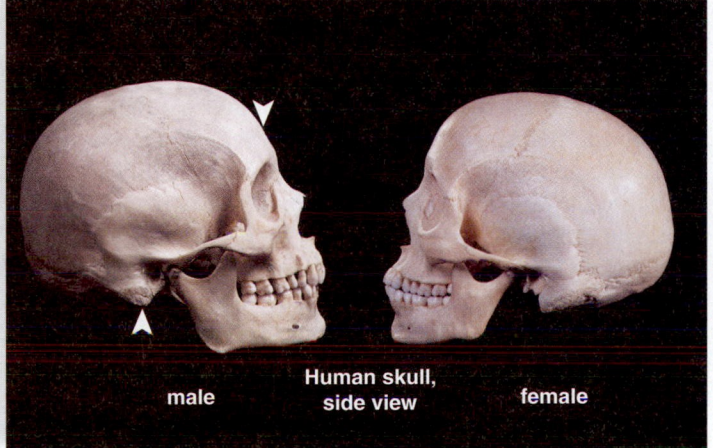

male | Human skull, side view | female

long bones are thicker and more dense, and the points of attachment for the muscles are bigger and more noticeable. The skull of a male tends to have a square chin, and the eyebrow ridges above the eye sockets are more prominent. Also, males have a larger mastoid process (the lump behind the ear).

Determining the ethnic origin of skeletal remains can be difficult because many people today have a mixed racial heritage. But again, the bones, especially the skull, offer clues. Individuals of African descent have a greater distance between the eyes, eye sockets that are roughly rectangular, and a large and prominent jaw. In Native Americans, the eye sockets are round, the cheekbones are prominent, and the palate is rounded. In Caucasians, the palate is U-shaped, and a suture line is apt to be visible. After gathering all such data, Dr. Bullock hopes to use missing persons reports to assign a specific name to the bones, which have revealed all.

In this chapter, we will survey the types of skeletons in the animal kingdom before concentrating on the bones and muscles of humans and how they move. In addition, the bones and muscles have many functions that contribute to homeostasis.

Forensic expert clears soil in mass grave in eastern Bosnia.

Diversity of Skeletons

Three types of skeletons—hydrostatic skeletons, exoskeletons, and endoskeletons—occur among animals. The endoskeleton of mammals provides a frame for the body and protects and assists several other systems of the body.

29.1 Animal skeletons can be hydrostatic, external, or internal

LEARNING OUTCOME

When you complete this section, you should be able to

1. Compare and contrast the three types of animal skeletons.

Skeletons serve as support systems for animals, providing rigidity, protection, and surfaces for muscle attachment.

Hydrostatic Skeleton In animals that lack a hard skeleton, a fluid-filled gastrovascular cavity or a fluid-filled coelom can act as a hydrostatic skeleton. A **hydrostatic skeleton** offers support and resistance to the contraction of muscles so that an animal can move. As an analogy, consider that a garden hose stiffens when filled with water, and that a water-filled balloon changes shape when squeezed at one end. Similarly, an animal with a hydrostatic skeleton can change shape and perform a variety of movements.

Hydras and flatworms (planarians) use their fluid-filled gastrovascular cavity as a hydrostatic skeleton. When epidermal cells in a hydra contract, the body or tentacles shorten rapidly. Planarians usually glide over a substrate with the help of muscular contractions that control the body wall and many cilia. Roundworms have a fluid-filled pseudocoelom and move in a whiplike manner when their longitudinal muscles contract. Earthworms are segmented and have septa that divide the coelom into compartments (Fig. 29.1A). Each segment has its own set of longitudinal and circular muscles and its own nerve supply, so each segment or group of segments may function independently. When circular muscles contract, the segments become thinner and elongated, just as a balloon would if you squeezed it. When longitudinal muscles contract, the segments become thicker and shorter, just as a balloon would if you pressed on it from both ends. By alternating circular muscle contraction and longitudinal muscle contraction and by using its setae to hold its position during contractions, the earthworm moves forward.

Even animals that have an exoskeleton or an endoskeleton move selected body parts by means of muscular hydrostats, meaning that fluid contained within certain muscles assists movement of that part. Muscular hydrostats are used by clams to extend their muscular foot and by sea stars to extend their tube feet. Spiders depend on them to move their legs, and moths depend on them to extend their long tubular feeding apparatus. In vertebrates, movement of an elephant's trunk involves a muscular hydrostat that allows the trunk to reach as high as about 7 m for a morsel of food, or to pull down a tree.

Exoskeleton Molluscs, arthropods, and vertebrates have rigid skeletons. The **exoskeleton** (external skeleton) of molluscs and arthropods protects and supports and provides a location for muscle attachment. Molluscs, such as snails and clams, use a thick, nonmobile calcium carbonate shell primarily for protection against the environment and predators. A mollusc's shell can grow as the animal grows.

The exoskeleton of arthropods, such as insects and crustaceans, is composed of chitin, a strong, flexible nitrogenous polysaccharide. This exoskeleton protects against wear and tear, predators, and drying out—an important feature for arthropods that live on land. The exoskeleton of arthropods is particularly suitable for terrestrial life in another way. The jointed and movable appendages allow flexible movements. To grow, however, arthropods must molt to rid themselves of an exoskeleton that has become too small. While molting, arthropods are vulnerable to predators.

old exoskeleton

Endoskeleton Both echinoderms and vertebrates have an **endoskeleton** (internal skeleton). The skeleton of a starfish (Fig. 29.1B) consists of plates of calcium carbonate embedded in the living tissue of the body wall. In contrast, the vertebrate endoskeleton is living tissue. Sharks and rays have skeletons composed only of cartilage. Other

FIGURE 29.1A The well-developed circular and longitudinal muscles of an earthworm push against a segmented, fluid-filled coelom.

circular muscles | longitudinal muscles | septa | fluid | setae

circular muscles contracted | longitudinal muscles contracted

circular muscles contract, and anterior end moves forward

longitudinal muscles contract, and segments catch up

circular muscles contract, and anterior end moves forward

vertebrates, such as bony fishes, amphibians, reptiles, birds, and mammals, have endoskeletons composed of bone and cartilage. An endoskeleton grows with the animal, and molting is not required. It supports the weight of a large animal without limiting the space for internal organs. An endoskeleton also offers protection to vital internal organs, but is protected by the soft tissues around it. Injuries to soft tissue are easier to repair than injuries to a hard skeleton. The vertebrate endoskeleton is also jointed, allowing for complex movements such as swimming, jumping, flying, and running.

FIGURE 29.1B
A starfish has an endoskeleton.

▶ **29.1 CHECK YOUR PROGRESS**

1. Identify a disadvantage for each type of skeleton.
2. Explain why an earthworm loses its cylindrical shape when it dies.

29.2 A skeleton serves many functions

LEARNING OUTCOME

When you complete this section, you should be able to

1. Describe how the skeletons of a crayfish and a human share five primary functions.

The jointed exoskeleton of a crayfish and the jointed endoskeleton of a human have certain basic functions in common. These are listed here:

The skeleton provides a frame for the body. Compare the skeleton of a crayfish (Fig. 29.2) with that of a human (see Fig. 29.3A) to appreciate that the shape of an animal is largely dependent on the shape of its skeleton. A skeleton also supports the body—the legs of a crayfish and the legs of a human support the entire body when the animal is standing.

The skeleton protects the internal organs. The carapace in a crayfish and the rib cage in humans protect the heart and other internal organs; the rostrum in a crayfish and the skull of a human protect the brain and sense organs such as eyes. The jointed abdominal skeletal plates of a crayfish protect the nerve cord, while this same function is supplied by the vertebrae in humans.

The skeleton assists digestion. Both crayfish and humans have jaws (i.e., mandibles and maxillae) and mouthparts (called the teeth in humans and the mouth appendages in crayfish) for chewing food. Chewing breaks food into pieces small enough to be swallowed and chemically digested. Without digestion, nutrients would not enter the body to serve as building blocks for repair and a source of energy for ATP production.

The skeleton stores calcium. The skeleton of a crayfish contains chitin but also calcium; the bones of humans also store calcium. Calcium ions play a major role in muscle contraction and nerve conduction. In humans, we know that the storage of Ca^{2+} in the bones is under hormonal control, and there is an interplay between calcium in the bones and calcium in the blood. When the blood level starts to fall, it is removed from bone so that the blood level is always near normal.

The skeleton is necessary for locomotion. Locomotion is efficient in crayfish and humans because they have a jointed skeleton for the attachment of muscles that move body parts. Freedom

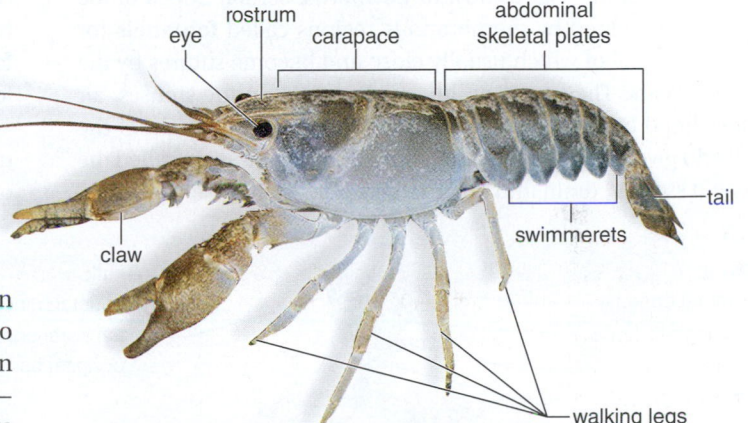

FIGURE 29.2 Exoskeleton of a crayfish.

of movement allows them to seek out and grasp prey (crayfish have claws and humans have flexible tool-using hands). A crayfish also has swimmerets attached to its abdomen for swimming, as is consistent with its life in the water.

The skeleton, in humans, assists all phases of respiration. Crayfish breathe by gills, but humans must ventilate their lungs. Prior to inhalation, the rib cage lifts up and out (see Fig. 33.5A), and the diaphragm moves down, expanding the chest. Now air automatically flows into the lungs, enabling oxygen to enter the blood, where it is transported to the tissues. As we shall see in section 29.5, some of the bones contain red bone marrow, which produces the red blood cells that transport oxygen. Although crayfish have no blood cells, they do have a respiratory pigment for transporting oxygen.

The skeleton, in humans, assists immunity. Red bone marrow produces not only the red blood cells but also the white blood cells. The white cells, which congregate in the lymphatic organs such as the lymph nodes shown in Figure 31.1A, are involved in defending the body against pathogens and cancerous cells.

▶ **29.2 CHECK YOUR PROGRESS**

1. List and describe ways in which the skeletons of crayfish and humans share the same functions.

Mammalian Skeleton

The bones of the mammalian skeleton occur in the midline (axial skeleton) and in the girdles and limbs (appendicular skeleton). Joints between the bones provide the flexibility needed in a land environment.

29.3 The bones of the axial skeleton lie in the midline of the body

LEARNING OUTCOME

When you complete this section, you should be able to

1. Describe the major regions and the bones within each region of the axial skeleton.

The **axial skeleton** consists of the bones in the midline of the body, and the appendicular skeleton consists of the limb bones and their girdles (Fig. 29.3A).

The Skull The cranium and the facial bones form the **skull,** which protects the brain (Fig. 29.3B). In newborns, certain bones of the cranium are joined by membranous regions called **fontanels** (or "soft spots"), all of which usually close and become sutures by the age of 2 years. The bones of the cranium contain the sinuses, air spaces lined by mucous membrane that reduce the weight of the skull and give a resonant sound to the voice. Two sinuses, called the mastoid sinuses, drain into the middle ear.

The major bones of the cranium have the same names as the lobes of the brain (see Fig. 27.4B). On the top of the cranium, the frontal bone forms the forehead, and the parietal bones make up the sides of the skull. Below the much larger parietal bones, each temporal bone has an opening that leads to the middle ear. In the rear of the skull, the occipital bone curves to form the base of the skull. At the base of the skull, the spinal cord passes upward through a large opening called the **foramen magnum** and becomes the brain stem.

Certain cranial bones contribute to forming the face. The sphenoid bones account for the flattened areas on each side of the forehead, which we call the temples. The frontal bone not only forms the forehead, but also has supraorbital ridges where the eyebrows are located. Glasses sit where the frontal bone joins the nasal bones.

The most prominent of the facial bones are the mandible, the maxillae, the zygomatic bones, and the nasal bones. The mandible, or lower jaw, is the only freely movable portion of the skull, and its

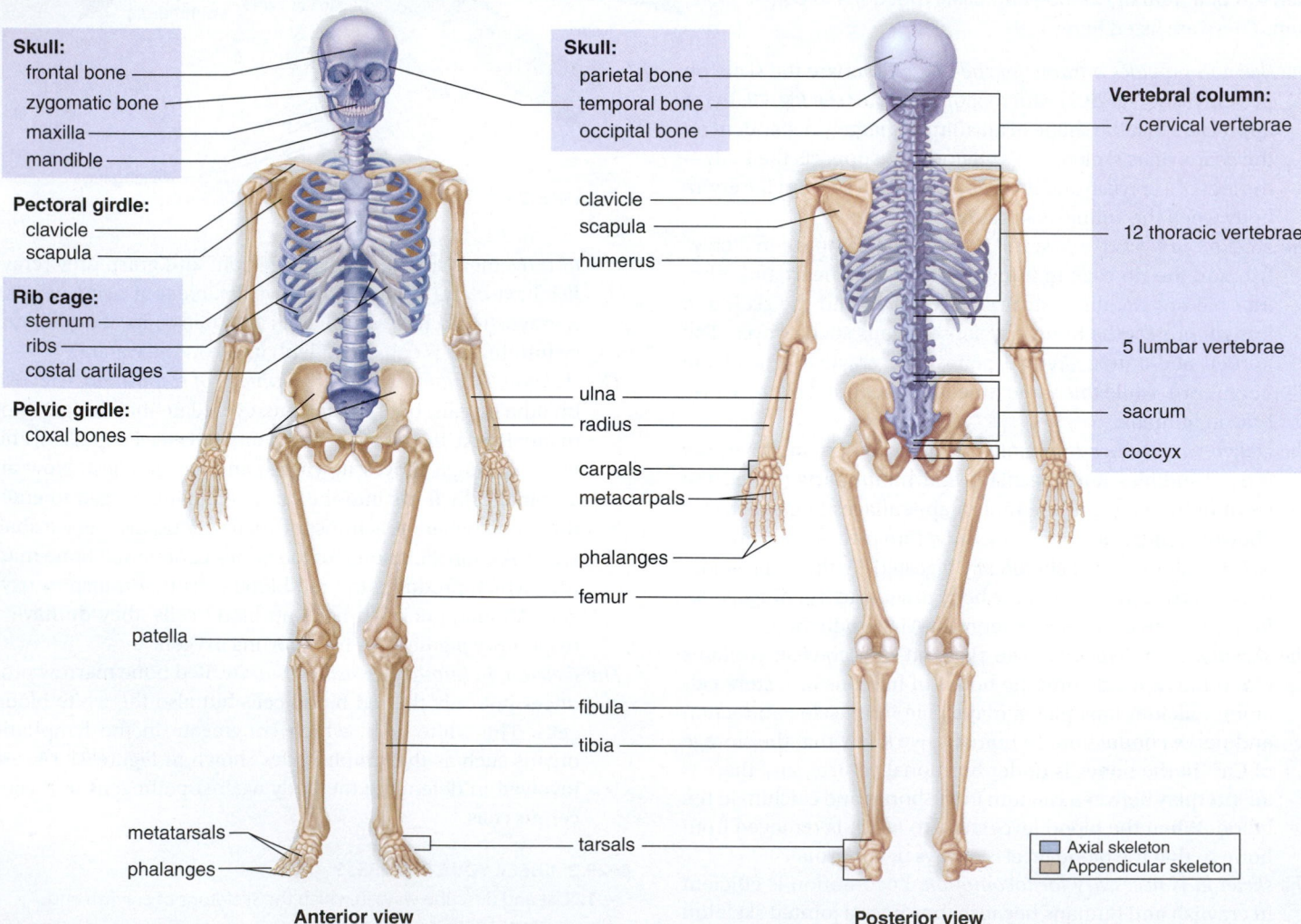

Skull:
- frontal bone
- zygomatic bone
- maxilla
- mandible

Pectoral girdle:
- clavicle
- scapula

Rib cage:
- sternum
- ribs
- costal cartilages

Pelvic girdle:
- coxal bones

patella

metatarsals
phalanges

Skull:
- parietal bone
- temporal bone
- occipital bone

clavicle
scapula
humerus

ulna
radius

carpals
metacarpals

phalanges

femur

fibula
tibia

tarsals

Vertebral column:
- 7 cervical vertebrae
- 12 thoracic vertebrae
- 5 lumbar vertebrae
- sacrum
- coccyx

Axial skeleton
Appendicular skeleton

Anterior view

Posterior view

FIGURE 29.3A The human skeleton.

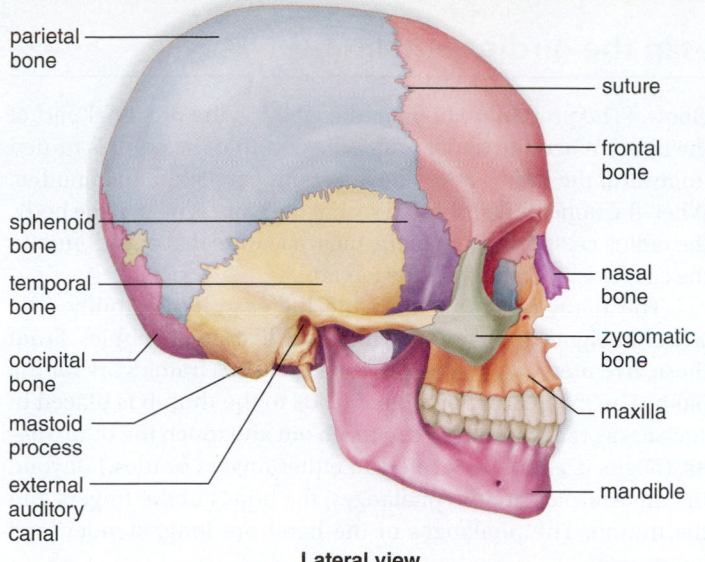

parietal bone

suture

frontal bone

sphenoid bone

temporal bone

nasal bone

occipital bone

zygomatic bone

mastoid process

maxilla

external auditory canal

mandible

Lateral view

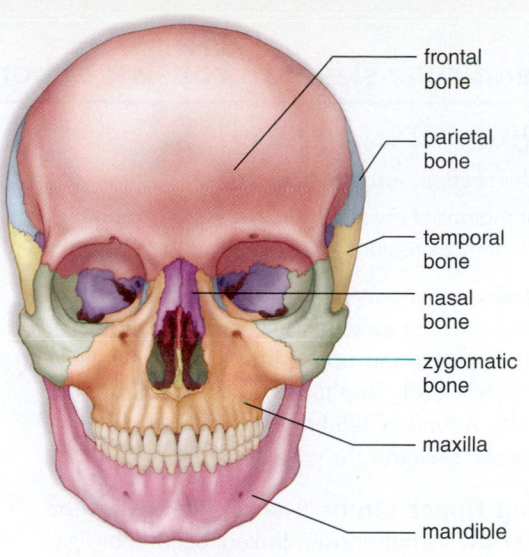

frontal bone

FIGURE 29.3B
Bones of the skull.

parietal bone

temporal bone

nasal bone

zygomatic bone

maxilla

mandible

Frontal view

action permits us to chew our food. It also forms the "chin." Tooth sockets are located on the mandible and on the maxillae, which form the upper jaw and a portion of the hard palate. The zygomatic bones are the cheekbone prominences, and the nasal bones form the bridge of the nose. Other bones make up the nasal septum, which divides the nose cavity into two regions. The nose is a mixture of bones, cartilage, and connective tissues.

MP3 The Skull

The Vertebral Column The head and trunk are supported by the **vertebral column,** which also protects the spinal cord and the roots of the spinal nerves. It is a longitudinal axis that serves either directly or indirectly as an anchor for all the other bones of the skeleton.

Twenty-four vertebrae make up the vertebral column (see Fig. 29.3A). The seven cervical vertebrae are located in the neck, and the twelve thoracic vertebrae are in the thorax. The five lumbar vertebrae are in the small of the back. The five sacral vertebrae are fused to form a single sacrum. The coccyx, or tailbone, is composed of several fused vertebrae. Normally, the vertebral column has four curvatures that absorb shock and also provide more resilience and strength for an upright posture than would a straight column. *Scoliosis* is an abnormal lateral (sideways) curvature of the spine. Two other well-known curvatures are hunchback (kyphosis) and swayback (lordosis), seen frequently in pregnant women.

Intervertebral disks, composed of fibrocartilage between the vertebrae, act as padding. They prevent the vertebrae from grinding against one another and absorb shock caused by movements such as running, jumping, and even walking. The presence of the disks allows the vertebrae to move as we bend forward, backward, and from side to side. Unfortunately, these disks become weakened with age and can herniate and rupture. Pain results if a disk presses against the spinal cord and/or spinal nerves. The body may heal itself, or the disk can be removed surgically. If removed, the vertebrae can be fused together, but this limits the flexibility of the body.

The Rib Cage The thoracic vertebrae are a part of the rib cage. The **rib cage** also contains the ribs, the costal cartilages, and the sternum, or breastbone (Fig. 29.3C).

There are twelve pairs of ribs. The upper seven pairs are "true ribs" because they attach directly to the sternum. The lower five pairs do not connect directly to the sternum and are called the "false ribs." Three pairs of false ribs attach by means of a common cartilage, and two pairs are "floating ribs" because they do not attach to the sternum at all (Fig. 29.3C).

The rib cage demonstrates how the skeleton is protective but also flexible. The rib cage protects the heart and lungs; yet it swings outward and upward upon inspiration and then downward and inward upon expiration.

MP3 The Vertebral Column and Thoracic Cage

▶ **29.3 CHECK YOUR PROGRESS**
   1. Identify the major regions of the axial skeleton and the internal organ(s) protected by each region.
   2. Identify which portion of the axial skeleton is most apt to suffer injury from daily wear and tear and explain.

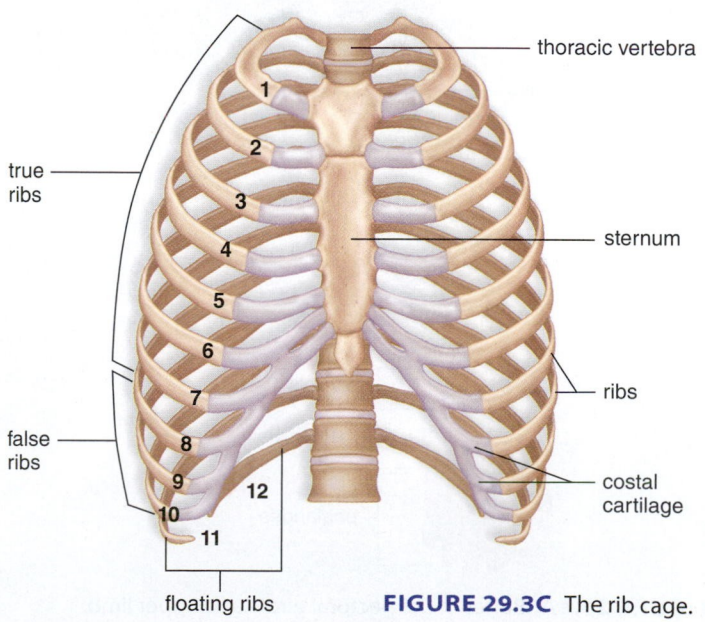

thoracic vertebra

true ribs

sternum

false ribs

ribs

costal cartilage

floating ribs

FIGURE 29.3C The rib cage.

LEARNING OUTCOME

When you complete this section, you should be able to

1. Identify the major regions of the appendicular skeleton and the bones within each of these regions.

The **appendicular skeleton** consists of the bones within the pectoral and pelvic girdles and the attached limbs (see Fig. 29.3A). The pectoral (shoulder) girdle and upper limbs are specialized for flexibility, but the pelvic girdle (hipbones) and lower limbs are specialized for strength. A total of 126 bones make up the appendicular skeleton.

MP3
The Appendicular Skeleton

Pectoral Girdle and Upper Limbs The components of the **pectoral girdle** (Fig. 29.4A) are only loosely linked together by ligaments. Each clavicle (collarbone) connects with the sternum in front and the scapula (shoulder blade) behind, but the scapula is largely held in place only by muscles. This allows it to freely follow the movements of the arm. The single long bone in the upper arm, the humerus, has a smoothly rounded head that fits into a socket of the scapula. The socket, however, is very shallow and much smaller than the head. Although this means that the arm can move in almost any direction, there is little stability. Therefore, this is the joint that is most apt to dislocate. The opposite end of the humerus meets the two bones of the forearm, the ulna and the radius, at the elbow. (The prominent bone in the elbow is the proximal part of the ulna.) When the upper limb is held so that the palm is turned frontward, the radius and ulna are about parallel to one another. When the upper limb is turned so that the palm is next to the body, the radius crosses in front of the ulna, a feature that contributes to the easy twisting motion of the forearm.

The many bones of the hand increase its flexibility. The wrist has eight carpal bones, which look like small pebbles. From these, five metacarpal bones fan out to form a framework for the palm. The metacarpal bone that leads to the thumb is placed in such a way that the thumb can reach out and touch the other digits. (Digits is a term that refers to either fingers or toes.) Beyond the metacarpals are the phalanges, the bones of the fingers and the thumb. The phalanges of the hand are long, slender, and lightweight.

Pelvic Girdle and Lower Limbs Two heavy, large coxal bones (hipbones) are joined at the pubic symphysis to form the **pelvic girdle** (Fig. 29.4B). The coxal bones are anchored to the sacrum, and together these bones form a hollow cavity called the pelvic cavity. The wider pelvic cavity in females accommodates childbearing. The weight of the body is transmitted through the pelvis to the lower limbs and then onto the ground. The largest bone in the body is the femur, or thighbone.

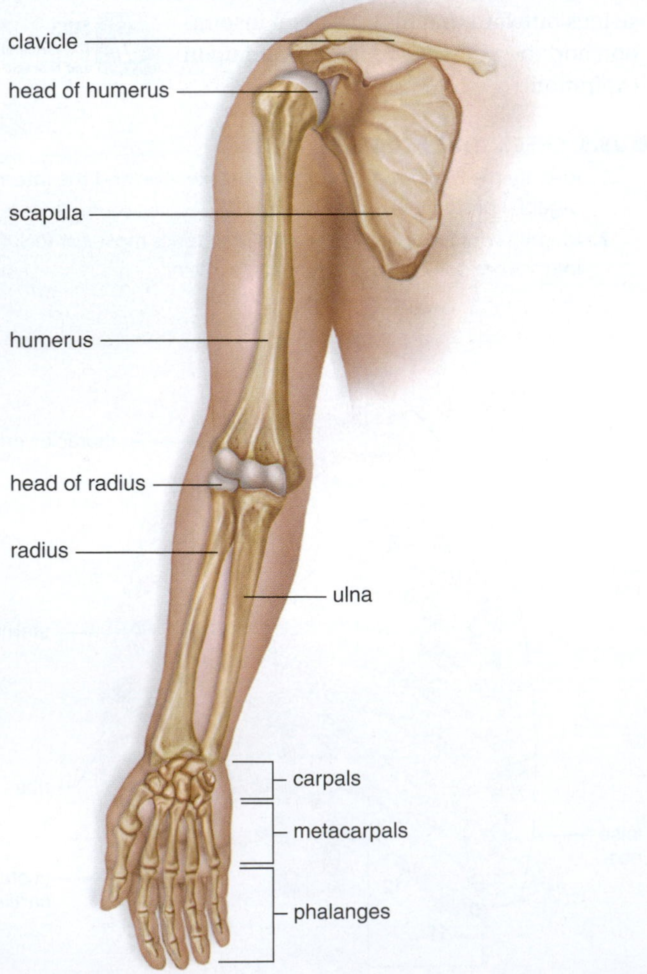

clavicle

head of humerus

scapula

humerus

head of radius

radius

ulna

carpals

metacarpals

phalanges

FIGURE 29.4A Bones of the pectoral girdle and upper limb.

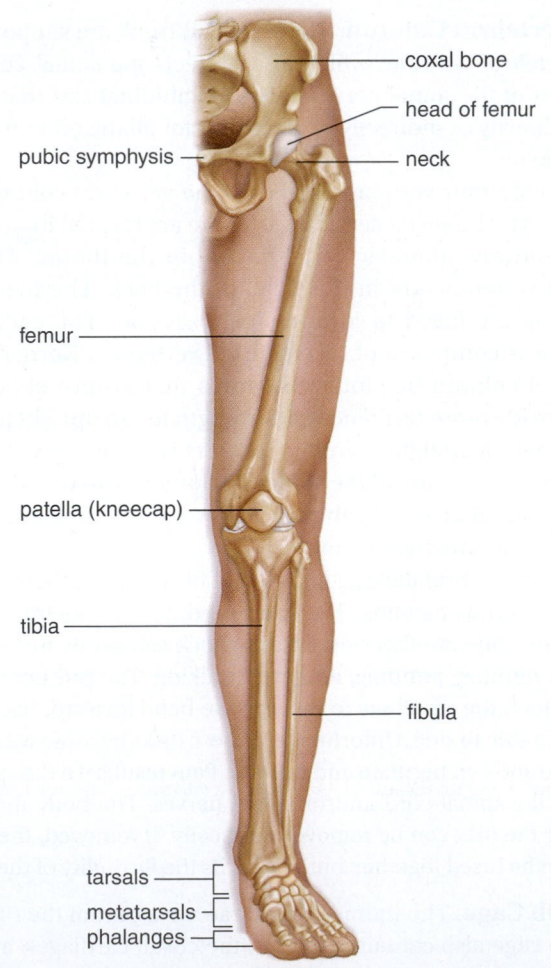

coxal bone

head of femur

pubic symphysis

neck

femur

patella (kneecap)

tibia

fibula

tarsals

metatarsals

phalanges

FIGURE 29.4B Bones of the pelvic girdle and lower limb.

Distal to the thigh, the larger of the two bones, the tibia, has a ridge we call the shin. Both of the bones of the leg have a prominence that contributes to the ankle—the tibia on the inside of the ankle and the fibula on the outside of the ankle. Although there are seven tarsal bones in the ankle and heel, only one tarsal bone receives the body's weight and passes it on to the heel and the ball of the foot. If you wear high-heeled shoes, the weight is thrown toward the front of your foot.

The metatarsal bones participate in forming the arches of the foot. There is a longitudinal arch from the heel to the toes and a transverse arch across the foot. These provide a stable, springy base for the body. If the tissues that bind the metatarsals together become weakened, "flat feet" are apt to result. The bones of the toes are called phalanges, just as are those of the fingers, but in the foot the phalanges are stout and extremely sturdy.

▶ **29.4 CHECK YOUR PROGRESS**
1. Identify the analogous bones (present in the same regions) of the arms and legs.
2. Give a reason why the hands and feet are shaped differently.

HOW LIFE CHANGES   *Application*

29A What Our Limbs Tell Us About Our Past

Although the human line of descent separated from that of apes several million years ago, we still retain evidence that a common ancestor had a brachiating mode of locomotion. A brachiator alternately uses its arms to reach up and its hands to swing from limb to limb in a tree. Although our arms are shorter than those of chimpanzees, which still depend on locomotion through trees, our arms are relatively long compared to those of other mammals. If we stand upright with arms to our sides, our fingers reach below our hips. By contrast, the forelimbs of a nonbrachiating mammal such as a dog or cat, if pressed back, do not reach as far. Also, in a brachiator, the second to the fifth fingers form a hook with which to grasp overhead branches. Without thinking it anything special, we use this same comfortable design to grip the handle of a suitcase carried at our side. The arm position changes, but the grip used is the same. Finally, in a running animal, such as a cat, the clavicle is reduced. But in ourselves and brachiators, the clavicle is a prominent structural element of the shoulder, serving to transfer the weight of the body to the arm (Fig. 29A).

In contrast, the design of our hindlimbs and pelvic girdle give evidence of compromises to our upright bipedal posture. Certainly in females but also in males, the pelvis is more bowl shaped than that of a chimpanzee. But widening the hips places the heads of the femurs far apart and outside the center line of the body weight. However, notice that the femurs are angled (red arrows) in humans to allow the limbs to swing directly beneath the body.

Our bipedal posture and pendulum-like leg motions also result in changes in foot design. Apes retain a grasping hindfoot with a projecting large toe. In humans, the toe is aligned with the other digits of the foot so that as the limbs swing beneath the body, they can be placed close to the line of travel without catching the projecting toe on the opposite leg. The human foot forms an arch, a way of broadening the base of support upon which the upper body stands. The arch adds bounce and helps lift the heel as we walk.

Adapted from Kenneth V. Kardong, *Vertebrates: Comparative Anatomy, Function, Evolution* (New York: McGraw-Hill Higher Education, 2009), p.347.

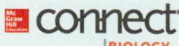 Explore the concepts through a variety of multimedia assets and question types.
www.mcgrawhillconnect.com

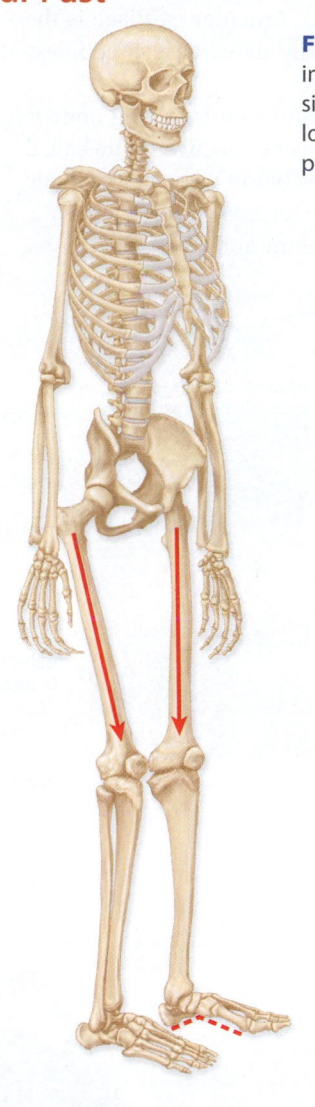
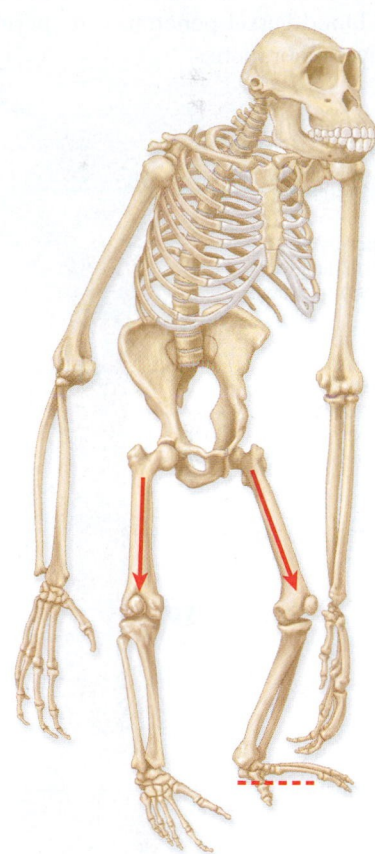

FIGURE 29A The upper limbs are shorter in humans than in apes, but they function similarly. In contrast, the pelvic girdle and lower limbs in humans facilitate an upright posture.

CONSIDER THESE QUESTIONS
1. Speculate on a possible evolutionary connection between a larger human brain and upright posture.
2. The knees of humans can bear more weight than can those of chimpanzees. What human conditions suggest that the knee is not fully adapted to bearing weight?

LEARNING OUTCOMES

When you complete this section, you should be able to

1. Describe the microscopic anatomy of the living tissues that make up a skeletal bone.

2. Describe the three types of joints that occur between bones.

As shown in Figure 29.5A, ① when a long bone such as the humerus is split open, the longitudinal section shows that it is not solid but has a cavity, called the medullary cavity, bounded at the sides by compact bone and at the ends by spongy bone. The cavity of a long bone usually contains yellow bone marrow, which stores fat. Beyond the spongy bone is a thin shell of compact bone and finally a layer of ② hyaline cartilage, called **articular cartilage** when it occurs at articulations (joints). Articular cartilage is the "teflon coating" for the bones; it normally allows easy, frictionless movement between the bones of a joint.

Except for the articular cartilage on its ends, a long bone is completely covered by a layer of fibrous connective tissue called the periosteum. This covering contains blood vessels, lymphatic vessels, and nerves. Note in Figure 29.5A how a blood vessel penetrates the periosteum and gives off branches.

MP3
Bone Structure

Compact bone makes up the shaft of a long bone. ③ It contains many osteons (also called Haversian systems), where **osteocytes** derived from osteoblasts lie in tiny chambers called **lacunae.** The lacunae are arranged in concentric circles around central canals that contain branches of blood vessels and nerves. The lacunae are separated by a matrix of collagen fibers and mineral deposits, primarily calcium and phosphorus salts, as also discussed in section 26. 3.

④ **Spongy bone** has numerous bony bars and plates separated by irregular spaces. Although lighter than compact bone, spongy bone is still designed for strength. Just as braces are used for support in buildings, the solid portions of spongy bone follow lines of stress. At the ends of long bones, the spaces in spongy bone are often filled with **red bone marrow,** a specialized tissue that produces blood cells. This is an additional way the skeletal system assists homeostasis. As you know, red blood cells transport oxygen, and white blood cells are part of the immune system, which fights infection.

Also, as shown in Figure 29.5A, a growth plate occurs near the end of a long bone. As long as a bone has a growth plate, it is capable of growing, because organized growth of bone in this region contributes to the length of the bone. The growth plate usually disappears when a person reaches maturity.

Animation
Bone Growth

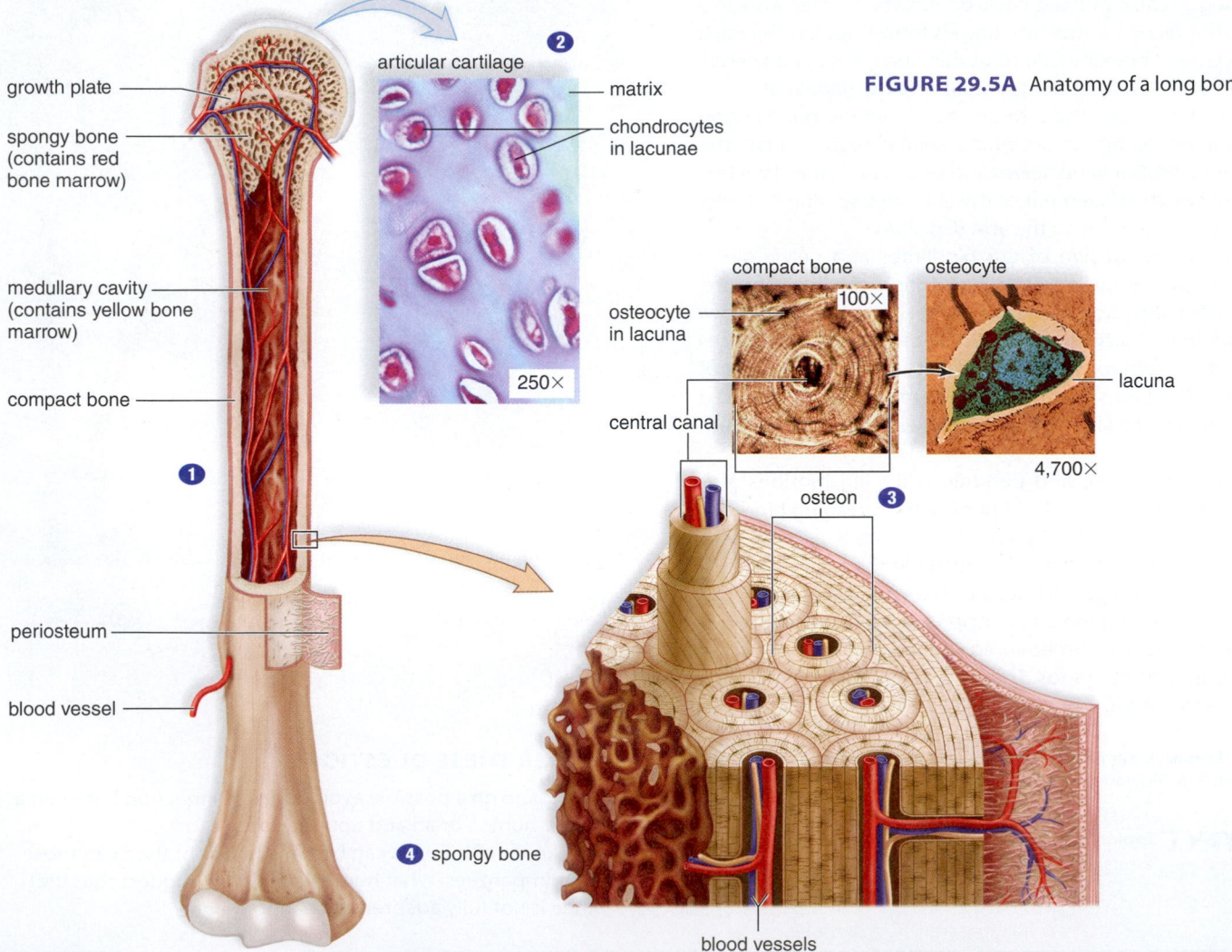

growth plate

spongy bone
(contains red
bone marrow)

medullary cavity
(contains yellow bone
marrow)

compact bone

periosteum

blood vessel

② articular cartilage

matrix

chondrocytes
in lacunae

250×

osteocyte
in lacuna

central canal

compact bone

100×

osteocyte

lacuna

4,700×

osteon ③

④ spongy bone

blood vessels

FIGURE 29.5A Anatomy of a long bone.

Joints Bones articulate at the joints, which are classified as fibrous, cartilaginous, or synovial. Each of the three types of joints has a different appearance:

- **Fibrous joints,** such as the sutures between the cranial bones, are immovable. The membranous "soft spots" in newborns are regions where the cranial bones will come together and become fibrous joints.
- **Cartilaginous joints,** which are connected by cartilage, tend to be slightly movable. The pubic symphysis, mentioned earlier, consists of fibrocartilage. The intervertebral disks are composed of fibrocartilage, and the ribs are joined to the rib cage by costal cartilages composed of hyaline cartilage.
- **Synovial joints** are freely movable (Fig. 29.5B) ❶ A synovial joint has a cavity lined with synovial membrane, which produces *synovial fluid.* Synovial fluid, which has an egg-white consistency, lubricates the joint. If the joint is stretched suddenly, the fluid does not immediately fill the joint, and in the meantime, the synovial membrane falls into the vacuum and a click is heard.

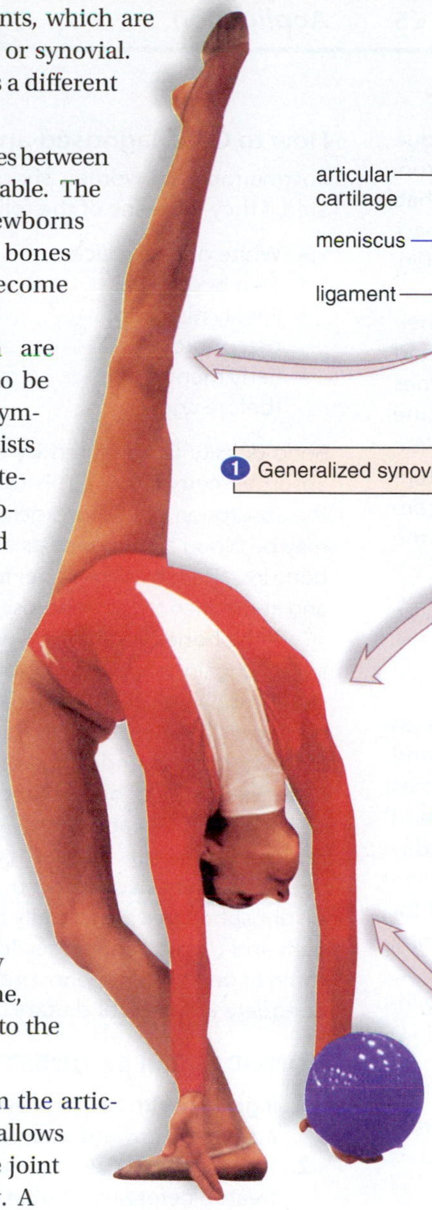

The absence of tissue between the articulating bones of a synovial joint allows them to be freely movable, but the joint has to be stabilized in some way. A synovial joint is stabilized by the joint capsule, a sleevelike extension of the periosteum of each articulating bone. Ligaments are fibrous bands that bind the two bones to one another and add even more stability. Tendons are fibrous tissue that connect muscle to bone and also help stabilize the joint.

The articulating surfaces of the bones are protected in several ways. First, the bones are covered by a layer of articular (hyaline) cartilage, described previously. Then, the **bursae,** which are fluid-filled sacs, ease friction between bone and overlapping muscles, or between skin and tendons. Inflammation of a bursa is called *bursitis.* **Menisci** are crescent-shaped pieces of cartilage in synovial joints that also ease friction between all parts of the joint. Injuries that involve the tearing of menisci are often called *torn cartilage.*

Two types of synovial joints are shown in Figure 29.5B: ❷ **Ball-and-socket joints,** found at the hips and shoulders, allow movement in all planes, even rotational movement. Adduction occurs when limbs are moved toward the midline of the body. Abduction occurs when limbs are moved away from the midline of the body. ❸ **Hinge joints,** such as the elbow and knee joints,

A gymnast depends on flexible joints.

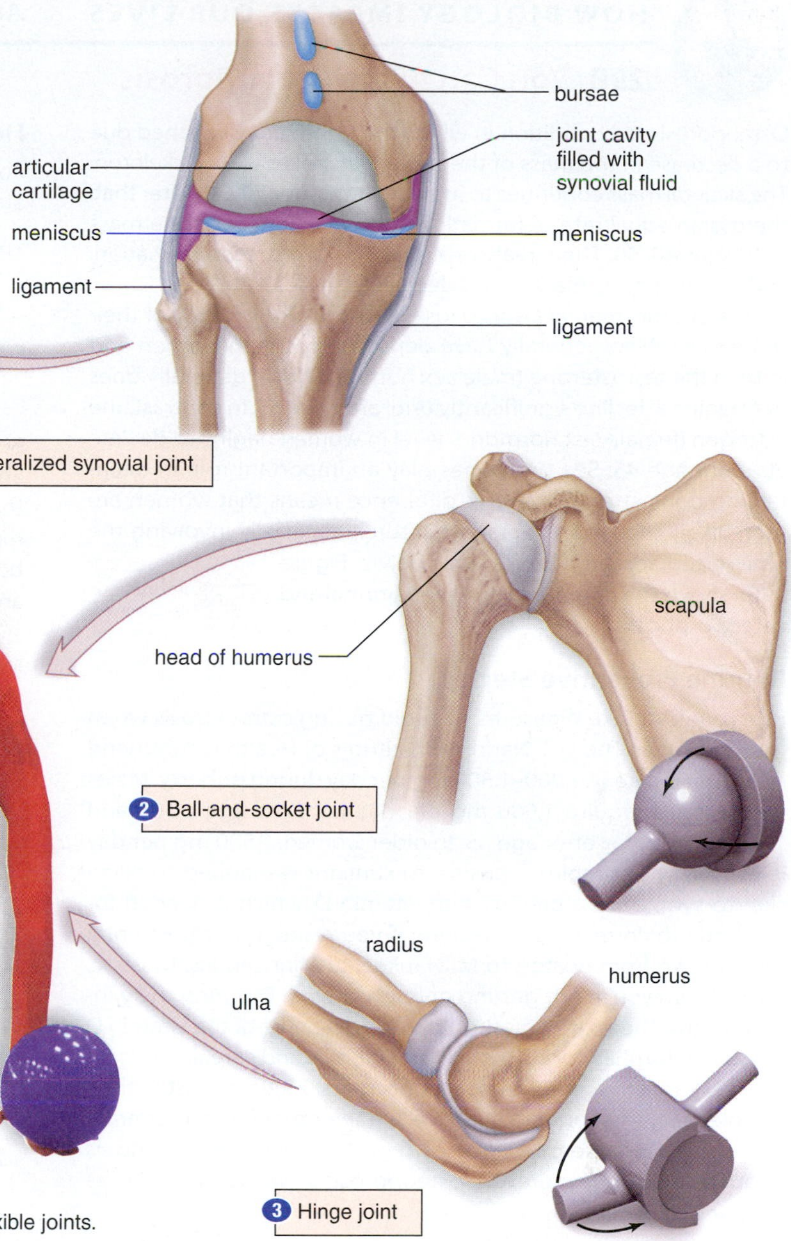

1 Generalized synovial joint

bursae
joint cavity filled with synovial fluid
articular cartilage
meniscus
ligament
meniscus
ligament

2 Ball-and-socket joint

head of humerus
scapula

3 Hinge joint

radius
ulna
humerus

FIGURE 29.5B Synovial joints.

largely permit movement up and down in one plane only, like a hinged door. Flexion occurs when the angle decreases, as when the forearm moves upward. Extension occurs when the angle increases, as when the forearm moves downward.

In our hands are three other types of synovial joints: saddle (one bone fits inside another), gliding (the bones slide against one another), and condyloid (the convex surface of one bone fits in a depression of the other).

▶ **29.5 CHECK YOUR PROGRESS**
1. Contrast the location and function of compact bone and spongy bone.
2. Distinguish between the function of ligaments and tendons at a synovial joint.

29B You Can Avoid Osteoporosis

Osteoporosis is a condition in which the bones are weakened due to a decrease in the mass of the bone that makes up the skeleton. The skeletal mass continues to increase until ages 20–30. After that, there is an equal rate of formation and breakdown of bone mass until ages 40–50. Then, reabsorption begins to exceed formation, and the total bone mass slowly decreases.

Over time, men are apt to lose 25% and women 35% of their bone mass. Men generally have denser bones than women and in men the testosterone (male sex hormone) level generally does not begin to decline significantly until after age 65. In contrast, the estrogen (female sex hormone) level in women begins to decline at about age 45. Sex hormones play an important role in maintaining bone strength, so this difference means that women are more likely than men to suffer fractures, especially involving the hip, vertebrae, long bones, and pelvis. Figure 29B*a* shows the difference between normal and osteoporotic bone.

Animation
Osteoporosis

Routine Preventive Steps

Everyone can take measures to avoid having osteoporosis when they get older. The U.S. National Institutes of Health recommend a calcium intake of 1,200–1,500 mg per day during puberty. Males and females require 1,000 mg per day until the age of 65 and 1,500 mg per day after age 65. In older women, 1,500 mg per day is especially desirable. Exposure to sunlight is required to allow skin to synthesize a precursor to vitamin D, which is needed for the body to correctly use calcium. If you reside on or north of a "line" drawn from Boston to Milwaukee, to Minneapolis, to Boise, chances are you're not getting enough vitamin D during the winter months. Therefore, you should take advantage of the vitamin D present in fortified foods such as low-fat milk and cereal.

Very inactive people lose bone mass 25 times faster than people who are moderately active. On the other hand, moderate weight-bearing exercise, such as regular walking or jogging, is another good way to prevent osteoporosis so we do not stoop as we age (Fig. 29B*b*).

How to Get Diagnosed and Treated

Postmenopausal women should have their bone density evaluated if they have any of the following risk factors:

- White or Asian race
- Thin body type
- Family history of osteoporosis
- Early menopause (before age 45)
- Smoking
- A diet low in calcium, or excessive alcohol consumption and caffeine intake
- Sedentary lifestyle

Bone density is measured by a method called dual energy X-ray absorptiometry (DEXA). This test measures bone density based on the absorption of photons generated by an X-ray tube. Soon there may be blood and urine tests to detect the biochemical markers of bone loss. Then it will be easier for physicians to screen older women and at-risk men for osteoporosis.

If the bones are thin, it is worthwhile to take all possible measures to gain bone density because even a slight increase can significantly reduce fracture risk. Although estrogen therapy does reduce the incidence of hip fractures, long-term estrogen therapy is rarely recommended for osteoporosis. Estrogen is known to increase the risk of breast cancer, heart disease, stroke, and blood clots. Other medications are available, however. Calcitonin, a thyroid hormone, has been shown to increase bone density and strength, while decreasing the rate of bone fractures. Also, the bisphosphonates are a family of nonhormonal drugs used to prevent and treat osteoporosis. To achieve optimal results with calcitonin or one of the bisphosphonates, patients should also receive adequate amounts of dietary calcium and vitamin D.

CONSIDER THESE QUESTIONS

1. If good health habits prevent osteoporosis, what should our attitude be toward people who have it?
2. What's the best way to make sure women are diagnosed and treated before they suffer a fracture due to osteoporosis?
3. Should insurance companies pay for preventable illnesses? Why or why not?

connect **BIOLOGY** Explore the concepts through a variety of multimedia assets and question types.

www.mcgrawhillconnect.com

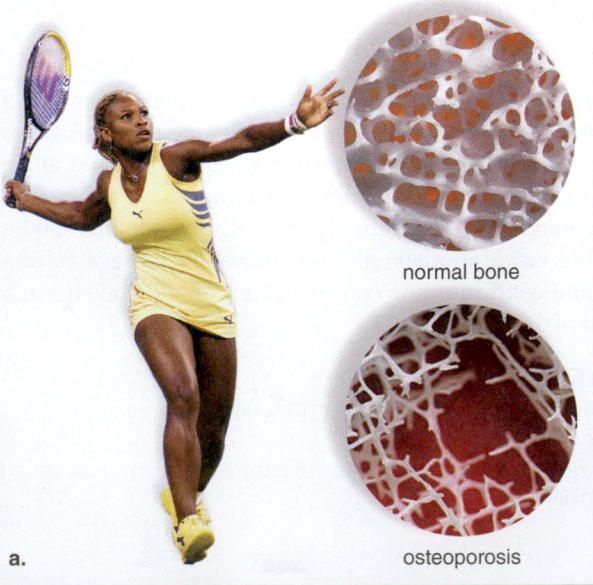

normal bone

osteoporosis

a.

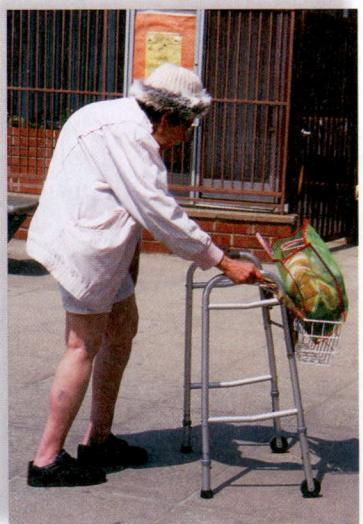

b.

FIGURE 29B a. Exercise when we are young can help prevent osteoporosis when we are older. **b.** Severely stooped posture is a sign of osteoporosis.

Vertebrate Skeletal Muscles

Of the three types of vertebrate muscle, skeletal is the only type attached to the bones. Its activity accounts for animals' ability to move about from place to place. In this section, we examine the anatomy and physiology of a whole skeletal muscle, and in the next section, we take a look at the anatomy and physiology of the skeletal muscle cell, called a muscle fiber.

29.6 Skeletal muscles primarily cause bones to move

LEARNING OUTCOMES

When you complete this section, you should be able to

1. Identify the general functions of skeletal muscles and state specific functions for selected muscles.
2. Understand the basic principles of whole muscle contraction.

As noted in Figure 26.4, smooth muscle is involuntary muscle found in the walls of internal organs. Cardiac muscle is involuntary and makes up the wall of the heart. Skeletal muscle can be moved voluntarily and makes up the nearly 700 skeletal muscles, which account for approximately 40% of the weight of an average human. Figure 29.6A illustrates several of the major skeletal muscles and their actions. The skeletal muscles perform many functions:

Skeletal muscles make bones move. Muscle contraction accounts not only for movement of the arms and legs but also for movement of the eyes, facial expressions, and breathing.

Skeletal muscles also support the body. Muscle contraction opposes the force of gravity and allows us to remain upright.

Skeletal muscles also help maintain a constant body temperature. Muscle contraction causes ATP to break down, releasing heat that is distributed about the body.

Skeletal muscle contraction also assists blood flow in cardiovascular veins. The pressure of skeletal muscle contraction keeps blood moving in cardiovascular veins.

Skeletal muscles also help protect internal organs and stabilize joints. Muscles pad the bones, and the muscular wall in the abdominal region helps protect the internal organs.

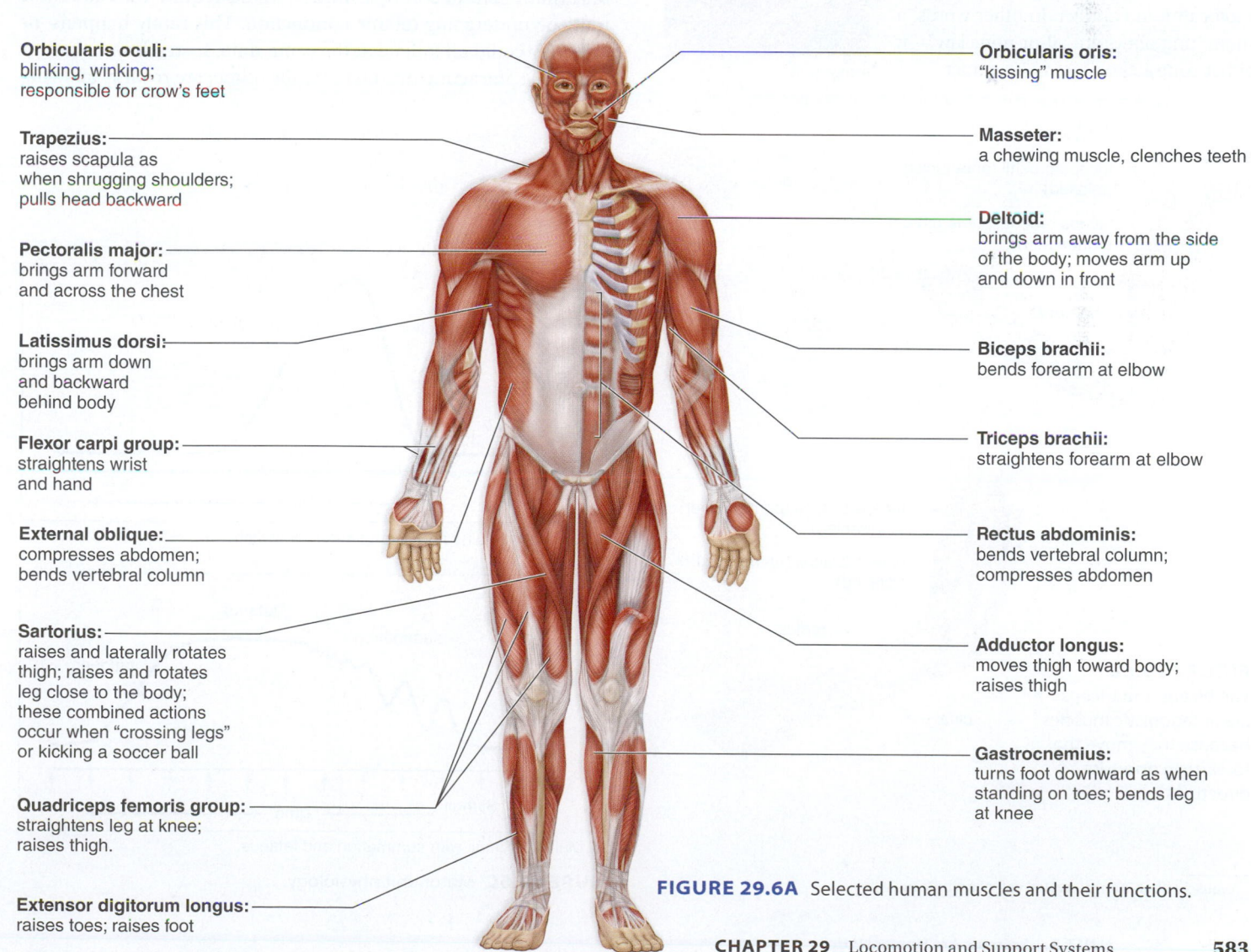

Orbicularis oculi: blinking, winking; responsible for crow's feet

Trapezius: raises scapula as when shrugging shoulders; pulls head backward

Pectoralis major: brings arm forward and across the chest

Latissimus dorsi: brings arm down and backward behind body

Flexor carpi group: straightens wrist and hand

External oblique: compresses abdomen; bends vertebral column

Sartorius: raises and laterally rotates thigh; raises and rotates leg close to the body; these combined actions occur when "crossing legs" or kicking a soccer ball

Quadriceps femoris group: straightens leg at knee; raises thigh.

Extensor digitorum longus: raises toes; raises foot

Orbicularis oris: "kissing" muscle

Masseter: a chewing muscle, clenches teeth

Deltoid: brings arm away from the side of the body; moves arm up and down in front

Biceps brachii: bends forearm at elbow

Triceps brachii: straightens forearm at elbow

Rectus abdominis: bends vertebral column; compresses abdomen

Adductor longus: moves thigh toward body; raises thigh

Gastrocnemius: turns foot downward as when standing on toes; bends leg at knee

FIGURE 29.6A Selected human muscles and their functions.

Antagonistic Pairs Skeletal muscles move the bones of the skeleton with the aid of bands of fibrous connective tissue called **tendons** that attach muscle to bone. In general, one muscle does most of the work of moving a bone, and that muscle is called a **prime mover.** When a muscle contracts, it shortens, and the tendon pulls on the bone. Therefore, muscles can only pull a bone; they cannot push it. Because of this, skeletal muscles must work in antagonistic pairs. If one muscle of an antagonistic pair *flexes* the joint and bends the limb, the other one *extends* the joint and straightens the limb. For example, the biceps brachii and the triceps brachii are antagonists; one bends the forearm, and the other straightens the forearm (Fig. 29.6B). If both of these muscles were to contract at once, the forearm would not move.

Motor Units A skeletal muscle has degrees of contraction because it is divided into motor units. A **motor unit** is composed of all the muscle fibers[1] under the control of a single motor axon. The axon has branches that terminate at the muscle fibers of a motor unit. Here, axon terminals release neurotransmitter molecules that cross a synapse, causing the motor unit to contract. We can liken a motor unit to a set of lights in a ceiling that is controlled by a single switch. A flip of the switch turns these lights on, much as a single axon causes its motor unit to contract. In other words, a motor unit obeys an "all-or-none law"—it either contracts or does not contract.

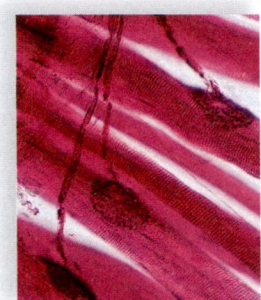
motor unit

The number of muscle fibers within a motor unit can vary. For example, in the ocular muscles that move the eyes, the innervation ratio is one motor axon per 23 muscle fibers, while in the gastrocnemius muscle of the leg, the ratio is about one motor axon per 1,000 muscle fibers. Thus, moving the eyes requires finer control than moving the legs.

When a motor unit is stimulated by a single stimulus, a contraction occurs that lasts only a fraction of a second. This response is called a **simple muscle twitch.** A muscle twitch is customarily divided into three stages: the latent period, or the period of time between stimulation and initiation of contraction; the contraction period, when the muscle shortens; and the relaxation period, when the muscle returns to its former length (Fig. 29.6C*a*). If a motor unit is given a rapid series of stimuli, it can respond to the next stimulus without relaxing completely. *Summation* is increased muscle contraction until maximal sustained contraction, called *tetanus,* is achieved (Fig. 29.6C*b*). Tetanus continues until the muscle fatigues due to depletion of energy reserves. Fatigue is apparent when a muscle relaxes even though stimulation continues.

A whole muscle typically contains many motor units, much as a ceiling might contain several sets of lights. For maximum lighting, all the switches are turned on. In the same way, as the intensity of nervous stimulation increases, more and more motor units in a muscle are activated. This phenomenon is known as *recruitment.* Maximum contraction of a muscle would require that all motor units be undergoing tetanic contraction. This rarely happens, or else they could all fatigue at the same time. Instead, some motor units are contracting maximally while others are resting, allowing

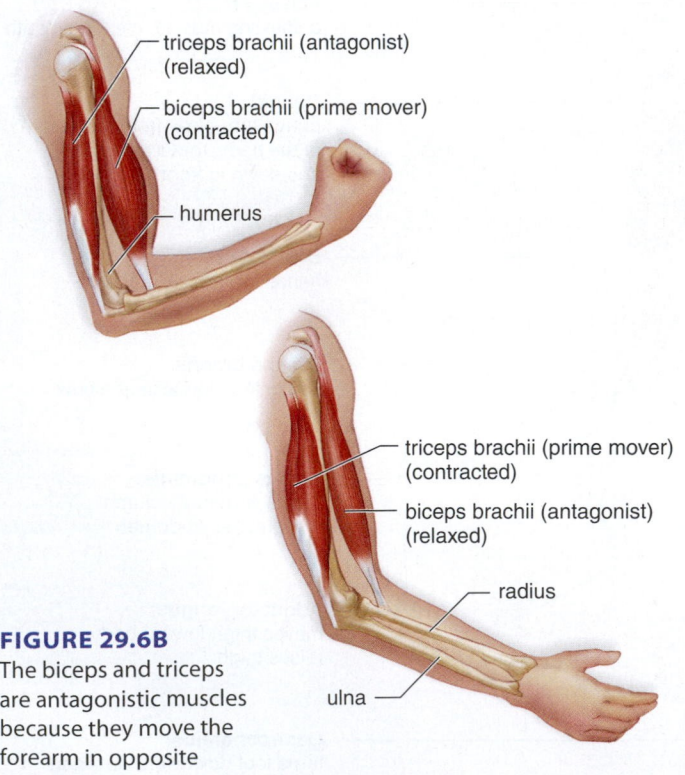

FIGURE 29.6B
The biceps and triceps are antagonistic muscles because they move the forearm in opposite directions.

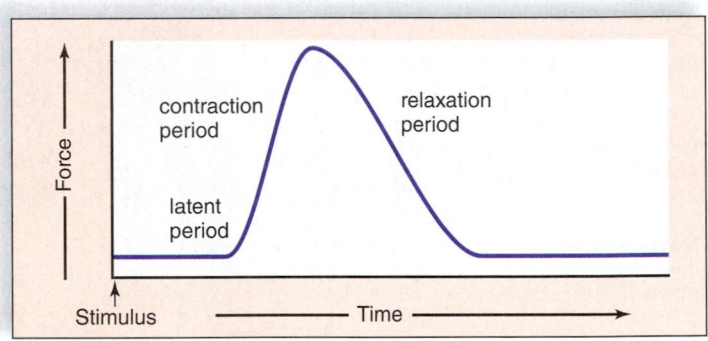

a. A single stimulus and a simple muscle twitch.

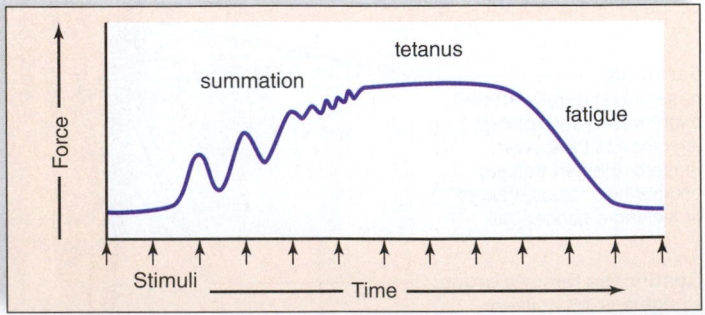

b. Multiple stimuli with summation and tetanus.

FIGURE 29.6C Motor unit physiology.

---

[1] A muscle fiber is an elongated cell of a whole muscle.

sustained contractions to occur. One desirable effect of exercise is to achieve good "muscle tone," which is dependent on muscle contraction. When some motor units are always contracted but not enough to cause movement, the muscle is firm and solid.

▶ **29.6 CHECK YOUR PROGRESS**
1. List the functions of skeletal muscles and give specific examples.
2. Explain why the force of muscle twitch can vary.

HOW SCIENCE PROGRESSES   *Application*

29C The Accidental Discovery of Botox

Several of the bacterial pathogens that cause serious human diseases such as cholera, diphtheria, tetanus, and botulism secrete potent toxins. Botulinum toxin, produced by the bacterium *Clostridium botulinum,* is an extremely lethal toxin. Less than a microgram of the purified toxin can kill an average size person and 4 kg would be enough to kill the entire world's population. Considering how lethal it is, it's doubtful that the scientists who discovered this toxin nearly 200 years ago would have ever thought it would be commonly used, grant you in a dilute form, for nonsurgical cosmetic purposes. Today, we know there are seven versions of the toxin produce by *Clostridium botulinum* and the one in common use is now called botulinum A or Botox.

As with many discoveries in science, a sequence of events led from considering the botulinum toxins as extremely lethal to using botulinum A for cosmetic purposes. In the 1820s, a German scientist named Justinus Kerner was able to prove that several people died because they had eaten spoiled sausage. Botulism, in fact, takes its name from *botulus,* the Latin word for sausage. A few decades after Kerner's discovery, a Belgian researcher, Emile Pierre van Ermengem, was able to show that *Clostridium botulinum* was the bacterium that produced the deadly toxin in sausage. By the 1920s, medical researchers at the University of California had obtained the toxin in pure form and this allowed them to determine that it prevented the release of acetylcholine from axon terminals of motor nerves. Botulinum A causes death by paralyzing the respiratory muscles so that breathing is impossible. These researchers reasoned, however, that a very dilute form of the toxin could prevent muscular spasms of facial muscles and vocal cords and also the crossing of eyes. In 1989 the FDA approved the use of dilute botulinum toxin A for these conditions.

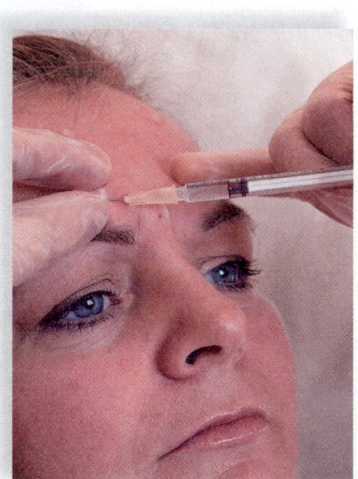

FIGURE 29C Treating wrinkles with Botox.

Then luck entered our story. A Canadian ophthalmologist, Jean Carruthers, had been using Botox to treat certain types of eye conditions when she noticed that some of her patients' wrinkles had subsided. One night during a family dinner, Dr. Carruthers shared this observation with her husband who was a dermatologist and he decided to see if a dilute form of the toxin, injected in the skin, would lessen the deep wrinkles of his patients. It worked! This husband and wife team went on to test the drug on many patients and themselves as well. They presented their findings at many scientific conferences and wrote research articles for various journals. At first the medical community was very skeptical but eventually came around to realizing that Botox is an effective treatment for wrinkles. Unfortunately, the Carruthers never thought to patent their discovery so they do not get a portion of the $1.3 billion in annual sales the drug earns for the company that did patent it.

By now, botulinum A, approved in 2002 by the FDA for the treatment of frown lines (Fig. 29C) has many more applications. For example in March 2010, the FDA approved the drug for the treatment of muscle stiffness in people with upper limb spasticity, and may approve it for as many as 90 other uses including migraine headaches. Perhaps the drug would have achieved its many applications even if Jean Carruthers hadn't noticed its effect on her patients' wrinkles, but no doubt progress would have been a lot slower. As the French microbiologist Louis Pasteur observed in 1854, "Chance favors the prepared mind," meaning that unforeseen observations can play just as important a role in scientific discoveries as many years of work by many investigators.

CONSIDER THESE QUESTIONS

1. What do you suppose is the treatment for botulism, considering that it is caused by a toxin?
2. Do you think companies should be allowed to legally patent a naturally occurring substance like botulinum toxin A? Why or why not?

 Explore the concepts through a variety of multimedia **BIOLOGY** assets and question types.

www.mcgrawhillconnect.com

LEARNING OUTCOMES

When you complete this section, you should be able to

1. Describe the microscopic structure of a muscle fiber.
2. Explain the sliding filament model and the sources of ATP for muscle fiber contraction.
3. Explain how nerves signal muscle fibers to contract.

While we are used to considering the contraction of whole muscles, we have just learned that muscle contraction actually involves motor units. Now, we wish to consider how each and every muscle fiber contracts. Because a muscle fiber has a slightly different structure from that of other cells, its parts are given special names. For example, the plasma membrane is called the **sarcolemma.** The sarcolemma of a muscle fiber forms a transverse (T)-tubule system. The T tubules penetrate, or dip down, into the fiber so that they come in contact—but do not fuse—with expanded portions of modified endoplasmic reticulum, called the **sarcoplasmic reticulum.** These expanded portions serve as storage sites for calcium ions (Ca^{2+}), which are essential for muscle contraction. Also present in a muscle fiber are many long, cylindrical organelles called **myofibrils,** which are the contractile portions of muscle fibers. In cross section, a myofibril contains many contractile units called **sarcomeres** (Fig. 29.7A). Each sarcomere lies between two visible boundaries called Z lines.

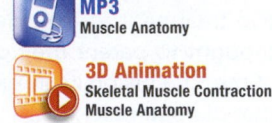

The electron microscope reveals that skeletal muscle striations are due to the placement of protein filaments in sarcomeres. A sarcomere contains thick filaments made up of **myosin** and thin filaments made up of **actin.** A myosin filament has many globular heads. An area in the middle of a sarcomere, called the H zone, has only myosin filaments. The actin filaments are composed of long strands of globular actin molecules twisted about one another. The actin filaments are attached to the Z lines.

Sliding Filament Model Figure 29.7A contrasts the appearance of a relaxed sarcomere with a contracted sarcomere. In a contracted sarcomere, the actin filaments are much closer to the center, and the H zone has all but disappeared. To achieve a contracted sarcomere, it is necessary for the actin filaments to slide past the myosin filaments. This occurs because the myosin heads pull the actin filaments toward the center of a sarcomere. When you play "tug of war," your hands grasp the rope, pull, let go, attach farther down the rope, and pull again. The myosin heads are like your hands—grasping, pulling, letting go, and then repeating the process. This model of muscle contraction is called the **sliding filament model.**

Figure 29.7B pertains to only one myosin head, but actually many myosin heads act in unison to achieve the contraction of a sarcomere. The cycle of events shown occurs over and over again, and with each cycle, the actin filaments move nearer the center of the sarcomere, until the H zone all but disappears. ATP provides the energy for muscle contraction in a way that is not obvious. Each myosin head has a binding site for ATP, and the heads have an enzyme that splits ATP into ADP and ⓟ. This activates the heads, making them ready to bind to actin. ADP and ⓟ remain on the myosin heads while the heads attach to actin, forming cross-bridges. Release of ADP and ⓟ causes the cross-bridges to bend sharply. This is the power stroke that pulls the actin filaments toward the middle of the sarcomere. When another ATP molecule binds, myosin detaches from actin. The cycle has begun again.

Rigor mortis is the stiffening of muscles that occurs in a dead body. It is often used to estimate the time of death when a recently deceased body is discovered. At temperatures of 21–24°C, rigor mortis begins within 1–3 hours; maximum rigidity is reached 10–12 hours after death. Stiffness persists for 24–36 hours, and

FIGURE 29.7A Skeletal muscle fiber structure and contraction.

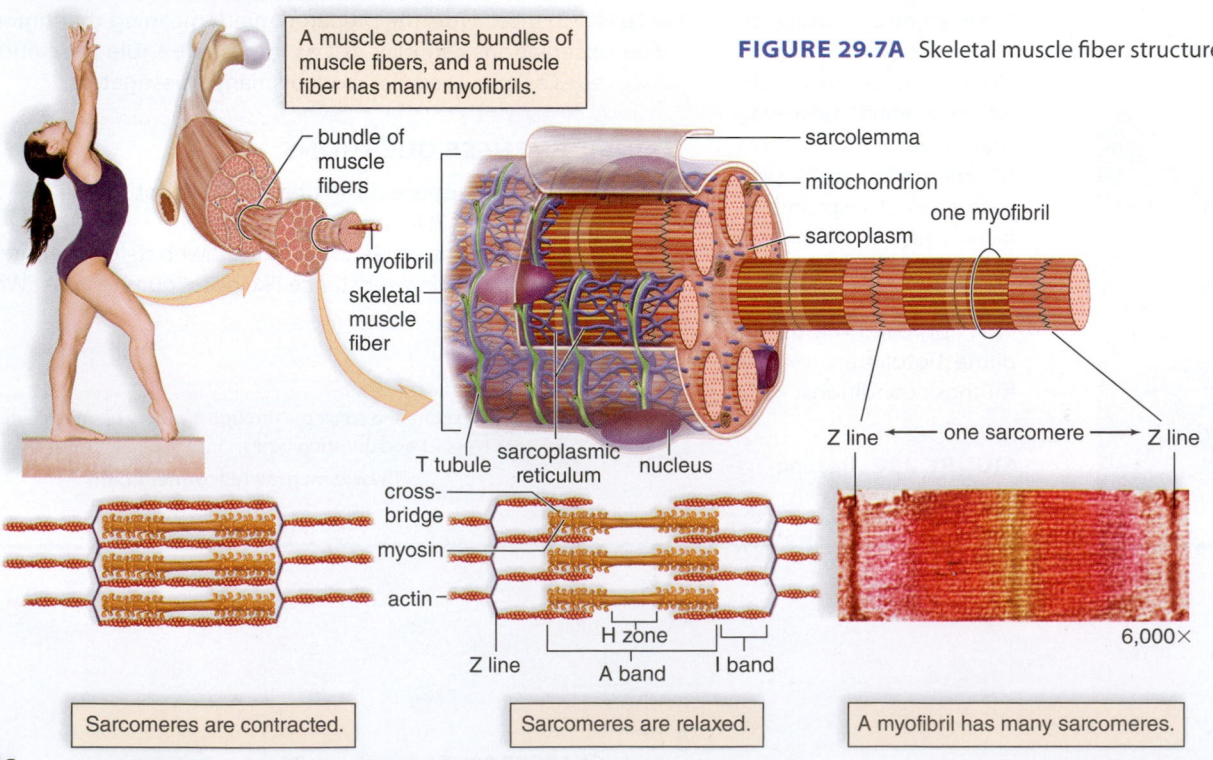

A muscle contains bundles of muscle fibers, and a muscle fiber has many myofibrils.

bundle of muscle fibers

myofibril

skeletal muscle fiber

sarcolemma

mitochondrion

one myofibril

sarcoplasm

T tubule

sarcoplasmic reticulum

nucleus

Z line ← one sarcomere → Z line

6,000×

cross-bridge

myosin

actin

Z line

H zone

A band

I band

Sarcomeres are contracted.

Sarcomeres are relaxed.

A myofibril has many sarcomeres.

FIGURE 29.7B Role of ATP in muscle contraction.

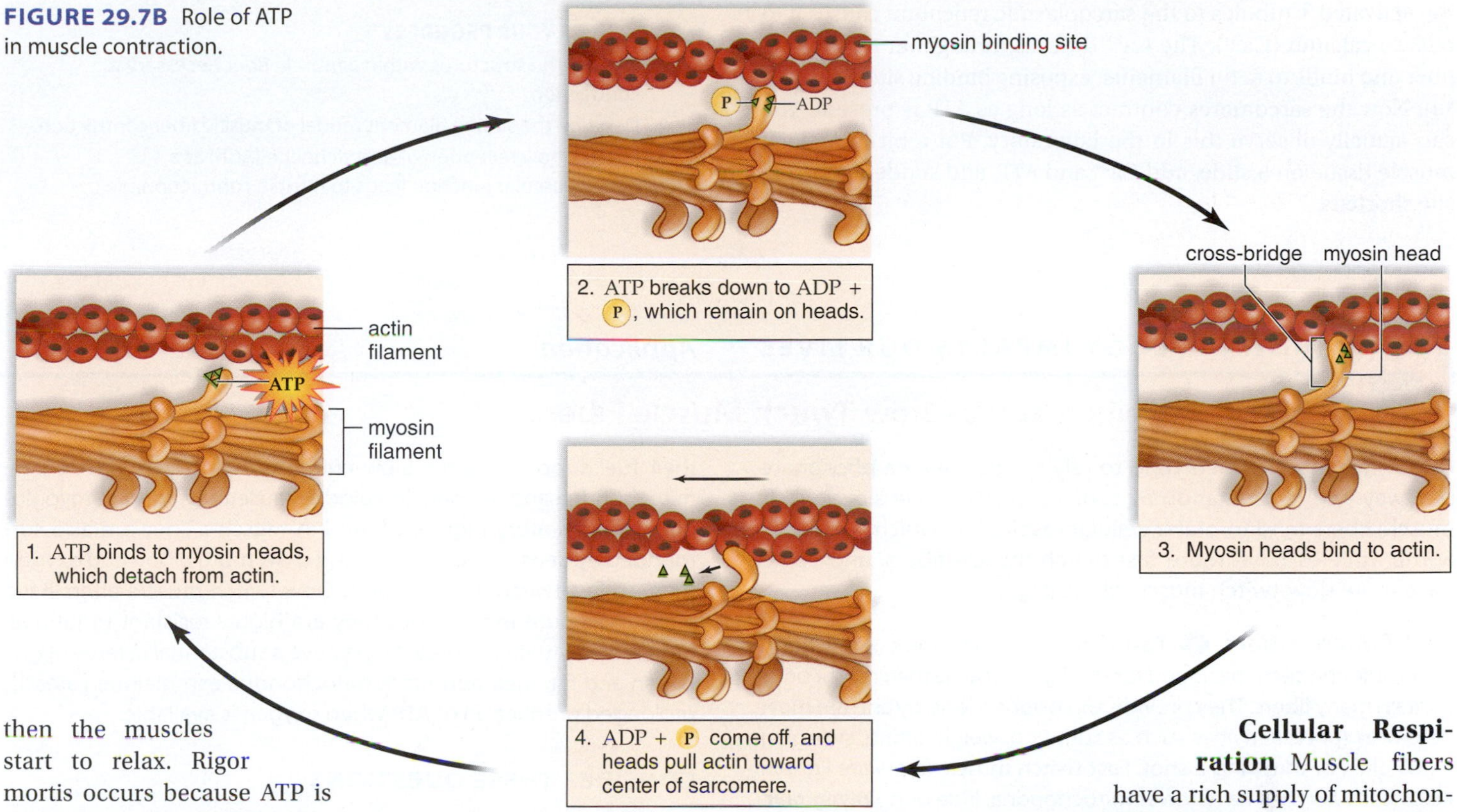

— myosin binding site

2. ATP breaks down to ADP + P, which remain on heads.

— actin filament

ATP

— myosin filament

1. ATP binds to myosin heads, which detach from actin.

cross-bridge   myosin head

3. Myosin heads bind to actin.

4. ADP + P come off, and heads pull actin toward center of sarcomere.

then the muscles start to relax. Rigor mortis occurs because ATP is needed in order for the myosin heads to detach from actin filaments. However, because ATP synthesis stops shortly after death, the myosin heads remain attached for a matter of hours, until deterioration sets in.

Animation
Sarcomere Contraction

3D Animation
Skeletal Muscle Contraction: Sliding Filament Model

Sources of ATP Muscle fibers store limited amounts of ATP, but they have three ways of acquiring more ATP for contraction once this supply has been used up: the creatine phosphate pathway, fermentation, and cellular respiration.

Creatine Phosphate (CP) Pathway Creatine phosphate is a molecule that contains a high-energy phosphate. Creatine phosphate is formed only when a muscle fiber is resting, and only a limited amount is stored. The simplest and most rapid way for muscle fiber to produce ATP is to transfer the high-energy phosphate from CP to ADP.

This reaction occurs in the midst of sliding filaments, and therefore this method of supplying ATP is the speediest energy source available to muscles. The CP pathway is used at the beginning of exercise and during short-term, high-intensity exercise that lasts less than 5 seconds.

Fermentation Fermentation, as you know, produces two ATP from the anaerobic breakdown of glucose to lactate (see Fig. 7.7). Fermentation, like the CP pathway, is fast-acting, but it results in the buildup of lactate, noticeable because it produces short-term muscle aches and fatigue upon exercising. Fermentation also results in **oxygen debt,** the oxygen required in part to complete the metabolism of lactate and restore cells to their original energy state.

Cellular Respiration Muscle fibers have a rich supply of mitochondria where cellular respiration supplies ATP, usually from the breakdown of glucose whenever oxygen is available.

Muscle Innervation Muscle fibers contract only because they are stimulated to do so by motor axons. A motor axon branches, and each branch terminates very close to a muscle fiber. This region, called a **neuromuscular junction,** contains a *synaptic cleft* (Fig. 29.7C). A nerve impulse traveling down an axon causes the axon terminals to release the neurotransmitter acetylcholine (ACh) (green). The sarcolemma of a muscle fiber contains receptors for ACh molecules, and when these molecules bind to the receptors, a muscle action potential begins. The muscle action potential travels down the T (transverse) tubules which are deep indentations of the sarcolemma. The close proximity of

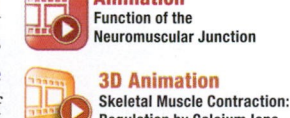

Animation
Function of the Neuromuscular Junction

3D Animation
Skeletal Muscle Contraction: Regulation by Calcium Ions

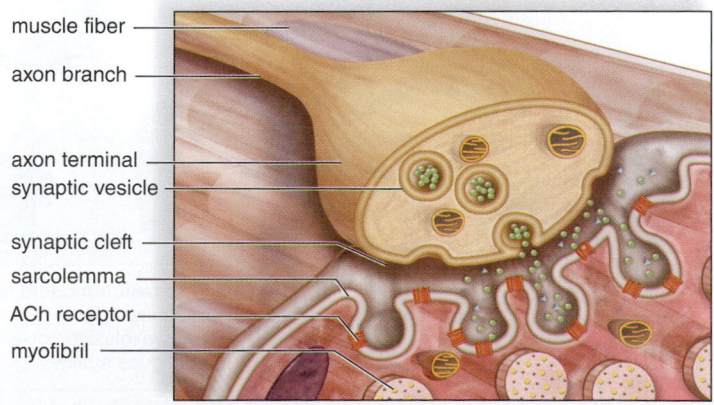

muscle fiber —
axon branch —

axon terminal —
synaptic vesicle —

synaptic cleft —
sarcolemma —
ACh receptor —
myofibril —

FIGURE 29.7C Neuromuscular junction (green = ACh).

the activated T tubules to the sarcoplasmic reticulum causes it to release calcium (Ca^{2+}). The Ca^{2+} diffuses throughout the muscle fiber and binds to actin filaments, exposing binding sites for myosin. Now the sarcomeres contract as long as ATP is present. You can actually observe this in the laboratory. Put a bit of skeletal muscle tissue on a slide, add Ca^{2+} and ATP, and suddenly the tissue shortens.

HOW BIOLOGY IMPACTS OUR LIVES *Application*

29D Fast-Twitch Versus Slow-Twitch Muscle Fibers

Fast-twitch muscle fibers tend to rely on the creatine phosphate pathway and fermentation as sources of ATP, while slow-twitch muscle fibers tend to prefer cellular respiration, which is aerobic. Some athletes have more fast-twitch muscle fibers, and some have more slow-twitch muscle fibers (Fig. 29D).

Fast-Twitch Fibers ❶ Fast-twitch muscle fibers are usually anaerobic and seem designed for strength because their motor units contain many fibers. They provide explosions of energy and are most helpful in sports activities such as sprinting, weight lifting, swinging a golf club, or throwing a shot. Fast-twitch muscle fibers are light in color because they have fewer mitochondria, little or no **myoglobin** (an oxygen-binding molecule found in muscle tissue), and fewer blood vessels than slow-twitch muscle fibers do. If fast-twitch, a muscle fiber can develop maximum tension more rapidly than if it is slow-twitch, and the maximum tension is greater. However, a dependence on anaerobic energy leaves a muscle fiber vulnerable to an accumulation of lactate, which causes it to fatigue quickly.

Slow-Twitch Fibers ❷ Slow-twitch muscle fibers have a steadier tug and more endurance, despite having more units with fewer cells. These muscle fibers are most helpful in sports such as long-distance running, biking, jogging, and swimming. Because they produce most of their energy aerobically, they tire only when their fuel supply is gone. Slow-twitch muscle fibers have many mitochondria and are dark in color because they contain myoglobin, the respiratory pigment found in muscles. They are also surrounded by dense capillary beds and draw more blood and oxygen than do fast-twitch muscle fibers. Slow-twitch muscle fibers have a low maximum tension, but they are highly resistant to fatigue. Because slow-twitch muscle fibers have a substantial reserve of glycogen and fat, their abundant mitochondria can maintain steady, prolonged production of ATP when oxygen is available.

CONSIDER THESE QUESTIONS

1. Our genetics determines our athletic abilities. Discuss this statement with reference to the composition of your muscles.
2. Suppose you wanted to be a weight lifter but you discovered you were better at sprinting than lifting weights. What would you do? Discuss your options.

connect
BIOLOGY Explore the concepts through a variety of multimedia assets and question types.

www.mcgrawhillconnect.com

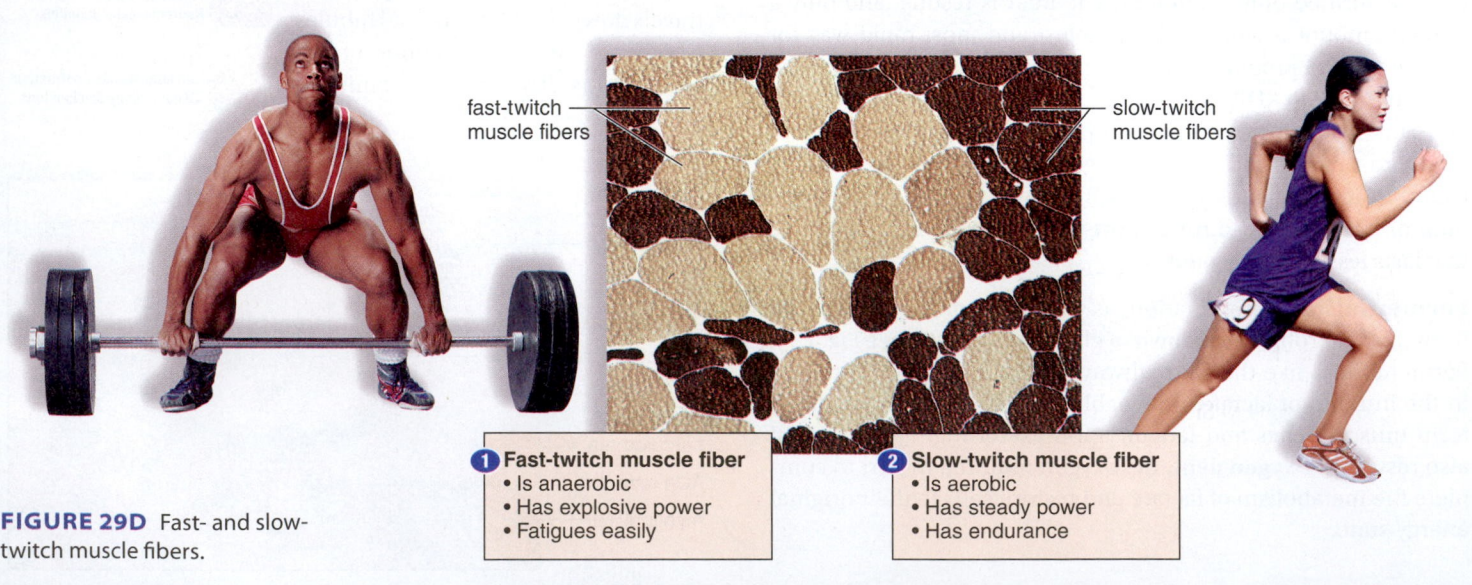

fast-twitch muscle fibers

slow-twitch muscle fibers

❶ **Fast-twitch muscle fiber**
• Is anaerobic
• Has explosive power
• Fatigues easily

❷ **Slow-twitch muscle fiber**
• Is aerobic
• Has steady power
• Has endurance

FIGURE 29D Fast- and slow-twitch muscle fibers.

THE CHAPTER IN REVIEW

CONNECTING THE CONCEPTS

This chapter gives us an opportunity to examine the body at both the macro and micro levels. The skeleton is easily observable at the macro level, and we can learn the names of the bones making up the axial and appendicular portions of the skeleton. Then we can consider the tissues of the bones and joints (compact bone, spongy bone, cartilage, and fibrous connective tissue). At this level, we can understand the injuries that occur when we misuse our joints.

Similarly, we can learn the names of various muscles and how they operate when we intentionally move our bones. Although the body has three types of muscles, this chapter concentrates on the skeletal muscles. We can't understand how skeletal muscles contract and move the bones until we study skeletal muscles at the cellular level. The theme "structure suits function" is observable at the macro level because whole muscles and bones have structures suitable to their functions. But this theme is even more observable at the micro level. A muscle fiber is suited to its task because it contains contractile organelles called myofibrils. Myofibrils contain the filaments (actin and myosin) that account for muscle contraction. Imagine the satisfaction of the electron microscopists and biochemists who first solved the riddle of muscle contraction and were able to explain to forensics specialists why rigor mortis occurs.

In Chapter 30, we continue our theme of homeostasis by studying the circulatory system, which is directly involved in homeostasis because blood and tissue fluid constitute the body's internal environment. When blood and tissue fluid remain relatively constant, so do the cells making up the body tissues—including those of the bones and muscles.

ANALYZE AND EVALUATE

1. Nerve and muscle cells are characteristics of animals and, therefore, were present in the first animals to evolve. Of what benefit are nerves and muscles to an animal?

2. Support the statement that "bones are diverse and specialized for particular functions." Support the same statement for muscles.

3. The jointed appendages of arthropods and vertebrates have allowed them to invade land. Why are jointed appendages an advantage on land?

SUMMARIZE

Diversity of Skeletons

29.1 Animal skeletons can be hydrostatic, external, or internal

- A **hydrostatic skeleton** is a fluid-filled gastrovascular cavity or coelom. An **exoskeleton** is a rigid external skeleton found in molluscs and arthropods. An **endoskeleton** is a rigid internal skeleton that protects the internal organs and is protected by soft tissues surrounding it; in vertebrates, the endoskeleton is jointed.

29.2 A skeleton serves many functions

- The skeleton is necessary to movement, protects internal organs, assists breathing, stores and releases calcium, and assists the immune and digestive systems.

Mammalian Skeleton

29.3 The bones of the axial skeleton lie in the midline of the body

- The **axial skeleton** consists of the **skull, vertebral column, rib cage,** sacrum, and coccyx: (1) The cranium and facial bones of the skull protect the brain. In newborns, the bones of the cranium are joined at **fontanels.** The **foramen magnum** is an opening at the base of the skull for the spinal cord. (2) The vertebral column, composed of vertebrae separated by shock-absorbing **intervertebral disks,** protects the spinal cord and nerves and anchors all other bones. (3) The rib cage, composed of the ribs, the costal cartilages, and the sternum, protects the heart and lungs.

29.4 The appendicular skeleton consists of bones in the girdles and limbs

- The **appendicular skeleton** contains (1) the bones of the **pectoral girdle** (shoulder) and the upper limbs, adapted for flexibility, and (2) the **pelvic girdle** (hipbones) and lower limbs, adapted for strength and support.

29.5 Bones and joints are composed of living tissues

- A long bone (e.g., humerus) has the following structures: (1) **Compact bone** at the sides contains osteons separated by a hard matrix. (2) The **osteocytes** are located in **lacunae.** (3) **Spongy bone** at the ends contains **red bone marrow.** (4) **Articular cartilage** covers the ends of a long bone. (5) The medullary cavity contains yellow bone marrow.

- **Fibrous joints** are immovable; **cartilaginous joints** are slightly movable; and synovial joints are freely movable.

- **Synovial joints** (e.g., ball-and-socket, hinge) are filled with synovial fluid, provide stability, and absorb shock. **Bursae** and **menisci** ease tension. **Ball-and-socket joints** allow movement in all planes, including rotation. **Hinge joints** allow movement in one direction only. Other types of synovial joints include saddle, gliding, and condyloid joints.

Vertebrate Skeletal Muscles

29.6 Skeletal muscles primarily cause bones to move

- Skeletal muscle supports the body, makes bones move, helps maintain a constant body temperature, assists blood movement in veins, helps protect internal organs, and stabilizes joints.

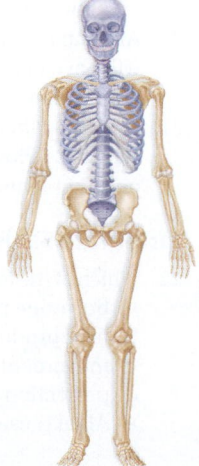

- Muscles are attached to bones by **tendons;** one muscle of a group is the **prime mover.** Muscles pull and do not push; therefore, they work in antagonistic pairs (e.g., biceps and triceps).
- A whole muscle has many **motor units.** Each unit is innervated by a single axon with many terminals. In the lab, stimulation causes a **simple muscle twitch,** which upon frequent stimulation exhibits summation and tetanus. Recruitment is activation of more units. Tone requires some units always contracting.

29.7 Muscles contract at the cellular level

- The plasma membrane of a muscle fiber is called the **sarcolemma,** and the endoplasmic reticulum is called the **sarcoplasmic reticulum. Myofibrils,** composed of units called **sarcomeres,** do the contracting. Sarcomeres contain **actin** and **myosin** filaments.
- The **sliding filament model** tells us that sarcomeres shorten when myosin heads bind to actin and pull the actin filaments past the myosin filaments. ATP is the energy source for muscle contraction. **Rigor mortis** occurs because muscle fibers lack the ATP needed for myosin heads to release actin filaments.
- A muscle fiber has three sources of ATP for contraction: (1) Creatine phosphate (CP) reacts with ADP to form ATP. (2) Fermentation (anaerobic) produces 2 ATP per glucose molecule. (3) Cellular respiration (aerobic) produces 36–38 ATP per glucose. Fermentation results in **oxygen debt.**
- Motor neurons cause muscle fibers to contract. At a **neuromuscular junction,** the neurotransmitter acetylcholine is received and the sarcoplasmic reticulum releases calcium ions. Calcium exposes myosin binding sites on actin and the sarcomeres contract.

TEST YOURSELF

Diversity of Skeletons

1. Unlike an exoskeleton, an endoskeleton
   a. grows with the animal.   c. is jointed.
   b. is composed of chitin.   d. protects internal organs.
2. _____ connect bone to bone, and _____ connect muscle to bone.
   a. Ligaments, ligaments   c. Ligaments, tendons
   b. Tendons, ligaments   d. None of these are correct.
3. Which of the following is not a function of the skeletal system?
   a. production of blood cells   d. storage of fat
   b. storage of minerals   e. production of body heat
   c. movement
4. All blood cells—red, white, and platelets—are produced by which of the following?
   a. yellow bone marrow   c. periosteum
   b. red bone marrow   d. medullary cavity
5. **THINKING CONCEPTUALLY** Skeletons have many fuctions. Which ones were least expected by you and why?

Mammalian Skeleton

6. A component of the appendicular skeleton is the
   a. rib cage.   c. femur.
   b. skull.   d. vertebral column.
7. Which of the following is not a bone of the appendicular skeleton?
   a. the scapula   c. a metatarsal bone
   b. a rib   d. the patella
8. This bone is the only movable bone of the skull.
   a. sphenoid   d. maxilla
   b. frontal   e. temporal
   c. mandible

For questions 9–15, match each bone to a location in the key. Answers can be used more than once.

KEY:
   a. arm (above forearm)   d. pelvic girdle
   b. forearm   e. thigh
   c. pectoral girdle   f. leg (below thigh)

9. Ulna
10. Tibia
11. Clavicle
12. Femur
13. Scapula
14. Coxal bone
15. Humerus
16. _____ occupies the _____.
    a. Cartilage, medullary cavity
    b. Marrow, foramen magnum
    c. Marrow, medullary cavity
    d. None of these are correct.
17. The deterioration of a synovial joint over time can cause
    a. arthritis.   c. a slipped disk.
    b. a sprain.   d. tendonitis.
18. Spongy bone
    a. contains osteons.
    b. contains red bone marrow, where blood cells form.
    c. lends no strength to bones.
    d. contributes to homeostasis.
    e. Both b and d are correct.
19. Which of these pairs is mismatched?
    a. slightly movable joint—vertebrae
    b. hinge joint—hip
    c. synovial joint—elbow
    d. immovable joint—sutures in cranium
    e. ball-and-socket joint—hip
20. After an examination, the doctor informs Isabella that she will have to have her baby by cesarean section. What skeletal abnormality is most likely?
21. **THINKING CONCEPTUALLY** The organization of the mammalian skeleton follows that of the central nervous system and the peripheral nervous system. How so?

Vertebrate Skeletal Muscles

22. Which of the following is not a function of the muscular system?
    a. hormone production
    b. heat production
    c. movement
    d. protection of internal organs
    e. All of these choices are functions of the muscular system.

23. The biceps and triceps are considered
    a. synergists.
    c. protagonists.
    b. antagonists.
    d. None of these are correct.
24. To increase the force of muscle contraction,
    a. individual muscle fibers must contract with greater force.
    b. motor units must contract with greater force.
    c. more motor units need to be recruited.
    d. All of these are correct.
    e. None of these are correct.
25. In a muscle fiber,
    a. the sarcolemma is connective tissue holding the myofibrils together.
    b. the sarcoplasmic reticulum stores calcium.
    c. both myosin and actin filaments have cross-bridges.
    d. there is a T system but no endoplasmic reticulum.
    e. All of these are correct.
26. The thick filaments of a muscle fiber are made up of
    a. actin.
    c. fascia.
    b. troponin.
    d. myosin.
27. A neuromuscular junction occurs between an axon terminal and a
    a. muscle fiber.
    d. sarcomere only.
    b. myofibril.
    e. Both a and d are correct.
    c. myosin filament.
28. Nervous stimulation of a muscle fiber
    a. occurs at a neuromuscular junction.
    b. involves the release of ACh.
    c. results in impulses that travel down the T system.
    d. causes calcium to be released from the sarcoplasmic reticulum.
    e. All of these are correct.
29. Which of the following statements about cross-bridges is false?
    a. They are composed of myosin.
    b. They bind to ATP after they attach to actin.
    c. They contain an ATPase.
    d. They split ATP before they attach to actin.
30. Which of these is the direct source of energy for muscle fiber contraction?
    a. ATP
    d. glycogen
    b. creatine phosphate
    e. Both a and b are correct.
    c. lactic acid

31. Myoglobin content is higher in _____ -twitch muscle fibers respiring _____.
    a. slow, aerobically
    c. slow, anaerobically
    b. fast, aerobically
    d. None of these are correct.
32. **THINKING CONCEPTUALLY** Why do myosin heads have to be attached to actin during the power stroke of muscle contraction?

GET INVOLVED

1. You work in a morgue and do frequent autopsies. You know that it is possible to watch very thin muscle tissue contract under the microscope. How would you test the statement that rigor mortis is due to lack of ATP?
2. Exercise physiologists tell us that people who exercise use less oxygen per unit time and rely on fermentation less than those who do not exercise. If you had access to a medical laboratory including oxygen tanks, how would you test this finding? If your results support it, what explanation is possible?
3. After viewing this animated exploration of the processes involved in the contraction of skeletal muscles, explain the role of calcium in muscle contraction.

3D Animation
Skeletal Muscle
Contraction

MEDIA STUDY TOOLS

mhhe.com/maderconcepts3

Enhance your study of this chapter with interactive study tools, practice tests, and engaging animations. Also, ask your instructor about the resources available through ConnectPlus, which includes LearnSmart, a personalized adaptive learning program, and a media-rich eBook.

30

Circulation and Cardiovascular Systems

BEFORE YOU BEGIN

Take a few minutes to recall

How the capillaries supply cells with oxygen and glucose (section 7.1)

How the cardiovascular system utilizes various tissues (sections 26.2–26.4)

How the cardiovascular system contributes to homeostasis (section 26.8)

Not All Animals Have Red Blood

Our blood is red, as you no doubt have witnessed after suffering a cut. We tend to think that most animals, whether vertebrates or invertebrates, are pretty much like ourselves. So, it comes as a surprise to learn that the blood of some invertebrates is green or blue, not red. The color of blood is dependent on the pigment that transports oxygen. The job of a respiratory pigment is to bind oxygen in areas of higher concentration (usually gas-exchange surfaces, such as lungs or gills) and to release it in areas of lower concentration, usually the tissues.

Vertebrates have red blood because their respiratory pigment, hemoglobin, is red when it is bound to oxygen. Hemoglobin is packaged inside blood cells, appropriately called red blood cells. Each subunit of hemoglobin consists of the protein globin plus an embedded heme group. The heme group contains an iron atom that binds to oxygen. When oxygen is attached to the iron, hemoglobin is red; when oxygen is not attached, hemoglobin is sort of purplish. The expression "blue-blooded" is used to describe royalty because, in days gone by, their pale, untanned skin allowed the blue-tinged oxygen-poor blood in their veins to show through.

An earthworm (an invertebrate) has red blood, not because it contains hemoglobin, but because it contains giant, free-floating blood proteins bound to many dozens, even hundreds, of iron-containing heme groups. However, other annelids—such as tube worms that live in the sea—have the respiratory pigment chlorocruorin. Chlorocruorin appears red when oxygenated, but green when deoxygenated!

All vertebrates, from pandas to lionfishes, have red blood.

Hemocyanin, the second most common oxygen-transporting pigment found among animals, uses copper-containing heme groups instead of iron-containing heme groups. Another big difference is that hemocyanin is dissolved in the blood rather than packaged in cells, as is the hemoglobin of vertebrates. Copper turns blue when oxygenated, so some invertebrates are truly blue-blooded. Hemocyanin is present in the blood of marine arthropods, such as lobsters and horseshoe crabs, and also in most molluscs, including squids. The heart of a giant squid pumps blue blood.

Which type of invertebrate is well known for having colorless blood with no respiratory pigment? The terrestrial insects, of course. They have no need of a respiratory pigment because little air tubes called tracheae take air directly to mitochondria just inside the muscle cells. The rapid delivery of oxygen-laden air to flight muscles is very adaptive because insects' mode of transportation on land is flying, which is energy-intensive.

The occurrence of respiratory pigments does not appear strongly connected to evolutionary relationships, so it is hard to find a reason why some animals have red, some blue, and some green blood. One idea is that the pH of the environment affects the type of respiratory pigment. Hemocyanin is an excellent oxygen carrier, but it is very sensitive to pH changes. A very slight change toward acidity can cause hemocyanin to unload too early. Hemoglobin is not as sensitive to pH changes, so it becomes the better choice when the respiratory pigment is exposed to a different pH in the lungs compared to the tissues. In humans, the pH of the tissues is slightly lower than that of the lungs. Why? Because when carbon dioxide combines with the water in plasma, the liquid part of blood, it forms carbonic acid.

In this chapter, we will study the circulatory systems of invertebrates and vertebrates in an evolutionary context. We will also investigate the composition of mammalian blood.

Tube worms have green deoxygenated blood.

Marine shellfish have blue blood.

The blood of an insect is colorless.

Invertebrate Circulatory Systems

To maintain homeostasis, cells need a continual supply of oxygen and nutrients. They also need to rid themselves of metabolic wastes. A system that speeds transport of materials to and from cells helps animals including invertebrates remain homeostatic and in good health.

30.1 A circulatory system serves the needs of cells

LEARNING OUTCOMES

When you complete this section, you should be able to

1. Explain why some invertebrates (e.g., hydras and planarians) can make do without a circulatory system.
2. Contrast the manner in which an earthworm transports oxygen to cells with the way a grasshopper transports oxygen.

All animals must move materials from one body location to another. Any animal system that uses a fluid to transport substances within vessels may be called a **circulatory system.** The fluid transports nutrients, gases, and metabolic wastes either in solution or bound to soluble compounds often within cells, such as blood cells. To appreciate the importance of a circulatory system, recall that cells require glucose and oxygen in order to carry out cellular respiration and must excrete carbon dioxide and other wastes in order to continue their existence. Only if these needs are met can cells maintain homeostasis and continue to exist.

Like any other system, a circulatory system is expected to be appropriate to the size, complexity, and lifestyle of the organism. Otherwise, it would not be advantageous to the organism.

Some Invertebrates Have No Circulatory System Cnidarians, such as hydras, and flatworms, such as planarians, do not have a circulatory system as we have defined it (Fig. 30.1A). Instead, they circulate external water within their gastrovascular cavity to move substances to and from their cells. The cells are either part of an external layer, or they line the gastrovascular cavity. In either case, each cell is exposed to water and can independently exchange gases and get rid of wastes. The cells that line the gastrovascular cavity are specialized to carry out digestion internally. They pass nutrient molecules to other cells by diffusion. In a planarian, a trilobed gastrovascular cavity branches throughout the small, flattened body. No cell is very far from one of the three digestive branches, so nutrient molecules can diffuse from cell to cell. Similarly, diffusion meets the respiratory and excretory needs of the cells. These organisms must remain small and/or flat so that nearly every cell can be exposed to external water for exchange of substances.

Some Invertebrates Use Their Coelom as a Circulatory System Recall that a coelom is a body cavity that houses the internal organs. This cavity contains coelomic fluid, which bathes the organs and can act as a hydrostatic skeleton. Some animals, such as pseudocoelomate invertebrates (i.e., roundworms), use the coelomic fluid of their body cavity for transport purposes, as do the coelomate echinoderms, such as sea stars.

Some Circulatory Systems Are Open Systems Two ways of circulating a fluid internally are seen among animals. These methods are called an **open circulatory system** and a **closed circulatory system.** Both systems utilize a transport fluid called blood within

one or more vessels and some sort of pumping device to keep blood flowing. The "blood" need not be red nor have any respiratory pigment (see "Not All Animals Have Red Blood" on pages 592–593). Open systems are utilized in animals that have a reduced coelom and whose organs and cavity are not lined by an internal membrane. Therefore, cells are better able to exchange substances directly with surrounding fluid. The open system is so-called because vessels transport blood (actually a mixture of blood and tissue fluid called **hemolymph**) to open places (termed sinuses) within the body cavity where it directly bathes the organs. The body cavity is appropriately called a **hemocoel.** Open circulatory systems are typical of arthropods, such as crayfish and insects, and molluscs such as clams but not cephalopods (i.e., squid).

For example, in most molluscs and arthropods, the heart pumps hemolymph containing the respiratory pigment hemocyanin via vessels into tissue spaces that are sometimes enlarged into saclike sinuses. Eventually, hemolymph drains back to the heart. In the grasshopper, an insect arthropod, the tubular

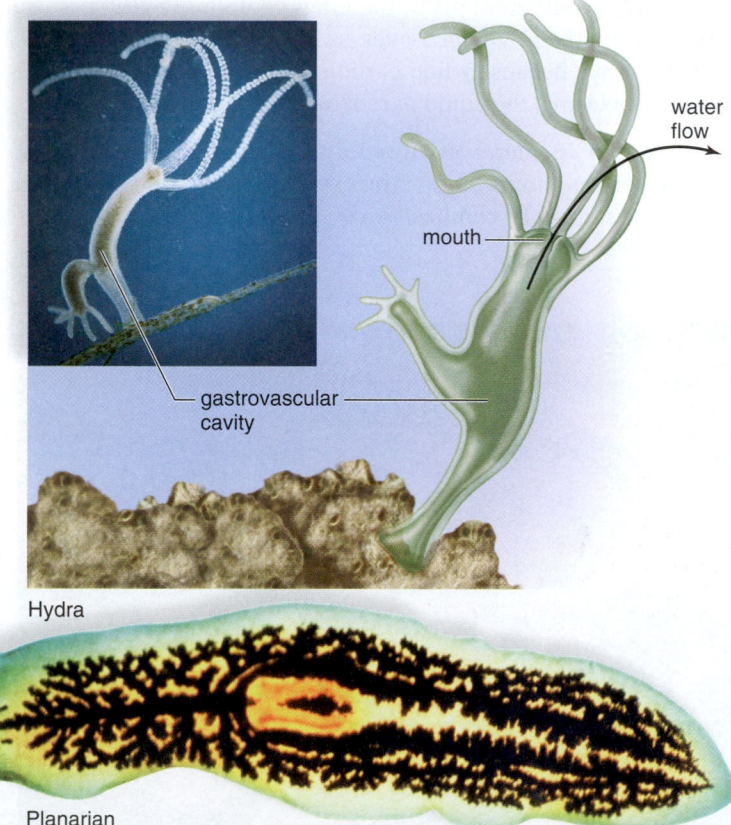

water flow

mouth

gastrovascular cavity

Hydra

Planarian

FIGURE 30.1A Invertebrates such as a hydra (above) and a planarian (below) have no circulatory system. Because they are small and/or flat, they can use their gastrovascular cavity to circulate substances to and from cells.

heart pumps hemolymph into a dorsal aorta, which empties into the hemocoel (Fig. 30.1B*a*). When the heart contracts, openings called ostia (sing., ostium) are closed; when the heart relaxes, the hemolymph is sucked back into the heart by way of the ostia. An open circulatory system has a slow delivery of oxygen and nutrients to cells but is appropriate for a clam, which is not quick acting. But an insect such as a grasshopper is quite active. A grasshopper has colorless blood and doesn't depend on its open circulatory system to deliver oxygen to its muscles. Instead, it has numerous little air tubes, called **tracheae,** that open to the outside and take oxygen-laden air directly to its flight muscles. Flying is an adaptation to life on land, and so is the use of tracheae to deliver oxygen to muscles.

The lower blood pressure and velocity in an open circulatory system may allow it to fulfill multiple functions. For example, in a clam the hemocoel functions in part as a hydrostat that expands its hatchet foot for digging. Also, in arthropods, the open circulatory system acts as a hydrostatic skeleton when its external skeleton undergoes molting.

Other Circulatory Systems Are Closed Systems Some invertebrates, such as annelids, and all the vertebrates have a circulatory system in which the blood never runs free and is always contained in blood vessels. Therefore, this is a closed system. The blood passes through a series of blood vessels from large to small until it reaches a capillary bed. Because capillaries have very thin walls that are a single cell-layer thick, they allow exchanges to take place by diffusion. Not only that, their collective cross-sectional area is large and this makes the blood flow slowly under reduced pressure facilitating exchange with the body's cells.

In animals that utilize a closed circulatory system, the coelom tends to be large and well developed. Organs including blood vessels are located in the coelom, but otherwise it plays no direct transport function.

It takes power to keep blood moving through a series of blood vessels. An earthworm has a series of contractile vessels that are often called hearts, and their activity keeps the blood moving. Vertebrates have a well-developed chambered heart as we will learn in "Evolution of a Four-Chambered Heart" on page 599. Even so, we can use the diagram that accompanies the earthworm in Figure 30.1B*b* to also represent a closed circulatory system in vertebrates. In vertebrates, the blood vessels that pump blood to a capillary bed are called arteries, and those that return blood to the heart are called veins.

If we were to examine the closed circulatory system of an earthworm in more detail, we would see that any one of the hearts pumps blood into a ventral blood vessel. Valves prevent the backward flow of blood. In every segment of the worm's body, the ventral blood vessel (an artery) branches into a lateral vessel. From a lateral vessel blood moves into capillaries, where exchanges with tissue fluid take place. Small veins then enter the dorsal blood vessel (a vein) and this blood vessel returns blood to the heart for repumping. The earthworm has red blood that contains a respiratory pigment akin to hemoglobin. The pigment is dissolved in the blood and not contained within cells. The earthworm has no specialized boundary, such as lungs, for gas exchange with the external environment. Gas exchange takes place across the body wall, which must always remain moist for this purpose.

▶ **30.1 CHECK YOUR PROGRESS**
1. Identify and explain why some animals are able to exist without a circulatory system.
2. Contrast the possible benefits of an open circulatory system with those of a closed system.

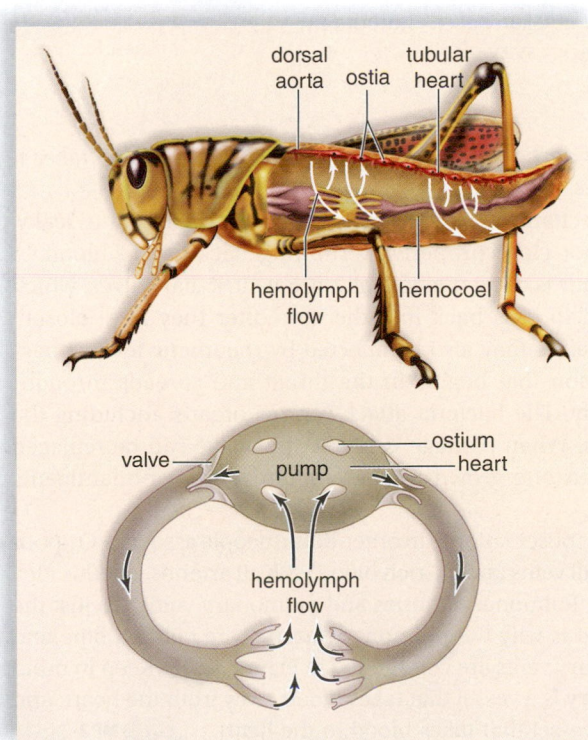

a.

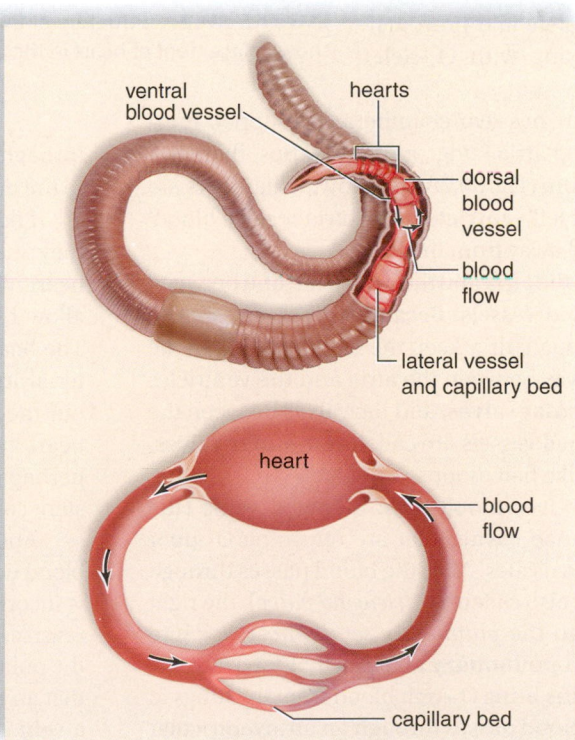

b.

FIGURE 30.1B
a. An open circulatory system occurs in a grasshopper. **b.** A closed circulatory system occurs in an earthworm.

Vertebrate Cardiovascular Systems

The cardiovascular system of vertebrates consists of a four-chambered heart and arteries (blood vessels) that take blood away from the heart and veins that bring blood to the heart. Arteries and veins are connected by capillaries, which are the smallest and most numerous of the blood vessels.

30.2 The mammalian heart is a double pump

LEARNING OUTCOMES

When you complete this section, you should be able to

1. Describe the anatomy of the human heart, including its chambers, valves, and attached blood vessels.

2. Explain how the right side of the heart contains O_2-poor blood and the left side contains O_2-rich blood.

3. Correlate the functions of the SA node and AV node with the heartbeat.

Mammals, like all vertebrates, have a closed circulatory system, which is called a **cardiovascular system** because it consists of a heart (*cardio*) and a system of blood vessels (*vascular*). The strong, muscular heart has four chambers.

A **septum** divides the heart into left and right sides. (Note that "right/left side of the heart" refers to how the heart is positioned in your body, not to the right side of a diagram.) The right side of the heart pumps O_2-poor blood to the lungs, and the left side of the heart pumps O_2-rich blood to the tissues. The septum is complete and prevents O_2-poor blood from mixing with O_2-rich blood.

Each side of the heart has two chambers. The upper, thin-walled chambers are called atria (sing., **atrium**)—thus, there is a right atrium and a left atrium (Fig. 30.2A). The lower chambers are the thick-walled right and left **ventricles.** The atria receive blood; the ventricles pump blood away from the heart.

Valves occur between the atria and the ventricles, and between the ventricles and attached vessels. Because these valves close after the blood moves through, they keep the blood moving in the correct direction. The valves between the atria and the ventricles are called the **atrioventricular valves,** and the valves between the ventricles and their attached vessels are called **semilunar valves,** because their cusps look like half moons.

The right atrium receives blood from attached veins (the **superior** and **inferior venae cavae**) that are returning O_2-poor blood to the heart from the tissues. After the blood passes through the atrioventricular valve (also called the *tricuspid valve*), the right ventricle pumps it through the *pulmonary semilunar valve* into the **pulmonary trunk** and **pulmonary arteries** that take it to the lungs. The **pulmonary veins** bring O_2-rich blood from the lungs to the left atrium. After this blood passes through an atrioventricular valve (also called the *bicuspid valve*), the left ventricle pumps it

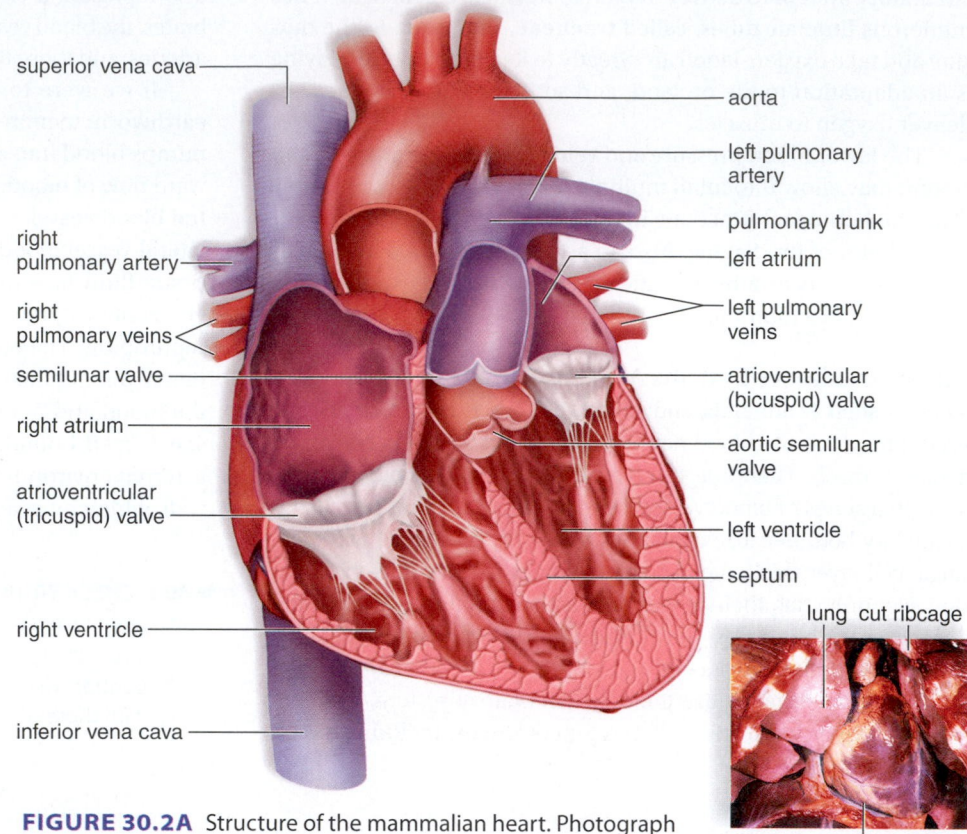

FIGURE 30.2A Structure of the mammalian heart. Photograph shows placement of heart in the thoracic cavity.

through the *aortic semilunar valve* into the **aorta,** which takes it to the tissues.

Like mechanical valves, the heart valves are sometimes leaky; they may not close properly, permitting backflow of blood. A heart murmur is often due to leaky atrioventricular valves, which allow blood to pass back into the atria after they have closed. The heart valves may also be affected by rheumatic fever, a bacterial infection that begins in the throat and spreads throughout the body. The bacteria attack various organs, including the heart valves. When damage is severe, the valve can be replaced perhaps with one grown in the laboratory from nonantigenic stem cells.

Another observation is in order. Some people associate O_2-poor blood with all veins and O_2-rich blood with all arteries, but this idea is incorrect: Pulmonary arteries and pulmonary veins are just the reverse. That is why the pulmonary arteries are colored blue and the pulmonary veins are colored red in Figure 30.2A. Keep in mind that an **artery** is a vessel that takes blood away from the heart, and a **vein** is a vessel that takes blood to the heart, regardless of the blood's oxygen content.

MP3
Heart Structure and Function

Heartbeat The average human heart contracts, or beats, about 70 times a minute, or 2.5 billion times in a lifetime. Each heartbeat lasts about 0.85 seconds, called the cardiac cycle, and can be divided into three phases (Fig. 30.2B):

1 The atria contract (while the ventricles relax); 0.15 seconds.

2 The ventricles contract (while the atria relax); 0.30 seconds.

3 All chambers rest; 0.40 seconds.

The term **systole** refers to contraction of the heart chambers, and the word **diastole** refers to relaxation of these chambers. Note that the heart is in diastole about 50% of the time. The short systole of the atria is appropriate, because the atria send blood only into the ventricles. It is the muscular ventricles that actually pump blood out into the cardiovascular system proper. The word *systole* used alone usually refers to the left ventricular systole. The volume of blood that the left ventricle pumps per minute into the systemic circuit is almost equivalent to the amount of blood in the body. During heavy exercise, the cardiac output can increase manyfold.

When the heart beats, the familiar "*lub-dub*" sound is heard as the valves of the heart close. The longer, lower-pitched *lub* is caused by vibrations of the heart when the atrioventricular valves

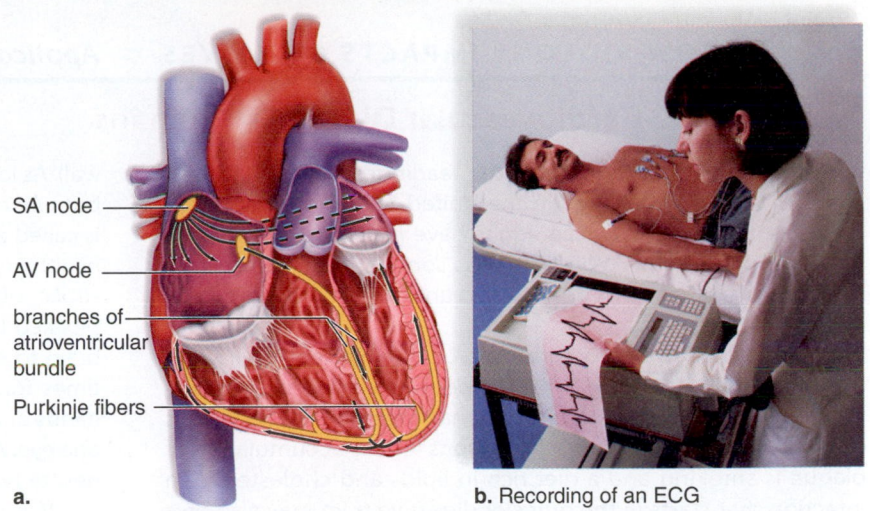

SA node

AV node

branches of atrioventricular bundle

Purkinje fibers

a.

b. Recording of an ECG

FIGURE 30.2C a. Conduction system of the heart results in (**b**) a normal ECG recording when it is working properly.

close due to ventricular contraction. The shorter, sharper *dub* is heard when the semilunar valves close due to back pressure of blood in the arteries. The **pulse** is a wave effect that passes down the walls of the arterial blood vessels following ventricular systole and can be felt at various points externally. The arterial pulse rate can be used to determine the heart rate, which is why taking your pulse is one of the first things a physician does during an examination.

The Cardiac Conduction System The rhythmic contraction of the heart is due to the **cardiac conduction system** (Fig. 30.2C*a*). Nodal tissue, which has both muscular and nervous characteristics, is a unique type of cardiac muscle. The SA (sinoatrial) node, located in the dorsal wall of the right atrium causes the atria to contract and initiates the heartbeat every 0.85 seconds. Therefore, the SA node is called the **cardiac pacemaker.** When the impulse reaches the AV (atrioventricular) node located in the base of the right atrium near the septum, the AV node signals the ventricles to contract by way of large fibers terminating in the smaller but more numerous Purkinje fibers. Although the beat of the heart is intrinsic, it is regulated by the nervous system, which can increase or decrease the rate.

An **electrocardiogram (ECG)** is a recording of the electrical changes that occur in the heart during a cardiac cycle. When an ECG is being taken, electrodes placed on the skin are connected by wires to an instrument that detects the heart's electrical changes (Fig. 30.2C*b*). As the heart contracts, the instrument prints out the ECG. An ECG can detect various types of abnormalities. Ventricular fibrillation (uncoordinated contractions) is of special interest because it can be caused by an injury or a drug overdose. It is the most common cause of sudden cardiac death in a seemingly healthy person. Once the ventricles are fibrillating, they have to be defibrillated by applying a strong electric current for a short period of time.

MP3 Cardiac Cycle

Animation Cardiac Cycle

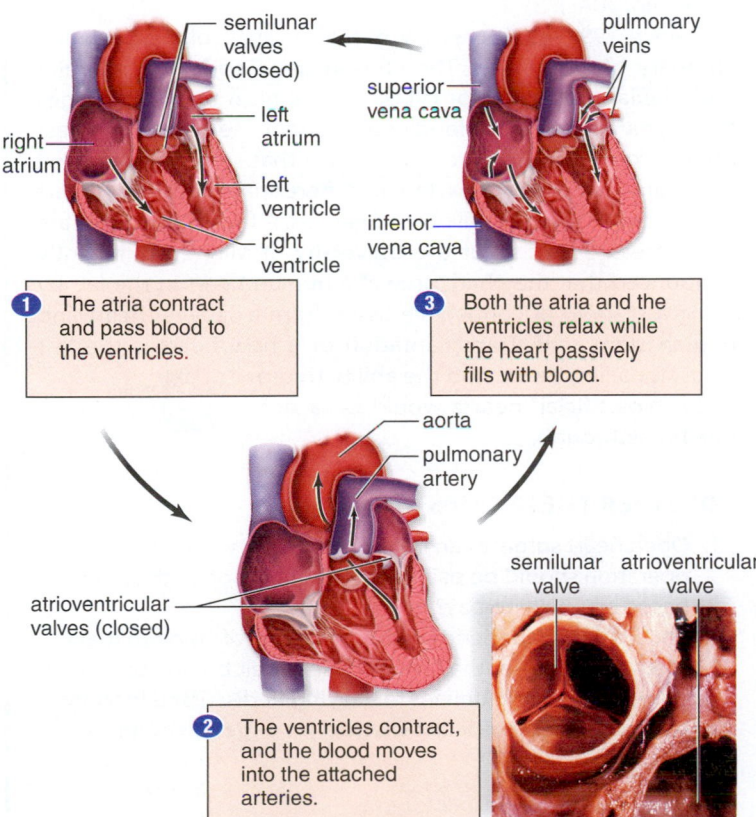

semilunar valves (closed)

pulmonary veins

superior vena cava

left atrium

right atrium

left ventricle

right ventricle

inferior vena cava

1 The atria contract and pass blood to the ventricles.

3 Both the atria and the ventricles relax while the heart passively fills with blood.

aorta

pulmonary artery

atrioventricular valves (closed)

semilunar valve

atrioventricular valve

2 The ventricles contract, and the blood moves into the attached arteries.

FIGURE 30.2B During the cardiac cycle, chambers fill with blood, contract, and then relax. The heartbeat sounds are due to closure of atrioventricular and semilunar valves.

▶ **30.2 CHECK YOUR PROGRESS**

1. Identify which artery carries O₂-poor blood and explain.
2. Contrast what causes the heart chambers to contract with what causes the sounds of the heartbeat.

30A   Cardiovascular Disease in Humans

Cardiovascular disease (CVD) is the leading cause of untimely death in Western countries. In the United States, it is estimated that about 33% of American adults have *hypertension,* which is high blood pressure. Hypertension is sometimes called a silent killer because it may not be detected until a stroke, heart attack, or an aneurysm occur.

Hypertension (a blood pressure reading higher than 130/90 in young adults and 140/95 in older people) is often seen in individuals who have atherosclerosis, a narrowing of blood vessels due to *plaque.* Chief among the reasons for the accumulation of plaque is smoking and a diet rich in lipids and cholesterol. An infection that starts in the gums or digestive tract may also be a cause. Atherosclerosis begins in early adulthood and develops progressively through middle age when symptoms appear.

An aneurysm or ballooning of a blood vessel, most often the aorta, occurs when atherosclerosis and high blood pressure weaken an arterial wall to the point that it may burst. Currently, if surgery is called for, the patient is given a stent to support the aorta or an artificial graft to replace a section of the aorta. In the future it may be possible to use a graft made in the laboratory by injecting a patient's cells inside an inert mold. In other instances, plaque causes a clot to form on the irregular arterial wall. As long as the clot remains stationary, it is called a thrombus, but when and if it dislodges and moves along with the blood, it is called an embolus. If thromboembolism is not treated, serious health problems can result. A cardiovascular incident, also called a *stroke,* often occurs when a small cranial arteriole bursts or is blocked by an embolus. Lack of oxygen causes a portion of the brain to die, and paralysis or death can result. A person is sometimes forewarned of a stroke by a feeling of numbness in the hands or the face, difficulty in speaking, or temporary blindness in one eye. A stroke is not limited to the elderly; in 2006, former Minnesota Twins baseball star Kirby Puckett died of a stroke at age 45.

If a coronary artery becomes partially blocked, the individual may first suffer from angina pectoris, characterized by a squeezing or burning sensation in the chest. When a coronary artery is completely blocked, a portion of the heart muscle dies due to lack of oxygen, and a myocardial infarction or *heart attack* occurs. Various procedures are available to correct this situation. During coronary bypass operations, often performed by robotic surgery (Fig. 30A*a*), a portion of a blood vessel from another part of the body is sutured, usually from the aorta to the coronary artery, past the point of obstruction (Fig. 30A*b*). Now blood flows normally from the aorta to the wall of the heart. In balloon angioplasty, a stent with an inner balloon is pushed into the blocked area. When the balloon inside the stent is inflated, it expands, locking itself in place (Fig. 30A*c*).

Video Heart Stem Cells

Since 1997, gene therapy has been available for clogged coronary blood vessels. The chosen gene codes for vascular endothelial growth factor (VEGF), a chemical that encourages the appearance of new blood vessels and these carry blood past the blocked ones. Research has shown that stem cells injected into damaged heart muscle will differentiate into new heart muscle cells as well as new blood vessels. A few clinical trials are in progress. Scientists at the University of Minnesota recently announced that they had "grown" a human heart in the lab by using a scaffold of connective tissue from a cadaver heart and human stem cells. Transplantation of a new heart is the last resort for ailing hearts and the ability to produce "bioartificial" hearts would be a definite breakthrough.

Video Cardiac Repair

CONSIDER THESE QUESTIONS

1. Open heart surgery can be expensive. Do you believe the operation should be paid for by insurance or by the patient based on ability to pay?
2. Should people depend on the medical profession to make them healthy after they have abused their bodies, or should they follow the guidelines for staying fit discussed in "How to Prevent Cardiovascular Disease" on page 603? Explain.

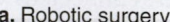

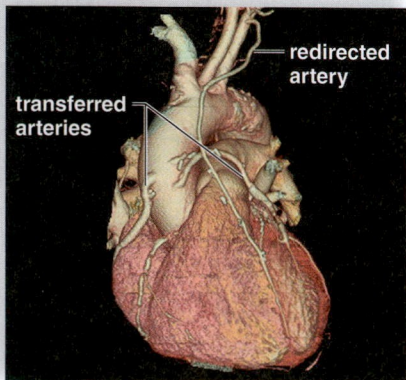

TV screen shows operation in progress.

Doctor manipulates instruments.

Movable arms hold instruments.

a. Robotic surgery

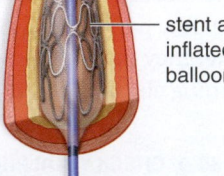

redirected artery

transferred arteries

b. Coronary bypass

stent and inflated balloon

c. Stenting

FIGURE 30A Treatment for a clogged coronary artery.

connect | BIOLOGY   Explore the concepts through a variety of multimedia assets and question types.

www.mcgrawhillconnect.com

30B Evolution of a Four-Chambered Heart

Two different types of circulatory pathways are seen among vertebrate animals. Fishes have a single-loop system in which the heart pumps the blood only to the gills. The other vertebrates have a two-circuit (double-loop) circulatory pathway. In the larger *systemic circuit,* the heart pumps blood to all parts of the body except the lungs; the smaller *pulmonary circuit* pumps blood to the lungs.

Fishes

In fishes, the heart has a single atrium and a single ventricle (Fig. 30B.1). The pumping action of the ventricle sends blood under pressure to the gills, where gas exchange with the external environment occurs. Fishes have an efficient means of respiration, and their blood is fully enriched with oxygen when it leaves the *gills,* the respiratory organ for aquatic organisms. But after passing through the gills, blood is no longer under pressure as it travels in the aorta to the rest of the body. This means the rate of oxygen delivery to the tissues is limited. However, fishes are ectotherms and have a reduced body temperature. A sluggish delivery of oxygen to the body proper is usually sufficient; the undulating movement of a fish's body helps move the blood back to the heart.

Amphibians and Reptiles

Amphibians were the first vertebrates to invade the land, and an evolutionary change in their cardiovascular system supports this change in their environment. A single ventricle pumps blood in the pulmonary circuit to the lungs, the respiratory organ of land vertebrates. It also pumps blood in the systemic circuit to the rest of the body (Fig. 30B.2).

Although both O_2-rich and O_2-poor blood enter a single ventricle, the two types of blood are kept somewhat separate because O_2-poor blood is pumped out of the ventricle to the lungs before O_2-rich blood enters and is pumped to the systemic circuit. Also, frogs and salamanders have a moist skin that helps recharge their blood with oxygen.

In some reptiles (e.g., lizards) the ventricle is partially divided, making it even less likely that O_2-rich and O_2-poor blood will mix in the single ventricle.

Birds and Mammals

In some reptiles (e.g., crocodiles and birds) and mammals, a septum completely separates the ventricles. This means that these animals have a complete separation of the pulmonary and systemic circuits (Fig. 30B.3). In birds and mammals, the right ventricle pumps blood under low pressure to the lungs, and the larger left ventricle pumps blood under high pressure to the rest of the body. Birds and mammals are endotherms, and they locomote well on land. Good oxygen delivery is needed for their active way of life and the maintenance of a warm internal temperature. These features may have helped them become the dominant animals on land today.

CONSIDER THESE QUESTIONS

1. Activity of the gene *TBX5* is needed for a septum to develop. In which animals would you expect to find expression of this gene during development?
2. Give examples to show that mammals exist in all parts of the globe. Argue that this is due to their efficient circulatory system.

 Explore the concepts through a variety of multimedia assets and question types.

www.mcgrawhillconnect.com

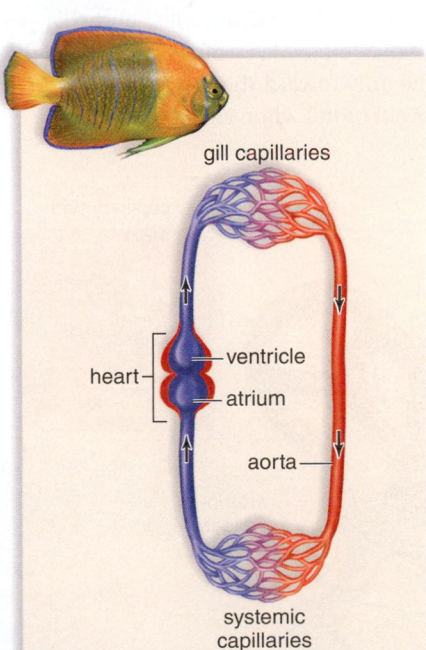

FIGURE 30B.1 A single-loop circulatory pathway occurs in fishes.

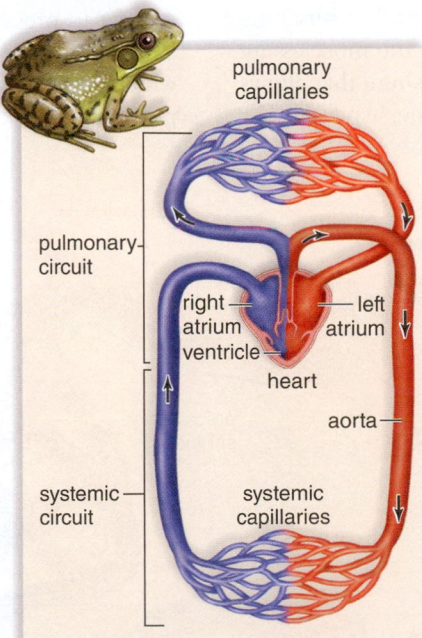

FIGURE 30B.2 A two-circuit pathway occurs in amphibians and most reptiles.

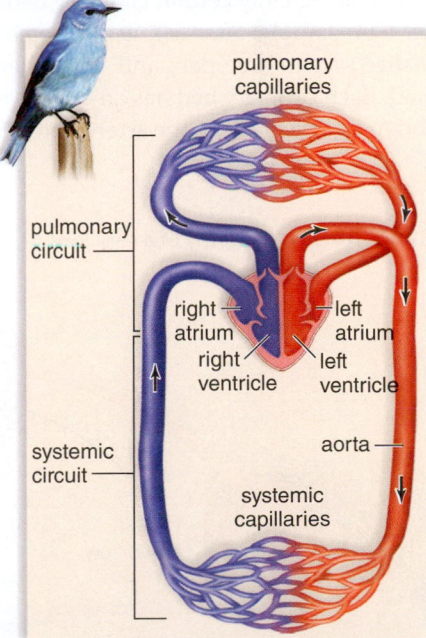

FIGURE 30B.3 Complete separation of pulmonary and systemic circuits occurs in birds, mammals, and some reptiles.

30.3 The structure of blood vessels is suited to their function

LEARNING OUTCOME

When you complete this section, you should be able to

1. Compare the structure, function, and movement of blood in arteries, veins, and capillaries.

The cardiovascular system has three types of blood vessels: arteries (and arterioles), which carry blood away from the heart to the capillaries; capillaries, which permit exchange of material with the tissues; and veins (and venules), which return blood from the capillaries to the heart (Fig. 30.3A).

MP3 Classification of Blood Vessels

Arteries have a much thicker wall than veins because of a well-developed middle layer consisting of smooth muscle and elastic tissue. The elastic tissue allows arteries to expand and accommodate the sudden increase in blood volume that results after each heartbeat. The well-developed smooth muscle prevents arteries from expanding too much.

Smaller arteries branch into a number of **arterioles,** which are just visible to the naked eye. The diameter of arterioles can be regulated by the nervous system. When arterioles are dilated, more blood flows through them, and when they are constricted, less blood flows. The constriction of arterioles can also raise blood pressure.

Arterioles branch into capillaries. **Capillaries** are extremely narrow—about 8–10 mm wide—and have thin walls composed of a single layer of *endothelium* (a simple squamous epithelium) with a basement membrane. The thin walls of a capillary facilitate capillary exchange. Although each capillary is small, collectively they form vast networks; their total surface area in humans is about 6,000 square meters. Because capillaries serve the cells, the heart and the other vessels of the cardiovascular system can be thought of as the means by which blood is conducted to and from the capillaries. Only certain capillary beds are open at any given time. For example, after eating, the capillary beds that serve the digestive system are open, and those that serve the muscles are closed. Each capillary bed has an arteriovenous shunt that allows blood to go directly from the arteriole to the venule, bypassing the

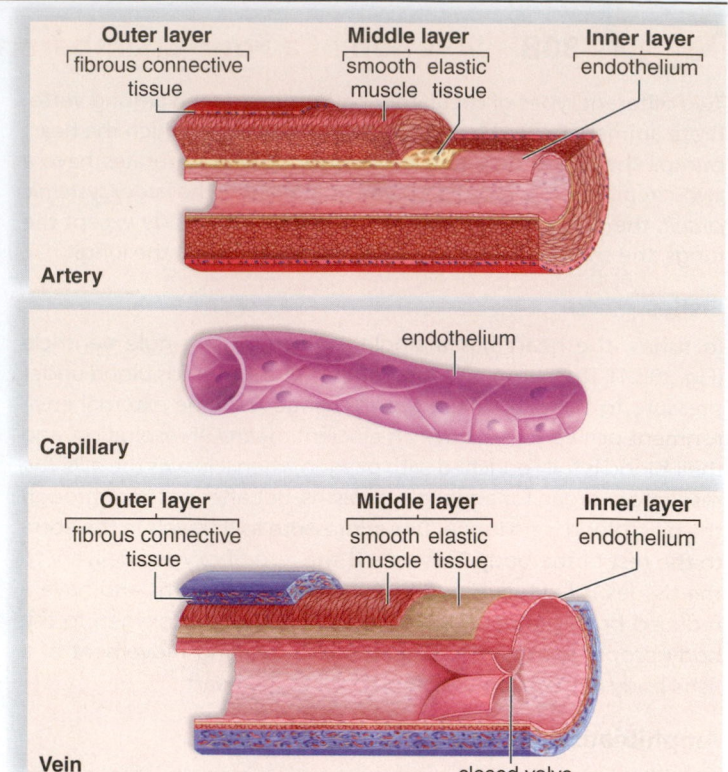

FIGURE 30.3A Types of blood vessels.

bed (Fig. 30.3B). Contracted precapillary sphincter muscles prevent the blood from entering the capillary vessels.

Veins and venules take blood from the capillary beds to the heart. First, the **venules** (small veins) drain blood from the capillaries; then they join to form a vein. The middle layer of a vein (and venule) is thinner than that of an artery. This makes them subject to pressure exerted by skeletal muscles, and this pressure helps move blood in the veins. Also, veins often have **valves,** which allow blood to flow only toward the heart when open and prevent the backward flow of blood when closed.

FIGURE 30.3B Anatomy of a capillary.

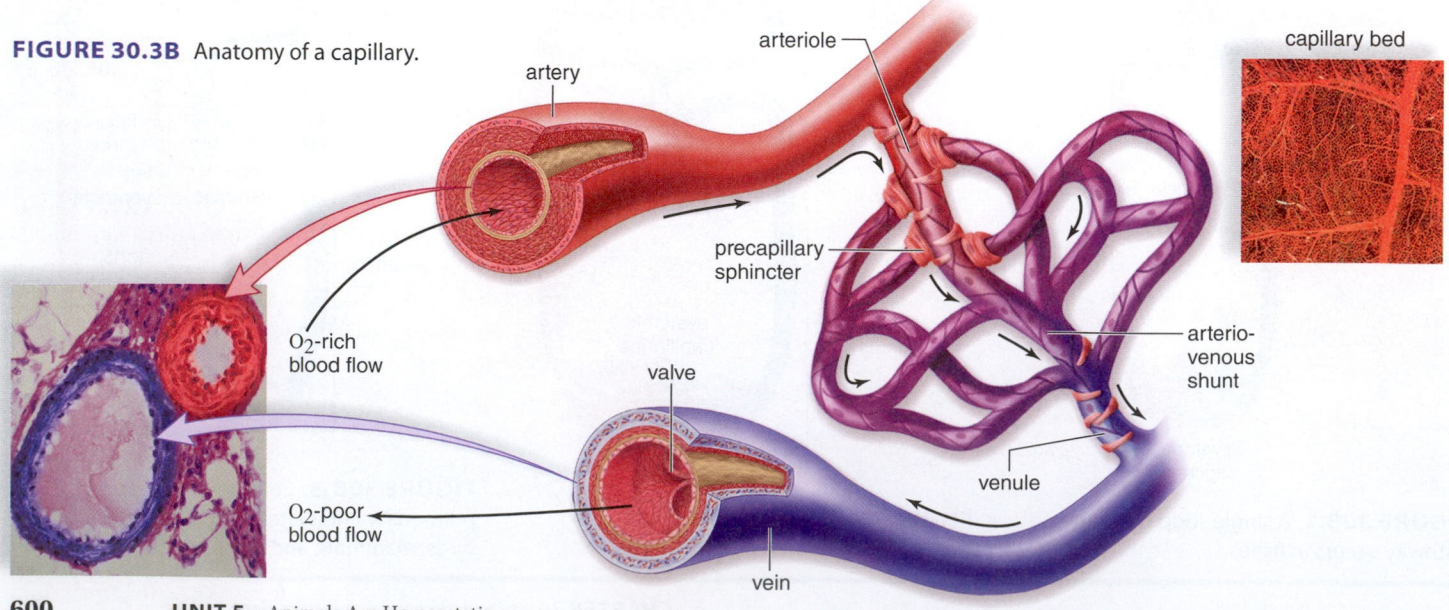

Blood Flow in Arteries When the left ventricle contracts, blood is forced into the aorta and other systemic arteries under pressure. **Systolic pressure** results from blood being forced into the arteries during ventricular systole, and **diastolic pressure** is the pressure in the arteries during ventricular diastole. Human **blood pressure** can be measured with a traditional mercury manometer or with a digital manometer. Skill is required to accurately use a mercury manometer, but a digital manometer usually requires no training. With both types of manometers, a pressure cuff determines the amount of pressure required to stop the flow of blood through an artery. Blood pressure is normally measured on the brachial artery of the upper arm, but digital manometers often use other parts of the body, such as the wrist. A blood pressure reading consists of two numbers—for example, 120/80—that represent systolic and diastolic pressures, respectively.

Blood pressure accounts for the flow of blood from the heart to the capillaries. As blood flows from the aorta into the various arteries, arterioles, and capillaries (Fig. 30.3C *top*) blood pressure falls. Also, the difference between systolic and diastolic pressure gradually diminishes. The fall of blood pressure and blood velocity in the capillaries may be related to the very high total cross-sectional area of the capillaries (Fig. 30.3C *bottom*). It has been calculated that if all the blood vessels in a human were connected end-to-end, the total distance would reach around the Earth at the equator two times. A large portion of this distance would be due to the quantity of capillaries. The slow movement

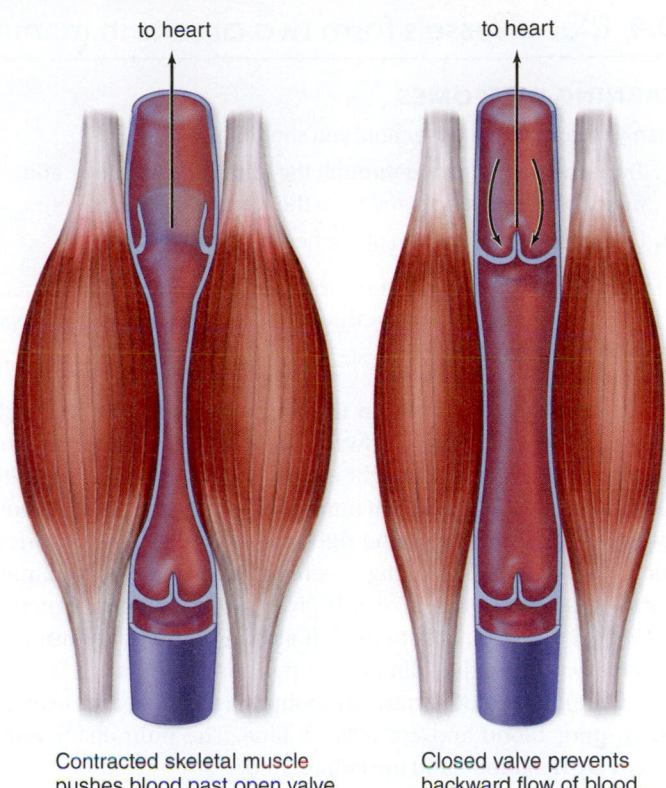

to heart   to heart

Contracted skeletal muscle   Closed valve prevents
pushes blood past open valve.   backward flow of blood.

FIGURE 30.3D How a valve affects the movement of blood in a vein.

of blood in the capillaries provides time for the gas exchange and nutrient-for-waste exchange that occur across capillary walls.

Blood Flow in Veins Blood pressure in the veins is low and cannot move blood back to the heart, especially from the limbs. Instead, venous return depends upon three factors: skeletal muscle contraction, the presence of valves in veins, and respiratory movements. When the skeletal muscles near veins contract, they put pressure on the collapsible walls of the veins and on the blood contained in these vessels. Veins, however, have valves that prevent the backward flow of blood, and therefore pressure from muscle contraction is sufficient to move blood through the veins toward the heart (Fig. 30.3D). When a person inhales, the thoracic pressure falls and the abdominal pressure rises as the chest expands. This also aids the flow of venous blood back to the heart because blood flows in the direction of reduced pressure. Blood velocity increases slightly in the venous vessels due to a progressive reduction in the cross-sectional area as small venules join to form veins.

MP3
Blood Flow and
Blood Pressure

Varicose veins, abnormal dilations in superficial veins, develop when the valves of the veins become weak and ineffective due to backward pressure of the blood. Crossing the legs or sitting in a chair so that its edge presses against the back of the knees can contribute to the development of varicose veins in the legs. Varicose veins of the anal canal are known as hemorrhoids.

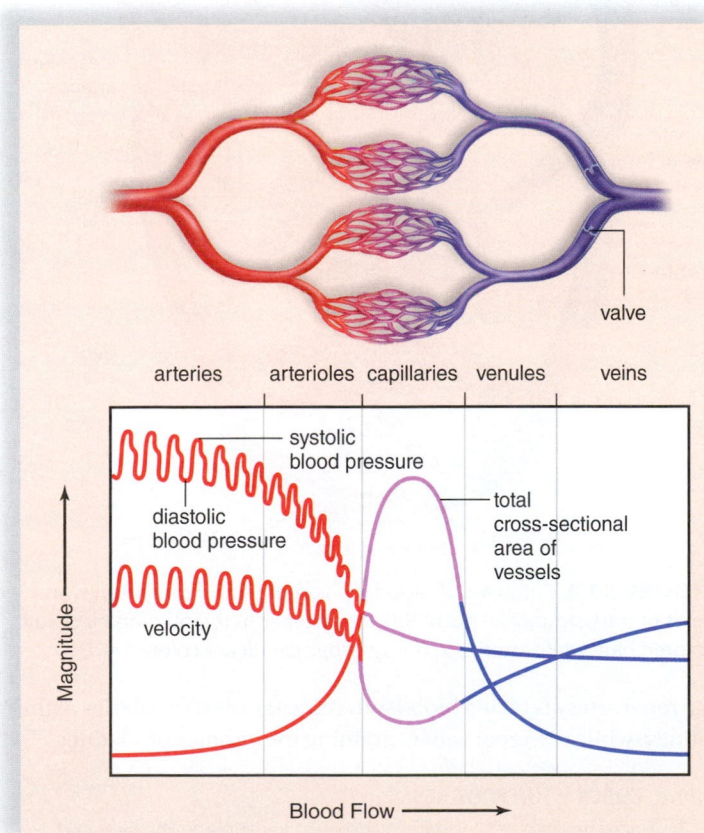

valve

arteries   arterioles   capillaries   venules   veins

systolic blood pressure

diastolic blood pressure

total cross-sectional area of vessels

velocity

Magnitude

Blood Flow

FIGURE 30.3C Velocity and blood pressure are related in part to the cross-sectional area of the blood vessels, depicted above.

▶ **30.3 CHECK YOUR PROGRESS**
1. Explain how the structure of arteries, veins, and capillaries is suited to their function.
2. Explain why a soldier who stands immobile for hours might faint.

30.4 Blood vessels form two circuits in mammals

LEARNING OUTCOMES

When you complete this section, you should be able to

1. Trace the path of blood through the human heart, lungs, and major blood vessels to and from the kidneys, head, and legs.
2. Understand the function of the hepatic portal system.

As exemplified in humans, the mammalian cardiovascular system includes two major pathways: the pulmonary circuit and the systemic circuit (Fig. 30.4).

The Pulmonary Circuit In the **pulmonary circuit,** the path of blood can be traced as follows. O_2-poor blood from all regions of the body collects in the right atrium and then passes into the right ventricle, which pumps it into the pulmonary trunk. The pulmonary trunk divides into the right and left pulmonary arteries, which carry blood to the lungs. As blood passes through pulmonary capillaries, carbon dioxide is given off and oxygen is picked up. O_2-rich blood returns to the left atrium of the heart, through pulmonary venules that join to form pulmonary veins.

Notice in Figure 30.4 that in the pulmonary circuit, arteries contain O_2-poor blood and are colored blue. The pulmonary veins contain O_2-rich blood and are colored red.

The Systemic Circuit In the **systemic circuit,** arteries contain O_2-rich blood and have a bright red color, but veins contain O_2-poor blood and appear dull red or, when viewed through the skin, blue. The aorta and the venae cavae (sing., **vena cava**) are the major blood vessels in the systemic circuit. To trace the path of blood to any organ in the body, you need only to start with the left ventricle and then mention the aorta, the proper branch of the aorta, the organ, and the vein returning blood to the vena cava, which enters the right atrium. For example, if you were tracing the path of the blood to and from the kidneys, you would mention the left ventricle, the aorta, the renal artery, the renal vein, the inferior vena cava, and the right atrium.

The coronary arteries (not shown in Fig. 30.4) are extremely important because they serve the heart muscle itself. Failure of the coronary arteries to perform this function results in a heart attack because the heart is not nourished by the blood in its chambers. The coronary arteries arise from the aorta just above the aortic semilunar valve. They lie on the exterior surface of the heart, where they branch into arterioles and then capillaries. In the capillary beds, nutrients, wastes, and gases are exchanged between the blood and the tissues. The capillary beds enter venules, which join to form the cardiac veins, and these empty into the right atrium.

A **portal system** begins and ends in capillaries. The **hepatic portal system** takes blood from the intestines to the liver. The liver, an organ of homeostasis, modifies substances absorbed by the intestines, removes toxins and bacteria picked up from the intestines, and monitors the composition of the blood. Blood leaves the liver by way of the hepatic vein, which enters the inferior vena cava.

Although Figure 30.4 gives the impression that only arteries occur on the left side of the body and only veins occur on the right side of the body, this is not the case. In fact, all parts of the body contain all three types of blood vessels. For example, the iliac artery and vein run side by side into each leg. Similarly, each kidney receives both a renal artery and a renal vein. For both kidneys,

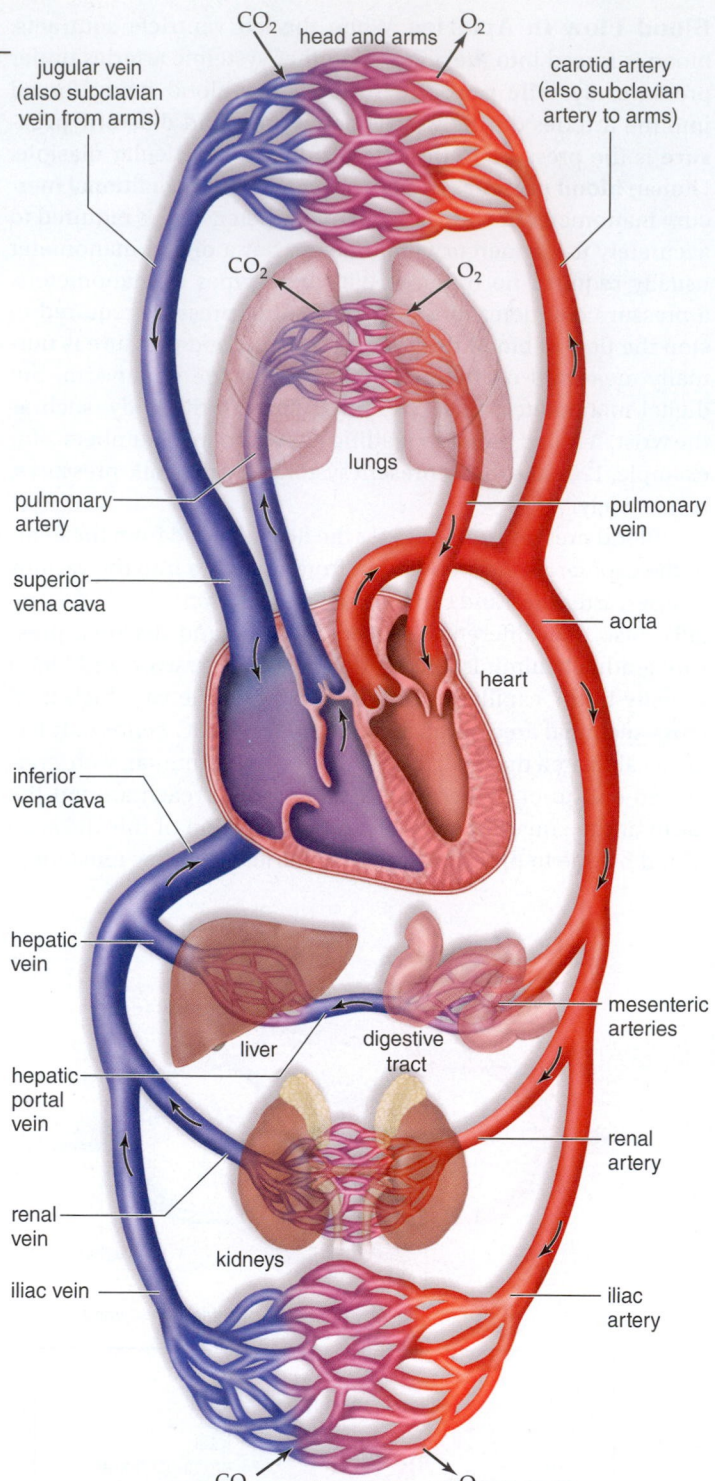

FIGURE 30.4 Pathways of blood in the most significant arteries and veins of the mammalian body. Why are arteries in the pulmonary circuit colored blue while arteries in the systemic circuit are colored red?

the renal artery is taking blood into the arterioles/capillaries of the kidney, while the renal vein is draining the venules of a kidney.

> **30.4 CHECK YOUR PROGRESS**
> 1. Trace the path of blood from the right ventricle to the liver and back to the right atrium.
> 2. Identify how the positioning of the hepatic portal system assists homeostasis.

30C How to Prevent Cardiovascular Disease

Many of us are predisposed to cardiovascular disease (CVD) due to factors beyond our control. Having a family history of heart attacks under age 55, being male, or being African American increases the risk. Other risk factors for CVD are related to our behavior.

The Don'ts

Smoking and Drug Abuse

When a person smokes, the drug nicotine, present in cigarette smoke, enters the bloodstream. Nicotine causes arterioles to constrict, including those that supply the heart itself. It also increases blood pressure, and the tendency of blood to clot. These factors may explain why about 20% of deaths from CVD are directly related to cigarette smoking.

Stimulants, such as cocaine and amphetamines, can cause an irregular heartbeat and lead to heart attacks even when using the drugs for the first time.

Alcohol abuse can destroy just about every organ in the body, the heart included. Based on several large research studies, however, the AHA notes that people who consume one or two drinks per day have a 30% to 50% reduced risk of CVD compared to nondrinkers. Importantly, because of the potential downsides of alcohol consumption, the AHA does not recommend that nondrinkers start using alcohol.

Obesity

People who are obese have more tissues that need to be supplied with blood. To meet this demand, the heart pumps blood out under greater pressure. Being overweight also increases the risk of type 2 diabetes, in which glucose damages blood vessels and makes them more prone to the development of plaque.

The Do's

Cholesterol Profile

Cholesterol is ferried by two types of plasma proteins, called LDL (low-density lipoprotein) and HDL (high-density lipoprotein). LDL (the "bad" lipoprotein) takes cholesterol from the liver to the tissues, and HDL (the "good" lipoprotein) transports cholesterol out of the tissues to the liver. When the LDL level in blood is high or the HDL level is too low, plaque accumulates on arterial walls (Fig. 30C).

Starting at age 20, all adults are advised to have their cholesterol levels tested at least every 5 years. Even in healthy individuals, an LDL level above 160 mg/100 ml and an HDL level below 40 mg/100 ml are matters of concern. Cholesterol-lowering medications are available for those who do not meet these minimum guidelines.

Healthy Diet

Eating foods high in saturated fat (red meat, cream, and butter) or trans-fats (most margarines, commercially baked goods, and deep-fried foods) increases LDL cholesterol. Replacement of these harmful fats with monounsaturated fats (olive and canola oil) and polyunsaturated fats (corn, safflower, and soybean oil) is recommended. Most nutritionists also suggest eating at least five servings of antioxidant-rich fruits and vegetables a day to protect against cardiovascular disease. Antioxidants protect the body from free radicals that can damage blood vessels.

The American Heart Association (AHA) recommends eating at least two servings of cold-water fish (e.g., halibut, sardines, tuna, and salmon) each week. These fish contain omega-3 polyunsaturated fatty acids that can reduce plaque.

Resveratrol

The "French paradox" refers to the observation that levels of CVD are relatively low in France, despite the common consumption of a high-fat diet. One possible explanation is that wine is frequently consumed with meals. In addition to its alcohol content, red wine contains an antioxidant called resveratrol. This chemical is mainly produced in the skin of grapes, so resveratrol is also found in grape juice, and supplements are available at health food stores.

Exercise

People who exercise are less apt to have cardiovascular disease. Exercise not only helps keep weight under control, but may also help minimize stress and reduce hypertension. And short bursts of exercise may be superior to longer sessions. In one study, as few as three 10-minute workout sessions a day reduced triglyceride levels in blood better than one 30-minute session.

Anxiety and Stress

Mental stress can increase the odds of a heart attack. Within an hour of a strong earthquake that struck near Los Angeles in 1994, 16 people died of sudden heart failure (compared to the average of about 4 per day). Over the next several days, the number of heart-related deaths declined, suggesting that emotional stress had triggered fatal complications in those who were already predisposed to them. Obviously, it is difficult to avoid earthquakes, but we can learn healthy ways to avoid and manage stress.

CONSIDER THESE QUESTIONS

1. Which of these recommendations would you be able to follow with no difficulty? Which would be more difficult? Why?

2. How is it possible to make young people realize they should take care of their bodies now so they have fewer health problems when they are older?

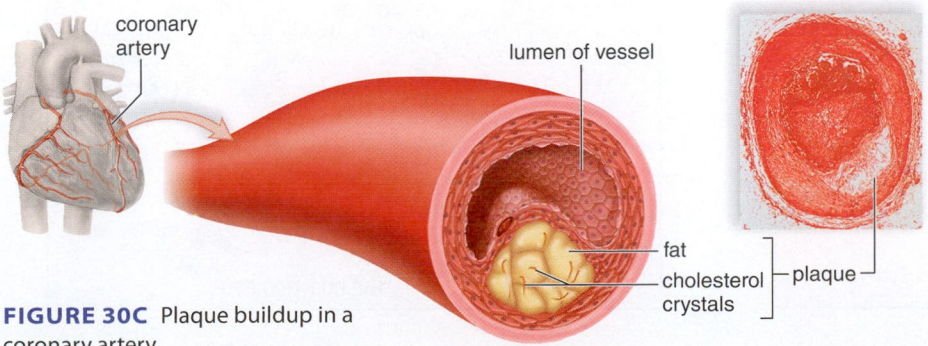

FIGURE 30C Plaque buildup in a coronary artery.

coronary artery

lumen of vessel

fat

cholesterol crystals

plaque

 Explore the concepts through a variety of multimedia assets and question types.

www.mcgrawhillconnect.com

Blood

The composition of blood is intimately associated with its functions, such as transport of substances, and blood clotting. At capillaries, blood's composition changes as gases and nutrients are exchanged for wastes. Transfusions of blood must be carefully monitored to prevent serious consequences.

30.5 Blood is a liquid tissue

LEARNING OUTCOMES

When you complete this section, you should be able to

1. List the functions of blood and of its various components, including plasma and blood cells.
2. Identify the major cellular and molecular events that result in a blood clot.

The numerous functions of **blood** include (1) transporting substances to and from the capillaries, where exchanges with tissue fluid take place; (2) helping defend the body against invasion by pathogens (e.g., disease-causing viruses and bacteria); (3) helping regulate body temperature; and (4) forming clots, preventing a potentially life-threatening loss of blood.

In humans, blood has two main portions: the liquid portion, called plasma, and the formed elements, consisting of various cells and platelets (Fig. 30.5A). The formed elements are manufactured continuously within the red bone marrow of certain bones, namely the skull, the ribs, the vertebrae, and the ends of the long bones.

MP3 General Functions and Composition of Blood

Plasma is composed mostly of water (90–92%) and proteins (7–8%), but it also contains smaller quantities of many types of molecules, including nutrients, wastes, and salts. The salts and proteins are involved in buffering the blood, effectively keeping the pH near 7.4. They also maintain the blood's osmotic pressure so that water has an automatic tendency to enter blood capillaries. Several plasma proteins (e.g., prothrombin and fibrinogen) are involved in blood clotting, and others transport large organic molecules in the blood. Albumin, the most plentiful of the plasma proteins, transports bilirubin, a breakdown product of hemoglobin. Lipoproteins transport cholesterol.

The **formed elements** are red blood cells, white blood cells, and platelets. Among the formed elements, **red blood cells** (**RBCs**), also called erythrocytes, transport oxygen. Red blood cells are small, biconcave disks that at maturity lack a nucleus and contain the respiratory pigment hemoglobin. There are 4–6 million red blood cells per mm^3 of whole blood, and each one of these cells contains about 250 million hemoglobin molecules. **Hemoglobin** contains iron, which combines loosely with oxygen; in this way, red blood cells transport oxygen. If the number of red blood cells is insufficient, or if the cells do not have enough hemoglobin, the individual suffers from *anemia* and feels tired and rundown.

Before they are released from the bone marrow into the blood, red blood cells lose their nuclei and begin to synthesize hemoglobin. After living about 120 days, red blood cells are destroyed, chiefly in the liver

Plasma	
Type	**Function**
Water (90–92% of plasma)	Maintains blood volume; transports molecules
Plasma proteins (7–8% of plasma)	Maintain blood osmotic pressure and pH
Globulins	Transport; fight infection
Fibrinogen	Blood clotting
Salts (less than 1% of plasma)	Maintain blood osmotic pressure and pH; aid metabolism
Gases (O_2 and CO_2)	Cellular respiration
Nutrients (lipids, glucose, and amino acids)	Food for cells
Wastes (urea and uric acid)	End product of metabolism; excretion by kidneys
Hormones	Aid metabolism

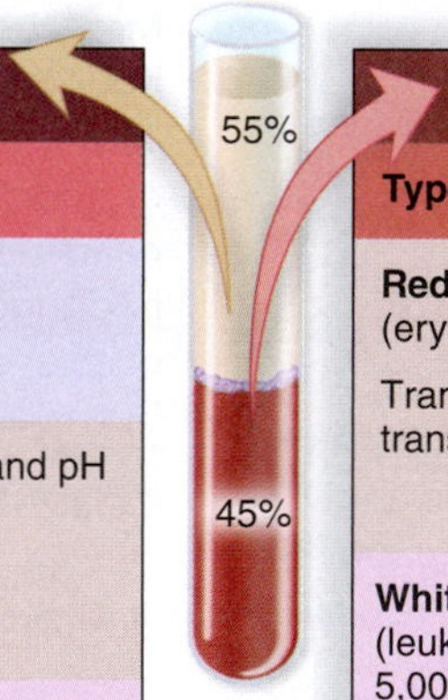

55%

45%

Formed Elements	
Type	**Number (per mm^3 blood)**
Red blood cells (erythrocytes) Transport O_2 and help transport CO_2	4 million–6 million
White blood cells (leukocytes) 5,000–11,000 Fight infection	Neutrophils 40–70% Lymphocytes 20–45% Monocytes 4–8% Eosinophils 1–4% Basophils 0–1%
Platelets (thrombocytes) Aid clotting	150,000–300,000

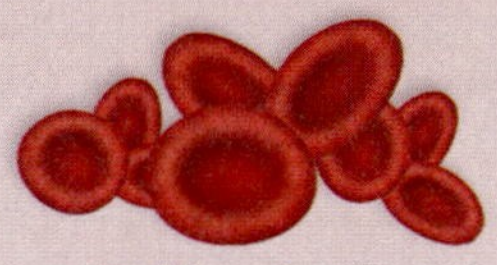

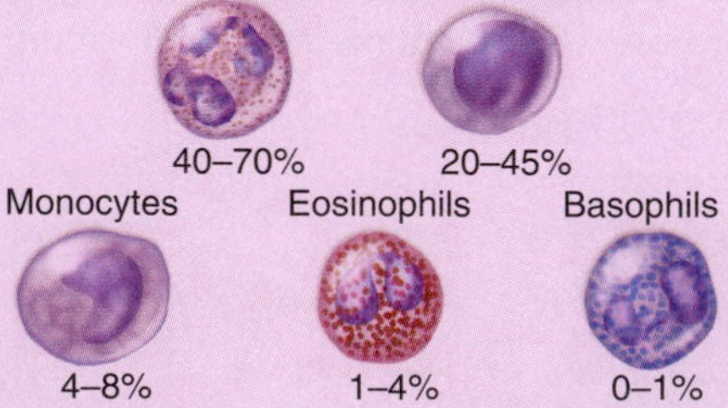

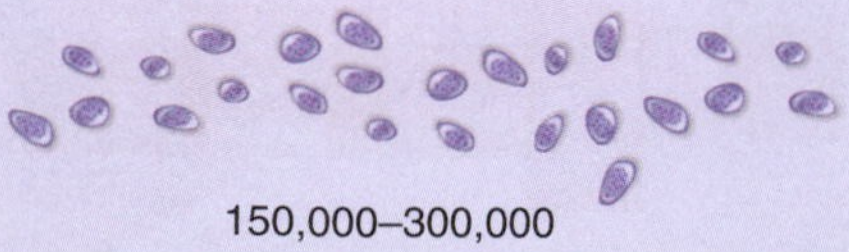

FIGURE 30.5A Composition of blood.

and the spleen, where they are engulfed by large phagocytic cells. When red blood cells are destroyed, hemoglobin is released. The iron is recovered and returned to the red bone marrow for reuse. The heme portion of the molecule undergoes chemical degradation and is excreted by the liver as bile pigments in the bile. The bile pigments are primarily responsible for the color of feces.

White blood cells (**WBCs**), also called leukocytes, help fight infections. White blood cells differ from red blood cells in several ways: They are usually larger and have a nucleus; they lack hemoglobin; and without staining, they appear translucent. With staining, white blood cells appear light blue unless they have granules that bind with certain stains. The white blood cells that have granules also have a lobed nucleus. The agranular leukocytes have a spherical or indented nucleus and no granules. There are approximately 5,000–11,000 white blood cells per mm³ of blood. Growth factors are available to increase the production of all white blood cells, and these are helpful to people with low immunity, such as AIDS patients.

Red blood cells are confined to the blood, but white blood cells are able to squeeze between the cells of a capillary wall. Therefore, they are found in tissue fluid, lymph, and lymphatic organs. When an infection is present, white blood cells greatly increase in number. Many white blood cells live only a few days—they probably die while engaging pathogens. Others live months or even years.

When microorganisms enter the body due to an injury, an *inflammatory response,* characterized by swelling, reddening, heat, and pain, occurs at the injured site. Damaged tissue releases kinins, which dilate capillaries, and histamines, which increase capillary permeability. White blood cells called **neutrophils,** which are amoeboid, squeeze through the capillary wall and enter the tissue fluid, where they phagocytize foreign material. White blood cells called **monocytes** appear and are transformed into macrophages, large phagocytizing cells that release white blood cell growth factors. Soon, the number of white blood cells increases explosively. A thick, yellowish fluid called pus contains a large proportion of dead white blood cells that have fought the infection.

Lymphocytes, another type of white blood cell, also play an important role in fighting infection. Lymphocytes called **T cells** attack infected cells that contain viruses. Other lymphocytes, called **B cells,** produce antibodies. Each B cell produces just one type of antibody, which is specific for one type of antigen. An **antigen,** which is most often a foreign protein but sometimes a polysaccharide, causes the body to produce an **antibody** to combine with the antigen. Antigens are present in the outer covering of parasites or in their toxins. When antibodies combine with antigens, the complex is often phagocytized by a macrophage. An individual is actively immune when a large number of B cells are all producing the antibody needed for a particular infection.

Blood Clotting Platelets result from fragmentation of large cells in the bone marrow called megakaryocytes. The blood contains 150,000–300,000 platelets per mm³. Figure 30.5B shows the process of blood clotting. ❶ When a blood vessel in the body is damaged, ❷ platelets clump at the site of the puncture and partially seal the leak. Platelets and the injured tissues release a clotting factor, called prothrombin activator, that converts *prothrombin* to thrombin. This reaction requires calcium ions (Ca²⁺). **Thrombin,** in turn, acts as an enzyme that severs two short amino acid chains from each *fibrinogen,* a plasma protein. These activated fragments then join end-to-end, forming long threads of **fibrin.** ❸ Fibrin threads wind around the platelet plug in the damaged area of the blood vessel and provide the framework for the clot. Red blood cells also are trapped within the fibrin threads; these cells make a clot appear red. A fibrin clot is present only temporarily. As soon as blood vessel repair is initiated, an enzyme called plasmin destroys the fibrin network and restores the fluidity of the plasma.

If blood is allowed to clot in a test tube, a yellowish fluid develops above the clotted material. This fluid, called serum, contains all the components of plasma, except fibrinogen. Common blood tests often measure the amount of a substance in the serum, rather than in the blood.

Hemophilia is a well-known, inherited clotting disorder. Due to the absence of a particular clotting factor, the slightest bump can cause internal bleeding. Bleeding into the joints damages cartilage, and reabsorption of bone follows. Bleeding into the muscles causes muscle atrophy, and bleeding into the brain can lead to death.

▶ **30.5 CHECK YOUR PROGRESS**
1. Explain why blood is a connective tissue and tell how each component contributes to homeostasis.
2. List the steps and explain the benefit of multisteps for blood to clot.

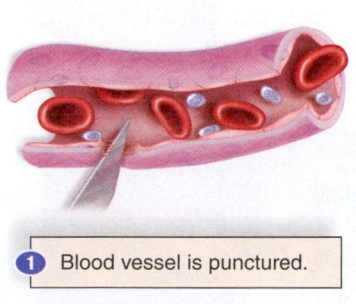

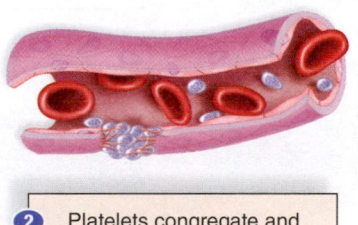

❷ Platelets congregate and form a plug.

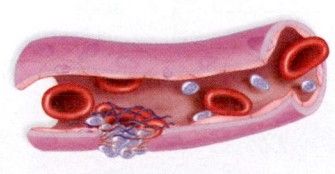

❸ Fibrin threads form and trap red blood cells.

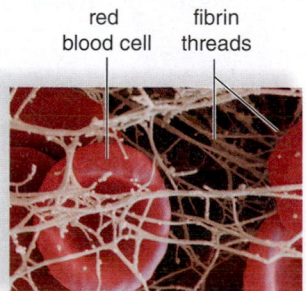

red blood cell   fibrin threads

❶ Blood vessel is punctured.

FIGURE 30.5B Blood clotting.

30.6 Exchanges between blood and tissue fluid occur at capillaries

LEARNING OUTCOMES

When you complete this section, you should be able to

1. Understand the importance of capillary exchange and the forces that are involved.
2. Explain the contribution of the lymphatic system to maintaining blood pressure.

Figure 30.6A illustrates capillary exchange between a systemic capillary and **tissue fluid,** the fluid between the body's cells. Blood that enters a capillary at the arterial end is rich in oxygen and nutrients, and it is under pressure created by the pumping of the heart. Two forces primarily control the movement of fluid through the capillary wall: blood pressure, which tends to cause water to move out of a capillary into the tissue fluid, and osmotic pressure, which tends to cause water to move from the tissue fluid into a capillary. At the arterial end of a capillary, blood pressure is higher than the osmotic pressure of blood (Fig. 30.6A, *left*). Osmotic pressure is created by the presence of salts and the plasma proteins. Because blood pressure is higher than osmotic pressure at the arterial end of a capillary, water exits a capillary at this end.

Midway along the capillary, where blood pressure is lower, blood pressure and osmotic pressure essentially cancel each other, and no net movement of water occurs. Solutes now diffuse according to their concentration gradient. Tissue fluid is always the area of lesser concentration of oxygen and nutrients because cells continually use them up. On the other hand, tissue fluid is always the area of greater concentration of carbon dioxide and wastes because cells generate wastes. Therefore, oxygen and nutrients (glucose and amino acids) diffuse out of the capillary, and carbon dioxide and other wastes diffuse into the capillary.

Red blood cells and almost all plasma proteins remain in the capillaries. The fluid and other substances that leave a capillary contribute to the tissue fluid. Because plasma proteins are too large to readily pass out of the capillary, tissue fluid tends to contain all the components of plasma, except much lesser amounts of protein.

At the venous end of a capillary, blood pressure has fallen to the point that osmotic pressure is greater than blood pressure,

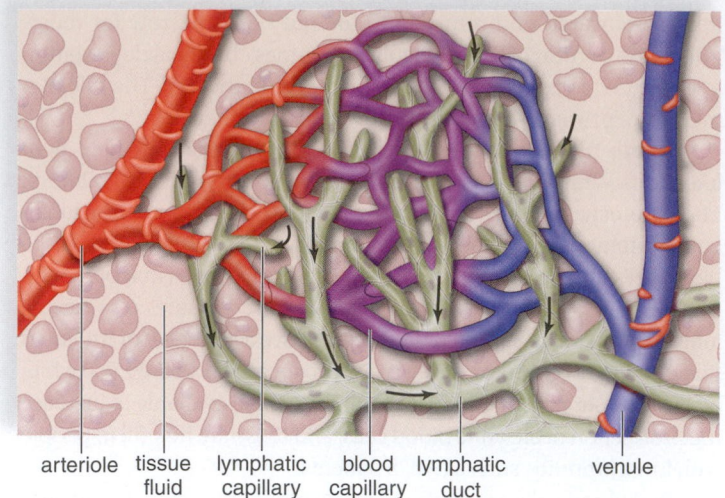

arteriole tissue lymphatic blood lymphatic venule
fluid capillary capillary duct

FIGURE 30.6B Lymphatic capillary beds lie near blood capillary beds.

and water tends to move into the capillary (Fig. 30.6A, *right*). Almost the same amount of fluid that left the capillary returns to it, although some excess tissue fluid is always collected by the lymphatic capillaries (Fig. 30.6B). **Tissue fluid** contained within lymphatic vessels is called **lymph.** Lymph is returned to the systemic venous blood when the major lymphatic vessels enter the subclavian veins in the shoulder region. Lymphatic vessels begin at lymphatic capillaries in the tissues and end at the subclavian veins. The one-way lymphatic vessels play a vital role in maintaining blood pressure by returning fluid to the cardiovascular system.

MP3
Capillary Exchange and Bulk Flow

Animation
Fluid Exchange Across the Walls of Capillaries

▶ **30.6 CHECK YOUR PROGRESS**

1. Support the proposal that the capillaries are the more significant component of the cardiovascular system.
2. Predict the effect on blood pressure if the protein content of the blood was greatly reduced (i.e., by malnutrition).

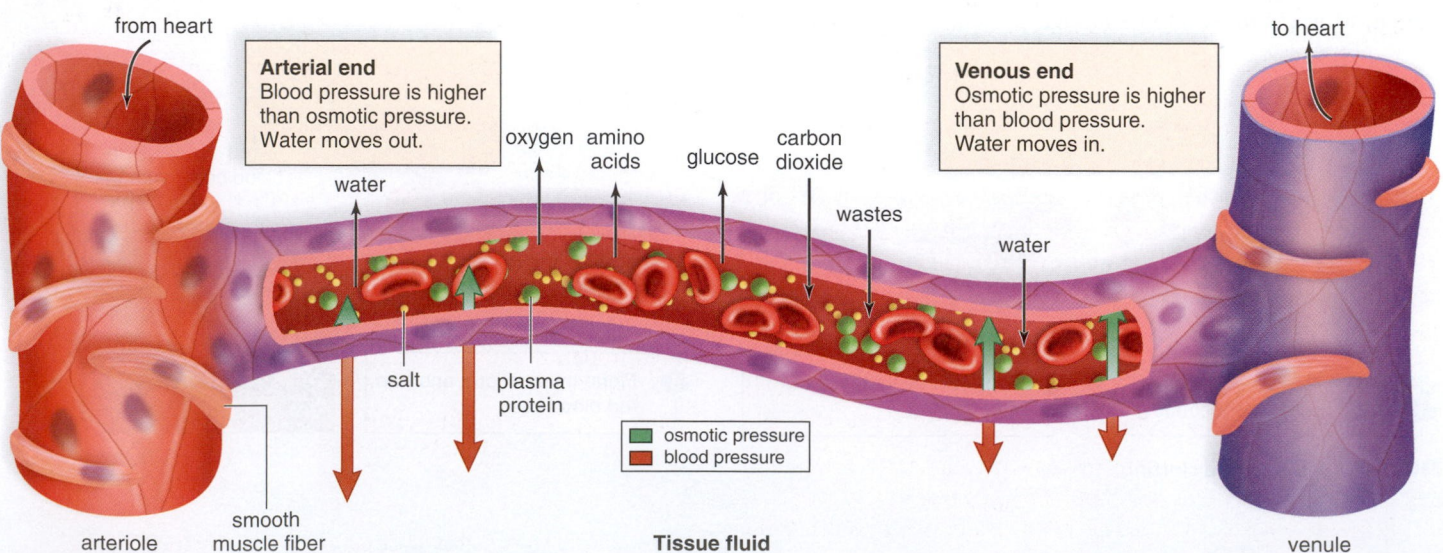

FIGURE 30.6A Capillary exchange in tissues.

30.7 Blood types must be matched for transfusions

Many early blood transfusions resulted in the illness and even death of some recipients. Eventually, it was discovered that only certain types of blood are compatible because red blood cell membranes carry specific proteins or carbohydrates that are antigens to blood recipients. As already defined, an antigen is a foreign molecule, usually a protein, that the body reacts to. Several groups of red blood cell antigens exist, the most significant being the ABO system. Clinically, it is very important that the blood groups be properly cross-matched to avoid a potentially deadly transfusion reaction. In such a reaction, the recipient may die of kidney failure within a week.

ABO System In the ABO system, the presence or absence of type A and type B antigens on red blood cells determines a person's blood type. For example, if a person has type A blood, the A antigen is on his or her red blood cells. This molecule is not an antigen to this individual, although it can be an antigen to a recipient who does not have type A blood.

The ABO system recognizes four types of blood: A, B, AB, and O. Within the plasma are antibodies to the antigens that are not present on the person's red blood cells. These antibodies are called anti-A and anti-B. This chart tells you what antibodies are present in the plasma of each blood type:

Blood Type	Antigen on Red Blood Cells	Antibody in Plasma
A	A	Anti-B
B	B	Anti-A
AB	A, B	None
O	None	Anti-A and anti-B

Because type A blood has anti-B and not anti-A antibodies in the plasma, a donor with type A blood can give blood to a recipient with type A blood (Fig. 30.7A). However, if type A blood is given to a type B recipient, clumping of red blood cells, or agglutination, occurs (Fig. 30.7B). Agglutination can cause blood to stop circulating in small blood vessels, and this leads to organ damage. Agglutination is also followed by hemolysis, or bursting of red blood cells, which if extensive, can cause the death of the individual.

Theoretically, which type of blood would be accepted by all recipients? Type O blood has no antigens on the red blood cells and is sometimes called the universal donor. A person with which blood type could receive blood from any other blood type? Type AB blood has no anti-A or anti-B antibodies in the plasma and is sometimes called the universal recipient. In practice, however, it is not safe to rely solely on the ABO system when matching blood. Instead, samples of the two types of blood are physically mixed, and the result is microscopically examined before blood transfusions are done.

Rh System Another important antigen in matching blood types is the Rh factor. Eighty-five percent of the U.S. population have this particular antigen on the red blood cells and are called Rh-positive. Fifteen percent do not have the antigen and are Rh-negative. Rh-negative individuals normally do not have antibodies to the Rh factor,

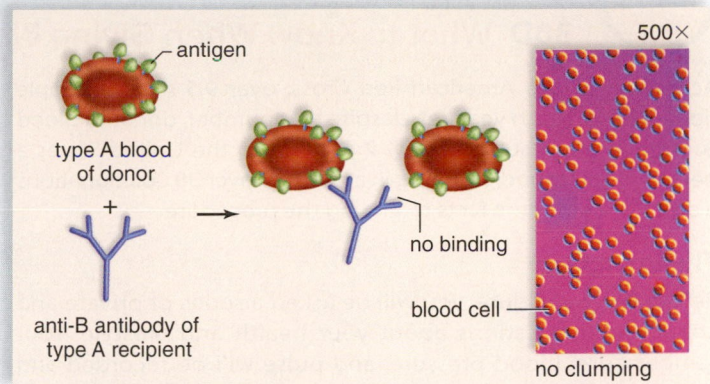

No agglutination

FIGURE 30.7A No agglutination occurs when the donor and recipient have the same blood type.

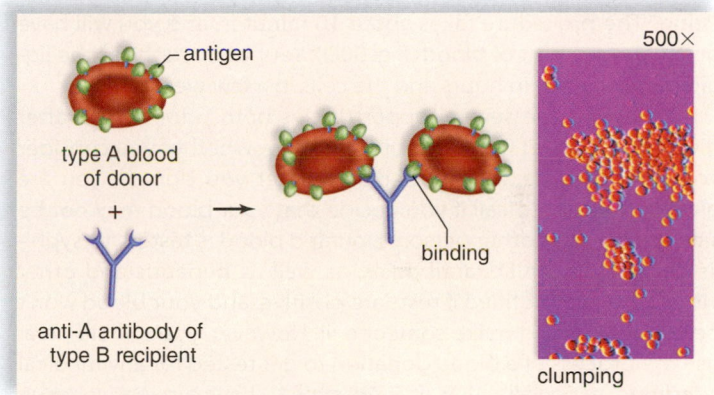

Agglutination

FIGURE 30.7B Agglutination occurs because blood type B has anti-A antibodies in the plasma.

but they may make them when exposed to the Rh factor. The designation of blood type usually also includes whether the person has or does not have the Rh factor on the red blood cells. This is done by attaching a minus (–) or plus (+) sign to the blood type. For example, some people have A-negative blood, which is symbolized as A⁻.

Erythroblastosis Fetalis During pregnancy, if the mother is Rh-negative and the father is Rh-positive, the child may be Rh-positive. The Rh-positive red blood cells may begin leaking across the placenta into the mother's cardiovascular system, because placental tissues normally break down before and at birth. Now, the mother produces anti-Rh antibodies. In this or a subsequent pregnancy with another Rh-positive baby, these antibodies may cross the placenta and destroy the child's red blood cells, a condition known as **erythroblastosis fetalis.** Nowadays, this problem is prevented by giving Rh-negative women an Rh immunoglobulin injection midway through the first pregnancy and no later than 72 hours after giving birth to an Rh-positive child. This injection contains anti-Rh antibodies that attack any of the baby's red blood cells in the mother's blood before she starts making her own antibodies.

▶ **30.7 CHECK YOUR PROGRESS**
1. Explain who can give blood to whom utilizing the ABO and Rh systems.

30D What to Know When Giving Blood

According to the American Red Cross, over 9.5 million people donate blood each year. Yet, despite that number, donated blood is often in short supply. Every 2 seconds in the United States a person needs blood, resulting in a need for over 38,000 donations a day. Here are some facts regarding the procedure.

The Procedure

Before the procedure, you will be asked a series of private and confidential questions about your health and lifestyle. Your temperature, blood pressure, and pulse will be recorded and a drop of your blood will be tested to ensure that you're not anemic.

The supplies used for your donation are sterile and are used only for you. You can't be infected with a disease when donating blood. When the actual donation is started, you may feel a brief "sting." The procedure takes about 10 minutes, and you will have given about a pint of blood (Fig. 30D). Your body replaces the liquid part (plasma) in hours and the cells in a few weeks.

You will have several opportunities both prior to and after giving blood to let Red Cross officials know whether you consider your blood to be safe. Immediately after you donate, you are given a number to call if you decide that your blood may not be safe to give to another person. Donated blood is tested for syphilis bacteria and AIDS antibodies, as well as hepatitis and other viruses. You are notified if tests are positive, and your blood won't be used if it could make someone ill. However, you should *never* use the process of a blood donation to get tested for any medical condition, especially AIDS. It is possible to have a negative result for AIDS antibodies and yet still spread the virus, because forming antibodies takes several weeks after exposure.

The Cautions

Some medications and medical conditions have waiting periods before you can donate blood. Before giving blood, you should inform the medical staff if you meet any of the conditions in the following list:

- You have recently had an infection or a fever.
- You have taken or are taking drugs that slow blood clotting. You should wait 48 hours after taking aspirin or aspirin-related drugs.
- You have had malaria, have taken drugs for malaria prevention, or have traveled to malaria-prone countries.
- You have a medical history of hepatitis or tuberculosis.
- You have been treated for syphilis or gonorrhea in the last 12 months.
- You have AIDS, have had a positive HIV test, or are at risk for getting an HIV infection due to one of the following:
  - You have ever injected illegal drugs.
  - You have taken clotting factor concentrates for hemophilia.
  - You have been given money or drugs for sex since 1977.
  - You had a sexual partner within the last year who did any of the above things.

For men:
- You had sex *even once* with another man since 1977, or had sex with a female prostitute within the last year.

For women:
- You had sex with a male or female prostitute within the last year, or had a male sexual partner who had sex with another man *even once* since 1977.

After the Procedure

Most people feel fine while they give blood and afterward, but a few donors have an upset stomach or feel faint or dizzy after donation. Resting, drinking fluid, and eating a snack usually help. Occasionally, bruising, redness, and pain occur at your donation site, so avoid strenuous exercise and lifting for a day or so. Very rarely, a person may have muscle spasms and/or suffer nerve damage. You should contact your physician if you have any concerns about potential side effects of donating blood.

For more information about blood donation, including eligibility requirements, visit the Red Cross website at www.redcrossblood.org.

CONSIDER THESE QUESTIONS

1. Why would individuals be restricted from donating if they have had a sexually transmitted disease in the past 12 months, even if they have been treated?
2. Are there other risk factors that you think should be considered that would exclude individuals from donating blood?

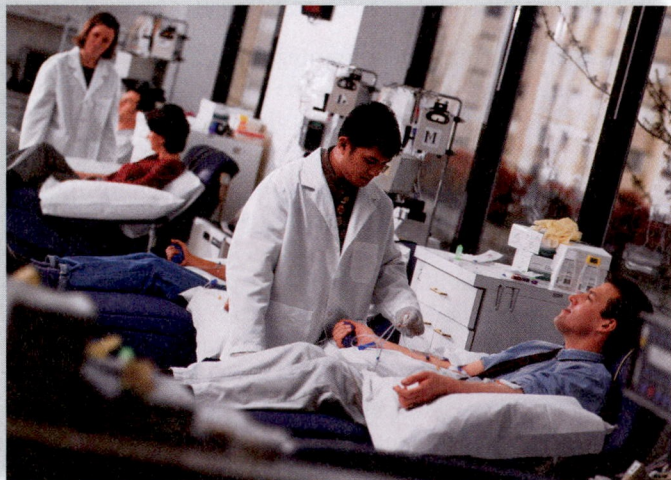

FIGURE 30D Donating blood can help save a life.

 **BIOLOGY** Explore the concepts through a variety of multimedia assets and question types.

www.mcgrawhillconnect.com

THE CHAPTER IN REVIEW

The cardiovascular system consists of the heart and the blood vessels, which can be likened to the streets of a town. Imagine that you have

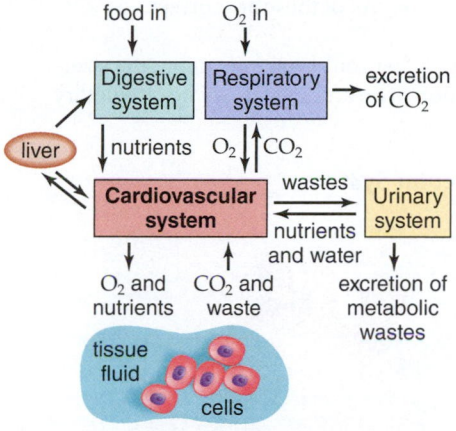

received birthday gifts and have decided to return some and pick up other items you prefer. In a similar way, the cardiovascular system picks up nutrients at the intestinal tract, exchanges carbon dioxide for oxygen at the lungs, and deposits metabolic wastes at the kidneys. Therefore, the composition of the blood and the tissue fluid stays relatively constant. Tissue fluid, which surrounds the cells, exchanges substances with the blood, and if the composition of the blood stays constant, tissue fluid is able to supply the cells with the nutrients and oxygen they require and also to receive their metabolic wastes. In this way, homeostasis, the relative constancy of the internal environment, is maintained.

The pumping of the heart merely helps keep the blood moving so that it can continue to service the cells. While we tend to think of the body in terms of organs, it is the cells of the organs that do the work of

keeping us alive, and only if the cells are well cared for do we remain healthy.

Body fluids make ideal culture media for the growth of infectious parasites, but these fluids often have ways to ward off an invasion. You already know that white blood cells are involved in these endeavors. Chapter 31 discusses how humans in particular are able to stay one step ahead of the microorganisms that want to take up residence in their bloodstream and cells.

ANALYZE AND EVALUATE

1. Show that, regardless of the type of circulatory system, the overall function remains the same.
2. Why do the veins require supplemental help in returning blood to the heart?
3. If proteins collected in tissue fluid, what would happen to blood pressure and blood flow? Explain.

SUMMARIZE

Invertebrate Circulatory Systems

30.1 A circulatory system serves the needs of cells

- An animal system that uses confined fluid to transport substances to and from cells can be called a **circulatory system.** Some invertebrates, such as cnidarians (e.g., hydra) and flatworms (e.g., planarians) do not have a circulatory system because they exchange materials with a fluid environment by diffusion. Other invertebrates have an open or a closed circulatory system: (1) **Open circulatory system: Hemolymph** (blood plus tissue fluid) is pumped into tissue spaces or a **hemocoel.** In molluscs and arthropods, the body cavity doubles as a hydrostatic skeleton. In insects, tracheae distribute oxygen to muscles. (2) **Closed circulatory system:** Blood is pumped into blood vessels as in earthworms and vertebrates.

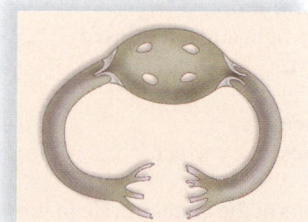

Open circulatory system

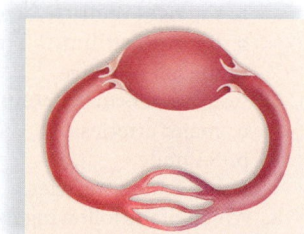

Closed circulatory system

Vertebrate Cardiovascular Systems

30.2 The mammalian heart is a double pump

- The mammalian heart is a part of a **cardiovascular system** (heart plus blood vessels). In the mammalian heart, a **septum**

separates the heart into right and left halves. Each side of the heart has an **atrium** and a **ventricle. Atrioventricular** and **semilunar valves** keep blood moving in the correct direction. **Arteries,** such as the **aorta, pulmonary trunk,** and **pulmonary arteries,** take blood away from the heart; **veins,** such as the **superior** and **inferior venae cavae** and the **pulmonary veins,** take blood to the heart.

- During the heartbeat, **systole** is contraction and **diastole** is relaxation of a heart chamber. First, the atria contract, and then the ventricles contract. The **pulse** results from expansion of the aorta when blood enters following ventricular contraction. The sounds of the heart are due to closure of valves. When the antrioventricular valves close, a "lub" is heard, and when the semilunar valves close, a "dub" is heard.

- The cardiac cycle is due to the **cardiac conduction system,** the SA node (**the cardiac pacemaker**) causes the atria to contract, and the AV node causes the ventricles to contract. An **electrocardiogram** (**ECG**) is a recording that depicts the electrical events of the **cardiac cycle.**

30.3 The structure of blood vessels is suited to their function

- **Blood pressure** in strong arteries and arterioles carries blood away from the heart. **Systolic pressure** is the pressure in arteries during ventricular systole. **Diastolic pressure** is the pressure in arteries during ventricular diastole. Blood flow in **venules** and veins is assisted by (1) skeletal muscle contraction pushing on weak veins and (2) lower thoracic pressure. Once blood in veins has passed a **valve,** it closes and blood cannot fall back.

- Capillary cross-sectional area causes blood pressure to be very low in **capillaries,** allowing time for diffusion across thin capillary walls.

30.4 Blood vessels form two circuits in mammals

- In the **pulmonary circuit,** pulmonary arteries take O_2-poor blood to the lungs; pulmonary veins return O_2-rich blood to the heart. In the **systemic circuit,** the left ventricle sends O_2-rich blood to the aorta; the **vena cava** takes O_2-poor blood back to the right atrium. A **portal system** begins and ends in capillaries. The **hepatic portal system** begins at the digestive tract and ends in the liver.

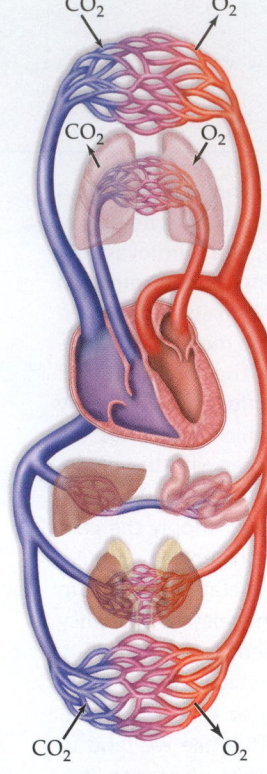

Blood

30.5 Blood is a liquid tissue

- **Blood,** which is composed of **plasma** and **formed elements** (red blood cells, white blood cells, and platelets), transports, defends against pathogens, helps regulate body temperature, and forms clots.

- Plasma is composed of mostly water and proteins along with some nutrients, wastes, and salts. **Red blood cells (RBCs)** contain **hemoglobin** for transport of oxygen. **White blood cells (WBCs)** including **neutrophils, monocytes,** and **lymphocytes** fight infection. Lymphocytes include **T cells** and **B cells.** T cells attack virus-infected cells. B cells produce **antibodies** in response to **antigens.**

- Blood clotting involves platelets. Platelets clump at the site of blood vessel damage, where they release clotting factors such as **thrombin;** long **fibrin** threads result and provide a framework for a blood clot. **Hemophilia** (bleeding that will not stop) is a clotting disorder.

30.6 Exchanges between blood and tissue fluid occur at capillaries

- At a capillary, water moves out at the arterial end due to blood pressure. Water moves in at the venous end due to osmotic pressure. Midway along the capillary, nutrients diffuse out and wastes diffuse in. Lymphatic capillaries collect excess **tissue fluid (lymph)** and return it to the cardiovascular system.

30.7 Blood types must be matched for transfusions

- ABO blood typing determines the presence or absence of A and B antigens on the surface of red blood cells. Clumping (agglutination) occurs when a corresponding antigen and antibody are put together. The Rh factor is another important antigen in matching blood types. **Erythroblastosis fetalis** occurs when the mother is Rh-negative and her child is Rh-positive.

TEST YOURSELF

Invertebrate Circulatory Systems

1. A circulatory system functions in
   - a. homeostasis.
   - b. transport of wastes.
   - c. transport of nutrients.
   - d. All of these are correct.
2. In insects with an open circulatory system, oxygen is taken to cells by
   - a. blood.
   - b. hemolymph.
   - c. tracheae.
   - d. capillaries.
3. Which one of these structures would you expect to be part of a closed, but not an open, circulatory system?
   - a. ostia
   - b. capillary beds
   - c. hemocoel
   - d. heart
   - e. All of these are correct.
4. **THINKING CONCEPTUALLY** What could you infer about the size, shape, and organization of an animal if it had no circulatory system?

Vertebrate Cardiovascular Systems

5. Label this diagram:

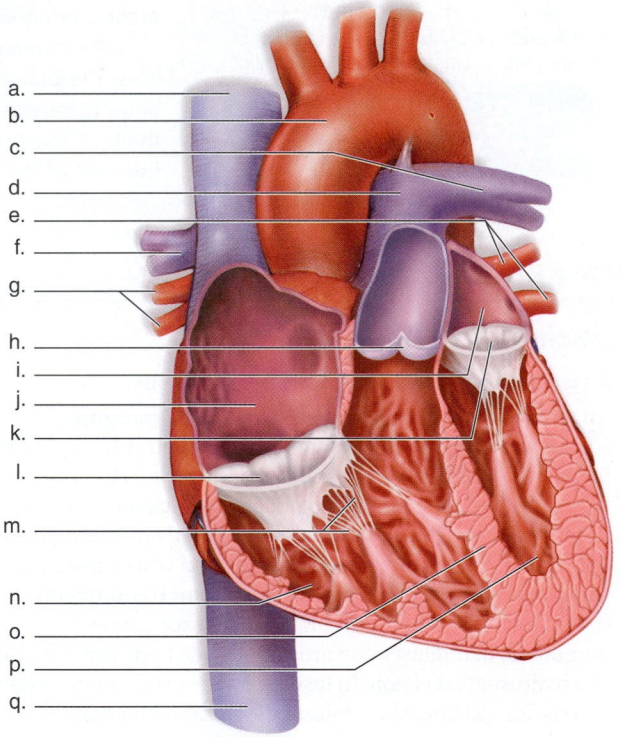

a. _____
b. _____
c. _____
d. _____
e. _____
f. _____
g. _____
h. _____
i. _____
j. _____
k. _____
l. _____
m. _____
n. _____
o. _____
p. _____
q. _____

6. In humans, blood returning to the heart from the lungs returns to
   - a. the right ventricle.
   - b. the right atrium.
   - c. the left ventricle.
   - d. the left atrium.
   - e. both the right and left sides of the heart.
7. Systole refers to the contraction of the
   - a. major arteries.
   - b. SA node.
   - c. atria and ventricles.
   - d. major veins.
   - e. All of these are correct.
8. Which of the following lists the events of the cardiac cycle in the correct order?
   - a. contraction of atria, rest, contraction of ventricles
   - b. contraction of ventricles, rest, contraction of atria
   - c. contraction of atria, contraction of ventricles, rest
   - d. contraction of ventricles, contraction of atria, rest

9. Which of the following statements is true?
   a. Arteries carry blood away from the heart, and veins carry blood to the heart.
   b. Arteries carry blood to the heart, and veins carry blood away from the heart.
   c. Arteries carry O_2-rich blood, and veins carry O_2-poor blood.
   d. Arteries carry O_2-poor blood, and veins carry O_2-rich blood.
10. Place the following blood vessels in order, from largest to smallest in diameter.
    a. arterioles, capillaries, arteries
    b. arteries, arterioles, capillaries
    c. capillaries, arteries, arterioles
    d. arterioles, arteries, capillaries
    e. arteries, capillaries, arterioles
11. Which of these factors has little effect on blood flow in arteries?
    a. heartbeat
    b. blood pressure
    c. total cross-sectional area of vessels
    d. skeletal muscle contraction
    e. the amount of blood leaving the heart
12. The best explanation for the slow movement of blood in capillaries is
    a. skeletal muscles press on veins, not capillaries.
    b. capillaries have much thinner walls than arteries.
    c. there are many more capillaries than arterioles.
    d. venules are not prepared to receive so much blood from the capillaries.
    e. All of these are correct.
13. **THINKING CONCEPTUALLY** Why is a closed circulatory system with a four-chambered heart consistent with the lifestyle of birds and mammals and their ability to live in cold climates?

Blood

14. Which association is incorrect?
    a. white blood cells—infection fighting
    b. red blood cells—blood clotting
    c. plasma—water, nutrients, and wastes
    d. red blood cells—hemoglobin
    e. platelets—blood clotting
15. Red blood cells
    a. reproduce themselves by mitosis.
    b. live for several years.
    c. continually synthesize hemoglobin.
    d. are destroyed in the liver and spleen.
    e. More than one of these are correct.
16. The last step in blood clotting
    a. is the only step that requires calcium ions.
    b. occurs outside the bloodstream.
    c. is the same as the first step.
    d. converts prothrombin to thrombin.
    e. converts fibrinogen to fibrin.

17. Label portions of these arrows (a–d) as either blood pressure or osmotic pressure:

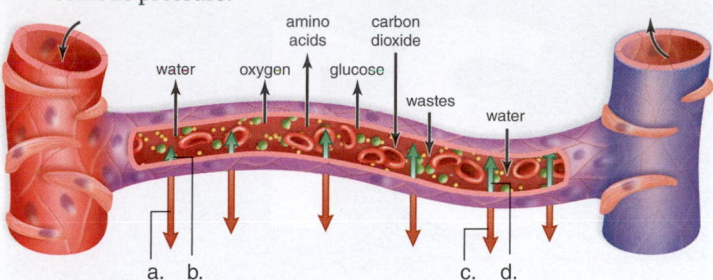

18. Which of the plasma proteins contributes most to osmotic pressure?
    a. albumin   c. erythrocytes
    b. globulins   d. fibrinogen
19. In the tissues, nutrients and _____ are exchanged for _____ and other wastes.
    a. blood, oxygen
    b. oxygen, carbon dioxide
    c. hemoglobin, tissue fluid
    d. None of these are correct.
20. **THINKING CONCEPTUALLY** Rita feels tired and run-down, and her physician diagnoses iron-deficiency anemia. The physician doesn't prescribe erythropoietin. Why not?

GET INVOLVED

1. What evidence from examining (a) dead vertebrates and (b) live vertebrates substantiates that their blood circulates?
2. You are a biochemist who decides to analyze the plasma composition of arterial versus venous blood. What data would be consistent with the function of capillaries? Explain.
3. After viewing this virtual lab, explain why frogs are not completely dependent on their lungs to oxygenate their blood and therefore frogs can make do with a 3-chambered heart.

Virtual Lab
Virtual Frog
Dissection

MEDIA STUDY TOOLS

mhhe.com/maderconcepts3

Enhance your study of this chapter with interactive study tools, practice tests, and engaging animations. Also, ask your instructor about the resources available through ConnectPlus, which includes LearnSmart, a personalized adaptive learning program, and a media-rich eBook.

31

Lymph Transport and Immunity

BEFORE YOU BEGIN

Take a few minutes to recall

The functions of plasma membrane proteins (section 5.1)
Functions of the lymphatic system (sections 26.3 and 26.7)
Structure and function of white blood cells (section 30.5)
How lymph forms and is transported (section 30.6)

HIV/AIDS: A Global Disaster

AIDS (acquired immunodeficiency syndrome) is caused by a virus known as the human immunodeficiency virus (HIV). HIV infects and reproduces in the cells of the immune system, the very cells that normally keep us free of disease. As the virus bursts forth, its host cell is destroyed. No wonder a person infected with HIV is unable to fight off the onslaught of viruses, fungi, and bacteria that attack the body every day.

At first a person infected with HIV suffers from weight loss, chronic fever, cough, diarrhea, swollen glands, and shortness of breath. Finally, without treatment, the person will have AIDS and develop one or more opportunistic infections. These are illnesses that can occur only because the immune system is so weak. The list is long but a common opportunistic infection is Kaposi sarcoma, a cancer of blood vessels caused by a herpes-virus. Reddish-purple, coin-sized spots and skin lesions occur with this type of cancer. *Pneumocystis* pneumonia, a fungal infection that causes the lungs to become useless as they fill with fluid and debris, is another common opportunistic illness.

An HIV infection is acquired by having sex with an infected person or by sharing a needle with an infected person. Therefore,

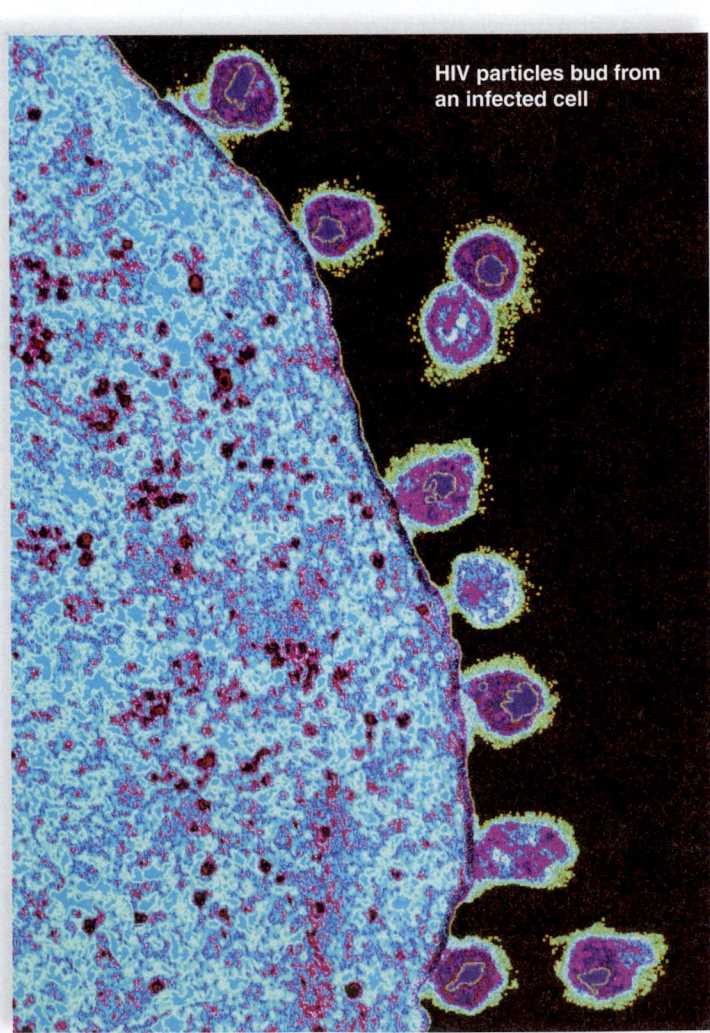

HIV particles bud from an infected cell

an HIV infection is clearly preventable. It is important to realize that HIV is not transmitted through casual contact in the workplace, schools, or social settings. A casual kiss, hugging, shaking hands; or toilet seats, doorknobs, dishes, drinking glasses, food, or pets cannot give you AIDS. Rather, to prevent an HIV infection, abstain from sexual contact or have sex only with the same uninfected person, or consistently and accurately use a condom when having sex.

AIDS victim: Kaposi sarcoma is evident

A person who does not follow these guidelines needs to be tested repeatedly for an HIV infection. Generally, an HIV test detects the presence of HIV antibodies in the blood, urine, or saliva. Antibodies are special proteins that the body produces when it is fighting an infection. If antibodies to HIV are detected, it means that the person is indeed infected with HIV. A problem is that most people only develop detectable HIV antibodies after, say, 6–12 weeks of infection. This means that in the meantime they may pass on the infection to a partner before they know they are infected. Hence, the requirement to always have sex with the same uninfected person or to always use a condom.

There is no cure for an HIV infection, but a treatment called highly active antiretroviral therapy (HAART) is usually able to stop HIV reproduction to such an extent that the virus becomes undetectable in the body. HAART utilizes a combination of drugs that interfere with the life cycle of HIV so that it can no longer reproduce additional viruses that go on to infect more immune cells. HAART uses a combination of drugs so that the virus is less likely to mutate and become resistant to the therapy. However, if HAART is discontinued the virus rebounds; therefore, the drugs must be continued indefinitely for the rest of a patient's life. In developing countries around the globe, millions of people cannot afford HAART therapy and are dying due to a lack of antiviral drugs and proper education about AIDS.

HIV/AIDS is a global catastrophe. Regions all over the world—Africa, Asia, Latin America, Europe, North America, and the Middle East—are suffering from the medical and also social, economical, and political impact of AIDS. In 2011, an estimated 35 million people worldwide were living with an HIV infection; 2.5 million more people get infected each year. Many more are HIV-positive but unaware of their infection. Since the beginning of the epidemic, nearly 30 million have died from AIDS, and 2 million additional people die from AIDS each year. Eighty percent of these AIDS deaths are among people ages 20–50. Young people ages 15–24 make up about 50% of all new HIV infections worldwide but in the United

States, those over 50 are the fastest growing group living with an HIV infection. One out of seven newly diagnosed people is now over 50. Women, who account for 50% of all people living with HIV, have left over 14 million AIDS orphans in Africa. As nations around the globe come together to fight the AIDS epidemic, there is hope that the "war on AIDS" will eventually be won as treatment and educational campaigns reach the hardest-hit regions of the world. Unfortunately, people might tend to forget about AIDS when it is no longer in the headlines. However, no war, natural disaster, or terrorist attack has ever killed more people than HIV/AIDS. AIDS is the worst and deadliest pandemic (global infection) that mankind has ever faced. And so far investigators have been unable to produce an effective vaccine; without it AIDS will continue to claim millions of lives throughout the world.

Understanding why AIDS patients are so sick gives us a whole new level of appreciation for the workings of a healthy immune system, the topic of this chapter. Homeostasis can only be maintained when the body is able to protect its own tissues and maintain its integrity.

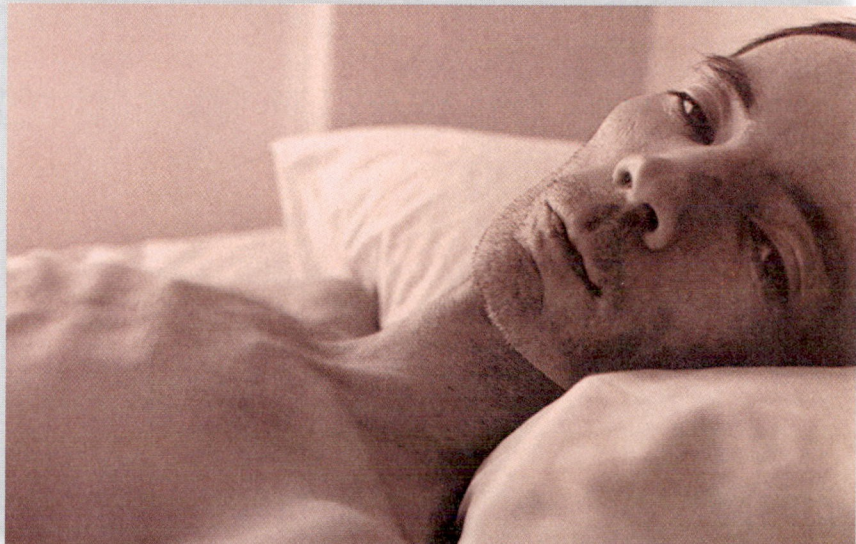

The health of a patient with an untreated HIV infection deteriorates until death occurs.

The Lymphatic System

The lymphatic system consists of lymphatic vessels and lymphatic organs. Lymphatic vessels transport lymph, and lymphatic organs produce and store cells that fight pathogens. Pathogens are disease-causing agents, such as viruses and bacteria.

31.1 Lymphatic vessels transport lymph

LEARNING OUTCOMES

When you complete this section, you should be able to

1. List four functions of the lymphatic system.
2. Describe the one-way transport of the lymphatic vessels.
3. State the chief functions of four lymphatic organs and three patches of lymphatic tissue.

The **lymphatic system,** which is closely associated with the cardiovascular system, has four main functions that contribute to homeostasis:

MP3
Lymphatic
System

- Lymphatic capillaries absorb excess tissue fluid and lymphatic vessels, called ducts, return it to the bloodstream.
- In the small intestines, lymphatic capillaries called **lacteals** absorb fats in the form of lipoproteins and lymphatic vessels transport them to the bloodstream.
- The lymphatic system is responsible for the production, maintenance, and distribution of lymphocytes which help defend the body against pathogens.

Lymphatic vessels form a one-way system that begins with lymphatic capillaries (Fig. 31.1A, *right*). Most regions of the body are richly supplied with lymphatic capillaries—tiny, closed-ended vessels. Lymphatic capillaries take up excess tissue fluid. The fluid inside lymphatic capillaries is called **lymph.** In addition to water and fat molecules, lymph contains the same ions, nutrients, gases, and proteins that are present in tissue fluid.

The lymphatic capillaries join to form lymphatic vessels that merge before entering one of two ducts: the thoracic duct or the right lymphatic duct. The construction of the larger lymphatic vessels is similar to that of cardiovascular veins, including the presence of valves. Skeletal muscle contraction forces lymph through lymphatic vessels, and the one-way valves prevent it from flowing backward.

The large lymphatic ducts help maintain blood volume and pressure when the right lymphatic duct empties its lymph into the right subclavian vein and the thoracic duct empties its lymph into the left subclavian vein (Fig. 31.1A, *left*).

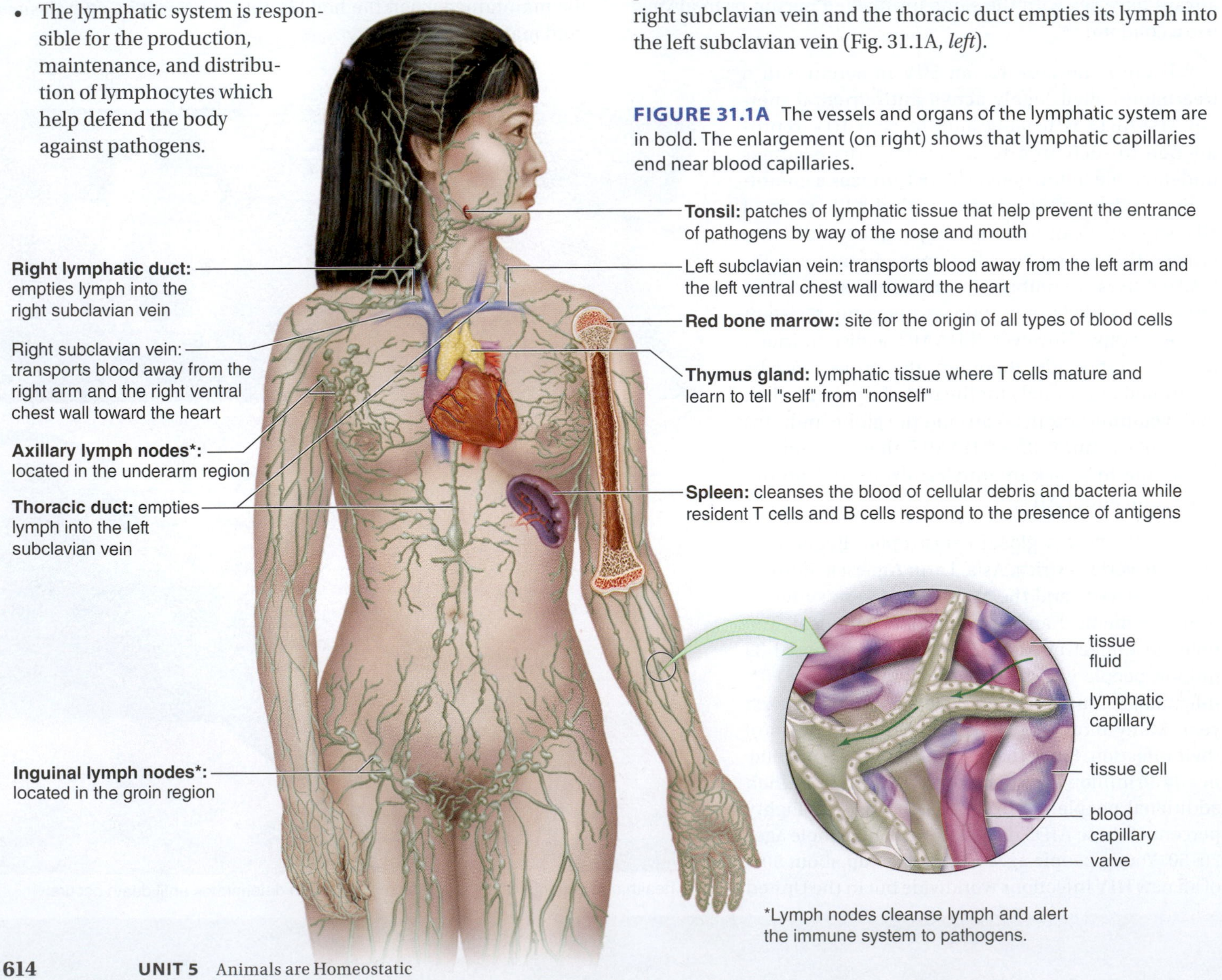

FIGURE 31.1A The vessels and organs of the lymphatic system are in bold. The enlargement (on right) shows that lymphatic capillaries end near blood capillaries.

Right lymphatic duct: empties lymph into the right subclavian vein

Right subclavian vein: transports blood away from the right arm and the right ventral chest wall toward the heart

Axillary lymph nodes*: located in the underarm region

Thoracic duct: empties lymph into the left subclavian vein

Inguinal lymph nodes*: located in the groin region

Tonsil: patches of lymphatic tissue that help prevent the entrance of pathogens by way of the nose and mouth

Left subclavian vein: transports blood away from the left arm and the left ventral chest wall toward the heart

Red bone marrow: site for the origin of all types of blood cells

Thymus gland: lymphatic tissue where T cells mature and learn to tell "self" from "nonself"

Spleen: cleanses the blood of cellular debris and bacteria while resident T cells and B cells respond to the presence of antigens

tissue fluid

lymphatic capillary

tissue cell

blood capillary

valve

*Lymph nodes cleanse lymph and alert the immune system to pathogens.

The **lymphatic (lymphoid) organs** are shown in Figure 31.1A and Figure 31.1B:

Red bone marrow is the site of stem cells that are ever capable of dividing and producing the various types of blood cells, including lymphocytes. Some of these become mature **B cells,** a major type of lymphocyte, in the bone marrow. In a child, most of the bones have red bone marrow, but in an adult, it is present only in the bones of the skull, the sternum (breastbone), the ribs, the clavicle (collarbone), the pelvic bones, the vertebral column, and the proximal heads of the femur and humerus.

The red bone marrow consists of a network of connective tissue fibers that supports the stem cells and their progeny—mature blood cells. These differentiated cells are poised to enter the bloodstream at sinuses filled with venous blood.

The soft, bilobed **thymus gland** is located in the thoracic cavity anterior and ventral to the heart (see Fig. 31.1A). Immature **T cells,** the other major type of lymphocyte, migrate from the bone marrow through the bloodstream to the thymus, where they mature. The thymus also produces thymic hormones, such as thymosin, that are thought to aid in the maturation of T cells. The thymus varies in size, but it is largest in children and shrinks as we get older. In the elderly, it is barely detectable. When well developed, it contains many lobules.

Lymph nodes are small (about 1–25 mm in diameter), ovoid structures occurring along lymphatic vessels. Lymph nodes are named for their location. For example, inguinal lymph nodes are in the groin, and axillary lymph nodes are in the armpits. Physicians often feel for the presence of swollen, tender lymph nodes in the neck as evidence that the body is fighting an infection.

A lymph node has many open spaces called sinuses. As lymph courses through the sinuses, it is filtered by macrophages, which engulf debris and pathogens. Many B cells and T cells are also present in and around the sinuses of a lymph node.

Unfortunately, cancer cells sometimes enter lymphatic vessels and congregate in lymph nodes. Therefore, when a person undergoes surgery for cancer, it is a routine procedure to remove some lymph nodes and examine them to determine whether the cancer has spread to other regions of the body.

The **spleen** is located in the upper left side of the abdominal cavity posterior to the stomach. Most of the spleen is red pulp that filters the blood. Red pulp consists of blood vessels and sinuses, where macrophages remove old and defective blood cells. Inside the red pulp of the spleen is white pulp that consists of little lumps of lymphatic tissue where B cells and T cells congregate.

The spleen's outer capsule is relatively thin, and an infection or a blow can cause the spleen to burst. Although the spleen's functions can be largely replaced by other organs, a person without a spleen is often slightly more susceptible to infections and may require antibiotic therapy indefinitely.

The **tonsils,** located in the pharynx; **Peyer patches,** located in the intestinal wall; and the vermiform **appendix,** attached to the cecum are patches of lymphatic tissue in the body. These structures, encounter pathogens and antigens that enter the body by way of the mouth.

This completes our discussion of the lymphatic system. The next part of the chapter begins our discussion of defenses against disease.

Animation
Lymphatic System

▶ **31.1 CHECK YOUR PROGRESS**

1. Summarize how the lymphatic system contributes to homeostasis.
2. Associate lymph nodes with a function of the lymphatic system. Explain this association.

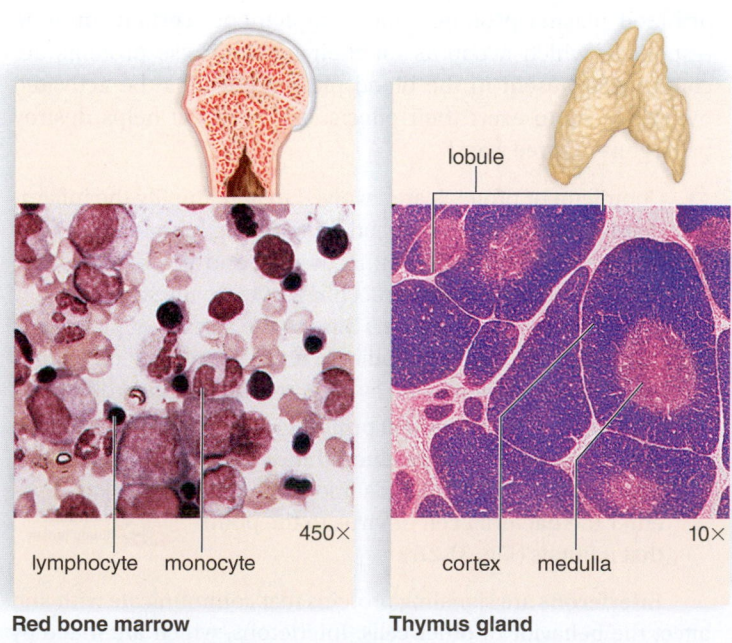

Red bone marrow — lymphocyte, monocyte — 450×

Thymus gland — lobule, cortex, medulla — 10×

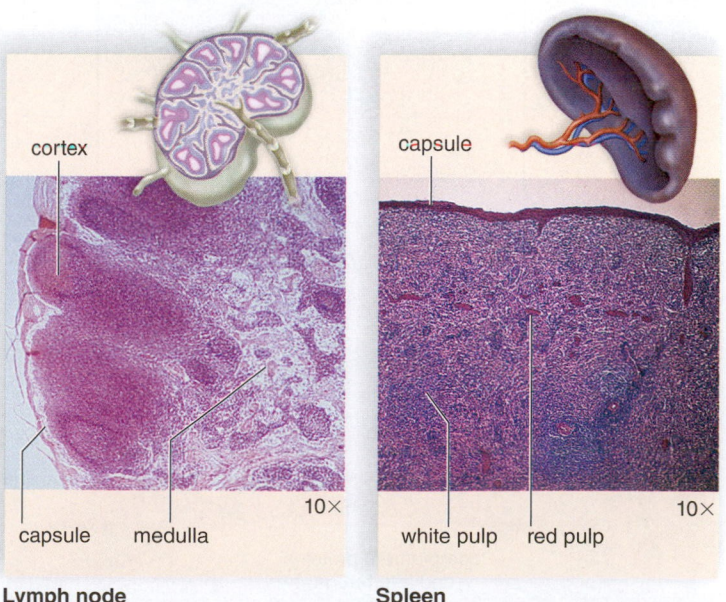

Lymph node — cortex, capsule, medulla — 10×

Spleen — capsule, white pulp, red pulp — 10×

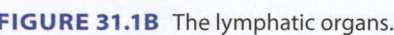

FIGURE 31.1B The lymphatic organs.

Innate Immunity

Immunity is the capability of resisting diseases and killing pathogens such as bacteria and viruses. Immunity has two major subdivisions: Innate immunity is our first line of defense while adaptive immunity is our second line of defense. First, we will discuss innate immunity including the inflammatory response and, then, later we will consider adaptive immunity.

31.2 Barriers to entry, protective proteins, and white blood cells are first responders

LEARNING OUTCOME

When you complete this section, you should be able to

1. Describe three mechanisms of innate immunity, including the tissues, cells, and chemicals involved.

We are constantly exposed to pathogens in our food and drink, the air we breathe, and the objects we touch, both living and inanimate. The warm temperature and constant supply of nutrients in our bodies make them an ideal place for pathogens to flourish. **Innate immunity** is so called because it is fully functional and immediately available without previous exposure to pathogens. Innate immunity is sometimes called nonspecific immunity because it consists of mechanisms that are useful for resisting or killing any pathogen. These mechanisms can be likened to a police force that is skilled at dealing with many different types of criminals.

Innate immunity involves the several defenses summarized in Figure 31.2A. They include barriers to entry such as the skin; protective proteins such as complement and interferons; phagocytes and natural killer cells; and the inflammatory response, which is

discussed in section 31.3. Innate immunity occurs automatically; however, no memory is involved—there is no recognition that the same intruder has been attacked before. Later, we will discuss adaptive defenses, which are directed against specific pathogens and do exhibit memory.

MP3 Barriers and Nonspecific Defenses

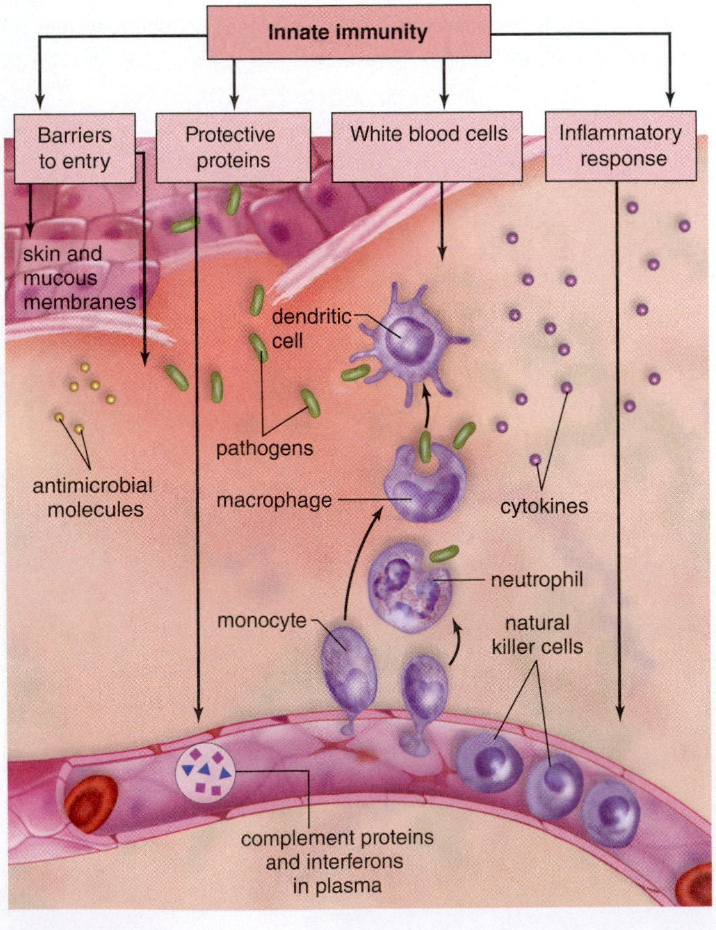

Barriers to Entry Barriers to entry by pathogens include nonchemical, mechanical barriers, such as the skin and the mucous membranes lining the respiratory, digestive, and urinary tracts. For example, the upper layers of our skin are composed of dead, keratinized cells that form an impermeable barrier. But when the skin has been injured, one of the first concerns is the possibility of an infection. We are also familiar with the importance of sterility before an injection is given. The injection needle and the skin must be free of pathogens. This testifies to the importance of skin as a defense against invasion by a pathogen.

The mucus of mucous membranes physically ensnares microbes. The upper respiratory tract is lined by ciliated cells that sweep mucus and trapped particles up into the throat, where they can be swallowed or expectorated (coughed out). In addition, the various bacteria that normally reside in the intestine and other areas, such as the vagina, prevent pathogens from taking up residence.

Barriers to entry also include antimicrobial molecules. Oil gland secretions contain chemicals that weaken or kill certain bacteria on the skin; mucous membranes secrete lysozyme, an enzyme that can lyse bacteria; and the stomach has an acidic pH, which inhibits the growth of many types of bacteria, or may even kill them.

Protective Proteins **Complement** is composed of a number of blood plasma proteins that "complement" certain immune responses, which accounts for their name. These proteins are continually present in the blood plasma but must be activated by pathogens to exert their effects. Complement helps destroy pathogens in three ways:

1. Complement proteins are involved in and amplify the inflammatory response because certain ones can bind to **mast cells** (a type of white blood cell in tissues) and trigger histamine release, and others can attract phagocytes to the scene.
2. Some complement proteins bind to the surface of pathogens already coated with antibodies, which ensures that the pathogens will be phagocytized by a neutrophil or a macrophage.
3. Certain other complement proteins join to form a **membrane attack complex** that produces holes in the surface of some bacteria and viruses. Fluids and salts then enter the bacterial cell or virus to the point that it bursts (Fig. 31.2B).

Animation Activation of Complement

Interferons are signaling proteins that communicate with and affect the behavior of other cells. Interferons, which are made by virus-infected cells, bind to the receptors of noninfected cells, causing them to produce substances that interfere with viral replication.

FIGURE 31.2A Overview of innate immunity.

Innate immunity

Barriers to entry | Protective proteins | White blood cells | Inflammatory response

skin and mucous membranes

dendritic cell

pathogens

antimicrobial molecules

macrophage

cytokines

neutrophil

monocyte

natural killer cells

complement proteins and interferons in plasma

Interferons, now available as a biotechnology product, are used to treat certain viral infections, such as hepatitis C.

Video
Antiviral Activity of Interferon

White Blood Cells Several types of white blood cells are effective against pathogens. **Neutrophils** are cells that can leave the bloodstream and phagocytize (engulf) bacteria in connective tissues. They have various other ways of killing bacteria also. For example, their granules release antimicrobial peptides called defensins. **Eosinophils** are phagocytic, but they are better known for mounting an attack against animal parasites such as tapeworms that are too large to be phagocytized. The two most powerful of the phagocytic white blood cells are **macrophages** and macrophage-derived **dendritic cells.** They engulf pathogens, which are then destroyed by enzymes when their endocytic vesicles combine with lysosomes. Dendritic cells are found in the skin; once they devour pathogens, they travel to lymph nodes, where they stimulate natural killer cells or lymphocytes. Macrophages are found in all sorts of tissues, where they voraciously devour pathogens and then stimulate lymphocytes to carry on adaptive immunity as discussed in the next part of the chapter.

Video
Neutrophils

Natural killer (NK) cells are large, granular lymphocytes that kill virus-infected cells and cancer cells by cell-to-cell contact. NK cells do their work while adaptive defenses are still mobilizing, and they produce cytokines that stimulate this type of immunity.

What makes NK cells attack and kill a cell? First, they normally congregate in the tonsils, lymph nodes, and spleen, where they are stimulated by dendritic cells before they travel forth. Then, NK cells look for a self-protein on the body's cells. As may happen, if a virus-infected cell or a cancer cell has lost its self-proteins, the NK cell

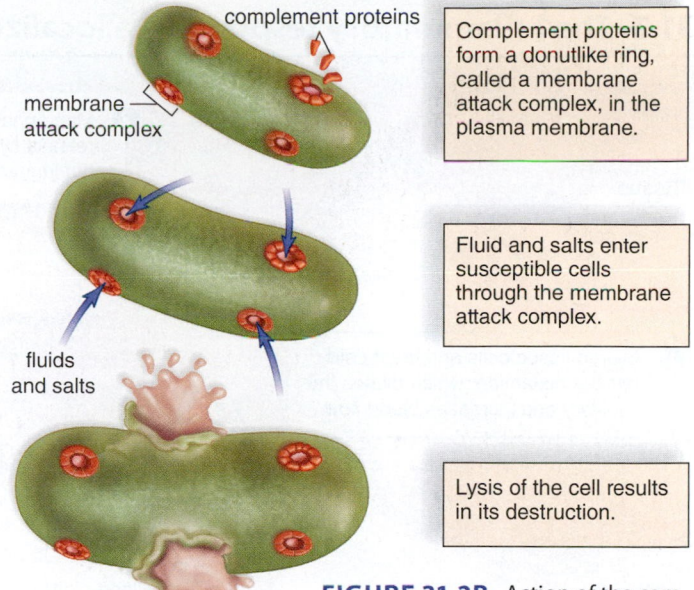

complement proteins

membrane attack complex

Complement proteins form a donutlike ring, called a membrane attack complex, in the plasma membrane.

Fluid and salts enter susceptible cells through the membrane attack complex.

fluids and salts

Lysis of the cell results in its destruction.

FIGURE 31.2B Action of the complement system against a bacterium.

kills it in the same manner used by cytotoxic T cells (see Fig. 31.7B). Unlike cytotoxic T cells, NK cells are not specific; they have no memory; and their numbers do not increase after stimulation.

▶ **31.2 CHECK YOUR PROGRESS**
   1. Explain the term innate immunity and how this term specifically pertains to barriers to entry, protective proteins, and white blood cells.

HOW LIFE CHANGES   *Application*

31A Evolution of the Immune System

A comparative study of organisms supports the hypothesis that immunity, the ability of an organism to destroy foreign cells, developed in stages. Protists such as an amoeba (Fig. 31A) have the ability to distinguish self from nonself, or else they would end up feeding on their own amorphous bodies. When cellular slime molds become multicellular, they even have sentinel cells specialized in engulfing bacteria and toxins. Both of these cells are reminiscent of macrophages, which are guided to their target cells by a chemoattractant. Many organisms, both invertebrate and vertebrate, have populations of phagocytic cells that patrol their internal tissues and fluids, keeping them free of foreign cells. Such an innate mechanism does not require the recognition of a specific invader as seen in adaptive immunity, so-called because it develops after exposure to a pathogen.

Scientists know that fruit flies (*Drosophila*) have plasma membrane receptors that combine with specific pathogenic microbes and thereby trigger an immune response. Therefore, it seems as if a form of adaptive immunity evolved among the invertebrates. Its success is borne out by the discovery that DNA sequences for these receptors are also found in an array of other organisms, even humans. But only jawed vertebrates evolved the ability to produce varied receptors that have the capability of recognizing most any foreign entity that appears on the scene. The ability to mount an effective response after exposure to an antigen allows vertebrates to develop long-lasting immunity to any new threat that evolves in the future.

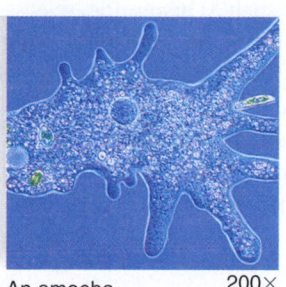

An amoeba   200×

FIGURE 31A Amoebas engulf their food into vacuoles that merge with lysosomes for digestion. Macrophages do the same when they engulf pathogens. It is hypothesized that macrophages and amoebas are related through evolution!

CONSIDER THESE QUESTIONS

1. Does it disturb you to hypothesize that macrophages are endosymbiotic amoebas patrolling our tissues?
2. Of what significance is it that we carry a DNA sequence for an immunoreceptor found in fruit flies?
3. Innate immunity that requires no prior exposure is much less complicated than adaptive immunity that requires exposure to a pathogen. Of what benefit is adaptive immunity?

 **connect**
BIOLOGY
Explore the concepts through a variety of multimedia assets and question types.

www.mcgrawhillconnect.com

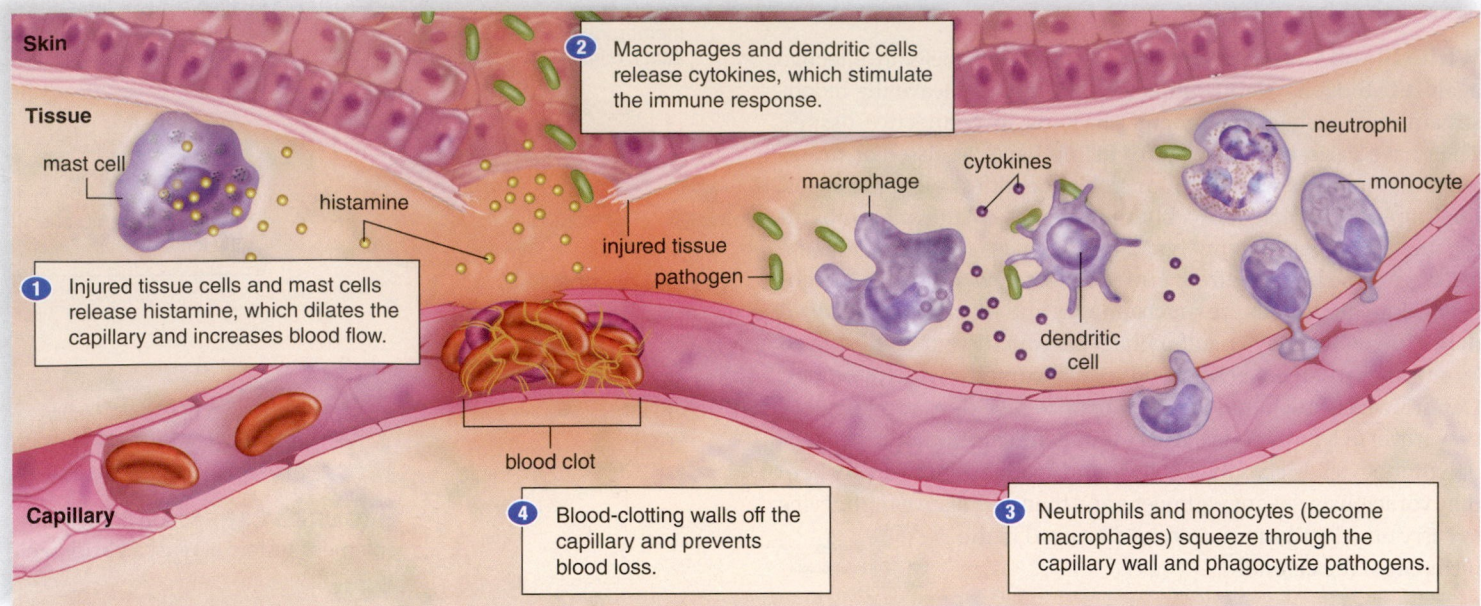

Skin

Tissue

mast cell

histamine

2 Macrophages and dendritic cells release cytokines, which stimulate the immune response.

macrophage

cytokines

neutrophil

monocyte

injured tissue

pathogen

dendritic cell

1 Injured tissue cells and mast cells release histamine, which dilates the capillary and increases blood flow.

blood clot

Capillary

4 Blood-clotting walls off the capillary and prevents blood loss.

3 Neutrophils and monocytes (become macrophages) squeeze through the capillary wall and phagocytize pathogens.

FIGURE 31.3A Inflammatory response.

LEARNING OUTCOME

When you complete this section, you should be able to

1. Summarize the events in the inflammatory response.

Whenever tissue is damaged by physical or chemical agents or by pathogens, a series of events occurs that is known as the **inflammatory response.**

An inflamed area has four outward signs: redness, heat, swelling, and pain. All of these signs are due to capillary changes in the damaged area. Figure 31.3A illustrates the participants in the inflammatory response. **1** Chemical mediators, such as **histamine,** released by damaged tissue cells and mast cells, cause the capillaries to dilate and become more permeable. Excess blood flow, due to enlarged capillaries, causes the skin to redden and become warm. Increased permeability of the capillaries allows proteins and fluids to escape into the tissues, resulting in swelling. The swollen area stimulates free nerve endings, causing the sensation of pain.

2 Macrophages also release colony-stimulating factors, cytokines that pass by way of the blood to the red bone marrow, where they stimulate the production and release of white blood cells, primarily neutrophils. As the infection is being overcome, some phagocytes die. These—along with dead tissue cells, dead bacteria, and living white blood cells—form pus, a whitish material. The presence of pus indicates that the body is trying to overcome an infection. **3** Migration of phagocytes, namely neutrophils and monocytes, also occurs during the inflammatory response. Neutrophils and monocytes are amoeboid and can change shape to squeeze through capillary walls and enter tissue fluid. Also present are dendritic cells, notably in the skin and mucous membranes, and macrophages, both of which are able to devour many pathogens and still survive (Fig. 31.3B).

The inflammatory response can be accompanied by other responses to the injury. **4** A blood clot can form to seal a break in a blood vessel.

Antigens, chemical mediators, dendritic cells, and macrophages move through the tissue fluid and lymph to the lymph nodes. There, B cells and T cells are activated to mount an adaptive defense to the infection.

Sometimes an inflammation persists, and the result is chronic inflammation that is often treated by administering drugs such as aspirin, ibuprofen, or cortisone. These medications act against the chemical mediators released by the white blood cells that promote inflammation.

We have now completed our discussion of innate defenses. The next part of the chapter discusses adaptive defenses.

 **Animation** Inflammatory Response

▶ **31.3 CHECK YOUR PROGRESS**

1. Explain what causes the redness, heat, swelling, and pain of the inflammatory response.

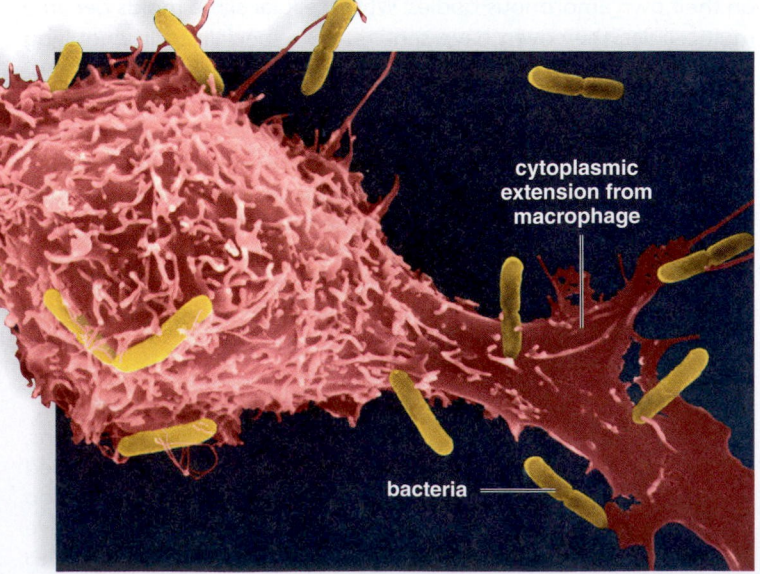

cytoplasmic extension from macrophage

bacteria

SEM 1,075×

FIGURE 31.3B A macrophage engulfing bacteria.

Adaptive Immunity

The body's second line of defense is called adaptive immunity, or sometimes specific immunity, because it consists of mechanisms that are specific to the particular disease. Certain white blood cells are involved in adaptive immunity and they can be likened to bounty hunters trained to seek out a very specific perpetrator.

31.4 Adaptive immunity targets a specific antigen

LEARNING OUTCOMES

When you complete this section, you should be able to

1. Distinguish between a foreign antigen and a self-antigen.
2. Distinguish between active and passive immunity.

When innate immunity has been inadequate to stem an infection, **adaptive immunity** comes into play. *First,* note that pathogens, such as bacteria, viruses, and transplanted tissues and organs, bear molecules called **foreign antigens.** They are called foreign antigens because the body does not produce them. Other antigens are termed **self-antigens** because the body itself produces them. (The self-proteins mentioned earlier are self-antigens.) It is unfortunate when the immune system reacts to the body's own pancreatic cells (causing diabetes mellitus) or to nerve fiber sheaths (causing multiple sclerosis), but fortunate when adaptive immunity can destroy the cancerous cells of a tumor.

Second, after recognizing foreign antigens, adaptive immunity can respond to them. Unlike innate immunity which occurs immediately, it usually takes 5–7 days to mount an adaptive response. *Third,* the cells involved in adaptive immunity remember the antigens they have met before. This is the reason, for example, that once we recover from the measles, we usually do not get the disease a second time. Under these circumstances the body is said to be immune to the disease.

Adaptive immunity can be active or passive. After a person is infected with a pathogen, such as a measles or chickenpox virus, **active immunity** develops naturally. However, active immunity is often induced when a person is well, so that a possible future infection will not take place. Individuals can be immunized with a vaccine to keep from being infected by a pathogen (Fig. 31.4A). The **vaccine** contains antigen that causes the body to develop

FIGURE 31.4B
Breast-feeding provides infants with antibodies.

specific **antibodies.** When an antibody combines with an antigen, the pathogen bearing that antigen is destroyed. The United States is committed to immunizing all children against the common childhood diseases; before children can enter school, they must show proof of immunization. Outbreaks of measles and mumps sometimes occur in college dormitories because students haven't been immunized or their immunity has worn off.

Passive immunity occurs when an individual is given prepared antibodies (immunoglobulins) to combat a disease. Because these antibodies are not produced by the individual's plasma cells, passive immunity is short-lived. For example, newborn infants are passively immune to some diseases because antibodies have crossed the placenta from the mother's blood. These antibodies soon disappear, however, so that within a few months, infants become more susceptible to infections. Breast-feeding prolongs the natural passive immunity an infant receives from its mother because antibodies are present in the mother's milk (Fig. 31.4B).

Even though passive immunity does not last, it is sometimes used to prevent illness in a patient who has been unexpectedly exposed to an infectious disease. Artificial passive immunity is used in the emergency treatment of rabies, measles, tetanus, diphtheria, botulism, hepatitis A, and snakebites. Usually, the patient receives a gamma globulin injection (serum that contains antibodies), extracted from a large, diverse adult population.

▶ **31.4 CHECK YOUR PROGRESS**
1. Explain why active immunity is longer lasting than passive immunity.

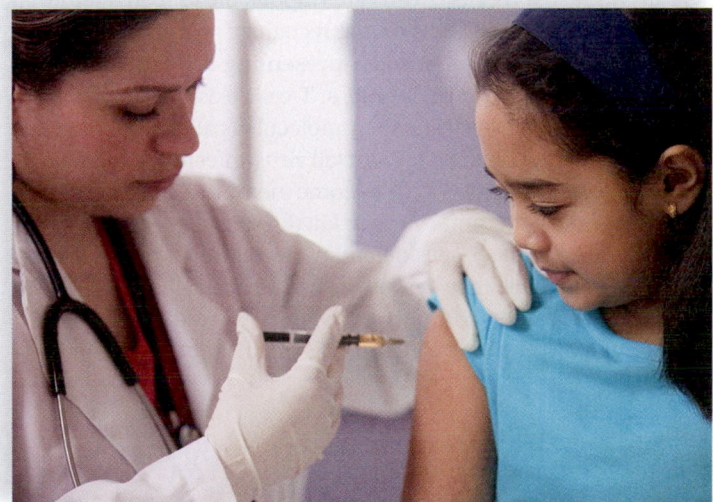

FIGURE 31.4A Vaccines immunize children against diseases.

LEARNING OUTCOME

When you complete this section, you should be able to

1. List the types of cells and their functions that are involved in adaptive immunity.

Adaptive immunity primarily depends on the two types of lymphocytes, called B cells and T cells. Both B cells and T cells are manufactured in the red bone marrow. *B* cells mature in *b*one marrow, but *T* cells mature in the *t*hymus. These cells are capable of recognizing antigens because they have specific **antigen receptors** that combine with antigens. B cells have **B-cell receptors** (**BCRs**), and T cells have **T-cell receptors** (**TCRs**) (Fig. 31.5). Each lymphocyte has receptors that will combine with only one type of antigen. If a particular B cell responds to an antigen—a molecule projecting from the bacterium *Streptococcus pyogenes*, for example—it does not react to any other antigen. It is often said that the receptor and the antigen fit together like a lock and a key. Remarkably, diversification occurs to such an extent during maturation that there are specific B cells and/or T cells for any possible antigen we are likely to encounter during a lifetime.

B cells are responsible for **antibody-mediated immunity** (Fig. 31.5). Once a B cell combines with an antigen, it gives rise to **plasma cells,** which produce specific antibodies. These antibodies can react to the same antigen as the original B cell. Therefore, an antibody has the same specificity as the BCR. Some progeny of the activated B cells become memory B cells, so-called because these cells always "remember" a particular antigen and make us immune to a particular illness, but not to any other illness.

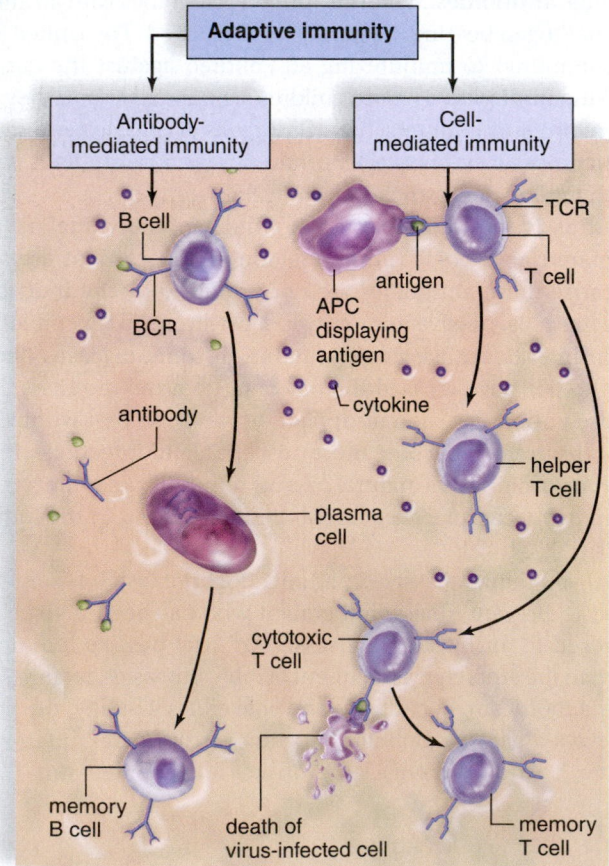

FIGURE 31.5 Overview of adaptive immunity.

TABLE 31.5		Cells Involved in Adaptive Immunity
Cell		**Functions**
Macrophages		Phagocytize pathogens; inflammatory response and specific defense
Dendritic cells		Phagocytize pathogens in skin; play a role in specific defense
Lymphocytes		Responsible for specific defense
B cells		Produce plasma cells and memory cells
Plasma cells		Produce antibodies to specific antigens
Memory B cells		Ready to produce antibodies in the future
Helper T cells		Regulate immunity
Cytotoxic T cells		Kill virus-infected and cancer cells
Memory T cells		Ready to attack infected cells in the future

In contrast to B cells, T cells are responsible for **cell-mediated immunity** (Fig. 31.5). T cells do not recognize an antigen until it is presented to them by an **antigen-presenting cell** (**APC**). Macrophages and dendritic cells are APCs. T cells exist as either **helper T cells,** which release stimulating molecules called **cytokines,** or **cytotoxic T cells,** which attack and kill virus-infected cells and cancer cells. Some cytotoxic T cells become memory T cells, ever ready to defend against the same virus or kill the same type of cancer cell again. Because B and T cells defend us from disease by specifically reacting to antigens, they can be likened to special forces that can attack selected targets without harming nearby residents (cells).

Significant immune cells are listed in Table 31.5 for easy reference.

▶ **31.5 CHECK YOUR PROGRESS**
1. Contrast the function of B cells and cytotoxic T cells.

31.6 Antibody-mediated immunity involves B cells

LEARNING OUTCOME

When you complete this section, you should be able to

1. Describe the clonal selection model and the production of antibodies.

When a *B-cell receptor* combines with an antigen the B cell undergoes **clonal selection** (Fig. 31.6A). ❶ The antigen is said to "select" the B cell that will clone. Also, at this time, cytokines secreted by helper T cells stimulate B cells to clone. ❷ Defense by B cells is called *antibody-mediated immunity* because most members of a clone become plasma cells that produce specific antibodies. It is also called humoral immunity because these antibodies are present in blood and lymph. (A *humor* is any fluid normally occurring in the body.) ❸ Some progeny of activated B cells become **memory B cells,** which are the means by which long-term immunity is possible. Once the threat of an infection has

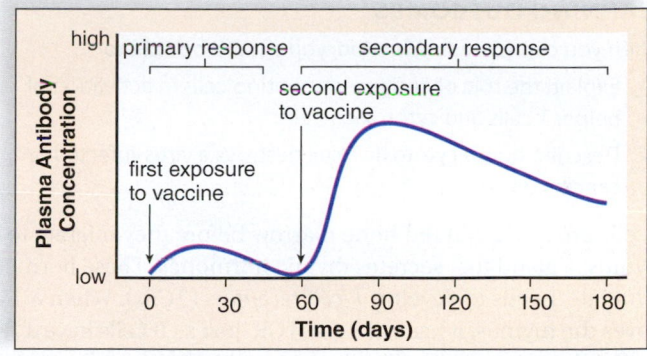

FIGURE 31.6B Antibody titers.

passed, the development of new plasma cells ceases, and those present undergo apoptosis.

Immunization, described in section 31.4, involves the use of vaccines to bring about clonal expansion, not only of B cells, but also of T cells. Traditionally, vaccines are the pathogens themselves, or their products, that have been treated so they are no longer virulent (able to cause disease). Vaccines against smallpox, polio, and tetanus have been successfully used worldwide. Today, it is possible to genetically engineer bacteria to mass-produce a protein from pathogens, and this protein can be used as a vaccine. This method was used to produce a vaccine against hepatitis B, a viral disease, and is being used to prepare a potential vaccine against malaria.

Animation Constructing Vaccines

Video Potato Vaccine

After a vaccine is given, it is possible to determine the antibody titer (the amount of antibody present in a sample of plasma). After the first exposure to a vaccine, a primary response occurs. For a period of several days, no antibodies are present; then the titer rises slowly, followed by first a plateau and then a gradual decline as the antibodies bind to the antigen or simply break down (Fig. 31.6B). After a second exposure, the titer rises rapidly to a plateau level much greater than before. The second exposure is called a "booster" because it boosts the antibody titer to a high level. The high antibody titer is expected to prevent disease symptoms when the individual is exposed to the antigen. Even years later, if the antigen enters the body, memory B cells quickly give rise to more plasma cells capable of producing the correct type of antibody.

Antibodies Another name for an antibody is **immunoglobulin (Ig).** The most typical antibody, called IgG, is a Y-shaped molecule with two arms. IgG has constant regions, where the sequence of amino acids is set, and variable regions, where the sequence of amino acids varies between IgGs. The variable regions become hypervariable at their tips and form antigen-binding sites. The shape of the variable region is specific for a particular antigen.

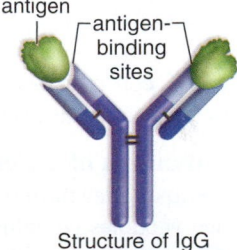
Structure of IgG

▶ **31.6 CHECK YOUR PROGRESS**

1. Explain why antibodies produced by a plasma cell against an HIV infection would not be effective against the virus that causes measles.

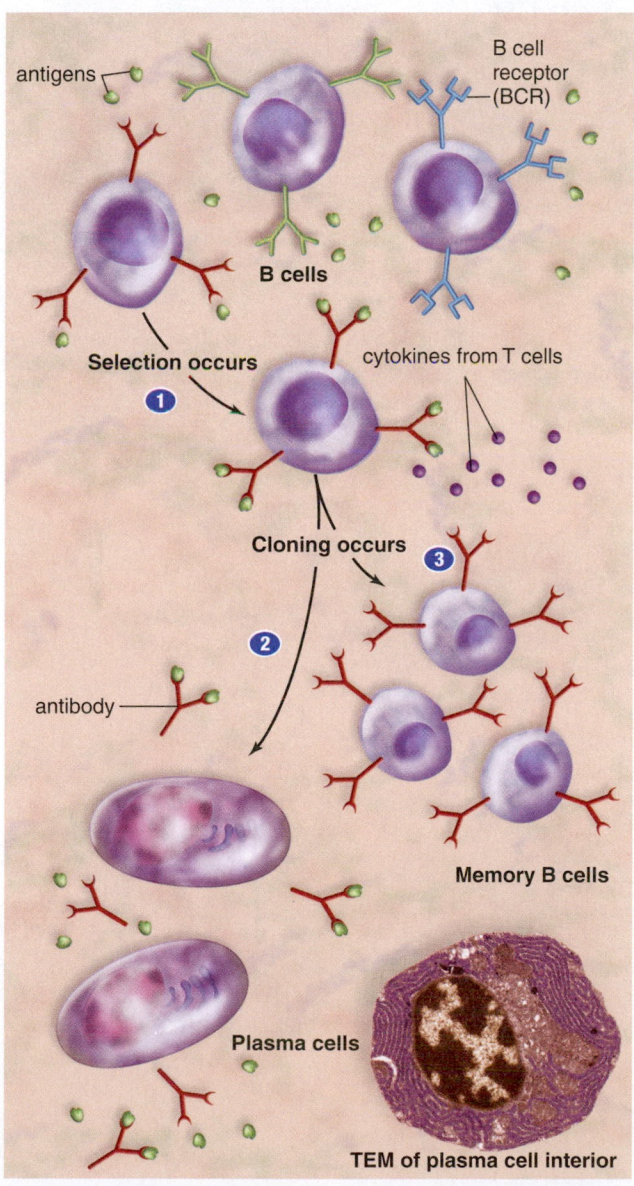

FIGURE 31.6A Clonal selection model as it applies to B cells.

LEARNING OUTCOMES

When you complete this section, you should be able to

1. Explain the role of antigen-presenting cells in activation of helper T cells and cytotoxic T cells.

2. Describe how a cytotoxic T cell destroys a virus-infected or cancer cell.

T cells are formed in red bone marrow before they migrate to the *t*hymus, a gland that secretes thymic hormones. These hormones stimulate T cells to develop *T-cell receptors* (*TCRs*). When a T cell leaves the thymus, it has a unique TCR, just as B cells have a BCR. Unlike B cells, however, T cells are unable to recognize an antigen without help. The antigen must be displayed to them by an antigen-presenting cell (APC), such as a dendritic cell or a macrophage. After phagocytizing a pathogen, APCs travel to a lymph node or the spleen, where T cells also congregate. In the meantime, the APC has broken the pathogen apart in a lysosome. A piece of the pathogen is then displayed in an **MHC** (**major histocompatibility complex**) **protein** on the cell's surface. MHC proteins are self-antigens because they mark cells as belonging to a particular individual and, therefore, make transplantation of organs difficult. MHC proteins differ by the sequence of their amino acids, and the immune system will attack as foreign any tissue that bears MHC antigens different from those of the individual.

In Figure 31.7A, the different types of T cells have specific TCRs represented by their different shapes and colors. ❶ A macrophage presents an antigen only to a T cell that has a TCR capable of combining with a particular antigen (colored green). ❷ A major difference between helper T cells and cytotoxic T cells is that helper T cells only recognize an antigen in combination with MHC class II molecules, while cytotoxic T cells only recognize an antigen in combination with MHC class I molecules. Notice in Figure 31.7A, a cytotoxic T cell binds to an antigen displayed in combination with an MHC I and undergoes cloning during which many copies of the cytotoxic T cell are produced. ❸ Some of the cloned cells become **memory T cells.**

❹ As the illness disappears, the immune response wanes, and active T cells become susceptible to apoptosis. As mentioned previously, apoptosis contributes to homeostasis by regulating the number of cells present in an organ, or in this case, in the immune system. When apoptosis does not occur as it should, T cell cancers (e.g., lymphomas and leukemias) can result. Also, in the thymus, any T cell that has the potential to recognize self-antigens and destroy the body's own cells undergoes apoptosis.

Functions of T Cells Cytotoxic T cells specialize in cell-to-cell combat. They have storage vacuoles containing perforins and storage vacuoles containing enzymes called granzymes (Fig. 31.7B*a*). ❶ After a cytotoxic T cell binds to a virus-infected cell or cancer cell, it releases perforin molecules, which perforate the plasma membrane, forming a pore. ❷ Cytotoxic T cells then deliver granzymes into the pore, and these cause the cell to undergo osmotic destruction and die. Once cytotoxic T cells have released the perforins and granzymes, they move on to the next target cell. Cytotoxic T cells are responsible for so-called cell-mediated immunity.

Animation
Cytotoxic T-Cell Activity
Against Target Cells

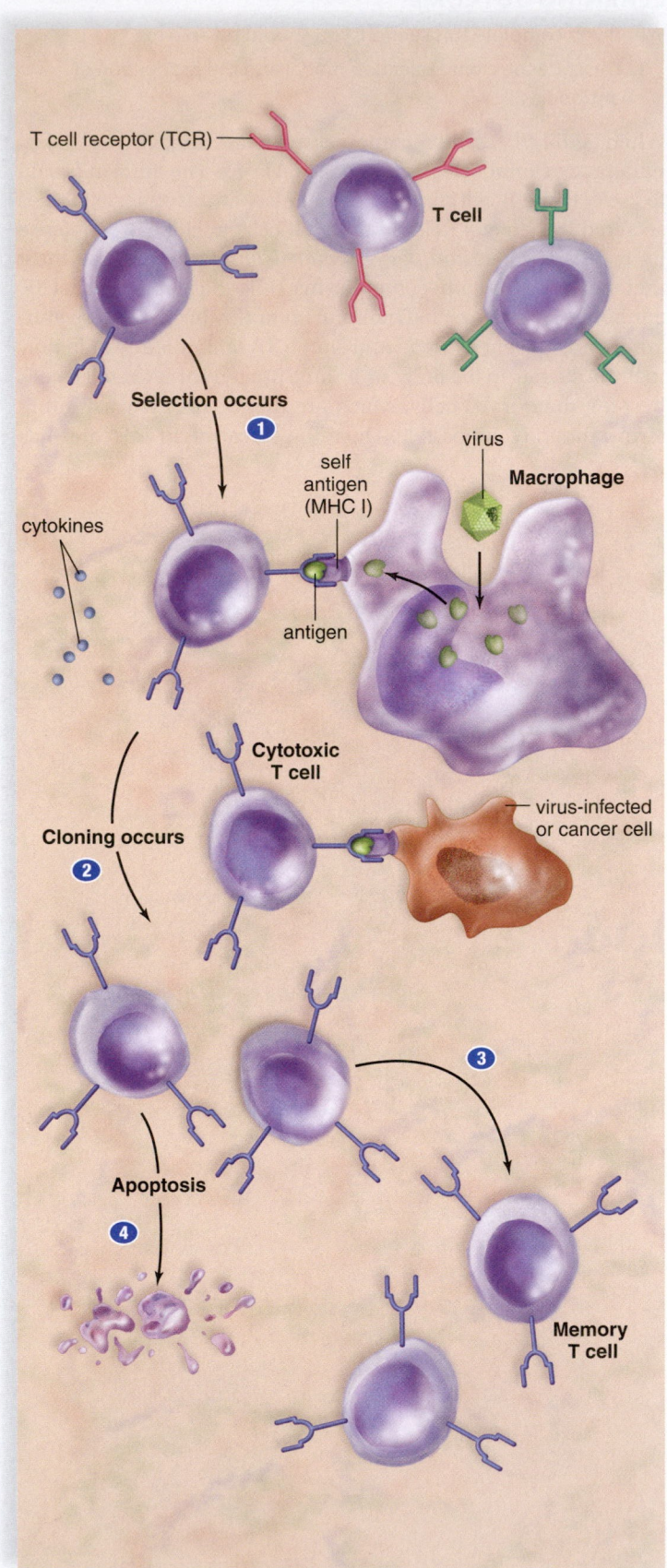

FIGURE 31.7A Clonal selection model as it applies to T cells.

Helper T cells play a critical role in coordinating innate defenses and adaptive defenses, including both cell-mediated immunity and antibody-mediated immunity. How do helper T cells perform this function? After a helper T cell is activated, it secretes cytokines. The cytokines attract neutrophils, natural killer cells, and macrophages to where they are needed. Cytokines stimulate phagocytosis of pathogens, as well as the clonal expansion of T cells and B cells.

Because more and more immune cells are recruited by helper T cells, the number of pathogens eventually begins to wane. (As discussed in the introduction, an HIV infection destroys helper T cells, and for this reason the individual is susceptible to other deadly infections.) Usually after an infection has passed, memory T cells are available. Like memory B cells, memory T cells are long-lived, and their number is far greater than the original number of T cells that could recognize a specific antigen. Therefore, when the same antigen enters the body later on, the reaction is so rapid that no illness is detectable.

Video T Lymphocyte

Cytokines and Cancer Therapy

The term cytokine simply means a soluble protein that acts as a signaling molecule. Cytokines allow immune cells to communicate with one another.

Because cytokines stimulate white blood cells, they have been studied as a possible adjunct therapy for cancer. We have already mentioned that interferons are cytokines. Interferons are produced by virus-infected cells, and they signal other cells of the need to prevent infection. Interferon has been investigated as a possible cancer drug, but so far it has been proven effective only in certain patients, and the exact reasons for this, as yet, cannot be discerned. Also, interferon has a number of side effects that limit its use.

Cytokines called interleukins are produced by white blood cells, and they act to stimulate other white blood cells. Scientists actively engaged in interleukin research believe that interleukins will soon be used to supplement vaccines, to treat chronic infectious diseases, and perhaps even to treat cancer. Interleukin antagonists may also prove helpful in preventing skin and organ rejection, autoimmune diseases, and allergies.

When, and if, cancer cells carry an altered protein on their cell surface, they should be attacked and destroyed by cytotoxic T cells (Fig. 31.7B*b*). Whenever cancer does develop, it is possible that the cytotoxic T cells have not been activated. In that case, interleukins might awaken the immune system and lead to the destruction of the cancer. In one technique being investigated, researchers first withdraw T cells from the patient and activate the cells by culturing them in the presence of an interleukin. The cells are then reinjected into the patient, who is given doses of interleukin to maintain the killer activity of the T cells.

Tumor necrosis factor (TNF) is a cytokine produced by macrophages that has the ability to promote the inflammatory response and to cause the death of cancer cells. Like the interferons and interleukins, TNF stimulates the body's immune cells to fight cancer. TNF also directly affects tumor cells, damaging them and the blood vessels within the tumor. Without an adequate blood supply, a cancerous tumor cannot thrive. However, researchers are still uncertain about exactly how TNF destroys tumors. Researchers have found that TNF therapy is most effective and least toxic when directed at a specific tumor site, rather than administered throughout. Clinical trials are under way.

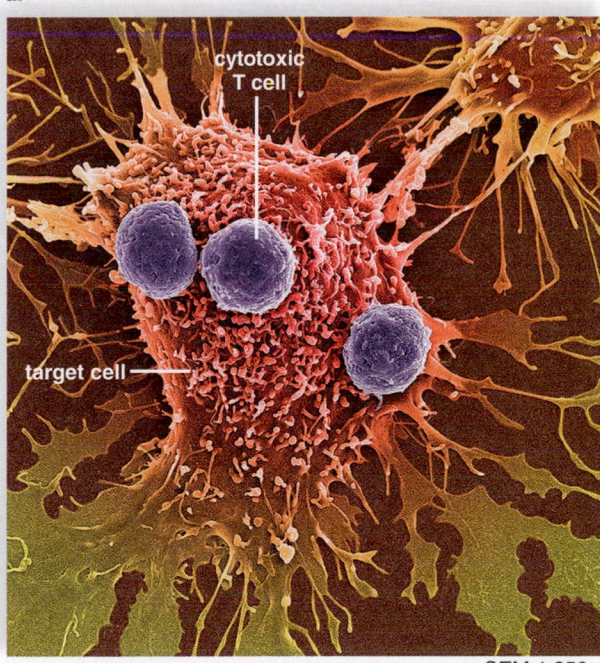

FIGURE 31.7B Cell-mediated immunity. **a.** How a T cell destroys a virus-infected cell or cancer cell. **b.** The scanning electron micrograph shows cytotoxic T cells attacking and destroying a cancer cell (target cell).

▶ **31.7 CHECK YOUR PROGRESS**
1. Explain the significance of #1–3 in Figure 31.7A.
2. Contrast the contribution of helper T cells and cytotoxic T cells to immunity.

31B Monoclonal Antibodies

Every plasma cell derived from a single B cell secretes antibodies against a specific antigen. These are called **monoclonal antibodies** because all of them are the same type. One method of producing monoclonal antibodies in vitro (outside the body in the laboratory) is depicted in Figure 31B. B cells were removed from an animal (usually mice are used) and exposed to a particular antigen. The B cell that responded to the antigen is a plasma cell that produces a particular antibody. ❶ This plasma cell is fused with a myeloma cell (malignant plasma cell that lives and divides indefinitely). The fused cell is called a hybridoma—*hybrid-* because it results from the fusion of two different cells, and *-oma* because one of the cells is a cancer cell. ❷ The first hybridoma gives rise to others and they all secrete large amounts of the specified monoclonal antibody, which recognizes only the antigen of interest.

Animation
Monoclonal Antibody
Production

Research Uses for Monoclonal Antibodies

The ability to quickly produce monoclonal antibodies in the laboratory has made them an important tool for academic research. Monoclonal antibodies are very useful because of their extreme specificity for only a particular molecule. A monoclonal antibody can be used to select out a specific molecule among many others, much like finding needles in a haystack. Now the molecule can be purified from all the others that are also present in a sample. In this way, monoclonal antibodies have simplified formerly tedious laboratory tasks, allowing investigators more time to focus on other priorities.

Medical Uses for Monoclonal Antibodies

Monoclonal antibodies also have many applications in medicine. For instance, they can now be used to make quick and certain diagnoses of various conditions. Today, a monoclonal antibody is used to signify pregnancy by detecting a particular hormone in the urine of a woman after she becomes pregnant. Thanks to this technology, pregnancy tests that once required a visit to a doctor's office and the use of expensive laboratory equipment can now be performed in the privacy and comfort of a woman's own home, at minimal expense.

Monoclonal antibodies can be used not only to diagnose infections and illnesses but also to fight them. Many bacteria and viruses possess unique proteins on their cell surfaces that make them easily recognized by an appropriate monoclonal antibody. When binding occurs with a certain monoclonal antibody but not another, a physician knows what type of infection is present. And because monoclonal antibodies can distinguish some cancer cells from normal tissue cells, they may also be used to identify cancers at very early stages when treatment can be most effective.

Finally, monoclonal antibodies have also shown promise as potential drugs to help fight disease. Respiratory syncytial virus (RSV), a common virus that causes serious respiratory tract infections in very young children, is now being successfully treated with a monoclonal antibody drug. The antibody recognizes a protein on the viral surface, and when it binds very tightly to the surface of the virus, the patient's own immune system can easily recognize the virus and destroy it before it has a chance to cause serious illness.

Other illnesses, such as cancer, are also being successfully treated with monoclonal antibodies. Because these antibodies

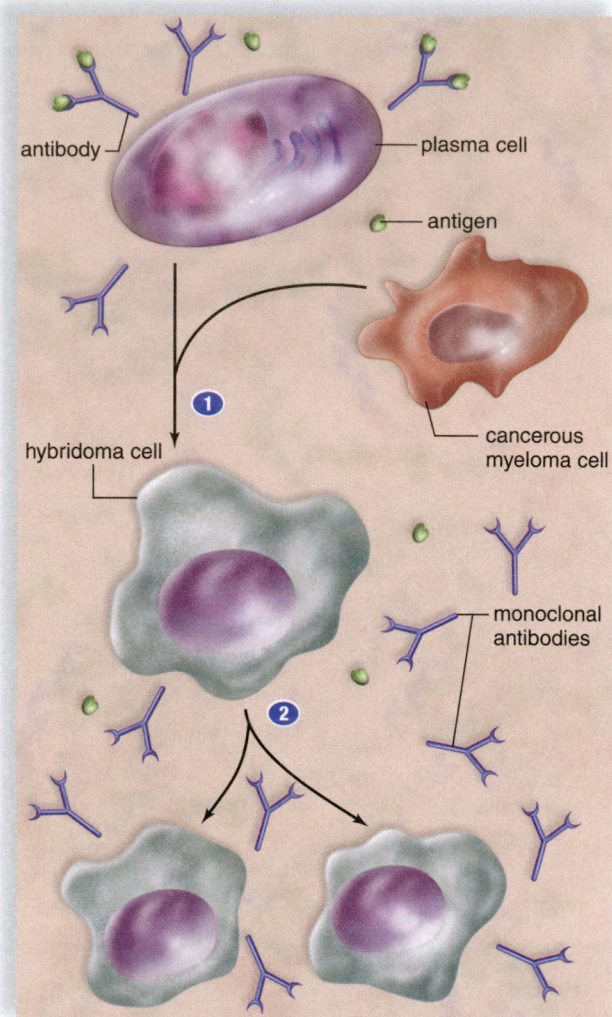

FIGURE 31B Production of monoclonal antibodies.

are able to distinguish between cancerous and normal cells, they have been engineered to carry radioisotopes or toxic drugs to tumors so that cancer cells can be selectively destroyed without damaging other body cells. In short, monoclonal antibodies are helping scientists in their research and physicians in their attempts to diagnose and cure patients.

CONSIDER THESE QUESTIONS

1. If monoclonal antibodies are made by cancer cells, why don't they cause cancer in the patient?
2. HeLa cells are a continuous line of cancer cells used in virology studies and derived from a tumor of a patient who died in 1951. The patient never knew that researchers were going to use her tumor cells in this way. Should she have been asked to give her consent? Should she have been paid for her contribution to science?

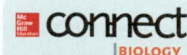 Explore the concepts through a variety of multimedia **|BIOLOGY** assets and question types.

www.mcgrawhillconnect.com

Immune System Failures

In certain instances, immunity works against the best interests of the body. For example, immune mechanisms can impede organ transplantation, cause autoimmune disorders, or lead to allergic reactions.

31.8 Autoimmune disorders can result in long-term illnesses

LEARNING OUTCOMES

When you complete this section, you should be able to

1. Explain the role of MHC antigens in tissue rejection.
2. Contrast several autoimmune diseases and describe the symptoms of each.

Certain organs, such as skin, the heart, and the kidneys, could be transplanted easily from one person to another if the body did not attempt to reject them. Rejection occurs because antibodies and cytotoxic T cells bring about the destruction of foreign tissues in the body. When rejection occurs, the immune system is correctly distinguishing between self and nonself.

Organ rejection can be controlled by carefully selecting the organ to be transplanted and administering immunosuppressive drugs. It is best if the transplanted organ has the same type of MHC antigens as those of the recipient, because cytotoxic T cells recognize foreign MHC antigens. Two well-known immunosuppressive drugs, cyclosporine and tacrolimus, both act by inhibiting the production of certain cytokines that stimulate cytotoxic T cells.

Xenotransplantation, the transplantation of animal tissues and organs into humans, is another way to solve the problem of organ scarcity. Genetic engineering can make pig organs less antigenic by removing the MHC antigens. The ultimate goal is to make pig organs as widely accepted as blood type O. Other researchers hope that tissue engineering, including the production of human organs by using stem cells, will one day do away with the problem of rejection. Scientists have recently grown a new heart in the laboratory using stem cells and researchers are now able to produce stem cells that contain the patient's DNA.

Autoimmune disorders are diseases that can be characterized by the failure of the immune system to distinguish between foreign antigens and the self-antigens that mark the body's own tissues. In an autoimmune disease, chronic inflammation occurs, and cytotoxic T cells or antibodies mistakenly attack the body's own cells as if they displayed foreign antigens.

Systemic lupus erythematosus (lupus) is a chronic autoimmune disorder characterized by the presence of antibodies to the nuclei of the body's cells. Lupus affects multiple tissues and organs, and is still very poorly understood. The symptoms vary somewhat, but most patients experience a characteristic skin rash (Fig. 31.8*a*), joint pain, and kidney damage. Lupus typically progresses to include many life-threatening complications.

Rheumatoid arthritis is a common autoimmune disorder that causes recurring inflammation in synovial joints (Fig. 31.8*b*). Complement proteins, T cells, and B cells all participate in deterioration of the joints, which eventually become immobile. This chronic inflammation gradually causes destruction of the delicate membrane and cartilage within the joint.

In myasthenia gravis, a well-understood autoimmune disease, antibodies attach to and interfere with the functioning of neuromuscular junctions. The result is muscular weakness.

The exact events that trigger an autoimmune disorder are not known. Some autoimmune disorders set in following a noticeable infection; for example, the heart damage following rheumatic fever is thought to be due to an autoimmune disorder triggered by the illness. Otherwise, most autoimmune disorders probably start after an undetected inflammatory response. However, the tendency to develop autoimmune disorders is known to be inherited in some cases, so there may be genetic causes as well.

Because little is known about the origin of autoimmune disorders, no cures are currently available. Regardless, most of these conditions can be managed over the long term with immunosuppressive drugs that control the various symptoms.

▶ **31.8 CHECK YOUR PROGRESS**
1. For patients needing new organs, why might immunosuppressive drugs become obsolete in the future?
2. Patients with an autoimmune disease are caught between a rock and a hard place. Explain.

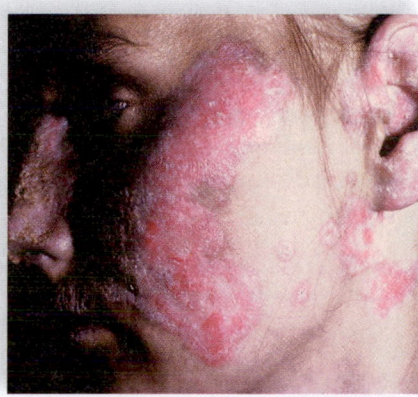

a. Systemic lupus characterized by a red rash on the face.

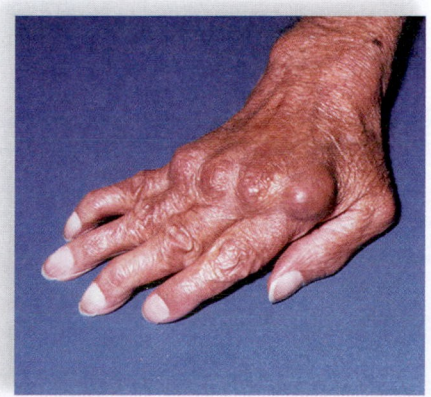

b. Rheumatoid arthritis.

FIGURE 31.8 Autoimmune diseases occur when the body's immune system attacks the body's cells as if they were foreign cells. The result is disfigurement, pain, and life-threatening illnesses.

31C Allergic Reactions

Allergies are hypersensitivities to substances, such as pollen, food, or animal hair, that ordinarily would do no harm to the body. The response to these antigens, called allergens, usually includes some degree of tissue damage.

An *immediate allergic response* can occur within seconds of contact with the antigen. The response is caused by an antibody known as IgE (Table 31C). IgE receptors are located in the plasma membrane of mast cells present in the tissues. After allergens attach to IgEs, they attach to their mast cell receptors and the mast cells release histamine and other substances that bring about the allergic symptoms (Fig. 31C). When pollen is an allergen, histamine combines with and stimulates the mucous membranes of the nose and eyes to release fluid, causing the runny nose and watery eyes typical of *hay fever*. If a person has *asthma*, the airways leading to the lungs constrict, resulting in difficult breathing accompanied by wheezing. When food contains an allergen, nausea, vomiting, and diarrhea result.

Drugs called antihistamines are used to treat allergies. These drugs compete for histamine receptors in the nose, eyes, airways, and lining of the digestive tract. In this way, histamine is prevented from binding and causing its unpleasant symptoms. However, antihistamines are only partially effective because mast cells release other molecules, in addition to histamine, that cause allergic symptoms.

Anaphylactic shock is an immediate allergic response that occurs because the allergen has entered the bloodstream. Bee stings and penicillin shots are known to cause this reaction in some individuals because both inject the allergen into the blood. Due to increased permeability of the capillaries by histamine, anaphylactic shock is characterized by a sudden and life-threatening drop in blood pressure.

People with allergies produce ten times more IgE than people without allergies. A new treatment using injections of monoclonal IgG antibodies for IgEs is currently being tested in individuals with severe food allergies. More routinely, injections of the allergen are given so that the body will build up high quantities of IgG antibodies. The hope is that these will combine with allergens received from the environment before they have a chance to combine with IgEs. Then the IgEs will not bind to their mast cell receptors.

A *delayed allergic response* is initiated by memory T cells at the site of allergen body contact. The allergic response is regulated by cytokines secreted by both T cells and macrophages. A classic example of a delayed allergic response is the skin test for tuberculosis (TB). When the test result is positive, the tissue where the antigen was injected becomes red and hardened. This indicates prior exposure to tubercle bacilli, the cause of TB. Contact dermatitis, which occurs when a person is allergic to poison ivy, jewelry, cosmetics, and many other substances that touch the skin, is another example of a delayed allergic response.

allergen

histamine and other chemicals

B cell

IgE antibodies

IgE receptor

plasma cell   mast cell

SEM of mast cell

FIGURE 31C An allergic reaction. An allergen attaches to IgE antibodies, and then IgEs cause mast cells to release histamine and other chemicals that are responsible for the allergic reaction.

CONSIDER THESE QUESTIONS

1. What might increase a person's tendency to develop an immediate allergic response?
2. Allergic responses are more common in developed as opposed to developing countries. Why might that be?

 Explore the concepts through a variety of multimedia assets and question types.

www.mcgrawhillconnect.com

TABLE 31C	Comparison of Immediate and Delayed Allergic Responses	
	Immediate Response	**Delayed Response**
Onset of Symptoms	Takes several minutes	Takes 1–3 days
Lymphocytes Involved	B cells	T cells
Immune Reaction	IgE antibodies	Cell-mediated immunity
Type of Symptoms	Hay fever, asthma, and many other allergic responses	Contact dermatitis (e.g., poison ivy)
Therapy	Antihistamine and adrenaline	Cortisone

CONNECTING THE CONCEPTS

The role of the lymphatic system in homeostasis cannot be overemphasized. Excess tissue fluid that is collected within lymphatic vessels is called lymph. Because this fluid is returned to the cardiovascular system when the large lymphatic ducts empty into the subclavian veins, blood volume and pressure are maintained. The lymphatic system is also intimately involved in immunity.

Levels of defense against invasion of the body by pathogens can be compared to the way we protect our homes. Homes usually have external defenses, such as a fence, a dog, or locked doors. Similarly, our skin and mucous membranes act as barriers to prevent pathogens from entering the blood and lymph. Like a home alarm system, if an invasion does occur, the immune system detects the intruder. First, innate immunity consisting of mechanisms, such as the complement system and phagocytosis by white blood cells, comes into play.

Then, adaptive immunity, consisting of mechanisms that allow B cells and T cells to respond to specific antigens, brings the infection to an end.

A strong connection exists between immunity and the nervous system in that cytokines can affect the nervous system. For example, researchers have learned that cytokines affect the brain's temperature control center so that a fever results. Also, cytokines bring about a feeling of sluggishness, sleepiness, and loss of appetite. These behaviors tend to make us take care of ourselves until we feel better.

In the next few chapters, we discuss other systems that, like the lymphatic system, interact directly with the cardiovascular system. Blood is refreshed because the digestive, respiratory, and urinary systems make exchanges with the external environment. The food we eat is digested into nutrients that enter the blood; gas exchange in the lungs permits the

removal of carbon dioxide from the blood and the entrance of oxygen into the blood; and the kidneys remove and excrete metabolic wastes taken from the blood. Homeostasis is critically dependent on these exchanges between blood and the external environment.

ANALYZE AND EVALUATE

1. HIV can be transmitted as a free virus and in an infected cell. Therefore, both antibody-mediated and cell-mediated responses are needed. Explain why both responses are needed.

2. A PCR HIV test detects the presence of the virus and not the presence of antibodies to HIV in the blood. What would be the benefits of this new type of testing?

3. Speculate on what has specifically gone wrong with innate and adaptive immunity when a person has an autoimmune disease such as rheumatoid arthritis.

SUMMARIZE

The Lymphatic System

31.1 Lymphatic vessels transport lymph

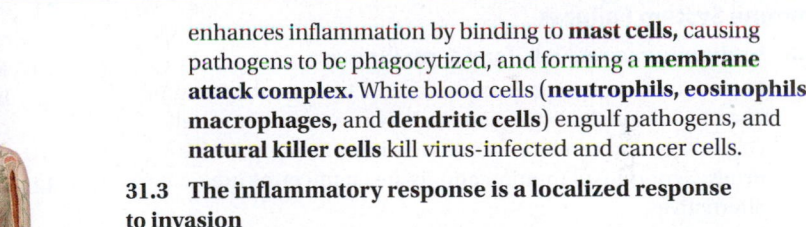

- The **lymphatic system** is composed of **lymphatic vessels,** which transport **lymph,** and various lymphatic organs. Lymphatic capillaries absorb excess tissue fluid, and those called **lacteals** absorb fats and then lymphatic vessels transport them to the bloodstream. The lymphatic system produces, maintains, and distributes lymphocytes, and helps defend the body against pathogens.

- Among the **lymphatic (lymphoid) organs,** the **red bone marrow** is where all blood cells are made and where **B cells** mature. The **thymus gland** is where **T cells** mature. **Lymph nodes** cleanse the lymph of pathogens and debris. In the **spleen,** blood is cleansed of pathogens and debris. **Tonsils, Peyer patches,** and the **appendix** are patches of lymphatic tissue that encounter pathogens and antigens.

Innate Immunity

31.2 Barriers to entry, protective proteins, and white blood cells are first responders

- **Innate immunity** is our first line of defense. Barriers to entry are the skin, mucous membranes, resident bacteria, and antimicrobial molecules. Protective proteins are **complement** (blood plasma proteins) and **interferons.** Complement

enhances inflammation by binding to **mast cells,** causing pathogens to be phagocytized, and forming a **membrane attack complex.** White blood cells (**neutrophils, eosinophils, macrophages,** and **dendritic cells**) engulf pathogens, and **natural killer cells** kill virus-infected and cancer cells.

31.3 The inflammatory response is a localized response to invasion

- The **inflammatory response** involves mast cells, which release **histamine** to increase capillary permeability. Redness, warmth, swelling, and pain result. Neutrophils and macrophages enter tissue fluid and engulf pathogens.

Adaptive Immunity

31.4 Adaptive immunity targets a specific antigen

- **Adaptive immunity** recognizes, responds to, and remembers **foreign antigens** as opposed to **self-antigens.**

- Adaptive immunity can be active or passive. **Active immunity** is long-lived; it develops naturally after infection or is induced by a **vaccine. Passive immunity** is short-lived; it occurs when a person is given **antibodies** or when a mother's antibodies are passed to her baby by breast-feeding.

31.5 Lymphocytes are directly responsible for adaptive immunity

- Each B cell and T cell has an **antigen receptor** for a specific antigen. B cells have **B-cell receptors (BCRs),** and T cells have **T-cell receptors (TCRs).** Altogether, B cells and T cells have antigen receptors for all possible antigens now and in the future. B cells are responsible for **antibody-mediated immunity.** After an antigen binds to a BCR, the B cell becomes an antibody-producing **plasma cell.** T cells are responsible for **cell-mediated immunity:** An **antigen-presenting cell** (**APC**) presents the antigen to a T cell. T cells exist as **helper T cells** and **cytotoxic T cells.**

31.6 Antibody-mediated immunity involves B cells

- After a B-cell receptor (BCR) combines with a specific antigen, the activated B cell undergoes **clonal selection,** with production of plasma cells and **memory B cells.**

Plasma cell

- **Immunization** involves the use of vaccines to bring about clonal expansion. An antibody (**immunoglobulin,** or **Ig**) is usually Y-shaped, with two binding sites for a specific antigen.

31.7 Cell-mediated immunity involves several types of T cells

- For a T-cell receptor (TCR) to recognize an antigen, the antigen must be presented to it by an antigen-presenting cell, along with an **MHC** (**major histocompatibility complex**) **protein.** Cytotoxic T cells recognize an antigen in combination with MHC class I. Helper T cells recognize an antigen in combination with MHC class II.

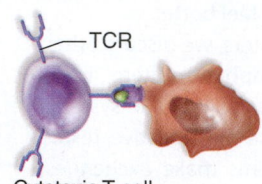

TCR

Cytotoxic T cell

- Cytotoxic T cells engage in cell-to-cell combat by releasing perforin (forms a hole) and granzymes (cause apoptosis of target cell). Helper T cells coordinate cell-mediated immunity and antibody-mediated immunity by releasing **cytokines. Memory T cells** jump-start an immune reaction to an antigen they recognize. Cytokines, which stimulate other white blood cells, are being used for cancer therapy.

Immune System Failures

31.8 Immunity can result in long-term illnesses

- Rejection of transplanted organs occurs because the immune system is correctly distinguishing between self and nonself. Organ rejection can be controlled by immunosuppressive drugs. Xenotransplantation and tissue engineering are alternatives.
- In **autoimmune disorders,** such as rheumatoid arthritis, myasthenia gravis, and lupus, the immune system mistakenly attacks the body's tissues.

TEST YOURSELF

The Lymphatic System

1. Which of the following is not a function of the lymphatic system?
   a. produces red blood cells
   b. returns excess fluid to the blood
   c. transports lipids absorbed from the digestive system
   d. defends the body against pathogens
2. Lymph nodes
   a. block the flow of lymph.
   b. contain B cells and T cells.
   c. decrease in size during an illness.
   d. filter blood.

3. B cells mature within the
   a. lymph nodes.   c. thymus.
   b. spleen.   d. bone marrow.
4. Which one is mismatched?
   a. red bone marrow—production of blood cells
   b. lymph nodes—cleaning of lymph
   c. spleen—maturation of all types of blood cells
   d. thymus—production of thymic hormones
5. Which of the following is a function of the spleen?
   a. produces T cells
   b. removes worn-out red blood cells
   c. produces immunoglobulins
   d. produces macrophages
   e. regulates the immune system
6. **THINKING CONCEPTUALLY** Why would you expect the lymphatic system to be a one-way system?

Innate Immunity

7. Mechanisms that protect the body against any infectious agent are called
   a. adaptive immunity.   c. barriers to entry.
   b. innate immunity.   d. immunity.
8. Complement
   a. is an innate defense mechanism.
   b. is involved in the inflammatory response.
   c. is a series of proteins present in the plasma.
   d. plays a role in destroying bacteria.
   e. All of these are correct.
9. Which cells will phagocytize pathogens?
   a. neutrophils   c. mast cells
   b. macrophages   d. Both a and b are correct.
10. _____ release histamines.
    a. Mast cells   c. Monocytes
    b. Neutrophils   d. All of these are correct.
11. Which of the following is not a goal of the inflammatory reactions?
    a. Bring more oxygen to the damaged tissues.
    b. Decrease the number of white blood cells in damaged tissues.
    c. Decrease blood loss from a wound.
    d. Prevent entry of pathogens into damaged tissues.
12. **THINKING CONCEPTUALLY** Explain why innate immunity would have evolved before adaptive immunity.

Adaptive Immunity

13. People who have had measles as children usually do not get the infection a second time because
    a. adults do not get chickenpox.
    b. the virus cannot enter the body twice.
    c. immune cells remember the virus and respond to it.
    d. the body has gained the ability to negate the effects of the virus.
14. Active immunity may be produced by
    a. having a disease.
    b. receiving a vaccine.
    c. receiving gamma globulin injections.
    d. Both a and b are correct.
    e. Both b and c are correct.

15. To contrast innate immunity with adaptive immunity, label this diagram:

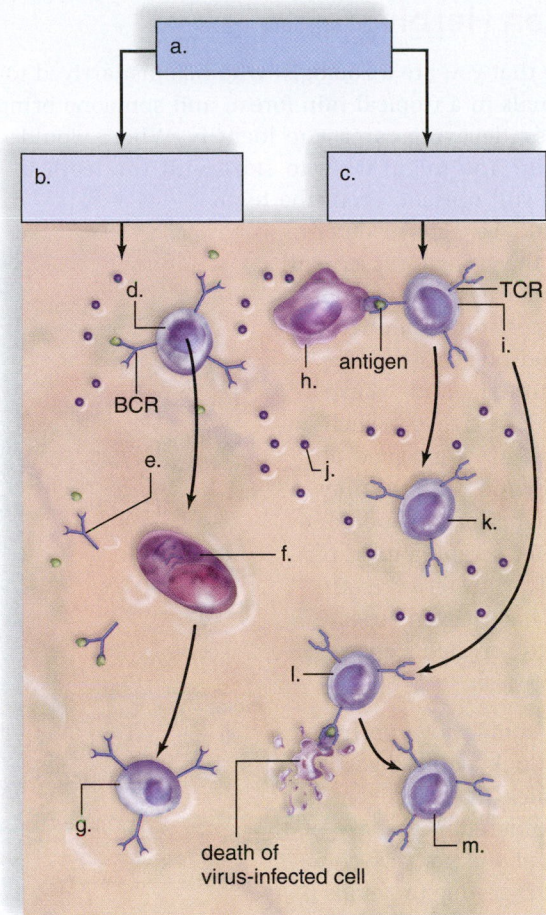

16. Which one of these does not pertain to B cells?
    a. have passed through the thymus
    b. have specific receptors
    c. are responsible for antibody-mediated immunity
    d. become plasma cells that synthesize and liberate antibodies
17. Unlike B cells, T cells
    a. require help to recognize an antigen.
    b. are components of adaptive immunity.
    c. are types of lymphocytes.
    d. contribute to homeostasis.
18. Which one of these pairs is mismatched?
    a. helper T cells—increase during an HIV infection
    b. cytotoxic T cells—active in tissue rejection
    c. macrophages—activate T cells
    d. memory T cells—long-living line of T cells
    e. T cells—mature in thymus
19. The clonal selection model says that
    a. an antigen selects certain B cells and suppresses them.
    b. an antigen stimulates the multiplication of B cells that produce antibodies against it.
    c. T cells select those B cells that should produce antibodies, regardless of antigens present.
    d. T cells suppress all B cells except the ones that should multiply and divide.
    e. Both b and c are correct.

20. During a secondary immune response,
    a. antibodies are made quickly and in great amounts.
    b. antibody production lasts longer than in a primary response.
    c. antibodies of the IgG class are produced.
    d. lymphocyte cloning occurs.
    e. All of these are correct.
21. Which of the following would not be a participant in cell-mediated immune defense?
    a. helper T cells   d. cytotoxic T cells
    b. macrophages   e. plasma cells
    c. cytokines
22. **THINKING CONCEPTUALLY** How are the B cells and T cells like bounty hunters?

Immune System Failures

23. Which of the following is not an example of an autoimmune disease?
    a. multiple sclerosis   d. systemic lupus erythematosus
    b. myasthenia gravis   e. rheumatoid arthritis
    c. contact dermatitis
24. **THINKING CONCEPTUALLY** People who die from an HIV infection succumb to an opportunistic infection. Why should that be?

GET INVOLVED

1. Revisit Figure 11.4D, p. 201, and formulate a hypothesis to account for the ability of lymphocytes to produce about 2 million different antibodies, despite having a limited number of genes for antibodies.
2. Design an experiment to test whether a drug known to suppress cell-mediated defense has an effect on antibody-mediated defense.

MEDIA STUDY TOOLS

mhhe.com/maderconcepts3

Enhance your study of this chapter with interactive study tools, practice tests, and engaging animations. Also, ask your instructor about the resources available through ConnectPlus, which includes LearnSmart, a personalized adaptive learning program, and a media-rich eBook.

32

Digestive Systems and Nutrition

BEFORE YOU BEGIN

Take a few minutes to recall

The organic components of cells (Chapter 3)

How all animals are heterotrophs that ingest their food (section 20.1)

The organs of the digestive system (section 26.7)

The contribution of the digestive tract and accessory organs to homeostasis (section 26.8)

How to Tell a Carnivore from an Herbivore

Imagine that you are a zoologist who has just arrived to study the animals in a tropical rain forest, and someone brings you a skinless, decaying carcass to identify. Where would you begin? You might want to start with the teeth if they are still present. From the teeth you would be able to determine whether the animal was a carnivore (meat eater), an herbivore (plant eater), or an omnivore (meat and plant eater). The long, curved, and sharp canines of a carnivore, such as a tiger, are very good at stabbing and killing prey. The front incisors are short and pointed for scraping meat off bones. The posterior teeth (molars) have jagged edges, similar to serrated knives, for slicing flesh.

You could move on from there to look at the digestive tract, which in a carnivore is short compared to the size of the animal and not very complicated. Digesting protein is a relatively easy matter, and that

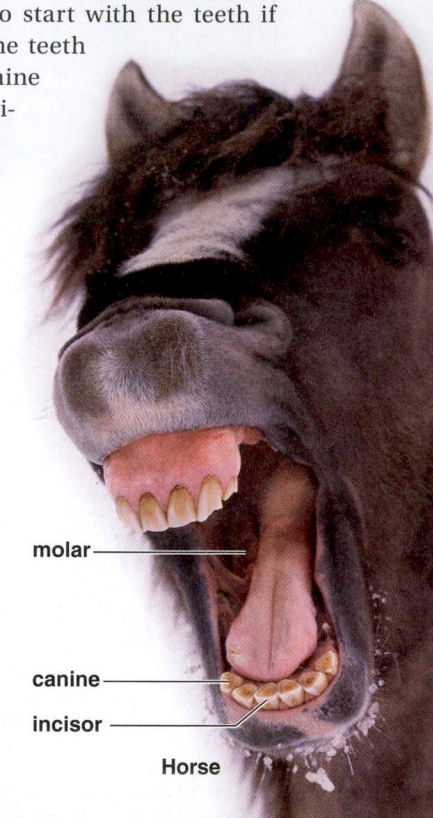

molar
canine
incisor
Horse

Tiger

incisor
canine
molar

is why carnivores tend to bolt their food and swallow it quickly, with little prep work before it enters a large, simple (single-chambered) stomach. Carnivores, such as wolves, do not eat all that often, so it is advantageous to have a large stomach they can fill up when they do eat. Digestion, which starts in the stomach, is finished in the short small intestine, where nutrients are absorbed. The large intestine (colon) of a carnivore is also simple and very short, as its only purpose is to absorb salt and water. The large intestine is cylindrical with a smooth appearance.

Herbivores, such as horses, zebras, and chimpanzees have a more varied digestive tract than carnivores because their diets are so different. Some eat only veggies, some only fruit, and some only nuts, for example. If they eat grass or leaves, the face is muscular in order to do a lot of chewing, and a muscular tongue moves the food around. The front incisors are broad and have clipped edges like a spade. The canines may be short, or long for defense, or absent. The molars, herbivores' most important teeth, have flat surfaces for grinding. Grinding breaks up the strong walls of plant cells, but a lot of digestion is still needed, and it begins in the mouth while the animal is still chewing.

Cows, deer, sheep, and giraffes are ruminants, with a large divided stomach where bacteria and protists are called upon to digest cellulose. Even so, the small intestine of all herbivores tends to be very long (10–12 times greater than body length), allowing space for digestion of plant material to continue in addition to absorption of nutrients. The large intestine is wider than the small intestine and also long enough to provide room for bacterial flora that live off the fiber that was not digested in the stomach or small intestine. In some herbivores (e.g., rabbits), the first section of the large intestine, called a cecum, is quite large, and serves as a primary or accessory place where microbes digest fiber. Some rabbits are known to re-ingest feces so that the nutrients can be absorbed by the small intestine the second time around.

Zebras

Omnivores are generalists. They have anterior teeth for holding and cutting, but also canines for stabbing and tearing and flat molars with sharp blades. The digestive tract proportions tend to match those of a carnivore more than those of an herbivore. Of course, a zoologist would have no trouble at all recognizing the teeth and digestive tract of a human, which many say are more like those of an herbivore than an omnivore. Why? Because we have flat molars (along with massive chewing muscles), and although we do not have a multi-divided stomach, we do have long intestines. Still, it is debatable whether our teeth and digestive tract are more like those of an herbivore or a carnivore. After all, some of our ancestors did have longer canine teeth—maybe not as long as those of a tiger, but certainly longer than those of humans today.

This chapter stresses the digestive traits that animals have in common and those that distinguish one group of feeders from another. It also describes the structure and function of the human digestive system, from the mouth to the large intestine. Finally, it discusses good nutrition and offers suggestions for a healthy diet.

Wolves

Rabbit

How Animals Acquire and Process Food

Animals must first acquire food and then digest it to nutrient molecules. They have evolved a number of different ways to obtain their food before breaking it down to molecules small enough to enter their bodies.

32.1 Animals have various ways to obtain food before processing occurs

LEARNING OUTCOMES

When you complete this section, you should be able to

1. Describe with examples four different feeding strategies.
2. List and discuss the four phases of digestion.

Animals have at least four different feeding strategies for ingesting food.

Bulk feeders bite off chunks of food, or even take in a whole organism, during one feeding. Humans are bulk feeders—they usually use utensils to cut up food or bring bite-sized pieces of food to the mouth. By contrast, the expandable jaws of snakes allow them to swallow prey whole, even if the prey is much larger than the snake's head. A great blue heron stands on the water's edge and uses its swordlike beak to capture a fish prior to engulfing it whole (Fig. 32.1A). Bulk feeders do not have to be carnivores. An elephant is an herbivore that feeds on tree foliage, bark, leaves, twigs, loose fruits, and tall grasses. Bulk feeders tend to eat *discontinuously,* meaning that they have definite periods of eating and not eating.

Filter feeders (also called suspension feeders) sift small food particles from the water. As do tube worms (see section 20.8), filter feeders often employ a ciliated surface to capture drifting particles from currents of water. Baleen whales are filter feeders that eat great quantities of small, shrimplike animals called krill. First the whale allows a large quantity of water to enter its open mouth. Then it lowers the baleen, a huge sieve made of keratin that hangs from the upper jaw (Fig. 32.1B). Finally, it uses its enormous tongue to force the water out, trapping the krill behind the baleen.

Substrate feeders live on, or in, the material they eat. For example, caterpillars (Fig. 32.1C) live and feed on leaves, while maggots live in and feed on dead and decaying tissue. Unlike the heron and baleen whale, caterpillars and maggots are *continuous feeders,* as are earthworms. They eat a small amount all the time. Earthworms plow through dirt using a powerful muscular pharynx to suck in whatever is in their path. After extracting the nutrients, the earthworm's digestive system expels the undigested residue at the surface.

Fluid feeders include parasites that live in a body fluid of a host, as well as external parasites, which have a way to tap into the host's vascular system. Recall that aphids (see section 23.4) use sharp mouthparts to reach the phloem sap of plants. Leeches have suckers and jaws that allow them to attach to and draw blood or hemolymph from a host. A vampire bat has fangs (actually pointed incisors) for piercing the skin of its host in order to get at its blood (Fig. 32.1D). A bat's saliva contains an anticoagulant called draculin, which runs into the bite; the bat then licks the flowing blood. Fluid feeders also include such pollinators as bees and hummingbirds, which feed on the nectar of flowers.

FIGURE 32.1A Great blue heron, a bulk feeder.

FIGURE 32.1B Baleen whale, a filter feeder.

FIGURE 32.1C Caterpillar, a substrate feeder.

FIGURE 32.1D Vampire bat, a fluid feeder.

Four Phases of Digestion Digestion contributes to homeostasis by providing the body's cells with the nutrients they need to continue living. An animal's digestive tract accomplishes the following processes: ingestion, digestion, absorption, and elimination, as shown by the model of a salamander in Figure 32.1E. It is possible to associate ingestion with a mouth and elimination with an anus, but the locations of digestion and absorption can differ somewhat, depending on the animal.

Ingestion is the act of taking in food during the process of feeding. *Digestion* is often divided into mechanical digestion and chemical digestion. Mechanical digestion can be accomplished by chewing in animals with suitable mouthparts, such as teeth, but in animals that lack these features, mechanical digestion can occur elsewhere. For example, earthworms and birds have a gizzard, where the grinding action of pebbles and sand accomplishes mechanical digestion.

Chemical digestion requires enzymes, the protein molecules that speed a reaction at body temperature. Enzymes are secreted by the digestive tract itself or by accessory glands that send their juices into the tract by way of ducts. Enzymes are specific and usually work in small steps. For example, it takes two steps to digest polysaccharides (carbohydrate polymers), polypeptides of proteins, and nucleic acids to molecules small enough to be absorbed across the epithelial lining of the digestive tract, as illustrated in Figure 32.1F. First, polysaccharides are digested to disaccharides, and then disaccharides are digested to monosaccharides. Similarly, the polypeptides of proteins are first digested to peptides, and then peptides are broken down to amino acids. The first step in the breakdown of nucleic acids produces nucleotides, and then nucleotides become pentose sugars, phosphate, and bases during the second step. Fats are not polymers, and it takes only one step to digest them to monoglycerides (glycerol and one fatty acid) and fatty acids.

Absorption occurs when nutrients are taken into the body, often into the bloodstream. In humans, the small intestine, where absorption occurs, should really be called the "long intestine"

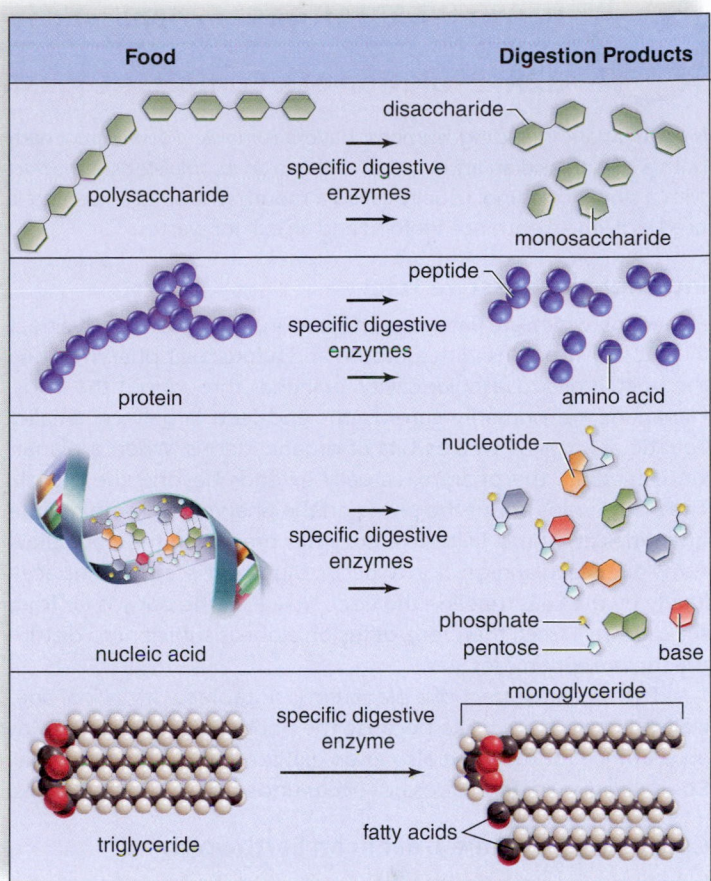

FIGURE 32.1F During chemical digestion, enzymes break down food. Polysaccharides are digested to monosaccharides, proteins to amino acids, nucleic acid to its components, and a triglyceride (e.g., fat) to a monoglyceride and fatty acids.

because of its length. The wall of the small intestine is convoluted and contains even smaller projections to increase the available surface area for absorption to take place. Any molecules that are not absorbed by the digestive tract cannot be used by the animal. They are the waste products that pass through the anus during *elimination* (defecation).

Even though all digestive tracts carry out the same four functions, there are modifications according to the animal's way of life. As we saw in the introduction to this chapter, carnivores have adaptations for eating meat, which is easily digested, and herbivores have modifications for eating plant material, which needs a lot of chewing and other processing to break up its cellulose cell walls. Humans, being omnivores, have modifications for eating both meat and plant material, but as we discussed, the human digestive system is more like that of an herbivore than a carnivore.

▶ **32.1 CHECK YOUR PROGRESS**

1. How might a carnivore obtain its food and which of the four phases of digestion are more easily accomplished in a carnivore?

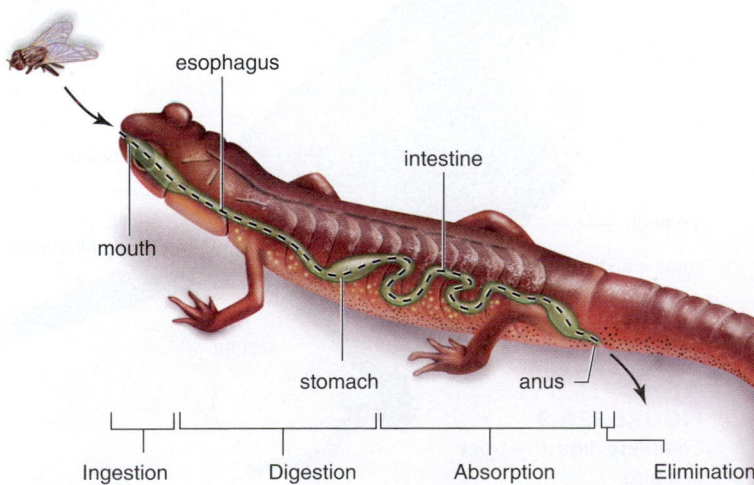

FIGURE 32.1E The four phases of digestion.

32A Evolution of a Complete Digestive Tract

Most animals, including humans, have a *complete digestive tract* with both a mouth and an anus, but a few have an *incomplete digestive tract* with a single opening, usually called a mouth. The single opening is used as both an entrance for food and an exit for wastes.

Incomplete Digestive Tract

Planarians, which are flatworms, have an incomplete digestive tract (Fig. 32A.1). It begins with a mouth and a muscular pharynx. Then the tract, a gastrovascular cavity, branches throughout the body. Planarians are primarily carnivorous and feed largely on smaller aquatic animals as well as bits of organic debris. When a planarian is feeding, the pharynx actually extends beyond the mouth. The body wraps about the prey, and the pharynx sucks up minute quantities at a time. Digestive enzymes present in the tract allow some extracellular digestion to occur. Digestion is finished intracellularly by the cells that line the tract. No cell in the body is far from the digestive tract; therefore, diffusion alone is sufficient to distribute the nutrient molecules.

The digestive tract of a planarian is notable for its lack of specialized parts. It is saclike because the pharynx serves not only as an entrance for food but also as an exit for undigestible material. Specialization of parts does not occur under these circumstances.

Complete Digestive Tract in an Earthworm

In contrast to planarians, earthworms, which are annelids, have a complete digestive tract (Fig. 32A.2). Earthworms are considered substrate feeders because they live underground and feed mainly on decayed organic matter in the soil. The muscular pharynx draws in food with a sucking action. Food then enters the *crop,* which is a storage area with thin, expansive walls. From there, food goes to the *gizzard,* which has thick, muscular walls that crush the food and pebbles that grind. Digestion is extracellular within an intestine. The surface area of digestive tracts is often increased for absorption of nutrient molecules, and in earthworms, this is accomplished by an intestinal fold called the *typhlosole*. Undigested remains pass out of the body at the anus.

Specialization of parts is obvious in the earthworm because the pharynx, the crop, the gizzard, and the intestine each have a particular function in the digestive process.

Complete Digestive Tract in Birds

Some birds, such as the great blue heron, are discontinuous bulk feeders, and they have a thin-walled crop, which serves as a storage area. This is useful because food may not always be available, and the presence of a predator may make it essential to "eat and run." Therefore, it seems adaptive that birds do not chew their food but instead have a gizzard where food is broken apart mechanically (Fig. 32A.3). The stomach, which secretes digestive juices, precedes the gizzard in birds. Following the gizzard are the intestines. Attached to the large intestine are two long ceca; a *cecum* is a blind (has no opening) sac. The ceca of an herbivorous bird, such as one that feeds on seeds, are well-developed because they are fermentation organs where microorganisms help digest cellulose. In all birds, absorption occurs in the small intestine, and elimination occurs at the *cloaca,* which receives undigestible material and urine from the bladder and also gametes from the testes or ovaries.

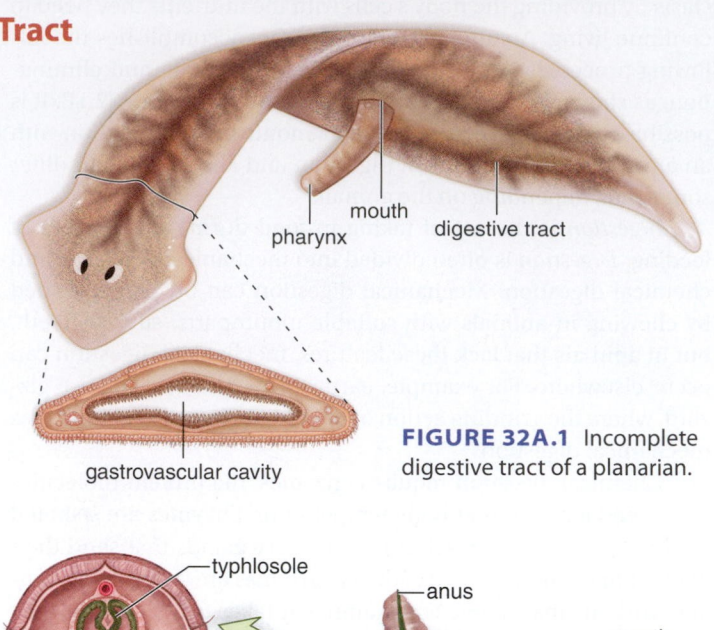

FIGURE 32A.1 Incomplete digestive tract of a planarian.

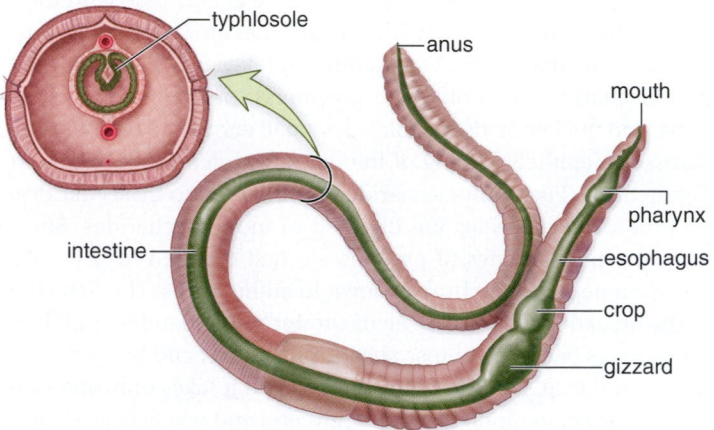

FIGURE 32A.2 Complete digestive tract of an earthworm.

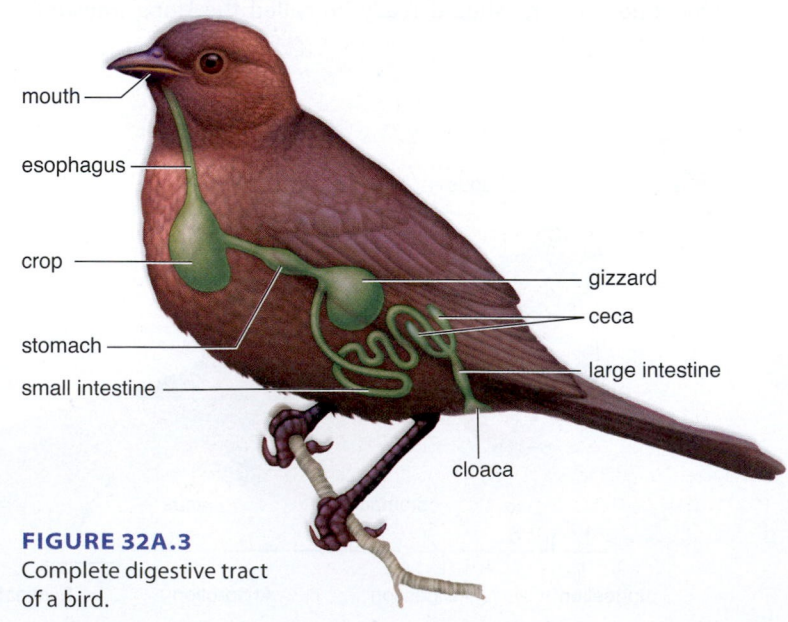

FIGURE 32A.3
Complete digestive tract of a bird.

Complete Digestive Tract in Mammalian Herbivores

The digestive tract of mammalian herbivores, such as cows and rabbits, is modified to provide a special compartment for microbes that can digest cellulose. No animal is able to produce an enzyme that can digest the cellulose of plant cell walls, but microbes do produce such enzymes. Mammalian herbivores can digest grass only by having a mutualistic relationship with these microorganisms. Mammalian herbivores called *ruminants,* such as cows, sheep, deer, and giraffes, have large divided stomachs with four chambers, one of which is called the rumen (Fig. 32A.4). ❶ The rumen is a large fermentation chamber that contains microbes and up to 50 gallons of undigested plant material. The location of the rumen at the entrance to the stomach allows partly digested food to be regurgitated and chewed again while the animal is at leisure. Because the animal "chews the cud," ❷ the rumen is able to do a better job when the food is re-swallowed, entering the rumen a second time. ❸ Now, the partially digested food moves through another chamber before ❹ it enters the true stomach and finally exits to the small intestine. The digestive tract of herbivores, in general, is long because the digestive action on cellulose also continues along the length of the tract.

Other mammalian herbivores, such as rabbits and horses, are not ruminants and instead make special use of a single, large cecum at the start of the large intestine (Fig. 32A.5*a*). The cecum contains microorganisms that digest cellulose. Because the cecum is located some distance from the stomach, these animals do not regurgitate their food for more chemical digestion. However, rabbits and hares do re-ingest their food. They produce two kinds of fecal pellets—a firm, dark variety and a larger, soft, light-colored kind, which is not dropped but is instead eaten directly from the anus. Horses do not re-ingest, and their feces contain some undigested and undigestible remains. The appendix attached to the cecum in rabbits contains lymphatic tissue where lymphocytes respond to antigens, as discussed in Chapter 31.

Recall that carnivores, such as foxes, dogs, cats, lions, and tigers, have a much shorter digestive tract than do herbivores. Their food is much easier to digest, and if they have a cecum, it is small in size (Fig. 32A.5*b*).

Carnivores do not have to spend much time mechanically chewing their food and can go hunting instead.

CONSIDER THESE QUESTIONS

1. The adaptations of an animal include its digestive system. For example, a bird has jaws but no teeth, whereas cows have both. Explain.
2. If you had your choice, would you rather be an herbivore or a carnivore? Explain the advantages of each way of life and your choice.
3. When humans arose in Africa, they were herbivores, but later on in Europe they were hunters of large animals. Explain why they were originally herbivores and what allowed them to take up the hunting way of life.

 Explore the concepts through a variety of multimedia assets and question types.
www.mcgrawhillconnect.com

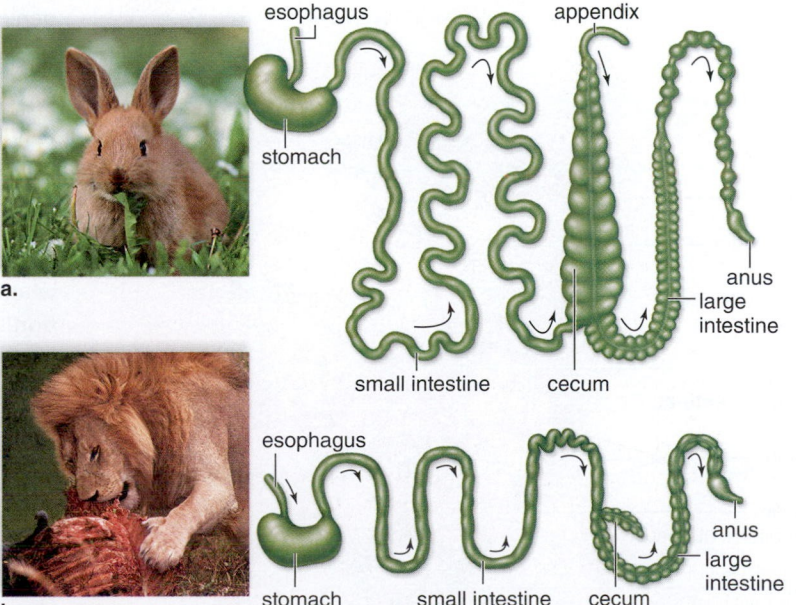

FIGURE 32A.5 Digestive tract of an herbivore compared to that of a carnivore.

Mammalian Organs of Digestion

Humans are omnivores and therefore exhibit features of both carnivores and herbivores. As you study the human digestive tract, determine whether each part of the tract is like that of a carnivore or an herbivore.

32.2 Some digestion occurs before food reaches the stomach

LEARNING OUTCOME

When you complete this section, you should be able to

1. Describe the anatomy and physiology of the mammalian mouth, pharynx, and esophagus.

The mouth is the first part of the complete digestive system of humans (Fig. 32.2A). The mouth serves multiple functions, from allowing us to enjoy the taste and texture of food to fighting infection. The roof of the mouth separates the nasal cavities from the oral cavity. The roof has two parts: an anterior (toward the front) **hard palate** and a posterior (toward the back) **soft palate** (Fig. 32.2B). The hard palate contains several bones, but the interior of the soft palate is composed entirely of muscle. The soft palate ends in a

MP3
Overview of the Digestive System

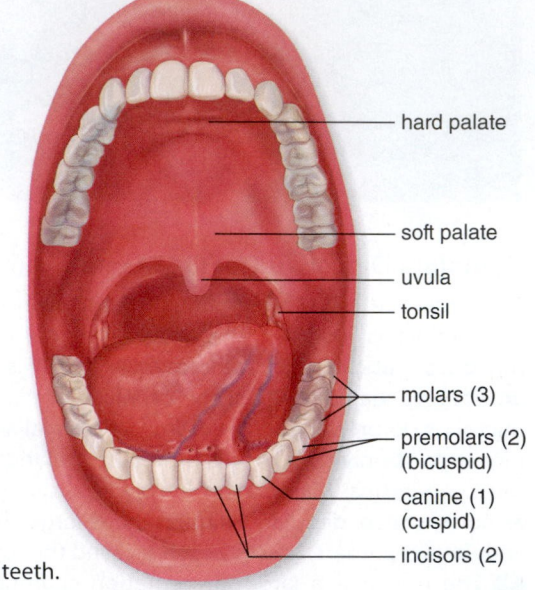

FIGURE 32.2B
Adult mouth and teeth.

finger-shaped projection called the uvula, which plays a role in snoring by obstructing the free flow of air through the respiratory passages. The tonsils are in the back of the mouth, on either side of the tongue, as well as in the nasopharynx, where they are called adenoids. The tonsils contain lymphatic tissue and help protect the body against infections. If the tonsils become inflamed, the person has tonsillitis. If tonsillitis recurs repeatedly, the tonsils may be surgically removed (called a tonsillectomy).

Salivary Glands Three pairs of **salivary glands** send juices (saliva), by way of ducts, to the mouth. Saliva has a neutral pH but is still one of the body's defenses against disease-causing pathogens because it contains antibacterial agents, including lysozyme. One pair of salivary glands lies at the sides of the face immediately below and in front of the ears. These glands swell when a person has the mumps, a disease caused by a viral infection. The glands have ducts that open on the inner surface of the cheek at the location of the second upper molar. Another pair of salivary glands lies beneath the tongue, and still another pair lies beneath the floor of the oral cavity. The ducts from these salivary glands open under the tongue. You can locate the openings if you use your tongue to feel for small flaps on the inside of your cheek and under your tongue. Saliva contains an enzyme called **salivary amylase** that begins the digestion of starch.

The tongue, which is composed of skeletal muscle, mixes the chewed food with saliva. It then forms this mixture into a mass, called a bolus, in preparation for swallowing.

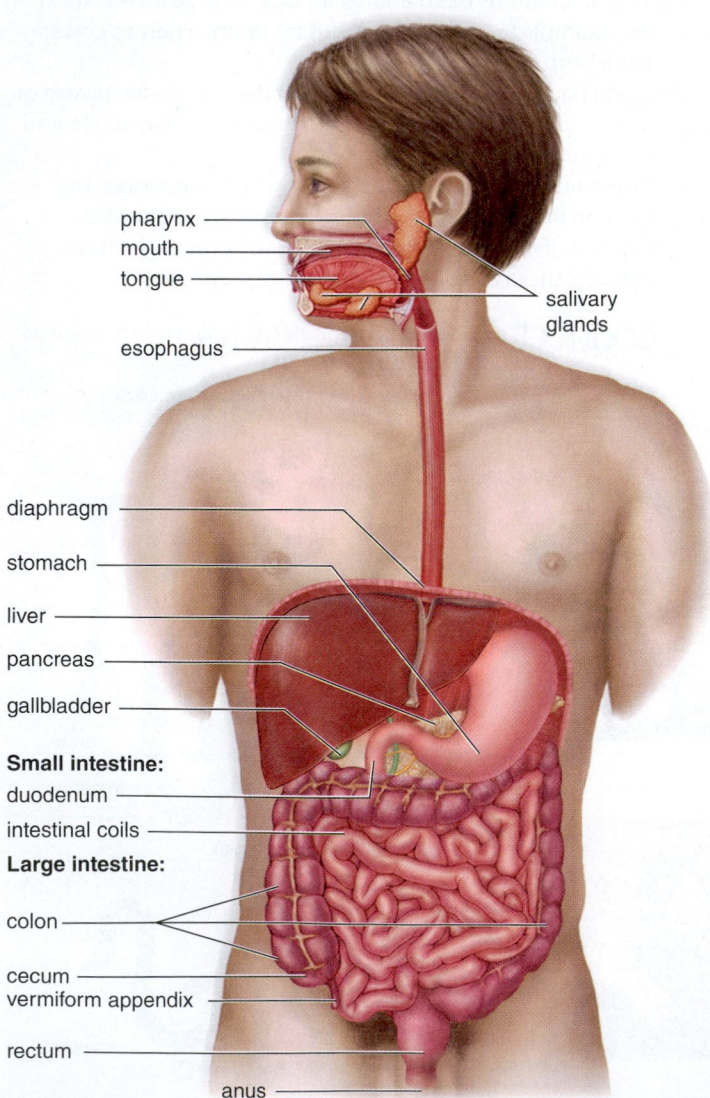

FIGURE 32.2A The human digestive tract.

The Teeth We use our teeth to chew food into pieces convenient for swallowing. During the first 2 years of life, the 20 smaller deciduous, or baby, teeth appear. These are eventually replaced by 32 adult teeth of four types (Fig. 32.2B). Anteriorly, four chisel-shaped incisors function for biting. Flanking them on either side are two pointed canine teeth, or cuspids, which help tear food. Laterally, four flat premolar teeth, or bicuspids, are used for grinding food. The most posterior are the six molar teeth, designed for crushing and grinding food. The third pair of molars, called the wisdom teeth, sometimes fail to erupt. If they push on the other teeth or cause pain, they can be removed by a dentist or an oral surgeon.

Tooth decay, or simply cavities, occurs when bacteria within the mouth metabolize sugar and give off acids, which erode teeth. Two measures can prevent tooth decay: eating a limited amount of sweets, and daily brushing and flossing the teeth. Fluoride treatments, particularly in children, can make the enamel stronger and more resistant to decay.

Brushing and flossing remove dental plaque, a sticky, colorless film that can cling to your teeth and line your gums. Bacteria accumulate in plaque and cause inflammation of the gums (gingivitis). Gingivitis can spread to the periodontal membrane, which lines the tooth sockets. A person then has periodontitis, characterized by loss of bone and loosening of the teeth so that extensive dental work may be required. Stimulation of the gums, as recommended by a dentist, can help control this condition.

The consequences of gum infection may not end with loss of teeth. Researchers are discovering that inflammation, including gum disease, may increase the risk of serious health problems, such as heart attack, stroke, poorly controlled diabetes, and pre-term labor.

The Esophagus The **esophagus** is a muscular tube that passes from the pharynx through the thoracic cavity and diaphragm into the abdominal cavity, where it joins the stomach (see Fig. 32.2A). The **pharynx** is a region that receives air from the nasal cavities and food from the mouth. The food passage and the air passage cross in the pharynx because the trachea (windpipe) is anterior to (in front of) the esophagus. As shown in Figure 32.2C, swallowing occurs in the pharynx. Swallowing is a reflex action performed

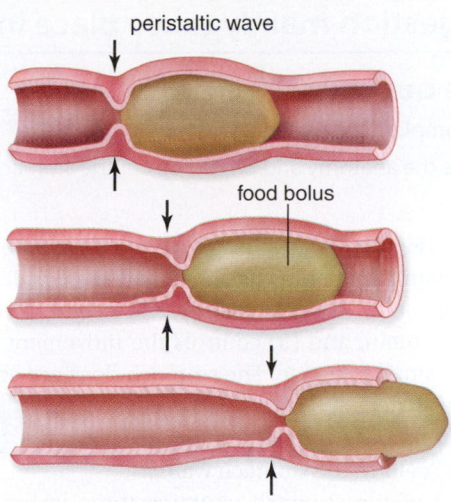

FIGURE 32.2D Peristalsis, a rhythmic progressive wave of muscular contraction, pushes the food bolus along the digestive tract.

automatically, without conscious thought. During swallowing, ❶ the soft palate moves back to close off the nasal cavities, and ❷ the trachea moves up under the **epiglottis,** a flap that now covers the **glottis** (opening to air passages). A food bolus then enters the ❸ esophagus because the air passages are blocked—we do not breathe when we swallow.

The esophagus plays no role in the chemical digestion of food. Its sole purpose is to move the food bolus from the mouth to the stomach. A rhythmic contraction called **peristalsis** pushes the bolus along the digestive tract, much as squeezing a tube of toothpaste pushes the toothpaste along (Fig. 32.2D). Peristalsis begins in the esophagus and continues in all the organs of the digestive tract.

The place where the esophagus enters the stomach is marked by a sphincter. **Sphincters** are muscles that encircle tubes and act as valves; tubes close when sphincters contract, and they open when sphincters relax. Relaxation of the sphincter allows food to pass into the stomach, while contraction prevents the acidic contents of the stomach from backing up into the esophagus. The occasional backup, or reflux, of stomach contents into the esophagus usually causes no difficulty because it can be neutralized by saliva, which is slightly basic. Some people, however, develop gastroesophageal reflux disease.

🎵 **MP3**
Oral Cavity, Esophagus, and Swallowing Reflex

Gastroesophageal Reflux Disease (GERD) GERD is apparent when refluxed stomach acid touches the lining of the esophagus and causes a burning sensation in the chest or throat, called heartburn or acid indigestion. Heartburn that occurs more than twice a week may be considered acid reflux disease.

Various medications—both prescription and nonprescription—are available to treat acid reflux. If these fail, surgery to strengthen the sphincter is available.

▶ **32.2 CHECK YOUR PROGRESS**
1. Identify the primary function of the mouth. Of the esophagus.
2. Account for why carbohydrate, not protein, digestion starts in the mouth of humans.

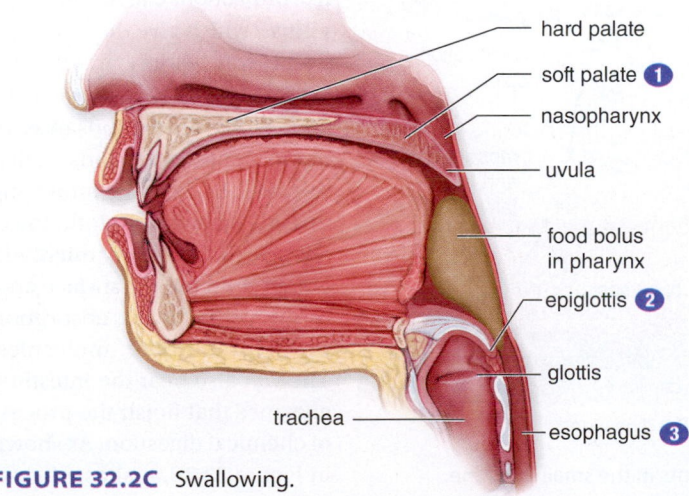

hard palate

soft palate ❶

nasopharynx

uvula

food bolus in pharynx

epiglottis ❷

glottis

trachea

esophagus ❸

FIGURE 32.2C Swallowing.

32.3 Digestion mainly takes place in the stomach and small intestine

LEARNING OUTCOME

When you complete this section, you should be able to

1. Describe the anatomy and physiology of the stomach and small intestine.

The *stomach* (Fig. 32.3A) is a thick-walled, J-shaped organ that is continuous with the esophagus above and the duodenum of the small intestine below. The stomach (1) stores food, (2) initiates the digestion of protein, and (3) controls the movement of digested food into the small intestine. The stomach does not absorb nutrients; however, it does absorb alcohol because alcohol is fat-soluble and can pass through the plasma membrane of cells in the stomach wall.

MP3 Stomach

The wall of the stomach contains three layers of smooth muscle—the oblique, circular, and longitudinal layers—that allow the stomach to stretch and to mechanically break down food into smaller fragments. The lining of the stomach has deep folds, called rugae, which disappear as the stomach fills to an approximate capacity of 1 l. The stomach lining also has millions of gastric pits, which are openings for gastric glands that produce gastric juice (Fig. 32.3A). Gastric juice contains a precursor to the enzyme **pepsin,** which digests protein, plus hydrochloric acid (HCl) and mucus. This mucus forms a coating that protects the stomach from self-digestion. HCl causes the stomach to have high acidity; a pH of about 2 is optimum for pepsin activity and is also beneficial because it kills most of the bacteria present in food. Although HCl does not digest food, it does break down the connective tissue of meat and activates pepsin, which requires an acid pH.

Animation Three Phases of Gastric Secretion

Normally, the stomach empties slowly in about 2–6 hours. When food leaves the stomach, it is a thick, soupy liquid called **chyme.** Chyme enters the small intestine in squirts by way of a sphincter, which acts like a valve, repeatedly opening and closing.

The **small intestine** is quite long—about 6 m in length (range: 4.6–9 m)—but it takes its name from its small diameter, which is about 2.5 cm. The first 25 cm of the small intestine is called

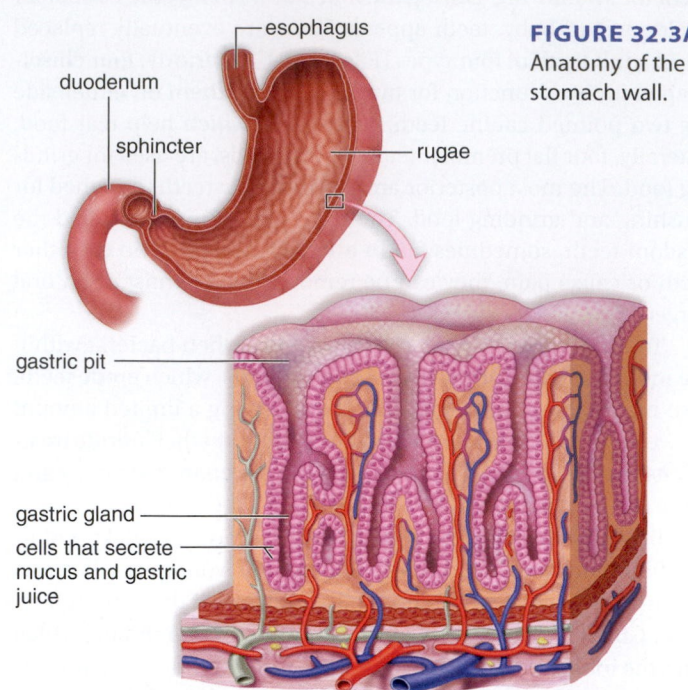

FIGURE 32.3A Anatomy of the stomach wall.

esophagus
duodenum
sphincter
rugae
gastric pit
gastric gland
cells that secrete mucus and gastric juice

the **duodenum.** A duct brings bile from the liver and gallbladder, and pancreatic juice from the pancreas, into the small intestine (see Fig. 32.5). **Bile** emulsifies fat—that is, it causes fat droplets to disperse in water. The intestine has a slightly basic pH because pancreatic juice contains sodium bicarbonate ($NaHCO_3$), which neutralizes chyme and provides an optimum pH for the enzymes in the small intestine. The enzymes in pancreatic juice (Fig. 32.3B) include **pancreatic amylase,** which digests starch to maltose; **trypsin,** which digests protein to peptides; **lipase,** which digests fat droplets to monoglycerides and fatty acids; and nucleases, which digest DNA and RNA to nucleotides.

Animation Enzymatic Action and the Hydrolysis of Sucrose

Certain anatomic features increase the surface area of the small intestine. The wall of the small intestine contains finger-like projections called villi (sing., **villus**), which give the intestinal wall a soft, velvety appearance (Fig. 32.3C). The outer cells of a villus have thousands of microscopic extensions called **microvilli.** Just as the projecting threads of a terry-cloth towel absorb water, the microvilli greatly increase the surface area of the villus for the absorption of small nutrient molecules. The villi also bear the intestinal enzymes that finish the process of chemical digestion. As shown in Figure 32.3B, maltase digests

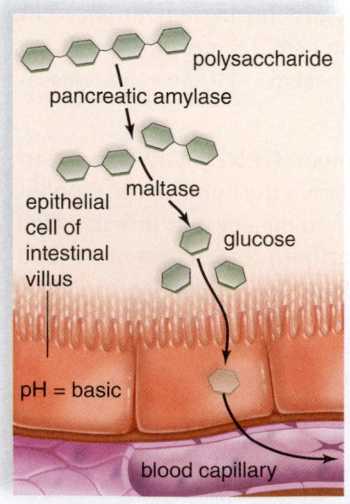

polysaccharide
pancreatic amylase
epithelial cell of intestinal villus
maltase
glucose
pH = basic
blood capillary

Carbohydrate digestion

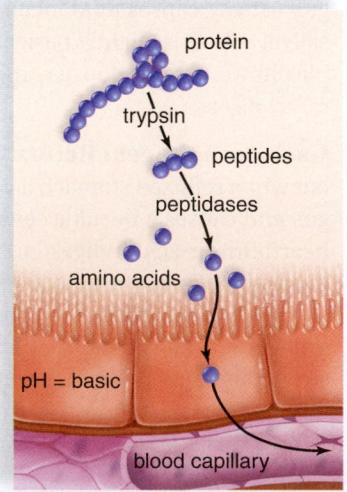

protein
trypsin
peptides
peptidases
amino acids
pH = basic
blood capillary

Protein digestion

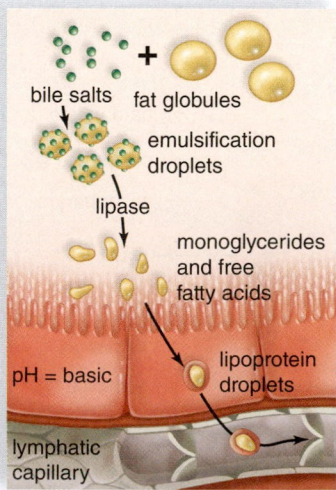

bile salts
fat globules
emulsification droplets
lipase
monoglycerides and free fatty acids
pH = basic
lipoprotein droplets
lymphatic capillary

Fat digestion

FIGURE 32.3B Digestion by pancreatic enzymes and absorption of nutrients in the small intestine.

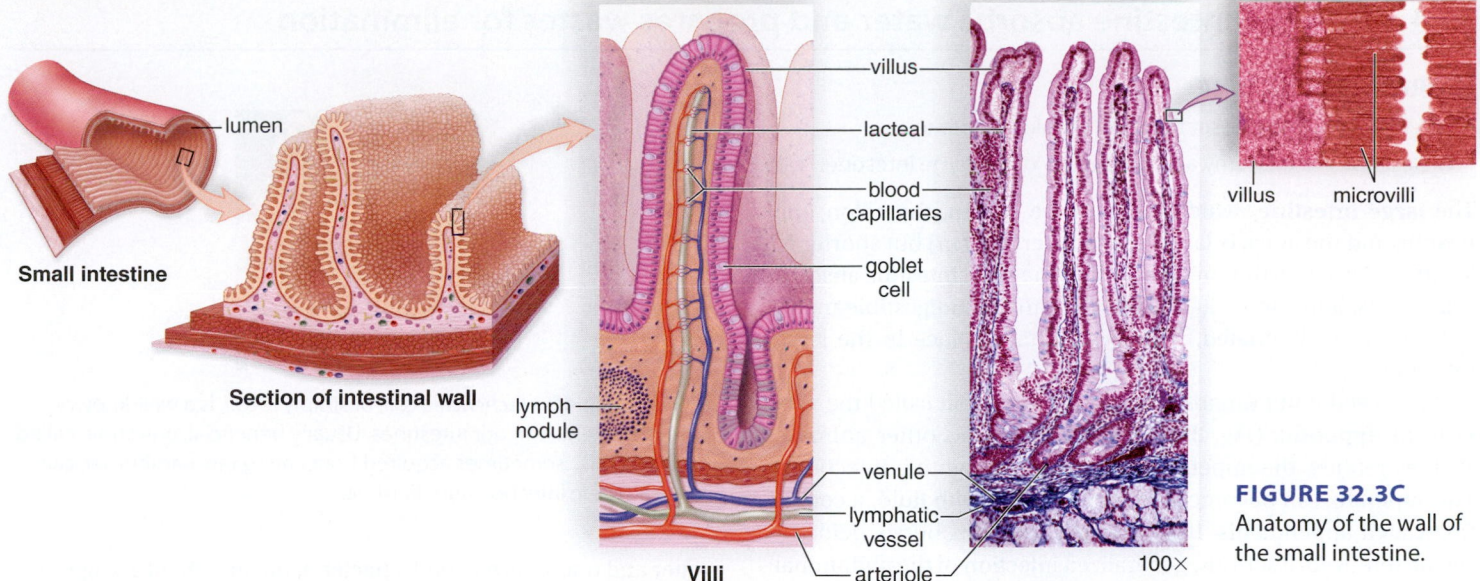

villus

lacteal

blood capillaries

goblet cell

villus   microvilli

FIGURE 32.3C
Anatomy of the wall of the small intestine.

lumen

Small intestine

Section of intestinal wall

lymph nodule

venule

lymphatic vessel

arteriole

Villi

100×

maltose to glucose, and peptidases break down peptides to amino acids. Also present are nucleotidase enzymes that digest nucleotides to their component parts.

Nutrient molecules are absorbed into the vessels of a villus, which contains blood capillaries and a lymphatic capillary called a **lacteal.** Sugars (digested from carbohydrates) and amino acids (digested from proteins) enter the blood capillaries of a villus. Monoglycerides and fatty acids (digested from fats) enter the epithelial cells of the villi, and within these cells they are joined and

packaged as lipoprotein droplets, which enter a lacteal (Fig. 32.3B *right*). Lacteals are a part of the lymphatic system, a one-way system of lymphatic vessels that return lymph to the bloodstream (see Fig. 31.1).

Animation
Absorption of Nutrients and Water

▶ **32.3 CHECK YOUR PROGRESS**
1. In what two organs are carbohydrates digested?
2. Argue that the small intestine contributes more to homeostasis than the other digestive organs.

HOW SCIENCE PROGRESSES   *Application*

32B The Cause of Ulcers: Bacteria!

In 1974, Dr. Barry Marshall, a brash, youthful physician trainee, made up his mind to pursue a research project with Dr. Robin Warren, a pathologist at the Royal Perth Hospital in Australia. Dr. Warren had observed an unusual spiral-shaped bacterium in the biopsy tissue of many ulcer patients. Extensive research identified the bacterium as a new, relatively unknown species, *Helicobacter pylori.*

In 1983, Marshall proposed at an international conference that the cause of ulcers was not stress, diet, or excess stomach acid as had been previously thought, but a bacterial infection caused by *H. pylori.* Initially, the idea was met with scorn and ridicule. How could any bacterium survive the harsh, acidic environment of the stomach? The theory that ulcers were caused by stress and diet was well ingrained. Volumes of studies had been published establishing the prevailing dogma, and millions of dollars had gone into developing drugs to inhibit excess stomach acid production.

Marshall and Warren soon found their novel theory in danger, because they were unable to culture the strange spiral bacteria in laboratory animals. Finally, in desperation and defiance, Dr. Marshall decided to use himself as a human guinea pig and quaffed a tube of *H. pylori.* About 1 week later, he began vomiting and suffering the painful symptoms of gastritis, or stomach inflammation. Medical tests confirmed this condition, and revealed that his stomach was teeming with the bizarre bacteria. He treated himself by taking a combination of antibiotics to kill the bacteria and

an acid-blocking drug to ease the symptoms. Within a few weeks, Dr. Marshall felt better. But more importantly, he had made his point. He challenged other researchers to prove his theory wrong, but before long, many studies indeed supported his theory.

Today, ulcer sufferers are no longer resigned to following a bland diet and looking for ways to reduce stress in their lives. Instead of simply treating the symptoms of an ulcer, doctors can now treat the cause of the condition with a simple round of antibiotics, coupled with an acid-reducing drug to allow the stomach lining to heal. For their innovative methodology to demonstrate the connection between *Helicobacter pylori* and ulcers, Dr. Marshall and Dr. Warren were awarded the 2005 Nobel Prize in Physiology or Medicine.

CONSIDER THESE QUESTIONS

1. Devise a controlled experiment that Dr. Marshall could have done.
2. Why would a controlled experiment be more reliable than the one Dr. Marshall did?

 Explore the concepts through a variety of multimedia
BIOLOGY assets and question types.

www.mcgrawhillconnect.com

LEARNING OUTCOME

When you complete this section, you should be able to

1. Describe the anatomy and physiology of the large intestine.

The **large intestine,** which includes the cecum, the colon, the rectum, and the anus, is larger in diameter (6.5 cm) but shorter in length (1.5 m) than the small intestine. The large intestine absorbs water, salts, and some vitamins. It also stores undigestible material until it is eliminated. No digestion takes place in the large intestine.

The cecum in humans has a small projection called the **vermiform appendix** (Fig. 32.4A). In humans and other animals, such as rabbits, the appendix plays a role in fighting infections. The appendix can become infected and filled with fluid, a condition called appendicitis. If an infected appendix bursts before it can be removed, a serious, generalized infection of the abdominal lining, called peritonitis, can result.

The **colon** is subdivided into the ascending, transverse, descending, and sigmoid colons (Fig. 32.4A). The sigmoid colon enters the **rectum,** the last 20 cm of the large intestine.

About 1.5 l of water enters the digestive tract daily as a result of eating and drinking. An additional 8.5 l enter the digestive tract each day carrying the various substances secreted by the digestive glands. About 95% of this water is absorbed by the small intestine, and much of the remaining portion is absorbed into the cells of the colon. If this water is not reabsorbed, diarrhea can lead to serious dehydration and ion loss, especially in children.

The large intestine has a large population of various bacteria that normally are nonpathogenic (Fig. 32.4B). These bacteria break down undigestible material, and they also produce some vitamins, such as vitamin K, which is necessary to blood clotting. The large intestine forms feces. Feces are normally three-quarters

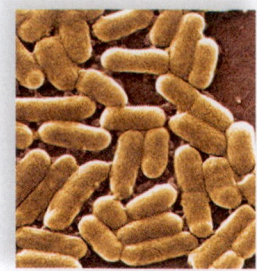

FIGURE 32.4B *Escherichia coli,* or simply *E. coli,* is a well-known type of bacterium in our intestines. Usually beneficial, one strain called *E. coli* 0157:H7, sometimes acquired from eating raw hamburger, can cause a severe infection and diarrhea.

water and one-quarter solids. Bacteria, dietary fiber (undigestible remains), and other undigestible materials make up the solid portion. Bacterial action on undigestible materials causes the odor of feces and also accounts for the presence of gas as discussed in "An Ecosystem in Your Large Intestine" on page 641. A breakdown product of bilirubin (a hemoglobin breakdown product) and the presence of oxidized iron cause the brown color of feces.

Video
Fat Microbes

Defecation, ridding the body of feces, is also a function of the large intestine. Peristalsis occurs infrequently in the large intestine, but when it does, feces are forced into the rectum. Feces collect in the rectum until it is appropriate to defecate. At that time, stretching of the rectal wall initiates nerve impulses to the spinal cord. Shortly thereafter, the rectal muscles contract and the anal sphincters relax. This allows the feces to exit the body through the anus. A person can inhibit defecation by contracting the external anal sphincter. Ridding the body of undigestible remains is another way the digestive system helps maintain homeostasis.

The term *fecal coliforms* refers to *E. coli* and its relatives in the large intestine. Water is considered unsafe by the Public Health Service for swimming when the coliform bacterial count reaches a certain number. A high count indicates that a significant amount of feces has entered the water. The more feces present, the greater the possibility that disease-causing bacteria are also present.

The colon is subject to the development of polyps, small growths arising from the epithelial lining. Polyps, whether benign or cancerous, can be removed surgically. Some investigators believe that dietary fat increases the likelihood of colon cancer. Dietary fat causes an increase in bile secretion, and it could be that intestinal bacteria convert bile salts to substances that promote the development of colon cancer. By contrast, dietary fiber absorbs water and adds bulk, thereby preventing constipation and facilitating the movement of substances through the intestine, as we learn in section 32.6.

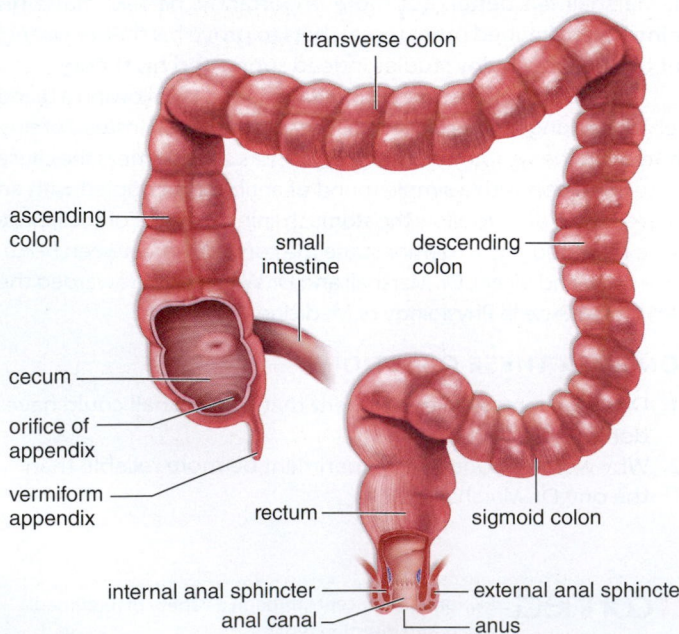

transverse colon

ascending colon

small intestine

descending colon

cecum

orifice of appendix

vermiform appendix

rectum

sigmoid colon

internal anal sphincter

external anal sphincter

anal canal

anus

FIGURE 32.4A The regions of the large intestine are the cecum, colon, rectum, and anal canal.

▶ **32.4 CHECK YOUR PROGRESS**

1. Identify how our intestines (small and large together) resemble those of an herbivore.

32C An Ecosystem in Your Large Intestine

Do you look at a bowl of ice cream or a glass of milk and think, "uh oh…" (Fig. 32C)? If you are lactose intolerant—like 75% of the world's adult population—the sight of dairy foods might make you think of bloating, flatulence, and even diarrhea. The problem is that your small intestine is not producing enough lactase to break down lactose, the sugar present in dairy products. When undigested lactose reaches the large intestine, the many bacterial populations that live there try to break it down, and this leads to your symptoms. The common symptoms of lactose intolerance vary between individuals, and their intensity depends in part on the unique combination of microbes living in each person's gut.

Your large intestine harbors cells from all three domains of life: eukarya, archaea, and bacteria. This small area can actually be considered an ecosystem, and its microbial life is more densely packed than in any other ecosystem on Earth. However, you might be surprised to know how little this microbial diversity in your gut has been studied—in fact, it is still an active area of research. Although our best current estimates are that the large intestine contains approximately 1,000 different species of bacteria, the exact types are not well characterized; even less detail is known about the varieties of archaea present. To further complicate the situation, each person typically has different types of microbes, occurring in different abundances. Beyond that, this microbial diversity may change over a person's lifetime, depending on health, diet, and other factors. Some researchers have suggested that future medical tests may provide a detailed profile of the types of microbes living in our guts, as a way of identifying current or developing health problems.

One type of archaean known to occur in some people's guts, *Methanobrevibacter smithii,* also lives in the intestines of cattle. As its name indicates, this archaean microbe makes methane, one of the gases responsible for climate change. (Interestingly, some researchers have suggested that climate change might be combatted by developing "anti-archaeal" drugs for cattle, to reduce their methane emissions.) Whether in humans or cows, that methane gas becomes part of normal flatulence—along with carbon dioxide and hydrogen gas produced by other microbes.

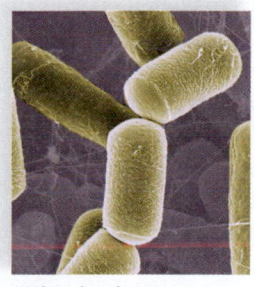

Methanobrevibacter smithii

On an average day, most of us produce and release 1–4 pints of gas. This amount, however, could be much greater for someone who is lactose intolerant. These excess quantities of gas are produced by the microbes—both bacteria and archaea—which can consume the undigested lactose sugars; the gas is a by-product of their metabolism. The other symptoms of lactose intolerance are influenced by exactly which microbes reside in your own internal ecosystem. For example, if you carry microbes that can use hydrogen gas for metabolism, the overall amount of gas produced in your gut will be lessened. If instead your body houses very few gas-absorbing bacteria or more methane-producing archaea, you may experience significant bloating and flatulence every time you eat lactose. Your intestine also houses

FIGURE 32C In North America, about 25% of adults show signs of decreased lactose digestion in the small intestine. Many of these are Asian, African, and Hispanic Americans.

other bacteria that ferment the lactose into organic compounds called short-chain fatty acids (SCFAs). However, some bacterial species make SCFAs faster than others. If your own personal microbes produce SCFAs more quickly than your body can absorb them, you may retain more water in your colon (through osmosis) and have diarrhea. Those lucky individuals who have the perfect balance of microbes have fewer unpleasant symptoms—in fact, some people who cannot break down lactose have no symptoms of lactose intolerance at all, thanks to their own internal ecosystem.

Studies have shown the best way to deal with lactose intolerance is to limit dairy products to about a cup per meal and to eat them with other foods so that digestion takes longer. Hard cheese and yogurt may be tolerated because lactose is often lost as hard cheeses are produced, and if the yogurt contains active bacteria, they digest the lactose and also release lactase for your use. Unfortunately, nondairy commercial products may still contain lactose. For example, if any of these are listed as an ingredient, the product will contain lactose: nonfat dry milk powder, whey, curds, milk by-products, or dry milk solids. If you need calcium in your diet, try calcium-fortified soy milk. Then, too, low-lactase milk and lactase pills are available for people who can find no other remedy.

CONSIDER THESE QUESTIONS

1. Antibiotic therapy can kill bacteria in your intestinal tract so that the proportions of bacteria are different than before. Why might this lead to lactose intolerance?
2. To control lactose intolerance, would you prefer to watch your diet or take pills? Explain your choice.

 **connect**
|BIOLOGY

Explore the concepts through a variety of multimedia assets and question types.

www.mcgrawhillconnect.com

Pancreas and Liver Are Vital Organs

The pancreas, liver, and gallbladder are accessory digestive organs. They secrete digestive juices, which enter the duodenum by way of ducts. The pancreas and the liver have several other functions that help maintain homeostasis.

32.5 Pancreas and liver contribute to chemical digestion

LEARNING OUTCOMES

When you complete this section, you should be able to

1. Describe several functions of the pancreas.
2. Describe several functions of the liver and discuss three liver illnesses.

Pancreas The **pancreas** lies deep in the abdominal cavity, behind the stomach. It is an elongated, somewhat flattened organ that has both endocrine and exocrine functions. As an endocrine gland, it secretes insulin and glucagon, hormones that help keep the blood glucose level within normal limits. In this chapter, however, we are interested in its exocrine function. Most pancreatic cells produce pancreatic juice, which contains sodium bicarbonate ($NaHCO_3$) and digestive enzymes for all types of food. Sodium bicarbonate neutralizes acidic chyme from the stomach. Pancreatic amylase digests starch, trypsin digests protein, and lipase digests fat, as stated in section 32.3.

Liver The **liver,** which is the largest gland in the body, lies mainly in the upper right section of the abdominal cavity, under the diaphragm (see Fig. 32.2A). The liver contains approximately 100,000 lobules that serve as its structural and functional units (Fig. 32.5). Between the lobules, a bile duct takes bile away from the liver; a branch of the hepatic artery brings O_2-rich blood to the liver; and a branch of the hepatic portal vein transports nutrients from the intestines to the liver. The central veins of the lobules drain blood from the liver and form the hepatic veins, which enter the inferior vena cava.

The liver has a number of important functions: (1) The liver acts as the gatekeeper to the blood. As blood in the hepatic portal vein from the intestines passes through the liver, the liver removes poisonous substances and detoxifies them. The liver has many metabolic functions. It removes and stores iron and the vitamins A, B_{12}, D, E, and K. (2) It manufactures the plasma proteins and helps regulate the quantity of cholesterol in the blood. (3) The liver maintains the blood glucose level at about 100 mg/100 ml (0.1%), even though a person eats intermittently. When insulin is present, any excess glucose in the blood is removed and stored by the liver as glycogen. Between meals, glycogen is broken down to glucose, which enters the hepatic veins. If the supply of glycogen is depleted, the liver converts glycerol (from fats) and also amino acids to glucose molecules. The conversion of amino acids to glucose requires the removal of amino groups. (4) Via a complex metabolic pathway, the liver then combines ammonia from the amino groups with carbon dioxide to form urea. **Urea** is the usual nitrogenous waste product from amino acid breakdown in humans; urea is excreted by the kidneys. (5) The liver produces bile, which is stored in the **gallbladder.** Bile has a yellowish-green color because it contains the bile pigment *bilirubin,* which is derived from the breakdown of hemoglobin, the red pigment of red blood cells. Bile contains bile salts, which emulsify fat in the small intestine. When fat is emulsified, it breaks up into droplets, providing a much larger surface area that can be acted upon by lipase, a digestive enzyme from the pancreas.

Liver Disorders When a person is jaundiced, the skin has a yellowish tint due to an abnormally large quantity of bile pigments in the blood. In hemolytic jaundice, red blood cells are broken down in abnormally large amounts in liver, spleen, and bone marrow; in obstructive jaundice, the bile duct is obstructed, or the liver cells are damaged. Obstructive jaundice often occurs when crystals of cholesterol precipitate out of the bile and form gallstones.

Jaundice can also result from a viral infection of the liver, called *hepatitis.* Hepatitis A is most often caused by eating contaminated food. Today, hepatitis B and C are mainly spread through needle sharing by drug addicts and the use of unsterilized tools by tattoo artists. These three types of hepatitis can also be spread by sexual contact.

Cirrhosis is a chronic liver disease in which the organ first becomes fatty, and then liver tissue is replaced by inactive fibrous scar tissue. Alcoholics often get cirrhosis, most likely due, at least in part, to the excessive amounts of alcohol the liver is forced to break down.

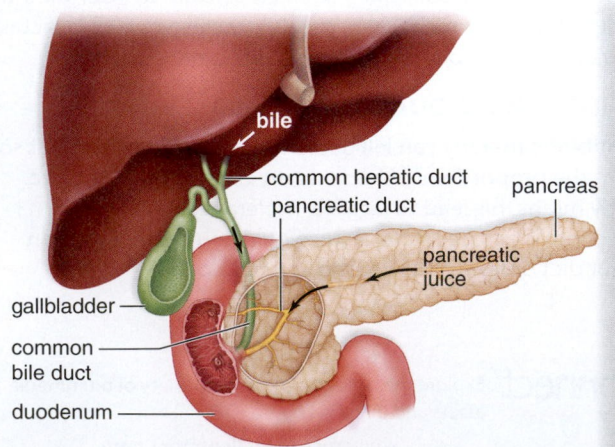

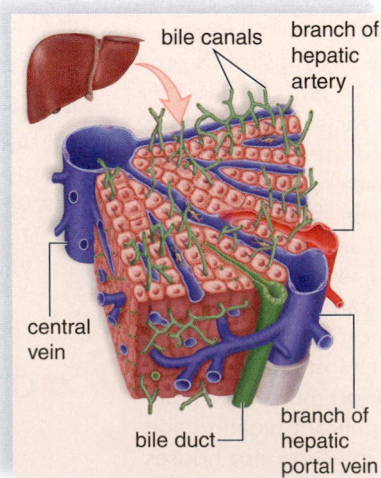

FIGURE 32.5 *Left,* relationship of liver, gallbladder, and pancreas. *Right,* portion of a liver lobule.

▶ **32.5 CHECK YOUR PROGRESS**

1. Identify what functions of the liver are so vital that we can't live without them.
2. When prescribed a medication, why can't you take it all at once and be done with it?

Nutrition

The vigilance of your immune system, the strength of your muscles, the circulation of your blood—all aspects of your physiology—depend on a balanced diet, one that supplies all the nutrients in the proper proportions necessary for a healthy, functioning body. Certain minerals and vitamins are also needed in order for the body to function properly.

 **Virtual Lab** Nutrition

32.6 Carbohydrates, lipids, and proteins supply nutrients

LEARNING OUTCOME

When you complete this section, you should be able to

1. Compare the benefits and drawbacks of carbohydrates, fats, and proteins in the diet.

Carbohydrates are present in food in the form of sugars, starch, and fiber. Fruits, vegetables, milk, and honey are natural sources of sugars. Glucose and fructose are monosaccharide sugars, and lactose (milk sugar) and sucrose (table sugar) are disaccharides. After being absorbed from the digestive tract into the bloodstream, all sugars are converted to glucose for transport in the blood and use by cells. Glucose is the preferred direct energy source in cells.

Plants store glucose as starch, and animals store glucose as glycogen. Good sources of starch are beans, peas, cereal grains, and potatoes. Starch is digested to glucose in the digestive tract, and any excess glucose is stored as glycogen. Although other animals likewise store glucose as glycogen in liver or muscle tissue (meat), little is left by the time an animal is eaten for food. Except for honey and milk, which contain sugars, animal foods do not contain carbohydrates.

Fiber includes various undigestible carbohydrates derived from plants. Food sources rich in fiber include beans, peas, nuts, fruits, and vegetables. Whole-grain products are also a good source of fiber, and are therefore more nutritious than food products made from refined grains. During *refinement,* fiber and also vitamins and minerals are removed from grains, so that primarily starch remains. For example, a slice of bread made from whole-wheat flour contains 3 g of fiber, while a slice of bread made from refined wheat flour contains less than 1 g of fiber.

Technically, fiber is not a nutrient for humans because it cannot be digested to small molecules that enter the bloodstream. Insoluble fiber, however, adds bulk to fecal material, which stimulates movement in the large intestine, preventing constipation. Soluble fiber combines with bile acids and cholesterol in the small intestine and prevents them from being absorbed. In this way, high-fiber diets may protect against heart disease. The typical American consumes only about 15 g of fiber each day; the recommended daily intake of fiber is 25 g for women and 38 g for men. To increase your fiber intake, eat whole-grain foods, snack on fresh fruits and raw vegetables, and include nuts and beans in your diet (Fig. 32.6A).

If you, or someone you know, has lost weight by following the Atkins or South Beach diet, you may think "carbs" are unhealthy and should be avoided. According to nutritionists, however, carbohydrates should supply a large portion of your energy needs. Evidence suggests that Americans are not eating the right kind of carbohydrates. In some countries, the traditional diet is 60–70% high-fiber carbohydrates, and these people have a low incidence of the diseases that plague Americans.

TABLE 32.6A	Reducing Dietary Sugar
To Reduce Dietary Sugar:	

1. Eat fewer sweets, such as candy, soft drinks, ice cream, and pastry.
2. Eat fresh fruits or fruits canned without heavy syrup.
3. Use less sugar—white, brown, or raw—and less honey and syrups.
4. Avoid sweetened breakfast cereals.
5. Eat less jelly, jam, and preserves.
6. Drink pure fruit juices, not imitations.
7. When cooking, use spices such as cinnamon, instead of sugar, to flavor foods.
8. Do not put sugar in tea or coffee.
9. Avoid potatoes and processed foods made from refined carbohydrates, such as white bread, rice, and pasta. These foods are immediately broken down to sugar during digestion.

Some nutritionists hypothesize that the high intake of foods that are rich in refined carbohydrates and fructose sweeteners processed from cornstarch may be responsible for the prevalence of obesity in the United States. Because certain foods, such as donuts, cakes, pies, and cookies, are high in both refined carbohydrates and fat, it is difficult to determine which dietary component is responsible for the current epidemic of obesity among Americans. In any case, they are empty-calorie foods that provide sugars but no vitamins or minerals. Table 32.6A tells how to reduce your sugar intake. Nutritionists also point out that consuming too much energy from any source contributes to body fat, which increases a person's risk of obesity and associated illnesses.

Lipids Like carbohydrates, **triglycerides** (fats and oils) supply energy for cells, but **fat** is stored for the long term in the body. Nutritionists generally recommend that people include unsaturated, rather than saturated, fats in their diets. Two unsaturated fatty acids (alpha-linolenic and linoleic acids) are *essential* in the

FIGURE 32.6A Fiber-rich foods.

diet. They can be supplied by eating fatty fish and by including plant oils, such as canola and soybean oils, in the diet. Growth delays and skin problems can develop in people whose diets lack these essential unsaturated fatty acids.

Animal-derived foods, such as butter, meat, whole milk, and cheeses, contain saturated fatty acids (Fig. 32.6B*a*). Plant oils contain unsaturated fatty acids; each type of oil has a particular percentage of monounsaturated and polyunsaturated fatty acids.

Cholesterol, a lipid, can be synthesized by the body. Cells use cholesterol to make various compounds, including bile, steroid hormones, and vitamin D. Plant foods do not contain cholesterol; only animal foods such as cheese, egg yolks, liver, and certain shellfish (shrimp and lobster) are rich in cholesterol. Elevated blood cholesterol levels are associated with an increased risk of cardiovascular disease, the number-one killer of Americans. A diet rich in cholesterol and saturated fats increases the risk of cardiovascular disease (see "Disorders Associated with Obesity" on page 647).

Statistical studies suggest that trans fatty acids (trans fats) are even more harmful than saturated fatty acids. Trans fatty acids arise when unsaturated oils are hydrogenated to produce a solid fat, as in shortening and some margarines. Trans fatty acids may reduce the function of the plasma membrane receptors that clear cholesterol from the bloodstream. Trans fatty acids are found in commercially packaged foods, such as cookies and crackers; in commercially fried foods, such as french fries; and in packaged snacks. Table 32.6B tells you how to cut back on harmful lipids in your diet.

Proteins Dietary **proteins** are digested to amino acids, which cells use to synthesize hundreds of cellular proteins. Of the 20 different amino acids, nine are *essential amino acids* that must be present in the diet. Children will not grow if their diets lack the essential amino acids. Eggs, milk products, meat, poultry, and most other foods derived from animals contain all nine essential amino acids and are considered "complete" or "high-quality" protein sources (see Fig. 32.6B*b*).

Foods derived from plants generally do not have as much protein per serving as those derived from animals, and each type of plant food generally lacks one or more of the essential amino acids. Therefore, most plant foods are "incomplete" or "low-quality" protein sources. Vegetarians, however, can still avoid animal sources of protein. To meet their protein needs, total vegetarians (vegans) may eat grains, beans, and nuts in various combinations. Also, tofu, soy milk, and other foods

TABLE 32.6B	Reducing Harmful Lipids

To Reduce Dietary Fat:

1. Choose poultry, fish, or dry beans and peas as a protein source.
2. Remove skin from poultry and trim fat from red meats before cooking; place meat on a rack while cooking so that fat drains off.
3. Broil, boil, or bake rather than frying.
4. Limit your intake of butter, cream, trans fats, shortening, and tropical oils (coconut and palm oils).*
5. Use herbs and spices to season vegetables instead of butter, margarine, or sauces. Use lemon juice instead of salad dressing.
6. Drink skim milk instead of whole milk, and use skim milk in cooking and baking.

To Reduce Dietary Cholesterol:

1. Eat white fish and poultry in preference to cheese, egg yolks, liver, and certain shellfish (shrimp and lobster).
2. Substitute egg whites for egg yolks in both cooking and eating.
3. Include soluble fiber in the diet. Oat bran, oatmeal, beans, corn, and fruits, such as apples, citrus fruits, and cranberries, are high in soluble fiber.

*Although coconut and palm oils are from plant sources, they are mostly saturated fats.

made from processed soybeans are complete protein sources. A balanced vegetarian diet is quite possible with a little planning.

According to nutritionists, protein should not supply the bulk of dietary calories. The average American eats about twice as much protein as he or she needs, and some people may be on a diet that encourages the intake of proteins, instead of carbohydrates, as an energy source. Also, bodybuilders should realize that excess amino acids are not always converted into muscle tissue. When amino acids are broken down, the liver removes the nitrogen portion (a process called *deamination*) and uses it to form urea, which is excreted in urine. The water needed for excretion of urea can cause dehydration when a person is exercising and losing water by sweating. High-protein diets can also increase calcium loss in the urine and encourage the formation of kidney stones. Furthermore, high-protein foods often contain a high amount of fat.

▶ **32.6 CHECK YOUR PROGRESS**
1. Which type of carbohydrate leads to poor health and which type leads to good health? Explain.
2. Which types of lipids lead to poor health and which types lead to good health? Explain.

a. Lipids   b. Proteins

FIGURE 32.6B a. Dietary lipids are needed only in limited amount. Lipids have health risks and can add unwanted calories to the diet. **b.** The diet should contain protein to ensure the intake of essential amino acids but a moderate amount is sufficient.

32.7 Minerals and vitamins have various roles in the body

LEARNING OUTCOME

When you complete this section, you should be able to

1. In general, discuss the need for minerals and vitamins in the diet.

The body needs about 20 elements called **minerals** for numerous physiological functions, including regulation of biochemical reactions, maintenance of fluid balance, and incorporation into certain structures and compounds. The body contains more than 5 g of each major mineral and less than 5 g of each trace mineral. Table 32.7A lists both the major and trace minerals, and gives their functions and food sources. It also tells the health effects of too little or too much intake.

Occasionally, individuals (especially women) do not receive enough iron, calcium, magnesium, or zinc in their diets. Menstruating adult females need more iron in the diet than males (18 mg compared to 10 mg). *Anemia,* characterized by a run-down feeling due to insufficient red blood cells, results when the diet lacks sufficient iron.

Many people take calcium supplements, as directed by a physician, to counteract *osteoporosis,* a degenerative bone disease that affects an estimated one-quarter of older men and one-half of older women in the United States.

Animation
Osteoporosis

One mineral that people consume too much of is sodium. The recommended amount of sodium intake per day is 2,400 mg, while the average American takes in 4,000–4,700 mg each day. About one-third of the sodium we consume occurs naturally in foods, another one-third is added during commercial processing, and the last one-third is added either during home cooking or at the table in the form of table salt.

TABLE 32.7A | Minerals

Mineral	Functions	Food Sources	Conditions Caused By: Too Little	Conditions Caused By: Too Much
Major Minerals				
Calcium (Ca^{2+})	Strong bones and teeth, nerve conduction, muscle contraction	Dairy products, leafy green vegetables	Stunted growth in children; low bone density in adults	Kidney stones; interferes with iron and zinc absorption
Phosphorus (PO_4^{3-})	Bone and soft tissue growth; part of phospholipids, ATP, and nucleic acids	Meat, dairy products, sunflower seeds, food additives	Weakness, confusion, pain in bones and joints	Low blood and bone calcium levels
Potassium (K^+)	Nerve conduction, muscle contraction	Many fruits and vegetables, bran	Paralysis, irregular heartbeat, eventual death	Vomiting, heart attack, death
Sodium (Na^+)	Nerve conduction, pH and water balance	Table salt	Lethargy, muscle cramps, loss of appetite	High blood pressure, calcium loss
Chloride (Cl^-)	Water balance	Table salt	Not likely	Vomiting, dehydration
Magnesium (Mg^{2+})	Part of various enzymes for nerve and muscle contraction, protein synthesis	Whole grains, leafy green vegetables	Muscle spasm, irregular heartbeat, convulsions, confusion, personality changes	Diarrhea
Trace Minerals				
Zinc (Zn^{2+})	Protein synthesis, wound healing, fetal development and growth, immune function	Meat, legumes, whole grains	Delayed wound healing, night blindness, diarrhea, mental lethargy	Anemia, diarrhea, vomiting, renal failure, abnormal cholesterol levels
Iron (Fe^{2+})	Hemoglobin synthesis	Whole grains, meat, prune juice	Anemia, physical and mental sluggishness	Iron toxicity disease, organ failure, eventual death
Copper (Cu^{2+})	Hemoglobin synthesis	Meat, nuts, legumes	Anemia, stunted growth in children	Damage to internal organs if not excreted
Iodine (I^-)	Thyroid hormone synthesis	Iodized table salt, seafood	Thyroid deficiency	Depressed thyroid function, anxiety
Selenium (SeO_4^{2-})	Part of antioxidant enzyme	Seafood, meat, eggs	Vascular collapse, possible cancer development	Hair and fingernail loss, discolored skin

Vitamins are organic compounds (other than carbohydrates, fats, and proteins, including amino acids) that regulate various metabolic activities and must be present in the diet. Table 32.7B lists the major vitamins and some of their functions, food sources, and associated disorders.

Although many people think vitamins can enhance health dramatically, prevent aging, and cure diseases such as arthritis and cancer, there is no scientific evidence that vitamins are "wonder drugs." However, vitamins C, E, and A are believed to defend the body against free radicals, and therefore they are termed **antioxidants.** These vitamins are especially abundant in fruits and vegetables, and nutritionists suggest that we eat about 4½ cups of fruits and vegetables per day. Assuming an adequate diet, skin cells normally contain a precursor cholesterol

molecule that is converted to vitamin D after UV exposure. Most milk today is fortified with vitamin D, which helps prevent the occurrence of rickets, characterized by defective mineralization of the skeleton.

Although many foods in the United States are now enriched, or fortified with vitamins, some individuals are still at risk for vitamin deficiencies, generally as a result of poor food choices. These include the elderly, young children, alcoholics, and low-income people.

Animation
B Vitamins

▶ **32.7 CHECK YOUR PROGRESS**

1. Explain why we should use less salt in our food.
2. Identify food sources that can be relied on to supply vitamins in the diet.

TABLE 32.7B	Vitamins			
Vitamin	**Functions**	**Food Sources**	**Conditions Caused By:**	
			Too Little	**Too Much**
Water-Soluble Vitamins				
Vitamin C	Antioxidant; collagen synthesis for capillaries, bones, and teeth	Citrus fruits, green leafy vegetables, tomatoes	Scurvy, delayed wound healing, infections	Gout, kidney stones, diarrhea
Thiamine (vitamin B₁)	Coenzyme needed for cellular respiration	Whole grains, legumes, and nuts	Beriberi, anemia, muscle weakness	Absorption of other vitamins prevented
Riboflavin (vitamin B₂)	Aids cellular respiration, including oxidation of protein and fat	Nuts, dairy products, whole grains, poultry, green leafy vegetables	Dermatitis, blurred vision, growth failure	Unknown
Niacin (nicotinic acid)	Coenzyme needed for cellular respiration	Peanuts, poultry, whole grains, green leafy vegetables	Pellagra, diarrhea, mental disorders	High blood sugar and uric acid, vasodilation
Folacin (folic acid)	Coenzyme needed for production of hemoglobin and DNA	Dark-green leafy vegetables, nuts, beans, whole grains	Megaloblastic anemia, spina bifida	May mask B₁₂ deficiency
Vitamin B₆	Coenzyme; synthesis of hormones and hemoglobin; CNS control	Whole grains, bananas, beans, poultry, nuts, green leafy vegetables	Rarely, convulsions, vomiting, seborrhea, muscle weakness	Insomnia, neuropathy
Pantothenic acid	Coenzyme A needed for oxidation of carbohydrates and fats	Nuts, beans, dark-green leafy vegetables, poultry, fruits, milk	Rarely, loss of appetite, mental depression, numbness	Unknown
Vitamin B₁₂	Coenzyme needed for synthesis of nucleic acids and myelin	Dairy products, fish, poultry, eggs, fortified cereals	Pernicious anemia	Unknown
Biotin	Coenzyme needed for metabolism of amino acids and fatty acids	Generally in foods; eggs	Skin rash, nausea, fatigue	Unknown
Fat-Soluble Vitamins				
Vitamin A	Antioxidant synthesized from beta-carotene; healthy eyes, skin, hair, and proper bone growth	Deep yellow/orange and green leafy vegetables, fruits, cheese, whole milk, butter, eggs	Night blindness, impaired growth of bones and teeth	Headache, nausea, hair loss, abnormal fetal development
Vitamin D	Steroid needed for development and maintenance of bones	Fortified milk, fish liver oil; exposure to sunlight	Rickets, bone decalcification and weakening	Calcification of soft tissues, diarrhea
Vitamin E	Antioxidant; prevents oxidation of vitamin A polyunsaturated fatty acids	Green leafy vegetables, fruits, nuts, and whole grains	Unknown	Diarrhea, headaches, fatigue, muscle weakness
Vitamin K	Synthesizes substances active in blood clotting	Green leafy vegetables, cauliflower	Easy bruising and bleeding	Interferes with anticoagulant drugs

32D   Disorders Associated with Obesity

Nutritionists point out that consuming too many Calories from any source contributes to body fat, which increases a person's risk of obesity and associated illnesses. (Obesity is defined as weighing 30% more than the ideal body weight for your height and body build.) Still, foods such as donuts, cakes, pies, cookies, and white bread, which are high in refined carbohydrates (starches and sugars), and fried foods, which are high in fat, may very well be responsible for the current epidemic of obesity among Americans. Also implicated is the lack of exercise because of a sedentary lifestyle. Type 2 diabetes and cardiovascular disease are often seen in people who are obese.  **Animation** Body Mass Index

Type 2 Diabetes

Diabetes mellitus is indicated by the presence of glucose in the urine. Glucose has spilled over into the urine because there is too high a level of glucose in the blood. Diabetes occurs in two forms. Type 1 diabetes is not associated with obesity. When a person has type 1 diabetes, the pancreas does not produce insulin (the hormone that causes cells to take up glucose), and the patient must have daily insulin injections. In contrast, children, and more often adults, with type 2 diabetes are usually obese and display impaired insulin production and instead have insulin resistance. In a person with insulin resistance, the body's cells fail to take up glucose, even when insulin is present. Therefore, the blood glucose level exceeds the normal level, and glucose appears in the urine.

The incidence of type 2 diabetes is increasing rapidly in most industrialized countries of the world. A healthy diet, increased physical activity, and weight loss have been shown to improve the insulin function in type 2 diabetics. How might a poor diet contribute to the occurrence of type 2 diabetes? Simple sugars in foods, such as candy and ice cream, immediately enter the bloodstream, as do sugars from the digestion of the starch within white bread and potatoes. When the blood glucose level rises rapidly, the pancreas produces an overload of insulin to bring the glucose level under control. Chronically high insulin levels apparently lead to insulin resistance, increased fat deposition, and a high blood fatty acid level. High fatty acid levels can lead to increased risk of cardiovascular disease. It is well worth the effort to control type 2 diabetes because all diabetics, whether type 1 or type 2, are at risk for blindness, kidney disease, and cardiovascular disease.

Cardiovascular Disease

In the United States, cardiovascular disease, which includes hypertension, heart attack, and stroke, is among the leading causes of death. Cardiovascular disease is often due to arteries blocked by plaque, which contains saturated fats and cholesterol. Cholesterol is carried in the blood by two types of lipoproteins: low-density lipoprotein (LDL) and high-density lipoprotein (HDL). LDL molecules are considered "bad" because they are like delivery trucks that carry cholesterol from the liver to the cells and to the arterial walls. HDL molecules are considered "good" because they are like garbage trucks that dispose of cholesterol. HDL transports cholesterol from the cells to the liver, which converts it to bile salts that enter the small intestine.

Consuming saturated fats, including trans fats, tends to raise LDL cholesterol levels, while eating unsaturated fats lowers LDL

FIGURE 32D Foods high in trans fats.

cholesterol levels. Beef, dairy foods, and coconut oil are rich sources of saturated fat. Foods containing partially hydrogenated oils (e.g., vegetable shortening and stick margarine) are sources of trans fats (Fig. 32D). Unsaturated fatty acids in olive and canola oils, most nuts, and cold-water fish (e.g., herring, sardines, tuna, and salmon) tend to lower LDL cholesterol levels. Furthermore, cold-water fish contain polyunsaturated fatty acids and especially *omega-3 unsaturated fatty acids*, which are believed to reduce the risk of cardiovascular disease. However, taking fish oil supplements to obtain omega-3s is not recommended without a physician's approval because excessive fatty acids can interfere with normal blood clotting.

The American Heart Association recommends limiting total cholesterol intake to 300 mg per day. This requires careful selection of the foods we include in our daily diets. For example, an egg yolk contains about 210 mg of cholesterol, which would be two-thirds of the recommended daily intake. Still, this doesn't mean eggs should be eliminated from a healthy diet, because the proteins in them are very nutritious; in fact, most healthy people can eat a couple of whole eggs each week without experiencing an increase in their blood cholesterol levels.

A physician can determine whether patients' blood lipid levels are normal. If a person's cholesterol and triglyceride levels are elevated, modifying the fat content of the diet, losing excess body fat, and exercising regularly can reduce them. If lifestyle changes do not lower blood lipid levels enough to reduce the risk of cardiovascular disease, a physician may prescribe special medications.

CONSIDER THESE QUESTIONS

1. What are some of the reasons a person might have an unhealthy diet, and what could be done about it?
2. Should certain foods carry warning labels, as cigarettes do, to forewarn the public?

 Explore the concepts through a variety of multimedia **BIOLOGY** assets and question types.

www.mcgrawhillconnect.com

32E Eating Disorders

People with eating disorders are dissatisfied with their body image. Social, cultural, emotional, and biological factors all contribute to the development of an eating disorder. For example, many women admire the extreme thinness of fashion models today. In a study at the University of Bath, two-thirds of participants favored thin models and were more likely to buy a product advertised by a thin model. In Madrid, steps have been taken to safeguard the health of fashion models: They must weigh a certain minimum according to their height, or they cannot participate in a fashion show.

Anorexia nervosa is a severe psychological disorder characterized by an irrational fear of getting fat that results in the refusal to eat enough food to maintain a healthy body weight. Anorexia nervosa patients think they are fat when actually they are thin (Fig. 32E.1). A self-imposed starvation diet is often accompanied by occasional binge-eating, followed by purging and extreme physical activity to avoid weight gain. Binges usually include large amounts of high-calorie foods, and purging episodes involve self-induced vomiting and laxative abuse. About 90% of the people suffering from anorexia nervosa are young women; an estimated 1 in 200 teenage girls is affected.

A person with *bulimia nervosa* binge-eats, and then purges to avoid gaining weight (Fig. 32E.2). The binge-purge cyclic behavior can occur several times a day. People with bulimia nervosa can be difficult to identify because their body weights are often normal and they tend to conceal their bingeing and purging. Women are more likely than men to develop bulimia; an estimated 4% of young women suffer from this condition.

Many obese people suffer from *binge-eating disorder*, which is characterized by episodes of overeating that are not followed by purging. Stress, anxiety, anger, and depression can trigger food binges. Any eating disorder, whether anorexia nervosa, bulimia nervosa, or simply binge eating, can lead to malnutrition, disability, and death.

Preoccupation with diet, bodybuilding activities, and body form characterize the disorder known as *muscle dysmorphia*. A person with this condition thinks his or her body is underdeveloped when actually it is extremely well developed (Fig. 32E.3). The person may spend hours in the gym each day, working out on muscle-strengthening equipment. Unlike anorexia nervosa and bulimia, muscle dysmorphia affects more men than women.

Regardless of the eating disorder, early recognition and treatment are crucial. Treatment usually includes psychological counseling and antidepressant medications.

CONSIDER THESE QUESTIONS

1. In what ways might our society be encouraging eating disorders?
2. Most likely you will not be able to reason with a person who has an eating disorder. What could you do to help?

 Explore the concepts through a variety of multimedia assets and question types.

www.mcgrawhillconnect.com

FIGURE 32E.1 Anorexia nervosa patients think they are fat, even though they are thin.

FIGURE 32E.2 Bulimia nervosa is characterized by occasional binge eating followed by purging.

FIGURE 32E.3 Muscle dysmorphia is also characterized by a distorted body image.

CONNECTING THE CONCEPTS

The human digestive system consists of the digestive tract with the mouth at one end and the anus at the other end. Food enters the mouth and is digested to small nutrient molecules, which enter the bloodstream at the small intestine. Undigestible residue (feces) exits at the anus.

Digestive juices enter the tract from the salivary glands and the pancreas. The gallbladder also sends bile to the tract for emulsification of fats. Bile is made in the liver, which makes a significant contribution to homeostasis by helping to keep the composition of the blood constant. For example, after eating and under the influence of insulin, the liver stores glucose from the digestive tract as glycogen; in between meals, it releases glucose to the blood. The liver also removes nitrogenous and other types of injurious molecules from the blood and metabolizes them to

excretory products that the bloodstream takes to the kidneys.

Nutrient molecules, such as amino acids, glucose, and vitamins, are absorbed from the small intestine into the blood. Fatty acids and glycerol enter the lacteals, a part of the lymphatic system. Body cells receive these molecules from blood and then build the body's own proteins and carbohydrates. In Chapter 33, we consider the contribution of the respiratory system to homeostasis.

ANALYZE AND EVALUATE

1. How does the digestive system help all the other systems of the body and in that way help maintain homeostasis?
2. Explain the position of the liver in the accompanying diagram.
3. Anti-obesity surgeries often reduce the size of the stomach. Why is this effective? What problems might arise?

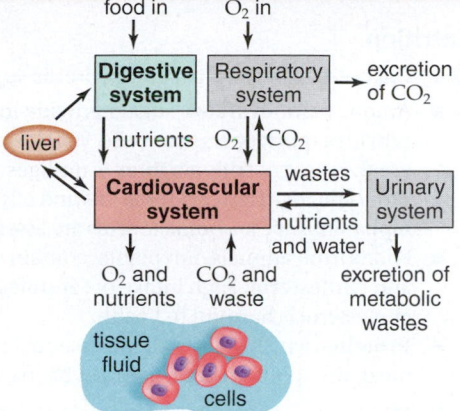

The digestive system receives bile from the liver and supplies nutrients to the cardiovascular system, which delivers them to the cells.

SUMMARIZE

How Animals Acquire and Process Food

32.1   Animals have various ways to obtain food before processing occurs

- A **bulk feeder** eats food in chunks or whole. A **filter feeder** collects and eats small particles from water. A **substrate feeder** lives on or in the food it eats. A **fluid feeder** consumes a fluid.

  Complete digestive tract

- A digestive system carries out these functions: (1) ingestion, the intake of food; (2) digestion, the breakdown of food; (3) absorption, the intake of nutrient molecules into the body, usually via the bloodstream; and (4) elimination, the removal of unabsorbed molecules from the body.

Mammalian Organs of Digestion

32.2   Some digestion occurs before food reaches the stomach

- Teeth chew food, **salivary glands** secrete saliva (contains **salivary amylase** for digesting starch) and the tongue forms a food bolus. The roof of the mouth consists of the **hard palate** and **soft palate.** During swallowing, the soft palate moves back to cover the nasal passages; the **epiglottis** covers the **glottis,** the opening for the air passages to the lungs. After swallowing, **peristalsis** begins in the esophagus. The **esophagus** is a muscular tube that passes from the **pharynx** to the stomach whose opening is guarded by a **sphincter.**

32.3   Digestion also takes place in the stomach and small intestine

- The stomach expands and stores food, which is churned and mixed with acidic gastric juices. It also secretes gastric juices containing **pepsin,** an enzyme that digests protein.
- In the **small intestine,** the duodenum receives **chyme** from the stomach, bile from the **gallbladder** and pancreatic juice from the **pancreas.** From the gallbladder, **bile** emulsifies fat and readies it for digestion by lipase. From the pancreas, **pancreatic amylase** digests carbohydrates, **trypsin** digests proteins, and **lipase** digests fats. Intestinal enzymes complete digestion to small nutrient molecules, which are absorbed at the **villi** (surface is increased by **microvilli**). Glucose and amino acids are absorbed into blood; lipid is absorbed at **lacteals.**

32.4   The large intestine absorbs water and prepares wastes for elimination

- The large intestine includes the cecum (from which projects the **vermiform appendix**), the **colon,** and the **rectum.** Besides water, the **large intestine** absorbs salts and some vitamins, and forms feces.

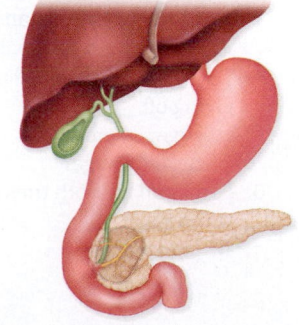

Pancreas and Liver Are Vital Organs

32.5 Pancreas and liver contribute to chemical digestion

■ The **liver** (1) detoxifies blood, (2) makes plasma proteins, (3) maintains the amount of glucose in the blood, (4) produces **urea,** and (5) produces bile, which is stored in the **gallbladder.** Liver disorders include jaundice, hepatitis, and cirrhosis.

Nutrition

32.6 Carbohydrates, lipids, and proteins supply nutrients

■ Among carbohydrates, sugars provide immediate and starch provides quick energy for cells. Whole-grain **carbohydrates** are the most nutritious; **fiber** is undigestible cellulose.

■ Among lipids, **triglycerides** in fats and oils provide stored energy. Alpha-linolenic and linoleic acids are essential fatty acids.

■ Foods from animals, not plants, contain saturated fats and **cholesterol.** High intake of saturated fats, trans fats, and cholesterol is harmful to health.

■ **Proteins,** which supply all the essential amino acids, include meat, fish, poultry, eggs, nuts, soybeans, and cheese.

32.7 Minerals and vitamins have various roles in the body

■ **Minerals** regulate biochemical reactions, maintain fluid balance, and are incorporated into structures and compounds. **Vitamins,** obtained from most foods, regulate metabolism and physiological development; vitamins C, E, and A are **antioxidants.**

TEST YOURSELF

How Animals Acquire and Process Food

1. The digestive system
   a. breaks down food into usable nutrients.
   b. absorbs nutrients.
   c. eliminates waste.
   d. All of these are correct.
2. Animals that feed discontinuously
   a. have digestive tracts that permit storage.
   b. are always filter feeders.
   c. exhibit extremely rapid digestion.
   d. have a nonspecialized digestive tract.
   e. usually eat only meat.
3. **THINKING CONCEPTUALLY** Argue that the digestive system is the most critical system to the life of an animal.

Mammalian Organs of Digestion

4. Which of these does not occur in the mouth?
   a. Amylase digests protein.
   b. Salivary glands secrete juices into the mouth.
   c. Chewing is mechanical digestion.
   d. Preparation for swallowing occurs.
5. Tooth decay is caused by bacteria metabolizing _____ and giving off _____.
   a. sugar, protein   c. acids, sugar
   b. protein, sugar   d. sugar, acids
6. Food and air both travel through the
   a. lungs.   c. larynx.
   b. pharynx.   d. trachea.

7. Peristalsis occurs
   a. from the mouth to the small intestine.
   b. from the beginning of the esophagus to the anus.
   c. only in the stomach.
   d. only in the small and large intestines.
   e. only in the esophagus and stomach.
8. Label this diagram:

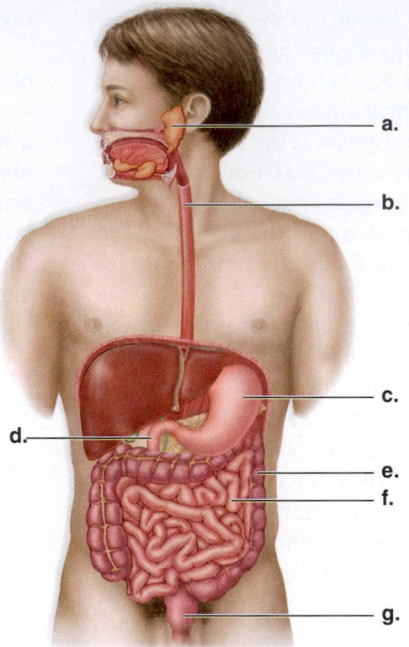

9. Which test tube would show the most digestion? Why?

Incubator

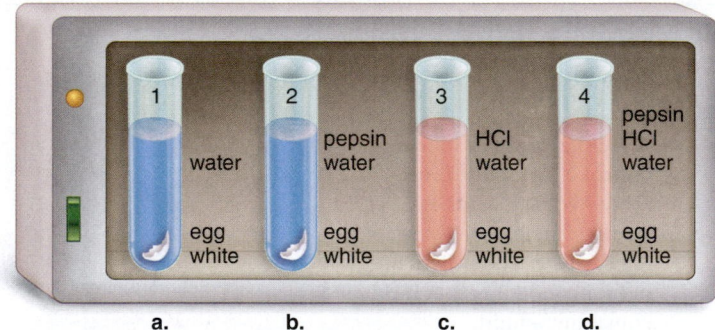

10. The stomach
    a. is lined with a thick layer of mucus.
    b. contains sphincter glands.
    c. has a pH of about 6.
    d. digests carbohydrates.
    e. More than one of these are correct.
11. Which of these is not a proper contrast between the stomach and the small intestine?
    a. digests protein—digests all large biomolecules
    b. stores food—absorbs food
    c. is very long—expands
    d. is connected to the esophagus—is connected to the large intestine

12. Bile
    a. is an important enzyme for the digestion of fats.
    b. cannot be stored.
    c. is made by the gallbladder.
    d. emulsifies fat.
    e. All of these are correct.
13. The lack of _____ activity would result in failure to maintain water balance.
    a. small intestinal   c. gallbladder
    b. large intestinal   d. stomach
14. The vermiform appendix
    a. is connected to the small intestine at the junction of the large intestine.
    b. plays a role in fighting infection.
    c. Both of these are correct.
    d. Neither of these is correct.
15. **THINKING CONCEPTUALLY** A ham sandwich contains what nutrients and is processed to what smaller molecules in which digestive organs?

Pancreas and Liver Are Vital Organs

16. Which organ has both an exocrine and an endocrine function?
    a. liver   c. pancreas
    b. esophagus   d. cecum
17. Which of these is not a proper contrast between the pancreas and the liver?
    a. secretes insulin—secretes bile
    b. sends digestive juices to the duodenum—sends toxins to the small intestine
    c. long and flat—composed of lobules
    d. sends a duct directly to the duodenum—first sends a duct to the gallbladder
18. Label this portion of a hepatic lobule:

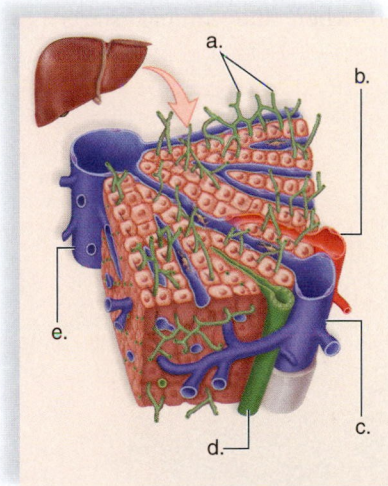

19. **THINKING CONCEPTUALLY** The liver is an accessory organ to the digestive system. To what other system could it also be considered an accessory organ?

Nutrition

For questions 20–24, choose the class of nutrient from the key that matches the description. Each answer may be used more than once.

KEY:
    a. carbohydrates   d. minerals
    b. lipids   e. vitamins
    c. proteins   f. water

20. Preferred source of direct energy for cells.
21. Include antioxidants.
22. Generally found in higher levels in animal sources than in plant sources.
23. An example is cholesterol.
24. Includes calcium, phosphorus, and potassium.
25. **THINKING CONCEPTUALLY** Explain why vegetarians need not be concerned that their tissues will contain plant proteins.

GET INVOLVED

1. (a) If you are testing the ability of pepsin to digest protein, what must your test tube contain? (b) What control will you use? Explain. (See section 32.3 and Test Yourself question 9.)
2. Findings from correlation studies, such as statistical studies that correlate saturated fats in the diet with increased chances of cardiovascular disease, often lead to medical decisions. What criteria would you use to judge correlation studies? (See "Disorders Associated with Obesity" on page 647.)
3. After viewing this virtual lab, relate the usual diet of a frog to the length of its intestines.  **Virtual Lab** Virtual Frog Dissection

MEDIA STUDY TOOLS

mhhe.com/maderconcepts3

Enhance your study of this chapter with interactive study tools, practice tests, and engaging animations. Also, ask your instructor about the resources available through ConnectPlus, which includes LearnSmart, a personalized adaptive learning program, and a media-rich eBook.

33

Gas Exchange and Transport in Animals

BEFORE YOU BEGIN

Take a few minutes to recall

The reason we breathe (section 7.1)

How the respiratory system contributes to homeostasis (sections 26.7–26.8)

Gas exchange at tissue capillaries (section 30.6)

Free-Diving Is Dangerous

Many aquatic mammals can dive to great depths and stay submerged for some time. Yet they breathe air, just as we do, and therefore can't breathe when they dive—no oxygen tank is provided! The northern elephant seal has been observed diving to a depth of 1,500 m. In comparison, a free-diving human can only dive to about 163 m. The Weddell seal typically dives to only 300–400 m, but stays submerged for up to 15 minutes. The limit for humans is about 3 minutes.

Let's look at how aquatic mammals, such as whales, seals, and dolphins, do it. First, they store oxygen before they dive. Blood doping (as when athletes get a blood transfusion before a competitive event) is natural to them because they have more blood cells and more blood per body weight than humans do. Not only that, their muscles are chock full of myoglobin, a respiratory pigment that specializes in keeping oxygen where it is most needed—namely, in the muscles.

A special *diving response* occurs when aquatic mammals dive: (1) The heart rate slows (called bradycardia) to about one-half to one-tenth the normal rate. (2) The peripheral blood vessels constrict, and the blood circulates to the heart and lungs only. (3) After the oxygen stored by myoglobin is used up, fermentation supplies ATP. The lactic acid produced is metabolized when the animal starts

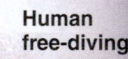

Human free-diving

Northern elephant seal

Weddell seal

Pilot whales

breathing again. (4) Finally, the spleen kicks in. The spleen acts as a storage area for red blood cells, in addition to cleansing the blood. As the water pressure increases during the dive, compression causes the spleen to release its supply of fully oxygenated red blood cells. This oxygen keeps the heart and brain going for a while longer.

Many humans want to free-dive—and they do. How far can we go in copying the aquatic mammals? Although we do not store oxygen to any great extent, it is helpful to warm up before diving into cold water. This makes sure the muscles are getting O_2-rich blood for as long as possible before the dive begins. Researchers also recommend that the mask not include the forehead—where cold receptors are located. Cold receptors help bring on the diving response that can also occur in humans. You can practice bringing on the four aspects of the diving response by holding your breath (called the apneic time) with your face submerged in cold water. If you can overcome the urge to breathe long enough, even the spleen will kick in. To get the spleen to discharge its red blood cells, the only requirement is to not breathe for a certain length of time. Researchers have found that it doesn't matter whether a person is in the water or standing on solid ground for this reaction to occur.

Aquatic mammals apparently don't suffer from the "bends" as we do. In humans, N_2 from air inhaled just before the dive enters the blood. Surfacing rapidly after diving causes the N_2 to bubble from the blood, bursting capillaries, even in the brain. It has been shown in the laboratory that the lungs of elephant seals collapse on the way down and do not reinflate until after the seals have ascended to a shallow depth. Researchers conclude that much of the nitrogen in the seals' lungs doesn't have a chance to enter the bloodstream as they submerge. Presumably, other deep marine mammals experience similar lung collapse while diving.

Free-diving can be very dangerous and severely limit your life expectancy, even if you train with an expert. Therefore, free-diving is not recommended for humans—they lack the adaptations that aquatic mammals have and are instead highly adapted to living on land. In this chapter, we will learn how animals in the water and on land breathe and transport gases to and from their cells.

Bottlenose dolphin

Gas-Exchange Surfaces

Respiratory systems bring external oxygen to a gas-exchange surface and take carbon dioxide from this surface to the exterior. In this first part of the chapter, we look at a variety of respiratory systems.

33.1 Respiration involves several steps

LEARNING OUTCOME

When you complete this section, you should be able to

1. Understand the three events that occur during respiration.

Respiration is the sequence of events that results in gas exchange between the body's cells and the environment. In terrestrial vertebrates, respiration includes these steps:

- **Ventilation** (i.e., breathing) includes inspiration (entrance of air into the lungs) and expiration (exit of air from the lungs).
- **External respiration** is gas exchange between the air and the blood within the lungs. In humans, red blood cells transport oxygen from the lungs to the tissues (Fig. 33.1, *top*).
- **Internal respiration** is gas exchange between the blood and the tissue fluid. In otherwords, the body's cells exchange gases with the tissue fluid. The blood then transports carbon dioxide to the lungs (Fig. 33.1, *bottom*).

Gas exchange takes place by the physical process of diffusion. For external respiration to be effective, the external gas-exchange region must be (1) moist, (2) thin, and (3) large in relation to the size of the body. Most complex animals have specialized external gas-exchange areas in gills or lungs where capillaries abound. Delivery of oxygen to the cells is promoted when the blood or the blood cells contain a respiratory pigment, such as hemoglobin. Thin capillary walls in the tissues make up the internal gas-exchange area.

Regardless of the particular external respiration surface and the manner in which gases are delivered to the cells, in the end, oxygen enters mitochondria, where cellular respiration takes place. Without ATP production life ceases.

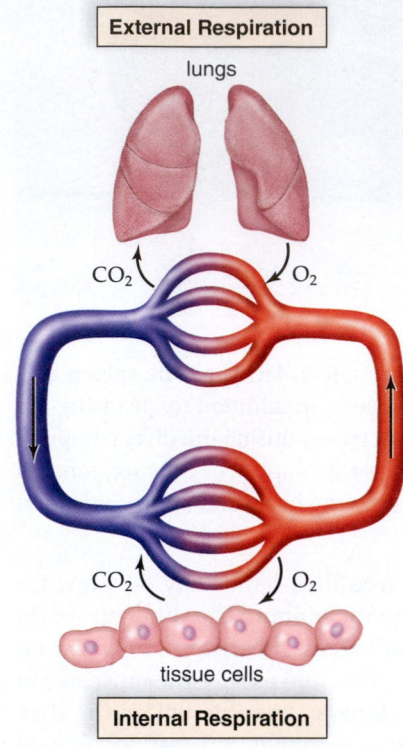

FIGURE 33.1 In humans, the lungs are the external surface where oxygen is exchanged for carbon dioxide.

External Respiration
lungs
CO_2 O_2
CO_2 O_2
tissue cells
Internal Respiration

▶ **33.1 CHECK YOUR PROGRESS**

1. When you hold your breath, which event does not occur during respiration?

HOW LIFE CHANGES   *Application*

33A Evolution of Gas-Exchange Surfaces

It is more difficult for animals to obtain oxygen from water than from air. Water fully saturated with air contains only a fraction of the amount of oxygen that would be present in the same volume of air. Also, water is more dense than air. Therefore, aquatic animals expend more energy carrying out gas exchange than do terrestrial animals. Fishes use as much as 25% of their energy output to respire, while terrestrial mammals use only 1–2% of their energy output for that purpose.

Hydras, which are cnidarians, and planarians, which are flatworms, have a large surface area compared to their size. This makes it possible for most of their cells to exchange gases directly with the environment. In hydras, the outer layer of cells is in contact with the external environment, and the inner layer can exchange gases with the water in the gastrovascular cavity

FIGURE 33A.1 External and internal respiration in a hydra.

(Fig. 33A.1). In planarians, the flattened body permits cells to exchange gases with the external environment.

The tubular shape of annelids (segmented worms) also provides a surface area adequate for external respiration. The earthworm, an annelid, is an example of a terrestrial invertebrate that uses its body surface for respiration because the capillaries come close to the

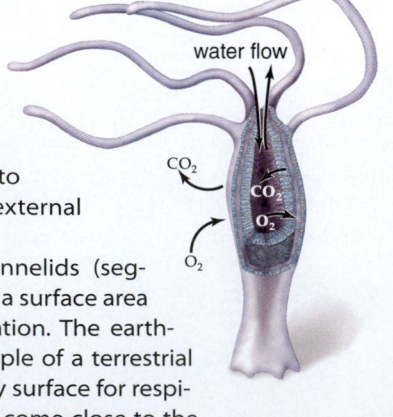

water flow
CO_2
CO_2
O_2
O_2

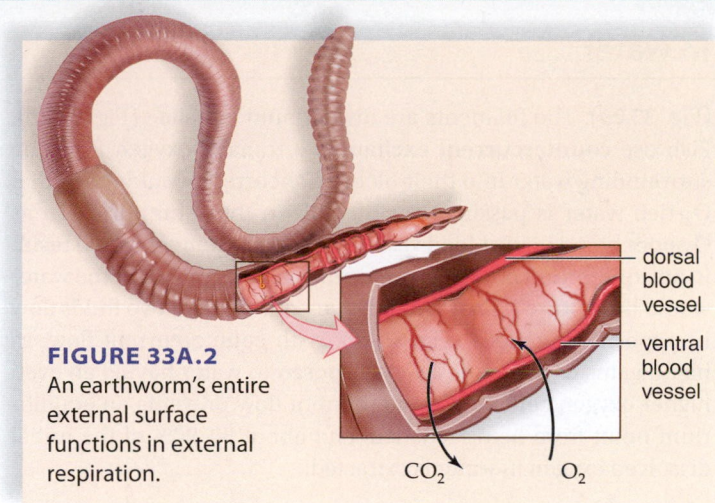

FIGURE 33A.2 An earthworm's entire external surface functions in external respiration.

dorsal blood vessel

ventral blood vessel

CO_2 O_2

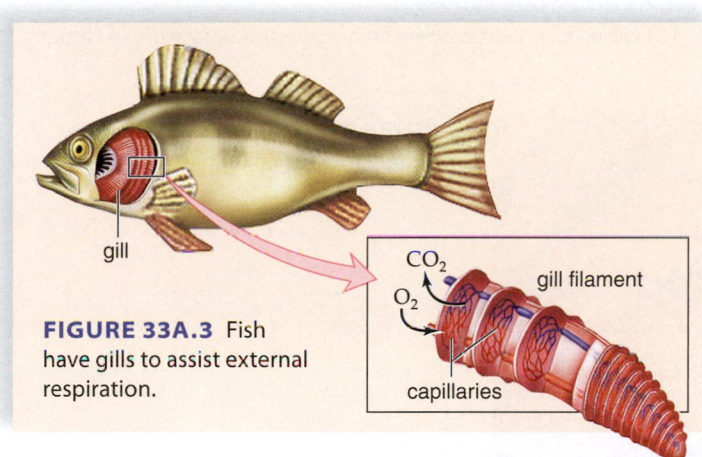

FIGURE 33A.3 Fish have gills to assist external respiration.

gill

CO_2 O_2 gill filament

capillaries

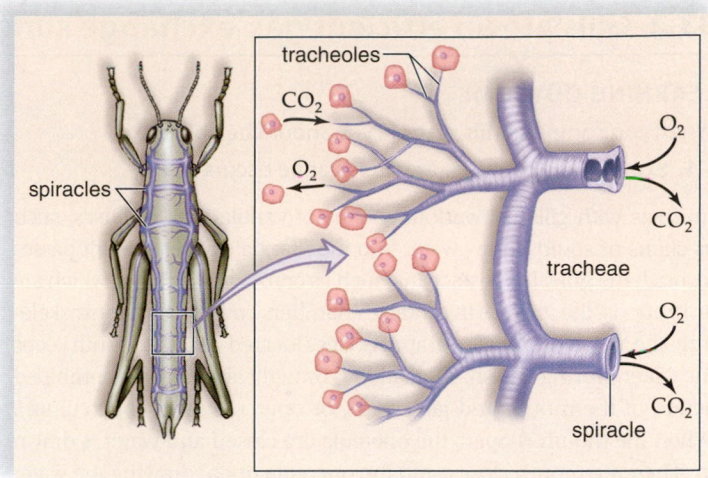

tracheoles

CO_2

spiracles

O_2

O_2

CO_2

tracheae

O_2

CO_2

spiracle

FIGURE 33A.4 Insects have a tracheal system that delivers oxygen directly to their cells.

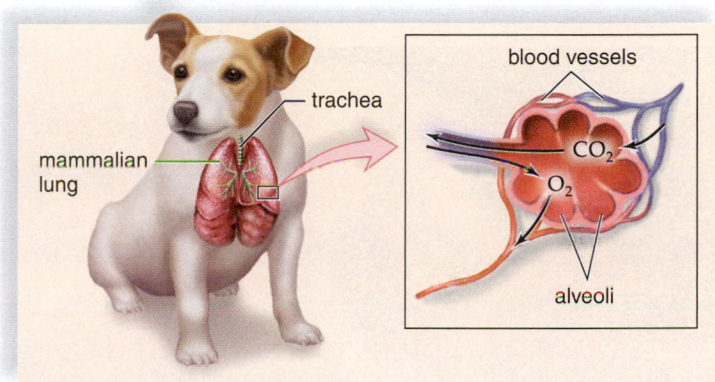

blood vessels

trachea

mammalian lung

CO_2

O_2

alveoli

FIGURE 33A.5 Vertebrates such as mammals have lungs with a large total external respiration surface.

surface (Fig. 33A.2). An earthworm keeps its body surface moist by secreting mucus and by releasing fluids from excretory pores. Further, the worm is behaviorally adapted to remain in damp soil during the day, when the air is driest. In addition to a tubular shape, aquatic polychaete worms have extensions of the body wall called parapodia, which are vascularized and used for gas exchange.

Aquatic invertebrates (e.g., clams and crayfish) and aquatic vertebrates (e.g., fish and tadpoles) have gills that extract oxygen from a watery environment. **Gills** are finely divided, vascularized outgrowths of the body surface or the pharynx (Fig. 33A.3). Various mechanisms are used to pump water across the gills, depending on the organism.

Insects have a system of air tubes called tracheae through which oxygen is delivered directly to the cells without entering the blood (Fig. 33A.4). Fluid occurs at the end of the smallest air tubes. Air sacs located near the muscles of the wings, legs, and abdomen act as bellows to help move the air into the tubes through spiracles, which are external openings.

Terrestrial vertebrates have **lungs,** which are vascularized outgrowths from the lower pharyngeal region (Fig. 33A.5). The tadpoles of frogs live in the water and have gills as external respiratory organs, but adult amphibians possess simple, saclike lungs. Most amphibians respire to some extent through the skin. In much the same way that fishes use their mouths to push water across the gills, amphibians use their mouths to push air into their lungs. Reptiles lack a diaphragm, but all mammals have a diaphragm, in addition

to a rib cage, to expand the lungs so that air comes rushing in (see Fig. 33.4a). The lungs of reptiles are somewhat divided, but the lungs of mammals are elaborately subdivided into small passageways and spaces (Fig. 33A.5). It has been estimated that the total surface area of human lungs is at least 50 times the skin's surface area. Air is a rich source of oxygen compared to water; however, it does have a drying effect on external respiratory surfaces. A human loses about 350 ml of water per day when the air has a relative humidity of only 50%. To keep the lungs from drying out, air is moistened as it moves through the passageways leading to the lungs.

CONSIDER THESE QUESTIONS

1. Exchange surfaces must be kept moist. How is this solved by the animals pictured in Figures 33A.1–5?
2. Can you breathe without involving your mouth? Fishes can't. Explain. Which other animals in Figures 33A.1–5 don't use their mouths to breathe?
3. Why do all animals need to exchange gases in order to live?

connect Explore the concepts through a variety of multimedia

BIOLOGY assets and question types.

www.mcgrawhillconnect.com

LEARNING OUTCOME

When you complete this section, you should be able to

1. Explain how countercurrent exchange occurs in gills.

Animals with gills use various means of ventilation. Molluscs, such as clams or squids, draw water into the mantle cavity, where it passes through the **gills.** In crustaceans, such as crabs and shrimps, which are arthropods, the gills are in thoracic chambers covered by the exoskeleton. The action of specialized appendages located near the mouth keeps the water moving. In fish, ventilation is brought about by the combined action of the mouth and gill covers, or opercula (sing., operculum). When the mouth is open, the opercula are closed and water is drawn in. Then the mouth closes, and the opercula open, drawing the water from the pharynx through the gill slits located between the gill arches.

The gills of bony fishes are outward extensions of the pharynx. On the outside of the gill arches, the gills are composed of filaments (Fig. 33.2a). The filaments are divided into lamellae (Fig. 33.2b). Fish use **countercurrent exchange** to transfer oxygen from the surrounding water into their blood. *Con*current would mean that O_2-rich water is passing over the gills in the same direction as O_2-poor blood in the blood vessels. This arrangement would result in an equilibrium point, at which only half the oxygen in the water would be captured. *Counter*current means that the two fluids flow in opposite directions (Fig. 33.2c). With countercurrent flow, as blood gains oxygen, it is always exposed to water having an even higher oxygen content. Countercurrent flow prevents an equilibrium point from being reached, and about 80–90% of the initial dissolved oxygen in water is extracted.

▶ **33.2 CHECK YOUR PROGRESS**
  1. Explain how countercurrent exchange assists uptake of O_2 in fishes.

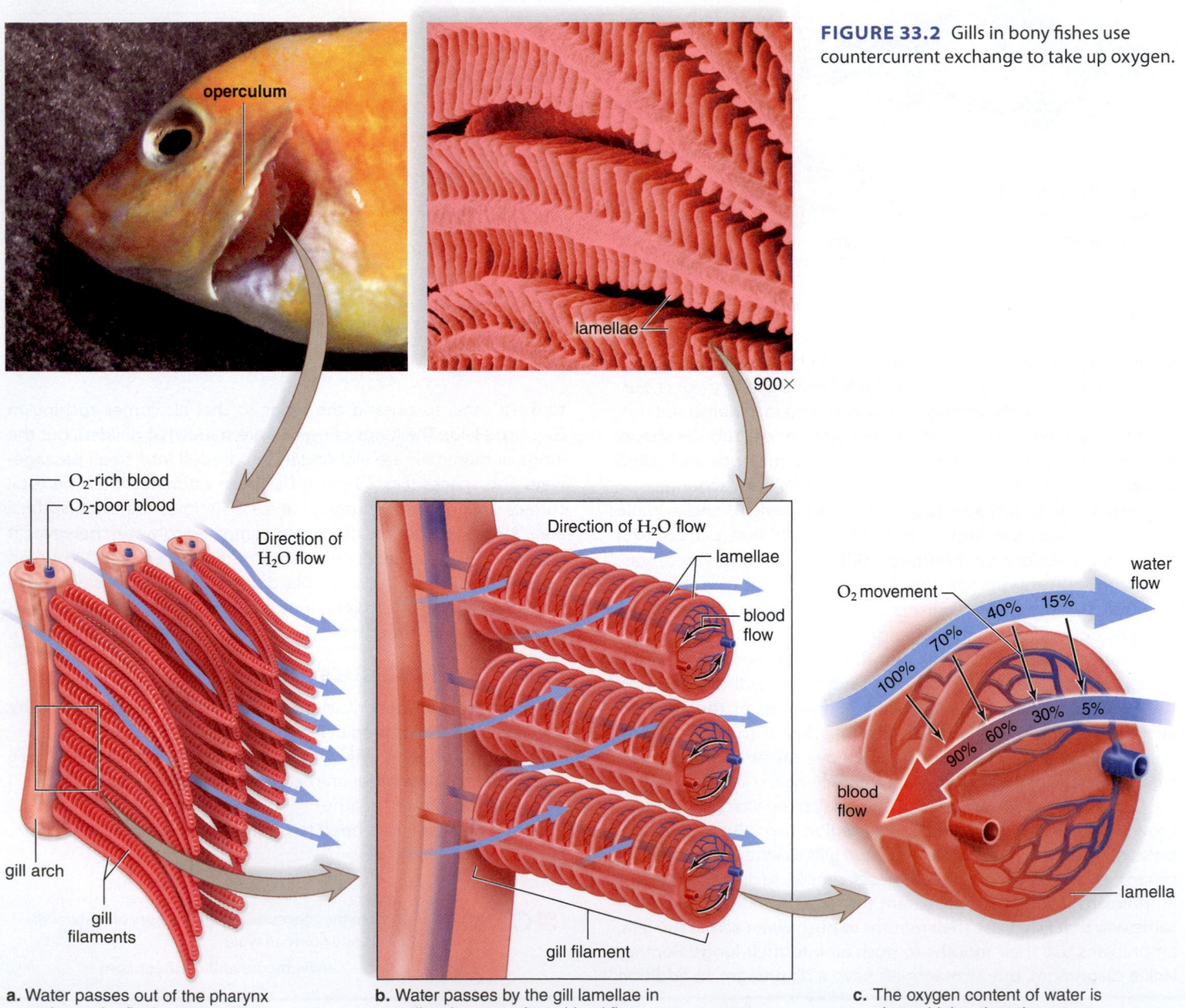

FIGURE 33.2 Gills in bony fishes use countercurrent exchange to take up oxygen.

operculum

lamellae

900×

O_2-rich blood
O_2-poor blood

Direction of H_2O flow

gill arch

gill filaments

a. Water passes out of the pharynx and over the finely divided gills.

Direction of H_2O flow

lamellae

blood flow

gill filament

b. Water passes by the gill lamellae in a direction opposite to blood flow.

O_2 movement

water flow

100% 70% 40% 15%
90% 60% 30% 5%

blood flow

lamella

c. The oxygen content of water is always higher than the oxygen content of the blood.

33.3 The tracheal system in insects permits direct gas exchange

LEARNING OUTCOME

When you complete this section, you should be able to

1. Explain how respiration occurs in insects that have a tracheal system.

Arthropods are coelomate animals, but the coelom is reduced and the internal organs lie within a cavity called the hemocoel because it contains hemolymph, a mixture of blood and lymph. Hemolymph flows freely through the hemocoel, making circulation in arthropods inefficient. Many insects are adapted for flight, and their flight muscles require a steady supply of oxygen. Insects overcome the inefficiency of their blood flow by having a respiratory system that consists of **tracheae,** tiny air tubes that take oxygen directly to the cells (Fig. 33.3). Tracheae have a single layer of cells supported by spiral thickenings of a cuticle lining. The tracheae branch into even smaller tubules called tracheoles, which also branch and rebranch until finally the air tubes are only about 0.1 μm in diameter. There are so many fine tracheoles that almost every cell is near one. Also, the tracheoles indent the plasma membrane so that they terminate close to mitochondria. Therefore, O_2 can flow more directly from a tracheole to mitochondria, where cellular respiration occurs. The tracheae also dispose of CO_2.

The smallest tracheoles are fluid-filled, but the larger tracheae contain only air and open to the outside by way of spiracles (Fig. 33.3). Usually, the spiracle has some sort of closing device that reduces water loss, and this may be why insects have no trouble inhabiting drier climates. It's been suggested that a tracheal system that has no efficient method to improve flow is sufficient only in small insects and tracheae actually expand and contract, thereby drawing air into and out of the system. And in fact, larger insects have still another mechanism to ventilate—keep the air moving in and out—their tracheae. Many larger insects have air sacs, which are thin-walled and flexible, located near major muscles. Contraction of these muscles causes the air sacs to empty, and relaxation causes the air sacs to expand and draw in air. This method is comparable to the way the human lungs expand to draw air into them. Otherwise, a tracheal system consisting of an expansive network of thin-walled tubes seems to be an entirely different mechanism of respiration from those used by other animals. Also, insects lack the efficient circulatory system of birds and mammals that is able to pump O_2-rich blood through arteries to all the cells of the body. This may be why insects remain small, despite Hollywood's attempts to make us think huge insects are possible.

A tracheal system is an adaptation to breathing air, and yet, some insect larval stages and even some adult insects live in the water. In these instances, the tracheae do not receive air by way of spiracles. Instead, diffusion of oxygen across the body wall supplies the tracheae with oxygen. Mayfly and stonefly nymphs have thin extensions of the body wall called tracheal gills where the tracheae are particularly numerous. This is an interesting adaptation because it dramatizes that tracheae function to deliver oxygen in the same manner as vertebrate blood vessels.

> ### 33.3 CHECK YOUR PROGRESS
> 1. Explain how insects accomplish ventilation and gas exchange.

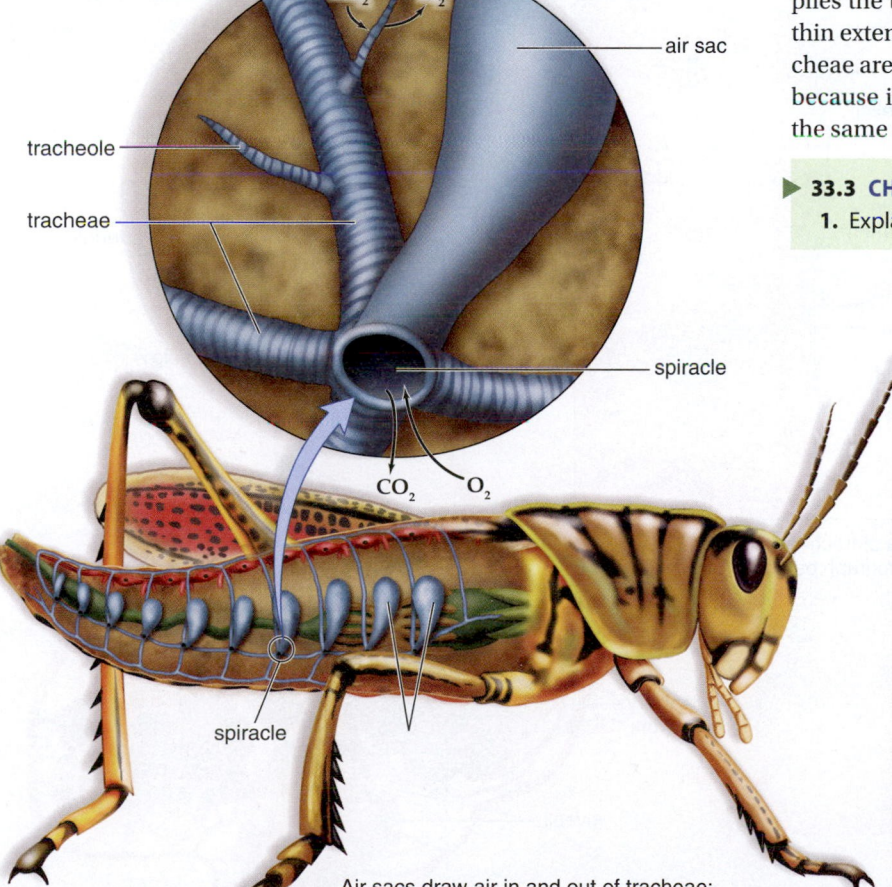

FIGURE 33.3 Tracheae in an insect.

CO₂   O₂

air sac

tracheole

tracheae

spiracle

CO₂   O₂

spiracle

Air sacs draw air in and out of tracheae; tracheoles take O_2 to cells and remove CO_2.

Tracheae and tracheoles are stiffened with bands of chitin.

LEARNING OUTCOME

When you complete this section, you should be able to

1. Name all the organs of the human respiratory system as you trace the path of air from the nose to the alveoli of the lungs.

The human respiratory system includes all of the structures that conduct air in a continuous pathway to and from the lungs (Fig. 33.4a). The lungs lie deep within the thoracic cavity, where they are protected from drying out.

As air moves through air passages to the lungs, it is filtered so that it is free of debris, warmed, and humidified. By the time the air reaches the lungs, it is at body temperature and saturated with water. In the nasal cavities (nose), hairs and cilia act as screening device. In the trachea and the bronchi, cilia beat upward, carrying mucus, dust, and occasional small bits of food that "went down the wrong way" into the throat, where the accumulation may be swallowed or expectorated (Fig. 33.4b).

MP3
Respiratory Structure and Function

Path of Air Air enters the body by way of the nasal cavities or the mouth. The hard and soft palates separate the nasal cavities from the mouth, but the air and food passages cross in the **pharynx** (throat). This may seem inefficient, and there is danger of choking if food accidentally enters the trachea; however, this arrangement does have the advantage of letting you breathe through your mouth in case your nose is plugged up. In addition, it permits more air intake during heavy exercise, when greater gas exchange is required.

Air passes from the pharynx through the **glottis,** an opening into the **larynx,** or voice box. At the edges of the glottis, embedded in mucous membrane, are the **vocal cords.** These flexible and pliable bands of connective tissue vibrate and produce sound when air is expelled past them through the glottis from the larynx.

The larynx can receive air at all times because it is held open by a complex of nine cartilages, among them the Adam's apple. Easily seen in many men, the Adam's apple resembles a small, rounded apple just under the skin in the front of the neck. When food is being swallowed, the larynx rises, and the glottis is closed by a flap of tissue called the **epiglottis.** A backward movement of the soft palate covers the entrance of the nasal passages into the pharynx. The food then enters the esophagus, which lies behind the larynx.

The larynx is part of the **trachea,** which conducts air to the **bronchi.** The bronchi enter the right and left lungs. Branching continues, eventually forming a great number of smaller passages called **bronchioles.** The trachea and the bronchi are held open by a series of C-shaped cartilaginous rings that do not completely meet in the rear, but as the bronchial tubes divide and subdivide, their walls become thinner, and rings of cartilage are no longer present. Each bronchiole terminates in an elongated space enclosed by a multitude of air pockets, or sacs, called **alveoli,** which make up the lungs (Fig. 33.4c). External gas exchange occurs between the air in the alveoli and the blood in the pulmonary capillaries.

▶ **33.4 CHECK YOUR PROGRESS**

1. Describe the path of carbon dioxide as it exits the body.

FIGURE 33.4 The human respiratory tract.

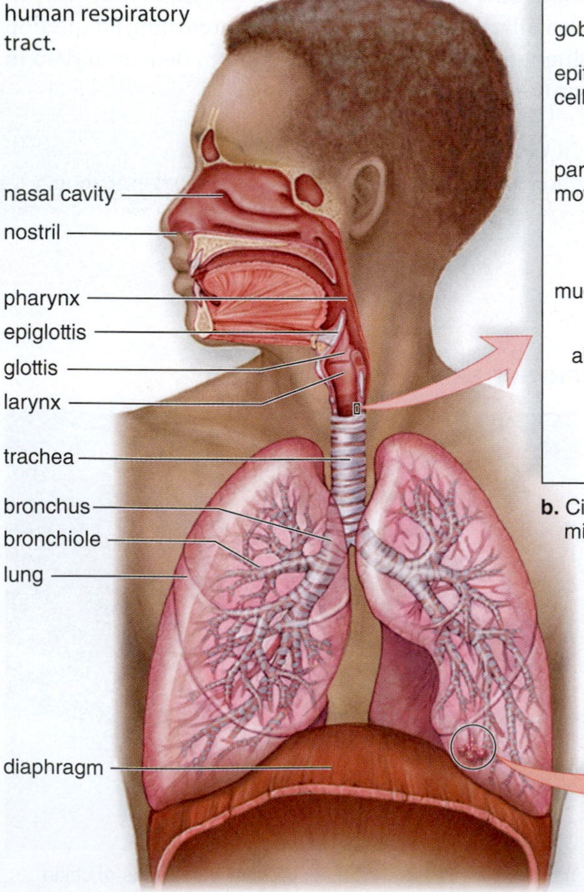

cilia

goblet cell

epithelial cell

particle movement

mucus

air

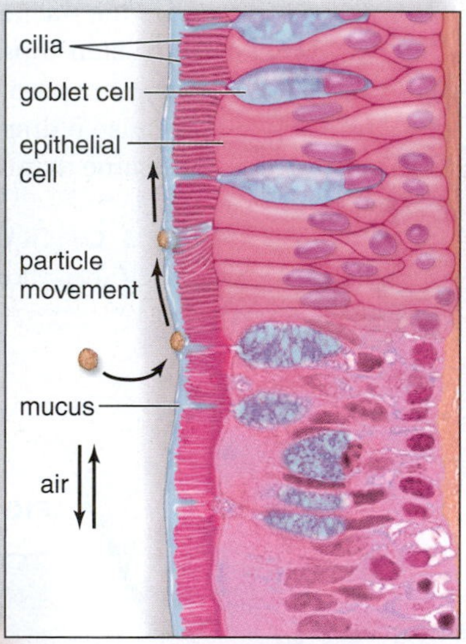

b. Cilia of tracheal lining (art above, micrograph below)

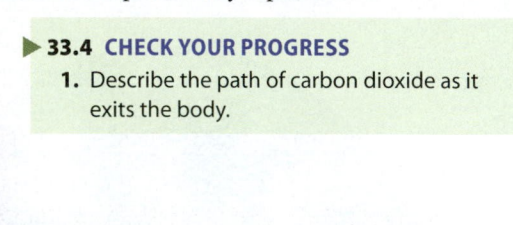

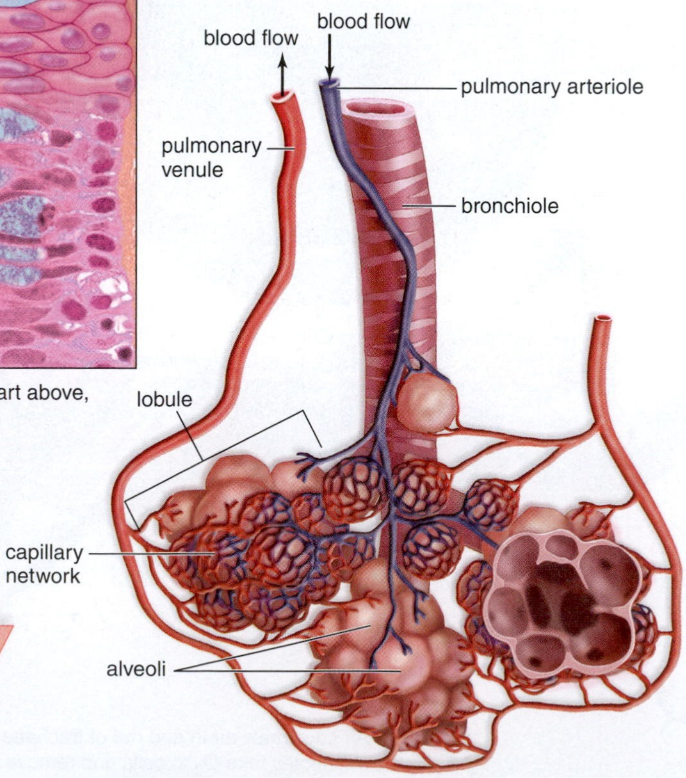

blood flow

blood flow

pulmonary arteriole

pulmonary venule

bronchiole

lobule

capillary network

alveoli

nasal cavity

nostril

pharynx

epiglottis

glottis

larynx

trachea

bronchus

bronchiole

lung

diaphragm

a. The path of air

c. Bronchiole and alveoli

33B   Questions About Tobacco, Smoking, and Health

Is There a Safe Way to Smoke?

No. All cigarettes can damage the human body. Any amount of smoke is dangerous. Cigarettes are perhaps the only legal product whose advertised and intended use—smoking—is harmful to the body and causes cancer.

Is Cigarette Smoking Really Addictive?

Yes. The nicotine in cigarette smoke causes addiction to smoking. Nicotine is an addictive drug (just like heroin and cocaine) for three main reasons: Small amounts make the smoker want to smoke more; smokers usually suffer withdrawal symptoms when they stop; and nicotine can affect the mood and nature of the smoker.

Does Smoking Cause Cancer?

Yes. Tobacco use accounts for about one-third of all cancer deaths in the United States. Smoking causes almost 90% of lung cancers (Fig. 33B). Smoking also causes cancers of the larynx (voice box), oral cavity, pharynx (throat), and esophagus, and contributes to the development of cancers of the bladder, pancreas, cervix, kidney, and stomach. Smoking is also linked to the development of some leukemias.

Why Do Smokers Have "Smoker's Cough"?

Cigarette smoke contains chemicals that irritate the air passages and lungs. When a smoker inhales these substances, the body tries to protect itself by producing mucus and stimulating coughing. Smoke also decreases the sweeping action of cilia that prevent impurities from reaching the lungs.

If you Smoke but Do Not Inhale, Is There Any Danger?

Yes. Wherever smoke touches living cells, it does harm. Even if smokers don't inhale, they are breathing the smoke secondhand and are still at risk for lung cancer. Pipe and cigar smokers, who often do not inhale, are at increased risk for lip, mouth, tongue, and several other cancers.

Does Cigarette Smoking Affect the Heart?

Yes. Smoking increases the risk of heart disease, which is the number-one cause of death in the United States. Smokers who have a heart attack are more likely to die within an hour of the attack than nonsmokers. Cigarette smoke can cause harm to the heart at very low levels, much lower than the amount that causes lung disease.

How Does Smoking Affect Pregnant Women and their Babies?

Smoking during pregnancy is linked to a greater chance of miscarriage, premature delivery, stillbirth, infant death, low birth weight, and sudden infant death syndrome (SIDS). Up to 10% of infant deaths would be prevented if pregnant women did not smoke. Dangerous chemicals in smoke enter the mother's bloodstream and then pass into the baby's body, preventing the baby from getting essential nutrients and oxygen for growth.

What are Some of the Short-Term and Long-Term Effects of Smoking Cigarettes?

Short-term effects include shortness of breath and a nagging cough, diminished ability to smell and taste, premature aging of the skin, and increased risk of sexual impotence in men. Smokers tend to tire easily during physical activity. Long-term effects include many types of cancer, heart disease, aneurysms, bronchitis, emphysema, and stroke. Smoking also contributes to the severity of pneumonia and asthma.

Are Chewing Tobacco and Snuff Safe Alternatives to Cigarette Smoking?

No. The juice from smokeless tobacco is absorbed directly through the lining of the mouth. This creates sores and white patches that often lead to cancer of the mouth. Smokeless tobacco users also greatly increase their risk of other cancers, including those of the pharynx. Other effects of smokeless tobacco include harm to teeth and gums.

CONSIDER THESE QUESTIONS

1. The health costs of smoking are borne by society. Do smokers still have a right to smoke if they want to?
2. Do you have any firsthand knowledge of the damage done by smoking? How so?
3. Do you blame the manufacturers of cigarettes and cigars for encouraging people to smoke, or do you blame friends and family members?

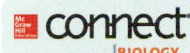 Explore the concepts through a variety of multimedia assets and question types.
www.mcgrawhillconnect.com

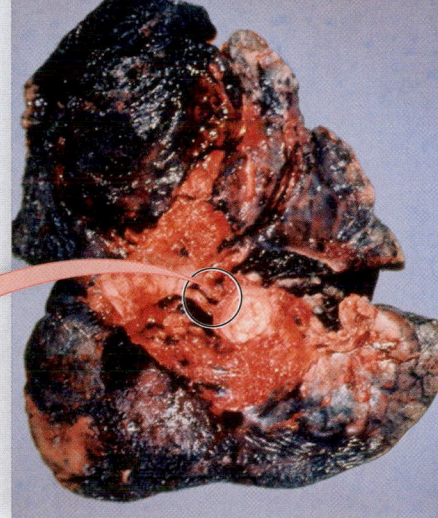

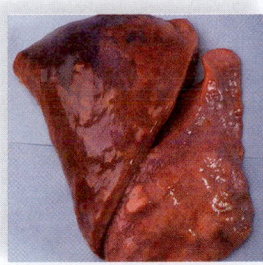

Normal lung of nonsmoker

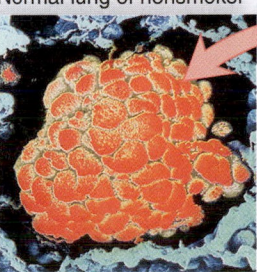

Cancerous tumor   Blackened lung of smoker

FIGURE 33B Compare the appearance of a normal lung (*upper left*) to that of a smoker (*right*). Smokers' lungs are always blackened by pollutants and often develop cancerous tumors (*lower left*).

Ventilation and Transport of Gases

Ventilation in terrestrial vertebrates is the process of getting oxygen to and from the lungs. During external respiration, gas exchange occurs in the lungs and, during internal respiration, gas exchange occurs in the tissues. Oxygen must be transported from the lungs to the tissues and carbon dioxide must be transported from the tissues to the lungs.

33.5 Breathing brings air into and out of the lungs

LEARNING OUTCOME

When you complete this section, you should be able to

1. Compare ventilation in humans with that in birds.

In terrestrial vertebrates, air moves into and out of the respiratory tract. Much as fishes use positive pressure to move water across the gills, amphibians use positive pressure to force air into the respiratory tract. With the mouth and nostrils firmly shut, the floor of the mouth rises and pushes the air into the lungs. Reptiles and mammals use negative pressure to move air into the lungs and positive pressure to move air out of the lungs. During **inspiration** (or inhalation), air moves into the lungs, and during **expiration** (or exhalation), air moves out of the lungs.

Reptiles have jointed ribs that can be raised to expand the lungs, but mammals have both a rib cage and a diaphragm. The **diaphragm** is a horizontal muscle that divides the thoracic cavity (above) from the abdominal cavity (below). During inspiration in mammals, the rib cage moves up and out, and the diaphragm contracts and moves down (Fig. 33.5A). As the thoracic (chest) cavity expands and lung volume increases, air flows into the lungs by way of the bronchi and bronchioles due to decreased air pressure in the thoracic cavity and lungs. This means inspiration is the active phase of breathing in reptiles and mammals.

During expiration in mammals, the rib cage moves down, and the diaphragm relaxes and moves up to its former position (Fig. 33.5B). No muscle contraction is required, making expiration the inactive phase of breathing in reptiles and mammals. During expiration, air flows out as a result of increased pressure in the thoracic cavity and lungs.

We can liken ventilation in reptiles and mammals to the function of a bellows, used to fan a fire. First, the handles of the bellows are pulled apart, decreasing the air pressure inside the bellows. This causes air to automatically flow into the bellows, just as air automatically enters the lungs because the rib cage moves up and out during inspiration. Then, when the handles of the bellows are pushed together, air automatically flows out because the air pressure increases inside the bellows. Similarly, air automatically exits the lungs when the rib cage moves down and in during expiration. The analogy is not exact, however, because no force required for the rib cage to move down, and inspiration is the only active phase of breathing. Forced expiration can occur if we so desire, however.

All terrestrial vertebrates, except birds, use a *tidal ventilation mechanism,* so-called because the air moves in and out by the same route. This means that the lungs of amphibians, reptiles, and mammals are not completely emptied before they are refilled during the next breathing cycle. Therefore, the air entering mixes with

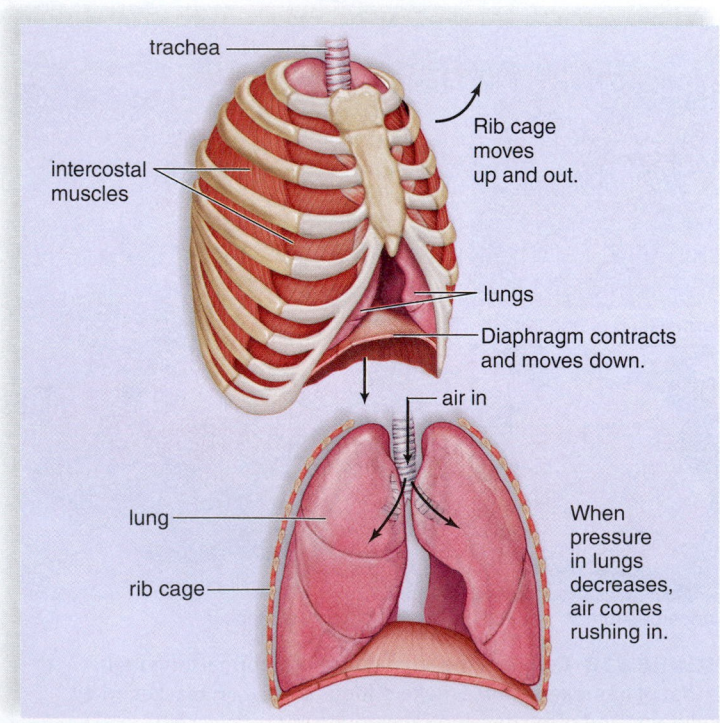

FIGURE 33.5A Inspiration.

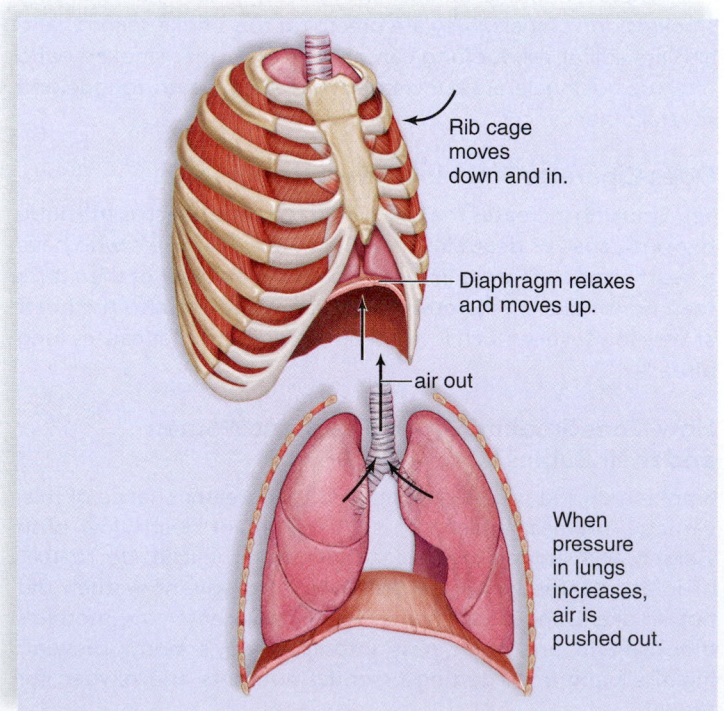

FIGURE 33.5B Expiration.

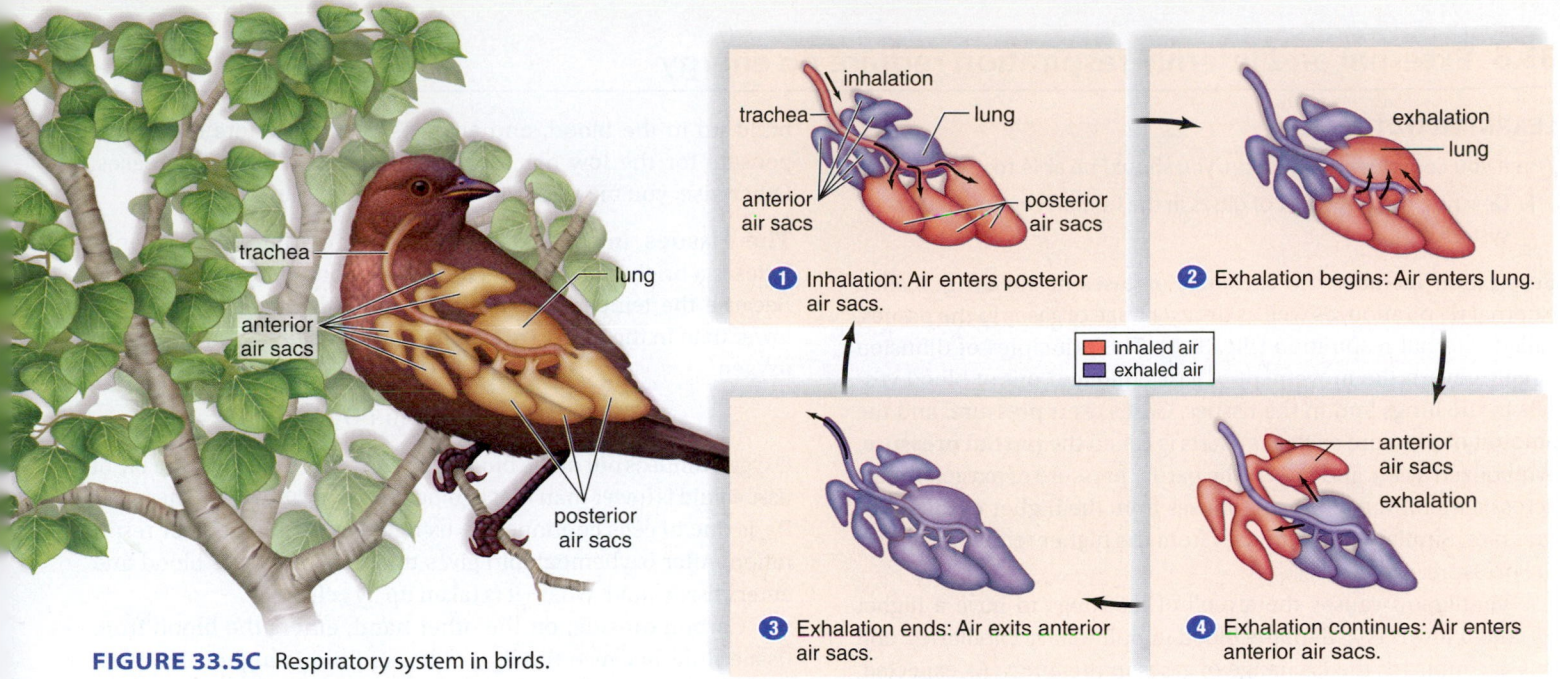

FIGURE 33.5C Respiratory system in birds.

1 Inhalation: Air enters posterior air sacs.

2 Exhalation begins: Air enters lung.

3 Exhalation ends: Air exits anterior air sacs.

4 Exhalation continues: Air enters anterior air sacs.

inhaled air
exhaled air

used air remaining in the lungs. While this does help conserve water, it also decreases gas-exchange efficiency. In contrast, birds use a *one-way ventilation mechanism* (Fig. 33.5C). 1 Incoming air is carried past the lungs by a trachea, which takes it to a set of posterior air sacs. 2 The air then passes forward through the finely divided lungs into a set of 3 anterior air sacs. 4 From here, it is finally expelled. Countercurrent exchange occurs because the air in lungs is flowing in the opposite direction to the flow of blood in the pulmonary capillaries. Notice that fresh, O_2-rich air never mixes with used air in the lungs of birds, thereby greatly improving gas-exchange efficiency.

Control of Breathing Normally, adults have a breathing rate of 12–20 ventilations per minute. The rhythm of ventilation is controlled by a **respiratory center** in the medulla oblongata of the brain. The respiratory center automatically sends out impulses by way of spinal nerves to the intercostal muscles of the rib cage and to the diaphragm (phrenic nerve) (Fig. 33.5D). Now inspiration occurs. When the respiratory center stops sending neuronal signals to the diaphragm and the rib cage, expiration occurs.

Although the respiratory center automatically controls the rate and depth of breathing, its activity can also be influenced by nervous input and chemical input. Following forced inhalation, stretch receptors in the alveolar walls initiate inhibitory nerve impulses that travel from the inflated lungs to the respiratory center. This stops the respiratory center from sending out nerve impulses.

The respiratory center is directly sensitive to the levels of hydrogen ions (H^+) in the blood. However, when carbon dioxide (CO_2) enters the blood, it reacts with water and releases hydrogen ions. In this way, CO_2 participates in regulating the breathing rate. When hydrogen ions rise in the blood and the pH decreases, the respiratory center increases the rate and depth of breathing. The chemoreceptors in the **carotid bodies,** located in the carotid arteries, and in the **aortic bodies,** located in the aorta, stimulate

the respiratory center during intense exercise due to a reduction in pH and also if and when arterial oxygen decreases to 50% of normal.

MP3
Control of Breathing

▶ **33.5 CHECK YOUR PROGRESS**

1. Explain the statement, "Inspiration is the only active phase of breathing in humans."
2. Discuss what causes the human respiratory center to increase the rate and depth of breathing.

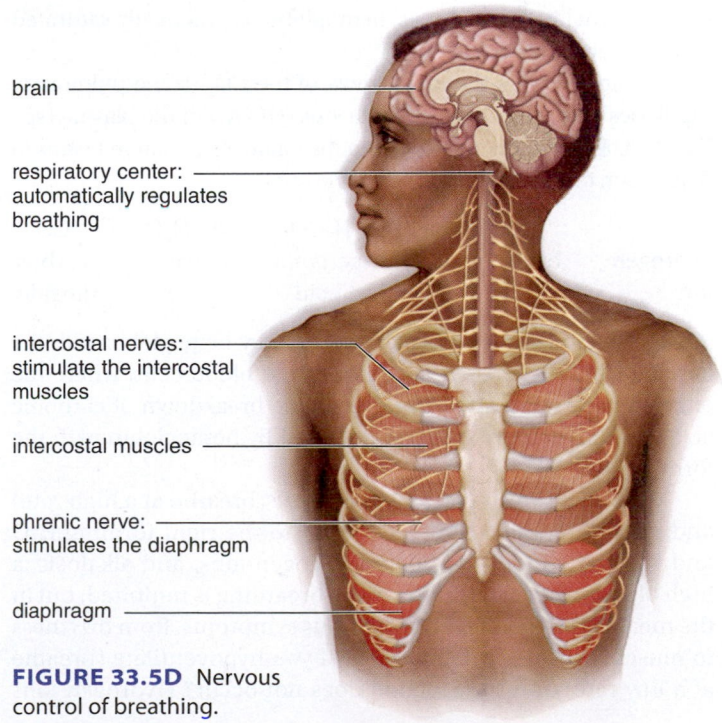

brain

respiratory center: automatically regulates breathing

intercostal nerves: stimulate the intercostal muscles

intercostal muscles

phrenic nerve: stimulates the diaphragm

diaphragm

FIGURE 33.5D Nervous control of breathing.

LEARNING OUTCOME

When you complete this section, you should be able to

1. Describe the transport of gases in the human cardiovascular system.

Respiration includes the exchange of gases in our lungs, called external respiration, as well as the exchange of gases in the tissues, called internal respiration (Fig. 33.6). The principles of diffusion largely govern the movement of gases into and out of blood vessels in the lungs and in the tissues. Gases exert pressure, and the amount of pressure each gas exerts is called the **partial pressure,** symbolized as P_{O_2} and P_{CO_2}. If the partial pressure of oxygen differs across a membrane, oxygen diffuses from the higher to the lower pressure. Similarly, CO_2 diffuses from the higher to the lower partial pressure.

Ventilation causes the alveoli of the lungs to have a higher P_{O_2} and a lower P_{CO_2} than the blood in pulmonary capillaries, and this accounts for the exchange of gases in the lungs. As expected, then, O_2 moves from the alveoli into the blood and CO_2 moves out of the blood into the alveoli. When blood reaches the tissues, cellular respiration in cells causes the tissue fluid to have a lower P_{O_2} and a higher P_{CO_2} than the blood in the systemic capillaries, and this accounts for the exchange of gases in the tissues. As predicted, O_2 moves from the blood into tissue fluid and then into cells; CO_2 moves from tissue fluid into the blood and is then transported to the lungs.

MP3
Gas Exchange

Animation
Changes in the Partial Pressure of Oxygen and Carbon Dioxide

The Lungs In the lungs, most oxygen entering the pulmonary capillaries from the alveoli combines with **hemoglobin (Hb)** in red blood cells (RBCs) to form **oxyhemoglobin** (see Fig. 33.6 ❶):

$$Hb + O_2 \longrightarrow HbO_2$$

deoxyhemoglobin   oxygen   oxyhemoglobin

At the normal P_{O_2} in the lungs, hemoglobin is practically saturated with oxygen.

As blood enters the lungs, most of the CO_2 in the pulmonary capillaries is carried as bicarbonate ions (HCO_3^-) in the plasma (see Fig. 33.6, ❷). As the free CO_2 from the following equation begins to diffuse out, this reaction is driven to the right:

$$H^+ + HCO_3^- \longrightarrow H_2CO_3 \longrightarrow H_2O + CO_2$$

hydrogen   bicarbonate   carbonic   water   carbon
ion   ion   acid   dioxide

Some of these hydrogen ions are donated by hemoglobin with the formula HbH^+. The reaction occurs in red blood cells, where the enzyme **carbonic anhydrase** speeds the breakdown of carbonic acid. A small amount of CO_2 is released by hemoglobin with the formula $HbCO_2$.

What happens if you hyperventilate (breathe at a high rate) and, therefore, push the reaction far to the right forming H_2O and CO_2? The blood has fewer hydrogen ions, and alkalosis, a high blood pH, results. In that case, breathing is inhibited, but in the meantime, you may suffer various symptoms, from dizziness to muscle spasms. What happens if you hypoventilate (breathe at a low rate) and the reaction does not occur? Hydrogen ions

build up in the blood, and acidosis occurs. Buffers may compensate for the low pH, and breathing most likely increases. Otherwise, you may become comatose and die.

The Tissues In the tissues, blood entering the systemic capillaries is a bright red color because RBCs contain oxyhemoglobin. Because the temperature in the tissues is higher and the pH is lower than in the lungs, oxyhemoglobin has a tendency to give up oxygen:

$$HbO_2 \longrightarrow Hb + O_2$$

Oxygen diffuses out of the blood into tissue fluid because the P_{O_2} of tissue fluid is lower than that of blood (see Fig. 33.6, ❸). The lower P_{O_2} is due to cells continuously using up oxygen in cellular respiration. After oxyhemoglobin gives up O_2, it leaves the blood and enters tissue fluid, where it is taken up by cells.

Carbon dioxide, on the other hand, enters the blood from tissue fluid because the P_{CO_2} of tissue fluid is higher than that of the blood. Carbon dioxide, produced continuously by cells, collects in tissue fluid. After CO_2 diffuses into the blood, it enters the red blood cells (see Fig. 33.6, ❹). A small amount of CO_2 combines with the protein portion of hemoglobin to form **carbaminohemoglobin** ($HbCO_2$). Most of the CO_2, however, is transported in the form of the **bicarbonate ion** (HCO_3^-). First, CO_2 combines with water, forming carbonic acid, and then this dissociates to a hydrogen ion (H^+) and HCO_3^-:

$$CO_2 + H_2O \longrightarrow H_2CO_3 \longrightarrow H^+ + HCO_3^-$$

carbon   water   carbonic   hydrogen   bicarbonate
dioxide   acid   ion   ion

Carbonic anhydrase also speeds this reaction. The HCO_3^- diffuses out of the red blood cells to be carried in the plasma.

The release of H^+ from this reaction could drastically change the pH of the blood. However, the H^+ is absorbed by the globin portions of hemoglobin. Hemoglobin that has combined with H^+ is called **reduced hemoglobin** and has the formula HbH^+. HbH^+ plays a vital role in maintaining the normal pH of the blood. Blood that leaves the systemic capillaries is a dark maroon color because red blood cells contain reduced hemoglobin.

Hemoglobin Each hemoglobin molecule contains four polypeptide chains, and each chain is folded around an iron-containing group called **heme.** It is actually the iron that forms a loose association with oxygen. Since there are about 250 million hemoglobin molecules in each red blood cell, each red blood cell is capable of carrying at least 1 billion molecules of oxygen. Unfortunately, carbon monoxide (CO) is an air pollutant that combines with hemoglobin more readily than does O_2, making Hb unavailable for O_2 transport. This is the reason that some homes are equipped with CO detectors.

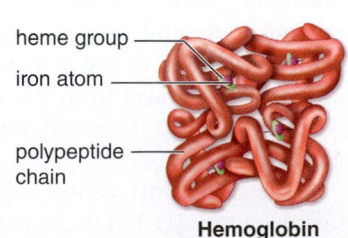

heme group
iron atom
polypeptide chain

Hemoglobin

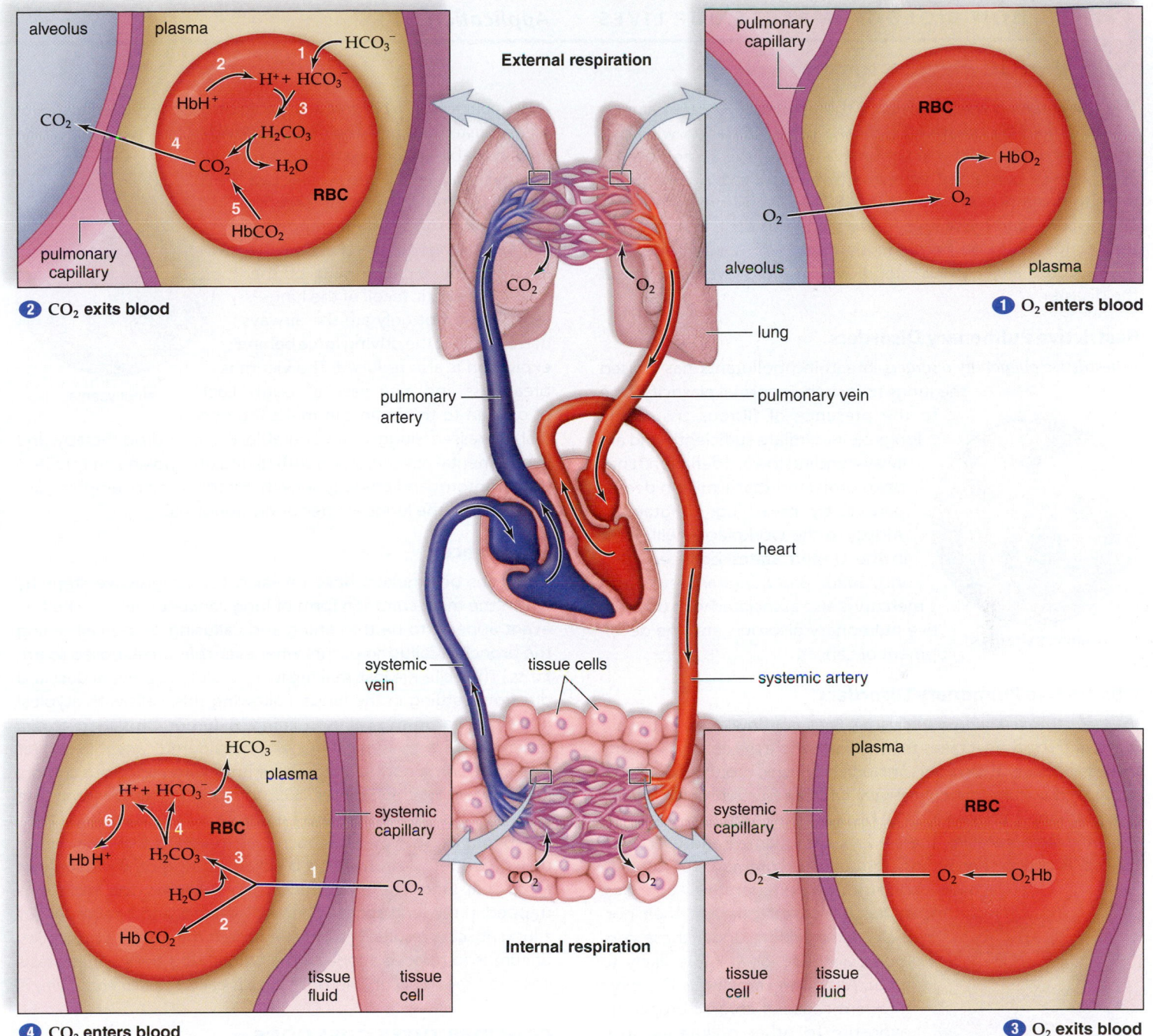

External respiration

alveolus **plasma**

HCO_3^-

1

2 $H^+ + HCO_3^-$

HbH^+ 3

CO_2 H_2CO_3

4 CO_2 H_2O

RBC

5

$HbCO_2$

pulmonary capillary

2 CO_2 **exits blood**

pulmonary capillary

RBC

HbO_2

O_2

O_2

alveolus **plasma**

1 O_2 **enters blood**

lung

pulmonary artery

pulmonary vein

CO_2 O_2

heart

systemic vein

tissue cells

systemic artery

plasma

HCO_3^-

plasma

$H^+ + HCO_3^-$ 5

6 4 **RBC**

HbH^+ H_2CO_3 3

H_2O 1 CO_2

2

$Hb CO_2$

systemic capillary

tissue fluid tissue cell

4 CO_2 **enters blood**

systemic capillary

Internal respiration

CO_2 O_2

plasma

RBC

O_2 O_2 O_2Hb

tissue cell tissue fluid

3 O_2 **exits blood**

FIGURE 33.6 During external respiration (*top*), CO_2 exits blood and enters an alveolus and O_2 enters blood from an alveolus. During internal respiration (*bottom*), O_2 exits blood and enters a cell while CO_2 enters blood and is transported away as a bicarbonate ion (HCO_3^-).

Hemoglobin is well adapted to being the oxygen carrier in mammals. The lungs have a lower acidity and a lower temperature than do the tissues. This environment is the optimum environment for the reaction $Hb + O_2 \longrightarrow HbO_2$. If hemoglobin didn't unload its oxygen it would be useless but the tissues have an environment suitable to bringing this about. The higher acidity and warmer temperature of the tissues causes hemoglobin to release its oxygen.

Hemoglobin has other attributes that serve mammals well. As you know, in the tissues, the blood tends to become acidic after it picks up CO_2 but hemoglobin helps prevent this from happening. When the heme portion of hemoglobin is not carrying oxygen, the globin (protein) portion combines with H^+ (forming HbH^+) and this helps buffer the blood. Hemoglobin will even combine with CO_2 (forming $HbCO_2$). Most of the CO_2, as stated, becomes the bicarbonate ion in red blood cells although the bicarbonate ion (HCO_3^-) is carried in the plasma.

Animation
Gas Exchange During Respiration

▶ **33.6 CHECK YOUR PROGRESS**
1. Explain how CO_2 is carried in the bloodstream to the lungs.

33C Respiratory Disorders

The self-cleaning mechanisms of the respiratory system can be overwhelmed by breathing polluted urban air laden with particles. The dust particles at Ground Zero after the September 11, 2001, attack contained asbestos from fire-proofing materials, mercury from fluorescent lightbulbs, and lead from computers. The rescuers and bystanders breathed in this dust, unaware that their lungs could be permanently damaged. A number of disorders could afflict thousands of New Yorkers even years after the disaster.

Restrictive Pulmonary Disorders

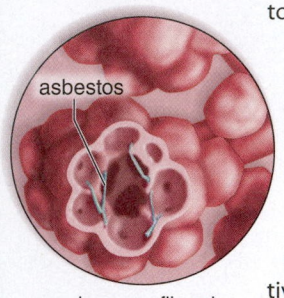

asbestos

pulmonary fibrosis

In *restrictive pulmonary disorders,* breathing pollutants has caused the lungs to lose their normal elasticity. Due to the presence of fibrous tissue, the lungs cannot inflate sufficiently and are always tending toward deflation. It has been projected that 1 million deaths caused by breathing pollutants—mostly in the workplace—will occur in the United States between 1990 and 2020. Breathing asbestos and mercury is also associated with obstructive pulmonary disorders and the development of cancer.

Obstructive Pulmonary Disorders

In *obstructive pulmonary disorders,* air does not flow freely in the airways, and the time it takes to inhale or exhale maximally is greatly increased. Several disorders, including chronic bronchitis, emphysema, and asthma, are collectively referred to as chronic obstructive pulmonary disease (COPD) because they tend to recur.

In *chronic bronchitis,* the airways are inflamed and filled with mucus. A cough that brings up mucus is common. The bronchi have undergone degenerative changes, including the loss of cilia and their normal cleansing action. Under these conditions, an infection is more likely to occur. Although smoking is the most frequent cause of chronic bronchitis, exposure to other pollutants, such as the dust at Ground Zero, can also cause this condition.

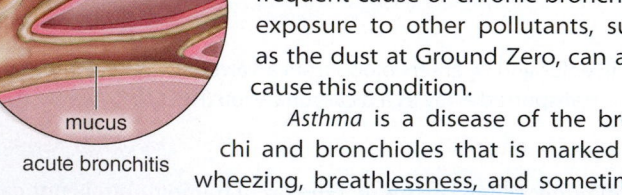

mucus

acute bronchitis

Asthma is a disease of the bronchi and bronchioles that is marked by wheezing, breathlessness, and sometimes a cough and expectoration of mucus. The airways are unusually sensitive to specific irritants, which can include a wide range of allergens such as pollen, animal dander, dust, tobacco smoke, and industrial fumes. When exposed to the irritant, the smooth muscle in the bronchioles undergoes spasms. Most asthma patients have some degree of bronchial inflammation that further reduces the diameter. Asthma is not curable, but it is treatable. Special inhalers can control the inflammation and possibly prevent an attack, while other types of inhalers can stop the muscle spasms should an attack occur.

Emphysema is a chronic, incurable disorder in which the alveoli are distended and their walls damaged, so that the surface area available for gas exchange is reduced. Emphysema, which we know can be caused by smoking, is often preceded by chronic bronchitis. Air trapped in the lungs leads to alveolar damage and noticeable ballooning of the chest. The elastic recoil of the lungs is reduced, so not only are the airways narrowed, but the driving force behind expiration is also reduced. The victim is breathless and may have a cough. Lack of oxygen to the brain can make the person feel depressed, sluggish, and irritable. Exercise, drug therapy, and supplemental oxygen, along with giving up smoking, may relieve the symptoms and possibly slow the progression of emphysema. Upon death, the lungs are decidedly abnormal.

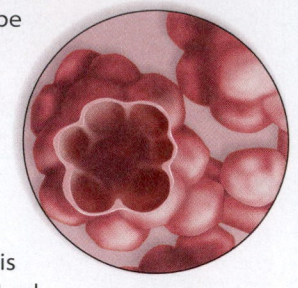

emphysema

Lung Cancer

Autopsies on smokers have revealed the progressive steps by which the most common form of lung cancer develops. The first event appears to be thickening and callusing of the cells lining the bronchi. (Callusing occurs whenever cells are exposed to irritants.) Then cilia are lost, making it impossible to prevent dust and dirt from settling in the lungs. Following this, cells with atypical nuclei appear in the callused lining. A tumor consisting of disordered cells with atypical nuclei is considered cancer in situ (at one location). A normal lung and a lung with cancerous tumors are shown in Figure 33B, page 659. A final step occurs when some of these cells break loose and penetrate other tissues, a process called metastasis. Now, the cancer has spread. The original tumor may grow until a bronchus is blocked, cutting off the supply of air to that lung. The entire lung then collapses, the secretions trapped in the lung spaces become infected, and pneumonia or a lung abscess results. The only treatment that offers a possibility of cure is to remove a lobe or the whole lung before metastasis has had time to occur. This operation is called a *pneumonectomy*.

CONSIDER THESE QUESTIONS

1. In what ways is smoking cigarettes and cigars the same as breathing 9/11 dust?
2. Create a scenario by which an organic chemical from cigarette smoke enters cells and affects a signaling pathway leading to cancer (see Fig. 12.8A).
3. Make a list of ordinary everyday activities a person with emphysema would have trouble performing.

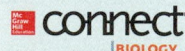

connect
| BIOLOGY

Explore the concepts through a variety of multimedia assets and question types.

www.mcgrawhillconnect.com

CONNECTING THE CONCEPTS

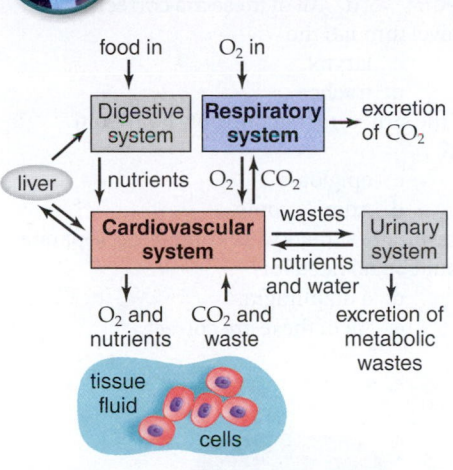

In mammals, the respiratory system consists of the respiratory tract with the nasal passages (or mouth) at one end and the lungs at the other end. Inspired air is 20% O_2 and 0.04% CO_2, while expired air is about 14% O_2 and 6% CO_2. Gas exchange in the lungs accounts for the difference in composition of inspired and expired air.

In the lungs, oxygen is absorbed into the bloodstream, and from there it is transported by red blood cells to the capillaries, where it exits and enters tissue fluid. On the other hand, carbon dioxide enters capillaries at the tissues and is transported largely as the bicarbonate ion to the lungs, where it is converted to carbon dioxide and exits the body. Diffusion alone accounts for gas exchange in the lungs, called external respiration, and for gas exchange in the tissues, called internal respiration. Energy is not needed, as gases follow their concentration gradients according to their partial pressures.

Internal gas exchange is extremely critical because cells use oxygen and release carbon dioxide as a result of cellular respiration, the process that generates ATP in cells. External gas exchange has the benefit of helping to keep the pH of the blood constant as required for homeostasis. When carbon dioxide exits, the blood pH returns to normal. In Chapter 34, we consider the contribution of the kidneys to homeostasis.

ANALYZE AND EVALUATE

1. The body's cells are dependent on a surface for exchanging gases with the external environment. Explain.
2. Excretion is ridding the body of a metabolic waste. How is the respiratory system involved in excretion? The digestive system doesn't excrete. Why not?
3. Most large animals have a respiratory pigment such as hemoglobin. What functions does hemoglobin perform for us?

SUMMARIZE

Gas-Exchange Surfaces

33.1 Respiration involves several steps

- The purpose of **respiration** is to deliver oxygen for cellular respiration. **Ventilation** is breathing. **External respiration** is gas exchange between the air and the blood in the lungs. **Internal respiration** is gas exchange between the blood and the tissue fluid.

33.2 Gills are an efficient gas-exchange surface in water

- In clams, squids, and fishes, water moves across **gills.** Fish use **countercurrent exchange** to transfer oxygen efficiently from water into their blood.

33.3 The tracheal system in insects permits direct gas exchange

- The **tracheae** branch into smaller tracheoles, which also branch until the smallest fluid-filled tracheoles indent the plasma membrane to terminate close to mitochondria. In some insects, ventilation involves air sacs that expand to draw in air.

33.4 The mammalian respiratory system utilizes lungs as a gas-exchange surface

- Air is warmed and humidified in the nose. The air and food passages cross in the **pharynx,** allowing us to breathe through

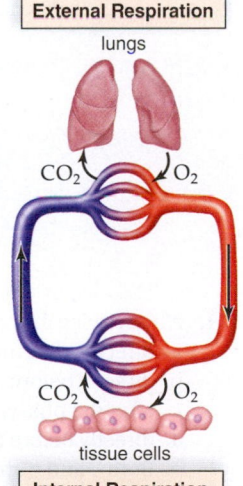

our mouths. The **glottis** when not covered by the **epiglottis** opens into the **larynx** (where **vocal cords** are located), and the **trachea** connects the larynx to the **bronchi.** Two bronchi lead to the right and left lungs and branch into **bronchioles.** Bronchioles end in **alveoli,** which make up the lungs in all terrestrial mammals.

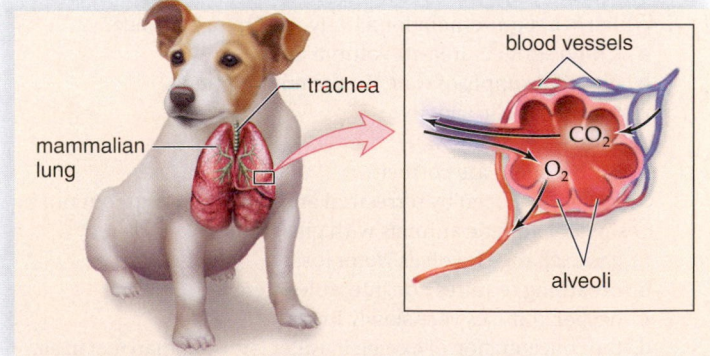

Ventilation and Transport of Gases

33.5 Breathing brings air into and out of the lungs

- In humans, **inspiration** consists of muscle contractions that lower the **diaphragm** and raise the ribs, followed by negative pressure that causes air to flow in. During **expiration,** the rib and diaphragm muscles relax, and air flows out due to increased pressure.

- All vertebrates but birds have a tidal ventilation system: New air entering the lungs mixes with used air in the lungs. Birds have a more efficient one-way ventilation system: Air moves from posterior air sacs through the lungs and anterior air sacs before exiting.
- Breathing is automatic. The **respiratory center** stimulates the intercostal muscles when inspiration occurs. Lack of stimulation causes expiration. A reduction in pH stimulates the respiratory center as well as certain chemoreceptors (**carotid bodies** and **aortic bodies**). Thereafter, the breathing rate increases.

33.6 External and internal respiration require no energy

- O_2 and CO_2 exert a pressure, called **partial pressure,** symbolized as P_{O_2} and P_{CO_2}. In the lungs, P_{CO_2} is higher in the blood than in the lungs, and P_{O_2} is higher in the lungs; CO_2 diffuses out of the blood into the lungs, and O_2 diffuses out of the lungs into the blood. In the tissues, P_{O_2} is higher in the blood, and P_{CO_2} is higher in the tissues; therefore, O_2 diffuses out of the blood, and CO_2 diffuses into the blood.
- In the lungs, the **heme** portion of **hemoglobin (Hb)** in red blood cells contains iron, which combines with O_2, forming HbO_2 (**oxyhemoglobin**). Carbon dioxide is carried as the **bicarbonate ion** (HCO_3^-). HCO_3^- combines with H^+ to form H_2CO_3, which breaks down to water and CO_2. This reaction is speeded by the enzyme **carbonic anhydrase.** CO_2 exits blood.
- In the tissues, HbO_2 gives up its O_2, which enters tissue cells. Hb combines with some CO_2 (**carbaminohemoglobin**). CO_2 combines with water, forming $H_2CO_3^-$, which becomes H^+ and HCO_3^-. Hb combines with some H^+ (**reduced hemoglobin**).

TEST YOURSELF

Gas-Exchange Surfaces

1. Internal respiration refers to
   a. the exchange of gases between the air and the blood in the lungs.
   b. the movement of air into the lungs.
   c. the exchange of gases between the blood and tissue fluid.
   d. cellular respiration, resulting in the production of ATP.
2. Which is a requirement for a body respiratory surface?
   a. a high surface-area-to-volume ratio
   b. plentiful supply of water in the environment
   c. low metabolic activity
   d. thin, moist body wall
   e. Both a and d are correct.
3. One problem faced by terrestrial animals with lungs, but not by freshwater aquatic animals with gills, is that
   a. gas exchange involves water loss.
   b. breathing requires considerable energy.
   c. oxygen diffuses very slowly in air.
   d. the concentration of oxygen in water is greater than that in air.
   e. All of these are correct.
4. Countercurrent exchange means that oxygen-_____ blood in the gills flows in the _____ direction as oxygen-rich water passing over the gills.
   a. rich, same   c. poor, same
   b. rich, opposite   d. poor, opposite

5. In which animal is the circulatory system not involved in gas transport?
   a. mouse   d. sparrow
   b. dragonfly   e. human
   c. trout
6. How is inhaled air modified before it reaches the lungs?
   a. It must be humidified.   c. It must be filtered.
   b. It must be warmed.   d. All of these are correct.
7. Food and air both travel through the
   a. lungs.   c. larynx.
   b. pharynx.   d. trachea.
8. What is the name of the structure that prevents food from entering the trachea?
   a. glottis   c. epiglottis
   b. septum   d. Adam's apple
9. If the digestive and respiratory tracts were completely separate in humans, there would be no need for
   a. swallowing.   d. a diaphragm.
   b. a nose.   e. All of these are correct.
   c. an epiglottis.
10. Label this diagram:

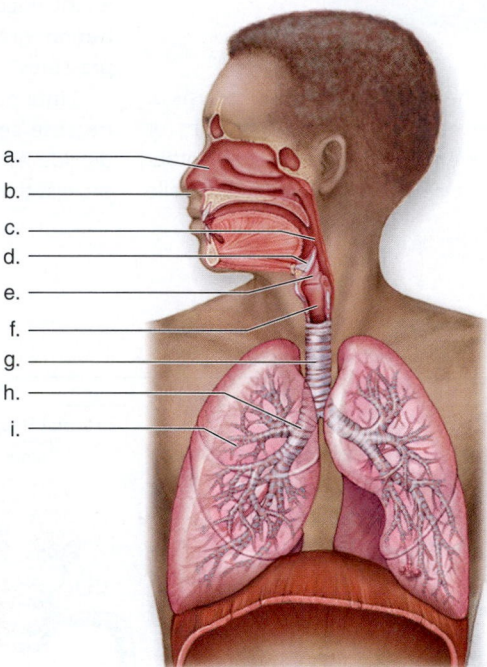

a. _____
b. _____
c. _____
d. _____
e. _____
f. _____
g. _____
h. _____
i. _____

11. When tracing the path of air in humans, you would list the trachea
    a. directly after the nose.
    b. directly before the bronchi.
    c. after the pharynx.
    d. directly before the lungs.
    e. Both a and c are correct.
12. Which of these statements is anatomically incorrect?
    a. The nose has two nasal cavities.
    b. The pharynx connects the nasal cavity and the mouth to the larynx.
    c. The larynx contains the vocal cords.
    d. The trachea enters the lungs.
    e. The lungs contain many alveoli.

For questions 13–17, match each description with a structure in the key.

KEY:

    a. pharynx   c. larynx   e. bronchi
    b. glottis   d. trachea   f. bronchioles

13. Branched tubes that lead from the bronchi to the alveoli
14. Reinforced tube that connects the larynx with the bronchi
15. Chamber behind the oral cavity and between the nasal cavities and larynx
16. Opening into the larynx
17. Divisions of the trachea that enter the lungs
18. **THINKING CONCEPTUALLY** Knowing that flight requires much oxygen, create a natural selection for flight scenario to explain why birds have such an efficient means of gas exchange.

Ventilation and Transport of Gases

19. Which animal breathes by positive pressure?
    a. fish   d. frog
    b. human   e. planarian
    c. bird
20. Label these drawings to describe the movement of the rib cage and diaphragm during inspiration and expiration:

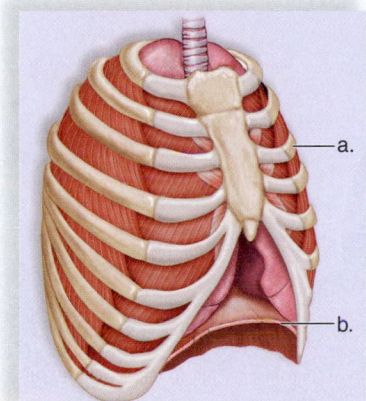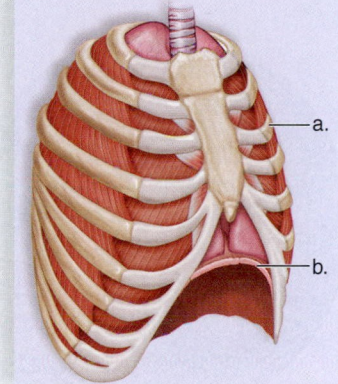

21. Air enters the human lungs because
    a. atmospheric pressure is lower than the pressure inside the lungs.
    b. atmospheric pressure is greater than the pressure inside the lungs.
    c. although the pressures are the same inside and outside, the partial pressure of oxygen is lower within the lungs.
    d. the residual air in the lungs causes the partial pressure of oxygen to be lower than it is outside.
22. Which of these statements correctly goes with expiration rather than inspiration?
    a. The rib cage moves up and out.
    b. The diaphragm relaxes and moves up.
    c. Pressure in the lungs decreases, and air comes rushing out.
    d. The diaphragm contracts and lowers.
23. Birds have more efficient lungs than humans because the flow of air
    a. is the same during both inspiration and expiration.
    b. never backs up as it does in human lungs.
    c. is not hindered by a larynx.
    d. enters birds' bones.
24. In humans, the respiratory control center
    a. is located in the medulla oblongata.
    b. controls the rate of breathing.
    c. is stimulated by hydrogen ion concentration.
    d. All of these are correct.

25. Which equation occurs in pulmonary capillaries?
    a. $Hb + O_2 \longrightarrow HbO_2$
    b. $H^+ + Hb \longrightarrow HHb$
    c. $CO_2 + H_2O \longrightarrow H_2CO_3$
    d. $H_2CO_3 \longrightarrow H^+ + HCO_3^-$
    e. More than one of these is correct.
26. Hemoglobin assists the transport of gases by
    a. combining with oxygen.
    b. combining with CO_2.
    c. combining with H^+.
    d. being present in red blood cells.
    e. All of these are correct.
27. **THINKING CONCEPTUALLY** Knowing that proteins are sensitive to pH changes, hypothesize why hemoglobin binds to oxygen in the lungs but releases it in the tissues.
28. Most CO_2 is carried in the blood
    a. as HCO_3^-.
    b. on hemoglobin.
    c. dissolved in plasma.
    d. Both a and c are correct.
29. The chemical reaction that converts carbon dioxide to a bicarbonate ion takes place in
    a. the blood plasma.
    b. red blood cells.
    c. the alveoli.
    d. the hemoglobin molecule.
30. The enzyme carbonic anhydrase
    a. affects the heart rate.
    b. is found in red blood cells.
    c. is active once carbon dioxide enters the blood.
    d. attaches carbon dioxide to hemoglobin.
    e. Both b and c are correct.
31. **THINKING CONCEPTUALLY** The human body is able to go without breathing for only minutes but can go without eating for days. Explain.

GET INVOLVED

1. Why might you hypothesize that the need to metabolize fat by a diabetic would lead to arterial blood with a higher pH, a lower CO_2 content, and a higher O_2 content than normal? How would you test your hypothesis?
2. You are a physician who has witnessed a riding accident and the rider is unconscious. Why might you immediately use mouth-to-mouth resuscitation until mechanical ventilation became available?

MEDIA STUDY TOOLS

mhhe.com/maderconcepts3

Enhance your study of this chapter with interactive study tools, practice tests, and engaging animations. Also, ask your instructor about the resources available through ConnectPlus, which includes LearnSmart, a personalized adaptive learning program, and a media-rich eBook.

34

Osmoregulation and Excretion

BEFORE YOU BEGIN

Take a few minutes to recall

The basics of osmosis (section 5.2)

Forces affecting capillary function (section 30.6)

The urinary system and its contribution to homeostasis (sections 26.7–26.8)

How various animals osmoregulate ("Evolution of Homeostasis" on p. 524)

Do Coral Reef Animals Regulate?

Most of us don't ever think about whether the beautiful animals in a coral reef are osmoregulating—that is regulating the quantity of ions, such as Na^+, K^+, Ca^{2+}, Mg^{2+}, or Cl^-, in their body fluids. Some invertebrates (sea anemones and sea stars, for example) do not osmoregulate. Their body fluids conform to the seawater around them, and they take what they get. Think of the energy they save. They're not pumping salt here and there to get their osmolarity just right. Remember osmosis? If a cell is in an isotonic solution and the concentration of salts is the same inside and out, there is no net movement of water. A cell in a hypotonic solution gains water, and a cell in a hypertonic solution loses water (see Fig. 5.2C). Invertebrate animals that live in the sea are isotonic to seawater. They have a lot of salts in their cells, but that is okay because their enzymes are adapted to this osmolarity.

Representatives from only a handful of the approximately 34 animal phyla have managed to colonize land. What's the problem? We know that an appropriate respiratory organ is needed for living on land—gills won't do, but lungs are great. And we know that legs

Barracuda

Coral reef fishes

are much better for locomoting on land than are flippers. But an animal also needs a good osmoregulatory organ, such as the vertebrate kidney. In addition, an animal must have various accessory ways to reduce water loss to the dry air, or it will never make it on land.

The blood plasmas of marine vertebrates, such as fishes and sea turtles, are hypotonic to seawater, and when these animals live in a coral reef, they tend to gain salt. The vertebrate kidney is much better at conserving water on land than pumping out extra salts in the marine environment. But the fishes in the coral reef get a little help from their gills. In marine environments, gills pump salts out; in fresh water, they do the opposite.

Sea turtles and seabirds also do just fine living in a marine environment. They have special salt glands that pump salts out. Ever see a green sea turtle cry? It's getting rid of salt by way of its lacrimal (tear) glands. With the exception of marine invertebrates, animals do spend a lot of energy regulating the osmolarity of their internal body fluids. Their enzymes demand it and won't function if the tonicity is not just right.

In this chapter, we learn how both aquatic and terrestrial animals excrete metabolic wastes, particularly nitrogenous wastes, and at the same time maintain their normal water-salt balance.

Christmas tree worms

Clownfish

Cup coral

Coral reef

Comparative Excretory Organs

Excretion is the elimination of metabolic wastes, such as nitrogenous wastes. The nitrogenous waste excreted by an animal is an adaptation to the environment. Invertebrates as well as vertebrates have excretory organs. Do not confuse excretion with defecation. Defecation is the elimination of digestive wastes at the anus; these molecules have never been a part of the metabolism in cells.

34.1 The nitrogenous waste product of animals varies according to the environment

LEARNING OUTCOME

When you complete this section, you should be able to

1. Contrast the advantages of excreting ammonia, urea, or uric acid in particular environments.

The breakdown of various molecules, including protein and nucleic acids, results in nitrogenous wastes. For simplicity's sake, we will limit our discussion to amino acid metabolism. When amino acids are broken down by the body to generate energy, or are converted to fats or carbohydrates, the amino groups ($-NH_2$) must be removed because they are not needed. Once the amino groups have been removed, they may be excreted from the body in the form of ammonia, urea, or uric acid, depending on the species. Removal of amino groups from amino acids requires a fairly set amount of energy. However, the amount of energy required to convert amino groups to ammonia, urea, or uric acid, and the amount of water needed to excrete them does differ, as indicated in Figure 34.1.

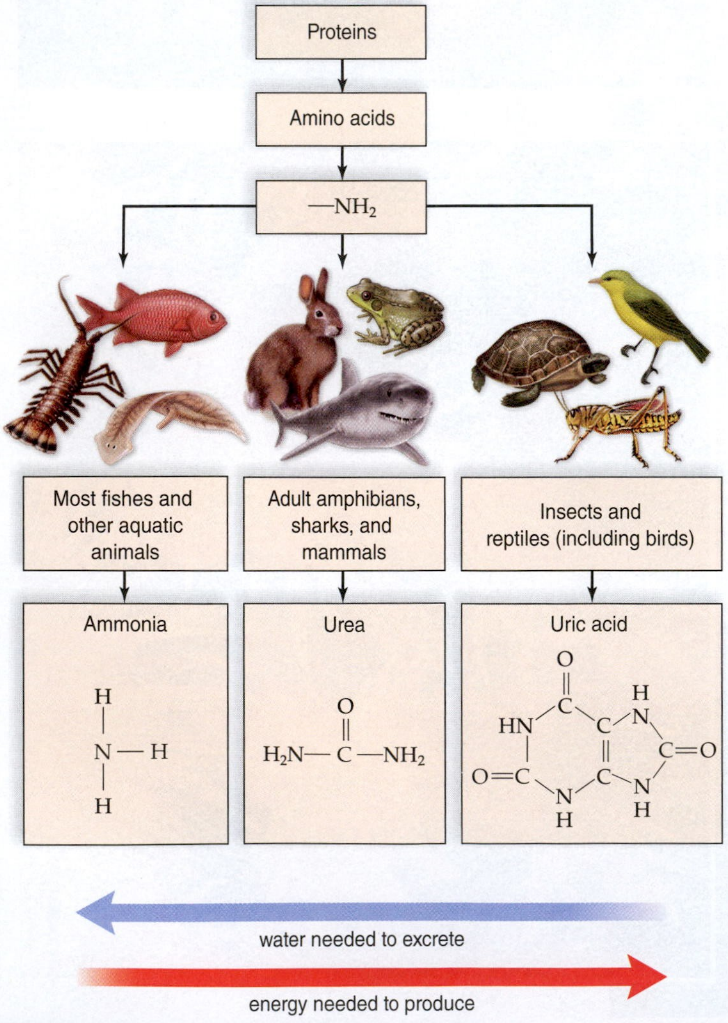

Ammonia Amino groups removed from amino acids immediately form **ammonia** (NH_3) by the addition of a third hydrogen ion. Little or no energy is required to convert an amino group to ammonia by adding a hydrogen ion. Ammonia is quite toxic and can be a nitrogenous excretory product if a good deal of water is available to wash it from the body. Ammonia is excreted by most fishes and other aquatic animals whose gills and skin surfaces are in direct contact with the water of the environment.

Urea Production of **urea** requires the expenditure of energy. Urea is produced in the liver by an energy-requiring metabolic pathway, known as the urea cycle. In this cycle, carrier molecules take up carbon dioxide and two molecules of ammonia, finally releasing urea. Urea is much less toxic than ammonia and can be excreted in a moderately concentrated solution. This allows body water to be conserved, an important advantage for terrestrial animals with limited access to water. Sharks, adult amphibians, and mammals usually excrete urea as their main nitrogenous waste.

Uric Acid Synthesis of **uric acid** requires a long, complex series of enzymatic reactions that use even more ATP than does urea synthesis. Uric acid is not very toxic, and it is poorly soluble in water. Therefore, it can be concentrated even more readily than can urea. Uric acid is routinely excreted by insects, reptiles, including birds. In reptiles and birds, a dilute solution of uric acid passes from the kidneys to the *cloaca*, a common reservoir for the products of the digestive, urinary, and reproductive systems. The cloacal contents can be refluxed into the large intestine, where water is reabsorbed. The white substance in bird feces is uric acid. Embryos of reptiles and birds develop inside completely enclosed shelled eggs. The production of insoluble, relatively nontoxic uric acid is advantageous for shelled embryos that store nitrogenous wastes inside the shell until hatching takes place. The advantage of water conservation seems to counterbalance the disadvantage of energy expenditure for uric acid production.

In general, it is possible to predict which metabolic waste product an animal excretes based on its anatomy and environment—but not always. For example, unlike most other birds, hummingbirds and sunbirds excrete more ammonia than uric acid. However, hummingbirds and sunbirds are fluid feeders, and as such, they have plenty of fluid available for excreting ammonia.

Most fishes and other aquatic animals	Adult amphibians, sharks, and mammals	Insects and reptiles (including birds)
Ammonia	Urea	Uric acid

water needed to excrete

energy needed to produce

FIGURE 34.1 Excretion of $-NH_2$.

▶ **34.1 CHECK YOUR PROGRESS**

1. Identify the advantage of excreting urea instead of ammonia or uric acid.
2. We usually think of evolution in terms of anatomical adaptations. What type of adaptation is described here?

34.2 Many invertebrates have organs of excretion

When you complete this section, you should be able to

1. Compare the organs of excretion in planarians, earthworms, and arthropods.

Most animals have tubular excretory organs that regulate the water-salt balance of the body and excrete metabolic wastes into the environment. Here we give three examples among the invertebrates.

Planarians The planarians are flatworms that live in fresh water. Their excretory organs consist of two strands of branching tubules that open to the outside of the body through excretory pores (Fig. 34.2, *top*). Located along the tubules are bulblike **flame cells,** each containing a cluster of beating cilia that looks like a flickering flame under the microscope. The beating of flame-cell cilia propels fluid through the excretory tubules and out of the body. The system is believed to function in ridding the body of excess water and excreting metabolic wastes.

Earthworms The body of an earthworm, an annelid, is divided into segments, and nearly every body segment has a pair of excretory structures called **nephridia.** Each nephridium is a tubule with a ciliated entrance and a pore that acts as an exit. (Fig. 34.2, *bottom*). As fluid from the coelom is propelled through the tubule by beating cilia, its composition is modified. For example, nutrient substances are reabsorbed and carried away by a network of capillaries surrounding the tubule. The urine of an earthworm contains metabolic wastes, salts, and water. Although the earthworm is considered a terrestrial animal, it excretes a very dilute urine. Each day, an earthworm may produce a volume of urine equal to 60% of its body weight. The excretion of ammonia is consistent with these data.

Arthropods Insects have a unique excretory system consisting of long, thin tubules called **Malpighian tubules** attached to the digestive tract. Uric acid is actively transported from the surrounding hemolymph into these tubules, and water follows a salt gradient established by active transport of K^+. Water and other useful substances are reabsorbed at the rectum, and uric acid leaves the body at the anus. Insects that live in water, or eat large quantities of moist food, reabsorb little water. Insects in dry environments reabsorb most of the water and excrete a dry, semisolid mass of uric acid.

The excretory organs of other arthropods are given different names, although they function similarly. In crustaceans (e.g., crabs, crayfish), nitrogenous wastes are generally removed by diffusion across the gills—in those species that have gills. Even so, crustaceans have excretory organs called *green glands* located in the ventral portion of the head region. Fluid, collecting within tubules from the surrounding hemolymph, is modified by the time it leaves the green gland. The absorption of salts into the tubules regulates the amount of urine excreted. In shrimp and pillbugs, excretory organs located in the maxillary segments are called *maxillary glands.* Spiders, scorpions, and other arachnids possess *coxal glands,* which are located near one or more appendages and used for excretion. Coxal glands are spherical sacs resembling annelid nephridia. Wastes are collected from the surrounding hemolymph and discharged through pores located at one to several pairs of appendages.

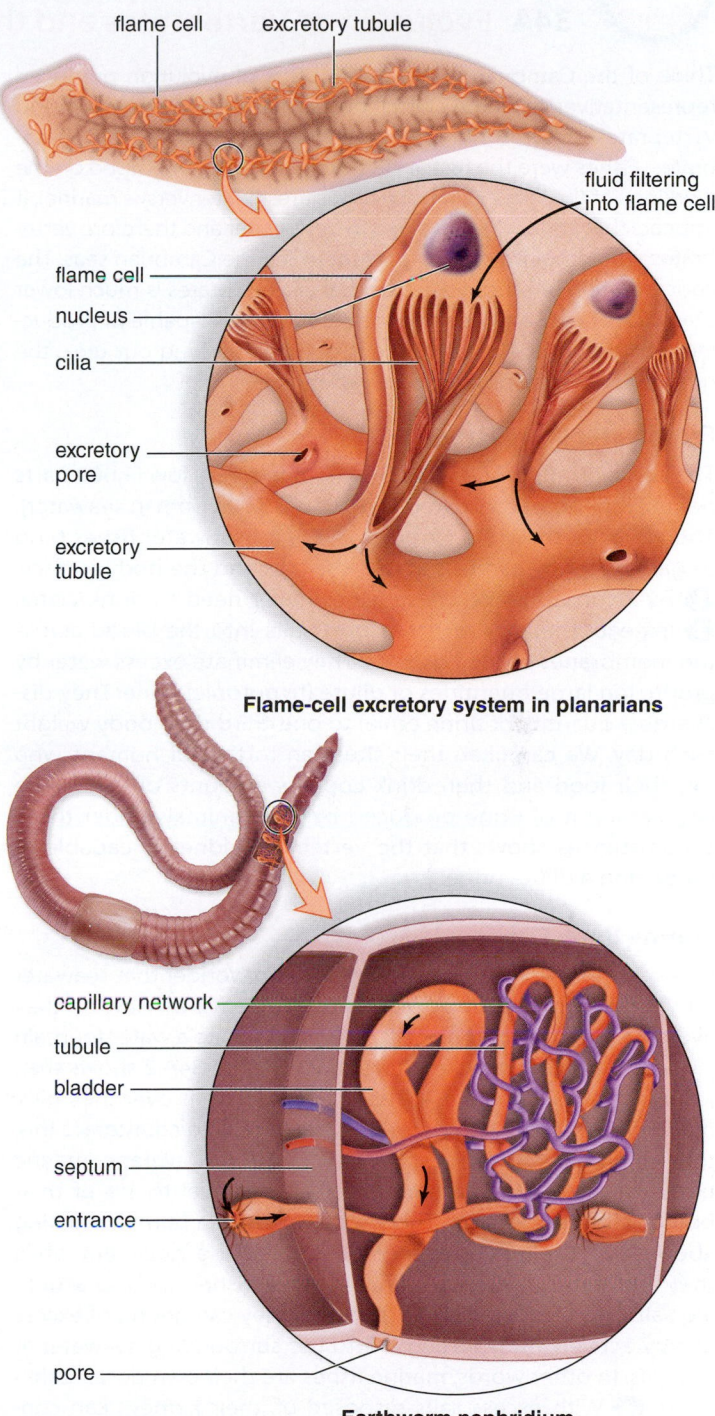

Flame-cell excretory system in planarians

Earthworm nephridium

FIGURE 34.2 Two examples of excretory organs in invertebrates.

In conclusion, invertebrates utilize tubules to rid the body of wastes and maintain a water-salt balance. On occasion, excretion also involves other organs, such as the rectum in insects and the gills in crayfish.

► **34.2 CHECK YOUR PROGRESS**

1. In Figure 34.2, how does the capillary network (associated with a nephridium) benefit an earthworm but is not needed in the flame-cell system of a planarian?

HOW LIFE CHANGES   *Application*

34A Evolution of Vertebrates and the Vertebrate Kidney

Think of the Cambrian seas when a burst of evolution produced representatives of all the animal phyla. That's when chordates and vertebrates arose, right? Chordates, yes—but perhaps not vertebrates. Fishes were the first vertebrates to evolve and based on the workings of the kidneys in bony fishes (freshwater versus marine), it appears that vertebrates evolved in fresh water and therefore vertebrates would not have been present in marine Cambrian seas. The concentration of salts in the blood of all vertebrates is much lower than that of seawater. Further, their kidneys are capable of producing a dilute urine. A dilute urine is good at washing out urea, the nitrogenous waste produced by vertebrates.

Freshwater Bony Fishes

The concentration of salts in fresh water is very low (1,000 parts per million [ppm] compared to 10,000–35,000 ppm in seawater). Therefore, as shown in Figure 34A.1, ❶ freshwater fishes tend to gain water by osmosis across the gills and the body surface. ❷ As a consequence, these fishes never need to drink water. ❸ Instead, they actively transport salts into the blood across the membranes of their gills. ❹ They eliminate excess water by producing large quantities of dilute (hypotonic) urine. They discharge a quantity of urine equal to one-third their body weight each day. We can liken their situation to that of humans who salt their food and then drink copious amounts of water. The large amount of urine produced by both animals under these circumstances shows that the vertebrate kidney is capable of producing a dilute urine.

Marine Bony Fishes

If vertebrates evolved in fresh water, it's no wonder that seawater is hypertonic to the blood plasma of marine fishes. How do they avoid dehydration of their cells? They can't go to a water fountain and drink fresh water like humans can. Figure 34A.2 shows that, just as you would expect, ❶ as seawater washes over their gills, marine bony fishes lose water by osmosis. ❷ To counteract this, they drink seawater almost constantly. On the average, marine bony fishes swallow an amount of water equal to 1% of their body weight every hour. This is equivalent to a human drinking about 700 ml of water every hour around the clock. But while they are getting water by drinking, these fishes are also acquiring salt. This is the rub. ❸ Fortunately, they can get rid of excess salt by actively transporting it into the surrounding seawater at the gills. In other words, marine fishes are their own desalination plant. ❹ With excess salts disposed of, their kidneys can conserve water, and marine fishes produce a scant amount of isotonic urine. The ability of the vertebrate kidney to produce a scant amount of urine is actually preadaptive to living on land because the land environment is dry.

A comparison of freshwater bony fishes with marine bony fishes does indeed suggest that freshwater bony fishes evolved first. Marine bony fishes would be far better off energetically speaking if their blood, like that of invertebrates, had the same concentration of salts as seawater. Why are their cells accustomed to a much lower concentration of salt (necessitating

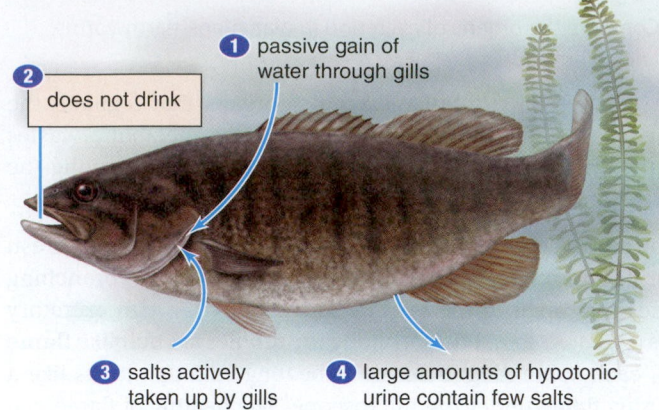

FIGURE 34A.1 Osmoregulation in freshwater bony fishes.

❶ passive gain of water through gills
❷ does not drink
❸ salts actively taken up by gills
❹ large amounts of hypotonic urine contain few salts

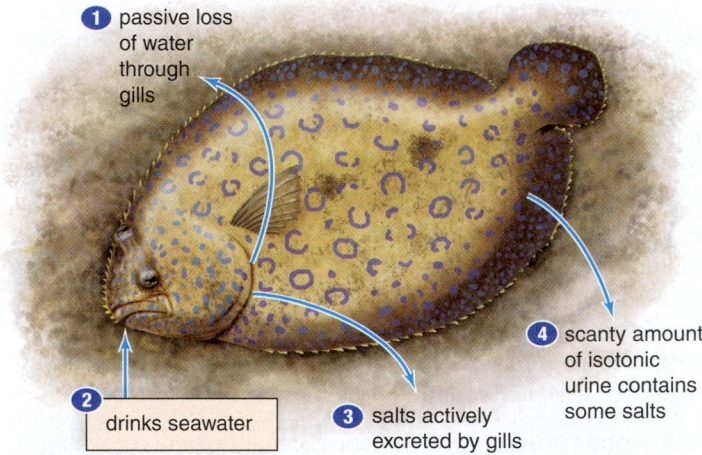

FIGURE 34A.2 Osmoregulation in marine bony fishes.

❶ passive loss of water through gills
❷ drinks seawater
❸ salts actively excreted by gills
❹ scanty amount of isotonic urine contains some salts

the pumping out of salt)? Probably because they evolved from freshwater fishes.

Cartilaginous Fishes

Just as with other vertebrates, the total concentration of the various ions in the blood of cartilaginous fishes is less than that in seawater. However, these fishes evolved a totally different method of dealing with the marine environment than did marine fishes. Vertebrates use urea as their nitrogenous waste, and cartilaginous fishes turned this to good use. Their blood plasma is nearly isotonic to seawater because they pump it full of urea, and this molecule gives their blood the same tonicity as seawater (Fig. 34A.3). Notice, then, that vertebrates can tolerate a high concentration of urea in their blood. Also, while kidneys do excrete nitrogenous wastes, they are primarily organs of osmoregulation. We humans are better able to tolerate urea in our blood than salts. Death from kidney failure is more likely due to the loss of osmoregulation than to the buildup of urea.

Like other marine vertebrates, cartilaginous fishes do have an auxiliary method of getting rid of salts. They have rectal glands that remove excess salt from the body, and in this way they do not solely depend on their kidneys. This is a take-home message: Vertebrates at sea need a way besides their kidneys to get rid of excess salt. The kidneys can't do it alone. Let's take another example.

Seagulls, Reptiles, and Mammals

Birds, reptiles, and mammals evolved on land, and they make use of the kidney's ability to conserve water when needed. However, some land animals have become secondarily adapted to living near or in the sea. They drink seawater and still manage to survive. Little is known about how whales, which are mammals, manage to get rid of extra salt, but we know that their kidneys are enormous. Other animals have been studied, and we have learned that marine reptiles, including seabirds have salt glands that may have later evolved into the tear glands of mammals. In the animals we will mention, salt glands pump out salt from blood plasma and leave behind much of the water, just as in a desalination plant.

In seabirds, salt-excreting glands are located near the eyes (Fig. 34A.4). The glands produce a salty solution that is excreted through the nostrils and moves down grooves on their beaks until it drips off. In marine turtles, the salt gland is more similar to a tear (lacrimal) gland, and in sea snakes, a salivary sublingual gland beneath the tongue gets rid of excess salt. The work of the gland is regulated by the nervous system. Osmoreceptors, perhaps located near the heart, are thought to stimulate the brain, which then orders the gland to excrete salt until the salt concentration in the blood decreases to a tolerable level.

This brings us to humans. The human kidney regulates the water-salt concentration of the blood and in most instances can produce scant to copious urine as needed. However, humans have no auxiliary gland to get rid of excess salt, and the kidneys are unable to excrete salt and other solutes such as urea without also excreting water. Therefore, we need access to fresh water in order to produce enough urine to wash excess salt from the body. If we drink seawater, our cells become dehydrated as they lose water to blood, which will have a higher concentration of salts than cytoplasm. If you get stranded at sea, don't drink seawater—it will only make you more thirsty (Fig. 34A.5).

CONSIDER THESE QUESTIONS

1. There are no amphibians that live in seawater. Revisit Figure 20.14B and explain how fossil remains are consistent with the hypothesis that vertebrates evolved in fresh water.
2. Amphibians produce both ammonia and urea as their nitrogenous waste product. Why is this consistent with their life history? Why would you predict that amphibians osmoregulate in the same way that freshwater fishes do?
3. Of all the animals discussed in this application, humans can best be compared to freshwater fishes. How so?

 Explore the concepts through a variety of multimedia assets and question types.

www.mcgrawhillconnect.com

FIGURE 34A.3 The blood of cartilaginous fishes, such as this shark, is isotonic to seawater.

Salt solution exits here.

Salt solution runs down beak here.

FIGURE 34A.4 Marine birds and reptiles are apt to have salt glands to pump excess salt.

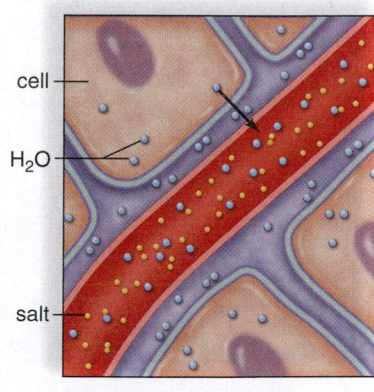

cell

H_2O

salt

water loss from cells to blood vessel

FIGURE 34A.5 Your kidneys are adapted to life on land, and you have to drink fresh water in order to wash excess salt from blood. If you drink seawater, your cells will lose water to a hypertonic blood (see arrow).

Mammalian Urinary System

Although the liver produces urea, the kidneys produce urine, a complex solution of nitrogenous wastes, salts, and water at a regulated pH. The other portions of the urinary system merely function to hold urine or pass it to the exterior.

34.3 The urinary system includes kidneys that contain tubules

LEARNING OUTCOME

When you complete this section, you should be able to

1. Describe the path of urine in humans and the microscopic anatomy of the mammalian kidney.

In mammals, the **kidneys** are bean-shaped, reddish-brown organs, each about the size of a fist. They are located in the lower back on either side of the vertebral column just below the diaphragm, where they are partially protected by the lower rib cage. The right kidney is slightly lower than the left kidney. While urea is made by the liver, ❶ **urine** is produced by the kidneys (Fig. 34.3A). Urine is conducted from the body by other organs in the urinary system. ❷ Each kidney is connected to a **ureter,** a duct that takes urine from the kidney to ❸ the **urinary bladder,** where it is stored until it is voided from the body through ❹ the single **urethra.** Urine is not found in any part of the body except these structures.

In males, the urethra passes through the penis, and in females, the opening of the urethra is ventral to (in front of) the vagina. There is no connection between the genital (reproductive) and urinary systems in females. But in males, the urethra also carries sperm during ejaculation. This is an either/or situation; either the male urethra carries sperm or it carries urine. It cannot do both at the same time.

Reptiles and birds have an elongated kidney resembling that of fishes and amphibians during development, but as adults, their kidney resembles that of mammals. In all vertebrates, except placental mammals, a duct from the kidney conducts urine to a *cloaca,* which you will recall is a common depository for undigestible remains, urine, and sex cells.

Macroscopic and Microscopic Anatomy of the Kidney

If a mammalian kidney is sectioned longitudinally, three major parts can be distinguished (Fig. 34.3B). The **renal cortex,** which is the outer region of a kidney, has a somewhat granular appearance. The **renal medulla** consists of six to ten cone-shaped renal pyramids that lie between the renal cortex and the renal pelvis. The **renal pelvis** is a hollow chamber where urine collects before it is carried to the bladder by the ureter.

A kidney stone, or renal calculus, is a hard granule of phosphate, calcium, protein, or uric acid that forms in the renal pelvis. Many pass out of the body unnoticed. However, larger, jagged stones can block the renal pelvis or ureter, causing intense pain and damage.

All vertebrate kidneys contain tiny tubules called **nephrons** that produce urine. The mammalian kidney is composed of over 1 million nephrons. Some nephrons are located primarily in the renal cortex, but others dip down into the renal medulla, as shown in Figure 34.3B, *right.* Each nephron is made of several parts (Fig. 34.3C). ❶ The blind end of a nephron is pushed in on itself to form a cuplike structure called the **glomerular capsule** (Bowman's capsule). The outer layer of the glomerular capsule is composed of squamous epithelial cells; the inner layer is composed of podocytes, specialized cells that allow easy passage of molecules.

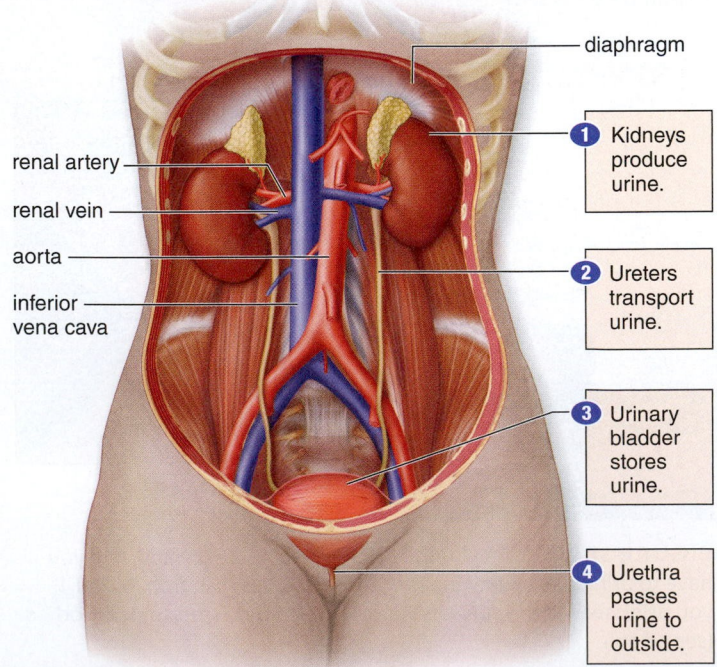

renal artery

renal vein

aorta

inferior vena cava

diaphragm

❶ Kidneys produce urine.

❷ Ureters transport urine.

❸ Urinary bladder stores urine.

❹ Urethra passes urine to outside.

FIGURE 34.3A The mammalian urinary system.

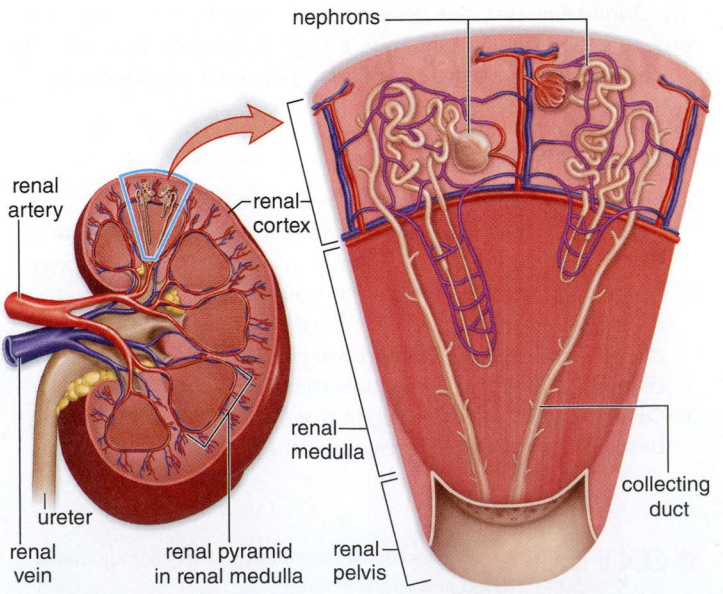

nephrons

renal artery

renal cortex

renal vein

ureter

renal pyramid in renal medulla

renal medulla

renal pelvis

collecting duct

FIGURE 34.3B Macroscopically (*left*), notice the renal cortex, renal medulla, and renal pelvis of the kidney. Microscopically (*right*), it's possible to see that the renal pyramids contain tubules called nephrons.

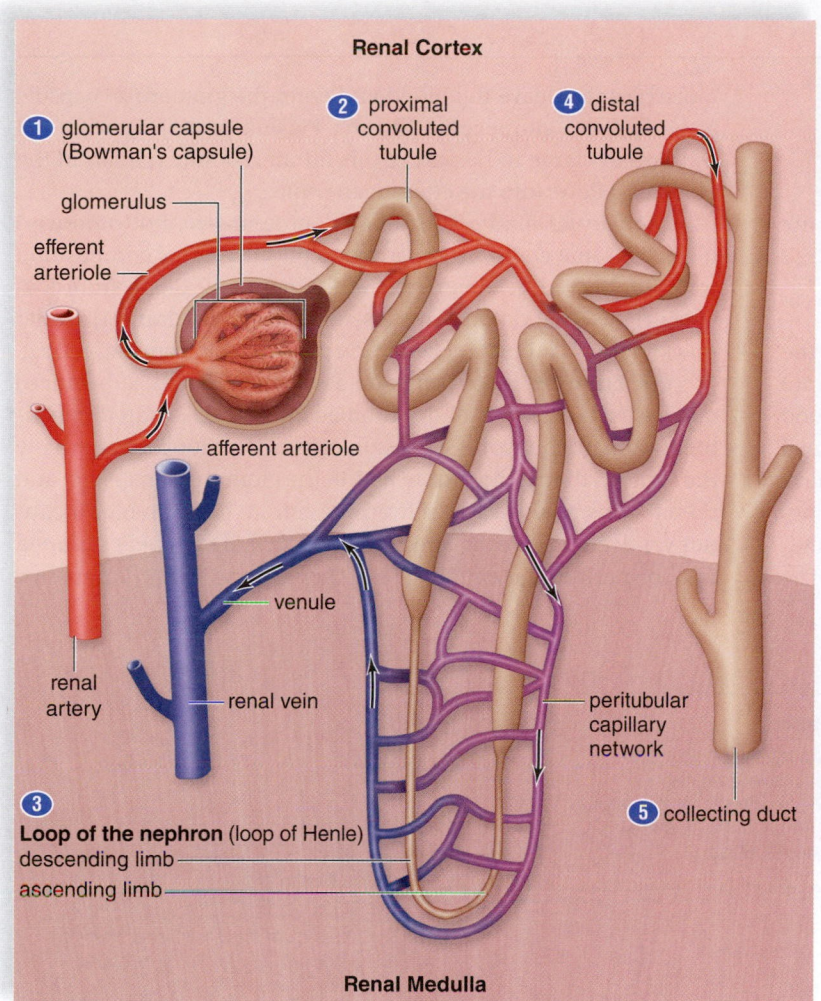

Anatomy of nephron

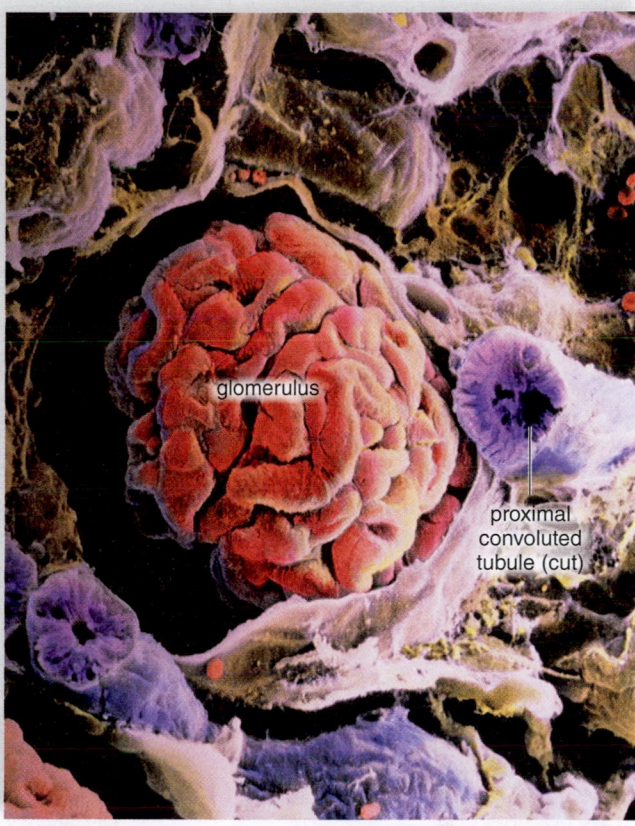

glomerulus

proximal convoluted tubule (cut)

SEM of glomerulus   469×

FIGURE 34.3C A nephron is made up of a glomerular capsule, the proximal convoluted tubule, the loop of the nephron, the distal convoluted tubule, and the collecting duct. You can trace the path of blood about the nephron by following the arrows. (MV = microvilli)

Leading from the glomerular capsule is a portion of the nephron known as ❷ the **proximal convoluted tubule,** which is lined by cells with many mitochondria and tightly packed microvilli that reabsorb nutrients, salts, and water. ❸ Next comes the **loop of the nephron** (loop of Henle), which has a descending limb and an ascending limb. ❹ The **distal convoluted tubule** follows after the loop of the nephron. Several distal convoluted tubules enter one collecting duct. ❺ **Collecting ducts** transport urine down through the renal medulla and deliver it to the renal pelvis. The many loops of the nephron and collecting ducts give the pyramids of the renal medulla a striped appearance. In mammals, these structures are essential to producing urine that is hypertonic to blood plasma.

Each mammalian nephron has an extensive blood supply (Fig. 34.3C). From the dorsal aorta, the renal artery leads to many afferent arterioles, one for each nephron. Each afferent arteriole

divides to form a capillary bed, the **glomerulus,** where liquid exits and enters the glomerular capsule. Reptiles and birds have a reduced glomerulus and, in this way, produce less urine for the excretion of uric acid. Blood from the glomerulus drains into an efferent arteriole, which subsequently branches into a second capillary bed around the tubular parts of the nephron. This capillary bed, called the **peritubular capillary network,** leads to venules that join to form the renal vein, a vessel that enters the inferior vena cava.

▶ **34.3 CHECK YOUR PROGRESS**

1. Distinguish between the function of ureters and that of the urethra.

2. Trace the path of fluid in a nephron from the glomerular capsule to the pelvis of the mammalian kidney.

LEARNING OUTCOME

When you complete this section, you should be able to

1. Understand the primary steps in urine formation.

Urine Formation Considering the several functions of the kidneys it is not surprising that the process of producing urine requires several steps and involves the entire nephron.

MP3 Overview of Urine Formation

As described in Figure 34.4, ❶ **glomerular filtration** is the movement of small molecules across the glomerular wall into the glomerular capsule as a result of blood pressure. When blood enters the glomerulus, blood pressure is sufficient to cause small molecules, such as water, nutrients, salts, and wastes, to move from the glomerulus to the inside of the glomerular capsule, especially because the glomerular walls are 100 times more permeable than the walls of most capillaries elsewhere in the body. The

molecules that leave the blood and enter the glomerular capsule are called the **glomerular filtrate.** Plasma proteins and blood cells are too large to be part of this filtrate, so they remain in the blood as it flows into the efferent arteriole.

Glomerular filtrate is essentially protein-free, but otherwise it has the same composition as blood plasma. If this composition were not altered in other parts of the nephron, death from loss of nutrients (starvation) and loss of water (dehydration) would quickly follow. Tubular reabsorption prevents this from happening.

❷ **Tubular reabsorption** takes place when substances move across the walls of the tubules into the associated peritubular capillary network. The osmolarity of the blood is essentially the same as that of the filtrate within the glomerular capsule, and therefore osmosis of water from the filtrate into the blood cannot yet occur. However, sodium ions (Na^+) are actively pumped into the peritubular capillary, and then chloride ions (Cl^-) follow

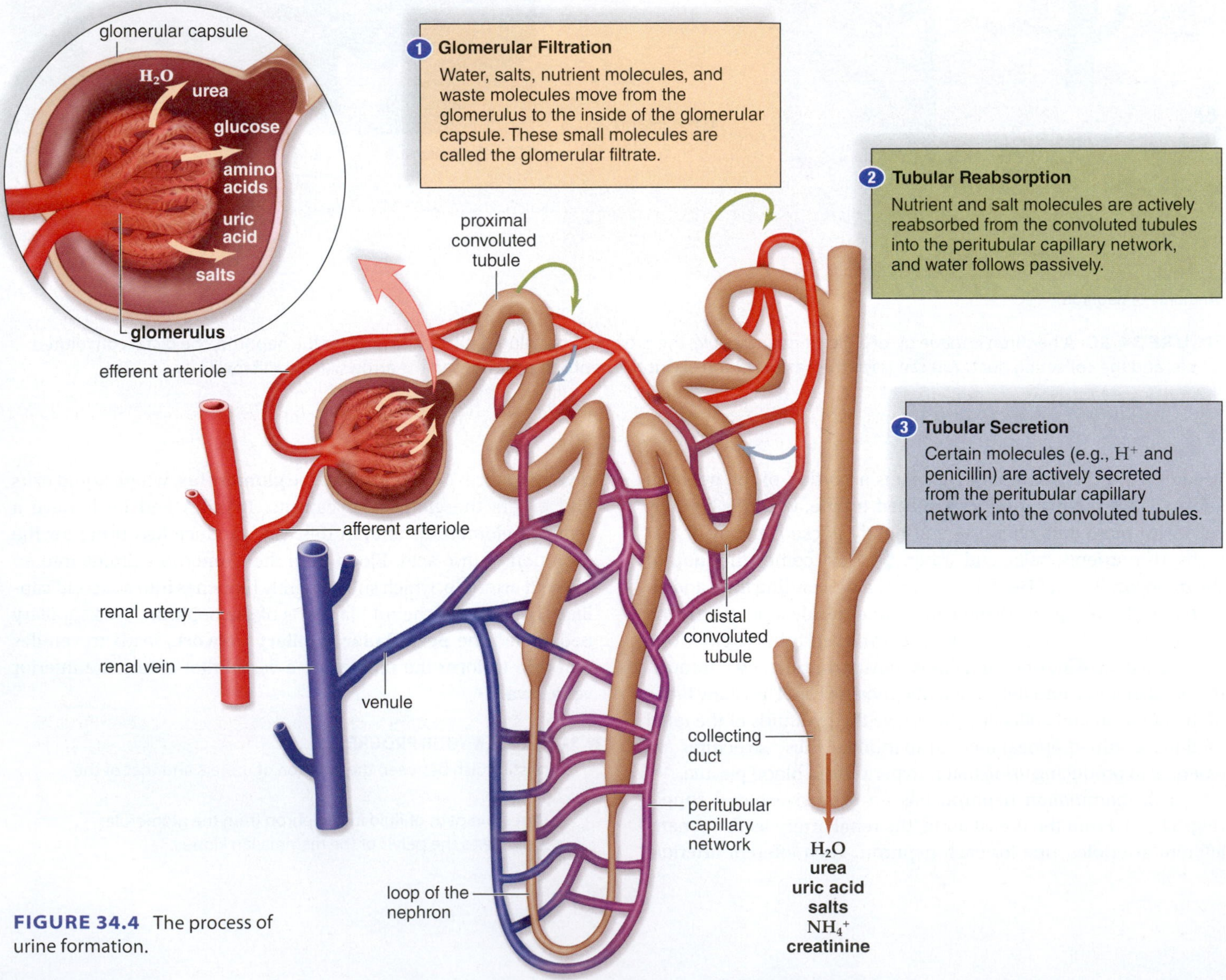

---

❶ Glomerular Filtration

Water, salts, nutrient molecules, and waste molecules move from the glomerulus to the inside of the glomerular capsule. These small molecules are called the glomerular filtrate.

❷ Tubular Reabsorption

Nutrient and salt molecules are actively reabsorbed from the convoluted tubules into the peritubular capillary network, and water follows passively.

❸ Tubular Secretion

Certain molecules (e.g., H^+ and penicillin) are actively secreted from the peritubular capillary network into the convoluted tubules.

glomerular capsule

H_2O
urea
glucose
amino acids
uric acid
salts

glomerulus

efferent arteriole

afferent arteriole

renal artery

renal vein

venule

proximal convoluted tubule

distal convoluted tubule

collecting duct

peritubular capillary network

loop of the nephron

H_2O
**urea
uric acid
salts
NH_4^+
creatinine**

FIGURE 34.4 The process of urine formation.

passively. Now the osmolarity of the blood is such that water moves passively from the tubule into the blood. About 60–70% of salt and water are reabsorbed at the proximal convoluted tubule.

Nutrients, such as glucose and amino acids, also return to the blood at the proximal convoluted tubule. This is a selective process, because only molecules recognized by carrier proteins in plasma membranes are actively reabsorbed. The cells of the proximal convoluted tubule have numerous microvilli, which increase the surface area, and numerous mitochondria, which supply the energy needed for active transport. Glucose is a molecule that ordinarily is reabsorbed completely because there is a plentiful supply of carrier molecules for it. However, if there is more glucose in the filtrate than there are carriers to handle it, glucose will exceed its renal threshold, or transport maximum. When this happens, the excess glucose in the filtrate appears in the urine. People with diabetes mellitus have an abnormally large amount of glucose in the blood and filtrate because the liver fails to store glucose as glycogen.

Urea is a substance that is passively reabsorbed from the filtrate. At first, the concentration of urea within the filtrate is the same as that in blood plasma. But after water is reabsorbed, the urea concentration is greater than that in the peritubular plasma. In the end, about 50% of the filtered urea is excreted.

3 **Tubular secretion** is the second way substances are removed from blood and added to tubular fluid. Substances such as uric acid, hydrogen ions, ammonia, creatinine, histamine, and penicillin are eliminated by tubular secretion. The process of tubular secretion helps rid the body of potentially harmful compounds that were not filtered into the glomerulus.

Urine which now enters the collecting duct and exits the body is available for urinalysis. "Urinalysis Can Detect Drug Use" on this page describes how drug detection is usually done today.

> **34.4 CHECK YOUR PROGRESS**
> 1. In your own words, state in sequence what happens during the three steps of urine formation.
> 2. Explain why you expect to see microvilli in the proximal convoluted tubule but not in the distal convoluted tubule.

HOW BIOLOGY IMPACTS OUR LIVES   *Application*

34B Urinalysis Can Detect Drug Use

Today, the use of urinalysis has expanded beyond medical applications to include forensic diagnosis of drug use. Screening for illegal drug use is now mandated by federal and state agencies as a condition of employment, and most private employers now require it as well. A court can order that urinalysis be done if drug abuse is involved in the commission of a crime. The National Collegiate Athletic Association requires all student athletes to undergo drug testing, as do many high school athletic programs.

Urinalysis is not used to screen for the drugs themselves, but for drug metabolites, the breakdown products of drugs that have been consumed or injected. Once in the body, drugs are metabolized by the liver and the breakdown products are filtered by the kidney. Thus, metabolites will be present in the urine of a drug abuser. Two types of techniques can detect metabolites. The first, a screening exam, involves placing a test strip into freshly voided urine. Strips are available that can test for five commonly abused drugs: marijuana, amphetamine, PCP, cocaine, and opiates such as heroin. Drug testing is also done to prevent athletes from taking performance-enhancing drugs, such as anabolic steroids or growth hormone. Lance Armstrong was recently stripped of his Tour de France medals for having used EPO (erythropoietin) and/or any one of several new versions of EPO. EPO is a hormone produced by the kidneys that stimulates stem cells in the bone marrow to produce more RBCs. So-called blood doping is any method of increasing the normal supply of RBCs for the purpose of delivering oxygen more efficiently, reducing fatigue, and giving an athlete a competitive edge.

Although urine-strip testing gives results within minutes, certain legal over-the-counter medications can produce false positives. Should the sample test positive, it can be immediately sent for a second, more sophisticated chemical analysis, such as gas chromatography. Tracking long-term drug use may require using hair samples in addition to urine samples, because drug metabolites can be incorporated into the hair.

The urine specimen must be properly collected to avoid possible tampering by the individual being tested. Specimens can be deliberately diluted with sink or toilet water or contaminated with any number of additives. Bleach, drain cleaner, soft drinks, and so on can be used,

Cyclist Lance Armstrong

as well as products specifically sold for the purpose of "helping to beat a drug screen." Drug abusers may also attempt to substitute the urine of a "clean" individual for their own. Tampering may be prevented by requiring that a witness be present at all times while the sample is being collected and stored. In addition, proper documentation must accompany any urine sample. Chain-of-custody forms record each step of the handling of a specimen, from collection to disposal. This provides proof of everything that happens to the specimen, and prevents the specimen from being rejected as evidence in court proceedings.

CONSIDER THESE QUESTIONS

1. Do you believe it's a private and not a legal matter whether people take drugs? If drug use is discovered, should the person go to jail or a rehabilitation facility? Explain.
2. Some athletes take drugs in order to win competitions. Can you rationalize this but still suggest ways to curtail these practices?

 Explore the concepts through a variety of multimedia assets and question types.

www.mcgrawhillconnect.com

Kidneys and Homeostasis

A kidney helps maintain homeostasis in four ways: (1) excretion of metabolic wastes, (2) maintenance of the water-salt balance, (3) maintenance of the acid-base balance and, therefore, the pH balance, and (4) hormone secretion.

34.5 The kidneys can concentrate urine to maintain water-salt balance

LEARNING OUTCOME

When you complete this section, you should be able to

1. Understand how the mammalian kidney maintains the water-salt balance.

Figure 34.5A depicts the process of regulating water-salt balance in the kidneys of mammals. **①** Along with nutrients, most of the water and salt (NaCl) present in the filtrate is reabsorbed across the wall of the proximal convoluted tubule. Reabsorption also occurs along the remainder of the nephron. If you drink little, excretion of a hypertonic urine is dependent on the reabsorption of water from the loop of the nephron and the collecting duct. During the process of reabsorption, water passes through water channels called **aquaporins.**

The long loop of the nephron, which typically penetrates deep into the renal medulla, is composed of a descending (going down) limb and an ascending (going up) limb. **②** Salt passively diffuses out of the lower portion of the ascending limb, but the upper, thick portion of the limb actively extrudes NaCl into the tissue of the outer renal medulla (Fig. 34.5A). Less and less NaCl is available for transport as fluid moves up the thick portion of the ascending limb. Because of these circumstances, there is an osmotic gradient within the tissues of the renal medulla: The concentration of NaCl is greater in the direction of the inner medulla. (Water cannot leave the ascending limb because the limb is impermeable to water, as indicated by the dark border in Figure 34.5A.)

The wide arrow to the far left in Figure 34.5A indicates that the medulla has an increasing concentration of solutes. This cannot be due to NaCl because active transport of NaCl does not start until fluid reaches the thick portion of the ascending limb. **③** Urea is believed to leak from the lower portion of the collecting duct, and this is the molecule that contributes to the high solute concentration of the inner medulla.

④ Because of the osmotic gradient within the renal medulla, water leaves the descending limb along its entire length. This is a countercurrent mechanism: As water diffuses out of the descending limb, the remaining fluid within the limb encounters an even greater osmotic concentration of solute; therefore, water continues to leave the descending limb from the top to the bottom.

When fluid enters the collecting duct from the distal convoluted tubule, it is isotonic to the cells of the renal cortex. Thus, to this point, the net effect of reabsorption of water and salt is the production of urine that has the same osmolarity as blood plasma. However, the fluid within the collecting duct also encounters the same osmotic gradient mentioned earlier. **⑤** Therefore, water diffuses out of the collecting duct into the renal medulla, and the urine within the collecting duct becomes hypertonic to blood plasma.

Antidiuretic hormone (ADH), released by the posterior lobe of the pituitary gland, plays a role in water reabsorption at the collecting duct. To understand the action of this hormone, consider its name. *Diuresis* means increased amount of urine, and *anti-diuresis* means decreased amount of urine. When ADH is present, more water is reabsorbed (blood volume and pressure rise), and a decreased amount of urine results. In practical terms, if an individual does not drink much water on a certain day, the posterior lobe of the pituitary releases ADH, causing more water to be reabsorbed and less urine to form. On the other hand, if an individual drinks a large amount of water and does not perspire much, ADH is not released. More water is excreted, and more urine forms. Diuretics, such as caffeine and alcohol, increase the flow of urine by interfering with the action of ADH.

Hormones Control the Reabsorption of Salt Usually, more than 99% of sodium (Na^+) filtered at the glomerulus is returned to the blood. Most sodium (67%) is reabsorbed at the proximal convoluted tubule, and a sizable amount (25%) is extruded by the ascending limb of the loop of the nephron. The rest is reabsorbed from the distal convoluted tubule and collecting duct.

Blood volume and pressure are regulated, in part, by salt reabsorption. When blood volume, and therefore blood pressure, is not sufficient to promote glomerular filtration, the kidneys secrete renin. **Renin** is an enzyme that changes angiotensinogen (a large plasma protein produced by the liver) into angiotensin I.

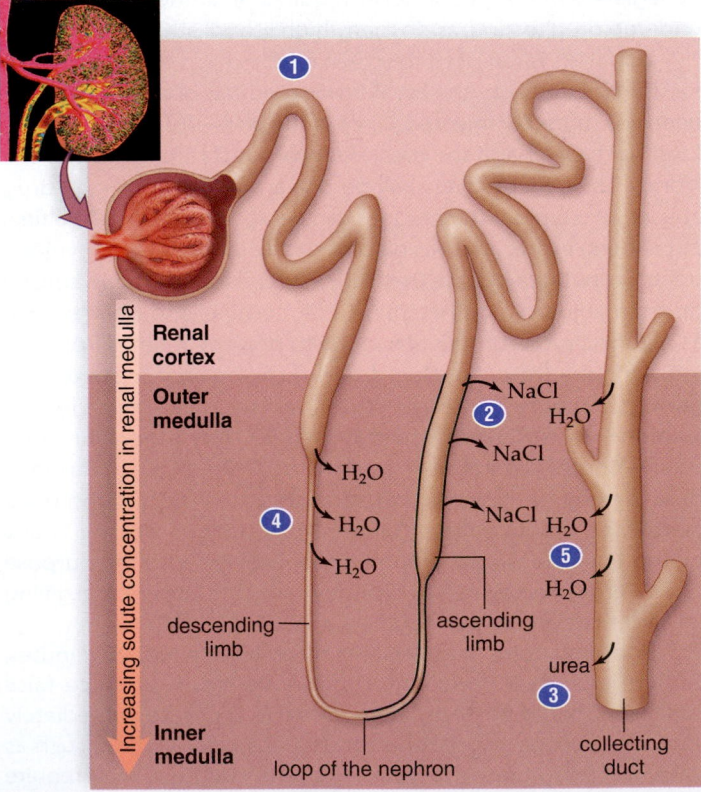

FIGURE 34.5A Regulation of water-salt balance in mammals.

Later, angiotensin I is converted to **angiotensin II,** a powerful vasoconstrictor that also stimulates the adrenal glands, which lie on top of the kidneys, to release aldosterone (Fig. 34.5B).

Aldosterone is a hormone that promotes the excretion of potassium ions (K^+) and the reabsorption of sodium ions (Na^+) at the distal convoluted tubule. The reabsorption of sodium ions is followed by the reabsorption of water. Therefore, blood volume and blood pressure increase.

Atrial natriuretic hormone (**ANH**) is a hormone secreted by the atria of the heart when cardiac cells are stretched due to increased blood volume. ANH inhibits the secretion of renin by the kidney and the secretion of aldosterone by the adrenal cortex. Its effect, therefore, is to promote the excretion of Na^+—that is, *natriuresis*. When Na^+ is excreted, so is water, and therefore blood volume and blood pressure decrease.

If you eat salty foods the hormone balance means that salt along with water will be excreted. Otherwise problems such as edema due to too much salt and water in the tissues, dehydration due to loss of water by cells, and increased blood pressure due to increased blood volume can arise.

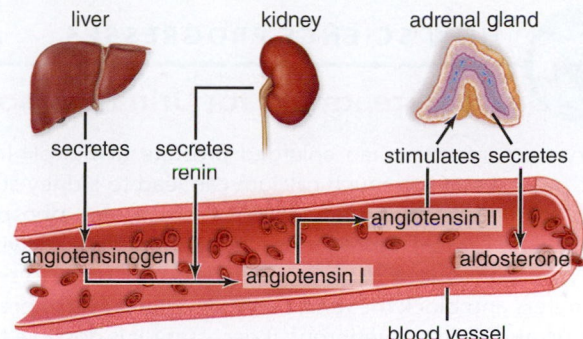

FIGURE 34.5B The renin-angiotensin-aldosterone system.

▶ **34.5 CHECK YOUR PROGRESS**
1. Use Figure 34.5A to describe how a nephron regulates the water-salt balance of the blood.
2. Describe how the hormones ADH, aldosterone, and ANH influence regulation of water-salt excretion.
3. Explain what might happen to metabolism inside cells if blood and tissue fluid were too salty. Explain your reasoning.

34.6 Lungs and kidneys maintain acid-base balance

LEARNING OUTCOME

When you complete this section, you should be able to

1. Explain how the bicarbonate buffer system, the lungs, and the kidneys are involved in regulating the acid-base balance of the blood.

The functions of proteins in the body—such as enzymes, hemoglobin, proteins of the electron transport chain, and others—are influenced by pH. Therefore, the regulation of pH is extremely important to good health.

The *bicarbonate ion* (HCO_3^-) *buffer system* works together with the breathing process to maintain the pH of the blood. Central to the mechanism is this reaction, which you last saw in section 33.6.

$$H^+ + HCO_3^- \rightleftharpoons H_2CO_3 \rightleftharpoons H_2O + CO_2$$

The excretion of carbon dioxide (CO_2) by the lungs helps keep the pH within normal limits, because when CO_2 is exhaled, this reaction is pushed to the right, and hydrogen ions (H^+) are tied up in water. Indeed, when blood pH decreases, chemoreceptors in the carotid bodies (located in the carotid arteries) and in aortic bodies (located in the aorta) stimulate the respiratory control center, and the rate and depth of breathing increase. On the other hand, when blood pH begins to rise, the respiratory control center is depressed, and the amount of bicarbonate ion increases in the blood.

As powerful as these mechanisms are, only the kidneys can rid the body of a wide range of acidic and basic substances. The kidneys are slower acting than the buffer/breathing mechanism, but they have a more powerful effect on pH. For the sake of simplicity, we can think of the kidneys as reabsorbing bicarbonate ions and excreting hydrogen ions as needed to maintain the normal pH of the blood. If the blood is acidic, hydrogen ions move from the blood into the tubule and are excreted. Bicarbonate ions are reabsorbed (Fig. 34.6). (If the blood is basic, hydrogen ions are not excreted,

capillary

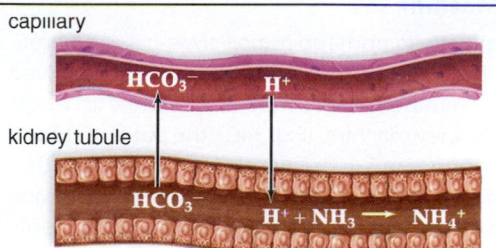

FIGURE 34.6 Regulation of blood pH involves movement of these ions as shown.

and bicarbonate ions are not reabsorbed.) The fact that urine is usually acidic (pH about 6) shows that usually an excess of hydrogen ions is excreted. Ammonia (NH_3), produced in tubule cells by the deamination of amino acids, provides a means of buffering these hydrogen ions in urine:

$$NH_3 + H^+ \longrightarrow NH_4^+$$

The presence of the ammonium ion (NH_4^+) is quite obvious in the diaper pail or kitty litter box by its odor.

Acidosis and Alkalosis The normal pH of arterial blood is around 7.4. Therefore, a person is said to have acidosis when the blood pH is below 7.34 and alkalosis when the blood pH is higher than 7.45. Abnormal breathing rates—too shallow or too rapid—can lead to acidosis and alkalosis, respectively. Alcohol consumption causes acidosis because alcohol is metabolized to acetic acid, which enters the blood. Similarly, diabetes can cause acidosis because of its effect on metabolism. Metabolic alkalosis is rare, but it can be caused by excessive intake of antacids or by an illness.

▶ **34.6 CHECK YOUR PROGRESS**
1. Use Figure 34.6 to describe how the kidneys regulate the pH of the blood.

34C Treatment for Urinary Disorders

Urinary tract infections, an enlarged prostate gland, pH imbalances, or intake of too much calcium can lead to kidney stones. Kidney stones are hard granules made of calcium, phosphate, uric acid, and protein. Kidney stones form in the renal pelvis and usually pass unnoticed in the urine flow. If they grow to several centimeters and block the renal pelvis or ureter, reverse pressure builds up and destroys nephrons. If necessary, it is possible to use shock waves to break up the stones into particles that can pass through the urinary tract.

Many types of illnesses, especially diabetes, hypertension, and recurrent infections, cause progressive renal disease and renal failure. One of the first signs of nephron damage is albumin, white blood cells, or even red blood cells in the urine. If damage is so extensive that more than two-thirds of the nephrons are inoperative, urea and other waste substances accumulate in the blood. This condition is called *uremia*. Although nitrogenous wastes can cause serious damage, the retention of water and salts is of even greater concern. The latter leads to edema, fluid accumulation in the body tissues, and eventually imbalance in the ionic composition of body fluids can lead to loss of consciousness and to heart failure.

Artificial Kidney Machine

Patients with renal failure can undergo *hemodialysis,* using an artificial kidney machine. Dialysis is defined as the diffusion of dissolved molecules through a semipermeable natural or synthetic membrane. In an artificial kidney machine (Fig. 34C), the patient's blood passes through a membranous tube, which is in contact with a dialysis solution, or *dialysate*. The pores in the synthetic membrane are small enough to allow only small molecules to pass through. Substances more concentrated in the blood diffuse into the dialysate, and substances more concentrated in the dialysate diffuse into the blood. The dialysate is continuously replaced to maintain favorable concentration gradients. In this way, the artificial kidney can be used either to extract substances from the blood, including waste products or toxic chemicals and drugs, or to add substances to the blood—for example, bicarbonate ions (HCO_3^-) if the blood

is acidic. In the course of a 3–6-hour hemodialysis treatment, from 50–250 g of urea can be removed from a patient, which greatly exceeds the amount excreted by normal kidneys. Therefore, a patient needs to undergo treatment only about twice a week.

Lab-Grown Bladders and Kidneys

Bladders grown in a lab have been successfully transplanted into a number of patients. The patient can contribute normal cells from his or her own diseased bladder, and then these cells are cultured in a lab to form a new bladder when grown on a collagen form shaped like a bladder. Eventually the new bladder is attached to the remaining portion of the patient's bladder. This alleviates any incontinence problems the patient previously experienced, and the risk of kidney damage is also decreased by lowering the pressure inside the bladder. Because the patient's own cells are used to grow a new bladder, there is no risk of rejection.

Using a combination of mouse embryonic kidney cells, stem cells, and growth factors, investigators have produced a kidney in the lab that can filter blood. Therefore, they are hopeful of one day producing working kidneys for transplant patients.

CONSIDER THESE QUESTIONS

1. Would you tend to treat a person who is on hemodialysis differently than a person who had a bladder replaced? Explain.
2. If hemodialysis is not successful, should the doctor, nurse, or technician suffer any consequences? Why or why not?
3. Should people who are on hemodialysis be hired, even though they have to be away from work a couple of afternoons a week? Explain.

 Explore the concepts through a variety of multimedia assets and question types.

www.mcgrawhillconnect.com

FIGURE 34C
An artificial kidney machine.

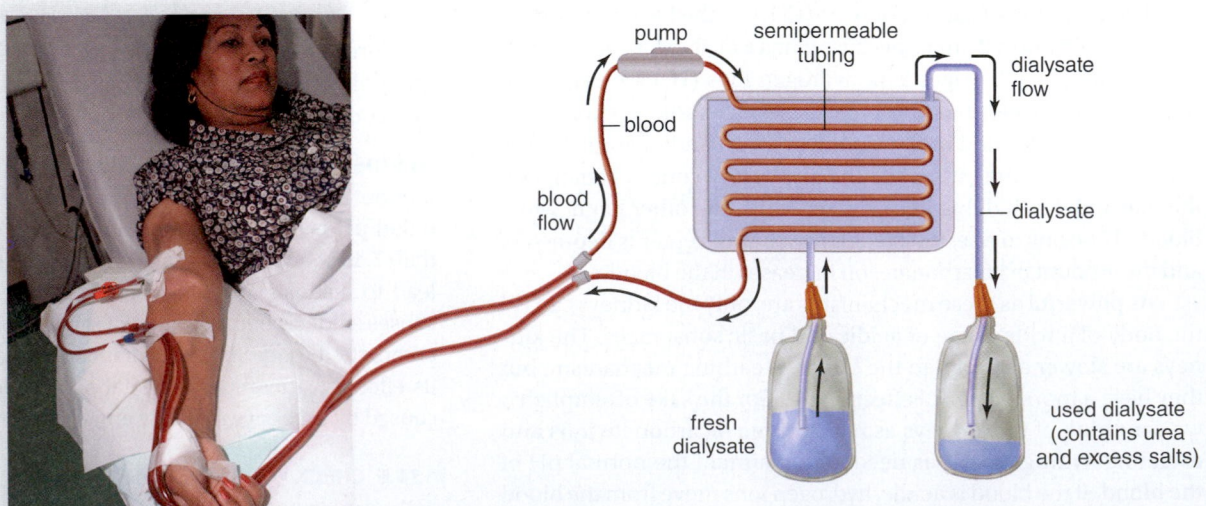

CONNECTING THE CONCEPTS

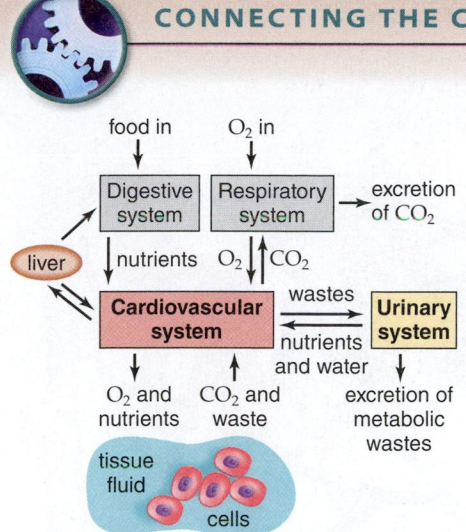

food in O_2 in

Digestive system → Respiratory system → excretion of CO_2

liver ↓ nutrients O_2 ↓↑ CO_2

Cardiovascular system wastes ⇄ Urinary system

nutrients and water

O_2 and nutrients CO_2 and waste   excretion of metabolic wastes

tissue fluid

cells

We have seen that the cardiovascular system works with the digestive and respiratory systems to maintain homeostasis as depicted in the accompanying illustration. Now we wish to consider the contribution of the urinary system. The kidneys are the chief regulators of the internal environment because they have ultimate control over what is removed and what is retained in the blood. They remove nitrogenous wastes such as urea (produced by the liver) and also uric acid. Even more importantly, the kidneys maintain the water-salt balance and the pH balance of the blood.

We have seen that excretion of CO_2 at the lungs and the retention of HCO_3^- at the kidneys help raise the pH. Also the kidneys can excrete H^+ as needed. On the other hand a basic pH can be brought under control by excreting HCO_3^- and retaining H^+ by the kidneys. The ion composition of the blood affects osmolarity and the workings of other body systems.

The ability of the vertebrate kidney to save water and produce a hypertonic urine was preadaptive to life on land where water is scarce. But when we drink a lot of water, why is the urine dilute? The general answer is that the kidneys are able to maintain the osmolarity of the blood. The specific answer is that ADH is not secreted from the posterior pituitary and water is lost. What about when we take in too much salt? Aldosterone is not secreted, and the kidneys excrete salt but also water. The problem is that the loss of water can cause the blood to become hypertonic to the body's cells, and unless we drink fresh water, we die from dehydration. Humans can't drink seawater because they have no way to get rid of excess salt without losing water. In contrast, marine fishes can drink sea water because they get rid of the excess salt across their gills.

ANALYZE AND EVALUATE

1. People who produce an insufficient amount of aldosterone have low blood pressure. Explain.
2. The systems of the body work together. Support this statement with reference to pH regulation of the blood.
3. Explain why you can't live without a kidney.

SUMMARIZE

Comparative Excretory Organs

34.1 The nitrogenous waste product of animals varies according to the environment

- Amino groups (—NH_2) removed from amino acids and nucleic acids are converted to ammonia, urea, or uric acid for **excretion.** Aquatic animals excrete **ammonia,** which requires water but no energy to produce. Sharks, adult amphibians, and mammals excrete **urea,** which requires energy to produce but less water to excrete. Reptiles and birds excrete **uric acid,** which requires the most energy to produce but the least water to excrete.

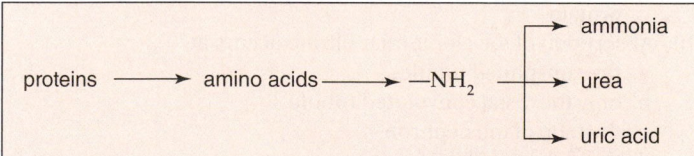

proteins → amino acids → —NH_2 → ammonia / urea / uric acid

34.2 Many invertebrates have organs of excretion

- Planarians have **flame cells,** earthworms have **nephridia,** and insects have **Malpighian tubules** for excretion of wastes.

Mammalian Urinary System

34.3 The urinary system includes kidneys that contain tubules

- Fishes and amphibians have elongated kidneys; the adult kidneys in reptiles and birds resemble those of mammals.

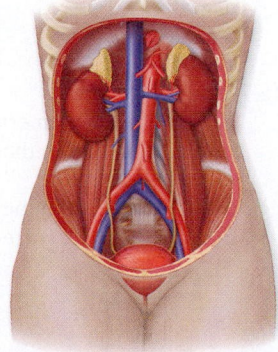

In mammals, the urinary system is composed of the **kidneys, ureters, urinary bladder,** and **urethra.** In most other vertebrates the ureters empty into a cloaca.

- The divisions of a kidney are the **renal cortex, renal medulla,** and **renal pelvis.** A **nephron** (kidney tubule) has a **glomerular capsule,** a **proximal convoluted tubule,** a **loop of the nephron,** a **distal convoluted tubule,** and a **collecting duct.** The blood supply of a nephron consists of an afferent arteriole, a **glomerulus,** an efferent arteriole, a **peritubular capillary network,** and venules.

34.4 The kidney tubules produce urine

- During **glomerular filtration,** small molecules move into the glomerular capsule and become a part of the **glomerular filtrate. Tubular reabsorption** occurs primarily at the proximal convoluted tubule and involves the reabsorption of most of the water, all of the nutrients, and other small molecules except urea. During **tubular secretion,** substances such as hydrogen ions, ammonia, creatinine, histamine, and penicillin are added to the filtrate.

Kidneys and Homeostasis

34.5 The kidneys can concentrate urine to maintain water-salt balance

- Most water and salts are reabsorbed at the proximal convoluted tubule. Water passes through **aquaporins.**
- In mammals, the loop of the nephron concentrates urine: In the ascending limb, the concentration of salt in the medulla increases, and urea leaks passively from the collecting duct. Water increasingly leaves the descending limb and collecting duct.
- Secretion of **antidiuretic hormone** (**ADH**) regulates water reabsorption from the distal convoluted tubule and collecting duct. Blood volume and pressure are regulated by water and salt absorption.
- The **renin**-angiotensin-aldosterone system which regulates salt reabsorption involves the liver, the kidneys, and the adrenal glands. **Angiotensin II** stimulates secretion of **aldosterone,** which promotes K^+ excretion and Na^+ absorption. Blood volume and pressure rise. **Atrial natriuretic hormone** (**ANH**) inhibits secretion of renin and aldosterone.

34.6 Lungs and kidneys maintain acid-base balance

- The lungs excrete carbon dioxide, a substance that tends to make the blood acidic. The kidneys excrete H^+ via formation of ammonium ions (NH_4^+) and reabsorption of bicarbonate ions, as needed to maintain blood pH: Acidosis is blood pH below 7.34, and alkalosis is blood pH above 7.45.

TEST YOURSELF

Comparative Excretory Organs

1. One advantage of urea excretion over uric acid excretion is that urea
   a. requires less energy than uric acid to form.
   b. can be concentrated to a greater extent.
   c. is not a toxic substance.
   d. requires no water to excrete.
   e. is a larger molecule.
2. Which of these pairs is mismatched?
   a. insects—excrete uric acid
   b. humans—excrete urea
   c. fishes—excrete ammonia
   d. birds—excrete ammonia
   e. All of these are correct.
3. Animals with which of these structures are most likely to excrete a semisolid nitrogenous waste?
   a. nephridia   d. flame cells
   b. Malpighian tubules e. All of these are correct.
   c. human kidneys
4. **THINKING CONCEPTUALLY** Of what benefit is a tubular shape to an excretory tubule? Explain your answer.

Mammalian Urinary System

5. Urine transport follows which order?
   a. kidneys, urethra, urinary bladder, ureter
   b. kidneys, ureter, urinary bladder, urethra
   c. urinary bladder, kidneys, ureter, urethra
   d. urethra, kidneys, ureter, urinary bladder

6. When tracing the path of glomerular filtrate, which structure is encountered just before the loop of the nephron?
   a. collecting duct
   b. distal convoluted tubule
   c. proximal convoluted tubule
   d. glomerulus
   e. renal pelvis
7. Label this diagram:

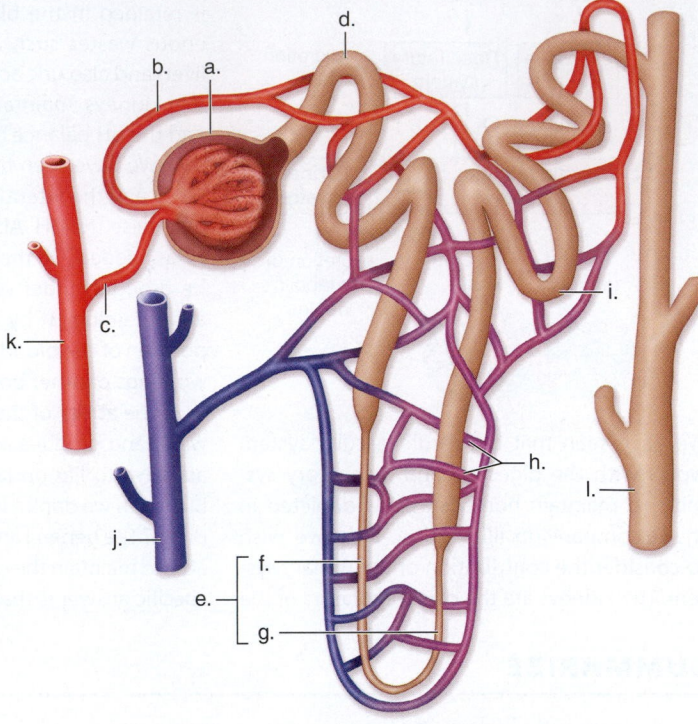

8. When tracing the path of blood, the blood vessel that follows the renal artery is called the
   a. peritubular capillary network.   d. renal vein.
   b. efferent arteriole.   e. glomerulus.
   c. afferent arteriole.
9. Which of these materials is not filtered from the blood at the glomerulus?
   a. water   d. glucose
   b. urea   e. sodium ions
   c. protein
10. Absorption of the glomerular filtrate occurs at
    a. the convoluted tubules.
    b. only the distal convoluted tubule.
    c. the loop of the nephron.
    d. the collecting duct.
11. Which of the following materials would not be maximally reabsorbed from the glomerular filtrate?
    a. water   d. urea
    b. glucose   e. amino acids
    c. sodium ions
12. In humans, water is
    a. found in the glomerular filtrate.
    b. reabsorbed from the nephron.
    c. in the urine.
    d. reabsorbed from the collecting duct.
    e. All of these are correct.

For questions 13–16, match the molecule to the correct process in the key. Answers can include more than one of these.

KEY:

    a. filtered
    b. reabsorbed
    c. tubular secreted
    d. appears normally in urine.

13. Water and Na$^+$
14. Urea
15. Glucose and amino acids
16. H$^+$
17. **THINKING CONCEPTUALLY** The concept that structure suits function is exemplified by the cells lining the proximal convoluted tubule. How so?

Kidneys and Homeostasis

18. Sodium is actively extruded from which part of the nephron?
    a. descending portion of the proximal convoluted tubule
    b. ascending portion of the loop of the nephron
    c. ascending portion of the distal convoluted tubule
    d. descending portion of the collecting duct
19. The function of a long loop of the nephron in the process of urine formation is
    a. reabsorption of water.   c. reabsorption of solutes.
    b. production of filtrate.   d. secretion of solutes.

For questions 20–22, match the hormone to the effect of its secretion. Choose more than one answer if correct.

KEY:

    a. excretion of Na$^+$
    b. excretion of H$_2$O
    c. reabsorption of Na$^+$
    d. reabsorption of H$_2$O

20. Aldosterone
21. ANH (atrial natriuretic hormone)
22. ADH (antidiuretic hormone)
23. The presence of ADH (antidiuretic hormone) causes an individual to excrete
    a. less salt.   d. more salt.
    b. less water.   e. Both a and c are correct.
    c. more water.
24. Which of these is not likely to cause a rise in blood pressure?
    a. aldosterone   c. renin
    b. antidiuretic hormone   d. atrial natriuretic hormone
    (ADH)   (ANH)

25. H$_2$CO$_3$ can be converted to
    a. H$^+$ + HCO$_3^-$.   c. Both a and b are correct.
    b. H$_2$O + CO$_2$.   d. Neither a nor b is correct.
26. To lower blood acidity,
    a. hydrogen ions are excreted, and bicarbonate ions are reabsorbed.
    b. hydrogen ions are reabsorbed, and bicarbonate ions are excreted.
    c. hydrogen ions and bicarbonate ions are reabsorbed.
    d. hydrogen ions and bicarbonate ions are excreted.
    e. urea, uric acid, and ammonia are excreted.
27. If a drug inhibits the kidneys' ability to reabsorb bicarbonate ions so that bicarbonate is excreted in the urine, the blood will become
    a. acidic.
    b. basic.
    c. first acidic and then basic.
    d. first basic and then acidic.
28. **THINKING CONCEPTUALLY** Knowing that osmosis plays a role in water reabsorption by the nephron, why would you recommend that people with high blood pressure reduce their salt intake?

GET INVOLVED

1. Suggest an anatomic study to substantiate that kangaroo rats are able to produce a very hypertonic urine. (See Fig. 26B on p. 524.)
2. You have built an artificial kidney but want to improve its ability to clear the blood of urea. What could you do? (See Fig. 34C on p. 680.)
3. After viewing the following virtual lab, answer this question. What does this virtual lab mean by the expression "liquid waste" and what molecules do you expect to find in a frog's "liquid waste"?

MEDIA STUDY TOOLS

mhhe.com/maderconcepts3

Enhance your study of this chapter with interactive study tools, practice tests, and engaging animations. Also, ask your instructor about the resources available through ConnectPlus, which includes LearnSmart, a personalized adaptive learning program, and a media-rich eBook.

35

Coordination by Hormone Signaling

BEFORE YOU BEGIN

Take a few minutes to recall

Receptor proteins in the plasma membrane (section 5.1)

The signal transduction pathway ("Evolution of the Plasma Membrane" on p. 91)

The location and function of the hypothalamus (section 27.5)

Type 2 diabetes ("Disorders Associated with Obesity" on p. 647)

Pheromones Among Us

Male and female peregrine falcons bow and "ee-chup," exchange food, and otherwise interact before they mate. Male elephant seals roar loudly, rear up on their foreflippers, and fight for territory and harems of females. Such courtship behaviors, no doubt, are strongly influenced by hormones.

But what about pheromones, each a mixture of small organic molecules wafting through the air? Pheromones are different from hormones because they are sent from one member of a species to another, and they modify the behavior and physiological processes of the recipient. If the receptors on the antennae of a male silkworm moth detect a few molecules of a sex attractant, the moth takes off immediately toward the female that released it. Social insects, such as ants and bees, use pheromones to keep each member of the society performing its particular function. What keeps fire ant workers working and the queen producing all those eggs? Pheromones.

To what degree do pheromones, in addition to hormones, affect the behavior of vertebrates, even mammals? Dr. Milos Novotny runs the Institute for Pheromone Research at Indiana University, and he says there is no doubt about their importance

Aggressive behavior

in mammalian behavior. He can show that in mice, pheromones trigger inter-male aggression and dominance, readiness for mating, onset of puberty, and communication of stress to other members of a colony. It's possible that pheromones are involved in all sorts of mammalian behaviors, including the social, maternal care, and migratory behaviors depicted on this page.

Courtship behavior

Like the mouse, humans apparently have an organ in the nose, called the vomeronasal organ (VNO), that can detect not only odors, but also pheromones. The neurons from this organ lead to the hypothalamus, the part of the brain that controls the release of so many hormones in the body, including the reproductive hormones. So, if we are receiving silent messages through the air from the other members of our species, could they be controlling our behavior, as occurs with the social insects? Well, maybe not exactly. It's possible that our ability to reason and think could override unaccountable urges.

Previous data have suggested that human physiological processes could possibly be controlled by pheromones. The dormitory effect, for example, refers to the ability of body odor to bring about synchronization of the menstrual cycle in women who live together. A few years ago, neurogeneticists at Rockefeller University and Yale University reported the isolation of a human gene, labeled *V1RL1*, which codes for a pheromone receptor. And other researchers, including David L. Berliner, have now identified a pheromone released by men that reduces premenstrual nervousness and tension in women. Perfumes containing the same chemical are now on sale! Berliner says, "Human beings communicate with each other with pheromones, just like any terrestrial animal. And they do it through the same organ (i.e., the nose) that all terrestrial animals have." Once received, pheromones may initiate nerve impulses that go to the hypothalamus. How the hypothalamus controls the secretion of hormones is one topic of this chapter. We also include a description of the endocrine system before studying the structure and function of the major glands that do its work.

Parental care

Social behavior

Migratory behavior

The Endocrine System

The nervous and endocrine systems work together to govern the activities of the internal organs. Whereas the nervous system uses neurotransmitters to signal muscles and glands, the endocrine system utilizes hormones to signal target organs (cells) throughout the body.

35.1 The endocrine and nervous systems work together

LEARNING OUTCOMES

When you complete this section, you should be able to

1. Compare the use of chemical signals by the nervous and endocrine systems.
2. List the major vertebrate endocrine glands, tell where they are located, and state their principal hormones.

The nervous system and the endocrine system work together to regulate the activities of other body systems (Table 35.1A). The nervous system sends nerve impulses along a nerve fiber directly to a target organ. Nerve impulses cause the terminals of axons to release a neurotransmitter, which binds to the plasma membrane receptors of muscles or glands. In Figure 35.1A, ❶ the neurotransmitter is binding to the receptors of smooth muscle fibers in the wall of an arteriole. Now the muscle fibers will contract and arteriole constriction will occur. The endocrine system uses the blood vessels of the cardiovascular system to send **hormones,** chemical messengers, to target organs. In Figure 35.1A, ❷ the target organ is the liver, and insulin from the pancreas is binding to receptors in the plasma membrane of liver cells. Now the liver will store glucose as glycogen.

Animation
Hormonal Communication

Axon conduction occurs rapidly, and so does diffusion of a neurotransmitter across a synapse. The rapid response of the nervous system is particularly useful if the stimulus is an external event that endangers our safety.

The **endocrine system** is composed of glands (Fig. 35.1B), which function more slowly than do nerves. It takes time for the blood to deliver hormones, and it takes time for cells to respond, but the effect is longer lasting. It is important to remember that only certain cells, called target cells, can respond to a hormone. For a cell to respond, the hormone and its plasma membrane receptor must bind together as a key fits a lock.

MP3
Endocrine System

TABLE 35.1	Comparison of Nervous and Endocrine Systems	
	Nervous System	**Endocrine System**
Composed Of	Neurons	Glands, usually
Delivery	Nerve impulse and neurotransmitter	Hormone
How Delivered	Axon and synapse	Bloodstream, usually
Target	Muscles and glands	Cells throughout body
Response	Rapid, short-lived	Slow, long-lasting
Controlled By	Negative feedback	Negative feedback, usually

Negative feedback, described in section 26.9, controls the activities of the nervous and endocrine systems. In the nervous system, a sensory receptor senses a change and signals the CNS, which brings about a corrective response. Once normalcy is restored, the sensory receptor no longer signals the CNS. In the endocrine system, a gland detects and responds to an abnormal blood level of a hormone or another substance by correcting that level. Once normalcy is restored, the gland becomes inactive.

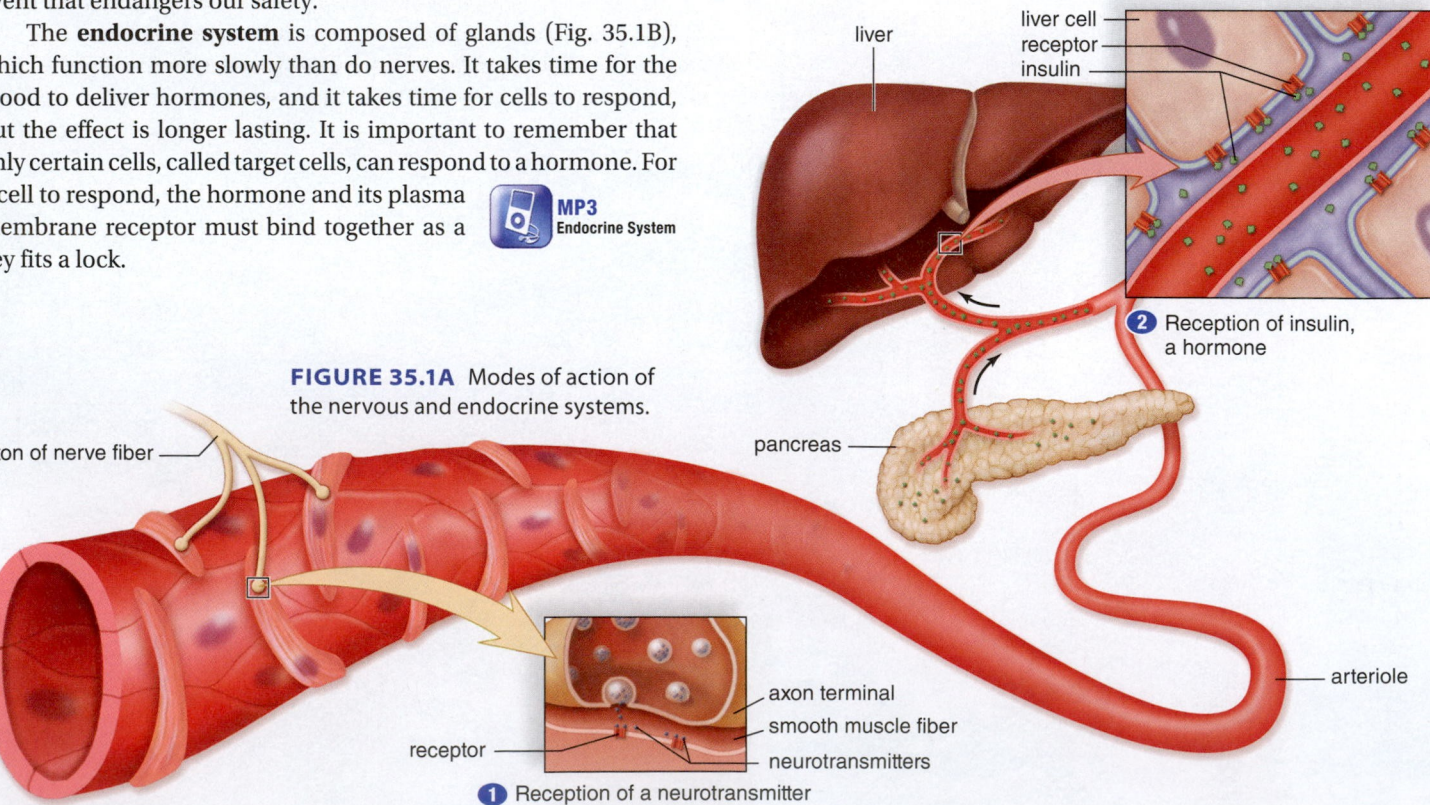

FIGURE 35.1A Modes of action of the nervous and endocrine systems.

axon of nerve fiber

liver

liver cell
receptor
insulin

❷ Reception of insulin, a hormone

pancreas

arteriole

axon terminal
smooth muscle fiber
receptor
neurotransmitters

❶ Reception of a neurotransmitter

HYPOTHALAMUS

Releasing and inhibiting hormones:
  regulate the anterior pituitary

PITUITARY GLAND

Posterior pituitary

Antidiuretic (ADH):
  water reabsorption by kidneys

Oxytocin: uterine contraction and milk
letdown

Anterior pituitary

Thyroid stimulating (TSH):
  stimulates thyroid

Adrenocorticotropic (ACTH):
  stimulates adrenal cortex

Gonadotropic (FSH, LH): egg and
  sperm production; sex hormone
  production

Prolactin (PL): milk production

Growth (GH): bone growth, protein
  synthesis, and cell division

THYROID

Thyroxine (T_4) and triiodothyronine
  (T_3): increase metabolic rate;
  regulate growth and development

Calcitonin: lowers blood calcium level

ADRENAL GLAND

Adrenal cortex

Glucocorticoids (cortisol):
  raises blood glucose level;
  stimulates breakdown of protein

Mineralocorticoids (aldosterone):
  reabsorption of sodium and
  excretion of potassium

Sex hormones: reproductive organs
  and bring about sex characteristics

Adrenal medulla

Epinephrine and norepinephrine:
  active in emergency situations;
  raise blood glucose level

GONADS

Testes

Androgens (testosterone):
  male sex characteristics

Ovaries

Estrogens and progesterone:
  female sex characteristics

FIGURE 35.1B Major glands of the human endocrine system are depicted along with the hormones they produce.

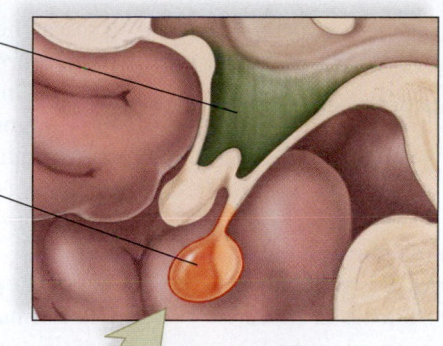

PINEAL GLAND

Melatonin: controls circadian
  and circannual rhythms

PARATHYROIDS

Parathyroid hormone (PTH):
  raises blood calcium level

parathyroid glands
(posterior surface
of thyroid)

THYMUS

Thymosins: production
  and maturation of T
  lymphocytes

PANCREAS

Insulin: lowers blood
  glucose level and
  promotes glycogen
  buildup

Glucagon: raises blood
  glucose level and
  promotes glycogen
  breakdown

testis
(male)

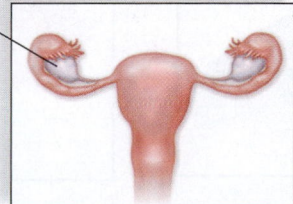

ovary (female)

▶ 35.1 CHECK YOUR PROGRESS

1. Compare the chemical signal used by
   the nervous system to that used by the
   endocrine system.
2. Identify the three hormonal glands
   found in, or close to, the brain.

LEARNING OUTCOMES

When you complete this section, you should be able to

1. Explain how hormones can be delivered to target organs and cells.
2. Contrast the manner in which peptide and steroid hormones affect cellular metabolism.

Although this chapter concentrates on mammals, various data tell us that invertebrates also produce hormones. It is well known that hormones control molting and metamorphosis in insects (Fig. 35.2A). Insulin–related peptides have been identified in insects and molluscs, suggesting that at least this hormone evolved early in the history of animal evolution. Hormones are active in all vertebrates but when identified may have a different function from that in humans. For example, the hormone prolactin inhibits metamorphosis in amphibians, stimulates skin pigmentation in reptiles, initiates incubation of eggs in birds, and stimulates milk production in mammals. Indeed, the study of hormone function across the spectrum of animal evolution is complicated!

This section emphasizes certain principles of hormone action:

1. Hormones are a means of communication between cells, between body parts, and even between individuals. Hormones affect the metabolism of **target cells.** Target cells must have a receptor specific to the hormone in order to receive it (Fig. 35.2B). Certain individuals with X and Y sex chromosomes and testosterone-producing testes in the abdominal cavity have the appearance of normal females because their body cells lack the receptor for testosterone! The condition is called androgen insensitivity.
2. Most hormones act at a distance between body parts, and in mammals they travel in the bloodstream from the glands that produced them to their target cells (Fig. 35.2B). Not all hormones act between body parts, however. Some are called local hormones because they affect cells in the area. Prostaglandins are a good example of a local hormone. After prostaglandins are produced, they are not carried elsewhere in the bloodstream. Instead, they diffuse to neighboring cells, sometimes promoting pain and inflammation. Growth factors are also local hormones that promote cell division and mitosis.
3. **Pheromones** are chemical signals that influence the behavior of other individuals, as discussed in the introduction to this chapter. As evidence that humans produce pheromones,

a researcher has isolated one released by men that reduces premenstrual nervousness and tension in women. Women who live in the same household often have menstrual cycles in synchrony. It could be that the armpit secretions of one menstruating woman are influencing the timing of the menstrual cycle in the other women of the household.

Video Sex and the Senses

Action of Peptide Hormones The term **peptide hormone** refers to hormones that are peptides, proteins, or glycoproteins. The hypothalamus, pituitary gland (both posterior and anterior), parathyroid, pancreas, thymus gland, pineal gland and the adrenal medulla produce peptide hormones. Epinephrine is a peptide hormone that leads to the breakdown of glycogen to glucose in muscle fibers, thereby providing energy for ATP production. Consult Figure 35.1B and note that the hormone epinephrine is produced by the adrenal medulla. It's the very same molecule that acts as a neurotransmitter in the sympathetic division of the autonomic system. Its ability to act as a hormone that promotes the breakdown of glycogen to glucose (Fig. 35.2C) is also consistent with its role in the fight-or-flight response.

Animation Peptide Hormone Action

In Figure 35.2C, note that after ❶ the hormone epinephrine binds to a receptor in the plasma membrane, ❷ the immediate effect is the activation of an enzyme that changes ATP to **cAMP** (**cyclic adenosine monophosphate**). ❸ The molecule cAMP activates an enzyme cascade, a series of enzymes that amplify the effect of the molecule. ❹ Finally, many molecules of glycogen are broken down to glucose, which enters the bloodstream.

Typical of a peptide hormone, epinephrine never enters the cell. Therefore, the hormone is called the **first messenger,** while cAMP, which sets the metabolic machinery in motion, is called the **second messenger.** To explain this terminology, let's imagine that the adrenal medulla is like the home office that sends out a courier (epinephrine) to a factory (the cell). The courier is the first messenger that doesn't have a pass to enter the factory, so when he arrives

FIGURE 35.2A An interplay of three hormones, unique to insects, determines whether their larvae molt or undergo pupation and metamorphosis.

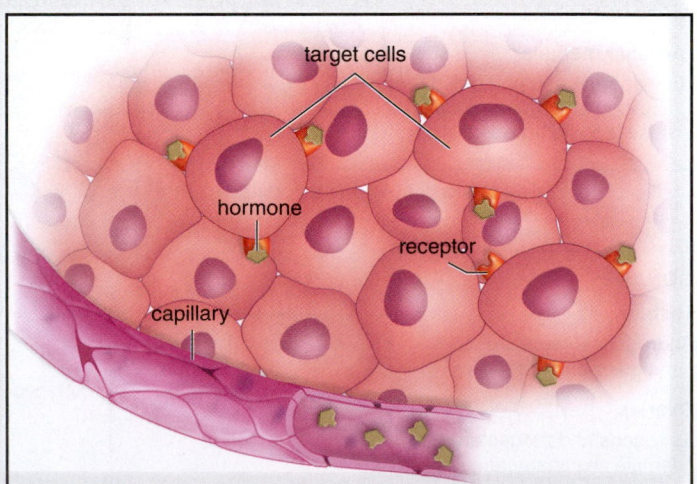

FIGURE 35.2B Cells that respond to hormones, called target cells, must have specific receptors to receive them. In mammals, most hormones are transported in the bloodstream to target cells.

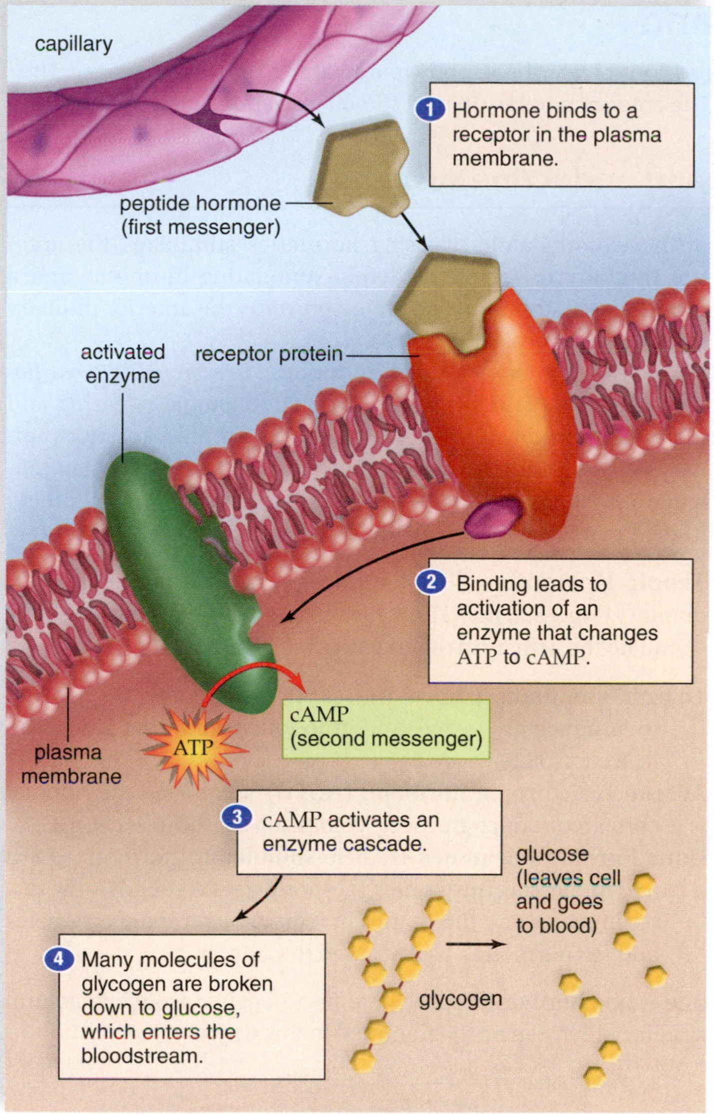
capillary
1 Hormone binds to a receptor in the plasma membrane.
peptide hormone (first messenger)
activated enzyme
receptor protein
2 Binding leads to activation of an enzyme that changes ATP to cAMP.
plasma membrane
ATP
cAMP (second messenger)
3 cAMP activates an enzyme cascade.
glucose (leaves cell and goes to blood)
4 Many molecules of glycogen are broken down to glucose, which enters the bloodstream.
glycogen

FIGURE 35.2C A peptide hormone (first messenger) binds to a receptor in the plasma membrane. Thereafter, cAMP (second messenger) forms and activates an enzyme cascade.

at the factory, he tells a supervisor through the screen door that the home office wants the factory to produce a particular product. The supervisor (cAMP, the second messenger) walks over and flips a switch that starts the machinery (the enzymatic pathway), and glucose is produced.

Animation Second Messengers

Action of Steroid Hormones **Steroid hormones** all have the same complex of four carbon rings because they are all derived from cholesterol (see section 3.6). Only the adrenal cortex and the gonads produce steroid hormones (see Fig. 35.1B). The sex hormones produced by the gonads are the most familiar of the steroid hormones. They are responsible for the development of the genitals and the secondary sex characteristics of the sexes. Unlike peptide hormones, which are broken down by digestive enzymes, steroid hormones, such as those in the birth control pill, can be taken by mouth.

Steroid hormones, and also thyroid hormone, which are amines, do not bind to plasma membrane receptors (Fig. 35.2D). (1) Instead, they are able to enter the cell because they are lipid soluble, as are the phospholipids of the plasma membrane. (2) Once inside, steroid hormones bind to receptors, usually in the nucleus, but sometimes in the cytoplasm. (3) Inside the nucleus, the hormone-receptor complex binds with DNA and activates transcription of certain portions of DNA. (4) Translation of messenger RNA (mRNA) transcripts at ribosomes results in enzymes and other proteins that can carry out a response to the hormonal signal. To continue our analogy, a steroid hormone is like a courier that has a pass to enter the factory (the cell). Once inside, it makes contact with the plant manager (DNA), who sees to it that the factory (cell) is ready to produce a product. Steroids act more slowly than peptides because it takes more time to synthesize new proteins than to activate enzymes already present in cells. Their action lasts longer, however.

Animation Mechanism of Steroid Hormone Action

▶ 35.2 CHECK YOUR PROGRESS

1. Explain why a second messenger is needed for most peptide hormones and why steroid hormones take longer to bring about a change in metabolism.

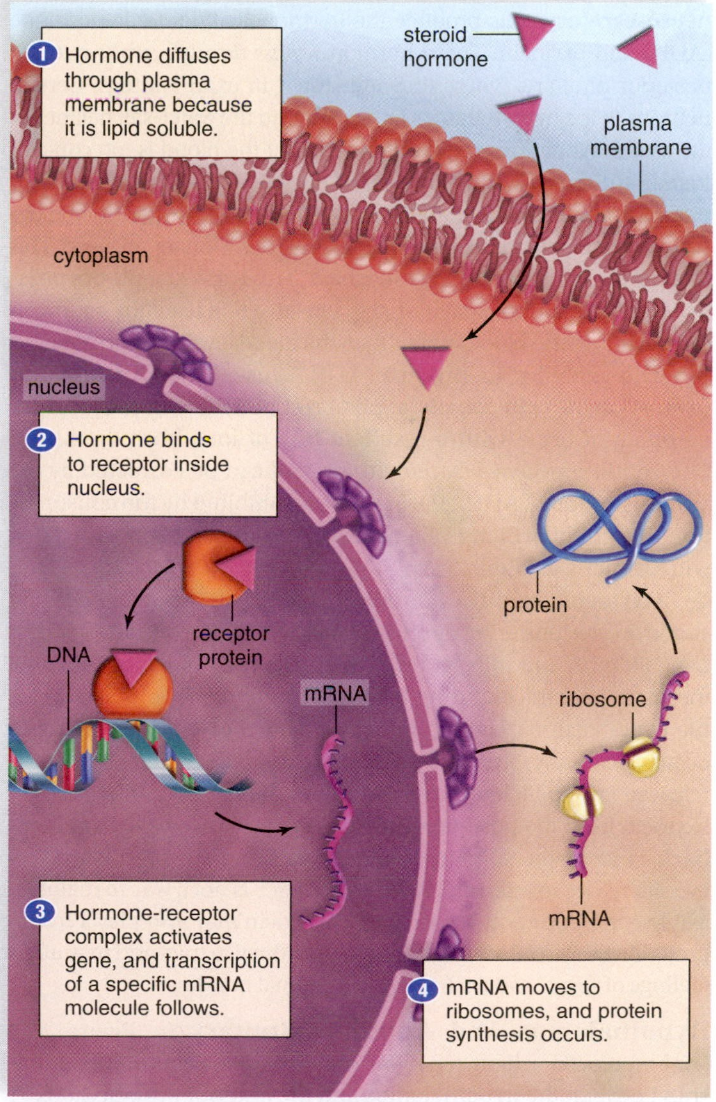
1 Hormone diffuses through plasma membrane because it is lipid soluble.
steroid hormone
plasma membrane
cytoplasm
nucleus
2 Hormone binds to receptor inside nucleus.
protein
DNA
receptor protein
mRNA
ribosome
mRNA
3 Hormone-receptor complex activates gene, and transcription of a specific mRNA molecule follows.
4 mRNA moves to ribosomes, and protein synthesis occurs.

FIGURE 35.2D A steroid hormone passes directly through the target cell's plasma membrane before binding to a receptor in the nucleus or cytoplasm. The hormone-receptor complex binds to DNA, and gene expression follows.

The Hypothalamus and the Pituitary Gland

The hypothalamus regulates the internal environment; for example, it helps control heartbeat, body temperature, and water balance. The hypothalamus also controls the posterior and anterior pituitary glands. The anterior pituitary in turn stimulates various other glands.

35.3 The hypothalamus is a part of the nervous and endocrine systems

LEARNING OUTCOMES

When you complete this section, you should be able to

1. Describe the function of the hormones secreted by the posterior and anterior pituitary glands.
2. Describe the three-tier relationship between the hypothalamus, the anterior pituitary, and the thyroid, adrenal cortex, and gonads.

A stalklike structure connects the hypothalamus to the **pituitary gland,** which is only about 1 cm in diameter. Figure 35.3 shows that the hypothalamus controls the glandular secretions of the **posterior pituitary** (*left*) and the **anterior pituitary** (*right*).

Hypothalamus and Posterior Pituitary In Figure 35.3 (*left,* colored blue), you can note that neurons in the hypothalamus called **neurosecretory cells** produce the hormones antidiuretic hormone (ADH) and oxytocin. These hormones *pass through axons* into the posterior pituitary, where they are stored in axon endings. Certain neurons in the hypothalamus are sensitive to the water-salt balance of the blood. When these cells determine that the blood is too concentrated, **antidiuretic hormone (ADH)** is released from the posterior pituitary. Upon reaching the kidneys, ADH causes water to be reabsorbed. As the blood becomes dilute, ADH is no longer released. This is another example of negative feedback. This is an example of a negative feedback control system that is sensitive to a resulting condition (osmolarity of the blood) rather than the blood level of a hormone.

The inability to produce ADH causes *diabetes insipidus* (watery urine), in which a person's body produces copious amounts of urine with a resultant loss of ions from the blood. The condition causes extreme thirst and can be corrected by the administration of ADH. ADH release is inhibited by the consumption of alcohol, which explains the frequent urination associated with drinking alcohol.

Oxytocin, the other hormone made in the hypothalamus and stored in the posterior pituitary, causes uterine contractions during childbirth and milk letdown when a baby is nursing. The more the uterus contracts during labor, the more nerve impulses reach the hypothalamus, causing oxytocin to be released until childbirth occurs. Similarly, the more a baby suckles, the more oxytocin is released until milk letdown occurs. In both instances, the release of oxytocin from the posterior pituitary is controlled by **positive feedback**—that is, the stimulus continues to bring about an effect that ever increases in intensity. Positive feedback is not a way to maintain stable conditions and homeostasis. Oxytocin may also play a role in propelling semen through the male reproductive tract and may affect feelings of sexual satisfaction and emotional bonding.

Hypothalamus and Anterior Pituitary In Figure 35.3 (*right,* colored a light tan) you can note the anatomic relationship between the hypothalamus and the anterior pituitary. Box #1 tells you that the hypothalamus controls the anterior pituitary by producing **hypothalamic-releasing hormones** and in some instances hypothalamic-inhibiting hormones. For example, one of the hypothalamic-releasing hormones stimulates the anterior pituitary to secrete a thyroid-stimulating hormone, and a hypothalamic-inhibiting hormone prevents the anterior pituitary from secreting prolactin.

As noted in box #2, these hormones are *secreted into cardiovascular veins*. In other words, the hypothalamic releasing and inhibiting hormones reach their target organ, the anterior pituitary, by the bloodstream as do most hormones. After a hypothalamic stimulatory hormone has stimulated the anterior pituitary, it secretes its hormones into the bloodstream (boxes #3 and #4).

Tropic Hormones Four hormones produced by the anterior pituitary (Fig. 35.3, *right*) are called tropic hormones because they stimulate the activity of other endocrine glands:

Thyroid-stimulating hormone (TSH) stimulates the thyroid to produce thyroxine and triiodothyronine, as discussed in section 35.5.
Adrenocorticotropic hormone (ACTH) stimulates the adrenal cortex to produce glucocorticoid, as discussed in section 35.7.
Gonadotropic hormones (follicle-stimulating hormone, FSH, and luteinizing hormone, LH) stimulate the gonads—the testes in males and the ovaries in females—to produce gametes and sex hormones, as discussed in Chapter 36.

The tropic hormones participate in a negative feedback control system, which can be diagrammed in this way:

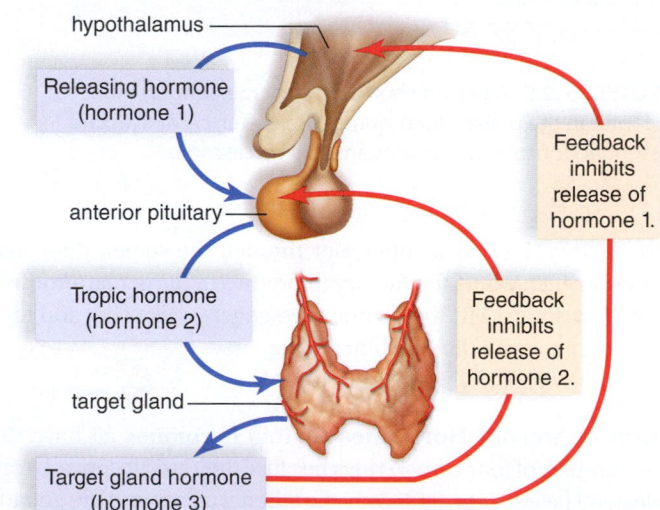

Like the negative feedback system mentioned on page 686, in this diagram the blood level of hormone is being controlled. What's unusual in this instance is that a negative feedback system involves two hormonal glands and three hormones. We can take as an example the blood level of thyroxine. In the diagram a low blood level of thyroxine would cause the hypothalamus to secrete a releasing hormone (hormone 1) and this hormone in turn would cause the anterior pituitary to secrete TSH (hormone 2). TSH stimulates

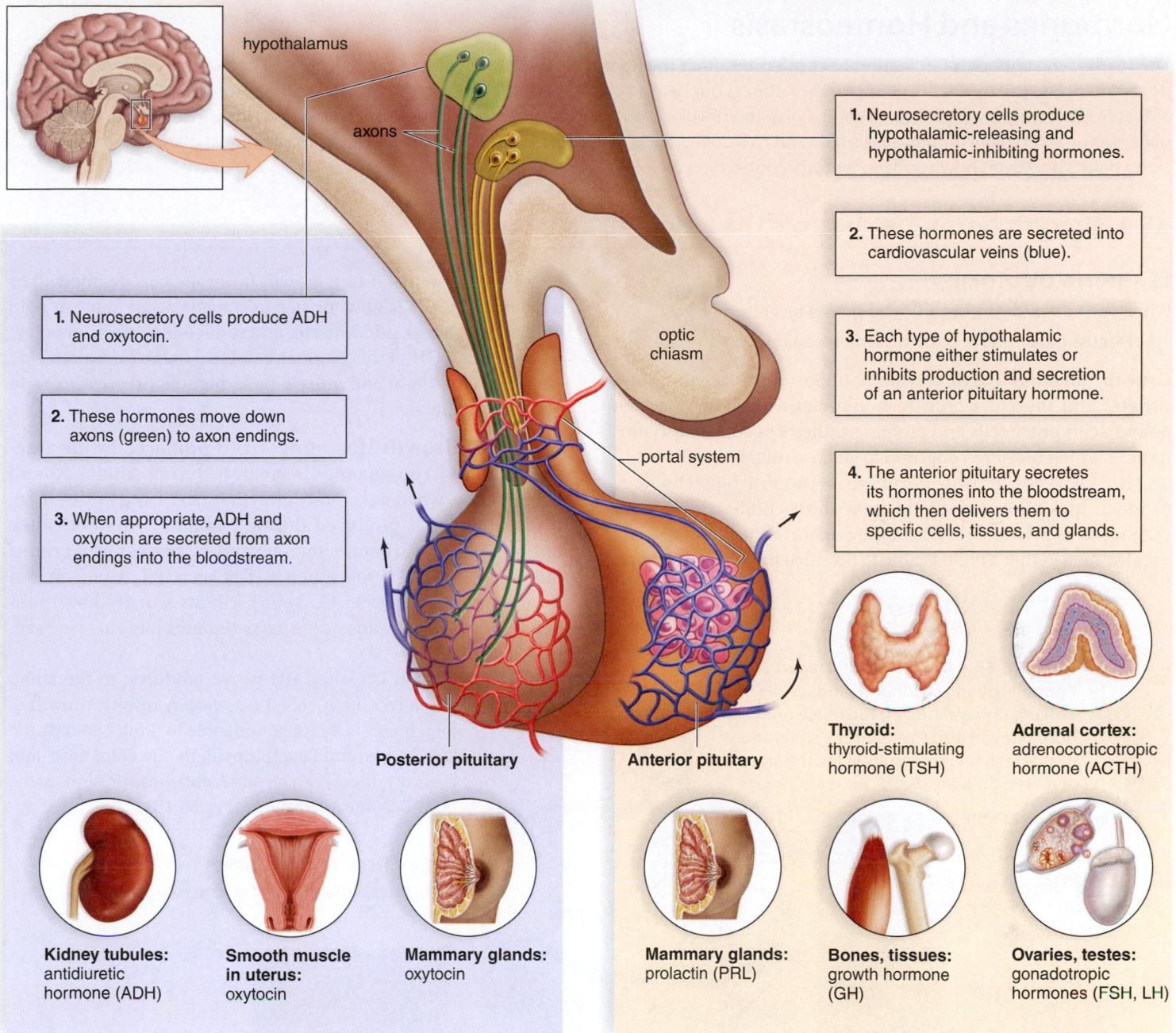

hypothalamus

axons

1. Neurosecretory cells produce hypothalamic-releasing and hypothalamic-inhibiting hormones.

2. These hormones are secreted into cardiovascular veins (blue).

optic chiasm

3. Each type of hypothalamic hormone either stimulates or inhibits production and secretion of an anterior pituitary hormone.

1. Neurosecretory cells produce ADH and oxytocin.

2. These hormones move down axons (green) to axon endings.

3. When appropriate, ADH and oxytocin are secreted from axon endings into the bloodstream.

portal system

4. The anterior pituitary secretes its hormones into the bloodstream, which then delivers them to specific cells, tissues, and glands.

Posterior pituitary

Anterior pituitary

Thyroid: thyroid-stimulating hormone (TSH)

Adrenal cortex: adrenocorticotropic hormone (ACTH)

Kidney tubules: antidiuretic hormone (ADH)

Smooth muscle in uterus: oxytocin

Mammary glands: oxytocin

Mammary glands: prolactin (PRL)

Bones, tissues: growth hormone (GH)

Ovaries, testes: gonadotropic hormones (FSH, LH)

FIGURE 35.3 The hypothalamus controls the pituitary gland, including both the posterior pituitary (*left,* blue) and the anterior pituitary (*right,* light tan).

the thyroid to secrete thyroxine (hormone 3). Thyroxine exerts feedback control over the activity of both the hypothalamus and the pituitary so that its blood level stays within normal range.

Nontropic Hormones Some hormones produced by the anterior pituitary are controlled by the hypothalamus and are called nontropic hormones because they do not affect other endocrine glands (Fig. 35.3, *right*).

Prolactin (**PRL**) causes the mammary glands in the breasts to develop and produce milk after childbirth. It also plays a role in carbohydrate and fat metabolism.

Growth hormone (**GH**) promotes skeletal and muscular growth. It stimulates the rate at which amino acids enter cells and

protein synthesis occurs. It also promotes fat metabolism, as discussed in section 35.4.

In addition, **melanocyte-stimulating hormone** (**MSH**) causes skin-color changes in vertebrates having melanophores, special skin cells that produce color variations. The concentration of this hormone in humans is very low.

▶ **35.3 CHECK YOUR PROGRESS**

1. Explain how the hypothalamus communicates with the endocrine system.

2. Identify what effect a rise in thyroxine would have on the secretions of the hypothalamus and the anterior pituitary.

Hormones and Homeostasis

The study of hormones is fascinating because they affect the appearance and/or health of the individual. For example, growth hormone produced by the anterior pituitary affects our height; oversecretion by the thyroid gland and the eyes bulge out; not enough aldosterone from the adrenal cortex and a nice tan develops but also life-threatening Addison disease. Too little insulin from the pancreas or an inability of the cells to respond to insulin results in diabetes, and finally melatonin from the pineal gland affects our sleep habits and even how soon we undergo puberty.

35.4 Growth hormone controls the height of an individual

LEARNING OUTCOME

When you complete this section, you should be able to

1. Discuss the effects of growth hormone on the body.

Growth hormone (GH), or somatotropic hormone, promotes skeletal and muscular growth. It stimulates the rate at which amino acids enter cells and protein synthesis occurs. It also promotes fat metabolism as opposed to glucose metabolism.

In the 1980s, growth hormone became a biotechnology product, making it possible to treat short children and those diagnosed as pituitary dwarfs. A growth hormone blood test can determine whether a child is able to produce the normal amount of growth hormone. If not, growth hormone can be injected as a medication. GH is now for sale as a medication that can help adults lose weight, add muscle, and reduce the effects of aging. Using human GH in this manner can have many undesired side effects, such as joint and muscle pain, high blood pressure, and diabetes mellitus.

Effects of Growth Hormone GH is produced by the anterior pituitary. The quantity is greatest during childhood and adolescence, when most body growth is occurring (35.4A, *left*). If too little GH is produced during childhood, the individual has **pituitary dwarfism,** characterized by perfect proportions but small stature. If too much GH is secreted, a person can become a giant (Fig. 35.4A, *right*). Giants usually have poor health, primarily because they have diabetes mellitus (see section 35.7).

On occasion, GH is overproduced in the adult, and a condition called *acromegaly* results. Growth of long bones is no longer possible in adults, so only the feet, hands, and face (particularly the chin, nose, and eyebrow ridges) can respond, and these portions of the body become overly large (Fig. 35.4B).

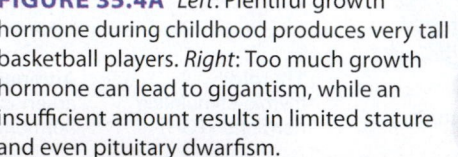

FIGURE 35.4A *Left*: Plentiful growth hormone during childhood produces very tall basketball players. *Right*: Too much growth hormone can lead to gigantism, while an insufficient amount results in limited stature and even pituitary dwarfism.

> **35.4 CHECK YOUR PROGRESS**
> 1. Explain the effects of acromegaly on the body.

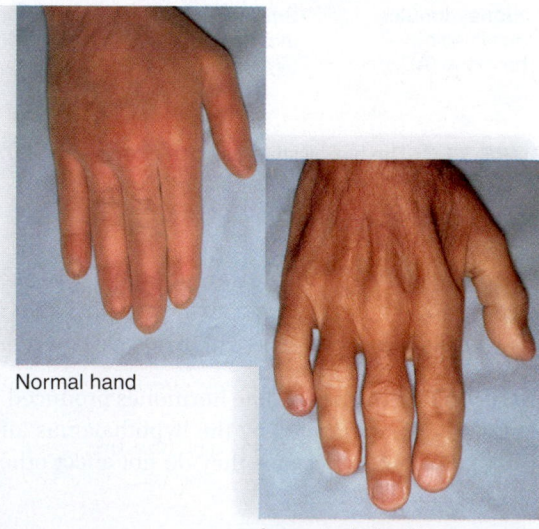

Normal hand

Acromegaly hand

FIGURE 35.4B Acromegaly is caused by overproduction of GH in the adult. It is characterized by enlargement of the bones in the face, the fingers (as shown here), and the toes as a person ages.

LEARNING OUTCOME

When you complete this section, you should be able to

1. Discuss the functions of the thyroid and parathyroids and disorders that can arise due to malfunctioning of these glands.

The **thyroid gland** is a two-lobed gland located in the neck; the four small **parathyroid glands** are embedded in the back of the thyroid (see Fig. 35.1B). The thyroid produces **triiodothyronine** (T_3), containing three iodine atoms and **thyroxine** (T_4), containing four iodine atoms. These hormones increase the metabolic rate of all cells so that they function at a faster rate.

Animation
Mechanism of Thyroxine Action

Hypothyroidism in an adult can result if the thyroid is unable to produce its hormones due to lack of iodine in the diet. Under these circumstances, the anterior pituitary increases its stimulation of the thyroid, which visibly enlarges and becomes a *goiter* (Fig. 35.5). The use of iodized salt in the United States prevents goiter formation. A pregnant woman should be sure to receive adequate iodine in order to prevent a condition called congenital hypothyroidism (cretinism). Children with this condition are short and stocky and often intellectually disabled.

Hyperthyroidism, also called *Graves disease*, results from oversecretion of the thyroid hormones. The eyes may protrude (exophthalmos) because of edema in the eye socket tissues and swelling of the muscles that move the eyes. The patient usually becomes nervous, and irritable, and suffers from insomnia. Removal or destruction of a portion of the thyroid by means of radioactive iodine is sometimes effective in curing the condition.

Calcium metabolism The body uses calcium for muscle contraction, neurotransmitter release, blood clotting, and other

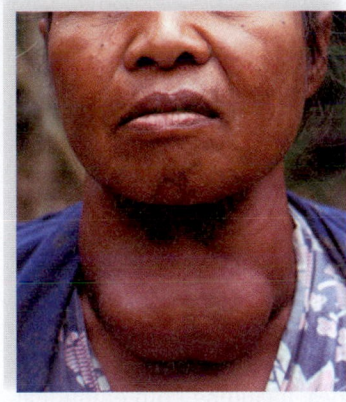

FIGURE 35.5 Simple goiter is due to a thyroid gland that enlarges because it is not receiving enough iodine.

vital activities. The thyroid produces **calcitonin,** a hormone that removes calcium from the blood and puts it into bones, and the parathyroid glands produce **parathyroid hormone** (**PTH**), which does the opposite. Calcitonin is an FDA approved drug for reducing bone loss in osteoporosis, a condition that can be worsened by PTH. Aside from directly raising the blood calcium level, PTH also promotes the reabsorption of calcium by the kidneys and activates vitamin D. Vitamin D stimulates the absorption of calcium from the intestine.

MP3
Calcium Homeostasis

▶ **35.5 CHECK YOUR PROGRESS**

1. Explain what makes the thyroid subject to goiters and how thyroid goiters can be prevented.
2. Explain how calcitonin and PTH are antagonistic hormones.

HOW LIFE CHANGES *Application*

35A Fish Gills and the Parathyroid Glands

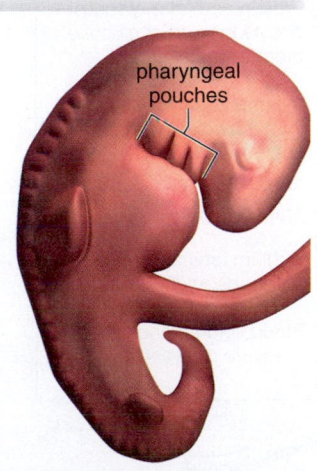

FIGURE 35A The parathyroid glands develop from pharyngeal pouches.

Observational data tell us that the parathyroids develop from embryonic pharyngeal pouches, the very same pouches that give rise to the gills of fish (Fig. 35A). This is good anatomic evidence that there is a link between the evolution of fish and tetrapods, which breathe by lungs and not gills! Now, researchers have discovered genetic evidence that makes the link stronger: Both gills and the parathyroid glands express a gene called *Gcm2*. Further, both structures require this gene's function for their formation! The conclusion is that the "fish" gene has been modified

through evolution to be critical to the formation of parathyroid glands also. Gills in fish carry out gas exchange, and they also maintain blood calcium levels by absorbing Ca^{2+} from the water. Most amazing, gills also possess PTH genes and generate active peptides. Thus, both fish gills and parathyroids are involved in regulating calcium blood levels, further substantiating why the parathyroid glands are in our neck. After all, they could have been located anywhere in the body.

CONSIDER THIS QUESTION

1. What makes the genetic evidence stronger than the anatomic evidence that fish gills and the parathyroid glands are related structures?

Explore the concepts through a variety of multimedia assets and question types.

www.mcgrawhillconnect.com

LEARNING OUTCOME

When you complete this section, you should be able to

1. Explain how the adrenal gland hormones work together to maintain homeostasis.

Two **adrenal glands** sit atop the kidneys (see Fig. 35.1B). Each gland is about 5 cm long and 3 cm wide and weighs about 5 g. Each adrenal gland consists of an inner portion called the **adrenal medulla** and an outer portion called the **adrenal cortex.** These portions, like the anterior pituitary and the posterior pituitary, have no physiological connection with one another. However, the hypothalamus exerts control over the activity of both portions of the adrenal glands. Stress of all types, including both emotional and physical trauma, prompts the hypothalamus to stimulate the adrenal glands.

Adrenal Medulla The hypothalamus initiates nerve impulses that travel by way of the brain stem, spinal cord, and sympathetic nerve fibers to the adrenal medulla, which then secretes its hormones (Fig. 35.6A, *left*). **Epinephrine** (adrenaline) and **norepinephrine** (noradrenaline) produced by the adrenal medulla rapidly bring about all the body changes that occur when an individual reacts to an emergency situation. The effects of these hormones are short-term. Epinephrine and norepinephrine accelerate the breakdown of glucose leading to the production of ATP,

trigger the mobilization of glycogen reserves in skeletal muscle, and increase the cardiac rate and force of contraction.

Adrenal Cortex The adrenal cortex secretes small amounts of male and female sex hormones in both sexes—that is, in the male, both male and female sex hormones are produced by the adrenal cortex, and in the female, both female and male sex hormones are produced by the adrenal cortex.

The adrenal cortex produces the **glucocorticoids** and the **mineralocorticoids** (Fig. 35.6A, *right*). These hormones provide a long-term response to stress. The hypothalamus, by means of corticotropin-releasing hormone, controls the anterior pituitary's secretion of ACTH, which in turn stimulates the adrenal cortex to secrete glucocorticoids. These hormones are involved in the three-tier relationship discussed in section 35.3.

Animation
Action of Glucocorticoid Hormone

Cortisol is a biologically significant glucocorticoid produced by the adrenal cortex. Cortisol raises the blood glucose level in at least two ways: (1) It promotes the breakdown of muscle proteins to amino acids, which the liver takes up from the bloodstream. The liver then breaks down these amino acids to form glucose, which enters the blood. (2) Cortisol promotes the metabolism of fatty acids rather than carbohydrates. Both of these actions spare glucose.

Cortisol also counteracts the inflammatory response, which leads to the pain and swelling of joints in arthritis and bursitis. The administration of cortisol in the form of cortisone aids these conditions because it reduces inflammation. Very high levels of glucocorticoids in the blood can suppress the body's defense system, including the inflammatory response that

FIGURE 35.6A Stress responses of the adrenal gland.

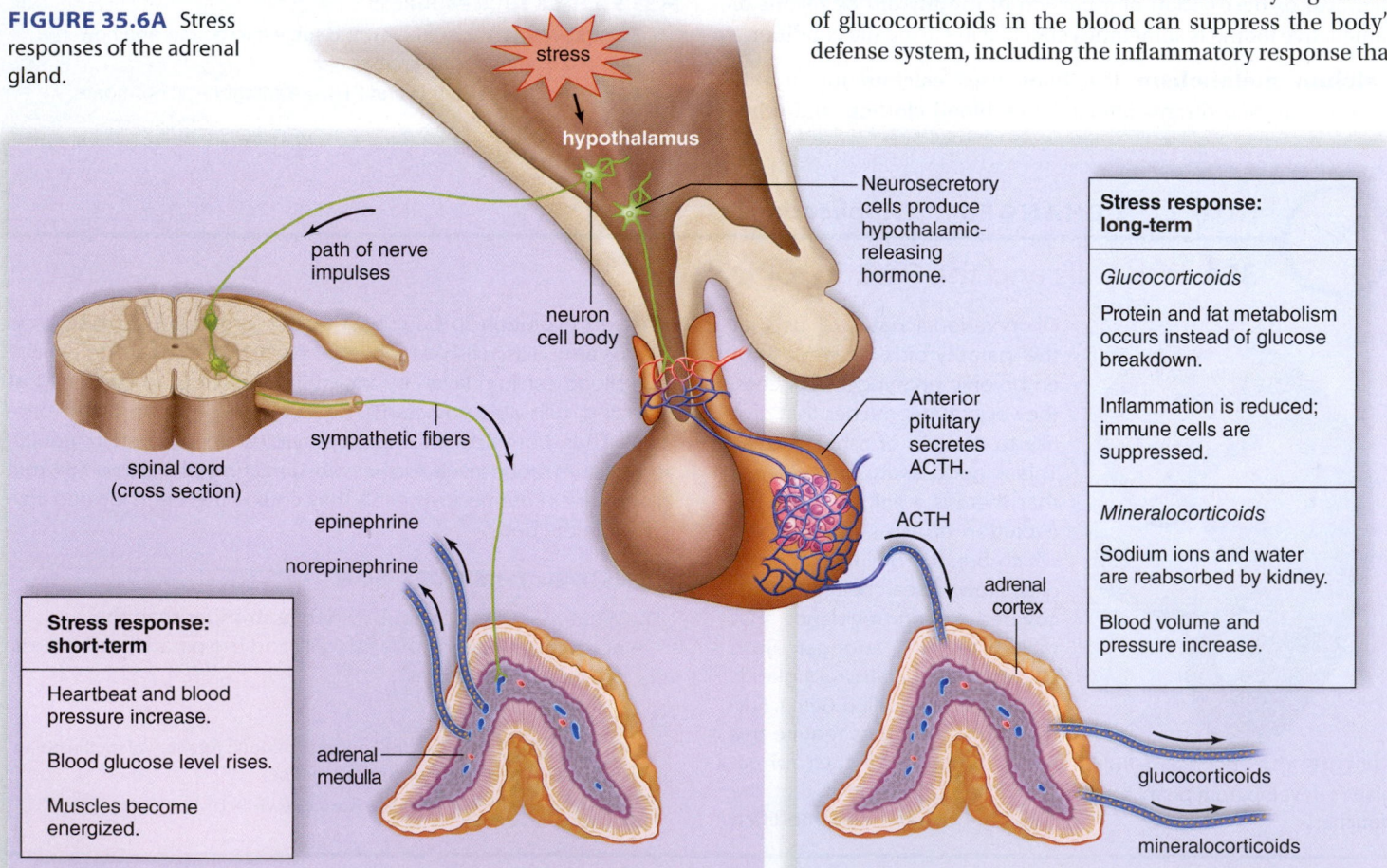

Stress response: short-term

Heartbeat and blood pressure increase.

Blood glucose level rises.

Muscles become energized.

Stress response: long-term

Glucocorticoids

Protein and fat metabolism occurs instead of glucose breakdown.

Inflammation is reduced; immune cells are suppressed.

Mineralocorticoids

Sodium ions and water are reabsorbed by kidney.

Blood volume and pressure increase.

occurs at infection sites. Cortisone and other glucocorticoids can relieve swelling and pain from inflammation, but by suppressing pain and immunity, they can also make a person highly susceptible to injury and infection.

Aldosterone (also discussed in Chapter 34) is the most important of the mineralocorticoids. Aldosterone primarily targets the kidney, where it promotes renal absorption of sodium (Na^+) and renal excretion of potassium (K^+).

The secretion of mineralocorticoids is not controlled by the anterior pituitary. When the blood sodium level, and therefore the blood pressure, are low, the kidneys secrete **renin.** Renin is an enzyme that converts the plasma protein angiotensinogen to angiotensin I, which later is changed to angiotensin II. Angiotensin II stimulates the adrenal cortex to release aldosterone (see Fig. 34.5B). The effect of this process, called the renin-angiotensin-aldosterone system, is to raise blood pressure in two ways: (1) Angiotensin II constricts the arterioles. (2) Aldosterone causes the kidneys to reabsorb Na^+. When the blood sodium level rises, water is reabsorbed. Then blood pressure increases.

Disorders When the blood level of adrenal cortex hormones is low, a person develops *Addison disease*. The presence of excessive

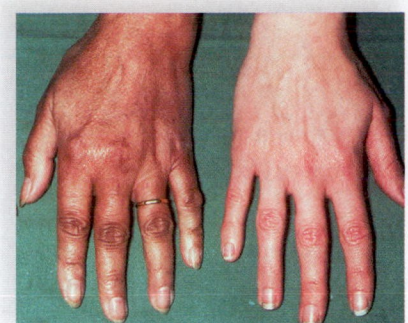

FIGURE 35.6B
Addison disease is characterized by a peculiar bronzing of the skin, as seen in the hand on the left.

amounts of ACTH causes a buildup of melatonin and bronzing of the skin (Fig. 35.6B). Without glucocorticoids, stress cannot be tolerated and even a mild infection can lead to death. The inability to produce aldosterone causes low blood pressure and possibly severe dehydration. *Cushing syndrome* due to excessive glucocorticoids is discussed next.

> **35.6 CHECK YOUR PROGRESS**
> 1. Associate an alarm pheromone with the release of a particular adrenal gland hormone. Explain.

HOW BIOLOGY IMPACTS OUR LIVES *Application*

35B Glucocorticoid Therapy Can Lead to Cushing Syndrome

When the level of adrenal cortex hormones is high due to hypersecretion, a person develops *Cushing syndrome*. Excess cortisol results in a tendency toward diabetes mellitus as muscle protein is metabolized and subcutaneous fat is deposited in the midsection. The trunk is obese, while the arms and legs remain normal in size. Excess aldosterone and reabsorption of sodium and water by the kidneys lead to a basic blood pH and hypertension (high blood pressure). The face is moon-shaped due to edema (Fig. 35B, *left*). Masculinization may occur in women because of excess adrenal male sex hormones. This condition, known as adrenogenital syndrome (AGS), is characterized by an increase in body hair, deepening of the voice, and beard growth in women.

Some people develop Cushing syndrome due to a tumor in the adrenal cortex or anterior pituitary. Normally, cortisol secretion fluctuates, with blood levels highest in the early morning and falling throughout the day. In people with Cushing syndrome, cortisol levels remain much the same throughout the day. Additional tests may help doctors identify the location of the tumor causing Cushing syndrome. Radiation therapy or surgery is necessary to remove the tumor.

Other individuals develop Cushing syndrome because they are taking glucocorticoid drugs. Glucocorticoids are often prescribed for medical conditions such as asthma, rheumatoid arthritis, and lupus. Glucocorticoid therapy is also used to suppress the immune system in order to prevent rejection of a transplanted organ. If glucocorticoid therapy in these instances leads to Cushing syndrome, the physician will try to reduce the dosage, while still effectively continuing treatment.

CONSIDER THESE QUESTIONS

1. Who is at fault when people are unaware that any kind of hormone therapy can have unpleasant side effects: the prescribing physician, a drug company, or the patients themselves?
2. Does it make you uncomfortable to know that you have to weigh the possible benefits of hormone therapy against the possible negative effects? Explain why or why not.

 **connect** |BIOLOGY Explore the concepts through a variety of multimedia assets and question types.
www.mcgrawhillconnect.com

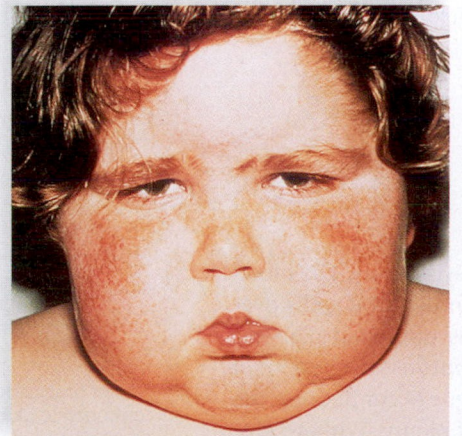
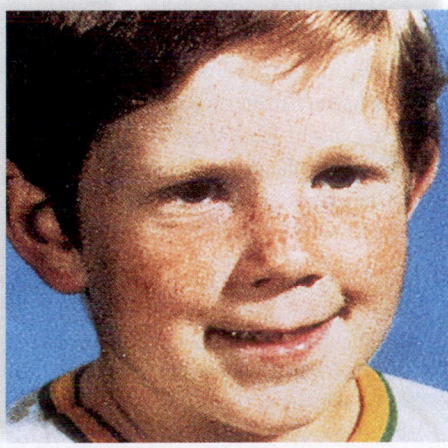

FIGURE 35B Patient with Cushing syndrome (*left*) before and (*right*) after treatment.

LEARNING OUTCOME

When you complete this section, you should be able to

1. Explain how the pancreatic hormones work together to maintain homeostasis.

The **pancreas** is a slender, pale-colored organ that lies transversely in the abdomen between the kidneys and near the duodenum of the small intestine (see Fig. 35.1B). The pancreas is rather lumpy in consistency and is composed of two types of tissue: exocrine and endocrine (Fig. 35.7A). Exocrine tissue produces and secretes digestive juices that travel by way of ducts to the small intestine. Endocrine tissue, called the **pancreatic islets** (islets of Langerhans), produces and secretes the antagonistic hormones insulin and glucagon directly into the blood. The majority of pancreatic tissues are exocrine in nature.

A negative feedback mechanism governs whether insulin or glucagon is secreted by the pancreas (Fig. 35.7B). **Insulin** is secreted when the blood glucose level is high, which usually occurs just after eating. Insulin stimulates the uptake of glucose by all cells, but especially those of the liver, muscles, and adipose tissue. In liver and muscle cells, glucose is then stored as glycogen. Therefore, insulin lowers the blood glucose level. Later, in muscle fibers glucose breakdown supplies energy for protein metabolism, and in adipose tissue cells, the glucose breakdown leads to fatty acids and glycerol for the formation of fat.

Glucagon is secreted from the pancreas, usually between meals, when the blood glucose level is low. The major target tissues of glucagon are the liver and adipose tissue. Glucagon stimulates the liver to break down glycogen to glucose and to use fat and protein in preference to glucose as energy sources. Also, adipose tissue cells break down fat to glycerol and fatty acids. The liver can later take these up and use them as substrates for glucose formation. In these ways, glucagon raises the blood glucose level.

Diabetes Twenty-four million people have now been diagnosed with *diabetes* in the United States. Heredity is a risk factor, so it is wise to investigate whether a close relative has diabetes. Also, the incidence of diabetes is higher in African Americans, Hispanics, and native Americans than in the general population. Obesity is an important risk factor as well. In particular, the following symptoms can indicate that a physician should check you for diabetes:

- Frequent urination, especially at night
- Unusual hunger and/or thirst
- Unexplained change in weight
- Blurred vision
- Sores that do not heal
- Excessive fatigue

It is relatively easy to explain some of these symptoms. The blood glucose level in a diabetic is too high, and much of it spills over into the urine. Unusual thirst accompanies untreated diabetes because glucose in the urine causes the nephrons to lose a lot of water by osmosis. Hunger and excessive fatigue occur because the cells are starving for glucose in the midst of plenty. Without glucose inside mitochondria, cells cannot produce ATP, the energy currency of cells.

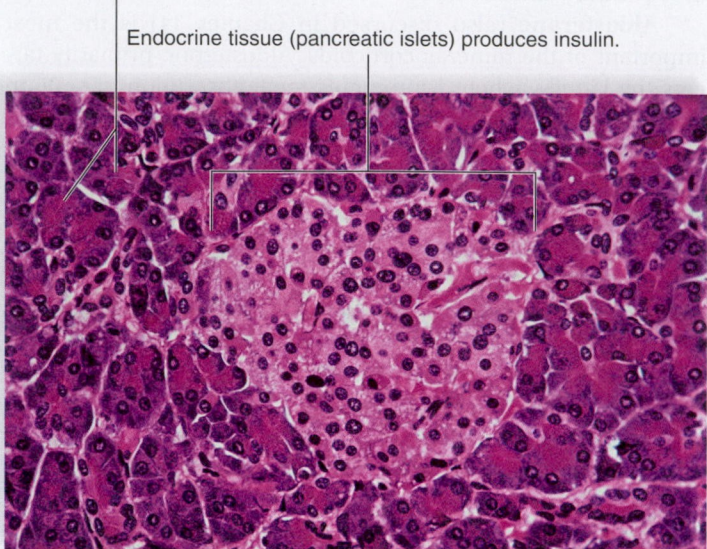

Exocrine tissue produces digestive juice.

Endocrine tissue (pancreatic islets) produces insulin.

FIGURE 35.7A This light micrograph shows that the pancreas has two types of cells. The exocrine tissue produces a digestive juice, and the endocrine tissue produces the hormones insulin and glucagon.

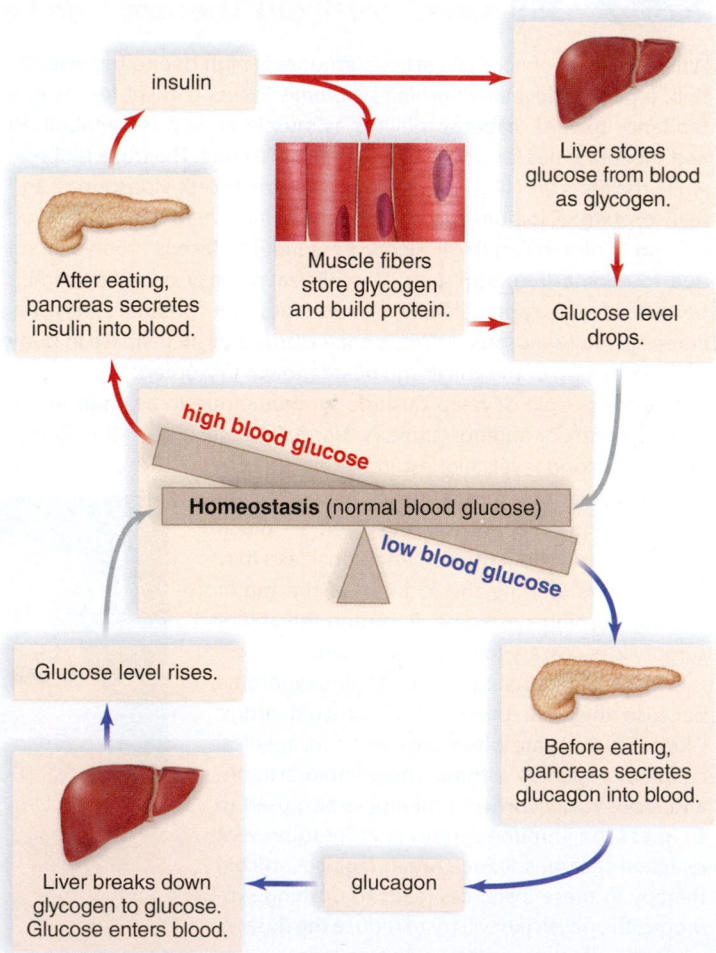

FIGURE 35.7B Regulation of blood glucose level.

A physician must first determine whether a patient has type 1 or type 2 diabetes. The number of people with type 2 diabetes, in particular, is rising at an alarming rate. By 2025, the incidence of type 2 diabetes is expected to be double what it is now.

In the nondiabetic, the hormone insulin secreted by the pancreas causes tissue cells, particularly muscle and liver cells, to take up glucose so that the blood sugar level remains within the normal range. In type 1 diabetes, the pancreas fails to secrete insulin because the cells that make insulin have died off. In type 2 diabetes, the pancreas usually doesn't secrete enough insulin, but the real problem is that the cells are resistant to insulin. Receptors in the plasma membrane don't bind insulin properly, and this causes the plasma membrane to have a reduced number of transporters for glucose (i.e., carrier proteins for glucose). Without an adequate number of transporters, glucose enters the cell in reduced amount and stays in the blood instead.

Type 1 diabetes usually occurs after a viral infection. The immune system gears up to fight the infection by killing off cells that are harboring the virus. When the infection is over, the immune system keeps on killing cells, this time the cells of pancreatic islets that produce insulin. Type 1 diabetes is clearly a self-inflicted disease, in the sense that the body itself brings it on. In a different way, *type 2 diabetes* is self-inflicted when it is caused by poor health habits. Type 2 diabetes was once considered an adult-onset disorder. Now, more and more children have it. Type 2 diabetes (not type 1) is associated with physical inactivity and excessive food intake leading to obesity, especially fat in the abdominal region.

Treatment People with diabetes must regularly check their blood sugar level, usually before meals and at bedtime. Today, automatic lances prick the finger, and computerized devices can record readings automatically. A normal fasting blood sugar level is between 70 and 110 mg/dl (milligrams per deciliter), but in diabetics, the blood sugar can rise to more than 126 mg/dl.

Patients with type 1 diabetes must have daily insulin injections, which today don't present too much of a problem because nonallergic insulin and sterile needles are readily available. Some diabetics have learned to use an insulin pump to better regulate their blood sugar level. The pump is worn outside the body, usually attached to a belt or waistband (Fig. 35.7C). Insulin is pumped from a reservoir through a tube inserted under the skin of the abdominal wall. Why can't insulin be taken by mouth? Because the digestive juices would digest insulin to amino acids.

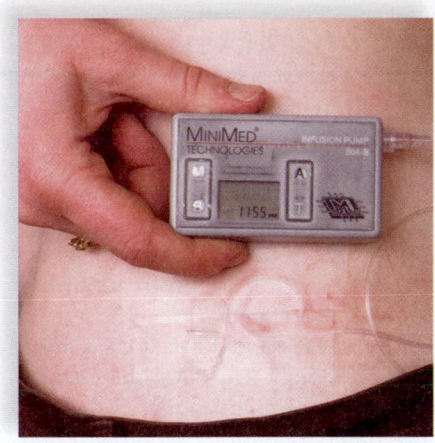

FIGURE 35.7C An insulin pump automatically delivers insulin in small amounts throughout the day.

Medication is available for type 2 diabetes, but a healthy diet and exercise can also help. Some intake of sugar with meals is fine, but these must be substituted for other carbohydrates, and not added on. Foods rich in complex carbohydrates, including dietary fiber, are preferred over easily digested foods such as many junk foods. Building up the muscles and using them regularly by, say, walking, riding a bike, or dancing uses up glucose and improves insulin efficiency. Walking, in particular, has been shown to increase the number of glucose transporters. All diabetics must work closely with a physician to tailor the correct regimen of medications, diet, and exercise for them.

It is well worth the effort to control diabetes because both types are associated with blindness, cardiovascular disease, and kidney disease. Nerve deterioration can lead to the inability to feel pain, particularly in the hands and feet. As a result, treatment of an infection may be delayed to the point that limb amputation is required.

Diabetes is definitely a disorder to avoid if at all possible. The very best course of action is to adopt a healthy lifestyle, the earlier in life the better. This is the way you can help your body maintain homeostasis, the relative constancy of the internal environment—in this case, a normal blood glucose level.

Animation
Blood Sugar Regulation in Diabetics

▶ **35.7 CHECK YOUR PROGRESS**
 1. Identify the types of foods that would cause insulin to be released after eating.

35.8 The pineal gland is involved in biorhythms

LEARNING OUTCOME

When you complete this section, you should be able to
 1. Discuss the function of the pineal gland.

The **pineal gland,** which is located in the brain (see Fig. 35.1B), produces the hormone **melatonin,** primarily at night. Melatonin is involved in a **circadian** sleep-wake cycle; normally we grow sleepy at night when melatonin levels increase and awaken once daylight returns and melatonin levels are low. Also, because light suppresses melatonin secretion, the duration of secretion is longer in the winter than in the summer. Daily 24-hour cycles such as this are called circadian rhythms, and circadian rhythms are controlled by an internal timing mechanism called a biological clock, as discussed in "The Hormone Melatonin" on page 698.

Animal research indicates that melatonin also regulates sexual development. In humans, researchers have noted that children whose pineal gland has been destroyed due to a brain tumor experience early puberty.

Video
Winter Moods

▶ **35.8 CHECK YOUR PROGRESS**
 1. Explain why you might darken the room if you want to take a nap during the day.

35C The Hormone Melatonin

The hormone melatonin is now sold as a nutritional supplement. Advertisements promote its use in pill form for sleep, aging, cancer treatment, sexuality, and more. It's possible that melatonin has some benefit in certain sleep disorders, but most physicians do not yet recommend its use because so little is known about the dosage requirements and possible side effects.

Melatonin is produced by the pineal gland in a daily cycle influenced by the seasonal length of the night (Fig. 35C). The change in duration of secretion serves as a biological signal in animals who change their reproductive habits, behavior, coat growth, and camouflage coloring according to the season. In general, however, melatonin secretion accompanies our natural sleep-wake cycle, and therefore it may control circadian rhythms also. (Any rhythm that occurs every 24 hours—such as retiring at 10 and rising at 6—is called a circadian [around a day] rhythm.)

All circadian rhythms seem to be controlled by an internal biological clock because they have a regular cycle even in the absence of environmental cues. In scientific experiments, humans have lived in underground bunkers where they never see the light of day. In a few of these people, the sleep-wake cycle drifts badly, but in most, the daily activity schedule is just about 25 hours. Usually, however, an individual's internal biological clock is reset each day by the environmental day-night cycle. If not, circadian rhythms can drift out of phase with the environmental day-night cycle.

Humans have a biological clock that acts as a pacemaker for circadian rhythms. The clock may reside in a cluster of neurons within the hypothalamus. These neurons, called the suprachiasmatic nucleus (SCN), undergo cyclical changes in activity and receive light from the eyes, which most likely resets the SCN. As the days get darker and darker during the fall and winter, some people become depressed, sometimes severely. Without a dose of early morning light, they find it difficult to keep going. This condition is known as seasonal affective disorder (SAD). People with SAD need to have their clock reset every day by daylight. A half-hour of simulated daylight from a portable light box first thing in the morning makes them feel operational again.

The SCN also controls the secretion of melatonin by the pineal gland, and in turn melatonin may quiet the operation of the neurons in the SCN. Research is still going forward to see if melatonin will be effective for circadian rhythm disorders such as SAD, jet lag, sleep phase problems, recurrent insomnia in the totally blind, and some other less common conditions. So-called jet lag occurs when you travel across several time zones and your biological clock is out of phase with local time. Jet-lag symptoms gradually disappear as your biological clock adjusts to the environmental signals of the local time zone. Many young people have a sleep phase problem because their circadian cycle lasts 25–26 hours. As they lengthen each day, they get out of sync with normal times for sleep and staying awake for school or work.

Clinical trials show that melatonin can shift circadian rhythms. Melatonin given in the afternoon shifts rhythms toward an earlier bedtime, while melatonin given in the morning shifts rhythms toward a later bedtime. For most people, the process was gradual:

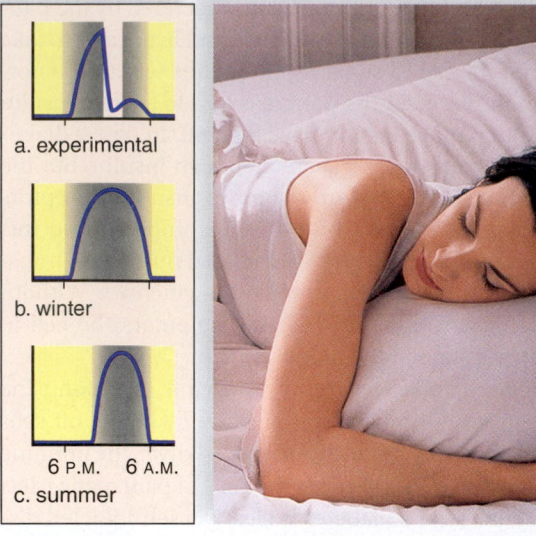

FIGURE 35C Melatonin production is greatest at night when we are sleeping. Light applied during an experiment suppresses melatonin production (**a**), so duration of production is longer in the winter (**b**) than in the summer (**c**).

The average rate of change was about an hour a day. Before you try melatonin, however, you might want to consider that this hormone is known to affect reproductive behavior in other mammals. You need to consider all possible side effects before deciding that melatonin therapy is worth the possible risks.

CONSIDER THESE QUESTIONS

1. Have you ever experienced any degree of seasonal affective disorder? Was there any reason it could have been a psychological rather than a chemical imbalance?
2. Are you more of a "morning person" or a "night person"? Is it possible to relate this difference in people to levels of melatonin secretion?
3. As with melatonin, some Internet sites proclaim the benefits of using growth hormone for weight loss, increased energy, and the like. However, there are also documented cases of diabetes and other problems resulting from its use, and some of the products have been contaminated or have been found to contain little or no growth hormone. Have you or would you ever take a supplement such as growth hormone to try to improve your overall health? Why or why not?

 Explore the concepts through a variety of multimedia assets and question types.

www.mcgrawhillconnect.com

CONNECTING THE CONCEPTS

In contrast to the nervous system, the endocrine system is not centralized but consists of several organs scattered throughout the body. The hormones secreted by endocrine glands travel through the bloodstream to reach their target cells. The metabolism of a cell changes when it has plasma membrane or nuclear receptors for that hormone. Still the nervous system and the endocrine system are structurally and functionally related. The hypothalamus, a portion of the brain, controls the pituitary, an endocrine gland. The hypothalamus even produces the hormones that are released by the posterior pituitary. Neurosecretory cells in the hypothalamus produce releasing hormones that control the activity of the anterior pituitary.

The nervous system is well known for bringing about an immediate response to environmental stimuli, as in the fight-or-flight reaction.

While doing so, it stimulates release of hormones from the adrenal medulla. In many instances, the chemical signals released by the nervous system, called neurotransmitters, help maintain homeostasis. Heart rate, breathing rate, and blood pressure are all controlled by the nervous system and thereby stay relatively constant. Hormones released by endocrine glands also help maintain homeostasis especially by keeping the levels of calcium, sodium, glucose, and other blood constituents within normal limits.

The endocrine system is slower acting than the nervous system and may regulate processes that occur over days or even months. Therefore, hormones secreted into the bloodstream control whole-body processes, such as growth and reproduction. In this chapter we mention several hormones that have to do with growth either directly or indirectly; in Chapter 36, we explore the activity of the hormones that influence reproduction.

ANALYZE AND EVALUATE

1. The fight-or-flight reaction is intensified when the adrenal glands release epinephrine. How does this show that the nervous system and the endocrine system work together?
2. Homeostasis is the maintenance of the internal environment within normal limits. Choose an organ of the endocrine system and show that it is involved in maintaining the internal environment through a negative feedback mechanism.
3. Years ago, the anterior pituitary was called the master gland. How might this be justified and, why is it better to call the hypothalamus a master gland?

SUMMARIZE

The Endocrine System

35.1 The endocrine and nervous systems work together

- The nervous system acts quickly because it utilizes axons that release neurotransmitters for stimulation of a target organ (e.g., a muscle). The **endocrine system** acts more slowly but its effects are longer lasting; its glands release **hormones** that travel in the bloodstream to target organs. Both systems utilize **negative feedback** mechanisms.
- Figure 35.1B lists the principal glands and hormones. Hormones affect blood osmolarity, blood glucose and calcium levels, metabolism, and the maturation and function of reproductive organs.

35.2 Hormones affect cellular metabolism

- Hormones are chemical signals that usually work within the body but can work between individuals as do **pheromones.** Hormones affect the metabolism of **target cells,** cells that have receptors to receive them. **Peptide hormones** are the **first messengers** received at a receptor in the plasma membrane. The **second messenger** (e.g., **cAMP,** or cyclic adenosine monophosphate)

changes the metabolism of the cell. **Steroid hormones** enter the cell and then nucleus, where they bind to a receptor. The complex stimulates a gene, resulting in a protein that carries out a response to the hormonal signal.

The Hypothalamus and the Pituitary Gland

35.3 The hypothalamus is a part of the nervous and endocrine systems

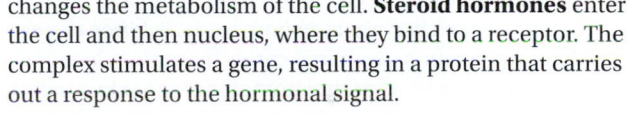

- **Neurosecretory cells** in the hypothalamus produce and send by way of axons the hormones released by the **posterior pituitary:** (1) **Antidiuretic hormone** (**ADH**) causes the reabsorption of water by the kidneys. (2) **Oxytocin** stimulates the uterus to contract during childbirth and promotes milk letdown while nursing. Unlike other hormones, oxytocin is controlled by **positive feedback.**
- Neurosecretory cells in the hypothalamus produce **hypothalamic-releasing hormones** and -inhibiting hormones that control the anterior pituitary. They reach the anterior pituitary by way of cardiovascular veins.
- The **anterior pituitary** produces nontropic and tropic hormones.
- Tropic hormones are **thyroid-stimulating hormone** (**TSH**), **adrenocorticotropic hormone** (**ACTH**), and the **gonadotropic hormones** (FSH and LH). The tropic hormones participate in a three-tier negative feedback system that controls their secretion. Nontropic hormones include **prolactin** (**PRL**), growth hormone (GH), and **melanocyte-stimulating hormone** (**MSH**).

Hormones and Homeostasis

35.4 Growth hormone controls the height of an individual

- **Growth hormone** (**GH**) promotes skeletal and muscular growth and the height of an individual. If too little GH is secreted, the individual can have **pituitary dwarfism**, and if too much GH is produced, the person can be a giant. Too much growth hormone in an adult leads to **acromegaly.**

35.5 The thyroid and parathyroids regulate metabolism

- The **thyroid gland** secretes **triiodothyronine** (T_3) and **thyroxine** (T_4), which increase the metabolic rate of all cells. Hypothyroidism is the cause of a *goiter.* Hyperthyroidism is characterized by nervousness and protruding eyes.
- **Calcitonin,** secreted by the thyroid gland, causes blood calcium levels to decrease. **Parathyroid glands** secrete **parathyroid hormone** (**PTH**), which causes the blood calcium level to increase.

35.6 The adrenal glands respond to stress

- Each **adrenal gland** has an **adrenal cortex** and an **adrenal medulla.** The adrenal medulla releases **epinephrine** and **norepinephrine,** which have short-term effects. The adrenal cortex secretes hormones that provide long-term response to stress: (1) **Glucocorticoids** (e.g., **cortisol**) regulate carbohydrate, protein, and fat metabolism, leading to increased blood glucose levels. (2) **Mineralocorticoids** (e.g., **aldosterone**) help regulate sodium and potassium excretion.

  adrenal medulla — adrenal cortex

- The adrenal cortex also releases a small amount of both male and female sex hormones in both sexes. *Addison disease,* due to undersecretion of adrenal cortex hormones, is characterized by intolerance to stress and bronzing of the skin. *Cushing syndrome,* due to oversecretion of adrenal cortex hormones, is characterized by diabetes mellitus and high blood pressure.

35.7 The pancreas regulates the blood sugar level

- In the **pancreas, pancreatic islet** cells release **insulin** after eating. Insulin causes all cells to take up glucose, thereby lowering blood glucose. Glucose is stored in liver and muscle fibers as glycogen. Islet cells release **glucagon** between meals, causing the breakdown of glycogen to glucose, and thereby increasing blood glucose.
- In **type 1 diabetes,** the pancreas fails to produce insulin, usually because pancreatic islet cells have died off following a viral infection. In **type 2 diabetes,** the pancreas produces at least some insulin, but cells are unable to respond and take up glucose.

35.8 The pineal gland is involved in biorhythms

- The **pineal gland** produces the hormone **melatonin.** Melatonin is involved in the **circadian** sleep-wake cycle and may also regulate sexual development.

TEST YOURSELF

The Endocrine System

1. The nervous system and the endocrine system differ in
   a. how their signals are delivered in the body.
   b. the speed of the body's response to their signals.
   c. the use of negative feedback in controlling their signaling.
   d. Both a and b are correct.
   e. a, b, and c are all correct.

2. Which hormones typically cross the plasma membrane?
   a. peptide hormones   c. Both a and b are correct.
   b. steroid hormones   d. Neither a nor b is correct.

3. Peptide hormones
   a. are received by a receptor located in the plasma membrane.
   b. are received by a receptor located in the cytoplasm.
   c. bring about the transcription of DNA.
   d. Both b and c are correct.

4. Steroid hormones are secreted by
   a. the adrenal cortex.   d. Both a and b are correct.
   b. the gonads.   e. Both b and c are correct.
   c. the thyroid.

5. Steroid hormones
   a. bind to a receptor located in the plasma membrane.
   b. cause the production of cAMP.
   c. activate the protein kinase.
   d. stimulate the production of mRNA.

For questions 6–10, match the hormone to the correct gland in the key.

KEY:
   a. pancreas   d. thyroid
   b. anterior pituitary   e. adrenal medulla
   c. posterior pituitary   f. adrenal cortex

6. Cortisol
7. Growth hormone (GH)
8. Oxytocin storage
9. Insulin
10. Epinephrine
11. **THINKING CONCEPTUALLY** Caffeine inhibits the breakdown of cAMP in the cell. Referring to Figure 35.2C, how would this influence a stress response brought about by epinephrine?

The Hypothalamus and the Pituitary Gland

12. Which of the following statements about the pituitary gland is incorrect?
   a. The pituitary lies inferior to the hypothalamus.
   b. Growth hormone and prolactin are secreted by the anterior pituitary.
   c. The anterior pituitary and posterior pituitary communicate with each other.
   d. Axons run between the hypothalamus and the posterior pituitary.

13. _____ is released through positive feedback and causes _____.
   a. Insulin, stomach contractions
   b. Oxytocin, stomach contractions
   c. Oxytocin, uterine contractions
   d. None of these are correct.

14. Complete this diagram by filling in blanks a–e:

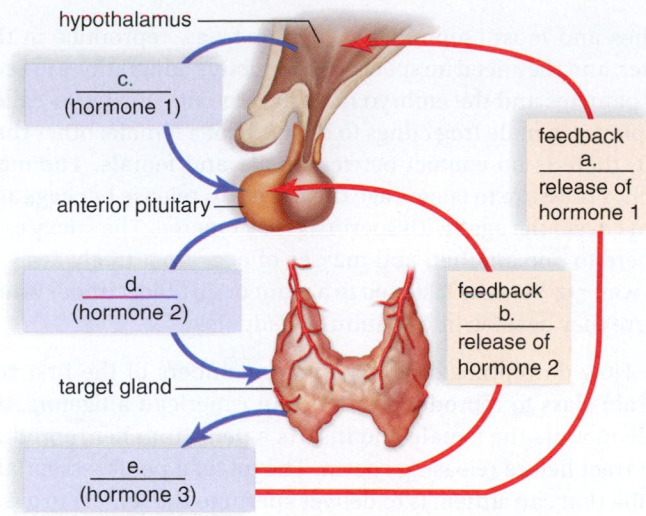

hypothalamus

_____
c.
(hormone 1)

anterior pituitary

_____
d.
(hormone 2)

target gland

_____
e.
(hormone 3)

feedback
a.
release of
hormone 1

feedback
b.
release of
hormone 2

15. The anterior pituitary controls the secretion(s) of
    a. both the adrenal medulla and the adrenal cortex.
    b. both the thyroid gland and the adrenal cortex.
    c. both the ovaries and the testes.
    d. Both b and c are correct.
16. Tropic hormones affect other endocrine tissues. Which of the following would be considered a tropic hormone?
    a. calcitonin   d. melatonin
    b. oxytocin   e. None of these are correct.
    c. glucagon
17. **THINKING CONCEPTUALLY** Use the figure in question 14 to suggest the possible causes for hypothyroidism when the body apparently lacks production of or sensitivity to hormones produced by the thyroid.

Hormones and Homeostasis

18. Growth hormone is produced by the
    a. posterior adrenal gland.
    b. posterior pituitary.
    c. anterior pituitary.
    d. kidneys.
    e. None of these are correct
19. Both the adrenal medulla and the adrenal cortex are
    a. endocrine glands.
    b. in the same organ.
    c. involved in our response to stress.
    d. All of these are correct.
20. The blood cortisol level does not control the secretion of
    a. hypothalamic-releasing hormone from the hypothalamus.
    b. adrenocorticotropic hormone (ACTH) from the anterior pituitary.
    c. mineralocorticoids from the adrenal cortex.
    d. All of these are correct.
21. Lack of aldosterone will cause a blood imbalance of
    a. sodium.   d. All of these are correct.
    b. potassium.   e. None of these are correct.
    c. water.

22. Glucagon causes
    a. use of fat for energy.
    b. glycogen to be converted to glucose.
    c. use of amino acids to form fats.
    d. Both a and b are correct.
    e. None of these are correct.
23. Diabetes is associated with
    a. too much insulin in the blood.
    b. a blood glucose level that is too high.
    c. blood that is too dilute.
    e. All of these are correct.
24. The difference between type 1 and type 2 diabetes is that
    a. for type 2 diabetes, insulin is produced but it is ineffective; type 1 results from lack of insulin production.
    b. treatment for type 2 involves insulin injections, while type 1 can be controlled, usually by diet.
    c. only type 1 can result in complications such as kidney disease, reduced circulation, or stroke.
    d. type 1 can be a result of lifestyle, and type 2 is thought to be caused by a virus or other agent.
25. Long-term complications of diabetes include
    a. blindness.   d. All of these are correct.
    b. kidney disease.   e. None of these are correct.
    c. circulatory disorders.
26. Which hormone and condition are mismatched?
    a. thyroxine—goiter
    b. parathyroid hormone—Addison disease
    c. cortisol—Cushing syndrome
    d. insulin—diabetes
27. PTH causes the blood level of calcium to _____, and calcitonin causes it to _____.
    a. increase, not change   d. decrease, increase
    b. increase, decrease   e. not change, increase
    c. decrease, also decrease
28. **THINKING CONCEPTUALLY** Negative feedback controls the secretion of calcitonin by the thyroid gland. Explain how it would work.

GET INVOLVED

1. Formulate a hypothesis to test the effect of calcitonin on osteoblasts (bone-forming cells) when the blood calcium level is high. (See section 35.5.)
2. How would you test the hypothesis that melatonin levels influence when we fall asleep and wake up?

MEDIA STUDY TOOLS

mhhe.com/maderconcepts3

Enhance your study of this chapter with interactive study tools, practice tests, and engaging animations. Also, ask your instructor about the resources available through ConnectPlus, which includes LearnSmart, a personalized adaptive learning program, and a media-rich eBook.

36

Reproduction and Development

BEFORE YOU BEGIN

Take a few minutes to recall

Reproduction is a characteristic of life (section 1.10)
The relationship between the hypothalamus, anterior
   pituitary, and gonads (section 35.3)
Mitosis and meiosis (Chapters 8–9)
The role of the *Hox* genes in development (section 15.7)
How the coelom forms in deuterostomes (Figure 20.2B)

How to Do It on Land

Fishes and most amphibians, such as frogs, reproduce in the water, and they need no special reproductive adaptations to keep the gametes and the embryo from drying out. During so-called amplexus, a male frog clings to the body of a female; other than that, there is no contact between male and female. The male applies pressure to encourage the female to release her eggs and then covers the eggs with sperm as they emerge. The sticky eggs adhere to one another, and masses of eggs float freely away in the water or become fastened to a plant or an object under water, where they develop into swimming tadpoles.

How do reptiles do it? They are members of the first vertebrate class to reproduce on land. In American alligators, the male mounts the female and inserts a penis into her reproductive tract before releasing sperm. The job of a penis, essentially a tube that can stiffen, is to deliver sperm to the female in a way that protects the sperm from drying out. The eggs of a female are protected deep inside her body. The female alligator lays her fertilized eggs, but they are protected by a shell, which keeps them from drying out. Inside the shell are a series of membranes that perform all sorts of services for the embryo, such as exchanging gases, collecting wastes, and providing a watery environment. That's right, alligators, and indeed all land vertebrates, still develop surrounded by water. In reptiles, the water is inside the shell. The evolution of the shelled egg was absolutely essential in order for vertebrates to reproduce on land. When baby alligators hatch, they are tiny beings only about 15–20 cm in length. But they are fully formed and able to survive on land, even though the mother may gently carry them inside her mouth to the water.

Frogs mating

gametes

How do mammals do it? Most mammals, including humans, copulate just like the alligators do. The male mounts the female and inserts a penis into her reproductive tract to release sperm. Copulation, even in humans, is simply a way to protect the sperm from drying out. No need to worry about the eggs of the female—they never leave her body.

One big difference between mammals and reptiles is that the mammalian embryo develops inside the uterus of the female, and so placental mammals do not lay eggs. But the embryo is still surrounded by water and by the same membranes. The outer membrane develops into the placenta, the point of contact between mother and offspring through which nutrient-for-waste exchange occurs. When a human is born, the "water breaks," and for the first time, the baby has to breathe air. A slap on the behind can make sure breathing begins.

Whales are mammals, and all sorts of evidence points to the conclusion that whales evolved from land animals. Even though whales are marine, the males do have a penis; they do insert it into the reproductive tract of the female; they do reproduce internally; and they are born fully able to breathe air. As soon as a whale is born, it can swim, breathe, produce clicking sounds, and nurse (because all mammals have mammary glands that produce milk for the newborn).

In this chapter, we discuss principles of animal reproduction, followed by the stages and processes of animal development.

Sea turtles mating

Kangaroos mating

Sea turtle hatching

Whale penises

703

How Animals Reproduce

The ways that animals reproduce are quite varied. By contrasting asexual and sexual reproduction along with development in water and on land, we can observe the general differences.

36.1 Both asexual and sexual reproduction occur among animals

LEARNING OUTCOMES

When you complete this section, you should be able to

1. Contrast asexual and sexual reproduction and explain the advantages of each.
2. Compare the strategy of reproduction in water with that on land.

Asexual Reproduction Several phyla of invertebrates, including free-living sponges, cnidarians, flatworms, annelids, and echinoderms, can reproduce asexually. In **asexual reproduction,** a single parent gives rise to offspring that are identical to the parent, unless mutations have occurred. The adaptive advantage of asexual reproduction is that organisms can reproduce rapidly and colonize a favorable environment quickly.

Sponges produce gemmules that develop into new individuals asexually. Cnidarians, such as hydras, can reproduce asexually by budding. A new individual arises as an outgrowth (bud) of the parent. Fragmentation, followed by regeneration, is well known among sponges and echinoderms. If a sea star is chopped up, it has the potential to regenerate into several new individuals. Also you can horizontally cut a planarian into as many as ten pieces in the laboratory and get ten new planarians. The parasitic flatworms reproduce asexually during certain stages of their complicated life cycle.

Several types of crustaceans, annelids, insects, fishes, lizards, and even some turkeys have the ability to reproduce parthenogenetically. During **parthenogenesis,** unfertilized eggs develop into complete but haploid individuals which is not disadvantageous as long as they are adapted to the present environment.

Sexual Reproduction The advantage of sexual reproduction is the production of an individual different from either parent. Usually during **sexual reproduction,** the egg of one parent is fertilized by the sperm of another. Animals most often produce gametes in specialized organs called **gonads.** Sponges are an exception to this rule because the collar cells lining the central cavity of a sponge give rise to sperm and eggs. Hydras and other cnidarians produce only temporary gonads in the fall, when sexual reproduction occurs. Animals in other phyla have permanent reproductive organs. The gonads are **testes,** which produce sperm, and **ovaries,** which produce eggs. The reproductive system also consists of a number of accessory structures, such as storage areas and ducts, that aid in bringing the gametes together.

Many aquatic animals, such as the invertebrates already mentioned and also fishes and amphibians, practice external fertilization and the egg develops into a **larva,** an immature stage (Fig. 36.1). Since the larva has a different lifestyle, it is able to use a different food source than the adult. In sea stars, the bilaterally symmetrical larva undergoes metamorphosis to become a radially symmetrical juvenile. Crayfish, on the other hand, do not have a larval stage; the egg hatches into a tiny juvenile with the same form

Video
Sea Urchin
Reproduction

FIGURE 36.1 Fishes, such as Atlantic salmon, release their eggs and sperm into fresh water, where they develop into larvae, which hide among the rocks.

as the adult. Some aquatic animals retain their eggs in various ways and release young able to fend for themselves. These animals are *ovoviviparous.* For example, oysters, which are molluscs, retain their eggs in the mantle cavity, and male seahorses, which are vertebrates, have a special brood pouch where the eggs develop.

In terrestrial vertebrates, males typically have a penis for depositing sperm into the vagina of females. In birds that lack a penis and vagina, a male transfers sperm to a female after placing his cloacal opening against hers. Following internal fertilization, reptiles lay a leathery-shelled egg containing **extraembryonic membranes** to serve the needs of the embryo and prevent drying out. One membrane surrounds an abundant supply of **yolk,** which is a nutrient-rich material. The shelled egg is a significant adaptation to the terrestrial environment. Birds lay and incubate hard-shelled eggs with extraembryonic membranes. They also care for their offspring until they are able to fly.

Among mammals, the duckbill platypus and the spiny anteater lay shelled eggs. In marsupials, immature newborns finish their development within a pouch where they are nourished on milk. The placental mammals are termed *viviparous,* because offspring develop inside the female's body until they can live independently. The **placenta** is a complex structure derived, in part, from the chorion, another of the reptilian extraembryonic membranes. The evolution of this type of placenta allows the developing young to internally exchange materials with the mother until they can function on their own. Viviparity represents the ultimate in caring for the unborn, and in placental mammals, the mammary glands of the mother continue to supply the nutrient needs of her offspring after birth.

▶ **36.1 CHECK YOUR PROGRESS**

1. Explain why it is advantageous for aphids, a type of insect, to reproduce asexually by parthenogenesis during the summer and reproduce sexually right before the onset of winter.

36A War Between the Sexes Results in Coevolution

Monogamy means always mating with the same partner, at least for a period of time (Fig. 36A). A few birds, such as swans, mate for life. This might be because they begin breeding in early spring and care for each brood for a full year, leaving limited time to seek a different mate. Goldfinches, on the other hand, are not monogamous. They don't begin breeding until late summer and need only about a month to raise their quickly maturing offspring. So, they have more time to find a new partner each year.

Mallard ducks are monogamous, except that males sometimes force themselves on other females. Researchers at Yale University believe this has led to a form of coevolution that can be called a sexual arms race. Unlike other birds that mate merely

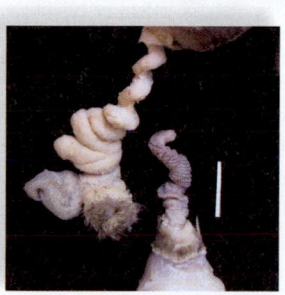

Corkscrew genitals of female (*left*) and male (*right*) mallard ducks.

by putting their cloacas together, water fowl, including mallards, have a penis. The penis is long and spirals in a counterclockwise direction. Might a long penis give sperm an advantage? After all, the egg is fertilized by the sperm that gets there first. But the story appears to be more complicated than that. The female oviduct that receives this penis is also long and it too spirals—but in the opposite direction to that of the penis. In addition, the oviduct has dead-end sacs. The clockwise spiral of the oviduct may make the entrance of a penis difficult, and the dead-end sacs may be false passages for deposit of sperm that will never reach the egg. The researchers speculate that these physical barriers disappear when females cooperate during mating. If so, the female seems to have evolved a way to prevent fertilization of her egg by any bird that is not her mate, and in that way she controls which bird will father her offspring.

DNA fingerprinting of offspring has shown that monogamy is the exception rather than the rule among animals. However, nonmonogamy is apt to correlate with sexual antagonism toward the opposite sex. William Rice and colleagues at the University of California describe an experiment in nonmonogamous fruit flies. Each generation, they selected the most sexually antagonistic males to mate, but let the females remain as they were. The males rapidly evolved increased reproductive success, but this male advantage resulted in reduced female survival. Death may have occurred because of increased remating with females or because male seminal fluid was ever more toxic to females. Therefore, it appears that only by allowing both sexes to evolve in relation to one another can a nonmonogamous species continue to exist.

Johanna Rönn from the University of Uppsala in Sweden studied seven species of seed beetle, known to have harmful penises covered in sharp spikes. The spikes anchor the penis—increasing the chances of reproductive success—but they also pierce the female during sex and injure her. Rönn hypothesized that as the males evolve more effective and spiny

genitalia, female beetles will evolve tougher integument to resist injury.

To test Rönn's hypothesis, researchers studied scanning electron micrographs of seed beetle penises. They counted the number of spines present in each species, and came up with a mathematical scale for ranking the harmfulness of the penises. They

Spiked penis of seed beetles.

also gave pictures of the genitalia to other biologists, who were novices in the area, and asked them to rank the genitalia in terms of seeming harmfulness. They found that, as the harmfulness of the penises increased, so did padding in the female reproductive tract. In this way, the females were more likely to survive and produce offspring. They speculate that this adaptation in males may benefit females by making their own male offspring more likely to successfully mate and pass on their genes. However, even minor imbalances in this sexual arms race can be very costly for females. In a sense, the resulting average sex organ phenotype seems to represent an evolutionary compromise between the interests of both sexes.

CONSIDER THESE QUESTIONS

1. Speculate why nonmonogamy rather than monogamy would promote a sexual arms race between the sexes.
2. Some investigators hypothesize that whenever males and females have different alleles, regardless of monogamy, an arms race would ensue. Why might that be?
3. Can you find any evidence of a sexual arms race among humans? If so, how would you go about testing your hypothesis?

 Explore the concepts through a variety of multimedia assets and question types.

www.mcgrawhillconnect.com

FIGURE 36A Parenting in goldfinches and in swans involves both parents.

Human Reproductive Systems

The male reproductive system produces sperm, which are delivered to the female by way of an erect penis. The female reproductive system produces an egg once a month. An egg is fertilized internally, and the resulting embryo also develops internally. In a nonpregnant female, the ovarian cycle drives the uterine cycle so that menstruation occurs once a month.

36.2 Testes are the male gonads

LEARNING OUTCOMES

When you complete this section, you should be able to

1. Identify the organs of the human male reproductive system and provide a function for each.
2. Describe the production of sperm and the functions of testosterone.

In the human male, the paired testes, which produce sperm, are suspended within the scrotum (Fig. 36.2A). The testes begin their development inside the abdominal cavity, but they descend into the scrotum as development proceeds. Normal sperm production is inhibited at body temperature; a slightly cooler temperature is required.

Sperm produced by the testes mature within the **epididymis,** a coiled tubule lying just outside each testis. Maturation seems to be required in order for the sperm to swim to the egg. Once the sperm have matured, they are propelled into the **vas deferens** by muscular contractions. Sperm are stored in both the epididymis and the vas deferens. When a male becomes sexually aroused, sperm enter first the ejaculatory duct and then the urethra, part of which is within the penis. (The urethra in males is part of both the urinary and the reproductive systems.)

The **penis** is a cylindrical organ that usually hangs in front of the scrotum. The penis has an enlarged tip normally covered by a layer of skin called the **foreskin.** Circumcision is the surgical removal of the foreskin. Three cylindrical columns of spongy, erectile tissue containing distensible blood spaces extend through the shaft of the penis. During sexual arousal, nervous reflexes cause an increase in arterial blood flow to the penis. This increased blood flow fills the blood spaces in the erectile tissue, and the penis, which is normally limp (flaccid), stiffens and increases in size. These changes are called an **erection.** Drugs have been developed to treat erectile dysfunction by increasing blood flow to the penis so that when a man is sexually excited, he can get and keep an erection.

Semen (seminal fluid) is a thick, whitish fluid that contains sperm and secretions from three types of glands. The **seminal vesicles** lie at the base of the bladder. As sperm pass from the vas deferens into the ejaculatory duct, these vesicles secrete a thick, viscous fluid containing nutrients for possible use by the sperm. Just below the bladder is the **prostate gland,** which secretes a milky basic fluid believed to activate or increase the motility of the sperm. In older men, the prostate gland may become enlarged, thereby constricting the urethra and making urination difficult. Also, prostate cancer is the most common form of cancer in men. Slightly below the prostate gland, on either side of the urethra, is a pair of small glands called **bulbourethral glands,** which secrete mucus that has a lubricating effect.

MP3
Male Reproductive Anatomy and Physiology

Male Orgasm If sexual arousal reaches its peak, ejaculation follows an erection. The first phase of ejaculation is called emission. During *emission,* the spinal cord sends nerve impulses via appropriate nerve fibers to the epididymis and vas deferens. Their subsequent motility causes sperm to enter the ejaculatory duct, whereupon the seminal vesicles, prostate gland, and bulbourethral glands release their secretions.

During the second phase of ejaculation, called *expulsion,* rhythmic contractions of muscles at the base of the penis and within the urethral wall expel semen in spurts from the opening of the urethra. These contractions are a part of male **orgasm,** the physiological and psychological sensations that occur at the climax of sexual stimulation. Following ejaculation, a male typically experiences a time, called the refractory period, during which stimulation does not bring about an erection.

Testes Anatomy and Function A testis is composed of compartments called lobules, each of which contains one to three tightly coiled **seminiferous tubules** (Fig. 36.2B). A seminiferous

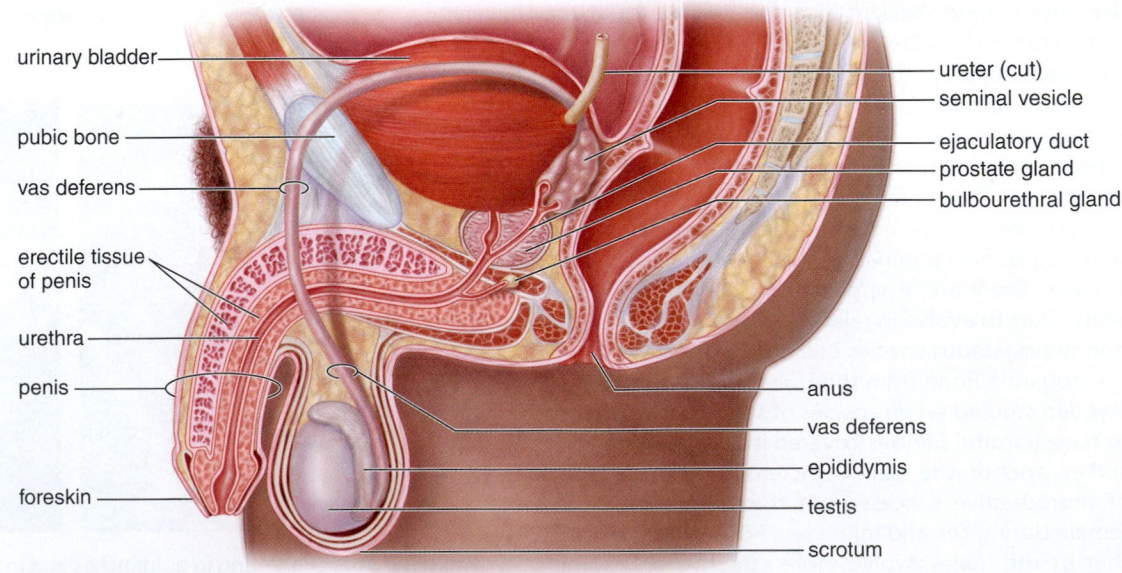

Labels: urinary bladder, pubic bone, vas deferens, erectile tissue of penis, urethra, penis, foreskin, ureter (cut), seminal vesicle, ejaculatory duct, prostate gland, bulbourethral gland, anus, vas deferens, epididymis, testis, scrotum

FIGURE 36.2A Male reproductive system.

tubule is packed with cells undergoing **spermatogenesis,** the production of sperm. Newly formed cells move away from the outer wall, increase in size, and undergo meiosis to become spermatids, which contain the haploid number of chromosomes. Spermatids then differentiate into sperm. Also present are **Sertoli cells,** which support, nourish, and regulate the production of sperm. A sperm has three distinct parts: a tail, a middle piece, and a head. The tail is a flagellum that allows a sperm to swim toward the egg, and the middle piece contains energy-producing mitochondria. The head contains a nucleus and is capped by a membrane-enclosed acrosome. The acrosome contains enzymes that assist a sperm in entering an egg during fertilization. The ejaculated semen of a normal human male contains 40 million sperm per milliliter, ensuring an adequate number for fertilization to take place. Fewer than 100 sperm ever reach the vicinity of the egg, however, and only one sperm normally enters an egg.

Animation
Spermatogenesis

Video
Human Sperm

Control of Testes Function The hypothalamus has ultimate control of the testes' sexual function because it secretes a hormone called gonadotropin-releasing hormone, or GnRH, that stimulates the anterior pituitary to produce two gonadotropic hormones. In males, **follicle-stimulating hormone** (**FSH**) promotes spermatogenesis in the seminiferous tubules. **Luteinizing hormone** (**LH**)

in males was formerly called interstitial cell–stimulating hormone (ICSH) because it controls the production of testosterone by the interstitial cells, which are scattered in the spaces between the seminiferous tubules.

Testosterone, the main sex hormone in males, is essential for the normal development and functioning of the male sexual organs. Testosterone is also necessary for the maturation of sperm. In addition, testosterone brings about and maintains the male **secondary sex characteristics** that develop at the time of **puberty,** the time of life when sexual maturity is attained.

Testosterone causes males to develop noticeable hair on the face, chest, and occasionally other regions of the body, such as the back. Testosterone also leads to the receding hairline and pattern baldness that occur in males. Testosterone is responsible for the greater muscular development in males, who are generally taller than females and have broader shoulders and longer legs relative to trunk length. The deeper voice of males compared to females is due to males having a larger larynx with longer vocal cords. Because the so-called Adam's apple is a part of the larynx, it is usually more prominent in males than in females.

▶ **36.2 CHECK YOUR PROGRESS**

1. Use Figure 36.2A to trace the path of sperm from the testes to the outside of the body.
2. Identify three ways testosterone affects the body of males.

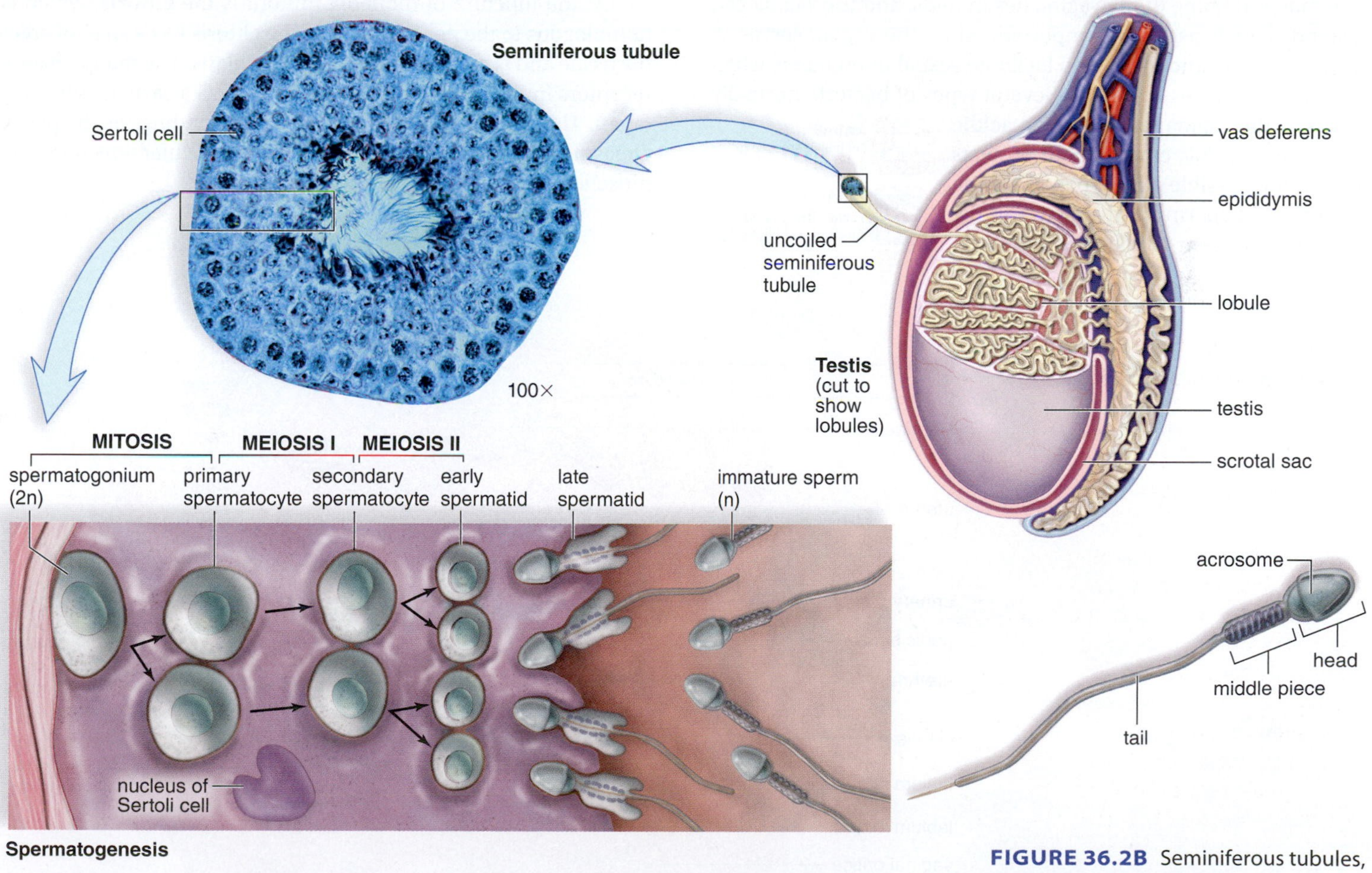

Seminiferous tubule

Sertoli cell

100×

| MITOSIS | MEIOSIS I | MEIOSIS II | | | |

spermatogonium (2n)

primary spermatocyte

secondary spermatocyte

early spermatid

late spermatid

immature sperm (n)

nucleus of Sertoli cell

Spermatogenesis

vas deferens

epididymis

uncoiled seminiferous tubule

lobule

Testis (cut to show lobules)

testis

scrotal sac

acrosome

head

middle piece

tail

FIGURE 36.2B Seminiferous tubules, where sperm are produced via the process of spermatogenesis.

LEARNING OUTCOME

When you complete this section, you should be able to

1. Identify the organs of the female reproductive system and provide a function for each.

In the human female, the **ovaries,** which produce one **oocyte** each month, lie in shallow depressions, one on each side of the upper pelvic cavity (Fig. 36.3A). The **oviducts,** also called uterine or fallopian tubes, extend from the ovaries to the uterus; however, the oviducts are not attached to the ovaries. Instead, they have fingerlike projections called **fimbriae** (sing., fimbria) that lie over the ovaries. When an oocyte bursts from an ovary during ovulation, it usually is swept into an oviduct by the combined action of the fimbriae and the beating of cilia that line the oviducts. Fertilization, if it occurs, normally takes place in an oviduct, and the developing embryo is propelled slowly by ciliary movement and tubular muscle contraction to the uterus. The **uterus** is a thick-walled muscular organ about the size and shape of an inverted pear. The narrow end of the uterus is called the **cervix.** The embryo completes its development after embedding itself in the uterine lining, called the **endometrium.** If, by chance, the embryo should embed itself in another location, such as an oviduct, a so-called ectopic pregnancy results.

A small opening in the cervix leads to the vaginal canal. The **vagina** is a tube at a 45-degree angle to the small of the back. The mucosal lining of the vagina lies in folds, and the vagina can distend. This is especially important when the vagina serves as the birth canal, and it can also facilitate sexual intercourse, when the vagina receives the penis. Several types of bacteria normally reside in the vagina and create an acidic environment. This environment protects against the possible growth of pathogenic bacteria, but sperm survive better in the basic environment provided by semen.

Animation
Female Reproductive System

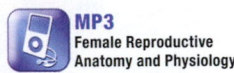
MP3
Female Reproductive Anatomy and Physiology

FIGURE 36.3B
Vulva.

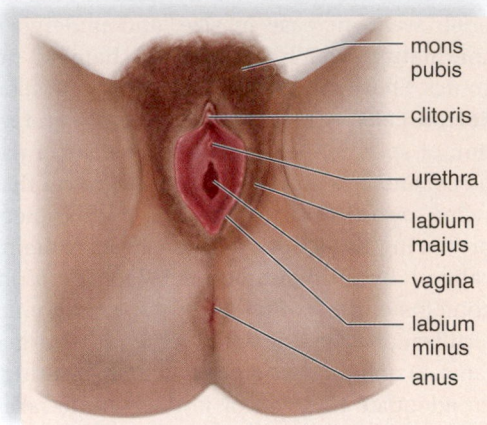

mons pubis

clitoris

urethra

labium majus

vagina

labium minus

anus

External Genital Organs and Orgasm In the female, the external genital organs are known collectively as the **vulva** (Fig. 36.3B). The mons pubis and two sets of skin folds called the labia minora (sing., labium minus) and labia majora (sing., labium majus) are on either side of the urethral and vaginal openings. Beneath the labia majora, pea-sized greater vestibular glands (Bartholin glands) open on either side of the vagina. They keep the vulva moist and lubricated during intercourse.

At the juncture of the labia minora is the **clitoris,** which is homologous to the penis in males. The clitoris has a shaft of erectile tissue and is capped by a pea-shaped glans. The many sensory receptors in the clitoris allow it to function as a sexually sensitive organ. The clitoris has twice as many nerve endings as the penis. Orgasm in the female is a release of neuromuscular tension in the muscles of the genital area, vagina, and uterus.

FIGURE 36.3A
Female reproductive system.

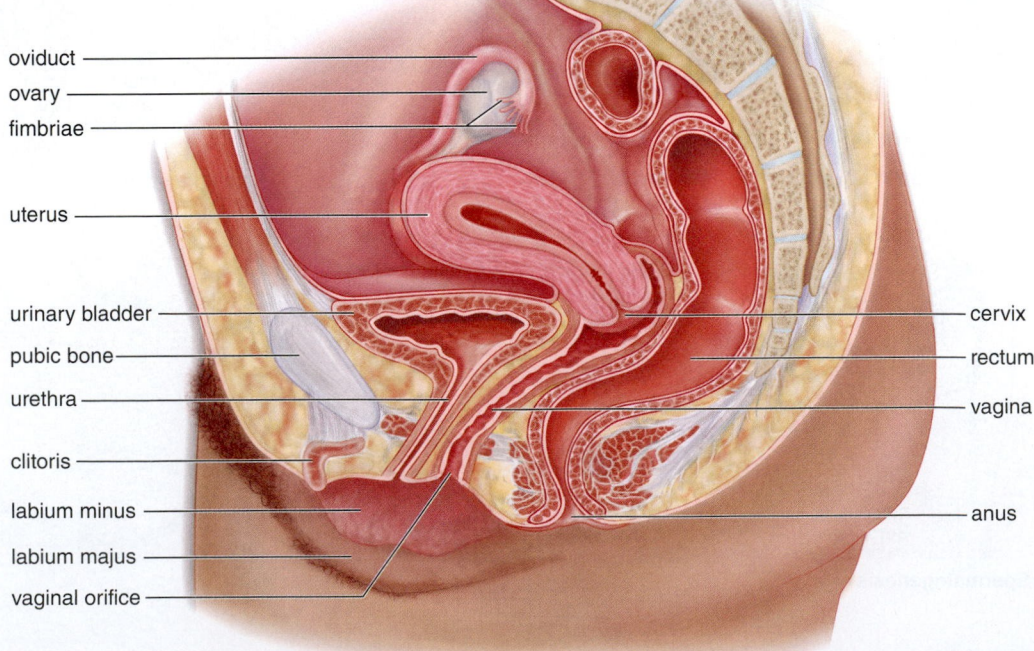

oviduct

ovary

fimbriae

uterus

urinary bladder

pubic bone

urethra

clitoris

labium minus

labium majus

vaginal orifice

cervix

rectum

vagina

anus

Oocyte Production An oocyte is produced as a **follicle** changes from a primary to a secondary to a vesicular (Graafian) follicle (Fig. 36.3C). ❶ Epithelial cells of a primary follicle surround a primary oocyte. ❷ Pools of follicular fluid surround the oocyte in a secondary follicle. ❸ In a vesicular follicle, a fluid-filled cavity increases to the point that the follicle wall balloons out on the surface of the ovary.

❹ The vesicular follicle bursts, releasing an oocyte surrounded by a clear membrane. This process is referred to as **ovulation.** For the sake of convenience, the released oocyte is often called an egg. ❺ Once a vesicular follicle has lost its oocyte, it develops into a **corpus luteum,** a glandlike structure.

During oocyte maturation **oogenesis,** a form of meiosis depicted in Figure 36.3D, is initiated and continues. The primary oocyte divides, producing two haploid cells. One cell is a secondary oocyte, and the other is called the first polar body. At ovulation, the secondary oocyte enters an oviduct, where it remains viable and capable of being fertilized for only 12–24 hours. The sperm survive for 24 and perhaps up to 72 hours. If **fertilization** occurs, a sperm enters the secondary oocyte, and then the oocyte completes meiosis. An egg with 23 chromosomes and a second polar body results. When the sperm nucleus unites with the egg nucleus, a zygote with 46 chromosomes is produced. ❻ If zygote

formation and pregnancy do not occur, the corpus luteum begins to degenerate after about 10 days.

Hormone Production The follicles and corpus luteum in the ovary also produce the female sex hormones, **estrogens** (collectively called estrogen) and **progesterone,** during the ovarian cycle. Estrogen, in particular, is essential for the normal development and functioning of the female reproductive organs. Estrogen is also largely responsible for the secondary sex characteristics in females, including body hair and fat distribution. In general, females have a more rounded appearance than males because of a greater accumulation of fat beneath their skin. Also, the pelvic girdle enlarges so that females have wider hips than males, and the thighs converge at a greater angle toward the knees. Both estrogen and progesterone are required for breast development as well.

Section 36.4 explains the hormonal relationship between the ovarian and uterine cycles.

▶ **36.3 CHECK YOUR PROGRESS**
1. Evaluate how the design of the female reproductive tract facilitates reproduction on land.
2. Identify two functions of an ovarian follicle.

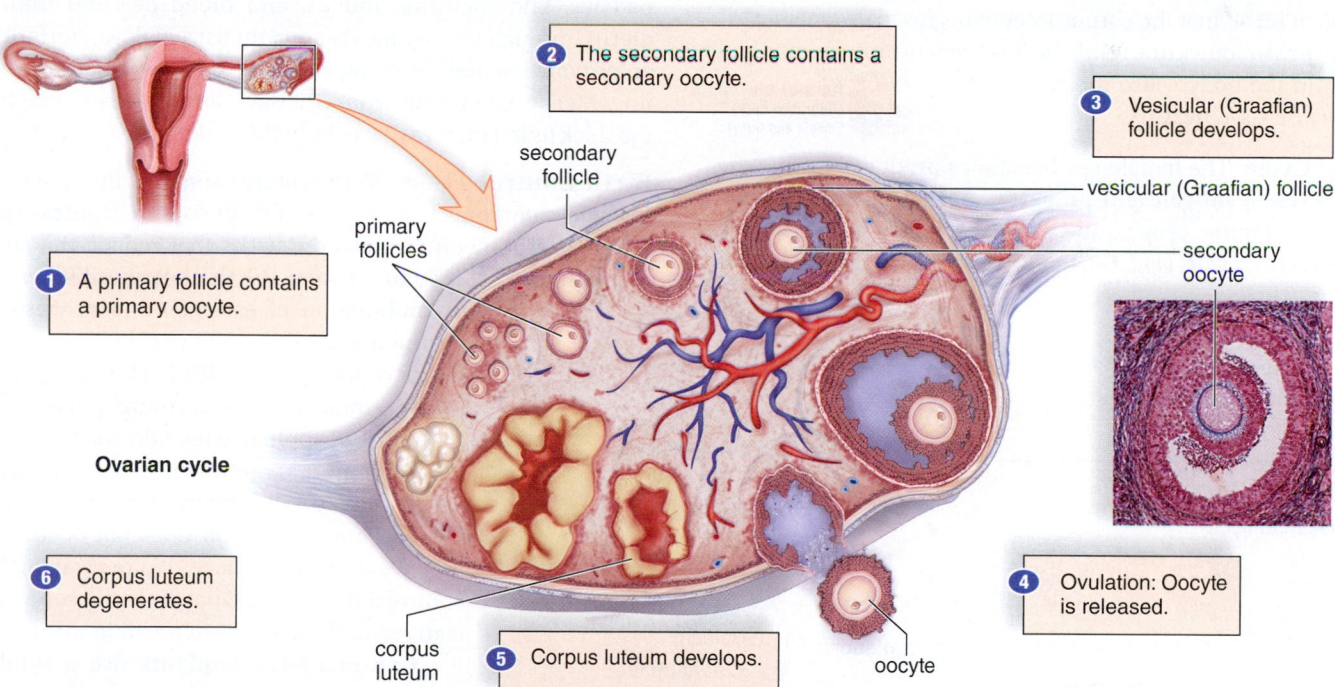

Ovarian cycle

The secondary follicle contains a secondary oocyte. ❷

Vesicular (Graafian) follicle develops. ❸

secondary follicle

vesicular (Graafian) follicle

primary follicles

secondary oocyte

❶ A primary follicle contains a primary oocyte.

❻ Corpus luteum degenerates.

Ovulation: Oocyte is released. ❹

corpus luteum

❺ Corpus luteum develops.

oocyte

FIGURE 36.3C Ovarian follicles, where eggs are produced. Over time, each follicle goes through the stages depicted.

FIGURE 36.3D Oogenesis.

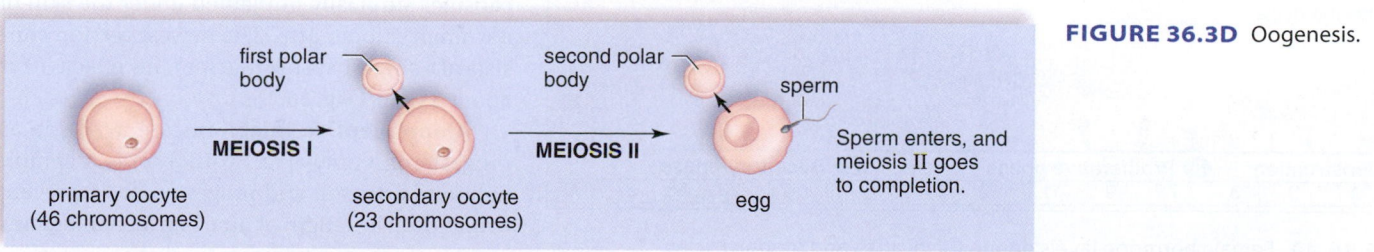

first polar body

second polar body

sperm

MEIOSIS I

MEIOSIS II

Sperm enters, and meiosis II goes to completion.

primary oocyte (46 chromosomes)

secondary oocyte (23 chromosomes)

egg

36.4 The ovarian cycle drives the uterine cycle

When you complete this section, you should be able to

1. List the stages of the ovarian and uterine cycles and explain what is occurring in each stage.
2. Describe various methods of birth control.

Ovarian Cycle The ovaries go through the same series of events, called the **ovarian cycle,** each month. In Figure 36.4A, the ovarian cycle begins with ❶ a **follicular phase,** during which the anterior pituitary produces follicle-stimulating hormone (FSH). FSH promotes the development of a follicle that secretes estrogen and some progesterone. As the estrogen level in the blood rises, it exerts feedback control over the anterior pituitary secretion of FSH so that the follicular phase comes to an end.

Presumably, the high level of estrogen in the blood causes the hypothalamus to suddenly secrete a large amount of gonadotropic-releasing hormone (GnRH). This leads to a surge in LH production by the anterior pituitary and to ❷ ovulation at about the fourteenth day of a 28-day cycle (Fig. 36.4A).

During the second half, or ❸ **luteal phase,** of the ovarian cycle, luteinizing hormone (LH) promotes the development of the corpus luteum, which primarily secretes progesterone. As the blood level of progesterone rises, it exerts feedback control over anterior pituitary secretion of LH so that the corpus luteum begins to degenerate. As the luteal phase comes to an end, the low levels of progesterone and estrogen in the body cause menstruation to begin, as discussed next.

▶ **Animation**
Maturation of the Follicle and Oocyte

Uterine Cycle The female sex hormones produced during the ovarian cycle (estrogen and progesterone) affect the endometrium of the uterus, causing the cyclical series of events known as the **uterine cycle** (Fig. 36.4A). Twenty-eight-day cycles are divided as follows:

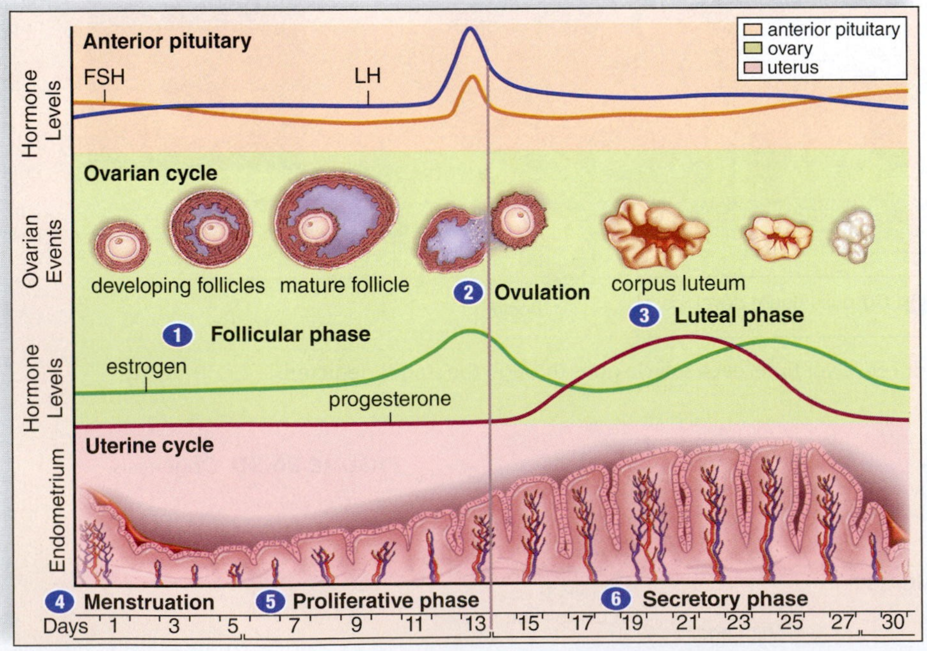

During *days 1–5*, the level of female sex hormones in the body is low, causing the endometrium to disintegrate and its blood vessels to rupture. A flow of blood and disintegrated tissue passes out of the vagina during ❹ **menstruation,** also known as the menstrual period.

During *days 6–13*, increased production of estrogen by an ovarian follicle causes the endometrium to thicken and to become vascular and glandular. This is called the ❺ **proliferative phase** of the uterine cycle.

Ovulation usually occurs on the fourteenth day of the 28-day cycle.

During *days 15–28*, increased production of progesterone by the corpus luteum causes the endometrium to double in thickness and the uterine glands to mature, producing a thick mucoid secretion. This is called the ❻ **secretory phase** of the uterine cycle. The endometrium is now prepared to receive the developing embryo. If pregnancy does not occur, the corpus luteum degenerates, and the cycle begins again.

During menstruation, arteries that supply the lining constrict, and the capillaries weaken. Blood spilling from the damaged vessels detaches layers of the lining, not all at once but in random patches. Endometrium, mucus, and blood descend from the uterus, through the vagina, creating menstrual flow. Fibrinolysin, an enzyme released by dying cells, prevents the blood from clotting. Menstruation lasts from 3–5 days, as the uterus sloughs off the thick lining that was 3 weeks in the making.

Birth Control Figure 36.4B features some of the most effective and commonly used means of birth control. **Contraception** makes use of medications and devices that reduce the chance of pregnancy. Oral contraception (**birth control pills**) contains a combination of estrogen and progesterone for the first 21 days, followed by seven days of inactive pills (Fig. 36.4B*h*). The estrogen and progesterone in the birth control pill or a patch (Fig. 36.4B*b*) applied to the skin effectively shuts down the pituitary production of both FSH and LH. Follicle development in the ovary is prevented. Since ovulation does not occur, pregnancy cannot take place. Women taking birth control pills or using a patch should see a physician regularly, because of possible side effects.

Contraceptive implants use a synthetic progesterone to prevent ovulation by disrupting the ovarian cycle. The older version of the implant consists of six match-sized, time-release capsules surgically implanted under the skin of a woman's upper arm. The newest version consists of a single capsule that remains effective for about 3 years (Fig. 36.4B*g*).

Contraceptive injections are available as progesterone only (Fig. 36.4B*c*) or as a combination of estrogen and progesterone to prevent ovulation. The length of time between injections can vary from one to several months.

FIGURE 36.4A Female hormone levels during the ovarian and uterine cycles.

An **intrauterine device (IUD)** is a small piece of molded plastic inserted into the uterus by a physician (Fig. 36.4B*a*). IUDs are believed to alter the environment of the uterus and oviducts so that fertilization probably will not occur. If fertilization should occur, implantation cannot take place.

The **diaphragm** is a soft latex cup with a flexible rim that lodges behind the pubic bone and fits over the cervix (Fig. 36.4B*d*). Each woman must be properly fitted by a physician, and the diaphragm can be inserted into the vagina no more than 2 hours before sexual relations. Also, it must be used with spermicidal jelly or cream and should be left in place at least 6 hours after sexual relations. The cervical cap is a minidiaphragm.

There has been renewed interest in barrier methods of birth control, because these methods offer some protection against sexually transmitted diseases. A **female condom,** now available, consists of a large polyurethane tube with a flexible ring that fits onto the cervix (Fig. 36.4B*e*). The open end of the tube has a ring that covers the external genitals. A **male condom** is most often a latex sheath that fits over the erect penis (Fig. 36.4B*f*). The ejaculate is trapped inside the sheath, and thus does not enter the vagina. When used in conjunction with a spermicide, protection is better than with the condom alone.

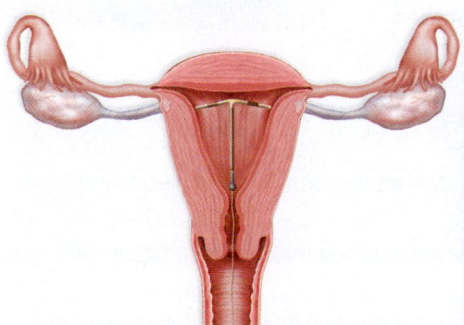

a. Intrauterine device placement

Intrauterine devices

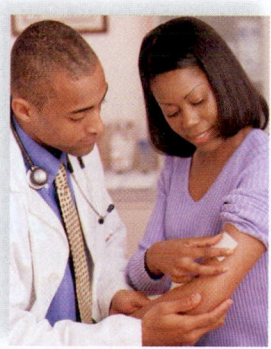

b. Hormone skin patch

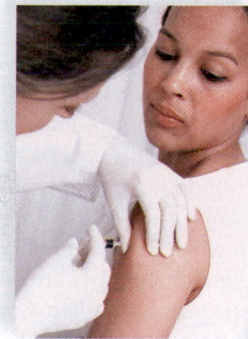

c. Hormone injection

d. Diaphragm and spermicidal jelly

e. Female condom

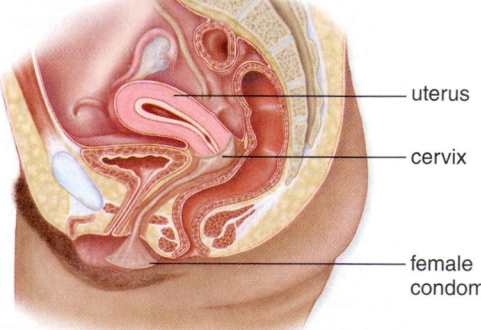

uterus

cervix

female condom

Female condom placement

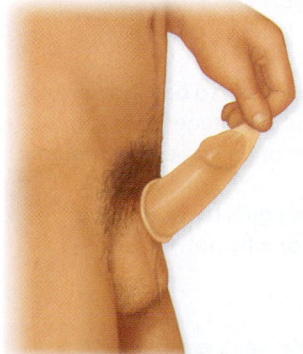

f. Male condom placement

Male condom

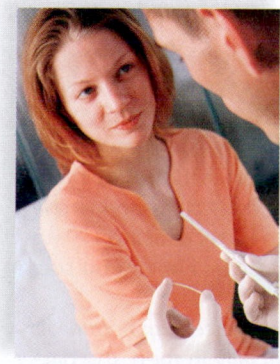

g. Hormone implant

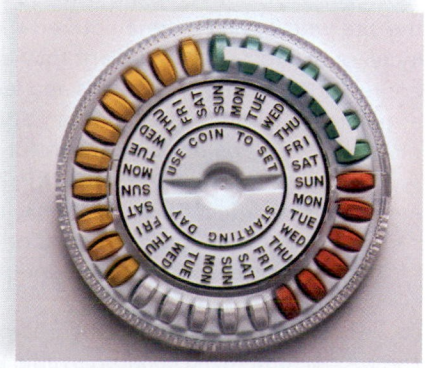

h. Oral contraception (birth control pills)

FIGURE 36.4B Birth control methods.

36B Sexual Activity Can Transmit Disease

A *sexually transmitted disease* (STD) is passed from one person to another during sexual relations. Prevention of STDs depends on consistently practicing safe sex, as described in Table 36B. We have already discussed AIDS (see pages 612–613), an STD caused by the human immunodeficiency virus (HIV), and there are many more STDs, caused by viruses, bacteria, or protozoans.

Genital herpes is caused by the *herpes simplex virus* (HSV). There are two types of HSV. HSV-1 usually affects the mouth, causing "fever blisters" or "cold sores." HSV-2 usually infects the genitals but sometimes the anus. In genital herpes, clusters of small blisters arise, rupture, and form painful sores. The sores heal in several days, but the virus persists in nearby nerve tissue and reemerges to cause periodic outbreaks of blisters and sores. Antibiotics are not effective against viral infections, and current antiviral drugs such as acyclovir (Zovirax) cannot cure herpes either. However, such drugs may reduce the number and duration of outbreaks an infected individual experiences.

Human papillomavirus (HPV) causes a common STD. Like HSV, different versions of HPV exist. Many types cause warts in areas such as the genitals or rectum that can be removed by freezing, burning with a laser, or surgery. However, a few types of HPV do not cause visible warts, but cause cervical cancer instead. Women aged 18–65 are encouraged to have an annual Pap smear, in which a sample of cells is removed from the cervix and examined for signs of cancer cells. In 2006, the FDA approved a vaccine against HPV, called Gardasil, which is now administered to both females and males. The vaccine prevents infection by the two most common cancer-causing types of HPV and the two types that most frequently cause genital warts.

Chlamydial infection, the most common STD in the United States, frequently occurs in the urethra or cervix, but may also exist in the rectum or throat. Usually, men with chlamydial infections experience painful urination and an abnormal discharge from the penis. Four out of five women infected with chlamydia do not experience any symptoms, although some report painful urination, vaginal bleeding, and/or abdominal pain. Chlamydial infection is a common cause of pelvic inflammatory disease (PID) in women. If a woman has a chlamydial infection, she can pass it to her baby during birth, resulting in eye infection or pneumonia. Antibiotics such as doxycycline are prescribed to treat the disease.

Syphilis is caused by the spirochete bacterium *Treponema pallidum.* An open sore called a chancre usually appears at or near the site where the bacteria entered the body. After the bacteria increase in number, the patient may experience a rash over the entire body or just on the hands and feet, mouth sores, fever, and a general ill feeling. After that, an apparent recovery may take place, with no obvious symptoms for several years. However, if syphilis is not eventually treated, the bones, heart, and/or brain are likely to be affected. If a woman with syphilis is pregnant, she can pass the bacterium to her child prior to birth, causing severe birth defects or even death. The antibiotic penicillin can be successfully used to cure syphilis.

Gonorrhea is caused by the bacterium *Neisseria gonorrhoeae;* can be called *Neisseria gonococcus* also. Especially in men, the bacteria may infect the urethra, causing painful urination and pus-filled urine. In women, cervical infection may go unnoticed for

TABLE 36B	Some Guidelines for Preventing the Spread of STDs

1. Abstain from sexual intercourse or develop a long-term monogamous (always the same person) relationship with a person who is free of STDs.

2. Refrain from having multiple sex partners or having a relationship with a person who has multiple sex partners.

3. Be aware that having sexual relations with an intravenous drug user is risky because the behavior of this group puts them at risk for certain sexually transmitted diseases.

4. Avoid anal intercourse because HIV has easy access through the lining of the rectum.

5. Always use a latex condom if your partner has not been free of STDs for the past 5 years.

6. Avoid oral sex because this may be a means of transmitting AIDS and other STDs.

7. Stop, if possible, the habit of injecting drugs, and if you cannot stop, at least always use a sterile needle.

several months. Meanwhile, the bacteria spread throughout the reproductive organs, often leading to PID. If a woman has gonorrhea, she may transmit the bacterium to her baby's eyes during vaginal birth; if untreated, blindness will result. Although penicillin has long been used to treat gonorrhea, resistant strains of *N. gonorrhoeae* have appeared in recent years.

Trichomoniasis is caused by the protozoan *Trichomonas vaginalis.* Usually the parasite infects the urethra in men and the vagina in women. Symptoms are more obvious in women. An infected woman (but not a man) typically notices a frothy, smelly, yellow-green discharge. Trichomoniasis in a pregnant woman may result in a premature and/or underweight baby. The infection is treated with the antiprotozoan drug metronidazole.

CONSIDER THESE QUESTIONS

1. Why is it evolutionarily beneficial for sex to be a powerful motivator? Why is it personally disadvantageous?
2. What are the social consequences of having a nontreatable sexually transmitted disease?
3. Do you approve of allowing young girls to receive the vaccine that prevents HPV? Why or why not?

 Explore the concepts through a variety of multimedia assets and question types.

www.mcgrawhillconnect.com

36C Reproductive Technologies Are Available to Help the Infertile

Infertility is the inability of a couple to achieve pregnancy after one year of regular, unprotected intercourse. The American Medical Association estimates that 15% of all couples are infertile. The cause of infertility can be attributed to the male (40%), the female (40%), or both (20%).

Sometimes the causes of infertility may be corrected by medical intervention so that couples can have children. If no obstruction is apparent and body weight is normal, it is possible for females to take fertility drugs, which are gonadotropic hormones that stimulate the ovaries and bring about ovulation. Such hormone treatments may cause multiple ovulations and multiple births.

When reproduction does not occur in the usual manner, many couples adopt a child. Others sometimes try one of the assisted reproductive technologies (ARTs) developed to increase the chances of pregnancy. In these cases, sperm and/or eggs are often retrieved from the testes and ovaries, and fertilization takes place in a clinical or laboratory setting.

Artificial Insemination by Donor (AID)

During artificial insemination, sperm are placed in the vagina by a physician. Sometimes a woman is artificially inseminated by her partner's sperm. This is especially helpful if the partner has a low sperm count, because the sperm can be collected over a period of time and concentrated so that the sperm count is sufficient to result in fertilization. Often, however, a woman is inseminated by sperm acquired from a donor. At times, a combination of partner and donor sperm is used.

A variation of AID is *intrauterine insemination* (*IUI*). In IUI, fertility drugs are given to stimulate the ovaries, and then the donor's sperm is placed in the uterus, rather than in the vagina.

If the prospective parents wish, sperm can be sorted into those believed to be X-bearing or Y-bearing to increase the chances of having a child of the desired sex. First, the sperm are dosed with a DNA-staining chemical. Because the X chromosome has slightly more DNA than the Y chromosome, it takes up more dye. When a laser beam shines on the sperm, the X-bearing sperm shine a little more brightly than the Y-bearing sperm. A machine sorts the sperm into two groups on this basis. Parents can expect about a 65% success rate for males and about 85% for females.

In Vitro Fertilization (IVF)

During IVF fertilization, the union of a sperm and an egg to form a zygote occurs in laboratory glassware (Fig. 36C). Ultrasound machines can now spot follicles in the ovaries that hold immature eggs; therefore, the latest method is to forgo fertility drugs and retrieve immature eggs by using a needle. The immature eggs are then brought to maturity in glassware before concentrated sperm are added. After about 2–4 days, the embryos are ready to be transferred to the uterus of the woman, who is now in the secretory phase of her uterine cycle. If desired, the embryos can be tested for a genetic disease, and only those found to be free of disease will be used. If implantation is successful, development is normal and continues to term.

Video
One Healthy Baby

Video
In Vitro Fertilization

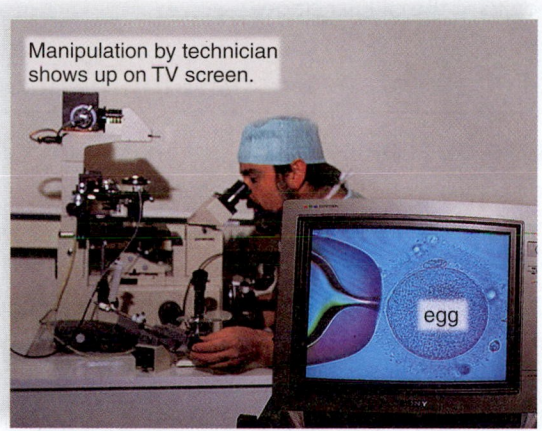
Manipulation by technician shows up on TV screen.

egg

FIGURE 36C Fertilization can occur in the laboratory.

Gamete Intrafallopian Transfer (GIFT)

Recall that the term *gamete* refers to a sex cell, either a sperm or an egg. Gamete intrafallopian transfer was devised to overcome the low success rate (15–20%) of in vitro fertilization. The method is exactly the same as for in vitro fertilization, except the eggs and the sperm are placed in the oviducts immediately after they have been brought together. GIFT has the advantage of being a one-step procedure for the woman—the eggs are removed and reintroduced all in the same time period. A variation on this procedure is to fertilize the eggs in the laboratory and then place the zygotes in the oviducts.

Surrogate Mothers

In some instances, women are contracted and paid to have babies. These women are called surrogate mothers. The sperm and even the egg can be contributed by the contracting parents.

Intracytoplasmic Sperm Injection (ICSI)

In this highly sophisticated procedure, a single sperm is injected into an egg. It is used effectively when a man has severe infertility problems.

If all the assisted reproductive technologies discussed here were employed simultaneously, it would be possible for a baby to have five parents: (1) sperm donor, (2) egg donor, (3) surrogate mother, (4) contracting mother, and (5) contracting father.

CONSIDER THESE QUESTIONS

1. Which of the potential parents listed in the last paragraph has the best claim to the child? Explain your reasoning.
2. Give the pros and cons of utilizing reproductive technologies. In your opinion do the benefits outweigh the drawbacks?
3. Suggest some legislation that might help regulate the use of reproductive technologies.

 **connect** | **BIOLOGY** | Explore the concepts through a variety of multimedia assets and question types.

www.mcgrawhillconnect.com

Early Developmental Stages and Processes

Development is divided into the cellular, tissue, and organ stages. Differentiation (specialization) of cells occurs during development, and it begins at the cellular stages.

36.5 Cellular stages of development precede tissue stages

LEARNING OUTCOME

When you complete this section, you should be able to

1. Describe the cellular stages of development from zygote to blastula.

The cellular stages of development are (1) cleavage resulting in a multicellular embryo and (2) formation of the blastula. **Cleavage** is cell division without growth. DNA replication and mitotic cell division occur repeatedly, and the cells get smaller with each division. In other words, cleavage increases only the number of cells; it does not change the original volume of the egg cytoplasm. As a model organism that demonstrates cleavage, we will use the lancelet, a chordate.

As shown in Figure 36.5, cleavage of a lancelet zygote is equal and results in uniform cells that form a **morula,** a ball of cells. The 16-cell morula resembles a mulberry and continues to divide, forming a blastula. A **blastula** is a hollow ball of cells having a fluid-filled cavity called a **blastocoel.** The blastocoel forms when the cells of the morula extrude Na^+ into extracellular spaces and water follows by osmosis. The water collects in the center, and the result is a hollow ball of cells.

Video
Blastocyst Formation

The zygotes of other animals, such as a frog, a chick, or a human, which are vertebrates, also undergo cleavage and form a morula. In frogs, cleavage is not equal because of the presence of yolk. When yolk is present, the zygote and embryo exhibit polarity, and the embryo has an animal pole and a vegetal pole. The animal pole of a frog embryo is deep gray in color because the cells contain melanin granules, and the vegetal pole is yellow because the cells contain yolk.

Similarly, all vertebrates have a blastula stage, but the blastula can look different from that of a lancelet. Birds lay a hard-shelled egg containing plentiful yolk. Because yolk-filled eggs do not participate in cleavage, the blastula is a layer of cells that spreads out over the yolk. The blastocoel is a space separating these cells from the yolk:

Chick blastula
(cross section)
blastocoel

The blastula of humans resembles that of the chick embryo, and yet this resemblance is not related to the amount of yolk because the human egg contains little yolk. Rather, this resemblance can be explained by the evolutionary history of these two animals. Because both birds and mammals are related to reptiles, all three groups develop similarly, despite a difference in the amount of yolk in their eggs.

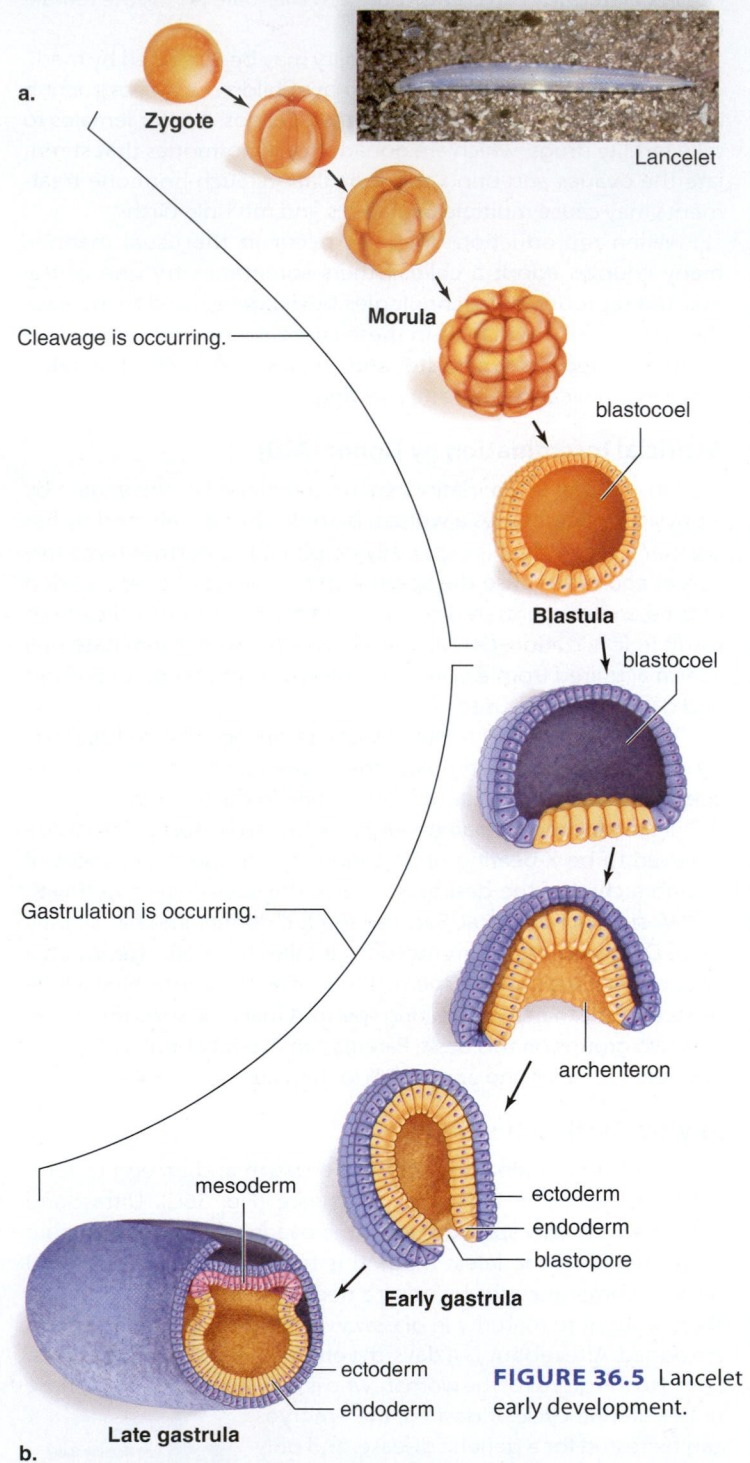

a.
Zygote

Lancelet

Cleavage is occurring.

Morula

blastocoel

Blastula

blastocoel

Gastrulation is occurring.

archenteron

mesoderm

ectoderm
endoderm
blastopore

Early gastrula

ectoderm
endoderm

Late gastrula

b.

FIGURE 36.5 Lancelet early development.

▶ **36.5 CHECK YOUR PROGRESS**

1. Explain how the cellular stages of development prepare for the tissue stages.
2. A lancelet develops in the water, and its zygote has minimal yolk; a bird embryo develops on land and has much yolk. Explain.

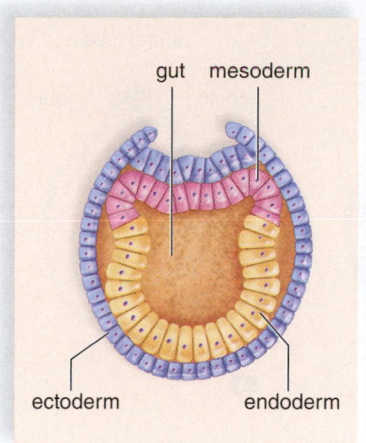

Lancelet late gastrula, c.s.

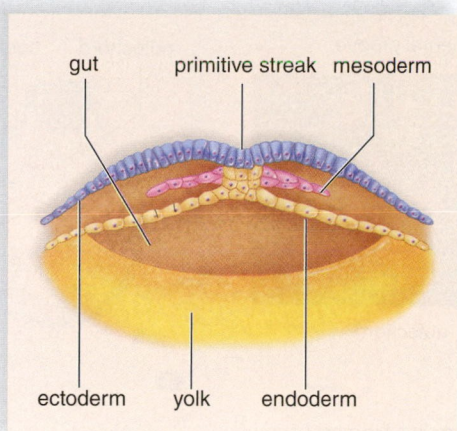

Chick late gastrula, c.s.

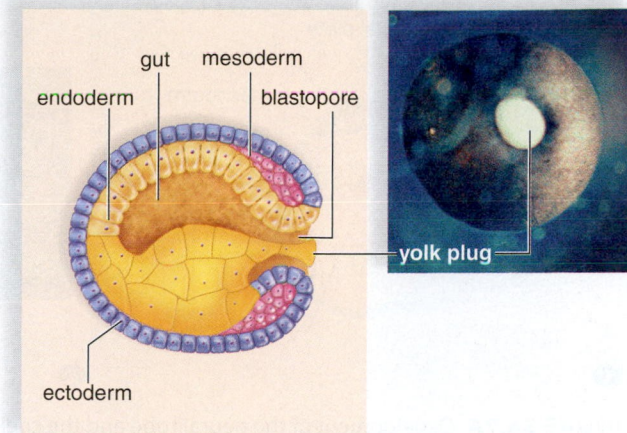

Frog late gastrula, l.s.

FIGURE 36.6 Comparative development of mesoderm. (c.s. = cross section, l.s. = longitudinal section)

LEARNING OUTCOME

When you complete this section, you should be able to

1. Explain the importance of gastrulation and the significance of germ layers.

The tissue stages of development are (1) the early gastrula and (2) the late gastrula. During early gastrulation, cells migrate into the blastocoel, creating a double layer of cells (see Fig. 36.5 *bottom*). Cells continue to migrate during other stages of development also, sometimes traveling quite a distance before reaching a destination, where they continue developing. Extracellular proteins and cytoskeletal elements allow migration to occur. As cells migrate, they "feel their way" by changing their pattern of adhering to extracellular proteins.

An early gastrula has two layers of cells. The outer layer is called the **ectoderm,** and the inner layer is called the **endoderm.** The endoderm borders a primitive gut. The pore, or hole, created by invagination is the **blastopore,** and in a lancelet, the blastopore eventually becomes the anus. **Gastrulation** is not complete until three layers of cells that will develop into adult organs are produced. In addition to ectoderm and endoderm, the late gastrula has a middle layer of cells called the **mesoderm** (see Fig. 36.5).

Figure 36.6 compares the lancelet, frog, and chick late gastrula stages. In the lancelet, mesoderm formation begins as outpocketings from the primitive gut. These outpocketings will grow in size until they meet and fuse, forming two layers of mesoderm. The space between them is the coelom. The coelom is a body cavity lined by mesoderm that contains internal organs. (In humans, the coelom becomes the thoracic and abdominal cavities of the body.)

In the frog, the cells containing yolk do not participate in gastrulation, and therefore they do not invaginate. Instead, a slitlike blastopore forms when the animal pole cells begin to invaginate from above, forming endoderm. Animal pole cells also move down over the yolk, to invaginate from below. Some yolk cells remain temporarily in the region of the blastopore, and are called the yolk plug. Mesoderm forms when cells migrate between the ectoderm and endoderm. Later, splitting of the mesoderm creates the coelom.

The chick egg contains so much yolk that endoderm formation does not occur by invagination. Instead, an upper layer of cells becomes ectoderm, and a lower layer becomes endoderm. Mesoderm arises as cells invaginate along the edges of a longitudinal furrow in the midline of the embryo. Because of its appearance, this furrow is called the *primitive streak*. Later, the newly formed mesoderm splits to produce a coelomic cavity.

Ectoderm, mesoderm, and endoderm are called the embryonic **germ layers.** No matter how gastrulation takes place, the result is the same—three germ layers form. It is possible to relate the development of future organs to these germ layers (Table 36.6).

▶ **36.6 CHECK YOUR PROGRESS**
1. Identify an organ that develops from each of the germ layers.

TABLE 36.6	Embryonic Germ Layers	
Embryonic Germ Layer	**Vertebrate Adult Structures**	
Ectoderm (outer layer)	Nervous system; epidermis of skin and derivatives of the epidermis (hair, nails, glands); tooth enamel, dentin, and pulp; epithelial lining of oral cavity and rectum	
Mesoderm (middle layer)	Muscular and skeletal systems; dermis of skin; cardiovascular system; urinary system; lymphatic system; reproductive system, including most epithelial linings; outer layers of respiratory and digestive systems	
Endoderm (inner layer)	Epithelial lining of digestive tract and respiratory tract, associated glands of these systems; epithelial lining of urinary bladder; thyroid and parathyroid glands	

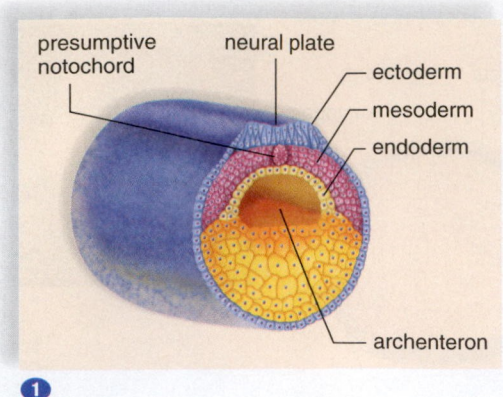
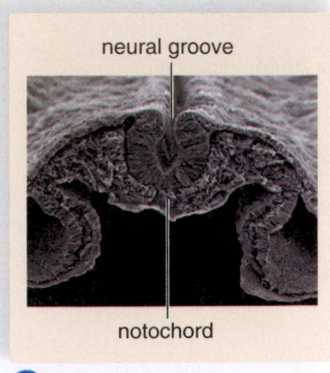
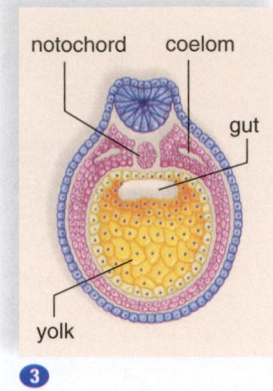
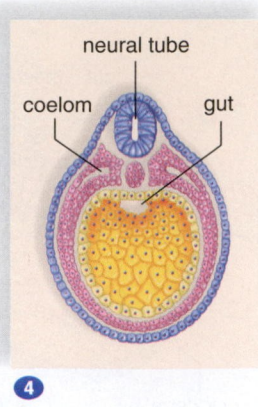

FIGURE 36.7A Development of the neural tube and the coelom in a frog embryo.

LEARNING OUTCOME

When you complete this section, you should be able to

1. Draw and identify the significance of each structure shown in Figure 36.7B.

The organs of an animal's body develop from the three embryonic germ layers. In this section, we explain this process, beginning with how the nervous system develops from the ectoderm.

The newly formed mesoderm cells lie along the main longitudinal axis of the animal and coalesce to form a *presumptive notochord*, so-called because these cells mark the location of the notochord. The notochord persists in lancelets, but in frogs, chicks, and humans, it is later replaced by the vertebral column (which explains why these animals are called vertebrates).

Figure 36.7A shows cross sections of frog development to illustrate the formation of the neural tube. The nervous system starts to develop from midline ectoderm located just above the presumptive notochord. ❶ At first, a thickening of cells, called the **neural plate,** appears along the dorsal surface of the embryo. ❷ Then neural folds develop on either side of a neural groove. ❸ The coelom appears, and ❹ the **neural tube** is complete. At this point, the embryo is called a **neurula.** Later, the anterior end of the neural tube develops into the *brain*, and the rest becomes the *spinal cord*. In addition, the neural crest is a band of cells that develops where the neural tube pinches off from the ectoderm. Neural crest cells migrate to various locations, where they contribute to the formation of skin and muscles as well as the adrenal medulla and the ganglia of the peripheral nervous system.

Midline mesoderm cells that did not contribute to the formation of the notochord now form two longitudinal masses of tissue. These two masses become blocked off into somites, which are serially arranged along both sides along the length of the notochord. Somites give rise to muscles associated with the axial skeleton and to the vertebrae. The serial origins of axial muscles and the vertebrae testify that vertebrates are segmented animals. Lateral to the somites, the mesoderm splits, forming the mesodermal lining of the coelom.

Figure 36.7B will help you relate the formation of the vertebrate structures and organs discussed in this section to the three embryonic layers of cells: the ectoderm, the mesoderm, and the endoderm. A primitive gut tube is formed by endoderm as the body itself folds into a tube. The heart, too, begins as a simple tubular pump formed from mesoderm. Organ formation continues until the germ layers have given rise to the specific organs listed in Table 36.6.

▶ **36.7 CHECK YOUR PROGRESS**
1. Describe how a notochord is different from a neural tube.

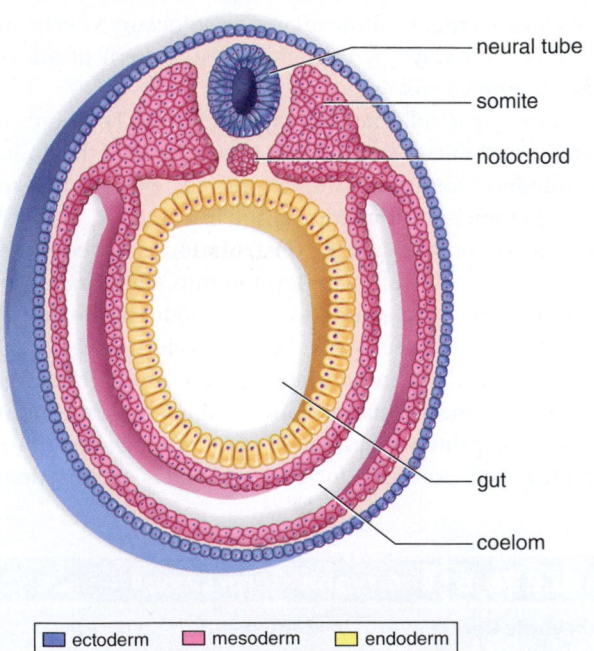

| ectoderm | mesoderm | endoderm |

FIGURE 36.7B Vertebrate embryo, cross section, at the neurula stage. The neural tube has formed and three germ layers are apparent. Each of the germ layers are associated with the development of particular organs (see Table 36.6). The somites give rise to the muscles and to the vertebrae, which replace the notochord in vertebrates.

36.8 Cellular differentiation begins with cytoplasmic segregation

LEARNING OUTCOME

When you complete this section, you should be able to

1. Explain the significance of cytoplasmic segregation.

Development requires growth, cellular differentiation, and morphogenesis. We have already seen that growth occurs when there is an increase in the number of cells, which then become larger. And we know that differentiation occurs when cells become specialized in structure and function—that is, a muscle cell looks and acts differently than a nerve cell. **Morphogenesis,** on the other hand, produces the shape and form of the body. One of the earliest indications of morphogenesis is cell movement. Later, morphogenesis includes **pattern formation,** the way that tissues and organs are arranged in the body. Apoptosis, or programmed cell death, first discussed in section 8.5, is an important part of pattern formation. In this section, we consider what event might precede cellular differentiation and morphogenesis.

Cytoplasmic Segregation Cellular differentiation begins long before we can recognize specialized types of cells. Ectodermal, endodermal, and mesodermal cells in the gastrula look quite similar, but they must be different because they develop into different organs. An examination of the frog's egg shows that it is not uniform (Fig. 36.8). ❶ The egg is polar and has both an anterior/posterior axis and a dorsal/ventral axis, which can be correlated with the **gray crescent,** a gray area that appears after the sperm fertilizes the egg. Hans Spemann, who received a Nobel Prize in 1935 for his extensive work in embryology, showed that ❷ if the gray crescent is divided equally by the first cleavage, each experimentally separated daughter cell develops into a complete embryo. ❸ However,

if the egg divides so that only one daughter cell receives the gray crescent, only that cell becomes a complete embryo. This experiment allows us to speculate that the gray crescent must contain particular chemical signals that are needed in order for development to proceed normally. We now know that an egg contains substances called *maternal determinants* that influence the course of development. As mitosis occurs, maternal determinants are parceled out, a process known as cytoplasmic segregation:

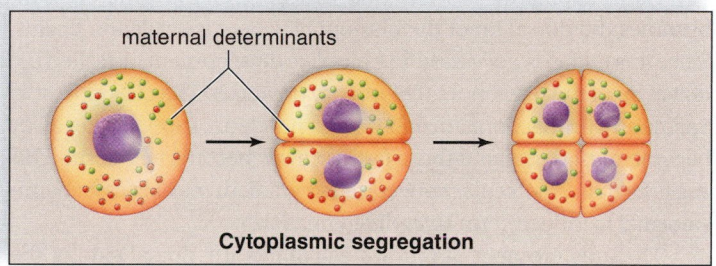

Cytoplasmic segregation

At one time, it was thought that the genes must be parceled out as development. Now we know that instead gene regulation accounts for the differentiation of cells. The first step toward gene regulation during development is cytoplasmic segregation. Cytoplasmic segregation helps determine how the various cells of the morula will develop.

▶ **36.8 CHECK YOUR PROGRESS**

1. A cell that does not receive a portion of the gray crescent never undergoes gastrulation. Therefore, development is halted at the cellular level. Explain.

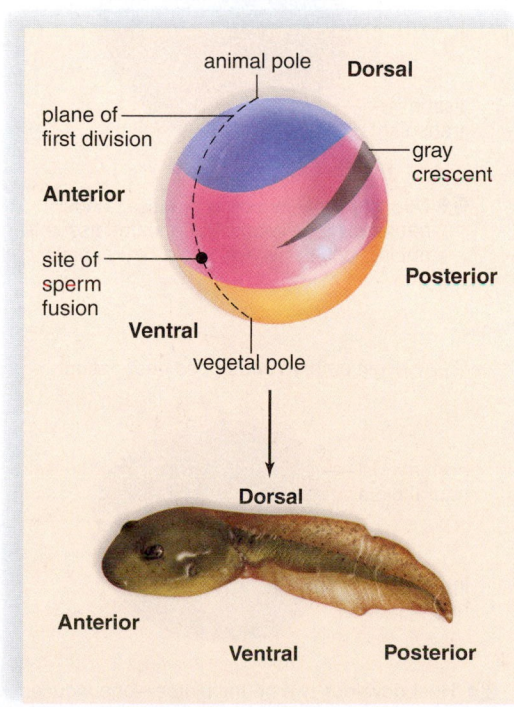

❶ Frog's egg is polar and has axes.

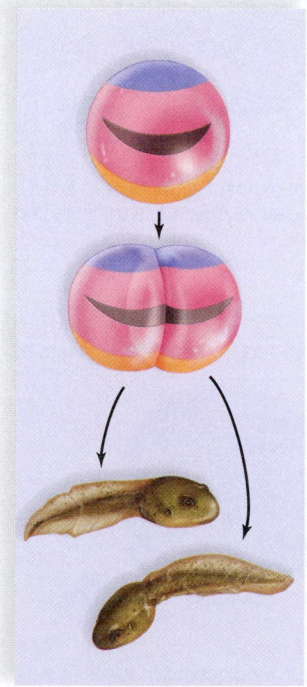

❷ Each cell receives a part of the gray crescent.

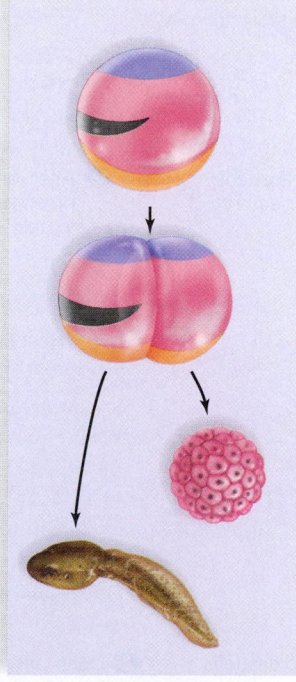

❸ Only the cell on the left receives the gray crescent.

FIGURE 36.8
Cytoplasmic influence on development.

36.9 Morphogenesis involves induction also

As development proceeds, specialization of cells and formation of organs are influenced not only by maternal determinants, but also by signals given off by neighboring cells. **Induction** is the ability of one embryonic tissue to influence the development of another tissue.

Hans Spemann showed that a frog embryo's gray crescent becomes the dorsal lip of the blastopore, where gastrulation begins. Since this region is necessary for complete development, he called the dorsal lip of the blastopore the primary organizer. The cells closest to Spemann's primary organizer become endoderm; those farther away become mesoderm; and those farthest away become ectoderm. This suggests that a molecular concentration gradient may act as a chemical signal to induce germ layer differentiation.

The gray crescent of a frog's egg marks the dorsal side of the embryo, where the mesoderm becomes notochord and ectoderm becomes the nervous system. In a classic experiment, Spemann and his colleague Hilde Mangold showed that presumptive (potential) notochord tissue induces the formation of the nervous system. Two experiments are described in Figure 36.9. In the first experiment, ❶ presumptive nervous system tissue, located just above the presumptive notochord, is cut out and ❷ transplanted to the belly region of the embryo; ❸ it does not form a neural tube. In the second experiment, ❶ presumptive notochord tissue is cut out and ❷ transplanted beneath what would be belly ectoderm; ❸ this ectoderm differentiates into neural tissue. Still other examples of induction are now known. In 1905, Warren Lewis studied the formation of the eye in frog embryos. He found that an optic vesicle, which is a lateral outgrowth of developing brain tissue, induces overlying ectoderm to thicken and become a lens. The developing lens, in turn, induces an optic vesicle to form an optic cup, where the retina develops. This suggests that induction, which involves signaling molecules, is the process that explains differentiation and morphogenesis.

Hox Genes Section 12.4 discusses the role of *Hox* (homeotic) genes in development. Most likely, signaling molecules during the process of induction turn on *Hox* genes that code for hox proteins, which are involved in morphogen gradients. A surprising finding has been that the hox proteins of so many different organisms, both invertebrate and vertebrate, contain the same homeodomain, a sequence of amino acids that binds to DNA and turns on other genes.

▶ **36.9 CHECK YOUR PROGRESS**
1. Briefly explain the experiment depicted in Figure 36.9.

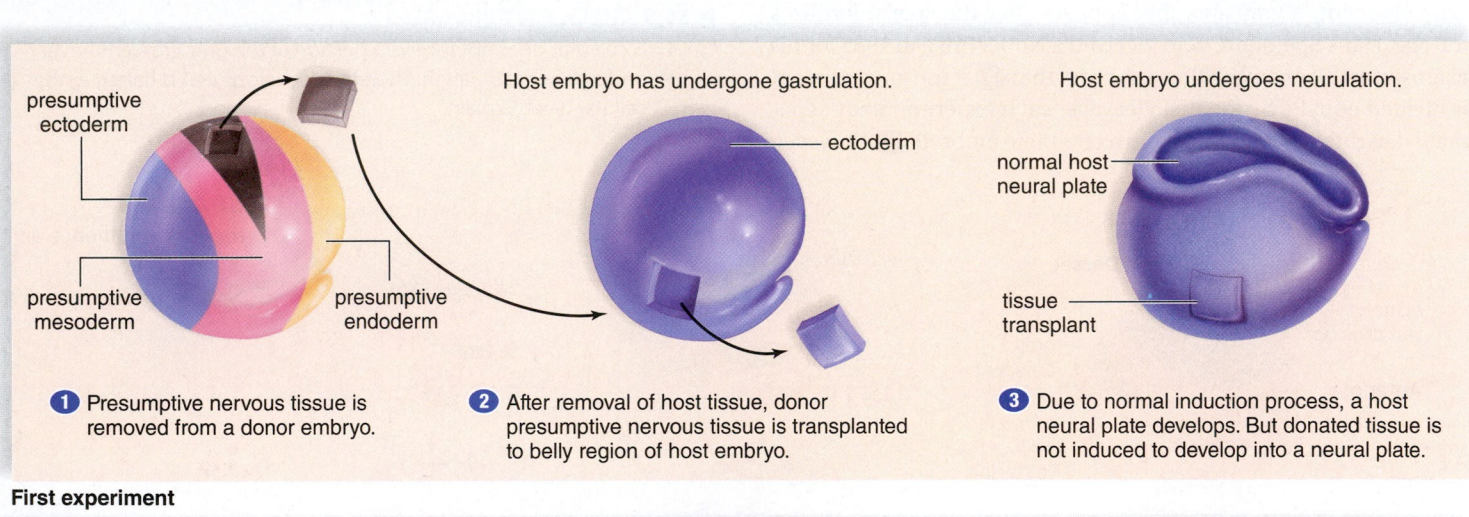

presumptive ectoderm

presumptive mesoderm

presumptive endoderm

Host embryo has undergone gastrulation.

ectoderm

Host embryo undergoes neurulation.

normal host neural plate

tissue transplant

❶ Presumptive nervous tissue is removed from a donor embryo.

❷ After removal of host tissue, donor presumptive nervous tissue is transplanted to belly region of host embryo.

❸ Due to normal induction process, a host neural plate develops. But donated tissue is not induced to develop into a neural plate.

First experiment

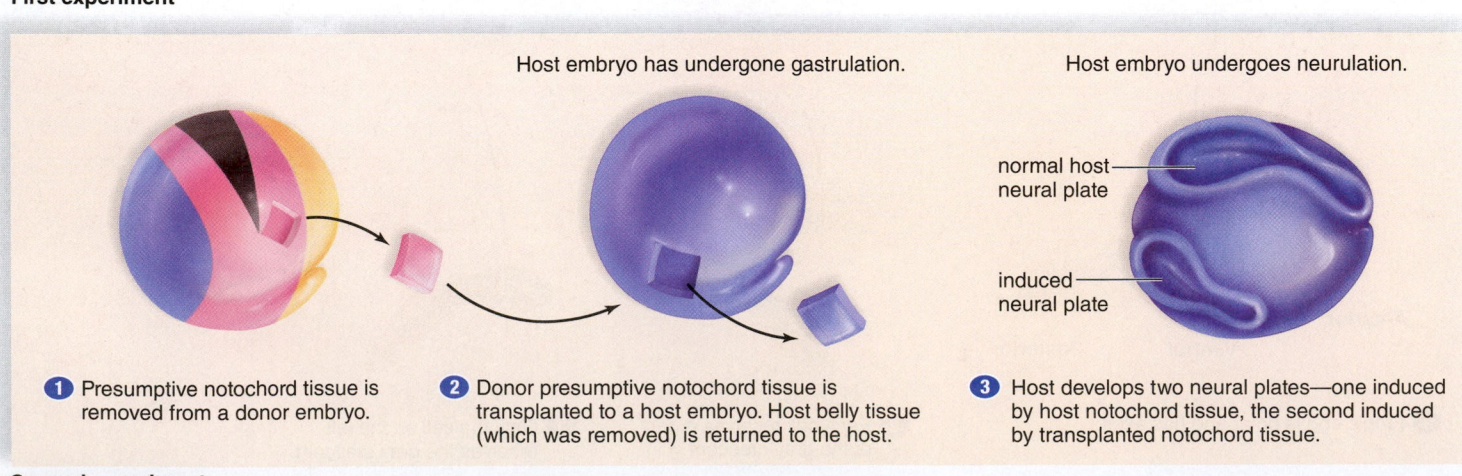

Host embryo has undergone gastrulation.

Host embryo undergoes neurulation.

normal host neural plate

induced neural plate

❶ Presumptive notochord tissue is removed from a donor embryo.

❷ Donor presumptive notochord tissue is transplanted to a host embryo. Host belly tissue (which was removed) is returned to the host.

❸ Host develops two neural plates—one induced by host notochord tissue, the second induced by transplanted notochord tissue.

Second experiment

FIGURE 36.9 Control of nervous system development.

Human Embryonic and Fetal Development

Although the extraembryonic membranes are not a part of the embryo, they play a significant role in human development. Human development is divided into embryonic and fetal stages. Fetal development ends with the birth of the newborn.

36.10 Extraembryonic membranes are critical to human development

LEARNING OUTCOME

When you complete this section, you should be able to

1. Contrast the function of extraembryonic membranes in chicks and in humans.

In humans, the length of time from conception (fertilization followed by **implantation**) to birth (parturition) is approximately nine months (266 days). It is customary to calculate the time of birth by adding 280 days to the start of the last menstrual period, because this date is usually known, whereas the day of fertilization is usually not known.

In humans, pregnancy, or gestation, is the time during which the mother carries the developing embryo. Biologists divide human development into embryonic development (months 1 and 2) and fetal development (months 3–9). During embryonic development, the major organs are formed, and during fetal development, these structures are refined.

Development can also be divided into trimesters, with each one characterized by specific developmental accomplishments. During the first trimester, embryonic and early fetal development occur. The second trimester is characterized by the development of organs and organ systems. By the end of the second trimester, the fetus appears distinctly human. In the third trimester, the fetus grows rapidly, and the major organ systems become functional. An infant born one month, or perhaps even two months, prematurely has a reasonable chance of survival.

Before we consider human development chronologically, we must understand the placement of **extraembryonic membranes.** Extraembryonic membranes are best understood by considering their function in reptiles and birds. In reptiles, these membranes made development on land possible. If an embryo develops in the water, the water supplies oxygen for the embryo and takes away waste products. The surrounding water prevents desiccation, or drying out, and provides a protective cushion. For an embryo that develops on land, all these functions are performed by the extraembryonic membranes.

In the chick, the extraembryonic membranes develop from extensions of the germ layers, which spread out over the yolk. Figure 36.10, *top*, shows the chick surrounded by the membranes. The **chorion** lies next to the shell and carries on gas exchange. The **amnion** contains the protective amniotic fluid, which bathes the developing embryo. The **allantois** collects nitrogenous wastes, and the **yolk sac** surrounds the remaining yolk, which provides nourishment.

Humans (and other mammals) also have extraembryonic membranes (Fig. 36.10, *bottom*). The chorion develops into the fetal half of the placenta, where nutrient-for-waste exchange occurs between the fetus and mother (see section 36.12). The yolk

sac, which lacks yolk, is the first site of blood cell formation; the blood vessels of the allantois become the umbilical blood vessels; and the amnion contains fluid to cushion and protect the embryo, which develops into a fetus. Therefore, the function of the membranes in humans has been modified to suit internal development, but their very presence indicates our relationship to birds and to reptiles. It is interesting to note that all chordate animals develop in water—either in bodies of water or surrounded by amniotic fluid within a shell or uterus.

▶ **36.10 CHECK YOUR PROGRESS**

1. Discuss how the extraembryonic membranes, which first evolved in reptiles, have been modified in mammals.
2. The presence of extraembryonic membranes in humans supports the evolutionary principle of common descent. How so?

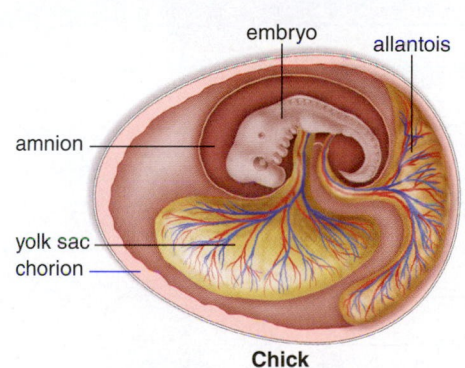

Chick

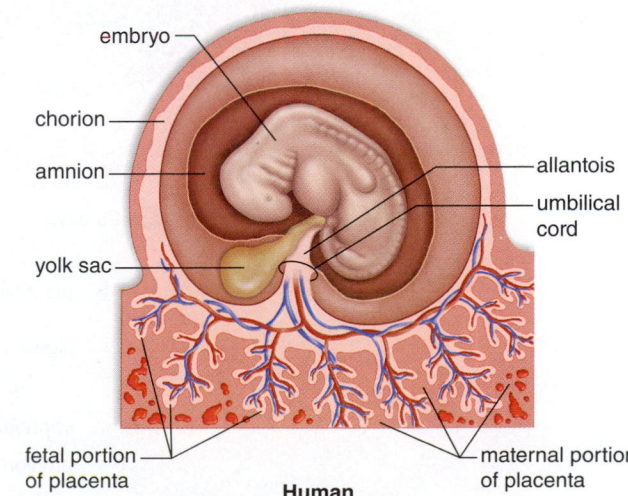

Human

FIGURE 36.10 Extraembryonic membranes.

36.11 Embryonic development involves tissue and organ formation

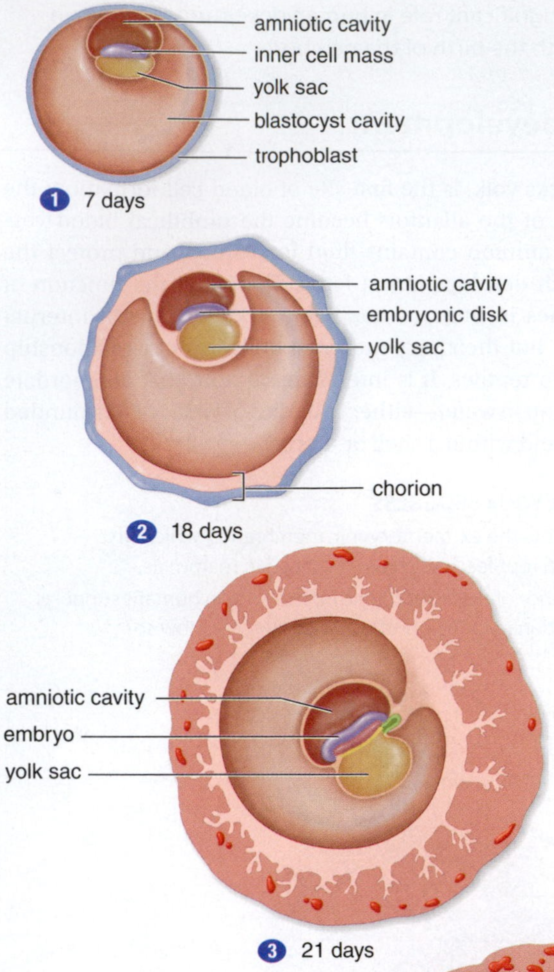

1 7 days

- amniotic cavity
- inner cell mass
- yolk sac
- blastocyst cavity
- trophoblast

2 18 days

- amniotic cavity
- embryonic disk
- yolk sac
- chorion

3 21 days

- amniotic cavity
- embryo
- yolk sac

4 25 days

- chorion
- amniotic cavity
- chorionic villi
- body stalk
- allantois
- yolk sac

5 35+ days

- placenta
- digestive tract
- chorionic villi
- amnion
- amniotic cavity
- umbilical cord

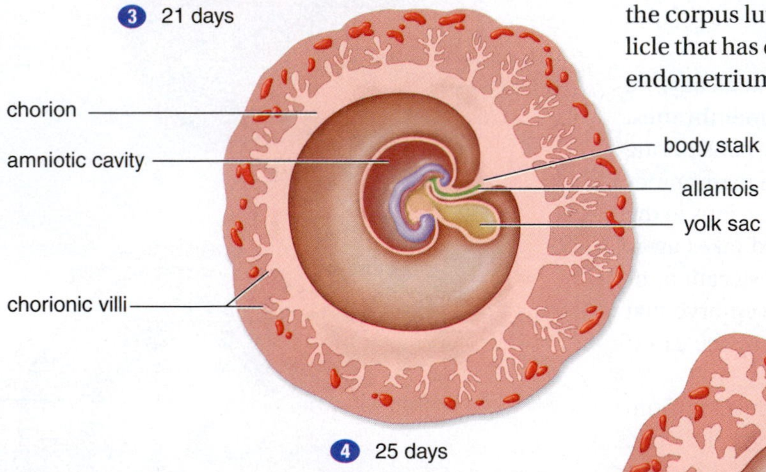

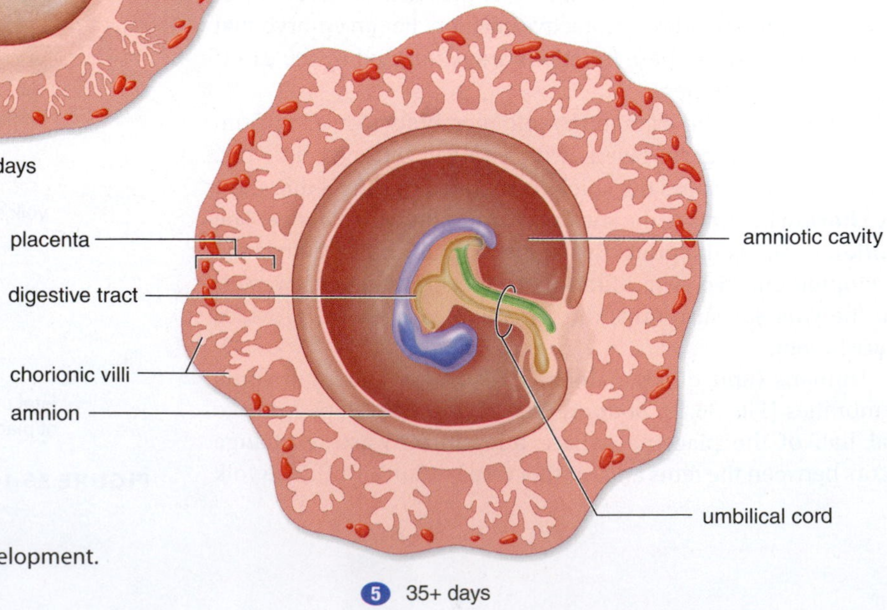

FIGURE 36.11A Overview of human embryonic development.

LEARNING OUTCOME

When you complete this section, you should be able to

1. Describe the events of human embryonic development, stressing the need for implantation.

Embryonic development includes the first 2 months of development.

Fertilization occurs in the upper third of the oviduct. Cleavage begins 30 hours after fertilization and continues as the embryo passes through the oviduct to the uterus. By the time the embryo reaches the uterus on the third day, it is a morula. The morula is not much larger than the zygote because, even though multiple cell divisions have occurred, there has been no growth of these newly formed cells. Around the seventh day, the morula is transformed into the blastocyst. In Figure 36.11A, **1** the **blastocyst** has a fluid-filled cavity called the amniotic cavity, a single layer of outer cells called the **trophoblast,** and an inner cell mass. The early function of the trophoblast is to provide nourishment for the embryo. The trophoblast will be a part of the *chorion,* one of the extraembryonic membranes. The inner cell mass eventually becomes the embryo, which develops into a fetus.

At about this time, the embryo begins the process of implanting in the wall of the uterus. The trophoblast secretes enzymes to digest away some of the tissue and blood vessels of the endometrium of the uterus. The embryo is now about the size of the period at the end of this sentence. The trophoblast begins to secrete **human chorionic gonadotropin** (**HCG**), the hormone that serves as the basis for the pregnancy test and maintains the corpus luteum past the time it normally disintegrates. (Recall that the corpus luteum is a yellow body formed in the ovary from a follicle that has discharged its secondary oocyte.) Because of this, the endometrium is maintained, and menstruation does not occur.

2 As the week progresses, the inner cell mass detaches itself from the trophoblast, and two more extraembryonic membranes form. In humans, the *yolk sac,* which forms below the inner cell mass, has no nutritive function as it does in chicks, but it is the first site of blood cell formation. However, the *amnion* and its cavity are where the embryo (and then the fetus) develops. In humans, amniotic fluid acts as an insulator against cold and heat and also absorbs shock, such as that caused by the mother exercising.

Gastrulation occurs during the second week. The inner cell mass now has flattened into the **embryonic disk,** composed of two layers of cells: ectoderm above and endoderm below. Once the embryonic disk elongates to form the primitive streak, the third germ layer, mesoderm, forms by invagination of cells along the streak. The trophoblast is reinforced by mesoderm and becomes the chorion. It is possible to relate the development of future organs to these germ layers (see Table 36.6).

3 Two important organ systems make their appearance during the third week as the embryonic disk becomes the embryo. The nervous system is the first organ system to be visually evident. At first, a thickening appears along the entire dorsal length of the embryo; then the neural folds appear. When the neural folds meet at the midline, the neural tube, which later develops into the brain and the nerve cord, is formed. After the notochord is replaced by the vertebral column, the nerve cord is called the spinal cord.

Although the embryo has already undergone quite a bit of development, the woman, as yet, probably doesn't even realize she is pregnant. Many authorities recommend that sexually active females always practice good health habits to protect a possible developing embryo from harm. This means avoiding alcohol, cigarettes, illegal drugs, and any medication not approved by a physician. Aside from a positive pregnancy test, the early symptoms of pregnancy are nausea, breast swelling and tenderness, and fatigue.

The second organ system to appear during the third week of development is the cardiovascular system as the heart begins to form. At first, there are right and left heart tubes; when these fuse, the heart begins pumping blood, even though the chambers of the heart are not fully formed. The veins enter posteriorly, and the arteries exit anteriorly from this largely tubular heart, but later the heart twists so that all major blood vessels are located anteriorly.

4 At 4 weeks, the embryo is barely larger than the height of this print. A bridge of mesoderm called the body stalk connects the caudal (tail) end of the embryo with the chorion, which has treelike projections called **chorionic villi.** The chorionic villi are a part of the placenta, the region of exchange between the mother and fetus. The fourth extraembryonic membrane, the *allantois,* is contained within this stalk, and its blood vessels become the umbilical blood vessels. **5** The head and the tail then lift up; the body stalk and the yolk sac fuse to become the **umbilical cord,** which connects the developing embryo to the placenta.

Little flippers called limb buds appear at the fifth week (Fig. 36.11B); later, the arms and the legs develop from the limb buds, and even the hands and the feet become apparent. At the same time—during the fifth week—the head enlarges, and the sense organs become more prominent. It is possible to make out the developing eyes, ears, and even the nose.

During the sixth to eighth weeks of development, the embryo becomes easily recognizable as human. Concurrent with brain development, the head achieves its normal relationship with the body as a neck region develops. The nervous system is developed well enough to permit reflex actions, such as a startle response to touch. At the end of this period, the embryo is about 38 mm long and weighs no more than an aspirin tablet; even so, all organ systems have been established.

▶ **36.11 CHECK YOUR PROGRESS**
  1. Relate human embryonic development to the cellular and tissue level development discussed earlier.

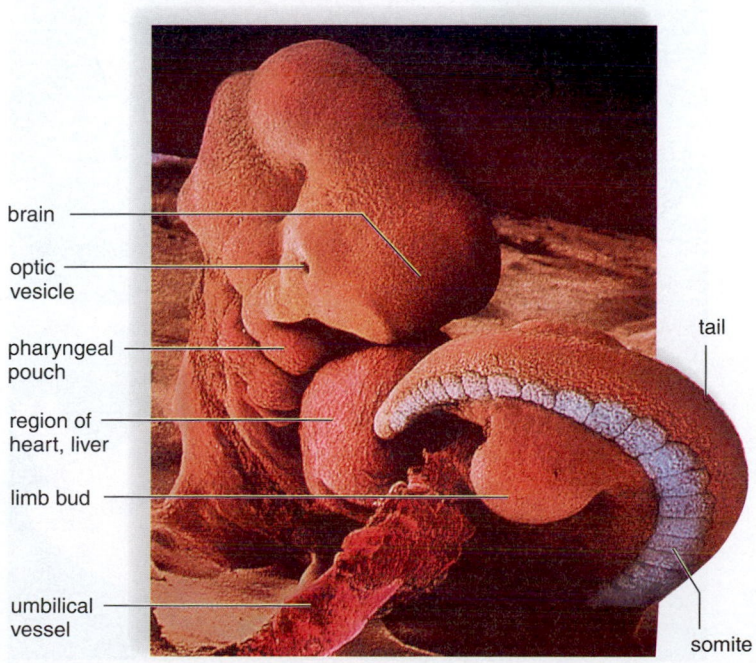

brain
optic vesicle
pharyngeal pouch
region of heart, liver
limb bud
umbilical vessel
tail
somite

FIGURE 36.11B Human embryo at fifth week.

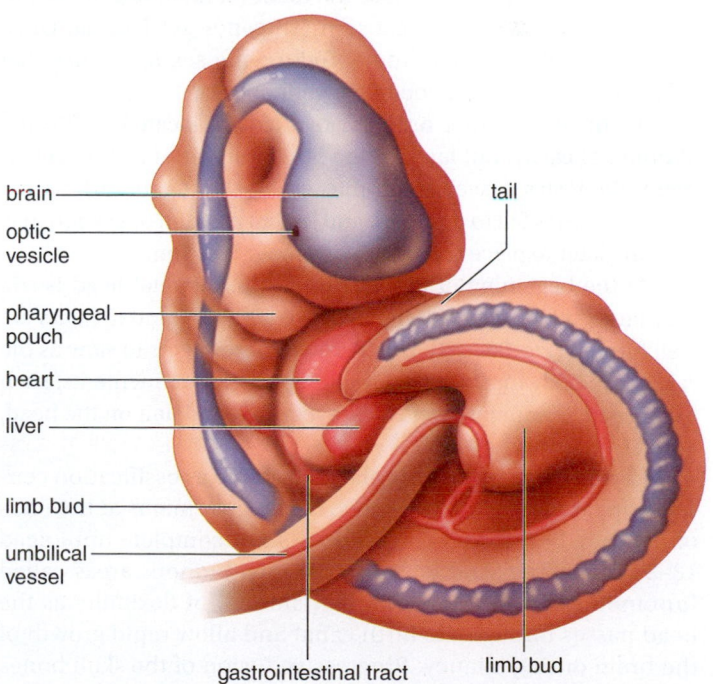

brain
optic vesicle
pharyngeal pouch
heart
liver
limb bud
umbilical vessel
tail
gastrointestinal tract
limb bud

LEARNING OUTCOME

When you complete this section, you should be able to

1. Describe the events of human fetal development, stressing the importance of the placenta.

Fetal development includes the third through the ninth months of development (Fig. 36.12A). Aside from an increase in weight, many of the physiological changes in the mother are due to the placental hormones that support fetal development. For example, progesterone decreases uterine motility, but smooth muscle relaxation also brings about the heartburn and constipation so often experienced by pregnant women.

The placenta is fully developed (see Fig. 36.11A **5**). The placenta has a fetal side contributed by the chorion, the outermost extraembryonic membrane, and a maternal side consisting of uterine tissues. The treelike chorionic villi are surrounded by maternal blood; yet maternal and fetal blood never mix because exchange always takes place across the villi. Carbon dioxide and other wastes move from the fetal side to the maternal side of the placenta, and nutrients and oxygen move from the maternal side to the fetal side. Simple diffusion and active transport are at work. The umbilical cord stretches between the placenta and the fetus. The umbilical blood vessels are an extension of the fetal circulatory system and simply take fetal blood to and from the placenta. Harmful chemicals can also cross the placenta. This is of particular concern during the embryonic period, when various structures are first forming. Each organ or part seems to have a sensitive period during which a substance can alter its normal development. For example, if a woman takes the drug thalidomide, a tranquilizer, between days 27 and 40 of her pregnancy, the infant is likely to be born with deformed limbs.

Sometime during the third month, it is possible to distinguish males from females. Researchers have discovered a series of genes on the chromosomes that cause the differentiation of gonads into testes. Ovaries develop because these genes are lacking. Once these have differentiated, they produce the sex hormones that influence the differentiation of the genital tract.

At this time, either testes or ovaries are located within the abdominal cavity, but later, in the last trimester of fetal development, the testes descend into the scrotal sacs (scrotum). Sometimes the testes fail to descend, and in that case, an operation may be done later to place them in their proper location.

At the beginning of the fourth month, the fetal head is still very large, the nose is flat, the eyes are far apart, and the ears are well formed (Fig. 36.12B). Head growth now begins to slow as the rest of the body increases in length. Epidermal refinements, such as fingernails, nipples, eyelashes, eyebrows, and hair on the head, appear.

Cartilage begins to be replaced by bone as ossification centers appear in most of the bones. Cartilage remains at the ends of the long bones, and ossification is not complete until aged 18–20 years. The skull has six large membranous areas called **fontanels,** which permit a certain amount of flexibility as the head passes through the birth canal and allow rapid growth of the brain during infancy. Progressive fusion of the skull bones causes the fontanels to close, usually by 2 years of age.

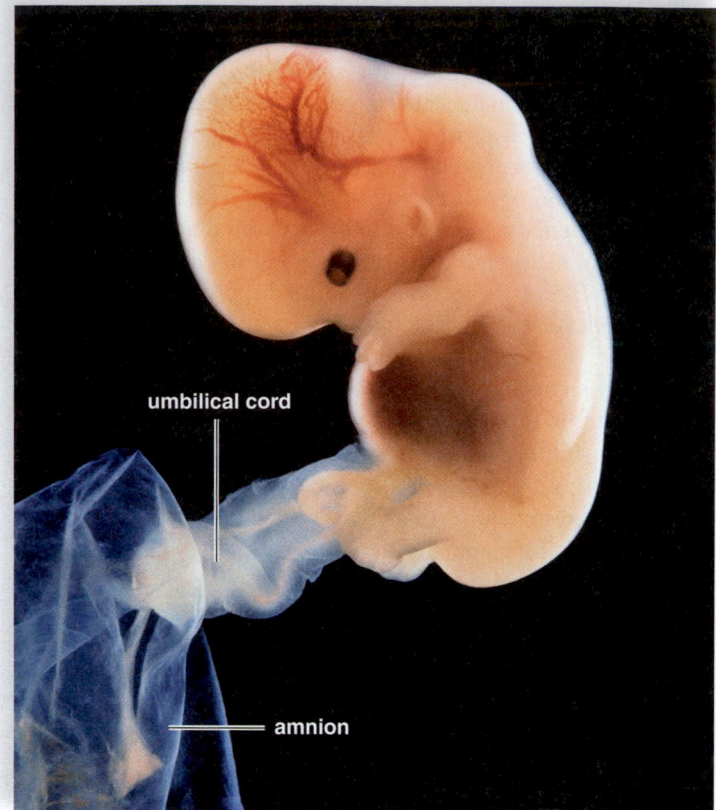

FIGURE 36.12A 9-week fetus.

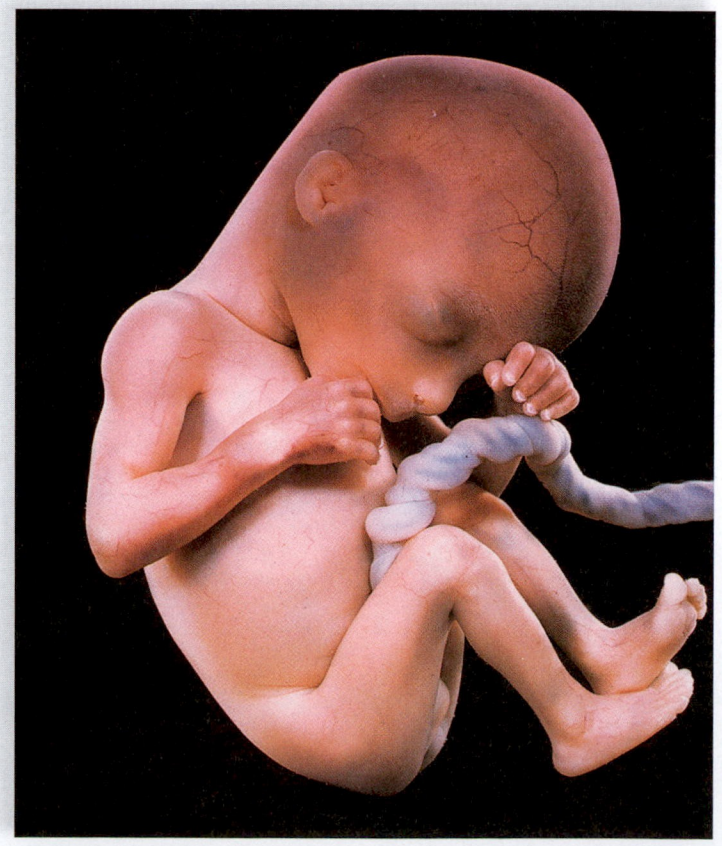

FIGURE 36.12B 16-week fetus.

TABLE 36.12 | Fetal Development

Time	Events for Baby
Third month	Gender can be distinguished by ultrasound. Fingernails appear.
Fourth month	Skeleton is visible. Hair begins to appear. Fetus is about 15 cm long and weighs about 170 g.
Fifth month	Protective cheesy coating, called vernix caseosa, begins to be deposited. Heartbeat can be heard.
Sixth month	Body is covered with fine hair called lanugo. Skin is wrinkled and reddish.
Seventh month	Testes descend into scrotum. Eyes are open. Fetus is about 31 cm long and weighs about 1.4 kg.
Eight month	Body hair begins to disappear. Subcutaneous fat begins to be deposited.
Ninth month	Fetus is ready for birth. It is about 52 cm long and weighs about 3.5 kg.

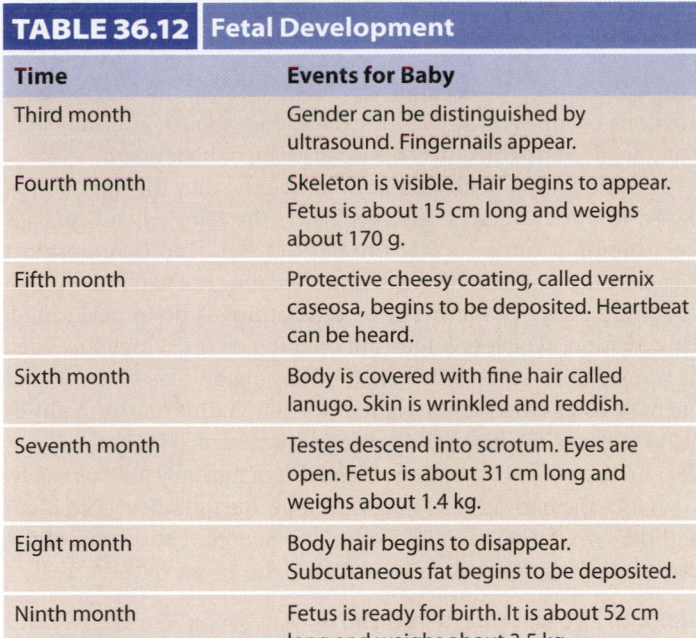

FIGURE 36.12C 24-week fetus.

During the fourth month, the fetal heartbeat is loud enough to be heard when a physician applies a stethoscope to the mother's abdomen, and fetal movement can be felt by women who have previously been pregnant. By the end of this month, the fetus is about 15 cm long and weighs about 170 g.

During the fifth through seventh months (Fig. 36.12C), fetal movement increases. What is at first only a fluttering sensation turns into kicks and jabs as the fetal legs grow and develop. The wrinkled, translucent, pink-colored skin is covered by a fine down called *lanugo.* This, in turn, is coated with a white, greasy, cheese-like substance called *vernix caseosa,* which probably protects the delicate skin from the amniotic fluid. The eyelids can now fully open.

At the end of the seventh month, the fetus's length has increased to about 31 cm, and the weight is about 1.4 kg. It is possible that, if born now, the baby will survive.

Of interest, the increasing size of the uterus during pregnancy contributes to an improvement in the mother's respiratory functions because breathing depth typically increases by 40%. The uterus comes to occupy most of the abdominal cavity, reaching nearly to the breastbone. This increase in size pushes the intestines out of the way and also widens the thoracic cavity. The fall in the mother's blood carbon dioxide level creates a favorable gradient for the flow of carbon dioxide from fetal blood to maternal blood at the placenta.

As the end of development approaches, the fetus usually assumes the fetal position, with the head bent down and in contact with the flexed knees. Eventually, the body rotates so that the head points toward the cervix. If the fetus does not turn, a breech birth (rump first) is likely. It is very difficult for the cervix to expand enough to accommodate the baby's body in this position, and asphyxiation of the baby is more likely to occur. Thus, a **cesarean section** (incision through the abdominal and uterine walls) may be prescribed for delivery of the fetus.

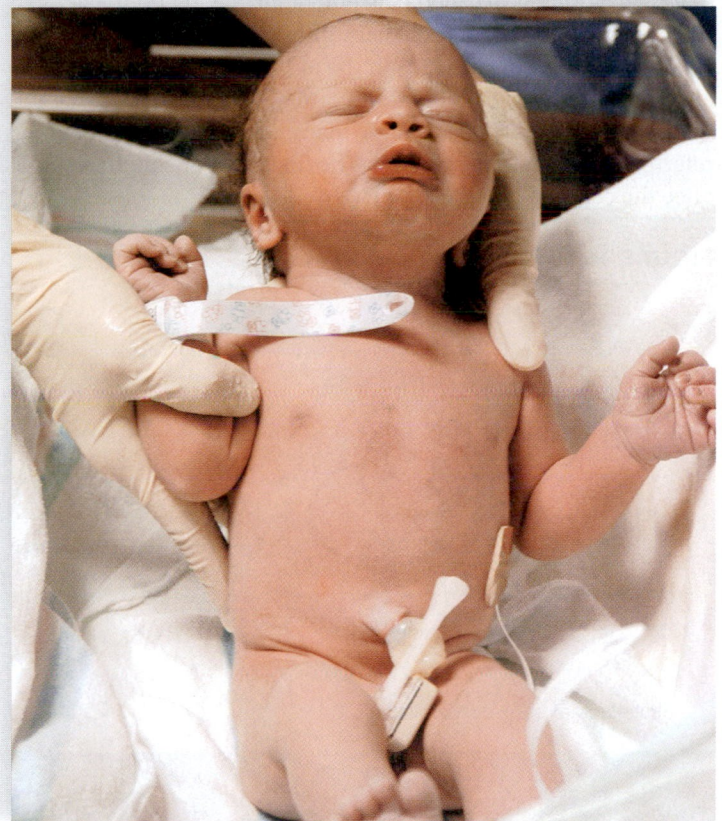

FIGURE 36.12D Newborn (40 weeks).

At the end of 9 months (Fig. 36.12D), the fetus is about 52 cm long and weighs about 3.4 kg. This latest weight gain is largely due to an accumulation of fat beneath the skin. Full-term babies have the best chance of survival.

▶ **36.12 CHECK YOUR PROGRESS**
1. Identify the exchanges that take place in the placenta so that the urinary, respiratory, and digestive systems in the fetus are not functioning.

LEARNING OUTCOME

When you complete this section, you should be able to

1. Divide the process of childbirth into three stages, stating the significance of each stage.

The uterus undergoes contractions throughout pregnancy. At first, these are light, lasting about 20–30 seconds and occurring every 15–20 minutes. Near the end of pregnancy, the contractions may become stronger and more frequent so that a woman thinks she is in labor. However, the onset of true labor is marked by uterine contractions that occur more frequently and last for 40 seconds or longer.

Prior to or at the first stage of childbirth, there may be a "bloody show" caused by expulsion of a mucous plug from the cervical canal. This plug prevents bacteria and sperm from entering the uterus during pregnancy. Childbirth has these three stages:

Stage 1 During the first stage of labor, the uterine contractions occur in such a way that the cervical canal slowly disappears as the lower part of the uterus is pulled upward toward the baby's head. This process is called effacement, or "taking up the cervix." With further contractions, the baby's head acts as a wedge to assist cervical dilation (Fig. 36.13a). If the amniotic membrane has not already ruptured, it is apt to do so during this stage, releasing the amniotic fluid, which leaks out of the vagina (an event sometimes referred to as "breaking water"). The first stage of childbirth ends once the cervix is dilated completely.

Stage 2 During the second stage of childbirth, the uterine contractions occur every 1–2 minutes and last about 1 minute each. They are accompanied by a desire to push, or bear down.

As the baby's head gradually descends into the vagina, the desire to push becomes greater. When the baby's head reaches the exterior, it turns so that the back of the head is uppermost (Fig. 36.13b). Because the vaginal orifice may not expand enough to allow passage of the head, an **episiotomy** is often performed. This incision, which enlarges the opening, is sewn together later. As soon as the head is delivered, the baby's shoulders rotate so that the baby faces either to the right or the left. At this time, the physician may hold the head and guide it downward, while one shoulder and then the other emerges. The rest of the baby follows easily.

Once the baby is breathing normally, the umbilical cord is cut and tied, severing the child from the placenta. The stump of the cord shrivels and leaves a scar, which is the navel.

Stage 3 The placenta, or **afterbirth,** is delivered during the third stage of childbirth (Fig. 36.13d). About 15 minutes after delivery of the baby, uterine muscular contractions shrink the uterus and dislodge the placenta. The placenta is then expelled into the vagina. As soon as the placenta and its membranes are delivered, the third stage of childbirth is complete.

▶ **36.13 CHECK YOUR PROGRESS**

1. Explain why the "afterbirth" is a part of the delivery of the newborn.

FIGURE 36.13 The three stages of birth.

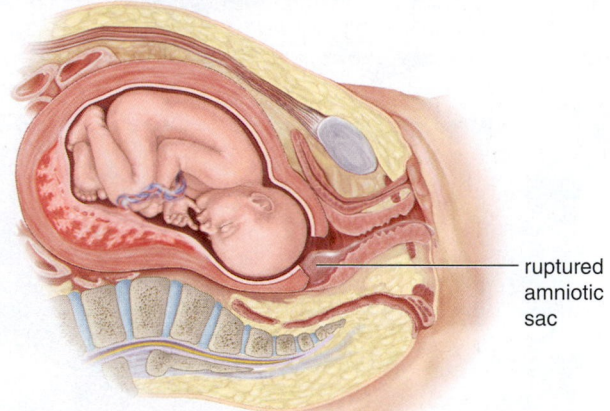

a. First stage of birth: cervix dilates

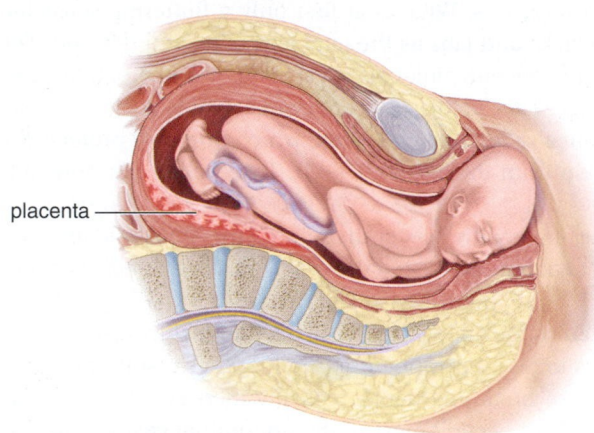

placenta

b. Second stage of birth: baby emerges

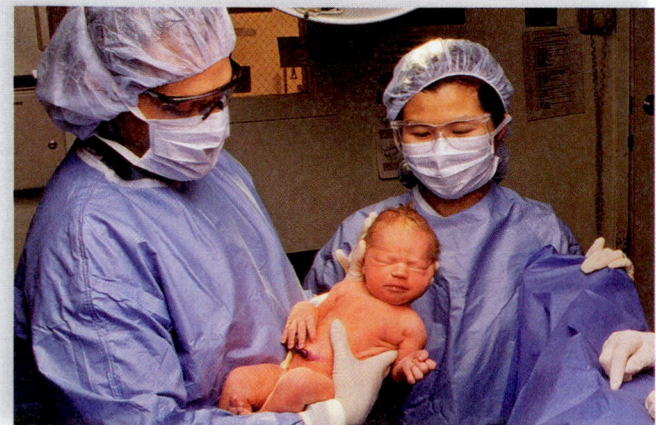

c. Baby has arrived

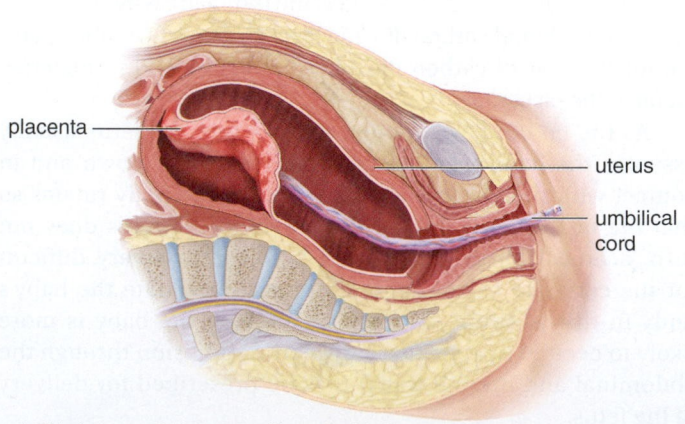

placenta

uterus

umbilical cord

d. Third stage of birth: expelling afterbirth

THE CHAPTER IN REVIEW

It's almost impossible not to consider evolutionary principles when studying development. Animals usually begin life as a zygote produced by the union of sperm and egg and then the zygote undergoes developmental stages that reflect the evolutionary history of animals. The early developmental stages of chordates parallel those of echinoderms, both of which are deuterostomes (see Fig. 20.2A). Then, too, all vertebrates have pharyngeal pouches which become gills in fish but become other structures in terrestrial vertebrates. The "add on" tendency of evolution is apparent as we study the development of animals.

The fact that animal development reflects the evolutionary history of animals was apparent to early investigators who could not have foreseen that developmental patterns are due to the passage of genes that control development. As we discussed in section 15.7 mutations in *Hox* genes (those master developmental genes that arose early in the history of life) can account for changes in development and therefore the evolution of new species during the history of life. In this chapter we have seen that investigators are piecing together exactly how cytoplasmic signaling molecules turn on genes during development. We last encountered signaling molecules when we

studied how steroid hormones turn on particular genes (see Fig. 35.2D).

ANALYZE AND EVALUATE

1. How is human reproduction an adaptation to life on land? In what way is human development still tied to an aquatic environment?
2. The presence of the same *Hox* genes (even if they have undergone mutations) in so many different organisms can show that all life forms are related. Explain why.
3. Why would a researcher decide to study the activity of genes during development rather than in the adult?

SUMMARIZE

How Animals Reproduce

36.1 Both asexual and sexual reproduction occur among animals

- In **asexual reproduction**, a single parent produces offspring that are genetically identical to the parent. During **parthenogenesis**, an egg develops into a complete individual.
- In **sexual reproduction**, the gametes are usually produced in **gonads**. The egg produced in the **ovary** of one parent is fertilized by the sperm produced in the **testes** of another. The offspring is genetically different from either parent.
- While many aquatic animals have an immature stage called a **larva**, land animals such as reptiles, including birds, develop in shelled eggs with a **yolk** and **extraembryonic membranes**.
- Ovoviviparous animals retain their eggs in the body until fully developed offspring hatch. Viviparous animals produce live young directly.
- Placental mammals modify the extraembryonic membranes for complete development in the body of the female where a **placenta** nourishes the offspring.

Human Reproductive Systems

36.2 Testes are the male gonads

- The human male reproductive system includes the testes and also the **epididymis, vas deferens,** and urethra. The **penis**, which has a **foreskin**, is the organ of sexual intercourse after **erection** occurs. **Semen** (**seminal fluid**) is composed of sperm and secretions from the **seminal vesicles, prostate gland,** and **bulbourethral glands.** Male **orgasm** results in ejaculation of semen from the penis.

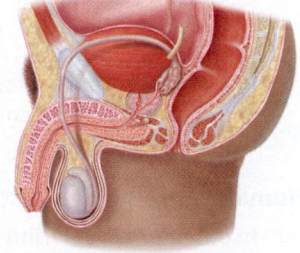

- The testes contain **seminiferous tubules,** which produce sperm during **spermatogenesis,** a process regulated by **Sertoli cells.** The testes also produce testosterone in interstitial cells. **Follicle-stimulating hormone** (**FSH**) stimulates spermatogenesis, and **luteinizing hormone** (**LH**) stimulates **testosterone** production. FSH and LH are produced by the anterior pituitary. Testosterone is the hormone that influences sex organ function and sperm maturation as well as the male **secondary sex characteristics** and **puberty.**

36.3 Ovaries are the female gonads

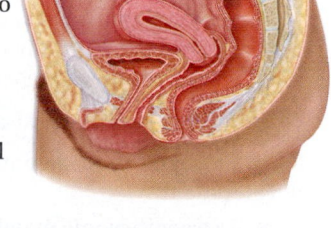

- The human female reproductive system includes the ovaries and also the **oviducts** ending in **fimbriae, uterus,** and **vagina.** The **cervix** is the narrow opening to the uterus, and the **endometrium** is its inner lining. The female external genital area, the **vulva,** includes the vaginal opening, **clitoris,** labia minora, and labia majora. Female orgasm culminates in uterine and oviduct contractions.
- **Oogenesis,** the production of an **oocyte** occurs as an ovarian **follicle** changes from a primary to a secondary to a vesicular follicle, which then releases the oocyte during **ovulation.** Following ovulation, the follicle becomes a **corpus luteum. Fertilization** of the egg occurs in an oviduct. The follicles also produce **estrogen** and **progesterone,** which maintain female sex characteristics.

36.4 The ovarian cycle drives the uterine cycle

- The **ovarian cycle** occurs as follows: (1) During the **follicular phase,** FSH causes maturation of a follicle that secretes estrogen and some progesterone. (2) During the **luteal phase,** LH promotes development of the corpus luteum, which secretes progesterone.

- The **uterine cycle** occurs as follows: (1) During days 1–5, **menstruation** occurs. (2) During days 6–13, the **proliferative phase,** estrogen causes the uterine lining to thicken. (3) During days 15–28, the **secretory phase,** progesterone prepares the uterine lining to receive the embryo.
- Numerous birth control methods are available. **Contraception** can involve taking **birth control pills,** the use of **contraceptive injections,** an **intrauterine device (IUD),** hormone **skin patch, diaphragm, female** and **male condoms,** or a **contraceptive implant.**

Early Developmental Stages and Processes

36.5 Cellular stages of development precede tissue stages

- **Cleavage** results in a multicellular embryo (a **morula**), and then a **blastula** forms. The blastula has a fluid-filled cavity called a **blastocoel.**

36.6 Tissue stages of development precede organ stages

- An embryo has three **germ layers: ectoderm, endoderm,** and **mesoderm.** During **gastrulation,** the first two layers arise by invagination, and the pore remaining is called the **blastopore.** Once the mesoderm forms by migration of cells, the embryo has three layers and is called a gastrula.

36.7 Organ stages of development occur after tissue stages

- Each germ layer develops into specific organs. The nervous system develops from ectoderm just above the notochord; the **neural plate** appears, and then neural folds become the **neural tube.** At the **neurula** stage, cross sections of all chordate embryos are similar in appearance.

36.8 Cellular differentiation begins with cytoplasmic segregation

- Growth refers to an increase in cell number and size. Cellular differentiation means that cells specialize in structure and function. Cellular differentiation begins with cytoplasmic segregation. For example, early on, the frog embryo has a visible **gray crescent.** During **morphogenesis,** the body takes shape, and **pattern formation** is evident.

36.9 Morphogenesis involves induction also

- **Induction** is the ability of one embryonic tissue to influence the development of another tissue. Signaling molecules turn on *Hox* genes that code for hox proteins, which may be involved in causing morphogen gradients.

Human Embryonic and Fetal Development

36.10 Extraembryonic membranes are critical to human development

- **Extraembryonic membranes (chorion, amnion, yolk sac, and allantois)** make internal development possible following **implantation.**

	chick	human
chorion	performs gas exchange	becomes part of placenta
amnion	holds amnionic fluid	holds amnionic fluid
yolk sac	contains nutrients	helps form blood cells
allantois	stores waste molecules	becomes umbilical blood vessel

36.11 Embryonic development involves tissue and organ formation

- Embryonic development spans the first two months of development, from fertilization through the acquisition of organ systems. During the first week, the **blastocyst** has a fluid-filled cavity and an outer layer of cells called the **trophoblast,** which will be part of the chorion. The trophoblast secretes **human chorionic gonadotropin (HCG),** the hormone that is the basis for the pregnancy test and that maintains the corpus luteum.
- During the second week, the **embryonic disk** has three germ layers. During the third week, the nervous system and the heart are forming; at 4 weeks, the **chorionic villi** and **umbilical cord** are present. At the end of 2 months, the embryo is recognizably human and all organ systems are present.

36.12 Fetal development involves refinement and weight gain

- **Fetal development** includes the third through the ninth months of development. Exchanges at the placenta supply the fetus with oxygen and nutrients and take away carbon dioxide and wastes. During the third and fourth months, the skeleton becomes ossified except for the **fontanels** of the skull. The fetus's sex is distinguishable. During the fifth through seventh months, fetal movement begins, and the fetus continues to grow and gain weight. As birth approaches, the fetus turns so that the head faces the cervix. If not, a **cesarean section** may be necessary.

36.13 Pregnancy ends with the birth of the newborn

- Stage 1: Uterine contractions begin; cervix dilates. Stage 2: Uterine contractions occur every 1–2 minutes; baby is born; umbilical cord is cut. An **episiotomy** may have been necessary. Stage 3: Uterine muscle contractions shrink the uterus and dislodge the placenta, or **afterbirth,** which is expelled.

TEST YOURSELF

How Animals Reproduce

1. Internal fertilization
   a. can prevent the drying out of gametes and zygotes.
   b. must take place on land.
   c. is practiced by humans.
   d. requires that males have a penis.
   e. Both a and c are correct.
2. The evolutionary significance of the placenta is that it allowed
   a. animals to retain their eggs until they hatched.
   b. offspring to be born in an immature state and then develop outside the mother.
   c. offspring to exchange materials with the mother while developing inside the mother.
   d. mammals to develop the ability to produce milk.
3. **THINKING CONCEPTUALLY** Some animals, including hydras, only reproduce sexually when environmental conditions are less than optimal. Explain.

Human Reproductive Systems

4. In human males, sterility results when the testes are located inside the abdominal cavity instead of in the scrotum because
   a. the sperm cannot pass through the vas deferens.
   b. sperm production is inhibited at body temperature.
   c. the sperm cannot travel the extra distance.
   d. digestive juices destroy the sperm.

5. Label this diagram of the male reproductive system:

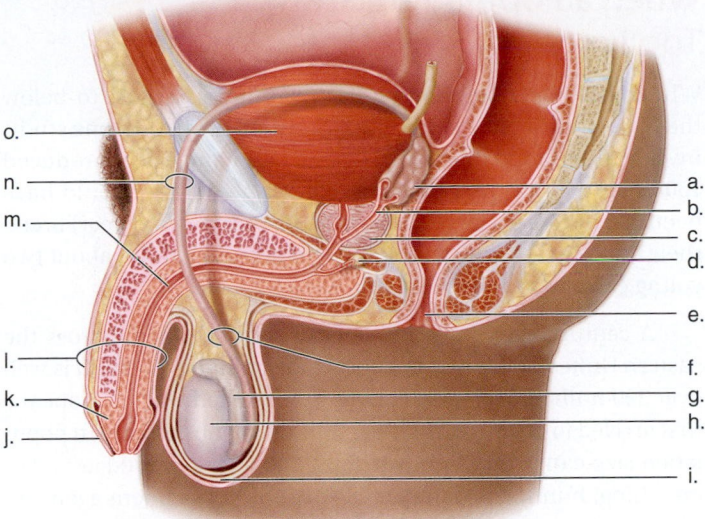

6. In the testes, the _____ produce(s) the sperm.
   a. seminiferous tubules   c. seminal vesicles
   b. bulbourethral glands   d. prostate gland
7. The release of the oocyte from the follicle is caused by
   a. a decreasing level of estrogen.
   b. a surge in the level of follicle-stimulating hormone.
   c. a surge in the level of luteinizing hormone.
   d. progesterone released from the corpus luteum.
8. Which of the following is not an event of the ovarian cycle?
   a. FSH promotes the development of a follicle.
   b. The endometrium thickens.
   c. The corpus luteum secretes progesterone.
   d. Ovulation of an egg occurs.
9. **THINKING CONCEPTUALLY** Giving men additional testosterone has been shown to decrease sperm production enough to be considered a successful form of birth control. Use negative feedback to find an explanation.

Early Developmental Stages and Processes

10. Gastrulation is influenced by the
    a. nervous system.   c. amount of yolk.
    b. environment.   d. presence of hormones.
11. Which of the germ layers is best associated with development of the heart?
    a. ectoderm   d. neurula
    b. mesoderm   e. All of these are correct.
    c. endoderm
12. Which of these stages is mismatched?
    a. cleavage—cell division
    b. blastula—gut formation
    c. gastrula—three germ layers
    d. neurula—nervous system
    e. Both b and c are mismatched.
13. Differentiation is equivalent to which term?
    a. morphogenesis   c. specialization
    b. growth   d. gastrulation
14. Which process refers to the shaping of the embryo and involves cell migration?
    a. cleavage   c. growth
    b. differentiation d. morphogenesis
15. Place the following terms in the correct order to describe development: blastula, cleavage, gastrula, morula, neurula, zygote.

16. **THINKING CONCEPTUALLY** Give an explanation for the observation that all chordates go through similar stages of development.

Human Embryonic and Fetal Development

17. Label this diagram and state a function for each structure.

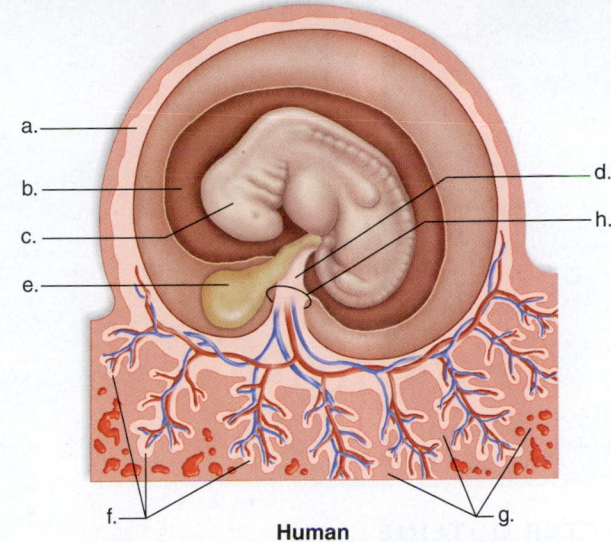

Human

18. In human development, which part of the blastocyst develops into a fetus?
    a. morula   d. chorion
    b. trophoblast e. yolk sac
    c. inner cell mass
19. In humans, the fetus
    a. has four extraembryonic membranes.
    b. has developed organs and is recognizably human.
    c. is dependent upon the placenta for excretion of wastes and acquisition of nutrients.
    d. All of these are correct.
20. **THINKING CONCEPTUALLY** Support the statement that the evolution of the placenta was just as important to the success of humans as was a mobile hand.

GET INVOLVED

1. State a hypothesis that would explain why prostate cancer is common in men. How would you test your hypothesis?
2. Mary has had several early miscarriages, so her doctor has prescribed progesterone therapy the next time she gets pregnant. Explain.

MEDIA STUDY TOOLS

mhhe.com/maderconcepts3

Enhance your study of this chapter with interactive study tools, practice tests, and engaging animations. Also, ask your instructor about the resources available through ConnectPlus, which includes LearnSmart, a personalized adaptive learning program, and a media-rich eBook.

37

Population Ecology

BEFORE YOU BEGIN

Take a few minutes to recall

The flow of energy through populations in "Life is Organized" on page 82
How populations adapt to their environments (section 14.2)
Microevolution in populations (section 14.7)

When a Population Grows Too Large

White-tailed deer, which live from southern Canada to below the equator in South America, are prolific breeders. In one study, investigators found that two male and four female deer produced 160 offspring in 6 years. Theoretically, the number could have been 300 because a large proportion of does (female deer) breed their first year, and once they start breeding, produce about two young each year of life.

A century ago, the white-tailed deer population across the eastern United States was less than half a million. Today, it is well over 200 million deer—even more than existed when Europeans first arrived to colonize America. This dramatic increase in population size can probably be attributed to a lack of predators. For one thing, hunting is tightly controlled by government agencies, and in some areas, it is banned altogether because of the danger it poses to the general public. Similarly, the natural predators of deer, such as wolves and mountain lions, are now absent from most regions. This can be traced to a large human population that fears large predators because they could possibly attack humans and domestic animals.

Buck

We like to see a mother with her fawns by the side of the road or scampering off into the woods with tails raised to show off the white underside. And we find it thrilling to see a large buck (male deer) with majestic antlers partially hidden in the woods. But the sad reality is that, in those areas where deer populations have become too large, the deer suffer from starvation as they deplete their own food supply. For example, after deer hunting was banned on Long Island, New York, the deer population quickly outgrew available food resources. The animals became sickly and weak and weighed so little that their ribs, vertebrae, and pelvic bones were visible through their skin.

Then, too, a very large deer population causes humans many problems. A homeowner is dismayed to see new plants decimated and evergreen trees damaged by munching deer. The economic damage that large deer populations cause to agriculture, landscaping, and forestry exceeds a billion dollars per year. Even more alarming, a million deer–vehicle collisions take place in the United States each year, resulting in over a billion dollars in insurance claims, thousands of human injuries, and hundreds of human deaths. Lyme disease, transmitted by deer ticks to humans, infects over 3,000 people annually. Untreated Lyme disease can lead to debilitating arthritic symptoms.

Deer overpopulation hurts not only deer and humans, but other species as well. The forested areas that are overpopulated by deer have fewer understory plants. Furthermore, the deer selectively eat certain species of plants, while leaving others alone. This can cause long-lasting changes in the number and diversity of trees in forests, leading to a negative economic impact on logging and forestry. The number of songbirds, insects, squirrels, mice, and other animals declines with an increasing deer population. It behooves us, therefore, to learn to manage deer populations. And the good news is that in some states, such as Texas, large landowners now set aside a portion of their property for a deer herd. They improve the nutrition of the herd and restrict the harvesting of young bucks, but allow the harvesting of does. The result is a self-sustaining herd that brings economic benefits—the landowners charge others for the privilege of hunting on their land.

In this chapter, we examine the general characteristics of populations. You will learn how the size, distribution, and age structure of a population can change over time and what factors influence populations. You will see that, like the deer in eastern North America, human populations too may suffer the consequences of overpopulation.

Doe running

Doe and fawn

Scope of Ecology

As previously shown in Figure 1.3A, the levels of biological organization extend beyond the organism to include the population, community, ecosystem, and biosphere. Ecology studies these higher levels of organization.

37.1 Ecology is studied at various levels

LEARNING OUTCOME

When you complete this section, you should be able to

1. Name and compare the ecological levels of study.

In 1866, the German zoologist Ernst Haeckel coined the word ecology from two Greek roots (*oikos,* home, and *-logy,* study of). He said that **ecology** is the study of the interactions of organisms with other organisms and with the physical environment. Haeckel also pointed out that ecology and evolution are intertwined because ecological interactions are selection pressures that result in evolutionary change, which in turn affects ecological interactions.

Ecology, like so many biological disciplines, is wide-ranging. An ecologist can study how the individual organism is adapted to its environment. For example, they might study how a fish is adapted to and survives in its **habitat** (the place where the organism lives) (Fig. 37.1). Most organisms do not exist singly; rather, they are part of a **population,** defined as all the organisms within an area belonging to the same species and interacting with the environment. At this level of study, ecologists are interested in factors that affect the growth and regulation of population size.

A **community** consists of all the various populations interacting at a locale. In a coral reef, there are numerous populations of algae, corals, crustaceans, fishes, and so forth. At this level,

ecologists want to know how interactions such as predation and competition affect the organization of the community.

An **ecosystem** encompasses a community of populations as well as the abiotic environment (e.g., the availability of sunlight for plants). Ecosystems rarely have distinct boundaries and are not totally self-sustaining. Usually, a transition zone called an *ecotone,* composed of a mixture of organisms from adjacent ecosystems, exists between ecosystems.

Video
Coral Reef Ecosystem

The **biosphere** encompasses the zones of the Earth's land, water, and air where organisms are found. Taking the global view, the entire biosphere is an ecosystem, a place where organisms interact among themselves and with the physical and chemical environments. These interactions help maintain ecosystems and, in turn, the biosphere.

► **37.1 CHECK YOUR PROGRESS**

1. Explain why all ecological levels are affected by the deer overpopulation discussed in "When a Population Grows Too Large" on page 728.

FIGURE 37.1
Ecological levels.

Organism ⟶ Population ⟶ Community ⟶ Ecosystem

Coral reef ecosystem

How Populations Change Over Time

A population is defined as all the members of a species living in the same locale at the same time. A population's *demographics* such as its density, distribution, and other characteristics discussed in this chapter shift over time. Researchers have identified three common survivorship curves that influence a population's life history pattern.

37.2 Density and distribution are aspects of population structure

LEARNING OUTCOME

When you complete this section, you should be able to

1. Explain density and contrast three patterns of population distribution.

Density Once population size has been estimated, it is possible to calculate the **population density,** which is the number of individuals per unit area. For example, the population density of the United States is estimated at 87 persons per square mile. Population density figures make it seem as if individuals are uniformly distributed, but this is often not the case. For example, we know full well that most people in the United States live in cities, where the number of people per unit area is dramatically higher than in the country. And even within a city, more people live in particular neighborhoods than others. Furthermore, such distributions can change over time. Therefore, basing ecological models solely on population density can lead to misleading results.

Distribution **Population distribution** is the pattern of dispersal of individuals across an area of interest. The availability of resources can affect where populations of a species are found. **Resources** are nonliving (abiotic) and living (biotic) components of an environment that support organisms. Light, water, space, mates, and food are some important resources for populations. **Limiting factors** are those environmental aspects that particularly determine where an organism lives. For example, trout live only in cool mountain streams, where the oxygen content is high, but carp and catfish are found in rivers near the coast because they can tolerate warm waters, which have a low concentration of oxygen. The timberline is the limit of tree growth in mountainous regions or in high latitudes because of low temperatures. The distribution of organisms can also be due to biotic factors. In Australia, the red kangaroo does not live outside arid inland areas because it is adapted to feeding on the grasses that grow there.

Three descriptions—*clumped, random*, and *uniform*—are often used to characterize observed patterns of distribution. Suppose you were to consider the distribution of a species across its full **range,** that portion of the globe where the species can be found. For example, red kangaroos live in Australia. On that scale, you would expect to find a clumped distribution because organisms are located in areas suitable to their adaptations. That is why red kangaroos live in grasslands, and catfish live in warm river water near the coast.

Within a smaller area, such as a single body of water or a single forest, the availability of resources again influences which distribution pattern is common for a particular population. A study of the creosote bush, a desert shrub (Fig. 37.2A), revealed that the distribution changed from clumped to random to uniform as the plants matured. As time passed, competition for belowground resources caused the distribution pattern to become uniform.

In another example, Cape gannet populations are clumped over their range, but in a nesting colony, the birds are uniformly distributed (Fig. 37.2B).

► **37.2 CHECK YOUR PROGRESS**

1. Explain what data are needed to calculate the density of a population.

Young, small shrubs

Medium shrubs

Larger shrubs

Clumped   Random   Uniform

Mature desert shrubs

FIGURE 37.2A Distribution patterns of the creosote bush.

FIGURE 37.2B Nesting colony of Cape gannets off the coast of New Zealand.

LEARNING OUTCOMES

When you complete this section, you should be able to

1. Explain the relationship between growth rate and biotic potential.

2. Describe three types of survivorship curves and contrast three age structure diagrams.

A population's annual growth rate is dependent upon the number of individuals born each year, the number of individuals that die each year, and annual immigration and emigration. Usually, it is possible to assume that immigration and emigration are equal and need not be considered in the calculation. Therefore, in a simple example, if the number of births is 30 per year, and the number of deaths is 10 per year per 1,000 individuals, the growth rate is 2.0%:

$$(30 - 10)/1,000 = 0.02 = 2.0\%$$

This population will grow because the number of births exceeds the number of deaths. On the other hand, if the number of deaths exceeds the number of births, the value of the growth rate is negative, and the population will shrink.

The **biotic potential** of a population is its highest possible growth rate when resources are unlimited. Whether the biotic potential is high or low depends primarily on the following factors:

1. Usual number of offspring per reproduction
2. Chances of survival until age of reproduction and until reproduction ceases
3. Age structure diagram; age reproduction begins
4. Length of time and how often an individual reproduces

For example, a pig population in which pig females produce many offspring that quickly mature to produce more offspring, has a much higher biotic potential than a rhinoceros population in which rhinoceros females produce only one or two offspring per infrequent reproductive event (Fig. 37.3A).

Survivorship Curves The population growth rate does not take into account that the individuals of a population are in different stages of their life span. **Cohort** is the term used to describe population members that are the same age and have the same chances of surviving. Some investigators study population dynamics and construct life tables that show how many members of a cohort are still alive after certain intervals of time. For example, Figure 37.3B*a* is a **life table** for Dall sheep. The cohort contains 1,000 individuals. The table tells us that after 1 year, 199 individuals have died. Another way to express this same statistic, however, is to consider that 801 individuals are still alive—have survived—after 1 year. **Survivorship** is the probability of cohort members surviving to particular ages.

If we plot the number surviving at each age, a survivorship curve is produced (Fig. 37.3B*b*). The results of such investigations show that each species has a particular survivorship curve. Three typical survivorship curves, numbered I, II, and III, are seen (Fig. 37.3B*b*). Mammals, represented here by the Dall sheep, usually have a type I survivorship curve; they survive well past the midpoint of the life span, and they do not die until near the end of the life span. On the other hand, the type III curve

FIGURE 37.3A Biotic potential is dependent on many factors, among them the number of offspring per reproductive event. Pigs, which produce many offspring that quickly mature to produce more offspring, have a much higher biotic potential than rhinoceroses, which produce only one or two offspring per infrequent reproductive event.

is typical of a population, such as oysters, in which most individuals will probably die very young. This type of survivorship curve occurs in many invertebrates, fishes, and humans in less-developed countries. In the type II curve, survivorship decreases at a constant rate throughout the life span; this pattern is typical of hydras, many songbirds, small mammals, and some invertebrates, for which death is usually unrelated to age.

Much can be learned about the life history of a species by studying its life table and the survivorship curve constructed from this table. For example, a type III survivorship curve indicates that since death probably comes early for most members, only a few live long enough to reproduce. How do you think the other two types of survivorship curves affect reproduction?

Other types of information are also derived from studying life tables. In the life table for a plant called blue grass, per capita seed production increases as plants mature, and then seed production drops off. The survivorship curve for blue grass shows that most individuals survive 6–9 months, and then the chances of survivorship diminish at an increasing rate.

Age Structure Diagrams When the individuals in a population reproduce repeatedly, several generations may be alive at any

Regulation of Population Size

Regulation of population size cannot be achieved by density-independent factors (e.g., natural disasters) but can be achieved by density-dependent factors (e.g., predators). The former only sporadically reduce population size.

37.5 Density-independent factors affect population size

LEARNING OUTCOME

When you complete this section, you should be able to

1. Describe examples of density-independent factors, telling how they relate to population size.

So far, we have observed that a population's particular density and growth pattern determine the population size. In addition, ecologists have long recognized that environmental interactions play an important role in population size. Abiotic environmental factors include droughts (lack of rain), freezes, hurricanes, floods, and forest fires. Any one of these natural disasters can cause individuals to die and lead to a sudden and catastrophic reduction in population size. However, such an event does not necessarily kill a larger percentage of a dense population compared to a less dense population. Therefore, an abiotic factor is usually a **density-independent factor,** meaning that the percentage of individuals killed remains the same regardless of the population size. In other words, the intensity of the effect does not increase with increased population size. The red line in Figure 37.5A shows that mortality percentage (percentage killed) remains the same, regardless of the density of the population.

An example of a density-independent factor is a drought on the Galápagos Islands that caused the population size of one of Darwin's finches (*Geospiza fortis*) to decline from 1,400 to 200 individuals. (The drought caused reduced availability of seeds this species ate.) This is a reduction of 86% of the original population size. Assuming no competition between members of the population, drought, in this instance, is acting as a density-independent factor. We can assume, then, that if the population began with 2,800 individuals, the population would reduce to 400 members, which is the same percentage reduction.

Natural disasters such as hurricanes and floods can have a drastic effect on a population and cause sudden and catastrophic reductions in population size. However, a flash flood does not necessarily kill a larger percentage of a dense population than a less dense population. Therefore, such a flood cannot be counted

Death rate = 60%

Death rate = 64%

FIGURE 37.5B A flood has a density-independent effect. At the top, a low-density population has a death rate of 3/5, or 60%, while at the bottom a high-density population has a similar death rate of 12/20, or 64%.

on to regulate population size, keeping it within the carrying capacity of the environment. Nevertheless, as with our first example, the larger the population, the greater the number of individuals probably affected. In the impact of a flash flood on a low-density population of mice living in a field (mortality rate of 3/5, or 60%) Figure 37.5B, *top,* is similar to the impact on a high-density population (mortality rate of 12/20, or 64%) (Fig. 37.5B, *bottom*).

FIGURE 37.5A
Percentage that die per density of population.

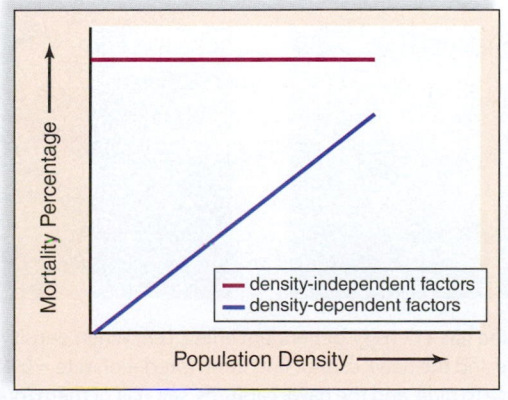

Mortality Percentage

— density-independent factors
— density-dependent factors

Population Density

▶ **37.5 CHECK YOUR PROGRESS**

1. List some examples of density-independent factors that could occur in a region overpopulated by deer.

LEARNING OUTCOME

When you complete this section, you should be able to

1. Explain examples of density-dependent factors, telling how they relate to population size.

Biotic factors tend to be **density-dependent factors.** The percentage of the population affected by these factors does increase as the density of the population increases (see the blue line in Fig. 37.5A). Competition, predation, and parasitism are all biotic factors that increase in intensity as the density increases. We will discuss these interactions between populations again in Chapter 38 because they influence community composition and diversity.

Competition occurs when members of the same species attempt to use needed resources (such as light, food, or space) that are in limited supply. As a result, not all members of the population have access to the resource to the degree necessary to ensure survival or reproduction. As a theoretical example of a density-dependent effect, let's consider a woodpecker population in which members have to compete for nesting sites. Each pair of birds requires a tree hole to raise offspring. In Figure 37.6A, *left*, if there are more holes than breeding pairs, each pair can have a hole in which to lay eggs and rear young birds. But if there are fewer holes than breeding pairs, each pair must compete to acquire a nesting site (Fig. 37.6A, *right*). Pairs that fail to gain access to holes will be unable to contribute new members to the population.

> **Video**
> Booby Chick Competition

As an actual example of competition, four male and 21 female reindeer (*Rangifer*) were released on St. Paul Island in the Bering Sea off Alaska in 1911. St. Paul Island was a completely undisturbed environment, with little hunting pressure and no predators. By 1939, the herd had grown exponentially to about 2,000 reindeer, which overgrazed the habitat and then abruptly declined to about eight animals by 1950.

> **Video**
> Hungry Reindeer

Predation occurs when one organism, the predator, eats another, the prey. In the broadest sense, predation can include not only animals such as lions, which kill zebras, but also filter-feeding blue whales, which strain krill from the ocean waters; parasitic ticks, which suck blood from their hosts; and even herbivorous deer, which browse on trees and bushes. The effect of predation on a prey population generally increases as the population grows more dense, because prey are easier to find when hiding places are limited. Consider a

> **Video**
> Harris Hawks

field inhabited by a population of mice. Each mouse must have a hole in which to hide to avoid being eaten by a hawk. If there are 100 mice, but only 98 holes, two mice will be left out in the open (Fig. 37.6B, *left*). It might be hard for the hawk to find only two mice in the field. If neither mouse is caught, then the predation rate is 0/2 = 0%. However, if there are 200 mice and only 100 holes, there is a greater chance that hawks will be able to find some of these 100 mice without holes (Fig. 37.6B, *right*). If half of the *exposed* mice are caught, the predation rate is 50/100 = 50%. Therefore, increasing the density of the available prey has increased the proportion of the population preyed upon.

In the next part of the chapter, we consider life history patterns.

▶ **37.6 CHECK YOUR PROGRESS**

1. List some examples of density-dependent factors that will come into play in a region overpopulated by deer.

FIGURE 37.6A Competition has a density-dependent effect. *Left:* When density is low, every woodpecker has a tree hole for raising young. *Right:* When density is high, some birds will not have a hole.

FIGURE 37.6B Predation has a density-dependent effect. *Left:* When density is low, only two mice cannot find a place to hide and the hawk cannot find them (predation rate = 0%). *Right:* When density is high, 100 mice are unable to hide, and the hawk captures, say, half of them (predation rate = 50%).

Life History Patterns

The life history of a species is based on the population demographics we have just studied as well as on the attributes that result in these demographics. The life history pattern can be used to predict the possibility of extinction.

37.7 Life history patterns consider several population characteristics

LEARNING OUTCOMES

When you complete this section, you should be able to

1. Contrast the characteristics of an opportunistic population with those of an equilibrium population.
2. List factors that help determine whether a population will become extinct.

Populations vary in terms of the number of births per reproductive event, the age at reproduction, the life span, and the probability of living the entire life span. Such particulars are part of a species' **life history,** and life histories often involve trade-offs. For example, each population is able to capture only so much of the available energy, and how this energy is distributed between its life span (short versus long), reproductive events (few versus many), care of offspring (little versus much), and so forth has evolved over the years. Natural selection shapes the final life history of individual species, and therefore it is not surprising that even related species, such as frogs and toads, may have different life history patterns if they occupy different types of environments.

Analysis of exponential and logistic population growth patterns suggests that some populations follow an opportunistic pattern and others follow an equilibrium pattern. An **opportunistic population** tends to live in a fluctuating and/or unpredictable environment. The population remains small until favorable conditions promote exponential growth. The members of the population are small in size, mature early, have a short life span, and provide limited parental care for a great number of offspring. Density-independent effects dramatically affect population size, which is large enough to survive an event that threatens to annihilate it. The population has a high dispersal capacity. Various types of insects and weeds are the best examples of opportunistic species (Fig. 37.7A), but there are others. A cod is a rather large fish, weighing up to 12 kg and measuring nearly 2 m in length—but the cod releases gametes in vast numbers, the zygotes form in the sea, and the parents make no further investment in developing offspring. Of the 6–7 million eggs released by a single female cod, only a few will become adult fish.

In contrast, some environments are relatively stable and predictable, allowing population size to remain fairly stable. **Equilibrium populations** exhibit logistic population growth, and the size of the population remains close to, or at, the carrying capacity. Resources are relatively scarce, and the individuals best able to compete—those with phenotypes best suited to the environment—have the largest number of offspring. Members allocate energy to their own growth and survival, and to the growth and survival of their few offspring. Therefore, they are fairly large, slow to mature, and have a relatively long life span (Fig. 37.7B). The size of equilibrium populations tends to be regulated by density-dependent effects. The best possible examples include long-lived plants (saguaro cacti, oaks, cypress, and pine), birds of prey (hawks and eagles), and large mammals (whales, elephants, bears, and gorillas).

Animation
r and k Strategies

Opportunistic Pattern

- Small individuals
- Short life span
- Fast to mature
- Many offspring
- Little or no care of offspring

FIGURE 37.7A Dandelions are opportunistic species.

Extinction is the total disappearance of a species or higher group. Which species, the dandelion or the mountain gorilla, is apt to become extinct? Because the dandelion matures quickly, produces many offspring at one time, and has seeds that are dispersed widely by wind, it can more easily withstand a local decimation than can the mountain gorilla.

▶ **37.7 CHECK YOUR PROGRESS**

1. Explain why the deer population is exploding, while other mammals (e.g., gorillas) are on the brink of extinction.

FIGURE 37.7B Bears are equilibrium species.

Equilibrium Pattern

- Large individuals
- Long life span
- Slow to mature
- Few and large offspring
- Much care of offspring

37A Adaptability of Small Populations

Small populations are typically very vulnerable and may often be considered threatened or endangered with extinction. Florida panthers, black-footed ferrets, California condors, whooping cranes, and leatherback turtles are examples of small populations (Fig. 37A). We usually find small populations in areas where there is a considerable amount of habitat destruction, overharvest of species, or a high level of introduced or invasive species. A considerable amount of time and money is spent each year trying to save or manage these populations. Therefore, it is important to consider if they have enough genetic diversity to adapt to future changes in the environment.

One major problem of small populations is inbreeding, or the mating of closely related individuals who may be heterozygous for the same disorder. In humans, hereditary conditions such as cystic fibrosis, hemophilia, and some forms of muscular dystrophy are passed on relatively rarely because the human population is large and the chances of both parents carrying the same recessive alleles are fairly low. But in a small population, matings between individuals with the same recessive alleles are more likely. Cheetahs, for example, are known to have limited genetic diversity, and today zoos take all possible steps to prevent inbreeding between closely related animals. Among other conditions, this should help prevent cheetahs from succumbing to the same infectious disease as has happened in the past.

How did the cheetah population—or how does any population—lose its genetic diversity? A review of population genetics and the Hardy-Weinberg principle illustrates how many endangered populations have lost their genetic diversity. The Hardy-Weinberg principle states that genetic diversity will remain the same from generation to generation unless, for example, genetic drift occurs. Genetic drift is a change in a population's usual allele frequencies due to chance alone. For example, the leatherback turtle population has now crashed due to overharvesting, accidental capture by fishermen, and destruction of nesting sites. Many members of the population have died, thus losing the opportunity to reproduce. Under these circumstances, gene pool allele frequencies have changed, and some alleles may have been lost. The loss of alleles can affect a population's ability to adapt to future changes, as the following example shows.

Let's refer to the peppered moth and industrial melanism to illustrate how the loss of genetic diversity can affect a population's ability to adapt to new environmental circumstances. Researchers have shown that a peppered moth population ordinarily contains mostly light-colored moths and a few dark-colored moths. When the natural vegetation becomes dark due to industrial pollutants, the dark-colored moths become more prevalent due to the ability of birds to find and eat mainly the light-colored moths. Suppose a natural disaster occurs and wipes out the few remaining light-colored moths, along with their alleles that code for light coloring. Will the population be able to adapt if pollution control measures are introduced and the vegetation returns to its natural color? Under these circumstances, birds will be able to see and eat dark-colored moths, but the population will be unable to return to its original coloration.

Because we now know that genetic diversity is important to the future adaptability of small populations, it is essential that biologists not only restore the habitat of endangered organisms, but also take steps to restore their genetic diversity. Only in this way can we increase the likelihood that a small population will be able to adapt to future changes in the environment.

CONSIDER THESE QUESTIONS

1. Some locations around the globe are called "hotspots" because they contain more diverse organisms than most other locations. Should we put our efforts into preserving hotspots rather than individual species?
2. Cheetahs are not faring well in their natural habitat, but they have been preserved in zoos. Is this okay, or should organisms only be preserved in their original natural environment?
3. Humans need to preserve the natural environments that provide us with food, absorb our wastes, and keep ecological cycles functioning. This being the case, should we preserve only enough species to keep natural environments in good health and forget about unique endangered species? Why or why not?

FIGURE 37A Leatherback turtles are sea turtles that once numbered about 115,000. The population size is now estimated to be as low as 26,000 because human activities have killed off many turtles or prevented them from reproducing. These huge animals can be 2.4 meters long and weigh 907 kg!

 Explore the concepts through a variety of multimedia
BIOLOGY assets and question types.

www.mcgrawhillconnect.com

Human Population Growth

The human population continues to increase, but demographics differ for the more-developed and the less-developed countries.

37.8 World population growth is exponential

LEARNING OUTCOMES

When you complete this section, you should be able to

1. Explain how the world population is still undergoing exponential growth.
2. Contrast the population growth and age distributions of the more-developed countries (MDCs) and the less-developed countries (LDCs).

The world's population has risen steadily to a present size of about 7.1 billion people (Fig. 37.8). Prior to 1750, the growth of the human population was relatively slow, but as more reproducing individuals were added, growth increased, until the curve began to slope steeply upward, indicating that the population was undergoing exponential growth. The number of people added annually to the world population peaked at about 87 million around 1990, and currently it is a little over 78 million per year. This is roughly equal to the current populations of Argentina, Ecuador, and Peru combined.

The potential for future population growth can be appreciated by considering the **doubling time,** the length of time it takes for the population size to double. Currently, the doubling time is estimated to be 50 years. This means that in 50 years, the world would need double the amount of food, jobs, water, energy, and so on just to maintain the present standard of living.

Many people are gravely concerned that the amount of time needed to add each additional billion persons to the world population has become shorter and shorter. The first billion didn't occur until 1800; the second billion was attained in 1930; the third billion in 1960; and thereafter, a billion was added every 12–15 years. Only when the number of young women entering the reproductive years is the same as those leaving those years can there be zero population growth, meaning that the birthrate equals the death rate, and population size remains steady. The world's population may level off at 7.7, 9.1, or 10.6 billion, depending on the speed at which the growth rate declines.

More-Developed and Less-Developed Countries The countries of the world can be divided into two groups. In the more-developed countries (**MDCs**), typified by countries in North America, Europe, Japan, and Australia, population growth is low, and most people enjoy a good standard of living. In the less-developed countries (**LDCs**), such as countries in Latin America, Africa, and Asia, population growth is expanding rapidly, and the majority of people live in poverty.

The MDCs doubled their populations between 1850 and 1950. This was largely due to a decline in the death rate, the development of modern medicine, and improved socioeconomic conditions. The decline in the death rate was followed shortly thereafter by a decline in the birthrate, so that populations in the MDCs experienced only modest growth between 1950 and 1975. This sequence of events (i.e., decreased death rate followed by decreased birthrate) is termed a **demographic transition.** Yearly growth of the MDCs has been decreasing and is now 0.1%.

Although the death rate in the LDCs began to decline steeply following World War II with the importation of modern medicine from the MDCs, the birthrate remained high. The yearly growth of the LDCs peaked at 2.5% between 1960 and 1965. Since that time, a demographic transition has occurred: The decline in the death rate slowed, and the birthrate fell. The yearly growth rate is now 1.4% with the largest proportion of this increase occurring in the very poorest of these countries. The LDCs as a whole may explode from 5.8 billion today to 8 billion in 2025. The LDCs experience a population momentum because they have more women entering the reproductive years than older women leaving them. Today, about 90% of the world's population lives in Asia (India and China), Africa, and Latin America.

▶ **37.8 CHECK YOUR PROGRESS**
1. Compare the population growth curve of the LDCs with that of the MDCs and suggest how to help an MDC lower its growth rate.

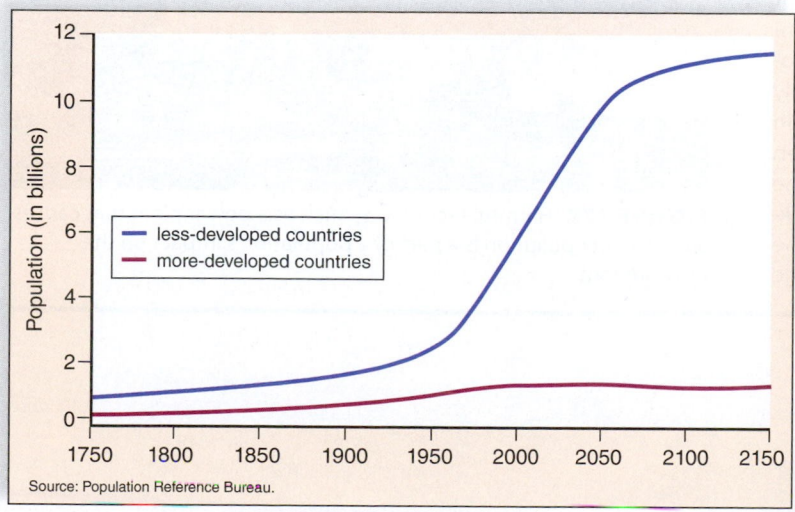

FIGURE 37.8 World population growth over time.

Living conditions in more-developed countries

Living conditions in less-developed countries

37B Sustainability of the U.S. Population

Is the United States population sustainable? *Sustainability* occurs when future generations are able to enjoy the same quality of life as the present generation. How can we judge when a population is sustainable? One way applies the following formula proposed by scientists Paul Ehrlich and John Holdren:

$$I = PAT$$

Impact = Population × Affluence × Technology

I = impact of the population on the environment
P = population size
A = affluence, the amount of consumption of goods and services per capita
T = technology, the degree to which inefficient and environmentally unsafe methods are used to produce goods and services

The less impact a population has on the environment, the more likely it will be sustainable. The United States could reduce its impact on the environment—for example, by having a zero population growth, reducing consumption, and improving the efficiency and cleanliness of technology. What is the possibility of stabilizing the U.S. population size? The United States is a more-developed country (MDC) with a growth rate of 0.5%, which is higher than many other MDCs. However, the fertility rate in the United States has been declining since 2007 when a slower economy resulted in a loss of jobs and many home foreclosures. The fertility rate is now 1.9, which is just about large enough to maintain the population at its current size. The continuing increase in United States population is due to immigration. Even a modest immigration of 1.1 million to 1.6 million per year would lead to a population of 399 million by 2050; the current population is 312 million.

The total impact of the U.S. population on the environment includes a high consumption of energy and minerals. The term consumption includes not only resource use, but also waste production (Fig. 37B). Consumption in and of itself is not necessarily a problem, but it becomes one when a resource is consumed in a negligent way. For example, fishing becomes undesirable when fisheries are overharvested and future yield is in jeopardy. The surface mining of coal and a new process called hydraulic fracking for oil create environmental problems from degrading the environment to polluting water supplies. Technology can come to the rescue, as when automobile engines are designed to reduce emissions or when we can use renewable energy sources, such as solar and wind which create less pollution than do fossil fuels. The aim is to manage our consumption of energy and minerals so that the natural environment can continue to provide for our needs without creating much waste. Manufacturing in the United States is up by 50% since 1980 but our energy consumption per person has not increased. Before we congratulate ourselves, however, we have to recognize that other countries have increased their energy consumption and assumed the burden of environmental pollution when we buy imported goods.

Several countries, such as India and China, are now moving from being less-developed to more-developed. Previously, their very large populations had minimal impact on the biosphere because of low consumption of resources, but this is no longer the case, because their level of consumption is increasing rapidly.

CONSIDER THESE QUESTIONS

1. Is a family in the United States acting selfishly when it has more than two children? Why or why not?
2. What benefits might arise if the U.S. population were to achieve sustainability? How might this affect the economy?
3. The United States has a high environmental impact, and yet we don't want other countries to have a similar impact. Should we reduce our environmental impact and expect other countries to do the same? Why or why not?

 Explore the concepts through a variety of multimedia assets and question types.

www.mcgrawhillconnect.com

FIGURE 37B Harmful technology such as a power plant that causes air and water pollution is a part of a population's impact on the environment.

THE CHAPTER IN REVIEW

CONNECTING THE CONCEPTS

Modern ecology began with descriptive studies by nineteenth-century naturalists. In fact, an early definition of the field was "scientific natural history." However, modern ecology has now grown from a simple descriptive field to an experimental, predictive science.

Much of the success in the development of ecology as a predictive science has come from studies of populations and the creation of models that examine how populations change over time. The simplest models are based on population growth when resources are unlimited. This results in exponential population growth, a type only rarely seen in nature. Pest species may exhibit exponential growth until they run out of resources. Because so few natural populations exhibit exponential growth, population ecologists realized they must incorporate resource

limitation into their models. The simplest models that account for limited resources result in logistic growth. Populations that exhibit logistic growth stop growing when they reach the environmental carrying capacity.

Many modern ecological studies are concerned with identifying the factors that limit population growth and set the environmental carrying capacity. A combination of careful descriptive studies, experiments done in nature, and sophisticated models has allowed ecologists to accurately predict which factors have the greatest influence on biotic potential, population growth, and life history patterns.

We continue this study in Chapter 38 as we see how the behavior of an organism is adapted to the environment. Adaptations to the environment, as you know, result in reproductive success, leading possibly to an increase

in population size, or at least maintenance of population size, in that environment.

ANALYZE AND EVALUATE

1. Discuss why a moderate population size is desirable and either a very large or a very small population size could be undesirable.
2. The U.S. population continues to increase, unlike those of other more-developed countries. Should we be concerned about how this affects consumption and possibly the sustainability of the population?
3. With reference to Figure 37.3C, discuss the statement that a population will continue to grow in size as long as more women are entering than leaving their reproductive years.

SUMMARIZE

Scope of Ecology

37.1 Ecology is studied at various levels

- **Ecology** is the study of the interactions of organisms with other organisms and with the physical environment. For example, an ecologist might study how an organism is adapted to its **habitat** and then go on to study life at various higher levels of organization including **population, community, ecosystem,** or the **biosphere.**

How Populations Change Over Time

37.2 Density and distribution are aspects of population structure

- **Population density** refers to the number of individuals per unit area and **population distribution** (whether clumped, random, or uniform) is the pattern of dispersal of individuals across a **range.**
- **Resources,** the components of the environment that support organisms, and **limiting factors,** such as the availability of particular nutrients, affect population density and distribution.

37.3 The growth rate results in population size changes

- Yearly birth and death rates mainly determine a population's growth rate. A population's **biotic potential** is its highest possible growth rate when resources are unlimited. Biotic potential is (1) affected by **survivorship,** the probability of **cohort** members surviving to a particular age as shown in a **life table,** and (2) dependent on the age structure as shown in an **age structure diagram,** which includes prereproductive, reproductive, and postreproductive age groups. A pyramid shape means the population will expand rapidly; a bell shape

indicates the population is stabilized; and an urn shape means the population will decline in size:

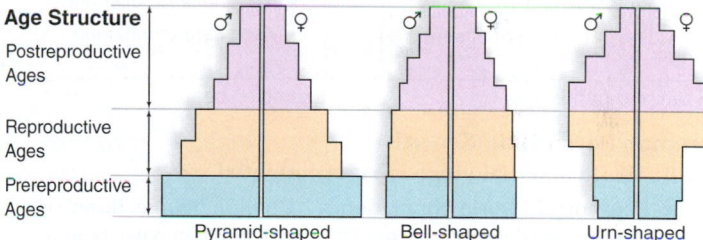

Pyramid-shaped   Bell-shaped   Urn-shaped

37.4 Patterns of population growth can be described graphically

- **Exponential growth,** which has a lag phase and an exponential growth phase, accelerates over time and results in a J-shaped curve.
- **Logistic growth,** which has an exponential growth phase, a deceleration phase, and a stable equilibrium phase, stabilizes when the **carrying capacity** has been reached, resulting in an S-shaped curve.

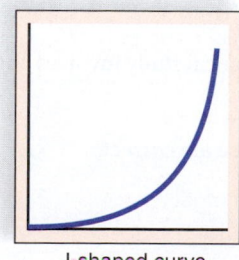

J-shaped curve

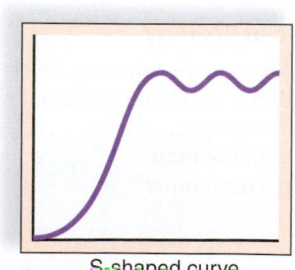

S-shaped curve

Regulation of Population Size

37.5 Density-independent factors affect population size

- Density-independent means that mortality (% killed) remains the same, regardless of density. **Density-independent factors** are abiotic factors that affect population size, such as droughts, freezes, hurricanes, floods, and forest fires.

37.6 Density-dependent factors affect large populations more

- Density-dependent means that mortality increases as the density of the population increases. **Density-dependent factors** are biotic factors that affect population size, such as competition, predation, and parasitism.

Life History Patterns

37.7 Life history patterns consider several population characteristics

- Population characteristics (life span, biotic potential, etc.) shape the population's **life history. Opportunistic populations** exhibit exponential growth; small individuals have a short life span and may die before reproducing. **Equilibrium populations** exhibit logistic growth; large individuals have a long life span, and most of their offspring survive to reproductive age. Opportunistic populations are adapted to a fluctuating environment, and equilibrium species are adapted to a relatively stable environment and more likely to experience **extinction:**

Opportunistic Pattern	Equilibrium Pattern
• Small individuals	• Large individuals
• Short life span	• Long life span
• Fast to mature	• Slow to mature
• Many offspring	• Few and large offspring
• Little or no care of offspring	• Much care of offspring

Example: dandelions

Example: bears

Human Population Growth

37.8 World population growth is exponential

- The current world population is 7.1 billion people. Based on the predicted **doubling time,** the world population will require double the amount of current resources in 50 years.
- **More-developed countries** (**MDCs**) are characterized by low population growth and generally a good standard of living. **Less-developed countries** (**LDCs**) are known for rapidly expanding populations and generally poor living conditions. Once the LDCs have passed through the **demographic transition** and deaths equal births, growth will stabilize.

TEST YOURSELF

Scope of Ecology

1. Which of the following levels of ecological study involves both abiotic and biotic components?
   a. organism
   b. population
   c. community
   d. ecosystem
   e. All of these are correct.

2. Place the following levels of organization in order, from lowest to highest.
   a. community, ecosystem, population, organism
   b. organism, community, population, ecosystem
   c. population, ecosystem, organism, community
   d. organism, population, community, ecosystem

How Populations Change Over Time

3. The distribution of the human population is
   a. variable.   c. random.
   b. clumped.   d. uniform.

4. If the human birthrate were reduced to 15 per 1,000 per year and the death rate remained the same (9 per thousand), what would be the growth rate?
   a. 9%   d. 0.6%
   b. 6%   e. 15%
   c. 10%

5. A population's maximum growth rate is also called its
   a. carrying capacity.   c. growth curve.
   b. biotic potential.   d. replacement rate.

6. Which of the following statements about a plant species is not relevant for determining its biotic potential?
   a. It produces 10 kg of mass per year.
   b. It produces its first flowers at 5 years of age.
   c. 50% of seedlings grow into mature plants.
   d. On average, 100 seedlings are produced by each plant every year.

7. If a population has a type I survivorship curve (most of its members live the entire life span), which of the following events would you also expect?
   a. a single reproductive event per adult
   b. most individuals reproduce
   c. sporadic reproductive events
   d. reproduction occurring near the end of the life span
   e. None of these are correct.

8. Exponential growth is best described by
   a. steep, unrestricted growth.
   b. an S-shaped growth curve.
   c. a constant rate of growth.
   d. growth that levels off after rapid growth.
   e. Both b and d are correct.

9. When the carrying capacity of the environment is exceeded, the population typically
   a. increases, but at a slower rate.
   b. stabilizes at the highest level reached.
   c. decreases.
   d. dies off entirely.

10. Label this diagram:

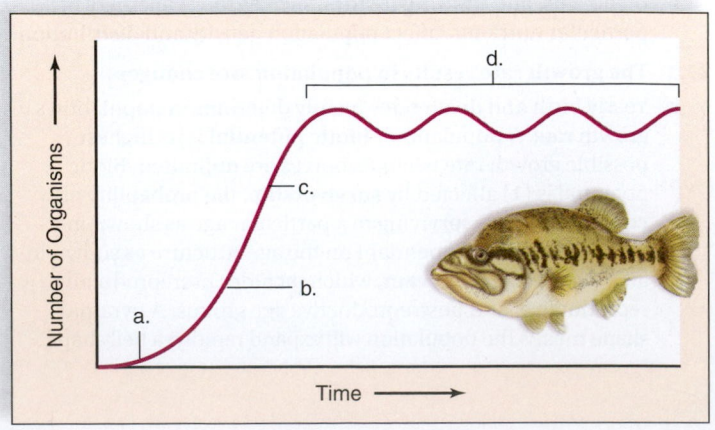

11. **THINKING CONCEPTUALLY** The range of a plant that reproduces asexually by runners would be smaller than one that reproduces by windblown seeds. Explain.

Regulation of Population Size

12. Which of these is a density-independent factor?
    a. competition
    b. predation
    c. weather
    d. resource availability

For questions 13–15, indicate the factor in the key exemplified by the scenario.

KEY:
    a. density-independent factor
    b. competition
    c. predation

13. A severe drought destroys the entire food supply of a herd of gazelle.

14. Only the swiftest coyotes are able to catch the limited supply of rabbits available as a food source. The remaining animals are not strong enough to reproduce.

15. Deer in a forest damage a dense thicket of oak saplings more severely than a few young oak trees.

16. Label each line on the following as representing density-independent factors or density-dependent factors:

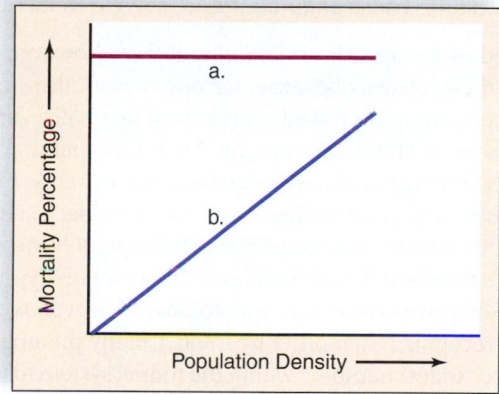

17. **THINKING CONCEPTUALLY** Does adaptation to the environment necessarily increase population size? Why or why not?

Life History Patterns

18. Which of the following is not a feature of an opportunistic life history pattern?
    a. many offspring
    b. little or no care of offspring
    c. long life span
    d. small individuals
    e. fast to mature

19. An equilibrium life history pattern does not include
    a. large individuals.
    b. long life span.
    c. few offspring.
    d. little or no care of offspring.

20. **THINKING CONCEPTUALLY** Under what conditions is it advantageous for a population to produce a large number of small, uncared-for offspring rather than large, well-cared-for offspring?

Human Population Growth

21. The human population
    a. is undergoing exponential growth.
    b. is not subject to environmental resistance.
    c. fluctuates from year to year.
    d. grows only if emigration occurs.
    e. All of these are correct.

22. The current doubling time for the world's population is 50 years. Therefore, if the population is 7.1 billion people today, what will it be in 100 years if the doubling rate does not change?
    a. 14.2 billion
    b. 28.4 billion
    c. 35.5 billion
    d. 81.0 billion

23. Decreased death rate followed by decreased birthrate has occurred in
    a. MDCs.
    b. LDCs.
    c. MDCs and LDCs.
    d. neither MDCs nor LDCs.

24. The overall reason the LDCs population is still expected to explode in the near future is the
    a. annual growth rate of the least developed countries.
    b. death rates are near zero.
    c. lack of medical care.
    d. the population is in the deceleration phase of exponential growth stage.

25. A pyramid-shaped age distribution means that
    a. the prereproductive group is the largest group.
    b. the population will grow for some time in the future.
    c. the country is more likely an LDC than an MDC.
    d. fewer women are leaving the reproductive years than entering them.
    e. All of these are correct

26. **THINKING CONCEPTUALLY** People in countries undergoing the most environmental stress reproduce the most. Why might that be?

GET INVOLVED

1. The right whale population remains dangerously small, despite many decades of complete protection. Formulate a hypothesis based on the four factors listed in section 37.3 to explain this observation. How would you test your hypothesis?

2. You are a river manager charged with maintaining the water flow through the use of dams so that trees, which have equilibrium life histories, can continue to grow along the river. What would you do?

MEDIA STUDY TOOLS

mhhe.com/maderconcepts3

Enhance your study of this chapter with interactive study tools, practice tests, and engaging animations. Also, ask your instructor about the resources available through ConnectPlus, which includes LearnSmart, a personalized adaptive learning program, and a media-rich eBook.

38

Behavioral Ecology

BEFORE YOU BEGIN

Take a few minutes to recall

Behavior between sexes helps reproductively isolate species (section 15.2)

How behavior helps identify evolutionary relationships in "Motherhood Among Dinosaurs" on page 298

Pheromone regulation of behavior in "Pheromones Among Us" on page 684

The dependence of behavior on the senses of organisms (section 28.1)

For the Benefit of All

A naked mole rat is a rodent, as are the more familiar mice, rats, hamsters, and gerbils. But mole rats are not as closely related to these rodents as they are to porcupines, guinea pigs, and chinchillas. Porcupines vary in size, but some are almost as large as beavers. They have a means of defense suited to their nocturnal lifestyle—quills. If a porcupine engages in combat, the quills become embedded in the skin—or even a vital organ—of the assailant. Porcupines lead a solitary life in forests, deserts, and grasslands. Chinchillas and guinea pigs are about half the size of rabbits. Chinchillas are known for their soft fur, which is much thicker than that of a rabbit. They live in the Andes mountains of South America where they are so agile that they can jump up to 152 cm above their heads. Chinchillas live in large colonies, but the social organization is not rigid and there are no dominant males or females. Guinea pigs live in open, grassy areas, and they seek shelter in naturally protected areas or burrows deserted by other animals. Guinea pigs are social and live in small groups consisting of a male, several females, and their young. Porcupines, chinchillas, and guinea pigs are usually nocturnal, preferring to be active in the dark. They feed on shrubs, grasses, tubers, or fruits. These adaptations are seen in mole rats.

Naked mole rats (*Heterocephalus glaber*), however, are different from their relatives because, for one reason, they have no fur and, indeed, are quite naked except for a few hairs. Also, a mole rat is tiny—only about 7.5 cm long. Mole rats spend almost their entire lives underground and are social to the extreme. In fact, their social behavior is more reminiscent of social insects, such as ants and bees, than most other mammals. Unfortunately, no one knows how much their behavior depends on the secretion of pheromones passed between members of the colony. However, we do know that they recognize each other by smell. Colony members visit the designated "toilet chamber" within the tunnel system to roll around in excrement and ensure that they smell as though they belong.

Porcupine

An average-sized mole rat colony contains about 75 animals. Naked mole rats are so dependent on being part of a colony that an individual kept in isolation will die. The mole rat society has a rigid hierarchical structure. At the top of the hierarchy is the queen, the only female who reproduces. At the bottom are workers of both sexes, who do not reproduce, but instead work tirelessly to help the queen and excavate long and intricate tunnels. When digging tunnels, they line up nose to tail and operate in conveyor-belt style. A digger at the front uses its teeth to break through new soil. Behind the first one, sweepers use their feet to whisk the dirt backwards. The last one in line kicks the dirt up onto the surface of the ground. While the workers dig, soldiers protect them from threats, such as predators and interloping mole rats from other colonies. Soldiers can be male or female and are physically larger than the workers.

The tunnels enable the mole rats to obtain the roots and tubers that they rely on for food and moisture. Without efficient tunnel digging, a mole rat can't live, and it takes several mole rats working together to efficiently dig a tunnel. Some researchers believe this explains why mole rats and some other burrowing animals are social.

The queen helps the colony not just by producing young but also by inspecting the tunnels. If the queen senses that more work needs to be done in a particular area of the tunnel system, she forces other colony members into that place.

Chinchillas

The queen keeps a harem of a few very close relatives with whom she mates. The result of this inbreeding is that all members of the colony share an unusually high degree of genetic similarity: About 80% of their genes are identical. This means that the members of the group see to the propagation of their genes when they help the rest of the colony survive. Interestingly, when the queen is pregnant, the mole rats—of both sexes—develop teats, even though only the queen produces milk to nurse the pups. Pheromones could be at work. The entire colony is devoted to the queen and her offspring. Once the pups are born, the queen delegates much of their care to workers. Such cohesion keeps the colony together.

Behavioral ecology, as discussed in this chapter, is dedicated to the principle that natural selection shapes behavior just as it does the anatomy and physiology of an animal. As we can see in mole rats, behavior is adapted to how and where animals live. The term behavioral ecology recognizes this phenomenon and serves as the title for this chapter. One day, studies will most likely determine how much of animal behavior is innate and how much is learned, a topic often called the "nature versus nurture" question.

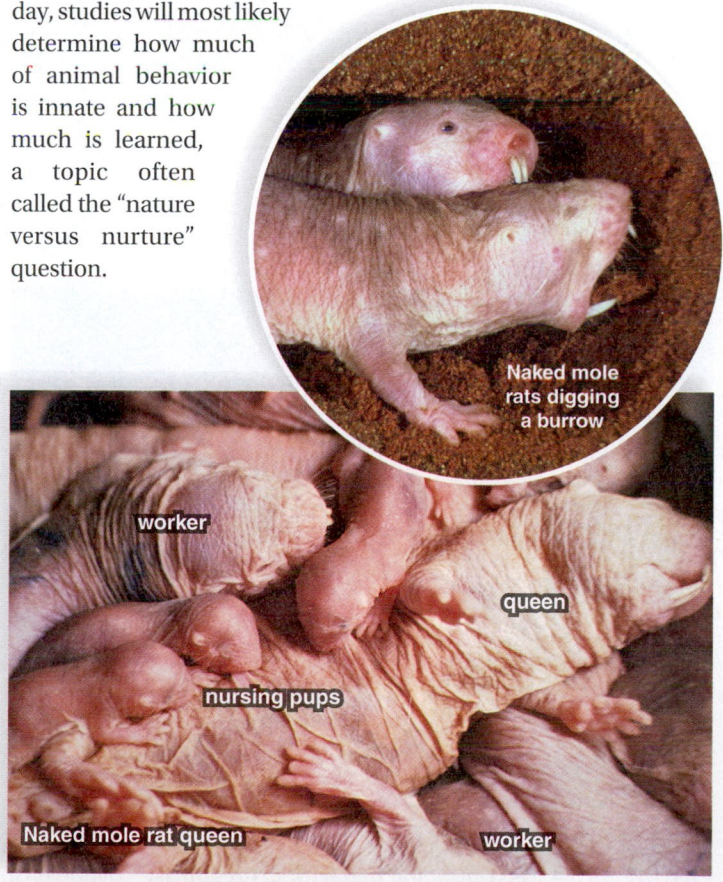
Naked mole rats digging a burrow

Guinea pigs

worker
queen
nursing pups
Naked mole rat queen
worker

Inheritance and Learning Influence Behavior

Behavior encompasses any action that can be observed and described. The "nature versus nurture" question asks to what extent our genes (nature) and environmental influences (nurture) affect behavior. The results of experiments support the hypothesis that most behaviors have, at least in part, a genetic basis.

38.1 Inheritance influences behavior

LEARNING OUTCOME

When you complete this section, you should be able to

1. Describe data showing that behavior is inherited.

An animal's anatomy and physiology determine what types of behavior are possible for that animal. Therefore, genes, which control anatomy and physiology, also control behavior. Experiments have been done to discover the degree to which genetics controls behavior, and numerous results support the hypothesis that behavior has a genetic basis. Apparently, genes influence the development of neural and hormonal mechanisms that control behavior.

Data Suggest Behavior Has a Genetic Basis Lovebirds are small, green and pink African parrots that nest in tree hollows. Several species of lovebirds in the genus *Agapornis* are closely related but build their nests differently (Fig. 38.1A). Peach-faced lovebirds, *Agapornis roseicollis,* cut somewhat shorter strips, and they carry them to the nest in a very unusual manner. They pick up the strips in their bills and then insert them into their rump feathers. In this way, they can carry several of these short strips with each trip to the nest. Fischer lovebirds, *Agapornis fischeri,* use their bills to cut large leaves (or in the laboratory, pieces of paper) into long strips. Then they carry one strip at a time in their bills to the nest, where they weave them with others to make a deep cup.

Researchers hypothesized that if the behavior for obtaining and carrying nesting material is inherited, hybrid lovebirds might show intermediate behavior. When the two species of birds were mated, the hybrid birds had difficulty carrying nesting materials. They cut strips of intermediate length and then attempted to tuck the strips into their rump feathers. However, they did not push the strips far enough into the feathers, and so the strips always came out as they walked or flew. Hybrid birds eventually (after about 3 years in this study) learned to carry the cut strips in their beak, but they still briefly turned their head toward their rump before flying off.

Peach-faced lovebird with nesting material in its rump feathers

Fischer lovebird with nesting material in its beak

FIGURE 38.1A Nest-building behavior in lovebirds. Hybrid offspring of these two types of lovebirds try but cannot tuck strips in their rump feathers (*left*), so they carry them in their beaks (*right*).

Similarly, mating experiments have been done with two populations of California garter snakes (Fig. 38.1B). Inland populations were aquatic and commonly fed underwater on frogs and fish. Coastal populations were terrestrial and fed mainly on slugs. In the laboratory, inland adult snakes refused to eat slugs while coastal populations readily did so. Matings between the two populations resulted in offspring with an intermediate incidence of slug acceptance. To determine this, reseachers did a clever experiment. When snakes eat, they use tongue flicks to recognize their prey. The tongue carries chemicals to an odor receptor in the roof of the mouth. The researchers dipped swabs in slug extract and recorded the number of tongue flicks for newborn hybrid snakes, and it was intermediate between newborn inland and coastal garter snakes. As with the lovebirds, these snake experiments suggest that behavior has a genetic basis.

Human twins, on occasion, have been separated at birth and raised under different environmental conditions. Studies of such twins show that they have similar food preferences and activity patterns, and they even select mates with similar characteristics. These twin studies lend support to the hypothesis that at least certain types of human behavior are primarily influenced by nature (i.e., the genes).

Data Show Behavior Has a Genetic Basis The experiments mentioned so far suggest that the nervous system is involved in behavior. Both the nervous and endocrine systems are responsible for coordinating body systems. Is the endocrine system also involved in behavior? Research studies indicate that it is. For example, the egg-laying behavior of the marine snail *Aplysia* involves a set sequence of movements. Following copulation, the animal extrudes long strings of egg cases—more than a million of them. The snail takes the egg case string in its mouth, covers it with mucus, waves its head back and forth to wind the string into an irregular mass, and attaches the mass to a solid object, such as a rock. Several years ago, scientists isolated and analyzed an egg-laying hormone (ELH) that can cause a snail to lay eggs even if it has not mated. ELH was found to be a small protein composed of 36 amino acids that diffuses into the circulatory system and excites the smooth muscle cells of the reproductive duct, causing them to contract and expel the egg string. Using recombinant DNA techniques, the investigators isolated the *ELH* gene. The gene's product turned out to be a protein with 271 amino acids. The protein can be cleaved into as many as 11 possible products, and ELH is one of these. ELH alone, or in conjunction with these other products, has been found to control all the components of egg-laying behavior in *Aplysia*.

In another study, investigators found that maternal behavior in mice was dependent on the presence of a gene called *fosB*. Normally, when a mother first inspects her newborn, various sensory information from her eyes, ears, nose, and touch receptors travel to the hypothalamus. This incoming information activates *fosB* alleles, and a particular protein is produced. The protein begins a process of activating cellular enzymes and other genes. The end result is a change in the neural circuitry within the hypothalamus, which manifests itself in good maternal behavior. Mice that lack good maternal behavior also lack *fosB* alleles, and the hypothalamus neither makes any of the products nor activates any of the enzymes and other genes that lead to good maternal behavior (Fig. 38.1C).

Inland garter snake   Coastal garter snake

FIGURE 38.1B Slug eating preferences. Inland garter snakes do not eat slugs but coastal garter snakes do eat slugs.

▶ **38.1 CHECK YOUR PROGRESS**
1. Identify specific genes that have been linked to behavior.
2. Determine what would happen to a mole rat colony if a mutation in the queen reduced the ability of the workers to cooperate in excavating tunnels.

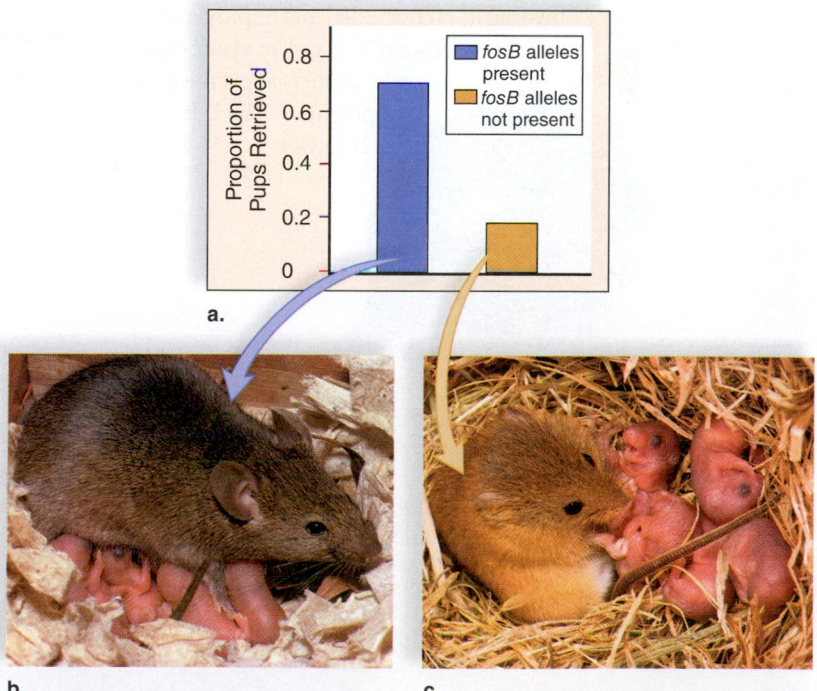

FIGURE 38.1C Maternal care in mice. Mice with *fosB* alleles spend time retrieving and crouching over their young, whereas mice that lack these alleles do not display these maternal behaviors.

38.2 Learning can also influence behavior

When you complete this section, you should be able to

1. Describe data showing that behavior can be learned.

Even though genetic inheritance (nature) serves as a basis for behavior, it is possible that environmental influences (nurture) also affect behavior. For example, behaviorists originally believed that some behaviors were **fixed action patterns (FAPs)** elicited by a sign stimulus. But then they found that many behaviors formerly thought to be FAPs improve with practice—that is, through learning. In this context, **learning** is defined as a durable change in behavior brought about by experience.

Learning in Birds Laughing gull chicks' begging behavior appears to be a FAP, because it is always performed the same way in response to the parent's red bill (the sign stimulus) (Fig. 38.2). ❶ A chick directs a pecking motion toward the parent's bill, grasps it, and strokes it downward. A parent can bring about the begging behavior by swinging its bill gently from side to side. After the chick responds, the parent regurgitates food onto the floor of the nest. If need be, the parent then encourages the chick to eat. This interaction between the chicks and their parents suggests that the begging behavior involves learning. To test this hypothesis, eggs were collected in the field and hatched in a dark incubator to eliminate visual stimuli before the test. For testing purposes, pictures of gull heads were painted on small cards, and each chick was allowed to make about a dozen pecks at the model. The chicks were returned to the nest, and then each was retested daily. ❷ The tests showed that on the average, only one-third of the pecks by a newly hatched chick strike the model. But 1 day after hatching, more than half of the pecks are accurate, and ❸ 2 days after hatching, the accuracy reaches a level of more than 75%. Investigators concluded that improvement in motor skills, as well as visual experience, strongly affect the development of chick begging behavior.

Imprinting **Imprinting** is considered a form of learning. Imprinting was first observed in birds when chicks, ducklings, and goslings followed the first moving object they saw after hatching. In the wild, this object is ordinarily their mother, but investigators found that birds can seemingly be imprinted on any object—a human or a red ball—if it is the first moving object they see during a *sensitive period* of 2–3 days after hatching. The term sensitive period (also called the critical period) means that the behavior develops only during this time.

A chick imprinted on a red ball follows it around and chirps whenever the ball is moved out of sight. Social interactions between parent and offspring during the sensitive period seem key to normal imprinting. For example, female mallards cluck during the entire time imprinting is occurring, and it could be that vocalization before and after hatching is necessary to normal imprinting.

Song Learning White-crowned sparrows sing a species-specific song, but the males of a particular region have their own dialect. Birds were caged in order to test the hypothesis that young white-crowned sparrows learn how to sing from older members of their species.

Three groups of birds were tested. Birds in the first group *heard no songs at all*. When grown, these birds sang a song, but it was not fully developed. Birds in the second group *heard tapes of white-crowns singing*. When grown, they sang in that dialect, as long as the tapes had been played during a sensitive period from about age 10–50 days. White-crowned sparrows' dialects (or other species' songs) played before or after this sensitive period had no effect on the birds. Birds in a third group did not hear tapes and instead were *given an adult tutor*. These birds sang a song of even a different species—no matter when the tutoring began—showing that social interactions apparently assist learning in birds.

► **38.2 CHECK YOUR PROGRESS**

1. Explain an experiment that could test whether worker mole rats have to learn (at least in part) to excavate tunnels.

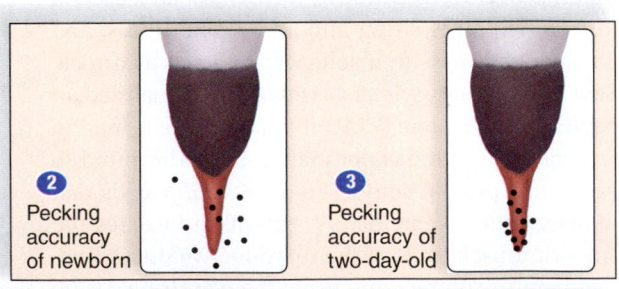

❷ Pecking accuracy of newborn   ❸ Pecking accuracy of two-day-old

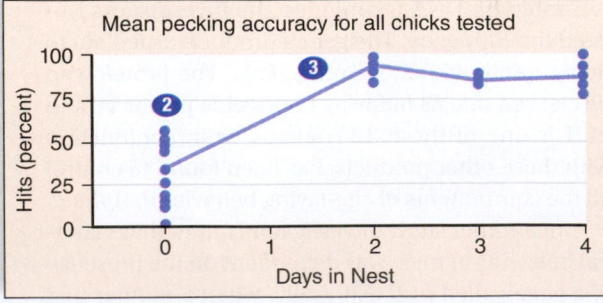

Mean pecking accuracy for all chicks tested

FIGURE 38.2 Begging behavior in laughing gulls improves with experience.

When you complete this section, you should be able to

1. Distinguish between classical conditioning, operant conditioning, insight learning, and habituation.

A change in behavior that involves an association between two events is termed **associative learning.** For example, birds that get sick after eating a monarch butterfly no longer prey on monarch butterflies, even though they may be readily available. Or the smell of fresh-baked bread may entice you, even though you have just eaten. If so, perhaps you associate the taste of bread with a pleasant memory, such as being at home. Examples of associative learning include classical conditioning and operant conditioning.

Classical Conditioning In **classical conditioning,** the presentation of two different types of stimuli (at the same time) causes an animal to form an association between them. The best-known laboratory example of classical conditioning was conducted by the Russian psychologist Ivan Pavlov (Fig. 38.3). Pavlov observed that dogs salivate when presented with food; in this case, ❶ food is an unconditioned stimulus that brings about ❷ an unconditioned response (saliva) (left-hand side of Fig. 38.3). Then Pavlov rang a bell whenever the dogs were fed. In time, the dogs began to salivate at the sound of the bell, without any food being present. As shown on the right-hand side of Figure 38.3, ❸ the sound of the bell is a conditioned stimulus that brings about ❹ the conditioned response—that is, salivation even though no food is present.

Classical conditioning suggests that an organism can be trained—that is, conditioned—to associate any response with any stimulus. Unconditioned responses occur naturally, as when salivation follows the presentation of food. Conditioned responses are learned, as when a dog learns to salivate when it hears a bell. Advertisements attempt to use classical conditioning to sell products. Why do commercials pair attractive people with a product being advertised? Advertisers hope that consumers will associate attractiveness with the product and that this pleasant association will cause them to buy it.

Some types of classical conditioning can be helpful. For example, psychologists have suggested that you hold children on your lap when reading to them. Why? Because they will associate a pleasant feeling with reading.

Operant Conditioning During **operant conditioning,** a stimulus-response connection is strengthened. Most people know that it is helpful to give an animal a reward, such as food or affection, when teaching it a trick. When we go to an animal show, it is quite obvious that trainers use operant conditioning. They present a stimulus, say, a hoop, and then give a reward (food) for the proper response (jumping through the hoop). The reward need not always be immediate. In latent operant conditioning, an animal makes an association without the immediate reward, as when squirrels make a mental map of where they have hidden nuts.

B. F. Skinner became well known for studying this type of learning in the laboratory. In the simplest type of experiment performed by Skinner, a caged rat happens to press a lever and is rewarded with sugar pellets, which it avidly consumes. Thereafter,

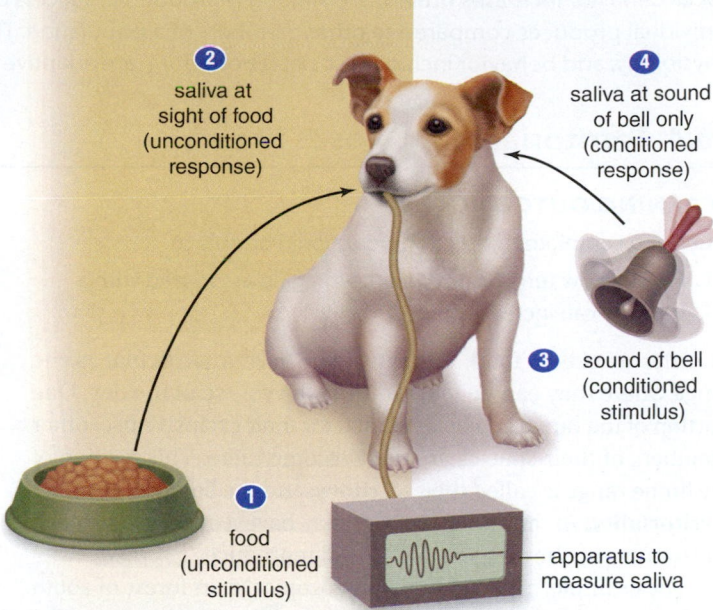

FIGURE 38.3 After classical conditioning, a dog salivates at the sound of a bell even though no food is present.

the rat regularly presses the lever whenever it wants a sugar pellet. In more sophisticated experiments, Skinner even taught pigeons to play Ping-Pong by reinforcing desired responses to stimuli. This technique can also be applied to child rearing. It's been suggested that parents who give positive reinforcement for good behavior will be more successful than parents who punish undesirable behavior.

Other Means of Learning In addition to the modes of learning already discussed, animals may learn through insight, imitation, and habituation. **Insight learning** occurs when an animal suddenly solves a problem without any prior experience with the problem. The animal appears to adapt prior experience to solving the problem. For example, chimpanzees have been observed stacking boxes to reach bananas in laboratory settings. Other animals, too, seem to be able to reason things out. In one experiment, ravens were offered meat attached to string hanging from a branch. It took several hours but eventually one raven reached down and with its beak pulled the string up over and over again, each time securing the string with its foot. Eventually, the raven was able to grab the meat.

Many organisms learn through observation and imitation. For example, Japanese macaques learn to wash sweet potatoes before eating them by imitating others.

Habituation occurs when an animal no longer responds to a repeated stimulus. Deer ignoring traffic as they graze on the side of a busy highway demonstrate habituation.

▶ **38.3 CHECK YOUR PROGRESS**
1. Identify the type of learning if an experimenter shows that older mole rat workers share food with younger workers who are new to tunnel excavation.

Behaviors That Increase Fitness

Social behavior increases **fitness,** the lifelong reproductive success of an individual—that is, the number of fertile offspring the individual produces compared to other members of a population. Through natural selection, all attributes of animals—anatomy, physiology, and behavior including societal behavior—are adaptive and increase an individual's fitness.

38.4 Territoriality increases fitness

LEARNING OUTCOME

When you complete this section, you should be able to

1. Explain how territoriality and the foraging associated with a territory can increase fitness.

In order to gather food, animals often have a particular home range where they can be found during the course of the day. One portion of the range can be defended for their exclusive use; other members of their species are not welcome there. This portion of the home range is called their **territory,** and the behavior is called **territoriality.** An animal's territory may have a good food source and/or may be the area in which it will reproduce.

For example, gibbons live in the tropical rain forest of south and southeast Asia. Normally, their home range can be covered in about 3–4 days. They are monogamous (always the same mate) and territorial. Territories are maintained by loud singing (Fig. 38.4A). Males sing just before sunrise, and mated pairs sing duets during the morning. Males, but not females, show evidence of fighting to defend their territory in the form of broken teeth and scars.

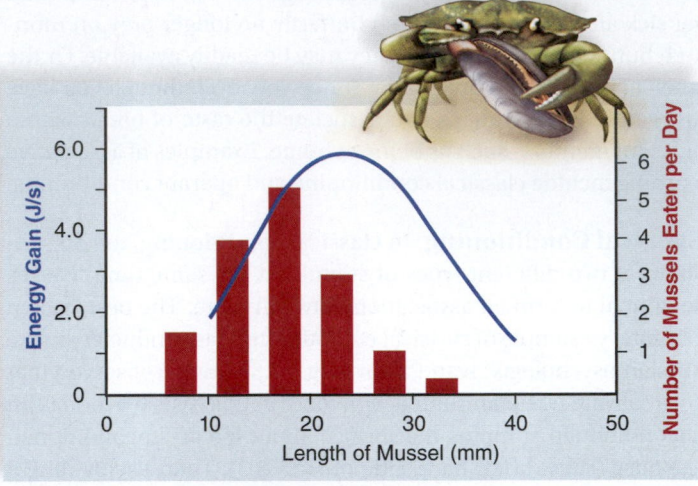

FIGURE 38.4B The shore crab, *Carcinus maenas,* prefers to open and eat intermediate-sized mussels. This size offers the most energy compared to the effort of opening the shell.

Obviously, defense of a territory has a certain cost; it takes energy to sing and fight off others. Also, you might get hurt. So, what is the adaptive benefit of being territorial? The chief benefits include ensuring a source of food, exclusive rights to one or more females, a place to rear young, and possible protection from predators. Cheetahs require a large territory in order to hunt for their prey, and therefore, they use urine to mark their territory (see Fig. 38.7A).

Territoriality is more likely to occur during times of reproduction. Seabirds have very large home ranges consisting of hundreds of kilometers of open ocean, but when they reproduce they become fiercely territorial. Each bird has a very small territory consisting of only a small patch of beach where it places its nest (see Fig. 37.2B).

Video
Cichlid Territoriality

Foraging for Food Food gathering is technically called foraging for food. One study showed that shore crabs eat intermediate-sized mussels because the net energy gain is more than if they eat larger-sized mussels (Fig. 38.4B). The large mussels take too much energy to open compared to the amount of energy they provide. The **optimal foraging model** states that it is adaptive for foraging behavior to be as energetically efficient as possible.

Even though it can be shown that animals that take in more energy are likely to have more offspring, animals must often deal with other factors as well, such as escaping from predation. If an animal is killed and eaten, it has no chance of having offspring. Animals often face trade-offs that cause them to modify their behavior so that they stop foraging for a while.

FIGURE 38.4A Siamang gibbons, *Hylobates syndactylus,* look alike and share the task of marking their territory by singing.

male

female

> **38.4 CHECK YOUR PROGRESS**
> 1. Evaluate whether foraging for food in a territory can increase fitness.

38.5 Sexual selection increases fitness

LEARNING OUTCOME

When you complete this section, you should be able to

1. Explain the role of female choice and male competition in sexual selection.

Sexual selection is a form of natural selection favoring features that increase an animal's chances of mating. In other words, these features are adaptive in that they lead to increased fitness—the chance of having more offspring. Because females produce a limited number of eggs in their lifetime and are generally the sex that provides the most parental care, it is adaptive for them to be choosy about their mates. If a female chooses a mate that passes on features that will eventually cause a male offspring to be chosen by females, her fitness has increased. This explains why male birds-of-paradise have such flamboyant plumage even though it increases the risk of being captured by a predator.

Females have a certainty that an offspring is theirs that males do not have. The best way for males to overcome this disadvantage is to have as many offspring as possible. However, competition is often required for males to gain access to females, and ornaments, such as antlers, can enhance a male's ability to compete. If physique alone does not cause a competitor to back off, bull elks resort to ramming each other with their antlers, a sure sign that they are competing with one another.

▶ **38.5 CHECK YOUR PROGRESS**
1. Determine the cost and the benefit of female choice to males.

HOW BIOLOGY IMPACTS OUR LIVES
Application

38A Sexual Selection Among Humans

A study of human mating behavior shows that humans are also subject to female choice and male competition. Of course, applying the principle of increased fitness to humans is chancy because we cannot be "preprogrammed" and are able to make conscious choices. That said, understanding the evolutionary basis of animal behavior can provide some interesting insights into why people behave the way they do.

Consider that a woman, by nature, must invest more in having a child than a man. After all, it takes 9 months to have a child, and pregnancy is followed by lactation during which a woman may nurse her infant. Men, on the other hand, need only contribute sperm during a sex act that may require only a few minutes. As a result, men are generally more available for mating than are women. Because more men are available, they necessarily have to compete for the privilege of mating. As in other animals, males tend to be larger and more aggressive than females, perhaps as a result of past sexual selection by females. But they may pay a price for their high level of testosterone, the energetic cost of male-male competition, and increased stress associated with finding mates. Male humans live on average 4 to 7 fewer years than females do.

David Buss, an evolutionary psychologist at the University of Texas, conducted studies of female mate preference across cultures in over 20 countries. His research, although somewhat controversial, suggests that the number-one trait females find attractive is financial success because it ensures that resources will be available to raise children (Fig. 38A). Only if her children survive will a female increase her fitness.

Other studies have shown that symmetry in facial and body features is also important to females. It could be that symmetry suggests to the female that the male has "good genes," which will help ensure the health of her children. In other animals, infection with parasites leads to asymmetry.

Just as women choose men who can provide resources, men prefer youthfulness and attractiveness in females, signs that their partner can provide them with children. One controversial study indicated that men ages 8–80 across four continents prefer

FIGURE 38A As exemplified by Queen Noor and the late King Hussein, women tend to prefer a husband who can support their children, and men tend to prefer to wed fertile younger women.

women with a WHR (waist-to-hip ratio) of 0.7. Previous research has shown that a WHR of 0.7 is optimal for conception; for each increase of 10% in WHR, the odds of conception during each ovulation event decrease by 30%. Men responding to questionnaires prefer physical attributes in females that biologists associate with a strong immune system and good health (e.g., symmetry), high estrogen levels (indicating fertility), and especially youthfulness. On average, men marry women 2.5 years younger than they are, but as men age, they tend to prefer women who are many years younger. Men are capable of reproduction for many more years than women. Therefore, by choosing younger women, older men can increase their fitness.

CONSIDER THESE QUESTIONS

1. Should evolutionary principles be applied to human reproductive behavior? Why or why not?
2. Do you have any evidence that human males compete for access to females and females choose their mates based on their financial success?

 Explore the concepts through a variety of multimedia assets and question types.

www.mcgrawhillconnect.com

38.6 Societies increase fitness

LEARNING OUTCOME

When you complete this section, you should be able to

1. Understand how living in a society can increase an individual's fitness.

The principles of evolutionary biology can be applied to the study of social behavior in animals, a discipline called **sociobiology.** Sociobiologists hypothesize that living in a society has a greater reproductive benefit than reproductive cost. A cost-benefit analysis can help determine if this hypothesis is supported.

Group Living Group living does have its benefits, including helping an animal avoid predators, rear offspring, and find food. A group of impalas is more likely than a solitary one to hear an approaching predator. Many fish moving rapidly in many directions might distract a would-be predator. Weaverbirds form giant colonies that help protect them from predators, and may also allow them to share information about food sources. Primates belonging to the same troop signal one another when they have found an especially bountiful fruit tree. Lions working together are able to capture large prey, such as zebras and buffalo. Pair bonding of trumpet manucodes helps the birds raise their young because, due to their particular food source, the female cannot rear as many offspring alone as she can with the male's help.

Group living also has its costs. When animals are crowded together into a small area, disputes can arise over access to the best feeding places and sleeping sites. A **dominance hierarchy** in which an animal with a higher rank receives benefits before a lower ranking animal is one way to keep order but it puts subordinates at a disadvantage. Among red deer, sons are preferred because if they become a harem master, sons result in a greater number of grandchildren. However, sons, being larger than daughters, need to be nursed more frequently and for a longer period of time. Subordinate females do not have access to enough food resources to adequately nurse sons, and therefore they tend to rear daughters, not sons. Still, like the subordinate males in a baboon troop, subordinate females in a red deer harem may be better off in terms of fitness if they stay with a group, despite the cost involved.

Living in close quarters also exposes individuals to illness and parasites that can easily pass from one animal to another. However, social behavior helps offset some of the proximity disadvantages. For example, baboons and other social primates invest much time in grooming one another, an activity that most likely helps them remain healthy. Likewise, humans have developed extensive medical care to help offset the health problems that arise from living in the densely populated cities of the world.

Altruism Altruism is a behavior that has the potential to decrease the fitness of an individual while benefiting the reproductive success of other members of the society. In order to determine whether altruism exists, we need to understand indirect selection. Genes are passed from one generation to the next in two quite different ways. The first way is direct: A parent passes a gene directly to an offspring. The second way is indirect: A relative that reproduces may pass the same gene to the next generation. *Direct*

Queen ant produces eggs.

Sterile nurses assist queen.

FIGURE 38.6A In an ant colony, only the queen reproduces. According to the idea of inclusive fitness, the nurses are being altruistic when they help the queen reproduce because they are closely related to the offspring.

selection is adaptive due to the reproductive success of an individual. *Indirect selection,* also called **kin selection,** is adaptive due to the reproductive success of the individual's relatives. Together, direct selection and indirect selection make up **inclusive fitness.** In other words, an individual's inclusive fitness includes its direct offspring and the offspring of its relatives. Let's look at examples of inclusive fitness among various societies.

In army ants, like many other insect societies, reproduction is often limited to only one pair, the queen and her mate, the only male in the colony. In addition, the society has three different sizes of sterile female workers. The smallest workers (3 mm), called the nurses, take care of the queen and larvae, feeding them and keeping them clean (Fig. 38.6A). The intermediate-sized workers, constituting most of the population, go out on raids to collect food. The soldiers (14 mm), with huge heads and powerful jaws, run along the sides and rear of raiding parties, protecting the column of ants from attack by intruders. Are the sterile workers being altruistic? Among social insects, the queen is diploid (2n), but her mate is haploid (n). If the queen has had only one mate, the sister workers are more closely related to each other. Sisters share, on average, 75% of their genes because they inherit 100% of their father's alleles. But their potential offspring would share, on average, only 50% of their genes. Therefore, a worker can achieve greater inclusive fitness by helping her mother (the queen) produce additional sisters than by directly reproducing. Under these circumstances, behavior that appears altruistic is more likely to evolve.

Kin selection can also occur among animals whose offspring receive only a half set of genes from both parents. Consider that your brother or sister shares 50% of your genes, your niece or nephew shares 25%, and so on. Therefore, the survival of two nieces (or nephews) is worth the survival of one sibling, assuming they both go on to reproduce.

Among chimpanzees in Africa, a female in estrus frequently copulates with several members of the same group, and the males make no attempt to interfere with each other's matings. How can they be acting in their own self-interest? Genetic relatedness appears to underlie their apparent altruism. Members of a group share more than 50% of their genes in common because members never leave the territory in which they are born.

Reciprocal Altruism In some bird species, offspring from a previous clutch of eggs may stay at the nest to help their parents rear the next batch of offspring. In a study of Florida scrub jays, the number of fledglings produced by an adult pair doubled when they had helpers. Mammalian offspring are also observed to help their parents. In Figure 38.6B, a meerkat is acting as a babysitter for its young sisters and brothers while their mother is away. Among jackals in Africa, solitary pairs managed to rear an average of 1.4 pups, whereas pairs with helpers reared 3.6 pups per breeding season.

Video Meerkat Warning Calls

What are the benefits of staying behind to help? First, a helper is contributing to the survival of its own kin. Therefore, the helper actually gains a fitness benefit. Second, a helper is more likely than a nonhelper to inherit a parental territory—including other helpers. Helping, then, involves making a minimal, short-term reproductive sacrifice in order to maximize future reproductive potential. Therefore, helpers at the nest are also practicing a form of **reciprocal altruism.** Reciprocal altruism also occurs in animals that are not necessarily closely related. In this case, an animal helps or cooperates with another animal with no immediate benefit. However, the animal that was helped will repay the debt at some later time. Reciprocal altruism usually occurs in groups of animals that are mutually dependent. Cheaters in reciprocal altruism are recognized and not reciprocated in future events. For example, reciprocal altruism occurs in vampire bats that live in the tropics. Bats returning to the roost after feeding share their blood meal with other bats in the roost. If a bat fails to share blood with one that previously shared blood with it, the cheater bat will be excluded from future blood sharing.

Vampire bat, *Desmodus rotundus*

Applications to Humans Sociobiologists interpret human behavior according to these same principles. Human infants are born helpless and have a much better chance of developing properly if both parents contribute to the effort. Perhaps this explains why the human female, unlike other primates, is continuously amenable to sexual intercourse. Under these circumstances, the male is more likely to remain and help care for the offspring. In any case, it has been suggested that parental love is selfish in that it promotes the likelihood that an individual's genes will be present in the next generation's gene pool.

FIGURE 38.6B A meerkat acting as a babysitter for younger siblings. Researchers point out that the helpful behavior of the older meerkat can lead to increased inclusive fitness.

Studies of other human cultures also lend themselves to sociobiological interpretations. Among primitive African tribes, one man often had several wives. This is advantageous not only to the male, but also to the females. In Africa, sources of protein are scarce, and this arrangement allows a female to nurse her baby longer ensuring it is well fed. By contrast, in remote rural regions of Nepal even today, brothers sometimes have the same wife. There, the environment is hostile, and two men are better able to provide the everyday necessities for one family. Since the men are brothers, they help each other look after common genes.

Some people object to interpreting human behavior based on evolutionary fitness, and they stress that modern human behavior need not be determined by relatedness, pointing out that people adopt and care for children who are not related to them.

The next part of the chapter reviews means of communication between animals living in social groups.

▶ **38.6 CHECK YOUR PROGRESS**
1. Describe with examples how altruistic behavior can be in the self-interest of the animal.

Animal Communication Suits the Environment

Animals that live in social groups communicate with one another by signaling. **Communication** is an action by a sender that may influence the behavior of a receiver. Types of communication include chemical, auditory, and tactile.

38.7 Communication with others involves the senses

LEARNING OUTCOME

When you complete this section, you should be able to

1. List various examples of communication and state possible advantages and disadvantages of each type.

Social behavior requires that animals communicate with one another. The communication can be purposeful, but it does not have to be. Bats send out a series of sound pulses and then listen for the corresponding echoes in order to find their way through dark caves and locate food at night. Some moths have the ability to hear these sound pulses, and they begin evasive tactics when they sense that a bat is near. Are the bats purposefully communicating with the moths? No; bat sounds are simply a cue to the moths that danger is near.

Chemical Communication Chemical signals have the advantage of being effective both night and day and from long distances, allowing protection to the communicator. The term **pheromone** designates chemical signals in low concentration that are passed between members of the same species (see the introduction to Chapter 35). Some animals are capable of secreting different pheromones, each with a different meaning. Female moths have special abdominal glands that secrete chemicals, which are detected downwind by receptors on male antennae. The antennae are especially sensitive, and this ensures that only male moths of the correct species (not predators) will be able to detect the pheromones.

> **Video**
> Bug Speak

Ants and termites mark their trails with pheromones. Cheetahs and other cats mark their territories by depositing urine, feces, and anal gland secretions at the boundaries (Fig. 38.7A). Klipspringers (small antelope) use secretions from a gland below the eye to mark twigs and grasses of their territory.

Auditory Communication Auditory (sound) communication has some advantages over other kinds of communication. It is faster than chemical communication, and it too is effective both night and day. Further, auditory communication can be modified not only by loudness but also by pattern, duration, and repetition. In an experiment with rats, a researcher discovered that an intruder can avoid attack by increasing the frequency with which it makes an appeasement sound.

Male crickets have calls, and male birds have songs for a number of different occasions. For example, birds may have one song for distress, another for courting, and still another for marking territory. Sailors have long heard the songs of humpback whales transmitted through the hull of a ship. Scientists have shown that the song has six basic themes, each with its own phrases, that can vary in length and be interspersed with sundry cries and chirps. The purpose of the song is probably sexual, serving to advertise

FIGURE 38.7A Chemical communication—a cheetah spraying urine to mark its territory.

the availability of the singer. Bottlenose dolphins have one of the most complex languages in the animal kingdom.

Language is the ultimate auditory communication. Only humans have the biological ability to produce a large number of different sounds and to put them together in many different ways. Nonhuman primates have different vocalizations, each with a definite meaning, such as when vervet monkeys give alarm calls (Fig. 38.7B). Although chimpanzees can be taught to use an artificial language, they never progress beyond the speech capability of a 2-year-old child. It has also been difficult to prove that chimps understand the concept of grammar or can use their language to reason. It still seems that humans possess a communication ability unparalleled by other animals.

> **Video**
> Meerkat Warning Calls

Visual Communication Visual signals are most often used by species that are active during the day. Contests between males make use of threat postures and possibly prevent outright fighting, a behavior that might result in reduced fitness. A male baboon displaying full threat is an awesome sight that establishes his dominance and keeps peace within the baboon troop. Hippopotamuses perform territorial displays that include mouth opening.

Many animals use complex courtship behaviors and displays. The plumage of a male Raggiana Bird of Paradise allows him to put on a spectacular courtship dance to attract a female, giving her a basis on which to select a mate. Territorial and courtship displays are exaggerated and always performed in the same way so that their meaning is clear.

Video Flirting Flies

FIGURE 38.7B **a.** Vervet monkeys, *Cercopithecus aethiops,* use auditory communication to alarm members of a troop. **b.** The sound patterns differ for an eagle (*left*) or a leopard (*right*).

FIGURE 38.7C Tactile communication: the waggle dance of bees.

Visual communication allows animals to signal others of their intentions without needing to provide any auditory or chemical messages. The body language of students during a lecture provides an example. Some students lean forward in their seats and make eye contact with the instructor. They want the instructor to know they are interested and find the material of value. Others lean back in their chairs and look at the floor or doodle. These students indicate they are not interested in the material. Teachers can use students' body language to determine whether they are effectively presenting the material and make changes accordingly.

Other human behaviors also send visual clues to others. The hairstyle and dress of a person or the way he or she walks and talks are ways to send messages to others. Some studies have suggested that women are more apt to dress in an appealing manner and be sexually inviting when they are ovulating. People who dress in black, move slowly, fail to make eye contact, and sit alone may be telling others that they do not want social contact. Psychologists have long tried to understand how visual clues can be used to better understand human emotions and behavior. Similarly, researchers are using body language in animals to suggest that they, as well as humans, have emotions (see "Do Animals Have Emotions?" on p. 756).

Tactile Communication Tactile communication occurs when one animal touches another. For example, laughing gull chicks peck at the parent's bill to induce the parent to feed them (see Fig. 38.2). A male leopard nuzzles the female's neck to calm her and to stimulate her willingness to mate. In primates, grooming—one animal cleaning the coat and skin of another—helps cement social bonds within a group.

Honeybees use a combination of communication methods, but especially tactile ones, to impart information about the environment. When a foraging bee returns to the hive, it performs a **waggle dance** that indicates the distance and the direction of a food source (Fig. 38.7C). As the bee moves between the two loops of a figure 8, it buzzes noisily and shakes its entire body in so-called waggles. Outside the hive, the dance is done on a horizontal surface, and the straight run indicates the direction of the food. Inside, it is dark, and other bees have to follow and touch the dancer in order to get the message. Inside, the angle of the straight run to that of the direction of gravity is the same as the angle of the food source with the sun. In other words, a 40° angle to the left of vertical means that food is 40° to the left of the sun.

Bees can use the sun as a compass to locate food because they have a biological clock, which allows them to compensate for the movement of the sun in the sky. A biological clock is an internal means of telling time. Today, we know that the "ticking" of the clock, in both insects and mammals (including humans), requires alterations in the expression of a gene called *period.*

▶ **38.7 CHECK YOUR PROGRESS**
  1. Determine by designing an experiment whether the singing of male birds to mark their territory increases fitness.

38B Do Animals Have Emotions?

Investigators are now interested in determining whether animals have emotions. When wolves reunite, they wag their tails to and fro, whine, and jump up and down; elephants vocalize—emit their "greeting rumble"—flap their ears, and spin about. Many young animals play with one another or even with themselves, as when dogs chase their own tails. On the other hand, upon the death of a friend or parent, chimps are apt to sulk, stop eating, and even die. Even people who rarely observe animals usually agree about what an animal must be feeling when it exhibits these behaviors (Fig. 38B).

In the past, scientists found it expedient to collect data only about observable behavior and to ignore the possible mental state of the animal. Why? Because emotions are personal, and no one can ever know exactly how another animal is feeling. B. F. Skinner, whose research method is described earlier in this chapter, regarded animals as robots that become conditioned to respond automatically to a particular stimulus. He and others never considered that animals might have feelings. But now, some scientists believe they have sufficient data to hypothesize that at least other vertebrates and/or mammals do have feelings, including fear, joy, embarrassment, jealousy, anger, love, sadness, and grief. And they believe that those who hypothesize otherwise should have to present the opposing data.

Perhaps it would be reasonable to consider the suggestion of Charles Darwin, the father of evolution, who said that animals are different in degree rather than in kind. This means that all animals can, say, feel love, but perhaps not to the degree that humans can. B. Würsig watched the courtship of two right whales. They touched, caressed, rolled side-by-side, and eventually swam off together. He wondered if their behavior indicated they felt love for one another. When you think about it, it is unlikely that emotions first appeared in humans with no evolutionary homologies in animals.

Iguanas, but not fish and frogs, tend to stay where it is warm. M. Cabanac has found that warmth makes iguanas experience a rise in body temperature and an increase in heart rate. These are biological responses associated with emotions in humans. Perhaps the ability of animals to feel pleasure and displeasure is a mental state that rises to the level of consciousness.

Neurobiological data support the hypothesis that other animals, aside from humans, are capable of enjoying themselves when they perform an activity such as play. Researchers have found a high level of dopamine in the brain when rats play, and the dopamine level increases even when rats anticipate the opportunity to play. Certainly even the staunchest critic is aware that many different species of animals have limbic systems and are capable of fight-or-flight responses to dangerous situations. Can we go further and suggest that animals feel fear even when no physiological response has yet occurred?

Laboratory animals may be too stressed to provide convincing data on emotions; we have to consider that emotions evolved under an animal's normal environmental conditions. This makes field research more useful. It is possible to fit animals with devices that

FIGURE 38B
Possible emotions in animals.

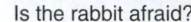
Is the rabbit afraid?   Is the young chimp comforted?

transmit information on heart rate, body temperature, and eye movements as they go about their daily routine. Such information will help researchers learn how animal emotions correlate with behavior, as they do in humans. One possible definition describes emotion as a psychological phenomenon that helps animals direct and manage their behavior.

M. Bekoff, who is prominent in the field of animal behavior, suggests:

> By remaining open to the idea that many animals have rich emotional lives, even if we are wrong in some cases, little truly is lost. By closing the door on the possibility that many animals have rich emotional lives, even if they are very different from our own or from those of animals with whom we are most familiar, we will lose great opportunities to learn about the lives of animals with whom we share this wondrous planet.[1]

[1]Bekoff, M. "Animal emotions: Exploring passionate natures," *Bioscience* 50:10, page 869 (October 2000).

CONSIDER THESE QUESTIONS

1. Does your personal experience with animals suggest they have emotions? Explain.
2. How might you behave differently toward animals, knowing that they may have emotions?
3. Would you be inclined to be a vegetarian if scientists conclude that animals do have emotions?

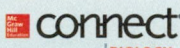 Explore the concepts through a variety of multimedia assets and question types.

www.mcgrawhillconnect.com

CONNECTING THE CONCEPTS

Birds build nests, dogs bury bones, cats chase quick-moving objects, and snakes bask in the sun. All animals, including humans, behave—they respond to stimuli, both from the physical environment and from other individuals of the same or different species. Experiments have shown that behaviors have a genetic basis, and yet behaviors can be modified by experience. Regardless, behaviorists use an evolutionary approach to generate hypotheses that can be tested to better understand how the behavior increases individual fitness (i.e., the capacity to produce fertile offspring more than other members of a population).

Behaviorists hypothesize that inheritance produces behavioral variations that are subject to natural selection, and therefore behavior will consist of adaptive responses to stimuli. Still, behaviorists concede that both nature and nurture determine behavior. Genetics determines, for example, that hawks hunt by using vision rather than smell, and that cats, not dogs, climb trees. Songbirds are born with the ability to sing, but which song or dialect they sing is strongly dependent on the songs they hear from their parents and siblings. A new chimpanzee mother naturally cares for her young, but she is a better mother if she has observed other females in her troop raising young.

Behaviorists can give us good examples that natural selection does affect behavior. Sexual selection, a form of natural selection, explains much about mating behavior. Males are apt to compete and females to choose because this type of behavior is most apt to lead to reproductive success. Also, there is survival value, after all, in the ability of male baboons to react aggressively when the troop is under attack. A cost-benefit analysis can help determine why a particular behavior that seemingly has a high cost (aggression can lead to death) actually has a benefit (increasing fitness) that outweighs the cost. A behavior such as territoriality or foraging has its trade-offs, but this behavior would not have evolved unless the benefit outweighed the cost. Why should older siblings help raise younger siblings? The evolutionary answer is that, through kin selection, older siblings are increasing their fitness. Otherwise, the behavior would not continue. Similarly, group living in which animals communicate must have benefits (e.g., availability of food) that outweigh the costs (e.g., infectious diseases).

The evolutionary approach to studying behavior has proved fruitful in helping us understand why birds sing their melodious songs, dolphins frolic in groups, and wondrous male birds-of-paradise display to females.

ANALYZE AND EVALUATE

1. See "The Theory of Natural Selection" on page 177 and give an example showing that behavior is subject to the process of natural selection.

2. Some people might object to applying the principles of evolution to behavior. Explain.

3. Do a cost-benefit analysis of some risky behavior. In evolutionary terms, what is the benefit?

SUMMARIZE

Inheritance and Learning Influence Behavior

38.1 Inheritance influences behavior

- **Behavior** is any action that can be observed and described.
- Lovebird experiments, garter snake experiments, and studies of human twins suggest that behavior has a genetic basis (nature).
- The gene that codes for a hormone responsible for *Aplysia* egg-laying behavior and a gene responsible for maternal behavior in mice have been discovered. These experiments show that behavior has a genetic basis.

38.2 Learning can also influence behavior

- Originally behaviorists believed that many behaviors were an inherited **fixed action pattern** (**FAP**) that could not be modified. Now we know that **learning** (nurture) can often cause a durable change in behavior. For example, in laughing gull chicks, visual experience and improvement in motor skills affect the development of begging behavior.

- **Imprinting** is a form of learning that allows baby birds to follow the first moving object they see during a sensitive period.

- Social interactions assist song learning in birds.

38.3 Associative learning links behavior to stimuli

- **Associative learning** is a change in behavior that involves an association between two events. In **classical conditioning** the presentation of two types of stimuli that occur at the same time (e.g., Pavlov's dogs) cause the animal to form an association between them. In **operant conditioning,** positive reinforcement is provided for desired behavior.

- Other means of learning include **insight learning,** which appears to have occurred when an animal solves a problem without prior experience and **habituation,** which has taken place when an animal no longer responds to a repeated stimulus.

Behaviors That Increase Fitness

38.4 Territoriality increases fitness
- **Fitness** is the lifetime reproductive success of an individual.
- **Territoriality,** which includes the behaviors needed to defend a **territory,** gives animals a secure environment in which to mate, raise young, and find food. Animals forage for food in a territory. The **optimal foraging model** states that it is adaptive for foraging behavior to be as energetically efficient as possible.

38.5 Sexual selection increases fitness
- Sexual selection is a form of natural selection that selects for traits that increase the chance of mating. Females produce few eggs and are expected to be selective about their mates. Males produce many sperm and are expected to compete to inseminate females.

38.6 Societies increase fitness
- **Sociobiology** applies the principles of evolutionary biology to the study of social behavior. Only if benefits (in terms of individual reproductive success) outweigh costs will societies come into existence.
- Group living has both advantages and disadvantages. **Dominance hierarchies** help organize interactions.
- **Altruism** is a behavior that benefits the reproductive success of another individual, but decreases the reproductive success of the altruist. Direct selection occurs when you are selected to reproduce; indirect selection, or **kin selection,** occurs when your kin are selected to reproduce. When animals act altruistically, they may be increasing their own **inclusive fitness.** **Reciprocal altruism** occurs when an animal helps another for its own future benefit.

Animal Communication Suits the Environment

38.7 Communication with others involves the senses
- **Communication** is a signal by a sender that affects the behavior of a receiver. Communication includes:
  1. Chemical communication involves **pheromones,** which are signals passed between members of a species.
  2. Auditory (sound) communication, like chemical communication, is possible both day and night.
  3. Language involves the spoken word, and visual communication includes posturing, body language, and eye contact.
  4. Tactile communication occurs via touch. Tactile communication occurs when bees do a **waggle dance** that tells the distance and direction of a food source.

TEST YOURSELF

Inheritance and Learning Influence Behavior

1. Behavior is
   a. any action that is learned.
   b. all responses to the environment.
   c. any action that can be observed and described.
   d. all activity that is controlled by hormones.
   e. unique to birds and humans.
2. Egg-laying hormone causes snails to lay eggs, implicating which system in this behavior?
   a. digestive   d. lymphatic
   b. endocrine   e. respiratory
   c. nervous

3. Which of the following is not an example of a genetically based behavior?
   a. Inland garter snakes do not eat slugs, while coastal populations do.
   b. One species of lovebird carries nesting strips one at a time, while another carries several.
   c. Human twins have similar activity patterns.
   d. Snails lay eggs in response to egg-laying hormone.
   e. Wild foxes raised in captivity are not capable of hunting for food.
4. How would the following graph differ if pecking behavior in laughing gulls were a fixed action pattern not influenced by learning?
   a. It would be a diagonal line with an upward incline.
   b. It would be a diagonal line with a downward incline.
   c. It would be a horizontal line.
   d. It would be a vertical line.
   e. None of these are correct.

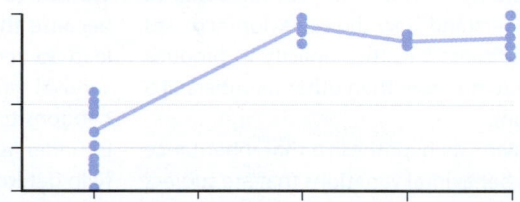

5. The benefits of imprinting (following and obeying the mother) generally outweigh the costs (following the wrong object instead of the mother) because
   a. an animal that has been imprinted on the wrong object can be reimprinted on the mother.
   b. imprinting behavior never lasts more than a few months.
   c. animals in the wild rarely imprint on anything other than their mother.
   d. animals that imprint on the wrong object generally die before they pass their genes on.
6. In white-crowned sparrows, social experience exhibits a very strong influence over the development of singing patterns. What observation led to this conclusion?
   a. Birds learned to sing only when they were trained by other birds.
   b. The window during which birds learn from other birds is wider than that during which they learn from tape recordings.
   c. Birds could learn different dialects only from other birds.
   d. Birds that learned to sing from a tape recorder could change their song when they listened to another bird.
7. Label this diagram:

8. Which of the following best describes classical conditioning?
   a. gradual strengthening of stimulus-response connections that are seemingly unrelated
   b. type of associative learning in which there is no contingency between response and reinforcer
   c. learning behavior in which an organism follows the first moving object it encounters
   d. learning behavior in which an organism exhibits a fixed action pattern from the time of birth
9. Using treats to train a dog to do a trick is an example of
   a. imprinting.
   b. tutoring.
   c. vocalization.
   d. operant conditioning.
10. Why does operant conditioning work? An animal
   a. is punished when it does not perform the behavior.
   b. associates the ringing of a bell with the behavior.
   c. is rewarded when it performs the behavior.
   d. is imprinted on the first moving object it sees.
11. **THINKING CONCEPTUALLY** How would you know that the laughing gull chicks in Figure 38.2 did exhibit learning but not a type of associative learning (see section 38.3)?

Behaviors That Increase Fitness

12. Male animals that protect a territory
   a. require more food and must travel longer distances.
   b. require more food in order to care for the young.
   c. are larger than females and prone to injury.
   d. must care for the young and travel longer distances.
13. This graph shows that shore crabs choose

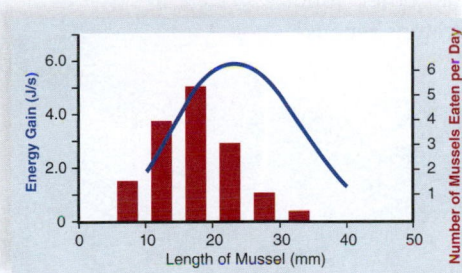

   a. the size mussel that provides the highest rate of net energy return.
   b. the largest mussels because they contain the most food.
   c. the smallest mussels because they are easier to catch.
14. **THINKING CONCEPTUALLY** Males typically compete for mates, and females typically choose between males for mates. Explain how these behaviors may have evolved.
15. Which of the following is not a benefit of group living?
   a. increased availability of food
   b. protection from predators
   c. increased success in rearing offspring
   d. protection from illness and parasites
16. Which of the following does not contribute to inclusive fitness?
   a. direct selection
   b. indirect selection
   c. kin selection
   d. Both b and c only are correct.
   e. All of these are correct.

17. On the average, how many genes do first cousins have in common?
   a. ⅛
   b. ½
   c. ¼
   d. ¹⁄₁₆
18. **THINKING CONCEPTUALLY** Some birds do not produce offspring for the first couple of years. Why would it be beneficial for them to help their parents reproduce or practice reciprocal altruism?

Animal Communication Suits the Environment

For questions 19–22, match the type of communication in the key with its description. Answers may be used more than once.

KEY:
   a. chemical communication
   b. auditory communication
   c. visual communication
   d. tactile communication

19. Aphids (insects) release an alarm pheromone when they sense they are in danger.
20. Male peacocks exhibit an elaborate display of feathers to attract females.
21. Ground squirrels give an alarm call to warn others of the approach of a predator.
22. Laughing gull chicks peck at a parent's bill to stimulate feeding.

GET INVOLVED

1. You have observed that New York rats keep their distance from Limburger cheese, which has a disagreeable odor, and do not approach to eat it, while Florida rats readily approach and eat the cheese. Design an experiment to test whether this behavior is genetically controlled.
2. Among meerkats, sentries stand guard on rocks and serve as lookouts while others feed. How would you decide whether sentry behavior is altruistic?
3. After viewing this virtual lab, come to a conclusion about the type of food preferred by mealworms. What particular experiments substantiate your conclusion?

Virtual Lab
Mealworm Behavior

MEDIA STUDY TOOLS

mhhe.com/maderconcepts3

Enhance your study of this chapter with interactive study tools, practice tests, and engaging animations. Also, ask your instructor about the resources available through ConnectPlus, which includes LearnSmart, a personalized adaptive learning program, and a media-rich eBook.

39

Community and Ecosystem Ecology

BEFORE YOU BEGIN
Take a few minutes to recall

"The Sun Drives Photosynthesis" (p. 102)
Energy flows and chemicals cycle in ecosystems (section 1.6)
Energy laws govern conversions of energy (section 5.5)
Significant interactions between populations (section 37.6)

Ridding the Land of Waste

If it weren't for scavengers and decomposers, dead organic matter would accumulate in deep piles so that you wouldn't even be able to step outside your door! But thanks to decomposition, that doesn't happen. As soon as an animal dies, all of the microorganisms in its gut set to work, and decomposition begins. The faint smell of decay very quickly attracts scavengers, which seek out dead animals and eat them.

One well-known scavenger is the vulture, a bird equipped with an excellent sense of smell that helps it locate an animal carcass from miles away. Vultures attract a lot of attention because they are rather large, usually black birds with bald heads devoid of feathers. In some rural mountainous areas of the world, humans rely on vultures for assistance with "sky burials," in which a human corpse is cut into small pieces and left for the scavenging birds to eat. In Tibet, this practice is necessary because the frozen ground is too hard to bury the dead, and wood is too scarce for cremation. This ritual is also based on an appreciation for nutrient recycling; the deceased is providing nutrients that ultimately sustain all organisms.

Besides vultures, other scavengers include beetles, earthworms, and some insects. We can't include carrion beetles and flies because they are attracted to decomposing animals not as a source of food but as a nutrient-rich place to lay their eggs. Dung beetles are not scavengers either, but they are of interest because they eat and raise their young in animal feces. They have no need to eat other foods or drink water, because

Vulture

Vultures assist in a sky burial.

feces provide them with complete nutrition. By helping recycle fecal nutrients back into the land and by improving the hygiene of livestock pastures, dung beetles provide $380 million worth of services to farmers in the United States every year.

Decomposers are not scavengers because they do not take into their bodies pieces of dead animals or plants, as scavengers do. Instead, they are saprotrophs that secrete their digestive juices into the environment and then absorb the nutrients. The most abundant and widespread saprotrophs in the world are microbes, predominantly bacteria and fungi. In order to compete with scavengers for the nutrients in decaying matter, microbes often secrete chemicals that are foul-smelling, distasteful, and/or toxic. According to biologist Dan Janzen, this is why fruits rot, seeds mold, and meat spoils. If you have ever avoided eating a moldy strawberry or drinking curdled milk, the microbes have won.

Decomposers perform a service for the entire biosphere when they recycle nutrients to plants, which produce food for themselves directly and for all other organisms indirectly. Also, microbial decomposers can anaerobically convert wood chips, animal (e.g., cow) feces, and all kinds of organic matter into biogas, which can be captured and used for heating, cooking, and generating electricity. In Rwanda, there are "poop-powered" prisons in which prisoners' feces are digested into biogas, and in San Francisco, biogas may soon be made from the huge quantity of pet feces produced within the city. Then, too, bacteria can clean up oil spills. Bioremediation scientists have shown that often the best

results can be achieved by letting bacteria decompose an oil spill. In the long run, bacteria do the best job in the shortest length of time.

Our society could also tap into the leftover high-energy compounds produced by decomposers. Your house could be heated, your food could be cooked, or your electronics could be powered using methane (biogas) obtained from decomposing matter!

In this chapter, we will look at how the populations in a community interact. Then we will consider how populations interact not only among themselves, but also with the physical environment in ecosystems.

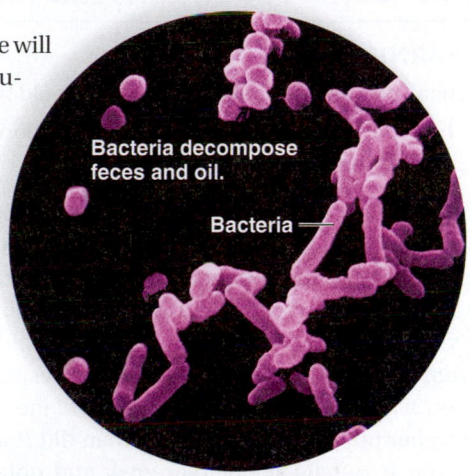
Bacteria decompose feces and oil.
Bacteria

Bacteria convert organic matter, including feces, to biogas in buildings called biodigesters.

Dung beetles rolling dung

Dung beetles assist in dung cleanup.

Before   After
Bacteria assist in oil cleanup.

Community Interactions

Step into a forest and look around. Most likely, you will notice populations of animals interacting with plants and other animals. Insects will be feeding on the leaves of trees, or visiting flowers or even grasses. If you are lucky, you might see a hawk dart from a tree on the edge of the forest and grab a rodent running through an adjoining meadow. All the populations interacting in the forest form a *community*. The populations interact particularly through competition, predation, and various symbiotic relationships (parasitism, commensalism, and mutualism).

39.1 Competition can lead to resource partitioning

LEARNING OUTCOME

When you complete this section, you should be able to

1. Understand and explain how competition leads to resource partitioning.

Competition is rivalry between populations for the same resources, such as light, space, nutrients, or mates. In the 1930s, G. F. Gause grew two species of *Paramecium* in one test tube containing a fixed amount of bacterial food. Although populations of each species survived when grown in separate test tubes, only one species, *Paramecium aurelia,* survived when the two species were grown together (Fig. 39.1A). *P. aurelia* acquired more of the food resource and had a higher population growth rate than did *P. caudatum.* Eventually, as the *P. aurelia* population grew and obtained an increasingly greater proportion of the food, the number of *P. caudatum* individuals decreased, and the population died out. This experiment and others helped ecologists formulate the **competitive exclusion principle,** which states that no two species can occupy the same niche at the same time. The **ecological niche** of an organism is the role it plays in its community, including its **habitat** (where the organism lives) and its interactions with other organisms and the environment. See Figure 39.7 for a more complete description of niche.

 **Video** Cichlid Specialization

Competition for resources does not always lead to localized extinction of a species. Multiple species can coexist in communities by partitioning resources. In another laboratory experiment using other species of *Paramecium,* Gause found that two species could survive in the same test tube if one species consumed bacteria at the bottom of the tube and the other ate bacteria suspended in solution. This **resource partitioning** decreased competition between the two species. What could have been one niche became two more-specialized niches due to species differences in feeding behavior.

When small and medium ground finches live on the same island of the Galápagos, their beak sizes differ because they feed on different-sized seed (Fig. 39.1B). When these ground finches live on different islands, the beak size does not differ and they feed on the same preferred range of seeds. Such so-called **character displacement** is often viewed as evidence that competition and resource partitioning do take place when closely related species share the same resource, such as seeds.

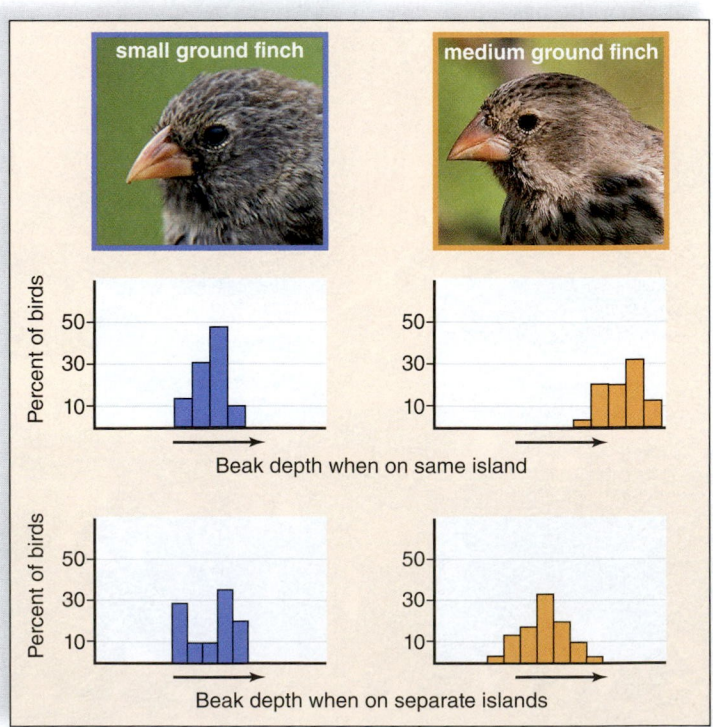 **Video** Finches Adaptive Radiation

The niche specialization that permits coexistence of multiple species can be very subtle. Species of warblers that live in North American forests are all about the same size, and all feed on budworms, a type of caterpillar found on spruce trees. Robert

FIGURE 39.1A Competition only occurs between two species of *Paramecium* when they are grown together.

Virtual Lab Population Biology

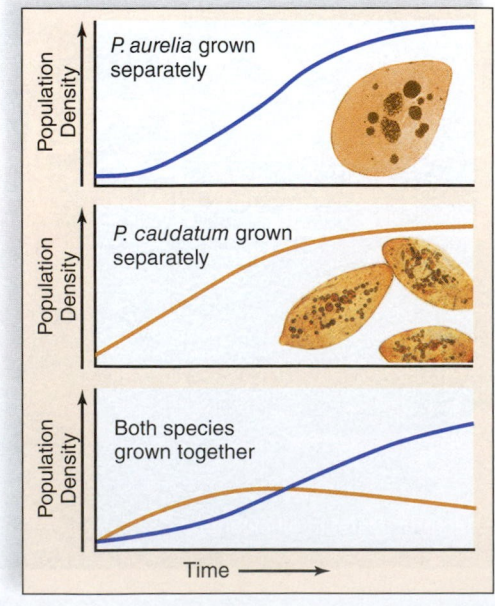

FIGURE 39.1B Character displacement occurs in finches when they coexist. Notice the similarity in beak depth when they are on separate islands.

MacArthur recorded the length of time each warbler species spent in different regions of spruce canopies to determine where each species did most of its feeding. He discovered that each species primarily used different parts of the tree canopy and, in that way, had a more specialized niche (Fig. 39.1C). As another example, consider that three types of birds—swallows, swifts, and martins—all eat flying insects and parachuting spiders. These birds even frequently fly in mixed flocks. But each type of bird has a different nesting site and migrates at a slightly different time of year.

In all these cases of niche specialization, we have merely assumed that what we observe today is due to competition in the past. Some ecologists are fond of saying that in doing so we have invoked the "ghosts of competition past." Are there any instances in which competition has actually been observed? Joseph Connell has studied the distribution of barnacles on the Scottish coast, where a small barnacle (*Chthamalus stellatus*) lives on the high part of the intertidal zone, and a large barnacle (*Balanus balanoides*) lives on the lower part (Fig. 39.1D). Why should that be, when their free-swimming larvae attach themselves to rocks at any point in the intertidal zone? The answer is that the faster-growing *Balanus* crowds out *Chthamalus* in the lower tidal zone, but cannot do so in the upper tidal zone because it is more susceptible to drying out than *Chthamalus*.

▶ **39.1 CHECK YOUR PROGRESS**

1. Identify the particular type of interaction and discuss the appropriate principle when warblers forage in different parts of a spruce tree.

Cape May warbler

Black-throated green warbler

Bay-breasted warbler

Blackburnian warbler

Yellow-rumped warbler

FIGURE 39.1C Niche specialization occurs among five species of coexisting warblers.

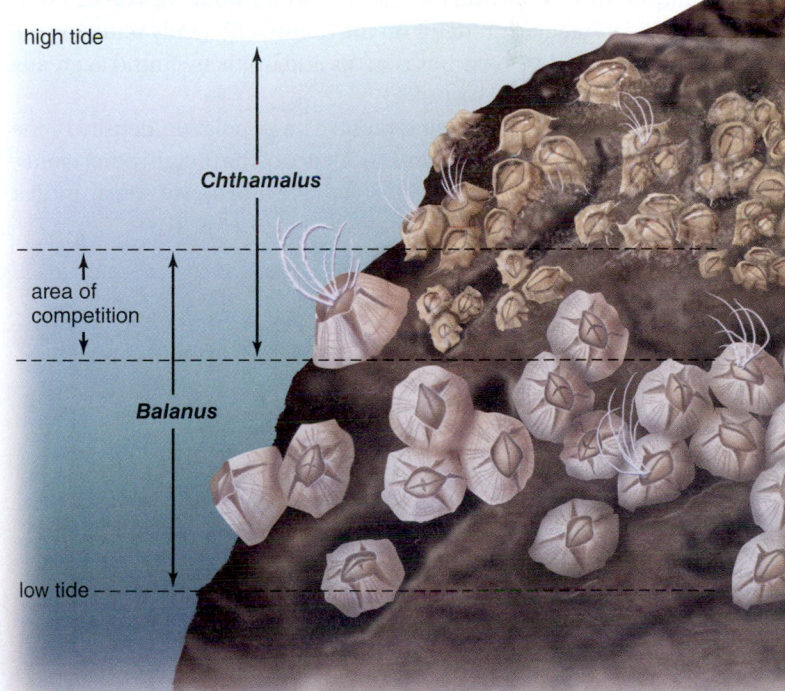

FIGURE 39.1D Competition occurs between two species of barnacles.

LEARNING OUTCOMES

When you complete this section, you should be able to

1. Differentiate predation from other interactions and discuss predator-prey population size dynamics.
2. List and discuss various types of prey defenses and give examples of two types of mimicry.

Predation occurs when one organism, called the predator, feeds on another, called the prey. In its broadest sense, predaceous consumers include not only animals such as lions, which kill zebras, but also filter-feeding blue whales, which strain krill from ocean waters; herbivorous deer, which browse on trees and shrubs; parasitic ticks, which suck blood from their victims; and *parasitoids,* which are wasps that lay their eggs inside the body of a host. The resulting larvae feed on the host, sometimes causing death. Parasitism can be considered a type of predation because one individual obtains nutrients from another (see section 39.3). Predation and parasitism are expected to increase the abundance of the predator and parasite at the expense of the abundance of the prey or host.

Video Harris Hawks

Predator-Prey Population Dynamics Do predators reduce the population density of prey? In another classic experiment, G. F. Gause reared the ciliated protozoans *Paramecium caudatum* (prey) and *Didinium nasutum* (predator) together in a culture medium. He observed that *Didinium* ate all the *Paramecium* and then died of starvation. In nature, we can find a similar example. When a gardener brought prickly-pear cactus to Australia from South America, the cactus spread out of control until millions of acres were covered with nothing but cactus. The cactuses were brought under control when a moth from South America, whose caterpillar feeds only on the cactus, was introduced. The caterpillar was a voracious predator on the cactus, efficiently reducing the cactus population. Now both cactus and moth are found at greatly reduced densities in Australia.

This raises an interesting point: The population density of the predator can be affected by the prevalence of the prey. In other words, the predator-prey relationship is actually a two-way street. In that context, consider that at first the biotic potential (maximum reproductive rate) of the prickly-pear cactus was maximized, but factors that oppose biotic potential came into play after the moth was introduced. And the biotic potential of the moth was maximized when it was first introduced, but the carrying capacity decreased after its food supply was diminished.

Sometimes, instead of remaining in a steady state, predator and prey populations first increase in size and then decrease. We can appreciate that an increase in predator population size is dependent on an increase in prey population size. But what causes a decrease in population size instead of the establishment of a steady population size? At least two possibilities account for the reduction: (1) Perhaps the biotic potential (reproductive rate) of the predator is so great that its increased numbers overconsume the prey, and then as the prey population declines, so does the predator population. (2) Perhaps the biotic potential of the predator is unable to keep pace with the prey, and the prey population overshoots the carrying capacity and suffers a crash. Now the predator population follows suit because of a lack of food. In either case, the result will be a series of peaks and valleys, with the predator population size lagging slightly behind that of the prey population.

A famous example of predator-prey cycles occurs between the snowshoe hare and the Canadian lynx, a type of small predatory cat (Fig. 39.2A). The snowshoe hare is a common herbivore in the coniferous forests of North America, where it feeds on terminal twigs of various shrubs and small trees. The Canadian lynx feeds on snowshoe hares but also on ruffed grouse and spruce grouse, two types of birds. Studies have revealed that the hare and lynx populations cycle regularly, as graphed in Figure 39.2A. Investigators at first assumed that the lynx brings about a decline in the hare population and that this accounts for the cycling. But others have noted that the decline in snowshoe hare abundance was accompanied by low growth and reproductive rates, which could be signs of a food shortage. Experiments were done to test whether factor 1, predation, or factor 2, lack of food, caused the decline in the hare population. The results suggest that both factors combined to produce a low hare population and the cycling effect.

FIGURE 39.2A

Predator-prey interaction between a snowshoe hare and a lynx.

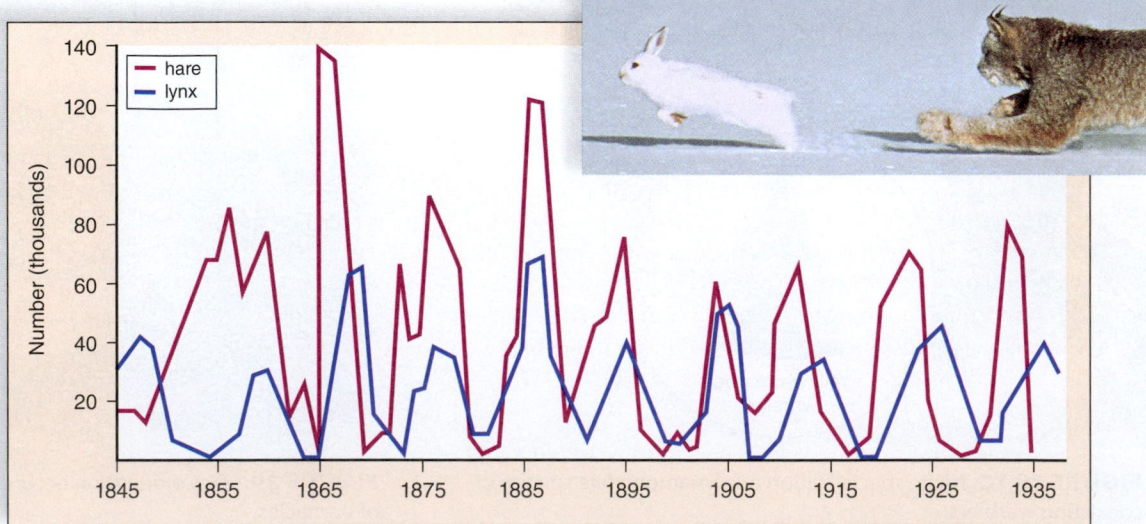

Camouflage

Warning coloration

eye

false head

Fright

FIGURE 39.2B Antipredator defenses.

Prey Defenses Prey defenses are mechanisms that thwart the possibility of being eaten by a predator. Prey species have evolved a variety of mechanisms that enable them to avoid predators, including heightened senses, speed, protective armor, protective spines or thorns, tails and appendages that break off, and chemical defenses. Consider, for the moment, only animals. There are a number of ways that a predator deceives its prey and that prey avoids capture by predators. One common strategy is **camouflage,** or the ability to blend into the background. Some animals have cryptic coloration that allows them to blend into their surroundings. For example, flounders can take on the same coloration as their background (Fig. 39.2B, *left*). Many examples of protective camouflage are known: Walking sticks look like twigs; katydids look like sprouting green leaves; some caterpillars resemble bird droppings; and some insects and moths blend into the bark of trees.

Video
Flounder
Camouflage

Video
Leaf-tailed
Gecko

Another common antipredator defense among animals is *warning coloration,* which tells the predator that the prey is potentially dangerous. As a warning to possible predators, poison-dart frogs are brightly colored (Fig. 39.2B, *middle*). Also, many animals, including caterpillars, moths, and fishes, possess false eyespots that confuse or startle another animal. Other animals have elaborate anatomic structures that cause *fright.* The South American lantern fly has a large false head with false eyes, making it resemble the head of an alligator (Fig. 39.2B, *right*). However, warning coloration and a fright defense are not always false. A porcupine certainly looks formidable, and for good reason. Its arrowlike quills have barbs that dig into the predator's flesh and penetrate even deeper as the enemy struggles after being impaled. In the meantime, the porcupine runs away.

Association with other prey is another common strategy that may help avoid capture. Flocks of birds, schools of fish, and herds of mammals stick together as protection against predators. Baboons that detect predators visually, and antelopes that detect predators by smell, sometimes forage together, gaining double protection against stealthy predators. The gazellelike springboks of southern Africa jump stiff-legged 2–4 m into the air when alarmed. Such a jumble of shapes and motions might confuse an attacking lion, allowing the herd to escape.

Mimicry Mimicry occurs when one species resembles another that possesses an overt antipredator defense. A mimic that lacks the defense of the organism it resembles is called a Batesian mimic (named for Henry Bates, who discovered the phenomenon). Once

an animal experiences the defense of the model, it remembers the coloration and avoids all animals that look similar. Figure 39.2C, *top row,* shows two insects (flower fly and longhorn beetle) that resemble a yellow jacket wasp but lack the wasp's ability to sting. Classic examples of Batesian mimicry include the scarlet kingsnake mimicking the venomous coral snake and the viceroy butterfly mimicking the foul-tasting monarch butterfly.

There are also examples of species that have the same defense and resemble each other. Many stinging insects—bees, wasps, hornets, and bumblebees—have the familiar black and yellow bands. Once a predator has been stung by a black and yellow insect, it is wary of that color pattern in the future. Mimics that share the same protective defense are called Müllerian mimics, after Fritz Müller, who suggested that this, too, is a form of mimicry. In Figure 39.2C, *bottom row,* the bumblebee is a Müllerian mimic of the yellow jacket wasp because both of them can sting.

▶ **39.2 CHECK YOUR PROGRESS**
1. Describe two factors that cause predator and prey populations to cycle in a predictable manner.

Flower fly

Longhorn beetle

Bumblebee

Yellow jacket

FIGURE 39.2C Mimicry: All of these insects have the same coloration.

39.3 Parasitism benefits one population at another's expense

LEARNING OUTCOME

When you complete this section, you should be able to

1. Explain with examples how parasitism is similar to predation.

Parasitism is similar to predation in that an organism, called the **parasite,** derives nourishment from another, called the **host.** Parasitism is one type of **symbiosis,** an association in which at least one of the species is dependent on the other (Table 39.3).

Viruses, such as HIV, that reproduce inside human lymphocytes are always parasitic, and parasites occur in all of the kingdoms of life as well. Bacteria (e.g., strep infection), protists (e.g., malaria), fungi (e.g., rusts and smuts), plants (e.g., mistletoe), and animals (e.g., tapeworms and fleas) all have parasitic members. While small parasites can be endoparasites (pinworms), larger ones are more likely to be ectoparasites (leeches), which remain attached to the exterior of the body by means of specialized organs and appendages. The effects of parasites on the health of the host can range from slightly weakening them to actually killing them over time. When host populations are at a high density, parasites readily spread from one host to the next, causing intense infestations and a subsequent decline in host density. Parasites that do not kill their host can still play a role in reducing the host's population density because an infected host is less fertile and becomes more susceptible to another cause of death.

In addition to nourishment, host organisms also provide their parasites with a place to live and reproduce, as well as a mechanism for dispersing offspring to new hosts. Many parasites have both a primary and a secondary host. The secondary host may be a vector that transmits the parasite to the next primary host. Usually both hosts are required in order to complete the life cycle. The association between parasite and host is so intimate that parasites are often specific and even require certain species as hosts.

TABLE 39.3	Symbiotic Relationships
Interaction	**Expected Outcome**
Parasitism	Abundance of parasite increases, and abundance of host decreases.
Commensalism	Abundance of one species increases, and the other is not affected.
Mutualism	Abundance of both species increases.

▶ **39.3 CHECK YOUR PROGRESS**
1. Explain why decomposers are not parasites.

39.4 Commensalism benefits only one population

LEARNING OUTCOME

When you complete this section, you should be able to

1. Contrast commensalism with other symbiotic relationships including parasitism.

Commensalism is a symbiotic relationship between two species in which one species is benefited and the other is neither benefited nor harmed.

Instances are known in which one species provides a home and/or transportation for the other species. For example, barnacles that attach themselves to the backs of whales and the shells of horseshoe crabs get both a home and transportation. It is possible, though, that the movement of the host is impeded by the presence of the attached animals, and therefore some scientists are reluctant to call these examples of commensalism.

Epiphytes, such as Spanish moss and some species of orchids and ferns, grow in the branches of trees, where they receive light, but they take no nourishment from the trees. Instead, their roots obtain nutrients and water from the air. Clownfishes live within the waving tentacles of sea anemones (Fig. 39.4). Because most fishes avoid the anemones' poisonous tentacles, clownfishes are protected from predators. Perhaps this relationship borders on mutualism, because the clownfishes may actually attract other fishes on which the anemone can feed.

Commensalism often turns out, on closer examination, to be an instance of either mutualism or parasitism. Cattle egrets are so-named because these birds stay near cattle, which flush out their prey—insects and other animals—from vegetation. The relationship becomes mutualistic when egrets remove ectoparasites from

FIGURE 39.4 A clownfish living among a sea anemone's tentacles.

the cattle. Remoras are fishes that attach themselves to the bellies of sharks by means of a modified dorsal fin acting as a suction cup. Remoras benefit by getting a free ride, and they also feed on a shark's leftovers. However, the shark benefits when remoras remove its ectoparasites. To some, it seems like wasted effort to try to classify symbiotic relationships into the three categories of parasitism, commensalism, and mutualism. Often, the amount of harm or good two species seem to do to one another depends on what the investigator chooses to measure.

▶ **39.4 CHECK YOUR PROGRESS**
1. Describe how you could determine that remoras are commensalistic, not parasitic, on a shark.

39A Coevolution Between Parasite and Host

Organisms in symbiotic associations, which include parasitism, commensalism, and mutualism (see Table 39.3), are especially prone to the process of coevolution. Coevolution is present when two species adapt in response to selective pressure imposed by the other. For example, the speed of a cheetah is probably selective for gazelles that are able to run away or jump high in the air to escape capture. The introduction to chapter 14, page 254, discussed how plants and their pollinators are adapted to one another.

The process of coevolution has been studied in the cuckoo, a social parasite that reproduces at the expense of other birds. A cuckoo lays its eggs in another bird species' nest, and when the cuckoo egg hatches, that species ends up caring for the cuckoo. It is odd to see a small bird feeding a cuckoo nestling several times its size. How did this strange behavior develop? Investigators discovered that in order to "trick" a host bird, the adult cuckoo must (1) lay an egg that mimics the host's egg, (2) lay its egg very rapidly (only 10 seconds are required) in the afternoon while the host is away from the nest, and (3) leave most of the eggs in the nest because hosts will desert a nest that has only one egg in it. (The cuckoo chick hatches first and is adapted to removing any other eggs in the nest [Fig. 39A].) It seems that the host birds may very well next evolve a way to distinguish the cuckoo from their own young.

Coevolution can take many forms. In the case of *Plasmodium,* a cause of malaria, the sexual portion of the life cycle occurs within mosquitoes (the vector), and the asexual portion occurs in humans. The human immune system uses surface proteins to detect pathogens, and *Plasmodium* has numerous genes for surface proteins. But it is capable of changing its surface proteins repeatedly, and in this way it stays one step ahead of the host's immune system. HIV has a similar capability, which has added to the difficulty of producing an AIDS vaccine.

The relationship between parasite and host can include the ability of parasites to seemingly manipulate the behavior of their hosts in self-serving ways. Ants infected with the lance fluke (but not those uninfected) mysteriously cling to blades of grass with their mouthparts. There, the infested ants are eaten by grazing sheep, and the flukes are transmitted to the next host in their life cycle. Similarly, snails of the genus *Succinea* are parasitized by worms of the genus *Leucochloridium.* As the worms mature, they invade the snail's eyestalks, making them resemble edible caterpillars. Birds eat the snails, and now the parasites release their eggs, and the embryos complete development inside the urinary tracts of the birds.

It used to be thought that as host and parasite coevolved, each would become more tolerant of the other because, if the opposite occurred, the parasite would soon run out of hosts. Parasites could first become commensal, or harmless to the host. Then, given enough time, the parasite and host might even become mutualists. In fact, the evolution of the eukaryotic cell by endosymbiosis is predicated on the supposition that bacteria took up residence inside a larger cell, and then the parasite and cell became mutualists.

However, this argument is too complex for some; after all, no organism is capable of "looking ahead" at its evolutionary fate. Rather, if an aggressive parasite could transmit more of itself in less time than a benign one, aggressiveness would be favored by natural selection. But other factors, such as the life cycle of the host, can determine whether aggressiveness is beneficial or not. For example, a benign parasite of newts will do better than an aggressive one. Why? Because newts take up solitary residence outside ponds in the forest for 6 years, and parasites have to wait that long before they are likely to meet up with another possible host.

CONSIDER THESE QUESTIONS

1. How is a predator-prey relationship similar to and different from a parasite-host relationship?
2. Compare the HIV virus to the H1N1 virus in order to show that the HIV virus is a highly successful parasite.
3. Coevolution between a pollinator and a flower and between a parasite and its host differs widely. How so?

 **connect** | BIOLOGY Explore the concepts through a variety of multimedia assets and question types.

www.mcgrawhillconnect.com

Cuckoo pushes out host's eggs after hatching first

FIGURE 39A Social parasitism in the cuckoo.

39.5 Mutualism benefits both populations

LEARNING OUTCOME

When you complete this section, you should be able to

1. Understand that mutualistic relationships are beneficial to the community as well as to the two species directly involved.

Mutualism is a symbiotic relationship in which both members benefit. As with other symbiotic relationships, it is possible to find numerous examples among all organisms. Bacteria that reside in the human intestinal tract acquire food, but they also provide us with vitamins, molecules we are unable to synthesize for ourselves. Termites would not be able to digest wood if not for the protozoans that inhabit their intestinal tracts and digest cellulose. Mycorrhizae are mutualistic associations between the roots of plants and fungal hyphae. The hyphae improve the uptake of nutrients for the plant, protect the plant's roots against pathogens, and produce plant growth hormones. In return, the plant provides the fungus with carbohydrates. Some sea anemones make their home on the backs of crabs. The crab uses the stinging tentacles of the sea anemone to gather food and to protect itself; the sea anemone gets a free ride that allows it greater access to food than other anemones. Lichens can grow on rocks because their fungal member conserves water and leaches minerals that are provided to the algal partner, which photosynthesizes and provides organic food for both populations. However, it's been suggested that the fungus is parasitic, at least to a degree, on the algae.

In tropical America, the bullhorn acacia tree is adapted to provide a home for ants of the species *Pseudomyrmex ferruginea* (see section 24.11, p. 489). Unlike other acacias, this species has swollen thorns with a hollow interior, where ant larvae can grow and develop. In addition to housing the ants, acacias provide them with food. The ants feed from nectaries at the base of the leaves and eat fat- and protein-containing nodules called Beltian bodies, found at the tips of the leaves. The ants constantly protect the plant from herbivores and other plants that might shade it because, unlike other ants, they are active 24 hours a day.

Video Thorn Tree Ants

The relationship between plants and their pollinators, mentioned previously, is a good example of mutualism. Perhaps the relationship began when herbivores feasted on pollen. The provision of nectar by the plant may have spared the pollen and, at the same time, allowed the animal to become an agent of pollination. By now, pollinator mouthparts are adapted to gathering the nectar of a particular plant species, and this species is dependent on the pollinator for dispersing pollen. The mutualistic relationships between flowers and their pollinators are examples of coevolution (see "Evolution of Insect Pollination," p. 499). As discussed, a striking example is the flower of the orchid *Ophrys apitera*. This flower resembles the body of a bumblebee, and its odor mimics the pheromones of a female bee. Therefore, male bees are attracted to the flower, and when they attempt to mate with it, they become covered with pollen, which they transfer to the next orchid flower. The orchid is dependent upon bees for pollination because neither wind nor other insects pollinate these flowers.

Video Pollinators

The outcome of mutualism is an intricate web of species interdependencies critical to the community. For example, in areas of the western United States, the branches and cones of whitebark

FIGURE 39.5A Clark's nutcrackers store and disperse the seeds of whitebark pine trees.

pine are turned upward, meaning that the seeds do not fall to the ground when the cones open. Birds called Clark's nutcrackers eat the seeds of whitebark pine trees and store them in the ground (Fig. 39.5A). Therefore, Clark's nutcrackers are critical seed dispersers for the trees. Also, grizzly bears find the stored seeds and consume them. Whitebark pine seeds do not germinate unless their seed coats are exposed to fire. When natural forest fires in the area are suppressed, whitebark pine trees decline in number, and so do Clark's nutcrackers and grizzly bears. When lightning-ignited fires are allowed to burn, or prescribed burning is used in the area, the whitebark pine populations increase, as do the populations of Clark's nutcrackers and grizzly bears.

Cleaning symbiosis is a symbiotic relationship in which crustaceans, fish, and birds act as cleaners for a variety of vertebrate clients. Large fish in coral reefs line up at cleaning stations and wait their turn to be cleaned by small fish that even enter the mouths of the large fish (Fig. 39.5B). Whether cleaning symbiosis is an example of mutualism has been questioned because of the lack of experimental data. If clients respond to tactile stimuli by remaining immobile while cleaners pick at them, then cleaners may be exploiting this response by feeding on host tissues, as well as on ectoparasites.

▶ **39.5 CHECK YOUR PROGRESS**

1. Identify the mutual benefit of protozoans living in the gut of a termite and the possible benefit to the community.

FIGURE 39.5B
Cleaning symbiosis occurs when small fish clean large fish.

Community Development

The composition and diversity of a community can change over time, a fact that is dramatically illustrated by ecological succession. Various models have been proposed to explain the changes that take place during succession. As we discuss in Chapter 40, climate dictates to a degree what type of community will develop, but other factors also influence the exact composition.

39.6 During ecological succession, community composition and diversity change

LEARNING OUTCOMES

When you complete this section, you should be able to

1. Discuss several models for the process of succession.
2. Compare and contrast the stages of primary and secondary succession.

A series of species replacements in a community following a disturbance is called **ecological succession.** Consider for example, the eruption of Mount St. Helens volcano on May 18, 1980, which cleared an area of 350 km. A few plants managed to survive the blast either by being covered with snow or by regrowing from underground roots. Plants such as fireweed (*Chamerion*) and prairie lupine (*Lupinus*) that specialized in colonizing disturbed areas joined these survivors. But only two types of trees, the lodgepole pine and the red alder have been able to take hold so far and it may take 500 years for the blast zone to become the rich forest it was before.

Scientists speak of *primary succession,* which occurs in areas where no soil is present, such as following a volcanic eruption or a glacial retreat and *secondary succession,* which begins in areas where soil is present. Figure 39.6A shows the stages of both primary and secondary succession to a large coniferous forest in central New York State. Secondary succession in the Mount St. Helens area is expected in general to follow this model.

Models of Succession In 1916, F. E. Clements proposed the *climax-pattern model* of succession, which suggests that succession in a particular area will always lead to the same type of community, called a **climax community.** Clements believed that climate, in particular, determines whether a desert, a type of grassland, or a particular type of forest results. This is the reason, he said, that coniferous forests occur in northern latitudes, deciduous forests in temperate zones, and tropical rain forests in the tropics. Secondarily, he believed that soil conditions might also affect the results. Shallow, dry soil might produce a grassland where a forest would otherwise be expected, or the rich soil of a riverbank might produce a woodland where a prairie would be expected.

Further, Clements believed that each stage facilitated the invasion and replacement by organisms of the next stage. As in the examples given, shrubs can't arrive until grasses have made the soil suitable for them. Each successive community prepares the way for the next, so that grass-shrub-forest development occurs sequentially. This is known as the *facilitation model* of succession.

Aside from the facilitation model, there is also an *inhibition model.* That model predicts that colonists hold onto their space and inhibit the growth of other plants until the colonists die or are damaged. Still another model, called the *tolerance model,* predicts that different types of plants can colonize an area at the same time. Sheer chance determines which seeds arrive first, and successional stages may simply reflect the length of time it takes species to mature. This alone could account for the grass-shrub-forest sequence that is often seen. The length of time it takes for trees to develop might give the impression that plant communities develop in a recognizable series, from the simple to the complex. But in reality, the models we have mentioned are not mutually exclusive, and succession is probably a complex process of these models at one time or another.

FIGURE 39.6A Primary succession begins on areas of bare rock. Secondary succession begins at the grass stage.

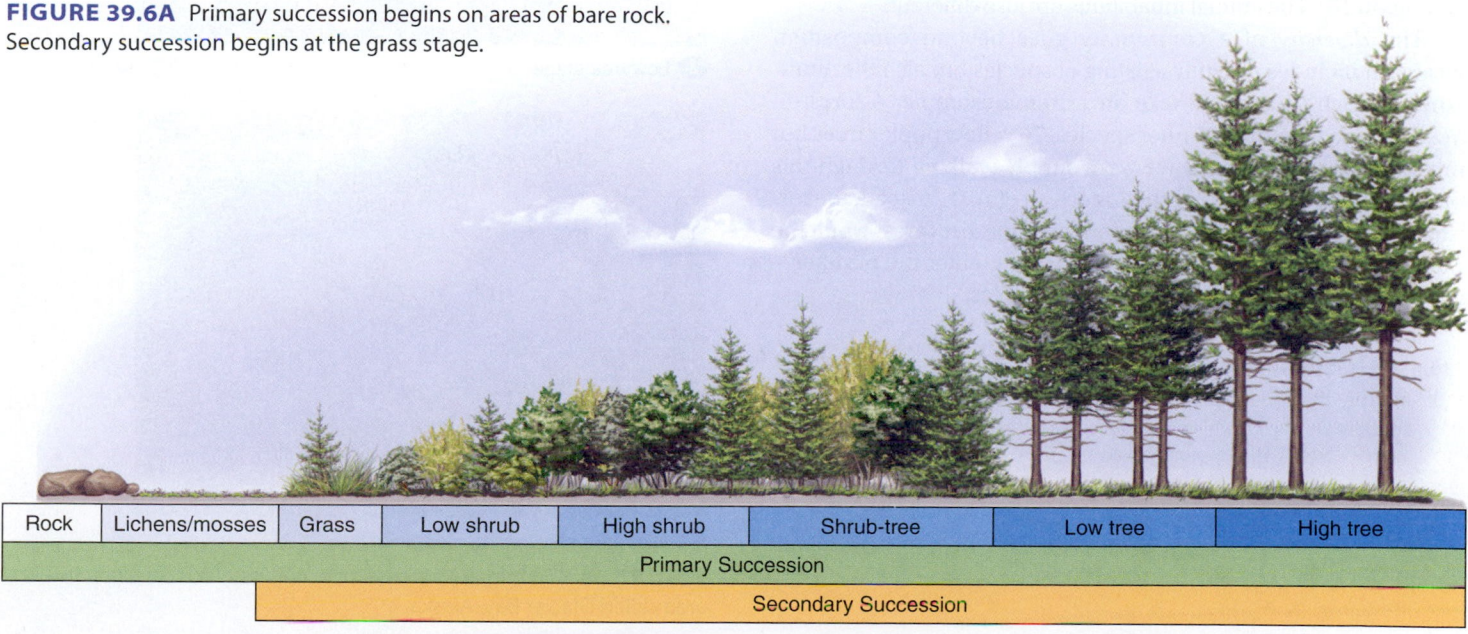

Rock	Lichens/mosses	Grass	Low shrub	High shrub	Shrub-tree	Low tree	High tree
				Primary Succession			
			Secondary Succession				

The Stages of Succession A portion of Alaska called Glacier Bay has become a laboratory for studying the stages of primary succession from bare rock to a climax community of spruce and hemlock trees (Fig. 39.6B). This description of primary succession utilizes the facilitation model:

1 Lichens and mosses too invade the rocky terrain left by a retreating glacier in Alaska. Lichens secrete acids that break down rocks and slowly create new soil; mosses help stabilize the soil and slowly the landscape prepares for the next invaders of the area.

2 Small bushes, such as *Dryas* plants, can now grow abundantly in the new soil because the bacteria in their root nodules are able to fix nitrogen, which so far is in short supply. The mats *Dryas* plants form further stabilize the soil for the possible growth of tree roots.

3 Fast-growing alder trees are also nitrogen-fixing plants, and their falling leaves create humus, the type of soil that holds water and contains microorganisms that enrich the soil. It's said that alder trees "nurse the seedlings of other trees."

4 The improved soil conditions allow the white spruce–Western hemlock community typical of Alaska to develop. These large trees grow to 200 ft and form a climax community that dominates the landscape.

Because Glacier Bay contains glaciers that have been retreating for different lengths of time, all four of these communities can be found in different areas of the bay. Most likely, all complex communities contain patches that are in various stages of secondary succession. If so, diversity is greater because each successional stage has its own mix of plants and animals.

Composition and Diversity When comparing successive communities, two fundamental characteristics that are examined are composition and diversity. The *composition* of a community is a thorough listing of the various species in a particular community. For example, pictorially, it is evident that broadleaved trees are numerous in a temperate deciduous forest (see Fig. 40.7), while succulent cacti and nonsucculent shrubs are numerous in some deserts (see Fig. 40.10). The animal inhabitants are also different.

The *diversity* of a community goes beyond composition because it includes not only a listing of species but also the abundance of each species. To take an extreme example: A forest in West Virginia has, among other species, 76 yellow poplar trees but only one American elm. If we were simply walking through this forest, we might miss seeing the American elm. If, instead, the forest had 36 poplar trees and 41 American elms, the forest would be more diverse. The greater the diversity, the greater the number— and the more even the distribution—of species.

▶ **39.6 CHECK YOUR PROGRESS**

1. Identify the sequence of stages that occur during the primary succession of a coniferous forest using the facilitation model.

lichen

1 Rock/lichen stage

Dryas plant

2 Shrub stage

3 Low tree stage

4 High tree stage: climax community

FIGURE 39.6B Primary succession occurs as glaciers retreat in an area called Glacier Bay, Alaska.

39B Preservation of Community Composition and Diversity

Would you expect larger coral reefs to have a greater number of species, called *species richness,* than smaller coral reefs? The area (space) occupied by a community can have a profound effect on its biodiversity. American ecologists Robert MacArthur and E. O. Wilson developed a general *model of island biogeography* to explain and predict how (1) distance from the mainland and (2) size of an island affect community diversity.

Imagine two new islands that, as yet, contain no species at all. One of these islands is near the mainland, and one is far from the mainland. Which island will receive more immigrants from the mainland? Most likely, the near one because it's easier for immigrants to get there. Similarly, imagine two islands that differ in size. Which island will be able to support a greater number of species? The large one, because its greater amount of resources can support more populations, while species on the smaller island may eventually face extinction due to scarce resources. MacArthur and Wilson studied the biodiversity on many island chains, including the West Indies, and discovered that species richness does correlate positively with island distance from mainland and island size. They developed a model of island biogeography that takes into account both factors. An equilibrium is reached when the rate of species immigration matches the rate of species extinction due to limited space. Notice in Figure 39B that the intersection (equilibrium) point results in higher species richness for an island near the mainland (high immigration) having a large size (low extinction). The equilibrium could be dynamic (new species keep arriving, and new extinctions keep occurring), or the composition of the community could remain steady unless disturbed.

Biodiversity

Conservationists note that the trends graphed in Figure 39B in particular apply to their work because humans often create preserved areas surrounded by farms, towns, and cities, or even water. For example, in Panama, Barro Colorado Island (BCI) was created in the 1910s when a river was dammed to form a lake. As predicted by the model of island biogeography, BCI lost species because it was a small island that had been cut off from the mainland. Among the species that became extinct were the top predators on the island, namely the jaguar, puma, and ocelot. Thereafter, medium-sized terrestrial mammals, such as the coatimundi, increased in number. Because the coatimundi is an avid predator of bird eggs and nestlings, soon there were fewer bird species on BCI, even though the island is large enough to support them.

The model of island biogeography suggests that the larger the conserved area, the better the chance of preserving more species. Is it possible to increase the amount of space without using more area? Two possibilities come to mind. If the environment has patches, it has a greater number of habitats—and thus greater diversity. As gardeners, we are urged to create patches in our yards if we wish to attract more butterflies and birds! One way to introduce patchiness is through stratification, the use of layers. Just as a high-rise apartment building allows more human families to live in an area, so can stratification within a community provide more and different types of living space for different species.

CONSIDER THESE QUESTIONS

1. Humans are apt to kill off large predators and introduce alien species to an area. What will happen to the normal composition and diversity of an area following such activities?
2. You are a member of a town board charged with deciding whether a subdivision can be built. The subdivision will separate one large natural area into two smaller areas. How might the information given here influence your decision?

 Explore the concepts through a variety of multimedia assets and question types.

www.mcgrawhillconnect.com

FIGURE 39B Large islands have a lower extinction rate than small islands (red curves). Islands near the mainland have more immigration than islands far from the mainland (blue curves). The equilibrium model of species richness is based on the intersection points (circles) between these curves because they predict the number of species on an island. To determine the relative number of species note the location of the intersection points from left to right (height of the intersection points is not relevant). As expected, a large island near the mainland will have the greatest number of species.

Dynamics of Ecosystems

When we study an ecosystem, we are concerned with the living community along with its physical environment. **Ecology** is the study of the interactions of populations with each other and with the physical environment. An ecosystem is characterized by energy flow and chemical cycling, as we first discussed in section 1.6, page 12.

39.7 Ecosystems have biotic and abiotic components

LEARNING OUTCOME

When you complete this section, you should be able to

1. Characterize the biotic components of a community according to their food source.

An ecosystem possesses both living (biotic) and nonliving (abiotic) components. The abiotic components include resources, such as sunlight and inorganic nutrients, and conditions, such as type of soil, water availability, prevailing temperature, and amount of wind. The biotic components of an ecosystem are influenced by the abiotic components, as when the force of the wind has affected the growth of a tree in Figure 39.7, *left*. Each biotic component has an *ecological niche* whose aspects are listed in Figure 39.7 for both plants and animals.

Biotic Components of an Ecosystem Among the biotic components of an ecosystem, **autotrophs** require only inorganic nutrients and an outside energy source to produce organic nutrients for their own use and indirectly for all the other members of a community. Photosynthetic organisms produce most of the organic nutrients for the biosphere. They are called **producers** because they produce food. Algae of all types contain chlorophyll and carry on photosynthesis in freshwater and marine habitats. Algae make up the phytoplankton, which are photosynthesizing organisms suspended in water. Green plants, such as trees in forests and corn plants in fields, are the dominant photosynthesizers on land.

 **Video** Plants

Heterotrophs need a preformed source of organic nutrients. They are called **consumers** because they consume food. **Herbivores** are animals that graze directly on plants or algae.

In terrestrial habitats, insects are small herbivores, and antelopes and bison are large herbivores. In aquatic habitats, zooplankton, which is composed of protozoans and tiny invertebrates, are small herbivores while some fishes, as well as manatees, are large herbivores. **Carnivores** feed on other animals; for example, birds that feed on insects are carnivores, and so are hawks that feed on birds. **Omnivores** are animals that feed on both plants and animals. Chickens, raccoons, and humans are omnivores.

Video Snake Eating

Scavengers, such as jackals and vultures (see chapter introduction), feed on the dead remains of animals and also on plants that have recently begun to decompose. **Detritus** refers to organic remains in the water and soil that are in the final stages of decomposition. Marine fan worms take detritus from the water, while clams take it from the substrate. Earthworms, some beetles, and termites feed on detritus in the soil. Bacteria and fungi, including mushrooms, are the **decomposers.** They use their digestive secretions to chemically break down dead organic matter, including animal wastes, in the external environment. Notice that decomposers are heterotrophs that produce detritus. Without decomposers, plants would be completely dependent only on physical processes, such as the release of minerals from rocks, to supply them with inorganic nutrients.

Video Decomposers

▶ **39.7 CHECK YOUR PROGRESS**

1. Explain why it is correct to say that autotrophs support all the biotic components (including decomposers) of an ecosystem.

Aspects of Niche for Plants

- Season of year for growth and reproduction

- Sunlight, water, and soil requirements

- Relationships with other organisms

- Effect on abiotic environment

Aspects of Niche for Animals

- Time of day for feeding and season of year for reproduction

- Habitat and food requirements

- Relationships with other organisms

- Effect on abiotic environment

FIGURE 39.7 Niche specifications of plants compared to animals.

39.8 Energy flow and chemical cycling characterize ecosystems

LEARNING OUTCOME

When you complete this section, you should be able to

1. Compare, with examples, energy flow through an ecosystem with chemical cycling within an ecosystem.

The diagram in Figure 39.8A illustrates that every ecosystem is characterized by two fundamental phenomena: energy flow and chemical cycling. Energy is lost from the biosphere, but inorganic nutrients are not. They recycle within and between ecosystems. Decomposers return inorganic nutrients directly to autotrophs, or they are imported or exported between ecosystems in global cycles.

Energy flow begins when producers absorb solar energy, and chemical cycling begins when producers take in inorganic nutrients from the physical environment. Thereafter, via photosynthesis, producers make organic nutrients (food) directly for themselves and indirectly for the other populations of the ecosystem. Energy flows through an ecosystem via photosynthesis because, as organic nutrients pass from one component of the ecosystem to another, as when an herbivore eats a plant or a carnivore eats an herbivore, a portion of those nutrients is used as an energy source. The rest of the energy dissipates into the environment as heat. Therefore, the vast majority of ecosystems cannot exist without a continual supply of solar energy.

Only a portion of the organic nutrients made by producers is passed on to consumers because plants use organic molecules to fuel their own cellular respiration. Similarly, only a small percentage of nutrients consumed by lower-level consumers, such as herbivores, is available to higher-level consumers, or carnivores. As Figure 39.8B demonstrates, a certain amount of the food eaten by an herbivore is never digested and is eliminated as feces. Metabolic wastes are excreted as urine. Of the assimilated energy, a large portion is used during cellular respiration for the production of ATP, and thereafter it becomes heat. Only the remaining energy, which is converted into increased body weight or additional offspring, becomes available to carnivores.

The elimination of feces and urine by a heterotroph, and indeed the death of all organisms, does not mean that organic nutrients are lost to the ecosystem; instead, they are made available to scavengers and decomposers. Decomposers convert the organic nutrients, such as glucose, back into inorganic chemicals, such as carbon dioxide and water, and release them to the soil or the atmosphere. Chemicals complete their cycle within an ecosystem when inorganic chemicals are absorbed by the producers from the atmosphere or from the soil.

The first law of thermodynamics states that energy cannot be created (or destroyed). This explains why ecosystems are dependent on a continual outside source of energy. The second law of thermodynamics states that, with every transformation, some energy is degraded into a less available form, such as heat. For example, because plants carry on cellular respiration, only 55% of the original energy absorbed by plants is available to an ecosystem.

▶ 39.8 CHECK YOUR PROGRESS

1. Explain what happens to its inorganic components, such as carbon dioxide and water, when glucose is metabolized by a decomposer.

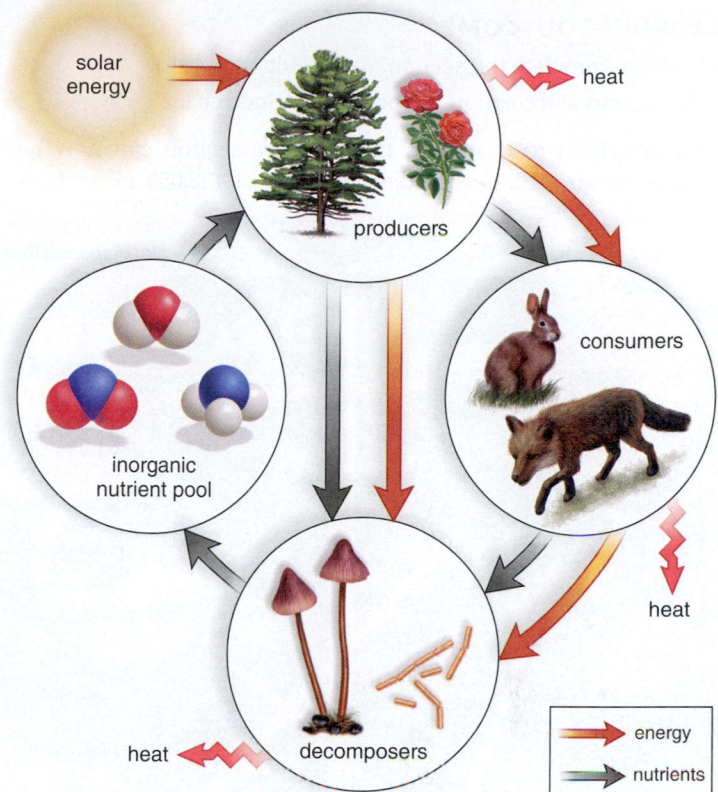

FIGURE 39.8A Energy flow and chemical cycling in an ecosystem.

FIGURE 39.8B Energy balances for an herbivore.

39.9 Energy flow involves food webs and food chains

LEARNING OUTCOME

When you complete this section, you should be able to

1. Contrast and compare a food web to a food chain.

The principles discussed in the previous section can now be applied to an actual ecosystem—a forest of 132,000 m² in New Hampshire. The various interconnecting paths of energy flow may be represented by a **food web,** a diagram that describes trophic (feeding) relationships. Figure 39.9, *top*, is a **grazing food web** because it begins with producers, specifically **1** the oak tree, herbs, and grass. **2** Insects, in the form of caterpillars, feed on leaves, while **3** mice, rabbits, and deer feed on plant material at

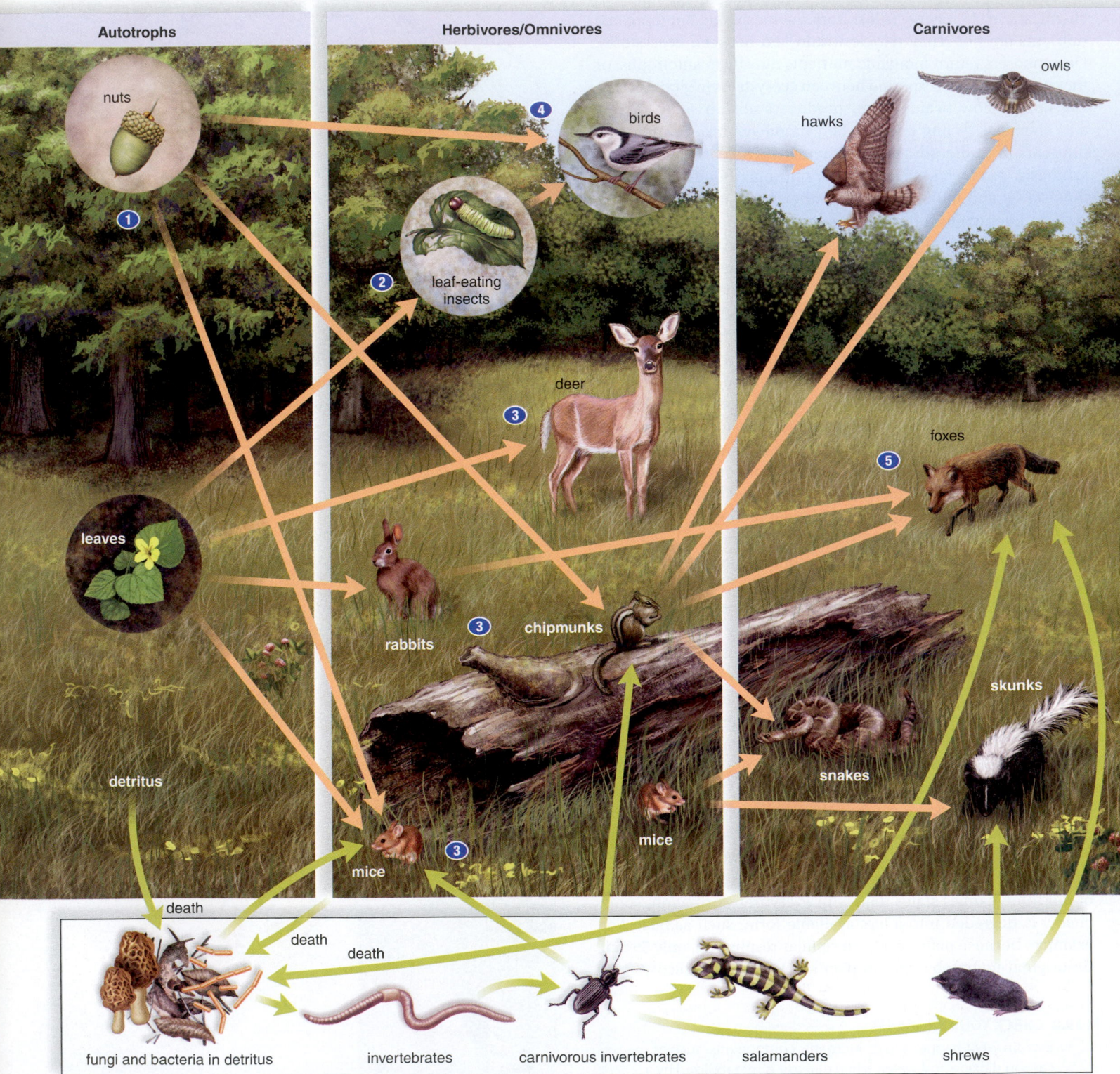

FIGURE 39.9 Grazing food web (*top*) and detrital food web (*bottom*).

or near the ground. Mice also feed on nuts, while ④ birds, collectively, are omnivorous, feeding on both nuts and caterpillars. Herbivores and omnivores all provide food for ⑤ a number of different carnivores.

Figure 39.9, *bottom*, is a **detrital food web,** which begins with detritus. Detritus is food for soil organisms such as earthworms. Earthworms are, in turn, eaten by carnivorous invertebrates, and these may be consumed by salamanders or shrews. Because the members of a detrital food web may become food for aboveground carnivores, the detrital and grazing food webs are connected.

We naturally tend to think that aboveground plants such as trees are the largest storage form of organic matter and energy, but this is not necessarily the case. In this particular forest, the organic matter lying on the forest floor and mixed into the soil contains over twice as much energy as the leaf matter of living trees. Therefore, more energy in a forest may be funneling through the detrital food web than through the grazing food web.

Trophic Levels The arrangement of the species in Figure 39.9 suggests that organisms are linked to one another in a straight line, according to feeding relationships, or who eats whom. Diagrams that show a single path of energy flow in an ecosystem are called **food chains.** For example, in the grazing food web, we could find this food chain:

$$\text{leaves} \longrightarrow \text{caterpillars} \longrightarrow \text{birds} \longrightarrow \text{hawks}$$

And in the detrital food web, we could find this food chain:

$$\text{detritus} \longrightarrow \text{earthworms} \longrightarrow \text{carnivores}$$

A **trophic level** (feeding level) is composed of organisms that occupy the same position within a food web or chain. In the grazing food web in Figure 39.9, *top*, going from left to right, the trees are producers (first trophic level), the first series of animals are primary consumers (second trophic level), and the next group of animals are secondary consumers (third trophic level). Energy moves from trophic level to level, but some is lost between the various levels because energy has been used to do biological work.

▶ **39.9 CHECK YOUR PROGRESS**
1. Compare the first link in a grazing food chain to that in a detrital food web.

39.10 Ecological pyramids are based on trophic levels

LEARNING OUTCOME

When you complete this section, you should be able to

1. Explain the shape of an ecological pyramid, whether considering biomass or energy.

An **ecological pyramid** is a graphic representation of the number of organisms, the biomass, or the relative energy content of the various trophic levels in an ecosystem. For example, Figure 39.10 shows the biomass content of each trophic level of a bog in Silver Springs, Florida. Data regarding the biomass or energy content of trophic levels place the producer trophic level at the base of the pyramid, and each succeeding trophic level, which has less biomass and energy, follows thereafter. In general, only about 10% of the energy of one trophic level is available to the next trophic level because of energy losses between trophic levels. Therefore, if an herbivore population consumes 1,000 kg of plant material, only about 100 kg is converted to herbivore tissue, 10 kg to first-level carnivores, and 1 kg to second-level carnivores. The so-called 10% rule of thumb explains why ecosystems have only a few trophic levels and why a food web can support only a few carnivores.

Ecological pyramids are helpful for explaining energy loss in an ecosystem, but they oversimplify energy flow. Most likely, a pyramid based on the number of organisms in each trophic level wouldn't work. For example, in Figure 39.9, each tree would contain numerous caterpillars, so there would be more herbivores than autotrophs! Similarly, in aquatic ecosystems, such as some lakes and open seas where algae are the only producers, the herbivores may have a greater biomass than the producers when they are measured, because the algae are consumed at a high rate. In any case, ecological pyramids based on grazing food webs don't account for energy that passes to decomposers. The energy content between the autotroph level and the herbivore level is disproportionate in Figure 39.10 because much of the energy in a bog flows through the detrital food web, not the grazing food web.

FIGURE 39.10 This ecological pyramid based on the biomass content of bog populations could also be used to represent an energy pyramid.

Top carnivores
1.5 g/m²

Carnivores
11 g/m²

Herbivores
37 g/m²

Autotrophs
809 g/m²

▶ **39.10 CHECK YOUR PROGRESS**
1. Determine what trophic level, if any, in an ecological pyramid accounts for decomposers and explain.

39.11 Chemical cycling includes reservoirs, exchange pools, and the biotic community

LEARNING OUTCOME

When you complete this section, you should be able to

1. Distinguish between the three basic components of all biogeochemical cycles.

The pathways by which chemicals circulate through ecosystems involve both living (biotic) and nonliving (geologic) components; therefore, they are known as **biogeochemical cycles.** In the next sections of this chapter, we describe three of the biogeochemical cycles: the phosphorus, nitrogen, and carbon cycles. A biogeochemical cycle may be sedimentary or gaseous. The phosphorus cycle is a sedimentary cycle; the chemical is absorbed from the soil by plant roots, passed to heterotrophs, and eventually returned to the soil by decomposers. The carbon and nitrogen cycles are gaseous, meaning that the chemical returns to and is withdrawn from the atmosphere as a gas.

Chemical cycling involves the components of ecosystems shown in Figure 39.11. A *reservoir* is a source normally unavailable to producers, such as the carbon present in calcium carbonate shells on ocean bottoms. An *exchange pool* is a source from which organisms do generally take chemicals, such as the atmosphere

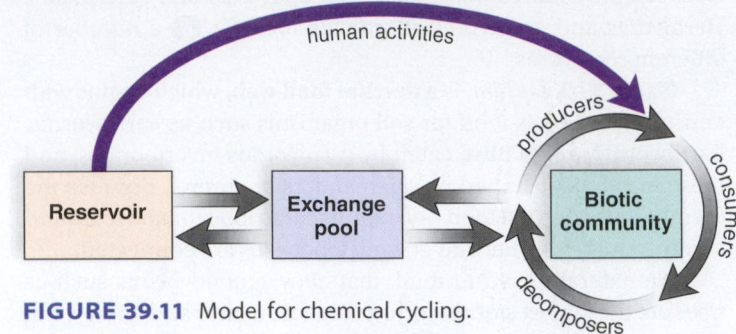

FIGURE 39.11 Model for chemical cycling.

or soil. Chemicals move along food chains in a *biotic community*, perhaps never entering an exchange pool.

Human activities (purple arrow) remove chemicals from reservoirs (or exchange pools) and make them available to the biotic community. In this way, human activities result in pollution because they upset the normal balance of nutrients for producers in the environment.

▶ **39.11 CHECK YOUR PROGRESS**

1. Identify the portion of Figure 39.11 that represents the abiotic environment.

39.12 The phosphorus cycle is sedimentary

LEARNING OUTCOME

When you complete this section, you should be able to

1. List and explain the major steps of the phosphorus cycle.

Figure 39.12 depicts the phosphorus cycle. ① Phosphorus, trapped in oceanic sediments, moves onto land after a geologic upheaval.

② On land, the very slow weathering of rocks places ③ phosphate ions (PO_3^- and HPO_4^{2-}) in the soil. ④ Some of these become available to plants, which use phosphate in a variety of molecules, including phospholipids, ATP, and the nucleotides that become a part of DNA and RNA. ⑤ Animals eat producers and incorporate some of the phosphate into their teeth, bones, and shells, which take many years to decompose. ⑥ However, eventually the death and decay of all organisms and also the decomposition of animal wastes make phosphate ions available to producers once again. Because the available amount of phosphate is already being used within food chains, phosphate is usually a limiting inorganic nutrient for plants—that is, the lack of it limits the size of populations in ecosystems.

⑦ Some phosphate naturally runs off into aquatic ecosystems, where algae acquire phosphate from the water before it becomes trapped in sediments. Phosphate in marine sediments does not become available to

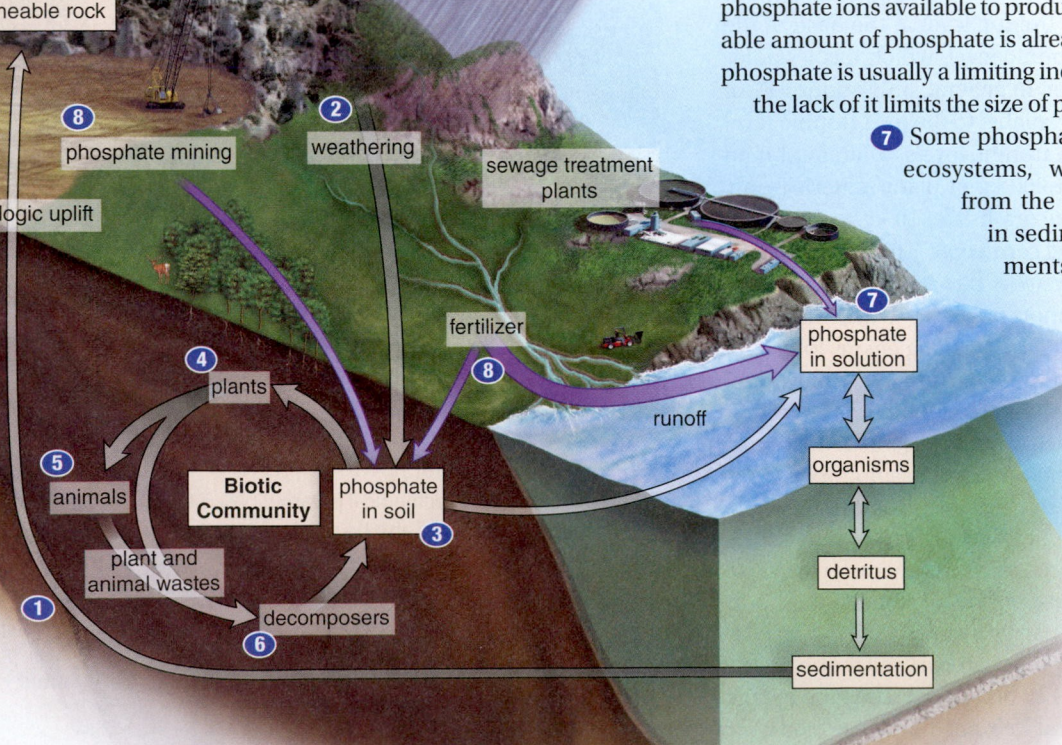

FIGURE 39.12
The phosphorus cycle.

producers on land again until a geologic upheaval exposes sedimentary rocks on land. Now, the cycle begins again.

Human Activities and the Phosphorus Cycle ❽ Humans boost the supply of phosphate by mining phosphate ores for producing fertilizer and detergents. Runoff of phosphate and nitrogen due to fertilizer use, animal wastes from livestock feedlots, and discharge from sewage treatment plants results in **eutrophication** (overenrichment) of waterways.

▶ **39.12 CHECK YOUR PROGRESS**
   1. Explain how the phosphate ions in animal bones become available to producers.

39.13 The nitrogen cycle is gaseous

LEARNING OUTCOME

When you complete this section, you should be able to
   1. List and explain the major steps of the nitrogen cycle.

Nitrogen gas (N_2) makes up about 78% of the atmosphere, but plants cannot use nitrogen in its gaseous form. Therefore, nitrogen can be a nutrient that limits the amount of growth in an ecosystem.

Ammonium (NH_4^+) Formation and Use In the nitrogen cycle, N_2 **(nitrogen) fixation** occurs when nitrogen gas (N_2) is converted to ammonium (NH_4^+), a form plants can use (Fig. 39.13). ❶ Some cyanobacteria in aquatic ecosystems and some free-living bacteria in soil are able to fix atmospheric nitrogen in this way. Other nitrogen-fixing bacteria live in nodules on the roots of legumes, such as beans, peas, and clover. They make organic compounds containing nitrogen available to the host plants so that the plant can form proteins and nucleic acids.

Nitrate (NO_3^-) Formation and Use Plants can also use nitrates (NO_3^-) as a source of nitrogen. The production of nitrates during the nitrogen cycle is called **nitrification**. ❷ Nitrification can occur in two ways: (1) Nitrogen gas (N_2) is converted to NO_3^- in the atmosphere when cosmic radiation, meteor trails, and lightning provide the high energy needed for nitrogen to react with oxygen. (2) Ammonium (NH_4^+) in the soil from various sources, including decomposition of organisms and animal wastes, is converted to NO_3^- by nitrifying bacteria in soil. Specifically, NH_4^+ (ammonium) is converted to NO_2^- (nitrite), and then NO_2^- is converted to NO_3^- (nitrate). ❸ During the process of assimilation, plants take up NH_4^+ and NO_3^- from the soil and use these ions to produce proteins and nucleic acids.

> 📹 **Video** Dung Beetles

Notice in Figure 39.13 that the subcycle involving the biotic community, which occurs on land and in the ocean, need not depend on the presence of nitrogen gas at all.

Formation of Nitrogen Gas ❹ **Denitrification** is the conversion of nitrate back to nitrogen gas, which then enters the atmosphere. Denitrifying bacteria living in the anaerobic mud of lakes, bogs, and estuaries carry out this process as a part of their own metabolism. In the nitrogen cycle, denitrification would counterbalance nitrogen fixation if not for human activities.

Human Activities and the Nitrogen Cycle ❺ Humans significantly alter the transfer rates in the nitrogen cycle by producing fertilizers from N_2—in fact, they nearly double the fixation rate. Fertilizer, which also contains phosphate, runs off into lakes and rivers and results in an overgrowth of algae and rooted aquatic plants. When the algae die off, enlarged populations of decomposers use up all the oxygen in the water, and the result is a massive fish kill.

Acid deposition occurs because nitrogen oxides (NO_x) and sulfur dioxide (SO_2) enter the atmosphere from the burning of fossil fuels. Both these gases combine with water vapor to form acids that eventually return to the Earth. Acid deposition has drastically affected forests and lakes in northern Europe, Canada, and the northeastern United States because their soils are naturally acidic and their surface waters are only mildly alkaline (basic). Acid deposition reduces agricultural yields and corrodes marble, metal, and stonework (see also p. 36).

▶ **39.13 CHECK YOUR PROGRESS**
   1. Define nitrate production by soil microbes.

FIGURE 39.13
The nitrogen cycle.

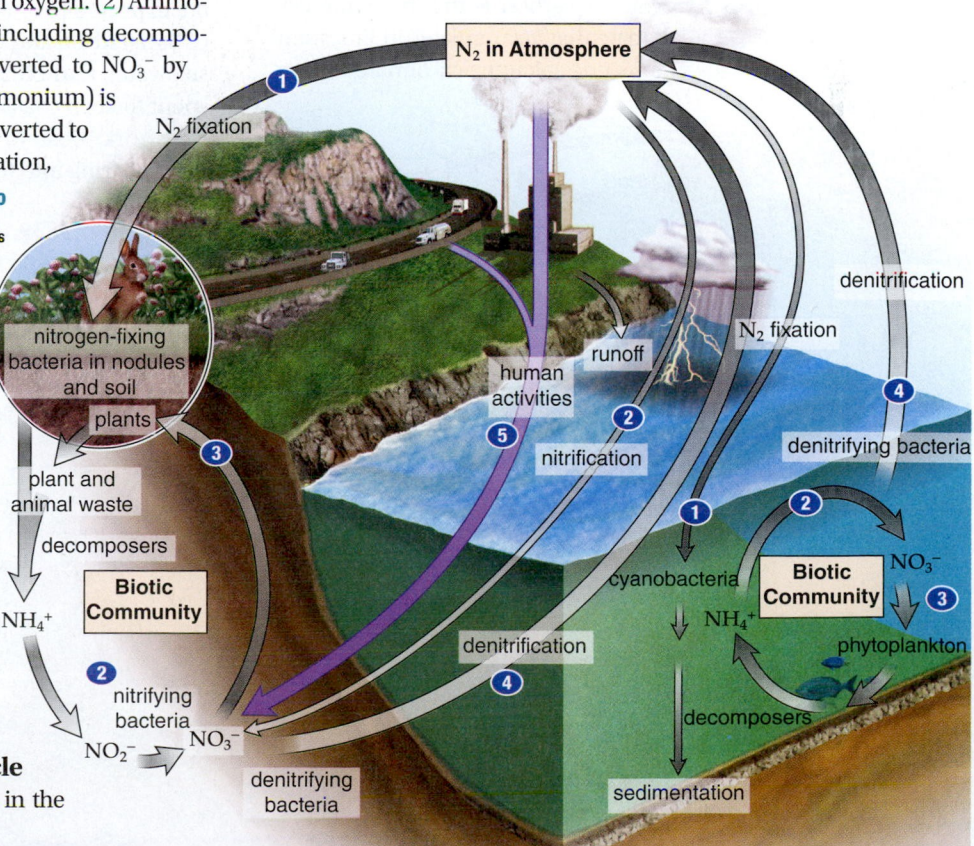

39.14 The carbon cycle is gaseous

LEARNING OUTCOME

When you complete this section, you should be able to

1. List and explain the major steps of the carbon cycle.

In the carbon cycle, organisms in both terrestrial and aquatic ecosystems exchange carbon dioxide (CO_2) with the atmosphere (Fig. 39.14). Therefore, the CO_2 in the atmosphere is the exchange pool for the carbon cycle. On land, plants take up CO_2 from the air, and through photosynthesis, they incorporate carbon into nutrients that are used by autotrophs and heterotrophs alike. **1** When organisms, including plants, respire, carbon is returned to the atmosphere as CO_2. **2** CO_2 then recycles to plants by way of the atmosphere.

In aquatic ecosystems, the exchange of CO_2 with the atmosphere is indirect. **3** Carbon dioxide from the air combines with water to produce bicarbonate ion (HCO_3^-), a source of carbon for algae that produce food for themselves and for heterotrophs. Similarly, when aquatic organisms respire, the CO_2 they give off becomes HCO_3^-. **4** The amount of bicarbonate in the water is in equilibrium with the amount of CO_2 in the air.

Reservoirs Hold Carbon Living and dead organisms contain organic carbon and serve as one of the reservoirs for the carbon cycle. **5** The world's biotic components, particularly trees, contain 800 billion metric tons of organic carbon, and an additional 1,000–3,000 billion metric tons are estimated to be held in the remains of plants and animals in the soil. **6** Ordinarily, decomposition of animals returns CO_2 to the atmosphere.

Some 300 MYA, plant and animal remains were transformed into coal, oil, and natural gas, the materials we call fossil fuels. Another reservoir for carbon is the inorganic carbonate that accumulates in limestone and in calcium carbonate shells.

Many marine organisms have calcium carbonate shells that remain in bottom sediments long after the organisms have died. Geologic forces change these sediments into limestone.

Human Activities and the Carbon Cycle **7** More CO_2 is being deposited in the atmosphere than is being removed, largely due to the burning of fossil fuels and the destruction of forests to make way for farmland and pasture. When we humans do away with forests, we reduce a reservoir and also the very organisms that take up excess carbon dioxide. Today, the amount of CO_2 released into the atmosphere is about twice the amount that remains in the atmosphere. Scientists hypothesize that much of this has been dissolving into the ocean.

Carbon dioxide, and other gases as well, are being emitted due to human activities. The other gases include nitrous oxide (N_2O) from fertilizers and animal wastes and methane (CH_4) from bacterial decomposition that takes place particularly in the guts of animals, in sediments, and in flooded rice paddies. These gases are known as **greenhouse gases** because, just like the panes of a greenhouse, they allow solar radiation to pass through but hinder the escape of infrared rays (heat) back into space. The greenhouse gases are contributing to an overall rise in the Earth's ambient temperature, a trend called **global climate change.** The global climate has already warmed about 0.6°C since the Industrial Revolution.

Video Global Warming

Scientists predict that, as the oceans warm, temperatures in the polar regions will rise to a greater degree than in other regions. If so, glaciers will melt, and sea level will rise, not only due to this melting but also because water expands as it warms. Increased rainfall is likely along the coasts, while dryer conditions are expected inland. Coastal agricultural lands, such as the deltas of Bangladesh and China, will be inundated with sea water, and billions of dollars will have to be spent to keep coastal cities such as New Orleans, New York, Boston, Miami, and Galveston from disappearing into the sea.

Video Warming Hurts Rice

Video Karoo Global Warming

> ▶ **39.14 CHECK YOUR PROGRESS**
> 1. Explain how global temperature would be affected if humans used renewable energy such as wind and solar as an energy source instead of fossil fuels.

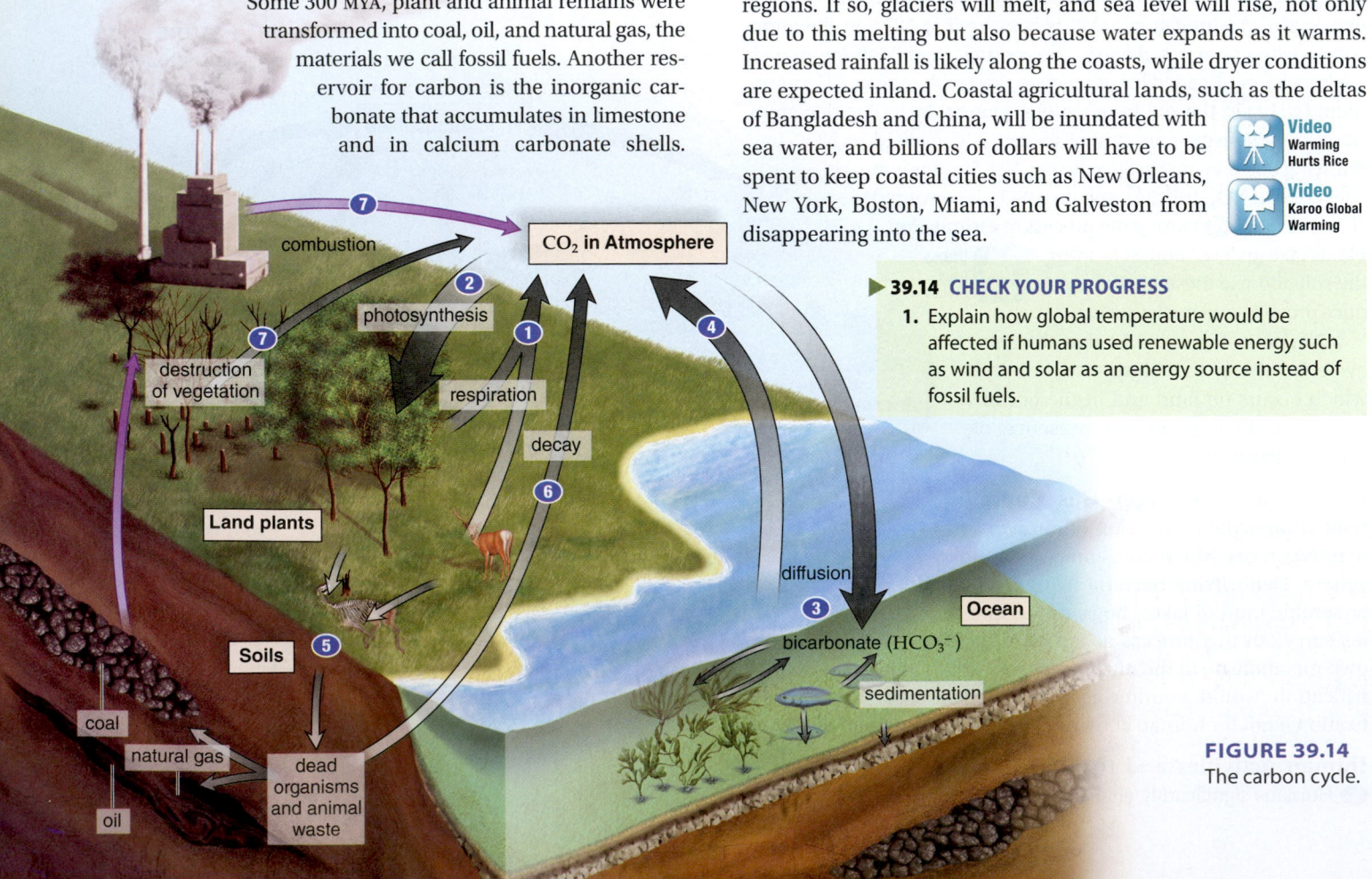

FIGURE 39.14
The carbon cycle.

THE CHAPTER IN REVIEW

CONNECTING THE CONCEPTS

Community ecology is concerned with how populations of different species interact with each other. Population size is influenced by negative interactions such as competition, predation, and parasitism. But positive interactions such as mutualism are also fairly common in nature (especially for plants) and are presumed to increase or maintain population sizes. Perhaps one of the most important discoveries about communities is that they are highly dynamic, meaning that the number of species, kinds of species, and sizes of populations within most communities are constantly changing. This dynamic quality is well demonstrated by the process of ecological succession.

Instead of studying the composition of diversity of communities, some ecologists concentrate on the movement of energy and nutrients through communities. The physical environment has a large influence on energy flow and chemical cycling within a community. Therefore, our study of communities must include the abiotic environment. Human activities also influence the operation of ecosystems. For example, burning fossil fuels and trees adds carbon dioxide to the atmosphere. Carbon dioxide and other greenhouse gases allow the sun's rays to pass through, but they absorb and reradiate heat back to the Earth. Therefore fossil fuel combustion may be leading to global warming. Transfer rates in both the phosphorus and nitrogen cycles are affected when we produce fertilizers and detergents. Nitrogen and phosphorus runoff causes eutrophication in aquatic ecosystems. Many human activities negatively impact the functioning of the biosphere, the largest ecosystem of all, and threaten the existence of all species, including our own. In Chapter 40, we take a look at the specific types of ecosystems that make up the biosphere.

ANALYZE AND EVALUATE

1. If comparing a community to your college, what would be the populations? Give examples to show that (a) the populations interact, (b) it would be best if the populations were balanced one to the other, and (c) the composition and diversity of the community change over time.

2. What evidence shows that the human population is part of an ecosystem in the biosphere? (See the introduction to Chapter 5 on p. 82) If the natural environment is "home" for the human population, should people work to preserve it? Why do many of our actions fail to preserve the natural environment?

SUMMARIZE

Community Interactions

39.1 Competition can lead to resource partitioning

- An **ecological niche** is the role an organism plays in its **community,** including both **habitat** and interactions.
- **Competition** is rivalry for the same resources (e.g., nesting sites, food). The **competitive exclusion principle** states that no two species can occupy the same niche at the same time. **Resource partitioning** decreases competition for resources between two species.
- Characteristics tend to diverge when similar species belong to the same community, a phenomenon known as **character displacement.**

39.2 Predator-prey interactions affect both populations

- **Predation** occurs when one organism feeds on another population. Predator-prey populations cycle: More predators/fewer prey result in fewer predators/more prey. Scarcity of food for prey also causes fewer predators. Prey defenses include senses, speed, protective body parts, and chemicals. Ways to deceive predators include **camouflage,** cryptic coloration, and **mimicry.**

39.3 Parasitism benefits one population at another's expense

- **Parasitism** occurs when a **parasite** derives nourishment from a host. Parasitism is a type of **symbiosis;** other types are commensalism and mutualism.

39.4 Commensalism benefits only one population

- Although **commensalism** benefits only one species, closer examination sometimes reveals more of a mutualistic or parasitic relationship.

39.5 Mutualism benefits both populations

- **Mutualism** results in an intricate web of species interdependencies critical to the community.

Community Development

39.6 During ecological succession, community composition and diversity change

- **Ecological succession** involves a series of species replacements in a community: (1) Primary succession occurs where there is no soil present. (2) Secondary succession occurs where soil is present and certain plant species can begin to grow.
- Various models for succession have been proposed. The facilitation model suggests that each stage facilitates the invasion and replacement by the next stage until the process results in a **climax community** typical of that area.

Dynamics of Ecosystems

39.7 Ecosystems have biotic and abiotic components

- Biotic components are living components: (1) **Autotrophs** are **producers.** (2) **Heterotrophs** are **consumers;** types of consumers include **herbivores, carnivores,** and **omnivores.** (3) **Scavengers** feed on dead remains of animals and plants. (4) **Detritus** is composed of the organic remains of decomposition found in water and soil. (5) **Decomposers** include heterotrophic bacteria and fungi.

decomposers

- Abiotic components are resources (e.g., sunlight, inorganic nutrients) and conditions (e.g., soil, water, temperature, wind).

39.8 Energy flow and chemical cycling characterize ecosystems

- Energy flows through ecosystems because as food passes from producers through consumers, each population makes energy conversions that result in a loss of usable energy in the form of heat.

producers

consumers

- Chemicals cycle because they pass from one population to the next until decomposers return them once more to the environment where producers can take them up again.

39.9 Energy flow involves food webs and food chains

- A **food web** is an interconnecting path of energy flow that describes trophic (feeding) relationships: (1) A **grazing food web** begins with a producer, such as an oak tree. (2) A **detrital food web** begins with detritus. (3) Grazing and detrital food webs are joined. A **trophic level** is composed of organisms that occupy the same feeding position within a food web.
- A **food chain** is a single path of energy flow.

39.10 Ecological pyramids are based on trophic levels

- An **ecological pyramid** is a graphic representation of the number of organisms, biomass, or energy content of trophic levels.

39.11 Chemical cycling includes reservoirs, exchange pools, and the biotic community

- **Biogeochemical cycles** may be sedimentary (phosphorus cycle) or gaseous (carbon and nitrogen cycles).
- Chemical cycling involves a reservoir, an exchange pool for an inorganic nutrient, and a biotic community.

inorganic nutrient pool

39.12 The phosphorus cycle is sedimentary

- Geologic upheavals move phosphorus from the ocean to land. Slow weathering of rocks returns phosphorus to the soil. Most phosphorus is recycled within a community, and phosphorus is a limiting nutrient except when humans add phosphate to fertilizers.

39.13 The nitrogen cycle is gaseous

- Plants cannot use nitrogen gas (N_2) from the atmosphere. During N_2 (**nitrogen**) **fixation**, N_2 converts to ammonium, making nitrogen available to plants. **Nitrification** is the production of nitrates, while **denitrification** is the conversion of nitrate back to N_2, which enters the atmosphere. Human activities increase transfer rates in biogeochemical cycles. Due to fossil fuel combustion, oxides (and sulfur dioxide) enter the atmosphere. There they combine with water vapor, and then return to Earth as **acid deposition.**

39.14 The carbon cycle is gaseous

- CO_2 in the atmosphere is an exchange pool for the carbon cycle; terrestrial and aquatic plants and animals exchange CO_2 with the atmosphere. Living and dead organisms serve as reservoirs for the carbon cycle because they contain organic carbon. Due to fossil fuel combustion, the levels of CO_2 and other **greenhouse gases** have risen in the atmosphere. Greenhouse gases contribute to **global climate change.**

TEST YOURSELF

Community Interactions

1. According to the competitive exclusion principle,
   a. one species is always more competitive than another for a particular food source.
   b. competition excludes multiple species from using the same food source.
   c. no two species can occupy the same niche at the same time.
   d. competition limits the reproductive capacity of species.
2. Resource partitioning pertains to
   a. niche specialization.
   b. character displacement.
   c. increased species diversity.
   d. the development of mutualism.
   e. All but d are correct.

For questions 3–7, indicate the type of interaction in the key that is described in each scenario.

KEY:
   a. competition   d. commensalism
   b. predation   e. mutualism
   c. parasitism

3. An alfalfa plant gains fixed nitrogen from the bacterial species *Rhizobium* in its root system, while *Rhizobium* gains carbohydrates from the plant.
4. Both foxes and coyotes in an area feed primarily on a limited supply of rabbits.
5. Roundworms establish a colony inside a cat's digestive tract.
6. A fungus captures nematodes as a food source.
7. An orchid plant lives in the treetops, gaining access to sun and pollinators, but not harming the trees.
8. **THINKING CONCEPTUALLY** You want to reintroduce a predator into an area. What concern might you have?

Community Development

9. The model of island biogeography is pertinent to
   a. only islands surrounded by water.
   b. explaining decreases in community composition and diversity.
   c. explaining the intermediate disturbance hypothesis.
   d. why exotics are such a problem today.
   e. All of these are correct.
10. Mosses growing on bare rock will eventually help create soil. These mosses are involved in _____ succession.
    a. primary   c. tertiary
    b. secondary
11. Assume that a field on a farm is allowed to return to its natural state. By chance, the field is first colonized by native grasses, which begin the succession process. This is an example of which model of succession?
    a. climax pattern   c. facilitation
    b. tolerance   d. inhibition
12. **THINKING CONCEPTUALLY** The description of succession in Glacier Bay, Alaska, utilizes the facilitation model. How is the description consistent with the facilitation model?

Dynamics of Ecosystems

13. The ecological niche of an organism
    a. is the same as its habitat.
    b. includes how it competes and acquires food.
    c. is specific to the organism.
    d. is usually occupied by another species.
    e. Both b and c are correct.

14. Label this diagram:

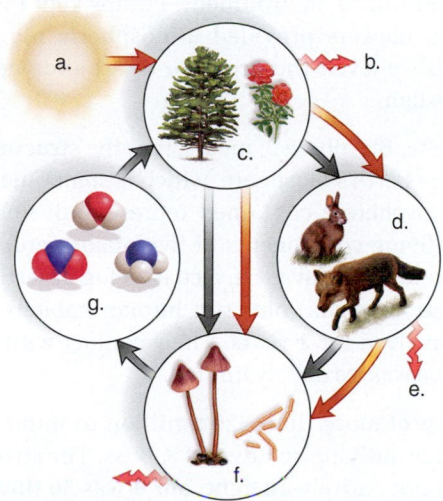

15. In what way are decomposers like producers?
    a. Either may be the first member of a grazing or a detrital food chain.
    b. Both produce oxygen for other forms of life.
    c. Both require nutrient molecules and energy.
    d. Both are present only on land.
    e. Both produce organic nutrients for other members of ecosystems.

16. When a heterotroph takes in food, only a small percentage of the energy in that food is used for growth. The remainder is
    a. not digested and eliminated as feces.
    b. excreted as urine.
    c. given off as heat.
    d. All of these are correct.
    e. None of these are correct.

17. During chemical cycling, inorganic nutrients are typically returned to the soil by
    a. autotrophs.   c. decomposers.
    b. detritivores.   d. tertiary consumers.

18. In a grazing food web, carnivores that eat herbivores are
    a. producers.   c. secondary consumers.
    b. primary consumers.   d. tertiary consumers.

19. Choose the statement that is true concerning this food chain:
    grass ⟶ rabbits ⟶ snakes ⟶ hawks
    a. Each predator population has a greater biomass than its prey population.
    b. Each prey population has a greater biomass than its predator population.
    c. Each population is omnivorous.
    d. Each population returns inorganic nutrients and energy to the producer.
    e. Both a and c are correct.
    f. Both a and b are correct.

20. Which of the following is a sedimentary biogeochemical cycle?
    a. carbon
    b. nitrogen
    c. phosphorus

21. Which of the following is not a component of the nitrogen cycle?
    a. proteins   d. photosynthesis
    b. ammonium   e. bacteria in root nodules
    c. decomposers

22. How do plants contribute to the carbon cycle?
    a. When plants respire, they release CO_2 into the atmosphere.
    b. When plants photosynthesize, they consume CO_2 from the atmosphere.
    c. When plants photosynthesize, they provide oxygen to heterotrophs.
    d. Both a and b are correct.

23. **THINKING CONCEPTUALLY** Show that the carbon cycle can be used to exemplify the model for chemical cycling. (See Fig. 39.11 and Fig. 39.14).

GET INVOLVED

1. As per Figure 15.1B, you observe three species of *Empidonax* flycatchers in the same general area, and you hypothesize that they occupy different niches. How could you substantiate your hypothesis?
2. In order to improve species richness, you decide to add phosphate to a pond. How might you determine how much phosphate to add in order to avoid eutrophication?
3. Open this virtual lab and place the plants and animals for one of the ecosystems according to the directions given. Explain why you placed the organisms as you did.

Virtual Lab
Model Ecosystems

MEDIA STUDY TOOLS

mhhe.com/maderconcepts3

Enhance your study of this chapter with interactive study tools, practice tests, and engaging animations. Also, ask your instructor about the resources available through ConnectPlus, which includes LearnSmart, a personalized adaptive learning program, and a media-rich eBook.

40

Major Ecosystems of the Biosphere

BEFORE YOU BEGIN

Take a few minutes to recall

That the biosphere is divided into ecosystems (Fig. 1.3A and section 1.6)

How climate affects the global distribution of animals in "Staying warm, staying cool" on page 510

The value of symbiotic relationships to the complexity of a biome (section 39.5)

The adaptations of aquatic animals includes osmoregulation in "Do coral reef animals regulate?" on page 668

Life Under Glass

Just a 20-minute drive north of Tucson, in the desert of Arizona, a futuristic glass structure emerges from the surrounding landscape. Called Biosphere 2, the structure was built to help establish a new field called ecological engineering and to investigate the interaction and evolution of ecosystems enclosed within a heavily subsidized environment. Pumps kept the hydrologic cycle going, blowers provided atmospheric movement, and huge "lungs" relieved the pressure that builds up in a glass-enclosed system.

The name, Biosphere 2, means that the structure was modeled after the Earth's biosphere, which is sometimes called Biosphere 1. Biosphere 2 contained representative organisms, as well as nonliving components of five major natural biomes (a tropical rain forest, a savanna, a coastal fog desert, a mangrove wetland, and an ocean) plus two human habitats where eight biospherians lived for 2 years. Their contact with Biosphere 1 where we live was extremely limited.

The cost of more than $200 million to build Biosphere 2 was funded by billionaire Edward P. Bass. The structure, when sealed, was not entirely airtight, but it was 30 times less leaky than the space shuttle, and thousands of times less leaky than a typical skyscraper. Ecosystems did not flourish as had been planned, but this gave information and insight into the workings of their analogs in nature. CO_2 concentrations soared, and O_2 concentrations dipped to below normal during the first years, indicating the importance of the world's oceans in absorbing CO_2 and the world's forests in producing O_2. Soil microbes and the building's concrete components also played a role in creating the imbalance of these gases in Biosphere 2. Some of the crops did not bear fruit because of pests, diseases,

Biosphere 2

© G. Kenny

782

Biosphere agriculture

Biosphere harvest
© A. Alling

Biosphere feast
© A. Alling

and the lack of pollinators. Oxygen was piped in, but many vertebrates died off. When cockroaches proliferated, toads and geckoes were introduced from the outside to eat them. The biospherians were stressed by living within a confined structure.

Still, the success of the Biosphere 2 project can be measured by the quality of the observations and extensive data now available to the scientific community. Biosphere 2, it turns out, was immensely valuable to many scientific disciplines, and it inspired quite a few scientific publications. Some papers presented computer models to describe the hydrologic balance and the heat versus humidity within the system. Other papers told of the changes that occurred in the rain forest, mangrove, ocean, and agronomic ecosystems in this CO_2-rich environment.

The overall ecological message is clear—we should appreciate and try to better understand the workings of the ecosystems within Biosphere 1. While Biosphere 2 was a start, more experimentation is needed before a self-sustaining habitat can be established on the Moon or on Mars. Biosphere 2 was only the first step toward achieving such a system.

Climate and the Biosphere

Weather consists of short-term changes in the atmosphere that affect temperature and rainfall whereas **climate** is the prevailing temperature and rainfall over a long period of time. The temperature and rainfall of a region are determined by (1) variations in solar radiation due to the curvature and tilt of the Earth as it orbits the sun and also by (2) the topography of an area and (3) the nearness and currents of the oceans.

40.1 Solar radiation and winds determine climate

LEARNING OUTCOME

When you complete this section, you should be able to

1. Discuss how temperature and rainfall determine climate.

The sun's rays have a direct affect on temperature. Because the Earth is a sphere, the sun's rays are more direct at the equator and more spread out at the polar regions (Fig. 40.1A, *left*). Therefore, the tropics are warmer than the temperate regions and the poles. However, the tilt of the Earth as it orbits the sun causes one pole or the other to be closer to the sun (except at the spring and fall equinoxes, when the sun aims directly at the equator), and this accounts for the seasons that occur in all parts of the Earth, but not at the equator (Fig. 40.1A, *right*). When the Northern Hemisphere is having winter, the Southern Hemisphere is having summer, and vice versa.

Air currents have a direct affect on rainfall. Because the Earth rotates on its axis daily and its surface consists of continents and oceans, an overall flow of warm and cold air currents are modified into three large circulation cells in each hemisphere (Fig. 40.1B, *large arrows*). At the equator, the sun heats the air and evaporates water. The warm, moist air rises, cools, and loses most of its moisture as rain. The greatest amounts of rainfall on Earth are near the equator. The rising air flows toward the poles, but at about 30° north and south latitude, it sinks toward the Earth's surface and reheats. As the air descends and warms, it is very dry, creating zones of low rainfall. The great deserts of Africa, Australia, and the Americas occur at these latitudes. At about 60° north and south latitude, the air rises and cools, producing additional zones of high rainfall. This moisture supports the great forests of the temperate zone. Part of this rising air flows equatorward, and part continues poleward, descending near the poles, which are zones of low precipitation.

The spinning of the Earth causes a curving pattern of the winds and ocean currents (Fig. 40.1B). Periods of calm called the *doldrums* occur at the equator. At about 30° latitude (below and above the equator), the winds blow from the east-southeast in the

FIGURE 40.1B

Wind circulation as air moves from the equator to the poles and back again.

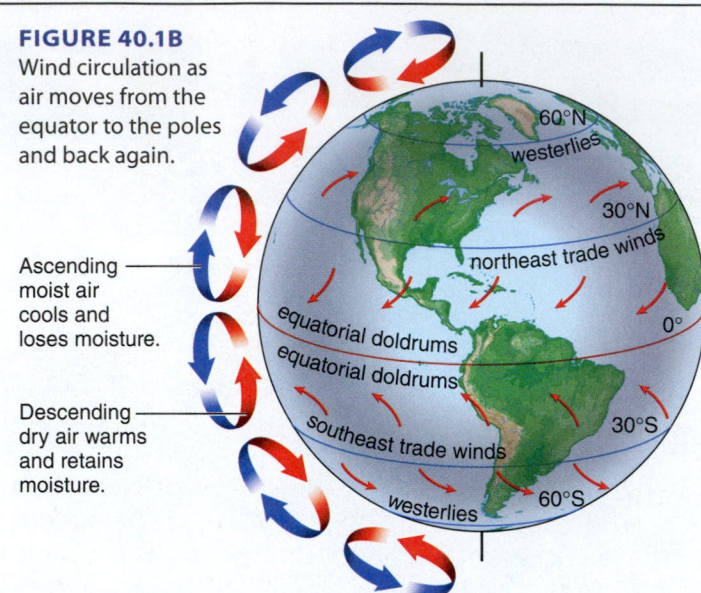

Ascending moist air cools and loses moisture.

Descending dry air warms and retains moisture.

Southern Hemisphere and from the east-northeast in the Northern Hemisphere (the east coasts of continents at these latitudes are wet). These are called trade winds because sailors depended on them to fill the sails of their trading ships. Between 30° and 60° north and south latitudes, strong winds, called the prevailing westerlies, blow from west to east.

▶ **40.1 CHECK YOUR PROGRESS**

1. Identify the particular solar and wind conditions (see Figs. 40.1A and 40.1B) that account for a warm, wet climate at the equator.

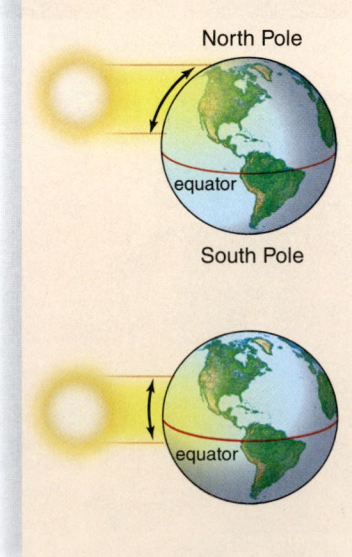

FIGURE 40.1A *Left*: Distribution of the sun's rays striking the Earth. *Right*: Distribution of solar energy as the Earth orbits the sun.

Distribution of sun's rays

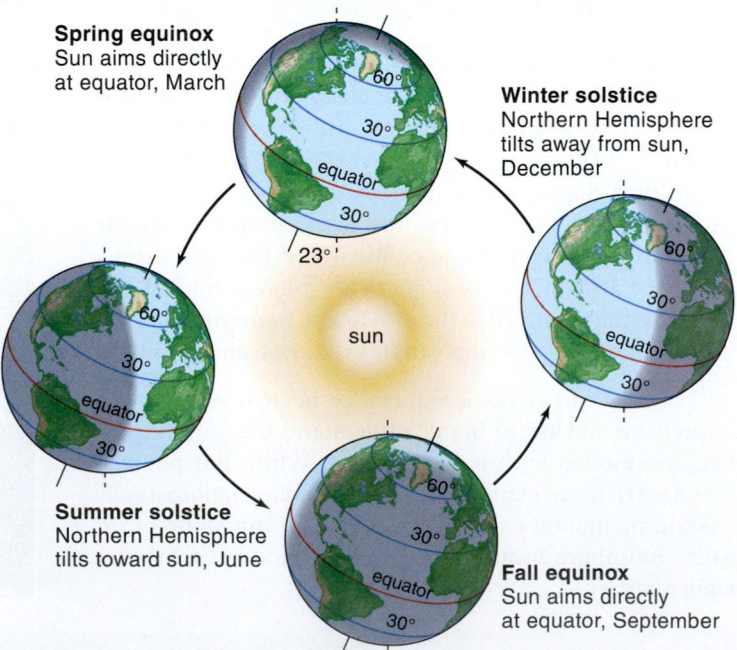

Spring equinox
Sun aims directly at equator, March

Winter solstice
Northern Hemisphere tilts away from sun, December

Summer solstice
Northern Hemisphere tilts toward sun, June

Fall equinox
Sun aims directly at equator, September

40.2 Topography and other effects also influence climate

When you complete this section, you should be able to

1. Discuss how surface features and nearby bodies of water affect rainfall.

Topography refers to the surface features of the land. Mountains are topographic features that affect climate, and therefore the distribution of ecosystems. For example, traveling from Florida to the North Pole, you might see first a subtropical forest, followed by a temperate deciduous forest, a coniferous forest, and tundra in that order because of the change in temperature. This same sequence is seen when ascending a mountain in the Northern Hemisphere (Fig. 40.2A). The coniferous forest of a mountain is called a montane coniferous forest, and the tundra near the peak of a mountain is called an alpine tundra.

Mountains also affect precipitation. As air blows up and over a coastal mountain range, it rises and releases its moisture as it cools. One side of the mountain, called the windward side, receives more rainfall than the other side, called the leeward side. On the leeward side, the air descends, picks up moisture, and produces clear weather (Fig. 40.2B). The difference between the windward side and the leeward side can be quite dramatic. In the Hawaiian Islands, for example, the windward side of a mountain receives more than 750 cm of rain a year, while the leeward side, which is in a **rain shadow,** gets on the average only 50 cm of rain and is generally sunny. In the United States, the western side of the Sierra Nevada Mountains is lush, while the eastern side is a semidesert.

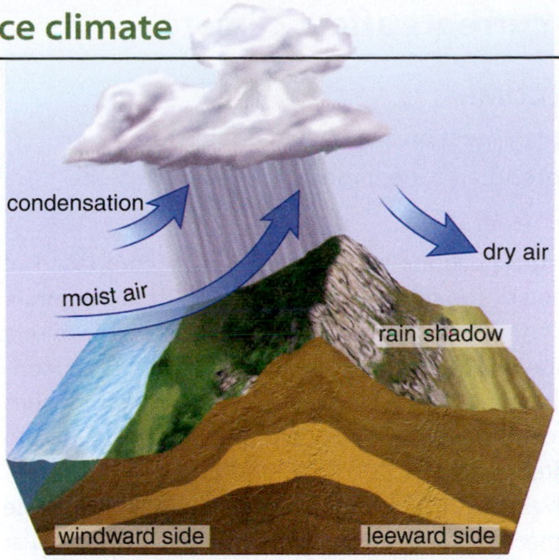

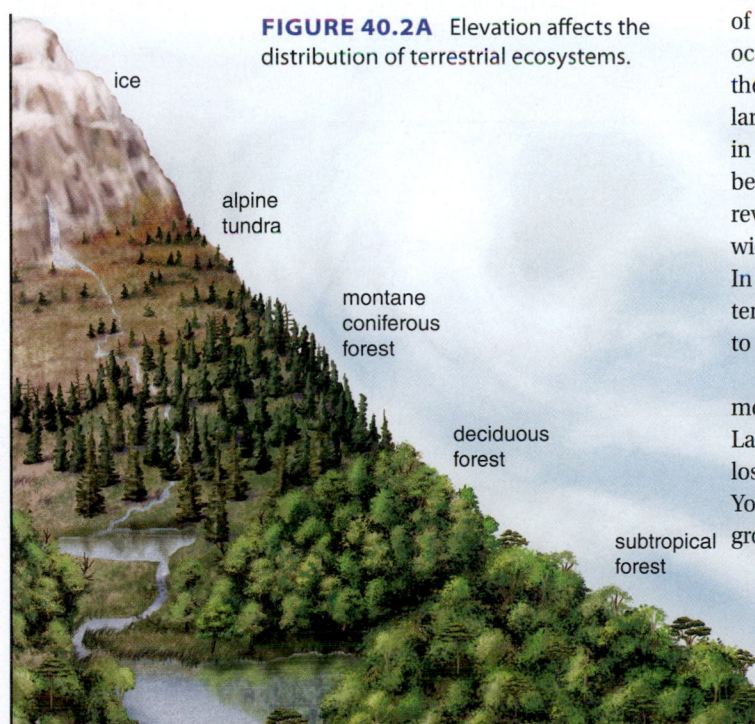

FIGURE 40.2B Formation of a rain shadow.

Nearby Bodies of Water The temperature of the oceans is more stable than that of the landmasses. Ocean water gains or loses heat more slowly than terrestrial environments do. This gives coasts a unique weather pattern that is not observed inland. During the day, the land warms more quickly than the ocean, and the air above the land rises. Then a cool sea breeze blows in from the ocean. At night, the reverse happens; the breeze blows from the land toward the sea.

India and some other countries in southern Asia have a **monsoon** climate, in which wet ocean winds blow onshore for almost half the year. During spring, the land heats more rapidly than the waters of the Indian Ocean, resulting in a temperature differential between the land and the ocean and a gigantic circulation of air: Warm air rises over the land, and cooler air comes in off the ocean to replace it. As the warm air rises, it loses its moisture, and the monsoon season begins. As just discussed, rainfall is particularly heavy on the windward side of hills. The town of Cherrapunji in northern India receives an average of 1,090 cm of rain per year because of its high altitude. By November, the weather pattern has reversed. The land is now cooler than the ocean; therefore, dry winds blow from the Asian continent across the Indian Ocean. In the winter, the air over the land is dry, the skies cloudless, and temperatures pleasant. The chief crop of India is rice, which starts to grow when the monsoon rains begin.

In the United States, people often speak of the "lake effect," meaning that in the winter, arctic winds blowing over the Great Lakes become warm and moisture-laden. As these winds rise and lose their moisture, snow begins to fall. Places such as Buffalo, New York, get heavy snowfalls due to the lake effect and have snow on the ground for an average of 90–140 days every year.

FIGURE 40.2A Elevation affects the distribution of terrestrial ecosystems.

ice

alpine tundra

montane coniferous forest

deciduous forest

subtropical forest

temperate deciduous forest

coniferous forest

tundra

ice

Increasing Altitude

Increasing Latitude

▶ **40.2 CHECK YOUR PROGRESS**

1. Explain the term "rain shadow."

40.3 Ocean currents affect climate

LEARNING OUTCOME

When you complete this section, you should be able to

1. Understand how ocean currents are formed and affect the continents.

The hydrosphere, the portions of the planet Earth composed of water, is also warmed by the sun, and that causes water to evaporate from the oceans. As moisture evaporates into the air, it carries along the heat of evaporation. When such warm, vapor-laden air continues to rise, the heat remains with the vapor until the air reaches an altitude where cooler temperature condenses the moisture into clouds and precipitation occurs.

Climate is driven by the sun, but the oceans play a major role in redistributing heat in the biosphere. Water tends to be warm at the equator and much cooler at the poles because of the distribution of the sun's rays, discussed earlier (see Fig. 40.1A). Air takes on the temperature of the water below, and warm air moves from the equator to the poles. In other words, the oceans make the winds blow. (Landmasses also play a role, but the oceans hold heat longer and remain cool longer during periods of changing temperature than do continents.)

When the wind blows strongly and steadily across a great expanse of ocean for a long time, friction from the moving air begins to drag the water along with it. Once the water has been set in motion, its momentum, aided by the wind, keeps it moving in a steady flow called a current. Because the ocean currents eventually strike land, they move in a circular path—clockwise in the Northern Hemisphere and counterclockwise in the Southern Hemisphere (Fig. 40.3). As the currents flow, they take warm water from the equator to the poles. One such current, called the Gulf Stream, brings tropical Caribbean water to the east coast of North America and the higher latitudes of western Europe. Without the Gulf Stream, Great Britain, which has a relatively warm temperature, would be as cold as Greenland. In the Southern Hemisphere, another major ocean current warms the eastern coast of South America.

Also in the Southern Hemisphere, a current called the Humboldt Current flows toward the equator. The Humboldt Current carries phosphorus-rich cold water northward along the west coast of South America. During a process called **upwelling,** cold offshore winds cause cold, nutrient-rich waters to rise and take the place of warm, nutrient-poor waters. In South America, the enriched waters cause an abundance of marine life that supports the fisheries of Peru and northern Chile. Birds feeding on these organisms deposit their droppings on land, where they are mined as guano, a commercial source of phosphorus.

When the Humboldt Current is not as cool as usual, upwelling of nutrients does not occur, stagnation results, the fisheries decline, and climate patterns change globally. This phenomenon, called an **El Niño–Southern Oscillation,** has a profound effect on weather; a severe El Niño affects the weather over three-quarters of the globe. This phenomenon is discussed in "Hurricane Patterns in the United States" on page 796.

> ▶ **40.3 CHECK YOUR PROGRESS**
>
> 1. Explain how and why the Gulf Stream is a warm ocean current that brings warmth to northern regions.

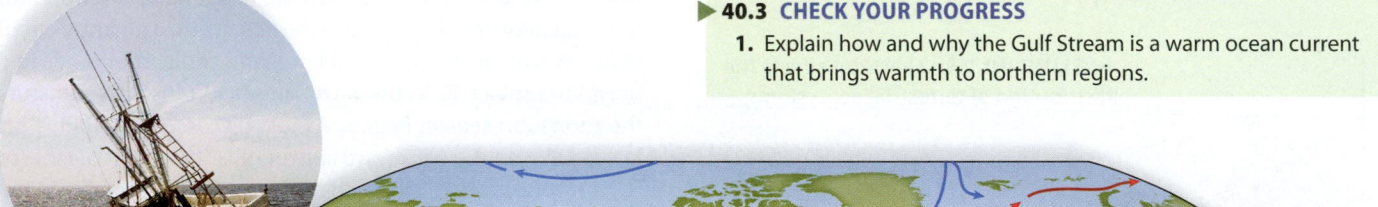

> Ocean currents can be helpful to sea life and humans, but they can also be deadly.

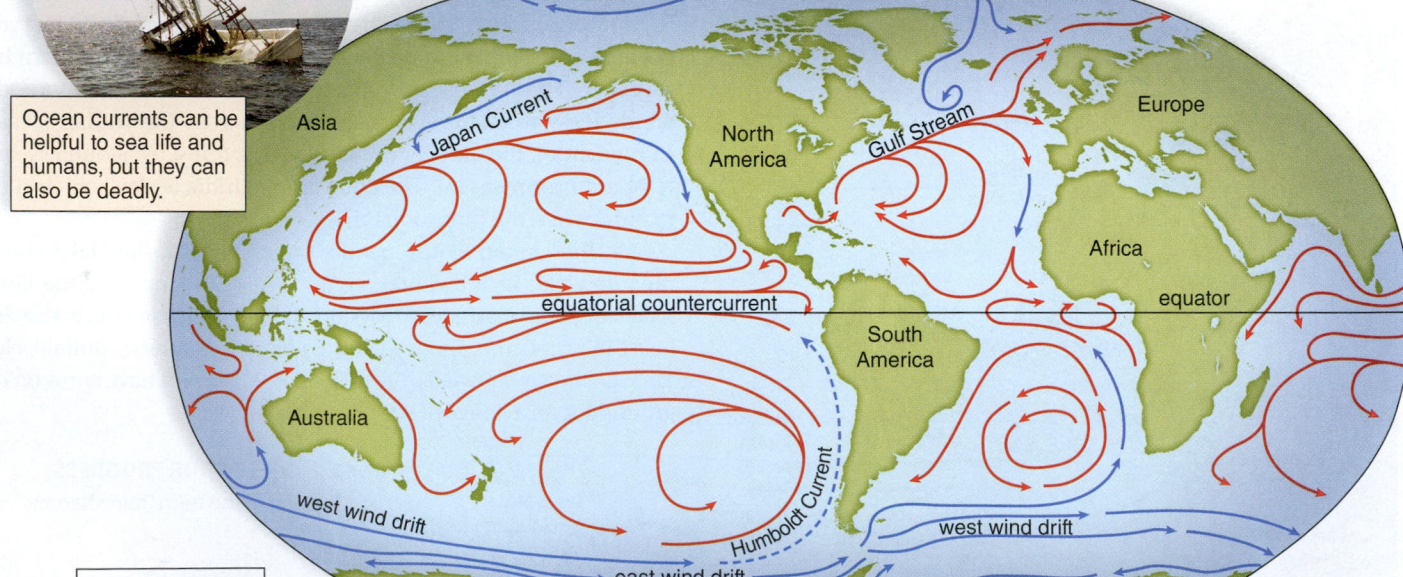

FIGURE 40.3 The arrows on this map indicate the locations and directions of the major ocean currents set in motion by the global wind circulation. By carrying warm water to cool latitudes (e.g., the Gulf Stream) and cool water to warm latitudes (e.g., the Humboldt Current), these currents have a major effect on the world's climates.

Terrestrial Ecosystems

Each major terrestrial ecosystem (sometimes called a **biome**) has a particular mix of plants and animals that are adapted to living under the prevailing environmental conditions.

40.4 Major terrestrial ecosystems are characterized by particular climates

LEARNING OUTCOME

When you complete this section, you should be able to

1. List and discuss in general the distribution of major terrestrial ecosystems.

As we have seen, climate consists of the prevailing temperature and rainfall of a region. When terrestrial ecosystems are plotted according to their climate, a particular distribution pattern results. The distribution of a number of terrestrial ecosystems is shown in Figure 40.4. However, we are going to concentrate on the selected ecosystems listed in Table 40.4. The table tells you about the climate but not about the organisms that are adapted to living in the ecosystem. The descriptions in the next few sections describe some of the major populations in each biome.

Animation Biomes

Be sure to note where the ecosystems listed in Table 40.4 are distributed. Even though Figure 40.4 shows definite demarcations, keep in mind that ecosystems gradually change from one type to the other. Also, although we will be discussing different ecosystems separately, we should remember that each one has inputs from and outputs to all the other terrestrial and aquatic ecosystems of the biosphere.

▶ **40.4 CHECK YOUR PROGRESS**

1. Why do the same types of ecosystems occur around the globe?

TABLE 40.4	Selected Terrestrial Ecosystems
Name	**Characteristic**
Tundra	Around North Pole; very cold (−12°C to −6°C); little rainfall (less than 25 cm/year); permafrost (permanent ice) year-round within a meter of surface.
Taiga (coniferous forest)	Large northern biome just below the Arctic Circle; temperature is below freezing for half the year; moderate precipitation (30–85 cm/year); long nights in winter and long days in summer.
Temperate deciduous forest	Eastern half of United States, Canada, Europe, and parts of Russia; four seasons of the year with hot summers and cold winters; ample precipitation (75–150 cm/year).
Grasslands	Savanna in Africa and temperate grassland elsewhere; hot in summer and cold in winter (U.S.); moderate precipitation; good soil for agriculture.
Tropical rain forest	Located near the equator in Latin America, Southeast Asia, and West Africa; warm (20°–25°C) and heavy precipitation (190 cm/year).
Desert	Northern and Southern Hemispheres at 30° latitude; hot (38°C) days and cold (7°C) nights; low precipitation (less than 25 cm/year).

FIGURE 40.4 Pattern of ecosystem distribution on land.

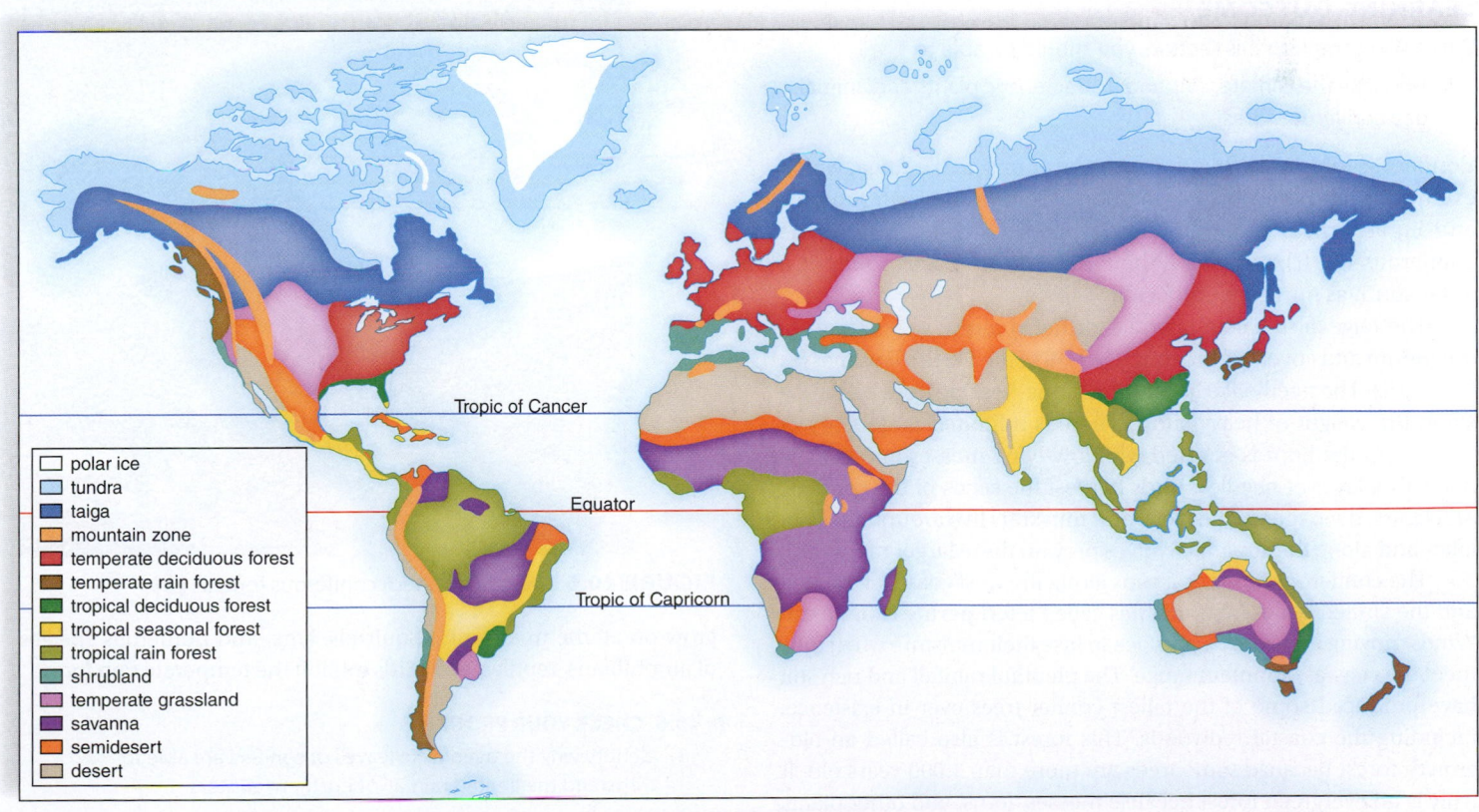

- ☐ polar ice
- ☐ tundra
- ☐ taiga
- ☐ mountain zone
- ☐ temperate deciduous forest
- ☐ temperate rain forest
- ☐ tropical deciduous forest
- ☐ tropical seasonal forest
- ☐ tropical rain forest
- ☐ shrubland
- ☐ temperate grassland
- ☐ savanna
- ☐ semidesert
- ☐ desert

Tropic of Cancer

Equator

Tropic of Capricorn

40.5 The tundra is cold and dark much of the year

The **Arctic tundra** ecosystem, which encircles the Earth just south of the ice-covered polar seas in the Northern Hemisphere, covers about 20% of the Earth's land surface (Fig. 40.5). (A similar ecosystem, called the alpine tundra, occurs above the timberline on mountain ranges.) The Arctic tundra is cold and dark much of the year. Its winters are extremely long, cold, and harsh, and its summers are short (6–8 weeks). Because rainfall amounts to only about 20 cm a year, the tundra could possibly be considered a desert, except that melting snow creates a landscape of pools and mires in the summer, especially because so little evaporates. Only the topmost layer of soil thaws; the **permafrost** beneath this layer is always frozen, and therefore drainage is minimal. The available soil in the tundra is nutrient-poor.

Trees are not found in the tundra because the growing season is too short, their roots cannot penetrate the permafrost, and they cannot become anchored in the boggy soil of summer. In the summer, the ground is covered with short grasses and sedges, as well as numerous patches of lichens and mosses. Flowers and dwarf woody shrubs, such as dwarf birch, seed quickly while there is plentiful sun for photosynthesis.

A few animals live in the tundra year-round. For example, the mouselike lemming stays beneath the snow; the ptarmigan, a grouse, burrows in the snow during storms; and the musk ox conserves heat

FIGURE 40.5 Tundra, the northern-most ecosystem.

because of its thick coat and short, squat body. Other animals that live in the tundra include snowy owls, lynx, voles, Arctic foxes, and snowshoe hares. In the summer, the tundra is alive with numerous insects and birds, particularly shorebirds and waterfowl that migrate inland. Caribou in North America and reindeer in Asia and Europe also migrate to and from the tundra, as do the wolves that prey upon them. Polar bears are common near the coastal regions.

▶ **40.5 CHECK YOUR PROGRESS**
1. Account for why only small, hardy plants grow in the tundra.

40.6 Coniferous forests are dominated by gymnosperms

Coniferous forests are found in three locations: in the **taiga,** which extends around the world in the northern part of North America and Eurasia; near mountaintops (where it is called a montane coniferous forest); and along the Pacific coast of North America, as far south as northern California.

The taiga, also called boreal (northern) forest, exists south of the tundra and covers approximately 11% of the Earth's landmasses (Fig. 40.6). The needlelike leaves of its cone-bearing trees can withstand the weight of heavy snow. There is a limited understory of plants, but the floor is covered with low-lying mosses and lichens beneath a layer of needles. Birds harvest the seeds of the conifers, and bears, deer, moose, beavers, and muskrats live around the cool lakes and along the streams. Wolves prey on these larger mammals.

The coniferous forest that runs along the west coasts of Canada and the United States is sometimes called a **temperate rain forest.** Winds moving in off the Pacific Ocean lose their moisture when they meet the coastal mountain range. The plentiful rainfall and rich soil have produced some of the tallest conifer trees ever in existence, including the coastal redwoods. This forest is also called an old-growth forest because some trees are more than 1,000 years old. It truly is an evergreen forest because mosses, ferns, and other plants

FIGURE 40.6 Taiga, a northern coniferous forest.

grow on all the tree trunks. Squirrels, lynx, and numerous species of amphibians, reptiles, and birds inhabit the temperate rain forest.

▶ **40.6 CHECK YOUR PROGRESS**
1. Identify why the needlelike leaves of conifers are able to withstand the limited rainfall of northern winters.

HOW LIFE CHANGES *Application*

40A Land of Beringia

Paleoecologists can use the rich fossil record to reconstruct ecological communities that exist no more. One such community called Beringia was located between present-day Siberia in northeastern Russia and Alaska in extreme northwestern North America. These two areas are now separated by the Bering Strait, an 80-km-wide body of water that connects the Arctic and Pacific Oceans. Over the long history of Earth, temperatures have fluctuated and sea levels have risen and fallen as part of natural processes. During the cold periods, large volumes of water were held in ice and snow. Consequently, there was less water in the oceans and sea levels dropped. Ecologists believe that during the last Ice Age 12,000–20,000 years ago, sea levels were 100–150 m lower than at present. The lower sea level exposed Beringia, a large stretch of land that formed a land bridge linking the two continents (Fig. 40A, *top*).

Paleoecologists have determined that Beringia was a cold, dry, open plain. There was likely little snowfall due to a "rain shadow" effect caused by the Siberian mountain ranges (see Fig. 40.2B). Fossils found on either side of the Bering Strait indicate that Beringia supported a steppe community with a flora dominated by grasses and numerous shrubs. Fossils have also shown that the fauna of Beringia included the large mammals of the Ice Age such as mammoths, steppe bison, and the scimitar cat (Fig. 40A, *bottom*). Beringia served as an important link between Asia and North America and provided a route for the exchange of plants, animals, fungi, and other life-forms that migrated between continents. For example, the alders (*Alnus* spp.) are shrubs and small trees that originated in Asia. They migrated through Beringia and spread throughout North America. Now, members of this genus are important components of many North American wetlands and stream-side communities. Many other genera show a similar history of migration from Asia into new areas where they became established and diversified. Possibly one of the most significant migrations through Beringia was that of humans. Ecologists believe that human populations from Asia migrated into Beringia and then into North America as the glaciers melted and made passage south possible. After the glaciers melted, sea levels rose, and the biologically important communities inhabiting this link between Asia and North America disappeared under the sea.

CONSIDER THESE QUESTIONS

1. Revisit Figure 16.2 and explain why all continents have the major groups of organisms, but particular types can vary between continents. If organisms such as the alders occur only on two continents what does it tell you?
2. The sequencing of DNA from a 4,000 year-old human found in Greenland showed that he came from Siberia. Speculate on how the man got to Greenland, which is east of North America. Could he have migrated by way of Beringia or did he travel between landmasses by sea?

FIGURE 40A *Top*: Beringia (purple) included a land bridge that is now under water. *Bottom*: Beringia existed when humans hunted big game such as mammoths.

 Explore the concepts through a variety of multimedia assets and question types.
www.mcgrawhillconnect.com

40.7 Temperate deciduous forests have abundant life

LEARNING OUTCOME

When you complete this section, you should be able to

1. Describe the climate, soil, and characteristic plants and animals of a temperate diciduous forest.

Temperate deciduous forests are found south of the taiga in eastern North America, eastern Asia, and much of Europe (Fig. 40.7). The climate in these areas is moderate, with relatively high rainfall (75–150 cm per year). The seasons are well defined, and the growing season ranges between 140 and 300 days. The trees, which include oak, beech, sycamore, and maple, have broad leaves and are termed deciduous trees; they lose their leaves in the fall and grow them in the spring. In southern temperate deciduous forests, evergreen magnolia trees can be found.

The tallest trees form a canopy, an upper layer of leaves that are the first to receive sunlight. Even so, enough sunlight penetrates to provide energy for another layer of trees, called understory trees. Beneath these trees are shrubs that may flower in the spring before the trees have put forth their leaves. Still another layer of plant growth—mosses, lichens, and ferns—resides beneath the shrub layer. This *stratification* provides a variety of habitats for insects and birds. Ground life is also plentiful. Squirrels, cottontail rabbits, shrews, skunks, woodchucks, and chipmunks are small herbivores. These and ground birds, such as turkeys, pheasants, and grouse, are preyed on by red foxes. White-tailed deer and black bears have been increasing in number for some time. In contrast to the taiga, amphibians and reptiles live in this ecosystem because the winters are not as cold. Frogs and turtles prefer an aquatic existence, as do the beavers and muskrats, which are mammals.

Autumn fruits, nuts, and berries serve as food for the winter. The leaves, after turning brilliant colors and falling to the ground, contribute to a rich layer of humus. The minerals within the rich

FIGURE 40.7 Temperate deciduous forest in the fall.

soil are washed far into the ground by spring rains, but the deep tree roots capture these minerals and bring them back up into the forest system again.

▶ **40.7 CHECK YOUR PROGRESS**

1. Explain why you would expect diverse plants and animals in a temperate deciduous forest.

40.8 Temperate grasslands have extreme seasons

LEARNING OUTCOME

When you complete this section, you should be able to

1. Describe the climate, soil, and characteristic plants and animals of a temperate grassland.

The **temperate grasslands** include the Russian steppes, the South American pampas, and the North American prairies (Fig. 40.8). In these grasslands, winters are bitterly cold, and summers are hot and dry. When traveling across the United States from east to west, temperate deciduous forest transitions into *tall-grass prairie* roughly along the border between Illinois and Indiana. The tall-grass prairie requires more rainfall than does the *short-grass prairie,* which occurs near deserts. Large herds of bison—estimated at hundreds of thousands—once roamed the prairies, as did herds of pronghorn antelope. Now, small mammals, such as mice, prairie dogs, and rabbits, typically live belowground, but usually feed aboveground. Hawks, snakes, badgers, coyotes, and foxes feed on these mammals. However, virtually all of these grasslands have been converted to agricultural lands because of their fertile soils. Before farming began, the grasses died off and decomposed and nutrients were added to the soil every year.

Video Tallgrass Prairie System

FIGURE 40.8 Temperate grassland in the summer.

▶ **40.8 CHECK YOUR PROGRESS**

1. Account for why grassland soil contains much organic matter.

40.9 Savannas have wet-dry seasons

LEARNING OUTCOME

When you complete this section, you should be able to

1. Describe the climate, soil, and characteristic plants and animals of a savanna.

Savannas occur in regions where a relatively cool dry season is followed by a hot rainy season. The largest savannas are in central and southern Africa; other savannas exist in Australia, Southeast Asia, and South America (Fig. 40.9). The savanna is characterized by large expanses of grasses with sparse populations of trees. The plants of the savanna have extensive and deep root systems that enable them to survive drought and fire. One tree that can survive the severe dry season is the thorny flat-topped acacia, which sheds its leaves during a drought. The African savanna supports the greatest variety and number of large herbivores of all the biomes (Fig. 40.9). Elephants and giraffes are browsers that feed on tree vegetation. Antelopes, zebras, wildebeests, water buffalo, and some rhinoceroses are grazers that feed on grasses. Any plant litter that is not consumed by grazers is attacked by a variety of small organisms, among them termites. Termites build towering nests in which they tend fungal gardens, their source of food. The herbivores support a large population of carnivores. Lions and hyenas sometimes hunt in packs, cheetahs hunt singly by day, and leopards hunt singly by night.

 **Video** Thorn Tree Ant

 **Video** Dung Beetles

FIGURE 40.9 The African savanna.

▶ **40.9 CHECK YOUR PROGRESS**

1. Account for why lions, as representative carnivores, occur in the African savanna.

40.10 Deserts have very low annual rainfall

LEARNING OUTCOME

When you complete this section, you should be able to

1. Describe the climate, soil, and characteristic plants and animals of a desert.

Deserts are usually found at latitudes of about 30° in both the Northern and Southern Hemispheres (Fig. 40.10). The winds that descend in these regions lack moisture, and the annual rainfall is less than 25 cm. Days are hot because lack of cloud cover allows the sun's rays to penetrate easily, but nights are cold because heat escapes easily into the atmosphere.

The Sahara Desert, which stretches all the way from the Atlantic coast of Africa to the Arabian peninsula, and a few other deserts have little or no vegetation. But most have a variety of plants. Desert plants are highly adapted to survive long droughts, extreme heat, and extreme cold. Adaptations to these conditions include thick epidermal layers, water-storing stems and leaves, and the ability to set seeds quickly in the spring. The best-known desert perennials in North America are the succulent, spiny-leafed cactuses, which have stems that store water and carry on photosynthesis.

Some animals are adapted to the desert environment. To conserve water, many desert animals are nocturnal or burrowing and have a protective outer body covering. A desert has numerous insects, which pass through the stages of development when there is rain. Reptiles, especially lizards and snakes, are perhaps the most characteristic group of vertebrates found in deserts, but running birds (e.g., the roadrunner) and rodents (e.g., the kangaroo

FIGURE 40.10 Desert with some vegetation.

rat) are also well known. Larger mammals, such as the kit fox, prey on the rodents, as do hawks.

▶ **40.10 CHECK YOUR PROGRESS**

1. Account for why desert animals are usually small.

40.11 Tropical rain forests are warm with abundant rainfall

LEARNING OUTCOME

When you complete this section, you should be able to

1. Describe the climate, soil, and characteristic plants and animals of a tropical rain forest.

In the **tropical rain forests** of South America, Africa, and the Indo-Malayan region near the equator, the temperature is always warm during the day (between 20° and 25°C), and rainfall is plentiful (a minimum of 190 cm per year). This may be the richest ecosystem, in terms of both number of different kinds of species and their abundance. The diversity of species is enormous—a 10-km^2 area of tropical rain forest may contain 1,500 species of flowering plants, including the trees.

A tropical rain forest has a complex structure, housing many levels of life, including the forest floor, the understory, and the canopy. The canopy filters out sunlight, and the plants of the forest floor, such as ferns, can tolerate minimal light. The understory consists of shorter trees that receive some light and bear epiphytes. **Epiphytes** are plants that grow on other plants but usually have roots of their own to absorb moisture and minerals leached from their hosts; others catch rain and debris in hollows produced by overlapping leaf bases. The most common epiphytes are related to pineapples, orchids, and ferns. The canopy, topped by the crowns of tall trees, is the most productive level of the tropical rain forest (Fig. 40.11A). Some of the broadleaf evergreen trees grow to 15–50 m or more. These tall trees often have trunks buttressed at ground level to prevent them from toppling over. Lianas, or woody vines, which encircle the tree as it grows, also help strengthen the trunk.

Although some animals live on the forest floor (e.g., pacas, agoutis, peccaries, and armadillos), most live in the trees (Fig. 40.11B). Insect life is so abundant that the majority of species

FIGURE 40.11A Levels of life in a tropical rain forest.

have not been identified yet. Termites play a vital role in decomposing woody plant material, and ants are everywhere, particularly in the trees. The various birds, such as hummingbirds, parakeets, parrots, and toucans, are often beautifully colored. Amphibians and reptiles are well represented by many types of frogs, snakes, and lizards. Lemurs, sloths, and monkeys are well-known primates that feed on the fruits of the trees. The largest carnivores are the big cats—the jaguars in South America and the leopards in Africa and Asia.

▶ **40.11 CHECK YOUR PROGRESS**

1. Account for why life in a tropical rain forest is extremely diverse.

Poison-dart frog,
Dendrobates histrionicus

Blue and gold macaw,
Ara ararauna

Thorny-devil katydid,
Panacanthus cuspidatus

Ocelot,
Felis pardalis

Brush-footed butterfly,
Anartia amalthea linnaeus

Lemur,
Lemur catta

Arboreal lizard,
Calotes calotes

FIGURE 40.11B Representative animals of the tropical rain forests of the world.

Aquatic Ecosystems

Freshwater aquatic ecosystems are extremely important to humans because they temporarily hold fresh water before it travels to the ocean. Marine ecosystems include those along the coast and within the ocean.

40.12 Fresh water flows into salt water

LEARNING OUTCOME

When you complete this section, you should be able to

1. Describe freshwater ecosystems, including oligotrophic lakes and eutrophic lakes.

Figure 40.12A shows how a freshwater ecosystem joins a saltwater ecosystem. Fresh water flows within **streams** and **rivers** and is contained, at least temporarily, in **lakes** and **ponds.** Mountain streams have cold, clear water that flows over waterfalls and rapids (Fig. 40.12A). ➊ Here, the clawed feet of a long-legged stonefly larva help it hold onto the stones in the streambed. The streams join to form a river that flows gently enough for ➋ trout to exist in occasional pools of oxygen-rich water. ➌ Carp are fish adapted to water that contains little oxygen and much sediment, as might be found at the bottom of a lake. The river meanders across broad, flat valleys, and finally empties into the ocean. At its mouth, the river divides into the many muddy channels of a delta. ➍ Blue crabs frequent such areas. Nearby, salt marshes, which are characterized by the presence of rushes, reeds, and other grasses, are extremely productive ecosystems. They provide food and habitats for fish, waterfowl, and other wildlife.

Video Thames River

In general, wetlands (lands that are wet some part of the year) directly absorb storm waters and also absorb overflows from lakes and rivers. In this way, they protect farms, cities, and towns from the devastating effects of floods. They also purify waters by filtering them and by diluting and breaking down toxic wastes and excess nutrients. Unfortunately, humans have the habit of filling in wetlands. Between the 1780s and the 1980s, approximately 53% of the original wetlands in the contiguous 48 states of the United States were lost. On average, this amounts to a loss of about 24.3 hectares* of wetlands per hour during that 200-year time span.

Oligotrophic lake

FIGURE 40.12B *Top:* Oligotrophic lakes are nutrient-poor and have limited algal growth. *Bottom:* Eutrophic lakes are nutrient-rich and have much algal growth.

Eutrophic lake

Lakes are often classified by their nutrient status (Fig. 40.12B). **Oligotrophic lakes** are nutrient-poor, having a small amount of organic matter and low productivity. **Eutrophic lakes** are nutrient-rich, having plentiful organic matter and high productivity. Eutrophic lakes are usually situated in naturally nutrient-rich regions, or are enriched by agricultural or urban and suburban runoff. Oligotrophic lakes can become eutrophic through large inputs of nutrients, a process called **eutrophication.**

Marine ecosystems are discussed in section 40.13.

▶ **40.12 CHECK YOUR PROGRESS**

1. Explain how it is possible to divide freshwater ecosystems into those that are flowing-water and those that are standing-water ecosystems.

*Hectares are a metric unit of measurement.

FIGURE 40.12A Freshwater ecosystems (e.g., stream, lake, and river) often adjoin wetland ecosystems (e.g., delta and salt marsh).

LEARNING OUTCOME

When you complete this section, you should be able to

1. Briefly describe coastal ecosystems (estuaries) and ocean ecosystems by stating their biotic and abiotic characteristics.

Coastal Ecosystems Border the Oceans Salt marshes, discussed previously, and also mudflats and mangrove swamps, featured in Figure 40.13A, are ecosystems that occur at a delta. Mangrove swamps develop in subtropical and tropical zones, while marshes and mudflats occur in temperate zones. These ecosystems are often designated as an **estuary.** So are coastal bays, fjords (an inlet of water between high cliffs), and some lagoons (a body of water separated from the sea by a narrow strip of land).

An estuary has a partially enclosed body of water where fresh water and seawater meet and mix as a river enters the ocean.

At mudflats migrating birds feed on small fish and shellfish.

Mangrove roots prevent erosion and serve as a nursery for sea life.

Rocky shore has diverse life in tide pools.

FIGURE 40.13A Coastal ecosystems.

Organisms living in an estuary must be able to withstand constant mixing of waters and rapid changes in salinity. But those organisms adapted to the estuarine environment find an abundance of nutrients. An estuary acts as a nutrient trap because the sea prevents the rapid escape of nutrients brought by a river. As the result of usually warm, calm waters and plentiful nutrients, estuaries are biologically diverse and highly productive.

Phytoplankton and shore plants thrive in nutrient-rich estuaries, providing an abundance of food and habitat for animals. It is estimated that nearly two-thirds of marine fishes and shellfish spawn and develop in the protective and rich environment of estuaries, making the estuarine environment the nursery of the sea. An abundance of larval, juvenile, and mature fish and shellfish attract a number of predators, such as reptiles, birds, and fishers of various types.

Rocky shores (Fig. 40.13A, *bottom*) and *sandy shores* are constantly bombarded by the sea as the tides roll in and out. The *intertidal zone* lies between the high- and low-tide marks (Fig. 40.13B). In the upper portion of the intertidal zone on a rocky shore, barnacles are glued so tightly to the stone by their own secretions that their calcareous outer plates remain in place, even after the enclosed shrimplike animal dies. In the midportion of the intertidal zone, brown algae, known as rockweed, may overlie the barnacles. Below the intertidal zone, macroscopic seaweeds, which are the main photosynthesizers, anchor themselves to the rocks by holdfasts.

On a sandy beach, organisms cannot attach themselves to shifting, unstable sands; therefore, nearly all the permanent residents dwell underground. They either burrow during the day and surface to feed at night, or they remain permanently within their burrows and tubes. Ghost crabs and sandhoppers (amphipods) burrow themselves above the high-tide mark and feed at night when the tide is out. Sandworms and sand (ghost) shrimp remain within their burrows in the intertidal zone and feed on detritus whenever possible. Still lower in the sand, clams, cockles, and sand dollars are found. A variety of shorebirds visit the beaches and feed on various invertebrates and fishes.

FIGURE 40.13B
Ocean ecosystems.

Oceans Shallow ocean waters (called the *euphotic zone*) contain a greater concentration of organisms than the rest of the sea (Fig. 40.13B). Here, **phytoplankton** (i.e., algae) is food not only for **zooplankton** (i.e., protozoans and microscopic animals) but also for small fishes. These attract a number of predatory and commercially valuable fishes. On the *continental shelf,* seaweed grows, even on outcroppings as the water gets deeper. Clams, worms, and sea urchins are preyed upon by sea stars, lobsters, crabs, and brittle stars.

Video Plankton Diversity

Coral reefs are areas of biological abundance just below the surface in shallow, warm, tropical waters (see Fig. 37.1). Their chief constituents are stony corals, animals that have a calcium carbonate (limestone) exoskeleton, and calcareous red and green algae. Corals provide a home for microscopic algae called *zooxanthellae*. The corals, which feed at night, and the algae, which photosynthesize during the day, are mutualistic and share materials and nutrients. The algae need sunlight, and this may be the reason coral reefs form only in shallow, sunlit waters.

Video Coral Reef Ecosystems

A reef is densely populated with life. The large number of crevices and caves provide shelter for filter feeders (sponges, sea squirts, and fanworms) and for scavengers (crabs and sea urchins). The barracuda, moray eel, and shark are top predators in coral reefs. There are many types of small, beautifully colored fishes. These become food for larger fishes, including snappers that are caught for human consumption.

Video Coral Reef Spawning

Most of the ocean lies within the **pelagic zone,** divided as noted in Figure 40.13C. The *epipelagic zone* lacks the inorganic nutrients of shallow waters, and therefore it does not have as high a concentration of phytoplankton, even though the surface is sunlit. Still, the photosynthesizers are food for a large assembly of zooplankton, which then become food for schools of various fishes. A number of porpoise and dolphin species visit and feed in the epipelagic zone. Whales too are mammals that live in this zone. Baleen whales strain krill (small crustaceans) from the water, and toothed sperm whales feed primarily on the common squid.

Video Ocean Fishing Ban

Animals in the deeper waters of the *mesopelagic zone* are carnivores, which are adapted to the absence of light, and tend to be translucent, red-colored, or even luminescent. There are luminescent shrimps, squids, and fishes, including lantern and hatchet fishes. Various species of zooplankton, invertebrates, and fishes migrate from the mesopelagic zone to the surface to feed at night.

The deepest waters of the *bathypelagic zone* are in complete darkness except for an occasional flash of bioluminescent light. Carnivores and scavengers inhabit this zone. Strange-looking fishes with distensible mouths and abdomens and small, tubular eyes feed on infrequent prey.

It once was thought that few animals exist on the *abyssal plain* (see Fig. 40.13B) because of the intense pressure and the extreme cold. Yet, many invertebrates survive there by feeding on debris floating down from the mesopelagic zone. Sea lilies (crinoids) rise above the seafloor; sea cucumbers and sea urchins crawl around on the sea bottom; and tube worms burrow in the mud.

The flat abyssal plain is interrupted by enormous underwater mountain chains called oceanic ridges. Along the axes of the ridges, crustal plates spread apart, and molten magma rises to fill the gap. At **hydrothermal vents,** seawater percolates through cracks and is heated to about 350°C, causing sulfate to react with

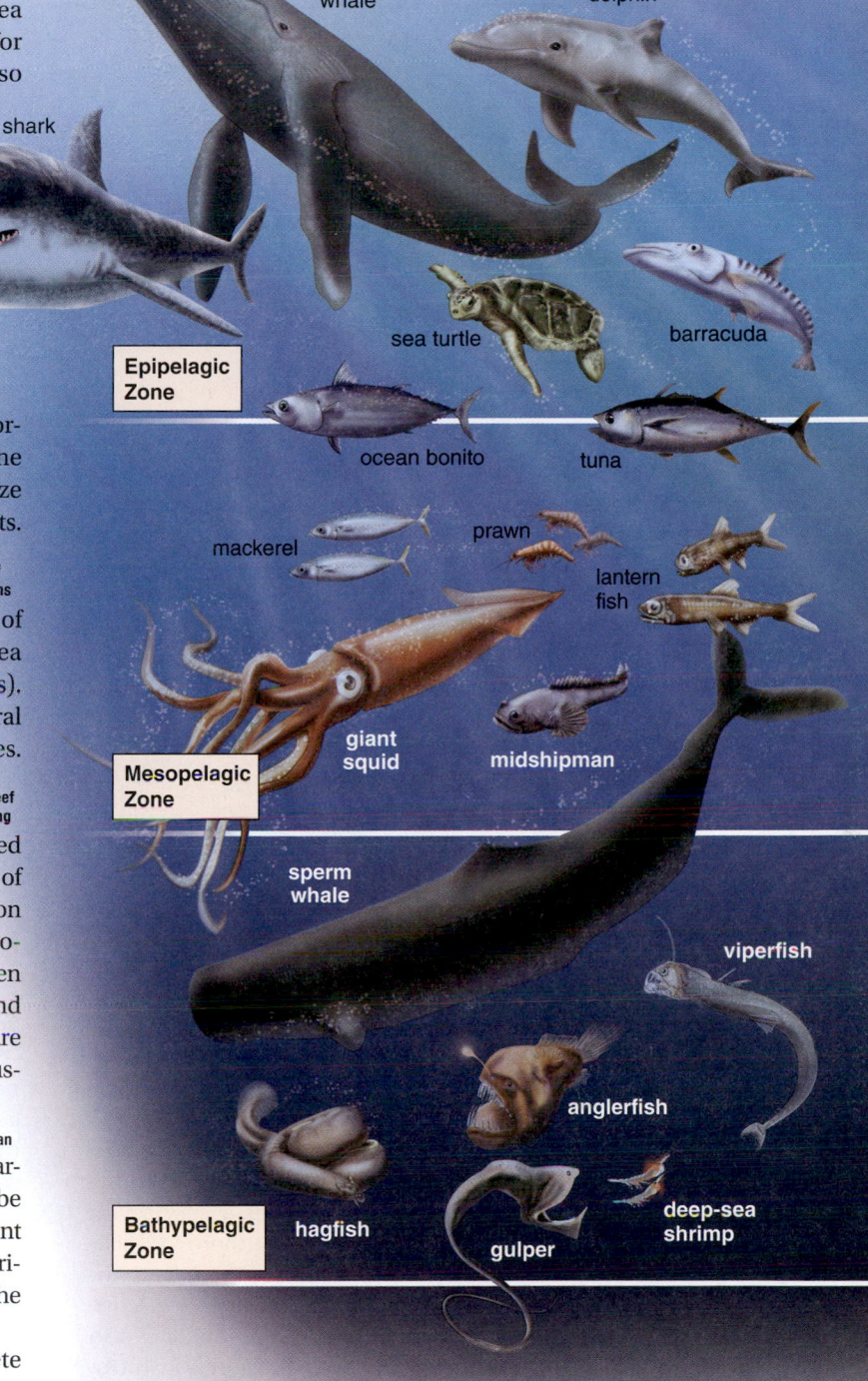

FIGURE 40.13C Ocean inhabitants in divisions of the pelagic zone.

water and form hydrogen sulfide (H_2S). Chemosynthetic bacteria that obtain energy from oxidizing hydrogen sulfide exist freely or mutualistically within the tissues of organisms. They are the start of food chains for an ecosystem that includes huge tube worms, clams, crustaceans, echinoderms, and fishes. This ecosystem can exist where light never penetrates because, unlike photosynthesis, chemosynthesis does not require light energy.

▶ **40.13 CHECK YOUR PROGRESS**
1. List and briefly describe three coastal ecosystems.
2. Account for why coral reefs, like tropical rain forests, have abundant life.

40B Hurricane Patterns in the United States

Normal conditions off the coast of South America are now called a La Niña, from Spanish meaning "the girl." The southeast trade winds normally move along the coast and turn west because of the Earth's daily rotation on its axis. As the winds drag warm ocean waters from east to west, there is an upwelling of nutrient-rich cold water from the ocean's depths, resulting in a bountiful Peruvian harvest of anchovies. When the warm ocean waters reach their western destination, the monsoons bring rain to India and Indonesia (Fig. 40B, *top*). Hurricanes are free to bring rain to the east coast of the United States.

El Niño

During an El Niño–Southern Oscillation, due to the change in barometric pressure, both the northeast and the southeast trade winds slacken. Upwelling no longer occurs, and the anchovy catch off the coast of Peru plummets. During a severe El Niño, from Spanish meaning "the boy," waters from the eastern Pacific never reach the west, and the winds lose their moisture in the middle of the Pacific instead of over the Indian Ocean (Fig. 40B, *bottom*). The monsoons fail, and drought occurs in India, Indonesia, Africa, and Australia. Harvests decline, cattle must be slaughtered, and famine is likely in highly populated India and Africa.

In addition, a backward movement of winds and ocean currents may even occur so that the waters warm to more than 14° above normal along the west coasts of the Americas. This is a sign that a very severe El Niño–Southern Oscillation has occurred, and the weather changes are dramatic in the Americas also. Southern California is hit by storms and even hurricanes, and the deserts of Peru and Chile receive so much rain that flooding occurs. A jet stream (strong wind currents) can carry moisture into Texas, Louisiana, and Florida, with flooding a near certainty. Or the winds can turn northward and deposit snow in the mountains along the West Coast so that flooding occurs in the spring. Some parts of the United States, however,

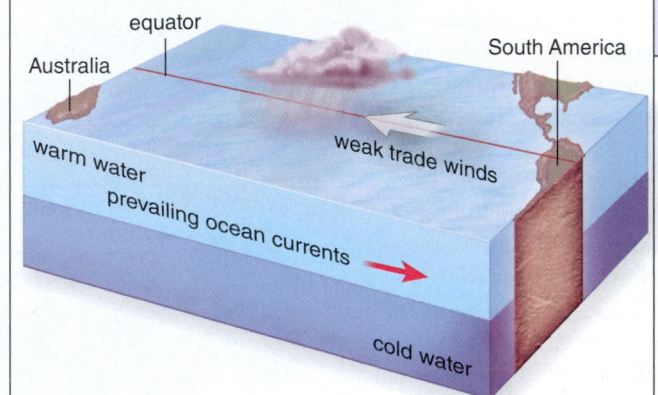

La Niña
• Barometric pressure is high over the southeastern Pacific.
• Upwelling off the west coast of South America brings cold waters to the surface.
• Hurricanes occur off the east coast of the United States.
• Monsoons associated with the Indian Ocean occur.

El Niño–Southern Oscillation
• Barometric pressure is low over the southeastern Pacific.
• Great ocean warming occurs off the west coast of the Americas.
• Hurricanes occur off the west coast of the United States.
• Monsoons associated with the Indian Ocean fail.

FIGURE 40B *Top:* La Niña. *Below:* El Niño.

benefit from an El Niño. The Northeast is warmer than usual, few if any hurricanes hit the East Coast, and there is a lull in tornadoes throughout the Midwest. Altogether, a severe El Niño affects the weather over three-quarters of the globe.

Eventually, an El Niño dies out, and normal conditions return. Since 1991, El Niños have varied in magnitude, and two record-breaking El Niños have occurred. Due to climate change, the severity of El Niños remains somewhat unpredictable.

CONSIDER THESE QUESTIONS

1. Ecologists have a saying: "Everything is connected to everything else." Give examples from (1) this application and (2) the carbon cycle (see section 39.14) that support this saying.

2. The monsoons must be very trying in India, and yet without them, the rice crop fails and people go hungry. Give some other examples in this or other countries that show humans are dependent on the natural environment.

Where will most rains occur? On the west coast or the east coast? Over the ocean or over land?

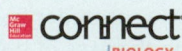

Explore the concepts through a variety of multimedia assets and question types.

www.mcgrawhillconnect.com

THE CHAPTER IN REVIEW

CONNECTING THE CONCEPTS

The Earth's diverse ecosystems have resulted from interactions between the biotic communities and the abiotic environment. Organisms have helped create the chemical and physical conditions of streams, lakes, and oceans. The soils of terrestrial ecosystems and the sediments of aquatic ecosystems are structured largely by the activities of organisms.

Over geologic time, the biosphere has been constantly changing. Changes in the sun's radiation output and in the tilt of the Earth's axis have altered the pattern of solar energy reaching the Earth's surface. Geologic processes have also modified conditions for life. The drifting of continents has changed the arrangement of landmasses and oceans. Mountain ranges have been thrust up or eroded. Through these changing conditions, life has been evolving, and the structure of the Earth's ecosystems has evolved as well. In the last few million years,

humans appeared and learned to exploit the Earth's ecosystems.

Modern humans have transformed vast areas of many of the terrestrial biomes into farmland, cities, highways, and other developments. Through our use of resources and release of pollutants, we have now become an agent of global importance. Yet, we still depend on the biodiversity that exists in the Earth's ecosystems, and on the interactions of other organisms within the biosphere. These interactions influence climate, patterns of nutrient cycling and waste processing, and basic biological productivity. The Earth's biotic diversity also provides enjoyment and inspiration to millions of people, who spend billions of dollars to visit coral reefs, deserts, rain forests, and even the Arctic tundra. Unfortunately, as we shall see in Chapter 41, many of the Earth's ecosystems may not survive in their present form for our descendants' benefit. Although most of us

value Earth's biodiversity, human activities are threatening many species with extinction.

ANALYZE AND EVALUATE

1. Associate an animal with (a) a temperate deciduous forest, (b) a tropical rain forest, and (c) the deep ocean. Name two characteristics of these animals that show they are adapted to where they live.
2. A mangrove swamp (see Fig. 40.13A, *middle*) can protect the coastline and lessen the effect of hurricanes. With this in mind, discuss whether it would behoove a human population following a devastating hurricane to restore a mangrove swamp instead of a housing development at the coast.
3. List five other negative effects of the human population on a natural area. Discuss several ways people can lessen these effects.

SUMMARIZE

Climate and the Biosphere

40.1 Solar radiation and winds determine climate

- Climate is the average yearly temperature and rainfall of a region. The distribution of solar energy and global wind patterns caused by the spherical Earth and the rotation of Earth around the sun affect temperature (seasons), amounts of rainfall, and how winds blow.

40.2 Topography and other effects also influence climate

- **Topography** affects temperature, and therefore mountains affect climate just as latitude does. Mountains in the path of winds affect rainfall. A **rain shadow** occurs on the leeward side of a mountain. Atmospheric circulations between the ocean and the landmasses influence regional climate conditions, accounting for why Asia has a **monsoon** climate.

40.3 Ocean currents affect climates

- Winds cause a steady flow of water; currents move clockwise in the Northern Hemisphere and counterclockwise in the Southern Hemisphere.
- The Gulf Stream brings tropical water to the eastern coast of North America and to western Europe. The Humboldt Current brings cold water north to western South America.
- **Upwelling** occurs when cold, nutrient-rich waters rise, resulting in plentiful fisheries.

Terrestrial Ecosystems

40.4 Major terrestrial ecosystems are characterized by particular climates

- The major terrestrial ecosystems, sometimes called **biomes,** are tundra (light blue), taiga (dark blue), temperate deciduous forest (red), grasslands (light and dark purple), desert (tan), and tropical rain forest (olive green). An ecosystem's climate consists of average yearly temperature and rainfall.

40.5 The tundra is cold and dark much of the year
- In the **Arctic tundra** (light blue), winters are long and summers are short. The **permafrost** is always frozen.

40.6 Coniferous forests are dominated by gymnosperms
- **Taiga** (dark blue), found south of the tundra, has less rainfall than other forests.
- **Temperate rain forests** (brown), occurring along the west coasts of Canada and the United States, have plentiful rain and rich soil.

40.7 Temperate deciduous forests have abundant life
- **Temperate deciduous forests** (red) occur south of the taiga in eastern North America, eastern Asia, and Europe; they are characterized by moderate climate, high rainfall, and seasons.

40.8 Temperate grasslands have extreme seasons
- The Russian steppes, South American pampas, and North American prairies are all **temperate grasslands** (light purple) known for very cold winters and hot, dry summers.

40.9 Savannas have wet-dry seasons
- The largest **savannas** (dark purple) exist in central and southern Africa; they are characterized by large expanses of grasses with scattered trees and many and varied large herbivores and carnivores.

40.10 Deserts have very low annual rainfall
- **Deserts** (tan) exist in the Northern and Southern Hemispheres at latitudes of about 30°; they are characterized by hot days, cold nights, and low annual rainfall.

40.11 Tropical rain forests are warm with abundant rainfall
- **Tropical rain forests** (olive green) exist in equatorial regions of South America, Africa, and the Indo-Malayan region; they are characterized by wide species diversity and abundance. The three levels of a rain forest are the forest floor (sparse vegetation), the understory (small plants that can live in shade, including **epiphytes**), and the canopy (the most productive level).

Aquatic Ecosystems

40.12 Fresh water flows into salt water
- Freshwater ecosystems include **lakes, ponds, streams, rivers,** and wetlands (marshes, swamps, and bogs). Lakes are classified as **oligotrophic lakes** if they are nutrient-poor with low productivity and **eutrophic lakes** if they are nutrient-rich with high productivity.

40.13 Marine ecosystems include those of the coast and the ocean
- Coastal ecosystems include **estuaries** and **intertidal zones.** Shallow ocean waters are sunlit and contain many inorganic nutrients for photosynthesizers, **phytoplankton,** and **zooplankton. Coral reefs** occur there.
- Deep ocean waters constitute the **pelagic zone,** which is divided into the epipelagic, mesopelagic, and bathypelagic zones. The abyssal plain is interrupted by **hydrothermal vents** where chemoautotrophic bacteria are at the start of a food chain that includes giant clams and tube worms.

TEST YOURSELF

Climate and the Biosphere

1. The seasons are best explained by
   a. the distribution of temperature and rainfall in biomes.
   b. the tilt of the Earth as it orbits the sun.
   c. the daily rotation of the Earth on its axis.
   d. the fact that the equator is warm and the poles are cold.
2. Which of these influences the location of a particular biome?
   a. latitude
   b. average annual rainfall
   c. average annual temperature
   d. altitude
   e. All of these are correct.
3. Label this diagram:

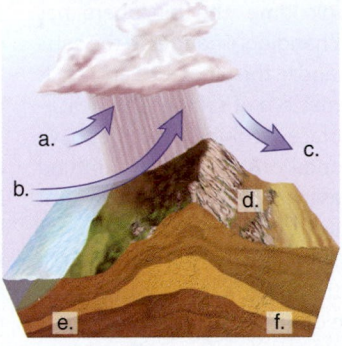

4. Whether ascending a mountain or traveling from the equator to the North Pole, you would observe terrestrial biomes in which order?
   a. coniferous forest, tropical rain forest, temperate deciduous forest, tundra
   b. tropical rain forest, coniferous forest, temperate deciduous forest, tundra
   c. tropical rain forest, temperate deciduous forest, coniferous forest, tundra
   d. tropical rain forest, coniferous forest, tundra, temperate deciduous forest
   e. tundra, temperate deciduous forest, tropical rain forest, coniferous forest
5. Which side of a mountain range lies in a rain shadow?
   a. windward
   b. leeward
6. A monsoon climate is produced
   a. when a landmass heats more rapidly than the ocean.
   b. when the ocean heats more rapidly than a landmass.
   c. in regions of the Earth that are tilting toward the sun.
   d. in regions of the Earth that are tilting away from the sun.
7. **THINKING CONCEPTUALLY** Rising temperatures at the equator are expected to intensify the effect of climate change in the temperate zone and at the poles. See section 40.1 and explain.

Terrestrial Ecosystems

8. Which of the following statements is not true of the Arctic tundra?
   a. Soil below the uppermost layer stays frozen year-round.
   b. Annual rainfall is low.
   c. Trees cannot grow because the growing season is too short.
   d. Large mammals are absent.

9. All of these phrases describe the tundra except
   a. low-lying vegetation.
   b. northernmost biome.
   c. short growing season.
   d. many different types of species.
10. Which biome is characterized by a coniferous forest with low average temperature and moderate rainfall?
    a. taiga
    b. savanna
    c. tundra
    d. tropical rain forest
    e. temperate forest
11. The major environmental factor that determines whether a prairie has tall grass or short grass is
    a. annual rainfall.
    b. mean annual high temperature.
    c. mean annual low temperature.
    d. soil nutrient levels.
    e. types of grazers present.
12. All of the following phrases describe a tropical rain forest except
    a. nutrient-rich soil.
    b. many arboreal plants and animals.
    c. canopy composed of many layers.
    d. broad-leaved evergreen trees.
13. Epiphytes are most likely to be found in
    a. deserts.
    b. tundra.
    c. grasslands.
    d. tropical rain forests.
    e. shrublands.
14. **THINKING CONCEPTUALLY** No matter the altitude, daytime temperatures are warm in the tropics but the nights are cool. Why might you expect to find desertlike conditions on the leeward side of mountains in the tropics?

Aquatic Ecosystems

15. Which of these is a function of a wetland?
    a. purifies water
    b. area where toxic wastes can be broken down
    c. helps absorb overflow and prevents flooding
    d. home for organisms that are links in food chains
    e. All of these are correct.
16. An oligotrophic lake
    a. is nutrient-rich.
    b. is cold.
    c. is likely to be found in an agricultural or urban area.
    d. has poor productivity.
17. Runoff of fertilizer and animal wastes from a large farm that drains into a lake would be the cause of which process?
    a. fall overturn
    b. eutrophication
    c. spring overturn
    d. upwelling
18. **THINKING CONCEPTUALLY** Why might a large population of decomposers in an oligotrophic lake cause it to have no fish?
19. In an ocean, the highest concentrations of phytoplankton would be found in the
    a. intertidal zone.
    b. euphotic zone.
    c. pelagic zone.
    d. All areas would have roughly equal concentrations.

20. Which zone of the oceanic province is completely dark?
    a. epipelagic zone
    b. mesopelagic zone
    c. bathypelagic zone
    d. euphotic zone
21. Energy for the food chain near hydrothermal vents comes from
    a. dead organisms that fall down from above.
    b. highly efficient photosynthetic phytoplankton.
    c. chemosynthetic bacteria.
    d. heat given off by the vents.
22. The directions of ocean currents are influenced by
    a. wind.
    b. landmasses.
    c. rivers.
    d. lunar cycles.
    e. Both a and b are correct choices.
23. The mild climate of Great Britain is best explained by
    a. the winds called westerlies.
    b. the spinning of Earth on its axis.
    c. Great Britain being a mountainous country.
    d. the flow of ocean currents.
24. Which of the following describes a change in weather conditions in the United States that can result from an El Niño–Southern Oscillation?
    a. fewer tornadoes in the Midwest
    b. less snow in the Northeast
    c. fewer hurricanes on the East Coast
    d. flooding in the south-central and southeastern regions
    e. All of these choices are correct.
25. **THINKING CONCEPTUALLY** Why may estuaries disappear if sea levels rise due to global warming?

GET INVOLVED

1. Global climate change might make it possible to test the hypothesis that climate determines the terrestrial ecosystem. How so?
2. Hypothesize why nearby logging, which causes dirt runoff in coastal areas, might interfere with the health of coral.

MEDIA STUDY TOOLS

mhhe.com/maderconcepts3

Enhance your study of this chapter with interactive study tools, practice tests, and engaging animations. Also, ask your instructor about the resources available through ConnectPlus, which includes LearnSmart, a personalized adaptive learning program, and a media-rich eBook.

41

Conservation Biology

Trouble in Paradise

Gardenia

Over 7 million tourists a year visit the Hawaiian Islands. Among the plentiful attractions are many species of native plants and animals found nowhere else on Earth. But sadly, about 63% of these species are at risk of extinction, and many others have already become extinct. What is causing such a tragedy?

Originally, the archipelago had around 10,000 native species of arthropods, 1,200 native species of land snails, 960 native species of flowering plants, and 115 native species of birds. (Because of Hawaii's distance from the mainland and the fact that most mammals cannot fly, it had only one native mammalian species: a bat.) Considering the impressive number of native species and the long period of time they have had to evolve in the tropical paradise, Hawaii has become an extraordinary place for biologists to come and study biodiversity.

Biologists were preceded by a long line of other people who came to live in Hawaii. Polynesians arrived more than 1,000 years ago, and Europeans came in the late 1700s. These people brought with them—both accidentally and intentionally—over 5,000 species of nonnative animals and plants. Mammals such as pigs, rats, cats, goats, and mongooses, as well as plants such as myrtle trees, Koster's curse (a weedy shrub), and fountain grass, are alien organisms that have caused much destruction in Hawaii.

Yellow hibiscus is the state flower of Hawaii.

Crested honeycreeper feeds on a Hawaiian myrtle.

The damage caused by feral pigs has gained a lot of attention in particular, and these animals are a good example of the problems created by alien species. The pigs voraciously eat all kinds of plants and animals, and create disturbed areas, allowing other aliens to come in. Kahili ginger, banana poke, and strawberry guava are examples of alien plants that thrive in pig-damaged areas. Furthermore, the water held in pig wallows allows mosquitoes to breed and spread avian malaria. Both the mosquitoes and the parasite that causes avian malaria have wreaked havoc on Hawaii's native bird populations.

Feral pig

Nonnative species are just part of the reason for the biodiversity crisis in Hawaii. Humans have depleted natural resources, created pollution, and overpopulated fragile coastal areas. Hawaii has some of the fastest-growing cities in the United States; the demand for fresh water is so high that frequent water-use restrictions exist in some areas; and almost a third of the landfills are at, or will soon reach, capacity. Hawaii is fast becoming a lost paradise with degraded and damaged vistas.

As you will see in this chapter, the first step toward a sustainable society in which resources, such as those of Hawaii, are protected for the present and future generations, is to recognize the damage that has been done. Then we must put our energies into correcting the activities that endanger the environment and enacting measures that will preserve it.

This finch-like honeycreeper behaves like a woodpecker.

In Waikiki, development stresses the coast.

Conservation Biology and Biodiversity

Conservation biology is an important part of achieving a **sustainable society,** one that conserves all resources for this generation and future generations. We should pay attention to preserving species, particularly in areas where biodiversity is the most complex. The direct value of biodiversity is observable in the services of individual wild species. The indirect value is evidenced by the many services provided by ecosystems.

41.1 Conservation biology is a practical science

LEARNING OUTCOMES

When you complete this section, you should be able to

1. Describe how conservation biology is an applied, goal-oriented, multidisciplinary field.
2. List and explain how biodiversity has four levels of concern.

Conservation biology is a new discipline that focuses on conserving natural resources for this generation and all future generations. Conservation biology is unique in that it is concerned with both developing scientific concepts and applying those concepts to the everyday world. A primary goal is the management of biodiversity for sustainable use by humans. To achieve this goal, conservation biologists are interested in, and come from, many subfields of biology that only now have been brought together into a cohesive whole:

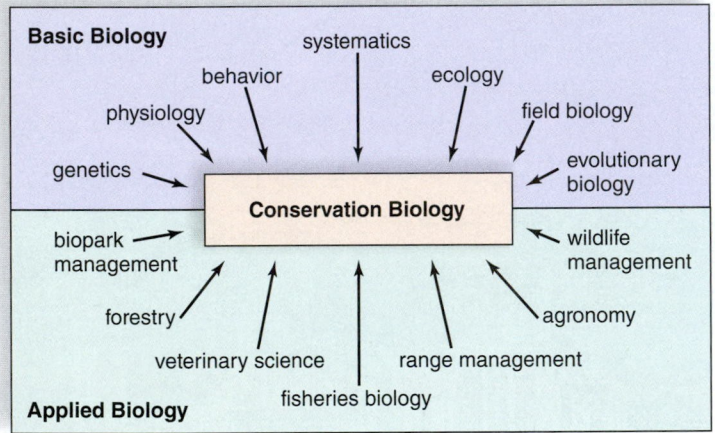

Like a physician, a conservation biologist must be aware of the latest findings, both theoretical and practical, and be able to use this knowledge to diagnose the source of trouble and suggest a suitable treatment. Often, it is necessary for conservation biologists to work with government officials at both the local and federal levels. Public education is another important duty of conservation biologists.

Conservation biology is unique among the life sciences because it supports the following ethical principles. (1) Biodiversity is desirable for the biosphere and, therefore, for humans. (2) Extinctions due to human actions are therefore undesirable. (3) The complex interactions in ecosystems support biodiversity and are desirable. (4) Biodiversity brought about by evolutionary change has value in and of itself, regardless of any practical benefit.

Conservation biology is often called a crisis discipline—never before in the history of the Earth are so many extinctions expected in such a short period of time. Estimates vary, but at least 10–20% of all species now living most likely will become extinct in the next 20–50 years, unless immediate action is taken. It is urgently important, then, that all citizens understand the concept of biodiversity, the value of biodiversity, the likely causes of present-day extinctions, and the potential consequences of reduced biodiversity.

To protect biodiversity, bioinformatics, the science of collecting and analyzing biological information, is applied. Throughout the world, molecular, descriptive, and biogeographical information about organisms is being collected. Eventually, this information will be used to help us understand and protect biodiversity.

Biodiversity At its simplest level, **biodiversity** is the variety of life on Earth. Scientists have estimated that between 10 and 50 million species may exist in all. If this is the case, many species are still to be found and described. Of the described species, nearly 1,200 in the United States and 19,000 worldwide are in danger of **extinction,** the total disappearance of a species or higher group. An **endangered species** is in peril of immediate extinction throughout all or most of its range. Examples of endangered species include the black lace cactus, armored snail, hawksbill sea turtle, California condor, West Indian manatee, and the snow leopard. **Threatened species** are organisms that are likely to become endangered in the foreseeable future. Examples of threatened species include the Navaho sedge, puritan tiger beetle, gopher tortoise, bald eagle, gray wolf, and Louisiana black bear.

To develop a meaningful understanding of life on Earth, we need to know more about species than their total number. Ecologists also study biodiversity as an attribute of three other levels of biological organization: genetic diversity, ecosystem diversity, and landscape diversity.

Genetic diversity includes the variations that occur among the members of a population. Populations with high genetic diversity are more likely to have some individuals that can survive a change in the structure of their ecosystem. For example, the 1846 potato blight in Ireland, the 1922 wheat failure in the Soviet Union, and the 1984 outbreak of citrus canker in Florida were all made worse by limited genetic variation among these crops. If a species' populations are quite small and isolated, that species is more likely to eventually become extinct because of a loss of genetic diversity. As organisms become endangered and threatened, they lose their genetic diversity.

Video
Cloning Endangered Species

Ecosystem diversity is dependent on the interactions of species at a particular locale. Although past conservation efforts frequently concentrated on saving particular charismatic species, such as the California condor, the black-footed ferret, or the spotted owl, this is a shortsighted approach. A better approach is to conserve species that play a critical role in an ecosystem. Saving an entire ecosystem can save many species, and the contrary is also true—disrupting an ecosystem threatens the existence of more than one species. For example, opossum shrimp, *Mysis relicta*, were introduced into Flathead Lake in Montana and its tributaries as food for salmon. But the shrimp ate so much zooplankton that there was, in the end, far less food for the salmon and ultimately for the grizzly bears and bald eagles as well (Fig. 41.1).

Video
Coral Reef Ecosystems

Video
Tallgrass Prairie Ecology

Landscape diversity involves a group of interacting ecosystems; within one landscape, for example, there may be plains, mountains, and rivers. Any of these ecosystems can be so fragmented that they are connected by only patches (remnants) or strips of land that allow organisms to move from one ecosystem to the other. Fragmentation of the landscape reduces reproductive capacity and food availability and can disrupt seasonal behaviors.

Distribution of Biodiversity Biodiversity is not evenly distributed throughout the biosphere; therefore, protecting particular areas will help save more species. Biodiversity is highest at the tropics, and it declines toward the poles whether

considering terrestrial, freshwater, or marine ecosystems. Also, for example, more species are present in the coral reefs of the Indonesian archipelago than in coral reefs west of this archipelago.

Some regions of the world are called **biodiversity hotspots** because they contain unusually large concentrations of species. Although hotspots harbor about 44% of all known higher plant species and 35% of all terrestrial vertebrate species, they are present in only about half of the Earth's landmass. The island of Madagascar, the Cape region of South Africa, Indonesia, the coast of California, and the Great Barrier Reef of Australia are all biodiversity hotspots.

One surprise of late has been the discovery that rain forest canopies and the deep-sea benthos have many more species than formerly thought. Some conservationists refer to these two areas as biodiversity frontiers.

The direct and indirect value of biodiversity is explained in the next part of the chapter.

▶ **41.1 CHECK YOUR PROGRESS**
1. List a number of applied disciplines that support conservation biology.
2. Explain why ecosystem-level conservation might be more important than species-level conservation.

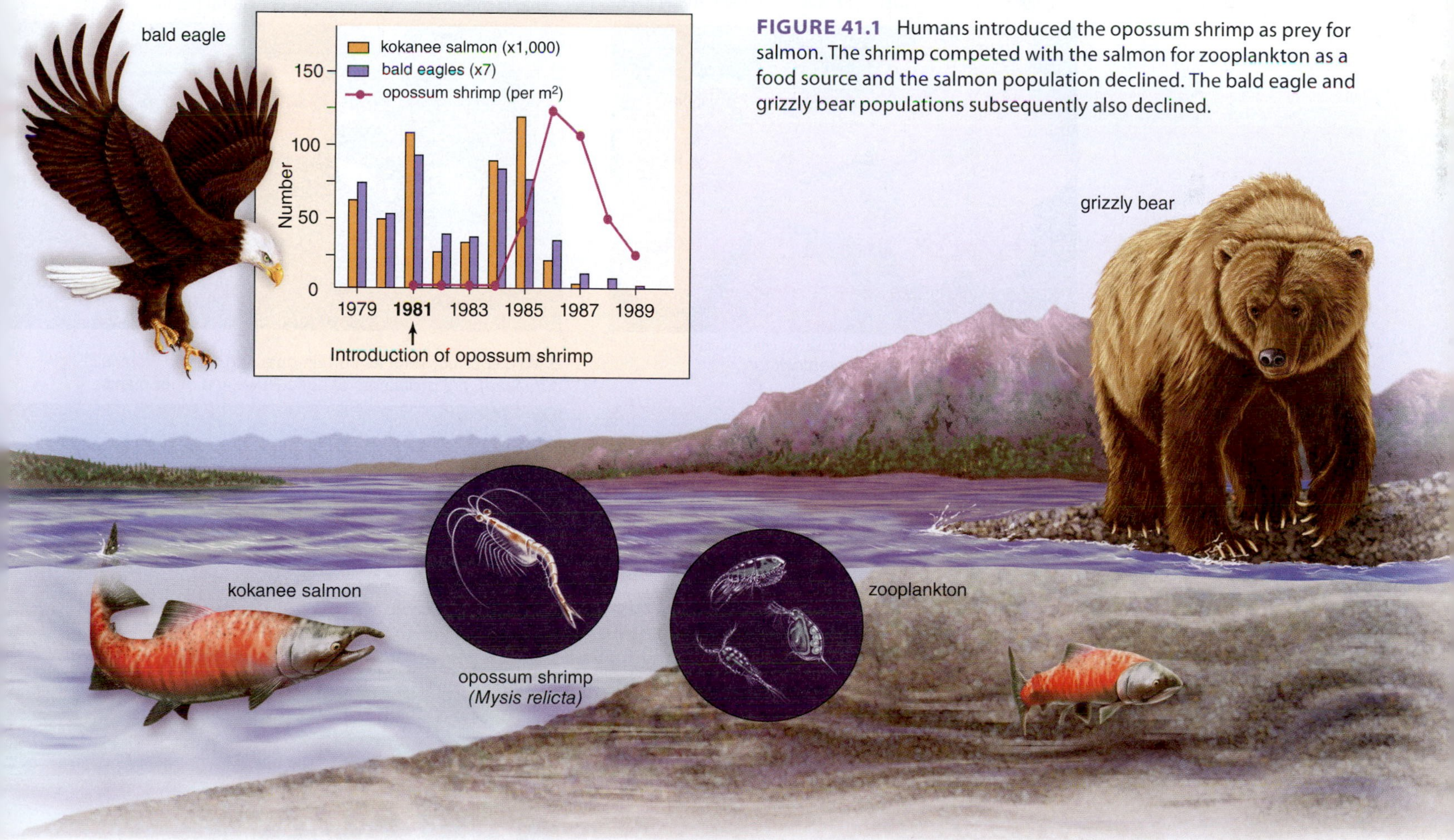

bald eagle

Legend:
- kokanee salmon (x1,000)
- bald eagles (x7)
- opossum shrimp (per m²)

Number

1979 **1981** 1983 1985 1987 1989

↑ Introduction of opossum shrimp

FIGURE 41.1 Humans introduced the opossum shrimp as prey for salmon. The shrimp competed with the salmon for zooplankton as a food source and the salmon population declined. The bald eagle and grizzly bear populations subsequently also declined.

grizzly bear

kokanee salmon

opossum shrimp (*Mysis relicta*)

zooplankton

LEARNING OUTCOME

When you complete this section, you should be able to

1. Explain the various direct values of biodiversity.

Conservation biology strives to reverse the trend toward the possible extinction of tens of thousands of organisms. To bring this about, it is necessary to make all people aware of the various ways that biodiversity has direct value and indirect value. Following are some of the ways that wildlife has direct value.

Wildlife Has Medicinal Value Most of the prescription drugs used in the United States, valued at over $200 billion, were originally derived from living organisms. The rosy periwinkle from Madagascar is an excellent example of a tropical plant that has provided us with useful medicines (Fig. 41.2a). Potent chemicals from this plant are now used to treat two forms of cancer: leukemia and Hodgkin disease. Researchers tell us that, judging from the success rate in the past, many additional types of drugs are yet to be found in tropical rain forests, and the value of this resource to society is probably over $150 billion.

You may already know that the antibiotic penicillin is derived from a fungus and that certain species of bacteria produce the antibiotics tetracycline and streptomycin. These drugs have been indispensable in the treatment of diseases, including sexually transmitted diseases such as gonorrhea and syphilis.

Leprosy is among the diseases for which there is, as yet, no cure. The bacterium that causes leprosy will not grow in the laboratory, but scientists have discovered that it grows naturally in the nine-banded armadillo (Fig. 41.2b). Having a source for the bacterium may make it possible to find a cure for leprosy. The blood of horseshoe crabs contains a substance called limulus amoebocyte lysate, which is used to ensure that medical devices such as pacemakers, surgical implants, and prosthetic devices are free of bacteria. Blood is taken from 250,000 crabs a year for extraction of this chemical, and then the crabs are returned to the sea unharmed.

Video
Indigenous Medicine
Good Poison

Wildlife Has Agricultural Value Crops such as wheat, corn, and rice are derived from wild plants that have been modified to be high producers. The same high-yield, genetically similar strains tend to be grown worldwide. When rice crops in Africa were being devastated by a virus, researchers grew wild rice plants from thousands of seed samples until they found one that contained a gene for resistance to the virus. They then used these wild plants in a breeding program to transfer the gene into high-yield rice plants. If this variety of wild rice had become extinct before it was discovered, rice cultivation in Africa might have collapsed.

FIGURE 41.2
Direct value of diverse wildlife.

a. Wild species, such as the rosy periwinkle, *Catharanthus roseus*, are sources of many medicines.

c. Wild species, including ladybugs, *Coccinella*, play a role in the biological control of agricultural pests.

b. Wild species, such as the nine-banded armadillo, *Dasypus novemcinctus*, play a role in medical research.

d. Wild species, such as certain bats (e.g., *Leptonycteris curasoae*), are pollinators of agricultural and other plants.

Biological pest controls—natural predators and parasites—are often preferable to chemical pesticides. When a rice pest called the brown planthopper became resistant to pesticides, farmers began to use natural brown planthopper enemies instead. The economic savings were calculated at well over $1 billion. Similarly, cotton growers in Cañete Valley, Peru, found that pesticides were no longer working against the cotton aphid because of resistance. Research identified natural predators, such as the ladybug, that cotton farmers are now using to an ever greater degree (Fig. 41.2c).

Most flowering plants are pollinated by animals, such as bees, wasps, butterflies, beetles, birds, and bats (Fig. 41.2d). The honeybee, *Apis mellifera,* has been domesticated, and it pollinates almost $10 billion worth of food crops annually in the United States. However, the danger of this dependency on a single species is exemplified by mites, which have now wiped out more than 23% of the commercial honeybee population in the United States. Where can we get resistant bees? From the wild, of course. The value of wild pollinators to the U.S. agricultural economy has been calculated at $4.1–$6.7 billion a year.

Video Pollinators

Wildlife Has Consumptive Use Value Humans have had much success cultivating crops, keeping domesticated animals, growing trees in plantations, and so forth. But so far, aquaculture, the growing of fish and shellfish for human consumption, has contributed only minimally to human welfare. Instead, most freshwater and marine harvests depend on the catching of wild animals, such as crustaceans (e.g., lobsters, shrimps, and crabs), mammals (e.g., whales), and fishes (e.g., trout, cod, tuna, and flounder) (Fig. 41.2e).

The environment provides a variety of other products that are sold in the marketplace worldwide, including wild fruits and vegetables, skins, fibers, beeswax, and seaweed. Also, by hunting and fishing, some people obtain their meat directly from the environment. In one study, researchers calculated that the economic value of the wild pig in the diet of native hunters in Sarawak, East Malaysia, was approximately $40 million per year.

Similarly, many trees in the natural environment are still felled for their wood. Researchers have calculated that a species-rich forest in the Peruvian Amazon is worth far more if the forest is used for fruit and rubber production than for timber production (Fig. 41.2f). Fruit and the latex needed to produce rubber can be brought to market for an unlimited number of years, whereas once the trees are gone, no more products can be harvested.

▶ **41.2 CHECK YOUR PROGRESS**
1. Explain why it is important to preserve aquatic biodiversity from a "consumptive" perspective.

e. Wild species, including many marine species, provide us with food.

f. Wild species, such as rubber trees, *Hevea,* can provide a product indefinitely if the forest is not destroyed.

LEARNING OUTCOME

When you complete this section, you should be able to

1. Explain the various indirect values of biodiversity.

Ecosystems perform many services for modern humans, who increasingly live in cities. The services discussed in this section are said to be indirect because they are pervasive and not easily discernible.

Biogeochemical Cycles Help Dispose of Waste The biodiversity within ecosystems contributes to the workings of the water, carbon, phosphorus, and nitrogen cycles. We are dependent on these cycles for fresh water, removal of carbon dioxide from the atmosphere, uptake of excess soil nitrogen, and provision of phosphate. When human activities upset the usual workings of biogeochemical cycles, the dire environmental consequences include the release of excess pollutants that are harmful to us. Technology is unable to artificially contribute to or create any of the biogeochemical cycles.

> **Video** Dung Beetles

As discussed in the introduction to Chapter 39, if not for decomposition, waste would soon cover the entire surface of our planet. We can build sewage treatment plants, but they are expensive, and few of them break down solid wastes completely to inorganic nutrients. It is less expensive and more efficient to water plants and trees with partially treated wastewater and let soil bacteria cleanse it completely. Biological communities are also capable of breaking down and immobilizing pollutants, such as heavy metals and pesticides, that humans release into the environment. A review of wetland functions in Canada assigned a value of $50,000 per hectare (or 10,000 m^2) per year to the ability of natural areas to purify water and take up pollutants.

> **Video** Decomposers

Natural Areas Provide Fresh Water, Prevent Soil Erosion, and Regulate Climate Few terrestrial organisms are adapted to living in a salty environment—they need fresh water. We can remove salt from seawater to obtain fresh water, but the cost of desalination is about four to eight times the average cost of fresh water acquired via the water cycle. Humans use fresh water in innumerable ways, including for drinking and for irrigating their crops. Freshwater ecosystems such as rivers and lakes also provide us with fish and other types of organisms for food.

> **Video** Thames River

Forests and other natural ecosystems exert a "sponge effect" (Fig. 41.3). The leaves of trees acquire water from the roots and then release it at a regular rate. The water-holding capacity of forests and wetlands reduces the possibility of flooding. The value of a marsh outside Boston, Massachusetts, has been estimated at $72,000 per hectare per year based solely on its ability to reduce floods. Forests release water slowly for days or weeks after the rains have ceased. Rivers flowing through forests in West Africa release twice as much water halfway through the dry season, and between three and five times as much at the end of the dry season, as do rivers bordered by coffee plantations.

Forested ecosystems naturally retain soil and prevent erosion. Due to deforestation, the Tarbela Dam in Pakistan is losing its storage capacity of 13.5 billion m^3 many years sooner than expected because silt is building up behind the dam. At one time, the Philippines were exporting $240.8 million worth of oysters, mussels, clams, and cockles each year. Now, silt carried down rivers following deforestation is smothering the mangrove ecosystem that serves as a nursery for the sea.

Globally, forests ameliorate the climate because they take up carbon dioxide. The leaves of trees use carbon dioxide when they photosynthesize, and the bodies of the trees store carbon. When trees are cut and burned, carbon dioxide is released into the atmosphere. Carbon dioxide makes a significant contribution to global warming, which is expected to be stressful for many plants and animals. If temperatures become warmer, only a small percentage of wildlife will be able to move northward to find weather suitable for them.

Ecotourism Is Enjoyed by Many In the United States, nearly 100 million people enjoy vacationing in a natural setting. To do so, they spend $4 billion each year on fees, travel, lodging, and food. Many tourists want to go sport fishing, whale watching, boat riding, hiking, birdwatching, and the like (Fig. 41.3). Others merely want to immerse themselves in the beauty and serenity of a natural environment.

▶ **41.3 CHECK YOUR PROGRESS**

1. List examples of indirect values of biodiversity.

FIGURE 41.3
Tourists (inset) love to visit natural ecosystems, such as this forest, which has indirect value because of its water-holding capacity and its ability to take up carbon dioxide.

Causes of Species Extinctions

Researchers have identified the major causes of extinction. They are, in order of significance, habitat loss, introduction of alien species, pollution, overexploitation, and disease. These causes can work together to reduce biodiversity.

41.4 Habitat loss is a major cause of wildlife extinctions

LEARNING OUTCOME

When you complete this section, you should be able to

1. Describe what causes loss of habitat in various ecosystems.

To stem the tide of extinction due to human activities, it is first necessary to identify the causes. Based on the records of 1,880 threatened and endangered wild species in the United States, habitat loss was involved in 85% of the cases (Fig. 41.4A). Other significant causes of extinction are introduction of alien species, pollution, overexploitation, and disease. In Figure 41.4A, the percentages add up to more than 100% because most species are imperiled for more than one reason. Macaws are a good example of a species in decline due to a combination of factors.

Habitat loss has occurred in all ecosystems, but concern has now centered on tropical rain forests and coral reefs because they are particularly rich in species. A sequence of events in Brazil offers a fairly typical example of how rain forest is converted to land uninhabitable for wildlife. The construction of a major highway first provided a way to reach the interior of the forest. Small towns and industries sprang up along the highway, and roads branching off the

Distant view

Close-up view

FIGURE 41.4B Destruction of a rain forest in Brazil.

main highway gave rise to even more roads. The result was fragmentation of the once immense forest. The government offered subsidies to anyone willing to take up residence in the forest, and people began to cut and burn trees in patches (Fig. 41.4B). Tropical soils contain limited nutrients, but when the trees are burned, nutrients are released that support a lush growth so that cattle can be grazed for about 3 years. Once the land has been degraded, farmers move on to another portion of the forest and start over again.

Loss of habitat also affects freshwater and marine biodiversity. Coastal degradation is mainly due to the large concentration of people living on or near the coast. Already 20% of coral reefs have been destroyed and show little likelihood of recovery; another 24% are under immediate danger of collapsing and 26% may collapse sometime in the future. Mangrove forests are also being decimated worldwide; 50% of all mangroves have disappeared in the past 50 years. Wetland areas, estuaries, and seagrass beds are also being rapidly destroyed by the actions of humans.

▶ **41.4 CHECK YOUR PROGRESS**

1. Identify the main cause of coastal loss of habitat, for example, in Hawaii.

FIGURE 41.4A Macaws, *Ara macao*, and other species are endangered for the reasons graphed here.

LEARNING OUTCOME

When you complete this section, you should be able to

1. List and discuss the ways humans have introduced alien species into ecosystems.

Ecosystems around the globe are characterized by unique assemblages of organisms that have evolved together in one location. Migrating to a new location is not usually possible because of barriers such as oceans, deserts, mountains, and rivers. Humans, however, have introduced **alien species,** nonnative members, into new ecosystems through the following means:

Colonization Europeans, in particular, brought various familiar species with them when they colonized new places. For example, the pilgrims brought the dandelion to the United States as a familiar salad green. In addition, they introduced pigs that have become feral, reverting to their wild state. In some parts of the continental United States, feral pigs are very destructive, just as they are in Hawaii.

Horticulture and agriculture Some aliens now taking over vast tracts of land have escaped from cultivated areas. Kudzu is a vine from Japan that the U.S. Department of Agriculture thought would help prevent soil erosion. The plant now covers much of the landscape in the South, including even walnut, magnolia, and sweet gum trees (Fig. 41.5A). The water hyacinth was introduced to the United States from South America because of its beautiful flowers. Today, it clogs waterways and diminishes natural diversity.

Accidental transport Global trade and travel accidentally bring many new species from one country to another. Researchers found that the ballast water released from ships into Coos Bay, Oregon, contained 367 marine species from Japan. The zebra mussel from the Caspian Sea was accidentally introduced into the Great Lakes in 1988. It now forms dense beds that squeeze out native mussels. Other organisms accidentally introduced into the United States include the Formosan termite, the Argentinian fire ant, and the nutria, a type of rodent.

Alien species can disrupt food webs. As mentioned earlier, opossum shrimp introduced into a lake in Montana added a trophic level that in the end meant less food for bald eagles and grizzly bears (see Fig. 41.1). Introduction of alien species, sometimes called exotic species, plays a role in nearly 50% of extinctions (see Fig. 41.4A).

Video Alien Invasion

Aliens on Islands Islands are particularly susceptible to environmental discord caused by the introduction of alien species. Islands have unique assemblages of native species that are closely adapted to one another and cannot compete well against aliens. Myrtle trees, *Myrica faya,* introduced into the Hawaiian Islands from the Canary Islands, are symbiotic with a type of bacterium that is capable of nitrogen fixation. This feature allows the species to establish itself on nutrient-poor volcanic soil, a distinct advantage in Hawaii. Once established, myrtle trees halt the normal succession of native plants on volcanic soil.

The brown tree snake has been inadvertently introduced onto a number of islands in the Pacific Ocean. The snake eats eggs,

FIGURE 41.5A Kudzu, a vine from Japan, has displaced many native plants in the southern United States.

FIGURE 41.5B Mongooses, introduced into Hawaii, prey on the native birds.

nestlings, and adult birds. On Guam, it has reduced ten native bird species to the point of extinction. On the Galápagos Islands, black rats have reduced populations of giant tortoises, while goats and feral pigs have changed the vegetation from highland forest to pampaslike grasslands and destroyed stands of cactus. In Australia, mice and rabbits have stressed native marsupial populations. In Hawaii, mongooses introduced to control rats also prey on native birds (Fig. 41.5B) and feral pigs continue to devastate forests. The pigs especially seem to prefer eating native species and their wallows allow other alien species to spread. Exterminating the pigs would be helpful but native peoples are opposed because they use them as a source of food. Fencing is being used to keep them out of national parks, however.

▶ **41.5 CHECK YOUR PROGRESS**

1. Identify why colonization introduces alien species to new ecosystems.

41.6 Pollution contributes to extinctions

LEARNING OUTCOME

When you complete this section, you should be able to

1. List and explain the ways humans have polluted the environment.

So far, we have discussed habitat loss and alien species as causes of extinction. Now, we will discuss pollution, which is a factor in 24% of extinctions (see Fig. 41.4A). In the present context, **pollution** can be defined as any environmental change that adversely affects the lives and health of organisms. Biodiversity is particularly threatened by the following types of environmental pollution:

Acid Deposition Both sulfur dioxide from power plants and nitrogen oxides in automobile exhaust are converted to acids when they combine with water vapor in the atmosphere. These acids return to Earth as either wet deposition (acid rain or snow) or dry deposition (sulfate and nitrate salts). Acid deposition causes trees to weaken and increases their susceptibility to disease and insects. Many lakes in the northern United States are now lifeless because of the effects of acid deposition.

Ozone Depletion The ozone shield is a layer of ozone (O_3) in the stratosphere, some 50 km above the Earth. The ozone shield absorbs most of the wavelengths of harmful ultraviolet (UV) radiation so that they do not strike the Earth. The cause of ozone depletion can be traced to chlorine atoms (Cl^-) that come from the breakdown of chlorofluorocarbons (CFCs). Severe ozone shield depletion can impair crop and tree growth and also kill plankton (microscopic plant and animal life) that sustain oceanic life. Due to an international agreement, manufacture of CFCs ceased in the United States in 1996, but the chemicals linger because they are chemically stable. The amount of depletion over Antarctica varies with weather conditions and the hole has actually decreased in 2012 to 1.1 million sq km compared to 18.5 million sq km 12 years ago.

Organic Chemicals Our modern society uses organic chemicals in all sorts of ways. Organic chemicals called nonylphenols are in products ranging from pesticides to dishwashing detergents, cosmetics, plastics, and spermicides. These chemicals mimic the effects of hormones and, in that way, most likely harm wildlife. Salmon are born in fresh water but mature in salt water. After investigators exposed young fish to nonylphenol, they found that 20–30% were unable to make the transition between fresh and salt water. Nonylphenols cause the pituitary to produce prolactin, a hormone that may prevent saltwater adaptation.

Global Climate Change Recall that certain gases, such as carbon dioxide and methane, are known as greenhouse gases because, just like the panes of a greenhouse, they allow solar radiation to pass through but hinder the escape of its heat back into space. Data collected around the world show a steady rise in the concentration of the various greenhouse gases due to the burning of fossil fuels and forests. A rise in greenhouse gases parallels a rise in global temperatures. The response of organisms to global climate change will be dramatic, as discussed in "Response of

FIGURE 41.6 *Top:* Normal coral reef. *Bottom:* Bleaching of a coral reef. A temperature rise of only a few degrees causes coral reefs to "bleach" and become lifeless. As the oceans warm and land recedes, coral reefs could move northward.

Organisms to Global Climate Change" on page 810. For example, the growth of corals is very dependent on mutualistic algae living in their walls. When the temperature rises by 4°C, corals expel their algae and are said to be "bleached" (Fig. 41.6).

Video Global Warming

Solid Waste Disposal Due to high consumption, the United States is one of the largest global generators of municipal wastes per capita. Between 1990 and 2005, Japan maintained its municipal waste generation at 405 kg per capita, but waste generation in the United States rose from 600–770 kg per capita. According to the Solid Waste Disposal Act of 1976, states are to maximize recycling and minimize waste, but many states have not yet complied. Plastic, which accounts for 90% of the solid waste now floating in the oceans, is a danger to wildlife. An estimated 1 million seabirds and 100,000 sea turtles die annually by choking on floating plastic objects or becoming entangled in them.

Eutrophication Lakes are also under stress due to overenrichment. When lakes receive excess nutrients due to runoff from agricultural fields and wastewater from sewage treatment, algae

begin to grow in abundance. An algal bloom is apparent as a green scum or excessive mats of filamentous algae. Upon death, the decomposers break down the algae, but in so doing, they use up oxygen. Thus a decreased amount of oxygen is available to fish, sometimes leading to a massive fish kill.

► **41.6 CHECK YOUR PROGRESS**

1. Explain why global climate change could cause the greatest loss of wildlife.

HOW LIFE CHANGES   *Application*

41A Response of Organisms to Global Climate Change

Climate change is occurring globally and all ecosystems on Earth are affected. In general the growing season is extending as the number of frost-free days increases; however, the preferred environmental conditions for certain species are shifting northward.

Biologists want to know how species are coping with the temperature and rainfall changes they are experiencing. Some species are able to shift their distributions poleward. For example, Edith's checkerspot butterfly (Fig. 41A.1), a North American species, has become rare in southern locations and is now found with greater frequency at northern locations. Similar range shifts are noted in European butterflies as well. A study of forest ecosystems in Europe found that several species are moving to higher elevations at a rate of approximately 29 m per decade. A study of Sierra Nevada birds showed that 48 of 53 bird species adjusted to climate change over the last century by moving to where their favored temperature and/ or rainfall conditions now exist. The Clark's nutcracker has moved upward to secure its favored temperature, while the Bullock's oriole is following its favored rainfall (Fig. 41A.2).The western bluebird shifted its range to achieve both suitable temperature and rainfall.

In addition to distribution changes, evolutionary effects are also expected. North American tree swallows (Fig. 41A.3) normally lay eggs in May, which indicates that the signal to start laying eggs is determined by temperature and day length. Over the past 40 years, researchers have documented that some members of this species are now laying their eggs 9 days earlier than usual. This behavior increases fitness and will be favored by natural selection. To the degree that it is genetically determined, more tree swallows will exhibit this behavior in the future. Similarly, wild female Soay sheep on the island of Hirta (Fig. 41A.4) have gotten smaller by 5% during the past two decades. Size is partly inherited, and up to now cold winters have favored large-sized sheep, which have better survival power during times of food scarcity. The population's phenotypic ratio is now changing, however, because milder winters do not select for large size, and small, weak youngsters have been able to survive.

The topic of species responses to climate change is not only interesting to ecologists and evolutionary biologists, but also has serious implications for public health. For example, consider a common and often unpopular insect, the mosquito. The life cycle of blood-sucking mosquitoes is strongly influenced by day length. During the long days of summer, they seek a blood meal to support their reproduction, but as the days shorten and temperatures cool, they hibernate. A study of the pitcher plant mosquito in North America found that most individuals also follow this day length–sensitive cycle, but a genetically distinct group continues to reproduce as the days shorten. As temperatures warm, this genetically distinct group is breeding longer and shifting its distribution northward, two adjustments that will spread its genes and increase its fitness. Pitcher plant mosquitoes feed on nectar, not blood, but many of their relatives do feed on blood. If these relatives also experience shifts in distribution and duration of reproduction, any diseases they carry could increase. Other diseases may also increase. For example, it's been determined that higher temperatures have favored the spread of the fungal infection that is causing the Harlequin toad, shown in Figure 41.8, to become extinct.

Video Karoo Global Warming

Video Warming Hurts Rice

CONSIDER THESE QUESTIONS

1. How might scientists use artificial selection to help species survive global climate change?
2. Even if a species could follow its favored temperature, why might it not survive in new surroundings?

connect |BIOLOGY Explore the concepts through a variety of multimedia assets and question types.

www.mcgrawhillconnect.com

FIGURE 41A.1 Edith's checkerspot butterfly, *Euphydryas editha.*

FIGURE 41A.2 Bullock's oriole, *Icterus bullockii.*

FIGURE 41A.3 Tree swallow, *Tachycineta ssp.*

FIGURE 41A.4 Soay sheep *Ovis aries.*

41.7 Overexploitation contributes to extinctions

When you complete this section, you should be able to

1. List and discuss ways humans have overexploited species.

Overexploitation occurs when the number of individuals taken from a wild population is so great that the population becomes severely reduced in number. Overexploitation accounts for about 17% of extinctions (see Fig. 41.4A). A positive feedback cycle ensues when the members of a small population are particularly prized. Poachers and members of criminal organizations collect and sell endangered and threatened species because it has become so lucrative. The overall international value of trading wildlife species is $20 billion, of which $8 billion is attributed to the illegal sale of rare species.

Declining species of mammals, such as the Siberian tiger, are still hunted for their hides, tusks, horns, or bones. Because of its rarity, a single Siberian tiger is now worth more than $500,000—its bones are pulverized and used as a medicinal powder. The horns of rhinoceroses become ornate carved daggers, and their bones are ground up to sell as a medicine. The ivory of an elephant's tusk is used to make art objects, jewelry, or piano keys. The fur of a Bengal tiger sells for as much as $100,000 in Tokyo.

The U.N. Food and Agricultural Organization tells us that humans have now overexploited 11 of 15 major oceanic fishing areas. Larger and more efficient fishing fleets are now decimating fishing stocks. Pelagic species such as tuna are captured by purse-seine fishing, in which a very large net surrounds a school of fish and is then closed like a drawstring purse. Up to thousands of dolphins that swim above schools of tuna are often captured and then killed in this type of net. Other fishing boats drag huge trawling nets, large enough to accommodate 12 jumbo jets, along the seafloor to capture bottom-dwelling fish (Fig. 41.7). Only large fish are kept; undesirable small fish and sea turtles are discarded, dying, back into the ocean. Other reptiles are sought for consumption. Collection and trade of terrestrial tortoises and freshwater turtles for food and other uses has surged in Asia over the past two decades and is now spreading

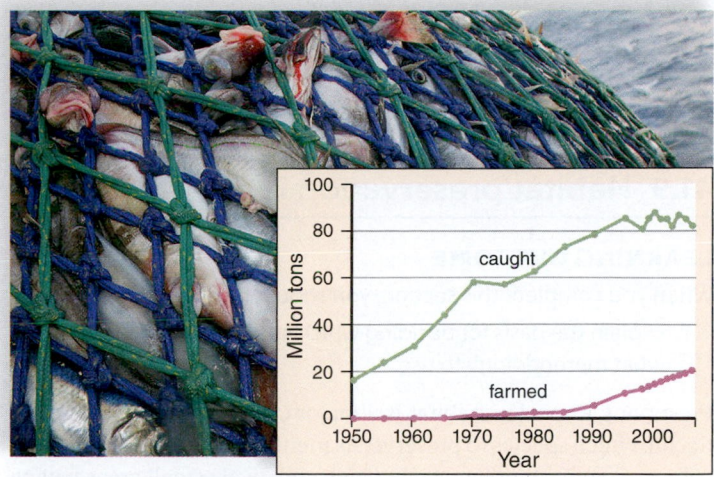

FIGURE 41.7 Huge nets catch massive numbers of fish. The graph above shows that the number of fish caught increased until 1995 and since then has remained fairly the same.

around the globe. These practices have wiped out populations in Asia and have brought others to the brink of extinction in a matter of years.

Video Ocean Fishing Ban

A marine ecosystem can be disrupted by overfishing, as exemplified on the U.S. West Coast. When sea otters began to decline in numbers, investigators found that they were being eaten by orcas (killer whales). Usually orcas prefer seals and sea lions to sea otters, but they began eating sea otters when few seals and sea lions could be found. What caused a decline in seals and sea lions? Their preferred food sources—perch and herring—were no longer plentiful due to overfishing. Ordinarily, sea otters keep the population of sea urchins, which feed on kelp, under control. But with fewer sea otters around, the sea urchin population exploded and decimated the kelp beds.

▶ **41.7 CHECK YOUR PROGRESS**
1. Explain with examples what is meant by overexploitation of species.

41.8 Disease contributes to extinctions

When you complete this section, you should be able to

1. Explain how pollution can lead to disease.

Scientists tell us the number of pathogens that cause diseases is on the rise, threatening human health as well as that of wildlife. Due

FIGURE 41.8 The Harlequin toad is near extinction due to a fungal pathogen.

to the encroachment of humans on their habitat and other general interventions, wildlife have been exposed to emerging diseases. For example, canine distemper was spread from domesticated dogs to lions in the African Serengeti, causing population declines. Avian influenza likely emerged from domesticated fowl (e.g., chicken) populations and could lead to the deaths of many wild birds.

Pollution can weaken organisms so that they are more susceptible to disease. Almost half of sea otter deaths along the coast of California are now due to infectious diseases. Pollution, most likely, plays a role in the worldwide decline of amphibians due to disease (Fig. 41.8).

The next part of the chapter discusses habitat preservation and restoration.

▶ **41.8 CHECK YOUR PROGRESS**
1. Develop a scenario that explains the worldwide loss of amphibians.

Achieving Species Preservation

To preserve species, it is necessary to preserve their habitat, or possibly restore a habitat. Conservation aims for sustainable development, which allows multiple uses of the land.

41.9 Habitat preservation is of primary importance

LEARNING OUTCOME

When you complete this section, you should be able to

1. Explain the basis for deciding which species to preserve and what methodology to use.

Preserving a species' habitat is a major concern, but first we must decide which species to preserve. As mentioned previously, the biosphere contains biodiversity hotspots, relatively small areas having a concentration of endemic (native) species not found anyplace else. In the tropical rain forests of Madagascar, 93% of the primate species, 99% of the frog species, and over 80% of the plant species are endemic to Madagascar. Preserving these forests and other hotspots will save a wide variety of organisms.

Keystone species are species that influence the viability of a community, although their numbers may not be excessively high. The extinction of a keystone species can lead to other extinctions and loss of biodiversity. For example, bats are designated a keystone species in tropical forests of the Old World. They are pollinators that also disperse the seeds of trees. When bats are killed off and their roosts destroyed, the trees fail to reproduce. The grizzly bear is a keystone species in the northwestern United States and Canada (Fig. 41.9A). Bears disperse the seeds of berries; as many as 7,000 seeds may be in one dung pile. Grizzlies also kill the young of many hoofed animals and thereby keep their populations under control. In addition, grizzlies are a principal mover of soil when they dig up roots and prey upon hibernating ground squirrels and marmots. Other keystone species are beavers in wetlands, bison in grasslands, alligators in swamps, and elephants in grasslands and forests.

Keystone species should not be confused with **flagship species,** which motivate us to preserve biodiversity purely because of our own emotional response. Flagship species include monarch butterflies, lions, tigers, dolphins, and the giant panda, which are not critical to an ecosystem in the same way keystone species are. The grizzly bear is a flagship species, but it is also a keystone species because of its critical role in an ecosystem. The grizzly bear population is actually a **metapopulation**, a population inadvertently fragmented by humans into several small, isolated populations. Originally, there were probably 50,000–100,000 grizzlies south of Canada, but this number has been reduced because communities have encroached on their home range. Now there are six virtually isolated subpopulations, totaling about 1,000 individuals.

Saving metapopulations sometimes requires determining which of the populations is the source and which are sinks. A **source population** is one that lives in a favorable area, and its birthrate is most likely higher than its death rate. Individuals from source populations move into **sink populations**, where the environment is not as favorable and where the birthrate equals the death rate at best. When trying to save the northern spotted owl, conservationists decided to prevent owls from leaving old-growth rain forests of the Pacific Northwest where they successfully reproduce (Fig. 41.9B). This decision proved beneficial in maintaining the populations.

Landscape Preservation May Be Necessary   To save grizzly bears, it is necessary to save diverse ecosystems that are at least connected by corridors. A landscape area called the Greater Yellowstone Ecosystem, where bears are free to roam, has now been defined. It contains over 400,000 hectares* in Yellowstone National Park; state lands in Montana, Idaho, and Wyoming; five different national forests; various wildlife refuges; and even private lands.

Landscape protection for grizzlies will be beneficial for other wildlife that share the same space. The last of the contiguous 48 states' harlequin ducks, bull trout, westslope cutthroat trout, lynx, pine martens, wolverines, mountain caribou, and great gray owls are found in areas occupied by grizzlies. Then, too, the grizzly range overlaps with 40% of Montana's vascular plants of special conservation concern.

▶ **41.9 CHECK YOUR PROGRESS**

1. Explain the need to preserve keystone species rather than flagship species.

---

*Hectares are a metric unit of measurement.

FIGURE 41.9B Old-growth forest, home of the northern spotted owl, *Strix occidentalis caurina.*

FIGURE 41.9A Landscape preservation will help grizzly bears, *Ursus arctos horribilis,* survive.

LEARNING OUTCOME

When you complete this section, you should be able to

1. Identify and explain the most useful procedures for habitat restoration.

Restoration ecology is a new subdiscipline of conservation biology that seeks scientific ways to return ecosystems to their former state. Three principles have so far emerged: (1) It is best to begin as soon as possible before remaining fragments of the original habitat are lost. These fragments are sources of wildlife and seeds from which to restock the restored habitat. (2) Once the natural history of the habitat is understood, it is best to use biological techniques that mimic natural processes to bring about restoration. This might take the form of using controlled burns to bring back grassland habitats, biological pest controls to rid the area of alien species, or bioremediation techniques to clean up pollutants. (3) The goal is **sustainable development,** the ability of an ecosystem to maintain itself while providing services to humans. We will use the Everglades ecosystem to illustrate these principles. "Emiquon Floodplain Restoration" on page 814 tells you about a restoration program in Illinois.

The Everglades The Everglades, located in southern Florida, is a vast sawgrass prairie, interrupted occasionally by hardwood tree islands. Within these islands, both temperate and tropical evergreen trees grow amongst dense and tangled vegetation. Mangroves are found along sloughs (creeks) and at the shoreline. The prop roots of red mangroves protect over 40 different types of juvenile fishes as they grow to maturity. During the wet season, from May to November, animals disperse throughout the region, but in the dry season, from December to April, they congregate wherever pools of water are found. Alligators are famous for making "gator holes," where water collects and fish, shrimp, crabs, birds, and a host of organisms survive until the rains come again. The Everglades once supported millions of large and beautiful birds, including herons, egrets, the white ibis, and the roseate spoonbill. Figure 41.10 shows various animals that live in the Everglades.

At the turn of the twentieth century, settlers began to drain land in central Florida to grow crops in the newly established Everglades Agricultural Area (EAA). A large dike was used to keep water in a large lake called Lake Okeechobee. The dike prevents water from overflowing its banks and moving slowly southward. Water is contained not only in the lake, but also in three so-called conservation areas established to the south of the lake. Water must be conserved to irrigate the farmland and to recharge the Biscayne aquifer (underground river), which supplies drinking water for the cities on the east coast of Florida. The Everglades National Park receives only water that is discharged artificially from a conservation area, and the discharge is scheduled according to the convenience of humans rather than the natural wet/dry season of southern Florida. Largely because of this, the Everglades is now dying, as witnessed by declining bird populations. The birds, which formerly numbered in the millions, now exist in only thousands.

A restoration plan has been developed that will sustain the Everglades ecosystem while maintaining the services society requires. The Everglades is to receive a more natural flow of water from Lake Okeechobee. This will require flooding the EAA and growing only crops such as sugarcane and rice that can tolerate these conditions. This has the benefit of stopping the loss of topsoil and preventing possible residential development in the area. There will also be an extended buffer zone between an expanded Everglades and the urban areas on Florida's east coast. The buffer zone will contain a contiguous system of interconnected marsh areas, detention reservoirs, seepage barriers, and water treatment areas. This plan is expected to stop the decline of the Everglades, while still allowing agriculture to continue and providing water and flood control to the eastern coast. Sustainable development will maintain the ecosystem indefinitely and still meet human needs.

Florida panther, *Puma concolor coryi*

American alligator, *Alligator mississippiensis*

White ibis, *Eudocimus albus*

Roseate spoonbill, *Ajaia ajaja*

Wood stork, *Mycteria americana*

FIGURE 41.10 A variety of animals make their home in the Everglades.

▶ **41.10 CHECK YOUR PROGRESS**

1. List and explain the three principles of habitat restoration.

41B Emiquon Floodplain Restoration

The Emiquon Floodplain Restoration Project at the confluence of the Spoon and Illinois Rivers involves the many organizations listed in Figure 41B.1 and takes in a complex system of wetlands and lakes adjacent to the two rivers.

In the early 1900s, the Illinois River was one of the most economically significant river ecosystems in the United States. It boasted the highest mussel abundance and productive commercial fisheries per mile of any stream in the United States. Each year during the seasonal flooding, the river would deposit nutrient-laden sediment onto the floodplain. This created a highly productive wetlands ecosystem. With the abundance of nutrients, a wide diversity of plant life was possible. This broad producer base supported a number of trophic levels as well as a wide diversity of food webs (Fig. 41B.2).

Throughout the twentieth century, modern cultures altered the natural ebb and flow of the Illinois River to better suit their needs. Levees were constructed, resulting in the isolation of nearly half of the original floodplain from the river. Native habitats were converted to agricultural land. Corn and soy beans soon replaced the variety of vegetation that were once a hallmark of the Emiquon floodplain. The river itself was also altered to better facilitate the export and import of a variety of commercial goods by boat. With the isolation of these areas from the seasonal flooding, a decrease in the amount of nutrients being deposited also occurred. This resulted in a continuous decline in the natural productivity of the floodplain.

Restoration Plan

The National Research Council has identified Emiquon as one of three large floodplain-river ecosystems in the United States that can be restored to some semblance of its original diversity and function. The floodplain restoration work at Emiquon is a key part of The Nature Conservancy's efforts to conserve the Illinois River.

The Emiquon Science Advisory Council is composed of over 40 scientists and managers who provide guidance for the restoration and managed connection between Emiquon's restored floodplain and the Illinois River. Once established, this connection would allow for the seasonal flooding that it previously experienced. This seasonal flooding would help naturalize the flow of the river and restore the cyclical process of flooding and drying out. Flooding would provide more nutrients for wetland plants as well as helping to improve the water quality. Access between the river and floodplain would once again be restored for various aquatic species, such as paddlefish and gar. Both species need a variety of habitats for reproduction and survival. The wetlands will in turn provide vital nutrients to the species that reside in the river. The Conservancy is working with the Illinois Natural History Survey, the University of Illinois, Springfield, and others to determine the baseline community diversity. These groups will monitor the community diversity to provide vital information about the progress of the restoration plan.

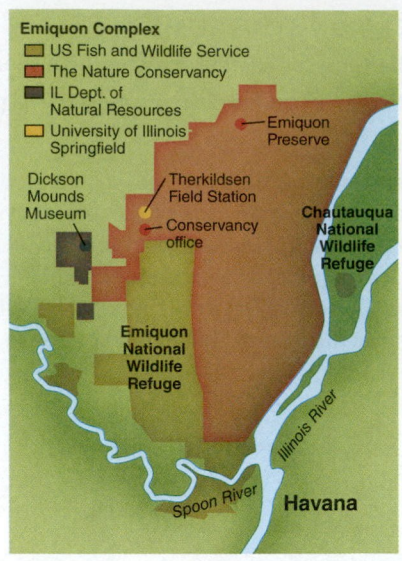

FIGURE 41B.1 The Emiquon Restoration Project calls for allowing the rivers to overflow their banks as they once did. The organizations listed are involved in the project because they have jurisdiction according to the various colors.

Local communities are investigating the potential economic benefits that can be gained as a result of the restoration. For example, perhaps more nature-based tourism can be offered at Emiquon and along the Illinois River. The knowledge gained from the research at Emiquon will potentially guide large floodplain-river restoration efforts around the world.

CONSIDER THESE QUESTIONS

1. In what ways might scientists assess whether the restoration plan was successful in its goal?
2. How might the Emiquon Floodplain Restoration plan be used as a model for other restoration projects?

connect | BIOLOGY Explore the concepts through a variety of multimedia assets and question types.

www.mcgrawhillconnect.com

Dragonfly nymph

White pelicans

Researcher inspects water plantain.

River otter

American paddlefish

FIGURE 41B.2
Wildlife of the Illinois River ecosystem. Many species in a number of animal phyla live in this ecosystem.

Achieving a Sustainable Human Society

A **sustainable society** will protect biodiversity and at the same time will be able to provide the same amount of goods and services for future generations as it does at present. In a sustainable society, energy sources will be renewable, water will be conserved, and agriculture will not harm the environment.

41.11 A sustainable society will preserve resources

LEARNING OUTCOME

When you complete this section, you should be able to

1. Identify characteristics of a sustainable society.

To achieve a sustainable society, resources such as clean air, water, an adequate amount of food, and living space for humans and wildlife will have to be preserved. This goal is not possible unless we carefully regulate our consumption of these resources today, taking into consideration that the human population is ever increasing.

A natural ecosystem can offer clues about how to make today's society sustainable. A natural ecosystem uses renewable solar energy, and its materials cycle through the various populations back to the producer once again. For example, coral reefs have been sustaining themselves for millions of years. At the same time, the reefs have provided sustenance to humans. The value of coral reefs has been assessed at over $300 billion a year. Their aesthetic value is immeasurable.

It is clear that if we want to develop a sustainable society, we too should use renewable energy sources and recycle materials. We should protect natural ecosystems such as coral reefs that help sustain our modern society. At least a quarter of the coral reefs exist close to the shores of an MDC (more-developed country), and the chances are good that these coral reefs are at least somewhat protected. Unfortunately, other coral reefs are threatened by unsustainable practices. The good news is that reefs are remarkably regenerative and will return to their former condition if left alone. The message of today's environmentalists is about what can be done to improve matters and make the environment sustainable (Fig. 41.11). There is still time to make changes and improvements.

Sustainability should be practiced in all areas of human endeavor, from agriculture to business. Efficiency is the key to sustainability. For example, an efficient car would be ultralight and gas thrifty. Efficient cars could be just as durable and speedy as the inefficient ones of today. Only through efficiency can we meet the challenges of limited resources and finances in the future.

▶ **41.11 CHECK YOUR PROGRESS**
1. Determine ways a sustainable society would preserve biodiversity.

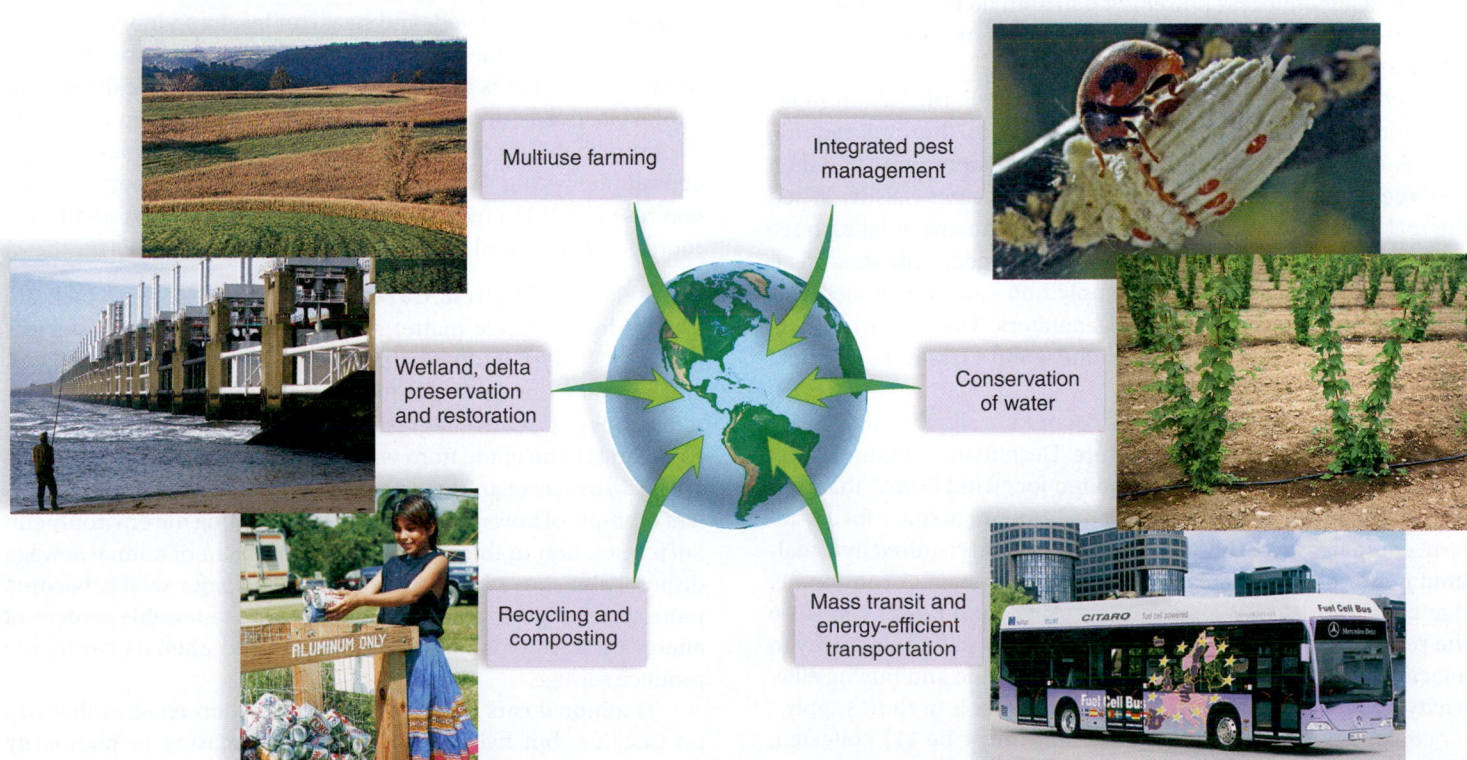

FIGURE 41.11 These activities are characteristic of a sustainable society. Arrows point inward to signify that these activities increase the carrying capacity of the Earth.

Multiuse farming

Integrated pest management

Wetland, delta preservation and restoration

Conservation of water

Recycling and composting

Mass transit and energy-efficient transportation

LEARNING OUTCOME

When you complete this section, you should be able to

1. Describe the current and future renewable sources of energy.

The goal of a sustainable society is to primarily use renewable resources and to increase the efficiency with which they are used. **Renewable resources** are capable of being naturally replenished; for example, solar energy is a renewable source of energy because the sun shines on the Earth every day without fail. In contrast, **nonrenewable resources** are in limited supply. Consider that the present supply of fossil fuels (oil, natural gas, and coal), which were formed many millions of years ago from the compressed remains of plants and animals, can run out. Much needs to be done to increase our reliance on renewable energy sources because in 2010, about 81% of the world's energy came from fossil fuels and 6% came from nuclear power, which is also a nonrenewable source. Fossil fuels cause air pollution, and nuclear power plants give off radioactive solid wastes that threaten our health.

Traditional Renewable Energy Sources Renewable energy sources, such as hydropower, geothermal energy, wind power, and solar energy, have been under development for some time (Fig. 41.12A).

Hydroelectric plants convert the energy of falling water into electricity. Worldwide, hydropower presently generates 16% of all the electricity utilized. Most of the presently available hydropower comes from enormous dams. Unfortunately, constructing these dams involved the destruction of ecosystems, and large dams are impractical for the reasons discussed in section 41.13. Small-scale dams that generate less power per dam, but do not have the same environmental impact, are believed to be the more environmentally responsible choice.

Geothermal energy occurs because the Earth has an internal source of heat. Elements such as uranium, thorium, radium, and plutonium undergo radioactive decay underground and then heat the surrounding rocks to hundreds of degrees Celsius. When the rocks are in contact with underground streams or lakes, huge amounts of steam and hot water are produced. This steam can be piped up to the surface to supply hot water for home heating or to run steam-driven turbogenerators. The California Geysers Project, for example, is one of the world's largest geothermal electricity–generating complexes.

Wind power is expected to account for a significant percentage of our energy needs in the future. Despite the common belief that a huge amount of land is required for "wind farms" that produce commercial electricity, the actual amount of space for a wind farm compares favorably to the amount of land required by a coal-fired power plant or a solar thermal energy system. A community that generates its own electricity by using wind power can solve the problem of uneven energy production by selling electricity to a local public utility when an excess is available and buying electricity from the same facility when wind power is in short supply.

Solar energy is diffuse energy that must be (1) collected, (2) converted to another form, and (3) stored if it is to compete with other available forms of energy. Solar energy plants use massive mirrors to track the sun and reflect the heat toward storage

Hydropower dams

Wind power

Solar panels on roof-top

Sun-tracking mirrors of a solar energy plant

FIGURE 41.12A Traditional sources of renewable energy.

tanks that drive a steam turbine. Passive solar heating of a house is successful when the windows of the house face the sun, the building is well insulated, and heat can be stored in water tanks, rocks, bricks, or some other suitable material. The use of photovoltaic (solar) cells is another way to tap the energy of the sun. In photovoltaic cells, one metal wafer absorbs solar energy and emits electrons that are collected by another wafer and passed on to appropriate wiring. Wind, solar, and geothermal energy production rose in 2013 in response to high energy prices for traditional sources of energy and growing state government support.

Biofuels and Electric Cars In the future, biofuels which are derived from organic matter may run power plants or your car (Fig. 41.12B, *left*). Biofuels such as ethanol, methanol, diesel gas, and methane are derived from plant material including agricultural crops (e.g., corn) or grasses (e.g., switchgrass) or algae. Biofuels can also be made from wastes such as wood chips or animal sewage. The use of animal sewage to produce biofuels is an excellent example of how recyling can avoid polluting the environment. Such a solution to the environmental problem of animal sewage disposal also shows how technology can help our society become more sustainable. Biofuels are derived from renewable sources of energy because crops and trees regrow and animals constantly produce sewage.

Traditional cars have internal combustion engines that run on gasoline, but hybrid cars that are increasing in popularity due to rising fuel costs run on both gasoline and electricity. The electricity is produced by a battery that is charged when the car is idle. Hybrid cars have improved mileage per gallon equivalent

FIGURE 41.12B Cars of the future may be powered by a biofuel or by electricity that recharges car batteries.

(MPGe), and purely electric cars may be capable in the future of 200 miles per gallon. The batteries of an electric car can be charged much as your cell phone is charged—by plugging it into an electrical outlet (Fig. 41.12B, *right*). Ideally, the power plants will be using a renewable source of energy such as those we have been discussing. Also, cars of the future may run on hydrogen. The car either burns hydrogen in an internal combustion engine, or a fuel cell produces electricity by combining hydrogen with oxygen. In either case, a car that burns hydrogen releases water and not carbon dioxide. Biofueled cars and hydrogen cars are now being produced and are available for sale.

▶ **41.12 CHECK YOUR PROGRESS**
  1. Determine the benefits for a society that makes use of renewable energy supplies.

41.13 Water sources should be conserved

LEARNING OUTCOME
When you complete this section, you should be able to
  1. Identify ways water can be conserved to protect the environment.

While people and wildlife need a constant supply of fresh water, at present most fresh water is utilized by industry and agriculture. Worldwide, 70% of all fresh water is used to irrigate crops! Although the needs of the human population overall do not exceed the renewable supply of water, this is not the case in certain regions of the United States and the world. When needed, humans increase the supply of fresh water by damming rivers and withdrawing water from aquifers. Dams have drawbacks: (1) Reservoirs behind the dam lose water due to evaporation and seepage into underlying rock beds. The amount of water lost sometimes equals the amount dams make available! (2) The salt left behind by evaporation and agricultural runoff increases salinity and can make a river's water unusable farther downstream. (3) Over time, dams hold back less water because of sediment buildup. Sometimes a reservoir becomes so full of silt that it is no longer useful for storing water. (4) The reduced water below the dam has a negative impact on the native wildlife.

Another mistaken practice is to meet freshwater needs by pumping vast amounts of water from aquifers, which are reservoirs just below, or as much as 1 km below, the ground's surface.

Aquifers hold about 1,000 times the amount of water that falls on land as precipitation each year. In the past 50 years, ground water depletion has become a problem in many areas of the world. Removal of water is causing land subsidence—that is, settling of the soil as it dries out. In California's San Joaquin valley, an area of more than 13,000 km² has subsided at least 30 cm due to groundwater depletion, and in the worst spot, the surface of the ground has dropped more than 9 m! Subsidence damages canals, buildings, and underground pipes.

Conservation of Water By 2025, two-thirds of the world's population may be facing serious water shortages. Agriculture could cut down on its use of water by planting drought- and salt-tolerant crops. Development of salt-tolerant traditional crops is already under way due to genetic engineering, as discussed in section 13.5. Using drip irrigation delivers more water to crops and saves about 50% over traditional methods while increasing crop yields as well (Fig. 41.13). Although the first drip systems were developed in 1960, they are currently used on less than 1% of irrigated land. At fault are government subsidies to farmers who irrigate in the usual manner. Industries could also cut their water needs by as much as one-half by reusing water and adopting other conservation measures. Power plants could use air rather than water for cooling purposes. The point is that we do have some leeway to conserve water in the future.

Much can also be done on the homefront. Domestically in MDCs, more water is usually used for bathing, flushing toilets, and watering lawns than for drinking and cooking. But houses could be equipped to recycle washing machine water and bath and sink water for reuse before it is discarded. Home yard irrigation should occur during dusk and dawn hours, as opposed to in the middle of the day when evaporation is at its highest. Purchasing and using water-saving toilets can save millions of gallons of water per year. Instead of running tap water to cool it, place a bottle of water in the refrigerator so that your drinking water is sure to be cool right away. You can go online to find many more ways to conserve water.

FIGURE 41.13 Drip irrigation saves water.

▶ **41.13 CHECK YOUR PROGRESS**
  1. List ways farmers and industry might conserve water instead of wasting it.

41.14 Agriculture can be more diverse

LEARNING OUTCOME

When you complete this section, you should be able to

1. Determine how agriculture can be improved to protect the environment.

Farmers today need to put greater emphasis on procedures that will make farming consistent with the goals of a sustainable society. Presently some farmers plant only a few genetic varieties. Because each crop is a monoculture (a genetically identical crop), a single destructive parasite or pathogen could cause huge crop losses. Instead, farmers need to practice polyculture, or the planting of several varieties of a crop, which will reduce the susceptibility of crops to pests or diseases (Fig. 41.14a). Polyculture also reduces the amount of herbicides necessary to kill weeds and can be used to replenish nutrients to topsoil.

Some farmers still rely on the heavy use of herbicides and pesticides, although biotechnology has produced plants that require lesser application of these products. Pesticides reduce soil fertility because they kill off beneficial soil organisms as well as pests. Pesticides also select for resistant insects, causing farmers to increase the amount of pesticide they use. Organic farms are increasing in number, and as mandated by the U.S. Department of Agriculture, they do not use synthetic herbicides or pesticides. Organic farming has become increasingly profitable in recent years because people are more willing to purchase organic produce, despite its higher cost compared to nonorganic produce. Health concerns surrounding pesticide use, as well as the desire for better-tasting food, have encouraged this trend. One way that organic and nonorganic farmers can eliminate the need for pesticides is by using integrated pest management, which advocates the growth of competitive beneficial insects and uses biological pest control methods (also called "biocontrol"). As discussed in section 41.2, biocontrol includes using natural predators, such as the ladybug beetle in Figure 41.14b.

Crop rotation can help reduce the use of nitrogen-containing fertilizers by farmers. When crop rotation is practiced, a nitrogen-fixing crop, notably a legume, which replenishes soil nutrition, is alternated with a crop such as corn that takes nitrogen from the soil. Multiuse farming techniques generally help increase the amount of organic matter and nutrients in the soil, as was confirmed by the experiment described in section 1.2.

In section 41.13, we discussed how farmers could reduce the amount of fresh water they use by planting drought-resistant plants and using drip irrigation. Once biofuels are readily available, farmers could use these renewable sources of energy to run heavy equipment including irrigation pumps and large machines to harvest crops. Another long-term goal might be to reduce the amount of animal husbandry because much of the grain produced in MDCs is used to feed livestock rather than humans. Eating grain instead of meat would feed more people, an important consideration because 925 million people go hungry every day.

Techniques such as contour farming (Fig. 41.14c) are available to reduce erosion and help minimize topsoil loss. Avoiding farming on steep slopes helps reduce erosion. Terrace farming involves converting steep slopes into steplike hills to minimize erosion. Farmers can plant "natural fences," such as rows of trees, around crops to prevent topsoil loss due to wind or other factors. These trees can also supply a useful product; some trees produce nuts and others are a good source of a particular chemical. Also, cover crops, which are often a mixture of legumes and grasses, help stabilize soil between rows of cash crops. Soil nutrients can be increased through composting, organic farming techniques, or other self-renewable methods. Finally, farmers should consider using precision farming (PF) techniques that help them micromanage the planting and harvesting of crops, while saving water and improving crop yields.

▶ **41.14 CHECK YOUR PROGRESS**

1. List and explain three improvements that can make farming more sustainable.

a. Polyculture

b. Biological pest control

c. Contour farming

FIGURE 41.14 A variety of methods can make farming more sustainable. **a.** Polyculture reduces the ability of one parasite to wipe out an entire crop and reduces the need to use an herbicide to kill weeds. This farmer has planted alfalfa between strips of corn, which also replenishes the nitrogen content of the soil (instead of adding fertilizers). Alfalfa, a legume, has root nodules that contain nitrogen-fixing bacteria. **b.** Instead of pesticides, it is possible to use a natural predator. Here, a ladybug is feeding on cottony-cushion scale insects on citrus trees. **c.** Contour farming with no-till conserves topsoil because water has less tendency to run off.

THE CHAPTER IN REVIEW

CONNECTING THE CONCEPTS

Our industrial societies can responsibly develop a sustainable society in which resources are preserved for the present and future generations. At the same time, biodiversity will be preserved. One goal of conservation biology is to protect, restore, and use biodiversity wisely because we now realize it is a resource of enormous economic value. If properly managed, sustainable yields of food and fiber can be obtained from many natural lands and waters. Modern genetic engineering technologies make the genes of millions of wild species available for breeding improved crops and domestic animals and for use as biological control agents. Enjoyment of nature can also enrich human life enormously.

If natural forests, grasslands, streams, lakes, and seas are degraded, human society must expend greater amounts of energy and materials to substitute for the benefits that biodiversity provides at no cost. Lost species, and ultimately lost ecosystems, cannot be replaced. To that end, the vision of conservation biology is a world where:

1. Leaders are committed to long-term environmental protection and to international leadership and cooperation in addressing the world's environmental problems, including the loss of biodiversity.
2. An environmentally literate citizenry has the knowledge, skills, and ethical values needed to protect biodiversity and achieve sustainable development.
3. Market prices and economic indicators reflect the full environmental and social costs of human activities.
4. A new generation of technologies contributes to the conservation of resources and the protection of the environment.
5. The landscape sustains natural systems, maximizes biological diversity, and uplifts the human spirit.
6. Human numbers are stabilized, all people enjoy a decent standard of living through sustainable development, and the global environment is protected for future generations. (Modified from the *Report of the National Commission on the Environment,* 1993.)

ANALYZE AND EVALUATE

1. Biotechnology has been offered as a way to improve crop yields to feed an ever-increasing world population. Do you agree with this solution? Offer and discuss an alternative one.
2. Why is it necessary to emphasize renewable energy sources in order to achieve a sustainable society? Do you agree with this emphasis? Why or why not?
3. Are you willing to pay extra for organic foods as a way to encourage farmers to reduce pesticide and herbicide use? Why would this be beneficial to society?
4. Do you agree that the size of the human population should be stabilized in order to achieve a sustainable society? What would that mean in terms of the number of children each couple can have?

SUMMARIZE

Conservation Biology and Biodiversity

41.1 Conservation biology is a practical science

- **Conservation biology** studies biodiversity with the goal of conserving natural resources and preserving biodiversity. **Biodiversity** refers to the variety of life on Earth.
- **Extinction** is the total disappearance of a species or higher group. **Endangered species** are at immediate risk of extinction. **Threatened species** are likely to soon become endangered.
- Biologists study biodiversity in terms of **genetic diversity, ecosystem diversity,** and **landscape diversity.** Diversity is highest at the tropics and declines toward the poles; **biodiversity hotspots** contain large concentrations of species.

41.2 The direct value of biodiversity is becoming better recognized

- Many prescription drugs were originally derived from living organisms, and still more drugs may be derived from rain forest species.
- Biodiversity can help protect crops against disease and save billions of dollars for a nation's agricultural economy.
- The environment produces wild fruits, vegetables, meats, and fish, and also sustainable products such as rubber.

41.3 The indirect value of biodiversity is immense

- Biodiversity contributes to the successful workings of the water, carbon, phosphorus, and nitrogen cycles.
- Freshwater ecosystems provide fresh water and fish; forests soak up and release water at a regular rate and take up carbon dioxide; and intact ecosystems retain soil.
- Ecotourists enjoy vacations in natural settings.

Causes of Species Extinctions

41.4 Habitat loss is a major cause of wildlife extinctions

- Habitat loss is the cause of 85% of threatened/endangered cases. Coastal degradation destroys freshwater and marine biodiversity:

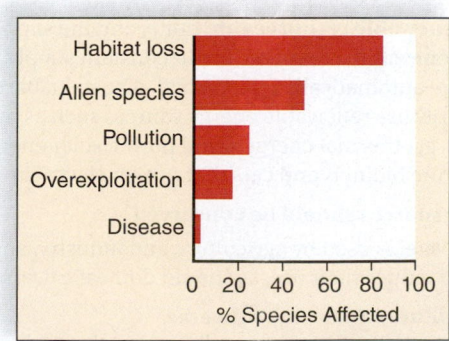

41.5 Introduction of alien species contributes to extinctions

- **Alien species** are nonnative species introduced through colonization, horticulture and agriculture, and accidental transport. Islands are particularly susceptible to disturbances by introduced aliens.

41.6 Pollution contributes to extinctions

- **Pollution** is any change in the environment that adversely impacts organisms. Forms of environmental pollution include acid deposition, eutrophication, ozone depletion, organic chemicals, climate change, and solid waste disposal.

41.7 Overexploitation contributes to extinctions

- **Overexploitation** is the removal of a number of individuals so that the population is severely reduced. The market for exotic plants and pets supports the legal and illegal trade of wild species. Declining species are still being hunted, and fisheries are being overfished.

41.8 Disease contributes to extinctions

- Pathogens are on the rise, subjecting wildlife to emerging diseases.

Achieving Species Preservation

41.9 Habitat preservation is of primary importance

- Biodiversity hotspots and keystone species must be preserved. **Keystone species** influence the viability of a community; their extinction can lead to other extinctions and loss of biodiversity. In contrast, **flagship species** engender an emotional response. Preserving keystone species requires landscape preservation, which will lead to preservation of other species as well.

- A **metapopulation** is subdivided into small, isolated populations, sometimes due to habitat fragmentation. A **source population** reproduces well most likely due to a favorable environment; a **sink population** has more deaths than births most likely due to an unfavorable environment.

41.10 Habitat restoration is sometimes necessary

- **Restoration ecology** seeks scientific ways to return ecosystems to their former state through the use of natural processes and with the goal of **sustainable development.**

Achieving a Sustainable Human Society

41.11 A sustainable society will preserve resources

- Like an ecosystem, a **sustainable society** preserves resources for use by the generations to come.

41.12 Energy sources should be renewable

- **Nonrenewable resources** are finite, and the supply can run out. **Renewable resources** are in constant supply because they are automatically replenished. A sustainable society will primarily use renewable energy sources such as hydroelectric plants, geothermal energy, wind power, solar energy, and in the future biofuels and cars that run on electricity alone.

41.13 Water sources should be conserved

- Most water is used by agriculture and industry, and both should reduce their use, as should domestic users.

41.14 Agriculture can be more diverse

- Agriculture should also use polyculture, biological pest control, and contour farming to be sustainable.

TEST YOURSELF

Conservation Biology and Biodiversity

1. Which of these tasks would not be within the realm of conservation biology?
   a. helping to manage a national park
   b. a government board charged with restoring an ecosystem
   c. writing textbooks and/or popular books about the value of biodiversity
   d. introducing endangered species back into the wild
   e. All of these are concerns of conservation biology.

2. Most likely, ecosystem performance improves
   a. the more diverse the ecosystem.
   b. as long as selected species are maintained.
   c. as long as species have both direct and indirect value.
   d. if extinctions are diverse.
   e. Both b and c are correct.

3. Biodiversity hotspots
   a. have few populations because the temperature is too hot.
   b. contain a large proportion of the Earth's species even though their area is small.
   c. are always found in tropical rain forests and coral reefs.
   d. are sources of species for the ecosystems of the world.
   e. All except a are correct.

4. Which of these associations does not show a contrast in the number of species?
   a. temperate zone—tropical zone
   b. hotspots—cold spots
   c. rain forest canopy—rain forest floor
   d. pelagic zone—deep-sea benthos

5. **THINKING CONCEPTUALLY** Draw an energy pyramid (see section 39.10) to show that if salmon ate only shrimp in Figure 41.1 the salmon population would get less energy than if they ate zooplankton.

6. The value of wild pollinators has been calculated to be $2.4 billion in California alone. What is the implication?
   a. Society could easily replace wild pollinators by domesticating various types of pollinators.
   b. Pollinators may be valuable, but that doesn't mean any other species won't provide us with valuable services also.
   c. If we did away with all natural ecosystems, we wouldn't be dependent on wild pollinators.
   d. Society doesn't always appreciate the services that wild species provide naturally and without any fanfare.
   e. All of these statements are correct.

7. Consumptive use value
   a. means we should think of conservation in terms of the long run.
   b. means we are placing too much emphasis on organisms that are useful to us.
   c. means some organisms, other than crops and farm animals, are valuable as products.
   d. is a type of direct value.
   e. Both c and d are correct.

8. The services provided to us by ecosystems are unseen. This means
   a. they are not valuable.
   b. they are noticed particularly when the service is disrupted.
   c. biodiversity is not needed in order for ecosystems to keep functioning as before.
   d. we should be knowledgeable about ecosystems and protect them.
   e. Both b and d are correct.

9. Which of the following is not a function that ecosystems can perform for humans?
   a. purification of water
   b. immobilization of pollutants
   c. reduction of soil erosion
   d. removal of excess soil nitrogen
   e. breakdown of heavy metals such as lead
10. Which of these is not an indirect value of a species?
    a. participates in biogeochemical cycles
    b. participates in waste disposal
    c. helps provide fresh water
    d. prevents soil erosion
    e. All of these are indirect values.

Causes of Species Extinctions

11. The most significant cause of the loss of biodiversity is
    a. habitat loss.
    b. pollution.
    c. alien species.
    d. disease.
    e. overexploitation.
12. Eagles and bears feed on spawning salmon. If shrimp are introduced that compete with salmon for food,
    a. the salmon population will decline.
    b. the eagle and bear populations will decline.
    c. only the shrimp population will decline.
    d. all populations will increase in size.
    e. Both a and b are correct.
13. Global climate change has nothing to do with
    a. habitat loss.
    b. introduction of alien species into new environments.
    c. pollution.
    d. overexploitation.
    e. Global climate change pertains to all of these.
14. Which of these is not expected because of global climate change?
    a. changes in the composition of ecosystems
    b. the bleaching and drowning of coral reefs
    c. rise in sea levels and loss of coastal wetlands
    d. pest preservation because cold weather reduces their population size
    e. All of these are expected.
15. Which of the following associations is not correct?
    a. excess nutrients—eutrophication
    b. carbon dioxide—ozone depletion
    c. sulfur dioxide—acid deposition
    d. methane—global climate change
16. **THINKING CONCEPTUALLY** Considering the food web diagram in Figure 39.9, why would you expect a predator, such as a mongoose, to prey on more than one type of organism?

Achieving Species Preservation

17. Why is a grizzly bear a keystone species existing as a metapopulation?
    a. Grizzly bears require many thousands of kilometers of preserved land because they are large animals.
    b. Grizzly bears have functions that increase biodiversity, but presently the population is subdivided into isolated subpopulations.
    c. When grizzly bears are preserved, so are many other types of species within a diverse landscape.
    d. Grizzly bears are a source population for many other types of organisms across several population types.
    e. All of these statements are correct.

18. A population in an unfavorable area with a high infant mortality rate would be a
    a. metapopulation.   c. sink population.
    b. source population.   d. new population.
19. The goal of restoration ecology is
    a. to return damaged ecosystems to their former states.
    b. to maximize direct values of ecosystems.
    c. sustainable development.
    d. Both a and c are correct choices.
    e. a, b, and c are all correct choices.

Achieving a Sustainable Human Society

20. A sustainable society would as much as possible
    a. use renewable energy sources.
    b. recycle materials.
    c. conserve water.
    d. All of these are correct.
21. Biofuels are made from
    a. both renewable and nonrenewable sources.
    b. any organic matter, including crops and waste materials.
    c. any pollutant in the environment.
    d. All of these are correct.
22. A sustainable agriculture will
    a. liberally irrigate crops.
    b. use limited amounts of pesticides, herbicides, and fertilizer.
    c. make sure a field contains only one species at a time.
    d. All of these are correct.
23. **THINKING CONCEPTUALLY** Name several actions an ordinary citizen can take to help preserve biodiversity.

GET INVOLVED

1. Hypothesize the effect of overharvesting on genetic diversity (see "Adaptability of Small Populations" on p. 738), even if the population is protected for a while and allowed to return to its normal level. How would you test your hypothesis?
2. A conservationist wants to rescue a species of songbirds from extinction by gathering a few eggs and raising a population that s/he can reintroduce into the wild. Criticize the plan.

MEDIA STUDY TOOLS

mhhe.com/maderconcepts3

Enhance your study of this chapter with interactive study tools, practice tests, and engaging animations. Also, ask your instructor about the resources available through ConnectPlus, which includes LearnSmart, a personalized adaptive learning program, and a media-rich eBook.

Appendix A | Answer Key

CHAPTER 1

Check Your Progress

1.1 1. Making observations is done to gather initial information about an interest concerning the natural world. Formulating a hypothesis helps to define what a study is about. Performing an experiment and making observations is done to test the hypothesis. Coming to a conclusion determines if the hypothesis is supported or not. 2. Cell theory: organisms are composed of cells which only come from other cells. Gene Theory: genes dictate the form, function, and behavior of organisms. Theory of Evolution: all organisms trace their ancestry to one source and are adapted to the environment. Theory of Homeostasis: organisms have an internal environment that is regulated. Theory of Ecosystems: organisms live in units of the biosphere where they interact. **1.2** 1. The test pots received either synthetic nitrogen fertilizer or a preplanting of pigeon peas. The control group received no fertilizer treatment. 2. The hypothesis was not supported and they formulated a new hypothesis. **1.3** 1. Levels of organization from cells to organism illustrate order. 2. New cells come from a division of previous cells, and cells must acquire nutrients and energy from outside themselves. **1.4** 1. DNA dictates what cells and organisms will be like and parental genes composed of DNA are passed on during reproduction. 2. DNA technology helps us understand cell structure and function; how organisms differ from one another; how organisms are related and evolved. **1.5** 1. In humans and other mammals, hormones regulate the blood level of various substances such as sugar; several types of animals bask in the sun to remain warm. 2. Each of these occurs in response to a stimulus, for example low blood sugar causes a hormonal response and a low body temperature causes an animal to move into the sun. **1.6** 1. As chemicals move (gray arrows) from one population to another they cycle and there is no loss but as energy moves from one population to the next (orange to red arrows) energy is lost and therefore energy cannot cycle. 2. Humans destroy ecosystems when they build houses, towns, cities and humans add pollutants to the environment. **1.7** 1. All organisms can trace their ancestry to the same source is the concept of common descent. A study of fossils, comparative anatomy and physiology, and development all support the theory of evolution. 2. An evolutionary tree shows how one species is related to another by way of common ancestors. **1.8** 1. The first domain gave rise to domains Bacteria and Archaea. Later, domain Eukarya split off from the archaeal line. 2. A scientific name tells the genus and specific epithet. 3. Fire ants would be classified in the domain Eukarya and the animal kingdom. **1.9** Scenario: In a population, a few fish can swim faster and escape predatory sharks. These fish have more offspring than slower fish, and with time, the entire population contains mostly fast swimmers. **1.10** The hawk has levels of organization; catches food for herself and offspring; remains homeostatic because she can respond to stimuli, lives in a semidesert ecosystem; and is adapted to flying.

Consider These Questions

Page 6: 1. No answer: The United States should not continue to export its current farming technology. Exporting technology that is known to be detrimental to groundwater and topsoil will eventually reduce agricultural yields, resulting in a food shortage. The United States would be better served to encourage sustainable farming practices like crop rotation to foster long-term success in agriculture. Yes answer: The United States should continue to export farming technology to other countries in order to support the global food market. As solutions to the long-term problems of these technologies arise, the U.S. should make these available to other countries as well. 2. Circumstances such as labor costs, profits, and marketing challenges might discourage a farmer from growing organically. These obstacles might be overcome through government subsidies for sustainable farming practices, sharing best practices through networking with successful organic farmers, and a worldwide movement toward eating primarily locally grown and seasonal produce. **Page 17:** 1. Yes, if we offer examples such as these: Practical problems in agriculture, medicine, and conservation can be solved with a knowledge of evolution. Because we know that natural selection drives the evolution of pesticide-resistant insects, changes to farming practices and pesticide use are possible. In medicine, awareness of the evolution of antibiotic-resistant bacteria allows patients and doctors alike to make changes to the way antibiotics are prescribed and used. In conservation efforts, knowledge of evolution can be used to make informed decisions regarding endangered species; directed evolution can also be used to select for organisms to clean up the environment. No: It is not possible unless people have open minds and will consider scientific findings. 2. No answer: The government should not interfere with our private lives. Yes answer: The government should do it can to protect the populace. 3. No answer: Sanitizers can prevent us from building up our natural immunity. Yes answer: Sanitizers offer another layer of protection.

Analyze and Evaluate

Page 18: 1. One example, that shows the relationship between two theories, is the link between ecosystems and evolution. The different species in an ecosystem are adapted to the biotic conditions (other organisms) in that ecosystem and abiotic (physical) conditions of that ecosystem. 2. Certain bacterial members of a population are resistant to the antibiotic; therefore they survive and reproduce more in the presence of the antibiotic than other members of the population. After several generations all members of the population are resistant.

Test Yourself

1. a; 2. d; 3. c; 4. Scientific theories arise due to innumerable observations and experimentation. 5.b; 6. c; 7. b; 8. c; 9. Cells are microscopic. 10. c; 11. d; 12. Each type organism has its own sequence of bases in its genes. 13. e; 14. e; 15. cellular; 16. d; 17. a; 18. d; 19. d; 20. A college campus has a location, as does an ecosystem. The populations of students, faculty, and administrators communicate with each other and the physical environment (the buildings). 21. c; 22. e; 23. d; 24. d; 25. c; 26. a; 27. c; 28. c; 29. e; 30. f; 31. c; 32. g 33. Evolution is related to all the other theories; for example, all organisms are cellular because their common ancestor was cellular.

Get Involved

1. Plant the same species of tomato plants in three large plots. All plots receive the same treatment, except plot 1, your control, receives no fertilizer; plot 2 receives the name-brand fertilizer in the same quantity as plot 3-which receives the generic brand. Measure the tomatoes from each plot and calculate the average size to determine which plot results in the largest tomatoes. 2. The experimental variable is the independent variable. The dependent variable (results) depends on the independent variable. 3. a. Bacteria don't die in sunlight when dye is present. b. Dye is protective against UV radiation. c. Experiment consists of exposing control and test groups to UV light. d. Hypothesis is not supported.

CHAPTER 2

Check Your Progress

2.1 1. Carbon, nitrogen, phosphorus, sulfur. 2. See Figure 2.2B, page 27, in text. 3. Calcium 40 has 20 neutrons and calcium 48 has 28 neutrons. **2.2** 1. Hydrogen has one shell, which is complete with two electrons; 2. Phosphorus (atomic # 15) has three shells with two electrons in the first shell; eight electrons in the second shell and five valence electrons in the outer shell. 3. Oxygen requires two more electrons for a completed outer shell. **2.3** 1.If sodium (Na) gives up an electron it will have a completed outer shell and if Chlorine (Cl) gains an electron it will have a completed outer shell. 2. H^+, OH^-. **2.4** 1. One end of the molecule is negative and the other end is positive because oxygen attracts electrons more than hydrogen does. 2. Yes, because electropositive hydrogens are attracted to either electronegative oxygen or nitrogen. **2.5** 1. Water molecules are cohesive (hydrogen bonding causes water molecules to stick together) and cohesive (hydrogen bonding cause waters molecules to stick to other polar molecules.) 2. Both people and plants use water for transport. See Figure 2.5, p. 31. **2.6** 1. The blocks of ice trap heat inside and prevent it from escaping to the environment. 2. The air loses heat as it causes water in the pad to evaporate. **2.7** Polarity makes the emulsifiers hydrophilic. **2.8** Water gains H^+ when acids are added to water and OH^- when bases are added to water. **2.9** Water is acidic; when the pH is 5.6 because it has more H^+. 2. When conditions are acidic, the bicarbonate ion combines with H^+ to form carbonic acid and when conditions are basic it gives up H^+ which combines with OH^- to form water.

Consider These Questions

Page 26: 1. Though controversy will undoubtedly arise over moral, monetary, and other issues, science must be free to ask and answer questions about the world in which we live. However, scientists should conduct research with minimal risks to their own safety and that of others. As technology advances, the restrictions placed upon research should be reevaluated to ensure their relevance. With careful and balanced monitoring, scientific endeavors should be supported and encouraged. 2. No answer: Safety should be paramount in experimentation and therefore seniors and people with preexisting conditions should not be permitted to engage in risky medical studies. Yes answer: People should be free to decide for themselves whether they want to risk participation. 3. Yes answer: Depending on the benefits that an experiment might possibly have, I would consider being a guinea pig. No answer: I would not be willing to serve as a guinea pig in experiments that may prove harmful to me. **Page 36:** 1. Yes, it concerns me. The changes in lakes due to acid rain kill fish and other wildlife and sometimes eliminate it altogether. Trees suffer as a result of acid rain's effects, becoming diseased or dying. If this continues, not only will the atmosphere's oxygen be affected, but the lumber industry will suffer. The food supply is at risk if plants as well as fish are negatively affected by acid rain. 2. Human beings must take some responsibility for what they do to the planet and make attempts to control their negative impact. Driving less could help to preserve lakes and the wildlife in them, forests, buildings made of limestone and marble, and the respiratory health of humans. Because of these factors, I would certainly be willing to drive less to help prevent acid rain. 3. Despite other stresses, environmental degradation should concern people at all times. If we do not preserve our planet and its ecosystems, our current way of life will become unsustainable and economic development will cease.

Analyze and Evaluate

Page 37: 1. Methane gas (CH_4) is formed when a carbon atom binds four hydrogen atoms. The bonds of methane are nonpolar covalent bonds in which electrons are shared equally between the atoms. Each bond in methane points to a corner of a tetrahedron. Water is formed when two hydrogen atoms bind one oxygen atom. The sharing of electrons is not entirely equal in this bent molecule. Oxygen, being more electronegative, attracts the electrons more than hydrogen does, and the result is a polar molecule. 2. The properties of water, which result from its structure, are critical to life. Cohesion, adhesion, high surface tension, high heat capacity, and high heat of vaporization all result from hydrogen bonding. These properties of water allow it to be a transport medium that also maintains the internal temperature of organisms. The difference in the density of frozen water compared to liquid water means that life is preserved under a layer of ice. Water being a polar molecule acts as a universal solvent that allows metabolism to occur.

Test Yourself

1. c; 2. b; 3. e; 4. d; 5. a; 6. c; 7. The characteristics of the elements recur; all atoms in the first group have one electron in the outer shell, for example. 8. c; 9. d; 10. Nitrogen needs three more electrons in the outer shell to be stable, and each hydrogen shares one electron. 11. a; 12. c; 13. b; 14. a; 15. b; 16. d; 17. d; 18. Gaining six electrons by sharing is not as likely as losing two electrons to become an ion. 19 a; 20. b; 21. c; 22. e; 23. Blood is transported in a tube (blood vessel), and water fills a tube due to its cohesive and adhesive properties. 24. a; 25. a; 26. a; 27. b; 28. c; 29. A bicarbonate buffer combines immediately with H^+ or OH^-, normalizing the blood pH. Respiration, by removing carbon dioxide from the blood, provides a slower response that decreases H_2CO_3 concentration. The kidneys excrete H^+ and provide the slowest of the three responses, but the kidneys also have the ability to cause the greatest overall change in the pH level.

Get Involved

1. Na^+Cl^- interrupts hydrogen bonding enough to prevent the formation of the ice lattice that forms during freezing. 2. Chemical behavior is dependent on the number of electrons in the outer shell, not the number of neutrons in the nucleus.

CHAPTER 3
Check Your Progress

3.1 1. Carbon can share with four different elements at a time; and bond with itself to form long chains; can form double bonds; hydrocarbons can double back to form a ring compound. 2. Carbohydrates, proteins, lipids and nucleic acids. 3. Testosterone and estrogen differ only by attached functional groups and yet they have a profound affect on the body. **3.2** 1. A dehydration reaction joins subunits together, and a hydrolysis reaction breaks them apart. 2. Starch releases glucose; protein releases amino acids; and a fat releases glycerol and fatty acids. **3.3** Glucose is a 6-carbon sugar used by all organisms as a quick source of energy; sucrose is a disaccharide used as transport sugar in plants for hydrolysis reactions. **3.4** All complex carbohydrates are sources of stored energy and are polymers of glucose; glycogen which is highly branched, and starch can be branched or not branched. The bonding between glucose subunits in cellulose, chitin and peptidoglycan makes them structural carbohydrates that are not digestible. Chitin and peptidoglycan have attached amino groups/peptide chains. **3.5** 1. Saturated fatty acids, which do not have double bonds between carbon atoms, contribute to plaque buildup in blood vessels but unsaturated fatty acids with double bonds are protective against plaque. **3.6** Cholesterol (which stabilizes animal cell plasma membranes) contains rings and both phospholipids (which forms the plasma membrane) and waxes (which have protective functions) contain long hydrocarbon chains. **3.7** 1. Globular proteins are enzymes present in all organisms, fibrous proteins are structural proteins, and plants do not use proteins as structural compounds. 2. All amino acids have an amino group and an acidic group and an R group. The R group is different for each amino acid. 3. The peptide bond is polar; the oxygen is partially negative and the hydrogen bond is partially positive. **3.8** Enzymes are globular and plant cells carry out many enzymatic reactions. Structural proteins are fibrous and plant cells use cellulose as a structural compound. **3.9** 1. The complementary DNA sequence is CTAGGT. 2. The sugar in RNA is ribose not deoxyribose as in DNA; uracil in RNA replaces thymine in DNA; RNA is single-stranded not double-stranded helix. 3. A mutation is an altered sequence of bases in DNA and an altered base sequence in DNA causes a change in the complementary base sequence in messenger RNA. **3.10** Glycogen is stored glucose and glucose leads to ATP buildup in cells. Without a continual supply of ATP, cells die.

Consider These Questions

Page 49: 1. Yes answer: Because of the increased instances of obesity, especially in children, it would be responsible for restaurants to decrease portion sizes. No answer: People should be able to eat responsibly on their own. Restaurants should not have to manage the diets of their customers. 2. Due to the implication of cornstarch-based sweeteners and other ingredients in manufactured foods, manufacturers should be required to limit the use of such ingredients in their foods. It is also the responsibility of the consumer, however, to intake cornstarch-based sweeteners and saturated fats in moderation. **Page 53:** 1. Yes answer: Studies should be encouraged as much as possible and considered worthwhile until evidence suggests otherwise. No answer: In these days of economic hardship, only studies judged by a panel of experts as worthwhile should be funded. 2. Yes answer: By inserting a gene

that allows tomato plants to thrive when watered with salty water, a genetic change has occurred within the organism. If the plants reproduce other genetically altered individuals, and a new population results, then evolution has occurred. No answer: This is an example of artificial evolution, and more appropriately, genetic engineering. Because the change was not the result of natural selection, it should not be termed evolution.

Analyze and Evaluate

Page 57: 1. (a) Energy storage and use: lipids, which contain many energy-storing bonds; carbohydrates, which are a source of quick energy in the form of glycogen and starches. (b) Genetic information storage: nucleic acids, like RNA and DNA which store genetic information in a sequence of nucleotides. (c) Ongoing activities of the cell: proteins, which provide structure and support in the cell, speed chemical reactions, transport molecules (i.e., oxygen), and which are used in defense and regulation; lipids, which make up membranes, serve as hormones, and protect plants and animals in the form of waxes. 2. Just as the alphabet has only 26 letters and yet can produce many different words, DNA contains only 4 bases but the particular order of these bases in each gene can vary. This means that each gene specifies a different sequence of amino acids in each type protein. 3. I would expect each cell type to have its own particular combination of proteins.

Test Yourself

1. c; 2. c; 3. b; 4. b; 5. Like carbon, silicon has four outer electrons and can form four covalent bonds, as in SiO_2, the main component of sand. Carbon has two, but silicon has three, shells of electrons. Because of its larger size, silicon rarely forms chains, nor does it bond to four different types of atoms. 6. b; 7. a; 8. d; 9. c; 10. d; 11. c; 12. c; 13. Cellulose chains can form fibers because they lie side-by-side and hydrogen bonds form between them. The branching observed in starch makes fiber formation more unlikely. 14. a; 15. b; 16. d; 17. d; 18. c; 19. The hydrophilic heads interact with fluids, allowing the hydrophobic tails to orient toward each other. 20. c; 21. c; 22. d; 23. c; 24. d; 25 The sequence of bases in DNA specifies the proteins of a cell but it is the proteins that control the structure and function of a cell. 26. e; 27. a; 28. c; 29. A nucleic acid has a sequence of nucleotides and protein has a sequence of amino acids.

Get Involved

1. a. Subject the seeds of temperate and tropical plants, for which you know the amount and kind of oil content, to a range of temperatures from above freezing to below freezing for an extended length of survival per type of plant. b. The presence of unsaturated oils in temperate plant seeds may be an adaptation to the environment. 2. Possible hypothesis: (1) The abnormal enzyme will not produce as much product per unit time as the normal enzyme. (2) The abnormal enzyme will have a different shape from the normal enzyme due to changes in organization.

CHAPTER 4
Check Your Progress

4.1 1. The human body is composed of cells, the basic units of life. New cells in the body arise from preexisting cells. 2. A high surface-area-to-volume ratio means that there will be enough surface area to allow for adequate exchanges between the environment and the contents of the cell. This makes the cell more efficient. **4.2** 1. Archaea and bacteria have a similar appearance but differ biochemically including their DNA/RNA. 2. Pili are appendages that attach to surfaces; ribosomes produce proteins that carry out metabolism; nucleoid contains chromatin, which controls the cell; plasma membrane is outer surface of cell where exchanges occur; cell wall and capsule are outer boundaries that protect the cell; flagella are appendages for movement in a fluid medium. **4.3** 1. Organelles allow a large cell to be efficient because they carry on specialized functions. 2. The presence of chloroplasts allows plants to photosynthesize; both types of cells require a continuous supply of ATP. **4.4** 1. The nuclear envelope has pores that allow substances to pass through. 2. The nucleolus produces the subunits of a ribosome. An RNA copy of gene moves from the nucleus to a ribosome; the protein made by the ribosome enters the ER. **4.5** 1. Ribosomes at RER produces proteins ; transport vesicles take them from RER to the Golgi apparatus; transport vesicles from Golgi take them to the plasma membrane. 2. Lysosomes contain hydrolytic enzymes that digest biomolecules and cell parts; transport vesicles only carry biomolecules from one organelle to another. The Golgi contains a series of saccules; possible each saccule specializes in producing particular proteins. **4.6** 1. The ER (produce) the Golgi (modify) and lysosomes (break down). **4.7** 1. The vacuole is filled with cell sap and pushes against the plasma membrane and cell wall to help maintain shape of cell. Cell cap contains pigments, excretory molecules and molecules toxic to insects, among other possible molecules. **4.8** 1. A peroxisome has an enzyme to break down hydrogen peroxide and often synthesizes and breaks down lipids. Lysosomes contain very caustic enzymes that breakdown large biomolecules and parts of the cell. **4.9** 1. Plants absorb solar energy and produce carbohydrates; mitochondria use the energy of carbohydrates to produce ATP. **4.10** 1. Cytoskeletal fibers (1) support the cell as when intermediate filaments support the nuclear envelope, and (2) allow cell parts to move as when microtubules become tracks for the movement of vesicles by a motor molecule. 2. Actin filaments contain two twisted strands of actin molecules. Bundles of actin filaments support cell structures. Microtubules are composed of 13 rows of tubulin. Microtubules help maintain the shape of the cell and serve as tracts for organelle movement. They also form the spindle which allows chromosomes to move during cell division. 3. Motor molecules use ATP as a source of energy that allows them change their position and create movement.

Consider These Questions

Page 64: 1. I would give a friend the opportunity to see for his/herself by taking him/her into a biological laboratory and allowing them to examine a cell, for example, microscopically. 2. Color enhancement of TEMs does not border on misrepresentation because it is often necessary to enhance images to view and analyze them more clearly. However, I would encourage a side-by-side comparison of colored and noncolored TEMs images to show what a cell actually looks like. **Page 72:** 1. Carbon is present in all organic molecules but sulfur is unique to amino acids. 2. Palade may have also found the labeled amino acids on free ribosomes in the cytoplasm. **Page 75:** 1. An "evolutionary advantage" means that the organism can capture more resources than other members of a population and therefore will produce more offspring than other members. 2. I would expect to find that both have (1) membrane-bound chlorophyll capable of absorbing light energy (2) centrally located DNA (3) a membrane boundary. A chloroplast has a double membrane because the outer one is derived from the plasma membrane of the host cell.

Analyze and Evaluate

Page 79: 1. Prokaryotes are smaller than eukaryotic cells but they have DNA and ribosomes. 2. The folding of the endoplasmic reticulum creates a large area for the attachment of ribosomes outside and presence of enzymes inside. The membranous nature of the membrane allows it to form transport vesicles. 3. The cytoskeleton allows the cell and its parts to move. The presence of microtubules in cilia and flagella and within the cytoplasm is consistent with this function.

Test Yourself

1. c; 2. a; 3. c; 4. d; 5. c; 6. d; 7. b; 8. c; 9. a; 10. The nuclear envelope has pores; transport vesicles; 11. c; 12. c; 13. e; 14. anatomical and experimental data; 15. d; 16. d; 17. a; 18. a;19. d; 20. Chloroplasts capture solar energy and form the carbohydrate broken down by mitochondria to produce ATP; 21. d; 22. b; 23. b; 24. c; 25. b; 26. a; 27. e; 28. d

Get Involved

1. Labeling RNA. You expect to find RNA in the nucleus, passing through a nuclear pore, and on RER. 2. Mitochondria have their own DNA and make their own proteins; therefore, they have RNA. The mitochondrial double membrane does not have any pores.

CHAPTER 5

Check Your Progress

5.1 1. Just as a mosaic is a pattern formed by adding decorative pieces so a membrane is formed by adding proteins. 2. Cell recognition proteins identify the cell; receptor proteins receive signaling molecules; enzymatic proteins carry out reactions; junction proteins join cells together **5.2** 1. Facilitated diffusion does not require energy because the substances traveling down its concentration gradient through a carrier protein. 2. In a hypotonic solution, a red blood cell is subject to bursting and in a hypertonic solution, it shrivels (crenates). **5.3** 1. Just as in active transport, the payment is like the use of ATP to change the shape of the carrier protein (turnstile) so that the substance can pass through. 2. Have the arrows proceed in the opposite direction. **5.4** 1. Both adjoin two cells and allow substances to pass through. 2. Carbohydrates make up the plant cell wall the ECM of animal cells contains proteins. **5.5** 1. Organisms need a supply of stored energy (potential energy) in the form of energy-rich molecules so they can exhibit kinetic energy (i.e. cell division, nerve conduction, muscle contraction) in cells. 2. A horse cannot create energy; therefore, it has to store energy in order to have a supply of ATP for muscle contraction (running away). **5.6.** 1. Just like money ATP is spent when a cell needs something (i.e. build a protein, nerve conduction) and just like small bills rather than big bills are convenient for spending money, ATP having a lot less energy than a glucose molecule is a convenient form of energy for cells to use. 2. Figure 5.3A illustrates that after a carrier protein breaks down ATP, the P attaches to it and this changes its shape so that a substance can enter the cell. Similarly, Figure 5.6C shows how P attached to myosin allows it to pull actin during muscle contraction. **5.7** 1. An enzymatic reaction cannot occur without an enzyme having a functional active site because this is where the substrate forms a complex with the enzyme. 2. Knowing the environmental conditions preferred by enzymes facilitates research (including medical research) and commercial applications 3. Enzyme inhibition is a major way the body has of regulating the amount of enzyme product to suit the needs of the cell. Unfortunately, deaths can occur due to enzyme inhibition caused by accidental or deliberate use of certain poisons by terrorists.

Consider These Questions

Page 89: 1. People could be convinced of the benefits of nutrition and exercise in the prevention of type 2 diabetes if shown case studies or other documentation of individuals with type 2 diabetes who have eliminated or controlled their disease through proper diet and fitness. 2. The extra cost of making sure the text is useable by a limited number of color-blind individuals and those with other visual impairments (i.e., poor eyesight) might cause the price of the textbook to go up. **Page 91:** 1. It seems reasonable that human and bacterial plasma membranes are the same when we

consider that all organisms are descended from an original ancestor. **Page 98:** 1. Even though poisons can be ingested accidentally they should not be banned. Poisons are used to preserve foods, control pests, eliminate weeds, and to preserve buildings. Proper storage and use is important when using poisons, as well as posting notice when dangerous substances such as rat poison are present. 2. Though the warning system may not be able to detect all poisons, it could save many lives if a system were in place to alert the public or a country's government of an impending attack.

Analyze and Evaluate

Page 99: 1. An enzyme serves to lower activation barriers by bringing reactants together within its active site. This is only possible due to the specific structure, or shape, of enzymes. The enzyme shape complements the shape(s) of the reactant(s), allowing them to interact chemically. 2. Polypeptide synthesis will not occur unless amino acids, enzymes, and the proper nucleic acids are present. Environmental pH and temperature must be appropriate. For any endergonic reaction to occur, ATP must be present to supply the necessary chemical energy. 3. Cell recognition glycoproteins, receptor proteins, and junction proteins found in the plasma membrane are all involved in cell-to-cell communication.

Test Yourself

1. c; 2. b; 3. b; 4. b; 5. c; 6. d; 7. c; 8. Through osmosis solutes retain the liquid portion of blood in the vessels, allowing the blood to flow. 9. a,b,c; 10. b, c, d; 11. d; 12. b, d; 13. d; 14. e 15. c; 16. a; 17. e; 18. d; 19. a; 20. Food in bulk is stored food in your pantry; glycogen is stored energy in the liver and muscles; meals provide only enough energy till you eat again; ATP is just enough energy for a reaction. 21. a; 22. e; 23. c; 24. b

Get Involved

1. Ecosystems need a source of energy because energy cannot be created. They need a continuous supply because with each energy transformation, heat is lost; eventually, all the energy taken in is lost to the system as heat. 2. You need to decide the proper enzyme versus substrate concentrations, type of glassware, amount of time needed for the reaction, how to vary the temperature and the pH, and how to test for the product.

CHAPTER 6

Check Your Progress

6.1 1. The two major parts of a chloroplast are the grana (thylakoids) where chlorophyll absorbs solar energy and the stroma where carbohydrate is produced. **6.2** 1. Water participates in the light reactions: it splits releasing electrons, hydrogen ions, and oxygen; carbon dioxide participates in the Calvin cycle reactions and is reduced to a carbohydrate (CH_2O); solar energy provides the energy for NADPH and ATP production which are used to reduce carbon dioxide during the Calvin cycle reactions. 2. Provides oxygen for us to breathe. Provides food for us and all other living things. **6.3** 1. Any and all pigments present have an opportunity to absorb solar energy from white light. **6.4** 1. A pigment complex absorbs solar energy so that all the energy becomes concentrated in the reaction center. **6.5** 1. The electron acceptor in PS II passes energized electrons to an electron transport chain. As the electrons pass down the chain, they release energy and this is used by some members of the chain to pump hydrogen ions across the membrane. A hydrogen ion gradient is established and the flow of H^+ down a ATP synthase complex provides the energy for ATP production. **6.6** 1. When the pigment complex of PSII absorbs solar energy, energized electrons leave the complex. Replacement electrons come from water which splits releasing oxygen. **6.7** NADP reductase is an enzyme that speeds the production of NADPH. The flow of H+ down a concentration gradient provides the energy for ATP synthase to produce ATP. **6.8** 1. ATP provides the energy and NADPH provides the hydrogen atoms to reduce carbon dioxide during the Calvin cycle reactions. 2. Sunlight is needed for photosynthesis which is the process that produces glucose. **6.9** Figure 6.9 shows how G3P, the product of the Calvin cycle reaction can be metabolized into the various molecules needed by a plant. **6.10** 1. In a C_4 plant the mesophyll cells are arranged concentrically around the bundle sheath cells and fixation of carbon dioxide results in a C_4 molecule that is pumped into bundle sheath cells where CO_2 enters the Calvin cycle. 2. Photorespiration does not occur in a C_3 plant when the weather is cool and moist and the Calvin cycle functions normally and directly after CO_2 uptake. 3. In a CAM plant, CO_2 is absorbed at night and released to the Calvin cycle during the day.

Consider These Questions

Page 110: 1. The evolution of oxygen-releasing photosynthesis is an example of the concept of "adding on" rather than starting over in that photosystem II (PS II), which is necessary for the splitting of water to release oxygen, evolved after photosystem I (PSI) but did not replace it. 2. Cyanobacteria can revert to a cyclic electron pathway, using hydrogen sulfide (H_2S) as a hydrogen source to reduce carbon dioxide like other autotrophic bacteria. Chloroplasts cannot revert to using H_2S in place of water. **Page 116:** 1. Preserving tropical rain forests is advantageous because it protects and preserves the habitats of many of the plants and animals that can be found only in these forests while contributing to stable global temperatures by absorbing much CO_2. 2. Countries without tropical rain forests can assist in preservation through contributing money to conservation efforts, publicly supporting the protection of these dynamic ecosystems, and refusing to consume or use products that result from deforestation.

Analyze and Evaluate

Page 117: 1.Cyanobacteria gave rise to chloroplasts once they were taken up by a pre-eukaryotic cell, making them significant in the history of life as an integral part in the evolution of photosynthetic eukaryotes. These organisms are also credited with introducing oxygen into the early atmosphere. 2. To produce a simplified diagram of the light reactions create a diagram based on Figure 6.6, which uses only words and arrows. To create a simplified diagram of the Calvin cycle reactions produce a diagram based on Figure 6.8 that uses the molecules listed in the upper box and all 5 steps. Stack your diagrams so that you can show how steps #7and #11 in Figure 6.6 connect to steps # 3, 4, 5 in Figure 6.8. 3. Photosynthesis put O_2 in the atmosphere. C_4 photosynthesis is more advantageous to organisms (rather than C_3 photosynthesis) when the atmosphere already contains significant amounts of O_2.

Testing Yourself

1. c; 2. d; 3. d; 4. c; 5. a. granum; b. thylakoid; c. O2; d. Calvin cycle; e. stroma; 6. Only plants produce organic food, and if animals only ate animals, food would run out. 7. a; 8. a; 9. a; 10 e; 11. d; 11. a; 12. a; 13. e; 14. b; 15. Solar energy is unable to participate directly in the chemical reactions that reduce carbon dioxide to carbohydrate, but ATP can do so. 16. d; 17. c; 18. a

Get Involved

1. The bacteria require oxygen and therefore the hypothesis that photosynthesis releases oxygen was supported. 2. It's adaptive for plants and for human vision to make use of the same best source of available light on planet Earth.

CHAPTER 7
Check Your Progress

7.1 1. NAD$^+$ and FAD are coenzymes that remove hydrogen atoms from glucose breakdown products and deliver electrons to the ETC. 2. The preparatory reaction and the Krebs cycle release carbon dioxide. **7.2** 1. During glycolysis two molecules of NAD$^+$ become NADH as oxidation occurs; two ATP are used to get the pathway started and four ATP are produced by substrate level ATP synthesis and therefore the net gain is two ATP. 2. A substrate must have a high-energy phosphate bond that can be passed to ADP in order to produce ATP **7.3** Pyruvate is oxidized by the removal of hydrogen atoms and this releases a CO_2 molecule. The reaction occurs twice per glucose molecule and therefore two CO_2 are made per glucose molecule. **7.4** The Krebs cycle produces 2 ATP and 6 NADH and 2 FAD per glucose molecule. This means that 24 ATP are produced per glucose molecule due to the Krebs cycle. **7.5** Some carries of the ETC pump H$^+$ into the intermembrane space. When the H$^+$ flows down its concentration gradient through an ATP synthase enzyme, ATP is produced. **7.6** 1. Outside the mitochondria only 2 ATP are produced during glycolysis and 2 are produced during the Krebs cycle. This compares to at least 30 ATP by the ECT. 2. The cytoplasmic portion evolved first because the early atmosphere had no oxygen which is utilized by mitochondria. **7.7** 1. During glycolysis NAD carries out oxidation of substrate molecules and delivers electrons to the respiratory chain; in fermentation NAD carries out oxidation and delivers hydrogen atoms to pyruvate which breaks down to two lactate molecules or two alcohol molecules with the release of CO_2. **7.8** Amino acids from protein, glucose from polysaccharides, and fatty acids from fats and oils can all be broken down to acetyl CoA and acetyl CoA can be the starting point for their synthesis

Consider These Questions

Page 132: 1. The discovery that bacteria in addition to yeast ferment may have seemed irrelevant at first but it led to the production of many products such as yogurt, pickles, and certain types of beer. 2. The products of this page do not negate that food is ultimately derived from plants. Fermentation is a process through which sugar is broken down in the absence of oxygen. Since sugar is produced by autotrophic organisms, namely plants, they play a large role in producing these products. **Page 134:** 1. Statistical information on the correlation between illness and exercise would substantiate the claim that exercise can help to prevent a wide range of ailments. Evidence from studies on diabetics and cancer patients, for example, that showed higher instances of these diseases in individuals that do not exercise regularly, would support this claim. 2. Some diseases are genetic or result from exposure; however, if evidence shows that exercise is useful in warding off diabetes, heart disease, or some forms of cancer, I would find it beneficial to prevent those ailments that are avoidable.

Analyze and Evaluate

Page 135: 1. The pre-eukaryotic cell must have been a fermenter living off abiotically produced organic molecules. Cellular respiration couldn't have evolved until oxygen entered the atmosphere due to the evolution of photosynthesis. The eukaryotic cell became capable of cellular respiration after it engulfed a prokaryote that could carry out cellular respiration. 2. Structural similarities often point to evolutionary ties. Both the chloroplasts and mitochondria have an internal membrane system where various complexes are located. They both have a semiliquid interior where enzymes metabolize. They both have an outer boundary that allows substances to pass into or out of the organelle .

Test Yourself

1. a; 2. c; 3. b; 4. Glucose breakdown begins with glycolysis; oxygen becomes water at the end of the ETC; carbon dioxide is produced by the prep reaction and Krebs cycle. ATPs are produced by glycolysis (2 ATP), Krebs cycle (2 ATP per two turns), and 32-34 ATP by the ETC. 5. a; 6. e; 7. d; 8. b; 9. b; 10. c; 11. d; 12. d; 13. b; 14. Inner membrane space serves as an area where hydrogen ions collect before passing through an ATP synthase complex; b. Matrix is location of preparatory reaction and Krebs cycle; c. Cristae contain electron transport chain and ATP synthase complex. 15. Only 39% of available energy becomes ATP and the rest becomes heat. 16. c; 17. d; 18. c; 19. b; 20. b; 21. c; 22. Chloroplasts and mitochondria were originally independent prokaryotes. Prokaryotes evolved from a common ancestor, which must have used an ETC.

Get Involved

1. Acidic because for ATP to be produced, H+ must flow through the ATP synthase complex. 2. The acetyl group that enters the Krebs cycle is oxidized to CO_2.

CHAPTER 8
Check Your Progress

8.1 1. Every new cell needs a complete copy of the organism's DNA. 2. G_1, S, G_2 (growth and synthesis) occur as the cell gets ready to divide; mitosis and cytokinesis occur when the cell does divide. **8.2** 1. The parent cell and the daughter cells have the same chromosomes: one long red and one long blue chromosome. 2. The daughter cells will also have one long red and one long blue chromosome. **8.3** 1. The duplicated chromosomes are positioned at the equator in such a way that the sister chromatids will separate one from the other. 2. Once separated the sister chromatids become daughter chromosomes. **8.4** A cell plate allows the new plant cell to build a new cell wall. Animal cells don't have a cell wall. **8.5** 1. G_1 checkpoint examines DNA for damage and G_2 checkpoint examines DNA for proper replication. If need be apoptosis occurs. 2. Apoptosis (cell death) ordinarily kills cells that could become cancerous. **8.6** A tumor indicates the cell cycle is out of control. A tumor contains cells that lack differentiation (specialization) and have abnormal nuclei.

Consider These Questions

Page 143: 1. If bacteria make a protein similar to tubulin it does strengthen the hypothesis that bacteria contributed to the evolution of the spindle apparatus. 2. Evolution must necessarily make use of the structures and molecules currently available. For example, the jaw evolved from the first two gill arches of jawless fishes; humans walk erect by standing on the hind legs of quadrupeds that preceded them. **Page 147:** 1. Yes answer: If I had lived 75 years and was facing a diminished quality of life including memory loss, the inability to maintain personal hygiene, or eat and drink, I would be willing to risk cancer in order to correct Alzheimer disease. Even though cancer has its own problems, I would not want to lose the memory of my family, friends, and myself as I lived my final years. No answer: Though the possibility of cancer is not absolute, I would not be willing to risk spending the last years of my life being treated for, or dying from, cancer in order to cure Alzheimer disease. I would seek other treatments to slow the progression of Alzheimer disease and attempt to maintain my quality of life for as long as it may last. 2. Yes answer: A scientist should be able to patent independent work because our society now recognizes that a person has a right to his or her intellectual work. No answer: A scientist should not be allowed to patent intellectual endeavors because it is the custom for scientists to work together and share knowledge for the benefit of society. **Page 150:** People who sunbathe, or drink or smoke cigarettes might be prone to follow the popular trend. They might feel that being one of a group engaged in these activities is worth it to them even though they may eventually die from them. You could give them examples of people who are popular even though they do not engage in these practices. 2. Because cancer does not occur immediately, the risk of cancer may seem unrealistic or too far into the future to be immediately relevant. However, having a friend or family member who has cancer, might make the illness more real to them.

Test Yourself

1. a; 2. a; 3.b; 4. d; 5. a; 6. c; 7. e; 8. c; 9. a. 10. b. 11. c; 12. a; 13. d; 14. c; 15. b; 16. c; 17. chromatid of chromosome or duplicated chromosome; b. centriole; c. spindle fiber or aster; d. nuclear envelope (fragment); 18.e; 19. a; 20. b; 21. b; 22. c; 23. a; 24. c; 25. b

Get Involved

1. No answer required. 2. Interphase takes the longest because it takes time for a cell to grow larger; produce new proteins and new organelles to form. 3. The cell cycle is out of control when a tissue is cancerous.

CHAPTER 9

Check Your Progress

9.1 1. Homologues are chromosomes that look alike and carry alleles for the same traits; the alleles can be different. Each homologue has sister chromatids when duplicated; the sister chromatids have exactly the same alleles. **9.2** 1. DNA replication results in two identical chromatids. 2. Homologues separate during meiosis: only the haploid number of centromeres is present in daughter nuclei. 3. Long red and long blue chromosomes are homologues and they separate during meiosis I. **9.3** 1. Nonsister chromatids can carry different alleles and crossing-over recombines the alleles. 2. All the mother chromosomes (e.g., red chromosomes) would go into one daughter cell and all the father (e.g., blue) chromosomes would go into the other daughter cell and the chromosomes would not be shuffled. **9.4** 1. Tetrad formation allows crossing over to occur and prepares the homologues for separation during meiosis I. 2. Anaphase I has duplicated chromosomes at the poles because when the homologues separate they are duplicated. **9.5** Replication allows mitosis to maintain the chromosomes and separation of homologues allows meiosis to reduce the chromosome number. **9.6** 1. If the homologues fail to separate during meiosis I or the sister chromatids go into the same daughter cell, an egg can receive one too many chromosomes. **9. 7** A person with Turner syndrome has only one X chromosome and a person with Klinefelter syndrome has one Y chromosome but more than one X chromosome. A person with Down syndrome is either XX or XY.

Consider These Questions

9A 1. You could consider asexual reproduction to be advantageous because it allows an organisms to quickly populate an area to which the organism is already adapted or you could consider a diploid adult to be advantageous because if one allele is nonfunctional the other one may be functional. 2. Yes, because the sex act has nothing to do with the point of sexual reproduction. No, because the sex act is usually required for sexual reproduction to take place. **9B** 1. Yes, It's possible that a trisomy individual might by chance produce a normal egg or sperm and an in vitro clinic could identify it and use it for reproductive purposes. No, because they will most likely produce an offspring that has a chromosome anomaly. 2. It depends on my symptoms. If I were like Chris Burke I would opt out of receiving therapy. If I had severe symptoms, I would want to try gene therapy. 9C Yes, special care and consideration should be given to a child with Klinefelter syndrome because

it will help them lead a normal life. No there are not enough resources available to give every person with a syndrome, special care and attention. 2. Yes, I would be able to treat a male with Klinefelter syndrome as a normal male because everyone has some defect or other even if it is just the necessity to wear glasses or a hearing aid; so we should not discriminate between people. No, I would not be able to treat them as a normal male because they do not have the sexual characteristics of a normal male.

Analyze and Evaluate

Page 167: 1. The proximity of the homologous chromosomes during synapsis allows for the exchange of genetic information that occurs through crossing-over. Synapsis also allows for the alignment of homologues during independent assortment. 2. In addition to crossing-over and independent assortment of chromosomes, the random fertilization that occurs through sexual reproduction results in increased variation among offspring. 3. As discussed in previous chapters, evolution is sometimes considered a process of "adding on" as opposed to starting over. Mitosis and meiosis being very similar events in which DNA is copied and passed on to a new cell, it is possible that a mitotic cell division gone awry was the first meiosis. For instance, a genetic defect occurred that allowed homologues to pair by mistake during mitosis. Then, crossing-over occurred between chromosomes, resulting in a new allele combination that benefited an organism. This "modified" mitosis which we call meiosis was passed on to the next reproductive generation.

Test Yourself

1. c; 2. a; 3. b; 4. The long chromosomes are homologues and the short chromosomes are homologues. The homologous chromosomes separate during meiosis I. If one daughter cell receives a red long chromosome, the other receives a blue long chromosome. The situation is similar for the short chromosomes. 5. c; 6. Homologues contain the same genes for the same traits but the exact genes are apt to vary. Not so for close relatives and if the genes are exact, separation of homologues and crossing-over will not introduce variations. 7. c; 8. b; 9. d; 10. d; 11. a; 12. e; 13. c; 14. a; 15. b; 16. d; 17. b; 18.c ; 19. a; 20. a; 21. a. 2n b. mitosis c. 2n d. 2n e. mitosis f. 2n g. fertilization h. n i n j. meiosis d; 22. d. 23. d; 24. d

Get Involved

1. Both parents because the egg had to be missing this chromosome and the sperm had to have two of this chromosome. 2. While theoretically possible, it would be practically impossible to find two gametes with the exact same genetic information that produced this individual.

CHAPTER 10

Check Your Progress

10.1 According to the blending model, all offspring would be intermediate in height. According to particulate model, offspring are not intermediate; they are either tall or short. **10.2** (1) a. RR b. Rr (2) a. all W; b. 1/2 R, 1/2 r; c. 1/2 T, 1/2 t; d. all T;(3) bb (4) 3 black rabbits: 1 white rabbit; 30 rabbits/120 are white. **10.3** 1. All offspring will have round seeds and purple flowers. 2. WS, Ws, wS, ws. **10.4** (1) 1 long wings, gray body, 1 long black, 1 short gray 1. short black ; (2) Tt x tt, tt. (3) 75%; (4) 9:3:3:1, 1/16. **10.5** Without careful records, the breeder may not know that all the individuals in row III are cousins who could carry the same recessive allele for, say, hip dysplasia. **10.6** (1) Each parent is heterozygous. (2) 25% **10.7** (1) The phenotypes of the potential offspring are: 1/4 are normal; 1/2 may suffer a heart attack as young adults; and 1/4 may suffer a heart attack in childhood. (2) Child: ii, mother: IAi; father: IAi, ii, IBi; **10.8** (1) In CF, faulty chloride channels

occur throughout the body and lead to problems in several organs, including the skin, lungs, and pancreas. (2) see Figure 10.8C on page 186. (3) Resistance is a polygenic trait with alleles on the X chromosome and chromosomes 2 and 3. (4) Diet is an environmental influence. 10.9 X^r Y; $X^r X^r$ 10.10 (1) mother $X^B X^b$, father X^bY, daughter $X^b X^b$; (2) $X^R X^r$, X^RY; (3) (b); 1:1.

Consider These Questions

Page 177: 1. The bulleted statements in the application show that the theory of evolution and the gene theory are compatible. When two theories are compatible it means that experimentation in one field supports the findings of another field as when two structures are supported by a wall they have in common. 2. During artificial selection humans and not the environment select which organisms will reproduce. 3. It worked because the phenotype is dependent upon the genotype. Darwin observed that the phenotype changes over time but Mendel deduced that the an organism has two factors (now called alleles) for each phenotypic trait. **Page 181:** 1. Testing either the 8-celled embryo or the egg prior to implantation is more acceptable than testing a child in utero and then terminating the pregnancy if the child tests positive for a genetic disorder. 2. IVF is acceptable to me, especially when used by people who struggle getting pregnant on their own. I do prefer testing the egg to testing the embryo because an egg will not produce a life without fertilization. Therefore, testing the egg and choosing not to use it cannot be considered termination of a life. However testing and discarding an embryo that tests positive for a genetic disorder can be considered a termination of life.

Analyze and Evaluate

Page 189: 1. Just as flipping a coin many times is more likely to give 50% heads and 50% tails, so counting many F2 plants is more likely to give a 3:1 ratio when the parent plants are heterozygous because the particular gametes that plants join with each fertilization is by chance. 2. Meiosis explains Mendel's laws. Mendel's law of segregation holds because the homologous pairs separate during meiosis I. Mendel's law of independent assortment holds because either homologous chromosome of each pair of homologues can face either pole during metaphase I of meiosis. 3. Mendelian genetics laid the groundwork to eventually understand how genes function. Because of its many applications to humans, Mendel's work allows us to understand the inheritance patterns of genetic disorders, including autosomal recessive and dominant disorders. Today, we can determine if parents are carriers for numerous genetic disorders and advise them of the chances they will pass on a particular disorder because of Mendel's laws.

Test Yourself

1. c; 2. The blending model of inheritance results in no variations, whereas Mendel's model does result in variations among the offspring. Evolution requires variations. 3. d; 4. c; 5. d; 6. a; 7. a; 8. b; 9. It shows that a chromosome (and a chromatid) has a sequence of alleles; 10.a; 11 b; 12. c; 13. c; 14. b;15. Egg testing because only eggs free of the faulty allele are fertilized. 16. e; 17.b; 18. a; 19. c; 20. e; 21. a; 22 d; 23.a; 24. $X^A X^a$

Get Involved

1. Cross it now with a fly that lacks the characteristic. Most likely, the fly is heterozygous and only a single autosomal mutation has occurred. Therefore, the cross will be Aa x aa with 1:1 results. If the characteristic disappears in males, cross F_1 flies to see if it reappears; it could be X-linked (see Fig. 9.16). 2. Give plants with a particular leaf pattern different amounts of fertilizer from none (your control) to overenriched, and observe the results. Keep other conditions, such as amount of water, the same for all.

CHAPTER 11

Check Your Progress

11.1 1. DNA contains deoxyribose, and the base T; RNA contains ribose and the base U instead of T. 2. The particular sequence of the nucleotide bases. **11.2** DNA is a double helix in all organisms. **11.3** 1. complementary base pairing between template strand and new strand. 2. Each cell has an exact copy of the organism's DNA and produces its own proteins. **11.4** 1. The genetic code is in triplet and each sequence of three bases in mRNA, called a codon, stands for a particular amino acid. 2. Transcription (production of an mRNA) is the first step toward gene expression. 3. It changes gene expression because it splices out certain introns and this affects what particular protein will be made during translation. **11.5** proline, tyrosine,arginine; **11.6** During elongation, tRNAs bring the amino acids to the ribosome in the order dictated by the order of the codons. **11.7** 1. The cell makes a protein by the process of transcription (DNA is a template for mRNA formation) and translation (tRNAs pair with mRNA codons)at a ribosome (rRNA). 2. The protein is transported by vesicle to the Golgi apparatus. **11.8** 1. The cell will be missing a protein because ATT codes for a stop 2. In a silent mutation the sequence of amino acids is not disturbed and with a frameshift mutation the sequence is completely different from normal. **11.9** 1. The duplication has two copies and the other parent donates a chromosome with a third copy. 2. If they inherited only the translocated short chromosome, a deletion would occur for *q* and *r*.

Consider These Questions

Page 199: 1. PCR should be used. Collection procedures should be standardized to eliminate the possibility of stray DNA entering the sample. Perhaps more than one lab technician could independently collect a sample and one of these technicians could be from an independent lab rather than a police lab. 2. Yes answer: Once a person has been convicted of a felony, it would make sense for any DNA fingerprints to remain on file in the event that another crime was committed. It would save time and money to already have this information readily available. No answer: Despite the fact that one crime was committed and a conviction made, there should be some degree of privacy with DNA fingerprinting, even for convicted felons. DNA fingerprinting can always be redone in the event of future crimes. 3. Yes answer: In the interest of science, public health and safety, PCR should be readily available. No answer: Government agencies like the Department of Health and Human Services could potentially regulate the use of PCR in instances where PCR would interfere with the privacy rights of an individual. **Page 210:** 1. Transposons replicate within the DNA without regard to the consequences. DNA can be considered selfish because just as the most successful members of a population have the most offspring, the most successful DNA passes itself to the most members of the next generation without regard to consequences. 2. It is possible that P elements produced some evolutionary advantage in *Drosophila melanogaster*, which was passed on via natural selection to spread into all populations of the species.

Analyze and Evaluate

Page 211: 1. The criteria for a genetic material: (1) it must be variable, accounting for the differences between species. DNA is variable. Differences in the base pairing combinations found in DNA molecules produce differences in proteins, which account for the variation between species; (2) it must be able to replicate, as DNA does, in a semiconservative process by which it duplicates itself; (3) it must be able to undergo mutations. DNA undergoes mutations, another factor that accounts for variability in organisms. DNA can undergo mutations during the replication process, due

to contact with external factors like radiation, or due to random events like jumping genes, or transposons, which in turn alter gene expression and cause many changes in organisms. 2. This suggests that protein-coding genes in specialized cells may be activate or inactive at different points in the cell's life. 3. Futuristic drugs might affect the early synthesis procedure, changing or eliminating the proteins to be produced by making changes in mRNA before it reaches the ribosomes.

Test Yourself

1. c; 2. e; 3. b; 4. a; 5. d; 6. c; 7. b; 8. DNA analogues prevent the formation of new DNA molecules, and each new virus requires a DNA molecule. 9. d; 10. b; 11. e; 12. a; 13. c; 14. b. 15. d; 16. d; 17. c; 18. b; 19. a. polypeptide b. amino acid c. tRNA d. anticodon e. codon f. ribosome; 20. c; 21. a. ACU CCU GAA UGC AAA; b. UGA GGA CUU ACG UUU c. threonine—proline—glutamate—cysteine—lysine; 22. genetic information, in the sequence of its bases; 23. b; 24. b; 25. c; 26. d; 27d; 28a; 29. c 30. Without mutations, new adaptations cannot occur and new life-forms cannot evolve.

Get Involved

1. Determine if any transposon base sequence occurs in the sequence for the neurofibromatosis gene. 2. Place the isolated gene in *Arabidopsis* cells and look for the mutation in cloned adult plants.

CHAPTER 12

Check Your Progress

12.1 When lactose is present, it combines with the active repressor and this inactivates the repressor. **12.2** Transcription factors are proteins that assist the binding of RNA polymerase to a promotor. Transcription enhancers are a region of DNA that bind to a transcription activator, a protein that activates transcription. **12.3** Euchromatin (loosely compacted) is available for transcription when DNA unpacking occurs. **12.4** A master developmental regulatory gene codes for transcription factors that turn on other regulatory genes so its effect can be extreme. **12.5** If splicing did not occur the mRNA would contain both introns and exons and look exactly like pre-mRNA. **12.6** The miRNAs inhibit translation of an mRNA molecule while siRNAs degrade the mRNA so it is not translated at all. **12.7** Translational control affects to what degree a protein is produced. For example when miRNA is still attached to an mRNA, translation does not occur. Posttranslational control determines if and how long a protein will be functional in a cell. For example if a protein is properly tagged it enters a proteasome and is broken down. **12.8** A mutation in a proto-oncogene causes overstimulation of the cell cycle. A mutation in a tumor suppressor gene means that the cell cycle with occur unimpeded even if the cells are abnormal.

Consider These Questions

Page 221: 1. Yes answer: The brain is a fascinating organ about which scientists, doctors, and everyday people have many questions. The more avenues we explore to help answer these questions, the more headway we might be able to make in matters involving the brain, such as Alzheimer disease or human intelligence. Evolutionary changes in the brain and its development might serve to teach us much more than we could ever imagine. No answer: Research should focus on the human brain alone and how it works. There is no guarantee that looking into the differences between human and ape brains will give us any relevant or useful information. 2. This reading tells us that our brain continues to develop after we are born and this accounts for its large size. It suggests that coordination between brain parts and the functioning of limbs was sacrificed to this enlargement. 3. Yes, it does make

sense. Initially, cells divide to make exact copies of themselves which serves to produce a larger organism. As the cell number grows, regulatory genes bring about differentiation of cell types so that they perform different functions. Without this specialization, cell division would simply create copy after copy of the original cell and development of an organism would not occur. **Page 225:** 1. We now know that epigenetic inheritance is just as important as genetic inheritance because it can affect whether a gene is active or not and the degree to which it is active. Researchers are working on drugs that will affect epigenetic inheritance. 2. Many levels of control means that genes can be finely controlled which is beneficial. It might be disadvantageous because more things can go wrong. **Page 228:** 1. Yes answer: By making a young person aware that they could carry a cancer gene, they might lead a healthier life to help ward off illness with exercise and a proper diet. An informed person may also decide, based upon knowing the probability, whether or not to have children and potentially pass the gene along. No answer: Making a young person aware that they could carry a cancer gene could potentially make the person's life miserable with paranoia and trips to the doctor for every ache and pain. A person may decide against having a family, participating in activities that expose them to sun or other carcinogens, and lead a depressing and anxious life. 2. Yes answer: Especially when there is a family history of cancer, the testing can help doctors catch and treat the cancer earlier, and advise the patient on the healthiest life possible to avoid the cancer. Decisions can also be made regarding reproduction and passing the alleles to children. No answer: It might cause a person to devote their lives to avoiding cancer and do nothing else.

Analyze and Evaluate

Page 229: 1. Many of the anatomical differences between apes and humans pertain to differences that might have arisen due to a change in gene regulation. For example differences in brain development and infant strength could be due to particular protein activities during development. 2. A knowledge of regulation can be used to understand the onset and progression of diseases, particularly those that result from genetic mutations that effect protein production. With an understanding of how regulation turns protein production "on" and "off" we might discover methods to artificially "flip the switch" to head off potentially detrimental changes in protein levels. 3. Genetic regulation is to cells as a teacher is to a classroom. The students (cells) need clear instructions on what tasks to perform (when protein production is needed) or chaos (disease) will result.

Testing Yourself

1. a. DNA; b. regulator gene; c. promoter; d. operator; e. active repressor; 2. b; 3. b; 4. b; 5. d; 6. b; 7. In both prokaryotes and eukaryotes DNA binding proteins are involved; in prokaryotes they turn off DNA and in eukaryotes they turn on genes. Eukaryotes have many more levels of control than do prokaryotes and RNA is involved in posttranscriptional control. 8. b; 9. b; 10. b; 11. Small RNS affect what the organism will be like by regulating what proteins will be made and/or the degree to which they will be active. 12. c; 13. e; 14. Transcription factors and small RNAs affect whether mRNA will be present in the cytoplasm for translation. 15. e; 16. d; 17 a; 18. c; 19. b; 20. Benefit is refinement of control; drawback is more chances of losing control; 21. b,c; 22. a,c; 23. a,d; 24. a,d; 25. c,d,a,b

Get Involved

1. You should also show that the genes having homeo-boxes result in similar developmental stages. 2. Culture normal cells in the presence of the pollutant and observe the results. Culture the same types of cells in the same manner but without the pollutant and observe the results.

CHAPTER 13

Check Your Progress

13.1 The nucleus is from a mature organism and the cytoplasm is from a fresh egg. **13.2** Both the human DNA and the plasmid were cleaved by the same restriction enzyme leaving the same exposed bases. Now the cut human DNA and the cut plasmid can be joined together. **13.3** Recombinant DNA technology is used to clone genes and PCR is used to clone a selected sequence of bases in DNA. PCR can confirm a viral infection, is now used in DNA fingerprinting, can help identify evolutionary relationships. **13.4** GM bacteria can (1) produce medicines that cure people of ills (2) produce vaccines that keep people from coming down with an infectious disease and (3) produce enzymes that help clean our clothes. (4) Clean up oil spills and other environmental pollutants. **13.5** GM crops have been produced that resist predaceous insects and/ or herbicides leading to increased yield. GM crops are now being produced that have greater nutrition value; soybeans can contain oleic acid, a monounsaturated fatty acid and golden rice contains provitamin A. **13.6** GM animals can have a gene for bovine growth hormone which causes them to grow larger, increasing yield and/or can produce a medicinal product in their milk. Mice that are model organisms can be given for genes human illnesses and this allows researchers to study and hopefully for these illnesses. **13.7** To cure SCID, bone marrow stem cells are removed from the bone marrow and infected with a virus that carries a normal gene for the enzyme adenosine deaminase, an enzyme that will allow the patient to produce white blood cells. Then the bone marrow stem cells are injected back into the patient. **13.8** PCR is done using a thermocycler and sequencing is done by an automated DNA sequencer. **13.9** Three types of noncoding DNA: introns, repetitive and unique. Repetitive DNA takes up the largest percentage of the genome and it regulates the activity of the coding genes. **13.10** Functional genomics studies variations in the genome (e.g., mutations and SNPs) to see how they affect the phenotype. Comparative genomics compares our genome to other organisms to determine their evolutionary relationship. **13.11** 1. Two mechanisms that increase the number of cellular proteins: (1)Alternative pre-mRNA splicing leads to various mature mRNA molecules and (2) proteins can be modified in the Golgi apparatus called posttranslational modification. 2. Much research has gone into studying the function of proteins and now scientists want to correlate their shape with their function. X-ray crystallography and nuclear magnetic resonance help them collect data. **13.12** 1. The Human Genome Project produced so much data that only modern computers and special software programs can deal with it properly. 2. Databases store the sequence of amino acids in proteins and their shape and function.

Consider These Questions.

Page 239: 1. Yes answer: I am in favor of banning GM crops until long-term studies have informed scientists on the long-term effects of consuming these foods. And I do approve of ecoterrorists burning crops and destroying labs that could potentially be harmful to the public. Getting rid of the crops and the labs they are genetically engineered in slows the spread of GM crops to the market and protects the public from what might create health problems in the long term. No answer: I do not favor a ban of GM crops. GM crops have increased yields, decreased losses due to herbicides and pests, and have been tested before hitting the shelf. I do favor the proper labeling of such crops and the decision to consume these or other crops being left up to the consumer. And I do not approve of ecoterrorism or the burning of crops and labs that could be useful in scientific research. Boycotting GM products or informing the public in peaceful displays would be desirable alternatives to

destroying crops and property. 2. Scientific studies that show that eating yellow rice does prevent childhood blindness and molecular studies that show the mode of effectiveness. **Page 248:** 1. If researchers know what genes are being expressed in a normal cell, they might be able to compensate for that gene or eventually be able to turn it on in an abnormal cell that is not expressing the gene. 2. Most drugs are proteins and a microarray tells us what genes are inactive and what proteins are not being produced in individuals with illnesses. Drugs can be developed to supply these proteins. **Page 250:** 1. Yes, I would have expected gene regulation to account for major evolutionary differences because the control of protein production results in the different characteristics we see between species. The types and amount of proteins being produced by an organism determine important attributes like appearance and development. It makes sense then, that control of protein production accounts for differences among the species. 2. Yes answer: If a study of my genome and the diseases prone to others with similar genetic makeup might lead to beneficial findings for other humans, I would be willing to participate in a comparative human study. No answer: The privacy of human subjects is at risk in comparative studies and I would prefer to preserve my privacy rather than participate in a study that exposes my genetic weaknesses. 3. Since we are not identical twins, I would expect my genome to differ slightly from my siblings as evidenced by our different characteristics.

Analyze and Evaluate

Page 251: 1. Biotechnology is correctly named because it takes what we know about genomics and utilizes developments in technology to artificially produce desired results for humans. For example, recombinant DNA technology produces drugs to cure illnesses and PCR can detect viral illnesses. 2. Biotechnology products can be (1) drugs to cure illnesses; (2) enzymes that improve the quality of food or clean up the environment; (3) commercial products that improve our lives or crops that need less pesticide or farm animals that grow faster or larger to provide food for humans. 3. Sequencing the human genome has led to the fields of functional and comparative genomics, and proteomics.

Test Yourself

1. a; 2. a; 3. d; 4. c,a,b,d; 5. c; 6. d; 7. begins: AATT; ends: TTAA; 8. Cloning can increase the number of desired organisms for human purposes; genes and segments for research. 9. a; 10. e; 11. Plants and animals can trace their ancestry to a common source, and therefore plants are able to express a human gene. 12. a; 13. c; 14. e; 15. Genetically modified stem cells pass on their modification to their offspring (more white blood cells) indefinitely. 16. b; 17. c; 18. Each gene performs only one function, so despite increased number of genes, only 30 gene functions would be available. 19. a; 20. d; 21. d; 22. c; 23. d; 24. a; 25. c

Get Involved

1. Use restriction enzymes to fragment the DNA of chromosome 10. Inject these fragments, one at a time, into groups of one-celled mouse embryos. Following development, see which mice have no tails. 2. An ex vivo study allows you to perfect your procedure in the lab and possibly avoid harm to the patient. 3. (1) Remove plasmid from parent bacteria and acquire donor DNA. (2) Use the same restriction enzyme to open plasmid and remove gene from DNA. (3) Use DNA ligase to splice gene into plasmid. (4) introduce plasmid into host bacteria.

CHAPTER 14

Check Your Progress

14.1 Scientists come to a conclusion after collecting data. Data are the evidence gained from making observations and doing experiments **14.2** 1. Members

of a population have variations. 2. Populations generally produce more offspring than the environment support. 3. Those that survive have advantageous traits. 4. The survivors pass on these traits to their offspring which thereby becomes the usual traits for the population. **14.3** 1. Common descent means that organisms are related (even back to the first organisms) by way of common ancestors. 2. *Archaeopteryx* has both reptilian and bird characteristics; *Ambulocetus natans* has traits in between those for water and land. **14.4** 1. For example, in birds and bats, the forelimb is adapted for flight; in the cat the phalanges aid running and also a soft landing; and in the horse, only one phalange became the hoof for running. 2. Both pigs and humans are vertebrates and vertebrates trace their ancestry to fish who breathe by using gills derived from pharyngeal pouches. **14.5** The separate continents have the same type environment to which the unrelated species are adapted. **14.6** The more related two species the more likely they will have the same sequence of bases in their DNA. 2. Master developmental genes turn on other regulatory genes so the end result can be different in different organisms. **14.7** Microevolution has occurred when gene pool frequencies change due to mutations, gene flow, nonrandom mating, genetic drift and natural selection. **14.8** "New variations" occur among the members of a population because of mutations and sexual recombination. **14.9** Nonrandom mating causes more individuals to be homozygous at certain locations. Gene flow causes more individual to be heterozygous at certain locations **14.10** 1. In small populations natural disasters are more likely to change the percentages of individuals with particular genes. 2. A bottleneck occurs because a natural disaster has killed off a large proportion of a population. A founder effect occurs when a small portion of a parent population starts a new subpopulation someplace else. **14.11** 1 and 2. If the peak phenotype range shifts in one direction directional selection has occurred. (Example, guppies become colorful if predation is absent) If the peak phenotype range narrows (birth weight becomes intermediate); British land snails are have coloring suitable to varied environments, stabilizing selection has occurred and if two peaks result, disruptive selection has occurred. **14.12** 1. Female choice can increase fitness when most females choose the same phenotype. Competition between males can involve fighting and larger males need plentiful food to stay alive. 2. The benefit of bright male plumage can be female choice but the cost can be attracting the attention of predators. Keeping a harem of females can mean more offspring but you may have to fight to prevent other males from taking over. **14.13** 1. Heterozygote advantage and ongoing mutations are two ways variation can be maintained in a population. 2. If the heterozygote is favored more and more members of the population will be heterozygous for sickle-cell disease.

Consider These Questions

Page 263: 1. The common occurrence of MRSA (methicillin-resistant *Staphylococcus aureus*) is an example of antibiotic resistance that many people now know about. Many patients think of antibiotics as a universal cure for all ailments and pressure doctors into prescribing them. However, physicians should stop overprescribing antibiotics and patients should be sure to take all doses that were prescribed. Also instead of increasing pesticide use or changing pesticides, farmers should set aside farmland that encourages the reproduction of nonresistant insects. If a farmer can increase the number of nonresistant insects on their land, lower levels of pesticides will continue to eliminate these insects. Current practices call for the increased use of pesticides and development of new ones as resistant insects rise in population. 2. Due to the extent of human involvement, drug development, design, and improvement is artificial selection for the best drugs, but not natural selection at work. **Page 268:** 1. A taste for sweet foods in humans, apes and monkeys benefits plants containing sweet proteins because the more these plants are ingested, the more widely the distribution of their seeds in the waste of humans and animals. The spreading of these seeds over larger areas and at high rates gives the plants a reproductive advantage over plants that are not eaten as much. 2. Humans are influencing the evolution of plants when they propagate them. When humans step in to produce more of one plant than another, for food or other purposes, they are artificially selecting for some characteristics. When humans genetically modify and breed plants, they certainly influence the plants' evolution. Changing something about the plant, which would occur only over time due to mutations and natural selection in nature, at least speeds the evolutionary process—that is, if the genetic modification made by humans would occur in nature at all. Any human involvement in plant reproduction or genetic composition influences evolution. The engineered plants humans create may evolve entirely differently than non-engineered plants would have. 3. Artificial selection can be harmful to plants and animals because it (1) reduces variation and (2) can make organisms prone to disease and (3) can select for traits that do not help the animal survive in the wild.

Analyze and Evaluate

Page 275: 1. Evolution via natural selection can be witnessed, particularly in organisms that have short life cycles and thus reproduce multiple generations over short periods of time. For example, over 5 years silent male field crickets increased from 10% to 90% of the population in a study conducted in Hawaii. The silent cricket had an advantage over the chirping cricket in its inability to be detected by a predator. This selective pressure caused silent crickets to survive to reproduce while chirping crickets were preyed upon. Other examples can be seen in the evolution of pesticide-resistant insects as a result of the increased use of pesticides and antibiotic-resistant bacteria due to the selective pressure of increased antibiotic use. 2. We expect evolution to have a genetic basis because genes control the phenotype of an individual including structural, metabolic, and behavioral traits. Thus genetic changes had to precede the observation of phenotypic changes. 3. Because evolution is not goal-oriented, it would be incorrect to say that bacteria became resistant to escape being killed by antibiotics. The mutations that allow bacteria to become resistant had to have occurred before they were exposed to the antibiotic. The environment "selected" the resistant individuals to have more offspring than the nonresistant individuals.

Test Yourself

1. b; 2. c; 3. c; 4. d; 5. Neither the occurrence of mutations nor the changing of environmental conditions is known ahead of time. 6. e; 7. d; 8. e; 9. e; 10. a; 11. d; 12. c; 13. b; 14. Their DNA base sequences were inherited from a common ancestor. 15. a; 16. a; 17. b; 18. c; 19. d; 20. b; 21. c; 22. Mutations are the raw material for evolution and natural selection results in adaptation to the environment.

Get Involved

1. If you know the genotype of the various colors, you could use the Hardy-Weinberg equation to calculate changes in gene pool frequencies. Otherwise, you could base sequence before and after the experiment to determine a change in the genome. 2. Yes; due to natural selection of boll weevils resistant to the insecticide.

CHAPTER 15

Check Your Progress

15.1 An advantage of the evolutionary species concept is it applies to sexually and asexually reproducing species and an advantage of the biological species concept is it

can be used when no differences in traits can be found. **15.2** Temporal prezygotic isolation means that species reproduce at different times of year/day. Different species of frogs mate at different times of the year. Hybrid sterility is a postzygotic isolation mechanism when a hybrid cannot reproduce. For example, a cross between a male horse and female donkey results in a sterile mule. **15.3** The green iguana could have swum from South America to the Galápagos Islands and also to Hispaniola . Here the subpopulations evolved into separate species that no longer reproduce with one another. **15.4** All the various finch species on the Galápagos are descended from a mainland species that invaded different islands and evolved separately. They are separate species because they cannot mate with one another. **15.5** Hybrid sterility. Crossing the plants results in a hybrid that cannot reproduce. **15.6** It needs to be determined if any of the skulls could be transitional links because gradualistic speciation is more apt to have transitional links. The skulls need to be dated because the punctuated equilibrium model assumes a more rapid space for speciation. **15.7** A change in regulatory genes can cause a major change in the phenotype immediately. **15.8** The tree would have only one trunk instead having many branches.

Consider These Questions

Page 288: 1. Yes answer: We should continue to convert corn to ethanol because corn does not need to be mined and does not pollute water supplies. No answer: We should not continue to convert corn to ethanol if it raises the cost of corn needed to feed humans and animals around the world. We must develop alternative methods for energy, particularly those that are more efficient and less disruptive than converting corn to ethanol. 2. Yes answer: We should eat less beef because cattle waste pollutes the water and eating beef causes circulatory problems in humans, and even obesity in our fast-food fed nation. No answer: We should eat beef from cattle that are range fed. It has less fat and the waste is disposed of by terrestrial organisms and it is does not pollute the water. **Page 290:** 1. Since fossil evidence can help to fill in blanks in evolutionary history, the fossils should be examined and evaluated in an attempt to figure out how they fit into the story of our planet. The scientific community should do its best to preserve such phenomena while objectively examining the history hidden inside. Eventually technology can be developed to meet the demands of dissecting such fragile fossil evidence. 2. Students should be presented objective information about the unexplained. This can encourage students to understand the progressive nature of science, along with how much we do not know about the world in which we live. Exposure to phenomena that are yet to be fully explained can help students understand the scientific process. These students may someday be scientists themselves, developing techniques or technology that will help us learn more about these phenomena.

Analyze and Evaluate

Page 295: 1. Biologists have made testable hypotheses about (1) the activity of *Hox* genes in many different species; (2) the effects of regulatory genes on developmental processes; (3) the need for geographic and reproductive isolation in order for a new species to arise; (4) the role of polyploidy in the origination of new plant species and (5) the role of gradualist and punctuated equilibrium models in speciation. 2. Paleontologists suggested the punctuated equilibrium model using data from the fossil record that sometimes indicated a fairly rapid appearance of new species. 3. Yes, the study of evolution is a scientific endeavor. In order to understand the process of evolution, scientists have formulated hypotheses and tested these hypothesis in the natural world. This has resulted in the collection of data and a conclusion that evolution does occur.

Natural selection can be used to explain observations in other fields of biology from genetics to behavior.

Test Yourself

1. c; 2. c; 3. b; 4. f; 5. a; 6. e; 7. h; 8. e; 9. c; 10. e; 11. Genetic differences maintained by reproductive isolation. 12. a. species 1; b. geographic barrier; c. genetic changes; d. species 2; e. genetic changes; f. species 3. 13. d; 14. b; 15. Allopatric speciation was possible on the Hawaiian Islands, but not on the Florida Keys. 16. b; 17. c; 18. d; 19. b; 20. b; 21. c; 22. b; 23. b; 24. b; 24. d; 25. *Hox* genes evolved early in the history of life. 26. d; 27. b

Get Involved

1. The evolutionary species concept allows you to trace the history of an organism in the fossil record, and the biological species concept provides a way to identify species without the need to examine them anatomically. 2. Their chromosomes are compatible, and the two species are very closely related. It's doubtful they should be considered different species.

CHAPTER 16

Check Your Progress

16.1 fishes: Paleozoic era, Cambrian period; amphibians: Paleozoic era, Devonian period; reptiles: Paleozoic era, Carboniferous period; mammals: Mesozoic era, Triassic period **16.2** At the start of the Mesozoic, the continents were joined and therefore the dinosaurs would be able to migrate to all the continents. **16.3** 1. Continental drift and meteorite collision with Earth. 2. Mass extinctions caused all other organisms to invade new territories. **16.4** 1. domain, kingdom, phylum, class, order, family, genus, and species 2. Genus and specific epithet 3. More inclusive categories (e.g., mammals) evolved before less inclusive categories (e.g., apes). **16.5** Sequencing of genes for rRNA is the basis for the three-domain system. Prokaryotes are in the domain Bacteria and the domain Archaea. Protists, plants, fungi, and animals are in the domain Eukarya. **16.6** Table summarizes the ancestral and derived traits of the species for which a cladogram will be constructed. An outgroup has at least one trait shared by all organisms in the table; the in group has shared derived traits; the oldest group has the most shared traits with the other groups and the last group(s) to evolve has the least number of shared traits with the other groups. **16.7** DNA sequence data are objective data and it's more readily available than fossil data.

Consider These Questions

Page 307: 1. In some instances the ability to readily identify any species of organism could be beneficial to society. In fields such as farming, education, and medicine this technology would prove particularly useful for identifying organisms that pose a threat to plants and/or humans. For example, a farmer, doctor, or student could differentiate poisonous or otherwise harmful insects or plant species that differ only subtly in appearance from nonharmful species. This could assist in medical diagnostics for rashes, insect or snake bites if some portion of the organism were available for identification. 2. The conversion of the CBOL methods to a criminal identification process would most likely require the isolation of definitive nucleotide sequences that differ between individuals or would otherwise require the sequencing of a person's entire genome. It could potentially work through inserting a sample of hair, blood, semen, or skin into a scanning device. Individuals could be required to have their DNA sequenced upon birth, arrest, or medical care such that a catalog of DNA could be used as a reference for comparison of the scanned genetic material, much like the catalog that exists for organisms that have been sequenced by scientists. 3. Yes answer: Scientists and inventors alike should be able to earn a living from the procedures and technology that they spend time developing, testing, and improving. No answer: Individuals should be compensated for their work but not profit from it because scientific knowledge should be available to everyone. **Page 310** 1. Cladistics is considered more objective than evolutionary systematics because it presents its data for all to see and no conclusions are subjective. This is why cladistics has most likely been welcomed and will continue to be well received by the scientific community. 2. Science should be willing to change while still retaining traditional methods that are also useful. We should particularly keep any traditional methods have been well tested. Innovation, however, propels science forward and is only natural in a field driven by hypotheses and experimentation.

Analyze and Evaluate

Page 313: 1. Yes answer: Putting all current organisms into a tree of life would put the vast nature of the world in perspective and would make me want to preserve all species now alive. No answer: The overwhelming nature of a complete tree of life might decrease my interest in the preservation of species by sheer over-stimulation. Seeing that so many species exist, some being very similar to one another might detract from the value of each species individually. 2. Yes answer: I do look forward this because determining the ancestry of humans all the way to the first living source is an exciting concept that seems to fit naturally into the human desire to understand who we are and where we came from. No answer: I do not look forward to tracing the ancestry of humans to the very first living source because it might cause me to realize that evolution does occur. 3. It does not surprise me that humans are related to all other organisms on the planet. An understanding of genetics and evolution helps explain why connections exist between all living organisms. Also, the shared characteristics of all living things help to support the undeniable interrelatedness of every organism, past and present.

Test Yourself

1. b; 2. d; 3. a; 4. c; 5. e; 6. e; 7.They evolved after the continents separated.8. a; 9. d; 10. d; 11. The three-domain system is based on differences/similarities in the sequencing of rRNA. This is backed up by differences in structure. 12 a; Ferns; b. produce seeds; c. naked seeds; d. needle-like leaves, Conifers; e. fan-shaped leaves, Gingkos; f. enclosed seeds; g. one embryonic leaf, Monocots; h. two embryonic leaves, Eudicots. 13. b; 14. d; 15. e; 16. b; 17 e; 18. b; 19. New and different structures arise due to DNA differences. 20. b; 21. a, b, c; 22. c, d, e; 23. b

Get Involved

1. The tree shows that all life-forms have a common source and how they are related, despite the occurrence of divergence, which gives rise to different groups of organisms. 2. The specialized environmental niche of these organisms is the same as it was when they first evolved.

CHAPTER 17

Check Your Progress

17.1 Capsid composed of protein and an inner core of nucleic acid. **17.2** In the lytic cycle new viruses are made and exit cell but in the lysogenic cycle the viral DNA enters the bacterial DNA. **17.3** Viruses enter cells by way of a receptor, some component of the cell membrane. **17.4** Retrovirus reproduction requires the enzyme reverse transcriptase and an animal virus does not require this enzyme because they have a DNA genome. b. Reverse transcriptase makes a DNA copy of the retrovirus's genetic material. **17.5** Prebiotic

Soup hypothesis: Atmospheric inorganic molecules combined in ocean (lightning can be energy source); Iron sulfur hypothesis: molten lava and nickel sulfides as catalyst caused ammonia to be converted to NH_3 at hydrothermal vents. **17.6** 1. Ribozymes are composed of RNA and they could have assisted RNA replication. 2. (a) Proteins (enzymes) allow a cell to metabolize and (b) genetic information (in the form of RNA or DNA) allows a cell to reproduce. 3. Aerobic cellular respiration requires oxygen. **17.7** The cell wall may be surrounded by a layer of polysaccharides; a well-organized layer is a capsule. **17.8** Steps: (1) DNA replication (2) Chromosome segregation (3) Cytokinesis 17.1 Gene transfer by transformation, gene transfer by conjugation and gene transfer by transduction. **17.9** Three ways prokaryotes can recombine their genetic material: (1) Gene transfer by transformation; (2) gene transfer by conjugation and (3) gene transfer by transduction. **17.10** Heterotrophs take in organic nutrients while autotrophs take in inorganic chemicals. Chemosynthetic bacteria oxidize inorganic chemicals to reduce carbon dioxide to an organic compound; photosynthetic bacteria use solar energy and hydrogen atoms to reduce carbon dioxide to organic compounds. **17.11** The methanogens live in swamps and marshes and couple the production of methane to the production of ATP; the halophiles live in high salt concentrations and depend on a pigment to synthesize ATP; the thermoacidophiles live In hot and acidic environments (e.g., hot springs) where they reduce sulfur to sulfides. **17.12** Get vaccinated; handle food safely, and wash hands frequently.

Consider These Questions

Page 321: 1. The special concerns associated with catching a newly discovered emergent disease include the potential of catching a new strain that cannot be treated or cured in the same manner as previous strains. A new disease may not be diagnosed and treated properly. 2. It is both a service and disservice for the media to publicize emergent diseases to the level that they do. It serves the public to be informed and take precautions to avoid infection. It does not serve the public to be worked into frenzy over worst-case scenarios that suggest the world may be coming to an end. 3. We are probably unnecessarily accusatory. A country cannot be exclusively blamed for being the origin of a disease due to factors that are largely out of the control of that country's people. It would benefit the entire world if we helped developing countries by campaigning for and helping to implement programs for clean water, good hygiene, and medical care in order to prevent the spread of disease. **Page 324:** 1. Hypothetical evolutionary stages between the first viruses and bacteria might include intermediate stages. (1) It would be necessary for the virus to independently carry out protein synthesis Since a virus already has either RNA and DNA, they would need only to acquire both at the same time. In fact some RNA viruses even now carry out reverse transcription in order to form DNA. A virus could also take for itself some of the DNA from a host cell. With the ability to produce enzymes, cellular respiration would be a possibility. Now the virus could live independently. (2) The virus would need to gain a plasma membrane and this may have happened by keeping a portion of the plasma membrane of the host cell. 2. If the viruses of today are degenerate forms of viruses that gave rise to cells, their parasitic lifestyle may have caused them to lose certain abilities out of disuse. In their degeneration, viruses of the past would have lost the ability to replicate independently, carry on metabolic activity independently, and thus to live outside of a host cell. 3. A parasite needs to produce a sufficient number of offspring so that some will go on to parasitize other hosts. Many offspring are expected to never find another host; the greater the number of offspring the more likely a new host will be infected.

Page 331: 1. No answer: If we discontinue flyovers, bioterrorists would certainly just find another method to attack with. We should take precautions to control air traffic during sports events and not discontinue activities in fear of possible attack. Yes answer: Bioterrorists might take advantage of the gathering of a large crowd and the fact that a flyover is not an abnormal event during a sporting event. Discontinuing flyovers may prevent a catastrophic attack using biological weapons. 2. Yes answer: Though the United States should not retaliate with biological weapons, it has a responsibility to retaliate and defend its people in the event of a bioterrorism attack. No answer: The United States should not retaliate and perpetuate a cycle of destruction using biological weapons, which pose a threat to all living things and has potentially unknown and far-reaching side effects. **Page 336:** 1. I would explain the unpleasant and dangerous symptoms of gonorrhea. The two most important points I would raise are that the risk of contracting HIV is increased when a person has gonorrhea, and that it is possible that some strains of gonorrhea will be incurable with antibiotics if this bacteria continues to evolve new strains, increasing the risk of permanent side effects when major organs are infected. 2. Children should be instructed about gonorrhea early, in late elementary to early middle school years. Just like other diseases, it is important to inform children about the risk of and precautions against sexually transmitted infection. The earlier the instruction takes place, the greater the chances that infection can be avoided instead of having to be treated. Teaching children to be responsible, even in matters that might make parents and other adults uncomfortable, provides them with the information they need to be safe and healthy.

Analyze and Evaluate

Page 337: 1. Not all bacteria cause disease. There are bacteria present in a healthy digestive system, in foods including yogurt and cheeses, and in food and beverage production processes like fermentation in bread and alcohol. 2. The first stages of the scenario described in figure 17.6 can be explained using the iron-sulfur world hypothesis through which thermal vents in the ocean could have heated water containing iron and nickel sulfides which then served as catalysts in the conversion of N_2 into NH_3. This conversion could have produced nutrient molecules capable of supporting life. Under the conditions of the thermal vents, inorganic molecules could have given rise to organic monomers, which could combine to form polymers like amino acids which form peptides in the presence of iron and nickel sulfides. Once this chemical evolution was complete, the raw materials for biological evolution were readily available for the formation of membranes, proteins, and genetic materials that are the building blocks of the cell. 3. Cyanobacteria first carried on photosynthesis. The endosymbiotic theory explains the evolution of photosynthetic eukaryotes by taking up cyanobacteria, organisms believed to be responsible for introducing oxygen into the early atmosphere and thus allowing the evolution of animals. Without the introduction of oxygen to the atmosphere, the world as we know it would not exist.

Test Yourself

1. c; 2. e; 3. c; 4. d; 5. b; 6. a. Lytic cycle: the virus is immediately produced and can go on to infect other cells. In life cycle b. Lysogenic cycle: the virus is being replicated along with the host cells and is protected from exposure to host defenses. 7. c; 8. c; 9. c; 10. a; 11. e; 12. b; 13. c; 14. c; 15. RNA can store genetic information and within the sequence of its base when it participates in the formation of proteins. RNA can act as an enzyme (e.g., ribozymes) and possibly could have replicated on its own. 16. a; 17. c; 18. b; 19. e; 20 a; 21. b; 22. a; 23. d; 24. b 25. Cyanobacteria do not have a nucleus or membranous organelles; their cell membrane contains peptidoglycan.

Get Involved

1. Viruses replicate inside human cells, and therefore, medications aimed at a virus can interfere with the workings of human cells. 2. The bacterium *E. coli* has the same genetic machinery as all other cells, including eukaryotic cells. They reproduce quickly and a large number can be kept in a small container. Since bacteria are haploid, mutations can be immediately observed.

CHAPTER 18

Check Your Progress

18.1 Photosynthetic organisms synthesize glucose in chloroplasts but transform the energy in glucose to ATP molecules in mitochondria. **18.2** Many protists are unicellular but their shapes can be completely different as in the various types of protozoans compared to the many types of algae. Protozoans are heterotrophic but some are predacious and some are parasitic. Algae are photosynthetic. **18.3** 1. A photoreceptor helps a photosynthetic *Euglena* find the light. 2. *Giardia* infection. **18.4** An amoeba uses pseudopods to engulf debris. 2. They both have tests (shells) that accumulate in the ocean. **18.5** Ciliates use trichocysts to capture prey. **18.6** Malaria is transmitted by a mosquito that can be killed by an insecticide. **18.7** Slime molds lives on the forest floor and look like a cytoplasmic mass covered by a slimy sheath. They engulf debris. The whitish filaments of water molds parasitize vegetables and animals on land and in the water. **18.8** Diatoms are covered by glass valves and dinoflagellates have cellulose plates that contain silicates. **18.9** *Fucus* is called rockweed because you find it along a rocky coast; the air bladders help hold the plant erect. **18.10** Molecular biologists rely on DNA/RNA sequencing to tell who is related to whom.

Consider These Questions

Page 345: 1. The theory of evolution has been supported by experimental evidence and by observation of fossil record, comparative anatomy, development, and chemistry of organisms. Any scientific endeavor has unanswered questions and evolutionists seek to refine the theory of evolution by studying new proposals. 2. The most reliable method for determining evolutionary lineages is a combination of both types of data. Using similarities and differences alone will not suffice because evolution is a variable process that can proceed "backward" at times, reverting organisms back to their former state under some conditions. To bolster the data taken from simply comparing protist groups structurally, we can use genetic comparison as well. The combination of both types of data creates a more reliable case for evolutionary lineages than either type of data alone. 3. In order to avoid mistaken conclusions based on observational data, scientists should record results precisely and avoid bias as much as possible. They should check their conclusions with other scientists and observations should be conducted repeatedly to confirm findings.

Analyze and Evaluate

Page 355: 1. Studying the first eukaryotes, the protists, with regard to their diversity of nutrition, reproduction, movement, and organization contributes to the study of biology in many ways. Protists can be used as models in the study of photosynthesis, examined for the role they play in the food chain, and studied for their medical impacts including malaria, dysentery, giardiasis, and Chagas disease. Through understanding the disease-causing ability of some protists, the large role others play in the food chain, or even the discrepancy in size between single-celled and multicellular protists, we gain an understanding of the diversity of life on our planet. 2. An evolutionary approach to relating the protists is a good one because it explores molecular data that can be analyzed more objectively than observational data.

By analyzing the genome of protists and comparing it to that of other organisms, we can determine the relationship between the first eukaryotes and organisms like plants and animals. 3. The term kingdom Protista is no longer used because of molecular evidence that many more kingdoms exist within the diverse group of protists and that protists are not closely related to one another. Some protists are more closely related to plants or animals than to each other. To place them into a kingdom together, then, is inaccurate. The term protist remains because it refers to an organism that is not a plant, animal, or fungi.

Test Yourself

1. d; 2. c; 3. a; 4. e; 5. Mitochondria were at one time free-living heterotrophic bacteria. 6. c; 7. b; 8. b; 9. c; 10. Flagella; trypanosomes cause disease, euglenoids contain chloroplasts. Pseudopods; amoeba sometimes cause disease, foraminiferans and radiolarians form tests. Cilia; ciliates such as paramecia are complex and have trichocysts and undergo conjugation. No means of locomotion; apicomplexans cause diseases. 11. Animals and protozoans share some features; they are both heterotropic and usually motile. Some protozoans are predaceous as are many animals. 12. b; 13. c; 14. Adaptation to similar habitats and ways of life makes organisms appear to be similar. 15. b;16. b; 17. d; 18. b; 19. a; 20. b; 21. a, c; 22. b, ; 23. a,d; 24. b; 25 b; 26. b; 27. b; 28. c; 29. d; 30. a; 31. e; 32. Diatoms have a two valve shell of silica and become diatomaceous earth. Dino-fl-agellates are responsible for "red tides." Red algae are variously structured seaweeds and are sources of agar and carrageenan. Brown algae are seaweeds harvested for food and are a source of alginate (algin). 33. Brown algae, diatoms, and water molds are traditionally separate groups based on mode of nutrition and/or structure, but they are grouped together in the Stramenopila. This means that their DNA shows they are closely related.

Get Involved

1. If single cells do not separate, and if each cell divides in a way that allows the cells to join end on end, the end result could be a filament. 2. Merozoites enter red blood cells, and if you knew by what process they enter red blood cells, you might be able to develop a way to stop them from doing so.

CHAPTER 19

Check Your Progress

19.1 1. Embryo protection, apical growth, vascular tissue, megaphylls, seeds, fruit. 2. Cell wall of cellulose, apical growth, plasmodesmata, haploid adult protects and passes nutrients to the zygote. 19.2 1. sporophyte produces spores; gametophyte produces gametes 2. The gametophyte (n) of a fern is an independent heart-shaped structure; seed plants have a female gametophyte located within the ovule that produces an egg and the male gametophyte which is a pollen grain. 19.3 1. All bryophytes lack vascular tissue and have a dominant gametophyte and produce windblown spores. 2. Disadvantage: flagellate sperm must swim to egg; advantage: windblown spores. 19.4 Presence of vascular tissue gave lycophytes true stem, root, and leaves (micropylls). Xylem has strong-walled dead cells that lend strength to stem. 19.5 1. Independent gametophyte lacks vascular tissue and flagellated sperm must swim to egg. 2. The sporophyte of both plants produce windblown spores; the gametophyte is dominant in a moss but the sporophyte is dominant in a fern. 19.6 Pine: Dominant sporophyte has vascular tissue and protects the female gametophyte in an ovule; hard-walled pollen grain (male gametophyte) carries sperm to carpel. Seeds disperse species Fern: Independent gameophyte lacks vascular tissue and flagellated sperm much swim to egg. Dominant sporophyte with vascular

tissue produces spores which disperse the species. 19.7 1. Flower attracts insects, which assist with pollination; ovule becomes embryo-containing seed retained inside ovary that becomes fruit carried by animals to new location. 2. Pollen sacs of stamen produce pollen grains that germinate on the stigma of carpel. Carpel has ovule-containing ovary where female gametophyte produces egg. Fertilization produces an embryo within a seed (develops from ovule) within a fruit. (develops from ovary). 19.8 1. Fungus is composed of hyphae that form a mycelium which sends out digestive enzymes into environment and absorbs nutrients. Lack of flagella and production of windblown spores is an adaptation to land. 2. Fungi are decomposers that return nutrients to plants; mycorrhizae are fungus-coated roots and lichens are fungus plus algae. Lichens are important soil formers. 3. Chytrids evolved first and gave rise to the zygospore fungi, which gave rise to the sac fungi and club fungi. 19.9 Black bread mold: + and —hyphae fuse and nuclei fusion leads to a zygospore where meiosis occurs. Upon germination aerial hyphae with sporangia at tip release spores. Mushroom: mushroom develops following fusion of hyphae; cap contains basidia where nuclei fusion, meiosis, and spore formation occur 2. Black bread mold: hyphae bear stalks that produce spores in sporangia; sac fungi: aerial hyphae produce chains of spores called conidia.

Consider These Questions

Page 372: 1. The plants present in an ancient swamp forest are the evolutionary predecessors of today's plants as well as the source of much of the energy that humans use in the form of coal. Because we depend on plants for oxygen, food, and the energy produced by decomposed plants from millions of years ago, we are dependent on the plants that gave rise to present-day plants. 2. Because the great swamp forest existed in what is now northern Europe, the Ukraine and the Appalachian mountains of the United States, I would predict that they occurred when the continents drifted and coalesced near the equator. Because the forest spanned what are now separate continents, they must have been present at a point when these continents came together and these regions were near one another in a warm and humid environment not found in all of these locations today. The equator was the perfect setting for the tall seedless vascular plants to thrive before falling into the rising swamp waters. **Page 377:** 1. Because it is possible that an infection could wipe out any or all of these three flowering grains, it is unfortunate that humans are so dependent upon only wheat, corn, and rice. To get people to eat more varied grains there would have to be an effort made to market alternative options at affordable prices. 2. In addition to the food they provide us, we are also dependent on plants because they are a source of oxygen, providing us with building materials, medicine, and fuel. We rely on plants for many things and thus could not live without them. **Page 382:** 1. Decomposition, nutrient cycling, fermentation (yeast), food (mushrooms), antibiotic production, pest control. 2. No. Many organisms, including fungi and bacteria can be considered "bad" or "good" depending on the specific organism in question. While some fungi cause disease that can harm humans, others are a source of food, nutrient cycling, antibiotics, or pest control. For instance, fungi used for pest control also cause disease, but they do so in pests that might harm crops that humans use as food sources. Thus, all disease causing fungi are not "bad" since some are used to human advantage. 3. Most plants can make their own food and for this reason are generally not parasitic. A parasite like the fungus that causes athlete's foot can be likened to a predator, such as a lion, because they are heterotrophic and must find and consume food. Predators cannot make their own food and must hunt for and consume food. 2. DNA is a selfish molecule, copying itself

Analyze and Evaluate

Page 383: 1. It is not a good strategy to have a separate and water-dependent gametophyte on land. It is a better strategy for the gametophyte to be protected by the sporophyte which has vascular tissue. It is also better to have the sperm transported to the egg in a way that does not require external water. 2. DNA is a selfish molecule, copying itself without regard for any other processes taking place in a cell or an organism. The energy expended by plants for reproduction is merely a means to perpetuate the plant's DNA; reproduction, after all, is meant to pass on the genetic information of an organism to later generations so that the species may live on. 3. Yes answer: Reproduction of land plants and animals can be compared, because the adaptations of animals and plants are alike as they moved from aqueous environments to dry land. Both flowering plants and humans are able to protect all stages of their life cycle from drying out, and internal fertilization are all good adaptations required for reproduction on land. No answer: I do not approve of comparing reproduction in plants to reproduction in humans. Plants often require intervention from other organisms or abiotic factors like wind or water in order to reproduce successfully, while the reproductive process in humans involves direct contact between a male and female.

Test Yourself

1. a; 2. a; 3.a; 4. e; 5. b; 6. Tall plants have better access to sunlight.; 7. a; 8. a; 9. c; 10. e; 11. b; 12. a. stigma; b. style; c. carpel; d. ovary; e. ovule; f. receptacle; g. sepal; h. petal; i. stamen; j. filament; k. anther. 13. a; 14. c; 15. e; 16. Pollen is less likely to be moved from the pollen cone to the seed cone of the same tree. 17. e; 18. e; 19. e; 20. e; 21. a; 22. e; 23. Mycorrhizal fungi are more likely to go along with the seedling in the native soil.

Get Involved

1. Only group (b) because mosses require a film of moisture in order for flagellated sperm to swim to an egg. 2. Because male moths attempt to mate with these flowers, they carry pollen only between flowers of this type

CHAPTER 20

Check Your Progress

20.1 1. Animals are multicellular heterotrophs that ingest their food; usually reproduce sexually; undergo developmental stages that produce specialized tissues and organs; and have muscles and nerves. 2. Animals evolved from a colonial flagellate, Tissues arose when some cells invaginate into the hollow ball of cells. 3. These terms refer to the fate of the embryonic blastopore. In protostomes the blastopore becomes the mouth and in deterostomes it becomes the anus. **20.2** 1. Type of symmetry, number of germ layers, and presence of segmentation. 2. In protostomes cleavage is spiral and determinate and in deuterostomes, cleavage is radial and indeterminate. In protostomes, the embryonic blastopore becomes the mouth. In deuterostomes, the blastopore becomes the anus. If a true coelom is present, in protostomes the coelom develops by splitting of the mesoderm. In deuterostomes, the coelom develops from an outpocketing of the gut. **20.3** Sponges lack symmetry, lack tissue layers and the only opening is an exit. **20.4** Both body plans have radial symmetry, incomplete digestive tract, two tissue layers, separated by mesoglea and the tissue level of organization. In the polyp body plan, the mouth is above and in the medusa the mouth is below. The polyp can reproduce asexually; the medusa reproduces sexually. **20.5** 1. Flatworms have three tissue layers body systems including a flame-cell system, ladderlike nervous system, a digestive tract and reproductive organs. 2. Tapeworms attach to the wall of the digestive tract. Flukes are named for their location in the body; there are liver flukes and blood flukes, for example. **20.6** A coelom provides a space

for organs to become specialized and have freedom of movement. If a skeleton is lacking, the coelom acts as a hydrostatic skeleton. **20.7** In gastropods (e.g., snails), the foot is flattened; in cephalopods (e.g., octopuses), the foot evolved into tentacles; and in bivalves (e.g., scallops), the foot projects from the shell. **20.8** 1. In annelids, segmentation is obvious: Rings encircle the body, septa divide the coelom, and paired nephridia and blood vessels occupy each segment. 2. Oligochaetes (earthworms) have few setae per segment, reside in soil, and do not have a well-developed head. Polychaetes (marine worms) have many setae per segment. They have jaws and sense organs, and are predaceous. Leeches have no setae per segment and are parasitic blood suckers. **20.9** Some roundworms are parasites, causing trichinosis (muscle infection), elephantiasis (lymph vessel infection), or other infections such as hookworm. **20.10** 1. Six characteristics of arthropods are jointed appendages, an exoskeleton of chitin that molts, segmentation, a well-developed nervous system including compound eyes, a variety of respiratory organs (insects use tracheae), and metamorphosis. 2.Among crustaceans, some (lobsters, crayfish, shrimp) have a cephalothorax in which the head bears sense organs and the thorax has five pairs of walking legs. 3. In insects, wings, and tracheae and presence of penis in males are adaptations to life on land. **20.11** Echinoderms, such as the sea star, sea urchin, sea cucumber, feather star, and brittle star, are marine animals with no head, brain, or segmentation. A water vascular system provides locomotion and helps carry out respiratory, excretory, and circulatory functions. **20.12** 1. During their life cycle, chordates have a notochord, a dorsal tubular nerve cord, pharyngeal pouches, and a postanal tail. 2. Tunicates have the four chordate characteristics as an embryo; gill slits are present in the adult. 3. Vertebrate evolution is marked by the evolution of vertebrae (jawless fishes) jaws (sharks), a bony skeleton (ray-finned fishes), lungs (lobe-finned fishes), limbs (amphibians) and the amniotic egg (reptile). **20.13** Ray-finned fishes have gills for respiration, streamline shape and fins for locomotion, swim bladder for buoyancy. **20.14** Metamorphosis allows the amphibian larval stage to exploit the aquatic environment while the adult exploits the terrestrial environment. The two stages do not compete for the same space and food source. **20.15** Sea turtles are actually reptiles that have lungs and lay leathery-shelled eggs on land when they reproduce. 2. Birds are adapted for flight. They have wings and air sacs in their bones which helps keep them airborne. **20.16** Monotremes (duckbill platypus, spiny anteater) lay a hard-shelled amniotic egg. Marsupials (kangaroo, opossum) have a pouch in which the newborn matures. Placental mammals (deer, dogs, cats, whales, humans) retain their offspring inside a uterus until birth.

Consider These Questions

Page 397: 1. Before molecular analysis, biologists had to use observational comparison of organisms to decide who was related to whom. By comparing structure, nutrition, and developmental information, among other characteristics, biologists did their best to determine relationships between organisms. 2. As seen in the example of the nemertine worm, anatomic data might lead scientists to believe that these organisms are closely related to flatworms; however, DNA data show a closer relationship to the annelids and mollusks. Because of the complex evolutionary relationship between organisms, it is best if anatomic and DNA data support the same conclusion. 3. Biologists should certainly go into the field to study organisms. Observation of an organism within its natural environment is extremely important in understanding behavioral characteristics, reproduction and nutrition, for example. Therefore, the study of organisms should take place in the field as well as in the lab. **Page 414:** 1. Scientists think that it is unlikely that

viruses will cross the species barrier; however, this has occurred in the past and may likely occur again due to the suppressed immune system of the human recipient along with the novelty of the animal disease(s) that may or may not be known human pathogens. Other unseen health consequences could result from the rapid and violent rejection of organs belonging to animal species that might lead to the death of the human recipient. There is also some concern over the genetic programming of humans versus animals: pigs, for example, age at a different rate than humans, presenting challenges in making any permanent transplants. 2. Yes answer: it is ethical to change the genetic makeup of vertebrates for use in drug or organ "manufacturing" so long as the animal is not harmed or killed in the process. No answer: It is not ethical to alter the vertebrate, use it as a factory or incubator, and then kill it in the process of harvesting the desired product. 3. It is possible that we are altering the relationship between humans and other vertebrates in a mutually detrimental way. The relationship between humans and the rest of the natural world should not be one of human exploitation and should avoid using vertebrates solely for our own purposes. If we slowly begin to alter every aspect of the planet for our own benefit, we may be forced to face consequences we never imagined possible.

Analyze and Evaluate

Page 415: 1. Ancestral protist, multicellularity, tissue layers, bilateral symmetry and 3 tissue layers and body cavity, deuterostome development, ancestral chordates, vertebrae, jaws, bony skeleton, lungs, limbs, amniotic egg, mammary glands, mammals, humans (16 steps). 2. The advantages of the parasitic way of life include the reliance of a parasite on its host for food, shelter, and a place to disperse offspring with little to no energy expended by the parasite. Parasites, specialized for their way of life, can also reproduce at higher rates (more quickly and in greater numbers) than their host. The disadvantages of the parasitic way of life include dealing with any harm done to the host from the parasite leeching nutrients and invading the host organism. Parasites reduce the "fitness" of their host organism while increasing their "fitness" via exploitation of the host. 3. The relationship or sharing of a common descent, between humans and echinoderms can teach us that over time, changes in organisms including specialization, adaptation, and mutation create new organisms that may appear to have little in common though they originate from some common place in evolutionary history.

Test Yourself

1. d; 2. e; 3. e; 4. a; 5. b; 6. a; 7. e; 8. d; 9. d; 10. b; 11. a; 12. Flatworms are small, have a large absorptive surface, and are hermaphroditic. 13. e; 14. c; 15. e.; 16. a; 17. e; 18. b; 19. e; 20. Transitional fossils show how evolution occurred.

Get Involved

1. As per Section 19.1, study whole sponges and determine how they acquire food, whether they move, how they reproduce, and what developmental stages they have. Microscopically, determine the structure and function of their cells. 2. a. DNA/RNA sequencing; b. The data mentioned in Section 15.7: Fossil record and homologies in anatomy and development. 3. You will not find a notochord nor vertebrae, gill slits, or a postanal tail in an earthworm. You will find a ventral solid nerve cord typical of invertebrates and not a dorsal hollow one typical of vertebrates.

CHAPTER 21

Check Your Progress

21.1 1.Primates (prosimians, monkeys, apes, humans) have mobile forelimbs and hindlimbs, stereoscopic vision, large, complex brains, and reduced reproductive rates. **21.2** Mammalian ancestor to common ancestor with prosimians ; common ancestor with all anthropoids; common ancestor with all hominoids; common ancestor with all apes; common ancestor with gorilla and chimpanzee; common ancestor with all hominins. **21.3** Brain: Ardi had a small brain and humans have a large brain (300- 350 cc compared to over 1,300 cc). In Ardi, the muzzle projects and the forehead is low with heavy eyebrow ridges; in humans the face is flat and the forehead is high with lighter eyebrow ridges; Ari has longer arms and the pelvis is more flared. Ardi has opposable toes and humans do not. Ardi is shorter than humans. **21.4** 1. In australopithecines, the brain was small (370-515 cc) but they walked erect. In *A. sediba*, a tree-climbing arm ends in a hand that can grasp tools. The modern-style ankle rests on a primitive heel. 2. *Homo erectus* advances: body proportions like that of a modern human, striding gait; large brain (about 1,000 cc); knowledge of fire; was able to migrate out of Africa. **21.5** Replacement model proposes that humans evolved in Africa from preceding species and then migrated to Asia and Europe where they replaced the archaic species living there (that evolved from *H. erectus*). This accounts for why there is only one species of modern humans. **21.6** Cro-Magnons: more sophisticated tool use leading to ability to kill large animals,; language and significant artistic achievements; they began agriculture and they lived in social groups. **21.7** Adaptations: (1) darker skin is protective against high UV intensity; lighter skin permits Vitamin D production (2) body type regulates heat—squat bodies and short appendages in cold climates conserve heat; lengthy limbs and appendages in warm climates allow the body to lose heat.

Consider These Questions

Page 428: 1. Culture is evident in almost all aspects of my daily life. Culture exists in traditions of cooking, eating, making a home, and using tools for cleaning, cooking, hanging a picture, creating written documents or drawing. 2. Aspects of my life that are influenced solely by biological inheritance have nothing to do with culture. For instance, I breathe, I eat, and I sleep because of the biology of being human. 3. Though our sources of food may be different than the hunter-gatherers, we still function in a similar manner, bringing nourishment back to a central location from a grocery store, farm, restaurant, etc. Also, it is still more customary for women to stay home to take care of the young and for men to go out and be aggressive in order to maintain the home base. **Page 431:** 1. Many U.S. citizens can trace their ancestry to forbearers in the various other continents. This means they are descended from immigrants. This does not involve a migration pattern. 2. Yes, as humans migrated from continent to continent, genetic drift (founder effect) would have occurred and may have contributed to the similarity between members of a common ethnicity, but natural selection would have also occurred to favor advantageous phenotypes for the new environment. For instance, close to the equator and exposed to hot sun, darker pigments in the skin would have been beneficial and thus evolutionarily favorable. 3. I think that migration patterns show that all humans are related to one another. Studies that examine mtDNA show relatedness of populations on separate continents. The founder effect helps explain how differences in appearance might result from decreased genetic variation in a region inhabited by a small population. Natural selection helps to explain evolution of groups with different skin color, hair color or texture, and eye color.

Analyze and Evaluate

Page 424: 1. See page 344, Fig. 18.1 and page 390, Fig. 20.2A and page 422 figure 21.2A. 2. I would suggest that biocultural evolution, "in which natural selection is influenced by cultural achievements rather than anatomic phenotype," is more influential today than human biological evolution because of modern inventions like technology and medicine. Humans now have the ability to adapt to the environment through human-made methods including treating illness, and moving from one climate to another. Populations in regions without technology to keep water clean and treat illness are less successful than groups with clean water and medical care. 3. Through education we can explain the necessity of preserving habitats of organisms other than ourselves, and explain how biodiversity is necessary to the preservation of our planet and all its inhabitants including humans. Education could be used to demonstrate what happens when small or large changes are made to an ecosystem.

Test Yourself

1.a; 2. c; 3. common ancestor for prosimians, anthropoids, hominoids, hominins; 4. a; 5. Ardi evolved before Lucy and yet it has fewer chimplike features than Lucy has; 6. d; 7. b; 8. b; 9. b; 10. d; 11. e; 12. b; 13. Darwinian evolution is dependent on genetic differences, and biocultural evolution is dependent on advances that cannot be inherited. The ability to learn is inherited and learning is necessary to biocultural evolution. 14. e; 15 a; 16. c; 17. d; 18. d; 19 a. *Homo erectus*; b. archaic humans; c. modern humans 20. d; 21. There is more variation within ethnic groups. This tells us that all humans are one species.

Get Involved

1. The benefits of bipedalism must have outweighed the cost or else bipedalism wouldn't have evolved. 2. Sequence the Neandertal genome using DNA from Neandertal bones, and compare the sequence to the human genome of today. Look for sequences present in both genomes.

CHAPTER 22

Check Your Progress

22.1 Stems support leaves and conduct materials to and from roots and leaves in their vascular tissue. Leaves are the chief organs of photosynthesis. Roots anchor a plant and absorb water and minerals from the soil. Roots often store the products of photosynthesis. **22.2** Monocots have flower parts in three and scattered vascular bundles in a stem. **22.3** 1. The epidermis has projections in a root called root hairs; has lenticels in a stem, and stomata in a leaf. 2. In vascular tissue, xylem, transports water and minerals from roots to leaves, and phloem often transports sugar from leaves to roots. **22.4** In a leaf, ground tissue which is called mesophyll exists between the upper and lower epidermis and receives water from leaf veins. In a herbaceous stem, ground tissue presents the cortex and pith receives water from vascular bundles. In a root the ground tissue, called in the cortex exists between the epidermis and the vascular cylinder from which it receives water. **22.5** Root apical meristem (protected by root cap) produces the primary meristems called protoderm, ground meristem, and procambium which respectively produce the epidermis, cortex, and vascular tissue of a root. These tissues allow a root to carry out its functions: absorption of water and storage of products of photosynthesis. Shoot apical meristem produces all the structures of a shoot including the leaves and the stem which supports the leaves as they carry on photosynthesis. **22.6** Wood is secondary xylem that builds up year after year in annual rings. A woody plant has some advantages (enough xylem and phloem to service a large shoot and root system and retention of a home site for the plant) but also some disadvantages (has to use energy to produce wood, offers a larger target for parasites and predators). **22.7** 1. In order to maintain homeostasis, organisms need to maintain their structure. Photosynthesis produces food for a plant and therefore provides a source of energy and building blocks both of which are needed to maintain the structure of a plant. 2. Regulating mechanisms are: closure of stomata by guard cells when plant is water-stressed; phloem transport from source to sink according to needs of plant; plant hormones: auxin promotes bending toward light; mutualistic relationships such as fungus roots; defense mechanisms include a waxy cuticle, thorns, the ability to produce toxins, and local cell death.

Consider These Questions

Page 441: 1. No answer: I do not find the consumption of an embryo, in the form of a grain, disturbing. Eating any plant is eating a living organism, and eating a steak, or chicken, or pork, is eating what used to be a living organism. Because nearly everything I consume used to be alive, eating grains, the embryos of plants, does not disturb me. Yes answer: I find it disturbing because of the word "embryo" and my tendency to equate that word with something immature that needs my help to become mature. 2. No answer: I would not mind eating only grains so that more people could be fed. Since grains provide more energy and cattle are polluting the environment, I would be happy to eat grains and not cattle. Grain can continue to be planted and harvested while cattle must be taken care of until old enough to be used as food. Yes answer: I would mind eating only grains so that more people could be fed. I like eating beef and want to continue to do so. People have been consuming cattle and other animals for some time and need to consider that our rapidly growing population needs all sorts of foods. **Page 449** 1. Trees provide beauty, shade, oxygen, clean air and water, fruit, nuts and many different wood products such as paper, furniture and housing. Also, literally thousands of products are made from wood including the medicine L-Dopa for treating Parkinsons disease, to the film for cameras. Cellulose fibers become rayon fabric and cellophane among other products. Bark is used for landscaping, and to generate electricity for paper and lumber mills. Wood pulping by-products are used in cleaning compounds, deodorants, hair spray, medicines and cosmetics for example. Because our modern civilization depends on trees for all these products, it is clear that future generations will depend on trees also.

Analyze and Evaluate

Page 453: 1. The presence of an epidermis and cuticle, stomata that open and close based upon the surrounding conditions, a root system, and a stem to carry water from the roots to leaves and flowers is evidence enough to convince me that the vegetative organs of plants have adapted to living on land. 2. A bush has more stems leading to its increased number of leaves, providing more vascular tissue for them. 3. Rhizomes provide an efficient way for plants to asexually multiply because each shoot represents an offspring of the parent plant.

Test Yourself

1. c; 2. c; 3. b; 4. c; 5. a; 6. Apical meristem is in the shoot tip and root tip where it produces cells that add to the length of the stem and the root. Axillary bud activity produces new branches (and flowers). 7. A cereal grain is a dry fruit that contains a seed. 8. b; 9. c; 10. c; 11. b; 12. a; 13. d; 14. e; 15. c; 16. d; 17. a; 18. d; 19. b; 20. b; 21. b; 22. They all consists of parenchyma cells that fill the interior of an organ. 23. Roots, stem and leaves would lack protection. Roots would lose their ability to absorb water by way of root hairs; stems and leaves should lose water across exposed cell walls. 24. e; 25. c; 26. b; 27. a; 28. e; 29. c; 30. b; 31. Apical meristem (shoot tip and root tip meristem) is responsible for primary growth and the result is increase in the length of stem and root. Vascular cambium is responsible for secondary growth and the result is increase in girth. Girth increases because secondary xylem builds up as annual rings (wood). 32. c; 33. Vessel

elements and sclerenchyma cells in xylem have very strong walls. 34. a. Broader expanse to collect sunlight; b. Prevents loss of water; c. Collects sunlight; d. Allows gas exchange; e. Allows carbon dioxide to enter and water to exit. 35. Photosynthesis allows a plant to produce the biomolecules and ATP it needs to maintain metabolism and its structure. 36. e; 37. d; 38. b; 39. a; 40. c; 41. e

Get Involved

1. Confirm that plasmodesmata do run between companion cells and sieve-tube members. Same as Palade (see page 72), use labeled amino acids to show that proteins pass by way of plasmodesmata from companion cells to sieve-tube members. 2. Zone of cell division slides should show small cells dividing; zone of elongation slides should show cells that are longer than the previous zone—no cell division is occurring; zone of maturation should show mature tissues and epidermal cells with root hairs.

CHAPTER 23

Check Your Progress

23.1 Vessel elements are nonliving cells with no cytoplasm and companion cells which carry on metabolism would not be an asset. **23.2** 1.Water is cohesive and sticks together due to hydrogen bonding; while adhesion to sides of vessels keeps the water from slipping back. 2. Evaporation of water through the stomata of a leaf creates a tension that pulls the water column from the roots to the leaves **23.3** Stomata close when turgor pressure decreases in guard cells due to the exit of K^+ and water. Also, the presence of light causes stomata to open, high CO_2 and the presence of the abscisic acid causes them to close. **23.4** Sugar is actively transported into phloem at a source, and water follows by osmosis. The resulting increase in pressure creates a flow, which moves water and sugar to a sink. **23.5** Macronutrients which are essential nutrients needed in large quantity by a plant include nitrogen, sulfur, and phosphorus those minerals required to produce protein and nucleic acid. **23.6** 1. If water and minerals take the apoplast route via cell walls, the Casparian strip causes them to enter endodermal cells before reaching xylem. 2. Topsoil contains humus and organisms that make nutrients available to plants. **23.7** Root nodules contain nitrogen-fixing bacteria that take up atmospheric nitrogen and incorporate it into amino acids that are available to the plant host while mycorrhizae (fungal roots) increase the root surface area and otherwise assist in water and mineral uptake.

Consider These Questions

Page 460: 1. Because of their remarkably long lives and great size, I would expect that redwoods are good competitors. Their shallow roots, capture water as it seeps into the ground from rainfall. Their thick fire-resistant bark allows them to survive a fire when other plants would burn. 2. I am certainly willing to eat new grains in order for farmers to increase biodiversity. Maintaining biodiversity is good for the planet as a whole. If simple changes like eating habits can assist in keeping our planet a suitable home for many organisms, I'd be happy to do so. **Page 463:** 1. Plants are a source of food, make oxygen, take in carbon dioxide, provide shelter, serve as building materials, clean up nitrate pollution, grow materials used for clothing like cotton, produce or aid in production of medicines, and serve to preserve soil. 2. Find some way to prevent pets and other animals from having access to these plants; actually it would be best to stop polluting the biosphere; then, we would not have to use plants to rid an area of the pollutant. 3. Society has to be willing to pay the cost of cleaning up the environment or else people and other living things may perish. I personally would be willing to pay a price for the rare and beneficial service of a plant

that could filter pollutants from soil or water to benefit the environment.

Analyze and Evaluate

Page 471: 1. Plants require inorganic nutrients from the environment in order to build organic molecules necessary for life. Carbohydrates are used to form their cell walls and to respire. 2. Plant cells absorb water and minerals through their root systems. The minerals are taken into the cells in their charged or ionic forms. The movement of ions across the plant cell plasma membrane requires active transport but it allows plants to concentrate minerals in their cells. When animals consume plants, they take in these concentrated minerals. 3. It is beneficial to have xylem and phloem in the same vascular bundle because in this way they are supported by the same sclerenchyma cells (in stems) and are surrounded by the same bundle shealth cells (in leaves).

Test Yourself

1. a; 2. Leaves must receive water for photosynthesis, and they produce sugar as a result of photosynthesis. 3. c; 4. d; 5 Atmospheric pressure cannot account for the ability of water to rise above 76 cm in a plant. However, transpiration does allow water to rise above 76 cm. 6. c; 7. d; 8. e; 9. c; 10. When water enters guard cells, the stomata open. 11.d; 12. b; 13. a; 14. e; 15. Water enters the first bulb by osmosis and this creates a pressure flow that moves the solute to the second bulb and beyond. 16. Phloem transport might stop because phloem transport is dependent on the entry of water after the active transport sugar at the source. 17. c; 18. d; 19. b; 20.d; 21. a; 22. b; 23. e; 24. c; 25. e; 26. c 27. If mycorrhizae fungi cannot live only, the relationship is mutualistic.

Get Involved

1. Due to diversity, some species may not be susceptible to the assault. Also, the susceptible plants may be protected by the unsusceptible. For example, insects may not find all the susceptible plants, or most of the available water will go to those plants that need it. 2. Divide a large number of identical plants into control and experimental groups. Both groups are to receive the same treatment, including all necessary nutrients, but the experimental group will not be given any calcium. It is expected that only the experimental group will suffer any ill effects. If only the control or if both groups do poorly, some unknown variable is affecting the results. 3. A fan (wind simulation) affects transpiration the most.

CHAPTER 24

Check Your Progress

24.1 Reception occurs when a receptor combines with the signal, which could be a plant hormone. Transduction occurs when the signal is amplified and transformed into one understood by the plant cell machinery. Cellular response occurs when a member of the transduction pathway actives an enzyme or a gene and the cell undergoes a change. **24.2** Reception of auxin activates a proton pump and the acidic conditions loosen cellulose leading to elongation of cells on the shady side. **24.3** After GA3 binds to receptors, a transduction pathway leads to transcription of gene for amylase which breaks down starch releasing sugars which the embryo uses to spur growth. **24.4** Three effects of cytokinins: (1) growth of plant tissues, seeds, and roots (2) growth and differentiation of cultured plant tissues, and (3) prevention of leaf senescence. **24.5** Abscisic converts buds to winter buds when a plant is temperature-stressed and closes stomata when a plant is water-stressed. **24.6** Once abscission has started, ethylene promotes leaf, fruit, or flower drop. Ethylene produces the flavor and smell of ripened fruit, weakens plant cell walls, and breaks down chlorophyll allowing

the plant to change color. **24.7** Positive geotropism occurs in roots when they bend toward the pull of gravity and negative geotropism occurs in stems when they bend away from the pull of gravity. Phototropism is bending toward the light and thigmotropism is movement in response to touch. **24.8** 1. Electrical charges which propagate through the lobes of the leaf and spring the trap could be due to a swelling of cells near the outer region of the lobes or could be due to a loss of water by cells near the midrib. 2. In prayer plants the sleep cycle changes to 26 hours when the plant is kept in constant dim light. **24.9** When daylight is present phytochrome is activated and turns on certain genes and then plant structures respond appropriately to the presence of light. **24.10** Cocklebur flowers when the night is longer than a critical length. When a flash of light interrupts the longer than critical length of night, the cocklebur does not flower. **24.11** Defenses against predators: leaf cell cuticle; secondary metabolites are poisonous or otherwise deter predators; The peptide system is produced when a plant is wounded by a predator and then other parts of the plant produce a proteinase inhibitor. Sometimes plants produce chemicals that attract the natural enemies of caterpillars attacking them. Acadia trees are well known to provide a home for ants that attack and sting herbivores.

Consider These Questions

Page 490: 1. Humans rely upon plants as a source of food and oxygen as well as filters to remove carbon dioxide from the air. Plants provide humans with the raw materials for shelter, a means to clean up nitrate pollution, grow materials used for clothing like cotton, produce or aid in production of medicines, and tools in soil preservation. 2. Some secondary metabolites produced by plants for defense, can inhibit cellular respiration and/or block DNA or RNA synthesis, making them useful in the fight against cancer.

Analyze and Evaluate

Page 491: 1. Population: Members of a plant population that opened and closed their stomata morning and night had an advantage over plants that did not do this. Organism: Sunlight acts as a stimulus that causes stomata to open during the day; the absence of sunlight causes stomata to close. Cellular: When K^+ enters guard cells, water follows, and stomata open. When K^+ exits guard cells water follows, and stomata close. Abscisic acid brings about closure of stomata in a plant that is water-stressed. 2. (1) Reception of the stimulus: red light, in day light, activates phytochrome. (2) Transduction of the stimulus: activated phytochrome moves into the nucleus, binding specific proteins, and activating specific genes. (3) Response to the stimulus: Flowering occurs. 3. Auxins: suppress lateral growth; Gibberellins: result in stem (internode) growth; Cytokinins: stimulate cell division; Abscisic acid: promotes dormancy; Ethylene: causes abscission of leaves; therefore reduces growth.

Test Yourself

1. c; 2. a; 3. d; 4. The first plant which responds to the administration of gibberellin is likely unable to produce it. The second plant which is unable to respond to the administration of gibberellin likely has a defective receptor; 5. c; 6. c; 7. d; 8. e; 9. d; 10. d; 11. b; 12. e; 13. a; 14. c; 15. Place the banana in a closed container with a ripened fruit. 16. d; 17. c; 18. d; 19. e; 20. a. active form of phytochrome, b. biological response, c. far-red light; d. red light 21. c; 22. b; 23. e; 24. a; 25. e; 26. In the presence of light, phytochrome helps turn on genes that code for enzymes; 27. b; 28. e; 29. The volatiles are pheromones which attract a specific carnivore.

Get Involved

1. Use a plant that tracks the sun as your experimental material. Make tissue slides to confirm the presence of a pulvinus, as in Figure 24.8. Apply ABA to live pulvinus

tissue under the microscope to test for the results described in Figure 24.8. 2. Shine a light underneath a plant growing on its side (see Fig. 24.7A, *upper left*). If the stem now curves down, the phototropic response is greater than the gravitropic response and your hypothesis is not supported.

CHAPTER 25

Check Your Progress

25.1 1. The male gametophyte is inside the pollen grain which protects it. The female gametophyte is inside the ovule in the ovary where it is protected. 2. The ovary at the base of the carpel contains ovules. An embryo sac which contains an egg cell develops within each ovule. **25.2** 1. The mature pollen grain contains the tube cell which develops into pollen tube and the generative cell which becomes two sperm that travel in the pollen tube to the embryo sac. 2. Double fertilization results in a 2n zygote which becomes the embryo and the 3N endosperm which becomes the stored food. 3. A flowering plant is adapted to reproduce on land because (1) pollination by wind or insect carries sperm to the stigma within the pollen grain which is the male gametophyte generation; (2) the female gametophyte is protected from drying out because it is retained inside an ovule until fertilization occurs. (3) the zygote is protected from drying out because it is located within a seed protected covered by fruit. **25.3** The embryo is derived from the zygote; the stored food is derived from the endosperm; and the seed coat is derived from the ovule wall. 2. Cotyledons are embryonic leaves that are present in seeds. Cotyledons contain nutrients derived from endosperm (in eudicots). **25.4** 1. Dry fruits, with a thin, and dry covering derived from the ovary are more apt to be windblown. Fleshy fruits, with juicy covering derived from the ovary and possibly other parts of the flower, are more apt to be eaten by animals. 2. An apple is derived from a single ovary that has several chambers. The flesh of an apple is largely from the receptacle. 3. Fruits promote the dispersal of seeds by wind or animals. **25.5** 1. Both types of seeds have the three main parts of a seed (embryo, stored food, seed coat); and the embryo has a hypocotyl and epicotyl but a bean seed has two cotyledons and a monocot seed has one cotyledon. 2. Eudicot seedlings have a hook shape, and monocot seedlings have a sheath to protect the first true leaves. **25.6** 1. Advantages to asexual reproduction include (1) the newly formed plant is often supported nutritionally by the parent plant until it is established (2) if the parent ideally suited to the environment the offspring will be as well, (3) Asexual reproduction is less complicated in that it doesn't require the fusion of two gametes. 2. For example stolons and rhizomes produce new shoots and roots; fruit trees produce suckers; and stem cuttings grow new roots and become a shoot system. **25.7** Tissue from leaves, meristem, and anthers can become whole plants in tissue culture.

Consider These Questions

Page 498: 1. A series of mutations are needed for a nonseed life cycle to evolved into the seed life cycle: (1) Sporophyte produces heterospores that develop into microscopic male and female gametophytes. (2) Female gametophyte is retained by sporophyte (3) Male gametophyte becomes pollen grain. (4) Following fertilization, the embryo sac becomes a seed. (5) Seed contains sporophyte embryo, food, and seed coat. 2. Humans are reproductively adapted to protect all stages of reproduction in a dry environment, much like seed plants. Human gametes are housed in the ovaries and testes and fertilization occurs internally. A fertilized egg implants in the uterus and is protected and nourished here until it is mature enough for birth and development outside the body of the mother. 3. Flowers and seed cones both bear ovules that develop into seeds. The flower produces both pollen and seeds but

the seed cone produces only seeds. Only the seeds of angiosperm are protected by fruit. **Page 499:** 1. Observe several different fl ower types and see if they are visited by only one type or several types of pollinators. 2. The efficiency of wind pollination is most likely low since gymnosperms produce a large amount of pollen.

Analyze and Evaluate

Page 507: 1. Within the seeds of angiosperms, the endosperm provides nutrients. 2. (1) Flowers have benefited from the use of animals in pollination. This has made flowers successful reproducers on land, producing more seeds. Animals have benefited from this relationship because they received food in the form of sugary nectar from plants. (2) Fruit produced by plants, eaten by animals, has aided flowering plants in seed dispersal. In this relationship, the animal also benefits from the nutrition received from the fruit of the flowering plant. 3. Flowering plants (but not ferns) produce (1) two types of spores, microspores and megaspores and (2) two types of gametophytes: male and female gametophytes.

Test Yourself

1. b; 2. b; 3. a; 4. e; 5. a. anther; b. filament; c. stamen; d. stigma; e. style; f. ovary; g. ovule; h. carpel. 6. d; 7. c; 8. b; 9. a; 10. e; 11. c; 12. d; 13. b; 14. d; 15. a; 16. A wind-pollinated plant produces more pollen because the method of pollen transfer is less efficient. 17. b; 18. a; 19. e; 20. c; 21. b; 22. d; 23. e; 24. The need for a period of cold weather helps ensure that the seeds germinate when the weather will be favorable to continued growth. 25. c; 26. e; 27. a; 28. When the environment is not changing or when male and female plants are not in close proximity.

Get Involved

1. Study (a) the anatomy of the wasp and flower, trying to determine if the mouthparts of the wasp are suitable for collecting nectar from this flower; (b) Study the appearance of the flower in sunlight/ultraviolet light to determine suitability to the vision of the wasp; and (c) Study the behavior of the wasp to see if it is compatible to that of the flower (i.e. flower is open when wasp is active). 2. Protoplasts can be made from leaf cells and then cultured to grow entire plants. These plants are expected to produce seeds that you can use to propagate the plant.

CHAPTER 26

Check Your Progress

26.1 1.Cell, tissue, kidney urinary system, organism. **26.2** 1.Squamous epithelium is a single layer of flat and thin cells that allows diffusion in the lungs and blood vessels. 2. Columnar: cells shaped like columns, tightly packed line a cavity such as small intestine and function in protection and absorption. Pseudostratified: Ciliated cells appear to be layered but are not line the trachea offer protection and sweep impurities toward throat. **26.3** 1. Loose fibrous: collagen and elastic fibers form a mesh in which fibroblast cells are trapped; occurs beneath skin and between organs functions in support. Compact bone: Osteocytes occur in concentric rings separated by a hard matrix containing mineral salts; make up the bones of the skeleton. 2. Blood: composed of cells and has a matrix, namely plasma. **26.4**. 1. Skeletal muscle: Striated cells with multinuclei occur in skeletal muscles and function in voluntary movement such as walking. Smooth muscle: spindle-shaped cells have no striations; occur in blood vessel walls and walls of digestive tract; function in involuntary movement such as peristalsis in the tract. 2. Smooth muscle occurs in the walls of internal organs and its functions occur below the level of consciousness. **26.5** 1. Dendrites of motor neuron are long branched extensions that receive signals from other neurons and conduct them to cell body; axon is a single long

extension that conducts signals to muscles or glands. 2. A long axon allows nerve impulses to travel from the brain and spinal cord to distant parts of the body. **26.6** 1. The epidermis of skin contains stratified squamous epithelium. The term stratified tells us that these squamous cells occur in layers; as the cells rise to the top layer they become keratinized. 2. Dead keratinized cells of stratified epithelium prevent loss of water and offer protection to the dermis of the skin. **26.7** For example, the respiratory system contains the lungs and tubes held open by incomplete cartilaginous rings that take air to the lungs where gas exchange occurs. In the lungs, oxygen enters the blood and goes to cells. Carbon dioxide exits the blood and is conducted out of the body. **26.8** 1. Homeostasis is the relative constancy of the internal environment; body cells cannot function well unless they exist in a relatively constant environment and the individual becomes ill. 2. The organs of the digestive system receive and digest food to small nutrient molecules that enter the blood. The cardiovascular system carries oxygen to the cells to the systems of the body which use it as they produce ATP. **26.9** When body temperature rises, the control center sends out nerve impulses that cause skin blood vessels to expand and sweat glands to become active. Negative feedback shuts down these responses when the body has cooled to 98.6° F (37°C). When body temperature falls below normal, the control center directs the blood vessels in skin to constrict and shivering may occur to raise the temp. Negative feedback shuts down these responses when the body has warmed to 98.6° F (37°C).

Consider These Questions

Page 519: 1.Young people should be more concerned with long-term health outcomes rather than social expectations because the former affects how long they will live. 2. The amount of vitamin D that is recommended depends on where a person lives and the color of their skin. Therefore, there is no recommendation that is suitable to all persons. **Page 524:** 1. For instance, the brain must communicate with the heart and lungs in order for the heart to pump blood and the lungs to exchange gases. These gases are transported in the blood to and from the lungs. 2. A terrestrial animal but not an aquatic one has the added burden of water conservation. 3. Our skin prevents water loss, our large intestine absorbs water and our kidneys can excrete a hypertonic urine if we do not drink enough that day. Blood volume is maintained so that tissue fluid can provide an aquatic environment for our cells.

Analyze and Evaluate

Page 523: 1. When the human body is cold, signals are sent to the bones and muscles to initiate movement which results in shivering. Shivering increases blood flow and body temperature. As soon as possible humans move to a more suitable environment to achieve homeostasis. 2. People who travel to the South Pole wear special clothing and take shelters that help maintain body temperature and prevent water loss. People who walk in space have a space suit that maintains their normal environment and provides them with the support systems they need (e.g., availability of oxygen). 3. No, because blood and tissue fluid are external to cells. Yes, because blood and tissue fluid are inside the body.

Test Yourself

1. c; 2. a; 3. a; 4. d; 5. e; 6. d; 7. b; 8. d; 9. The dendrites offer wide surface area for the reception of stimuli and the long axon is suitable for the transmission of stimuli. 10. a, c, g; 11. b, d, e; 12. b, c, f; 13. a; 14. b; 15. c; 16. Epithelial cells are exposed to mutagens (agents that cause mutations) in the environment. Also, high rate of cell division means that spontaneous mutations may occur that lead to cancer. 17. e; 18. c; 19. The epidermis is composed of stratified epithelium, which provides an impenetrable barrier to invasion by microorganisms.

20. a; 21. b; 22. e; 23. d; 24. e; 25. c; 26. a; 27. a; 28. e; 29. Shivering is due to muscle contraction which gives off heat. 30. You would expect the cardiovascular system to have a pump (the heart) to move a liquid (blood) through tubular vessels.

Get Involved

1. Examine the tissue visually, trying to determine the particular organ before preparing microscope slides so that you can compare the slides you have made to known tissue slides. 2. Test two groups: (1) People who visit tanning salons, say two or more times a week. (2) People who never visit tanning salons. Find out how many people in each group have been treated for skin cancer. Compare the percentages and determine if the difference is significant.

CHAPTER 27

Check Your Progress

27.1 1. Integration is the summing up of information from various neurons; it occurs in the CNS where signals are received. 2. Motor neurons, sensory neurons, interneurons. Cell bodies and dendrites of sensory and motor neurons are in the CNS; axons are in the PNS; interneurons are completely within the CNS. **27.2** The work of the pump establishes the resting potential of -70: more sodium (Na$^+$) is outside the neuron than inside and more potassium (K$^+$) is inside the axon than outside. During the action potential first Na$^+$ passes into the axon and this results in depolarization as the membrane potential rises to +40 mV and then K$^+$ exits the axon and repolarization occurs as the membrane potential returns to -70 mV. **27.3** 1. When the nerve impulse reaches a synapse, Ca$_2$$^+$ enters and synaptic vesicles move to the presynaptic membrane where they release a neurotransmitter. The neurotransmitter diffuses across the synaptic cleft and is received by receptors at the post synaptic membrane. Depending on the amount of Na$^+$ that enters the neuron the receiving neuron fires or doesn't fire. 2. Many neurons synapse with a cell body. Because of synapses a receiving neuron receives both inhibitory and excitatory signals and the summing of these signals is integration. **27.4** 1. The cerebrum receives sensory input and originates motor output. Most important to humans the cerebrum is associated with reasoning; planning, speech, emotions, and problem solving. The cerebrum can override primitive emotions so that we act rationally even though we are in a rage or very fearful, for example. 2. Look at Figure 27.4B, and note parietal lobe (receives sensory information from skin), occipital lobe (receives sensory information from eyes); temporal lobe (receives sensory information from ears); frontal lobe (final integration and sends out motor impulses). **27.5** 1. The cerebrum and the cerebellum help us play baseball because the cerebrum receives input such as the speed of the ball and the cerebellum fine-tunes the cerebrums motor output so we can hit the ball. Two parts of the brain most involved in homeostasis are The hypothalamus (regulates hunger sleep thirst, body temperature and water balance) and the medulla oblongata (regulates heartbeat, breathing, and blood pressure) are the two parts of the brain most involved in homeostasis. 2. Nerve impulses from the spinal cord reach the thalamus which sends them to the cerebrum as appropriate. A malfunctioning RAS could result in a comatose state (cerebrum receives no signals) or a super excitatory state (cerebrum constantly receives signals about everything) and inability to sleep. **27.6** The limbic system is necessary to our survival because it is involved in memory storage including for example, any previous dangers we have been exposed to and in that way we can avoid them in the future. 2. Hippocampus receives information regarding stored memories and passes them to the frontal lobe and the amygdala adds emotion (particularly fear) to events and memories.

27.7 1. A nerve is composed of many axons with their myelin sheaths organized in bundles. 2. Cranial nerves come off the brain (take information from major sense receptors and motor muscles to face) and spinal nerves come off the spinal cord. (take information from sensory receptors in skin and tissues and motor muscles to muscles of the trunk and limbs). **27.8** Touching a hot stove results in a reflex arc: impulses from sensory receptors in skin travel to the spinal cord and pass to interneurons which sends them to the brain and to a motor neuron. The motor neurons commands the muscles to withdraw your hand and in the meantime impulses reach the brain and you realize what's going on. The reflex arc is quick acting because it does not involve the brain. **27.9** The sympathetic system is associated with fight and flight and the parasympathetic with rest and digest. Therefore the sympathetic results in an increased heart beat and increased breathing rate and the parasympathetic in a decreased heart beat and slower breathing rate. **27.10** Both methamphetamine and cocaine intensify the effect of dopamine a neurotransmitter that creates a sense of well-being.

Consider These Questions

Page 537: 1. Both animals have a central and peripheral nervous system. Planarians have a ladderlike nervous system with two nerve cords connected to the cerebral ganglia and extend to the posterior end of the body. Transverse nerves connect the nerve cords and the ganglia to the eye spots. The human nervous system is composed of a brain, and spinal cord from which branches a network of nerves throughout the body. 2. The arthropod nerve cord is ventrally placed while the vertebrate nerve cord is dorsally placed. 3. No, the structure and function of the nervous system in all mammals is about the same. **Page 541:** 1. The body is free to allocate energy to strengthening neural circuits. 2. REM sleep might be a time of preparation for deep sleep much like in the cell cycle the cell prepares for DNA replication during the G$_1$ stage. 3. More neural circuits can be devoted to the problem and this helps solve the problem.

Analyze and Evaluate

Page 548: 1. The nervous system is composed of two parts, the brain and spinal cord (central nervous system) and the nerves and ganglia (peripheral nervous system). The central nervous system receives information from the body, integrates it , and sends instructions to the body's muscles and glands via the nerves of the peripheral nervous system. 2. Neurons are the functional units of both the CNS and the PNS; therefore, every thought, feeling, and emotion we have and action we take is dependent on the nerve impulse and transmission across a synapse. 3. The limbic system influences motor output by its communication with the forebrain.

Test Yourself

1. d; 2. c; 3. b; 4. a; 5. a; 6 b;7. b; 8. New characters often arise by modifications of previously evolved characters. 9.c 10. c 11 b; 12. Myelination enables signals to travel quickly down an axon, which helps motor skills. 13. a; 14. c; 15. d; 16. a; 17. e; 18. c; 19. Learning requires the formation for new associations and this is mirrored in the brain by the formation of new synapses. 20. a central canal; b. gray matter; c. white matter; d. interneuron; e. dorsal root ganglion; f. cell body of motor neuron; 21. a; 22. b; 23. d; 24. a; 25. d; 26. d; 27.d; 28. d; 29. They affect the limbic system and either promote or decrease the action of a particular neurotransmitter.

Get Involved

1. Administer a medication that interferes with the reception of norepinephrine at a synapse. The patient may not respond properly to a real danger. 2. Severed sensory neurons are still releasing neurotransmitters in the spinal cord, resulting in messages to the brain that are interpreted as pain in the limb.

CHAPTER 28

Check Your Progress

28.1 Chemoreceptors (taste and smell); pain receptors (pain); electromagnetic receptors ; (radiation such as visual light, Earth's magnetic field); mechanoreceptors (sound waves, pressure waves for hearing, balance, and somatic senses) 2. stimulus, sensory receptor, brain, vision occurs in occipital lobe. **28.2** 1. Tongue (taste buds contain elongated taste cells with microvilli sensitive to chemicals in solution); olfactory cells with olfactory cilia sensitive to odor molecules. 2. Olfactory cells initiate nerve impulses that travel to the brain and eventually the frontal lobe which is receiving messages from other parts of the brain also. Integration leads to decision to investigate and nerve impulses go out to appropriate muscles. **28.3** Outer coat is sclera which includes the cornea which assists in focusing. Choroid (middle later) absorbs stray light rays. Retina inner layer contains rods and cones for vision. 2. Only when ciliary muscle contracts does the lens round up for close vision. **28.4** 1. Rods located in all areas of retina are very sensitive to light and result in nondistinct vision in dim light; cones located in fovea centralis require bright light but result in color and sharp vision. 2. Cones have a 1:1or somewhat greater relationship with bipolar cells where as rods have a 150:1 relationship. **28.5** 1. Tympanic membrane (located between auditory canal and middle ear) receives sound waves and engages the ossicles. Oval window (located between stapes and inner ear) receives magnified sound waves from stapes and initiates pressure waves in vestibular canal of the cochlea. 2. The organ of Corti contains hair cells on the basilar membrane. When the membrane vibrates due to pressure waves in tympanic canal of the cochlea, the vibration causes the cilia of the hair cells embedded in the tectorial membrane to bend and this leads to nerve impulses going to brain. 3. If the stereocilia disappear, hearing is impossible because they initiate the nerve impulses that go to the brain resulting in our ability to hear. **28.6** The receptors for rotational balance are in the ampullae of the semicircular canals. The ampullae contain hair cells with stereocilia embedded in a cupula. When the head rotates, the cupula is displaced, bending the stereocilia. Thereafter, nerve impulses travel in the vestibular nerve to the brain which interprets them. The receptors for gravitational balance are in the utricle and the saccule of the vestibule. When the head bends, otoliths are displaced, causing the membrane to sag and the stereocilia to bend. Thereafter, nerve impulses are generated and go the brain which interprets them. **28.7** Proprioceptors detect the degree of muscle contraction, the stretch of tendons, and the movement ligaments. This information goes to the brain which then knows where the limbs are located even if our eyes are closed. 2. To maintain balance, we need proprioceptors, touch receptors, photoreceptors and even pain receptors upon occasion.

Consider These Questions

Page 559: 1. Damage to the retina will cause loss of vision; other damages may be repaired. For example, inflammation of the conjunctivae, may cause some blurring of vision but if treated , vision will improve. 2. Yes answer: My vision is important to me for my safety as well as enjoyment of the world around me. No answer: I would rather enjoy life as I like and be willing to take the consequences. **Page 560:** 1. This is reasonable because of cultural influences that make a person seem disabled if they wear glasses or that make them they are "nerds." However, I think it is wise to disregard any stigma and do what is best for oneself. 2. Everyone should weigh the consequences of their actions and after being well informed, do what is best for them. **Page 564:** 1. I would rather to participate in rock concerts and exercise classes with the aid of earplugs to preserve my hearing. 2. We should not be annoyed because it could be that they did protect their hearing and their hearing loss is due to no fault of their own. **Page 565:** 1. It shows that evolution makes use of the structures available to it rather than inventing something new. 2. The inner ear of mammals evolved from our predecessors, water-dwelling animals. As such, it would make evolutionary sense that the inner ear of mammals makes use of fluid pressure waves, as our aquatic ancestors would have. 3. Mammals evolved on land and use vision as well as sound to detect the presence of other organisms. Sound waves are amplified by the presence of three ossicles surrounded by air in the middle ear.

Analyze and Evaluate

Page 569: 1. Humans' keenest senses are probably hearing and vision, which would have been adaptive for a primate dwelling on land and seeking food and shelter and protection in trees. 2. Parents use touch, vision, and sound to bond with their children. Birds use the same senses. 3. I would give up taste as a weight control mechanism.

Test Yourself

1. e; 2. c; 3. The brain requires input from sensory receptors in order to produce sensations about the world at large. 4. e; 5. See Fig. 28.3A; 6. For a near object the lens is rounded; 7. d; 8. c; 9. c; 10. c; 11. b; 12. e; 13. d; 14. c; 15. Both procedures can correct an inability to focus properly; The first with an artificial lens and the second by changing the shape of the cornea. 16. See Fig. 28.5A; 17. a; 18. a; 19 c; 20. c; 21. c; 22. c; 23. b; 24. a; 25. b 26. Rather than visual information, a blind person uses information from proprioceptors in joints and tendons and touch receptors for example in the buttocks to know they are sitting in a chair.

Get Involved

1. One possible answer: The size of the auditory cortex is larger in those who have perfect pitch. Test the pitch ability of subjects, and then stimulate the brain directly to determine the size of the auditory cortex. 2. LASIK surgery only corrects the shape of the cornea in order to achieve 20/20 vision.

CHAPTER 29

Check Your Progress

29.1 1. hydrostatic : movement is limited to changes in body shape; exoskeleton: must be shed periodically; endoskeleton: covered by living tissue and cannot protect from drying out. 2.Dehydration occurs and hydroskeleton disappears. **29.2** Bones assist breathing by lungs; store calcium needed by nerves and muscles; red bone marrow produces red blood cells needed by circulatory system and white blood cells needed by immune system; assist digestive system by chewing food; protect all internal organs belonging to other systems. **29.3** 1. Skull protects the brain; rib cage protects the lungs; vertebral column protects the spinal cord; 2. Vertebral disks between the vertebrae wear out and can herniate and rupture and this causes pain when they press against the spinal cord. **29.4** 1. Humerus of the forearm and femur of the thigh are analogous long strong bones; Ulna of arm and fibula of leg are both narrow thin bones; the radius of the arm and the tibia of the leg are the larger bones of these parts. 2. The hands manipulate tools while the feet are adapted for walking and balance. **29.5** 1. Compact bone makes up the shaft of a long bone and provides strength; spongy bone located at ends of long bones contains red bone marrow where blood cells are produced. 2. Ligaments hold bones together and form the capsule if a joint; tendons attach muscles to bones.

29.6 1. Support the body (e.g., quadriceps femoris group); make bones move (e.g., biceps brachii brings forearm up); maintain constant body temperature (all muscles give off heat); assist blood flow (all muscles contract against veins); protect internal organs (rectus abdominis protects the intestines). **29.7** 1. Muscle fiber has a sarcolemma surrounding contents including sarcoplasmic reticulum and several myofibrils containing actin and myosin filaments organized into sarcomeres. 2. When a muscle fibers contracts myosin heads bind to actin filaments and they slide past the myosin filaments. 3. After ACh is received at the sarcolemma, an action potential travels down its T tubules and the sarcoplasmic reticulum releases Ca^{2+}. Calcium exposes binding sites on actin filaments for myosin and contraction follows.

Consider These Questions

Page 579: 1. A larger human brain and upright posture might be evolutionarily connected because as walking upright evolved, increased brain size was also needed to ensure the coordination and balance required for upright posture. 2. Arthritis, tendonitis, ligament tears and other injuries suggest that the knee is not fully adapted to bearing weight. Also, the knees are subject to injury when playing sports. **Page 582:** 1. We should give people the benefit of the doubt because knowledge about osteoporosis is relatively new some people are more prone to the disease than others. 2. Screening for osteoporosis should be part of any type of physical exam. 3. Payment for health insurance should be on a sliding scale dependent on the health habits of the individual. Smoking can be detected by chronic bronchitis; lack of exercise can be detected by their musculature. **Page 585:** 1. If applied in time, antitoxin can be used. 2. Yes answer: If a company develops how it can be used they should be able to patent the product. No answer: If the substance is naturally occurring, it should be available to all. **Page 588:** 1. The answer depends on personal experience. 2. I would immediately switch to being a runner (avoiding cross-country) and give up any ideas of being a weight lifter.

Analyze and Evaluate

Page 589: 1. Animals are heterotrophs and have to seek their food. Nerves and muscles assist animals in pursuing their way of life such as being a predator or escaping predation. 2. To take an example, the pelvis and femur meet in a ball and socket, a synovial joint that allows for the rotational mobility of the hip, while the fibrous joints of the skull are immovable. These immovable joints have an important function since the bones of the skull must remain intact in order to protect the brain. There are three types of muscles: (1) Skeletal muscles are striated for strength and voluntary contraction when needed to move bones. (2) Cardiac muscle is striated for strength, but involuntary, ensuring that the heart continues to beat. (3) Smooth muscle is nonstriated and involuntary, ensuring that the muscles of the digestive tract always perform peristalsis. 3. Jointed appendages allow arthropods and vertebrates to perform flexible movements needed for walking, running, jumping on land.

Test Yourself

1. a; 2. c; 3. e; 4. b; 5. Least expected would be the production of all blood cells. 6. c; 7. b; 8. c; 9. b; 10. f; 11. c;12. e; 13. c; 14. d; 15. a; 16. c; 17. a; 18. e; 19. b; 20. Pelvic girdle is too small for a normal delivery.21. The axis of the nervous system is the spinal cord and brain; the axis of the skeleton is the vertebral column; the peripheral nervous system consists of the nerves; the appendicular skeleton consists of the bones of the limbs. Unique to the skeletal system is the rib cage and the girdles which don't compare to anything in the nervous system. 22. a; 23. b; 24. c; 25. b; 26. d; 27. a; 28. e; 29. b; 30. a; 31. a ; 32. Unless they were attached, myosin

couldn't cause the actin filaments to move thereby shortening sarcomeres.

Get Involved

1. Remove muscle tissue from a corpse in rigor mortis, slice it thin. While watching under the microscope, flood your slide with ATP and necessary ions to see if muscle relaxes and then contracts. 2. Acquire two test groups: aerobics instructors and confirmed couch potatoes. Oxygen tanks supply their only air, while they are running on a treadmill. Those who routinely exercise have more mitochondria than those who do not exercise, accounting for why the first group uses less oxygen and has less lactate (lactic acid) in their blood. 3. Calcium attaches to actin filaments exposing the binding sites for myosin.

CHAPTER 30

Check Your Progress

30.1 1. A flat aquatic animal with a gastrovascular cavity can use its water environment to move materials directly to cells and these substances can enter and exit by diffusion. 2. In an open system substances need only diffuse through the hemolymph into cells. However, the animal is unable to increase or decrease distribution and velocity of hemolymph flow because hemolymph is too widespread. A closed system allows an animal to have control over delivery to tissues because a heart pumps blood into a vascular system. A closed system can rapidly take oxygen to where it is currently needed. **30.2** 1. Pulmonary artery carries O2-poor blood from the heart to the lungs. 2. Cardiac conduction system consisting of the SA node and the AV node cause the heart to contract while closure of first the atrioventricular and then the semilunar valves causes the heart sounds. **30.3** 1. Arteries can expand (when the heart contracts and sends blood into them under pressure) and constrict (to maintain pressure) because they have a thick middle layer consisting of elastic fibers and smooth muscle; veins have a thin middle layer and are weak allowing the skeletal muscles to push on them and move blood past a valves; capillaries have only a wall consisting of endothelial cells and this allows substances to diffuse out of them. 2. Without movement the blood is unable to move in veins to the heart. **30.4** 1. Right ventricle—pulmonary trunk—pulmonary artery—lungs, pulmonary vein—left atria—left ventricle –aorta—mesenteric arteries—digestive tract—hepatic portal vein—liver—hepatic vein—inferior vena cava—right atria. 2. The liver monitors the blood and adjusts out intake of nutrients so that the balances of substances in the blood stay within normal range. **30.5** 1. Blood is a connective tissue because it contains cells (red and white) in a extracellular matrix (plasma). 2. First, blood vessel is punctured, platelets aggregate and convert prothrombin to thrombin. Thrombin causes fibrinogen to become fibrin threads. **30.6** 1. Exchange only occurs at the capillaries and without exchange, cells die. 2. Capillaries are unable to reabsorb water and blood pressure lowers. **30.7** The same type blood can also give and receive blood from the same type; AB is the universal recipient because it has no antibodies and O is the universal donor because it has no antigens.

Consider These Questions

Page 598 1. Insurance answer: I believe that open heart surgery should be paid for by insurance, especially in cases where the need for the surgery is a genetic condition or birth defect. Patient answer: I believe the operation should be paid for by the patient based on ability to pay to help offset the cost of insurance premiums, especially in cases where cardiac health has suffered due to poor diet, lack of exercise, or drug abuse. 2. Ideally, people would follow the guidelines for staying fit and would not need to rely on the medical profession to make them healthy after they've abused their bodies. **Page 599:** 1. Most likely reptiles (bird and crocodiles) and mammals express this gene because their heart has a septum, which completely separates the ventricles. 2. The efficient closed circulatory system of mammals permits them to maintain homeostasis in various environments. These examples are from Chapter 40: Polar ice—polar bear; tundra—musk ox; coniferous forest—brown bear; temperate deciduous forest—skunk; temperate grassland—coyote; savanna—zebra; desert; lizard; tropical rain forest; lemur; freshwater lakes; otter; ocean—whale. **Page 603:** 1. Recommendations not to smoke or abuse drugs would be easier for me to follow than recommendations for diet and exercise. Food and laziness are easier for me to buy into than cigarettes and drugs. 2. I think education including the opportunity to see the effects of poor diet, lack of exercise and/or effects of drug use is the best method to make for making young people realize that they should take care of their bodies. **Page 608:** 1. These precautions prevent the possibility of passing on a sexually transmitted disease by way of blood. 2. The blood concentration of cholesterol and lipoproteins in the blood in the donor's blood should be considered because a high level of cholesterol and an imbalance of HDL versus LDL can lead to cardiovascular disease.

Analyze and Evaluate

Page 609: 1. Circulatory systems always have exchange surfaces with the external environment so they can keep the internal environment constant. For example, aquatic animals have gills and terrestrial animals have lungs where gas exchange occurs. They also exchange materials with the cells because their ultimate function is to serve the needs of cells. 2. When the blood is returning to the heart, it lacks the blood pressure provided by the pumping of the heart and venous blood may be traveling against gravity to move from the extremities back to the heart. 3. Blood pressure and therefore blood flow would be reduced due to the movement of water into tissue fluid.

Test Yourself

1.d; 2. c; 3. b; 4. Most likely the animal may be flat, have a small size and a gastrovascular cavity. 5. See Fig. 30.2A; 6. d; 7. c; 8. a; 9. c; 10. b; 11. d; 12. c; ; 13. Efficient delivery of oxygen to the muscles allows birds and mammals to have an active lifestyle and generate heat to maintain a warm body temperature. 14. b; 15. d; 16. e; 17. a. blood pressure b. osmotic pressure c. blood pressure d. osmotic pressure; 18. b; 19. a; 20 Erythropoietin increases the number of red blood cells, and Rita's problem is lack of iron in her diet.

Get Involved

1. (1) By dissecting animals, you will see three different types of blood vessels; the valves in the heart directed toward the arteries and in the veins valves directed toward the heart. (2) A deep cut to a vertebrate limb draws bright, red arterial blood under pressure; pressing on a vein causes it to expand on the far side. This could only be if the blood circulates. 2. The amount of amino acids, sugar, and oxygen is higher in arterial blood and the amount of bicarbonate ion is higher in venous blood. Nutrients and oxygen leave a capillary and carbon dioxide enters a capillary midway along its length. When carbon dioxide enters the capillary, the main portion becomes the bicarbonate ion. The moist skin of a frog acts as an organ of respiration. As a result, frogs are not completely dependent on their lungs for respiration and are not hindered by a 3-chambered heart.

CHAPTER 31

Check Your Progress

31.1 Lymphatic system absorbs excess tissue fluid; absorb fats from small intestine; lymphatic organs help defend body by producing and discriminating white blood cells. 2. Lymph nodes contain many lymphocytes that help defend the body. **31.2** Innate immunity occurs without preparation; skin and low pH of digestive tract is always present; complement proteins are always active and white blood cells are always being produced. **31.3** Redness and heat are caused by the movement of blood into the area; swelling is due to loss of fluid from capillaries; and pain is due to stimulation of pain receptors. **31.4** Active immunity occurs when the administration of a vaccine or an infection causes the body to be able to produce their own antibodies now and in the future. Since passive immunity is the reception of someone else's antibodies and when they are gone, they're gone. **31.5** When a B cell reacts to an antigen it becomes a plasma cell that produces antibodies, When an antigen is presented to a T cell, it is able to seek out and destroy a virus-containing cell on contact. **31.6** Antibodies fit their antigens like a hand fits a glove; therefore HIV antibody cannot combine with a measles virus. **31.7** 1. (1) A T cell whose TCR can combine with the antigen presented by a macrophage is selected to undergo cloning. (2) Each of these T cells is capable of binding to and destroying s virus-infected cell. (3) Some of these activated T cells become memory T cells. 2. Helper T cells release cytokines which stimulate other immune cells; cytotoxic T cells destroy virus-containing cells. **31.8** 1. Tissue-engineering is producing organs in the lab. 2. The immune system attacks the body when an autoimmune disease occurs; however the patient needs an immune system to keep from getting infections.

Consider These Questions

Page 617: 1. No it does not disturb me because I am well aware that we are related to all organisms through symbiotic relationships and evolution. 2. This shows that immunity began in organisms ancestral to us and then was passed on to future generations. 3. It's specific to the pathogen. **Page 624:** 1. Cancer is caused by an agent that can affect our genes. Monoclonal antibodies do not activate transduction pathways or activate genes. 2. Yes, the patient should have been asked for her consent and should have been compensated. Even now her heirs could be compensated. Page 626: 1. Contact with an allergen over a period of time can lead to an allergic response. 2. The presence of IgG antibodies can prevent an allergic response. Good hygiene in developed countries has resulted in fewer types of antibodies in people's blood and tissues.

Analyze and Evaluate

Page 627: 1. An antibody-mediated response is needed to identify and destroy the virus once it is in the blood. If the viral DNA has integrated itself into the infected cell's genome, then cell-mediated immunity is also needed. During cell-mediated immunity, cytotoxic T cells destroy a cell infected with a virus. 2. The PCR test is positive faster because it detects the present of the virus whereas the antibody test is not positive until the immune system has produced antibodies against the HIV virus. You get faster and more accurate results with the PCR test. 3. Following a viral infection a body cell might mistakenly display a viral antigen. Natural killer cells (innate immunity) attack these body cells. Also, an APC cell detects the antigen and presents it to a T cell. (adaptive immunity). The T cell attacks the body cell even though it is no longer infected.

Test Yourself

1.a; 2. b; 3. d; 4. c; 5. b; 6. It collects excess tissue fluid at the blood capillaries and returns it to the subclavian veins of the cardiovascular system. 7. b; 8. e; 9. d; 10. a;11. b; 12. Innate immunity is less complicated than adaptive immunity 13. c; 14. d; 15. See Fig. 31.5; 16. a; 17. a; 18 a; 19. b; 20 e; 21. e; 22. B and T cells are specific to the agent that has harmed the body. 23. c; 24. Their immune system is impaired because HIV brings about the death of helper T cells.

Get Involved

1. Hypothesis: each type of antibody is coded for by a different sequence of exons from the same gene or genes. 2. Control group is vaccinated against a specific disease and then the pathogen is administered. They are expected to remain well. The test group is vaccinated against a specific disease and then is administered the drug plus the pathogen. If the drug suppresses antibody-mediated defense, the test group will become ill. Repeat the procedure for other diseases.

CHAPTER 32

Check Your Progress

32.1 Carnivores are predators that attack and kill other animals. Digestion (both mechanical and chemical) of meat is more easily accomplished than the digestion of carbohydrates. **32.2** 1. Mouth: chew food; esophagus: conduct food bolus to the stomach; 2. Salivary glands contain salivary amylase, an enzyme for carbohydrate digestion. This might be adaptive in a herbivore whose main source of food is fiber-rich plant material. **32.3** 1. Carbohydrates: mouth and small intestine 2. The small intestine digests all types of food to nutrients that are absorbed into the body; therefore it contributes more than other digestive system organs to homeostasis. **32.4** Like a herbivore whose food is hard to digest our intestines are very long. **32.5** 1. The liver is a gatekeeper to what nutrients enter the bloodstream and the liver detoxifies drugs. 2. The liver breaks down the medicine for excretion, so it is necessary to keep taking it in order to maintain a certain level in the body. **32.6** 1. Low-fiber, refined carbohydrates lead to poor health; high-fiber, whole-grain carbohydrates lead to good health. 2. Fats containing saturated fatty acids and/or trans fatty acids, in particular, lead to poor health. Oils containing unsaturated fatty acids lead to good health. **32.7** 1. Salts increase the osmolarity of blood and cause more water to be absorbed by the kidneys, leading to hypertension. 2. Whole grains, fruits, and vegetables, in general, supply vitamins in the diet.

Consider These Questions

Page 635: 1. Birds are adapted for flight and a beak is light because it does not have teeth. The food a bird eats whether a seed or a fish does not need to be chewed. Cows are large animals that have few predators; they can take their time eating grass which requires much chewing and digestion. 2. I would rather be an herbivore so that I would not have to hunt and kill in order to eat. I could spend all my time eating and digesting my food. A carnivorous lifestyle does have its benefits as well: I could eat large meals high in protein and spend less time digesting my food. 3. When humans first evolved, they were not tool makers and didn't have the means to hunt and kill animals. As the brain increased in size, humans made tools and learned to use fire. When they migrated to Europe the weather turned cold, and because of their spears and greater intelligence, men could cooperate to hunt and kill large animal and use their skins for clothing. **Page 639:** 1. To do a controlled experiment, Dr. Marshall needed two groups of volunteers; the test group would ingest a sample of *Heliobacter pylori* and the control group would ingest a sample that does not contain *Heliobacter pylori*. Each group is later tested for the presence of ulcers. 2. The absence of ulcer in the control group supports the hypothesis that *Heliobacter pylori* is the cause of ulcer in the test group. **Page 641:** 1. Antibiotic therapy might kill off the bacteria that can help you digest lactose. 2. Pro-pill answer: I would prefer to take a pill that supplies enzymes because I love milk and other dairy products. If the intolerance could be managed with a simple pill I would take it over excluding some of my favorite foods from my diet. Pro-watch diet answer: I would prefer to manage my diet for lactose intolerance because I don't think a pill should be the answer to every ailment, and the pill may have side effects. **Page 647:** 1. A person might have an unhealthy diet due to family dietary habits, lack of money, or a poor knowledge of a balanced diet. Schools could provide healthy foods in the dining room and should teach students the essentials of good nutrition. 2. Yes answer: Warning labels should be used because the government should protect us so we do not make poor food choices. No answer: Warning labels should not be used because education is a better protective measure than warning labels. The government should not be involved in every choice we make during our lifetimes. **Page 648:** 1. Our society admires thinness and to avoid being seen as overweight, a person might develop an eating disorder. Advertisements encourage children to eat high-calorie foods and that could start them on the road to obesity. 2. You could offer emotional support for any underlying issue that may contribute to the disorder.

Analyze and Evaluate

Page 649: 1. The digestive system provides the nutrients that allow all the systems of the body to produce ATP and synthesize the molecules needed to maintain structure. 2. The placement of the liver between the digestive system and the circulatory system stresses that the liver monitors the quantity and purity of the molecules received from our food; it breaks down poisonous molecules and stores excess glucose. 3. Reducing the size of the stomach can reduce food intake by making us feel full sooner. Malnutrition can arise.

Test Yourself

1. d; 2. a; 3. Life is sustained by a source of energy, and the digestive system provides the nutrients that provide energy to animals. 4. a; 5. d; 6. b; 7. b; 8. a. salivary glands b. esophagus c. stomach d. duodenum e. large intestine f. small intestine g. colon; 9. d because it contains the right enzyme, the right pH, and the right substrate; 10. a; 11. c; 12. d; 13. b; 14. b; 15. A ham sandwich contains protein which is digested to amino acids in the stomach and small intestine and carbohydrates digested to glucose in the mouth and in the small intestine, and fat which is digested to glycerol and fatty acids in the small intestine. 16 c; 17. b; 18. a. bile canals, b. hepatic artery, c. hepatic portal vein, d. bile duct e. central vein; 18. see Fig. 32.5, right; 19. The liver could also be considered an accessory organ to the circulatory system because it produces the plasma proteins and clotting factors. 20 a; 21.e; 22. c; 23. b; 24. d; 25. The source of amino acids is of no consequence because the DNA of each cell specifies the types of proteins for that cell.

Get Involved

1. Pepsin, HCL, substrate (e.g., piece of cooked egg white), water. Omit the pepsin. If digestion still occurs, pepsin may not be the cause of digestion. 2. The use of a control group and a large number of participants makes the correlation more certain. 3. Carnivores, such as frogs, have short intestines.

CHAPTER 33

Check Your Progress

33.1 Ventilation does not occur. **33.2** A counter flow mechanism ensures that blood is always exposed to a higher O_2 concentration; therefore, O_2 is continually taken up by blood. **33.3** Larger insects have a means of ventilating the tracheae, but external respiration (exchange of gases with incoming air), and internal respiration (exchange of gases in the tissues) comprise a single event in insects. **33.4** lungs, bronchiole, bronchus, trachea, larynx, glottis, pharynx, nasal cavity. **33.5** 1. Muscle contraction moves the rib cage up and out and then inspiration occurs. No contraction is necessary to lower the rib cage. 2. A higher than normal level of H+ increases the rate and depth of breathing. **33.6** Most CO_2 is transported as the bicarbonate

ion but some combines with hemoglobin forming carbaminohemoglobin.

Form Your Opinion

Page 654: 1. (1) A hydra lives in an aquatic environment, which ensures a moist exchange surface. (2) Earthworms keep themselves moist by living in moist soil. (3) Fish live in aquatic environments and thus have moist exchange surfaces. (4) Insects have fluid filled tracheoles, which aid in exchange. (5) Mammals use the internal environment to keep exchange surfaces moist. 2. Fishes must use the motion of their mouths to power water over the gills in order receive oxygen from water. Humans can breathe through their noses; insects, earthworms and hydras also do not use their mouths to breathe. 3. Cellular respiration requires O_2 and gives off CO_2. **Page 659:** 1. Yes answer: Smokers have a right to smoke, but should confine their smoking to non-public places since secondhand smoke is a serious threat to anyone near a smoker. No answer: Smokers do not have a right to smoke and smoking should be outlawed because nonsmokers should not have to pay for the health costs of smoking. 2. Smokers suffer from bronchitis and often lung cancer and other types of cancer. 3. Cigarette manufacturers should be held responsible for providing cigarettes to the world, while friends and family members of smokers might bear some blame if they do not do their best to prevent smoking in their families. **Page 660:** 1. Both activities bring pollutant irritants into the lungs and can cause conditions like chronic bronchitis and lung cancer. 2. An organic chemical could stimulate or inhibit a transduction pathway that ends with a transcription factor or activator that regulates gene activity. The end result of this altered gene regulation could be cancer. 3. Walking, household chores, climbing or descending a stairway, and heavy lifting would be troublesome for a person with emphysema.

Analyze and Evaluate

Page 665: 1. The lungs have external exchange surfaces because air from the external environment enters the lungs. 2. The respiratory system excretes CO_2 which arises due to cellular metabolism. The digestive system rids the body of nondigestible remains which have never been a part of the body. 3. Hemoglobin (1) transports O_2 to the cells, (2) helps transport CO_2 to the lungs, (3) combines with H+ and thereby helps buffer the blood.

Test Yourself

1. c; 2. e; 3. a; 4. b; 5. b; 6. d; 7. b; 8. c; 9. c; 10. See Fig. 33.4; 11. b; 12. d; 13. f; 14. d; 15. a; 16. b; 17. e; 18. Birds with a more efficient means of gas exchange had more offspring than those less efficient systems because they better adapted to flying. As generations passed birds became better and better adapted for flight. 19. d; 20. left: a. rib cage up and out; b. diaphragm moves down; right: a. rib cage down and in; b. diaphragm moves up. 21. b; 22. b; 23. b; 24. d; 25. a; 26. e; 27. The shape of hemoglobin changes when the pH changes from near neutral in the lungs to slightly acidic in the tissues, and this causes it to unload its oxygen. 28. d; 29. b; 30. e; 31. The body has a limited capacity to store oxygen and has a better ability to store energy.

Get Involved

1. Fat metabolism results in acids that enter the bloodstream; a lower pH stimulates the respiratory center and causes increased breathing, which lowers the CO_2, but raises the O_2 blood level. Test the blood of a diabetic group and a normal group for these blood levels and compare the results of the tests. 2. A severed spinal cord prevents the medulla oblongata from communicating with the rib cage and diaphragm via the phrenic nerve and intercostal nerves.

CHAPTER 34

Check Your Progress

34.1 Urea is not as toxic as ammonia, and it does not require as much water to excrete; it takes less energy to prepare urea than uric acid. 2. physiological adaptation **34.2** The capillary system reabsorbs water from the nephridium; the planarian has no need to reabsorb water because it lives in fresh water. **34.3** 1. The ureters take urine from the kidneys to the bladder and the urethra take urine from bladder to the exterior. 2. Glomerular capsule, proximal convoluted tubule, loop of nephron, distal convoluted tubule, collecting duct, pelvis. **34.4** 1. First step in urine formation: Glomerular filtration: blood is filtered and all but proteins and red blood cells enter glomerulus. Second step: Tubular reabsorption: primarily water, nutrients, and salts are reabsorbed. Third step: Tubular secretion: certain molecules are added to urine and pH is adjusted as needed. 2. Microvilli aid in reabsorption. **34.5** 1. Most water and salt are reabsorbed at the proximal convoluted kidney; due to an active extrusion of salt by ascending limb an osmotic gradient exists in the medulla and this draws water out of the ascending limb along its entire length. 2. ADH affects the permeability of the collecting duct, so that more water is reabsorbed and less urine forms; aldosterone promotes the excretion of K+ and the reabsorption of Na+ so that more water is reabsorbed in the distal convoluted tubule. ANH inhibits the secretion of aldosterone (and also renin). 3. Cells would become dehydrated and enzymatic reactions would most likely stop. **34.6** Fig. 34.6 shows how the kidneys regulate the pH of the blood: If the blood is basic, bicarbonate is excreted and not reabsorbed and H+ is not excreted. Therefore more H+ ions will be available to combine with OH- and form water. If the blood is acidic, H+ is excreted and ammonia combines with H+ to buffer it.

Consider These Questions

Page 673: 1. Evolution would have followed this pathway: freshwater fishes; to lobe-finned fishes in small bodies of fresh water on land; to amphibians that could locomote on land but had to return to fresh water to reproduce. 2. Excretion of ammonia is consistent with the early life of amphibians in the water; excretion of urea is consistent with their later life on land. You would predict that amphibians osmoregulate as their ancestors (freshwater fishes) did. 3. Humans osmoregulate by regulating the amount of salt and water in their blood as do freshwater fishes in that both have no way other than the kidneys to rid the body of salt. **Page 677:** 1 Drug use is a private matter unless it impinges on the rights of others or causes harm to others. Jail might scare some people to give up drugs, however, addictions should be treated as illnesses and treated safely with medical supervision and with the support of drug/alcohol counselors. 2. To use drugs to acquire an advantage in a competition is impinging on the rights of others and therefore it should not be condoned. However, you can understand why people take drugs in order to win competitions. The only way to curtail the practice is to take away the medals etc. won by competitors who take drugs. **Page 680:** 1. I might treat a dialysis patient that has to spend hours of the week in the hospital for hemodialysis like a "sick" person who needs special care. People have so many organ replacements today that I probably would not treat a person with a replaced bladder differently. 2. Except in cases of neglect or malpractice, I do not think that doctors, nurses, or technicians should be held responsible for unsuccessful hemodialysis. 3. If a person is able to perform the duties associated with a job and needs time off for medical treatment, some provisions should be made so they can be hired.

Analyze and Evaluate

Page 681: 1. Aldosterone promotes the reabsorption of Na^+ which promotes reabsorption of water and maintenance of blood pressure. Therefore insufficient aldosterone would cause low blood pressure. 2. Blood pH is regulated through the carbonate buffer system in the blood and by the excretory and respiratory systems. The carbonate buffer system combines with acid or base to keep the pH constant; to raise pH the respiratory system excretes carbon dioxide and the kidneys excrete H^+ and reabsorb bicarbonate. The kidneys can also do the opposite to lower the pH when necessary. 3. The kidneys regulate the composition of the blood by excreting metabolites, and they regulate the salt-water balance so that the tonicity and volume of blood stays constant.

Test Yourself

1. a; 2. d; 3. b; 4. A tubular shape provides an expanded surface area for reabsorption of substances into the blood or hemolymph and for the excretion of substances into the tubule. The tubular shape also offers a way for a excretory fluid to travel to the exterior. 5. b; 6. c; 7. See Fig. 34.3C; 8. c; 9. c; 10. a; 11. d; 12. e; 13.a,b,d; 14. a, b, d; 15. a, b; 16 b,c; 17. These cells reabsorb most of the contents of the nephron and need increased surface area and energy to better pump molecules back into the blood. 18. b; 19. a; 20. c,d; 21. a,b; 22. d; 23. b; 24. d; 25. c; 26. a; 27. a; 28. The presence of salt in the blood causes more water to be passively reabsorbed, and the resulting increase in blood volume contributes to high blood pressure

Get Involved

1. A microscopic study of their kidneys should reveal that they have a reduced glomerulus and a very long loop of the nephron. 2. Use the pump to increase the pressure of the blood passing through the tubing, increase the length of the tubing, and increase the rapidity with which the dialysis fluid passes through the apparatus. 3. liquid waste = urine; urea would be present in a frog's liquid waste.

CHAPTER 35
Check Your Progress

35.1 1. The nervous system uses neurotransmitters while the endocrine system uses hormones as chemical signals. 2. Hypothalamus is a part of the brain; pituitary is attached to hypothalamus by a stalk, and pineal gland is embedded in brain. **35.2** Peptide hormones are received by plasma membrane receptors; steroid hormones cause genes to be expressed. **35.3** 1. The hypothalamus produces ADH and oxytocin secreted by posterior pituitary; it controls the anterior pituitary by secreting hypothalamic-releasing or inhibiting hormones 2. Negative feedback inhibits the hypothalamus and the anterior pituitary. **35.4** Acromegaly occurs only in adults and affects only the bones that can still grow: feet, hands, face. **35.5** When there is not enough iodine in the diet, the anterior pituitary keeps stimulating the inactive thyroid and a goiter develops, Iodine in the diet prevents goiter formation. 2. Calcitonin causes calcium to enter the bones and PTH causes calcium to exit the bones. **35.6** Epinephrine because it gets the body ready for fight or flight. **35.7** sugar-containing foods. **35.8** Melatonin level increases when it becomes dark.

Consider These Questions

Page 698: 1. Sleeping in a room without a window seemed to bring on a degree of seasonal affective disorder for me. Without morning light to rouse me from sleep, I found myself sleeping later and later into the day. Though I attributed the change to a shift in melatonin, it could have been psychological also. 2. Night owl: I am a night person that likes to sleep in and go to bed in the early morning hours. This could be attributed to secretion of melatonin in the morning. Early bird: I am a morning person that likes to get up with the sunrise and get to bed early. This could be attributed to secretion of melatonin in the afternoon. 3. I would be hesitant to take a supplement such as growth hormone due to the associated risks and reported cases of diabetes and other problems. I would employ conventional methods of eating a healthy diet, exercising, and getting plenty of rest in order to improve my overall health.

Analyze and Evaluate

Page 699: 1. Stress is a stimulus that causes the hypothalamus to send nerve impulses (nervous system involvement) to the adrenal gland which then releases epinephrine (endocrine system involvement). 2. Example: Low blood calcium is a stimulus that causes the parathyroid glands to secrete parathyroid hormone and this hormone brings about a rise in blood calcium level. The increase in blood calcium does away with the original stimulus and the parathyroid glands no longer release parathyroid hormone. 3. Biologists originally called the anterior pituitary gland the master gland because the anterior pituitary gland secretes many different hormones and some of these control the secretion of other endocrine glands. Later, biologists discovered that the hypothalamus secretes hormones that control the secretion of the pituitary gland and therefore it is the master gland.

Test Yourself

1. d; 2. b; 3. a; 4. d; 5. d; 6. f; 7. b; 8. c; 9. a; 10. e; 11. Caffeine would increase the stress effect of epinephrine because cAMP, which drives the effect, would be slower to break down. 12. c; 13. c; 14. a; inhibits, b. inhibits, c. releasing hormone, d. tropic hormone, e. target gland hormone; 15. d; 16. e; 17. Hypothyroidism could be caused by an absence of or decrease of hypothalamus releasing hormone or thyroid stimulating hormone from the anterior pituitary. It could also be due to the inability of the thyroid to produce its hormones. 18. c; 19. d; 20. c; 21. d; 22. b; 23. b; 24. d; 25. d; 26. b; 27. b; 28. High blood calcium causes the thyroid to secrete calcitonin, which leads to the uptake of calcium by bone. Calcium level drops and thyroid is no longer stimulated to release calcitonin.

Get Involved

1. Calcitonin, being a peptide hormone, stimulates the metabolism of osteoblasts to form bone utilizing calcium. 2. Use two groups of volunteers; one group consists of the "night owls" and the other group are "early birds." Collect blood samples from all volunteers when they are typically sleepy and when they typically wake up. If the hypothesis is supported, the melatonin level rises earlier at night and lowers earlier in the day for the early birds.

CHAPTER 36
Check Your Progress

36.1 Asexual reproduction makes many copies of an already successful organism; sexual reproduction produces new phenotypes one of which may be successful in changed environment. **36.2** 1. Testis, epididymis, vas deferens, ejaculatory duct, urethra. 2. Broad shoulders, muscle strength, narrow hips, beard, deep voice. **36.3** 1. The vagina receives the penis the sperm are deposited inside the genital tract, and the offspring develops in the uterus. 2. To produce an oocyte and to produce estrogen and progesterone **36.4** 1. FSH causes a follicle to develop and LH brings about ovulation and maintains the corpus luteum. 2. The measures that rely on the administration of (a) hormone(s) prevent ovulation. **36.5** 1. The blastula is a hollow ball of cells that permits invagination to form the germ layers. 2. In the water, a zygote develops quickly into a larval form that can take food from the water. On land, it takes more time for a chick to develop and yolk provides the needed nutrients. **36.6** Ectoderm: nervous system; mesoderm: heart and blood vessels; endoderm: digestive tract. **36.7** A notochord is a supporting rod and a neural tube is a hollow tube that gives rise to the brain and spinal cord. **36.8** Distribution of the gray crescent is a form of cytoplasmic segregation. Cytoplasmic segregation determines how the various cells of the morula will develop. **36.9** Presumptive nervous tissue will not develop into a nervous system unless presumptive notochord is beneath it. **36.10** To be suitable to internal development; the chorion becomes a part of the placenta; the yolk sac produces blood cells, blood vessels of the allantois become umbilical vessels. **36.11** Cleavage occurs after fertilization occurs in the oviduct; the morula arrives in the uterus; the blastocyst that embeds itself has an inner cell mass which becomes the embryonic disk with two germs layers. Mesoderm forms as in the chick. Future development relates to the germ layers. **36.12** Gas exchange occurs at the placenta; waste for nutrients exchange also occurs, and therefore the urinary, respiratory, and digestive systems are not needed. **36.13** 1. The afterbirth consists of the placenta and the extraembryonic membranes that served the fetus so it could develop internally.

Consider These Questions

Page 705: 1. When the relationship is monogamous, males are fairly assured that the offspring is theirs. Females exercise choice by selecting as mates, males that appear to evolutionarily fit. When the relationship is not monogamous, males become sexually aggressive and females become defensive and only let down their defenses when they choose to. 2. Each sex wants their particular alleles to be passed on to the next generation. 3. Traditionally, before marriage males chase females and females withhold sexual favors. This might cause males to become deceptive and make the female believe they are interested in marriage when they are not interested. It might make the female be the first to break off a relationship that does not seem to be headed toward marriage. To test the hypothesis, I would take a survey of females to determine how many broke off a relationship because marriage did not seem in the offing. **Page 712:** 1. It is evolutionary beneficial for sex to be a powerful motivator in order to perpetuate a species. It is personally disadvantageous because of the risk of sexually transmitted infections such as those discussed in this passage. Women who reproduce earlier than planned may have to give up plans for a career or will find it more difficult to pursue a career while caring for children. 2. Social consequences that might result from having a non-treatable sexually transmitted disease (STD) range from economical to interpersonal. Treatment for STDs and lost work time represent an economic consequence. Socially, infertility and the risk of transmission to a partner could make a person less desirable as a long-term mate. 3. Yes answer: I do approve of allowing young girls to receive the HPV vaccine. I think the benefits of decreasing the risk of cervical cancer are enormous and it is no different than a vaccine to protect a child from any other disease. No answer: I do not approve of allowing girls as young as 9 to receive the HPV vaccine. I think that a girl should be presented with this option at an age deemed appropriate by a parent or physician in the early teen years. **Page 713:** 1. sperm donor- because the donor is the genetic parent of the child. egg donor—because the donor is the genetic parent of the child. Surrogate mother—because she carries a child and is ultimately responsible for its being born a healthy, viable child. contracting mother—because of legal agreement to care for the child produced from the pregnancy of the surrogate mother. contracting father—because of legal agreement to care for the child produced from the pregnancy of the surrogate mother. 2. Pros: successful pregnancy, overcoming reproductive challenges. Cons: multiple failed attempts at pregnancy, cost of procedures, pain surrounding hormone injection, legal issues surrounding surrogates or sperm/ egg donations, emotional wear and tear. For couples who have struggled with reproducing and are forced to utilize these technologies to produce an offspring, the benefits outweigh the drawbacks. 3. Legislation should regulate which of the "5 potential parents" listed actually does have the best claim to the child. Decisions should be made beforehand; for example, whether or not to make a sperm or egg donor known to a child later in life. In addition, the use of reproductive technology should be restricted to avoid cases like the "Octo-mom" where a large single parent family occurs; thus creating the risk that the children will be neglected or the single parent will be unable to support all of the children.

Analyze and Evaluate

Page 725: 1. The human male reproductive organs are adapted to the production and storage of sperm in the temperature-regulated scrotum, secretion of protective fluids for the safe transfer of mature sperm, as well as direct deposit of seminal fluids into the female reproductive organs. Internal development of a fetus in amniotic fluid shows that human development is still tied to an aquatic environment. 2. Animals go through the same developmental stages and this would not develop similarly if they were not related and descended from a common ancestor. 3. Many genes are active only during development and they get turned off as the organism matures.

Test Yourself

1. e; 2. c; 3. Sexual reproduction produces variations among the offspring and one of these may be phenotypes that can deal with the new conditions. 4. b; 5. See Fig. 36.2A; 6. a; 7. c; 8. b; 9. A high blood testosterone level shuts down the anterior pituitary production of gonadotropic hormones, reducing sperm production. 10. c; 11. b; 12. b; 13. c; 14. d; 15. zygote, cleavage, morula, blastula, gastrula, neurula. 16. All vertebrates can trace their ancestry to fishes. 17. a. chorion (gas exchange), b. amnion (protection), c. embryo, d. allantois (umbilical blood vessels), e. yolk sac (blood cell formation) f. and g. (placenta is area of exchange with mother); h. umbilical cord 18. c; 19. d; 20. The placenta allows the female to be more active and therefore allowed mammals to invade new areas such as trees. Many of our successful traits such as flexible hand are the result of being arboreal.

Get Involved

1. Testosterone causes increased cell division in the prostate. Microscopically, observe the effect of testosterone on cells taken from the prostate. 2. Progesterone maintains the uterine lining, and if the placenta doesn't begin producing it when it should, a woman could lose the embryo embedded in the lining.

CHAPTER 37

Check Your Progress

37.1 When a population explodes, it affects the statistics of all the other levels of an ecosystem. Predators, if there are any should increase in number and prey (be an animal or plant would decrease in number, for example. **37.2** Population density is determined by the average number of organisms per square mile such as "the density of clams in the Bay is 3000 per square mile." **37.3** A immediate reduction in reproductive ages would change the survivor ship to a type III curve in Fig. 37.3Bb. **36.4** 1. Exponential growth would occur until competition for resources slowed the population growth and it was at the carrying capacity of the environment. 2. The fish populations enter a lag phase

and it takes years before they will again experience exponential growth. **37.5** Severe natural disasters (such as hurricanes) and weather-related food shortages are two examples density-independent factors that could severely affect a deer population. **37.6** Competition for food and habitat resources are two examples density dependent factors that could affect a deer population. **37.7** Deer are exploding because they have no predators and are able to live in the vicinity of people. Gorillas prefer a restricted environment that is shrinking and they form small populations. **37.8** 1. The population growth curve for the LDCs is sharply exponential compared to that of the MDCs. The use of birth control, education of girls, and providing the necessary capital for women to start a business can reduce the fertility of women as measured by the number of children per woman.

Consider These Questions

Page 738 : 1. In efforts to protect individual species, we pay special attention to preserving their habitats. As such, protecting "hotspots" would essentially be a means to protect many organisms and their homes. However, if "hotspots" are not home to any endangered species, we should still preserve individual species that inhabit other locations. 2. Yes answer: Though it is a controversial issue, I support the preservation of cheetahs in zoos if it can help increase the population. Continued monitoring of genetic diversity is essential. Perhaps if cheetahs can be bred in zoos for a period of time and later strategically released onto protected lands, I can see the use of zoos as beneficial. No answer: I think that efforts should be made to preserve cheetahs in their natural environment and not in the captivity of a zoo where they are unable to run freely and hunt food, skills that are characteristic of cheetahs in the wild. 3. Although it is important for our own needs to preserve specific species and natural environments, we should not neglect unique endangered species. Minor changes in ecosystems can have widespread and unpredictable effects. The loss of unique endangered species might alter our planet in ways we did not anticipate. We should make every effort to preserve as much of the natural environment as possible. **Page 740:** 1. Yes answer: Considering that our current way of living and reproducing is not sustainable, families in the U.S. are acting selfishly when they have more than two children. This is selfish behavior because it has little regard for the future and the impact we are making on our planet that will be detrimental, and possibly preventative, of generations of human beings to come. No answer: I believe families in the U.S. that have more than two children are not acting selfishly if they make efforts to have minimal impact on the environment. It is more important, no matter the size of families, that people in the U.S. pay attention to energy use and exploitation of resources in order to have less negative impact on the planet. 2. Sustainability would provide a quality of life for future generations that is equal to the current conditions. This would help the economy by providing a steady source of materials for production, thus stabilizing prices in the long run. 3. The United States should hold itself, along with other countries, to the same standards with regard to environmental impact. More developed countries tend to have a greater impact on the environment and should be keenly aware of the predicament this puts the planet in for the future. The U.S. should certainly reduce its environmental impact while expecting other countries to do the same. The benefits for the planet and the future of humanity depend upon our responsible management of resources and preservation of the planet we inhabit today.

Analyze and Evaluate

Page 741: 1. A moderate population size in desirable because it strikes a balance between the negative aspects of very large and very small populations. Very

large populations overexploit the resources of their environment, and might become extinct due to lack of resources. Very small populations lose genetic variability and run the risk of extinction from disease associated with inbreeding, infectious diseases, or extreme environmental changes (flood, drought, etc.). 2. We should definitely be concerned about how the growing U.S. population affects consumption and sustainability of our population. If we persist in a trend of increased population without regard to the environmental impact we are making, we run the risk of not only running out of natural resources, but also destroying ecosystems due to pollution and overdevelopment. 3. As demonstrated in Figure 37.3C, more women entering than leaving their reproductive years means that births will continue to increase population over time. A population that consists of more young members than old lends itself to growth that is not balanced out by deaths. In populations with a balance between older and younger members, stabilization occurs.

Test Yourself

1. d; 2. d; 3. b; 4. d; 5. b; 6. a; 7. b; 8. a; 9. c; 10. See Fig. 37.4B. 11. The offspring of a plant that reproduces by runners would remain near the parent. The offspring of a plant that reproduces by windblown seeds could be taken far away from the parent. 12. c; 13. a; 14. b; 15. c; 16. a. density independent factor, b. density dependent factor; 17. The asexually reproduced plant has to remain in the vicinity of its parent but the windblown seed can be taken far away from its parent. 18. c; 19. d; 20. When the environment is unstable due to density independent and density-dependent factors, a few of the many small, uncared for offspring might have a better chance of dispersal to a favorable environment. 21. a; 22. a; 23. c; 24. a; 25. e. 26. An unsafe environment leads to thinking that the more children you have the more likely a few will survive.

Get Involved

1. You might hypothesize that the right whale has only one offspring per reproduction and chances of death before maturity are good. The right whale begins reproducing well after maturity, and reproduces infrequently. To test your hypothesis, you would have to observe a captive population or tag individuals in the wild and observe them from a distance. 2. Determine the original normal flow of the river and maintain the flow as close to normal as possible.

CHAPTER 38

Check Your Progress

38.1 1. Gene for egg-laying behavior hormone in Aplysia was isolated and its protein product controls egg-laying behavior. The gene fosB has been found to control maternal behavior in mice. 2. Because all members of the colony have the same mother, they would all inherit the gene, and the colony would die out as tunnel excavating ceased. **38.2** Observe a mole rat in its colony from the time of birth and see if efficiency in building tunnels improves over time. **38.3** Operant conditioning because the younger workers would be rewarded for tunnel work. **38.4** A territory provides an exclusive place to forage for food and the amount of food should be greater because of decreased amount of competition. **38.5** Female choice causes males to compete which can be detrimental to their life and health however if they are chosen they often have exclusive rights to mate with a desirable female thereby increasing their fitness. **38.6** Altruistic behavior usually includes some self-interest. For example, when an offspring helps its parents raise young, it is ensuring some of its own genes are passed on. **38.7** Raise birds that are unable to perform this behavior and then compare the fitness of birds that sing to mark their territory with those that do not.

Consider These Questions

Page 751: 1. Yes answer: Evolutionary principles should be applied to human reproductive behavior. Like other animals, humans demonstrate sexual selection and the mating behavior of both males and females indicates a preference towards selecting the partner with the optimal fitness. No answer: I believe evolutionary principles should not be applied to human reproductive behavior, since our complex culture influences reproduction in ways that are not possible in other animals. For example, selection of males based on financial security is unique to humans. In addition, our technological advances can help "less fit" individuals survive and reproduce. 2. I have observed many examples that support male competition for females and female choice based on financial stability. On average, men marry and have children with younger women. Traditional dating behavior follows a trend of a man impressing a woman in order to "earn the privilege" of mating and reproduction. **Page 756:** 1. My experience with dogs leads me to suggest that dogs do feel affection for their owners, anger demonstrated through aggression, excitement for food and play, shame when admonished for undesirable behavior, and mourning at the permanent loss of a playmate (another animal) or human companion. 2. With regard to pets, I would take care not to provoke aggression, and reward them and praise them for desirable behaviors. In nature, I would also refrain from provoking aggressive or territorial behavior in an animal, bearing in mind that the emotions of wild animals are tied closely to instincts. 3. Yes answer: The conclusion that animals do have emotions makes them seem closer to human beings, thus making eating an animal nearly cannibalistic in nature. Though I tend to associate emotions more with pets than animals raised for food, I would reconsider eating meat with evidence that all animals display emotions. No answer: I would continue to be an omnivore even with conclusions that animals have emotions. I think that animal emotions differ from human emotions and are tied largely to instinctual response to stimuli.

Analyze and Evaluate

Page 757: 1. Natural selection favors the successful competitor for territory while the losing animal(s) may not survive to reproduce without having a territory. 2. The idea that behavior is largely response to stimulus, and that some responses yield increased fitness might be offensive to people who prefer to think of behavior as learned and correlated with cognitive choice. 3. Crickets make calls to attract mates. This behavior is risky, and can lead to predation. However, the benefits of attracting a mate have overcome the risk evolutionarily, since natural selection has perpetuated populations of crickets that make calls for mates.

Test Yourself

1. c; 2. b; 3. e; 4. c; 5. c; 6. b; 7. See Fig. 37.3; 8. a; 9. d; 10. c; 11. Associative learning needs an accompanying stimulus and there was no other stimulus aside from acquiring food. 12. c; 13. a; 14. Because females produce few gametes during their lifetime, they place an emphasis on quality of offspring. Because males produce many gametes all the time, they place an emphasis on quantity of offspring. 15. d; 16. e; 17. a; 18. They would be increasing their inclusive fitness when they help their parents at a time they cannot be reproducing themselves. Reciprocal altruism would ensure that they will successfully reproduce when the time comes. 19 a; 120. c; 21. b; 22. d

Get Involved

1. Mate the two types of rats and test how the offspring react to limburger cheese. If they show an intermediate response, such as being willing to approach the cheese but still not eating it, then the behavior may be genetically controlled. 2. Observe sentry behavior more closely. Recent observations have shown that sentries are the first ones to reach safety when a predator is spotted, and meerkats only serve as sentries after they have eaten.3. The mealworms ate the bran flakes readily; however, it would be hard to extrapolate from so few experiments.

CHAPTER 39

Check Your Progress

39.1 Competition leads to resource partitioning allowing several types of warbles to forage in the same spruce tree. **39.2** Predation and lack of food causes the prey population to decline followed by the predator population. **39.3** Parasites feed on a living host; decomposers consume a dead one. **39.4** Study the feeding behavior of a remora when not attached to a shark. If it can survive independently, it is more likely commensalistic with the shark. **39.5** The protozoans gain a home and the termites gain nutrients they would be unable to acquire on their own. The benefit to the community is a new source of food. **39.6** Lichens and mosses change rocks to soil allowing small shrubs to invade an area. The shrubs enrich the soil allowing nitrogen fixing trees to take hold. Leaves from trees create water-holding humus that permit seeds of large trees such as spruce to germinate and these trees are a climax community. **39.7** Autotrophs are producers of food for themselves and heterotrophs and when they die they become food for decomposers. **39.8** The inorganic chemicals are taken up by plants and the cycling of chemicals in an ecosystem begins again. **39.9** The first link in a grazing food web is a producer that provides food for a herbivore and the first link in a detrital food chain is detritus, provided by bacteria and fungi of decay. **39.10** None; ecological pyramids do not include decomposers. The energy or biomass attributable to decomposers is in the bodies of the other levels. **39.11** The reservoir and the exchange pool are abiotic. **39.12** Decomposers gradually break down the bones making phosphate ions and other chemicals available. **39.13** Nitrification. **39.14** It would decrease greenhouse emissions and reduce the threat of a rise in global temperature.

Consider These Questions

Page 767: 1. In both the predator-prey and parasitehost relationships, one of the individuals (predator and parasite) benefits more than the other (prey and host). The difference is that in a predatorprey relationship, the death of prey is guaranteed, while the death of a host is not always the case in parasitism. 2. HIV cannot currently be effectively vaccinated against, and even if caught early, remains incurable. H1N1 can be vaccinated against and is also treatable in certain stages of illness. HIV is a retrovirus capable of producing DNA from its RNA within a host cell, thus taking over the host entirely over time. 3. A pollinator coevolves with a flower such that the flower evolves to attract the pollinator and the pollinator adapts to feed from the flower (i.e. long proboscis when the flower is deep). This coevolution takes place in the same "direction" the flower to be pollinated and the pollinator in order to feed. Coevolution between parasite and host differs in that hosts evolve under selective pressure to avoid being infected by parasites, while parasites evolve under pressures to evade host defenses. The parasite and host are evolving in different "directions"- the host to avoid the parasite and the parasite to overcome attempts to be avoided. **Page 771:** 1. Food chains will be altered; killing off large predators or the introduction of alien species will create new predator-prey relationships, potentially exterminating some prey. Each minor change to the area has trickle down effects that touch not only the animals, but also plants as well as abiotic resources.

2. Small areas, cut off from former connections lose species, altering predator-prey relationships and causing other potentially disruptive changes throughout an environmental system.

Analyze and Evaluate

Page 779: 1. Three populations would be students, faculty, administrators: (1) Students interact with faculty for the exchange of information. Administrators interact with faculty for record keeping, financial exchanges etc. (2) Faculty must not be too small to meet the needs of students and the number of administrators, must not be too many or else the community is notsustainable financially. (3) A larger student body causes a larger faculty but this is financially sustainable because each students pays tuition. 2. The human population is dependent on natural resources and people should definitely work to preserve the natural environment in order to maintain the resources needed for life. Many of our actions fail to preserve the natural environment because we exploit natural resources, alter the habitats of many organisms, and create large amounts of waste at a rate unmatched by our efforts to replenish and repair the damage we do.

Test Yourself

1. c; 2. e; 3. e; 4. a; 5. c; 6. b; 7. d; 8. Will the area be able to meet the niche requirements of this predator? (see Fig. 39.7); 9. b; 10. a; 11. b; 12. It follows the utilization model because each community prepares the way for the next community. For example, lichens and mosses are soil formers for small bushes and they stabilize the soil for the growth of trees; the first types of trees enrich the soil for the final community consisting of a white spruce-Western hemlock community. 13. e; 14. See Fig. 38.8A; 15. c; 16. d; 17. c; 18. c; 19. b; 20. c; 21. d; 22. d; 23. The reservoirs include coal, natural gas, oil, and limestone (calcium carbonate.) The exchange pool is the atmosphere and the biotic community includes the organisms.

Get Involved

1. Observe the birds carefully to see if they differ in ways suggested by Figure 38.7. 2. Measure and fill a large container with water from the pond. Add measured amounts of phosphate slowly over several days or months, and when you see growth, calculate the amount of phosphate you need for the pond.

CHAPTER 40

Check Your Progress

40.1 The sun shines directly on the equator and moisture laden winds rise and lose their moisture. **40.2** As winds rise above mountains they lose their moisture so that there is little moisture left for the other (shadow) side of the mountain. **40.3** The ocean currents are due to wind patterns. Those that begin in the tropics like the Gulf Stream bring warmth to in the Northern hemisphere. **40.4** The same climate occurs per latitude around the globe. No matter what continent for example, the climate is warm and moist at the equator. **40.5** Limited rainfall, limited sun, and the presence of permafrost prevent larger plants from living in the tundra. **40.6** Their needle-like leaves do not lose moisture like a large flat leaf does. **40.7** Solar energy and rainfall are sufficient to allow the growth of trees and the stratified forests provide food and niches for many different type of organisms. **40.8** The yearly growth and death of grasses over many years leads to rich organic soil. **40.9** The African savanna supports great herds of herbivores and they provide food for carnivorous lions. **40.10** Small animals can hide better in limited vegetation and more easily burrow to get in out of the sun during day and stay warm during cold nights. **40.11** Much solar energy and rainfall allow trees to flourish; the stratified forests and many types of organisms provide many more niches than other ecosystems. 40.12 Fresh water ecosystems consisting of rivers and streams which have flowing water; and lakes and ponds which have standing water. 40.13 1. Mangrove roots prevent erosion and serve as a nursery for sea life; at mudflats migrating birds feed on small fish and shellfish. Rocky shore has diverse life in tide pools. 2. Coral reefs occur in warm, sun lit waters and the corals and many organisms there provide plentiful food and niches for organisms.

Consider These Questions

Page 789: 1. The major groups of organisms can be found on all continents because the present-day land masses used to be a supercontinent. Types of organisms can vary between continents depending upon when the animals evolved relative to the separation of the continents. 2. Because of the age of the man, he most likely traveled across sea or ice but did not make his way to Greenland from Siberia via Beringia since this land bridge is believed to have been exposed during the last ice age 12,000 to 20,000 years ago. **Page 796:** 1. Everything is connected to everything else. The effect of the barometric pressure over the Southeast Pacific and the Indian Ocean upon weather on the coasts of the United States is an example of this connection. Also on land, pollution that enters the atmosphere from smoke stacks is distributed to cities, towns and ecosystems far away. 2. Prior to the completion of a dam in 1970, the annual flooding of the Nile River left behind fertile silt capable of sustaining agriculture in Egypt. A small flood or no flood at all would result in famine for this area. The annual flood was a dangerous event, but without it, Egypt's farming would have been impossible.

Analyze and Evaluate

Page 797: 1. (a) Squirrels: small herbivorous animals which are behaviorally adapted to collect and store foods (like nuts) in order to have nourishment through the winter. The thick fur coats of squirrels are also adaptations to colder winters in the temperate deciduous forest. (b) Sloths: slow moving within the trees, these animals are adapted to expend very little energy which coincides with the slow digestion of fruit and leaves carried out by their specially adapted digestive system. Since sloths use very little energy, they do not need to eat much. (c) Squid: have a very large eye for hunting in the extremely low light levels of the deep sea as well as the ability to maneuver through water via jet propulsion, making them mostly fast-moving creatures. 2. Due to the protection offered by the mangrove swamp, it would certainly benefit a human population to restore a mangrove swamp over a housing development that would likely see damage or devastation shortly after it was rebuilt. Hurricane damage can be avoided if we resist the urge to build homes as close to the shore as possible, and instead guard our shores with protective barriers such as mangrove forests. 3. Five negative effects of the human population on the oceans: Pollution from developed areas including sewage, toxic chemicals, insecticides, and detergents invades the ocean and is carried by the tides throughout the ocean where they can invade the habitats of animals, or even be consumed by them. Burning of fossil fuels has increased acidity of rain and lakes to the detriment of forests and freshwater inhabitants. Litter and improperly disposed trash is

trapping and being eaten by animals, killing them. Dredging the ocean floor with heavy nets and chains as a method of fishing disturbs the ocean floor habitat and kills many non-target fish/animals in the process (by-catch). Drilling for oil sometimes results in oil spills and causes long-term damage to the oceans, killing animals and destroying shoreline habitats.

Test Yourself

1. b; 2. e; 3. See Fig. 39.2B; 4. c; 5. b; 6. a; 7. Hot air from the equator moves toward the poles. 8. d; 9. d; 10. a; 11. a; 12. a; 13. d; 14. The atmosphere loses its moisture on the windward side. 15. e; 16. d ; 17. b; 18.Decomposers use up any available oxygen, leaving none for the fish. 19 b; 20. c; 21 c; 22 e; 23. d; 23. e; 25. Sea level will be too deep for previous estuaries to exist, and new ones may not develop farther inland due to development of coastal regions.

Get Involved

1. You will be able to see if a rising global temperature affects the distribution of biomes in the biosphere. 2. Coral houses microscopic algae and if the dirty water blots out the sun, the algae will die.

CHAPTER 41

Check Your Progress

41.1 1. Forestry, veterinary science fisheries biology, range management, agronomy (crop and soil management) wildlife management. 2. Ecosystem-level conservation has the potential to save a large number of species instead of just one. **41.2** Reduced aquatic biodiversity would limit the amount and types of seafood we consume. **41.3** Natural areas provide fresh water, prevent soil erosion and regulate climate. **41.4** A large number of people living and building structures on or near the coast. **41.5** People who colonize bring the animals and plants from their former homes to their new homes. **41.6** Global climate change because it will affect so many species. **41.7** Overfishing, killing rare species for a body part, collecting species from the wild to sell to others. **41.8** First their habitat becomes polluted and then they succumb to disease. **41.9** Other species in an ecosystem depend on keystone species; flagship species simply appeal to humans. **41.10** Begin as soon as possible; mimic natural processes; develop the ability of the ecosystems to maintain itself while providing services to human beings. **41.11** Primarily by preserving ecosystems and areas of species richness. **41.12** Preservation of natural resources and less pollution of the environment. **41.13** Famers can grow salt tolerant plants and otherwise use drip irrigation; industry can use air instead of water for cooling purposes. **41.14.** Use other means of controlling pests rather than pesticides such as polyculture and biological pest control; use drip irrigation to water plants, saving water; use legumes to fertilize land instead of artificial fertilizers.

Consider These Questions

Page 810: 1. Scientists could use artificial selection to breed members of species that display adaptive traits or behaviors that make them tolerable of climate change in order to help these species survive, essentially fast forwarding what might occur via natural selection over a

longer period of time. 2. Though the temperature might be suitable for survival, the new surroundings could be home to predators not previously encountered by a species and may not contain an adequate food supply. The ecosystem in which the species is adapted to living is suitable due to many factors, not temperature alone. **Page 814:** 1. The restoration plan will be judged by how closely the community comes to resemble the original community in Emiquon floodpain 2. If the plan produce the expected data showing good restoration, the plan and the data could be made available on the web to assist other restoration projects.

Analyze and Evaluate

Page 819: 1. Instead of increasing the manipulation of plants and animals via biotechnology, I would rather see the needs of the growing population met with an alternative method. For example, if the human population was to shift to growing and consuming locally, fostering crops suitable to the environment in which they are grown rather than the most popular crops that yield the best profits, we could use nature to our advantage to feed the growing population. 2. Emphasis on renewable energy is necessary in order to achieve sustainability because nonrenewable energy sources are quickly being depleted and cannot ever be replaced. I agree with this emphasis. It is important for people to understand that the sources of energy we are accustomed to using cannot feasibly be utilized long term and will, sooner than later, run out. 3. Yes, I am willing to pay extra for organic foods. The societal benefits of reduced pesticide and herbicide use include reduction of pollution of water and air and decreased side effects on non-target populations of animals/insects. 4. Yes answer: I agree that the human population should stabilize and that each couple should only replace itself because resources are finite. No answer: I do not agree that human population should stabilize and believe technology will always find a way to increase the availability of resources.

Test Yourself

1. e; 2. a; 3. b; 4. b; 5. The added shrimp pushes the salmon to a higher place in the food pyramid, indicating that less energy is now available to them. 6. d; 7. e; 8. e; 9. e; 10. e; 11. a; 12. e; 13. d; 14. e; 15. b; 16. If predators prey on more than one food source, it is more likely they will always be able to find food. 17. b; 18. c; 19. e; 20. d; 21. b; 22. b 23 To help preserve biodiversity, restrict the use of pesticides and herbicides and also purchase organically grown foods. Because global warming is a threat to biodiversity drive a fuel-efficient vehicle, conserve electricity in your home, and purchase "green power". Try to cut back on consumption and buy only Ecolabeled products; when buying wood products, for example, look for the Forest Stewardship Council (FSC) logo. This indicates the wood comes from sustainably managed forests.

Get Involved

1. Overharvesting reduces genetic diversity due to the bottleneck effect. As with the cheetah, determine how many loci are now homozygous. 2. Besides having produced a population with limited genetic diversity, none of the problems that brought the species to near extinction (see Sections 40.5–40.9) have been solved.

Glossary

A

abscisic acid (ABA) Plant hormone that causes stomata to close and initiates and maintains dormancy.

abscission Dropping of leaves, fruits, or flowers from a plant.

accessory fruit Fruit, or an assemblage of fruits, whose fleshy parts are derived from tissues other than the ovary (e.g., strawberry).

acetylcholine (ACh) (uh-seet-ul-koh-leen) Neurotransmitter active in both the peripheral and central nervous systems.

acetylcholinesterase (AChE) (uh-seet-ul-koh-luh-nes-tuh-rays) Enzyme that breaks down acetylcholine bound to postsynaptic receptors within a synapse.

acid Molecules tending to raise the hydrogen ion concentration in a solution and to lower its pH numerically.

actin (ak-tin) One of two major proteins of muscle; makes up thin filaments in myofibrils of muscle fibers. (See myosin.)

actin filament Cytoskeletal filaments of eukaryotic cells composed of the protein actin; also refers to the thin filaments of muscle fibers.

action potential Electrochemical changes that take place across the axomembrane; the nerve impulse.

active site Region on the surface of an enzyme where the substrate binds and where the reaction occurs.

active transport Use of a plasma membrane carrier protein and energy to move a substance into or out of a cell from lower to higher concentration.

adaptation An organism's modification in structure, function, or behavior suitable to the environment.

adaptive radiation Evolution of several species from a common ancestor into new ecological or geographical zones.

addiction Physiological and psychological need for a habit-forming drug.

adenosine Portion of ATP and ADP that is composed of the base adenine and the sugar ribose.

adenosine diphosphate (ADP) (ah-den-ah-seen dy-fahs-fayt) Nucleotide with two phosphate groups that can accept another phosphate group and become ATP.

adenosine triphosphate (ATP) (ah-den-ah-seen try-fahs-fayt) Nucleotide with three phosphate groups. The breakdown of ATP into ADP + P makes energy available for energy-requiring processes in cells.

adhesion Attachment of cells, as when water adheres to the vessel walls of plants.

adipose tissue Connective tissue in which fat is stored.

adrenal cortex (uh-dree-nul kor-teks) Outer portion of the adrenal gland; secretes mineralocorticoids, such as aldosterone, and glucocorticoids, such as cortisol.

adrenal gland (uh-dree-nul) An endocrine gland that lies atop a kidney, consisting of the inner adrenal medulla and the outer adrenal cortex.

adrenal medulla (uh-dree-nul muh-dul-uh) Inner portion of the adrenal gland; secretes the hormones epinephrine and norepinephrine.

adrenocorticotropic hormone (ACTH) (uh-dree-noh-kawrt-ih-koh-troh-pik) Hormone secreted by the anterior lobe of the pituitary gland that stimulates activity in the adrenal cortex.

adult stem cells Cells in a mature body that have the ability to divide; found in red bone marrow.

aerobic Phase of cellular respiration that requires oxygen.

afterbirth Placenta and the extraembryonic membranes, which are delivered (expelled) during the third stage of birth.

age structure diagram In demographics, a display of the age groups of a population; a growing population has a pyramid-shaped diagram.

aggregate fruit Fruit developed from several separate carpels of a single flower.

aldosterone (al-dahs-tuh-rohn) Hormone secreted by the adrenal cortex that decreases sodium and increases potassium excretion; raises blood volume and pressure.

alga Type of protist that carries on photosynthesis; unicellular forms are a part of phytoplankton, and multicellular forms are called seaweed.

alien species Nonnative species that migrate or are introduced by humans into a new ecosystem; also called exotics.

allantois (uh-lan-toh-is) Extraembryonic membrane that contributes to the formation of umbilical blood vessels in humans.

allele (uh-leel) Alternative form of a gene; alleles occur at the same locus on homologues.

allergy Immune response to substances that usually are not recognized as foreign.

allopatric speciation Origin of new species between populations that are separated geographically.

alloploidy Polyploid organism that contains the genomes of two or more different species.

alternation of generations Life cycle, typical of plants, in which a diploid sporophyte alternates with a haploid gametophyte.

alternative mRNA splicing Variation in pre-mRNA processing resulting in different mRNAs and different protein products.

altruism Social interaction that has the potential to decrease the lifetime reproductive success of the member exhibiting the behavior.

alveolus (pl., alveoli) Air sac of a lung.

amino acid Organic molecule having an amino group and an acid group, which covalently bonds to produce peptide molecules.

ammonia Colorless gas that has a penetrating odor and is soluble in water.

amnion (am-nee-ahn) Extraembryonic membrane that forms an enclosing, fluid-filled sac.

amniotic egg Egg that has an amnion, as seen during the development of reptiles, birds, and mammals.

amoeboid Cell that moves and engulfs debris with pseudopods.

amphibian Member of a class of vertebrates that includes frogs, toads, and salamanders; they are still tied to a watery environment for reproduction.

ampulla Expansion at the end of each semicircular canal that houses the receptors for rotational balance.

amygdala (uh-mig-duh-luh) Portion of the limbic system that functions to add emotional overtones to memories.

anabolism Metabolic process by which larger molecules are synthesized from smaller ones; anabolic metabolism.

anaerobic Growing or metabolizing in the absence of oxygen.

analogous structure Structure that has a similar function in separate lineages but differs in anatomy and ancestry.

ancestral character Structural, physiological, or behavioral trait that is present in a common ancestor and all members of a group.

anemia (uh-nee-mee-uh) Inefficiency in the oxygen-carrying ability of blood due to a shortage of hemoglobin.

aneuploid Individual whose chromosome number is not an exact multiple of the haploid number for the species.

angiogenesis (an-jee-oh-jen-uh-sis) Formation of new blood vessels, an event that occurs to promote the enlargement of a tumor.

angiotensin II Hormone produced from angiotensinogen (a plasma protein) by the kidneys and lungs; raises blood pressure.

animal Multicellular, heterotrophic organism belonging to the animal kingdom.

annelid Member of a phylum of invertebrates that contains segmented worms, such as the earthworm and the clam worm.

annual ring Layer of wood (secondary xylem) usually produced during one growing season.

anterior pituitary (pih-too-ih-tair-ee) Portion of the pituitary gland that is controlled by the hypothalamus and produces six types of hormones, some of which control other endocrine glands.

anther In flowering plants, pollen-bearing portion of stamen.

anthropoid Group of primates that includes monkeys, apes, and humans.

antibody (an-tih-bahd-ee) Protein produced in response to the presence of an antigen; each antibody combines with a specific antigen.

anticodon (an-tih-koh-dahn) Three-base sequence in a transfer RNA molecule base that pairs with a complementary codon in mRNA.

antidiuretic hormone (ADH) (an-tih-dy-uh-ret-ik) Hormone secreted by the posterior pituitary that increases the permeability of the collecting ducts in a kidney.

antigen (an-tih-jun) Foreign substance, usually a protein or a polysaccharide, that stimulates the immune system to produce antibodies.

antigen-presenting cell (APC) Cell that displays the antigen to the cells of the immune system so they can defend the body against that particular antigen.

antigen receptor Receptor proteins in the plasma membrane of immune system cells whose shape allows them to combine with a specific antigen.

antioxidant Substances, such as vitamins C, E, and A, which defend the body against free radicals.

aorta (ay-or-tuh) Major systemic artery that receives blood from the left ventricle.

aortic body Sensory receptor in the aortic arch sensitive to the O_2, CO_2, and H^+ content of the blood.

apical dominance Influence of a terminal bud in suppressing the growth of axillary buds.

apical meristem In vascular plants, masses of cells in the root and shoot that reproduce and elongate as primary growth occurs.

apoptosis (ap-uh-toh-sis, ahp-) Programmed cell death involving a cascade of specific cellular events leading to death and destruction of the cell.

appendicular skeleton (ap-un-dik-yuh-lur) Portion of the skeleton forming the pectoral girdles and upper extremities and the pelvic girdle and lower extremities.

aquaporin Protein membrane channel through which water can diffuse.

arachnid Group of arthropods that contains spiders and scorpions.

arboreal Living in trees.

Archaea One of the three domains of life; contains prokaryotic cells that often live in extreme habitats and have unique genetic, biochemical, and physiological characteristics; its members are sometimes referred to as archaea.

archaic human Regionally diverse descendants of *H. erectus* that lived in Africa, Asia, and Europe; considered by some to be a separate species.

archegonium Egg-producing structure, as in the moss life cycle.

Arctic tundra Biome that encircles the Earth just south of ice-covered polar seas in the Northern Hemisphere.

Ardi Fossilized remains of *Ardipithecus ramidus,* a hominin that lived 4.4 MYA.

ardipithecine One of several species of *Ardipithecus,* a genus that contains humanlike hominins and lived some 4–5 MYA.

arteriole (ar-teer-ee-ohl) Vessel that takes blood from an artery to capillaries.

artery Vessel that takes blood away from the heart to arterioles; characteristically possesses thick, elastic, muscular walls.

arthropod Member of a phylum of invertebrates that contains, among other groups, crustaceans and insects that have an exoskeleton and jointed appendages.

articular cartilage (ar-tik-yuh-lur) Hyaline cartilaginous covering over the articulating surface of the bones of synovial joints.

artificial selection Change in the genetic structure of populations due to selective breeding by humans.

ascus Fingerlike sac where ascospores are produced during sexual reproduction of sac fungi.

asexual reproduction Reproduction that requires only one parent and does not involve gametes.

associative learning Acquired ability to associate two stimuli or a stimulus and a response.

assortative mating Individuals tend to mate with those that have the same phenotype as themselves with respect to certain characteristics.

aster Short, radiating fibers about the centrioles at the poles of a spindle.

atom Smallest particle of an element that displays the properties of the element.

atomic mass Mass of an atom equal to the number of protons plus the number of neutrons within the nucleus.

atomic number Number of protons within the nucleus of an atom.

atomic symbol One or two letters that represent the name of an element— e.g., H stands for a hydrogen atom.

ATP synthase The enzyme that makes use of a hydrogen ion gradient and produces ATP during photosynthesis and cellular respiration.

atrial natriuretic hormone (ANH) (ay-tree-ul nay-tree-yoo-ret-ik) Hormone secreted by the heart that increases sodium excretion and, therefore, lowers blood volume and pressure.

atrioventricular valve Valve located between the atrium and the ventricle.

atrium (ay-tree-um) One of the upper chambers of the heart, either the left atrium or the right atrium, that receives blood.

australopithecine One of several species of *Australopithecus*, a genus that contains the first generally recognized hominids.

Australopithecus africanus Hominid that lived between 3.6 and 3 MYA; e.g., Lucy, discovered at Hadar, Ethiopia, in 1974.

autoimmune disorder Disorder that results when the immune system mistakenly attacks the body's own tissues.

autonomic system (aw-tuh-nahm-ik) Branch of the peripheral nervous system that has control over the internal organs; consists of the sympathetic and parasympathetic divisions.

autoploidy Polyploid organism that contains a duplicated genome of the same species.

autosome (aw-tuh-sohm) Any chromosome other than the sex chromosomes.

autotroph Organism that can capture energy and synthesize organic nutrients from inorganic nutrients.

auxins A group of plant hormones regulating growth, particularly cell elongation; most often indoleacetic acid (IAA).

avian influenza Bird flu caused by a virus that is able to spread from birds to humans.

axial skeleton (ak-see-ul) Portion of the skeleton that supports and protects the organs of the head, the neck, and the trunk.

axillary bud Bud located in the axil of a leaf.

axon (ak-sahn) Elongated portion of a neuron that conducts nerve impulses, typically from the cell body to the synapse.

B

Bacteria One of the three domains of life; contains prokaryotic cells that differ from archaea because they have their own unique genetic, biochemical, and physiological characteristics.

bacteriophage Virus that infects bacteria.

bark External part of a tree, containing cork, cork cambium, and phloem.

Barr body Dark-staining body (discovered by M. Barr) in the nuclei of female mammals that contains a condensed, inactive X chromosome.

basal body Cytoplasmic structure that is located at the base of and may organize cilia or flagella.

basal nuclei (bay-sul) Subcortical nuclei deep within the white matter that serve as relay stations for motor impulses and produce dopamine to help control skeletal muscle activities.

base Molecules tending to lower the hydrogen ion concentration in a solution and raise the pH numerically.

basement membrane Layer of nonliving material that anchors epithelial tissue to underlying connective tissue.

basidium Clublike structure in which nuclear fusion, meiosis, and basidiospore production occur during sexual reproduction of club fungi.

B cell Lymphocyte that matures in the bone marrow and, when stimulated by the presence of a specific antigen, gives rise to antibody-producing plasma cells.

B-cell receptor (BCR) Molecule on the surface of a B cell that binds to a specific antigen.

behavior Observable, coordinated responses to environmental stimuli.

beneficial nutrient In plants, element that is either required or enhances the growth and production of a plant.

bicarbonate ion Ion that participates in buffering the blood; the form in which carbon dioxide is transported in the bloodstream.

bilateral symmetry Body plan having two corresponding or complementary halves.

bile Secretion of the liver that is temporarily stored and concentrated in the gallbladder before being released into the small intestine, where it emulsifies fat.

binary fission Bacterial reproduction into two daughter cells without utilizing a mitotic spindle.

binomial nomenclature The scientific name of a species that consists of two parts, the genus name and the specific species name, for example, *Apis mellifera*.

biocultural evolution Phase of human evolution in which cultural events affect natural selection.

biodiversity Variety of life within an ecosystem, biome, or biosphere.

biodiversity hotspot Region of the world that contains unusually large concentrations of species.

biogeochemical cycle (by-oh-jee-oh-kem-ih-kul) Circulating pathway of elements such as carbon and nitrogen involving exchange pools, storage areas, and biotic communities.

biogeography Study of the geographical distribution of organisms.

biological clock Internal mechanism that maintains a biological rhythm in the absence of environmental stimuli.

biological species concept The concept that defines species as groups of populations that have the potential to interbreed and are reproductively isolated from other groups.

biome One of the biosphere's major terrestrial communities, characterized by certain climatic conditions and particular types of plants.

biomolecules Organic molecules common to organisms: carbohydrates, proteins, fats, and nucleic acids.

biosphere (by-oh-sfeer) Zone of air, land, and water at the surface of the Earth in which living organisms are found.

biotechnology Use of a natural biological system to produce a product or achieve an end desired by humans; may involve using recombinant DNA technology.

biotechnology product Product created by using biotechnology techniques.

biotic potential Maximum reproductive rate of an organism, given unlimited

resources and ideal environmental conditions.

bipedalism Walking erect on two feet.

bird Endothermic reptile that has feathers and wings, is often adapted for flight, and lays hard-shelled eggs.

bivalve Type of mollusc with a shell composed of two valves; includes clams, oysters, and scallops.

blade Broad, expanded portion of a plant leaf that may be single or compound.

blastocoel Fluid-filled cavity of a blastula.

blastocyst (blas-tuh-sist) Early stage of human embryonic development that consists of a hollow, fluid-filled ball of cells.

blastopore Opening into the primitive gut formed at gastrulation.

blastula Hollow, fluid-filled ball of cells occurring during animal development prior to gastrula formation.

blind spot Region of the retina lacking rods or cones where the optic nerve leaves the eye.

blood pressure Force of blood pushing against the inside wall of an artery.

bone Connective tissue having protein fibers and a hard matrix of inorganic salts, notably calcium salts.

bottleneck effect Cause of genetic drift; occurs when a majority of genotypes are prevented from participating in the production of the next generation as a result of a natural disaster or human interference.

brain stem Portion of the brain consisting of the medulla oblongata, pons, and midbrain.

bronchi (sing., bronchus) Two major divisions of the trachea leading to the lungs.

bronchioles (brahng-kee-ohlz) Smaller air passages in the lungs that begin at the bronchi and terminate in alveoli.

brown algae Marine photosynthetic protists with a notable abundance of xanthophyll pigments; this group includes well-known seaweeds of northern rocky shores.

bryophyte Member of one of three phyla of nonvascular plants—the mosses, liverworts, and hornworts.

buffer Substance or group of substances that tends to resist pH changes of a solution, thus stabilizing its relative acidity and basicity.

bulbourethral gland (bul-boh-yoo-ree-thrul) Either of two small structures located below the prostate gland in males; each adds secretions to semen.

bulk feeder Animal that eats relatively large pieces of food.

bulk transport Movement of elements in an organism in large amounts.

bundle sheath Sheath located around the veins of a leaf; formed from tightly packed cells.

C

C₃ photosynthesis Type of photosynthesis in which the first stable product following carbon dioxide fixation is a 3-carbon compound produced by the Calvin cycle.

C₄ photosynthesis Type of photosynthesis in which the first stable product is a 4-carbon molecule that releases carbon dioxide to the Calvin cycle.

calcitonin (kal-sih-toh-nin) Hormone secreted by the thyroid gland that increases the blood calcium level.

calorie (kcal) Amount of heat energy required to raise the temperature of 1 g of water 1°C.

Calvin cycle reactions In photosynthesis. the reactions that use ATP and NADPH from the light reactions to reduce CO₂ and produce an organic molecule.

Cambrian explosion Sudden appearance in the fossil record of most major groups of complex animals around 530 MYA.

CAM photosynthesis Crassulacean-acid metabolism; plant fixes carbon dioxide at night to produce a C₄ molecule that releases carbon dioxide to the Calvin cycle during the day.

capillary (kap-uh-lair-ee) Microscopic vessel connecting arterioles to venules; exchange of substances between blood and tissue fluid occurs across its thin walls.

capsid Protein coat or shell that surrounds a virion's nucleic acid.

capsule Gelatinous layer surrounding the cells of cyanobacteria and certain bacteria.

carbaminohemoglobin Hemoglobin carrying carbon dioxide.

carbohydrate Class of organic compounds that includes monosaccharides, disaccharides, and polysaccharides.

carbonic anhydrase (kar-bahn-ik an-hy-drays, -drayz) Enzyme in red blood cells that speeds the formation of carbonic acid from the reactants water and carbon dioxide.

carcinogenesis (kar-suh-nuh-jen-uh-sis) Development of cancer.

cardiac conduction system System of specialized cardiac muscle fibers that conduct impulses from the SA node to the chambers of the heart, causing them to contract.

cardiac cycle One complete cycle of systole and diastole for all heart chambers.

cardiac muscle Striated, involuntary muscle found only in the heart.

cardiovascular system Organ system in which blood vessels distribute blood under the pumping action of the heart.

carnivore (kar-nuh-vor) Consumer in a food chain that eats other animals.

carotenoid Yellow or orange pigment that serves as an accessory to chlorophyll in photosynthesis.

carotid body (kuh-raht-id) Structure located at the branching of the carotid arteries; contains chemoreceptors sensitive to the O₂, CO₂, and H⁺ content in blood.

carpel Ovule-bearing unit that is a part of a pistil.

carrier Heterozygous individual who has no apparent abnormality but can pass on an allele for a recessively inherited genetic disorder.

carrying capacity Maximum number of individuals of any species that can be supported by a particular ecosystem on a long-term basis.

cartilage Connective tissue in which the cells lie within lacunae separated by a flexible proteinaceous matrix.

Casparian strip Layer of impermeable lignin and suberin bordering four sides of root endodermal cells; prevents water and solute transport between adjacent cells.

catabolism Metabolic process that breaks down large molecules into smaller ones; catabolic metabolism.

cell Smallest unit that displays the properties of life; always contains cytoplasm surrounded by a plasma membrane.

cell body Portion of a neuron that contains a nucleus and from which dendrites and an axon extend.

cell cycle Repeating sequence of cellular events that consists of interphase, mitosis, and cytokinesis.

cell-mediated defense Specific mechanism of defense in which T cells destroy antigen-bearing cells.

cell plate Structure across a dividing plant cell that signals the location of new plasma membranes and cell walls.

cell theory One of the major theories of biology; states that all organisms are made up of cells and that cells come only from preexisting cells.

cellular respiration Metabolic reactions that use the energy primarily from carbohydrates but also from fatty acid or amino acid breakdown to produce ATP molecules.

cellular slime mold Free-living amoeboid cells that feed on bacteria and yeasts by phagocytosis and aggregate to form a plasmodium that produces spores.

cellulose (sel-yuh-lohs, -lohz) Polysaccharide that is the major complex carbohydrate in plant cell walls.

cell wall Structure that surrounds a plant, protistan, fungal, or bacterial cell and maintains the cell's shape and rigidity.

central nervous system (CNS) Portion of the nervous system consisting of the brain and spinal cord.

central vacuole In a plant cell, a large, fluid-filled sac that stores metabolites. During growth, it enlarges, forcing the primary cell wall to expand and the cell surface-area-to-volume ratio to increase.

centriole (sen-tree-ohl) Cellular structure, existing in pairs, that possibly organizes the mitotic spindle for chromosomal movement during mitosis and meiosis.

centromere (sen-truh-meer) Constriction where sister chromatids of a chromosome are held together.

centrosome Central microtubule organizing center of cells. In animal cells, it contains two centrioles.

cephalization Having a well-recognized anterior head with a brain and sensory receptors.

cephalopod Type of mollusc in which a modified foot develops into the head region; includes squid, cuttlefish, octopus, and nautilus.

cerebellum (ser-uh-bel-um) Part of the brain located posterior to the medulla oblongata and pons that coordinates skeletal muscles to produce smooth, graceful motions.

cerebral cortex (suh-ree-brul, ser-uh-brul kor-teks) Outer layer of cerebral hemispheres; receives sensory information and controls motor activities.

cerebral hemisphere One of the large, paired structures that together constitute the cerebrum of the brain.

cerebrospinal fluid Fluid found in the ventricles of the brain, in the central canal of the spinal cord, and in association with the meninges.

cerebrum (sair-uh-brum, suh-ree-brum) Main part of the brain consisting of two large masses, or cerebral hemispheres; the largest part of the brain in mammals.

cervix (sur-viks) Narrow end of the uterus, which projects into the vagina.

character Any structural, chromosomal, or molecular feature that distinguishes one group from another.

chemiosmosis Ability of certain membranes to use a hydrogen ion gradient to drive ATP formation.

chemoreceptor (kee-moh-rih-sep-tur) Sensory receptor that is sensitive to chemical stimuli—for example, receptors for taste and smell.

chemosynthetic Organism able to synthesize organic molecules by using carbon dioxide as the carbon source and the oxidation of an inorganic substance (such as hydrogen sulfide) as the energy source.

chitin Strong but flexible nitrogenous polysaccharide found in the exoskeleton of arthropods.

chlorophyll, chlorophyll *a*, chlorophyll *b* Green pigment that absorbs solar energy and is important in algal and plant photosynthesis; occurs as chlorophyll *a* and chlorophyll *b*.

chloroplast Membrane-enclosed organelle in algae and plants with chlorophyll-containing membranous thylakoids; where photosynthesis takes place.

cholesterol One of the major lipids found in animal plasma membranes; makes the membrane impermeable to many molecules.

chordate Member of the phylum Chordata, which includes lancelets, tunicates, fishes, amphibians, reptiles, birds, and mammals; characterized by a notochord, dorsal tubular nerve cord, pharyngeal gill pouches, and a postanal tail at some point in the life cycle.

chorion (kor-ee-ahn) Extraembryonic membrane that contributes to placenta formation.

chorionic villi (kor-ee-ahn-ik vil-eye) Treelike extensions of the chorion that project into the maternal tissues at the placenta.

chorionic villus sampling (CVS) Removal of cells from the chorionic villi portion of the placenta. Karyotyping is done to determine if the fetus has a chromosomal abnormality.

choroid (kor-oyd) Vascular, pigmented middle layer of the eyeball.

chromatin (kroh-muh-tin) Network of fine threads in the nucleus that are composed of DNA and proteins.

chromosomal mutation Any mutation that affects chromosome structure.

chromosome (kroh-muh-som) Chromatin condensed into a compact structure.

chyme Thick, semiliquid food material that passes from the stomach to the small intestine.

ciliary body (sil-ee-air-ee) Structure associated with the choroid layer that contains ciliary muscle and controls the shape of the lens of the eye.

ciliary muscle Within the ciliary body of the vertebrate eye, the ciliary muscle controls the shape of the lens.

ciliate Complex unicellular protist that moves by means of cilia and digests food in food vacuoles.

cilium (pl., cilia) (sil-ee-um) Short, hairlike projection from the plasma membrane, occurring usually in large numbers.

circadian rhythm Regular physiological or behavioral event that occurs on an approximately 24-hour cycle.

clade Taxon or other group consisting of an ancestral species and all of its descendants, forming a distinct branch on a phylogenetic tree.

cladistics Method of systematics that uses shared derived characters to place organisms in clades and construct cladograms.

cladogram In cladistics, a branching diagram that shows the relationship among species in regard to their shared derived characters.

classical conditioning Type of learning whereby an unconditioned stimulus that elicits a specific response is paired with a neutral stimulus so that the response becomes conditioned.

cleavage Cell division without cytoplasmic addition or enlargement; occurs during the first stage of animal development.

cleavage furrow In animal cells, an indentation that encircles the cell to divide the cytoplasm during cytokinesis.

climate Weather condition of an area, including especially prevailing temperature and average/yearly rainfall.

climax community In ecology, community that results when succession has come to an end.

clonal selection Concept that an antigen selects which lymphocyte will undergo clonal expansion and produce more lymphocytes bearing the same type of antigen receptor.

closed circulatory system Blood is confined to vessels and is kept separate from the interstitial fluid.

cnidarian Invertebrate in the phylum Cnidaria existing as either a polyp or a medusa with two tissue layers and radial symmetry.

cochlea (kohk-lee-uh, koh-klee-uh) Portion of the inner ear that resembles a snail's shell and contains the spiral organ, the sense organ for hearing.

codominance Inheritance pattern in which both alleles of a gene are equally expressed.

codon Three-base sequence in messenger RNA that causes the insertion of a particular amino acid into a protein or termination of translation.

coelom (see-lum) Embryonic body cavity lying between the digestive tract and body wall that becomes the thoracic and abdominal cavities.

coenzyme (koh-en-zym) Nonprotein organic molecule that aids the action of the enzyme to which it is loosely bound.

coevolution Interaction of two species such that each influences the evolution of the other.

cofactor Nonprotein adjunct required by an enzyme in order to function; many cofactors are metal ions, others are coenzymes.

cohesion Clinging together of water molecules.

cohesion-tension model Explanation for upward transport of water in xylem, based upon transpiration-created tension and the cohesive properties of water molecules.

cohort Group of individuals having a statistical factor in common, such as year of birth, in a population study.

coleoptile Protective sheath that covers the young leaves of a seedling.

collecting duct Duct within the kidney that receives fluid from several nephrons; the reabsorption of water occurs here.

colon (koh-lun) The major portion of the large intestine, consisting of the ascending colon, the transverse colon, and the descending colon.

colonial flagellate hypothesis The proposal first put forth by Haeckel that protozoans descended from colonial protists; supported by the similarity of sponges to flagellated protists.

columnar epithelium Type of epithelial tissue with cylindrical cells.

commensalism Symbiotic relationship in which one species is benefited, and the other is neither harmed nor benefited.

common ancestor Ancestor held in common by at least two lines of descent.

communication Signal by a sender that influences the behavior of a receiver.

community Assemblage of populations interacting with one another within the same environment.

compact bone Type of bone that contains osteons consisting of concentric layers of matrix and osteocytes in lacunae.

companion cell Cell associated with sieve-tube members in the phloem of vascular plants.

comparative genomics Study of genomes through a comparison of their coding and noncoding DNA sequences.

competitive exclusion principle Theory that no two species can occupy the same niche.

competitive inhibition Form of enzyme inhibition whereby the substrate and inhibitor are both able to bind to the enzyme's active site; each complexes with the enzyme. Only when the substrate is at the active site will product form.

complement Collective name for a series of enzymes and activators in the blood, some of which may bind to antibody and may lead to rupture of a foreign cell.

complementary base pairing Hydrogen bonding between particular bases. In

DNA, thymine (T) pairs with adenine (A), and guanine (G) pairs with cytosine (C); in RNA, uracil (U) pairs with A, and G pairs with C.

complex carbohydrates Mixture of carbohydrates that must be digested to release sugars; preferably also contain cellulose that acts as roughage in the diet.

compound Substance having two or more different elements, united chemically in fixed ratio.

compound light microscope Consists of a two-lens system, one above the other, to magnify an object.

concentration gradient Gradual difference in the distribution of a substance from a high amount to a low amount or vice versa.

conclusion Statement following an experiment as to whether the results support the hypothesis.

cone cells In vertebrates, a type of light-sensitive nerve cell in the retina noted for color and detail discrimination.

conifer Member of a group of cone-bearing gymnosperm plants that includes pine, cedar, and spruce trees.

conjugation Transfer of genetic materials from one cell to another.

conjunctiva Delicate membrane that lines the eyelid protecting the sclera.

connective tissue Type of tissue characterized by cells separated by a matrix that often contains fibers.

conservation biology Scientific discipline that seeks to understand the effects of human activities on species, communities, and ecosystems and to develop practical approaches to preventing the extinction of species and the destruction of ecosystems.

consumer Organism that feeds on another organism in a food chain; primary consumers eat plants, and secondary consumers eat animals.

control group Sample that goes through all the steps of an experiment but lacks the factor or is not exposed to the factor being tested; used as a standard against which experimental results are checked.

coral reef Area of biological abundance in warm, shallow tropical waters on and around coral formations.

cork cambium Lateral meristem that produces cork.

corm Underground, upright plant stem where food is stored, usually in the form of starch.

cornea (kor-nee-uh) Transparent, anterior portion of the outer layer of the eyeball.

corpus luteum (kor-pus loot-ee-um) Yellow body that forms in the ovary from a follicle that has discharged its secondary oocyte; secretes progesterone and some estrogen.

cortex In plants, ground tissue surrounded by the epidermis and vascular tissue in stems and roots; in animals, outer layer of an organ such as the cortex of the kidney or the adrenal gland.

cortisol (kor-tuh-sawl) Glucocorticoid secreted by the adrenal cortex that responds to stress on a long-term basis; reduces inflammation and promotes protein and fat metabolism.

cotyledon Seed leaf for the embryo of a flowering plant; provides nutrient molecules for the developing plant before photosynthesis begins.

countercurrent flow Fluids flow side-by-side in opposite directions, as in the exchange of fluids in the kidneys.

coupled reaction Reaction that occurs simultaneously; one is an exergonic reaction that releases energy, and the other is an endergonic reaction that requires energy in order to occur.

covalent bond (coh-vay-lent) Chemical bond in which atoms share one pair of electrons.

cranial nerve Nerve that arises from the brain.

cristae Short, fingerlike projections formed by the folding of the inner membrane of mitochondria.

Cro-Magnon Common name for the first fossils to be designated *Homo sapiens*.

crossing-over Exchange of segments between nonsister chromatids of a tetrad during meiosis.

crustacean Member of a group of marine arthropods that contains, among others, shrimps, crabs, crayfish, and lobsters.

cuboidal epithelium Type of epithelial tissue with cube-shaped cells.

Cushing syndrome (koosh-ing) Condition resulting from hypersecretion of glucocorticoids; characterized by thin arms and legs and a "moon face," and accompanied by high blood glucose and sodium levels.

cuticle In plants, a waxy, noncellular layer on the outer wall of epidermal cells.

cyanobacteria Photosynthetic bacteria that contain chlorophyll and release oxygen; formerly called blue-green algae.

cycad Type of gymnosperm with palmate leaves and massive cones; cycads are most often found in the tropics and subtropics.

cyclic adenosine monophosphate (cAMP) (sy-klik, sih-klik) ATP-related compound that acts as the second messenger in peptide hormone transduction; it initiates activity of the metabolic machinery.

cytochrome Any of several iron-containing protein molecules that serve as electron carriers in photosynthesis and cellular respiration.

cytokine (sy-tuh-kyn) Type of protein secreted by a T lymphocyte that stimulates cells of the immune system to perform their various functions.

cytokinesis (sy-tuh-kyn-ee-sus) Division of the cytoplasm following mitosis and meiosis.

cytokinin Plant hormone that promotes cell division; often works in combination with auxin during organ development in plant embryos.

cytoplasm (sy-tuh-plaz-um) Contents of a cell between the nucleus and the plasma membrane that contains the organelles.

cytoskeleton Internal framework of the cell, consisting of microtubules, actin filaments, and intermediate filaments.

cytotoxic T cell (sy-tuh-tahk-sik) T lymphocyte that attacks and kills antigen-bearing cells.

D

data Facts or pieces of information collected through observation and/or experimentation.

daughter cell Cell that arises from a parental cell by mitosis or meiosis.

day-neutral plant Plant whose flowering is not dependent on day length, e.g., tomato and cucumber.

deamination Removal of an amino group ($-NH_2$) from an amino acid or other organic compound.

deciduous Plant that sheds its leaves annually.

decomposer Organism, usually a bacterium or fungus, that breaks down organic matter into inorganic nutrients that can be recycled in the environment.

dehiscent Anther, fruit, or other plant structure that opens to permit the release of reproductive bodies inside.

dehydration reaction Chemical reaction resulting in a covalent bond and the loss of a water molecule.

delayed allergic response Allergic response initiated at the site of the allergen by sensitized T cells, involving macrophages and regulated by cytokines.

deletion Change in chromosome structure in which the end of a chromosome breaks off, or two simultaneous breaks lead to the loss of an internal segment; often causes abnormalities (e.g., cri du chat syndrome).

demographic transition Due to industrialization, a decline in the birthrate following a reduction in the death rate so that the population growth rate is lowered.

denatured (denaturation) Loss of an enzyme's normal shape so that it no longer functions; caused by a less than optimal pH or temperature.

dendrite (den-dryt) Branched ending of a neuron that conducts signals toward the cell body.

dendritic cell Antigen-presenting cell of the epidermis and mucous membranes.

denitrification Conversion of nitrate or nitrite to nitrogen gas by bacteria in soil.

dense fibrous connective tissue Type of connective tissue containing many collagen fibers packed together and found in tendons and ligaments, for example.

density-dependent factor Biotic factor, such as disease or competition, that affects population size according to the population's density.

density-independent factor Abiotic factor, such as fire or flood, that affects population size independent of the population's density.

deoxyribose Pentose sugar found in DNA that has one less hydroxyl group than ribose.

depolarization Loss in polarization, as when a nerve impulse occurs.

derived character Structural, physiological, or behavioral trait that is present in a specific lineage and is not present in the common ancestor for several lineages.

dermis Region of skin that lies beneath the epidermis.

desert Ecological biome characterized by a limited amount of rainfall; deserts have hot days and cool nights.

detrital food web (dih-tryt-ul) Complex pattern of interlocking and crisscrossing food chains that begins with detritus.

detritus Partially decomposed organic matter derived from tissues and animal wastes.

deuterostome A coelomate animal in which the second embryonic opening is associated with the mouth; the first embryonic opening, the blastopore, is associated with the anus.

diaphragm (dy-uh-fram) Dome-shaped horizontal sheet of muscle and connective tissue that divides the thoracic cavity from the abdominal cavity.

diastole (dy-as-tuh-lee) Relaxation period of a heart chamber during the cardiac cycle.

diastolic pressure (dy-uh-stahl-ik) Arterial blood pressure during the diastolic phase of the cardiac cycle.

diatom Golden-brown alga with a cell wall having two parts, or valves; significant part of phytoplankton.

diencephalon (dy-en-sef-uh-lahn) Portion of the brain in the region of the third ventricle that includes the thalamus and hypothalamus.

differentially permeable Ability of plasma membranes to regulate the passage of substances into and out of the cell,

allowing some to pass through and preventing the passage of others.

diffusion The net movement of dissolved molecules or other particles from a region of higher concentration to a region of lower concentration.

digestive system Organ system that includes the mouth, esophagus, stomach, small intestine, and large intestine (colon), which receives food and digests it into nutrient molecules. Also has associated organs: teeth, tongue, salivary glands, liver, gallbladder, and pancreas.

dihybrid cross Single genetic cross involving two different traits, such as flower color and plant height.

dinoflagellate Photosynthetic unicellular protist with two flagella, one whiplash and the other located within a groove between protective cellulose plates; a significant part of phytoplankton.

diploid (2n) Cell condition in which two of each type of chromosome are present in the nucleus.

diploid life cycle Life cycle, typical of animals, in which meiosis occurs only during formation of gametes so that the individual is always diploid.

directional selection Outcome of natural selection in which an extreme phenotype is favored, usually in a changing environment.

disaccharide (dy-sak-uh-ryd) Sugar that contains two units of a monosaccharide (e.g., maltose).

disruptive selection Outcome of natural selection in which the two extreme phenotypes are favored over the average phenotype, leading to more than one distinct form.

DNA (deoxyribonucleic acid) Nucleic acid polymer produced from covalent bonding of nucleotide subunits that contain the sugar deoxyribose; the genetic material of nearly all organisms.

DNA fingerprinting The use differences in the number of short tandem repeats units in DNA to identify particular individuals.

DNA ligase (ly-gays) Enzyme that links DNA fragments; used during production of recombinant DNA to join foreign DNA to vector DNA.

DNA microarrays Thousands of different DNA fragments (probes) arranged in an array (grid); used to detect and measure gene expression.

DNA polymerase During replication, an enzyme that joins the nucleotides complementary to a DNA template.

DNA replication Synthesis of a new DNA double helix prior to mitosis and meiosis in eukaryotic cells and during prokaryotic fission in prokaryotic cells.

domain The primary taxonomic group above the kingdom level; all living organisms may be placed in one of three domains.

dominant allele (uh-leel) Allele that exerts its phenotypic effect in the heterozygote; it masks the expression of the recessive allele.

dormancy In plants, a cessation of growth under conditions that seem appropriate for growth.

dorsal root ganglion (gang-glee-un) Mass of sensory neuron cell bodies located in the dorsal root of a spinal nerve.

double fertilization In flowering plants, one sperm joins with polar nuclei within the embryo sac to produce a 3n endosperm nucleus, and another sperm joins with an egg to produce a zygote.

double helix Double spiral; describes the three-dimensional shape of DNA.

duodenum (doo-uh-dee-num) First part of the small intestine where chyme enters from the stomach.

duplication Change in chromosome structure in which a particular segment is present more than once in the same chromosome.

E

ecdysozoan Protostome characterized by periodic molting of the exoskeleton. Includes the roundworms and the arthropods.

echinoderm Group of marine animals that includes sea stars, sea urchins, and sand dollars; characterized by radial symmetry and a water vascular system.

ecological niche Role an organism plays in its community, including its habitat and its interactions with other organisms.

ecological pyramid Pictorial graph based on the biomass, number of organisms, or energy content of various trophic levels in a food web—from the producer to the final consumer populations.

ecological succession The slow orderly progression of changes in community composition that takes place through time.

ecology Study of the interactions of organisms with other organisms and with the physical and chemical environment.

ecosystem (ek-oh-sis-tum, ee-koh-) Biological community together with the associated abiotic environment; characterized by energy flow and chemical cycling.

ecosystem diversity Variety of species in a particular locale, dependent on the species interactions.

ectoderm Outermost primary tissue layer of an animal embryo; gives rise to the nervous system and the outer layer of the integument.

ectotherm Organism having a body temperature that varies according to the environmental temperature.

electrocardiogram (ECG) (ih-lek-troh-kar-dee-uh-gram) Recording of the electrical activity associated with the heartbeat.

electromagnetic receptor Sensory receptor that detects energy of different wavelengths, such as electricity, magnetism, and light.

electron Negative subatomic particle, moving in an energy level around the nucleus of an atom.

electronegativity Ability of an atom to attract electrons toward itself in a chemical bond.

electron shell Concentric energy levels in which electrons orbit.

electron transport chain (ETC) Passage of electrons along a series of membrane-associated carrier molecules from a higher to lower energy level; the energy released is used for the synthesis of ATP.

element Substance that cannot be broken down into substances with different properties; composed of only one type of atom.

El Niño–Southern Oscillation Warming of water in the Eastern Pacific equatorial region such that the Humboldt Current is displaced with possible negative results, including reduction in marine life.

embryo A multicellular developmental stage that follows cell division of the zygote.

embryonic disk Stage of embryonic development following the blastocyst stage that has two layers; one layer will be endoderm, and the other will be ectoderm.

embryo sac Female gametophyte of flowering plants that produces an egg cell.

emphysema (em-fih-see-muh) Degenerative lung disorder in which the bursting of alveolar walls reduces the total surface area for gas exchange.

endangered species A species that is in peril of immediate extinction throughout all or most of its range (e.g., California condor, snow leopard).

endergonic reaction Chemical reaction that requires an input of energy; opposite of exergonic reaction.

endocrine system Organ system involved in the coordination of body activities; uses hormones as chemical signals secreted into the bloodstream.

endocytosis Process by which substances are moved into the cell from the environment by phagocytosis (cellular eating) or pinocytosis (cellular drinking); includes receptor-mediated endocytosis.

endoderm Innermost primary tissue layer of an animal embryo that gives rise to the linings of the digestive tract and associated structures.

endodermis Plant root tissue that forms a boundary between the cortex and the vascular cylinder.

endomembrane system A collection of membranous structures involved in transport within the cell.

endometrium Mucous membrane lining the interior surface of the uterus.

endoplasmic reticulum (ER) (en-duh-plaz-mik reh-tik-yuh-lum) System of membranous saccules and channels in the cytoplasm, often with attached ribosomes.

endoskeleton Protective internal skeleton, as in vertebrates.

endosperm In angiosperms, the 3n tissue that nourishes the embryo and seedling and is formed as a result of a sperm joining with two polar nuclei.

endospore Spore formed within a cell; certain bacteria form endospores.

endosymbiotic theory Possible explanation of the evolution of eukaryotic organelles by phagocytosis of prokaryotes.

endotherm Animal that maintains a constant body temperature independent of the environmental temperature.

energy Capacity to do work and bring about change; occurs in a variety of forms.

energy of activation Energy that must be added in order for molecules to react with one another.

enhancer DNA sequence that increases the level of transcription when a transcription activator binds to it.

entropy Measure of disorder or randomness.

enzyme (en-zym) Organic catalyst, usually a protein, that speeds a reaction in cells due to its particular shape.

eosinophil (ee-oh-sin-oh-fill) White blood cell containing cytoplasmic granules that stain with acidic dye.

epidermal tissue Exterior tissue, usually one cell thick, of leaves, young stems, roots, and other parts of plants.

epidermis In plants, tissue that covers roots, leaves, and stems of a nonwoody organism; in animals, the outer protective region of the skin.

epididymis (ep-uh-did-uh-mus) Coiled tubule next to the testes where sperm mature and may be stored for a short time.

epiglottis (ep-uh-glaht-us) Structure that covers the glottis during the process of swallowing.

epinephrine (ep-uh-nef-rin) Hormone secreted by the adrenal medulla in times of stress; adrenaline.

epiphyte Plant that takes its nourishment from the air because its placement among other plants gives it an aerial position.

episiotomy (ih-pee-zee-aht-uh-mee) Surgical procedure performed during childbirth in which the opening of the vagina is enlarged to avoid tearing.

epithelial tissue Type of tissue that lines hollow organs and covers surfaces; epithelium.

equilibrium population Population whose members exhibit logistic population growth and whose size remains at or near the carrying capacity. Its members are large in size, slow to mature, have a long life span, have few offspring, and provide much care to offspring (e.g., bears, lions).

erection Increase in blood flow to the penis during sexual arousal, causing the penis to stiffen and become erect.

E site In a ribosome, the place where a spent tRNA exits the ribosome.

esophagus (ih-sahf-uh-gus) Muscular tube for moving swallowed food from the pharynx to the stomach.

essential nutrient In plants, substance required for normal growth, development, or reproduction.

estrogen (es-truh-jun) Female sex hormone that helps maintain sex organs and secondary sex characteristics.

estuary Portion of the ocean located where a river enters and fresh water mixes with salt water.

euchromatin Chromatin that is extended and accessible for transcription.

eudicot Abbreviation of eudicotyledon. Flowering plant group; members have two embryonic leaves (cotyledons), net-veined leaves, vascular bundles in a ring, flower parts in fours or fives and their multiples, and other characteristics.

euglenoid Flagellated and flexible freshwater unicellular protist that usually contains chloroplasts and has a semirigid cell wall.

Eukarya One of the three domains of life, consisting of organisms with eukaryotic cells and further classified into the kingdoms Protista, Fungi, Plantae, and Animalia.

eukaryotic cell (eukaryote) Type of cell that has a membrane-enclosed nucleus and membranous organelles.

eutrophication Enrichment of water by inorganic nutrients used by phytoplankton. Often, overenrichment caused by human activities leads to excessive bacterial growth and oxygen depletion.

eutrophic lake Lake containing many nutrients and decaying organisms, often tinted green with algae.

evolution Descent of organisms from common ancestors with the development of genetic and phenotypic changes over time that make them more suited to the environment.

evolutionary species concept Every species has its own evolutionary history, which is partly documented in the fossil record.

evolutionary (phylogenetic) tree Diagram that shows how groups of organisms are related by way of common ancestors.

excretion Elimination of metabolic wastes by an organism at exchange boundaries such as the plasma membrane of unicellular organisms and excretory tubules of multicellular animals.

exergonic reaction Chemical reaction that releases energy; opposite of endergonic reaction.

exocytosis Process in which an intracellular vesicle fuses with the plasma membrane so that the vesicle's contents are released outside the cell.

exon A segment of a gene that codes for a protein.

exoskeleton Protective external skeleton, as in arthropods.

experiment Test of an experimental variable for the purpose of collecting data.

experimental variable A value that is expected to change as a result of an experiment; represents the factor being tested by the experiment.

expiration (ek-spuh-ray-shun) Act of expelling air from the lungs; also called exhalation.

exponential growth Growth at a constant rate of increase per unit of time; can be expressed as a constant fraction or exponent.

external respiration Exchange of oxygen and carbon dioxide between alveoli and blood.

extinction Total disappearance of a species or higher group.

extracellular matrix (ECM) Materials secreted by animal cells that form a complex network which supports the cells and allows them to communicate.

extraembryonic membrane (ek-struh-em-bree-ahn-ik) Membrane that is not a part of the embryo but is necessary to the continued existence and health of the embryo.

F

F$_1$ generation In genetics, the first (filial) generation of offspring.

F$_2$ generation In genetics, the second (filial) generation of offspring.

facilitated diffusion Passive transfer of a substance into or out of a cell along a concentration gradient by a process that requires a carrier.

facultative anaerobe Prokaryote that is able to grow in either the presence or the absence of gaseous oxygen.

FAD Flavin adenine dinucleotide; a coenzyme of oxidation-reduction that becomes FADH$_2$ as oxidation of substrates occurs, and then delivers electrons to the electron transport chain in mitochondria during cellular respiration.

fat Organic molecule that contains glycerol and fatty acids and is found in adipose tissue.

fatty acid Molecule that contains a hydrocarbon chain and ends with an acid group.

female gametophyte In seed plants, the gametophyte that produces an egg; in flowering plants, an embryo sac.

fermentation Anaerobic breakdown of glucose that results in a gain of two ATP and end products such as alcohol and lactate.

fern Member of a group of plants that have large fronds; in the sexual life cycle, the independent gametophyte produces flagellated sperm, and the vascular sporophyte produces windblown spores.

fertilization Union of a sperm nucleus and an egg nucleus, which creates a zygote.

fetal development Period of development from the ninth week through birth.

fiber Structure resembling a thread; also, plant material that is undigestible.

fibroblast Cell in connective tissues that produces fibers and other substances.

filament End-to-end chains of cells that form as cell division occurs in only one plane; in plants, the elongated stalk of a stamen.

filter feeder Method of obtaining nourishment by certain animals that strain minute organic particles from the water in a way that deposits them in the digestive tract.

fimbria (pl., fimbriae) (fim-bree-uh) Finger-like extension from the oviduct near the ovary.

first messenger Chemical signal such as a peptide hormone that binds to a plasma membrane receptor protein and alters a cell's metabolism because a second messenger is activated.

fitness Ability of an organism to reproduce and pass its genes to the next fertile generation; measured against the ability of other organisms to reproduce in the same environment.

fixed action pattern Innate behavior pattern that is stereotyped, spontaneous, independent of immediate control, genetically encoded, and independent of individual learning.

flagellum (pl., flagella) (fluh-jel-um) Slender long extension that propels a cell through a fluid medium.

flame cell Found along excretory tubules of planarians; functions in propelling fluid through the excretory canals and out of the body.

flatworm Unsegmented worm lacking a body cavity; phylum Platyhelminthes.

fluid feeder Animal that gains needed nutrients by sucking nutrient-rich fluids from another living organism.

fluid-mosaic model Model for the plasma membrane based on the changing location and pattern of protein molecules in a fluid phospholipid bilayer.

follicle (fahl-ih-kul) Structure in the ovary that produces a secondary oocyte and the hormones estrogen and progesterone.

follicle-stimulating hormone (FSH) Hormone secreted by the anterior pituitary gland that stimulates the development of an ovarian follicle in a female or the production of sperm in a male.

follicular phase First half of the ovarian cycle, during which the follicle matures and much estrogen (and some progesterone) is produced.

fontanel (fahn-tun-el) Membranous region located between certain cranial bones in the skull of a fetus or infant.

food chain The order in which one population feeds on another in an ecosystem, from detritus (detrital food chain) or producer (grazing food chain) to final consumer.

food web In ecosystems, complex pattern of interlocking and crisscrossing food chains.

foramen magnum (fuh-ray-mun mag-num) Opening in the occipital bone of the vertebrate skull through which the spinal cord passes.

foraminiferan Member of the phylum Foraminifera bearing a calcium carbonate test with many openings through which pseudopods extend.

foreign antigen Organism does not produce this type of antigen.

foreskin Skin covering the glans penis in uncircumcised males.

formed element Constituent of blood that is either cellular (red blood cells and white blood cells) or at least cellular in origin (platelets).

fossil Evidence of usually an extinct species that has been preserved in the Earth's crust.

fossil record History of life recorded from remains from the past.

founder effect Cause of genetic drift due to colonization by a limited number of individuals who, by chance, have different gene frequencies than the parent population.

fovea centralis Region of the retina consisting of densely packed cones; responsible for the greatest visual acuity.

frameshift mutation Alteration in a gene due to deletion of a base, so that the reading "frame" is shifted; can result in a nonfunctional protein.

frond Leaf of a fern.

fruit Flowering plant structure consisting of one or more ripened ovaries that usually contain seeds.

functional genomics Study of DNA function at the genomic level; involves the study of many genes simultaneously and the use of microarrays.

functional group Specific cluster of atoms attached to the carbon skeleton of organic molecules that enters into reactions and behaves in a predictable way.

fungus (pl., fungi) Saprotrophic decomposer; the body is made up of filaments called hyphae that form a mass called a mycelium.

G

G_0 stage In the cell cycle, a period of time that occurs should a cell leave the cycle during the G_1 stage before committing to the complete cycle.

G_1 stage In the cell cycle, a period of time during which the cell grows in size; includes the G_1 checkpoint when cells commit to completing the cycle.

G_2 stage In the cell cycle, a period of time after the S stage and before the M stage during which growth includes organelle duplication.

G3P In photosynthesis, a Krebs cycle molecule that is the starting point for many types of organic molecules produced by plants, including glucose and starch. In cellular respiration, the molecule that occurs after glucose is split during glycolysis.

gallbladder Organ attached to the liver that stores and concentrates bile.

gamete (ga-meet, guh-meet) Haploid sex cell; the egg or a sperm, which join in fertilization to form a zygote.

gametophyte Haploid generation of the alternation-of-generations life cycle of a plant; produces gametes that unite to form a diploid zygote.

ganglion Collection or bundle of neuron cell bodies usually outside the central nervous system.

gap junction Junction between animal cells that provides a passageway for intercellular transport.

gastropod Mollusc with a broad flat foot for crawling (e.g., snails and slugs).

gastrovascular cavity In animals with an incomplete digestive tract, a cavity that serves for digestion of food and transport of oxygen and nutrients to body cells.

gastrula Stage of animal development in which germ layers have formed, at least in part, by invagination.

gene Unit of heredity existing as alleles on the chromosomes; in diploid organisms, typically two alleles are inherited—one from each parent.

gene cloning Production of one or more copies of the same gene.

gene flow Sharing of genes between two populations through interbreeding.

gene locus Specific location of a particular gene on homologues.

gene pharming Use of a genetically modified organism to produce a commercial medical product.

gene pool Total of all the genes of all the individuals in a population.

gene theory Concept that organisms contain coded information dictating their form, function, and behavior.

gene therapy Correction of a detrimental mutation by the addition of normal DNA and its insertion into a genome.

genetically engineered Alteration of genomes for medical or industrial purposes.

genetically modified organism (GMO) Organism that carries the genes of another organism as a result of DNA technology.

genetic code Universal code that specifies protein synthesis in the cells of all organisms. Each codon consists of three letters standing for the DNA nucleotides that make up one of the 20 amino acids found in proteins.

genetic drift Mechanism of evolution due to random changes in the allelic frequencies of a population; more likely to occur in small populations or when only a few individuals of a large population reproduce.

genetic mutation Alteration in chromosome structure or number and also an alteration in a gene due to a change in DNA composition.

genetic profile Gene expression in an individual or a cell as detected by the use of a microarray.

genome Full set of genetic information for a species or a virus.

genotype (jee-nuh-typ) Genes of an individual for a particular trait or traits; often designated by letters, for example, *BB* or *Aa*.

genus One of the categories, or taxa, used by taxonomists to group species; contains those species that are most closely related through evolution.

geologic timescale History of the Earth since the beginning of time divided into eras, periods, and epochs based in part on the fossil record.

germinate Beginning of growth of a seed, spore, or zygote, especially after a period of dormancy.

germ layer Primary tissue layer of a vertebrate embryo—namely, ectoderm, mesoderm, or endoderm.

gibberellin Plant hormone promoting increased stem growth; also involved in flowering and seed germination.

gills Respiratory organ in most aquatic animals; in fish, an outward extension of the pharynx.

gland Epithelial cell or group of epithelial cells that are specialized to secrete a substance.

glomerular capsule (gluh-mair-yuh-lur) Double-walled cup that surrounds the glomerulus at the beginning of the nephron.

glomerular filtration Movement of small molecules from the glomerulus into the glomerular capsule due to the action of blood pressure.

glomerulus (gluh-mair-uh-lus, gloh-mair-yuh-lus) Cluster; for example, the cluster of capillaries surrounded by the glomerular capsule in a nephron, where glomerular filtration takes place.

glottis (glaht-us) Opening for airflow in the larynx.

glucagon (gloo-kuh-gahn) Hormone secreted by the pancreas that causes the liver to break down glycogen and raises the blood glucose level.

glucocorticoid (gloo-koh-kor-tih-koyd) Type of hormone secreted by the adrenal cortex that influences carbohydrate, fat, and protein metabolism; see cortisol.

glucose (gloo-kohs) Six-carbon sugar that organisms degrade as a source of energy during cellular respiration.

glycerol Three-carbon carbohydrate with three hydroxyl groups attached; a component of fats and oils.

glycogen (gly-koh-jun) Storage polysaccharide that is composed of glucose molecules joined in a linear fashion but having numerous branches.

glycolipid Lipid in plasma membranes that bears a carbohydrate chain attached to a hydrophobic tail.

glycolysis Anaerobic breakdown of glucose that results in a gain of two ATP.

glycoprotein Protein in plasma membranes that bears a carbohydrate chain.

Golgi apparatus (gohl-jee) Organelle, consisting of saccules and vesicles, that processes, packages, and distributes molecules about or from the cell.

gonad (goh-nad) Organ that produces gametes; the ovary produces eggs, and the testis produces sperm.

gonadotropic hormone (goh-nad-uh-trahp-ic, -troh-pic) Chemical signal secreted by the anterior pituitary that regulates the activity of the ovaries and testes; principally, follicle-stimulating hormone (FSH) and luteinizing hormone (LH).

granum (pl., grana) Stack of chlorophyll-containing thylakoids in a chloroplast.

gravitational balance Maintenance of balance when the head and body are motionless.

gravitropism Growth response of roots and stems of plants to the Earth's gravity; roots demonstrate positive gravitropism, and stems demonstrate negative gravitropism.

gray crescent Gray area that appears in an amphibian egg after being fertilized by the sperm; thought to contain chemical signals that turn on the genes that control development.

gray matter Nonmyelinated axons and cell bodies in the central nervous system.

grazing food web Complex pattern of interlocking and crisscrossing food chains that begins with populations of autotrophs serving as producers.

green algae Members of a diverse group of photosynthetic protists; contain chlorophylls *a* and *b* and have other biochemical characteristics like those of plants.

greenhouse gases Gases involved in the reradiation of solar heat toward the Earth, sometimes called the greenhouse effect.

ground tissue Tissue that constitutes most of the body of a plant; consists of parenchyma, collenchyma, and sclerenchyma cells that function in storage, basic metabolism, and support.

growth hormone (GH) Substance secreted by the anterior pituitary; controls the size of an individual by promoting cell division, protein synthesis, and bone growth.

guard cell One of two cells that surround a leaf stoma; changes in the turgor pressure of these cells cause the stoma to open or close.

gymnosperm Type of woody seed plant in which the seeds are not enclosed by fruit and are usually borne in cones, such as those of the conifers.

H

H1N1 Emerging virus that causes a flu commonly called swine flu.

habitat Place where an organism lives and is able to survive and reproduce.

habituation Simplest form of learning, in which an animal learns not to respond to irrelevant stimuli.

halophile Type of archaean that lives in extremely salty habitats.

haploid (n) (hap-loyd) The n number of chromosomesomes—half the diploid number; the number characteristic of gametes, which contain only one set of chromosomes.

haploid life cycle Life cycle, typical of protists and fungi, in which meiosis occurs immediately after zygote formation so that the individual is always haploid.

Hardy-Weinberg principle Law stating that the gene frequencies in a population remain stable if evolution does not occur due to nonrandom mating, selection, migration, and genetic drift.

heat Type of kinetic energy; captured solar energy eventually dissipates as heat in the environment.

helper T cell T lymphocyte that secretes cytokines that stimulate all kinds of immune system cells.

heme Iron-containing portion of a hemoglobin molecule.

hemocoel Residual coelom found in molluscs and arthropods that is filled with hemolymph.

hemodialysis (he-moh-dy-al-uh-sus) Cleansing of blood by using an artificial membrane that causes substances to diffuse from blood into a dialysis fluid.

hemoglobin (Hb) (hee-muh-gloh-bun) Iron-containing pigment in red blood cells that combines with and transports oxygen.

hemolymph Circulatory fluid that is a mixture of blood and tissue fluid; seen in animals that have an open circulatory system, such as molluscs and arthropods.

hemophilia Most common of the severe clotting disorders, caused by the absence of a blood clotting factor.

hepatic portal system Pathway of blood flow between intestinal capillaries and liver capillaries.

herbaceous Nonwoody stem.

herbivore (ur-buh-vor) Primary consumer in a grazing food chain; a plant eater.

hermaphrodite Animal having both male and female sex organs.

herpes simplex virus (HSV) Virus that causes genital herpes, a sexually transmitted disease.

heterochromatin Highly compacted chromatin that is not accessible for transcription.

heterocyst Cyanobacterial cell that synthesizes a nitrogen-fixing enzyme when nitrogen supplies dwindle.

heterotroph Organism that cannot synthesize organic molecules from inorganic nutrients and therefore must take in organic nutrients (food).

heterozygote advantage Situation in which individuals heterozygous for a trait have a selective advantage over those who are homozygous; an example is sickle-cell disease.

heterozygous Possessing unlike alleles for a particular trait.

hinge joint Type of joint that allows movement as a hinge does, such as the movement of the knee.

hippocampus (hip-uh-kam-pus) Portion of the limbic system where memories are stored.

histamine (his-tuh-meen, -mun) Substance, produced by basophils in blood and mast cells in connective tissue, that causes capillaries to dilate.

histone Protein molecule responsible for packing chromatin.

HIV provirus Viral DNA that has been integrated into host cell DNA.

homeobox 180-nucleotide sequence located in all homeotic genes.

homeostasis (hoh-mee-oh-stay-sis) Maintenance of normal internal conditions in a cell or an organism by means of self-regulating mechanisms.

Hox **gene** Gene that controls the overall body plan by controlling the fate of groups of cells during development.

hominin Classification category that includes chimpanzees, humans, and species very closely related to humans.

homologous chromosome (hoh-mahl-uh-gus, huh-mahl-uh-gus) See homologue.

homologous structure Structure that is similar in two or more species because of common ancestry.

homologue One of a pair of chromosomes of the same kind located in a diploid cell; one copy of each pair of homologues comes from each gamete that formed the zygote; also called homologous chromosome.

homozygous Possessing two identical alleles for a particular trait.

hormone Chemical signal produced in one part of an organism that controls the activity of other parts.

human chorionic gonadotropin (HCG) (kor-ee-ahn-ik goh-nad-uh-trahp-in, -troh-pin) Hormone produced by the chorion that functions to maintain the uterine lining.

Human Genome Project (HGP) Initiative to determine the complete sequence of the human genome and to analyze this information.

human papillomavirus (HPV) Virus that causes genital warts, a common sexually transmitted disease; also linked to cervical cancer.

humus Decomposing organic matter in the soil.

Huntington disease Genetic disease marked by progressive deterioration of the nervous system due to deficiency of a neurotransmitter.

hyaline cartilage Cartilage whose cells lie in lacunae separated by a white translucent matrix containing very fine collagen fibers.

hydrogen bond Weak bond that arises between a slightly positive hydrogen atom of one molecule and a slightly negative atom of another or between parts of the same molecule.

hydrogen ion (H⁺) Hydrogen atom that has lost its electron and therefore bears a positive charge also called a proton.

hydrolysis reaction (hy-drahl-ih-sis re-ak-shun) Splitting of a compound by the addition of water, with the H^+ being incorporated in one fragment and the OH^- in the other.

hydrophilic (hy-druh-fil-ik) Type of molecule that interacts with water by dissolving in water and/or forming hydrogen bonds with water molecules.

hydrophobic (hy-druh-foh-bik) Type of molecule that does not interact with water because it is nonpolar.

hydroponics Technique for growing plants by suspending them with their roots in a nutrient solution.

hydrostatic skeleton Fluid-filled body compartment that provides support for muscle contraction resulting in movement; seen in cnidarians, flatworms, roundworms, and segmented worms.

hydrothermal vent Hot springs in the seafloor along ocean ridges where heated seawater and sulfate react to produce hydrogen sulfide; here, chromosynthetic bacteria support a community of varied organisms.

hydroxide ion (OH⁻) One of two ions that results when a water molecule dissociates; it has gained an electron, and therefore bears a negative charge.

hypersensitive response (HR) Plants respond to pathogens by selectively killing plant cells to block the spread of the pathogen.

hypertension Elevated blood pressure, particularly the diastolic pressure.

hypertonic solution Higher solute concentration (less water) than the cytoplasm of a cell; causes cell to lose water by osmosis.

hypha (pl., hyphae) Filament of the vegetative body of a fungus.

hypothalamic-releasing hormone One of several hormones produced by the hypothalamus that stimulates the secretion of an anterior pituitary hormone.

hypothalamus (hy-poh-thal-uh-mus) Part of the brain located below the thalamus that helps regulate the internal environment of the body and produces releasing factors that control the anterior pituitary.

hypothesis (hy-pahth-ih-sis) Supposition that is formulated after making an observation; it can be tested by obtaining more data, often by experimentation.

hypotonic solution Lower solute (more water) concentration than the cytoplasm of a cell; causes cell to gain water by osmosis.

I

immediate allergic response Allergic response that occurs within seconds of contact with an allergen; caused by the attachment of the allergen to IgE antibodies.

immune system All the cells in the body that protect the body against foreign organisms and substances and also cancerous cells.

immunity Ability of the body to protect itself from foreign substances and cells, including disease-causing agents.

immunization (im-yuh-nuh-zay-shun) Use of a vaccine to protect the body against specific disease-causing agents.

immunoglobulin (Ig) (im-yuh-noh-glahb-yuh-lin, -yoo-lin) Globular plasma protein that functions as an antibody.

implantation Attachment and penetration of the embryo into the lining of the uterus (endometrium).

imprinting Learning to make a particular response to only one type of animal or object.

inclusive fitness Fitness that results from personal reproduction and from helping nondescendant relatives reproduce.

incomplete dominance Inheritance pattern in which the offspring has an intermediate phenotype, as when a red-flowered plant and a white-flowered plant produce pink-flowered offspring.

indehiscent Remaining closed at maturity, as are many fruits.

independent assortment Homologues segregate independently of each other during meiosis so that the gametes contain all possible combinations of chromosomes.

induced fit model Change in the shape of an enzyme's active site that enhances the fit between the active site and its substrate(s).

induction Ability of a chemical or a tissue to influence the development of another tissue.

industrial melanism Increased frequency of a darkly pigmented (melanic) form in a population when predators more easily

see and capture the lightly pigmented form because it is more visible against vegetation that has turned dark due to industrial pollution.

inflammatory response Tissue response to injury that is characterized by redness, swelling, pain, and heat.

initiation During transcription, initiation is the step whereby transcription begins with the start codon AUG.

inner ear Portion of the ear consisting of a vestibule, semicircular canals, and the cochlea where equilibrium is maintained and sound is transmitted.

insect Member of a group of arthropods in which the head has antennae, compound eyes, and simple eyes; the thorax has three pairs of legs and often wings; and the abdomen has internal organs.

insight learning Ability to apply prior learning to a new situation without trial-and-error activity.

inspiration (in-spuh-ray-shun) Act of taking air into the lungs; also called inhalation.

insulin (in-suh-lin) Hormone secreted by the pancreas that lowers the blood glucose level by promoting the uptake of glucose by cells and the conversion of glucose to glycogen by the liver and skeletal muscles.

integration Summing up of excitatory and inhibitory signals by a neuron or by some part of the brain.

integumentary system Organ system consisting of skin and various organs, such as hair, that are found in skin.

intercalated disks Region that holds adjacent cardiac muscle fibers together and appears as dense bands at right angles to the muscle striations.

interferon (in-tur-feer-ahn) Antiviral agent produced by an infected cell that blocks the infection of another cell.

interkinesis Period of time between meiosis I and meiosis II during which no DNA replication takes place.

intermediate filaments Ropelike assemblies of fibrous polypeptides in the cytoskeleton that provide support and strength to cells; so called because they are intermediate in size between actin filaments and microtubules.

intermembrane space Space between the inner and outer membranes of a mitochondrion where hydrogen ions collect prior to passing through an ATP synthase complex.

internal respiration Exchange of oxygen and carbon dioxide between blood and tissue fluid.

interneuron Neuron located within the central nervous system that conveys messages between parts of the central nervous system.

internode In vascular plants, the region of a stem between two successive nodes.

interphase Portion of the cell cycle that includes the G_1, S, and G_2 stages but not the mitotic stage.

intron Segment of a gene that does not code for a protein.

inversion Change in chromosome structure in which a segment of a chromosome is turned around 180 degrees; this reversed sequence of genes can lead to altered gene activity and abnormalities.

invertebrate An animal without a serial arrangement of vertebrae.

invertebrate chordate Chordate in which the notochord is never replaced by the vertebral column.

ion (eye-un, -ahn) Charged particle that carries a negative or positive charge.

ionic bond (eye-ahn-ik) Chemical bond in which ions are attracted to one another by opposite charges.

isomers Molecules with the same molecular formula but a different structure, and therefore a different shape.

isotonic solution Solution that is equal in solute concentration to that of the cytoplasm of a cell; causes cell to neither lose nor gain water by osmosis.

isotope (eye-suh-tohp) One of two or more atoms with the same atomic number but a different atomic mass due to the number of neutrons.

J

jawless fish Type of fish that has no jaws; includes today's hagfishes and lampreys.

K

karyotype (kar-ee-uh-typ) Duplicated chromosomes arranged by pairs according to their size, shape, and general appearance.

keystone species Species whose activities significantly affect community structure.

kidneys Paired organs of the vertebrate urinary system that regulate the chemical composition of the blood and produce a waste product called urine.

kilocalorie Caloric value of food; 1,000 calories.

kinetic energy Energy associated with motion.

kingdom One of the categories used to classify organisms; the category below domain.

kin selection Indirect selection; adaptation to the environment due to the reproductive success of an individual's relatives.

Klinefelter syndrome Condition caused by the inheritance of XXY chromosomes.

Krebs cycle Cycle of reactions in mitochondria that begins with citric acid; it breaks down an acetyl group as CO_2, ATP, NADH, and $FADH_2$ are given off; also called the citric acid cycle.

L

lacteal (lak-tee-ul) Lymphatic vessel in an intestinal villus; it aids in the absorption of lipids.

lacuna Small pit or hollow cavity, as in bone or cartilage, where a cell or cells are located.

lake Body of fresh water, often classified by nutrient status, such as oligotrophic (nutrient-poor) or eutrophic (nutrient-rich).

lancelet Invertebrate chordate with a body that resembles a lancet and has the four chordate characteristics as an adult.

land fungi Fungi (zygospore, sac, and club) that live on land and reproduce by producing windblown spores.

landscape diversity Variety of habitat elements within an ecosystem (e.g., plains, mountains, and rivers).

large intestine Last major portion of the digestive tract, extending from the small intestine to the anus and consisting of the cecum, the colon, the rectum, and the anal canal.

larva Immature form in the life cycle of some animals; it sometimes undergoes metamorphosis to become the adult form.

larynx (lar-ingks) Cartilaginous organ located between the pharynx and the trachea that contains the vocal cords; also called the voice box.

Law of Independent Assortment Mendel's second law of heredity, stating that each pair of factors (called alleles today) assorts (separates) independently of other pairs during the production of gametes , and therefore all possible combinations of factors can occur in the gametes.

Law of Segregation Mendel's first law of heredity, stating that factors (called alleles today) for the same trait segregate from each other in the production of gametes.

leaf Lateral appendage of a stem, highly variable in structure, often containing cells that carry out photosynthesis.

learning Relatively permanent change in behavior that results from practice and experience.

legume Plant with root nodules containing bacteria able to fix atmospheric nitrogen.

lens Clear, membranelike structure found in the eye behind the iris; brings objects into focus.

lenticel Structure in the bark of woody plants that permits gas exchange between the interior of a plant and the external atmosphere.

less-developed country (LDC) Country that is becoming industrialized; typically, population growth is expanding rapidly, and the majority of people live in poverty.

lichen Symbiotic relationship between certain fungi and algae, in which the fungi possibly provide inorganic food or water and the algae provide organic food.

life history Adaptations in characteristics that influence an organism's biology, such as how many offspring it produces, its survival, and factors such as age and size that determine its reproductive maturity.

ligament Tough cord or band of dense fibrous connective tissue that joins bone to bone at a joint.

light reactions In photosynthesis, the reactions in which light energy is captured and used to produce ATP and NADPH.

lignin A complex organic compound that makes plant cell walls more rigid; an important component of wood.

limbic system Association of various brain centers, including the amygdala and hippocampus; governs learning and memory and various emotions, such as pleasure, fear, and happiness.

limiting factor Resource or environmental condition that restricts the abundance and distribution of an organism.

linkage group Alleles of different genes that are located on the same chromosome and tend to be inherited together.

Linnaean classification Use of traditional categories (domain, kingdom, phylum, class order, family, and genus) to group organisms according to anatomical and genetic homologies.

lipase Fat-digesting enzyme secreted by the pancreas.

lipid (lip-id, ly-pid) Class of organic compounds that tends to be soluble only in nonpolar solvents such as alcohol; includes fats and oils.

liposome Droplet of phospholipid molecules formed in a liquid environment.

liver Large, dark red internal organ that produces urea and bile, detoxifies the blood, stores glycogen, and produces the plasma proteins, among other functions.

lobe-finned fish Type of fish with limblike fins.

logistic growth Population increase that results in an S-shaped curve; growth is slow at first, steepens, and then levels off due to environmental resistance.

long-day plant Plant that flowers when day length is longer than a critical length (e.g., wheat, barley, clover, spinach).

loop of the nephron (nef-rahn) Portion of the nephron lying between the proximal convoluted tubule and the distal convoluted tubule that functions in water reabsorption.

loose fibrous connective tissue Tissue composed mainly of fibroblasts widely separated by a matrix containing collagen and elastic fibers.

luteinizing hormone (LH) Hormone produced by the anterior pituitary gland that stimulates the development of the corpus luteum in females and the production of testosterone in males.

lymph (limf) Fluid, derived from tissue fluid, that is carried in lymphatic vessels.

lymphatic (lymphoid) organ Organ other than a lymphatic vessel that is part of the lymphatic system; includes lymph nodes, tonsils, spleen, thymus gland, and bone marrow.

lymphatic system Organ system consisting of lymphatic vessels and lymphatic organs that transports lymph and lipids and aids the immune system.

lymphatic vessel Vessel that carries lymph.

lymph node Mass of lymphatic tissue located along the course of a lymphatic vessel.

lymphocyte (lim-fuh-syt) Specialized white blood cell that functions in specific defense; occurs in two forms— T lymphocyte and B lymphocyte.

lysogenic cycle Bacteriophage life cycle in which the virus incorporates its DNA into that of a bacterium; occurs preliminary to the lytic cycle.

lysosome (ly-suh-sohm) Membrane-enclosed vesicle that contains hydrolytic enzymes for digesting macromolecules.

lytic cycle Bacteriophage life cycle in which the virus takes over the operation of the bacterium immediately upon entering it and subsequently destroys the bacterium.

M

macroevolution Large-scale evolutionary change, such as the formation of new species.

macronutrient Essential element needed in large amounts for plant growth, such as nitrogen, calcium, or sulfur.

macrophage (mak-ruh-fayj) Large phagocytic cell derived from a monocyte that ingests microbes and debris.

malaria Serious infectious illness caused by the parasitic protozoan *Plasmodium*. Malaria is characterized by bouts of chills and high fever that occur at regular intervals.

male gametophyte In seed plants, the gametophyte that produces sperm; a pollen grain.

Malpighian tubule Blind, threadlike excretory tubule near the anterior end of an insect's hindgut.

mammal Homeothermic vertebrate characterized especially by the presence of hair and mammary glands.

marsupial Member of a group of mammals bearing immature young nursed in a marsupium, or pouch (e.g., kangaroo and opossum).

mass extinction Episode of large-scale extinction in which large numbers of species disappear in a few million years or less.

mast cell Cell to which antibodies attach, causing it to release histamine, thus producing allergic symptoms.

master developmental genes Genes that regulate the transcription of other genes so that development can proceed normally.

matrix (may-triks) Unstructured semifluid substance that fills the space between cells in connective tissues or inside organelles.

matter Anything that takes up space and has mass.

mechanoreceptor (mek-uh-noh-rih-sep-tur) Sensory receptor that responds to mechanical stimuli, such as that from pressure, sound waves, or gravity.

medulla oblongata (muh-dul-uh ahb-lawng-gah-tuh) Part of the brain stem that is continuous with the spinal cord; controls heartbeat, blood pressure, breathing, and other vital functions.

megaphyll In plants, a leaf that has several to many veins connecting it to the vascular tissue of the stem; most plants have megaphylls.

megaspore One of the two types of spores produced by seed plants; develops into a female gametophyte (embryo sac).

meiosis, meiosis I, meiosis II (my-oh-sis) Type of nuclear division that occurs as part of sexual reproduction, in which the daughter cells receive the haploid number of chromosomes in varied combinations.

melanocyte-stimulating hormone (MSH) Substance that causes melanocytes to secrete melanin in lower vertebrates.

melatonin (mel-uh-toh-nun) Hormone, secreted by the pineal gland, that is involved in biorhythms.

memory Capacity of the brain to store and retrieve information about past sensations and perceptions; essential to learning.

memory B cell Forms during a primary immune response but enters a resting phase until a secondary immune response occurs.

memory T cell T cell that differentiates during an initial infection and responds rapidly during subsequent exposure to the same antigen.

meninges (sing., meninx) (muh-nin-jeez) Protective membranous coverings about the central nervous system.

meningitis Condition that refers to inflammation of meninges that cover the brain and spinal cord.

meniscus (pl., menisci) (muh-nis-kus,-kee, -sy) Cartilaginous wedges that separate the surfaces of bones in synovial joints.

meristem Undifferentiated, embryonic tissue in the active growth regions of plants.

mesoderm Middle primary tissue layer of an animal embryo that gives rise to muscle, several internal organs, and connective tissue layers.

mesoglea In animals with only two tissue layers, a transparent, jellylike packing material that occurs between the ectoderm and the endoderm.

mesophyll Inner, thickest layer of a leaf consisting of palisade and spongy mesophyll; the site of most photosynthesis.

messenger RNA (mRNA) Type of RNA formed from a DNA template that bears coded information for the amino acid sequence of a polypeptide.

metabolic pathway Series of linked reactions, beginning with a particular reactant and terminating with an end product.

metabolic pool Metabolites that are the products of and/or substrates for key reactions in cells, allowing one type of molecule to be changed into another type, such as carbohydrates converted to fats.

metabolism All of the chemical reactions that occur in a cell.

metapopulation Population subdivided into several small, isolated populations due to habitat fragmentation.

metastasis (muh-tas-tuh-sis) Spread of cancer from the place of origin throughout the body; caused by the ability of cancer cells to migrate and invade tissues.

methanogen Type of archaean that lives in oxygen-free habitats, such as swamps, and releases methane gas.

MHC (major histocompatibility complex) protein Protein marker that is a part of cell-surface markers anchored in the plasma membrane, which the immune system uses to identify "self."

microevolution Change in gene frequencies between populations of a species over time.

micronutrient Essential element needed in small amounts for plant growth, such as boron, copper, and zinc.

microphyll In plants, a leaf that has only one vein connecting it to the vascular tissue of the stem; lycophytes in particular have microphylls.

microRNA (miRNA) Type of small RNA that may bind to mRNA and thereby regulate its activity following transcription.

microspores One of two types of spores produced by seed plants; develops into a male gametophyte (pollen grain).

microtubule (my-kro-too-byool) Small, cylindrical structure that contains 13 rows of the protein tubulin surrounding an empty central core; present in the cytoplasm, centrioles, cilia, and flagella.

microvillus Cylindrical process that extends from an epithelial cell of a villus of the intestinal wall and serves to increase the surface area of the cell.

midbrain Part of the brain located below the thalamus and above the pons; contains reflex centers and tracts.

middle ear Portion of the ear consisting of the tympanic membrane, the oval and round windows, and the ossicles; where sound is amplified.

mimicry Superficial resemblance of two or more species; a mechanism that avoids predation by appearing to be noxious.

mineral Naturally occurring inorganic substance containing two or more elements; certain minerals are needed in the diet.

mineralocorticoid (min-ur-uh-loh-kor-tih-koyd) Type of hormone secreted by the adrenal cortex that regulates salt and water balance, leading to increases in blood volume and blood pressure.

mitochondrion (mite-oh-KAHN-dree-uhn) Membrane-enclosed organelle in which ATP molecules are produced during the process of cellular respiration.

mitosis (my-toh-sis) Type of cell division in which daughter cells receive the exact chromosomal and genetic makeup of the parent cell; occurs during growth and repair.

model Stand-in for an experimental subject that is not available for experimentation.

model of island biogeography Model developed by ecologists Robert MacArthur and E. O. Wilson to explain the effects of distance from the mainland and size of an island on its diversity.

molecule Union of two or more atoms of the same element; also, the smallest part of a compound that retains the properties of the compound.

molting Periodic shedding of the exoskeleton in arthropods.

monoclonal antibody One of many antibodies produced by a clone of hybridoma cells that all bind to the same antigen.

monocot Abbreviation of monocotyledon. Flowering plant group; among other characteristics, members have one embryonic leaf, parallel-veined leaves, and scattered vascular bundles.

monohybrid cross Single genetic cross involving only one trait, such as flower color.

monomer Alternate term for subunit of a macromolecule called a polymer.

monosaccharide (mahn-uh-sak-uh-ryd) Simple sugar; a carbohydrate that cannot be decomposed by hydrolysis (e.g., glucose).

monosomy One less chromosome than usual.

monotreme Egg-laying mammal (e.g., duckbill platypus and spiny anteater).

monsoon Climate in India and southern Asia caused by wet ocean winds that blow onshore for almost half the year.

more-developed country (MDC) Country that is industrialized; typically, population growth is low, and the people enjoy a good standard of living.

morphogenesis Emergence of shape in tissues, organs, or entire embryo during development.

morula Spherical mass of cells resulting from cleavage during animal development prior to the blastula stage.

moss Type of bryophyte.

motor neuron Nerve cell that conducts nerve impulses away from the central nervous system and innervates effectors (muscles and glands).

motor unit Motor neuron and all the muscle fibers it innervates.

M stage In the cell cycle, the period of time during which mitosis occurs to produce daughter cells.

multifactorial trait Trait or illness determined by several genes and the environment.

multiple alleles (uh-leelz) Inheritance pattern in which there are more than two alleles for a particular trait; each individual has only two of all possible alleles.

multiple fruit Cluster of mature ovaries produced by a cluster of flowers, as in a pineapple.

muscular system System of muscles that produces movement within the body and movement of its limbs; principal components are skeletal muscle, smooth muscle, and cardiac muscle.

muscular tissue Type of tissue composed of fibers that can shorten and thicken.

mutate To undergo a permanent genetic change.

mutation Alteration in chromosome structure or number; also, alteration in a gene due to a change in DNA composition.

mutualism Symbiotic relationship in which both species benefit in terms of growth and reproduction.

mycelium Mass of hyphal filaments composing the vegetative body of a fungus.

mycorrhizae Mutually beneficial symbiotic relationship between a fungus and the roots of vascular plants.

myelin sheath (my-uh-lin) White, fatty material, derived from the membrane of Schwann cells, that forms a covering for nerve fibers.

myofibril (my-uh-fy-brul) Contractile portion of muscle fibers that contains a linear arrangement of sarcomeres and shortens to produce muscle contraction.

myosin (my-uh-sin) One of two major proteins of muscle; makes up thick filaments in myofibrils of muscle fibers. (See actin.)

N

N$_2$ (nitrogen) fixation Process whereby free atmospheric nitrogen is converted into compounds, such as ammonium and nitrates, usually by bacteria.

NAD$^+$ Nicotinamide adenine dinucleotide; coenzyme of oxidation-reduction that accepts electrons and hydrogen ions to become NADH + H$^+$ as oxidation of substrates occurs. During cellular respiration, NADH carries electrons to the electron transport chain in mitochondria.

natural killer (NK) cell Lymphocyte that causes an infected or cancerous cell to burst.

natural selection Mechanism resulting in adaptation to the environment due to differential reproductive success.

negative feedback Mechanism of homeostatic response by which the response to a stimulus is inhibited because the stimulus has been negated.

nematocyst In cnidarians, a capsule that contains a threadlike fiber whose release aids in the capture of prey.

nephridia Segmentally arranged, paired excretory tubules of many invertebrates, as in the earthworm.

nephron (nef-rahn) Microscopic kidney unit that regulates blood composition by glomerular filtration, tubular reabsorption, and tubular secretion.

nerve Bundle of nerve fibers outside the central nervous system.

nerve fiber Axon; conducts nerve impulses away from the cell. Nerve fibers are classified as either myelinated or unmyelinated, based on the presence or absence of a myelin sheath.

nerve net Diffuse, noncentralized arrangement of nerve cells in cnidarians.

nervous system Organ system consisting of the brain, spinal cord, and associated nerves that coordinates the other organ systems of the body.

nervous tissue Tissue that contains nerve cells (neurons), which conduct impulses, and neuroglia, cells that support, protect, and provide nutrients to neurons.

neural plate Region of the dorsal surface of the chordate embryo that marks the future location of the neural tube.

neural tube Tube formed by closure of the neural groove during development. In vertebrates, the neural tube develops into the spinal cord and the brain.

neuroglia Nonconducting nerve cells that are intimately associated with neurons and function in a supportive capacity.

neuron Nerve cell that characteristically has three parts: dendrites, cell body, and axon.

neurotransmitter Chemical stored at the ends of axons that is responsible for transmission across a synapse.

neurula The early embryo during the development of the neural tube from the neural plate, marking the first appearance of the nervous system; the next stage after the gastrula.

neutron (noo-trahn) Neutral subatomic particle, located in the nucleus and having a weight of approximately one atomic mass unit.

neutrophil (noo-truh-fill) Granular leukocyte that is the most abundant of the white blood cells; first to respond to infection.

nitrification Process by which nitrogen in ammonia and organic molecules is oxidized to nitrites and nitrates by soil bacteria.

node In plants, the place where one or more leaves attach to a stem.

node of Ranvier (rahn-vee-ay) Gap in the myelin sheath around a nerve fiber.

noncompetitive inhibition Form of enzyme inhibition by which the inhibitor binds to an enzyme at a location other than the active site; while at this site, the enzyme shape changes, the inhibitor is unable to bind to its substrate, and no product forms.

nondisjunction Failure of homologues or daughter chromosomes to separate during meiosis I and meiosis II, respectively.

nonpolar covalent bond Bond in which the sharing of electrons between atoms is fairly equal.

nonrandom mating Mating among individuals on the basis of their phenotypic similarities or differences, rather than randomly.

nonrenewable resource Resource that is finite and cannot be replenished by a natural means at the same rate it is being consumed.

nonvascular plants Land plants (i.e., bryophytes) that have no vascular tissue and therefore are low-lying and generally found in moist locations.

norepinephrine (NE) (nor-ep-uh-nef-rin) Neurotransmitter of the postganglionic fibers in the sympathetic division of the autonomic nervous system; also, a hormone produced by the adrenal medulla.

notochord Cartilaginous-like supportive dorsal rod in all chordates sometime in their life cycle; replaced by vertebrae in vertebrates.

nuclear envelope Double membrane that surrounds the nucleus and is connected to the endoplasmic reticulum; has pores that allow substances to pass between the nucleus and the cytoplasm.

nuclear pore Opening in the nuclear envelope that permits the passage of proteins into the nucleus and ribosomal subunits out of the nucleus.

nucleic acid Polymer of nucleotides; both DNA and RNA are nucleic acids.

nucleoid An irregularly shaped region in the prokaryotic cell that contains its genetic material.

nucleolus (noo-klee-uh-lus, nyoo-) Dark-staining, spherical body in the cell nucleus that produces ribosomal subunits.

nucleosome In the nucleus of a eukaryotic cell, a unit composed of DNA wound around a core of eight histone proteins, giving the appearance of a string of beads.

nucleotide Subunit of DNA and RNA consisting of a 5-carbon sugar bonded to a nitrogen-containing base and a phosphate group.

nucleus (noo-klee-us, nyoo-) Membrane-enclosed organelle that contains chromosomes and controls the structure and function of the cell.

O

obligate anaerobe Prokaryote unable to grow in the presence of free oxygen.

observation Step in the scientific method by which data are collected before a conclusion is drawn.

octet rule States that an atom other than hydrogen tends to form bonds until it has eight electrons in its outer shell; an atom that already has eight electrons in its outer shell does not react and is inert.

oil Substance, usually of plant origin and liquid at room temperature, formed when a glycerol molecule reacts with three fatty acid molecules.

olfactory cell (ahl-fak-tuh-ree) Modified neuron that is a sensory receptor for the sense of smell.

oligotrophic lake Lake with few nutrients, usually very blue.

omnivore (ahm-nuh-vor) Organism in a food chain that feeds on both plants and animals.

oncogene (ahng-koh-jeen) Cancer-causing gene.

oocyte Immature egg that is undergoing meiosis; upon completion of meiosis, the oocyte becomes an egg.

oogenesis (oh-uh-jen-uh-sis) Production of an egg in females by the process of meiosis and maturation.

open circulatory system Arrangement of internal transport in which blood bathes the organs directly, and there is no distinction between blood and interstitial fluid.

operant conditioning Learning that results from rewarding or reinforcing a particular behavior.

operator In an operon, the sequence of DNA that binds tightly to a repressor, and thereby regulates the expression of structural genes.

operon Group of structural and regulating genes that function as a single unit.

opportunistic population Population demonstrating a life history pattern in which members exhibit exponential population growth. Its members are small in size, mature early, have a short life span, produce many offspring, and provide little or no care to offspring (e.g., dandelions).

opposable thumb Fingers arranged in such a way that the thumb can touch the fingertips of all four fingers.

organ Combination of two or more different tissues performing a common function.

organelle (or-guh-nel) Small membranous structure in the cytoplasm having a specific structure and function.

organic chemistry The study of carbon compounds; chemistry of the living world.

organism Individual living thing.

organ of Corti Structure in the vertebrate inner ear that contains auditory receptors; also called the spiral organ.

organ system Group of related organs working together.

orgasm Physiological and psychological sensations that occur at the climax of sexual stimulation.

osmosis (ahz-moh-sis, ahs-) Diffusion of water through a selectively permeable membrane.

ossicle (ahs-ih-kul) One of the small bones of the middle ear—malleus, incus, and stapes.

osteocyte (ahs-tee-uh-syt) Mature bone cell located within the lacunae of bone.

otolith One of several calcium carbonate granules associated with receptors for gravitational balance; in vertebrates, located in the utricle and saccule.

outer ear Portion of the ear consisting of the pinna and the auditory canal.

ovarian cycle (oh-vair-ee-un) Monthly follicle changes occurring in the ovary that control the level of sex hormones in the blood and the uterine cycle.

ovary In animals, the female gonad, the organ that produces eggs, estrogen, and progesterone; in flowering plants, the base of the pistil that protects ovules and, along with associated tissues, becomes a fruit.

overexploitation Occurs when the number of individuals taken from a wild population is so great that the population becomes severely reduced in numbers.

oviduct Tube that transports oocytes to the uterus; also called a uterine tube.

ovulation (ahv-yuh-lay-shun, ohv-) Release of a secondary oocyte from the ovary; if fertilization occurs, the secondary oocyte becomes an egg.

ovule In seed plants, the structure in which the megaspore becomes an egg-producing female gametophyte; it develops into a seed following fertilization.

oxidation Loss of one or more electrons from an atom or molecule; in biological systems, generally the loss of hydrogen atoms.

oxygen debt Amount of oxygen needed to metabolize lactate, a compound that accumulates during vigorous exercise.

oxyhemoglobin (ahk-see-hee-muh-gloh-bin) Compound formed when oxygen combines with hemoglobin.

oxytocin (ahk-sih-toh-sin) Hormone released by the posterior pituitary that causes contraction of the uterus and milk letdown.

P

p53 Protein coded for by the p53 gene that halts the cell cycle when DNA mutates and is in need of repair.

pain receptor Sensory receptor that is sensitive to chemicals released by damaged tissues or excess heat or pressure stimuli.

paleontologist Individual who studies fossils and the history of life.

paleontology Study of fossils that results in knowledge about the history of life.

palisade mesophyll Layer of tissue in a plant leaf containing elongated cells with many chloroplasts.

pancreas (pang-kree-us, pan-) Internal organ that produces digestive enzymes and the hormones insulin and glucagon.

pancreatic amylase Enzyme that digests starch to maltose.

pancreatic islets (islets of Langerhans) Masses of cells that constitute the endocrine portion of the pancreas.

parasite Species that is dependent on a host species for survival, usually to the detriment of the host species.

parasitism Symbiotic relationship in which one species (the parasite) benefits in terms of growth and reproduction to the detriment of the other species (the host).

parasympathetic division That part of the autonomic system that is active under normal conditions; uses acetylcholine as a neurotransmitter.

parathyroid gland (par-uh-thy-royd) Gland embedded in the posterior surface of the thyroid gland; it produces parathyroid hormone.

parenchyma cell Plant tissue composed of the least-specialized of all plant cells; found in all organs of a plant.

parent cell Cell that divides to form daughter cells.

parthenogenesis Development of an egg cell into a whole organism without fertilization.

partial pressure Pressure exerted by each gas in a mixture of gases.

pattern formation Positioning of cells during development that determines the final shape of an organism.

pectoral girdle (pek-tur-ul) Portion of the skeleton that provides support and attachment for an arm; consists of a scapula and a clavicle.

pedigree Graphic representation of matings and offspring over multiple generations for a particular genetic trait.

pelagic zone Open portion of the sea.

pelvic girdle Portion of the skeleton to which the legs are attached; consists of the coxal bones.

penis External organ in males through which the urethra passes; also serves as the organ of sexual intercourse.

pepsin Enzyme secreted by gastric glands that digests proteins to peptides.

peptide Two or more amino acids joined together by covalent bonding.

peptide bond Type of covalent bond that joins two amino acids.

peptide hormone Type of hormone that is a protein, a peptide, or derived from an amino acid.

peptidoglycan Unique molecule found in bacterial cell walls.

pericycle Layer of cells surrounding the vascular tissue of roots; produces branch roots.

periderm Protective tissue that replaces epidermis; includes cork and cork cambium.

peripheral nervous system (PNS) (puh-rif-ur-ul) Nerves and ganglia that lie outside the central nervous system.

peristalsis (pair-ih-stawl-sis) Wavelike contractions that propel substances along a tubular structure, such as the esophagus.

peritubular capillary network (pair-ih-too-byuh-lur) Capillary network that surrounds a nephron and functions in reabsorption during urine formation.

permafrost Permanently frozen ground, usually occurring in the tundra, a biome of arctic regions.

peroxisome Enzyme-filled vesicle in which fatty acids and amino acids are metabolized to hydrogen peroxide that is broken down to harmless products.

petal A flower part that occurs just inside the sepals; often conspicuously colored to attract pollinators.

petiole Part of a plant leaf that connects the blade to the stem.

Peyer patches Lymphatic organs located in the small intestine.

P generation In genetics, the parental generation.

phagocytize To ingest extracellular particles by engulfing them, as amoeboid cells do.

phagocytosis (fag-uh-sy-toh-sis) Process by which amoeboid-type cells engulf large substances, forming an intracellular vacuole.

pharynx (far-ingks) Portion of the digestive tract between the mouth and the esophagus that serves as a passageway for food and also for air on its way to the trachea.

phenotype (fee-nuh-typ) Visible expression of a genotype—for example, brown eyes or attached earlobes.

pheromone Chemical signal released by an organism that affects the metabolism or influences the behavior of another individual of the same species.

phloem Vascular tissue that conducts organic solutes in plants; contains sieve-tube elements and companion cells.

phloem sap Solution of sugars, nutrients, and hormones found in the phloem tissue of a plant.

phospholipid (fahs-foh-lip-id) Molecule that forms the bilayer of the cell's membranes; has a polar, hydrophilic head bonded to two nonpolar hydrophobic tails.

phospholipid bilayer Comprises the plasma membrane; each polar, hydrophilic head is bonded to two nonpolar, hydrophobic tails; contains embedded proteins.

photoperiodism Relative lengths of daylight and darkness that affect the physiology and behavior of an organism.

photorespiration Action of enzyme rubisco, which catalyzes the oxidation of RuBP, releasing CO_2; this reverses carbon fixation and can reduce the yield of photosynthesis.

photosynthesis Process occurring usually within chloroplasts whereby chlorophyll-containing organelles trap solar energy to reduce carbon dioxide to carbohydrate.

photosystem I (PSI) and photosystem II (PSII) Photosynthetic unit where solar energy is absorbed and high-energy electrons are generated; contains a pigment complex and an electron acceptor.

phototropism Growth response of plant stems to light; stems demonstrate positive phototropism.

pH scale Measurement scale for hydrogen ion concentration.

phylogeny Evolutionary history of a group of organisms.

phylum One of the categories, or taxa, used by taxonomists to group species; the taxon above the class level.

phytochrome Photoreversible plant pigment that is involved in photoperiodism and other responses of plants such as etiolation.

phytoplankton Part of plankton containing organisms that photosynthesize, releasing oxygen to the atmosphere and serving as food producers in aquatic ecosystems.

phytoremediation The use of plants to restore a natural area to its original condition.

pili Threadlike appendages that allow bacteria to attach to surfaces and to each other.

pineal gland (pin-ee-ul, py-nee-ul) Endocrine gland located in the third ventricle of the brain; produces melatonin.

pinocytosis Process by which vesicle formation brings macromolecules into the cell.

pioneer species First species to colonize an area devoid of life.

pith Parenchyma tissue in the center of some stems and roots.

pituitary gland Endocrine gland that lies just inferior to the hypothalamus; consists of the anterior pituitary and the posterior pituitary.

placenta Organ formed during the development of placental mammals from the chorion and the uterine wall; allows the embryo, and then the fetus, to acquire nutrients and rid itself of wastes; produces hormones that regulate pregnancy.

placental mammal Member of the mammalian subclass characterized by the presence of a placenta during the development of an offspring.

planarian Free-living flatworm with a ladderlike nervous system.

plankton Free-floating, mostly microscopic aquatic organisms.

plant hormone Chemical signal that is produced by various plant tissues and coordinates the activities of plant cells.

plant tissue culture Process of growing plant cells in the laboratory.

plaque (plak) Accumulation of soft masses of fatty material, particularly cholesterol, beneath the inner linings of the arteries.

plasma (plaz-muh) Liquid portion of blood; contains nutrients, wastes, salts, and proteins.

plasma cell Cell derived from a B cell that is specialized to mass-produce antibodies.

plasma membrane Membrane surrounding the cytoplasm that consists of a phospholipid bilayer with embedded proteins; functions to regulate the entrance and exit of molecules from the cell.

plasmid (plaz-mid) Self-replicating ring of accessory DNA in the cytoplasm of bacteria.

plasmodesmata (sing., plasmodesma) In plants, cytoplasmic strands that extend through pores in the cell wall and connect the cytoplasm of two adjacent cells.

plasmodial slime mold Free-living mass of cytoplasm that moves by pseudopods on a forest floor or in a field, feeding on decaying plant material by phagocytosis; reproduces by spore formation.

plasmolysis Contraction of the cell contents due to the loss of water.

platelet Cell fragment that is necessary to blood clotting; also called a thrombocyte.

pleiotropy Inheritance pattern in which one gene affects many phenotypic characteristics of the individual.

point mutation Alteration in a gene due to a change in a single nucleotide; results of this mutation vary.

polar body In oogenesis, a nonfunctional product; two to three meiotic products are of this type.

polar covalent bond Bond in which the sharing of electrons between atoms is unequal.

pollen grain In seed plants, the sperm-producing male gametophyte.

pollen sacs In flowering plants, the portions of the anther where microspore mother cells undergo meiosis to produce microspores.

pollen tube In seed plants, a tube that forms when a pollen grain lands on the stigma and germinates. The tube grows, passing between the cells of the stigma and the style to reach the egg inside an ovule, where fertilization occurs.

pollination In seed plants, the delivery of pollen to the vicinity of the egg-producing female gametophyte.

pollution Any environmental change that adversely affects the lives and health of organisms.

polygenic inheritance Pattern of inheritance in which a trait is controlled by several allelic pairs; each dominant allele contributes to the phenotype in an additive and like manner.

polymer Macromolecule consisting of covalently bonded subunits; for example, a polypeptide is a polymer of subunits called amino acids.

polymerase chain reaction (PCR) (pahl-uh-muh-rays, -rayz) Technique that uses the enzyme DNA polymerase to produce millions of copies of a particular piece of DNA.

polypeptide Polymer of many amino acids linked by peptide bonds.

polyploid Having a chromosome number that is a multiple greater than twice that of the monoploid number.

polyploidy Condition in which one or more entire sets of chromosomes is added to the diploid genome.

polyribosome (pahl-ih-ry-buh-sohm) String of ribosomes simultaneously translating regions of the same mRNA strand during protein synthesis.

polysaccharide (pahl-ee-sak-uh-ryd) Polymer made from sugar subunits; the polysaccharides starch and glycogen are polymers of glucose subunits.

pond Freshwater basin, smaller than a lake.

pons (pahnz) Portion of the brain stem above the medulla oblongata and below the midbrain; assists the medulla oblongata in regulating the breathing rate.

population Organisms of the same species occupying a certain area.

population density The number of individuals per unit area or volume living in a particular habitat.

portal system Pathway of blood flow that begins and ends in capillaries, such as the portal system located between the small intestine and the liver.

positive feedback Mechanism in which the stimulus initiates reactions that lead to an increase in the stimulus.

posterior pituitary Portion of the pituitary gland that stores and secretes oxytocin and antidiuretic hormone produced by the hypothalamus.

postzygotic isolating mechanism Anatomical or physiological difference between two species that prevents successful reproduction after mating has taken place.

potential energy Stored energy as a result of location or spatial arrangement.

predation Interaction in which one organism (the predator) uses another (the prey) as a food source.

preparatory (prep) reaction Reaction that oxidizes pyruvate with the release of carbon dioxide; results in acetyl CoA and connects glycolysis to the Krebs cycle.

prezygotic isolating mechanism Anatomical or behavioral difference between two species that prevents the possibility of mating.

primary growth In plants, growth that originates in the apical meristems of the shoot and root; causes the plant to increase in length.

primary motor area Area in the frontal lobe where voluntary commands begin; each section controls a part of the body.

primary somatosensory area (soh-mat-uh-sens-ree, -suh-ree) Area dorsal to the central sulcus where sensory information arrives from skin and skeletal muscles.

primate A type of mammal adapted to living in trees; includes prosimians, monkeys, apes, and hominins.

prime mover Muscle most directly responsible for a particular movement.

prions Misfolded proteins that cause other proteins to also become misfolded; cause of mad cow disease and other rare diseases.

producer Photosynthetic organism at the start of a grazing food chain that makes its own food (e.g., green plants on land and algae in water).

proembryo Smaller portion of divided sporophyte embryo that, after dividing repeatedly, becomes the embryo of a plant.

progesterone (proh-jes-tuh-rohn) Female sex hormone that helps maintain sex organs and secondary sex characteristics.

prokaryotic cell (prokaryote) Organism that lacks the membrane-enclosed nucleus and membranous organelles typical of eukaryotes.

prolactin (PRL) (proh-lak-tin) Hormone secreted by the anterior pituitary that stimulates the production of milk from the mammary glands.

proliferative phase Phase of the uterine cycle in which there is increased production of estrogen, causing the endometrium to thicken.

promoter In an operon, a sequence of DNA where RNA polymerase binds prior to transcription.

proprioceptor Sensory receptor that responds to changes in muscle or tendon tension.

prosimian A primate of a group that includes lemurs and tarsiers, and may resemble the first primates that evolved.

prostate gland (prahs-tayt) Gland located around the male urethra below the urinary bladder; adds secretions to semen.

protease Enzyme that digest proteins.

proteasome A large, cylindrical cellular structure that contains proteases and digests tagged proteins following translation.

protein Molecule consisting of one or more polypeptides.

proteome Collection of proteins resulting from the translation of genes into proteins.

proteomics The study of all proteins in an organism.

protist Member of the kingdom Protista.

protobiont In biological evolution, a possible cell forerunner that became a cell once it could reproduce.

proton Positive subatomic particle, located in the nucleus and having a weight of approximately one atomic mass unit.

proto-oncogene (proh-toh-ahng-koh-jeen) Normal gene that can become an oncogene through mutation.

protostome A coelomate animal in which the first embryonic opening (the blastopore) is associated with the mouth.

protozoan Heterotrophic, unicellular protist that moves by flagella, cilia, or pseudopodia, or is immobile.

pseudocoelom A body cavity lying between the digestive tract and the body wall that is incompletely lined by mesoderm.

pseudogene Gene copy that is nonfunctional due to a mutation.

pseudopod Cytoplasmic extension of amoeboid protists; used for locomotion and engulfing food.

pseudostratified ciliated columnar epithelium Appearance of layering in some epithelial cells when, actually, each cell touches a baseline and true layers do not exist.

P site In a ribosome, the place where a tRNA carrying a peptide is bound to mRNA.

puberty Period of life when secondary sex changes occur in humans; marked by the onset of menses in females and sperm production in males.

pulmonary artery (pool-muh-nair-ee, pul-) Blood vessel that takes blood away from the heart to the lungs.

pulmonary circuit Circulatory pathway that consists of the pulmonary trunk, the pulmonary arteries, and the pulmonary veins; takes O_2-poor blood from the heart to the lungs and O_2-rich blood from the lungs to the heart.

pulmonary trunk Large blood vessel that divides into the pulmonary arteries; takes blood away from the heart to the lungs.

pulmonary vein Blood vessel that takes blood from the lungs to the heart.

pulse Vibration felt in arterial walls due to expansion of the aorta following ventricular contraction.

Punnett square Grid used to calculate the expected results of simple genetic crosses.

pupil (pyoo-pul) Opening in the center of the iris of the eye.

pyruvate End product of glycolysis; its further fate, involving fermentation or entry into a mitochondrion, depends on oxygen availability.

R

radial symmetry Body plan in which similar parts are arranged around a central axis, like spokes of a wheel.

rain shadow Leeward side (side sheltered from the wind) of a mountainous barrier, which receives much less precipitation than the windward side.

ray-finned fish Group of bony fishes with fins supported by parallel bony rays connected by webs of thin tissue.

receptacle Area where a flower attaches to a floral stalk.

receptor-mediated endocytosis Selective uptake of molecules into a cell by vacuole formation after they bind to specific receptor proteins in the plasma membrane.

recessive allele (uh-leel) Allele that exerts its phenotypic effect only in the homozygote; its expression is masked by a dominant allele.

reciprocal altruism The trading of helpful or cooperative acts, such as helping at the nest, by individuals—the animal that was helped will repay the debt at some later time.

recombinant DNA (rDNA) DNA that contains genes from more than one source.

recombinant DNA technology Use of DNA that contains genes from more than one source, often to produce genetically modified organisms.

rectum (rek-tum) Terminal end of the digestive tube between the sigmoid colon and the anus.

red algae Marine photosynthetic protists with a notable abundance of phycobilin pigments; include coralline algae of coral reefs.

red blood cell (RBC) Formed element that contains hemoglobin and carries oxygen from the lungs to the tissues; erythrocyte.

red bone marrow Blood-cell-forming tissue located in the spaces within spongy bone.

redox reaction Oxidation-reduction reaction; one molecule loses electrons (oxidation) while another molecule simultaneously gains electrons (reduction).

reduced hemoglobin Hemoglobin molecule that is carrying hydrogen ions derived from carbonic acid.

reduction Chemical reaction that results in addition of one or more electrons to an atom, ion, or compound. Reduction of one substance occurs simultaneously with oxidation of another.

reflex Automatic, involuntary response of an organism to a stimulus.

reflex action An action performed automatically, without conscious thought (e.g., swallowing).

refractory period (rih-frak-tuh-ree) Time following an action potential when a neuron is unable to conduct another nerve impulse.

regulatory gene In an operon, a gene that codes for a protein that regulates the expression of other genes.

renewable resource Resource that is replenished by a natural means at the same rate or faster than it is consumed.

renin (ren-in) Enzyme released by the kidneys that leads to the secretion of aldosterone and a rise in blood pressure.

repetitive DNA Sequence of DNA nucleotides that is repeated several times in a genome.

repolarization Recovery of a neuron's polarity to the resting potential after the neuron ceases transmitting impulses.

repressor In an operon, protein molecule that binds to an operator, preventing transcription of structural genes.

reproduce To produce a new individual of the same kind.

reproductive cloning Production of an organism that is genetically identical to the original individual.

reproductive system Organ system that contains male or female organs and specializes in the production of offspring.

reptile Member of a class of terrestrial vertebrates characterized by internal fertilization, scaly skin, and an egg with a leathery shell; includes snakes, lizards, turtles, and crocodiles and birds.

resource partitioning Mechanism that increases the number of niches by apportioning the supply of a resource such as food or living space between species.

respiration Sequence of events that results in gas exchange between the cells of the body and the environment.

respiratory center Group of nerve cells in the medulla oblongata that send out nerve impulses on a rhythmic basis, resulting in involuntary inspiration on an ongoing basis.

respiratory system Organ system consisting of the lungs and tubes that bring oxygen into the lungs and take carbon dioxide out.

resting potential Polarity across the plasma membrane of a resting neuron due to an unequal distribution of ions.

restoration ecology Subdiscipline of conservation biology that seeks ways to return ecosystems to their former state.

restriction enzyme Bacterial enzyme that stops viral reproduction by cleaving viral DNA; used to cut DNA at specific points during production of recombinant DNA.

retina (ret-n-uh, ret-nuh) Innermost layer of the eyeball that contains the rod cells and the cone cells.

retrovirus RNA virus containing the enzyme reverse transcriptase that carries out RNA-to-DNA transcription.

reverse transcriptase Enzyme that transcribes RNA into DNA.

rhizome Rootlike underground stem.

rhodopsin (roh-dahp-sun) Light-absorbing molecule in rod cells and cone cells that contains a pigment and the protein opsin.

ribose Pentose sugar found in RNA.

ribosomal RNA (rRNA) (ry-buh-soh-mul) Type of RNA found in ribosomes where protein synthesis occurs.

ribosome (ry-buh-sohm) RNA and protein in two subunits; site of protein synthesis in the cytoplasm.

ribozyme Enzyme that carries out mRNA processing.

rigor mortis Contraction of muscles at death due to lack of ATP.

river Freshwater channel that flows eventually to the oceans.

RNA (ribonucleic acid) (ry-boh-noo-klee-ik) Nucleic acid produced from covalent bonding of nucleotide subunits that contain the sugar ribose; occurs in three forms: messenger RNA, ribosomal RNA, and transfer RNA.

rod cell Light-sensitive nerve cell found in the vertebrate retina; sensitive to very dim light; responsible for "night vision."

root The usually belowground portion of a plant that anchors it and serves as the major point of entry for water and minerals.

root cap In plants, the tissue that protects the growing tips of roots.

root hair Extension of a root epidermal cell that increases the surface area for the absorption of water and minerals.

root nodule Structure on a plant root that contains nitrogen-fixing bacteria.

root system Includes the main root and any and all of its lateral (side) branches.

rough ER (RER) Membranous system of tubules, vesicles, and sacs in cells; has attached ribosomes.

roundworm Member of the phylum Nematoda, having a cylindrical body with a complete digestive tract and a pseudocoelom; some forms are free-living in water and soil; many are parasitic.

RuBP carboxylase Enzyme that is required for carbon dioxide fixation (atmospheric CO_2 attaches to RuBP) in the Calvin cycle.

S

saccule (sak-yool) Saclike cavity in the vestibule of the inner ear; contains sensory receptors for gravitational equilibrium.

salivary amylase (sal-uh-vair-ee am-uh-lays, -layz) Secreted from the salivary glands; the first enzyme to act on starch.

salivary gland Gland associated with the mouth that secretes saliva.

salt Compound produced by a reaction between an acid and a base.

saltatory conduction Movement of nerve impulses from one neurolemmal node to another along a myelinated axon.

saprotroph Organism that secretes digestive enzymes and absorbs the resulting nutrients back across the plasma membrane.

sarcolemma (sar-kuh-lem-uh) Plasma membrane of a muscle fiber; also forms the tubules of the T system involved in muscular contraction.

sarcomere (sar-kuh-mir) One of many units, arranged linearly within a myofibril, whose contraction produces muscle contraction.

sarcoplasmic reticulum (sar-kuh-plaz-mik rih-tik-yuh-lum) Smooth endoplasmic reticulum of skeletal muscle fibers; surrounds the myofibrils and stores calcium ions.

saturated fatty acid Fatty acid molecule that lacks double bonds between the atoms of its carbon chain.

savanna Terrestrial biome that is a grassland in Africa, characterized by few trees and a severe dry season.

scanning electron microscope Beam of electrons scans over a specimen point by point and builds up an image on a fluorescent screen.

scavenger Animal that specializes in the consumption of dead animals.

Schwann cell Cell that surrounds a fiber of a peripheral nerve and forms the myelin sheath.

sclera (skleer-uh) White, fibrous, outer layer of the eyeball.

sclerenchyma cell Plant tissue composed of cells with heavily lignified cell walls; functions in support.

seaweed Multicellular forms of red, green, and brown algae found in marine habitats.

secondary growth In vascular plants, an increase in stem and root diameter made possible by cell division of the lateral meristems.

secondary metabolite Molecule not directly involved in growth, development, or reproduction of an organism; in plants, these molecules, which include nicotine, caffeine, tannins, and menthols, can discourage herbivores.

secondary sex characteristic Trait that is sometimes helpful but not absolutely necessary for reproduction and is maintained by the sex hormones in males and females.

second messenger Chemical signal such as cyclic AMP that causes the cell to respond to the first messenger—a hormone bound to a plasma membrane receptor.

secretory phase Phase of the uterine cycle in which increased production of progesterone causes the endometrium to double in thickness, producing a thick, mucoid secretion.

seed Mature ovule that contains an embryo with stored food enclosed in a protective coat.

seed coat In seed plants, the outer layer of the ovule, which becomes a relatively impermeable barrier to protect the dormant embryo and stored food.

seedless vascular plants Land plants, such as lycophytes and ferns, which have vascular tissue but do not produce seeds. Reproduction involves flagellated sperm and production of windblown spores.

seed plant Land plant whose reproduction involves the production of seeds. Reproduction in seed plants is fully adapted to living on land because all reproductive structures are protected from drying out. Includes gymnosperms and angiosperms.

segmentation Repetition of body parts as segments along the length of the body; seen in annelids, arthropods, and chordates.

selective agent Environmental factor that affects the ability of an organism to survive and produce fertile offspring.

self-antigen Antigen that is produced by an organism.

semen (seminal fluid) (see-mun) Thick, whitish fluid consisting of sperm and secretions from several glands of the male reproductive tract.

semicircular canal (sem-ih-sur-kyuh-lur) One of three tubular structures within the inner ear that contain sensory receptors responsible for the sense of rotational equilibrium.

semiconservative replication Duplication of DNA resulting in two double helix molecules, each having one parental and one new strand.

semilunar valve (sem-ee-loo-nur) Valve resembling a half moon located between the ventricles and their attached vessels.

seminal vesicle (sem-uh-nul) Convoluted, saclike structure attached to the vas deferens near the base of the urinary bladder in males; adds secretions to semen.

seminiferous tubule (sem-uh-nif-ur-us) Long, coiled structure contained within chambers of the testis; where sperm are produced.

senescence Sum of the processes involving aging, decline, and eventual death of a plant or plant part.

sensation Conscious awareness of a stimulus due to a nerve impulse sent to the brain from a sensory receptor by way of sensory neurons.

sensory adaptation The phenomenon in which a sensation becomes less noticeable once it has been recognized by constant repeated stimulation.

sensory neuron Nerve cell that transmits nerve impulses to the central nervous system after a sensory receptor has been stimulated.

sensory receptor Structure that receives either external or internal environmental stimuli and is a part of a sensory neuron or transmits signals to a sensory neuron.

sepal Outermost, sterile, leaflike covering of the flower; usually green in color.

Sertoli cell Type of cell in seminiferous tubules with FSH receptors; helps nourish and support developing sperm.

sessile Describes an animal that tends to stay in one place.

severe acute respiratory syndrome (SARS) Respiratory syndrome caused by a virus that emerged in China and spread around the world.

sex chromosome Chromosome that determines the sex of an individual; in humans, females have two X chromosomes, and males have an X and a Y chromosome.

sexually transmitted disease (STD) Illness communicated primarily or exclusively through sexual contact.

sexual reproduction Reproduction involving meiosis, gamete formation, and fertilization; produces offspring with chromosomes inherited from each parent with a unique combination of genes.

sexual selection A type of differential reproduction that results from variable success in obtaining mates.

shoot system Aboveground portion of a plant consisting of the stem, leaves, and flowers.

short-day plant Plant that flowers when day length is shorter than a critical length (e.g., cocklebur, poinsettia, and chrysanthemum).

sieve-tube member Member that joins with others in the phloem tissue of plants as a means of transport for nutrient sap.

simple goiter (goy-tur) Condition in which an enlarged thyroid produces low levels of thyroxine.

simple muscle twitch Contraction of a whole muscle in response to a single stimulus.

single nucleotide polymorphism (SNP) Site present in at least 1% of the population at which individuals differ by a single nucleotide. These can be used as genetic markers to map unknown genes or traits.

sink In the pressure-flow model of phloem transport, the location (roots) from which sugar is constantly being removed. Sugar will flow to the roots from the source.

sink population Population that is found in an unfavorable area where at best the birthrate equals the death rate; sink populations receive new members from source populations.

sister chromatid One of two genetically identical chromosomal units that are the result of DNA replication and are attached to each other at the centromere.

skeletal muscle Striated, voluntary muscle tissue that comprises skeletal muscles; also called striated muscle.

skeletal system System of bones, cartilage, and ligaments that works with the muscular system to protect the body and provide support for locomotion and movement.

skin Outer covering of the body; can be called the integumentary system because it contains organs such as sense organs.

skull Bony framework of the head, composed of cranial bones and the bones of the face.

sliding filament model An explanation for muscle contraction based on the movement of actin filaments in relation to myosin filaments.

slime mold Protists that decompose dead material and feed on bacteria by phagocytosis; silences (inactivates) a chosen mRNA.

small interfering RNA (siRNA) Type of small RNA that combines with a complex and thereafter silences (inactivates) a chosen mRNA.

small intestine In vertebrates, the portion of the digestive tract that precedes the large intestine; in humans, consists of the duodenum, jejunum, and ileum.

small RNA (sRNA) RNAs of limited length that function in various ways within the nucleus to regulate gene expression following transcription.

smooth ER (SER) Membranous system of tubules, vesicles, and sacs in eukaryotic cells; lacks attached ribosomes.

smooth muscle Nonstriated, involuntary muscle tissue found in the walls of internal organs.

sociobiology Application of evolutionary principles to the study of social behavior of animals, including humans.

sodium-potassium pump Carrier protein in the plasma membrane that moves sodium ions out of and potassium ions into cells; important in nerve and muscle fibers.

soft palate (pal-it) Entirely muscular posterior portion of the roof of the mouth.

soil Accumulation of inorganic rock material and organic matter that is capable of supporting the growth of vegetation.

soil profile Vertical section of soil from the ground surface to the unaltered rock below.

solute Substance that is dissolved in a solvent, forming a solution.

solution Fluid (the solvent) that contains a dissolved solid (the solute).

solvent Liquid portion of a solution that serves to dissolve a solute.

somatic system That portion of the peripheral nervous system containing motor neurons that control skeletal muscles.

sorus (pl., sori) Dark spot on the underside of fern fronds that is a collection of spore-producing structures.

source In the pressure-flow model of phloem transport, the location (leaves) of sugar production. Sugar will flow from the leaves to the sink.

source population Population that can provide members to other populations of the species because it lives in a favorable area, and the birthrate is most likely higher than the death rate.

speciation Origin of new species due to the evolutionary process of descent with modification.

species Group of similarly constructed organisms capable of interbreeding and producing fertile offspring; organisms that share a common gene pool; the taxon at the lowest level of classification.

species richness Number of species in a community.

specific epithet In the binominal system of taxonomy, the second part of an organism's name; it may be descriptive.

spermatogenesis (spur-mat-uh-jen-ih-sis) Production of sperm in males by the process of meiosis and maturation.

sphincter (sfingk-tur) Muscle that surrounds a tube and closes or opens the tube by contracting and relaxing.

spinal cord Part of the central nervous system; the nerve cord that is continuous with the base of the brain plus the vertebral column that protects the nerve cord.

spinal nerve Nerve that arises from the spinal cord.

spindle apparatus Microtubule structure that brings about chromosome movement during nuclear division.

spleen Large, glandular organ located in the upper left region of the abdomen; stores and purifies blood.

spongy bone Porous bone found at the ends of long bones where red bone marrow is sometimes located.

spongy mesophyll Layer of tissue in a plant leaf containing loosely packed cells, increasing the amount of surface area for gas exchange.

sporangium (pl., sporangia) Structure that produces spores.

spore Asexual reproductive or resting cell capable of developing into a new organism without fusion with another cell, in contrast to a gamete.

sporophyte Diploid generation of the alternation-of-generations life cycle of a plant; produces haploid spores that develop into the haploid generation.

squamous epithelium Type of epithelial tissue that contains flat cells.

S stage In the cell cycle, the period of time during which DNA replication occurs so that the chromosomes are duplicated.

stabilizing selection Outcome of natural selection in which extreme phenotypes are eliminated and the average phenotype is conserved.

stamen In flowering plants, the portion of the flower that consists of a filament and an anther containing pollen sacs where pollen is produced.

starch Storage polysaccharide found in plants that is composed of glucose molecules joined in a linear fashion with few side chains.

statolith Sensors found in root cap cells that cause a plant to demonstrate gravitropism.

stem Usually the upright, vertical portion of a plant that transports substances to and from the leaves.

stereoscopic vision Vision characterized by depth perception and three-dimensionality.

steroid (steer-oyd) Type of lipid molecule having a complex of four carbon rings; examples are cholesterol, progesterone, and testosterone.

steroid hormone Type of hormone that has a complex of four carbon rings but different side chains from other steroid hormones.

stigma In flowering plants, portion of the pistil where pollen grains adhere and germinate before fertilization can occur.

stolon Stem that grows horizontally along the ground and may give rise to new plants where it contacts the soil—e.g., the runners of strawberry plants.

stoma (pl., stomata) Small opening between two guard cells on the underside of leaf epidermis through which gases pass.

stratified As in the outer layer of skin, having several layers.

stratum (pl., strata) Ancient layer of sedimentary rock; results from slow deposition of silt, volcanic ash, and other materials.

stream Freshwater channel, smaller than a river.

striated Having bands; in cardiac and skeletal muscle, alternating light and dark crossbands produced by the distribution of contractile proteins.

strobilus Terminal cluster of specialized leaves that bear sporangia.

stroma Fluid within a chloroplast that contains enzymes involved in the synthesis of carbohydrates during photosynthesis.

structural gene Gene that codes for an enzyme in a metabolic pathway.

style Elongated, central portion of the pistil between the ovary and stigma.

subcutaneous layer A sheet that lies just beneath the skin and consists of loose connective and adipose tissue.

substrate Reactant in a reaction controlled by an enzyme.

substrate feeder Organism that lives in or on its food source.

substrate-level ATP synthesis Process in which ATP is formed by transferring a phosphate from a metabolic substrate to ADP.

surface-area-to-volume ratio Ratio of a cell's outside area to its internal volume.

survivorship Probability of newborn individuals of a cohort surviving to particular ages.

sustainable development Management of an ecosystem so that it maintains itself while providing services to humans.

sustainable society Interactive group of individuals who provide for their needs in a way that will allow future generations to enjoy the same standard of living.

symbiosis Relationship that occurs when two different species live together in a unique way; it may be beneficial, neutral, or detrimental to one and/or the other species.

sympathetic division The part of the autonomic system that usually promotes activities associated with emergency (fight-or-flight) situations; uses norepinephrine as a neurotransmitter.

sympatric speciation Origin of new species in populations that overlap geographically.

synapse (sin-aps, si-naps) Junction between neurons consisting of the presynaptic (axon) membrane, the synaptic cleft, and the postsynaptic (usually dendrite) membrane.

synapsis (sih-nap-sis) Pairing of homologues during prophase I of meiosis I.

synaptic cleft (sih-nap-tik) Small gap between presynaptic and postsynaptic membranes of a synapse.

syndrome Group of symptoms that appear together and tend to indicate the presence of a particular disorder.

synovial joint (sih-noh-vee-ul) Freely movable joint in which two bones are separated by a cavity.

syphilis (sif-uh-lis) Sexually transmitted disease caused by the bacterium *Treponema pallidum* that, if untreated, can lead to cardiac and central nervous system disorders.

systemic circuit Blood vessels that transport blood from the left ventricle and back to the right atrium of the heart.

systemin In plants, an 18-amino-acid peptide that is produced by damaged or injured leaves and leads to the wound response.

systole (sis-tuh-lee) Contraction period of the heart during the cardiac cycle.

systolic pressure (sis-tahl-ik) Arterial blood pressure during the systolic phase of the cardiac cycle.

T

taiga Terrestrial biome that is a coniferous forest extending in a broad belt across northern Eurasia and North America.

tandem repeat Sequence of DNA nucleotides that is repeated many times in a row.

taxon Group of organisms that fills a particular classification category.

taxonomy Branch of biology concerned with identifying, describing, and naming organisms.

T cell Lymphocyte that matures in the thymus. Cytotoxic T cells kill antigen-bearing cells outright; helper T cells release cytokines that stimulate other immune system cells.

T-cell receptor (TCR) Molecule on the surface of a T cell that can bind to a specific antigen fragment in combination with an MHC molecule.

temperate deciduous forest Forest found south of the taiga; characterized by deciduous trees such as oak, beech, and maple; moderate climate; relatively high rainfall; stratified plant growth; and plentiful ground life.

temperate rain forest Coniferous forest—e.g., the forest running along the west coast of Canada and the United States—characterized by plentiful rainfall and rich soil.

template (tem-plit) Pattern or guide used to make copies; parental strand of DNA serves as a guide for the production of daughter DNA strands, and DNA also serves as a guide for the production of messenger RNA.

tendon Strap of fibrous connective tissue that connects skeletal muscle to bone.

terminal bud Bud that develops at the apex of a shoot.

territoriality Marking and/or defending a particular area against invasion by another species member.

territory Area occupied and defended exclusively by an animal or group of animals; often used for the purpose of feeding, mating, and caring for young.

testcross Cross between an individual with the dominant phenotype and an individual with the recessive phenotype. The resulting phenotypic ratio indicates whether the dominant phenotype is homozygous or heterozygous.

test group Group that participates in an experiment and is exposed to the experimental variable.

testis (pl., testes) Male gonad that produces sperm and the male sex hormones.

testosterone (tes-tahs-tuh-rohn) Male sex hormone that helps maintain sexual organs and secondary sex characteristics.

tetrad Homologues, each having sister chromatids that are joined during meiosis.

thalamus (thal-uh-mus) Part of the brain located in the lateral walls of the third ventricle that serves as the integrating center for sensory input; it plays a role in arousing the cerebral cortex.

theory of ecosystems Concept that organisms are members of populations that interact with each other and with the physical environment at a particular locale.

theory of evolution Concept that all organisms have a common ancestor, but each is adapted to a particular way of life.

thermoacidophile Type of archaean that lives in hot, acidic, aquatic habitats, such as hot springs or near hydrothermal vents.

thermoreceptor Sensory receptor that is sensitive to changes in temperature.

thigmotropism In plants, unequal growth due to contact with solid objects, as the coiling of tendrils around a pole.

threatened species Species that is likely to become endangered in the foreseeable future (e.g., bald eagle, gray wolf, Louisiana black bear).

threshold Electrical potential level (voltage) at which an action potential or nerve impulse is produced.

thrombin (thrahm-bin) Enzyme that converts fibrinogen to fibrin threads during blood clotting.

thylakoid Flattened sac within a granum whose membrane contains chlorophyll and where the light reactions of photosynthesis occur.

thylakoid space Inner compartment of the thylakoid.

thymus gland Lymphatic organ, located along the trachea behind the sternum, involved in the maturation of T lymphocytes. Secretes hormones called thymosins, which aid the maturation of T cells and perhaps stimulate immune cells in general.

thyroid gland Endocrine gland in the neck that produces several important hormones, including thyroxine, triiodothyronine, and calcitonin.

thyroid-stimulating hormone (TSH) Substance produced by the anterior pituitary that causes the thyroid to secrete thyroxine and triiodothyronine.

thyroxine (T4) (thy-rahk-sin) Hormone secreted from the thyroid gland that promotes growth and development; in general, it increases the metabolic rate in cells.

tight junction Junction between animal cells that seals the cells to one another.

tissue Group of similar cells that perform a common function.

tissue culture Process of growing tissue artificially, usually in a liquid medium in laboratory glassware.

tissue fluid Fluid that surrounds the body's cells; consists of dissolved substances that leave the blood capillaries by filtration and diffusion.

topography Surface features of the Earth.

torpedo stage Stage of development of a sporophyte embryo; embryo has a torpedo shape, and the root and shoot apical meristems are present.

totipotent Cell that has the full genetic potential of the organism, including the potential to develop into a complete organism.

toxin Substance produced by a bacterium that has a poisonous effect on the body and causes illness.

trachea (tray-kee-uh) In birds and mammals, passageway that conveys air from the larynx to the bronchi; also called the windpipe.

tracheae In insects, air tubes located between the spiracles and the tracheoles.

tracheid In flowering plants, type of cell in xylem that has tapered ends and pits through which water and minerals flow.

tract Bundle of myelinated axons in the central nervous system.

transcription Process whereby a DNA strand serves as a template for the formation of mRNA.

transcription activator Protein that initiates transcription by RNA polymerase and thereby starts the process that results in gene expression.

transcription factor In eukaryotes, protein required for the initiation of transcription by RNA polymerase.

transduction Exchange of DNA between bacteria by means of a bacteriophage.

trans fats Fats known to cause cardiovascular disease because they contain partially hydrogenated fatty acids.

transfer RNA (tRNA) Type of RNA that transfers a particular amino acid to a ribosome during protein synthesis; at one end, it binds to the amino acid, and at the other end it has an anticodon that binds to an mRNA codon.

transformation Taking up of extraneous genetic material from the environment by bacteria.

transitional fossil A fossil that bears a resemblance to two groups that in the present day are classified separately.

translation Process whereby ribosomes use the sequence of codons in mRNA to produce a polypeptide with a particular sequence of amino acids.

translation repressor protein In the cytoplasm, one of a number of proteins that attach to an mRA and prevent it from binding to a ribosome.

translocation A type of chromosomal mutation that occurs when a chromosomal segment moves from one chromosome to another nonhomologous chromosome, leading to abnormalities (e.g., Down syndrome).

transmission electron microscope Similar to the scanning electron microscope, but the image is colored by a computer.

transpiration Plant's loss of water to the atmosphere, mainly through evaporation at leaf stomata.

transport vesicle Vesicle formed in the ER that carries proteins and lipids to the Golgi apparatus.

transposon DNA sequence capable of randomly moving from one site to another in the genome.

trichocyst Found in ciliates; contains long, barbed threads useful for defense and capturing prey.

trichome Outgrowth of the epidermis, such as a hair or a thorn.

trichomoniasis Sexually transmitted disease caused by the parasitic protozoan *Trichomonas vaginalis*.

triglyceride (trih-glis-uh-ryd) Neutral fat composed of glycerol and three fatty acids.

triplet code Each sequence of three nucleotide bases in the DNA of genes stands for a particular amino acid.

triploid endosperm In flowering plants, nutritive storage tissue that is derived from an egg uniting with polar nuclei during double fertilization.

trisomy One more chromosome than usual.

trochophore larva Independent motile feeding stage in the development of the trochozoa; a trochozoan recognized by two bands of cilia around its middle.

trochozoan Type of protostome that produces a trochophore larva, which has two bands of cilia around its middle.

trophic level Feeding level of one or more populations in a food web.

tropical rain forest Biome near the equator in South America, Africa, and the Indo-Malay regions; characterized by warm weather, plentiful rainfall, a diversity of species, and mainly tree-living animal life.

tropism In plants, a growth response toward or away from a directional stimulus.

trypsin Protein-digesting enzyme secreted by the pancreas.

tuber Enlarged, short, fleshy underground stem—e.g., potato.

tubular reabsorption Movement of primarily nutrient molecules and water from the contents of the nephron into blood at the proximal convoluted tubule.

tubular secretion Movement of certain molecules from blood into the distal convoluted tubule of a nephron so that they are added to urine.

tumor (too-mur) Cells derived from a single mutated cell that has repeatedly undergone cell division; benign tumors remain at the site of origin, and malignant tumors metastasize.

tumor suppressor gene Gene that codes for a protein that ordinarily suppresses cell division; inactivity can lead to a tumor.

tunicate Type of primitive invertebrate chordate.

turgor movement In plant cells, pressure of the cell contents against the cell wall when the central vacuole is full.

turgor pressure The internal pressure inside a plant cell, resulting from osmotic intake of water, that presses its plasma membrane tightly against the cell wall, making the cell rigid.

Turner syndrome Condition caused by the inheritance of a single X chromosome.

tympanic membrane (tim-pan-ik) Located between the outer and middle ear where it receives sound waves; also called the eardrum.

U

umbilical cord Cord connecting the fetus to the placenta through which blood vessels pass.

uniformitarianism Belief espoused by James Hutton that geological forces act at a continuous, uniform rate.

unique noncoding DNA DNA that does not code for a protein and whose unknown function may be different from that of other noncoding DNA.

unpacking Prior to transcription, the transformation of heterochromatin to euchromatin.

unsaturated fatty acid Fatty acid molecule that has one or more double bonds between the atoms of its carbon chain.

upwelling Upward movement of deep, nutrient-rich water along coasts; it replaces surface waters that move away from shore when the direction of prevailing winds shifts.

urea Main nitrogenous waste of terrestrial amphibians and most mammals.

uremia High level of urea nitrogen in the blood.

ureter (yoor-uh-tur) One of two tubes that take urine from the kidneys to the urinary bladder.

urethra (yoo-ree-thruh) Tubular structure that receives urine from the bladder and carries it to the outside of the body.

uric acid Main nitrogenous waste of insects, reptiles, and birds.

urinary bladder Organ where urine is stored before being discharged by way of the urethra.

urinary system Organ system consisting of the kidneys and urinary bladder; rids the body of nitrogenous wastes and helps regulate the water-salt balance of the blood.

urine Liquid waste product made by the nephrons of the vertebrate kidney through the processes of glomerular filtration, tubular reabsorption, and tubular secretion.

uterine cycle (yoo-tur-in, -tuh-ryn) Monthly occurring changes in the characteristics of the uterine lining (endometrium).

uterus (yoo-tur-us) Organ located in the female pelvis where the fetus develops; also called the womb.

utricle (yoo-trih-kul) Saclike cavity in the vestibule of the inner ear that contains sensory receptors for gravitational equilibrium.

V

vaccine Antigens prepared in such a way that they can promote active immunity without causing disease.

vacuole Membrane-enclosed sac, larger than a vesicle; usually functions in storage and can contain a variety of substances. In plants, the central vacuole fills much of the interior of the cell.

vagina Organ that leads from the uterus to the vestibule and serves as the birth canal and organ of sexual intercourse in females.

valence shell Outer shell of an atom.

valve Membranous extension of a vessel or the heart wall that opens and closes, ensuring one-way flow.

vascular cambium In vascular plants , a cylindrical sheath of meristematic cells, the division of which produces secondary phloem and secondary xylem and allows the plant to increase in girth.

vascular cylinder In dicot roots, a core of tissues surrounded by the endodermis, consisting of vascular tissues and pericycle.

vascular plants Land plants that have xylem and phloem. Xylem transports water and helps support an erect stem.

vascular tissue Transport tissue in plants consisting of xylem and phloem.

vector (vek-tur) In genetic engineering, a means to transfer foreign genetic material into a cell (e.g., a plasmid).

vein Vessel that takes blood to the heart from venules; characteristically has nonelastic walls.

vena cava Large systemic vein that returns blood to the right atrium of the heart in tetrapods; either the superior or inferior vena cava.

ventilation Process of moving air into and out of the lungs; also called breathing.

ventricle (ven-trih-kul) Cavity in an organ, such as a lower chamber of the heart or the ventricles of the brain.

venule (ven-yool, veen-) Vessel that takes blood from capillaries to a vein.

vermiform appendix Small, tubular appendage that extends outward from the cecum of the large intestine.

vertebral column (vur-tuh-brul) Series of joined vertebrae that extends from the skull to the pelvis.

vertebrate Chordate in which the notochord is replaced by a vertebral column.

vessel element Cell that joins with others to form a major conducting tube found in xylem.

vestibule (ves-tuh-byool) Space or cavity at the entrance of a canal, such as the cavity that lies between the semicircular canals and the cochlea.

vestigial structure Remains of a structure that was functional in some ancestor but is no longer functional in the present-day organism.

villus (pl., villi) (vil-us) Small, fingerlike projection of the inner small intestinal wall.

viroids Naked strands of RNA that cause diseases in plants by directing the plant cell to produce more viroids.

visual accommodation Ability of the eye to focus at different distances by changing the curvature of the lens.

vitamin Essential requirement in the diet, needed in small amounts. Vitamins are often part of coenzymes.

vocal cord Fold of tissue within the larynx; creates vocal sounds when it vibrates.

vulva External genitals of the female that surround the opening of the vagina.

W

waggle dance Figure-eight dance performed by honeybees to indicate locations of nectar sources.

water column In plants, water molecules joined together in xylem from the leaves to the roots.

water molds Filamentous organisms having cell walls made of cellulose; typically decomposers of dead freshwater organisms, but some are parasites of aquatic or terrestrial organisms.

water vascular system Series of canals that takes water to the tube feet of an echinoderm, allowing them to expand.

wax Sticky, solid, waterproof lipid consisting of many long-chain fatty acids usually linked to long-chain alcohols.

white blood cell (WBC) Leukocyte, of which there are several types, each having a specific function in protecting the body from invasion by foreign substances and organisms.

white matter Myelinated axons in the central nervous system.

wood Secondary xylem that builds up year after year in woody plants and becomes the annual rings.

X

xenotransplantation Use of animal organs, instead of human organs, in human transplant patients.

xylem Vascular tissue that transports water and mineral solutes upward through the plant body; it contains vessel elements and tracheids.

xylem sap Solution of inorganic nutrients moved from a plant's roots to its shoots through xylem tissue.

Y

yeast Unicellular fungus that has a single nucleus and reproduces asexually by budding or fission, or sexually through spore formation.

yolk Dense nutrient material in the egg of a bird or reptile.

yolk sac Extraembryonic membrane that encloses the yolk of birds; in humans, it is the first site of blood cell formation.

Z

zone of cell division In plants, the part of the young root that includes the root apical meristem and the cells just posterior to it; cells in this zone divide every 12–36 hours.

zone of elongation In plants, the part of the young root that lies just posterior to the zone of cell division; cells in this zone elongate, causing the root to lengthen.

zone of maturation In plants, the part of the root that lies posterior to the zone of elongation; cells in this zone differentiate into specific cell types.

zooplankton Part of plankton containing protozoans and other types of microscopic animals.

zygospore Thick-walled resting cell formed during sexual reproduction of zygospore fungi; meiosis and spore formation occur upon germination.

zygote (zy-goht) Diploid cell formed by the union of sperm and egg; the product of fertilization.

Credits

LINE ART

Chapter 23

Figs. 23A.1, 23A.2: Courtesy of G. David Tilman of the University of Minnesota.

Chapter 39

Fig. 39.1a: Data from G.F. Gause, *The Struggle for Existence,* 1934, Williams & Wilkins Company, Baltimore, MD. **Fig. 39.2a:** Data from D.A. MacLulich, *Fluctuations in the Numbers of the Varying Hare (Lepus americanus),* University of Toronto Press, Toronto, 1937, reprinted 1974.

Chapter 41

Fig. 41.1: Redrawn from "Shrimp Stocking, Salmon Collapse, and Eagle Displacement" by C.N. Spencer, B. R. McClelland, and J. A. Stanford, *Bioscience, 41*(1): 14–21. Copyright © 1991 American Institute of Biological Sciences.

PHOTOGRAPHS

Photo Researcher: Evelyn Jo Johnson

Chapter 1

Openers (mound): © S.B. Vinson–Texas A&M University; (ants, eggs, colony): Courtesy of the U.S. Department of Agriculture (USDA); (fire ant stinger): © Alex Wild/Visuals Unlimited; (ant stings): © Scott Camazine/Science Source; 1.1A (left): © Holger Winkler/Corbis; 1.1A (center): © Peter Dazeley/Getty Images; 1.1A (right): © Alexis Rosenfeld/Science Source; 1.1B: © Getty Images/Digital Vision RF; p. 6 (nodules): © Wally Eberhart/Visuals Unlimited; p. 6 (buying vegetables): © Eric Glenn/Getty Images; 1.2a (all): Courtesy Jim Bidlack; 1.3Ba: © Masahiro Nakano/a. collection RF/Getty RF; 1.3Bb: © Peter Stanley/Getty Images; 1.3Bc: © Ezhilan A (Kaviyarasu) (microbeauty. blogspot.com); 1.4 (bacteria): © Dr. Dennis Kunkel/Phototake; 1.4 (*Paramecium*): © M. Abbey/Visuals Unlimited; 1.4 (morel): © Corbis RF; 1.4 (sunflower): © Dave Thompson/Life File/Getty RF; 1.4 (snow goose): © Winfried Wisniewski/Getty RF; 1.5A: © George D. Lepp/Corbis; 1.5B: © Dorling Kindersley/Getty Images; 1.7A–B (both): © Joe Tucciarone; 1.8B: © Ralph Robinson/Visuals Unlimited; 1.8C: © A.B. Dowsett/SPL/Science Source; 1.8D (protist): © Michael Abbey/Visuals Unlimited; 1.8D (plant): © Pat Pendarvis; 1.8D (fungi): © Tinke Hamming/Ingram Publishing RF; 1.8D (animal): © Corbis RF; 1.9B: © Photodisc/Getty RF; p. 17 (agriculture): © Heather Winters/Brand X Pictures/Getty RF; p. 17 (medicine): © George Disario/Corbis; p. 17 (conservation): © Lawrence Berkeley Nat'l Lab/Roy Kaltschmidt, photographer; p. 18 (hawk and young): © John Cancalosi/Getty Images.

Chapter 2

Openers (Earth): © Image created by Reto Stockli, Nazmi El Saleous, and Marit Jentoft-Nilsen, NASA-GSFC; (waterfall): © Scenics of America/PhotoLink/Getty RF; (egg, sperm): © Nestle/Petit Format/Science Source; (fetus): © Brand X Pictures/PunchStock RF; (planets): © The International Astronomical Union/Martin Kommesser; 2.1A: © Michael & Patricia Fogden/Minden Pictures; 2A.1: © Biomed Commun./Custom Medical Stock Photo; 2A.2 (left): © Mazzlota et al./Science Source; 2A.2 (right): Courtesy National Institutes of Health; 2A.3: © SPL/Science Source; 2.3A (crystals): © Evelyn Jo Johnson; 2.3A (salting food): © PM Images/Getty RF; p. 31 (water strider): © Matti Suopajarvi/mattisj/Getty RF; 2.5 (human): © Image Source/Getty RF; 2.5 (tree): © Paul Davies/Alamy; 2.6C: © Andrew Woodley/Alamy; 2B (smokestacks): © Charles O'Rear/Corbis; 2B (lake): © Roger Evans/Science Source; 2B (trees): © Frederica Georgia/Science Source; 2B (statue): © Ray Pfortner/Getty Images.

Chapter 3

Openers (strawberries): © Corbis RF; (raspberries): © C Squared Studios/Getty RF; (boy in tree): © Inti St. Clair/Getty RF; (woman eating fruit): © Design Pics/Monkey Business RF; (supermarket vegetables): © Image Source/Age Fotostock RF; (fawn): © Alaska Stock RF; 3.1A (cactus): © Brand X Pictures/PunchStock; 3.1A (crab): © Ingram Publishing/Alamy RF; 3.1A (bacterium): © H. Pol/CNRI/SPL/Science Source; 3.1C (both): © Corbis RF; 3.1C: © Corbis RF; 3.2B (man): © Hill Street Studios/Getty RF; 3.2B (carbohydrates, lipids, protein foods): © The McGraw-Hill Companies, Inc./John Thoeming, photographer; 3.3A: © Steve Bloom/Taxi/Getty; 3.4 (woman): © Image Source/Corbis RF; 3.4 (glycogen): © Don W. Fawcett/Science Source; 3.4 (starch): © Jeremy Burgess/SPL/Science Source; 3.4 (cellulose): © Science Source/J.D. Litvay/Visuals Unlimited; 3.6: © Claire Higgins/Getty Images; 3A: © Digital Vision/Getty RF; 3.7Aa: © Comstock/PunchStock RF; 3.7Ab: © P. Motta & S. Correr/Science Source; 3.7Ac: © Duomo/Corbis; 3B: © Brian D. Baer; p. 53 (red flower): © Maryellen Baker/Botanica/Getty Images; p. 53 (pink flower): © James Steinberg/Science Source; 3.9A (DNA model): © Photodisk Red/Getty RF; p. 54 (girl, rabbit): © Radius Images/Corbis RF; 3.19c (both): © Eye of Science/Science Source.

Chapter 4

Openers (Hooke's sketches): © Science VU/Zeiss/Visuals Unlimited; (nerve cells): © Dr. Dennis Kunkel/Visuals Unlimited; (bacteria): © David Scharf/SPL/Science Source; (Euglena): © T. E. Adams/Visuals Unlimited; (root cells): © Eye of Science/Science Source; 4.1A (lilac): © Geoff Bryant/Science Source; 4.1A (leaf cells): Courtesy Ray F. Evert/University of Wisconsin Madison; 4.1B (rabbit): © Jeremy Woodhouse/Getty RF; 4.1B (intestine cells): Courtesy O. Sabatakou and E. Xylouri-Frangiadaki; 4Aa (LM): © Stephen Durr; 4Ab (TEM): © M. Schliwa/Visuals Unlimited; 4Ac (SEM): © Dr. Richard Kessel & Dr. Gene Shih/Visuals Unlimited; 4.4A (nucleus): © David M. Phillips/Science Source; 4.4A (pores TEM): Courtesy Ron Milligan/Scripps Research Institute; 4.5A: © R. Bolender & D. Fawcett/Visuals Unlimited; 4.5B: Courtesy Charles Flickinger, from *Journal of Cell Biology* 49:221-226, 1971, Fig. 1 page 224; 4.5C: Courtesy Daniel S. Friend; 4.7: © Newcomb/Wergin/Biological Photo Service; 4.8: © S.E. Frederick & E.H. Newcomb/Biological Photo Service; 4.9 (animal): © Kevin Schafer/Getty Images; 4.9 (mitochondrion): Courtesy Dr. Keith Porter; 4.9 (chloroplast): © Dr. Jeremy Burgess/Science Source; 4.10A (actin): © M. Schliwa/Visuals Unlimited; 4.10A (intermediate, microtubules): © K.G. Murti/Visuals Unlimited; 4.10C (sperm): © Y. Nikas/Science Source; 4.10C (flagellum, basal body): © William L. Dentler/Biological Photo Service.

Chapter 5

Openers (cheetah running): © Steve Bloom/Getty Images; (cattle): © Peter Adams/Getty Images; (cheetah with kill): © PhotoAlto/PunchStock RF; (hot gasses): © John Chumack/Science Source; (sunrise): © Teri Dixon/Getty RF; (people): © Horizon International Images Limited/Alamy; 5.2C (top left): © Dr. Stanley Flegler/Visuals Unlimited; 5.2C (top: hypotonic, hypertonic) : © David M. Phillips/Science Source; 5.2C (bottom: isotonic, hypotonic): © Dwight Kuhn; 5.2C (bottom right): © Ed Reschke; 5.3Ba: © SPL/Science Source; 5Ab: Courtesy Cystic Fibrosis Foundation; 5.4A: © E.H. Newcomb/Biological Photo Service; 5.4B: Courtesy Camillo Peracchia, M.D.; 5.5A (left and center): © Patrik Giardino/Corbis; 5.5A (right): © Joe McBride/Corbis; 5.5B (both): © Tim Graham/Alamy; 5.6B: © Banana Stock/Jupiter Images RF; 5.6C (muscle cell): © CNRI/Science Source; 5.7A (left): © Tony Freeman/PhotoEdit; 5.7A (right): © David R. Frazier Photolibrary, Inc./Alamy; 5.7D: © James Watt/Visuals Unlimited; 5.7E: © Wayne R. Bilenduke/Getty Images; 5C: © AFP/Getty Images.

Chapter 6

Openers (lake): © Michiaki Omori/a.collectionRF/Getty RF; (algae): © Ed Reschke/Getty Images; 6.1A (Euglena): © T. E. Adams/Visuals Unlimited; 6.1A (trees): © Brand X Pictures/PunchStock RF; 6.1A (sunflower): © Corbis RF; 6.1A (kelp): © Chuck Davis/Stone/Getty Images; 6.1A (cyanobacteria): © Sherman Thomas/Visuals Unlimited; 6.1A (diatoms): © Ed Reschke; 6.1A (moss): © Steven P. Lynch; 6.1B: © Dr. George Chapman/Visuals Unlimited; p. 107 (autumn leaves): © Brand X Pictures/PunchStock RF; 6A.2 (top): © Biophoto Associates/Science Source; 6A.2 (bottom): © Michael Abbey/Visuals Unlimited; 6.9: © Creatas/Punchstock RF; 6.10A: © Pixtal/age fotostock RF; 6.10C: Courtesy USDA/Doug Wilson, photographer; 6.10D: © S. Alden/PhotoLink/Getty RF; 6B: © Fridmar Damm/ZEFA/Corbis.

Chapter 7

Openers (snakes): © NHPA/Superstock; (octopus): © Marevision/Getty Images; (bacteria): © Dr. Linda M. Stannard, University of Cape Town/Science Source; (swamp walk): © Edward F. Nelson/TravelingTed. com; 7.1A: © Peter Cade/Getty Images; 7.3 (left): © Scott Camazine/Science Source; 7.3 (right): © Dr. Donald Fawcett and Dr. Porter/Visuals Unlimited; 7.5C: © BananaStock/PunchStock RF; p. 131 (grapes): © Corbis RF; p. 132 (wine, cheese): © The McGraw Hill Companies, Inc./John Thoeming, photographer; p. 132 (yogurt, pickles): © The McGraw Hill Companies, Inc./

16.5 (Paramecium): © M. Abbey/Visuals Unlimited; 16.5 (Acetabularia): © Linda L. Sims/Visuals Unlimited; 16.5 (Flower): © Ed Reschke; 16.5 (Mushroom): © IT Stock/ age fotostock RF; 16.5 (Wolf): © Art Wolf/Stone/Getty Images; 16A.1 (both fish): © Steve Bly/Alamy RF; 16A.2: © Lars Klove/The New York Times/Redux.

Chapter 17

Openers (virus capsid): © Lee Simon/Science Source; (Syphilis bacteria): © Science Source; (H. pylori): © Dr. Linda M. Stannard, University of Cape Town/ Science Source; (E. coli): © David Scharf/SPL/Science Source; (anthrax): © Dr. Gary Gaugler/Science Source; (SARS, circle): © Science VU/Visuals Unlimited; (SARS virus budding): © Eye of Science/Science Source; (nodules): © Dr. Jeremy Burgess/Science Source; (before/after inset): © Science VU/Visuals Unlimited; (bioremediation): © Accent Alaska.com/Alamy; 17.1A: © Dr. Hans Gelderblom/Visuals Unlimited; 17.1B: © K.G. Murti/Visuals Unlimited; 17.3A: © Brad Mogen/ Visuals Unlimited; 17.3B: © AY Images/Anna Yu/Getty RF; 17A.1: © AFP/Getty Images; 17A.2: © Vassil Donev/ epa/Corbis; p. 325 (hydrothermal vent): © Ralph White/ Corbis; 17.6B: © Science VU/Visuals Unlimited; 17.6C: Courtesy Dr. David Deamer; 17.7B (cocci): © Dr. David M. Phillips/Visuals Unlimited; 17.7B (bacilli): © Dr. Dennis Kunkel/Visuals Unlimited; 17.7B (spirilla): © Dr. Gary D. Gaugler/Phototake; 17.8: © CNRI/SPL/ Science Source; p. 329 (endospore): © Alfred Pasieka/ SPL/Science Source; 17C: © AP Images/Lambert; 17.10A: © Science VU/Visuals Unlimited; 17.10B (marsh): © Michael P. Gadomski/Science Source; 17.10B (bacteria): © Dr. Dennis Kunkel/Visuals Unlimited; 17.10C (Gloeocapsa): © Michael Abbey/ Science Source; 17.10C (Anabaena): © Philip Sze/ Visuals Unlimited; 17.10C (Oscillatoria): © Tom Adams/ Visuals Unlimited; 17.11A (swamp): © altrendo nature/ Getty Images; 17.11A (inset): © Ralph Robinson/Visuals Unlimited; 17.11B (Great Salt Lake): © John Sohlden/ Visuals Unlimited; 17.11B (inset): From J.T. Staley, et al., Bergey's Manual of Systematic Bacteriology, Vol. 13, © 1989 Williams and Wilkins Col, Baltimore. Prepared by A.L. Usted Photography by Dept. of Biophysics, Norwegian Institute of Technology; 17.11C (geysers): © Jeff Lepore/Science Source; 17.11C (inset): Courtesy Dennis W. Grogan, University of Cincinnati; 17D.1a: Reproduced by permission of Dr. Marc Steben; 17D.1b: Courtesy Ira Abrahamson, M. D.; 17D.2a: © Michael Abbey/Visuals Unlimited; 17D.2b: © Carolina Biological Supply/Phototake.

Chapter 18

Openers (mosquito): © CDC/PHIL/Corbis; (malaria infected blood): © Dr. Gopal Murti/Science Source; (sandfly): © Sinclair Stammers/Science Source; (Leishmaniasis): © BSIP/Science Source; 18.2 (foraminiferan): © Astrid & Hanns-Frieder Michler/ Science Source; 18.2 (ciliate): © Eric Grave/Science Source; 18.2 (dinoflagellate): © D.P. Wilson/Science Source; 18.2 (green alga): © Linda L. Sims/Visuals Unlimited; 18.2 (red alga): © Daniel V. Gotschall/ Visuals Unlimited; 18.2 (diatoms): © M.I. Walker/ Science Source; 18.2 (protozoan): © Michael Abbey/ Visuals Unlimited; 18A.2 (tree): © Pat Canova/Alamy; 18.3: © Michael Abbey/Visuals Unlimited; 18.4B (inset): © NHPA/Superstock; 18.4B (cliffs): © Stockbyte/ Getty RF; 18.4C: © Dr. Richard Kessel & Dr. Gene Shih/Visuals Unlimited; 18.5A: © Carolina Biological Supply Company/Phototake; 18.5B: © Ed Reschke/ Getty Images; 18.5C: © Eric Grave/Science Source; 18.7 (plasmodium): © Carolina Biological Supply Company/ Visuals Unlimited; 18.7 (sporangia): © V. Duran/Visuals Unlimited; p. 350 (water mold): © James Richardson/ Visuals Unlimited; 18.8A: © Dr. Ann Smith/Science Source; 18.8B: © Biophoto Associates/Science Source; 18.9A: © Steven P. Lynch; 18.9B: © D.P Wilson/Eric & David Hosking/Science Source; 18.10B: © M.I. Walker/

Science Source/Science Source; 18.10C (top): © John D. Cunningham/Visuals Unlimited; 18.10C (lower right): © Carolina Biological Supply/Visuals Unlimited; 18.10D: © Evelyn Jo Johnson; 18.10E (left): © Dr. John D. Cunningham/Visuals Unlimited; 18.10E (right): © Kingsley Stern.

Chapter 19

Openers (flytrap with fly top, circle): © Jerome Wexler/ Visuals Unlimited; (flytrap flower, sundew plant): © Barry Rice/Visuals Unlimited; (flytrap plant): © David Sieren/Visuals Unlimited; (sundew leaf, circle): © Dr. Jeremy Burgess/Science Source; (pitcher plant with bugs, circle): © David Sieren/Visuals Unlimited; (pitcher plants, tall): © Milton H. Tierney, Jr./Visuals Unlimited; (pitcher plant flowers): © Jeffrey Lepore/ Science Source; 19.1B (Chara): © Natural Visions/ Alamy; 19.1B (Coleochaete): © T. Mellichamp/Visuals Unlimited; 19.2Ca-b: © Ed Reschke; 19.2D (left): © Kingsley Stern; 19.2D (right): © Andrew Syred/SPL/ Science Source; 19.3A (hornwort): © Steven P. Lynch; 19.3A (liverwort): © Harold Taylor/Getty Images; 19.3A (moss): © Nigel Cattlin/Science Source; 19.3B (top): © Peter Lilja/Getty Images; 19.3B (bottom): © Steven P. Lynch; 19.4A: Courtesy Hans Steur, the Netherlands; 19.5B: © Gerald & Buff Corsi/Visuals Unlimited; 19.5C: © Carolina Biological Supply Company/Phototake; 19.5D (cinnamon fern): © James Randklev/Getty Images; 19.5D (hart's tongue): © Organics image library/Alamy RF; 19.5D (maidenhair): © Jeff Foott/ Getty Images; 19.5E: © Matt Meadows/Getty Images; 19.6A (ginkgo): © Kingsley Stern; 19.6A (Gnetophyte): © Evelyn Jo Johnson; 19.6A (cycad): © D. Cavagnaro/ Visuals Unlimited; 19.6A (conifer): © D. Giannechini/ Science Source; 19.6B: © Carolina Biological Supply Company/Phototake; 19A (fossil fern): © Sinclair Stammers/SPL/Science Source; p. 373 (Amborella): Courtesy Stephen McCabe/Arboretum at University of California Santa Cruz; 19B.1 (wheat): © Creatas Images RF; 19B.1 (corn plants): © Corbis RF; 19B.1 (ear of corn): © Dorling Kindersley/Getty RF; 19B.1 (rice grains): © Dex Image/Getty RF; 19B.1 (rice plants): © Corbis RF; 19B.2a: © moodboard/Corbis RF; 19B.2b: © Courtesy USDA, David Nance, photographer; 19B.2c: © Brand X Pictures/Jupiterimages RF; 19B.2d: © Photodisc Blue/Getty RF; 19.8A: © Gary R. Robinson/Visuals Unlimited; p. 379 (mycorrhizae): © Dana Richter/ Visuals Unlimited; 19.8C (fungus): © Jeff Lepore/ Science Source; 19.8C (spores): © Dr. Richard Kessel & Dr. Gene Shih/Visuals Unlimited; 19.8D (foliose): © Ed Reschke/Getty Images; 19.8D (fruticose): © Stephen Sharnoff/Visuals Unlimited; 19.9A: © Ed Reschke/Getty Images; 19.9B (cup fungi): © Felix Labhardt/Getty RF; 19.9B (morel): © Robert Marien/Corbis RF; 19.9Ca–b: © David Philips/Visuals Unlimited; 19.9D (mushroom): © Biophoto Assoc./Science Source; 19.9D (shelf fungi): © Inga Spence/Alamy; 19.9D (puffball): © L. West/ Science Source; 19C.1: © Kingsley Stern; 19C.2a: © John Hadfield/SPL/Science Source; 19C.2b: © CMSP/Getty Images; 19C.2c: Courtesy of the Centers for Disease Control and Prevention.

Chapter 20

Openers (flying fox bat): © Austin J. Stevens/Animals Animals; (vampire bat): © Michael Fogden/Animals Animals; (vampire bat feeding): © Oxford Scientific/ Animals Animals; (bat eating cactus): © Dr. Merlin D. Tuttle/Bat Conservation International/Science Source; (gnome fruit bat): © Andrew M. Snyder/Getty RF; 20.1A (adult frog): © Dwight Kuhn; 20.1A (top row, all embryonic stages): © Carolina Biological Supply Company/Phototake; 20.1A (bottom row, all): © Dwight Kuhn; 20.3: © Andrew J. Martinez/Science Source; 20.4A (sea anenome): © Azure Computer & Photo Services/ Animals Animals; 20.4A (coral): © Ron & Valerie Taylor/ Bruce Coleman/Photoshot; 20.4A (man-of-war): © Stephen Frink Collection/Alamy; 20.4A (jellyfish):

© Frederic Pacorel/Getty Images; 20.4B: © Carolina Biological Supply Company/Visuals Unlimited; 20.5A: © Tom E. Adams/Visuals Unlimited; 20.5B (left): © Dennis Kunkel Microscopy, Inc./Phototake; 20.5B (right): © Science VU/Visuals Unlimited; 20.5C: © SPL/Science Source; 20A.1 (mudflats): © LOOK Die Bildagentur der Fotografen GmbH/ Alamy; 20A.2 (students): © Purestock/Getty RF; 20A.3 (ribbon worm): © Dr. Cleveland P. Hickman, Jr.; 20.7A (radula): © Kjell Sandved/Butterfly Alphabet; 20.7B (snail): © Georgette Douwma/Science Source; 20.7B (nudibranch): © Kenneth W. Fink/Bruce Coleman/ Photoshot; 20.7B (nautilus): © Maximilian Weinzierl/ Alamy; 20.7B (octopus): © Photographer's Choice/ Superstock RF; 20.7B (mussels): © Michael Lustbader/ Science Source; 20.7B (scallop): © ANT Photo Library/ Science Source; 20.8Ba: © James H. Carmichael; 20.8Bc: © St. Bartholomews Hospital/SPL/Science Source; 20.9a: © Lauritz Jensen/Visuals Unlimited; p. 401 (Trichinella): © James Solliday/Biological Photo Service; p. 401 (man): © Vanessa Vick/The New York Times/ Redux; 20.10B (caterpillar): © Creatas/Punchstock RF; 20.10B (pupa): © PBNJ Productions/Getty RF; 20.10B (metamorphosis, emergence, butterfly): © Creatas/ Punchstock RF; 20.10C (crab): © Michael Lustbader/ Science Source; 20.10C (barnacles): © Kjell Sandved/ Butterfly Alphabet; 20.10C (crayfish): © DEA/C. Galasso/ Getty Images; 20.10C (copepod): © Kim Taylor/Bruce Coleman/Photoshot; 20.10D (centipede): © David M. Dennis/Animals Animals; 20.10D (millipede): © Stuart Wilson/Science Source; 20.10E (scorpion): © Tom McHugh/Science Source; 20.10E (horseshoe crab): © Jana R. Jirak/Visuals Unlimited; 20.10E (spider): © John Serrao/Science Source; 20.10F (mealybug): © AgStock Images, Inc./Alamy; 20.10F (grasshopper): © Chris Mattison/Frank Lane Picture Agency/Corbis; 20.10F (dragonfly): © Thomas Shahan/Getty Images; 20.10F (louse): © Alastair Macewen/Getty Images; 20.10F (wasp): © Johnathan Smith; Cordaiy Photo Library/Corbis; 20.10F (butterfly): © McDonald Wildlife Photography/Animals Animals; 20.10F (beetle): © George Grall/Getty Images; 20.10F (housefly): © L. West/Bruce Coleman/Photoshot; 20.10F (leafhopper): © Whitehead Images/Alamy RF; 20.11 (sea cucumber): © Alex Kerstitch/Visuals Unlimited; 20.11 (sea urchin): © Biophoto Associates/Science Source; 20.11 (sand dollar): © Andrew J. Martinez/Science Source; 20.11 (feather star): © Borut Furlan/Getty Images; 20.11 (brittle star): © Diane R. Nelson; 20.11 (sea lily): © Georgette Douwma/Science Source; 20.12Ba: © Heather Angel/ Natural Visions; 20.12Bb: © Amar and Isabelle Guillen-Guillen Photography/Alamy; 20.13 (lamprey): © Heather Angel/Natural Visions; 20.13 (shark): © James Watt/ Animals Animals; 20.13 (bony fish): © Ron & Valerie Taylor/Bruce Coleman/Photoshot; 20.14Aa: © Justus de Cuveland/Getty RF; 20.14Ab: © Suzanne L. Collins & Joseph T. Collins/Science Source; 20.15A (turtles): © Michael Patrick O'Neill/Alamy; 20.15A (Gila monster): © Joe McDonald/Visuals Unlimited; 20.15A (rattlesnake): © Joel Sartorie/National Geographic/Getty Images; 20.15A (Tuatara): © Nathan W. Cohen/Visuals Unlimited; 20.15A (alligator): © OS21/PhotoDisc/Getty RF; 20.15B (downstroke): © Mike Black Photography/ Getty RF; 20.15B (upstroke): © Mark Strevens/Getty RF; 20.15C (eagle, flamingo): © Creatas/PunchStock RF; 20.15C (woodpecker): © Joe McDonald/Corbis; 20.15C (vulture): © Robert Comport/Animals Animals; 20.15C (cardinal): © Jeremy Woodhouse/Getty RF; 20.16Aa, 20.15Ac: © Fritz Prenzel/Animals Animals; 20.16Ab (opossum), 20.16B (deer): © Stephen J. Krasemann/ Getty Images; 20.16B (lioness): © John Downer/Getty Images; 20.16B (human): © Comstock/Punchstock RF; 20.16B (killer whale): © 2011 Tory Kallman/Getty RF; p. 413 (pangolin): © Nigel J. Dennis/Science Source; 20Ba: © Mark Smith/Science Source; 20Bb: © Allan Friedlander/SuperStock; 20Bc: © Account Phototake/ Phototake.

Chapter 21

Openers (Ardi hand, teeth, foot): © HO/Reuters/Corbis; 21.1A (lemur): © Art Wolfe/Getty Images; 21.1A (tarsier): © David Tipling/Getty Images; 21.1A (monkey): © Paul Souders/Corbis; 21.1A (baboon): © St. Meyers/Okapia/Science Source; 21.1A (orangutan): © Tim Davis/Science Source; 21.1A (gibbon): © Hans & Judy Beste/Animals Animals; 21.1A (chimpanzee): © Martin Harvey/Getty Images; 21.1A (gorilla): © Creatas/PunchStock RF; 21.1A (humans): © Tim Davis/Science Source; 21.2Bb: © Carolina Biological Supply Company/Visuals Unlimited; 21.3A (*A. ramidus*): © Richard T. Nowitz/Science Source; 21.3A (*A. afarensis*): © Scott Camazine/Alamy; 21.3A (*A. africanus*): © Philippe Plailly/Science Source; 21.3A (*H. habilis*): © Kike Calvo VWPics/Superstock; 21.3A (*H. sapiens*): © Kenneth Garrett/Getty Images; 21.4A (footprints): © John Reader/Science Source; 21.4A (Lucy): © Dan Dreyfus and Associates; p. 426 (*A. sediba* hand): Image created by Peter Schmid, courtesy Lee R. Berger and the University of the Witwatersrand; 21.4B (left): © Gallo Images/Corbis; 21.4B (center): © Brett Eloff/Lee Berger/University of the Witwatersrand; 21.4B (right): © Deco Images/Alamy; 21.4C: © Science VU/Visuals Unlimited; 21.5A: © Mark Thiessen/National Geographic Stock; 21.6A: © Tomas Abad/agefotostock/SuperStock; 21.7: © PhotoDisc/Getty RF; p. 433 (*A. ramidus*): © Richard T. Nowitz/Science Source; p. 433 (*A. afarensis*): © Scott Camazine/Alamy; p. 434 (*H. habilis*): © Kike Calvo VWPics/Superstock; p. 434 (*H. sapiens*): © Kenneth Garrett/Getty Images.

Chapter 22

Openers (switchgrass, baler, power plant): © Dr. David I. Bransby; (biofuel pump): Photo by Bensinger, Charles and Renewable Energy Partners of New Mexico, NREL 13531; 22.1B (cactus): © Patti Murray/Animals Animals; 22.1B (cucumber): © Michael Gadomski/Science Source; 22.1B (flytrap): © Steven P. Lynch; 22.1C (taproot): © Jonathan Buckley/Getty Images; 22.1C (fibrous root): © The McGraw-Hill Companies Inc./Evelyn Jo Johnson, photographer; 22A (rice plants): © Corbis RF; 22A (rice grain head): © Dex Image/Getty RF; 22A (wheat): © Earl Roberge/Science Source; 22A (ear of corn): © Doug Wilson/Corbis; 22A (corn plants): © Adam Hart-Davis/SPL/Science Source; 22A (barley plants): © Sundell Larsen/Getty RF; 22A (barley grains): © C. Sherburne/Photolink/Getty RF; 22.3Aa: © J.R. Waaland/Biological Photo Service; 22.3Ab: © Evelyn Jo Johnson; 22.3Ac: © Kingsley Stern; 22.3B (all): © Biophoto Associates/Science Source; 22.3Ca: © J.R. Waaland/Biological Photo Service; 22.3D: © George Wilder/Visuals Unlimited; 22.4Bb: © Ed Reschke; 22.4Bc-d: © Carolina Biological Supply Company/Phototake; 22.5Ab: Courtesy Ray F. Evert/University of Wisconsin Madison; 22.5B (both): © J.R. Waaland/Biological Photo Service; 22B.1 (left): © Dr. William E. Stein; 22B.1 (center, right): © New York State Museum, Albany, NY; 22B.2: © Richard T. Nowitz/Corbis; 22.7B: Courtesy Ray F. Evert/University of Wisconsin Madison; 22.7C (both): © Jeremy Burgess/SPL/Science Source; 22.7D: © D. H. Marx/Visuals Unlimited.

Chapter 23

Openers (man and Sequoia): © Grant Faint/Getty Images; (car and Sequoia): © D. Falconer/PhotoLink/Getty RF; (tall Sequoia): © Stephen Sharnoff/National Geographic Stock; (man climbing Sequoia): © Mark Moffett/Minden Pictures/Corbis; 23.1B (wood): © N.C. Brown Center for Ultrastructure Studies, SUNY, College of Environmental Science & Forestry, Syracuse, NY; 23.1B (vessel elements, end wall): © John N. A. Lott/Biological Photo Service; 23.1C: © Dr. Richard Kessel & Dr. Gene Shih/Visuals Unlimited; 23.2A: © Kallista Images/Getty Images; 23.2B (tree): © Paul Davies/Alamy; 23.3B (both): © Jeremy Burgess/SPL/Science Source; 23B: Courtesy Gary Banuelos/Agriculture Research Service/USDA; 23.4A (top): © Bruce Iverson/

SPL/Science Source; 23.4A (bottom): From M.H. Zimmerman "Movement of Organic Substances in Trees" in SCIENCE 133 (13) January 1961 page 667, Fig. 3 page 73, © 1961 AAAS; 23.5A: © Dorling Kindersley/Getty Images; 23.5B (all): Courtesy Mary E. Doohan; 23.6Ba: © Carolina Biological Supply Company/Phototake; p. 470 (dodder): © Kevin Schafer/Corbis; 23.7A (root nodules): © Dwight Kuhn; 23.7A (cell): © E.H. Newcomb & S.R. Tardon/Biological Photo Service; 23.7B (plants, top): © Science VU/R. Roncadori/Visuals Unlimited; 23.7B (mycorrhizae, circle): © Dana Richter/Visuals Unlimited; 23.7C (sundew plant): © Barry Rice/Visuals Unlimited; 23.7C (sundew leaf, prey): © Dr. Jeremy Burgess/Science Source.

Chapter 24

Openers (lettuce, normal): © Michael Thompson/Animals Animals; (lettuce, bolted): © Ros Drinkwater/Alamy; (Jeffrey pine, windswept): © Kent and Donna Dannen/Science Source; (Jeffrey pine, straight): © M. & C. Photography/Getty Images; (both maple trees): © Gary Vestal/Getty Images; p. 477 (apical dominance): Courtesy Prof. Malcolm B. Wilkins; 24.3Aa: © Robert E. Lyons/Visuals Unlimited; 24.3Ab: © Sylvan Whittwer/Visuals Unlimited; 24.4 (1): Courtesy Prof. Dr. Hans-Ulrich Koop, from Plant Cell Reports, 17:601-604; 24.4 (2, 3, 4): Courtesy Alan Darvill and Stefan Eberhard, Complex Carbohydrate Research Center, University of Georgia; 24.5A: © Evelyn Jo Johnson; 24.5B: Courtesy Dr. Donald R. McCarty, University of Florida; 24.6A (both), 24.7A (negative): © Kingsley Stern; 24.7A (positive): © McGraw-Hill Companies, Inc./Evelyn Jo Johnson photographer; 24.7A (sedimentation): © BioPhot; 24.7B: © Maryann Frazier/Science Source; 24.7D: © John D. Cunningham/Visuals Unlimited; 24.8A (both): © John Kaprielian/Science Source; p. 484 (venus flytrap): © Steven P. Lynch; 24.8B (top, both): © Tom McHugh/Science Source; 24.8B (both): © Takahisa Hirano/Minden Pictures; 24.9B (both): © Nigel Cattlin/Visuals Unlimited; 24.11A (alfalfa plant bug): Courtesy U.S. Department of Agriculture/Agricultural Research Service, Photo by Scott Bauer; 24.11A (fungus): © Kingsley Stern; 24.11A (caterpillar): © The McGraw-Hill Companies Inc./Kevin Cavanagh, photographer; 24.11A (butterfly): © Getty Images/Photographers Choice RF; p. 489 (leaf spot): © Nigel Cattlin/Science Source; p. 489 (mutualism): © Kevin Schafer/Alamy; 24A: © Goldberg Diego/Corbis Sygma; p. 493 (tracking light): © Kim Taylor/Bruce Coleman/Photoshot.

Chapter 25

Openers (chipmunk): © ASO FUJITA/amana images RF; (songbird): © Stockbyte RF; (duck): © Jennifer Loomis/Animals Animals; (dog): © Kallista Images/Getty Images; 25.1Ca: © Marg Cousens/Getty Images; 25.1Cb: © Pat Pendarvis; 25.1D (staminate): © Adam Hart-Davis/SPL/Science Source; 25.1D (carpellate): Courtesy USDA/Doug Wilson, photographer; 25Ba: © Creatas/PunchStock RF; 25Bb: © Stephen P. Parker/Science Source; 25Bc: © Glenn Bartley/Getty Images; 25Bd: © David Dennis/Animals Animals; 25Be: © Charles Melton/Visuals Unlimited/Corbis; 25.2B (left): © 2007 John Bolivar/Custom Medical Stock; 25.2B (right): © Microworks Color/Phototake; 25.2C (both): © Heather Angel/Natural Visions; 25.2D (right): © Dwight Kuhn; 25.3 (proembryo, globular, heart): Courtesy Dr. Chun-Ming Liu; 25.3 (torpedo): © Biology Media/Science Source; 25.3 (mature embryo): © Jack Bostrack/Visuals Unlimited; 25.4 (maple seed): © James Mauseth; 25.4 (strawberries): © Corbis RF; 25.4 (raspberries): © C Squared Studios/Getty RF; 25.4 (pineapple): © BJ Miller/Biological Photo Service; 25.5A: © Ed Reschke; 25.5B: © Ed Reschke/Getty Images; 25.6: © G.I. Bernard/Animals Animals; 25.7Aa (all): Courtesy Prof. Dr. Hans-Ulrich Koop, from Plant Cell Reports, 17:601-604; 25.7B: © Kingsley Stern; p. 507 (embryo sac): © Ed Reschke.

Chapter 26

Openers (dragonfly): © Anders Blomqvist/Getty Images; (chimpanzee): © NHPA/Superstock; (iguana): © Digital Vision Ltd. RF; (rainforest): © Steven P. Lynch; (frog): © Comstock/Punchstock RF; (tortoise): © Digital Vision/PunchStock RF; (caribou): © Johnny Johnson/Animals Animals; (polar bears): © Wayne R. Bilenduke/Getty Images; (penguins): © Martin Ruegner/Getty RF; 26.2 (all): © Ed Reschke; 26.3A (loose connective, adipose): © The McGraw-Hill Companies, Inc., Al Telser, photographer; 26.3A (dense fibrous): © McGraw-Hill Higher Education, Dennis Strete, Photographer; 26.3A (hyaline, compact bone), 26.4 (skeletal, cardiac): © Ed Reschke; 26.4 (smooth): © McGraw-Hill Education, Dennis Strete, Photographer; 26.5: © Ed Reschke; 26.6B (skin): © John D. Cunningham/Visuals Unlimited; 26.6B (woman): © Fancy Photography/Veer RF; 26Aa-b: © Dr. P. Marazzi/Science Source; 26Ac: © James Stevenson/SPL/Science Source; 26B: © Bob Calhoun/Bruce Coleman/Photoshot.

Chapter 27

Openers (clam): © Andrew J. Martinez/Science Source; (sea urchins): © Randy Morse/Animals Animals; (octopus): © Bruce Watkins/Animals Animals; (dolphin): © Gerard Lacz/Animals Animals; (grasshopper): © Masterfile RF; (trout): © Tom & Pat Leeson/Science Source; (crab): © Michael Lustbader/Science Source; (student): © BananaStock/JupiterImages RF; 27.1B: © M.B. Bunge/Biological Photo Service; 27.1D: © doc-stock/Visuals Unlimited; 27.3Ba: © Science VU/Lewis-Everhart-Zeevi/Visuals Unlimited; 27.4A: © Colin Chumbley/Science Source/Science Source; 27B: © Evelyn Jo Johnson; 27.10A: © Vol. 94 PhotoDisc/Getty RF; 27.10B: © Reuters/Corbis.

Chapter 28

Openers (bee): © Steven P. Lynch; (flower mosaic): © James Gould; (squid): © Jeff Rotman/Science Source; (woman): © Ingram Publishing RF; (zebra): © Ingrid Van Den Berg/Animals Animals; 28.1A: © Denis Scott/Corbis; 28.1B: © David M. Dennis/Animals Animals; 28.1C: © Dr. Merlin D. Tuttle/Bat Conservation International/Science Source; 28.2B (all): © Omikron/SPL/Science Source; p. 557 (ice cream): © Coneyl Jay/Science Source; p. 559 (cataract): © Sue Ford/Science Source; 28.4: © Frank S. Werblin, Ph.D.; p. 561 (color blindness test): © Steve Allen/Getty RF; 28.5B: © P. Motta/SPL/Science Source; 28Ca (both): Courtesy Dr. Yeohash Raphael, the University of Michigan, Ann Arbor; 28.6B: © Myrleen Ferguson Cate/PhotoEdit; 28.6C: © Dr. David Furness, Keele University/Science Source.

Chapter 29

Openers (child, adult skulls): © 2007 Educational Images Ltd./Custom Medical Stock Photo; (fracture): © Scott Camazine/Phototake; (femur): © Dr. Fred Hossler/Visuals Unlimited; (pelvis): © L. Bassett/Visuals Unlimited; (male and female skulls): © Ralph Hutchings/Visuals Unlimited; (forensics): © Fehim Demir/epa/Corbis; p. 574 (exoskeleton): © Michael Fogden/Animals Animals; 29.1B: © Dynamic Graphics Group/PunchStock RF; 29.2: © DEA/C. Galasso/Getty Images; 29.5A (articular cartilage, compact bone): © Ed Reschke; 29.5A (osteocyte): © Biophoto Associates/Science Source; 29.5B (gymnast): © Gerard Vandystadt/Science Source; 29Ba (tennis player): © Susan Mullane/NewSport/Corbis; 29Ba (normal bone): © Perennou Nuridsany/Science Source; 29Ba (osteoporosis): © Ed Reschke/Getty Images; 29Bb: © Bill Aaron/PhotoEdit; p. 584 (motor unit): © Victor B. Eichler, Ph.D. 29C: © Steve Lindridge/Alamy; 29.7A (gymnast): © Corbis RF; 29.7A (myofibril): © Biology Media/Science Source; 29D (left): © Lawrence Manning/Corbis; 29D (center): © G. W. Willis/Visuals Unlimited; 29D (right): © Corbis RF.

Index

iPS (induced pluripotent stem) cells, 147, 147*f*
Iridium, and mass extinctions, 303
Iris (of eye), 558, 558*f*, 569
Iris (plant), rhizomes of, 505, 505*f*
Iron, 24
  deficiency, 645*t*
  excess, 645*t*
  food sources of, 645*t*
  functions in body, 645, 645*t*
  of hemoglobin, 662, 662*f*
  in red blood cells, 604–605
Iron-sulfur world hypothesis, 325, 338
Island biogeography, model of, 771, 771*f*
Isoetes. See Quillworts
Isomer(s), 43, 57
Isotonicity, 668
Isotonic solution, 87, 87*f*, 99
Isotope(s), 25, 37
  radioactive, 25
    medical uses of, 26, 26*f*
IUD. *See* Intrauterine device (IUD)
IUI. *See* Intrauterine insemination (IUI)
IVF. *See* In vitro fertilization (IVF)
Ivory, 811

J

Jacob, François, 216
Janzen, Dan, 761
Jasmonic acid, 489, 489*f*
Jaundice, 642
  hemolytic, 642
  obstructive, 642
Jaw(s), 575
  of clam worm, 400, 400*f*
  evolution of, among fishes, 407–408
  of fish, 407
Jeffrey pine, phenotypic plasticity of, 475*f*
Jellyfish, 15, 390, 392, 394, 394*f*
  bioluminescent, 233*f*
  nerve nets of, 536
Jet lag, 698
Jet stream, 796
Joint(s), 581, 581*f. See also* Articulation(s)
  ball-and-socket, 581, 581*f*, 589
  cartilaginous, 581, 589
  condyloid, 581
  fibrous, 581, 589
  forensic examination of, 572
  gliding, 581
  hinge, 581, 581*f*, 589
  saddle, 581
  synovial, 581, 581*f*, 589
Joint capsule, 581
Jojoba, 490
Jordan, Karl, 255
Jugular vein(s), 602*f*
Junction proteins, 85, 85*f*, 99–100
Juniper(s), 370

K

Kangaroo(s), 412
  reproduction of, 703*f*
Kangaroo rat, homeostasis in, 524, 524*f*
Kaposi sarcoma, in HIV-infected (AIDS) patients, 612, 613*f*
Karyotype, 156, 156*f*, 167
  in Down syndrome, 165*f*, 166
Katydid(s), 765
Kelp(s), 104*f*, 352
Keratin, 50, 76, 247*t*, 513, 518
Kerner, Justinus, 585
Ketamine, 547
Ketone, 43*t*
Keystone species, 812, 820
Kidney(s), 512, 512*f*, 521, 602, 602*f*, 681
  and acid–base balance, 679, 682
  anatomy of

macroscopic, 674, 674*f*
  microscopic, 674–675, 674*f*
  artificial, 680, 680*f*
  of birds, 674
  functions of, 674, 674*f*, 681–682
  and homeostasis, 522, 522*f*, 678–679, 682
  human, 673, 673*f*, 674, 674*f*
    anatomy of, 674, 674*f*
  lab-grown, 680
  mammalian, 673–674, 674*f*
  in osmoregulation, 669
  of reptiles, 673–674
  urine production in, 681
  vertebrate, evolution of, 672
  and water-salt balance, 678–679, 682
  of whales, 673
Kidney stones, 644, 645*t*–646*t*, 674, 680
Killer whale(s), 413*f. See also* Orca(s)
Kilocalorie(s), 92, 100
Kinesin(s), 76, 77*f*
Kinetic energy, 92, 92*f*, 100
Kinetin, 489
Kinetochore(s), 143–144, 143*f*
Kinetochore fibers, 143, 143*f*
Kingdom(s), 14, 19, 304, 313
Kingdom Animalia, 14*t*, 15
Kingdom Fungi, 15, 378
Kingdom Plantae, 14*t*, 15
Kingsnake, 765
Kinins, 605
Kinocilium, 567, 567*f*
Kin selection, 752–753, 757–758
Kissing bug, 341
Kiwi(s), 411
Klinefelter syndrome, 166, 168
Klipspringers, chemical communication by, 754
Knee, human, and ardipith, comparison of, 425
Kneecap. *See* Patella
Knee-jerk reflex, 568, 568*f*
Knop, Wilhelm, 466
Koala, 412, 412*f*
Kohlrabi, 258, 258*f*
Krebs cycle, 133, 133*f*, 134–136
  in cellular respiration, 123, 123*f*, 126, 126*f*, 127, 127*f*
  energy benefit of, 127
  inputs of, 127
  outputs of, 127
  steps in, 127, 127*f*
Krill, 403, 632, 795
Kudzu, 808, 808*f*
Kurosawa, Ewiti, 478
Kyphosis, 577

L

Labial palps, of grasshopper, 404*f*
Labia majora, 708, 708*f*
Labia minora, 708, 708*f*
Labium, of grasshopper, 404*f*
Labor, stages of, 724, 724*f*
Labrum, of grasshopper, 404*f*
lac operon, 216–217, 216*f*, 229
Lactase, 641
Lactate, production of, in fermentation, 131, 131*f*
Lacteal(s), 614, 627, 639, 649
Lactic acid, metabolism, in diving response, 652–653
Lactic acid bacteria, 132
Lactobacillus, and beer production, 132
Lactobacillus bulgaricus, 132
Lactose, 45, 132, 643
  dietary sources of, 641
Lactose intolerance, 45, 641, 641*f*
Lacunae (sing., lacuna)
  in bone, 514*f*, 515, 580, 580*f*, 589
  in cartilage, 514*f*, 515
Ladybug(s), 818, 818*f*
  in pest control, 804*f*, 805
Lagena, 565, 565*f*

Lagomorphans, 413
Lagoon(s), 794
Lag phase
  of exponential growth, 734, 734*f*
  of logistic growth, 734, 734*f*
Lake(s), 798
  acid rain and, 36, 36*f*
  eutrophic, 793, 793*f*, 798
  oligotrophic, 793, 793*f*, 798
Lake effect, 785
Lamarck, Jean-Baptiste de, 256
Lamellae
  of gill filaments, 656, 656*f*
  of spider's book lungs, 403
Laminaria, 352
Laminarin, 352
Lamprey, 408, 408*f*
Lance fluke, 767
Lancelets, 392, 406, 406*f*, 416
  early development of, 714, 714*f*, 715
  evolution of, 407*f*
  late gastrula, 714*f*, 715, 715*f*
Landscape diversity, 803, 819
Landscape preservation, 812
Language, 754
  Cro-Magnons and, 430
La Niña, 796, 796*f*
Lantern fish, 795, 795*f*
Lantern fly, 765, 765*f*
Lanugo, 723
Large intestine, 649
  anatomy, 640, 640*f*
  of birds, 634, 634*f*
  of carnivores, 631, 635*f*
  ecosystem in, 641
  functions of, 640
  of herbivores, 631, 635*f*
  human, 636*f*
  physiology, 640
Larrea tridentata. See Creosote bush
Larva (pl., larvae), 725
  of aquatic animals, 704, 704*f*
  of arthropods, 402
  of crabs, 402
  echinoderm, 405
  of fire ant, 3, 3*f*
  of frog, 388, 388*f*
  insect, 402, 402*f*, 688, 688*f*
  of planarians, 395
  of tapeworm, 396, 396*f*
  Trichinella, 401, 401*f*
  trochophore, 391, 395, 395*f*, 415
  tunicate, 406
Larynx, 658, 658*f*, 665
  amphibian, 409
LASIK surgery, 560
Lateral line, of fishes, 408, 565, 565*f*
Lateral nerve cord, of *Ascaris,* 401*f*
Lateral sulcus, 539*f*
Lateral ventricle, 538, 538*f*
Latex, 376–377, 377*f*
Latissimus dorsi muscle, 583*f*
Laughing gulls
  learning in, 748, 748*f*
  tactile communication in, 755
Laurasia, 302, 302*f*
Law of conservation of energy, 93, 93*f*
Law of independent assortment, 176, 189
Law of segregation, 174, 178, 189
Laws of probability, 179
Laws of thermodynamics, 92–93, 100
LDCs. *See* Less-developed countries (LDCs)
LDL. *See* Low-density lipoproteins (LDL)
Leaf (pl., leaves), plant. *See also* Megaphylls
  anatomy of, 450, 450*f*
  blade of, 438*f*, 439, 453
  of C$_3$ plant, 114, 114*f*
  of C$_4$ plant, 114, 114*f*
  embryonic. *See* Cotyledon(s)
  of eudicot, 450, 450*f*

of flowering plants, 438–439, 438*f*, 453
  immature, of eudicot embryo, 501*f*, 504, 504*f*
  internal structure of, 445*f*
  of lycophytes, 366, 366*f*
  monocot vs eudicot, 440, 440*f*
  photosynthesis in, 105, 105*f*. *See also* Photosynthesis
  of pines, 370
  and plant homeostasis, 451*f*, 454
  sessile, 439
  sleep movements of, 485, 485*f*
  structure of, 105, 105*f*
  sun vs shade, of trees, 475
  tissues of, 444, 444*f*
  transport in, 458–459, 458*f*–459*f*
  water transpiration from, 461, 461*f*
  wound responses in, 489, 489*f*
Leaf curl fungi, 380, 382
Leafhopper, 404*f*
  and plant viruses, 320
Leaflets, of ferns, 368, 368*f*
Leaf primordia (sing., primordium), 447, 447*f*
Leaf scar, of plants, 447, 447*f*
Leaf veins, 105, 105*f*, 440, 440*f*, 443, 453–454
  of C$_4$ plant, 114, 114*f*
  of eudicot leaf, 444, 445*f*, 450, 450*f*
  patterns of, 440, 440*f*
Leafwing butterfly, 214*f*
Learning
  associative, 749, 757
  and behavior, 748, 748*f*, 757
  definition of, 748
  insight, 749, 757
  limbic system and, 542
Least flycatcher, 281, 281*f*
Leatherback turtles, 738, 738*f*
Leech(es), 392, 400, 400*f*, 632, 766
Leg(s). *See also* Walking legs
  of arthropods, 402, 402*f*
  of crustaceans, 402*f*, 403, 403*f*
  embryology of, 721
  of frog, 388, 388*f*
  insect, 402, 402*f*, 688, 688*f*
  number of, *Hox* genes and, 293
  of tetrapods, 409, 409*f*
Legume
  definition of, 6
  root nodules of, 6, 6*f*
Leishmaniasis, 341, 346
  cutaneous, 341
  visceral, 341
Lemur, 413, 792*f*
  evolution of, 422*f*
Lemus catta. See Ring-tailed lemur
Lens, of eye, 421, 558, 558*f*, 559, 569
Lens(es), corrective, 560, 560*f*
Lenski, Richard, 53
Lenticels, 442, 442*f*, 448, 448*f*, 454
Leprosy, research on, armadillo and, 804, 804*f*
Less-developed countries (LDCs), population growth in, 739, 739*f*, 740, 742
Lethargus, in *Caenorhabditis elegans,* 541
Lettuce, phenotypic plasticity, 474, 475*f*
Leucochloridium, 767
Leuconostoc, 132
Leukemia(s), 149
Leukocytes (white blood cells). *See* White blood cells
Lewis, Warren, 718
Lewontin, Richard, 432
LH. *See* Luteinizing hormone (LH)
Lianas, 792, 792*f*
Lichen(s), 333, 338, 353, 378–379, 379*f*, 384, 768
  crustose, 379*f*
  foliose, 379*f*
  fruticose, 379*f*
Licht, Louis, 463

Life
  characteristics of, 18
  extraterrestrial, 23
  origin of, 325–327, 338
  six types of atoms basic to, 24
  water and, 22–23
Life cycle
  alternation of generations, 157, 162, 162*f*, 361–363, 362*f*, 383, 496, 496*f*, 498, 507
  definition of, 162
  diploid, 162, 162*f*
  of flowering plants, 374–375, 374*f*–375*f*, 500, 500*f*
  haploid, 162, 162*f*
Life history
  definition of, 737, 742
  patterns of, 737, 742
Life table, 732, 733*f*, 741
Ligament(s), 515, 581
Liger(s), 278–279, 278*f*
Light micrographs, 64
Light microscope(s), visual range of, 62–63, 63*f*
Light reactions, 106, 106*f*, 107–111, 117
  noncyclic electron pathway during, 109, 109*f*, 110, 118
Lignin, 366, 383, 442–443
Lilac, 62, 62*f*
Limb(s)
  development of, 293
  lower, bones of, 578–579, 578*f*
  upper, bones of, 578, 578*f*
  vertebrate, evolution of, 407
Limb buds, 293, 721, 721*f*
Limbic system, 542, 542*f*, 546, 549
  and olfaction, 557
Limestone, 36
  acid rain and, 36
  foraminaferan, 347
Limiting factors, 731
Limulus amoebocyte lysate, 804
Limulus polyphemus. See Horseshoe crab
Linkage group, 175, 189
Linnaeus, Carolus, 304
Linnean classification, 304–305, 305*f*, 313
  vs cladistics, 310, 310*f*
Linnean systematics, 304–306, 313
Linoleic acid, 643–644, 650
Lion(s), 413*f*
  as flagship species, 812
Lipase, 638, 638*f*, 642, 649
Lipid(s), 42, 47–48, 57
  blood levels of, 647
  dietary, 643–644, 644*f*
  in food, 44, 44*f*
  functions of, 48
  subunits of, 44*t*
  synthesis of, 70, 79
Lipoprotein(s), 604. *See also* High-density lipoproteins (HDL); Low-density lipoproteins (LDL)
  synthesis of, 70
Liposome(s), 327, 327*f*, 338
  in gene therapy, 241
Little bluestem, 460, 460*f*
Liver, 602, 602*f*, 650
  anatomy of, 642, 642*f*
  and carbohydrate metabolism, 45
  disorders of, 642
  embryology of, 721*f*
  functions of, 642
  glycogen storage in, 46
  and homeostasis, 522, 522*f*
  human, 636*f*
  lobules of, 642, 642*f*
Liver cell(s), gene expression in, 244, 244*f*
Liverworts, 364–365, 364*f*, 383
  characteristics of, 361
  evolution of, 360*f*
Lizard(s), 13, 410, 410*f*

Metric System

Unit and Abbreviation	Metric Equivalent	Approximate English-to-Metric Equivalents		Units of Temperature
Length				
nanometer (nm)	$= 10^{-9}$ m (10^{-3} μm)			
micrometer (μm)	$= 10^{-6}$ m (10^{-3} mm)			
millimeter (mm)	$= 0.001$ (10^{-3}) m			
centimeter (cm)	$= 0.01$ (10^{-2}) m	1 inch	= 2.54 cm	
		1 foot	= 30.5 cm	
meter (m)	$= 100$ (10^{2}) cm	1 foot	= 0.30 m	
	$= 1,000$ mm	1 yard	= 0.91 m	
kilometer (km)	$= 1,000$ (10^{3}) m	1 mi	= 1.6 km	
Weight (mass)				
nanogram (ng)	$= 10^{-9}$ g			
microgram (μg)	$= 10^{-6}$ g			
milligram (mg)	$= 10^{-3}$ g			
gram (g)	$= 1,000$ mg	1 ounce	= 28.3 g	
		1 pound	= 454 g	
kilogram (kg)	$= 1,000$ (10^{3}) g		= 0.45 kg	
metric ton (t)	$= 1,000$ kg	1 ton	= 0.91 t	
Volume				
microliter (μl)	$= 10^{-6}$ l (10^{-3} ml)			
milliliter (ml)	$= 10^{-3}$ liter	1 tsp	= 5 ml	
	$= 1$ cm^3 (cc)	1 fl oz	= 30 ml	
	$= 1,000$ mm^3			
liter (l)	$= 1,000$ ml	1 pint	= 0.47 liter	
		1 quart	= 0.95 liter	
		1 gallon	= 3.79 liter	
kiloliter (kl)	$= 1,000$ liter			

Temperature scale (°F / °C):

212° — 210 / 100 — 100°
160° — 160 / 70 — 71°
134° / 131° — 130 / 57°
105.8° — 110 / 40 — 41°
98.6° — 100 / 37°
56.66° — 60 / 13.7°
32° — 30 / 0 — 0°
-40 / -40

°C	°F	
100	212	Water boils at standard temperature and pressure.
71	160	Flash pasteurization of milk.
57	134	Highest recorded temperature in the United States, Death Valley, July 10, 1913.
41	105.8	Average body temperature of a marathon runner in hot weather.
37	98.6	Human body temperature.
13.7	56.66	Human survival is still possible at this temperature.
0	32.0	Water freezes at standard temperature and pressure.

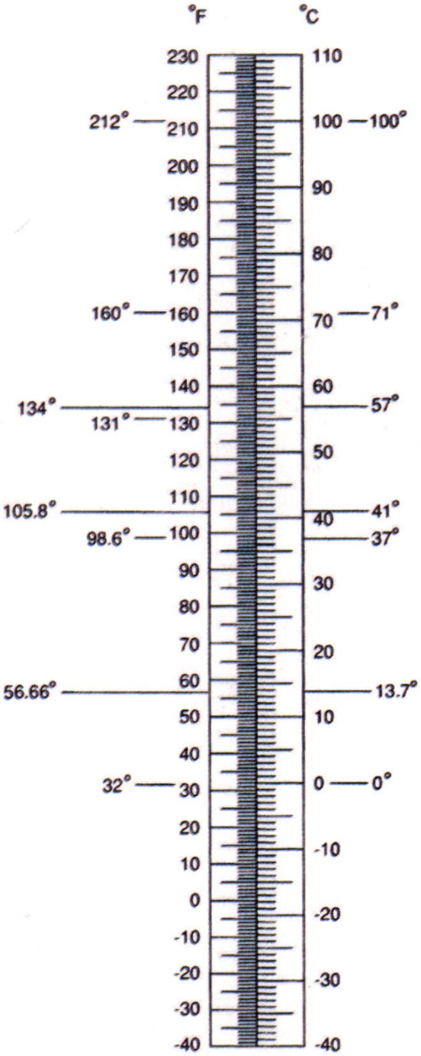

To convert temperature scales:

$$°C = \frac{(°F - 32)}{1.8}$$

$$°F = 1.8\,(°C) + 32$$